中国科学院院士文集

贺贤士论文选集

《贺贤士论文选集》编委会 编

ZHEJIANG UNIVERSITY PRESS
浙江大学出版社

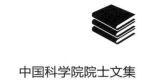

前言

贺贤土，理论物理学家，中国科学院院士。1962 年毕业于浙江大学物理系理论物理专业。1962 年 11 月进入北京应用物理与计算数学研究所工作。1986 年 6 月至 1987 年 12 月为美国马里兰大学高级访问学者。自 1988 年起任北京应用物理与计算数学研究所研究员，1988—1991 年任研究所科技委副主任，1991—1997 年任副所长。1995 年当选为中国科学院院士。1993—2001 年先后任国家高技术研究发展计划（863 计划）直属惯性约束聚变主题秘书长、第二任首席科学家。

2001—2006 年先后任中国科学院数理学部常委、副主任、主任，以及中国科学院学部主席团成员和执行委员会成员。1999—2009 年任浙江大学理学院院长和宁波职业技术学院院长。自 2007 年起任北京大学应用物理与技术中心主任。曾兼任科技部 863 计划顾问组成员、国家自然科学奖评审委员会委员、总装备部科技委兼职委员等职。

贺贤土院士长期从事国家重大科学工程研究和基础科学研究。

1963—1986 年，我国核武器物理研究和设计。在我国原子弹、氢弹以及中子弹的突破性研究和核试验任务中，完成了大量开拓性的研究工作和内部报告。

1988 年至今，从事激光驱动惯性约束聚变（inertial confinement fusion，ICF）研究。曾作为秘书长和首席科学家，领导了 863 计划直属惯性约束聚变主题，取得了多项科学和技术重大成果。在他的领导下，我国在十分薄弱的研究基础上建成了独立自主的 ICF 研究体系，为之后被纳入“国家中长期科学和技术发展规划”点火计划奠定了重要基础。

1980 年至今，完成了大量基础科学研究工作。他在十分繁忙的国家任务中，始终抽出时间坚持基础科学研究，在深度科学认知的基础上完成他所肩负的国家任务。2002 年，他因年龄关系不再担任一线领导职务，这使他有更多时间和精力进行基础研究。他在所从事的大科学工程中提炼了大量基础问题，并结合快速发展的高能量密度物理的相关前沿科学问题，领导团队进行了深入研究，在斑图（pattern，这里指携带物质质量、动量、能量的一种空间结构）动力学和时空混沌、等离子体湍流、自生磁场产生机制、强场带电粒子加速、强场原子电离、激光等离子体相互作用、温稠密物质特性、流体力学不稳定性和可压缩流体湍流、惯性约束聚变点火靶物理等方面，取得了一系列具有国际影响力的成果，发表了约 300 篇学术论文。除了早期个人研究成果之外，他指导的博士研究生和博士后的研究成果以及与合作者的研究成果同样非常丰硕。

50 多年来，贺贤土院士服从国家需要，不断改变研究方向。从核武器科学、ICF 物理到基础科学研究，虽然学科跨度很大，但他都出色地完成了研究任务。

贺贤土院士是惯性聚变科学界和高能量密度物理学界的国际著名科学家，是这些领域中一些国际会议的发起者、主席和科学顾问，享有很高的国际声誉。他多次被邀请在惯性聚变科学与应用国际会议（International Conference on Inertial Fusion Sciences and Applications，IFSA）、国际原子能机构聚变能源大会（IAEA Fusion Energy Conference）等重大国际会议上做大会邀请报告和综合报告，积极宣传我国惯性聚变和高能量密度物理研究进展，使我国相关研究成果得到国际广泛认可并在国际上占有一席之地。他参与了许多国际合作，并共同创建了亚洲核聚变与等离子体协会等国际学术组织。

贺贤土院士曾获国家自然科学奖二等奖一项，国家科学技术进步奖一等奖、二等奖各一项，部委级奖八项。1992 年获光华科技基金奖一等奖，2000 年获何梁何利奖，2001 年获国家 863 计划突出贡献先进个人奖。

今年是贺贤土院士从事科学研究 55 周年。我们选取他已发表的 150 篇学术论文（发表于 1980—2017 年）进行编辑出版，以飨读者。由于保密原因，他所从事的有关核武器物理研究和设计的论文未被收录在本选集中。

《贺贤土论文选集》编委会

2017 年 4 月 28 日

目录

激光驱动惯性约束聚变

斑图动力学和时空混沌

等离子体自生磁场

激光等离子体相互作用和粒子加速

A 电子共振加速

D 快点火

流体力学不稳定性和可压缩流体湍流

温稠密物质

惯性约束聚变（inertial confinement fusion，ICF）是实现可控热核聚变能源的主要途径之一，同时在国防和基础科学研究等方面有重要应用。它是国际上持续关注的重大科技前沿问题，其研究水平是一个国家科技实力的重要标志和国防实力的重要体现。目前激光是 ICF 的主要驱动源，激光驱动惯性约束聚变也称激光聚变。

1988 年，贺贤土被任命为北京应用物理与计算数学研究所领导，专门负责激光聚变物理研究，从此开始了他奋斗半生的 ICF 事业。他积极推动了我国 ICF 研究进入国家高技术研究发展计划（863 计划）和国家中长期科学和技术发展规划，领导论证并制定了我国 ICF 总体发展规划，是我国 ICF 事业的领军科学家。在他的领导下，我国 ICF 研究团队完成了激光器建造的关键单元技术攻关和神光（SG）系列部分激光器的建造，建立了靶物理研究的数值模拟大型系列程序平台，首次实现我国间接和直接两种驱动方式的热核聚变并测得高产额中子数，发展了精密诊断和精密靶制备技术等，建成了我国独立自主的 ICF 研究体系。

贺贤土院士在 ICF 点火靶物理研究中，也为我国 ICF 发展做出了重大贡献。在他的领导下，我国 ICF 研究走了一条与美国不同的发展道路。我国靶物理研究的激光器建造在万焦耳和百万焦耳量级之间加上了中间一步：①从当时的几千焦耳级能量 SG-Ⅱ发展到几万焦耳级的 SG-Ⅱ升级和 SG-Ⅲ原型；②建造 20 万～40 万焦耳的 SG-Ⅲ；③建造百万焦耳级的 SG-Ⅳ点火装置。20 万～40 万焦耳的 SG-Ⅲ激光器这一中间环节的物理规律研究为我国 ICF 点火研究大大降低了科学上靶物理规律过大倍数外推造成的风险。最近美国国家点火装置（National Ignition Facility，NIF）实验点火失败表明，通过计算机模拟 ICF 靶物理规律定标外推近百倍是不可信的。

与此同时，贺贤土院士坚持走创新道路，与合作者完成了大量开拓创新的研究成果。他早年提出的从较低温度点火发展到高温燃烧的体点火模型被国际同行评论为重要的发现（key discovery）。在激光等离子体相互作用、非局域电子热传导等方面，其团队也做出了许多创新工作。美国 NIF 点火实验失败后，贺贤土院士总结经验，先后提出了六孔球腔二次冲击间接驱动模型和间接—直接混合驱动模型等新型点火途径，并领导团队开展了大量数值模拟和关键物理问题分解实验，取得了一系列创新成果，得到了国际同行的广泛关注。其混合驱动点火模型能在较低激光能量下实现高增益聚变，被国际同行评论为“中国人的点火模型”。

PHYSICAL REVIEW E VOLUME 50, NUMBER 1 JULY 1994

Modified formula of nonlocal electron transport in a laser-produced plasma

Y. Xu
Institute of Applied Physics and Computational Mathematics, P.O. Box 8009, Beijing 100088, People's Republic of China

X. T. He
Institute of Applied Physics and Computational Mathematics, P.O. Box 8009, Beijing, People's Republic of China and China Center of Advanced Science and Technology (World Laboratory), P.O. Box 8730, Beijing 100080, People's Republic of China.

(Received 28 October 1993)

A nonlocal heat-transport formula for electrons is derived to include the terms associated with the electrostatic potential and $\partial/\partial v(f_0,f_1)$ in the Fokker-Planck (FP) equation. Then the FP equation for a strongly inhomogeneous plasma is solved. It is found that the behavior of the electron thermal conductivity at a large temperature gradient is considerably affected by the electrostatic field, and the thermal conductivity $\kappa/\kappa_{\rm SH}$ for electrons scales as $1/k$ in a large temperature gradient k when there exists a non-negligible electrostatic field, where $\kappa_{\rm SH}$ is the Spitzer-Härm heat coefficient.

PACS number(s): 52.25.Fi, 52.50.Jm

I. INTRODUCTION

It has been found in laser fusion experiments in the region of the steeped temperature and density gradient near the critical surface of the laser produced plasma that the electron thermal conductivity is far less than that predicted by the Spitzer-Härm value obtained from classical theory. Since this phenomenon is found in many experiments measuring properties of a laser-produced plasma, much effort [1–7] has been made to obtain an analytical formula which can give the best approximation of the results given by the numerical simulation. The most successful of these works is that of Albritton *et al.* [2]. They derived a formula (AWBS) for nonlocal electron transport by directly solving the Fokker-Planck (FP) equation in a simplified form. There are also some modified models of AWBS [3–7].

Despite the success of AWBS, there exist some problems that need further study. Comparing the results obtained from the formula of AWBS and the numerical simulation, Epperlein and Short [3] pointed out that the results of AWBS deviate from that of the numerical simulation in the large temperature gradient, and the reason for that deviation is that the theory of AWBS is too delocalized, while that of Spitzer-Härm is too localized.

Neglecting the electrostatic field is one possible way to cause the excess delocalization of AWBS. When the temperature gradient is steep, slow electrons inside the steeped region cannot compensate for the departure of fast electrons outside. In order to satisfy the condition of quasineutrality, an electrostatic field must occur. This electrostatic field directs electrons to the outside of the steeped region, and increases with increasing position and drops to zero in the corona region. The non-negligible electrostatic field associated with the large temperature gradient makes the simplified FP equation of the AWBS model inadequate in the large temperature gradient.

In this paper we will discuss the contribution of the electrostatic field on electron transport in the steeped region in a laser-produced plasma. The electrostatic field is determined self-consistently by the quasineutrality condition. In Sec. II, we give a modified formula of electron distribution function, from which we obtain a modified formula for particle and energy flux. The discussion and conclusion are given in Sec. III.

II. MODIFIED ELECTRON DISTRIBUTION AND ELECTRON THERMAL CONDUCTIVITY

We start with the FP equation for electrons:

$$\frac{\partial f}{\partial t}+\mu v\frac{\partial f}{\partial x}-\frac{eE}{m}\left[\mu\frac{\partial f}{\partial v}+\frac{1-\mu^2}{v}\frac{\partial f}{\partial \mu}\right]=\left[\frac{\partial f}{\partial t}\right]_c , \tag{1}$$

where $\mu=v_x/v$, v_x is the x component of velocity v, and $[\partial f/\partial t]_c$ is the FP collision term

$$\left[\frac{\partial f}{\partial t}\right]_c=\frac{1}{v^2}\frac{\partial}{\partial v}\left\{v^2\left[\frac{D_1}{2}\frac{\partial f}{\partial v}+Cf\right]\right\}+\frac{D_2}{2v^2}\frac{\partial}{\partial \mu}\left[(1-\mu^2)\frac{\partial f}{\partial \mu}\right] , \tag{2}$$

where C is a constant for slowing down, and D_1 and D_2 represent the coefficients for diffusion in velocity space and angular space, respectively. Because energy is transported at predominantly high electron velocities, we use the high-velocity asymptotic form of these coefficients [8].

To solve the electron kinetic equation, one expands the distribution function into spherical harmonics, i.e.,

$$f(\mathbf{r},\mathbf{v},t)=\sum_{l=0}^{N} f_l(x,v,t)P_l(\mu) ,$$

where P_l is the l-order Legendre polynomial.

Numerical calculation [4] has found that although the high-order terms $(f_2,f_3,\ldots)$ are not negligible, their influence on f_1 (which describes heat flow and current) and f_0 (temperature and density) is small. Hence we take the kinetic equation curtailed in $N=1$, and have

$$\frac{v}{3}\left[\frac{\partial}{\partial x}-\frac{eE}{mv}\left[\frac{\partial}{\partial v}+\frac{2}{v}\right]\right]f_1=\frac{v^2}{2\lambda_e}\frac{\partial}{\partial v}\left[\frac{T}{mv}\frac{\partial f_0}{\partial v}+f_0\right], \tag{3}$$

$$v\left[\frac{\partial}{\partial x}-\frac{eE}{mv}\frac{\partial}{\partial v}\right]f_0=-\frac{2v}{\lambda_{90}}f_1, \tag{4}$$

where E is the electric field, λ_e is the stopping length of electrons defined as $\lambda_e=T^2/4\pi ne^4(Z+1)^{1/2}\ln\Lambda$, and

$$\lambda_{90}=(mv^2)^2/2\pi ne^4(Z\ln\Lambda_{ei}+\Lambda_{ee})$$

is the scattering mean free path, while Z is the ion charge in the plasma, T is the electron temperature, and Λ the Coulomb logarithm.

In order to derive Eqs. (3) and (4) we have assumed that all temporal variations are slow.

The number of electrons carrying the heat flux is much smaller than that of background electrons. Therefore we approximate the parallel diffusion term in Eq. (3) by the local Maxwellian distribution function $f_0\sim f_{MB}=n(m/2\pi T)^{3/2}\exp[-\epsilon/T]$, where $\epsilon=\frac{1}{2}mv^2+e\phi$ is the electron energy and ϕ is the electrostatic potential, so that Eqs. (3) and (4) can be rewritten as

$$\frac{v}{3}\left[\frac{\partial}{\partial x}-\frac{eE}{mv}\left[\frac{\partial}{\partial v}+\frac{2}{v}\right]\right]f_1=\frac{v^2}{2\lambda_e}\frac{\partial}{\partial v}[f_0-f_{MB}], \tag{5}$$

$$v\left[\frac{\partial}{\partial x}-\frac{eE}{mv}\frac{\partial}{\partial v}\right]f_0=-\frac{2v}{\lambda_{90}}f_1. \tag{6}$$

In other works [5,6], the second terms in the left-hand sides of Eqs. (5) and (6) are neglected. Those terms may have an important influence on the state of nonlocal models.

Substituting Eq. (6) into Eq. (5), we have

$$\epsilon^3\left[\frac{\partial^2 f_0}{\partial\xi^2}+\tilde{\lambda}_s eE\left[\frac{2}{T}-\frac{3}{\epsilon}\right]\frac{\partial f_0}{\partial\xi}\right]+\frac{\partial f_0}{\partial\epsilon}=-\frac{f_{MB}}{T}, \tag{7}$$

where $\xi=x/\tilde{\lambda}_s$ and $\tilde{\lambda}=\lambda/(mv^2)^2$, and $\tilde{\lambda}_s=(2\tilde{\lambda}_e\tilde{\lambda}_{90}/3)^{1/2}$. $\epsilon\gg e\phi$ is exploited everywhere by letting $e\phi\to 0$ except under differentiation by ξ.

Since $\tilde{\lambda}_s eE=-\partial e\phi/\partial\xi$ and $\epsilon\gg e\phi$, we can rewrite Eq. (7) as

$$(\epsilon+e\phi)^3\left[\frac{\partial^2 f_0}{\partial\xi^2}+\tilde{\lambda}_s eE\frac{2}{T}\frac{\partial f_0}{\partial\xi}\right]+\frac{\partial f_0}{\partial\epsilon}=-\frac{f_{MB}}{T}. \tag{8}$$

In order to transform Eq. (8) into a parabolical-type equation, we make a transformation $y=y(\xi)$, and have

$$\frac{\partial f_0}{\partial\xi}=\frac{\partial y}{\partial\xi}\frac{\partial f_0}{\partial y}$$

and

$$\frac{\partial^2 f_0}{\partial x^2}=\frac{\partial^2 y}{\partial\xi^2}\frac{\partial f_0}{\partial y}+\left[\frac{\partial y}{\partial\xi}\right]^2\frac{\partial^2 f_0}{\partial y^2}.$$

Substituting into Eq. (8), we find that when

$$\frac{\partial^2 y}{\partial\xi^2}+\tilde{\lambda}_s eE\frac{2}{T}\frac{\partial y}{\partial\xi}=0, \tag{9}$$

Eq. (8) becomes a parabolic-type equation:

$$(\epsilon+e\phi)^3\left[\frac{\partial y}{\partial\xi}\right]^2\frac{\partial^2 f_0}{\partial y^2}+\frac{\partial f_0}{\partial\epsilon}=-\frac{f_{mB}}{T}. \tag{10}$$

Integrating Eq. (9) over ξ, we have

$$\frac{\partial y}{\partial\xi}=C_1\exp\left[-\int d\xi\,\tilde{\lambda}_s eE\frac{2}{T}\right],$$

where C_1 is an integer. When the electric field is negligible, $y\to\xi$ and $\partial y/\partial\xi\to 1$, we have $C_1=1$.

$$\frac{\partial y}{\partial\xi}=\exp\left[-\int d\xi\,\tilde{\lambda}_s eE\frac{2}{T}\right], \tag{11}$$

$$y=\int d\xi\exp\left[-\int d\xi\,\tilde{\lambda}_s eE\frac{2}{T}\right]. \tag{12}$$

From Eq. (11), we find that when modified terms due to the electrostatic field are negligible, Eq. (8) is reduced to the results of Albritton *et al.* [2]. But when the electrostatic field is non-negligible, y will deviate quickly from ξ along with an increase of the electrostatic field E or of the temperature gradient.

Solving Eq. (10), we obtain the formula for the electron distribution function:

$$f_0=\int dy(\xi')\int_\epsilon^\infty d\epsilon'\frac{\exp\left[-\dfrac{[y(\xi)-y(\xi')]^2}{[\epsilon'+e\phi(\xi')]^4\left[\dfrac{\partial y}{\partial\xi'}\right]^2-[\epsilon+e\phi(\xi)]^4\left[\dfrac{\partial y}{\partial\xi}\right]^2}\right]}{\left\{\pi\left[[\epsilon'+e\phi(\xi')]^4\left[\dfrac{\partial y}{\partial\xi'}\right]^2-[\epsilon+e\phi(\xi)]^4\left[\dfrac{\partial y}{\partial\xi}\right]^2\right]\right\}^{1/2}}\frac{f_{MB}(\xi',\epsilon')}{T(\xi')}. \tag{13}$$

The fluxes of the particle and energy are defined as

$$\begin{Bmatrix}\Gamma\\ Q\end{Bmatrix}=-16\pi\int_{-e\phi}^{\infty}d\epsilon\times\begin{Bmatrix}1\\ \epsilon\end{Bmatrix}\times\epsilon^3\frac{\tilde{\lambda}_{90}}{3m^2}\frac{\partial f_0}{\partial\xi}. \tag{14}$$

Substituting Eq. (13) into Eq. (14), and we write $\partial f_0/\partial\xi$ as

$$\frac{\partial f_0}{\partial \xi}=\int dy(\xi')\int_{\epsilon}^{\infty} d\epsilon' K(\xi,\xi',\epsilon,\epsilon')\frac{f_{\mathrm{MB}}(\xi',\epsilon')}{T(\xi')}\frac{\partial}{\partial \xi'}\frac{\exp\left[-\frac{[y(\xi)-y(\xi')]^2}{\epsilon'^4\left[\frac{\partial y}{\partial \xi'}\right]^2-\epsilon^4\left[\frac{\partial y}{\partial \xi}\right]^2}\right]}{\left\{\pi\left[\epsilon'^4\left[\frac{\partial y}{\partial \xi'}\right]^2-\epsilon^4\left[\frac{\partial y}{\partial \xi}\right]^2\right]\right\}^{1/2}},$$

where

$$K(\xi,\xi',\epsilon,\epsilon')=\frac{\left[2[y(\xi)-y(\xi')]+\epsilon^4\left[1-2\frac{[y(\xi)-y(\xi')]^2}{\left[\epsilon'^4\left[\frac{\partial y}{\partial \xi'}\right]^2-\epsilon^4\left[\frac{\partial y}{\partial \xi}\right]^2\right]^2}y''(\xi)\right]\right]y'(\xi)}{\left[-2[y(\xi)-y(\xi')]+\epsilon'^4\left[1-2\frac{[y(\xi)-y(\xi')]^2}{\left[\epsilon'^4\left[\frac{\partial y}{\partial \xi'}\right]^2-\epsilon^4\left[\frac{\partial y}{\partial \xi}\right]^2\right]^2}y''(\xi')\right]\right]y'(\xi')},$$

where y' and y'' are the first and second differentials of y. When $E\to 0$, $K(\xi,\xi',\epsilon,\epsilon')\to -1$. And when $\epsilon \gg e\phi$, the $e\phi$ only brings in a second-order modification. Thus we replace $\epsilon+e\phi$ with ϵ in the expressions above and below, and integrate by parts in ξ' to cast the spatial derivative to $f_{\mathrm{MB}}(\xi',\epsilon')/T(\xi')$. Thus the particle and heat flux Γ, Q are expressed as

$$\begin{aligned}\begin{Bmatrix}\Gamma\\ Q\end{Bmatrix}=&-16\pi\int_{-e\phi}^{\infty} d\epsilon \times \begin{Bmatrix}1\\ \epsilon\end{Bmatrix}\times \epsilon^3\frac{\tilde{\lambda}_{90}}{3m^2}\\ &\times\int dy(\xi')\int_{\epsilon}^{\infty} d\epsilon' \frac{\exp\left[-\frac{[y(\xi)-y(\xi')]^2}{\epsilon'^4\left[\frac{\partial y}{\partial \xi'}\right]^2-\epsilon^4\left[\frac{\partial y}{\partial \xi}\right]^2}\right]}{\left\{\pi\left[\epsilon'^4\left[\frac{\partial y}{\partial \xi'}\right]^2-\epsilon^4\left[\frac{\partial y}{\partial \xi}\right]^2\right]\right\}^{1/2}}\\ &\times\frac{f_{\mathrm{MB}}(\xi',\epsilon')}{T^2(\xi')}\left[\frac{\epsilon}{T}\frac{\partial T}{\partial x}-\frac{\partial e\phi}{\partial x}-\frac{5}{2}\frac{\partial T}{\partial x}+\frac{T}{n}\frac{\partial n}{\partial x}+\frac{\partial}{\partial x}K(\xi,\xi',\epsilon,\epsilon')\right],\end{aligned}\tag{15}$$

where

$$\begin{aligned}\frac{\partial}{\partial x}K(\xi,\xi',\epsilon,\epsilon')=&\frac{2y'(\xi)y'(\xi')+\frac{4\epsilon^4[y(\xi)-y(\xi')]y'(\xi)y'(\xi')y''(\xi)}{-[\epsilon^4y'(\xi)^2]+\epsilon'^4y'(\xi')^2}+\frac{4\epsilon^4\epsilon'^4[y(\xi)-y(\xi')]^2y'(\xi)y'(\xi')y''(\xi)y''(\xi')}{[-(\epsilon^4y'(\xi)^2)+\epsilon'^4y'(\xi')^2]^2}}{-2[y(\xi)-y(\xi')]y'(\xi')+\epsilon'^4y'(\xi')y''(\xi')-\frac{2\epsilon'^4[y(\xi)-y(\xi')]^2y'(\xi')y''(\xi')}{-(\epsilon^4y'(\xi)^2)-\epsilon'^4y'(\xi')^2}}\\ &+\left[-2[y(\xi)-y(\xi')]y'(\xi)+\epsilon^4y'(\xi)y''(\xi)-\frac{2\epsilon^4[y(\xi)-y(\xi')]^2y'(\xi)y''(\xi)}{-(\epsilon^4y'(\xi)^2)+\epsilon'^4y'(\xi')^2}\right]\\ &\times\left[\frac{2y'(\xi')^2-2[y(\xi)-y(\xi')]y''(\xi')+\frac{4\epsilon'^4[y(\xi)-y(\xi')]y'(\xi')^2y''(\xi')}{-(\epsilon^4y'(\xi)^2)-\epsilon'^4y'(\xi')^2}+\epsilon'^4y''(\xi')^2}{\left[-2[y(\xi)-y(\xi')]y'(\xi')+\epsilon'^4y'(\xi')y''(\xi')-\frac{2\epsilon'^4[y(\xi)-y(\xi')]^2y'(\xi')y''(\xi')}{-(\epsilon^4y'(\xi)^2)-\epsilon'^4y'(\xi')^2}\right]^2}\right.\\ &+\frac{-\frac{4\epsilon'^8(y(\xi)-y(\xi'))^2y'(\xi')^2y''(\xi')^2}{[-(\epsilon^4y'(\xi)^2)-\epsilon'^4y'(\xi')^2]^2}-\frac{2\epsilon'^4(y(\xi)-y(\xi'))^2y''(\xi')^2}{-(\epsilon^4y'(\xi)^2)-\epsilon'^4y'(\xi')^2}}{\left[-2[y(\xi)-y(\xi')]y'(\xi')+\epsilon'^4y'(\xi')y''(\xi')-\frac{2\epsilon'^4[y(\xi)-y(\xi')]^2y'(\xi')y''(\xi')}{-(\epsilon^4y'(\xi)^2)-\epsilon'^4y'(x')^2}\right]^2}\\ &\left.+\frac{+\epsilon'^4y'(\xi')y^{(3)}(\xi')-\frac{2\epsilon'^4[y(\xi)-y(\xi')]^2y'(\xi')y^{(3)}(\xi')}{-(\epsilon^4y'(\xi)^2)-\epsilon'^4y'(\xi')^2}}{\left[-2[y(\xi)-y(\xi')]y'(\xi')+\epsilon'^4y'(\xi')y''(\xi')-\frac{2\epsilon'^4[y(\xi)-y(\xi')]^2y'(\xi')y''(\xi')}{-(\epsilon^4y'(\xi)^2)-\epsilon'^4y'(\xi')^2}\right]^2}\right].\end{aligned}\tag{16}$$

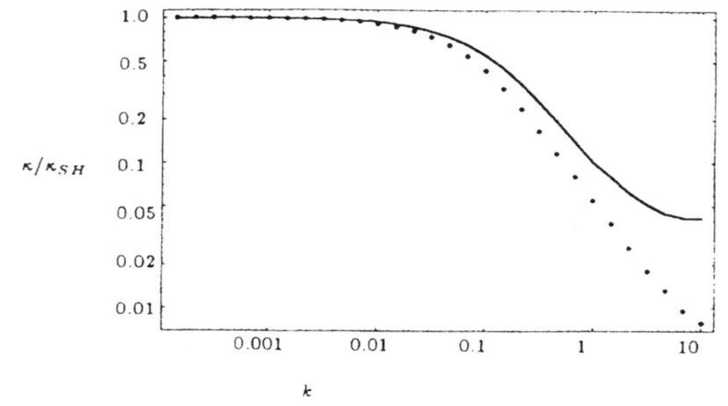

FIG. 1. The electron thermal conductivity $\kappa/\kappa_{\rm SH}$ vs k with a temperature form $T_c\exp[k(x/L-1)]$ at the point $x=L$, when $R_t=0.001$. The dots and the line correspond to the modified formula and that given by Albritton *et al.*

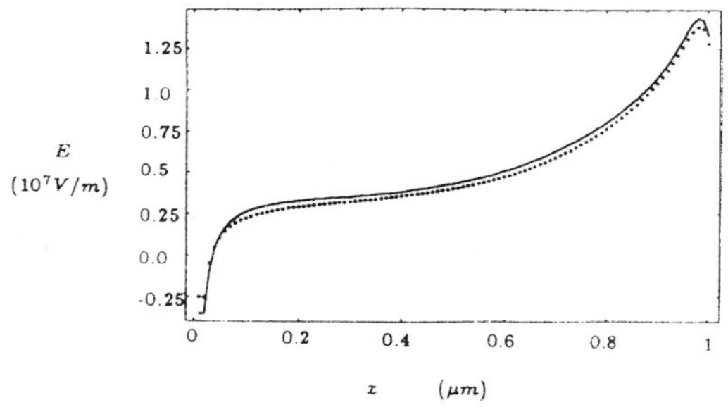

FIG. 3. The electrostatic field E vs position x with a temperature gradient $k\sim10$ at the point $x=L$ calculated with the modified formula. The dots and the line correspond to the formula of Albritton *et al.* and our modified formula, respectively.

Equation (15) expresses Γ and Q as implicit functions of electric field E. The electric field is determined by the requirement of quasineutrality, $n=NZ$, where N is the ion density. In this case, the electric field is determined by the vanishing of the particle flux $\Gamma(E)=0$. However, there are two unknown parameters, electron density $n(x)$ and temperature $T(x)$, in Eq. (15). In principle, $n(x), T(x)$ can be determined from an implicit equation

$$\begin{Bmatrix} n \\ T \end{Bmatrix} = \int_{-e\phi}^{\infty} d\epsilon \times \begin{Bmatrix} 1 \\ \epsilon \end{Bmatrix} \times \epsilon^{1/2} f_0 \ .$$

III. RESULTS AND CONCLUSION

The electrostatic field determined from the quasineutrality condition in a laser-produced plasma can be written as $eE=-T\partial[\ln(xT^{\gamma})]/\partial x$ [2]; here γ is a parameter that is determined by total configuration and temperature gradient. When nT^{γ} is not a constant, but varies with position, an electrostatic field occurs.

First we discuss the effect of the electrostatic field on nonlocal electron transport in different temperature gradients by assuming the form of temperature $T(x)$ and density $n(x)$. Without loss generality, we assume that the plasma temperature has a form of $T_c\exp[k(x/L-1)]$, where T_c is the critical temperature, L is a scale length of temperature gradient, and $nT^{\gamma}=C_2T^{R_t}$. Here C_2 and R_t are constants that will be given later. Thence the relation of electrostatic field to $\partial T/\partial x$ is $E=-(R_t/e)\partial T/\partial x$, and Eq. (12) becomes

$$y=-\frac{L}{2k\tilde{\lambda}_s R_t}\exp\left[-\frac{2kxR_t}{L}\right] . \tag{17}$$

We take $L=100\ \mu$m in our calculation. Substituting the above assumptions into Eq. (15), using the condition $\Gamma(E)=0$, we determine the profile of electron thermal conductivity.

In the calculation we find that $\partial K/\partial x$ is a small modification that can be neglected.

Figure 1 shows that an electrostatic field will modify the formula of Albritton *et al.* at the large temperature gradient. Fitting the curve, the $\kappa/\kappa_{\rm SH}$ profile satisfies the relation

$$\kappa/\kappa_{\rm SH}=\frac{1}{1+16k} ,$$

where $\kappa_{\rm SH}=640\sqrt{2\pi}\varepsilon_0^2 k(kT)^{5/2}\varepsilon_e\delta_e/Ze^4\sqrt{m}\ \ln\Lambda_e$ is the Spitzer-Härm heat coefficient [9], and ε_e, δ_e are two

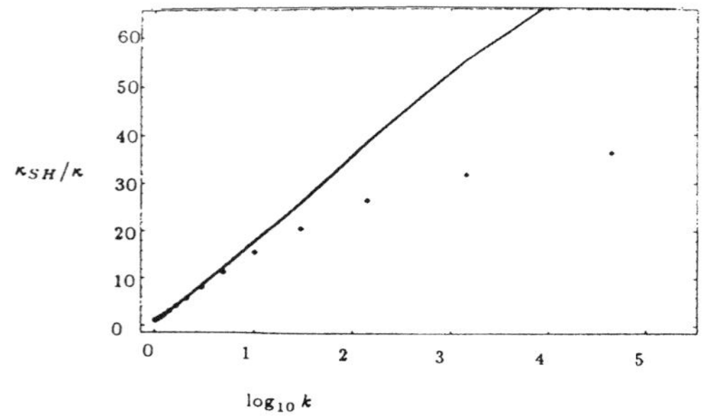

FIG. 2. The electron thermal conductivity $\kappa_{\rm SH}/\kappa$ vs k with a temperature form $T_c\exp[k(x/L-1)]$ at the point $x=L$ calculated with the modified formula, when $R_t=0.001$ or 0.000 01, respectively. The dots and the line correspond to $R_t=0.000\,0001$ or 0.001.

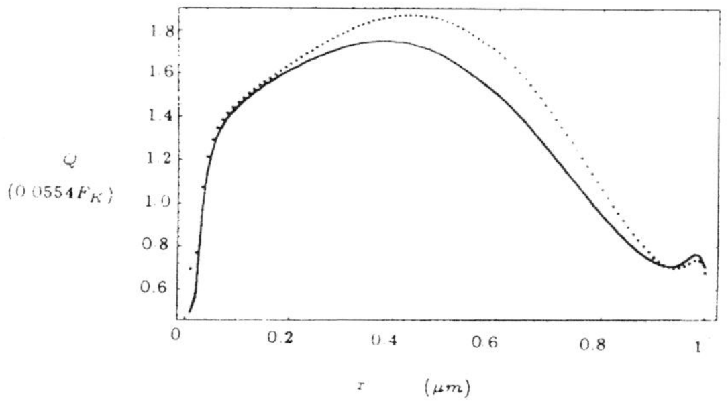

FIG. 4. The heat flux Q vs position x with a temperature gradient $k\sim10$ at the point $x=L$ calculated with the modified formula. The dots and the line correspond to the formula of Albritton *et al.* and our modified formula, respectively. Here $F_K=4.29\times10^{-25}n/m^{0.5}$ J m/s.

modified factors given by Spitzer, while the κ/κ_{SH} profile obtained from the formula given by Albritton *et al.* shows it to be a quadratic function of k.

It can be seen from Fig. 2 that, when the electrostatic field is non-negligible ($R_t=0.0001$), κ/κ_{SH} scales as $1/k$ at large k, and when the electrostatic field is negligible ($R_t=0.00001$), κ/κ_{SH} scales as $1/k^2$ at large k.

Second, we determine the electrostatic field self-consistently from quasineutrality $\Gamma(E)=0$, and

$$\begin{Bmatrix} n \\ T \end{Bmatrix} = \int_{-i\phi}^{\infty} d\epsilon \times \begin{Bmatrix} 1 \\ \epsilon \end{Bmatrix} \times \epsilon^{1/2} f_0 \; .$$

Then we use the value of E obtained to calculate the heat flux. Figure 3 shows the electrostatic field versus x, and Fig. 4 the heat flux inside the steeped region, when the initial test temperature gradient is $k \sim 10$. It can be seen from Figs. 3 and 4 that although the electrostatic field obtained from our formula is a little different from that of Albritton *et al.* the heat flux has a non-negligible modification.

Finally, we summarize our results. By including terms associated with an electric field in the FP equation, we have a modified formula of nonlocal electron transport. The effect of the electrostatic field on the electron thermal conductivity in various temperature gradients is calculated. We find that the electron thermal conductivity scales as $1/k$ at large k when a small electrostatic field exists, and hence the electrostatic field will have a non-negligible modification on the formula for nonlocal electron transport in a laser-produced plasma.

ACKNOWLEDGMENT

This work is supported by the National Science Foundation of China under Grant No. 19305008, and is partly supported by CAEP Foundation HD9317.

[1] J. F. Luciani, P. Mora, and J. Virmont, Phys. Rev. Lett. **51**, 1664 (1983).

[2] J. R. Albritton, E. A. Williams, I. B. Bernstein, and K. P. Swartz, Phys. Rev. Lett. **57**, 1887 (1986).

[3] E. M. Epperlein and R. W. Short, Phys. Fluids B **3**, 3092 (1991).

[4] J. P. Matte and J. Virmont, Phys. Rev. Lett. **49**, 1936 (1982).

[5] Arshad M. Mirza, G. Murtaza, and M. S. Qaisar, Phys. Scr. **42**, 85 (1990).

[6] Arshad M. Mirza and G. Murtaza, Phys. Rev. A **41**, 4460 (1990).

[7] F. Minotti and C. Ferro Fontan, Phys. Fluids B **2**, 1725 (1990).

[8] A. R. Bell, Phys. Fluids **28**, 2007 (1985).

[9] L. Spitzer and R. Härm, Phys. Rev. **89**, 977 (1953).

PHYSICAL PROCESSES OF VOLUME IGNITION AND THERMONUCLEAR BURN FOR HIGH-GAIN INERTIAL CONFINEMENT FUSION

X.T.He and Y.S.Li
Institute of Applied Physics and Computational Mathematics
P.O.Box 8009, Beijing 100088, China

ABSTRACT

Of late years, the volume ignition for inertial confinement fusion(ICF) has received great attention. In this paper, physical processes of volume ignition and burn from local thermodynamic equilibrium (LTE) transiting to the non-LTE are described, threshold values for the volume ignition are given, the property of the critical point that determines the maximum ion temperature in the burn process, and the quenching process of the thermonuclear system are discussed. Finally, a model is numerically computed, the above processes are well demonstrated, and the high gain is reached.

I. INTRODUCTION

So far there are essentially two kinds of ignition models for research on high gain ICF. The central spark ignition(CSI) model [1][2] is a most active one, for this model the central hot spot of deuterium-tritium (DT) fuel inside a pellet is first ignited with high ion temperature Ti $\cong 55(10^6\mathrm{K})$, and also high compressed density ρ and area density ρR for sufficient burn, where R is a radius for fuel. However, it requires to meet the harsh terms, such as, the precision implosion symmetry, the R-T instability and mix etc.

On the other hand, the volume ignition(VI) model[3], in which the DT fuel in pellet is compressed and ignited in its entirety, may considerably relax requirements for CSI to certain extent if the ablator of pellet is elaborately designed to meet the conditions of both inertial layer and thermal barrier.

In this paper ,we shall discuss the physical processes of volume ignition and burn in detail, to our knowledge, which are lack of investigation so far. The ignition for the VI model is performed under the regime of a local thermal equilibrium (LTE) state, then, the self-sustaining burn of DT fuel in the LTE state is going on until the volume burn in the non-local equilibrium (non-LTE) state begins. Where by the non-LTE, we mean that the x-rays are in dynamic distribution, and electron and ion temperatures are in strong separated. This way from LTE ignition to non-LTE burn considerably decreases the ignition temperature and ρR values comparing to the CSI model, also reduces the requirement of the implosion symmetry to be not so overcritical, and the mixture of the hot spot with "cold " main fuel, which is a seriously harmful factor to ignition, is not present in the VI model.

In order to discuss the system behavior in the LTE stage, we observe the entropy evolution. Ignition, i.e., the first critical point in LTE, indeed, takes place in a near minimum entropy state; then the LTE burn is bifurcated to non-LTE state before the system evolves to maximum entropy, that is, the second critical point.

Once the system evolves to the non-LTE stage, there exist new two critical points. The first one corresponds to non-LTE ignition which is the same as occurring in the hot spot for the CSI model. Then the unstable growth from this fixed point rapidly transits to a transient stationary state, i.e., the second critical point in non-LTE, in which the maximum ion temperature of this system is achieved, and finally the system will tend to the quenching process.

The paper is organized as follows: volume ignition and burn in LTE are discussed in section II, volume ignition and burn in non-LTE are investigated in section III, and the numerical investigation and discussion are given in final section.

II. VOLUME IGNITION AND BURN IN LTE STATE

1. Compression and increment entropy

It is convenience to use entropy to describe the evolution in implosion process. Assumed that temperature T is approximately homogeneous in space due to fast heat conduction in DT fuel, and fluid density ρ is substituted by an average compression ratio $\sigma = \rho/\rho_0$, ρ_0 is an initial density, in an ideal gas the equation for entropy can be solved as follows:

$$\frac{\Delta S}{MCv} = \ln[\frac{T}{T_4}(\frac{\sigma}{4})^{-(\gamma-1)}] + \frac{aT_4^4}{3\rho_0 CvT_4}(\frac{T}{T_4})^3(\frac{\sigma}{4})^{-1}[1-(\frac{T_4}{T})^3\frac{\sigma}{4}] \tag{1}$$

where ΔS is the subsystem entropy variation between T and T_4 states, T_4 is the temperature as σ=4 behind shock, $\gamma = 5/3$, Cv is the specific heat, a is a radiation constant, and M is the DT mass.

On the other hand, the rate of the subsystem entropy can also be expressed in the following:

Using the assumption of the temperature gradient small, we approximately write

$$\frac{dS}{dt} + \dot{J}_s = \frac{\dot{Q}}{T} \tag{2}$$

where $\dot{Q} = \int_M \dot{q}\, dm$. $\dot{q}$ is the heat rate from thermonuclear reaction for unit mass,

$\dot{J}_S = \frac{\dot{F}}{T}$ is the rate of the entropy flux, and

$$\dot{F} = -4\pi R_b^2 K \frac{\partial T}{\partial R}\Big|_{R=Rb} \quad (3)$$

is the radiative flux into the pellet shell from DT fuel, and R_b is the boundary radius for DT fuel, K is a coefficient for the radiation heat conduction.

From Eq.(2) we obtained

$$\Delta S = S(T) - S(T_4) = -\Delta J_s + \Delta Q_s \quad (4)$$

with

$$\Delta J_S = \int_{t_4}^{t} \frac{\dot{F}}{T} dt \quad (5)$$

and

$$\Delta Q_S = \int_{t_4}^{t} \frac{\dot{Q}}{T} dt \quad (6)$$

Notice that the radiation pressure comparing with the fluid pressure can be neglected in the ignition case, from Eq.(1) one obtained the relation

$$T = T_4 \left(\frac{\sigma}{4}\right)^{\gamma-1} \exp\left\{\frac{\Delta S}{MCv}\right\} \quad (7)$$

where ΔS is determined by Eq.(4). ΔS<0 always holds true due to $\Delta Qs < \Delta Js$ until the ignition is performed, therefore, ΔS is approximately the ratio of the radiation flux to the internal energy. In order to ignite to be successful, the internal energy $MCvT >> (\Delta S)T$ has to be required, thus, it is convenient to rewrite Eq.(4) in the following form

$$T = T_4 \left(\frac{\sigma}{4}\right)^{(\gamma-1)(1-\alpha)} \quad (8)$$

where $\alpha = \frac{1}{(\gamma-1)} \frac{|\Delta Js|}{MCv \ln\frac{\sigma}{4}} < 1$ for $\ln\frac{\sigma}{4} > 1$.

Our purpose is to gain the condition α<<1, so that the compression is approximately adiabatic with temperature T increasing, it is that to appropriately increase the fuel mass M and to considerably reduced the entropy flux ΔJ are necessary. Because the mean free path for radiation or its coefficient K depends on the ablator property, the pellet ablator must skillfully be designed to reduce the ΔJs loss. Here the self-adjustable effect on ΔJs and T must be stressed as seen from Eqs.(7) and (8). In fact, for high Z ablator, $K\sim T^n$, usually n=4-5, as a result, ΔJs depends on T sensitively, thus T and ΔJs are restricted each other. Therefore, ΔJs is impossible to be getting large, it guarantees T to be unceasingly increased due to doing work. In addition, the entropy increment from thermonuclear heat can also partially offset against ΔJs before ignition.

All these effects may ensure $\frac{\Delta S}{MCv} << 1$, the numerical calculation has shown that $\alpha << 1$ is easily to achieve for high Z shell, and α changes slowly with σ increasing in the extent of our interesting parameters.

2. Ignition definition and its physical nature

In the process of doing work to the fusion fuel from imploding the thermonuclear reaction is always going on due to temperature rising. What is the ignition conception in such a process? and what is the ignition conditions? We shall discuss them starting from competitions between the deposited energy rate from thermonuclear reaction $\dot{Q}$ and the lost energy rate of the radiation heat conduction to ablator $\dot{F}$. In the early stage, $\dot{F}$ is naturally great than $\dot{Q}$, however, $\dot{Q}$ increases rapidly with work or temperature due to self-adjustable effect and $\dot{Q} \sim T^{5-6}$ in lower temperature. Once the state $\dot{F}=\dot{Q}$ is achieved the ignition is coming, the thermonuclear reaction begins to carry on self-sustainly.

We arrive from Eq.(2) that the ignition state just is a state of the minimum entropy, $\frac{dS}{dt}=0$,in the fuel subsystem, because the system experiences the evolution from $\dot{Q}<\dot{F}$ to $\dot{Q}>\dot{F}$

Next we proceed to discuss the ignition condition. According to $\dot{Q}=\dot{F}$ we have

$$0.92\times10^4 \xi n_T^2 Q_0 \frac{\rho^2 R_b}{\beta_0} = \frac{T^4}{<\sigma\upsilon>} \tag{9}$$

approximately to write

$$-K\frac{\partial T}{\partial R}\Big|_{Rb} \approx \beta_0 \sigma_0 T^4 \tag{10}$$

where ξ= the ratio of particle number for deuterium and tritium.

n_T= tritium number per gram, in $10^{24}/g$ unit

$<\sigma\upsilon>$ is the DT reaction rate, in $10^{-24}cm^3/\mu s$ unit. Q_0 is the effective α particle energy including the neutron deposited energy in Mev unit. ρ in g/cm^3, R in cm, time in μs. In Eq.(9) and below, we shall also adopt the following units: T in $10^6\ {}^0K$, energy in 10^{12} erg, thus Stefan-Boltzman constant σ_0 =58 in our units. Unless otherwise state, hereafter the units for these quantities are omitted.

$\beta_0 \approx 5\%$ in a factor of 2 is an adjustable parameter determined by the numerical simulation of the pellet model, and depend on T insensitively. The curve $\rho^2 R_b \sim \frac{T^4}{<\sigma\upsilon>}$ varying with T is plotted in Fig.1. We can see from Fig.1 that there is a minimum point of the curve at T about 35, as $<\sigma\upsilon> \sim T^n$,$n>4$ in lower temperature and n<4 in higher temperature. Usually to take ignition temperature $T_{ig} \approx 15-25$, and

$10^{-4}T^{3/2} << \rho R_b < 1$, $\rho \approx 20-25$ perhaps is reasonable, so that the ignition state is in lower region of the curve.

3. Volume burn

In order to prove the self-sustaining burn after the ignition we discuss the unstable burn. For convenience, we assume that the compression ratio σ is fixed, and expand the entropy at ignition temperature point

$$S(T,\sigma)=S(T_{ig},\sigma)+S'(T_{ig},\sigma)\,\Delta T+\ldots\ldots \quad (11)$$

Using Eq.(7) and expanding $\dot{Q}$ and $\dot{F}$ at T_{ig}, we have the linearized equation

$$\frac{d\Delta T}{dt}=\frac{T_{ig}}{MCv}[(\frac{\dot{Q}}{T})'_{Tig}-(\frac{\dot{F}}{T})'_{Tig}]\Delta T \quad (12)$$

Bearing $\dot{Q}\sim T^{5-6}$ and $K\sim T^{4-5}$ in mind, even if $\frac{d\sigma}{dt}=0$, one obtained at ignition point

$$\frac{d\Delta T}{dt}>0 \quad (13)$$

It is verified that the thermonuclear reaction can be going on self-sustainly if the system is ignited. With temperature increasing and fuel depleting, ΔS changes from positive to negative due to ΔJ_s growing to be great than ΔQ_S, thus the second critical point in LTE is reached in this process, and the temperature is in the maximum $T\max \approx 120-150$, it can be obtained using $\dot{Q}=\dot{F}$ again and also can clearly be seen from the curve in Fig.1.

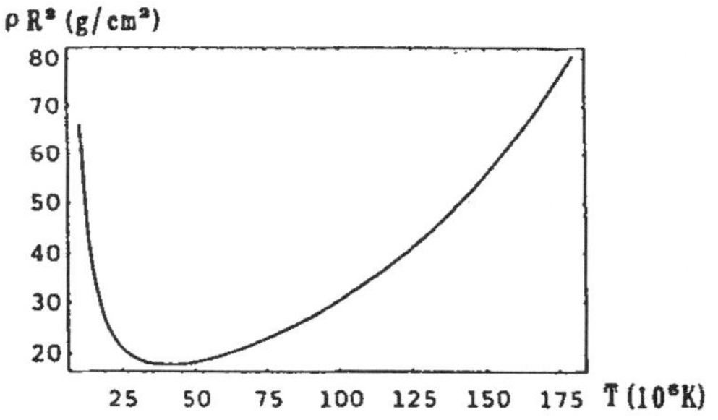

Fig. 1 Area density ρR^2 varys with temperature T for $\zeta=1$, $Q_0=3.5$Mev and $\beta_0=3\%$

III. VOLUME IGNITION AND BURN IN NO-LTE

Before T_{max} is coming, in fact, the violent burn in LTE would be bifurcated into the non-LTE state, when the emission rate of the spontaneous bremsstralung is equal to the deposited energy rate from thermonuclear reaction. Here in our time scale by the non-LTE means that the temperatures for electrons and ions are considerably separated, and X-rays are in non-Planck dynamic distribution .

The evolution of the system is determined by competition between various processes, we can write the governing equations in the thermonuclear reaction system as follows.

Energy equation for electron(e)

$$Cve\frac{dTe}{dt}+Pe\frac{d\frac{1}{\rho}}{dt}=\frac{1}{\rho R^2}\frac{\partial}{\partial R}(R^2 Ke\frac{\partial Te}{\partial R})+\frac{\rho\gamma_{ei}}{Te^{3/2}}(Ti-Te)-W_R+\beta W_T \qquad (14)$$

Energy equation for ion(i)

$$Cvi\frac{dTi}{dt}+Pi\frac{d\frac{1}{\rho}}{dt}=\frac{1}{\rho R^2}\frac{\partial}{\partial R}(R^2 Ki\frac{\partial Ti}{\partial R})-\frac{\rho\gamma_{ei}}{Te^{3/2}}(Ti-Te)+(1-\beta)W_T+Wn \qquad (15)$$

where $C_{v\alpha}, P_\alpha$ and K_α are the specific heat, the pressure, the conductivity for α = e,i respectively, $\frac{\gamma_{ei}}{Te^{3/2}}$ is a coefficient related to e - i coulomb collision, W_R is the dissipative rate of unit mass for the bremsstralung and compton energy, W_T and W_n are the heat rate of unit mass for charged particles and neutrons respectively, and β is the fraction of charged particle transferring energy to electrons.

To simplify investigation, as an approximation, without loss of generality, we have to assume that X-rays are approximately a Planck distribution in an equivalent temperature Tr. This approximation has been discussed in detail.[4] Therefore, we have equation for T_r as follows

$$\frac{4aT_r^3}{\rho}\frac{dTr}{dt}+\frac{4}{3}aT_r^4\frac{\partial}{\partial t}(\frac{1}{\rho})=\frac{1}{\rho R^2}\frac{\partial}{\partial R^2}(R^2 Kr\frac{\partial Tr}{\partial R})+W_R \qquad (16)$$

Eqs.(14), (15) and (16) coupled other equations for fluid mechanics, fuel depletion and neutron transport are our starting point to discuss the evolution of the fusion system.

The detailed expressions for $W_t, W_r, W_n, \gamma_{ei}, K_\alpha\ (\alpha=e,i,r)$ etc. in Eqs.(14)-16) will be given in another paper.

1. Temperature threshold to begin non-LTE burn

Eqs.(14)-(15) and (16) have at least two kinds of critical points. For the first one, ignoring X-rays and heat conduction, one obtained the threshold expression for temperature to ignite the non-LTE burn from Eqs.(14)- (15)

$$A_b=(1+\xi)^2\sqrt{T}=As\,\xi<\sigma\upsilon>Q_0$$

where $T_e \approx T_i \approx T$ due to the dominate coulomb collision and the temperature separation just to begin.

Taking ξ =1, we found the threshold temperature for the non-LTE ignition as follows:

For DT, Tc≈ 35 - 55, corresponding to $Q_0=17.6-3.5$,

For DD, Tc ≈336 -522, corresponding to $Q_0=7.3-2.5$,

It is shown that the threshold Tc for DD is too high to provide it usually. On the other hand, we observe that Tc for DT, which is much lower than that for DD, is proportional to about $\xi^{1/3.5}$ and ξ>>1 because $<\sigma\upsilon>\sim T_i^4$ for 25<Ti<100. Therefor, if a few tritium are mixed up in deuterium, we would achieve the purpose to burn deuterium which are plenty on the earth.

2. Volume burn in non-LTE

What is going on once to achieve the threshold temperature? We commence to discuss the evolution of the system.

a. Unstable excitation of non-LTE burn

To discuss the instability at the first critical point, neglecting the term of heat conduction and work in Eqs.(14)-(15) and taking Tr=0, we have the linerized equations in the form $\delta T_\alpha = T_\alpha - T_c, \alpha = e,i$. From the characteristic equation, it is easily to prove that the solution is a saddle point, therefore, there exists the growth behavior at $T_i = T_e = T_c$. The detailed discussion will be given otherwise.

Once this state is achieved, any small perturbation can lead to temperature rapid raise and separation, and to vigorous burn. Because the spontaneous bremsstrahlung emission on an average is to loss electron energy, and β decreases with Te increasing, as a result, the temperature separation evolves always toward $T_i > T_e$ direction. The unstability is clearly due to $<\sigma\upsilon>\sim T_i^{4.5}$, but spontaneous bremsstrahlung $\sim\sqrt{Te}$.

b. Transient stationary state

The system rapidly evolves with vigorous temperature growth and separation, $T_i > T_e$, and eventually it will be restrained and stopped by dissipative processes including fluid mechanics expansion and inverse compton scattering, which depend on pellet feature. Then the second critical point for non-LTE, which corresponds to the maximum temperature for ions, is coming.

From Eqs.(14) - (16), the maximum ion temperature Tim and simultaneous electron temperature Tem varying with parameters are given in Table I and II. where the local deposition for α-particles with 3.5Mev has been assumed. For high enough T_e, the range of α-particles, which is $10^{-4}T_e^{3/2}$, will be great than size R, $T_{\alpha m}(\alpha = e,i)$ will be reduced to a certain extent, however, the shell in high ρ and ρR will also prevent the α -particle escape from fuel system.

Table I. for the optically thin system, T_r=0, Q_0=3.5

ξ	1.0		9.9	
ρR	T_{im}	T_{em}	T_{im}	T_{em}
0.5	1581	758	1119	666
1.0	2300	1060	1654	940
3.0	4250	1825	3029	1620

Table II. for the optically thick system, ξ = 1.0

T_r	80		100	
Q_0	3.5	7.0	3.5	7.0
T_{im}	1930	4165	1050	2190
T_{em}	748	1045	453	619

We can see from Table II that $T_{\alpha m}$ depends on T_r sensitively, where T_r, indeed , is determined by radiation from inner wall of shell.

c. Steady burn stage

When the energy dissipation rate exceeds the heat rate, Ti begins to descend and the temperature separation is reduced. However, for a fuel system in which the size R is great than that of 14.1 Mev neutron free path the heating rate from neutron slow-down in transport process will approximately be steady in a slow-down duration, the fuel will further be burnt.

d. Burn decay

After steady burn, the system tends to a quench, and eventually, the LTE state will be formed again.

In the case of high compression, for the high Z ablator if the fuel system size $R > C_S\,\tau$, where Cs is a sound speed and τ is duration of thermonuclear reaction, the quench is mainly determined by compton dissipation, whereas in the opposite case, the quench process is essentially governed by the fluid expansion.

If neglecting the electron heat conduction, it is easy to find from Eq.(14) the compton scattering is dominant in the following inequality

$$Tr \geq [\frac{3\Gamma_e u}{A_c^{(0)} n_e}(\frac{\rho}{a_0})^{1/3}]^{1/4} \quad \text{, for } Te >> Tr \tag{17}$$

where the bremsstrahlung process has been neglected and $p_e \dfrac{d\frac{1}{\rho}}{dt} \approx 3\Gamma e T_e a^{-1/3} \rho^{1/3} u$

due to $\rho R^3 = \dfrac{3}{4\pi} M_{DT} = a_0$. For $\xi=1, M_{DT}=5\times10^{-3}g, \rho=30, u=20$, one obtained Tr≥33.

We now discuss the burn decay passed through Tim point. Neglecting bremsstrahlung process and heat conduction for electrons and ions, and regarding Tr as a parameter for convenience, after tedious operation we arrive the solution of the linearized equation for $\delta T_\alpha, \alpha = i,e$, at (T_{im}, T_{em}) point,

where T_{im}=1390, T_{em}=748, W_K=0, we concluded that at T_{im} point there are two negative roots that correspond to node point. It shows that burn decay takes place in the DT system. The detailed discussion will be shown otherwise.

Decay time τ_d for Te is a pproximately

$$\tau_d(\mu s) \approx \min\{\frac{Cve}{A_c n_e T_r^4}, \frac{Cve}{3\Gamma_e u(\frac{\rho}{a_0})^{1/3}}\}$$

for $C_{ve}=33$, $Tr=50$, $u=20$, $\rho=30$, $\tau_d(compton) \approx 0.1ns$

IV. NUMERICAL RESULTS AND DISCUSSIONS

In order to illustrate the physical processes of volume ignition and burn, a pellet is roughly designed to simulate the behavior. The DT fuel system is analylogus to the CSI model with a large cavity where the initial density $\rho_0 = 0.01 g/cm^3$ and initial radius $R_0 = 5000 \mu m$, and a main fuel layer where $\rho_0 = 0.15 g/cm^3$ and thickness $\Delta R = 80 \mu m$. The pellet shell is a sandwich with Au, Al, C_8H_8 and Au from inner to outer, where Al as an ablator is to enhance the imploding pressure in inner shell Au, the radiation energy 12 MJ with duration 20 ns and in Gaussian distribution is put to the C_8H_8 layer where the deposited energy from ion beams or laser beams is simulated.

The numerical simulation is performed in the case of no mixing, first of all, the three-temperature code CFJ-3T is used to compute the LTE evolution of the system and to investigate the LTE ignition, then, the code RH-5 in which the x-rays dynamics distribution satisfied F-P equation, the temperature separation for electrons and ions, fluid mechanics, fuel depletion, neutron transport etc. are involved is switched on to compute the non-LTE evolution.

The results show that at the ignition instantaneity the fuel sphere is clearly seen in two region structure yet, the LTE ignition is carried out in main fuel region where $T_{ig} \approx 20, \rho \approx 18, \rho\Delta R \approx 0.30$, however, for the central region, $T \approx 100$, $\rho \approx 4.2$, $\rho\Delta R \approx 0.12$ where $\rho\Delta R$ does not satisfy the non-LTE burn condition, in addition, the "cold" gold inner shell where $\rho \approx 300$ and $\rho\Delta R \approx 3$, plays an important role to preserve temperature in fusion fuel and as a inertial layer. The non-LTE threshold $T_i \approx 55$ is also observed at time $t \approx 0.29ns$ that the zero point is taken in ignition instantaneity.

The evolution and competition among thermonuclear heating and energy loss due to the dissipative processes are given in Fig. 2. With the lapse of time, the heating rate W_T that depends on $<\sigma v>$ rapidly increases, and then decreases because $<\sigma v>$ is of a maximum at T_i about 700. Meanwhile, the dissipative processes including W_R for inverse Compton scattering and bremsstrahlung emission, W_{ei} for coulomb collision of electrons and ions, and W_K for fluid expansion, also increase a little later. The transient stationary balance is observed at $t \approx 0.56ns$ as shown in Fig.2, and the maximum ion temperature in outer fuel is $T_{im} \approx 1770$, where the electron temperature $T_e \approx 450$, and radiation temperature $T_r \approx 70$, the time dependence of temperatures is plotted in Fig. 3, in which the local deposition of α-particles with 3.5Mev is assumed. At that time, $\rho R \approx 0.6$ for DT system. The heating rate from neutron transport is unimportant, the quenching process from Fig.2, indeed, is mainly determined by W_R in which later the main contribution is from inverse Compton scattering in our model as shown in computation because the gold ablator is in high density and optically thick, therefore, large amount of X-rays from the spontaneous bremsstrahlung emission of high T_e fuel are absorbed in a thin layer of inner gold shell, and then are coverted into lower energy X-rays and reemitted into the DT fuel system to increase the inverse Compton scattering and rapidly decrease T_e and thus T_i through e-i collision.

The energy rate and energy yield from thermonuclear reaction are plotted in Fig.4 and Fig.5. It is shown that the energy yield in final state is 857 MJ, thus, the gain is G=71 and the fuel depletion is $\eta_b = \xi \dfrac{\rho R}{\rho R + 6}$ with $\xi > 1$ due to the confinement effect of the high compressed A_u shell.

In the above numerical demonstration , the pellet shell is imagined as a an ablator for radiative ablation drive, it makes drive energy utilization to be not economic. A new pellet design for the VI model which may considerably reduce the requirement for drive energy will be discussed otherwise.

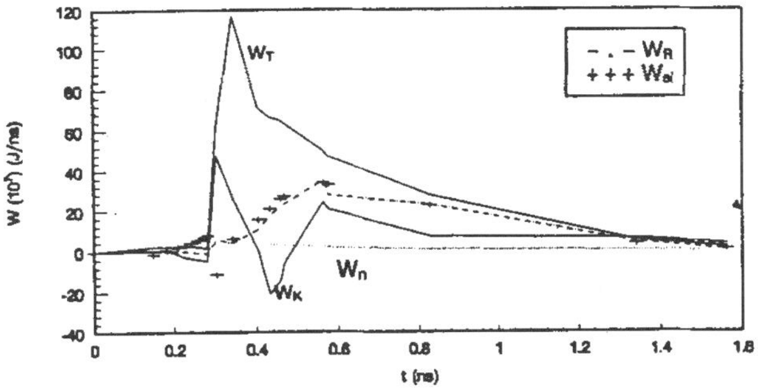

Fig. 2 Various process competitions with time

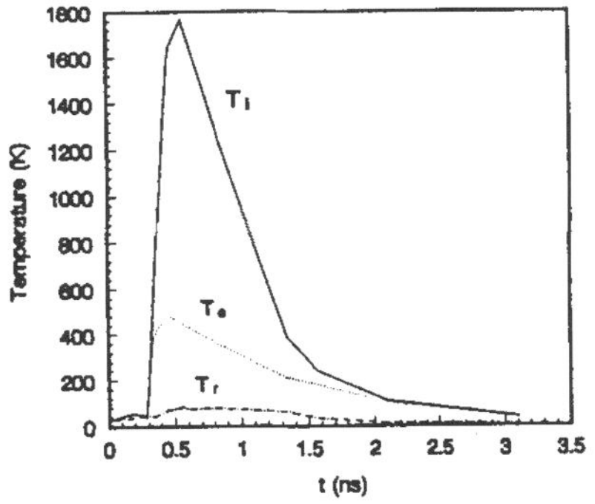

Fig. 3 Temperature T_α (α = i, e, γ) varys with time

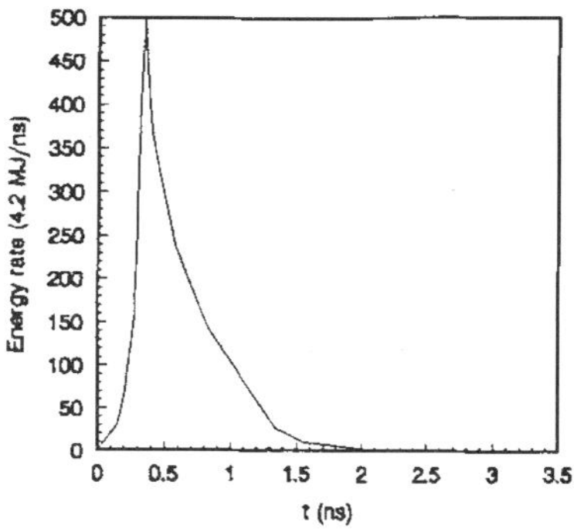

Fig. 4 Thermonuclear energy rate varys with time

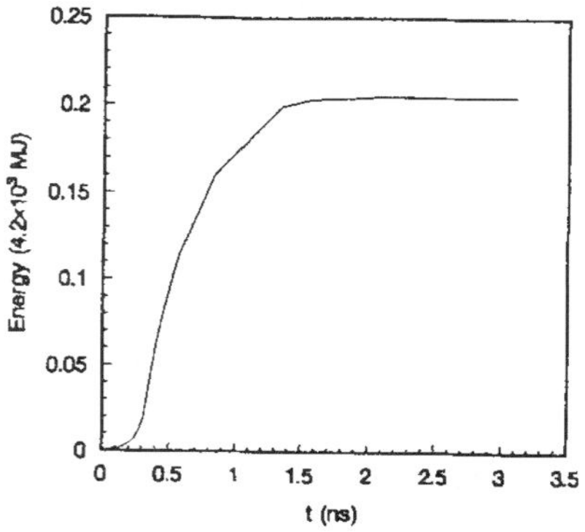

Fig. 5 Thermonuclear energy varys with time

Acknowledgment

Thanks Dr. Gao yaoming for typing manuscript. This work was performed under the National Inertial Confinement Fusion Project.

References

1. E.Storm, J.D.Lindl, E.M.Campbell et al., Progress in Laboratory High Gain ICF, Prospects for the future, the Paper on the Eighth Session of International Seminars On Nuclear War, Erice Sicily, Italy(1988)

2. T.Q.Chang, X.T.He and M.Yu, Laser Plasma Interaction and Related Plasma Phenomena 9, 565(1991)

3. H.Hora and P.S.Ray, Z.Naturforsch 33A, 890(1978);
H.Hora, Z.Naturforsch 42A, 1293(1987)

4. He Xiantu (X.T.He), Shen Naiwan, Liu Jun and Zhang Tianshu, Kexue TongBao (A monthly Journal of Science) 29, 1460(1984)

第7卷第3期 强激光与粒子束 Vol.7,No.3
1995年8月 HIGH POWER LASER AND PARTICLE BEAMS Aug, 1995

高增益靶体点火的动力学过程分析

李沄生 高耀明 贺贤土 于 敏

(北京应用物理与计算数学研究所，北京8009信箱，100088)

摘 要 针对中心热斑点火和热核传播燃烧（简称中心点火）的高增益靶设计理论要求很高和技术上的难度很大，我们通过阻热层的设计，在高增益靶的整体点火和燃烧（简称体点火）设计上首先实现挡光系统的非稳定温度脱离，由平衡点火和燃烧逐步过渡到非平衡点火和燃烧，从而将驱动能量降到15MJ左右，实现了体点火和体燃烧，并达到了可观的增益。

关键词 体点火 高增益靶 热核点火 热核燃烧

ABSTRACT A good heat-resistance layer and a good inert layer as the shell of DT are used to make volume ignition and volume burn (VIVB) realistic. An unstable non-LTE state and the transition from LTE ignition and burn to non-LTE ignition and burn is discussed, and in this way the volume ignition and volume burn design is achieved and high gain fusion is obtaind.

KEY WORDS volume ignition, hign gain, thermonuclear ignition, thermonuclear burn

0 引 言

E.Storm等人阐述中心点火高增益靶设计的物理和技术难点[1]主要是等熵压缩和脉冲调制；界面不稳定性和混合；高度的均匀性和对称性要求；高能离子对靶球的破坏。中心点火的优势是可以将点火能量降低。1978年，H.Hora等人提出了体点火的设想[2,3]，贺贤土等人证明了通过氘氚在平衡态下整体点火和燃烧，在一定条件下过渡到整体非平衡点火和燃烧的必然性，从而证明了体点火在科学上的合理性和可行性[4]。

体点火可以消除中心点火几乎所有的难题。合并了点火区和燃烧区，消除冷热区之间的混合问题；只要求靶球收缩比约为15，无需严格的波形调制和等熵压缩，也没有长脉冲带来的高能离子的影响。最后，由于不存在很小的点火热斑，对驱动过程的均匀性和对称性的要求也可降低。体点火的难点是点火能量需要增加。

体点火如果考虑直接非平衡点火，则需要的点火能量约是中心点火能量的10倍左右。本文通过阻热层的设计，找到了在挡光氘氚系统中，首先实现平衡点火，经由非稳定温度脱离情况下的燃烧，再逐步过渡到非平衡点火和燃烧的技术途径，可以把5mg的氘氚的点火能量降低到1.5MJ以下。

1 热核反应动力学过程分析

热动平衡条件下的能量平衡方程

1994年9月21日收到原稿，1994年11月15日收到修改稿。
国家自然科学基金资助项目（19377204）。

$$\frac{dE}{dt}=W_T-F-W \tag{1}$$

讨论氘氚系统的温度 (10^6K)、密度 (g/cm^3) 分布为均匀的情况，E (MJ) 是氘氚区的内能，W_T、F 和 W (MJ/ns) 分别是单位时间内氘氚区热核反应能、氘氚边界辐射能流和压力作功。

$$W_T=A_s\rho\xi n_T^2<\sigma v>Q_0 \tag{2}$$

$$F=-\frac{4\pi R^2}{m_{DT}}K\frac{\partial T}{\partial R}\Big|_{R=R_D}\approx\frac{3}{\rho R}\beta_0(T,\rho')\sigma_0 T^4 \tag{3}$$

$$W=\frac{3}{\rho R}pu \tag{4}$$

$[1-\beta_0(T,\rho')]$ 是氘氚外壳阻热层的等效辐射反射率，ρ' 是金的密度，对于金外壳，可以得到拟合公式

$$\beta_0=5.39\rho'^{-0.587}T^{-0.527} \tag{5}$$

p 和 u 的估计比较复杂，与驱动方式和靶结构有关。这里简单取 $p=\Gamma\rho'T$ (ρ'、T 分别是驱动层的密度和温度)，推进层的最大速度 $u^{max}\approx40$。

非热动平衡的三温能量平衡方程

$$\begin{cases}\frac{dE_i}{dt}=W_{T_i}-W_i-F_i-\omega_{ei}\\ \frac{dE_e}{dt}=W_{T_e}-W_e-F_e+\omega_{ei}-\omega_R\\ \frac{dE_r}{dt}=W_{T_r}-W_r-F_r+\omega_R\end{cases} \tag{6}$$

e、i、r 分别表示电子、离子、光子各分量。各物理量的详细表达式可见文献[4]。电子和光子之间的能量交换率

$$\omega_R=\omega_b+\omega_c \tag{7}$$

$$\begin{cases}\omega_b=A_b\rho(1+\xi)\frac{n_e n_T}{T_e^{1/2}}\{T_e-T_r[\Psi(1+\frac{T_r}{T_e})-\Psi(1)]\}\\ \omega_c=A_c n_e T_r^4(T_e-T_r)\end{cases} \tag{8}$$

ω_b 是电子的轫致能耗，ω_c 是电子的康普顿能耗，Ψ 是 Psi 函数[5]。不难看出，将 (6) 中的三式相加即得 (1) 式。

由 (1) 式可得热动平衡时压缩升温条件是

$$-W-F\geq0 \tag{9}$$

在压缩阶段，推进层向氘氚区推进作功，温度较低时，辐射流 F 很小，氘氚区的温度将不断升高。随着温度升高，氘氚区向推进层的辐射流损耗 F 也不断增大；最后将导致 (9) 式的反转。当 (9) 式取等号时，系统将达到平衡压缩升温的最高温度

$$T=\left(\frac{\Gamma\rho u^{max}}{\beta_0\sigma_0}\right)^{1/3} \tag{10}$$

对于 $u^{max}\approx40$，可得系统由于压缩可达到的最高温度随 ρR 的变化，这时有 $T=\left(\frac{46.4\rho}{\beta_0}\right)^{1/3}$，见表1。

表 1　氘氚系统压缩过程的最大温度

Table 1　Maximum temperatures in process of compressing DT

m_{DT}	ρR	0.5	1.0	1.5	2.0	3.0
5mg	T	34.0	51.8	66.2	78.9	100.8
	T'	10.2	14.4	17.7	20.4	25.0
1mg	T	24.6	37.4	47.8	56.9	72.8
	T'	7.8	11.9	13.5	15.6	19.2

表中 T'是氘氚透明系统（$\beta_0=1.0$ 时）所能达到的最高温度，T 是考虑阻热层后的系统所达到的最高温度，T 比 T'高得多，因此，在体点火的设计中，阻热层是必需的，也是至关重要的。

对于热核系统的点火条件，首先给出直接非平衡点火的阈值条件。所谓直接非平衡点火，是指在透明系统中，辐射温度近似地可以认为 $T_r=0$ 时，氘氚系统的离子电子温度达到阈值而点火。由（6）式的头两式相加，忽略电子离子作功和热传导时，系统电子离子的升温条件是

$$W_T-\omega_R\geqslant 0 \tag{11}$$

由于 $T_r=0$，有 $\omega_c=0$，以及

$$\omega_b=A_b\rho\ (1+\xi)n_e n_T T_e^{1/2} \tag{12}$$

（11）式取等号可得直接非平衡点火条件（$\xi=1$，$\varepsilon=1$，$T_r=0$）

$$T_c=48.79 \tag{13}$$

若能创造条件，使氘氚系统的温度升到 T_c 以上，氘氚系统便能实现自持燃烧。问题是要将 5mg 氘氚（这是高增益靶所要求的装量，考虑 30% 的燃耗，烧掉 1.5mg 的氘氚，可获得约 500MJ 的能量输出）加热到 T_c 需要 50MJ 的驱动能量。

由方程（1）可得在热动平衡条件下氘氚系统的点火条件是

$$W_T-F\geqslant 0 \tag{14}$$

此即

$$\frac{\rho R}{3}\ W_T-\beta_0\sigma_0 T^4\geqslant 0 \tag{15}$$

由（15）式可见，在考虑平衡点火条件时，是使热核反应能超过辐射能流损耗，而不是象在非平衡点火条件中那样，是使热核反应能超过轫致能耗。平衡点火阈值温度依赖于点火时刻系统的 ρR 和 β_0，在能源给定的情况下，可以通过改变推进层的厚度在一定程度上调整氘氚区的温度和密度；在体点火设计中，由于有较好的惰层，系统要求的ρR 值可以比中心点火设计中低很多，取 1.0 左右就足够了。为了使 α 粒子不致大量漏失，$\rho R>0.3$ 是必需的。

在压缩阶段，系统是靠外界作功而升高温度的；W_T 很小，这时有 $W_T-F<0$。随着系统温度升高，W_T 越来越大，当满足条件（14）时，系统靠本身的热核反应放能就能使温度升高，这时系统便实现了点火而进入自持热核反应阶段。

由（14）式可以求出系统在不同的 ρR 值时的点火阈值温度或点火阈值能量。

定义

$$W_T'=\frac{W_T}{\rho}$$

则 W_T' 是一个与 ρ 无关，只与 T 有关的量。由（15）式得

$$\rho^2 R \geqslant \frac{3\beta_0(T, \rho')\sigma_0 T^4}{W_T'}$$

对于 $\beta_0(T, \rho') = 1$ 的透明系统，当 $T = 40.3$ 时，不难求出

$$\left(\frac{3\sigma_0 T^4}{W_T'}\right)_{\min} \approx 1640 \tag{16}$$

由（15）式可得，对于 $m_{DT} = 5\text{mg}$ 的氘氚透明系统，有 $(\rho R)_{\min} = 5.03$；对于 $m_{DT} = 1\text{mg}$ 的氘氚透明系统，有 $(\rho R)_{\min} = 3.64$。可见氘氚透明系统的平衡点火阈值太高，很难达到。

考虑阻热层后，可求出系统相应的 ρR 的最小值：在用金作阻热层材料，取 $\rho' = 350$ 时，对于 $m_{DT} = 5\text{mg}$ 的氘氚系统，有 $(\rho R)_{\min} = 1.09$；对于 $m_{DT} = 1\text{mg}$ 的氘氚系统，有 $(\rho R)_{\min} = 0.79$。若取 $\rho' = 1350$，则对于 $m_{DT} = 5\text{mg}$ 的氘氚系统，有 $(\rho R)_{\min} = 0.798$；对于 $m_{DT} = 1\text{mg}$ 的氘氚系统，有 $(\rho R)_{\min} = 0.579$。可见对于 $\rho R = 1.0$ 的氘氚系统，要实现平衡点火是有可能的；对具有更小的 ρR 值的氘氚系统，实现平衡点火就比较困难了。对不同的 ρR 值的平衡点火阈值温度和相应的点火能量见表 2。

表 2 氘氚系统的平衡点火阈值温度和点火能量

Table 1 Thresholds of temperature and energy for LTE ignition

m_{DT}	ρR	1.10	1.50	1.80	2.00	3.00
5mg	T_i	53.69	20.70	16.01	14.15	9.66
	E_{ig}/MJ	2.68	1.04	0.801	0.708	0.483
1mg	T_i	51.22	23.97	13.46	10.35	7.59
	E_{ig}/MJ	0.512	0.240	0.136	0.104	0.076

以氘氚装量为 5mg 为例，由表1，对于 $\rho R = 1.5$ 的情况，作功加热可使系统达到 47.8 百万度，而从表 2 可知，这时系统要求的点火温度是 20.70 百万度；因此，这种系统的平衡点火是可能的。

实际上，氘氚系统在达到非平衡点火温度阈值之前，由于系统本身的热核放能，必然存在电子、光子的温度脱离现象（离子和电子温度是相等的）。由于系统存在温度脱离现象，便可降低对系统 ρR 的要求。

在微聚变领域内，由于驱动能量的限制，ρR 一般都不大；为了实现平衡燃烧向非平衡点火过渡，还必须考虑在温度低于非平衡点火阈值温度时就存在的温度脱离（即 $T_i \approx T_e > T_r$）的现象。由于这种温度脱离有赖于辐射温度，所以称之为不稳定温度脱离。

由于在点火之前，系统的温度较低，可进一步忽略掉作功项和离子、电子流项，可得系统升温条件

$$\begin{cases} W_T - \omega_R \geqslant 0 \\ -F_r + \omega_R \geqslant 0 \end{cases} \tag{17}$$

取等号便可求得系统的点火温度阈值。由（17）的第一式可得 T_e（或 T_i）与 T_r 的拟合公式

$$\begin{cases} T_r = T_e & T_e \leqslant 20 \\ T_r = -0.0002125\,T_e^3 + 0.00448\,T_e^2 + 0.9841\,T_e & 20 < T_e \leqslant 30 \\ T_w = -0.000931\,T_e^3 + 0.037\,T_e^2 + 0.6583\,T_e & 30 < T_e \leqslant 50 \end{cases} \tag{18}$$

由 (17)式的第二式可得点火温度和点火能量 (见表 3)。

表 3　氘氚系统三温状态下的点火阈值温度和点火能量 ($\rho' = 350$)

Table 3　Thresholds of temperature and energy for non－LTE ignition

m_{DT}	ρR	0.5	1.0	1.5	2.0	3.0
5mg	T_i	42.35	32.73	20.16	14.15	9.66
	E_{ig}/MJ	2.12	1.64	1.01	0.71	0.48
1mg	T_i	38.84	22.58	13.64	10.35	7.59
	E_{ig}/MJ	0.388	0.226	0.136	0.104	0.0759

由表 3 和表 2 的比较可见，非稳定温度脱离是一个很重要的效应。与平衡点火的差别在于 (8)式中的 Ψ 函数项对轫致辐射能耗的减弱作用。其实质就是低频光子对轫致辐射的抑制作用。

如前所述，对于热动平衡情况，当 $\rho R < 1.0$ 时，实现平衡点火是困难的。这主要是因为辐射温度太高，辐射能流损失太大。在考虑了温度脱离效应后，由于辐射能流损失取决于辐射温度，而能源取决于离子温度 (与电子温度相等)，故即使系统的 ρR 低到 0.5，也存在点火温度阈值，只不过已经非常接近透明系统的非平衡点火阈值温度了。由表 3 可见，提高氘氚系统的ρR 值，对点火是很重要的。

由上述分析不难看出，氘氚球外壳起着推进层、惰层和阻热层的三重作用，在利用激光烧蚀或辐射烧蚀的设计中，还应考虑它的烧蚀层的作用，或称第四个作用，所以，如何设计好氘氚球外壳是非常关键的。

要提高外壳的阻热作用，就必须选取高 Z 材料，并提高阻热层的密度，从而减小 β_0 的值；要提高外壳的惰层作用，除提高惰层的密度外，还要增加惰层的质量；提高外壳的推进层作用，也就是提高推进速度，除提高压力外，可增加腔体外壳与推进层的质量比和增大形状因子。但是，提高压力要以提高驱动能量为代价，增大形状因子要以增加界面不稳定性和混合为代价。如何平衡得失，选取一组恰当的值，正是设计的难点。将来还必须结合实验加以解决。

我们设计的目的是要尽可能地降低系统的点火阈值温度，为此就要选取好的阻热层和提高氘氚的 ρR 值。

对于热核燃烧条件，热动平衡条件氘氚裸球的燃耗是

$$f = \frac{\rho R}{\rho R + \theta} \tag{19}$$

其中

$$\theta = \frac{4\sqrt{\Gamma T}}{\xi n_T^{(0)} \langle \sigma v \rangle} \tag{20}$$

有惰层时燃耗公式为

$$f = \frac{\zeta}{\zeta + \theta} \tag{21}$$

其中 ζ 定义如下

$$\zeta=\rho R+\eta\rho'\Delta R \tag{22}$$

η 与惰层的密度 ρ 和面密度 ρR 有关，对密度的关系比较弱，这里主要给出对面密度 ρR 的依赖关系。由数值模拟可得拟合公式

$$\eta=2.127\,(\rho'\Delta R)^{-0.4873} \tag{23}$$

对于 $\rho R=1.0$, $\rho'\Delta R=2.5$, 有 $\eta=1.361$, 可求得 $\zeta=4.4$。由表4可见，考虑惰层燃烧有较大的提高，但平衡燃耗即使考虑惰层也是很低的，要有可观的燃耗，必须创造条件，使系统由平衡燃烧过渡到非平衡燃烧。

对于非热动平衡下的燃烧条件，这时公式(18)照样适用，但由于温度大幅度脱离，$<\sigma v>$ 随离子温度增加而大幅度增加，θ 便小于平衡态下的值，使得氘氚的燃耗大幅度提高。

$$\theta=\frac{4\sqrt{\Gamma_e T_e+\Gamma_i T_i}}{\xi n_T^{(0)}<\sigma v>} \tag{24}$$

T_i 和 T_e 的关系可令(6)式中的第二式等于零求得拟合公式

$$\begin{cases} T_e=\dfrac{T_i}{-1.090\times10^{-5}T_i^2+0.0175T_i-0.0476} & 98.0\leqslant T_i<860.0 \\ T_e=0.1395T_i-7.238\times10^{-8}T_i^2 & 860.0\leqslant T_i<2000 \end{cases} \tag{25}$$

由此求得燃耗见表5。考虑非平衡温度脱离之后，由于氘氚系统的离子温度可以远远高于平衡温度（从数值模拟结果可以得到，氘氚系统的平均离子温度可达到800～1000百万度），加上惰层的作用，致使氘氚系统的燃耗可达到30%～40%。

由于我们的氘氚系统是保温系统，辐射温度一般比较高（约70百万度），康普顿能耗比较大，影响了氘氚系统的温度进一步上升，同时也缩短了系统的燃烧时间宽度，从而限制了燃烧的进一步提高。在考虑光子非平衡分布效应后，康普顿能耗将大幅度下降，可以进一步提高氘氚系统的燃烧，这一点已经超出了本文的讨论范围了。

表 4 氘氚系统的平衡燃耗 ($\rho R=1.0$)

Table 4 Fuel depletion of LTE for DT

T	20	50	80	100	120
$<\sigma v>$	0.143	7.78	38.9	75.6	124.0
θ	8483	246.5	62.4	35.9	24.0
f'	0.000118	0.00404	0.0158	0.0271	0.040
f	0.000519	0.0175	0.0659	0.109	0.155

表 5 氘氚系统的非平衡燃耗

Table 5 Fuel depletion of non-LTE for DT

T_i	100	150	200	500	780	1000	1500
$<\sigma v>$	75.6	210.3	362.7	779.6	827.2	814.2	746.7
θ	32.41	13.32	8.63	5.94	6.92	7.95	10.62
f'	0.030	0.0698	0.104	0.144	0.126	0.112	0.0861
f	0.143	0.288	0.385	0.476	0.438	0.404	0.337

f or f' is the fuel depletion with (without) inert layer

2 数值模拟结果

表 6 高增益靶模型（形状因子f_a=250）

Table 6 Model of high gain target

target	0 (DT)	0.500 (Au)	0.502 (LiH)	0.600 (Au) 0.620
ρ_0	0.01	19.13	0.7923	19.13
m /mg	5.24	120.7	297.0	1789.0

表 7 低增益点火靶模型（形状因子f_a= 200）

Table 7 Model of low gain target

target	0 (DT)	0.300 (Au)	0.3015 (LiH)	0.400 (Au) 0.620
ρ_0	0.01	19.13	0.7923	19.13
m/mg	1.13	32.62	121.4	394.3

由表 6、7 示出的高增益、低增益靶模型的力学压缩过程为：

	高增益靶	低增益靶		高增益靶	低增益靶
燃料收缩比 f_c	13.3	14.9	推进层最大平均密度 $\overline{\rho}'_{max}$	353.8	338.0
燃料最大平均密度 ρ_{DT}^{max}	23.74	33.05	推进层最小厚度 ΔR_{min}	0.00715	0.00915
燃料最小半径 R_{min}	0.03748	0.02014	推进层最大面密度 $(\rho'\Delta R)_{max}$	2.53	2.42
燃料最大面密度 $(\rho R)_{max}$	0.890	0.664	加源区的最大压强 $p_{max}/10^{11}$Pa	149	130
燃料最高温度 T^{max}	55.0	60.0	推进层的最大推进速度 u_f^{max}	25.7	20.0
推进层最大密度 ρ'_{max}	1932	1246	流体力学效率 η_h	13.0%	8.2%

靶模型的热核反应过程为：

输入的驱动能量 E_z/MJ	13.06	4.54	氘氚燃耗 f	33.6%	31.2%
点火温度 T_f	35.0	40.0	系统增益 G	45.34	26.1
中子产额 G/ 个	0.21×10^{21}	0.42×10^{20}	最高离子温度 T_i^{max}	1274	1040
热核反应放能 Y/MJ	529.2	118.44	最高光子温度 T_r^{max}	65	65

低增益模型和高增益模型的主要差别在于氘氚装量、驱动能量不同；从而导致各个物理量在数量上有差别。但有大致相同的物理图象和动作过程，如图 1 表示高增益和低增益模型氘氚区的温度随时间变化曲线。由图1可见，电子、离子温度只在热核反应激烈进行的很短时间内有1～2倍的脱离。电子、光子温度存在较长时间的脱离，由于存在较好的阻热层，辐射温度较高，脱离度也不大。

图 2 表示高、低增益模型的热核反应放能率（B），热传导能流率（F）和推进层向氘氚区的作功（W）随时间的变化曲线。由图2可见，高增益模型的放能率更高，热核燃烧的时间宽度更大，热核点火的时间更晚；同时可以看出，热核熄火的主要因素是流体力学膨胀（在 W 曲线的峭壁处熄火）。

图 3 表示高、低增益模型 B、W、F 三条离子温度的变化曲线。图中 B、F 曲线的交点就是系统的点火点。由图可见，高增益模型的点火温度约为 35，低增益模型约为40。W、F 曲线的交点近似表示系统压缩所能达到的最高温度（由于这时热核反应放能很少，对系统的温度影响不大）。由图可见，高增益模型能达到 55 左右，低增益模型是 60 左右。

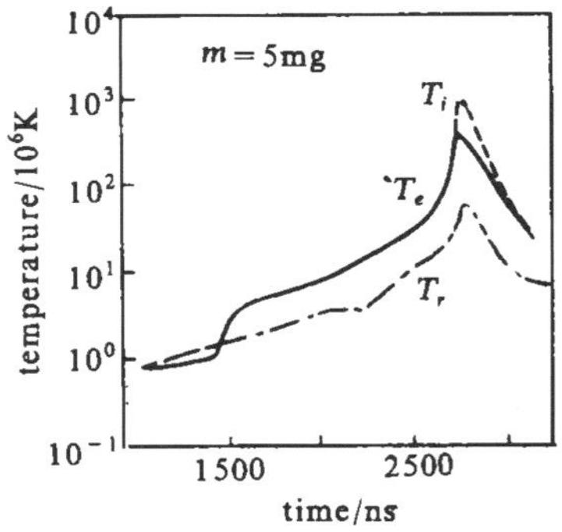

a. high gain target

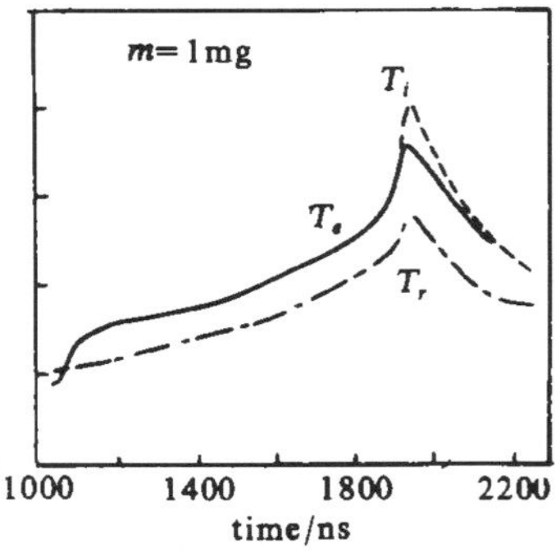

b. low gain target

Fig.1 Temperatures in DT *vs* time

图 1 氘氚区温度随时间变化曲线

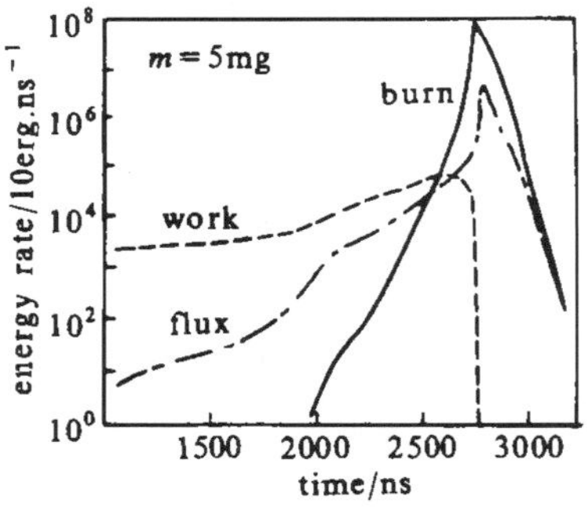

a. high gain target

Fig.2 *B F W vs* time

图 2 *B*、*F*、*W* 随时间变化曲线

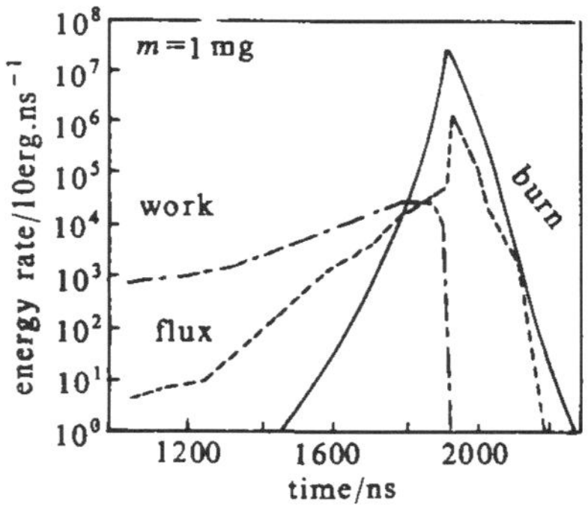

b. low gain target

Fig.2 *B F W vs* time

图 2 *B*、*F*、*W* 随时间变化曲线

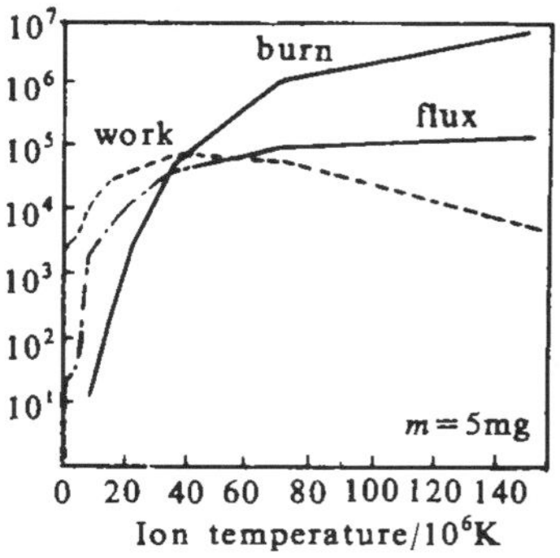

a. high gain target

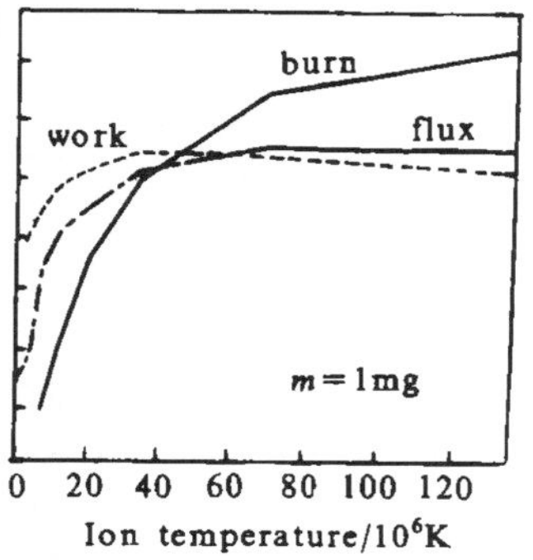

b. low gain rarget

Fig.3 *B*、*F*、*W vs* ion temperature

图 3 *B*、*F*、*W* 随离子温度变化曲线

3 结 论

目前只讨论了三温温度脱离的非平衡情况，尚未涉及光子非平衡分布效应的影响。只讨论了氘氚的压缩过程、点火过程和燃烧过程，尚未讨论输入能量的过程和驱动过程。数值模拟研究也是初步的，尚未进行模型的优化选择；只讨论了气体靶球，尚未讨论冷冻靶的情况。尽管如此，从本文的结果已不难得到：

体点火的原理及其采用的靶球结构，基本上避免了中心点火原理和靶球结构中的一系列困难问题；体点火原理的实现需要比较多的驱动能量，在采用了较好的阻热层后，可在比中心点火原理要求的驱动能量略多一些的条件下，实现挡光氘氚系统的点火和燃烧；模型中采用了 LiH 作驱动层材料，只能用离子束驱动。如用激光驱动，就只能用烧蚀增压原理，估计流体力学效率将降低，增益也会降低。离子束驱动器的效率可达 30%，就可适当降低对靶球增益的要求，这样约50 左右的增益将是合适的；在进一步考虑光子的非平衡效应后，氘氚燃耗将进一步提高。这一点将在今后的工作中予以阐述；作为一种可能的技术途径，我们不仅要研究中心点火原理，同时也应对体点火原理的研究引起充分的注意。

参考文献

1　Storm E, Lindl J D, Campbell E M . Progress in Laboratory High Gain ICF, Prospects for the Future, the Paper on the Eighth Session of International Seminars on Nuclear War, Erice Sicily, Italy, 1988

2　Hora H and Ray P S, *Z.Naturforsch*, 1978, **33A**: 890

3　Hora H, *Z.Naturforsch*, 1987, **42A**: 1293

4　X. T. HE, Y. S. Li and M. Yu .Physics of Volume Thermonuclear Ignition, and Burn for High－Gain ICF, Invited Paper at 11th International Workshop on Laser Interaction and Related Plasma Phenomena, Monterey, CA, 25－30, 1993

5　Jahnke Emde Losch. Tables of higher functions, Sixth edition, Revised by Friedrich Losch, Mcgranw－Hill book Company. inc., 1960, 11

THE KINETIC PROCESS ANALYSIS OF VOLUME THERMONUCLEAR IGNITION IN HIGH GAIN INERTIAL CONFINEMENT FUSION

Li Yunsheng, Gao Yaoming, He Xiantu, Yu Min

(*Institute of Applied Physics and Computational Mathematics, P.O.Box* 8009, *Beijing* 100088)

The central ignition and propagation burn (CIPB) is a great achievement for many year's research in LLNL, it is also the best theory of target design until now. Recent years the volume ignition and volume burn (VIVB) has begun to receive a lot attention and investigation because of the great difficulties in CIPB.

The direct non－local thermodynamic equilibrium (non－LTE) ignition (the threshold temperature is about 5keV) in VIVB demands too much ignition energy to practise it. In this paper a good heat－resistance layer and a good inert layer as the shell of DT are used to make VIVB realistic. An unstable non－LTE state and the transition from LTE ignition and burn to non－LTE ignition and burn is discussed, and in this way the driving energy can be lowered to 15MJ or so. The VIVB design is achieved and high gain fusion is obtained. In this paper, the threshold conditions for temperature－increading by compression, and for ignition and burn are given, Finally, a high gain and a low gain models for VIVB by computational simulation are given in this paper.

A Moderate-Gain ICF Model From LTE Ignition to non-LTE Burn

X.T.He ,Y.S.Li and M.Yu

Institute of Applied Physics and Computational Mathematics, P.O.Box 8009, Beijing 100088, China

Abstract A moderate - gain indirect-driven ICF model from LTE volume ignition to non-LTE burn is presented, a high compressed glass tamper, which consists of a capsule glass shell that has not been ablated, is required to enhance the DT burn. The numerical simulation shows that the energy yield of 18.1MJ from DT fuel of 0.2mg is achieved when a shaped laser beam of about 1.25MJ is invested in a hohlraum, a enhanced factor of 2.3 for the DT fuel burn up fraction is obtained.

1. INTRODUCTION

At present one concerned about ignition and moderate-gain(<20,say) burn in inertial confinement fusion (ICF) research on National Ignition Facility.

For an indirect-driven ICF, the investigation for above purpose has been shown (1),(2) more recently. The typical ablators of the capsules used on the above work are made up of *Br*-doped *CH* or *Cu*-doped *Be* , both them are in a same inner radius of 0.95mm and 0.16mm thickness. The deuterium-tritium (DT) ice layer neighboring ablator is of a mass of 0.2 mg, and DT gas with density 0.3mg/cm^3 is filled in a central region of the capsule. Laser energy of 1.6 MJ in a pulse of 17 ns is delivered into a hohlraum to generate a spatial uniform radiation filed with peak temperature about 3.5-4.0 $MK(MK \equiv 10^6 K)$ surrounding the capsule, the clean yield is calculated about 5 MJ. It is easy to estimate from the radiative ablation depth that the ablator of 0.16mm thickness is essentially exhausted during the laser pulse, the central hot spot ignition in extremely high density with a marginal ignition temperature 55 MK obtained from the balance between the emission rate of the spontaneous bremsstrahlung and the heating rate of the thermonuclear reaction seems to be required, this is so called non-local -thermal-equilibrium (non-LTE) ignition model.

In this paper, we suggest an alternative model that the local thermal equilibrium (LTE) ignition, which is in a lower ignition temperature comparing to non-LTE ignition, is required, and after short sustaining LTE burn the DT system rapidly transits to the non-LTE burn as discussed in our previous work (3). The capsule configuration is similar to that of the non-LTE ignition model as shown in (1) and (2). However, the shell of the capsule is thickened so that a portion of this shell is ablated to provide an imploding energy and the rest, which has not been ablated in a laser duration, is a high compressed "cold" layer and acts first as a pusher and later as a tamper to confine the high pressure DT plasma and further to enhance the fuel depletion. The material of the shell is carefully chosen so that it is optically thick until LTE ignition is coming, and optically thin during non-LTE burn. Here, the material of the shell is required to be a low Z one or doped a few high Z elements. Such a system is essentially robust, i.e., the central hot ignition is absent, in-flight aspect ratio is far less than 25, and the requirement of the imploding symmetry may be relaxed to a certain extent.

2. PHYSICAL CONSIDERATIONS

The LTE volume ignition in the lower temperature, and the enhanced confinement effect on DT plasma using the properly thickened shell in lower Z material are the essential points of our model. The physical considerations are discussed below.

a. LTE ignition

The LTE ignition occurs in a condition that the heating rate of the thermonuclear reaction balances with the losing rate of the radiative heat conduction that the flux goes into the shell of the capsule from DT fuel , then the transition from LTE to non-LTE happened in burn process. The threshold values of the LTE ignition of

DT system under the optimized implosion conditions are given in (3), that is, the temperature about 20-25MK and areal density $10^{-4}T^{3/2} << \rho R < 1$, where fluid density ρ usually is about 20-50 g/cm^3, R in centimeter is the radius of the compressed DT fuel, temperature T is in unit of MK.

Under these conditions , the internal energy for the DT fuel of 0.2 mg is about 0.05MJ. If laser energy of one megajoule is delivered into hohlraum, as long as 5% laser energy is invested to DT fuel, it may trigger a sustained DT burn. For moderate-gain ICF, the target possesses only DT fuel of a few tenths of 1 mg, therefore, the LTE volume ignition may reduce the requirement of laser energy as well.

b. Shell design

As above mentioned, the shell of the capsule must be required to have the properties that it is optically thick for LTE ignition and optically thin for non-LTE burn. Our work has shown that the glass shell or doped a few high Z elements is able to approximately meet this requirement.

At LTE ignition, if the imploding kinetic energy is converted into the internal energy of the high compressed "cold" layer (called tamper) and DT fuel, the energy conservation reads:

$$\frac{1}{2} M_1 \upsilon_0^2 = M_1 \alpha E_S + m C_\upsilon T_{ig} \tag{1}$$

where υ_0 is an imploding velocity, α is a modified factor close to one due to radiation heat-conduction from DT to tamper, M_1 and E_S are mass and internal energy for the tamper respectively, m, C_υ, T_{ig} are mass, specific heat, ignition temperature for DT fuel, respectively.

We write $m = \frac{4\pi(\rho R)_{ig}^3}{3P_{ig}^2} C_{ig}^4$, where the pressure $P_{ig} = C_{ig}^2 \rho_{ig}$, C_{ig} sound speed for DT fuel. For high compressed "cold" glass tamper, pressure $P_S \propto \rho_S^{5/3}$, internal energy of the unit mass $E_S \propto \rho_S^{2/3}$, where ρ_S is the tamper density.

Assume that the mass of the tamper is an approximate invariant in a main implosion stage to invest energy to the DT fuel, and regard M_1, E_S as a function of $P_s = P_{ig}$, we discuss the minimal energy for the LTE volume ignition, or the minimal mass of the tamper as υ_0 is given.

Fix C_{ig} and $(\rho R)^3_{ig}$, and let $\frac{dM_1}{dP_{ig}} = 0$, from Eq.(1) and equation of state for glass,

we have $\alpha E_s = \frac{5}{12}\upsilon_0^2$, and thus $\frac{m}{M_1} = \frac{\upsilon_0^2}{12C_\upsilon T_{ig}}$. If the fuel mass is $m=0.2mg$ and $\upsilon_o = 2.5\times 10^7\ cm/s$, we obtain the tamper mass $M_1 \approx 9mg$.

The ablated mass of the shell is roughly estimated below. The ablative depth from asymptotic similar solution(4) for glass shell is

$$X_R(cm) \approx 0.5\times 10^{-3}\rho^{-1.2}T^{2.9}\sqrt{t} \qquad (2)$$

where the glass density ρ is in g/cm^3, the ablation temperature T in MK, the ablation duration t in ns.

Taking the effective boundary temperature $T\approx 3.5MK$, normal glass density $\rho = 2.5\,g/cm^3$, the ablative duration t=6ns, for the outer radius of 1.55mm, the ablated mass $M_2 \approx 4\pi R^2\rho X_R \approx 11mg$. Thus, the total shell mass $M = M_1 + M_2 \approx 20mg$, which is 100 fold mass of the DT fuel.

C. Enhancement of fuel depletion

It is well known that the DT fuel burn up fraction in the case of exhausted ablator and the ion temperature about 200 *MK* is approximately (5)

$$\eta_0 = \frac{A_0}{A_0+6} \qquad (3)$$

where $A_0 = \rho R$ is an areal density for DT fuel.

However, when there exists the high compressed "cold" tamper, the burning duration τ is approximately determined by the relation $\frac{1}{2}\frac{P_{DT}}{(\rho\Delta R)}\tau^2 = \frac{R}{3}$, and

$\frac{\tau}{\tau_o} = \sqrt{\frac{6(\rho\Delta R)_t}{(\rho\Delta R)_{DT}}}$, where $(\rho\Delta R)_t$ and $(\rho R)_t$ are the areal density for DT fuel and tamper respectively, and P_{DT} is the pressure of the DT fuel. As a result, Eq.(3) should be modified by

$$\eta = \frac{A}{A+6} \qquad (4)$$

where $A = A_0\xi$, $\xi \equiv \sqrt{\frac{6(\rho\Delta R)_t}{(\rho\Delta R)_{DT}}}$. In general, $\xi > 1$ as shown in numerical simulation.(6)

As a result, when there exists the high compressed tamper, the DT depletion may considerably enhanced.

3. NUMERICAL RESULTS AND DISCUSSIONS

We consider a target for demonstration of the moderate-gain ICF from LTE volume ignition to non-LTE burn.

For calculation convenience, the hohlraum is simulated into a spherical cavity with a gold case, the gas of $He + H_2$ with density of $1\ mg/cm^3$ is filled inside.

The capsule in the hohlraum is constructed of a glass shell with outer radius of 1.55mm and 0.48mm thickness, a cryogenic DT fuel layer of 0.2mg and DT gas with density $10^{-4}\ g/cm^3$ filled in the central region.

The one-dimensional and three-temperature code CFJ is used to compute the system evolution. A laser beam of 1.25*MJ* with about 6 ns prepulse and 1.5 *ns* main pulse is invested in the hohlraum, and a radiation filed with peak temperature 4.50*MK* is produced surrounding the capsule to drive an imploding process.

The computed results are as follows. The LTE volume ignition occurs at time t=11.95*ns*, the internal energy of the DT fuel is 0.043*MJ*, thus, the averaged temperature $T_i \approx T_e \approx T_\gamma = 23MK$, where T_i, T_e and T_γ are the temperature for ion, electron and x-ray, respectively;$(\rho R)_{DT} \approx 0.41\ g/cm^2$ with the radius of 0.082*mm*. For the tamper, with $(\rho\Delta R)_t \approx 1.19\ g/cm^2$; About 14*mg* of the glass shell has been ablated and the rest as a tamper is about 11mg. The in-flight aspect ratio is about 2.7 which is far less than 25-35, as a result, the hydrodynamic instability is expected to be unimportant.

After short LTE burn, at 12.1ns the maximal areal density of $(\rho R)_{DT} \approx 0.8\ g/cm^2$ reaches and the non-LTE burn begins; at 12.36 ns the maximal ion temperature $T_{i\,max} \approx 804MK$ and $T_e \approx 278MK$ and $T_\gamma \approx 33.8MK$ arrive.

Finally, 18.1*MJ* energy yield (gain G=14.5) is released, thus, the DT burn up fraction is $\eta \approx 0.27$. This is very well expressed by Eq.(4) using the maximal areal density $(\rho R)_{DT} \approx 0.8\, g/cm^2$ for DT fuel and simultaneously $(\rho R)_t \approx 1.38\, g/cm^2$ is taken in account, we obtained $\xi = 3.2$, thus, η is enhanced to a factor of 2.3 due to confinement effect from tamper.

This work was supported in part by the foundation of the National Hi-Tec ICF Committee and by the National Natural Science foundation in China.

REFERENCES

1. M.Cray, "Inertial confinement fusion at Los Alamos National Laboratory", invited talk at *China-Us-Japan Workshop on Laser Plasma and Drivers*, October 17-21, 1994, Beijing, China.

2. J.D.Lindl, "Review of indirect drive ICF", Presented to *The Am.Phys.Society and Div. of Plasma Society*, November 7-11,1994, Minneapolis, MN, USA.

3. X.T.He and Y.S.Li, "Physical processes of volume ignition and thermonuclear burn for high-gain ICF", *Laser Interaction and Relation Plasma Phenomena*, AIP Conference Proceedings 318, G.H.Miley, ed., AIP Press, New York, 1994, pp.334-344.

4. X.T.He, T.Q.Chang and M.Yu, "X-ray conversion in high gain radiation drive ICF", *Laser Interaction and Related Plasma Phenomena*, H.Hora and G.H.Miley, eds.,(Plenum Press, New York), **9**,553(1991).

5. R.E.Kidder, "Energy gain of laser-compressed pellets: a simple model calculation", *Nuclear Fusion* **163**,405(1976).

6. Y.S.Li,X.T.He,Y.M.Gao and M.Yu, "The Kinetic Process Analysis of Volume Thermonuclear Ignition in High Gain Inertial Confinement Fusion", *High power Laser and Particle Beams* (to be published in 1995).

The Volume Ignition for ICF Ignition Target

Y.S.Li X.T.He M.Yu

Institute of Applied Physics and Computational Mathematics
P.O.Box 8009, Beijing 100088, China

Abstract Compared with central model, volume ignition has no hot spot, avoids the mixing at the hot-cold interface, the α-particle escaping, and the high convergence, greatly reduces the sharp demanding for uniformity. In laser indirect driving, from theoretical estimation and computational simulation, we have proved that using a tamper with good heat resistance, the DT fuel can be ignited in LTE at ~3KeV and then evolves to the non-LTE ignition at >5KeV. In this case, 1MJ radiation energy in the hohlraum could cause near 10MJ output for a pellet with 0.2mg DT fuel. We have compared results with and without α-particle transport, it shows that in the condition of $\rho R>0.5g/cm^2$ of DT fuel, both have the same results. For the system with $\rho R\approx0.5$ g/cm^2 we can use α-particle local deposition scheme. The non-uniformly doped tamper with density $\rho\approx1$-5g/cc can reduce mixing due to the small convergence ratio. The input energy is deposited in DT and tamper during the implosion, we try to reduce the tamper energy by changing the ratio of CH and doped Au and the thickness of the tamper.

1. INTRODUCTION

Following the work of [1], [2] and [3], we continue to design the shell of the capsule, to exam the influence of the α-particle escaping from the DT system and to emphasize the non-LTE effect during the ignition process.

As being previously pointed out, the key points for volume ignition are that the shell must be both a good tamper and a good heat-resistance layer in order to decrease ignition energy, to enhance the nuclear reaction and to keep the high radiation temperature to reach the LTE ignition condition. For this purpose we have to use high Z material (say, Au), but at the mean time, it increases the aspect ratio and the density gradient between DT and shell, and as a result, raises the mixing risk. To reduce these factors, we use CH non-uniformly doped with Au as a tamper in stead of using Au as a tamper.

Secondly, we take into consideration of the α-particle escaping, compared it's results at deferent ρR between local deposit and transport theories. For central spark ignition it is a big factor to raise the ignition temperature due to the small ρR of the hot spot, but for volume ignition the issue is greatly reduced because the total ρR reaches 0.4~0.5 g/cm^2 .

CP406, *Laser Interaction and Related Plasma Phenomena*: 13th International Conference,
edited by G. H. Miley and E. M. Campbell

Thirdly, we investigate the effect of the Non-LTE phenomena during the ignition process as the temperature rising from 10 to 30 MK, the Non-LTE effect cause the split between the material temperature Te (and Ti≈Te) and radiation temperature Tr, even if the radiation energy is mostly kept inside the capsule.

2. α-PARTICLE ESCAPING

For a ignition spot which area density ρR is smaller than 0.3 g/cm^2 [4], the α-particles will escape from the ignition area and the ignition temperature will be affected seriously. The computational simulation of the ignition process using the α-particle transport codes and comparison of the results of the neutron yield between the transport and local deposit of the α-particle energy are shown in table 1.

Table 1. The neutron yield at deferent ρR for transport (N) and local deposit (N')

ρR	0.05	0.10	0.15	0.20	0.25	0.30	0.40	0.50
N	0.18(14)	0.10(16)	0.28(17)	0.21(17)	0.12(19)	0.67(20)	0.91(20)	0.96(20)
N'	0.14(14)	0.27(15)	0.97(20)	0.96(20)	0.97(20)	0.97(20)	0.97(20)	0.97(20)

From table 1, it is obvious that for a capsule with ρR≥0.4, the energy output (or neutron yield) is the same, but for a capsule with ρR<0.3, the output by local deposit is much larger than by the transport theory. For our capsule with ρR≈0.4~0.5, we can use the local deposit theory to simulate the nuclear reaction process without causing much error.

3. NON-LTE PHENOMENA IN IGNITION PROCESS

As the DT gas is compressed and the temperature is rising, there is a temperature split between ion and radiation and the temperature at the center is higher than at the edge (see Fig 1). If taking the work done by the pusher into consideration the temperature split will be bigger. This two effects will reduce the ignition temperature, because the radiation energy loss and the electron energy loss only depend on the radiation and electron temperature at the edge of the DT area.

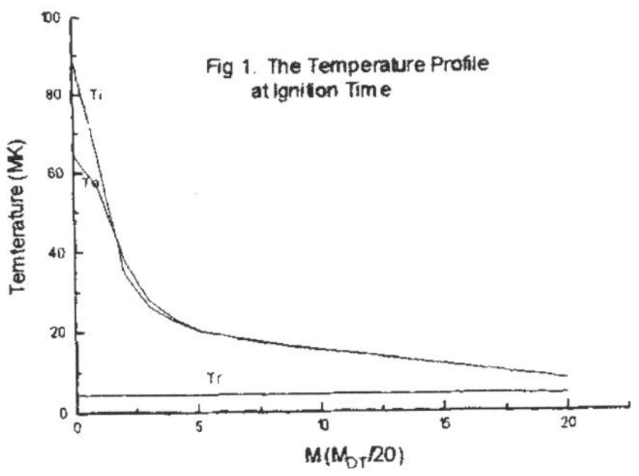

Fig 1. The Temperature Profile at Ignition Time

4. SHELL DESIGN

(1) A Good Heat Resistance layer.

We describe the heat resistance property by β_0 [3],

$$F_r = \frac{3}{\rho R}\beta_0\sigma_0 T^4 \qquad (1)$$

Here F_r is the energy flux from DT to the shell, $\sigma_0 T^4$ is the radiation energy of the black body. Smaller the β_0 is, better the heat resistance of the shell will be. We can improve these properties through doping CH with some high Z materials (say, Au). And in order to reduce the flight aspect ratio, we control the doping percentage of Au, and so change the density of the shell as shown in fig 2.

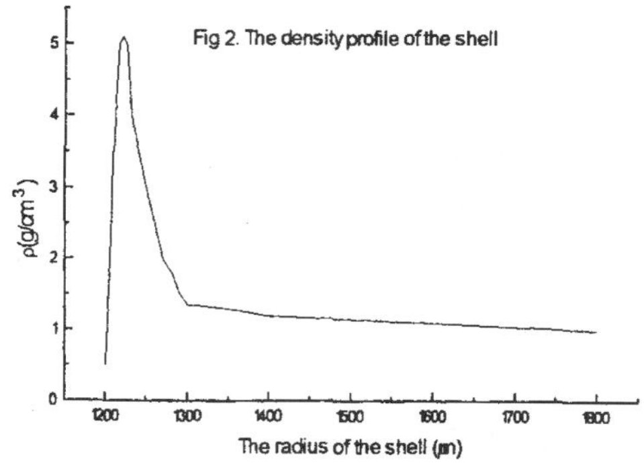

Fig 2. The density profile of the shell

(2) A Good Tamper Layer.

It is obvious that a good tamper can enhance the nuclear reaction and increase the depletion of nuclear burn as pointed out in [2].

$$\xi = \sqrt{\frac{6(\rho\Delta R) + \rho R}{\rho R}} \qquad (2)$$

$$\eta = \frac{\xi\rho R}{\xi\rho R + 6.0} \qquad (3)$$

Here η is the depletion of the nuclear burn.

In the central ignition model, the high density DT outside the central spot acts as a tamper. The energy deposit in tamper should be included in the ignition energy. Compared with TD tamper, the High Z tamper need less energy for the same area density ρR. We can change the thickness of the tamper and the doping percentage to control the energy deposited in the tamper. We can properly shape the energy source curve to get suitable density and temperature in DT.

5. NUMERICAL SIMULATION AND CONCLUSION

As we discussed above, we use ~0.2mg DT in a capsule of 1.2mm radius and 0.6mm thick tamper of CH non-uniformly doped with Au (see fig 2), for input radiation energy of ~1MJ, we obtained ~10MJ thermonuclear output energy and the gain of ~10. We use DT gas of density 0.028g/cm^3 to reduce the sharp demanding of the DT ice manufacture and moderately shaping the source energy curve (see fig 3) to get suitable temperature and density of DT gas. At ignition point, the ignition energy is about 0.054MJ in total including the tamper energy of 0.009MJ and the average temperature of DT is about 23MK. The ablation energy is ~0.31MJ. So the hydrodynamic efficiency is about 17% and the implosion efficiency is about 31%, much higher than in central ignition scheme. The profile of the DT temperature at maximum burning point is shown in fig 4.

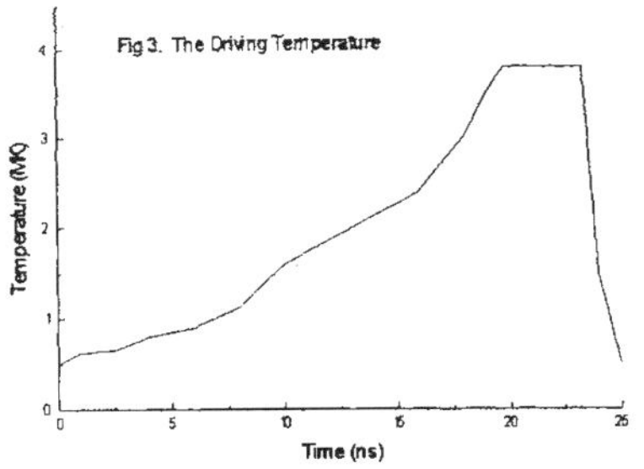

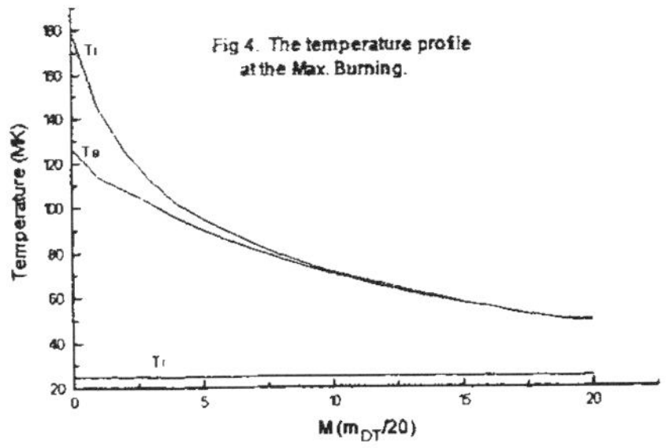

6. REFERENCE

1. X. T. He and Y. S. Li, "Physical processes of volume ignition and thermonuclear burn for high-gain ICF", Laser Interaction and Related Plasma Phenomena, AIP Conference Proceedings 318, G. H. Miley, AIP Press, New York, 1994, pp. 334-344.

2. X. T. He, Y. S. Li and M. Yu, "A moderate-gain ICF model From LTE ignition to non-LTE Burn", Laser Interaction and Related Plasma Phenomena, AIP Conference Proceedings 369, Sadao Nakai and G. H. Miley, AIP Press, New York, 1995, pp. 321-326.

3. Y.S. Li, X. T. He, Y. M. Gao and M. Yu, "The kinetic process analysis of volume thermonuclear ignition in high gain inertial confinement fusion" (Chinese Version), High Power Laser and Particle Beams, Vol. 7, No. 3 Aug., 1995, pp. 325-333.

4. John D. Lindl, Robert L. McCrory and E. Michael Campbell, "Progress Toward Ignition and Burn Propagation in Inertial Confinement Fusion", Physics Today, September 1992, pp. 32-40.

ELSEVIER

Fusion Engineering and Design 44 (1999) 57–60

Fusion Engineering and Design

Status and progress in the Chinese ICF program

X.T. He*, X.M. Deng, D.Y. Fan, X.M. Zhang, Z.Q. Lin, N.Y. Wang, Z.J. Zheng, J.R. Liu

National Hi-Tech ICF Committee of China, Office: P.O. Box 8009-55, Beijing 100088, P.R. China

Abstract

The Chinese ICF program is aimed towards inertial fusion energy in the 21st century and other applications. In this presentation, driver developments involving solid state lasers, i.e. Shenguang series, and the gas laser, i.e. KrF excimer laser, are presented; the theoretical and experimental studies for target physics, the equipment development for diagnostics, and the target fabrication are described; the achievements of ICF research in the past few years are mentioned. Precision physics is the basic point in ICF research of target physics in China. And the prospects for the Chinese ICF program are encouraging.

Keywords: Chinese ICF program; Driver developments; Equipment; Target fabrication

1. Introduction

The goals of the Chinese program for inertial confinement fusion (ICF) is towards inertial fusion energy in the 21st century and other applications.

For driver development in ICF research, during 1987–1994, the solid state laser facility called Shenguang-I (SG-I) or LF-12, which has an output laser energy of 2 kJ at wavelength $\lambda = 1.053$ μm with two beams, was operating. Fundamental research in ICF physics was conducted on SG-I which was retired in 1994. In addition, there is another solid state laser called the Xingguang (XG) facility for resolution research of ICF physics.

In order to further develop the Chinese ICF program to organize ICF research in China, an overall plan was completed in 1992, and the National Hi-Tech ICF Committee was authorized by the Chinese government in 1993.

The new solid state lasers, Shenguang series, have been planned and developed. The SG-II began construction in 1994, and is operating since the summer of 1997. The project to design a new generation facility, SG-III, which will be serving ICF physics research before ignition, is expected to be constructed in 2004. The ignition facility, SG-IV, will commence in 2010.

The KrF excimer laser, named Heaven-1 (H-1), is one of the candidates for drivers in inertial

* Corresponding author.

0920-3796/99/$ - see front matter
PII: S0920-3796(98)00281-6

fusion energy in China, and is operating and is being developed.

Furthermore, short pulse technology and drivers are also advanced.

For target physics research, achievements have been made in the past few years, and the theoretical and experimental research, the diagnostics, the target fabrication, etc. are stressed in precision physics.

2. The drivers for ICF and other applications

2.1. Solid state laser XG-II

The XG-II is located at the Southwest Institute of Nuclear Physics and Chemistry in Mianyang, Sichuan province and was constructed for research of some physical processes in ICF and also for exploration of X-ray laser, etc.

This laser facility with one beam has an output laser energy of 260 J at wavelength $\lambda = 1.053$ μm and 130 J at $\lambda = 0.35$ μm.

Some physical experiments involving laser plasma interaction, radiative hydrodynamics behavior, laser absorption and X-ray conversion in plasma, etc. have been conducted on this facility.

It is planned to further upgrade the XG-II to higher energy and performance in the future.

2.2. Solid state laser SG-II

The SG-II [1] is located at the National Laboratory on High Power Laser and Physics in Shanghai, and is the largest solid state laser in recent years in China and began operating in June 1997.

The purpose of constructing this facility is mainly for research of basic ICF target physics, and also for investigation of X-ray laser, etc. This is the main facility before the SG-III operation in 2004.

The SG-II with eight beams has a total output laser energy of about 3.0 kJ at $\lambda = 0.35$ μm and of 6 kJ at $\lambda = 1.053$ μm, pulse duration $\tau_p = 0.1–2$ ns, and has two chambers for ICF and x-ray laser, respectively. The laser bay and chamber room are in a clean environment. New technologies, such as the coaxial double-pass for 2×2 array disk amplifiers with aperture of 200 mm, loss modulation longitudinal mode oscillator, pulse intensity stabilizer, pulse shaper using time-space-time converting, etc. have been adopted in SG-II.

The precision work for SG-II began in 1996 [2], and is expected to be completed in 1999. The power imbalances of 10% (rms) in peak power and 20% (rms) in foot power, the energy imbalance of 5% (rms) for indirect drive ICF and the pointing precision of 20 μm required in a hohlraum with diameters of 1.2–2 mm are the first priority; the illumination non-uniformity $< 8\%$ (rms) will also be considered later.

2.3. Solid state laser SG-III

This facility will be used to investigate and grasp target physics for both direct-drive and indirect-drive ICF except for ignition physics, and to scale up to ignition parameters in China in the future. To meet those requirements, the SG-III design has an output laser energy of 60 kJ with 60 beams at $\lambda = 0.35$ μm; pulse duration 1–3 ns. The arbitrary modulation capability temporally for each beam and the 60-beam power imbalance $< 5\%$ (rms) in peak power and $< 10\%$ (rms) in foot power required are to be considered.

For the physical purpose of indirect-drive ICF, the temperature asymmetry on the capsule in a hohlraum has to be controlled in about 1% (rms), and thus the six rings for laser beam incident on the inner wall of the hohlraum and pointing precision 30 μm for a hohlraum with diameters 1.6–2.4 mm are required.

For the physical purpose of direct-drive ICF, the irradiation non-uniformity for a laser beam incident on a capsule surface has to be limited.

The architecture for SG-III will adopt many advanced technologies involving multi-pass main amplifier with aperture diameter 280 mm, higher performance Nd glass with platinum-free, higher damage threshold coating, etc.

Two steps in the design of SG-III have been considered. The first step is to construct a two-beam prototype with the output laser energy of 2.0–3.0 kJ at $\lambda = 0.35$ μm, which is a synthesized test bed for test of the synthetical performance in the integrated technical route and for demonstration and check of critical element technologies in engineering, and to provide the experience and to reduce the risk for the SG-III project. For this step, the primary design has been passed, and the conceptional design was completed in 1997. After the engineering design the two-beam prototype will be constructed in 2001.

For the second step, based on the above work, the SG-III will be constructed in about 2004.

The development and the preparation for the advanced critical technologies and the important components for SG-III, in fact, began in 1993. The high performance phosphate glass with platinum-free has been produced in a crucible with 15 l, and will be extended to the crucible with 40 l this year. The coating technology for diameter of 300 mm with higher damaged threshold has been performed, and the efficiency of the flash light has been improved, etc. Their performances will be tested on SG-II and XG-II. The equipment and the capabilities for optical fabrication are being planned and prepared.

2.4. KrF excimer laser H-1

The H-1 [3] has been developed and advanced at the China Institute of Atomic Energy in Beijing since 1993. This laser, which is a pulsed power system and has an output laser energy of 300 J with wavelength $\lambda = 0.248$ μm, is pumped by electron beams from double sides, and was further developed into an eight-beam MOPA system with angle multiplexer in 1997. And the six-beam laser energy of 300 J incident on the target with the compressed pulse duration 23 ns and divergent angle 0.3 mrad will be achieved this year.

In order to improve the beam quality for physical purposes, using a KrF system to pump a Raman amplified system, the output laser system of 75 J with a pulse duration 5 ns and a further reduced divergent angle 0.03 mrad will be achieved before 2000.

2.5. Short pulse laser

For ICF purposes, the laser with TW and pulse duration < 1 ps is being developed at the Southwest Institute of Nuclear Physics and Chemistry, and the relevant technologies for short pulse high power laser are being investigated at the National Laboratory on Higher Power Laser and Physics and other institutions in China.

3. Achievements, precision physics, prospects of ICF research in China

The achievements for ICF research have been made in the past few years. A series of physical experiments have been conducted on SG-I, especially, the properties of a radiation field converted by $\lambda = 1.053$ μm laser in a hohlraum were investigated in detail. Stimulated Raman scattering (SRS), stimulated Brillouin scattering (SBS), hot electrons, etc. in a hohlraum have been measured. The neutron yields from a thermonuclear reaction in indirect-drive ICF were first observed in 1990, and were repeated in 1991 [4]. The hydrodynamic behavior was studied as well.

Since SG-I retired in 1994, the fundamental experiments have been carried out on XG-II using $\lambda = 0.35$ μm laser, some important results are achieved and compared with that on SG-I, such as, the detailed angle distributions of laser absorption and conversion, the X-ray conversion bands and secondary emission in laser-produced plasma, the stopping up for diagnostic hole, etc. are measured in detail, and a mismatch related to SRS energy and hot electrons is observed.

A series of experimental techniques for third harmonics laser drive ICF are investigated and developed. The equipments and instruments with higher precision for physical diagnostics are used and prepared.

For target technology, the hohlraum and the microballoon shells with polystyrene or glass can be produced, the filling system for deuterium-tritium gas is used, the target materials involving lower density polystyrene foam are prepared, various coating techniques are developed, characterization for targets is measured, and the machines for target fabrication are available.

For theoretical study, a large number of numerical simulations for target physics have been carried out under the parameters for the Shenguang series. Some important codes in two dimensions have been made and developed. The ignition physics have been studied and explored.

The basic point for research of target physics in China is to advance precision physics and to emphasise new ideas, thus, the precision for drivers, experimental diagnostics, target fabrication and numerical simulations are known and have begun.

The prospects of the Chinese ICF program are encouraging. China will enter a new milestone later due to SG-III construction, the investigation on this facility puts the Chinese program in a critical stage towards ignition physics, and the SG-IV operation to begin in about 2015 will reach this goal in China.

Acknowledgements

It gives me pleasure to acknowledge the contributions of many participants in the Chinese ICF program. I especially would like to mention the teams at the National Laboratory on High Power Laser and Physics, Southwest Institute of Nuclear Physics and Chemistry, Institute of Applied Physics and Computational Mathematics, and China Institute of Atomic Energy.

References

[1] D.Y. Fan, X.M. Deng, Shenguang-II project: a 6 kJ laser fusion driver, Proceedings of IAEA TCM on Drivers for ICF, 1994, p. 89.

[2] Z. Lin, et al., SG-II laser elementary research and precision SG-II program, IAEA TCM on Drivers and Ignition Facilities for Inertial Fusion, Osaka University, Japan, 1997.

[3] Ma. Weiyi, et al., Chin. J. Lasers B5 (1996) 485.

[4] Z.J. Zheng, D.Y. Tang, P. Xie, High Power Laser Part. Beams 3 (1991) 286.

PHYSICS OF PLASMAS VOLUME 9, NUMBER 11 NOVEMBER 2002

Laser hohlraum coupling efficiency on the Shenguang II facility

Tieqiang Chang,[1] Yongkun Ding,[2] Dongxian Lai,[1] Tianxuan Huan,[2] Shaoping Zhu,[1] Zhijian Zheng,[2] Guangyu Wang,[1] Yongmin Zheng,[2] Xiantu He,[1] Wenbing Pei,[1] Qingsheng Duan,[1] Weiyan Zhang,[2] Tinggui Feng,[1] Guangnan Chen,[1] and Peijun Gu[1]
[1]*Institute of Applied Physics and Computational Mathematics, Beijing 100088, People's Republic of China*
[2]*Center for Laser Fusion Physics, China Academy of Engineering Physics, Sichuan Mianyang 621900, People's Republic of China*

(Received 8 May 2002; accepted 23 August 2002)

Recently, hohlraum experiments were performed at the Shenguang-II (SG-II) laser facility [Lin *et al.*, Chin. J. Lasers **B10**, Suppl. IV6 (2001)]. The measured maximum radiation temperature was 170 eV for the standard hohlraum and 150 eV for a 1.5-scaled one. This paper discusses the radiation temperature and laser hohlraum coupling efficiency in terms of a theoretical model [Phys. Plasmas **8**, 1659 (2001)] and numerical simulation. A 2D laser–hohlraum coupling code, LARED-H [Chin. J. Comput. Phys. **19**, 57 (2002)], gives a satisfactory coincidence with the measured time-resolved radiation temperature. Upon fitting the time-resolved curve, the theoretical model obtains the hohlraum coupling efficiency and, furthermore, the parameter $n+s$ for the hohlraum wall material (Au) can be determined simultaneously, where n, s are the power exponents of temperature for the radiation Rosseland mean-free path and specific internal energy, respectively.
 [DOI: 10.1063/1.1516781]

I. INTRODUCTION

For indirect drive laser fusion, a high-Z hohlraum (cavity) is crucial in which the high-temperature radiation field is generated to drive the fusion capsule to implode. The hohlraum is usually cylindrical and has a hole at each end for laser beams to enter (LEH). The laser beams strike a part of the hohlraum inner wall surface and heat it to produce soft x rays. The hohlraum radiation temperature is very important for driving the capsule, since the radiation ablation pressure is very sensitive to it. In order to increase the radiation temperature in a given laser condition, the hohlraum should be designed to increase the laser coupling efficiency, which depends on the laser x-ray conversion efficiency, wall ablation, and radiation energy loss through the LEH. Here, some key points are: (1) To minimize the radiation energy loss, the LEH should be as small as possible, so long as it does not obstruct the laser beams going through; (2) The design should avoid the laser light escaping out of the hohlraum due to reflection and deflection as far as possible, once it enters; and (3) To control the nonlinear laser plasma interaction, such as SRS (stimulated Raman scattering), SBS (stimulated Brillouin scattering), and hot electron generation, to an acceptable level (for example, SRS and hot electron energy should be less than 5% and SBS 10%). The hohlraum experiments have been performed at many laser facilities, especially on Nova.[1–4] Recently, the experiments were also performed at Shenguang-II (SG-II).[5] This paper will concisely introduce the SG-II hohlraum experimental results on radiation temperature in Sec. II and the 2D numerical simulations in Sec. III. The investigation of the coupling efficiency with the aid of an analytic model is described in Sec. IV, and Sec. V is a brief summary.

II. HOHLRAUM EXPERIMENTS AT THE SG-II LASER FACILITY

The SG-II laser has eight beams of $3\omega_0$ light with total energy of 2000 J reaching the target surface when operating with pulse duration of about 1 ns. The laser pulse is a ladder-shaped one, as shown in Fig. 1. The standard hohlraum target made of Au is designed as a cylindrical cavity of 1400 μm long, 800 μm in diameter, and the two laser-entrance holes located at the ends of the cylindrical hohlraum have the diameter of 380 μm. The eight beams of light enter into the hohlraum at 45° with the cylindrical axis, four beams each side. Besides the standard hohlraum, a 1.5-scaled hohlraum is also designed, but with the LEH unchanged. The hohlraum radiation temperature is measured by an array of x-ray diodes (ten energy channels x-ray spectrometer; the detectable photon energy is from 100 eV to about 1.5 keV and time resolution is about 150 ps) collimated at 30° with the hohlraum axis for collecting the radiation flux from the LEH. The SRS scattered light was carefully investigated. We separately measured its backscattered part, which was collected by the laser focus lens, and the outside-lens part. The angular distribution of the outside-lens part was detected by using an array of separated photon diodes equipped with a bandpass filter of 0.4–0.6 μm light, and the backscattered part was detected by a large-aperture calorimeter equipped with the same bandpass filter. The total SRS scattered light energy, then, can be obtained by integration, and was about 1% of the laser energy for the SG-II standard hohlraum. Similarly, SBS scattered light was detected, but with a different filter for the 0.35 ± 0.02 μm light to pass. The measured SBS light energy was about 8% of the laser energy. The hot electron energy was also estimated to be less than 3%. The absorption efficiency for the standard hohlraum was 0.9 on average. The measured maximum radiation temperature is 170 eV for the

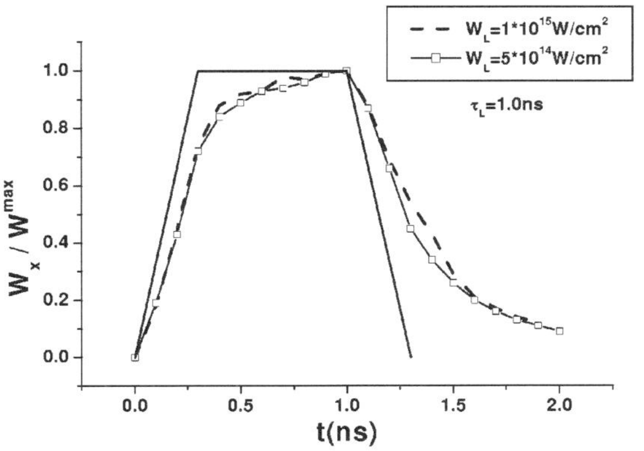

FIG. 1. SG-II Laser pulse and the laser-produced radiation pulse for two laser intensities.

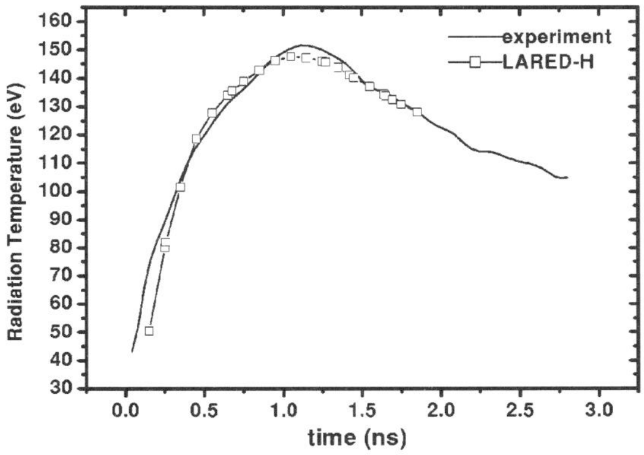

FIG. 3. LARED-H simulation result for SG-II 1.5-scaled hohlraum radiation temperature (shot no. 49).

standard hohlraum and 150 eV for the 1.5-scaled one. Figures 2 and 3 show the measured time-dependent radiation temperatures.

To the end, the hohlraum radiation temperature is determined by the available laser energy. The laser energy of SG-II is much lower than Nova, GekkoXII,[6] and Phebus,[7] so the radiation temperature of SG-II is, of course, correspondingly lower. However, the laser coupling efficiencies of SG-II hohlraums, which depends on laser beam quality and target design, are not low, as can be seen in Fig. 4, the diagram of $\ln T_r \sim \ln(P_L \tau^{1/2}/A)$, where T_r is the hohlraum radiation temperature, P_L the laser power, τ the pulse duration, and A the hohlraum wall area. The data for SG-II standard and 1.5-scaled hohlraums lie on the same line with the other results obtained at the facilities of higher laser energy.

III. LARED-H SIMULATION

The numerical simulation is carried out in terms of the code LARED-H.[8] LARED-H is a 2D non-LTE (nonlocal thermal equilibrium) laser hohlraum coupling code, in which the hydrodynamic equations with separated electron–ion and radiation temperatures are considered. The laser propagation and energy deposition are calculated by using the ray-tracing method. In order to calculate the nonequilibrium x-ray emission, the population rate equations dPn/dt are solved for principal quantum energy levels of Au ions, coupled with the hydrodynamic equations. The calculated time-resolved curves of the radiation temperature for the standard and 1.5-scaled hohlraums are compared with the measured ones in Figs. 2 and 3, and the coincidences are satisfactorily within the experimental error of about $\pm 7\%$. In fact, the measured radiation temperature is an averaged one, since the array of x-ray diodes spectrometer collects the radiation energy emitted by the whole plasma located in the spectrometer line of sight. Thanks to the low plasma density in the hohlraum, the measured radiation temperature is in reality the one at the hohlraum wall. However, the numerical simulation shows that the wall temperature varies from element to element along the path of the radiation penetration. In order to get the

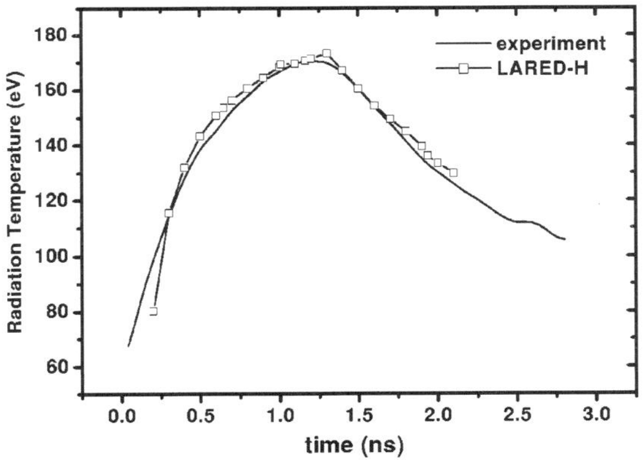

FIG. 2. LARED-H simulation result for SG-II standard hohlraum radiation temperature.

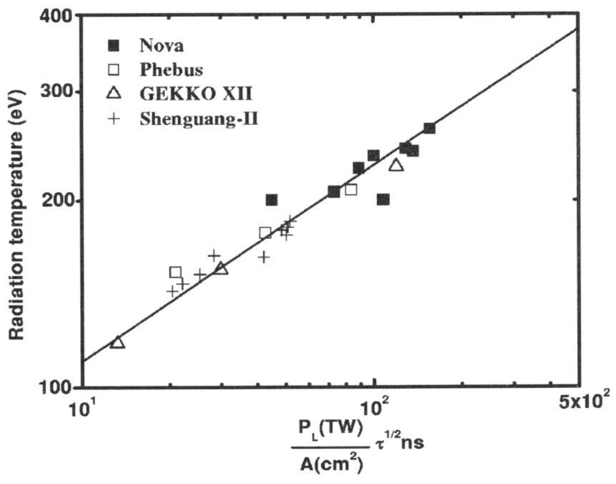

FIG. 4. Hohlraum radiation temperatures obtained at some laser facilities (the data on Nova, GekkoXII, and Phebus can be found, for example, in Ref. 11).

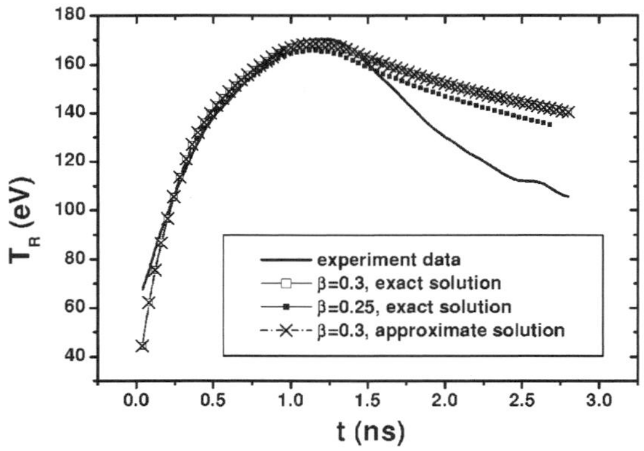

FIG. 5. Comparison of the radiation temperature between the measured and the analytic model for SG-II standard hohlraum.

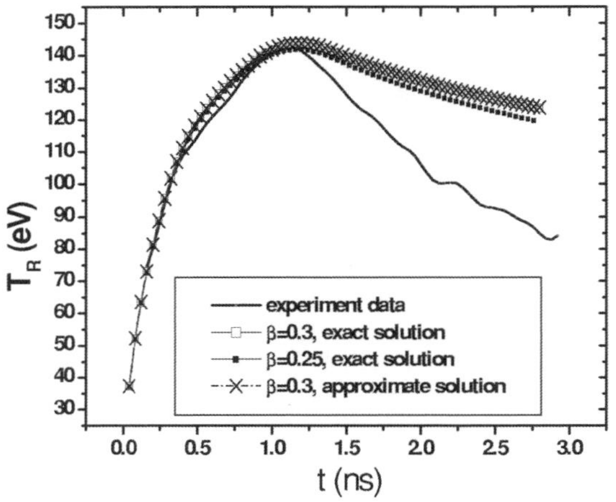

FIG. 6. Comparison of the radiation temperature between the measured and the analytic model for SG-II 1.5-scaled hohlraum (shot no. 72).

radiation temperature from numerical simulation to compare with the measured one, we solve the radiation transport equation

$$\frac{dI(\nu,s,t)}{ds}=j\nu(B\nu-I(\nu,s,t)),\tag{1}$$

where I is the radiation flux, ν, s are the photon frequency and the distance along the path of the spectrometer line of sight, respectively. $B\nu$, $J\nu$ are the Planckian distribution (multiplied by $c/4\pi$) and absorption coefficient of radiation with frequency ν in the hohlraum plasma, and

$$B\nu=\frac{2h\nu^3}{c^2}\frac{1}{e^{(h\nu/kT_R)}-1}.\tag{2}$$

Equation (1) has a solution as

$$I(\nu,s,t)=\int^s ds'\, j\nu B\nu \exp[-\int_{s'}^{s}ds''J\nu].\tag{3}$$

Taking s as the detecting point, using the $J\nu$ values of our database, and integrating the photon frequency, we obtain the radiation flux at time t and the average radiation temperature can be determined in the usual way. From Figs. 2 and 3 we can see that the coincidence between the experiment data and numerical simulation is satisfactory.

We also fit the radiation temperature by a simpler method, using only one-dimensional numerical codes. In this way, we must first obtain the time behavior of the converted x ray for a given laser pulse. It can be done by using a 1D laser-target coupling code and the radiation pulses so obtained are plotted in Fig. 1. Then, by means of the obtained time behavior and an assumed radiation energy, we calculate the radiation ablation process of the hohlraum wall by using another 1D code, which is a pure radiation-hydrodynamic one without laser coupling. Of course, in the ablation calculation the cylindrical geometry with the real hohlraum diameter and wall inner surface must be guaranteed. We change the radiation energy until an appropriate one is selected to give the measured time-resolved radiation temperature. This method is successful to fit the SG-II hohlraum experimental data, giving a primary and approximate result.

IV. LASER HOHLRAUM COUPLING EFFICIENCY

We also calculated the hohlraum radiation temperature by the analytic formula that we obtained in Refs. 9 and 10. The formula gives

$$T_R(t)\approx 3.21\{t^{-1/2}\int_0^t dt'\exp[-\phi(t,t',0.5)]\omega(t')\}^{\beta},\tag{4}$$

with

$$\phi(t,t',\alpha)\equiv\frac{(3.21)^{1/\beta}\alpha_H\sigma b^{4\beta-1}}{1-\xi-4\beta(\alpha-\xi)}\times[t^{[1-\xi-4\beta(\alpha-\xi)]}-t'^{[1-\xi-4\beta(\alpha-\xi)]}].\tag{5}$$

In Eq. (4) α has been taken to be 0.5, as the scaling relation required,[9,10] and

$$\omega(t)=P_{\rm rad}(t)/S-(4\beta-1)\alpha_H\sigma b^{4\beta-1}\times[Z_0(t)\ln(Z_o(t)/\overline{Z}(t))/t^{[4\beta(0.5-\xi)+\xi]}],\tag{6}$$

$$Z_0(t)=(3.21)^{1/\beta}\int_0^t\frac{P_{\rm rad}(t')}{S}\exp\{-\phi(t,t',0.5)\}dt',\tag{7}$$

and $\overline{Z}\equiv bt^{\xi}$

$$P_{\rm rat}(t)=\eta_0 t^{\nu}P_L(t).\tag{8}$$

Equation (4) is an approximate solution of the integral equation of T_R. β is defined as $2/(n+s+4)$ and n, s are the exponents of temperature of the Rosseland mean-free path and specific internal energy, respectively, $P_{\rm rad}$, P_L, the x-ray and laser power, and $\eta_0 t^{\nu}$ the laser-x-ray conversion efficiency. α_H is the ratio between the LEH area and the whole hohlraum wall area, $\alpha_H=S_{\rm LEH}/S$ and σ is the Stefan–Boltzmann constant (for the meaning of other notation, see Ref. 9). When β equals 0.25, the solution becomes a exact one, and then

$$T_R=3.21\left[\frac{1}{S\sqrt{t}}\int_0^t dt'\,P_{\rm rad}(t')e^{-2\lambda\alpha_H(\sqrt{t}-\sqrt{t'})}\right]^{0.25},\tag{9}$$

TABLE I. Maximum $T_R(10^6$ K) calculated for Nova hohlraums by the analytic model.

Holhraum size	β	$\eta_0=0.75$	$\eta_0=0.70$	$\eta_0=0.65$	$\eta_0=0.60$
$\phi=1.6$ mm	0.25	0.36 (3.10)	3.30 (3.05)	3.24 (3.03)	
	0.30	3.39 (**3.13**)	3.32 (3.07)	3.24 (3.0)	
$\phi=1.2$ mm	0.25		3.81 (3.34)	3.74 (3.28)	3.67 (3.21)
	0.30		3.94 (3.45)	3.86 (**3.38**)	3.77 (3.30)

with $\lambda=(3.21)^4\sigma$. It should be noticed that the above formulas are valid only in the time interval from 0 to the peak time of the radiation temperature.

As is well known, the laser-x-ray conversion efficiency cannot be directly measured. Usually, it can only be obtained by comparing the experimental radiation temperature with numerical simulation or some theoretical models. Our method is to select the parameters η_0, ν, and β, and compare the above analytic solution with the measured time-resolved radiation temperature in the interval from 0 to peak time. The processes will continue until the coincidence is satisfactory. The two kinds of hohlraum target mentioned above are selected to be compared. We find that $\beta=0.3$ is appropriate and consistent with some other results, for example the Nova results. Also, $\nu=0.12$ is proper. However, η_0 differ from each other: 0.6 for the standard hohlraum and 0.75 for the 1.5 scaled one. This is reasonable: The larger the size, the less the laser intensity on the hohlraum wall, hence, higher absorption and x-ray conversion efficiencies, and the less laser energy loss escaping out of the cavity. With these parameters the coincidences between experimental and theoretical hohlraum radiation temperature are rather good up to its maximum value. Beyond the maximum the theory ceases to be valid. The results are shown in Figs. 5 and 6, in which the exact solution of the integral equation is also presented, and it is seen that the exact solution and its approximate one [Eq. (4)] give almost the same results.

Using the given formula, we also calculated the hohlraum radiation temperature and the coupling efficiency for some Nova hohlraums (square laser pulse of 1 ns).[1–3] The hohlraums to be compared are scale 1 and a smaller one. The calculated and Ref. 1-given values are listed in Table I. Nova laser and hohlraum parameters: laser power $P_L=0.3$ (10^{14} W), $\tau=1$ ns pulse length (square shape), $\eta=\eta_0 t^{0.12}$, cylindrical diameter $\phi=1.6$ mm, 1.2 mm (for details see Refs. 1–3). The measured maximum radiation temperatures are $T_R=3.13(10^6$ K), 3.36, and the coupling efficiencies are 0.75 and 0.65, respectively.

The figures in parentheses are the corrected values due to the radiation loss through LEHs, whereas the ones outside the parentheses are the corresponding values without considering the loss correction and are obtained from Eq. (4) with $\alpha_H=0$ or directly from the integral equation of the radiation temperature[9] for square laser pulse. The values in bold give the ones nearest to Nova experiment data. The comparison leads to the following conclusions: (1) the theoretical model is accurate enough to get the Ref. 1-given values of x-ray conversion efficiency; and (2) including the correction due to radiation energy loss escaping from the LEH is important in determining the coupling efficiency.

In addition, due to the inevitable error bar, the reproducibility of the experimental data from shot to shot is not perfect. Figures 7 and 8 give the values based on the analytical model for different shots of the SG-II standard and 1.5-scaled hohlraum experiment, respectively.

V. BRIEF SUMMARY

The laser hohlraum coupling efficiency is a key parameter for increasing the radiation temperature. Although the SG-II has only 2 kJ of laser energy on target, the coupling efficiency is relatively high thanks to the good quality of the laser beams and the proper target design. The radiation temperature is detected from the radiation flux out of the LEH, which is more reliable than that from a diagnostic hole.[4] We

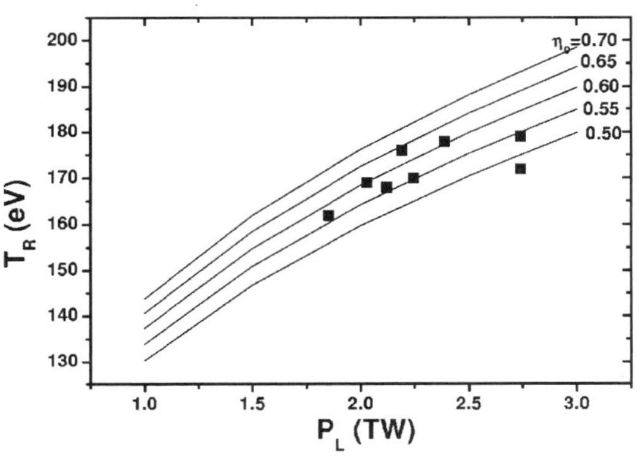

FIG. 7. Laser hohlraum coupling efficiency of SG-II standard hohlraums.

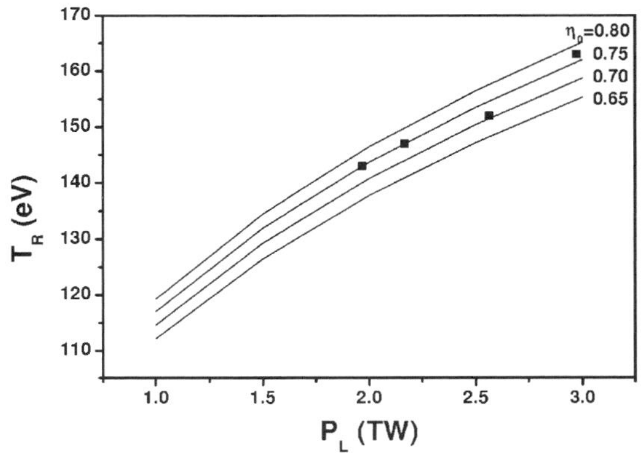

FIG. 8. Laser hohlraum coupling efficiency of SG-II 1.5-scaled hohlraums.

have used an analytic model and 2D numerical simulation to investigate the hohlraum radiation temperature, and the coincidence between them and the experimental data is satisfactory. Furthermore, the parameter $n+s$ for gold material can be determined simultaneously, where n, s are the power exponents of temperature for the radiation Rosseland mean-free path and specific internal energy, respectively. We have also discussed the laser x-ray conversion efficiency for SG-II hohlraums in detail. The good accuracy of the theoretical model with the experimental data guarantees that the analysis of the coupling efficiency by the method is appropriate.

ACKNOWLEDGMENTS

This work was performed under the auspices of the National High-Tech Inertial Confinement Fusion Committee in China, the National Natural Science Foundation of China (No 19735002), and the National Key Laboratory on Plasma (No 51480030101ZW0902).

[1] R. L. Kauffman, L. J. Suter, C. B. Darrow *et al.*, Phys. Rev. Lett. **73**, 2320 (1994).
[2] L. J. Suter, R. L. Kauffman, C. B. Darrow *et al.*, Phys. Plasmas **3**, 2057 (1996).
[3] J. Lindl, Phys. Plasmas **2**, 3933 (1995).
[4] E. Dattolo, L. Suter, M-C. Monteil *et al.*, Phys. Plasmas **8**, 260 (2001).
[5] L. Zunqi, W. Shiji, F. Dianyuan *et al.*, Chin. J. Lasers **B10**, Suppl. IV6 (2001).
[6] H. Nishimura, H. Shiraga, H. Takabe *et al.*, *Fourteenth International Conference on Plasma Physics and Controlled Nuclear Fusion Research, Wurzburg, Germany, 1992*, Nuclear Fusion Research, 1992 (International Atomic Energy Agency, Vienna, 1993), 3, 97.
[7] B. Canaud, M. Vandenboomgaerde, F. Chaigneau *et al.*, in *Inertial Fusion Sciences and Applications 99, Bordeaux, 12–17 September 1999*, edited by C. Labaune, W. J. Hogan, and K. A. Tanaka (Elsevier, Amsterdam, 2000), p. 261.
[8] Q. Duan, T. Chang, W. Zhang *et al.*, Chin. J. Comput. Phys. **19**, 57 (2002).
[9] T. Chang and G. Wang, Phys. Plasmas **8**, 1659 (2001).
[10] W. Pei, T. Chang, G. Wang *et al.*, Phys. Plasmas **6**, 3337 (1999).
[11] J. D. Lindl, *Inertial Confinement Fusion* (AIP, New York, 1998).

Inertial fusion advance toward ignition and gain (Summary talk)

X. T. He

Institute of Applied Physics and Computational Mathematics, P. O. Box 8009, Beijing 100088, P. R. China
Email: xthe@iapcm.ac.cn

Abstract

In this article that summarized some papers submitted to the 21th IAEA Fusion Energy Conference, the recent most important advancements in inertial confinement fusion are reported. For the traditional central hot spot ignition scheme, National Ignition Campaign to establish the scientific basis for the first ignition demonstration in indirect-drive implosion on National Ignition Facility is defined. Key scientific and technological questions to scale up ignition demonstration are being experimentally verified on the OMEGA and the Z/ZR facility using D_2 and DT cryogenic fuel targets. For the alternative ignition scheme-fast ignition, significant progress of target physics has been made in recent two years, the enhancement of the hot electron flux and the reduction of the divergent angle of the electron beam found in the cone target experiments highlight the latest meaningful findings. In addition, the fusion science driven by heavy ion beams and the Z-pinch is advanced.

PACS number: 52.57.-z, 52.57.Kk, 52.57.Fg, 52.57.Bc, 52.58.Hm, 52.58.Lq

1 Introduction

Inertial confinement fusion (ICF) parallel to magnetic confinement fusion (MCF) is an alternative approach to gain fusion energy. After a long way of more than forty years, now ICF is going on the eve of the ignition. The first experimental demonstration of a single capsule (filled mixture of deuterium-tritium fusion fuel) performed ignition, obviously, is the most important milestone on the way to inertial fusion energy (IFE) based on ICF.

The National Ignition Facility (NIF) in the US and the Laser Megajoule (LMJ) in France will be the first two laser drivers to serve for ignition campaign in future 4-6 years. The construction of the NIF that has laser energy output of 1.8 MJ at the wavelength of 0.35 μm is on the schedule to be completed in 2009 and experiments with targets on it begin in 2008 [Sangster,OV/5-2]. The first ignition experiment to begin in 2010 is expected in the indirect-driver implosion on the NIF based on a central hot spot scheme [1], in which the physical and technological explorations have been conducted for more than forty years. The direct-drive ignition demonstration on NIF based on the central ignition scheme will be done after the indirect-drive ignition. Although the central ignition scheme is requested severely to control the spherical symmetry of the imploding capsule and the technological accuracy, the recent progress in the fields of target physics, target fabrication etc., toward robust design, has provided a confidence to be successful.

In France, the LMJ is designed to deliver laser energy of about 2 MJ at the wavelength of 0.35 μm to target and will be gradually commissioned from early 2011, the first fusion experiments will begin in 2012 [2].

In China, a similar program to perform the ignition demonstration in 2020 is advancing. The construction of the SG-IV laser facility is planned, which has laser energy output of 1.5 MJ at the wavelength of 0.35 μm. The construction of the SG-III, which has laser energy output of 200 kJ at the wave length of 0.35 μm and serves for the scaled ignition, is underway and will be finished in 2010 [Zhang,OV/6-1].

Different from the central ignition scheme, alternative ignition concept, fast ignition, separated from imploding fusion fuel compression is also being actively explored. Its advantages are clearly in reducing the energy demands on drivers by a large margin and in relaxing conditions of the imploding symmetry.

In Japan, under the condition of the GEKK-II coupled with a lower energy petawatt (PW) laser, recent experiments at ILE of Osaka University have achieved the encouraging results [3]. However, to scale up fast ignition, it has to be checked by the tens kJ PW laser experiments probably in 1-3 years late. In order to further demonstrate the fast heating efficiency and the ignition temperature of 5-10 keV, the projects of FIREX (Fast Ignition Realization Experiment) in Japan [Miyanaga, IF/P5-2] [Mima,OV/5-1] and of OMEGA-EP (Experimental performance) in the US [Sangster,OV/5-2], and of SG-IIU-PW (SG-II upgrading coupled with PW laser) in China [Zhang,OV/6-1], are being executed. The similar projects in Europe are also planned. Major efforts are toward realization of fast ignition IFE with high gain.

In addition to laser drivers, the Z-pinch facility with wire-array and the heavy ion driver are also developed.

Twenty four reports are submitted to the 21st IAEA Fusion Energy Conference 2006, among them, there are three overview presentations [OV/5-1, OV/5-2, OV/6-1], five presentations[IF/1-1, IF/1-2R(a-c), IF/1-3], fourteen posters [IF/P5-(1-14)], and two fusion technology posters [FT/P5-39, FT/P5-40], however, the present paper based on the summary talk delivered at conference is impossible to exhaust all of them.

1.1 Central ignition scheme

In laboratory, ICF occurs in a deuterium and tritium (DT) fuel filled capsule that undergoes an implosion dynamics driven by drivers, which here and below are focused on laser driver unless otherwise specified.

The direct-drive way refers to the laser directly irradiating the capsule surface, while the indirect-drive one means x-rays to ablate the capsule surface, where x-rays are produced by laser incident on the inner wall of a cylindrical hohlraum. The central hot spot ignition scheme refers to that during the imploding process the main fuel layer in the capsule is adiabatically compressed to extremely high density, say $300g/cm^3$, and the ignition hot spot (the areal density of about $0.3g/cm^2$ and the temperature of about 10keV) is generated in the central region of the imploded capsule.

1.1.1 Indirect-drive implosion

In the US, the first ignition experiment on the NIF is expected through the indirect-drive way to demonstrate the propagation of a thermonuclear burn wave from the ignition spot to the main fuel layer. The National Ignition Campaign (NIC) is to establish the scientific basis for an ignition demonstration campaign on the NIF, the goal of the NIC is to define the initial ignition target designs, to experimentally validate the key physics, to develop the complex cryogenic target handling system, to demonstrate the fabrication of an indirect-drive target to specification, and to define and implement the appropriate target diagnostics [Sangster,OV/5-2].

Advances have been made in indirect-drive target designs with emphasis on developing more efficient hohlraum coupling and capsules that ignite with less absorbed energy. The indirect-drive ignition-point design is based on many years of experimental and theoretical work, which is shown in Fig. 1 [Sangester, OV/5-2]. The capsule with radius of 1.0 mm inside hohlraum is made up of Be ablator, DT ice and DT gas fill. The hohlraum is a cylinder filled with low pressure helium gas and has a diameter of 5.1 mm and a length of 9.1 mm, laser beams in 2 rings are incident on inner wall of the hohlraum to generate x-ray emission. The hohlraum design uses a high-Z mixture (cocktail) wall to increase opacity and reduce absorption while simulations indicate the improved hydrodynamic stability through the use of a radially varying Cu dopan in the Be ablator. Also, end shields have been added, which reduce losses through the laser entrance holes and allow for a shorter hohlraum. Experiments are also beginning to validate the hydrodynamic effects resulting from a fuel filled tube perturbation, to demonstrate the performance of the point design Be-Cu ablator and to mitigate hohlraum energy losses due to laser plasma interaction.

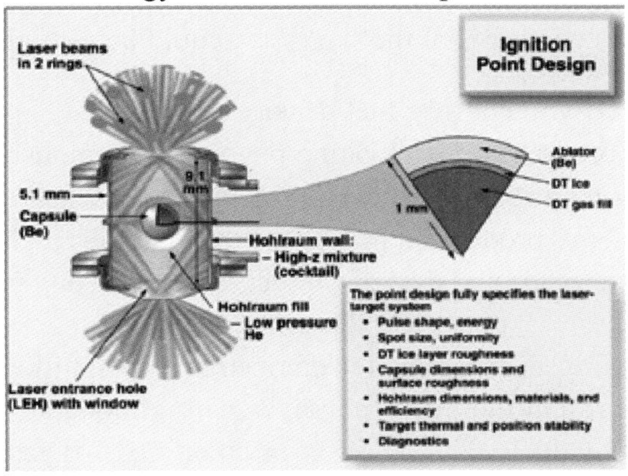

Fig.1 The indirect-drive ignition-point design [Sangester, OV/5-2]

In fact, two capsule designs with graded-doped Be or CH ablators have been developed that ignite with about 1.0 MJ laser energy. Such ablators improve the hydrodynamic stability and manage the x-ray preheat. Each ablator has advantages and issues for target performance and manufacturing [1].

The challenge now is to fabricate the indirect-drive point-design target to specification, such as, for the Be capsule, the outer surface finish and the ice-surface smoothness of the DT layer are closed to specification, the low mode isotherm control for cryogenic hohlraum is demonstrated [Sangster, OV/5-2].

Four integrated experiment teams are developing the requirements for the campaigns leading up to ignition. Laser beam qualities, hohlraum and capsule performance, and diagnostics development and calibration for overall ignition campaign-plan integrated implosion involved ignition failure check, are being developed. The scaled ignition implosions conducting on the 60-beam, 30-kJ OMEGA laser facility will validate much of the target physics required for successful ignition experiments that begin in 2010.

Much work has shown the high performance of ice DT targets. For examples, the power spectrum varying with mode number shows that the smoothness of the inner DT ice is about 0.91 μm rms, which is similar to direct-drive implosion about 1.0 μm rms; a uniform layer with a spherically symmetric temperature gradient across the DT (or D_2) ice is controlled to the thermal gradients of mK precision; for transparent ablator, the smoothness of the inner ice surface is measured using optical shadowgraphy and the radius varying with angle is analyzed to get the Fourier power spectrum Pn; a 3D representation of the inner ice surface can be constructed from multiple views; more than half of the DT capsules created to date have produced layers with sub-1 μm rms roughness; for indirect-drive ignition point design, the DT layers in Be shells at the vapor density of 0.3 mg/cc can meet the NIF smoothness standard for modes ≥ 10, which requires the ice temperature to be about 18.3 K, etc.

Target physics with cryogenic DT implosions is investigated on OMEGA. The neutron averaged areal density $\langle \rho R \rangle_n$ is greater than 100mg/cm^2 for cryogenic D_2 implosion, further analysis is underway to infer a $\rho R(t)$ by convolving the neutron emission rate with the measured proton spectrum; the peak areal density $\langle \rho R \rangle_{peak}$ may be inferred by using core self emission to backlight the fuel shell, 1D simulations suggest that the $\langle \rho R \rangle_{peak}$ could be as high as 180 to 190 mg/cm^2, 2D simulations are expected shortly to confirm fuel density estimation.

The US inertial fusion program is also looking beyond the demonstration of ignition on the NIF. A large, multi-institutional program is developing high gain target designs and the technology needed for inertial fusion energy production [4]. These technologies include high rep-rate DPPSL and KrF lasers, final optic materials and designs, target fabrication and injection, and first wall protection.

In China, it is performing a program to demonstrate the ignition and the propagation of a thermonuclear burn wave through the main fuel layer with a moderate gain for central ignition in about 2020 and with a high gain for fast ignition in about 2015 if fast ignition is workable. The SG-II laser driver with eight beams and routine energy output of $\geq$6kJ (1ω) and $\geq$3kJ (3ω) has provided 3000 shots for experiments of target physics since 2000; The LARED package that consists of 1D-3D codes to serve for numerical simulations of target physics is developed.

In the past years, the radiation generation and transport in hohlraum were studies on the SG-II. The evolution of the interface between capsule shell (doped sulphur in the inner wall) and fuel compression (doped argon in fuel deuterium) through x-ray-drive implosion on SG-II was observed using backlighting image and pin-hole (ϕ 5 μ m) image with resolution 6 μ m, and the interface convergence from initial diameter 300 μm to stagnation 20 μm was measured. To investigate indirect-drive implosion symmetry, the deformed profiles of capsules varying with different hohlraum length were experimentally observed on SG-II, and LARED simulations reproduce the measuring results [Zhang, OV/6-1]. Under the condition of the SG-III that has laser energy of 200kJ (3ω), the scaled indirect-drive ignition physics was studied by LARED simulations, and will be investigated by experiments in 2012.

The ignition implosion driven by the SG-IV, which has the top laser energy of 1.5MJ, will be demonstrated in 2020. The preliminary capsule design with DT fuel has been completed, as shown in Fig.2, it is feasible to achieve the fusion yield of 12.7 MJ by indirect-drive implosion in a radiation temperature of 300eV with delivering laser energy of 1.0 MJ (main pulse 4 ns) in hohlraum.

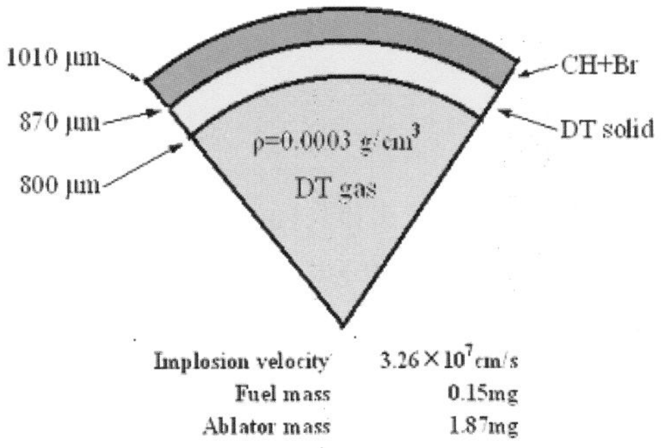

Fig. 2 Preliminary capsule design for indirect-drive ignition in China [Zhang, OV/6-1].

The measurement of the Hugoniot curve on SG-II gave a maximal pressure of 2.7Tpa with high precision for gold material, in terms of the Al-Au impedance-matched method. It would provide a basis to design hohlraum target.

In addition, the advanced diagnostic techniques and the relatively integrated system, and the precise target fabrications are coordinately developed as well.

1. 1. 2 Direct drive implosion

The direct-drive ignition due to much better energy coupling is the baseline approach for laser-based IFE. The approach to ignition and the physics of cryogenic capsule implosions are being validated by imploding direct-drive cryogenic D_2 and DT-filled capsules on OMEGA. Experiments are measuring the sensitivity of implosion performance to parameters such as the surface roughness of the frozen fuel layer, the adiabat of the fuel, the driver symmetry and energy absorbed by the capsule. Direct-drive ignition demonstration on the NIF is expected when the areal density of the central hot spot reaches $0.3g/cm^2$ and a temperature of about 10 keV same as the indirect-drive ignition. The cryogenic targets being imploded on the OMEGA are predicted to achieve the total areal densities in excess of $0.2g/cm^2$ and the temperature of about 2-3 keV.

In order to perform direct-drive ignition experiment on NIF, LLE at Rochester University has developed a new concept for direct-drive experiments in the indirect-drive configuration. These designs, called Polar Direct Drive (PDD), are illustrated in Fig.3 [Sangster, OV/5-2], where half of the beams are repointed toward the midplane on the capsule.

The high density compression by direct-drive implosion, in fact, was achieved in the end of the 1980s at ILE of Osaka University in Japan, over 600 times initial solid density was obtained on GEKKO-XII by the laser implosion of a hollow shell pellet of deuterated and tritiated plastic (CDT) [5] [Mima, OV/5-1], it was confident of advancing inertial fusion.

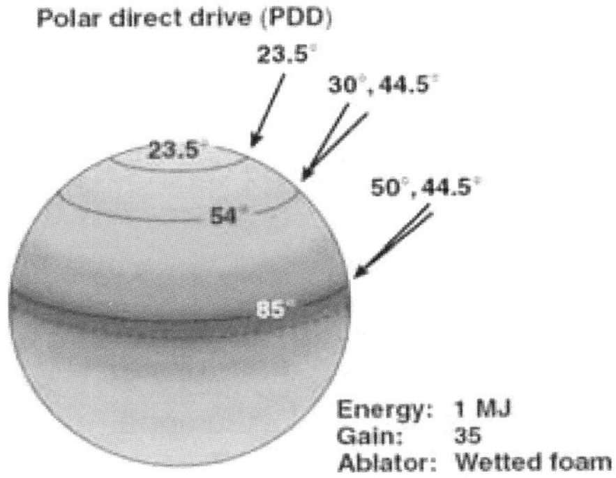

Fig. 3 Polar illumination geometry for initial direct drive ignition [Sangster, OV/5-2]

2. Fast ignition scheme

Fast ignition concept is that the capsule filled DT fuel firstly undergoes an implosion compression by long pulse (nanoseconds) laser and then an ignition hot spot is generated by the separated PW (picoseconds) laser in the edge of the core where the DT density is maximal. It can significantly reduce the demand on driver energy and relax restrictions of capsule symmetry in implosion process.

The recent experiments first demonstrate that fast ignition is a viable scheme. Impressive progress with a novel target, which a hollow gold cone was inserted into a deuterated polystyrene (CD) shell and a PW laser beam was through the hollow cone incident on the compressed CD shell to heat a hot spot, has been achieved. Using the above cone attached target (cone-shell target), experiments on GEKKO-XII with laser energy of 1.2 kJ coupled with the PW laser energy of about 60J have shown that the PW laser heating efficiency (laser energy converted into the internal energy of the hot spot) is estimated about (15-30)%, the divergent angle of the electron beam is less than 20° (FWHM) from the heated region on the back of the $200\mu m$ Al touched the tip of the cone, which is less than that of a plane target ($20^\circ - 30^\circ$) by use of 40TW/0.5ps laser [3], and the neutron yield enhancement from a thermonuclear reaction is about three orders of magnitude at 0.5 PW laser compared to that without heating laser [6]. In order to scale up ignition target, it is necessary to extend the above experiments to larger laser energy. Immediately, the experiments of the cryogenic cone-shell targets were conducted on the OMEGA without the PW laser for indirect- drive and direct-drive implosion to test the collapse predictions. And much work was also devoted to an understanding of the heating efficiency and the complex interaction of the PW laser with the relativistic plasma.

2.1 Implosion studies of cryogenic cone shell targets

A cryogenic cone-shell target has been used to further study the imploding compression experiment on OMEGA [7]. X-ray drive targets were driven exactly as a 1/2 mm diameter of the central ignition shell in a scale hohlraum using 14 kJ drive power. The reentrant cone (opening angle 70°, the tip 40 μm from the center of the shell) was notched outside the shell to prevent an excessive x-ray flux on the cone near the shell. The direct-drive targets had a 1 mm diameter shell, and the cone base blocked one set of beams entirely. The shell was driven with 11 kJ from 15 half power and 20 full power beams.

The experimental results, as shown in Figs. 4 and Table 1, show that the cone seen in indirect-drive experiments appears considerably blunted compared to its original shape. The central gas (simulations show a density of $10-20g/cm^3$, heated to 400eV, and moving at $\sim 10^7$ cm/s for an initially empty shell; the density is doubled and the velocity is about halved for a filled shell) is flowing directly toward the tip, eventually punching it in (in simulations that collapse starts later for a filled shell). On that basis these shells achieve $\geq 50\%$ of 1D; A 2D LASNEX simulation shows that a substantial part of the core gas can escape (the cone tip "punctures" the target) in a reentrant cone target. Thus, experimentally, these fast ignition targets should perform better than central ignition shells on this measure.

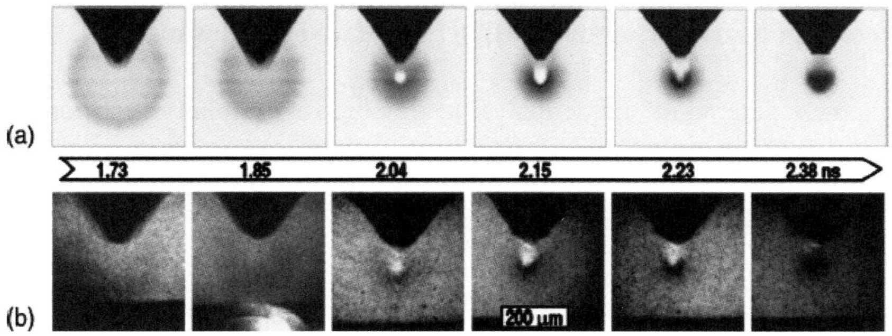

Fig.4 Sequence of direct drive collapse image (a) simulation in LASNEX and (b) experiment shot [7].

	Experiment	Simulation	
	Cone shell	Cone shell	Sphere
ρR(mg/cm^2)	100/60–80 (±20%) (x-ray/laser drive)	130/200	130/150

Tab. 1 Maximal areal density from OMEGA experiments and LASNEX simulations [7].

Under FIREX program in Japan [Mima, OV/5-1], the cone-shell target implosion, which has performed on the GEKKO-XII and will conduct on heating laser LFEX (Laser for Fast Ignition Experiment, 10 kJ/10ps/1.06 μm) in late 2007, has been simulated by 2D radiation hydrodynamic code PINOCO, which is one of the Fast Ignition Integrated Interconnecting Codes (FI^3). The shape of imploded core plasma in experiments is well reproduced in the simulation. It is shown that the tip of the cone is damaged by the jet produced by the stagnation of imploding fuel shell. The PINOCO simulation results also show that the plasma density is as high as the maximum density of spherical implosion and the areal density of the core will reach 0.45 g/cm^2 and 1.2 g/cm^2 respectively in the FIREX-I, and –II. It is found that implosion of cone target is relatively insensitive to the hydrodynamic instabilities in comparison with spherical target. A planar cryogenic foam layer was irradiated with GEKKO-XII laser recently, it is observed that a deuterium layer is compressed and accelerated by ablation pressure as expected in simulations. The GEKKO-XII also irradiated a foam cryogenic deuterium layer with the PW laser to investigate the heat transport, DD neutron from a cryogenic foam layer target has been measured. The neutron yield of cryogenic target is compared with CD plastic target, which shows heat transport is strongly inhibited on a surface of a cryogenic DD layer.

The non-spherical implosion with initial perturbation on the target surface to estimate the effect of Rayleigh-Taylor (RT) instability [Nagatomo, IF/P5-4] is also discussed. The cone shell target (CH-DT shell with gold cone, radius 250 μm) and laser condition (4.5 kJ, Gaussian pulse) are similar to the FIREX-I experiment. Results show that even though there is RT instability, a high density core plasma is formed as well as the non-perturbed case. These results give the evidence of the robustness of non-spherical implosion which is a favorable fact for fast ignition.

In China, under the SG-IIU condition, the LARED simulation of the cone shell target also shows that the RT instability from P_4 perturbation can insignificantly affect the fuel density and temperature distribution [Zhang, OV/6-1].

2. 2 Hot electron generation, acceleration, transport in relativistic plasma

Hot electrons, which are generated in interaction of the PW laser with relativistic plasma, accelerate and transport in the underdense plasma, where the plasma density is less than γn_c, γ is the relativistic factor and $n_c \sim 10^{21}/cm^3$. The hot electron propagation separates from the laser beam when the latter begins hole boring, then hot electrons enter the overdense ($>\gamma n_c$) plasma, and penetrate a depth about $50 \sim 100\mu m$ and finally deposit their energy of about $20kJ$ in the compressed fuel core with the extremely high density (say, about $300g/cm^3$), where an ignition spot with the radius about $20\mu m$ and the temperature about $12keV$ is formed [8].

In the underdense plasma, hot electrons undergo an acceleration by the ponderomotive force and the nonlinear electric fields and a collimation by spontaneously magnetic fields as high as over 100MG generated by the nonlinear current induced by the PW laser of intensity $I = 10^{19} W/cm^2$ [9]. Experiments, theoretical models and numerical simulations have justified the existence of such strong magnetic fields.

In the overdense plasma, the return current from the background plasma compensates the relativistic current $\gamma n_c c \sim 10TA/cm^2$, it ensures that the hot electrons go forward. However, hot electrons propagated in the overdense plasma undergo a scattering not only by the background electrons, but also by the Ohm's electric field induced by the net current, it leads to the electron beam spot with a divergent angle about 40° (FWHM) driven by the linearly polarized laser [10] though the electron beam is pinched by an azimuthally magnetic field induced by the axially net current. Meanwhile, the hot electron beam suffers from energy decay by Ohm's electric field to lose its partial energy.

For fast ignition, the beam divergent angle and the heating efficiency of PW laser converting into internal energy of the ignition spot, have to be essentially concerned. Especially, the studies of the integrated experiments and simulations are essentially important

2.2.1 Enhancement of the laser brightness and the electron flux in the guiding cone

Recently, experiments [Kodama, IF/1-2Rb] have shown the evidence that the cone target comparing with a plane target can significantly enhance laser beam propagation since the wall of the cone would reflect and refocus the lost laser light into the cone tip, which was observed by temporally resolved measurement of the UV emission from the rear side of the target. The peak brightness for the cone targets is 2-3 times higher than that for the plane target.

Energetic electrons, generated by the enhanced laser at the side wall and the tip of the cone, could be guided along the cone wall as a result of the source of the return current, and are significantly affected by self-generated electromagnetic fields surrounding the cone. The energetic electrons are pushed outside by the magnetic field and return into the cone by the sheath fields. This field balance results in the collimation and confinement of the electrons in the cone device. Experimental results show that the divergence of the electron beam is $20^{o}\pm3^{o}$ (FWHM) for the cone target and 30° (FWHM) for the planar target. The number of the generated electrons in the cone is estimated to be 1.7 to 3 times larger than that of plane target, which is consistent with the UV emission experimentally.

The collimation of energetic electrons was also examined as shown in Fig. 5 [Kodama, IF/1-2Rb]. The relative energy flux of the electrons in the Al solid cylinder and solid taper attached to a 10 μm Al plate was measured and compared with that in a 200 μm thick plane target. The MeV electrons generated at the front flat surface propagated through the shaped target to the rear plane target. Optical emission due to the energetic electrons from the rear side of the plane target was obtained. A good collimation of the electrons, which is dependent on the target shape, is clearly seen on the emission images as shown in Fig. 5. The peak flux of the emission was enhanced in the cylinder by a factor of about 2 and in the taper by a factor of 4.5 as compared with that in the block target. It shows that such plasma devices guide and collimate the high density electrons similar manner to the control of light with a taper shaped optic-block collimator. More efficient heating of core plasmas would be expected as a way to control charged particles at high flux (TA/cm^2) with an enormous devices energy density in a fashion analogous to light control with conventional optical devices.

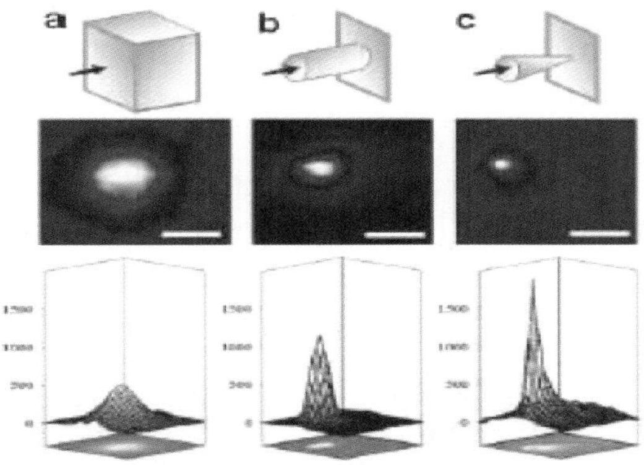

Fig.5 The relative energy flux of electrons in the 200 μm thick planar target (a), the Al solid cylinder (b) and the solid taper (c) attached to a 10 μm Al plate [Kodama,IF/1-2Rb].

2.2.2 Hot electron propagation along target surface

Experiments have evidenced through x-ray image that fast electrons produced by the 20TW laser target interaction can be self-organized at the plane target surface to form a jet emission and transport along the surface when laser incident angle is greater than 70°, as shown in Fig.6, [11] [Zhang, OV/6-1]. It means that the fast electrons will be guided and focused as it propagates along the cone wall. Static electromagnetic fields are self-induced around the surface, the surface electrons are confined by these fields and their motion is dependent on laser intensities, laser polarization, target materials and plasma properties in surface. 2D simulations justify experimental results.

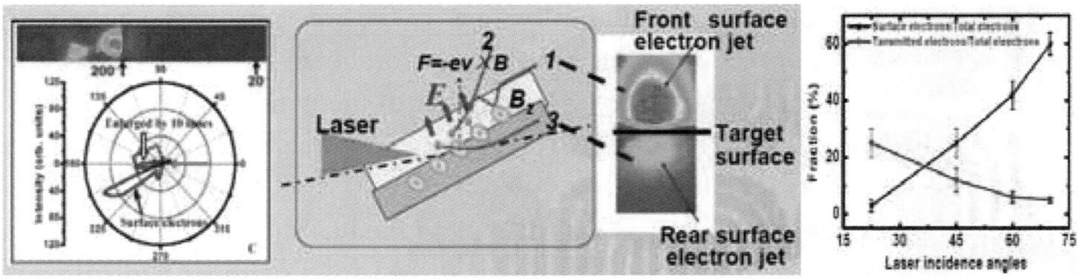

Fig.6 Electron transport along target surface by experiments and simulations [Zhang, OV/6-1].

In another experimental group, hot electron transport along target surface is also observed at a laser intensity $3\times10^{18}W/cm^2$ [Tanaka, IF/1-2Rc]. Fig.10 shows the experimental results for three different laser intensities varying up to $3\times10^{18}W/cm^2$. The dotted lines indicate the laser axis, target surface, and the laser specular reflection directions from the top to the bottom. When the laser intensity is lower [Fig.7(a)-(b)], the MeV electrons are mainly directed toward the laser specular reflection direction. However, the electrons appear clearly to the target surface direction at the highest laser intensity [Fig. 7(c)]. The results are fully supported by the PIC simulation and show that the relativistic electrons created along the surface are bound by a strong magnetic field of the order of 30 MG generated by both hot electrons and the cold return electrons in the vicinity of the target surface. This behavior promises that the relativistic electrons created within the cone will be attracted toward the tip in the 3D configuration and contribute to the highly efficient fast heating of compressed cores. Those hot electrons departing from the target are subject to a huge potential formation on the rear side of the target. The effect on the electron behavior was studied by irradiating plane targets with and without pre-formed plasma on the target rear side. The results are compared with the PIC simulation. It is shown that the potential formation on the target rear is controlled with the plasma formation which is caused by the return current within the plasma to compensate the hot electron current. This understanding will lead to an accurate target design of fast ignition integrated experiment planned within a few years with multi-kJ PW laser systems.

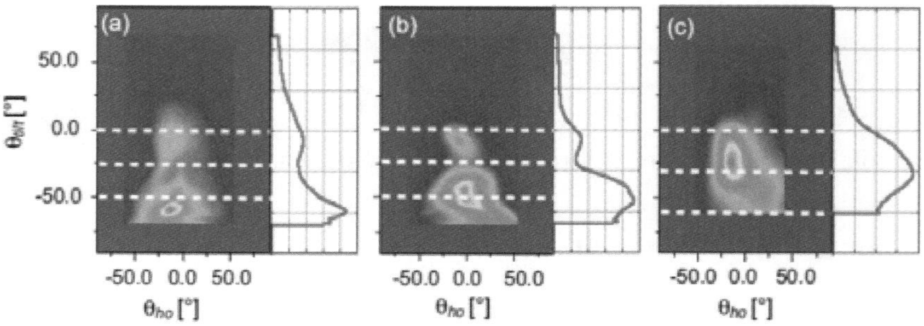

Fig.7 Laser intensity dependence of electron emission angle with plasma condition at (a) $1\times10^{17}W/cm^2$, (b) $1\times10^{18}W/cm^2$, and (c) $3\times10^{18}W/cm^2$. The horizontal lines represent the laser axis, target surface, and specular direction planes from the top to down. The red line shows intensity profiles along the vertical axis at 0° at the horizontal [Tanaka,IF/1-2Rc].

2.2.3 Fuel preheating and the PW laser heating efficiency

In the previous experiments, the deuterated plastic shells as surrogate target for fusion fuel lead to extremely high radiation loss due to carbon atoms [3-6]. Under the FIREX program, cryogenically cooled liquid deuterium targets are used instead of previous CD targets, two essential issues, i.e., the fuel preheating by hot electrons generated from laser plasma interaction during imploding compression and the PW laser heating efficiency have been explored [Azechi,IF/1-1].

For the fuel preheating investigation, a low density (0.34 g/cc) plastic foam with 80 μm thickness as a sustainer of liquid deuterium was irradiated by a long pulse laser light at an intensity of 10^{14} W/cm^2 with wavelength 0.5 μm ,the foam layer is attached with a solid density (1 g/cc) plastic with 5 μm thickness as an ablator. To suppress the preheating, the same target but with a thin gold layer between the plastic ablator and the foam layer was also used. The gold layer was designed to be thin enough (0.05 μm) to prevent potential hydrodynamic instabilities between two layers. The preheating temperature on the rear target surface was conventionally measured from blackbody radiation in a UV-visible light region. Since the shock heated temperature can be controlled by laser pulse shape, the primary concern is the preheat temperature just before the shock arrival. It has been found that the target preheat level is about the Fermi temperature (5 eV) corresponding to liquid density deuterium. This preheating level may further be reduced by the insertion of the gold layer. In any case, the moderate preheat implies that the energy required for compression is close to that of perfectly degenerated Fermi gas. It appears therefore that the concern of preheat in deuterium targets was somewhat groundless [Azechi,IF/1-1].

For the PW laser heating efficiency in deuterium targets, which is (15-30)% in the previous experiments that the cone-shell (CD) target was used, it will be addressed in the coming experimental campaign. Recent preliminary experiments have injected the PW laser onto a liquid deuterium target with 100 μm thickness sandwiched by two plastic layers with 7 μm thickness. To study hot electron transport in the target, experiments have observed Kα line emission generated by hot electron bombardment with Cl atoms that were doped into the plastic foam, a clear DD neutron signal from the liquid deuterium target was measured. These results indicate that hot electrons, indeed, transport through the plastic layer and deposit their energy in the deuterium target [Azechi, IF/1-1].

2.2.4 Electron energy spectra

Recently, the 2D large scale simulation developed in FI^{3}code is executed successfully. The electron energy spectra from cone target and plane target are shown in Fig.8a-b [Nagatomo,IF/P5-4]. The spectra from cone targets indicated that there are three electron acceleration processes dominant in laser cone interaction. The first is the electron acceleration and transport at the sidewall via surface magnetic field, which results in electrons for middle temperature (1.8 MeV in the figure). The second is ponderomotive acceleration at the cone tip by intensified laser pulse due to cone guiding, which generates high energy electrons fitted with highest temperature (5.0MeV). And the third appears when there exists steep density gradient and the current density is too high to be current neutralized, a static field is induced to decelerate high energy electrons and to heat up bulk electrons to the temperature of sub-MeV. These three acceleration processes lead to electron spectrum with three-temperature. Since the cone targets have additional parameters such as cone angle and scale length at the sidewall comparing to plane targets, it is difficult to derive general comparison on absorption rate. But in above simulations the absorption rate is about 40% for cone target which is twice higher than that for plane target (Fig. 8b), thus it is possible to say that cone targets can effectively convert laser energy to electrons. These differences are important, because it was observed that the energy deposition in core plasma is very sensitive to these absorption rate and electron spectra by integrated simulation.

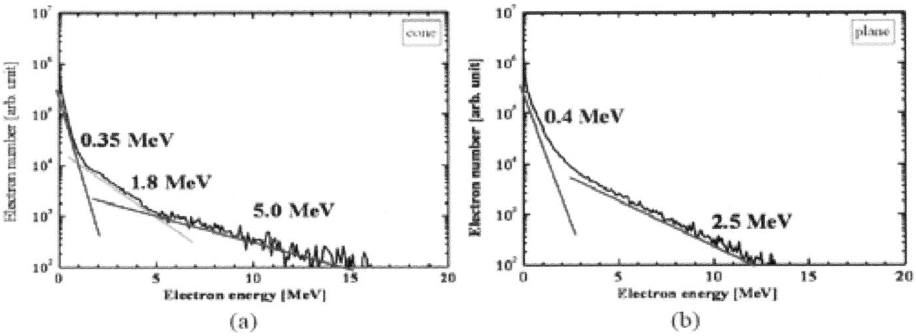

Fig.8 Electron spectrum of cone target (a) and electron spectrum of plane target (b) respectively [Nagatomo,IF/P5-4].

2.2.5 Electron isochoric heating

Studies and modeling of the laser accelerated electron sources, the transport of energy by electrons and the consequent isochoric heating have advanced considerably but there is still no well established modeling capability to enable extrapolation from small scale experiments to full scale fast ignition. The processes are complex and challenging from both experimental study and modeling aspects. Recent work has studied two limiting cases as shown in Fig.9 a)-d), which address specific issues in electron transport and offer good opportunities for comparison with modeling [Mackinnon,IF/1-2Ra].

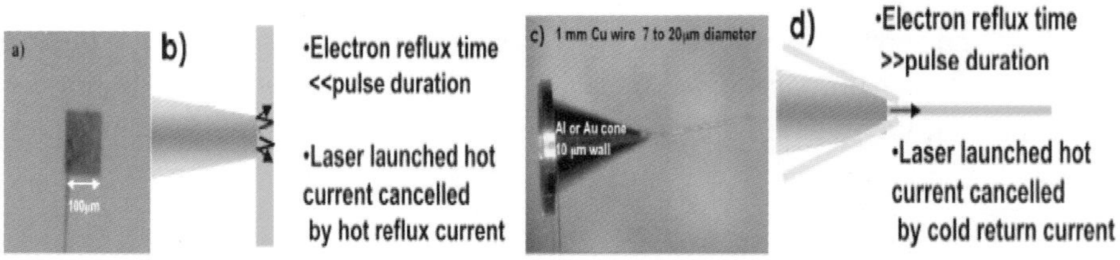

Fig.9 Targets for electron transport studies. a, b) low mass foil which maximizes hot electron refluxing. c, d) wire that has no refluxing [Mackinnon, IF/1-2Ra].

(1) Thin foils of small area constrain hot electrons to hot reflux between the surfaces in a short time compared to the laser pulse duration so that the laser launched hot current is always approximately cancelled by the hot reflux current, thus eliminating the effect of Ohmic heating by the return current of cold electrons, which is a dominant effect in initially cold solid targets in the absence of refluxing. The isochoric phase of target heating is measured by 256eV x-ray imaging, the results show that electron thermal conduction equalized front/back temperature during the XUV emission in thin targets and that 12% of the laser energy is converted to thermal energy in the target.

(2) In the opposite limit a hollow cone couples the laser to a long thin wire and provides a situation where there is 100% compensation of the laser launched hot electron injection into the wire cancelled by the cold electron return current, thus maximizing Ohmic effects. For this target, thermal emissions were imaged in the XUV at 256 eV and energetic electron current from 8 keV $Cu-K_\alpha$ spectroscopy. LASNEX modeling of the $Cu-K_\alpha$ XUV emission gives a maximum

thermal temperature of 350 eV, while the $Cu-K_{\alpha}$ line width gives a temporal mean temperature of 160 eV. The electron propagation along the wire is shown by the $Cu-K_{\alpha}$ imaging system to have a 1/e attenuation length of $\sim 100\mu m$. Both the temperature and propagation length results are in reasonable agreement with LASNEX model.

Electron propagation along the wire is also observed by the $Cu-K_{\alpha}$ imaging system to have much longer range component ($l>$ 1mm) about 2% of the peak.

2.3 Proton beam fast ignition

Protons offer an alternative means of isochoric heating for fast ignition with very different physical constraints [Mackinnon,IF/1-2Ra]. Much work has investigated proton acceleration by PW laser interaction with plasma, mainly numerical simulations, such as, recent LARED- HFPIC simulations, show that protons with energy of 3 MeV can be generated by the PW laser with intensity of $3x10^{20}$ w/cm^2 interacting with a 4 μm thickness CH target, and sharp electric fields of 30 GV/cm are observed in both the front and the rear of target, see Fig.10 [Zhang, OV/6-1].

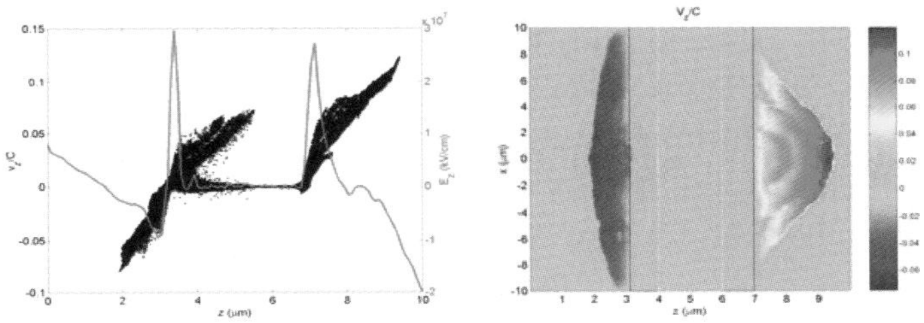

Fig.10 Electric field and velocity distribution of mono-energetic of proton beam observed in the rear of CH target by 3D HF-PIC simulation (2D) [Zhang, OV/6-1].

The requirement for fast ignition is to deliver the PW laser about 20 kJ to a less than 40 μm diameter spot with a proton axial temperature of about 3MeV. The conversion efficiency from the PW laser energy to protons should exceed 15% assuming the beam is focused within the spot.

For proton fast ignition (Fig.11a), the cone geometries must protect the proton source foil from rear surface plasma formation induced by the implosion but it should not cause Molieré scattering outside the required $<40\ \mu m$ hot spot. The source foil should be thick enough to protect its rear surface from pre-pulse shock modification but thin enough to allow adiabatic energy loss to acceleration of protons to dominate over collisional energy loss for the refluxing electrons. The laser irradiation should produce sufficiently high temperature electrons uniformly across the source foil to make collisional losses relatively insignificant and to result in a sheath axial development that is spatially uniform giving radial proton focusing to a spot size $<40\ \mu m$. The laser pulse length should be short enough to limit edge effects on the sheath to avoid significant loss of well focused protons. The protons must also deliver their energy in a shot time; because of their velocity spread, the proton generating foil must be quite close to the core, such as <2 mm depending on details. A prototype of such a proton source was built, as seen in Fig.11b, to test its capabilities and performance on OMEGA and GEKKO-XII. The experiment demonstrated that focused heating of a Cu foil target (Fig.11c,d) inside a cone could be achieved but was far from optimized.

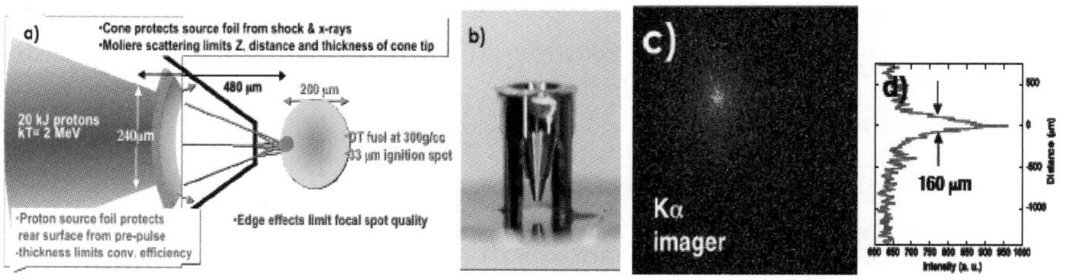

Fig.11 a) Requirements for proton ignition structure, b) test structure $Cu-K_{\alpha}$, c) image and d) line-out of spot heated by focused protons [Mackinnon, IF/1-2Ra].

The study using PW class lasers at the RAL Vulcan, ILE GEKKO and LLNL Titan facilities showed that proton focusing was significantly aberrated in the experiments where a relatively small laser focal spot produced a central maximum in the sheath extension giving radial components to the acceleration.

Conversion efficiency to protons of energy >3MeV is also a key research area in proton fast ignition. To date conversion efficiency has not exceeded 10% of the laser energy. Analytic methods and numerical simulations were used to study the conversion efficiency, such as, recent 1D numerical modeling with the hybrid PIC code LSP are beginning to show promising results, namely greater than 50% conversion of electron energy to proton energy >3MeV. These theoretical predictions are currently being experimentally tested with cryogenic target layers.

The PW laser beam at the RAL Vulcan, delivering typically 0.5 PW, 350J pulses of 0.7ps was used and the Titan laser system, 350J, 0.5ps focused to an irradiance of $3\times10^{20}W/cm^2$ will be used for proton conversion efficiency measurements. Conversion efficiency modeling suggests that efficiencies higher than so far obtained and meeting the need for proton fast ignition should be possible with suitable design optimization.

The other way for the proton fast ignition is the use of the plasma blocks with ballistic focusing geometry [Hora,IF/P5-14] where very high proton current densities are possible while a minor thermal expansion increases the thickness of the accelerated layer before the PW laser interacting with the pre-compressed thousand times solid dense DT plasma. This increase of the layer thickness offers the realization of the proton fast igniter.

2.4. Impact fast ignition

Instead of the charged particle heating, recently a totally new ignition scheme, impact fast ignition, is proposed [Murakami,IF/P5-3].Such a target is composed of a spherical DT shell coated with an ablator and a hollow conical target (a fragmental spherical shell with DT and ablator) that impacts onto the spherical shell, as seen in Fig.12. The main idea is to accelerate the impact shell to collide against the main fuel. Upon the collision, shock waves generated at the contact surface transmit in two opposite directions heating the fuels to produce an igniting hot spot. The impact shell itself becomes the igniter by directly converting its kinetic energy into the thermal energy rather than by boosting the main fuel heating as in the particle (electrons or ions) drive fast ignition. Target structure of the impact ignition target overlapped with the maximum compression fuel. The ignition temperature Ts is one of the crucial parameters to be specified for gain estimation, and should strongly depend on the process how the igniter is generated. Because the compression and heating result from rather complex hydrodynamic process through the spherically converging implosion dynamics, the ignition temperature required for the impact fast ignition scheme is still an open question. The igniting temperature Ts is determined by implosion velocity being $(1.1-1.5)\times10^{8}cm/s$, where Ts = 5-10 keV is assumed. To achieve an ultra-high velocity of the order of $10^{8}cm/s$, preliminary experiments have been conducted very recently at GEKKO-II with a high laser intensity of the order of $10^{15}W/cm^{2}$. Two different types of planar targets, i.e., the polystyrene with mass density of $1.06g/cm^{3}$ and the Br-doped plastic target with mass density of $1.35g/cm^{3}$ with variation of thickness of 12 - 20 μm, were used. For the latter target, highly stable acceleration was expected owing to a newly found physical effect that substantially suppresses the growth of the RT instability, in which a double-ablation structure plays a crucial role. Indeed, a maximum velocity of about $6.5\times10^{7}cm/\sec$ has been experimentally observed with a 14 μm thick Br-doped plastic target Fig.12. The 2D hydrodynamic simulation has demonstrated such key velocity for the impact shell implosion as the velocity $\sim10^{8}cm/\sec$, the compressed density $\sim300-400g/cm^{3}$, and the temperature $\sim5keV$. Those results have shown the feasibility and the potential of this impact fast ignition scheme.

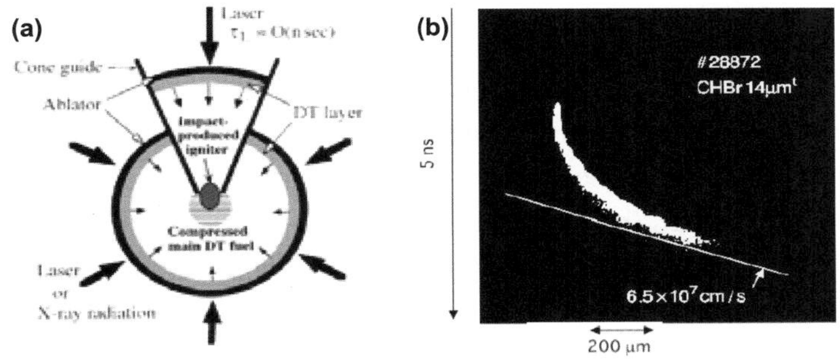

Fig.12 A new ignition scheme-impact fast ignition. a) A hollow conical target impacts onto the spherical shell to produce an ignition spot. B) Preliminary experiment on plane target was shown [Murakami,IF/P5-3].

In addition, a jet impact fast ignition concept was also discussed [Velarde,IF/P5-5]. The collision of jets, produced by a special configuration in the implosion process, converts their kinetic energy into thermal energy of the nuclear fuel, which is expected to produce ignition under proper design. Recently, 2D radiation transport code (ARWEN) is used to simulate both the jet production and the interaction with the compressed target, and low Z materials for the jets are applied to improve the target design.

2.5 Fast ignition experiment program

In Japan, the scientific feasibility of fast ignition will be fully examined in the FIREX program. In order to make the program flexible, two phases are divided. The purpose of the first phase (FIREX-I) is to demonstrate fast heating of a fusion fuel up to the ignition temperature of 5-10 keV. The heating laser for this program is a high energy PW (10 kJ/1 ps) laser [Miyanaga,IF/P5-2] that is currently under construction. The first experiment of FIREX-I will start in FY2007, followed by fully integrated experiments until the end of FY2010. If subsequent FIREX-II will start as proposed, the ignition and burn will be demonstrated in parallel to that at the NIF in the US and at the LMJ in France, providing a scientific database of both central and fast ignition [Azechi,IF/1-1].

In the US, a high-energy PW capability is nearing completion at LLE of Rochester University. The OMEGA-EP will deliver two short-pulse, multi-PW, multi-kJ beams to the OMEGA target chamber. Experiments in target physics to perform the scaled fast ignition will begin in 2008. The combined facilities will provide an integrated platform for the validation of fast ignition relevant physics using ignition relevant targets,. Based on the OMEGA-EP scaled ignition to be successful, the NIF may not only serve for the central hot spot ignition and but also for high gain fast ignition.

In China, the SG-II is upgrading to laser energy of 18 kJ for third harmonics (SG-IIU), and the PW laser beam with laser energy of about 1.5kJ, which is based on the present laser beam with laser energy of 4.5kJ and pulse duration of 3ns, is underway. The SG-IIU-PW will provide the integrated study of the implosive compression, and will perform the scaled fast ignition in 2008. Then, if fast ignition is workable the SG-III coupled with PW laser beams with laser energy of tens kJ will demonstrate fast ignition in 2015. Finally, the SG-IV will serve for high gain IFE.

In Europe, recent information shows that it is planning to perform a HiPER project [12], a panel of over 40 scientists from 9 countries has developed the laser facility design over the past 18 months, along with a detailed project plan for the next phase. It is estimated that the HiPER laser fusion facility system required to achieve ignition goals would consist of long pulsed laser $\sim 200kJ$, combined with PW laser $\sim 70kJ$.

3. Heavy ion fusion advance

Heavy ion beam is a leading alternative (non-laser) driver candidate because of its efficiency. To realize heavy ion fusion, intense beams of heavy ions (such as $A>80$) must be accelerated to multi-GeV kinetic energies (several megajoules total), temporally compressed to duration $\sim 10ns$, and focused onto a series of small (a few millimeters) targets, each containing a spherical capsule of fusion fuel. Such intense beams represent a significant extension beyond current state-of-the-art space-charge dominated beams.

Much progress has been made in the past two years, in experiment, theory and simulation, toward the essentially questions on heavy ion fusion science and high energy density physics for the US program. Main results are listed below [Simon, IF/P5-11].

Longitudinal and transverse beam compression: The Neutralized Transport Experiment (NTX) demonstrated transverse beam compression with density enhancement by a factor greater than 100, and the Neutralized Drift Compression Experiment (NDCX) showed longitudinal beam compression by a factor of 50. In both experiments, extensive 3D simulations, using LSP, were carried out, and the agreement with experiments was excellent. A 3D kinetic model for longitudinal compression was developed, and a class of exact solutions for the problem was obtained.

Beam target interaction: It is shown that the target temperature uniformity can be maximized if the ion energy at target corresponds to the maximum in the energy loss rate dE/dX. Ions of moderate energy (about a few to tens of MeV) may be used. The energy must be deposited in times much shorter than the hydrodynamic expansion time ($\sim ns$ for metallic foams at 0.01 to 0.1 times of solid density). Hydrodynamic simulations have confirmed that uniform conditions with temperature variations of less than a few per cent can be achieved.

High brightness transport: Unwanted electrons can lead to deleterious effects for high brightness ion beam transport. The electron accumulation in quadrupole and solenoid beam transport systems is being studied. In parallel with the experimental campaign, a new approach to large time-step advancement of electron orbits, as well as a comprehensive suite of models for electrons, gas, and wall interactions have been developed and implemented in the 3D beam dynamics PIC code WARP. If sufficient electrons are accumulated within the beam, severe distortion of the beam phase space can result. Simulations of this effect have reproduced the key features observed in the experiments.

Beam production and acceleration: The merging-beamlet injector experiment recently completed demonstrates the feasibility of a compact, high-current injector for heavy ion fusion drivers. The measured unnormalized transverse emittance (phase space area) of 200-250 mm-mrad for the merged beam met fusion driver requirement. These measurements are in good agreement with the PIC simulations using WARP3D. The physical design of a short pulse injector suitable for warm density matter (WDM) studies was completed. A new concept for acceleration, the Pulse Line Ion Accelerator (PLIA), offers the potential of a very low cost accelerator for WDM studies. The experimental verification of the predicted PLIA beam dynamics was shown. Measured energy gain, longitudinal phase space, and beam bunching are in good agreement with WARP3D simulations.

Computational models and simulator experiments: The pioneering merger of Adaptive Mesh Refinement and PIC methods underlies much of the recent success of WARP3D. BEST, the Beam Equilibrium Stability and Transport code was optimized for massively parallel computers and applied to studies of the collective effects of 3D bunched beams and the temperature-anisotropy instability. Space-charge-dominated beam physics experiments relevant to long-path accelerators were carried out.

Advances in the last two years will enable the first heavy ion beam target interaction experiments to begin in 2008 in the US [Sangster,OV/5-2].

4. Z-pinch for inertial fusion

Using pulsed power driven Z pinch to drive targets capable of ICF provides an alternative and complementary technology to the laser based approach.

In US, the ZR project is upgrading the performance of Sandia's "Z"-pinch facility. The ZR device and the Z-Beamlet are underway [Sangster,OV/5-2]. The ZR will be of the current increased from 19MA to 26 MA, 2x increase in diagnostic access, 2x increase in short-rate capability, $100-200ns$ pulses for ICF/Z-pinches and $100-300ns$ pulses for equation of state experiments. The Z-Petawatt project is also upgrading the capability of Sandia's Z-Beamlet laser facility, that is, power increased from 2 TW to 4 PW, x-ray backlighter energies to 40 keV and integrated fast ignition experiments with peak deuterium fuel $\rho R \sim 0.8g/cm^2$. The upgraded Z and Z-PW facilities on schedule will begin operation in 2007.

In Russia, a project to construct a generator with the current of 50 MA to investigate the physics of Z-pinch array compression is being carried out [Grabovsky, IF/1-3]. This generator is able to provide an x-ray pulse with energy above 10 MJ. Such a generator known as "Baikal" will be possible to investigate the ignition of a DT target. The development of the "Baikal" facility is being performed on the module prototype of the facility, i.e., the "Mol" facility, which is under construction now at SRC RF TRINITI. The "Mol" was aimed at developing three steps of shortening the duration of the output electric pulse. The fast Z-pinch experiments aimed at the IFE are conducted at the Kurchatov Institute, a series of issues are studied on the S-300 pulsed power machine [Bakshaev,IF/P5-12].

In China, Z-pinch device with about ten MA peak load current is under the way of construction [Zhang, OV/6-1]. Simulations to design the device are being conducted.

4. Target fabrication for fast ignition

The RF-PF shell with a diameter of $300\mu m$ has been successfully produced, and the conical laser guide is partially inserted into the shell, as seen in Fig.13 [Mima, OV/5-1]. To directly supply the fuel, the gas or liquid feeder with 30/10 μm outer/inner tip diameters is connected to the shell.

Then, the liquid fuel is fed into the shell through the feeder and is soaked up by the foam under capillarity, finally, the fuel is solidified. Thus, an ideal cone-shell target would be foamed. There are subjects for the foam target development involving fabrication of the foam shell and assembling the three parts to foam the target. The foam fabrication method has to be modified to realize ultra low density of ~10 mg/cm^3. Fuel layering in the practical foam target is going to be studied. Uniform fuel layer formation and fine fuel-quantity control are expected to be challenging.

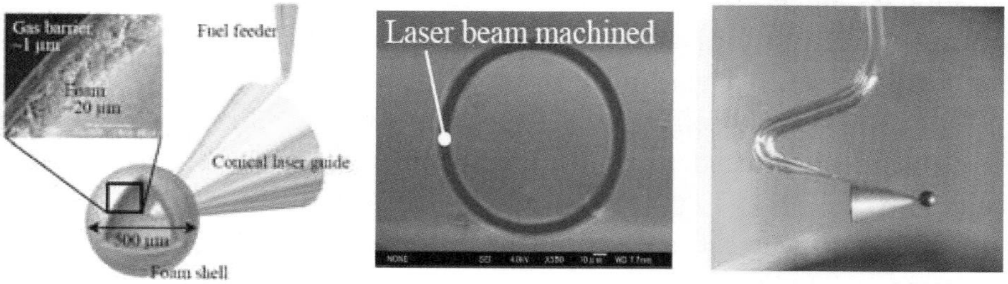

Fig.13 The foam cryogenic target for the FIREX project. [Mima, OV/5-1]

VII Conclusions

Significant progress in laser based inertial fusion energy is advancing toward laboratory-based thermonuclear ignition demonstration through imploding spherical cryogenic capsules of the frozen DT mixture. The first ignition experiment beginning in 2010 is expected in the indirect-drive implosion based on NIF that is on schedule to be completed in 2009. For this purpose, National Ignition Campaign as a comprehensive effort to establish the scientific basis for the ignition campaign on NIF is defined. Key scientific and technological questions relevant to ignition are being experimentally verified on the OMEGA and the Z/ZR facility. The capsule physics of the direct-drive ignition to be realized later after the indirect-drive ignition is also advanced, and the scaled ignition targets that meet the ignition specification for inner ice smoothness have been imploded on the OMEGA. The following ignition demonstration based on central hot spot ignition scheme will be in France in 2012.

In the same time, an alternative ignition scheme, fast ignition, is extensively explored. The projects of FIREX in Japan, OMEGA-EP in US, SG-IIU-PW in China and others in Europe, are being performed to understand the scaled fast ignition physics. It was greatly inspired by the achievement of integrated experiments using the cone-shell targets driven by lower laser energy. The physics of the fast heating ignition spot is further studied and significant progress is made in recent two years, the enhancement of hot electron flux and the reduction of the divergent angle of the electron beam by using the cone target highlight the latest meaningful findings. However, the heating efficiency of the PW laser energy with tens kJ converting into the internal energy of the ignition hot spot is still waiting for check.

In addition, the fusion physics driven by the heavy ion beam and the Z-pinch driver is also actively advanced.

List of papers on ICF submitted to the 21th IAEA Fusion Energy Conference 2006:

OV/5-1, Mima, K., Recent Progress on FIREX Project and Related Fusion Researches at ILE, Osaka.

OV/5-2, Sangster,T. C., Overview of Inertial Fusion Research in the United States.

OV/6-1, Zhang, W. Y., Status of Inertial Energy Program in China.

IF/1-1, Azechi, H, Compression and Fast Heating of Liquid Deuterium Targets in FIREX Program

IF/1-2Ra, Mackinnon, A. J., Studies of electron and proton isochoric heating for fast Ignition.

IF/1-2Rb, Kodama, R., Plasma Photonic Devices for Fast Ignition Concept in Laser Fusion Research

IF/1-2Rc, Tanaka, K., Relativistic Electron Generation and Its Behaviors Relevant to Fast Ignition.

IF/1-3, Grabovski, E. V., Radiating Z-pinch Investigation and “Baikal” Project For ICF.

IF/P5-1, Iwamoto, A., Development of the Foam Cryogenic Target for the FIREX Project.

IF/P5-2, Miyanaga, N., Development of 10-kJ PW Laser for the FIREX-I Program.

IF/P5-3, Murakami, M., Hyper-velocity Acceleration of Foil Targets for Impact Fast Ignition.

IF/P5-4, Nagatomo, H., Fast Ignition Integrated Interconnecting Code (FI3)-Integrated Simulation and Element Physics.

IF/P5-5, Velarde, P., Recent results on fast ignition jet impact scheme.

IF/P5-6, Koresheva, E. R., Reactor-scaled cryogenic target formation: mathematical modeling and experimental results.

IF/P5-7, Perlado, J. M., Advances in Target Design and Materials Physics for IFE at DENIM.

IF/P5-8 ·Pesme, D., Modelling the Nonlinear Saturation of the Parametric Instabilities Generated by Laser-Plasma Interaction.

IF/P5-9 ·Rudraiah N., Effects of Laser Radiation, Nanostructured Porous Lining and Electric Field on the Control of Rayleigh-Taylor Instability at the Ablative Surface of the IFE Target.

IF/P5-10 Khaydarov, R.T., Investigation of characteristics of laser source of ions from the targets of different densities.

IF/P5-11 Yu, S. S., Heavy-Ion-Fusion-Science: Summary of U.S. Progress.

IF/P5-12 Kingsep, A.S., Fast Z-Pinch Experiments at the Kurchatov Institute Aimed at the Inertial Fusion Energy.

IF/P5-13 Zhang, Y., Linear analysis of sheared flow profiles and compressibility effects on Z-pinch MRT instability.

IF/P5-14 Hora, H., Fast Ingnition by Laser Driven Particle Beams of Very High Intensity.

FT/P5-39 Norimatsu,·T., Conceptual Design of Laser Fusion Reactor KOYO-F Based on Fast Ignition Scheme.

FT/P5-40 Kawanaka,·J., New Concept of Laser Fusion Energy Driver Using Cryogenic Yb:YAG Ceramics.

References

[1] Moses E.I. et al 2006 J. Phys. IV France **133** 9

[2] Besnard D. 2006 J. Phys. IV France **133** 47
[3] Kodama R. Et al 2001 Nature **412** 798
[4] Lindl J. D. 1995 Phys. Plasmas **11** 339
[5] Azechi H. et al 1991 Laser Part. Beam **9** 193
[6] Kodama R et al 2002 Nature **418** 933
[7] Stephens R. B. et al 2005 Phys. Plasmas **12** 056312
[8] Atzeni S. 1999 Phys. Plasmas **6** 3316
[9] Zheng C. Y. 2005 Phys. Plasmas **12** 044505
[10] Stephens Y. et al 2004 Phys. Rev. E**69** 066414
[11] Li Y. .T. et al 2006 Phys. Rev. Lett..**96** 165003
[12] http://www.hiper-laser.org/public/fusion.htm

J. Phys. IV France **133** (2006) 89–94

DOI: 10.1051/jp4:2006133017

Status of inertial confinement fusion research in China

X.T. He[1] and W.Y. Zhang[1]

[1] *National Hi-Tech Inertial Confinement Fusion Committee,*
Office Address: PO Box 8009-55, Beijing 100088, China

Abstract. The goal of the first milestone of the inertial confinement fusion (ICF) program in China is fusion ignition and plasma burning in about 2020. Under the program, in the past years, the target physics research achieved great progress; SG-II has been operating with high quality since 2000 and SG-III prototype began operating in 2005, and the support technologies for laser drivers are developed and improved; precise diagnostic techniques are developed and relatively integrated system is set up; precise target fabrications are coordinately developed.

1. INTRODUCTION

The ICF program under National Hi-Tec Inertial Confinement Fusion Committee in China has been performed since 1993, the key institutions to join the program are the Laser Fusion Research Center of China Academy of Physics Engineering, Institute of Applied Physics and Computational Mathematics, National Laboratory on High Power Laser and Physics, China Institute of Atomic Energy, and also Institute of Physics of Chinese Academy of Sciences (CAS) and Shanghai Institute of Fine Optics and Mechanics of CAS, etc. The mission of the program is to study and understand target physics toward ignition and plasma burning, hence, to coordinately develop driver technologies and architectures, precision measurements and relevant diagnostic systems, and precision target fabrications.

This paper briefly reports the latest advances and the future plan for ICF research and development in China.

2. DEVELOPMENT OF DRIVERS

For the development of drivers involving lasers and Z-pinch, solid state laser facilities (Shenguang–SG series) are the first priority.

After SG-I retired in 1994, SG-II that has eight beams and routine energy output of $\geq 6\,\text{kJ}$ (1ω) and $\geq 3\,\text{kJ}$ (3ω), is operating in high beam quality and has provided about thousands shots for experiments since 2000 [1]. This facility is the first successful development of two sets of double-pass and co-axial array main amplifier with the aperture of $\phi\,200\,\text{mm}$ for each beam. In addition to eight beams of SG-II, a new laser beam (called the ninth beam), in Fig.1, with laser energy output of 4.5 kJ (1ω). and 2.2 kJ (2ω), pulse duration 3ns, began operating in the end of 2004 and is coupling with SG-II to be used as a diagnostic beam, such as x-ray backlighting with camera, etc.

SG-IIU that is an upgraded SG-II and has laser energy output of 18 kJ (3ω), pulse duration 3 ns, will be operating in 2007.

The ninth beam, in addition to being the diagnostic beam, will be set up as a petawatt laser (PW laser) with laser energy of 1-2 kJ, pulse duration of 1-2 picosecond, and will be operative in 2007.

SG-III prototype with eight beams (Fig.2), which is one of the eight bundles of SG-III and delivers an energy of 15-20 kJ (3ω) serving for both technological platform of SG-III construction and physical experiment [2], began operating in 2005. The facility has used the advanced technologies, such as the fiber optics front-end, etc.

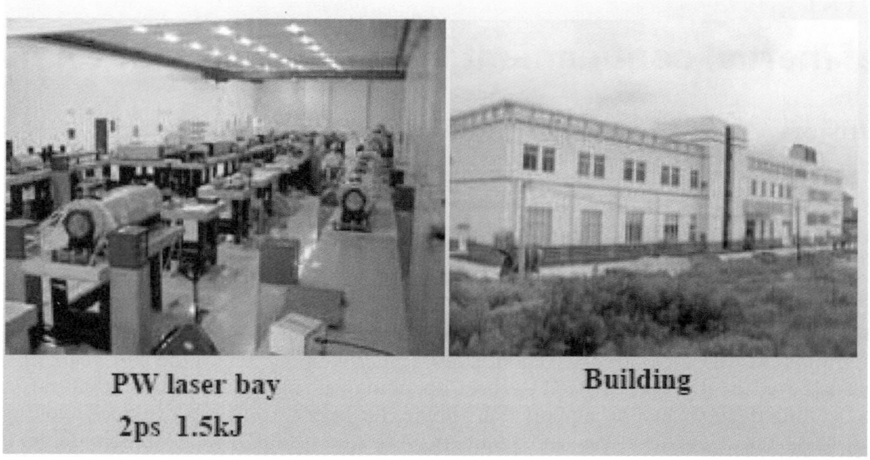

Figure 1. The ninth laser beam.

Figure 2. SG-III prototype.

SG-III laser facility to be used to investigate target physics before ignition for both direct-driven and indirect-driven ICF, will be operating in about 2010. The design has been completed and the support technologies are developed. The facility has 64 beams (eight bundles) and laser energy output of 150-200 kJ (3ω) for pulse duration about 3 ns. If fast ignition is workable, SG-III will couple with a PW laser of tens kJ to demonstrate fast ignition.

The project to architect the SG-IV, National Ignition Driver (NID), that has laser energy output of about MJ serving for the central ignition with moderate gain and fast ignition with high gain, has been authorized, and the architecture of SG-IV will be completed in about 2020.

In addition, KrF excimer laser facility and Z-Pinch device are developed as well in China.

3. THE LATEST ADVANCES IN TARGET PHYSICS

(1) The LARED family consists of the 2D and 3D codes that are continuously being improved and checked by experiments on SG-II, and has been applied to investigate hohlraum physics, radiation transport, direct-drive and indirect-drive implosion dynamics, hydrodynamic instabilities, and laser-plasma interactions.

(2) Direct-drive implosion experiments on SG-II have obtained many important data involving the capsule dynamics framing image and encoding image of the α particles escaping from the compressed fuel region (Fig.3), etc.

(3) Indirect-drive implosion experiments on SG-II have supplied many important results involving hohlraum behavior, fuel core framing and self-backlighting images, etc. In particular, we have precisely measured the dynamic process of the interface between capsule shell (doped sulphur thin film on the inner wall) and fuel region (doped argon) using x-ray backlighting and pin-hole images, see Fig.4. The LARED simulations reproduced the experimental results of the capsule implosion compression in hohlraum, the experimental profiles of the imploding fuel core agree with numerical simulations, see Fig.5.

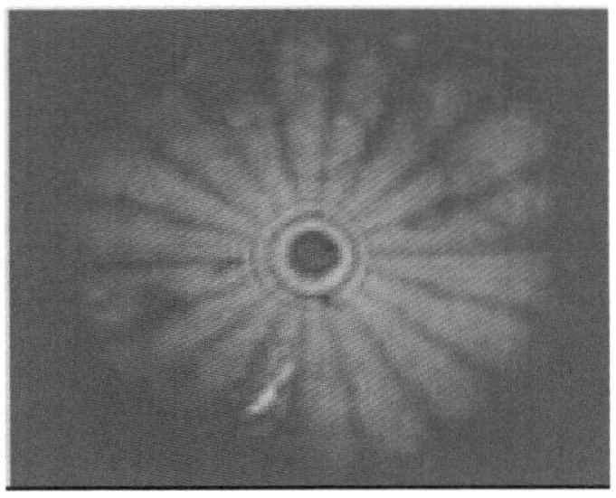

Figure 3. α particle encoding image.

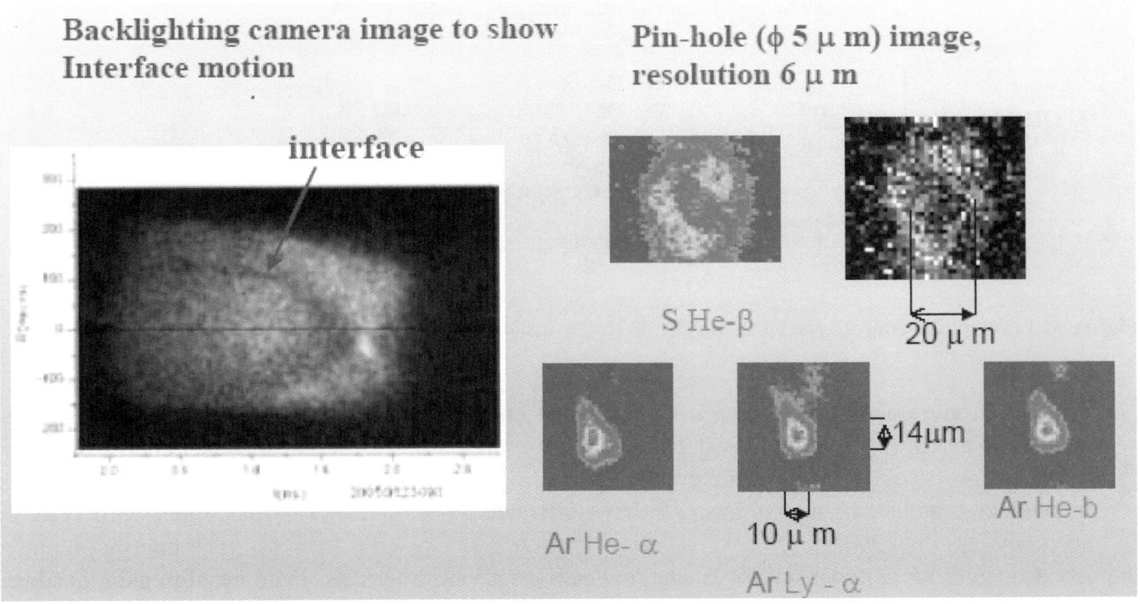

Figure 4. Dynamic process of interface between inner wall of capsule shell and fuel region.

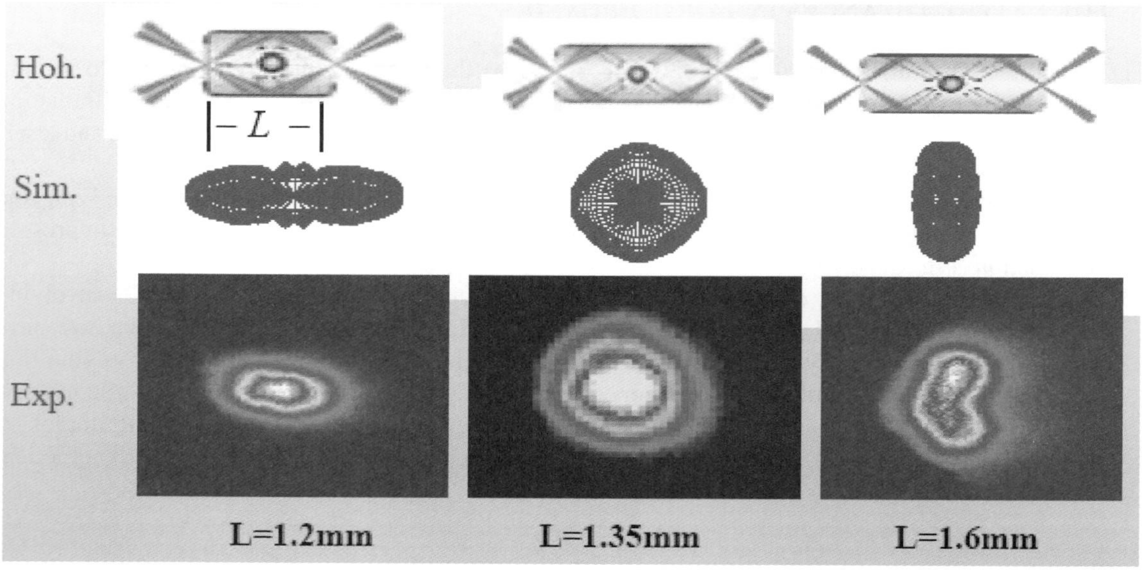

Figure 5. Experimental and theoretical comparison of indirect-driven fuel core profiles.

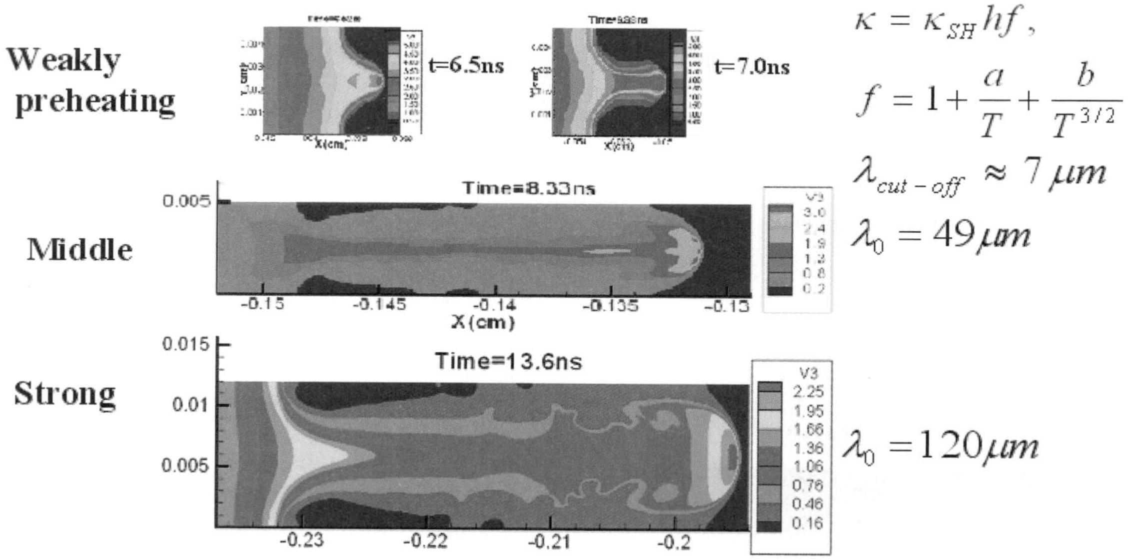

Figure 6. Long jet generated by ablative Rayleigh-Taylor instability.

(4) The radiation transport experiments have measured the transport time, energy spectrum structures, and heat wave propagation in low density foam.

(5) Equation of state (EOS) in high pressure is precisely measured using SG-II laser to irradiate impedance-matched planar targets. We have obtained EOS of 23 Mbar for copper with velocity accuracy less than 2%.

(6) Ablation hydrodynamic instability in the preheating case is numerically simulated by LARED codes, it is shown that the moderate preheating can adequately suppress the Kelvin-Helmholtz instability, as a result, the ablative Rayleigh-Taylor instability drives the formation of the long jets, in Fig.6 [3].

(7) The collimation and the stabilization of the relativistic electron beam (REB) in PW laser interaction with relativistic plasma are one of the critical issues for fast ignition. Numerical simulations using 3D-PIC (one of LARED family) show that the circularly polarized (CP) laser interaction with dense plasma generates both axial and azimuthal magnetic fields. The azimuthal field causes a pinch effect on REB and the axial field plays a trapping role that leads to REB having a small divergent angle, such as 2–3° for laser intensity of 10^{19} w/cm^2. On the other hand, the linearly polarized (LP) laser interaction with dense plasma generates only azimuthally magnetic field that leads to the REB having a large divergent angle, such as about 30–40° for laser intensity 2×10^{19} w/cm^2. Moreover, in the intense LP laser beam interaction with relativistic plasma, the transverse spot of the electric current exhibits an ellipse-like structure that is different from the cylindrical structure for the CP laser, therefore, when the laser beam began boring hole in the dense plasma and the REB separates from the LP laser beam to penetrate overdense plasma, the ellipse-like structure of the REB may cause the increase of the growth rate of Weibel instability in overdense plasma. As a result, the CP laser serving for fast ignition may be better than the LP laser [4–7].

4. DIAGNOSTIC TECHNIQUES AND SYSTEMS

We have diagnostic techniques in many ways and have set up an integrated diagnostic system. The latest advances in the diagnostic techniques are dedicated to measuring the density near critical surface using the interferogram of the soft x-ray laser at wavelength 13.9 nm produced by the nickel-like silver in hot plasma, and to obtaining plasma temperature using Thomson scattering method, non-local heat flow and the plasma expanding velocity using Thomson scattering combined with numerical calculation [8].

5. TARGET FABRICATIONS

In addition to direct-driven capsules with glass shell and multi-layer plastic shell, the hohlraum targets in various shapes and sizes have been used to experimentally study radiation transport, opacity, indirect-drive implosion dynamics, etc.

6. FINAL REMARKS

In the past years, a series of experiments on SG-II have obtained many important achievements to understand target physics; the numerical simulations using LARED family, which are continuously improved and checked by experiments, can reproduce many experimental results and predict many new physical phenomena, and thus have provided the design parameters to construct SG-IIU, SG-III prototype and SG-III.

SG-IIU deriver will serve for experiments in target physics beginning in 2007, and will couple with PW laser to investigate hot spot behavior of fast ignition.

SG-III prototype that began operating in 2005 will provide the use of both technological platform for SG-III architecture and physical experiment.

SG-III driver will be completed in about 2010 and be coupled with tens kJ PW lasers for fast ignition demonstration.

The ignition project was authorized. The architecture of SG-IV laser facility, National Ignition Driver, with laser energy output of about MJ for central ignition with moderate gain and fast ignition with high gain is being planned. China is moving forward to the goal of ignition and plasma burning in about 2020 and it could be earlier in about 2015 if fast ignition is favorable.

Acknowledgments

The authors would like to express their appreciation for the many people and the relevant institutions to diligently work under the ICF program in China. The work was supported by the National Hi-Tech Inertial Confinement Fusion Committee of China.

References

[1] X. T. He et al., "Inertial fusion energy research progress in China", in 6th Symposium On Current Trends in International Fusion Research: A Review, 2005, www.physicsessays.com
[2] W. Zheng, "Status of prototype for Shenguang-III laser facility," presentation at this conference.
[3] X. T. He, "Jet-like long spike of ablative Rayleigh-Taylor instability", in 19th International Conference on Numerical Simulation of Plasma and 7th Asia Pacific Plasma Theory Conference, July 12-15, 2005, Nara, Japan.
[4] C. Y. Zheng et al., Phys. Plasmas **12**, 044505 (2005).
[5] B. Qiao et al., Phys. Plasmas **12**, 083102 (2005).
[6] B. Qiao et al., Phys. Plasmas **12**, 053104 (2005).
[7] H. Liu, X. T. He, and S. G. Chen, Phys. Rev. E **69**, 066409 (2004).
[8] Q. Z. Yu et al., Phys. Rev. E**71**, 046407 (2005).

Eur. Phys. J. D **44**, 227–231 (2007)
DOI: 10.1140/epjd/e2007-00005-1

THE EUROPEAN
PHYSICAL JOURNAL D

Inertial fusion research in China

X.T. He[a] and W.Y. Zhang

National Hi-Tech Inertial Confinement Fusion Committee, P.O. Box 8009-55, Beijing 100088, P.R. China

Received 3rd January 2006 / Received in final form 28 July 2006
Published online 17 January 2007 –

Abstract. The goal of the first milestone of the inertial fusion program in China is to reach fusion ignition and plasma burning in about 2020. Under the program, in the past years, the inertial fusion physics research achieved great progress; the laser facilities and the support technologies for laser drivers are advanced; the advanced diagnostic techniques are developed and the relatively integrated system is set up; the precise target fabrications are coordinately developed.

PACS. 52.38.Kd Laser-plasma acceleration of electrons and ions – 52.38.Fz Laser-induced magnetic fields in plasmas

1 Introduction

The inertial fusion program under National Hi-Tec Inertial Confinement Fusion Committee in China has been performed since 1993, the key institutions to join the program are the Laser Fusion Research Center of China Academy of Physics Engineering; Institute of Applied Physics and Computational Mathematics; National Laboratory on High Power Laser and Physics; China Institute of Atomic Energy. In addition, Institute of Physics of Chinese Academy of Sciences (CAS) and Shanghai Institute of Fine Optics and Mechanics of CAS, and some universities join the program as well. The mission of the program is to study and understand inertial fusion physics and to develop the drivers, the goal is toward ignition and plasma burning in about 2020, for this purpose, hence, to coordinately develop driver technologies and architectures, precision measurements and relevant diagnostic systems, and precision target fabrications.

The emphases of this paper are given on the latest advances of inertial fusion physics and drivers, and the future plan for inertial fusion research and development in China.

2 Development of drivers

For the development of drivers involving lasers and Z-pinch, solid state laser facilities (Shenguang–SG series) are put to the first priority.

After SG-I was decommissioned in 1994, SG-II (Fig. 1a) that has eight beams and routine energy output of ≥6 kJ (1ω) and ≥3 kJ (3ω), is operating in high beam quality and has provided about thousands shots for experiments since 2000 [1]. This facility is the first successful development of two sets of double-pass and co-axial array main amplifier with the aperture of $\phi = 200$ mm for each beam.

Fig. 1. (a) SG-II laser bay and target chamber, (b) the ninth laser beam to be PW laser.

In addition to eight beams of SG-II, a new laser beam (named ninth beam) in Figure 1b, with laser energy output of 4.5 kJ (1ω) and 2.2 kJ (2ω), pulse duration 3 ns, began operating in the end of 2004 and is coupling with SG-II to be used as a diagnostic beam, such as X-ray backlighting, etc.

SG-IIU that is an upgraded SG-II and has laser energy output of 18 kJ (3ω), pulse duration 3 ns, will be operating in 2007.

[a] e-mail: xthe@iapcm.ac.cn

Fig. 2. SG-III prototype.

The ninth beam will also be set up as a petawatt laser (PW laser) with laser energy of 1–2 kJ, pulse duration of 1–2 picosecond, and be operating in 2007.

SG-III prototype with eight beams (Fig. 2), which is one of the eight bundles of SG-III and is of laser energy output of 15–20 kJ (3ω) serving for both technological platform of SG-III construction and physical experiment [2], began operating in 2005. The facility uses the most advanced technologies, such as a whole optical fiber front end and whole solid state laser pumping, etc.

SG-III laser facility will be used to investigate target physics before ignition for both direct-drive and indirect-drive ICF and will be operating in about 2010. The design has been completed and the support technologies are advanced. The facility has 64 beams (eight bundles) and laser energy output of 150–200 kJ (3ω) for a pulse duration of about 3 ns. If fast ignition is workable, SG-III will couple with PW laser of tens kJ to demonstrate fast ignition.

The project to architect SG-IV, National Ignition Driver (NID), that has laser energy output of about 1–1.5 MJ serving for the central ignition with moderate gain and fast ignition with high gain, has been authorized, and the architecture of SG-IV will be completed in about 2020.

In addition, KrF excimer laser facility and Z-Pinch device are developed as well in China.

3 The latest advances in target physics

(1) In the past ten years, the LARED series of codes that consists of the 2D and 3D codes has been developed and is currently checked by experiments on SG-II. The series cover 2D code for hydrodynamics and 3D code for radiation transport; 2D codes for direct-drive and indirect-drive implosion dynamics, ignition and burning propagation; 2D and 3D codes for hydrodynamic instabilities and implosion dynamics; 2D and 3D for particle simulation; and others. It has been applied to investigate hohlraum physics, radiation transport, direct-drive and indirect-drive implosion dynamics, hydrodynamic instabilities, thermonuclear ignition and plasma burning and laser-plasma interactions, and so on.

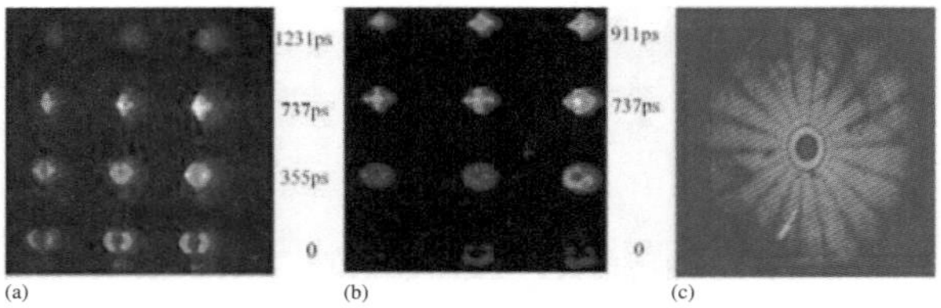

Fig. 3. (a) and (b) Capsule implosion dynamic framing, (c) particle encoding image.

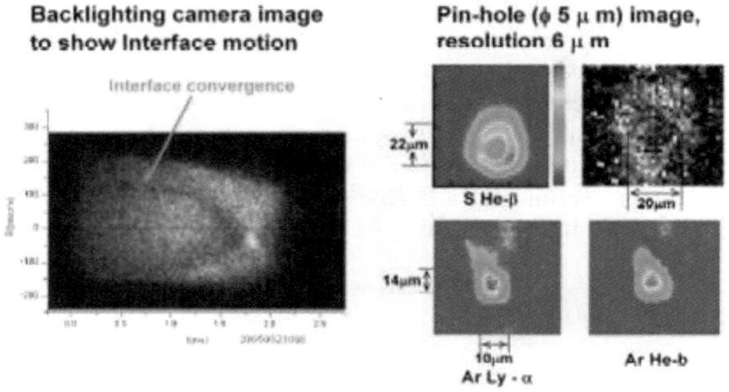

Fig. 4. Dynamic process of interface between inner wall of capsule shell and fuel region.

(2) Implosion dynamics studies on SG-II facility and numerical simulations using LARED series.

Direct-drive implosion experiments on SG-II have obtained many important data involving the capsule dynamics framing image and encoding image of the particles escaping from the compressed fuel region, as seen in Figure 3, etc.

Indirect-drive implosion experiments on SG-II have observed many important results involving hohlraum behavior, fuel core framing and self-backlighting images, etc. In particular, we have precisely measured the dynamic process of the interface between capsule shell (doped sulphur thin film on the inner wall) and fuel region (doped argon) using X-ray backlighting and pin-hole images with the resolution of 6 μm, see Figure 4, it contributes to our understanding of implosion dynamics and checking of LARED series.

The LARED simulations reproduced the indirect-drive experimental results of the capsule implosion compression, and the experimental profiles of the imploding fuel core, which vary with the length L of the hohlraum, agree with numerical simulations, see Figure 5.

(3) The radiation transport experiments have measured the transport time, energy spectrum structures, and heat wave propagation in the low density foam.

(4) Equation of state (EOS) in high pressure is precisely measured using SG-II laser to irradiate aluminium-copper impedance-matched planar targets (Fig. 6a) and 23 Mbar for copper with velocity accuracy less than 2% is

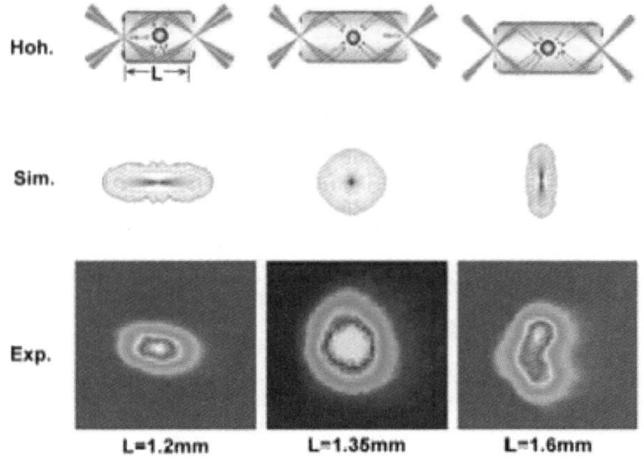

Fig. 5. Experimental and theoretical comparison of the fuel core profiles in indirect-drive implosion.

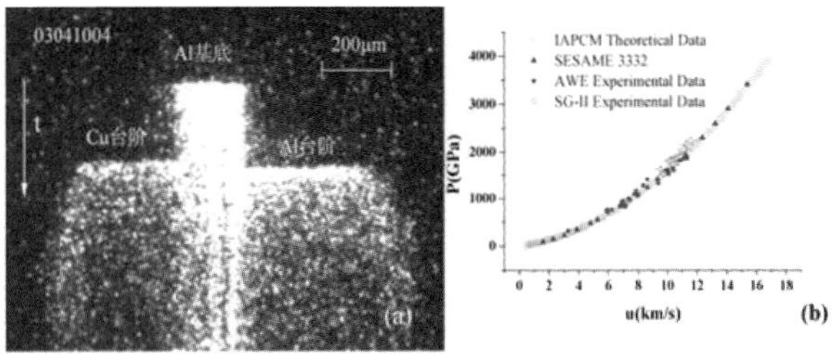

Fig. 6. Plots for measurements of EOE: (a) aluminium-copper impedance-matched target shock wave recorded by streak camera with velocity accuracy less than 2%, (b) experimental and theoretical curves.

obtained. The experimental results on SG-II and the theoretical simulations are shown in Figure 6b, where other experiments are also compared.

(5) Hydrodynamic instabilities may occur at an interface between two fluids with different densities when a perturbation exists. Ablation hydrodynamic instability occurs at a sharp ablation front (Fig. 7a), which is formed when a laser beam is incident on a target surface where the nonlinear heat conduction penetrates into over critical density. At that time, a small temperature tongue, i.e., preheating, exists ahead of the sharp ablation front due to X-ray emission from high temperature plasma and/or nonlocal heat conduction. We write the heat conductivity in the form of $\kappa = \kappa_{sh} h f$, where κ_{sh} is Spitzer-Harm conductivity, h is the flux limitation, $f = 1 + a/T + b/T^{3/2}$ and temperature T in units of 10^6 K, $b = 1$ and $c = 0.3$ (weak preheating), $b = 10$ and $c = 1$ (moderate), $b = 100$ and $c = 10$ (strong). Numerical simulations by LARED codes show that the weak preheating can adequately suppress the Kelvin-Helmholtz (KH) instability and thus leads to the spike formation without mushroom structure, however, the weak preheating cannot suppress the harmonic mode that results in the rupture of the spike due to in-

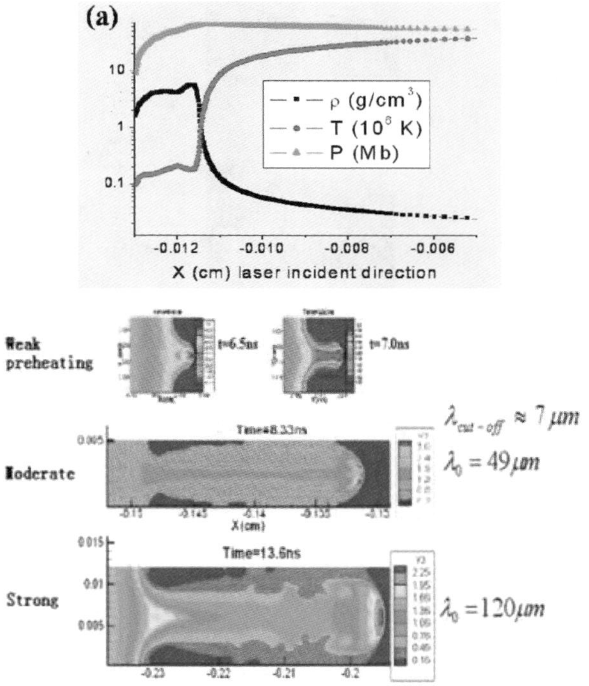

Fig. 7. (a) Background: a typical simulated 1D profiles of density, temperature and pressure nearby the ablative front. There may have a small temperature tongue from non-local heat conduction at ablative front. (b) Long jet generated by ablative Rayleigh-Taylor (RT) instability.

teraction of the master mode with the harmonic mode. Enhancing the preheating still further, the moderate preheating can adequately suppress both KH instability and harmonic instability, as a result, the ablative Rayleigh-Taylor instability drives the formation of the long jets [3], see Figure 7b. It may contribute to understand some processes in inertial fusion and astrophysics.

(6) Fast ignition investigation by numerical simulations.

The implosion dynamics of the cone target is investigated by numerical simulations using LARED-S code under SG-IIU condition. The laser energy of 10 kJ is incident on the surface of the cone target capsule. After a laser pre-pulse power rising linearly from $W_0(t = 0) = 0$ to $W_1(t = 3.45 \text{ ns}) = 0.54 \times 10^{11}$ W/cm^2, the laser power rises linearly from W_1 to $W_2 = 10^{13}$ W/cm^2 at $t_2 = 4.2$ ns, and then keeps a constant W_2 till $t_3 = 4.9$ ns, at last, reduces to zero at $t_4 = 5.2$ ns. The capsule of the cone target has a CD shell with density of 1.0 g/cc, and the initial radius of the capsule is 325 μm. The cone makes an outer angle of 60 degree and the cone wall consists of copper.

The numerical results show that during implosion process the maximal density of the CD shell reaches over 1000 g/cc (Fig. 8a) and jets-like due to nonlinear RT instability are observed (10b). The implosion dynamics in

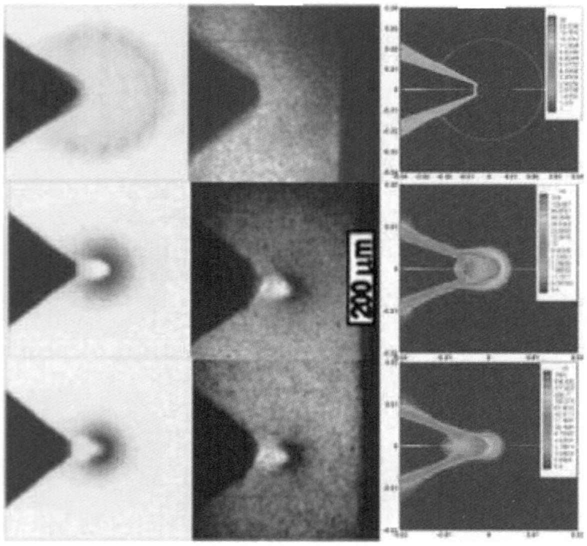

(a) (b) (c)

Fig. 8. Implosion dynamics of the cone target in different times from top to bottom: (a) experiment on OMEGA, (b) numerical simulation by LASNEX, (c) numerical simulation by LARED-S.

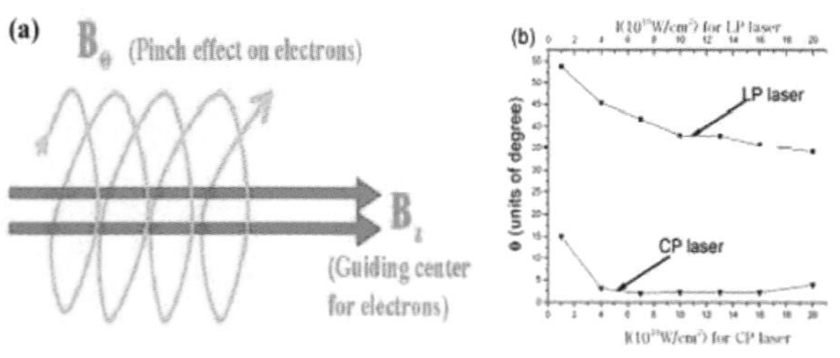

Fig. 9. Collimation of electrons: (a) magnetic field effects, (b) divergent angles.

our case is compared with experiment on OMEGA laser facility and numerical simulation by LASNEX, and the profiles agree very well.

The collimation and the stabilization of the relativistic electron beam (REB) in PW laser interaction with relativistic plasma are one of the critical issues for fast ignition. Numerical simulations using 3D-PIC (one of LARED series) show that the circularly polarized (CP) laser interaction with dense plasma generates both axially and azimuthally magnetic fields, the azimuthal field causes a pinch effect on REB and the axial field plays a trapping role that leads to REB having a small divergent angle, such as 2–3° for laser intensity of 10^{19} W/cm^2. On the other hand, the linearly polarized (LP) laser interaction with dense plasma generates only azimuthally magnetic field that leads to the REB having a large divergent angle, such as about 30–40° for laser intensity 2×10^{19} W/cm^2, as shown in Figure 9. Moreover, the transverse spot of

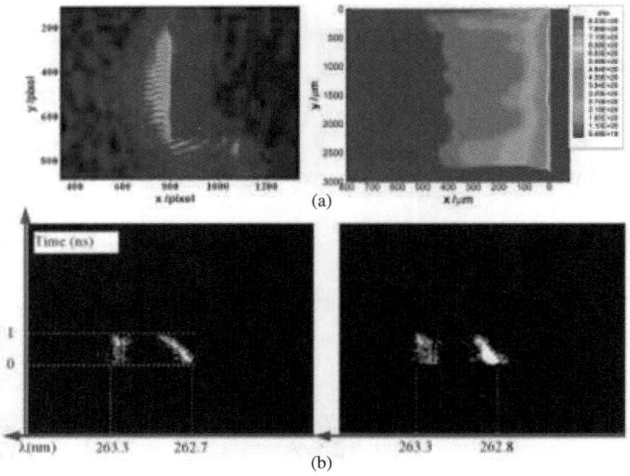

Fig. 10. (a) The dynamic interferogram image (left) of soft X-ray laser with wavelength 13.9 nm and the reconstructed electronic density of CH plasma near critical surface at the time of 1 ns after the peak of the drive laser intensity (right). (b) The patterns from Thomson scattering experiment.

the electric current exhibits an ellipse-like structure as the REB penetrating dense plasma deeply, therefore, when the REB separating from the LP laser beam penetrates the overdense plasma where the REB deposits energy to form a hot spot for fast ignition, such a profile of the REB may cause the increase of the growth rate of Weibel instability to compare with CP laser [4–7].

4 Diagnostic techniques and systems

We apply diagnostic techniques in many ways and set up a relatively integrated diagnostic system to measure physical experiments, such as, using the ninth beam to measure implosion dynamics, using the gated multi-framing camera and Wolter X-ray microscopes to diagnose radiation transport and imploding behavior, and so on.

The latest advances in the diagnostic techniques are using the dynamic Mach-Zehnder interferogram image of the saturated soft X-ray laser at wavelength 13.9 nm, which is produced by the nickel-like silver in hot plasma driven by Nd glass laser, to measure the plasma density (Fig. 10a), and also using Thomson scattering experiments (Fig. 10b) combined with other methods to obtain plasma temperature, non-local heat flow and the plasma expanding velocity from the pattern structure of the experimental pictures on SG-II [8].

5 Target fabrications

In addition to direct-drive capsules with glass shell and multi-layer plastic shell, the hohlraum targets in various shapes have been applied to experimentally study radiation transport, opacity, indirect-drive implosion dynamics, etc. (Fig. 11).

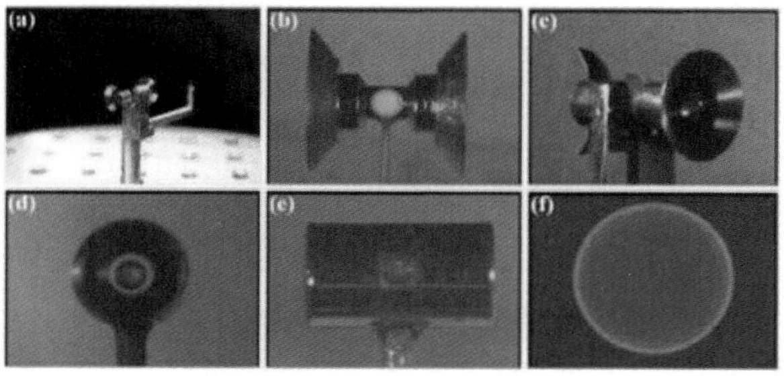

Fig. 11. Various configurations of the indirect-drive targets: (a) opacity target with two-point backlighting, (b) radiation transport target, (c) ablation target with two-point backlighting, (d, e) cylindrical hohlraum target (capsule inside), about $\phi = 800 \times 1600$ μm for SG-II experiments, (f) multi-layer (PS-PVA-CH) capsules, about $\phi = 300$–500 μm.

6 Summary and conclusion

In the past years, many important achievements were obtained in a series of experiments on SG-II to understand target physics; the numerical simulations using LARED series of codes can reproduce many experimental results and predict many new physical phenomena and provided the design parameters to construct SG-IIU, SG-III prototype and SG-III.

SG-IIU driver will serve for experiments in target physics beginning in 2007, and will couple with PW laser to investigate hot spot behavior of fast ignition.

SG-III prototype that began operating in 2005 will provide the use of both the technological platform for SG-III architecture and the physical experiment.

SG-III driver will be completed in about 2010 and be coupled with tens kJ PW lasers for fast ignition demonstration.

The ignition project was authorized. The architecture of SG-IV laser facility, National Ignition Driver, with laser energy output of about MJ for central ignition with moderate gain and fast ignition with high gain is being planned. China is going to be toward the goal of ignition and plasma burning in about 2020 and may advance to carry out it in about 2015 if fast ignition is favorable.

The authors would like to express their appreciation for the many people and the relevant institutions to diligently work under the ICF program in China. The work was supported by the National Hi-Tech Inertial Confinement Fusion Committee of China.

References

1. X.T. He et al., Inertial fusion energy research progress in China, in *6th Symposium On Current Trends in International Fusion Research: A Review*, 2005, `www.physicsessays.com`
2. W. Zheng, "Status of prototype for Shenguang-III laser facility," presentation at IFSA05
3. X.T. He, Jet-like long spike of ablative Rayleigh-Taylor instability, in *19th International Conference on Numerical Simulation of Plasma and 7th Asia Pacific Plasma Theory Conference*, July 12-15, 2005, Nara, Japan
4. C.Y. Zheng et al., Phys. Plasmas **12**, 044505 (2005)
5. B. Qiao et al., Phys. Plasmas **12**, 083102 (2005)
6. B. Qiao et al., Phys. Plasmas **12**, 053104 (2005)
7. H. Liu, X. T. He, S.G. Chen, Phys. Rev. E **69**, 066409 (2004)
8. Q.Z. Yu et al., Phys. Rev. E **71**, 046407 (2005)

APPLIED PHYSICS LETTERS **93**, 011101 (2008)

Twenty times lower ignition threshold for laser driven fusion using collective effects and the inhibition factor

H. Hora,[1,a)] B. Malekynia,[2] M. Ghoranneviss,[2] G. H. Miley,[3] and X. He[4]
[1]*Department of Theoretical Physics, University of New South Wales, Sydney 2052, Australia*
[2]*Plasma Physics Research Center, Science and Research Branch, Islamic Azad University, Tehran-Poonak 14835-159, Iran*
[3]*Department of Nuclear, Plasma and Radiological Engineering, University of Illinois, Urbana, Illinois 61801, USA*
[4]*Institute of Applied Physics and Computational Mathematics, Beijing 100088, People's Republic of China*

(Received 2 April 2008; accepted 17 June 2008; published online 7 July 2008)

Hydrodynamic analysis for ignition of inertial fusion by Chu [Phys. Fluids **15**, 413 (1972)] arrived at extremely high thresholds of a minimum energy flux density E^* at 4×10^8 J/cm^2 which could be provided, e.g., by spark ignition. In view of alternative schemes of fast ignition, a re-evaluation of the early analysis including later discovered collective stopping power and the inhibition factor results in a 20 times lowering of the threshold for E^*. © *2008 American Institute of Physics.* [DOI: 10.1063/1.2955839]

In order to produce a fusion flame by an incident radiation flux density E^* in deuterium tritium of higher than solid-state density, hydrodynamic calculations arrived at necessary extremely high energy flux densities, e.g., at a threshold of $E_t^*=4\times10^8$ J/cm^2 for solid density deuterium tritium (DT).[1,2] One way to reach these conditions was the scheme of spark igntion[3] where it was clarified from exceedingly detailed numerical studies, that a spherically irradiated DT plasma had to have an about 7 keV hot core of modest density for producing a fusion flame into an outer DT shell of low temperature and very high compression for reaching gains of 200 more fusion energy than irradiated laser energy. The reaction in the core follows volume ignition and resulted in a value above E_t^*.[4]

In view of later developments of fast ignition[5] and further advanced modifications by block ignition,[6,7] re-examination of the early hydrodynamic computations was indicated in view of later discovered phenomena. One of them is the inhibition factor E^* for reduction of thermal conduction and the other is the effect reducing stopping lengths of the generated alphas in the fusion plasma due to the collective effects. The following hydrodynamic theory is rather limited and does not include the interpenetration effects of particles from hot into cold plasmas[8] which may be arriving at different results by using the particle in cell (PIC) techniques[9] as applied to this problem, but the question about the collision processes and the stopping lengths may still not be finalized for the present results.

In order to arrive at the best possible understanding how the effects observed since 1972 led to differences to the earlier results, the hydrodynamic equations are as close as possible following Chu[1] with appropriate modifications. The change due to the inhibition factor F^* for reduced thermal conduction in the interfaces between the hot and cold plasmas were discussed in detail before Ref. 10, where F^* was discovered experimentally and an explanation by electric double layers seemed to be sufficient while the reduction of thermal conduction by the spontaneous magnetic fields in the laser produced plasmas seemed to be of minor influence. The following results include additionally the collective effects for the stopping length of the DT fusion produced alphas. In contrast to the binary collisions of the alphas with electrons, Gabor[11] concluded a stronger collective interaction with the whole electron cloud in a Debye sphere. An experiment for very intense 2 MeV electrons[12] could be explained completely by the collective effect.[13] The drastic difference for the stoping length for plasma temperatures above 100 eV, for binary collisions used by Chu following the Winterberg approximation [see Eq. (7) of Ref. 1] for the stopping length R_{BB} of the binary Bethe–Bloch theory

$$R_{BB}\propto T^{3/2} \tag{1}$$

was shown in Fig. 6 of Ref. 14, where the collective interaction[11,15] resulted in a very weak dependence of the range R on the temperature T with very much shorter stopping lengths for solid-state density DT

$$R=0.01-1.7002\times10^{-4}T\ \text{cm}\quad(\text{temperature } T \text{ in keV}). \tag{2}$$

To be as close as possible to the analysis of Chu,[1] the following hydrodynamic equations are used. The equations of continuity and reactions $(D+T\rightarrow\alpha+n)$ may be combined to yield as equations of mass conservation

$$\frac{\partial\rho}{\partial t}+\frac{\partial}{\partial x}(\rho u)=0 \tag{3}$$

and

$$\frac{\partial Y}{\partial t}+u\frac{\partial Y}{\partial x}=W, \tag{4}$$

where ρ is the mass density, u is the mass velocity, and Y is the fraction of material burned, defined by

$$Y=(n_\alpha+n_n)/(n_D+n_T+n_\alpha+n_n).$$

W is the reaction rate function, given by

[a)]Electronic mail: h.hora@unsw.edu.au.

$$W=\tfrac{1}{2}n(1-Y)^2\langle\sigma v\rangle.$$

It is obvious that Eq. (3) is the same as the mass conservation equation, due to the small percentage ($\sim 0.35\%$) of the mass transformed into energy. In the equation for Y, the n's are the particle densities, and the subscripts are for the different particle species. In the equation for W, the n stands for the total number density of the ions.

The equation of motion expressing the conservation of momentum is

$$\frac{\partial u}{\partial t}+u\frac{\partial u}{\partial x}=-\rho^{-1}\frac{k}{m_i}\frac{\partial}{\partial x}[\rho(T_i+T_e)]+\rho^{-1}\frac{\partial}{\partial x}\left[(\mu_i+\mu_e)\frac{\partial u}{\partial x}\right], \tag{5}$$

in which the pressure and viscosity terms are included. $\mu_{i,e}$ are the viscosity coefficients whose values are taken to be

$$\mu_{i,e}=\frac{0.406 m_{i,e}{}^{1/2}(kT_{i,e})^{5/2}}{e^4\ln\Lambda},$$

where $\ln\Lambda$ is the usual Spitzer logarithm.

The ion and electron temperature equations expressing the conservation of energies

$$\frac{\partial T_i}{\partial t}+u\frac{\partial T_i}{\partial x}=-\frac{2}{3}T_i\frac{\partial u}{\partial x}+\frac{2m_i}{3k\rho}\mu_i\left(\frac{\partial u}{\partial x}\right)^2+\frac{2m_i}{3k\rho}\frac{\partial}{\partial x}\left(K_i\frac{\partial T_i}{\partial x}\right)+W_i+\frac{T_e-T_i}{\tau_{ei}} \tag{6}$$

and

$$\frac{\partial T_e}{\partial x}+u\frac{\partial T_e}{\partial x}=-\frac{2}{3}T_e\frac{\partial u}{\partial x}+\frac{2m_i}{3k\rho}\mu_e\left(\frac{\partial u}{\partial x}\right)^2+\frac{2m_i}{3k\rho}\frac{\partial}{\partial x}\left(K_e\frac{\partial T_e}{\partial x}\right)+W_e+\frac{T_i-T_e}{\tau_{ei}}-A\rho T_e^{1/2} \tag{7}$$

were included on the right-hand side using the pressure, viscosity, conductivity, thermonuclear energy generation, equilibration terms, and energy transfer terms W_1 and W_2 following Chu.[1] The last term on the right-hand side of Eq. (7) is the bremsstrahlung term.

For the following reported computations, the bremsstrahlung is based on the electron temperature T_e working with Eq. (15) of Chu[1] with the maximum at $x=0$; thus,

$$W_i+W_e=A\rho T_e^{1/2}+\tfrac{8}{9}(k/m_i)(1/aT_e^{1/2})+\tfrac{2}{9}(T_e/t). \tag{8}$$

Equation (8) is a little different from Eq. (20) of Chu[1] where $T_i=T_c$ is assumed while the following computations with the collective stopping has to be for general temperatures. The α particles are assumed to deposit their energy in the plasma. They have a mean free path for solid-state density DT according to Eq. (2) for the collective effect to distinguish from Eq. (1) used by Chu.[1] For the calculation of the collective effect, a term to the right-hand side of Eq. (8) was added

$$W_i+W_e=A\rho T_e^{1/2}+\tfrac{8}{9}(k/m_i)(1/aT_e^{1/2})+\tfrac{2}{9}(T_e/t)+P, \tag{9}$$

where P is the thermonuclear heating rate per unit time obtained from the burn rate and the fractional alpha particle deposition:

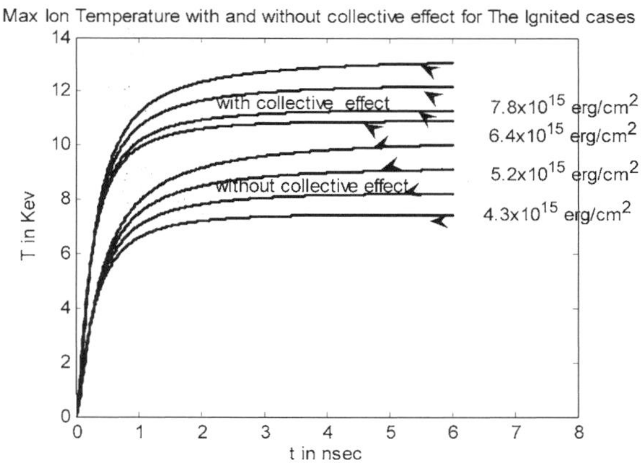

FIG. 1. Characteristics of the plasma temperature T on time t similar to Fig. 2 of Chu (Ref. 1) with the energy flux density E^* in erg/cm^2 with and without the collective effect for the stopping of the alphas from the DT reaction. Transition into constant temperature on time for 4.3×10^8 J/cm^2 corresponds to the threshold for ignition nearly identical with the computation of Chu (Ref. 1) without collective effect and all without inhibition factor.

$$P=\rho\phi E_\alpha f, \tag{10}$$

$$\phi=\frac{dW}{dt}=\frac{d}{dt}\left(\tfrac{1}{2}n(1-Y)^2\langle\sigma v\rangle\right), \tag{11}$$

where E_α=3.5 MeV and f is the fraction of alpha particle energy absorbed by electrons or ions, which has been given by

$$f_i=\left(1+\frac{32}{T_e}\right)^{-1}\ \text{and}\ f_e=1-f_i. \tag{12}$$

For the equations after Eq. (8) the temperatures of the electrons and of the ions were used to be equal T as used in Eq. (2) for the following numerical evaluations.

Ignition can best seen from Fig. 2 of Chu[1] showing the temperature on time t in the irradiated plasma. The present work without the collective effect is shown in Fig. 1. At an energy flux density $E^*=4.3\times 10^8$ J/cm^2, the plasma temperature T merges into a constant time dependence exactly what was the result of Chu, see Fig. 2 of Chu.[1] This E^* is the threshold E_t^* because lower values of E^* result in a decrease of T on t and higher values show an increase of T on t. Figure 1 reports also the results with the collective effect obviously resulting in higher plasma temperatures.

The calculation with the inclusion of the inhibition factor was based on experiments where E^* was measured to be between 30 and 100.[10] The theoretical value is determined by the thermal conductivity within the double layer, given by that of the ions K_i and not by that of the electrons K_e. Classical theory without magnetic fields arrives at

$$K_i=K_e(m_e/m_i)^{1/2} \tag{13}$$

determined by the electron mass m_e and ion mass m_i. For DT the value E^* is then 67.5. Using this inhibition factor E^* in the hydrodynamic computations, the plots similar to Fig. 1 with and without collective effect are given in Fig. 2. This shows that the threshold for ignition with the collective effect and with the inhibition factor is at a value

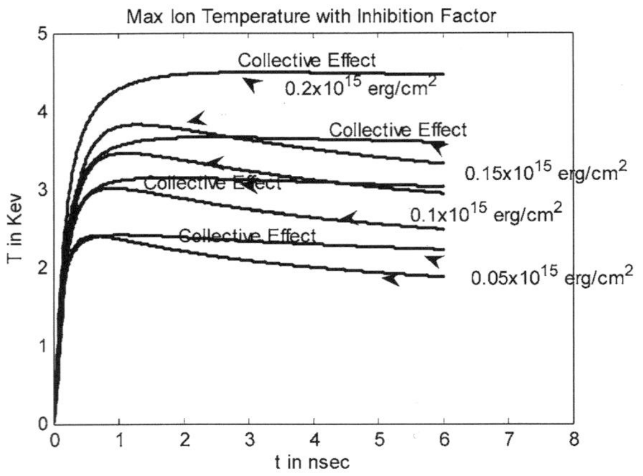

FIG. 2. Same as Fig. 1 with inhibition factor and with and without the collective effect for stopping of the alphas demonstrating ignition at 2×10^7 J/cm^2.

$$E_t^* = 2 \times 10^7 \text{ J/cm}^2. \quad (14)$$

This is a reduction from the initial value of Chu[1] by a factor 21.5. A separate analysis without collective effects was treated in several details only with the inhibition factor.[10]

The value E^* in Eq. (14) is still extremely high, and it will be a question whether conditions of the block ignition with many petawatt-picosecond laser pulses[7] could be within realistic conditions. In view of the initially mentioned interpenetration problems,[8,9] the final decision may still be open. When the next measurements with spark ignition may be gained from experiments like that of the National Ignition Facility NIF,[16] it may be of interest whether the here reported re-evaluation of the hydrodynamic model apart from other theories based on PIC or other techniques may lead to a clarification of the processes involved.

Though the here presented results were initiated by a modification of fast ignition for laser fusion with laser pulses of picosecond duration and one dimensional plane geometry interaction,[7] the connection with nanosecond laser pulse interaction for central spark ignition[17] may be seen from Chu's[1] threshold $E^* = 4 \times 10^8$ J/cm^2. The computations for the spark ignition with the hydrodynamically equivalent physics[17] were possible with the absence of alpha-particle deposition. The re-calculation of the spark ignition[4] had shown that the energy flux density E^* from the spark for ignition of the outer low temperature and high compression mantle was not much higher than the value of Chu. It should be noted that the reaction of the spark was found to fit[4] with volume ignition,[18–20] where the alpha reheat based on the collective stopping was essential. Other measurements of reduced stopping lengths are known from research on heavy ion driven laser fusion.[21] This is the reason why the here reported change of the ignition threshold may lead also to phenomena to be explored with the next studies of spark ignition with NIF (Ref. 16) or similar experiments.

The authors from the Plasma Physics Research Center of IAU (B.M. and M.G.) gratefully acknowledge support and encouragement for fusion energy research by the Coordinated Research Program (CRP) No. 34774 of the International Atomic Energy Agency (IAEA) in Vienna. Special thanks by one of the authors (H.H.) are due to several discussions at the International Center of Theoretical Physics in Trieste/Italy with Professor Reza Amrollahi (Amir Kabir University, Tehran) favoring this project.

[1] M. S. Chu, Phys. Fluids **15**, 413 (1972).
[2] J. L. Bobin, in *Laser Interaction and Related Plasma Phenomena*, edited by H. Schwarz and H. Hora (Plenum, New York, 1974), Vol. 4B, pp. 465–494.
[3] J. L. Nuckolls, in *Inertial Nuclear Confinement Fusion: A Historical Approach by its Pioneers*, edited by G. Velarde and N. Carpinetro-Santamaria (Foxwell & Davies, London, 2007), p. 1.
[4] H. Hora, H. Azechi, Y. Kitagawa, K. Mima, M. Murakami, S. Nakai, K. Nishihara, H. Takabe, C. Yamanaka, M. Yamanaka, and T. Yamanaka, J. Plasma Phys. **60**, 743 (1998).
[5] M. Tabak, J. Hammer, M. N. Glinski, W. L. Kruer, S. C. Wilks, J. Woodworth, E. M. Campbell, M. D. Perry, and R. Mason, Phys. Plasmas **1**, 1626 (1994).
[6] H. Hora, Laser Part. Beams **25**, 37 (2007).
[7] H. Hora, J. Badziak, M. N. Read, Y. Li, T. Liang, Y. Cang, H. Liu, M. Zheng, J. Zhang, F. Osman, G. H. Miley, W. Zhang, X. He, H. Peng, S. Glowacz, S. Jablonski, J. Wolowski, Z. Sklandanowski, K. Jungwirth, K. Rohlena, and J. Ullschmied, Phys. Plasmas **14**, 072701-1 (2007).
[8] H. Hora, Atomkernenerg./Kerntech. **42**, 7 (1983).
[9] S. C. Wilks, W. L. Kruer, M. Tabak, and A. B. Langdon, Phys. Rev. Lett. **69**, 1383 (1992).
[10] M. Ghoranneviss, B. Malekynia, H. Hora, G. H. Miley, and X. He, Laser Part. Beams **26**, 105 (2008).
[11] D. Gabor, Proc. R. Soc. London, Ser. A **213**, 72 (1953).
[12] J. R. Kearns, W. C. Rogers, and J. G. Clark, Bull. Am. Phys. Soc. **17**, 629 (1972).
[13] E. Bagge and H. Hora, Atomkernenergie **24**, 143 (1974).
[14] J. Stepanek, in *Laser Interaction and Related Plasma Phenomena*, edited by H. Schwarz, H. Hora, M. Lubin, and B. Yaakobi (Plenum, New York, 1981), Vol. 5, p. 341.
[15] P. S. Ray and H. Hora, in *Laser Interaction and Related Plasma Phenomena*, edited by H. Schwarz and H. Hora (Plenum, New York, 1977), Vol. 5A, p. 341.
[16] E. I. Moses, G. H. Miller, and R. L. Kauffman, J. Phys. IV **133**, 9 (2006).
[17] J. D. Lindl, P. Amend, R. L. Berger, S. G. Glendinning, S. H. Glenzer, S. W. Hahn, R. L. Kauffman, O. L. Landen, and L. J. Suter, Phys. Plasmas **11**, 339 (2004).
[18] H. Hora and P. S. Ray, Z. Naturforsch. A **33A**, 890 (1978).
[19] R. C. Kirkpatrick and J. A. Wheeler, Nucl. Fusion **21**, 389 (1981).
[20] G. H. Miley, H. Hora, F. Osman, P. Evans, Y. Cang, and P. Toups, Laser Part. Beams **23**, 453 (2005).
[21] D. H. H. Hoffmann, A. Blazevic, P. Ni, P. Rosmej, M. Roth, N. A. Tahir, A. Tauschwitz, S. Udera, D. Vaentsov, K. Weyrich, and Y. Maron, Laser Part. Beams **23**, 47 (2005).

第 58 卷 第 5 期 2009 年 5 月
1000-3290/2009/58(05)/3255-07

物 理 学 报
ACTA PHYSICA SINICA

Vol.58, No.5, May, 2009
©2009 Chin. Phys. Soc.

神光Ⅲ原型 1 ns 激光驱动黑腔辐射温度实验研究*

李三伟[1]† 易荣清[1] 蒋小华[1] 何小安[1] 崔延莉[1] 刘永刚[1] 丁永坤[1] 刘慎业[1]
蓝 可[2] 李永升[2] 吴畅书[2] 古培俊[2] 裴文兵[2] 贺贤土[2]

1)(中国工程物理研究院激光聚变研究中心，绵阳 621900)
2)(北京应用物理与计算数学研究所，北京 100088)

(2008 年 7 月 7 日收到；2008 年 7 月 30 日收到修改稿)

在神光Ⅲ原型装置上，利用八束三倍频激光注入金柱腔靶，开展了光学高温计与软 X 射线能谱仪测量辐射温度的比对研究；利用功率平衡关系式，分析了神光Ⅲ原型 1 ns 内爆标准腔、1 ns 输运腔的辐射温度与激光功率的关系，得到了两种腔靶 X 射线转换效率(耦合效率)约为 50%—55%.

关键词：辐射温度，冲击波，黑腔靶，激光靶耦合效率
PACC: 5225，5240，5270L

1. 引 言

利用高功率激光器、强流粒子束和 Z-箍缩发生器等大型装置可以在实验室产生极高能量密度的物质，从而允许我们详细地探索逼近天体物理学系统状态下的物理现象，提供关于流体力学混合、冲击现象、辐射流、辐射不透明度、高马赫数射流、状态方程、相对论等离子体等方面的实验数据，这些数据可以校验超新星研究、惯性约束聚变(ICF)和国防应用中使用的数值模拟程序.在这些研究中常常需要建立类似黑体辐射的高温 X 射线辐射源[1,2].

在 ICF 研究中通常依靠黑腔把入射激光转换成 X 射线，X 射线辐射在腔壁约束下，经过多次吸收和再发射，把能量输运到整个空腔内表面并产生光学厚的高温等离子体环境，该环境使辐射场在空间均匀化，并渐近热力学平衡，X 射线能谱逼近黑体辐射的 Planck 谱.然后利用可控高温 X 射线辐射源对高能量密度物理(HEDP)的一些基本环节进行实验研究[3—6]，例如辐射输运，辐射流体力学，流体力学不稳定性；核反应动力学等等.因此高温黑腔辐射源研究是内爆物理、辐射输运和流体力学不稳定性等研究的基础.

高温黑腔辐射源特性一方面依赖于激光参数，另一方面依赖于黑腔结构，当新的激光器——神光Ⅲ原型于 2006 年初步建成后，即激光参数初步确定后，需要通过一系列的激光腔靶耦合实验优化激光打靶参数与黑腔靶构形和尺寸的匹配，建立多种应用实验所需要的高温黑腔辐射源.

辐射温度是黑腔辐射源最重要的特征物理量，是我们了解激光腔靶耦合物理过程以及辐射源应用实验设计的基础.目前辐射温度的测量通常采用两种方法：一种方法是通过上下激光注入孔或诊断孔用软 X 射线能谱仪(SXS)直接测量腔内辐射温度[7—10]；另一种方法是利用光学高温计(SOP)测量冲击波图像，由冲击波速度与辐射温度定标关系式给出腔内等效辐射温度[3,11,12].2007 年度在神光Ⅲ原型实验中首次开展了不同实验方法测量辐射温度的一致性研究，由于开展了三台软 X 射线能谱仪同时测量的比对、光学条纹相机全屏扫描非线性的精密标定、以及辐射温度测量不确定度的深入分析等细致工作，使我们对辐射温度有了更进一步的认识.

2. 激光打靶条件和实验安排

在本实验中设计了两种柱型腔靶.一种为 1 ns 输运腔：腔直径 $\Phi 1.0$ mm，腔长为 2.1 mm，腔两端激光注入孔(LEH)为 $\Phi 0.7$ mm，腔壁材料为 35 μm 厚

* 国家重点基础研究发展计划(批准号：2007CB814802)资助的课题.
† 通讯联系人. E-mail: lisanwei@sohu.com, or lisanwei@mail.ru

的金，在柱腔壁上中间位置开矩形诊断孔 400 μm × 600 μm，孔上贴台阶 Al 样品，Al 基底厚度为 35 μm，台阶处厚度为 50 μm，靶上设计了专门屏蔽杂散光的屏蔽筒，其长度有 16 mm 和 32 mm 两种，锥角 30°，其目的是避免散射光入射到冲击波样品上进入光学条纹相机的视场范围而给测量带来干扰．另外一种靶为 1 ns 内爆标准腔：腔直径 Φ1.0 mm，腔长为 1.7 mm，LEH 为 Φ0.7 mm，腔壁材料为 20 μm 厚的金，在柱腔壁上中间位置开圆形诊断孔 Φ0.4 mm．腔两端屏蔽片尺寸 Φ3.0 mm，锥形部分长 1.5 mm．

由于神光Ⅲ原型装置上剩余基频、二倍频光要进入靶室，这些杂散光辐照到柱腔外壳和屏蔽片上产生的 X 射线辐射要干扰辐射温度（其大小由腔内辐射流强度确定）测量，根据 2006 年度磨合实验结果可知，当屏蔽片和柱腔外壳涂上 15 μm 厚的 CH 层与不涂 CH 层相比，杂散光产生的 X 射线信号强度下降到 1/5—1/10，为了降低杂散光对辐射温度测量的影响，所有腔靶屏蔽片和腔体外壳均涂 CH 层，CH 涂层厚度大于 15 μm．另外输运腔长度要比内爆标准柱腔要长 400 μm，其目的是为了降低激光第一打击点产生的高能 X 射线对冲击波样品的辐照，降低预热效应[13]，同时腔长增大，改善了辐射波在样品中产生冲击波的平面性．

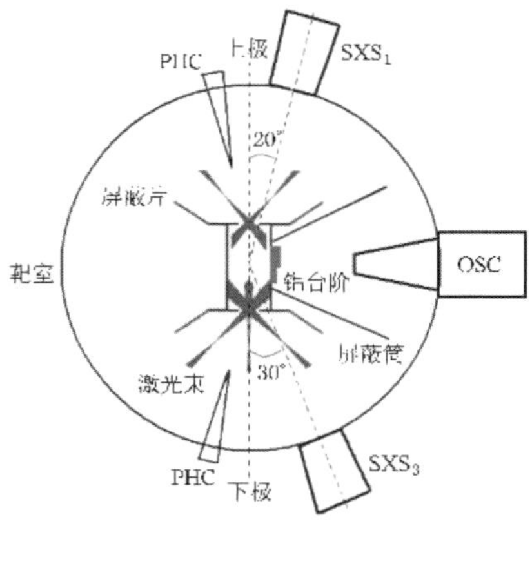

图 1 靶结构和实验探测器排布示意图

靶结构和实验探测器排布示意图如图 1 所示，腔靶位于真空靶室中央，腔轴上下竖直放置，为了便于说明在示意图中腔靶尺寸人为放大．神光Ⅲ原型八束三倍频激光分别从上、下两个方向相对柱腔轴线成 45°角从腔靶注入口进入腔内，激光聚焦注入，上、下各四束激光焦点重合，焦点分别位于上、下激光注入孔中心，上、下两环激光光路错开 45°，激光波长 0.35 μm，脉宽 1 ns，每束激光能量 300 J—1 kJ，靶面上激光焦斑为棱形状，短对角线长度约 300 μm，长对角线约 420 μm，靶面激光功率密度 4.5×10^{14} W/cm^2—1.5×10^{15} W/cm^2．

本实验主要诊断设备为扫描光学高温计[14—16]（SOP）和三台新设计加工的 15 道软 X 射线能谱仪（SXS）．SOP 由成像系统（Cassegrain 望远镜，放大倍数为 10）和 979 光学条纹相机（OSC）组成，其时间分辨为 5 ps，空间分辨为 10 μm，SOP 安装在靶室水平面上沿着腔靶方形诊断口方向测量台阶 Al 样品背侧可见光（300—500 nm）信号的时空分辨图像，从而给出冲击波传播速率．SXS_1 号谱仪位于靶室上半球，与腔轴成 20°的位置，通过激光上注入孔观测腔内时间分辨辐射温度，由于激光注入孔 Φ0.7 mm 偏大，SXS_1 号谱仪通过两个激光注入孔少部分看穿了，腔内软 X 射线发射以面发射为主，因此测得的辐射温度偏低一点．SXS_2 号谱仪位于靶室水平面上通过诊断孔观测腔内时间分辨辐射温度．SXS_3 号谱仪位于靶室下半球，与腔轴成 30°的位置，通过下激光注入孔观测腔内时间分辨辐射温度，SXS_3 号谱仪测得的辐射流基本上等效于辐照到腔中心靶丸上的辐射能流．软 X 射线能谱仪每道由 X 射线滤片、掠入射 X 射线平面镜和真空 X 射线二极管（XRD）构成，通过滤片的低能截止和平面镜的高能截止把 X 射线进行能谱分割成一段能区测量，SXS 测谱范围为 50 eV—5 keV．另外在靶室上、下半球近极轴位置各安装一台针孔相机（PHC）对注入孔 X 射线发射进行时间积分空间成像监测，考察激光注入情况．

3．腔靶不同方位辐射温度理论模拟

在间接驱动惯性约束聚变实验研究中，由冲击波速度通过定标关系确定辐射温度是一种常用的技术手段．在 Nova 激光装置上通过 PS22 脉冲激光注入腔内辐射驱动冲击波实验获得了 Al 样品中冲击波速度 D 与腔内峰值辐射温度 T_R 的定标关系[4,12]

$$T_R(\mathrm{eV}) = 0.0126(D/\mathrm{cm/s}))^{0.63} = 17.8(D/\mathrm{km/s}))^{0.63}. \qquad (1)$$

我们的研究结果表明，冲击波速度与辐射温度定标关系式是与辐射驱动波形相关的．因此，针对神光Ⅲ原型装置的辐射驱动波形，我们利用 RDMG 程序及原子参数数据库通过数值模拟给出了冲击波速

度与辐射温度定标关系式

$$T_R(eV) = 0.00919(D/cm/s))^{0.647} = 15.787(D/km/s))^{0.647}. \quad (2)$$

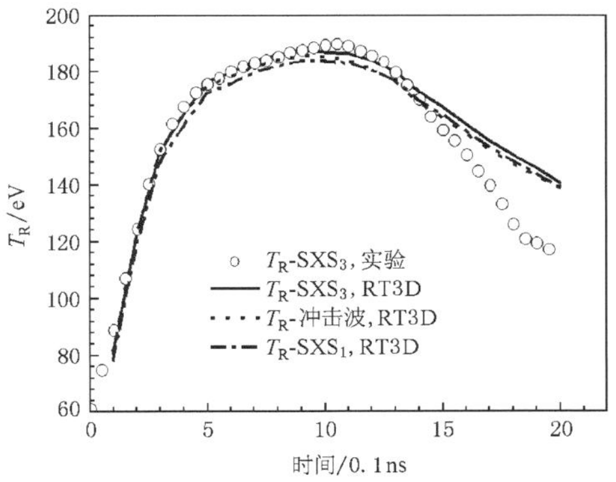

图 2　对应于 SXS 观测方向由 RT3D 模拟给出的时间分辨辐射温度曲线

同时利用 RT3D 程序对位于不同方位安装的软 X 射线能谱仪所测的辐射温度和通过冲击波样品所测的辐射温度进行了细致的模拟，并与冲击波样品区入射流所定义的辐射温度进行了对比分析.通过冲击波速度得到的辐射温度是样品区感受到的峰值辐射温度，由从黑腔进入样品的入射能流所决定；而通过 SXS 谱仪所测得的辐射温度则由黑腔壁再发射流所决定，与其安装方位，即观测角度有关.根据 RT3D 给出的图像，SXS_1 谱仪能完整地看到一个激光光斑区以及两个光斑的一部分，同时可以看到激光下注入孔的一小部分；SXS_2 谱仪测得的是黑腔腰部的再发射流，它只能看到一小部分光斑区；SXS_3 谱仪基本上能完整地看到两个光斑区.

针对第 54 发激光波形和 1 ns 输运腔结构，对应于不同 SXS 观测方向由 RT3D 模拟给出的时间分辨辐射温度曲线如图 2 所示，实线为 SXS_3 观测方向由 RT3D 模拟给出的时间分辨辐射温度曲线；点画线为 SXS_1 观测方向模拟结果；虚线为由入射流定义的腔壁平均辐射温度模拟结果，即由冲击波速度确定的辐射温度对应于此曲线峰值；空心圆圈为 SXS_3 谱仪测得的辐射温度时间波形.由图 2 的数值模拟结果看到，三台谱仪中，SXS_3 谱仪测得的辐射温度最高，而 SXS_2 谱仪的最低，这是由于观测方位不同导致的；由冲击波速度确定的峰值辐射温度介于 SXS_1 与 SXS_2 之间，但更接近 SXS_3 方位的辐射温度，约低 2 eV.

4. 实验结果与分析

4.1. 光学条纹相机全屏扫描速度标定

由于条纹相机中晶体管的老化将导致斜坡扫描电压发生畸变，从而改变条纹相机的扫描速度.为了提高冲击波速度测量精度，需要对相机不同时刻、不同空间位置的扫描速度进行精密标定.

改变延迟箱的延迟时间，并利用扫描电路的触发晃动得出不同位置的扫描条纹图.将时间轴人为划分成若干小段，然后根据扫描条纹图及序列脉冲时间间隔求得每一小段时间间隔内的平均扫描速度.将多次测得的扫描速度加权处理后得到这一区域的平均扫描速度.

通过一组扩束器将主激光扩束后照亮整个相机狭缝，得狭缝扫描像，再截取不同位置的扫描图进行处理即可获得对应的扫描速度.

实验前利用短脉冲 YAG 激光器（光源波长 526 nm，能量 6 mJ，脉宽 30 ps）配 200 ps 光学标准具对 979 光学条纹相机全屏扫描非线性进行了仔细标定，CCD 像素为 512 × 512，5 ns 档标定结果如图 3 所示，横轴为扫描方向荧光屏上像素（pixel）坐标，纵轴为荧光屏不同空间位置每个单元像素点的平均扫描速度.5 ns 档和 10 ns 档初始时刻的扫描速度明显慢于其他时刻，扫描方向为从右向左.这说明由于条纹相机的老化，相机扫描速度在不同时刻差别较大，采用平均扫描速度和扫速非线性进行数据处理已不适用.

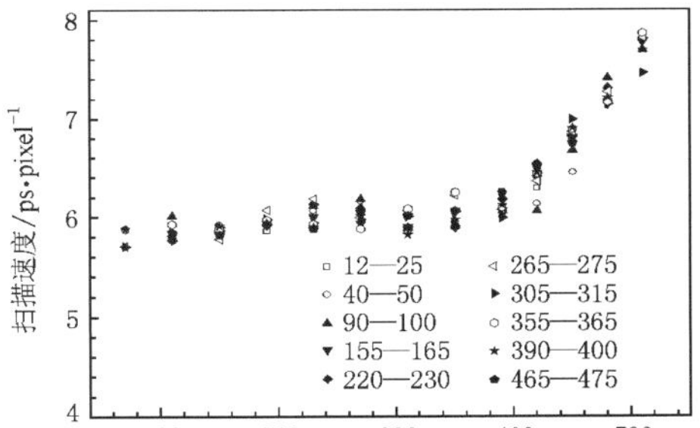

图 3　979 光学条纹相机全屏扫描非线性标定结果（5 ns 档）

4.2. 光学高温计测量辐射温度

由冲击波速度确定辐射温度，其原理是将标准材料 Al 样品放置在腔壁诊断口上，当 X 射线辐射在

低 Z 材料 Al 样品中被吸收时，将产生向内传播的冲击波，首先在 Al 基底中传播，然后冲击波进入 Al 台阶区（已知厚度差 Δd），利用 SOP 测量出台阶区冲击波的渡越时间 Δt，从而给出 Al 样品中冲击波传播速度 $D=\Delta d/\Delta t$，再通过冲击波速度与驱动冲击波辐射场的辐射温度 T_R 的定标关系求出等效辐射温度.

图 4 给出了第 23 发获得的台阶 Al 样品冲击波发光图像. 图像上半部份对应 Al 台阶，厚度 49.677 μm，下半部分对应 Al 基片，厚度 34.154 μm，图像横向为时间扫描方向，纵向为空间分辨方向. 图 4 右边给出了第 23 发样品背侧可见光强度沿时间方向扫描曲线，经过去本底归一化处理后得出冲击波上升沿约 80 ps，位于 979 相机荧光屏上不同空间位置的冲击波图像按标定结果图 3 分段处理，即扫描速度分段计算，考虑冲击波图像空间位置后，冲击波速度 42.16 km/s，根据关系式(2)式由冲击波速度推得腔内等效辐射温度 177.7 eV.

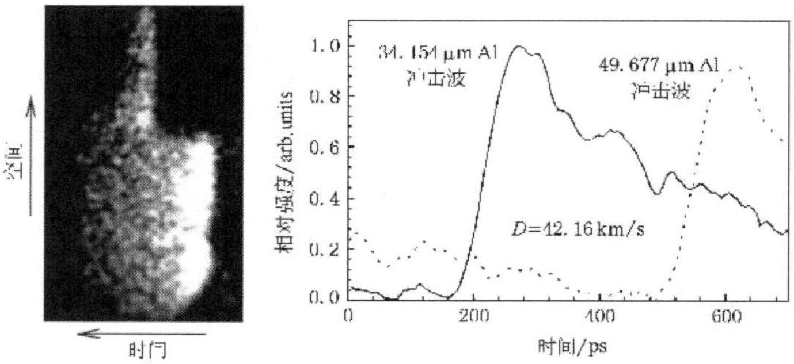

图 4　第 23 发 Al 台阶样品背侧冲击波图像和可见光强度沿时间方向扫描图

由冲击波关系式

$$D = C_0 + \lambda u, \tag{3}$$

$$P = \rho_0 D u, \tag{4}$$

得到

$$P = \rho_0 D(D - C_0)/\lambda, \tag{5}$$

式中 D 为冲击波速度，u 为波阵面后的粒子速度，P 为冲击压力，ρ_0 为样品常态下的密度，C_0，λ 为样品的冲击绝热参数，测得冲击波速度后，可求得样品中的冲击波压力. 对于 Al 样品[17]，$C_0=5.25$ km/s，$\lambda=1.39$，$\rho_0=2.79$ g/cm^3，由第 23 发得出 Al 样品中冲击波速度 $D=42.2$ km/s，对应的冲击波压力 $P=3.1\times10^{12}$ Pa.

冲击波速度测量误差主要来源有样品台阶高度的测量误差及样品表面粗糙度的影响；样品背侧冲击波发光时间拟合误差，包括条纹相机的噪声，样品非均匀性以及冲击波空间的非平面性；条纹相机的分辨率及标定误差；冲击波的衰减（即不稳定性）引起的误差，以及预热引起的样品密度变化带来的误差. 实验中对冲击波速度测量不确定度进行了仔细分析，台阶厚度采用光学干涉仪测量，台阶厚度的相对不确定度约 2%；考虑光学条纹相机时间分辩、成像系统空间分辩、以及狭缝宽度影响后，台阶样品处冲击波渡越时间相对不确定度约 5.5%；冲击波速度测量相对不确定度约 5.9%；根据(2)式得出辐射温度相对不确定度约 4%.

4.3. 软 X 射线能谱仪测量辐射温度

为了考察软 X 射线能谱议测量结果的一致性，实验期间三台 15 道 SXS 同时安装在靶室上半球，与腔轴均成 20°的位置，三台谱议视场完全一致，三台 SXS 测得的辐射温度曲线基本一致，SXS_1 谱仪获得的辐射温度约高一点，约 4 eV，实验前 SXS_2 谱仪上的 XRD 探测器在北京同步辐射源上标定过，因此其余两台谱仪实验数据按 SXS_2 谱仪归一. 正式实验期间 SXS_3 谱仪位于靶室下半球，与腔轴成 30°的位置，通过下激光注入孔观测腔内时间分辨辐射温度.

在美国 Nova 和 Omega 装置上通常采用从激光注入口测量辐射通量而不采用通过中平面诊断孔测量辐射温度[4]. 通过激光注入口方向测量腔内辐射温度，其优点在于注入口尺寸比中平面诊断口大得多，可以不考虑诊断孔缩孔效应，而且注入口附近的稀薄等离子体通过激光加热温度较高，对腔内发射的 X 射线更透明. 通过激光注入口获得的辐射通量包含了激光加热光斑和激光未辐照腔壁再发射的通量，选择恰当的观察方向能够等效于辐照到腔中心靶丸上的辐射能流.

图 5 给出了典型的 1 ns 内爆标准腔靶辐射温度时间波形.在激光脉冲（激光能量 7.7 kJ，脉宽 1 ns）作用期间，SXS_3 谱仪获得的激光注入孔辐射温度时间过程 T_R（LEH）和 SXS_2 谱仪获得的诊断孔辐射温度时间过程 T_R（DH）与激光脉冲（Laser）基本是同步的，说明在激光脉冲结束前腔内没有出现明显的等离子体堵腔效应.

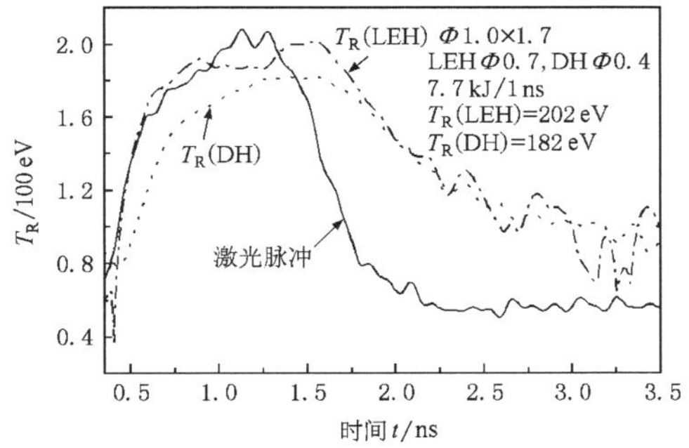

图 5　1 ns 内爆标准腔靶辐射温度时间波形

目前 SXS 谱仪数据处理采用积分辐射流[18]的方法给出辐射温度，积分能区取 50 eV—4.1 keV.积分区间高能截止点选择越低，给出的辐射温度越高，当高能截止点大于 4.1 keV 时，给出的辐射温度趋向稳定.

时间分辨辐射温度不确定度的主要来源有 XRD 探测器灵敏度的不确定度、平面镜反射率的不确定度、平面镜安装角度的不确定度、滤片透过率的不确定度、探测器的立体角因子的不确定度、信号电缆和衰减头的衰减系数不确定度、示波器的测量不确定度以及数据处理不确定度等.目前 X 射线辐射功率测量不确定度约 30.5%，辐射温度相对不确定度 7.5%.

表 1　SOP 测温和 SXS 测温比较（输运腔）

发次号	激光总能量/J	$T_{R\text{-SXS}}$/eV	D/km·s^{-1}	$T_{R\text{-Shock}}$/eV
8-27-18	5363	183 ± 14	43.7 ± 2.6	181.8 ± 6.9
8-28-23	4500	175 ± 13	42.2 ± 2.5	177.7 ± 6.7
8-29-27	4717	176 ± 13	41.7 ± 2.5	176.4 ± 6.7

考虑冲击波图像空间位置后，同一发打靶 SOP 测温结果 $T_{R\text{-Shock}}$ 和 SXS_3 测温结果 $T_{R\text{-SXS}}$ 基本一致，即两种测量原理完全不同的设备得到的结果基本一致，见表 1 所示，与理论模拟结果相符合，说明目前辐射温度测量是可信的.

4.4. 1 ns 腔靶辐射温度定标关系

在固定激光功率条件下辐射温度主要依赖于腔壁漏失和 X 射线转换效率.图 6 和图 7 分别给出了神光Ⅲ原型 1 ns 标准腔、1 ns 输运腔辐射温度与激光功率之间的定标关系.图中不同 X 射线转换效率 η 曲线按功率平衡关系式[4]

$$\eta P_L = (A_w(1-\alpha) + A_h)\sigma T_R^4 \qquad (6)$$

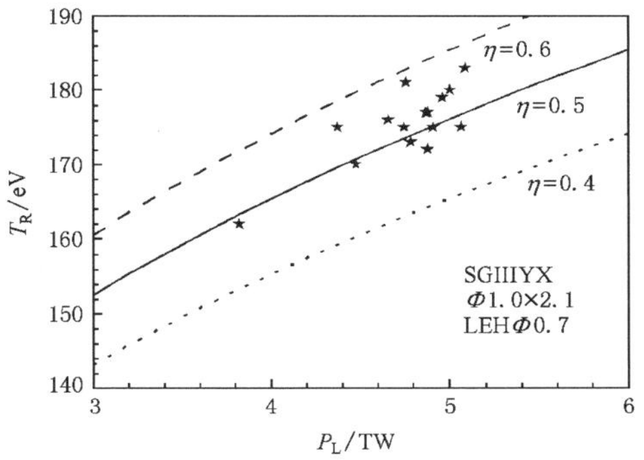

图 6　神光Ⅲ原型 1 ns 输运腔辐射温度与激光功率定标关系

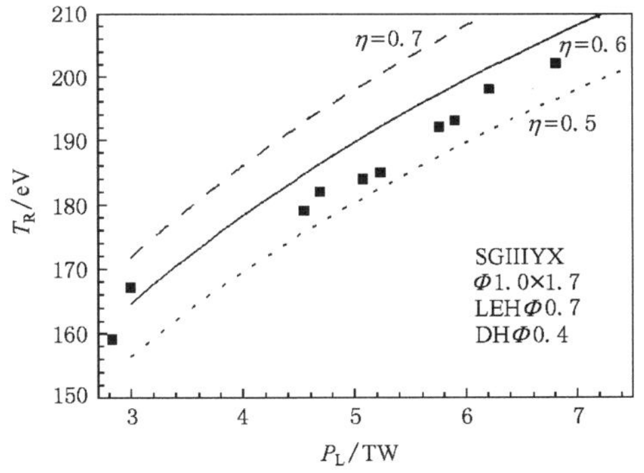

图 7　神光Ⅲ原型 1 ns 标准腔辐射温度与激光功率定标关系

计算给定，其中 T_R 是腔内辐射温度，P_L 为激光功率，η 为激光—X 射线转换效率，A_w 为腔壁内表面积，A_h 为腔靶开口部分面积.α 为腔壁材料 X 射线反照率，是辐射温度、激光脉宽的隐形函数，$\sigma = 1.03 \times 10^5$ W/(cm^2·eV^4)为斯提芬玻尔兹曼常量.我们利用 RDMG 程序的数值模拟结果给出了金材料的反照率 $\alpha = 1 - 0.34 T_R^{-0.65} \cdot \tau^{-0.41}$，这与文献[19]给出的结果是一样的.在图 6 中黑五星和图 7 中黑方块符

号分别为 SXS_3 谱仪获得的 1 ns 输运腔、1 ns 标准腔靶辐射温度实验数据，由图 6 和图 7 得出神光Ⅲ原型 1 ns 输运腔、1 ns 内爆标准腔 X 射线转换效率（耦合效率）分别约为 50%和 55%.

5. 结　　论

本文研究了不同实验方法测量辐射温度的一致性.实验数据分析表明在同一发打靶实验中，光学高温计测温结果和软 X 射线能谱仪测温结果基本一致，与理论模拟结果相符合.在此基础上获得了神光Ⅲ原型 1 ns 内爆标准腔、1 ns 输运腔辐射温度定标关系，给出了 1 ns 输运腔、1 ns 内爆标准腔 X 射线转换效率分别约为 50%和 55%.本文描述的 1 ns 激光驱动黑腔辐射温度实验数据给未来辐射烧蚀、内爆物理、辐射输运等实验研究奠定了良好基础.

由于目前神光Ⅲ原型激光束质量原因，腔靶两端激光注入孔（目前直径 Φ0.7 mm）偏大，腔内辐射流漏失严重，辐射温度偏低，同时腔内辐射场 P_2 不对称性增大，靶丸内爆压缩度变低.考虑激光注入条件、激光能量耦合效率和黑腔对称性后激光注入孔直径通常为腔直径的二分之一，对神光Ⅲ原型典型黑腔而言，腔直径为 Φ1 mm，因此较理想的激光注入孔直径为 Φ0.5 mm，即光束质量有待改进.

神光Ⅲ原型黑腔物理实验是在中国工程物理研究院激光聚变研究中心三部、二部、五部、四部、一部、以及北京应用物理与计算数学研究所四室相关人员的协作下共同完成的，在此向他们表示感谢！同时对有关领导、专家的指导帮助，以及对所各职能部门的组织协调作用表示感谢！

[1] Drake R P 2006 *High Energy Density Physics*（New York: Springer）p335

[2] Campbell E M, Hogan W J 1999 *Plasma Phys. Control. Fusion* **41** B39

[3] Lindl J D 1995 *Phys. Plasmas* **2** 3933

[4] Lindl J D, Amendt P, Berger R L, Glendinning S G, Glenzer S H, Haan S W, Kauffman R L, Landen O L, Suter L J 2004 *Phys. Plasmas* **11** 339

[5] Cavailler C 2005 *Plasma Phys. Control. Fusion* **47** B389

[6] Rothman S D, Evans A M, Horsfield C J, Graham P, Thomas B R 2002 *Phys. Plasmas* **9** 1721

[7] Sun K X, Huang T X, Ding Y K, Yi R Q, Jiang S E, Cui Y L, Tang X Q, Chen J S, Zhang B H, Zheng Z J 2002 *Acta Phys. Sin.* **51** 1750 (in Chinese)［孙可熙、黄天暄、丁永坤、易荣清、江少恩、崔延莉、汤小青、陈久森、张保汉、郑志坚 2002 物理学报 **51** 1750］

[8] Sun K X, Jiang S E, Yi R Q, Cui Y L, Ding Y K, Liu S Y 2006 *Acta Phys. Sin.* **55** 68 (in Chinese)［孙可熙、江少恩、易荣清、崔延莉、丁永坤、刘慎业 2006 物理学报 **55** 68］

[9] Campbell K M, Weber F A, Dewald E L, Glenzer S H, Landen O L, Turner R E, Waide P A 2004 *Rev. Sci. Instrum.* **75** 3768

[10] Dewald E L, Landen O L, Suter L J, Schein J, Holder J, Campbell K, Glenzer S H, McDonald J W, Niemann C, Mackinnon A J, Schneider M S, Haynam C, Hinkel D, Hammel B A 2006 *Phys. Plasmas* **13** 056315

[11] Kauffman R L, Kornblum H N, Phillion D W, Darrow C B, Laslnskl B F, Suter L J, Theissen A R, Wallace R J, Ze F 1995 *Rev. Sci. Instrum.* **66** 678

[12] Kauffman R L, Suter L J, Darrow C B, Kilkenny J D, Kornblum H N, Montgomery D S, Phillion D W, Rosen M D, Theissen A R, Wallace R J, Ze F 1994 *Phys. Rev. Lett.* **73** 2320

[13] Qi L Y, Jiang X H, Zhao X W, Li S W, Zhang W H, Li C G, Zheng Z J, Ding Y K 2000 *Acta Phys. Sin.* **49** 492 (in Chinese)［祁兰英、蒋小华、赵雪薇、李三伟、张文海、李朝光、郑志坚、丁永坤 2000 物理学报 **49** 492］

[14] Jiang S E, Li W H, Sun K X, Jiang X H, Liu Y G, Cui Y L, Chen J S, Ding Y K, Zheng Z J 2004 *Acta Phys. Sin.* **53** 3424 (in Chinese)［江少恩、李文洪、孙可煦、蒋小华、刘永刚、崔延莉、陈久森、丁永坤、郑志坚 2004 物理学报 **53** 3424］

[15] Oertel J A, Murphy T J, Berggren R R 1999 *Rev. Sci. Instrum.* **70** 803

[16] Olson R E, Leeper R J, Nobile A, Oertal J A, Chandler G A, Cochrane K, Dropinski S C, Evans S, Haan S W, Kaae J L, Knauer J P, Lash K, Mix L P, Nikroo A, Rochau G A, Rivera G, Russell C, Schroen D, Sebring R J, Tanner D L, Turner R E, Wallace R J 2004 *Phys. Plasmas* **11** 2778

[17] Peng X S, Li S W, An Z, Jiang X H, Liu Y G 2007 *High Power Laser and Particle Beams* **19** 741 (in Chinese)［彭晓世、李三伟、安　竹、蒋小华、刘永刚 2007 强激光与粒子束 **19** 741］

[18] Fehl D L, Stygar W A, Chandler G A, Cuneo M E, Ruiz C L 2005 *Rev. Sci. Instrum.* **76** 103504

[19] Hammer J H, Rosen M D 2003 *Phys. Plasmas* **10** 1829

Experimental study of radiation temperature for gold hohlraum heated with 1 ns, 0.35 μm lasers on SG-Ⅲ prototype laser facility*

Li San-Wei[1)†]　Yi Rong-Qing[1)]　Jiang Xiao-Hua[1)]　He Xiao-An[1)]　Chui Yan-Li[1)]
Liu Yong-Gang[1)]　Ding Yong-Kun[1)]　Liu Shen-Ye[1)]
Lan Ke[2)]　Li Yong-Sheng[2)]　Wu Chang-Shu[2)]　Gu Pei-Jun[2)]　Pei Wen-Bing[2)]　He Xian-Tu[2)]
1)(*Research Center of Laser Fusion, Chinese Academy of Engineering Physics, Mianyang*　621900, *China*)
2)(*Beijing Institute of Applied Physics and Computation Mathematics, Beijing*　100088, *China*)
(Received 7 July 2008; revised manuscript received 30 July 2008)

Abstract

Experimental measurement of radiation temperature by a streaked optical pyrometer and a soft X-ray spectrometer viewing through the laser entrance hole are performed on SG-Ⅲ prototype laser facility. It was found that the two methods compares well. Using the power balance relation, the laser-hohlraum coupling efficiency for $\phi 1.0$ mm × 1.7 mm hohlraum and $\phi 1.0$ mm × 2.1 mm hohlraum is around 50%—55%.

Keywords: radiation temperature, shock wave, hohlraum, laser-hohlraum coupling efficiency
PACC: 5225, 5240, 5270L

* Project supported by the National Basic Research Program of China (Grant No. 2007CB814802).
† Corresponding author. E-mail: lisanwei@sohu.com, or lisanwei@mail.ru

Laser and Particle Beams (2010), *28*, 421–427.

doi:10.1017/S026303461000042X

An initial design of hohlraum driven by a shaped laser pulse

KE LAN, PEIJUN GU, GUOLI REN, XIN LI, CHANGSHU WU, WENYI HUO, DONGXIAN LAI, AND XIAN-TU HE
Institute of Applied Physics and Computational Mathematics, Beijing, People's Republic of China
(RECEIVED 24 March 2010; ACCEPTED 1 June 2010)

Abstract

In this paper, the plasma-filling model was extrapolated to the case of a hohlraum driven by a shaped laser pulse, and this extended model was used to obtain an initial design of the hohlraum size. A density criterion of $n_e = 0.1$ was used for designing hohlraums which have low plasma filling with maximum achievable radiation. The method was successfully used to design a half hohlraum size with a three-step laser pulse on SGIII prototype and a U hohlraum size with shaped laser pulse for ignition. It was shown that the extended model with the criterion can provide reasonable initial design of a hohlraum size for optimal designing with a two-dimensional code.

Keywords: Hohlraum design; Inertial fusion; Plasma filling

INTRODUCTION

Since the hohlraum target plays a key role in the study of indirect drive inertial fusion study, the experimental and theoretical target design, and simulation continuously attract extensive interests and efforts worldwide (Aleksandrova *et al.*, 2008; Atzeni & Meyer-Ter-vehn, 2004; Borisenko *et al.*, 2008; Bret & Deutsch, 2006; Chatain *et al.*, 2008; Deutsch *et al.*, 2008; Eliezer *et al.*, 2007; Foldes & Szatmari, 2008; Hoffmann, 2008; Holmlid *et al.*, 2009; Hora, 2007; Imasaki & Li, 2007; Koresheva *et al.*, 2009; Li *et al.*, 2010; Ramis *et al.*, 2008; Rodriguez *et al.*, 2008; Strangio *et al.*, 2009; Winterberg, 2008; Yang *et al.*, 2008). The optimal size of a hohlraum is a trade-off between the requirements for energy and power, and the need for symmetry and acceptable plasma filling. Hence, a series of theory for target study have been developed, such as the hohlraum coupling efficiency theory, capsule radiation uniformity theory (Lindl, 2004), and plasma-filling model (Schneider *et al.*, 2006). In order to design a suitable hohlraum driven by a given laser source, we often use a two-dimensional (2D) hydrodynamic code to select the optimum hohlraum size by simulation. However, what is the initial size of the hohlraum that we should put into 2D simulation for further optimization? As known, the optimization searching process may be extremely laborious and computationally expensive if without a suitable initial value. In this paper, we will present a method for giving a suitable initial hohlraum size by using an extrapolated plasma-filling model, which was developed under the condition of a constant input laser power (Schneider *et al.*, 2006). In the following text, we will first extend the plasma-filling model to the condition of shaped laser pulse. Second, with this extended model, we will give an initial size of a half Au hohlraums, which is driven by a three-step laser pulse on SGIII prototype laser facility. This half hohlraum is used to test the optical diagnostics VISAR on SGIII prototype, and a low plasma filling with maximum achievable radiation is required in the hohlraum. Third, we will use our 2D simulation code Laser with Atomic, Radiation and Electron hydroDynamic code for Hohlraum physics study (LARED-H) (Duan *et al.*, 2002; Pei, 2007) to search the optimal size of the half hohlraum by simulation. The initial half hohlraum size obtained from above extended model will be used as an initial input. As it will be shown, the optimum hohlraum size obtained by simulation is close to its initial design. In the fourth part, we further use the extended plasma-filling model to obtain an initial design of the U hohlraum under laser pulse for ignition. Finally, we will present a summary.

Address correspondence and reprint requests to: Xin Li, Institute of Applied Physics and Computational Mathematics, P.O. Box 8009-14, Beijing, 100088, People's Republic of China. E-mail: li_xin@iapcm.ac.cn

PLASMA-FILLING MODEL FOR THE HOHLRAUM DRIVEN BY A SHAPED PULSE

In this section, we first relate the radiation temperature T_r to input laser power P and time τ by power balance (Sigel *et al.*,

1988), and then extend the plasma-filling model (Dewald *et al.*, 2005; McDonald *et al.*, 2006; Schneider *et al.*, 2006) to the case of a hohlraum driven by a shaped laser pulse.

In a hohlraum driven by a laser pulse, the radiation temperature T_r in steady-state conditions is related to input laser power P *via* the power balance:

$$\eta P = \sigma T_r^4\{(1-\alpha)A_W + A_H\}, \tag{1}$$

where η is the laser-to-X-ray coupling efficiency, A_W is the hohlraum wall area, A_H is the hohlraum hole area, α is the albedo of wall, and σ is the Stefan-Boltzmann constant. Under a radiation that is proportional to time, the albedo at time τ can be expressed as:

$$\alpha = 1 - H/(T_r^{\gamma}\tau^{\beta}), \tag{2}$$

where H, γ, and β are parameters related to radiation temporal profile.

We can approximately relate T_r to P and τ from Eqs. (1) and (2) by defining a hohlraum geometrical factor:

$$f_H = 1 + \frac{A_H}{(1-\tilde{\alpha})A_W}, \tag{3}$$

in which $\tilde{\alpha}$ is an average albedo of wall over time. Inserting Eq. (3) into Eq. (1), we have:

$$\eta P \approx (1-\alpha)\, f_H A_W \sigma T_r^4. \tag{4}$$

Then inserting Eq. (2) into Eq. (4), we obtain:

$$T_r = DP^E\tau^F, \tag{5}$$

where

$$D \approx \left[\eta\left(Hf_H\sigma\, A_W\right)^{-1}\right]^{\frac{1}{4-\gamma}}, \tag{6}$$

$$E \approx \frac{1}{4-\gamma}, \tag{7}$$

$$F \approx \frac{\beta}{4-\gamma}. \tag{8}$$

If an appropriate $\tilde{\alpha}$ is given, then radiation temperature obtained in this way is very close to the accurate solution of Eqs. (1) and (2). When there is a capsule inside the hohlraum or there is a sample on the hohlraum wall, the hohlraum geometrical factor can be redefined by taking the area of capsule or sample into consideration.

In the plasma-filling model (Dewald *et al.*, 2005; McDonald *et al.*, 2006; Schneider *et al.*, 2006), the laser beam channel is characterized by laser power, electron density n_e, electron temperature T_e, and ionization state Z_h of the hot plasma inside laser channel. The radiation ablated volume surrounding the laser channel is characterized by the radiation temperature T_r and electron density Zn_i, in which Z is the ionization state and n_i is the ion density of the radiation volume. Note that all densities are expressed in unit of the critical density. It is assumed that the laser power per unit length absorbed *via* inverse bremsstrahlung in the laser hot channel equals to the power per unit length conducted out of the channel by thermal conduction. The filling time for the hot laser channel to reach density n_e can be obtained by the pressure balance equation for the laser channel and the surrounding plasma, and the equations relating Z_h to T_e and Z to T_r.

Now we extend this plasma-filling model to a case in which a hohlraum is driven by a shaped laser pulse with high contrasts (>1) between different steps. This condition is widely used in inertial fusion research to produce a time-dependent radiation to assure a nearly isentropic compression of capsule inside the hohlraum and achieve thermonuclear ignition. The maximum radiation temperature in the hohlraum is produced in the last step of laser pulse. We focus on the Au hohlraum here and assume that the differences of T_r between steps are large and the mass of wall material ablated in each step of radiation is much larger than that in its former step. Neglecting the influence of former radiation steps on ablation, the mass of wall ablated in each step can be calculated independently. However, in each step the wall albedo increases and contributes to the sum of the ablated mass, which eventually influences the plasma-filling time in the hohlraum.

Assuming that the laser pulse has J steps, we use P_j to denote the incident laser power, τ_j is the time width, $T_{r,j}$ is the radiation temperature, $T_{r,j} \propto (t - \sum_{j'=1}^{j-1}\tau_{j'})^{k_j}$, and α_j is the albedo of the jth step ($j = 1, \ldots, J$). The coupling efficiency from laser to X-ray is assumed to be the same for each step, and then Eqs. (1) and (2) for the jth step become:

$$\eta P_j = \sigma {T_{r,j}}^4\{(1-\alpha_j)A_W + A_H\}, \tag{9}$$

$$\alpha_j = 1 - H_j/({T_{r,j}}^{\gamma_j}\tau^{\beta_j}). \tag{10}$$

The corresponding hohlraum geometrical factor is:

$$f_{H,j} = 1 + \frac{A_H}{(1-\tilde{\alpha}_j)A_W}, \tag{11}$$

where $\tilde{\alpha}_j$ is an average albedo of the jth step. Same as above, we have:

$$T_{r,j} = D_j P_j^{E_j}\tau_j^{F_j}, \tag{12}$$

in which

$$D_j = \eta\,(H_j f_{H,j}\sigma A_W)^{-1}, \tag{13}$$

$$E_j = \frac{1}{4-\gamma_j}, \tag{14}$$

$$F_j = \frac{\beta_j}{4-\gamma_j}. \tag{15}$$

The mass ablated in the jth step of radiation in Au can be expressed as (Lindl, 1995):

$$m_j(\mathrm{g/cm^2}) = 2.3\times10^{-7}(T_{r,j}/\mathrm{eV})^{1.86} \times (3.44k_j+1)^{-0.46}(\tau_j/ns)^{0.54+1.86k_j}. \tag{16}$$

Thus, the total ablated mass is $\sum_{j=1}^{J} m_j$. We therefore have the material density ρ and the ion density n_i in the hohlraum:

$$\rho = \frac{S_{abl}}{V_{hohl}} \times \sum_{j=1}^{J} m_j, \tag{17}$$

$$n_i = 1.07\times10^3 \times \frac{\lambda_L^2}{A} \times \frac{S_{abl}}{V_{hohl}} \sum_{j=1}^{J} m_j, \tag{18}$$

where S_{abl} ($\mathrm{cm^2}$) is the effective area of wall being ablated, V_{hohl}($\mathrm{cm^3}$) is the hohlraum volume, and A is the atomic mass of the hohlraum material. Considering hydrodynamic losses and coronal radiative losses from the laser beam entrance hole of the hohlraum, we take $S_{abl} = S_{hohl} - S_{LEH}$, where S_{hohl} is the inner area of the hohlraum, and S_{LEH} is the area of the laser entrance hole (LEH). For an Au hohlraum driven by a 3ω shaped laser pulse, we have

$$n_i = 0.67 \times \frac{S_{abl}}{V_{hohl}} \sum_{j=1}^{J} m_j. \tag{19}$$

As we mentioned above, the maximum temperature in the hohlraum will be obtained in the last step of laser pulse, i.e., $T_{r,J} = D_J\, P_J^{E_J}\tau_J^{F_J}$. Hereafter, we neglect the index J for the last step. As in Schneider *et al.* (2006), considering the power balance between the laser power per unit length being absorbed *via* inverse bremsstrahlung in the laser hot channel, and that exiting the channel *via* thermal conduction, as well as the pressure balance between the laser channel and surrounding plasma, we obtain:

$$P(\mathrm{TW}) = 3.1\times10^{-16}\times n_i^{4.6} n_e^{-6.6}(T_r/eV)^{6.67}(1-n_e)^{0.5}. \tag{20}$$

Inserting $T_r = DP^E\tau^F$ into the above expression and solving for n_e, we obtain the average density in the last laser step:

$$n_e(1-n_e)^{-\frac{1}{13.2}} = 4.47\times10^{-3}\times D^{1.01}P^{1.01E-\frac{1}{6.6}}\tau^{1.01F}n_i^{0.697}. \tag{21}$$

Usually, $n_e < <1$ and $(1-n_e)^{-\frac{1}{13.2}} \sim 1$, so Eq. (21) can be rewritten as:

$$n_e \approx 4.47\times10^{-3}\times D^{1.01}\times P^{1.01E-\frac{1}{6.6}}\tau^{1.01F}n_i^{0.697}. \tag{22}$$

Furthermore, we can obtain the average electron temperature T_e in laser hot channel by using the pressure balance 1000 n_eT_e (keV) $= Zn_iT_r$ (eV) and the scaling $Z = 3.15T_r^{0.45}$ (Lindl, 1995):

$$T_e(\mathrm{keV}) = 3.15\times10^{-3}n_e^{-1}n_iT_r^{1.45}. \tag{23}$$

When the plasma filling becomes serious, the laser absorption region shifts far from the hohlraum wall, and more, the hydrodynamic loss and the thin coronal radiative loss from LEH increase rapidly. The density $n_e = 0.1$ is usually used as a threshold that prevents the laser from propagating into the hohlraum due to absorption. For the models concerned in this paper, we use $n_e = 0.1$ as a criterion for giving initial size of a hohlraum in which a low plasma filling with maximum achievable radiation is required. Note that this is different from that used in Dewald *et al.* (2005), in which $\lambda_{IB} = 0.7R_{\mathrm{LEH}}$ is taken as a criterion for a hot hohlraum. Here, R_{LEH} is the LEH radius and λ_{IB} is the inverse bremsstrahlung absorption length.

INITIAL DESIGN OF A HALF HOHLRAUM UNDER A SHAPED LASER PULSE ON SGIII PROTOTYPE

In this Section, we use the extended plasma filling model to give an initial design of size of a half hohlraum driven by a 6 kJ three-step laser pulse with 1:4:16 contrasts at 0.35 μm on SGIII prototype laser facility. As it is explained above, this half hohlraum is used to test the optical diagnostics VISAR on SGIII prototype. A low plasma filling with maximum achievable radiation is required in the hohlraum. Eight laser beams from 45° angle of incidence cone simultaneously irradiate the hohlraum from one side. The power profile of the laser beams are shown in Figure 1. The durations of the three steps are 1.5 ns, 1 ns, and 0.6 ns, respectively. The ratio of hohlraum length L to radius R is taken as 1.7, i.e., $L/R = 1.7$. The radius of LEH is required to be 350 μm, and the laser-to-X-ray coupling efficiency in half hohlraum is around 70% on SGIII prototype.

In order to use the extended plasma-filling model to design hohlraum size under a given laser drive, firstly we need to know the temporal profile of the radiation in hohlraum and the time-dependent albedo of the wall in order to obtain coefficients of k_j, H_j, γ_j, β_j, and $\tilde{\alpha}_j$ in Eqs. (10), (11), and (16).

From our simulation, for cases we concern, the radiation profile coefficient k_j is mainly decided by laser profile if the plasma filling is not serious. Meanwhile, the albedo profile coefficients, H_j, γ_j, and β_j, are mainly decided by k_j. The average albedo $\tilde{\alpha}_j$ is decided by maximum T_r in every step, but only sensitive to T_r in the first step when T_r is relatively low. Moreover, our study reveals that the radiation ablation behavior of the wall obtained from 2D simulation

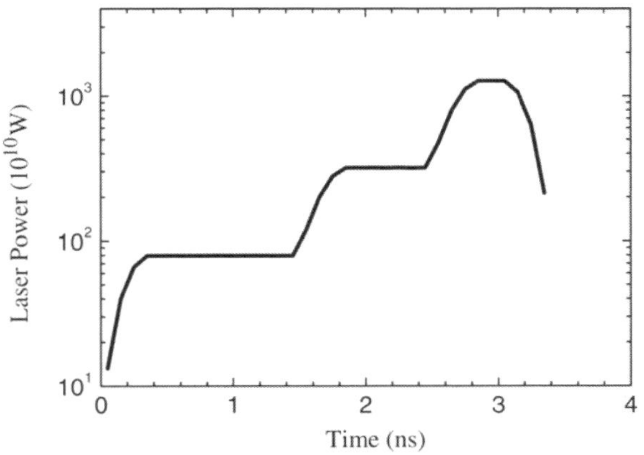

Fig. 1. Shaped laser pulse on SGIII prototype.

is very similar to that from one-dimensional simulation. The same result was also obtained in Suter *et al.* (1996).

Hence, we can first use our 2D code LARED-H to obtain a primary radiation in a half hohlraum with a reasonable supposed size under the given laser pulse, and then use one-dimensional code radiation hydrodynamic code of multi-groups (RDMG) (Feng *et al.*, 1999; Li *et al.*, 2010) to obtain the wall albedo under the radiation. Figure 2 shows the temporal T_r from LARED-H by taking R as 850 mm and L as 1450 mm, and the corresponding temporal albedo from RDMG.

Now, from the radiation and albedo obtained above, we can extract the coefficients by fitting: $k_j = 0.19$, 0.13, 0.12, $H_j = 8.5$, 7, 6.27, $\gamma_j = 0.66$, 0.66, 0.66, $\beta_j = 0.54$, 0.35, 0.54, for $j = 1$, 2, and 3, respectively. In addition, we take the average albedo of the three steps as: $\tilde{\alpha}_1 = 0.6$, $\tilde{\alpha}_2 = 0.7$, $\tilde{\alpha}_3 = 0.75$.

Inserting the above coefficients into Eqs. (12) and (20), we obtain the variations of n_e and T_r as R, as shown in Figure 3.

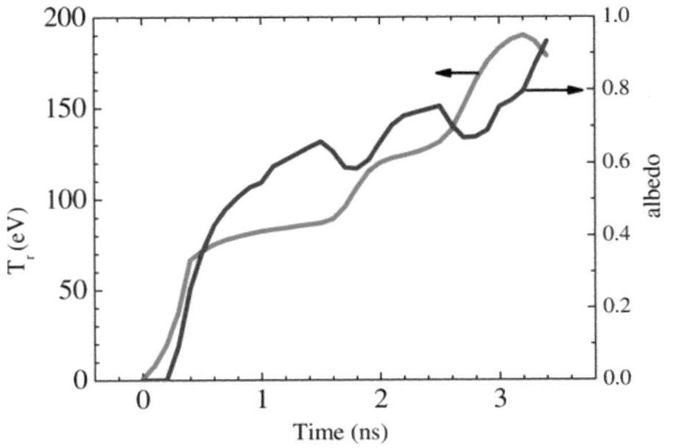

Fig. 2. (Color online) Temporal T_r (red line) in Au hohlraum from LARED-H and albedo (blue line) from RDMG. The hohlraum size is taken as $R = 850$ μm and $L = 1450$ μm.

Here, η is taken as 70%. As it is indicated, both n_e and T_r increases as R decreases. It means that the radiation is stronger and plasma filling is more serious in a smaller hohlraum. Thus, the criterion $n_e = 0.1$ gives the initial design of the half hohlraum as $R = 690$ μm with $T_r = 210$ eV.

2D SIMULATION RESULTS BY LARED-H

In the last Section, we obtained $R = 690$ μm as the initial design of the half hohlraum by using the extended plasma filling model. In order to find out an optimum design of the hohlraum size, we now use LARED-H to simulate three models of hohlraum with radii that are close to the initial design: $R = 600$, 650, and 700 μm. A total of 6 kJ of laser energy entered the half hohlraum from its LEH, and 10% laser energy is deducted after 2 ns due to laser plasma interaction loss from LEH.

In Figure 4, it shows spatial distribution of n_e in the hohlraum at 3 ns for the three models. As shown, the region of $n_e \leq 0.1$ is very small in the hohlraum of $R = 600$ μm, while relatively large in $R = 700$ μm hohlraum. In Figure 5, it presents the decay of the laser power in laser beam center along its transferring direction in the hohlraum. As indicated, the laser power decays to 37% of its initial value when it transfers to half of the hohlraum radius for $R = 600$ μm model, decays to 56% for $R = 650$ μm, and 77% for $R = 700$ μm. It means that most part of the laser has been absorbed by plasmas when it reaches half of its way to the hohlraum wall for $R = 600$ μm model, and hence it is not our optimum model. In Figure 6, it gives T_r in the hohlraum for the three models, obtained by post processing along the line about 30° from the hohlraum axis through LEH. As shown, the radiation temperature arrives, its maximum at around 3.2 ns, and it is about 233 eV, 221 eV, and 211 eV for the three models, respectively.

In order to get the maximum achievable radiation in a low plasma filling half hohlraum, the model of $R = 650$ μm is selected as the optimum design for the experiment. The

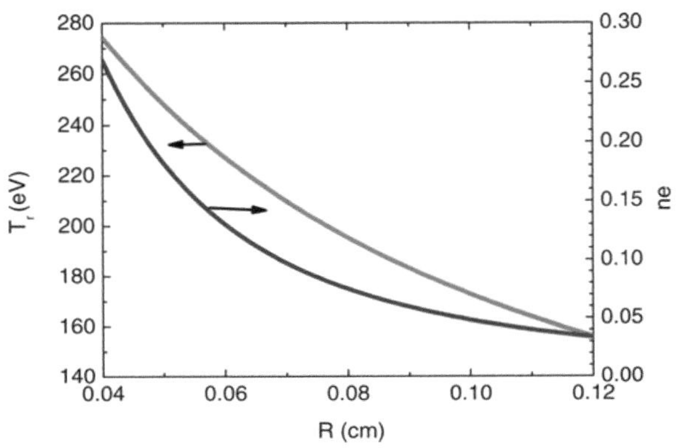

Fig. 3. (Color online) Variations of T_r (red line) and n_e (blue line) as R from Eqs. (12) and (20).

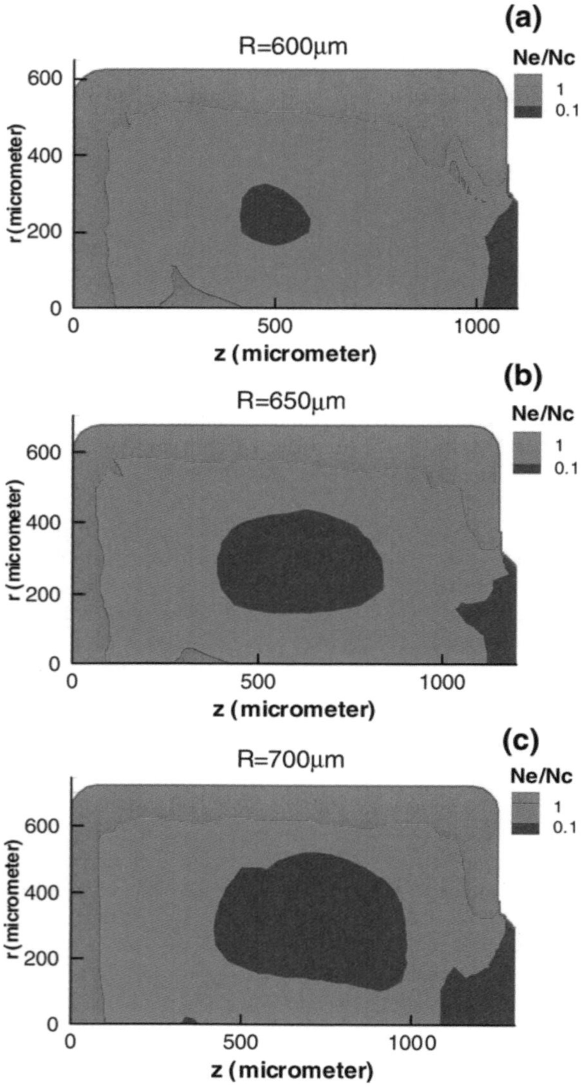

Fig. 4. (Color online) Spatial distribution of N_e/N_c in half hohlraum at 3 ns from LARED-H for the three models.

simulation results will be compared with the observations in a follow-up paper with experimental data we are collecting.

INITIAL DESIGN OF 300 eV IGNITION TARGETS OF NIF

In last section, we have used the extended plasma-filling model to obtain an initial design of a half hohlraum size to produce the maximum achievable radiation with low plasma filling under a given shaped laser pulse. In fact, the extended model can also be used to give an initial design of both the hohlraum size and the laser energy to produce a required radiation. Here, we use it to give an initial design

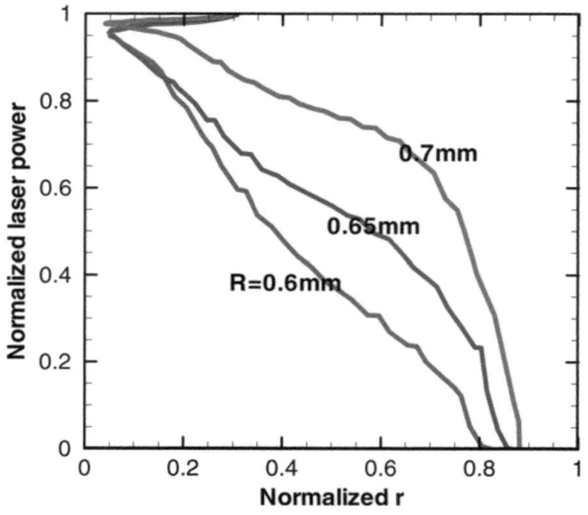

Fig. 5. (Color online) Decay of the laser power at 3 ns in laser beam center along its transferring direction in hohlraum for the three models. The abscissa is the radial position at the laser center and is normalized to hohlraum radius. The ordinate is the laser power at the laser center and is normalized to power at the center of LEH.

of the U hohlraum size and laser energy for producing the typical 300 eV ignition radiation designed for NIF (Hinkel *et al.*, 2008), although ignition hohlraum is filled with low-pressure gas while the plasma-filling model is made for an empty hohlraum. We take the contrasts between consecutive laser steps from Hinkel *et al.* (2008). The ratio of the hohlraum length to radius is taken as 3.62 and the coupling efficiency η is 75%.

In Figure 7, we present contour lines of $T_r = 300$ eV and $\eta_{e=0.1}$ in the plane of the hohlraum length and the laser energy for U hohlraum under the 300 eV ignition radiation drive. The intersection of $T_r = 300$ eV and $\eta_{e=0.1}$ gives the

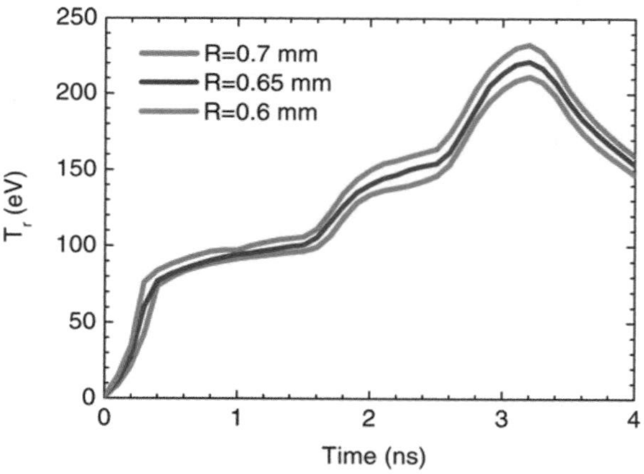

Fig. 6. (Color online) Postprocessed calculated T_r along the line about 30° from the hohlraum axis through LEH at 3 ns for the three models.

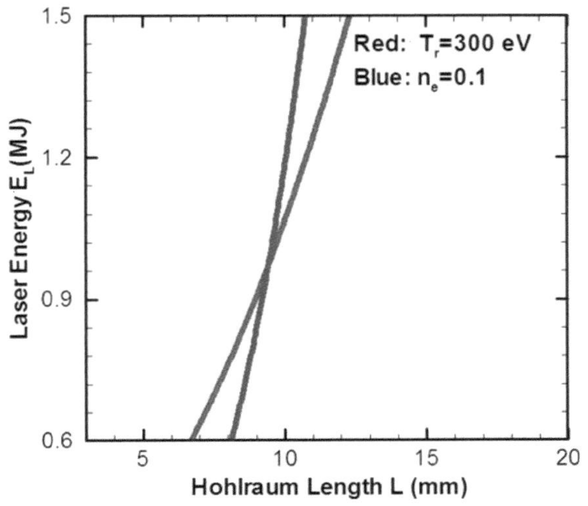

Fig. 7. (Color online) Contour lines of T_r = 300 eV (red line) and n_e = 0.1(blue line) in L/E_L plane for U hohlraum 300 eV ignition design.

initial design of 9.3 mm long hohlraum with 0.95 MJ laser energy, which is close to the 300 eV candidate hohlraum design for the 2009 ignition campaign on NIF (Hinkel *et al.*, 2008).

SUMMARY

In conclusion, the extended plasma-filling model can be used to give a reasonable initial design of the hohlraum size under a given laser pulse, or give a initial design of both the hohlraum size and laser energy to produce a required radiation. In particular, we can also have initial design of hohlraum with different geometries by this method, such as rugby-ball shaped hohlraum (Vandenboomgaerde *et al.*, 2007).

ACKNOWLEDGMENT

This work was performed under National Basic Research Program of China (973 program, No. 2007CB814800).

REFERENCES

Aleksandrova, I.V., Belolipeskiy, A.A., Koresheva, E.R. & Tolokonnikov, S.M. (2008). Survivability of fuel lasers with a different structure under conditions of the environmental effects: Physical concepts and modeling results. *Laser Part. Beams* **26**, 643–648.

Atzeni, S. & Meyer-Ter-Vehn, J. (2004). *The Physics of Inertial Fusion*. Oxford: Oxford Science Press.

Borisenko, N.G., Bugrov, A.E., Burdonskiy, I.N., Fasakhov, I.K., Gavrilov, V.V., Goltsov, A.Y., Gromov, A.I., Khalenkov, A.M., Kovalskii, N.G., Merkuliev, Y.A., Petryakov, V.M., Putilin, M.V., Yankovskii, G.M. & Zhuzhukalo, E.V. (2008). Physical processes in laser interaction with porous low-density materials. *Laser Part. Beams* **26**, 537–543.

Bret, A. & Deutsch, C. (2006). Density gradient effects on beam plasma linear instabilities for fast ignition scenario. *Laser Part. Beams* **24**, 269–273.

Chatain, D., Perin, J.P., Bonnay,. P., Bouleau, E., Chichoux, M., Communal, D., Manzagol, J., Viargues, F., Brisset, D., Lamaison, V. & Paquignon, G. (2008). Cryogenic systems for inertial fusion energy. *Laser Part. Beams* **26**, 517–523.

Deutsch, C., Bret, A., Firpo, M.C., Gremillet, L., Lefebvre, E. & Lifschitz, A. (2008). Onset of coherent electromagnetic structures in the relativistic electron beam deuterium-tritium fuel interaction of fast ignition concern. *Laser Part. Beams* **26**, 157–165.

Dewald, E.L., Suter, L.J., Landen, O.L., Holder, J.P., Schein, J., Lee, F.D., Campbell, K.M., Weber, F.A., Pellinen, D.G., Schneider, M.B., Celeste, J.R., Mcdonald, J.W., Foster, J.M., Niemann, C.J. Mackinnon, A., Glenzer, S.H., Young, B.K., Haynam, C.A., Shaw, M.J., Turner, R.E., Froula, D., Kauffman, R.L., Thomas, B.R., Atherton, L.J., Bonanno, R.E., Dixit, S.N., Eder, D.C., Holtmeier, G., Kalantar, D.H., Koniges, A.E., Macgowan, B.J., Manes, K.R., Munro, D.H., Murray, J.R., Parham, T.G., Piston, K., Van Wonterghem, B.M., Wallace, R.J., Wegner, P.J., Whitman, P.K., Hammel, B.A. & Moses, E.I. (2005). Radiation-driven hydrodynamics of high-Z hohlraums on the national ignition facility. *Phys. Rev. Lett.* **95**, doi: 10.1103/PhysRevLett.95.215004.

Duan, Q., Chang, T., Zhang, W., Wang, G., Wang, C. & Zhu, S. (2002). Two-dimensional numerical simulation of laser irradiated gold disk targets. *Chinese J. Comp. Phys.* **19**, 57–61.

Eliezer, S., Murakami, M. & Val, J.M.M. (2007). Equation of state and optimum compression in inertial fusion energy. *Laser Part. Beams* **25**, 585–592.

Feng, T.G., Lai, D.X. & Xu, Y. (1999). An artificial-scattering iteration method for calculating multi-group radiation transfer problem. *Chinese J. Comput. Phys.* **16**, 199–205.

Foldes, I.B. & Szatmari, S. (2008). On the use of KrF lasers for fast ignition. *Laser Part. Beams* **26**, 575–582.

Hinkel, D.E., Callahan, D.A., Langdon, A.B., Langer, S.H., Still, C.H. & Williams, E.A. (2008). Analyses of laser-plasma interactions in National Ignition Facility ignition targets. *Phys. Plasmas* **15**, 056314.

Hoffmann, D.H.H. (2008). Intense laser and particle beams interaction physics toward inertial fusion. *Laser Part. Beams* **26**, 295–296.

Holmlid, L., Hora, H., Miley, G. & Yang, X. (2009). Ultrahigh-density deuterium of Rydberg matter clusters for inertial confinement fusion targets. *Laser Part. Beams* **27**, 529–532.

Hora, H. (2007). New aspects for fusion energy using inertial confinement. *Laser Part. Beams* **25**, 37–45.

Imasaki, K. & Li, D. (2007). An approach to hydrogen production by inertial fusion energy. *Laser Part. Beams* **25**, 99–105.

Koresheva, E.R., Aleksandrova, I.V., Koshelev, E.L., Nikitenko, A.I., Timasheva, T.P., Tolokonnikov, S.M., Belolipetskiy, A.A., Kapralov, V.G., Sergeev, V.T., Blazevic, A., Weyrich, K., Varentsov, D., Tahir, N.A., Udrea, S. & Hoffmann, D.H.H. (2009). A study on fabrication, manipulation and survival of cryogenic targets required for the experiments at the Facility for Antiproton and Ion Research: FAIR. *Laser Part. Beams* **27**, 255–272.

Li, X., Lan, K., Meng, X., He, X., Lai, D. & Feng, T. (2010). Study on Au + U + Au sandwich hohlraum wall for ignition targets. *Laser Part. Beams* **28**, doi:10.1017/S0263034609990590.

Lindl, J.D. (1995). Development of the indirect-drive approach to inertial confinement fusion and the target physics basis for ignition and gain. *Phys. Plasmas* **2**, 3933.

Lindl, J.D., Amendt, P., Berger, R.L., Glendinning, S.G., Glenzer, S.H., Haan, S.W., Kauffman, R.L., Landen, O.L. & Suter, L.J. (2004). The physics basis for ignition using indirect-drive targets on the National Ignition Facility. *Phys. Plasmas* **11**, 339–491.

Mcdonald, J.W., Suter, L.J., Landen, O.L., Foster, J.M., Celeste, J.R., Holder, J.P., Dewald, E.L., Schneider, M.B., Hinkel, D.E., Kauffman, R.L., Atherton, L.J., Bonanno, R.E., Dixit, S.N., Eder, D.C., Haynam, C.A., Kalantar, D.H., Koniges, A.E., Lee, F.D., Macgowan, B.J., Manes, K.R., Munro, D.H., Murray, J.R., Shaw, M.J., Stevenson, R.M., Parham, T.G., Van Wonterghem, B.M., Wallace, R.J., Wegner, P.J., Whitman, P.K., Young, B.K., Hammel, B.A. & Moses, E.I. (2006). Hard X-ray and hot electron environment in vacuum hohlraums at the National Ignition Facility. *Phys. Plasmas* **13**, doi: 10.1063/1.2186927.

Pei, W. (2007). The construction of simulation algorithms for laser fusion. *Commun. Comput. Phys*. **2**, 255–270.

Ramis, R., Ramirez, J. &Schurtz, G. (2008). Implosion symmetry of laser-irradiated cylindrical targets. *Laser Part. Beams* **26**, 113–126.

Rodriguez, R., Florido, R., Gll, J.M., Rubiano, J.G., Martel, P. & Minguez, E. (2008). RAPCAL code: A flexible package to compute radiative properties for optically thin and thick low and high-Z plasmas in a wide range of density and temperature. *Laser Part. Beams* **26**, 433–448.

Schneider, M.B., Hinkel, D.E., Landen, O.L., Froula, D.H., Heeter, R.F., Langdon, A.B., May, M.J., Mcdonald, J., Ross, J.S., Singh, M.S., Suter, L.J., Widmann, K. & Young, B.K. (2006). Plasma filling in reduced-scale hohlraums irradiated with multiple beam cones. *Phys. Plasmas* **13**, doi: 10.1063/1.2370697.

Sigel, R., Pakula, R., Sakabe, S. & Tsakiris, G.D. (1988). X-ray generation in a cavity heated by 1.3 or 0.44 mm laser light III Comparison of the experimental results with theoretical predictions for x-ray confinement. *Phys. Rev. A* **38**, 5779–5785.

Strangio, C., Caruso, A. & Aglione, M. (2009). Studies on possible alternative schemes based on two-laser driver for inertial fusion energy applications. *Laser Part. Beams* **27**, 303–309.

Suter, L.J., Kauffman, R.L., Darrow, C.B., Hauer, A.A., Kornblum, H., Landen, O.L., Orzechowski, T.J., Phillion, D.W., Porter, J.L., Powers, L.V., Richard, A., Rosen, M.D., Thiessen, A.R. & Wallace, R. (1996). Radiation drive in laser-heated hohlraums. *Phys. Plasmas* **3**, 2057.

Vandenboomgaerde, M., Bastian, J., Casner, A., Galmiche, D., Jadaud, J.-P., Laffite, S., Liberatore, S., Malinie, G. & Philippe, F. (2007). Prolate-spheroid ("rugby-shaped") hohlraum for inertial confinement fusion. *Phys. Rev. Lett.* **99**, doi: 10.1103/PhysRevLett.99.065004.

Winterberg, F. (2008). Lasers for inertial confinement fusion driven by high explosives. *Laser Part. Beams* **26**, 127–135.

Yang, H., Nagai, K., Nakai, N. & Norimatsu, T. (2008). Thin shell aerogel fabrication for FIREX-I targets using high viscosity (phloroglucinol carboxylic acid)/formaldehyde solution. *Laser Part. Beams* **26**, 449–453.

EPJ Web of Conferences **59**, 01009 (2013)
DOI: 10.1051/epjconf/20135901009

Advances in the national inertial fusion program of China

X.T. He[a], W.Y. Zhang and the Chinese ICF team

Institute of Applied Physics and Computational Mathematics, Beijing 100088, China

Abstract. The planned inertial confinement fusion (ICF) ignition in China in around 2020 is to be accomplished in three steps. The first is carrying out target physics experiments in the existing laser facilities SG-II, SG-IIIP and SG-IIU (operating in 2012) of output energy 3-24 kJ at 3ω. Results have been obtained for better understanding the implosion dynamics and radiation transport. Recent studies include efficiency of radiation generation, hydrodynamic instabilities, shock waves in cryogenic targets, opacity measurements using kJ lasers, etc. Hydrodynamic codes (the LARED series) have been developed and experimentally verified with over 5000 shots, and are applied to investigating target physics and ignition target design. For fast ignition, a large number of experiments and numerical simulations have led to improved understanding relevant to target design, hot electron transport, collimation by the spontaneous magnetic fields in overdense plasmas, etc. In addition to the SG-II, SG-IIU and SG-IIIP, the SG-III laser facility with energy of 200–400 kJ at 3ω shall operate in 2014 and be used for advanced target physics research. In the last step, the 1.5 MJ SG-IV laser facility still under design will be used to investigate ignition and burning.

1. INTRODUCTION

The national research institutes such as the Research Center of Laser Fusion (RCLF), the Institute of Applied Physics and Computational Mathematics (IAPCM), the National Laboratory on High Power Laser and Physics (NLHPLP), as well as many universities and other institutions are participating in the Inertial Confinement Fusion (ICF) Program of China. The program involves target physics (theory and experiment) and target fabrication, high-power lasers, and precision diagnostics. The roadmap towards ignition in around 2020 is being carried out in three steps. In the first step, target physics research and target design are based on the existing tens-kJ laser facilities in China. In the second step, pre-ignition target design and advanced target physics research, based on the SG-III laser facility, are crucial as the basis for the ignition target design in the coming six years, which should identify and solve the essential science and technology issues, to reduce risks, and to scale up the design to the ignition target. In the last step, ignition is to be demonstrated on the SG-IV laser ignition facility.

2. DEVELOPMENT OF LASER DRIVERS

The SG-II in operation has 8 beams with output energy of 3kJ (here and below, unless otherwise stated the laser energy is for 3ω) and pulse duration of 1 ns. Since 2000 it has provided over 3500 shots for experiments. The SG-IIU (SG-II upgrade) under construction for target physics research has an output energy of 24 kJ for the 3 ns pulse and should be operating in 2012. The petawatt (PW) laser with pulse duration of 1–5 picoseconds and output laser energy of 1 kJ at 1ω should also be operating in 2012. The SG-IIU combined with the PW laser will serve the fast ignition research.

[a]e-mail: xthe@iapcm.ac.cn

Figure 1. (a) SG-III laser bay. (b) The bottom of the target chamber, the operating-beam guide is in blue color.

An SG-III prototype (SG-IIIP) has energy output of 15 kJ with pulse duration of 3 ns. Since 2006 it has completed over 1000 shots for experiments.

The SG-III facility (Figure 1a) under construction has 48 beams at output laser energy of 200–400 kJ with pulse duration of 3–5 ns. The first bundle (8 beams) was installed in 2010 and the second bundle in 2010–2011. One beam (Figure 1b) with laser energy of 16 kJ at 1ω has been working since June 2011 for tests. Full laser energy output is expected in 2013 [1].

The SG-IV ignition facility is still being designed. The amount of laser energy required for ignition will be determined by scaling up the results from the target physics research on the SG-III. The SG-IV facility should be completed in around 2020.

3. IMPROVED SIMULATION OF TARGET PHYSICS WITH LARED AND JASMIN

JASMIN has an adaptive structure mesh application infrastructure. It builds a bridge among the LARED package, the parallel computers and the users. It allows for using non-structured multi-block mesh for complex geometries and Lagrangian mesh across material interfaces for dealing with large deformation, etc. It also improved indirect-drive simulations involving complicated geometries, multi-material and large deformations, parallel computing with several thousand processors.

The LARED series is a numerical code package for simulating direct- and indirect-drive ICF, and has been verified with about 5000 shots in experiments. A LARED-integrated code involving 2D hydrodynamics, radiation transport and multi-group diffusion, 3D ray tracing in the hohlraum, thermal conductivities for electrons and ions, and plasma burning, is also developed. Combined with JASMIN, LARED greatly improves simulations of indirect-drive ICF, and is being used in advanced research on target physics, especially of the ignition target.

4. ADVANCES IN TARGET PHYSICS EXPERIMENTS (SELECTED EXAMPLES)

M-band fraction. A novel flat-response x-ray detector for photon energies in the range 0.1–4 keV has been applied to measure the radiation temperature T and M-band emission from the hohlraum gold wall in the SG-II facility. For the M-band measurement, the system consists of the filter system 720 nm B and 3.9 μm Sc and 34 μm Sc (transmitting area 18%), as well as Al cathode XRD. The signal $V = \int R(E)S(E,t)dEd\Omega = RS$, where $S(E,\Omega)$ is the spectrum of x-ray energy E and solid angle Ω, and R is a response function that approaches to a constant. Thus, we can obtain the measured x-ray flux $S = \int S(E,t)dEd\Omega$ when R is calibrated. An M-band fraction of $\sim$6% was measured for laser energy of 2.1kJ (MSE $\sim$ 5% for 1.4–4.4 keV) at the SG-II facility [2].

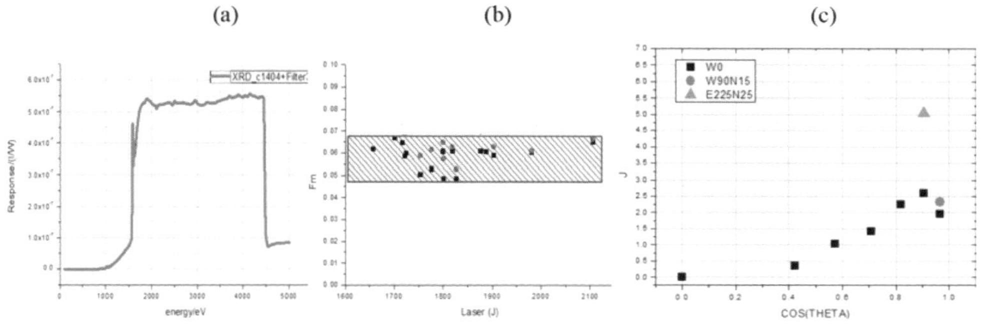

Figure 2. M-band emission from gold wall in the hohlraum as measured on the SG-II. (a) Response function R vs x-ray energy. (b) M-band fraction vs laser energy. (c) Angular distribution of the M band.

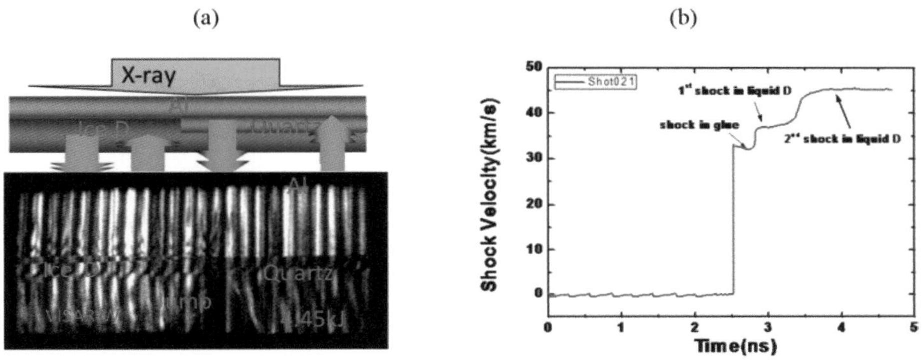

Figure 3. (a) The jumps of the first and second shocks in the iced deuterium can be clearly observed. (b) The velocity jumps are shown as the first one arrived and the second one catches up with the first.

Shock waves. Shocks are measured by VISAR (shock velocity resolution ~2%) on the SG-IIIP using cryogenic targets. A 4.45 kJ laser is incident on the inner wall of the hohlraum and the radiation ablates the Al foil and drives shock waves. The first shock propagates directly to the iced deuterium and the second one first penetrates the quartz and then into the iced deuterium. The second shock catches up with the first, as shown in Figure 3(a). The velocities of the two shocks vary with time, the shock passes through the glue interface between the Al and the quartz, causing a velocity change, as shown in Figure 3(b).

Opacity measurements. A new target configuration is used to measure opacity. The target is between two *separated* parts of the hohlraum and has a clean environment, where the scattering light and expanded plasmas from hohlraum are blocked by foam (Figure 4a). Thus, smaller hohlraum and lower laser energy can be used to achieve higher-temperature opacities without pollution. We have obtained Al opacity with 95 eV radiation temperature at 1.5 kJ laser energy. The Au, Fe, and Ge opacities measured at 85 eV radiation temperature (for $\rho \approx 20\,\mathrm{mg/cm^3}$) and 1.2 kJ laser energy are shown in Figures 4(b)–(c). For measuring opacity of the dense target we have used x-ray scattering, as shown in Figures 4(d)–(f) [3].

Measurements of the equation of state (EOS). Pressures of about 110 Mbar for Al and about 230 Mbar for Au are obtained from the measured shock velocity (SOP) by the step targets connected with the half-hohlraun. A laser of 6 kJ from SG-IIIP is incident on a small-size hohlraum of $\varnothing 700\mu\mathrm{m} \times 585\,\mu\mathrm{m}$. Simulations give the M-band fraction of ~25% in such a hohlraum, where the energy from SRS has been deducted. The equivalent radiation temperature of $T = 270$–290 eV is obtained from

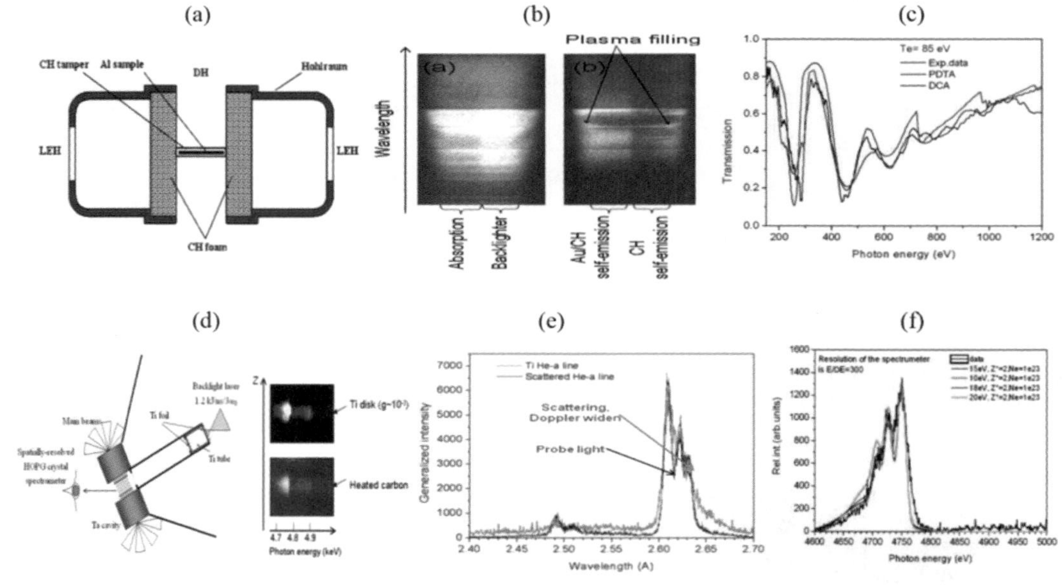

Figure 4. Measurements of opacity. (a) Clean hohlraum target, (b) absorption, backlight and self-emission spectrum measured using a time-gated grating spectrometer, (c) atomic physics calculation for opacity: DCA–detailed configuration model, PDTA–pseudo-detailed model, (d)–(e) bulk heating of carbon foam by Ta M-band x-ray, (f) comparison of experimental data and fits for various T_e.

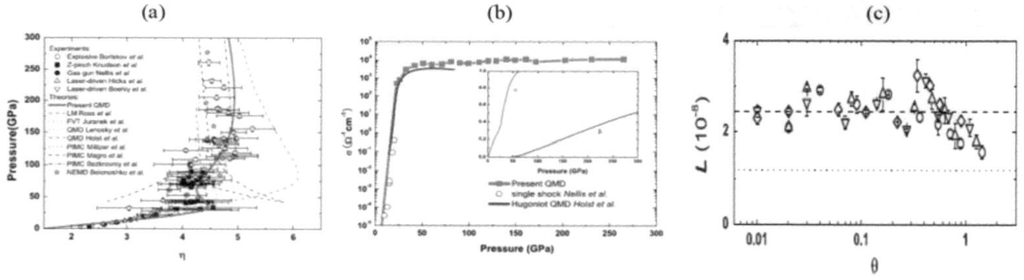

Figure 5. (a) Hugoniot curve for D for <3 GPa, (b) electric conductivity σ vs pressure, (c) Lorentz number vs θ, for densities of 400(O), 480(Δ), 600(∇), 800(◇) g/cm^3.

the scaling relation $T = \delta D\gamma$, where δ and γ are adjustable parameters [4]. EOS for deuterium (D) for pressure less then 3 Mbar is simulated by QMD and OFMD coDEs. Results (red) by QMD [5] show maximal compressed ratio $\eta = 4.5$ around 40 GPa and η =4.95 about 180 GPa due to D2 dissociation (α, red) and D ionization (β, black) as shown in Figure 5(a)–(b), and agree with the experiments. In the process of adiabatically compression of ICF, the fusion fuel and its reaction products are warm dense matter (WDM) under extremely high density and pressure, and are in partial degenerate states that have not a satisfactory theory to deal with. We have to apply massively parallel computing to investigate EOS. A Lorentz number of $L = \kappa/\sigma T$, where κ and σ are thermal and electric conductivities, respectively, is a constant of 2.44×10^{-8} in the full degenerate state and 1.18×10^{-8} in the non-degenerate state, which can be used to identify full degeneracy or not [6]. Numerical simulations show that for densities of 400–800 g/cm^3 the full degenerate theory is applicable for $\theta = T/T_F \leq 0.35$, and the partial degenerate theory for $0.35 < \theta < 2$, where T_F is the Fermi temperature (see Figure 5(c)).

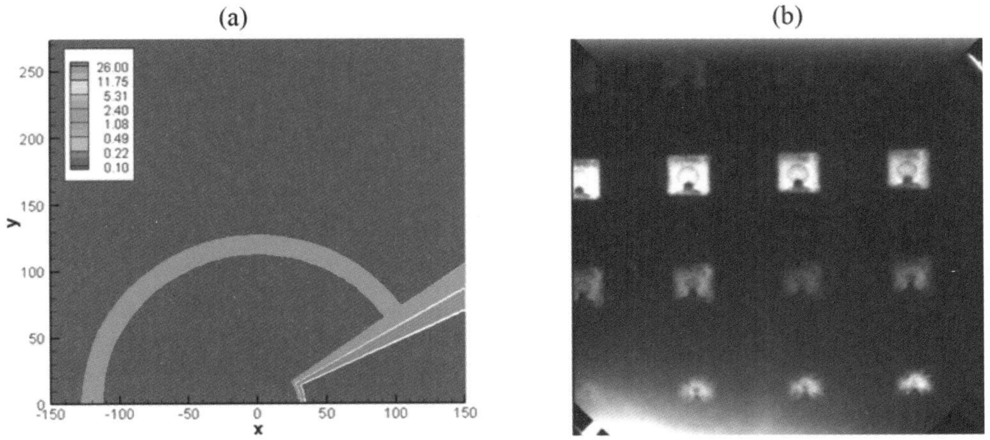

Figure 6. Simulation and experiment on fast ignition. (a) Surface of Au cone is coated by CD material, (b) indirect-drive implosion experiment on SG-II using Au cone-CD shell ($r \sim 250\ \mu$m) with hohlraum size $\varnothing 1000\ \mu\text{m} \times 1350\ \mu$m.

Fast ignition. Fast ignition is investigated by simulations and experiments. Implosion dynamics for the Au cone-CD shell targets shows that Au without coating is mixed into the CD plasma, but the mixing is suppressed with CD coating on the surface of the Au cone. It also shows that a guiding filament in double-cone tip generates spontaneous magnetic field that collimates the hot electron beam [7].

5. REMARKS

Since 2000, target physics research has been made through a series of experiments in the SG-II and SG-IIIP facilities, as well as extensive numerical simulations using the LARED-JASMIN codes. The ignition physics is also being advanced. Beginning 2012 the SG-IIU driver will serve experiments on the target physics, and will be coupled with the PW laser to investigate hotspot formation in fast ignition. The SG-III driver will be completed in about 2014. Target physics research on the SG-III is an important intermediate step towards ignition, and should find and solve relevant issues in the science and technology of eventual ignition. The 1.5 MJ SG-IV driver for direct ignition and high-gain fast ignition is in the design stage. China expects to attain moderate/high gain ignition and burning in about 2020.

References

[1] W. G. Zheng et al., *Progress of the SG-III solid-state laser facility*, presentation at IFSA2011, Bordeaux, France

[2] Z. C. Li et al., Rev. Sci. Instrum. **82**, 106106 (2011)

[3] Jiyan Zhao et al., Phys. Rev. E **79**, 016401 (2009); Yang Zhao, et al., Rev. Sci. Instrum. **80**, 043505 (2009); Y. Xu et al., Phys. Plasmas **14**, 052701 (2007)

[4] Yongsheng Li et al., Phys. Plasmas **18**, 022701(2011)

[5] Cong Wang et al., J. Appl. Phys. **108**, 044909 (2010)

[6] Cong Wang et al., Phys. Rev. Lett. **106**, 145002 (2011)

[7] Hong-bo Cai et al., Phys. Rev. E **83**, 036408 (2011); C. T. Zhou et al., Optics Lett. **36**, 924 (2011)

PHYSICS OF PLASMAS 20, 082707 (2013)

Preheat of radiative shock in double-shell ignition targets

J. W. Li,[1,2,a)] W. B. Pei,[2] X. T. He,[1,2] J. H. Li,[2] W. D. Zheng,[2] S. P. Zhu,[2] and W. Kang[1]
[1]*Key Lab of High Energy Density Physics Simulation, Center for Applied Physics and Technology, Peking University, Beijing 100871, China*
[2]*Institute of Applied Physics and Computational Mathematics, P.O. Box 8009, Beijing 100094, China*

(Received 14 May 2013; accepted 4 August 2013; published online 28 August 2013)

For the double-shell ignition target, the nonuniform preheat of the inner shell by high-energy x rays, especially the M-band line radiation and L-shell radiation from the Au hohlraum, aggravates the hydrodynamic instability that causes shell disruption. In this paper, for the first time, we propose another preheating mechanism due to the radiative shock formed in the CH foam, and also confirm and validate such preheat of radiative shock by numerical results. We also give an estimate of the improved double-shell in which the CH foam is replaced by the metallic foam to mitigate the hydrodynamic instabilities, and find that the radiative shock formed in the metallic foam produces a much stronger radiation field to preheat the inner shell, which plays a role in better controlling the instabilities. In double-shells, the preheat of radiative shock, as a potential effect on the instabilities, should be seriously realized and underlined. © *2013 AIP Publishing LLC*. [http://dx.doi.org/10.1063/1.4818970]

I. INTRODUCTION

Double-shell targets, as an additional design for demonstrating and exploring ignition on the NIF,[1–3] consist of two concentric shells, a low-Z ablator (outer) shell surrounding a high-Z (inner) shell kept in place by a low density foam. The inner shell, referred to as the pusher, is filled with high density fuel at room temperature. The outer shell acts as an efficient absorber of hohlraum-generated x rays, which magnifies and transfers the acquired energy density upon spherical convergence and subsequent collision with the inner shell. Compared to the cryogenic single-shell ignition target,[4–6] the double-shell target has its expected benefits including non-cryogenic deuterium-tritium (DT) fuel preparation, lower ignition temperature ($\sim$4 keV), a lower implosion velocity, and a relatively low characteristic fuel convergence ($\sim$10) for a relaxed symmetry requirement.[1,2,7]

Despite the expected advantages in the double-shells, the hydrodynamic instabilities such as Rayleigh-Taylor (RT) and Richtmyer-Meshkov (RM) are especially critical due to the existence of several interfaces. Perturbations on the surface of the pusher (the inner shell) have been found to be particularly troublesome, which can be a feed-through to the inner surface of the shell. The outer surface is RM unstable at time of shell collision and may subsequently become RT unstable, possibly disrupting the inner shell. What is more, the preheat due to high-energy x rays, especially the M-band line radiation and L-shell (>8 keV) radiation from the Au hohlraum, which can freely pass through the outer shell and ablate the outer surface of the inner shell, will also result in the instabilities because of the nonuniform inner shell preheat.[1,2,8] Besides that, it is also found that those high-energy photons can promote a large outward expansion of the pusher material, which is subsequently recompressed by the converging outer shell. During this recompression phase, the low density foam pushes onto the higher density ablated gold setting up the classic conditions for the RT instability, another pathway for the RT instability that causes shell disruption.[2,9,10]

To control the instabilities, reducing the inner shell preheat by the energetic x-rays flux asymmetry component is very important. An improved ablator acting as an efficient absorber of high-energy hohlraum x rays is proposed.[9] It seems to be a good method to reduce the inner shell preheat only if there is high-energy x rays preheating it. Actually, there may exist other preheating mechanism to preheat the inner shell, and a new preheating by radiative shock is first proposed in the paper, which also plays a role in promoting the inner shell expanding outwardly and affecting the hydrodynamic instabilities. The radiative shock is formed in the low-density foam rather than the outer shell (ablator). Even though the improved ablator can affectively absorb the high-energy x rays, the inner shell will still be preheated by the radiative shock. Such preheating mechanism should be seriously realized and paid much attention to, as well as the preheat by the high-energy x rays in designing the double-shell target.

It is known that if a shock is so strong that the radiative energy flux or the radiation pressure plays an essential role in the dynamics, the shock will become radiating.[11,12] Unless the material in the pre-shock is completely transparent, it will be preheated by the radiation emitted from the post-shock to a higher temperature and pressure before the shock arrives because the radiation runs much faster than the shock; when the shock arrives, the material been preheated by the radiation will be heated again by the shock; then cooled by the emission of part of its energy as radiative flux. It is likely for a strong shock to become radiative when it propagates through a low-density material.[13] In the double-shell target, when the converging radiatively driven shock enters the low-density foam (10 mg/cc), it always has a velocity larger than 100 km/s, so it will become radiative easily. For the radiative shock, another radiative heat wave is

a)li_jiwei@iapcm.ac.cn

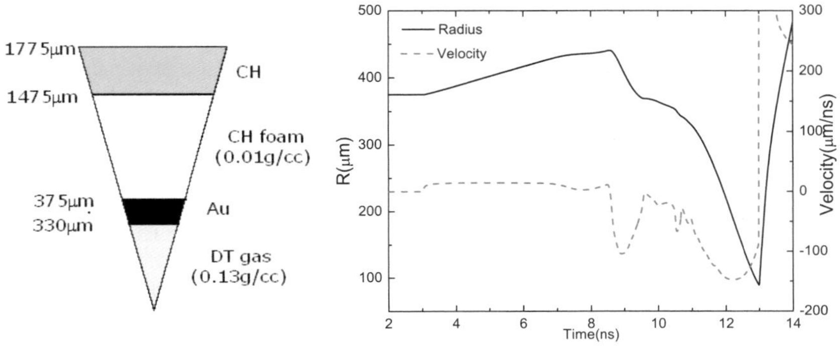

FIG. 1. Schematic diagram of the double-shell ignition target design (left), the trajectory (black solid line) and velocity (red dashed line) history of the outer surface of the inner Au shell (right) for the double-shell target driven by a 215 eV x ray source.

formed due to the radiation from the shocked foam in the post-shock, and it will run ahead of the shock front and preheat the material in the pre-shock. When the radiative heat wave passes through the low-density foam, it will also preheat (ablate) the outer surface of the inner shell before the shock arrives. The preheat (ablation) of the unstable Au/foam interface by the radiative shock, together with high-energy x rays, plays a role in the hydrodynamic instabilities.

The paper is organized as follows. In Sec. II, we give an analysis of the preheat by radiative shock in the double-shell target and validate its existence by numerical results. In Sec. III, we demonstrate the role of the preheat in mitigating the hydrodynamic instabilities, especially for the improved double-shell target, the radiative shock formed in the metallic foam produces a stronger preheat of the inner shell, which can further suppress the instabilities.

II. PREHEAT OF RADIATIVE SHOCK IN DOUBLE SHELLS

Here we demonstrate a double-shell ignition target design as shown in the left frame of Fig. 1, and the numerical results are obtained by the 1D radiation hydrodynamics code RDMG.[14] In the simulation, the radiation transport equation is treated with a multi-group Sn method, and the electron and ion temperature are by the thermal conduction with a flux limiter. The double-shell ignition target is driven by a 215 eV x ray drive, Planckian in frequency to disregard the inner shell preheat by the M-band line radiation and L-shell radiation. The right frame in Fig. 1 shows the trajectory and velocity history of the outer surface of the inner Au shell, indicating that by about t ∼ 3.0 ns, the shell begins to move outwardly; and about t ∼ 8.6 ns, it has been ablated to a radius 440 μm from its initial radius 375 μm. At a later time, the expanding shell is recompressed by the CH foam setting up the conditions for the RT instability.[9] Obviously, the inner shell preheat is responsible for the perturbation of the Au/CH foam interface. Here the drive is Planckian in frequency, so it seems that there is only some high-frequency photons, which can pass freely through the ablator to preheat the inner shell according to previous work.[1,2,9]

Now we examine the preheating radiation and electron temperature at the outer surface of the inner Au shell as shown in Fig. 2. We can see that at the time of beginning, before the shock driven by the ablation pressure passes through the ablator, the preheat is very weak with an electron temperature less than 10 eV because there is only the high-frequency photons of the x ray passing freely through the ablator to preheat the inner shell; after time t ∼ 5.6 ns, there is a rapid increase of the preheating radiation and electron temperature, resulting in a large outward expansion of the shell, which is a consequence of the further preheat by radiative shock. Actually at time t ∼ 3.5 ns, the shock driven by the ablation pressure enters the CH foam (10 mg/cc) and becomes radiative as the radiative flux plays a role in the dynamics, then another radiative heat wave emitted by the post-shock rather than the x ray drive runs ahead of the shock front and preheats the material in the pre-shock, which passes through the CH foam and begins to ablate the outer surface of the inner shell at time t ∼ 5.6 ns. Here it is obvious that the preheat due to the radiative shock plays an important role in promoting the shell's outward expansion. By the numerical results, we observe that, besides the high-energy photons, a radiative shock formed in the low-density CH foam will also preheat the inner shell, which has never been specifically pointed out.

As shown in Fig. 3, at time t ∼ 7.0 ns, when the shock has entered the CH foam and becomes radiative, the converging radiative shock has reached a velocity 170 km/s.

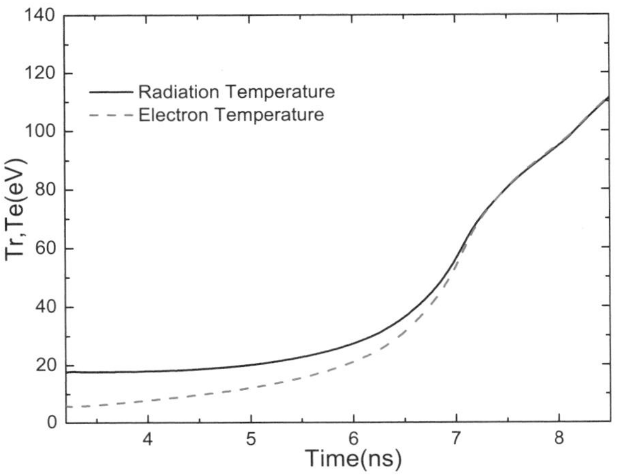

FIG. 2. The preheating radiation (black solid line) and electron (red dashed line) temperature at the outer surface of the inner Au shell due to preheat for the double-shell target.

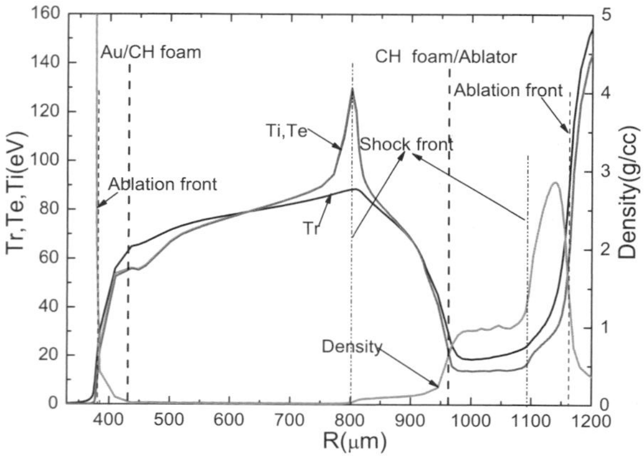

FIG. 3. Radiative shock formed in the CH foam in the double-shell ignition target. Radiation (black), electron (red) and ion (blue) temperature and density (green) for the double-shell target at time t ~ 7.0 ns when the shock driven by the ablation front has entered the CH foam. Before the shock front, another radiative heat wave driven by the post-shock has ablated the inner Au shell.

There exists a Zel'dovich spike (ion and electron temperature) located at the shock front.[11] The radiative heat wave emitted from the post-shock has propagated supersonically through the CH foam upstream only with an optical depth ~0.1, much faster than the shock and ahead of the shock front ~420 μm, and obviously, it has ablated the outer surface of the inner Au shell to a great extent. Here, the ion and electron temperature in the post-shock decrease toward the ablator to maintain a continuous pressure profile across the foam/ablator interface. At the time, two ablation fronts appear in the double-shell target as shown in Fig. 3: one is in the ablation zone driven by the x ray source, and the other in the inner Au shell driven by the post-shock. After being ablated by the radiative shock, the outer surface of the inner shell is heated with an electron temperature 56 eV, much higher than the preheating result by high-frequency photons of the x-ray. Correspondingly, the inner shell suffers a larger outward expansion due to the further preheat of radiative shock. It is shown that the shock driven by the ablation pressure has entered the CH foam, but there is also another shock moving inwardly in the ablator, which is the consequence of the first shock successively reflected by the CH foam/ablator interface and the ablation front.[1]

To further validate and confirm the preheat of radiative shock in the double-shell target, we artificially make the shock propagating through the CH foam non-radiative. Here, in the code, the equations of radiation hydrodynamics are solved with the three-temperature diffusion model, so there is no existence of the high-frequency photons preheating the inner shell because all the photons move at a moderate speed. When the shock enters the low-density CH foam, we artificially suppress the radiation transport by infinitely increasing the opacity of the foam, then the shock is non-radiative, and there is no radiative heat wave ahead of the shock front to preheat the inner Au shell. The left frame in Fig. 4 shows the numerical results for the double-shell target where the shock in the foam is radiative (upper frame) and artificially non-radiative (lower frame), respectively. Contrastingly, the artificial non-radiative shock does not emit a radiative heat wave to preheat the inner shell, so the inner shell suffers little perturbation as shown in the right frame in Fig. 4 where the inner shell does not begin to move inwardly until the shock arrives; on the other hand, the radiative shock preheats the inner shell to a higher temperature and promotes it expanding outwardly significantly. The radiative shock formed in the CH foam does preheat the inner shell, another potential effect on the implosion and hydrodynamic instabilities, and it should be attached importance to in the double-shells.

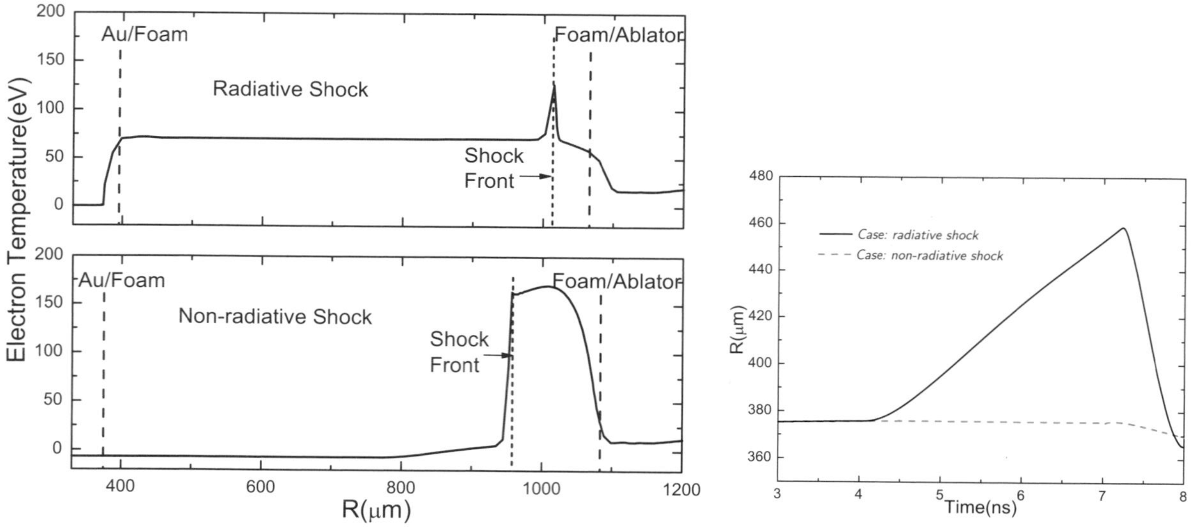

FIG. 4. Comparison of the electron temperature (left) between the radiative shock (upper) and artificial non-radiative shock (lower) in the CH foam at time t ~ 5.0 ns, and the corresponding trajectory history of the outer surface of the inner Au shell (right) for the double-shell target. When the shock enters the CH foam, the opacity is artificially increased to prevent the shock from radiating.

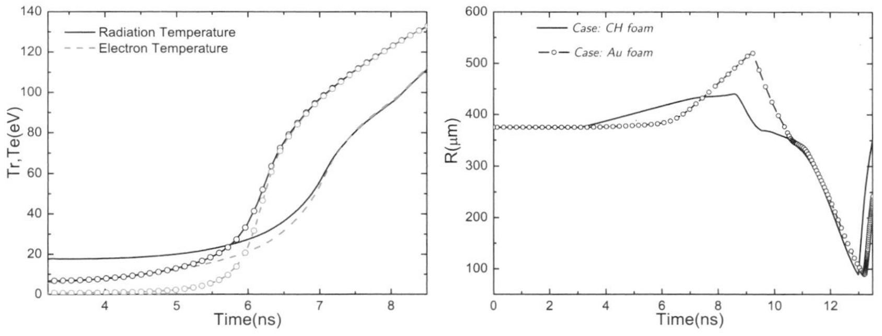

FIG. 5. The preheating radiation (black solid line) and electron (red dashed line) temperature at the outer surface of the inner shell (left), trajectory history of the outer surface of the inner shell (right) for the double-shell target. The different curves correspond to the targets filled with the CH foam (line) and the Au foam (line with circle), respectively.

III. MITIGATION OF RT INSTABILITY BY THE PREHEAT OF RADIATIVE SHOCK

It has been indicated that the radiative shock depends on the material in which the shock propagates besides the shock's velocity,[13] so the preheat of radiative shock may perform differently for different kinds of foam. Recently, an improved target design is proposed with low-density (<100 mg/cc) metallic foams rather than the CH foams for structural support of the inner shell and hydrodynamic instability mitigation, and the instability mitigation is attributed to the reduction of the Atwood number at the inner shell/foam interface.[2,9] Maybe it is the preheat of the radiative shock in the metallic foam that plays such a role. In the simulation, we simply substitute the Au foam (20 mg/cc) for the CH foam in the double-shell target to study the preheat by radiative shock formed in the metallic foam. The left frame in Fig. 5 shows the preheating radiation and electron temperature at the outer surface of the inner shell due to preheat for two targets with the CH foam and the Au foam, respectively. By contrasting, we can see that at the time of beginning, the Au foam can suppress the inner shell preheat and retard the shell's outward expansion because of its opacity which can both effectively absorb the high-frequency photons of the x ray and delay the preheat by the radiative shock and its harder equation of state (EOS) to retard the shell's moving; at a later time when the radiative heat wave driven by the post-shock, or by the shocked foam begins to ablate the inner shell, the shell experiences a quick increase in radiation and electron temperature at its outer surface, resulting in a larger outward expansion as shown in the right frame in Fig. 5 due to the stronger radiative shock in the high-Z materials.

However, the performance of radiative shock may depend on the material properties of the foam, such as the EOS and ionization.[15] To study these effects, we simulate the double-shell targets filled with the CH foam (10 mg/cc) and the Au foam (10 mg/cc), in which we also artificially manipulate the EOS of CH foam, just identical to that of gold material, and the outward expansion of the inner shell due to preheat as shown in Fig. 6. By comparing the two targets filled with the CH foams with its real and artificial EOS, respectively, we can see that at an earlier time when there is only the high-energy x rays preheating the shell, the harder artificial EOS of the foam can retard the shell moving outwardly, but after time t ~ 5.6 ns when the radiative shock plays a role in preheating the inner shell, the EOS can help produce a stronger radiative shock to preheat the inner shell to a larger outward expansion; while by comparing the targets with the artificial CH foam and the gold foam, which are of the same EOS, we can see that, after time t ~ 5.6 ns, the ionization of gold can also help produce a much stronger radiative shock to preheat the inner shell. The EOS and ionization of the foam can affect the radiative shock in the foam and its preheat of the inner shell. What is more, the density of the

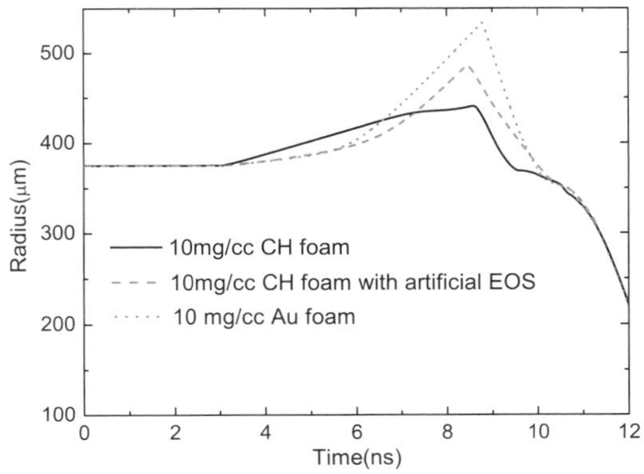

FIG. 6. The trajectory history of the outer surface of the inner shell due to preheat for the double-shell targets. The different curves correspond to the targets filled with the CH foam (black solid line), the CH foam with artificial EOS (red dashed line), and the Au foam (green dotted line), respectively.

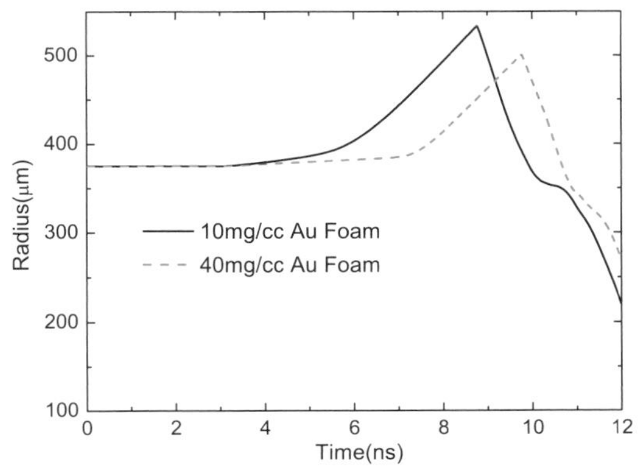

FIG. 7. The trajectory of the outer surface of the inner shell due to preheat for the double-shell targets filled with the Au foam of 10 and 40 mg/cc densities, respectively.

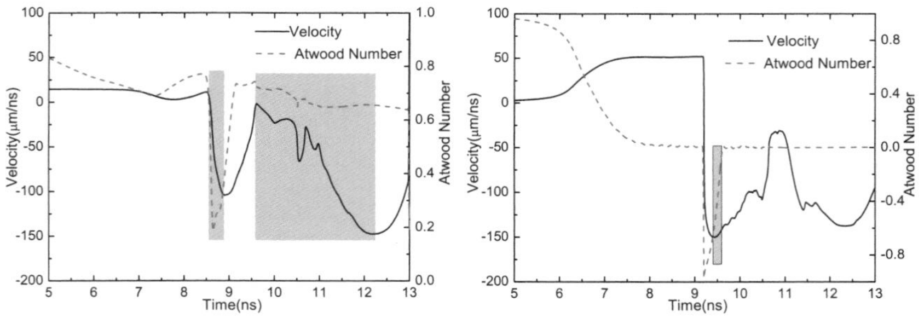

FIG. 8. The velocity (black solid line) and Atwood number (red dashed line) history at the Au/foam interface for the double-shell targets filled with the 10 mg/cc CH foam (left) and the 20 mg/cc Au foam (right). The gray zones correspond to the RT growth episodes.

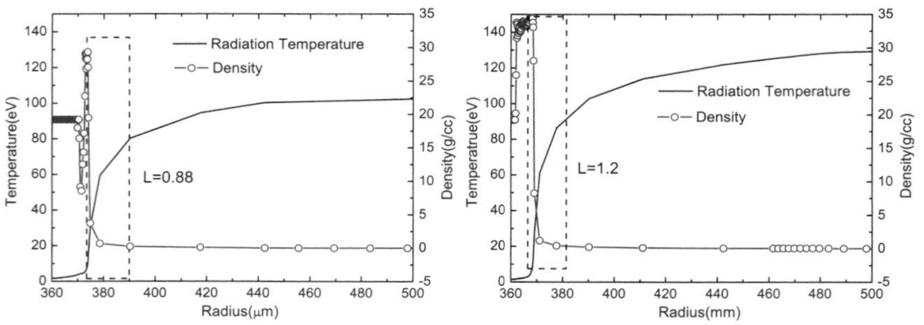

FIG. 9. The density scale length at the ablation front inside the inner shell due to radiative shock for the double-shell targets filled with the 10 mg/cc CH foam (left) and 20 mg/cc Au foam (right) at time $t \sim 8.0$ ns, respectively.

foam may also affect the preheat and the expansion of the inner shell, and the denser foam may play a role in restricting the inner shell from expanding outwardly. As shown in Fig. 7 where the two targets filled with the gold foam of different densities, because of its opacity and its density, the 40 mg/cc gold foam renders the inner shell expanding at a later time than the 10 mg/cc foam does. Anyway, the role of the higher density in retarding the shell's moving counterbalances that of the preheat in driving the shell's outward moving outwardly, then the inner shell suffers a less outward expansion.

The preheat of radiative shock promotes the inner shell an outward expansion, and the move of the interface renders the effective reduction of the Atwood number at the interface as shown in Fig. 8, which plays a role in mitigating the hydrodynamic instabilities. For the target filled with CH foam, we can see that the RT instability occurs when the acceleration is directed inwardly as shown in the shadow regions in the left frame in Fig. 8: during the first stage when the shock passes the interface, the Atwood number is significantly reduced with the preheat due to the radiative shock; during the second stage, the number is ~ 0.7, less than the initial value 1 without any movement of the inner shell, correspondingly the RT instabilities at the interface is alleviated to some extent. While for the improved target filled with the metallic foam, because of the stronger preheat by radiative shock, the Atwood number at the Au/foam has been reduced close to zero before the expanded Au shell is recompressed with a consequence of almost no RT instability occurring at the Au/foam interface during the inward acceleration stages as shown in the right frame in Fig. 8. Because of the stronger preheat of the radiative shock formed in the metallic foam, the so-called new pathway for the RT instability at the Au/foam interface may not be so serious as has been indicated.[9]

On the other hand, the subsonic ablation of the inner shell by the radiative shock also imposes another ablative RT instability at the ablation front inside the shell. As has been indicated, the ablatively RT instability can be stabilized by finite density gradient due to the ablation.[16,17] Figure 9 shows the density scale length at the ablation front inside the inner shell at time $t \sim 8.0$ ns for the targets filled with the CH (10 mg/cc) and gold (20 mg/cc) foam, respectively. By comparing, we can see that the stronger radiative shock produced in the gold foam can result in a longer density scale length and can better control the ablatively RT instability at the ablation front inside the sell. Anyway, with the stronger ablation of the unstable Au/foam interface by the radiative shock in the metallic foam, the RT instability is better controlled by both finite density gradient and ablative stabilization.[16,18] Similarly, the unstable foam/CH interface is also stabilized by the ablation of radiative shock.[18]

IV. CONCLUSION

In conclusion, we have proposed another inner shell preheating mechanism by radiative shock or shocked matter in the double-shell target besides the high-energy x rays, especially the M-band line radiation and L-shell radiation from the Au hohlraum, also confirmed and validated the existence of such preheating mechanism by the numerical results. The effect of the inner shell preheating on the hydrodynamic instabilities is analyzed and studied. Especially for the improved double-shell target filled with the metallic foam, the preheat of the unstable Au/foam interface by the stronger radiative shock formed in the metallic foam plays a role in better controlling the instabilities. The potential effect of radiative shock should be seriously realized in double shells.

ACKNOWLEDGMENTS

The work was supported by the National Natural Science Foundation of China under Grant Nos. 11175027 and 11275028.

[1]P. Amendt, J. D. Colvin, R. E. Tipton, D. E. Hinkel, M. J. Edwards, O. L. Landen, J. D. Ramshaw, L. J. Suter, W. S. Varnum, and R. G. Watt, Phys. Plasmas **9**, 2221 (2002).

[2]P. Amendt, C. Cerjan, A. Hamza, D. E. Hinkel, J. L. Milovich, and H. F. Robey, Phys. Plasmas **14**, 056312 (2007).

[3]P. Amendt, J. Milovich, L. J. Perkins, and H. Robey, Nucl. Fusion **50**, 105006 (2010).

[4]J. D. Lindl, Phys. Plasmas **2**, 3933 (1995).

[5]J. D. Lindl, P. Amendt, R. L. Berger, S. G. Glendinning, S. H. Glenzer, S. W. Haan, R. L. Kauffman, O. L. Landen, and L. J. Suter, Phys. Plasmas **11**, 339 (2004).

[6]S. W. Haan, J. D. Lindl, D. A. Callahan, D. S. Clark, J. D. Salmonson, B. A. Hammel, L. J. Atherton, R. C. Cook, M. J. Edwards, S. Glenzer, A. V. Hamza, S. P. Hatchett, M. C. Herrmann, D. E. Hinkel, D. D. Ho, H. Huang, O. S. Jones, J. Kline, G. Kyrala, O. L. Landen, B. J. MacGowan, M. M. Marinak, D. D. Meyerhofer, J. L. Milovich, K. A. Moreno, E. I. Moses, D. H. Munro, A. Nikroo, R. E. Olson, K. Peterson, S. M. Pollaine, J. E. Ralph, H. F. Robey, B. K. Spears, P. T. Springer, L. J. Suter, C. A. Thomas, R. P. Town, R. Vesey, S. V. Weber, H. L. Wilkens, and D. C. Wilson, Phys. Plasmas **18**, 051001 (2011).

[7]H. F. Robey, P. A. Amendt, J. L. Milovich, H.-S. Park, A. V. Hamza, and M. J. Bono, Phys. Rev. Lett. **103**, 145003 (2009).

[8]W. S. Varnum, N. D. Delamater, S. C. Evans, P. L. Gobby, J. E. Moore, J. M. Wallace, and R. G. Watt, Phys. Rev. Lett. **84**, 5153 (2000).

[9]J. L. Milovich, P. Amendt, M. Marinak, and H. Robey, Phys. Plasmas **11**, 1552 (2004).

[10]P. A. Amendt, H. F. Robey, H.-S. Park, R. E. Tipton, R. E. Turner, J. L. Milovich, M. Bono, R. Hibbard, H. Louis, and R. Wallace, Phys. Rev. Lett. **94**, 065004 (2005).

[11]Y. B. Zel'dovich and Y. P. Raizer, *Physics of Shock Waves and High-Temperature Hydrodynamics Phenomena* (Academic Press, New York, 1966).

[12]D. Mihalas and B. W. Mihalas, *Foundations of Radiation Hydrodynamics* (Academic Press, New York, 1984).

[13]R. P. Drake, Phys. Plasmas **14**, 043301(2007).

[14]T. Feng, D. Lai, and Y. Xu, Chin. J. Comput. Phys. **16**, 199 (1999) (in Chinese).

[15]R. P. Drake, *High Energy Density Physics: Fundamentals, Inertial Confinement Fusion and Experimental Astrophysics* (Springer, New York, 2006).

[16]J. D. Kilkenny, S. G. Glendinning, S. W. Haan, B. A. Hammel, J. D. Lindl, D. Munro, B. A. Remington, S. V. Weber, J. P. Knauer, and C. P. Verdon, Phys. Plasmas **1**, 1379 (1994).

[17]S. Atzeni and M. Temporal, Phys. Rev. E **67**, 057401 (2003).

[18]C. M. Huntington, C. C. Kuranz, R. P. Drake, A. R. Miles, S. T. Prisbrey, H.-S. Park, H. F. Robey, and B. A. Remington, Phys. Plasmas **18**, 112703 (2011).

PHYSICS OF PLASMAS 21, 100705 (2014)

A wedged-peak-pulse design with medium fuel adiabat for indirect-drive fusion

Zhengfeng Fan,[1] X. T. He,[1,2] Jie Liu,[1,2] Guoli Ren,[1] Bin Liu,[1] Junfeng Wu,[1] L. F. Wang,[1,2] and Wenhua Ye[1,2]
[1]*Institute of Applied Physics and Computational Mathematics, Beijing 100088, China*
[2]*Center for Applied Physics and Technology, Peking University, Beijing 100871, China*

(Received 16 July 2014; accepted 8 October 2014; published online 17 October 2014)

In the present letter, we propose the design of a wedged-peak pulse at the late stage of indirect drive. Our simulations of one- and two-dimensional radiation hydrodynamics show that the wedged-peak-pulse design can raise the drive pressure and capsule implosion velocity without significantly raising the fuel adiabat. It can thus balance the energy requirement and hydrodynamic instability control at both ablator/fuel interface and hot-spot/fuel interface. This investigation has implication in the fusion ignition at current mega-joule laser facilities. © *2014 AIP Publishing LLC.*
[http://dx.doi.org/10.1063/1.4898682]

In central hot-spot ignition of inertial confinement fusion (ICF),[1–3] a spherical capsule containing deuterium-tritium (DT) fusion fuel is imploded to a much smaller size (than its initial one) to achieve high density and high temperature in the central hot spot, triggering main fuel burn, and producing significant thermonuclear energy gain. According to the current point design of indirect drive at the National Ignition Facility (NIF),[4] ~1.6 MJ laser energy at a peak power of 410 TW is delivered into a 1 cm-long and 5.44 mm-diameter cylindrical hohlraum, generating x-rays with a peak temperature of 300 eV. One-dimensional (1D) simulations show that the x-rays drive the DT fuel to a peak implosion velocity of ~370 km/s to achieve a temperature of $T_i \geq 5$ keV in the central hot spot with an areal density of $\rho R_{hotspot} \sim 0.3$ g/cm^2, surrounded by a dense main-fuel layer with $\rho R_{fuel} \sim 1.5$ g/cm^2. Finally, ~20 MJ thermonuclear energy is expected to be gained.

In the above scheme, from the least energy consideration, the compressing process is designed to be adiabatic using pulse-shaping techniques with a low-foot drive, so that the fuel adiabat, defined as the ratio of the fuel pressure to the ideal Fermi-degenerate pressure at the same density, is low. Unfortunately, during the capsule implosion of low adiabat, the hydrodynamic instability is found to be much more prominent than expected. It leads to hot-spot deformation, main-fuel-layer distortion, and more severely, ablator mixing into the hot spot.[4–6] In the low-adiabat ignition experiments on NIF, the implosion velocity has achieved ~350 km/s, which is ~95% of the ignition requirement of 370 km/s,[7] and the areal density of the fuel has reached 1.3 g/cm^2, which is ~84% of the ignition requirement of 1.5 g/cm^2,[8] but the highest neutron yield of 7×10^{14} (Ref. 9) is much below the goal of ignition. In the symcap and ignition experiments with Ge-doped CH ablator, bright spots have been observed within the hot spot in broadband, gated x-ray implosion images, indicating that the ablator material has mixed into the hot spot.[10] In the experiments with silicon-doped ablator, mixing of the ablator material into the hot spot has been quantified from the measured elevated absolute x-ray emission of the hot spot.[9]

Based on the hypotheses that the ablation-front-driven instability was responsible partly for the observed degraded yield performance, high-foot high-adiabat implosion experiments have been tested on NIF.[11] High-foot implosions have high ablation velocities and large density gradient scale lengths and are more resistant to ablation-front Rayleigh-Taylor instability (RTI) induced mixing of ablator material into the DT hot spot. Recent experiments demonstrated that the neutron yield did increase as expected: the neutron yield has achieved a record high of 9.6×10^{15}.[12]

Even though great progress has been made with the high-foot design, there are still a lot of works need to be done for achieving fusion ignition. The minimum (shell kinetic) energy for ignition[13] is expressed as the fuel adiabat α,[14] (mass-averaged) fuel implosion velocity V_{imp}, and (maximum) ablation pressure P_a. Typical ignition specific values for them are 1.4, 370 km/s, and 140 Mbars. As the fuel adiabat increases from a nominal value of 1.4 to a high value of 2.8,[15] with the other two parameters kept unchanged, the minimum energy would increase by a factor of ~6,[16] requiring a capability far beyond the current laser facilities. Therefore, keeping the fuel adiabat within a reasonable value and at the mean time making the hydrodynamic instability controllable become very essential for the ignition target design, calling for investigation from theoretical side urgently.

In this letter, we propose the design of a wedged-peak pulse at the late stage of the indirect drive. First, the wedged peak grows gradually creating a compression wave which compresses the fusion fuel without significantly raising the fuel adiabat, and the fuel adiabat is kept at a reasonable value; second, the compression wave merges as they travel into the hot spot forming a strong inward enhancement shock (ES) wave or compression wave which prohibits the reflection of the outward rebounding shock (from the spherical center) at the interface of the hot spot and main-fuel layer, thus stabilizing the deceleration stage hydrodynamic instability.[17]

In our design, the fuel implosion velocity is moderately increased to 400 km/s, and the ablation pressure could be raised to ~200 Mbars. Then, by assuming a constant energy conversion of ~1.6% from laser energy to capsule shell kinetic energy,[4] the laser energy requirement is ~1.6 MJ for achieving ignition with a medium fuel adiabat of ~2.3. Considering ~20% uncertainty, the laser energy required would probably be ~2.0 MJ, which is within the capability of current laser facilities.

In order to demonstrate the design of the wedged-peak pulse, a cryogenic DT fusion capsule is adopted, as illustrated in Fig. 1(a). The capsule contains a 176 μm-thick CH ablator with density of 1.0 g/cm^3, a 80 μm-thick DT fusion fuel layer with density of 0.25 g/cm^3, and a fuel-gas-filled cavity with a radius of 870 μm and density of 0.3 mg/cm^3. The ablator mass is ~2.4 mg and the fuel mass is ~0.21 mg.

The new design of the radiation temperature pulse with the wedged-peak pulse is plotted in Fig. 1(b) with red solid line. By comparison, a low-foot and a high-foot radiation temperature pulses are also presented, as plotted, respectively, with blue dashed line and green dashed-dotted line. The low-foot pulse contains four steps such that four successive shocks satisfying the Munro criteria[18] are generated, and the fuel adiabat is ~1.4 at maximum implosion velocity (note that for the low-foot and high-foot pulses, the ablator thickness is 160 μm instead of 176 μm). The high-foot temperature pulse adopted to imitate the NIF high-foot drive reduces the pulse steps from four to three, and in the first pulse the drive temperature is increased so that the fuel adiabat is increased to ~2.8 at maximum implosion velocity. The wedged-peak-pulse design improves the main pulse of the high-foot scheme, and a wedged peak is applied at the late time of the capsule implosion so that the fuel adiabat at maximum implosion velocity is reduced to ~2.3. Figure 1(c) compares the time histories of the fuel adiabat for the three pulses.

The wedged peak is designed according to the following three principles. First, the radiation-drive pressure at the end of the peak pulse should be obviously larger than that at the beginning of the peak pulse. A reasonable requirement is $P_{end}/P_{beginning} \sim 2$; as $P \propto T_r^{3.5}$, we set $T_{r,beginning} = 285$ eV and $T_{r,end} = 340$ eV. Second, the slope of the wedged peak is realizable, namely, the slope should not be too large; according to the results of NIF experiments,[16] a rate of 33 eV/ns is realizable. Third, the wedged peak should be carefully timed such that the resultant enhancement shock/compression wave encounters the outward rebounding shock near the interface of the hot spot and the main-fuel layer so as to mitigate the hydrodynamic instability there.

Simulations of the capsule implosions are first performed in 1D using the radiation-hydrodynamics code LARED-S,[19] which is multi-dimensional, massively parallel, and Eulerian mesh based. The code includes laser ray tracing, multi-group radiation discussion, plasma hydrodynamics, electron and ion thermal conduction, atomic physics, nuclear reaction, alpha particle transport, and the quotidian equations of state.[20] In the radial direction, totally 2000

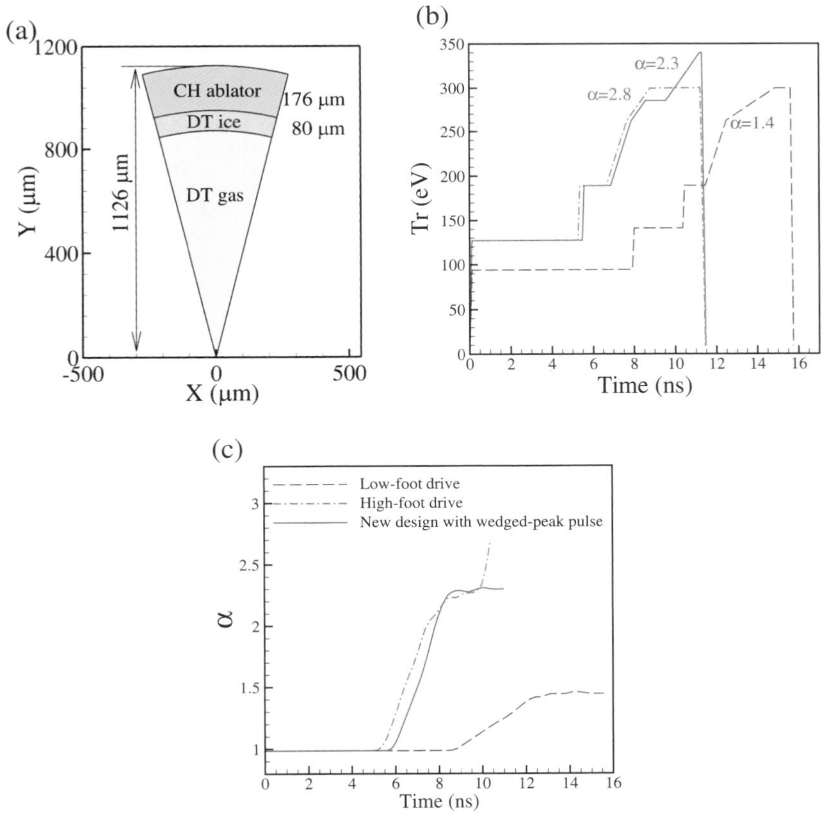

FIG. 1. (a) Schematic of the ignition capsule; (b) radiation-drive temperature pulses, with red solid line plotting the new design with the wedged-peak pulse, blue dashed line plotting the pulse for the low-adiabat drive, and green dashed-dotted line plotting the pulse for the high-adiabat drive; and (c) time histories of the fuel adiabat for the three pulses.

TABLE I. Parameters characterizing the three types of drives, namely, the new design with the wedged-peak pulse, the high-foot pulse, and the low-foot pulse.

Description	New design with peak-pulse	High-foot	Low-foot
Peak T_r (eV)	340	300	300
E_{abs} (kJ)	175	156	143
Ablator thickness (μm)	176	160	160
α	2.3	2.8	1.4
Convergence ratio	30	28	35
Fuel ρR (g/cm^2)	1.2	1.0	1.4
Fusion yield (MJ)	11	3	18

meshes are applied and the target shell region is locally refined with minimum mesh size of 0.05 μm so as to guarantee the convergence of the results.

Table I presents a brief summary of the parameters characterizing the three types of drives, namely, the new pulse design with the wedged-peak pulse, the high-foot pulse, and the low-foot pulse. The capsule absorbed energy, for the new design with peak pulse, is ~175 kJ, about 20% higher than the value of ~143 kJ for the low-foot drive. The implosion velocity reaches 400 km/s, 5% higher than that of the low-foot drive. We observe that with the design of the wedged-peak pulse, the fuel adiabat averaged at maximum implosion velocity is ~2.3, ~20% lower than that of the high-foot drive. As a result, the fuel compression is enhanced and the fuel ρR is increased as compared with the high-foot pulse. Consequently, the fusion yield increases by a factor of about four. From the comparisons, we can conclude that 1D performance of the new design with the wedged-peak pulse is better than the high-foot pulse and comparable to that of low-foot pulse.

Of great importance, the new design with the wedged-peak pulse takes the advantage of low growth of hydrodynamic instability with a medium fuel adiabat. In the high-foot drive, the ablation-front-driven hydrodynamic instability is reduced by high ablation velocities and large density gradient scale lengths. However, little attention is paid for controlling the deceleration-phase hydrodynamic instability, which takes place at the hot spot and main fuel interface. Aside from reducing the ablation-front-driven hydrodynamic instability, our new design creates an enhancement shock wave or compression wave, which is utilized to reduce the instability growth at the hot spot and main fuel interface.

The late-time implosion diagram is shown in Fig. 2 for the fusion capsule, in both cases with and without the wedged-peak pulse. The red thick line plots the trajectory of the merged shock (MS) of the shocks created by the first three radiation temperature steps. If the wedged-peak pulse is not presented, as shown in Fig. 2(b), the MS would encounter the incoming main-fuel layer and reflect off at the hot spot and main fuel interface, resulting in deceleration of the main fuel; soon after the MS's first reflection, the pressure of the low-density hot spot exceeds that of the heavier main-fuel layer, causing an opposition of the pressure gradient and density gradient, and thus leading to significant RTI growth at the hot spot and main fuel interface. However, as

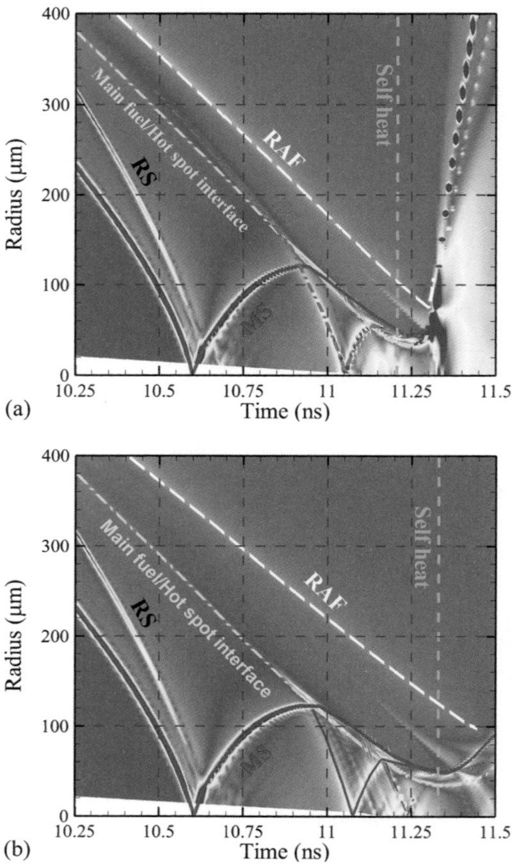

FIG. 2. The late-time implosion diagram ($|\nabla u|$) of the fusion capsule, (a) with and (b) without the wedged-peak pulse. The white dashed line and the yellow dashed-dotted line plot the movements of the radiation ablation front (RFA) and the interface of the hot spot and main fuel, respectively; while the pink dashed-dotted line and the red thick line show the trajectories of the enhancement shock/compression wave (ES) and the merged shock (MS) of the former shocks created by the first three radiation temperature steps, respectively; the reflected shock (RS) from the ablation front is marked with RS.

the wedged-peak pulse is applied, the resultant ES/compression wave, as plotted with pink dashed-dotted line in Fig. 2(a), encounters and overtakes the MS so that the pressure of the (low-density) hot spot remains lower than that of the (heavier) main-fuel layer for an additional 20 ps. As the pressure of the hot spot exceeds the main fuel, the hot spot temperature and pressure have achieved higher values so that within a short period of ~190 ps, ~130 ps shorter than the case without the wedged-peak pulse, the hot spot begins to selfheat. With a delayed, shorter deceleration and an enhanced ablative stabilization, the deceleration-phase RTI would be obviously stabilized.

In order to validate the above picture of instability control, two-dimensional simulations are carried out to investigate the perturbation growth. In order to save computation time, the simulations are done on wedges with angles $\theta \in [\pi/2 - \pi/L, \pi/2 + \pi/L]$, where L is the perturbation mode number. Within the wedges, the initial perturbations, in the form

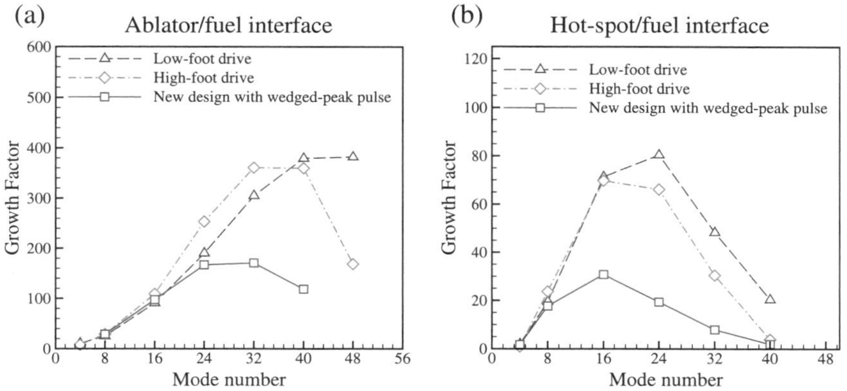

FIG. 3. Comparisons of the growth factors of hydrodynamic instabilities between the new design with the wedged-peak pulse and the low-foot and high-foot drives. (a) Growth factors to the ablator and fuel interface and (b) growth factors to the hot spot and main fuel interface at ignition times.

of cosine waves, are placed on the ablator outer surface. Fifty uniform angular meshes are applied to resolve the perturbations. Three cases are compared: the low-foot and high-foot drives and the new design with the wedged-peak pulse. The concept of growth factor, defined as the ratio of the perturbation amplitude at concerned time and specific interface to the initial amplitude at the ablator outer surface, is adopted to quantify the perturbation growth.

During the capsule acceleration stage, the ablation-front-drive instabilities, in both the high-foot drive and the new design, are reduced by enhanced ablation velocities and enlarged density gradient scale lengths, which are more effective on higher mode perturbations. In our new design, the ablative stabilization effect is more stronger than the high-foot design due to a higher drive temperature and larger effective perturbation wave numbers.[21] Reduced ablation-front-drive instability decreases the seeds for the subsequent perturbation growth at the ablator and fuel interface. Figure 3(a) compares the ignition-time[22] growth factors to the ablator and fuel interface for the new design with the wedged-peak pulse and the low-foot and high-foot drives.[23] For the modes with $L \geq 24$, the growth factors are obviously reduced in the new design, and, as the mode number gets larger, the reduction becomes stronger.

With the prohibition of the shock reflection at the hot spot and main fuel interface, the new design with the wedged-peak pulse also obviously reduces the growth factors to the hot spot and main fuel interface for the $L \geq 8$ modes, compared with the low-foot and high-foot drive pulses, as shown in Fig. 3(b). The maximum growth factor of 78 in the low-foot drive at $L = 24$ reduces to 31 at $L = 16$ with a reduction of ~60%. Figures 4(a) and 4(b) compare the density contours at ignition times between the new design with the wedged-peak pulse and the low-foot drive, for the $L = 16$ mode with an initial roughness of 35 nm. The bubble and spike structures at the hot spot and main fuel interface are much smaller in the new design. This is important for reducing the hot-spot deformation and mixing with the cold main fuel layer, so as to guarantee a successful hot-spot formation and increase the robustness of hot-spot ignition.

Simulations with the LARED-JC code indicate that the wedged-peak pulse, raising to a maximum of 340 eV at a speed of ~30 eV/ns, is achievable with a (cylinder) hohlraum driven by a laser facility with energy of ~2.0 MJ. Moreover, our scheme is also robust with respect to the maximum radiation temperature. In the simulations, the slope of the wedged-peak pulse decreases and the launch time remains fixed. The fusion yield, as shown in Fig. 5, decreases smoothly as the

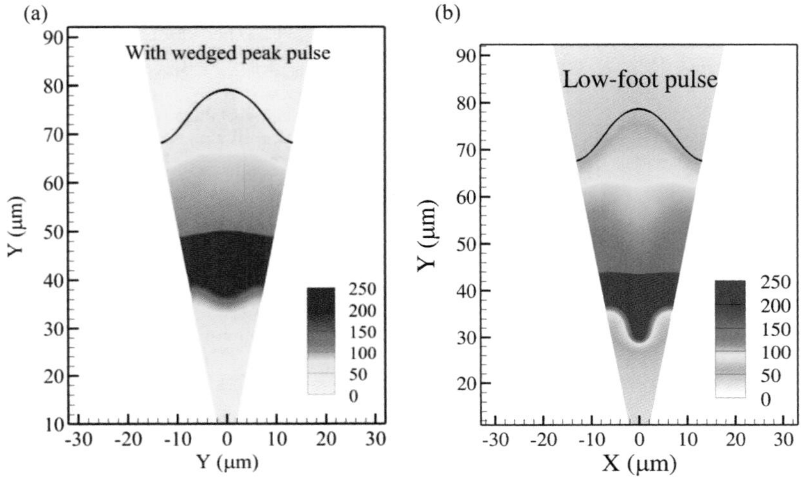

FIG. 4. Density contours for a $L = 16$ mode perturbation with an initial roughness of 35 nm (a) for the new design with the wedged-peak pulse and (b) for the low-foot pulse, at ignition times. Black curves represent the ablator and fuel interfaces.

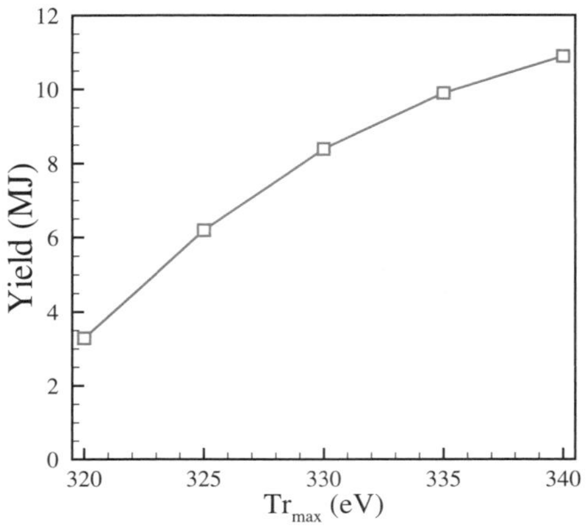

FIG. 5. Fusion yield versus the maximum radiation temperature of the wedged-peak pulse.

maximum radiation temperature drops, and we do not observe apparent cliff as the maximum radiation temperature drops to 320 eV.

In summary, we propose a new design of a wedged-peak pulse at the late stage of indirect drive. With the wedged-peak pulse, the fuel adiabat is kept at reasonable value such that the 1D performance of the capsule is better than that of the high-adiabat implosions. Meanwhile, the hydrodynamic instabilities, at both ablator/fuel interface and hot-spot/main-fuel interface, are significantly reduced compared with low-adiabat or even high-adiabat implosions. Our above investigation has implication in the fusion ignition at current mega-joule laser facilities.

This work has been supported by the National ICF program of China, the National Basic Research Program of China (Grant No. 2013CB34100), the National High Technology Research and Development Program of China (Grant No. 2012AA01A303), and the National Natural Science Foundation of China (Grant Nos. 11475032, 91130002, 11274026 and 11205010).

[1] J. H. Nuckolls, L. Wood, A. Thiessen, and G. B. Zimmerman, "Laser compression of matter to super-high densities: Thermonuclear (CTR) applications," Nature (London) **239**, 139 (1972).
[2] J. D. Lindl, *Inertial Confinement Fusion* (Springer, New York, 1998).
[3] S. Atzeni and J. Meyer-ter-Vehn, *The Physics of Inertial Fusion* (Clarendon Press, Oxford, 2004).
[4] S. W. Haan, J. D. Lindl, D. A. Callahan, D. S. Clark, J. D. Salmonson, B. A. Hammel, L. J. Atherton, R. C. Cook, M. J. Edwards, S. Glenzer *et al.*, Phys. Plasmas **18**, 051001 (2011).
[5] D. S. Clark, S. W. Haan, and J. D. Salmonson, Phys. Plasmas **15**, 056305 (2008).
[6] D. S. Clark, D. E. Hinkel, D. C. Eder, O. S. Jones, S. W. Haan, B. A. Hammel, M. M. Marinak, J. L. Milovich, H. F. Robey, L. J. Suter, and R. P. J. Town, Phys. Plasmas **20**, 056318 (2013).
[7] D. A. Callahan, N. B. Meezan, S. H. Glenzer, A. J. MacKinnon, L. R. Benedetti, D. K. Bradley, J. R. Celeste, P. M. Celliers, S. N. Dixit, T. Döppner *et al.*, Phys. Plasmas **19**, 056305 (2012).
[8] V. A. Smalyuk, L. J. Atherton, L. R. Benedetti, R. Bionta, D. Bleuel, E. Bond, D. K. Bradley, J. Caggiano, D. A. Callahan, D. T. Casey *et al.*, Phys. Rev. Lett. **111**, 215001 (2013).
[9] T. Ma, P. K. Patel, N. Izumi, P. T. Springer, M. H. Key, L. J. Atherton, L. R. Benedetti, D. K. Bradley, D. A. Callahan, P. M. Celliers *et al.*, Phys. Rev. Lett. **111**, 085004 (2013).
[10] S. P. Regan, R. Epstein, B. A. Hammel, L. J. Suter, J. Ralph, H. Scott, M. A. Barrios, D. K. Bradley, D. A. Callahan, C. Cerjan *et al.*, Phys. Plasmas **19**, 056307 (2012).
[11] O. A. Hurricane, D. A. Callahan, D. T. Casey, P. M. Celliers, C. Cerjan, E. L. Dewald, T. R. Dittrich, T. Döppner, D. E. Hinkel, L. F. Berzak Hopkins, J. L. Kline, S. Le Pape, T. Ma, A. G. MacPhee, J. L. Milovich, A. Pak, H.-S. Park, P. K. Patel, B. A. Remington, J. D. Salmonson, P. T. Springer, and R. Tommasini, Nature **506**, 343–348 (2014).
[12] See https://lasers.llnl.gov/news/status/2014/january for further details of the experiments.
[13] M. C. Herrmann, M. Tabak, and J. D. Lindl, Phys. Plasmas **8**(5), 2296 (2001).
[14] Conceptually, the adiabat is defined as the pressure at given density divided by the Fermi degenerate pressure at the same density. We here define the adiabat α as mass-weighted-averaged adiabat of the main fuel. At each specific position, $\alpha_{sp} = P_{1000}/\mathrm{PF}$, where P_{1000} is the pressure at 1000 g/cm^3 on the isentrope line and P_F is the ideal Fermi degenerate pressure for DT at 1000 g/cm^3.
[15] T. R. Dittrich, O. A. Hurricane, D. A. Callahan, E. L. Dewald, T. Döppner, D. E. Hinkel, L. F. Berzak Hopkins, S. Le Pape, T. Ma, J. L. Milovich, J. C. Moreno, P. K. Patel, H.-S. Park, B. A. Remington, and J. D. Salmonson, Phys. Rev. Lett. **112**, 055002 (2014).
[16] J. Lindl, O. Landen, J. Edwards, E. Moses, and NIC Team, Phys. Plasmas **21**, 020501 (2014).
[17] Z. F. Fan, M. Chen, Z. S. Dai, H. B. Cai, S. P. Zhu, W. Y. Zhang, and X. T. He, e-print arXiv:1303.1252 [physics.plasm-ph].
[18] D. H. Munro, P. M. Celliers, G. W. Collins, D. M. Gold, L. B. Da Silva, S. W. Haan, R. C. Cauble, B. A. Hammel, and W. W. Hsing, Phys. Plasmas **8**, 2245 (2001).
[19] W. Pei, Commun. Comput. Phys. **2**, 255 (2007).
[20] R. M. More, K. H. Warren, D. A. Young, and G. B. Zimmerman, Phys. Fluids **31**, 3059 (1988).
[21] As the wedged-peak pulse launches at the late stage of indirect drive, the capsule has been imploding. At this time, the effective perturbation wave number defined as the ratio of the mode number to the capsule radius gets larger.
[22] Here, the ignition time is defined as the time when the hot spot starts to selfheat, namely, when the hot-spot thermonuclear heating equals its energy losses by thermal conduction and radiation leak.
[23] During the capsule pulse-shaping stage, we observe that the ablative Richtmyer-Meshkov instability grows much more for the high-foot drive than for the low-foot drive when the mode number is less than 40. As a result, we see in Fig. 3(a) that the growth factors (to the ablator and fuel interface) are slightly larger for the high-foot drive than for the low-foot drive when the mode number is less than 40. However, as the mode number gets larger, the high-foot drive greatly reduces the growth factors to the ablator and fuel interface as compared with the low-foot drive.

CHIN. PHYS. LETT. Vol. 31, No. 4 (2014) 045201

Design of an Indirect-Drive Pulse Shape for ~1.6 MJ Inertial Confinement Fusion Ignition Capsules *

WANG Li-Feng(王立锋)[1,2], WU Jun-Feng(吴俊峰)[1], YE Wen-Hua(叶文华)[1,2**], FAN Zheng-Feng(范征锋)[1], HE Xian-Tu(贺贤土)[1,2**]

[1] *Institute of Applied Physics and Computational Mathematics, Beijing 100094*
[2] *HEDPS, Center for Applied Physics and Technology, Peking University, Beijing 100871*

(Received 26 February 2014)

We present a design of indirect-drive pulse shape for inertial confinement fusion ignition capsules using laser energy ~1.6 MJ with a moderate gain (~10) on the Shenguang IV laser facility. The trade-off fuel compression (pressure) for resistance to the hydrodynamic instability (HI) in the recent high-foot (HF) implosion campaign [Dittrich T R et al Phys. Rev. Lett. 112 (2014) 055002] is recovered. The proposed design modifies the "main" pulse shape, which features a decompression-recompression step for the fuel shell resulting in higher areal density than that of the "simple" HF design, and thereby approaches the conditions required for ignition avoiding at the expense of more laser energy while holding the HI under control.

PACS: 52.35.Py, 52.57.Fg, 52.38.Mf DOI: 10.1088/0256-307X/31/4/045201

In the central ignition indirect-drive (ID) inertial confinement fusion (ICF) scheme, a typical target consists of a spherical thin shell that contains a deuterium-tritium (DT) ice layer filled with a DT gas that is accelerated inward by x-rays generated from lasers interacting with a high-Z cavity (holhraum).[1] The x-rays ablate the outer shell, leading to a rocket-like reaction force, spherically imploding the shell at a velocity of $\geq$350 μm/ns. The spherically convergent imploding causes the pressure of the central DT gas to rise, resulting in a strong reflected shock wave. The shell is then decelerated, both converting its kinetic energy into hotspot internal energy (i.e. heating the hotspot) and assembling a dense shell $\geq$1000 g/cc. Ignition is approached as the hotspot central temperature reaches $\geq$4 keV and the hotspot areal density ρR approximately achieves α-particle stopping range $\rho R \geq 0.3\,\mathrm{g/cm^2}$. The desired ignition will occur if the hotspot is confined long enough for its temperature to bootstrap by α-heating to several tens of keV, which needs the fuel $\rho R \geq 1.0\,\mathrm{g/cm^2}$.

It is still not so clear how to achieve central ignition with a moderate gain G using laser energy $E_{\rm L} \sim 2.0$ MJ on the Shenguang (SG) IV laser facility[2] which will be built in China. Here, we focus on shedding new light on this issue.

In order to achieve extreme conditions on implosion velocities and areal densities, the National Ignition Campaign (NIC) on the National Ignition Facility (NIF) adopts a low-foot (LF), high-convergence spherical implosion which requires precise control of the laser pulse and the target fabrication. Unfortunately, the progress towards ignition of the NIC is not satisfactory. The NIC experiments have demonstrated ~90% of the required peak fuel implosion velocity ($V_{\rm imp}$) and ~85% of the required fuel ρR, but not at the same time.[3] Moreover, the pressures of the hotspot are 2–3 times below predictions. The NIC experiments indicate that the ablation-front instability is more aggressive than the original prediction.[4] The recent experiment shows that the ablation-front Rayleigh–Taylor (RT) growth is directly responsible for the hydrodynamic instability (HI) mixing of the ablator material with dopants into the hotspot.[5] In fact, the complex turbulent-like flow forms in the NIC LF-implosions, which has been observed in simulations.[6]

The so-called "adiabat" α (ratio of pressure to the Fermi-degenerate pressure at peak velocity) measuring fuel entropy is a commonly used parameter in the ICF community characterizing the fuel compression. In laser fusion, the fuel compression (entropy) is sensitive to the preheating caused by, e.g., shock, hot electrons, high-energy photons, and high-energy ions, etc. Our previous work indicates that the preheating can obviously reduce the peak density $\rho_{\rm p}$ (which means that the ablation-front Atwood number $A_{\rm T}$ decreases with increasing α since $A_{\rm T} \propto \rho_{\rm p}$) and clearly enhance the density gradient scale length $L_{\rm m}$ of the ablation front.[7] Furthermore, the large values of the α on the ablation-front increase the ablation velocity $v_{\rm a}$ since $v_{\rm a} \propto \alpha^{3/5}$. The linear growth of ablative RT perturbations can be described by

$$\gamma = \sqrt{A_{\rm T} kg/(1 + kL_{\rm m})} - \beta k v_{\rm a},$$

where k is the perturbation wave number, g is the acceleration, and β depends fundamentally on the flow parameters, usually it is 1–3.[8,1] From this evaluation, it is concluded that the RT growth can be effectively suppressed by the adiabat (preheating). Additionally, it is noteworthy that in direct-drive (DD) ICF target designs, the laser-induce adiabat shaping theory

*Supported by the National Natural Science Foundation of China under Grant Nos 11275031, 11105013, 11205017, 11274026, and the National Basic Research Program of China under Grant No 2013CB834100.
**Corresponding authors. Email: ye_wenhua@iapcm.ac.cn; xthe@iapcm.ac.cn

has been developed by Goncharov and Anderson *et al.* to quench the RT growth while maintaining the one-dimensional (1-D) implosion performance.[9]

Shock preheating has been adopted to inhibit ablation-front RT growth in the recent high-foot (HF) implosion campaign on the NIF.[10] Since the fuel α is mainly set by the 1st-shock strength, the radiation temperature $T_{\rm r}$ of the foot is increased in comparison with that used in the NIC LF-implosions, which has exhibited high yield-over-clean (YOC) and low hotspot mix (<100 ng). As is expected, the RT growth is significantly reduced.[11] However, since the HF-implosion trades off compression for resistance to the HI, the fuel ρR is only $\sim$0.8 g/cm^2. Capsule simulations using the radiation-hydrocode RDMG indicate that for a 300-eV isotropic x-ray drive, the laser energy needed becomes as high as $E_{\rm L} \sim 6.0$ MJ when the $V_{\rm imp} = 370$ μm/ns, $\alpha = 3$, and ignition threshold factor[12] ITF = 3 are fixed.

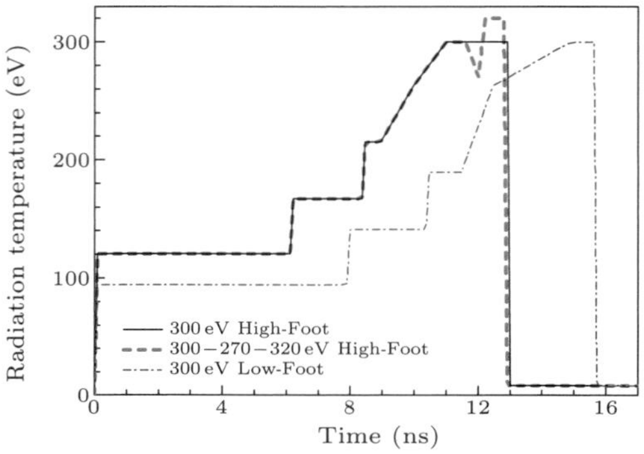

Fig. 1. Radiation temperature versus time for HF and LF drives.

Since the $E_{\rm L}$ needed via the simple HF-implosion achieving ICF ignition cannot be easily satisfied, a new avenue towards ICF ignition with more resistance to HI on the SG-IV using $E_{\rm L} \leq 2$ MJ should be explored. Lowering the ignition energy, a larger kinetic pressure is desired to create a hotspot with smaller volume but higher pressure (density). For the ID scheme, we have, approximately $V_{\rm imp} = 10^7\sqrt{T_{\rm r}}\ln(m_0/m)$, where m_0 is the shell initial mass and m the final mass.[1] When $T_{\rm r}$ and m_0 are fixed, the $V_{\rm imp}$ increases with decreasing m. The m is strongly connected to shell thickness (ΔR). However, a small ΔR may lead to undesirable shell break-up by RT growth. In view of the above facts, increasing $V_{\rm imp}$ will bring an increasing risk of RT-induced shell break-up, and hence increasing the density of the shell seems to be a more practicable alternative. As previously mentioned, the HI is out of control in LF-implosions owing to larger $A_{\rm T}$, smaller $L_{\rm m}$, and smaller $v_{\rm a}$ of the ablation front. By contrast, the RT growth is reduced noticeably in HF-implosions.[11] For that purpose, a new x-ray drive is designed in the present study, which implements stability via HF high-adiabat acceleration. In order to maximize the kinetic energy of the shell under the condition of control over the HI mixing, in the acceleration stage, we first accelerate the shell to a suitable velocity, reduce the $T_{\rm r}$ to decompress the shell secondly, and then improve the $T_{\rm r}$ to compress the shell again.

In the present work, the capsule-only radiation hydrodynamic code LARED-S[13] employing the multi-group diffusion approximation is used to perform 1-D and 2-D simulations. The adopted target outer radius is 1140 μm. The CH-ablator has an initial thickness of 190 μm with a density of 1.0 g/cc. The 80 μm-thick DT-ice layer has an initial density of 0.25 g/cc. The DT-ice layer interior is filled with DT-gas with a density of 3×10^{-4} g/cc.

This study follows the recent 3-shock HF pulse shape.[10] Shock timing has been determined according to Munro's coalescence criterion.[14] To compress the fuel by shock, the pressure difference is needed. From Ref. [1], for ID implosions, the shell-ablation pressure increases approximately as $T_r^{3.5}$. However, for the holhraum design, it is very difficult to create a high radiation temperature (e.g. $T_{\rm r} \sim 350$ eV) owing to laser-plasma interactions. For this purpose, we has redesigned the "main" pulse shape as shown in Fig. 1 (a typical LF drive is also plotted for comparison). In our new design, the T_r begins to decrease at 11.6 ns, arrives at 270 eV at 12.0 ns, and then increases rapidly up to 320 eV at 12.2 ns. The effects of this novel design are discussed in the following.

To perform simulations in a reasonable time and track the "steep surfaces" throughout the implosion, the target shell is locally refined along the radial direction with the zone of a fine grid around the ablation front. In that zone, the uniform mesh size is 0.05 μm. The number of grid points used in the radial direction is 2000. Representative results from 1-D calculations are listed in Table 1. For comparison, the 1-D performance for an LF implosion is also listed. The LF-implosion used a capsule similar to that adopted here but with a 160-μm-thick ablator.

Table 1. Parameters characterizing the designs from 1-D calculations (α-particle transport and energy deposition are included).

	HF	New HF	LF
Nominal $E_{\rm L}$ (MJ)	1.6	1.6	1.5
Expected yield (MJ)	2.8	15.7	16.9
$\rho_{\rm p}$ at $T_{\rm r,end}$ (g/cc)	13.0	15.6	25
R at $T_{\rm r,end}$ (μm)	210	217	148
Fuel $V_{\rm peak}$ (μm/ns)	367	369	382
α at $V_{\rm peak}$	2.14	2.07	1.55
R at $V_{\rm peak}$ (μm)	145	140	130
$\rho_{\rm p}$ at $V_{\rm peak}$ (g/cc)	23	28	30
CH remaining (%)	7.3	6.8	4.0
Fuel ρR (g/cm^2)	1.20	1.26	1.43

Before proceeding further with the analysis, let us mention that the radius and velocity of the shell at 11.6 ns (decompression starting time) are $R \sim 610$ μm and $u \sim 240$ μm/ns, respectively. The comparisons between the "simple" and "new" HF-implosions are illustrated in Figs. 2 (pressure profiles) and 3 (density profiles).

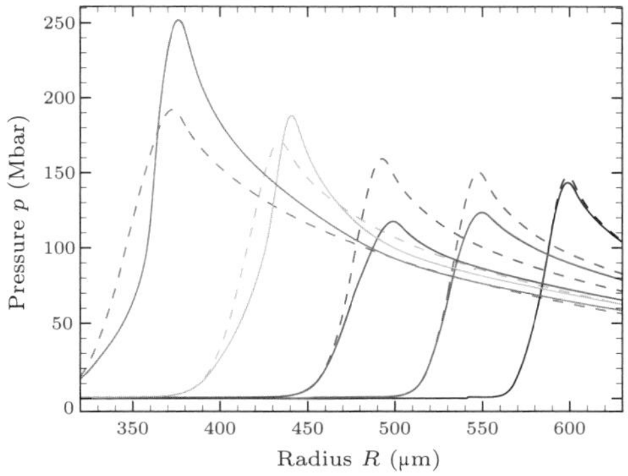

Fig. 2. Pressure profiles at 11.7 (black), 11.9 (red), 12.1 (blue), 12.3 (cyan), and 12.5 ns (magenta), respectively, for simple (dashed lines) and new (solid lines) HF designs.

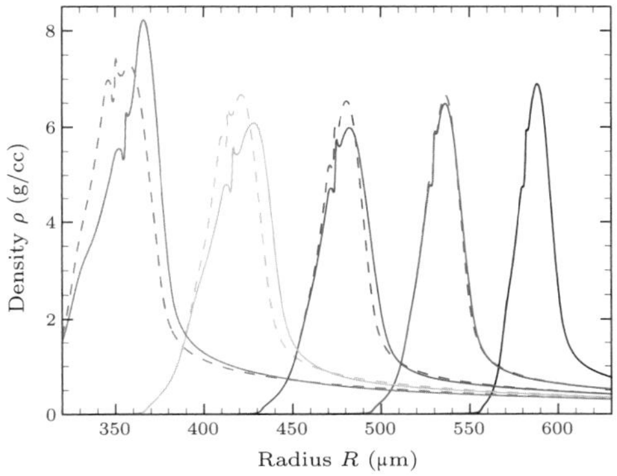

Fig. 3. Density profiles at 11.7 (black), 11.9 (red), 12.1 (blue), 12.3 (cyan), and 12.5 ns (magenta), respectively, for simple (dashed lines) and new (solid lines) HF designs.

It can be seen from Fig. 2 that the shell-ablation pressure of the new design, P, first decreases dramatically during the decompression regime (~11.6–12.2 ns), and then increases rapidly in the later recompression regime, but the P of the simple design grows slowly all the while due to the spherical convergence effects. At t = 11.6, 12.1 and 12.3 ns, the P of the simple (new) design are 147 (147), 159 (117), and 170 (188) Mbar, respectively. When convergence effects are ignored, the ratio of P driven by 300 eV and 270 eV is 1.37 and that driven by 270 eV and 320 eV is 0.60 ($P \propto T_r^{3.5}$). According to the temporal evolution of pressure profiles (see Fig. 2), we can assume that the P values at 11.6, 12.1, and 12.3 ns are driven by 300, 270, and 320 eV, respectively. The ratio of P at 11.6 ns (12.1 ns) to that at 12.1 ns (12.3 ns) is $P_{11.6}/P_{12.1} = 1.26$ ($P_{12.1}/P_{12.3} = 0.62$). For simplicity, we remove the convergence effects by referring to the P from the simple design, and then the ratios become $P_{11.6}/P_{12.1} = 1.36$ and $P_{12.1}/P_{12.3} = 0.66$, respectively, which are consistent with the estimates mentioned above.

As can be seen in Fig. 3, the density profiles are obviously affected by the main pulse shape. On the one hand, the typical width of the shell from the new design becomes thick in comparison with that from the sample one. The full width at half maximum (FWHM) of the shell from the new design at t = 12.3 ns is ~41 μm, while the FWHM from the simple design at same time is ~35 μm. One can expect that the new design is more resistant to the "feedthrough" from the outer ablation-front to the shell's inner interface (which is subject to the RT growth in the subsequent deceleration phase). On the other hand, the ρ_p (L_m) of the new design is smaller (larger) than that of the simple design in the decompression regime, while it then becomes larger (smaller) than that in the subsequent recompression regime, as depicted in Fig. 3. At $t \sim 12.3$ ns, a "recompression" shock wave passes through the dense shell. Thus, it is reasonable to idealize the physical quantities at $t = 12.2$ ns and $t = 12.4$ ns as unshocked and post-shocked conditions by the recompression shock wave, respectively. Based on the well-known shock relations

$$\frac{\rho_2}{\rho_1} = \frac{p_2(\gamma+1)+p_1(\gamma-1)}{p_1(\gamma+1)+p_2(\gamma-1)},$$

where the unshocked and post-shocked fluids are designated by the subscripts 1 and 2, respectively, we obtain $\rho_{12.4}/\rho_{12.2} \simeq 1.29$ for the polytropic index $\gamma = 5/3$. In fact, the ratio of ρ_p at $t = 12.2$ ns to that at $t = 12.4$ ns is $\rho_{12.4}/\rho_{12.2} \sim 1.19$. Removing the convergence effects by referring to the ρ_p of the simple design, the ratio becomes $\rho_{12.4}/\rho_{12.2} \sim 1.13$. The increased densities of the new design are very encouraging, since they are in agreement with the theoretical predictions. In addition, this effect on the HI will be evaluated later.

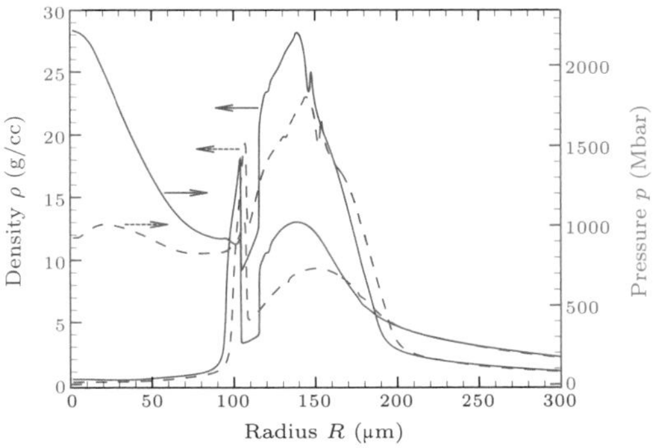

Fig. 4. Density and pressure profiles at peak velocity for simple (dashed lines) and new (solid lines) HF designs.

Figure 4 exhibits the density and pressure profiles at peak velocity of the simple and new designs, respectively. The ρ_p and P of the shell in the simple (new) design are 23 (28) g/cc and 730 (1020) Mbar, respectively. The pressure at $R \sim 0$ of the simple and new designs are 920 and 2220 Mbar, respectively. As can be seen in Fig. 4, the fuel compression (pressure) significantly gains in our new design.

In the following discussion, we present stability analysis results from 2-D calculations. In the 2-D simulation, the mesh setup in the radial direction is kept the same as the earlier 1-D calculation. The computational domain for the angle is $[\pi/2-\pi/L, \pi/2+\pi/L]$ and the number of uniformly spaced grid points used is 40. Single-mode outer surface perturbation in the form $\delta R=\eta_0 P_L(\cos\theta)$ is initiated in our simulations, where $P_L(\cos\theta)$ is the Legendre mode with L being the mode number, and η_0 is the initial perturbation amplitude. To make the evolution of perturbations within the linear growth regime, the η_0 in our calculation has been well adjusted.

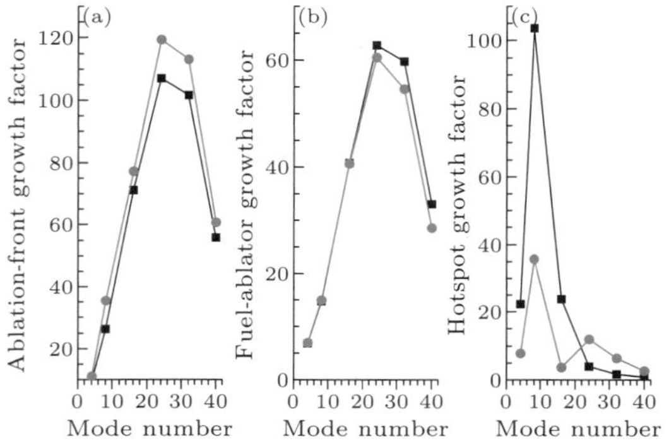

Fig. 5. Growth factors for perturbations seeded on the outer surface. (a) Growth at the ablation-front and (b) perturbations as they grow to the fuel-ablator interface, at peak velocity. (c) Perturbations as they grow to the hotspot surface at stagnation time ($G\sim 1$). The square and circle represent the simple and new HF-implosions, respectively.

Figure 5 displays the growth factor (the ratio of final to initial amplitude) as a function of mode number at the ablation front, the fuel-ablator interface, and the hotspot surface, respectively, for simple and new HF designs. Firstly, the maximum GF at the ablation front of the simple (new) design is GF $\sim$ 110 (120) at $L\sim 24$. Secondly, the maximum GF at the ablator-fuel interface of the simple (new) design is GF $\sim$ 64 (60) at $L\sim 24$. Finally, the maximum GF at the hotspot surface of the simple (new) design is GF $\sim$ 104 (36) at $L\sim 8$. It can be seen from Fig. 5 that the stability of the new LF-implosions at the ablation front does not decline seriously, but the ones at the ablator-fuel interface and the hotspot surface are improved in comparison to that of the simple HF case.

In conclusion, based on the experimental results of the NIC and the recent HF-implosions on the NIF, we have proposed a novel ID pulse shape for high-adiabat ICF ignition capsules with a moderate gain ($G\sim 10$) on the SG-IV laser facility using $E_{\rm L}\sim 1.6$ MJ. In spite of being insensitive to 2-D and 3-D perturbations demonstrating performance close to 1-D predictions[11] in the simple HF-implosions on the NIF, its ultimate performance is limited at the expense of fuel compression (pressure) and potential gain. The path to ignition via the simple HF-implosion requires a higher implosion velocity and a higher convergence ratio. However, these variations will definitely make the capsule implosion less tolerant of the HI mixing. Otherwise, the design presented here takes advantage of the stability in the HF-implosion, but recovers the compression (pressure) by altering the "main" drive pulse shape, and hence avoids the expense of requiring more laser energy. In addition, the idea of adiabat shaping[9] as mentioned above can be introduced in the present ID pulse shape to further increase the α on the ablation-front to damp the RT growth but decrease that on the shell inner surface for higher compression (gain), which will be pursued in future investigations.

The authors wish to thank Jianfa Gu, Zhensheng Dai, Yongsheng Li, Changli Liu, Fengjun Ge, Wudi Zheng and many others at IAPCM and CAPT for their support of this work.

References

[1] Lindl J D et al 2004 *Phys. Plasmas* **11** 339
[2] He X T and Zhang W Y 2007 *Eur. Phys. J.* D **44** 227
[3] Callahan D A et al 2012 *Phys. Plasmas* **19** 056305
Robey H F et al 2012 *Phys. Rev. Lett.* **108** 215004
Edwards M J et al 2013 *Phys. Plasmas* **20** 070501
[4] Goncharov V and Hurricane O 2012 *Panel 3: Implosion Hydrodynamics* LLNL-TR-562014
[5] Regan S P et al 2013 *Phys. Rev. Lett.* **111** 045001
[6] Clark D S et al 2013 *Phys. Plasmas* **20** 056318
[7] Ye W et al 2002 *Phys. Rev.* E **65** 057401
Ye W H et al 2010 *Phys. Plasmas* **17** 122704
Wang L F et al 2012 *Phys. Plasmas* **19** 100701
Wang L F et al 2010 *Phys. Plasmas* **17** 122706
[8] Bodner S 1974 *Phys. Rev. Lett.* **33** 761
H Takabe et al 1985 *Phys. Fluids* **28** 3676
[9] Goncharov V N et al 2003 *Phys. Plasmas* **10** 1906
Anderson K et al 2003 *Phys. Plasmas* **10** 4448
[10] Hurricane O A 2013 *APS DPP Meeting: High-Foot Implosion Campaign* ed Dittrich T R
Hurricane O A 2014 *Phys. Rev. Lett.* **112** 055002
Park H S et al 2014 *Phys. Rev. Lett.* **112** 055001
Hurricane O A 2014 *Nature* **506** 343
[11] Smalyuk V 2013 *APS DPP Meeting: Hydrodynamic Instability Growth and Mix Experiments at the National Ignition Facility*
[12] Herrmann M C et al 2001 *Nucl. Fusion* **41** 99
[13] Fan Z et al 2012 *Europhys. Lett.* **99** 65003
[14] Munro D H et al 2001 *Phys. Plasmas* **8** 2245

PHYSICS OF PLASMAS 21, 010704 (2014)

High flux symmetry of the spherical hohlraum with octahedral 6LEHs at the hohlraum-to-capsule radius ratio of 5.14

Ke Lan,[1,2] Jie Liu,[1,2] Dongxian Lai,[1] Wudi Zheng,[1] and Xian-Tu He[1,2]
[1]*Institute of Applied Physics and Computational Mathematics, Beijing, 100088, China*
[2]*Center for Applied Physics and Technology, Peking University, Beijing, 100871, China*

(Received 30 November 2013; accepted 10 January 2014; published online 23 January 2014)

We propose a spherical hohlraum with octahedral six laser entrance holes at a specific hohlraum-to-capsule radius ratio of 5.14 for inertial fusion study, which has robust high symmetry during the capsule implosion and low backscatter without supplementary technology. To produce an ignition radiation pulse of 300 eV, it needs 1.5 MJ absorbed laser energy in such a golden octahedral hohlraum, about 30% more than a traditional cylinder. Nevertheless, it is worth for a high symmetry and low backscatter. The proposed octahedral hohlraum is also flexible and can be applicable to diverse inertial fusion drive approaches.
[http://dx.doi.org/10.1063/1.4863435]

The hohlraum is crucial for the inertial fusions of both indirect drive[1–3] and the hybrid indirect-direct drive proposed recently.[4] In hohlraum design, the hohlraum shape, size, and the number of Laser Entrance Hole (LEH) are optimized to balance tradeoffs among the needs for capsule symmetry, the acceptable hohlraum plasma filling, the requirements for energy and power, and the laser plasma interactions. Among many requirements, the energy coupling and flux symmetry are of most concerned. A higher energy coupling will economize the input energy and increase the fusion energy gain. More importantly, a very uniform flux from the hohlraum on the shell of capsule is mandatory because a small drive asymmetry of 1%[2] can lead to the failure of ignition. Actually, the small flux asymmetry will be magnified during the compression process due to the varied kinds of instabilities and results in a serious hot-cold fuel mixture that can dramatically lessen the temperature or density of the hot spot for ignition. Various hohlraums with different shapes have been proposed and investigated, such as cylinder hohlraum,[1,2] rugby hohlraum,[5,6] and elliptical hohlraum.[7] These hohlraums are elongated with a length-to-diameter ratio greater than unity and have cylindrically symmetry with two LEHs on the ends. Among all above hohlraums, the cylindrical hohlraums are used most often in inertial fusion studies and are chosen as the ignition hohlraum on NIF,[8] though it breaks the spherical symmetry and leads to cross coupling between the modes. Intuitively, spherical hohlraum has the feature of the most symmetry compared to other geometric shapes. In fact, the spherical hohlraums with 4 LEHs or 6 LEHs have been investigated both theoretically[9,10] and experimentally[11,12] for many years, and all these investigations have shown the advantage of a spherical hohlraum in high uniformity. However, two sets of laser beams are needed for a 4-LEH spherical hohlraum in order to minimize the flux asymmetry by varying the relative power, thus a 4-LEH spherical hohlraum has no superiority over a traditional cylindrical hohlraum on this point. For the 6-LEH spherical hohlraums, a wide range of the hohlraum-to-capsule radius ratio from 2.5 to 7 was used in the studies, but an optimum ratio has not been addressed. As we know, the radius ratio is a crucial parameter which strongly concerns the flux symmetry and hohlraum energetics in an ignition target design.

In this letter, we investigate the spherical hohlraum with octahedral 6 LEHs from the theoretical side, addressing the most important issues of the flux symmetry and laser energy. We find a golden hohlraum-to-capsule radius ratio of 5.14 at which the flux asymmetry can be reduced to about 0.1%. We call the hohlraum with the golden ratio as golden octahedral hohlraum. From our study, a golden octahedral hohlraum has high symmetry and low backscatter without supplementary technologies.

For convenience, we consider that octahedral hohlraum has two poles and an equator though it is round. In the octahedral hohlraum, there are six LEHs, one at per pole and four along the equator coordinately. In the hohlraum system, we define θ as polar angle and ϕ as azimuthal angle. We use R_H to denote the hohlraum radius, R_C the capsule radius, and R_L the LEH radius. The laser beams are clustered in quads of four beams. We assume the quad shape at LEH to be a circle and use R_Q to denote the quad radius at LEH. Here, each quad through a LEH is characterized by θ_L and ϕ_L, where θ_L is the opening angle that the quad makes with the LEH normal direction and ϕ_L is the azimuthal angle about the normal of the LEH. For the golden octahedral hohlraums, the laser entering each LEH is in one cone with the same θ_L, thus a smaller LEH size is allowable, compared to the big LEH size of a cylindrical hohlraum which needs multilaser cones entering each LEH to adjust the symmetry. The relative fluxes of the laser spot, the hohlraum wall, and LEH are denoted as F_{spot}, F_{wall}, and F_{LEH}, respectively. Usually, we take F_{spot}:F_{wall}:F_{LEH} = 2:1:0, unless declaring. Fig. 1 shows the scenography of the octahedral hohlraum with six LEHs and laser spot of 48 quads and its pattern in the θ/ϕ plane, by taking $R_H/R_C = 5.1$, $R_C = 1.1$ mm, $R_L = 1$ mm, $R_Q = 0.3$ mm, and $\theta_L = 55°$. From our calculation, the flux asymmetry is about 0.1% on a capsule of 1.1 mm radius.

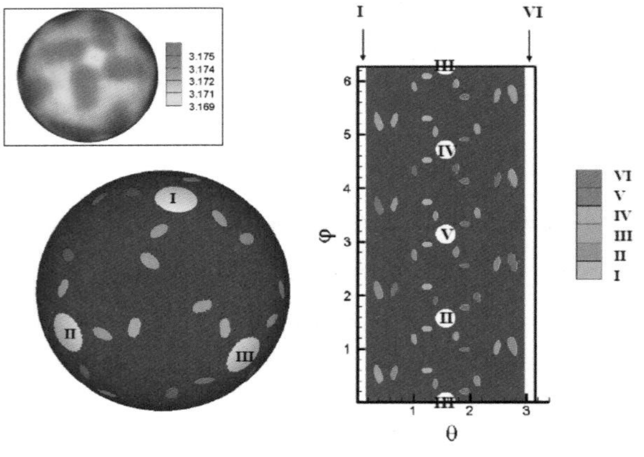

FIG. 1. Scenography of octahedral hohlraum with six LEHs (white color) and laser spot of 48 quads (red color) on the left-hand side and its pattern in the θ/ϕ plane on the right-hand side, by taking $R_H/R_C = 5.1$, $R_C = 1.1$ mm, $R_L = 1$ mm, $R_Q = 0.3$ mm, and $\theta_L = 55°$. Here, the LEHs are numbered and the laser spots are colored according to its LEH number. At the top left corner, it is the scenography of the flux distribution on capsule.

We first use a simple model to prove that a golden hohlraum-to-capsule radius ratio exists for an octahedral hohlraum at which the flux asymmetry can reach its minimum. In this simple model, the LEHs are treated as negative sources, and the wall and laser spots are treated as a homogeneous background by neglecting their flux difference. Only considering the negative effect of LEH on capsule, we present in Fig. 2(a) schematic of a capsule inside an octahedral hohlraum. The capsule is concentric with the octahedral hohlraum with their center at point O. On capsule, there are two kinds of points which see LEH most different, such as points A and B in the figure. The normal of point A is in the same direction as that of the LEH centered on point M, while the normal of point B has equal angles with that of three LEHs, centered on points L, M, and N, respectively. Hence, we can study the flux asymmetry on capsule by comparing the irradiation on points A and B.

The flux irradiated on a capsule point is mainly decided by the solid angle of the source opened to that point and the angle of the connecting line with respect to the normal of the capsule point. By denoting the LEH area as S, the solid angle of LEH M seen by point A is $d\Omega_A = S/(R_H - R_C)^2$, and the solid angle of LEH N seen by point B is $d\Omega_B = S\cos\alpha/l^2$. Here, l is the length of line BN, and α is the angle of BN with respect to the normal of N. We assume that $R_L \ll R_H$, and thus solid angles $d\Omega_A$ and $d\Omega_B$ are infinitesimal. We use β to denote the angle between the normal of N and the normal of B, then the angle of BN with respect to the normal of B is $\alpha + \beta$. Note that both LEH M and LEH L open the equal solid angle to point B as LEH N. Then the quality of the illumination on the capsule can be quantified approximately by $f = 0.5 \times [3d\Omega_B \times \cos(\alpha+\beta) - d\Omega_A]/d\Omega_A$. From Fig. 2, we have the following geometrical relationships: $tg\beta = \sqrt{2}$, $tg\alpha = R_C\sin\beta/(R_H - R_C\cos\beta)$ and $l \times \cos\alpha = R_H - R_C\cos\beta$. Then, we obtain

$$f = 0.5 \times \left[3\cos^3\alpha \times \cos(\alpha+\beta) \times \left(\frac{\frac{R_H}{R_C} - 1}{\frac{R_H}{R_C} - \cos\beta} \right)^2 - 1 \right]. \quad (1)$$

The variation of $|f|$ as R_H/R_C is presented in Fig. 3. As shown, $|f|$ reaches its minimum at $R_H/R_C = 5.14$. It predicts the emergence of the minimum flux asymmetry at the golden hohlraum-to-capsule radius ratio.

To certify the above theoretical prediction, we further exploit the view factor model[5] to calculate the radiation flux on the shell of the capsule numerically. We define ratio $|\Delta F/\langle F\rangle|$ in which $\Delta F = 0.5 \times (F_{max} - F_{min})$ and $\langle F\rangle$ is the average value of flux F upon the capsule. The black solid line shown in Fig. 3 is variation of $|\Delta F/\langle F\rangle|$ as R_H/R_C on the capsule shown in Fig. 2, which is inside an octahedral hohlraum with $R_L = 1$ mm, $R_Q = 0.5$ mm, and $\theta_L = 55°$. As indicated, an asymmetry minimum do exist for $|\Delta F/\langle F\rangle|$ at $R_H/R_C = 5.1$, quite close to the simple model. Using $F(\mathbf{P})$ to denote the total flux F at point $\mathbf{P}(\theta, \phi)$ on capsule, the asymmetry of flux on capsule can be expanded as $F(\mathbf{P}) = \sum_{l=0}^{\infty}\sum_{m=-l}^{l} a_{lm}Y_{lm}(\theta,\phi)$, where $Y_{lm}(\theta,\phi)$ are the spherical harmonics defined in quantum mechanics and a_{lm} is spherical

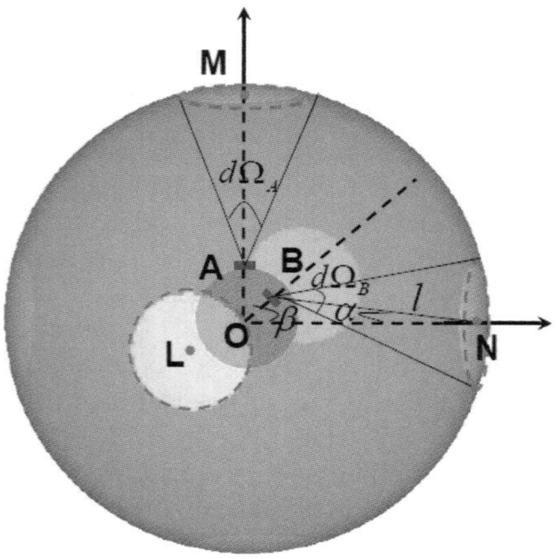

FIG. 2. Schematic of a capsule inside an octahedral hohlraum.

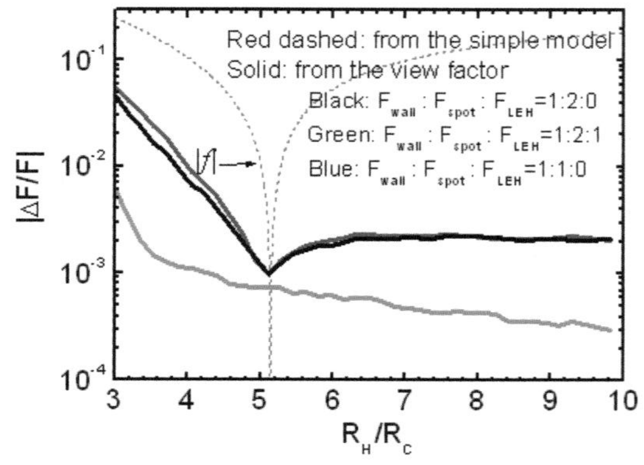

FIG. 3. Variations of $|f|$ as R_H/R_C from the simple model (red line) and $|\Delta F/F|$ from the view factor model (black solid line).

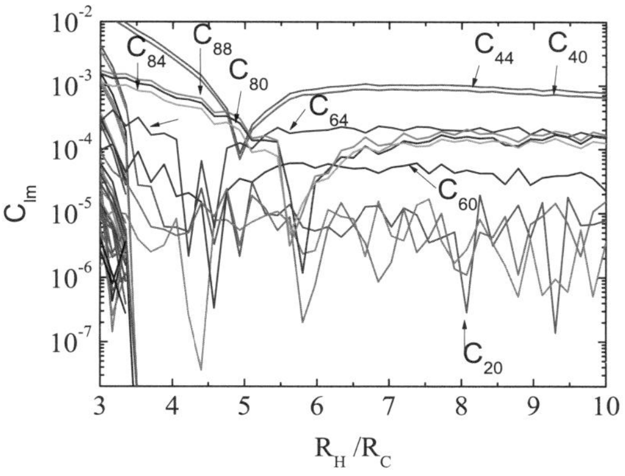

FIG. 4. Variations of C_{lm} as R_H/R_C.

harmonic decomposition. We further define $C_{l0} = a_{l0}/a_{00}$ and $C_{lm} = 2a_{lm}/a_{00}$ for $m > 0$. Shown in Fig. 4 is variations of C_{lm} as R_H/R_C for the same model in Fig. 3. As shown, C_{40} and C_{44} dominate the capsule flux asymmetry, except around the golden ratio where the asymmetry is dominated by C_{80}, C_{84}, and C_{88} with values much smaller than 0.1%. Notice that C_{2m} is on noise level and can be thoroughly neglected inside an octahedral hohlraum, quite different from the case inside a cylindrical geometry. The minimums of C_{4m} at $R_H/R_C = 5$, C_{6m} at around $R_H/R_C = 4$, and C_{8m} at around $R_H/R_C = 6$ are due to the asymmetry smoothing factor on capsule inside a concentric spherical hohlraum.[1,2,5] According to Sec. 9.4.3 in Ref. 1, the smoothing factor depends on mode number l but not on the directional mode number m because the choice of the direction of the polar axis is arbitrary for spherical symmetry. Here, it is worth to mention the 2LEH cylindrical and 4LEH spherical hohlraums in which the symmetries are dominated, respectively, by Y_{2m} and Y_{3m}, while the smoothing factors of Y_{2m} and Y_{3m} are much less reduced, especially at $R_H/R_C \leq 5$.

In order to distinguish the asymmetry contributions from LEH and laser spot, we use the view factor model to calculate the asymmetry of spherical hohlraums with only octahedral 6LEHs and only 48 quads, respectively. As shown in Fig. 3, the asymmetry contributed by the LEHs is larger than that by the spots, and the asymmetry minimum is thoroughly due to the six LEHs. That is why R_H/R_C of the minimum asymmetry from the simple model agrees so well with that from view factor model. In fact, it is reasonable that the LEH dominates the asymmetry inside an octahedral hohlraum, because the LEH number is much less than the laser spot number and also the large numbers of laser spots distribute much more homogeneously on hohlraum wall. Because the contribution of the laser spots to capsule asymmetry is small, the asymmetry is insensitive to the movements and other nonlinear properties of the laser spots. This is one of the superiorities of the octahedral hohlraum, quite different from the case for the elongated cylindrical hohlraums with two LEHs on the ends.

According to our calculations, R_H/R_C of the minimum asymmetry has small deviation from 5.14 under different LEH-to-hohlraum radius ratio and different arrangement of laser beams. We call $R_H/R_C = 5.14$ as the golden ratio of the octahedral hohlraum. Here, it is worth to mention the pioneer experiments on 6LEH spherical hohlraum fielded at the ISKRA-5 facility with 12 laser beams[12] in which the ratio is taken as $R_H/R_C = 7$. Obviously, it cost much more energy than the golden ratio to generate the same radiation.

Notice that $|\Delta F/\langle F\rangle|$ is smaller than or around 0.2% at $R_H/R_C \geq 4.7$, which means that the golden octahedral hohlraum has high symmetry not only at the early stage of capsule implosion when R_H/R_C becomes a little small due to wall plasma expansion, spot motion, and expansion of the outer layers of the capsule but also during the stages of capsule inward acceleration and ignition when R_H/R_C becomes very large. Thus, the golden octahedral hohlraum has a robust high symmetry.

As we know, stimulated backscatter of the laser beam, including Brillouin scattering and Raman scattering, is function of the beam intensity, of the density, temperature, and velocity fields along the path of the beams in the hohlraum and the beam path length.[3,8,13] Therefore, whether filling gas or vacuum, the backscatter of a hohlraum is strongly connected with the plasma density, beam overlap, and cross beam energy transfer. Thus, the golden octahedral hohlraum also has superiority on low backscatter because of following reasons. First, the spherical octahedral hohlraums contain only one cone at each LEH, so the issues of beam overlapping and crossed-beam transfer do not exist; thus, the backscatter can be remarkable decreased. Second, the volume of a golden octahedral hohlraums is 125 times of that capsule volume, more than 2 times of that of the traditional cylindrical hohlraum which is about 50 times of the capsule volume. Thus, the plasma filling inside such an octahedral hohlraum is obviously lower than inside a cylinder, which further benefits a low backscatter. With a lower backscatter, the golden octahedral hohlraums have a higher laser absorption efficiency than the cylinders. However, a quantitative assessment of the potential laser plasma instabilities (LPI) effects remains a significant uncertainty, and it needs experiments.

Of course, it needs more laser energy to drive a larger hohlraum for producing same radiation. As an example, we compare the laser energies required for generating an ignition radiation pulse of 300 eV inside a golden octahedral hohlraum and a traditional cylinder. We consider the ignition target recently designed for NIF[8] and use the expended plasma-filling model[7] to calculate the required laser energy and the plasma filling inside the hohlraums. According to Ref. 8, the cylindrical uranium hohlraums with dimensions of 5.75 mm in diameter and 9.4 mm in length are used for a deuterium-tritium (DT) capsule of $R_C = 1.13$ mm. To have same LEH area, we take the $R_L = 1.732$ mm for cylinder and $R_L = 1$ mm for the golden octahedral hohlraum. For simplicity, both kinds of hohlraums are vacuum. Here, we do not consider the backscatter and assume that the conversion efficiency from laser to x-ray is 87% for both hohlraums. From our calculation, it needs 1.5 MJ absorbed laser energy by using the golden octahedral hohlraum with $n_e = 0.067$ and 1.1 MJ by using the cylinder with $n_e = 0.094$. Here, n_e is electron density in unit of the critical density, and the plasma filling criterion is $n_e = 0.1$. Obviously, it needs more than 30% laser energy by using a

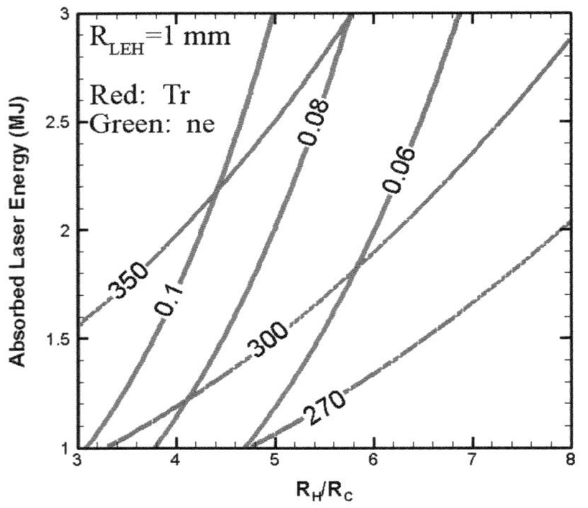

FIG. 5. Initial design of laser energy and R_H/R_C to produce a 300 eV ignition radiation pulse[8] by using an octahedral hohlraum.

golden octahedral hohlraum. Nevertheless, it is worth to spend some more laser energy for a high symmetry and a low backscatter. Shown in Fig. 5 is the initial design for an octahedral hohlraum. It is worth to mention that the energetics is sensitive to R_L. Approximately, the absorbed laser energy required by the golden octahedral hohlraum to produce above ignition radiation pulse is (in unit of MJ) 1.5 $+1.06(R_L/mm-1)$ for $1 \leq R_L \leq 1.8mm$.

Now we discuss the laser arrangement and constraints of octahedral hohlraum. We define the hohlraum pole axis as z axis. Axis x is defined by the centers of two opposite LEHs on equator, and y is defined by the other two. We name the LEHs centered on z axis as LEH1 and LEH6, on y axis as LEH2 and LEH4, and on x axis as LEH3 and LEH5. Each quad through a LEH is characterized by θ_L and ϕ_L. There is only one cone in our design, so all quads coming from the six LEHs have the same θ_L. The chooses of θ_L and ϕ_L are not only related to the ratios of R_H/R_C, R_L/R_C, R_Q/R_C but also interactional. There are three constraints which govern the quad arrangement. First, the lasers cannot hit to the opposite half sphere in order to have a short transfer distance inside hohlraum for suppressing the increase of LPI, which limits the opening angle $\theta_L > 45°$. Second, the laser cannot enter the hohlraum at a very shallow angle in order to avoid absorbing by blowoff from the wall and making unclearance of the hole. Third, a laser beams cannot cross and overlap with other beams. In Fig. 1, we take $\theta_L = 55°$.

We use N_Q to denote the quad number per LEH. The quads come in each LEH coordinately around LEH axis at the azimuthal angles of $\phi_{L0} + k \times 360°/N_Q$ $(k = 1, ..., N_Q)$. Here, ϕ_{L0} is azimuthal angle deviated from x axis in the xy plane for LEH1 and LEH6, from x axis in the xz plane for LEH2 and LEH4, and from y axis in the yz plane for LEH3 and LEH5. From the geometrical symmetry, we have $0° < \phi_{L0} < 360°/2N_Q$. In order to avoid overlapping between laser spots and transferring out of neighbor LEHs, we usually take ϕ_{L0} around $360°/4N_Q$. In Fig. 1, we take $\phi_{L0} = 11.25°$.

In summary, we have proposed a spherical hohlraum with octahedral 6LEHs at a golden hohlraum-to-capsule radius ratio of 5.14 for inertial fusion study, which has high symmetry on capsule and low backscatter without supplementary technology. This novel spherical hohlraum design has important implications for laser inertial fusion, and it can be implemented on the Omega laser and will be conducted on SG laser facilities in near future.

The authors wish to acknowledge the beneficial help from Dr. Yiqing Zhao, Dr. Zhengfeng Fan, Professor Xiaomin Zhang, Professor Jürgen Meyer-ter-Vehn, Professor Grant Logan, and Professor Min Yu. This work was supported by the National Fundamental Research Program of China (Contact Nos. 2013CBA01502 and 2013CB83410) and CAEP (Contact No. 2013A0102002).

[1]S. Atzeni abd and J. Meyer-ter-Vehn, *The Physics of Inertial Fusion* (Oxford Science, Oxford, 2004).
[2]J. D. Lindl, Phys. Plasmas **2**, 3933 (1995).
[3]S. W. Haan, J. D. Lindl, D. A. Callanhan, D. S. Clark, J. D. Salmonson, B. A. Hammel, L. J. Atherton, R. C. Cook, M. J. Edwards, S. Glenzer, A. V. Hamza, S. P. Hatchett, M. C. Herrmann, D. E. Hinkel, D. D. Ho, H. Huang, O. S. Jones, J. Kline, G. Kyrala, O. L. Landen, B. J. MacGowan, M. M. Marinak, D. D. Meyerhofer, J. L. Milovich, K. A. Moreno, E. I. Moses, D. H. Munro, A. Nikroo, R. E. Olson, K. Peterson, S. M. Pollaine, J. E. Ralph, H. F. Robey, B. K. Spears, P. T. Springer, L. J. Suter, C. A. Thomas, R. P. Town, R. Vesey, S. V. Weber, H. L. Wilkens, and D. C. Wilson, Phys. Plasmas **18**, 051001 (2011).
[4]Z. Fan, M. Chen, Z. Dai, H.-B. Cai, S.-P. Zhu, W. Y. Zhang, and X. T. He, e-print arXiv:1303.1252[physics.plasm-ph]; X. T. He, the plenary presentation at IFSA 8, Nara, Japan, September 9–13, 2013.
[5]A. Caruso and C. Strangio, Jpn. J. Appl. Phys., Part 1 **30**, 1095 (1991).
[6]P. Amendt, C. Cerjan, A. Hamza, D. E. Hinkel, J. L. Milovich, and H. F. Robey, Phys. Plasmas **14**, 056312(2007).
[7]K. Lan, P. Gu, G. Ren, C. Wu, W. Huo, D. Lai, and X. He, Laser Part. Beams **28**, 421 (2010); K. Lan, D. Lai, Y. Zhao, and X. Li, *ibid.* **30**, 175 (2012).
[8]J. L. Kline, D. A. Callahan, S. H. Glenzer, N. B. Meezan, J. D. Moody, D. E. Hinkel, O. S. Jones, A. J. MacKinnon, R. Bennedetti, R. L. Berger, D. Bradley, E. L. Dewald, I. Bass, C. Bennett, M. Bowers, G. Brunton, J. Bude, S. Burkhart, A. Condor, J. M. Di Nicola, P. Di Nicola, S. N. Dixit, T. Doeppner, E. G. Dzenitis, G. Erber, J. Folta, G. Grim, S. Lenn, A. Hamza, S. W. Hann, J. Heebner, M. Henesian, M. Hermann, D. G. Hicks, W. W. Hsing, N. Izumi, K. Jancaitis, O. S. Jones, D. Kalantar, S. F. Khan, R. Kirkwook, G. A. Kyrala, K. LaFortune, O. L. Landen, L. Lain, D. Larson, S. Le Pape, T. Ma, A. G. MacPhee, P. A. Michel, P. Miller, M. Montincelli, A. S. Moore, A. Nikroo, M. Nostrand, R. E. Olson, A. Pak, H. S. Park, M. B. Schneider, M. Shaw, V. A. Smalyuk, D. J. Strozzi, T. Suratwala, L. J. Suter, R. Tommasini, R. P. J. Town, B. Van Wonterghem, P. Wegner, K. Widmann, C. Widmayer, H. Wilkens, E. A. Williams, M. J. Edwards, B. A. Remington, B. J. MacGowan, J. D. Kikenny, J. D. Lindl, L. J. Atherton, S. H. Batha, and E. Moses, Phys. Plasmas **20**, 056314 (2013).
[9]M. Murakami, Nucl. Fusion **32**, 1715 (1992).
[10]D. W. Phillion and S. M. Pollaine, Phys. Plasmas **1**, 2963 (1994).
[11]J. M. Wallace, T. J. Murphy, N. D. Delamater, K. A. Klare, J. A. Oertel, G. R. Magelssen, E. L. Lindman, A. A. Hauer, P. Gobby, J. D. Schnittman, R. S. Craxton, W. Seka, R. Kremens, D. Bradley, S. M. Pollaine, R. E. Turner, O. L. Landen, D. Drake, and J. J. MacFarlane, Phys. Rev. Lett. **82**, 3807 (1999).
[12]S. A. Bel'kov, A. V. Bessarab, O. A. Vinokurov, V. A. Gaidash, N. V. Zhidkov, V. M. Izgorodin, G. A. Kirillov, G. G. Kochemasov, A. V. Kunin, D. N. Litvin, V. M. Murugov, L. S. Mkhitar'yan, S. I. Petrov, A. V. Pinegin, V. T. Punin, N. A. Suslov, and V. A. Tokarev, JEPT Lett. **67**, 171 (1998).
[13]M. D. Rosen, H. A. Scott, D. E. Hinkel, E. A. Williams, D. A. Callahan, R. P. J. Town, L. Divol, P. A. Michel, W. L. Kruer, L. J. Suter, R. A. London, J. A. Harte, and G. B. Zimmerman, High Energy Density Phys. **7**, 180 (2011).

PHYSICS OF PLASMAS **21**, 052704 (2014)

Octahedral spherical hohlraum and its laser arrangement for inertial fusion

Ke Lan,[1,2] Xian-Tu He,[1,2] Jie Liu,[1,2] Wudi Zheng,[1] and Dongxian Lai[1]
[1]*Institute of Applied Physics and Computational Mathematics, Beijing 100088, China*
[2]*Center for Applied Physics and Technology, Peking University, Beijing 100871, China*

(Received 18 February 2014; accepted 21 April 2014; published online 21 May 2014)

A recent publication [K. Lan *et al.*, Phys. Plasmas **21**, 010704 (2014)] proposed a spherical hohlraum with six laser entrance holes of octahedral symmetry at a specific hohlraum-to-capsule radius ratio of 5.14 for inertial fusion study, which has robust high symmetry during the capsule implosion and superiority on low backscatter without supplementary technology. This paper extends the previous one by studying the laser arrangement and constraints of octahedral hohlraum in detail. As a result, it has serious beam crossing at $\theta_L \le 45°$, and $\theta_L = 50°$ to $60°$ is proposed as the optimum candidate range for the golden octahedral hohlraum, here θ_L is the opening angle that the laser quad beam makes with the Laser Entrance Hole (LEH) normal direction. In addition, the design of the LEH azimuthal angle should avoid laser spot overlapping on hohlraum wall and laser beam transferring outside hohlraum from a neighbor LEH. The octahedral hohlraums are flexible and can be applicable to diverse inertial fusion drive approaches. This paper also applies the octahedral hohlraum to the recent proposed hybrid indirect-direct drive approach. © *2014 AIP Publishing LLC.*
[http://dx.doi.org/10.1063/1.4878835]

I. INTRODUCTION

The study and design of a hohlraum target are essentially important in inertial fusion because it seriously influences the capsule symmetry and the hohlraum energetics, which are the most important issues of inertial fusion.[1–3] Furthermore, it also decides the laser beam arrangement and the geometrical configuration of a laser facility to be designed, which usually costs billions of dollars for ignition goal. In hohlraum design, the hohlraum shape, size, and the number of Laser Entrance Hole (LEH) are determined by a variety of trade-offs, such as the capsule symmetry, the hohlraum energetics, and the laser plasma instability (LPI). Among these requirements, the flux symmetry is of most concerned[4] because it is imperative that the shell of capsule remain spherical up to the point of ignition. Typical capsule convergence ratios are 25 to 45, so that drive asymmetry can be no more than 1%,[2] a demanding specification on hohlraum design. Of course, the flux symmetry is strongly connected with the hohlraum geometrical structure. Up to now, various hohlraums with different shapes have been proposed and investigated theoretically and experimentally, such as cylinder hohlraum,[1,2] rugby hohlraum,[5–10] elliptical hohlraum,[11] and spherical hohlraums with 4 LEHs[12,13] or 6 LEHs.[14,15] Among all these hohlraums, the cylindrical hohlraums are used most often in inertial fusion studies and are chosen as the ignition hohlraum on The National Ignition Facility (NIF).[2,3,16,17] Because the cylindrical structure breaks the spherical symmetry and both Legendre modes P_2 and P_4 can be issues, so the time-varying flux symmetry is controlled by using two rings per side in the hohlraum and by adjusting the power ratio between the two rings (the approach called as beam phasing) on NIF.[2,3] However, the laser beams injected at multi cones from same LEH may overlap and cross and will arouse nonlinear phenomena and complicated laser plasma interaction issues, which further changes the capsule symmetry and thus, a three-wavelength scheme is used to maintain the required symmetry.[18] Obviously, a strict agreement of simulations with experiments is a prerequisite for effectively using the beam phasing technology and the three-wavelength scheme to keep the symmetry.

In a previous publication,[19] we proposed a spherical hohlraum with 6 LEHs of octahedral symmetry at a specific hohlraum-to-capsule radius ratio of 5.14, which has robust high symmetry during the capsule implosion. We call this specific radius ratio as golden radius ratio and call the 6 LEH spherical hohlraum with this ratio as the golden octahedral hohlraum. In addition to the robust high symmetry, the golden octahedral hohlraum also has superiority on low backscatter without supplementary technology because of two reasons below. First, the octahedral hohlraums contain only one cone at each LEH, so all laser beams are the same and the issue of crossed-beam transfer does not exist. Also, the laser beam crossing inside hohlraum can be avoided by taking a suitable injection angle of the laser beams, which will be discussed in this paper. In contrast, for a cylindrical ignition hohlraum,[20] it has four different laser cones entering together at each LEH, with three times the laser beam number of the octahedral hohlraum, so the crossed-beam transfer cannot be avoided either at LEH or inside the cylindrical hohlraum. Second, the volume of a golden octahedral hohlraum is more than 2 times of that of a traditional cylindrical hohlraum, with a much lower plasma filling density than the latter, and this benefit suppressing the increase of LPIs. The third superiority of an octahedral hohlraum is its low possibility of generating plasma jet which may preheat capsule and arouse asymmetry, because the laser spots are near-homogenous distributed on its wall with relatively long distance between them. In contrast, the last spots on cylindrical hohlraum wall must be arranged close enough to form two rings to get a cylindrical symmetry, and so the plasma jet cannot be avoided inside a vacuum hohlraum.[21] Certainly, the absorbed laser energy required by a golden octahedral hohlraum is higher than a traditional cylinder because of its larger hohlraum surface area than the latter. According to our previous results, it needs about 1.5 MJ absorbed laser energy to generate a 300 eV ignition radiation pulse[17] inside such a golden octahedral hohlraum, about 30% more than a traditional cylinder. Nevertheless, it is worth to exchange some laser energy for a high and robust symmetry and much less LPI.

In fact, the initial hohlraum-to-capsule radius ratio mainly influences the initial asymmetry on capsule inside an octahedral hohlraum, and it can keep a high symmetry as the radius ratio becomes larger and larger during the capsule implosion.[19] We can therefore choose an octahedral hohlraum with radius ratio of 4 when one considers that the tolerant asymmetry for ignition is 1%, then the required absorbed laser energy for generating the 300 eV ignition pulse reduces to 1.18 MJ, almost the same as that required by a traditional cylinder. However, we mainly focus on the golden octahedral hohlraum in this extension paper.

In this paper, we extend the results of our previous publication by discussing the concept of golden octahedral hohlraum and studying the laser arrangement and constraints of octahedral hohlraum in detail. We also apply the golden octahedral hohlraum to the recent proposed Hybrid Indirect-Direct drive approach (HID).[22] In this work, we use the view factor model to investigate the laser arrangement and the capsule asymmetry of golden octahedral hohlraum, without considering the hohlraum dynamics, the laser-plasma interaction and the capsule implosion physics, also without beam imbalance, target fabrication imperfection and any instabilities. Of course, an integrated 3D simulation is needed for our future detail study and design on ignition target. Nevertheless, the symmetry of the golden octahedral hohlraum from the integrated 3D simulation should not be far from the view factor model because of its high and robust symmetry, at least not so far as that of a cylindrical hohlraum.

The organization of this paper is follows. Section II discusses the concept of golden octahedral hohlraum in detail and presents the symmetry inside a tetrahedral hohlraum for comparison. Section III details the laser arrangement and constraints of the golden octahedral hohlraum. The octahedral hohlraums are flexible and can be applicable to diverse inertial fusion drive approaches. As an example, Sec. IV applies the golden octahedral hohlraum to HID and gives the laser energy required by the indirect-drive phase of HID. Section V presents a summary.

II. OCTAHEDRAL HOHLRAUM

In this section, we discuss the concept of golden octahedral hohlraum in detail. From our previous publication,[19] the more important asymmetries inside an octahedral hohlraum are due to its holes (only 6) not to the laser hot spots (there are more than enough, in a near homogeneous distribution on hohlraum wall), and a key point is that the dominating spherical harmonics Y_{4m} produced by the holes cancels at the golden radius ratio of 5.14. Thus, there is a high symmetry inside the golden octahedral hohlraums.

As an example, we consider a golden octahedral hohlraum with 192 laser beams for a capsule of 1.1 mm radius which is designed for a 300 eV radiation drive.[17] Denoting the hohlraum radius as R_H and the capsule radius as R_C, we have $R_H = 5.14\ R_C$ for a golden octahedral hohlraum. The laser beams are clustered in quads of four beams, so we have 48 quads in our model. Assuming the quad waist is a circle, we use R_Q to denote its waist radius and take $R_Q = 0.3$ mm. Each quad through a LEH is characterized by θ_L and ϕ_L, where θ_L is the opening angle that the quad makes with the LEH normal direction and ϕ_L is the azimuthal angle about the normal of the LEH. The LEH radius R_L of a hohlraum is not only determined by θ_L and R_Q but also determined by the radiation pulse length and temperature. By assuming that the LEH boundary moves inward is δ under radiation, then it requires $R_L > \delta + R_Q/\cos\theta_L$ in order to avoid laser absorbing by blowoff from the wall and making unclearance of the LEH. To simulate δ under the 300 eV ignition radiation pulse of 20 ns width and four steps,[17] we use our one dimensional multi-group radiation transfer hydrodynamic code RDMG[23] in which the multi-group radiation transfer equation is resolved with S-N discrete coordinate scheme and the energy coupling of radiation with matter is resolved by the matrix operator splitting method. Under the 300 eV ignition radiation, the critical surface of Au material moves about 290 μm from RDMG. Notice that radiation at LEH is usually a little lower than at capsule, so the LEH closure of the ignition hohlraum designed to produce the same radiation pulse on capsule should be smaller than this result. According to our discussions in the next section, we take $\theta_L = 55°$. Hence, it requires $R_L > 0.813$ mm. In this model, we take $R_L = 1$ mm. Shown in Fig. 1 is scenography of the golden octahedral hohlraum with 6 LEHs and 48 quad beams and in Fig. 2 the scenography of the 48 quad laser spot distribution on the octahedral hohlraum wall, with $R_C = 1.1$ mm, $R_H/R_C = 5.14$, $R_L = 1$ mm, $R_Q = 0.3$ mm, and $\theta_L = 55°$.

For an octahedral hohlraum, it is an octahedron when one connects the 6 LEH centers by straight lines. In addition,

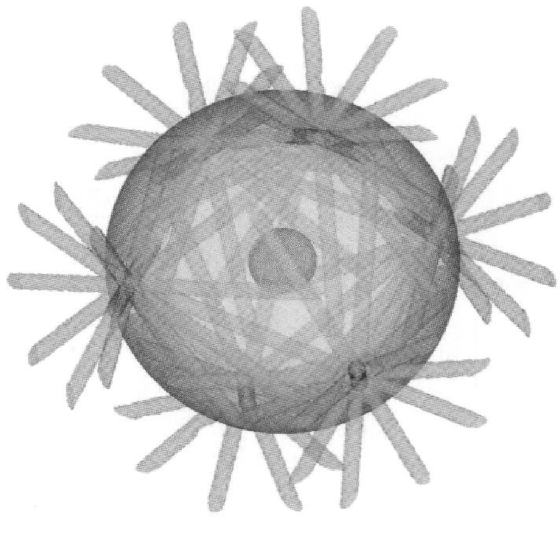

FIG. 1. Scenography of the golden octahedral hohlraum with 6 LEHs, 48 laser quad beams and centrally located fusion capsule.

FIG. 2. Scenography of the 48 laser quad spot distribution on the wall of the golden octahedral hohlraum. Hereafter, we use orange color for hohlraum, white color for LEHs, and red color for laser spots on hohlraum wall.

it separates the spherical hohlraum into 8 coordinate parts with the same laser spot distribution on each when one connects the 6 LEH centers by the shortest curves along the sphere surface. From Ref. 19, the asymmetry from the six LEHs goes to zero at the golden radius ratio and it is the laser spots from these 8 coordinate parts which dominates the asymmetry on capsule.

The relative fluxes of the laser spot, the hohlraum wall and the LEHs are denoted as F_{spot}, F_{wall}, and F_{LEH}, respectively. In the view factor model, we take F_{spot}: $F_{wall} = 2{:}1$, according to the experimental results[2] and our 2D simulations from our two-dimensional radiation hydrodynamic code LARED.[24] Shown in Fig. 3 is the scenography of the flux distribution on a capsule of 1.1 mm radius inside above golden octahedral hohlraum and the corresponding pattern in the θ/ϕ plane. We define ratio $|\Delta F/\langle F\rangle|$ as flux asymmetry on capsule in which $\Delta F = 0.5 \times (F_{max} - F_{min})$ and $\langle F\rangle$ is the average value of flux F upon the capsule. As shown, the flux asymmetry is about 0.1% for this model. From the view factor model, the dominating Y_{4m} tends to zero at $R_H/R_C = 5.14$ and it is Y_{8m} which dominates the capsule asymmetry inside a golden octahedral hohlraum.[19]

For a capsule inside a golden octahedral hohlraum, the 48 laser spots look like stars distributed on a round sky. Caused by the plasma evolution, the spot movement and the capsule implosion, all "stars" are either closer or farer together to the capsule during the laser pulse, but nevertheless it always keeps a round sky for the capsule and has a high symmetry. We therefore pointed that the golden octahedral hohlraum has a robust high symmetry in our previous publication.[19] This is quite different for a 2-LEH cylinder hohlraum. For the latter, the sky of the capsule is a cylinder and the asymmetry on capsule is strongly connected to the length-to-radius ratio of the cylindrical emission sky, while this ratio is changed with plasma evolution and the spot movement, and it becomes larger and larger during the laser pulse. Thus, the capsule asymmetry inside a cylinder is strongly time-dependent and needs to be controlled with the inner/outer cone ratio for the cylindrical hohlraums on NIF.

As a comparison, it is also interesting to present the flux symmetry of capsule inside a tetrahedral hohlraum, which is a spherical hohlraum with four LEHs located at the vertices of a tetrahedron. The concept of tetrahedral hohlraum was first proposed by Phillion and Pollaine,[12] and the first experiment with this kind of hohlraum was done at OMEGA by Wallace *et al.*[13] In that experiment, the tetrahedral hohlraums of $R_H/R_C = 2.56$–4 were used, with multi laser cones entering from each LEH. Here, we consider a tetrahedral hohlraum with only one cone of six quad beams from each LEH, and take $R_C = 1.13$ mm, $R_L = 1$ mm, $R_Q = 0.3$ mm, and $\theta_L = 55°$ in our model. Shown in Fig. 4 is a variation of $|\Delta F/\langle F\rangle|$ as R_H/R_C for this model. The inset plots are scenography of the tetrahedral hohlraum, the LEH and spot pattern in θ/ϕ plane and scenography of the flux distribution on capsule. As shown, $|\Delta F/\langle F\rangle|$ inside a tetrahedral hohlraum decreases as R_H/R_C, and different from the octahedral hohlraum,[19] it does not have a specific hohlraum-to-capsule ratio at which a minimum asymmetry exists. This is a reasonable

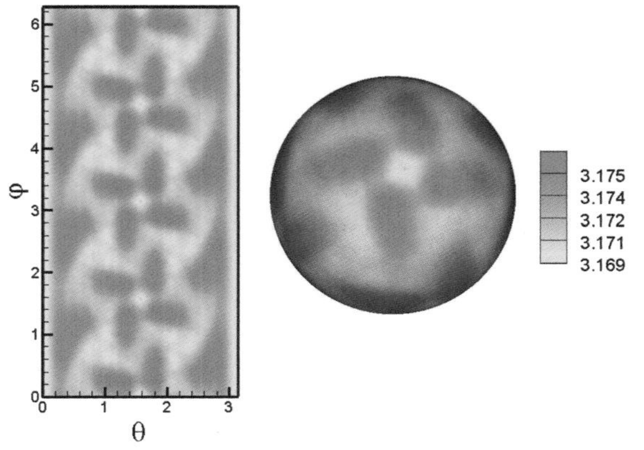

FIG. 3. Scenography of the flux distribution on a capsule of 1.1 mm radius inside the octahedral hohlraum shown in Fig. 1 on the left-hand side and its pattern in the θ/ϕ plane on the right-hand side. Notice that $|\Delta F/\langle F\rangle|$ is about 0.1%.

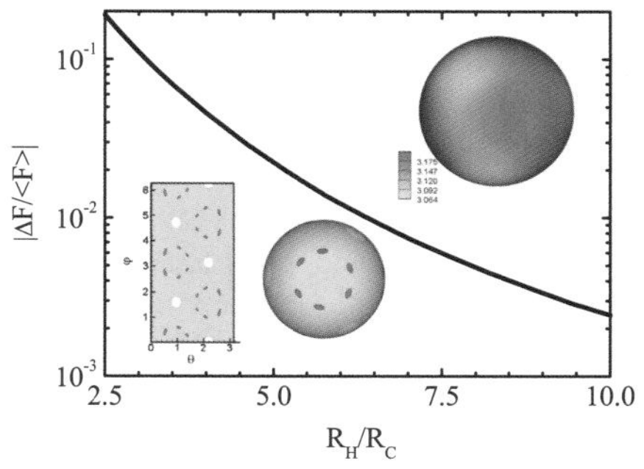

FIG. 4. Variation of $|\Delta F/\langle F\rangle|$ as R_H/R_C on capsule inside a tetrahedral hohlraum. The inset plots at the low left corner are scenograph of the LEH and spot pattern in θ/ϕ plane (left) and the tetrahedral hohlraum (right), at the top right corner the scenography of the flux distribution on capsule. In this model, $R_C = 1.13$ mm, $R_L = 1$ mm, $R_Q = 0.3$ mm, $\theta = 55°$, and $N_Q = 6$. For the inset plots, $R_H/R_C = 5$.

result for a tetrahedral hohlraum in which the capsule flux asymmetry is mainly contributed by the four LEHs and dominated by the spherical harmonics Y_{32}. According to Refs. 1 and 2, the asymmetry smoothing factor of Y_{3m} on capsule inside a concentric spherical hohlraum has no zero-point, and more, Y_{3m} is much less reduced as the increase of R_H/R_C, especially at $R_H/R_C < 5$. This result can be proved also by using the simple model proposed in Ref. 19. From Fig. 4, $|\Delta F/\langle F\rangle|$ inside a tetrahedral hohlraum decreases to 1% at $R_H/R_C = 6.4$. In contrast, $|\Delta F/\langle F\rangle|$ inside an octahedral hohlraum decreases to 1% at $R_H/R_C = 3.7$ and reaches its minimum of 0.1% at $R_H/R_C = 5.14$. Obviously, a very big tetrahedral hohlraum and corresponding high laser energy are needed to obtain the high flux symmetry required by fusion ignition, if without an assistant technology. Nevertheless, the requirement of laser energy higher than 2 MJ is not realistic at present laser facilities. Thus, a technology of dynamical compensation by using two sets of laser beams was proposed in Ref. 12 in order to decrease the asymmetry inside a tetrahedral hohlraum. However, the dynamical compensation strongly depends on the numerical simulations, while the agreement of the simulations with the observations is a serious issue.

III. LASER ARRANGEMENT AND CONSTRAINTS

The laser arrangement is of crucial important in hohlraum design, which not only strongly influences the capsule symmetry, laser-plasma coupling, laser-plasma instabilities, hydrodynamic instability and mix on capsule but also decides the geometrical configuration of laser facility and laser building. In this section, we discuss the laser arrangement and constraints of the golden octahedral hohlraum. We define the hohlraum pole axis as z axis. Axis x is defined by the centers of two opposite LEHs on equator, and y is defined by the other two. We name the LEHs centered on z axis as LEH1 and LEH6, on y axis as LEH2 and LEH4, and on x axis as LEH3 and LEH5. Each quad through a LEH is characterized by θ_L and ϕ_L. There is only one cone in our design, so all quads coming from the six LEHs have the same θ_L. We use N_Q to denote the quad number per LEH. The quads come in each LEH coordinately around LEH axis at the azimuthal angles of $\phi_{L0} + k \times 360°/N_Q$ $(k = 1, ..., N_Q)$. Here, ϕ_{L0} is azimuthal angle deviated from x axis in the xy plane for LEH1 and LEH6, from x axis in the xz plane for LEH2 and LEH4, and from y axis in the yz plane for LEH3 and LEH5. From the geometrical symmetry, we have $0° < \phi_{L0} < \phi_{LM}$, here $\phi_{LM} = 360°/2N_Q$.

The choose of ϕ_{L0} should avoid laser spot overlapping on hohlraum wall and laser beam transferring outside hohlraum from a neighbor LEH. Shown in Fig. 5 is schematic of laser spot overlapping on hohlraum wall and laser transferring outside from LEH. The determination of ϕ_{L0} is strongly connected with θ_L, R_H, R_L, R_Q, and N_Q. For $\theta_L \geq 50°$, we usually take $\phi_{L0} = 0.5\phi_{LM}$. For the model in Fig. 1, we take $\phi_{L0} = 11.25°$.

The choose of θ_L is somewhat a little complicated and several constraints must be taken into consideration. First, a high symmetry must be maintained. From the view factor

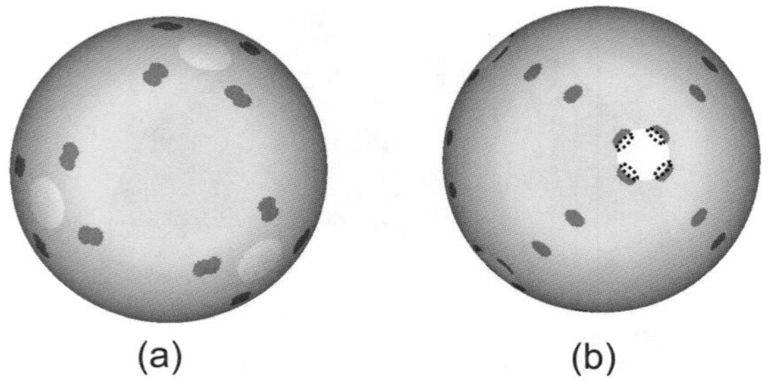

FIG. 5. Schematics of laser spot overlapping on hohlraum wall at $\phi_{L0} = 24.5°$ and $\theta = 45°$ (a), and laser transferring outside of hohlraum from LEH at $\phi_{L0} = 10°$ and $\theta = 45°$ (b). In (b), the part of laser energy inside the dashed black curves transfers outside from a neighbor LEH and is lost.

model, the capsule asymmetry at θ_L is almost the same as that at $90° - \theta_L$ inside an octahedral hohlraum, except the laser spot shapes on hohlraum wall are a little different. Shown in Fig. 6 is variation of $|\Delta F/\langle F\rangle|$ as θ_L. As indicated, $|\Delta F/\langle F\rangle|$ varies from about 0.1% to 0.27% when θ_L is between 30° and 60° and its peak happens at around 45°. This result is understandable, because the positive sources of laser spot and the negative sources of LEH partially cancel each other out, and the cancelling is more effective when the two kinds of sources are closer. In this model, ϕ_{L0} is taken as 11.25°, and it has the phenomena of laser beam transferring outside of hohlraum when θ_L is taken between 42° and 48°. Thus, the range of (42°, 48°) should be avoided for θ_L at $\phi_{L0} = 11.25°$. From Fig. 6, it has a high symmetry of around 0.12% at $\theta_L \leq 35°$ or $\theta_L \geq 55°$. Second, laser beams cannot cross and overlap with other beams in order to avoid arousing nonlinear phenomena and complicated laser plasma interaction issues. From our calculations, it has beam crossing inside the golden octahedral hohlraum at $\theta_L < 50°$, especially serious at $\theta_L \leq 45°$. Shown in Fig. 7 is the case for $\theta_L = 35°$. Notice that each quad beam crosses three beams for this case. In contrast, there is not any beam crossing at $\theta_L \geq 50°$. Presented in Fig. 8 is the case for $\theta_L = 55°$. In fact, the laser transfer distance inside hohlraum is shorter at a larger θ_L, which is also good for suppressing the increase of LPI. However, the laser cannot enter the hohlraum at a very shallow angle in order to avoid absorbing by blowoff

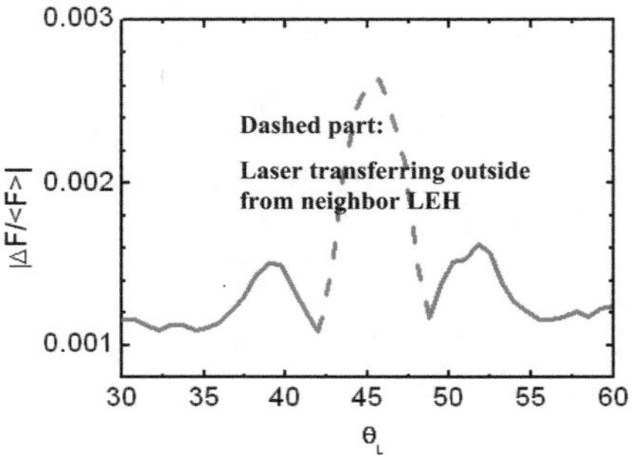

FIG. 6. Variation of $|\Delta F/\langle F\rangle|$ as θ_L for the capsule of $R_C = 1.13$ mm inside the golden octahedral hohlraum of $R_H = 5.14\ R_C$, $R_Q = 0.3$ mm, $R_L = 1$ mm, and $\phi_{L0} = 11.25°$. Notice the dashed part, in which range the laser quad beams transfer outside of hohlraum from neighbor LEHs at $\phi_{L0} = 11.25°$.

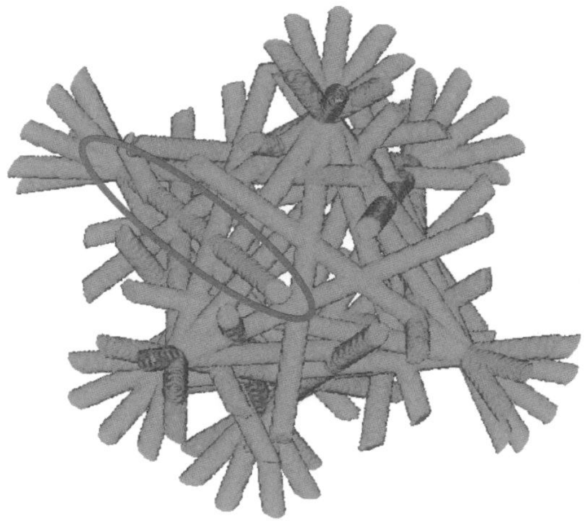

FIG. 7. Schematic of 48 quad beams injected at $\theta_L = 35°$ into the golden octahedral hohlraum presented in Fig. 1. Here, the hohlraum and capsule are omitted. Notice that the quad beam inside the red circle crosses other three beams, and in fact this is the same for all beams because they are equal inside the octahedral hohlraums.

from the wall and making unclearance of the hole. This is also the third constraint for choosing θ_L. From the geometry, it requires $\cos\theta_L \geq \max(\frac{R_L}{R_H}, \frac{R_Q}{R_L-\delta})$. For model in Fig. 1, it requires $\theta_L < 64°$. However, the upper limitation on θ_L is also strongly connected with the laser pulse, including its energy, power, and shape, and should be determined by future experiments. As a result, we propose $\theta_L = 50°–60°$ as the optimum candidate for the golden octahedral hohlraum.

Here, it is worth to mention that the shortest distance from a laser quad beam to the capsule surface is $R_H \times \sin\theta_L - R_C$, so it is 2.6 R_C at $\theta_L = 35°$ and 3.2 R_C at $\theta_L = 55°$ inside the golden octahedral hohlraum, obviously longer than the distance from the inner laser beam to the capsule surface inside a traditional cylindrical hohlraum on NIF. For the latter, the hohlraum-to-capsule radius ratio is usually taken as 2.6 to 2.8, so the distance from its inner laser cone to the capsule surface is about 1.6 to 1.8 R_C. Thus, the ablated plasmas from capsule has less influence on the laser transferring inside the golden octahedral hohlraum than that inside a cylinder, and this is beneficial to generating a clean hohlraum environment for the capsule and also a clean pass for the laser transport.

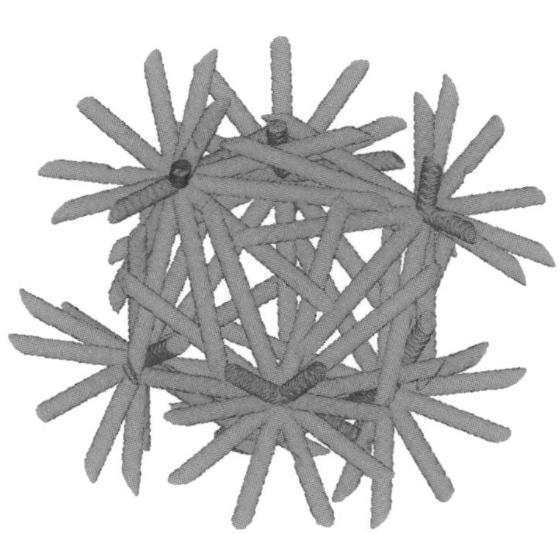

FIG. 8. Schematic of 48 quad beams injected at $\theta_L = 55°$ into the golden octahedral hohlraum presented in Fig. 1. Here, the hohlraum and capsule are omitted. Notice that there is not any beam crossing for this case.

IV. INITIAL DESIGN FOR HID

The proposed octahedral hohlraum is flexible and can be applicable to diverse inertial fusion drive approaches such as indirect and hybrid indirect-direct drives. As an application, we design a golden octahedral hohlraum for the HID scheme. In the HID scheme,[22] the fuel capsule is first compressed by indirect-drive x rays, and then by both indirect-drive x rays and direct drive lasers. According to the HID model, a four-step radiation pulse with the fourth step of 260 to 270 eV and 1.7 ns is required for a capsule with radius $R_C = 850\,\mu$m. In this section, we first give an estimation of the absorbed laser energy required by the indirect-drive phase of HID by using the expended plasma-filling model, and then study the flux asymmetry on HID capsule inside octahedral hohlraum by using the view factor model.

The plasma-filling model was first developed to calculate the plasma filling time for the hot laser channel to reach electron density inside a hohlraum driven by a constant input laser power,[25,26] and then was extended to the case of a hohlraum driven by a shaped laser pulse and is applied to the initial design of a hohlraum for generating a required radiation drive.[11,27] Shown in Fig. 9 is the initial design of laser energy and R_H/R_C by using the extended plasma-filling model. Here, R_L is also taken as 1 mm. The two semi-empirical criterions used here are: $n_e \leq 0.1$ related to the threshold for laser plasma instability,[25] and $n_{IB} \equiv (l/\sqrt{2})/(\lambda_{IB}) \leq 1$ related to the inverse bremsstrahlung absorption length λ_{IB}.[28] Here, n_e is electron density in unit of the critical density and l is the

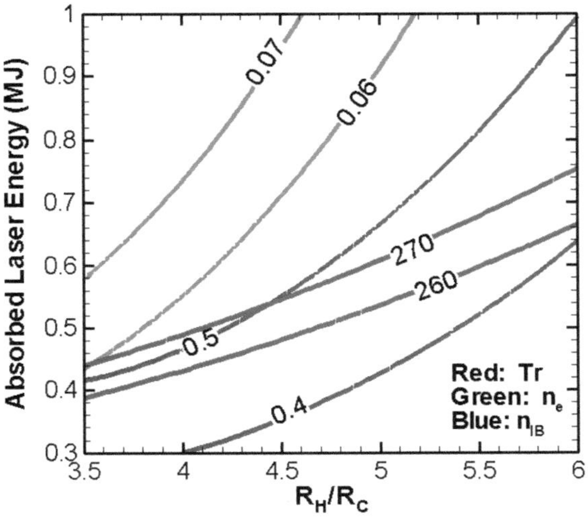

FIG. 9. Initial design of laser energy and R_H/R_C to produce the required radiation for the HID capsule inside an octahedral hohlraum. Red lines are contours of $T_r = 260$ eV and 270 eV and green lines are contours of $n_e = 0.06$ and 0.07, and blue lines are contours of $n_{IB} = 0.4$ and 0.5.

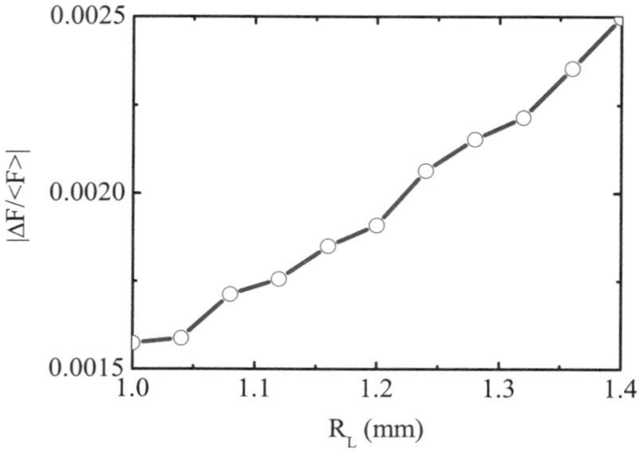

FIG. 10. Variation of $|\Delta F/\langle F\rangle|$ as R_L for the 0.85 mm capsule of HID model inside the golden octahedral hohlraum. Here, $R_H=5.14\ R_C$, $R_Q=0.3$ mm, $\theta_L=55°$, $\phi_{L0}=22.5°$, and $N_Q=4$. Notice that it is not a smooth line because of numerical noise.

transfer distance of laser beam inside a spherical hohlraum. As shown, it needs absorbed laser energy of 0.5 to 0.6 MJ to produce a 260 to 270 eV radiation pulse inside a golden octahedral hohlraum with n_e and n_{IB} well meeting the criterions. Of course, we can also use a smaller octahedral hohlraum with $R_H/R_C=4$ if the tolerant asymmetry is 1% for HID, then the absorbed laser energy required by the indirect phase of HID is reduced to 0.4 to 0.5 MJ.

The dependence of the capsule asymmetry on R_L, R_Q, θ_L, and the relative flux of laser spot to hohlraum wall is important for choosing the optimum design of hohlraum. Using the view factor model, we study the variations of $|\Delta F/\langle F\rangle|$ as these quantities for the HID model. The results indicate that $|\Delta F/\langle F\rangle|$ is far smaller than 1% at the golden ratio in the ranges of R_L, R_Q, and θ_L concerned in our model. Shown in Fig. 10 is variation of $|\Delta F/\langle F\rangle|$ as R_L for the HID capsule inside the golden octahedral hohlraum.

V. SUMMARY

In summary, we have extended our previous work by discussing the concept of golden octahedral hohlraum and studying the laser arrangement and constraints of octahedral hohlraum in detail. As a result, we propose $\theta_L=50°$ to $60°$ as the optimum candidate range for the golden octahedral hohlraum. We also applied the golden octahedral hohlraum to the HID scheme, and it shows that the absorbed laser energy of 0.5 to 0.6 MJ is needed for the indirect-drive phase of HID. Up to now, what we use in our study are analytic models of the view factor model and the expended plasma-filling model. In fact, the integrated 3D simulation is necessary for future detail study and design of an ignition target. In particularly, we have little experience on existing facilities for the octahedral hohlraums. Therefore, abundant experiments should be done with the octahedral hohlraums on hohlraum energetics and capsule implosion physics, including laser-plasma coupling, hohlraum dynamics, x-ray generation and transport, flux asymmetry, hydrodynamic instability and mix on capsule. Especially, the effects of laser plasma instabilities under different laser arrangements should be experimentally invested for determining a suitable θ_L for the octahedral hohlraum. These experiments are being designed on our SGIII laser facility and will begin in near future.

ACKNOWLEDGMENTS

The authors wish to acknowledge the beneficial discussions with Dr. Guoli Ren and Professor Min Yu. This work was supported by the National Fundamental Research Program of China (Contact Nos. 2013CBA01502 and 2013CB83410) and the Fundamental Research Program of CAEP (Contact No. 2013A0102002).

[1]S. Atzeni and J. Meyer-ter-Vehn, *The Physics of Inertial Fusion* (Oxford Science, Oxford, 2004).
[2]J. D. Lindl, Phys. Plasmas **2**, 3933 (1995).
[3]S. W. Haan, J. D. Lindl, D. A. Callanhan, D. S. Clark, J. D. Salmonson, B. A. Hammel, L. J. Atherton, R. C. Cook, M. J. Edwards, S. Glenzer, A. V. Hamza, S. P. Hatchett, M. C. Herrmann, D. E. Hinkel, D. D. Ho, H. Huang, O. S. Jones, J. Kline, G. Kyrala, O. L. Landen, B. J. MacGowan, M. M. Marinak, D. D. Meyerhofer, J. L. Milovich, K. A. Moreno, E. I. Moses, D. H. Munro, A. Nikroo, R. E. Olson, K. Peterson, S. M. Pollaine, J. E. Ralph, H. F. Robey, B. K. Spears, P. T. Springer, L. J. Suter, C. A. Thomas, R. P. Town, R. Vesey, S. V. Weber, H. L. Wilkens, and D. C. Wilson, Phys. Plasmas **18**, 051001 (2011).
[4]O. A. Hurricane, D. A. Callahan, D. T. Casey, P. M. Celliers, C. Cerjan, E. L. Dewald, T. R. Dittrich, T. Döppner, D. E. Hinkel, L. F. Berzak Hopkins, J. L. Kline, S. Le Pape, T. Ma, A. G. MacPhee, J. L. Milovich, A. Pak, H.-S. Park, P. K. Patel, B. A. Remington, J. D. Salmonson, P. T. Springer, and R. Tommasini, Nature **506**, 343 (2014).
[5]A. Caruso and C. Strangio, Jpn. J. Appl. Phys., Part 1 **30**, 1095 (1991).
[6]P. Amendt, C. Cerjan, A. Hamza, D. E. Hinkel, J. L. Milovich, and H. F. Robey, Phys. Plasmas **14**, 056312 (2007).
[7]M. Vandenboomgaerde, J. Bastian, A. Casner, D. Galmiche, J.-P. Jadaud, S. Laffite, S. Liberatore, G. Malinie, and F. Philippe, Phys. Rev. Lett. **99**, 065004 (2007).
[8]A. Casner, D. Galmiche, G. Huser, J.-P. Jadaud, S. Liberatore, and M. Vandenboomgaerde, Phys. Plasmas **16**, 092701 (2009).
[9]H. F. Robey, P. Amendt, H.-S. Park, R. P. J. Town, J. L. Milovich, T. Döppner, D. E. Hinkel, R. Wallace, C. Sorce, D. J. Strozzi, F. Phillippe, A. Casner, T. Caillaud, O. Landoas, S. Liberatore, M.-C. Monteil, F. Sguin, M. Rosenberg, C. K. Li, R. Petrasso, V. Glebov, C. Stoeckl, A. Nikroo, and E. Giraldez, Phys. Plasmas **17**, 056313 (2010).
[10]F. Philippe, A. Casner, T. Caillaud, O. Landoas, M. C. Monteil, S. Liberatore, H. S. Park, P. Amendt, H. Robey, and C. Sorce, Phys. Rev. Lett. **104**, 035004 (2010).
[11]K. Lan, D. Lai, Y. Zhao, and X. Li, Laser Part. Beams **30**, 175 (2012).
[12]D. W. Phillion and S. M. Pollaine, Phys. Plasmas **1**, 2963 (1994).
[13]J. M. Wallace, T. J. Murphy, N. D. Delamater, K. A. Klare, J. A. Oertel, G. R. Magelssen, E. L. Lindman, A. A. Hauer, P. Gobby, J. D. Schnittman, R. S. Craxton, W. Seka, R. Kremens, D. Bradley, S. M. Pollaine, R. E. Turner, O. L. Landen, D. Drake, and J. J. MacFarlane, Phys. Rev. Lett. **82**, 3807 (1999).
[14]M. Murakami, Nucl. Fusion **32**, 1715 (1992).
[15]S. A. Bel'kov, A. V. Bessarab, O. A. Vinokurov, V. A. Gaidash, N. V. Zhidkov, V. M. Izgorodin, G. A. Kirillov, G. G. Kochemasov, A. V. Kunin, D. N. Litvin, V. M. Murugov, L. S. Mkhitar'yan, S. I. Petrov, A. V. Pinegin, V. T. Punin, N. A. Suslov, and V. A. Tokarev, JEPT Lett. **67**, 171 (1998).
[16]D. A. Callahan, N. B. Meezan, S. H. Glenzer, A. J. MacKinnon, L. R. Benedetti, D. K. Bradley, J. R. Celeste, P. M. Celliers, S. N. Dixit, T. Doppner, E. G. Dzentitis, S. Glenn, S. W. Haan, C. A. Haynam, D. G. Hicks, D. E. Hinkel, O. S. Jones, O. L. Landen, R. A. London, A. G. MacPhee, P. A. Michel, J. D. Moody, J. E. Ralph, H. F. Robey, M. D. Rosen, M. B. Schneider, D. J. Strozzi, L. J. Suter, R. P. Town, K. L. J. Atherton, G. A. Kyrala, J. L. Kline, R. E. Olsen, D. Edgell, S. P. Regan, A. Nikroo, H. Wilkins, J. D. Kikenny, and A. S. Moore, Phys. Plasmas **19**, 056305 (2012).
[17]J. L. Kline, D. A. Callahan, S. H. Glenzer, N. B. Meezan, J. D. Moody, D. E. Hinkel, O. S. Jones, A. J. MacKinnon, R. Bennedetti, R. L. Berger, D.

Bradley, E. L. Dewald, I. Bass, C. Bennett, M. Bowers, G. Brunton, J. Bude, S. Burkhart, A. Condor, J. M. Di Nicola, P. Di Nicola, S. N. Dixit, T. Doeppner, E. G. Dzenitis, G. Erber, J. Folta, G. Grim, S. Lenn, A. Hamza, S. W. Hann, J. Heebner, M. Henesian, M. Hermann, D. G. Hicks, W. W. Hsing, N. Izumi, K. Jancaitis, O. S. Jones, D. Kalantar, S. F. Khan, R. Kirkwook, G. A. Kyrala, K. LaFortune, O. L. Landen, L. Lain, D. Larson, S. Le Pape, T. Ma, A. G. MacPhee, P. A. Michel, P. Miller, M. Montincelli, A. S. Moore, A. Nikroo, M. Nostrand, R. E. Olson, A. Pak, H. S. Park, M. B. Schneider, M. Shaw, V. A. Smalyuk, D. J. Strozzi, T. Suratwala, L. J. Suter, R. Tommasini, R. P. J. Town, B. Van Wonterghem, P. Wegner, K. Widmann, C. Widmayer, H. Wilkens, E. A. Williams, M. J. Edwards, B. A. Remington, B. J. MacGowan, J. D. Kikenny, J. D. Lindl, L. J. Atherton, S. H. Batha, and E. Moses, Phys. Plasmas **20**, 056314 (2013).

[18]P. Michel, L. Divol, R. P. J. Town, M. D. Rosen, D. A. Callahan, N. B. Meezan, M. B. Schneider, G. A. Kyrala, J. D. Moody, E. L. Dewald, K. Widman, E. Bond, J. L. Kline, C. A. Thomas, S. Dixit, E. A. Williams, D. E. Hinkel, R. L. Berger, O. L. Landen, M. J. Edwards, B. J. MacGowan, J. D. Lindl, C. Haynam, L. J. Suter, S. H. Glenzer, and E. Moses, Phys. Rev. E **83**, 046409 (2011).

[19]K. Lan, J. Liu, D. Lai, W. Zheng, and X. He, Phys. Plasmas **21**, 010704 (2014).

[20]J. D. Lindl, O. Landen, J. Edwards, Ed. Moses, and NIC Team, Phys. Plasmas **21**, 020501 (2014).

[21]C. K. Li, F. H. Séguin, J. A. Frenje, M. J. Rosenberg, H. G. Rinderknecht, A. B. Zylstra, R. D. Petrasso, P. A. Amendt, O. L. Landen, A. J. Mackinnon, R. P. J. Town, S. C. Wilks, R. Betti, D. D. Meyerhofer, J. M. Soures, J. Hund, J. D. Kilkenny, and A. Nikroo, Phys. Rev. Lett. **108**, 025001 (2012).

[22]Z. Fan, M. Chen, Z. Dai, H.-b. Cai, S.-p. Zhu, W. Y. Zhang, and X. T. He, e-print arXiv:1303.1252[physics.plasm-ph]; X. T. He, in Proceedings of the Plenary Presentation at International Farming System Association 8, September 9–13, 2013, Nara, Japan.

[23]T. Feng, D. Lai, and Y. Xu, Chin. J. Comput. Phys. **16**, 199 (1999); Y. Li, W. Y. Huo, and K. Lan, Phys. Plasmas **18**, 022701 (2011).

[24]W. Y. Huo, G. Ren, K. Lan, X. Li, C. Wu, Y. Li, C. Zhai, X. Qiao, X. Meng, D. Lai, W. Zheng, P. Gu, W. Pei, S. Li, R. Yi, T. Song, X. Jiang, D. Yang, S. Jiang, and Y. Ding, Phys. Plasmas **17**, 123114 (2010).

[25]E. L. Dewald, L. J. Suter, O. L. Landen, J. P. Holder, J. Schein, F. D. Lee, K. M. Campbell, F. A. Weber, D. G. Pellinen, M. B. Schneider, J. R. Celeste, J. W. Mcdonald, J. M. Foster, C. J. Niemann, A. Mackinnon, S. H. Glenzer, B. K. Young, C. A. Haynam, M. J. Shaw, R. E. Turner, D. Froula, R. L. Kauffman, B. R. Thomas, L. J. Atherton, R. E. Bonanno, S. N. Dixit, D. C. Eder, G. Holtmeier, D. H. Kalantar, A. E. Koniges, B. J. Macgowan, K. R. Manes, D. H. Munro, J. R. Murray, T. G. Parham, K. Piston, B. M. Van Wonterghem, R. J. Wallace, P. J. Wegner, P. K. Whitman, B. A. Hammel, and E. I. Moses, Phys. Rev. Lett. **95**, 215004 (2005).

[26]J. W. Mcdonald, L. J. Suter, O. L. Landen, J. M. Foster, J. R. Celeste, J. P. Holder, E. L. Dewald, M. B. Schneider, D. E. Hinkel, R. L. Kauffman, L. J. Atherton, R. E. Bonanno, S. N. Dixit, D. C. Eder, C. A. Haynam, D. H. Kalantar, A. E. Koniges, F. D. Lee, B. J. Macgowan, K. R. Manes, D. H. Munro, J. R. Murray, M. J. Shaw, R. M. Stevenson, T. G. Parham, B. M. Van Wonterghem, R. J. Wallace, P. J. Wegner, P. K. Whitman, B. K. Young, B. A. Hammel, and E. I. Moses, Phys. Plasmas **13**, 032703(2006).

[27]K. Lan, P. Gu, G. Ren, X. Li, C. Wu, W. Huo, D. Lai, and X. He, Laser Part. Beams **28**, 421 (2010).

[28]J. Dawson, P. Kaw, and B. Green, Phys. Fluids **12**, 875 (1969).

PHYSICS OF PLASMAS 22, 122707 (2015)

Laser imprint reduction for the critical-density foam buffered target driven by a relatively strong foot pulse at early stage of laser implosions

J. W. Li,[1,2,a)] W. Kang,[1] X. T. He,[1,2] J. H. Li,[2] and W. D. Zheng[2]
[1]*Key Lab of High Energy Density Physics Simulation, Center for Applied Physics and Technology, Peking University, Beijing 100871, China*
[2]*Institute of Applied Physics and Computational Mathematics, P. O. Box 8009, Beijing 100094, China*

(Received 9 July 2015; accepted 2 December 2015; published online 18 December 2015)

In order to reduce the effect of laser imprint in direct-drive ignition scheme a low-density foam buffered target has been proposed. This target is driven by a laser pulse with a low-intensity foot at the early stage of implosion, which heats the foam and elongates the thermal conduction zone between the laser absorption region and ablation front, increasing the thermal smoothing effect. In this paper, a relatively strong foot pulse is adopted to irradiate the critical-density foam buffered target. The stronger foot, near $1 \times 10^{14}\,W/cm^2$, is able to drive a radiative shock in the low-density foam, which helps smooth the shock and further reduce the effect of laser imprint. The radiative shock also forms a double ablation front structure between the two ablation fronts to further stabilize the hydrodynamics, achieving the similar results to a target with a high-Z dopant in the ablator. 2D analysis shows that for the critical-density foam buffered target irradiated by the strong foot pulse, the laser imprint can be reduced due to the radiative shock in the foam and an increased thermal smoothing effect. It seems viable for the critical-density foam buffered target to be driven by a relatively strong foot pulse with the goal of reducing the laser imprint and achieving better implosion symmetry in the direct-drive laser fusion. © *2015 AIP Publishing LLC.*
[http://dx.doi.org/10.1063/1.4938037]

I. INTRODUCTION

For direct-drive laser fusion[1–3] in inertial confinement fusion (ICF), very high levels of irradiation uniformity are required to achieve ignition and high gain. Even with the optical smoothing schemes that are available, such as the use of random phase plates, phase zone plates, kinoform phase plates, and smoothing by spectral dispersion, there still remains an issue of laser nonuniformity at very early times. This has been called the "laser imprint" problem.[4] The laser imprint or the single beam nonuniformity, possibly the most important source of nonuniformity, can seed Rayleigh-Taylor (RT) instability in the shell of an imploding capsule, resulting in shell breakup and disruption of the implosion. Therefore, it is of crucial importance to control laser imprint at the early stage for the successful implosion of laser-driven ICF targets.

To reduce the laser imprint, it has been suggested to cover the target shell with a low-density foam layer, which is also called foam-buffered target.[5–8] The foam-buffered target can help mitigate the initial laser imprint and prevent a cryogenic layer from melting by thermal insulation,[9] and it is also important that the foam can decrease the absolute levels of stimulated Raman scattering (SRS) and stimulated Brillouin scattering (SBS).[10] For the critical or overcritical density foam, the laser energy is deposited nearby the critical density, and a thermal conduction layer forms between the critical surface and the ablation front in which thermal smoothing may reduce the initial seed amplitude. The foam layer, after being ionized by laser-driven shock,[1,8,9,11] external x-ray pulse,[6] or the x-ray flash produced by the high-Z overcoat[6,10,12–15] can become a uniform plasma to elongate the thermal conduction zone by stretching the space between the target and the absorption region. Correspondingly, the laser inhomogeneity is better smoothed according to the "cloudy-day" model.[16] In the foam-buffered target with shock-preheated foam layer,[1,11] it has been concluded that the properties of the shocks in the foam layer have important effects on the implosion stability.[9] Traditionally, in laser-driven fusion, a low-intensity foot pulse is used to drive a weak shock that first compresses the fuel in order to prevent an entropy increase and achieve a high compression.

However, if the low-density foam-buffered target is irradiated by a high-intensity foot pulse at the early stage, the strong shock driven by the pulse may become radiating when it propagates through the foam,[17–19] and the radiative properties of the shock will help stabilize the hydrodynamics to further reduce the laser imprint. For a radiative shock, a radiation precursor from the shocked material can preheat the foam ahead of the shock front, generating a more uniform plasma. This upstream plasma smooths the rippled shock and decreases the susceptibility of the foam/ablator interface to RT instability. In this paper, a critical-density foam-buffered target driven by a relatively strong foot pulse has been proposed, which not only maximizes the thermal conduction zone, but also makes it easier for the laser-driven shock to become radiating with a purpose of reducing the laser imprint and improving the implosion symmetry.

The paper is organized as follows. In Sec. II, we overview the "cloudy-day" model to explain why the low-density

[a)]li_jiwei@iapcm.ac.cn

foam layer can reduce the laser imprint; in Sec. III the properties of the radiative shock in the critical-density foam driven by the strong pulse are analyzed; in Sec. IV the double ablation front (DAF) structure and formation of a plateau between the fronts as a result of the radiative shock are studied; in Sec. V 2D analysis of the foam-buffered target is performed, showing that the laser imprint is effectively reduced for the critical-density foam-buffered target driven by a relatively strong pulse.

II. OVERVIEW OF "CLOUDY-DAY" MODEL

When the laser irradiates the target, it is absorbed typically at a critical surface, and a thermal conduction layer forms between the critical surface and the ablation front. The thermal conduction can effectively smooth the laser-induced perturbations in a target over a larger thermal conduction distance D_{ac}. To illustrate the thermal smoothing effect, the "cloudy-day" model[16] has been proposed, which assumes that all perturbed quantities $\tilde{\Phi}$ satisfy $\nabla^2\tilde{\Phi} = 0$, and it predicts that the perturbations decay exponentially away from the critical surface. Thus, the perturbation with wave number k is reduced by a factor $e^{-kD_{ac}}$ from the critical surface to the ablation front.

It is realized that a larger thermal conduction distance D_{ac} is beneficial for reducing the laser imprint. The foam with a much lower density has a higher nonlinear conductivity,[13,20] which can maximize thermal conduction distance[6,21] for a better smoothing effect, so a critical-density foam is proposed to coat the target.

III. PROPERTIES OF RADIATIVE SHOCK IN THE FOAM

The typical irradiance of the laser foot incident onto foam-buffered targets in direct-drive fusion is $10^{13}\,W/cm^2$ to irradiate the foam-buffered target in direct-drive laser fusion. After the laser energy is deposited into the low-density foam, the laser-driven shock has a velocity less than 100 km/s, and its radiative properties are negligible, and the foam ahead of the shock front suffers no preheating by any x-ray precursor. As the pulse intensity increases to $5\text{–}10 \times 10^{13}\,W/cm^2$, the laser-driven shock can reach a velocity greater than 150 km/s so that the radiation from the post-shock plays a role in the dynamics, and the shock becomes a radiative shock. In the radiative shock, an x-ray precursor from the post-shock preheats the foam ahead of the shock, which can improve the thermal smoothing effect compared with the case without an x-ray precursor ahead of the shock front.[7] In our simulation, a 1.0 g/cc CH planar target coated with a 30 mg/cc foam layer (critical density for the 3ω laser) is irradiated by a flat-topped laser pulse, and the thicknesses of the CH and the foam layer are 117 μm and 200 μm, respectively. In the simulation, the targets were irradiated with laser pulses of 2×10^{13} and $10^{14}\,W/cm^2$. We can see that the shock driven by the strong pulse with a velocity of 180 km/s has become radiating as shown in Fig. 1. The radiation from the post-shock supersonically propagates through the foam, and preheats the foam ahead of the shock front with the electron temperature 30–90 eV without any density perturbation, so the foam's scale lengths, absolute temperature, and uniformity ahead of the laser-induced shock increase to smooth the rippled shock and reduce the laser imprint.

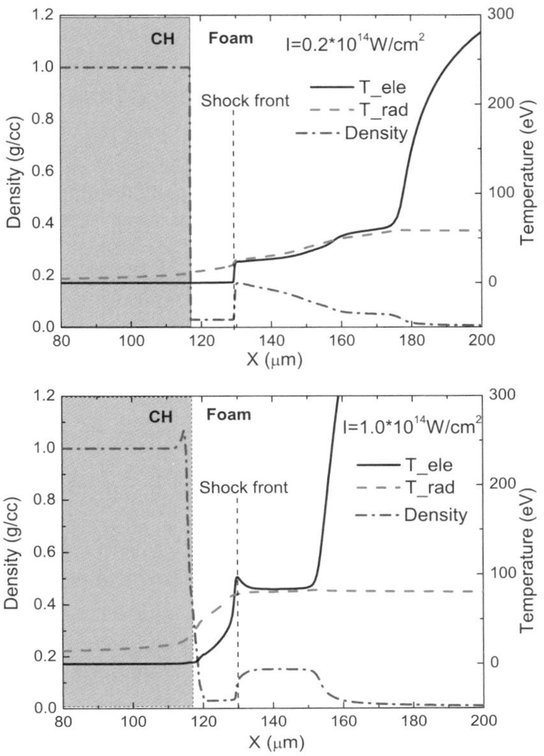

FIG. 1. Radiative shock formed in the foam for the critical-density foam buffered target driven by a weak pulse $I = 2.0 \times 10^{13}\,W/cm^2$ (upper frame) and a relatively strong pulse $I = 1.0 \times 10^{14}\,W/cm^2$ (bottom frame). The electron (black solid curve) and radiation (red dashed curve) temperature and density (blue dash-dotted curve).

Potentially, the CH/foam interface is classically RT unstable because of the opposing pressure and density gradients at the interface.[1] However, if a relatively strong pulse is adopted to drive the foam-buffered target, the RT instability at the interface can be partially suppressed. As shown in Fig. 1, the radiative heat wave from the radiative shock has reached the CH/foam interface, and the wave becomes subsonic to ablate the CH layer off towards the foam with a longer density scale length at the interface compared with the non-radiative shock case, then the potential RT instability at the interface can be stabilized by both finite density gradient and ablative stabilization due to the radiation preheating.[22–24]

The radiative properties of the shock depend on the foam's density. For a lower density foam, it is much more likely for a strong shock to become radiating.[19,22] The shock launched by the laser will travel faster through a lower density foam, generating a greater flux of radiation and more effectively preheating the foam upstream of the shock. Therefore, it is proposed that the target should be coated with a foam layer of critical density and the foam-buffered target should be driven by a relatively strong foot pulse in laser-driven fusion, which can both maximize the thermal conduction distance as has been discussed in Sec. II and produce a stronger radiative shock to further reduce the laser imprint.

IV. FORMATION OF THE PLATEAU BETWEEN THE TWO ABLATION FRONTS

After the shock becomes radiating, a radiative ablation front is formed ahead of the electron ablation front, and the two energy transport mechanisms—the electron heat flux and the radiative energy flux—drive a DAF structure[25–27] in Fig. 1. After the shock enters the dense CH layer, it is not radiative any more, as shown in Fig. 2. Meanwhile the high temperature of the shocked foam continues ablating the CH layer to form another radiative heat wave ahead of the electron ablation front, and the DAF structure survives for a time.

As the radiation ablates the CH layer, there also appears a piston-like plateau with a density of 0.2 g/cc and 60–70 eV electron and radiation temperature in equilibrium between the two ablation fronts, as shown in Fig. 2. At the time interval δt, the accumulated mass of the plateau is given by $\delta m = (\dot{m}_{ar} - \dot{m}_{ae})\delta t$, and the expanding length of the plateau by $\delta R = \frac{1}{\rho_a}\left(\dot{m}_{ar} - \frac{\rho_a}{\rho}\dot{m}_{ae}\right)\delta t$, where ρ_a is the density at the radiation ablation front, ρ the density of the plateau between the two ablation fronts, and $\dot{m}_{ar}$ and $\dot{m}_{ae}$ the mass ablation rate of radiation and laser, respectively. The plateau density ρ is always much smaller than the density ρ_a, and the mass increases faster than the distance between the two ablation fronts; therefore, the density continues increasing to form a higher density plateau. The formation of the plateau can stabilize the hydrodynamics to further reduce the laser imprint.[25–29]

V. MITIGATION OF LASER IMPRINT FOR THE FOAM-BUFFERED TARGET

To study the effect of foam on the laser imprint for the critical-density foam buffered target driven by a high-intensity pulse, we also give a 2D analysis of a spherical CH target with the radiation hydrodynamic code LARED-S,[30] which is multi-dimensional, massively parallel, and Eulerian mesh based, and 2000 meshes with minimum grid size of 0.05 μm. The code includes laser ray-tracing, multi-group radiation diffusion (20 groups), radiation hydrodynamics, electron and ion heat conduction, atomic physics, nuclear reaction, α particle transport, and quotidian equation of state (QEOS).

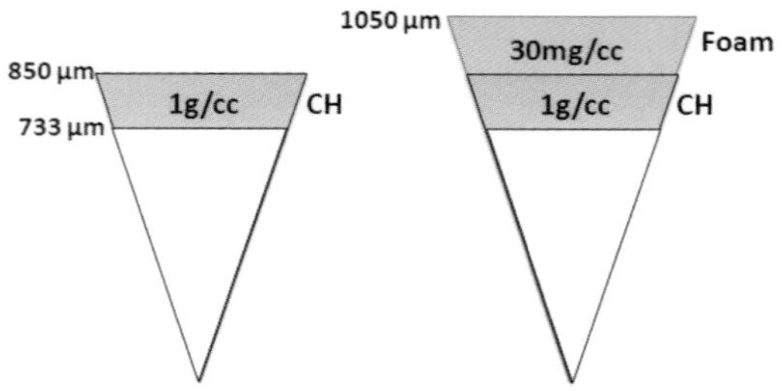

FIG. 3. Schematic of the bare CH target and the foam-buffered CH target. The foam layer has a density of 30 mg/cc, the critical-density for the 3ω laser.

When the nonuniform laser irradiates a smooth target surface and deposits its energy there, spatial nonuniformities at the critical surface induced by the laser cause modulation in the ablation pressure and, in turn, generate hydrodynamic perturbations in the target. When the laser-driven rippled shock passes across the CH layer, the rear surface of the layer will become unstable; therefore, the foam-induced smoothing can be studied through the shock breakout from the shell and the distortion of the shell's rear surface.[20,31] For the direct-drive laser implosion, the low-order modes or the long wavelengths of laser nonuniformity, mainly from the single-beam nonuniformity, are always dangerous to the implosion symmetry, so here we numerically evaluate the target performance under the influence of applied Legendre laser-illumination perturbations of low-order modes $\ell \leq 20$. In the simulation, both the bare and foam-buffered target in Fig. 3 are analyzed.

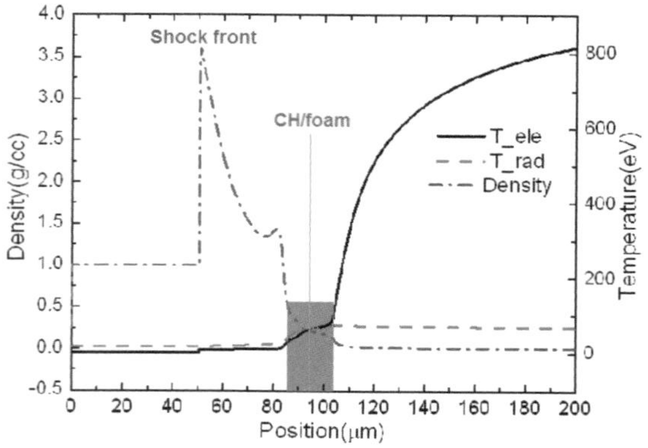

FIG. 2. A piston-like plateau between the two ablation fronts at time $t \sim 2.4$ ns for the critical-density foam buffered target driven by a strong pulse $I = 1.0 \times 10^{14}$ W/cm^2. The electron (black solid curve) and radiation (red dashed curve) temperature, and the plateau (gray region) is of 0.2 g/cc density and 60–70 eV electron and radiation temperature in equilibrium.

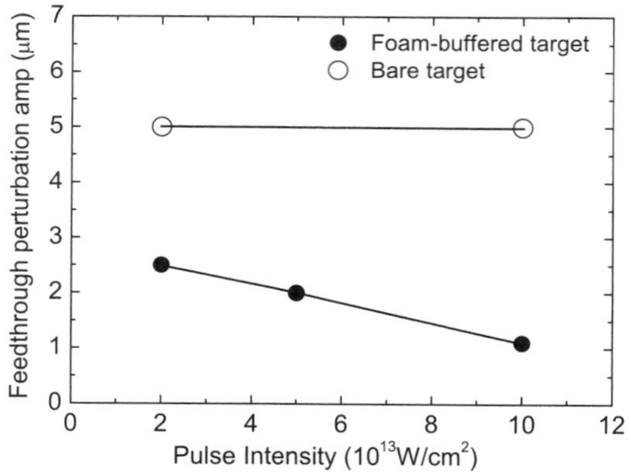

FIG. 4. Perturbation of the rear surface of CH layer brought by the rippled laser-driven shock for the bare and the foam-buffered target driven by foot pulse of intensity $I = 0.2$, 0.5, 1.0×10^{14} W/cm^2 with perturbation of mode $\ell = 8$, respectively.

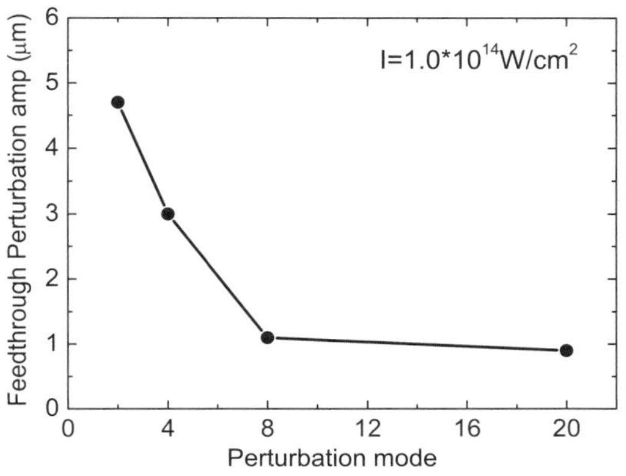

FIG. 5. Perturbation of the rear surface of CH layer brought by the rippled shock at shock breakout time for critical-density foam-buffered target, driven by the relatively strong pulse $I = 1.0 \times 10^{14}\,\mathrm{W/cm^2}$ with perturbation of modes $\ell \leq 20$.

First, we study the laser imprint for the bare CH spherical target, which has a radius of 850 μm with the CH layer 117 μm in the left frame in Fig. 3, and the target is driven by a weak-intensity pulse of $I = 2.0 \times 10^{13}\,\mathrm{W/cm^2}$ with a 20% perturbation (mode $\ell = 8$). At the time of the laser-driven rippled shock breakout, the perturbed rear surface has an amplitude of 5 μm as shown in Fig. 4, mainly feedthrough by the nonuniform laser.

Comparably, we also study the laser imprint for the target coated by a 200 μm critical-density foam, as in the right frame of Fig. 3. The foam-buffered target is driven by the laser with a range of intensities from 0.2 to $1.0 \times 10^{14}\,\mathrm{W/cm^2}$. For the weak pulse $I = 0.2 \times 10^{14}\,\mathrm{W/cm^2}$, at the time of the shock breakout, the shell's rear surface suffers an amplitude of 2.5 μm as shown in Fig. 4, only half of that for the bare target. The laser-driven shock propagating through the low-density foam is non-radiating, and the foam reduces the laser imprint by increasing the thermal conduction zone. As the laser intensity increases so that the laser-driven shock becomes radiating and preheats the foam ahead of the shock front, it can be seen that the laser imprint is more effectively reduced with a perturbation amplitude of only 1.1 μm as in Fig. 4 for the pulse intensity of $I = 1.0 \times 10^{14}\,\mathrm{W/cm^2}$.

It has been studied that the relatively strong pulse can increase the ablation velocity to enhance the ablative stabilization of the RT growth at the ablation front,[32] but there seems no reduction of the laser imprint if a strong pulse $I = 1.0 \times 10^{14}\,\mathrm{W/cm^2}$ is used to irradiate the bare target as in Fig. 4. In our simulation, the pulse intensity is constant, and the ablation front travels at constant velocity without RT growth during the time, so the laser imprint suffers no amplification by the RT instability at the ablation front.

As has been indicated by "Cloudy-day" model,[16] the perturbation of lower-order mode (long wavelength) is less smoothed than the higher-order one. For the same foam-buffered target, the perturbations imposed on the target of different modes perform differently in the laser imprint reduction. In our simulation of the spherical target, perturbations decay as $e^{-\ell d/R}$ where the foam's depth $d = 200$ μm and the initial radius of the foam-buffered target $R = 1050$ μm, which indicates that the perturbations with modes $\ell < 5$ (perturbation wavelength larger than the foam length) are less reduced by the thermal smoothing effect. Here we study the foam-buffered target driven by the pulse of $I = 1.0 \times 10^{14}\,\mathrm{W/cm^2}$ with perturbations in the range of $\ell = 2$–20. When the laser-driven rippled shock passes across the CH shell, for lower-order modes $\ell = 2$ and $\ell = 4$, the shell's inner surface suffers a much larger amplitude of 4.7 and 3 μm, respectively. Further, simulations of a higher order mode, $\ell = 20$, result in a feedthrough amplitude of 0.9 μm, smaller than the 1.1 μm for the $\ell = 8$ mode, as shown in Fig. 5.

Interestingly, we can also see that the phase of the laser-driven shock with the perturbation of mode $\ell = 20$ is reversed when it passes across the inner surface of the CH layer as shown in the left frame of Fig. 6 because the shock's traveling distance is larger than the perturbation wavelength.[33–35] If we reduce the foam's length to 100 μm so that the shock's traveling distance is smaller than the perturbation wavelength, the laser-driven shock will not be reversed, as show in the right frame of Fig. 6. However, the reduction of the laser imprint is not so effective with the perturbation amplitude of 2 μm because of the shorter foam's thickness, consistent with the "Cloudy-day" model.

In direct-drive laser fusion, the low-intensity foot pulse is used because it can prevent an entropy increase and achieve a high areal density of the fuel for a high gain. When a relatively strong foot is used, as proposed here, a stronger shock compresses the fuel layer and the entropy increase cannot be limited to a low value. For the foam-buffered target

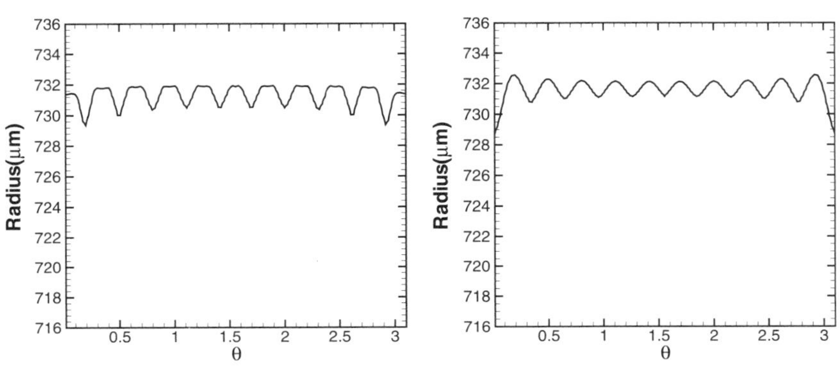

FIG. 6. Perturbation of the rear surface of CH layer brought by the rippled shock breakout for the foam-buffered target with 200 μm (left frame) and 100 μm (right frame) foam, driven by a relatively strong pulse $I = 1.0 \times 10^{14}\,\mathrm{W/cm^2}$ with perturbation of mode $\ell = 20$.

driven by the strong pulse $I = 1.0 \times 10^{14}\,\mathrm{W/cm^2}$, it is estimated that the adiabat of the CH layer is 50% higher than that for the target driven by the weak pulse $I = 0.2 \times 10^{14}\,\mathrm{W/cm^2}$ in our simulation, and the effect of the yield needs to be further studied for the actual DT target driven by the foot and the main pulse. To achieve a high gain fusion for the foam-buffered target, the adiabat shaping[36–38] with a strong foot pulse may be adopted to both reduce the laser imprint effectively and compress the fuel substantially.

VI. CONCLUSIONS

In conclusion, we analyzed the reduction of the laser imprint for the foam-buffered target driven by a high-intensity pulse and suggested a critical-density foam layer to coat the target because it both maximized the thermal conduction distance and made it easier for a strong shock to become radiating. The radiative shock driven by the high-intensity foot pulse produced an x-ray precursor to preheat the foam ahead of the shock front, which smoothed the rippled shock and controlled the potential RT instability occurring at foam/ablator interface to further reduce the laser imprint. There also appeared the DAF structure and a piston-like plateau between the two fronts to further stabilize the hydrodynamics.

2D analysis indicated that the foam can effectively reduce the laser imprint of a smaller perturbation amplitude of the shell's inner surface, which was induced by the laser-driven rippled shock breakout. The laser imprint was further reduced as the pulse intensity increases and the laser-driven shock becomes radiating. It was also seen that the higher-mode perturbations are effectively reduced. It seems viable that the critical-density foam buffered target can be driven by an adiabat shaping laser with a high-intensity foot pulse at the early stage, which helps reduce the laser imprint and stabilize the hydrodynamics while compresses the fuel substantially to produce a high gain.

ACKNOWLEDGMENTS

We would like to thank the referee for useful comments and suggestions that made the paper more clear. The work was supported by the National Natural Science Foundation of China under Grant Nos. 11175027, 11275028, 11371065, and 11475032.

[1]S. E. Bodner, D. G. Colombant, J. H. Gardner, R. H. Lehmberg, S. P. Obenschain, L. Phillips, A. J. Schmitt, J. D. Sethian, R. L. McCrory, W. Seka, C. P. Verdon, J. P. Knauer, B. B. Afeyan, and H. T. Powell, Phys. Plasmas **5**, 1901 (1998).
[2]S. E. Bodner, D. G. Colombant, A. J. Schmitt, and M. Klapisch, Phys. Plasmas **7**, 2298 (2000).
[3]P. W. McKenty, V. N. Goncharov, R. P. J. Town, S. Skupsky, R. Betti, and R. L. McCrory, Phys. Plasmas **8**, 2315 (2001).
[4]D. H. Kalantar, M. H. Key, L. B. DaSilva., S. G. Glendinning, J. P. Knauer, B. A. Remington, F. Weber, and S. V. Weber, Phys. Rev. Lett. **76**, 3574 (1996).
[5]T. Afshar-rad, M. Desselberger, M. Dunne, J. Edwards, J. M. Foster, D. Hoarty, M. W. Jones, S. J. Rose, P. A. Rosen, R. Taylor, and O. Willi, Phys. Rev. Lett. **73**, 74 (1994).
[6]M. Desselberger, M. W. Jones, J. Edwards, M. Dunne, and O. Willi, Phys. Rev. Lett. **74**, 2961 (1995).
[7]M. Dunne, M. Borghesi, A. Iwase, M. W. Jones, R. Taylor, O. Willi, R. Gibson, S. R. Goldman, J. Mack, and R. G. Watt, Phys. Rev. Lett. **75**, 3858 (1995).
[8]Andrew J. Schmitt, A. L. Velikovich, J. H. Gardner, C. Pawley, S. P. Obenschain, Y. Aglitskiy, and Y. Chan, Phys. Plasmas **8**, 2287 (2001).
[9]T. Takeda, K. Mima, T. Norimatsu, H. Nagatomo, and A. Nishiguchi, Phys. Plasmas **10**, 2608 (2003).
[10]O. Willi, L. Barringer, A. Bell, M. Borghesi, J. Davies, R. Gaillard, A. Iwase, A. MacKinnon, G. Malka, C. Meyer, S. Nuruzzaman, R. Taylor, C. Vickers, D. Hoarty, P. Gobby, R. Johnson, R. G. Watt, N. Blanchot, B. Canaud, H. Croso, B. Meyer, J. L. Miquel, C. Reverdin, A. Pukhov, and J. Meyer-ter-Vehn, Nucl. Fusion **40**, 537 (2000).
[11]J. H. Gardner, A. J. Schmitt, J. P. Dahlburg, C. J. Pawley, S. E. Bodner, S. P. Obenschain, V. Serlin, and Y. Aglitskiy, Phys. Plasmas **5**, 1935 (1998).
[12]R. G. Watt, D. C. Wilson, R. E. Chrien, R. V. Hollis, P. L. Gobby, R. J. Mason, R. A. Kopp, R. A. Lerche, D. H. Kalantar, B. MacGowan, M. B. Nelson, T. Phillips, P. W. McKenty, and O. Willi, Phys. Plasmas **4**, 1379 (1997).
[13]R. J. Mason, R. A. Kopp, H. X. Vu, D. C. Wilson, S. R. Goldman, R. G. Watt, M. Dunne, and O. Wiili, Phys. Plasmas **5**, 211 (1998).
[14]R. G. Watt, J. Duke, C. J. Fontes, P. L. Gobby, R. V. Hollis, R. A. Kopp, R. J. Mason, D. C. Wilson, C. P. Verdon, T. R. Boehly, J. P. Knauer, D. D. Meyerhofer, V. Smalyuk, R. P. J. Town, A. Iwase, and O. Willi, Phys. Rev. Lett. **81**, 4644 (1998).
[15]A. N. Mostovych, D. G. Colombant, M. Karasik, J. P. Knauer, A. J. Schmitt, and J. L. Weaver, Phys. Rev. Lett. **100**, 075002 (2008).
[16]M. H. Emery, J. H. Orens, J. H. Gardner, and J. P. Boris, Phys. Rev. Lett. **48**, 253 (1982).
[17]Y. B. Zel'dovich and Y. P. Raizer, *Physics of Shock Waves and High-Temperature Hydrodynamics Phenomena* (Academic Press, New York/London, 1966).
[18]D. Mihalas and B. W. Mihalas, *Foundations of Radiation Hydrodynamics* (Academic Press, New York, 1984).
[19]R. P. Drake, Phys. Plasmas **14**, 043301 (2007).
[20]D. Batani, W. Nazarov, T. Hall, T. Löwer, M. Koenig, B. Faral, A. Benuzzi-Mounaix, and N. Grandjouan, Phys. Rev. E **62**, 8573 (2000).
[21]R. Betti, V. N. Goncharov, R. L. McCrory, P. Sorotokin, and C. P. Verdon, Phys. Plasmas **3**, 2122 (1996).
[22]J. W. Li, W. B. Pei, X. T. He, J. H. Li, W. D. Zheng, S. P. Zhu, and W. Kang, Phys. Plasmas **20**, 082707 (2013).
[23]C. M. Huntington, C. C. Kuranz, R. P. Drake, A. R. Miles, S. T. Prisbrey, H.-S. Park, H. F. Robey, and B. A. Remington, Phys. Plasmas **18**, 112703 (2011).
[24]D. G. Colombant, S. E. Bodner, A. J. Schmitt, M. Klapisch, J. H. Gardner, Y. Aglitskiy, A. V. Deniz, S. P. Obenschain, C. J. Pawley, V. Serlin, and J. L. Weaver, Phys. Plasmas **7**, 2046 (2000).
[25]S. Fujioka, A. Sunahara, K. Nishihara, N. Ohnishi, T. Johzaki, H. Shiraga, K. Shigemori, M. Nakai, T. Ikegawa, M. Murakami, K. Nagai, T. Norimatsu, H. Azechi, and T. Yamanaka, Phys. Rev. Lett. **92**, 195001 (2004).
[26]J. Sanz, R. Betti, V. A. Smalyuk, M. Olazabal-Loume, V. Drean, V. Tikhonchuk, X. Ribeyre, and J. Feugeas, Phys. Plasmas **16**, 082704 (2009).
[27]V. Drean, M. Olazabal-Loumé, J. Sanz, and V. T. Tikhonchuk, Phys. Plasmas **17**, 122701 (2010).
[28]C. Yañez, J. Sanz, M. Olazabal-Loumé, and L. F. Ibaéz, Phys. Plasmas **18**, 052701 (2011).
[29]C. Yañez, J. Sanz, and M. Olazabal-Loumé, Phys. Plasmas **19**, 062705 (2012).
[30]Z. F. Fan, X. T. He, J. Liu, G. L. Ren, B. Liu, J. F. Wu, L. F. Wang, and W. H. Ye, Phys. Plasmas **21**, 100705 (2014).
[31]R. Ishizaki, K. Nishihara, H. Sakagami, and Y. Ueshima, Phys. Rev. E **53**, 5592 (1996).
[32]T. J. B. Collins and S. Skupsky, Phys. Plasmas **9**, 275 (2002).
[33]R. Ishizaki and K. Nishihara, Phys. Rev. Lett. **78**, 1920 (1997).
[34]V. N. Goncharov, S. Skupsky, T. R. Boehly, J. P. Knauer, P. McKenty, V. A. Smalyuk, R. P. J. Town, O. V. Gotchev, R. Betti, and D. D. Meyerhofer, Phys. Plasmas **7**, 2062 (2000).
[35]J. O. Kane, H. F. Robey, B. A. Remington, R. P. Drake, J. Knauer, D. D. Ryutov, H. Louis, R. Teyssier, O. Hurricane, D. Arnett, R. Rosner, and A. Calder, Phys. Rev. E **63**, 055401(R) (2001).
[36]J. P. Knauer, K. Anderson, R. Betti, T. J. B. Collins, V. N. Goncharov, P. W. McKenty, D. D. Meyerhofer, P. B. Radha, S. P. Regan, T. C. Sangster, V. A. Smalyuk, J. A. Frenje, C. K. Li, R. D. Petrasso, and F. H. Séguin, Phys. Plasmas **12**, 056306 (2005).

[37]T. J. B. Collins, J. A. Marozas, R. Betti, D. R. Harding, P. W. McKenty, P. B. Radha, S. Skupsky, V. N. Goncharov, J. P. Knauer, and R. L. McCrory, Phys. Plasmas **14**, 056308 (2007).

[38]V. N. Goncharov, T. C. Sangster, P. B. Radha, R. Betti, T. R. Boehly, T. J. B. Collins, R. S. Craxton, J. A. Delettrez, R. Epstein, V. Yu. Glebov, S. X. Hu, I. V. Igumenshchev, J. P. Knauer, S. J. Loucks, J. A. Marozas, F. J. Marshall, R. L. McCrory, P. W. McKenty, D. D. Meyerhofer, S. P. Regan, W. Seka, S. Skupsky, V. A. Smalyuk, J. M. Soures, C. Stoeckl, D. Shvarts, J. A. Frenje, R. D. Petrasso, C. K. Li, F. Séguin, W. Manheimer, and D. G. Colombant, Phys. Plasmas **15**, 056310 (2008).

PHYSICS OF PLASMAS 23, 082706 (2016)

A hybrid-drive nonisobaric-ignition scheme for inertial confinement fusion

X. T. He,[1,2,3,4,a)] J. W. Li,[1,2,3] Z. F. Fan,[1] L. F. Wang,[1,2,3] J. Liu,[1,2,3] K. Lan,[1,2,3] J. F. Wu,[1] and W. H. Ye[1,2,3]
[1]*Institute of Applied Physics and Computational Mathematics, P. O. Box 8009, Beijing 100094, China*
[2]*Center for Applied Physics and Technology, HEDPS, Peking University, Beijing 100871, China*
[3]*IFSA Collaborative Innovation Center of MoE, Shanghai Jiao-Tong University, Shanghai 200240, China*
[4]*Institute of Fusion Theory and Simulation, Zhejiang University, Hangzhou 310027, China*

(Received 24 February 2016; accepted 1 August 2016; published online 18 August 2016)

A new hybrid-drive (HD) nonisobaric ignition scheme of inertial confinement fusion (ICF) is proposed, in which a HD pressure to drive implosion dynamics increases via increasing density rather than temperature in the conventional indirect drive (ID) and direct drive (DD) approaches. In this HD (combination of ID and DD) scheme, an assembled target of a spherical hohlraum and a layered deuterium-tritium capsule inside is used. The ID lasers first drive the shock to perform a spherical symmetry implosion and produce a large-scale corona plasma. Then, the DD lasers, whose critical surface in ID corona plasma is far from the radiation ablation front, drive a supersonic electron thermal wave, which slows down to a high-pressure electron compression wave, like a snowplow, piling up the corona plasma into high density and forming a HD pressurized plateau with a large width. The HD pressure is several times the conventional ID and DD ablation pressure and launches an enhanced precursor shock and a continuous compression wave, which give rise to the HD capsule implosion dynamics in a large implosion velocity. The hydrodynamic instabilities at imploding capsule interfaces are suppressed, and the continuous HD compression wave provides main *pdV* work large enough to hotspot, resulting in the HD nonisobaric ignition. The ignition condition and target design based on this scheme are given theoretically and by numerical simulations. It shows that the novel scheme can significantly suppress implosion asymmetry and hydrodynamic instabilities of current isobaric hotspot ignition design, and a high-gain ICF is promising. *Published by AIP Publishing.*
[http://dx.doi.org/10.1063/1.4960973]

I. INTRODUCTION

In a central hotspot-ignition scheme of inertial confinement fusion (ICF),[1,2] a layered capsule with a frozen deuterium-tritium (DT) fuel shell and an outer shell of ablator is imploded; the central hotspot is compressed and heated to reach ignition condition. A thermonuclear burn wave from the ignited hotspot propagates outward towards the high compressed fuel shell, resulting in the self-sustaining burn and the fusion energy gain. Two main implosion approaches, direct drive (DD)[3,4] and indirect drive (ID),[5] have been proposed. In the DD approach, lasers directly focus on the layered capsule ablator surface to drive implosion dynamics and hotspot ignition. The shock ignition[4] is a DD scheme that is extensively investigated so far. In this scheme, the implosion lasers first drive the shock to compress main fuel and generate a central hotspot where the shock experiences reflections at the interface between hotspot and the compressed fuel; the ignition lasers with a short pulse (<1 ns) then provide an ignitor shock to collide the rebounded shock from hotspot to perform hotspot ignition.[4,6] In the ID approach, the layered capsule is placed inside a high-Z hohlraum and lasers irradiate on an inner wall of the hohlraum to generate radiation (soft x-ray) that ablates and heats the capsule ablator surface, where a high temperature and pressure plasma are formed, resulting in the capsule implosion dynamics to highly compress the fuel and heat the hotspot performing ignition. So far, the ID approach with an assembled target of a cylindrical hohlraum and a layered DT capsule inside has been performed in recent experiments on the National Ignition Facility (NIF).[7–11] However, to perform the spherical-symmetric capsule implosion is quite difficult due to radiation nonuniformity at imploding capsule surface even if lasers in two rings incident upon the inner wall of the cylindrical hohlraum are applied to tune radiation uniformity. On the other hand, the implosion shock running inside the hotspot experiences reflections back and forth from the center to an interface between the hotspot and the high compressed fuel (below, called the hotspot interface) till ignition. It results in the fact that each reflection exerts a force from a light fluid (the hotspot) to an inward-accelerating heavy fluid (the high-compressed density fuel) that is soon decelerated, and consequently, hotspot distortion, hydrodynamic instabilities,[2,12] and mix severely occur since there always exist spatial roughness at capsule interfaces and non-ideal spherical symmetry of the implosion shock. Meanwhile, these hydrodynamic instabilities are further aggravated by those from near the radiation ablation front (RAF) via feedthrough. In addition, the peak ablation pressure $\sim$100 Mbar (Ref. 7) in NIF experiments seems to be lower, resulting in maximal implosion velocity less than 380 km/s (Ref. 7) and difficulty in implosion ignition. Although recent high-foot experiments[7–10] on

[a)]xthe@iapcm.ac.cn

NIF have achieved hotspot ignition, no self-sustaining burn occurs due to the hotspot distortion and cold fuel mix. It has exhausted the existing NIF energy of $\sim$2 MJ. In addition, in the high-foot experiments, the DT fuel only arrived at the lower areal density ($<1.0\,g/cm^2$), which is hard to perform burn with gain.

In this paper, we proposed a new hotspot ignition scheme via hybrid drive (HD) that combined ID radiation and DD lasers using an assembled target of a spherical hohlraum (SH) with six laser entrance holes (LEHs) and octahedral symmetry and a layered DT capsule inside[13] for ICF, schematically shown in Figs. 1(a) and 1(b). The SH,[14–17] instead of the conventional cylindrical hohlraum, is applied to ensure the radiation in spherical-harmonic uniformity and the capsule implosion symmetry, and the ID laser beams incident upon the inner wall of the SH for each LEH are in a single ring; therefore, the SH can be designed with no laser beam overlap inside, resulting in laser plasma interaction (LPI) lower comparing with the cylindrical hohlraum.[11] The capsule inside the SH is first compressed by the ID radiation and then is further compressed by the HD pressure to perform the HD nonisobaric ignition of the hotspot when the HD implosion is providing the large enough *pdV* work to the hotspot rather than the conventional isobaric ignition near stagnation. The further explanation is in the following. The ID radiation ablating capsule ablator [CH ablator, Fig. 1(b)] with a radiation ablation front (RAF) produces a large-scale corona plasma due to large ID mass ablation rate,[2] where there exists a DD laser critical surface far from the RAF. The DT fuel shell is first compressed adiabatically by the successively intensive four shocks driven by the ID ablation pressure from the four-step temperature [Fig. 1(c), black line] at the CH ablator [Fig. 1(b)] surface, and a merged shock (MS) synthesized by the four shocks near the inner surface of the DT fuel enters the hotspot that is first compressed and heated, and rebounds from the hotspot center to the hotspot interface. On the other hand, DD lasers with a flat-top pulse in last duration [red line in Fig. 1(c)] via six LEHs enter the large-scale corona plasma formed by the ID ablated CH ablator and are absorbed in a region up to the critical surface where a supersonic electron-thermal-wave is launched and propagated inward towards the RAF far from the critical surface, and decays across the ID corona plasma where the wave energy is deposited due to electron-ion collision. Once the supersonic electron-thermal-wave slows down to sonic in the large-scale corona plasma, an electron ablation precursor shock and a continuous electron compression wave with high electron pressure are generated. This electron compression wave with an electron ablation front (EAF), like a snowplow, continually drives the ID corona plasma that is piled up into a higher plasma density ahead of the EAF. Meanwhile, the CH is continuously ablated by the ID temperature, and thus, a high plasma density plateau between EAF and RAF with a large width L is formed, generating a HD pressurized plateau with the ID temperature, which is analytically discussed in Sec. III. An enhanced HD precursor shock (ES) and a continuous HD compression wave are launched by the pressurized plateau where the peak HD plasma pressure is far larger than that in conventional ID and DD. The ES enters the ID compressed capsule and rapidly arrives at the hotspot interface and collides with the MS that is stopped before first reflecting there; therefore, the hydrodynamic instabilities and asymmetry caused by the MS reflections at the hotspot interface are suppressed. The hydrodynamic instabilities at the RAF are significantly mitigated as well due to $kL = \ell L/R_A \gg 1$ even if in low mode ℓ, where k is a perturbed wave number and R_A is a radius of the ID ablation front.

The continuous HD compression wave via the hotspot interface provides large *pdV* work to the hotspot that is further compressed and heated, resulting in the HD nonisobaric ignition before stagnation with a large margin soon after the ES running in the hotspot first reflects near the hotspot interface, where the hotspot distortion and hydrodynamic instabilities due to the MS reflections like in NIF experiments[7] or shock ignition[4] are suppressed. In the present model, the averaged CH corona plasma density piled is $\rho \sim 5\,g/cc$ within a width $L \sim 100\,\mu m$ obtained from simulations, and the ID peak temperature is $T_{max} \sim 270\,eV$ [Fig. 1(b)], thus the peak HD pressure $P = \Gamma \rho T$ reaches $\sim$800 Mbar $\gg$ ablation pressure P_a in the conventional ID and DD. Such a high pressure is generated only via increasing the corona plasma density ρ and is hard to appear in ID and DD where the ablation pressure increases via increasing temperature. The peak

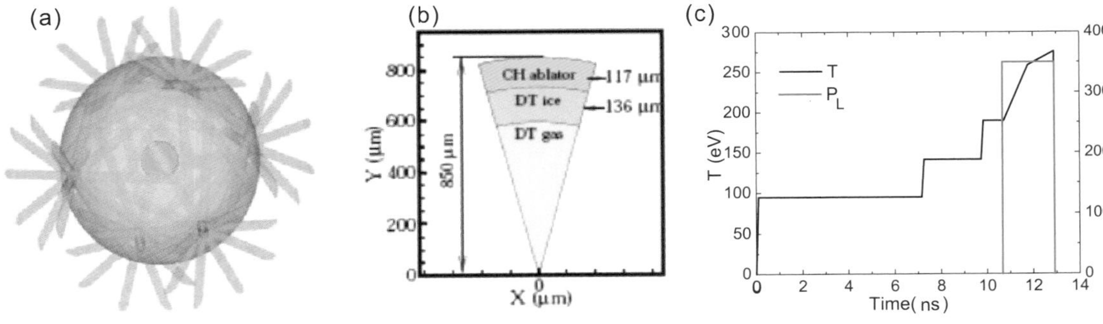

FIG. 1. Configurations of the SH and layered capsule, and the evolution of the ID temperature and DD laser power. (a) The schematic plot of the SH with six laser entrance holes (LEHs) and octahedral symmetry; (b) the capsule structure: CH ablator (the outer radius of 0.85 mm), DT ice layer, DT gas cavity; and (c) the radiation temperature T(eV, black) with four steps and the steady DD laser power $P_L = 350\,TW$ (red) in the pulse duration of last 2.2 ns.

ablation pressure in NIF experiments is only $P_a \sim 100\,\mathrm{Mbar}$ with peak ID temperature $\sim 300\,\mathrm{eV}$ (Refs. 7 and 8), and no supersonic electron-ablation wave is generated. In the DD approach, a typical one is the shock ignition,[4,6] and the electron ablation pressure $P_{EA} \sim 300\,\mathrm{Mbar}$ (Ref. 6) with electron temperature $T_e \sim 8.8\,\mathrm{keV}$ is requested. The space between the electron ablation front at the ablated ablator surface and the critical surface formed by the implosion lasers is far less than that in HD due to the DD ablation mass rate far less than the ID under the present ICF parameters so that the supersonic electron-ablation wave caused by the ignitor lasers fast arrives at the imploding capsule surface without formation of the continuous compression wave and the snowplow effect.[6] However, the ID role in the HD is very important because it not only drives the first symmetric implosion compression by the uniform radiation field in the SH but also provides a large enough space ($\sim 300\,\mu\mathrm{m}$) between RAF and EAF to make the supersonic electron thermal-wave slowing down to the electron compression wave.

II. HYBRID-DRIVE NONISOBARIC IGNITION CONDITION

The HD nonisobaric ignition condition for such a HD implosion dynamics is discussed in the following. The rate of the total internal energy E in the hotspot varies from $dE/dt < 0$ to $dE/dt > 0$ during implosion dynamics. The hotspot ignition condition can be calculated by using $dE/dt = 0$. Assume that the temperature T and density ρ in the hotspot are homogeneous, we obtain an equation for the ignition condition (α-particle self-heat beginning)[2,5] for an assembled DT system under spherical symmetry geometry in the form

$$g(T,\xi)(\rho r_h)^2 + A_W \Gamma T u(\rho r_h) = A_S \frac{T^{7/2}}{\ln\Lambda}, \tag{1}$$

Eq. (1) is a power balance and indicates a relationship between the areal density $\rho r_h(\mathrm{g/cm^2})$ and temperature $T(\mathrm{keV})$, where r_h is the hotspot radius. The first and second terms in left hand side of Eq. (1) are the net energy deposition rate in the hotspot and the power pu (p pressure and u velocity) imposed at the hotspot interface, respectively, and the first term in right hand side of Eq. (1) is the loss rate of the electron thermal-flux via the hotspot interface. In Eq. (1),

$$g(T,\xi) = A_Q n_T^2 \xi(1-\varphi)\langle\sigma v\rangle_{DT}\varepsilon_0 f_\alpha - A_B n_T^2 (1+\xi)^2\sqrt{T} \tag{2}$$

is the net increase rate ($g > 0$) of α-particle energy deposition from thermonuclear reaction [the first term in the right side of Eq. (2)] taking off electron bremsstrahlung emission energy (the second term), where $f_\alpha = 1 - \frac{1}{4\theta_\alpha} + \frac{1}{160\theta_\alpha^3}$ with $\theta_\alpha = \rho r_h/\rho\ell_\alpha$ is a fraction of the α-particle energy deposited into the hotspot[18] and $\rho\ell_\alpha = \frac{0.025T^{5/4}}{1+0.0082T^{5/4}}$ is the α-particle range at densities $\rho = (10-100)\mathrm{g/cc}$,[2,7] and a fraction of bremsstrahlung x-ray energy absorbed in the hotspot has been neglected. $g(T,\xi) = 0$ is a boundary between the α-particle energy deposition rate in thermonuclear reaction and the electron energy emission loss rate, where $\xi = n_D/n_T$ is a ratio of the deuterium particle number n_D to the tritium n_T. The multi-deuterium system is very important for ICF because the ion temperature during burning process can far exceed the deuterium-deuterium ignition threshold, in which the tritium is generated, and thus, the initial assembled tritium is saved.

We take $\xi = 1$ and the burn-up fraction $\phi = 0$ in Eqs. (1) and (2) hereafter. From $g = 0$, we can get the minimal ignition temperature $T_{im} \sim 4.3\,\mathrm{keV}$, the ignition temperature T must be greater than T_{im}. In Eq. (1), $p = \Gamma\rho T$ is the pressure and $\Gamma = 766[10^{12}\,\mathrm{erg/g\cdot keV}]$ is the pressure coefficient; u is the inward velocity in cm/s at the hotspot interface; $\ln\Lambda$ is the Coulomb logarithm; $\langle\sigma v\rangle_{DT}$ in $10^{-18}\,\mathrm{cm^3/s}$ is the DT thermonuclear reaction rate; and ε_0 in MeV is the α-particle energy. The coefficients in Eqs. (1) and (2): $A_Q = 1.6\times10^{12}[10^{30}\,\mathrm{erg/g^2}]$, $A_B = 5.8\times10^5[10^{12}\,\mathrm{erg\cdot cm^3/(g^2\cdot s\cdot keV^{1/2})}]$, and $A_W = 3$, $A_S = 2.7\times10^8\,[10^{12}\,\mathrm{erg/(s\cdot cm\cdot keV^{7/2})}]$.

The numerical solutions of Eq. (1) are plotted in Fig. 2. The isobaric ignition occurs at stagnation when the pdV work $4\pi\int r_h^2 pudt$ to the hotspot during implosion dynamics has been exhausted.[2] The isobaric ignition threshold curve $(T, \rho r_h)$ is shown in Fig. 2 (black solid-line). Eq. (1) is a quadric if $f_\alpha(\theta_\alpha)$ in Eq. (2) is regarded as independent of ρr_h since $\theta_\alpha = \rho r_h/\rho\ell_\alpha \sim 1$. An analytical expression for isobaric ignition from Eq. (1) is given approximately as

$$\rho r_h = \sqrt{\frac{A_s T^{7/2}/\ln\Lambda}{g}} \equiv f_2(T). \tag{3}$$

Eq. (3) is quite close to the solution of Eq. (1) as shown in Fig. 2 (black curve). The well known isobaric ignition threshold ($\rho r_h \sim 0.35\,\mathrm{g/cm^2}, T = 5\,\mathrm{keV}$) is obtained from Fig. 2 or Eq. (1).

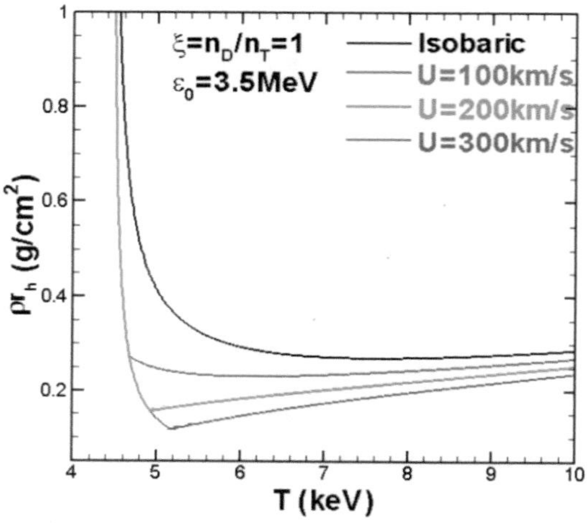

FIG. 2. Hot-spot ignition condition Eq. (1): the areal density ρr_h versus temperature T for $\xi = 1$ and α-particle energy $\varepsilon_0 = 3.5\,\mathrm{MeV}$. The curves show the isobaric ignition (u = 0, black solid line), the nonisobaric ignition with velocities $u(\mathrm{km/s}) = 100$ (red line), 200 (green line, $h \sim 0.2$ for $T \sim 5\,\mathrm{keV}$), and 300 (blue line, $h \sim 0.08$ for $T \sim 5\,\mathrm{keV}$).

The nonisobaric ignition occurs when the implosion dynamics is providing pdV work to the hotspot rather than at stagnation, and the thresholds $(T, \rho r_h)$ for different u (power pu) are plotted in Fig. 2. It is seen from Eq.(3) and Fig. 2 that under the condition of the same T the ρr_h for nonisobaric ignition is always less than that in the isobaric ignition, which shows that the nonisobaric ignition occurs in the convergence lower than that in the isobaric ignition and is beneficial to suppression of hydrodynamic instabilities. The HD ignition occurs at the time when implosion dynamics is providing the large enough pdV work to the hotspot. An approximate expression for HD nonisobaric ignition can be obtained as well if $f_\alpha(\theta_\alpha)$ is regarded as independent of ρr_h, and we have

$$\rho r_h = \frac{A_s T^{7/2}/\ln\Lambda}{A_w \Gamma T u} \equiv f_3(T, u), \tag{4}$$

if

$$\frac{4gA_s T^{7/2}/\ln\Lambda}{A_w^2 \Gamma^2 T^2 u^2} \equiv h(T, u) \ll 1, \tag{5}$$

where $h \sim T^{4.5}/u^2$ if taking $g \sim \langle\sigma v\rangle_{DT} \sim T^3$ and $\ln\Lambda$ independent of T. Eq. (4) gives $\rho r_h \sim 0.15\,\mathrm{g/cm^2}$ for the ignition temperature $T \sim 5\,\mathrm{keV}$ ($\ln\Lambda \sim 2$) and the hotspot interface velocity $u = u_{ig} = 200\,\mathrm{km/s}$ for $h \sim 0.2 \ll 1$). It means that the implosion velocity V_{im} of the high compressed fuel, which is close to the interface velocity u prior to ignition, has to be much larger than u_{ig}. We shall see in Sec. V that the maximal implosion velocity $V_{im} \geq 400\,\mathrm{km/s}$ is requested to meet the performance of the HD nonisobaric ignition. Such an implosion velocity is impossible to be provided by the conventional implosion dynamics driven by the low ablation pressure $\sim$100 Mbar (Refs. 7 and 8) as the NIF did. Therefore, the HD approach we proposed is necessary for HD nonisobaric ignition.

III. SNOWPLOW MODEL AND HD PRESSURIZATION

We now discuss the formation of the HD pressurized plateau with high density between the EAF and RAF. Using a snowplow model, we study how the high density is formed from the corona plasma that is driven and piled up by the electron compression wave when the supersonic electron-ablation wave slows down to an electron isothermal-sonic $C_s = (\Gamma_e T_e)^{1/2}$ at time $t = t_s$, where T_e and Γ_e are the electron temperature and pressure coefficient. Assume that the areal plasma density (mass per unit area) at time t accumulated in front of the EAF (here and below, the EAF means the EAF of the electron compression wave, unless otherwise specified) is $\mu(\tau) = \mu_0(0) + \mu_1(\tau)$ with $\tau = t - t_s$, where $\mu_0(0) = \int_{R_{RA}(0)}^{R_{EA}(0)} \rho dR$ is the areal density between the radius $R_{EA}(\tau = 0)$ in the EAF and radius $R_{RA}(\tau = 0)$ in the RAF, $\mu_1(\tau) = \int_0^\tau \dot{m} d\tau$ is the areal density of the ID ablated mass of the ablator at the RAF, where the mass ablation rate $\dot{m} \sim T^3$.[19] The motion of the corona plasma pushed by the EAF can be described by the snowplow equation:

$$\frac{d}{d\tau}\left(\mu R^2 \frac{dR}{d\tau}\right) = -R^2[P_{EA}(\tau) - P(\tau)], \tag{6}$$

where P_{EA} is a high pressure ahead of the electron compression wave and $P = \Gamma\rho T$ is the corona plasma pressure. Taking initial conditions: $R(\tau = 0) = R_{EA}(0)$ and $dR/d\tau(\tau = 0) \approx 0$, the radius R_{EA} of the EAF is obtained from Eq. (6), i.e., $R_{EA}(\tau) = R_{EA}(0) - \int_0^\tau \frac{d\tau_1}{\mu(\tau_1)R^2(\tau_1)} \left\{\int_0^{\tau_1} d\tau_2 R^2(\tau_2)[P_{EA}(\tau_2) - P(\tau_2)]\right\}$. By considering that the velocity at the RAF is approximately the imploding velocity, we can write the plateau width between the EAF and the RAF in the form

$$\Delta R_{ER}(\tau) = \Delta R_{ER}(0) + \int_0^\tau d\tau_1 \left\{ V_{im} - \frac{1}{\mu_1(\tau_1)R^2(\tau_1)} \times \int_0^{\tau_1} d\tau_2 R^2(\tau_2)\left[P_{EA}(\tau_2) - P(\tau_2)\right] \right\}, \tag{7}$$

where $\Delta R_{ER}(\tau) = R_{EA}(\tau) - R_{RA}(\tau)$, $\Delta R_{ER}(0) = R_{EA}(0) - R_{RA}(0)$ and R_{EA} is the radius of the EAF and R_{RA} is the radius of the RAF. In early stages, P_{EA} driven by the steady DD lasers is close to a constant, at the EAF, $P_{EA}/P = \Gamma_e T_e/\Gamma T \gg 1$, where the electron temperature T_e is far greater than the radiation temperature T, and $V_{im} \sim \sqrt{T}$[19] at the RAF is less than the driving velocity by P_{EA}; therefore, $V_{im} < \frac{1}{\mu_1(\tau_1)} \int_0^\tau d\tau_2 P_{EA}(\tau_2) \sim C_S = (\Gamma_e T_e)^{1/2}$. As a result, the snowplow pushes the density plateau width to be contracted, i.e., $\Delta R_{ER}(\tau) < \Delta R_{ER}(0)$, and thus, the averaged plasma density $\langle\rho\rangle \sim \mu_0/\Delta R_{ER}(\tau)$ goes up within the density plateau. At the later time, $P = \Gamma\rho T \gg P_{EA}$ due to T and ρ rising, while T_e decreasing within the $\Delta R_{ER}(\tau)$, and thus, $\Delta R_{ER}(\tau) > \Delta R_{ER}(0)$. However, noticing that $\mu_1(\tau) \sim \int_0^\tau T^3 d\tau, V_{im} \sim T^{1/2}$ and $P/\mu \sim T^{-2}$; as a result, the averaged corona plasma density $\langle\rho\rangle \sim \mu(\tau)/\Delta R_{ER}(\tau)$ between the EAF and the RAF increases rapidly within $\Delta R_{ER}(\tau)$, where an extremely high pressure $P = \Gamma\rho T$ far greater than the ID ablation pressure P_a is produced. Thus, the analytical discussion showed that the steady DD lasers combined with the ID radiation can drive the high corona plasma density between the RAF and EAF and give rise to the HD pressurization plateau. It is such a plateau that drives the ES and the continuous compression wave toward the implosion capsule with a high implosion velocity to perform the HD nonisobaric ignition of the hotspot.

IV. TARGET CONFIGURATION: SPHERICAL HOHLRAUM AND LAYERED CAPSULE

The assembled target structure and the energy source to drive implosion dynamics are schematically plotted in Fig. 1. The capsule inside the SH is made up of a CH ablator with an outer diameter of 1.7 mm, a cryogenic DT fuel layer with mass of 0.19 mg, and a cavity filled DT gas with density about 0.3 mg/cc, as shown in Fig. 1(b). The SH with 6 LEHs filled an electron density of $0.05n_c$ inside, where n_c is the critical density of electron number, has a diameter of 8.4 mm with a diameter of 2.2 mm for each LEH [Fig. 1(a)]. The area ratio of the LEHs to the hohlraum is $\sim$10%. The ID temperature is of the four-step temporal distribution from

100 eV to peak 270 eV [Fig. 1(c)] in a total pulse duration of 12.9 ns, corresponding to the ID laser energy of ~0.6 MJ. The steady DD lasers that are of a flat-top power of $P_L = 350$ TW [red color in Fig. 1(c)] imposed in last duration of 2.2 ns and possess laser energy of 0.77 MJ enter the SH via 6 LEHs with a light-waist diameter of 1.7 mm for each LEH, which has a space of 0.25 mm between LEH and light waist to ensure lasers passing the LEHs without being blocked in total pulse duration of 12.9 ns.[20] Calculations by view factor show that in our case the radiation flux asymmetry of $|\Delta F/F| \sim 0.07\%$ for the initial diameter ratio 4.94 of the SH to the capsule, mainly from the spherical harmonics Y_{44} and Y_{40}[17] [detailed in Fig. 4(a) below] without Legendria P_2 that only appears in cylindrical hohlraum. It is of better radiation symmetry and higher coupling of lasers to target compared with the cylindrical hohlraum. The 6-LEHs octahedral-symmetry SH as well is better than the 4-LEHs tetrahedral SH in which the radiation flux asymmetry as low as 1% (rms) was experimentally observed on OMEGA.[21]

V. ONE-DIMENSIONAL SIMULATION RESULTS

The radiation hydrodynamics code LARED-S with ray tracing[22] is applied to simulate one- and two-dimensional implosion dynamics involving ignition and fuel burn. The one-dimensional results of implosion dynamics are shown in Fig. 3, where the radii of the interfaces (fronts) of the imploding capsule varying with time are indicated. The radiation ablation front RAF [white solid line in Fig. 3(a)] at the ablating CH surface converges rapidly during implosion dynamics. At $t \sim 11.4$ ns, the merged shock MS appears in the vicinity of the inner surface of the DT fuel layer and enters the hotspot that is first compressed and heated. At $t \sim 12.95$ ns, the MS rebounded from the hotspot center reaches the hotspot interface. On the other hand, beginning at $t \sim 10.7$ ns, the DD lasers enter the SH and are absorbed in a range up to the critical density n_c at the radius of 960 μm [white-dashed line in Fig. 3(a)] in the ID corona plasma, where the electron temperature T_e rises rapidly near the critical surface. The supersonic

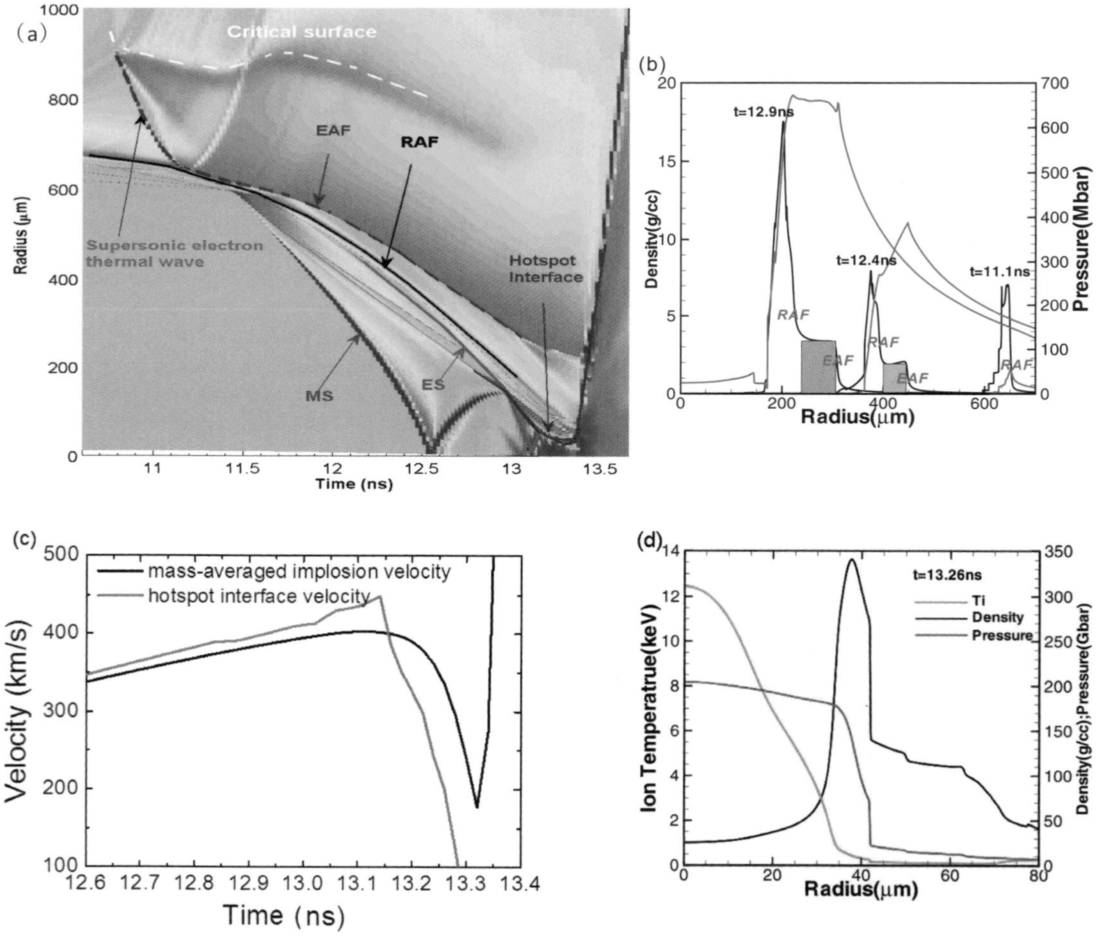

FIG. 3. Implosion physics of the HD approach. (a) The radii of interfaces (fronts) in the capsule implosion process versus time: the DD laser critical surface (white-dashed line), the electron ablation front EAF (dark-gray-dashed line), the radiation ablation front RAF (white-solid line), the enhanced shock ES (green line), the merged shock MS (blue line), the hotspot boundary (black line); (b) the ID ablation pressure of ~50 Mbar (red color) at $t \sim 11.1$ ns before the EAF arriving; the high density and pressurized plateaus (gray region) between the EAF and RAF at $t \sim 12.4$ ns and 12.9 ns, respectively; (c) implosion velocity V_{im} and hotspot interface velocity u vary with time; (d) ion temperature (green), density (black), and pressure (red) at $t \sim 13.26$ ns (ignition time).

electron thermal-wave propagation in the ID corona plasma is shown in Fig. 3(a). The electron ablation shock and the follow-up electron compression wave are launched when the supersonic thermal-wave slows down to a sound speed of $\sim$250 km/s at the radius $R_{EA} \sim 660\,\mu$m ($\sim$20 μm away from the radius R_{RA} of the RAF) at $t \sim 11.1$ ns, where the electron ablation pressure is $P_{EA} \sim 240$ Mbar far greater than the ID ablation pressure of $\sim$50 Mbar [Fig. 3(b)]. Such an electron compression wave with an electron ablation front EAF, like the snowplow, drives the corona plasma and piles up it into a high density pressurized plateau between the EAF and RAF, where the averaged density is of $\sim$2.0 g/cc within the plateau width $\Delta R_{ER} \sim 45\,\mu$m at $t \sim 12.4$ ns and $\sim 3.3 g/cc$ within $\Delta R_{ER} \sim 65\,\mu$m at $t \sim 12.9$ ns [gray region in Fig. 3(b)], the HD peak pressure is of $\sim$400 Mbar and $\sim$680 Mbar, respectively (red curves in Fig. 3(b)), and the ID temperature is $260 \sim 270$ eV there during this period [Fig. 2(c)]. Such a HD plateau predicted in analytical discussion in Sec. III plays a role of a high-pressure piston that launches the ES and the continuous compression wave towards the imploding capsule. The ES fast arrives at the hotspot interface at the radius of $\sim$150 μm at $t \sim 12.95$ ns and collides with the MS that just arrives there and is stopped.

Then, the ES and the follow-up continuous compression wave accelerate the hotspot interface, where the maximal interface velocity reaches $u_m \sim 450$ km/s [Fig. 3(c)] and results in large pdV work $4\pi \int r_h^2 pudt$ to the hotspot that is further compressed and heated, resulting in rising of ion temperature and pressure rapidly. The HD nonisobaric ignition ($\alpha-$particle self-heating beginning) occurs at $t \sim 13.26$ ns soon after the ES rebounded from the hotspot center reaches the hotspot interface. At the ignition time, the hotspot possesses itself of the mass averaged ion temperature $T_i \sim 5.1$ keV and the averaged areal density $\rho r_h \sim 0.15$ g/cm^2 (the ignition threshold is quite close to the analytical prediction in Sec. II), the pressure $\sim$200 Gbar, the pdV work $\sim$4 kJ [2/3 from the ES and the continuous HD compression wave, the interface velocity u decreases from maximum value $\sim$450 km/s to $\sim$200 km/s in Fig. 3(c)], and fusion energy of $\sim$25 kJ is released (1/5 deposited in the hotspot), the convergence ratio (the initial outer radius of capsule to the final radius of the hotspot) is $\sim$25 lower than in isobaric ignition, as shown in Fig. 3(d). Meanwhile, the maximal HD pressure at the RAF is $\sim$800 Mbar (far larger than the ID maximal ablation pressure $\sim$100 Mbar in NIF experiments[7,8]), the maximal implosion velocity averaged by fuel mass achieves $\sim$400 km/s [Fig. 3(c)]. After ignition, the inward velocity u at the hotspot interface, indeed, is still large enough to be able to further supply the pdV work to the ignited hotspot till stagnation ($u=0$), which provides a margin of $\sim$2 kJ ($\sim$33%, the ratio of the margin to the total work) for the ignited hotspot. Until stagnation, the isobaric pressure of $\sim$1000 Gbar in the ignited hotspot is observed, the system has released the fusion yield of $\sim$200 kJ ($\sim$35 kJ deposited in the hotspot), and the areal density of the fuel is 1.5 g/cm^2. Finally, the DT fuel has the burn-up fraction of $\sim$24%, and the fusion yield of 15MJ under total laser energy of 1.37 MJ is achieved. In addition, if we artificially shortened the DD laser pulse duration from 2.2 ns to 1.5 ns, i.e., the DD laser energy reduces from 0.77 MJ to 0.525 MJ by 25%, the one-dimensional results show that the HD pressure is $\sim$350 Mbar, the peak implosion velocity is over 390 km/s, and the yield is 7 MJ by total laser energy of 1.125 MJ. It indicates that this model has better robustness.

VI. TWO-DIMENSIONAL SIMULATION RESULTS

The two-dimensional implosion dynamics is simulated by 2D LARED-S code, and the results are discussed as follows.

A. ID asymmetry influence on hotspot ignition

First, we look at the asymmetry influence on the hotspot ignition due to the radiation flux asymmetry $|\Delta F/F| \sim 0.07\%$ initially at the RAF of the ablating CH surface from the spherical harmonics Y_{44} and Y_{40},[17] as seen in Fig. 4(a) by view factor calculation, which are the Legendria P_4 in 2-dimensional simulations. The simulation results at ignition time show that a deviation of the hotspot shape from spherical symmetry at the hotspot interface during convergent implosion dynamics is locally amplified to $P_4/P_0 \sim +10\%$ [Fig. 4(b)] in the top of a rhomb-like profile [Fig. 4(b)], where $P_0 \sim 33\,\mu$m is an averaged hotspot radius. The spatial change of ion temperature T_i [Fig. 4(b)] from the hotspot center to border is very close to the one-dimensional simulation [Fig. 3(c)], and the averaged $\rho r_h = 0.15$ g/cm^2 is the same as one-dimensional results. The ratio of two-dimension to one-dimension for the fusion yield Y caused by the radiation asymmetry

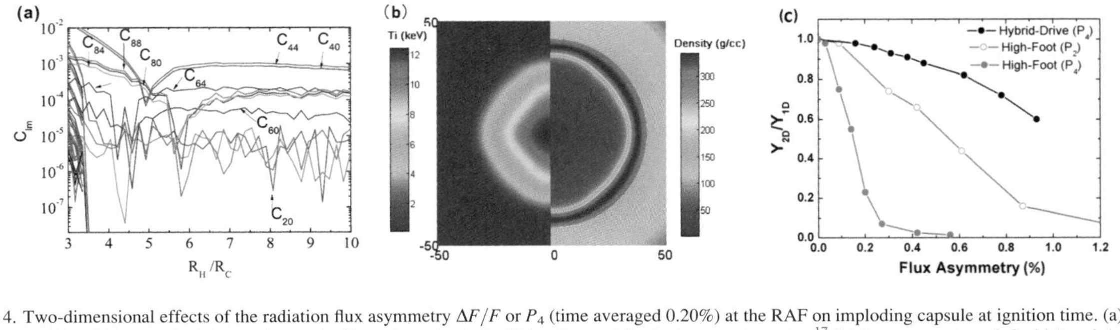

FIG. 4. Two-dimensional effects of the radiation flux asymmetry $\Delta F/F$ or P_4 (time averaged 0.20%) at the RAF on imploding capsule at ignition time. (a) The coefficient C_{lm} of the spherical harmonics vs R_H/R_C, where R_H is the SH radius and R_C is the capsule radius;[17] (b) ion temperature (left-side) and density (right-side) for hotspot and fuel; (c) Y_{2D}/Y_{1D} vs flux asymmetry P_4 for HD (black), P_2 and P_4 for high-foot drive (red).

$|\Delta F/F| \sim 0.2\%$ [Fig. 4(a)] at the RAF is $Y_{2D}/Y_{1D} \sim 97.7\%$, obviously the asymmetry influence can be neglected. We also see from Fig. 4(c) that the reduction of the fusion yield is 60% even if $|\Delta F/F| \sim 1.0\%$ at the RAF, while in the NIF high-foot experiments, $Y_{2D}/Y_{1D} \sim 10\%$ and ~ 0 if P_2 and P_4 are 1% [Fig. 4(c), red color], respectively.

B. Electron thermal wave smoothing in corona plasma

The DD lasers entering the SH through six LEHs to form six separate cylinder-like profiles with each light-waist diameter $D_{LWD} = 1.7$ mm propagate in the ID corona plasma, and the lasers intersect and overlap in a spherical surface of a diameter of ~2.4 mm, where the laser energy nonuniformity would take place. 2D LARED-S code with 3D ray-tracing was applied to investigate nonuniformity. Results showed that the laser nonuniformity $\delta I/I$ ($t \sim 10.9$ ns) is $Y_{44} = -2.6\%$, $Y_{4-4} = -2.6\%$, and $Y_{40} = -4.2\%$ with the root mean square ~5.5%, and the electron flux nonuniformity is reduced to ~0.1% by thermal smoothing when the supersonic thermal wave propagating in the ID corona plasma slows down to sonic from the critical surface ($R_{cr} \sim 960\,\mu$m) to the sonic position ($R_{EA} \sim 660\,\mu$m) and time from 10.7 ns to 11.1 ns, as shown in Fig. 5. Then, it is further smoothed from sonic point to the radiation ablation front (RAF). Therefore, during DD phase, the asymmetry of electron flux at the RAF is smaller than or comparable to that of the radiation flux in our model.

Laser nonuniformity also arises from power imbalance of beam to beam and mistiming, which are related to the laser facility structure, and is around 5% for main pulse (~3.5 ns, like in NIF) and mistiming is ~30 ps. In our HD scheme, this nonuniformity of DD lasers without prepulse (flat-top pulse of 2.2 ns) is at most 5%; total nonuniformities at most 8% from beam to beam and mistiming are reduced to ~0.67% at critical surface if there is a laser facility with 144 beams for DD and others for ID. In addition, the nonuniformity is also caused by CBET due to beam to beam energy crossing and is $\lesssim 1\%$ (Ref. 23) near critical surface in our model where the radius ratio $\sim 0.8 - 1.0$ of laser beam cluster to critical surface. Therefore, those nonuniformities are less than 5.5% from lasers intersecting and overlapping.

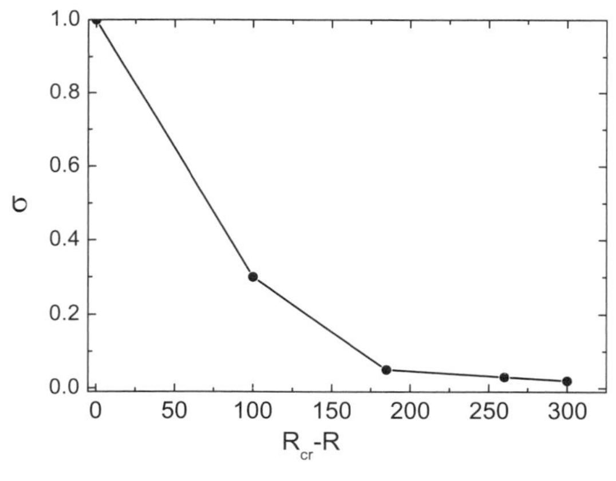

FIG. 5. The ratio σ of the asymmetry of electron thermal flux at the sonic point R_{EA} to that at critical surface R_{cr} induced by the nonuniform lasers as the supersonic electron thermal wave propagates from the critical surface to sonic position with a significant thermal smoothing effect.

C. Significant suppression of hydrodynamic instability

We now discuss the hydrodynamic instabilities[12] during capsule implosion, involving Rayleigh-Taylor instability (RTI) caused by a gravity exerted at an interface with different density, Richtmyer-Meshkov instability (RMI) caused by a shock across an interface, and Kelvin-Helmoltz instability (KHI) caused by shear motion in two sides at interface. They may severely degrade capsule implosion compression and bring about material mix from heavy fluid to light one. Here, we put the emphasis on the RTI[12] since it is more severe than others in the deceleration stage prior to ignition. The linear growth rate for RTI with ablation effect is $\gamma(k) = (Agk/(1+kL))^{1/2} - \beta k V_a$,[2] where $A = (\rho_1 - \rho_2)/(\rho_1 + \rho_2)$ is the Atwood number with fluid density $\rho_1 > \rho_2$, g is a deceleration, $k = 2\pi/\lambda$ is the perturbation wave number (wavelength λ), L is a characteristic density scale length at the ablation front, V_a is an ablation velocity, and β is an adjustable parameter in the range 1–3. The perturbation wavelength $\lambda = 2\pi R/\ell$ must be greater than the cut-off wavelength λ_c getting from $\gamma(k) = 0$, where R is the interface radius and the mode number $\ell \approx kR$. To define a growth factor (GF) as a ratio of the final amplitude to the initial perturbation is beneficial to discussing hydrodynamic instabilities, where $\mathrm{GF} = \exp[\int \gamma_{eff}(t)dt]$ is the form and γ_{eff} is an effective growth rate, and the greater GF, the more severe hydrodynamic instabilities affect implosion dynamics and mix.

2D simulations for HD show that the growth factors (GFs) varying with modes ℓ till the ignition time (red curves in Fig. 6) are significantly suppressed to compare with the ID high-foot isobaric ignition experiments on NIF[7,8] (black curves in Fig. 6). For comparison with the high-foot results in NIF experiments, both GFs are simulated by 2D LARED-S code. We see that at the radiation ablation front RAF and the fuel-ablator interface, the maximal GFs both are ~75 for the HD far less than 360 [Fig. 6(b)] and 200 [Fig. 6(c)] for the high-foot drive, respectively. Most essential is the GF at the hotspot interface [Fig. 6(a)], where the maximal GF for the HD is only ~10 far less than ~70 for the high-foot drive, and it is shown that the hydrodynamic instabilities for HD affect the hotspot ignition quite weakly. We now discuss why the GFs in our HD approach are significantly suppressed? Below, we use the linear growth rate γ instead of γ_{eff} to give an insight into the GFs.

First, look at the RTI at the RAF with radius $R = R_A \sim 120\,\mu$m (Table I) in ignition time, where the characteristic density scale length L is the pressurized plateau width $L = \Delta R_{ER} \sim 105\,\mu$m with the averaged density $\rho \sim 5$ g/cc formed by the snowplow as seen in Fig. 3(b), the linear growth rate $\gamma = (Ag/L)^{1/2} - \beta(\ell/R)V_a$ due to $kL = \ell L/R \gg 1$ for all $\ell > 5$ modes is significantly reduced, resulting in strong stabilization effect, which is absent in any full ID or DD implosion dynamics.

At the hotspot interface, GF ~ 10 is extremely small, it may attribute to the deceleration g small because the ES

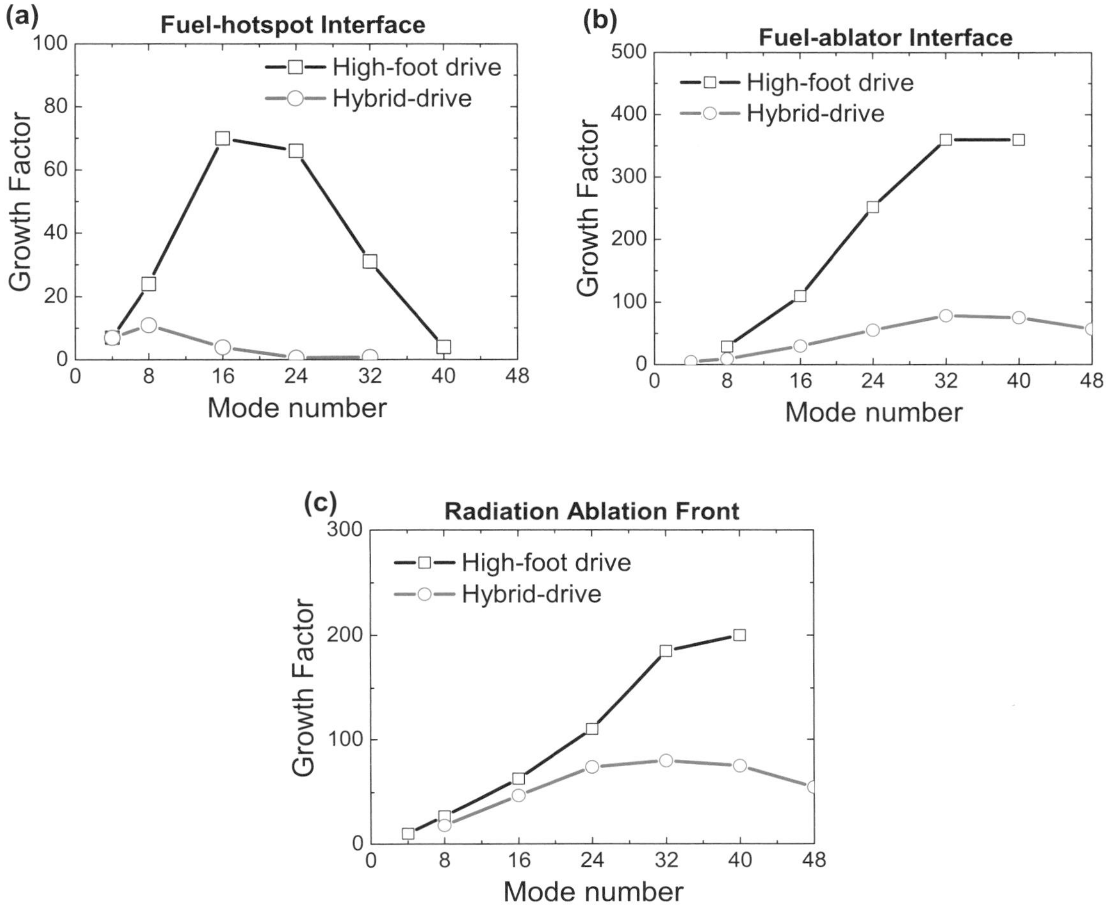

FIG. 6. Comparisons of the GFs of hydrodynamic instabilities (till ignition time) between the HD capsule and the ID capsule for perturbations initially seeded at the ablator outer surface, and the 2-dimensional capsule profile at the ignition time. Density contours for a perturbation with a mode $\ell = 16$ and an initial roughness of 350A. (a)-(c) All red curves for present HD and all black curves for the high-foot indirect drive in NIF experiments: (a) at the fuel-hotspot interface, (b) at fuel/ablator interface, and (c) at the radiation ablation front (RAF).

stops the MS reflections there and the hotspot is ignited before stagnation, and also to an ignorable feedthrough effect from the RAF due to a decaying factor $\sim e^{-k\Delta}$,[24] where $\Delta = \Delta R_F + \Delta R_A + \Delta R_{ER} \sim 182\,\mu$m (Table I) and $k\Delta = \ell\Delta/R_F > 1$ for modes $\ell > 3$ and hotspot radius $R_F \sim 33\,\mu$m.

At the fuel-ablator interface, the perturbation growth mainly occurs in an acceleration process and is from the RMI in the cold fluid, where the radius is $R_F \sim 41\,\mu$m and the thickness of the remaining ablator is $\Delta R_A \sim 79\,\mu$m, thus

TABLE I. The estimated interface radii R and the layer thickness ΔR at 13.26 ns (ignition time), both R and ΔR in the units of μm.

Radius R[a]				Thickness ΔR[b]		
R_H	R_F	R_A	R_E	ΔR_F	ΔR_A	ΔR_{ER}
33	41	120	225	8	79	105

[a] R_H, R_F, R_A, and R_E are the radii of hotspot, fuel, radiation ablation front, and electron ablation front, respectively.
[b] ΔR_F, ΔR_A, and ΔR_{ER} are the thickness of fuel layer, ablator, and pressurized plateau in the averaged density of $5g/cc$, respectively.

the feedthrough from the RAF results in the same stabilization effect as at the RAF, and the thick ΔR_{ER} also avoids the remaining ablator ruptures. As a result, all aforementioned effects make GFs in HD approach be unable to be developed, especially the thick width ΔR_{ER} of the pressurized plateau plays a pivotal role.

D. Laser plasma interaction

Laser plasma interaction (LPI) in ignition target is concerned for two reasons: the laser energy degradation by instabilities that are mainly from the stimulated Raman scattering (SRS), stimulated Brillouin scattering (SBS), and two-plasmon decay (TPD), and energetic electrons generated by these instabilities, which might preheat the imploding fuel core to degrade its compression. In our HD system, the ID laser beams of total energy of 0.6 MJ during 12.9 ns pulse pass six LEHs into the SH with each LEH only in one ring of 55° and laser power of ~28 TW. It is similar to the outer ring in NIF high-foot experiments with total laser energy of ~1.9 MJ into two LEHs of cylindrical hohlraum.[7] If the laser intensity to pass each LEH in present SH same as this in NIF, we deduce the LPI close to the level of NIF experiments, where the total backscatter mainly is from SBS and remains less than 5% of the total incident ID laser energy.[7] Recently, we performed experiments on SG-III of 48 laser beams, and the backscattering of 5.9% (3.5%)[25] without (with) phase plates was measured. In addition, the laser energy loss of ~6% mainly from SBS was measured on the OMEGA experiments,[21] where lasers were incident upon a SH with 4 LEHs. As a result, the ID laser energy coupling to the target is likely greater than 90% in our HD system. We refer to the NIF experiments to estimate the preheating effect, where the energetic electrons with kinetic energy >100 keV are absorbed in about 0.2 mg DT ice (preheat) in an upper bound of (5±3)J,[26] for total laser energy of 1.37 MJ. Therefore, in our HD system, the preheating DT energy by energetic electrons >120 keV is estimated at most (2.1±1.3)J for the ID laser energy of 0.6 MJ, and thus, the effect of fuel preheating on the adiabat is negligible.

Now, we discuss the LPI from DD lasers that enter the SH through six LEHs and interact with the ID corona plasma, where laser intensity is over 1.0 PW/cm^2 (PW/cm^2 $= 10^{15}$ W/cm^2) and it may give rise to LPI more severe than in ID. It is seen from the laser power at $t = 11.2$ ns in Fig. 7(a) that lasers are weakly absorbed by inverse bremsstrahlung till a quarter critical density at a radius of 1.3 mm in corona plasma and are remained at 80 TW near the critical surface at the radius of 0.85 mm, where the intensities are in a range from ~1.0 PW/cm^2 to ~1.5 PW/cm^2. The LPI with intensities over 1.0 PW/cm^2 is a quite challenging issue in the study of laser-drive ignition target, and much work was achieved in shock ignition.[6,29] The threshold curves depending on density scale length L_n[27] are plotted in Fig. 7(b) where the electron temperature $T_e \sim 6$ keV. Both SRS and TPD generate only near or below a quarter critical density $n_c/4$, where, in our case, the density scale length is $L_n \sim 320\,\mu$m and the intensity is ~1.4 PW/cm^2. The absolute SRS is easily to be excited as seen in Fig. 7(b) and saturates fast at level of picoseconds by PIC simulations, but the convective SRS is impossible to appear in $L_n < 350\,\mu$m due to strong Landau damping of the electron plasma wave under high $T_e \sim 6$ keV. The SBS appears in a larger region below critical density and is dominant, where density scale length $L_n \sim 250\,\mu$m (n_c) in laser intensity ~1.0 PW/cm^2 and ~320 μm ($n_c/2$) in maximal intensity ~1.7 PW/cm^2. Recent experiments performed on SG-III prototype[28] showed the backward scattering fraction of ~$(10-13)$% without phase plates for intensity of 2 PW/cm^2 (3ω, 1 ns). We as well as refer to the backscattering experiments on OMEGA,[29] where the backscattering fraction of ~10% mainly from SBS without phase plates for intensity of ~1.5 PW/cm^2, as well as discussed by Batani.[6] The thresholds of the TPD are ~0.4 PW/cm^2 for absolute instability and 0.7 PW/cm^2 for convective at $L_n \sim 320\,\mu$m, and increase with electron temperature T_e. The laser energy converted to total electron energy via TPD saturates at a level of 1% beginning at the overlapped laser intensity ~0.35 PW/cm^2 by the experiments[30] on OMEGA, where DD lasers were incident upon a long-scale length CH plasma with $L_n = 400\,\mu$m and $T_e \sim 3.5$ keV at $n_c/4$. However, hot electron temperature T_h increases with laser intensity and is extrapolated to $T_h > 120$ keV as the intensity is ~1.4 PW/cm^2. How much laser energy is converted to the hot electrons >120 keV is still unclear to be further investigated. The hot electron range in the CH plasma is $\rho R(\mathrm{g/cm^2}) \approx 0.589 \times 10^{-5} E_{\mathrm{keV}}^{1.661}$ for $E_{\mathrm{keV}} > 2$ keV (Ref. 31) and is 0.016 g/cm^2 for energy $E \sim 120$ keV. In the present model, the simulations by LARED-S showed that at $t \sim 11.1$ ns the ID implosion dynamics has compressed the remaining CH layer to reach an areal density $\rho R \sim 0.02$ g/cm^2, as seen in Fig. 3(b), larger than that for the range of electrons ~120 keV; it means that these electrons are stopped in the compressed remaining CH ablator, and the preheating DT effect may not be severe.

The CBET of the DD lasers occurs in vicinity of critical surface in ID corona plasma, and it reduces the laser absorption and hydrodynamic efficiency of implosion in the HD scheme. So far, several methods have been proposed to significantly reduce the CBET, which involves high-Z dopant

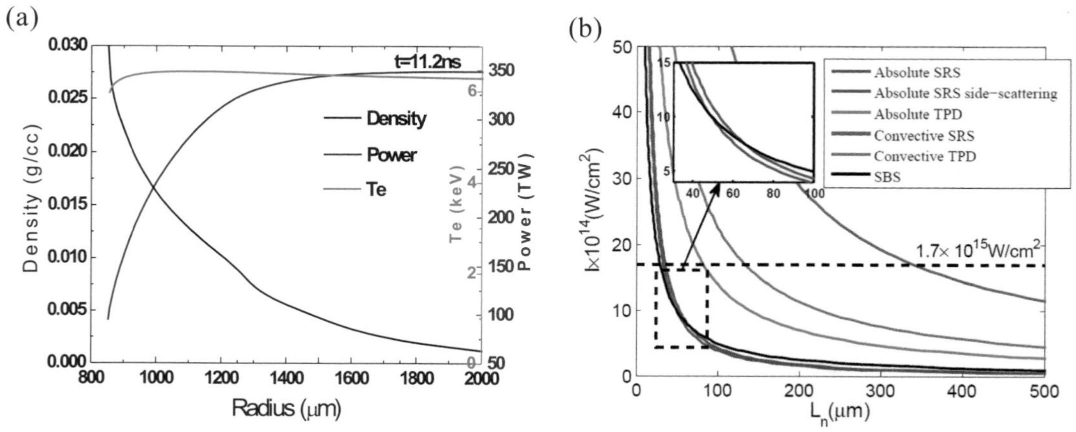

FIG. 7. Parameters related to the LPI and thresholds. (a) The propagating laser power P_L, density ρ, and electron temperature T_e vary with the radius of the SH and capsule obtained from simulations by one-dimensional LARED-S with 3D ray tracing; (b) the thresholds of the LPI for $T_e = 6$ keV vs the density scale L_n and intensity by PIC simulations.

CH ablator to increase laser absorption, dynamic narrow beams (zooming), and multicolor lasers to promote coupling efficiency.[32]

The self-focusing due to laser speckle is easily to take place due to the lower critical power from the ponderomotive effect, which is $P_s = 34T_{\mathrm{keV}}(n_c/n_e)(1 - n_c/n_e)^{1/2}$MW (Ref. 6) and is far less than the power of hundreds TW in HD case. It may be easily smoothed as performed in experiments.[33,34] The long-scale ID corona plasma may help laser speckle smoothing as well. The LPI with intensity over 1.0 PW/cm^2 has been concerned in shock ignition and much work was achieved.

The LPI is one of the critical issues of laser energy loss and implosion dynamics. So far, we are lacking in proper workable models and experimental results close to ignition scale for the lasers of intensities over 1 PW/cm^2 interacting with corona plasma; therefore, further investigations in theory and experiment are urgent.

VII. DISCUSSION AND CONCLUSIONS

In summary, we have proposed the HD nonisobaric ignition scheme for inertial confinement fusion. The layered fuel capsule inside the octahedral symmetric spherical hohlraum is compressed first by the ID, then by the DD combined with the ID in the last 2.2 ns. The HD pressure to drive implosion dynamics increases via increasing density, which is different from the conventional ablation pressure via increasing electron or radiation temperature in the DD and ID approaches. Theory and simulations show that a snowplow effect piles up the ID corona plasma into a higher density plateau and produces high HD pressure far greater than the ID ablation pressure, and the ID implosion dynamics is significantly improved. The enhanced shock and continuous compression wave for ignition are driven by the HD high pressure with peak value of ~800 Mbar, and the maximal imploding velocity reaches ~400 km/s demanded for the HD nonisobaric ignition of the hotspot. The ES stops the ID shock reflections at the hotspot interface and suppresses the hydrodynamic instabilities and asymmetry there, and the large plateau width of ~100 μm at the RAF mitigates the hydrodynamic instability and the feedthrough to the hotspot interface. Thus, two-dimensional influence may be significantly improved. The continuous compression wave provides mainly large enough *pdV* work to the hotspot that is further compressed and heated, resulting in the HD nonisobaric ignition in a lower convergence ratio of 25. Finally, the fusion energy of 15 MJ under total laser energy of 1.37 MJ is achieved.

These results showed that laser energy at NIF level is able to drive the HD ignition and burn gain if new laser facilities, involving NIF modified configuration, are provided. In addition, the present HD scheme is a physical concept model in which the conclusions based on theoretical analysis, 1D and 2D simulations, and recent NIF experiences must be checked and confirmed by 3D simulations and experiments in ignition scales, involving HD pressure generation, nonuniformity, and asymmetry (especially, power imbalance of beam to beam related to laser facilities), LPI, CBET, etc.

ACKNOWLEDGMENTS

We would like to thank C. Y. Zheng, H. B. Cai, Z. S. Dai, W. Y. Huo, Z. J. Liu, L. Hao, and M. Chen at IAPCM for their providing some data, F. Q. Li at LFRC for laser beam arrangement discussions, and C. K. Li at MIT and J. Fernandez at LANL for their encouragements. This work was performed under the auspices of the Nature Science Foundation of China (Grant Nos. 11475032, 11175027, 11274026, 11575033, and 11205010), the Foundation of President of Chinese Academy of Engineering Physics (Grant Nos. 2014-1-040 and 201402037), and the CAEP-FESTC (Grant No. R2014-05010-01).

[1] J. Nuckolls, L. Wood, A. Thiessen, and G. Zimmerman, Nature (London) **239**, 139 (1972).
[2] S. Atzeni and J. Meyer-ter-Vehn, *The Physics of Inertial Fusion: Beam Plasma Interaction, Hydrodynamics, Dense Plasma Physics* (Clarendon Press, Oxford, 2004).
[3] S. E. Bodner, D. G. Colombant, J. H. Gardner, R. H. Lehmberg, S. P. Obenschain, L. Phillips, A. J. Schmitt, J. D. Sethian, R. L. McCrory, W. Seka, C. P. Verdon, J. P. Knauer, B. B. Afeyan, and H. T. Powell, Phys. Plasmas **5**, 1901 (1998).
[4] R. Betti, C. D. Zhou, K. S. Anderson, L. J. Perkins, W. Theobald, and A. A. Solodov, Phys. Rev. Lett. **98**, 155001 (2007).
[5] J. Lindl, Phys. Plasmas **2**, 3933 (1995).
[6] D. Batani, S. Baton, A. Casner, S. Depierreux, M. Hohenberger, O. Klimo, M. Koenig, C. Labaune, X. Ribeyre, C. Rousseaux, G. Schurtz, W. Theobald, and V. T. Tikhonchuk, Nucl. Fusion **54**, 054009 (2014).
[7] O. A. Hurricane, D. A. Callahan, D. T. Casey, P. M. Celliers, C. Cerjan, E. L. Dewald, T. R. Dittrich, T. Döppner, D. E. Hinkel, L. F. Berzak Hopkins, J. L. Kline, S. Le Pape, T. Ma, A. G. MacPhee, J. L. Milovich, A. Pak, H.-S. Park, P. K. Patel, B. A. Remington, J. D. Salmonson, P. T. Springer, and R. Tommasini, Nature **506**, 343 (2014).
[8] T. R. Dittrich, O. A. Hurricane, D. A. Callahan, E. L. Dewald, T. Döppner, D. E. Hinkel, L. F. Berzak Hopkins, S. Le Pape, T. Ma, J. L. Milovich, J. C. Moreno, P. K. Patel, H.-S. Park, B. A. Remington, and J. D. Salmonson, Phys. Rev. Lett. **112**, 055002 (2014).
[9] H.-S. Park, O. A. Hurricane, D. A. Callahan, D. T. Casey, E. L. Dewald, T. R. Dittrich, T. Döppner, D. E. Hinkel, L. F. Berzak Hopkins, S. Le Pape, T. Ma, P. K. Patel, B. A. Remington, H. F. Robey, and J. D. Salmonson, Phys. Rev. Lett. **112**, 055001 (2014).
[10] T. Döppner, D. A. Callahan, O. A. Hurricane, D. E. Hinkel, T. Ma, H.-S. Park, L. F. Berzak Hopkins, D. T. Casey, P. Celliers, E. L. Dewald, T. R. Dittrich, S. W. Haan, A. L. Kritcher, A. MacPhee, S. Le Pape, A. Pak, P. K. Patel, P. T. Springer, J. D. Salmonson, R. Tommasini, L. R. Benedetti, E. Bond, D. K. Bradley, J. Caggiano, J. Church, S. Dixit, D. Edgell, M. J. Edwards, D. N. Fittinghoff, J. Frenje, M. Gatu Johnson, G. Grim, R. Hatarik, M. Havre, H. Herrmann, M. Izumi, S. F. Khan, J. L. Kline, J. Knauer, G. A. Kyrala, O. L. Landen, F. E. Merrill, J. Moody, A. S. Moore, A. Nikroo, J. E. Ralph, B. A. Remington, H. F. Robey, D. Sayre, M. Schneider, H. Streckert, R. Town, D. Turnbull, P. L. Volegov, A. Wan, K. Widmann, C. H. Wilde, and C. Yeamans, Phys. Rev. Lett. **115**, 055001 (2015).
[11] J. L. Kline, D. A. Callahan, S. H. Glenzer, N. B. Meezan, J. D. Moody, D. E. Hinkel, O. S. Jones, A. J. MacKinnon, R. Bennedetti, R. L. Berger, D. Bradley, E. L. Dewald, I. Bass, C. Bennett, M. Bowers, G. Brunton, J. Bude, S. Burkhart, A. Condor, J. M. Di Nicola, P. Di Nicola, S. N. Dixit, T. Doeppner, E. G. Dzenitis, G. Erbert, J. Folta, G. Grim, S. Glenn, A. Hamza, S. W. Haan, J. Heebner, M. Henesian, M. Hermann, D. G. Hicks, W. W. Hsing, N. Izumi, K. Jancaitis, O. S. Jones, D. Kalantar, S. F. Khan, R. Kirkwood, G. A. Kyrala, K. LaFortune, O. L. Landen, L. Lagin, D. Larson, S. Le Pape, T. Ma, A. G. MacPhee, P. A. Michel, P. Miller, M. Montincelli, A. S. Moore, A. Nikroo, M. Nostrand, R. E. Olson, A. Pak, H. S. Park, J. P. Patel, L. Pelz, J. Ralph, S. P. Regan, H. F. Robey, M. D. Rosen, J. S. Ross, M. B. Schneider, M. Shaw, V. A. Smalyuk, D. J. Strozzi, T. Suratwala, L. J. Suter, R. Tommasini, R. P. J. Town, B. Van Wonterghem, P. Wegner, K. Widmann, C. Widmayer, H. Wilkens, E. A. Williams, M. J. Edwards, B. A. Remington, B. J. MacGowan, J. D. Kilkenny, J. D. Lindl, L. J. Atherton, S. H. Batha, and E. Moses, Phys. Plasmas **20**, 056314 (2013).

[12]G. Taylor, Proc. R. Soc. London, Ser. A **201**, 192 (1950).
[13]X. T. He, Plenary talk at the IFSA2013, Nara, Japan, also in Proceedings of IFSA 2013; X. T. He, Z. F. Fan, J. W. Li, J. Liu, K. Lan, J. F. Wu, L. F. Wang, and W. H. Ye, e-print arXiv:1503.08979v1 [physics.plasm-ph] (2015).
[14]J. D. Schnittman and R. S. Craxton, Phys. Plasmas **3**, 3786 (1996).
[15]G. R. Bennett, J. M. Wallace, T. J. Murphy, R. E. Chrien, N. D. Delamater, P. L. Gobby, A. A. Hauer, K. A. Klare, J. A. Oertel, R. G. Watt, D. C. Wilson, W. S. Varnum, R. S. Craxton, V. Yu. Glebov, J. D. Schnittman, C. Stoeckl, S. M. Pollaine, and R. E. Turner, Phys. Plasmas **7**, 2594 (2000).
[16]S. A. Bel'kov, A. V. Bessarab, O. A. Vinokurov, V. A. Gaidash, N. V. Zhidkov, V. M. Izgorodin, G. A. Kirillov, G. G. Kochemasov, A. V. Kunin, D. N. Litvin, V. M. Murugov, L. S. Mkhitar'yan, S. I. Petrov, A. V. Pinegin, V. T. Punin, N. A. Sulsov, and V. A. Tokarev, JEPT Lett. **67**, 171 (1998).
[17]K. Lan, J. Liu, D. Lai, W. Zheng, and X.-T. He, Phys. Plasmas **21**, 010704 (2014); K. Lan, X.-T. He, J. Liu, W. Zheng, and D. Lai, *ibid.* **21**, 052704 (2014).
[18]O. N. Krokhin and V. B. Rozanov, Sov. J. Quantum Electron. **2**, 393 (1973).
[19]R. Olson, G. Rochau, O. Landen, and R. Leeper, Phys. Plasmas **18**, 032706 (2011).
[20]S. A. MacLaren, M. B. Schneider, K. Widmann, J. H. Hammer, B. E. Yoxall, J. D. Moody, P. M. Bell, L. R. Benedetti, D. K. Bradley, M. J. Edwards, T. M. Guymer, D. E. Hinkel, W. W. Hsing, M. L. Meezan, A. S. Moore, and J. E. Ralph, Phys. Rev. Lett. **112**, 105003 (2014).
[21]J. M. Wallace, T. J. Murphy, N. D. Delamater, K. A. Klare, J. A. Oertel, G. R. Magelssen, E. L. Lindman, A. A. Hauer, P. Gobby, J. D. Schnittman, R. S. Craxton, W. Seka, R. Kremens, D. Bradley, S. M. Pollaine, R. E. Turner, O. L. Landen, D. Drake, and J. J. MacFarlane, Phys. Rev. Lett. **82**, 3807 (1999).
[22]Z. F. Fan, X. T. He, J. Liu, G. L. Ren, B. Liu, J. F. Wu, L. F. Wang, and W. H. Ye, Phys. Plasmas **21**, 100705 (2014).
[23]D. H. Froula, I. V. Igumenshchev, D. T. Michel, D. H. Edgell, R. Follett, V. Yu. Glebov, V. N. Goncharov, J. Kwiatkowski, F. J. Marshall, P. B. Radha, W. Seka, C. Sorce, S. Stagnitto, C. Stoeckl, and T. C. Sangster, Phys. Rev. Lett. **108**, 125003 (2012).
[24]L. F. Wang, H. Y. Guo, J. F. Wu, W. H. Ye, J. Liu, W. Y. Zhang, and X. T. He, Phys. Plasmas **21**, 122710 (2014).
[25]SG-III experiments: 48-beam lasers with total energy $86\,\mathrm{kJ}/3\omega/3\,\mathrm{ns}$ incident upon the SH of diameter 3.6 mm with a capsule of diameter 0.96 mm inside via 2-LEHs (each diameter of 1.2 mm) in a 55° with the SH radius.
[26]T. Döppner, C. A. Thomas, L. Divol, E. L. Dewald, P. M. Celliers, D. K. Bradley, D. A. Callahan, S. N. Dixit, J. A. Harte, S. M. Glenn, S. W. Haan, N. Izumi, G. A. Kyrala, G. LaCaille, J. K. Kline, W. L. Kruer, T. Ma, A. J. MacKinnon, J. M. McNaney, N. B. Meezan, H. F. Robey, J. D. Salmonson, L. J. Suter, G. B. Zimmerman, M. J. Edwards, B. J. MacGowan, J. D. Kilkenny, J. D. Lindl, B. M. Van Wonterghem, L. J. Atherton, E. I. Moses, S. H. Glenzer, and O. L. Landen, Phys. Rev. Lett. **108**, 135006 (2012).
[27]C. Z. Xiao, Z. J. Liu, C. Y. Zheng, and X. T. He, Phys. Plasmas **23**, 022704 (2016).
[28]Our recent experiments on SG-III prototype using a laser of intensity $2\times10^{15}\,\mathrm{W/cm^2}$ $(3\omega,\ 1\,\mathrm{ns})$ incident upon a 200 μm foam put on CH plasma target.
[29]W. Theobald, R. Nora, M. Lafon, A. Casner, X. Ribeyre, K. S. Anderson, R. Betti, J. A. Delettrez, J. A. Frenje, V. Yu. Glebov, O. V. Gotchev, M. Hohenberger, S. X. Hu, F. J. Marshall, D. D. Meyerhofer, T. C. Sangster, G. Schurtz, W. Seka, V. A. Smalyuk, C. Stoeckl, and B. Yaakobi, Phys. Plasmas **19**, 102706 (2012).
[30]D. H. Froula, B. Yaakobi, S. X. Hu, P-Y. Chang, R. S. Craxton, D. H. Edgell, R. Follett, D. T. Michel, J. F. Myatt, W. Seka, R. W. Short, A. Solodov, and C. Stoeckl, Phys. Rev. Lett. **108**, 165003 (2012).
[31]S. Z. Wu, C. T. Zhou, S. P. Zhu, H. Zhang, and X. T. He, Phys. Plasmas **18**, 022703 (2011).
[32]I. V. Igumenshchev, W. Seka, D. H. Edgell, D. T. Michel, D. H. Froula, V. N. Goncharov, R. S. Craxton, L. Divol, R. Epstein, R. Follett, J. H. Kelly, T. Z. Kosc, A. V. Maximov, R. L. McCrory, D. D. Meyerhofer, P. Michel, J. F. Myatt, T. C. Sangster, A. Shvydky, S. Skupsky, and C. Stoeckl, Phys. Plasmas **19**, 056314 (2012).
[33]S. Depierreux, C. Labaune, D. T. Michel, C. Stenz, P. Nicolaï, M. Grech, G. Riazuelo, S. Weber, C. Riconda, V. T. Tikhonchuk, P. Loiseau, N. G. Borisenko, W. Nazarov, S. Hüller, D. Pesme, M. Casanova, J. Limpouch, C. Meyer, P. Di-Nicola, R. Wrobel, E. Alozy, P. Romary, G. Thiell, G. Soullié, C. Reverdin, and B. Villette, Phys. Rev. Lett. **102**, 195005 (2009).
[34]Max Karasik, J. L. Weaver, Y. Aglitskiy, J. Oh, and S. P. Obenschain, Phys. Rev. Lett. **114**, 085001 (2015).

First Investigation on the Radiation Field of the Spherical Hohlraum

Wen Yi Huo (霍文义),[1] Zhichao Li (李志超),[2,3] Yao-Hua Chen (陈耀桦),[1] Xuefei Xie (谢旭飞),[2] Ke Lan (蓝可),[1,3,4] Jie Liu (刘杰),[1,3,4] Guoli Ren (任国利),[1] Yongsheng Li (李永升),[1] Yonggang Liu (刘永刚),[2] Xiaohua Jiang (蒋小华),[2] Dong Yang (杨冬),[2,3] Sanwei Li (李三伟),[2] Liang Guo (郭亮),[2] Huan Zhang (章欢),[2] Lifei Hou (侯立飞),[2] Huabing Du (杜华冰),[2] Xiaoshi Peng (彭晓世),[2] Tao Xu (徐涛),[2] Chaoguang Li (李朝光),[2] Xiayu Zhan (詹夏宇),[2] Guanghui Yuan (袁光辉),[2] Haijun Zhang (张海军),[2] Baibin Jiang (蒋柏斌),[2] Lizhen Huang (黄丽珍),[2] Kai Du (杜凯),[2,3] Runchang Zhao (赵润昌),[2] Ping Li (李平),[2] Wei Wang (王伟),[2] Jingqin Su (粟敬钦),[2,3] Yongkun Ding (丁永坤),[2,3] Xian-Tu He (贺贤土),[1,3,4] and Weiyan Zhang (张维岩)[5]

[1]*Institute of Applied Physics and Computational Mathematics, Beijing 100088, China*
[2]*Research Center of Laser Fusion, Chinese Academy of Engineering Physics, Mianyang 621900, China*
[3]*Collaborative Innovation Center of IFSA, Shanghai Jiao Tong University, Shanghai 200240, China*
[4]*Center for Applied Physics and Technology, Peking University, Beijing, 100871, China*
[5]*Chinese Academy of Engineering Physics, Mianyang 621900, China*

(Received 18 February 2016; revised manuscript received 6 June 2016; published 7 July 2016)

The first spherical hohlraum energetics experiment is accomplished on the SGIII-prototype laser facility. In the experiment, the radiation temperature is measured by using an array of flat-response x-ray detectors (FXRDs) through a laser entrance hole at four different angles. The radiation temperature and M-band fraction inside the hohlraum are determined by the shock wave technique. The experimental observations indicate that the radiation temperatures measured by the FXRDs depend on the observation angles and are related to the view field. According to the experimental results, the conversion efficiency of the vacuum spherical hohlraum is in the range from 60% to 80%. Although this conversion efficiency is less than the conversion efficiency of the near vacuum hohlraum on the National Ignition Facility, it is consistent with that of the cylindrical hohlraums used on the NOVA and the SGIII-prototype at the same energy scale.

DOI: 10.1103/PhysRevLett.117.025002

Indirect drive inertial confinement fusion (ICF) uses high-Z cavities or hohlraums to convert intense laser power or charged particle beams into soft x rays. These x rays are used to produce an ablation drive that compresses the D-T fuel capsule placed at the center of the hohlraum, driving it to ignition and burn [1,2]. In hohlraum physics studies, hohlraum energetics and radiation symmetry are the two crucial but ambivalent contents. A successful ignition hohlraum design must balance the trade-off between energetics and radiation symmetry. Recent experiments on the National Ignition Facility (NIF) indicate that radiation asymmetry is one of the main obstacles in achieving ignition [3–6]. Based on this understanding, Lan *et al.* proposed the octahedral spherical hohlraum as a candidate for the ignition hohlraum [7]. It has been shown that the octahedral spherical hohlraums [7,8] have natural superiority in maintaining high radiation symmetry during the whole capsule implosion process and are much more insensitive to the random variations [9]. Then, the energetics of the spherical hohlraum is the other important issue concerned with the ignition hohlraum design.

The conversion efficiency of the hohlraum, that is the ability of the hohlraum to convert incident lasers into soft x rays, is a key issue in hohlraum energetics studies. The ignition threshold factor [10–12], a measure of how close the experiments are to ignition, scales approximately as implosion velocity to the eighth power. The implosion velocity depends directly on the hohlraum drive. Hence, how many radiation fluxes can be produced by a hohlraum for a given laser energy is always the first issue of concern in hohlraum studies. There have been a great deal of experiments performed on the NOVA [13,14], OMEGA [15], SGIII-Prototype [16–18], and NIF [19–21] in order to investigate the conversion efficiencies of the cylindrical hohlraums with 2 laser entrance holes (LEHs). In these experiments, the hohlraum radiation flux was mainly measured by a broadband x-ray spectrometer through a LEH at a specific angle relative to the hohlraum axis, and the hohlraum conversion efficiency is calculated by using the measured flux [22].

How to characterize the radiation field of the octahedral spherical hohlraums with six LEHs is an important but difficult issue for energetics study since there is not any experiment studying the octahedral hohlraum energetics. However, there are great numbers of experimental data of the cylindrical hohlraum energetics. In order to explore the characterization of the radiation field of the octahedral spherical hohlraum on a future laser facility, we performed the first energetics experiment of the spherical hohlraum with 2 LEHs adopting the experimental experience of the cylindrical hohlraums on the SGIII prototype. For the first time, the hohlraum radiation flux is measured by an array of flat-response x-ray detectors (FXRD) [23] through a LEH at different angles with respect to the hohlraum axis. The radiation temperature and the M-band fraction inside the hohlraum are determined by using the shock wave technique [17].

The experiments are conducted on the SGIII-prototype laser facility consisting of eight laser beams with 800 J per beam at 0.35 μm, and the eight laser beams simultaneously irradiate the hohlraum from two ends at an incidence cone of 45° angle. The laser beams are smoothed by using the continuous phase plates (CPPs). The CPP produces a nearly circular laser spot at the hohlraum wall with radius of about 250 μm, and the laser intensity is about 4×10^{14} W/cm^2. In the experiments, a constant power (flattop) laser pulse of 1 ns with 150 ps rising and falling times is used. It should be noted that the four laser beams from one LEH have a 45° difference in the azimuthal direction with those from the other LEH on the SGIII prototype. So the laser spots nearly form a ring on the waist of the spherical hohlraum. The Full Aperture Backscatter System and Near Backscatter System are used to measure the backscatter fraction of the incident laser due to the stimulated Brillouin scattering and the stimulated Raman scattering. The experimental results indicate that the total reflected energy due to laser plasma interactions is less than 3% of the total incident energy.

Two types of spherical hohlraums are used in the experiments to investigate the energetics difference between them. One is the spherical hohlraum with plane LEHs, and the other is the spherical hohlraum with cylindrical LEHs [7,24]. The gold (Au) vacuum hohlraums have a wall thickness of 25 μm, a radius of 850 μm, and a LEH radius R_L of 400 μm. For the spherical hohlraum with cylindrical LEHs, the radius of the cylindrical LEH outer ring is taken as $1.5R_L = 600$ μm. The experimental results indicate that there are no intrinsic differences between these two types of hohlraums. In the following, we will not present the experimental results and relevant analyses of the two hohlraums separately. An array of FXRDs is used to measure the soft x-ray radiation flux streaming out of a LEH at four different angles with respect to the hohlraum axis. The four angles are 20°, 30°, 45°, and 55°, respectively. The FXRD is a combination of a windowless x-ray detector and a compound filter [23]. In some shots, the shock wave technique with two witness materials, aluminum and titanium, are used to determine the radiation temperature and the fraction of Au M-band flux (> 1.6 keV) inside the hohlraum [17]. And, we use a streaked optical pyrometer (SOP) to measure the shock velocities in aluminum and titanium. In these shots with shock wave witness materials, seven of eight laser beams are used to drive the hohlraums. The M-band fraction is also determined by the M-band flux observed by a filtered M-band x-ray detector (MXRD) [25], which is placed at the same place with the FXRD at 20°. In Fig. 1, we present the schematic of the experimental setup, the arrangement of the main diagnostic devices, and the view fields of the FXRDs.

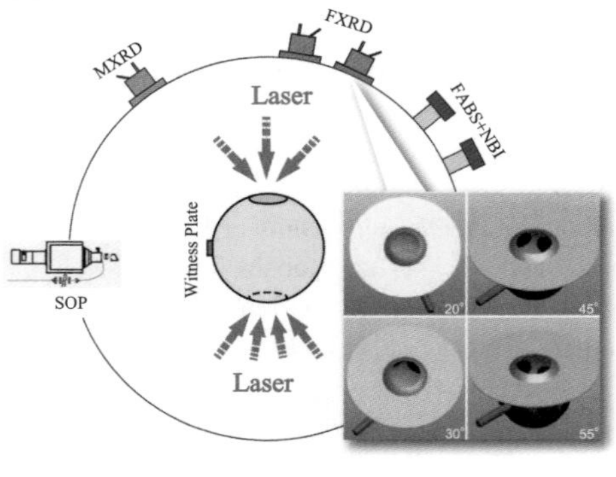

FIG. 1. Schematic view of the experimental setup, the arrangement of the main diagnostic devices and the view fields of the FXRDs at the four different angles.

The experiments are modeled with the two-dimensional radiation hydrodynamic code LARED [16]. LARED uses a flux-limited model with a flux limiter based on the Spitzer-Härm electron conduction of $f_e = 0.1$. In the LARED code, the sophisticated opacity and equation of state databases are used for the plasmas in the local thermodynamic equilibrium (LTE) state. The non-LTE state plasmas are described by the ideal gas EOS and the relativistic Hartree-Fock-Slater self-consistent average atom model OPINCH [26], which involves the processes of free-free, free-bound, and bound-bound transitions.

Figure 2 compares the temporal histories of radiation temperature measured by the FXRDs with the simulated results for one experimental shot with laser energy 5.4 kJ. It is clear that the simulation agrees well with the observed

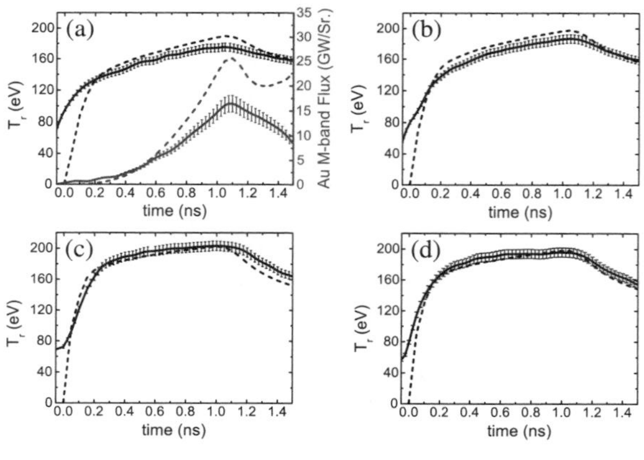

FIG. 2. The temporal behaviors of radiation temperature measured by the FXRDs (black solid lines) and calculated by the LARED code (black dashed lines) at the four different angles. (a) 20°, (b) 30°, (c) 45°, and (d) 55°. The blue solid and dashed lines in (a) are the measured and simulated M-band fluxes at 20°.

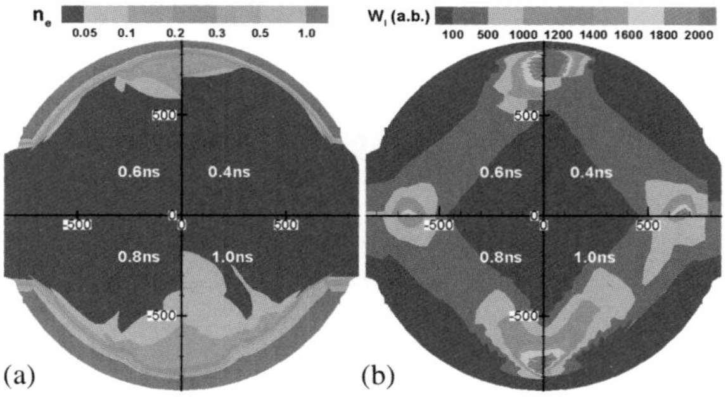

FIG. 3. The simulated spatial distributions of the normalized electron density n_e (a) and the laser deposition W_l (b) inside the spherical hohlraum at different times.

flux of the FXRDs at 45° and 55°. Since there are laser spots, hohlraum wall, and corona plasmas in the view fields of the FXRDs at 45° and 55°, the agreement indicates that the code can predict the radiation flux from the hybrid region well [15,16,27]. However, the simulated radiation flux is higher than the measured data of 20° and 30°. As indicated in Fig. 1, there is no laser spot in the view field of the FXRD at 20°. A portion of the radiation flux measured by the FXRD at 20° comes from the reemission region of the hohlraum wall, and others come from the hot Au bubble plasmas [28] entering into the view field of the FXRD. The Au bubble plasmas are from the laser spots and contribute the photons with higher energies to the total flux observed by the FXRD at 20°. Because the LARED code properly predicts the measurement results of the FXRDs at 45° and 55°, the discrepancy between the simulated radiation flux and the observed results of the FXRDs at 20° and 30° may be due to the overprediction on the movement of the hot Au bubble from laser spots into the hohlraum interior. In Fig. 3(a), we present the spatial distributions of the electron density normalized to the critical density inside the spherical hohlraum at different times. Here, $n_e = N_e/N_c$, N_e is the electron density, and N_c is the critical electron density for the laser with wavelength of 0.35 μm. Shown in Fig. 3(b) is the laser deposition. It is obvious that there are dense and hot plasmas entering into the view field of the FXRDs at 20° and 30°. This also leads to the overprediction of the Au M-band flux observed at 20°. As shown in Fig. 2(a), the simulated Au M-band flux is obviously higher than the experimental observation after 0.6 ns.

Shown in Fig. 4(a) are the simulated peak radiation temperatures as a function of the view angles compared to the measurements of the FXRDs at the four different angles. The peak radiation temperatures are taken from the simulation with laser energy 5.4 kJ. In order to make the comparison, all the measured peak temperatures are scaled to that of 5.4 kJ. In Fig. 4(b), we present the simulated M-band fraction at peak radiation temperature against the experimental M-band fraction determined by the FXRD at 20°. Although the simulated angular distribution of peak radiation temperature is consistent with the observation, there is relatively large quantitative discrepancy between simulation and experiment. The simulated temperature of 45° is close to the measured result, while the simulated temperatures of 20°, 30°, and 55° are higher than the measured results. In addition, the simulated M-band fraction is slightly higher than the experimental result obtained by the FXRD and MXRD at 20°. Considering that the simulated radiation flux of 20° is higher than the measurement, this result indicates that the code may obviously overpredict the movement of the Au bubble plasmas and the M-band emissions of the corona plasmas.

The radiation temperature and M-band fraction inside the hohlraum determined by the shock wave technique are higher than the measurement obtained by the FXRDs, as shown in Fig. 5. The radiation temperature inside the hohlraum is about 210 eV with drive laser energy of 5.5 kJ, and the M-band fraction is about 15%. The witness plate of the shock wave is placed on the waist of the spherical hohlraum, so the radiation flux from the laser spots which are also located at the waist of the hohlraum contributes largely to the flux incident on the interior surface of the shock witness. Because the x-ray emission, especially the M-band emission, from laser spots is obviously higher than

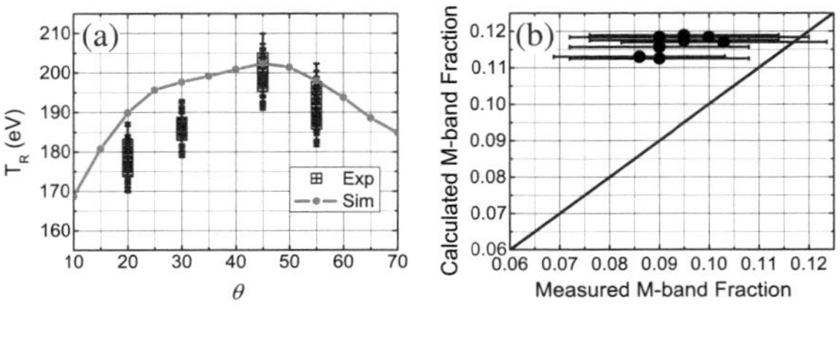

FIG. 4. (a) The simulated peak radiation temperatures as a function of the view angle (the red line with solid circle) compared to the measured results (the black squares). (b) The simulated M-band fraction at peak radiation temperature against the experimental M-band fraction determined by the FXRD and MXRD at 20°.

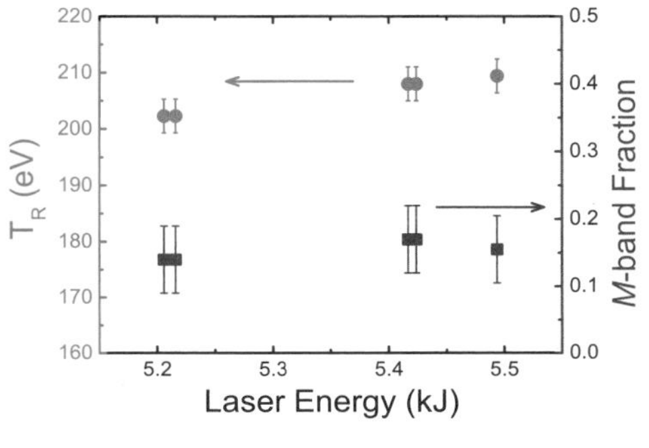

FIG. 5. The radiation temperature (red solid circles) and M-band fraction (blue solid squares) inside the spherical hohlraum determined by the shock wave technique.

the hohlraum wall, the radiation temperature and M-band fraction determined by the shock wave technique are higher than the measurement results of the FXRDs.

The hohlraum conversion efficiency, which characterizes the ability of a hohlraum to convert a laser into soft x rays, is one of the fundamental problems in indirect drive ICF. The ability of any hohlraum to convert lasers into soft x rays can be modeled by a source-sink equation,

$$\eta_{CE}(E_L - E_{BS}) = E_{\text{wall}} + E_{\text{LEH}}. \quad (1)$$

Here, η_{CE} is the laser to x-ray conversion efficiency, E_L is the laser input energy, E_{BS} is the laser energy scattered out of the hohlraum via the laser plasma interactions, E_{wall} and E_{LEH} are radiation losses into the hohlraum wall and out of the LEHs, respectively. Because E_{wall} and E_{LEH} cannot be measured directly in the experiments, it depends heavily on the radiation hydrodynamic code in order to obtain the real hohlraum conversion efficiency. If the radiation hydrodynamic code cannot reproduce all of the experimental observations, it is quite difficult to obtain the real hohlraum conversion efficiency.

In order to evaluate the conversion efficiency, we present the energy partition as a function of time extracting from the simulation, as shown in Fig. 6(a). According to the simulation, the converted x-ray energy is mainly absorbed by the hohlraum wall, and the simulated conversion efficiency η_{sim} is about 80%. However, the simulated conversion efficiency η_{sim} is not the real hohlraum conversion efficiency due to the discrepancy between simulation and experiment. In order to find out the real hohlraum conversion efficiency, we define the nominal conversion efficiency as

$$\eta_N = (E_{\text{sim}}/E_{\text{exp}})\eta_{\text{sim}}. \quad (2)$$

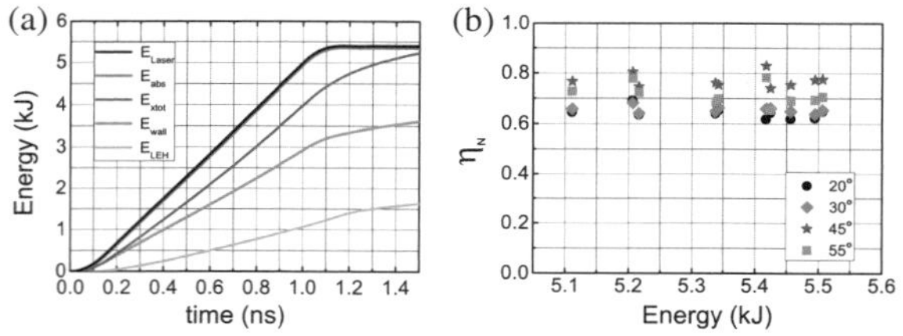

FIG. 6. (a) The time-dependent sums of incident laser energy E_{laser} (black line), absorbed laser energy E_{abs} (red line), total converted x-ray energy E_{xtot} (blue line), x-ray energy absorbed by the hohlraum wall E_{wall} (green line), and the x-ray energy escaping from the LEHs E_{LEH} (cyan line). (b) The nominal conversion efficiencies of the spherical hohlraum for the different experimental shots obtained by using the observed radiation temperatures of the FXRDs. The black circles, red diamonds, blue stars, and green squares are the nominal conversion efficiencies calculated by using the radiation temperatures measured by the FXRDs at 20°, 30°, 45°, and 55°, respectively.

Here, E_{sim} is the input laser energy in the simulation, $E_{\text{exp}} = E_L - E_{BS}$ is the measured laser energy coupled the hohlraum in the experiment. The nominal conversion efficiency is generally obtained by adjusting the input laser energy so that the simulated radiation temperature is consistent with the measured temperature at a specific angle [16]. If the simulation code can reproduce all of the observed flux at different angles for one experimental shot, the nominal conversion efficiency would be the real hohlraum conversion efficiency. However, one would obtain different nominal conversion efficiencies even for the same shot if the code cannot reproduce all of the observed radiation flux at different angles. In this case, the nominal conversion efficiencies would be a range of the real conversion efficiency. In Fig. 6(b), we present the nominal conversion efficiencies of the spherical hohlraum for different experimental shots by using the radiation temperature measured by the FXRDs at the four different angles. As indicated in Fig. 6(b), the nominal conversion efficiency is not unique for one experimental shot if the observed radiation temperatures of the FXRDs at different angles are used. As a result, we cannot obtain the real conversion efficiency but a range of the hohlraum conversion efficiency. And the conversion efficiency of the spherical hohlraum used in the experiment is in the range from 60% to 80%.

Of course, there are some external reasons for the discrepancies between the experimental observations and simulations, such as the three-dimensional effects and the power fluctuations in laser beams. The variations in hohlraum conversion efficiencies of a specific angle for different experimental shots reflect the fluctuations in power of laser beams. And the difference between the nominal conversion efficiencies of different angles for the same shot specifies the discrepancy between the simulation and experiment. The experiments on the NIF show that there is drive deficit in the gas filled hohlraum [22,29], that is the simulation code overpredicts the incident flux on the capsule. Both our investigation and the experiments on the NIF indicate that the real hohlraum conversion efficiency is difficult to be determined if the simulation code cannot reproduce all of the experimental observations.

To conclude, we have accomplished the first spherical hohlraum energetics experiment on the SGIII-prototype laser facility. The radiation temperature is measured by using an array of FXRDs through a LEH at different angles for the first time. In addition, the radiation temperature and M-band fraction inside a spherical hohlraum are determined by the shock wave technique. The experimental results show that the radiation temperatures measured by the FXRDs at different angles are different and depend on the FXRDs' view field. For the spherical hohlraum used in the experiment, the radiation temperature is about 197 eV with laser energy about 5.5 kJ. Because the shock wave witness plate is facing to the laser spots, the radiation temperature and M-band fraction determined by the shock wave technique are slightly higher than the measurement results of the FXRDs.

The hohlraum conversion efficiency is generally calculated by comparing the measured hohlraum radiation temperature with the simulation result. In the experiment, we have measured the hohlraum radiation temperature at four different angles. Because the simulation code cannot reproduce all the measurement results, we find that the real hohlraum conversion efficiency is quite difficult to be determined. As a result, we obtain different nominal conversion efficiencies when comparing the measurement results at different angles with the simulation results. Therefore, what we obtained is not an accurate conversion efficiency but a range of hohlraum conversion efficiency. According to the experimental results, the conversion efficiency of the vacuum spherical hohlraum is in the range from 60% to 80%. Although this conversion efficiency is less than the conversion efficiency of the near vacuum hohlraum on the NIF [30], it is consistent with that of the cylindrical hohlraums used on the NOVA [14] and the SGIII-prototype [16,18] at the same energy scale. The difference of the conversion efficiency between the vacuum spherical hohlraum on the SGIII-prototype and the near vacuum hohlraum on the NIF may be due to the difference in the laser pulse and energy scale.

This work is supported by the National Natural Science Foundation of China (Grants No. 11405011, No. 11475033 and No. 11305159).

[1] J. D. Lindl, Phys. Plasmas **2**, 3933 (1995).
[2] S. Atzeni and J. Meyer-ter-Vehn, *The Physics of Inertial Fusion* (Oxford Science, Oxford, 2004).
[3] D. A. Callahan *et al.*, Phys. Plasmas **19**, 056305 (2012).
[4] R. P. J. Town *et al.*, Phys. Plasmas **21**, 056313 (2014).
[5] O. A. Hurricane *et al.*, Phys. Plasmas **21**, 056314 (2014).
[6] D. S. Clark, M. M. Marinak, C. R. Weber, D. C. Eder, S. W. Haan, B. A. Hammel, D. E. Hinkel, O. S. Jones, J. L. Milovich, P. K. Patel, H. F. Robey, J. D. Salmonson, S. M. Sepke, and C. A. Thomas, Phys. Plasmas **22**, 022703 (2015).
[7] K. Lan, J. Liu, D. X. Lai, W. D. Zheng, and X. T. He, Phys. Plasmas **21**, 010704 (2014); K. Lan, X. T. He, J. Liu, W. D. Zheng, and D. X. Lai, *ibid.* **21**, 052704 (2014); K. Lan and W. D. Zheng, *ibid.* **21**, 090704 (2014).
[8] S. A. Bel'Kov *et al.*, Laser Part. Beams **17**, 591 (1999).
[9] W. Y. Huo, J. Liu, Y. Q. Zhao, W. D. Zheng, and K. Lan, Phys. Plasmas **21**, 114503 (2014).
[10] D. S. Clark, S. W. Haan, and J. D. Salmonson, Phys. Plasmas **15**, 056305 (2008).
[11] J. F. Gu, Z. S. Dai, Z. F. Fan, S. Y. Zou, W. H. Ye, W. B. Pei, and S. P. Zhu, Phys. Plasmas **21**, 012704 (2014).
[12] S. W. Haan *et al.*, Phys. Plasmas **18**, 051001 (2011).
[13] R. L. Kauffman, L. J. Suter, C. B. Darrow, J. D. Kilkenny, H. N. Kornblum, D. S. Montgomery, D. W. Phillion, M. D. Rosen, A. R. Theissen, R. J. Wallace, and F. Ze, Phys. Rev. Lett. **73**, 2320 (1994).
[14] L. J. Suter, A. A. Hauer, L. V. Powers, D. B. Ress, N. Delameter, W. W. Hsing, O. L. Landen, A. R. Thiessen, and R. E. Turner, Phys. Rev. Lett. **73**, 2328 (1994).
[15] C. Decker, R. E. Turner, O. L. Landen, L. J. Suter, P. Amendt, H. N. Kornblum, B. A. Hammel, T. J. Murphy, J. Wallace, N. D. Delamater, P. Gobby, A. A. Hauer, G. R. Magelssen, J. A. Oertel, J. Knauer, F. J. Marshall, D. Bradley, W. Seka, and J. M. Soures, Phys. Rev. Lett. **79**, 1491 (1997).
[16] W. Y. Huo, G. L. Ren, K. Lan, X. Li, C. S. Wu, Y. S. Li, C. L. Zhai, X. M. Qiao, X. J. Meng, D. X. Lai, W. D. Zheng, P. J. Gu, W. B. Pei, S. W. Li, R. Q. Yi, T. M. Song, X. H. Jiang, D. Yang, S. E. Jiang, and Y. K. Ding, Phys. Plasmas **17**, 123114 (2010).
[17] W. Y. Huo *et al.*, Phys. Rev. Lett. **109**, 145004 (2012).
[18] H. S. Zhang, D. Yang, P. Song, S. Y. Zou, Y. Q. Zhao, S. W. Li, Z. C. Li, L. Guo, F. Wang, X. S. Peng, H. Y. Wei, T. Xu, W. D. Zheng, P. J. Gu, W. B. Pei, S. E. Jiang, and Y. K. Ding, Phys. Plasmas **21**, 112709 (2014).
[19] N. B. Meezan *et al.*, Phys. Plasmas **17**, 056304 (2010).
[20] J. L. Kline *et al.*, Phys. Rev. Lett. **106**, 085003 (2011).
[21] R. E. Olson, L. J. Suter, J. L. Kline, D. A. Callahan, M. D. Rosen, S. N. Dixit, O. L. Landen, N. B. Meezan, J. D. Moody, C. A. Thomas, A. Warrick, K. Widmann, E. A. Williams, and S. H. Glenzer, Phys. Plasmas **19**, 053301 (2012).
[22] J. D. Moody *et al.*, Phys. Plasmas **21**, 056317 (2014).
[23] Z. C. Li, X. H. Jiang, S. Y. Liu, T. X. Huang, J. Zheng, J. M. Yang, S. W. Li, L. Guo, X. F. Zhao, H. B. Du, T. M. Song, R. Q. Yi, Y. G. Liu, S. E. Jiang, and Y. K. Ding, Rev. Sci. Instrum. **81**, 073504 (2010).
[24] W. Y. Huo *et al.*, Mat. Rad. Extr. **1**, 2 (2016);K. Lan *et al.*, *ibid.* **1**, 8 (2016).
[25] L. Guo, S. W. Li, J. Zheng, Z. C. Li, D. Yang, H. B. Du, L. F. Hou, Y. L. Cui, J. M. Yang, S. Y. Liu, S. E. Jiang, and Y. K. Ding, Meas. Sci. Technol. **23**, 065902 (2012).
[26] F. J. D. Serduke, E. Minguez, S. J. Davidson, and C. A. Iglesias, J. Quant. Spectrosc. Radiat. Transfer **65**, 527 (2000).
[27] A. S. Moore, T. M. Guymer, J. Morton, B. Williams, J. L. Kline, N. Bazin, C. Bentley, S. Allan, K. Brent, A. J. Comley, K. Flipp, J. Cowan, J. M. Taccetti, Katie Mussack-Tamashiro, D. W. Schmidt, C. E. Hamilton, K. Obrey, N. E. Lanier, J. B. Workman, and R. M. Stevenson, J Quant. Spectrosc. Radiat. Transf. **159**, 19 (2015).
[28] M. B. Schneider, S. A. MacLaren, K. Widmann, N. B. Meezan, J. H. Hammer, B. E. Yoxall, P. M. Bell, L. R. Benedetti, D. K. Bradley, D. A. Callahan, E. L. Dewald, T. Döppner, D. C. Eder, M. J. Edwards, T. M. Guymer, D. E. Hinkel, M. Hohenberger, W. W. Hsing, M. L. Kervin, J. D. Kilkenny *et al.*, Phys. Plasmas **22**, 122705 (2015).
[29] S. A. MacLaren, M. B. Schneider, K. Widmann, J. H. Hammer, B. E. Yoxall, J. D. Moody, P. M. Bell, L. R. Benedetti, D. K. Bradley, M. J. Edwards, T. M. Guymer, D. E. Hinkel, W. W. Hsing, M. L. Kervin, N. B. Meezan, A. S. Moore, and J. E. Ralph, Phys. Rev. Lett. **112**, 105003 (2014).
[30] L. F. Berzak Hopkins *et al.*, Phys. Rev. Lett. **114**, 175001 (2015).

PHYSICS OF PLASMAS 23, 010703 (2016)

Ignition conditions relaxation for central hot-spot ignition with an ion-electron non-equilibrium model

Zhengfeng Fan,[1,2,a)] Jie Liu,[1,2,3] Bin Liu,[1] Chengxin Yu,[1] and X. T. He[1,3]
[1]*Institute of Applied Physics and Computational Mathematics, Beijing 100088, China*
[2]*Center for Fusion Energy Science and Technology, Chinese Academy of Engineering Physics, Beijing 100088, China*
[3]*Center for Applied Physics and Technology, Peking University, Beijing 100871, China*

(Received 29 October 2015; accepted 4 January 2016; published online 22 January 2016)

Fusion ignition experiments on the National Ignition Facility have demonstrated >5 keV hot spot with ρR_h lower than 0.3 g/cm^2 [Döppner *et al.*, Phys. Rev. Lett. **115**, 055001 (2015)]. We present an ion-electron non-equilibrium model, in which the hot-spot ion temperature is higher than its electron temperature so that the hot-spot nuclear reactions are enhanced while energy leaks are considerably reduced. Theoretical analysis shows that the ignition region would be significantly enlarged in the hot-spot ρR-T space as compared with the commonly used equilibrium model. Simulations show that shocks could be utilized to create and maintain non-equilibrium conditions within the hot spot, and the hot-spot ρR requirement is remarkably reduced for achieving self-heating.
 [http://dx.doi.org/10.1063/1.4940315]

Inertial confinement fusion[1–3] aims at assembling and heating an equimolecular mixture of deuterium-tritium (D-T) fusion fuel to high densities and temperatures, so that the D-T fuel reacts strongly within a short period (~100–200 ps) before it disassembles, producing considerable thermonuclear energy. In laser-driven central hot-spot ignition, a spherical shell of cryogenic D-T fuel, coated by a low-Z ablator, is imploded by direct laser beams or by indirect x-ray radiation converted from laser beams to a high velocity, so that the fuel is highly compressed under the spherical convergent effect, and an ignition hot spot with a much lower density and a much higher temperature is formed in the center, triggering main fuel burn. Commonly, ions and electrons in the central hot spot are considered to be in equilibrium.[3,4] Under this consideration, it is required that the hot-spot areal density, $\rho R_h \geq\sim 0.3$ g/cm^2, and the hot-spot temperature, $T_h \geq\sim 5$ keV.

Recent experiments have made great progress toward ignition. The indirect drive experiments, done on the National Ignition Facility (NIF),[5] the world's largest and most powerful laser system in Livermore, California, have made the following key advances. The laser energy coupling to the gas-filled hohlraum reaches 84% over a wide range of laser parameters from 350 to 520 TW and from 1.2 to 1.9 MJ.[6] The hohlraum radiation temperature reaches ~320 eV, which exceeds the 300-eV point-design goal. The fuel compression achieves ~1.3 g/cm^2, and the hot-spot symmetry meets the ignition specification.[7,8] Especially, the NIF high-foot high-adiabat experiments have demonstrated that the energy produced through fusion reactions exceeds the energy delivered to D-T fusion fuel during capsule implosion.[9–11] Meanwhile, recent direct-drive experiments demonstrated the generation of gigabar spherical shock on the OMEGA laser facility.[12] Nevertheless, the hot-spot areal densities achieved in the NIF experiments are lower than 0.3 g/cm^2,[2,8,11,13,14] and moreover, the capsule suffers from severe drive asymmetries and hydrodynamic instabilities.[15,16] Under these circumstances, it has special significance to lower the ignition threshold.

In this letter, an ion-electron non-equilibrium model is proposed for relaxing the central hot-spot ignition conditions. According to this model, a higher hot-spot ion temperature relative to a lower electron temperature enhances the hot-spot nuclear reactions and reduces its energy leaks from the electron thermal conduction and electron bremsstrahlung, and as a result the ignition region is significantly enlarged in the ρR-T space of the hot spot. The following content consists of two parts. First, theoretical results of our ion-electron non-equilibrium model are discussed based on both hot-spot self-heating analysis and hot-spot burn-and-ignition dynamics. Second, simulations of an ignition capsule are presented to demonstrate improvements by creating ion-electron non-equilibrium conditions in the central hot spot.

The hot-spot-main-fuel configuration is shown in Fig. 1. The hot spot, with a uniform density of ρ_h and a radius of R_h, is surrounded by a highly compressed main fuel layer. D-T thermonuclear reactions occur within the hot spot, and a fraction of the main fuel enters the hot spot via mass ablation. The propagation of the burn wave through the main fuel is not included because in that case the capsule has already been ignited. Initial conditions are investigated at the stagnation time when the inward kinetic energy of the capsule has been exhausted and the fuel compression has reached the maximum. Thereafter, the hot spot burns, and an outward shock is created due to the pressure imbalance between the hot spot and the main fuel. Meanwhile, the hot spot expands approximately at the post-shock velocity u_h. We emphasize that the hot spot has separate ion and electron temperatures T_i and T_e, which is the key mechanism for relaxing the hot-spot ignition conditions. In order to compare with the equilibrium case, we define the corresponding equilibrium temperature as $T_h = (T_i + T_e)/2$ and an non-equilibrium factor as $f = T_i/T_h$.

a)fan_zhengfeng@iapcm.ac.cn

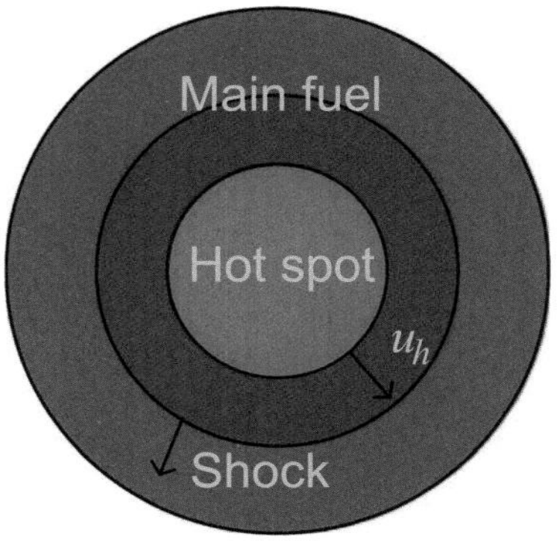

FIG. 1. Schematic of the hot-spot-main-fuel configuration.

In the hot spot, the major energy source comes from energy deposition of thermonuclear products (majorly, alpha particles), while the energy losses are caused by radiation loss, thermal conduction, and outward mechanical work (PdV work) to the main fuel. A successful ignition requires self-sustaining thermonuclear reactions within the hot spot until the D-T fuel disassembles. A simple way to describe the ignition is the self-heating condition, i.e., the alpha-particle heating power per unit volume greater than or equal to that of the radiation loss and electron conduction

$$W_\alpha - W_r - W_e \geq 0, \tag{1}$$

where $W_\alpha = A_\alpha \rho_h^2 \langle\sigma v\rangle(T_i) f_\alpha$, $W_r = A_r \rho_h^2 T_e^{1/2}$, and $W_e = 3A_e T_e^{7/2}/R_h^2/\ln\Lambda$; $\langle\sigma v\rangle(T_i)$ is the D-T reactivity and the Hively expression[17] is used, f_α is the fraction of alpha-particle energy deposited within the hot spot, $\ln\Lambda$ is the Coulomb logarithm, and A_α, A_r, and A_e are three constants as given in Ref. 3. By approximating f_α as unity, we have

$$\rho_h R_h \geq \left(\frac{3A_e}{\ln\Lambda} \frac{(2-f)^{7/2} T_h^{7/2}}{A_\alpha \langle\sigma v\rangle(fT_h) - A_r(2-f)^{1/2} T_h^{1/2}}\right)^{1/2}, \tag{2}$$

where T_i has been substituted by fT_h. The denominator at the right side of Eq. (2) should be non-negative, which determines the minimal hot-spot temperature required for ignition

$$T_m \simeq \frac{T_{m,0}(2-f)^{0.13}}{f^{1.18}}, \tag{3}$$

where $T_{m,0} = 4.3$ keV is the value for the equilibrium case. Table I presents the minimal hot-spot temperatures for different non-equilibrium factors ranging from 0.7 to 1.4. Clearly, the minimal hot-spot temperature decreases dramatically as f increases. For f greater than unity, the required temperatures are much smaller than the classical value of 4.3 keV. Taking $f = 1.2$ for example, it turns to be 3.4 keV, which is ~22% lower than the classical value. Figure 2 shows the hot-spot self-heating conditions in the ρR-T space for different non-equilibrium factors ranging from 0.9 to 1.3. In Fig. 2, the up-right part enclosed by each line denotes the region for hot-spot self-heating. When the non-equilibrium factor is less than unity, this region shrinks greatly in the ρR-T space. Whereas on the other side ($f > 1$), this region is significantly enlarged. It is worth of noting that the relaxation time between ions and electrons is about several picoseconds as $\rho_h \in [50, 100]$ g/cm^3 and $T_e \in [4, 10]$ keV, and there may be doubt about the validity of our model. However, in the following hot-spot ignition-and-burn dynamics model, where the energy transfer between ions and electrons is included, we observe that after ~2–3 ps the self-heating conditions for the equilibration case are satisfied. Therefore, in the non-equilibrium regime where the hot-spot ion temperature is greater than its electron temperature, the conditions required for the hot-spot self-heating is relaxed.

TABLE I. Minimum hot-spot temperatures for different non-equilibrium factors.

f	0.7	0.8	0.9	1.1	1.2	1.3	1.4
T_m (keV)	6.8	5.8	5.0	3.8	3.4	3.0	2.7

The above discussion on the hot-spot self-heating is static because it is focused on a specific time. Now we discuss the hot-spot ignition-and-burn dynamics. The hot-spot energy equations for both the ions and electrons are written separately as

$$C_{v,i}\rho_h \frac{dT_i}{dt} = W_\alpha f_{\alpha i} - W_{ie} - W_{m,i}, \tag{4}$$

$$C_{v,e}\rho_h \frac{dT_e}{dt} = W_\alpha f_{\alpha e} + W_{ie} - W_{m,e} - W_r - W_e, \tag{5}$$

where $C_{v,i}$ and $C_{v,e}$ are the specific heat respectively for ions and electrons, and an ideal equation of state is used; $f_{\alpha i} \simeq T_e/(T_e + 32)$ and $f_{\alpha_e} \simeq 32/(T_e + 32)$ are the fractions of alpha-particle energy deposited respectively to them with T_e and T_i in unit of keV.[18] The energy exchange power (per unit volume) between ions and electrons is given by $W_{ie} = A_{\omega,ei}\rho_h^2 \ln\Lambda_{ei}(T_e - T_i)/T_e^{3/2}$ with $A_{\omega,ei} = 5.6 \times 10^{14}$ kJ cm^3 g^{-2} s^{-1} keV$^{1/2}$. $W_{m,i} = 3P_i u_h/R_h$ and $W_{m,e} = 3P_e u_h/R_h$, respectively, represent the mechanical power of ions and electrons. The evolution of the hot spot mass depends on the mass ablation rate off the main fuel layer. At a main-fuel density of ~1000 g/cm^3, the mean free paths of radiation

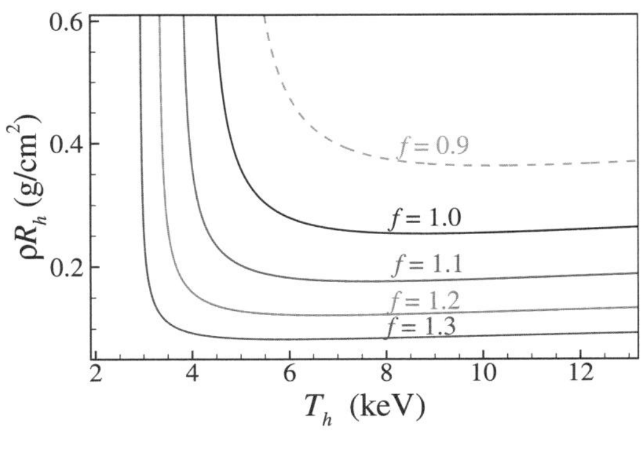

FIG. 2. Hot-spot self-heating conditions for different non-equilibrium factors.

x-rays, electrons, and charged alpha particles are all short enough that the energy leaks from the hot spot can be recycled back by mass ablation. Then we have

$$\frac{d\rho_h V_h}{dt} = \frac{[W_r + W_e + W_\alpha(1-f_\alpha)/f_\alpha]V_h}{C_{v,i}T_i + C_{v,e}T_e}, \tag{6}$$

where V_h is the hot-spot volume. The hot-spot radius expands approximately at the post-shock speed, which is defined by

$$\frac{dR_h}{dt} = u_h. \tag{7}$$

In this letter, the dynamics of the hot-spot ignition and burn process is solved by numerical integrations of equations (4)–(7). When the main-fuel layer is thick enough, whether the hot-spot ignites or not depends only on the initial values of T_h, ρ_h, R_h and the non-equilibrium factor f. Here, we do a parameter scanning for $T_h \in [3, 10]$ keV, $\rho_h \in [10, 120]$ g/cm^3, and $R_h \in [5, 65]$ μm. And the ignition is defined by whether the hot-spot ion temperature could achieve 64.2 keV at which the D-T reactivity reaches the maximum or by whether the hot-spot energy gain exceeds 500.[19] The outcome is demonstrated in Fig. 3(a), where the ignition conditions (color surfaces) for different non-equilibrium factors are presented. For the non-equilibrium factors larger than unity, the ignition conditions are significantly relaxed in the ρ-R-T space of the hot spot.

To give an intuitive comprehension, we address two specific comparisons on the hot-spot ignition and burn dynamics. In the first comparison, ρ_h, T_h, and R_h are set to be 120 g/cm^3, 8 keV, and 30 μm, respectively, and we compare a non-equilibrium case ($f = 1.2$) with the case of equilibrium ($f = 1.0$), as shown in Fig. 3(b). Initial conditions for T_i and T_e are 9.6 keV and 6.4 keV when f is 1.2, and both of them are 8 keV when $f = 1.0$. Obviously, the hot-spot ion and electron temperatures, as well as the energy gain, grow much faster in the non-equilibrium case ($f = 1.2$) than that in the case of equilibrium due to the enhanced nuclear fusion reactions and reduced energy leaks from the hot spot. In the second comparison, ρ_h, T_h, and R_h are set to be 90 g/cm^3, 5 keV, and 30 μm, respectively, as shown in Fig. 3(c). Initial conditions are chosen for T_i and T_e as 6.0 keV and 4.0 keV when f is 1.2, and both are 5 keV when f is unity. In the equilibrium case ($f = 1.0$), the hot-spot energy deposition is less than its energy leaks, resulting in the quick decrease of both the ion and electron temperatures, and the hot-spot energy gain is almost two orders less than that required for ignition. Whereas in the non-equilibrium case ($f = 1.2$), the hot spot ignites. First, the hot-spot ion temperature decreases slightly due to the energy exchange between ions and electrons. Meanwhile, the electron temperature grows rapidly. After a short while (~3.4 ps), the electron temperature catches up with the ion temperature at 5.3 keV (6% higher than the initial value of 5.0 keV) and from then on both the ion and electron temperatures grow quickly.

In Fig. 4(a), we redraw the ignition conditions (blue dots) in the hot-spot ρR-T space for a special case of $f = 1.1$.

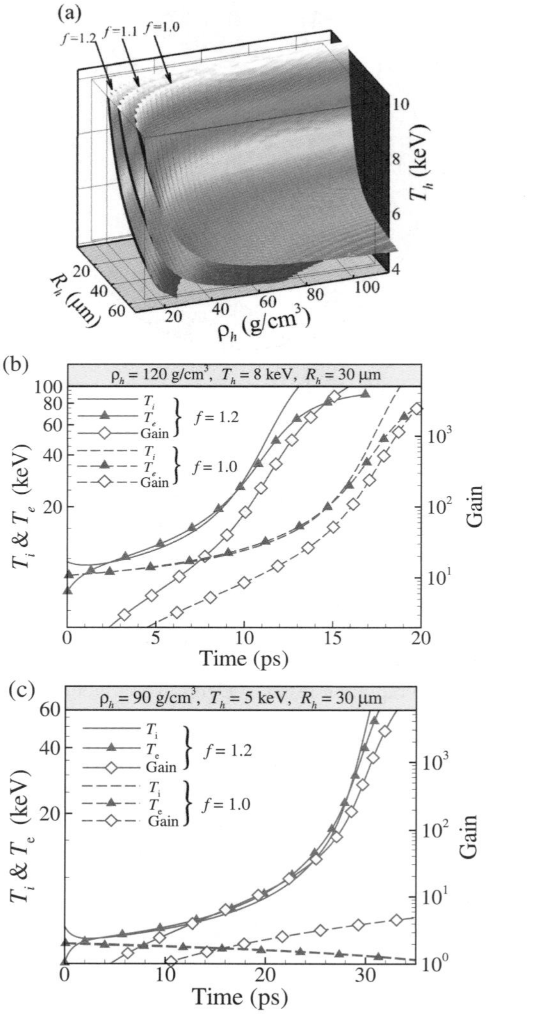

FIG. 3. Results of the hot-spot ignition-and-burn dynamics. (a) Ignition conditions for different non-equilibrium factors in the ρ-R-T space; (b) and (c) comparisons of two non-equilibrium cases with the corresponding equilibrium cases on the hot-spot ignition-and-burn dynamics.

It is seen that the threshold for ignition do not coincide strictly for different hot-spot densities and sizes even if the ρR_h is the same. However, the divergence is not significant, and in this letter we use numerical fittings to obtain averaged ignition conditions. Numerical fittings indicate that the hot-spot ignitions could be fitted using the formula

$$\rho R_h \geq \frac{a - b/T_h}{T_h - T_m}, \tag{8}$$

where a and b are parameters dependent on the non-equilibrium factor f and T_m could be obtained from (3). In Fig. 4(a), the fitting result plotted with a red solid line is compared with the original result for the case of $f = 1.1$. Table II presents a and b parameters for varied non-equilibrium factors. In Fig. 4(b), the fitting results are plotted in curves for varied non-equilibrium factors. Comparison of these curves strongly

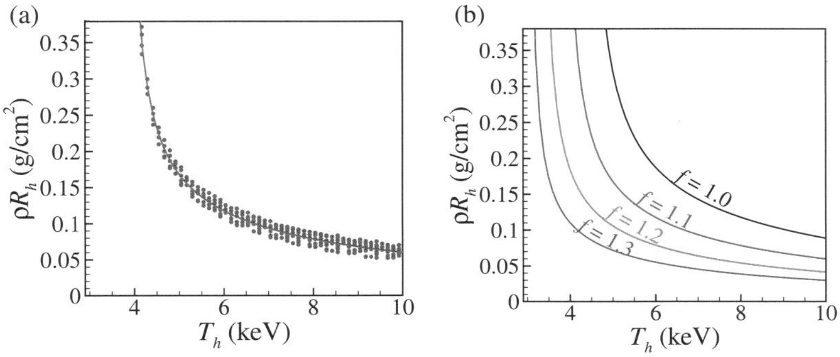

FIG. 4. Fitting results of the hot-spot ignition conditions in the ρR-T space. (a) Comparisons of the fitting result with the original result for $f = 1.1$ and (b) fitting results for varied non-equilibrium factors.

TABLE II. Coefficients in formula (8).

f	0.9	1.0	1.1	1.2	1.3
a	1.18	0.79	0.55	0.39	0.28
b	4.82	2.83	1.71	1.08	0.67

TABLE III. Hot-spot statistics at the self-heating time for the three radiation drives in Fig. 5(b).

Pulse	f	ρR_h (g/cm^2)	T_h (keV)
265–335 eV	1.03	0.19	4.01
275–335 eV	1.04	0.16	3.98
295–335 eV	1.08	0.13	3.96

suggests that the conditions required for hot-spot ignition is relaxed in the non-equilibrium regime when the hot-spot ion temperature is greater than its electron temperature.

One way to realize non-equilibrium central hot-spot ignition is to use properly timed shock(s), where the thermal energy is majorly deposited into ions. Experiments on the OMEGA laser facility have shown that the hot-spot ions and electrons are in a non-equilibrium state at the end of shock burn.[20] In ignition target designs, the shocks created in the pulse-shaping phase merge into a much stronger one, which would create the non-equilibrium states in the central hot spot. However, as the hot-spot satisfies self-heating conditions (or ignitions conditions), the energy relaxation between ions and electrons makes their temperatures close to each other. In this letter, we utilize a secondary shock to maintain the temperature separation between ions and electrons.[21,22] As an example, an indirect-drive ignition capsule is shown in Fig. 5(a), and three well-tuned radiation drives with wedged-peak pulses[22] are shown in Fig. 5(b). The wedged-peak pulses, beginning, respectively, at 265 eV, 275 eV, and 295 eV, peaking at 335 eV, are used to create the secondary shocks. Simulations have been performed with LARED-S which is a multi-dimensional, massively parallel, and Eulerian mesh based radiation hydrodynamic code. Our simulation incorporates the physics of hydrodynamics, electron and ion thermal conduction, radiation, nuclear reaction, and alpha particle transport. Especially electrons and ions have separate temperatures, and the energy transfer between them is described by the Brysk-Spitzer model.[23] Table III presents a brief summary of the hot-spot statistics. The non-equilibrium factor (*f*), obtained by mass-weighted average of the hot spot non-equilibrium (T_i/T_h), increases from 1.03 to 1.08 as the beginning of the wedged-peak pulse is raised from 265 eV to 295 eV, and correspondingly the hot-spot areal density (ρR_h) and burn-weighted average temperature (T_h) required for hot-spot self-heating relaxes. Particularly, the hot-spot areal density in the 295–335 eV pulse is reduced by about 30% compared with the 265–335 eV pulse.

In summary, this letter presents an ion-electron non-equilibrium model for relaxing the central hot-spot ignition conditions in inertial confinement fusion. Both the hot-spot self-heating analysis and a more comprehensive hot-spot dynamics model show that the ignition region would be significantly enlarged in the hot-spot ρR-T

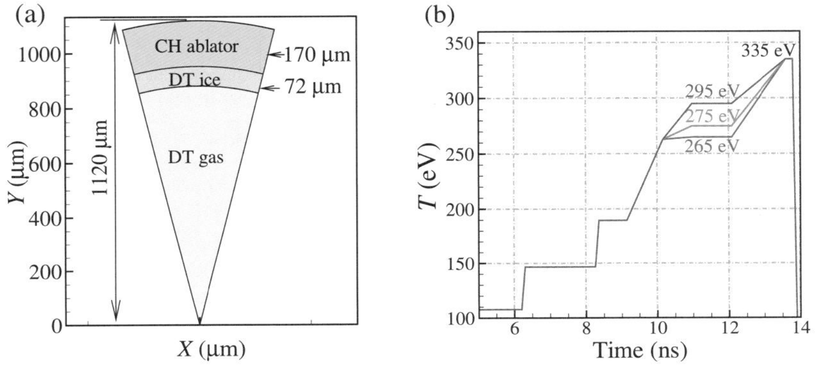

FIG. 5. (a) Sketch of the ignition capsule and (b) three radiation drives with wedged-peak pulses.

space. Simulations show that a secondary shock could be utilized to raise the hot-spot non-equilibrium, and the hot-spot ρR_h requirement is remarkably reduced for achieving hot-spot self-heating. This ion-electron non-equilibrium model poses a promising approach for improving the inertial fusion ignition performed at current mega-joule laser facilities.

This work has been supported by the Foundation of President of Chinese Academy of Engineering Physics (Grant Nos. 201402037 and 201401040), the National Natural Science Foundation of China (Grant Nos. 11475032 and 91130002), the National High Technology Research and Development Program of China (Grant No. 2012AA01A303), the CAEP-FESTC (Grant No. R2014-0501-01), and the National Basic Research Program of China (Grant No. 2013CB34100).

[1]J. H. Nuckolls, L. Wood, A. Thiessen, and G. B. Zimmerman, "Laser compression of matter to super-high densities: Thermonuclear (CTR) applications," Nature (London) **239**, 139 (1972).
[2]J. D. Lindl, *Inertial Confinement Fusion* (Springer, New York, 1998).
[3]S. Atzeni and J. Meyer-ter-Vehn, The Physics of Inertial Fusion (Clarendon Press, Oxford, 2004).
[4]P. Y. Chang, R. Betti, B. K. Spears, K. S. Anderson, J. Edwards, M. Fatenejad, J. D. Lindl, R. L. McCrory, R. Nora, and D. Shvarts, "Generalized Measurable Ignition Criterion for Inertial Confinement Fusion," Phys. Rev. Lett. **104**, 135002 (2010).
[5]E. I. Moses, R. N. Boyd, B. A. Remington, C. J. keane, and R. Al-Ayat, Phys. Plasmas **16**, 041006 (2009).
[6]J. L. Kline, D. A. Callahan, S. H. Glenzer, N. B. Meezan, J. D. Moody, D. E. Hinkel, O. S. Jones, A. J. MacKinnon, R. Bennedetti, R. L. Berger, D. Bradley, E. L. Dewald, I. Bass, C. Bennett, M. Bowers, G. Brunton, J. Bude, S. Burkhart, A. Condor, J. M. Di Nicola, P. Di Nicola, S. N. Dixit, T. Doeppner, E. G. Dzenitis, G. Erbert, J. Folta, G. Grim, S. Glenn, A. Hamza, S. W. Haan, J. Heebner, M. Henesian, M. Hermann, D. G. Hicks, W. W. Hsing, N. Izumi, K. Jancaitis, O. S. Jones, D. Kalantar, S. F. Khan, R. Kirkwood, G. A. Kyrala, K. LaFortune, O. L. Landen, L. Lagin, D. Larson, S. Le Pape, T. Ma, A. G. MacPhee, P. A. Michel, P. Miller, M. Montincelli, A. S. Moore, A. Nikroo, M. Nostrand, R. E. Olson, A. Pak, H. S. Park, J. P. Patel, L. Pelz, J. Ralph, S. P. Regan, H. F. Robey, M. D. Rosen, J. S. Ross, M. B. Schneider, M. Shaw, V. A. Smalyuk, D. J. Strozzi, T. Suratwala, L. J. Suter, R. Tommasini, R. P. J. Town, B. Van Wonterghem, P. Wegner, K. Widmann, C. Widmayer, H. Wilkens, E. A. Williams, M. J. Edwards, B. A. Remington, B. J. MacGowan, J. D. Kilkenny, J. D. Lindl, L. J. Atherton, S. H. Batha, and E. Moses, "Hohlraum energetics scaling to 520 TW on the National Ignition Facility," Phys. Plasmas **20**, 056314 (2013).
[7]V. A. Smalyuk, L. J. Atherton, L. R. Benedetti, R. Bionta, D. Bleuel, E. Bond, D. K. Bradley, J. Caggiano, D. A. Callahan, D. T. Casey, P. M. Celliers, C. J. Cerjan, D. Clark, E. L. Dewald, S. N. Dixit, T. Döppner, D. H. Edgell, M. J. Edwards, J. Frenje, M. Gatu-Johnson, V. Y. Glebov, S. Glenn, S. H. Glenzer, G. Grim, S. W. Haan, B. A. Hammel, E. P. Hartouni, R. Hatarik, S. Hatchett, D. G. Hicks, W. W. Hsing, N. Izumi, O. S. Jones, M. H. Key, S. F. Khan, J. D. Kilkenny, J. L. Kline, J. Knauer, G. A. Kyrala, O. L. Landen, S. Le Pape, J. D. Lindl, T. Ma, B. J. MacGowan, A. J. Mackinnon, A. G. MacPhee, J. McNaney, N. B. Meezan, J. D. Moody, A. Moore, M. Moran, E. I. Moses, A. Pak, T. Parham, H.-S. Park, P. K. Patel, R. Petrasso, J. E. Ralph, S. P. Regan, B. A. Remington, H. F. Robey, J. S. Ross, B. K. Spears, P. T. Springer, L. J. Suter, R. Tommasini, R. P. Town, S. V. Weber, and K. Widmann, "Performance of high-convergence, layered DT implosions with extended-duration pulses at the National Ignition Facility," Phys. Rev. Lett. **111**, 215001 (2013).
[8]J. Lindl, O. Landen, J. Edwards, E. Moses, and NIC Team, "Review of the National Ignition Campaign 2009–2012," Phys. Plasmas **21**, 020501 (2014).
[9]T. R. Dittrich, O. A. Hurricane, D. A. Callahan, E. L. Dewald, T. Döppner, D. E. Hinkel, L. F. Berzak Hopkins, S. Le Pape, T. Ma, J. L. Milovich, J. C. Moreno, P. K. Patel, H.-S. Park, B. A. Remington, J. D. Salmonson, and J. L. Kline, "Design of a high-foot high-adiabat ICF capsule for the National Ignition Facility," Phys. Rev. Lett. **112**, 055002 (2014).
[10]H.-S. Park, O. A. Hurricane, D. A. Callahan, D. T. Casey, E. L. Dewald, T. R. Dittrich, T. Döppner, D. E. Hinkel, L. F. Berzak Hopkins, S. Le Pape, T. Ma, P. K. Patel, B. A. Remington, H. F. Robey, J. D. Salmonson, and J. L. Kline, "High-adiabat high-foot inertial confinement fusion implosion experiments on the National Ignition Facility," Phys. Rev. Lett. **112**, 055001 (2014).
[11]O. A. Hurricane, D. A. Callahan, D. T. Casey, P. M. Celliers, C. Cerjan, E. L. Dewald, T. R. Dittrich, T. Döppner, D. E. Hinkel, L. F. Berzak Hopkins, J. L. Kline, S. Le Pape, T. Ma, A. G. MacPhee, J. L. Milovich, A. Pak, H.-S. Park, P. K. Patel, B. A. Remington, J. D. Salmonson, P. T. Springer, and R. Tommasini, "Fuel gain exceeding unity in an inertially confined fusion implosion," Nature **506**, 343–348 (2014).
[12]R. Nora, W. Theobald, R. Betti, F. J. Marshall, D. T. Michel, W. Seka, B. Yaakobi, M. Lafon, C. Stoeckl, J. Delettrez, A. A. Solodov, A. Casner, C. Reverdin, X. Ribeyre, A. Vallet, J. Peebles, F. N. Beg, and M. S. Wei, "Gigabar spherical shock generation on the OMEGA laser," Phys. Rev. Lett. **114**, 045001 (2015).
[13]T. Döppner, D. A. Callahan, O. A. Hurricane, D. E. Hinkel, T. Ma, H.-S. Park, L. F. Berzak Hopkins, D. T. Casey, P. Celliers, E. L. Dewald, T. R. Dittrich, S. W. Haan, A. L. Kritcher, A. MacPhee, S. Le Pape, A. Pak, P. K. Patel, P. T. Springer, J. D. Salmonson, R. Tommasini, L. R. Benedetti, E. Bond, D. K. Bradley, J. Caggiano, J. Church, S. Dixit, D. Edgell, M. J. Edwards, D. N. Fittinghoff, J. Frenje, M. Gatu Johnson, G. Grim, R. Hatarik, M. Havre, H. Herrmann, N. Izumi, S. F. Khan, J. L. Kline, J. Knauer, G. A. Kyrala, O. L. Landen, F. E. Merrill, J. Moody, A. S. Moore, A. Nikroo, J. E. Ralph, B. A. Remington, H. F. Robey, D. Sayre, M. Schneider, H. Streckert, R. Town, D. Turnbull, P. L. Volegov, A. Wan, K. Widmann, C. H. Wilde, and C. Yeamans, "Demonstration of high performance in layered deuterium-tritium capsule implosions in uranium hohlraums at the national ignition facility," Phys. Rev. Lett. **115**, 055001 (2015).
[14]J. D. Moody, D. A. Callahan, D. E. Hinkel, P. A. Amendt, K. L. Baker, D. Bradley, P. M. Celliers, E. L. Dewald, L. Divol, T. Döppner, D. C. Eder, M. J. Edwards, O. Jones, S. W. Haan, D. Ho, L. B. Hopkins, N. Izumi, D. Kalantar, R. L. Kauffman, J. D. Kilkenny, O. Landen, B. Lasinski, S. LePape, T. Ma, B. J. MacGowan, S. A. MacLaren, A. J. Mackinnon, D. Meeker, N. Meezan, P. Michel, J. L. Milovich, D. Munro, A. E. Pak, M. Rosen, J. Ralph, H. F. Robey, J. S. Ross, M. B. Schneider, D. Strozzi, E. Storm, C. Thomas, R. P. J. Town, K. L. Widmann, J. Kline, G. Kyrala, A. Nikroo, T. Boehly, A. S. Moore, and S. H. Glenzer, "Progress in hohlraum physics for the National Ignition Facility," Phys. Plasmas **21**, 056317 (2014).
[15]R. P. J. Town, D. K. Bradley, A. Kritcher, O. S. Jones, J. R. Rygg, R. Tommasini, M. Barrios, L. R. Benedetti, L. F. Berzak Hopkins, P. M. Celliers, T. Döppner, E. L. Dewald, D. C. Eder, J. E. Field, S. M. Glenn, N. Izumi, S. W. Haan, S. F. Khan, J. L. Kline, G. A. Kyrala, T. Ma, J. L. Milovich, J. D. Moody, S. R. Nagel, A. Pak, J. L. Peterson, H. F. Robey, J. S. Ross, R. H. H. Scott, B. K. Spears, M. J. Edwards, J. D. Kilkenny, and O. L. Landen, "Dynamic symmetry of indirectly driven inertial confinement fusion capsules on the National Ignition Facility," Phys. Plasmas **21**, 056313 (2014).
[16]D. S. Clark, M. M. Marinak, C. R. Weber, D. C. Eder, S. W. Haan, B. A. Hammel, D. E. Hinkel, O. S. Jones, J. L. Milovich, P. K. Patel, H. F. Robey, J. D. Salmonson, S. M. Sepke, and C. A. Thomas, "Radiation hydrodynamics modeling of the highest compression inertial confinement fusion ignition experiment from the National Ignition Campaign," Phys. Plasmas **22**, 022703 (2015).
[17]T. Döppner, D. A. Callahan, O. A. Hurricane, D. E. Hinkel, T. Ma, H.-S. Park, L. F. Berzak Hopkins, D. T. Casey, P. Celliers, E. L. Dewald, T. R. Dittrich, S. W. Haan, A. L. Kritcher, A. MacPhee, S. Le Pape, A. Pak, P. K. Patel, P. T. Springer, J. D. Salmonson, R. Tommasini, L. R. Benedetti, E. Bond, D. K. Bradley, J. Caggiano, J. Church, S. Dixit, D. Edgell, M. J. Edwards, D. N. Fittinghoff, J. Frenje, M. Gatu Johnson, G. Grim, R. Hatarik, M. Havre, H. Herrmann, N. Izumi, S. F. Khan, J. L. Kline, J. Knauer, G. A. Kyrala, O. L. Landen, F. E. Merrill, J. Moody, A. S. Moore, A. Nikroo, J. E. Ralph, B. A. Remington, H. F. Robey, D. Sayre, M. Schneider, H. Streckert, R. Town, D. Turnbull, P. L. Volegov, A. Wan, K.

Widmann, C. H. Wilde, and C. Yeamans, "Demonstration of high performance in layered deuterium-tritium capsule implosions in uranium hohlraums at the national ignition facility," Phys. Rev. Lett. **115**, 055001 (2015).

[18]G. S. Fraley, E. J. Linnebur, R. J. Mason, and R. L. Mose, "Thermonuclear burn characteristics of compressed deuterium-tritium microspheres," Phys. Fluids **17**(2), 474–489 (1974).

[19]The hot-spot energy gain is defined as the ratio of its fusion energy to its initial energy. Usually, the energy coupling from laser to the final hot spot is ~2–3‰, and a hot-spot energy gain of 500 means that the fusion energy is equivalent to the laser energy.

[20]J. R. Rygg, J. A. Frenje, C. K. Li, F. H. Séguin, and R. D. Petrasso, "Electron-ion thermal equilibration after spherical shock collapse," Phys. Rev. E **80**, 026403 (2009).

[21]R. Betti, C. D. Zhou, K. S. Anderson, L. J. Perkins, W. Theobald, and A. A. Solodov, "Shock ignition of thermonuclear fuel with high areal density," Phys. Rev. Lett. **98**, 155001 (2007).

[22]Z. Fan, X. T. He, J. Liu, G. Ren, B. Liu, J. Wu, L. Wang, and W. Ye, "A wedged-peak-pulse design with medium fuel adiabat for indirect-drive fusion," Phys. Plasmas **21**, 100705 (2014).

[23]H. Brysk, "Electron-ion equilibration in a partially degenerate plasma," Plasma Phys. **16**, 927–932 (1974).

8th International Conference on Inertial Fusion Sciences and Applications (IFSA 2013) IOP Publishing
Journal of Physics: Conference Series **688** (2016) 012029 doi:10.1088/1742-6596/688/1/012029

The updated advancements of inertial confinement fusion program in China

X. T. He[1*] with ICF teams in China

[1]Institute of Applied Physics and Computational Mathematics, Beijing, 100094, China

E-mail: *xthe@iapcm.ac.cn

Abstract.

The recent achievements of ICF research in China are reviewed. The constructions of laser facilities of SG-III and SG-IIUP are completed in this year and the full energy output operation will be in 2014. The target physics studies involving numerical simulations of a new ignition scheme, which is proposed to enhance implosion velocity and suppress hydrodynamic instability and distortion at interface between hot spot and main fuel, and experimental results (a few selected examples) are presented.

1. Introduction

The goal of the first milestone of the inertial confinement fusion (ICF) program in China is to perform the ignition and plasma burning in laboratory around 2020. The national research institutes, such as the Research Centre of Laser Fusion (RCLF), the Institute of Applied Physics and Computational Mathematics (IAPCM), the National Laboratory on High Power Laser and Physics (NLHPLP), as well as many universities and other institutions are participating in the Program. Since 1993, based on this program, target physics studies involving theories (models and simulations) and experiments (measurement techniques), establishments of precise diagnostic systems, constructions of laser facilities (SG-series) and technology development, and precise target fabrications have been coordinately developed, and the great achievements have been gained. The pre-ignition target design and advanced target physics research, based on the SG-III laser facility, are crucial as the basis for the ignition target design in the coming years, which should identify and solve the essential science and technology issues, to reduce risks, and to scale up the design to the ignition target. Finally, we will demonstrate the ICF ignition on the designing SG-IV laser ignition facility.

2. Development of laser drivers

The SG-series of laser facilities were constructed and/or are being developed to serve for ICF research in China. The SG-II laser facility, which has 8 beams and energy output of $3kJ/3\omega$ with pulse duration of 1ns and is supplemented by a ninth laser beam with energy output of $5.2kJ/1\omega$ $(3kJ/3\omega)$ and pulse duration of $2-3ns$ as a backlighter, has provided over 4000 shots related to the ICF experiments since 2000. The SG-III Prototype (SG-IIIP), which has 8 beams and laser energy output of $15kJ/3\omega$ and pulse duration of 3ns, has been operating for physical experiments over 2000 shots since 2006, and a new laser beam as its backlighter is being prepared. Also, the SG-IIU is a combined laser system which is made up of the upgrading SG-II

(8 beams, $24kJ/3\omega$, 3ns) and a PW laser ($\sim 1kJ$, $3-5ps$) to be operating in 2014. The SG-III with 48 beams (6 bundles) and the output laser energy of $250kJ/3ns/3\omega$ and $400kJ/5ns/3\omega$ has been completed in 2013, see Fig.1. All bundles are now operating for experiments and full energy operation and physical shots will be in 2015. The ignition facility (SG-IV) is being designed to serve for ignition demonstration around 2020 in China.

Figure 1. The SG-III laser system with 48 beams (6 bundles) and the output laser energy of $250kJ/3ns/3\omega$ and $400kJ/5ns/3\omega$ has been completed in early 2013.

3. A hybrid indirect-direct drive ignition scheme by a target of spherical hohlraum and layered fuel capsule

For the conventional indirect-drive hotspot ignition scheme of ICF [1], the ignition occurs near stagnation (isobaric) time. For such scheme, there exist at least two issues in implosion dynamics, i.e., strong rarefaction wave due to the ablated plasma expanding towards vacuum at the ablator surface leads to severe reduction of the ablative pressure and implosion velocity; the mix from Rayleigh-Taylor instability (RTI) and asymmetry amplification (distortion) due to the implosion shock multiple-reflection at the interfaces of hotspot/main fuel and others resulting in deceleration, may lead to ignition difficulty.

To solve the issues, we proposed a new ignition scheme of indirect-direct hybrid drive for ICF: a layered fuel capsule inside a spherical hohlraum (SH) is compressed first by indirect-drive (ID) x-ray ablation with laser energy of about 0.5MJ and pulse duration of 12.7 ns, and then by intense direct-drive (DD) lasers with laser energy of about 0.85MJ and pulse duration of last $2ns$ (figure 2c, green colour). The SH, which has a radius of 3.5 mm with six laser entrance holes (LEFs) and octahedral symmetry (figure. 2a), is designed by using plasma-filling model under conditions of the specific indirect-drive (ID) radiation temperature (figure 2c, red colour) and the filling density. The non-uniformity of about a few thousandth on the capsule surface, which has total deuterium-tritium (DT) fuel mass of 0.19 mg and an initial outer radius of $850\mu m$

(figure 2b), is obtained by view factor code simulations if the radius ratio of SH to capsule is greater than 5, the incident angles of laser beams are between $50^o - 60^o$, the radii of the laser bundle and LEH are properly chosen.

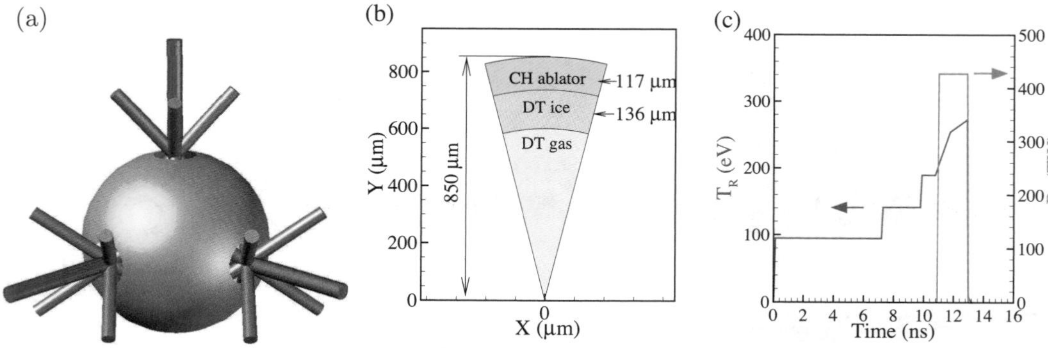

Figure 2. (a) The SH configuration; (b) capsule cross section; (c) indirect-drive radiation temperature (red) and direct-drive laser power (green).

The implosion dynamics of the capsule is investigated by using LARED-S code. The 1D simulation results are plotted in Figure 3. The ablator (CH material with initial density of $1.0g/cm^3$) of the capsule is ablated by 4 radiation- temperature steps with a top $Tr = 270eV$ (figure 2c, red colour), a radiation ablation front (RAF) appears in the ablated CH surface. Meanwhile four shocks are generated and propagates in the capsule that is imploded and compressed. Beginning at time $t = 10.7ns$, the DD lasers with intensity of $3.4 \times 10^{15} W/cm^2$ are imposed at a critical surface, where is $950\mu m$ away from the capsule centre, in the corona region. A supersonic electron thermal wave with an electron ablation front (EAF) is formed there and runs from the critical surface towards the RAF about $650\mu m$ away from the capsule centre. The thermal wave gradually slows down to a sonic/subsonic one due to the energy deposition in corona during propagation, followed a DD shock wave is driven. The four shocks driven by ID and the DD shock merge into a merged shock (MS) near inner wall of the main fuel. When the MS runs into the DT gas filling cavity a hot spot involving the DT gas and a part of innermost main fuel is formed and compressed by the MS, then the MS rebounds from the capsule centre and moves towards the interface of the hot spot and main fuel.

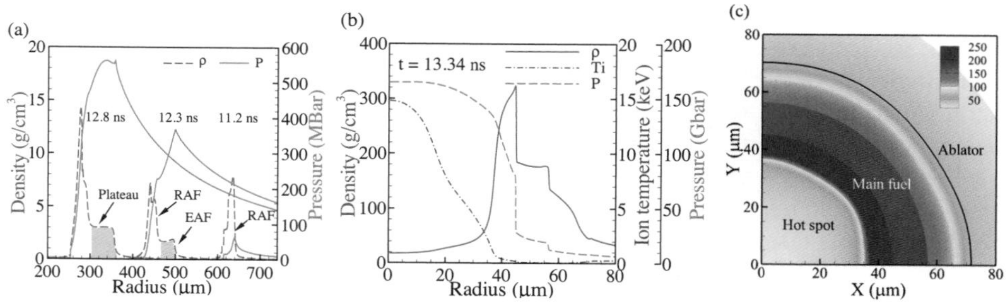

Figure 3. (a) and (b) are the density (red), pressure (green), ion temperature (blue) vs radii of the imploding capsule in different time, and (c) is the 2D profile (hot spot, main fuel, ablator) at ignition time.

Meanwhile a compression wave with the EAF followed the DD shock, likes a snowplow, drives the ablated CH of density $\rho \ll 1.0g/cm^3$ in the corona, which is gradually piled into a higher CH density plateau between the EAF (sonic) and RAF(subsonic). At $t = 12.3ns$, the plateau has a spatial width of $\sim 40\mu m$ (figure 3a, grey), the density $\rho \sim 2.0g/cm^3$ and the maximal radiation temperature $Tr \sim 270eV$, and thus, the peak pressure $P \sim 380$ Mbar; at t=12.8 ns, the plateau has a spatial width of $\sim 60\mu m$, the density $\rho \sim 3.0g/cm^3$ and the peak pressure $P \sim 570$ Mbar, it is far greater than ID ablation pressure of $\sim$ tens Mbar as shown at t=11.2 ns in figure 3a.

Such a plateau likes an extreme-high pressure piston to push out an enhanced shock (ES). The ES fast runs in the imploding capsule. At $t \sim 13.02$ ns the ES reaches the interface of hot spot/main fuel and collides with the MS that just arrives there from the hot spot centre, thus the MS is stopped to reflect at the hot spot interface. Then, the ES runs in the hot spot that is further compressed and heated. At $t \sim 13.34$ns, the non-isobaric ignition (particle heating rate equals to electron cooling rate) occurs before stagnation when the ES rebounded from the hot spot centre arrives near the interface of the hot spot/main fuel in a lower convergent ratio of 25 (figure 3b). As a result, the RTI, mix, and the interface distortion are prevented due to no deceleration phase happened. Finally, the fusion energy of 17.4 MJ and a gain of 13 for total laser energy of 1.35MJ are achieved. This hybrid scheme is robust, even if the laser energy is reduced to 1.1MJ the yield still has 16MJ.

The 2D simulation results are investigated by using LARED-S. The non-uniformity from DD laser irradiation at critical surface is suppressed by the thermal smoothing effect [2] during the supersonic electron thermal wave propagation from the critical surface ($\sim 950\mu m$) to the RAF ($\sim 650\mu m$) of the implosion capsule. The growth factors (GFs) of hydrodynamic instabilities varying with mode l for hybrid scheme are shown in figure 4, and compared with the traditional 1D one. The hybrid GF at interface of hot spot/main fuel is suppressed as shown in figure 4a, while the GFs at interfaces of ablator/main fuel and RAF are about 220 and 150 (figure 4b and 4c), respectively, which are lower than that for the traditional 1D scheme as seen in figure 4 (green). The simulation results also showed that the profile of imploding core at ignition time is quite close to symmetry and 2D effects seem to be neglectable. Therefore, the yield in 2D simulations is same as 1D result.

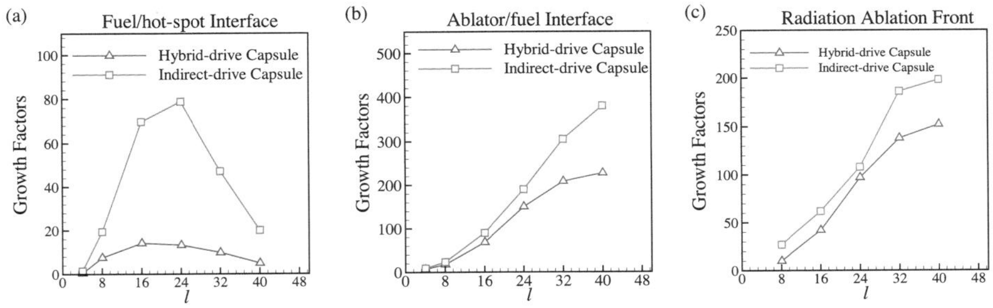

Figure 4. The growth factors of hydrodynamic instabilities vs mode l for hybrid scheme (red) to compare with the traditional 1D one (green). At the interface: (a) hot spot/main fuel, (b) ablator/main fuel, (c) radiation ablation front (RAF).

The expected efficiency of the laser energy coupling to the SH for ID phase is over 90% higher than the cylindrical and rugby hohlraum, and is $\sim 20\%$ for the DD lasers of intensity of $3.4 \times 10^{15} W/cm^2$ [3] with a relatively low hot electron temperature of $\sim 30keV$, which is easily stopped by the compressed surplus ablator.

8th International Conference on Inertial Fusion Sciences and Applications (IFSA 2013) IOP Publishing
Journal of Physics: Conference Series **688** (2016) 012029 doi:10.1088/1742-6596/688/1/012029

4. Physical experiments (selected examples)

4.1. Hohlraum test experiments with the first two bundle beams of SG-III

The hohlraum test experiments reported here were performed at the SG-III laser facility with the first two bundles of beams. In these experiments we compared the radiation temperature Tr of the Au hohlraums that were heated with the different incident angle lasers. We use two kinds of hohlraum in these experiments: hohlraum A is $1.4mm$ in diameter, $2.1mm$ in length, and has LEH diameters of $1.0mm$; hohlraum B is $2.4mm$ in diameter, $3.4mm$ in length, and has LEH diameters of $1.6mm$. The hohlraum A was heated by 16 beams using a 3 ns flattop laser pulse with a total energy of up to 80kJ and 45kJ, and The hohlraum B was heated with a 1ns flattop laser pulse with a total energy of up to 16kJ. Temperature Up16 (Up42, Up64) was diagnosed from upper-half-chamber with an angle of 16° (42°, 64°) between observation direction and hohlraum axis; Down20 (Down42) was diagnosed from down-half-chamber with an angle of 20° (42°) between observation direction and hohlraum axis. The temperature differences are shown in figure 5.

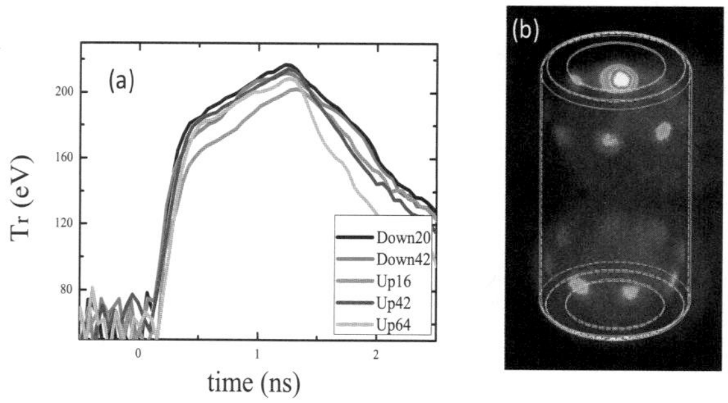

Figure 5. Hohlraum temperature experiment on SG-III laser system with only two bundles of beams.

4.2. Target physics–cryogenic capsule experiments

Cryogenic deuterium-deuterium (DD) capsule direct-drive experiments are carried on SG-IIIP. The target assembly (figure 6a) is made up of a copper cylinder clamped by two cooling arms; the plastic capsule cooled by helium gas to below the deuterium triple point; supported and filled with D2 gas by fill tube of $10\mu m$ in diameter; ice thickness is uniformed by tuning the temperature difference between two cooling arms; inserted into target chamber on a six-dimensional adjustable platform. The experiments are carried out with a polar direct drive mode: beam pointing is optimized for best irradiation uniformity. The laser has a pulse shape of two steps with power contrast of $1:10$ to minimize the energy loss on the sealing film. The experiments showed that the cryogenic capsule (ice layer of $115\mu m$) implosion achieved DD neutron yield of 4.7×10^6, while warm plastic capsule implosion achieved neutron yield of 7.8×10^7. As shown in Fig. 6, the target alignment accuracy in cryogenic condition is demonstrated, the technique for fuel areal density is validated, and the implosion dynamics is measured.

4.3. Monochromatic imaging for implosion measurement

Monochromatic imaging (MI) based on spherically bent crystals (BCs) was developed and used to explore implosion dynamics. Imploding ablator performance was measured on SGII using this MI system. Here, 5th order of reflection of the mica spherical crystal was used and the measured photon energy $E_{photon} = 3.14keV$ (Pd). The detrimental long-range spatial structure from backlighter non-uniformities are absent using monochromatic imaging system. The experimental results showed that crystal resolution is smaller than $5\mu m$, system resolution (involving streak

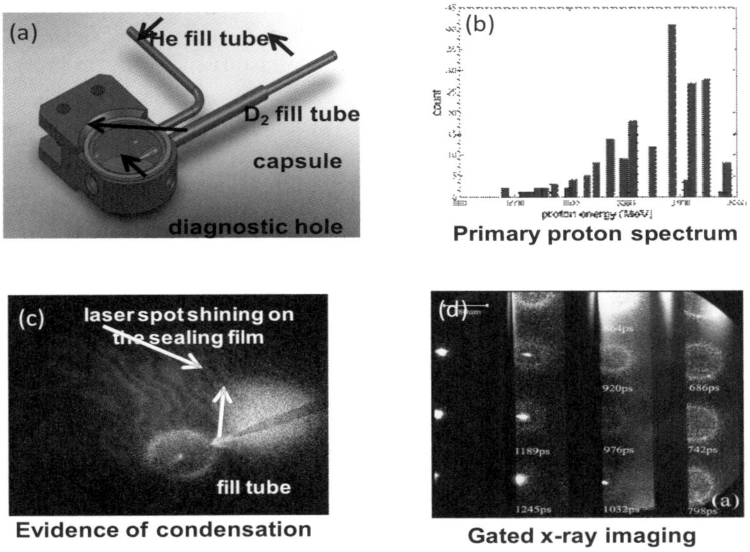

Figure 6. Preliminary experimental results of cryogenic deuterium-deuterium (DD) capsule direct-drive experiments on SG-IIIP: (a) cryogenic capsule involving He and D2 filling tubes, diagnostic hole; (b) primary proton spectrum; (c) evidence of condensation; (d) gated x-ray imaging.

camera) is about $10\mu m$. The uncertainty in this experiment is that the implosion velocity is about 10% and the residual mass is about 15%, see Fig. 7.

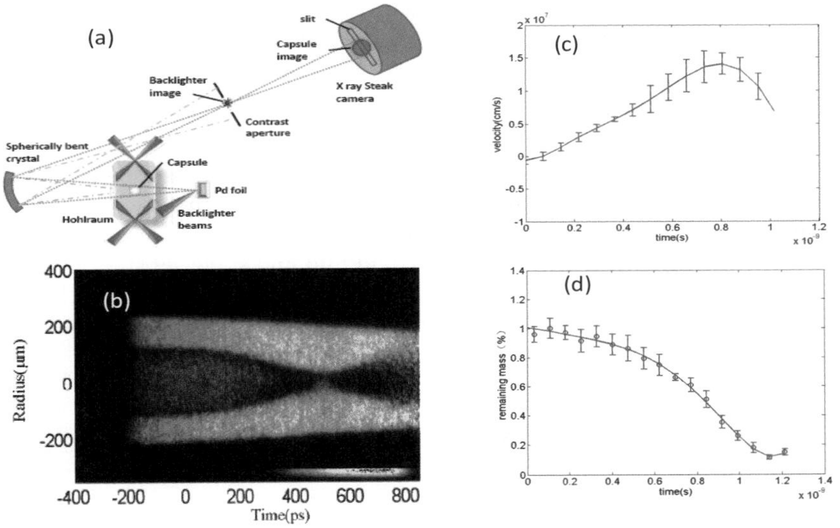

Figure 7. (a) A schematic of the experimental setup for the imploding process measurement; (b) 3.14keV monochromatic streaked radiograph; (d) Implosion velocity; and (d) Residual mass.

4.4. K-shell photoabsorption edge of strongly coupled matter

Warm dense matter (WDM) is a non-ideal plasma with full and partial degenerate and exhibits the continuum lowing of efficient ionization potential, phase transition, pressure ionization, etc., it is very important to measure WDM properties for target simulations. Here K-edge shift of Potassium Chloride (KCl) of $Ek = 2.8keV$ is investigate experimentally. A sandwich sample (KCl) target (figure 8a) is compressed (figure 8c) by shocks driven by x rays from "dog bone" hohlraum (figure 8), then short pulse backlighting (130ps laser from SG-II) is applied to get the time-resolved absorption spectroscopy from the compressed KCl sample (figure 8b). The red shift of K edge up to 11.7 eV and broadening of 15.2 eV from backlighting spectra in the maximal compression, resulting in K-shell ionization energy increase and continuum lowering, are clearly observed, and it shows that the WDM KCl is in partial degeneracy [4].

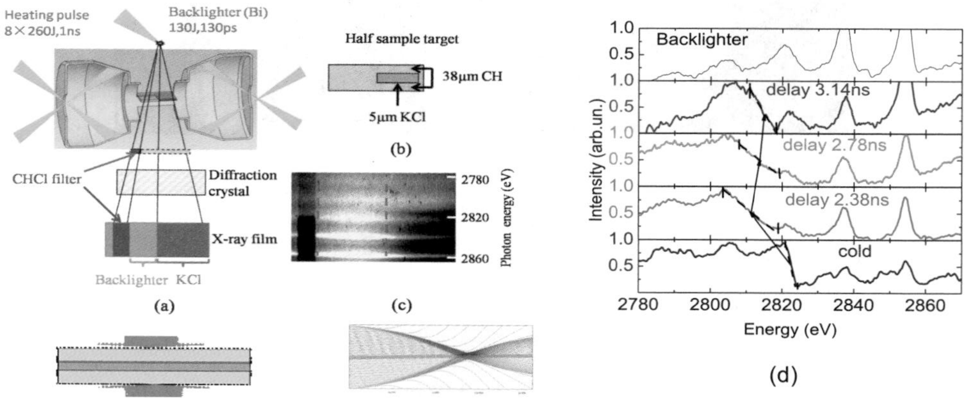

Figure 8. The "dog-bone" hohlraum schematic for partial degeneracy experiments: (a) the sandwich sample (KCl) target, (b) typical spectral image, (c) the shock compressed sample (KCl), (d) the time-resolved absorption spectroscopy from the compressed KCl sample.

Final remarks

In summary, we are performing the ignition target design to improve the conventional indirect central ignition target and to explore new ignition schemes involving hybrid drive, and as well will provide physical data for the SG-IV ignition driver design. Target physics research on SG-III is a quite important step towards ignition, and would find and solve substantial issues in science and technology before transiting to ignition. The SG-IV ignition facility with laser energy of around 1.5MJ for third harmonics is being designed. China is going to be toward moderate/high gain ignition and burning goal around 2020.

References

[1] J.D. Lindl *et al.*, Phys. Plasmas **11**, 339(2004).
[2] M. Keskinen, Phys.Rev.Lett. **103**, 055001(2009).
[3] W. Theobald *et al.*, Phys. Plasmas **19**, 102706(2009).
[4] Y. Zhao *et al.*, Phys.Rev.Lett. **111**, 155003 (2013).

PRL 118, 165001 (2017) PHYSICAL REVIEW LETTERS week ending 21 APRIL 2017

Neutron Generation by Laser-Driven Spherically Convergent Plasma Fusion

G. Ren,[1] J. Yan,[2] J. Liu,[1,3,4,*] K. Lan,[1] Y. H. Chen,[1] W. Y. Huo,[1] Z. Fan,[1] X. Zhang,[2] J. Zheng,[2] Z. Chen,[2] W. Jiang,[2] L. Chen,[2] Q. Tang,[2] Z. Yuan,[2] F. Wang,[2] S. Jiang,[2] Y. Ding,[2] W. Zhang,[1,3] and X. T. He[1,3]

[1]*Institute of Applied Physics and Computational Mathematics, Beijing 100088, China*
[2]*Research Center of Laser Fusion, Chinese Academy of Engineering Physics, Mianyang 621900, China*
[3]*Center for Applied Physics and Technology, Peking University, Beijing 100871, China*
[4]*Collaborative Innovation Center of IFSA, Shanghai Jiao Tong University, Shanghai 200240, China*
(Received 14 December 2016; published 19 April 2017)

We investigate a new laser-driven spherically convergent plasma fusion scheme (SCPF) that can produce thermonuclear neutrons stably and efficiently. In the SCPF scheme, laser beams of nanosecond pulse duration and 10^{14}–10^{15} W/cm^2 intensity uniformly irradiate the fuel layer lined inside a spherical hohlraum. The fuel layer is ablated and heated to expand inwards. Eventually, the hot fuel plasmas converge, collide, merge, and stagnate at the central region, converting most of their kinetic energy to internal energy, forming a thermonuclear fusion fireball. With the assumptions of steady ablation and adiabatic expansion, we theoretically predict the neutron yield Y_n to be related to the laser energy E_L, the hohlraum radius R_h, and the pulse duration τ through a scaling law of $Y_n \propto (E_L/R_h^{1.2}\tau^{0.2})^{2.5}$. We have done experiments at the ShengGuangIII-prototype facility to demonstrate the principle of the SCPF scheme. Some important implications are discussed.

DOI: 10.1103/PhysRevLett.118.165001

Introduction.—Neutron sources have important applications in many fields of science and engineering including the nuclear fuel cycle, neutron activation analysis, mineral and petroleum exploration, etc. Among the many ways to produce neutrons, which are usually based on high-energy accelerators and fission reactors through isotope disintegration and spallation, the laser-driven neutron sources are of particular interest and have attracted much attention recently [1–4].

The laser cluster interactions (LCIs) use an intense (10^{16}–10^{18} W/cm^2) femtosecond laser to strip off peripheral electrons from deuterium-tritium (DT) clusters of 10^3–10^5 atoms, forming DT ion clusters with positive charges, triggering a Coulomb explosion which accelerates the DT ions to an energy of several million electron volts (MeV). The collisions between the high-energy ions result in nuclear fusion reactions. The LCI approach can be implemented in a tabletop size, but the neutron yield is low, around 10^7 per laser pulse [1,2]. With a laser of picosecond pulse and higher intensity (10^{20}–10^{21} W/cm^2), the laser target interaction approach (LTI) [3] can accelerate deuterons to 100 MeV energies that are dumped in low-Z converter (Be) targets to produce neutrons via the nuclear reaction $^9\mathrm{Be}(d,n)$. This approach has produced a high neutron flux of 10^{10}/sr. However, the neutron spectrum is distributed widely from a few MeV to 100 MeV.

On the other hand, for the laser-driven implosions of a DT capsule by direct drive [5–7] and indirect drive [4], the neutron yield can reach as high as 5×10^{13} on the 30 kJ Omega laser [8] and 9×10^{15} on the 1.8 MJ National Ignition Facility (NIF) laser [9,10], respectively. Nevertheless, these inertial implosion schemes require a high convergence ratio (~30), and the implosions are often degraded by hydrodynamic instabilities [11–14] and laser plasma instabilities [15] as well as radiation asymmetry [16]. Other interesting approaches include the laser-driven exploding pusher target [17], shock ignition [18] and laser-assisted magnetized liner inertial fusion [19,20], etc.

In this Letter, we investigate a new laser-driven scheme of spherically convergent plasma fusion (SCPF) that can generate a thermonuclear fusion neutron efficiently and robustly. Theoretically, we have modeled the ion temperature and density of converging hot plasmas and derived an explicit scaling law of the neutron yield with respect to laser parameters and target radius. In particular, we have performed experiments to demonstrate the principle of SCPF at the ShengGuangIII-prototype (SGIII-proto) facility. In the experiment, we have obtained a maximum deuterium-deuterium (DD) fusion neutron yield of 3.5×10^9 with a 6.3 kJ laser. The ion temperature is about 7 keV. The observed dependence of the neutron yields on the laser and target parameters agrees with our theoretical prediction. We further discuss the optimization of the SCPF scheme and some important implications.

SCPF principle and scaling law.—In the SCPF scheme, we suppose multiple laser beams of nanosecond (ns) duration and 10^{14}–10^{15} W/cm^2 intensities uniformly irradiate and ablate the thermonuclear fuel layer (~20 μm) lined inside a spherical hohlraum [Fig. 1(a)]. The fuel layer expands and implodes towards the center [Fig. 1(b)], and eventually the hot fuel plasmas converge and stagnate at the central region of the sphere [Fig. 1(c)], converting most of

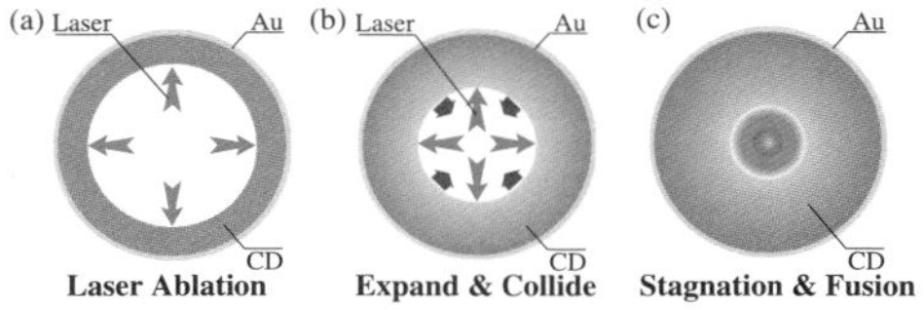

FIG. 1. Schematic plot of laser-driven spherically convergent plasma fusion processes.

their kinetic energy to the internal energy, raising the ion temperature to around 10 keV, and causing thermonuclear fusion reactions to occur.

For the nanosecond laser ablation of the low-Z fuel layer, such as deuterated polystyrene (CD), a laser-driven stationary ablation model for a spherical target can be properly exploited [21]. The mass ablation rate $\dot{m}$ and ablating plasma temperature T_{ab} can be deduced as $\dot{m} \propto I_L^{0.76}$ and $T_{ab} \propto I_L^{2/3}$, respectively, where I_L denotes the laser intensity. Meanwhile, the plasmas expand at a velocity of v_i related to the acoustic speed of the ablated hot plasmas, $v_i \propto T_{ab}^{1/2} \propto I_L^{1/3}$.

The ablating plasmas expand, collide, and stagnate at the sphere center. At stagnation, the kinetic energy is converted into ion thermal energy, making the ion temperature T_i increase dramatically, i.e., $T_i \propto v_i^2 \propto I_L^{2/3}$; here, we assume approximately that for the ablated plasmas the ions expand adiabatically [22].

The total neutron yield equals the nuclear reaction rate integrated over the plasma volume and confinement time, i.e., $Y_n = \frac{1}{2} n_D^2 \langle \sigma v \rangle_{\mathrm{DD}} V \Delta t$, where the plasma number density $n_D = N/V$ with $N \sim \dot{m} S_{\mathrm{spot}} \tau$, where S_{spot} is the total area of the laser spots that is proportional to πR_h^2, R_h is the radius of the sphere hohlraum, and τ is the laser pulse length. The average reactivity of deuterium on deuterium is proportional to T_i^2 near 10 keV. V is the burn plasma volume of radius $R_c \propto R_h$ according to the pressure balance principle, and Δt is the fusion burn time, which is proportional to R_c / v_{shock} with $v_{\mathrm{shock}} \propto \sqrt{T_i}$ the speed of the outward propagating shock after stagnation. We finally obtain [22]

$$Y_n \propto \left(\frac{E_L}{R_h^{1.2} \tau^{0.2}} \right)^{2.5} = D^{2.5}, \tag{1}$$

where $D = E_L / R_h^{1.2} \tau^{0.2}$ served as the scaling parameter for the SCPF and E_L is the laser energy which is related to the intensity through $E_L = I_L \tau S_{\mathrm{spot}}$.

The above deductions indicate that the SCPF neutron yield is proportional to the incident laser energy to the power of 2.5. For the laser-driven capsule implosion, the scaling is $Y_n \propto E_L{}^3$ without alpha-particle self-heating; it rises to $Y_n \propto E_L{}^{4-5.8}$ with alpha-particle self-heating [27] and is expected to rise further with the realization of ignition and burn. In contrast, for the LTI scheme the yield scaling is roughly $Y_n \propto E_L$ [3], and for the LCI scheme the yield scaling is roughly $Y_n \propto E_L{}^2$ [2].

A few other related fusion concepts based on the spherical configuration have been discussed previously. The spherically convergent ion focus [28] uses electrostatic confinement where ions are accelerated towards a spherical cathode. The fusion occurs mainly by beam-beam or beam-target interactions, and therefore the fusion reactions are not fully thermonuclear and, consequently, may be limited to low gains.

The plasma-jet-driven magneto-inertial fusion (PJMIF) uses a spherical array of plasma guns to generate plasma jets which merge and form an imploding plasma liner to compress preheated magnetized target plasmas to thermonuclear condition [29,30]. Both PJMIF and our SCPF make use of the center-oriented plasma motion and spherical convergence to amplify the imploding power density of the converging plasmas by several orders of magnitude. However, in our SCPF scheme, because the lasers are capable of delivering a power density orders of magnitude higher than the plasma guns, the initial power density of the ablated plasmas in the SCPF can be orders of magnitude higher than the plasma jets produced in PJMIF.

Experiment setup.—To demonstrate the SCPF principle, we have performed experiments at the SGIII-proto laser facility. The SGIII-proto laser is a neodymium glass laser operating at triple frequency with the wavelength of 351 nm. It has eight laser beams with four beams on each end, outputting 8 kJ energy at maximum and a pulse duration from 1 to 3 ns. We fabricate spherical hohlraums out of gold with a shell thickness of 25 μm and two laser entrance holes (LEHs) at the poles. The gold spherical hohlraum has an inner diameter of 1700 μm. The diameter of the LEHs is 1000 μm. We line the wall of the gold hohlraum with CD of 40 μm in depth. We also fabricate some smaller hohlraums with an inner diameter of 1500 μm.

The experimental setup is shown in Fig. 2. The laser beams are aligned in two conical rings and enter the hohlraum through the LEHs. The beam angle with respect to the vertical hohlraum axis is 45°. Usually, the diameter of the laser spots is 500 μm with the smoothing technique of the continuous phase plate (CPP). We also conduct experiments in which the laser spot size is 200 μm without using the CPP. We use one scintillator (Sc.D1 in Fig. 2) to count the neutron yield, another one (Sc.D2 in Fig. 2) to measure the neutron bang time, and a neutron time of flight (nTOF) measurement to infer the DD ion temperature. We use a gated x-ray image (GXI) camera to capture time-resolved x-ray pictures of the plasma converging process. The x-ray flux flowing out of the LEH is measured using an absolutely calibrated flat-response x-ray diode (FXRD). Laser injection is monitored by using a pinhole x-ray camera (PHC). Laser backscatterings are measured by a full

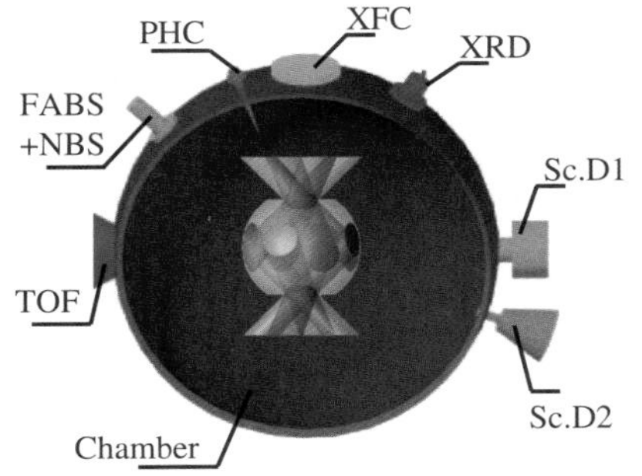

FIG. 2. Illustration of the experimental setup, and the to-scale side-on view of the two-LEH spherical hohlraum with laser beams.

aperture backscattering system (FABS) and a near back-scattering system (NBS) and to be less than 10% in our experiments. Note that for the low-Z CD hohlraum the radiation temperature is as low as 130 eV, which cannot affect the plasma dynamics significantly.

Experimental results and discussion.—The GXI camera measuring x-ray emission above 2 keV is positioned to image emissions through the LEH to investigate the plasma converging processes. Clear and time-resolved pictures are successfully caught as shown in Fig. 3(a). In each image in Fig. 3(a), the outer ring is the image of the x-ray emission from the edge of the LEH. We will address the central portion of the image that is caused by the emissions emanating from the converging plasmas. Clear bright emissions are observed from 0.8 to 1.0 ns. We believe these are emitted by colliding high-speed plasma jets squeezed out by adjacent laser ablated bubbles [31], which soon fade away because of the rapid electron energy loss due to the bremsstrahlung emission and plasma rarefaction. Since the x-ray emissions appear at 0.8 ns, we estimate the

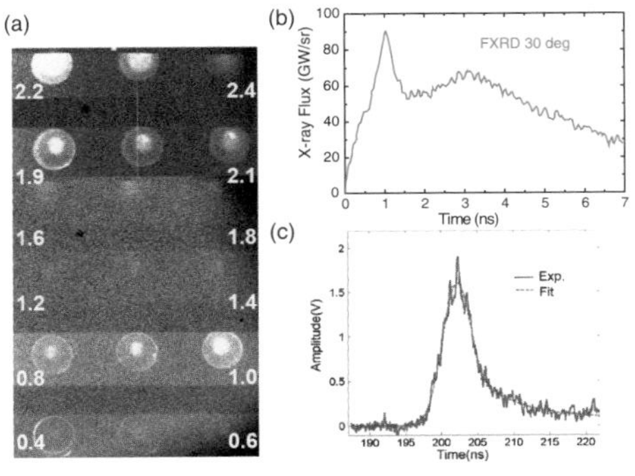

FIG. 3. In the panel, (a) is the GXI images of the convergent plasma emission above 2 keV with the time pinpointed, and the gate time is 70 ps; (b) is the temporal x-ray flux measured by FXRD with an energetic range from 100 eV to 4.4 keV from LEH with 30° to the axis; (c) is the nTOF line with a fitting curve.

velocity of the plasma jets to be about 1000 km/s, which is about 4 times the sound speed [32,33]. From 1.0 to 1.8 ns, the images become dark at the center. Note that the neutron peak emission time or the bang time is in between, i.e., about 1.5 ns. From 1.9 to 2.4 ns, brighter and larger emissions are seen again. We believe the second round of bright emissions are the x rays emitted by the converged wall plasmas ablated by the lasers.

In Fig. 3(b), we observe the spectrally integrated x-ray flux measured by FXRD through the LEH at an angle of 30° to the axis. From this direction, the FXRD views emissions from both the hohlraum wall area and the central region. The x-ray flux rises quickly during the first 1 ns when the laser beams are on, and it is contributed mostly by the emissions from the ablation of the hohlraum wall by the lasers. After the lasers are turned off at 1.0 ns, the x-ray flux drops quickly. At around 2.0 ns, the flux begins to rise slowly again and reaches a second peak at 3.3 ns. After 3.3 ns, the x-ray flux drops slowly and lasts for several nanoseconds. The second flux hump is associated with the emission of converging plasmas; it stands for a longer time and fades away slower.

In Fig. 3(c), the nTOF signal is fitted reasonably well by assuming a thermonuclear DD reaction of plasma with an ion temperature of 6.6 keV, confirming our theory of convergent plasma thermalization and concomitant thermo-nuclear fusion rather than other fusion mechanisms. The above thermalized plasma fusion picture is consistent with our estimated ion equilibration time of the order of a few picoseconds [22].

The neutron bang time of 1.5 ns is about 0.4 ns earlier than the second round of hard x-ray emission in GXI and 1.8 ns earlier than the FXRD second emission peak, reflecting intrinsically the electron-ion relaxation and non-equilibrium energy transfer in the processes of plasma convergence and subsequent outward shock propagation [22]. The bubble implosion speed is estimated to be around 500 km/s calculated by the inner radius of the hohlraum (810 μm) divided by 1.5 ns transit time. When the plasmas converge into the sphere center at about 1.5 ns, the ion temperature reaches its peak value of about 7 keV due to kinetic-thermal energy conversion. At this time, the plasmas have an apparent accumulation at the central region yielding a maximum neutron generation.

The 0.4 ns time delay of the hard x-ray emission to the neutron bang time is due to the fact that the ions are soon thermalized in a few picoseconds while the electron temperature rise is through an ion-electron relaxation of ~0.7 ns time scale, with considering typical SCPF plasma parameters of $T_e = 2$ keV, $T_i = 7$ keV, and $n_i = 1.5 \times 10^{21}/\mathrm{cm}^3$ according to our simulations [22]. The electron temperature remains low until 1.9 ns, leading to the dark image recorded by GXI. From 1.9 to 2.3 ns, the electron temperature rises to a certain level above 2 keV due to ion-electron energy relaxation, and the GXI images can be seen clearly. After 2.3 ns, the electron energy loss through bremsstrahlung emission leads to the fadeaway of GXI images, while the FXRD signals can still increase until 3.3 ns. This is because the FXRD of broadly spectrum-integrated x-ray flux are not so sensitive to the electron temperature, and the FXRD view field is large and can receive emissions from both the central and the peripheral plasmas. The peripheral low-density plasmas are compressed and heated to consecutively emit x rays when swept by the outgoing shock [22]. The shock effect and the time delay between neutron and x-ray emissions also emerged in capsule implosion experiments [24].

Our radiation hydrodynamic simulations with the LARED code [25] corroborate the above thermalization and relaxation pictures [22]. The nonequilibrium plasmas with a higher ion temperature and a lower electron temperature are desirable for nuclear fusion, because it increases the fusion reactivity and at the same time reduces energy losses through electron bremsstrahlung emission [34].

The neutron yields are measured by a plastic scintillator. The highest neutron yield is 3.5×10^9 achieved using a smaller hohlraum with a diameter of 1500 μm and an input laser energy of 6.3 kJ. It is worth mentioning that the neutron yields are remarkably stable. Three consecutive shots are fired with the 1700 μm hohlraum and a laser energy of 6.1, 5.9, and 5.8 kJ. The corresponding neutron yields are 3.2×10^9, 2.8×10^9, and 2.9×10^9, respectively. With all the experimental variations, such as the laser shot-to-shot energy variation and the target fabrication imperfections, the robustness in neutron yields is clearly evident. According to the nTOF measurement, the ion temperatures are typically around 6–8 keV, which are apparently higher than the 3–5 keV seen in NIF inertial implosions [35].

A total of 11 laser shots are fired, and neutron yield data are collected. In Fig. 4, the red stars show the neutron yield produced by a hohlraum with a diameter of 1700 μm and a laser spot diameter of 500 μm. The number of laser beams is adjusted to give the required total laser energies. The red

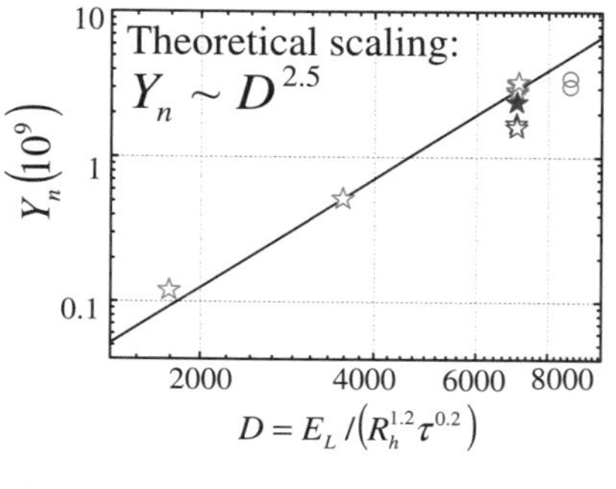

FIG. 4. The neutron yield data versus the scaling parameter. The error bars of the data are about 10%–15%, approximately equivalent to the vertical size of the varied scatters. E_L in units of J, R_h in units of mm, and τ in units of ns. Different scatters represent varied parameters; refer to the text for details.

circles represent the data with a smaller hohlraum of 1500 μm diameter. The hollow blue stars in Fig. 4 are yields obtained with a hohlraum of 1700 μm diameter and smaller laser spots of 200 μm diameter. Figure 4 shows that the neutron yield data are in agreement with the theoretical scaling law.

Two laser spot sizes with diameters of 500 and 200 μm are used to study the neutron yield dependency on spot sizes. As shown in Fig. 4, with the other target and laser parameters remaining identical, the yields with big spots (red stars) are about 50% higher than those with small spots (blue hollow stars). Though the plasma temperature with big spots is lower than that with small spots, the ablated mass is larger, giving rise to a larger fuel density at convergence and a longer confinement time.

We intentionally move the laser pointing location to form two distinct geometrical configurations to investigate yield dependency on geometric symmetry. The first configuration consists of one ring of laser spots on the equator. The second configuration has two rings of spots on each side of the equator with a distance of 600 μm between two rings. The first configuration gives the neutron yield of 1.7×10^9 and ion temperature of 6.4 keV (blue hollow stars in Fig. 4), while the second configuration gives 2.4×10^9 with an ion temperature of 7.6 keV (blue solid star in Fig. 4). Our interpretation is that, with one equatorial ring of laser spots, the colliding plasmas will eventually form a more cylindrical shape of hot dense plasmas (2D convergence), while in the second situation the colliding plasmas will form a relatively more spherical shape of hot dense plasma (3D convergence). In the latter case, the higher-dimensional convergence leads to a higher stagnation density and a higher ion temperature. For this reason, the octahedral spherical hohlraum [36,37] with six LEHs is a better choice for the SCPF scheme, because with the six-LEH the convergent plasmas are more spherically symmetric than that with two-LEH used in our experiment.

The choice of fuel material for the ablation layer is crucial for the SCPF. The neutron yield of different materials depends not only on their reactivity but also on the average atomic number A and average nuclear charge number Z through $Y_n \propto (A/Z)^{1/3}(1+Z)^{-1.5}A^{-0.5}$ [22]. For cryogenic DT, we expect to achieve a higher neutron yield of $Y_{\rm DT} = 12 r_{\sigma v} Y_{\rm CD}$ with $r_{\sigma v} = \langle \sigma v \rangle_{\rm DT} / \langle \sigma v \rangle_{\rm DD}$ the ratio of DT fusion reactivity to DD fusion reactivity. It indicates that the 3.5×10^9 DD neutron yield obtained in the current experiment is equivalent to the 4.2×10^{12} DT neutron. Moreover, according to the scaling law of the SCPF, one might expect to achieve $\sim 3.0 \times 10^{13}$ DT neutrons on the 30 kJ Omega laser with the spherical hohlraum diameter of 3.4 mm and a duration of 1 ns and $\sim 3.2 \times 10^{17}$ DT neutrons on the 1.8 MJ NIF laser with the spherical hohlraum diameter of 5.6 mm and a duration of 3 ns. The fusion energy can even balance or exceed the input energy with a laser energy larger than 3.3 MJ and a properly designed SCPF target.

In summary, the concept of laser-driven spherical convergent plasma fusion is proposed and investigated, and its principle has been demonstrated experimentally at the SGIII-prototype laser facility. It is shown that the SCPF scheme is efficient, robust, and insensitive to experimental variations, and the neutron yield can be predicted by our simple scaling law. The SCPF concept has important implications in neutron source and proton diagnostics. It can be further optimized by shaping the laser pulse or raise the convergent plasma temperature with fast heating by directly driven lasers [38,39], etc. Related works are ongoing.

This work is supported by the National Basic Research Program of China (2013CB834100), National Natural Science Foundation of China under Grants No. 11475027, No. 11674034, and No. 11475032.

*liu_jie@iapcm.ac.cn

[1] T. Ditmire, J. W. G. Tisch, E. Springate, M. B. Mason, N. Hay, R. A. Smith, J. Marangos, and M. H. R. Hutchinson, Nature (London) **386**, 54 (1997).
[2] T. Ditmire, J. Zweiback, V. P. Yanovsky, T. E. Cowan, G. Hays, and K. B. Wharton, Phys. Plasmas **7**, 1993 (2000).
[3] M. Roth *et al.*, Phys. Rev. Lett. **110**, 044802 (2013).
[4] J. D. Lindl, P. Amendt, R. L. Berger, S. Gail Glendinning, S. H. Glenzer, S. W. Haan, R. L. Kauffman, O. L. Landen, and L. J. Suter, Phys. Plasmas **11**, 339 (2004).
[5] R. L. McCrory *et al.*, Phys. Plasmas **15**, 055503 (2008).
[6] V. N. Goncharov *et al.*, Phys. Rev. Lett. **104**, 165001 (2010).
[7] T. J. B. Collins *et al.*, Phys. Plasmas **19**, 056308 (2012).
[8] S. P. Regan *et al.*, Phys. Rev. Lett. **117**, 025001 (2016).
[9] O. A. Hurricane *et al.*, Nature (London) **506**, 343 (2014).
[10] T. Döppner *et al.*, Phys. Rev. Lett. **115**, 055001 (2015).
[11] L. Rayleigh, Proc. London Math. Soc. **14**, 170 (1883); G. I. Taylor, Proc. R. Soc. A **201**, 192 (1950).
[12] R. D. Richtmyer, Commun. Pure Appl. Math. **13**, 297 (1960); E. E. Meshkov, Fluid Dyn. **4**, 101 (1972).
[13] H. Helmholtz, Philos. Mag. **36**, 337 (1868); W. Thomson, Philos. Mag. **42**, 362 (1871).
[14] T. Ma *et al.*, Phys. Rev. Lett. **111**, 085004 (2013).
[15] R. K. Kirkwood *et al.*, Plasma Phys. Controlled Fusion **55**, 103001 (2013).
[16] R. H. H. Scott *et al.*, Phys. Rev. Lett. **110**, 075001 (2013).
[17] M. J. Rosenberg *et al.*, High Energy Density Phys. **18**, 38 (2016).
[18] R. Betti and O. A. Hurricane, Nat. Phys. **12**, 717 (2016).
[19] S. A. Slutz, M. C. Herrmann, R. A. Vesey, A. B. Sefkow, D. B. Sinars, D. C. Rovang, K. J. Peterson, and M. E. Cuneo, Phys. Plasmas **17**, 056303 (2010).
[20] M. R. Gomez *et al.*, Phys. Rev. Lett. **113**, 155003 (2014).
[21] S. Atzeni and J. Meyer-ter-vehn, *The Physics of Inertial Fusion* (Clarendon, Oxford, 2004).
[22] See Supplemental Material at http://link.aps.org/supplemental/10.1103/PhysRevLett.118.165001 for the detailed deduction of the scaling law, discussion of relaxation rate, the time scales for neutron and x-ray emissions, and some hydrodynamic simulations, which includes Refs. [21,23–26].
[23] J. Wesson, *Tokamaks* (Clarendon, Oxford, 2003).
[24] S. Le Pape *et al.*, Phys. Rev. Lett. **112**, 225002 (2014).
[25] Z. Fan, S. Zhu, W. Pei, W. Ye, M. Li, X. Xu, J. Wu, Z. Dai, and L. Wang, Europhys. Lett. **99**, 65003 (2012).
[26] E. L. Dewald *et al.*, Phys. Rev. Lett. **95**, 215004 (2005).
[27] O. A. Hurricane *et al.*, Nat. Phys. **12**, 800 (2016).
[28] T. A. Thorson, R. D. Durst, R. J. Fonck, and A. C. Sontag, Nucl. Fusion **38**, 495 (1998).
[29] Y. C. F. Thio, J. Phys. Conf. Ser. **112**, 042084 (2008).
[30] Y. C. F. Thio, C. E. Knapp, R. C. Kirkpatrick, R. E. Siemon, and P. J. Turchi, J. Fusion Energy **20**, 1 (2001).
[31] S. H. Glenzer *et al.*, Phys. Plasmas **6**, 2117 (1999).
[32] C. K. Li *et al.*, Phys. Rev. Lett. **102**, 205001 (2009).
[33] C. K. Li *et al.*, Science **327**, 1231 (2010).
[34] Z. Fan, J. Liu, B. Liu, C. Yu, and X. T. He, Phys. Plasmas **23**, 010703 (2016).
[35] A. L. Kritcher *et al.*, Phys. Plasmas **23**, 052709 (2016).
[36] K. Lan, J. Liu, D. Lai, W. Zheng, and X.-Tu He, Phys. Plasmas **21**, 010704 (2014).
[37] K. Lan *et al.*, Matter Radiat. Extremes **1**, 8 (2016).
[38] Y. Kitagawa *et al.*, Phys. Rev. Lett. **108**, 155001 (2012).
[39] Y. Mori *et al.*, Phys. Rev. Lett. **117**, 055001 (2016).

Supplemental Material

To manuscript titled:

"Neutron generation by laser-driven spherically convergent plasma fusion"

By authors: G. Ren, J. Yan, J. Liu, et al

In this Supplemental material, we present the detailed deduction of the scaling law and discuss the relaxation, thermalization issues of convergent plasma and related FXRD, GXI and neutron diagnosis. We also present some preliminary one-dimensional hydrodynamic simulations to validate the physical pictures of SCPF.

1 Deduction of the scaling law

We start with the basic neutron yield formula as follows,

$$Y_n = \frac{1}{2} n_D^2 \langle \sigma v \rangle_{DD} V \Delta t, \tag{1}$$

where the deuteron ion number density $n_D = N_D/V$ with N_D the ion number, V the burn plasmas volume, and $\langle\sigma v\rangle_{DD}$ the mean reactivity of deuterium on deuterium.

We then have,

$$V = \frac{4\pi}{3}R_c^3, \tag{2}$$

with R_c the radius of the converged plasmas sphere.

The average reactivity is proportional to T_i^2 near 10 keV (T_i is the ion temperature of the converged plasma), i.e.,

$$\langle\sigma v\rangle \propto T_i^2, \tag{3}$$

The streaming plasma stagnates at the central region forming a high temperature plasma region with the pressure of more than one hundred Mbars, and subsequently it invokes an outgoing shock wave in the peripheral plasmas of relatively low density. The shock will decay in expanding radially outwards. This picture is clearly demonstrated in our following hydrodynamic simulations. According the above picture, we can estimate the fusion burn time to be proportional to the converged plasma sphere radius divided by the shock wave speed which is proportional to the sound speed, i.e.,

$$\Delta t \propto R_c/C_T \propto R_c/T_i^{0.5}. \tag{4}$$

Combine the above equations(2), (3), (4) with (1), we have

$$Y_n \propto N_D^2 T_i^{1.5} R_c^{-2}. \tag{5}$$

The convergent hot plasmas originate from the laser ablating CD fuel lined inside the gold spherical hohlraum , the total deuteron ion number N_D can be written as,

$$N_D \sim \dot{m} S_{spot} \tau \propto I_L^{0.76} S_{spot} \tau, \tag{6}$$

where the mass ablation rate $\dot{m} \propto I_L^{0.76}$, S_{spot} is the total area of the laser spots, and τ is the laser duration.

With the stationary ablation assumption [1], the mass ablation rate $\dot{m}$ and ablating plasmas temperature T_{ab} can be deduced as, $\dot{m} \propto I_L^{0.76}, T_{ab} \propto I_L^{2/3}$. Meanwhile, the plasmas expand at a velocity of v_i related to the acoustic speed of the ablated hot plasmas, $v_i \propto T_{ab}^{1/2} \propto I_L^{1/3}$. I_L denotes the laser intensity taking the form,

$$I_L = E_L/S_{spot}/\tau, \tag{7}$$

where, E_L is total laser energy.

In our SCPF scheme, laser beams are supposed to uniformly irradiate the fuel layer lined inside a spherical hohlraum of radius R_h, implying that,

$$S_{spot} \sim 4\pi R_h^2. \tag{8}$$

The ablating plasmas expand, collide and stagnate at the sphere center. At stagnation, the kinetic energy is converted into ion thermal energy, making the ion temperature T_i increase dramatically. With the ion adiabatic expansion assumption which allows us to approximately omit the thermal energy of the ions before stagnation, we get that

$$T_i \propto v_i^2 \propto I_L^{2/3}. \tag{9}$$

Combining the relations of (6), (7), (8) , (9) with (5), we have

$$Y_n \propto E_L^{2.52}/(R_h^{1.04} R_c^2 \tau^{0.52}), \tag{10}$$

The relation between R_c and R_h can be deduced with considering the pressure balance principle, that is, the thermal pressure of the converged plasma equates the kinetic pressure of the converging plasmas, i.e.,

$$P_c = \rho v^2. \tag{11}$$

here,P_c is the converged plasma's thermal pressure, ρ is the density of the streaming plasma, v is the stream velocity. We have omitted the thermal pressure of the streaming plasma which is small compared to the kinetic pressure.

The expressions of the thermal pressure and kinetic pressure read $P_c \sim \frac{NkT}{R_c^3}$ and $\rho v^2 \sim \frac{N}{R_h^3} v^2$, respectively. Here N is the total particle number of the laser ablated plasmas. On the other hand, as discussed in the main text, both T and v^2 are linearly proportional to the temperature T_{ab} of the ablated plasmas. Taking above considerations into equation (11), we can reasonably assume that ,

$$R_c \propto R_h \tag{12}$$

Substitute (12) to (10), we then finally get following approximate scaling law,

$$Y_n \propto E_L^{2.5} / (R_h^{3.0} \tau^{0.5}). \tag{13}$$

The neutron yield of different materials depends not only on their reactivity but also on the average atomic number A and average nuclear charge number

Z. One can readily obtain the dependency on A and Z as $Y_n \propto (A/Z)^{1/3}(1+Z)^{-1.5}A^{-0.5}$, with considering following relations [1] , $N_D \propto \dot{m}S_{spot}\tau/Am_p$, $\dot{m} \propto \rho_c^{2/3}I_L^{0.76}$, and $T_i \propto v_i^2/\Gamma \propto I_L^{2/3}\rho_c^{-2/3}Am_p/(1+Z)$, here m_p is the mass of a proton, $\rho_c = (N_c/Z)Am_p \propto A/Z$ is the critical mass density and $\Gamma = (1+Z)/(Am_p)$.

2 Relaxation rate of the convergent plasmas

First we do some calculations for the plasmas equilibration time by adopting the following formula from Wesson (2003) [2],

$$\tau_{ee} = 1.09 \times 10^{-11}\frac{T_e^{3/2}}{n_i Z_i^2 ln\Lambda_e}, \tag{14}$$

$$\tau_{ii} = 6.6 \times 10^{-10}\frac{A^{1/2}T_i^{3/2}}{n_i Z_i^4 ln\Lambda_i}, \tag{15}$$

$$\tau_{ei} = 0.99 \times 10^{-8}\frac{AT_e^{3/2}}{n_i Z_i^2 ln\Lambda_e}, \tag{16}$$

where T_e and T_i are eletron and ion temperature in units of keV , respectively, n_i is the ion density in units of $10^{21}/cm^3$, A is the mass number of the ion, and Z_i the ionization state, $ln\Lambda$ is the Coulomb logarithm and set as 10. $\tau_{ee}, \tau_{ii}, \tau_{ei}$ are the equilibration time of electron, ion and electron-ion in units of second, respectively.

In the SCPF scheme, laser beams of nanosecond pulse duration and $10^{14\sim15}W/cm^2$ intensity uniformly irradiate the fuel layer lined inside a spherical hohlraum. The fuel layer is ablated and heated to expand inwards. Eventually the hot fuel plasmas converge, collide, merge and stagnate at the central region, converting most of their kinetic energy to thermal energy, forming a thermonuclear fusion fireball. When the ablated plasmas expands at high speed toward the center, most of the kinetic energy of the plasmas is carried by the ions. When the convergent plasmas collide and stagnate, the kinetic energy of plasmas is converted into the ion thermal energy at first. Subsequently, by ion-electron relaxation, some of the ion thermal energy is transferred to the electrons. The typical converged CD plasma parameters are $T_e = 2keV$ and $T_i = 7keV$, $n_i = 1.5 \times 10^{21}/cm^3$. In the above (14), (15) and (16) expressions, we choose the average quantities for the CD mixed plasmas as $A = \langle A \rangle = 7$, $Z_i^2 = \langle Z_i^2 \rangle$ = 18.5, $Z_i^4 = \langle Z_i^4 \rangle$= 648.5, we get the e-quilibration times as $\tau_{ee} = 0.11ps$, $\tau_{ii} = 3.3ps$, $\tau_{ei} = 0.7ns$. The ion equilibration time is of the order of a few picoseconds which tells us that the ions will be soon thermalized to equilibrium state at stagnation while the ion-electron equilibration will take a few nanoseconds.

The above relaxation times can help us to understand the different time scales for the neutron and x-ray emissions.

3 Discussions on the time scales for the neutron and x-ray emissions

Neutron yield production per volume per time takes the form of $n_D^2\langle\sigma v\rangle_{DD} \propto n_D^2 T_i^2$, indicating that the fusion rate is sensitive to both ion density and ion temperature, while the x-ray Bremsstrahlung emission is scaled as $n_e^2 T_e^{0.5}$, which is more dependent on electron density and relatively less on electron temperature.

For the Bremsstrahlung emission, we have spectral power emission per volume as following,

$$\eta_\nu = \eta_0 \frac{Z_i^3 \rho^2}{T_e^{1/2} A^2} exp(-h\nu/T_e), \tag{17}$$

The power emission per volume of the total spectra and the GXI spectra are integration of (17) as below, in units of W/cm^3

$$P_{br}^{XRD} = 1.76 \times 10^{17} \frac{Z_i^3 \rho^2}{T_e^{1/2} A^2}, \tag{18}$$

$$P_{br}^{GXI} = 1.76 \times 10^{17} \frac{Z_i^3 \rho^2}{T_e^{1/2} A^2} exp(-T_G/T_e), \tag{19}$$

where $T_G = 2$ keV, and since the FXRD records the x-ray spectrally expanding from 100 eV to 4.4 keV, we approximately integrate the spectra from 0 to infinity. In the above two integrations, we also assume that the both the response functions are constant over their spectral ranges. The constant response is appropriate for the FXRD, but is not necessarily true in the case of GXI. However, it is enough to qualitatively evaluate how the GXI images brightness depends on electron temperature.

The above analysis indicates that the hard x-ray signals (collected by GXI) are much more sensitive to the plasma electron temperature than the FXRD signals.

In our experiments, FXRD measures the spectrally integrated x-ray flux through the LEH at an angle of 30° to the axis. From this direction, the FXRD views both emissions from the hohlraum wall and emissions from the central converging plasmas. The x-ray flux rises quickly during the first 1 ns when the laser beams are on, and it is contributed mostly by the emissions from the ablation of hohlraum wall by the lasers.

We see in the GXI images of the earlier emission at 0.8 ns to 1.0 ns, which are believed to be the fast moving plasma jet with higher electron temperatures yet much smaller amount of plasmas. The jet converging can be easily seen in the GXI images for its higher electron temperature, but for the much smaller amount of plasmas, the neutron production is trivial compared to the neutron production at bang time of 1.5 ns.

The neutron bang time of 1.5 ns is about 0.4 ns earlier than the second round hard x-ray emission in GXI and 1.8 ns earlier than the FXRD second emission peak, reflecting intrinsically the electron-ion relaxation and nonequilibrium energy transfer in the processes of plasmas convergence and subsequent outgoing shock propagation. When the plasmas converge into the sphere center at about 1.5 ns, the ion temperature reaches its peak value of about 7 keV due to kinetic-thermal energy conversion. At this time, the plasmas have an apparent accumulation at central region yielding a maximum neutron generation.

The 0.4 ns time delay of the hard x-ray emission to the neutron bang time is due to that the ions are soon thermalized in a few picoseconds while electron temperature rise needs a time scale of nanosecond through ion-electron relaxation. The electron temperature keeps low until 1.9 ns leading to the dark image recorded by GXI. From 1.9 ns to 2.3 ns, the electron temperature rises to a certain level above 2 keV due to ion-electron energy relaxation and the GXI images can be seen clearly. After 2.3 ns, the energy loss through Bremsstrahlung emission leads to the fadeaway of GXI images, while the FXRD signals can still increase until 3.3 ns. This is because the FXRD of broadly spectrum-integrated x-ray flux are not so sensitive to the electron temperature, and the FXRD at an angle of 30° to the axis can view the emissions from both the central and the peripheral plasmas. The peripheral plasmas of low density are compressed and heated to consecutively emit x-ray when swept by the outgoing shockwave. The shock effect and the time delay between neutron production and hard x-ray emission are also observed in recent capsule implosion experiment [3].

From the above discussions on relaxation, thermalization issues of convergent plasma and related FXRD, GXI diagnosis and neutron measurement, we can see that the different time scales for the neutron (recorded by scintillator and nTOF)and x-ray emissions (recorded by FXRD and GXI) coincide and crosscheck each other that certify the physical principle of SCPF.

4 Hydrodynamic simulations

We use LARED-S code to simulate one-dimensional SCPF hydrodynamic processes. LARED-S is a Eulerian, multi-material, radiation hydrodynamic code, it includes plasma hydrodynamics, multi-group radiation diffusion, electron and ion thermal conduction, laser-ray tracing, etc [4]. In our simulations, we adopt the average ion $A = 7$ and the average ionization state $Z_i = 3.5$. The inner radius of CD shell is $R_h = 800\ \mu m$ and the shell width is 100 $\mu\ m$. We apply 1 ns square laser pulse of energy of 5.1 kJ to irradiate uniformly the inside of CD shell, the corresponding laser intensity is of $6.3 \times 10^{13}\ W/cm^2$.

Figure S1 (a), (b) and (c) show the distributions of density, ion temperature and electron temperature of the laser ablated plasmas at t=0.5 ns, 1.0 ns, and 1.6 ns, respectively. In Figure S1 (a), the laser ablated plasmas are expanding inward. It shows that the laser deposits energy near critical density surface and the heated plasmas are expanding fast toward the sphere center with a rapid density rarefaction. On the opposite direction, an outgoing shock wave is invoked by laser ablation pressure, creating a sharp density peak in the CD layer. Note that since the laser heating is through inverse Bremsstrahlung so that the electrons expand almost isothermally ($\sim$ 1.3 keV) while the ions expand most likely in an adiabatical way in which the ion temperature decays quickly to about 0.3 keV at a distance of 400 μm. Figure S1 (b) shows the distributions at the moment close to stagnation. The hot plasmas pile up in the central region, forming a density plateau with density about 30 mg/cm^3 with electron density of about 1.0 N_c (critical density). The ion temperature rises to a high level of about 10 keV due to kinetic-thermal energy conversion. At this time, the electron temperature also increases to about 3.3 keV through ion-electron thermal relaxation. Figure S1 (c) shows that the ion and electron temperatures tend to equilibrate. The fast falling of the ion temperature leads to the quenching of thermonuclear reaction. Even with ion-electron energy transfer, the electron temperature still decrease because of electron energy loss due to the Bremsstrahlung emission.

Figure S2 shows the temporal evolution of the pressure and indicates clearly the SCPF processes, including the ablated plasma expansion, stagnation, and the following outward decaying shock waves.

The above one-dimensional LARED-S simulations clearly demonstrate the important SCPF processes of ablation, expansion, stagnation, outgoing shock and related thermalization and relaxtion, and qualitatively certificate the physical pictures of SCPF. Quantitatively, the numerical simulations predict higher ion temperature, faster expansion velocity and higher plasma density compared to experiments. This is because that in experiments, only limited number of laser beams are available so that the plasmas dynamics are 3D rather than ideal 1D spherical convergence. Other 3D effects such as the jets squeezed out by adjacent laser ablated bubbles also can not reproduced by the present 1D simulation. The dimension effects will decrease the energy conversion efficiency and the density of the converged plasmas, for instance, we also extend to make some preliminary 2D simulations which give the density of the converged plasmas of only 0.5-0.7 N_c. Detailed discussions of our modeling results will be presented in a later paper.

On the other hand, hohlraum plasma filling might prevent the incident laser from propagating in the hohlraum and will reduce the energy coupling efficiency significantly. It is an important problem in the high-Z hohlraums in which high temperature radiation ablation might lead to severe plasma filling [5]. While in our

SCPF of low-Z case the radiation temperature is very low and the hohlraum filling becomes less important. From our 2D simulations, we clearly see that in the low-Z plasmas the incident lasers can smoothly propagate to the hohlraum wall. Therefore, the previous model needs to be modified for the low-Z hohlraum case. Details of the plasma filling in SCPF will be investigated in our future work.

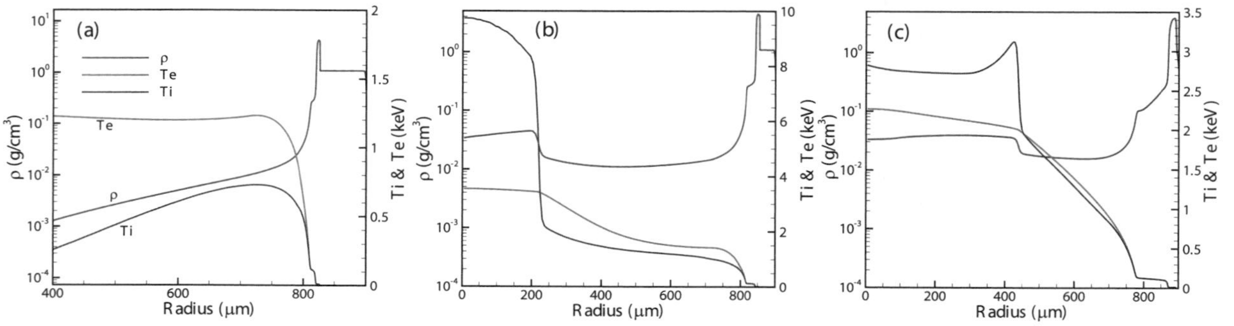

Fig. S1 (Color online) The evolution of the SCPF plasmas in simulations of using the LARED-S code. Figure (a), (b) and (c) correspond to time of 0.5 ns, 1.0 ns and 1.6 ns, respectively.

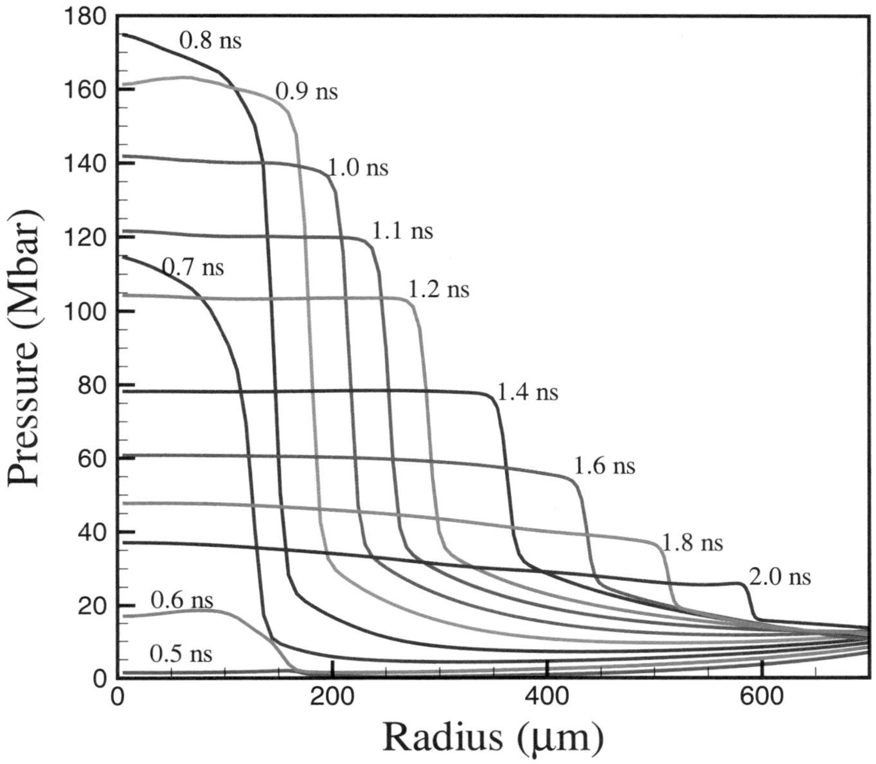

Fig. S2 (Color online) The temporal evolution of the pressure of the SCPF plasmas in the simulations of LARED-S code, indicating the plasma converging, stagnation and the outward decaying shock waves.

1. S. Atzeni and J. Meyer-ter-vehn, *The Physics of Inertial Fusion*, (Clarendon Oxford, 2004).

2. Wesson J.(2003). Tokamaks. Clarendon Press, Oxford.

3. S. Le Pape, et al., Phys. Rev. Lett. 112, 225002 (2014).

4. Z. F. Fan et al., Europhys. Lett. 99, 65003 (2012).

5. E. L. Dewald, L. J. Suter, O. L. Landen, et al., Phys. Rev. Lett. 95, 215004 (2005).

斑图动力学和时空混沌

斑图（pattern）是指非线性相互作用产生的携带物质质量、动量和能量的一种空间结构（例如，天空中的云图、宇宙中的星团、海洋中的波浪、流体中的涡旋等），它是系统动量和能量传输的载体，对系统演化起着关键性作用。

贺贤土院士发展了斑图动力学理论，是最早提出把传统的斑图选择（selection）和斑图形成（formation）研究扩展为斑图动力学研究的学者之一。早在20世纪80年代，他首次从电磁波与等离子体作用过程获得类孤立子（相干斑图）解，并发现斑图时间随机演化行为。

贺贤土院士发现了近可积哈密顿系统初始相干（或规则）斑图相互作用可以演化成时间随机、空间不规则的时空混沌及其演化途径。若干初始相干或规则斑图通过非线性相互作用，会自发地演化到统计意义上的时空混沌或“湍流”状态。这些研究成果在非线性物理和统计物理领域引起了国际同行的广泛兴趣，被评论为“发现了近可积系统时空混沌”。此后，国际上此类研究工作也陆续大量开展。

第 31 卷 第 10 期
1982 年 10 月

物 理 学 报
ACTA PHYSICA SINICA

Vol. 31, No. 10
Oct., 1982

等离子体中大幅波与低频振荡粒子非线性相互作用效应

贺 贤 土
(北京 8009 信箱)

1981 年 11 月 2 日收到

提 要

本文从 Vlasov-Maxwell 方程组出发，保留高频场到四级项，导得了纵、横波大幅高频场与低频振荡粒子拍频作用的动力学方程组. 结果表明，在空间一维情况下，对亚声波，高阶大幅高频场的存在使形成 Soliton 的有质动力凝聚作用削弱，而对超声波则更不存在 Soliton 解；在三维情况下，似乎使 Caviton 形成变慢.

一、引 言

等离子体中不稳定性发展的后阶段，常常伴随着增大了振幅的高频波与低频振荡粒子的非线性拍频作用. 例如，波与粒子的非线性散射产生了场能在长波长区的积累[1]，导致了调制不稳定性[2]开始，其结果是增大了振幅的场，改变了波的传播特性，产生了明显的非线性色散效应. 波与低频振荡粒子的这种非线性耦合可能导致强湍流过程的开始[3]. 在文献 [4] 中，我们从 Vlasov 方程和电磁场的 Maxwell 方程组出发，讨论了两个高频场与低频振荡粒子的拍频作用，导得了一组包含阻尼的有质动力作用下广义的 Zakharov 方程组，它可以导致 Soliton 和 Caviton 等现象的出现[5]. 但是，随着场振幅的增大，应当考虑更高阶场与粒子运动的耦合，并研究它们对二级场产生效应的影响. 本文仍然从 Vlasov-Maxwell 方程组出发，取自洽场到四级项，导得了低频振荡粒子与四级场非线性作用的粒子微观分布和宏观流体密度的动力学方程组. 结果表明，无论纵波或横波，四级场的出现都可明显影响二级场的非线性作用，使一维 Soliton 粒子密度抽空只能具有比初始密度小得多的量级，而使超声一维 Soliton 却变得更加不稳定；对于三维 Caviton 运动，四级场的作用可能使 Caviton 形成变慢.

二、Fourier 表象中粒子微观速度分布与大振幅波的耦合

为了叙述方便，我们重复文献 [4] 中的某些考虑和表示. 引入单位偏振矢量 $\boldsymbol{e}_k^\sigma$，$\boldsymbol{k}$ 为波矢，σ 为波符号. 对纵波 $\sigma = l$，$\boldsymbol{e}_k^l = \boldsymbol{k}/|\boldsymbol{k}|$；对横波 $\sigma = t$，$\boldsymbol{e}_k^t$ 已对两个偏振方向求和. 写 F 空间的电场矢量为

$$\boldsymbol{E}(k) = \sum_{\sigma=l,t} \boldsymbol{e}_{\boldsymbol{k}}^{\sigma} E^{\sigma}(k). \tag{1}$$

这里 $k = \{\omega, \boldsymbol{k}\}$ 为频率 ω 和波矢 $\boldsymbol{k}$ 的缩写，以下同. 对于每个 σ 波 $\boldsymbol{k}$ 有两个传播方向，我们约定

$$\boldsymbol{e}_{-\boldsymbol{k}}^{\sigma} = -\boldsymbol{e}_{\boldsymbol{k}}^{\sigma}, \quad E^{\sigma}(-k) = -E^{\sigma}(k)^{*}. \tag{2}$$

* 为复数记号. 现在可把电场力与 Lorentz 力写成

$$\boldsymbol{G}(k) = \sum_{\sigma} \boldsymbol{e}_{\boldsymbol{k}}^{\prime\sigma} E^{\sigma}(k), \tag{3}$$

其中

$$\boldsymbol{e}_{\boldsymbol{k}}^{\prime\sigma} = \left(1 - \frac{\boldsymbol{k}\cdot\boldsymbol{v}}{\omega}\right)\boldsymbol{e}_{\boldsymbol{k}}^{\sigma} + \frac{\boldsymbol{k}\boldsymbol{v}\cdot\boldsymbol{e}_{\boldsymbol{k}}^{\sigma}}{\omega}, \tag{4}$$

$\boldsymbol{v}$ 为粒子速度，显然

$$\boldsymbol{e}_{\boldsymbol{k}}^{\prime\sigma}\cdot\boldsymbol{v} = \boldsymbol{e}_{\boldsymbol{k}}^{\sigma}\cdot\boldsymbol{v}. \tag{5}$$

有了上述表示后，就可从 Vlasov-Maxwell 方程组得到 F 表象中高频 (h) 和低频 (d) 的粒子速度分布函数及场的表达式如下[4]:

粒子速度分布函数的低频部分

$$\delta F_{\alpha}(q) = \frac{e_{\alpha}}{i m_{\alpha}(\Omega - \boldsymbol{q}\cdot\boldsymbol{v})}\left\{\boldsymbol{G}_{d}(q)\cdot\frac{\partial F_{0\alpha}}{\partial \boldsymbol{v}} + \int_{(d)} \boldsymbol{G}_{h}(k_1)\cdot\frac{\partial f_{\alpha}(k_2)}{\partial \boldsymbol{v}}\cdot\delta^4(q - k_1 - k_2)\frac{dk_1 dk_2}{(2\pi)^4}\right\}. \tag{6}$$

上式中已略去了两个低频波拍频产生的新低频波项，因为只感兴趣高频波拍频为低频的效应. (6) 式中右端第一项为低频场力作用使初态分布 $F_{0\alpha}$ 变化对粒子低频扰动 $\delta F_{\alpha}(q)$ 的贡献，第二项为高频分布与高频场作用对 δF_{α} 的贡献. 这里低频分布函数已用了表示

$$F_{\alpha}(q) = F_{0\alpha}(\boldsymbol{v})(2\pi)^4\delta^4(q) + \delta F_{\alpha}(q), \tag{7}$$

其中 $\delta F_{\alpha}(q) \equiv \delta F_{\alpha}(q, \boldsymbol{v})$, $f_{\alpha}(k) \equiv f_{\alpha}(k, \boldsymbol{v})$, $\alpha = i$ (离子), e (电子). m_{α} 为粒子质量，e_{α} 为电荷，$q = \{\Omega, \boldsymbol{q}\}$ 为低频波的频率和波矢，k_1, k_2 为高频波缩写，$dk = d^3k d\omega$, $dq = d^3q d\Omega$, $\delta^4(k - k_1 - k_2) = \delta(\omega - \omega_1 - \omega_2)\delta(\boldsymbol{k} - \boldsymbol{k}_1 - \boldsymbol{k}_2)$. 积分号 $\int_{(d)}$ 表示只取高频向低频拍频过程.

粒子的高频分布

$$f_{\alpha}(k) = \frac{e_{\alpha}}{i m_{\alpha}(\omega - \boldsymbol{k}\cdot\boldsymbol{v})}\left\{\boldsymbol{G}_{h}(k)\cdot\frac{\partial}{\partial \boldsymbol{v}}F_{0\alpha}(\boldsymbol{v}) + \int \boldsymbol{G}_{h}(k_1)\cdot\frac{\partial \delta F_{\alpha}(q)}{\partial \boldsymbol{v}} \times\delta^4(k - k_1 - q)\frac{dk_1 dq}{(2\pi)^4} + \int \boldsymbol{G}_{d}(q)\cdot\frac{\partial f_{\alpha}(k_1)}{\partial \boldsymbol{v}}\delta^4(k - k_1 - q)\frac{dk_1 dq}{(2\pi)^4}\right\}. \tag{8}$$

这里已略去了两个高频波拍频产生的更高频波项，因为只感兴趣频率接近等离子体频率的一类高频波. (8) 式中右端第一项来自高频场与初态分布粒子相互作用对高频分布的贡献，第二、三两项分别为高频场与低频振荡粒子作用以及低频场与高频粒子作用引起高频振荡粒子分布的变化.

由电磁场 Maxwell 方程，得低频场和高频场表达式分别为

$$\varepsilon_d^{\sigma_q}(q)E_d^{\sigma_q}(q) = -\frac{4\pi}{\Omega}\sum_{\alpha,\sigma_1}\frac{e_\alpha^2}{m_\alpha}\int_{(d)}\frac{\boldsymbol{e}_q\cdot\boldsymbol{v}}{\Omega-\boldsymbol{q}\cdot\boldsymbol{v}+i\delta}\,\boldsymbol{e}_1'\cdot\frac{\partial}{\partial\boldsymbol{v}}f_\alpha(k_2)E_h^{\sigma_1}(k_1)$$
$$\cdot\,\delta^4(q-k_1-k_2)\frac{dk_1dk_2d^3v}{(2\pi)^4}, \tag{9}$$

$$\varepsilon_h^{\sigma}(k)E_h^{\sigma}(k) = -\frac{4\pi}{\omega}\sum_{\alpha}\frac{e_\alpha^2}{m_\alpha}\int\frac{\boldsymbol{e}_k\cdot\boldsymbol{v}}{\omega-\boldsymbol{k}\cdot\boldsymbol{v}+i\delta}\Big[\sum_{\sigma_1}\boldsymbol{e}_1'\cdot\frac{\partial}{\partial\boldsymbol{v}}\delta F_\alpha(q)E_h^{\sigma_1}(k_1)$$
$$+\sum_{\sigma_q}\boldsymbol{e}_q'\cdot\frac{\partial f_\alpha(k_1)}{\partial\boldsymbol{v}}E_d^{\sigma_q}(q)\Big]\delta^4(k-k_1-q)\frac{dk_1dq}{(2\pi)^4}d^3v. \tag{10}$$

(9), (10) 式右边分别是低频和高频非线性电流部分，被积函数中 $\delta\to 0_+$，积分线路在复平面上沿实轴上岸向下绕过极点，为了方便，以后将省略 $i\delta$ 不写. 式中还应用了简写 $\boldsymbol{e}_k'\equiv\boldsymbol{e}_k'^{\sigma}$, $\boldsymbol{e}_1'\equiv\boldsymbol{e}_{k_1}'^{\sigma}\cdots$ 等等，以下同.

对 $\sigma=l$，高频纵波线性介质函数为

$$\varepsilon_h^l(k) = 1+\sum_{\alpha=i,e}\frac{\omega_{p\alpha}^2}{k^2}\int\frac{\boldsymbol{k}\cdot\dfrac{\partial f_{0\alpha}}{\partial\boldsymbol{v}}}{\omega-\boldsymbol{k}\cdot\boldsymbol{v}}d^3v$$
$$=1-\sum_{\alpha=i,e}\left(\frac{\omega_{p\alpha}^2}{\omega^2}+\frac{3k^2v_{T_\alpha}^2}{\omega_{p\alpha}^2}+\cdots\right)+\operatorname{Im}\varepsilon_h^l(k). \tag{11}$$

对 $\sigma=t$，高频横波线性介质函数为

$$\varepsilon_h^t(k)+\frac{c^2k^2}{\omega^2}\doteq 1-\frac{\omega_{pe}^2}{\omega^2}. \tag{12}$$

(11) 和 (12) 式中已令 $F_{0\alpha}=n_0f_{0\alpha}$，其中 $n_{0e}=n_{0i}=n_0$ 为初始单位体积粒子数. $v_{T_\alpha}^2=T_\alpha/m_\alpha$，$T_\alpha$ 为粒子温度，以能量为单位. $\omega_{p\alpha}^2=4\pi n_0e_\alpha^2/m_\alpha$. 对于 $\varepsilon_d^l(q)$ 和 $\varepsilon_d^t(q)$ 亦有类似表达式.

从 (8) 和 (9) 式可知，低频场 $E_d^{\sigma_q}(q)$ 系两个高频场拍频产生的次级场，因此，粒子低频分布 (6) 式最低阶也是高频场的二级拍频效应，其次才是四级场、六级场…效应. 在文献 [4] 中处理时，仅保留了二级场的有质动力作用，所以对低频分布的贡献完全来自有质动力与初态粒子分布的相互作用. 但是，当场幅进一步增大时，四级场的拍频效应不能忽略，这时除了波场对初态分布的作用外，还进一步作用于粒子速度分布的扰动态，导致了粒子在速度空间的扩散过程. 因此，叠代过程必须保留四级场的存在. 展开粒子高频分布至场的三级项，得

$$f_\alpha(k) = \frac{e_\alpha}{im_\alpha(\omega-\boldsymbol{k}\cdot\boldsymbol{v})}\Big\{n_0\sum_{\sigma_k}\boldsymbol{e}_k\cdot\frac{\partial f_{0\alpha}}{\partial\boldsymbol{v}}E_h^{\sigma_k}(k)+\sum_{\sigma_1}\int\boldsymbol{e}_1'\cdot\frac{\partial\delta F_\alpha(q)}{\partial\boldsymbol{v}}E_h^{\sigma_1}(k_1)$$
$$\times\,\delta^4(k-k_1-q)\frac{dk_1dq}{(2\pi)^4}+\sum_{\beta\sigma_q\sigma_1\sigma_2\sigma_3}n_0\omega_{p\beta}^2\frac{e_\alpha e_\beta}{m_\alpha m_\beta}\iint_{(d)}\boldsymbol{e}_q'\cdot\frac{\partial}{\partial\boldsymbol{v}}\frac{\boldsymbol{e}_1'\cdot\dfrac{\partial f_{0\alpha}}{\partial\boldsymbol{v}}}{\omega_1\boldsymbol{k}_1\cdot\boldsymbol{v}}$$
$$\times\frac{H_\beta^{\sigma_q\sigma_2\sigma_3}(q,k_2,k_3)}{\Omega\varepsilon_d^{\sigma_q}(q)}E_h^{\sigma_1}(k_1)E_h^{\sigma_2}(k_2)E_h^{\sigma_3}(k_3)\delta^4(k-k_1-q)\delta^4$$
$$\times\,(q-k_2-k_3)\frac{dk_1dk_2dk_3dq}{(2\pi)^8}\Big\}. \tag{13}$$

同样,取至四级场可得低频场

$$E_d^{\sigma q}(q) = -\sum_{\alpha\sigma_1\sigma_2} \frac{\omega_{p\alpha}^2 e_\alpha}{im_\alpha} \left\{ \int_{(d)} \frac{H_\alpha^{\sigma q\sigma_1\sigma_2}(q, k_1, k_2)}{\Omega \epsilon_d^{\sigma q}(q)} E_k^{\sigma_1}(k_1) E_k^{\sigma_2}(k_2) \right.$$
$$\times \delta^4(q - k_1 - k_2) \frac{dk_1 dk_2}{(2\pi)^4} + \frac{1}{n_0} \int_{(d)} \frac{Z_\alpha^{\sigma q\sigma_1\sigma_2}(q, k_1, k_2 + q', k_2, q')}{\Omega \epsilon_d^{\sigma q}(q)}$$
$$\times E_k^{\sigma_1}(k_1) E_k^{\sigma_2}(k_2) \delta^4(q - k_1 - k_2 - q') \frac{dk_1 dk_2 dq'}{(2\pi)^8}$$
$$+ \sum_{\beta\sigma_{q'}\sigma_3\sigma_4} \omega_{p\beta}^2 \frac{e_\alpha e_\beta}{m_\alpha m_\beta} \int_{(d)} \frac{K_\alpha^{\sigma q\sigma_1\sigma_{q'}\sigma_2}(q, k_1, k_2 + q', q', k_2)}{\Omega \epsilon_k^{\sigma q}(q)}$$
$$\times \frac{H_\beta^{\sigma q'\sigma_3\sigma_4}(q', k_3, k_4)}{\Omega' \epsilon_d^{\sigma q'}(q')} E_k^{\sigma_1}(k_1) E_k^{\sigma_2}(k_2) \cdot E_k^{\sigma_3}(k_3) E_k^{\sigma_4}(k_4)$$
$$\left. \times \delta^4(q - k_1 - k_2 - q') \delta^4(q' - k_3 - k_4) \frac{dk_1 \cdots dk_4 dq'}{(2\pi)^{12}} \right\}. \tag{14}$$

(13) 及 (14) 式中

$$H_\alpha^{\sigma q\sigma_1\sigma_2}(q, k_1, k_2) = \int \frac{\boldsymbol{e}_q \cdot \boldsymbol{v}}{\Omega - \boldsymbol{q} \cdot \boldsymbol{v}} \mathscr{S}_\alpha^{\sigma_1\sigma_2}(k_1, k_2, \boldsymbol{v}) d^3\boldsymbol{v}, \tag{15}$$

$$\mathscr{S}_\alpha^{\sigma_1\sigma_2}(k_1, k_2, \boldsymbol{v}) \equiv \frac{1}{2}\left(\boldsymbol{e}_1' \cdot \frac{\partial}{\partial \boldsymbol{v}} \frac{\boldsymbol{e}_2' \cdot \frac{\partial f_{0\alpha}}{\partial \boldsymbol{v}}}{\omega_2 - \boldsymbol{k}_2 \cdot \boldsymbol{v}} + 1 \longleftrightarrow 2 \right); \tag{16}$$

$$Z_\alpha^{\sigma q\sigma_1\sigma_2}(q, k_1, k_2+q', k_2, q') = \int \frac{\boldsymbol{e}_q \cdot \boldsymbol{v}}{\Omega - \boldsymbol{q} \cdot \boldsymbol{v}} \mathscr{R}_\alpha^{\sigma_1\sigma_2}(k_1, k_2 + q', k_2, q', \boldsymbol{v}) d^3 v, \tag{17}$$

$$\mathscr{R}_\alpha^{\sigma_1\sigma_2}(k_1, k_2 + q', k_2, q', \boldsymbol{v}) \equiv \frac{1}{2}\left(\boldsymbol{e}_1' \cdot \frac{\partial}{\partial \boldsymbol{v}} \frac{1}{\omega_2 + \Omega' - (\boldsymbol{k}_2 + \boldsymbol{q}') \cdot \boldsymbol{v}} \boldsymbol{e}_2' \right.$$
$$\left. \times \frac{\partial \delta F_\alpha(q')}{\partial \boldsymbol{v}} + 1 \longleftrightarrow 2 \right); \tag{18}$$

$$K_\alpha^{\sigma q\sigma_1\sigma_{q'}\sigma_2}(q, k_1, k_2 + q', q', k_2) = \int \frac{\boldsymbol{e}_q \cdot \boldsymbol{v}}{\Omega - \boldsymbol{q} \cdot \boldsymbol{v}} \mathscr{T}_\alpha^{\sigma_1\sigma_{q'}\sigma_2}(k_1, k_2 + q', q', k_2, \boldsymbol{v}) d^3 v, \tag{19}$$

$$\mathscr{T}_\alpha^{\sigma_1\sigma_{q'}\sigma_2}(k_1, k_2 + q', q', k_2, \boldsymbol{v}) \equiv \frac{1}{2}\left(\boldsymbol{e}_1' \cdot \frac{\partial}{\partial \boldsymbol{v}} \frac{1}{\omega_2 + \Omega' - (\boldsymbol{k}_2 + \boldsymbol{q}') \cdot \boldsymbol{v}} \boldsymbol{e}_{q'}' \right.$$
$$\left. \times \frac{\partial}{\partial \boldsymbol{v}} \frac{\boldsymbol{e}_2' \cdot \frac{\partial f_{0\alpha}}{\partial \boldsymbol{v}}}{\omega_2 - \boldsymbol{k}_2 \cdot \boldsymbol{v}} + 1 \longleftrightarrow 2 \right). \tag{20}$$

H_α, Z_α 与 K_α 各自对k_1, k_2 对称. 此外,可以证明 H_α 尚具有下列性质:

$$\frac{1}{\Omega} H_\alpha^{\sigma q\sigma_1\sigma_2}(q, k_1, k_2) = \frac{1}{\omega_2} H_\alpha^{\sigma_2\sigma q\sigma_1}(k_2, q, -k_1). \tag{21}$$

(21) 式只对 $\Omega - \boldsymbol{q} \cdot \boldsymbol{v} = 0$ 无奇点条件下成立.

在以下的书写中,除特殊情况外,将省去 H_α, Z_α, K_α 和 ϵ_d 上角标波符号不写. 现在写出低频分布 (6) 式的进一步形式为(至四级场)

$$\delta F_\alpha(q)=\frac{e_\alpha}{m_\alpha(\Omega-\boldsymbol{q}\cdot\boldsymbol{v})}\sum_{\sigma_1\sigma_2}n_0\int_{(d)}\Big[\sum_{\beta\sigma_{\boldsymbol{q}}}\omega_{p\beta}^2\frac{e_\beta}{m_\beta}\boldsymbol{e}'_{\boldsymbol{q}}\cdot\frac{\partial f_{0\alpha}}{\partial\boldsymbol{v}}\frac{H_\beta(q,k_1,k_2)}{\Omega\epsilon_d(q)}$$
$$-\frac{e_\alpha}{m_\alpha}\mathscr{S}_\alpha(k_1,k_2,\boldsymbol{v})\Big]E_{h}^{\sigma_1}(k_1)E_{h}^{\sigma_2}(k_2)\delta^4(q-k_1-k_2)\frac{dk_1dk_2}{(2\pi)^4}$$
$$+\frac{e_\alpha}{m_\alpha(\Omega-\boldsymbol{q}\cdot\boldsymbol{v})}\Big\{\sum_{\sigma_1\sigma_2}\iint_{(d)}\Big[\sum_{\beta\sigma_{\boldsymbol{q}}}\omega_{p\beta}^2\frac{e_\beta}{m_\beta}\boldsymbol{e}'_{\boldsymbol{q}}$$
$$\times\frac{\partial f_{0\alpha}}{\partial\boldsymbol{v}}\frac{Z_\beta(q,k_1,k_2+q',k_2,q')}{\Omega\epsilon_d(q)}-\frac{e_\alpha}{m_\alpha}\boldsymbol{e}'_1$$
$$\times\frac{\partial}{\partial\boldsymbol{v}}\frac{1}{\omega_2+\Omega'-(\boldsymbol{k}_2+\boldsymbol{q}')\cdot\boldsymbol{v}}\boldsymbol{e}'_2\cdot\frac{\partial\delta F_\alpha(q)}{\partial\boldsymbol{v}}\Big]E_{h}^{\sigma_1}(k_1)E_{h}^{\sigma_2}(k_2)$$
$$\times\delta^4(q-k_1-k_2-q')\frac{dk_1dk_2dq'}{(2\pi)^8}+\sum_{\gamma\sigma_{\boldsymbol{q}'}}\sum_{\sigma_1\sigma_2\sigma_3\sigma_4}\omega_{p\gamma}^2\frac{e_\gamma}{m_\gamma}$$
$$\times\iiint_{(d)}\int_{(d')}\Big[\sum_{\beta\sigma_{\boldsymbol{q}}}\frac{\omega_{p\beta}^2}{4\pi m_\beta}\boldsymbol{e}'_{\boldsymbol{q}}\cdot\frac{\partial f_{0\alpha}}{\partial\boldsymbol{v}}\frac{K_\beta(q,k_1,k_2+q',q',k_2)}{\Omega\epsilon_d(q)}$$
$$-\frac{\omega_{p\alpha}^2}{4\pi m_\alpha}\mathscr{T}_\alpha(k_1,k_2+q',q',k_2)\Big]\cdot\frac{H_\gamma(q',k_3,k_4)}{\Omega'\epsilon_d(q')}$$
$$\times E_{h}^{\sigma_1}(k_1)E_{h}^{\sigma_2}(k_2)E_{h}^{\sigma_3}(k_3)E_{h}^{\sigma_4}(k_4)\delta^4(q-k_1-k_2-q')\delta^4(q'-k_3-k_4)$$
$$\times\frac{dk_1\cdots dk_4dq'}{(2\pi)^{12}}\Big\}.\tag{22}$$

(22) 式右端第一项为二级场效应,其余均为四级场贡献.

现在进一步化简 (22) 式. 低频波 $\sigma_{\boldsymbol{q}}$ 可以是迴旋波[1)]和纵波, 在以下讨论中, 为方便只取 $\sigma_{\boldsymbol{q}}$ 为纵波 (例如离子声波), 这时 $\boldsymbol{e}_{\boldsymbol{q}}=\boldsymbol{q}/|\boldsymbol{q}|$. 同时, 由于离子质量 m_i 远大于电子 m_e, 在 (22) 式的低频分布中可略去离子响应,即取 $\alpha=e$. 这样得到

$$\delta F_e(q)=\frac{1}{\Omega-\boldsymbol{q}\cdot\boldsymbol{v}}\Big\{\sum_{\sigma_1\sigma_2}\int_{(d)}B_e(q,k_1,k_2,\boldsymbol{v})E_{h}^{\sigma_1}(k_1)E_{h}^{\sigma_2}(k_2)\delta^4(q-k_1-k_2)\frac{dk_1dk_2}{(2\pi)^4}$$
$$+\frac{e^2}{m_e^2}\sum_{\sigma_1\sigma_2}\iint_{(d)}\Big[\omega_{pe}^2\boldsymbol{e}_{\boldsymbol{q}}\cdot\frac{\partial f_{0e}}{\partial\boldsymbol{v}}\frac{Z_e(q,k_1,k_2+q',k_2,q')}{\Omega\epsilon_d(q)}$$
$$-\boldsymbol{e}'_1\cdot\frac{\partial}{\partial\boldsymbol{v}}\frac{1}{\omega_2+\Omega'-(k_2+\boldsymbol{q}')\cdot\boldsymbol{v}}\boldsymbol{e}'_2\cdot\frac{\partial}{\partial\boldsymbol{v}}\delta F_e(q')\Big]E_{h}^{\sigma_1}(k_1)E_{h}^{\sigma_2}(k_2)$$
$$\times\delta^4(q-k_1-k_2-q')\cdot\frac{dk_1dk_2dq'}{(2\pi)^4}+\sum_{\sigma_1\sigma_2\sigma_3\sigma_4}$$
$$\times\iiint_{(d)}\int_{(d')}C_e(q,k_1,k_2+q',k_2,q',k_3,k_4,\boldsymbol{v})E_{h}^{\sigma_1}(k_1)E_{h}^{\sigma_2}(k_2)$$
$$\times E_{h}^{\sigma_3}(k_3)E_{h}^{\sigma_4}(k_4)\delta^4(q-k_1-k_2-q')\delta^4(q'-k_3-k_4)\frac{dk_1\cdots dk_4dq'}{(2\pi)^{12}}\Big\},\tag{23}$$

式中

1) 例如：当有自发磁场.

$$B_e(q, k_1, k_2, \boldsymbol{v}) = \frac{\omega_{pe}^4}{4\pi m_e} \boldsymbol{e}_q \cdot \frac{\partial f_{0e}}{\partial \boldsymbol{v}} \frac{H_e(q, k_1, k_2)}{\Omega \varepsilon_d(q)} - \frac{\omega_{pe}^2}{4\pi m_e} \mathscr{S}_e(k_1, k_2, \boldsymbol{v}), \tag{24}$$

$$C_e(q, k_1, k_2+q', q', k_3, k_4, \boldsymbol{v}) = \frac{\omega_{pe}^6}{(4\pi m_e)^2 n_0} \left\{ \boldsymbol{e}_q \cdot \frac{\partial f_{0e}}{\partial \boldsymbol{v}} \omega_{pe}^2 \frac{K_e(q, k_1, k_2+q', q', k_2)}{\Omega \varepsilon_d(q)} - \mathscr{T}_e(k_1, k_2+q', q', k_2) \right\} \frac{H_e(q', k_3, k_4)}{\Omega' \varepsilon_d(q')}, \tag{25}$$

其中 B_e 对脚标 1，2，C_e 对脚标 3，4 各为对称.

前面已经指出，积分号 $\int_{(d)}$ 只是对两个相反符号的频率及波矢发生作用，也即只感兴趣频率接近 ω_{pe} 的两个高频波拍频为低频波的过程. 这样，从现在起就可把高频波的求和号表示为对每两个相反符号波求和，即

$$\sum_{\sigma_1\sigma_2} \to \sum_{\substack{\{\sigma_1^+\sigma_2^-\} \\ \{\sigma_1^-\sigma_2^+\}}}, \quad \sum_{\sigma_1\sigma_2\sigma_3\sigma_4} \to \sum_{\substack{\{\sigma_1^+\sigma_2^-\} \\ \{\sigma_1^-\sigma_2^+\}}} \sum_{\substack{\{\sigma_3^+\sigma_4^-\} \\ \{\sigma_3^-\sigma_4^+\}}}.$$

经过这样处理后，利用关系式(2)，便可写(22)式为

$$\begin{aligned}
\delta F_e(q) = \frac{1}{\Omega - \boldsymbol{q}\cdot\boldsymbol{v}} \Big\{ & -2\int B_e(q, k_1, -k_2) E_h^{\sigma_1^+}(k_1) E_h^{\sigma_2^-}(k_2)^* \delta^4(q-k_1+k_2) \frac{dk_1 dk_2}{(2\pi)^4} \\
& - \frac{e^2}{m_e^2} \int \Big[2\omega_{pe}^2 \boldsymbol{e}_q \cdot \frac{\partial f_{0e}}{\partial \boldsymbol{v}} \frac{1}{\Omega \varepsilon_d(q)} Z_e(q, k_1, -k_2+q', -k_2, q') \\
& - 2\mathscr{R}_e(k_1, -k_2+q', -k_2, q') \Big] E_h^{\sigma_1^+}(k_1) E_h^{\sigma_2^-}(k_2)^* \delta^4(q-k_1+k_2-q') \\
& \times \frac{dk_1 dk_2 dq'}{(2\pi)^8} + 4\int C_e(q, k_1, -k_2+q', -k_2, q', k_3, -k_4, \boldsymbol{v}) \\
& \times E_h^{\sigma_1^+}(k_1) E_h^{\sigma_2^-}(k_2)^* E_h^{\sigma_3^+}(k_3) E_h^{\sigma_4^-}(k_4)^* \cdot \delta^4(q-k_1+k_2-q') \\
& \times \delta^4(q'-k_3+k_4) \frac{dk_1 \cdots dk_4 dq'}{(2\pi)^{12}} \Big\}.
\end{aligned} \tag{26}$$

现在处理高频场方程. 利用(13)和(14)式，取(10)式到四级场，可得

$$\begin{aligned}
\varepsilon_h^\sigma(k) E_h^\sigma(k) = & -\frac{1}{\omega} \sum_{\sigma_1} \frac{\omega_{pe}^2}{n_0} \int \frac{\boldsymbol{e}_k \cdot \boldsymbol{v}}{\omega - \boldsymbol{k}\cdot\boldsymbol{v}} \boldsymbol{e}_1' \cdot \frac{\partial \delta F_e(q)}{\partial \boldsymbol{v}} E_h^{\sigma_1}(k_1) \\
& \times \delta^4(k-k_1-q) \frac{dk_1 dq}{(2\pi)^4} d^3v - \frac{1}{\omega} \sum_{\sigma_5\sigma_6} \omega_{pe}^4 \frac{e^2}{m_e^2} \\
& \times \int_{(d)} \Big\{ \sum_{\sigma_1} \int \tilde{H}_e(k, q, k_1) E_h^{\sigma_1}(k_1) \delta^4(k-k_1-q) dk_1 \\
& + \sum_{\sigma_2} \frac{1}{n_0} \int \tilde{Z}_e(k, q, k_2+q', k_2, q') E_h^{\sigma_2}(k_2) \delta^4 \\
& \times (k-k_2-q-q') \frac{dk_2 dq'}{(2\pi)^4} + \sum_{\sigma_2\sigma_3\sigma_4} \omega_{pe}^2 \frac{e^2}{m_e^2} \\
& \times \iint_{(d)'} \tilde{K}_e(k, q, k_2+q', q', k_2) \frac{H_e(q', k_3, k_4)}{\Omega' \varepsilon_d(q')} E_h^{\sigma_2}(k_2) E_h^{\sigma_3}(k_3) E_h^{\sigma_4}(k_4) \\
& \times \delta^4(k-k_2-q-q') \delta^4(q'-k_3-k_4) \frac{dk_2 dk_3 dk_4}{(2\pi)^8} \Big\} \frac{H_e(q, k_5, k_6)}{\Omega \varepsilon_d(q)}
\end{aligned}$$

$$\times E_{h}^{\sigma_5}(k_5)E_{h}^{\sigma_6}(k_6)\cdot\delta^4(q-k_5-k_6)\frac{dk_5dk_6dq}{(2\pi)^8}-\frac{1}{\omega}\sum_{\sigma_1}\omega_{pe}^4\frac{e^2}{m_e^2}$$

$$\times\int\frac{\widetilde{H}_e(k,q,k_1)}{\Omega\epsilon_d(q)}E_h^{\sigma_1}(k_1)\cdot\left\{\sum_{\sigma_2\sigma_3}\frac{1}{n_0}\int_{(d)}Z_e(q,k_2,k_3+q',k_3,q')\right.$$

$$\times E_h^{\sigma_2}(k_2)E_h^{\sigma_3}(k_3)\delta^4(q-k_2-k_3)\frac{dk_2dk_3dq'}{(2\pi)^8}$$

$$+\sum_{\sigma_2\sigma_3\sigma_4\sigma_5}\omega_{pe}^2\frac{e^2}{m_e^2}\int_{(d)}\int_{(d')}K_e(q,k_2,k_3+q',q',k_2)$$

$$\times\frac{H_e(q',k_4,k_5)}{\Omega'\epsilon_d(q')}E_h^{\sigma_2}(k_2)E_h^{\sigma_3}(k_3)\cdot E_h^{\sigma_4}(k_4)E_h^{\sigma_5}(k_5)\delta^4(q-k_2-k_3-q')$$

$$\left.\times\delta^4(q'-k_4-k_5)\frac{dk_2\cdots dk_5dq'}{(2\pi)^{12}}\right\}\delta^4(k-k_1-q)\cdot\frac{dk_1dq}{(2\pi)^4},\tag{27}$$

式中

$$\widetilde{H}_e(k,q,k_1)=\int\frac{\boldsymbol{e}_k\cdot\boldsymbol{v}}{\omega-\boldsymbol{k}\cdot\boldsymbol{v}}\boldsymbol{e}_q\cdot\frac{\partial}{\partial\boldsymbol{v}}\frac{\boldsymbol{e}_1'\cdot\dfrac{\partial f_{0e}}{\partial\boldsymbol{v}}}{\omega_1-\boldsymbol{k}_1\cdot\boldsymbol{v}}d^3v,\tag{28}$$

$$\widetilde{K}_e(k,q,k_2+q',q',k_2)=\int\frac{\boldsymbol{e}_k\cdot\boldsymbol{v}}{\omega-\boldsymbol{k}\cdot\boldsymbol{v}}\boldsymbol{e}_q\cdot\frac{\partial}{\partial\boldsymbol{v}}\frac{1}{\omega_2+\Omega'-(\boldsymbol{k}_2+\boldsymbol{q}')\cdot\boldsymbol{v}}\boldsymbol{e}_{q'}$$

$$\times\frac{\partial}{\partial\boldsymbol{v}}\frac{\boldsymbol{e}_2'\cdot\dfrac{\partial f_{0e}}{\partial\boldsymbol{v}}}{\omega_2-\boldsymbol{k}_2\cdot\boldsymbol{v}}d^3v,\tag{29}$$

$$\widetilde{Z}_e(k,k_2+q',k_2,q')=\int\frac{\boldsymbol{e}_k\cdot\boldsymbol{v}}{\omega-\boldsymbol{k}\cdot\boldsymbol{v}}\boldsymbol{e}_q\cdot\frac{\partial}{\partial\boldsymbol{v}}\frac{\boldsymbol{e}_2'\cdot\dfrac{\partial\delta F_e(q')}{\partial\boldsymbol{v}}}{(\omega_2+\Omega')-(\boldsymbol{k}_2+\boldsymbol{q}')\cdot\boldsymbol{v}}d^3v,\tag{30}$$

$\widetilde{H}_e$, $\widetilde{K}_e$ 及 $\widetilde{Z}_e$ 没有对称性.

为了化简 (27) 式，将对 (23) 式作速度积分以消去 (27) 式等号右边最后一个大括号项. 注意到

$$\epsilon_{d\alpha}(q)=\frac{\omega_{p\alpha}^2}{|\boldsymbol{q}|}\int\frac{\boldsymbol{e}_q\cdot\dfrac{\partial f_{0\alpha}}{\partial\boldsymbol{v}}}{\Omega-\boldsymbol{q}\cdot\boldsymbol{v}}d^3v+1\quad\alpha=i,e,$$

得到

$$n_e(q)=\int\delta F_e(q)d^3v=-\frac{e^2}{m_e^2}\frac{|\boldsymbol{q}|}{\Omega}\frac{\epsilon_{di}(q)}{\epsilon_d(q)}\left\{\sum_{\sigma_1\sigma_2}n_0\int_{(d)}H_e(q,k_1,k_2)\right.$$

$$\times E_h^{\sigma_1}(k_1)E_h^{\sigma_2}(k_2)\delta^4(q-k_1-k_2)\frac{dk_1dk_2}{(2\pi)^4}$$

$$+\sum_{\sigma_1\sigma_2}\iint_{(d)}Z_e(q,k_1,k_2+q',k_2,q')\cdot E_h^{\sigma_1}(k_1)E_h^{\sigma_2}(k_2)$$

$$\times\delta^4(q-k_1-k_2-q')\frac{dk_1dk_2dq'}{(2\pi)^8}+\sum_{\sigma_1\sigma_2\sigma_3\sigma_4}\frac{\omega_{pe}^4}{4\pi m_e}$$

$$\times\iiint_{(d)}\int_{(d')}K_e(q,k_1,k_2+q',q',k_2)\frac{H_e(q',k_3,k_4)}{\Omega'\epsilon_d(q')}E_h^{\sigma_1}(k_1)E_h^{\sigma_2}(k_2)E_h^{\sigma_3}(k_3)E_h^{\sigma_4}(k_4)$$

$$\times \delta^4(q-k_1-k_2-q')\delta^4(q'-k_3-k_4)\frac{dk_1\cdots dk_4dq'}{(2\pi)^{12}}\Big\}. \tag{31}$$

用(31)式代入(27)式,整理后得

$$\begin{aligned}
\epsilon_h^\sigma(k)E_h^\sigma(k) = &-\frac{1}{\omega}\sum_{\sigma_1}\frac{\omega_{pe}^2}{n_0}\int\Big[\int\frac{\boldsymbol{e}_h\cdot\boldsymbol{v}}{\omega-\boldsymbol{k}\cdot\boldsymbol{v}}\boldsymbol{e}_1'\cdot\frac{\partial\delta F_e(q)}{\partial\boldsymbol{v}}d^3v\\
&-\omega_{pe}^2\frac{\widetilde{H}_e(k,q,k_1)}{|\boldsymbol{q}|\epsilon_{di}(q)}n_e(q)\Big]\cdot E_h^{\sigma_1}(k_1)\delta^4(k-k_1-q)\frac{dk_1dq}{(2\pi)^4}\\
&-\frac{1}{\omega}\sum_{\sigma_1\sigma_2\sigma_3}\omega_{pe}^2\frac{e^4}{n_0m_e^2}\iint_{(d)}\int_{(d')}\widetilde{Z}_e(k,q,k_1+q',k_1,q')\\
&\times\frac{H_e(q,k_2,k_3)}{\Omega\epsilon_d(q)}E_h^{\sigma_1}(k_1)E_h^{\sigma_2}(k_2)E_h^{\sigma_3}(k_3)\delta^4(k-k_1-q-q')\\
&\times\delta^4(q-k_2-k_3)\frac{dk_1dk_2dk_3dq'dq}{(2\pi)^{12}}-\frac{1}{\omega}\sum_{\sigma_1\cdots\sigma_5}\omega_{pe}^6\frac{e^4}{m_e^4}\\
&\times\iint_{(d)}\int_{(d')}\widetilde{K}_e(k,q;k_1+q',q',k_1)\frac{H_e(q,k_4,k_5)}{\Omega\epsilon_d(q)}\\
&\times\frac{H_e(q',k_2,k_3)}{\Omega'\epsilon_d(q')}E_h^{\sigma_1}(k_1)E_h^{\sigma_2}(k_2)E_h^{\sigma_3}(k_3)E_h^{\sigma_4}(k_4)\\
&\times E_h^{\sigma_5}(k_5)\delta^4(k-k_1-q-q')\delta^4(q'-k_2-k_3)\\
&\times\delta^4(q-k_4-k_5)\frac{dk_1\cdots dk_5dqdq'}{(2\pi)^{16}}.
\end{aligned}\tag{32}$$

经过对向低频拍频波求和后,注意到 H_e 的对称性,从(32)式得到

$$\begin{aligned}
\epsilon_h^\sigma(k)E_h^\sigma(k) = &-\frac{1}{\omega}\sum_{\sigma_1}\frac{\omega_{pe}^2}{n_0}\int\Big[\int\frac{\boldsymbol{e}_h\cdot\boldsymbol{v}}{\omega-\boldsymbol{k}\cdot\boldsymbol{v}}\boldsymbol{e}_1'\cdot\frac{\partial}{\partial\boldsymbol{v}}\delta F_e(q)d^3v\\
&-\omega_{pe}^2\frac{\widetilde{H}_e(k,q,k_1)}{|\boldsymbol{q}|\epsilon_{di}(q)}n_e(q)\Big]E_h^{\sigma_1}(k_1)\delta^4(k-k_1-q)\frac{dk_1dq}{(2\pi)^4}\\
&+\frac{2}{\omega}\sum_{\sigma_1}\omega_{pe}^4\frac{e^2}{n_0m_e^2}\cdot\iint\widetilde{Z}_e(k,q,k_1+q',k_1,q')\frac{H_e(q,k_2,-k_3)}{\Omega\epsilon_d(q)}\\
&\times E_h^{\sigma_1}(k_1)E_h^{\sigma_2^+}(k_2)E_h^{\sigma_3^-}(k_3)^*\delta^4(k-k_1-q-q')\\
&\times\delta^4(q-k_2+k_3)\frac{dk_1dk_2dk_3dqdq'}{(2\pi)^{12}}-\frac{4}{\omega}\sum_{\sigma_1}\omega_{pe}^6\frac{e^4}{m_e^4}\\
&\times\iint\widetilde{K}_e(k,q,k_1+q',q',k_1)\cdot\frac{H_e(q,k_4,-k_5)}{\Omega\epsilon_d(q)}\\
&\times\frac{H_e(q',k_2,-k_3)}{\Omega'\epsilon_d(q')}E_h^{\sigma_1}(k_1)E_h^{\sigma_2^+}(k_2)E_h^{\sigma_3^-}(k_3)^*E_h^{\sigma_4^+}(k_4)E_h^{\sigma_5^-}(k_5)^*\\
&\times\delta^4(k-k_2-q-q')\cdot\delta^4(q-k_2+k_3)\\
&\times\delta^4(q-k_4+k_5)\frac{dk_1\cdots dk_5dqdq'}{(2\pi)^{16}}.
\end{aligned}\tag{33}$$

三、时空表象中动力学方程形式

方程 (26) 和 (33) 是 F 表象中粒子分布函数与高频场耦合的闭合方程组，现在我们着手在不同近似下把它们变换为时空表象中动力学方程组.

1. 二级场近似下的动力学方程组

这时方程 (26) 只保留大括号内头一项，二级场拍频形成有质动力，驱动着粒子低频振荡密度和分布函数的变化. 注意到非线性色散关系和非线性作用占优势过程中，高频波线性 Landau 阻尼不重要，这样可略去高频波虚部项，同时，在跃迁概率中高频波频率的 Doppler 加宽效应亦可取低阶近似. 经这样处理后，关系式 (15) 和 (21) 便为

$$\begin{aligned} H_e(q, k_1, k_2) &= \frac{\Omega}{\omega_2} H_e(k_2, q, -k_1) \\ &= \frac{1}{2}\frac{\Omega}{\omega_2}\left[\frac{|\boldsymbol{q}|}{\omega_{pe}^2}\frac{\boldsymbol{e}_1\cdot\boldsymbol{e}_2}{\omega_2}(\varepsilon_{de}(q)-1) + o\left(\frac{\boldsymbol{k}\cdot\boldsymbol{v}}{\omega}\right)_{1,2}\right]. \end{aligned} \tag{34}$$

于是 (24) 式便为

$$\begin{aligned} B_e(q, k_1, -k_2) = -\frac{\omega_{pe}^2}{4\pi m_e}\Bigg[&\frac{|\boldsymbol{q}|\boldsymbol{e}_q\cdot\frac{\partial f_{0e}}{\partial \boldsymbol{v}}(\varepsilon_{de}(q)-1)}{2\omega_2^2\varepsilon_d(q)}(\boldsymbol{e}_1\cdot\boldsymbol{e}_2) \\ &+ \mathscr{S}_e(k_1, -k_2, \boldsymbol{v})\Bigg]. \end{aligned} \tag{35}$$

(34) 式中 $\Omega/\omega \ll 1$, $\boldsymbol{k}\cdot\boldsymbol{v}/\omega \ll 1$. 精确到同样阶

$$\begin{aligned} \mathscr{S}_e(k_1, -k_2) &= \frac{1}{2}\left[\boldsymbol{e}_1'\cdot\frac{\partial}{\partial \boldsymbol{v}}\frac{\boldsymbol{e}_2'\cdot\frac{\partial f_{0e}}{\partial \boldsymbol{v}}}{\omega_2-\boldsymbol{k}_2\cdot\boldsymbol{v}} - \boldsymbol{e}_2'\cdot\frac{\partial}{\partial \boldsymbol{v}}\frac{\boldsymbol{e}_1'\cdot\frac{\partial f_{0e}}{\partial \boldsymbol{v}}}{\omega_1-\boldsymbol{k}_1\cdot\boldsymbol{v}}\right] \\ &= \frac{1}{2}\Bigg[\frac{\boldsymbol{e}_1'}{\omega_2}\cdot\frac{\partial}{\partial \boldsymbol{v}}\boldsymbol{e}_2\cdot\frac{\partial f_{0e}}{\partial \boldsymbol{v}} + \frac{\boldsymbol{e}_1'\cdot\boldsymbol{e}_2'}{(\omega_2-\boldsymbol{k}_2\cdot\boldsymbol{v})^2}\boldsymbol{k}_2 \\ &\quad \cdot\frac{\partial f_{0e}}{\partial \boldsymbol{v}} + \frac{\boldsymbol{e}_2\cdot\boldsymbol{v}}{\omega_2(\omega_2-\boldsymbol{k}_2\cdot\boldsymbol{v})}\boldsymbol{e}_1'\cdot\frac{\partial}{\partial \boldsymbol{v}}\boldsymbol{k}_2\times\frac{\partial f_{0e}}{\partial \boldsymbol{v}} - 1\longleftrightarrow 2\Bigg] \\ &= \frac{1}{2}\left[\frac{\Omega-\boldsymbol{q}\cdot\boldsymbol{v}}{\omega_1\omega_2}\boldsymbol{e}_1\cdot\frac{\partial}{\partial \boldsymbol{v}}\boldsymbol{e}_2\cdot\frac{\partial}{\partial \boldsymbol{v}}f_{0e} - \frac{\boldsymbol{e}_1\cdot\boldsymbol{e}_2}{\omega_2^2}|\boldsymbol{q}|\boldsymbol{e}_q\cdot\frac{\partial f_{0e}}{\partial \boldsymbol{v}}\right] \\ &\quad + o\left(\frac{\boldsymbol{k}\cdot\boldsymbol{v}}{\omega}\right)_{1,2}^2. \end{aligned} \tag{36}$$

前面已经提到，我们只感兴趣频率接近等离子体频率的一类高频，所以可取 $\omega_1 \approx \omega_2 = \omega_{pe}$，于是在二级场拍频近似下，(26) 式便为

$$\begin{aligned} \delta F_e(q) = \frac{1}{4\pi m_e(\Omega-\boldsymbol{q}\cdot\boldsymbol{v})}\int\Bigg[&-|\boldsymbol{q}|\boldsymbol{e}_q\cdot\frac{\partial f_{0e}}{\partial \boldsymbol{v}}\frac{\varepsilon_{di}(q)}{\varepsilon_d(q)}(\boldsymbol{e}_1\cdot\boldsymbol{e}_2) \\ &+ (\Omega-\boldsymbol{q}\cdot\boldsymbol{v})\boldsymbol{e}_1\cdot\frac{\partial}{\partial \boldsymbol{v}}\boldsymbol{e}_2\cdot\frac{\partial}{\partial \boldsymbol{v}}f_{0e}\Bigg] \\ &\times E_h^{\sigma_1^+}(k_1)E_h^{\sigma_2^-}(k_2)^*\delta^4(q-k_1+k_2)\frac{dk_1dk_2}{(2\pi)^4}. \end{aligned} \tag{37}$$

下面我们专门研究低频波为离子声波的情况．如果不考虑电子、离子的阻尼效应，则

$$\frac{\epsilon_{di}(q)}{\epsilon_d(q)} \doteq -\frac{q^2v_s^2}{\Omega^2-q^2v_s^2} \qquad 对\quad qv_{T_i}\ll\Omega\ll qv_{T_e};$$

$$\doteq\frac{T_e}{T_e+T_i} \qquad 对\quad \Omega\ll qv_{T_i}. \tag{38}$$

这里 $v_s=(T_e/m_i)^{1/2}$ 为离子声速度．

(37) 式两边乘 $(\Omega-\boldsymbol{q}\cdot\boldsymbol{v})$ 因子，然后作 Fourier 逆变换，注意到

$$\delta F_e(\boldsymbol{x},\boldsymbol{v},t)=\int\frac{d\Omega}{2\pi}\int\frac{d^3q}{(2\pi)^3}\delta F_e(q)e^{i(\boldsymbol{q}\cdot\boldsymbol{x}-\Omega t)},$$

$$\boldsymbol{E}^\sigma(\boldsymbol{x},t)=\int\frac{d\omega}{2\pi}\int\frac{d^3k}{(2\pi)^3}\boldsymbol{e}_{\boldsymbol{k}}E_k^\sigma(k)e^{i(\boldsymbol{k}\cdot\boldsymbol{x}-\omega t)}, \tag{39}$$

便得到在 $qv_{T_i}\ll\Omega\ll qv_{T_e}$ 近似下时空表象中粒子低频分布函数动力学方程

$$\left(\frac{\partial^2}{\partial t^2}-v_s^2\nabla^2\right)\left(\frac{\partial}{\partial t}+\boldsymbol{v}\cdot\nabla\right)\left(F_e(\boldsymbol{x},\boldsymbol{v},t)-\frac{\partial}{\partial\boldsymbol{v}}\cdot\mathbf{D}^\sigma\cdot\frac{\partial}{\partial\boldsymbol{v}}F_{0e}\right)$$

$$=-\frac{v_s^2}{8\pi m_en_0}\nabla^2\left(\frac{\partial F_{0e}}{\partial\boldsymbol{v}}\cdot\nabla|\boldsymbol{E}^\sigma(\boldsymbol{x},t)|^2\right). \tag{40}$$

这里

$$F_{0e}(\boldsymbol{v})=n_0f_{0e}(\boldsymbol{v}),\quad F_e(\boldsymbol{x},\boldsymbol{v},t)=F_{0e}(\boldsymbol{v})+\delta F_e(\boldsymbol{x},\boldsymbol{v},t), \tag{41}$$

$$\mathbf{D}^\sigma=\frac{1}{8\pi n_0m_e}(\boldsymbol{E}^\sigma\boldsymbol{E}^{\sigma*})_s, \tag{42}$$

$$(\boldsymbol{E}^\sigma\boldsymbol{E}^{\sigma*})_s=2\boldsymbol{E}^{\sigma+}(\boldsymbol{E}^{\sigma-})^*, \tag{42'}$$

$\mathbf{D}^\sigma$ 为对称张量．(40) 和 (42) 式中已用了关系式

$$\boldsymbol{E}^\sigma=\boldsymbol{E}^{\sigma+}+(\boldsymbol{E}^{\sigma-})^*. \tag{43}$$

在 $\Omega\ll qv_{T_i}$ 的静态近似下，(37) 式便为

$$\left(\frac{\partial}{\partial t}+\boldsymbol{v}\cdot\nabla\right)\left(F_e(\boldsymbol{x},\boldsymbol{v},t)-\frac{\partial}{\partial\boldsymbol{v}}\cdot\mathbf{D}^\sigma\cdot\frac{\partial}{\partial\boldsymbol{v}}F_{0e}\right)$$

$$=\frac{T_e}{8\pi n_0m_e(T_i+T_e)}\frac{\partial F_{0e}}{\partial\boldsymbol{v}}\cdot\nabla|\boldsymbol{E}^\sigma|^2. \tag{44}$$

取 F_{0e} 为 Maxwell 分布，稳定态下 (44) 式就为

$$F_e(\boldsymbol{x},\boldsymbol{v})=\frac{\partial}{\partial\boldsymbol{v}}\cdot\mathbf{D}^\sigma\cdot\frac{\partial}{\partial\boldsymbol{v}}F_{0e}-\frac{|\boldsymbol{E}^\sigma|^2}{8\pi n_0(T_i+T_e)}F_{0e}(\boldsymbol{v}). \tag{45}$$

(40) 或 (44) 式就是二级场近似下低频粒子分布函数(包含初态)所服从的动力方程．这里，除了有质动力与等离子体背景粒子分布函数相互作用驱动 $\boldsymbol{F}_e$ 的时空变化外，还附加了二级场作用下背景粒子在速度空间扩散过程的贡献．

当 Ω 靠近 qv_{T_i} 或 qv_{T_e} 端时，离子或电子的 Landau 阻尼起重要作用，本文不作进一步讨论，详见文献 [4]．

保留 (33) 式到二级场，得高频场方程

$$\epsilon_h^\sigma(k)E_h^\sigma(k)=-\frac{1}{\omega}\sum_{\sigma_1}\frac{\omega_{pe}^2}{n_0}\int\left[\frac{\boldsymbol{e}_k\cdot\boldsymbol{v}}{\omega-\boldsymbol{k}\cdot\boldsymbol{v}}\boldsymbol{e}_1'\cdot\frac{\partial\delta F_e(q)}{\partial\boldsymbol{v}}d^3v-\frac{\omega_{pe}^2\widetilde{H}_e(k,q,k_1)}{|\boldsymbol{q}|\epsilon_{di}(q)}n_e(q)\right]$$
$$\times E_h^{\sigma_1}(k_1)\delta^4(k-k_1-q)\frac{dk_1dq}{(2\pi)^4}. \tag{46}$$

由(28)式可知，方程右边第二项中被积函数较第一项至少小一个因子 $\boldsymbol{k}_1\cdot\boldsymbol{v}/\omega_1$，故可略去．于是就有

$$\epsilon_h^\sigma(k)E_h^\sigma(k)=\frac{\omega_{pe}^2}{\omega^2}\sum_{\sigma_1}\int(\boldsymbol{e}_k\cdot\boldsymbol{e}_1)\frac{\delta F_e(q)}{n_0}E_h^{\sigma_1}(k_1)\delta^4(k-k_1-q)\frac{dk_1dq}{(2\pi)^4}d^3v. \tag{47}$$

利用(11)和(12)式对上式作 Fourier 逆变换后得高频场方程为

$$\frac{\partial^2\boldsymbol{E}^\sigma}{\partial t^2}-(u^\sigma)^2\nabla^2\boldsymbol{E}^\sigma=-\omega_{pe}^2\boldsymbol{E}^\sigma-\frac{\omega_{pe}^2\boldsymbol{E}^\sigma}{n_0}\int\delta F_e(\boldsymbol{x},\boldsymbol{v},t)d^3v, \tag{48}$$

$$\begin{aligned}u^\sigma&=c\ (\text{光速})\quad 对\quad \sigma=t;\\&=\sqrt{3}\,v_{T_e}\quad 对\quad \sigma=l.\end{aligned} \tag{49}$$

(40)或(44)与(48)式就为二级场近似下的闭合方程组．

如果对(37)式进行速度积分，便得

$$n_e(q)=-\frac{|\boldsymbol{q}|^2}{4\pi m_e\omega_{pe}^2}\frac{\epsilon_{di}(q)}{\epsilon_d(q)}(\epsilon_{de}(q)-1)\int(\boldsymbol{e}_1\cdot\boldsymbol{e}_2)E_h^{\sigma_1^+}(k_1)E_h^{\sigma_2^-}(k_2)^*$$
$$\times\delta^4(q-k_1+k_2)\frac{dk_1dk_2}{(2\pi)^4}. \tag{50}$$

对(50)式作 Fourier 逆变换，展开 $\epsilon_{di}(\epsilon_{de}-1)/\epsilon_d$ 因子，便可得到不同近似下密度时空变化方程，此方程与高频场(47)式耦合构成了广义的 Zakharov 方程组[4]．

2. 四级场近似下的动力学方程组

先求扰动密度 $n_e(q)$ 的方程．对(26)式进行速度积分后得到

$$n_e(q)=\frac{\omega_{pe}^2}{2\pi m_e}\frac{|\boldsymbol{q}|}{\Omega}\frac{\epsilon_{di}(q)}{\epsilon_d(q)}\left\{\int H_e(q,k_1,-k_2)E_h^{\sigma_1^+}(k_1)E_h^{\sigma_2^-}(k_2)^*\right.$$
$$\times\delta^4(q-k_1+k_2)\frac{dk_1dk_2}{(2\pi)^4}+\frac{1}{n_0}\int Z_e(q,k_1,-k_2+q',-k_2,q')$$
$$\times E_h^{\sigma_1^+}(k_1)E_h^{\sigma_2^-}(k_2)^*\delta^4(q-k_1+k_2-q')\frac{dk_1dk_2dq'}{(2\pi)^8}$$
$$-\frac{\omega_{pe}^4}{2\pi n_0m_e}\int K_e(q,k_1,-k_2+q',q',-k_2)\frac{H_e(q',k_3,-k_4)}{\Omega'\epsilon_d(q')}$$
$$\times E_h^{\sigma_1^+}(k_1)E_h^{\sigma_2^-}(k_2)^*E_h^{\sigma_3^+}(k_3)E_h^{\sigma_4^-}(k_4)^*$$
$$\left.\times\delta^4(q-k_1+k_2-q')\delta^4(q'-k_3+k_4)\frac{dk_1\cdots dk_4dq'}{(2\pi)^{12}}\right\}. \tag{51}$$

这里 Z_e 见(17)式，其中

$$2\mathscr{R}_e(k_1,-k_2+q',-k_2,q')=\boldsymbol{e}_1'\cdot\frac{\partial}{\partial\boldsymbol{v}}\frac{1}{\omega_2-\Omega'-(\boldsymbol{k}_2-\boldsymbol{q}')\cdot\boldsymbol{v}}\boldsymbol{e}_2'\cdot\frac{\partial}{\partial\boldsymbol{v}}\delta F_e(q')$$
$$-\boldsymbol{e}_2'\cdot\frac{\partial}{\partial\boldsymbol{v}}\frac{1}{\omega_1+\Omega'-(\boldsymbol{k}_1+\boldsymbol{q}')\cdot\boldsymbol{v}}\boldsymbol{e}_1'\cdot\frac{\partial}{\partial\boldsymbol{v}}\delta F_e(q')$$

$$= \left(\boldsymbol{e}_1' \cdot \frac{\partial}{\partial \boldsymbol{v}} \frac{1}{\omega_2 - \boldsymbol{k}_2 \cdot \boldsymbol{v}} \boldsymbol{e}_2' \cdot \frac{\partial}{\partial \boldsymbol{v}} \delta F_e(q') - 1 \longleftrightarrow 2 \right)$$

$$+ \left(\boldsymbol{e}_1 \cdot \frac{\partial}{\partial \boldsymbol{v}} \frac{\Omega' - \boldsymbol{q}' \cdot \boldsymbol{v}}{(\omega_2 - \boldsymbol{k}_2 \cdot \boldsymbol{v})^2} \boldsymbol{e}_2 \cdot \frac{\partial}{\partial \boldsymbol{v}} \delta F_e(q') + 1 \longleftrightarrow 2 \right) + o\left(\frac{\Omega' - \boldsymbol{q}' \cdot \boldsymbol{v}}{(\omega - \boldsymbol{k} \cdot \boldsymbol{v})_{1,2}} \right)^2$$

$$= \frac{\Omega - \Omega' - (\boldsymbol{q} - \boldsymbol{q}') \cdot \boldsymbol{v}}{\omega_1 \omega_2} \boldsymbol{e}_1 \cdot \frac{\partial}{\partial \boldsymbol{v}} \boldsymbol{e}_2 \cdot \frac{\partial}{\partial \boldsymbol{v}} \delta F_e(q') - \frac{\boldsymbol{e}_1 \cdot \boldsymbol{e}_2}{\omega_{pe}^2} (\boldsymbol{q} - \boldsymbol{q}') \cdot \frac{\partial}{\partial \boldsymbol{v}} \delta F_e(q')$$

$$- \left(\frac{\boldsymbol{e}_1 \cdot \boldsymbol{q}' \boldsymbol{e}_2 \cdot \frac{\partial}{\partial \boldsymbol{v}} \delta F_e(q')}{(\omega_2 - \boldsymbol{k}_2 \cdot \boldsymbol{v})^2} + 1 \longleftrightarrow 2 \right) + o\left(\frac{\Omega' - \boldsymbol{q}' \cdot \boldsymbol{v}}{(\omega - \boldsymbol{k} \cdot \boldsymbol{v})_{1,2}} \right)^2$$

$$= \frac{\Omega + \Omega' - (\boldsymbol{q} + \boldsymbol{q}') \cdot \boldsymbol{v}}{\omega_{pe}^2} \boldsymbol{e}_1 \cdot \frac{\partial}{\partial \boldsymbol{v}} \boldsymbol{e}_2 \cdot \frac{\partial}{\partial \boldsymbol{v}} \delta F_e(q') - \frac{\boldsymbol{e}_1 \cdot \boldsymbol{e}_2}{\omega_{pe}^2} (\boldsymbol{q} - \boldsymbol{q}') \cdot \frac{\partial \delta F_e(q')}{\partial \boldsymbol{v}}$$

$$- \frac{1}{\omega_{pe}^2} \left(\boldsymbol{e}_1 \cdot \boldsymbol{q}' \boldsymbol{e}_2 \cdot \frac{\partial}{\partial \boldsymbol{v}} \delta F_e(q') + 1 \longleftrightarrow 2 \right) + o\left(\frac{\Omega' - \boldsymbol{q}' \cdot \boldsymbol{v}}{(\omega - \boldsymbol{k} \cdot \boldsymbol{v})_{1,2}} \right)^2. \tag{52}$$

为了进一步求得 Z_e 具体表达式，(52) 式中的 $\delta F_e(q')$ 以 (26) 式中的二级场部分代入，即应用 (37) 式，经过对 Maxwell 分布背景粒子积分，取 $\Omega' \ll |\boldsymbol{q}'| v_{T_e}$ 近似，便得

$$Z_e(q, k_1, -k_2 + q', -k_2, q') = \frac{\Omega}{8\pi T_e |\boldsymbol{q}| v_{T_e}^2 \omega_{pe}^2} \int \Big\{ -(\boldsymbol{e}_3 \cdot \boldsymbol{e}_4) \frac{A(1,2)}{q^2} \frac{\varepsilon_{di}(q')}{\varepsilon_d(q')}$$

$$- 2B(1,2,3,4) + (\boldsymbol{e}_1 \cdot \boldsymbol{e}_2)(\boldsymbol{e}_3 \cdot \boldsymbol{e}_4) \frac{\varepsilon_{di}(q')}{\varepsilon_d(q')} \left[q^2 \lambda_e^2 (\varepsilon_{de}(q) - 1) - \frac{\boldsymbol{q} \cdot \boldsymbol{q}'}{q^2} \right]$$

$$+ 2(\boldsymbol{e}_1 \cdot \boldsymbol{e}_2)(\boldsymbol{q} - \boldsymbol{q}') \cdot \boldsymbol{q} \frac{\boldsymbol{e}_3 \cdot \boldsymbol{q} \boldsymbol{e}_4 \cdot \boldsymbol{q}}{q^4} + \frac{1}{q^2} (\boldsymbol{e}_1 \cdot \boldsymbol{q} \boldsymbol{e}_2 \cdot \boldsymbol{q}' + 1 \longleftrightarrow 2)$$

$$\times \left(\boldsymbol{e}_3 \cdot \boldsymbol{e}_4 \frac{\varepsilon_{di}(q')}{\varepsilon_d(q')} + \frac{2}{q^2} \boldsymbol{e}_3 \cdot \boldsymbol{q} \boldsymbol{e}_4 \cdot \boldsymbol{q} \right) \Big\} E_h^{\sigma_3^+}(k_3) E_h^{\sigma_4^-}(k_4)^*$$

$$\times \delta^4(q' - k_3 + k_4) \frac{dk_3 dk_4}{(2\pi)^4}, \tag{53}$$

其中

$$A(1,2) \equiv (\boldsymbol{e}_1 \cdot \boldsymbol{q} \boldsymbol{e}_2 \cdot \boldsymbol{q}' + 1 \longleftrightarrow 2) - 2\boldsymbol{e}_1 \cdot \boldsymbol{q} \boldsymbol{e}_2 \cdot \boldsymbol{q} \frac{\boldsymbol{q} \cdot \boldsymbol{q}'}{q^2}, \tag{54}$$

$$B(1,2,3,4) \equiv A(1,2) \boldsymbol{e}_3 \cdot \boldsymbol{q} \boldsymbol{e}_4 \cdot \boldsymbol{q} + \begin{pmatrix} 1 \longleftrightarrow 3 \\ 2 \longleftrightarrow 4 \end{pmatrix}, \tag{55}$$

$$\lambda_e = \left(\frac{T_e}{4\pi n_0 e^2} \right)^{1/2}. \tag{56}$$

关于 K_e 的计算比 Z_e 更复杂，但是比较方程 (51) 式右边第三项和第二项（其中 δF_e 已用表达式 (37) 代入）的被积函数后，很容易看到前者多了一个 $\boldsymbol{q} \cdot \boldsymbol{v}/\omega_{pe}$ 因子，故在四级场近似下 $n_e(q)$ 方程只取右边头两项就够了.

下面在空间为一维情况下讨论从 F 表象到二维时空表象的变换. 这时 (53) 式中 $A = B = 0$，对 $\sigma = t$，

$$Z_e(q, k_1, -k_2 + q', -k_2, q') = \frac{\Omega}{8\pi T_e q_x v_{T_e}^2 \omega_{pe}^2} \int \frac{\varepsilon_{di}(q')}{\varepsilon_d(q')} \left[q_x^2 \lambda_e^2 (\varepsilon_{de}(q) - 1) - \frac{q_x'}{q_x} \right]$$

$$\times E_h^{t_3^+}(k_3)E_h^{t_4^-}(k_4)^*\delta^2(q'-k_3+k_4)\frac{dk_3dk_4}{(2\pi)^2}. \tag{57}$$

这里 $q=\{\Omega, q_x\}$，$k=\{\omega, k_x\}$，$dk=d\omega dk_x$，$\delta^2(q'-k_3+k_4)=\delta(\Omega'-\omega_3+\omega_4)\delta(q'_x-k_{3x}+k_{4x})$，$E_h^t(k)$ 为垂直于 x 轴的 F 表象中量.

对 $\sigma=l$，

$$Z_e(q,k_1,-k_2+q',-k_2,q')=\frac{\Omega}{8\pi T_e q_x v_{T_e}^2\omega_{pe}^2}\int\left\{\frac{\varepsilon_{di}(q')}{\varepsilon_d(q')}\left[q_x^2\lambda_e^2(\varepsilon_{de}(q)-1)+\frac{q'_x}{q_x}\right]+2\left(1+\frac{q'_x}{q_x}\right)\right\}E_h^{l_3^+}(k_3)E_h^{l_4^-}(k_4)^*\delta^2(q'-k_3+k_4)\frac{dk_3dk_4}{(2\pi)^2}. \tag{58}$$

这里的 $E_h^l(k)$ 为平行于 x 轴的 F 表象中量.

于是由 (51) 式可得四级场截断下的动力学方程，对 $\sigma=t$，

$$n_e(q)=-\frac{1}{4\pi T_e}q_x^2\lambda_e^2(\varepsilon_{de}(q)-1)\frac{\varepsilon_{di}(q)}{\varepsilon_d(q)}\left\{\int E_h^{t_1^+}(k_1)E_h^{t_2^-}(k_2)\times\delta^2(q-k_1+k_2)\frac{dk_1dk_2}{(2\pi)^2}-\frac{1}{4\pi n_0T_eq_x^2\lambda_e^2(\varepsilon_{de}(q)-1)}\times\int\frac{\varepsilon_{di}(q')}{\varepsilon_d(q')}\left[q_x^2\lambda_e^2(\varepsilon_{de}(q)-1)-\frac{q'_x}{q_x}\right]E_h^{t_1^+}(k_1)E_h^{t_2^-}(k_2)^*\times E_h^{t_3^+}(k_3)E_h^{t_4^-}(k_4)^*\delta^2(q-k_1+k_2-q')\times\delta^2(q'-k_3+k_4)\frac{dk_1\cdots dk_4dq'_x}{(2\pi)^8}\right\}. \tag{59}$$

当阻尼不重要时，$q_xv_{T_i}\ll\Omega\ll q_xv_{T_e}$，这时

$$n_e(q)=\frac{1}{4\pi T_e}\frac{q_xv_s^2}{\Omega^2-q_x^2v_s^2}\left\{q_x\int E_h^{t_1^-}(k_1)E_h^{t_2^-}(k_2)^*\delta^2(q-k_1+k_2)\frac{dk_1dk_2}{(2\pi)^2}-\frac{1}{4\pi n_0T_e}\int(q_x-q'_x)\cdot\frac{q_x'^2v_s^2}{q_x'^2v_s^2-\Omega'^2}E_h^{t_1^+}(k_1)E_h^{t_2^-}(k_2)^*E_h^{t_3^+}(k_3)E_h^{t_4^-}(k_4)^*\times\delta^2(q-k_1+k_2-q')\delta^2(q'-k_3+k_4)\frac{dk_1\cdots dk_4dq'}{(2\pi)^8}\right\}. \tag{60}$$

变换到二维时空表象，得

$$\left(\frac{\partial^2}{\partial t^2}-v_s^2\frac{\partial^2}{\partial x^2}\right)n_e(x,t)=\frac{1}{8\pi m_i}\frac{\partial}{\partial x}\left\{\frac{\partial}{\partial x}|E^t(x,t)|^2-\left[\frac{\partial}{\partial x}(|E^t(x,t)|^2\psi_t(x,t))-|E^t(x,t)|^2\frac{\partial}{\partial x}\psi_t(x,t)\right]\right\},$$

即有

$$\left(\frac{\partial^2}{\partial t^2}-v_s^2\frac{\partial^2}{\partial x^2}\right)n_e(x,t)=\frac{1}{8\pi m_i}\frac{\partial}{\partial x}\left\{\left(1-\frac{\psi_t(x,t)}{8\pi n_0T_e}\right)\frac{\partial}{\partial x}|E^t(x,t)|^2\right\}; \tag{61}$$

其中

$$\psi_t(x,t)=2\int\frac{1}{\left(1-\left(\frac{v_\phi}{v_s}\right)^2\right)}E_h^{t_3^+}(k_3)E_h^{t_4^-}(k_4)^*e^{i(k_3-k_4)_xx-i(\omega_3-\omega_4)t}\frac{dk_3dk_4}{(2\pi)^4},$$

$$v_\phi = \frac{\Omega'}{q'_x}, \quad q'_x = (\boldsymbol{k}_3 - \boldsymbol{k}_4)_x, \quad \Omega' = \omega_3 - \omega_4. \tag{62}$$

在超声情况下 $\psi_t < 0$；亚声情况下 $\psi_t > 0$，在弱亚声时，$\frac{\Omega'}{q'_x} \ll v_s$，$\psi_t = |E^t|^2$.

上述表达式中的 E^t 见(43)式表示.

当相速度 v_ϕ 不随频率和波数变化时(例横波 Soliton)，(61)式便为

$$\left(\frac{\partial^2}{\partial t^2} - v_s^2 \frac{\partial^2}{\partial x^2}\right) n_e(x,t) = \frac{1}{8\pi m_i} \frac{\partial^2}{\partial x^2} \times \left\{|E^t|^2 \left(1 - \frac{|E^t|^2}{\left[1 - \left(\frac{v_\phi}{v_s}\right)^2\right] 16\pi n_0 T_e}\right)\right\}. \tag{63}$$

在 $\Omega \ll q_x v_{T_i}$ 静态近似下，(51)式变换结果为

$$n_e(x) = -\frac{|E^t|^2}{8\pi(T_e + T_i)}\left(1 - \frac{|E^t|^2}{16\pi n_0 (T_e + T_i)}\right). \tag{64}$$

对 $\sigma = l$，再一次略去 Landau 阻尼，(51)式变为

$$n_e(q) = \frac{1}{4\pi T_e} \frac{q_x v_s^2}{(\Omega^2 - q_x^2 v_s^2)} \left\{ q_x \int E_h^{l_1^+}(k_1) E_h^{l_2^-}(k_2)^* \delta^2(q - k_1 + k_2) \frac{dk_1 dk_2}{(2\pi)^2} - \frac{1}{4\pi n_0 T_e} \int (q_x + q'_x)\left(2 + \frac{q'^2_x v_s^2}{q'^2_x v_s^2 - \Omega'^2}\right) E_h^{l_1^+}(k_1) E_h^{l_2^-}(k_2)^* E_h^{l_3^+}(k_3) E_h^{l_4^-}(k_4)^* \times \delta^2(q - k_1 + k_2 - q') \delta^2(q' - k_3 + k_4) \frac{dk_1 \cdots dk_4 dq'_x}{(2\pi)^8} \right\}. \tag{65}$$

变换到二维时空表象，即有

$$\left(\frac{\partial^2}{\partial t^2} - v_s^2 \frac{\partial^2}{\partial x^2}\right) n_e(x,t) = \frac{1}{8\pi m_i} \frac{\partial}{\partial x} \left\{ \left(1 - \frac{\psi_e(x,t)}{8\pi n_0 T_e}\right) \frac{\partial}{\partial x} |E^l(x,t)|^2 - \frac{|E^l(x,t)|^2}{4\pi n_0 T_e} \frac{\partial}{\partial x} \psi_l(x,t) \right\}. \tag{66}$$

这里

$$\psi_l(x,t) = 2\int \beta E_h^{l_3^+}(k_3) E_h^{l_4^-}(k_4)^* \exp[i(\boldsymbol{k}_3 - \boldsymbol{k}_4)_x x - i(\omega_3 - \omega_4)t] \frac{dk_3 dk_4}{(2\pi)^4}. \tag{67}$$

$$\beta = \frac{3 - 2\left(\frac{v_\phi}{v_s}\right)^2}{1 - \left(\frac{v_\phi}{v_s}\right)^2} < 0 \quad 当 \quad 1 < \left(\frac{v_\phi}{v_s}\right)^2 < \frac{3}{2};$$

$$> 0, \quad 其它区间. \tag{68}$$

当相速度不随频率和波数变化时(例纵波 Soliton)，

$$\psi_l(x,t) = \beta |E^l(x,t)|^2. \tag{69}$$

这时方程(66)就为

$$\left(\frac{\partial^2}{\partial t^2} - v_s^2 \frac{\partial^2}{\partial x^2}\right) n_e(x,t) = \frac{1}{8\pi m_i} \frac{\partial^2}{\partial x^2} \left\{ |E^l(x,t)|^2 \left(1 - \frac{3\beta |E^l(x,t)|^2}{16\pi n_0 T_e}\right)\right\}. \tag{70}$$

在 $\Omega \ll q_x v_{T_i}$ 静态近似下，得

$$n_e(x) = -\frac{|E^l(x)|^2}{8\pi(T_e+T_i)}\left(1-\frac{3(2T_i+3T_e)}{16\pi n_0 T_e(T_i+T_e)}|E^l(x)|^2\right). \tag{71}$$

当有阻尼时介质函数的极化项必须保留虚部，这里不作进一步讨论.

在四维时空表象情况下，利用 (53) 式变换会出现结构复杂的项，为了简单起见，假定参与两个高频波拍频的第三个低频波其频率和波数远较新生成的波小，即 $\Omega'\ll\Omega$, $|\boldsymbol{q}'|\ll|\boldsymbol{q}|$，这时 (53) 式成为

$$Z_e = \frac{\Omega(\boldsymbol{e}_1\cdot\boldsymbol{e}_2)}{8\pi T_e|\boldsymbol{q}|v_{T_e}^2\omega_{pe}^2}\int\left\{(\boldsymbol{e}_3\cdot\boldsymbol{e}_4)q^2\lambda_e^2(\varepsilon_{de}(q)-1)\frac{\varepsilon_{di}(q')}{\varepsilon_d(q')}+2\frac{\boldsymbol{e}_3\cdot\boldsymbol{q}\,\boldsymbol{e}_4\cdot\boldsymbol{q}}{q^2}\right\}$$
$$\times E_h^{\sigma_3^+}(k_3)E_h^{\sigma_4^-}(k_4)^*\delta^4(q-k_3+k_4)\frac{dk_3dk_4}{(2\pi)^4}. \tag{72}$$

这样，在忽略阻尼情况下，扰动密度空间三维动力学方程便为

$$\left(\frac{\partial^2}{\partial t^2}-v_s^2\nabla^2\right)n_e(\boldsymbol{x},t)=\frac{1}{8\pi m_i}\nabla\cdot\left\{\left(1-\frac{\psi_\sigma(\boldsymbol{x},t)}{8\pi n_0T_e}\right)\nabla|\boldsymbol{E}^\sigma(\boldsymbol{x},t)|^2\right.$$
$$\left.-\frac{1}{8\pi n_0T_e}(|\boldsymbol{E}^\sigma|^2\nabla\psi_\sigma+2\nabla\cdot[(\boldsymbol{E}^\sigma\boldsymbol{E}^{\sigma*})_s|\boldsymbol{E}^\sigma|^2])\right\}, \tag{73}$$

$$\psi_\sigma(\boldsymbol{x},t)=2\int\frac{1}{1-\left(\dfrac{v_\phi}{v_s}\right)^2}(\boldsymbol{e}_3\cdot\boldsymbol{e}_4)\boldsymbol{E}_h^{\sigma_3^+}(k_3)\boldsymbol{E}_h^{\sigma_4^-}(k_4)^*\exp$$
$$\times[i(\boldsymbol{k}_3-\boldsymbol{k}_4)\cdot\boldsymbol{x}-i(\omega_3-\omega_4)t]\frac{dk_3dk_4}{(2\pi)^8}. \tag{74}$$

在静态近似下，得到

$$\nabla n_e=-\frac{1}{8\pi(T_i+T_e)}\left\{\nabla\left[|\boldsymbol{E}^\sigma|^2\left(1-\frac{|\boldsymbol{E}^\sigma|^2}{8\pi n_0(T_i+T_e)}\right)\right]\right.$$
$$\left.-\frac{1}{4\pi n_0T_e}\nabla\cdot[(\boldsymbol{E}^\sigma\boldsymbol{E}^{\sigma*})_s|\boldsymbol{E}^\sigma|^2]\right\}. \tag{75}$$

关于四级场近似下高频场方程，从 (33) 式可以看到，除了右边方括号内第一项外，其余项的速度空间被积函数中至少多一个 $\boldsymbol{k}\cdot\boldsymbol{v}/\omega_{pe}$ 因子，故可以忽略. 因此，高阶场近似下高频场方程形式完全同二级场的 (47) 式一样，所不同的是这时 n_e 中包含了四级场项. 这样，方程 (47) 与二维方程 (63) 或 (64)，(70) 或 (71) 以及四维方程 (73) 或 (75) 构成了相应条件下的各个闭合方程组.

现在讨论粒子速度分布所服从的方程，仍从方程 (26) 出发，并略去右边最后一项. 为了得到简洁表达式，利用方程 (51) 代入 (26) 式以消去含 Z_e 的项，这样得到

$$\delta F_e(q)=\frac{-1}{\Omega-\boldsymbol{q}\cdot\boldsymbol{v}}\left\{\frac{\omega_{pe}^2\boldsymbol{q}\cdot\dfrac{\partial f_{0e}}{\partial\boldsymbol{v}}}{|\boldsymbol{q}|^2\varepsilon_{di}(q)}\int\delta F_e(q)d^3v-\right.$$
$$\frac{\omega_{pe}^2}{2\pi m_e}\int\mathscr{S}_e(k_1,-k_2,\boldsymbol{v})\;E_h^{\sigma_1^+}(k_1)E_h^{\sigma_2^-}(k_2)^*$$
$$\times\delta^4(q-k_1+k_2)\frac{dk_1dk_2}{(2\pi)^4}-\frac{\omega_{pe}^2}{2\pi n_0m_e}\int\mathscr{R}_e(k_1,-k_2+q',-k_2,q')$$

$$\times E_h^{\sigma_1^+}(k_1)E_h^{\sigma_2^-}(k_2)^*\delta^4(q-k_1+k_2-q')\cdot\frac{dk_1dk_2dq'}{(2\pi)^8}\Big\}. \tag{76}$$

应用关系式(36)和(53)，注意到 $qv_{T_i}\ll\Omega\ll qv_{T_e}$ 时，$\varepsilon_{di}=-\omega_{pi}^2/\Omega^2$，在忽略阻尼条件下，(76)式变换到时空表象后得

$$\begin{aligned}\nabla^2\left(\frac{\partial}{\partial t}+\boldsymbol{v}\cdot\nabla\right)F_e(\boldsymbol{x},\boldsymbol{v},t)&=-\frac{m_i}{m_e}\frac{\partial^2}{\partial t^2}\left(\frac{\partial F_{0e}}{\partial\boldsymbol{v}}\cdot\frac{\nabla n_e(\boldsymbol{x},t)}{n_0}\right)\\&+\nabla^2\left\{\left(\frac{\partial}{\partial t}+\boldsymbol{v}\cdot\nabla\right)\frac{\partial}{\partial\boldsymbol{v}}\cdot\boldsymbol{D}^\sigma\cdot\frac{\partial}{\partial\boldsymbol{v}}\nabla F_e(\boldsymbol{x},\boldsymbol{v},t)\right.\\&+\frac{\partial}{\partial\boldsymbol{v}}F_e(\boldsymbol{x},\boldsymbol{v},t)\cdot\nabla\frac{|\boldsymbol{E}^\sigma|^2}{8\pi n_0m_e}+\frac{\partial}{\partial\boldsymbol{v}}\cdot\boldsymbol{D}^\sigma\cdot\frac{\partial}{\partial\boldsymbol{v}}\frac{\partial}{\partial t}F_e(\boldsymbol{x},\boldsymbol{v},t)\\&\left.+\left(\frac{\partial}{\partial\boldsymbol{v}}\cdot\boldsymbol{D}^\sigma\cdot\frac{\partial}{\partial\boldsymbol{v}}\nabla F_e\right)\cdot\boldsymbol{v}+\frac{\partial}{\partial\boldsymbol{v}}\cdot(\boldsymbol{D}^\sigma+\boldsymbol{D}^{\sigma*})\cdot\nabla F_e\right\}.\end{aligned} \tag{77}$$

这里 F_e 和 $\boldsymbol{D}^\sigma$ 的表达式见(41)和(42)式.

(77)式稳定态为

$$\begin{aligned}\boldsymbol{v}\cdot\nabla F_e=\boldsymbol{v}\cdot\nabla\frac{\partial}{\partial\boldsymbol{v}}\cdot\boldsymbol{D}^\sigma\cdot\frac{\partial}{\partial\boldsymbol{v}}F_e+\frac{\partial}{\partial\boldsymbol{v}}F_e\cdot\nabla\frac{|\boldsymbol{E}^\sigma|^2}{8\pi n_0m_e}\\+\left(\frac{\partial}{\partial\boldsymbol{v}}\cdot\boldsymbol{D}^\sigma\cdot\frac{\partial}{\partial\boldsymbol{v}}\nabla F_e\right)\cdot\boldsymbol{v}+\frac{\partial}{\partial\boldsymbol{v}}\cdot(\boldsymbol{D}^\sigma+\boldsymbol{D}^{\sigma*})\cdot\nabla F_e.\end{aligned} \tag{78}$$

可见 F_e 的变化包含着速度空间的扩散过程.

四、拉氏函数和守恒量

我们讨论二维时空表象中横波或纵波有质动力作用下的密度方程(63),(64)和(70),(71)的拉氏密度函数 $\mathscr{L}$ 和守恒量. 首先讨论静态近似下结果. (64)或(71)式代入(48)式后得高频场方程 ($\sigma=l,t$)

$$\begin{aligned}\frac{\partial^2E^\sigma}{\partial t'^2}-(u^\sigma)^2\frac{\partial^2}{\partial x'^2}E^\sigma=-\omega_{pe}^2E^\sigma+\omega_{pe}^2E^\sigma\frac{|E^\sigma|^2}{8\pi n_0(T_i+T_e)}\\\times\left(1-\frac{|E^\sigma|^2}{16\pi n_0(T_e+T_i)}\right).\end{aligned} \tag{79}$$

分离 E^σ 的缓变部分和高频振荡部分，令

$$E^\sigma(x',t')=\tilde{\mathscr{E}}^\sigma(x',t')e^{-i\omega_{pe}t'}. \tag{80}$$

这里 $\tilde{\mathscr{E}}^\sigma(x',t')$ 为随时间缓慢变化振幅，代入方程(79)，注意到

$$\omega_{pe}\tilde{\mathscr{E}}^\sigma\gg\frac{\partial}{\partial t'}\tilde{\mathscr{E}}^\sigma,$$

略去时间的二阶导数项，得到包含四级场作用下的高阶非线性 Schrödinger 方程

$$\begin{aligned}i\frac{\partial}{\partial t}\mathscr{E}^\sigma(x,t)+\frac{\partial^2}{\partial x^2}\mathscr{E}^\sigma(x,t)=-\mathscr{E}^\sigma(x,t)|\mathscr{E}^\sigma(x,t)|^2\\\times(1-|\mathscr{E}^\sigma(x,t)|^2).\end{aligned} \tag{81}$$

(81)式中已用了无量纲关系

$$\mathscr{E}^{\sigma}=\frac{\tilde{\mathscr{E}}_{\sigma}}{\sqrt{16\pi n_0(T_i+T_e)}},\quad t=\omega_{pe}t',\quad x=\sqrt{\frac{2\omega_{pe}^2}{(u^{\sigma})^2}}x'. \tag{82}$$

注意，现在空间座标的标度与波的种类有关.

很容易写出(81)式的拉氏密度函数为

$$\mathscr{L}^{\sigma}=\frac{1}{2}i(\mathscr{E}^{\sigma *}\mathscr{E}_t^{\sigma}-\mathscr{E}^{\sigma}\mathscr{E}_t^{\sigma *})-|\mathscr{E}_x^{\sigma}|^2+\frac{1}{2}|\mathscr{E}^{\sigma}|^4-\frac{1}{3}|\mathscr{E}^{\sigma}|^6. \tag{83}$$

这里 $\mathscr{E}_t^{\sigma}=\frac{\partial\mathscr{E}^{\sigma}}{\partial t}$，$\mathscr{E}_x^{\sigma}=\frac{\partial\mathscr{E}^{\sigma}}{\partial x}$.

现在给出方程(81)式的三个运动积分.

由四度流守恒：$\frac{\partial j_{\mu}}{\partial x_{\mu}}=0$，$j_{\mu}=-i\left(\frac{\partial\mathscr{L}}{\partial\mathscr{E}_{\mu}}\mathscr{E}-\frac{\partial\mathscr{L}}{\partial\mathscr{E}_{\mu}^{*}}\mathscr{E}^{*}\right)$，$\mu=1,\ 2$，$x_{\mu}=x_{\mu}(t,x)$，$\mathscr{E}_{\mu}=\mathscr{E}_{\mu}(\mathscr{E}_t,\mathscr{E}_x)$，得粒子数守恒

$$N=\int j_0dx=\int|\mathscr{E}^{\sigma}|^2dx=\text{const}. \tag{84}$$

由动量-能量张量密度守恒 $\frac{\partial T_{\mu\nu}}{\partial x_{\nu}}=0$，$T_{\mu\nu}=\mathscr{L}\delta_{\mu\nu}-\mathscr{E}_{\mu}^{(i)}\frac{\partial\mathscr{L}}{\partial\mathscr{E}_{\nu}^{(i)}}$，$\mathscr{E}^{(i)}=\{\mathscr{E},\ \mathscr{E}^{*}\}$，得动量守恒

$$P=\int T_{10}dx=\frac{1}{2}i\int(\mathscr{E}^{\sigma}\mathscr{E}_x^{\sigma *}-\mathscr{E}^{\sigma *}\mathscr{E}_x^{\sigma})dx. \tag{85}$$

能量守恒

$$G=-\int T_{00}dx=\int\left(|\mathscr{E}_x^{\sigma}|^2-\frac{1}{2}|\mathscr{E}^{\sigma}|^4+\frac{1}{3}|\mathscr{E}^{\sigma}|^6\right)dx. \tag{86}$$

在非静态近似下，这时$\mathscr{L}$中包含着密度 n_e 与场的相互作用项. 先由(48)式写出高频场方程

$$\frac{\partial^2E^{\sigma}}{\partial t'^2}-(u^{\sigma})^2\frac{\partial^2}{\partial x'^2}E^{\sigma}=-\omega_{pe}^2E^{\sigma}\left(1+\frac{\tilde{n}_e}{n_0}\right). \tag{87}$$

为了以后应用方便，这里已把 n_e 改写为 $\tilde{n}_e$. 采用表示式(80)后，得

$$\frac{i}{\omega_{pe}}\frac{\partial\tilde{\mathscr{E}}^{\sigma}}{\partial t'}+\frac{(u^{\sigma})^2}{2\omega_{pe}^2}\frac{\partial^2}{\partial x'^2}\tilde{\mathscr{E}}^{\sigma}=\frac{\tilde{\mathscr{E}}^{\sigma}}{2}\frac{\tilde{n}_e}{n_0}. \tag{88}$$

其次把方程(63)和(70)统一写成

$$\left(\frac{\partial^2}{\partial t'^2}-v_s^2\frac{\partial^2}{\partial x'^2}\right)\tilde{n}_e(x',t')=\frac{1}{8\pi m_i}\frac{\partial^2}{\partial x'^2}\times\left\{|\tilde{\mathscr{E}}^{\sigma}|^2\left(1-\frac{\gamma_{\sigma}}{(1-V_{\phi}^2)}\frac{|\tilde{\mathscr{E}}^{\sigma}|^2}{16\pi n_0T_e}\right)\right\}. \tag{89}$$

这里

$$V_{\phi}=v_{\phi}/v_s, \tag{90}$$

$$\gamma_{\sigma}=1\quad 对\quad \sigma=t;$$
$$=\frac{3(2T_i+3T_e)}{T_i+T_e}\quad 对\quad \sigma=l. \tag{91}$$

引入表示：$\xi_\sigma = \frac{v_s}{u^\sigma}$，$t = 2\omega_{pe}\xi_\sigma^2 t'$，$x = \frac{2\omega_{pe}\xi_\sigma^2}{v_s} x'$，

$$n_e = \frac{\tilde{n}_e}{4n_0\xi_\sigma^2}, \quad |\mathscr{E}^\sigma|^2 = \frac{|\tilde{\mathscr{E}}^\sigma|^2}{32\pi n_0\xi_\sigma^2 T_e}. \tag{92}$$

(88)和(89)式就分别写成为

$$i\frac{\partial\mathscr{E}^\sigma}{\partial t} + \frac{\partial^2}{\partial x^2}\mathscr{E}^\sigma = \mathscr{E}^\sigma n_e, \tag{93}$$

$$\left(\frac{\partial^2}{\partial t^2} - \frac{\partial^2}{\partial x^2}\right) n_e = \frac{\partial^2}{\partial x^2}\left\{|\mathscr{E}^\sigma|^2\left(1 - \frac{\Gamma_\sigma}{1 - V_\phi^2}|\mathscr{E}^\sigma|^2\right)\right\}, \tag{94}$$

其中

$$\Gamma_\sigma = 2\gamma_\sigma\xi_\sigma^2. \tag{95}$$

对于(93)和(94)式我们只能构筑一个近似的拉氏密度函数

$$\mathscr{L}^\sigma = \frac{1}{2}i(\mathscr{E}^{\sigma*}\mathscr{E}_t^\sigma - \mathscr{E}^\sigma\mathscr{E}_t^{\sigma*}) - |\mathscr{E}_x^\sigma|^2 + \frac{1}{2}|\mathscr{E}^\sigma|^4 - \frac{\Gamma_\sigma}{3(1 - V_\phi^2)}|\mathscr{E}^\sigma|^6 - \varphi_t|\mathscr{E}^\sigma|^2\left(1 - \frac{\Gamma_\sigma}{(1 - V_\phi^2)}|\mathscr{E}^\sigma|^2\right) + \frac{1}{2}(\varphi_t^2 - \varphi_x^2). \tag{96}$$

这里

$$\varphi_t = n_e + |\mathscr{E}^\sigma|^2\left(1 - \frac{\Gamma_\sigma}{1 - V_\phi^2}|\mathscr{E}^\sigma|^2\right), \tag{97}$$

$$\varphi_{tt} = \varphi_{xx} + |\mathscr{E}^\sigma|^2\left(1 - \frac{\Gamma_\sigma}{1 - V_\phi^2}|\mathscr{E}^\sigma|^2\right). \tag{98}$$

(96)式可以准确地给出方程(94)，但是在导得(93)式时多了高阶项

$\frac{2\Gamma_\sigma}{1 - V_\phi^2}|\mathscr{E}^\sigma|^2(n_e + |\mathscr{E}^\sigma|^2)\mathscr{E}^\sigma - 2\frac{\Gamma_\sigma^2}{(1 - V_\phi^2)^2}|\mathscr{E}^\sigma|^6\mathscr{E}^\sigma$，除了第二项超过方程本身精度自然可忽略外，对四级场部分，在稳定态下取 $n_e = -|\mathscr{E}^\sigma|^2$，第一项也可近似消去.

从(96)式可以得到类似于(84)—(86)式的三个守恒量.

五、结论和讨论

1. 从方程(48)，(63)，(70)或(93)，(94)可以看到，四级场的出现在某些情况下将明显改变作为有质动力的二级场作用. 例如考虑以下解：

$\xi = x - V_\phi t$，这里 V_ϕ 取用(90)式特定形式. 再令

$$\mathscr{E}^\sigma(x, t) = \mathscr{E}_0^\sigma(\xi)e^{igx - i\Theta t}, \quad \mathscr{E}_0 \text{ 为实数}, \tag{99}$$

$$n_e(x, t) = n_e(\xi). \tag{100}$$

由(93)式就得到

$$n_e(\xi) = -\frac{(\mathscr{E}_0^\sigma)^2}{1 - V_\phi^2}\left[1 - \frac{\Gamma_\sigma}{(1 - V_\phi^2)}(\mathscr{E}_0^\sigma)^2\right]. \tag{101}$$

(94)式则为

$$\frac{1}{2}(\mathscr{E}_{0\xi})^2 + b_0\mathscr{E}_0^4 - a_0\mathscr{E}_0^6 - c_0\mathscr{E}_0^2 = 0, \tag{102}$$

$$a_0 = \frac{\Gamma_\sigma}{6(1 - V_\phi^2)^2}, \quad b_0 = \frac{1}{4(1 - V_\phi^2)}, \quad c_0 = \frac{1}{2}\left(\frac{V_\phi^2}{4} - \Theta\right). \tag{103}$$

方程(102)是不难解的，但我们直接从相平面上势

$$V = -a_0\mathscr{E}_0^6 + b_0\mathscr{E}_0^4 - c_0\mathscr{E}_0^2, \tag{104}$$

就可以看到，有了四级场后仍然存在 Soliton 解，但是比起二级场来却有了明显的变化．在 $V_\phi < 1$ 亚声条件下，当 $\frac{\Gamma_\sigma}{1 - V_\phi^2}(\mathscr{E}_0^\sigma)^2 = \frac{\gamma_\sigma}{1 - V_\phi^2}\left(\frac{|E^\sigma|^2}{16\pi}\Big/ n_0T_e\right)$ 趋向于 1 时，n_e 趋向于 0，这时四级场的作用完全抵消了二级场效应，使一维 Soliton 不可能形成．只有在 $\frac{\Gamma_\sigma}{1 - V_\phi^2}(\mathscr{E}_0^\sigma)^2 \ll 1$ 时，(101)式右边第二项可忽略，才有通常的 Soliton 出现，这说明一维稳定结构解只出现在早期调制不稳定性场幅还较小的发展阶段或 $V_\phi \ll 1$ 时，随着场幅的增加或在 V_ϕ 接近 1 时，高阶场的作用使非线性凝聚作用开始消失，直至消灭 Soliton．所以四级场的作用促使密度抽空效应只能具有 $\frac{n_e}{n_0} < 1$ 的大小．

另一方面，从色散关系看，方程(81)的色散关系为

$$\Omega_\sigma = \frac{q^2(\mu^\sigma)^2}{2\omega_{pe}} - \omega_{pe}\frac{|E^\sigma|^2}{16\pi n_0(T_e + T_i)}\left(1 - \frac{|E^\sigma|^2}{16\pi n_0(T_e + T_i)}\right). \tag{105}$$

这里四级场的作用总是抑制着二阶场的非线性色散性，使波的传播在高场能密度时接近线性色散关系特征．

2．从方程(73)和(74)，可以预料四级场的出现将阻碍 Caviton 的形成过程．由于含 $(\boldsymbol{EE}^*)_s$ 项的出现，也不见得出现太强凝聚力而加速"坍塌"．

3．在阻尼效应不能忽略情况下，$\varepsilon_d(q)$ 和 $\varepsilon_{d\alpha}(q)$ 必须保留虚部，这时，在密度方程中就出现阻尼项，详细处理见文献[4]．

感谢于敏教授的有益讨论．

参　考　文　献

[1] R. C. Davidson, Methods in nonlinear plasma theory, (1972).
[2] Y. H. Ichikawa and T. Taniuti, *J. Phys. Soc. Japan*, **34**(1973), 513.
[3] V. N. Tsytovich, *Comments on Plasma Physics and Controlled Fusion* (*part E*), **2**(1976), 127.
S. G. Thornhill and D. ter Haar, *Phys. Reports*, **43c**(1978), 45.
[4] 贺贤土，核聚变，**1**(1980)，245.
[5] В. И. Захаров, *ЖЭТФ*, **62**(1972), 1945.

NON-LINEAR EFFECT ON THE LARGE AMPLITUDE WAVES INTERACTION WITH PARTICLES OF LOW FREQUENCY OSCILLATION IN PLASMA

He Xian-tu
(*P. O. Box* 8009, *Beijing, China*)

ABSTRACT

In this paper, retaining high-frequency fields up to fourth order, the kinetic and hydrodynamic equations for the beat frequency interaction between longitudinal, transverse high-frequency fields and particles in low-frequency oscillation were derived from Vlasov-Maxwell equations. The results show that, in the case of one dimension, the large amplitude high order fields in sub-acoustic waves weaken the ponderomotive force which causes solitons. In three dimensions, it possibly makes the formation of cavitons slower.

第32卷 第5期 物 理 学 报 Vol. 32, No. 5
1983年5月 ACTA PHYSICA SINICA May, 1983

等离子体中调制不稳定性和波包的坍缩过程

贺贤土
（中国科学院原子能研究所）

1982年4月26日收到

提 要

本文给出了等离子体中由动力学理论所得到的形成强湍流的高频和低频非线性电流的表达式．根据文献[2]给出的包含有质动力、自生磁场和它们的阻尼效应的方程组，我们推广了 Kono 等人关于调制不稳定性的分析，得到了在各种情况下由 Langmuir 波或横波 pump 波激发的纵波和横波的增长率与参量的关系式．最后，讨论了波包的坍缩动力学问题，把非线性 Schrödinger 方程的坍缩讨论推广到密度和场耦合的方程组的情况．

本文继续文献[1,2]的讨论，研究等离子体中长波长 Langmuir 波或横波扰动的调制不稳定性（MI）以及其发展后期波包的非线性坍缩过程．我们知道，当长波能量超过阈值以后，且它们的波数 $|\boldsymbol{k}_0|\leqslant k^*\equiv\frac{1}{3}\mu^{1/2}\lambda_e^{-1}$ （$\mu=m_e/M$——电子离子质量比，λ_e 为电子 Debye 长度）时，MI 通过自调制激发波数大于 $|\boldsymbol{k}_0|$ 的波，其结果是被激发波的振幅变得足够大，两个高频波拍频成为低频波的非线性作用促使低频振荡粒子密度有强的扰动，激发非线性有旋电流，产生低频自生磁场，这时波场的能量便会在局部空间集中．为了维持平衡，通常在能量聚积当地电子（从而离子）密度必须被排空．这种能量集中和密度排空运动具有准粒子特性，即准粒子数、动量和能量守恒．在一维情况下数值模拟表明，MI 可以发展成为一个个几乎是互相独立的尖峰结构[3]，可能对应稳定态时的 soliton 运动．就物理上来说，它是波包运动在空间一维方向上的弥散与非线性凝聚力之间竞争的结果；在多维情况下，波包运动在几个方向上同时与非线性凝聚力之间处于平衡状态成为困难．在 MI 发展后期非线性凝聚效应愈来愈强，最后导致运动的不平衡收缩．根据守恒关系[4,5]，可以预言会出现波包的坍缩．一旦波包不稳定收缩，最终可能出现奇点．但是，实际过程则是收缩到 Debye 长度大小时就会迅速被阻尼而导致粒子的急剧加热．这种不稳定性及其以后发展过程对于研究强湍流运动的形成和发展具有重要的理论意义．事实上，从 1965 年 Vedenov, Rudakov[6] 工作开始一直到现在，这方面研究文章很多，其中包括近几年 Tsytovich, Thornhill 以及 Harr[7,8] 等人的评论性文章．正如上面所提到的，在 MI 过程中除了两个高频波向低频波拍频激发离子声涨落的放大以外，还可以通过两个高频波交叉拍频激发低频振荡电子非线性电流的有旋分量从而激发低频磁场，其具体形式的给出以及它对 MI 过程影响的研究，则是最近一年左右时间的工作[2,9,10]，虽然实验的发现和

无碰撞磁场机制的提出是七十年代的事情．研究表明，在一维情况下自生磁场不起作用，但是二维和三维情况下它有助于 MI 过程中坍缩的发生[10]．

本文目的是给出我们从 Vlasov-Maxwell 体系出发导得的各种非线性电流形式和描述强湍流运动(包括自生磁场效应)的动力学方程组[1,2]，以及从它们出发进一步讨论 MI 发展及其坍缩问题．这里我们扩展了文献[10]关于色散关系的讨论，得到了包含密度、磁场阻尼效应的单色波的色散方程．对纵、横 pump 波的扰动，近似求解了色散关系；关于坍缩问题，我们推广了迄今为止只对非线性 Schrödinger 方程的讨论到较为普遍包含场幅和密度扰动耦合的方程组情况．

一、波与粒子相互作用的动力学方程组

如上所述，在 MI 后期高频有限幅波与背景粒子的扰动分布相互作用激发了高频和低频非线性电流，而这些非线性电流又进一步驱动着高频和低频波的传播，以及促进了有限幅波与低频粒子的耦合关系．在最低阶非线性近似下，产生的高频非线性电流有如下两种类型[1,2]：第一种系高频场与粒子的低频振荡相互作用产生的低频电子（或离子）的定向运动电流．如果略去高频波非线性 Landau 阻尼和频率 $\omega \simeq \omega_{pe} \gg \boldsymbol{k}\cdot\boldsymbol{v}$ 项（这里，ω_{pe} 为电子等离子体频率，$\boldsymbol{v}$ 为粒子速度)，就可得到这种电流的 Fourier 表象

$$\boldsymbol{j}_h^{(1)}(k)=\frac{i\omega_{pe}^2}{4\pi\omega}\sum_{\sigma_1}\int\boldsymbol{E}_h^{\sigma_1}(k_1)\frac{n_e(q)}{n_0}\,\delta^4(k-k_1-q)\,\frac{dk_1dq}{(2\pi)^4}. \tag{1}$$

这里 $k=\{\omega,\ \boldsymbol{k}\}$，$q=\{\Omega,\ \boldsymbol{q}\}$ 分别为高、低频四度动量，$dk=d^3k\,d\omega$，$dq=d^3q\,d\Omega$，$\boldsymbol{E}_h^{\sigma}(k)$ 为高频 σ 波（$\sigma=l,\ t$）的场强，$n_e(q)$ 为低频电子扰动密度．

第二种高频电流由粒子高频振荡与低频电场相互作用所产生．它主要表现为高频电场与低频自生磁场相互作用形成的电子螺旋运动的非线性电流

$$\boldsymbol{j}_h^{(2)}(k)=\frac{\omega_{pe}e}{4\pi m_e\omega c}\sum_{\sigma_1}\int\boldsymbol{E}_h^{\sigma_1}(k_1)\times\boldsymbol{B}(q)\delta^4(k-k_1-q)\,\frac{dk_1dq}{(2\pi)^4}, \tag{2}$$

其中 e 为电子电荷，c 为光速．

同样近似下，又可得到两类低频电流，它们分别为高频场与电子高频振荡相互作用拍频成为低频波时激发的纵向电流 $\boldsymbol{j}_d^{(1)}(q)$ 和螺旋运动电流 $\boldsymbol{j}_d^{(2)}(q)$，其表达式为

$$\boldsymbol{j}_d^{(1)}(q)=\frac{\Omega e}{4\pi m_e\omega_{pe}^2}(\varepsilon_e^l(q)-1)\boldsymbol{q}\sum_{\sigma_1^+\sigma_2^-}\int\boldsymbol{E}_h^{\sigma_1^+}(k_1)$$
$$\cdot\boldsymbol{E}_h^{\sigma_2^-}(k_2)\delta^4(q-k_1+k_2)\,\frac{dk_1dk_2}{(2\pi)^4}, \tag{3}$$

其中

$$\varepsilon_\alpha^l(q)=1+\frac{\omega_{p\alpha}^2}{|\boldsymbol{q}|}\int\frac{\boldsymbol{e}_d^l\cdot\dfrac{\partial F_{0\alpha}}{\partial\boldsymbol{v}}d^3v}{\Omega-\boldsymbol{q}\cdot\boldsymbol{v}}, \tag{4}$$

$$\varepsilon_d^l(q)=\sum_{\alpha=i,e}\varepsilon_\alpha^l(q)-1 \tag{5}$$

为纵波线性介电函数，$F_{0\alpha}$ 为背景粒子 Maxwell 速度分布函数（归一化），$\sigma^{\pm}$ 表示以四度动量 $\{\pm\omega, \pm\boldsymbol{k}\}$ 振荡的波，$\boldsymbol{e}_d^l = \boldsymbol{q}/|\boldsymbol{q}|$.

$$\boldsymbol{j}_d^{(2)}(q) = \frac{e}{4\pi m_e\omega_{pe}} \sum_{\sigma_1^+\sigma_2^-} \int [\boldsymbol{q} \times (\boldsymbol{E}_h^{\sigma_1^+}(k_1) \times \boldsymbol{E}_h^{\sigma_2^-}(k_2)^*)]\delta^4(q - k_1 + k_2)\frac{dk_1dk_2}{(2\pi)^4}. \tag{6}$$

由(1),(2),(3),(6)式就可得到高、低频场方程分别为

$$\varepsilon_h^\sigma(k)\boldsymbol{E}_h^\sigma(k) = -\frac{4\pi i}{\omega} \sum_v \boldsymbol{j}_h^{(v)}(k), \tag{7}$$

$$\varepsilon_d^\sigma(q)\boldsymbol{E}_d^\sigma(q) = -\frac{4\pi i}{\Omega} \sum_v \boldsymbol{j}_d^{(v)}(q). \tag{8}$$

为了使 (7), (8) 式闭合，尚需给出低频自生磁场 $\boldsymbol{B}$ 和电子扰动密度 n_e 表达式. 在 $\omega \simeq \omega_{pe} \gg \Omega$ 或 $\boldsymbol{q}\cdot\boldsymbol{v}$ 近似下，可得低频电子扰动密度

$$n_e(q) = -\frac{1}{4\pi m_e}\frac{\varepsilon_i^l(q)}{\varepsilon_d^l(q)}(\varepsilon_e^l(q) - 1)\frac{|\boldsymbol{q}|^2}{\omega_{pe}^2}\sum_{\sigma_1^+\sigma_2^-}\int \boldsymbol{E}_h^{\sigma_1^+}(k_1) \cdot \boldsymbol{E}_h^{\sigma_2^-}(k_2)^*\delta^4(q - k_1 + k_2)\frac{dk_1dk_2}{(2\pi)^4}. \tag{9}$$

至于低频磁场，由 Faraday 感应定律得

$$\boldsymbol{B}(q) = \frac{c}{\Omega}\boldsymbol{q} \times \boldsymbol{E}_d^\sigma(q), \tag{10}$$

它是(8)式中低频螺旋运动电流激发结果.

如果被激发低频波的相速度 $\Omega/|\boldsymbol{q}|$ 处在电子离子热速度之间，则变换到时、空表象时(7),(9),(10)式就成为

$$\frac{\partial^2}{\partial t^2}\boldsymbol{E} + c^2\nabla \times (\nabla \times \boldsymbol{E}) - 3v_{T_e}^2\nabla\nabla \cdot \boldsymbol{E} = -\omega_{pe}^2\left(1 + \frac{n_e}{n_0}\right)\boldsymbol{E} + \frac{i\omega_{pe}e}{m_ec}\boldsymbol{E} \times \boldsymbol{B}, \tag{11}$$

$$\left(\frac{\partial^2}{\partial t^2} + \Gamma_e\frac{\partial}{\partial t} - v_s^2\nabla^2\right)n_e(\boldsymbol{x}, t) = \nabla^2\frac{|\boldsymbol{E}|^2}{8\pi M}, \tag{12}$$

$$\Gamma_e\frac{\partial}{\partial t}n_e \equiv -(2\pi)^{-3/2}\mu^{1/2}v_s\nabla^2\frac{\partial}{\partial t}\int\frac{n_e(\boldsymbol{x}', t)}{|\boldsymbol{x}' - \boldsymbol{x}|^2}d^3x', \tag{13}$$

$$\nabla \times \nabla \times \boldsymbol{B} = \frac{e}{2im_e\omega_{pe}c}\nabla \times [\nabla \times (\boldsymbol{E} \times \boldsymbol{E}^*)] - \frac{\omega_{pe}^2}{(2\pi)^{3/2}v_{T_e}c^2}\frac{\partial}{\partial t}\int\frac{\boldsymbol{B}(\boldsymbol{x}', t)}{|\boldsymbol{x} - \boldsymbol{x}'|^2}d^3x', \tag{14}$$

其中 $\boldsymbol{E}(\boldsymbol{x}, t)$ 包括纵波 $(\nabla \times \boldsymbol{E} = 0)$ 和横波 $(\nabla \cdot \boldsymbol{E} = 0)$ 之和，并已省去下标 "h" 记号；$v_{T_e} = (T_e/m_e)^{1/2}$，$T_e$ 为电子温度(能量单位)；$v_s = (T_e/M)^{1/2}$；$\mu = m_e/M$.

(13)式和(14)式中等号右端第二项分别为电子扰动密度和磁场的电子 Landau 阻尼项，这里略去了离子的阻尼效应，详见文献[2].

应当提到，当波的相速度处在 v_{T_e} 和 v_{T_i}(离子热速度)之间时 (11)—(14) 式是比较完善的方程组. 早在 1972 年 Захаров[11] 首先导得了 $\boldsymbol{B} = 0$ 无阻尼的方程组(11)—(12)

形式，其后 Tsytovich, Rudakov, Harr 和 Thornhill[12,9,10] 又综合讨论了这组方程的 MI 发展行为，并把它们用于强湍流理论研究．但是所有这些，除了文献［12］中部分讨论以外，都建立在双流体矩方程模型基础上进行的．因此，微观作用的细致图象不能充分反映，Landau 阻尼等效应自动消失，且推导过程所作的近似物理意义也很不明确．我们从 Vlasov-Maxwell 微观理论体系出发，重新导得包含阻尼、无磁场的方程组 (11)—(12)[1]，并接着又导得了包含磁场的方程组(11)—(14)[2]．关于单一的(14)形式的磁场方程，Kono[9] 等人于 1980 年发表了他们的结果，但是他们的阻尼项与我们结果差了一个号．最近，我们又见到他们的另一篇文献[10]讨论了(11)—(14)式的动力学行为．但是他们仍然用双流体力学模型结果进行处理，所以密度的阻尼项不能出现，这就与磁场方程的阻尼项发生不自洽．由于没有密度的阻尼效应，在波包坍缩后期通过非线性 Landau 阻尼把场能转化为粒子能就不可能，并要出现奇点．

为了能方便求解(11)—(14)式，通常把高频场 $\boldsymbol{E}$ 分解为缓变振幅 $\boldsymbol{\phi}$ 和高频振荡因子两部分，以便使所得的振幅方程具有与 n_e 和 $\boldsymbol{B}$ 相同的时标，

令
$$\boldsymbol{E}=\frac{1}{2}(\boldsymbol{\phi}e^{-i\omega_{pe}t}+\text{c. c.}). \tag{15}$$

如果略去 $\boldsymbol{\phi}$ 的二次时间导数项，并引入如下替换：

$$\frac{2}{3}\omega_{pe}\mu t\to t,\ \frac{2}{3}\frac{\sqrt{\mu}}{\lambda_e}\boldsymbol{x}\to\boldsymbol{x},\ \frac{3n_e}{4\mu n_0}\to n,$$
$$\frac{\sqrt{3}\boldsymbol{\phi}}{(64\pi\mu n_0T_e)^{1/2}}\to\boldsymbol{\varepsilon},\ \frac{3\boldsymbol{B}}{(64\pi\mu^2n_0m_ec^2)^{1/2}}\to\boldsymbol{B},\quad \alpha^2=\frac{c^2}{3v_{T_e}^2}, \tag{16}$$

(11)—(14)式就成为如下无量纲的方程组：

$$i\frac{\partial}{\partial t}\boldsymbol{\varepsilon}-\alpha^2\nabla\times(\nabla\times\boldsymbol{\varepsilon})+\nabla(\nabla\cdot\boldsymbol{\varepsilon})=n\boldsymbol{\varepsilon}-i(\boldsymbol{\varepsilon}\times\boldsymbol{B}), \tag{17}$$

$$\left(\frac{\partial^2}{\partial t^2}-\nabla^2\right)n=\nabla^2|\boldsymbol{\varepsilon}|^2+\frac{\xi_0}{2\pi^2}\nabla^2\frac{\partial}{\partial t}\int\frac{n(\boldsymbol{x}',t)}{|\boldsymbol{x}'-\boldsymbol{x}|^2}d^3x', \tag{18}$$

$$\nabla^2\boldsymbol{B}=i\beta^2\nabla\times[\nabla\times(\boldsymbol{\varepsilon}\times\boldsymbol{\varepsilon}^*)]+\frac{\chi_0}{2\pi^2}\frac{\partial}{\partial t}\int\frac{\boldsymbol{B}(\boldsymbol{x}',t)}{|\boldsymbol{x}'-\boldsymbol{x}|^2}d^3x', \tag{19}$$

其中

$$\beta^2=\frac{v_{T_e}^2}{c^2},\ \xi_0=\frac{\pi\mu^{1/2}}{(2\pi)^{1/2}},\ \chi_0=\frac{9}{4}\frac{\beta^2\pi}{\mu^{1/2}(2\pi)^{1/2}}. \tag{20}$$

在 Landau 阻尼可忽略情况下，(19)式变为

$$\boldsymbol{B}=-i\beta^2(\boldsymbol{\varepsilon}\times\boldsymbol{\varepsilon}^*), \tag{21}$$

(17)，(18)式就成为

$$i\frac{\partial}{\partial t}\boldsymbol{\varepsilon}-\alpha^2\nabla\times(\nabla\times\boldsymbol{\varepsilon})+\nabla(\nabla\cdot\boldsymbol{\varepsilon})=n\boldsymbol{\varepsilon}-\beta^2\boldsymbol{\varepsilon}\times(\boldsymbol{\varepsilon}\times\boldsymbol{\varepsilon}^*), \tag{22}$$

$$\left(\frac{\partial^2}{\partial t^2}-\nabla^2\right)n=\nabla^2|\boldsymbol{\varepsilon}|^2. \tag{23}$$

二、有限振幅单色波的色散关系

方程组(17)—(19)或(22)—(23)都具有调制不稳定性（MI）和三波不稳定性解，后者发生在 $|\boldsymbol{k}_0| > k^* = \frac{1}{3}\frac{\mu^{1/2}}{\lambda_e}$ 区域．这里我们只感兴趣 MI 问题．为了简单起见，只讨论波谱较窄的波包近似单色波问题．这相应于 MI 的早期发展过程；当 MI 后期强湍流形成波包有显著宽度时，这里的结果需要作一些修正．

关于方程(22)，(23)的 Langmuir 波 MI 稳定性分析，较详细的可见文献[8]；在有磁场情况下，Kono[10] 等人的工作最近刚见到．正如前面所提到的，他们只是在有磁场阻尼而无密度阻尼情况下，讨论了纵 pump 波激发纵波的色散关系，而我们这里将用方程(17)—(19)——包含磁场和密度 Landau 阻尼——去讨论纵和横 pump 波激发纵波和横波的情况．

设初始未扰动的波为

$$\boldsymbol{\varepsilon}_0^\sigma e^{ik_0x},\ n = 0,\ B = 0,\ \sigma = l,\ t,$$
$$k_0x \equiv \boldsymbol{k}_0\cdot\boldsymbol{x} - \omega_0 t. \tag{24}$$

式中 $\boldsymbol{\varepsilon}_0^\sigma = \boldsymbol{e}_{0\sigma}\varepsilon_0^\sigma$，分别为 $\boldsymbol{e}_{0l} = \boldsymbol{k}_0/|\boldsymbol{k}_0|$ 的 Langmuir 波或 $\boldsymbol{e}_{0t}\perp\boldsymbol{e}_{0l}$ 频率接近 ω_{pe} 的横波．将上式代入(17)式后，(24)式满足

$$\omega = \begin{cases} k_0^2, & \sigma = l; \\ \alpha^2k_0^2, & \sigma = t. \end{cases} \tag{25}$$

现在讨论初始单色波(24)式扰动后场和粒子密度的稳定性问题．设场和扰动密度的形式为

$$\boldsymbol{\varepsilon} = \sum_{\sigma=l,t}\boldsymbol{\varepsilon}_0^\sigma e^{ik_0x} + \sum_{\sigma=l,t}(\boldsymbol{\varepsilon}_\sigma e^{iqx} + \boldsymbol{\varepsilon}_\sigma^+ e^{-iqx})e^{ik_0x}. \tag{26}$$

这里
$$\boldsymbol{\varepsilon}_\sigma = \boldsymbol{e}_\sigma\varepsilon_\sigma,\ \boldsymbol{\varepsilon}_\sigma^+ = \boldsymbol{e}_\sigma^+\varepsilon_\sigma^*,\ \boldsymbol{e}_l = \frac{\boldsymbol{q}+\boldsymbol{k}_0}{|\boldsymbol{q}+\boldsymbol{k}_0|},\ \boldsymbol{e}_l^+ = \frac{\boldsymbol{q}-\boldsymbol{k}_0}{|\boldsymbol{q}-\boldsymbol{k}_0|}, \tag{27}$$

$\boldsymbol{e}_t$ 和 $\boldsymbol{e}_t^+$ 分别与 $\boldsymbol{e}_l$ 和 $\boldsymbol{e}_l^+$ 垂直，$qx \equiv \boldsymbol{q}\cdot\boldsymbol{x} - \Omega t$.

$$\begin{pmatrix} n \\ \boldsymbol{B} \end{pmatrix} = \begin{pmatrix} \delta n \\ \delta\boldsymbol{B} \end{pmatrix}(e^{iqx} + e^{-iqx}) = 2\begin{pmatrix} \delta n \\ \delta\boldsymbol{B} \end{pmatrix}\cos qx. \tag{28}$$

为了使方程(17)—(19)闭合解，(17)式必须被分成 $\boldsymbol{\varepsilon}$ 和 $\boldsymbol{\varepsilon}^*$ 两个方程．把方程组线性化后，取含扰动因子 $e^{i(q+k_0)x}$ 的 $\boldsymbol{\varepsilon}_l$, $\boldsymbol{\varepsilon}_t$ 和 $e^{-i(q-k_0)x}$ 的 $(\boldsymbol{\varepsilon}_l^+)^*$, $(\boldsymbol{\varepsilon}_t^+)^*$ 部分，注意到 $(\boldsymbol{q}, \boldsymbol{k}_0)$ 平面上的夹角关系：$\boldsymbol{e}_l\cdot\boldsymbol{e}_{0l} = \cos\theta_+$, $\boldsymbol{e}_l^+\cdot\boldsymbol{e}_{0l} = \cos\theta_-$ 和 $\boldsymbol{B}$ 垂直 $(\boldsymbol{q}, \boldsymbol{k}_0)$ 平面，可以得到如下色散方程：

$$\begin{vmatrix} A_+^\sigma & 0 & 0 & 0 & \varepsilon_0^\sigma b_+^\sigma & -\varepsilon_0^\sigma a_+^\sigma \\ 0 & A_-^\sigma & 0 & 0 & \varepsilon_0^{*\sigma} b_-^\sigma & \varepsilon_0^{*\sigma} a_-^\sigma \\ 0 & 0 & B_+^\sigma & 0 & \varepsilon_0^\sigma a_+^\sigma & \varepsilon_0^\sigma b_+^\sigma \\ 0 & 0 & 0 & B_-^\sigma & \varepsilon_0^{*\sigma} a_-^\sigma & -\varepsilon_0^{*\sigma} b_-^\sigma \\ \varepsilon_0^{*\sigma} b_+^\sigma & \varepsilon_0^\sigma b_-^\sigma & \varepsilon_0^{*\sigma} a_+^\sigma & \varepsilon_0^\sigma a_-^\sigma & \left(i\Omega\xi_0 + \dfrac{\Omega^2 - q^2}{q^2}\right) & 0 \\ -\varepsilon_0^{*\sigma} a_+^\sigma & \varepsilon_0^\sigma a_-^\sigma & \varepsilon_0^{*\sigma} b_+^\sigma & -\varepsilon_0^\sigma b_-^\sigma & 0 & \beta^{-2}\left(\dfrac{i\Omega\chi_0}{q^2} - 1\right) \end{vmatrix} = 0. \tag{29}$$

(29)式中取电场沿 $\boldsymbol{e}_\sigma$, $\boldsymbol{e}_\sigma^+$ 方向分量,磁场沿垂直于 $(\boldsymbol{q}, \boldsymbol{k}_0)$ 平面方向分量. 行列式中

$$A_\pm^\sigma = k_\sigma^2 \pm \Omega - (\boldsymbol{q} \pm \boldsymbol{k}_0)^2,$$
$$B_\pm^\sigma = k_\sigma^2 \pm \Omega - \alpha^2(\boldsymbol{q} \pm \boldsymbol{k}_0)^2; \tag{30}$$

$$a_\pm^\sigma = \begin{cases} \sin\theta_\pm, & \sigma = l, \\ \cos\theta_\pm, & \sigma = t; \end{cases}$$

$$b_\pm^\sigma = \begin{cases} \cos\theta_\pm, & \sigma = l, \\ -\sin\theta_\pm, & \sigma = t; \end{cases} \tag{31}$$

$$k_\sigma = \begin{cases} k_0, & \sigma = l, \\ \alpha k_0, & \sigma = t. \end{cases} \tag{32}$$

方程(29)即为

$$\beta^{-2}\left(\frac{i\Omega\chi_0}{q^2} - 1\right)\left(i\Omega\xi_0 + \frac{\Omega^2 - q^2}{q^2}\right) - |\boldsymbol{\varepsilon}_0^\sigma|^2 \left[\left(i\Omega\xi_0 + \frac{\Omega^2 - q^2}{q^2}\right)D_1^\sigma + \beta^{-2}\left(\frac{i\Omega\chi_0}{q^2} - 1\right)D_2^\sigma\right] + |\boldsymbol{\varepsilon}_0|^4 \left\{D_1^\sigma D_2^\sigma - \left[a_+^\sigma b_+^\sigma\left(\frac{1}{A_+^\sigma} - \frac{1}{B_+^\sigma}\right) - a_-^\sigma b_-^\sigma\left(\frac{1}{A_-^\sigma} - \frac{1}{B_-^\sigma}\right)\right]^2\right\} = 0, \tag{33}$$

其中

$$D_1^\sigma = \frac{(a_+^\sigma)^2}{A_+^\sigma} + \frac{(a_-^\sigma)^2}{A_-^\sigma} + \frac{(b_+^\sigma)^2}{B_+^\sigma} + \frac{(b_-^\sigma)^2}{B_-^\sigma},$$
$$D_2^\sigma = \frac{(b_+^\sigma)^2}{A_+^\sigma} + \frac{(b_-^\sigma)^2}{A_-^\sigma} + \frac{(a_+^\sigma)^2}{B_+^\sigma} + \frac{(a_-^\sigma)^2}{B_-^\sigma}. \tag{34}$$

求解(33)式是十分复杂的,但可以作简化讨论. 由于 pump 波具有长波长特性,我们可以只讨论 $|\boldsymbol{q}| \gg |\boldsymbol{k}_0|$ 情况,这时 $\cos\theta_\pm \simeq \cos\theta$, $\sin\theta_\pm \simeq \sin\theta$, $A_\pm^\sigma \simeq \pm\Omega - q^2$, $B_\pm^\sigma \simeq \pm\Omega - \alpha^2 q^2$. 考虑到量纲化后 $\Omega/|\boldsymbol{q}| \to \Omega/|\boldsymbol{q}|v_s$, 而一般情况下 $\alpha^2 \gg 1$, 所以(33)式中含因子 $(\Omega^2 - \alpha^2 q^4)^{-2}$ 项可以略去,这样(33)式可得到简化. 下面首先讨论 pump 波为纵波 $\sigma = l_0$ 的情况,这时

$$D_1^{l_0} = 2q^2[\sin^2\theta(\Omega^2 - q^4)^{-1} + \alpha^2\cos^2\theta(\Omega^2 - \alpha^4 q^4)^{-1}]$$
$$D_2^{l_0} = 2q^2[\cos^2\theta(\Omega^2 - q^4)^{-1} + \alpha^2\sin^2\theta(\Omega^2 - \alpha^4 q^4)^{-1}]. \tag{35}$$

于是(33)式便成为

$$\Omega^6\frac{(1 + \chi_0\xi_0)}{q^2} - \Omega^4[1 + q^2(1 + \chi_0\xi_0)(1 + \alpha^4) - 2|\boldsymbol{\varepsilon}_0|^2\beta^2 C_\alpha^2] + \Omega^2\{q^4(1 + \alpha^4) + \alpha^4 q^6(1 + \chi_0\xi_0) - 2|\boldsymbol{\varepsilon}_0|^2 q^2[\beta^2 C_\alpha^2 + \alpha^2\beta^2 q^2 S_\alpha^2 + S_\alpha^2] - 4|\boldsymbol{\varepsilon}_0|^4\beta^2\sin^2\theta\cos^2\theta\} - i\Omega\left\{\left(\frac{\Omega^2 - q^2}{q^4}\chi_0 - \xi_0\right)(\Omega^2 - q^4)(\Omega^2 - \alpha^4 q^4) - 2|\boldsymbol{\varepsilon}_0|^2[\beta^2 q^2\xi_0(\Omega^2 C_\alpha^2 - \alpha^2 q^4 S_\alpha^2) + \chi_0(\Omega^2 S_\alpha^2 - \alpha^2 q^4 C_\alpha^2)]\right\} + 2|\boldsymbol{\varepsilon}_0|^2\alpha^2 q^6(\beta^2 S_\alpha^2 + C_\alpha^2) - \alpha^4 q^8 - 4\alpha^2\beta^2|\boldsymbol{\varepsilon}_0|^4 q^4(\alpha^2\sin^2\theta\cos^2\theta + \cos^4\theta + \sin^4\theta) = 0, \tag{36}$$

其中

$$C_\alpha^2 = \sin^2\theta + \alpha^2\cos^2\theta, \quad S_\alpha^2 = \cos^2\theta + \alpha^2\sin^2\theta. \tag{37}$$

(36)式中 pump 波 $\boldsymbol{\varepsilon}_0$ 已省写了上角标 l_0.

下面分两种情况讨论(36)式解:

1. 考虑 l_0-l 激发，即纵 pump 波激发沿 $\boldsymbol{q}$ 方向偏振的波，这时去掉含 α 因子的项. (36)式有纯增长解，令 $\Omega=i\gamma$，因为量纲化后 $\gamma/|\boldsymbol{q}|\to\gamma/|\boldsymbol{q}|v_s\ll 1$，保留 γ^4 以下的项，从(36)式就得如下二次方程:

$$\gamma^2[1+q^2(1+\chi_0\xi_0)-2|\boldsymbol{\varepsilon}_0|^2\beta^2\sin^2\theta]+\gamma[q^2(\chi_0+\xi_0q^2)-2|\boldsymbol{\varepsilon}_0|^2(\beta^2q^2\xi_0\sin^2\theta+\chi_0\cos^2\theta)]+q^4-2|\boldsymbol{\varepsilon}_0|^2q^2(\beta^2\sin^2\theta+\cos^2\theta)-4|\boldsymbol{\varepsilon}_0|^4\beta^2\sin^2\theta\cos^2\theta=0. \tag{38}$$

要使(38)式有增长解其阈值满足方程

$$2|\boldsymbol{\varepsilon}_0|^2q^2(\beta^2\sin^2\theta+\cos^2\theta)+4|\boldsymbol{\varepsilon}_0|^4\beta^2\sin^2\theta\cos^2\theta-q^4=0, \tag{39}$$

它与阻尼常数无关. 当 $|\boldsymbol{\varepsilon}_0|^2\ll 1$ 时，(39)式变为

$$|\boldsymbol{\varepsilon}_0|_s^2=\frac{q^2}{2(\beta^2\sin^2\theta+\cos^2\theta)}. \tag{40}$$

(39)式对于 $\theta=0$,

$$|\boldsymbol{\varepsilon}_0|_s^2=\frac{q^2}{2}, \tag{41}$$

还原量纲后即

$$\frac{|\boldsymbol{E}_0|^2}{8\pi n_0T_e}=3q^2\lambda_e^2.$$

对于 $\theta=\frac{\pi}{2}$,

$$|\boldsymbol{\varepsilon}_0|_s^2=\frac{q^2}{2\beta^2}, \tag{42}$$

还原量纲后即 $\frac{|\boldsymbol{E}_0|^2}{8\pi n_0T_e}=3\frac{q^2\lambda_e^2}{\beta^2}$. 显见 l_0-l 激发阈值随 θ 增大而增加. 由于一般情况下 $\beta^2\ll 1$，所以激发 $\boldsymbol{q}\perp\boldsymbol{k}_0$ 波的阈值要很大才行.

当 $|\boldsymbol{\varepsilon}_0|^2\gg 1$ 时，对 $2\theta\neq n\pi(n=0,1,2,\cdots)$ 阈值便为

$$|\boldsymbol{\varepsilon}_0|_s^2=\frac{q^2}{\beta|\sin 2\theta|}. \tag{43}$$

从二次方程(38)很容易求得增长率

$$\gamma=\gamma(|\boldsymbol{\varepsilon}_0|^2,\ T_e,\ |\boldsymbol{q}|,\theta,\ \chi_0,\ \xi_0) \tag{44}$$

的明显表达式. 当阻尼为 0 时，$\xi_0=\chi_0=0$，得

$$\gamma^2=[2|\boldsymbol{\varepsilon}_0|^2q^2(\beta^2\sin^2\theta+\cos^2\theta)+4|\boldsymbol{\varepsilon}_0|^4\beta^2\sin^2\theta\cos^2\theta-q^4](1+q^2-2|\boldsymbol{\varepsilon}_0|^2\beta^2\sin^2\theta)^{-1}. \tag{45}$$

由于 $\beta^2\ll 1$，足够大的 pump 波能量能使分母大于 0. 当 $|\boldsymbol{\varepsilon}_0|^2\ll 1$ 时，极大增长率

$$\gamma_{max}(\theta)=|\boldsymbol{\varepsilon}_0|^2(\beta^2\sin^2\theta+\cos^2\theta), \tag{46}$$

相应的波数

$$q_{max}(\theta)=(|\boldsymbol{\varepsilon}_0|^2)^{1/2}(\beta^2\sin^2\theta+\cos^2\theta)^{1/2}. \tag{47}$$

这时(45)式退化为文献[10]的结果．显然，$\theta=0$ 时 $\gamma_m(0)$ 最大，这时(46)，(47)式退化为文献[8]的无磁场的结果．

这里，我们看到 MI 发展结果形成了类似"铁饼"状空间结构，$\boldsymbol{q}/\!/\boldsymbol{k}_0$ 发展最快，至 $\boldsymbol{q}\perp\boldsymbol{k}_0$ 时几乎停止增长．

2．l_0-t 激发，即纵 pump 波激发偏振方向垂直 $\boldsymbol{q}$ 的横波．考虑到 $\alpha^2\gg1$，(36)式中略去 γ 的高阶项后，得到如下的二次方程：

$$\gamma^2\left[q^4(1+\alpha^4)+\alpha^4q^6(1+\chi_0\xi_0)-2|\boldsymbol{\varepsilon}_0|^2q^2\left(\beta^2C_\alpha^2+\frac{q^2}{3}S_\alpha^2+S_\alpha^2\right)\right.$$
$$\left.-4|\boldsymbol{\varepsilon}_0|^4\beta^2\sin^2\theta\cos^2\theta\right]+\gamma\left[\alpha^4q^6(\chi_0+\xi_0q^2)-2|\boldsymbol{\varepsilon}_0|^2q^4\left(\frac{\xi_0}{3}q^2S_\alpha^2+\chi_0\alpha^2C_\alpha^2\right)\right]$$
$$-2|\boldsymbol{\varepsilon}_0|^2q^6\left(\frac{S_\alpha^2}{3}+\alpha^2C_\alpha^2\right)+\alpha^4q^8+\frac{4}{3}|\boldsymbol{\varepsilon}_0|^4q^4(\alpha^2\sin^2\theta\cos^2\theta+\cos^4\theta$$
$$+\sin^4\theta)=0. \tag{48}$$

(48)式有增长解的阈值为

$$|\boldsymbol{\varepsilon}_0|_s^2=\frac{3}{4}q^2\left(\frac{S_\alpha^2}{3}+\alpha^2C_\alpha^2\right)(\alpha^2\sin^2\theta\cos^2\theta+\cos^4\theta+\sin^4\theta)^{-1}$$
$$-\frac{3}{4}q^2\left[\left(\frac{S_\alpha^2}{3}+\alpha^2C_\alpha^2\right)^2(\alpha^2\sin^2\theta\cos^2\theta+\cos^4\theta+\sin^4\theta)^{-2}\right.$$
$$\left.-\frac{4}{3}\alpha^4(\alpha^2\sin^2\theta\cos^2\theta+\cos^4\theta+\sin^4\theta)^{-1}\right]^{1/2}. \tag{49}$$

对于 $\theta=0-\frac{\pi}{2}$ 变化，很易证明，根号内总是大于零且阈值随 θ 增加而增加．由于 $\alpha^2\gg1$，$\beta^2\ll1$，在 $|\boldsymbol{\varepsilon}_0|^2$ 相当宽的变化范围内，γ^2 系数也总大于零，所以(48)式的 $\gamma>0$ 增长解总是存在．下面分两种情况讨论：

1) 当 $\theta\approx0$ 时，注意到 $\alpha^2\gg1$，$\beta^2\ll1$，写(48)式为

$$a\gamma^2+b\gamma+c=0, \tag{50}$$

其中

$$a=\alpha^4q^4(1+q^2)-\frac{(8+2q^2)}{3}q^2|\boldsymbol{\varepsilon}_0|^2 \quad (\text{已略去 } \chi_0\xi_0\ll1 \text{ 项}),$$
$$b=\alpha^4q^6(\chi_0+\xi_0q^2)-2|\boldsymbol{\varepsilon}_0|^2q^4\left(\chi_0\alpha^4+\frac{\xi_0}{3}q^2\right),$$
$$c=\frac{4}{3}|\boldsymbol{\varepsilon}_0|^4q^4-2|\boldsymbol{\varepsilon}_0|^2\alpha^4q^6+\alpha^4q^8. \tag{50'}$$

(50)式的阈值与 l_0-l 激发一样为

$$|\boldsymbol{\varepsilon}_0|_s^2\doteq\frac{q^2}{2}. \tag{51}$$

当 pump 波能量增加到 $|\boldsymbol{\varepsilon}_0|^2<\frac{3}{8}\alpha^4q^2$ 时，总有 $a>0$，$c<0$．因此，在相当宽的 pump 波能区内可以出现纯增长解．当阻尼不存在时，如 $|\boldsymbol{\varepsilon}_0|^2\ll\frac{3}{8}\alpha^4q^2$，增长率为

$$\gamma^2 \simeq \left[2|\boldsymbol{\varepsilon}_0|^2q^2 - q^4 - \frac{4}{3}\alpha^{-4}|\boldsymbol{\varepsilon}_0|^4\right](1+q^2)^{-1}, \tag{52}$$

极大增长率为

$$\gamma_{\max} = [2(1+|\boldsymbol{\varepsilon}_0|^2)(1+2a_0)^{1/2} - 2(1+2a_0)]^{1/2}(1+2a_0)^{-1/4}, \tag{53}$$

$$a_0 \equiv |\boldsymbol{\varepsilon}_0|^2 + \frac{2}{3}\alpha^{-4}|\boldsymbol{\varepsilon}_0|^4. \tag{53'}$$

相应于 $\gamma_{\max}$ 的波数为

$$q_{\max} \equiv [(1+2a_0)^{1/2} - 1]^{1/2}. \tag{54}$$

当 $|\boldsymbol{\varepsilon}_0|^2 \ll 1$ 时，

$$\gamma_{\max} = |\boldsymbol{\varepsilon}_0|^2, \tag{55}$$

$$q_{\max} = (|\boldsymbol{\varepsilon}_0|^2)^{1/2}. \tag{56}$$

这与文献[13]的结果相同，且与 $l_0 - l$ 激发相类似.

2) 当 $\theta \approx \frac{\pi}{2}$ 时，对 $\alpha^2 \gg 1$，$\beta^2 \ll 1$，(48) 式变为

$$\gamma^2\left[\alpha^4q^4(1+q^2) - \left(\frac{6+2q^2}{3}\right)\alpha^2q^4|\boldsymbol{\varepsilon}_0|^2\right] + \gamma\left[\alpha^4q^6(\chi_0+\xi_0q^2) - 2|\boldsymbol{\varepsilon}_0|^2\alpha^2q^4\left(\chi_0 + \frac{\xi_0}{3}q^2\right)\right] - \frac{8}{3}|\boldsymbol{\varepsilon}_0|^2\alpha^2q^6 + \alpha^4q^8 + \frac{4}{3}|\boldsymbol{\varepsilon}_0|^4q^4 = 0, \tag{57}$$

阈值为

$$|\boldsymbol{\varepsilon}_0|_s^2 = \frac{\alpha^2q^2}{2}. \tag{58}$$

把 (57) 式写成 (50) 式形式后，我们看到，增加 pump 波能量出现下列情况：由于在 $|\boldsymbol{\varepsilon}_0|^2 < \alpha^2/2$ 以前恒有 $a > 0$，仅当 $|\boldsymbol{\varepsilon}_0|^2$ 处于从 $\frac{1}{2}\alpha^2q^2$ 到 $\frac{3}{2}\alpha^2q^2$ 的窄的区域内，$c < 0$ 有 $\gamma > 0$ 增长解外，其余区域 $b^2/a < 4c$ 无纯增长解.

从(49)式和上述结果我们看到，与 $l_0 - l$ 激发一样，阈值随 θ 增加而变得很大. 无阻尼时，

$$\gamma^2 = \alpha^{-4}\left(\frac{8}{3}\alpha^2q^2|\boldsymbol{\varepsilon}_0|^2 - \alpha^4q^4 - \frac{4}{3}|\boldsymbol{\varepsilon}_0|^4\right)\left[1 - 2\alpha^{-2}|\boldsymbol{\varepsilon}_0|^2 + \left(1 - \frac{2}{3}\alpha^{-2}|\boldsymbol{\varepsilon}_0|^2\right)q^2\right]^{-1}. \tag{59}$$

极大增长率为

$$\gamma_{\max} = \left\{\frac{2d_0^2}{(1-2\alpha^{-2}|\boldsymbol{\varepsilon}_0|^2)}\left[\left(1 + \frac{4}{3}\frac{|\boldsymbol{\varepsilon}_0|^2}{d_0\alpha^2}\right)(1+2b_0)^{1/2} - (1+2b_0)\right]\right\}^{1/2}(1+2b_0)^{-1/4}, \tag{60}$$

$$b_0 = \frac{2}{3}\frac{|\boldsymbol{\varepsilon}_0|^2}{\alpha^2d_0^2}(2d_0 + \alpha^{-2}|\boldsymbol{\varepsilon}_0|^2),\quad d_0 = (1 - 2\alpha^{-2}|\boldsymbol{\varepsilon}_0|^2)\left(1 - \frac{2|\boldsymbol{\varepsilon}_0|^2}{3\alpha^2}\right)^{-1}, \tag{60'}$$

$$q_{\max} = d_0^{1/2}[(1+2b_0)^{1/2} - 1]^{1/2}. \tag{61}$$

当 $\alpha^{-2}|\boldsymbol{\varepsilon}_0|^2 \ll 1$ 时，

$$\gamma_{\max} = \frac{2}{3}\alpha^{-2}|\boldsymbol{\varepsilon}_0|^2, \tag{62}$$

$$q_{\max} = 2\alpha^{-1}\left(\frac{|\boldsymbol{\varepsilon}_0|^2}{3}\right)^{1/2}. \tag{63}$$

可见，与 l_0-l 激发一样，增长率随 θ 增加而变小. 至 $\theta=\pi/2$ 时，比 $\theta=0$ 小一个 $\alpha^{-2}=3\beta^2\ll 1$ 因子，而波长大了一个 α 因子.

同样，我们看到，l_0-t 激发具有与 l_0-l 类似的“铁饼”状结构.

关于 t_0 激发的情况类似 l_0 激发的讨论，只是把 $\cos^2\theta$ 与 $\sin^2\theta$ 互换一下，这里不再赘述.

三、守恒量和坍缩动力学

本节将先给出方程(22)，(23)的守恒量，然后讨论 MI 发展后期波包坍缩动力学问题. 这里我们将目前为止见到的文献中，只对非线性 Schrödinger 方程的坍缩讨论，推广到方程组(22)，(23)的情况. 从文献[2]得到拉氏密度函数

$$\mathscr{L} = \frac{1}{2}i\left(\boldsymbol{\varepsilon}^*\cdot\frac{\partial\boldsymbol{\varepsilon}}{\partial t}-\boldsymbol{\varepsilon}\cdot\frac{\partial\boldsymbol{\varepsilon}^*}{\partial t}\right)-\frac{\alpha^2}{2}|\nabla\times\boldsymbol{\varepsilon}|^2-|\nabla\cdot\boldsymbol{\varepsilon}|^2-\frac{\partial\varphi}{\partial t}|\boldsymbol{\varepsilon}|^2$$
$$+\frac{1}{2}\left[|\boldsymbol{\varepsilon}|^4+\beta^2|\boldsymbol{\varepsilon}\times\boldsymbol{\varepsilon}^*|^2+\left(\frac{\partial\varphi}{\partial t}\right)^2-(\nabla\varphi)^2\right], \tag{64}$$

其中 φ 在形式上相当于流体力学流势，即 $\boldsymbol{u}=-\nabla\varphi$；$\varphi$ 满足 $\frac{\partial\varphi}{\partial t}=n+|\boldsymbol{\varepsilon}_0|^2$. 从(64)式可得能量-动量张量密度

$$T_{\mu\nu}=\mathscr{L}\delta_{\mu\nu}-\Phi^{\gamma}_{,\mu}\frac{\partial\mathscr{L}}{\partial\Phi^{\gamma}_{,\nu}} \tag{65}$$

的具体形式. 这里

$$\Phi^{\gamma}=\Phi\left(\boldsymbol{\varepsilon},\boldsymbol{\varepsilon}^*;\frac{\partial\boldsymbol{\varepsilon}}{\partial t},\frac{\partial\boldsymbol{\varepsilon}^*}{\partial t};\frac{\partial\varphi}{\partial t},\nabla\varphi\right), \tag{66}$$

$$\Phi^{\gamma}_{,\mu}=\left(\frac{\partial\Phi^{\gamma}}{\partial x},\cdots\right). \tag{67}$$

当 $\mathscr{L}$ 平移不变时，则有守恒方程

$$\frac{\partial T_{\mu 0}}{\partial t}+\frac{\partial}{\partial x_k}T_{\mu k}=0\quad k=1,2,3;\ x_k=x,y,z. \tag{68}$$

上式取 $\mu=0$ 就是能量守恒方程；取 $\mu=k$ 时得线动量守恒方程

$$\frac{\partial\boldsymbol{p}}{\partial t}+\nabla\cdot\mathbf{T}=0, \tag{69}$$

其中

$$p_k=T_{k0}=\sum_j\frac{1}{2}i\left(\varepsilon_j\frac{\partial}{\partial x_k}\varepsilon_j^*-\varepsilon_j^*\frac{\partial}{\partial x_k}\varepsilon_j\right)+nu_k, \tag{70}$$

$$T_{ij}=\left(\frac{\partial\varepsilon_j}{\partial x_i}\nabla\cdot\boldsymbol{\varepsilon}^*+\text{c.c.}\right)+\alpha^2\left[\sum_k\frac{\partial\varepsilon_k}{\partial x_i}\left(\frac{\partial\varepsilon_k^*}{\partial x_j}-\frac{\partial\varepsilon_j^*}{\partial x_k}\right)+\text{c.c.}\right]$$

$$+u_iu_j-\frac{1}{2}\delta_{ij}\left\{\nabla\cdot(\boldsymbol{\varepsilon}\nabla\cdot\boldsymbol{\varepsilon}^*+\text{c. c.})+\beta^2|\boldsymbol{\varepsilon}\times\boldsymbol{\varepsilon}^*|^2+|\boldsymbol{\varepsilon}|^4-\left(\frac{\partial\varphi}{\partial t}\right)^2\right.$$
$$\left.+(\nabla\varphi)^2+\alpha^2|\nabla\times\boldsymbol{\varepsilon}|^2-\frac{\alpha^2}{2}[\boldsymbol{\varepsilon}\cdot(\nabla\times(\nabla\times\boldsymbol{\varepsilon}))+\text{c. c.}]\right\}\ (i,j\neq 0). \tag{71}$$

于是得线动量和能量守恒关系

$$\boldsymbol{P}=\int\boldsymbol{p}d^3\boldsymbol{x}=\text{const.}, \tag{72}$$

$$H=-\int T_{00}d^3x=\int\left\{\frac{\alpha^2}{2}|\nabla\times\boldsymbol{\varepsilon}|^2+|\nabla\cdot\boldsymbol{\varepsilon}|^2+n|\boldsymbol{\varepsilon}|^2-\frac{1}{2}[\beta^2|\boldsymbol{\varepsilon}\times\boldsymbol{\varepsilon}^*|^2\right.$$
$$\left.-n^2-u^2]\right\}d^3x=\text{const.}. \tag{73}$$

另外，从复场规范不变性得四度流

$$j_\mu=-i\left(\frac{\partial\mathscr{L}}{\partial\Phi^\gamma_{,\mu}}\Phi^\gamma-\frac{\partial\mathscr{L}}{\partial\Phi^{*\gamma}_{,\mu}}\Phi^{*\gamma}\right). \tag{74}$$

守恒方程为

$$\frac{\partial j_0}{\partial t}+\nabla\cdot\boldsymbol{j}=0. \tag{75}$$

这里

$$\boldsymbol{j}=i(\boldsymbol{\varepsilon}\nabla\cdot\boldsymbol{\varepsilon}^*-\boldsymbol{\varepsilon}^*\nabla\cdot\boldsymbol{\varepsilon})+i\alpha^2[\boldsymbol{\varepsilon}\times(\nabla\times\boldsymbol{\varepsilon}^*)-\boldsymbol{\varepsilon}^*\times(\nabla\times\boldsymbol{\varepsilon})], \tag{76}$$

$$j_0=|\boldsymbol{\varepsilon}|^2. \tag{77}$$

(75)式表示

$$\int j_0d^3x=\int|\boldsymbol{\varepsilon}|^2d^3\boldsymbol{x}=N=\text{const.}, \tag{78}$$

N 为准粒子数.

有了上述守恒量，就可以讨论方程组 (22)，(23) 中波包 $\boldsymbol{\varepsilon}$ (或 n) 随时间的坍缩情况. 定义波包的中心座标 $\langle\boldsymbol{x}\rangle$，平均$\langle\cdots\rangle$系按归一化的准粒子数密度分布 $|\boldsymbol{\varepsilon}|^2/N$ 作为权重进行，这样

$$\langle\boldsymbol{x}\rangle=\int\boldsymbol{x}\frac{|\boldsymbol{\varepsilon}|^2}{N}d^3\boldsymbol{x}. \tag{79}$$

波包的均方根宽度为 $\langle\delta\boldsymbol{x}^2\rangle^{1/2}$，其平方值

$$\langle\delta\boldsymbol{x}^2\rangle=\langle\boldsymbol{x}^2\rangle-\langle\boldsymbol{x}\rangle^2. \tag{80}$$

从(75)式得

$$\frac{\partial\langle\boldsymbol{x}\rangle}{\partial t}=\frac{\boldsymbol{J}}{N}=\text{const.}, \tag{81}$$

其中

$$\boldsymbol{J}\equiv\int\boldsymbol{j}d^3x. \tag{82}$$

对(80)式求二次时间导数得

$$\frac{\partial^2}{\partial t^2}\langle\delta\boldsymbol{x}^2\rangle=\frac{\partial^2}{\partial t^2}\langle\boldsymbol{x}^2\rangle-2\left(\frac{\boldsymbol{J}}{N}\right)^2, \tag{83}$$

其中

$$\frac{\partial^2}{\partial t^2}\langle\boldsymbol{x}^2\rangle=-\frac{\partial}{\partial t}\int\frac{\boldsymbol{x}^2}{N}\nabla\cdot\boldsymbol{j}d^3x. \tag{84}$$

下面为了方便，只讨论 Langmuir 波情况，这时 $\nabla\times\boldsymbol{\varepsilon}=0$，于是(70)式为

$$\boldsymbol{p}=\frac{1}{2}i[(\boldsymbol{\varepsilon}\cdot\nabla)\boldsymbol{\varepsilon}^*-(\boldsymbol{\varepsilon}^*\cdot\nabla)\boldsymbol{\varepsilon}]+n\boldsymbol{u}. \tag{85}$$

(85)式由(70)式把 $\boldsymbol{\varepsilon}$ 表示为势的梯度并交换算子 $\frac{\partial}{\partial x_j}\frac{\partial}{\partial x_k}$ 后而得到，对于 Langmuir 波，

$$\boldsymbol{j} = i\nabla\times(\boldsymbol{\varepsilon}\times\boldsymbol{\varepsilon}^*) + 2\boldsymbol{p} - 2n\boldsymbol{u}. \tag{86}$$

所以

$$\frac{\partial^2}{\partial t^2}\langle \boldsymbol{x}^2\rangle = \frac{4}{N}\int \mathrm{Tr}\mathbf{T}d^3x + 2\frac{\partial}{\partial t}G(t), \tag{87}$$

其中

$$G(t) = -\frac{2}{N}\int \boldsymbol{x}\cdot\boldsymbol{u}nd^3x, \tag{88}$$

$$\int \mathrm{Tr}\mathbf{T}d^3x = 2H - \frac{N}{2}[(D-2)I(t) + DF(t)], \tag{89}$$

$$H = \int\left\{|\nabla\cdot\boldsymbol{\varepsilon}|^2 + n|\boldsymbol{\varepsilon}|^2 - \frac{1}{2}[\beta^2|\boldsymbol{\varepsilon}\times\boldsymbol{\varepsilon}^*|^2 - n^2 - u^2]\right\}d^3x, \tag{90}$$

$$I(t) = \frac{1}{N}\int[\beta^2|\boldsymbol{\varepsilon}\times\boldsymbol{\varepsilon}^*|^2 - 2n|\boldsymbol{\varepsilon}|^2 - n^2]d^3x, \tag{91}$$

$$F(t) = \frac{1}{N}\int u^2d^3x, \tag{92}$$

D 为空间维数．在静态近似下，$\boldsymbol{u}=0$，$n=-|\boldsymbol{\varepsilon}|^2$，于是(89)式变为

$$\int \mathrm{Tr}\mathbf{T}d^3x = 2H_s - \frac{N}{2}(D-2)I_s(t), \tag{93}$$

其中

$$H_s = \frac{1}{2}\int[2|\nabla\cdot\boldsymbol{\varepsilon}|^2 - |\boldsymbol{\varepsilon}|^4 - \beta^2|\boldsymbol{\varepsilon}\times\boldsymbol{\varepsilon}^*|^2]d^3x, \tag{94}$$

$$I_s = \frac{1}{N}\int[|\boldsymbol{\varepsilon}|^4 + \beta^2|\boldsymbol{\varepsilon}\times\boldsymbol{\varepsilon}^*|^2]d^3x. \tag{95}$$

这时就退化为文献[10]中非线性 Schrödinger 方程的结果．

一般情况下，方程(22)，(23)中波包宽度平方值的变化由(83)式得

$$\langle\delta\boldsymbol{x}^2\rangle = A_0t^2 + B_0t + C_0 - 2\int_0^t dt'\int_0^{t'}[(D-2)I(t'') + DF(t'')]dt'' - 2\int_0^t G(t')dt', \tag{96}$$

其中 $A_0 = \frac{4H}{N} - \left(\frac{\boldsymbol{J}}{N}\right)^2$ 为运动积分常数，B_0，C_0 为积分常数．我们知道，在 MI 发展过程中亚声情况下 $n<0$．因此，如果 MI 发展后期场幅有了足够增大使非线性凝聚项超过空间弥散项的作用，即有

$$\int\left\{|\nabla\cdot\boldsymbol{\varepsilon}|^2 + n|\boldsymbol{\varepsilon}|^2 - \frac{1}{2}[\beta^2|\boldsymbol{\varepsilon}\times\boldsymbol{\varepsilon}^*|^2 - n^2]\right\}d^3x < 0, \tag{97}$$

于是

$$2H - \frac{N}{2}F < 0, \tag{98}$$

而

$$I(t) > 0, \tag{99}$$

所以(96)式等号右端

$$A_0t^2 - 2\int_0^t dt'\int_0^{t'}dt''[(D-2)I(t'') + DF(t'')] < 0 \quad (\text{对 } D\geqslant 2). \tag{100}$$

故 t 增大时，在有限的时间内 t 的二次方项起主导作用，对 $D \geqslant 2$，$\langle \delta \boldsymbol{x}^2 \rangle$ 收缩变得足够小，出现波包的坍缩运动．实际情况是，当 $\langle \delta \boldsymbol{x}^2 \rangle^{1/2}$ 接近几个 Debye 长度时，方程 (22)，(23)不再适用，必须由包含非线性 Landau 阻尼效应的方程(17)—(19) 来描述，这时阻尼项起主要作用，波能迅速耗散加热等离子体背景粒子．由于一般情况下 $\beta^2 \ll 1$，磁场的作用是不重要的，但是它的存在却起了加快坍缩过程的作用．

参　考　文　献

[1]　贺贤土，核聚变，**1**(1980)，245.
[2]　贺贤土，物理学报，**32**(1983)，325.
[3]　D. R. Nicholson and M. V. Goldman. *Phys. Fluids*, **21** (1978). 1766.
[4]　E. A. Кузнецов, *ЖЭТФ*, **66** (1974), 2037.
[5]　M. V. Goldman and D. R. Nicholson, *Phys. Rev. Lett.*, **41**(1978), 406.
[6]　A. A. Vedenov and L. I. Rudakov, *Sov. Phys. Doklady*, **9**(1965), 1073.
[7]　L. I. Rudakov and V. N. Tsytovich, *Phys. Reports*, **40C**(1978), 3.
[8]　S. G. Thornhill and D. ter Harr, *Phys. Reports*, **43C**(1978), 45.
[9]　M. Kono *et al.*, *Phys. Lett. A.*, **77**(1980), 27.
[10]　M. Kono *et al.*, *J. Plasma Physics*, **26**(1981), 123.
[11]　B. И. Захаров, *ЖЭТФ*, **62** (1972), 1745.
[12]　V. N. Tsytovich, Theory of Turbulent Plasma, Consultants Bureau, New York, (1977).

THE MODULATION INSTABILITY AND THE COLLAPSE PROCESS OF WAVE PACKET IN PLASMA

HE XIAN-TU
(*Institute of Atomic Energy, Academia Sinica*)

ABSTRACT

In this paper, the expressions of the high-frequency and low-frequency nonlinear currents that leads to strong turbulence formation in plasma were found by using Kinetic theory. According to the system of equations in the reference [2] which included pondermotive force, self-generated magnetic field and their damping effects, we generalized the modulation instability analysis given by Kono et al. and obtained the expressions of growth rate parameter etc. in various cases while the longitudinal and transverse waves were excited by Langmuir or transverse pump waves. Finally, we discussed the collapse dynamics of wave packet and extended the collapse argument on nonlinear Schrödinger equation to the case of the density and field coupling equation system.

大幅 Langmuir Soliton 及其 Liapunov 稳定性

刘之景 贺贤土
(中国科学技术大学近代物理系,合肥) (中国科学院原子能研究所,北京)

迄今为止,等离子体中的 Soliton 问题的研究,大部分是在弱非线性理论基础上进行的,例如文献[1,2]只考虑二阶高频场作用.文献[3,4]虽然讨论了强非线性和四阶高频场作用,但未给出这种情形下的孤立子解的形式. 本文目的是求解四阶高频场作用的 Langmuir 波拍频产生的 Soliton 的解析形式及其稳定性证明.

从 Vlasov-Maxwell 方程组出发,可以求得四阶场作用下的动力学方程组,在静态近似下,得到一维高阶非线性 Schrödinger 方程的无量纲形式[4]:

$$i\frac{\partial E}{\partial t}+\frac{\partial^2 E}{\partial x^2}=-E|E|^2(1-g|E|^2), \tag{1}$$

式中 $g=0,1$ 分别对应二阶场和四阶场的非线性 Schrödinger 方程.(1)式所用的无量纲关系为

$$E=\frac{\tilde{E}}{\sqrt{16\pi n_0(T_i+T_e)}},\quad t=\omega_{pe}t',\quad x=\sqrt{\frac{2}{3}}\frac{x'}{\lambda_{D_e}}. \tag{2}$$

电场 E' 包含慢变振幅 $\tilde{E}$ 和快变位相因子 $\exp(-i\omega_{pe}t')$ 两部分,而且 $\omega_{pe}\tilde{E}\gg\frac{\partial\tilde{E}}{\partial t}$.

在低频相速度远小于离子热速度的静态近似下,无量纲密度为

$$n(x,t)=-2|E|^2(1-r|E|^2). \tag{3}$$

这里 r 值取为

$$r=\begin{cases}3(2T_i+3T_e)/T_e & (\text{四阶场}),\\ 0 & (\text{二阶场}).\end{cases} \tag{4}$$

利用熟知的方法,可求得(1)式的稳态解为

$$E_0^2=\frac{4\alpha_0^2}{A\,\mathrm{ch}2\alpha_0\eta+1}, \tag{5}$$

式中

$$A=\left(1-\frac{16}{3}\alpha_0^2 g\right)^{1/2}>0,\quad \alpha_0^2=\frac{v^2}{4}-\Omega,\quad \eta=x-vt. \tag{6}$$

这里 v,Ω 分别为速度和频率.当 $\eta\to\pm\infty$ 时,$E_0^2\to 0$. 显然(5)为 Soliton 解.当 $g=0$ 时,(5)式退化为二阶场的非线性 Schrödinger 方程的解

$$E_0=\sqrt{2}\,\alpha_0 S\mathrm{ech}\alpha_0\eta. \tag{7}$$

四阶场作用下的 Soliton 的最大高度和宽度为

$$E_{0\max}^2\simeq 2\alpha_0^2\left(1+\frac{4}{3}g\alpha_0^2\right),\quad \Delta\simeq\frac{1}{2\alpha_0}\ln(17+12\sqrt{2})+\frac{2\sqrt{2}}{3}g\alpha_0. \tag{8}$$

有质动力使密度下陷,其最大值

本文 1983 年 5 月 30 日收到.

$$|n|_{\max} \simeq 4\alpha_0^2\left[1-\left(2r-\frac{4}{3}g\right)\alpha_0^2\right]. \tag{9}$$

对于二阶场的非线性 Schrödinger 方程,其相应量 ($g=0$, $r=0$) 为

$$E^2_{0(2)\max}=2\alpha_0^2,\quad \Delta_{(2)}=\frac{1}{2\alpha_0}\ln(17+12\sqrt{2}),\quad |n_{(2)}|_{\max}=4\alpha_0^2. \tag{10}$$

注意：宽度Δ是在 $E^2_{0\max}$ 下降一半时求得的. 所以四阶场的作用使 Soliton 振幅增大，波形加宽，密度下陷减弱.

下面我们来证明 Soliton 解(5)式的稳定性. 从物理上来说，关于稳定性的 Liapunov 理论就是指任何一种稳态都保持自由能最小. 构造的 Liapunov 泛函 L，它具有如下特征(参见文献[5]):

(i) 对于靠近原始态的扰动 f 来说，$L(f)$ 有上下界;(ii) L 是非递增的. 即 $dL/dt \leqslant 0$.

我们首先来寻求具有上述特征的 Liapunov 泛函,对于一个可逆系统，Liapunov 泛函可以由运动常数组成,而方程(1)的三个运动积分为

$$N=\int EE^*dx,\quad P=\frac{i}{2}\int\left(E\frac{\partial E^*}{\partial x}-C.\ C.\right)dx,\quad H=\int\left(\frac{\partial E}{\partial x}\frac{\partial E^*}{\partial x}-\frac{1}{2}(EE^*)^2+\frac{1}{3}(EE^*)^3\right)dx. \tag{11}$$

N、P、H. 分别表示粒子数、动量、能量守恒量. 于是 Liapunov 泛函可写为

$$L=L_1-L_0,$$

L_0 为零阶分布量,利用拉格朗日乘子法,可得

$$L_1=\left(\frac{v^2}{2}-\Omega\right)N-vP+H. \tag{12}$$

我们构造的 L，因为由运动常数组成，条件(ii)即 $dL/dt\leqslant 0$ 是显然满足的；对于条件(i)，还要求 L 对扰动的泛数存在上下界,上界可用通常的方法得到，对于下界，文献[5]证明了它存在的判据为

$$\frac{\partial}{\partial\alpha_0^2}\int dxE_0^2>0. \tag{13}$$

对于四阶场的 Soliton 解(5)式,求得

$$\int_{-\infty}^{+\infty}dxE_0^2=\sqrt{3}\ln\frac{a+1}{a-1}, \tag{14}$$

式中
$$a=\left(\frac{1+A}{1-A}\right)^{1/2}.$$

于是
$$\frac{\partial}{\partial\alpha_0^2}\int dxE_0^2=\frac{2}{\alpha_0\left(1-\frac{16}{3}\alpha_0^2\right)}>0,$$

从而证明了四阶场 Langmuir 波拍频产生的 Soliton 解是稳定的.

参 考 文 献

[1] Zakharov, V. E., *Zh. Eksp. Teor. Fiz.*, **62** (1972). 1745 [Sov. Phys. ——JETP, **35** (1972), 908].
[2] 贺贤土,物理学报, **32**(1983), 325.
[3] Schamel, H., Yu, M. Y. & Sukla, P. K., *Phys. Fluids*, **20** (1977), 8: 1286.
[4] 贺贤土,物理学报, **31**(1982), 10: 317.
[5] Leadke, E. W. & Spatschek, K. H., *Phys. Fluids*, **23** (1980), 1: 44.

Vol. 29 No. 11 KEXUE TONGBAO November 1984

DYNAMICAL BEHAVIOR OF THE PHOTONS IN HIGH TEMPERATURE PLASMA

HE XIANTU (贺贤土), SHEN NAIWAN (沈乃万),
LIU JUN (刘 钧) AND ZHANG TIANSHU (张天树)
(*Institute of Applied Physics and Computational Mathematics, Beijing*)
Received April 22, 1983.

I. INTRODUCTION

The photons play quite an important role in the relaxation process which takes place in the fully ionized system at high temperature. Because the photons interacting with the electrons almost dominate the evolution of the system, it is very important to study the behavior of the photons and its role in the process of tending to equilibrium. However, in order to study this process in detail, it is necessary to solve the relativistic, multi-group and nonlinear integral-differential equation of the photons. But this is a complex and time-consuming procedure. To simplify the equation, Cooper expanded the scattering integral into Fokker-Planck form (abbreviated as F—P) and studied it numerically in 1971[1]. However, the accurate calculation of relativistic second order moment of the photons energy change due to the Compton scattering indicates that in low and high energy region the results obtained by accurate computation are different from Cooper's results (accurate to the same order of F—P approximation). At the same time, Cooper compared his F—P result with that obtained by Monte-Carlo method. He also found that in high electron temperature region the F—P approximation method is not valid. We notice that Cooper only treated pure Compton scattering problem. Thus, the equilibrium solution of the photons equation is only B—E distribution with an arbitrary constant. However, in an actual system, a number of photons are emitted continually in bremsstrahlung process. Therefore the number of photons are not a constant any longer. The unique equilibrium solution of a photon equation including bremsstrahlung radiation term is only the Planck distrbution.

To improve these problems, we have solved directly a relativistic photon transport equation including bremsstrahlung term by numerical method.

II. THE BASIC EQUATION

The basic equation is

$$\frac{df_\nu(t)}{dt} = \int_0^\infty \left[\exp\left(\frac{\nu'-\nu}{T_e}\right) f_{\nu'}(t)(1+f_\nu(t)) - f_\nu(t)(1+f_{\nu'}(t))\right] S(\nu,\nu',T_e)d\nu' + B_\nu(t), \quad (1$$

where ν is the frequency of the photon (in unit of energy), $f_\nu(t)$ is the photons number in one quantum state. Both $f_\nu(t)$ and $B_\nu(t)$ are the functions of the frequency ν and the time t;

$$B_\nu(t) = ck'_\nu(f_{\nu p}(T_e) - f_\nu(t)) \tag{2}$$

is not the rate of the bremsstrahlung radiation. The k'_ν in Eq. (2) is the effective absorption coefficient, c is the velocity of photon

$$f_{\nu p}(T_e) = \left(\exp\left(\frac{\nu}{T_e}\right) - 1\right)^{-1}, \tag{3}$$

where T_e is the electron temperature (in unit of energy). The scattering kernel $S(\nu, \nu', T_e)$ in Eq. (1) is

$$S(\nu, \nu', T_e) = 2\pi C N_e \int_{-1}^{1} \left\langle \frac{d^3\sigma}{d\nu' d\Omega'} \right\rangle d\mu', \tag{4}$$

where N_e and $\frac{d^3\sigma}{d\nu' d\Omega'}$ are the free electron number density and the scattering differential cross section in the laboratory frame of reference, respectively. The latter is obtained by the Lorentz transformation from the electron rest frame to the laboratory frame. $\langle \cdots \rangle$ denotes the average over the relativistic Maxwellian distribution. By the complex operating[2], one can get

$$\left\langle \frac{d^3\sigma}{d\nu' d\Omega'} \right\rangle = \frac{m_e^2 r_0^2 e^{-X_{\min}}}{\nu \sqrt{y_1} N_e(T_e)} \int_0^\infty d\zeta e^{-\zeta} \int_{-1}^{1} \frac{\Delta}{\sqrt{1-\xi^2}} d\xi, \tag{5}$$

where m_e, r_0 are the electron rest mass and the classical electron radius, respectively.

$$N_e(T_e) = 4\pi \int_0^\infty e^{-X} \beta E^2 dx, \tag{6}$$

$$\beta = P/E, \tag{7}$$

where P, E are the momentum and the energy of a electron, respectively.

$$\Delta = 2 + \frac{\rho^2}{1-\rho}\left(1 - 2\omega + \frac{\rho^2\omega^2}{1-\rho}\right), \tag{8}$$

$$\rho = \frac{\nu'(1-\mu)}{E(1-\beta\mu_a)}, \qquad \omega = \frac{m_e^2}{\nu\nu'(1-\mu')}, \tag{9}$$

where μ, μ' are the cosine of a scattering angle of the incident and the scattered photons respectively μ_a is the cosine of the angle between incident photon and incident electron. $X_{\min}$ in Eq. (5) is defined as

$$X_{\min} = \frac{1}{2T_e}\left\{\nu'\left[1 + \sqrt{y_1(2\omega+1)}\right] - 2m_e - \nu\right\}, \tag{10}$$

where y_1 is

$$y_1 = \left(\frac{\nu}{\nu'}\right)^2 + 1 - 2\left(\frac{\nu}{\nu'}\right)\mu'. \tag{11}$$

The Compton and bremsstrahlung energy loss rates are defined by Eq. (1), i. e.

$$A\int(1)\nu^3 d\nu = \frac{dE_R}{dt} = \dot{W}_C + \dot{W}_B, \tag{12}$$

where A is a constant that depends on the dimension.

III. The Difference Scheme

Eq. (1) can be rewritten as

$$\frac{df_g}{dt} = \sum_{l=1}^{K} G_{gl} + y_g(f_{gp} - f_g), \tag{13}$$

where K is the total group number,

$$G_{gl} = \left[\exp\left(\frac{\Delta\nu_l}{T_e}\right) f_l(1 + f_g) - f_g(1 + f_l)\right] S_{gl}\Delta\nu_l \tag{14}$$

$$S_{gl} = \frac{1}{\Delta\nu_g \Delta\nu_l}\int_{\Delta\nu_g}\int_{\Delta\nu_l} S(\nu, \nu', T_e) d\nu' d\nu \tag{15}$$

$$y_g = N_e K'_\nu,\ g,\ l = 1,\ 2, \cdots\cdots K, \tag{16}$$

with

$$\Delta\nu_g = \nu_{g+1} - \nu_g,\ \Delta\nu_l = \nu_{l+1} - \nu_l.$$

In numerical computation, we used the semi-implicit and semi-explicit method, i.e. the f_g in G_{gl} uses the value at the nth rank, while the f_g in the bremsstrahlung term uses the value at the $(n + 1)$th rank. Thus Eq. (13) can be written as

$$f_g^{n+1} = \left\{f_g^n + \left[\sum_{l=1}^{K-1} \frac{1}{2}\,(G_{gl}^n + G_{gl+1}^n) + y_g f_{gp}\right] \Delta t\right\} \Big/ (1 + y_g \Delta t).$$

IV. The Main Results and Simple Analysis

In the numerical calculation, the photons energy and the cosine of a scattering angle are divided into 110 and 40 intervals, respectively. The initial distribution is given by the Planck function in terms of the effective photon temperature T_r (considering T_r as a parameter). We have assumed T_e to be constant and $T_e \gg T_r$. The minimum time step used is $10^{-8}\mu s$. The error of the energy conservation is less than 0.5%. We computed the relaxation behavior of the photons in a pure deuterium system at different densities and temperature in detail. The main results obtained can be summarized as follows.

(1) *f_ν as the function of ν* (Fig.1). When $T_e > 10$ keV, the low frequency part of the f_ν first approaches the equilibrium distribution by the inverse-brems strahlung absorption (absorption coefficient $\sim \nu^{-3}$), then the high frequency part of the f_ν approaches the equilibrium distribution by the Compton scattering through which the runaway photons from the bremsstrahlung absorption process are transformed into the high frequency photons. For the intermediate frequency $(\nu < T_e)$, f_ν approaches the equilibrium distribution at last. Therefore, in this frequency region, the profile of the energy spectrum displays an apparent concavity. The computation results show that the concavity removes continuously toward lower frequency region

with increasing time, and becomes narrower and narrower until it disappears at last. By this time, in entire frequency region, the photon energy distribution is a Planck function in terms of the electron temperature.

Clearly, with decreasing T_e the concavity appears in lower frequency region, when $T_e < 10$ keV it has fallen into a frequency region where the bremsstrahlung absorption are the most important. In this case the concavity disappears completely. The reason for forming the concavity may be further explained by investigating pure bremsstrahlung spectrum and pure Compton-transform spectrum. For pure bremsstrahlung spectrum (excluding Compton process) its low frequency part rapidly approaches equilibrium spectrum by the strong absorption, but in somewhat higher frequency region the absorption becomes weak rapidly (because $K'_\nu \sim \nu^{-3}$), therefore, its higher frequency part (as ν increases, it decreases rapidly, Fig. 1 c) may be far from the equilibrium spectrum in longer time. For pure Compton process (excluding bremsstrahlung process), the photon number in the system does not change, the Compton scattering only transforms low ν photons into high ν photons. Therefore, the Compton transforming spectrum tends to the flatness, thus its high frequency part rapidly approaches equilibrium distribution (Fig. 1d). When two processes proceed simultaneity, in intermediate frequency region the f_ν then displays the big concavity (Fig. 1e).

(2) *Sub-steady state "plateau"*. When $T_e > 10$ keV, for the different electron number density $\dot{W}_C$ and $\dot{W}_B$ nearly do not depend on time in longer time, thus the so-called "sub-steady state plateau" of $\dot{W}_C$ and W_B appears. This phenomenon may be explained very well by the $f_\nu \sim \nu$ relation in (1). Because the big concavity appears in $\nu \leqslant T_e$ region, in this frequency region f_ν is far from $f_{\nu p}$ (T_e). Thus the $\dot{W}_B$ is nearly determined by the spontaneous bremsstrahlung radiation of the electrons, (spontaneous radiation rate $\propto T_e$) therefore, the $\dot{W}_B$ hardly vary with time before the concavity disappears. While for $\dot{W}_C$, as the background photons and the runaway photons in the bremsstrahlung absorption process are very few in the early period, therefore, as photon number increases, it has a rising front edge (time $\approx t$), since the photons are produced in the unvaried bremsstrahlung radiation rate, the net Compton scattering rate by which the photons escaping absorption with the fixed speed are transformed, tending quickly to the kinetic balance, then the $\dot{W}_C$ displays a "plateau". When f_ν approaches $f_{\nu p}$ (T_e), the "plateau" disappears. As in the case described in (1), when $T_e < 10$ keV the "plateau" disappears, too.

When $N_e \approx 4 \times 10^{24}/\text{cm}^3$; the computation results show that for 10keV $< T_e$ $<$100keV $\dot{W}_C$ is larger than $\dot{W}_B$ by several to above ten times. In this case the time of the rising front edge is about 0.8 ns and hardly vary with T_e, but the relaxation time rapidly varies with T_e. These phenomena have very important meaning to investigating the relaxation process of a system.From above, we may reach a conclusion that even though without the leakage of photons, the deuteriun system remains in nonequilibrium state with longer time as well, showing that for the inertia confinement DT pellet compressed by a laser beams and for similar other thermonuclear system, by using the so-called three-temperature method, $\dot{W}_C$ obtained is too large and the relaxation time is too short. Thus the electron and ion temperature becomes

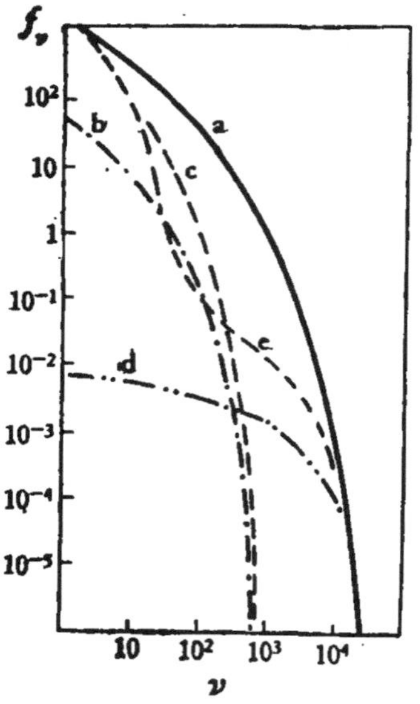

Fig. 1. Diagram illustrating schematically the relation between f_ν and ν.
a—the equilibrium distribution $f_{\nu p}(T_e)$;
b—the initial distribution $f_{\nu p}(T_r)$;
c—the spectrum distribution for pure bremsstrahlung radiation;
d—the spectrum distribution for pure Compton scattering case;
e—the non-equilibrium distribution at some time including both bremsstrahlung and Computon scattering processes.
ν in constant × erg.

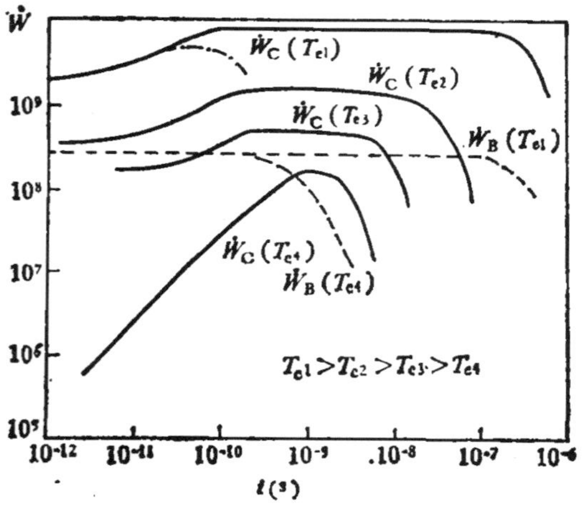

Fig. 2. Schematic curves showing the dependence of $\dot{W}_C$ and $\dot{W}_B$ on a time t. ($\dot{W}$ in constant × erg/(se×cm³)). Th dot-dash line is $\dot{W}_C$ (T_{e1}) due to pure Compton process.

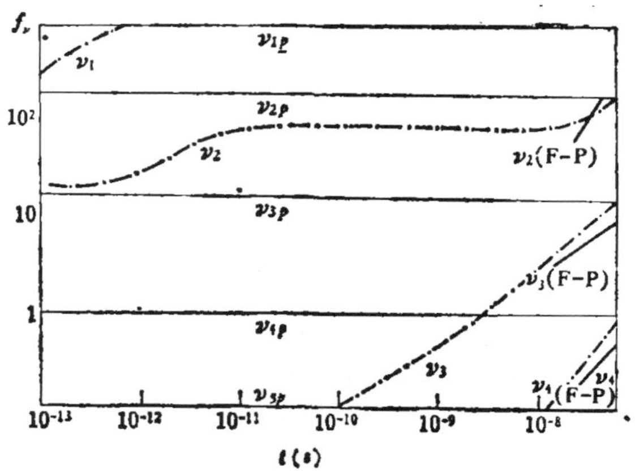

Fig. 3. Graph showing relaxation time of photons in different frequencies.
The horizontal lines denote the equilibrium distribution at the frequency ν_{ip} (subscript i denotes the certain group number and subscript p means "equilibrium"). The dot-dash lines show how f_ν at the ν_{ip} varies with time, and the solid lines denote corresponding results from F—P approximation method.

evidently low in the thermonuclear burn process.

In low and high frequency regions, f_ν obtained here is obviously different from that obtained by using multi-group F—P approximation. (Fig. 3).

This shows that in F—P approximation it is not enough to have only $\Delta\nu^2$ terms to achieve a good result. However, for $\dot{W}_C$, etc. quantities integrate over ν, because the energy contribution of the photons is not essential in the two ends of frequency region, the results obtained by accurate computation are very close to the results obtained by approximation method.

References

[1] Cooper, G. E., *Phys. Rev.*, **D3** (1971), 10:2312.
[2] Stone, S. & Nelson, R. G., UCRL—14918—T, 1966.

第35卷 第3期　　物 理 学 报　　Vol. 35, No. 3
1986年3月　　ACTA PHYSICA SINICA　　Mar., 1986

高频波激发低频磁场和离子声波的重整化强湍动理论

贺 贤 土
(应用物理与计算数学研究所，北京)

1983年10月17日收到；1985年1月22日收到修改稿

提 要

本文给出了高温等离子体中高频波激发低频磁场和离子声波强湍动过程的重整化理论，以便改善通常的弱非线性处理方法．从 Vlasov-Maxwell 方程组出发，在 Fourier 表象中得到了包含"最发散"和"次发散"效应互相耦合的高频和低频传播子重整化方程组，从而获得了高、低频振荡粒子重整化分布函数和场的耦合关系．

在"最发散"重整化近似下，我们求解了高低频传播子方程组，得到了展开到 ν_e（高频湍动场能密度与等离子体热能密度之比）一次方的近似解和重整化介电函数等表达式．然后，在 Fourier 逆变换下导得了我们所要的时空表象中重整化强湍动方程组．

最后，作为一个说明重整化作用的例子，在一维稳态下求解了孤立子的形式．

在文献[1, 2]中，从微观动力学理论而不是传统的流体动力学框架出发，我们导得了纵、横高频波拍频激发离子声波和低频磁场波，并包含自洽的非线性 Landau 阻尼的强湍动方程组，同时讨论了它们的一些重要特性，从而在较为普遍情况下推广了 Thornhill 和 Haar[3] 以及 Rudakov 和 Tsytovich[4] 等人于1978年提出的 Langmuir 波强湍动理论．

但遗憾的是上述所有结果都是建立在二次高频场拍频为低频波的弱非线性基础上进行．为了改善这种结果，我们曾讨论过四次高频场的拍频修正[5]，表明了四次高频场可能削弱二次非线性场能在局部空间的凝聚作用．为了使强湍动理论包含更高次的非线性效应．本文将从 Vlasov-Maxwell 方程组出发尝试用重整化方法去导得高频湍动场和低频波相互作用相干性的动力学方程组，把 Horton 等人[6]研究 Vlasov-Poisson 方程组的重整化方案推广应用到我们研究的问题上来，即把高次微扰久期项中"最发散"和"次发散"图形收集在一起，得到一组包含湍动碰撞频率的非线性传播子方程．接着把重整化后的粒子分布函数和传播子方程分成高、低频耦合关系，并在"最发散"近似下求解了高、低频传播子方程组，得到了展开到 ν_e 一次方项的高、低频重整化传播子方程近似解和相应介电函数表达式．

对有关方程作 Fourier 逆变换后，便可得到时空表象中高频电场、低频自生磁场、离子声波相互耦合的重整化强湍动方程组．最后还求得了方程组的孤立子形式解．

一、Vlasov-Maxwell 方程组的重整化形式

如果以电场 $\boldsymbol{E}(k)=\boldsymbol{e}_k^{\sigma}E^{\sigma}(k)$ 取代文献[6]中 Poisson 场 $\boldsymbol{k}\phi_k$，就可得到我们情况下的重整化传播子 $G_\alpha(k,\boldsymbol{v})$ 所服从的方程

$$G_\alpha(k,\boldsymbol{v})=\{\Theta-\boldsymbol{k}\cdot\boldsymbol{v}+i\Gamma_\alpha(k,\boldsymbol{v})\}^{-1}. \tag{1}$$

这里

$$\Gamma_\alpha(k,\boldsymbol{v})=-i\left(\frac{e_\alpha}{m_\alpha}\right)^2\sum_{k_1}|E^{\sigma}(k_1)|^2\frac{\partial}{\partial\boldsymbol{v}}\cdot\boldsymbol{e}_{k_1}^{\prime\sigma}G_\alpha(k-k_1,\boldsymbol{v})\boldsymbol{e}_{k_1}^{\sigma}\cdot\frac{\partial}{\partial\boldsymbol{v}} \tag{2}$$

为湍动碰撞频率算子；$k\equiv\{\Theta,\boldsymbol{k}\}$ 为频率 Θ 和波矢 $\boldsymbol{k}$ 的四维矢量；m_α，e_α 分别为粒子质量和电荷，$\alpha=i$，e；ϕ_k 为静电势；σ 为波型符号；重复角标表示求和.

$$\boldsymbol{e}_k^{\prime\sigma}\equiv\left(1-\frac{\boldsymbol{k}\cdot\boldsymbol{v}}{\Theta}\right)\boldsymbol{e}_k^{\sigma}+\frac{\boldsymbol{e}_k^{\sigma}\cdot\boldsymbol{v}}{\Theta}\boldsymbol{k}, \tag{3}$$

$\boldsymbol{e}_k^{\sigma}$ 为 σ 波单位偏振矢量. 显然 $\boldsymbol{e}_k^{\prime\sigma}\cdot\boldsymbol{v}=\boldsymbol{e}_k^{\sigma}\cdot\boldsymbol{v}$.

方程(1)只包含了最重要的"最发散"和"次发散"的久期图形求和部分. 有了重整化传播子，就可写出重整化粒子分布函数和场方程. 我们感兴趣具有长波长湍动特性的高频波拍频为低频波过程，这时高频波(包括纵波和横波)的频率都接近电子等离子体频率，于是从 Vlasov-Maxwell 方程组取到二次高频场拍频近似就可得到高、低频耦合的重整化关系式如下：

高频振荡粒子分布函数

$$\begin{aligned}f_\alpha(k,\boldsymbol{v})=&-i\frac{e_\alpha}{m_\alpha}G_\alpha(k,\boldsymbol{v})\Big\{E_h^{\sigma}(k)\boldsymbol{e}_k^{\prime}\cdot\partial/\partial\boldsymbol{v}F_{0\alpha}\\&+\int\boldsymbol{e}_1^{\prime}\cdot\partial/\partial\boldsymbol{v}\delta F_\alpha(q,\boldsymbol{v})E_h^{\sigma_1}(k_1)\,\delta^4(k-k_1-q)\frac{d^4k_1d^4q}{(2\pi)^4}\\&-i\frac{e_\alpha}{m_\alpha}\int\boldsymbol{e}_q^{\prime}\cdot\frac{\partial}{\partial\boldsymbol{v}}G_\alpha(k_1,\boldsymbol{v})\boldsymbol{e}_1^{\prime}\cdot\frac{\partial}{\partial\boldsymbol{v}}F_{0\alpha}(\boldsymbol{v})\,E_h^{\sigma_1}(k_1)\\&\times E_d^{\beta}(q)\delta^4(k-k_1-q)\frac{d^4k_1d^4q}{(2\pi)^4}\Big\},\end{aligned} \tag{4}$$

低频振荡粒子分布函数

$$\begin{aligned}\delta F_\alpha(q,\boldsymbol{v})=&-i\frac{e_\alpha}{m_\alpha}G_\alpha(q,\boldsymbol{v})\Big\{E_d^{\beta}(q)\boldsymbol{e}_q^{\prime}\cdot\frac{\partial}{\partial\boldsymbol{v}}F_{0\alpha}-i\frac{e_\alpha}{m_\alpha}\\&\times\int_{(d)}\boldsymbol{e}_1^{\prime}\cdot\frac{\partial}{\partial\boldsymbol{v}}G_\alpha(k_2,\boldsymbol{v})\boldsymbol{e}_2^{\prime}\cdot\frac{\partial}{\partial\boldsymbol{v}}F_{0\alpha}E_h^{\sigma_1}(k_1)E_h^{\sigma_2}(k_2)\\&\times\delta^4(q-k_1-k_2)\frac{d^4k_1d^4k_2}{(2\pi)^4}\Big\},\end{aligned} \tag{5}$$

高频传播子方程

$$\left\{\omega-\boldsymbol{k}\cdot\boldsymbol{v}-i\frac{\partial}{\partial\boldsymbol{v}}\cdot\mathbf{D}_\alpha(k,\boldsymbol{v})\cdot\frac{\partial}{\partial\boldsymbol{v}}\right\}G_\alpha(k,\boldsymbol{v})=1. \tag{6}$$

这里

$$\mathbf{D}_\alpha(k,\boldsymbol{v})=i\left(\frac{e_\alpha}{m_\alpha}\right)^2\int \boldsymbol{e}_1'\boldsymbol{e}_1'|E_h^{\sigma_1}(k_1)|^2 G_\alpha(q,\boldsymbol{v})\delta^4(k-k_1-q)\frac{d^4k_1d^4q}{(2\pi)^4} \tag{7}$$

与均匀稳态高频湍动谱有关．(7)式中已用了四维盒子体积趋向∞时 $\sum_k \to \int\cdots\frac{d^4k}{(2\pi)^4}$ 关系式．

低频传播子方程

$$\left\{\Omega-\boldsymbol{q}\cdot\boldsymbol{v}-i\frac{\partial}{\partial\boldsymbol{v}}\cdot\mathbf{D}_\alpha(q,\boldsymbol{v})\cdot\frac{\partial}{\partial\boldsymbol{v}}\right\}G_\alpha(q,\boldsymbol{v})=1, \tag{8}$$

而

$$\mathbf{D}_\alpha(q,\boldsymbol{v})=i\left(\frac{e_\alpha}{m_\alpha}\right)^2\int_{(d)}\boldsymbol{e}_1'\boldsymbol{e}_1'|E_h^{\sigma_1}(k_1)|^2G_\alpha(k_2,\boldsymbol{v})\delta^4(q-k_1-k_2)\frac{d^4k_1d^4k_2}{(2\pi)^4} \tag{9}$$

同样与均匀稳态高频湍动谱有关．

在方程组(5)—(9)中，$F_{0\alpha}$ 为背景分布函数，以后取为温度为 T_α 的 Maxwell 分布，$\int F_{0\alpha}(\boldsymbol{v})d^3v=n_0$，$n_0$ 为背景粒子数密度；$k\equiv\{\omega,\boldsymbol{k}\}$ 和 $q\equiv\{\Omega,\boldsymbol{q}\}$ 分别表示高低频)波的四维波矢，并记 $d^4k\equiv d\omega d^3k$，$\boldsymbol{e}_1'\equiv\boldsymbol{e}_{k_1}'^{\sigma}$，$\boldsymbol{e}_q'\equiv\boldsymbol{e}_q'^{\beta}$ 等等．方程组中，$E_h^\sigma(k)$ 和 $E_d^\beta(q$ 分别表示高低频标量电场；积分号 $\int_{(d)}$ 表示积分只作用于高频波拍频成为低频波过程，即积分只取 $\{-k_1,+k_2\}$ 和 $\{+k_1,-k_2\}$两组情况．

现在通过重整化电流从电磁场 Maxwell 方程组就可得到高频场方程如下：

$$\varepsilon_h^\sigma(k)E_h^\sigma(k)=S_1^\sigma(k)+S_2^\sigma(k), \tag{10}$$

其中

$$S_1^\sigma(k)=-\sum_\alpha\frac{\omega_\alpha^2}{n_0\omega}\int \boldsymbol{e}_k\cdot\boldsymbol{v}G_\alpha(k,\boldsymbol{v})\boldsymbol{e}_1'\cdot\frac{\partial}{\partial\boldsymbol{v}}\delta F_\alpha(q,\boldsymbol{v})E_h^{\sigma_1}(k_1)$$
$$\times\delta^4(k-k_1-q)\frac{d^4k_1d^4q}{(2\pi)^4}d^3v, \tag{11}$$

$$S_2^\sigma(k)=i\sum_\alpha\frac{\omega_\alpha^2}{n_0\omega}\frac{e_\alpha}{m_\alpha}\int H_\alpha^{\sigma\beta\sigma_1}(k,q,k_1)E_h^{\sigma_1}(k_1)E_d^\beta(q)$$
$$\times\delta^4(k-k_1-q)\frac{d^4k_1d^4q}{(2\pi)^4}, \tag{12}$$

这里

$$\omega_\alpha^2=\frac{4\pi n_0e_\alpha^2}{m_\alpha},$$

$$H_\alpha^{\sigma\beta\sigma_1}(k,q,k_1)=\int \boldsymbol{e}_k\cdot\boldsymbol{v}G_\alpha(k,\boldsymbol{v})\boldsymbol{e}_q'\cdot\frac{\partial}{\partial\boldsymbol{v}}G_\alpha(k_1,\boldsymbol{v})\boldsymbol{e}_1'$$
$$\cdot\frac{\partial}{\partial\boldsymbol{v}}F_{0\alpha}(\boldsymbol{v})d^3v. \tag{13}$$

低频电场方程为

$$\varepsilon_d^\beta(q)E_d^\beta(q)=i\sum_\alpha\frac{\omega_\alpha^2}{n_0\Omega}\frac{e_\alpha}{m_\alpha}\int_{(d)}H_\alpha^{\beta\sigma_1\sigma_2}(q,k_1,k_2)E_h^{\sigma_1}(k_1)E_h^{\sigma_2}(k_2)$$
$$\times\delta^4(q-k_1-k_2)\frac{d^4k_1d^4k_2}{(2\pi)^4}, \tag{14}$$

其中

$$H_\alpha^{\beta\sigma_1\sigma_2}(q, k_1, k_2) = \frac{1}{2}\int \boldsymbol{e}_q\cdot\boldsymbol{v}G_\alpha(q, \boldsymbol{v})\left[\boldsymbol{e}_1'\cdot\frac{\partial}{\partial\boldsymbol{v}}G_\alpha(k_2, \boldsymbol{v})\boldsymbol{e}_2'\cdot\frac{\partial}{\partial\boldsymbol{v}}F_{0\alpha} + 1\longleftrightarrow 2\right]d^3v. \tag{15}$$

交换下标 1 与 2 互为对称.

在(10)和(14)式中,重整化高低频色散响应函数分别为

$$\varepsilon_h^\sigma(k) = 1 - \frac{c^2k^2}{\omega^2}\delta_{\sigma t} + \sum_\alpha\frac{\omega_\alpha^2}{n_0\omega}\int \boldsymbol{e}_k\cdot\boldsymbol{v}G_\alpha(k, \boldsymbol{v})\,\boldsymbol{e}_k'\cdot\frac{\partial}{\partial\boldsymbol{v}}F_{0\alpha}(\boldsymbol{v})d^3v, \tag{16}$$

$$\varepsilon_d^\beta(q) = 1 - \frac{c^2q^2}{\Omega^2}\delta_{\beta t} + \sum_\alpha\frac{\omega_\alpha^2}{n_0\Omega}\int \boldsymbol{e}_q\cdot\boldsymbol{v}G_\alpha(q, \boldsymbol{v})\boldsymbol{e}_q'\cdot\frac{\partial}{\partial\boldsymbol{v}}F_{0\alpha}(\boldsymbol{v})d^3v. \tag{17}$$

这里 $\varepsilon_h^\sigma(k) + \frac{c^2k^2}{\omega^2}\delta_{\sigma t}$ 为介电函数, $\delta_{\sigma t} = \begin{cases}1 & \sigma = t\ (\text{横波}),\\ 0 & \sigma = l\ (\text{纵波}).\end{cases}$

最后,写出低频磁场方程

$$\boldsymbol{B}(q) = \frac{c}{\Omega}\boldsymbol{q}\times\boldsymbol{E}_d^t(q), \tag{18}$$

c 为光速.

(4)—(18)式一起构成了 Fourier 表象中重整化强湍动方程组. 显然, 当 G_α 变成自由传播子 $G_\alpha^{(0)}$ 时就退化为文献[1]中结果.

二、传播子方程的近似解

与一般的重整化结果不同,这里高低频传播子是互相耦合的非线性方程组,无法严格求解. 为简化问题,下面只研究"最发散"项组成的重整化解,这时张量系数 $\mathbf{D}$ 中的 G_α 变成自由传播子,从而(6)和(8)式变成互相独立的线性方程,于是对均匀稳态的高频湍动谱作些近似假定可望获得方程组的近似解.

下面只讨论高频场是 Langmuir 波情况,但是被激发的低频波可以是纵场和横场,后者将导致低频自生磁场的产生. 对于高频横场可仿此讨论.

1. 低频传播子方程的近似解

方程(8)的 Green 函数方程为

$$\left\{\Omega - \boldsymbol{q}\cdot\boldsymbol{v} - i\frac{\partial}{\partial\boldsymbol{v}}\cdot\mathbf{D}_\alpha(q, \boldsymbol{v})\cdot\frac{\partial}{\partial\boldsymbol{v}}\right\}G_\alpha(q, \boldsymbol{v}; \boldsymbol{v}') = \delta(\boldsymbol{v} - \boldsymbol{v}'). \tag{19}$$

利用(19)式的解可得算子作用关系式

$$G_\alpha(q, \boldsymbol{v})\Phi_\alpha(q, \boldsymbol{v}) = \int G_\alpha(q, \boldsymbol{v}; \boldsymbol{v}')\Phi_\alpha(q, \boldsymbol{v}')d^3v', \tag{20}$$

其中 $\Phi_\alpha(q, \boldsymbol{v})$ 表示(5)式右端大括号内量. 在"最发散"重整化条件下,张量扩散系数为

$$\mathbf{D}_\alpha(q, \boldsymbol{v}) = i\left(\frac{e_\alpha}{m_\alpha}\right)^2\int \boldsymbol{E}_+(k_1)\boldsymbol{E}_-^*(k_1)\left\{\frac{\delta^4(q + k_1 - k_2)}{\omega_2 - \boldsymbol{k}_2\cdot\boldsymbol{v} + i0_+}\right.$$

$$
-\frac{\delta^4(q-k_1+k_2)}{\omega_2-\boldsymbol{k}_2\cdot\boldsymbol{v}-i0_+}\Big\}\frac{d^4k_1d^4k_2}{(2\pi)^4}
$$

$$
\simeq -i\left(\frac{e_\alpha}{m_\alpha}\right)^2\frac{(\Omega-\boldsymbol{q}\cdot\boldsymbol{v})}{\omega^2}\int \boldsymbol{e}_1\boldsymbol{e}_1|E(k_1)|^2\frac{d^4k_1}{(2\pi)^4}. \tag{21}
$$

这里 $\Omega=\omega_1-\omega_2$，$\boldsymbol{q}=\boldsymbol{k}_1-\boldsymbol{k}_2$，$e_k'\to\boldsymbol{e}_k=\dfrac{\boldsymbol{k}}{|\boldsymbol{k}|}$；同时 $\boldsymbol{E}(k)=\boldsymbol{E}_+(k)+\boldsymbol{E}_-(k)$，$\boldsymbol{E}_\pm(k)\equiv\boldsymbol{E}(\pm k)$，显然 $\boldsymbol{E}(k)\boldsymbol{E}^*(k)=2\boldsymbol{E}_+(k)\boldsymbol{E}_\pm^*(k)$，Langmuir 波场的上、下角标已被省略；(21)式中已用了 $\omega_1\simeq\omega_e\gg\Omega$，$\boldsymbol{k}\cdot\boldsymbol{v}$ 近似.

注意到长波长区弱湍动高频谱在与粒子作用时所表现出来的各向同性化特性[3,7]，假定由调制不稳定性发展的强湍动早期高频湍动场能谱接近各向同性化分布，这样，

$$
\mathbf{D}_\alpha(q,\boldsymbol{v})\simeq -i\mathbf{I}(\Omega-\boldsymbol{q}\cdot\boldsymbol{v})a_\alpha, \tag{22}
$$

其中

$$
a_\alpha=v_\alpha^2\frac{\omega_\alpha^2}{\omega_e^2}\nu_\alpha,\quad \nu_\alpha=\int|E(k_1)|^2\frac{d^4k_1}{(2\pi)^4}\Big/4\pi n_0T_\alpha, \tag{23}
$$

$\mathbf{I}$ 为单位张量．于是，方程(19)简化为

$$
\Big\{\Omega-\boldsymbol{q}\cdot\boldsymbol{v}+a_\alpha\boldsymbol{q}\cdot\frac{\partial}{\partial\boldsymbol{v}}-a_\alpha(\Omega-\boldsymbol{q}\cdot\boldsymbol{v})\frac{\partial}{\partial\boldsymbol{v}}
\cdot\frac{\partial}{\partial\boldsymbol{v}}\Big\}G_\alpha(q,\boldsymbol{v};\boldsymbol{v}')=\delta(\boldsymbol{v}-\boldsymbol{v}'), \tag{24}
$$

(24)式是自伴的，所以有对易关系 $G_\alpha(q,\boldsymbol{v};\boldsymbol{v}')=G_\alpha(q,\boldsymbol{v}';\boldsymbol{v})$，并满足 $\boldsymbol{v},\boldsymbol{v}'\to\infty$ 时，$G_\alpha(q,\boldsymbol{v};\boldsymbol{v}')\to 0$ 的条件.

方程(24)的解为

$$
\begin{aligned}
G_\alpha(q,\boldsymbol{v};\boldsymbol{v}')=&\frac{-i}{(2\pi)^3|\boldsymbol{q}|}\int d\xi_\perp\int_{-\infty}^{\infty}d\xi_\parallel\int_0^\infty d\tau\exp\Big\{\frac{i}{|\boldsymbol{q}|}(\Omega-\boldsymbol{q}\cdot\boldsymbol{v}')\tau\\
&+i\xi_\perp(\boldsymbol{v}_\perp-\boldsymbol{v}_\perp')+i\xi_\parallel(v_\parallel-v_\parallel')-\frac{1}{2}\\
&\times\ln[1+a_\alpha(\xi_\perp^2+(\xi_\parallel+\tau)^2)]-\frac{1}{2}\ln[1+a_\alpha(\xi_\parallel^2+\xi_\perp^2)]\Big\}.
\end{aligned} \tag{25}
$$

对于 Maxwell 分布背景等离子体，只有 $|\boldsymbol{v}|\sim v_\alpha$ 的 G_α 对以后计算才是重要的．由于 $|\boldsymbol{\xi}|\sim 1/v_\alpha$，因此，当 $\nu_\alpha<1$ 时，主要贡献来自 $a_\alpha\xi^2<1$ 项．展开(25)式的对数部分，取到一级项积分后便得

$$
\begin{aligned}
G_\alpha(q,\boldsymbol{v};\boldsymbol{v}')=&\frac{-i}{8|\boldsymbol{q}|(\pi a_\alpha)^{3/2}}\int_0^\infty d\tau\exp\Big\{-\frac{a_\alpha}{4}\tau^2+\frac{i}{|\boldsymbol{q}|}\\
&\times\Big[\Omega-\frac{(\boldsymbol{v}+\boldsymbol{v}')}{2}\cdot\boldsymbol{q}\Big]\tau-\frac{(\boldsymbol{v}-\boldsymbol{v}')^2}{4a_\alpha}\Big\}.
\end{aligned} \tag{26}
$$

显然，$\nu_\alpha\to 0$ 时，自然退化到自由传播子 Green 函数解 $G_\alpha(q,\boldsymbol{v};\boldsymbol{v}')=\dfrac{\delta(\boldsymbol{v}-\boldsymbol{v}')}{\Omega-\boldsymbol{q}\cdot\boldsymbol{v}'}$.

下面给出几个有用的关系式：

$$
\int G_\alpha(q,\boldsymbol{v};\boldsymbol{v}')d^3v'=\frac{-i}{|\boldsymbol{q}|}\int_0^\infty d\tau\exp\Big\{\frac{-a_\alpha}{2}\tau^2+\frac{i}{|\boldsymbol{q}|}(\Omega-\boldsymbol{q}\cdot\boldsymbol{v})\tau\Big\}
$$

$$= G_{0\alpha}(q, \boldsymbol{v}) = G^{(0)}(q, \boldsymbol{v}) + \frac{a_\alpha q^2}{(\Omega - \boldsymbol{q}\cdot\boldsymbol{v})^3} + o(\nu_\alpha^2), \tag{27}$$

$$\int \boldsymbol{e}_{\boldsymbol{q}}^{\mathrm{l}} \cdot \boldsymbol{v}' G_\alpha(q, \boldsymbol{v}; \boldsymbol{v}') d^3v' = \frac{1}{|\boldsymbol{q}|} \{\Omega G_{0\alpha}(q, \boldsymbol{v}) - 1\}, \tag{28}$$

$$\int \boldsymbol{e}_{\boldsymbol{q}}^{\mathrm{t}} \cdot \boldsymbol{v}' G_\alpha(q, \boldsymbol{v}; \boldsymbol{v}') d^3v' = \boldsymbol{e}_{\boldsymbol{q}}^{\mathrm{t}} \cdot \boldsymbol{v} G_{0\alpha}(q, \boldsymbol{v}). \tag{29}$$

(28)式可由方程(24)严格求得,而(29)式利用(26)式得到.

于是,(17)式的低频纵波介电函数为 $\varepsilon_d^{\mathrm{l}}(q) = {}^{(0)}\varepsilon_d^{\mathrm{l}}(q) + {}^{(r)}\chi_e^{\mathrm{l}}$, 这里 ${}^{(0)}\varepsilon_d^{\mathrm{l}}$ 为通常没有重整化的部分,而

$$^{(r)}\chi_e^{\mathrm{l}}(q) = a_e \frac{\omega_e^2}{n_0} \int \frac{\boldsymbol{q} \cdot \dfrac{\partial F_{0e}}{\partial \boldsymbol{v}}}{(\Omega - \boldsymbol{q}\cdot\boldsymbol{v})^3} d^3v + o(\nu_e^2) \tag{30}$$

为重整化电子极化部分. 由于 $a_i \ll a_e$, 得到(30)式时离子重整化效应已被略去.

低频横波的色散响应函数为 $\varepsilon_d^{\mathrm{t}}(q) = {}^{(0)}\varepsilon_d^{\mathrm{t}}(q) - \dfrac{c^2q^2}{\Omega^2} + {}^{(r)}\chi_e^{\mathrm{t}}(q)$, ${}^{(0)}\varepsilon_d^{\mathrm{t}}(q)$ 为没有重整化部分,而重整化电子极化部分为

$$\begin{aligned} ^{(r)}\chi_e^{\mathrm{t}}(q) &= a_e \frac{q^2\omega_e^2}{n_0\Omega} \int \frac{\boldsymbol{e}_{\boldsymbol{q}}^{\mathrm{t}} \cdot \boldsymbol{v}\boldsymbol{e}_{\boldsymbol{q}}'^{\mathrm{t}} \cdot \dfrac{\partial}{\partial \boldsymbol{v}} F_{0e}(\boldsymbol{v})}{(\Omega - \boldsymbol{q}\cdot\boldsymbol{v})^3} d^3v \\ &= - a_e \frac{q^2\omega_e^2}{n_0\Omega} \int_{-\infty}^{\infty} \frac{F_{0e}(v)}{(\Omega - 8v)^3} dv + o(\nu_e^2). \end{aligned} \tag{31}$$

这样, 在 $v_i \ll \dfrac{\Omega}{|\boldsymbol{q}|} \ll v_e$ 相速度区, Landau 阻尼可忽略,就有

$$\varepsilon_d^{\mathrm{l}}(q) = \frac{1}{q^2\lambda_e^2\Omega^2} [\Omega^2(1 - \nu_e) - q^2v_s^2] + o(\nu_e^2), \tag{32}$$

$$\varepsilon_d^{\mathrm{t}}(q) = - \frac{1}{q^2\lambda_e^2\Omega^2} [\Omega^2(1 - \nu_e) + q^2v_s^2 + c^2\lambda_e^2q^4] + o(\nu_e^2). \tag{33}$$

这里 $\lambda_\alpha = \left(\dfrac{T_\alpha}{4\pi n_0 e_\alpha^2}\right)^{1/2}$ 为 Debye 长度 $(\alpha = i, e)$, $v_s = \left(\dfrac{T_e}{m_i}\right)^{1/2}$.

2. 高频传播子方程近似解

方程(6)的 Green 函数方程为

$$\left\{\omega - \boldsymbol{k}\cdot\boldsymbol{v} - i\frac{\partial}{\partial \boldsymbol{v}} \cdot \mathbf{D}_\alpha(k, \boldsymbol{v}) \cdot \frac{\partial}{\partial \boldsymbol{v}}\right\} G_\alpha(k, \boldsymbol{v}; \boldsymbol{v}') = \delta(\boldsymbol{v} - \boldsymbol{v}'). \tag{34}$$

在"最发散"近似下,注意到高频波 $(\boldsymbol{k}\cdot\boldsymbol{v}/\omega)_{1,2} \ll 1$ 及 $\sum\limits_{l(\pm k)} \boldsymbol{E}^\sigma(k)\boldsymbol{E}^{\sigma*}(k) = \boldsymbol{E}(k)\boldsymbol{E}^*(k)$, 即有

$$\begin{aligned} \mathbf{D}_\alpha(k, \boldsymbol{v}) &= i\left(\frac{e_\alpha}{m_\alpha}\right)^2 \int \frac{\boldsymbol{e}_1\boldsymbol{e}_1 |E(k_1)|^2 \delta^4(k - k_1 - q)}{\Omega - \boldsymbol{q}\cdot\boldsymbol{v} + i0_+} \frac{d^4k_1 d^4q}{(2\pi)^8} \\ &= \mathbf{D}_\alpha^{\mathrm{r}}(k, \boldsymbol{v}) + \mathbf{D}_\alpha^{\mathrm{nr}}(k, \boldsymbol{v}), \end{aligned} \tag{35}$$

其中

$$\mathbf{D}_\alpha^{rc}(k, \boldsymbol{v}) = \pi\left(\frac{e_\alpha}{m_\alpha}\right)^2\int \boldsymbol{e}_1\boldsymbol{e}_1|E(k_1)|^2\delta(\omega-\omega_1-\boldsymbol{k}\cdot\boldsymbol{v}+\boldsymbol{k}_1\cdot\boldsymbol{v})\frac{d^4k_1}{(2\pi)^4} \tag{36}$$

为共振张量扩散系数，而

$$\mathbf{D}_\alpha^{nr}(k, \boldsymbol{v}) = i\left(\frac{e_\alpha}{m_\alpha}\right)^2\mathscr{P}\int\frac{\boldsymbol{e}_1\boldsymbol{e}_1|E(k_1)|^2}{\omega-\omega_1-(\boldsymbol{k}-\boldsymbol{k}_1)\cdot\boldsymbol{v}}\frac{d^4k_1}{(2\pi)^4} \tag{37}$$

为非共振张量扩散系数，$\mathscr{P}$ 为主值积分符号.

当高频湍动谱接近各向同性时，

$$\mathbf{D}_\alpha^{rc}(k, \boldsymbol{v}) \simeq \frac{\pi}{4v}\int\frac{d^4k_1}{(2\pi)^4}\frac{|E(k_1)|^2}{|\boldsymbol{k}_1|}\{(\hat\theta\hat\theta)+\hat\varphi\hat\varphi)(1-\zeta^2)+\hat v\hat v2\zeta^2\}, \tag{38}$$

$$\begin{aligned}\mathbf{D}_\alpha^{nr}(k, \boldsymbol{v}) = {}& \frac{i}{4v}\left(\frac{e_\alpha}{m_\alpha}\right)^2\int\frac{d^4k_1}{(2\pi)^4}\frac{|E(k_1)|^2}{|\boldsymbol{k}_1|}\mathscr{P}\int_{-1}^{1}[(\hat\theta\hat\theta+\hat\varphi\hat\varphi)(1-\mu'^2)\\ &+2\hat v\hat v\mu'^2](\mu'-\zeta)^{-1}d\mu',\end{aligned} \tag{39}$$

式中 $\zeta=\dfrac{\boldsymbol{k}\cdot\boldsymbol{v}-\Omega}{|\boldsymbol{k}_1|v}$，$\Omega=\omega-\omega_1$；$\hat v$，$\hat\theta$，$\hat\varphi$ 表示球坐标三个轴的单位矢量. 注意到

$$\begin{aligned}\mathscr{P}\int_{-1}^{1}\frac{x^2dx}{x-\zeta} &= 2\zeta+\zeta^2\left(\int_{-1}^{\zeta-\Delta}\frac{dx}{x-\zeta}+\int_{\zeta+\Delta}^{1}\frac{dx}{x-\zeta}\right)\\ &= 2\zeta+\zeta^2\ln\frac{|1-\zeta|}{|1+\zeta|},\end{aligned} \tag{40}$$

其中Δ为奇点附近半圆的半径，△→0.

在粒子与高频场有明显作用情况下，$\mu=\dfrac{\boldsymbol{k}\cdot\boldsymbol{v}}{|\boldsymbol{k}|v}<1$，$\zeta<1$，于是

$$(1-\zeta^2)\ln\frac{|1-\zeta|}{|1+\zeta|}-2\zeta\simeq-4\zeta.$$

同时 $\mathbf{D}_\alpha^{nr}$ 及 $\mathbf{D}_\alpha^{rc}$ 中 $\hat v\hat v$ 方向分量可忽略. 这样，对方位角 φ 平均后，方程(34)近似为

$$\left\{\omega-|\boldsymbol{k}|v\mu-i\frac{\partial}{\partial\mu}\frac{(1-\mu^2)}{v^2}(D_\alpha^{rc}+D_\alpha^{nr})\frac{\partial}{\partial\mu}\right\}G_\alpha(k, \mu; \mu')=\delta(\mu-\mu'), \tag{41}$$

其中

$$G_\alpha(k, \boldsymbol{v}; \boldsymbol{v}')=G_\alpha(k, \mu; \mu')\frac{\delta(v-v')}{2\pi v'^2}, \tag{42}$$

$$\frac{1}{v^2}(D_\alpha^{nr}+D_\alpha^{rc})=\Lambda_\alpha-i\frac{4}{\pi}\Lambda_{1\alpha}\mu, \tag{43}$$

$$\begin{pmatrix}\Lambda_\alpha\\ \Lambda_{1\alpha}\end{pmatrix}=\frac{\pi}{4v^3}\left(\frac{e_\alpha}{m_\alpha}\right)^2\int\frac{|E(k_1)|^2}{|\boldsymbol{k}_1|}\begin{pmatrix}1\\ \dfrac{\boldsymbol{k}}{|\boldsymbol{k}_1|}\end{pmatrix}\frac{d^4k_1}{(2\pi)^4}. \tag{44}$$

令(44)式中 $|\boldsymbol{k}_1|\sim|\boldsymbol{k}|$，即有 $\Lambda_\alpha=\Lambda_{1\alpha}\simeq\dfrac{\pi}{4}\dfrac{\omega_\alpha^2}{|\boldsymbol{k}|\lambda_\alpha}\left(\dfrac{v_\alpha}{v}\right)^3\nu_\alpha.$　　(45)

记 $x=|\boldsymbol{k}|v\mu$ 及 $b_\alpha=\Lambda_\alpha^2k^2v^2$，便可写(41)式为

$$\left\{\omega-x-ib_\alpha\frac{\partial^2}{\partial x^2}\right\}G_\alpha(k, x; x')=|\boldsymbol{k}|v\delta(x-x'). \tag{46}$$

(46)式是自伴的，其解为

$$G_{\alpha}(k, x; x') = \frac{-i|\boldsymbol{k}|v}{2\sqrt{\pi b_{\alpha}}}\int_0^{\infty}\frac{d\tau}{\sqrt{\tau}}\exp\left\{i\left(\omega - \frac{x+x'}{2}\right)\tau - \frac{b_{\alpha}}{12}\tau^3 - \frac{(x-x')^2}{4b_{\alpha}\tau}\right\}. \tag{47}$$

显然，交换 x 与 x' 是互为对称的.

下面给出与高频 Green 函数积分量有关的几个式子：

$$\int_{-1}^{1} G_{\alpha}(k, x; x')d\mu' \simeq -i\int_0^{\infty} d\tau \exp\left\{i(\omega - \boldsymbol{k}\cdot\boldsymbol{v})\tau - \frac{b_{\alpha}}{3}\tau^3\right\} \equiv G_{0\alpha}(k, \boldsymbol{v}). \tag{48}$$

得到上式时仍设 $\nu_e < 1$，并取积分 $\int_{\sigma_-}^{\sigma_+} e^{-\sigma^2}d\sigma \simeq \sqrt{\pi}$，这里 $\sigma_{\pm} = \frac{1}{2\sqrt{b_{\alpha}\tau}}(\pm kv - x + ib_{\alpha}\tau^2)$.

$$\int_{-1}^{1} \mu' G_{\alpha}(k, x; x')d\mu' = \frac{\omega}{|\boldsymbol{k}|v}\left\{G_{0\alpha}(k, \boldsymbol{v}) - \frac{1}{\omega}\right\}. \tag{49}$$

$b_{\alpha} \to 0$ 时，(48)式即为自由传播子 $G^{(0)}(k, \boldsymbol{v}) = 1/(\omega - \boldsymbol{k}\cdot\boldsymbol{v})$，一般情况下，

$$G_{0\alpha}(k, \boldsymbol{v}) = G^{(0)}(k, \boldsymbol{v}) + i\frac{c_{\alpha}\nu_{\alpha}}{v(\omega - \boldsymbol{k}\cdot\boldsymbol{v})^4} + o(\nu_e^2). \tag{50}$$

这里 $c_{\alpha} = \frac{\pi}{2}|\boldsymbol{k}|\lambda_{\alpha}\omega_{\alpha}^3\nu_{\alpha}$.

现在写出 Langmuir 波的重整化介电函数具体形式. 由(16)式，$\varepsilon_h^1(k) = {}^{(0)}\varepsilon_h^1(k) + {}^{(r)}\chi_e^1$，其中无重整化时的介电函数为

$${}^{(0)}\varepsilon_h^1(k) \simeq 1 - \frac{\omega_e^2}{\omega^2} - 3k^2\lambda_e^2 + i\sqrt{\frac{\pi}{2}}\frac{1}{|\boldsymbol{k}|\lambda_e}e^{-\frac{1}{2k^2\lambda_e^2}}, \tag{51}$$

而重整化电子极化部分为

$${}^{(r)}\chi_e^1(k) = i\frac{\pi}{2}\frac{\omega_e^3 v_e^3}{|\boldsymbol{k}|\lambda_e n_0}\nu_e\int\frac{\boldsymbol{k}\cdot\frac{\partial F_{0e}}{\partial \boldsymbol{v}}d^3v}{v(\omega - \boldsymbol{k}\cdot\boldsymbol{v})^4} + o(\nu_e^2). \tag{52}$$

因为 $\omega \gg |\boldsymbol{k}|\nu_e$，立刻得到

$${}^{(r)}\chi_e^1(k) \simeq -i\frac{4}{3}\sqrt{2\pi}\left(\frac{\omega_e}{\omega}\right)^5|\boldsymbol{k}|\lambda_e\nu_e + o\left(\nu_e^2, \left(\frac{\boldsymbol{k}\cdot\boldsymbol{v}}{\omega}\right)^2\right). \tag{53}$$

由于高频传播子湍动碰撞频率中只考虑了共振项作用，所以重整化结果相当于阻尼效应. 当 $\frac{8}{3}k^2\lambda_e^2\nu_e > \exp\left(-\frac{1}{2k^2\lambda_e^2}\right)$ 时，重整化效应超过了线性 Landau 阻尼，因为 $|\boldsymbol{k}|\lambda_e \ll 1$，所以这一条件在 $\nu_e \ll 1$ 时也很好满足.

三、Langmuir 湍动场激发低频自生磁场和离子声波的重整化方程组

1. 高、低频重整化非线性电流跃迁系数的计算

在背景等离子体粒子为 Maxwell 分布情况下重整化高频传播子(50)式可近似地写为

$G_{0e}(k, \boldsymbol{v}) \simeq \frac{1}{\omega - \boldsymbol{k}\cdot\boldsymbol{v}}\left[1 + i\frac{\pi}{2}k\lambda_e\nu_e\right] + o(\nu_e^2)$，(27)式比较低频传播子 $G_{0e}(q, \boldsymbol{v}) \simeq \frac{1}{\Omega - \boldsymbol{q}\cdot\boldsymbol{v}}\left[1 + \frac{q^2v_e^2}{(\Omega - \boldsymbol{q}\cdot\boldsymbol{v})^2}\nu_e\right] + o(\nu_e^2)$ 可知，前者重整化部分远较后者小，故在计算低频重整化非线性电流跃迁系数时(15)式中可取 $G_{0e}(k, \boldsymbol{v}) \simeq \frac{1}{\omega - \boldsymbol{k}\cdot\boldsymbol{v}}$，由此得

$$H_e^{\beta ll}(q, k_1, k_2) = \int \boldsymbol{e}_q^\beta \cdot \boldsymbol{v} G_e(q, \boldsymbol{v}; \boldsymbol{v}')\mathscr{L}_e(k_1, k_2, \boldsymbol{v}')d^3v'd^3v, \tag{54}$$

其中

$$\mathscr{L}_e(k_1, k_2, \boldsymbol{v}) = \frac{1}{2}\left[\boldsymbol{e}_1\cdot\frac{\partial}{\partial\boldsymbol{v}}\frac{1}{\omega_2 - \boldsymbol{k}_2\cdot\boldsymbol{v}}\boldsymbol{e}_2\cdot\frac{\partial}{\partial\boldsymbol{v}}F_{0e}(\boldsymbol{v}) + 1\longleftrightarrow 2\right]. \tag{55}$$

对 $\beta = l$，

$$H_e^{lll}(q, k_1, k_2) = {}^{(0)}H_e^{lll}(q, k_1, k_2) + {}^{(r)}H_e^{lll}(q, k_1, k_2). \tag{56}$$

这里[1]，

$$\begin{aligned}{}^{(0)}H_e^{lll}(q, k_1, k_2) &= \frac{\Omega}{\omega_2}\,{}^{(0)}H_e^{lll}(k_2, q, k_1)\\ &= \frac{\Omega|\boldsymbol{q}|n_0}{2\omega_e^2\omega_2^2}\boldsymbol{e}_1\cdot\boldsymbol{e}_2[{}^{(0)}\varepsilon_e^l(q) - 1] + o\left(\frac{\boldsymbol{k}\cdot\boldsymbol{v}}{\omega}\right)_{1,2}\end{aligned} \tag{57}$$

为非重整化部分，${}^{(0)}\varepsilon_e^l(q) = 1 + \frac{\omega_e^2}{n_0q^2}\int\frac{\boldsymbol{q}\cdot\frac{\partial F_{0e}}{\partial\boldsymbol{v}}}{\Omega - \boldsymbol{q}\cdot\boldsymbol{v}}d^3v$，而重整化部分

$$\begin{aligned}{}^{(r)}H_e^{lll}(q, k_1, k_2) &= \nu_e q^2v_e^2\frac{\Omega}{2|\boldsymbol{q}|}\int\left\{\frac{(\Omega - \boldsymbol{q}\cdot\boldsymbol{v})}{\omega_1\omega_2}\boldsymbol{e}_1\cdot\frac{\partial}{\partial\boldsymbol{v}}\boldsymbol{e}_2\right.\\ &\quad\left.\cdot\frac{\partial}{\partial\boldsymbol{v}}F_{0e} + \frac{\boldsymbol{e}_1\cdot\boldsymbol{e}_2}{\omega_2^2}\boldsymbol{q}\cdot\frac{\partial F_{0e}}{\partial\boldsymbol{v}}\right\}(\Omega - \boldsymbol{q}\cdot\boldsymbol{v})^{-3}d^3v\\ &\quad + o\left(\nu_e^2, \left(\frac{\boldsymbol{k}\cdot\boldsymbol{v}}{\omega}\right)^2_{1,2}\right) = \frac{\nu_e}{2}\frac{\Omega n_0}{\omega_e^4q^2\lambda_e^2|\boldsymbol{q}|}\{2\boldsymbol{e}_1\cdot\boldsymbol{q}\boldsymbol{e}_2\cdot\boldsymbol{q} - \boldsymbol{e}_1\cdot\boldsymbol{e}_2q^2\}\\ &\quad + o(\nu_e^2),\end{aligned} \tag{58}$$

后一等式是在 $v_i \ll \Omega/|\boldsymbol{q}| \ll v_e$ 条件下得到．由于这时 ${}^{(0)}\varepsilon_e^l(\boldsymbol{q}) - 1 \simeq 1/q^2\lambda_e^2$，比较(57)和(58)式可知，重整化修正约为一个 ν_e 因子．

对 $\beta = t$，利用(29)后(54)式可写成

$$H_e^{tll}(q, k_1, k_2) = {}^{(0)}H_e^{tll}(q, k_1, k_2) + {}^{(r)}H_e^{tll}(q, k_1, k_2), \tag{59}$$

其中

$${}^{(0)}H_e^{tll}(q, k_1, k_2) = \frac{n_0}{2\omega_1\omega_2^2}\boldsymbol{e}_q^t\cdot[\boldsymbol{q}\times(\boldsymbol{e}_1\times\boldsymbol{e}_2)] + o\left(\frac{\boldsymbol{k}\cdot\boldsymbol{v}}{\omega}\right)^{[1]}_{1,2} \tag{60}$$

是导致自生磁场产生的低频非线性电流激发系数；而重整化系数 ${}^{(r)}H_e^{tll}(q, k_1, k_2) = \nu_e q^2v_e^2\int\frac{\boldsymbol{e}_q^t\cdot\boldsymbol{v}}{(\Omega - \boldsymbol{q}\cdot\boldsymbol{v})^3}\mathscr{L}_e(k_1, k_2, \boldsymbol{v})d^3v + o(\nu_e^2) = -\frac{\nu_e\Omega n_0}{\omega_e^2q^2v_e^2}\boldsymbol{e}_q^t\cdot[\boldsymbol{e}_1\boldsymbol{e}_2\cdot\boldsymbol{q} + 1\longleftrightarrow 2] + o(\nu_e^2).$ (61)

得到(61)式后一等式时同样用了 $v_i \ll \Omega/|\boldsymbol{q}| \ll v_e$ 条件．比较(60)与(61)式可见，后者约小一个因子 $\frac{\Omega\omega_e}{q^2v_e^2}\nu_e$．

下面给出重整化高频跃迁系数的计算．由(13)式得

$$H_e^{l\beta l}(k, q, k_1) = {}^{(0)}H_e^{l\beta l}(k, q, k_1) + {}^{(r)}H_e^{l\beta l}(k, q, k_1), \tag{62}$$

其中

$$^{(0)}H_e^{l\beta l}(k,q,k_1)=\int\frac{\boldsymbol{e}_k\cdot\boldsymbol{v}}{\omega-\boldsymbol{k}\cdot\boldsymbol{v}}\left[\left(1-\frac{\boldsymbol{q}\cdot\boldsymbol{v}}{\Omega}\right)\boldsymbol{e}_q^\beta+\frac{\boldsymbol{e}_q^\beta\cdot\boldsymbol{v}}{\Omega}\boldsymbol{q}\right]$$

$$\cdot\frac{\partial}{\partial\boldsymbol{v}}\cdot\frac{\boldsymbol{e}_1\cdot\dfrac{\partial F_{0e}}{\partial\boldsymbol{v}}}{\omega_1-\boldsymbol{k}_1\cdot\boldsymbol{v}}d^3v=\frac{n_0}{\omega\omega_1\Omega}\boldsymbol{e}_k\cdot[\boldsymbol{e}_1\times(\boldsymbol{q}\times\boldsymbol{e}_q^\beta)]$$

$$+o\left(\frac{\boldsymbol{k}\cdot\boldsymbol{v}}{\omega}\right).\tag{63}$$

显然，$^{(0)}H_e^{lll}(k,q,k_1)=0$. 重整化部分

$$^{(r)}H_e^{l\beta l}(k,q,k_1)=-i\frac{\omega_1}{|\boldsymbol{k}_1|v_e}c_e(k_1)\int\frac{\boldsymbol{e}_k\cdot\boldsymbol{v}}{\omega-\boldsymbol{k}\cdot\boldsymbol{v}}\boldsymbol{e}_q'$$

$$\cdot\frac{\partial}{\partial\boldsymbol{v}}\frac{F_{0e}(\boldsymbol{v})}{v(\omega_1-\boldsymbol{k}_1\cdot\boldsymbol{v})^4}d^3v+i\frac{\omega}{|\boldsymbol{k}|}c_e(k)v_e\int\frac{1}{v(\omega-\boldsymbol{k}\cdot\boldsymbol{v})^4}\boldsymbol{e}_q'$$

$$\cdot\frac{\partial}{\partial\boldsymbol{v}}\frac{\boldsymbol{e}_1\cdot\dfrac{\partial}{\partial\boldsymbol{v}}F_{0e}}{(\omega_1-\boldsymbol{k}_1\cdot\boldsymbol{v})}d^3v+o(\nu_e^2).\tag{64}$$

于是有

$$^{(r)}H_e^{lll}(k,q,k_1)=i\sqrt{\frac{\pi}{2}}\frac{n_0\nu_e}{\omega_e^2v_e}\left[\boldsymbol{e}_q^l\cdot\boldsymbol{e}_k-\frac{1}{3}\cos\theta_0\right]+o\left(\nu_e^2,\frac{\boldsymbol{k}\cdot\boldsymbol{v}}{\omega}\right),\tag{65}$$

$$^{(r)}H_e^{ltl}(k,q,k_1)=i\sqrt{\frac{\pi}{2}}\frac{n_0\nu_e}{\omega_e^2v_e}\left[\boldsymbol{e}_q^t\cdot\boldsymbol{e}_k-\frac{2}{3}\sin\theta_0\right]+o\left(\nu_e^2,\frac{\boldsymbol{k}\cdot\boldsymbol{v}}{\omega}\right),\tag{66}$$

这里 $\boldsymbol{q}=\boldsymbol{k}-\boldsymbol{k}_1$，$\cos\theta_0=(q^2+k_1^2-k^2)/2|\boldsymbol{k}_1||\boldsymbol{q}|$. (64)式中 $c_e(k)$ 见(50)式. 比较(63)和(66)式可见，高频跃迁系数重整化修正比低频的小一个因子 $\dfrac{\Omega}{|\boldsymbol{q}|v_e}\nu_e\ll1$.

2. Langmuir 波激发低频磁场和离子声强湍动过程的重整化方程组

首先给出高频场方程形式. 由于 $\nu_e<1$ 时高频跃迁系数的重整化修正较小，因此在计算高频非线性电流时略去了它的贡献，只在高频线性介电函数中保留了重整化效应. 这样，略去了离子非线性电流后，(11)式成为

$$S_1(k)=-\frac{\omega_e^2}{n_0\omega}\int\boldsymbol{e}_k\cdot\boldsymbol{v}G_e(k,\boldsymbol{v};\boldsymbol{v}')\boldsymbol{e}_1\cdot\frac{\partial}{\partial\boldsymbol{v}'}\delta F_e(q,\boldsymbol{v}')E(k_1)\delta^4(k-k_1-q)$$

$$\times d^3vd^3v'\frac{d^4k_1d^4q}{(2\pi)^4}\simeq\frac{\omega_e^2}{\omega^2}\int\frac{n_e(q)}{n_0}\boldsymbol{e}_k\cdot\boldsymbol{E}(k_1)\delta^4(k-k_1-q)$$

$$\times\frac{d^4k_1d^4q}{(2\pi)^4}+o\left(\frac{\boldsymbol{k}\cdot\boldsymbol{v}}{\omega}\right),\tag{67}$$

其中 $n_e(q)=\int\delta F_e(q,\boldsymbol{v})d^3v$ 为低频振荡电子密度.

注意到(63)式后，(12)式成为

$$S_2(k)=\frac{-i\omega_ee}{\omega^2m_e\Omega}\sum_\beta\int\boldsymbol{e}_k\cdot\{\boldsymbol{E}(k_1)\times(\boldsymbol{q}\times\boldsymbol{E}_d^\beta(q))\}\delta^4(k-k_1-q)$$

$$\times \frac{d^4k_1d^4q}{(2\pi)^4}+o\left(\frac{\boldsymbol{k}\cdot\boldsymbol{v}}{\omega}\right)=\frac{-i\omega_e e}{\omega^2 m_e c}\int \boldsymbol{e_k}\cdot\{\boldsymbol{E}(k_1)\times\boldsymbol{B}(q)\}$$

$$\times \delta^4(k-k_1-q)\frac{d^4k_1d^4q}{(2\pi)^4}+o\left(\frac{\boldsymbol{k}\cdot\boldsymbol{v}}{\omega}\right). \tag{68}$$

应用(51)和(53)式，忽略线性 Landau 阻尼，在(10)式两边乘上 $\boldsymbol{e_k}\cdot\boldsymbol{e_k}=1$ 后，得到高频场的方程如下：

$$\left\{\omega^2-\omega_e^2-3k^2v_e^2-i\frac{4}{3}\sqrt{2\pi}|\boldsymbol{k}|\lambda_e\nu_e\omega_e^2\right\}\boldsymbol{E}(k)$$

$$=\omega_e^2\int\left\{\frac{n_e(q)}{n_0}\boldsymbol{E}(k_1)-\frac{ie}{m_e\omega_e c}\boldsymbol{E}(k_1)\times\boldsymbol{B}(q)\right\}\delta^4(k-k_1-q)$$

$$\times\frac{d^4k_1d^4q}{(2\pi)^4}+o\left(\nu_e,\frac{\boldsymbol{k}\cdot\boldsymbol{v}}{\omega}\right). \tag{69}$$

上式右端已忽略了重整化小项.

为了使(69)式闭合，尚需给出 $n_e(q)$ 和 $\boldsymbol{B}(q)$ 的方程，为此，先研究低频电场方程(14). 略去离子非线性电流项，注意到(57)，(58)式及 $H_e^{\beta ll}(q,k_1,k_2)$ 对下标 1,2 的对称性，我们得到

$$E_d^\beta(q)=\frac{i2\omega_e^2e}{n_0m_e\Omega\varepsilon_a^\beta(q)}\int H_e^{\beta ll}(q,k_1,-k_2)E_+(k_1)E_-^*(k_2)\delta^4(q-k_1+k_2)\frac{d^4k_1d^4k_2}{(2\pi)^4}. \tag{70}$$

这里用了关系式 $\boldsymbol{e}_{-\boldsymbol{k}}=-\boldsymbol{e_k}$，$E(-k)=-E_-^*(k)$.

对低频纵波，$\beta=l$，应用(57)和(58)式并在(70)式两边作用 $\boldsymbol{e}_{\boldsymbol{q}}^l=\frac{\boldsymbol{q}}{|\boldsymbol{q}|}$，便有

$$\boldsymbol{E}_d^l(q)=\frac{-ie\boldsymbol{q}}{m_e\omega_e^2\varepsilon_d^l(q)}\int\left\{(\varepsilon_e^l(q)-1)\boldsymbol{E}_+(k_1)\cdot\boldsymbol{E}_-^*(k_2)\right.$$

$$\left.+2\frac{v_e^2}{q^2\lambda_e^2q^2}\boldsymbol{q}\cdot\boldsymbol{E}_+(k_1)\boldsymbol{q}\cdot\boldsymbol{E}_-^*(k_2)\right\}\delta^4(q-k_1+k_2)$$

$$\times\frac{d^4k_1d^4k_2}{(2\pi)^4}+o(\nu_e^2), \tag{71}$$

其中 $\varepsilon_e^\beta(q)={}^{(0)}\varepsilon_e^\beta(q)+{}^{(r)}\chi_e^\beta(q)$. (71)式也可写成

$$\boldsymbol{E}_d^l(q)=\frac{-ie\boldsymbol{q}}{m_e\omega_e^2\varepsilon_d^l(q)}\int\left\{({}^{(0)}\varepsilon_e^l(q)-{}^{(r)}\chi_e^l(q)-1)\boldsymbol{E}_+(k_1)\cdot\boldsymbol{E}_-^*(k_2)\right.$$

$$\left.+2\frac{v_e}{q^2\lambda_e^2q^2}\boldsymbol{q}\cdot[(\boldsymbol{q}\times\boldsymbol{E}_+(k_1))\times\boldsymbol{E}_-^*(k_2)]\right\}$$

$$\times\delta^4(q-k_1+k_2)\frac{d^4k_1d^4k_2}{(2\pi)^4}. \tag{72}$$

对低频横波，$\beta=t$，(70)式两边作用 $\boldsymbol{e}_{\boldsymbol{q}}^t\cdot\boldsymbol{e}_{\boldsymbol{q}}^t=1$ 后，应用(60)和(61)式便有

$$\boldsymbol{E}_d^t(q)=\frac{-ie}{m_e\omega_e\Omega\varepsilon_d^t(q)}\int\left\{\boldsymbol{q}\times(\boldsymbol{E}_+(k_1)\times\boldsymbol{E}_-^*(k_2))-\frac{2\nu_e\Omega\omega_e}{q^2v_e^2}\right.$$

$$\left.\times[\boldsymbol{E}_+(k_1)\boldsymbol{q}\cdot\boldsymbol{E}_-^*(k_2)+\boldsymbol{E}_-^*(k_2)\boldsymbol{q}\cdot\boldsymbol{E}_+(k_1)]\right\}$$

$$\times\delta^4(q-k_1+k_2)\frac{d^4k_1d^4k_2}{(2\pi)^4}+o\left(\nu_e^2,\frac{\boldsymbol{k}\cdot\boldsymbol{v}}{\omega}\right). \tag{73}$$

于是自生磁场便为

$$B(q)=\frac{-iec}{m_e\omega_e\Omega^2\varepsilon_d^t(q)}\int q\times\Big\{q\times(E_+(k_1)\times E_-^*(k_2)) - \frac{2\nu_e\Omega\omega_e}{q^2v_e^2}[E_+(k_1)q\cdot E_-^*(k_2)+E_-^*(k_2)q\cdot E_+(k_1)]\Big\} \times\delta^4(q-k_1+k_2)\frac{d^4k_1d^4k_2}{(2\pi)^4}+o\left(\nu_e^2,\frac{k\cdot v}{\omega}\right). \tag{74}$$

现在给出低频振荡电子密度的表达式. 从(5)式得

$$n_e(q)=\int\delta F_e(q,v)d^3v=i\frac{e}{m_e}E_d^\beta(q)\int G_e(q,v;v')e_q'^\beta \cdot\frac{\partial}{\partial v}F_{0e}(v')d^3v'd^3v-\frac{e^2}{m_e^2}\int_{(d)}I_e(q,k_1,k_2)E(k_1)E(k_2) \times\delta^4(q-k_1-k_2)\frac{d^4k_1d^4k_2}{(2\pi)^4}, \tag{75}$$

其中

$$I_\alpha(q,k_1,k_2)=\int G_\alpha(q,v;v')e_1\cdot\frac{\partial}{\partial v'}G_\alpha(k_2,v';v'')e_2 \cdot\frac{\partial}{\partial v''}F_{0\alpha}(v'')d^3v''d^3v'd^3v=\int G_{0\alpha}(q,v)e_1 \cdot\frac{\partial}{\partial v}G_\alpha(k_2,v;v')e_2\cdot\frac{\partial}{\partial v'}F_{0\alpha}(v')d^3v'd^3v. \tag{76}$$

注意到(28)式,即有

$$I_\alpha(q,k_1,k_2)=\frac{|q|}{\Omega}H_\alpha^{lll}(q,k_1,k_2), \tag{77}$$

同时由于

$$\int G_\alpha(q,v;v')e_q'^\beta\cdot\frac{\partial}{\partial v'}F_{0\alpha}(v')d^3v'd^3v=\frac{|q|}{\Omega}\int e_q^l\cdot vG_\alpha(q,v;v')e_q'^\beta \cdot\frac{\partial}{\partial v'}F_{0\alpha}(v')d^3vd^3v' =\begin{cases}\dfrac{n_0|q|}{\omega_e^2}(\varepsilon_\alpha^l(q)-1) & \text{对 } \beta=l,\\ 0 & \text{对 } \beta=t,\end{cases} \tag{78}$$

代入(75)式后便得到

$$n_e(q)=2\frac{e^2|q|}{m_e^2\Omega}\frac{{}^{(0)}\varepsilon_i^l(q)}{\varepsilon_\alpha^l(q)}\int H_e^{lll}(q,k_1,-k_2)E_+(k_1)E_-^*(k_2) \times\delta^4(q-k_1+k_2)\frac{d^4k_1d^4k_2}{(2\pi)^4}=-\frac{q^2}{4\pi m_e\omega_e^2}\frac{{}^{(0)}\varepsilon_i^l}{\varepsilon_d^l(q)} \times\int\Big\{(\varepsilon_e^l(q)-1)E_+(k_1)\cdot E_-^*(k_2)+\frac{2\nu_e}{q^2\lambda_e^2q^2}q \cdot E_+(k_1)q\cdot E_-^*(k_2)\Big\}\delta^4(q-k_1+k_2)\frac{d^4k_1d^4k_2}{(2\pi)^4}. \tag{79}$$

从上式可知，低频横波或自生磁场不直接影响低频振荡粒子密度的变化，它只通过对高频场的反作用产生间接效应.

现在对(69)式进行逆 Fourier 变换，应用关系式$-i\Omega \to \frac{\partial}{\partial t}$ 及 $i\boldsymbol{q} \to \nabla$，便得到高频场方程

$$\left(\frac{\partial^2}{\partial t^2} - 3v_e^2\nabla^2\right)\boldsymbol{E}(\boldsymbol{x}, t) - i\frac{4}{3}(2\pi)^{-3/2}\lambda_e v_e \omega_e^2 \nabla^2 \int \frac{E(\boldsymbol{x}, t)}{|\boldsymbol{x} - \boldsymbol{x}'|^2} d^3x'$$
$$= -\omega_e^2\left(1 + \frac{n_e(\boldsymbol{x}, t)}{n_0}\right)\boldsymbol{E}(\boldsymbol{x}, t) + \frac{i\omega_e e}{m_e c}\boldsymbol{E} \times \boldsymbol{B}(\boldsymbol{x}, t). \tag{80}$$

这里重整化产生的修正项，表示了湍动碰撞对高频场的作用.

注意到(32)式及相应相速度范围内 ${}^{(0)}\varepsilon_i^l(q) = -\omega_i^2/\Omega^2$，$\varepsilon_e^l(q) - 1 \doteq \frac{1}{q^2\lambda_e^2}(1 - \nu_e)$，(79)式两边乘上 $(\Omega^2 - q^2v_s^2 - \Omega^2\nu_e)$ 因子，逆变换后便可得 $n_e(q)$ 方程，这时出现了

$$\int \boldsymbol{q} \cdot \boldsymbol{E}_+(k_1)\boldsymbol{q} \cdot \boldsymbol{E}_-^*(k_2)\delta^4(q - k_1 + k_2)e^{i(\boldsymbol{q}\cdot\boldsymbol{x} - \Omega t)}\frac{d^4k_1d^4k_2d^4q}{(2\pi)^8}$$
$$= -\{\boldsymbol{E}_-^*(x, t) \cdot \nabla(\nabla \cdot \boldsymbol{E}_+) + \boldsymbol{E}_+(\boldsymbol{x}, t) \cdot \nabla(\nabla \cdot \boldsymbol{E}_-^*)$$
$$+ \nabla \cdot \boldsymbol{E}_+\nabla \cdot \boldsymbol{E}_-^* + \nabla\boldsymbol{E}_+ : \nabla\boldsymbol{E}_-^*\}$$
$$= -\frac{1}{2}\{\nabla^2|\boldsymbol{E}|^2 + |\nabla \cdot \boldsymbol{E}|^2 - \nabla\boldsymbol{E} : \nabla\boldsymbol{E}^*\}. \tag{81}$$

得到上式时已用了 Langmuir 波是有势场的性质以及关系式 $\boldsymbol{E}(k)\ \boldsymbol{E}^*(k) = 2\boldsymbol{E}_+(k)\ \boldsymbol{E}_-^*(k)$. 于是时空表象中 n_e 的方程为

$$\left[(1 - \nu_e)\frac{\partial^2}{\partial t^2} - v_s^2\nabla^2\right]n_e(\boldsymbol{x}, t) = \frac{(1 + \nu_e)}{8\pi m_i}\nabla^2|\boldsymbol{E}(\boldsymbol{x}, t)|^2$$
$$+ \frac{\nu_e}{4\pi m_i}[|\nabla \cdot \boldsymbol{E}|^2 - \nabla\boldsymbol{E} : \nabla\boldsymbol{E}^*]. \tag{82}$$

同样，(74)式两边乘以(33)式，注意到

$$\int \boldsymbol{q} \times \{\boldsymbol{E}_+(k_1)\boldsymbol{q} \cdot \boldsymbol{E}_-^*(k_2) + \boldsymbol{E}_-^*(k_2)\boldsymbol{q} \cdot \boldsymbol{E}_+(k_1)\}e^{i(\boldsymbol{q}\cdot\boldsymbol{x} - \Omega t)}$$
$$\times \delta^4(q - k_1 + k_2)\frac{d^4k_1d^4k_2d^4q}{(2\pi)^8} = \frac{1}{2}\{\boldsymbol{E}^* \times \nabla(\nabla \cdot \boldsymbol{E}) + c.c.\}, \tag{83}$$

便得自生磁场方程

$$\left[(1 - \nu_e)\frac{\partial^2}{\partial t^2} - (\lambda_e^2c^2\nabla^2 - v_s^2)\nabla^2\right]\boldsymbol{B}(\boldsymbol{x}, t)$$
$$= -\frac{ce}{2m_e\omega_e}\Big\{i\lambda_e^2\nabla^2\nabla \times [\nabla \times (\boldsymbol{E} \times \boldsymbol{E}^*)]$$
$$+ \frac{2\nu_e}{\omega_e}\frac{\partial}{\partial t}[\boldsymbol{E} \times \nabla(\nabla \cdot \boldsymbol{E}^*) + c.c.]\Big\}. \tag{84}$$

通常，条件 $|\boldsymbol{q}|\lambda_e \gg v_s/c$ 及 $|\boldsymbol{q}|\lambda_e \gg \Omega/c|\boldsymbol{q}| = \frac{\Omega}{|\boldsymbol{q}|v_e}\beta_e$，$\beta_e = v_e/c$ 很容易成立，所以(84)式又可写成

$$\nabla^4 \boldsymbol{B}(\boldsymbol{x}, t) = \frac{e}{2\lambda_e^2 c m_e \omega_e}\Big\{ i\lambda_e^2 \nabla^2 \nabla \times [\nabla \times (\boldsymbol{E} \times \boldsymbol{E}^*)] + \frac{2\nu_e}{\omega_e}\frac{\partial}{\partial t}[\boldsymbol{E} \times \nabla(\nabla \cdot \boldsymbol{E}^*) + c.c.]\Big\}. \tag{85}$$

方程 (80),(82) 和 (84) 组成了时空表象中描述强湍动过程闭合的重整化方程组. 当$\nu_e \to 0$ 时,上述方程组自然退化到文献[1]中导得的无 Landau 阻尼时的形式. 为了包含 Landau 阻尼效应,处理是简单的,完全可以搬用文献[1]的结果.

在 $\Omega \ll |\boldsymbol{q}| v_i$ 静态近似下,也容易得到重整化方程组的形式. 这时

$$\varepsilon_d^l(q) = {}^{(0)}\varepsilon_d^l(q) + {}^{(r)}\chi_e^l(q) \simeq \frac{1}{q^2\lambda_e^2}\left[\frac{T_e + T_i}{T_i} - \nu_e\right] \qquad (\nu_e \gg q^2\lambda_e^4), \tag{86}$$

$$\frac{{}^{(0)}\varepsilon_i^l(q)(\varepsilon_e^l(q) - 1)}{\varepsilon_d^l(q)} \simeq \frac{1}{q^2\lambda_i^2}\,\frac{(1 - \nu_e)}{\left(\dfrac{T_e + T_i}{T_i} - \nu_e\right)}, \tag{87}$$

$$\varepsilon_d^t(q) = {}^{(0)}\varepsilon_d^t(q) + {}^{(r)}\chi_e^t(q) - \frac{c^2q^2}{\Omega^2} \simeq -\frac{1}{q^2\lambda_e^2\Omega^2} \times \left[c^2q^4\lambda_e^2 + \left(\frac{T_e + T_i}{T_i} - \nu_e\right)\Omega^2\right]. \tag{88}$$

于是时空表象中,低频振荡粒子密度和自生磁场方程分别为

$$-\nabla^2 n_e(\boldsymbol{x}, t) = \frac{(1 + \nu_e)}{8\pi(T_e + T_i)}\nabla^2|\boldsymbol{E}(\boldsymbol{x}, t)|^2 + \frac{\nu_e}{4\pi(T_e + T_i)} \times [|\nabla \cdot \boldsymbol{E}|^2 - \nabla \boldsymbol{E} : \nabla \boldsymbol{E}^*], \tag{89}$$

$$\left(\frac{T_e + T_i}{T_i} - \nu_e\right)\frac{\partial^2}{\partial t^2}\boldsymbol{B}(\boldsymbol{x}, t) - \lambda_e^2 c^2 \nabla^4 \boldsymbol{B}(\boldsymbol{x}, t) = \frac{-ce}{2m_e\omega_e}\Big\{ i\lambda_e^2 \nabla^2 \nabla \times [\nabla \times (\boldsymbol{E} \times \boldsymbol{E}^*)] + \frac{2\nu_e}{\omega_e}\frac{\partial}{\partial t} \times [\boldsymbol{E} \times \nabla(\nabla \cdot \boldsymbol{E}^*) + c.c.]\Big\}. \tag{90}$$

当获得(85)式的类似条件满足时,(90)式便成为(85)式, 这时 $n_e(\boldsymbol{x}, t)$ 和 $\boldsymbol{B}(\boldsymbol{x}, t)$ 中的时间变量均以隐参量形式出现.

四、一维稳态解

为了求解方便,需要把方程组(80)、(82) 和 (84) 中的双时标关系变换到同一时间尺度. 分离出高频场中的快振荡因子,就可得到慢时标的振幅方程,令

$$\boldsymbol{E} = \frac{1}{\sqrt{2}}(\boldsymbol{\varepsilon}' e^{-i\omega_e t} + \boldsymbol{\varepsilon}'^* e^{i\omega_e t}), \tag{91}$$

略去 $\boldsymbol{\varepsilon}'$ 的二次时间导数项,(80)式即为

$$\left(i\frac{\partial}{\partial t} + \frac{3v_e^2}{2\omega_e}\nabla^2\right)\boldsymbol{\varepsilon}' + i\,\frac{4}{3}(2\pi)^{-3/2}\nu_e v_e \nabla^2 \int \frac{\boldsymbol{\varepsilon}'(\boldsymbol{x}', t)}{|\boldsymbol{x} - \boldsymbol{x}'|^2}\, d^3x'$$

$$= \frac{\omega_e}{2} \frac{n_e}{n_0} \boldsymbol{\varepsilon}' - \frac{ie}{2m_e c} \boldsymbol{\varepsilon}' \times \boldsymbol{B}. \tag{92}$$

对(92),(82)和(84)式引入如下无量纲变换：

$$\frac{2}{3} \omega_e \frac{m_e}{m_i} t \to t, \quad \frac{2}{3} \sqrt{\frac{m_e}{m_i}} \frac{\boldsymbol{x}}{\lambda_e} \to \boldsymbol{x}, \quad \frac{3}{4} \frac{m_i}{m_e} \frac{n_e}{n_0} \to n,$$

$$\frac{\sqrt{3m_i}\, \boldsymbol{\varepsilon}'}{(64\pi m_e n_0 T_e)^{1/2}} \to \boldsymbol{\varepsilon}, \quad \frac{3m_i \boldsymbol{B}}{(64\pi n_0 m_e^3 c^2)^{\frac{1}{2}}} \to \boldsymbol{B}, \tag{93}$$

上述方程就成为如下无量纲形式：

$$i \frac{\partial}{\partial t} \boldsymbol{\varepsilon} + \nabla^2 \boldsymbol{\varepsilon} + i\gamma' \nabla^2 \int \frac{\boldsymbol{\varepsilon}(\boldsymbol{x}', t)}{|\boldsymbol{x} - \boldsymbol{x}'|^2} d^3x' = n\boldsymbol{\varepsilon} - i\boldsymbol{\varepsilon} \times \boldsymbol{B}, \tag{94}$$

$$\left[(1 - \nu_e) \frac{\partial^2}{\partial t^2} - \nabla^2\right] n = (1 + \nu_e)\nabla^2 |\boldsymbol{\varepsilon}|^2 + 2\nu_e[|\nabla \cdot \boldsymbol{\varepsilon}|^2 - \nabla\boldsymbol{\varepsilon}:\nabla\boldsymbol{\varepsilon}^*], \tag{95}$$

$$\left\{(1 - \nu_e) \frac{\partial^2}{\partial t^2} - \left[\left(\frac{2}{3\beta_e}\right)^2 \nabla^2 - 1\right]\nabla^2\right\}\boldsymbol{B} = -i \frac{4}{9} \nabla^2\nabla \times [\nabla \times (\boldsymbol{\varepsilon} \times \boldsymbol{\varepsilon}^*)]$$

$$- \frac{4}{3} \nu_e \frac{\partial}{\partial t} [\boldsymbol{\varepsilon} \times \nabla(\nabla \cdot \boldsymbol{\varepsilon}^*) + c.c.]. \tag{96}$$

这里 $\gamma'_e = \frac{4}{3(2\pi)^2} \sqrt{\frac{2\pi m_i}{m_e}}\, \nu_e$. 与(85)式相对应,(96)式还可由下式很好代替,即

$$\nabla^4 \boldsymbol{B} = i\beta_e^2 \nabla^2 \nabla \times [\nabla \times (\boldsymbol{\varepsilon} \times \boldsymbol{\varepsilon}^*)] + 3\beta_e^2 \nu_e \frac{\partial}{\partial t} [\boldsymbol{\varepsilon} \times \nabla(\nabla \cdot \boldsymbol{\varepsilon}^*) + c.c.]. \tag{97}$$

在一维平板几何下,方程组成为

$$i \frac{\partial}{\partial t} \varepsilon + \frac{\partial^2}{\partial x^2} \varepsilon - \gamma_e \frac{\partial}{\partial x} \varepsilon = n\varepsilon, \tag{98}$$

$$\left[(1 - \nu_e) \frac{\partial^2}{\partial t^2} - \frac{\partial^2}{\partial x^2}\right] n = (1 + \nu_e) \frac{\partial^2}{\partial x^2} |\boldsymbol{\varepsilon}|^2, \tag{99}$$

其中 $\gamma_e = \frac{2}{3}\sqrt{\frac{2\pi m_i}{m_e}}\, \nu_e$. 得到(98)式时已应用了三维变为一维的关系式,即

$$\nabla^2 \frac{1}{|\boldsymbol{x} - \boldsymbol{x}'|^2} \to i \frac{(2\pi)^2}{2} \delta(y - y')\delta(z - z') \frac{\partial}{\partial x} \delta(x - x').$$

引入行波变量 $\xi = x - ut$ 及函数关系 $n = n(\xi)$, $\varepsilon = \varepsilon_0(\xi)\exp(iqx - i\Omega t)$, 这里 u, q 和 Ω 均为参量,于是由(99)式得

$$n(\xi) = -\frac{(1 + \nu_e)\varepsilon_0^2(\xi)}{1 - (1 - \nu_e)u^2}. \tag{100}$$

由于 ν_e 是正定的,所以与无重整化时一样,方程组 (98)—(99) 具有孤立子解只能在 $u < \sqrt{1/(1 - \nu_e)}$ 亚声条件下,这时从(98)式得到

$$\frac{d^2\varepsilon_0}{d\xi^2} - \alpha_0^2\varepsilon_0 + \beta_0^2\varepsilon_0^3 = 0, \tag{101}$$

其中要求 $\alpha_0^2 = \frac{\gamma_e^2 q}{2q - u} + q^2 - \Omega > 0$；$\beta_0^2 = \frac{1 + \nu_e}{1 - (1 - \nu_e)u^2} > 0$. (101) 式解为熟知的形式 $\varepsilon_0(\xi) = \varepsilon_{00}\mathrm{sech}\alpha_0\xi$， (102)

振幅

$$\varepsilon_{00} = \sqrt{2}\,\frac{\alpha_0}{\beta_0} = \left\{2\left[\frac{\gamma_e^2 q + (q^2 - \Omega)(2q - u)}{2q - u}\right] \times \left[\frac{1 - (1 - \nu_e)u^2}{1 + \nu_e}\right]\right\}^{\frac{1}{2}}. \tag{103}$$

满足 $\alpha_0^2 > 0$ 和 $\beta_0^2 > 0$ 条件的解 (100) 和 (102) 式就是重整化后方程的孤立子解. 当 $\nu_e \to 0$ 时，它们退化为通常 Zakharov 方程的双参量孤立子形式：$q = \frac{u}{2}$，$\alpha_0^2 \to q^2 - \Omega$，$\beta_0^2 \to 1/(1 - u^2)$. 这样重整化作用使孤立子的宽度和高度发生了明显的变化.

在静态近似下，方程组(80)，(85)和(89)的无量纲替换基本上同(93) 式，只把 $\boldsymbol{\varepsilon}'$ 的变换改为

$\frac{\sqrt{3m_i}\,\boldsymbol{\varepsilon}'}{(64\pi m_e n_0(T_e + T_i))^{\frac{1}{2}}} \to \boldsymbol{\varepsilon}$，这样，(80) 式形式上仍为(94)式而密度和磁场方程分别为

$$-\nabla^2 n = (1 + \nu_e)\nabla^2|\boldsymbol{\varepsilon}|^2 + 2\nu_e[|\nabla\cdot\boldsymbol{\varepsilon}|^2 - \nabla\boldsymbol{\varepsilon}:\nabla\boldsymbol{\varepsilon}^*], \tag{104}$$

$$\nabla^4\boldsymbol{B} = i\beta_e'^2\nabla^2\nabla\times[\nabla\times(\boldsymbol{\varepsilon}\times\boldsymbol{\varepsilon}^*)] + 3\beta_e'^2\nu_e\frac{\partial}{\partial t}[\boldsymbol{\varepsilon}\times\nabla(\nabla\cdot\boldsymbol{\varepsilon}^*) + c.c.], \tag{105}$$

其中 $\beta_e'^2 = (T_e + T_i)/m_e c^2$.

这些方程组同样有孤立子解，且当 $\nu_e \to 0$ 时，它们就成为通常的非线性 Schrödinger 方程结果.

参 考 文 献

[1] 贺贤土，物理学报，**32**(1983)，325.
[2] 贺贤土，物理学报，**32**(1983)，627.
[3] S. G. Thornhill and D. ter Haar, *Phys. Reports*, **43C** (1978), 43.
[4] L. I. Rudakov and V. N. Tsytovich, *Phys. Reports*, **40C** (1978), 1.
[5] 贺贤土，物理学报，**31**(1982)，1317.
[6] W. Horton, Jr. and Duk-In Choi, *Phys. Reports*, **49** (1979), 273.
[7] V. N. Tsytovich, Theory of Turbulent Plasma, Consultants Bureau, New York, (1977).

THE RENORMALIZED STRONG TURBULENCE THEORY FOR THE LOW-FREQUENCY MAGNETIC FIELD AND THE ION ACOUSTIC WAVE EXCITED BY HIGH-FREQUENCY WAVE

HE XIAN-TU

(Institute of Applied Physics and Computational Mathematics, Beijing)

ABSTRACT

In this paper, the renormalized turbulence theory for the lowfrequency magnetic field and the ion acoustic wave in the high temperature plasma is developed in order to improve the usual weak nonlinear-approach. From Vlasov-Maxwell equations, the coupled renormalized equations of the high and low frequency propagator, with the "most divergence" and "secondary divergence" effects included, are derived in the Fourier representation. Thus, we obtain the coupled relation of the renormalized partical distribution function and field for high and low-frequency oscillation.

Under "most divergence" renormalization approximation, the propagator equations for high and low-frequency are solved. Expanding to the order of ν_e—the ratio of the energy density of the high-frequency turbulence field to the thermal energy density of the plasma particle, the approximate solutions for the propagator and the expressions for the renormalized dielectric function are obtained. Then, by performing Fourier inverse transformation, the renormalized strong turbulence equations are derived in the spacetime representation.

Finally, as an example which shows the renormalized effects, under onedimensional and stationary approximation, the analytical form for the soliton is solved.

第36卷 第2期 物 理 学 报 Vol. 36, No. 2
1987年2月 ACTA PHYSICA SINICA Feb., 1987

等离子体中粒子与 cavitons 相互作用

贺 贤 土
(应用物理与计算数学研究所)

1986年4月14日收到

提 要

从 Vlasov-Poisson 方程组出发,我们导出了电子慢变分布函数 δF_e, 慢变场 $\boldsymbol{E}_s$ 和快变场 $\boldsymbol{E}$ 之间耦合的动力学方程组,并在一维稳态下解析求解了这一方程组,从而获得了粒子分布函数与 soliton 作用关系;进一步讨论了高维下 δF_e, $\boldsymbol{E}_s$, $\boldsymbol{E}$ 塌缩运动和它们的多个 cavitons 行为.

我们知道,当湍动态的L波(Langmuir波)波包在 Fourier 空间展宽到一定大小时,且能量达到了 $|E|^2/8\pi n_0 T_e \gtrsim (\Delta k \lambda_e)^2$ 后,在长波长区的湍动母体上可以激发短波方向相干性的局域运动不稳定性,随后发展成为 solitons(一维稳态)或 cavitons(高维塌缩)运动,并向新的湍动态过渡.这里 $|E|^2/8\pi$ 为L波场能; T_e, λ_e 分别为电子温度和 Debye 长度; Δk 为波包宽度; n_0 为背景粒子数密度.

应用 Захаров 方程能够很好预言上述系统的调制不稳定性,获得L波场和低频振荡粒子数密度扰动随空、时变化关系.但遗憾的是它们只是流体力学框架下的结果,无法给出波被波本身势阱中捕获的细致情况,更无法了解势阱中粒子微观速度分布,而这两者对了解粒子与波相互作用,包括粒子在波中的加速情况是十分重要的. 本文将从 Vlasov-Poisson 方程出发,应用导得 Захаров 方程时同样的近似[1],获得了粒子分布函数,低频场与L波之间耦合方程组,并且在一维稳态下解析求得这组方程的解,从而获得了粒子在 solitons 中的微观分布行为和低频场的变化情况,进而还讨论了它们在高维下塌缩性质.

一、粒子分布函数与高、低频场耦合方程组

把 Vlasov 分布函数分成 $f_\alpha + \delta F_\alpha + F_{0\alpha}$ 三部分,其中 f_α, δF_α 分别为高频和低频振荡部分, $F_{0\alpha}$ 为背景部分;并考虑到L波(以下也称高频波)二次非线性频率差拍过程,从 Vlasov 方程得到高频分布形式解[1]

$$f_\alpha(k, \boldsymbol{v}) = \frac{e_\alpha}{i m_\alpha(\omega - \boldsymbol{k}\cdot\boldsymbol{v})}\left\{\boldsymbol{E}(k)\cdot\frac{\partial F_{0\alpha}}{\partial v} + \int \boldsymbol{E}(k_1)\right.$$
$$\left.\cdot\frac{\partial \delta F_\alpha(q, \boldsymbol{v})}{\partial \boldsymbol{v}}\,\delta^4(k - k_1 - q)\,\frac{d^4k_1 d^4q}{(2\pi)^4}\right.$$

$$+\int \boldsymbol{E}_s(q)\cdot\frac{\partial f_\alpha(k_1)}{\partial \boldsymbol{v}}\delta^4(k-k_1-q)\frac{d^4k_1d^4q}{(2\pi)^4}\Big\}. \tag{1}$$

这里 $\boldsymbol{E}_s(q)$ 为二次非线性差拍激发的低频场，一般情况下它可以是纵、横波，但为方便我们只考虑纵场情况。$k\equiv\{\boldsymbol{k},\omega\}$ 表示高频波四维波矢；同样 $q\equiv\{\boldsymbol{q},\Omega\}$ 为低频波四维波矢。m_α，e_α 分别为粒子质量和电荷。$\alpha=\mathrm{e}$，i。$F_{0\alpha}$ 满足

$$\int F_{0\alpha}(\boldsymbol{v})d^3v=n_{0\alpha}=n_0.$$

同样可得低频分布

$$\delta F_\alpha(q,\boldsymbol{v})=\frac{e_\alpha}{im_\alpha(\Omega-\boldsymbol{q}\cdot\boldsymbol{v})}\Big\{\boldsymbol{E}_s(q)\cdot\frac{\partial F_{0\alpha}}{\partial\boldsymbol{v}}+\int_{(s)}\boldsymbol{E}(k_1)\cdot\frac{\partial f_\alpha(k_2,\boldsymbol{v})}{\partial\boldsymbol{v}}\delta^4(q-k_1-k_2)\frac{d^4k_1d^4k_2}{(2\pi)^4}\Big\}. \tag{2}$$

(2) 式中已略去低频波二次非线性拍频过程，显然它们是高频场的四次非线性效应。积分号 $\int_{(s)}$ 表示只取差拍过程。

为了以下运算方便，引入一些简易表示：令高频场

$$\boldsymbol{E}(k)=\boldsymbol{e}_{\boldsymbol{k}}E(k),\quad \boldsymbol{E}(k_1)=\boldsymbol{e}_1E(k_1),$$

这里

$$\boldsymbol{e}_{\boldsymbol{k}}=\boldsymbol{k}/|\boldsymbol{k}|,\quad \boldsymbol{e}_1=\boldsymbol{e}_{\boldsymbol{k}_1};$$

显然

$$\boldsymbol{e}_{-\boldsymbol{k}}=-\boldsymbol{e}_{\boldsymbol{k}},\quad E^-(k)\equiv E(-k)=-E^{-*}(k),$$

而

$$E^+(k)\equiv E(+k).$$

对于低频场 $\boldsymbol{E}_s(q)=\boldsymbol{e}_{\boldsymbol{q}}E_s(q)$ 也有类似表示。现在把(1)和(2)式代入 Poisson 方程，得到高、低频场方程分别为

$$\varepsilon(k)E(k)=-4\pi\sum_\alpha\frac{e_\alpha^2}{m_\alpha|\boldsymbol{k}|}\int\frac{1}{\omega-\boldsymbol{k}\cdot\boldsymbol{v}}\Big\{\boldsymbol{e}_1\cdot\frac{\partial\delta F_\alpha(q,\boldsymbol{v})}{\partial\boldsymbol{v}}E(k_1)+\boldsymbol{e}_{\boldsymbol{q}}\cdot\frac{\partial f_\alpha(k_1,\boldsymbol{v})}{\partial\boldsymbol{v}}E_s(q)\Big\}\delta^4(k-k_1-q)\frac{d^4k_1d^4q}{(2\pi)^4}d^3v, \tag{3}$$

$$\varepsilon_s(q)E_s(q)=-4\pi\sum_\alpha\frac{e_\alpha^2}{m_\alpha|\boldsymbol{q}|}\int_{(s)}\frac{1}{\Omega-\boldsymbol{q}\cdot\boldsymbol{v}}\boldsymbol{e}_1\cdot\frac{\partial f_\alpha(k_2,\boldsymbol{v})}{\partial\boldsymbol{v}}E(k_1)\cdot\delta^4(q-k_1-k_2)\frac{d^4k_1d^4k_2}{(2\pi)^4}d^3v, \tag{4}$$

其中

$$\varepsilon(k)=1+\sum_\alpha\frac{\omega_\alpha^2}{n_0|\boldsymbol{k}|}\int\frac{\boldsymbol{e}_{\boldsymbol{k}}\cdot\dfrac{\partial F_{0\alpha}}{\partial\boldsymbol{v}}}{\omega-\boldsymbol{k}\cdot\boldsymbol{v}}d^3v \tag{5}$$

为熟知的L波介电函数，ω_α 为 $\alpha=\mathrm{e}$，i 的等离子体频率。$\varepsilon_s(q)$ 是低频介电函数。

若只取到二次高频场拍频，这时(3)式中 f_α 只需取 (1) 式右端第一项就够。现在若 ε 的实部取到 k^2 项以及(3)式右端速度积分时略去 $\dfrac{\boldsymbol{k}\cdot\boldsymbol{v}}{\omega}$ 项，这样，(3) 式两边乘上 ω^2，

并作双边 Fourier 逆变换后，不难得到空时表象中高频场 $\boldsymbol{E}$ 的方程为

$$\frac{\partial^2}{\partial t^2}\boldsymbol{E}-3v_e^2\nabla^2\boldsymbol{E}=-\frac{\omega_e^2}{n_0}\left(n_0+\int\delta F_e(\boldsymbol{x},\boldsymbol{v},t)d^3v\right)\boldsymbol{E}+i\sqrt{\frac{\pi}{2}}\frac{1}{v_e\lambda_e^2}\int\frac{d^4k}{(2\pi)^4}\frac{\omega^3}{k^3}$$
$$\cdot\exp\left\{-\frac{\omega^2}{2k^2v_e^2}+i\boldsymbol{k}\cdot\boldsymbol{x}-i\omega t\right\}\cdot\boldsymbol{E}(\boldsymbol{k},\omega), \tag{6}$$

其中等号右边第二项为线性 Landau 阻尼项来自 ε 虚部，$v_\alpha=\sqrt{T_\alpha/m_\alpha}$ 为 $\alpha=$ e,i 粒子热速度. (6) 式中只留下扰动密度 $n_e=\int\delta F_e d^3v$ 为未知，但为了研究粒子微观分布，我们将保留 δF_e 的形式，并设法导得 δF_e 满足的方程. (1) 式代入 (2) 式后，取到高频场二次非线性拍频，我们得到

$$i(\Omega-\boldsymbol{q}\cdot\boldsymbol{v})\delta F_\alpha(q,\boldsymbol{v})=\frac{e_\alpha}{m_\alpha}\left\{\boldsymbol{E}_s(q)\cdot\frac{\partial F_{0\alpha}}{\partial\boldsymbol{v}}+2i\frac{e_\alpha}{m_\alpha}\int h_\alpha(k_1,-k_2,\boldsymbol{v})\right.$$
$$\left.\times E^+(k_1)E^{-*}(k_2)\delta^4(k-k_1+k_2)\frac{d^4k_1d^4k_2}{(2\pi)^4}\right\}, \tag{7}$$

其中

$$h_\alpha(k_1,k_2,\boldsymbol{v})=\frac{1}{2}\left\{\boldsymbol{e}_1\cdot\frac{\partial}{\partial\boldsymbol{v}}\frac{\boldsymbol{e}_2\cdot\dfrac{\partial F_{0\alpha}}{\partial\boldsymbol{v}}}{\omega_2-\boldsymbol{k}_2\cdot\boldsymbol{v}}+1\longleftrightarrow 2\right\}. \tag{8}$$

由于已取了差拍，积分号下指标 s 便可省去.

进一步化简 (8) 式，得

$$h_\alpha(k_1,-k_2,\boldsymbol{v})=\frac{1}{2}\left\{\frac{\Omega-\boldsymbol{q}\cdot\boldsymbol{v}}{(\omega_1-\boldsymbol{k}_1\cdot\boldsymbol{v})(\omega_2-\boldsymbol{k}_2\cdot\boldsymbol{v})}\boldsymbol{e}_1\cdot\frac{\partial}{\partial\boldsymbol{v}}\boldsymbol{e}_2\cdot\frac{\partial}{\partial\boldsymbol{v}}\right.$$
$$\left.+\boldsymbol{e}_1\cdot\boldsymbol{e}_2\left(\frac{\boldsymbol{k}_2}{(\omega_2-\boldsymbol{k}_2\cdot\boldsymbol{v})^2}-\frac{\boldsymbol{k}_1}{(\omega_1-\boldsymbol{k}_1\cdot\boldsymbol{v})^2}\right)\right\}\cdot\frac{\partial F_{0\alpha}}{\partial\boldsymbol{v}}$$
$$=\frac{1}{2\omega_e^2}\left\{(\Omega-\boldsymbol{q}\cdot\boldsymbol{v})\boldsymbol{e}_1\cdot\frac{\partial}{\partial\boldsymbol{v}}\boldsymbol{e}_2\cdot\frac{\partial}{\partial\boldsymbol{v}}F_{0\alpha}-\boldsymbol{e}_1\cdot\boldsymbol{e}_2\boldsymbol{q}\cdot\frac{\partial F_{0\alpha}}{\partial\boldsymbol{v}}\right\}$$
$$+o\left(\frac{\boldsymbol{k}\cdot\boldsymbol{v}}{\omega}\right)^2_{1,2}. \tag{9}$$

上式代入 (7) 式并作 Fourier 逆变换后，得到空时表象中的方程 ($\alpha=$ e)

$$\left(\frac{\partial}{\partial t}+\boldsymbol{v}\cdot\nabla\right)\left\{\delta F_e(\boldsymbol{x},\boldsymbol{v},t)-\frac{\partial}{\partial\boldsymbol{v}}\cdot\mathbf{D}_e\cdot\frac{\partial}{\partial\boldsymbol{v}}F_{0e}(\boldsymbol{v})\right\}$$
$$=\frac{e}{m_e}\left\{\frac{e}{2m_e\omega_e^2}\nabla(\boldsymbol{E}\cdot\boldsymbol{E}^*)+\boldsymbol{E}_s(\boldsymbol{x},t)\right\}\cdot\frac{\partial F_{0e}}{\partial\boldsymbol{v}}, \tag{10}$$

其中，

$$\mathbf{D}_e\doteq\frac{e^2}{2m_e^2\omega_e^2}\boldsymbol{E}\boldsymbol{E}^*,$$

$$\begin{pmatrix}\boldsymbol{E}\cdot\boldsymbol{E}^*\\ \boldsymbol{E}\boldsymbol{E}^*\end{pmatrix}=2\iint\frac{d^4k_1d^4k_2}{(2\pi)^4}\begin{pmatrix}\boldsymbol{E}^+(k_1)\cdot\boldsymbol{E}^{-*}(k_2)\\ \boldsymbol{E}^+(k_1)\boldsymbol{E}^{-*}(k_2)\end{pmatrix}\cdot\exp\{i(\boldsymbol{k}_1-\boldsymbol{k}_2)\cdot\boldsymbol{x}-i(\omega_1-\omega_2)t\}. \tag{11}$$

为了使方程(10)及(6)式闭合，现在需要给出 $\boldsymbol{E}_s$ 的空时方程(1)代入(4)式，并取二次拍频，得到

$$E_s = -i\frac{(\varepsilon_e(8)-1)}{\varepsilon_s(q)}\frac{e}{n_0 m_e \omega_e^2}q\int E^+(k_1)\cdot E^{*-}(k_2)\delta^4(q-k_1+k_2)\frac{d^4k_1d^4k_2}{(2\pi)^4}. \quad (12)$$

上式已略去离子非线性电流的贡献.

注意到

$$\frac{\varepsilon_e(q)-1}{\varepsilon_s(q)} \doteq \begin{cases} \Omega^2\left(\Omega^2 - q^2v_s^2 + i\sqrt{\dfrac{\pi}{2}}|q|\Omega v_s\sqrt{\mu}\right)^{-1} & \text{对 } v_i \ll \dfrac{\Omega}{|q|} \ll v_e, \\ \dfrac{T_i}{T_e+T_i} & \text{对 } \dfrac{\Omega}{|q|} \ll v_i. \end{cases} \quad (13)$$

这里,

$$\varepsilon_e(q) = 1 + \frac{\omega_e^2}{q^2n_0}\int\frac{q\cdot\dfrac{\partial F_{0e}}{\partial v}}{\Omega - q\cdot v}, \quad v_s = \sqrt{\frac{T_e}{m_i}}$$

为离子声速度, $\mu = m_e/m_i$. 从 (12) 式得到低频纵场的空时方程为

$$\left(\frac{\partial^2}{\partial t^2} + \Gamma_e\frac{\partial}{\partial t}\right)E_s(x, t) - v_s^2\nabla^2 E_s = -\frac{e}{2\omega_e^2 m_e}\frac{\partial^2}{\partial t^2}\nabla(E\cdot E^*), \quad (14)$$

其中 Landau 阻尼项

$$\Gamma_e\frac{\partial}{\partial t}E_s = v_s\sqrt{\frac{\mu\pi}{2}}\frac{\partial}{\partial t}\int|q|E_s(q, t)\cdot\exp(iq\cdot x)\frac{d^3q}{(2\pi)^3}$$

$$= -(2\pi)^{-3/2}\sqrt{\mu}\, v_s\frac{\partial}{\partial t}\int\frac{E_s(x', t)}{|x-x'|^2}d^3x'. \quad (15)$$

(14)式从表明像低频密度方程一样,低频场 E_s 是由有质动力 $\nabla(E\cdot E^*)$ 驱动所形成,并以离子声速度 v_s 进行传播.

方程组 (6), (10) 及 (14) 虽然是闭合方程组,但是由于方程 (6) 具有与其它方程不同的时标,不便于求解, 为此, 把高频场 E 分解为缓变振幅 ε 和快振荡因子两部分. 令

$$E = \frac{1}{\sqrt{2}}(\varepsilon e^{-i\omega_e t} + \text{c.c.}). \quad (16)$$

注意到 $\omega_e\varepsilon \gg \dfrac{\partial}{\partial t}\varepsilon$, 略去 ε 的两次时间导数后,方程 (6) 式便为

$$i\left(\frac{\partial}{\partial t} + \gamma_e\right)\varepsilon + \frac{3v_e^2}{2\omega_e}\nabla^2\varepsilon = \frac{\omega_e}{2n_0}\varepsilon\int\delta F_e(x, v, t)d^3v, \quad (17)$$

其中线性 Landau 阻尼项

$$\gamma_e\varepsilon = \sqrt{\frac{\pi}{2}}\frac{1}{2v_e^2\lambda_e}\int\frac{d^3kd\omega}{(2\pi)^4}\frac{\omega^3}{k^3}\varepsilon(k, \omega)\cdot\exp\left(-\frac{\omega^2}{2k^2v_e^2} + ik\cdot x\right). \quad (18)$$

(10), (14)、(17) 式是我们讨论问题的出发点. 由于 (16), (10) 和 (14) 式中的

$$E\cdot E^* = \text{Re}\,\varepsilon\cdot\varepsilon = |\varepsilon|^2.$$

二、Soliton 中低频场和粒子速度分布函数

方程组 (10), (14) 和 (17) 是包含 Landau 阻尼效应的非线性偏微分方程组. 为了

了解粒子在 caviton 中与波场的相互作用，我们将在 1 + 1 维无阻尼情况下求这组方程的稳态解. 引入行波坐标

$\xi = x - ut$，其中 u 是波包平均速度，这时低频纵场 (14) 式的稳态解为

$$E_s(\xi) = \frac{u^2}{v_s^2 - u^2}\frac{e}{2m_e\omega_e^2}\frac{\partial}{\partial\xi}|\varepsilon|^2. \tag{19}$$

解 (19) 式已取 $\xi \to \pm\infty$ 时，E_s 和 ε 及它们的导数 $\longrightarrow 0$.

同样，从方程 (10) 得到低频振荡电子分布函数稳态解为

$$\delta F_e(\xi, v) = \frac{|\varepsilon|^2}{8\pi n_0 T_e}\left\{\frac{v_s^2}{(v_s^2 - u^2)(v - u)} + \frac{\partial}{\partial v}\right\}\frac{\partial F_{0e}(v)}{\partial v}. \tag{20}$$

这里 $F_{0e}(v)$ 取为电子的 Maxwell 分布. 对(20)式速度积分后，得到慢变电子扰动密度

$$\frac{n_e}{n_0} = \frac{1}{n_0}\int_{-\infty}^{\infty}\delta F_e(\xi, v)dv = \begin{cases} -\dfrac{|\varepsilon|^2}{8\pi n_0 T_e}\dfrac{(1 - u_e^2)}{(1 - u_s^2)} & \text{对 } u_e \ll 1, \\ \dfrac{|\varepsilon|^2}{8\pi n_0 T_e}\dfrac{1}{u_e^2(u_s^2 - 1)} & \text{对 } u_e \gg 1. \end{cases} \tag{21}$$

对 (20) 式速度积分时，在 v 接近 u 处有奇点，积分线路与求介电函数时一样，取主值积分即得 (21) 式. 这里 $u_e \equiv u/v_e$，$u_s \equiv u/v_s$. 对 $u_e \ll 1$，在 $u_s < 1$ 亚声区粒子数密度扰动发生凹陷，有 Soliton 解，这正是 Захаров 方程粒子数密度的稳态解；而在 $1 < u_s \ll 1/\sqrt{\mu}$ 超声区，不出现密度扰动凹陷，是波被波捕获的不稳定区，稳态解没有意义. 对 $u_e \gg 1$，即波包平均速度远大于电子热速度，我们不感兴趣.

最后，我们求解 $\gamma_e = 0$ 的方程 (17)，表示行波坐标系的量为

$$\varepsilon(x, t) = \varepsilon(\xi)\exp(iq_0x - i\Omega_0 t), \tag{22}$$

其中 $\varepsilon(\xi)$ 为模量. (22) 式代入 (17) 式，应用 (21) 式亚声密度 n_e，虚部方程给出了 $q_0 = \frac{1}{3\lambda_e}u_e$，而实部方程则为

$$\frac{d^2\varepsilon(\xi)}{d\xi^2} - \alpha_0^2\varepsilon(\xi) + \beta_0^2\varepsilon(\xi)^3 = 0. \tag{23}$$

这里

$$\alpha_0^2 = \frac{1}{9\lambda_e^2}\left(u_e^2 - 6\frac{\Omega_0}{\omega_e}\right),$$

$$\beta_0^2 = \frac{1}{24\pi n_0 T_e \lambda_e^2}\frac{(1 - u_e^2)}{(1 - u_s^2)}. \tag{24}$$

如果 α_0^2，$\beta_0^2 > 0$，即 $u_s < 1$ 和 $u_e > 6\Omega_0/\omega_e$，(23) 式的解便为熟知的

$$\varepsilon(\xi) = \varepsilon_0 \mathrm{Sech}\,\alpha_0\xi, \tag{25}$$

其中 $\varepsilon_0 = \sqrt{2}\,\alpha_0/\beta_0$. (25)式也是通常 Захаров 方程中 L 波场振幅的稳态解形式，它是双参量 Ω_0, u 的 Soliton 解，显然，$\xi \to \pm\infty$ 时，$\varepsilon(\xi)$ 及 $\frac{d\varepsilon(\xi)}{d\xi} \to 0$.

有了 (25) 式便可写出低频场 E_s 和低频分布 δF_e 的具体形式. 从 (19) 式得到

$$E_s(\xi) = -A_0 \mathrm{sech}^2\alpha_0\xi \tanh\alpha_0\xi. \tag{26}$$

这里

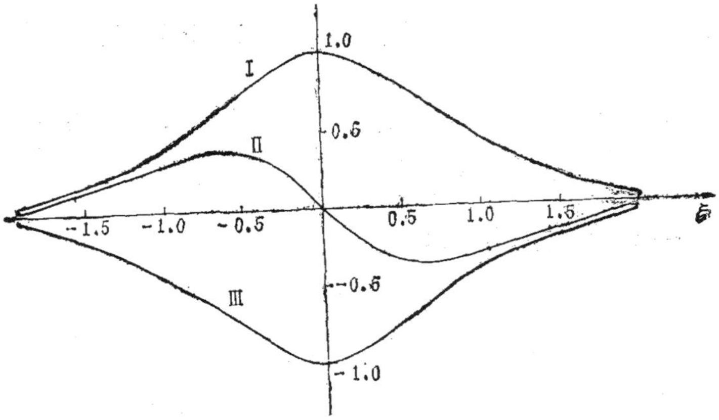

图 1 $|\varepsilon|^2$, E_s 及 n_e 随 ξ 变化

曲线 I $|\varepsilon|^2/\varepsilon_0^2$; 曲线 II $\frac{8\pi n_0 T_e}{\varepsilon_0^2}\left(\frac{(1-u_s^2)e}{2u_s^2\alpha_0 T_e}\right)E_s$; 曲线 III $\frac{1-u_s^2}{1-u_e^2}\,\frac{8\pi n_0 T_e}{\varepsilon_0^2}$

$$A_0=\frac{u_s^2}{1-u_s^2}\,\frac{2\alpha_0 T_e}{e}\left(\frac{\varepsilon_0^2}{8\pi n_0 T_e}\right). \tag{27}$$

图 1 给出了 $|\varepsilon|^2$, n_e/n_0 及 E_s 随 ξ 的变化关系，可见 $E_s(\xi)$ 以 $\xi=0$ 为对称点，由负值 $(\xi>0)$ 到正值 $(\xi<0)$ 反极性地改变. 由于 $\xi\to\pm\infty$, E_s 及其导数均趋向于零，显然，E_s 亦是一种 Soliton 运动，它的谷和峰位置在 $\alpha_0\xi=\pm 0.7$ 点，峰谷间距 $\Delta\xi\doteq 1.4\alpha_0^{-1}$. 在这两点 $|\varepsilon|^2$ 已从 $\alpha_0\xi=0$ 的极大值降到相对比 0.625 左右.

Soliton 中低频粒子流体运动将取决于 E_s，热压力和有质动力三者作用结果，在相消后剩余力 $-\left(\frac{e}{m_e}E_s+\frac{c_s^2}{n_0}\frac{\partial n_e}{\partial x}-\frac{F_p}{n_0 m_e}\right)$ 作用下粒子发生加速运动. 这里有质动力 $F_p\doteq -\frac{1}{8\pi}\frac{\partial|\varepsilon|^2}{\partial x}$，等离子体声速 $c_s=\sqrt{\Gamma T_e}$，$\Gamma=3$.

现在讨论 $\delta F_e(\xi,v)$ 随 ξ, v 的变化关系. 利用解 (25) 式后，(20) 式可写成

$$\delta F_e(\xi,v)=\frac{\varepsilon_0^2}{8\pi n_0 T_e}\operatorname{sech}^2\alpha_0\xi\left\{\tilde v^2-\left(1+\frac{\tilde v}{(\tilde v-u_e)(1-u_s^2)}\right)\right\}F_{0e}(\tilde v), \tag{28}$$

其中无量纲速度

$$\tilde v=\frac{v}{v_e},\quad F_{0e}(\tilde v)=\frac{n_0}{\sqrt{2\pi}}\cdot\exp\left(-\frac{\tilde v^2}{2}\right).$$

从方程 (10) 可以知道，在二次高频场作用下，δF_e 的空、时变化是由于有质动力和低频场 E_s 所驱动. 因为 E_s 也是以有质动力为驱动源，所以方程 (10) 的解，即这里的 (28) 式直接由 $|\varepsilon|^2$ 的 Soliton 场及与背景粒子速度变化有关两部分组成. 因此 δF_e 的空时行为完全是 Soliton 运动. 而保持空间的对称结构.

$\delta F_e(\xi,\tilde v)$ 随 $\tilde v$ 的分布行为如上(见图 1).

在 $\tilde v=0$ 点有一个负的极小值 $-\frac{1}{\sqrt{2\pi}}\frac{|\varepsilon|^2}{8\pi n_0 T_e}$，可以看到速度在 $-\tilde v_0<\tilde v<v_0\doteq\left(1+\frac{1}{1-u_s^2}\right)^{1/2}$ 区的粒子在 E_s 和有质动力作用下，一部分跑出 Soliton 铃形空间区，另一

部分加速到其它速度区域．前者正是发生扰动密度 $n_e < 0$ 凹陷的原因；后者导致了 $\tilde{v}$ 轴上有两个极大点 $\tilde{v}_{m_1} = \pm\left(3 + \dfrac{1}{1-u_s^2}\right)^{1/2}$，它近似与场无关，这里已用了 $\tilde{v}_{m_1} \gg u_e = \sqrt{\mu}u_s$ 的假定．最后，δF_e 形成了以 $\tilde{v}=0$ 为对称轴的图 2 的双峰分布形状，峰值 $\delta F_e(\pm\tilde{v}_{m_1}) = 2\dfrac{\varepsilon_0^2}{8\pi n_0 T_e}\operatorname{sech}^2\alpha_0\xi F_{0e}(\pm\tilde{v}_{m_1})$．显见在 $-\tilde{v}_0<\tilde{v}<\tilde{v}_0$ 区扰动分布 $\delta F_e(\tilde{v})<0$，而对于 $\tilde{v}\gg 1$区 $\delta F_e \sim \dfrac{|\varepsilon|^2}{8\pi n_0 T_e}\tilde{v}^2\cdot\exp\left(-\dfrac{\tilde{v}^2}{2}\right)$ 形状，比指数减少缓慢．

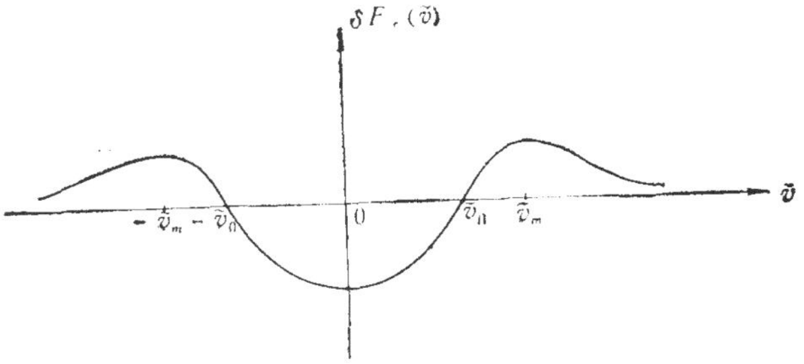

图 2　扰动分布函数 $\delta F_2(\tilde{v})$ 随 $\tilde{v}$ 的变化关系

在调制相互作用时间尺度上平均的慢变分布函数便为

$$F_e(\xi,\ \tilde{v}) = \left\{1 + \frac{|\varepsilon|^2}{8\pi n_0 T_e}\left[\tilde{v}^2 - \left(1 + \frac{\tilde{v}}{(\tilde{v}-u_e)(1-u_s^2)}\right)\right]\right\} F_{0e}(\tilde{v}). \qquad (29)$$

(29) 式在 $-\tilde{v}_0 < \tilde{v} < \tilde{v}_0$ 区低于背景粒子分布，显然在 $\tilde{v}=0$ 中心附近

$$F_e(\xi,\ \tilde{v}) = \left\{1 - \frac{|\varepsilon|^2}{8\pi n_0 T_e}\right\} F_{0e}(\tilde{v}),$$

峰值为

$$F_e(\xi,\ 0) = \frac{n_0}{\sqrt{2\pi}}\cdot\left\{1 - \frac{|\varepsilon|^2}{8\pi n_0 T_e}\right\}.$$

当场能密度不太小且满足 $\left(\dfrac{|\varepsilon|^2}{8\pi n_0 T_e}\right)^{-1} < \tilde{v}_{m_1}^2$ 时，$F_e(\xi,\ \tilde{v})$ 随 $\tilde{v}$ 变化亦有双峰结构，其极大值发生在 $\tilde{v}_{m_2} \doteq \left(\tilde{v}_{m_1}^2 - \dfrac{1}{A_1}\right)^{1/2}$，这里 $A_1 = |\varepsilon|^2/8\pi n_0 T_e$；对于很低的场能密度 $A_1 \ll 1$，从而 $\tilde{v}_{m_1}^2 < 1/A_1$，这时虽然 $\delta F_e(\tilde{v})$ 有双峰结构，但叠加背景分布后，主要仍取决于 F_{0e} 随 $\tilde{v}$ 变化的单峰结构．

当 $\tilde{v}\gg 1$ 时，$F_e(\xi,\ \tilde{v}) = \left(1 + \dfrac{|\varepsilon|^2}{8\pi n_0 T_e}\tilde{v}^2\right)\cdot\exp(-\tilde{v}^2/2)$，表明了粒子在 Soliton 内受到加速，高能粒子多于 Maxwell 分布．在 (29) 式等号两边乘上 $\dfrac{1}{2}mv^2$ 积分后，得粒子动能

$$E_k = \frac{1}{2}n_0 T_e\left\{1 + \frac{1}{n_0}\int_{-\infty}^{\infty}\tilde{v}^2\delta F_e d\tilde{v}\right\} \approx \frac{1}{2}n_0 T_e\left\{1 + \frac{|\varepsilon|^2}{8\pi n_0 T_e}\left(2 - \frac{1}{1-u_s^2}\right)\right\}. \qquad (30)$$

(30)表明了 Soliton 中粒子平均地获得或减少能量取决于亚声波包平均速度 $u_s(<1)$ 的

大小．当 $u_s < 1/\sqrt{2}$ 时，粒子发生了等效加热作用，有效温度

$$T_{eff} = T_s\left\{1 + \frac{|\varepsilon|^2}{8\pi n_0 T_e}\left(2 - \frac{1}{1 - u_s^2}\right)\right\}.$$

三、关于慢变分布函数、慢变场塌缩运动讨论

在讨论 δF_e 及 $\boldsymbol{E}_s$ 的塌缩运动之前，我们首先讨论 Захаров 方程的一些性质．如果从导得 δF_e，$\boldsymbol{E}_s$ 和 $\boldsymbol{\varepsilon}$ 的原始方程(7)，(12)及(3)式出发，在二级高频场拍频近似下，Fourier 逆变换后不难获得所需要的 Захаров 方程

$$i\left(\frac{\partial}{\partial t} + \gamma_e\right)\boldsymbol{\varepsilon} + \frac{3v_e^2}{2\omega_e}\nabla^2\boldsymbol{\varepsilon} = \frac{\omega_e}{2n_0}n_e\boldsymbol{\varepsilon}, \tag{31}$$

$$\left(\frac{\partial^2}{\partial t^2} + \Gamma_e\frac{\partial}{\partial t}\right)n_e - v_s^2\nabla^2 n_e = \frac{1}{8\pi m_i}\nabla^2(\boldsymbol{E}\cdot\boldsymbol{E}^*). \tag{32}$$

这个方程组已有很多重要性质被揭露． Захаров 和他的合作者首先证明了[2-4]类似球层结构的 Caviton 是不稳定的，具有类似冲击波自相似聚心运动规律，调制不稳定产生的高维扰动能够导致波包塌缩运动，或者伴随声辐射．

对于保守系统，从方程(31)和(32)可以得到各种守恒关系和守恒量． 引入无量纲量

$$\frac{2}{3}\frac{\sqrt{\mu}}{\lambda_e}\boldsymbol{x} \to \boldsymbol{\xi},\ \frac{2}{3}\omega_e\mu t \to \tau,\ \frac{\sqrt{3}\,\boldsymbol{\varepsilon}}{(32\pi\mu n_0 T_e)^{1/2}} \to \boldsymbol{\varepsilon},$$

(31)和(32)式可以无量纲化．于是准粒子数守恒给出

$$N = \int|\boldsymbol{\varepsilon}(\xi, \tau)|^2 d^3\xi \tag{33}$$

即

$$|\boldsymbol{\varepsilon}(\xi, \tau)| \sim \xi^{-3/2}. \tag{34}$$

这表明(17)式的自相似解形式为[1]

$$|\boldsymbol{\varepsilon}| = \frac{1}{\tau}\phi\left(\frac{\xi}{\tau^{2/3}}\right), \qquad \int\phi(\xi)d^3\xi = 1. \tag{35}$$

因此，当 Caviton 向球心塌缩时，$\phi(\xi) \to \infty$，对 $\tau < 1$，就有 $|\boldsymbol{\varepsilon}| \to \infty$．

大量数值解都肯定了[5-8]Захаров 方程具有 Cavitons 行为，并表明了有多个 Cavitons 解的性质．事实上，调制不稳定性使原始一个大尺度波包破碎成为一系列小尺度波包之和，正是它们在调制相互作用发展过程中形成了很多 Cavitons．

在低频波相速度小于离子热速度情况下，Захаров 方程退化为非线性 Schrödinger 方程．Goldman 和 Nicholson[9]利用维里定理证明了非线性 Schrödinger 方程所描述的波包平均平方宽度随时间变化关系为

$$\langle\delta\boldsymbol{x}^2\rangle = At^2 + Bt + C - (D-2)\int_0^t dt'\int_0^{t'} dt'' I(t''), \tag{36}$$

其中

$$I(t) = \int|\varepsilon|^4 d^D\boldsymbol{x},$$

D 为空间维数，$A<0$ 为运动常数．因此，当 $D\geqslant 2$ 时，随着时间 t 的增大，波包总是不断地塌缩．

上述有趣的结论虽然是从 Захаров 方程组得出，但是前面已经提到，Захаров 方程组中的密度方程(32)是由 δF_e 方程(7)和 $\boldsymbol{E}_s$ 方程(12)耦合而来，因此，方程(10)，(14)和(17)与 Захаров 方程是等价的．这样，利用 Захаров 方程求得的 $\boldsymbol{\varepsilon}$ 从(14)和(17)式就可自然求得 δF_e 和 $\boldsymbol{E}_s$，并立刻可知 $\boldsymbol{\varepsilon}$ 的塌缩性质必然使 δF_e 和 $\boldsymbol{E}_s$ 也具有塌缩行为．这一点还可从一维行波解结果(26)和(28)式直接看出来，由于 $\varepsilon(\xi)$ 具有 Soliton 性质，δF_e 和 E_s 的解直接与 $\varepsilon(\xi)$ 有关而出现了 Soliton 行为．

参　考　文　献

[1] 贺贤士，物理学报，**32**(1983)，325.
[2] В. Е. Захаров, *ЖЭТФ*, **62**(1972), 1745.
[3] Л, М. Дегтяре и пр., *ЖЭТФ*, **67**(1974), 533.
[4] А. Г. Литвак, Г. М. Фраймаи, Л. Д. Юнаковский. *Письма в МЭТФ*, **19**(1974), 23.
[5] Л. М. Дегтярев, В. Е. Захаров, Динамика Ленгмюровского коллапса, 第二次等离子体理论国际会议，基辅，(1974).
[6] D. R. Nichclson, *Physica Scripta*, **27**(1983), 77.
[7] N. R. Pereira, R. N. Sudan, J. Denavit, *Phys. Fluids*, **20**(1977), 936.
[8] D. R. Nicholson and M. V. Goldman, *Phys. Fluids*, **21**(1978), 1766.
[9] M. V. Goldman and D. R. Nicholson, *Phys. Rev. Lett.*, **41**(1978), 406.

THE INTERACTION OF PARTICLES WITH CAVITONS IN PLASMA

HE XIAN-TU

(Institute of Applied Physics and Computational Mathematics, P. O. Box 8009, *Beijing)*

ABSTRACT

Starting from Vlasov-Doisson equations, the coupled dynamics equations for slowly varying electron distribution function δF_e, slowly varying field $\boldsymbol{E}_s$ and fast varying field $\boldsymbol{E}$ are derived and the system of above equations is analytically solved. Thus, the interactions of the particles distribution functions with solitons are revealed. Furthermore, the collapse motions of δF_e $\boldsymbol{E}_s$ $\boldsymbol{E}$ for higher dimensions and their many cavitons behavior are discussed.

PHYSICAL REVIEW A VOLUME 43, NUMBER 4 15 FEBRUARY 1991

Relative diffusion and formation of clumps in phase space of electrons in localized Langmuir fields

Shao-ping Zhu and X. T. He
Center for Nonlinear Studies, Institute of Applied Physics and Computational Mathematics, P.O. Box 8009, Beijing, China
(Received 12 February 1990; revised manuscript received 4 September 1990)

A different relative diffusion coefficient for electrons interacting with coherent, localized Langmuir wave packets is proposed. It is shown that the coherent, localized wave packets can drive the formation of clumps in phase space, and only when the relative velocity is zero does the lifetime of clumps diverge logarithmically, with the relative positions of the constituents getting closer to one another.

In this paper we report that trajectories of electrons interacting with the localized, coherent Langmuir wave packets exhibit an effect of the strong two-point correlation. These wave packets are important in many aspects of laser fusion, space plasma, and turbulent plasma, etc. Thus the clump phenomena in phase space, as proposed by Dupree in 1972,[1] can appear in the case of the spatially inhomogeneous and coherent fields.

Write the localized, coherent fields in the form of a soliton structure:[2]

$$E(x,t)=E_0\operatorname{sech}(gx)\cos(k_0x-\omega_p t)\ , \tag{1}$$

where $g=\beta E_0/\sqrt{8\pi n k_B T}$, β is a parameter, E_0 amplitude, k_0 Langmuir wave number, T electron temperature, n electron number density, and $\omega_p=(4\pi n e^2/m_e)^{1/2}$, where m_e is electron mass.

Extending sech(gx) to whole space with a periodic length L, and expressing it in terms of wave packets, the equations for electron trajectories in phase space are written in the dimensionless form as follows:

$$\begin{aligned}\frac{dx_i(t)}{dt}&=v_i(t)\ ,\\ \frac{dv_i(t)}{dt}&=-\sum_{l=-l_0}^{l_0}E_l\cos[k_l x_i(t)-t],\quad i=1,2\end{aligned}\tag{2}$$

where $k_l=k_0+(2l\pi/L)$, $l=0,\pm1,\pm2,\ldots,\pm l_0$, $k_0=\pi/L$; E_l is in units of $\sqrt{8\pi n k_B T}$, space is in units of the Debye length $\lambda_e=(T_e/4\pi n e^2)^{1/2}$, time is in units of $1/\omega_p$.

It is shown that if the overlapping criterion is satisfied,[3] the periodically trapping trajectories of the electrons in phase space may lead to stochastic processes that form a stochastic region in phase space, accompanied by the existence of small regular islands in this stochastic region.[2,4] Thus the stochastic motion may lead to strong correlations of neighboring electrons in phase space, because relative diffusion is greatly reduced, and so may tend to enable clumps to form in phase space under localized and coherent fields.

Introducing the relative and barycentric motion coordinates

$$\begin{aligned}r&=x_1-x_2,\quad R=x_1+x_2\ ,\\ u&=v_1-v_2,\quad V=v_1+v_2\ ,\end{aligned}\tag{3}$$

from Eq. (2) we can write the following equations for the above-noted coordinate system in the form

$$\begin{aligned}\frac{dr}{dt}&=u,\quad \frac{du}{dt}=\sum_n 2E_n\sin(\tfrac{1}{2}k_nR-t)\sin(\tfrac{1}{2}k_nr)\ ,\\ \frac{dR}{dt}&=V,\quad \frac{dV}{dt}=-\sum_n 2E_n\cos(\tfrac{1}{2}k_nR-t)\cos(\tfrac{1}{2}k_nr)\ .\end{aligned}\tag{4}$$

According to the definition of the diffusion coefficient, we obtain the relative diffusion coefficient

$$D_-(R,r,V,u)=\frac{1}{2t}\langle[u(t)-u(0)]^2\rangle\ , \tag{5}$$

and the absolute diffusion coefficient

$$D_{ii}(v_i,x_i)=\frac{1}{2t}\langle[v_i(t)-v_i(0)]^2\rangle\ , \tag{6}$$

where the symbol $\langle\ \rangle$ denotes an ensemble average.

Next we express Eqs. (5) and (6) in a specific form. First of all, to integrate Eqs. (2), we obtain the trajectories for electrons as follows:

$$\begin{aligned}x_i(t)&=x_i^0+v_i^0t\\ &\quad-\int_0^t dt'(t-t')\sum_n E_n\cos[k_nx_i(t')-t']\ ,\\ v_i(t)&=v_i^0-\int_0^t dt'\sum_n E_n\cos[k_nx_i(t')-t']\ .\end{aligned}\tag{7}$$

Substituting Eqs. (7) into Eq. (5), and transforming to $t_1=t'$, $t_2=t_1-t''$, Eq. (5) can be rewritten in the form

$$\begin{aligned}D_-=\frac{1}{t}\int_0^t dt_1\int_0^{t_1}dt_2\sum_{n,m}E_nE_m\langle\{&\cos[k_nx_1(t_1)-t_1]\cos[k_mx_1(t_1-t_2)-(t_1-t_2)]\\ &-\cos[k_nx_1(t_1)-t_1]\cos[k_mx_2(t_1-t_2)-(t_1-t_2)]\}+(1\leftrightarrow2)\rangle\ ,\end{aligned}\tag{8}$$

where (1↔2) represents the preceding terms with the indices 1 and 2 interchanged.

We now express the first integrand term inside the braces in the exponent form; then, we obtain

$$\langle \cos[k_n x_1(t_1)-t_1]\cos[k_m x_1(t_1-t_2)-(t_1-t_2)]\rangle$$
$$=\tfrac{1}{2}(\cos\{k_0(R^0+r^0)-2[1-\tfrac{1}{2}k_0(V^0+u^0)]t_1+[1-\tfrac{1}{2}k_{-n}(V^0+u^0)]t_2\}$$
$$\times\delta_{n,-m}\exp\{-\tfrac{1}{2}\langle[k_n\Delta x_1(t_1)+k_{-n}\Delta x_1(t_1-t_2)]^2\rangle\}$$
$$+\cos\{[1-\tfrac{1}{2}k_n(V^0+u^0)]t_2\}\delta_{n,m}\ \exp\{-\tfrac{1}{2}\langle[k_n\Delta x_1(t_1)-k_n\Delta x_1(t_1-t_2)]^2\rangle\}) , \qquad (9)$$

where the cumulant expansion approximation has been used, and $\Delta x_i(t)\equiv x_i(t)-x_i^0-v_i^0 t$.

Treating the rest of Eq. (8) in a way similar to Eq. (9), then, we obtain

$$D_-=\frac{1}{2t}\int_0^t dt_1\int_0^{t_1}dt_2\sum_{i=1}^{2}\sum_n E_n^2(\cos[\xi_1+(-1)^{i-1}\eta_1]\exp\{-\tfrac{1}{2}\langle[k_n\Delta x_i(t_1)+k_{-n}\Delta x_i(t_1-t_2)]^2\rangle\}$$
$$-\cos[\xi_1+(-1)^{i-1}\eta_2]\exp\{-\tfrac{1}{2}\langle[k_n\Delta x_i(t_1)+k_{-n}\Delta x_i(t_1-t_2)$$
$$-(-1)^{i-1}k_{-n}\Delta r(t_1-t_2)]^2\rangle\}$$
$$+\cos[\xi_2+(-1)^{i-1}\eta_3]\exp\{-\tfrac{1}{2}\langle[k_n\Delta x_i(t_1)-k_n\Delta x_i(t_1-t_2)]^2\rangle\}$$
$$-\cos[\xi_2+(-1)^{i-1}\eta_4]\exp\{-\tfrac{1}{2}\langle[k_n\Delta x_i(t_1)-k_n\Delta x_i(t_1-t_2)$$
$$+(-1)^{i-1}k_n\Delta r(t_1-t_2)]^2\rangle\}) , \qquad (10)$$

where

$$\Delta r(t)=r(t)-r^0-u^0t,\quad \xi_1=k_0R^0-2(1-\tfrac{1}{2}k_0V^0)t_1+(1-\tfrac{1}{2}k_{-n}V^0)t_2 ,$$
$$\eta_1=k_0r^0+k_0u^0t_1-\tfrac{1}{2}k_{-n}u^0t_2,\quad \eta_2=(k_0-k_{-n})r^0+(k_0-k_{-n})u^0t_1+\tfrac{1}{2}k_{-n}u^0t_2,\quad \xi_2=(1-\tfrac{1}{2}k_nV^0)t_1 ,$$
$$\eta_3=-\tfrac{1}{2}k_nu^0t_2,\quad \eta_4=-k_nr^0-k_nu^0t_1+\tfrac{1}{2}k_nu^0t_2 . \qquad (11)$$

In order to calculate the cumulant expansion in Eq. (10), we assume that the trajectory stochasticity in velocity space can be described by a Wiener process.[5] Then the probability density of finding a particle at time t with velocity increment Δv_i, where $\Delta v_i(t=0)=0$, is given by

$$P(\Delta v_i,t)=\frac{1}{\sqrt{4\pi D_{ii}t}}\exp\left[-\frac{(\Delta v_i)^2}{4D_{ii}t}\right]\quad \text{for } t>0 . \qquad (12)$$

Using the Wiener average instead of an ensemble average, we have

$$\langle[k_n\Delta x_i(t_1)-k_n\Delta x_i(t_1-t_2)]^2\rangle=\tfrac{2}{3}D_{ii}k_n^2t_2^2(3t_1-2t_2) , \qquad (13)$$

$$\langle[k_n\Delta x_i(t_1)+k_{-n}\Delta x_i(t_1-t_2)]^2\rangle=\tfrac{2}{3}D_{ii}\{k_0(4t_1^3-6t_2t_1^3+3t_2t_1)+2k_0(k_n-k_0)[t_1^3-(t_1-t_2)^3]$$
$$+(k_n-k_0)^2t_2^2(3t_1-2t_2)\} , \qquad (14)$$

$$\langle[k_n\Delta x_i(t_1)+k_{-n}\Delta x_i(t_1-t_2)-(-1)^{i-1}k_{-n}\Delta r(t_1-t_2)]^2\rangle$$
$$=\langle[k_n\Delta x_i(t_1)+k_{-n}\Delta x_i(t_1-t_2)]^2\rangle-2(-1)^{i-1}\langle[k_n\Delta x_i(t_1)+k_{-n}\Delta x_i(t_1-t_2)]k_{-n}\Delta r(t_1-t_2)\rangle$$
$$+k_{-n}^2\langle[\Delta r(t_1-t_2)]^2\rangle , \qquad (15)$$

$$\langle[k_n\Delta x_i(t_1)-k_n\Delta x_i(t_1-t_2)+(-1)^{i-1}k_n\Delta r(t_1-t_2)]^2\rangle$$
$$=\langle[k_n\Delta x_i(t_1)-k_n\Delta x_i(t_1-t_2)]^2\rangle+2(-1)^{i-1}\langle[k_n\Delta x_i(t_1)-k_n\Delta x_i(t_1-t_2)]k_n\Delta r(t_1-t_2)\rangle+k_n^2\langle[\Delta r(t_1-t_2)]^2\rangle . \qquad (16)$$

In Eqs. (15) and (16) there exist the terms $\langle[\Delta r(t)]^2\rangle$ and $\langle\Delta x_i(t_1)\Delta r(t_2\le t_1)\rangle$. It is clear that $\langle|\Delta r(t)|^2\rangle$ is proportional to D_-, so is $\langle\Delta x_i(t)\Delta r(t_2\le t)\rangle$, but $\langle\Delta x_i(t_1)\Delta x_i(t_1-t_2)\rangle$ is proportional to D_{ii}. In our case, D_- is much less than D_{ii} when r^0 and u^0 are smaller, so the contribution of $\langle|\Delta r(t)|^2\rangle$ and $\langle\Delta x_i(t)\Delta r(t_2\le t)\rangle$ to D_- in Eq. (10) is very small. For the convenience of discussions, we calculate them under the unperturbed trajectory approximation, and obtain

$$\langle[\Delta r(t)]^2\rangle=\sum_{i=1}^{2}\sum_{n}E_n^2\{F(2k_0x_i^0,-(1-k_nv_i^0),-(1-k_{-n}v_i^0),t,t)+F(0,1-k_nv_i^0,-(1-k_nv_i^0),t,t)$$
$$-F(2k_0x_i^0-(-1)^{i-1}k_{-n}r^0,-(1-k_nv_i^0),-[1-k_{-n}v_i^0+(-1)^{i-1}k_{-n}u^0],t,t)$$
$$+F(-(-1)^{i-1}k_nr^0,1-k_nv_i^0,-[1-k_nv_i^0+(-1)^{i-1}k_nu^0],t,t)\}\ , \tag{17}$$

$$\langle\Delta x_i(t_1)\Delta r(t_2\le t_1)\rangle=\sum_{n}E_n^2\{F(2k_0x_i^0,-(1-k_nv_i^0),-(1-k_{-n}v_i^0),t_1,t_2)+F(0,1-k_nv_i^0,-(1-k_nv_i^0),t_1,t_2)$$
$$-F(2k_0x_i^0-(-1)^{i-1}k_{-n}r^0,-(1-k_nv_i^0),-[1-k_{-n}v_i^0+(-1)^{i-1}k_{-n}u^0],t_1,t_2)$$
$$+F(-(-1)^{i-1}k_nr^0,1-k_nv_i^0,-[1-k_nv_i^0+(-1)^{i-1}k_nu^0],t_1,t_2)\}\ , \tag{18}$$

where function F is defined as follows:

$$F(a,b,c,t_1,t_2)=\frac{\cos a-\cos(a+bt_1)-\cos(a+ct_2)+\cos(a+bt_1+ct_2)}{2b^2c^2}-\frac{(bt_1+ct_2)\sin a}{2b^2c^2}$$
$$+\frac{t_1\sin(a+ct_2)}{2bc^2}+\frac{t_2\sin(a+bt_1)}{2b^2c}-\frac{t_1t_2\cos a}{2bc}\ . \tag{19}$$

We find from Eq. (10) that D_- vanishes only for $r^0=0$, $u^0=0$; however, if u^0 is a nonzero constant, D_- is not equal to zero even if r^0 vanishes. Because $\eta_2-\eta_1=-k_{-n}[r^0+u^0(t_1-t_2)]$, $\eta_4-\eta_3=-k_n[r^0+u^0(t_1-t_2)]$ and $t_1\ge t_2$, we are sure that D_- is of minimum when $r^0>0$, $u^0<0$ or $r^0<0$, $u^0>0$.

By using Eq. (7), the absolute diffusion coefficient in Eq. (6) can be easily rewritten in the form

$$D_{ii}=\frac{1}{2t}\int_0^t dt_1\int_0^{t_1}dt_2\sum_n E_n^2(\cos[2k_0x_i^0-2(1-k_0v_i^0)t_1$$
$$+(1-k_{-n}v_i^0)t_2]\exp\{-\tfrac{1}{2}\langle[k_n\Delta x_i(t_1)+k_{-n}\Delta x_i(t_1-t_2)]^2\rangle\}$$
$$+\cos[(1-k_nv_i^0)t_2]\exp\{-\tfrac{1}{2}\langle[k_n\Delta x_i(t_1)-k_n\Delta x_i(t_1-t_2)]^2\rangle\})\ . \tag{20}$$

With numerical solutions of Eq. (4), according to the definition of Eq. (5), we write the absolute and the relative diffusion coefficients to be computed in the form

$$D_{ii}^p=\frac{1}{2\tau}\langle[v_i(\tau)-v_i(0)]^2\rangle\ ,$$
$$D_-^p=\frac{1}{2\tau}\langle[u(\tau)-u(0)]^2\rangle\ , \tag{21}$$

where

$$\tau=\frac{1}{2}\left[\frac{L}{v_1}+\frac{L}{v_2}\right]$$

is the approximate time for one pass through the periodic structure. The ensemble average is computed by averaging over a set of particles with different start times but with the same initial velocity and positions.

The spectrum of the wave packet is presented in Fig. 1 for parameters $E_0^2/8\pi nk_BT=0.53$, $L=36$, and $\beta=1/\sqrt{3}$, where the overlapping conditions are satisfied except for the modes $n=0,-1$. Figures 2 and 3 show the values of D_- given by Eqs. (10), (4), and (21), where the factors $\langle[\Delta r(t)]^2\rangle$ and $\langle\Delta x_i(t_1)\Delta r(t_2\le t_1)\rangle$ in Eq. (10) have been neglected to economize computing time. The fact that D_- and D_-^p in Figs. 2 and 3 are almost identical ob-

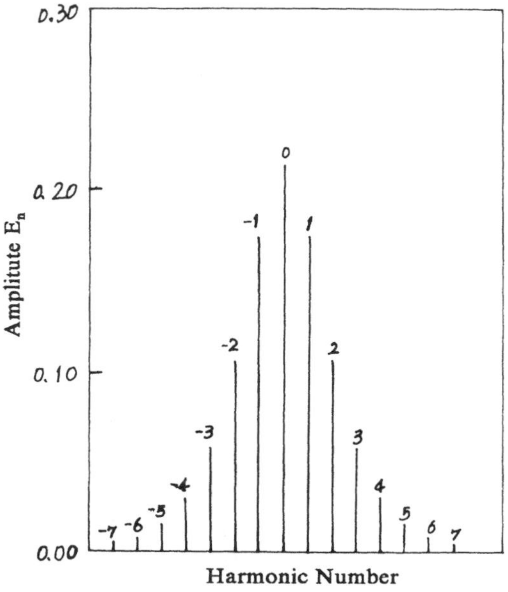

FIG. 1. Spectrum of the Langmuir wave packet.

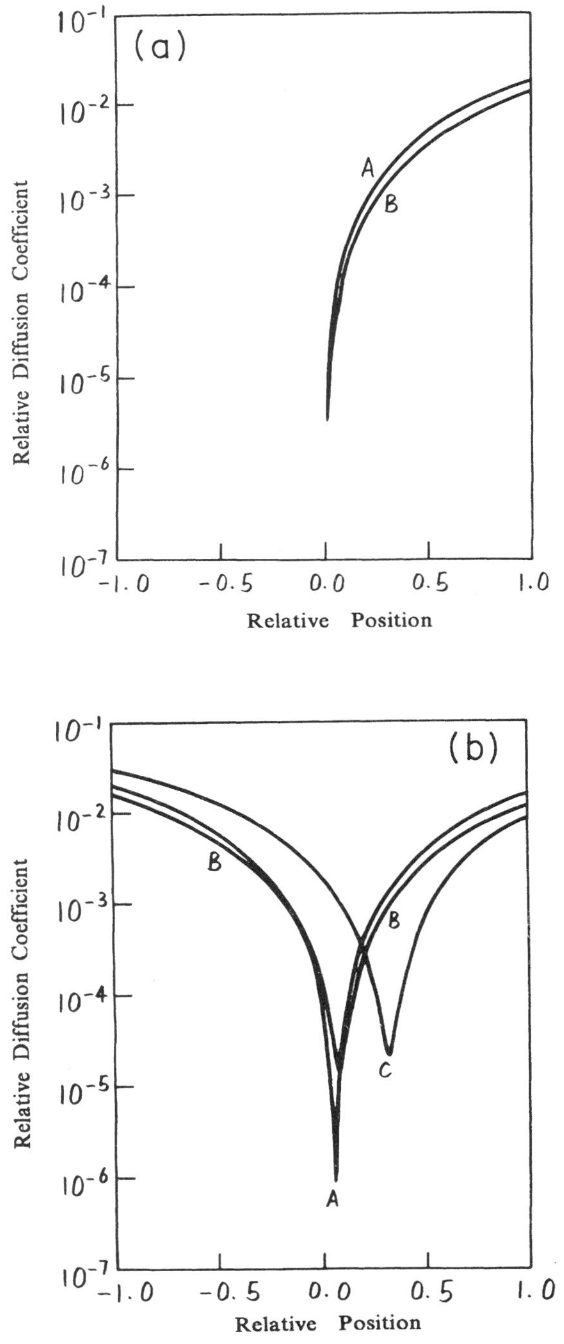

FIG. 2. (a) Relative diffusion coefficients vs the relative positions when $u^0=0$. Curve A given by numerical results; curve B by Eq. (10). (b) Relative diffusion coefficients vs the relative positions. Curves A and C given by numerical results corresponding to $u^0=-0.01,-0.05$, respectively; curve B by Eq. (10) when $u^0=-0.01$.

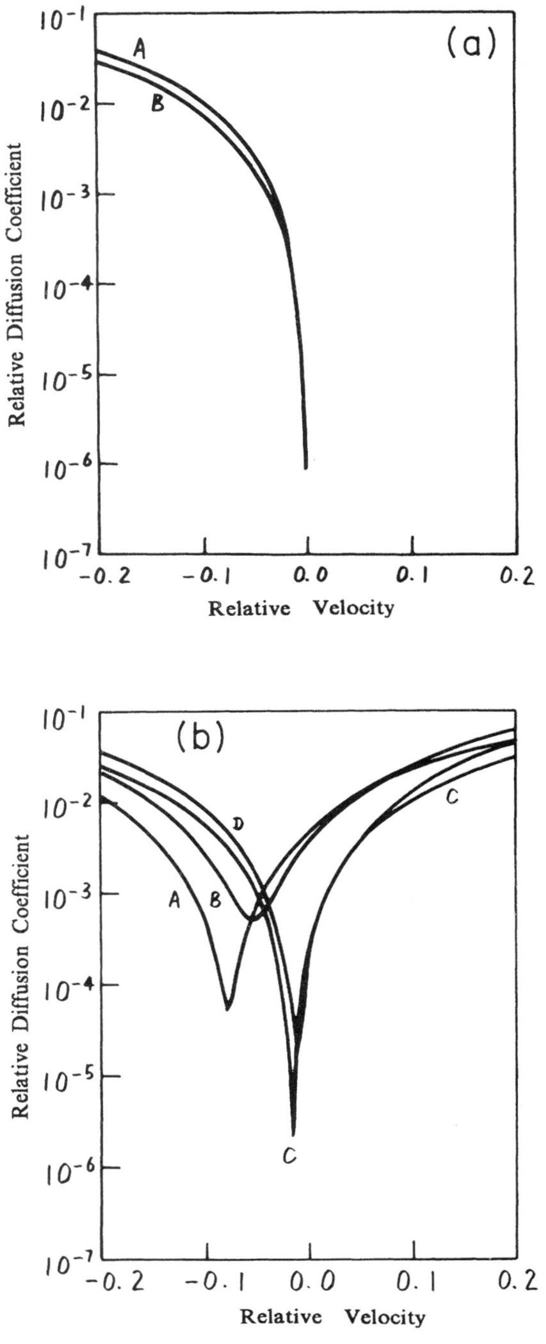

FIG. 3. (a) Relative diffusion coefficients vs the relative velocities when $r^0=0$. Curve B given by numerical results; curve A by Eq. (10). (b) Relative diffusion coefficients vs the relative velocities. Curves A and C given by numerical results corresponding to $r^0=0.1,0.5$, respectively; curves B and D by Eq. (10) corresponding to $r^0=0.1,0.5$, respectively.

viously supports the above approximation to be reasonable.

One can see from Figs. 2 and 3 that D_- of Eq. (10) and D_-^p of Eq. (21) are almost identical, therefore the numerical computation is reliable. Our results show that for $u^0=0$ D_- decreases when r^0 decreases, which is in keeping with that proposed by Dupree,[1] who gave the relative diffusion coefficient as follows:

$$D_-^d=\frac{q^2}{m_e^2}\int_0^\infty dt\sum_{k,\omega}|E_{k,\omega}|^2\exp[i(kv-\omega)t-\tfrac{1}{3}k^2Dt^3]\times(1-\cos kr^0)\ , \qquad (22)$$

where D is the absolute diffusion coefficient. However, our results show a dependence of D_- on u^0 as well, which is just like r^0, whereas D_-^d of Eq. (22) is not dependent on u^0.

As we expected, D_-^p given by numerical simulation has a nonzero minimum when $r^0+\frac{1}{2}\tau u^0=0$, as shown in Figs. 2 and 3. The fact that D_- has a nonzero minimum means that the lifetime of clumps is finite. It is easy to understand, because D_{11}, D_{22}, and D_- are non-negative, and D_- approaches $D_{11}+D_{22}$ when $|r^0|\to\infty$ or $|u^0|\to\infty$, so D_- varying with r^0 or u^0 must have a minimum. Thus the relative diffusion is greatly reduced close to D_- minima, and the two-point correlation is considerably enhanced.

Analyzing the D_-^p given by Eqs. (4) and (21), we find that the value for $u^0=0$ coincides with the following formula very well:

$$D_-^p=\tfrac{1}{2}k_c^2\langle r^2(\tau)\rangle(D_{11}+D_{22})\ , \qquad (23)$$

where the characteristic wave number $k_c=0.157$ with the parameters in Figs. 2 and 3. Under the condition of $u^0=0$, and defining the lifetime of clumps τ_{cl}

$$k_c^2\langle r^2(\tau_{cl})\rangle=1\ , \qquad (24)$$

Fig. 4 shows the dependence of τ_{cl} on the relative positions. We find that the results can be represented very well by

$$\tau_{cl}=18.27-6.1\ln r^0\ , \qquad (25)$$

τ_{cl} diverges logarithmically as r^0 vanishes.

We also give the coherent time which is defined as

$$\tau_c=\left[\frac{2}{k_c^2(D_{11}^p+D_{22}^p)}\right]^{1/3}\ ,$$

and Fig. 4 shows $\tau_{cl}\gg\tau_c$ for small r^0,u^0.

Clumps can behave like a single discrete particle and are considered to enhance the radiation emission, resistivity, etc.; the above-mentioned discussions assure us that clumps can be formed in this case.

Figure 5 shows the values of D_-^p for different r^0 at different times. Curve A is from particles passing through one periodic structure, and curve B is from particles passing through two periodic structures. The two

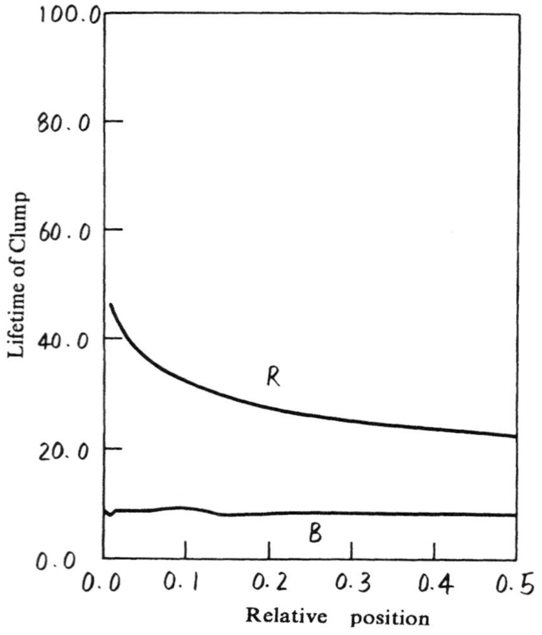

FIG. 4. The lifetime of clumps vs the relative positions when $u^0=0$. Curve R presents the lifetime of clumps, and curve B shows the coherent time.

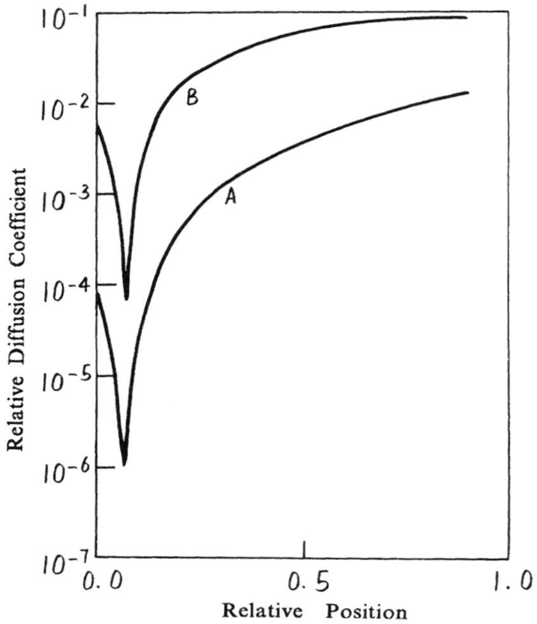

FIG. 5. The relative diffusion coefficients vs the relative positions at different time. Curves A,B corresponding to one pass, two passes, respectively.

curves have the same shape, and are minimized at the same position.

In addition, we find from numerical computation that the values of the relative diffusion coefficient D_{-}^{p}, to a certain extent, are the varying function of barycentric position R^0 and velocity V^0, and the profiles of D_{-}^{p} for different R^0 and V^0 are similar to each other.

In conclusion we have found a general form of the relative diffusion coefficient for electrons moving in Langmuir field with soliton structure. The behavior of the relative diffusion coefficient substantially differs from the prediction by Dupree. We also show that the coherent, localized wave packets can drive the formation of clumps in phase space. With the numerical solutions of the equations of motion, we verified that only when the relative velocity is zero does the lifetime of clumps indeed diverge logarithmically, with the relative positions of the constituents getting closer to one another.

We thank Professor S. G. Chen and Dr. M. S. Li for helpful discussions. This work was supported by the National Natural Science Foundation of China Grant No. 1880109.

[1]T. H. Dupree, Phys. Fluids. **15**, 334 (1972).

[2]W. Rozmus and J. C. Samson, Phys. Lett. A **126**, 263 (1988).

[3]J. A. Krommes, in *Handbook of Plasma Physics,* edited by A. A. Galeev and R. N. Sudan (North-Holland, Amsterdam, 1984), Vol. 2, p. 183.

[4]X. T. He and Y. Tan, Bull. Am. Phys. Soc. **34**, 2162 (1989).

[5]A. Salat, Phys. Fluids. **31**, 1499 (1988).

贺贤士论文选集
258

PHYSICAL REVIEW A VOLUME 46, NUMBER 5 1 SEPTEMBER 1992

Basic dynamic properties of the high-order nonlinear Schrödinger equation

Cangtao Zhou, X. T. He, and Shigang Chen
Center for Nonlinear Studies, Institute of Applied Physics and Computational Mathematics, P.O. Box 8009, Beijing 100 088, China
(Received 26 November 1991)

The dynamic behavior of the nonlinear Schrödinger equation (NSE) with an additional high-order Hamiltonian perturbation that displays a nonlinear interaction between Langmuir waves and electrons in plasmas is studied. It is shown that periodic solutions, solitary waves, and recurrence exist, which are also inherent in the cubic NSE. A more important phenomenon, stochasticity, has been found in such a continuum Hamiltonian dynamic system. This phenomenon illustrates that nonintegrability arises from the high-order Hamiltonian perturbation. In addition, our numerical results also show that recurrence, which sensitively depends on initial conditions, is only a special feature of the dynamic system.

PACS number(s): 47.10.+g, 05.45.+b, 52.35.Ra

I. INTRODUCTION

The cubic nonlinear Schrödinger nonlinear equation (NSE)

$$iE_t+E_{xx}+|E|^2E=0\ , \tag{1}$$

which is known to be completely integrable by the inverse scattering transform (IST) [1], is one of the basic evolution models for nonlinear waves in various branches of physics. A class of periodic solutions and solitons can be obtained by IST, and much work on Langmuir solitons in plasmas, optical solitons, magnetic solitons, etc., is stimulated [2,3]. The solitons developed by modulational instabilities keep their spatially coherent structures and temporally periodic evolutions.

In particular, it has been found that the unstable modulations to the uniform solution would first grow at an exponential rate as predicted by Benjamin and Feir [4] and that eventually the solution would demodulate and return to a near-uniform state. The energy in the system, which is initially confined to a few low modes, would spread to many higher modes due to the nonlinear interaction, but would eventually regroup into the original modes [5,6]. The well-known Fermi-Pasta-Ulam (FPU) recurrence [7] in connection with deep water waves has been verified experimentally by Lake *et al.* [8] and Yuen and Lake [9]. The simple and complex recurrence phenomena were also discussed numerically by Yuen and Ferguson [10] and analytically by Stiassnie and Korszynski [11].

In recent years, attention has been focused on analyzing the behavior of high-order NSE's involving quintic terms [12–15]. In Langmuir plasmas, the beat-frequency interaction between the large-amplitude parts of high-frequency fields and particles can occur in the evolutive later stages of plasma instability. In Ref. [14], one of us (X.T.H.) derived the dynamic equation involving quintic fields from Vlasov-Maxwell equations. Under the static approximation, the high-order NSE can be written as the following dimensionless form:

$$iE_t+E_{xx}+|E|^2E-g|E|^4E=0\ , \tag{2}$$

where $E(X,t)$ is the slowly varying complex amplitude of high-frequency electric fields of plasmons, and g is the coupling constant of high-frequency fields with electrons, which depends on electron temperature and density, etc. The dimensionless density is [15]

$$n(X,t)=-2|E|^2(1-r|E|^2)\ , \tag{3}$$

where r is taken as

$$r=\begin{cases}3(2T_i+3T_e)/T_e & \text{for Eq. (2)}\\ 0 & \text{for Eq. (1)}\ .\end{cases} \tag{4}$$

Here T_i and T_e are the temperature of the ion and electron, respectively.

For Eq. (2), a solitary-wave solution was obtained [15] and recurrence was also discussed numerically [12]. It is natural to ask whether periodic solutions can exist and whether integrability will be broken down under high-order Hamiltonian perturbation. To our knowledge, there is no systematic treatment of this problem.

In Sec. II a periodic solution and a solitary wave are obtained. The properties of the solitary wave are discussed from the point of review of physics. The linearized stability analysis is given in Sec. III. The nonintegrable behavior is analyzed in Sec. IV. In Sec. V we simply discuss the relationship of recurrence and stochasticity. Some conclusions are given in the final section.

II. PERIODIC SOLUTION AND SOLITON

Equation (2) can also be derived from the Lagrangian density

$$L=(i/2)(E^*E_t-EE_t^*)-|E_x|^2+\tfrac{1}{2}|E|^4-(g/3)|E|^6\ .$$

The invariance of the equation under phase shift, space, and time translations leads to the existence of the following invariants: The quasiparticle number

$$N=\int|E|^2dX\ , \tag{5}$$

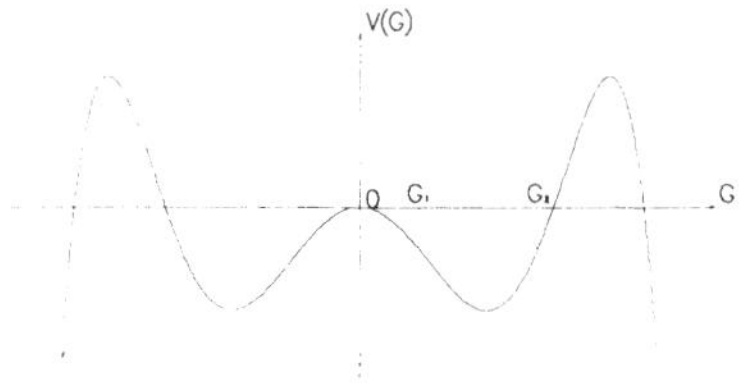

FIG. 1. The pseudopotential function in the regions of $0<g<3/16\alpha^2$.

the momentum

$$P=\tfrac{1}{2}i\int(EE_x^*-E_xE^*)dX\ , \tag{6}$$

and energy

$$H=\int(|E_x|^2-\tfrac{1}{2}|E|^4+\tfrac{1}{3}g|E|^6)dX\ . \tag{7}$$

Defining $E(X,t)=G(X,t)e^{i\Phi(X,t)}$ with G and Φ both real functions, we have equations for G and Φ,

$$G_t+2G_x\Phi_x+G\Phi_{xx}=0\ , \tag{8}$$

$$-G\Phi_t+G_{xx}-G\Phi_x^2+G^3-gG^5=0\ . \tag{9}$$

Also, let

$$G(X,t)=G(\xi),\quad \xi=X-Vt,\quad \Phi=qX-\Omega t\ . \tag{10}$$

Combining Eqs. (8)–(10), we get

$$q=V/2 \tag{11}$$

and

$$\frac{1}{2}\left[\frac{dG}{d\xi}\right]^2+V(G)=H_0\ . \tag{12}$$

Obviously, Eq. (12) is the energy integral of a classical particle with unit mass, H_0 represents the pseudoenergy of a particle, and $V(G)$ is the pseudopotential

$$V(G)=-\tfrac{1}{6}gG^6+\tfrac{1}{4}G^4-\tfrac{1}{2}\alpha^2G^2\ , \tag{13}$$

where $\alpha^2=\frac{1}{4}V^2-\Omega$.

Considering $0<g<3/16\alpha^2$, a plot of the pseudopotential function is given in Fig. 1. A particle lies in the region of pseudoenergy $H_0<0$, and starting from the left point $G=G_1$ will go to the right-hand side of the well, reflect from the place $G=G_2$, and return to $G=G_1$, making a cycle. This particle orbit corresponds to a periodic solution. If a particle lies in the origin point (say $H_0=0$) initially, its orbit corresponds to a very special solution in G, a single-pulse solution called a solitary wave (also called a soliton in an integrable system).

Supposing $H_0\le 0$ and considering

$$\tfrac{1}{3}gG^6-\tfrac{1}{2}G^4+\alpha^2G^2+2H_0=0\ , \tag{14}$$

we can find three positive real roots $G^2=a_1$, a_2, and a_3, and assume $0\le a_1<a_2<a_3$.

Thus, the solution of Eq. (12) can be expressed with the elliptic integrals of the first kind,

$$G(X,t)=\left[\frac{a_2(a_3-a_1)-a_3(a_2-a_1)\mathrm{sn}^2(\sqrt{g/3}M\xi,k)}{(a_3-a_1)-(a_2-a_1)\mathrm{sn}^2(\sqrt{g/3}M\xi,k)}\right]^{1/2}, \tag{15}$$

where

$$M=[a_2(a_3-a_1)]^{1/2}\ , \tag{16}$$

$$k=\left[\frac{a_3(a_2-a_1)}{a_2(a_3-a_1)}\right]^{1/2}. \tag{17}$$

Hence, the periodic solution of Eq. (2) is

$$E(X,t)=G(X,t)e^{i[(V/2)X-(V^2/4-\alpha^2)t]}\ . \tag{18}$$

If $H_0=0$, we can get

$$a_1=0,\quad a_2=\frac{3[1-(1-\frac{16}{3}g\alpha^2)^{1/2}]}{4g}\ ,\qquad a_3=\frac{3[1+(1-\frac{16}{3}g\alpha^2)^{1/2}]}{4g} \tag{19}$$

and

$$k=1,\quad M=\alpha\sqrt{3/g}\ . \tag{20}$$

Thus, a solitary-wave solution can be obtained

$$E(X,t)=\frac{2\alpha}{[(1-\frac{16}{3}g\alpha^2)^{1/2}\cosh(2\alpha\xi)+1]^{1/2}}\times e^{i[(V/2)X-(V^2/4-\alpha^2)t]}\ . \tag{21}$$

The solitary solution was obtained by Liu and He using the test function method [15]. For Eq. (21), we define $E_0=E(0,0)$ and obtain

$$\alpha^2=\frac{E_0^4}{2}\left[\frac{1}{E_0^2}-\frac{2g}{3}\right],$$

hence the condition for the soliton wave satisfied is $0\le g<3/2E_0^2$. In the case of $g=0$, the periodic solution for cubic NSE (1) is

$$E(X,t)=\sqrt{2}\alpha\,\mathrm{dn}(\alpha\xi,k)e^{i[(V/2)X-(V^2/4-\alpha^2)t]}\ , \tag{22}$$

and the solitary-wave solution (21) can be reduced to

$$E(X,t)=\sqrt{2}\alpha\,\mathrm{sech}(\alpha\xi)e^{i[(V/2)X-(V^2/4-\alpha^2)t]}\ , \tag{23}$$

which is the so-called Langmuir soliton [16].

The physical significance of the solitary wave is very clear. From Eq. (21), we obtain the height and width of

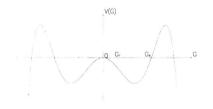

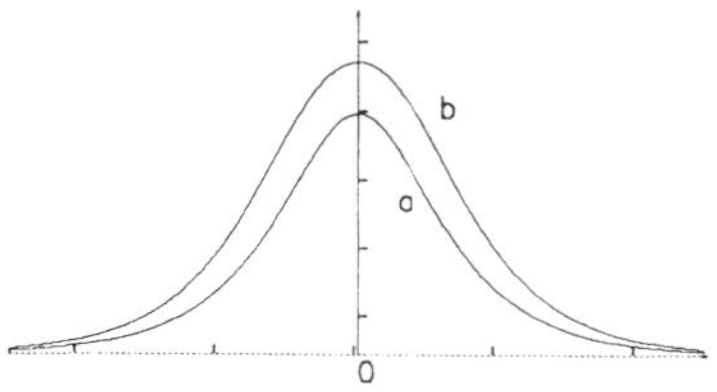

FIG. 2. Solitary-wave solution (21) with $\alpha=1.0$. Curve a corresponds to $g=0$. Curve b corresponds to $g=0.1625$.

the solitary wave,

$$E_{0\max}\approx\sqrt{2}\alpha(1+\tfrac{4}{3}g\alpha^2)\ , \tag{24}$$

$$\Delta\approx\frac{1}{2\alpha}\ln(17+12\sqrt{2})+\frac{2\sqrt{2}}{3}g\alpha\ . \tag{25}$$

The pondermotive force creates density depletion, and the maximum density can be obtained from Eq. (3),

$$|n|_{\max}\approx4\alpha^2[1-(2r-\tfrac{4}{3}g)\alpha^2]\ . \tag{26}$$

Obviously, the effects of high-order nonlinear terms enhance the amplitude and width of the solitary wave, and weaken the density depletion as compared with those for the case of $g=0$ (see Fig. 2).

III. LINEARIZED STABILITY ANALYSIS

In the preceding section, we only obtain a special periodic solution and a solitary wave. Are there more general periodic solutions and many-solitary solutions? Here, we do not discuss this problem. In this section we further qualitatively analyze the dynamic properties of Eq. (2) and consider the homogeneous solution,

$$E_s(t)=E_0e^{it}\ , \tag{27}$$

where E_0 are

$$E_0=0\ ,\qquad E_0=\pm\left[\frac{(1\pm\sqrt{1-4g}\,)}{2g}\right]^{1/2}\ . \tag{28}$$

The linear growth rate as a function of an unstable wave number K obtained from Eq. (2) is [17]

$$\gamma(K)=K[2E_0^2(1-2gE_0^2)-K^2]^{1/2} \tag{29}$$

for $0<|K|<E_0[2(1-2gE_0^2)]^{1/2}$. The unstable wave number corresponding to the maximum instability mode is $K_{\max}=E_0(1-2gE_0^2)^{1/2}$.

Defining

$$E(X,t)=E_s(t)+\delta E(X,t) \tag{30}$$

and linearizing Eq. (2), we have

$$\begin{bmatrix} i\partial_t+\partial_{xx}+L(|E_0|^2) & h(t) \\ h^*(t) & -i\partial_t+\partial_{xx}+L(|E_0|^2)\end{bmatrix}\begin{bmatrix}\delta E\\ \delta E^*\end{bmatrix}=0\ , \tag{31}$$

where L and h are expressed in the following form:

$$L(|E_0|^2)=2|E_0|^2-3g|E_0|^4\ ,\qquad h(t)=E_s^2(t)(1-2g|E_0|^2)\ . \tag{32}$$

If we consider the periodic boundary conditions $E(-L/2)=E(L/2)$, the eigenfunction of Eq. (31) can be chosen as

$$\begin{bmatrix}\delta E\\ \delta E^*\end{bmatrix}=\begin{bmatrix}\epsilon e^{i\lambda t}\\ \epsilon^* e^{-i\lambda^* t}\end{bmatrix}\cos(KX)\ , \tag{33}$$

where ϵ,ϵ^* are small parameters. Here, we choose the wave number of modulational instabilities to the uniform solution as $K=K_{\max}$, and obtain

$$\lambda=1\pm i \tag{34}$$

for cubic NSE (1) as $E_0=\pm1$, and

$$\lambda=1\pm i(-1+4g+\sqrt{1-4g}\,)/2g \tag{35}$$

for high-order NSE (2) as $E_0=\pm[(1-\sqrt{1-4g}\,)/2g]^{1/2}$ with $0\le g<\frac{1}{4}$.

To display the numerical results in later sections, we construct the phase-space diagrams (A,A_t) at $X=0$ as done by Moon [18], which are defined as

$$A(0,t)=|E(0,t)|-E_0\ ,\qquad A_t(0,t)=\frac{dA(0,t)}{dt}\ . \tag{36}$$

For phase space (A,A_t), it is easily seen that $E_0=\pm1$ for cubic NSE (1), and

$$E_0=\pm\left[\frac{1-\sqrt{1-4g}}{2g}\right]^{1/2}$$

with $0\le g<\frac{1}{4}$ for Eq. (2) correspond to the hyperbolic fixed points, respectively. The linearized stability analysis also shows that $E_0=0$ for Eqs. (1) and (2) corresponds to the elliptic fixed point. For Eq. (2), the other fixed points, i.e.,

$$E_0=\pm\left[\frac{1+\sqrt{1-4g}}{2g}\right]^{1/2}$$

for all parameter values g, and

$$E_0=\pm\left[\frac{1\pm\sqrt{1-4g}}{2g}\right]^{1/2}$$

for $g>\frac{1}{4}$ correspond to the elliptic points. Because the cubic NSE is integrable, therefore, the trajectories in phase space (A,A_t) correspond to the periodic recurrence solutions. Of course, the different trajectories should be associated with the different initial conditions. The origin is a saddle point.

On the other hand, although the recurrence phenomena have also been found in Eq. (2) [12], the integrable problem has not been discussed. In the following section, we further analyze the integral behavior numerically.

IV. NONINTEGRABILITY

For the cubic NSE (1), the integrability had been corroborated by IST. However, more general NSE's, e.g., higher dimensions or noncubic nonlinearity, do not possess integrability, and many interesting phenomena can appear. For example, a collapse of nonlinear Langmuir waves, when the spatial dimension d and the coefficient p in nonlinear term $|E|^pE$ satisfies $dp>4$, can occur [19]. This important process solves the old problem of small-K condensation in weak turbulent theory. In the case of one dimension, the integrability of systems will, in general, break down if other physical perturbations, such as driven damping [18,20–22], force dissipation [23,25], and nonlinear inhomogeneous media [26,27] are taken into account. A rich variety of temporal chaos and spatial coherence, which suggests that a low-dimensional chaotic attractor exists in an infinite-dimensional system, has been observed [28–31]. In the presence of Hamiltonian perturbations, however, the problem becomes extremely difficult for general initial conditions since the system does not reduce to a finite-dimensional system as that in the case of dissipative perturbations.

In the following discussions, we choose as the initial condition

$$E(X,0)=E_0+0.1e^{i\theta}\cos(K_{\max}X)\ , \tag{37}$$

where

$$E_0=\left[\frac{1-\sqrt{1-4g}}{2g}\right]^{1/2} .$$

Obviously, it reduces to the initial conditions as considered by Moon as $g=0$ [18]. In numerical discussions, the periodic boundary conditions and the split-step method [32] have been used. The periodic length is taken as $L=2\pi/K_{\max}$.

According to Eqs. (31)–(35), we obtain

$$\frac{\delta E}{\delta E^*}=\pm i \tag{38}$$

for our dynamic models. Combining Eq. (37), we see that the unstable manifold of the hyperbolic point corresponds to $\theta=45°$ *and* $\theta=225°$ for available values of g. In the case of $g=0$, however, our numerical results show that a homoclinic orbit (HMO) would appear as θ being a bit larger than 45° and the hyperbolic point would deviate from the origin, which results from the modulational amplitude in Eq. (37) being not too small. We find that the period of recurrence for $\theta=45.225°$ is much larger than that for $\theta=0°$ *and* $\theta=90°$ (see Figs. 3 and 6). The trajectory corresponding to $\theta=45.225°$ is near the HMO. In the presence of small perturbation, the periodic trajectories may shift from the HMO to adjacent trajectories whose periods are sensitive to the size of the displacement [24]. The field propagation can then appear to be chaotic. The small perturbation may lead to irregular HMO crossings [18,28].

In numerical experiments, we also find that the phase trajectories for $0°\le\theta<45.225°$ are the periodic motions around an elliptic fixed point and those for $45.225°<\theta\le 90°$ are the periodic motions around two elliptic fixed points and a hyperbolic fixed point, but do not cross over the hyperbolic fixed point. The largest distance for trajectory deviating from the hyperbolic point is that corresponding to $\theta=90°$ [Fig. 6(a)].

Taking $-g|E|^4E$ as a Hamiltonian perturbation, we expect that the irregular motions would easily appear nearby HMO if the integrability of the system is indeed broken down. Figures (4 and 5) clearly illustrate the chaotic behavior of the amplitude of fields. As shown in the figures for the A_t-A phase space, irregular HMO

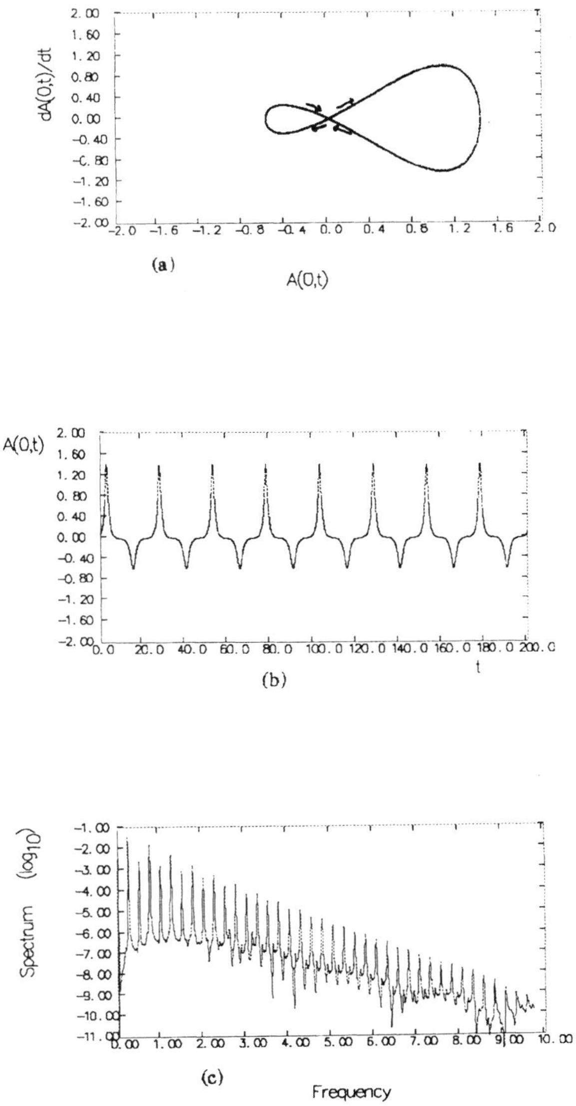

FIG. 3. Solution of Eq. (1) with $\theta=45.225°$. (a) Phase trajectories, where arrows indicate the directions of motion. (b) The time evolution of the amplitude of fields. (c) The Fourier spectrum of $A(0,t)$ with respect to time.

crossings have been observed. These demonstrate the presence of stochastic motions for complicated dynamics. The time evolutions for the amplitude of fields also exhibit the irregular motions, and the power spectra clearly indicate the broadband structures and noiselike spectra being typical of chaotic time evolutions. These phenomena show that the high-order Hamiltonian perturbation in Eq. (2) leads to chaos. As far as plasma turbulence is concerned, the cubic NSE describes the nonlinear interaction between the Langmuir wave and ion-acoustic wave under the subsonic regime, and finite soliton solutions can be obtained by IST. Solitons developed by parameteric instabilities keep their spatially coherent structures and temporally periodic evolutions. The velocity of field propagation remains constant [Fig. 6(c)]. In long-time evolution processes, integrability would be broken down by other physical effects, for example, damping, dissipation, and high-order nonlinearity. In this paper, our dynamic model shows that Langmuir wave fields are no longer periodic evolutions even without damping and dissipative effects, but still keep their spatially localized structures. Figures 4(d) and 5(d)

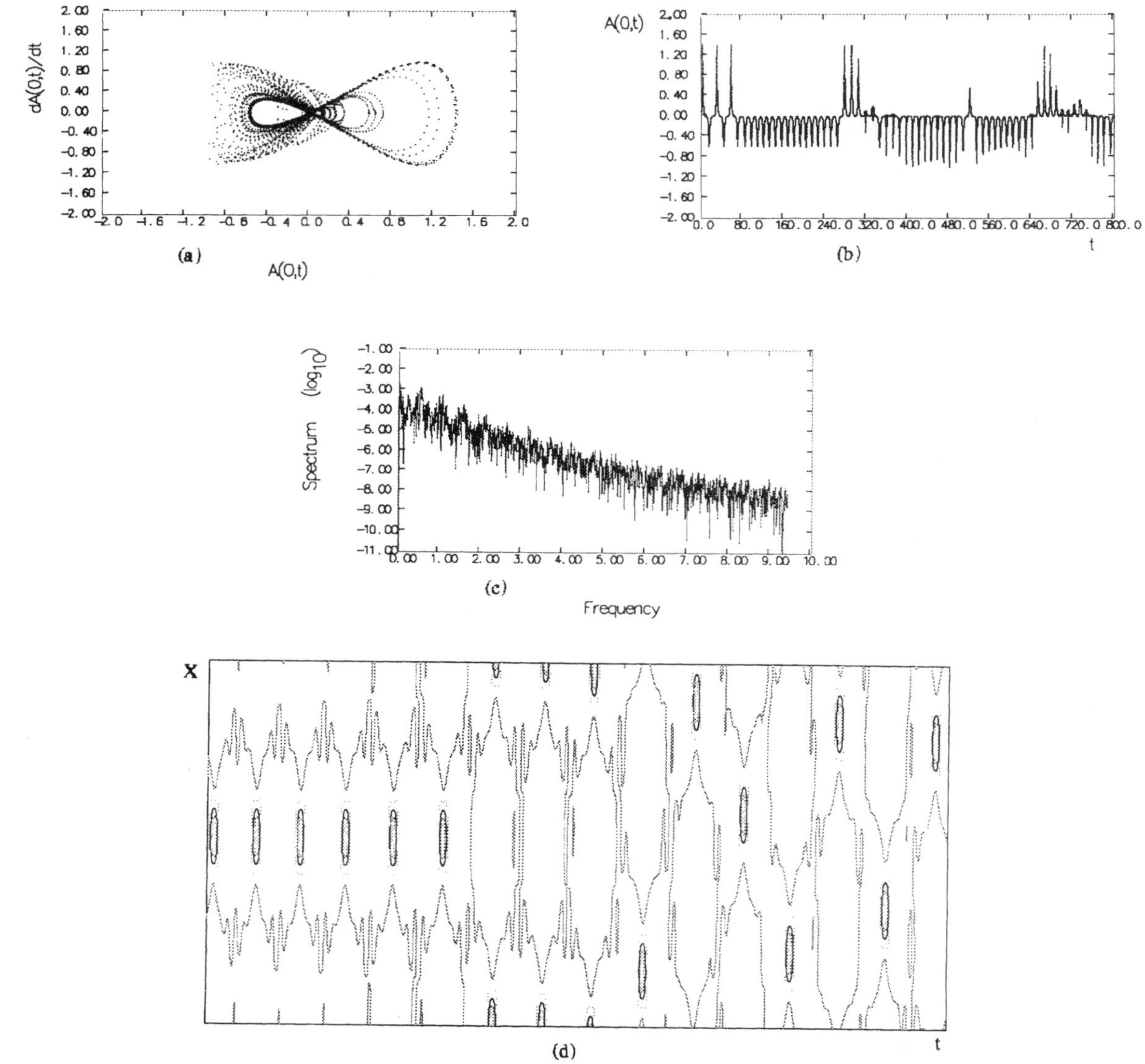

FIG. 4. Chaotic trajectories of Eq. (2) with $g=0.001$ and $\theta=45.225°$. (a) Phase motions in the phase space. Notice the irregular HMO crossings. (b) The time evolution of the amplitude of fields. (c) The Fourier spectrum of $A(0,t)$ indicates the broadband structure and noiselike spectrum being typical of chaotic time evolution. (d) Contours of $|E(X,t)|^2=\text{const}$, where $t=550-700$ and $X=-L/2\sim L/2$.

indicate that due to the high-order nonlinear effects the propagative velocity of the localized structures is not constant.

V. RECURRENCE AND STOCHASTICITY

In the Introduction, we simply accounted for the recurrence of the cubic NSE, which is associated with complete integrability. For Eq. (2), Cloot, Herbst, and Weideman [12] qualitatively proved the existence of bound solutions by use of the invariant (5) and (7). Their numerical results displayed the recurrence of solutions. Here, we further analyze the relationship between recurrence and stochasticity. For the initial condition (37), we expect that recurrence should easily appear nearby those trajectories corresponding to $\theta=90°$; Figure 7 illustrates the type of recurrence, where $g=0.2$ and $\theta=90°$. The phase orbit corresponds to a periodic motion of three-cycle, which can also be verified in Fig. 7(b) from the well-formed structures of the time evolution of field amplitude for long-time prediction. The FPU recurrence phenomenon is displayed in Fig. 7(c), where three kinds of spatially localized structures appear and the velocity of field propagation is invariable. Comparing Figs. 6 and 7, we observe that the recurrence period for $g=0$ is about five times as long as that for $g=0.2$.

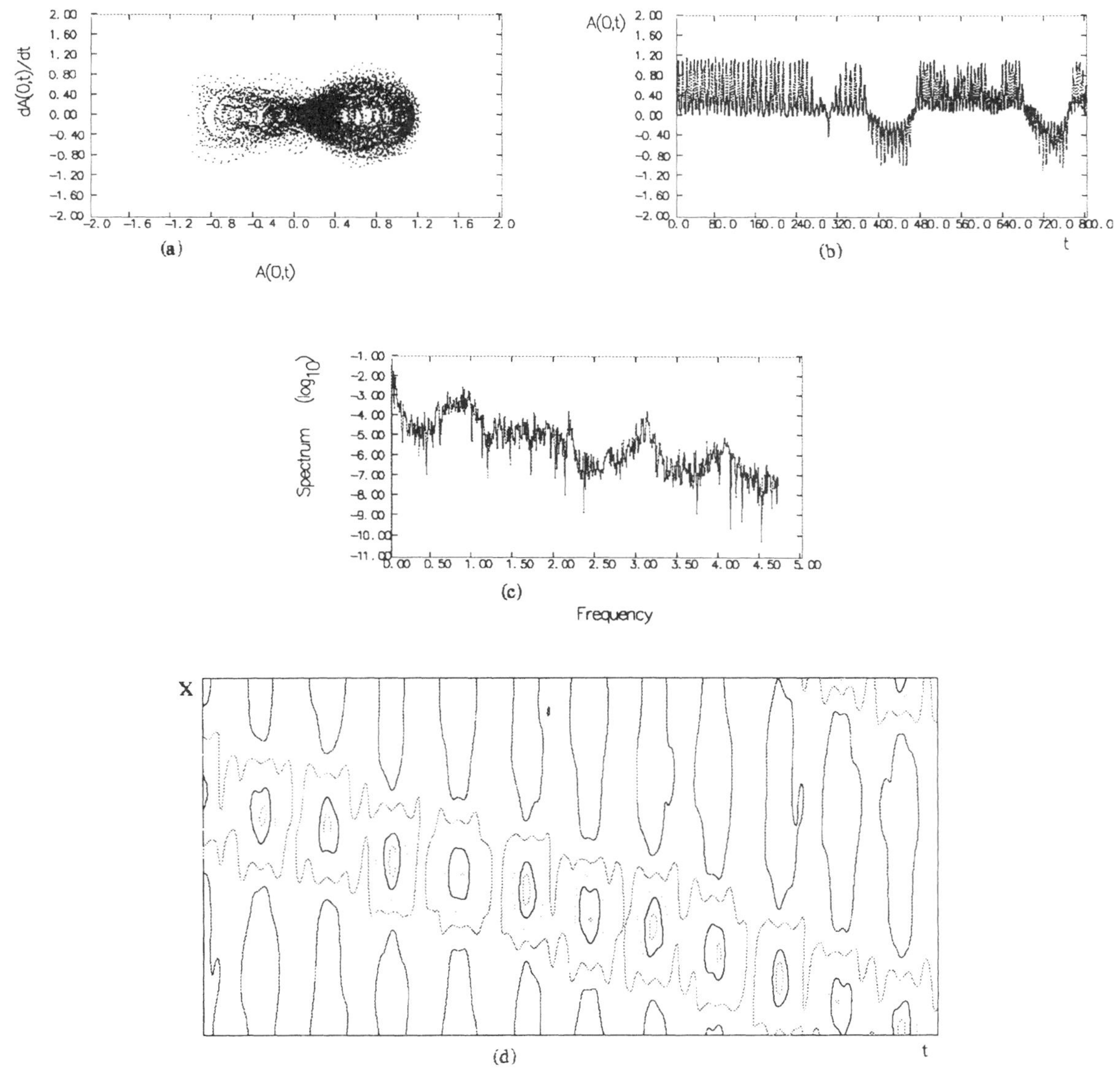

FIG. 5. Chaotic trajectories of Eq. (2) with $g=0.1$ and $\theta=45.225°$. (a), (b), and (c) as before, and (d) corresponds to $t=400-560$.

If we choose the different parametric values g and change the initial phase, we observe that the chaotic behavior appears as shown in Table I. These phenomena demonstrate that the recurrence sensitively depends on the initial conditions and the parametric values. The results in Table I express that the recurrence only is the special feature and the stochasticity should be the main one for the dynamic system (2). The fact that these completely different dynamic behaviors occur is very interesting and important. Obviously, the presence of recurrence is not necessarily associated with complete integrability. Our recent numerical investigations seem to indicate that there may be cases in which both effects, recurrence and

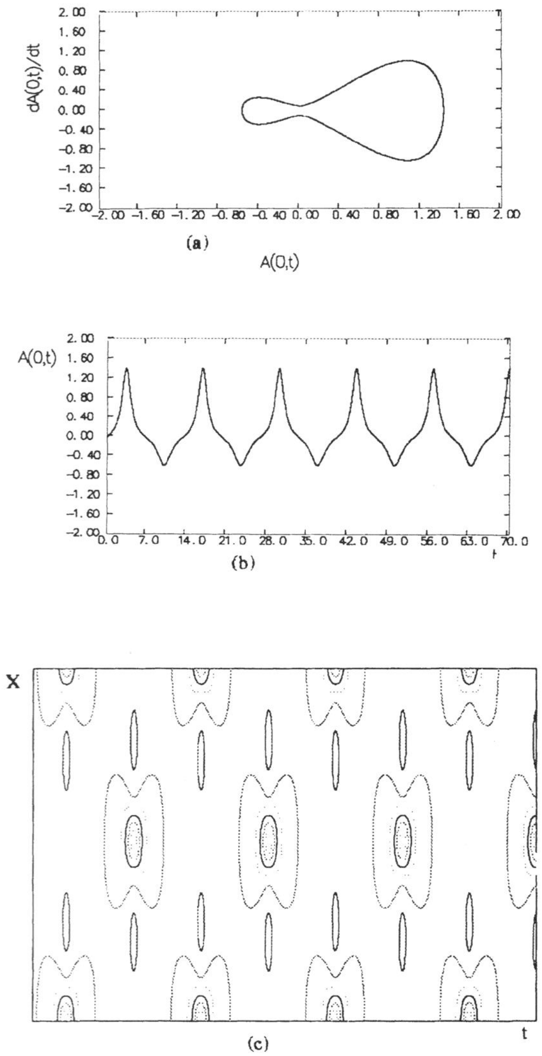

FIG. 6. solution of Eq. (1) with $\theta=90°$. (a) and (b) as illustrated in Fig. 3. (c) Contours of $|E(X,t)|^2=$const, where $t=0-50$.

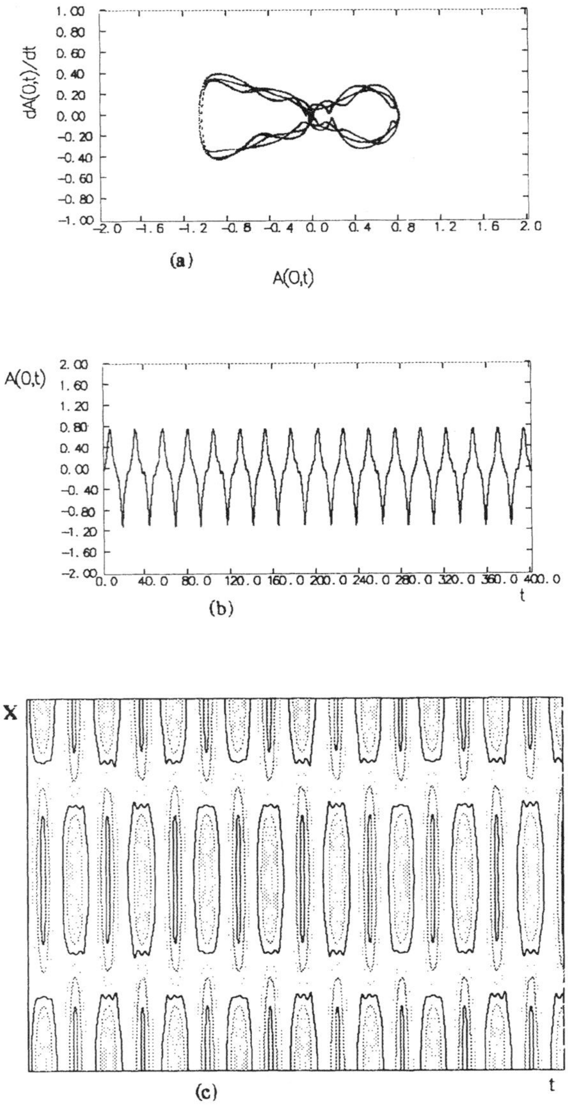

FIG. 7. Solution of Eq. (2) with $g=0.2$ and $\theta=90°$. (a), (b), and (c) as indicated in Figs. 6, where (c) corresponds to $t=0-200$. Note that the well-formed structures in (b) and (c) show that the period of recurrence is about five times as long as that in Figs. 6(b) and 6(c).

TABLE I. The letter C represents chaos and R represents recurrence. — shows that there is no numerical result.

g / θ (deg)	0.001	0.05	0.1	0.2	0.245	0.249
45	—	C	C	C	—	—
45.225	C	—	C	C	R	C
90	—	C	C	R	C	C

stochasticity, are simultaneously important for the dynamic system (2).

VI. CONCLUSIONS

A systematic analysis on the dynamic properties of high-order NSE (2) has been given. One of the most important conclusions is that the high-order Hamiltonian perturbation (i.e., quintic nonlinear effects) may lead to the breakdown of NSE integrability. For plasma physics, our dynamic model can be derived from Vlasov-Maxwell equations. The cubic NSE describes the nonlinear interaction between the Langmuir wave and the ion-acoustic wave in the subsonic regime, where the interaction of the second-order fields is only considered. In the initial stages of field evolutions, many physical phenomena, such as solitons developed by modulational stabilities, can be explained in terms of this model. In strong turbulent plasma theory, however, the cubic NSE is no longer valid. The other physical effects, for example, damping and dissipation, have to be considered. These effects would lead to stochasticity. In particular, the high-order field interaction has to be dealt with in the evolutive later stages of plasma instability. Although our dynamic model only involves the fourth-order field interaction, it is shown that this high-order Hamiltonian perturbation may also result in the presence of chaos. The field evolution from coherence to turbulence is temporally chaotic but keeps their spatially localized structures. The propagative velocity of the localized structures is variable where the mechanism is due to high-order nonlinear effects.

For a finite-dimensional dynamic system, of course, it is well known that small nonintegrable perturbations can lead to chaos [33]. Some important properties, such as a route to temporal chaos by double periodic bifurcation and irregular HMO crossings, etc, are also observed in the infinite-dimensional dissipative systems [18,30,34]. Our numerical results have also shown that the irregular HMO crossings occur in the presence of high-order Hamiltonian perturbation. Possible theoretical work on the HMO chaos may be continued in terms of Melnikov's method [34–36]. Owing to the complication consisting of a scattering spectrum for the cubic NSE with periodic boundary, however, it may be very tedious and difficult work.

On the other hand, a periodic solution and a solitary-wave solution have been obtained analytically. The parameteric values of periodic solutions can be given in virtue of pseudoenergy H_0. It is shown that the effects of high-order nonlinear terms enhance the amplitude and width of the solitary wave, and weaken the density depletion as compared with those for the cubic NSE. The recurrence solutions could be associated with a class of periodic solutions, but our results on recurrence do not correspond to our periodic solution. The reason is that the periodic solution in Sec. II is a traveling-wave solution, but the recurrence solution is derived from the initial-value problem with periodic boundary condition. Such a recurrence behavior which sensitively depends on the initial conditions is a general feature of Hamiltonian systems but is not necessarily associated with complete integrability.

It is worthwhile for us to analyze the detailed relationship between recurrence and stochasticity. In subsequent study, we expect that some theoretical work can be obtained. In addition, we plan to discuss some more interesting phenomena, for example, critical value g_c, energy spectrum, etc.

ACKNOWLEDGMENTS

We wish to acknowledge useful discussions with Professor B. L. Gou about the problem of the integrability of NSE, and we are grateful to S. P. Zhu for helping in numerical discussions. One of us (C.T.Z.) would like to thank Tan Yu for his stimulated discussions on various issues relating to this paper. This research has been supported by the National Natural Science Foundation of China, Grant No. 1880109.

[1] V. E. Zakharov and A. B. Shabat, Zh. Eksp. Teor. Fiz. **61**, 118 (1972) [Sov. Phys. JETP **34**, 62 (1972)].
[2] M. J. Ablowitz and J. F. Ledik, Stud. Appl. Math. **55**, 213 (1976).
[3] *Soliton*, edited by R. K. Bullough and P. J. Caudrey (Springer-Verlag, Berlin, 1980).
[4] T. B. Benjamin and J. E. Feir, J. Fluid. Mech. **27**, 417 (1967).
[5] A. Thyagaraja, Phys. Fluids **22**, 2093 (1979).
[6] A. Thyagaraja, Phys. Fluids **24**, 1972 (1981).
[7] E. Fermi, T. Pasta, and S. Ulam, *Collected Papers of Enrico Fermi*, edited by E. Segre (University of Chicago, Chicago, 1965), p. 978.
[8] B. M. Lake, H. C. Yuen, H. Rungaldier, and W. E. Ferguson, J. Fluid. Mech. **83**, 49 (1977).
[9] H. C. Yuen and B. M. Lake, Phys. Fluids **18**, 956 (1975).
[10] H. C. Yuen and W. E. Ferguson, Phys. Fluids **21**, 1275 (1978).
[11] M. Stiassnie and U. I. Kroszynski, J. Fluid. Mech. **116**, 207 (1982).
[12] A. Cloot, B. M. Herbst, and J. A. C. Weideman, J. Comput. Phys. **86**, 127 (1990).
[13] B. J. Lemesnrier, G. Papanicollaon, C. Sulem, and P. L. Sulem, Physica D **31**, 79 (1988).
[14] X. T. He, Acta Phys. **31**, 1317 (1982) (in Chinese).
[15] Z. J. Liu and X. T. He, Kexue Tong Bao. **29**, 1328 (1984).
[16] S. G. Thornhill and D. ter Haar, Phys. Rep. **43**, 79 (1978).
[17] M. S. Cramer and L. T. Watson, Phys. Fluids **27**, 821 (1984).
[18] T. H. Moon, Phys. Rev. Lett. **64**, 412 (1990).
[19] K. H. Spatschek, H. Pietsch, E. W. Leadk and Th. Eckermann, in *Nonlinear and Turbulent Processes in Physics,*

edited by V. E. Bar'yakhtar *et al.* (World Scientific, Singapore, 1989), Vol. 1, p. 10.
[20] K. Nozaki and N. Bekki, Phys. Rev. Lett. **50**, 1226 (1983); J. Phys. Soc. Jpn. **54**, 2363 (1985); Physica D **21**, 381 (1986).
[21] K. J. Blow and N. J. Doran, Phys. Rev. Lett. **52**, 526 (1984).
[22] K. H. Spatschek, H. Pietsch, E. W. Leadk, and Th. Erckermann, in *Singular Behavior and Nonlinear Dynamics,* edited by St. Pnevmatikos *et al.* (World Scientific, Singapore, 1989), Vol. 2, p. 555.
[23] T. H. Moon and M. V. Goldman, Phys. Rev. Lett. **53**, 1821 (1984).
[24] N. Nakhmediev, D. R. Heatley, G. I. Stegeman, and E. M. Wright, Phys. Rev. Lett. **65**, 1423 (1990).
[25] K. Nozaki and N. Bekki, Phys. Lett. **102A**, 383 (1984).
[26] W. Shyu, P. N. Guzdar, H. H. Chen, Y. C. Lee, and C. S. Liu, Phys. Lett. A **147**, 49 (1990).
[27] O. Larroche and D. Pesme, Phys. Fluids B **2**, 1751 (1990).
[28] A. B. Bishop, K. Fisser, P. S. Lomdahl, W. C. Kerr, M. B. Williams, and S. E. Trullinger, Phys. Rev. Lett. **50**, 1095 (1983).
[29] See also K. H. Spatshek, H. Pietsch, E. W. Leadke, and Th. Erckermann, in *Nonlinear and Turbulent Processes in Physics* (Ref. [19]), p. 978.
[30] F. Kh. Abudallaev, Phys. Rep. **179**, 1 (1989).
[31] K. Nozaki, Physics D **23**, 369 (1986).
[32] T. R. Taha and M. J. Ablowitz, J. Comput. Phys. **55**, 202 (1984).
[33] A. J. Lichtenberg and M. A. Lieberman, *Regular and Stochastic Motion* (Springer-Verlag, Berlin, 1984).
[34] M. Taki, K. H. Spatschek, J. C. Fernandez, R. Grauer, and G. Reinisch, Physica D **40**, 65 (1989).
[35] P. Holmes, in *Dynamical Systems and Turbulence, Warwick 1980,* edited by A. Dold and B. Echmann (Springer-Verlag, Berlin, 1981), p. 164.
[36] J. Cuckenheimer and P. Homles, *Nonlinear Oscillations, Dynamical Systems and Bifurcations of Vector Fields* (Springer-Verlag, Berlin, 1983).

CHINESE PHYS. LETT. Vol.9, No.8(1992)415

Correlation Function of Clumps for Electrons in Localized Coherent Langmuir Fields*

ZHOU Cangtao, HE Xiantu, ZHU Shaoping
Center for Nonlinear Studies, Institute of Applied Physics and Computational Mathematics, Beijing 100088

(Received 3 January 1992)

A correlation function of electrons interacting with localized coherent Langmuir wave packets is obtained by particle simulation. It is shown that the peak behavior of the correlation function for sufficiently small relative coordinates r^0 and u^0 in phase space also occurs in a spatially inhomogeneous system, and the neighboring particles in phase space are correlated with each other for a lifetime exceeding the coherent time. Our numerical results display the clump effects, which are consistent with previous theoretical prediction of the relative diffusion behavior.

PACS: 52.35.Mw, 05.45.+b

In previous paper[1] we have obtained a new relative diffusion coefficient for electrons interacting with localized coherent Langmuir wave packets which drive the trajectory stochasticity for electrons. It is revealed that clumps can occur in an inhomogeneous system. In order to investigate the clump behavior in more detail, one usually tries to discuss the correlation function through a renormalization of the two-point correlation from Vlasov equation.[2,3] For an inhomogeneous plasma, however, the renormalized turbulence technique is no longer valid, therefore it is very difficult to obtain an equation for correlation function related to the clump effects. On the other hand the particle simulation is also used.[4,5] In this letter we measure the correlation function by the particle simulation. We consider a model for electrons interacting with localized coherent Langmuir wave packets as discussed by Rozmus *et al.*[6,7] The electric field E, which can be obtained from the Zakharov equation or the cubically nonlinear Schrödinger one, is

$$E(X,t) = E_0 \mathrm{sech}(gX)\cos(\omega_p t) \quad , \tag{1}$$

where $g = \frac{1}{\lambda_D\sqrt{3}}\frac{E_0}{\sqrt{8\pi n_0 k_B T}}$, other notations are standard.

Extending sech(gX) into whole space with a periodic length L, we have the equations of motion for electrons in the dimensionless form

$$\frac{d^2X_i(t)}{dt^2} = -\sum_n E_n\cos(k_nX_i(t)-t) \quad , \tag{2}$$

where $k_n = \frac{2\pi n}{L}$, $n = 0, \pm 1, \cdots, \pm N$, E_n is in unit of $\sqrt{8\pi n_0 k_B T}$, space and time are in units of Debye length and plasma oscillating frequency, respectively. Even though the Eq. (2) is relatively simple, the phase-space trajectories of the accelerated electrons

*Supported by the National Natural Science Foundation of China No. 1880109.

can be extremely complicated. Even the low amplitude, slow modes are also likely to satisfy with the overlapping criterion.[8] Figure 1 shows that irregular motion, trapped orbits, secondary resonances, and the KAM surface must simultaneously be dealt with for such a system.

The clump effects can be illustrated by the relative diffusion coefficient, which is defined as following

$$D_-^u(R,r,U,u)=\frac{1}{2\tau_d}\langle(u(t)-u^0)^2\rangle \quad , \tag{3}$$

where $\tau_d=\frac{1}{2}(\frac{L}{V_1^0}+\frac{L}{V_2^0})$, and r,u,R,U are the relative and barycentric motion coordinates.

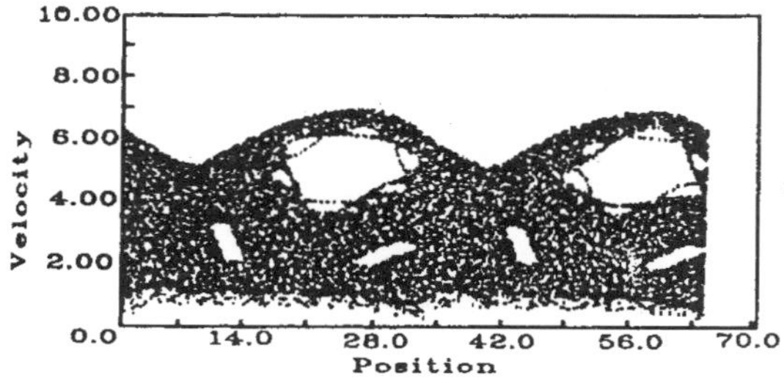

Fig. 1: Poincare surface of section plot, based on numerical solution of the equation of motion (2).

In the numerical experiments, the ensemble average is approximately computed by averaging over a set of particles with different start times but the same initial velocities and positions. Figure 2 shows that the relative diffusion is considerably slowing down when the relative motion trajectories are much closed, which accounts for the small scale correlation behaviour. In order to discuss the clump effects, we further study the two-point correlation function by using the particle simulated technique. We assume that the initial electron distribution function consists of the low-frequency oscillating electrons in the form[9,10]

$$F(X,V,0)=f_0(V)\left\{1+\left[\sum_n E_n\cos(k_nX)\right]^2\left(V^2-2\right)\right\} \quad , \tag{4}$$

where $f_0(V)$ is a Maxwellian distribution. The initial distribution function $F(X,V,0)$ at $X=0$ has the symmetric double-peak structure, which results from particles interacting with low-frequency fields and ponderomotive potentials.[9] We choose parameters $(E_0^2)/(8\pi n_0k_BT)=0.4$, $L=32.0$, and $N=9$, and assume that the initial particles are localized in a soliton envelope, i.e., $X_i\in[-16,16]$, and $V_i\in[-3,3]$. The initial size of each cell is taken as $(\Delta X,\Delta V)$=(0.1,0.1).

To measure the correlation function $\langle\delta f(1)\delta f(2)\rangle=\langle\delta f(X_1,V_1,t)\delta f(X_2,V_2,t)\rangle$ for small r^0 and small u^0 we define the fluctuation of the distribution function as: $\delta f(i)=\frac{N_i-\langle N_i\rangle}{\langle N_i\rangle}$. The correlation function can be expressed as

$$\langle\delta f(1)\delta f(2)\rangle=\frac{1}{m}\sum_{i=1}^{m}\left[\delta f(1)\delta f(2)\right]_i \quad , \tag{5}$$

where m represents the experimental ensemble average times. Here, the whole calculation is repeated 50 times, and the correlation function is measured at $t=60$ due to the limitation of computing time, where 60 is much greater than coherent time. To analyse the strong correlation property of clumps, we simultaneously calculate the coherent correlation function $\langle \delta f^{(c)}(1) \delta f^{(c)}(2) \rangle$, which is obtained by dint of the method of unperturbed particle orbits, and correlation function for electron clumps $C(1,2) = \langle \delta f(1) \delta f(2) \rangle - \langle \delta f^{(c)}(1) \delta f^{(c)}(2) \rangle$ is defined.

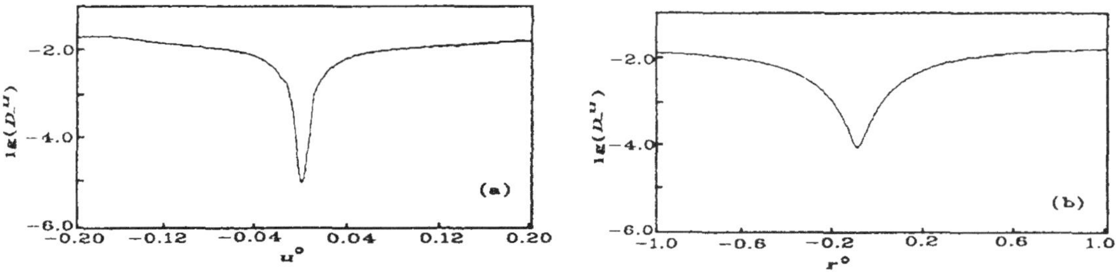

Fig. 2: Relative diffusion coefficients.
(a) Corresponds to $r^0=0.1$. (b) Corresponds to $u^0=0.01$.

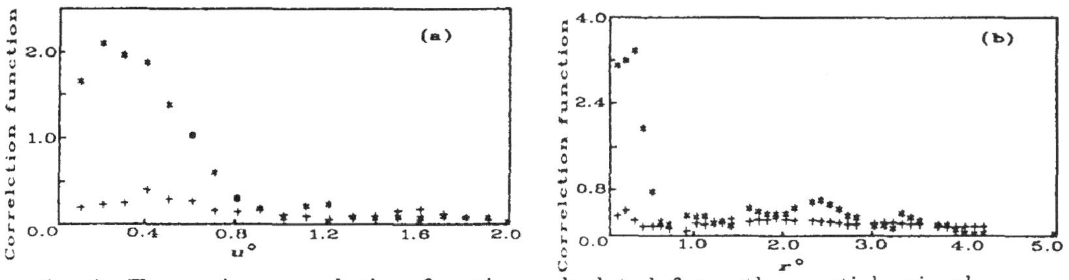

Fig. 3: Two-point correlation function calculated from the particle simulation with $X_1=125.0$, $V_1=2.0$. The asterisk represents the correlation function $\langle \delta f(1) \delta f(2) \rangle$. The plus shows the coherent part $\langle \delta f^{(c)}(1) \delta f^{(c)}(2) \rangle$. (a) Corresponds to $r^0=0.4$. (b) Corresponds to $u^0=0.4$.

Figure 3 is the main numerical result. Comparing the correlation function $\langle \delta f(1) \delta f(2) \rangle$ with $\langle \delta f^{(c)}(1) \delta f^{(c)}(2) \rangle$, we observed that the peak behavior of $\langle \delta f(1) \delta f(2) \rangle$ only occurs for very small r^0 and u^0. So far as coherent part $\langle \delta f^{(c)}(1) \delta f^{(c)}(2) \rangle$ is concerned, it is nearly a smooth function for r^0 and u^0. Obviously, the whole correlations are much greater than coherent parts for sufficiently small r^0 and u^0. With r^0 or u^0 increasing, $\langle \delta f(1) \delta f(2) \rangle$ tends to $\langle \delta f^{(c)}(1) \delta f^{(c)}(2) \rangle$. These phenomena account for the so-called clump effects. Such clumps will behave like a single large discrete particle or macroparticle, that is, each particle in a sufficiently small cell nearly sees the same force.

In Ref. 1, we got that the clump lifetime τ_{cl} diverges logarithmically as r^0 vanishes, and found that τ_{cl} is much greater than the coherent time τ_c for small r^0 and u^0. From current numerical results, we can see that the correlation length for clumps is nearly less than λ_D, which is qualitatively consistent with previous theoretical predictions and numerical results. In simulation, we also find that the peak behaviour of the correlation function depends on the barycentric positions in phase space, i. e., $\langle \delta f(1) \delta f(2) \rangle$ depends on the diffusion in velocity space as well as that in configuration space. It is clearly different from that in spatially homogeneous plasmas.

In the work discussed by Dupree *et al.*[2–5] they only treated the spatially homogeneous or weak inhomogeneous systems, where the initial stochastic electric E was considered. To current research, however, the formation of clumps in phase space arises from the intrinsic stochasticity which is due to electrons interacting with localized coherent Langmuir wave packets as shown in Fig. 1. So far, we may clearly have come to a conclusion that clumps can indeed occur in such a spatially inhomogeneous plasma. Though the theoretical work for discussing the clump correlation functions is very difficult, it is a rather important and interesting subject, specifically, because of clumps having been proposed to be an important mechanism in anomalous transport.[11]

We wish to acknowledge useful discussions with Professor S. G. Chen.

References

[1] S. P. Zhu and X. T. He, Phys. Rev. A 47 (1991) 1988.
[2] T. H. Dupree, Phys. Fluids, 18 (1975) 1167.
[3] T. Boutros-Ghali and T. H. Dupree, Phys. Fluids, 24 (1981) 1839.
[4] B. H. Hui and T. H. Dupree, Phys. Fluids, 18 (1975) 235.
[5] H. R. Bermain, D. J. Tetrealt and T. H. Dupree, Phys. Fluids, 26 (1983) 2437.
[6] W. Rozmus, J. C. Samson and A. A. Offenberger, Phys. Lett. A 126 (1988) 263.
[7] W. Rozmus and J. C. Samson, Phys. Fluids, 31 (1988) 2904.
[8] B. V. Chirikov, Phys. Rep. 52 (1979) 263.
[9] X. T. He, Chin. J. Phys. 8 (1988) 9.
[10] X. T. He, *The strong turbulence theory of the interaction of cavities and clumps in Vlasov plasma*, Plasma Preprint, UMLPF 87-10, University of Maryland.
[11] T. H. Dupree, Phys. Rev. Lett. 25 (1970) 789.

PHYSICAL REVIEW A VOLUME 46, NUMBER 6 15 SEPTEMBER 1992

Dynamic properties and the two-point correlation function for electrons in localized coherent Langmuir fields

Cangtao Zhou, Shao-ping Zhu, and X. T. He
Center for Nonlinear Studies, Institute of Applied Physics and Computational Mathematics, P.O. Box 8009, Beijing 100088, China
(Received 12 September 1991; revised manuscript received 30 March 1992)

The dynamic properties for electrons in the localized coherent Langmuir wave fields have been investigated numerically. The Poincaré map and the absolute diffusion coefficient account for the existence of regular islands embedded within stochastic regions of phase space. The relative diffusion behavior illustrates that the localized coherent Langmuir wave packets can lead to the formation of a microscale correlation for electrons in phase space. The time evolution of relative quantities further shows that such a microscale correlation has the characteristic of phase-space particle-density granulations, although the electric fields are regular. Finally, the two-point correlation function has been measured by use of the particle simulation technique. That the strong peak behavior of the correlation function can occur only for sufficiently small r^0 and u^0 has been displayed by the simulated results. All these phenomena indicate that the microscale correlation effects (defined also as "clumps") indeed exist in our dynamic system.

PACS number(s): 52.35.Mw, 52.65.+z, 05.45.+b

I. INTRODUCTION

An important subject in nonlinear plasma physics is the study of the dynamic properties of particles due to the acceleration of charged particles by localized coherent Langmuir wave packets. In a previous paper [1] we have obtained a relative diffusion coefficient for electrons interacting with coherent wave fields. It is revealed that the coherent fields can derive the formation of the microscale correlation. The lifetime of the microscale correlation diverges logarithmically when the relative positions become closer.

On the other hand, many results had also been obtained in the study of acceleration of electrons by localized fields. Fuchs *et al.* [2] used a quasilinear diffusion model to describe particles interacting periodically with coherent wave packets. Colunga, Luciani, and Mora [3] studied the acceleration of electrons by the fields produced during resonance absorption of laser light. Rozmus and co-workers [4,5] and one of the present authors [6] discussed the particle dynamics. Their theoretical model for the absolute diffusion coefficient accounts for the existence of large adiabatic islands, embedded within the stochastic region of the phase space.

In turbulence plasma, a very important phenomenon, clumps, in which particles are granulated by random electric fields, has been investigated extensively [7–13]. Dupree [8] has pointed out that the microscale random phase-space granulations or clumps arise because the Vlasov equation preserves phase-space density along particle orbits. The densities at neighboring phase-space points may originally have been widely separated in phase space. The granulations are due to the mixing of "fluids" of different densities that do not interpenetrate owing to Liouville's theorem.

In Ref. [1], we reported that the coherent Langmuir wave fields may lead to the microscale correlation, where the characteristic is similar to that of clump effects, and defined this microscale correlation as "clumps." The main purpose in this paper is to study the dynamic properties of electrons interacting with the localized coherent Langmuir wave packets, systematically. In particular, we want to know whether the features of clumps discussed by Dupree and others can indeed be exhibited in our dynamics systems, even where the electric fields interacting with electrons are regular.

In Sec. II the theoretical model is discussed. The absolute diffusion, the relative diffusion, etc., dynamic properties, have been discussed in Secs. III and IV, respectively. In Sec. V the two-point correlation function has been measured by the particle simulation technique. Finally some conclusions are given in Sec. VI.

II. DESCRIPTION OF LOCALIZED COHERENT LANGMUIR WAVE PACKETS

The coupled nonlinear equations, known as Zakharov equations [14], describe the nonlinear interaction between a high-frequency Langmuir wave and a low-frequency ion acoustic wave by the ponderomotive force. Under the subsonic regime, Zakharov equations can be reduced to the cubically nonlinear Schrödinger equation. The electric field $E(X,t)$ that may be obtained from a stationary solution of the Zakharov equations or the cubically nonlinear Schrödinger one can be expressed in the following form:

$$E(X,t)=E_0\operatorname{sech}(gX)\cos(\omega_p t) , \quad (1)$$

where

$$g=\frac{1}{\sqrt{3}\lambda_D}\frac{E_0}{\sqrt{8\pi n_e k_B T}} ,$$

$$\lambda_D=\left[\frac{k_B T}{4\pi e^2 n_e}\right]^{1/2} ,$$

$$\omega_p=\left[\frac{4\pi e^2 n_e}{m_e}\right]^{1/2} ,$$

E_0 is the amplitude, T the electron temperature, n_e the electron number density, and m_e the electron mass.

Consider a model for describing electrons interacting with the localized coherent Langmuir wave packets, as discussed by Rozmus and co-workers [4,5]. The wave packets in each structure are of the form of expression (1). Extending sech(gx) with a periodic length L, we have the equations of motion for the test particles in the dimensionless form

$$\frac{dX_i(t)}{dt}=V_i ,$$
$$\frac{dV_i(t)}{dt}=-\sum_{n=-N}^{N} E_n \cos[K_n X_i(t)-t] , \tag{2}$$

where $K_n=2\pi n/L$, E_n is in units of $\sqrt{8\pi n_e k_B T}$, and space and time are in the Debye length λ_D and $1/\omega_p$, respectively. In the following numerical discussions, we choose parameters $E_0^2/8\pi n_e k_B T=0.4$, $L=32$. In order to exhibit the characteristic of the soliton envelope (1) in each periodic structure in terms of the Fourier modes E_n, we truncate the summation in Eq. (2) as $E_{\pm N}/E_0\sim(10^{-3})$. The soliton structure and the spectrum of the wave packets are presented in Fig. 1.

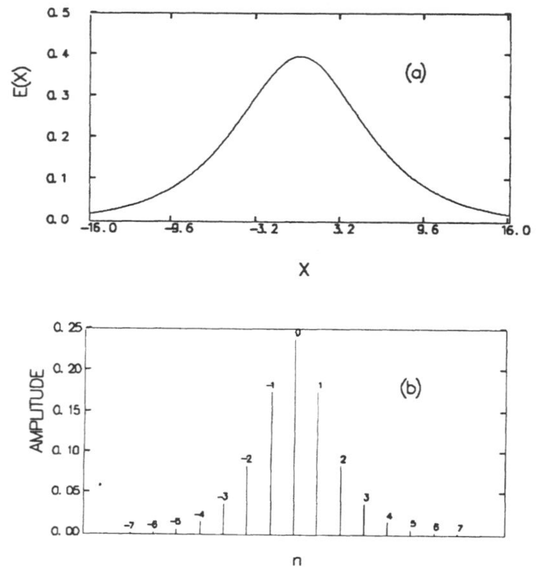

FIG. 1. (a) Soliton envelope. (b) Spectrum of the wave packet.

III. ABSOLUTE DIFFUSION

The numerical solution of Eq. (2) is depicted in the phase-space plots in Fig. 2(a), which is produced by mapping the particles velocities and positions at intervals of $T_p=2\pi$. Figure 2(a) clearly illustrates that the stochastic regions are not uniformly filled with the phase space. Several areas with the trapped orbits, characterized by regular trajectories, exist within the stochastic region.

According to the resonant overlapping criterion [15,16],

$$\tfrac{1}{2}(\Delta V_n+\Delta V_{n+1})>\left|\frac{\omega_p}{K_n}-\frac{\omega_p}{K_{n+1}}\right| , \tag{3}$$

where $\Delta V_n=4(E_n/K_n)^{1/2}$ corresponds to the width of the unperturbed trapping region of the nth mode. We find that all modes except $n=0$ satisfy the overlapping criterion. The trapping regions of the $n=0$ mode are separated from the stochastic bands by intervals with regular untrapped orbits. The regular islands embedded in the stochastic regions are the first-order resonances related to the overlapping criterion [17]. The influence of the high-order resonances can also be observed in Fig. 2(a). Thus the irregular motion, trapped orbits, secondary resonance, and the Kolmogorov-Arnold-Moser (KAM) surfaces must simultaneously be considered. The chaotic motion of electrons arises from the coherent Langmuir wave packets. The absolute diffusion of the random particles can be measured as follows:

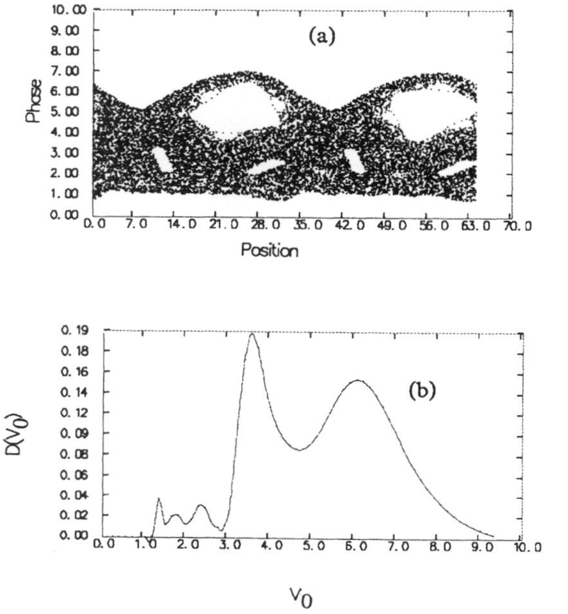

FIG. 2. (a) Poincaré surface of the section plot, based on numerical solutions to the equations of motions (2). (b) Absolute diffusion coefficient.

$$D(V_0)=\frac{1}{2\tau_d}\langle [V(t)-V_0]^2\rangle \ , \tag{4}$$

where $\tau_d=L/V_0$ is the approximated time for one pass through the periodic structures, and the symbol $\langle\ \rangle$ denotes an ensemble average. The ensemble average in all the numerical discussions is computed by averaging over a set of particles with different start times but with the same initial velocities and positions.

Figure 2(b) shows the numerical results for the absolute diffusion coefficient calculated from Eq. (4). The stable islands in the stochastic region [see Fig. 2(a)] have also been displayed in Fig. 2(b). When the particle's velocities are close to zero, particles experience a ponderomotive potential and are trapped by this potential well. The diffusion coefficient near zero velocity is zero. The diffusion increases in the stochastic region and decreases near the stable islands. With the increase of the initial velocity (which depends on the energy in the system), the separation layer forms in phase space and the diffusion coefficient also tends to zero.

IV. RELATIVE DIFFUSION PROPERTIES

The absolute diffusion coefficient only displays the dynamic behavior of one particle in phase space. To describe the features of the neighbor particles, we introduce the relative and barycentric motion coordinates

$$r=X_1-X_2\ , \quad R=X_1+X_2\ , \tag{5}$$

$$u=V_1-V_2\ , \quad U=V_1+V_2\ .$$

The equations of the relative motions become

$$\frac{dr}{dt}=u\ , \quad \frac{du}{dt}=\sum_n 2E_n\sin\left[\frac{K_nR}{2}-t\right]\sin\left[\frac{K_nr}{2}\right]\ , \tag{6}$$

$$\frac{dR}{dt}=U\ , \quad \frac{dU}{dt}=-\sum_n 2E_n\cos\left[\frac{K_nR}{2}-t\right]\cos\left[\frac{K_nr}{2}\right]\ .$$

First, we discuss the relative diffusion behaviors and define the relative diffusion coefficient as

$$D_-^u(R,r,U,u)=\frac{1}{2\tau_d}\langle [u(t)-u^0]^2\rangle \tag{7}$$

for the velocity space, and

$$D_-^r(R,r,U,u)=\frac{1}{2\tau_d}\langle [r(t)-r^0]^2\rangle \tag{8}$$

for the configuration space, where $\tau_d=\frac{1}{2}(L/V_1^0+L/V_2^0)$. In the previous study [1], we only considered the relative diffusion in velocity space; here, we also do the relative diffusion in configuration space. From Figs. 3 and 4 we see that D_-^i. $(i=u,r)$ decreases for small r^0 with decreasing u^0. In other words, it shows that the relative diffusion slows down considerably when the relative motion trajectories become closer and closer. These numerical results for the relative diffusion coefficient indicate the existence of the microscale correlation. The fact that D_-^i has a nonzero minimum means that the lifetime

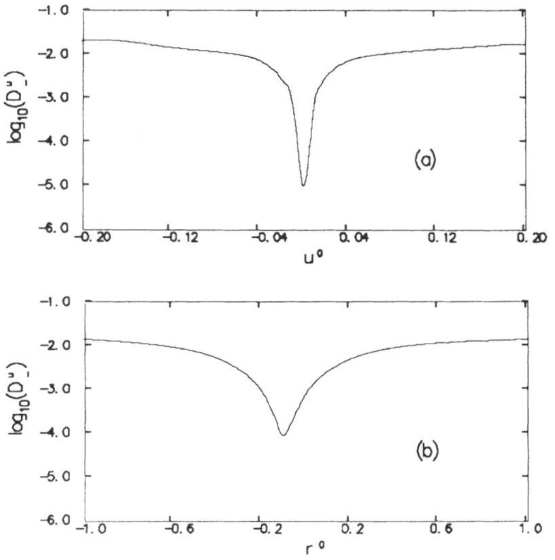

FIG. 3. Relative diffusion coefficient in velocity space. (a) Relative diffusion coefficient vs the relative velocities when $r^0=0.01$. (b) Relative diffusion coefficient vs the relative coordinates with $u^0=0.01$.

is finite. The lifetime τ_{cl} (also called clump lifetime) may be obtained from [8–10]

$$K_c^2\langle r^2(\tau_{cl})\rangle=1\ , \tag{9}$$

where K_c is the characteristic wave number. For our dy-

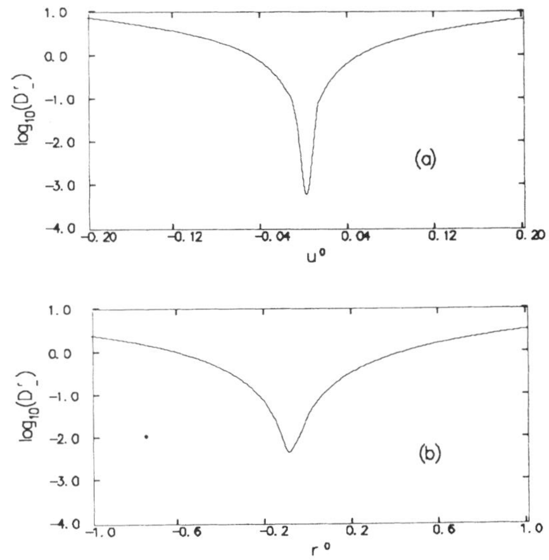

FIG. 4. Relative diffusion coefficient in configuration space. (a) Relative diffusion coefficient vs the relative velocities with $r^0=0.01$. (b) Relative coefficient vs the relative coordinates with $u^0=0.01$.

namic system we have [1]

$$\tau_{cl}=a-b\ln r^{0}\ , \tag{10}$$

where a,b are positive constants. We find that τ_{cl} diverges logarithmically as r^0 vanishes, and is much greater than the coherent time [1]

$$\tau_c=\left[\frac{2}{K_c^2(D_1+D_2)}\right]^{1/3}, \tag{11}$$

where D_1 and D_2 are the absolute diffusion coefficients for two neighboring particles.

The lifetime may also be analyzed in terms of the time evolutions of the relative quantities. In the stochastic regime, the neighboring particles are of the exponential separation type [17]. For plasma turbulence, Dupree [7–9] showed that the exponential separation trajectories occur over most of the clump scale. Misguich and Balescu [18] even illustrated that the relative diffusion of charged particles in turbulent electric fields can be divided into three time regions, i.e., a short-time regime (of order τ_c), a trajectory renormalization regime, and a long-time regime. The clump behavior is associated with the trajectory renormalization regime. In the second region, there exists a universal behavior for the relative evolution, that is,

$$\langle r^2(t)\rangle \sim e^{t/\tau_0}\ , \tag{12}$$

where τ_0 corresponds to the diffusion time scale. They pointed out that this exponential behavior is due to an extrinsic turbulence field [18]. This extrinsic stochasticity leads to a finite clump lifetime.

In Fig. 5(a), we give the time evolution of the neighboring particles. Obviously, the exponential separation also exists in our dynamic system, where chaotic behavior results from an intrinsic stochasticity. The evolution of the separation of neighboring trajectories permits a determination of the lifetime of microscale correlation. From Fig. 5(a), we observe that $\ln\langle r^2(t)\rangle \sim t$ as $t<70$. On the other hand, Fig. 5(b) indicates that the relative velocity of neighboring particles basically remains invariable as $t<70$. In a sense, the neighboring particles in sufficiently small phase-space cells suffer the same force, which is just the property of clumps.

Another characteristic quantity that can account for the clump effects is $K_\xi=\langle \xi^4\rangle/\langle \xi^2\rangle^2$. When a random quantity $\xi(t)$ is a Gaussian process, the magnitude of the fourth cumulant $\langle \xi^4\rangle-3\langle \xi^2\rangle^2$ vanishes. The kurtosis of K_ξ is associated with the clump lifetime [19]. Pettini *et al.* pointed out that the clump time scale (at which the peaks of K_ξ occur) can be related to the Kolmogorov-Sinai (KS) entropy. Figure 6 clearly displays a strong peak behavior around the lifetime of microscale correlation. With the evolution of time, K_r tends to a constant.

Now, we simply summarize our numerical results and give a basically physical picture. For homogeneous or weak inhomogeneous plasma, the stochastic electric fields leads to the chaotic diffusion of particles. When regions of different phase-space density are mixed by the fluctuating electric fields or magnetic fields, the microscale random phase-space granulations or clumps are produced [7–13]. For our dynamic system, the particle stochasticity arises from the localized coherent Langmuir wave packets, where the phase velocities of the modes satisfy the overlapping criterion. The numerical results show that the microscale correlation also exists in our dynamic system. Now we can conclude that such a microscale correlation has indeed the characteristic of clumps. In order to illustrate this feature, we further investigate the statistical behavior, where the two-point correlation function will be measured by the particle simulation technique.

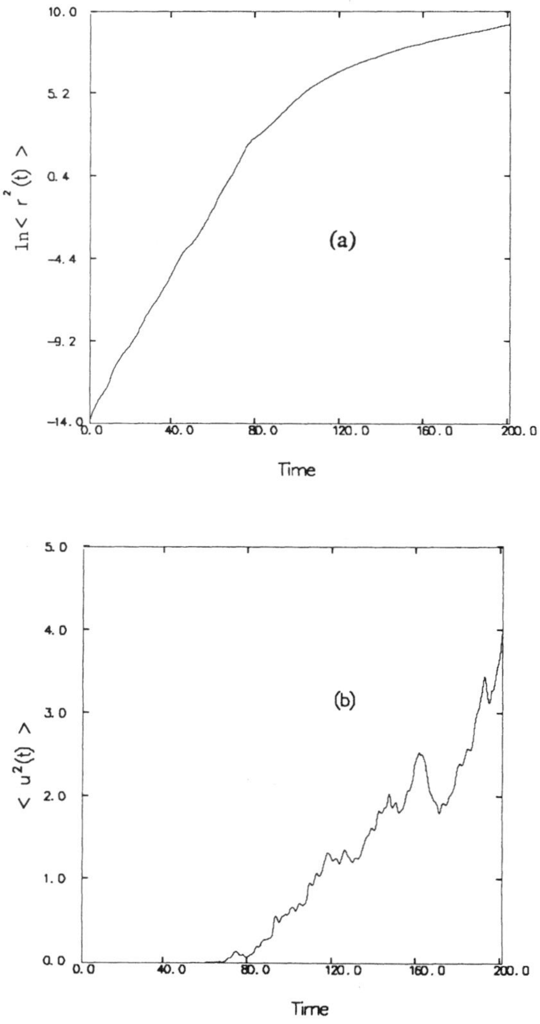

FIG. 5. (a) Time evolution of $\langle r^2(t)\rangle$. (b) Time evolution of $\langle u^2(t)\rangle$. The parameter values correspond to $r^0=0.001$ and $u^0=0.0002$.

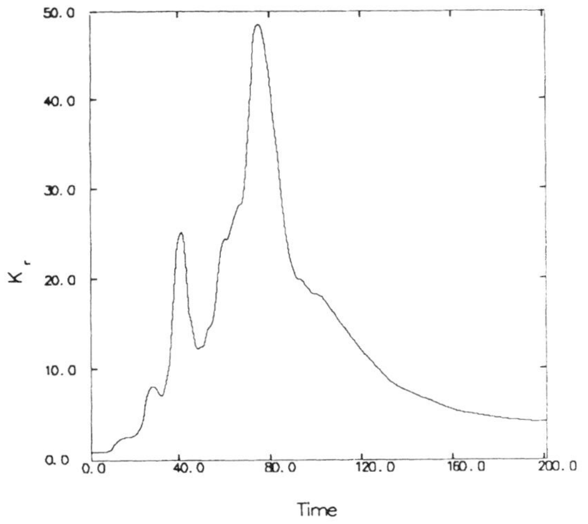

FIG. 6. Time evolution of K_r, where the kurtosis is associated with the lifetime of the microscale correlation.

V. TWO-POINT CORRELATION FUNCTION

To finish the particle simulation, we must choose an initial background distribution function. In the work discussed by Hui and Dupree, they only involved the spatial homogeneous or weak inhomogeneous plasmas. Thus an initial Maxwellian distribution was considered. For our system, the initial Maxwellian distribution is obviously not available due to the existence of spatial structure for electric fields.

In nonlinear plasma theory, the fact is that the soliton fields developed by modulational instability can well be explained in terms of Zakharov equations or the cubically nonlinear Schrödinger equation. However, the particle distribution function trapped by wave fields cannot be given by these equations. On the other hand, if we divide the Vlasov distribution function into three parts: δf, δF, and F_0, where δf is a fast varying fluctuation and corresponds to a high-frequency (HF) oscillation, and δF is a slowly varying coherent part and corresponds to a low-frequency (LF) oscillation. Also, the electric fields should include the HF field E_f and the LF field E_g. Considering the second-order beat frequency interaction between the LF field and the HF Lagmuir wave, one of us had derived the equations of the HF field and of the slowly varying electron distribution function by use of the Vlasov-Poisson equation [20]. The HF Langmuir field obtained from the above ideal is completely consistent with that obtained from the Zakharov equations or the cubically nonlinear Schrödinger equation. The LF oscillation electron distribution function consisting of δF and F_0 can be approximately expressed into the following form [20]:

$$F(X,V,0)=F_0(V)\left[1+\frac{|E(X)|^2}{8\pi n_e k_B T}(V^2-2)\right], \tag{13}$$

where V is dimensionless velocity and

$$F_0(V)=\frac{1}{\sqrt{2\pi}}e^{-V^2/2} \tag{14}$$

is a Maxwellian distribution. In simulation, we have replaced electric fields $E(X)$ by the Fourier modes structure.

It is clear that the distribution function remains positive as $0<E_0^2/8\pi n_e k_B T<0.5$. From Fig. 7, we observe that the initial distribution function $F(X,V,0)$ at $X=0$ has a symmetrical double-peak structure, which results from particles interacting with low-frequency fields and ponderomotive potential [20]. In addition, this structure would be destroyed with the particle trajectory stochasticity. He [21] had shown that spatially inhomogeneity plays a dominant role in a certain range of velocity for clump formation. These particles would run out of the soliton and carry the interaction information between particles and waves, i.e., cause a rearrangement of the average distribution function.

In the study of clumps, usually people try to discuss the correlation function through a renormalization of the two-point correlation from the Vlasov equation [9–11] or using a particle simulation technique [12,13]. For our inhomogeneous plasmas; however, the renormalized turbulent technique is no longer valid. Hence, we consider the two-point correlation function that exhibits the mi-

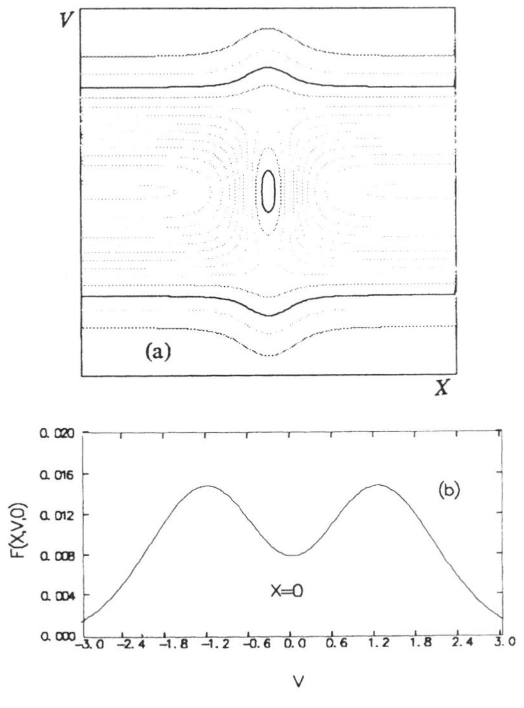

FIG. 7. (a) Contours of the initial electron distribution function $F(X,V,0)$. (b) Plot of the initial electron distribution function $F(X,V,0)$ with $X=0$.

croscale correlation behavior by the particle simulation technique. In simulation, we assume that the initial electrons are localized in a soliton envelope, i.e., $X_i \in [-16, 16]$, $V_i \in [-3, 3]$. The initial size of each cell is taken as $(\Delta X, \Delta V) = (0.1, 0.1)$.

The particles in the ith cell can be calculated by integrating

$$N_i(t=0) = n_0 \int_{X_1 - \Delta X/2}^{X_i + \Delta X/2} \int_{V_i - \Delta V/2}^{V_i + \Delta V/2} F(X_i', V_i', 0) dX_i' dV_i' . \tag{15}$$

To measure the correlation function $\langle \delta f(X_1, V_1, t) \delta f(X_2, V_2, t) \rangle$ for small $r^0 = X_1 - X_2$ and $u^0 = V_1 - V_2$, we define the fluctuation δf of the distribution function as $\delta f_i = N_i - \langle N_i \rangle / \langle N_i \rangle$. The simulated correlation function can be expressed as

$$\langle \delta f(1) \delta f(2) \rangle = \frac{1}{m} \sum_m [\delta f(1) \delta f(2)]_i , \tag{16}$$

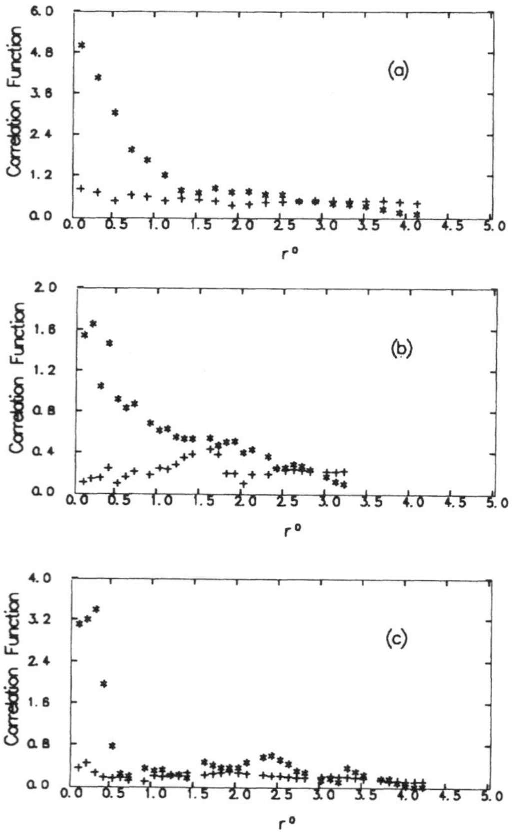

FIG. 8. Two-point correlation function calculated from the particle simulation vs r^0 with $u^0 = 0.4$. The asterisk represents the correlation function $\langle \delta f(1) \delta f(2) \rangle$. The plus shows the coherent part $\langle \delta f^{(c)}(1) \delta f^{(c)}(2) \rangle$. (a) $X_1 = 100.0$, $V_1 = 2.25$; (b) $X_1 = 100.0$, $V_1 = 1.25$; (c) $X_1 = 125.0$, $V_1 = 2.0$.

where m represents the experimental ensemble average times. The whole calculation is repeated 50 times. From the numerical results given in the above section, we have known that the microscale correlation time (or so-called clump lifetime) is finite. Here, we measure the correlation function at $t = 60$. The particles at $t = 60$ can be obtained by way of tracing the orbit of each particle lying in the initial cells. In addition, we simultaneously calculate the coherent correlation function $\langle \delta f^{(c)}(1) \delta f^{(c)}(2) \rangle$, which is obtained by use of the method of unperturbed particle orbits, and define $C(1,2) = \langle \delta f(1) \delta f(2) \rangle - \langle \delta f^{(c)}(1) \delta f^{(c)}(2) \rangle$ as the correlation function of clumps.

Figures 8 and 9 are the simulated numerical results. Comparing the correlation function $\langle \delta f(1) \delta f(2) \rangle$ with $\langle \delta f^{(c)}(1) \delta f^{(c)}(2) \rangle$, we observe that the peak behavior of $\langle \delta f(1) \delta f(2) \rangle$ only occurs for very small r^0 and u^0. As far as the coherent part $\langle \delta f^{(c)}(1) \delta f^{(c)}(2) \rangle$ is concerned, it is nearly a smooth function for r^0 and u^0; that is, $\langle \delta f^{(c)}(1) \delta f^{(c)}(2) \rangle$ does not have peak phenomena as r^0 and u^0 become smaller and smaller. Obviously, the whole correlation is much greater than that for coherent parts with sufficiently small r^0 and u^0. With r^0 and u^0 increasing, $\langle \delta f(1) \delta f(2) \rangle$ tends to $\langle \delta f^{(c)}(1) \delta f^{(c)}(2) \rangle$.

As to the relative diffusion [see Figs. 3(b) and 4(b)], we know that the correlation length of neighboring particles is less than λ_D. The same conclusion can be given in terms of the current simulation results (see Fig. 8). In addition, we find in simulation that the peak behavior of the correlation function depends on the barycentric positions in phase space, i.e., is associated with the diffusion in ve-

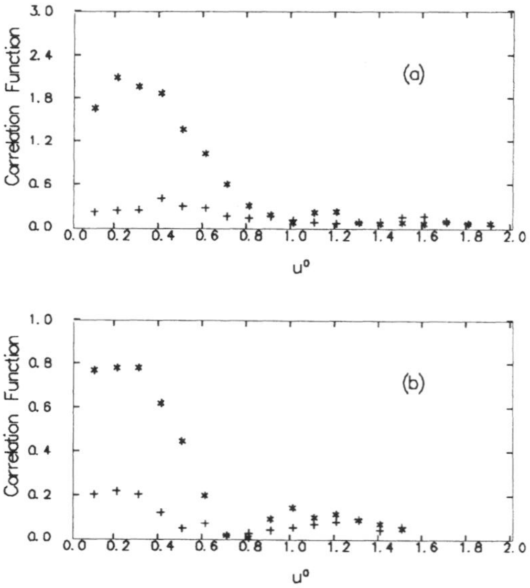

FIG. 9. Two-point correlation function calculated from the particle simulation vs u^0 with $r^0 = 0.4$, the symbols are as indicated in Fig. 8. (a) $X_1 = 125.0$, $V_1 = 2.0$; (b) $X_1 = 100.0$, $V_1 = 1.5$.

locity space as well as that in configuration space. Although the theoretical work for discussing the two-point correlation function of the inhomogeneous system has not been obtained as yet, we can conclude from current numerical results that the equation of the two-point correlation function would also contain terms to describe the relative diffusion behavior in configuration space, which is different from that discussed by Dupree and others [8–11] for spatially homogeneous plasmas. In their equation, only the relative velocity diffusion coefficient is exhibited.

VI. CONCLUSIONS

A numerical simulation of the dynamic properties for electrons in localized coherent Langmuir fields has been given. It is shown that these coherent wave packets lead to particle stochastic diffusion. The coexistence of irregular motions, trapped orbits, secondary resonances, and the KAM surfaces can be illustrated by the Poincaré mapping and the absolute diffusion coefficient. In plasma turbulence, the random fields derive particle trajectory stochasticity. The mixing of "fluids" of different densities would lead to the formation of microscale random phase-space granulations or clumps. These microscale structures may be an important mechanism in anomalous transport [7].

For our dynamic system, on the other hand, the intrinsic stochasticity is due to the phase velocities of the modes satisfying the overlapping criterion. Liouville's theorem also preserves the different densities from interpenetrations. When neighboring particles lie in the sufficiently small cell, however, they may be subject to the same force and behave like a single large discrete particle or macroparticle during a finite time. The basic properties for such a "macroparticle" can be described by the relative diffusion of neighboring particles and the two-point correlation function. From the numerical results, we find that the microscale phase-space granulations for electrons can also occur due to the interaction of the coherent Langmuir wave packets. The time evolutions of relative quantities further show that the lifetime of the microscale correlation is finite. In particular, the strong correlation behavior for sufficiently small r^0 and u^0 has been exhibited by our particle simulation.

Finally, we should mention the main difference between our results and those obtained by some authors. In the work of Rozmus and co-workers [4,5], they only discussed the one-particle diffusion. Although the resonance structures can well be explained in terms of their theoretical and numerical work, the dynamics behaviors of neighboring particles has not been analyzed. In our previous work [1], we analytically gave the relative diffusion coefficient and lifetime of the microscale correlation by considering the Wiener ensemble average [22], but the time evolution of relative quantities and the two-point correlation function were not discussed. In addition, Dupree and co-workers [7–13] were only involved with the spatial homogeneous or weak inhomogeneous plasmas. Their theoretical or numerical work is of great significance in the study of plasma turbulence and the mechanism of anomalous transport. In our current work, we systematically investigate the dynamics behaviors of one particle and two particles, while, on the other hand, the spatial inhomogeneous plasmas are dealt with. It is also worth noting that the interaction of particles with coherent wave packets is the central problem of plasma turbulence and laser plasmas, etc. We expect that our analysis, therefore, can prove to be useful in understanding the basically physical picture.

ACKNOWLEDGMENTS

We would like to acknowledge useful discussions with Professor S. G. Chen. This work has been supported by the National Natural Science Foundation of China, Grant No. 1880109.

[1] Shao-ping Zhu and X. T. He, Phys. Rev. A **43**, 1988 (1991).
[2] V. Fuchs, V. Krapchev, A. Ram, and A. Bers, Physica **14D** 141 (1985).
[3] M. Colunga, J. F. Luciani, and P. Mora, Phys. Fluids **29**, 3407 (1986).
[4] W. Rozmus, J. C. Samson, and A. A. Offenberger, Phys. Lett. A **126**, 263 (1988).
[5] W. Rozmus and J. C. Samson, Phys. Fluids **31**, 2904 (1988).
[6] X. T. He and Y. Tan, Bull. Am. Phys. Soc. **34**, 7162 (1988).
[7] T. H. Dupree, Phys. Rev. Lett. **25**, 789 (1970).
[8] T. H. Dupree, Phys. Fluids **15**, 334 (1972).
[9] T. H. Dupree, Phys. Fluids **18**, 1167 (1975).
[10] T. Boutros-Ghali and T. H. Dupree, Phys. Fluids **24**, 1839 (1981).
[11] R. Balescu and J. M. Misguich, in *Statistical Physics and Chaos in Fusion Plasmas,* edited by C. W. Horton, Jr. and L. E. Reichl (Wiley, New York, 1984), p. 295.
[12] B. H. Hui and T. H. Dupree, Phys. Fluids **18**, 235 (1975).
[13] H. R. Bermain, D. J. Tetrealt, and T. H. Dupree, Phys. Fluids **26**, 2437 (1983).
[14] D. R. Nicholson, *Introduction to Plasma Theory* (Wiley, New York, 1983).
[15] G. M. Zaslavsky, *Chaos in Dynamic Systems* (Harwood, London, 1985).
[16] R. Z. Segdeev, D. A. Usikov, and G. M. Zaslavsky, *Nonlinear Physics* (Harwood, London, 1988).
[17] A. J. Lichtenberg and M. A. Lieberman, *Regular and Stochastic Motion* (Springer-Verlag, New York, 1983).
[18] J. M. Misguich and R. Balescu, Plasma. Phys. **25**, 289 (1982).
[19] M. Pettini, A. Vulpiani, J. H. Misguich, M. D. Leener, J. Orban, and R. Balescu, Phys. Rev. A **38**, 344 (1988).
[20] X. T. He, Chin. Phys. **8**, 9 (1988).
[21] H. T. He (unpublished).
[22] A. Salat, Phys. Fluids **31**, 1499 (1988).

PHYSICAL REVIEW A VOLUME 45, NUMBER 8 15 APRIL 1992

Pattern form and homoclinic structure in Zakharov equations

Y. Tan, X. T. He, S. G. Chen, and Y. Yang
Center for Nonlinear Studies, Institute of Applied Physics and Computational Mathematics, P.O. Box 8009, Beijing, People's Republic of China
(Received 21 October 1991)

The relations between the homoclinic structure and spatial coherent pattern in Zakharov equations (ZE's) are discussed. Our results present Kolmogorov-Arnold-Moser curves and homoclinic crossing for ZE's, which exhibit the property of a near-integrable system, and Hamiltonian chaos in the ZE's is revealed.

PACS number(s): 47.10.+g, 05.45.+b, 52.35.Ra

It is well known that Zakharov equations (ZE's) are the most popular model to describe strong Langmuir turbulence in a plasma, a subject which has been studied extensively. But until now, the study for the complex behavior of the ZE's is mainly limited in the dissipation case because the study of Hamiltonian chaos is difficult, and the discussion of the integrable problem of the ZE's is subtle. In our previous work [1], it is shown that the modulation intabilities belong to a class of periodic solutions of the ZE's. The interest in such periodic solutions is connected with the study of chaos for the ZE's that describe homoclinic structure (HS) in pattern dynamics. This is an active research area recently [2].

HS in the nonlinear Schrödinger equation and the sine-Gordon equation has been studied in recent years [2,3]. The ZE's are more complex than they are. In order to understand the origin of strong Langmuir turbulence, it is very important to study the HS of the ZE's. In this Brief Report, we are investigating the HS related to a typical pattern of the ZE's. Our results present Kolmogorov-Arnold-Moser (KAM) curves and homoclinic crossings that indicate the existence of Hamiltonian chaos; this is a well-known phenomenon in a near-integrable finite-degree-freedom system [4].

We consider the ZE's in the one-dimensional case [5]:

$$i\partial_t E+\partial_x^2 E=nE\ ,\qquad \partial_t^2 n-\partial_x^2 n=\partial_x^2|E|^2\ , \tag{1}$$

where $E(x,t)$ is a slowly varying envelope of the high-frequency electric field and $n(x,t)$ is a low-frequency quasineutral density perturbation. In the linearized form, Eqs. (1) lead to the following dispersion relation:

$$(\omega^2-k^2)(\omega^2-k^4)=2E_0^2k^4\ , \tag{2}$$

where E_0 is an amplitude of the homogeneous Langmuir field, k is a wave number of the disturbance, and the time dependence is proportional to $\exp(-i\omega t)$. If k lies in the range

$$0<k<\sqrt{2}E_0\equiv k_c \tag{3}$$

the system will evolve from a spatially homogeneous state to a localized structure, which is known as modulation instability [6].

Equations (1) have at least two kinds of fixed points (FP's)

$$(E,n)=O$$

the (0,0) FP, and

$$(E,n)=Q$$

the $(E_0,0)$ FP, where E_0 is a real constant. The stability of these FP's can be examined by the linearized Eqs. (1) and solving the eigenvalue equation. Here we consider a simple case that only one wave number k_0 in the system satisfies Eq. (3), say, according to the condition

$$k_0\equiv 0.95k_c\ . \tag{4}$$

In this case, around the FP's, E and n are eigenfunctions of a spatial operator ∂_x^2 which has the form

$$\partial_x^3 E=-k_0^2E\ ,\qquad \partial_x^2 n=-k_0^2 n\ . \tag{5}$$

The analyses of the stability of the FP's for Eqs. (1) show that O is a center, and Q is the saddle and center in their own subspace. Combining the above facts and the periodic solutions of Eqs. (1), we believe that the HS exists in the ZE's. In order to study it we do a local analysis for the FP Q in detail. We divide the complex variable E of Eqs. (1) into real and imaginary components

$$E=v+iu\ .$$

Now we rewrite Eqs. (1) as

$$\partial_t v=-\partial_x^2 u+nu\ ,\qquad \partial_t u=\partial_x^2 v-nv\ ,\qquad \partial_t n=m\ ,\qquad \partial_t m=\partial_x^2 n+\partial_x^2(u^2+v^2)\ . \tag{6}$$

Combining Eqs. (5), we have the linearized form in a neighborhood U of Q:

$$\dot{\mathbf{w}} = Df(Q)\mathbf{w} \ , \tag{7}$$

where $Df(Q)$ is the Jacobian matrix at Q which has the form

$$Df(Q) = \begin{pmatrix} 0 & k_0^2 & 0 & 0 \\ -k_0^2 & 0 & -E_0 & 0 \\ 0 & 0 & 0 & 1 \\ -2E_0k_0^2 & 0 & -k_0^2 & 0 \end{pmatrix}$$

and

$$\mathbf{w} = \begin{pmatrix} v \\ u \\ n \\ m \end{pmatrix} ,$$

the eigenvalue equation $\|Df(Q) - \lambda I\| = 0$ gives eigenvalues

$$\begin{aligned} &\lambda_1 = \{\tfrac{1}{2}k_0^2[(1-k_0^2)^2 + 8E_0^2]^{1/2} - (k_0^2+1)\}^{1/2} \equiv \lambda_0 \ , \\ &\lambda_2 = -\lambda_0 \ , \\ &\lambda_3 = i\{\tfrac{1}{2}k_0^2[(1-k_0^2)^2 + 8E_0^2]^{1/2} + (k_0^2+1)\}^{1/2} \equiv i\omega_0 \ , \\ &\lambda_4 = -i\omega_0 \ , \end{aligned} \tag{8}$$

where λ_0 and ω_0 are real if Eq. (4) is satisfied. In our case, the solution of Eq. (7) can be written as follows:

$$\mathbf{f} = C_1 e^{\lambda_0 t}\mathbf{f}_1 + C_2 e^{-\lambda_0 t}\mathbf{f}_2 + C_3 e^{i\omega_0 t}\mathbf{f}_3 + C_4 e^{-i\omega_0 t}\mathbf{f}_4 \ ,$$

where C_i, $i = 1,2,3,4$ are arbitrary constants, and $\mathbf{f}_0$ are eigenvectors which have the form

$$\mathbf{f}_i = \begin{pmatrix} k_0^2 \\ \lambda_i \\ \dfrac{-(k_0^4+\lambda_i^2)}{E_0} \\ \dfrac{-\lambda_i(k_0^4+\lambda_i^2)}{E_0} \end{pmatrix} \quad \text{for } i - 1,2,3,4 \ . \tag{9}$$

Obviously the FP Q is the saddle and the center in the (f_1,f_2) and (f_3,f_3) respective subspace.

A very important thing is that the periodic solutions for the ZE's, which are known as the Fermi-Pasta-Ulam (FPU) recurrence phenomena [7], are the phase shifts [1]. Then the HS exists only in phase space $(\rho, \partial_t\rho, n, \partial_t n)$, where $\rho = (\mathrm{Re}E)^2 + (\mathrm{Im}E)^2$, rather than $(\mathrm{Re}E, \mathrm{Im}E, n, \partial_t n)$. Next, we construct a phase space which is beneficial to investigate HS. Substituting E by $\sqrt{\rho}e^{i\phi}$ we have another linear form at FP Q for the ZE's

$$\dot{y} = Ty \ , \tag{10}$$

where

$$y = \begin{pmatrix} \rho \\ \mu \\ n \\ m \end{pmatrix}, \quad \mu \equiv \partial_t\rho, \quad m = \partial_t n \ ,$$

and

$$T = \begin{pmatrix} 0 & 1 & 0 & 0 \\ -k_0^4 & 0 & -2E_0^2k_0^2 & 0 \\ 0 & 0 & 0 & 1 \\ -k_0^2 & 0 & -k_0^2 & 0 \end{pmatrix} .$$

The local structure of Eq. (10) is qualitatively similar to Eqs. (6). We divide the subspace spanned by the eigenvectors into two classes, the saddle subspace

$$E_S = \mathrm{span}(y_1, y_2)$$

and the center subspace,

$$E_C = \mathrm{span}(y_2, y_4) \ ,$$

where y_1, y_2 from Eq. (10) are the eigenvectors whose eigenvalues have negative and positive real parts and y_3, y_4 from Eq. (10) are those whose eigenvalues have zero real parts. We will put global information of the system into those subspaces in the following manner:

$$y_j - s_{j1}\rho + s_{j2}\partial_t\rho + s_{j3}n + s_{j4}\partial_t n \tag{11}$$

for $j = 1,2,3,4$ where s_{ji} are the elements of matrix S which satisfies

$$STS^{-1} = \begin{pmatrix} T_1' & 0 \\ 0 & T_2' \end{pmatrix} ,$$

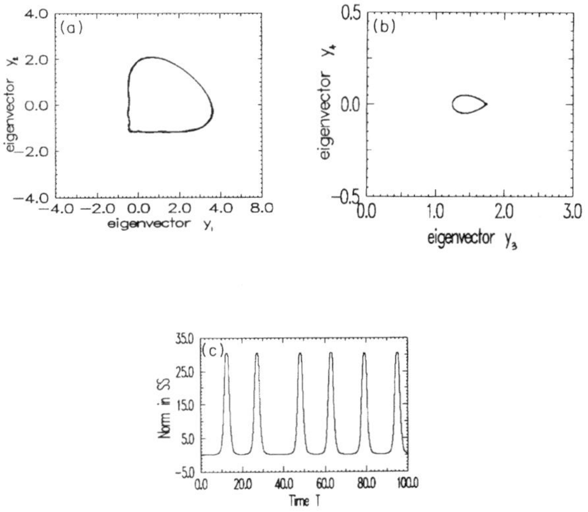

FIG. 1. $C_1 = 1.0$, $C_2 = 1.0$. (a) Phase trajectory in the saddle subspace. Notice this orbit forms the KAM curves. (b) Phase trajectory in the center subspace. Motion in this subspace is excited by the motion in the saddle subspace. (c) Time evolution of the norm $(y_1^2 + y_2^2)$ in the saddle subspace.

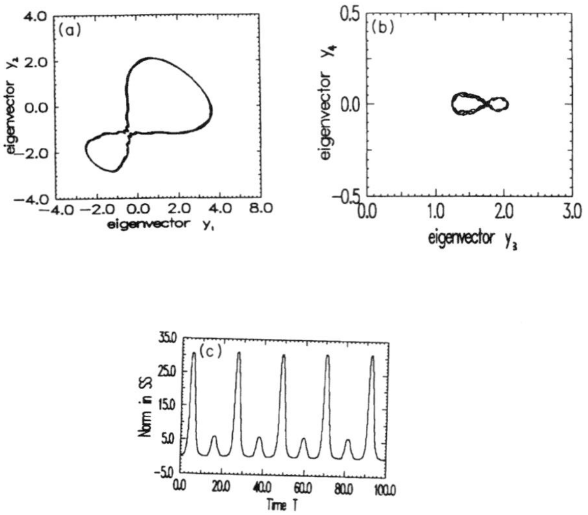

FIG. 2. The same as in Fig. 1, but the adjustable constants are $C_1=1.0$, $C_1=-1.0$.

where T'_1 and T'_2 are the 2×2 matrix, respectively. Numerical solution of Eqs. (1) was obtained with periodic boundary conditions. The spatial length L is chosen in such a way that $k_0L=2\pi$, and the number of grids is 64. The initial values are given as

$$\psi=\psi_0+(C_1\mathbf{f}_1+C_2\mathbf{f}_2)E_0\cos(k_0x)/500.0\ ,$$

where

$$\psi_0=\begin{pmatrix}\mathrm{Re}E\\ \mathrm{Im}E\\ n\\ m\end{pmatrix}_{t=0}=\begin{pmatrix}E_0\\ 0\\ 0\\ 0\end{pmatrix},\quad E_0=2.0\ ;$$

$\mathbf{f}_1$ and $\mathbf{f}_2$ are given by Eq. (9) and C_1,C_2 are adjustable constants. The dynamic variables are recorded according to Eq. (11) at $x=0$. Our initial conditions ensure that phase trajectories will locally lie in saddle subspace (SS) E_S. The long-time evolution of Eqs. (1) is obtained with C_1 and C_2 given in the following manner.

Case (i). $C_1=1.0$, $C_2=1.0$, the long-time evolution of the system is displayed in SS E_S and center subspace (CS) E_C, which are shown in Figs. 1(a) and 1(b), respectively.

Case (ii). $C_1=1.0$, $C_2=-1.0$, the same long-time evolution is shown in Fig. 2. Summing up Figs. 1 and 2, we see that there are two kinds of recurrence motions, one-loop and double-loop orbits in SS E_S. This property is the same as particle motion in a double-well potential with different level sets. Clearly, this behavior exhibits the HS in the ZE's. Around the FP Q, the SS E_S and CS E_C are invariant subspaces. Beyond local analysis, it is not true. Figures 1(b) and 2(b) show that there is a low level excitation in CS E_S because the motion in SS E_S and in CS E_C influence each other. This coupling typically leads to the creation of "chaotic" orbits for the

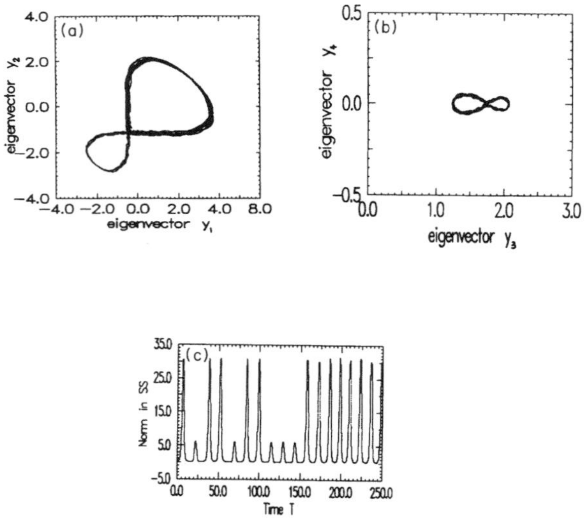

FIG. 3. $C_1=1.0$, $C_2=-0.1$. (a) Phase trajectory in the saddle subspace. Notice irregular homoclinic crossings. (b) Phase trajectory in center subspace. (c) Time evolution of the norm $(y_1^2+y_2^2)$ in the saddle subspace. Notice the irregular period corresponds to homoclinic crossings.

homoclinic orbits [8]. This is why the ZE's are the near-integrable system. The phase trajectories in Figs. 1(a) and 2(a) are not broken up by the addition of the coupling. These orbits form the KAM curves in the SS E_S the same as a finite dimensional system [4].

Now we discuss the case that the orbit is closing to the homoclinic orbit.

Case (iii). $C_1=1.0$, $C_2=-0.1$, the long-time evolution is displayed in Fig. 3. Comparing Fig. 3(c) to 2(c) and 1(c), it is shown that the time period of orbit in case (iii) is longer than that in cases (i) and (ii). It is easy to understand because the orbits in case (iii) are closing homoclinic orbits whose periods are infinite in relation to cases (i) and (ii). Under the influence of motion in CS E_C Fig. 3(a) shows that the motion in case (iii) is complex in SS E_S. The evolution of the system can now fall "on both sides" of the saddle point—one route corresponding to a rotation and the other to a libration. The system must again and again follow one route or the other. It leads to undetermined consequences. This feature which is known as homoclinic crossings is presented in Fig. 3(c). That indicates the existence of a Hamiltonian chaos for the ZE's.

The FPU recurrence is a very interesting phenomenon that exists in an integrable system, such as nonlinear Schrödinger equation, the Kerteweg–de Vries equation, etc. [9]. Recently, Akhmedive *et al.* studied the FPU problem in a near-integrable system [10]. They conclude that a near-integrable system can lead to pseudorecurrence rather than exact recurrence. This view is not quite right because there exist KAM curves. In our problem cases (i) and (ii) lead to an exact recurrence and case (iii) leads to pseudorecurrence. A more detailed investigation will be presented later.

In conclusion, although there are no inertial manifolds

to reduce the dimension in the Hamiltonian system, coherent spatial structure can lead to the degeneracy of degrees of freedom of the system. Some properties of finite dimensional systems, such as KAM curves and homoclinic crossings, are discovered in the ZE's as they relate to typical pattern formation. FPU recurrence in a near-integrable system is also discussed. That Hamiltonian chaos exists in the ZE's is very interesting for a better understanding of Langmuir turbulence.

This work was supported by Chinese Natural Science Foundation Grant No. 1880109.

[1] Y. Tan, X. T. He, and Y. Yang (unpublished).

[2] A. R. Bishop, D. W. McLaughlin, M. G. Forest, and E. A. Overman II, Phys. Lett. A **127**, 335 (1988); H. T. Moon, Phys. Rev. Lett. **64**, 412 (1990).

[3] A. R. Bishop, D. W. McLaughlin, M. G. Forest, and E. A. Overman II, Physics D **23**, 293 (1986).

[4] M. A. Lichtenberg and M. A. Lieberman, *Regular and Stochastic Motion* (Springer, New York, 1983).

[5] V. E. Zakharov and A. B. Shabat, Zh. Eksp. Teor. Fiz. **61**, 118 (1971) [Sov. Phys.—JETP **34**, 62 (1972)].

[6] T.B. Benjamin and J. E. Feir, J. Fluid. Mech. **27**, 417 (1967).

[7] R. Fermi, J. Pasta, and S. Ulam, in *Collected Papers of Enrico Fermi* (University of Chicago Press, Chicago, 1955), p. 978.

[8] S. Wiggins, *Global Bifuractions and Chaos—Analytical Methods* (Springer, Berlin, 1988).

[9] H. C. Yuen and W. E. Ferguson, J. Fluid. Mech. **21**, 1275 (1978); T. Taniuti and K. Nishihara, *Nonlinear Waves* (Pitman, Boston, 1983).

[10] N. N. Akhmedive, D. D. Heatley, G. I. Stegeman, and E. M. Wright, Phys. Rev. Lett. **65**, 1423 (1990).

Commun. Theor. Phys. 19(1993)231–242
© International Academic Publishers
Vol. 19, No. 2

Chaotic Behavior of One-Dimensional Nonlinear Schrödinger Equations Involving High Order Saturable Nonlinear Terms[1]

Cang-Tao ZHOU and X.T. HE
Center for Nonlinear Studies, Institute of Applied Physics and Computational Mathematics, P.O. Box 8009, Beijing 100088, China

(Received April 15, 1992)

Abstract

The non-integrable behavior in one-dimensional generalized nonlinear Schrödinger equations involving high order saturable nonlinearities, which govern the dynamic behavior in a class of physical systems, is investigated numerically. The numerical results illustrate that the high order saturable nonlinearities would lead to chaos, where the irregular homoclinic orbit (HMO) crossings have been observed in phase space. On the other hand, we show that the periodic solutions and solitary waves may exist in such non-integrable continuum Hamiltonian dynamic systems.

I. Introduction

The generalized nonlinear Schrödinger equation (NSE)

$$iE_t + E_{xx} + F(|E|^2)E = 0\,, \tag{1}$$

in which the potential F is a differential smooth real function, is one of the basic evolution models for nonlinear waves in various branches of physics, such as nonlinear optics, magnetic media, hydrodynamics and plasma physics.

The Lagrangian density corresponding to Eq. (1) is

$$L = \frac{i}{2}(E^* E_t - EE_t^*) - |E_x|^2 + f(|E|^2)\,, \tag{2}$$

where $f(|E|^2) = \int_0^{|E|^2} F(s)ds$. The invariance of the equation under phase shift, space and time translations leads to the existence of the following invariants: the quasiparticle number

$$N = \int |E|^2 dx\,, \tag{3}$$

the momentum

$$P = \frac{i}{2}\int (E_x E^* - EE_x^*)dx\,, \tag{4}$$

and the energy

$$H = \int [|E_x|^2 - f(|E|^2)]dx\,. \tag{5}$$

Some particular types of $F(|E|^2)$ in Eq. (1) have been obtained due to different physical backgrounds. For example[1–8]

$$\text{(I)}\quad F(|E|^2) = |E|^{2\sigma}\,, \tag{6}$$

$$\text{(II)}\quad F(|E|^2) = |E|^2 - g|E|^4\,, \tag{7}$$

[1]The project supported by National Natural Science Foundation of China under Grant No. 19175039.

$$\text{(III)} \quad F(|E|^2) = \frac{|E|^2}{1+g|E|^2}, \tag{8}$$

$$\text{(IV)} \quad F(|E|^2) = \frac{1}{2g}(1 - e^{-2g|E|^2}). \tag{9}$$

For model (I), a collapse of nonlinear wave[2,6], when the spatial dimension (D) and the coefficient σ satisfy $\sigma \cdot D > 2$, can occur. In nonlinear plasmas, this important process solves the problem of small-K condensation in weak turbulent theory. As $\sigma = 1$, the one-dimensional cubic NSE posses the integrability. A class of periodic solutions and solitons can be obtained by inverse scattering transform (IST). Solitons developed by modulational instabilities keep their spatially coherent structures and temporally periodic evolutions. In particular, the well-known Fermi-Pasta-Ulam (FPU) recurrence in connection with deep water waves had been verified experimentally by Lake *et al.*[9] and Yuen *et al.*[10]. The simple and complex recurrence phenomena were discussed numerically by Yuen and Ferguson[11] and analytically by Stiassnie and Korszynski[12].

For model (II), the fifth order correction to cubic term arises from the nonlinear interaction between Langmuir waves and electrons[8]. The coupling nonlinear equations, known as Zakharov equations, describe the nonlinear interaction between high frequency Langmuir wave and ion-acounstic wave by the ponderomotive force. However, the beat frequency interaction between the large amplitude parts of high frequency fields and particles can occur in the evolutive stages of plasma instability. Under the static approximation, HE[8] had obtained an NSE where the potential $F(|E|^2)$ can be expressed by Eq. (7) in terms of Vlasov-Maxwell equations. For this model, Cloot *et al.*[13] qualitatively proved the existence of bound solutions by making use of the invariants (3) and (5). Their numerical results displayed recurrence of solutions.

In the one-dimensional case, a periodic solution and a solitary wave solution are found analytically[14]. It is shown that the effects of the quintic order nonlinear term enhance the amplitude and width of solitary and weaken the density depletion as compared with those for cubic NSE. A more important phenomenon, stochasticity, has been discussed numerically[14]. We show that the recurrence behavior, which sensitively depends on the initial condition and parametric value g, is only a special feature of dynamic system and is not necessarily associated with complete integrability.

For models (III) and (IV), the focusing singularities can occur in two or three dimensions. The conditions of blow-up had been discussed analytically[1]. For model (III), Akhmediek *et al.*[7] studied the pseudorecurrence in two-dimensional modulation instability. They showed that the approximated recurrence to an initial homogeneous field can appear, and this pseudorecurrence arises only for a restricted range of spatial modulational frequency. Model (IV) is considered as a useful model in a homogeneous unmagnetized plasma[15]. When the phase velocity of the slow plasma oscillation is much smaller than the ion thermal velocity, one can obtain the adiabatic (quasi-static) electron density under the quasi-neutral approximation

$$n_e = n_0 e^{-|E|^2}. \tag{10}$$

Combining the coupling equation which exhibits the slowly varying complex amplitude $E(X,t)$ interaction with the low-frequency plasma motion, one can easily obtain Eq. (1) involving nonlinear terms (9).

The main purpose of this paper is to discuss the following questions: Whether can the high order saturable nonlinearities lead to integrability broken down? What are the temporal evolutive behaviors and the spatial patterns? Whether may there be periodic solutions and solitary waves? These problems are very interesting in physics and mathematics. In Sec. II we qualitatively get the linearized stability analysis. The numerical results that display the nonintegrable behavior are discussed in Sec. III. The existence of periodic solutions and solitary waves is analysed in Sec. IV and some conclusions are summarized in the final section.

II. Linearized Stability Analysis

In a previous paper[14], we investigated the NSE involving quintic (7). Here, we turn our attention to analyzing models (III) and (IV). The dynamic equations can be rewritten as

$$iE_t + E_{xx} + \frac{|E|^2 E}{1+g|E|^2} = 0\,, \tag{11}$$

$$iE_t + E_{xx} + \frac{1}{2g}(1 - e^{-2g|E|^2})E = 0\,. \tag{12}$$

Consider a homogeneous solution

$$E_s(t) = E_0 e^{it}\,, \tag{13}$$

where E_0 satisfies

$$E_0[1 - F(|E_0|^2)] = 0\,, \tag{14}$$

then, we have

$$E_0 = 0\,, \quad E_0 = \pm\sqrt{\frac{1}{1-g}} \tag{15}$$

for Eq. (11) with $g \neq 1$, and

$$E_0 = 0\,, \quad E_0 = \pm\sqrt{\frac{1}{2g}\ln(1-2g)^{-1}} \tag{16}$$

for Eq. (12) with $0 \leq g < 1/2$.

Defining

$$E(X,t) = E_s(t) + \delta E(X,t) \tag{17}$$

and linearizing Eq. (1), we get

$$\begin{pmatrix} i\partial_t + \partial_{xx} + L_i(|E_0|^2) & h_i(t) \\ h_i^*(t) & -i\partial_t + \partial_{xx} + L_i(|E_0|^2) \end{pmatrix} \begin{pmatrix} \delta E \\ \delta E^* \end{pmatrix} = 0\,, \tag{18}$$

where L_i and h_i depend on different models. They can be expressed as follows:

$$L_1(|E_0|^2) = \frac{2|E_0|^2 + g|E_0|^4}{(1+g|E_0|^2)^2}\,, \quad h_1 = \frac{E_s^2(t)}{(1+g|E_0|^2)^2} \tag{19}$$

for Eq. (11), and

$$L_2(|E_0|^2) = \frac{1}{2g}[1 - (1 - 2g|E_0|^2)e^{-2g|E_0|^2}]\,, \quad h_2(t) = E_s^2(t)e^{-2g|E_0|^2} \tag{20}$$

for Eq. (12).

If we consider the periodic boundary conditions, the eigenfunction of Eq. (18) can be chosen as

$$\begin{pmatrix} \delta E \\ \delta E^* \end{pmatrix} = \begin{pmatrix} \epsilon e^{i\lambda t} \\ \epsilon^* e^{-i\lambda^* t} \end{pmatrix} \cos(KX)\,, \tag{21}$$

where ϵ, ϵ^* are small parameters, and K is the wave number of modulational instabilities to the uniform solution. The linear growth rate as a function of an unstable wave number K is[16]

$$\gamma(K) = K[E_0 F'(E_0) - K^2]^{1/2} \tag{22}$$

for $0 < K < K^* = \sqrt{E_0 F'(E_0)}$, where the prime denotes differentiation. The unstable wave number corresponding to the maximum instability mode is $K_{\max} = \sqrt{[E_0 F'(E_0)]/2}$. In Eq. (21), we choose $K = K_{\max}$ and obtain

$$\lambda = 1 \pm i(1-g) \tag{23}$$

for Eq. (11) as $E_0 = \pm\sqrt{1/(1-g)}$ with $0 \leq g < 1$,

$$\lambda = 1 \pm i\left|\frac{1-2g}{2g}\ln(1-2g)\right| \tag{24}$$

for Eq. (12) as $E_0 = \pm\sqrt{1/2g\ln(1-2g)^{-1}}$ with $0 \leq g < 1/2$.

We further construct the phase space diagrams (A, A_t), which are defined as

$$A(X,t) = |E(X,t)| - E_0\,, \quad A_t(X,t) = \frac{d|E(X,t)|}{dt}\,. \tag{25}$$

For $|\delta E|$, obviously, we can conclude that $E_0 = \pm 1$ for cubic NSE, $E_0 = \pm\sqrt{1/(1-g)}$ with $0 \leq g < 1$ for Eq. (11) and $E_0 = \pm\sqrt{(1/2g)\ln(1-2g)^{-1}}$ with $0 \leq g < 1/2$ for Eq. (12) correspond to the hyperbolic fixed points, respectively. The linearized stability analysis also shows that $E_0 = 0$ for all the models considered in this paper corresponds to the elliptic fixed point.

III. Stochasticity

In numerical experiments, we choose the initial condition as

$$E(X,t) = E_0 + \epsilon e^{i\theta}\cos(K_{\max}X)\,, \tag{26}$$

where ϵ is a real parameter, $E_0 = \sqrt{1/(1-g)}$ for Eq. (11) and $E_0 = \sqrt{(1/2g)\ln(1-2g)^{-1}}$ for Eq. (12). Equation (26) reduces to the initial condition as considered by Moon for $g = 0$[17].

In numerical discussions, the periodic boundary condition has been used, where periodic length is taken as $L = 2\pi/K_{\max}$. The conventional split-step method[25] has been improved in order to increase the time accuracy. The number of modes for fast Fourier transformation is 2^n, and n is chosen from 5 to 8 and the time steps are chosen from 0.0001 to 0.0005, which depend on the accuracy of the conversed quantities being preserved to the order of 10^{-8}.

According to Eqs. (18), (23) and (24), we obtain

$$\frac{\delta E}{\delta E^*} = \pm i\,. \tag{27}$$

Combining Eq. (26), we see that the unstable manifold of the hyperbolic fixed point corresponds to $\theta = 45°$ and $\theta = 225°$ for availe value of g.

In the case of $g = 0$, however, our numerical results show that an HMO would appear as θ being a bit larger than $\theta = 45°$ and the hyperbolic point would deviate from the origin, which results from the modulational amplitude ϵ being not too small in our computer experiments. The trajectories in phase space are associated with space positions. The trajectory in Fig. 1 corresponds to that for $X = L/2$ and $\theta = 45.225°$.

In the presence of small perturbations, the periodic trajectories may shift from the HMO to adjacent ones whose periods are sensitive to the size of the displacement. In the case of one dimension, the integrability of systems will, in general, break down if other physical perturbations, such as driven damping[17-20], force dissipation[21,22] and nonlinear inhomogeneous media[23,24] are taken into account. A rich variety of temporal chaos and spatial coherence, which suggests that a low-dimensional chaotic attractor exists in an infinite-dimensional system, has been observed. In the presence of Hamiltonian perturbations, for example, the quintic correction, the chaotic behavior has also been found[14].

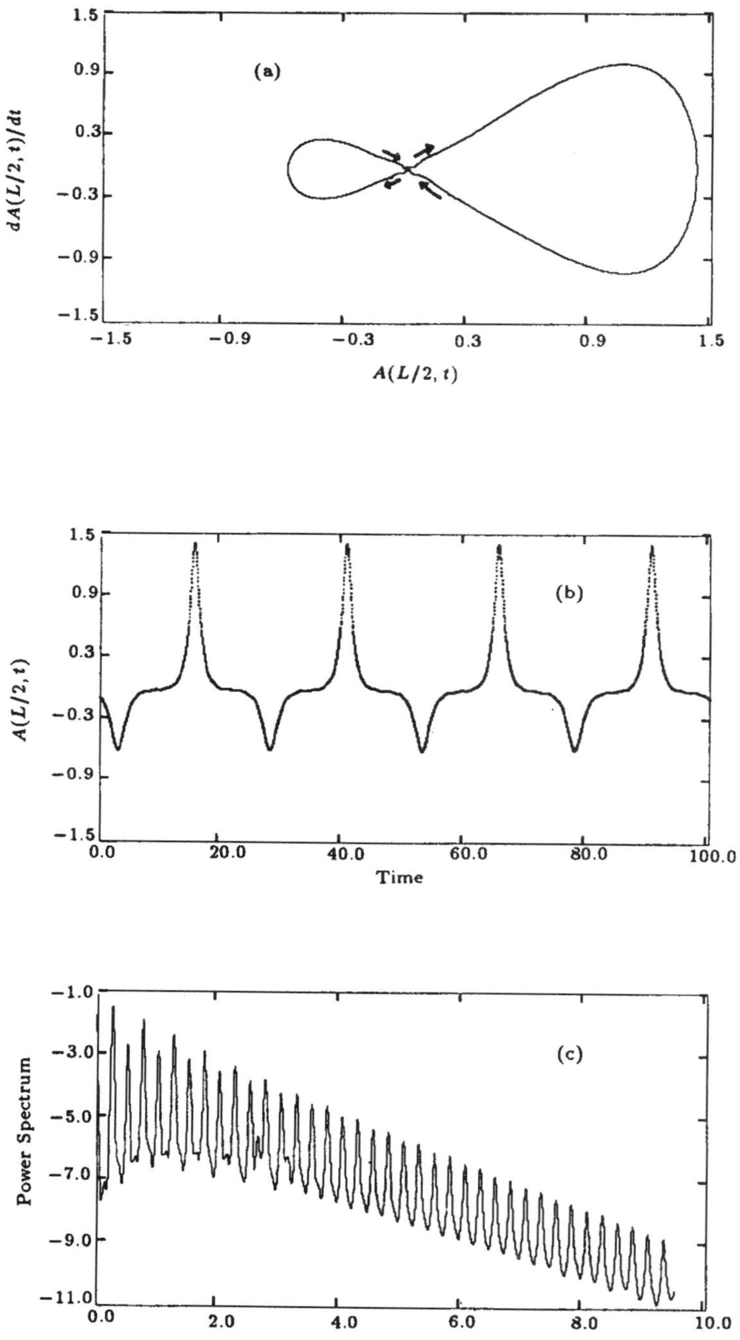

Fig. 1. Solution of cubic NSE with $\theta = 45.225°$. (a) Phase trajectories, where arrows indicate the directions of motions. (b) The time evolutions of the amplitude of fields. (c) The Fourier spectrum of $A(L/2, t)$ with respect to time.

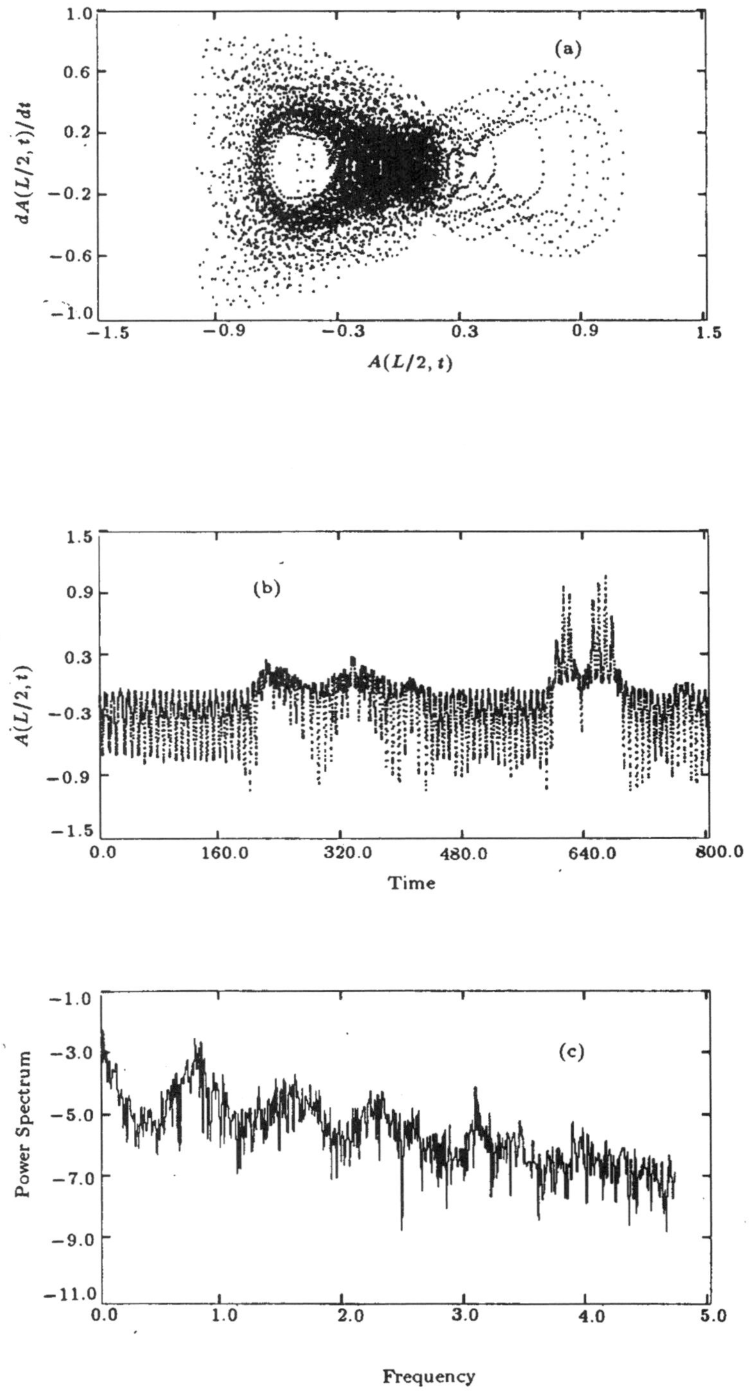

Fig. 2. Chaotic trajectories of Eq. (11) with $g = 0.1$ and $\theta = 45.225°$. **(a) Phase motion** in the phase space. Notice the irregular HMO crossings. (b) The time evolutions of the amplitude of fields. (c) The Fourier spectrum of $A(L/2, t)$ indicates the broadband structure and noise-like spectrum.

Taking the high order corrections to cubic terms in Eqs. (11) and (12) as Hamiltonian perturbations, we expect that irregular motion easily appears in nearby HMO. Comparing

Figs. 2 and 3 with Fig. 1, we observe that the completely different phenomena occur in such dynamic systems. As shown in Figs. 2a and 3a, the irregular HMO crossings in phase space $A_t - A$ are observed, which demonstrate that the presence of the HMO is a potential source of stochastic behavior. The time evolutions of the amplitude of fields $|E|$ display the stochastic motion. The power spectra clearly indicate the noise-like characteristic being typical of chaotic time evolutions.

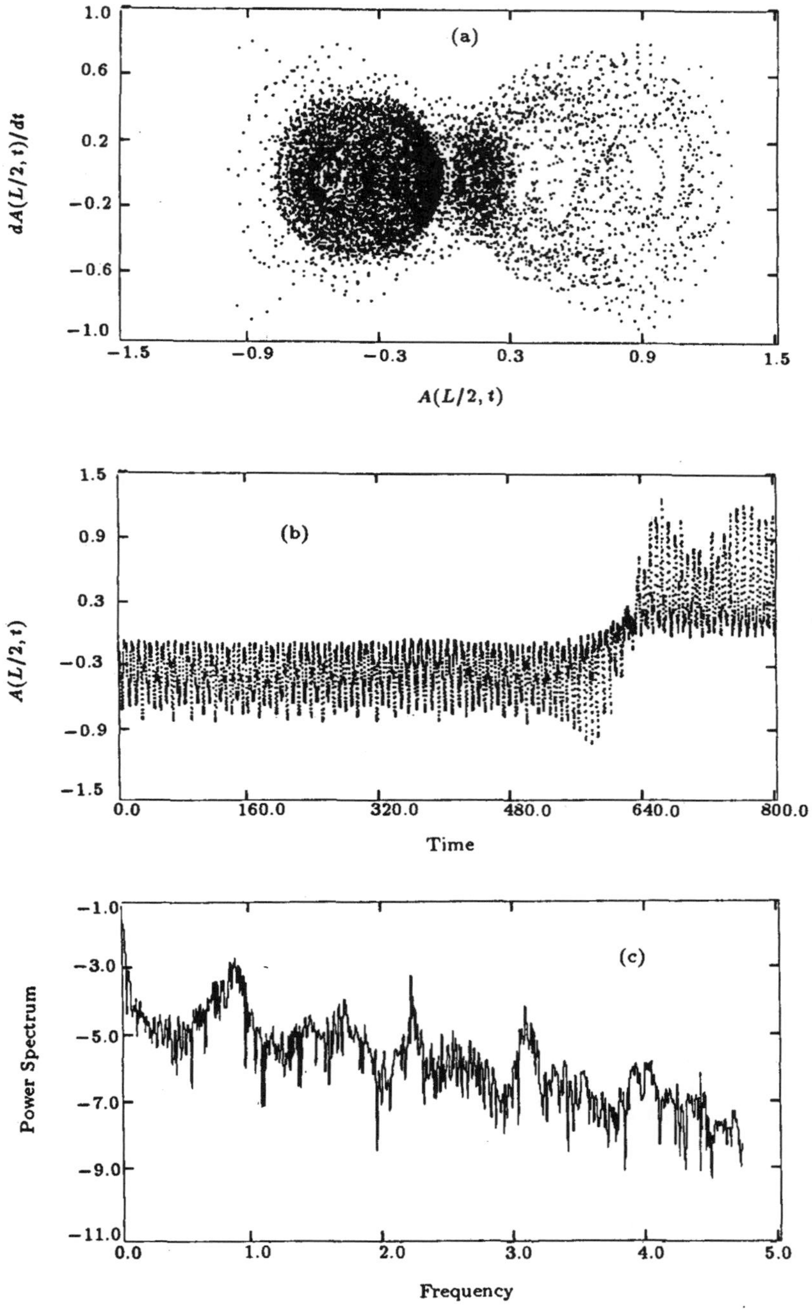

Fig. 3. Chaotic trajectories of Eq. (12) with $g = 0.1$ and $\theta = 45.225°$. (a), (b) and (c) are the same as those indicated in Fig. 2.

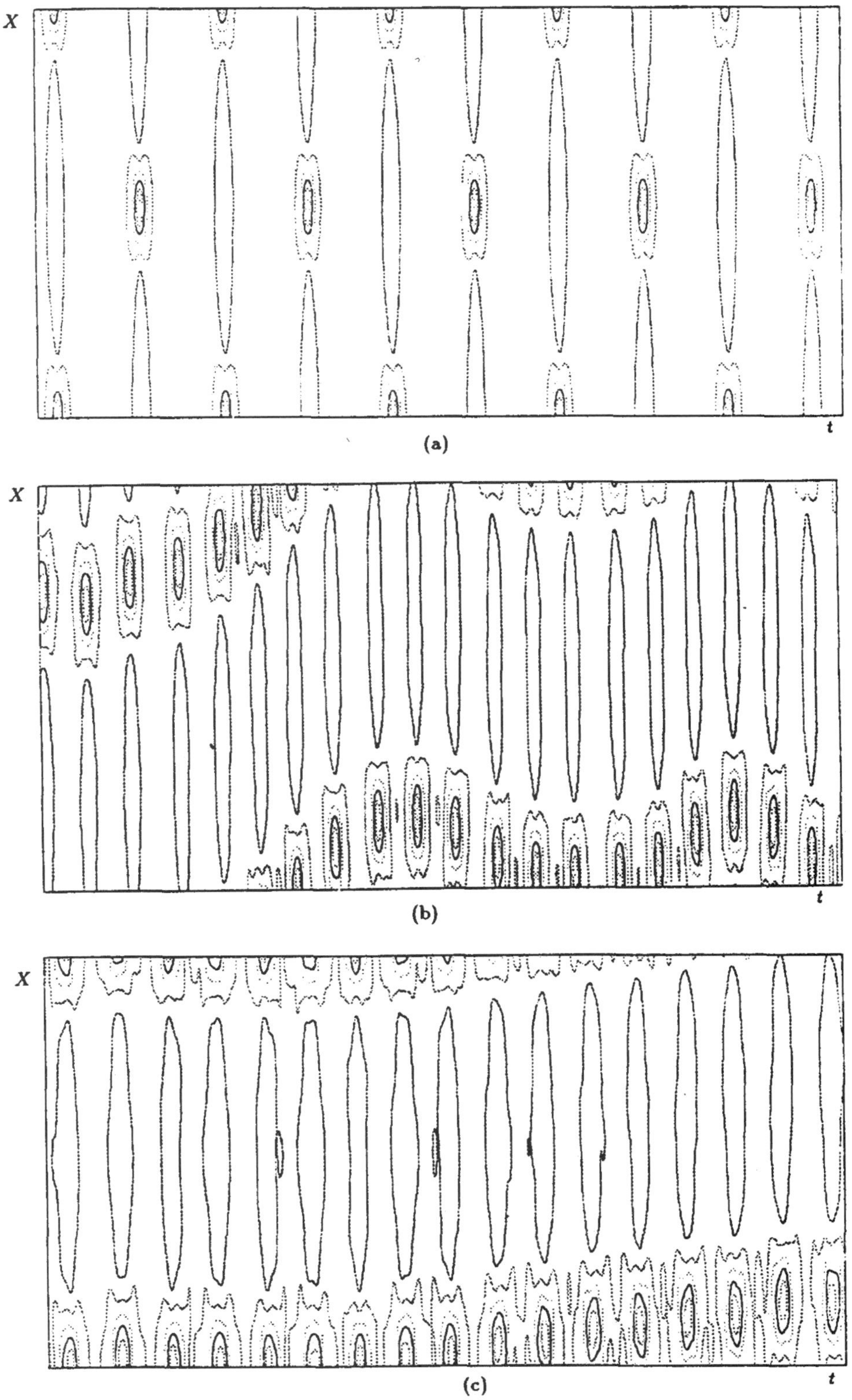

Fig. 4. Contours of $|E(X,t)|^2$ =const. (a) corresponds to the recurrence solution of cubic NSE with $t = 0 \sim 120$; (b) corresponds to the solution of Eq. (11) with $t = 200 \sim 360$; (c) corresponds to the solution of Eq. (12) with $t = 200 \sim 360$.

Figure 4 shows that recurrence, which is inherent in cubic NSE, has indeed disappeared for the NSE involving the saturable nonlinearities. For the cubic NSE, the well-known FPU recurrence is exhibited in Fig. 4a, where the field is of temporally periodic evolutions and spatially coherent structures. The propagative velocity of the coherent structures keeps constant. For high order saturable Eqs. (11) and (12), however, figures 4b and 4c illustrate that the field propagations are temporally stochastic but still keep their spatially localized structures. The temporal behaviors are more obvious in Figs. 2b and 3b. From the figures, also, we see that the high order saturable nonlinear effects may lead to the presence of different localized structures. In the further numerical experiments, we can show that the energy in systems, which is initially confined to the lowest mode, would spread to many higher modes due to the nonlinear instability, but would not regroup into the original lowest mode. On the other hand, we find that the first two modes dominate the large portion of energy in systems, which leads to the spatially localized structures being kept. The energy confined in the higher modes evolves stochastically with time and results in the existence of various spatial patterns of the localized structures, i.e., recurrence disappears. In addition, we find that the irregular behavior of phase of field propagation can also occur when the amplitude of fields oscillates stochastically.

IV. Periodic Solutions and Solitary Wave Solutions

In Sec. III, we have numerically discussed the stochasticity of dynamic systems (11) and (12). Here, we further analyse the properties of solutions.

For Eq. (1), we look for a solution in the form of

$$E(X,t) = G(X,t)e^{i\Phi(X,t)} . \tag{28}$$

Thus equation (1) becomes

$$-G\Phi_t + G_{xx} - G\Phi_x^2 + F(G^2)G = 0 , \tag{29}$$

$$G_t + G\Phi_{xx} + 2G_x\Phi_x = 0 . \tag{30}$$

Letting

$$G(X,t) = G(\zeta), \quad \zeta = X - Vt , \quad \Phi(X,t) = qX - \Omega t , \tag{31}$$

then we have

$$q = \frac{V}{2} \tag{32}$$

and

$$\frac{\partial^2 G}{\partial \zeta^2} - \alpha^2 G + F(G^2)G = 0 , \tag{33}$$

where $\alpha^2 = V^2/4 - \Omega$.

Multiplying both sides of Eq. (33) by $dG/d\zeta$, and integrating once, we obtain

$$\frac{1}{2}\left(\frac{dG}{d\zeta}\right)^2 + V(G) = H_0 , \tag{34}$$

where H_0 is the integral constant which represents the pseudoenergy, and the pseudopotential $V(G)$ is

$$V(G) = \int F(G^2)GdG - \frac{\alpha^2}{2}G^2 . \tag{35}$$

The expression $V(G)$ is dependent on the different models, that is

$$V(G) = \frac{1}{4}G^4 - \frac{\alpha^2}{2}G^2 \tag{36}$$

for cubic NSE,

$$V(G) = -\frac{g}{6}G^6 + \frac{1}{4}G^4 - \frac{\alpha^2}{2}G^2 \tag{37}$$

for model (II),

$$V(G) = -\frac{1}{2g^2}\ln(1+gG^2) + \left(\frac{1}{2g} - \frac{\alpha^2}{2}\right)G^2 \tag{38}$$

for Eq. (11), and

$$V(G) = \frac{1}{8g^2}(e^{-2gG^2} - 1) + \left(\frac{1}{4g} - \frac{\alpha^2}{2}\right)G^2 \tag{39}$$

for Eq. (12).

Here, we mainly discuss the properties of dynamic equations (11) and (12). The pseudopotential functions (38) and (39) are plotted in Fig. 5. Clearly, the curves are similar to those for cubic NSE. If we consider Eq. (34) to be energy integral of a classical particle with unit mass, the bound solutions may be found as $H_0 < 0$. Suppose a particle to lie in the potential well, then the particle motion is periodic and the orbit corresponds to a periodic solution. As $H_0 = 0$, the orbit corresponds to a very special solution in G, that is, a single pulse solution called a solitary, where the periodic length tends to infinity. From the review of point of physics, the saturable nonlinearities tend to steepen the wave whereas the wave dispersion tends to spread it out. Therefore, it is possible to obtain a stationary solution if these two effects exactly balance each other. According to the above analysis, an explicit periodic solution and a solitary have been obtained for model (II)[14]. The periodic solution is

$$G(X,T) = \left[\frac{a_2(a_3-a_1) - a_3(a_2-a_1)\mathrm{sn}^2(\sqrt{g/3}M\zeta,k)}{(a_3-a_1)-(a_2-a_1)\mathrm{sn}^2(\sqrt{g/3}M\zeta,k)}\right]^{1/2}, \tag{40}$$

where a_1, a_2, a_3 are three roots of the following equation, and assume $0 \le a_1 < a_2 < a_3$,

$$\frac{g}{3}a^3 - \frac{1}{2}a^2 + \alpha^2 a + 2H_0 = 0\,, \tag{41}$$

and

$$M = \sqrt{a_2(a_3-a_1)}\,, \tag{42}$$

$$k = \left[\frac{a_3(a_2-a_1)}{a_2(a_3-a_1)}\right]^{1/2} \tag{43}$$

As $H_0 = 0$, we have

$$a_1 = 0\,, \quad a_2 = \frac{3(1-\sqrt{1-16g\alpha^2/3})}{4g}\,, \quad a_3 = \frac{3(1+\sqrt{1-16g\alpha^2/3})}{4g} \tag{44}$$

and

$$k = 1\,, \quad M = \alpha\sqrt{3/g}\,. \tag{45}$$

Hence, a solitary wave solution becomes

$$G(\zeta) = \frac{2\alpha}{\sqrt{(1-16g\alpha^2/3)^{1/2}\cosh(2\alpha\zeta)+1}}\,. \tag{46}$$

If $g = 0$, equation (46) reduces to

$$G(\zeta) = \sqrt{2}\alpha\,\mathrm{sech}(\alpha\zeta)\,, \tag{47}$$

which is the so-called Langmuir soliton.

For potential functions (38) and (39), however, as a result of their complication, the explicitly analytical solutions have not been obtained yet. For Eq. (12), the region of existence of envelope solitary had been analysed by Shukal[15]. In this paper, we do not try to find their explicitly periodic solutions, but we can conclude that there may exist some special periodic solutions and solitary wave for dynamic systems (11) and (12). The effects of the high order saturable nonlinearities will make the amplitude and width of Langmuir electric field envelope solitary and the density cavity change.

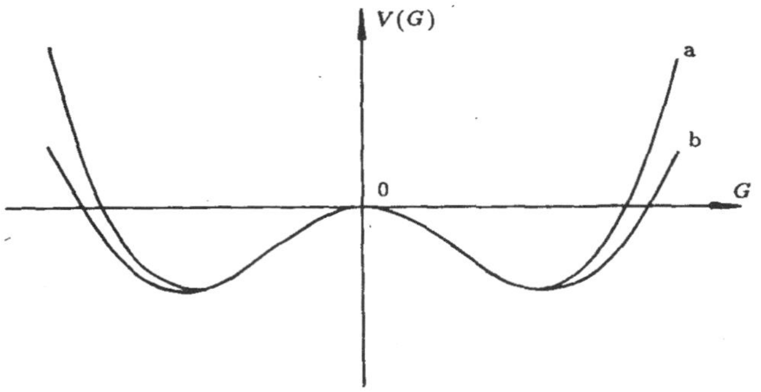

Fig. 5. The pseudopotential functions (a) corresponds to Eq. (38) and (b) corresponds to Eq. (39), where $\alpha = 1$ and $g = 5/16$.

In general, the recurrence solutions correspond to a class of periodic solutions, but the periodic solution discussed in this section is not associated with recurrence. The reason is that the periodic solution discussed in this section is a traveling solution, and the recurrence solution is derived from the initial value problem with periodic boundary condition. In Ref. [14], we had pointed out that recurrence which sensitively depends on the initial condition is only a special feature for such continuum non-integrable Hamiltonian dynamic systems.

V. Conclusions and Discussions

For cubic NSE, the integrability has been verified by finding Lax-pair, and a class of periodic solutions and solitons can be obtained by means of IST. Many physical phenomena can be explained due to the properties of solitons and recurrence of solutions. The solitons developed by modulational instabilities keep their spatially coherent structures and temporally periodic evolutions. When the dissipation or damping is considered, the field evolution is stochastic, which is well known for a finite dimensional system.

For our dynamic systems, the integrability will also be broken down due to high order saturable nonlinear Hamiltonian perturbation. It is shown that these nonlinear effects can lead to the presence of chaos, where the field evolutions are temporally chaotic but keep their spatially localized structures. However, the periodic solutions and envelope solitary could simultaneously exist in such non-integrable continuum Hamiltonian dynamic systems. These features illustrate that the existence of periodic solutions and solitary wave solutions is not necessarily associated with complete integrability. The fact that the completely different dynamic behaviors exist is the important properties of these dynamic systems.

As far as various branches of physics are concerned, the presence of stochastic wave fields depends on the evolutive processes of systems. In other words, the lowest order nonlinear term $|E|^2E$ is the predominative nonlinear mechanism in the early stages of field evolutions. Some important phenomena, such as coherent structures and the collision of solitons, can availably be explained by making use of the cubic NSE. In the later stages of field evolutions, other physical effects have to be considered. For example, the effects of Landau damping of the high frequency waves become quite significant in plasma turbulence. On the other hand, the non-linear interaction, such as the fourth field interaction and other nonlinear saturable mechanism should be involved in the processes of plasma instabilities. These nonlinear saturable effects make rich dynamic behaviors occur. In a sense, these complicated dynamic phenemona would probably be the characteristic of wave propagation. Finally, we expect that more important theoretical work on HMO chaos may be furthered by way of Melnikov's method[26,27], but it might be a very tedious and difficult work. The reason is that the scattering spectrum of cubic NSE with periodic boundary, hence its HMO in phase space, is quite complicated.

Acknowledgments

We wish to acknowledge stimulated discussions with Prof. S.G. CHEN on various issues relating to this paper. One of us (ZHOU) is grateful to Tan YU for helping in numerical discussions.

References

[1] J.J. Rasmussen and K. Rypdal, Physica Scripta **33**(1986)481; K. Rypdal and J.J. Rasmussen, Physica Scripta **33**(1986)498.

[2] B.J. Lemesurier, G. Papanicollaou, C. Sulem and P.L. Sulem, Physica **D31**(1988)78.

[3] D. Anderson and M. Bonnedal, Phys. Fluids **22**(1979)105.

[4] V.G. Makhankov, Physica Scripta **20**(1979)558.

[5] D.W. Mclaughin, G.G. Papanicolaou, C. Sulem and P.L. Sulem, Phys. Rev. **A34**(1986)1200.

[6] K.H. Spatschek, H. Pietsch, E.W. Laedke and Th. Erckermann, in *"Nonlinear and Turbulent Processes in Physics"*, eds. V.E. Bar'yakhtar *et al.,* World Scientific, Singapore **1**(1989)10.

[7] N.N. Akhmediev, D.R. Heatley, G.I. Stegeman and E.M. Wright, Phys. Rev. Lett. **65**(1990)1423.

[8] X.T. He, Acta. Physica **31**(1982)1317 (in Chinese).

[9] B.M. Lake, H.C. Yuen, H. Rungalier and W.E. Ferguson, J. Fluid Mech. **83**(1977)49.

[10] H.C. Yuen and W.E. Ferguson, Phys. Fluids **18**(1975)956.

[11] H.C. Yuen and W.E. Ferguson, Phys. Fluids **21**(1978)1275.

[12] M. Stiassnie and U.I. Kroszynski, J. Fluid. Mech. **116**(1982)207.

[13] A. Cloot, B.M. Herbst and J.A.C. Weideman, J. Comput. Phys. **86**(1990)127.

[14] Cangtao Zhou, X.T. He and Shigang Chen, Phys. Rev. **A46**(1992)2277.

[15] P.K. Shukla, *"Solitons in Plasma Physics in Nonlinear Waves"*, ed. by L. Debnath, Cambridge Univ. Press, London (1983) Chapt. 11.

[16] M.S. Cramer and L.T. Watson, Phys. Fluids **27**(1984)821.

[17] T.H. Moon, Phys. Rev. Lett. **64**(1990)412.

[18] K. Nozaki and N. Bekki, Phys. Rev. Lett. **50**(1983)1226; J. Phys. Soc. Jap. **54**(1985)2362; Physica **21D**(1986)381.

[19] K.J. Blow and N.J. Doran, Phys. Rev. Lett. **52**(1984)526.

[20] K.H. Spatschek, H. Pietsch, E.W. Leadk and Th. Erckermann, in *"Singular Behavior and Nonlinear Dynamics"*, eds. St. Pnevmatikos *et al.,* World Scientific, Singapore **2**(1989)555.

[21] T.H. Moon and M.V. Goldman, Phys. Rev. Lett. **53**(1984)1821.

[22] K. Nozaki and N. Bekki, Phys. Lett. **102A**(1984)383.

[23] W. Shyu, P.N. Guzdar, H.H. Chen, Y.C. Lee and C.S. Liu, Phys. Lett. **147A**(1990)49.

[24] O. Larroche and D. Pesme, Phys. Fluids **B2**(1990)1751.

[25] T.R. Taha and M.J. Ablowitz. J. Comput. Phys. **55**(1984)202.

[26] F.Kh. Abdullaev, Phys. Rep. **179**(1989)1.

[27] J. Cuckenheimer and P. Homles, *"Nonlinear Oscillations, Dynamical Systems and Bifurcations of Vector Fields"*, Springer-Verlag, Berlin (1983).

Commun. Theor. Phys. 20(1993)353–358
© International Academic Publishers
Vol. 20, No. 3

Relative Diffusion and Two-Point Microscale Correlation in Phase Space of Electrons in Localized Langmuir Fields[1]

Shao-Ping ZHU and X.T. HE
Center for Nonlinear Studies
Institute of Applied Physics and Computational Mathematics, Beijing 100088, China

(Received March 18, 1992; Revised April 28, 1992)

Abstract

We discuss both relative spatial diffusion coefficient D_-^r and relative velocity diffusion coefficient D_-^u for electrons interacting with coherent, localized Langmuir wave packets. We find that D_-^r and D_-^u have nearly the same positions in which they get a nonzero minimum for small relative position r and relative velocity u and the diffusion coefficients are greatly reduced close to minima. We also find that the position in which D_-^u has a minimum shifts when r and u vary, while the position for D_-^r does not move. So we show that the coherent, localized wave packets can drive the formation of strong two-point microscale correlation in phase space only when r and u are small. We also discuss the lifetime of clump.

In the past two decades, much attention has been paid to strong two-point microscale correlation in phase space of particles due to relative diffusion, the so-called clump from Vlasov-Poisson equations proposed firstly by Dupree[1] in 1972. Following Dupree's idea, Balescu *et al.* discussed it in detail from particle trajectory theory[2]. It is believed that clump can behave as a single discrete particle and enhance the radiation and emission, resistivity, etc. Up to now, almost all the study on clump is for homogeneous plasma system and assume that the variations of the driving electric fields are stochastic[1,2]. However, in recent years, Rozmus *et al.*[3] and one of the authors[4] have found that in phase space the trajectories of electrons moving in the regular electric fields are also stochastic due to Chirikov resonance[5]. It implies that trajectories between two points in phase space can be very close and lead to strong two-point microscale correlation because at this moment the electric fields in two points are close to be equal to each other. In the previous paper[6], we considered the so-called clump phenomena in regular, localized Langmuir plasma system. Reference [6] shows that the localized, coherent wave packets can drive the formation of clumps which behave as the one defined by Dupree[1] essentially in phase space according to the relative velocity diffusion coefficients. But in such an inhomogeneous system with coherent structure, it is necessary to study the spatially relative diffusion in order to clarify whether clump can appear in the system or not further. In this paper, we discuss both spatially relative diffusion coefficients and relative velocity diffusion coefficients. Our results do show that the clump in phase space can be driven by the localized, coherent wave packets only when relative velocity and position are small.

Write the localized, coherent fields in the form of a soliton structure[3]

$$E(x,t)=\sqrt{2}E_0\,\mathrm{sech}\,(gx)\cos(k_0x-\omega_pt), \tag{1}$$

where $g=\beta E_0/\sqrt{8\pi nk_BT}$, β is a parameter, E_0 is the amplitude, k_0 is the Langmuir wave number, other notations are standard.

[1]The project partially supported by National Natural Science Foundation of China through Grant No. 1880109.

Analogy to Ref. [6], we have the dimensionless equations of motion for electrons

$$\begin{aligned} \frac{dr(t)}{dt} &= u\,, & \frac{du(t)}{dt} &= \sum_n 2E_n \sin\left[\frac{1}{2}k_n R(t) - t\right] \sin\left[\frac{1}{2}k_n r(t)\right], \\ \frac{dR(t)}{dt} &= V, & \frac{dV(t)}{dt} &= -\sum_n 2E_n \cos\left[\frac{1}{2}k_n R(t) - t\right] \cos\left[\frac{1}{2}k_n r(t)\right], \end{aligned} \tag{2}$$

where r, u and R, V are relative and barycentric motion coordinates respectively, $k_n = k_0 + (2n\pi/L)$, $n = 0, \pm1, \pm2, \ldots, \pm n_0$, $k_0 = \pi/L$, L is a periodic length.

It is shown that the trajectories in phase space of electrons moving in one wave packet are periodic, and the periodically trapping trajectories of electrons interacting with many wave packets which satisfy overlapping criterion may lead to stochastic processes. So the trajectories of electrons form a stochastic region in phase space[3,4], accompanied by the existence of small regular islands in this region. Thus the stochastic motion may lead to strong correlations of neighboring electrons in phase space because relative diffusion is greatly reduced, and so may tend to enable clumps to form in phase space under localized and coherent fields.

Equations (2) are strongly nonlinear. It is difficult to solve them analytically. We solve Eqs. (2) numerically by fourth-order Runge-Kutta method with time step $\Delta t = 0.1$. With numerical solutions of Eqs. (2), according to the definition of the diffusion coefficients, we write the absolute diffusion coefficients to be computed in the form of

$$D^v_{ii} = \frac{1}{2\tau}\langle[v_i(\tau) - v_i]^2\rangle, \qquad D^x_{ii} = \frac{1}{2\tau}\langle[x_i(\tau) - x_i]^2\rangle, \tag{3}$$

where v_i and x_i are the initial velocity and position respectively, and the relative diffusion coefficients in the form of

$$D^u_- = \frac{1}{2\tau}\langle[u(\tau) - u]^2\rangle, \qquad D^r_- = \frac{1}{2\tau}\langle[r(\tau) - r]^2\rangle, \tag{4}$$

where u and r are the initial relative velocity and position respectively, $\tau = \frac{1}{2}(\frac{L}{v_1} + \frac{L}{v_2})$ is the approximate time for one passing through the periodic structure. The ensemble average is computed by averaging over a set of particles with different start time but with the same initial velocity and position.

The spectrum of the wave packet is presented in Fig. 1 for parameters $E_0^2/(8\pi n k_B T) = 0.53$, $L = 72$ and $\beta = 1/(2\sqrt{3})$, where the overlapping conditions are satisfied except for the modes $n = 0, -1$.

We can see from Fig. 2 and Fig. 3 that D^u_- and D^r_- have a nonzero minimum. The fact that D_- has a nonzero minimum means that the lifetime of clumps is finite. It is easy to understand, because D_{11}, D_{22} and D_- are nonnegative, and D_- approaches $D_{11} + D_{22}$ when $|r| \to \infty$ or $|u| \to \infty$, so D_- varying with r or u must have a minimum. Thus, the relative diffusion is greatly reduced close to D_- minima, and the two-point correlation is considerably enhanced. As shown in Fig. 2 and Fig. 3, D^u_- (solid curve) and D^r_- (dotted curve) have nearly the same positions in which they get a nonzero minimum for small r and u. But the position in which D^u_- has a minimum shifts when r or u increases, while the position for D^r_- does not move. To be more exact, D^u_- has a nonzero minimum when $r + \tau u/2 = 0$. This fact means that clump can form only for small r and u. In fact, for this inhomogeneous system, it is believed that the two-point correlation function $C(r, u)$ satisfies the following equation[1,7]:

$$\frac{\partial C}{\partial t} + u\frac{\partial C}{\partial r} - \frac{\partial}{\partial r}\left(D^r_- \frac{\partial C}{\partial r}\right) - \frac{\partial}{\partial u}\left(D^u_- \frac{\partial C}{\partial u}\right) = S(r, u)\,. \tag{5}$$

As r and u vanish, the source term S varies slowly, so the function C has peak behavior only when both D^u_- and D^r_- decrease very greatly.

贺贤土论文选集
296

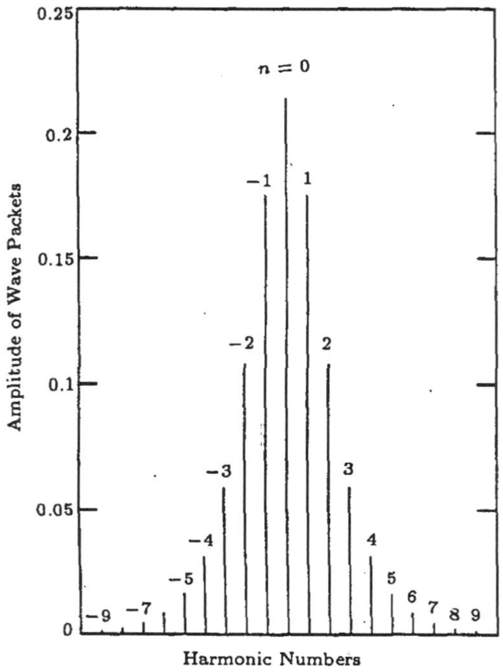

Fig. 1. The spectrum of the Langmuir wave packet.

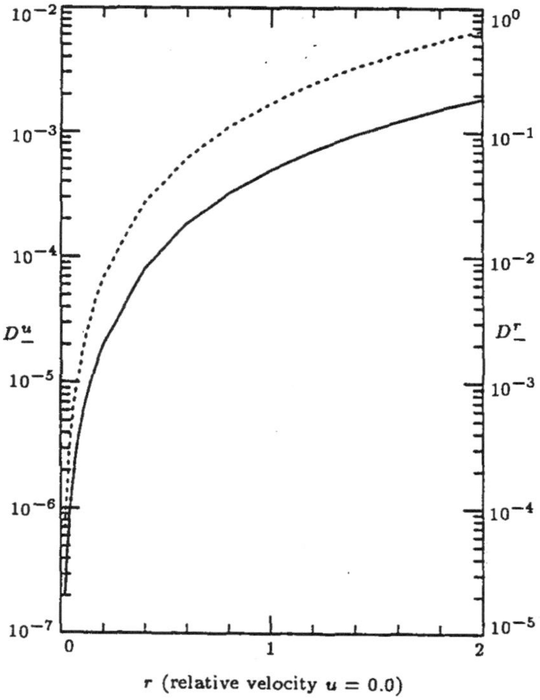

Fig. 2a. The relative diffusion coefficients vs. the relative positions for $u = 0.00$.

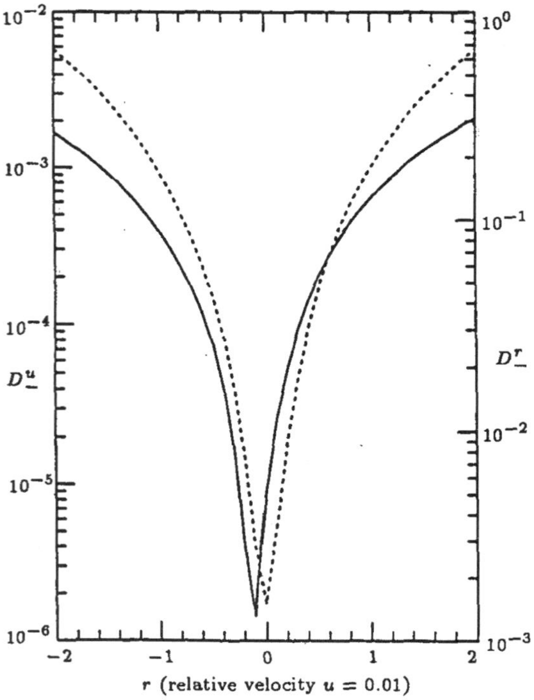

Fig. 2b. The relative diffusion coefficients vs. the relative positions for $u = 0.01$.

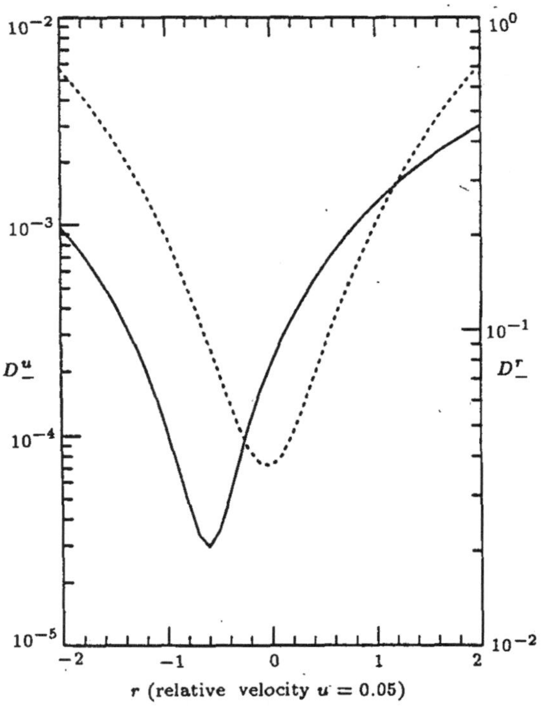

Fig. 2c. The relative diffusion coefficients vs. the relative positions for $u = 0.05$.

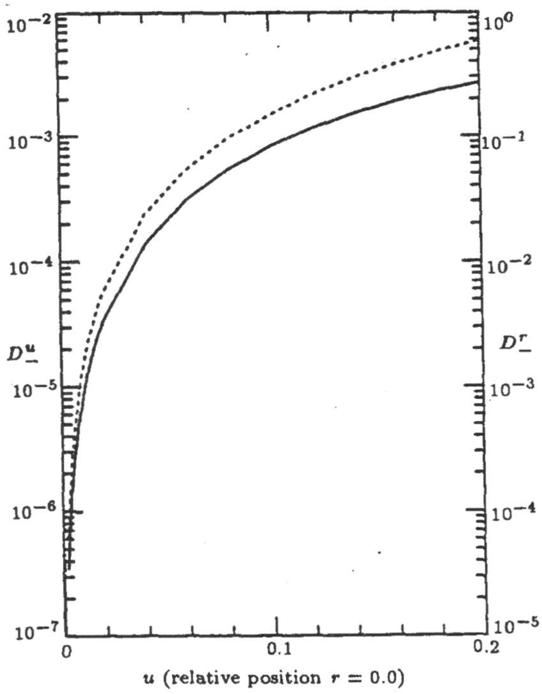

Fig. 3a. The relative diffusion coefficients vs. relative velocities for $r = 0.00$.

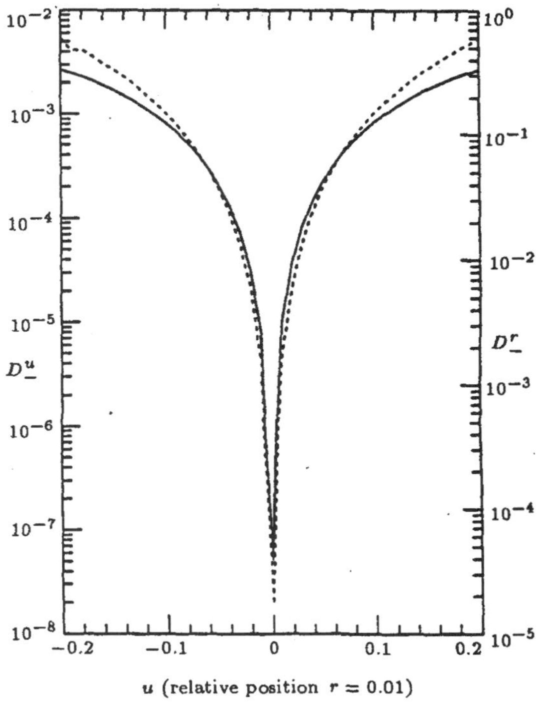

Fig. 3b. The relative diffusion coefficients vs. relative velocities for $r = 0.01$.

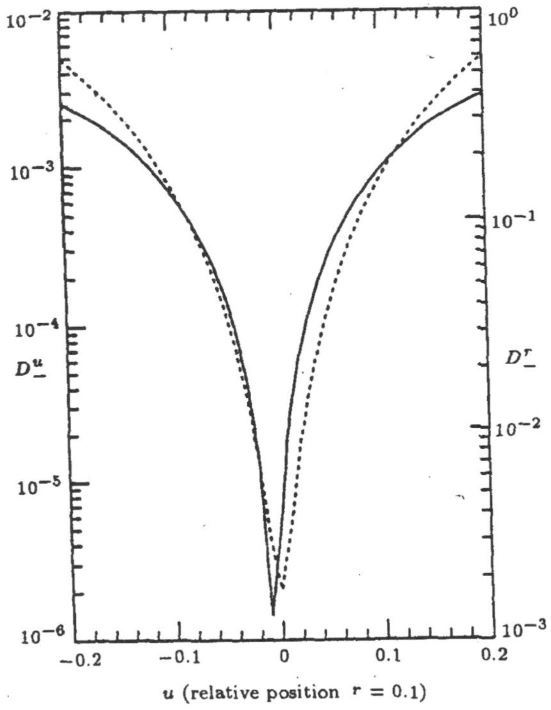

Fig. 3c. The relative diffusion coefficients vs. relative velocities for $r = 0.10$.

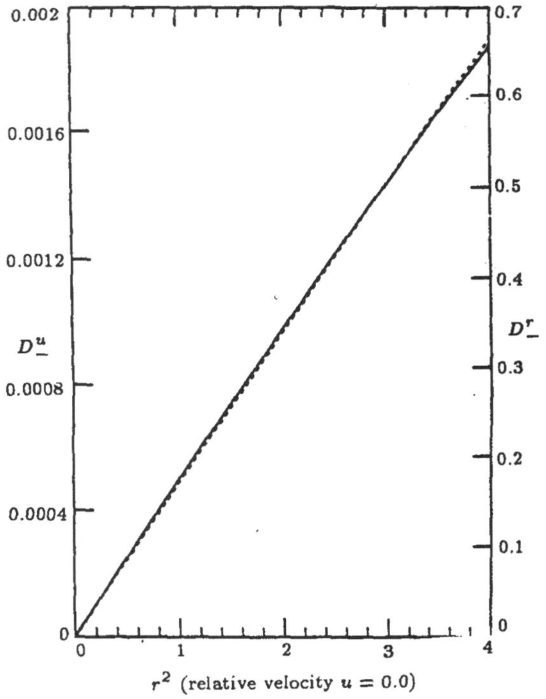

Fig. 4. The relative diffusion coefficients vs. the square of relative positions for $u = 0.0$. The solid curve — D^u_-, the dotted curve — D^r_-.

Analyzing the D^u_- and D^r_- given by numerical simulation, we find that their values for $u = 0$ coincide with the following formulae very well

$$D^u_- = \frac{1}{2}k_u^2\langle r^2(\tau)\rangle(D^v_{11} + D^v_{22}), \tag{6}$$

$$D^r_- = \frac{1}{2}k_r^2\langle r^2(\tau)\rangle(D^x_{11} + D^x_{22}), \tag{7}$$

where k_u and k_r are the characteristic wave numbers, $k_u = 0.1012$ and $k_r = 0.015$ with the parameters in Fig. 2 and Fig. 3. This behavior is shown in Fig. 4 and Fig. 5.

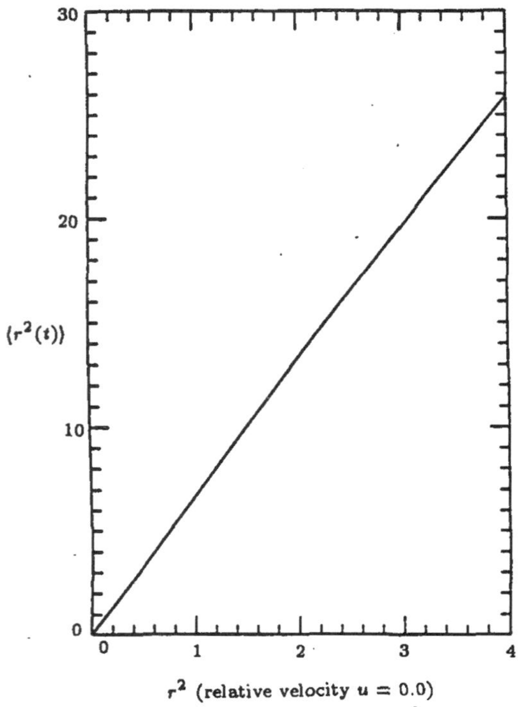

Fig. 5. $\langle r^2(t)\rangle$ vs. the square of relative positions when $u = 0.0$ and $t = \tau$.

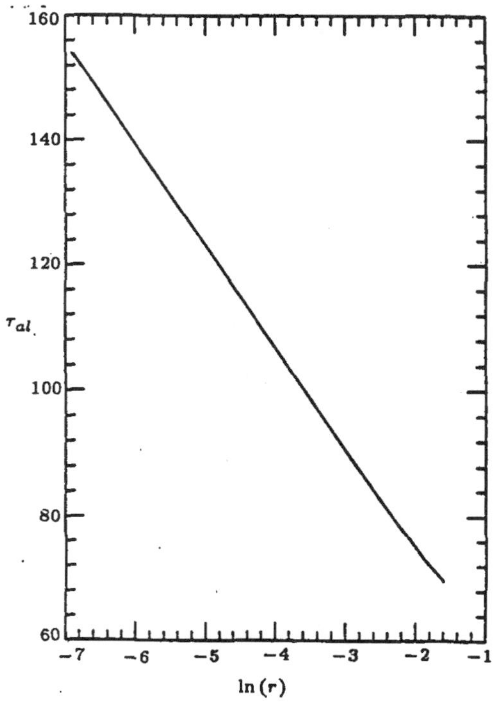

Fig. 6. The lifetime of clumps vs. the relative positions when $u = 0.0$.

From Eqs. (6) and (7), we can get two characteristic times, we define the smaller one as lifetime of clumps τ_{cl}, that is

$$\max(k_r^2, k_u^2)\langle r^2(\tau_{cl})\rangle = 1, \tag{8}$$

figure 6 shows the dependence of τ_{cl} on the relative positions. We find that the results can be represented very well by

$$\tau_{cl} = A - B\ln r, \tag{9}$$

where A and B are constants, in our case $A = 41.91$ and $B = 16.24$. τ_{cl} diverges logarithmically as r vanishes.

We also give the coherent time which is defined as[8]

$$\tau_c = \left[\frac{2}{k_u^2(D^v_{11} + D^v_{22})}\right]^{1/3}.$$

Our results show that $\tau_{cl} \gg \tau_c$ for small r and u.

In conclusion we have studied the relative diffusion coefficients of electrons moving in the localized, coherent Langmuir fields. Our results show that the localized, coherent wave packets can drive the formation of clumps in phase space only when the relative position r and velocity u are small because relative spatial and velocity diffusion coefficients have different positions where they have nonzero minimum for large r and u. With the numerical solutions of the equations of motion, we found that the lifetime of clump is much larger than coherent time (Fig. 6) and only when the relative velocity is zero does the lifetime of clump indeed diverge logarithmically, with the relative positions of the constituents getting closer to one another.

Acknowledgments

One of the authors (Shao-Ping ZHU) would like to thank the International Centre for Theoretical Physics (Trieste, Italy) for hospitality during his visit from 23 May to 2 July 1991. We would particularly thank the referee for good suggestions.

References

[1] T.H. Dupree, Phys. Fluids **15**(1972)334.
[2] M. Pettini *et al.*, Phys. Rev. **A38**(1988)344.
[3] W. Rozmus and J.C. Samson, Phys. Lett. **A126**(1988)263.
[4] X.T. HE and Y. TAN, Bull. Am. Phys. Soc. **34**(1989)2162.
[5] B.V. Chirikov, Phys. Rep. **52**(1979)492.
[6] Shao-Ping ZHU and X.T. HE, Phys. Rev. **A43**(1991)1988.
[7] T.H. Dupree, Phys. Fluids **17**(1974)100.
[8] J.A. Krommes, in "*Handbook of Plasma Physics*", ed. A.A. Galeev and R.N. Sudan, North Holland, Amsterdam, **2**(1984)183.

J. Phys. A: Math. Gen. 26 (1993) 4123–4133. Printed in the UK

Spatiotemporal complexity of the cubic–quintic nonlinear Schrödinger equation

X T He and Cang-tao Zhou

CCAST (World Laboratory), PO Box 8730, Beijing, 100080, People's Republic of China and Institute of Applied Physics and Computational Mathematics, PO Box 8009, Beijing, 100088, People's Republic of China

Received 15 April 1992, in final form 25 November 1992

Abstract. The integrability of the cubic–quintic nonlinear Schrödinger equation is investigated numerically. By analytically studying the linear stability of the homogeneous state and numerically solving such a continuum Hamiltonian dynamic system, we show that the quintic nonlinear term leads to a spatiotemporal complexity of wave fields, and illustrate that this behaviour is associated with the stochastic partition of energy in Fourier modes. In addition, we show the presence of the stochastic motion is due to the homoclinic orbit crossings.

1. Introduction

The cubic–quintic nonlinear Schrödinger equation (NSE)

$$i\partial_t E + \partial_{xx} E + |E|^2 E - g|E|^4 E = 0 \tag{1}$$

has recently received much attention due to its physical application, where g is a parameter. In nonlinear optics, the refractive index is expanded to the second order in electric field E, the well known cubic NSE is obtained. However, Aciolic *et al* [1] experimentally measured the optical susceptibilities of CdS_xSe_{1-x} doped glasses using a spatially resolved phase-matched multiwave mixing technique, and showed that fifth-order nonlinear effects should also be considered. In many-nucleon systems, usually, collision processes of heavy ion are described on classical or quasi-classical levels [2]; it is shown that in the quasi-classical limit the nuclear hydrodynamics equations with the Skyrme forces can be reduced to the non-relativistic cubic–quintic model for the corresponding choice of variable [3]. In Langmuir plasmas, the beat frequency interaction between the large amplitude parts of high-frequency Langmuir fields and ion-acoustic waves can occur in the later stages of evolution of plasma instability. The fourth field interaction is an important mechanism [4].

For the cubic–quintic NSE (1), on the other hand, some theoretical work has been obtained analytically and numerically. Puskharov *et al* [5] obtained solitary wave solutions. Cowan *et al* [6] showed numerically these are not solitons, but behave like quasi-solitons. Gagnon *et al* [7, 8] presented a large set of exactly analytic travelling-wave solutions for such a model. Zhou *et al* [9] also show that a special periodic solution can be reduced to a solitary wave as pseudo-energy being zero. The effects of the quintic nonlinear term enhance the amplitude and width of the solitary wave as compared with that for the cubic Langmuir soliton. Cloot *et al* [10] qualitatively

0305-4470/93/164123+11$07.50

proved the existence of bounded solutions by use of the invariants for quasi-particle number and energy. Their numerical results displayed the well known Fermi-Pasta-Ulam (FPU) recurrent phenomenon. The recurrence is only a special feature for such a system. In high space dimensions, the Lie symmetry group was analysed by Gagnon and Winternitz [11, 12]. They obtained a class of group-invariant solutions. Lemesnrier *et al* [13] pointed out that the focusing singularities can occur in two- or three-dimensions. Also, the stable particle-like solutions may exist in this dynamic system [3]. Even in the case of one-dimension, however, (1) for $g \neq 0$ does not belong to the class of integrable nonlinear evolution equations [14], which means that it cannot be solved by the inverse-scattering method and solitons and multi-solitons are not to be expected [7]. Many new and very interesting dynamic properties of the system, such as the stochastic propagation of wavefields, can appear [9]. In this paper, we theoretically and numerically show that the stochasticity results from the irregular homoclinic orbit (HMO) crossings, and we report that the quintic Hamiltonian perturbation to the cubic NSE can lead to the spatiotemporal complexity of fields. The mechanism of these complicated dynamics is also analysed in terms of energy spectra.

2. Homoclinic orbit crossings

In the case of one-dimension, the integrability of the cubic NSE can be verified by finding the Lax pair. A class of periodic solutions and solitons can be obtained by inverse-scattering transform. The solitons developed by modulational instabilities to a homogeneous solution keep their spatially coherent structures and temporally periodic evolutions. In order to analyse the dynamic properties of the cubic–quintic NSE (1), we consider a homogenous solution being

$$E_s(t) = E_0\,\mathrm{e}^{\mathrm{i}t} \tag{2}$$

where E_0 satisfies with

$$E_0 = 0 \pm \sqrt{\frac{1 \pm \sqrt{1-4g}}{2g}}\,. \tag{3}$$

If we represent the solution of (1) in the form

$$E(X, t) = E_s + \delta E(X, t) \tag{4}$$

and linearize around E_s, we have

$$\begin{bmatrix} \mathrm{i}\partial_t + \partial_{xx} + L(|E_0|^2) & h(t) \\ h^*(t) & -\mathrm{i}\partial_t + \partial_{xx} + L(|E_0|^2) \end{bmatrix} \begin{bmatrix} \delta E \\ \delta E^* \end{bmatrix} = 0 \tag{5}$$

where L and h are expressed as following form

$$\begin{aligned} L(|E_0|^2) &= 2|E_0|^2 - 3g|E_0|^2 \\ h(t) &= E_s^2(t)(1-2g|E_0|^2). \end{aligned} \tag{6}$$

In the further analyses, the unstable wavenumber and the linear growth rate must be defined. Considering the homogeneous solution $E_s(t)$ to be modulated by the very small perturbation, we also represent the solution of (1) in the form [17]

$$E(X, t) = E_0\,\mathrm{e}^{\mathrm{i}t} + \delta E_+\,\mathrm{e}^{\mathrm{i}(KX+\Omega t)} + \delta E_-\,\mathrm{e}^{-\mathrm{i}(KX+\Omega^* t)} \tag{7}$$

where δE_+ and δE_- are all much smaller than E_0. According to the linearizing analysis, we can obtain the linear disperse relation as follows

$$\Omega^2 = -K^2[2E_0^2(1-2gE_0^2) - K^2]. \tag{8}$$

When $0<|K|<K^* = E_0[2(1-2gE_0^2)]^{1/2}$, Ω is an imaginary number and is defined as $\Omega = \mathrm{i}\gamma$, where γ is called as the linear growth rate of modulational instability. Thus, the linear growth rate as a function of an unstable wavenumber K is

$$\gamma(K) = K[2E_0^2(1-2gE_0^2) - K^2]^{1/2} \tag{9}$$

for $0<|K|<K^* = E_0[2(1-2gE_0^2)]^{1/2}$. If we consider the periodic boundary conditions, the eigenfunction of (5) can be chosen as

$$\begin{bmatrix} \delta E \\ \delta E^* \end{bmatrix} = \begin{bmatrix} \varepsilon \mathrm{e}^{\mathrm{i}\lambda t} \\ \varepsilon^* \mathrm{e}^{-\mathrm{i}\lambda^* t} \end{bmatrix} \cos(KX) \tag{10}$$

where ε, ε^* are the small parameters.

It is convenient for the following numerical experiments that we take $K = K_{\max} = E_0(1-2gE_0^2)^{1/2}$, which corresponds to the maximum instability mode.

Hence, we easily get

$$\lambda = 1 \pm \mathrm{i}\frac{-1+4g+\sqrt{1-4g}}{2g} \tag{11}$$

for $E_0 = \pm\{[1-(1-4g)^{1/2}]/2g\}^{1/2}$ with $0 \leqslant g < \frac{1}{4}$. Thus, the amplitude of the eigenfunction of the linearizing operator can be written as

$$|\delta E(X, t)| = c_1 \cos(K_{\max}X)\,\mathrm{e}^{At} + c_2 \cos(K_{\max}X)\,\mathrm{e}^{-At} \tag{12}$$

where c_1 and c_2 are the small real parameters, and $A = [-1+4g+(1-4g)^{1/2}]/2g$. If we construct phase-space $(|E(X, t)|, |E(X, t)|_t)$, obviously, $E_0 = \pm\{[1-(1-4g)^{1/2}]/2g\}^{1/2}$ with $0 \leqslant g < \frac{1}{4}$ correspond to the hyperbolic fixed points. It is shown that $E_0 = \pm\{[1+(1-4g)^{1/2}]/2g\}^{1/2}$ and $E_0 = 0$ correspond to the elliptic points by means of the same process.

In the numerical experiment, on the other hand, we choose the initial condition as

$$E(X, 0) = E_0 + \varepsilon \mathrm{e}^{\mathrm{i}\theta} \cos(K_{\max}X) \tag{13}$$

here ε is the small real parameter, $E_0 = \{[1-(1-4g)^{1/2}]/2g\}^{1/2}$, and (13) is called as simply initial condition for the cubic NSE [15]. Considering (12) and (13), we have that the unstable manifolds for possessing the saddle point $(|E_0|, 0)$ correspond to $\theta = 45°$ and $\theta = 225°$, and the stable manifolds correspond to $\theta = 135°$ and $\theta = 315°$, as sketched in figure 1. In the numerical processes, the periodic length of system is taken as $L = 2\pi/K_{\max}$, and the splitting-time-step spectral method [16] has been improved in order to increase the time accuracy. The available modes for fast Fourier transform and the time-step are considered, which depend on the accuracy of the conserved quantities being preserved to 10^{-8}.

In order to analyse the dynamic properties of the cubic–quintic NSE (1), first, we simply discuss the cubic NSE. For $g = 0$, the integrability can be illustrated in terms of the exactly periodic recurrent motion in phase-space. Moon [17] experimentally guessed the fixed point (1, 0) could be a saddle point, which has been confirmed in the above linearized analysis. In addition, the orbit to possess the hyperbolic fixed point (1, 0) should correspond to the homoclinic one. As far as a finite dimensional dynamic system is concerned, the stable and unstable orbits for the hyperbolic fixed point would smoothly be joined to each other if the unperturbed system is taken to be integrable. For a Hamiltonian perturbation the orbits generically intersect transversely, leading

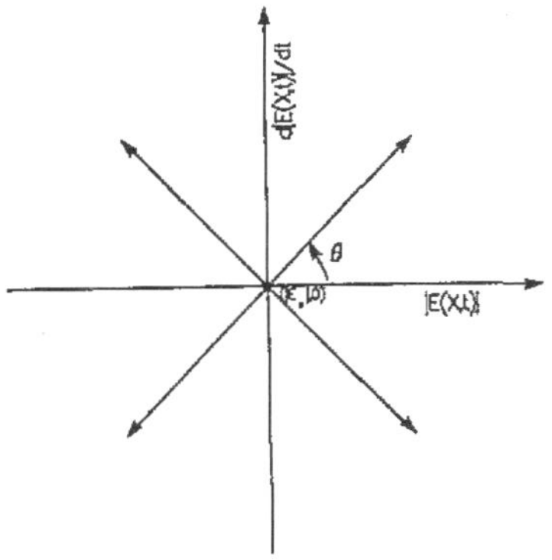

Figure 1. Local phase flows of saddle point $(|E_0|, 0)$, which are obtained in terms of (12).

to an infinite number of homoclinic points and chaotic motion [18]. For our continuum Hamiltonian system, the integrability of the cubic NSE is also illustrated in terms of such an idea. Making use of the results obtained in the linearized stability analysis, we choose the initial parametric values as $g = 0$, $E_0 = 1$ and $\theta = 45°$. From figure 2(a), we observe that the stable manifold $W^{(s)}$ smoothly joins with the unstable manifold $W^{(u)}$. The orbit to possess the saddle point $(1, 0)$ corresponds to the homoclinic one. However, when system is added in the quintic Hamiltonian perturbation ($g \neq 0$), we find that stable and unstable manifolds in phase-space do not smoothly join together (see figure 2(b)), which illustrates that the current system is near integrable. On the other hand, we note that there are not an infinite number of homoclinic points in figure 2(b). In fact, our phase-space is only the projection of a high-dimensional space, where the information of phase for wave fields is not contained. Therefore, there may exist some difference between our HMO crossings with those in a finite-dimensional Hamiltonian system. To further analyse the chaotic behaviour, we must discuss the long-time evolution of wave fields.

3. Spatiotemporal complicated dynamics

In section 2, we have shown that the irregular HMO crossings would appear when the quintic Hamiltonian perturbation is considered. To analyse the long-time behaviour, we also deal with cases of $g = 0$ and $g \neq 0$.

For the cubic NSE, the integrability can be displayed from figure 3. Figure 3(a) indicates the amplitude of fields is periodically temporal evolution. The periodic behaviour can also be illustrated by the plot of power spectrum (figure 3(c)), where the foundational frequency is about $\omega_0 = 0.549$. The exactly periodic recurrent solution is exhibited in figure 3(b, d). These phenomena account for the integrability of the cubic NSE.

As $g \neq 0$, however, a completely different dynamic behaviour is shown in figure 4. Taking into account the quintic Hamiltonian perturbation, we find that the periodic behaviour disappears. The temporal evolution for the amplitude of fields in figure 4(a) indicates the typically chaotic characteristic. As shown in figure 4(b), the irregular HMO crossings are very clear. A continuous non-periodic spectrum which obviously accounts for the chaotic characteristic is described in figure 4(c). These behaviours

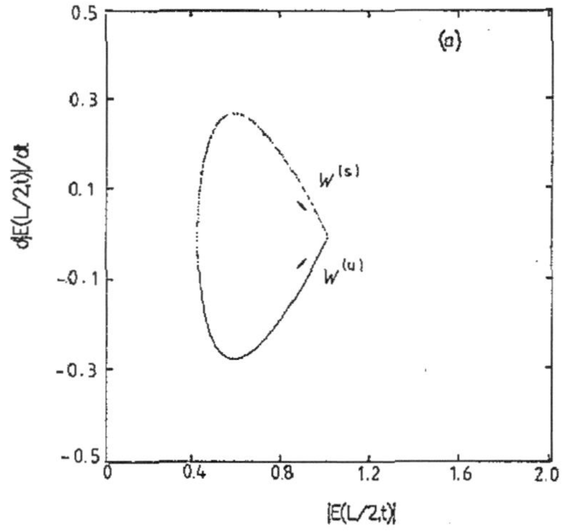

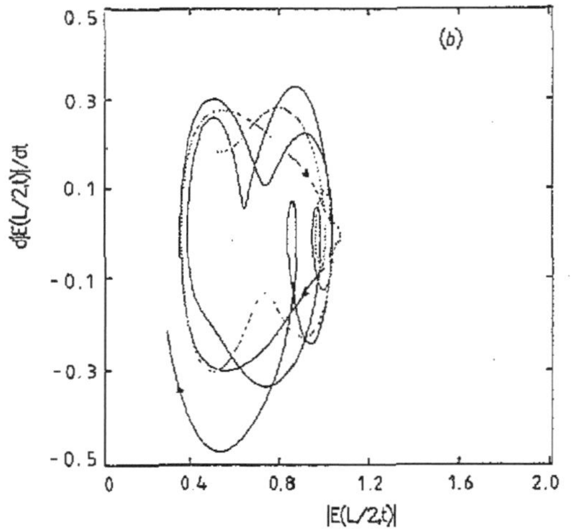

Figure 2. Stable ($W^{(s)}$) and unstable ($W^{(u)}$) manifolds for the hyperbolic fixed point $(|E_0|, 0)$, where solid line is computed with $t>0$, and dotted line is computed with $t<0$. (a) Integrable cubic NSE; the orbits smoothly join. (b) Non-integrable cubic–quintic NSE; the stable and unstable orbits intersect.

show that the quintic Hamiltonian perturbation leads to chaos. It is noted from figures 3 and 4 that the quintic nonlinear term in (1) behaves not completely like a general dissipative perturbation. Some more complicated dynamic phenomena cannot be explained only in terms of the non-integrable perturbation, such as the 'foundational' frequency (corresponding to the first maximum peak) in figure 4(c) is about $\omega_0 = 0.84$. But such a complicated behaviour will not be discussed in this paper. From figure 4(d), in addition, we observe that the spatially localized structures are still kept in the propagative processes of fields, but are considerably irregular. The presence of these irregular spatial patterns shows the recurrence broken down. Thus, the spatiotemporal complicated behaviour is realized in the present dynamic system.

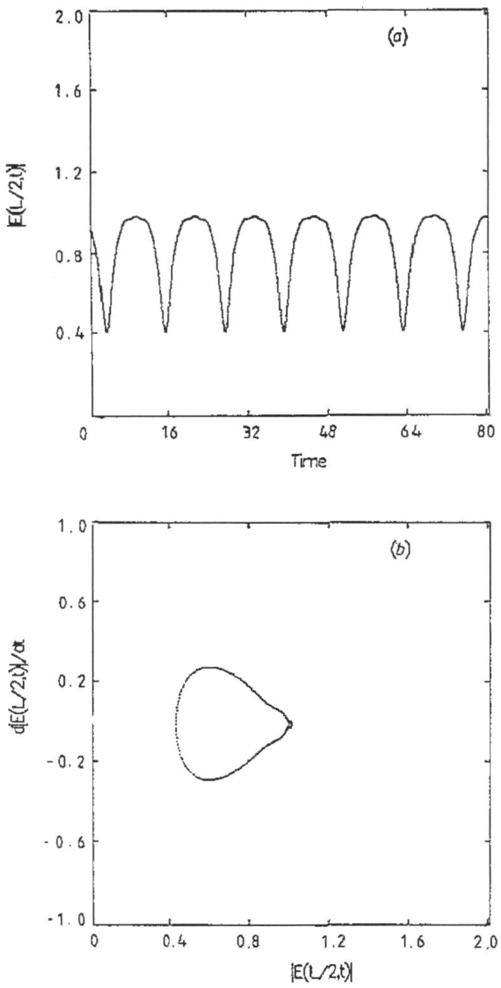

Figure 3. Solution of the cubic NSE with $\theta = 45°$ and $\varepsilon = 0.1$: (*a*) the periodic evolution for the amplitude of fields; (*b*) the phase-space orbit; (*c*) the power spectrum, where the foundational frequency $\omega_0 \approx 0.549$; and (*d*) contours of $|E(X, t)|^2 = \text{constant}$, where $t = 0 \sim 100$.

To analyse the mechanism that leads to spatiotemporal complexity, we further investigate the processes that the energy in Fourier modes evolves with time. For (1), we define energy of system as

$$H = \int |E|^2 \, dX. \tag{14}$$

In Fourier space, it can be rewritten as

$$H = \sum_K H_K = \sum_K |E_K|^2. \tag{15}$$

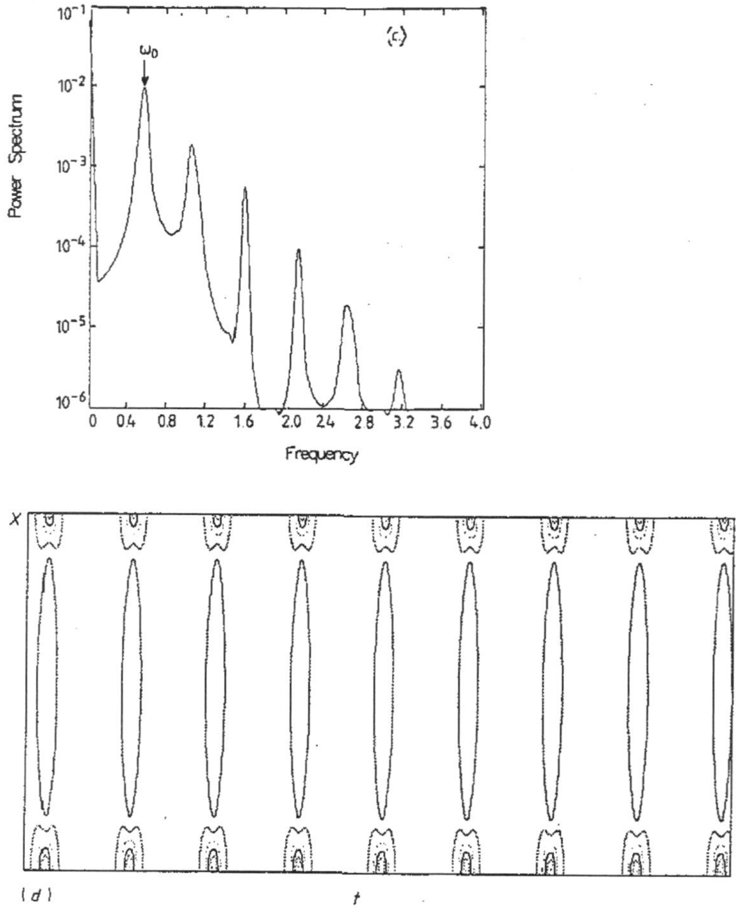

Figure 3. (continued)

Here K stands for the kth Fourier mode. The time evolutions of the amplitude for fields in Fourier space, which also correspond to the evolutions of energy in Fourier modes, i.e., $H_K = |E_K|^2$, are shown in figure 5. Obviously, the large part of energy in system is lain in the first and the second modes. For the case of $g = 0$, the time evolutions of energy in all modes is periodic, which is consistent with periodic recurrent solution (see figure 3). The amplitude of soliton structures is dominated by the total energy of system, the width and patterns of coherent structures are associated with energy partition in the high modes. Therefore, the evolution of fields is temporally periodic and spatially coherent. For the case of $g \neq 0$, but, figure 5 shows that energy in the system, which is initially confined to the lowest mode, would spread to many higher modes due to the nonlinear instability, but would not regroup into the original lowest mode. Because the initial energy is added to the maximum modulational instability mode, the main energy would be controlled by the finite modes in the processes of evolutions. We see from figure 5 that energy contained in the first two modes dominates the spatially localized structures being kept. The energy contained in higher modes,

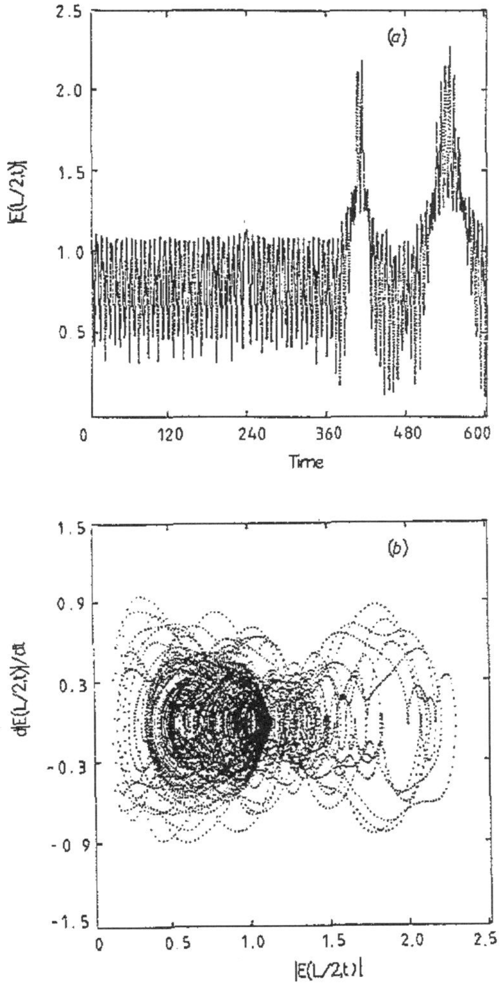

Figure 4. Solution of the cubic-quintic NSE with $g=0.1$, $\theta=45°$ and $\varepsilon=0.1$: (*a*) chaotic evolution for the amplitude of fields; (*b*) the phase-space trajectory illustrates the irregular HMO crossings; (*c*) a continuous noise-like power spectrum; and (*d*) contours of $|E(X,t)|^2=\text{constant}$, where $t=550\sim600$.

for example the clearly stochastic behaviours in the third and the fourth modes, leads to the presence of various spatial patterns of the localized structures. The spatio-temporal complexity as shown in figure 4(*d*) would be associated with the stochastic evolutions of energy lain in all modes.

4. Conclusions and discussions

The essentially important characteristic of the cubic-quintic NSE (1) has been discussed theoretically and numerically. We show that the quintic Hamiltonian perturbation may

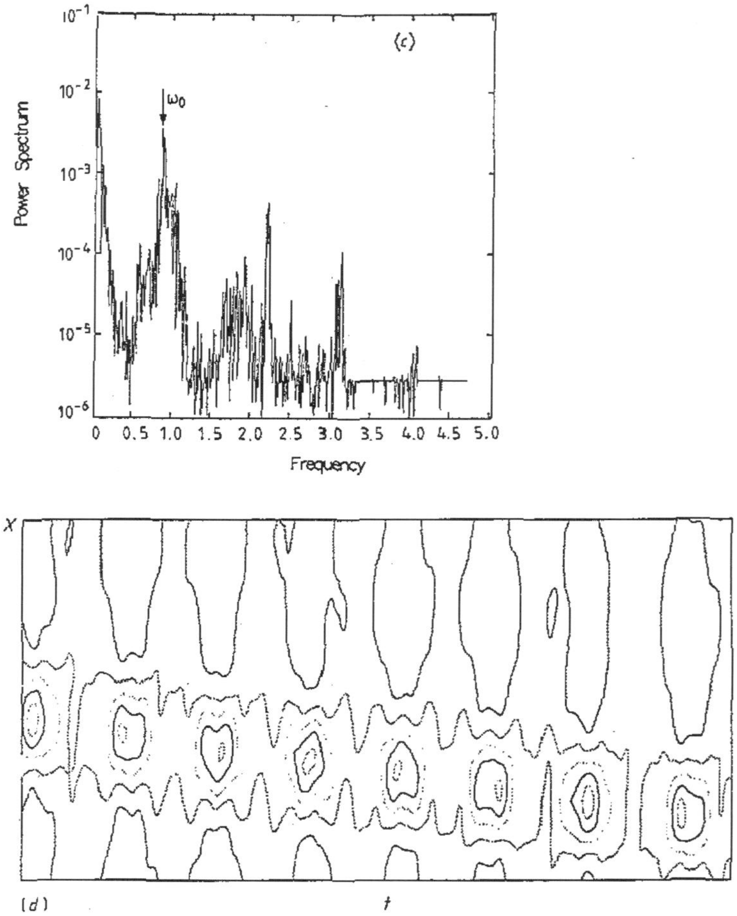

Figure 4. (continued)

lead to the integrability of system broken down. The propagation of fields is temporally stochastic but keeps their spatially localized patterns that developed by modulational instability to a homogeneous solution. The spatiotemporal complexity is due to the irregular HMO crossings. We also show that these complicated dynamic behaviours are associated with the fact that energy contained in Fourier modes stochastically evolves with time. In particular, the stochasticity of energy in higher modes would lead to the presence of various irregularly spatial structures.

Our current results are of significance in physics. In plasma turbulence, for example, the cubic NSE describes the nonlinear interaction between Langmuir wave and ion-acoustic wave under the subsonic regime, where the second-order fields are considered. In the evolutive initial stage, some physical phenomena can well be explained in terms of the cubic NSE. In the nonlinear strong turbulent stage, however, the cubic NSE may be no longer valid. The other physical effects, such as damping and dissipation, etc, would lead to the presence of the complicated dynamics of Langmuir fields. On the other hand, the high order field interaction has to be considered in the evolutive later

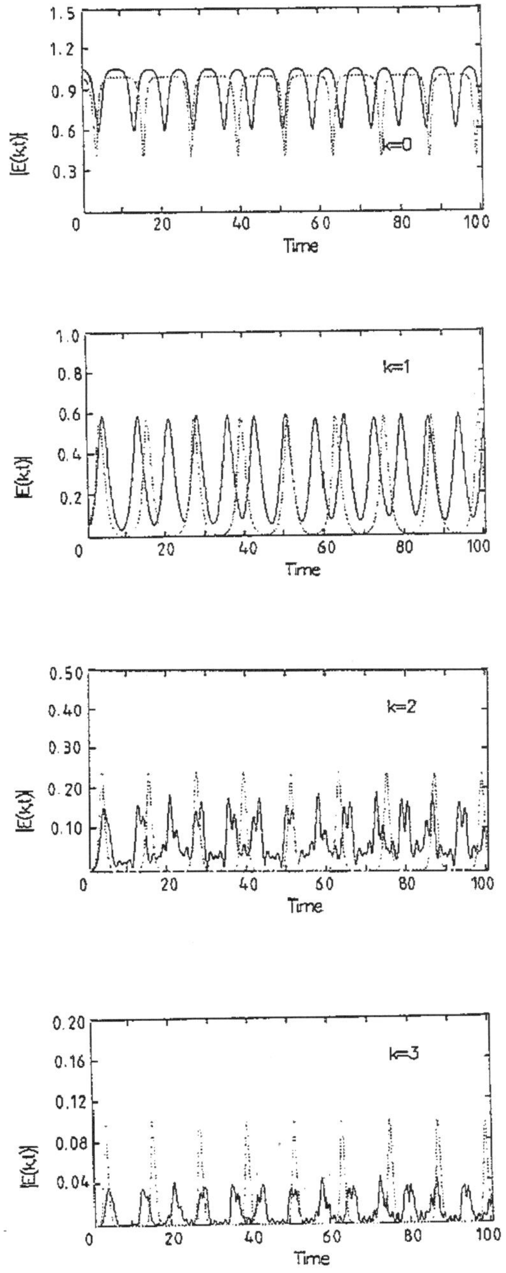

Figure 5. The time evolution of the amplitude for fields in Fourier space (also corresponding to the energy in Fourier modes, i.e., $H_K = |E(K)|^2$) for (1) with $\theta = 45°$, $\varepsilon = 0.1$, where dotted curves correspond to solutions of the cubic NSE and solid curves correspond to solutions of cubic-quintic NSE with $g = 0.1$, respectively.

stage of plasma instability with the increase of the amplitude of fields. Although the dynamic model (1) only involves in the fourth-order field interaction, it is shown that the quintic Hamiltonian corrections may also drive the chaotic evolutions of Langmuir fields. In addition, our research would also be interesting in nonlinear optics, etc.

Finally, we should mention the main difference between our work and those discussed by some authors. For the study of the spatiotemporal complexity in Langmuir turbulence, Doolen *et al* [19] first investigated the driven dissipative Zakharov equations numerically. Their studies demonstrated the co-existence of coherent spatial structures—cavitons—and temporal chaos. After their work, a set of dynamic models have been discussed in studying the process of coherence to turbulence. But, much work deals with the driven damping, force dissipation and nonlinear inhomogeneous media [20]. A rich variety of spatiotemporal complexity, which suggests that a low-dimensional chaotic attactor exists in an infinite-dimensional system, has been observed [21]. For the continuum Hamiltonian system, however, the problem becomes extremely difficult for general initial conditions since the system does not reduce to a finite-dimensional system as that in the case of dissipative perturbations though the existence of the solitary waves could slow down the dimensions of system. Therefore, more care is needed to investigate chaos of the continuum Hamiltonian system.

Acknowledgments

We wish to acknowledge stimulated discussions with Professor Chen Shigang on various issues relating to this paper. This work has been supported by the National Natural Science Foundations of China No. 1880109 and No. 19175039.

References

[1] Acioli L H, Gomes A S L and Rios Leit J R 1988 *Appl. Phys. Lett.* **53** 1799
[2] Djolos R V, Kartavenko V G and Permyakov V G 1981 *Yader Fiz.* **34** 144
[3] Makhankov V G 1989 *Soliton Phenomenology* (Dordrecht: Reidel)
[4] He X T 1982 *Acta Physica Sinica* **31** 1317 (in Chinese)
[5] Pushkarov Kh I, Pushkarov D I and Tomov I V 1979 *Opt. Quantum Electron* **11** 471
[6] Cowan S, Enns R H, Rangnekar S S and Sanghera S S 1986 *Can. J. Phys.* **64** 311
[7] Gagnon L 1989 *J. Opt. Soc. Am.* **A6** 1477
[8] Gagnon L and Winternitz P 1989 *Phys. Lett.* **134A** 276
[9] Zhou C T, He X T and Chen S G 1992 *Phys. Rev.* A **46** 2277
[10] Cloot A, Herbst B M and Weideman J A C 1990 *J. Comput. Phys.* **86** 127
[11] Gagnon L and Winternitz P 1988 *J. Phys. A: Math. Gen.* **21** 1493; 1989 *J. Phys. A: Math. Gen.* **22** 469
[12] Gagnon L and Winternitz P 1989 *Phys. Rev.* A **39** 296
[13] Lemesnrier B J, Papanicollaon G, Sulem C and Sulem P L 1988 *Physica* D **31** 79
[14] Calogero F and Degasperis 1982 *Spectral Transform and Solitons I* (Amsterdam: North-Holland)
[15] Taha T R and Ablowitz M J 1984 *J. Comput. Phys.* **55** 202
[16] Yuen H C and Ferguson W E 1978 *Phys. Fluids* **21** 1275
[17] Moon T H 1990 *Phys. Rev. Lett.* **64** 412
[18] Lichtenberg A J and Lieberman M A 1983 *Regular and Stochastic Motion* (New York: Springer-Verlag)
[19] Doolen A, DuBois D F and Rose H A 1983 *Phys. Rev. Lett.* **51** 335
[20] Spatschek K H, Pietsch H, Leadke E W and Erckermann Th 1989 *Nonlinear and Turbulent Processes in Physics*, eds Bar'yakhar V E *et al* (Singapore: World Scientific) p 978, see also their refs
[21] Nozaki K 1986 *Physica* D **23** 369

CHINESE PHYS. LETT. Vol.10, No.6(1993)321

Controlling Chaos Using Modified Lyapunov Exponents*

TAN Yi, HE Xiantu (X. T. He)
CHEN Shigang (S. G. Chen)
Center for Nonlinear Studies,
Institute of Applied Physics and Computational Mathematics,
Beijing 100088

(Received 15 March 1993)

We suggested a method using a small perturbation to modify Lyapunov exponents of the unstable periodic orbit for controlling chaos. The benefit of this control scheme is studied and the effect of noise is discussed. Based on these studies, a more practical control method is recommended, and some experiments can be better understanded.

PACS: 05.45.+b, 03.20.+i

In 1990 Ott, Grebogi and Yorke[1] (OGY) proposed a new way to achieve controlling a chaotic dynamical system in terms of stabilizing one of many unstable periodic orbits that are embedded in the chaotic attractor, through only the application of small, carefully computed perturbation to a system parameter. If we stayed in the OGY's description, however, the OGY control method would be a numerical curiosity that requires experimentally unattainable precision. Recently, some experiments[5,6] showed that even if imprecise measurements of the eigenvalues and eigenvectors are used, chaotic behavior of system can also successfully be tamed. In order to provide a solid theoretical ground for these experiments, to reexamine OGY's work is necessary. In this letter, we use a small perturbation to modify the Lyapunov exponents of the periodic orbit instead of the OGY's scheme for controlling chaos. The interesting results can better explain the experimental work,[5] and thus more practical control method is recommended.

For simplicity, in this letter we only discuss a discrete dynamical system . The same results can be extended to other cases. Let system depend on some accessible parameters $p \in (p_0 - \delta p_{\max}, p_0 - \delta p_{\max})$ with maximal possible parameter $\delta p_{\max}$, i.e., $\boldsymbol{\xi}_{n+1} = \boldsymbol{F}(\boldsymbol{\xi}_n, p)$, where $\boldsymbol{\xi} \in \boldsymbol{R}^n$ and $p \in \boldsymbol{R}$, and also let $\boldsymbol{\xi}_{\mathrm{f}} = \boldsymbol{F}(\boldsymbol{\xi}_{\mathrm{f}}, p_0)$ denote unstable periodic orbit embedded in chaotic attractor, which one wants to stabilize, then around this orbit we have a linearized form

$$\delta\boldsymbol{\xi}_{n+1} = \boldsymbol{A}\cdot\delta\boldsymbol{\xi}_n + \boldsymbol{Q}\delta p_n \ , \tag{1}$$

where $\delta\boldsymbol{\xi}_{n+1} = \boldsymbol{\xi}_n - \boldsymbol{\xi}_{\mathrm{f}}$, $\delta p = p - p_0$, $\boldsymbol{A} = \partial\boldsymbol{F}/\partial\boldsymbol{\xi}_n(\boldsymbol{\xi}_{\mathrm{f}}, p_0)$, and $\boldsymbol{Q} = \partial\boldsymbol{F}/\partial p(\boldsymbol{\xi}_{\mathrm{f}}, p_0)$. Write δp_n in the general feedback form $\delta p_n = \boldsymbol{Z}\cdot\delta\boldsymbol{\xi}_n$, now the Eq. (1) becomes

$$\delta\boldsymbol{\xi}_{n+1} = (\boldsymbol{A} + \boldsymbol{Q}\boldsymbol{Z})\cdot\delta\boldsymbol{\xi}_n \ . \tag{2}$$

Thus, one can choose the suitable value of $\boldsymbol{Z}$ in above equation such that the norm of all engenvalues of matrix $\boldsymbol{A} + \boldsymbol{Q}\boldsymbol{Z}$ is less than 1. Notice that the Lyapunov exponents can directly be given by the eigenvalues of the matrix of linearized system for periodic orbits, then unstable periodic orbit which we want to stabilize can be stabilized by choosing the suitable values of $\boldsymbol{Z}$. Of course, we stress that we achieve control by making use of only a small perturbation in an accessible system parameter. In this case, giving maximal possible perturbation $\delta p_{\max}$, the control is only activated if the condition $|\boldsymbol{Z}\cdot\delta\boldsymbol{\xi}_n| \le \delta p_{\max}$ is satisfied.

*Supported by the grand of Chinese National Basic Research Project "Nonlinear Science" and the National Natural Science Foundation of China, under Grant No. 19175039.

As an example, we now use Henon chaotic attractor to proceed to discussion. Our purpose is to illustrate what value of $\boldsymbol{Z}$ can be used to achieve control and how can we utilize the advantage in so doing for practical application.

The Henon map depending on control parameter p is $x_{n+1} = 1 - ax_n^2 + y_n + p$, $y_{n+1} = bx_n$, where $b = 0.3$ and $a = 1.4$ are taken, and the control parameter p can vary by a small amount near zero. For $p = 0$ it is known that there exists a chaotic attractor in this dynamical system and many unstable periodic orbits embedded in this attractor.[7] Now let us stabilize the period-1 orbit in Henon chaotic attractor $(x_f(p = 0), y_f(p = 0))$. Near (x_f, y_f) and $p_o = 0$, the linearization of Henon map is expressed in the form of Eq. (1), and Eq. (2) reads as follows

$$\boldsymbol{\delta\xi}_{n+1} = \boldsymbol{B} \cdot \boldsymbol{\delta\xi}_n , \tag{3}$$

where

$$B = \boldsymbol{A} + \boldsymbol{QZ}, \quad \boldsymbol{A} = \begin{pmatrix} -2ax_f & 1 \\ b & 0 \end{pmatrix}, \quad \boldsymbol{Q} = \begin{pmatrix} 1 \\ 0 \end{pmatrix}, \quad \boldsymbol{Z} = (z_1, z_2) .$$

Thus, the suitable values of $\boldsymbol{Z}$ to achieve control must be determined from relations $\| \lambda_+ \| < 1$ and $\| \lambda_- \| < 1$, here λ_+ and λ_- satisfy equation $\| \boldsymbol{B} - \lambda \boldsymbol{I} \| = 0$. Calculating gives that the available values of $\boldsymbol{Z}$ in this example are confined inside the region which is surrounded by the following three lines, i.e.,

$$z_2 < \frac{1}{b}(1 - z_1 + 2ax_f) - 1, \quad z_2 > -\frac{1}{b} - 1, \quad z_2 < \frac{1}{b}(1 + z_1 - 2ax_f) - 1 .$$

This region is shown in Fig.1. Within this region, any value of $\boldsymbol{Z}$ can be used to achieve control chaos. We name this region as controllable region. As a numerical example, now arbitrarily taking $\boldsymbol{Z} = (z_1, z_2) = (2.0, 0.0)$ with $\delta p_{\max} = 0.4$, we stabilize the period-1 orbit to achieve control as shown in Fig.2.

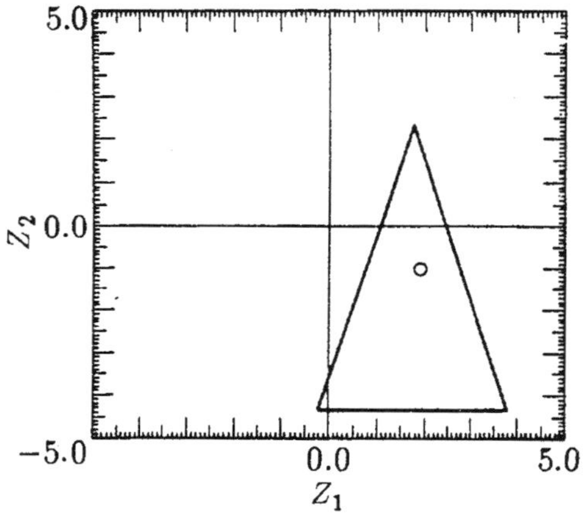

Fig. 1: Controllable region for controlling period-1 orbit in Henon chaotic attractor. Notes that the circle corresponds to the OGY control parameter.

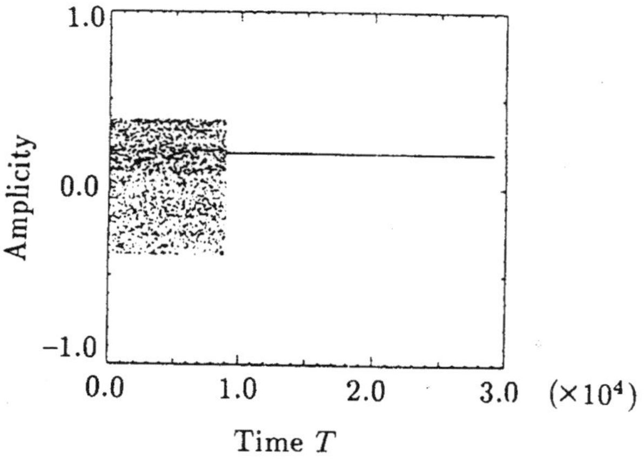

Fig. 2: Time-series data (amplitude vs iterations) for controlling period-1 orbit, control was initiated after iteration 9000.

Let us now answer the question why the experiments[5,6] can achieve control even if using imprecise measurements of the eigenvalues and eigenvectors in OGY control scheme. From general point of view, in fact, the OGY control scheme appeared in our scheme as a special case. In the above mentioned example, it is found that setting control parameter $z_1 = -\lambda_u$, $z_2 = -1$, where $\lambda_u = -ax_f - \sqrt{a^2x_f^2 + b}$, the OGY control formula can be obtained from Eq. (2). We have marked this OGY control parameter point in the parameter space as shown in Fig. 1. For most dynamical system the OGY control parameter point, like the above example showed in Fig. 1, is near the center of the controllable region in the parameter space because the control formula (1) modifies the Lyapunov exponents of unstable periodic orbits in such a way that one is the eigenvalue of stable manifold of unstable periodic orbit and others are zero (here the Lyapunov exponents are not in logrithme sense). Thus, making change not to be large for the OGY control parameter still keeps the value of control parameter in the controllable region, i.e., it is impossible to destroy the condition of achieving control. This is why the experiments[5,6] can be successful even if the imprecise measurements of the eigenvalues and eigenvectors are used to achieve control in the OGY control scheme.

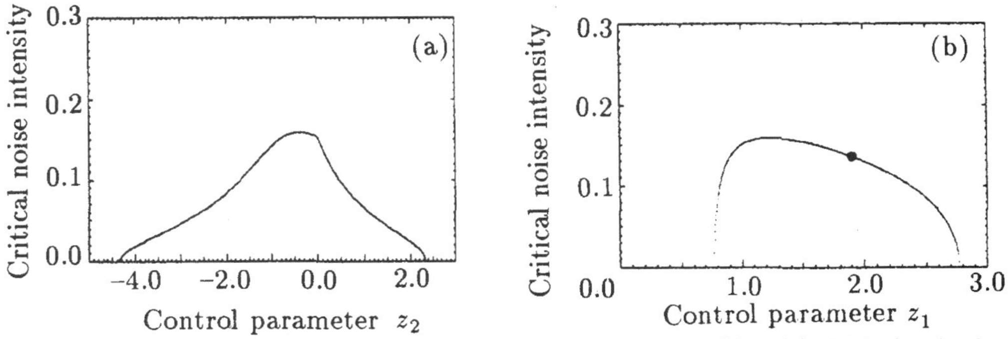

Fig. 3: Critical noise intensity for controlling period-1 orbit. (a) Critical noise intensity on the line $z_1 = 2ax_f$, (b) critical noise intensity on the line $z_2 = -1$. Notes that the dot corresponds to the critical noise intensity of OGY control.

With regard to possible applications of the controllable region, we investigate the effect of noise. In order to quantitatively characterize the ability of antinoise of every point which lies in the controllable region in parameter space, we introduce the conception of critical noise intensity ϵ_c which is defined as that the control chaos can not hold for infinite time if the noise intensity is greater than the critical one. Here, we should add noise terms $\epsilon\delta_{xn}$ and $\epsilon\delta_{yn}$ to the right-hand sides of Henon map, as done by OGY.[1] The symbol ϵ is the noise intensity and the random quantities δ_{xn} and δ_{yn} satisfy the following relations: $<\delta_{xn}>=<\delta_{yn}>=<\delta_{xn}\delta_{yn}>=0$; $<\delta_{xn}\delta_{xn'}>=<\delta_{yn}\delta_{yn'}>=0$ for $n \neq n'$; $<\delta_{xn}^2>=<\delta_{yn}^2>=1$, and have a Gaussian probability density. From these relations, one can compute $\sqrt{<\delta x_n^2(\epsilon, n \to \infty)>}$ and $\sqrt{<\delta y_n^2(\epsilon, n \to \infty)>}$, and then the critical noise intensity ϵ_c for the control parameter point (z_1, z_2) is obtained from equation $\delta p_{\max} = |z_1|\sqrt{<\delta x_n^2(\epsilon_c,\ n \to \infty)>} + |z_2|\sqrt{<\delta y_n^2(\epsilon_c,\ n \to \infty)>}$. In our example the critical noise intensity ϵ_c is evaluted along two lines $z_1 = 2ax_f$ and $z_2 = -1$. The results are given as follows:

$$\epsilon_c = \frac{\sqrt{1 - b^2(1+z_2)^2}}{|z_1|\sqrt{1+(1+z_2)^2} + |z_2|\sqrt{1+b^2}}\delta p_{\max}\ , \quad \text{for} \quad z_1 = 2ax_f \tag{4}$$

and

$$\epsilon_c = \frac{\sqrt{1 - (z_1 - 2ax_f)^2}}{|z_1| + |z_2|\sqrt{1 + b^2 - (z_1 - 2ax_f)^2}}\delta p_{\max}\ , \quad \textbf{for} \quad z_2 = -1\ , \tag{5}$$

which are shown in Figs. 3(a) and 3(b). Where $\delta p_{\max}$ is a maximal possible perturbation, x_f is the x coordinate of the fix point in phase space, a and b are parameters appearing in Henon map, and $(z_1,\ z_2)$ is the control parameter point in parameter space. The critical noise intensity for OGY control parameter is also shown in Fig. 3(b). The Figs. 3(a) and 3(b) show that most of the controllable regions are available comparing with OGY's for achieving control in real system(there always is noise in real system). Now we have got important conclusions that there exist controllable region and most of this region have good ability of antinoise. Although these conclusions is obtained from the special example, it does not lose the generality. Thus, we describe a new feasible control method here. The controllable region can be detected by adjusting the instruments. Then the detecting method to achieve control is so simple that we need only localizing one of fix points which we want stabilizing, and then adjust instrument on this point to find the controllable region . This method will be very useful for stabilized high periodic orbits or controlling chaos in high dimension system. As an example, we use the detecting method to stabilize period-2 orbit which embbeded in Henon chaotic attractor. The control equations are $x_{n+1} = 1 - ax_n^2 + y_n$, $y_{n+1} = bx_n$, $x_{n+2} = 1 - ax_{n+1}^2 + y_{n+1} + \theta_p$, $y_{n+2} = bx_{n+1}$, where p is control parameter varying around zero and has the form: $p = z_1(x_{n+1} - x_f) + z_2(y_{n+1} - y_f)$, where $(x_f,\ y_f)$ is one of two fix points, and θ is a step function such that $\theta = 0$ for $p > p_{max}$ and $\theta = 1$ for $p < p_{max}$, where p_{max} is the maximal possible perturbation. We detect the controllable region along the line $z_2 = 0$, and find that $-3.25 < z_1 < -1.75$ is the controllable region on the line $z_2 = 0.0$(detecting resolution is 0.25). We arbitrarily choose one point in this region , say $z_1 = -2.0$, and then combine $z_2 = 0.0$ with it to construct the perturbation $p = z_1(x_{n+1} - x_f) + z_2(y_{n+1} - y_f) = -2.0(x_{n+1} - x_f)$. Given $p_{max} = 0.4$, using the value of p to monitor system, when the system falls into a narrow strip $|p| \leq p_{max}$ stabilizing period-2 orbit is successfully achieved. The control results are shown in Fig. (4).

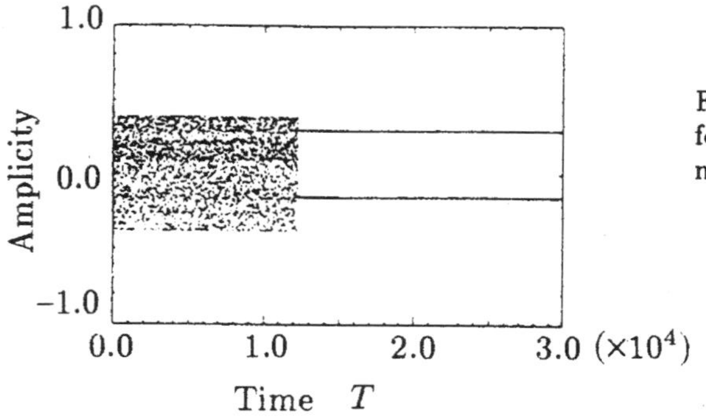

Fig. 4: Time-series data (amplitude vs iterations) for controlling period-2 orbit using detecting method. Control was initiated after iteration 12000.

In conclusion, we have investigated possible extension of the OGY control scheme. As a example, Henon system is studies in detail. From this special example, we extract some general results. Based on these results, some experiments can be better understanded, and a new feasible control method is given. We believe that the experimenters can get more convenience from our discussion.

Note: After Mailing our paper, we found a recent work published on Physica D 58, 165 (1992), which has the similar idea.

References

[1] E. Ott, C. Grebogi and J. A. Yorke, Phys. Rev. Lett. 64 (1990) 1196.
[2] For example, A. Hubler, Helv. Phys. Acta. 62 (1989) 343.
[3] N. H. Packard *et al.*, Phys. Rev. Lett. 45 (1980) 712.
[4] F. Takens, in *Dynamical Systems and Turbulence*, edited by D. Rand and L. S. Young (Springer-Verlag, Berlin, 1981) p. 230.
[5] W. L. Ditto *et al.*, Phys. Rev. Lett. 65 (1990) 3211.
[6] Rajarshi Roy *et al.*, Phys. Rev. Lett. 68 (1992) 1259; Phys. Rev. Lett. 68 (1990) 1.
[7] Robert L. Devaney, *An introduction to Chaotic Dynamical System* (Menlo park: Benjamin-Cummings, 1986).

PHYSICAL REVIEW E VOLUME 50, NUMBER 5 NOVEMBER 1994

Pattern structures on generalized nonlinear Schrödinger equations with various nonlinear terms

Cangtao Zhou,[1,2] X. T. He,[1,2] and Tianxing Cai[2]
[1] *China Center of Advanced Science and Technology (World Laboratory), P.O. Box 8730, Beijing 100080, China*
[2] *Laboratory of Computational Physics and Center for Nonlinear Studies,*
*Institute of Applied Physics and Computational Mathematics, P.O. Box 8009, Beijing 100088, China**
(Received 7 June 1994)

The spatiotemporal characteristics on a series of the generalized nonlinear Schrödinger equations (NSE's) in one- and two-dimensional space have been systematically discussed. The current investigations show that the high order Hamiltonian perturbation can lead to the destruction of the coherent structures and the formation of the spatiotemporally complicated patterns. The route from the coherent patterns to the complicated ones experiences a quasiperiodic route. For two-dimensional problems, the numerical experiments illustrate that the singular solution, spatiotemporal chaos, and pseudorecurrence can appear in the same NSE with the increase of the saturable nonlinear effects, where the spatial patterns and the energy partition of the system in the Fourier modes exhibit completely different dynamic behaviors. In particular, the symmetric destruction of the spatial patterns is associated with the energy in the nondiagonal modes.

PACS number(s): 52.35.Mw, 47.10.+g, 05.45.+b, 52.40.Db

I. INTRODUCTION

The generalized nonlinear Schrödinger equation (NSE)

$$iE_t + \nabla^2 E + F(|E|^2)E = 0, \tag{1.1}$$

in which the potential F is a differential smooth real function, is one of the basic evolution models for nonlinear waves in various branches of a wave train in conservative, disperse systems. Early applications of the NSE were in the context of nonlinear optics where it described the propagation of light beams in nonlinear media [1]. Also, it has been applied to gravity waves on deep water, for which the predicted modulational instability and envelope soliton formation have been clearly demonstrated experimentally [2]. In plasma physics, it was first derived for nonlinear hydromagnetic waves by Taniuti and Washimi using the reductive perturbation method [3].

For the one-dimensionally cubic NSE, the Lax pair has been found by Zakharov and Shabat and the inverse scattering transform (IST) is well established [4]. The infinite integrals of motion and N-soliton solutions, which are associated with integrability, can also be obtained by IST [5]. For the case of high-dimensional space, Rasmussen and Rypdal [6] showed that focusing singularities can occur in two- and three-dimensional space, and gave the condition for the singular solution analytically.

For Eq. (1.1), the Lagrangian density is

$$L = \frac{i}{2}(E^* E_t - E E_t^*) - |\nabla E|^2 + f(|E|^2), \tag{1.2}$$

where $f(|E|^2) = \int_0^{|E|^2} F(s)ds$. According to the Norther theorem, we can obtain the following invariants: the quasiparticle number

$$N = \int |E|^2 d\mathbf{X}, \tag{1.3}$$

the momentum

$$P = \frac{i}{2}\int (E_{\mathbf{X}} E^* - E E_{\mathbf{X}}^*) d\mathbf{X}, \tag{1.4}$$

the energy

$$H = \int [|E_{\mathbf{X}}|^2 - f(|E|^2)] d\mathbf{X}, \tag{1.5}$$

and the other invariants, for example, angular momentum in the three-dimensional space and the moment of inertia, etc. [6].

Taking the different physical backgrounds into account, some particular types of $F(|E|^2)$ in Eq. (1.1) have frequently been applied. For example [7–28],

$$\text{(i)}\ F(|E|^2) = |E|^{2q}, \tag{1.6}$$

$$\text{(ii)}\ F(|E|^2) = |E|^2 - g|E|^4, \tag{1.7}$$

$$\text{(iii)}\ F(|E|^2) = \frac{|E|^2}{1 + g|E|^2}, \tag{1.8}$$

$$\text{(iv)}\ F(|E|^2) = \frac{1}{2g}(1 - e^{-2g|E|^2}). \tag{1.9}$$

For model (1.6), a collapse of the nonlinear wave [7,8], when the spatial dimension (D) and the coefficient q sat-

*Mailing address.

isfy $qD \geq 2$, can occur. In nonlinear plasmas, this important process solves the problem of small-K condensation in weak turbulent theory. When $q = 1$, the one-dimensional cubic NSE possesses integrability. Solitons developed by modulational instabilities keep their spatially coherent structures and temporally periodic evolutions. In particular, the well-known Fermi-Pasta-Ulam (FPU) recurrence in connection with deep water waves has been verified experimentally by Lake, Rungaldier, and Ferguson [9] and Yuen *et al.* [10]. The simple and complex recurrence phenomena were discussed numerically by Yuen and Ferguson [11] and analytically by Stiassnie and Kroszynski [12].

For model (1.7), the fifth order correction to the cubic term arises from the nonlinear interaction between Langmuir waves and electrons [13]. The coupling nonlinear equations, known as Zakharov equations, describe the nonlinear interaction between the high frequency Langmuir wave and the ion-acoustic wave by ponderomotive force. However, a beat frequency interaction between the large amplitude parts of high frequency fields and particles can occur in the nonlinear strong turbulence stages. Under the static approximation, He [13] has obtained a NSE where the potential $F(|E|^2)$ can be expressed by Eq. (1.7) in terms of Vlasov-Maxwell equations. In many-nucleon systems, usually, collision processes of heavy ions are described on classical or quasiclassical levels [14]; it was shown that in the quasiclassical limit the nuclear hydrodynamics equations with the Skyrme forces can be reduced to the nonrelativistic cubic-quintic model for the corresponding choice of variable [15]. For this model, on the other hand, some theoretical work has been obtained analytically and numerically. Puskharov *et al.* [15] obtained solitary wave solutions. Cowan *et al.* [16] numerically showed that these are not solitons, but behave like quasisolitons. Gagnon and Winternitz *et al.* [17,18] presented a large set of exactly analytical traveling wave solutions. Zhou *et al.* [19] also showed that a special periodic solution can be reduced to a solitary wave when the pseudoenergy is zero. The effects of the quintic nonlinear term enhance the amplitude and width of the solitary wave as compared with those for the cubic Langmuir soliton. Cloot *et al.* [20] qualitatively proved the existence of bound solutions by making use of the invariants (1.3) and (1.5). Their numerical results displayed recurrence of solutions. In high space dimensions, the Lie symmetry group was analyzed by Gagnon and Winternitz [21]. They obtained a class of group-invariant solutions.

For model (1.8), Akhmediek *et al.* [22] studied the pseudorecurrence in two-dimensional instability. They showed that an approximated recurrence to an initial homogeneous field can appear, and this pseudorecurrence arises only for a restricted range of the spatially modulational frequency. Mclaughin *et al.* [23] gave the condition for the singular solution.

Model (1.9) is considered as a useful model in homogeneous unmagnetized plasmas [24] and laser-produced plasmas [25–28]. When the phase velocity of the slow plasma oscillation is much smaller than the ion thermal velocity, one can obtain the adiabatic (quasistatic) electron density under the quasineutral approximation: $n_e = n_0 e^{-|E|^2}$. Combining the coupling equation which exhibits the slowly varying complex amplitude $E(\mathbf{X}, t)$ interacting with the low-frequency plasma motion, one can easily obtain Eq. (1.1) involving nonlinear terms (1.9) [24]. In laser plasmas, the light beam filamentation for this model due to the ponderomotive force was extensively studied. Max [25] showed that the self-focusing solution becomes a periodic oscillatory phenomenon, rather than a catastrophic process due to the exponential nonlinearity. Lam *et al.* [26] illustrated that the self-trapped beams are stable. Kaw *et al.* [27] showed that an electromagnetic wave interacting with a plasma is subject to instabilities leading to filamentation. In previous work [28], we showed that the stochastic propagation of beams can accompany the production of the oscillating filamentation instability.

We note that, while previous theoretical and numerical work [7–28] for discussing models (1.6)–(1.9) has concentrated on the solitary wave solutions, the singular solutions, and pseudorecurrence, the problem of the spatiotemporal patterns in one- and high-dimensional space appears to have received very little attention. The main purpose of this paper is to discuss the following questions. Can we choose an available method that completely describes the dynamic behaviors of the spatiotemporal patterns in one- and high-dimensional space? What are the characteristics and patterns of evolution? What is the mechanism for the formation of patterns? These problems are very interesting and important in physics and mathematics. In Sec. II, we constitute the phase space and the initial condition by considering linear analyses. In Sec. III, we describe the integrability and the pattern structures for the one-dimensionally cubic NSE. The spatiotemporal chaos and pattern dynamics in one-dimensional space and two-dimensional space are discussed in Sec. IV and Sec. V, respectively. Some conclusions are summarized in the final section.

II. CHOICE OF INITIAL CONDITION AND CONSTITUTION OF PHASE SPACE

A. Linearized analyses

For the continuum Hamiltonian system (1.1), the dynamic description is dependent on the choice of the initial condition. In particular, we cannot give a reasonable explanation if we take an arbitrary initial condition. Therefore, it is necessary that we choose an available initial condition and constitute a reasonable phase space in order to discuss the spatiotemporally evolutive phenomena of this system. For simplicity, in this work, we only deal with the developing behavior of an initial homogeneous state due to the modulational instability. We assume a homogeneous solution for Eq. (1.1) as $E_s(t) = E_0 e^{i\omega_s t}$, where E_0 satisfies

$$E_0[1 - F(|E_0|^2)] = 0. \tag{2.1}$$

Simultaneously, we define

$$E(\mathbf{X},t) = E_s(t) + \delta E(\mathbf{X},t), \tag{2.2}$$

and linearize Eq. (1.1), and yield

$$\begin{pmatrix} i\partial_t + \nabla^2 + L_i(|E_0|^2) & h_i(t) \\ h_i^*(t) & -i\partial_t + \nabla^2 + L_i(|E_0|^2) \end{pmatrix} \times \begin{pmatrix} \delta E \\ \delta E^* \end{pmatrix} = 0, \tag{2.3}$$

where

$$L_i(|E_0|^2) = F(|E_0|^2) + F'(a)|_{a=|E_0|^2}|E_0|^2 (a \equiv |E|^2), \tag{2.4}$$

$$h_i(t) = F'(a)|_{a=|E_0|^2} E_s^2(t).$$

On the other hand, we can obtain the relation of the linear growth rate with the instability wave number by using the standard method [29,30]

$$\gamma(K_A) = K_A[E_0F'(E_0) - K_A^2]^{1/2}, \tag{2.5}$$

where $K_A \equiv K$ for the case of one dimension and $K_A = \sqrt{\mathbf{K}_\perp \cdot \mathbf{K}_\perp} = \sqrt{K_X^2 + K_Y^2}$ for the case of two dimensions, and K, K_X, and K_Y are the wave numbers of the modulational instabilities to the uniform solution in the one-dimensional space and the two-dimensional space, respectively.

Considering the periodic boundary condition, for Eq. (2.3), we can define the eigenfunction as

(i) the case of one dimension:

$$\begin{pmatrix} \delta E \\ \delta E^* \end{pmatrix} = \begin{pmatrix} \epsilon e^{i\lambda t} \\ \epsilon^* e^{-i\lambda^* t} \end{pmatrix} \cos(KX); \tag{2.6}$$

(ii) the case of two dimensions:

$$\begin{pmatrix} \delta E \\ \delta E^* \end{pmatrix} = \begin{pmatrix} \epsilon e^{i\lambda t} \\ \epsilon^* e^{-i\lambda^* t} \end{pmatrix} \cos(K_X X)\cos(K_Y Y), \tag{2.7}$$

where ϵ and ϵ^* are small parameters. In this work, we choose the wave number to be that corresponding to the maximum growth rate, that is, $K_{\max} = \sqrt{\frac{1}{2}E_0F'(E_0)}$ for the one-dimensional space and $K_{X,\max} = K_{Y,\max} = \frac{1}{2}\sqrt{E_0F'(E_0)}$ for two-dimensional space.

Inserting Eq. (2.6) or Eq. (2.7) into Eq. (2.3), we obtain the equation of eigenvalue:

$$\lambda^2 - 2\omega_s\lambda - A^2 + 2\omega_s A - E_0^2 = 0, \tag{2.8}$$

where $A = -K_A^2 + L_i(|E_0|^2)$.

B. Initial conditions

According to Eq. (2.2) and Eq. (2.6) or Eq. (2.7), we can choose the initial condition in the numerical experiments to be

(i) the case of one dimension:

$$E(X,0) = E_0 + \epsilon e^{i\theta}\cos(KX); \tag{2.9}$$

(ii) the case of two dimensions:

$$E(X,Y,0) = E_0 + \epsilon e^{i\theta}\cos(K_X X)\cos(K_Y Y). \tag{2.10}$$

C. Phase space

For the continuum Hamiltonian system (1.1), an available choice of the phase space is also a key problem to analyze the dynamic behaviors due to the existence of the infinite degrees of freedom. From Eq. (2.8), we note that E_0 would correspond to the saddle point if there were a pair of conjuge complex roots for eigenvalue λ, that is, $\lambda = \alpha \pm i\beta$. To further illustrate this problem, we should discuss some models in detail. Without loss of generality, we assume $\omega_s \equiv 1$ and finally obtain the following.

(i) For model (1.6) (the case of $q = 1$),

$$L_1 = 2|E_0|^2, \quad h_1 = E_s^2(t),$$

$$K_{\max} = E_0, \quad K_{X,\max} = K_{Y,\max} = \frac{1}{\sqrt{2}}E_0, \tag{2.11}$$

$$\lambda = 1 \pm i,$$

for $E_0 = 1$.

(ii) For model (1.7),

$$L_2 = 2|E_0|^2 - 3g|E_0|^4, \quad h_2 = E_s^2(t)(1 - 2g|E_0|^2),$$

$$K_{\max} = E_0(1 - 2gE_0^2)^{1/2}, \quad K_{X,\max} = K_{Y,\max} = \frac{1}{\sqrt{2}}K_{\max}, \tag{2.12}$$

$$\lambda = 1 \pm i\frac{-1 + 4g + \sqrt{1-4g}}{2g},$$

for $E_0 = \sqrt{\frac{1-\sqrt{1-4g}}{2g}}$ with $0 \le g < 1/4$.

(iii) For model (1.8),

$$L_3 = \frac{2|E_0|^2 + g|E_0|^4}{(1 + g|E_0|^2)^2}, \quad h_3 = \frac{E_s^2(t)}{(1 + g|E_0|^2)^2},$$

$$K_{\max} = \frac{E_0}{1 + gE_0^2}, K_{X,\max} = K_{Y,\max} = \frac{1}{\sqrt{2}}K_{\max}, \tag{2.13}$$

$$\lambda = 1 \pm i(1 - g),$$

for $E_0 = \sqrt{\frac{1}{1-g}}$ with $0 \le g < 1$.

(iv) For model (1.9),

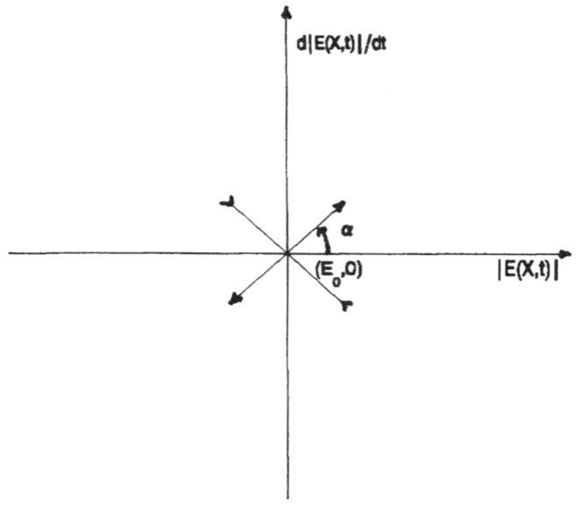

FIG. 1. Local phase flows of saddle point $(E_0, 0)$, which are obtained in terms of Eqs. (2.13) and (2.15), and where $\alpha = 1 - g$.

$$L_4 = \frac{1}{2g}[1 - (1 - 2g|E_0|^2)e^{-2g|E_0|^2}],$$

$$h_4 = E_s^2(t)e^{-2g|E_0|^2},$$

$$K_{\max} = E_0 e^{-gE_0^2}, \quad K_{X,\max} = K_{Y,\max} = \frac{1}{\sqrt{2}}K_{\max}, \tag{2.14}$$

$$\lambda = 1 \pm i\frac{1-2g}{2g}\ln(1 - 2g),$$

for $E_0 = \sqrt{\frac{1}{2g}\ln(1 - 2g)^{-1}}$ with $0 \le g < 1/2$.

Therefore, we have that (a) in the case of one dimension

$$|\delta E| = c_1\cos(K_{\max}X)e^{\beta t} + c_2\cos(K_{\max}X)e^{-\beta t},$$

$$\frac{d|\delta E|}{dt} = c_1\beta\cos(K_{\max}X)e^{\beta t} - c_2\beta\cos(K_{\max}X)e^{-\beta t}; \tag{2.15}$$

(b) in the case of two dimensions

$$|\delta E| = c_1\cos(K_{X,\max}X)\cos(K_{Y,\max}Y)e^{\beta t} + c_2\cos(K_{X,\max}X)\cos(K_{Y,\max}Y)e^{-\beta t},$$

$$\frac{d|\delta E|}{dt} = c_1\beta\cos(K_{X,\max}X)\cos(K_{Y,\max}Y)e^{\beta t} - c_2\beta\cos(K_{X,\max}X)\cos(K_{Y,\max}Y)e^{-\beta t}; \tag{2.16}$$

and

$$\left.\frac{\delta E}{\delta E^*}\right|_{t=0} = \pm i. \tag{2.17}$$

Thus we can construct the phase space as $(|\delta E|, d|\delta E|/dt)$ or $(|E|, d|E|/dt)$. Obviously, the point $(E_0, 0)$ in phase space $(|E|, d|E|/dt)$ corresponds to the hyperbolic fixed point. Comparing initial condition Eq. (2.9) or Eq. (2.10) with Eq. (2.17), we easily determine that the unstable manifolds for possessing the hyperbolic fixed point correspond to $\theta = 45°$ and $225°$, and the stable manifolds correspond to $\theta = 135°$ and $315°$ (see Fig. 1).

III. THE INTEGRABILITY OF THE ONE-DIMENSIONALLY CUBIC NSE

To better discuss the pattern dynamics of Eq. (1.1), we should simply describe the dynamic characteristics of the well-known cubic NSE. For convenience, we write the cubic NSE in an obvious form:

$$i\partial_t E + \partial_{XX}E + |E|^2 E = 0. \tag{3.1}$$

In Sec. II, we have shown that $(1, 0)$ in the phase space corresponds to a saddle point. If we take $\theta = 45°$ in the initial condition (2.9), the orbit that possesses the saddle point $(1, 0)$ corresponds to the homoclinic orbit (HMO) [19,31]. If we consider $\theta \neq 45°$, then the orbit should deviate from the HMO. Considering the integrability of the cubic NSE, on the other hand, we have that these orbits must be the exact recurrent solutions [11].

In the following, we turn to the numerical experiments. Here, the standard splitting-step spectral method [32] has been improved in order to increase the accuracy of conserved quantities. For the one-dimensional problem, the periodic length of the system is taken as $L = 2\pi/K_{\max}$. For the two-dimensional problem that will be discussed in Sec. V, the periodic lengths of the system are chosen as $L_X = 2\pi/K_{X,\max}$ and $L_Y = 2\pi/K_{Y,\max}$, respectively.

A. Homoclinic connection

In the initial condition, we take $\theta = 45°$ and $225°$, respectively. To determine the stable and unstable manifolds for the hyperbolic fixed point $(1, 0)$, we trace the unstable manifold by considering $t > 0$ and measure the stable manifold by simulating $t < 0$. Figure 2 is our computational result, which is traced at $X = 0$. From Fig. 2, we observe that the stable manifold $W^{(s)}$ smoothly joins with the unstable manifold $W^{(u)}$, that is, $W^{(s)}(\mathbf{X}^*) = W^{(u)}(\mathbf{X}^*)$ $[\mathbf{X}^* \equiv (1, 0)]$. This structure is called a homoclinic connection. The orbit that possesses the saddle point $(1, 0)$ corresponds to the HMO. As far as a finite dimensional dynamic system is concerned, the stable and unstable orbits for the hyperbolic fixed point would be smoothly joined to each other if the unperturbed system was taken to be integrable. For the cubic NSE, it is illustrated in Fig. 2 that such an idea is still effective for us to analyze the integrability of the continuum Hamiltonian system. Of course, almost any nonintegrable perturbation may destroy the connection. In the following section, we will again discuss such a problem for the saturable NSE (1.1).

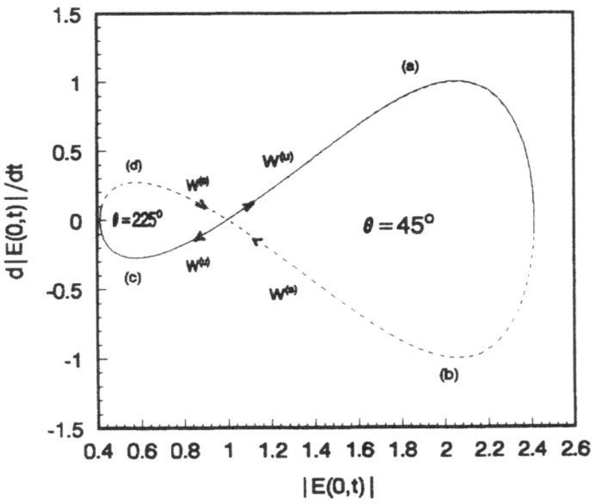

FIG. 2. Stable $W^{(s)}$ and unstable $W^{(u)}$ manifolds for the hyperbolic fixed point $(E_0, 0)$ with $\theta = 45°$ and $225°$, where the solid curve is computed with $t > 0$, and the dashed curve is computed with $t < 0$.

B. Recurrence

1. *Periodic trajectory*

Because of the integrability of the cubic NSE, its solutions can be displayed in terms of a class of periodic solutions. For our initial condition (2.9), Yuen and Ferguson [11] showed that the evolution of fields should be "simple." In other words, the solution developed by modulational instability corresponds to an exactly periodic solution (see Fig. 3).

It is noted from Fig. 3 that different trajectories are exhibited when we vary the initial phase θ. In numerical simulations, we find that the value of the amplitude for fields is larger than the value 1 as seen in Fig. 3(a) for $0 \leq \theta < 45°$, and the value periodically oscillates within the regions of 0 and 2.6 as shown in Fig. 3(b) for $45° < \theta \leq 90°$. For phase-space trajectories, Fig. 3(c) exhibits a clearer picture. When $\theta = 45°$, the trajectory that possesses the hyperbolic fixed point $(1, 0)$ is of the periodic motion around an elliptic fixed point. When $\theta = 0°$, the trajectory that deviates from the hyperbolic fixed point $(1, 0)$ periodically moves around the same elliptic fixed point as that for $\theta = 45°$. When $\theta = 90°$, however, the trajectory is of the periodic motion around two elliptic fixed points and the hyperbolic point $(1, 0)$. For these pictures, of course, the perturbation parameter ϵ in the initial condition should be quite small. If ϵ is not small, there could be some difference between theoretical results and numerical ones for our determining the region of θ.

2. *Spatially coherent patterns*

As mentioned above, we understand that the integrability of the one-dimensionally cubic NSE can be de-

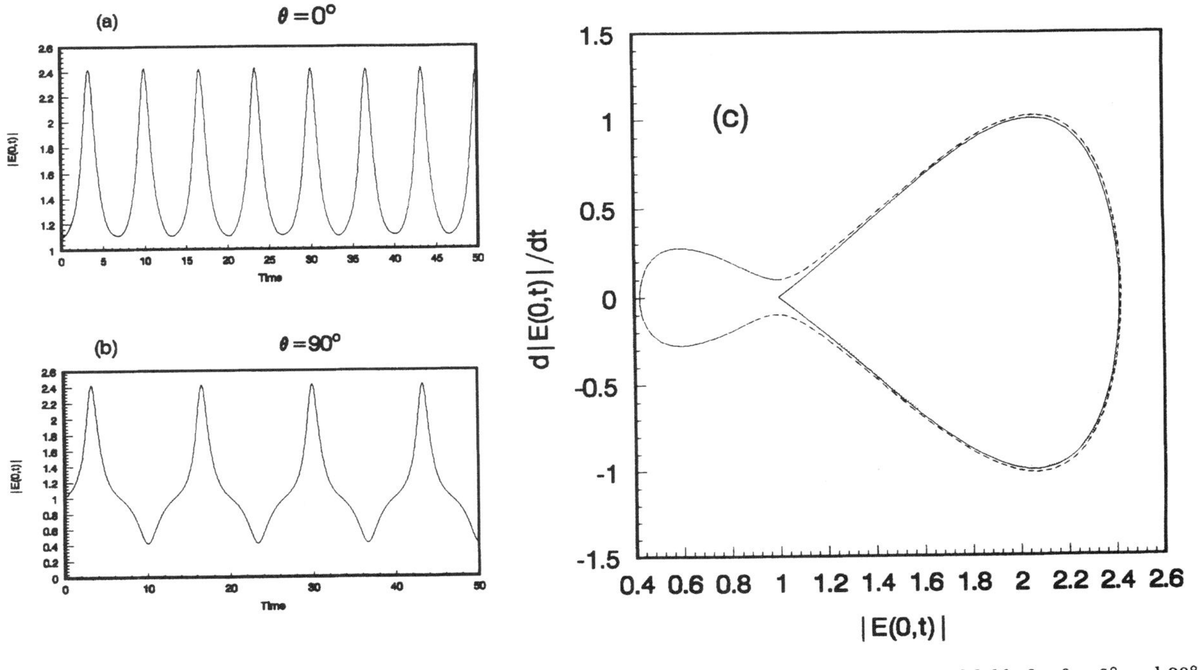

FIG. 3. Solutions of the one-dimensionally cubic NSE. (a) and (b) correspond to the amplitude of fields for $\theta = 0°$ and $90°$; (c) is the phase-space trajectories, where the dotted line corresponds to $\theta = 0°$, the solid line corresponds to $\theta = 45°$, and the dash line is $\theta = 90°$.

scribed in terms of strictly periodic trajectories. As a matter of fact, such an integrability is also associated with the well-known FPU recurrence. For our simply recurrent initial condition, the evolution of the spatially coherent patterns as shown in Fig. 4 exhibits an exactly periodic recurrent solution.

Particularly, Fig. 4(a) and Fig. 4(b) represent two types of patterns. In other words, a structure similar to that plotted in Fig. 4(a) will appear for $0 \leq \theta \leq 45°$, and one similar to that described in Fig. 4(b) will be formed for $45° < \theta \leq 90°$. Comparing Fig. 3 with Fig. 4, we easily understand that the choice process of these patterns depends on the phase-space manifolds. If the trajectory only recurs around an elliptic point [Fig. 3(a)], then pattern Fig. 4(a) will be formed. If the trajectory moves around two elliptic points and the hyperbolic fixed point $(1,0)$ [Fig. 3(b)], then the pattern Fig. 4(b) will be produced. Considering the Hamiltonian perturbation that is added in the cubic NSE, in the following sections, one can observe that the stochastic choice of these patterns is associated with the irregular HMO crossings.

3. Evolution of energy spectrum

To display this recurrence, we further measure the evolution of the energy contained in the Fourier modes. In Fourier space, we define the energy of the system as

$$H = \sum_n H_{K_n} = \sum_n |E_{K_n}|^2. \tag{3.2}$$

In our experiment, the initial energy is added to the mode $K_{\max}$. From Fig. 5, we observe that a large part of the energy in the system lies in the first and second modes. The temporal evolution of the energy in all modes is periodic, which is consistent with the periodic recurrence solution [Figs. 3 and 4]. In a sense, the unstable modulation to the uniform solution would first grow at an exponential rate as predicted by Benjamin and Feir [33], but eventually the solution would demodulate and return to a near-uniform state.

On the other hand, the periodically evolving process of the energy in Fourier modes is associated with the spatially coherent structures. That is, the amplitude of soliton structures is dominated by the total energy of the system, while the width and the pattern of the coherent structures are associated with the energy partition in the high modes. Thus, solitons developed by modulational instability keep their spatially coherent patterns and temporally periodic evolutions. The integrability of the one-dimensionally cubic NSE can also be described in terms of the exactly periodic recurrence in the energy space.

IV. SPATIOTEMPORAL CHAOS IN ONE-DIMENSIONAL SPACE

In the case of one dimension, the Hamiltonian of the NSE (1.1) can be rewritten as

$$H = H_0 + H_1, \tag{4.1}$$

where

$$H_0 = \int (|E_X|^2 - \tfrac{1}{2}|E|^4) dX, \tag{4.2}$$

$$H_1 = \int [\tfrac{1}{2}|E|^4 - f(|E|^2)] dX. \tag{4.3}$$

In Sec. III, we obviously illustrate that the integrability of Hamiltonian H_0 can be described in terms of the homoclinic connection of the periodic trajectories. Is the Hamiltonian perturbation H_1 integrable or nonintegrable? As yet, one has still not referred to a quite satisfactory theory to deal with these problems, especially on the integrability of the continuum Hamiltonian system. Generally speaking, there may exist finite solitons for integrable systems, such as the Langmuir soliton in the cubic NSE. However, the existence of a solitary

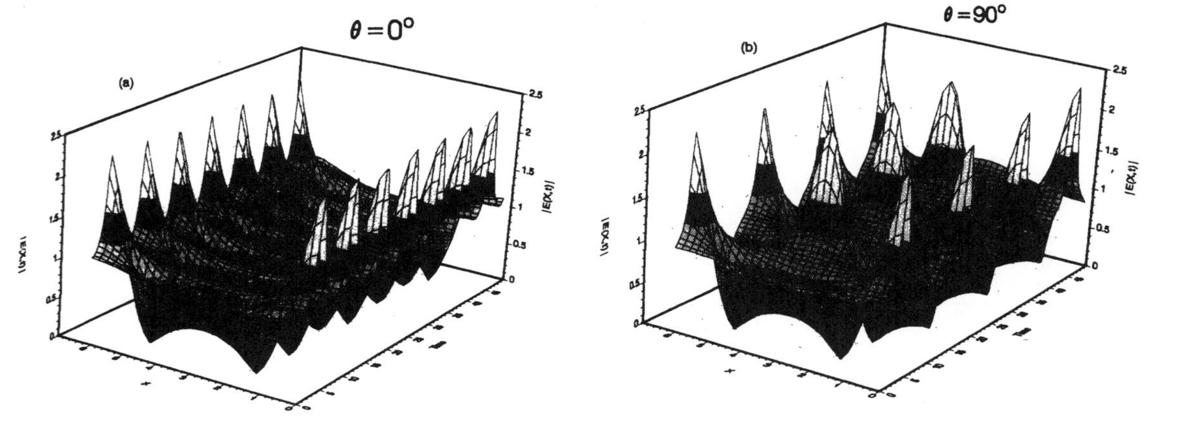

FIG. 4. The coherent structures of the one-dimensionally cubic NSE; (a) and (b) correspond to $\theta = 0°$ and $90°$, respectively.

cannot show the system to be integrable unless we can constitute a Lax pair. For example, there exists a standard Langmuir solitary for the model (1.6) [19], but the propagation of fields displays a typically chaotic characteristic. Here we numerically discuss the integrablility problem of the NSE (1.1).

A. Irregular HMO crossings

Taking into account Eq. (1.1) and Eqs. (1.6)–(1.9), we know that these NSE's correspond to the integrable equations when $g = 0$, and the stable and unstable manifolds that possess the saddle point $(1, 0)$ are smoothly joined to each other. Making use of the results obtained in linear stability analyses, we also choose the initial parameters as $\theta = 45°$ and $g = 0.1$, and measure their characteristics of the stable and unstable manifolds for models (1.7)–(1.9), respectively.

From Fig. 6, we obviously observe that the stable and unstable manifolds in the phase space do not smoothly join together, which illustrates that these saturable NSE's are nearly integrable [31,34]. Comparing our simulation results (Fig. 6) with those in finite dimensional Hamiltonian systems [34], it is also noted that there are not an infinite number of homoclinic points in Fig. 6. As for the finite dimensional system, the fact that one transverse homoclinic point implies an infinity

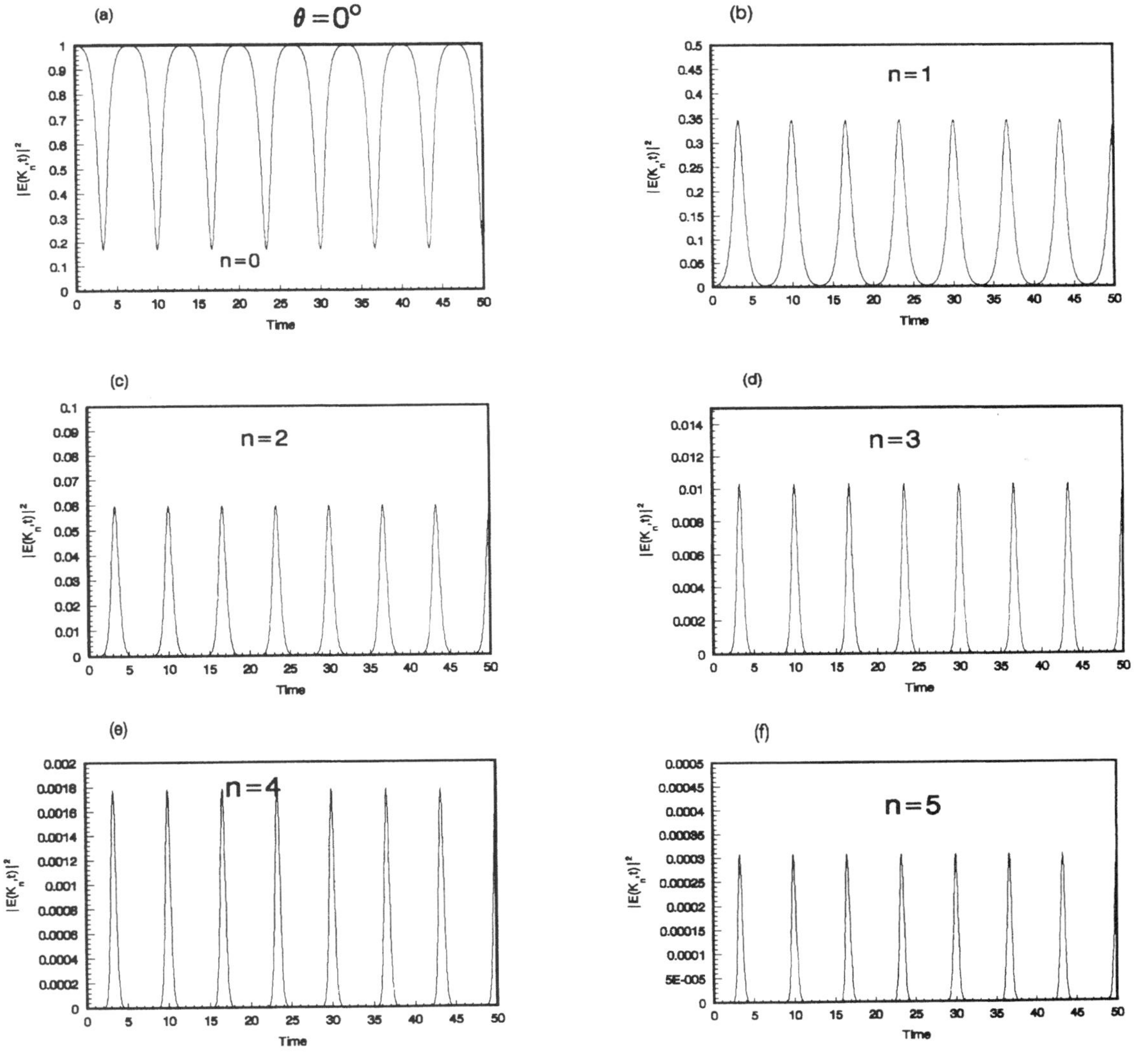

FIG. 5. The energy evolution in the first six Fourier modes for the one-dimensionally cubic NSE.

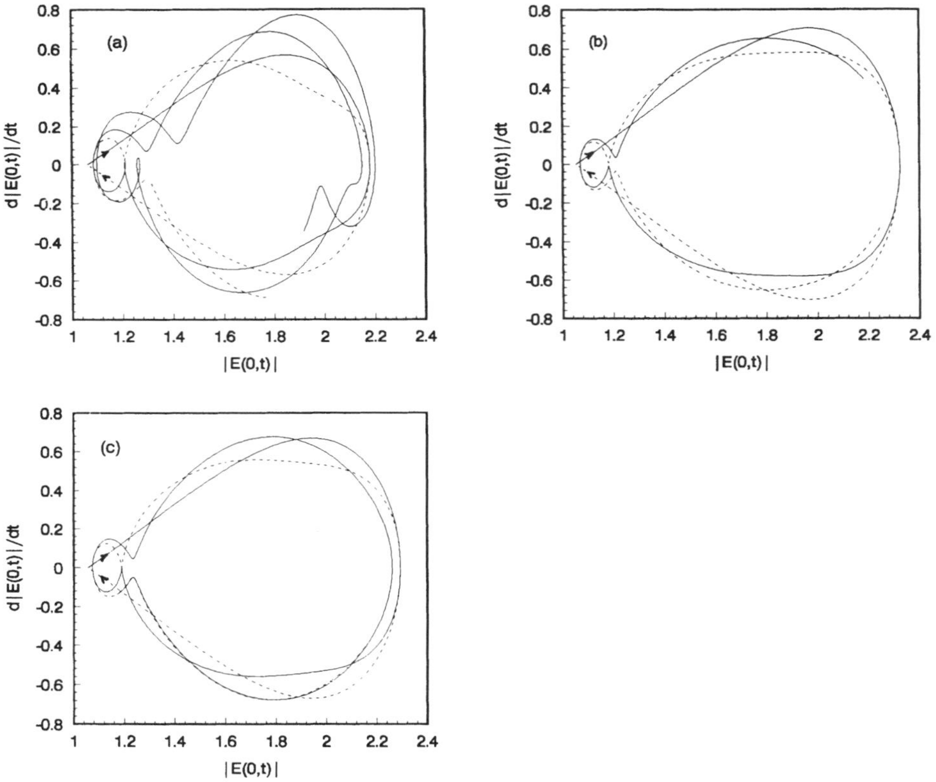

FIG. 6. Stable and unstable manifolds for the hyperbolic fixed point $(E_0, 0)$ with $\theta = 45°$, where (a), (b), and (c) correspond to the models (1.7), (1.8), and (1.9), respectively.

of homoclinic points leads to extreme stretching and folding of the manifolds. However, we notice that our phase space $(|E(X,t)|, d|E(X,t)|/dt)$ is only the projection of a high-dimensional space, where the information about phase for wave fields is not considered. Therefore, there may exist some difference between our HMO crossings and those in a finite dimensional Hamiltonian system in which infinite homoclinic points appear.

We also observe from Fig. 6(a)–Fig. 6(c) that their topologic structures are basically consistent. In a sense, the Hamiltonian perturbation H_1 (4.3) destroys the homoclinic connection and leads to the destruction of the integrability.

B. Spatiotemporal chaos

1. Quasi-periodic route to chaos

As stated above, we know that the high order Hamiltonian perturbation can result in the formation of irregular HMO crossings. We recall that the infinite periodic homoclinic states act, under perturbation, as sources of sensitivity (including chaos). Therefore, we should further discuss the chaotic characteristics and route.

We do not give a detailed description on all dynamic models as mentioned in Sec. I. As a specific example, we only deal with the model (1.8), which is rewritten as

$$i\partial_t E + \partial_{XX} E + \frac{|E|^2}{1 + g|E|^2} E = 0. \qquad (4.4)$$

In numerical processes, we choose the initial parameters as $\epsilon = 0.1$, $\theta = 1.0005\pi/4$, and vary g.

After many experiments, we find that the route from the exactly periodic solution to chaos is a quasiperiodic route as shown in Fig. 7. From the figure, we see that the periodic oscillation of amplitude for fields is broken down with the increase of g; in particular, the stochastic behavior occurs as displayed in Fig. 7(d) for $g = 0.1$. For the phase-space trajectories, an exactly recurrent motion for the cubic NSE ($g = 0$) is given in Fig. 7(a) and Fig. 7(a′). We also note that the phase trajectories in Fig. 7(a′) only move around an elliptic fixed point, and are not completely consistent with those discussed in Sec. III (B). This is because the initial parameter ϵ is not quite small. For $g = 0.0005$, the figures [Fig. 7(b) and Fig. 7(b′)] of phase space behave like the quasirecurrent solution where the trajectory is not the exactly periodic one. For $g = 0.005$, Figs. 7(c)–7(c″) show that more subharmonics appear. Comparing Fig. 7(b) with Fig. 7(c), we see that the amplitude of the wave fields for the case of $g = 0.005$ is more irregular; especially the trajectories of the phase space Fig. 7(c′) illustrate that the KAM tori become thicker but are not completely destroyed. Figure 7(c″) shows that more frequencies are produced due to the increase of the saturable nonlinearity. For $g = 0.1$, the irregular homoclinic orbit crossings are also drawn in Fig. 7(d′). To describe the route, a better technique is to measure the power spectrum. The foundation frequency of the periodic solution for the cubic NSE is labeled in Fig. 7(a″). Also, Fig. 7(d″) indicates the noiselike characteristic typical of chaotic time evolution. It is particularly noted that Fig. 7(b″) exhibits a quasiperiodic spectrum [34,35], where two incommensurate frequencies $\{\omega_0, \omega_1\}$ appear, and $\omega_2 = \omega_0 - \omega_1$. In the numerical simulations, we find that such a standard spectrum is sensitively dependent on the initial parameters, such as θ and g. In general, more subharmonics [as described in Fig. 7(c″)] may be observed with variation of the parameters [36]. Hence we can conclude the route is as follows: *period* (coherent structures) $\{\omega_0\}$ → *quasiperiod* $\{\omega_0, \omega_1\}$ (pseudorecurrence) → *subharmonics* $(\omega_0, \omega_1, \ldots, \omega_n)$ → *chaos* (spatiotemporal complexity).

Speaking in terms of physics, the Hamiltonian perturbation

$$H_1 = \int \left\{ \frac{1}{2}|E|^4 - \left[\frac{1}{g}|E|^2 - \frac{1}{g^2}\ln(1 + g|E|^2) \right] \right\} dX \tag{4.5}$$

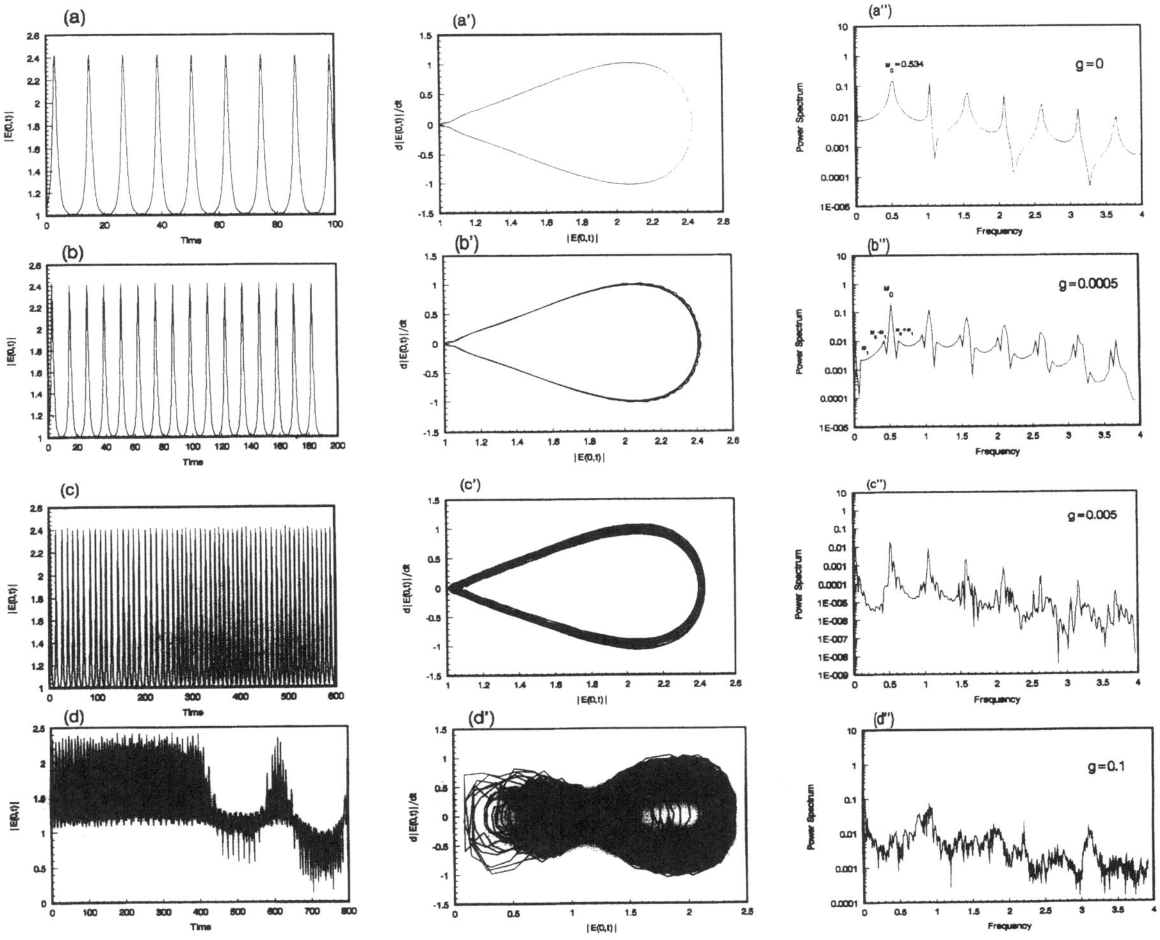

FIG. 7. The solutions of Eq. (4.4) with $\epsilon = 0.1$ and $\theta = \frac{1.0005\pi}{4}$, where a fixed position ($X = 0$) is traced. (a), (b), (c), and (d) correspond to the temporal evolution of amplitude for fields; (a′), (b′), (c′), and (d′) are the structures of phase space, noting that (d′) indicates the irregular homoclinic orbit crossings; (a″), (b″), (c″), and (d″) represent the power spectra corresponding to (a), (b), (c), and (d), respectively. In particular, (b′) exhibits the quasiperiodic spectrum.

drives the formation of stochastic fields and the destruction of recurrence. When $g = 0$, the trajectory is periodic and the spatial patterns are coherent due to the integrability of the system. When g is quite small, a periodic trajectory becomes the quasiperiodic one. With further increase of g, the nonintegrable Hamiltonian perturbation H_1 drives the KAM tori and the coherent structures to be completely broken down.

2. *Spatially complicated patterns*

We have understood that nonintegrable perturbation leads to the chaotic propagation of wave fields. On the other hand, we observe from Fig. 8 that the solitonlike structures are still kept but are quite irregular. The appearance of these complicated patterns means that the exactly recurrent structures disappear. In fact, the stochastic choice of these patterns is associated with the irregular HMO crossings. Comparing Fig. 7(d′) with Fig. 8, one can easily understand that these two similar kinds of patterns as described in Fig. 4(a) and Fig. 4(b) may also be formed under the current parameters. In addition, these irregular patterns depend on the partition of the energy in the Fourier modes. For the coherent structures (Fig. 4), the energy in the Fourier modes represents homogeneous decay. For these spatiotemporally complicated patterns (Fig. 8), however, the partition of the energy in the Fourier modes represents inhomogeneous decay [35]. To better analyze the mechanism by which complicated patterns are formed, we should also discuss the processes of evolution of the energy contained in the Fourier modes.

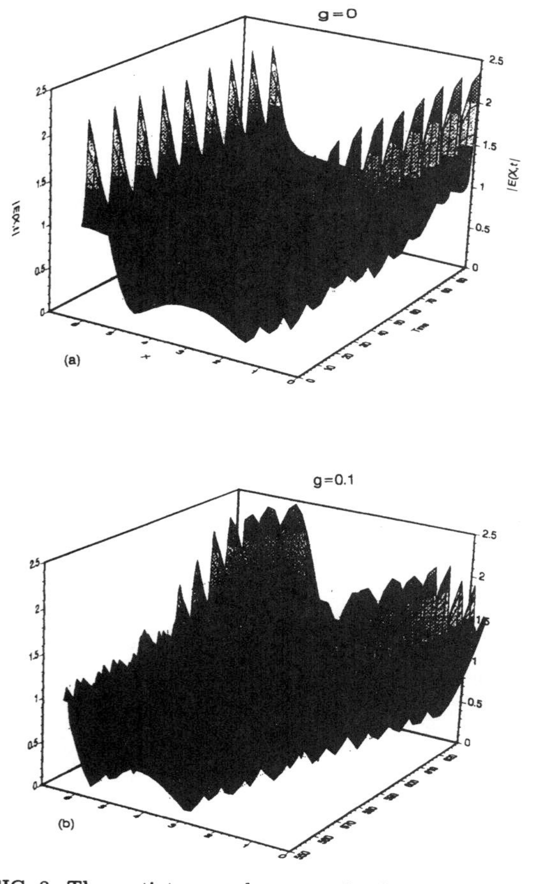

FIG. 8. The spatiotemporal patterns for the saturable NSE (4.4) with $\epsilon = 0.1$ and $\theta = \frac{1.0005\pi}{4}$, but the value g changed. (a) Coherent patterns with $g = 0$; (b) spatiotemporally complicated patterns with $g = 0.1$.

3. *Stochastic evolution of energy spectrum*

The evolution of energy in the first four Fourier modes is described in Fig. 9. Obviously, a large part of the energy in the system still lies in the low Fourier modes. As for the first two modes, the evolution of the energy behaves like a quasiperiodic motion, which makes sure the spatially localized structures (Fig. 8) are still kept. However, the evolution of the energy in the high modes exhibits a stochastic behavior, which illustrates spatial patterns that are very irregular as seen in Fig. 8. Physically, the energy in the system, which is initially confined to the master mode, would spread to many slave harmonic modes because of the nonlinear interaction, but would not regroup into the original lowest mode. The stochastic evolution of the energy in the Fourier modes, in which the slave modes interact with the master one, leads to the formation of spatiotemporal chaos.

V. PATTERN STRUCTURES IN TWO-DIMENSIONAL SPACE

In this section, we turn to discussing the dynamic characteristics of the two-dimensional NSE. Also we only deal with the model (1.8) which is clearly written as follows:

$$iE_t + \nabla^2 E + \frac{|E|^2}{1+g|E|^2}E = 0. \qquad (5.1)$$

For this model, Akhemediev *et al.* [22] showed numerically that pseudorecurrence can appear in two-dimensionally modulational instability. Moreover, the singular solution can also exist [23]. As we know, however, spatiotemporal dynamics in two-dimensional space has not been systematically studied as yet, especially the spatial patterns and the chaotic characteristics.

Here, we further investigate the behaviors of the solution in terms of the modulational instability in two-dimensional space. First, we write the linear growth rate (2.5) as the following obvious form:

$$\gamma(K_\perp) = K_\perp \left[\frac{2gE_0^2}{(1+gE_0^2)^2} - K_\perp^2\right]^{1/2}, \qquad (5.2)$$

where $K_\perp = \sqrt{K_X^2 + K_Y^2}$, which is sketched in Fig. 10.

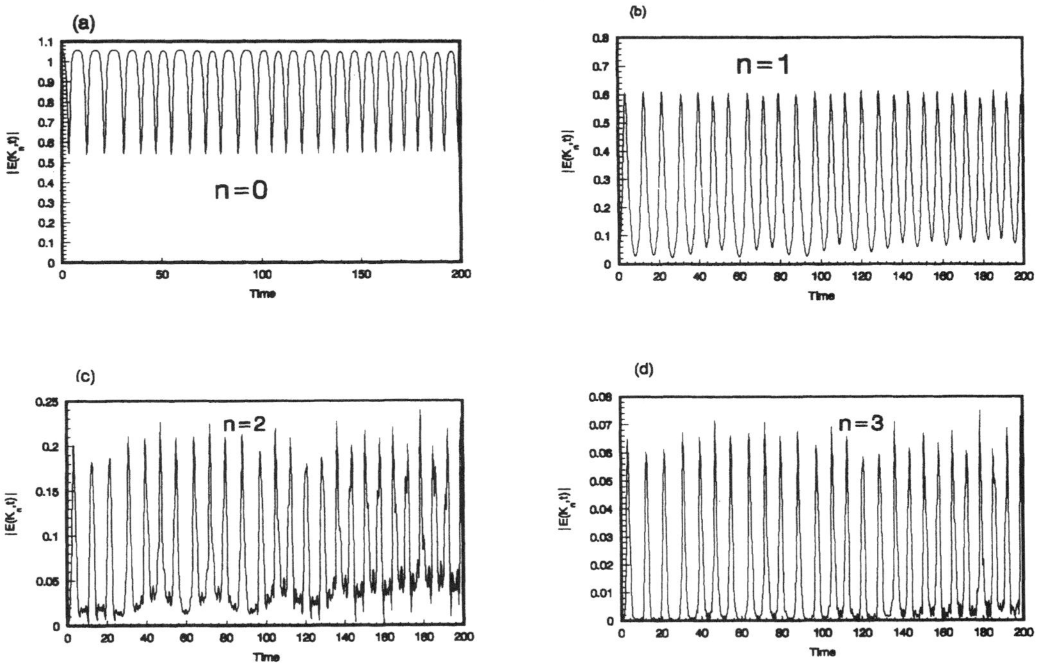

FIG. 9. The energy evolution in the first four Fourier modes for the NSE (4.4), where the parametric values correspond to those in Fig. 7(d).

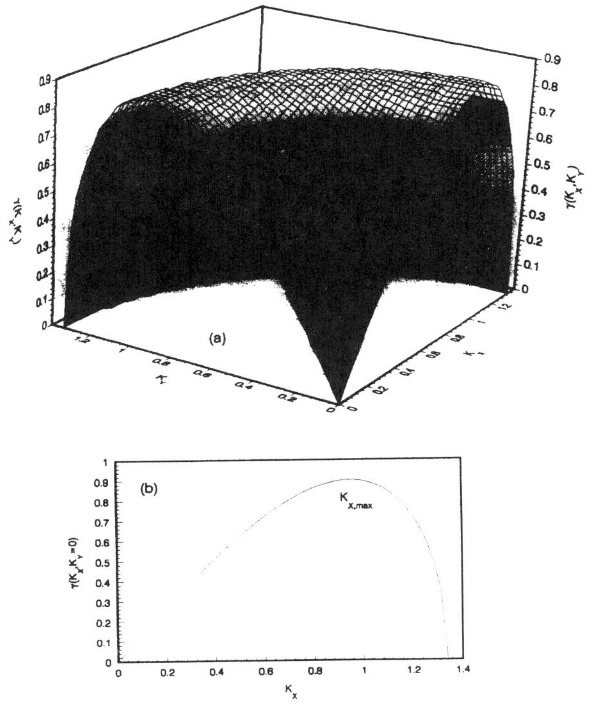

FIG. 10. (a) The linear growth rate vs K_X and K_Y. (b) The growth rate vs K_X with $K_Y = 0$.

A. Irregular HMO crossings

From Eqs. (2.16) and (2.17), we know that $(E_0, 0)$, where $E_0 = [1/(1-g)]^{1/2}$ with $0 \leq g < 1$, corresponds to a saddle point in the phase space $(|E(X,Y,t)|, d|E(X,Y,t)|/dt)$. The stable and unstable manifolds that possess the saddle point also correspond to $\theta = 45°, 225°$ and $\theta = 135°, 315°$, respectively. As discussed in Sec. IV, here, we choose the parametric values as $\theta = 45°$, $g = 0.1$, and trace the trajectories at the fixed space point $X = 0$ and $Y = 0$. From Fig. 11, we observe that the phase-space trajectories are completely similar to that in the case of one dimension as shown in Fig. 6. Therefore, we can say that the formation of chaos in the current two-dimensional space is also due to irregular HMO crossings. Of course, we cannot think that such a HMO chaos in the saturable NSE (5.1) only arises from the nonlinear perturbation. This is because the cubic NSE in two-dimensional space is also not integrable. In the following analyses, we may understand that the pseudorecurrence appears if the saturable nonlinear effects are very strong. With the decrease of the saturable effects, the irregular HMO crossings lead to the destruction of pseudorecurrence and the appearance of stochasticity for fields.

B. Blowup, chaos, and pseudorecurrence

In the following simulations, we choose parametric values as $\epsilon = 0.1$ and $\theta = \frac{1.0005\pi}{4}$, and vary the g value.

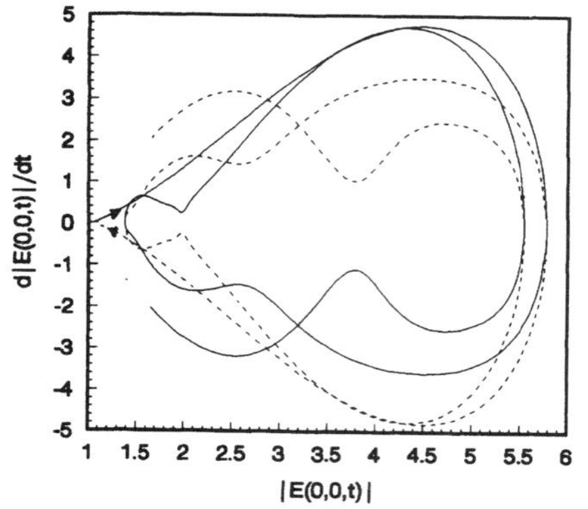

FIG. 11. Stable and unstable manifolds for the hyperbolic fixed point $(E_0, 0)$ in the two-dimensional NSE (5.1); it is noted that this structure is similar to that in the case of one dimension, which illustrates the characteristic of HMO chaos.

1. Singular solution

To clearly study the features of the dynamic system (5.1), we first discuss characteristics of the two-dimensionally cubic NSE. As $g = 0$, Eq. (5.1) corresponds to the critical NSE. The solution should be the singular one, where the time of blowup is finite [23] and the field intensities exhibit catastrophic processes.

In order to describe the temporally evolutive characteristics in more detail, we trace the trajectory at a fixed position, i.e., $X = 0, Y = 0$. Figure 12(a) exhibits the time evolution of amplitude for fields, where the solution is rapidly developed from the initial value 1 to the larger value within a finite time that is associated with the time of blowup as seen from Fig. 13. Owing to the rapid change of amplitude with time, its slope must be very large. Therefore, Fig. 12(b) indicates that the trajectory in phase space is not bounded within a small area.

2. Chaotic solution

Taking into account $g = 0.1$, we observe that the amplitude stochastically oscillates as shown in Fig. 12(c). The field E_0 is developed due to the modulational instability from the initial value E_0 to a finite amplitude toward saturation, then oscillates down because of the saturable nonlinear force (ponderomotive force), and so on. For the trajectory, its manifolds exhibit irregular HMO crossings, where the trajectory as seen in Fig. 12(d) is bounded with a finite region of phase space. Thus our simulation indeed verifies the prediction given by Akhemediev *et al.* [22] on HMO chaos.

3. Pseudorecurrent solution

A more interesting phenomenon is observed in Fig. 12(e)–Fig. 12(h). With the increase of parameter g, we find that pseudorecurrence appears. As plotted in Fig. 12(e) and Fig. 12(g), the amplitude of the fields behaves like a periodiclike oscillation. These phase trajectories are basically regular, where the motion is smooth in small regions of phase space as described in Fig. 12(f) and Fig. 12(h). In other words, the KAM tori are not completely broken down when the parameter g comes close to 1. In the work of Akhemediev *et al.* [22], they chose $g = 1$ and also observed such a pseudorecurrent behavior.

C. Spatial patterns

In the following, we describe the pattern structures corresponding to these three kinds of solutions as stated in the above. For the case of $g = 0$, we observe from Fig. 13(a)–Fig. 13(d) the developing processes of the initial field (2.10) due to the modulation instability. Figure 13(c) and Fig. 13(d) correspond to the singular solutions. Comparing Fig. 13(b) with Fig. 13(c), we understand that the time of blowup lies in the regions of $t^* = 1.2 - 2.0$. With the evolution of time, the solution is rapidly contracted and the energy of the system is concentrated on some small spots. In laser-produced plasmas, these processes describe filamentation of laser beams [25–27,37] as the time variable t is replaced by the propagative distance Z. In plasma turbulence, they describe the Langmuir collapse due to the ion-acoustic perturbation [8], which is influenced by ponderomotive force in the NSE (1.1).

For the case of the chaotic solution, the numerical experiments show that the singular solution does not occur where the saturable nonlinear effects prevent blowup of the solution. Speaking in terms of physics, the saturable nonlinear force prevents the collapse of Langmuir waves or the concentration of energy. We find from Fig. 14 that different patterns are formed. In particular, the spatial symmetry is broken down with the development of wave fields as sketched in Fig. 14(b), which are different from those in the initial stages of evolution where the spatial symmetry is kept as shown in Fig. 14(a). For these structures, we will now give a description in terms of the energy spectrum.

For the case of pseudorecurrent solutions, we see from Fig. 15 where $g = 0.8$ that the spatial structures at $t = 50$ are different from those at $t = 200$; however, the symmetry of the solution is basically kept. In addition, we note from Fig. 16 where $g = 0.98$ that the spatial structure at $t = 50$ is basically consistent with that at $t = 200$.

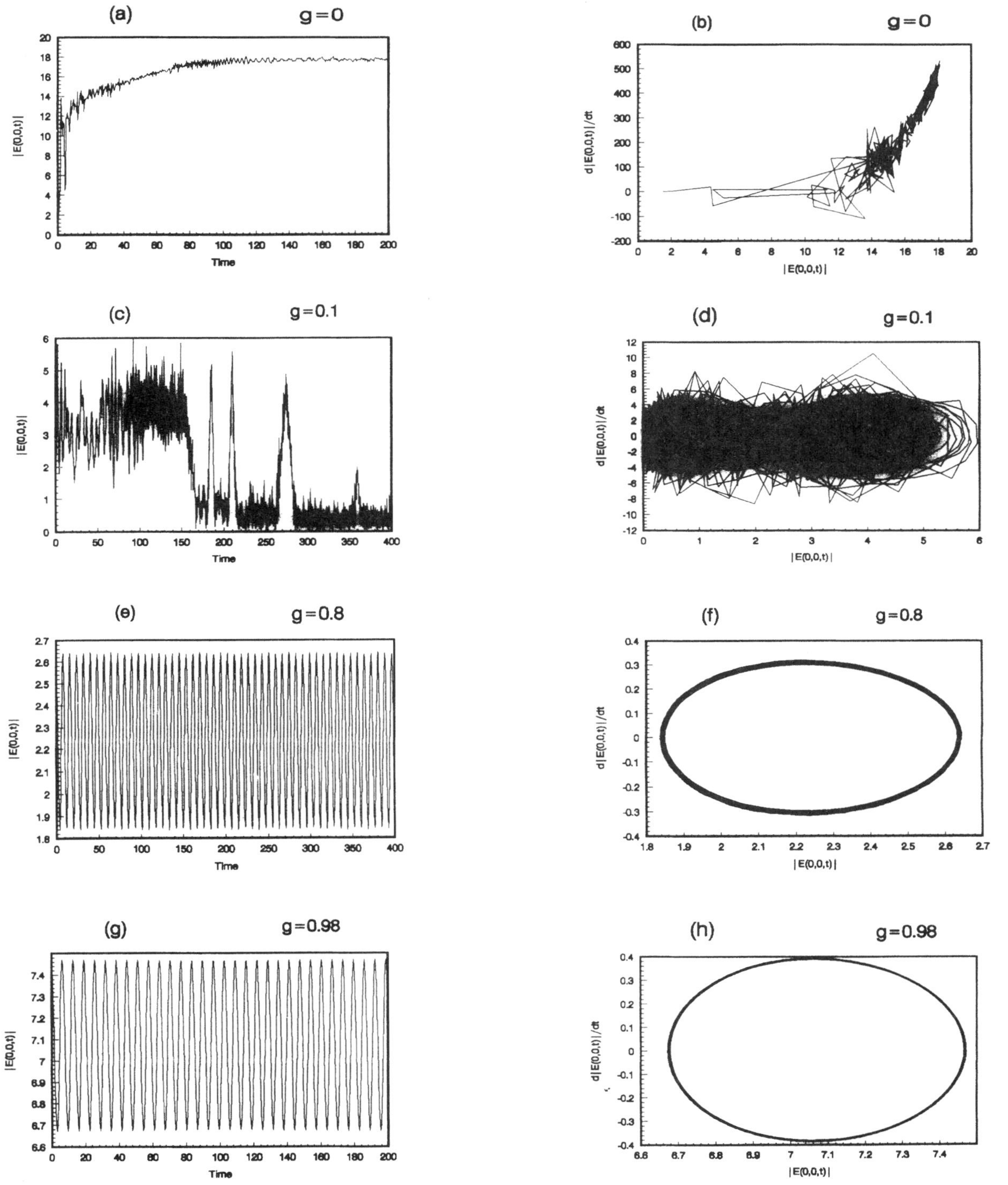

FIG. 12. Solutions of Eq. (5.1). (a), (c), (e), and (g) are the amplitude of fields, (b), (d), (f), and (h) the phase-space trajectory; (a) and (b) correspond to the singular solution with $g = 0$, (c) and (d) correspond to the chaotic solution with $g = 0.1$, and (e), (f) and (g), (h) correspond to the pseudorecurrent solutions with $g = 0.8$ and $g = 0.98$, respectively.

Taking into account these pictures for the case of pseudorecurrence, therefore, we can say that the symmetry of the solution is not basically broken down and the singular solution is completely prevented by the saturable nonlinearity.

D. Analyses of the energy spectrum

To illustrate clearly the mechanism that leads to the formation of these different complicated patterns, we also measure the evolution of the energy in the Fourier modes, and define

$$H = \sum_{K_{n,m}} |E(K_n, K_m, t)|^2, \tag{5.3}$$

where $K_{n,m} = \{K_n, K_m\}$ represents the nth Fourier mode in X space and the mth in Y space. The evolution of energy in Fourier modes, i.e., $H_{n,m} = |E_{K_n,K_m}|^2$, is shown in Figs. 17–19. Figure 17 corresponds to the case of $g = 0$, where we observe that the energy in the diagonal modes decays within a very short time and then stochas-

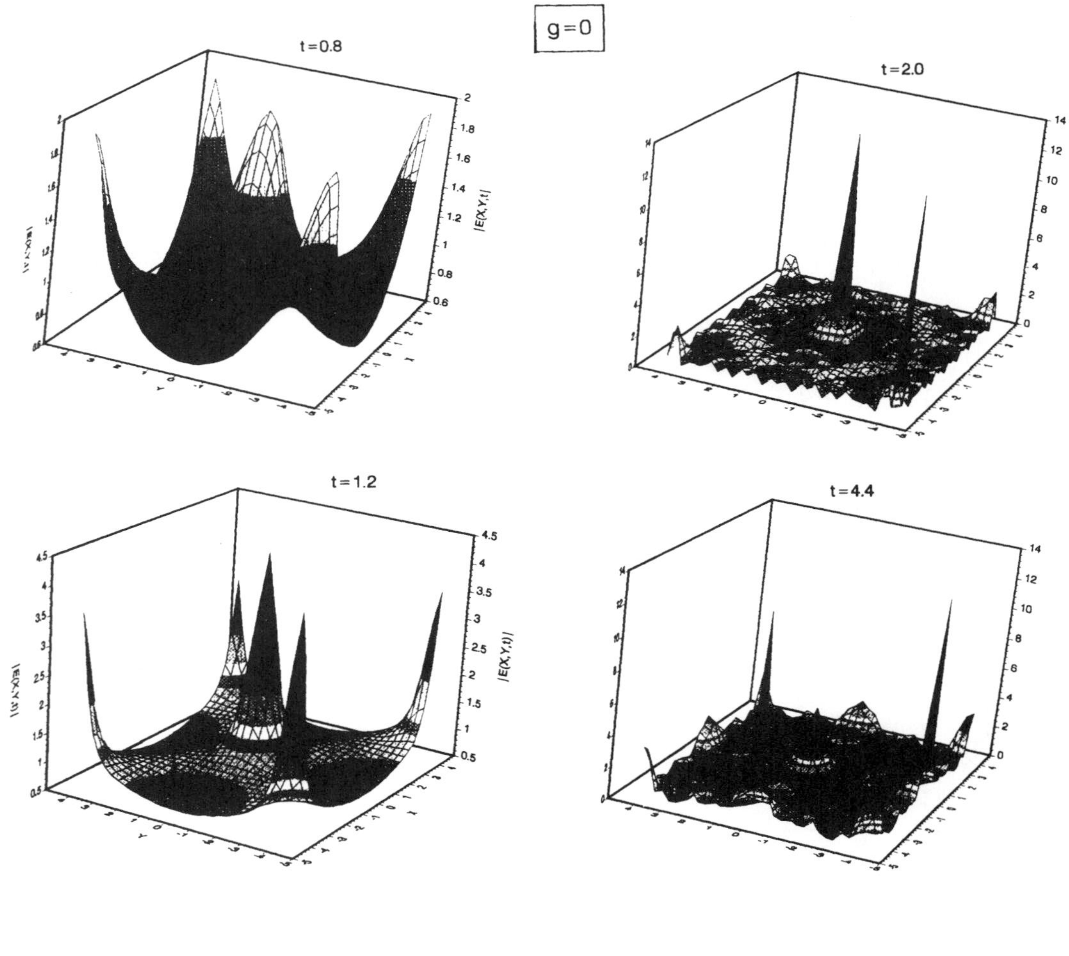

FIG. 13. The spatial patterns of the blowup processes with $g = 0$.

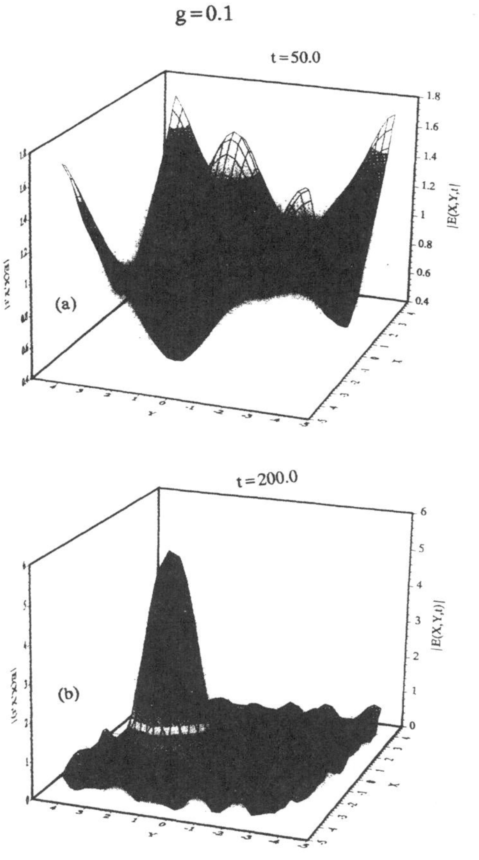

FIG. 14. The spatial patterns of spatiotemporal chaos with $g = 0.1$. (a) the symmetric structures, (b) the nonsymmetric structures.

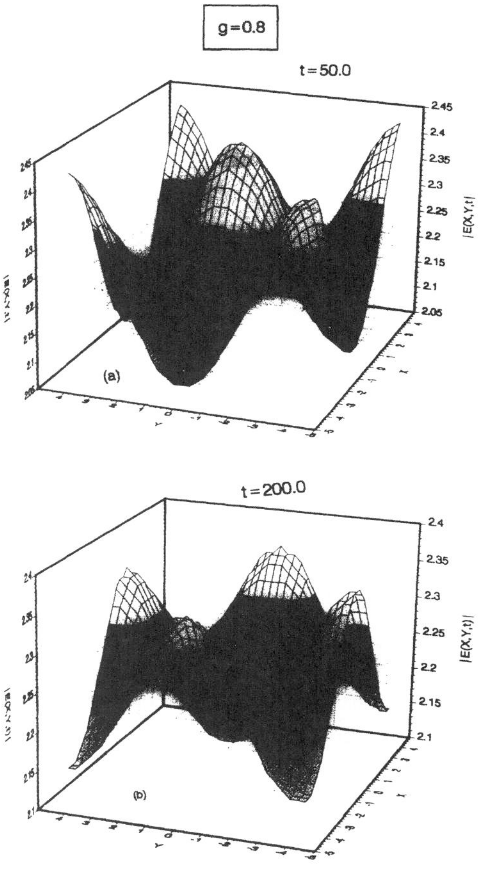

FIG. 15. The spatial patterns of the pseudorecurrent solution with $g = 0.8$.

tically oscillates. Also, the energy in the nondiagonal modes increases within a short time. In other words, time to reach the blowup may be associated with the evolution of the energy in the Fourier modes. When the blowup occurs, the energy in the main modes is decreased. In fact, the energy in these modes would determine the structures of the system. If the energy in the diagonal modes is very large, then the width of the wave packets would be large enough, which indicates that the blowup does not occur. With the development of time, the energy in the diagonal modes decreases and that in the nondiagonal modes increases. At this time, the energy in the system would be concentrated on some small spots, which shows that the singular solution occurs. Moreover, the nonsymmetric patterns may appear [see Fig. 13(c) and Fig. 13(d)] as soon as the energy in the nondiagonal modes is quite large.

It is noticed from Fig. 18 that an important phenomenon is exhibited. When $t < 60$, the energy in the diagonal modes experiences a stochastic oscillation. Simultaneously, their largest amplitude is near equal though the amplitude is actually stochastic. In particular, the energy in the nondiagonal modes equals zero for $t < 60$. When $t > 60$, the energy in the diagonal modes is decreased, and that in the nondiagonal modes is increased but still has the stochastic characteristic. In addition, we find that a large part of the energy lies in the main mode (K_{00}), where the initial energy is injected. According to these analyses, a clear conclusion is that the spatially nonsymmetric structures are mainly dependent on the energy in the nondiagonal modes. That is, the nonsymmetry should be associated with a stochastic partition of the energy in the Fourier modes, especially in the nondiagonal modes. Comparing Figs. 13 and 14 with Figs. 17 and 18, also, we conclude that the concentration of energy in the system depends on the distribution of the

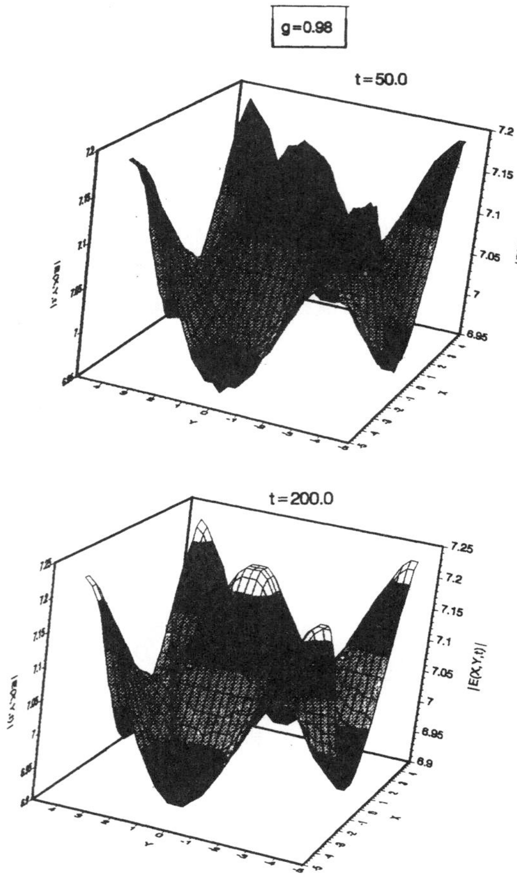

FIG. 16. The spatial patterns of the pseudorecurrent solution with $g = 0.98$.

energy in the Fourier modes. That is, the less energy is partitioned in the diagonal modes or the more in the nondiagonal modes, the higher the height of the spatial patterns is or the smaller the width.

Now, we discuss the spectrum evolution corresponding to the pseudorecurrent solution. From Fig. 19, we see that the energy in the main mode K_{00} compared with the other modes is very large and oscillates within a small regime. The evolution of the energy in the diagonal modes exhibits a periodiclike motion. In particular, the energy in mode K_{22} indicates the energy is not exactly FPU recurrent, but the partition of the energy in this mode is not irregular, which actually illustrates the pseudorecurrent feature. On the other hand, we find from the numerical experiments that the energy in the nondiagonal modes is nearly zero, which shows that the spatial structures developed by the modulational instability are symmetric, as seen in Fig. 15.

VI. SUMMARIES AND DISCUSSIONS

We have systematically described the spatiotemporal characteristics of a series of generalized NSE's in one- and two-dimensional space. Our investigations show that a high order Hamiltonian perturbation can lead to the destruction of the coherent structures and the formation of spatiotemporally complicated patterns. The route from coherent patterns to complicated ones is quasiperiodic. For two-dimensional problems, our numerical experiments illustrate that the singular solution, spatiotemporal chaos, and pseudorecurrence can appear for the same NSE with the increase of the saturable nonlinear effects, where the spatial patterns and the energy partition of the system in the Fourier modes exhibit completely different dynamic behaviors. We emphasize in particular that the symmetric destruction of the spatial patterns is associated with the energy in the nondiagonal modes. These results should be of significance in the study of pattern dynamics in plasma waves and other nonlinear waves.

As far as various branches of physics are concerned, the presence of the stochastic wave fields depends on the evolution processes of systems. In other words, the lowest order nonlinear term $|E|^2E$ is the predominant nonlinear mechanism in the evolving initial stages. Some important phenomena, such as coherent structures and Langmuir wave collapse, etc., can be reasonably explained by making use of the cubic NSE. In the strongly nonlinear stages, other physical effects could play an important role. For example, the effects of Landau damping of the high frequency waves become quite significant in plasma turbulence. The energy dissipation should be considered [38,39]. On the other hand, nonlinear interactions, such as the fourth field interaction and other nonlinear saturable mechanisms, should be involved in the processes of plasma instabilities. These nonlinear saturable effects make rich dynamic behaviors occur. In a sense, these complicated dynamic phenomena would probably be the characteristic of wave propagation.

We should mention that our results are also important in laser fusion studies. Considering the process of ponderomotive filamentation, under the static approximation, one studies the characteristics of ponderomotive filamentation in terms of the NSE's (1.6)–(1.9) [25–28,37] when the time variable t is replaced by the spatial variable Z. It is shown that the focusing becomes a periodic oscillation, rather than a catastrophic process due to the saturable effects. In the previous work, it was illustrated that there exists competition between SBS and filamentation instability [25,26]. Taking into account filamentation produced by perturbations or nonuniformities in light that cause (or are caused by) local changes in the dielectric constant, index of refraction, or the medium, we easily understand that these inhomogeneities may be enlarged due to mode-mode interactions, where the transverse instability as described in this work or the self-modulation as discussed by Max *et al.* [40] can broaden the frequency spectrum of the incident laser. For these instabilities, as a matter of fact, there is no reason to separate them in the nonlinear regime. They can coexist and affect each other. Of course, SBS is related to the resonant interaction of a three-wave coupling, whereas self-focusing and filamentation instabilities involve nonresonant ion density disturbances. As far as the stochastic instabilities are concerned, the mode-mode resonant interaction [41] may also drive the chaotic propagation of light. The current investigations show that an initial homogeneous beam may develop into incoherent light, where the symmetry of the laser could be broken down and the center of the beams would be shifted off axis due to the influence of the ponderomotive force. Of course, the recent modulation should be the spatial modulation in the transverse direction, because the starting equations (1.6)–(1.9) only allow for transverse inhomogeneities when the time variable t is replaced by the axial variable Z.

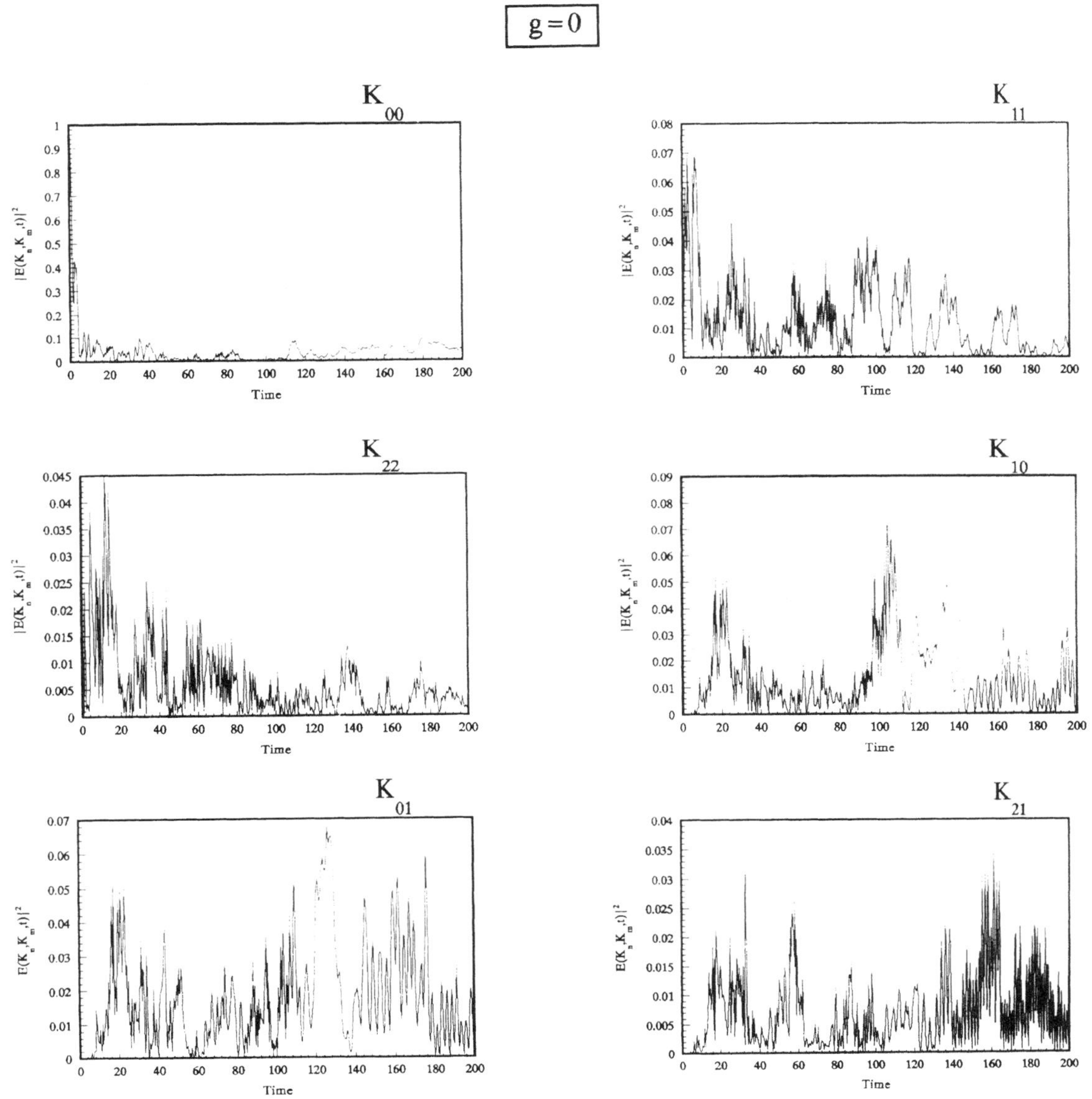

FIG. 17. The evolution of the energy in the Fourier modes with $g = 0$.

Finally, we point out that our work only deals with the initial homogeneous state. It is shown that the presence of the HMO crossings is a potential source of stochastic behavior. We expect that more important theoretical work on HMO chaos in the continuum Hamiltonian system can be furthered in future study. We also note that it is an interesting and important piece of research for us to consider the initial inhomogeneous state, which could be more difficult for theoretical analysis as in this work. We think that the key problem should still be the constitution of a suitable phase space for these continuum Hamiltonian systems so that one can readily describe their dynamic characteristics in terms of integrable and nonintegrable equations.

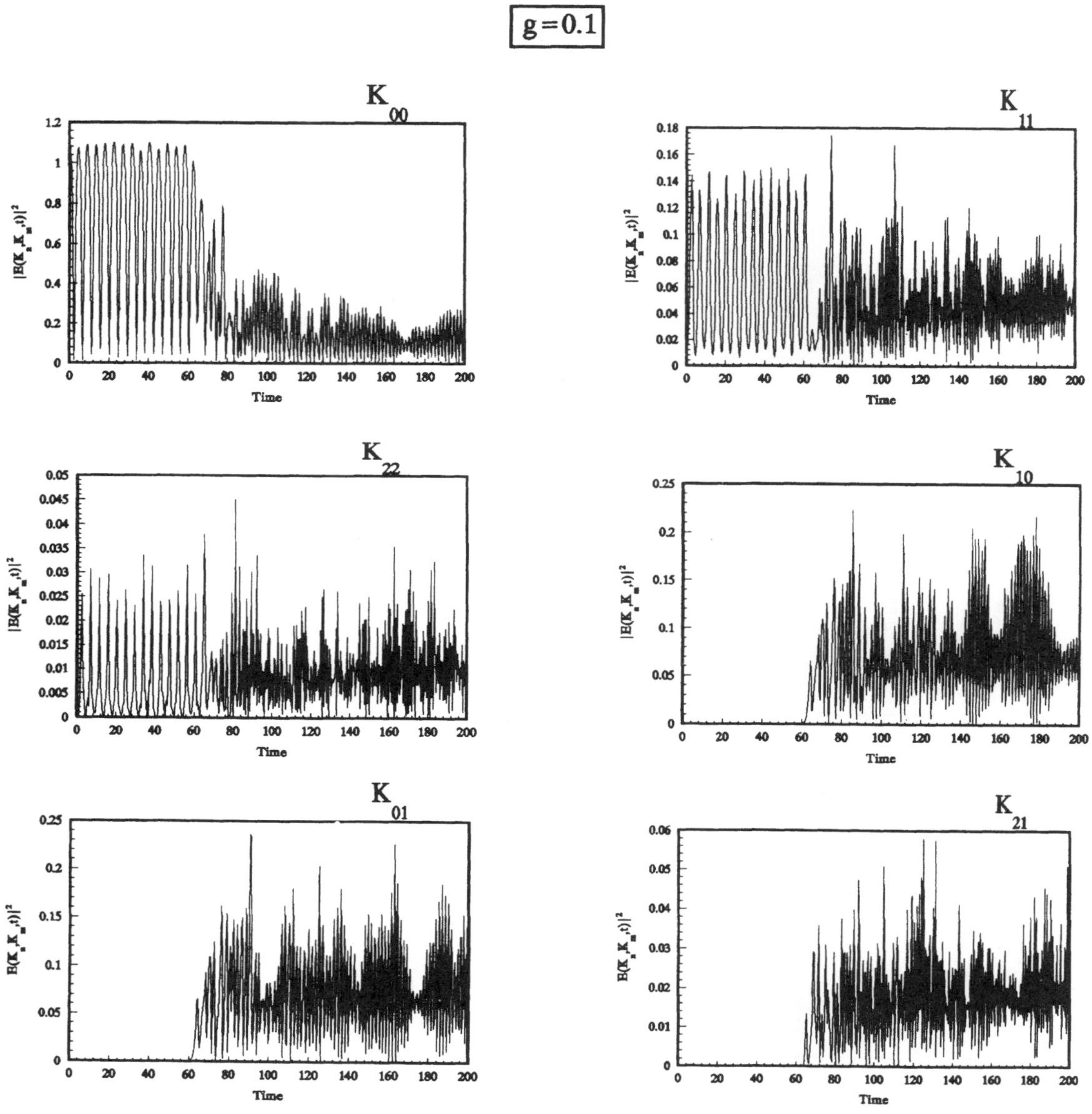

FIG. 18. The evolution of the energy in the Fourier modes with $g = 0.1$.

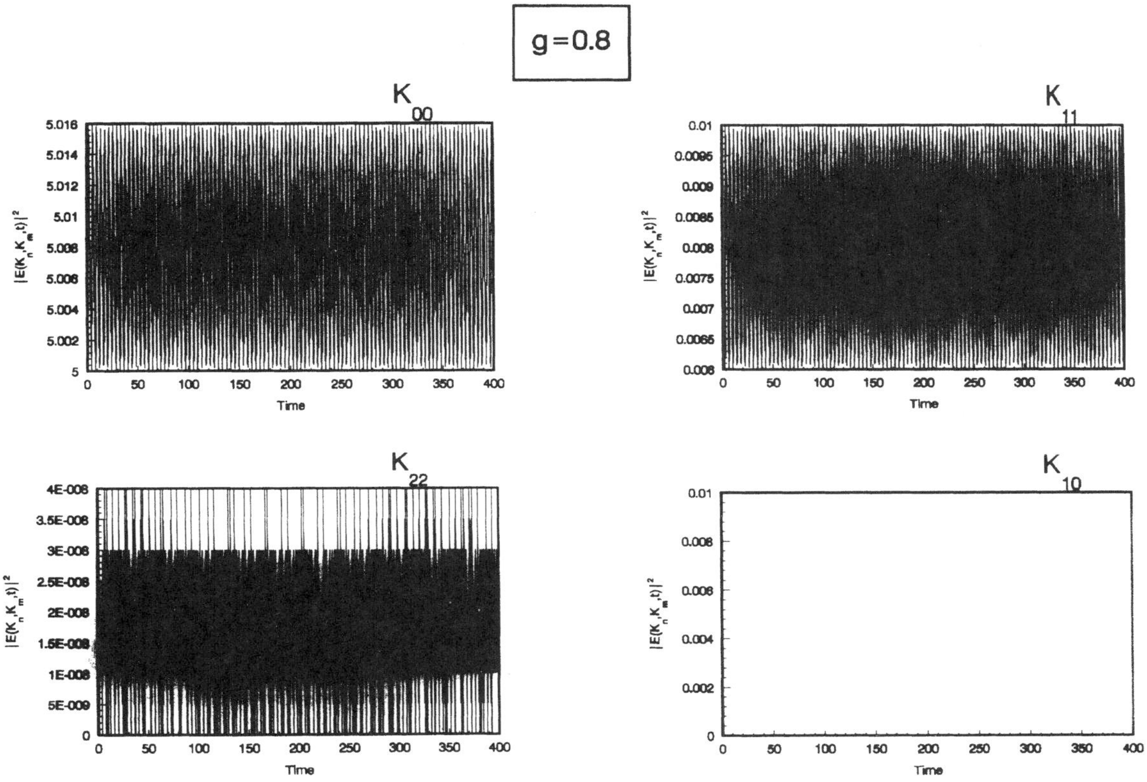

FIG. 19. The evolution of the energy in the Fourier modes with $g = 0.8$.

ACKNOWLEDGMENTS

One of us (C. T. Zhou) wishes to acknowledge useful discussions on filamentation instability in the NSE with Professor T. W. Johnston, and is grateful to Professor S. G. Chen, Professor J. L. Zhang, and Professor L. J. Shen for helping in theoretical analyses and numerical discussions. This work is supported by the CAEP foundation, Grant No. HD9317 (C.T.Z), and in part by the Natural Science Foundation of China (NSFC) and Nonlinear NSFC.

[1] P. L. Kelley, Phys. Rev. Lett. **15**, 1005 (1965).
[2] M. J. Alblowitz and H. Segur, J. Fluid Mech. **93**, 691 (179).
[3] T. Taniuti and H. Washimi, Phys. Rev. Lett. **21**, 209 (1968).
[4] V. E. Zakharov and A. B. Shabat, Zh. Eksp. Teor. Fiz. **61**, 118 (1972) [Sov. Phys. JETP. **34**, 62 (1972)].
[5] V. E. Zakharov, S. V. Manakov, S. Novikov, and L. P. Pitayevsky, *Introduction to the Theory of Solitons* (Plenum, New York, 1983).
[6] J. J. Rasmussen and K. Rypdal, Phys. Scr. **33**, 481 (1986).
[7] V. G. Makhankov, *Soliton Phenomenology* (Reidel, Dordrecht, 1989).
[8] M. V. Goldman, Rev. Mod. Phys. **54**, 709 (1984).
[9] B. M. Lake, H. C. Rungaldier, and W. E. Ferguson, J. Fluid Mech. **83**, 49 (1977).
[10] H. C. Yuen, B. M. Lake, and W. E. Ferguson *The Significance of Nonlinearity in the Natural Science* (Plenum, New York, 1977).
[11] H. C. Yuen and W. E. Ferguson, Jr., Phys. Fluids **21**, 1275 (1978).
[12] M. Stiassnie and U. I. Kroszynski, J. Fluid Mech. **116**, 207 (1982).
[13] X. T. He, Acta Phys. Sin. **30**, 1415 (1981).
[14] R. V. Djolos, V. G. Kartavenko, and V. G. Permyakov, Yad. Fiz. **34**, 1444 (1981) [Sov. J. Nucl. Phys. **34**, 800 (1981)].
[15] K. I. Pushkarov, D. I. Pusharov, and I. V. Tomov, Opt. Quantum Electron. **11**, 471 (1979).

[16] S. Cowan, R. H. Enns, S. S. Rangnekar, and S. S. Sanghera, Can. J. Phys. **64**, 311 (1986).
[17] L. Gagnon and P. Winternitz, J. Phys. A **21**, 1489 (1988).
[18] L. Gagnon and P. Winternitz, J. Phys. A **22**, 469 (1989).
[19] C. T. Zhou, X. T. He, and S. G. Chen, Phys. Rev. A **46**, 2277 (1992).
[20] A. Cloot, B. M. Herbst, and J. A. C. Weideman, J. Comput. Phys. **86**, 127 (1990).
[21] L. Gagnon and P. Winternitz, Phys. Rev. A **39**, 296 (1989).
[22] N. N. Akhemediev, D. R. Heatley, G. I. Stegement, and E. M. Wight, Phys. Rev. Lett. **65**, 1423 (1990).
[23] (a) D. W. Mclaughin, G. Papanicolaou, C. Sulem, and P. L. Sulem, Phys. Rev. A **34**, 1200 (1986); (b) B. J. Lemsurier, G. Papanicolaou, C. Sulem, and P. L. Sulem, Physica D **31**, 78 (1988).
[24] P. K. Shukla, in *Solitons in Plasma Physics in Nonlinear Waves*, edited by L. Debnath (Cambridge University Press, London, 1983), Chap. 11.
[25] C. E. Max, Phys. Fluids **19**, 74 (1976).
[26] J. F. Lam, B. Lippmann, and F. Tappert, Phys. Fluids **20**, 1176 (1977).
[27] P. Kaw, G. Schmidt, and T. Wilcox, Phys. Fluids **16**, 1522 (1973).
[28] C. T. Zhou and X. T. He (unpublished).
[29] D. R. Nicholson, *Introduction to Plasma Theory* (Wiley, New York, 1983).
[30] M. S. Cramer and L. T. Watson, Phys. Fluids **27**, 821 (1990).
[31] J. Guckenheimer and P. Holms, *Nonlinear Oscillations, Dynamical Systems, and Bifurcations of Vector Fields* (Springer-Verlag, New York, 1983).
[32] (a) T. R. Taka and M. A. Ablowitz, J. Comput. Phys. **55**, 203 (1984); (b) M. D. Feit *et al.*, J. Opt. Soc. Am. B **5**, 639 (1988).
[33] T. B. Benjamin and J. E. Feir, J. Fluid Mech. **27**, 417 (1967).
[34] A. J. Lichtenberg and M. A. Lieberman, *Regular and Stochastic Motion* (Springer-Verlag, Berlin, 1984).
[35] C. T. Zhou and X. T. He, Phys. Rev. E **49**, 4417 (1994).
[36] C. T. Zhou, Ph.D thesis, Institute of Applied Physics and Computational Mathematics, Beijing, 1992.
[37] T. W. Johnston (unpublished).
[38] H. T. Moon, Phys. Rev. Lett. **64**, 412 (1990); see also the references therein.
[39] K. Nozaki and N. Bekki, (a) Phys. Rev. Lett. **50**, 1226 (1983); (b) Phys. Lett. **102A**, 383 (1984).
[40] C. E. Max, J. Arons, and A. B. Langdon, Phys. Rev. Lett. **33**, 209 (1974).
[41] R. Z. Segdeev, D. A. Usikov, and G. M. Zaslavsky, *Nonlinear Physics From the Pendulum to Turbulence and Chaos* (Harwood Academic, Chur, Switzerland, 1988).

Modern Physics Letters B, Vol. 8, No. 13 (1994) 833–840

QUASIPERIODIC AND CHAOTIC BEHAVIOR ON ZAKHAROV EQUATIONS

C. Y. ZHENG, X. T. HE, and Y. TAN
Institute of Applied Physics and Computational Mathematics
Box 30, P.O. Box 8009, Beijing, P.R. China

Received 7 April 1994

It is first shown from the modulation instability of the conserved Zakharov equations (ZE) that spatiotemporal chaos is formed in the different regions of the unstable wave numbers. We find that chaos sets in when two homoclinic structures are present and then many and continual homoclinic crossings occur throughout the time series. It is shown that the presence of a homoclinic structure is a potential source of both temporal chaos and spatial patterns.

Solutions of nonlinear evolution equation often exhibit rich patterns in space and time which can have both coherent and chaotic components. Significant progress is being made in understanding such "space-time complexity."[1,2] But the study of the complex behavior is mainly limited in the dissipative case because the study of Hamiltonian chaos is difficult. In this letter, we investigate the feature of spatial temporal chaos by the use of the conserved Zakharov equations (ZE) in the dimensionless form of one-dimensional geometry, which are written as follows:[3]

$$\begin{aligned} i\partial_t E + \partial_x^2 E &= nE\,, \\ \partial_t^2 n - \partial_x^2 n &= \partial_x^2 |E|^2\,, \end{aligned} \tag{1}$$

where $E(x,\,t)$ is a slowly varying envelope of the high frequency electric field and $n(x,\,t)$ is a low frequency quasineutral density perturbation. In the linearized form, Eqs. (1) lead to the following dispersion relation:

$$(\omega^2 - k^2)(\omega^2 - k^4) = 2E_0^2 k^4\,, \tag{2}$$

where E_0 is an amplitude of the homogeneous Langmuir field, k is a wave number of the disturbance, and the time dependence is proportional to $\exp(-i\omega t)$. If k lies in the range

$$0 < k < \sqrt{2}E_0 \equiv k_c\,, \tag{3}$$

PACS Nos.: 47.10.+g, 05.45.+b, 52.35.Ra.

the system will evolve from a spatially homogeneous state to a spatial localized structure, which is known as modulation instability.[4]

Equations (1) have at least two kinds of fixed points (FP):

$$(E,\, n) = \mathrm{O} : (0,\, 0)$$

and

$$(E,\, n) = \mathrm{Q} : (E_0,\, 0)\,,$$

where E_0 is a real constant. The stability of these FP can be examined by the linearized Eqs. (1) and by solving the eigenvalue equation. Here, we consider a simple case that only one wave number k in the system satisfies Eq. (3), where according to the previous work,[5] O is the center and Q is the saddle. In order to investigate the global behavior, the split time step method[6] was used to compute Eqs. (1) with periodic boundary conditions. The spatial length L is chosen in such a way that $kL = 2\pi$ and the number of grids is 64. The initial values are given as

$$\psi = \psi_0 + (C_1\mathbf{f}_1 + C_2\mathbf{f}_2)E_0\cos(kx)/500.0\,,$$

where

$$\psi_0 = \begin{pmatrix} \mathrm{Re}\,E \\ \mathrm{Im}\,E \\ n \\ \partial_t n \end{pmatrix} = \begin{pmatrix} E_0 \\ 0 \\ 0 \\ 0 \end{pmatrix}, \quad E_0 = 2.0\,,$$

and $\mathbf{f}_i$ are eigenvectors of linearized Eqs. (1), which have the form

$$\mathbf{f}_i = \begin{pmatrix} k_0^2 \\ \lambda_i \\ \dfrac{-(k_0^4 + \lambda_i^2)}{E_0} \\ \dfrac{-\lambda_i(k_0^4 + \lambda_i^2)}{E_0} \end{pmatrix}, \quad i = 1,\, 2,\, 3,\, 4\,, \tag{4}$$

with λ_i being an eigenvalue. $(f_1,\, f_2)$ are the eigenvectors whose eigenvalues have positive and negative real parts, so our initial condition ensures that trajectories in phase space will lie locally in a saddle subspace. C_1, C_2 are adjustable constants where $C_1 = C_2 = 1$ that corresponds to the most stable orbit far from HMO has been chosen in our case. The conserved quantities are preserved to the order of 10^{-6} under computing.

The Fermi–Pasta–Ulam (FPU) recurrence is a very interesting phenomenon that has been shown to exist in integrable systems, such as the nonlinear Schrödinger equation, the KdV equation, etc.[7] In this paper, we study the FPU problem and the

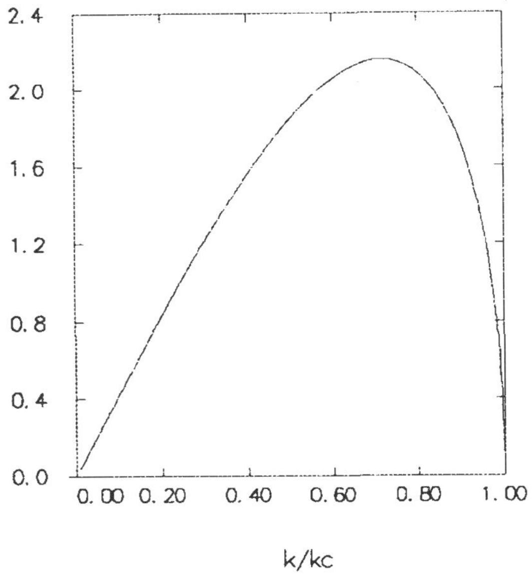

Fig. 1. The dependence of the growth rate on k.

relationship between recurrence and stochasticity in this near integrable ZE system numerically.

Numerical results were obtained for different spatial modulation frequencies in which the growth rate is positive (Fig. 1). When the variables $E(x, t)$ and $n(x, t)$ evolve from almost flat initial conditions, the most modulationally unstable mode evolves into a set of steepened localized structures.

To display the results, we choose to construct the phase space diagrams $(E, \partial_t E)$ at $x = 0$. Figure 2 shows the result obtained for $k = 0.98k_c$. Due to the initial conditions (i.e. $C_1 = C_2 = 1$), the trapped motion within one lobe of symmetric HMO is formed. The corresponding KAM torus is the most stable of all the trajectories. This behavior can also be detected from the power spectrum, which is shown in Fig. 2(b). We observed one frequency $\omega_1 = 0.4148$ and, of course, its harmonics. Clearly, this behavior exhibits an exact recurrence.

When decreasing the parameter k, the following route to temporal chaos was observed. First, as shown in Fig. 3, for $k = 0.95k_c$, ω_1 increased to 0.5732 and an incommensurate frequency component $\omega_2 = 0.1157$, and the linear combination of $\omega_{1,2}$ appeared. This should correspond to quasiperiodic recurrent motion.

Figure 4 shows the numerical result for $k = 0.93k_c$. At this time, in addition to $\omega_1 = 0.6471$ and the incommensurate frequency $\omega_2 = 0.2324$, the subharmonics of ω_1 appears, which shows the doubling of the original fixed point. In fact, when $k = 0.9295k_c$, the quasiperiodic behavior disappears, and only subharmonics exists with the further enhanced spikes, as seen in Fig. 4(c). So at a narrow interval of k, we observe windows of subharmonic locking.

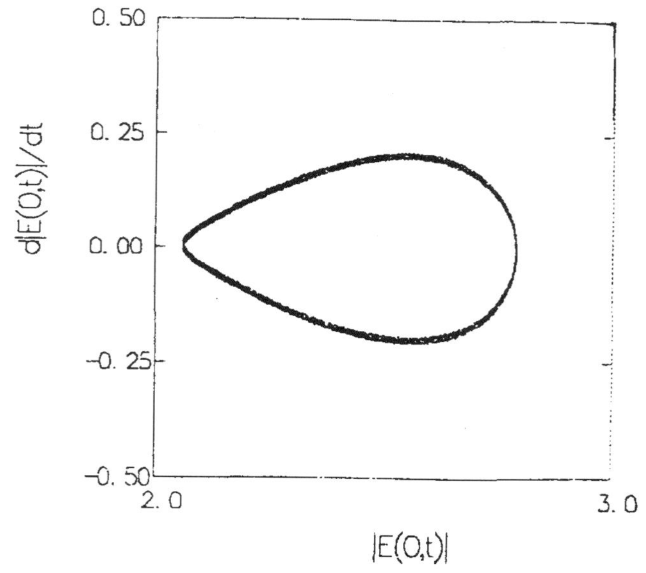

(a)

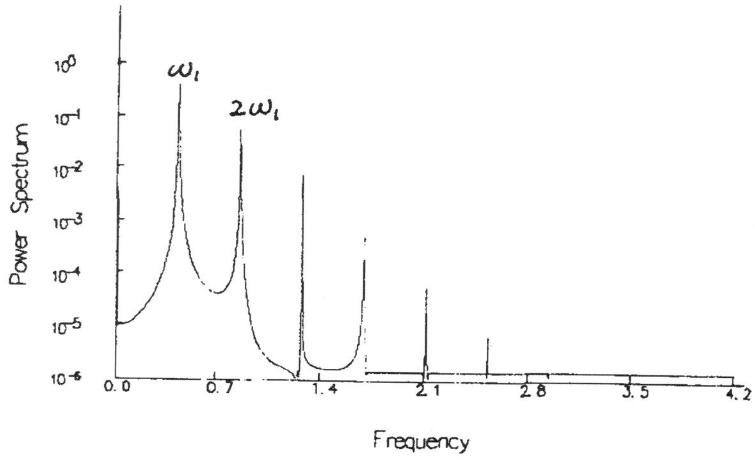

(b)

Fig. 2. (a) Manifolds of phase space from the saddle point $(E_0, 0)$. (b) Power spectrum for $k = 0.98k_c$.

A further decrease of k leads to chaos, as shown in Fig. 5 ($k = 0.92k_c$). Then, the homoclinic connection is destroyed and an intersection with transversal homoclinic points appears, so the stretching and folding of the manifolds should be the key mechanism for chaos and the phase space will be completely filled with increasing time. At this time, the power spectrum with a broadband noiselike structure occurs.

The temporal chaos for the continuum Hamiltonian system can also be checked by the Lyapunov exponents, which measure the exponential rate of separation of two solutions with nearby initial conditions. In Fig. 6, we summarize results for the Lyapunov exponent L vs k, and we note that L increases with decreasing k, which

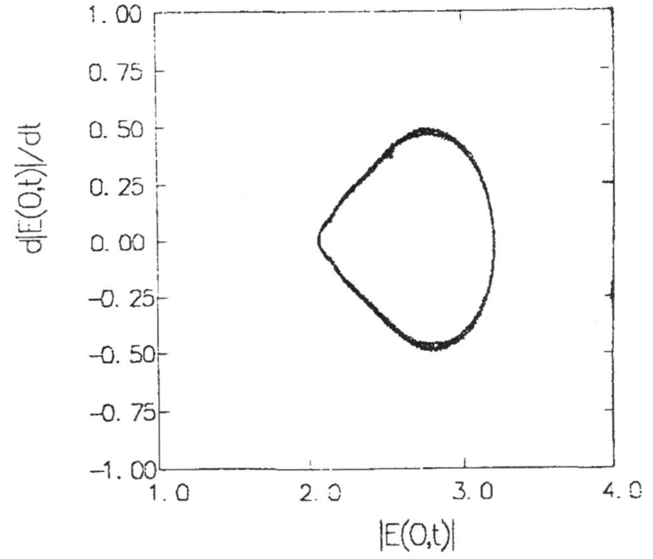

(a)

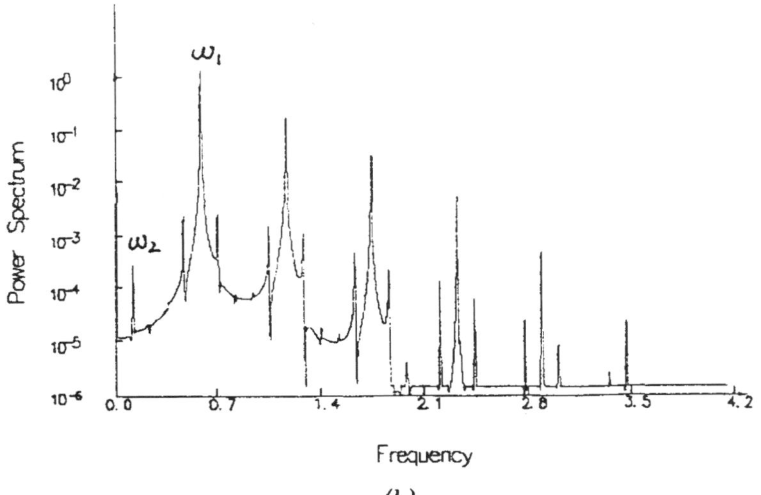

(b)

Fig. 3. The same as Fig. 2, but $k = 0.95k_c$.

means that stochasticity is enhanced in this system. And it is also apparent that for the unstable wave number k, there exists a bifurcation point $k_b \leq 0.93k_c$, which introduces a completely different and distinguished behavior into the system.

Thus, the evolution from quasiperiod to quasiperiod and subharmonics to subharmonics to chaos is found in our system. We would explain the reason in terms of the concept of harmonic oscillators if one regards each frequency as an oscillator.

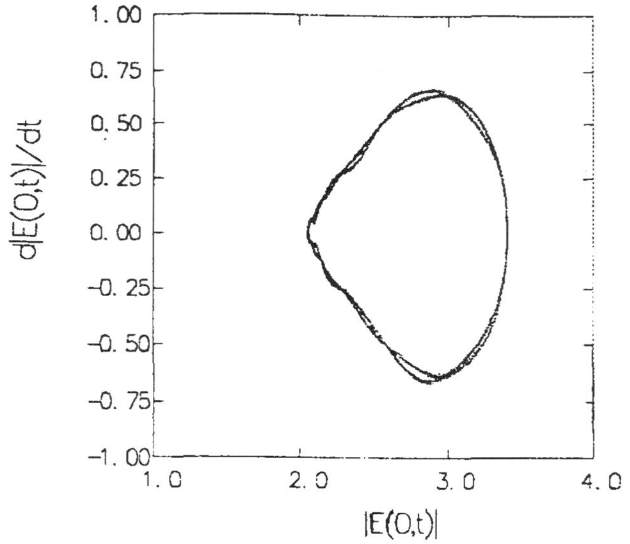

(a)

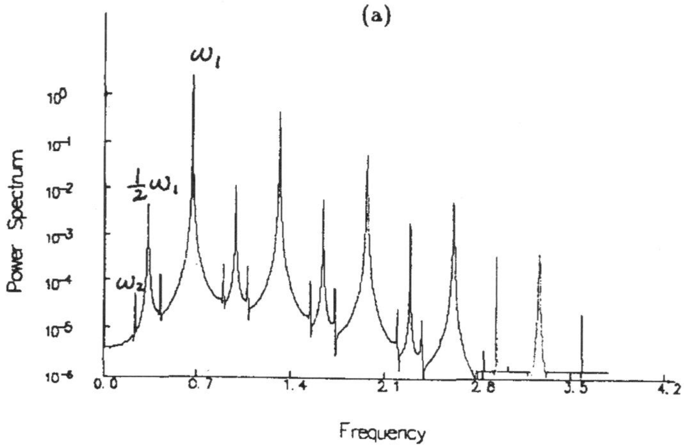

(b)

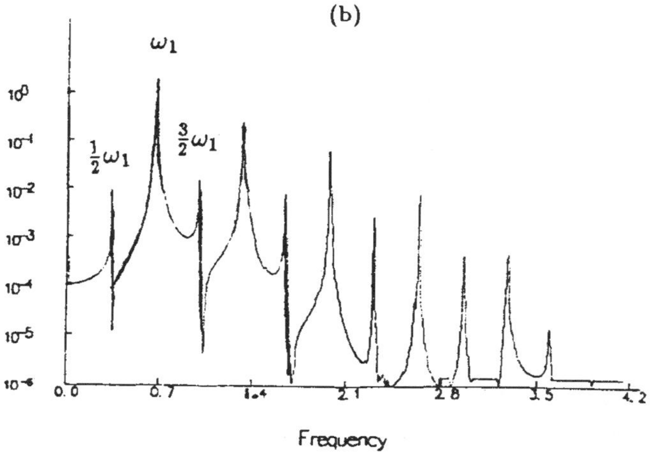

(c)

Fig. 4. (a) Manifolds of phase space. (b) Power spectrum for $k = 0.93k_c$. (c) Power spectrum for $k = 0.9295k_c$.

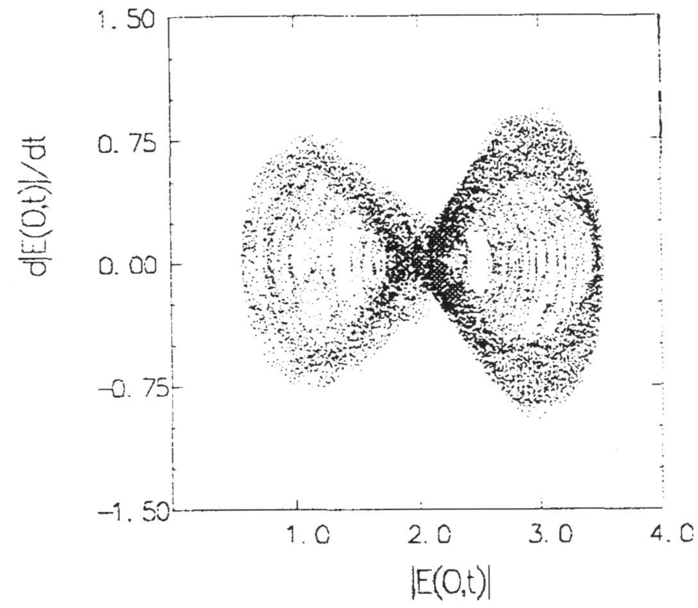

(a)

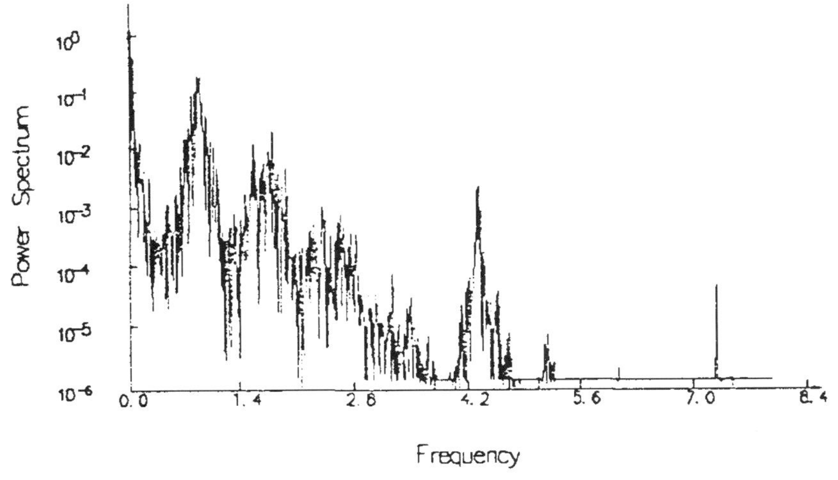

(b)

Fig. 5. The same as Fig. 2, but $k = 0.92k_c$.

When k is 0.98, as seen in Fig. 2(b), the discrete spikes are isolated from each other. However, when k approaches k_b, the growth rate does increase, and ω_2, $\omega_1 \pm \omega_2$, and the subharmonics make frequency intervals get narrower, as shown in Figs. 3(b) and 4(b), and eventually the oscillator overlapping for $k < k_b$ occurs because nonlinear interaction of Langmuir wave with ion acoustic wave is enhanced.

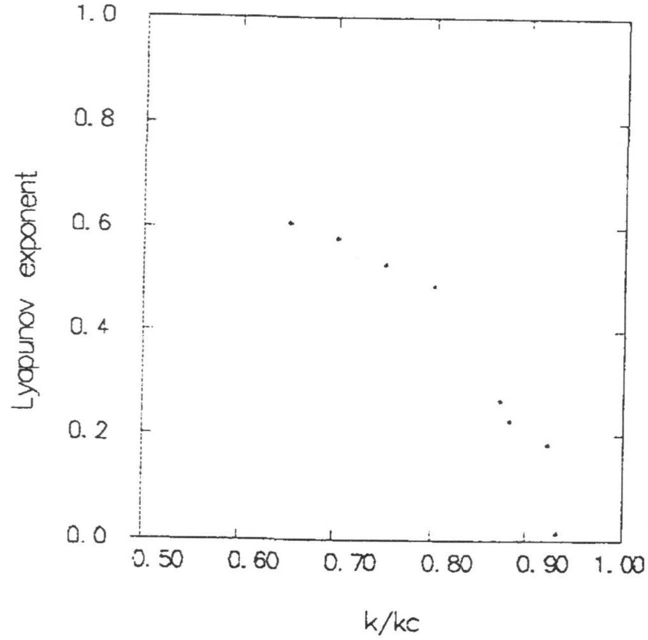

Fig. 6. Lyapunov exponent vs k.

For our investigations on the continuum Hamiltonian dynamic systems, some explicit conclusions can be given. The propagation of fields can be temporal and periodic keeping the spatially coherent structures in a certain parameter regime. Outside of this parameter range, the spatiotemporal complexity appears and the FPU-like recurrence is broken down. The route from the quasiperiod route to the subharmonics to spatial temporal chaos for Eqs. (1) was found. These results should be of great significance in the study of chaos in nonlinear wave propagation.

Acknowledgement

This work was supported by the Chinese Natural Science Foundation No. 19175039.

References

1. A. R. Bishop, D. W. McLaughlin, M. G. Forest, and E. A. Overman II, *Phys. Lett.* **A127**, 335 (1988); H. T. Moon, *Phys. Rev. Lett.* **64**, 412 (1990).
2. A. R. Bishop, D. W. McLaughlin, M. G. Forest, and E. A. Overman II, *J. Phys.* **D23**, 293 (1986).
3. V. E. Zakharov and A. B. Shabat, *Sov. Phys. JETP* **34**, 62 (1972).
4. T. B. Benjamin and J. E. Feir, *J. Fluid Mech.* **27**, 417 (1967).
5. Y. Tan, X. T. He, S. G. Chen, and Y. Yang, *Phys. Rev.* **A45**, 6109 (1992).
6. T. R. Taha and M. J. Ablowitz, *J. Comput. Phys.* **55**, 202 (1984).
7. H. C. Yuen and W. E. Ferguson, *J. Fluid Mech.* **21**, 1275 (1978); T. Taniuti and K. Nishihara, *Nonlinear Waves* (Pitman, Boston, 1983).

PHYSICAL REVIEW E VOLUME 49, NUMBER 5 MAY 1994

Spatial chaos and patterns in laser-produced plasmas

Cangtao Zhou and X. T. He
China Center of Advanced Science and Technology (World Laboratory), P.O. Box 8730, Beijing 100080, China
and Laboratory of Computational Physics and Center for Nonlinear Studies,
*Institute of Applied Physics and Computational Mathematics, P.O. Box 8009, Beijing 100088, China**
(Received 20 July 1993)

Spatial chaos and patterns of laser beams in plasmas are investigated in terms of a saturable nonlinear Schrödinger equation, theoretically and numerically. The linear analysis shows that the homoclinic orbit crossings may exist in phase space, which is also verified in our numerical experiments. In particular, current research illustrates that the spatial chaos and complicated patterns of light wave interacting with plasmas arise from the high order saturable nonlinear effects. The complicated patterns are associated with the stochastic partition of energy contained in Fourier modes. In addition, some physical significance for our work has been discussed.

PACS number(s): 52.35.Mw, 47.10.+g, 05.45.+b, 52.40.Db

I. INTRODUCTION

In laser-plasma interaction, considering the Maxwell equations and the motion equations of an electron, one can obtain the wave equation for the light electric fields [1,2],

$$2ik_0c^2\frac{\partial E}{\partial z}+c^2\nabla_\perp^2E+\omega_{p0}^2\frac{\delta n}{n_0}E=0\ , \qquad (1.1)$$

and the fluid equation of the electron density,

$$\frac{\partial^2}{\partial t^2}l_n n-c_s^2\nabla_\perp^2 l_n n=\nabla_\perp^2 c_s^2+\frac{Ze^2}{4m_e m_i}\nabla_\perp^2|E|^2\ , \qquad (1.2)$$

where $c_s=\sqrt{(ZT_e+T_i)/m_i}$ is the acoustic velocity, c is the light velocity, $n\equiv n_0-\delta n$, n_0 is the unperturbed electron density, k_0 is the laser wave number, ω_{p0} is the plasma frequency, and T_e,T_i and m_e,m_i represent the electron and ion temperature and mass, respectively.

In the static approximation, Eqs. (1.1) and (1.2) are reduced to the nonlinear Schrödinger equation (NSE) [3–6]

$$2ik_0c^2\frac{\partial E}{\partial z}+c^2\nabla_\perp^2E+\omega_{p0}^2(1-e^{-\beta|E|^2})E=0\ , \qquad (1.3)$$

where $\beta=e^2/4m_e\omega_{p0}^2T_e(1+T_i/ZT_e)$. Equation (1.3) obviously describes the case where a light beam propagates in a steady state, but varies in space. In other words, the ponderomotive force completely dominates ion inertia, and must be balanced by pressure forces instead.

For the dynamic model (1.3), the self-focusing, self-trapped, and laser beam filamentations due to the ponderomotive force in plasmas were extensively studied. Max [3] showed that the self-focusing becomes a periodic oscillatory phenomenon, rather than a catastrophic process due to the exponential nonlinearity. Lam, Lippmann, and Tappert [4] illustrated that the self-trapped beams are stable. Kaw, Schmidt, and Wilcox [5] showed that an electromagnetic wave interacting with a plasma is subject to instabilities leading to filamentation. Johnston [6] numerically discussed the fast and slow filamentations in plasmas.

Considering many physical applications, here we rewrite Eq. (1.3) as a more generally one-dimensional dimensionless form,

$$i\frac{\partial E}{\partial z}+\frac{\partial^2E}{\partial x^2}+\frac{1}{2g}(1-e^{-2g|E|^2})E=0\ , \qquad (1.4)$$

where g is the parameter. When $g=0$, Eq. (1.4) becomes the well-known cubic NSE, which is integrable due to the existence of the Lax pair. A class of periodic solutions and solitons can be obtained by the inverse scattering transform.

Although the self-focusing and filamentation phenomena have been extensively discussed, as we know, another important dynamic behavior for Eq. (1.3) or Eq. (1.4) to describe, the laser beam propagation (spatial chaos) has not been studied in detail. Our main purpose here is to investigate the spatial patterns of laser beams interacting with plasmas. In addition, the integrable problem of Eq. (1.4) is discussed. In Sec. II, we simply analyze the periodic solution of Eq. (1.4) and give the linear discussions. The spatial chaos and patterns are described in Secs. III and IV, respectively, and some summaries and discussions are given in the final section.

II. QUALITATIVE ANALYSES

A. Traveling-wave solution

For Eq. (1.4), we assume

$$E(x,z)=G(\xi)e^{i[(v/2)x-\Omega Z]}\ , \qquad (2.1)$$

where $\xi=x-vz$. Inserting Eq. (2.1) into Eq. (1.4) and integrating once, we yield

$$\frac{1}{2}\left[\frac{dG}{d\xi}\right]^2+V(G)=H_0\ , \qquad (2.2)$$

*Mailing address.

where H_0 is the integration constant that represents the pseudoenergy, and the pseudopotential $V(G)$ is

$$V(G)=\frac{1}{8g^2}(e^{-2gG^2}-1)+\left[\frac{1}{4g}-\frac{\alpha^2}{2}\right]G^2 , \qquad (2.3)$$

where $\alpha^2=v^2/4-\Omega$, which is plotted in Fig. 1. Equation (2.2) can be considered as an energy integral of a classical particle with unit mass. We know from the classical mechanics that the particle motion is periodic if a particle lies in the potential well ($H_0<0$), and the solution corresponds to a periodic one. As $H_0=0$, the solution corresponds to a solitary, where the strength and the characteristic width of the laser beams can be given when we obtain such a solution [7]. However, the explicitly analytical solitary has not been obtained as yet, which is due to the complication of the potential function (2.3).

B. Linearized analysis

On the other hand, a simpler homogeneous solution can be written as

$$E_s(z)=E_0e^{iz} , \qquad (2.4)$$

where

$$E_0=0 , \quad \pm\left[\frac{1}{2g}l_n(1-2g)^{-1}\right]^{1/2} \qquad (2.5)$$

for $0\leq g<\frac{1}{2}$. We assume the initial state to be E_0. Now let us analyze the developed behavior for such an initial state. Defining

$$E(x,z)=E_s(z)+\delta E(x,z) \qquad (2.6)$$

and linearizing Eq. (1.4) as done in Ref. [8], we easily obtain

$$\left.\frac{\delta E}{\delta E^*}\right|_{z=0}=\pm i \qquad (2.7)$$

and

$$|\delta E(x,z)|=c_1e^{Az}\cos(K_{\max}x)+c_2e^{-Az}\cos(K_{\max}x) \qquad (2.8)$$

for $E_0=\sqrt{(1/2g)l_n(1-2g)^{-1}}$ with $0\leq g<\frac{1}{2}$,

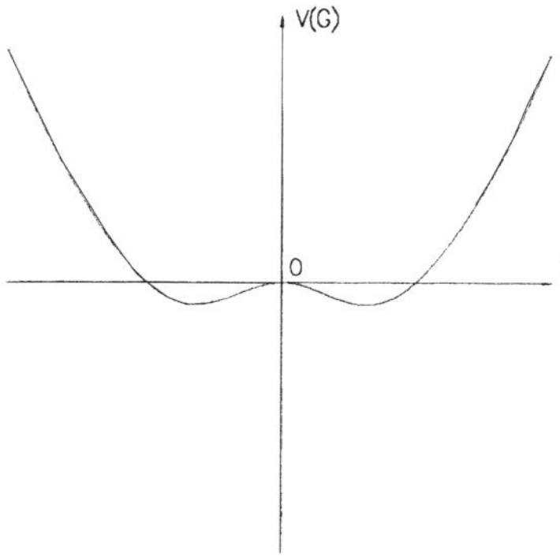

FIG. 1. The pseudopotential function (2.3).

where c_1 and c_2 are the small parameters, $A=[(1/2g)-1]l_n(1-2g)$, and $K_{\max}$ represents the maximum instability wave number. If we construct the phase space as $(|E(x,z)|,d|E(x,z)|/dz)$, then $[\sqrt{(1/2g)l_n(1-2g)^{-1}},0]$ corresponds to a saddle point in phase space. Owing to the integrability of the cubic NSE, on the other hand, the orbits that pass the saddle point (1,0) must be the homoclinic orbits (HMO's) [9].

III. SPATIAL CHAOS AND ITS DESCRIPTIONS

From steady-state analysis, we know the linear developed behavior of the homogeneous solution. A further discussion will be finished by using the numerical experiments.

According to the results obtained in Sec. II B, on the other hand, we take the initial condition as

$$E(x,0)=E_0+\epsilon e^{i\theta}\cos(K_{\max}x) , \qquad (3.1)$$

where ϵ is a real parameter and $E_0=\sqrt{(1/2g)l_n(1-2g)^{-1}}$. In numerical processes, the periodic length of the system is taken as $L=2\pi/K_{\max}$, and the standard splitting-step spectral method [10] has been improved in order to increase the accuracy, which depends on the accuracy of conserved quantities being preserved to 10^{-8}.

In order to analyze the chaotic properties of the saturable NSE (1.4), we first discuss the behavior of the HMO crossings. In Sec. II B, we show that $(E_0,0)$ [where $E_0=\sqrt{(1/2g)l_n(1-2g)^{-1}}$] corresponds to a saddle point in phase space. Comparing Eqs. (2.6) and (2.7) with the initial condition (3.1), we easily determine that the unstable manifolds for possessing the saddle point $(E_0,0)$ correspond to $\theta=45°$ and 225°, and the stable manifolds correspond to $\theta=135°$ and 315°.

As far as a finite dimensional dynamic system is concerned, the stable and unstable orbits for the hyperbolic fixed point would be smoothly joined to each other if the unperturbed system were taken to be integrable. For a Hamiltonian perturbation, the orbits generically intersect transversely, leading to an infinite number of homoclinic points and chaotic motion [11]. According to this idea, we also deal with our continuum Hamiltonian system. As $g=0$, the stable and unstable manifolds that possess the saddle point (1,0) should be smoothly joined to each other due to the integrability of the cubic NSE. Making use of the results obtained in the linear stability analysis, we choose the initial parameters as $\epsilon=0.0001$, $\theta=45°$, and consider two cases of $g=0$ and $g=0.1$, respectively. From Fig. 2(a), we observe that the stable manifold $W^{(s)}$ smoothly joins with the unstable manifold $W^{(u)}$. The orbit that possesses the saddle point (1, 0) corresponds to the homoclinic one. As $g\neq0$, however, we find that the stable and unstable manifolds in phase space do not smoothly join together as shown in Fig. 2(b), which illustrates that the current system is near integrable. It is also noted that there are not an infinite number of homoclinic points in Fig. 2(b). In fact, our phase space $(|E(x,z)|,d|E(x,z)|/dz)$ is only the projection of a high-dimensional space, where the information of phase for wave fields is not considered. Therefore, there may exist some difference between our HMO crossings with those in a finite-dimensional Hamiltonian system in which the infinite homoclinic points appear.

To analyze the spatial chaos, we further discuss the propagative behavior of wave fields. Here, we choose an initial position that lies in the nearby saddle point $(E_0,0)$, that is, $\epsilon=0.1$, $\theta=45.225°$. For $g=0$, the integrability for the cubic NSE can be seen from our previous work [8], where the amplitude of wave fields propagates periodically. The exactly periodic recurrent solution is also exhibited in Fig. 3(a). For $g=0.1$, however, a completely different dynamic behavior is shown in Fig. 3(b). We observe that the localized structures are still kept in the propagative processes of wave fields. The presence of these complicated patterns shows the coherence broken down. To better analyze the propagative characteristic of wave fields, we trace a fixed position in x space. As shown in Fig. 4, the typically chaotic characteristic has been described. As for wave fields, the amplitude, $|E(x,z)|$ experiences the stochastic oscillation at sufficiently far distances [see Fig. 4(b)]. In particular, the irregular HMO crossings as exhibited in Fig. 4(a) are very clear. A continuous nonperiodic spectrum is described in Fig. 4(c). From these figures, we believe that the wave fields are stochastic propagation, which arises from the Hamiltonian perturbation

$$H_1=\int \left\{ \frac{1}{4}|E|^4-\frac{1}{2g}\left[|E|^2+\frac{1}{2g}(e^{-2g|E|^2}-1)\right]\right\}dx \ . \tag{3.2}$$

To illustrate the route to spatial chaos, we fix parameters $\theta=45.225°$, $\epsilon=0.1$ and vary g. From Fig. 5, we find the foundational frequency $\omega_0=0.2513$ for $g=0$. When $g=0.0002$, the wave field still seems to propagate with periodic behavior, but some small peaks in the power spectrum appear and the base frequencies are not uniquely defined (see Table I). When $g=0.0008$, in particular, more peaks are produced (but still countable), although the solution does not recur within the finite distance. In

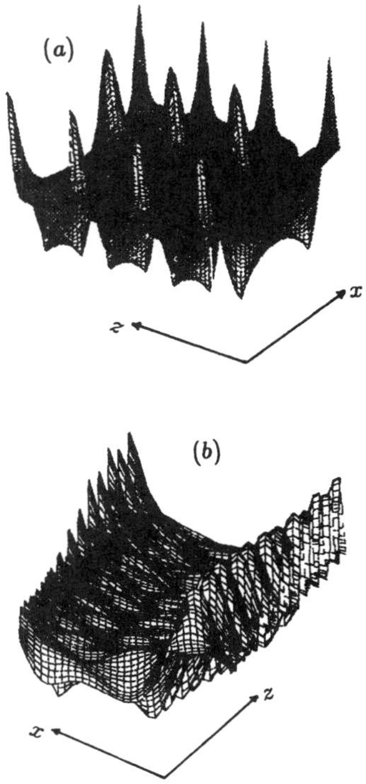

FIG. 3. Envelope amplitude of Eq. (1.4), with $\epsilon=0.1$ and $\theta=45.225°$. (a) $g=0$. (b) $g=0.1$.

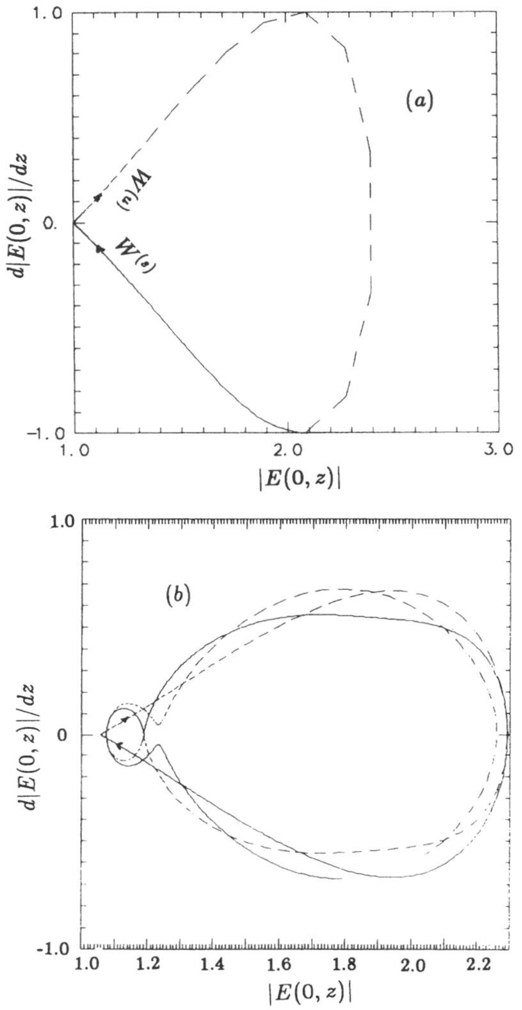

FIG. 2. Stable ($W^{(s)}$) and unstable ($W^{(u)}$) manifolds for the hyperbolic fixed point $[\sqrt{(1/2g)l_n(1-2g)^{-1}},0]$, where the point line is computed with $z>0$ and the solid line is computed with $z<0$. (a) Integrable cubic NSE; the orbits smoothly join. (b) Nonintegrable saturable NSE (1.4) with $g=0.1$.

a sense, these solutions are called quasiperiodic solutions [12]. To demonstrate this conclusion, we should discuss the definition of a quasiperiodic solution. As a matter of fact, a quasiperiodic solution is one that can be expressed as a countable sum of periodic functions with the following two properties: (i) it is linear independent; and (ii) it forms a finite integral base for the frequency. From Fig. 5 and Table I, we see that the countable frequencies do exist. In special cases, we find that the phase-space trajectory, the wave form of $|E(x,z)|$, and the spectrum of the wave form are completely similar to those cases reported in Ref. [12], in which the quasiperiodic solutions of the Ver Pol equation are described. Of course, owing to the impossibility of determining whether a measured value is rational or irrational numerically, a spectrum that appears to be quasiperiodic may actually be periodic with an extremely long periodic solution (see Fig. 5 for $g=0.0002$). On the other hand, with the increase of parameter g, the continuous power spectrum reveals the chaotic behavior (see Fig. 5, in the case of $g=0.01$, and Fig. 4). In addition, we also measure the maximum Lyapunov exponent. It is seen from Fig. 6 that the maximum Lyapunov exponent is positive for the large parts of parametric g. Therefore, the spatial chaos is the main characteristic of the current Hamiltonian system.

Here, we simply illustrate the forming mechanism for the appearance of our spatial chaos. From the standpoint of nonlinear dynamics, the base frequency of the cubic NSE is unique (ω_0) for our parameter due to the integrability of the system. However, the nonintegrable perturbation H_1 (3.2) will make the frequency shift, that is, $\omega_i=\omega_0+\Delta\omega_i$. When the parameter g is quite small, there exists finite countable frequencies. With the increase of g, the oscillatory overlapping of the modes could occur [13], which leads to the formation of the stochastic layer and the appearance of the continuous power spectrum.

IV. COMPLICATED PATTERNS AND THEIR CHARACTERISTICS

In Sec. III we have shown that nonintegrable perturbation leads to the chaotic propagation of wave fields. Also, the complicated patterns have been described in Fig. 3(b), in which we observe the solitarylike structures to be quite irregular. In fact, these irregular patterns are associated with the partition of energy in Fourier modes. For the cubic NSE, the energy in Fourier modes is homogeneous decay, which leads to the formation of coherent

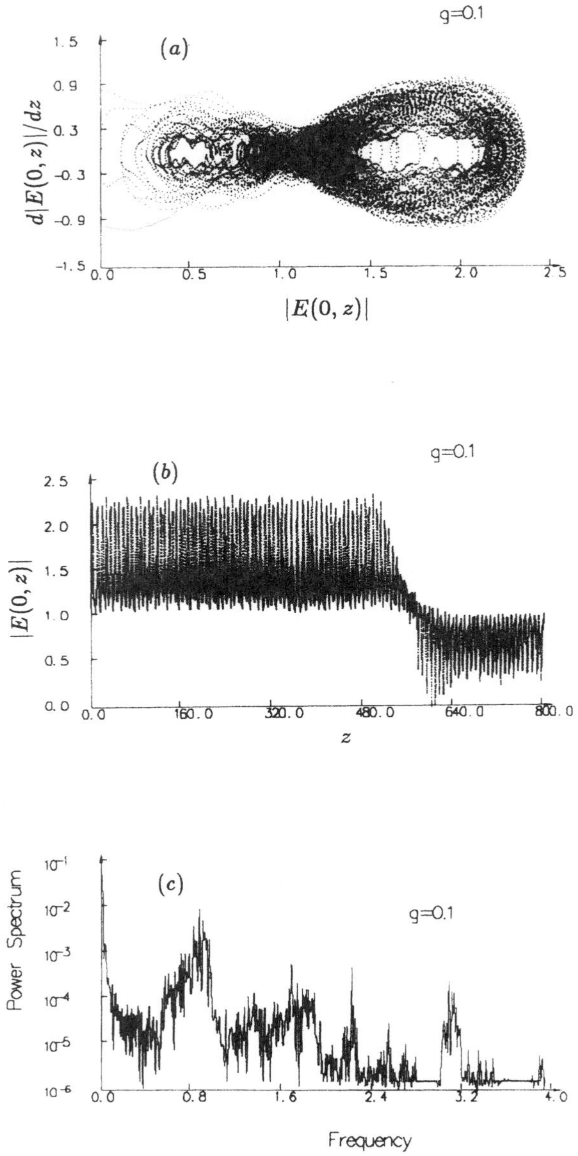

FIG. 4. (a) Phase trajectories illustrate the irregular HMO crossings. (b) Propagation of the field amplitude. (c) Continuous power spectrum shows the chaotic characteristic.

TABLE I. Frequencies corresponding to the high peaks, where p represents the nth high peak. Note that the base frequency for $g=0$ equals 0.251 32, and is about 0.251 33 for $g=0.0002$. However, the "base" frequency for $g=0.0005$ and 0.0008 is not unique, which illustrates that their solutions are quasiperiodic.

g P	1	2	3	4	5	6	7	8	9	10
0.0	0.251 32	0.502 65	0.753 98	1.005 30	1.256 63	1.507 92	1.759 29	2.010 61	2.261 94	2.513 27
0.0002	0.251 33	0.502 65	0.769 69	1.005 31	1.256 64	1.492 256	1.743 584	1.994 921	2.246 238	2.497 566
0.0005	0.235 62	0.486 95	0.722 56	0.989 60	1.240 92	1.476 55	1.712 16	1.979 21	2.214 82	2.497 56
0.0008	0.282 74	0.644 02	1.005 31	1.209 51	1.507 96	1.853 55	2.151 99	2.224 78	2.560 39	2.701 77

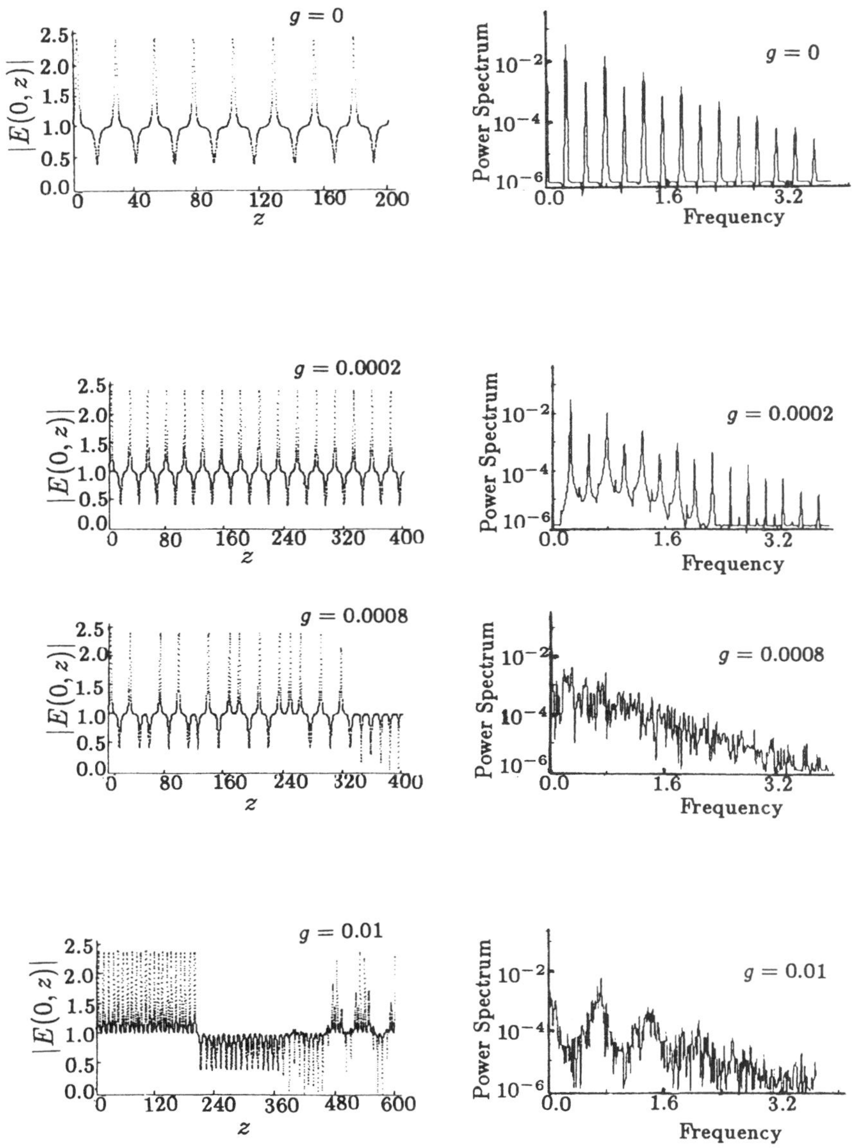

FIG. 5. The periodic solution, quasiperiodic solutions, and chaotic solutions with $\epsilon=0.1$ and $\theta=45.225°$. (a) Propagation of the field amplitude. (b) Power spectrum.

structures. For the saturable NSE (1.4), however, Fig. 7 shows that the partition of energy in Fourier modes is inhomogeneous decay, which results in the localized structures becoming considerably irregular.

To analyze the mechanism that leads to the appearance of the complicated patterns, we further investigate the evolutive process of energy in Fourier modes. For Eq. (1.4), we define the energy of the system as

$$H=\int |E|^2 dx \ . \tag{4.1}$$

In Fourier space, it can be rewritten as

$$H=\sum_n H_{K_n}=\sum_n |E_{K_n}|^2 \ . \tag{4.2}$$

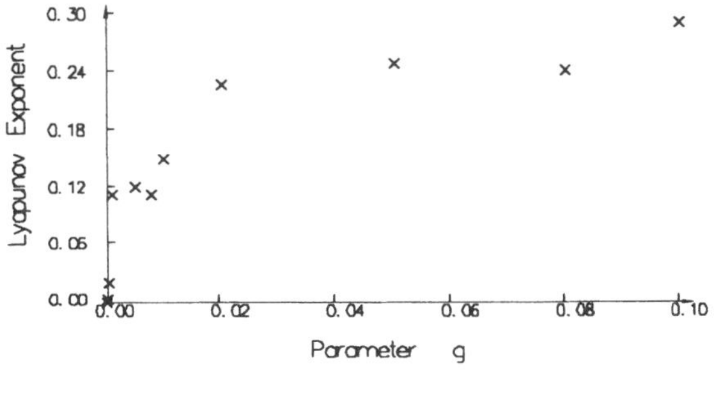

FIG. 6. Lyapunov exponent computed from the series $(|E(0,z)|\sim z)$,

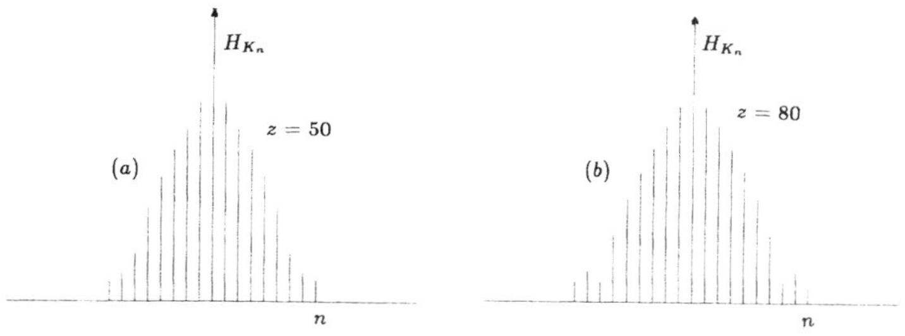

FIG. 7. Energy contained in Fourier modes with the differently propagative distance.

Here K_n stands for the nth Fourier mode. The propagative processes of the amplitude for wave fields in Fourier space, which also correspond to the evolutions of energy in Fourier modes, i.e., $H_{K_n}=|E_{K_n}|^2$, are shown in Fig. 8. Obviously, the large part of the energy in the system lies in the low Fourier modes. For $g=0$, the propagation of energy in all modes is periodic [14], which is consistent with the periodic recurrent solution [Fig. 3(a)]. The amplitude of the soliton structures is dominated by the total energy of system, while the width and pattern of coherent structures are associated with the energy partition in the high modes. For $g\neq 0$, however, Fig. 8 shows that the energy in the system, which is initially confined to the master mode, would spread to many slaved harmonic modes because of the nonlinear interaction, but would not regroup into the original lowest mode. The irregular

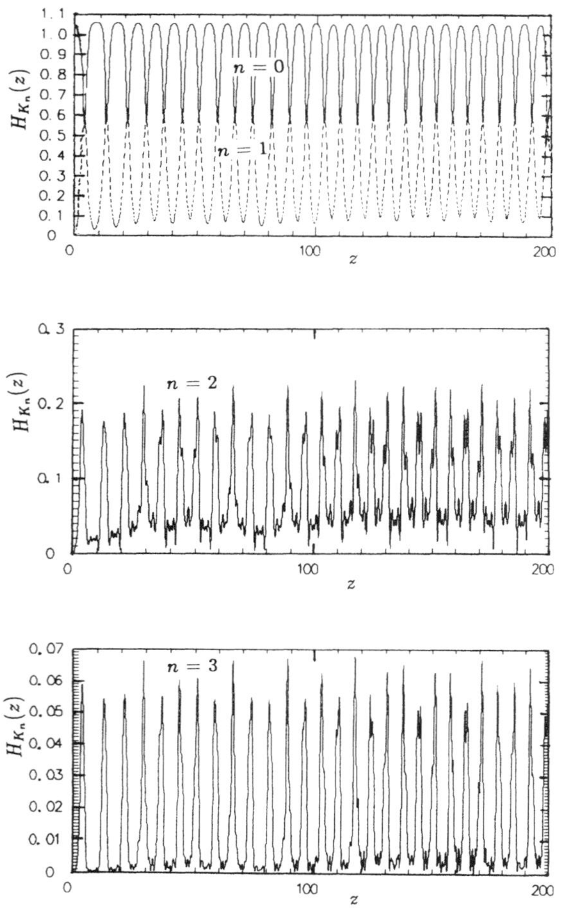

FIG. 8. Propagation of energy in the first four Fourier modes.

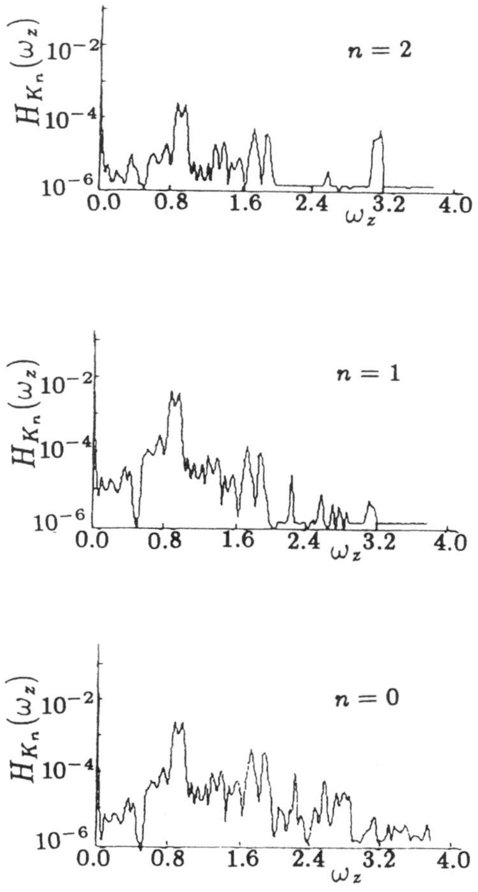

FIG. 9. The Fourier spectrum of energy ($H_{K_n}(\omega_z)\sim\omega_z$).

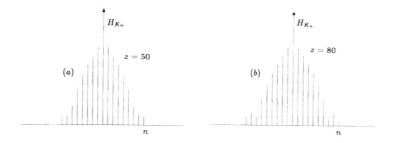

oscillation in the Fourier modes, in which the slaved modes interact with the master one, leads to the formation of the complicated patterns. On the other hand, we note from Fig. 9 that the power spectrum for energy in various modes is not exhibited as a clearly chaotic noise spectrum. In a sense, such a phenomenon illustrates the existence of solitarylike structures, although the well-formed structures are very irregular. In a word, the pattern dynamics is associated with the propagation of energy deposited in all Fourier modes.

V. SUMMARIES AND DISCUSSIONS

From our theoretical analysis and numerical discussions, we can summarize the main conclusions as follows. (i) The periodic solution and solitary wave solution may exist in our continuum Hamiltonian dynamic system, and the higher order saturable nonlinear effects, compared with those for the cubic term could make the amplitude and width of the light wave change. (ii) The analytical and numerical results show that nonintegrable Hamiltonian perturbation H_1 derives from the presence of chaos, where the wave field evolutions are spatially chaotic but the localized structures are still kept. (iii) The formation of the complicated patterns is associated with the stochastic partition of energy contained in Fourier modes. (iv) The route from coherent structures to complicated patterns is a quasiperiodic one.

In this research, our dynamic model, the saturable nonlinear NSE, is based on the static approximation. Speaking in terms of physics, the conditions for such an approximation to be valid are that the macroscopic length scale L must be much larger than the Debye length, macroscopic velocities must be small compared with the sound speed c_s, and macroscopic time scales must be long compared with L/c_s [3]. One should consider the light velocity to be far larger than the acoustic velocity. In fact, the response of density to the light fields is only exhibited by the nonlinear force (ponderomotive force), which corresponds to the exponential nonlinear term in Eq. (1.3). From the current investigation, we understand that such a response would lead to the stochastic propagation of light beams. On the other hand, this model has been assumed, where, for a beam pulse length much larger than the transverse spot size, the electron density is determined by the balance of ponderomotive force and electrostatic force only in the radial direction [15]. Generally speaking, the simple case of describing the light beam propagation is considered to be the axisymmetric cylindrical coordinates with $\nabla_\perp=(1/r)(\partial/\partial r)r(\partial/\partial r)$, where r is the radial variable of laser spot. Here we only deal with the case where the transverse spot consists of one variable x, which is obviously different from the real physical picture. Taking more generally physical and mathematical applications into account, however, we think that the current work is still of significance for our understanding of the propagative behavior of the light wave.

If we consider that the z variable in model (1.4) is equivalent to the evolutive time t, obviously this work exhibits the spatiotemporally chaotic dynamics of fields in the continuum Hamiltonian dynamic system, which still remains open because of the complexity of the system in the infinite degree of freedom. On the other hand, the dynamic model (1.4) can also describe the developed process of Langmuir turbulence when the variable z is replaced by the time variable t [16]. In the approximation of the weak or cubic nonlinearity, the formation of the coherent structures (solitons) can be well explained in terms of the modulational instability. But, the saturable nonlinearity has to be considered in the evolutive latter stage of plasma instability with the increase of field intensity. Our investigation also shows that the high order Hamiltonian perturbation may drive the formation of the pattern dynamics for Langmuir fields.

At last, we should mention the main difference between our work and those works that discuss the spatiotemporally complicated patterns for NSE. Until fairly recently, one studied the NSE with the driven damping, force dissipation, or nonlinear inhomogeneous media [17,18]. A rich variety of complicated patterns, which suggested that a low-dimensional chaotic attractor existed in an infinite-dimensional system, were observed. For the continuum Hamiltonian system, however, the problem becomes extremely difficult for the generally initial conditions since the system cannot be reduced to a finite-dimensional one, such as that found in the case of the dissipative perturbation, though the existence of the solitary waves could slow down the dimensions of the system. In addition, Akhmediev *et al.* only discussed the pseudorecurrence in the two-dimensional modulational instability [19], and our previous work [8] on the chaotic dynamics for the cubic-quintic NSE only described the chaotic trajectories in phase space. Here, we refer to a better description of the complicated patterns in a continuum Hamiltonian system.

One of us (C.T.Z.) wishes to acknowledge stimulating discussions with Professor S. G. Chen, and is grateful to Professor J. L. Zhang and Mr. Y. Tan for helping in numerical discussions. This work is supported by the National Natural Science Foundation of China, and in part by the CAEP foundation, Grant No. HD9317 (C.T.Z.).

[1] A. J. Schmitt, Phys. Fluids, **31**, 3079 (1988).
[2] R. L. Berger, B. F. Lasinski, T. B. Kaiser, E. A. Williams, A. B. Langdom, and B. I. Cohen, ICF Q. Rep. **2**, 77 (1992).
[3] C. E. Max, Phys. Fluids **19**, 74 (1976).
[4] J. F. Lam, B. Lippmann, and F. Tappert, Phys. Fluids **20**, 1176 (1977).
[5] P. Kaw, G. Schmidt, and T. Wilcox, Phys. Fluids **16**, 1522 (1973).
[6] T. W. Johnston (unpublished).
[7] G. Schmidt and W. Horton, Comments Plasma Phys. Controlled Fusion **9**, 85 (1985).
[8] C. T. Zhou, X. T. He, and S. G. Chen, Phys. Rev. A **46**, 2277 (1992).
[9] C. T. Zhou and X. T. He, Chin. Phys. Lett. **9**, 569 (1992).

[10] T. R. Taha and M. J. Ablowtz, J. Comp. Phys. **55**, 202 (1984).
[11] A. J. Lichtenberg, and M. A. Lieberman, *Regular and Stochastic Motion* (Springer-Verlag, Berlin, 1983).
[12] T. S. Perker and L. O. Chua, *Practical Numerical Algorithms for Chaotic Systems* (Springer-Verlag, Berlin, 1989).
[13] R. Z. Sagdeev, D. A. Usikov, and G. M. Zaslavsky, *Nonlinear Physics From the Pendulum to Turbulence and Chaos* (Harwood Academic, Chur, Switzerland, 1988).
[14] H. C. Yuen and W. E. Ferguson, Phys. Fluids **21**, 1275 (1978).
[15] G. Z. Sun, E. Ott, Y. C. Lee, and P. Guzdar, Phys. Fluids **30**, 526 (1987).
[16] P. K. Shukla, *Solitons in Plasma Physics in Nonlinear Waves,* edited by L. Debnath (Cambridge University Press, London, 1983), Chap. 11.
[17] K. H. Spatschek, H. Pietsch, E. W. Leadke, and Th. Erchermann, *Nonlinear and Turbulent Processes in Physics,* edited by V. E. Bar'yakhtar *et al.* (World Scientific, Singapore, 1989), p. 978, and references therein.
[18] T. H. Moon, Phys. Rev. Lett. **64**, 412 (1990).
[19] N. N. Akhmediev, D. R. Heatley, G. I. Stegeman, and E. M. Wright, Phys. Rev. Lett. **65**, 1423 (1990).

Modern Physics Letters B, Vol. 8, No. 24 (1994) 1503–1510

SPATIOTEMPORAL PATTERNS IN LANGMUIR PLASMAS

CANGTAO ZHOU and X. T. HE
Laboratory of Computational Physics and Center for Nonlinear Studies, Institute of Applied Physics and Computational Mathematics, P.O. Box 8009, Beijing 100088, China

Received 24 May 1994

The constitutions of the phase space, stochasticity, and the complicated patterns of Langmuir fields are investigated in terms of a two-dimensional cubic-quintic nonlinear Schrödinger equation. The numerical results obviously illustrate that the quintic nonlinearity leads to the production of the complicated patterns. The mechanism to form these spatial patterns is also analyzed by measuring the spectrum of energy in Fourier space. It is shown that the complicated patterns are associated with the complexity of trajectory in phase space and the stochastic partition of energy in Fourier modes.

In Langmuir turbulence, the interaction between Langmuir waves and particles can be described in terms of the following equations[1]:

$$\frac{\partial \delta F_\alpha}{\partial t} + \mathbf{V} \cdot \nabla_X \delta F_\alpha + \frac{e_\alpha}{m_\alpha} [\mathbf{G}_s \cdot \nabla_V \delta F_\alpha + (\mathbf{G}_h \cdot \nabla_V \delta f_\alpha)_s] = 0 \,, \tag{1}$$

$$\frac{\partial \delta f_\alpha}{\partial t} + \mathbf{V} \cdot \nabla_X \delta f_\alpha + \frac{e_\alpha}{m_\alpha} [\mathbf{G}_s \cdot \nabla_V \delta f_\alpha + \mathbf{G}_h \cdot \nabla_V \delta F_\alpha + (\mathbf{G}_h \cdot \nabla_V \delta f_\alpha)_h] = 0 \,, \tag{2}$$

$$\nabla \cdot E(\mathbf{X}, t) = 4\pi \sum_{\alpha=i,e} e_\alpha \int [\delta f_\alpha(\mathbf{X}, \mathbf{V}, t) + \delta F_\alpha(\mathbf{X}, \mathbf{V}, t)] \, d^3V \,, \tag{3}$$

$$\mathbf{G} \equiv \sum_{\sigma=l,t} e^\sigma E^\sigma(\mathbf{X}, t) \,, \tag{4}$$

where δf_α and δF_α are the fast and slow distribution functions, E_h and E_s are the fast and slow fields, $\alpha = i, e$, and h and s stand for the fast and slow time variables, respectively.

Considering the second order beat frequency interaction between Langmuir wave and ion-acoustic wave, one can obtain the well-known Zakharov equations or the cubic NSE under the static approximation. Generally speaking, the high order beat frequency interactions, such as the fourth order fields, the sixth order fields, etc.,

PACS Nos.: 05.45.+b, 52.35Mw.

have to be considered with the increase of energy of the Langmuir fields. When we only involve the fourth order interaction, Langmuir wave can be described in the following dimensionless equation in the adiabatic limit[1-3]:

$$iE_t + \nabla^2 E + |E|^2 E - g|E|^4 E = 0\,, \tag{5}$$

where E is the slowly varying envelope of Langmuir wave and g is the coupling constant for high frequency fields with electrons.

In the one-dimensional space, there exists an analytical solitary wave solution for Eq. (5),[3] but the beat frequency interaction between the Langmuir wave and ion-acoustic wave would lead to the breakdown of the coherent structures of Langmuir waves and formation of the spatiotemporal chaos.[4]

In addition, for Eq. (5), Lemesurier *et al.*[5] investigated numerically and theoretically the nature of the focusing singularities in two and three dimensions, and described their universal properties. Also, Rasmussen and Rypdal[6] showed that the quintic nonlinearity would prevent blow-up of the solution, but may lead to multifocusing. We know, however, that the pattern dynamics has not been studied in detail as yet. Here, we attempt to discuss the pattern dynamics and the spatiotemporal chaos of Eq. (5) in the two-dimensional space as done for analyzing these in the one-dimensional case.[4]

A key problem that describes the spatiotemporally evolutive phenomena of the Hamiltonian dynamic system (5) is also to choose the available initial condition and constitute the phase space. Here, we define the homogeneous solution of Eq. (5) which is similar to that done for analyzing the one-dimensional space[3] as $E_s(t) = E_0 e^{i\omega_s t}$, where $E_0 = \sqrt{(1-\sqrt{1-4\omega_s g})/2g}$, and let $E(X, Y, t) = E_s(t) + \delta E(X, Y, t)$ which is inserted into Eq. (5). Thus we have

$$\begin{pmatrix} i\partial_t + \nabla^2 + L(|E_0|^2) & h(t) \\ h^*(t) & -i\partial_t + \nabla^2 + L(|E_0|^2) \end{pmatrix} \begin{pmatrix} \delta E \\ \delta E^* \end{pmatrix} = 0\,, \tag{6}$$

where $\nabla^2 = \partial^2/\partial X^2 + \partial^2/\partial Y^2$, $L(|E_0|^2) = 2E_0^2 - 3gE_0^4$, and $h(t) = E_s^2(t)(1-2gE_0^2)$.

On the other hand, we may assume $E(X, Y, t) = E_0 e^{i\omega_s t} + \delta E_+ e^{i(\mathbf{K}\cdot\mathbf{X}+\Omega t)} + \delta E_- e^{-i(\mathbf{K}\cdot\mathbf{X}+\Omega^* t)}$, where $\delta E_+ \ll E_0$, $\delta E_- \ll E_0$. Then, we have the linear dispersion relation $\Omega^2 = -K^2[2E_0^2(1-2gE_0^2) - K^2]$, where $K = \sqrt{\mathbf{K}\cdot\mathbf{K}}$. If the condition $0 < |K| < E_0\sqrt{2(1-2gE_0^2)}$ is satisfied, Ω is an imaginary number and we define $\Omega = i\gamma$ (γ is called the linear growth rate of modulation instability). Thus the relation for the linear growth rate with the instability wave number can be written as $\gamma(K) = K[2E_0^2(1-2gE_0^2) - K^2]^{1/2}$.

Consider the periodic boundary of Eq. (5); the eigenfunction of Eq. (6) can be defined as

$$\begin{pmatrix} \delta E \\ \delta E^* \end{pmatrix} = \begin{pmatrix} \epsilon e^{i\lambda t} \\ \epsilon^* e^{-i\lambda^* t} \end{pmatrix} \cos(K_{X,\max} X)\cos(K_{Y,\max} Y)\,, \tag{7}$$

where ϵ and ϵ^* are the small parameters, $K_{X,\max} = K_{Y,\max} = (\sqrt{2}/2)E_0(1 - 2gE_0^2)^{1/2}$ is the wave number of the maximum modulational instabilities to the uniform solution, and the periodic length is taken as $L_X = 2\pi/K_{X,\max}$ and $L_Y = 2\pi/K_{Y,\max}$. Hence, we can choose the initial condition in the numerical experiments as

$$E(X, Y, 0) = E_0 + \epsilon e^{i\theta} \cos(K_{X,\max}X) \cos(K_{Y,\max}Y) . \tag{8}$$

Simultaneously, we have from Eqs. (6) and (7)

$$|\delta E| = c_1 e^{At} \cos(K_{X,\max}X) \cos(K_{Y,\max}Y) + c_2 e^{-At} \cos(K_{X,\max}X) \cos(K_{Y,\max}Y) , \tag{9}$$

where $A = (-1 + 4g\omega_s + \sqrt{1 - 4g\omega_s})/2g$, and

$$\left.\frac{\partial E}{\partial E^*}\right|_{t=0} = \pm i . \tag{10}$$

Therefore, we can construct the phase space as $(|\delta E|, |\delta E|_t)$ or $(|E|, |E|_t)$. Obviously, the point $(E_0, 0)$ in the phase space $(|E|, |E|_t)$ corresponds to the hyperbolic fixed point. Comparing Eq. (8) with Eq. (10), we see that the unstable manifolds to possess the hyperbolic fixed point correspond to $\theta = 45°$ and $\theta = 225°$, and the stable manifolds correspond to $\theta = 135°$ and $\theta = 315°$. Following this analysis, we[4] completely verify that the nonintegrable perturbation drives the system to form the homoclinic chaos.[7,8]

Now, we carry on with the numerical experiments. Here, we choose the initial parameters as $\omega_s = 0.8$, $\epsilon = 0.1$, and $\theta = 1.1\pi/4$. In the following numerical processes, the standard splitting-time method is improved to increase the accuracy of the conserved quantities.

We know for $g = 0$ that Eq. (5) corresponds to the critical NSE ($q \cdot d = 4$; q is the coefficient in nonlinear term $|E|^q E$ and d is the spatial dimension). An expression of the singular solution can analytically be obtained[5,9] and the blow-up time corresponds to a finite value t^*.

To analyze the pattern dynamics of Eq. (5), we also consider the case of $g = 0$. The spatial patterns for the fixed time are plotted in Fig. 1, which obviously correspond to the blow-up solutions. The blow-up time lies in the region of $t^* = 1.0$–2.0, where the singular solution is developed due to the modulation instability to the homogeneous solution. As g is large enough, numerical experiments show that such a blow-up phenomenon would not occur, where the negative quintic nonlinear effects prevent the blow-up of solution.[6]

Taking $g = 0.3$ into account, we analyze the evolutive behavior of fields. From Fig. 2(c), we see that the amplitude of the fields experiences the stochastic oscillation (also can be exhibited from the power spectrum), which illustrates the nonintegrability of the equation. In fact, the two-dimensional cubic NSE and the one-dimensional cubic-quintic NSE (5) do not belong to the classes of the integrable continuum equations[3] (see Fig. 2(a)). As far as the phase space is concerned, the homoclinic chaos which is similar to that reported in Ref. 4 is also observed as shown in Fig. 2(d), where the solution shifts from a homoclinic orbit to adjacent periodic

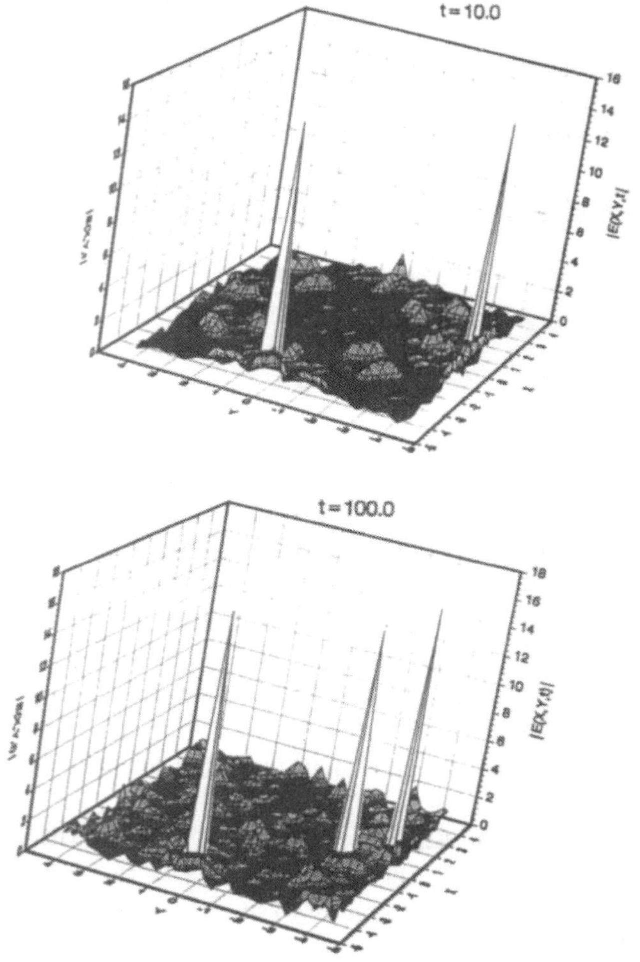

Fig. 1. The blow-up solutions of the two-dimensional cubic NSE with $t = 10$ and $t = 100$.

orbits due to the nonintegrable perturbation.[4,7,8] For the case of $g = 0$, the slope of $|E(X, Y, t)|$ is very large (see Fig. 2(b)) due to the blow-up of the fields, where the trajectory of phase space is not bounded within a small phase space as displayed in Fig. 2(b). On the other hand, the rapidly oscillatory behavior of the amplitude for fields or the irregular trajectory in the phase space illustrates the complexity of the spatial patterns as seen in Figs. 1 and 3.

As for the spatial patterns, Fig. 3 obviously illustrates that the quintic nonlinearity prevents the blow-up behavior. In other words, the collapse of Langmuir waves does not occur due to the influence of the higher order beat frequency interaction between the Langmuir waves and ionacoustic waves. However, the different patterns are formed, which may be because of the competition of mode–mode.[10] In

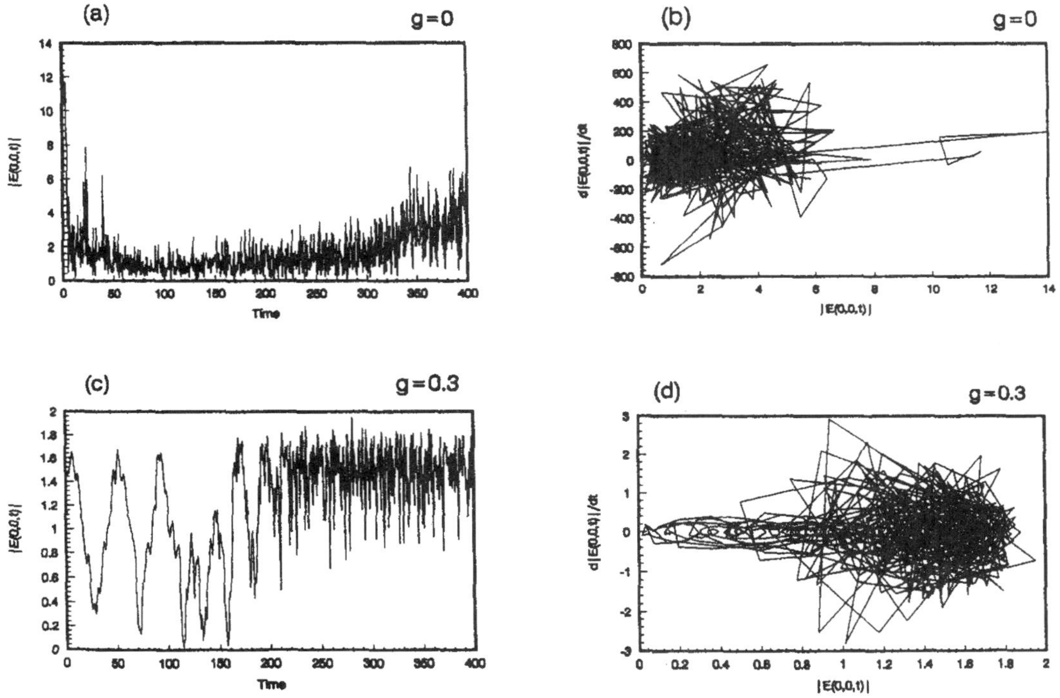

Fig. 2. Plots of the temporal evolution of the amplitude and of the phase space for Eq. (5).

particular, we observe from Fig. 3(a) that the spatial symmetry is kept in the initial evolutive stages of the Langmuir waves. With the development of waves, such a symmetry may be broken down (see Fig. 3(b); a clearer picture has been observed in the plot of contours); of course, this depends on the partitioned processes of energy in the Fourier modes. Speaking in terms of physics, the formation of these different patterns results from the modulation instability to the initial homogeneous state.

To describe the mechanism that leads to the formation of these complicated patterns more clearly, we further measure the evolution of energy in the Fourier modes, and define $H = \sum_{K_{n,m}} |E(K_n, K_m, t)|^2$, where $K_{n,m} = \{K_n, K_m\}$ represents the nth Fourier mode in X space and the mth in Y space. The evolution of energy in Fourier modes, i.e. $H_{n,m} = |E_{K_n,K_m}|^2$, is shown in Fig. 4. From the figure, we observe that the large part of energy lies in the main mode (K_{00}), where the initial energy is injected. For these diagonal modes $n = m$, their evolutive behaviors exhibit the stochastic oscillation, which results in the break-down of recurrence and the appearance of the complicated patterns. In particular, the energy in the nondiagonal modes ($n \neq m$) is associated with the symmetry of patterns. We find that the energy in these nondiagonal modes is nearly zero as $t < 170$, but would be developed as $t > 170$, and can be compared with those corresponding to the

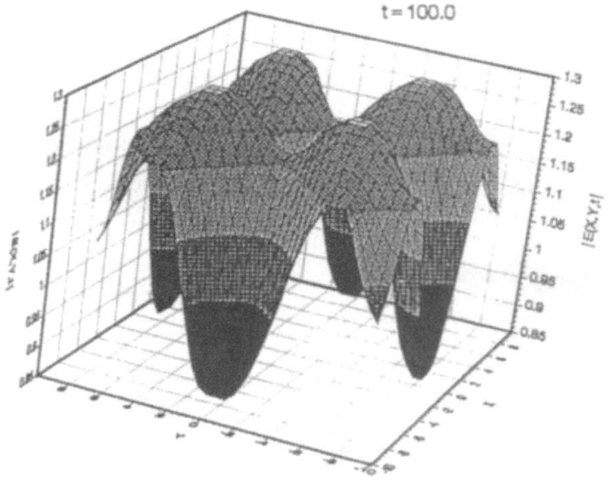

(a)

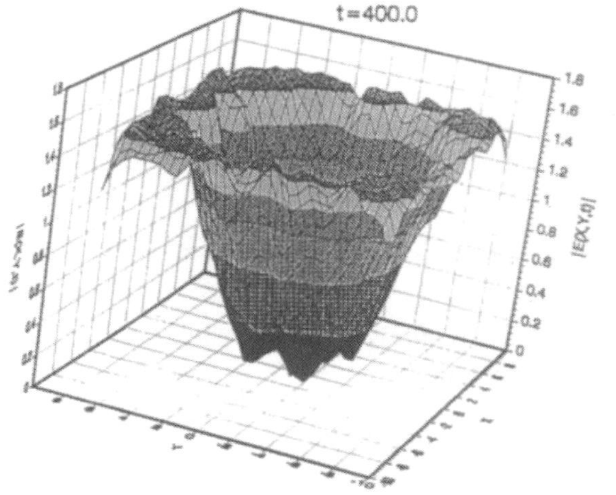

(b)

Fig. 3. The spatial patterns of the Langmuir fields for the fixed time. (a) The symmetric structures for $t = 100$. (b) The nonsymmetric structures for $t = 400$.

diagonal modes; that is, the nonsymmetric phenomenon should be dependent on the stochastic partition of energy in the Fourier modes, especially in the nondiagonal modes.

It is also noted from Fig. 3(b) that the symmetry of the spatial patterns is not completely destroyed, which can also be explained in terms of the partitive process of energy contained in the nondiagonal modes; that is, $|E(K_{nm})|^2 \sim |E(K_{mn})|^2$, $n \neq m$, as $t > 170$ though the fact is that there is some small difference for their values.

In conclusion, a homogeneous state in the Langmuir plasmas would be developed into the spatiotemporal complicated state and not the blow-up state due to the

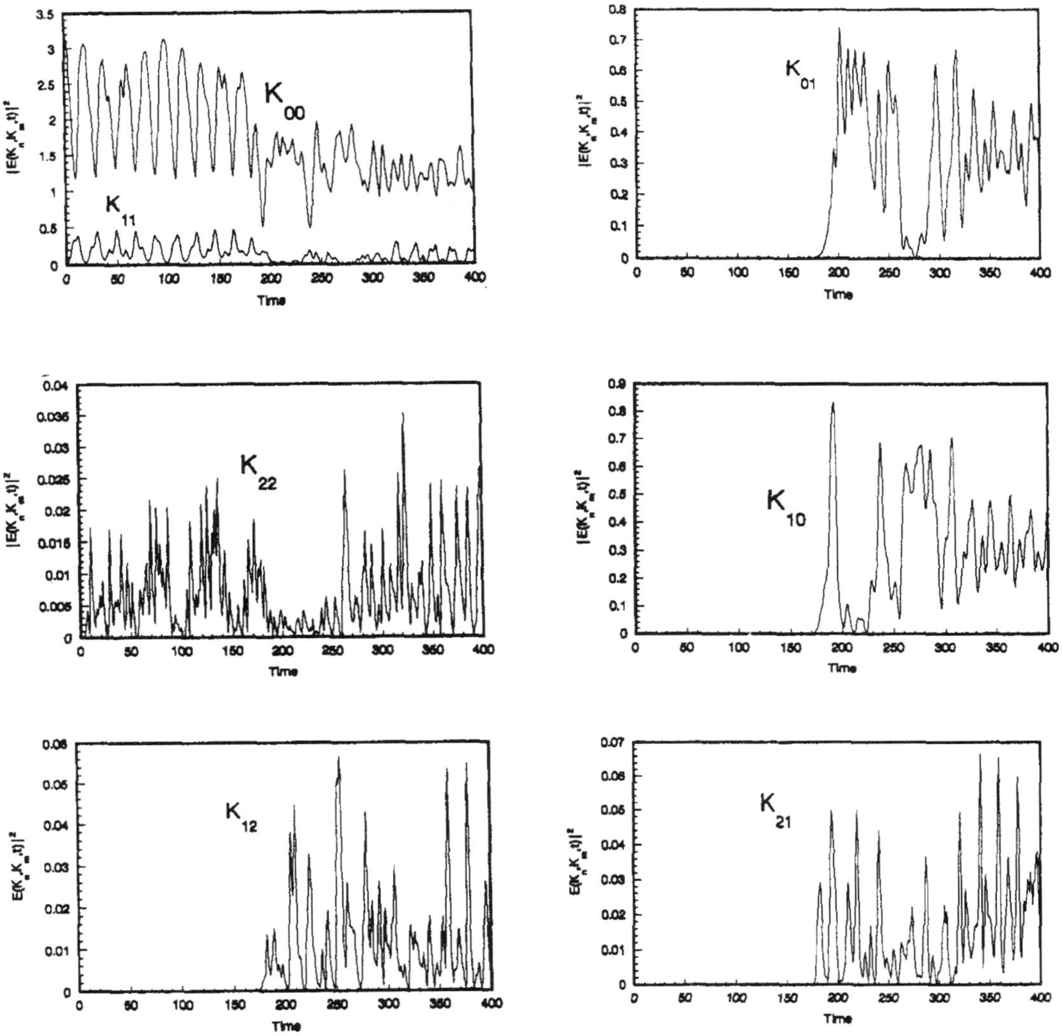

Fig. 4. The evolution of energy in the Fourier modes with $g = 0.3$.

modulation instability if the high order (such as the fourth, sixth order, etc.) beat frequency interactions are considered, where the different spatial patterns would appear stochastically. Simultaneously, the symmetric or nonsymmetric structures depend on the partition of energy in the Fourier modes. Therefore, such a homogeneous state cannot be evolved into a completely developing turbulent one. These results should be of significance in the study of pattern dynamics in plasma waves or other nonlinear waves.

Acknowledgements

One of us (Zhou) would like to thank Profs. S. G. Chen, J. L. Zhang, and L. J. Shen for useful discussions on various problems. This work is supported by the foundation in the China Academy of Engineering Physics under grant No. HD9317 (C. T. Zhou) and partially by the Nature Science Foundation of China (NSFC) and Nonlinear NSFC.

References

1. X. T. He, *Acta Phys. Sinica* **31**, 1317 (1982), in Chinese.
2. K. Nozaki, *Physica* **D23**, 369 (1986).
3. C. T. Zhou, X. T. He, and S. G. Chen, *Phys. Rev.* **A46**, 2277 (1992).
4. X. T. He and C. T. Zhou, *J. Phys.* **A26**, 4123 (1993).
5. B. J. Lemsurier, G. Papanicolaou, C. Sulem, and P. L. Sulem, *Physica* **D31**, 78 (1988).
6. I. Rasmussen and K. Rypdal, *Phys. Scripta* **33**, 481 (1986).
7. T. H. Moon, *Phys. Rev. Lett.* **64**, 412 (1990).
8. N. N. Akhemediev, D. R. Heatley, G. I. Stegement, and E. M. Wight, *Phys. Rev. Lett.* **65**, 1423 (1990).
9. V. E. Zakharov, *Sov. Phys. JETP* **35**, 908 (1972).
10. R. Z. Sagdeev, D. A. Usikov, and G. M. Zaslavsky, *Nonlinear Physics from the Pendulum to Turbulence and Chaos* (Harwood, 1988).

Physica Scripta. Vol. 50, 415–418, 1994

Stochastic Diffusion of Electrons in Evolutive Langmuir Fields

Cangtao Zhou and X. T. He

Laboratory of Computational Physics and Center for Nonlinear Studies, Institute of Applied Physics and Computational Mathematics, P.O. Box 8009, Beijing 100088, P. R. China

Received December 13, 1993; accepted March 30, 1994

Abstract

The stochastically diffusive processes of electrons in the spatiotemporally complicated Langmuir fields, which are directly simulated by the cubic-quintic nonlinear Schrödinger equation (NSE), and solitary envelope fields are investigated, respectively. It is shown that the diffusion of the low velocity particles is a very important mechanism for energy being transformed from the region low velocity to the region of high velocity.

The study of anomalous transport of charge particles has recently received extensive attention in laser-produced plasmas [1]. Up to now, one has mainly concentrated on discussing electrons heated by the soliton-like envelope fields which are the solutions of the cubic NSE or Zakharov equations [2–4]. All the numerical and theoretical results reveal a consistent conclusion, that is, the coherent Langmuir wave packets can drive the formation of particle orbit stochasticity, which arises from resonant overlapping.

In Langmuir turbulence, on the other hand, the spatiotemporally complicated structures are also exhibited by the Zakharov equations [5] or NSE [6]. The previous investigations demonstrated the coexistence of the spatially irregular localized patterns and temporal chaos. An interesting problem is how to diffuse electrons if we inject electrons into these spatiotemporally complicated fields. Our main purpose in this letter is to study the motion of electrons coupling with the cubic-quintic NSE.

For the Vlasov plasmas, considering the second order beat frequency interaction between Langmuir wave and ion-acoustic wave, one can obtain the well-known Zakharov equations or the cubic NSE under a static approximation. Generally speaking, the high order beat frequency interaction, such as the fourth order fields, the sixth order fields, etc., have to be considered with the increase of energy of Langmuir fields. When we only involve in the fourth order interaction, a Langmuir wave can be described in the following dimensionless equation under the adiabatic limit [7]

$$\mathrm{i}\frac{\partial}{\partial t}E + \frac{\partial^2}{\partial X^2}E + |E|^2E - g|E|^4E = 0 \tag{1}$$

where E is the slowly varying envelope of the Langmuir wave, and g is the coupling constant between the high frequency fields and electrons. Here, the dimensionless units used in eq. (1) for time, distance and electric field have been defined as,

$$[t] = \frac{1}{\omega_{\mathrm{pe}}}, \quad [X] = \sqrt{\tfrac{3}{2}}\lambda_{\mathrm{D}}, \quad [E] = \sqrt{32\pi n_0 T_{\mathrm{e}}}, \tag{2}$$

where ω_{pe}, λ_{D}, n_0 and T_{e} correspond to the plasma frequency, electron Debye wavelength, electron density, and electron temperature, respectively. For electrons moving in the Langmuir field ($\Psi(X, t) = \frac{1}{2}[E(X, t)\mathrm{e}^{-\mathrm{i}\omega_{\mathrm{pe}}t} + c.c.]$), the equation of motion can be written as

$$\frac{\mathrm{d}^2X(t)}{\mathrm{d}t^2} = -\frac{4}{\sqrt{3}}\,\mathrm{Re}\,E(X, t)\,\mathrm{e}^{-\mathrm{i}\omega_{\mathrm{pe}}t}] \tag{3}$$

where the dimensionless units [eq. (2)] have been used. In Fourier space, eq. (3) becomes

$$\frac{\mathrm{d}^2X}{\mathrm{d}t^2} = -\frac{4}{\sqrt{3}}\,\mathrm{Re}\left[\sum_{n=-\infty}^{\infty} E_{K_n}(t)\,\mathrm{e}^{\mathrm{i}(K_nX - \omega_{\mathrm{pe}}t)}\right]. \tag{4}$$

Here, $E_{K_n}(t)$ is the solution of eq. (1).

In order to obtain $E_{K_n}(t)$, we discuss the dynamic feature of Langmuir fields which are determined by eq. (1). Defining the homogeneous solution of eq. (1) as

$$E_{\mathrm{s}}(t) = E_0\,\mathrm{e}^{\mathrm{i}\omega_{\mathrm{s}}t} \tag{5}$$

where

$$E_0 = \sqrt{\frac{1 - \sqrt{1 - 4g\omega_{\mathrm{s}}}}{2g}},$$

and letting

$$E(X, t) = E_{\mathrm{s}}(t) + \delta E(X, t) \tag{6}$$

we have

$$\begin{pmatrix} \mathrm{i}\partial_t + \partial_{XX} + L(|E_0|^2) & h(t) \\ h^*(t) & -\mathrm{i}\partial_t + \partial_{XX} + L(|E_0|^2) \end{pmatrix} \times \begin{pmatrix} \delta E \\ \delta E^* \end{pmatrix} = 0 \tag{7}$$

where

$$L(|E_0|^2) = 2E_0^2 - 3gE_0^4, \quad h(t) = E_{\mathrm{s}}^2(t)(1 - 2gE_0^2). \tag{8}$$

Considering the periodic boundary condition, we can choose the eigenfunction as

$$\begin{pmatrix} \delta E \\ \delta E^* \end{pmatrix} = \begin{pmatrix} \varepsilon\,\mathrm{e}^{\mathrm{i}\lambda t} \\ \varepsilon^*\,\mathrm{e}^{-\mathrm{i}\lambda^* t} \end{pmatrix} \cos(K_{\max}X) \tag{9}$$

where ε and ε^* are small parameters, $K_{\max} = E_0(1 - 2gE_0^2)^{1/2}$ is the wave number of maximum modulational instability to the uniform solution [8], and the periodic length is taken as $L = 2\pi/K_{\max}$. Hence, we can choose the initial condition in the numerical experiments as

$$E(X, 0) = E_0 + \varepsilon\,\mathrm{e}^{\mathrm{i}\theta}\cos(K_{\max}X). \tag{10}$$

On the other hand, we have from eqs (5)–(9) that

$$|\delta E| = c_1\cos(K_{\max}X)\mathrm{e}^{At} + c_2\cos(K_{\max}X)\mathrm{e}^{-At} \tag{11}$$

and

$$\left.\frac{\delta E}{\delta E^*}\right|_{t=0} = \pm\mathrm{i} \tag{12}$$

where $A = (-1 + 4g\omega_s\sqrt{1 - 4g\omega_s})/2g$. If we construct the phase space $(|\delta E(X, t)|, |\delta E(X, t)|_t)$, obviously, (0, 0) is the hyperbolic fixed point. Comparing eq. (10) with eq. (12), we also have that the unstable manifolds to possess the hyperbolic fixed point correspond to $\theta = 45°$ and $\theta = 225°$, and the stable manifolds correspond to $\theta = 135°$ and $\theta = 315°$.

In the simulation, we take the parameter values $\theta = 45.225°$, $g = 0.1$ and $\omega_s = 0.8$, and consider the split-time-step method. The detailed numerical results were reported in Refs [8], in which we pointed out that the high order Hamiltonian perturation to the cubic NSE will lead to the breakdown of coherence and the appearance of stochasticity for Langmuir wave fields. The mechanism of stochasticity is associated with the irregular homoclinic orbit crossings. In particular, the formation of the spatiotemporal chaos depends on the stochastic evolution of energy contained in Fourier modes. Figure 1 displays the time behavior of the first six modes. As $g \neq 0$, the energy in the system, which initially confined to the lowest mode, would spread to many higher modes due to the nonlinear instability, but would not regroup into the original lowest modes. It is the stochastic evolution of Fourier modes that leads to the presence of various irregular patterns of the localized structures (Fig. 2).

Now, we consider that electrons are heated by such spatiotemporally complicated fields. The motion of electrons is described by using eq. (4) and the absolute diffusion of electrons is measured as follows

$$D(V_0) = \frac{1}{2\tau_d}\langle (V(t) - V_0)^2 \rangle \qquad (13)$$

where $\tau_d = L/V_0$ is the approximate time for one pass through a wave packet of the periodic length L, and the symbol $\langle \cdots \rangle$ denotes an ensemble average. The ensemble average in numerical discussions is computed by averaging over a set of particles with different start times but with the same initial velocities and positions.

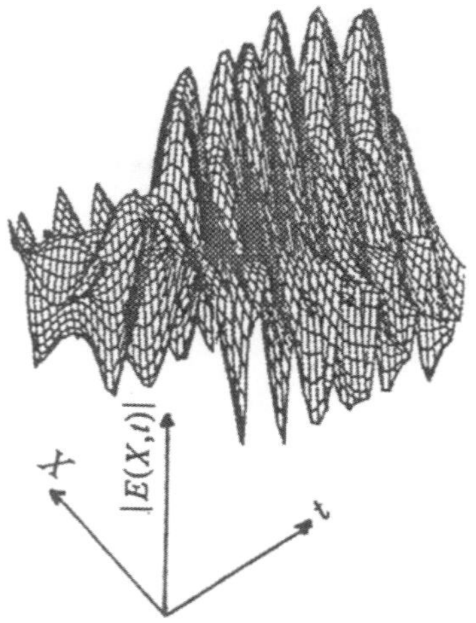

Fig. 2. The temporal evolution of the spatially localized structures for eq. (1).

Figure 3 is our simulated results. We note that the first peak near $1.9V_{th}$ for diffusion coefficient $D(V_0)$ is high enough, which may be compared with the second high peak appeared within $(4–5)V_{th}$. This phenomenon illustrates that the diffusion of the low velocity particles in the current fields may be quite strong, which cannot be explained in terms of the quasilinear theory [9].

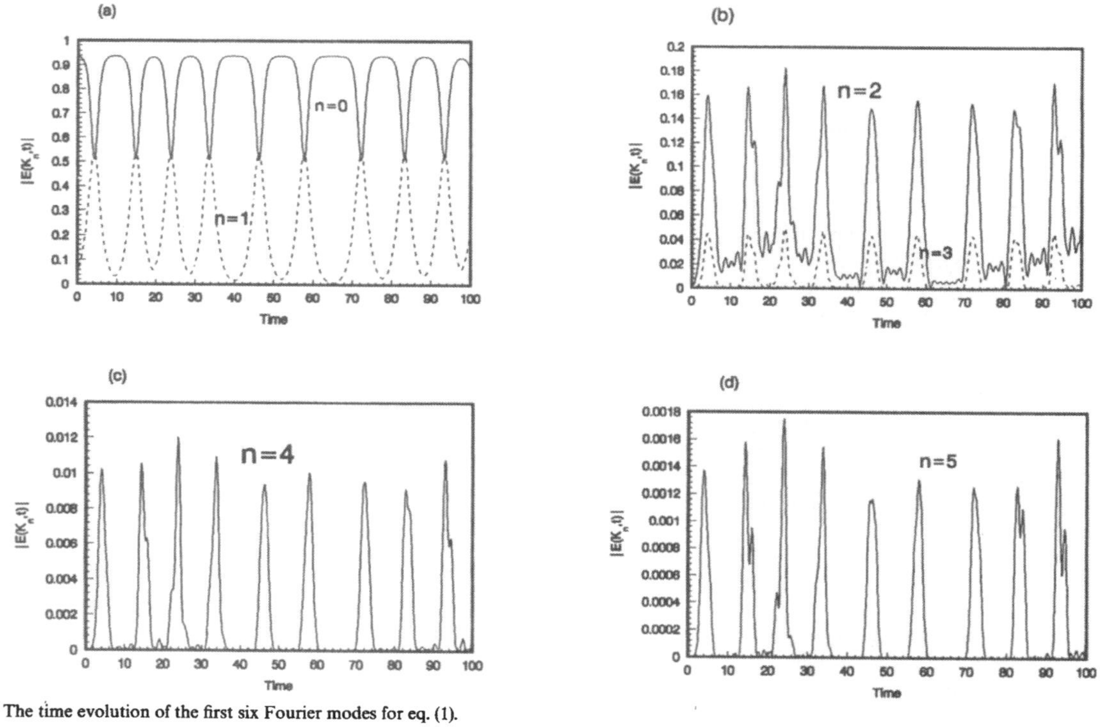

Fig. 1. The time evolution of the first six Fourier modes for eq. (1).

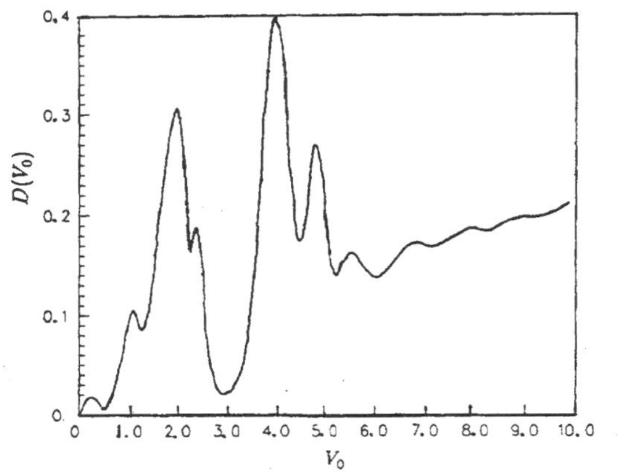

Fig. 3. The absolute diffusion coefficient of particles in the spatiotemporal complicated fields.

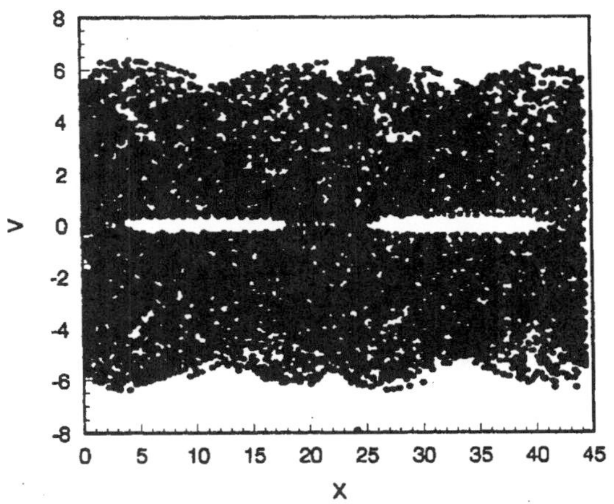

Fig. 4. The Poincaré surface of section plot, based on numerical solutions to the equation of motion (15).

To qualitatively analyze such a diffusive feature, we also discuss the diffusion of particles heated by the coherent localized Langmuir packets. For eq. (1), a solitary envelope is [8]

$$E(X)=\frac{2\alpha}{\sqrt{(1-\frac{16}{3}g\alpha^2)^{1/2}\cosh(2\alpha X)+1}}. \qquad (14)$$

Defining $E_0 = E(0)$, we have $\alpha = E_0^2/\sqrt{2}[(1/E_0^2)-(2g/3)]^{1/2}$. In the further analysis on particle dynamics, we suppose that the electric field in each periodic structure is of the form of a solitary envelope, eq. (14), then the equation of motion can be rewitten as, according to eqs (3) and (4),

$$\frac{d^2X}{dt^2}=-\frac{4}{\sqrt{3}}\sum_{n=-N}^{N}E_n\cos(K_nX-\omega_{pe}t) \qquad (15)$$

where E_n is the amplitude of the n-th Fourier mode, $K_n = 2\pi n/sl$ and sl is the periodic length. For the dynamic eq. (15), Rozmus *et al.* [2–4] have shown that stochasticity will occur if the resonant overlapping criterion

$$\tfrac{1}{2}(\Delta V_n+\Delta V_{n+1}) > \left|\frac{\omega_{pe}}{K_n}-\frac{\omega_{pe}}{K_{n+1}}\right| \qquad (16)$$

is satisfied, where $\Delta V_n = 4(E_n/K_n)^{1/2}$ corresponds to the width of the unperturbed trapping region of the n-th mode, and the upper and lower boundaries for stochastic region in phase space

$$V_h=\max\left(\frac{\omega_{pe}}{K_n}+2\sqrt{\frac{E_n}{K_n}}\right)$$

and

$$V_l=\min\left|\frac{\omega_{pe}}{K_n}-2\sqrt{\frac{E_n}{K_n}}\right|,$$

respectively. In numerical discussions, we choose parameters $sl = 3L$, which preserves $E_n > 0$, and truncate the summation in eq. (15) as $E_{\pm N}/E_0 = O(10^{-3})$. Poincaré mapping in Fig. 4 illustrates the formation of the large region of interconnected stochasticity. In the numerical process, we also find that the stochastic particles cannot transit from the region of positive velocity to the region of negative velocity with the decrease of Langmuir field intensity as discussed by Rozmus [3].

As far as the absolute diffusion coefficient is concerned, we observe from Fig. 5 that $D(V_0)$ near zero velocity is zero, and increases in the stochastic region. In fact, when the particle's velocities are less than V_l, particles experience a ponderomotive potential and are trapped by this potential well. Therefore, these particles cannot diffuse toward the stochastic region. As the particle's velocity is larger than V_l, the first order resonance drives the formation of stochasticity of particle orbit, and the stochastic diffusion occurs. In particular, the second nonlinear force (ponderomotive one) has important influence on the diffusion of the low velocity particles.

Finally, we turn to discuss the difference for the diffusion coefficient as exhibited in both Fig. 3 and Fig. 5. Considering structures of Langmuir fields as refered in eq. (1), we can rewrite eq. (3) as follows

$$\frac{d^2X}{dt^2}=-\frac{e}{m}[E^{(c)}(X,t)+\tilde{E}(X,t)] \qquad (17)$$

where $E^{(c)}$ corresponds to the coherent part of the fields and $\tilde{E}$ corresponds to the incoherent part of the fields – a stochastic perturbation. If $E^{(c)}$ is of the soliton-like structure, it

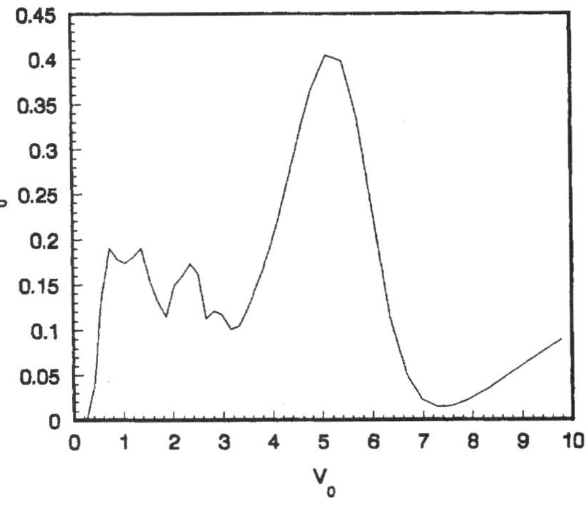

Fig. 5. The absolute diffusion coefficient of particles in the solitary envelope field (14).

can be expanded in terms of Fourier modes. Obviously, the coherent fields may drive the particle trajectory stochasticity as soon as the resonant overlapping is satisfied. For eq. (14), on the other hand, the ponderomotive potential is

$$\phi_{\text{pon}}(X=0)=\frac{4\alpha^2}{(1-\frac{16}{3}g\alpha^2)^{1/2}+1}. \quad (18)$$

If the condition

$$\tfrac{1}{2}mV_m^2<\phi_{\text{pon}}(X=0) \quad (19)$$

is satisfied, then electrons are trapped by the potential well, and the diffusion coefficient is zero as $V \leqslant V_m(V_m \simeq V_l)$. When the stochastic field $\tilde{E}$ is added to the coherent field, however, the ponderomotove potential eq. (18) will be broken down and the high order resonant structures and KAM tori may easily be destroyed, which leads to the diffusion of the low velocity particles increasing as shown in Fig. 3. For the spatiotemporally evolutive fields, therefore, the diffusive behavior of the particles is more complicated due to the influence of ponderomotive force and stochastic one.

In summary, the diffusion of the low velocity particle is an important mechanism for anomalous transport, which may lead to the formation of a non-quasilinear tail for distribution function.

Acknowledgements

One of us (Zhou) would thank Professors S. G. Chen, L. J. Shen and J. L. Zhang for useful discussions on various problems. This work is supported by NSFC NO.19175039 and NNSFC, and in part by a foundation in CAEP, Grant No. HD9317, held by C. T. Zhou.

References

1. Rubenchik, A. M. and Zakharov, V. E., in: "Handbook of Plasma Physics 3" (North-Holland 1989), p. 335.
2. Rozmus, W. and Samson, J. C., Phys. Fluids **31**, 2904 (1988).
3. Rozmus, W. and Goldstein, P. P., Phys. Rev. **A38**, 5745 (1988).
4. Zhou, C. T., Zhu, S. P. and He, X. T., Phys. Rev. **A46**, 3486 (1992).
5. Doolen, A., DuBois, D. F. and Rose, H. A., Phys. Rev. Lett. **51**, 335 (1983).
6. Moon, T. H., Phys. Rev. Lett. **64**, 412 (1990).
7. He, X. T., Acta Physica Sinica **31**, 1317 (1982) (in Chinese).
8. Zhou, C. T., He, X. T. and Chen, S. G., Phys. Rev. **A46**, 2277 (1992).
9. Spatschek, K. H., "Theoretische Plasmaphysik" (B. G. Teubner, Stuttgart 1990).

VOLUME 74, NUMBER 1 PHYSICAL REVIEW LETTERS 2 JANUARY 1995

Spatiotemporal Chaos in The Regime of the Conserved Zakharov Equations

X. T. He[1,2] and C. Y. Zheng[2]

[1]*Chinese Center of Advanced Science and Technology (World Laboratory), P.O. Box 8730, Beijing 100080, People's Republic of China*

[2]*Center for Nonlinear Studies and Laboratory of Computational Physics, Institute of Applied Physics and Computational Mathematics, P.O. Box 8009, Beijing 100088, People's Republic of China**

(Received 26 August 1993)

It is first shown from the modulation instability of the conserved Zakharov equations that regular soliton patterns with a periodic sequence in space and time, and spatiotemporal chaos with irregular localized patterns, are formed in the different regions of the unstable wave numbers. The route from quasiperiodic to the coexistence of quasiperiodic and subharmonic, then from quasiperiodic and subharmonic to subharmonic, and furthermore to spatiotemporal chaos evolving from regular spatial patterns, is found. The formation mechanism of irregularly localized patterns is also revealed.

PACS numbers: 47.10.+g, 52.35.Mw

Spatiotemporal chaos in a continuum Hamiltonian system is an important subject used to investigate the long-term behavior of a volume preserving system with an infinite number of degrees of freedom, which leads to the soliton pattern structure. To our knowledge, so far, little work has been done on this problem in plasmas.

In the present Letter, we investigate the pattern dynamics and the feature of spatiotemporal chaos evolving from spatial patterns first using the conserved one-dimensional Zakharov equations (ZE's) [1],

$$\begin{aligned} i\partial_t E + \partial_x^2 E &= nE, \\ \partial_t^2 n - \partial_x^2 n &= \partial_x^2 |E|^2. \end{aligned} \tag{1}$$

Here $E(x,t)$ is a slowly varying envelope of the electric field, and $n(x,t)$ is an ion-acoustic density.

So far the ZE's are the most extensively studied model used to describe strong turbulence in plasmas. This system has three conserved quantities, i.e., the quasiparticle number, the momentum, and the total energy of the system.

The linearized stability analysis of the perturbation $\exp[i(k'x - \omega t)]$ from a spatially homogeneous field E_0 for Eqs. (1) gives the growth rate of the modulational instability (MI) as [2]

$$\gamma = \frac{k'}{\sqrt{2}}\left\{[(1 - k'^2)^2 + 8E_0^2]^{1/2} - (1 + k'^2)\right\}^{1/2}, \tag{2}$$

for $0 < k' < \sqrt{2}E_0$.

It was realized the localized patterns are formed from MI, and the single soliton for such Hamiltonian systems has analytically been obtained [1].

Equations (1) are a near-integrable system [2] and have at least two kinds of fixed points: a center at (0,0) and a saddle at $(E_0, 0)$. In a simple case wherein only one specified unstable wave number k', which satisfies Eq. (2), is considered, we solve the linearized Eqs. [1] in initial values $(E_0, 0)$ and obtain the eigenvalues

$$\lambda_1 = \left\{\tfrac{1}{2}k'^2[(1 - k'^2)^2 + 8E_0^2]^{1/2} - (k'^2 + 1)\right\}^{1/2} \equiv \lambda_0,$$
$$\lambda_2 = -\lambda_0,$$
$$\lambda_3 = i\left\{\tfrac{1}{2}k'^2[(1 - k'^2)^2 + 8E_0^2]^{1/2} + (k'^2 + 1)\right\}^{1/2} \equiv i\omega_0,$$
$$\lambda_4 = -i\omega_0,$$

and the eigenfunctions $f = \sum_{i=1}^{4} c_i e^{\lambda_i} f_i$, where $f_i = (k'^2, \lambda_i, -g_i, -\lambda_i g_i)$, c_i is coefficient, and $g_i = k'^2 + \lambda_i/E_0$. Obviously the saddle is in (f_1, f_2) subspace.

We now discuss numerical simulations of Eqs. (1) to investigate the global behavior and choose the initial condition to adjoin the saddle, which at $t = 0$ a very small spatial inhomogeneity is added on the spatial homogeneous state, $\Psi_0 = (E_0, 0, 0, 0)$, as follows [2]:

$$\Psi = \Psi_0 + (c_1 f_1 + c_2 f_2) E_0 \cos(k'x)/500, \tag{3}$$

where $\Psi_0 = (\mathrm{Re}E, \mathrm{Im}E, n, \partial_t n)_{t=0} = (E_0, 0, 0, 0)$, and factor $\frac{1}{500}$ is only to show that this is a very small perturbation adjusted for convenient calculations. Our initial condition ensures that the trajectories in phase space will locally lie in a saddle subspace. The case $c_1 = c_2 = 1$, which corresponds to the most stable orbit far from the homoclinic (HMO) [2], has been chosen in the discussion below.

The split-step method for E with a preestimate for density perturbation n is first used to compute Eqs. (1) with the periodic boundary condition in which the spatial grid L is chosen, so that $k'L = 2\pi$, and the number of grids is 64. During the computation the conserved quantities are preserved to the order of 10^{-6}.

It is noticed from the numerical results that by varying with $k = k'/\sqrt{2}E_0$, one finds a bifurcation point at $k_c \leq 0.9295$, which divides the system into completely different dynamic behaviors both temporally and spatially.

For $k_c < k \leq 1$, the long-term behavior of manifolds in phase space is only a single loop starting from and returning to the same saddle $(E_0, 0)$ to form the Kol'mogorov-Arnol'd-Moser torus as seen in Fig. 1(a). The envelope amplitude of the electric field is performing an exactly periodic oscillation [Fig. 1(b)]. Simultaneously, the power spectrum, as seen in Fig. 1(c), is of an isolated discrete distribution with equal frequency interval, which shows a sharp peak for $k = 0.98$ at the fundamental frequency

 0031-9007/95/74(1)/78(4)$06.00

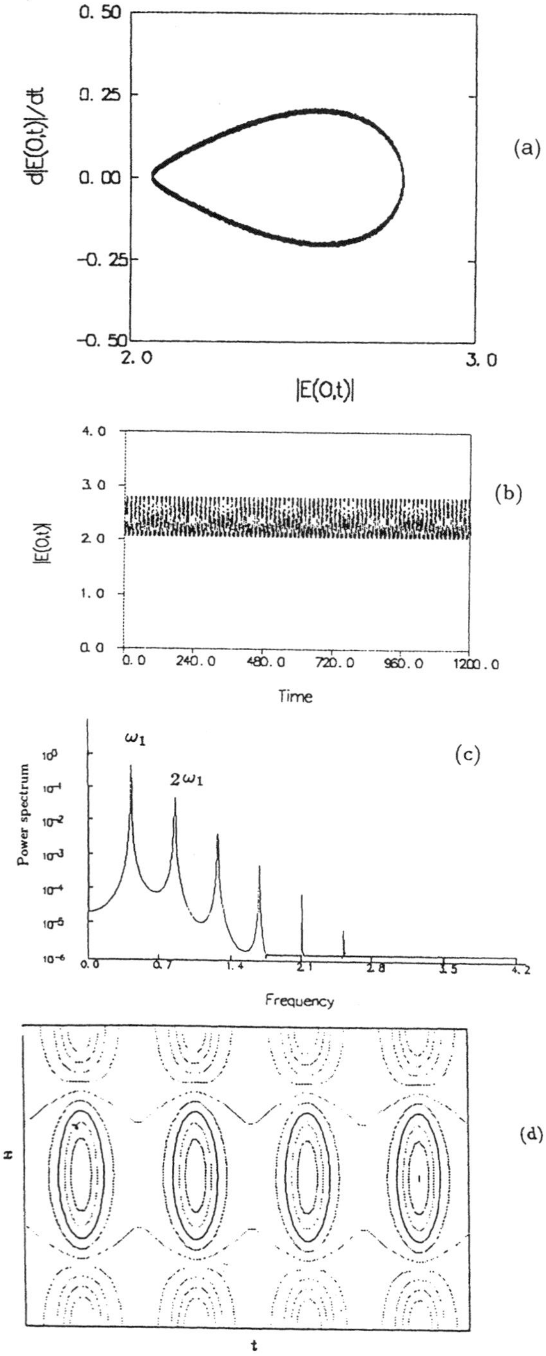

FIG. 1. Manifolds of phase space from the saddle point $(E_0, 0)$ at $x = 0$ for $k = 0.98$, (b) envelope amplitude $|E(x,t)|$ versus time for $k = 0.98$, (c) power spectrum for $k = 0.98$, and (d) contours of $|E(x,t)| = \text{const}$ for $k = 0.98$, $t = 500-550$.

$\omega_1 = 0.4188$ and is excited from the spatially homogeneous state and nth ($n = 2, 3, 4, \ldots$) harmonics of ω_1. However, the intensities of the harmonic spikes are extremely weak compared with the fundamental one. These features show that there exists an exactly periodic recurrence, like the Fermi-Past-Ulam (FPU) [3] process. For the spatial characteristics, we see from Fig. 1(d) that regularly localized patterns with a spatially and temporally periodic sequence are formed in the evolution of this system. These solutions propagate undisturbed with a constant speed. We observed that each one of these solutions is of symmetry and rapidly exponential decay in two sides around peak amplitude and is similar to single soliton solution of Eqs. (1) in the form $E(x,t) \sim \text{sech}\,\alpha\xi$, where α^{-1} is the half-width and $\xi = x - ut$, u is a constant speed [1]. Thus, the ZE's system for $k_c < k < 1$, indeed, is executing a regular motion with coherent patterns. When k is reduced to 0.93, the motion in a two-cycle [Fig. 2(a)], which corresponds to localized patterns with slightly different structures [Fig. 2(b)], emerges, and the pseudorecurrence appears.

When $k < k_c$, the evolution of trajectories in phase space with the long lapse of time exhibits clearly irregular homoclinic orbit crossings filling up substantial portraits of phase space as shown in Fig. 3(a). At the same time,

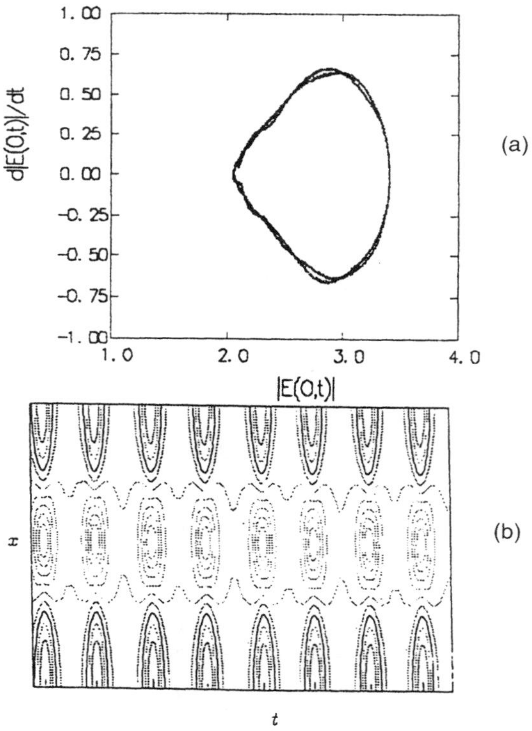

FIG. 2. Manifolds of phase space for $k = 0.93$, and (b) contours of $|E(x,t)| = \text{const}$ for $k = 0.93$, $t = 350-425$.

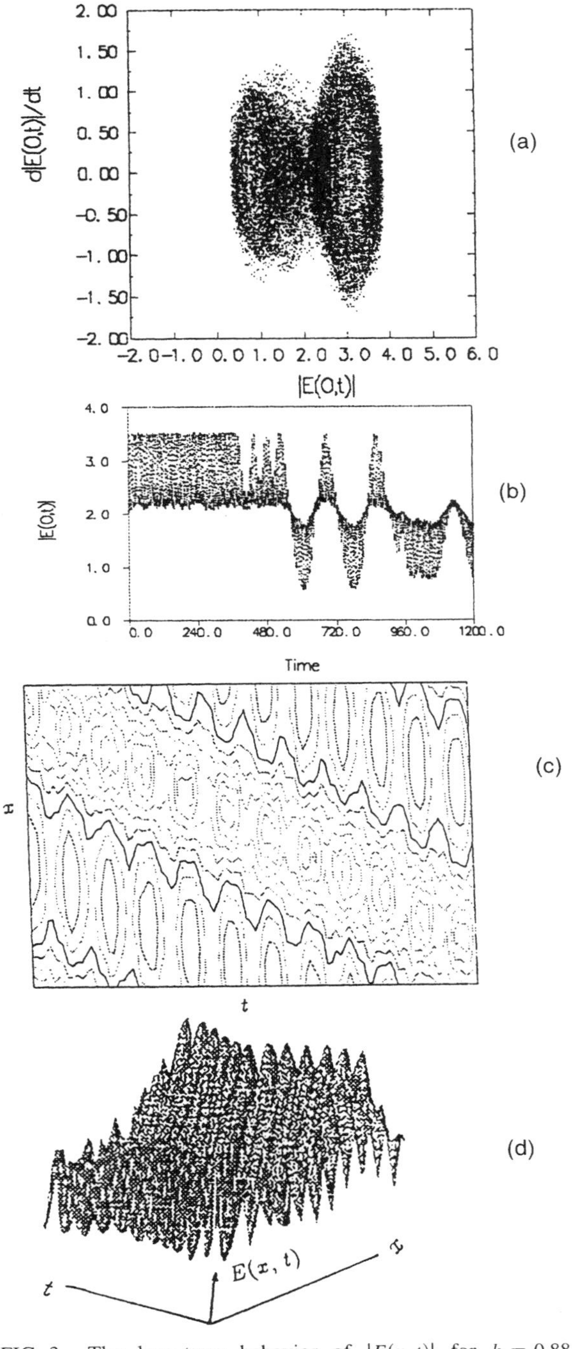

FIG. 3. The long-term behavior of $|E(x,t)|$ for $k = 0.88$. (a) Manifolds of phase space at $x = 0$, (b) envelope amplitude $|E(x = 0, t)|$ versus time, (c) contours of $|E(x,t)| = \text{const}$, $t = 500-550$, and (d) spatially localized structure, $t = 500-550$.

amplitude of the electric field experiences a quasiperiodic oscillation and exhibits stochastic behavior as seen in Fig. 3(b). That the contours for $E(x,t) = \text{const}$ in Fig. 3(c) are distorted indicates the existence of many irregularly localized patterns, which are still kept in the propagating process with stochastic speeds as shown in Fig. 3(d). The above feature illustrates the evolution of the spatial pattern that follows the temporal chaos.

Furthermore, the largest Lyapunov exponent with the definition

$$L_e = \lim_{t\to\infty} \lim_{D(t=0)\to 0} \left\{ t^{-1} \ln\left[\frac{D(t)}{D(t=0)} \right] \right\}$$

is positive for $k < k_c$, as seen in Fig. 4, where $D(t) = |E_1(0,t) - E_2(0,t)|$, $E_1(0,t)$, and $E_2(0,t)$ denote the $E(x,t)$ from two neighboring points in phase space at $x = 0$. Thus, the temporal chaos associated with the spatial patterns for the continuum Hamiltonian system is clearly observed.

In order to understand the evolutive route to spatiotemporal chaos, we observe the structure of the power spectrum. When $k = 0.95$, ω_1 increases to 0.5732, and an incommensurate frequency component $\omega_2 = 0.1157$ and the linear combination $\omega_1 \pm \omega_2$ appear simultaneously in Fig. 5(a). When k is reduced to 0.93, in addition to $\omega_1 = 0.6471$, $\omega_2 = 0.2324$, and $\omega_1 \pm \omega_2$, where their spikes of spectrum are weakened, the subharmonics $(2n - 1)\omega_1/2$, $n = 1, 2, 3, \ldots$, appear in Fig. 5(b). When $k = 0.9295$, the quasiperiodic behavior disappears, and only subharmonics exist with the further enhanced spikes, as seen in Fig. 5(c). Furthermore, to reduce k, the power spectrum with a broad-band noiselike structure occurs in Fig. 5(d). Thus, the evolution from quasiperiodic to quasiperiodic and subharmonic, then from quasiperiodic and subharmonic to subharmonic, furthermore, to chaos is found in our system. For the different c_1 and c_2, the above route is also observed.

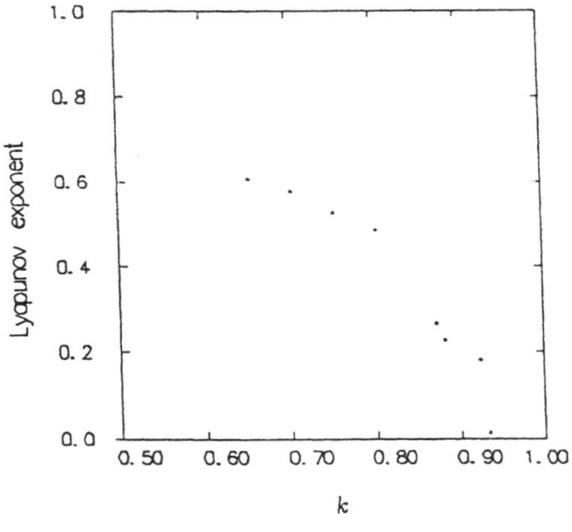

FIG. 4. Lapunov exponent versus k.

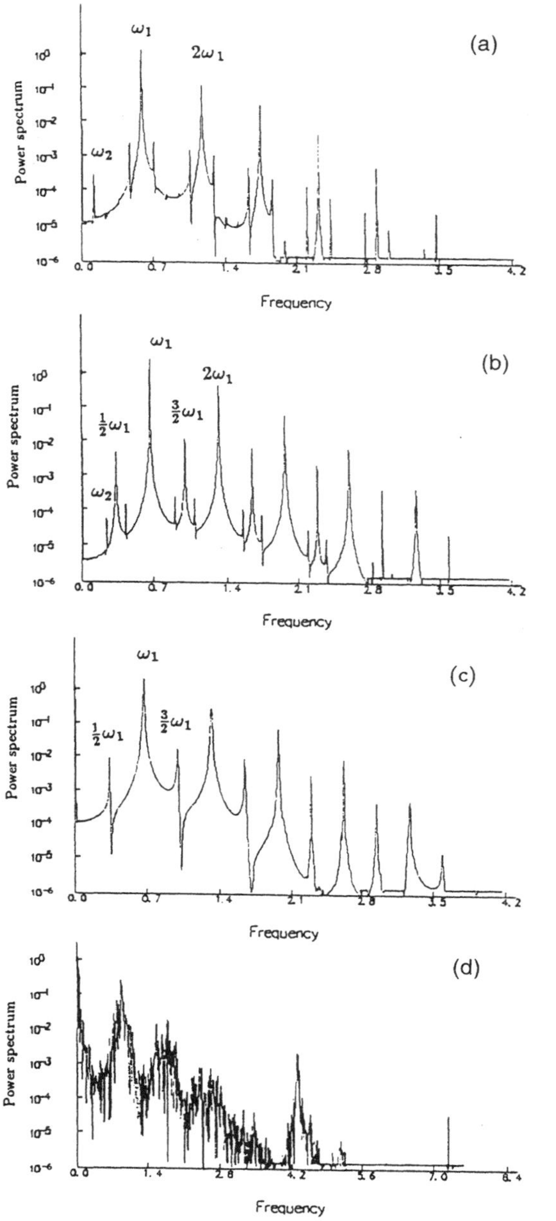

FIG. 5. Power spectrum. (a) $k = 0.95$, (b) $k = 0.93$, (c) $k = 0.9295$, (d) $k = 0.92$.

We will explain the reason in terms of the conception of harmonic oscillators by regarding each frequency as an oscillator. When k is 0.98, as seen in Fig. 1(c), the discrete spikes are isolated. However, when k approaches to k_c, the growth rate does increase, and ω_2, $\omega_1 \pm \omega_2$ and subharmonics make frequency intervals become narrow [see Figs. 5(a)–5(c)] and eventually oscillator overlapping for $k < k_c$ occurs because of the enhanced nonlinear interaction of the Langmuir wave with the ion-acoustic wave.

In order to further investigate the chaotic property of this system, we also computed the autocorrelation function [4] and found that the rapid decay of the correlator and the large Lyapunov exponents (Fig. 4) represent a very strong chaotic property for system, Eqs. (1). These justify the nature of the mixing of trajectories in phase space. Thus, there exists a Kol'mogorov-Sinai entropy [5].

To comprehend the formation of the irregular spatial structure, we investigate the energy shared by each of Fourier modes for $E(x,t) = \sum_n E_n(t)e^{ik_nx}$, where $k_n = 2\pi n/L$, $E_n = E_{k_n}$, and E_n versus time. This is similar to that in Ref. [6]. The spatially localized structure, indeed, can be described only in terms of finite modes. In our simulations, only eleven modes, i.e., $n = 5$, are included within the accuracy of 10^{-6} for conserved quantities, because the intensities of the higher Fourier modes have rapidly decreased. Energy initially confined to the master mode ($n = 1$), whose intensity increases with k decreasing and is excited from the spatially homogeneous state by MI, would extend to high harmonic slave modes. These modes are periodic in the sense of the Fermi-Pasta-Ulam (FPU)-like recurrence for $k_c < k$ but are unstable and stochastic for $k < k_c$. In the latter case, the interaction of slave modes with the master one causes the latter to be unstable due to the reaction of slave modes, and transferring more and more energy from the master to high harmonics at a fast rate. Thus, the FPU-like recurrence breaks. It leads to the formation of spatially irregular patterns, which are influencing temporal chaos.

In summary, the dynamics of regular spatial patterns with a periodic sequence in time and space are shown, The route from the quasiperiod route to subharmonics to spatiotemporal chaos for Eqs. (1) is first found. Features of spatiotemporal chaos with the coexistence of temporal chaos and spatially irregular patterns are observed.

We would like to acknowledge the helpful discussions of Professor S. G. Chen, Dr. W. M. Yang, and Dr. Y. Tan. This work is supported by National Natural Science Foundation of China (NNSFC) under Grant No. 19175039 and Nonlinear Science Foundation of China.

*Mailing address.

[1] V. E. Zakharov, Zh. Eksp. Teor. Fiz. **62**, 1745 (1972).
[2] Y. Tan, X. T. He, S. G. Chen, and Y. Yang, Phys. Rev. A **45**, 6109 (1992).
[3] R. Fermi, J. Pasta, and S. Ulam, *Collected Papers of Enrico Fermi* (Chicago University Press, Chicago, 1955), p. 978.
[4] G. M. Zaslavsky, *Chaos in Dynamic Systems* (Harwood, London, 1985).
[5] B. V. Chirikov, Phys. Rep. **52**, 265 (1979).
[6] X. T. He and C. T. Zhou, J. Phys. A **26**, 4123 (1993).

Characterization of Pattern Formation from Modulation Instability in the Cubic Schrödinger Equation *

LONG Tao(龙滔)[1], HE Xian-tu(贺贤土)[2,3]
[1] Graduate School, China Academy of Engineering Physics, Beijing 100088
[2] Institute of Applied Physics and Computational Mathematics, Beijing 100088
[3] Institute of Theoretical Physics, Chinese Academy of Sciences, Beijing 100080

(Received 1 April 1998)

The study of pattern dynamics in a Hamiltonian system(HS) having an infinite number of degree of freedom is very difficult due to the absence of attractors in such system. In this letter, we propose a useful method that only a few representative manifolds in phase space are investigated, and it can be used to reveal the pattern formation of HS. The conserved cubic Schrödinger equation is discussed. Although a special model is chosen, this method can be applied to more general case such as the near-integrable HS.

PACS: 47.10.+g, 05.45.+b, 52.35.Mw

There exists a region still needing much research, that is, the coherent structure in complex, spatially extended system. The visible pattern forms are the coherent structure extended in space. Cross *et al.*[1] have reviewed the mechanism of pattern formation in general. However, the systems that they discussed are essentially dissipative. For pattern dynamics in infinite Hamiltonian system, things become difficult. For such a system, different initial parameters will lead to different manifolds for lack of attractors. In this paper, we try to introduce a particular method to study the pattern formation in integrable Hamiltonian system, and the cubic nonlinear Schrödinger equation(NSE) is employed as a model.

The cubic NSE is a universal equation describing the evolution of wave envelops in a dispersive weakly nonlinear medium (see, for example, Taniuti[2]). It finds many applications in physics, such as plasma physics,[3] nonlinear optics,[4] hydrodynamic description of nuclear matter,[5] etc. The theoretical works about NSE have been proceeded for many years. Zakharov *et al.*[6] studied its integrability, Gagnon *et al.*[7] considered the properties of NSE's generalized form, and so on.

We take NSE in dimensionless form as follows:

$$\mathrm{i}\,\partial_t E + \partial_{xx} E + (|E|^2 - |E_0|^2)E = 0, \tag{1}$$

where E_0 is a constant. This form is different from the original one in a gauge transformation.

Now we consider how patterns form and evolute when the fixed point E_0, which manifests a homogeneous spatiotemporal state, is disturbed by an unstable perturbation.

For this purpose, we set

$$E = u + \mathrm{i}\,v, \tag{2}$$

and Eq. (1) gives

$$\begin{cases} \partial_t u = -\partial_{xx} v - (u^2 + v^2 - E_0^2)v, \\ \partial_t v = \partial_{xx} u + (u^2 + v^2 - E_0^2)u, \end{cases} \tag{3}$$

linearized at the fixed point E_0:

$$\begin{cases} \partial_t \delta u = -\partial_{xx}\delta v, \\ \partial_t \delta v = \partial_{xx}\delta u + 2E_0^2 \delta u. \end{cases} \tag{4}$$

For simplicity, we take the perturbation function as $\cos(kx)$. Denoting

$$k_0^2 = 2E_0^2, \tag{5}$$

then Eqs.(4) become

$$\begin{cases} \mathrm{d}_t \delta u = k^2 \delta v, \\ \mathrm{d}_t \delta v = -k^2 \delta u + k_0^2 \delta u. \end{cases} \tag{6}$$

The eigenvalues of Eqs.(6) are

$$\lambda_{1,2} = \pm k\sqrt{k_0^2 - k^2}, \tag{7}$$

obviously, when $0 < k < k_0$, a linear unstability appears and k is an unstable wavenumber at this time. The fixed point E_0 acts as a saddle point.

For convenience, we construct a phase plane $(|E(x,t)|, \partial_t |E(x,t)|)_{x=x_0}$. It corresponds to a Poincare section, where the coordinate is fixed at $x = x_0$. The eigenvalues in this phase plane are also λ in Eq. (7). In fact , the transformation

$$(u, v) \longrightarrow (|E| = \sqrt{u^2 + v^2}, \partial_t |E| = \frac{u\partial_t u + v\partial_t v}{\sqrt{u^2 + v^2}}) \tag{8}$$

is a diffeomorphisms if $E_0 \neq 0$.

* Supported in part by the National Climbing Project Nonlinear Science from State Science and Technology Commision of China.

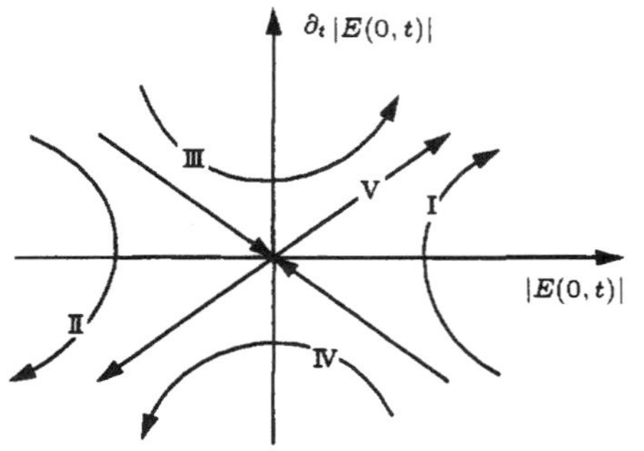

Fig. 1. Classification of orbits in phase plane.

From the linearized theory of dynamic system, orbits in phase plane, where the original is $(E_0, 0)$ and $x = 0$ is chosen, may be devided into five kinds as shown in Fig. 1, where the fifth kind is homoclinic orbits(HMO). For Eq. (1), the classification of orbits in the above phase plane may serve as a guide. That is, we can classify the patterns according to initial perturbations which lead to different representative orbits in the phase plane. These considerations will obviously simplify the complicated patterns evolution in the Hamiltonian system.

In our numerical experiments about Eq.(1), we choose the initial conditions as

$$\Psi = \Psi_0 + \epsilon E_0 (c_1 f_1 + c_2 f_2) \cos(kx), \tag{9}$$

where

$$\Psi = \begin{pmatrix} u \\ v \end{pmatrix}, \quad \Psi_0 = \begin{pmatrix} E_0 \\ 0 \end{pmatrix}, \tag{10}$$

and ϵ is the small real parameter. The eigenfunctions f_1 and f_2 in the phase plaue are

$$f_i = \begin{pmatrix} k^2 \\ \lambda_i \end{pmatrix}, \quad i = 1, 2. \tag{11}$$

In the numerical calculation, the values of parameters are $E_0 = 1.0$, $\epsilon = 0.002E_0$, and $k/k_0 = 0.6$. And we have used the split-step method, which can preserve the accuracy of the conserved quantities of 10^{-7}. The periodic boundary conditions are implied. We choose the spatial length L to satisfy $kL = 2\pi$. The number of grids is 512 and the time step for calculation is 10^{-4}, they work efficiently to avoid the occurrence of numerical induced chaos in NSE.[8]

From numerical simulation, we found five kinds of representative evolution patterns of Eq. (1) in the following manner.

Case I: $c_1 = 1.0$ and $c_2 = 1.0$. This corresponds to the first kind of representative orbit I in the phase plane. From the contour of the patterns as shown in Fig. 2, it is seen that a regularly spatiotemporal coherent structure with pattern centroid at $x = \pm nL, n = 0, 1, 2, \cdots$, is constructed. In this case, we name this pattern as type-I.

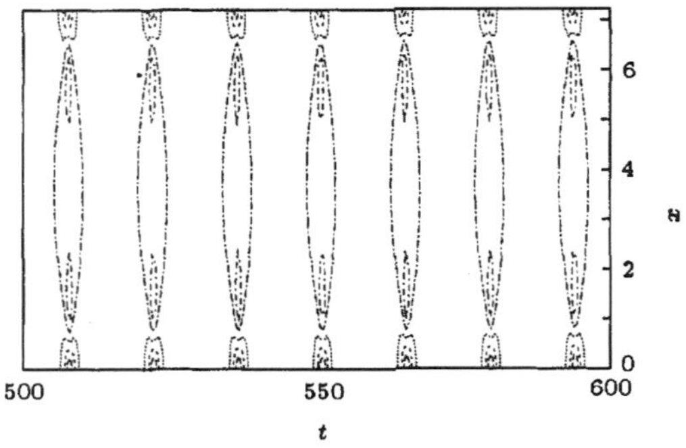

Fig. 2. Spatiotemporal structure of type-I pattern with $c_1 = 1.0, c_2 = 1.0$, and $k/k_0 = 0.6$.

Case II: $c_1 = -1.0$ and $c_2 = -1.0$. This pattern with pattern centroid at $x = \pm(n + 1/2)L, n = 0, 1, 2, ...$, is named as type-II and corresponds to the second kind of representative orbit II. Figure 3 shows its spatiotemporal structure. This pattern is also regular.

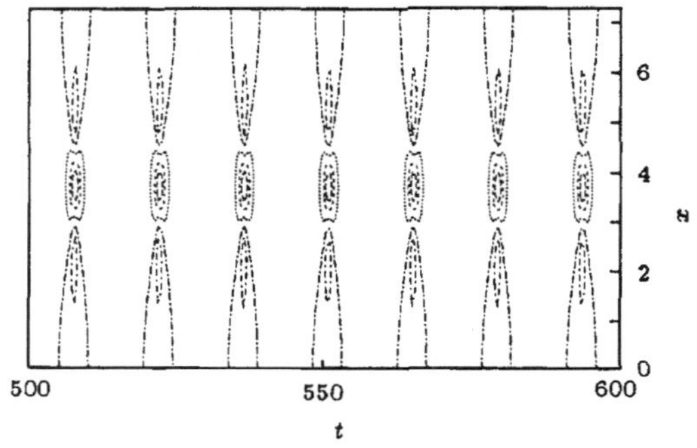

Fig. 3. Spatiotemporal structure of type-II pattern with $c_1 = -1.0, c_2 = -1.0$, and $k/k_0 = 0.6$.

Case III: $c_1 = 1.0$ and $c_2 = -1.0$. This is pattern type-III. The spatiotemporal structure is shown in Fig. 4. It is seen that this pattern is an alternatively combination of the types-I and II. When the orbit moves in the right side of the phase plane as shown in Fig. 1, the contour is like type-I, and in the left side of the phase plane, it is like type-II.

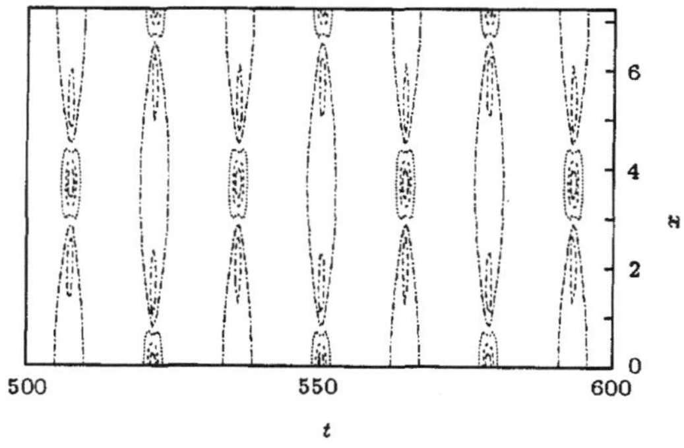

Fig. 4. Spatiotemporal structure of type-III pattern with $c_1 = -1.0, c_2 = -1.0$, and $k/k_0 = 0.6$.

Case IV: $c_1 = -1.0$ and $c_2 = 1.0$. This pattern, type-IV Fig. 5, is similar to pattern type-III, but the corresponding representative orbit moves in the opposite direction contrasting with that of type-III pattern.

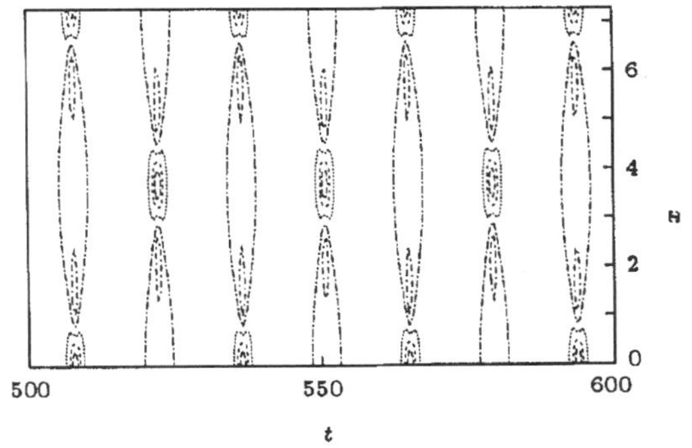

Fig. 5. Spatiotemporal structure of type-IV pattern with $c_1 = -1.0, c_2 = 1.0$, and $k/k_0 = 0.6$.

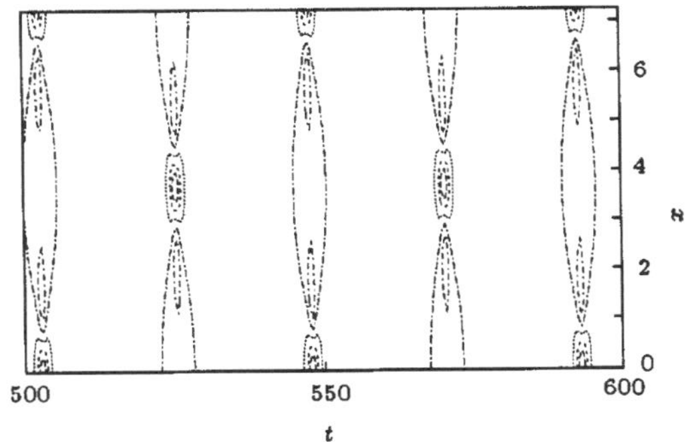

Fig. 6. Spatiotemporal structure of type-V pattern with $c_1 = 1.0, c_2 = 0.0001$, and $k/k_0 = 0.6$.

Case V: $c_1 = 1.0$ and $c_2 = 0.0001$. The spatiotemporal structure of pattern type-V is ploted in Fig. 6. It corresponds to orbits near HMO, the contour motion is regular too. However, we should mention that in a near-integrable system, these orbits in phase plane will appear in HMO crossings[9], and the contour motion occurs in temporal chaos, but spatial structure is still coherent and regular if the system size L is smaller than the correlation length ξ.[10]

In conclusion, we have proposed a useful description for stduying an extended integrable Hamiltonian system, that is, we divided the orbits of the system into five representative kinds. From investigating these representative orbits, we can get some insight into such complicated system. As an example, we study the cubic nonlinear Schrödinger equation. We found that regular patterns can exist in some region of parameter. Although we have chosen a typical example, this method can be easily applied to more general cases such as near-integrable Hamiltonian system if it has local saddle structure. More investigation will be made in our next works.

We thank Dr. Y. Tan and Dr. C. Y. Zheng for valuable discussion.

REFERENCES

1 M. C. Cross and P. C. Hohenberg, Rev. Mod. Phys. 65 (1993) N3.
2 T. Taniuti, Prog. Theor. Phys. Suppl. 55 (1974) 1.
3 N. Asano, T. Taniuti and N. Yajima, J. Math. Phys. 10 (1968) 2020.
4 R. I. Chiao, E. Garmire and C. H. Townes, Phys. Rev. Lett. 13 (1964) 479.
5 E. F. Hefter, S. Raha and R. M. Weiner, Phys. Rev. C32 (1985) 2201.
6 V. E. Zakharov and L. D. Faddeev, Func. Anal. Appl. 5 (1971) 280.
7 L. Gagnon and P. Winternitz, Phys. Lett. A 134 (1989) 276.
8 M. J. Ablowitz and B. M. Herbst, SIAM J. Appl. Math. 50 (1990) 339.
9 H. T. Moon, Phys. Rev. Lett. 64 (1990) 412.
10 P. Hohenberg and M. Shraimar, Physica, D 37 (1989) 109.

Modern Physics Letters B, Vol. 12, No. 22 (1998) 943–954

PATTERN FORMATION AND COMPETITION IN THE REGIME OF THE CONSERVED CUBIC-QUINTIC NONLINEAR SCHRÖDINGER EQUATION

TAO LONG[†] and X. T. HE[*,‡,§]

[*] *CCAST (World Laboratory), P.O. Box 8730, Beijing 100080, China*
[†] *Graduate School, China Academy of Engineering Physics, P. O. Box 8009, Beijing, 100088, China*
[‡] *Institute of Applied Physics and Computational Mathematics, P. O. Box 8009, Beijing 100088, China*
[§] *Institute of Theoretical Physics, Academia Sinica, P. O. Box 2735, Beijing 100080, China*

Received 3 April 1998

In this letter, the pattern formation, pattern competition and spatiotemporal chaos are first investigated in the regime of cubic-quintic nonlinear Schrödinger equation in a near-integrable Hamiltonian continuum system using a few important representative manifolds in phase space. We find that the regular pattern competition can exist in some regions of the controlled parameters, and the spatiotemporal chaos may result from stochastic competition of the patterns. This conclusion may be generalized to the systems which have the saddle structure.

PACS Number(s): 47.10.+g, 05.45.+b, 52.35.Mw, 83.50.Ws

1. Introduction

The existence of coherent structure in complex, spatially extended system has been known for long time.[1] The visible pattern forms are the coherent structure extended in space. Cross and Hohenberg[2] have reviewed the mechanism of pattern formation in general. However, the systems that they discussed are essentially dissipative. For a near-integrable Hamiltonian continuum system, which is very important to comprehend the connotation of statistical physics in the sense of the intrinsic stochasticity involving the origination of turbulence etc., there still have many things to do. The difficulty to deal with such system is that the finite number of manifolds in phase space, which correspond to the infinite number of initial parameters, must be considered. In this paper, we try to use a particular method to investigate the pattern formation and competition, which are very interesting phenomena of common occurrence in various fields related to extended near-integrable system, where a cubic-quintic nonlinear Schrödinger equation (c-q NSE) is introduced as a model.

The c-q NSE originated from many interesting physical systems, such as nonlinear optics, hydrodynamics, plasma physics, especially when the higher order effects are becoming important. In recent years, c-q NSE has received great attention. Much work has been done analtyically and numerically. Aciolic *et al.*[3] have shown that in nonlinear optics, the fifth-order nonlinear effects should be considered in some cases. He *et al.*[4] and Puskharov *et al.*[5] obtained solitary wave solutions. Cowan *et al.*[6] showed numerically these are not solutions, but behave like quasi-solitons. Gagnon *et al.*[7,8] presented a large set of exactly analytic travelling wave solutions for such a model. He *et al.*[9] investigate the integrability of c-q NSE numerically, and recently Y. Yang *et al.*[10] discussed the peak point displacement for intense laser beam propagation in plasma using c-q NSE, and so on.

C-q NSE can be written in dimensionless form as follows

$$i\partial_t E + \partial_{xx} E + (|E|^2 - g|E|^4)E = 0\,, \tag{1}$$

where g is a small parameter. It is shown that the term $g|E|^4E$ causes the near-integrability.[9]

This paper is organized as follows. In part 2 we give the necessary stability analysis for discussing pattern formation and competition of c-q NSE. Then we do numerical calculation in part 3. The last part is the conclusion of our investigation.

2. Linear Stability Analysis

Now we discuss the linear stability of solutions of Eq. (1) near Stokes wave

$$E_s(t) = E_0 e^{it}\,, \tag{2}$$

where E_0 satisfies

$$E_0 = 0, \pm\sqrt{\frac{1 \pm \sqrt{1-4g}}{2g}}\,. \tag{3}$$

Let

$$E(x,t) = \bar{E}(x,t)e^{it}\,. \tag{4}$$

we obtain an equation about $\bar{E}$ from Eq. (1) as follows

$$i\partial_t \bar{E} + \partial_{xx}\bar{E} + (|\bar{E}|^2 - g|\bar{E}|^4 - E_0^2 + gE_0^4)\bar{E} = 0\,. \tag{5}$$

Then we consider the stability of solutions of Eq. (5) at the fixed point E_0. For simplicity, we will denote $\bar{E}$ as E below.

Let

$$E(x,t) = E_0 + \delta E(x,t) \tag{6}$$

and linearized around E_0, we have

$$\begin{bmatrix} i\partial_t + \partial_{xx} + E_0^2 - gE_0^4 & E_0^2 - gE_0^4 \\ E_0^2 - gE_0^4 & -i\partial_t + \partial_{xx} + E_0^2 - gE_0^4 \end{bmatrix} \begin{bmatrix} \delta E \\ \delta E^* \end{bmatrix} = 0\,. \tag{7}$$

October 14, 1998 12:2 WSPC/147-MPLB 0068

If we consider the periodic boundary conditions, the eigenfunction of Eq. (7) can be chosen as

$$\begin{bmatrix} \delta E \\ \delta E^* \end{bmatrix} = \begin{bmatrix} \beta e^{\lambda t} \\ \beta^* e^{\lambda t} \end{bmatrix} \cos kx \,, \tag{8}$$

where β, β^* are the small complex conjugate parameters.

The eigenvalues of Eq. (7) is easily got as follows

$$\lambda = \pm k\sqrt{k_0^2 - k^2} \,, \tag{9}$$

where

$$k_0^2 = E_0^2(1 - gE_0^2) \,. \tag{10}$$

Obviously, if $k_0^2 > 0$ and $k < k_0$ a linear unstability appears and k is unstable wavenumber at this time. With the expression of E_0, it is easy to find that

$$E_0 = \pm\sqrt{\frac{1 - \sqrt{1-4g}}{2g}} \,, \tag{11}$$

with $0 \le g < 1/4$, corresponding to hyperbolic points;

$$E_0 = 0, \pm\sqrt{\frac{1 + \sqrt{1-4g}}{2g}} \,, \tag{12}$$

with $0 \le g < 1/4$, corresponding to elliptic points.

Without loss of generality, we will study the local structure of hyperbolic point

$$E_{+h} = \sqrt{\frac{1 - \sqrt{1-4g}}{2g}} \,. \tag{13}$$

as schematically shown in Fig. 1.

The local saddle structure affords an important description for a system governed by a nonlinear partial differential equation. Usually, it is difficult to study the near-integrable Hamiltonian continuum system, in which the volume in phase space is preserved and the system has infinite numbers of degrees of freedom, and there are no attractors, that is, each initial condition corresponds to each orbit in phase space. As known from the theory of dynamic system, orbits in the saddle subspace can be divided into five kinds. Figure 1 gives out one representative orbit for each kind of orbits schematically, where the fifth kind is very near homoclinic orbits (HMO). For the original system which is controlled by a nonlinear partial differential equation, the classification of orbits in local saddle subspace can serve as a guide. That is, we can classify the patterns according to initial perturbations which lead to different representative orbits in the subspace. These considerations will obviously simplify the complicated patterns evolution due to no attractors in the Hamiltonian system. If we investigate these five kinds of representative orbit, we can get some insight into the pattern formation, pattern competition, the stochasticity and the spatiotemporal choas of the system. It will be shown in the next section in details from numerical simulation of Eq. (5).

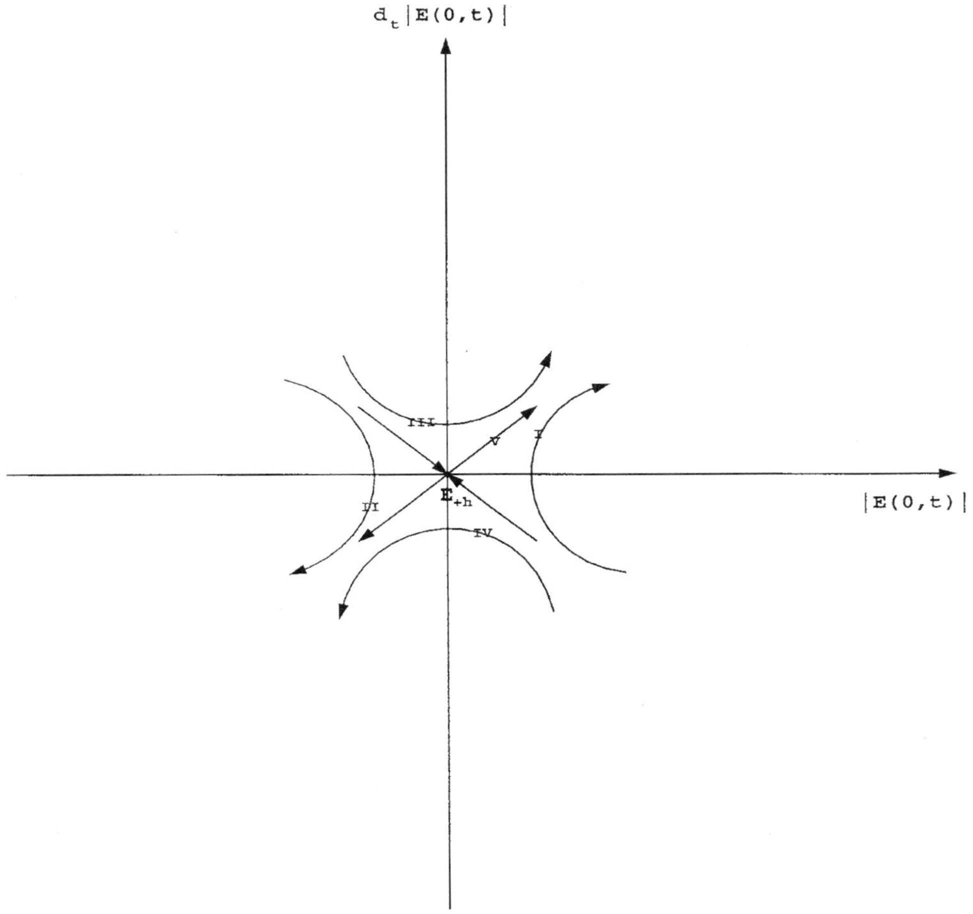

Fig. 1. The classification of orbits in phase space $(|E(0,t)|, d_t|E(0,t)|)$.

3. Numerical Results

It is convenient for numerical simulation to use the real and imaginary parts instead of E and E^*. We give the initial condition in the saddle subspace as follows

$$\Psi = \Psi_0 + \epsilon E_0(c_1 f_1 + c_2)\cos kx\,, \tag{14}$$

where

$$E = u + iv \tag{15}$$

and

$$\Psi = \begin{pmatrix} u \\ v \end{pmatrix}, \qquad \Psi_0 = \begin{pmatrix} E_0 \\ 0 \end{pmatrix} \tag{16}$$

ϵ is a small real parameter, $E_0 = \{[1-(1-4g)^{1/2}]/2g\}^{1/2}$ as mentioned before. The eigenfunction components f_1 and f_2 are in the saddle subspace

$$f_i = \begin{pmatrix} k^2 \\ \pm k\sqrt{k_0^2 - k^2} \end{pmatrix}, \qquad i = 1, 2\,. \tag{17}$$

We first choose a very small parameter $g = 0.00015$ to show the integrability of the system. The other parameters are $E_0 \sim 1.0$, $\epsilon = 0.002E_0$ and the initial unstable wavenumber $k/k_0 = 0.6$, where $k_0 = 1.414$ depends on g as shown in Eq. (10). In the numerical simulation, we have used the split-step method, which can preserve the accuracy of the conserved quantities being 10^{-8}. The periodic boundary conditions are implied, the spatial length L is chosen to satisfy $kL = 2\pi$. The number of grids are 512 and the time step for calculation is 10^{-4}, they work efficiently to avoid the occurrence of numerical induced chaos in NSE.[11]

The results are discussed as follows.

For $c_1 = 1.0$, $c_2 = 1.0$, the numerical result indicates this is a bowel orbit far inside HMO. Its evolution is the most stable one among the first kind of orbits, and we choose it as the representative of this kind of orbits. The long term behavior of the system is shown in Figs. 2(a) to 2(d), respectively. The manifold in phase space $(|E|, \partial_t|E|)$ with the lapse of time is always a single loop on the right side of $\mathbf{E}_{+\mathbf{h}}$ to form a KAM tours as seen in Fig. 2(a). Correspondently, the envelope of E is performing an exactly periodic oscillation [Fig. 2(b)]. In order to discuss its space-time characters, we draw a 3-dimensional picture [Fig. 2(c)] and its contour [Fig. 2(d)], all these pictures manifest that a regular spatiotemporal coherent structure is constructed. We name this kind of pattern type-I.

For $c_1 = -1.0$, $c_2 = -1.0$, this pattern is named type-II. Figure 3(a) shows it corresponds to orbits which are somewhere similar to those of type-I. We see from the initial perturbations Eq. (14) that the type-I and II have opposite signs of c_1 and c_2, and the manifolds evolve in a contrary direction. So one may expect this pattern has some relations with pattern-I. In fact, the space-time constructions [Fig. 3(c)] and its contour [Fig. 3(d)] express this relation clearly that type-II is different with type-I in half space period translation. However, these two patterns are not just different in a x-translation. The representative orbits of these patterns are essentially different in the sense that they can not go across saddle point topologically from one to another before chaos is appearing and expanding.

For $c_1 = 1.0$, $c_2 = -1.0$, this is the type-III. It corresponds to the third kind of orbits in phase space $(|E|, \partial_t|E|)$, which are outer of HMO as shown in Fig. 4(a). The contour in Fig. 4(d) and the $|E(x,t)|$ varying with x, t in Fig. 4(c) show that these structures for the patterns are alternative combination of the type-I and type-II as seen in Figs. 4(c) and 4(d), it is clearly that when the manifold moves in the right side of the saddle, the behavior looks like the type-I; whereas motion in left side of saddle, the behavior is like the type-II.

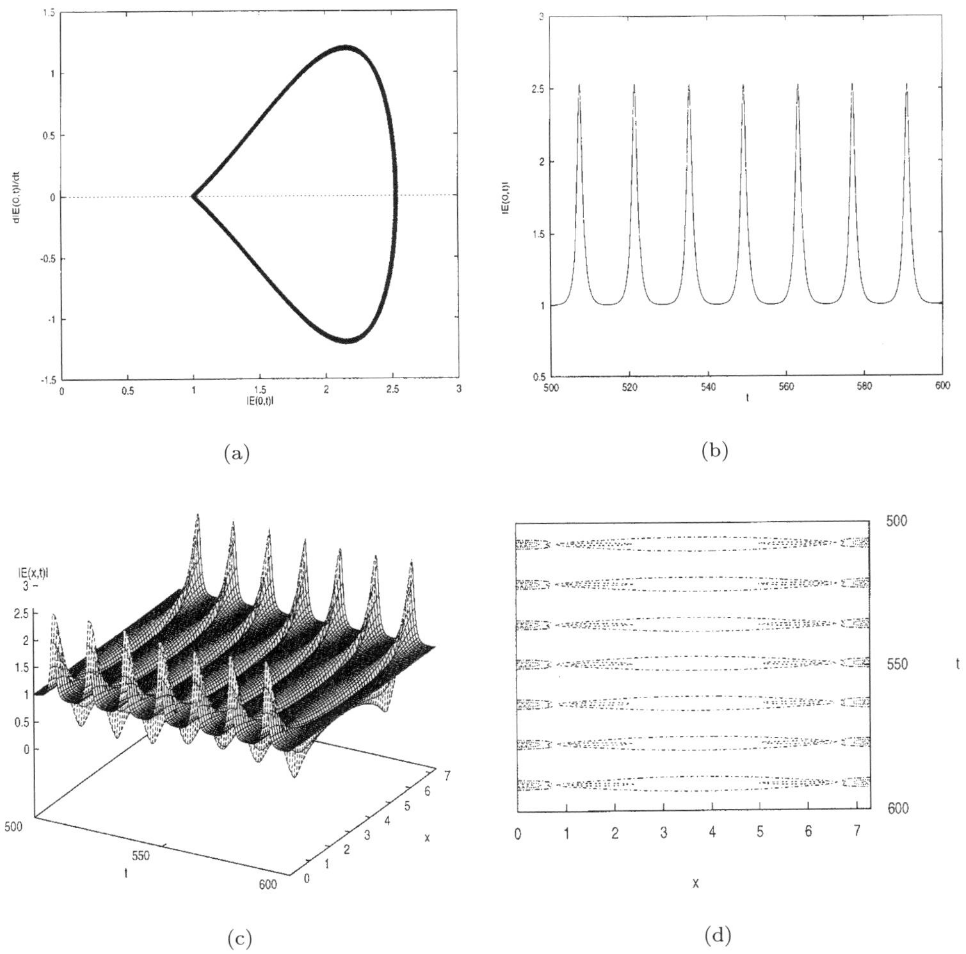

Fig. 2. The figure descriptions of pattern type-I with $c_1 = 1.0$, $c_2 = 1.0$ and $k/k_0 = 0.6$ when $g = 0.00015$.

For $c_1 = -1.0$, $c_2 = -1.0$. This pattern, type-IV, is all the same with type-III pattern except that the corresponding representative orbit moves in the opposite direction contrasting with that of type-III pattern.

For $c_1 = 1.0$, $c_2 = -0.001$. It corresponds to the orbits near HMO. The spatiotemporal construction of this type-V is plotted in Figs. 5(a)–(d). It is different from types I, II, III, IV and is irregular in space-time. Homoclinic crossings as observed in Fig. 5(a) indicate a spatiotemporal choas state in system.

The above results show that among types I–IV, as shown in Figs. 2(c), 2(d), 3(c), 3(d), 4(c), 4(d), one type pattern appearing in space, such as $x = 4$, are preferred over the other types, and thus pattern competition due to varying c_1 and

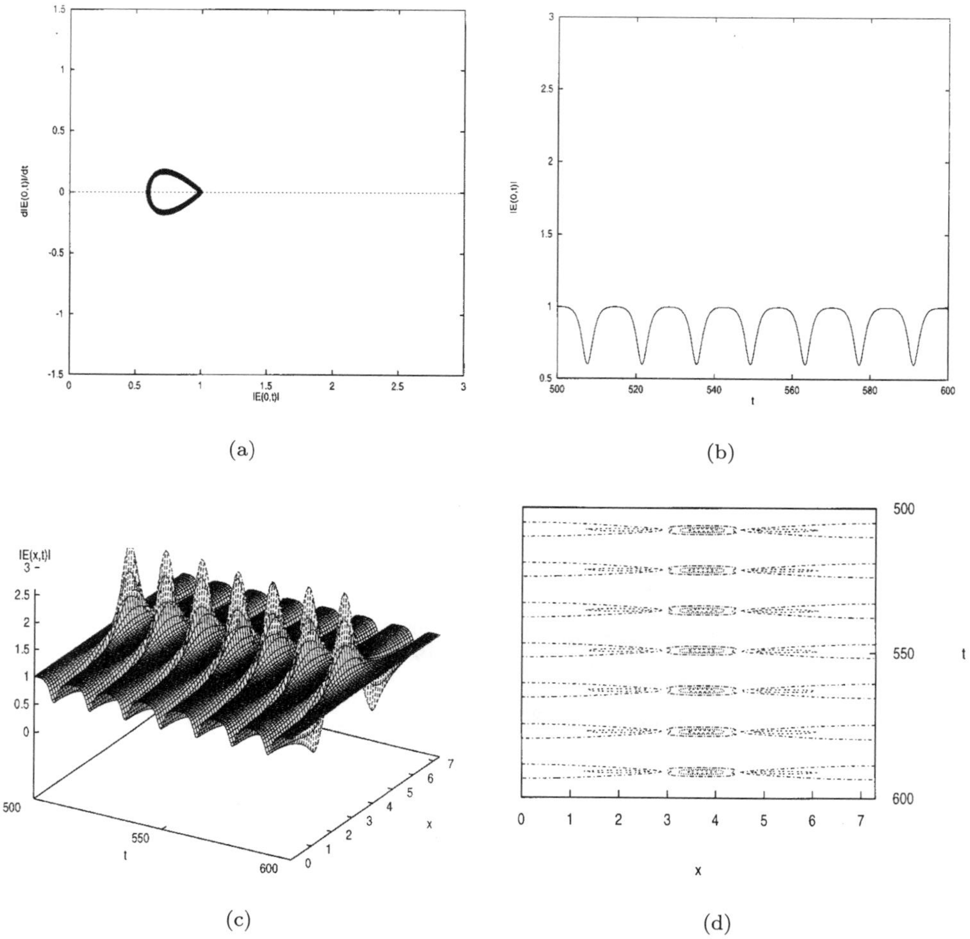

Fig. 3. The figure descriptions of pattern type-II with $c_1 = -1.0$, $c_2 = -1.0$ and $k/k_0 = 0.6$ when $g = 0.00015$.

c_2 is shown. Such a system that only the manifolds in the neighborhood of HMO are stochastic and others are regular is called weakly spatiotemporal chaos.

Now, further to increase g, i.e., k_0 to have a greater perturbation away from saddle E_{+h}. In addition to type-V, type-III and IV are also stochastic, the manifolds occur in HMO crossings like Fig. 5(a) and thus the destroyed region of KAM tori is extended. Here the stochastic competition between types I and II is seen. This is a transient choas state for the system. If we further increase the perturbations to be far away from saddle, for $g = 0.0015$, the manifolds, which correspond to the most stable orbits $c_1 = c_2 = 1.0$ and $c_1 = c_2 = -1.0$, become stochastic and HMO crossings occurred as shown in Figs. 6(a) and 7(a). And the pattern motion as shown in Figs. 6(c) and (d) and Figs. 7(c) and (d), and $|E(x,t)|$ versus time are stochastic

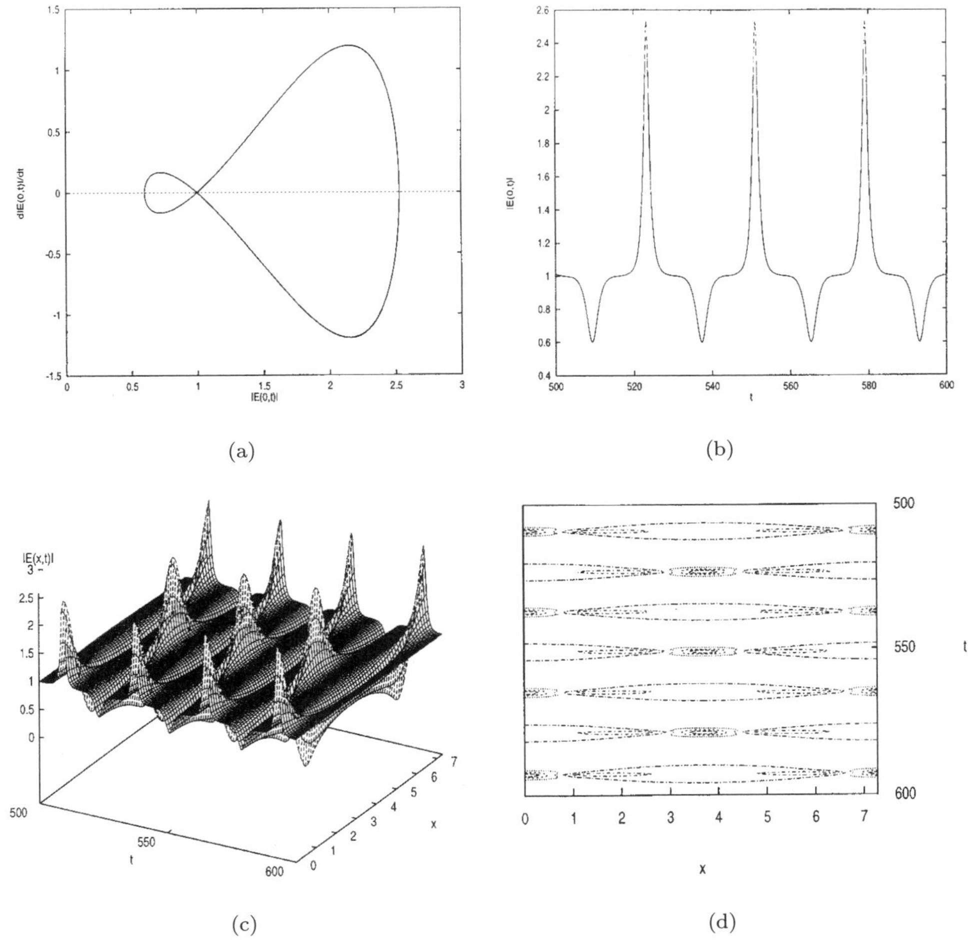

Fig. 4. The figure descriptions of pattern type-III with $c_1 = 1.0$, $c_2 = -1.0$ and $k/k_0 = 0.6$ when $g = 0.00015$.

as well. At present, the whole system is of stochasticity, such a system is called strong spatiotemporal chaos. These phenomena reveal that stochastic competition among pattern types switches on spatiotemporal chaos. From our calculation, we find that g_c, the critical value of perturbation parameter g, which switches the system from weak chaos state to strong chaos state, is about 0.0007. So there exist some regular patterns for $0 < g < 0.0007$, and for $g > 0.0007$ the system is in strong spatiotemporal chaos state.

Obviously, the critical value g_c is associated with a given k. For different k, g_c also different. Here we give some results of g_c with corresponding k (see Table 1), which can be found through numerical experiment.

Table 1. The critical value g_c versus k.

k	g_c
$0.5k_0$	0.00004
$0.6k_0$	0.0007
$0.7k_0$	0.0056
$0.8k_0$	0.0149
$0.9k_0$	0.245

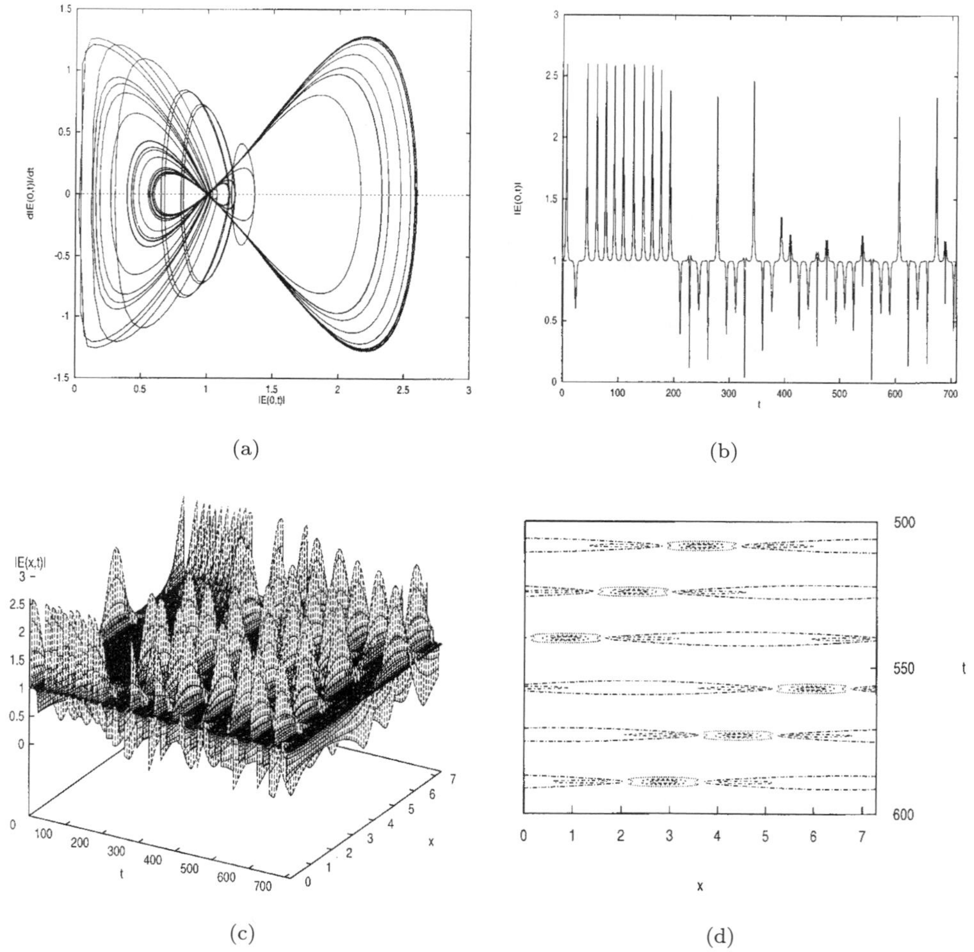

Fig. 5. The figure descriptions of pattern type-V with $c_1 = 1.0$, $c_2 = -0.0001$ and $k/k_0 = 0.6$ when $g = 0.00015$.

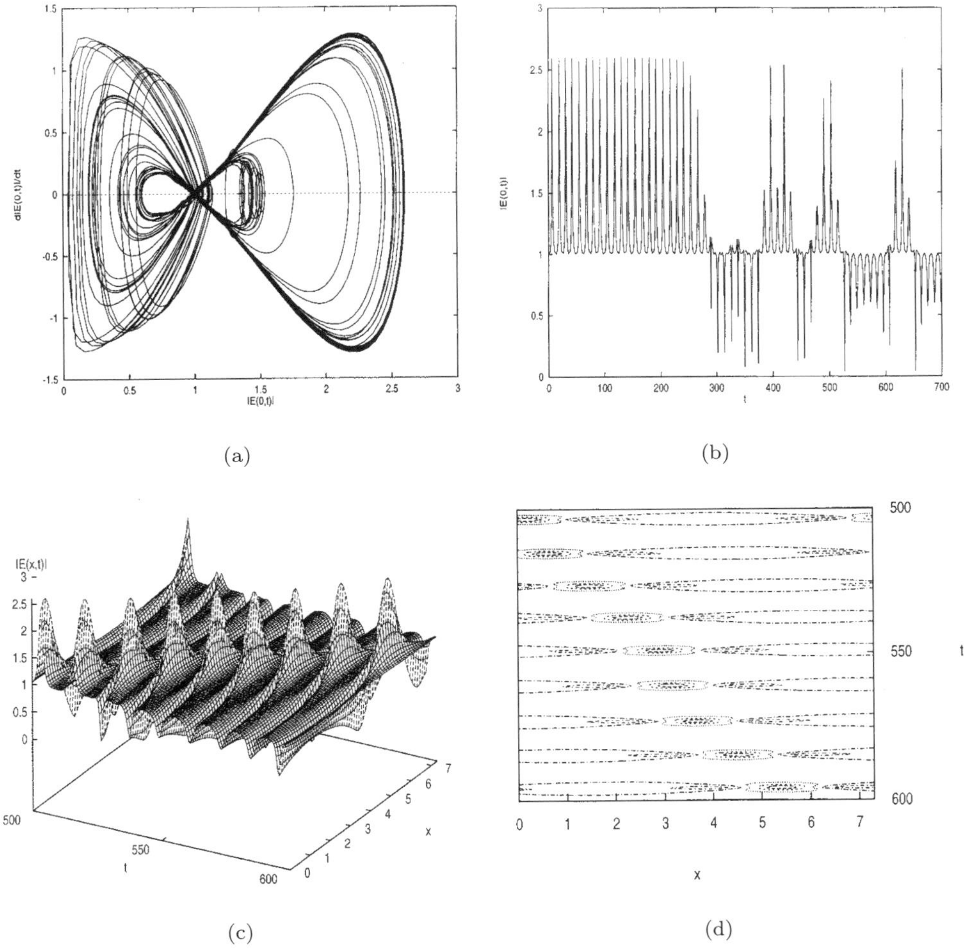

Fig. 6. The figures which show the manifolds corresponding to the most stable orbits $c_1 = c_2 = 1.0$ are becoming irregular when $g = 0.0015$, $k/k_0 = 0.6$.

4. Conclusion

We introduced an important description to reveal the complex properties for the extended near-integrable Hamiltonian continuum system in the regime of c-q NSE. Through investigating five kinds of representative orbit, the regular and stochastic motions in different parameters and the transition from regularity to stochasticity due to pattern competition are shown. The stochastic competition switching on spatiotemporal chaos is revealed, and weak and strong chaos are demonstrated. Although we have chosen a special example, the conclusion may be generalized to more general cases in which particle-like moves in the double well, i.e., there exists homoclinic orbit. More investigations will be made in our next works.

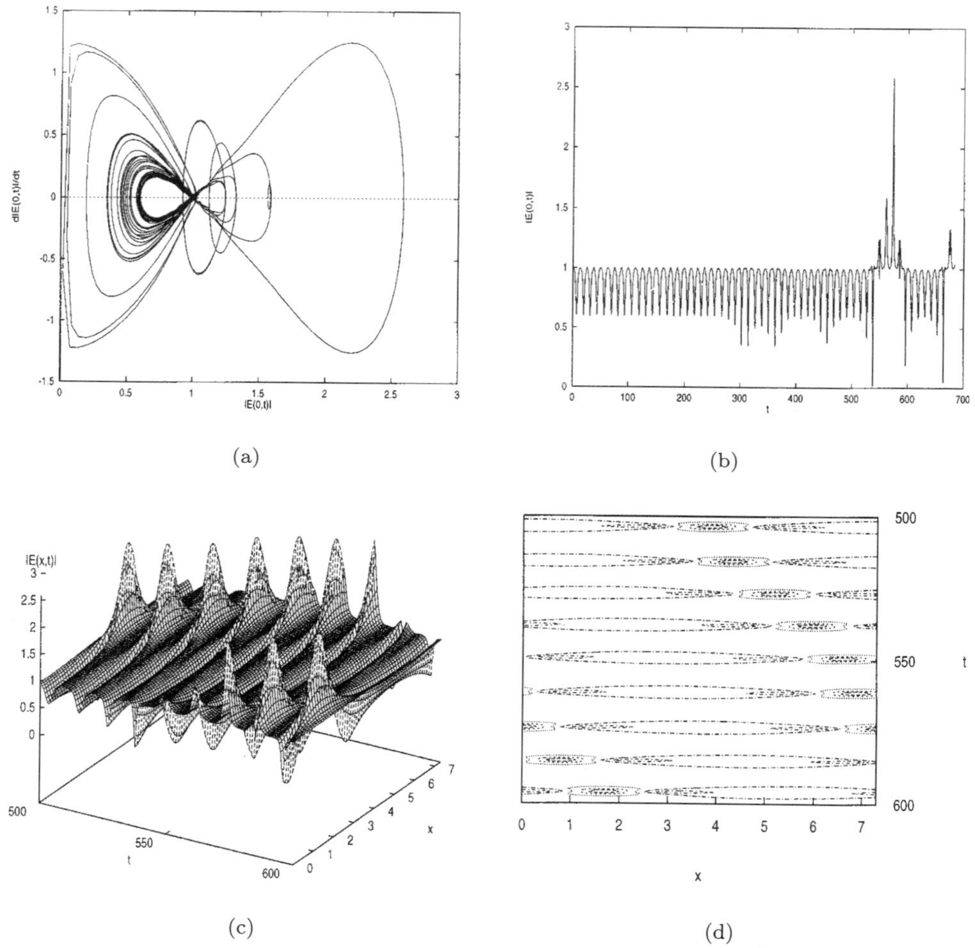

Fig. 7. The figures which show the manifolds corresponding to the most stable orbits $c_1 = c_2 = -1.0$ are becoming irregular when $g = 0.0015$, $k/k_0 = 0.6$.

Acknowledgments

This work was supported partially by the National climbing project "Nonlinear Science" and by the National Natural Science Foundation of China. We thank Dr. Y. Tan and Dr. C. Y. Zheng for valuable discussion.

References

1. A. A. Townsend, *The Structure of Turbulent Shear Flow* (The University Press, Cambridge, U.K., 1956).
2. M. C. Cross and P. C. Hohenberg, *Rev. Mod. Phys.* **65**, 3 (1993).
3. L. H. Acioli, A. S. Gomes and J. R. Rios Leit, *Appl. Phys. Lett.* **53**, 1779 (1988).
4. X. T. He and Z. Liu, *Chinese Science Bulletin* **29**, 1328 (1984).

5. Kh. I. Pushkarkov, D. I. Pushkarov and I. V. Tomov, *Opt. Quan. Electr.* **11**, 471 (1979).
6. S. Cowan, R. H. Enns, S. S. Rangnekar and S. S. Sanghere, *Can. J. Phys.* **64**, 311 (1986).
7. L. Gagnon, *J. Opt. Soc. Am.* **A6**, 1477 (1989).
8. L. Gagnon and P. Winternizt, *Phys. Lett.* **134A**, 276 (1989).
9. X. T. He and C. T. Zhou, *J. Phys. A: Math. Gen.* **26**, 4123 (1993).
10. Y. Yang, Y. Tan, W. Y. Zhang and C. Y. Zheng, *Phys. Lett.* **A216**, 247 (1996).
11. M. J. Ablowitz and B. M. Herbst, *SIAM J. Appl. Math.* **50**, 339 (1990).

ELSEVIER
Physica A 288 (2000) 338–343
PHYSICA A
www.elsevier.com/locate/physa

Pattern dynamics and spatiotemporal chaos in the conserved Zakharov equations

X.T. He[a,b,*], C.Y. Zheng[a], Shao-ping Zhu[a]
[a]Institute of Applied Physics and Computational Mathematics, P.O. Box 8009, Beijing 100088, People's Republic of China
[b]Department of Physics, Zhejiang University, Hangzhou 310027, People's Republic of China

Abstract

It is shown that many coherent solitary patterns can be excited and saturated from unstable harmonic modes through modulational instability in the conserved Zakharov equations. The evolution of solitary patterns may experience three states: spatiotemporal coherence, temporal chaos but spatially partial coherence, and spatiotemporal chaos that is caused by collisions and fusions among patterns with strong ion-sound emission, the energy in the system is strongly redistributed, it may switch on the onset of weak turbulence in plasma. © 2000 Elsevier Science B.V. All rights reserved.

PACS: 47.10.+g; 52.35. Mw

Keywords: Solitary pattern; Collision and fusion; Spatiotemporal chaos

It is shown that as Langmuir waves couple with ion-sound waves in plasma, a spatially modulational perturbation imposed on a spatially uniform field can cause the modulational instability (MI) and sequently drive the solitary pattern formation through nonlinear saturation [1].

In the recent years, it is found that the evolution of solitary patterns excited by the master mode and unstable harmonic modes from MI can transit to temporal chaos (TC) that is coexistent with solitary patterns and spatiotemporal chaos (STC) that is characterized by the coexistence of TC with incoherent patterns in a near-integrable Hamiltonian system [2–4]. However, how to transit to TC and STC from the pattern evolution in the above system is unclear. In this paper we investigate the relevant

* Correspondence address: Institute of Applied Physics and Computational Mathematics, P.O. Box 8009, Beijing 100088, People's Republic of China. Fax: +86-10 62010108.

PII: S0378-4371(00)00431-3

issues based on one-dimensional conserved Zakharov equations (ZEs) in the form [1]

$$i\partial_t E + \partial_x^2 E = nE ,$$
$$\partial_t^2 n - \partial_x^2 n = \partial_x^2 |E|^2 , \tag{1}$$

where $E(x,t)$ is a slowly varying envelope of Langmuir electric field, and $n(x,t)$ is an ion-sound density.

So far the ZEs are extensively studied as a model describing strong plasma turbulence, so the exploration of a transition from pattern evolution to STC (sometime called weak turbulence) based on ZEs contributes to the comprehension of the onset of turbulence.

The linearized stability analysis of the perturbation $\exp[i(kx-\omega t)]$ from a spatially homogeneous field E_0 for Eqs. (1) gives the growth rate of the MI as [2]

$$\gamma = kE_0\{[(1-2E_0^2k^2)^2 + 8E_0^2]^{1/2} - (1+2E_0^2k^2)]\}^{1/2} , \tag{2}$$

where k is normalized by $\sqrt{2}E_0$ and $0<k<1$. It is easy to prove that Eqs. (1) have at least two kinds of fixed points: a center at $(0,0)$ and a saddle at $(E_0,0)$.

It is realized that pattern formation is due to MI. In the present paper Eqs. (1) are integrated numerically to investigate the global behavior. The initial condition to adjoin the saddle is chosen, that is, a very small spatial inhomogeneity is added at $t=0$ on the spatially homogeneous state $\Psi_0 = (\operatorname{Re}E, \operatorname{Im}E, n, \partial_t n)_{t=0} = (E_0, 0, 0, 0)$ as follows:

$$\Psi = \Psi_0 + (c_1 f_1 + c_2 f_2)E_0 \cos(kx)/500 , \tag{3}$$

where f_1 and f_2 are the components of the eigenvector in the saddle subspace for the linearized Eqs. (1), c_1 and c_2 are coefficients, and the factor $\frac{1}{500}$ is only to show that a very small perturbation is adjusted for convenient calculation. The initial condition Eq. (3) ensures that the manifolds in phase space will locally lie in a saddle subspace.

Obviously, as $\frac{1}{2} \leqslant k \leqslant 1$ only one unstable wave number with a spatially modulational scale $L = 2\pi k^{-1}$ can be included. The unstable master mode is excited and saturated to form the spatially periodic solitary pattern that is coherent or spatially regular. However, in the case of $k < \frac{1}{2}$, there may exist many solitary patterns with spatially modulational length $l_m = L/m$, where $m=1$ is for master mode and $m = 2, 3, \ldots, M$ are for harmonic modes, here M less than unstable wave numbers $N = [k^{-1}]$ is due to pattern selection. As a result, the envelope E can be written as follows:

$$E(x,t) = \sum_{m=1}^{M} E_m(t)e^{imkx} + \sum_{m=M+1}^{\infty} E_m(t)e^{imkx} , \tag{4}$$

where the first term on the right-hand side of Eq. (4) comes from the master mode and unstable harmonic modes, and the second term is due to the nonlinear interaction.

We now discuss the pattern dynamics, TC and STC. The split-step method for E with a pre-estimate for density perturbation n is used to integrate Eqs. (1) with the periodic boundary condition in which the largest length L in space is chosen, so that $kL = 2\pi$. The number of grids is 2048 and $E_0 = 2.0$ is taken below. The conserved quantities are preserved to the order of 10^{-8} during the computation.

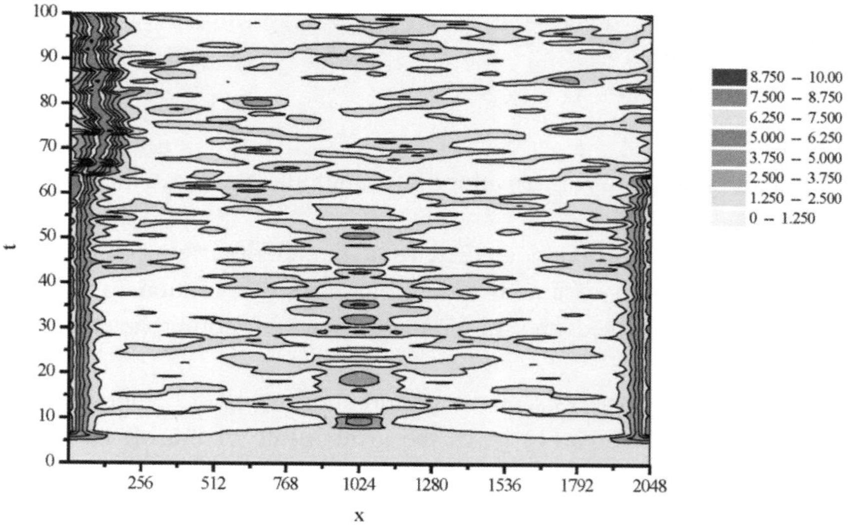

Fig. 1. The contour for $k = 0.24$. Single solitary pattern at $x = \pm mL$, $m = 0, 1, 2, \dots$ is excited by the master mode. Although the harmonic modes are excited to form solitary patterns as well, disappear due to ion-acoustic emission after a short time. The system is in a coexistance with TC and the master pattern.

It is shown [4] that there exists a critical wave number k_c in the case $\frac{1}{2} \leqslant k \leqslant 1$ for each set of c_1 and c_2, and the motion of the coherent solitary patterns is the temporal recurrence (periodic) or the pseudo-recurrence (quasi-periodic) as the unstable wave number is in $k_c < k < 1$. It is also known [4] that as $\frac{1}{2} \leqslant k \leqslant k_c$ the motion of the centre of the coherent solitary pattern exhibits stochastic behavior with a velocity v in the form [5]

$$\frac{\mathrm{d}v}{\mathrm{d}t} = \frac{2\partial_x n}{1 + \frac{8}{3}c^2(1 + 5v)/(1 - v^2)^4}\,, \tag{5}$$

where c is a constant and n is the ion-sound perturbation. The coexistence of TC and coherent solitary structure patterns occurs for the case of $\frac{1}{2} \leqslant k \leqslant k_c$, and it has been pointed out [2] that TC is caused by resonant overlapping. In the present paper, unless otherwise mentioned, without loss of generality, $c_1 = c_2 = 1.0$ is chosen. For different set of c_1 and c_2, only the initial position of the solitary pattern is shifted in space.

The rich phenomena are further observed by varying k, where k is less than $\frac{1}{2}$. As only a few harmonic waves, such as $k = 0.24$ (3 harmonics), are included, the $2k$ harmonic is excited early, but disappears after a short time due to ion-acoustic emission as shown in Fig. 1, hence only one solitary pattern from the master mode to form one pattern train varying with x and t. As a result, similar to $\frac{1}{2} \leqslant k \leqslant k_c$, only TC with the coherent solitary pattern train appears. As k is reduced to 0.12, Fig. 2 shows that five solitary-pattern trains from the master mode and harmonic modes appear. The stochastic motion of the trains leads to the collision between neighbouring solitary patterns due to the spatial interval getting narrower, and the collision results in a fusion of two coherent solitary patterns into a new incoherent pattern with the strengthened amplitude and the narrower width, the strong ion-sound emission $I(x,t) \gg |E(x,t)|^2$,

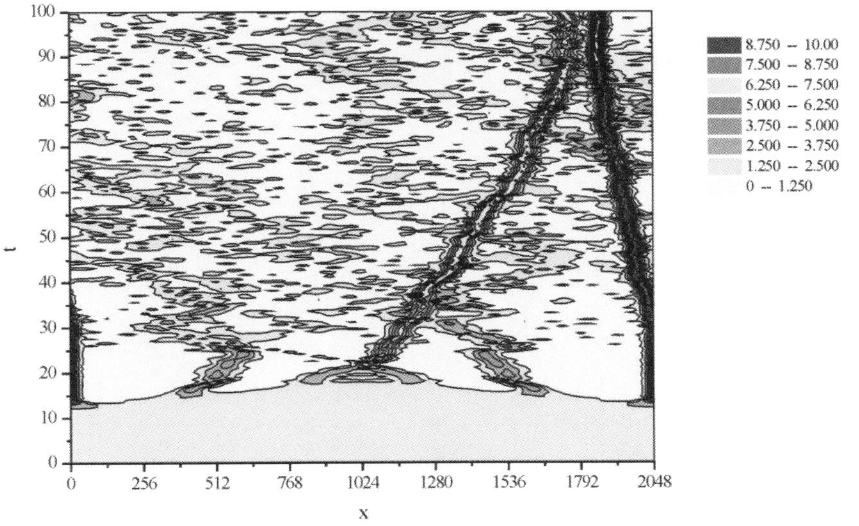

Fig. 2. The contour for $k = 0.12$. Five solitary patterns at $x = \pm(m + m'/5)L$, $m = 0, 1, 2, \ldots, m' = 1, 2, 3, 4$, are excited. At $t \simeq 20$, two solitary patterns initially at $x \simeq 800$ and 1200, respectively, collide and fuse into a new incoherent pattern accompanied strong ion-sound emission, then at $t \simeq 40$ the latter collides with the solitary pattern initially at $x = 1600$ to fuse into another new incoherent pattern.

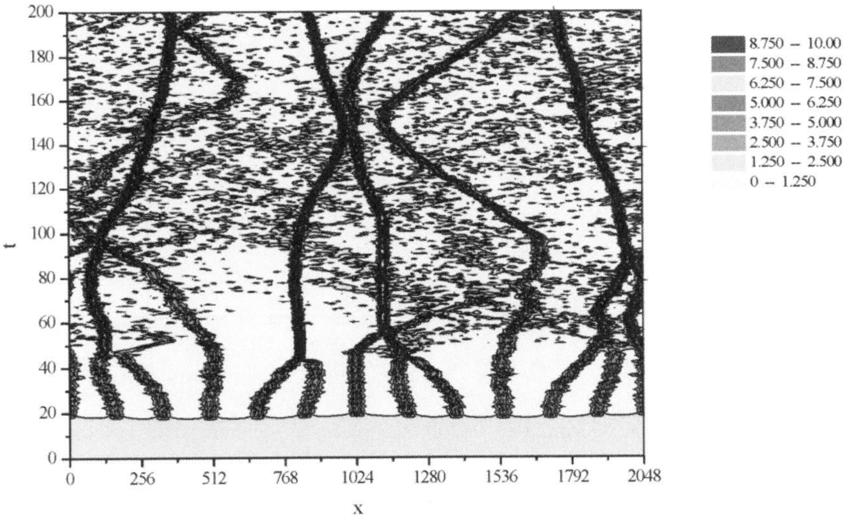

Fig. 3. The contour for $k = 0.049$. 12 solitary patterns at $x \simeq \pm L(m + m'/12)L$, $m' = 1, 2, \ldots, 11$, are excited and saturated. The first collision happened during $t = 60$–70, to fuse into four incoherent patterns from eight solitary patterns, accompanied strong ion-sound emission. With the lapse of long time, a few incoherent patterns are formed from collisions and fusions, the ion-sound emission spreads throughout, STC emerges.

where the ion-sound density $n = -|E|^2 + I$, is observed simultaneously, and the solitary patterns are seriously distorted. Thus, STC is emerging.

When k is further reduced, such as $k = 0.049$, 12 solitary pattern trains are excited and saturated. Soon collisions and fusions among some trains take place and the new

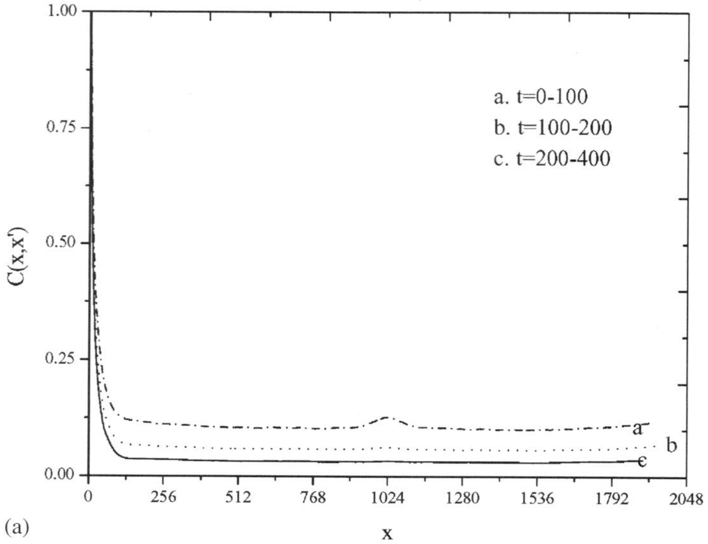

(a)

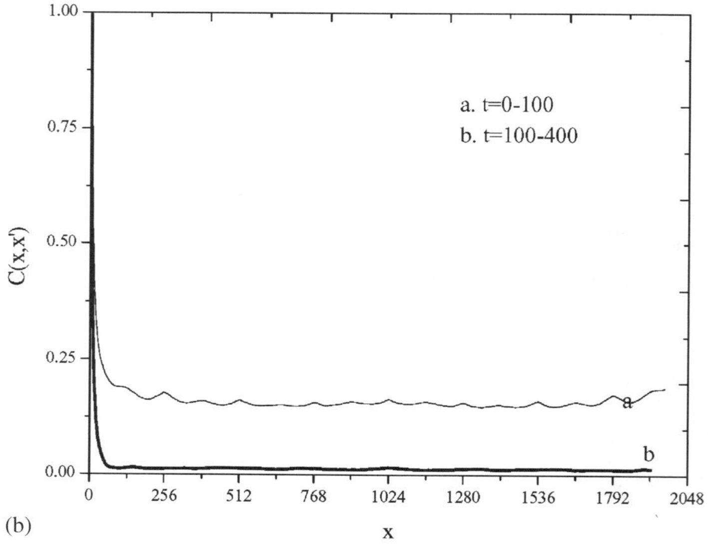

(b)

Fig. 4. (a) $C(x,x')$ for $k = 0.16$. During $t \simeq 0$–200, the partially coherent state is shown. After $t \geqslant 200$, $C(x,x')$, indeed, is of the behavior of the exponential decay $e^{-x/\xi}$, the system is in STC state. (b) $C(x,x')$ for $k = 0.08$. After $t \geqslant 100$, the behavior of the exponential decay is clearly shown, and the correlation length $\xi \simeq 30$, the system is in STC state.

incoherent pattern trains are formed accompanying strong ion-sound emission, then these incoherent patterns collide with others again and again with the lapse of long time. Finally, the early 12 solitary pattern trains are fused into a few incoherent patterns in stochastic motion with the greater part of the energy in system, as shown in Fig. 3, meanwhile many stable modes ($k > 1$) are also excited to share the energy. The system energy in the STC state is transfered from one place to the others in collision and fusion processes. Thus, the pattern evolution (collision, fusion, distortion) and the ion-sound emission can cause weak turbulence if there initially exist many excitable lengths. In fact, finally the dissipative effect will play an important role after a time greater than the damping time, the incoherent patterns will gradually disappear as the damping rate is included in Eqs. (1).

In order to further comprehend STC, we now discuss the behavior of the spatial two-point correlation function $C(x,x') = \langle(|E(x,t)| - \langle E\rangle)(|E(x',t)| - \langle|E|\rangle)\rangle$ [6], where $\langle\cdots\rangle$ presents the average over time. If $C(x,x')$ manifests the exponential decay behavior $e^{-x/\xi}$, a useful concept of correlation length $\xi \ll L$ (the system size) can be introduced. For $k = 0.52$ the numerical result shows that only single solitary pattern train with the stochastic motion appears, $C(x,x')$ is not exactly exponential decay and the correlation length can not be defined. It means that the solitary pattern is coherent, however, the Lyapunov exponents are positive [2], so the system is in the TC state. For $k = 0.16$, Fig. 4a shows that $C(x,x')$ remains a partially coherent effect due to rare collision at the early stage of $t = 0$–100, but the system evolutes gradually into STC after $t = 400$. For $k = 0.08$, $C(x,x')$ is plotted in Fig. 4b, and similar results are observed as in Fig. 4a, pattern trains experience a partially coherent stage and then transit from TC to STC. These results imply that the collision and the fusion among pattern trains cause the STC state, i.e., weak turbulence. It is natural that the second critical point k_{cs} exists, where the transition from TC to STC occurs. In the present model k_{cs} is less than 0.5 but greater than 0.16.

In summary, many coherent solitary patterns can be excited and saturated from unstable harmonic modes through modulational instability in the conserved Zakharov equations. The evolution of solitary patterns experiences three states: spatiotemporal coherence; temporal chaos but spatially partial coherence; spatiotemporal chaos that is caused by collisions and fusions among patterns with strong ion-sound emission. As a result, the energy in the system is strongly redistributed. If there initially exist many excitable lengths, the pattern evolution may switch on the onset of weak turbulence in plasma.

Acknowledgements

This work is supported by National Natural Science Foundation of China under Grant No. 19735002 and National Basic Research Project "Nonlinear Science" in China.

References

[1] V.E. Zakharov, Zh. Eksp. Teor. Fiz. 62 (1972) 1745.
[2] X.T. He, C.Y. Zheng, Phys. Rev. Lett. 74 (1995) 78.
[3] F.B. Rizzato, G.I. de Oliveira, R. Erichsen, Phys. Rev. E 57 (1998) 2776.
[4] X.T. He, C.Y. Zheng, J. Kor. Phys. Soc. (Proc. Suppl.) 31 (1997) s139.
[5] F.Kh. Abdullaev, Phys. Rep. 179 (1989) 1.
[6] P.C. Hohenberg, M. Shraiwar, Physica D 37 (1989) 109.

PHYSICAL REVIEW E **66**, 037201 (2002)

Harmonic modulation instability and spatiotemporal chaos

X. T. He,[1,2] C. Y. Zheng,[1] and S. P. Zhu[1]
[1]*Institute of Applied Physics and Computational Mathematics, P.O. Box 8009, Beijing 100088, China*
[2]*Department of Physics, Zhejiang University, Hangzhou 310027, China*
(Received 31 January 2002; published 18 September 2002)

It is shown from the conserved Zakharov equations that many solitary patterns are formed from the modulational instability of unstable harmonic modes that are excited by a perturbative wave number. Pattern selection in our case is discussed. It is found that the evolution of solitary patterns may appear in three states: spatiotemporal coherence, chaos in time but the partial coherence in space, and spatiotemporal chaos. The spatially partial coherent state is essentially due to ion-acoustic wave emission, while spatiotemporal chaos characterized by its incoherent patterns in both space and time is caused by collision and fusion among patterns in stochastic motion. So energy carried by patterns in the system is redistributed.

DOI: 10.1103/PhysRevE.66.037201 PACS number(s): 05.45.Jn, 89.75.Kd, 47.54.+r, 52.35.Mw

Pattern formation in spatially extended nonequilibrium systems has become a major frontier area in science [1]. The existence of spatiotemporal chaos (STC) characterized by its extensive, incoherent (irregular) pattern dynamics in both space and time significantly enriches the study of pattern formation. STC has been discussed numerically and experimentally in many physical systems such as dissipative hydrodynamics [1–3], optics [4], liquid crystals [5], etc. However, the mechanism leading to STC still remains unclear. Recently, transition to STC-like phenomena with coexisting temporal chaos (TC) and spatially partial coherence (SPC) from solitary patterns excited by a perturbative wave number (the master mode) and unstable harmonic modes in a plasma system through the modulational instability of a spatially homogeneous background field has been observed in numerical simulations [6,7]. In this paper, we consider the evolution of patterns in space and time using the well-known Zakharov equations. It is found that the evolution of solitary patterns may undergo the following states: spatiotemporal coherence, coexistence of TC and SPC, as well as STC. Collision and fusion among solitary patterns in the stochastic motion are the main cause of STC.

The conserved one-dimensional Zakharov equations (ZEs) in dimensionless form [8]

$$i\partial_t E+\partial_x^2 E=nE,$$

$$\partial_t^2 n-\partial_x^2 n=\partial_x^2|E|^2 \qquad (1)$$

are used as a model for studying the evolution of patterns. Here $E(x,t)$ is a slowly varying envelope of the Langmuir wave electric field and $n(x,t)$ is the ion density perturbation. The ZEs and their derivatives have been widely used to describe laser-plasma interaction, strong turbulence in plasmas [8], as well as many other physical phenomena. The study of the transition to STC based on the ZEs may therefore contribute to the understanding of the onset of plasma turbulence as well as the evolution of patterns in various branches of science.

Linear stability analysis of the perturbation $\exp[i(kx-\omega t)]$ of a spatially homogeneous field E_0 for Eqs. (1) gives the growth rate of the modulational instability as [6]

$$\gamma=E_0k[\sqrt{(1-2E_0^2k^2)^2+8E_0^2}-(1+2E_0^2k^2)]^{1/2}, \qquad (2)$$

for $0<k<1$. Here the unstable wave number k has been renormalized by $\sqrt{2}E_0$. Equations (1) constitute a near-integrable Hamiltonian system and have at least two types of fixed points: one (0,0) is a center and the other (E_0,0) a saddle.

To investigate the global behavior, we add a small spatial inhomogeneity at $t=0$ on the spatially homogeneous saddle state $\Psi_0=(\mathrm{Re}\,E,\mathrm{Im}\,E,n,\partial_t n)_{t=0}=(E_0,0,0,0)$ as follows:

$$\Psi_{t=0}=\Psi_0+(c_1f_1+c_2f_2)E_0\cos(kx)/500, \qquad (3)$$

where f_1 and f_2 are the components of the eigenvector in the saddle subspace for the linearized Eqs. (1) with the constant coefficients c_1, c_2, and the factor 1/500 is to emphasize that the perturbation is very small. The initial condition ensures that the manifolds in the phase space will locally lie in a saddle subspace.

Spatial modulation from the unstable wave number results in a local accumulation of the wave field E and a depletion of the ion density there. Since $n<0$, the latter leads to an increase of the index of refraction $\eta=(1-\omega_{pe}^2/\omega_0^2)^{1/2}$, where ω_0 and $\omega_{pe}=[4\pi(n_0+n)e^2/m_e]^{1/2}$ are the wave and the electron plasma frequencies, respectively, n_0 is the average plasma density. Thus the wave energy is further enhanced in the cavity. When the unstable wave number $k\leqslant 1$, corresponding to a small growth rate γ, the waves are trapped by the ion density cavity, and a solution of Eqs. (1) is $n=-|E|^2/(1-u^2)\sim\mathrm{sech}^2(\alpha\xi)$, where $\xi=x-ut$ and the sound speed satisfies $u\ll 1$ [8]. That is, the localized structure propagates below the ion sound speed. The ZEs are derived under the quasineutrality assumption. When space charge effects are included the solution still has a soliton structure (the dressed soliton) but with additional small spatial oscillations for larger ξ. When the frequency shift vanishes and $u\ll 1$, the dressed soliton degenerates to the simple soliton here [9].

As k decreases, the stationary state gradually disintegrates. Ion-acoustic wave emission [$I(x,t)$] becomes important. The solution of Eqs. (1) may be approximately expressed as $n(x,t) \simeq -|E(x,t)|^2 + I(x,t)$ [7], and the system is more unstable. This results in a distortion of the solitary pattern structure.

As $k<1/2$, the master mode k can in principle result in the excitation of $N-1$ unstable harmonic modes, where $N=[k^{-1}]$. Many solitary patterns may be formed from the modulation lengths $l_m = L/m$, where $m=1,2,3,\ldots,M$, $l_1 = L = 2\pi k^{-1}$ is for the master mode, and the others are for the unstable harmonic modes. The envelope E can be written as

$$E(x,t)=\sum_{m=1}^{M} E_m(t)\exp(imkx)+\sum_{m=M+1}^{\infty} E_m(t)\exp(imkx), \tag{4}$$

where the first term on the right-hand side of Eq. (4) comes from the master mode and unstable harmonic modes, and $M<N-1$ is from pattern selection. The second term on the right-hand side of Eq. (4) is from nonlinear interaction.

In the discussion below, for convenience the normalized time-averaged spatial two-point correlation function $\Gamma(x-x')=\Gamma(r)$ [1] and the temporal correlation function (or the Lyapunov exponent) will be used to demonstrate the spatiotemporal features of pattern evolution in the system. As discussed in Refs. [1] and [10,11], if $\Gamma(r)$ behaves like $\sim\exp(-r/\xi)$ for $r\to\infty$ and $\xi>L$ the system is spatially coherent (regular), or $\Gamma(r)=1$, where ξ is the correlation length and L is the system size. In the opposite limit $\xi \ll L$, or $\Gamma(r)\to 0$, the system is spatially incoherent (irregular). However, as we shall see below, in many situations there exists an intermediate state that is SPC, or $0<\Gamma(r)<1$.

We now proceed with the numerical solution of Eqs. (1) and (3) to consider the pattern dynamics and STC. The split-step method for E with a preestimate for the density perturbation n is used to integrate Eq. (1) with periodic boundary conditions. The largest length L from the master mode ($kL=2\pi$) is chosen. The number of grids is 2048 for L and $E_0=2.0$ is taken. In the computation, the conserved quanties are preserved to order 10^{-8}. Furthermore, unless otherwise stated, we set $c_1=c_2=1.0$ (far from homoclinic orbit [10]) without loss of generality.

It has been shown [6] that for $1/2 \leqslant k \leqslant 1$, for each set of c_1 and c_2 there exists a critical wave number k_c. Since $k_c<k<1$, the motion of the solitary patterns is temporal recurrent (periodic) or pseudorecurrent (quasiperiodic), and that of the solitary pattern is STC. Since $1/2 \leqslant k \leqslant k_c$, the motion of the center of the solitarylike pattern, whose initial peak is located at $x=\pm mL$, $m=0,1,2,\ldots$, exhibits stochastic behavior [10]. The amplitude of the pattern oscillates and its width varies temporally. The system is in TC, as shown by the positive Lyapunov exponent, and the spatial behavior is of SPC due to ion-acoustic wave emission. It has been shown [6] that resonant overlapping may be the cause for TC.

New phenomena are observed first for $k<1/2$ ($N>1$). First, a few harmonic modes are excited. For example, for $k=0.23$ the solitary pattern from the master mode, or master pattern, is first formed, Then patterns from the harmonic modes begin to appear. Two points are of interest: first, during the pattern formation, ions are repelled from the regions

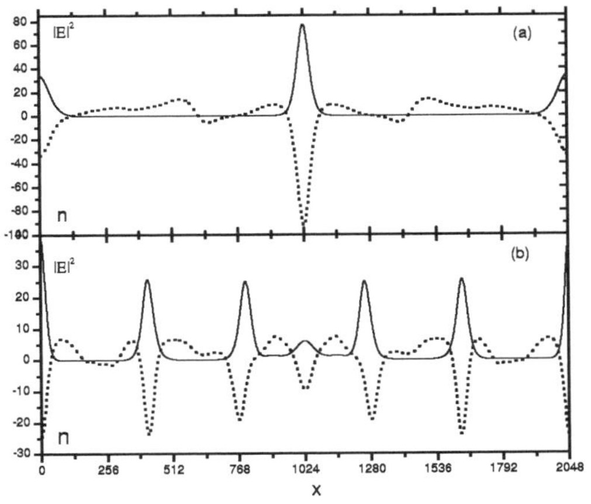

FIG. 1. $|E(x,t)|^2$ (solid line) and $n(x,t)$ (dotted line) vs x at $t=10$. The positive ion perturbation density n is distributed among the patterns formed. This shows pattern selection in the process of the pattern formation. (a) $k=0.23$, (b) $k=0.11$.

of the master mode and some harmonic patterns by the wave ponderomotive force and form spatial distributions with $n>0$ (dotted line of Fig. 1). These density humps, consisting of ions ponderomotively expelled by the master and harmonic wave fields, move stochastically on either side of the patterns. The waves cannot be trapped because of a corresponding decrease of the index of refraction. Second, for the case studied, no initial perturbation at $X/L=1/4$ and 3/4 appears. Finally, only one harmonic pattern is later formed at $X/L=1/2$. Thus, pattern selection leads to less than $N-1$ observed harmonic patterns. As shown in Fig. 2, actually only one master pattern and one harmonic pattern eventually move stochastically in the form of pattern trains. As a result, similar to the regime $1/2 \leqslant k \leqslant k_c$, the system is in TC and SPC.

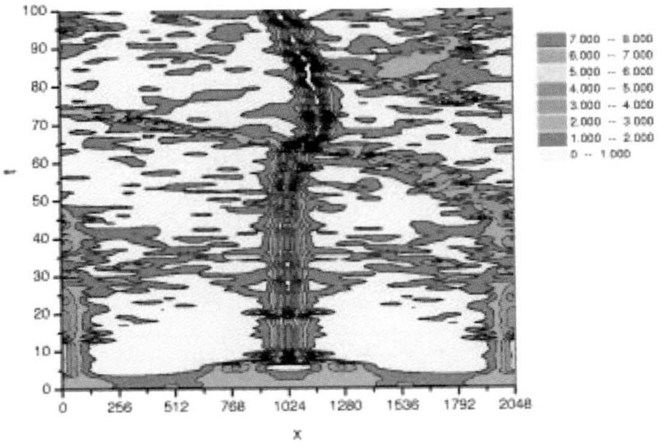

FIG. 2. (Color online) Contours of $|E(x,t)|=\text{const}$ for $k=0.23$. The solitary master pattern, initially peaked at $x=\pm mL(m=0,1,2,\ldots)$, is formed by the master mode. The pattern selection leads to only one harmonic pattern, initially peaked at $x=L/2$. The system is in the coexistent TC and SPC state.

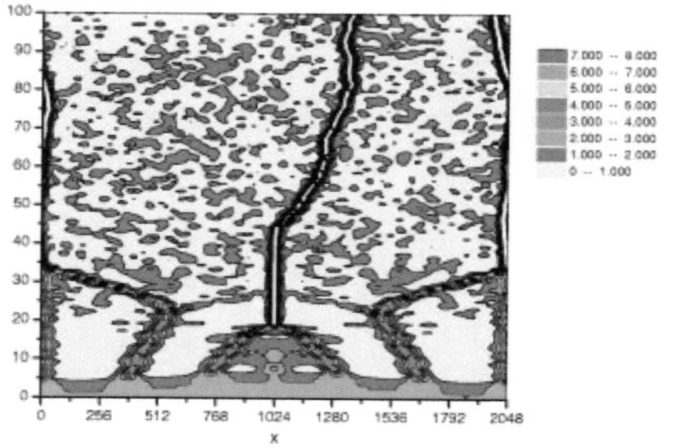

FIG. 3. (Color online) Contours of $|E(x,t)|=\mathrm{const}$ for $k=0.11$. Five solitary patterns, initially peaked at $x\sim\pm(m+m'/5)L$, $m=0,1,2,\ldots$, $m'=1$, 2, 3, 4 are formed. Two collisions, at $t\simeq20$ and 33, respectively, occurred. The system is still in the coexistence of TC and STC.

As k reduces to 0.11, pattern selection leads to only five solitary patterns [Fig. 1(b)], formed initially from the master and harmonic modes (Fig. 3). At $t\simeq20$, two solitary patterns, initially peaked at $x\simeq768$ and 1280, respectively, collide and fuse into a new pattern with a strengthened amplitude and narrower width. At the same time, strong ion-acoustic wave emission $I(x,t)>|E(x,t)|^2$ is observed everywhere and the pattern structures are distorted. At $t\simeq33$, the other patterns, initially peaked at 384 and 1664, collide with the master pattern and fuse into another new one. After two collisions, the four-harmonic patterns produce one new pattern that coexists with the distorted master pattern. However, as seen in Fig. 4, case (1), the spatial correlation function is still in SPC, i.e., $0<\Gamma(r)<1$. This implies that a few collisions do not seem to be enough to cause STC. The coherence in the system is partially retained, so that TC and SPC still coexist in the system.

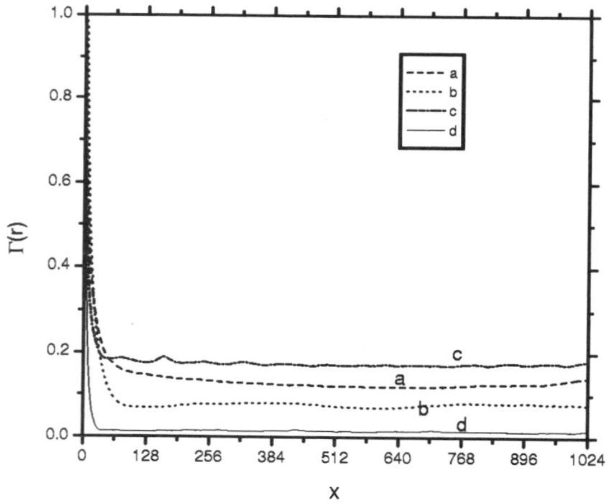

FIG. 4. The spatial correlation function $\Gamma(r)$ vs r: (1) for $k=0.11$, and (a) $t=0-100$, (b) $t>100$; (2) for $k=0.047$, and (c) $t=0-100$ (the system is in SPC), (d) $t>100$ (the system is in STC), with the correlation length $\xi\simeq10\ll2048$.

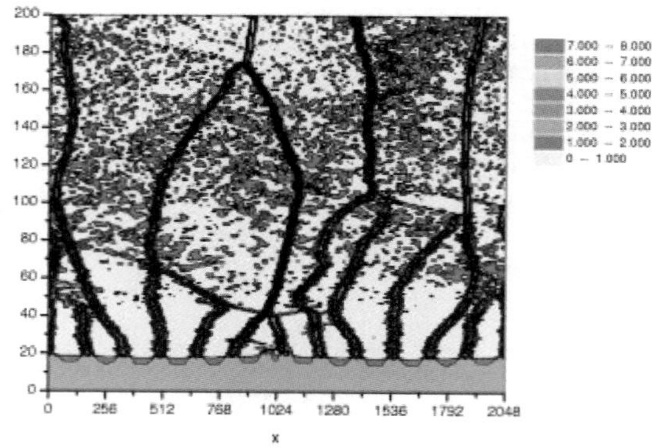

FIG. 5. (Color online) Contours of $|E(x,t)|=\mathrm{const}$ for $k=0.047$. Twelve solitary patterns, initial peaks at $x\sim\pm(m+m'/12)L$, $m=0,1,2,\ldots$, $m'=1,2,\ldots,11$ are formed. After some time, several incoherent patterns are fused from the original twelve solitary patterns by collisions. Ion-acoustic wave emission also occurs everywhere and the STC state emerges.

We now consider the case when many unstable modes are excited and saturated to form initially twelve solitary pattern trains by pattern selection (Fig. 5). The first collision occurs at $t\simeq40$. The four patterns, initially at $x\simeq170$, 340, 820, and 1100, respectively, are fused into two new patterns, accompanied by strong ion-acoustic wave emission. At $t\simeq90-110$, collisions again take place among some of the pattern trains, and new incoherent patterns accompanied by strong ion-acoustic wave emission appear. At longer times, the new incoherent patterns collide with each other repeatedly. The patterns are also much distorted and the original twelve solitary pattern trains are finally fused into a few incoherent patterns. For $t>100$, the spatial correlation function $\Gamma(r)\rightarrow0$ and the system tends to STC. The correlation length $\xi\simeq10\ll L=2048$ is shown in Fig. 4, case (2). A certain amount of the system energy is carried by the incoherent patterns as well as many stable high-harmonic modes with shorter wavelengths ($k>1$) excited through nonlinear interaction. The system energy is thus spatially redistributed in the process of pattern collision and fusion. Therefore, if initially there exist many unstable modulation lengths to form patterns, collision and fusion of many pattern trains can lead to the STC state. There should then exist a critical point k_{cs} where STC transition from TC can occur. In our model, k_{cs} is less than 0.11 but larger than 0.08.

In summary, solitary patterns may be formed by the harmonic modes excited by a spatially modulational master mode. If only the master pattern ($1/2<k<1$) exists, the system may be in either the STC state or the TC and SPC state, depending on the modulation wave number [6]. However, as $k<1$, pattern selection restricts the number of the harmonic patterns formed to be less than that of the unstable harmonic modes. If a few harmonic patterns coexist with the master pattern, the system will still be in TC and SPC. However, if many harmonic patterns occur in the early phase, the system may experience the SPC state. With the lapse of time collision and fusion among patterns occur frequently and several new incoherent patterns are formed. The system enters the STC state. Energy is transferred to the stable high-harmonic modes with shorter wavelengths, and transport occurs in the system. In addition, as seen in Fig. 4, the more the patterns formed, the faster is the dynamic transition to STC. This implies that the excitation of a sufficient number of spatially harmonic modulation lengths is the main condition to cause STC.

This work is supported by National Natural Science Foundation of China under Grant No. 19735620 and the Special Funds for Major State Basic Research Projects.

[1] M.C. Cross and P.C. Hohenberg, Rev. Mod. Phys. **65**, 851 (1993).
[2] H.W. Xi, J.P. Gunton, and J. Vinals, Phys. Rev. Lett. **71**, 2030 (1993); Y. Hu, E. Ecke, and G. Ahlers, *ibid.* **74**, 391 (1995); H.W. Xi, X.J. Li, and J.D. Gunton, *ibid.* **78**, 1046 (1997); K.M.S. Bajaj, J. Liu, B. Naberhuis, and G. Ahlers, *ibid.* **81**, 806 (1998).
[3] B.J. Gluckman, P. Marcq, J. Bridger, and J.P. Gollub, Phys. Rev. Lett. **71**, 2043 (1993).
[4] F.T. Arecchi, G. Giacomelli, P.L. Rammazza, and S. Riesidor, Phys. Rev. Lett. **65**, 253 (1990); J.V. Moloney and A. Newell, Physica D **44**, 1 (1990); S.W. Morris, E. Bodenschatz, D.S. Cannel, and G. Ahlers, Phys. Rev. Lett. **71**, 2026 (1998).
[5] S. Rudroff and I. Rehberg, Phys. Rev. E **55**, 2742 (1997).
[6] X.T. He and C.Y. Zheng, Phys. Rev. Lett. **74**, 78 (1995).
[7] F.B. Rizzato, G.I. deOliveira, and R. Erichsen, Phys. Rev. E **57**, 2776 (1998).
[8] V.E. Zakharov, Zh. Eksp. Teor. Fiz. **62**, 1745 (1972) [Sov. Phys. JETP **35**, 908 (1972)].
[9] P. Deeskow *et al.*, Phys. Fluids **30**, 2703 (1987).
[10] X.T. He and C.Y. Zheng, J. Korean Phys. Soc. **31**, s139 (1997).
[11] P.C. Hohenberg and M. Shraiwar, Physica D **37**, 109 (1989).

CHAOS 16, 033113 (2006)

A note on chaotic unimodal maps and applications

C. T. Zhou and X. T. He
Institute of Applied Physics and Computational Mathematics, P. O. Box 8009, Beijing 100088, People's Republic of China
M. Y. Yu
Institut für Theoretische Physik I, Ruhr-Universität Bochum, D-44780 Bochum, Germany
L. Y. Chew
Division of Physics and Applied Physics, School of Physical and Mathematical Sciences, Nanyang Technological University, Nanyang Avenue, Singapore 639798
X. G. Wang
Department of Physics, National University of Singapore, 21 Lower Kent Ridge Road, Singapore 119077

(Received 1 April 2006; accepted 6 June 2006; published online 30 August 2006)

Based on the word-lift technique of symbolic dynamics of one-dimensional unimodal maps, we investigate the relation between chaotic kneading sequences and linear maximum-length shift-register sequences. Theoretical and numerical evidence that the set of the maximum-length shift-register sequences is a subset of the set of the universal sequence of one-dimensional chaotic unimodal maps is given. By stabilizing unstable periodic orbits on superstable periodic orbits, we also develop techniques to control the generation of long binary sequences. © *2006 American Institute of Physics*. [DOI: 10.1063/1.2218048]

Chaotic pseudorandom number sequences are especially relevant to code optimization in an asynchronous direct sequence code-division multiple-access system. It is hoped that chaotic codes are of potential application for the third- or fourth-generation mobile communication systems. Some previous research has showed that truncated and quantized chaotic sequences allow performances that are superior to those of classical sequences. In this work, we are interested in directly applying chaotic periodic orbits (stable or unstable) for binary sequence generators. We first consider the simplest one-dimensional unimodal map, and find that the set of the maximum-length shift-register sequences is a subset of the set of its corresponding kneading sequences. With such a conclusion, we have established the link between chaotic periodic sequences and conventional maximum-length shift-register sequences, or *m* sequences. Moreover, we apply the word-lift technique of symbolic dynamics of one-dimensional chaotic unimodal maps to investigate the limiter controller. We analytically obtain control parameters for stabilizing an arbitrarily long unstable periodic sequence. The results can be relevant to understanding chaos-based sequences and securing data transmission, as well as analysis and manipulation of noisy data from laboratory experiments and astrophysical observations.

I. INTRODUCTION

Since Lorenz[1] discovered the phenomenon of deterministic chaos, the theory and application of chaos have undergone rapid development,[2] especially in the past decade. Chaos theory is about finding the underlying order in apparently random conditions or data.[3–7] The most important property of chaotic systems is their sensitive dependence on the initial conditions: a small change in the latter can drastically change the long-term behavior of the system. Chaos has applications in many fields, including science, medicine, finance, as well as engineering. It is also natural to make use of the properties of chaotic systems in providing security in modern communications.

Communication security can be realized using two strategies: cryptography or spreading spectrum system.[8] For example, advanced encryption standard, data encryption standard, and public-key cryptosystem are commonly used encryption standards,[9,10] while *m*, Gold, and Kasami sequences are popular pseudorandom (PN) codes for spread-spectrum systems.[11,12] While *m* sequences are the longest sequences that can be generated by a linear feedback shift-register (LFSR) of a given length, Gold sequences are constructed by taking a pair of specially selected *m* sequences, called preferred *m* sequences, and forming the modulo-2 sum of the two sequences for each of the cyclically shifted versions of one sequence relative to the other. Kasami sequences are constructed by decimating an *m* sequence.

Besides these well-known PN codes, there are other binary sequences, especially that based on chaos,[13–15] that have been investigated for spread-spectrum communications.[16] Chaos is a state of nonlinear dynamical systems. It has the essential characteristics of randomness, which enables the generation of noiselike chaotic waveforms that possess the property of low probability of detection (LPD). In addition, with chaotic dynamical system being extremely sensitive to small changes in the initial conditions, it also has the property of low probability of intercept (LPI) of the communication signals. Thus, by employing chaos instead of the conventional PN codes to the spread-spectrum system, one can achieve not only LPD but also LPI. Furthermore, a system that sensitively depends on the initial state is also capable of suppressing interference from jamming and multipath, and can be used as a multiple-access device, with the potential of accommodating a larger number of users.

Since chaos is of ergodic and "Lebesgue spectrum" nature,[17] it is not too surprising that it exhibits such features as both LPD and LPI. However these properties are mainly based on the assumption that a chaotic system can generate an ideal scrambling/spreading code consisting of an *infinite* sequence (related exactly to a real-value system) of random codes, an unrealistic task for finite digital hardware. Thus, in practice usually periodic PN codes are used.[17–19] On the other hand, from a topological point of view chaos should consist of an infinite number of unstable periodic orbits (UPOs). It is then natural to replace the m sequences with these chaotic binary UPO sequences in direct-sequence spread-spectrum communication systems. Although substantial knowledge already exists in this area,[15,17–19] some crucial points still need to be resolved. One of the most important is to clarify the properties of both the UPO and the conventional PN code, and to design a practical periodic PN sequence generator based on simple chaotic systems. In this paper we apply symbolic dynamics of 1D unimodal maps to analyze the relation between the m and U sequences [the latter is also known as the universal or kneading sequence, or superstable periodic orbits (SPOs) (Refs. 20 and 21)], and determine the control parameters of arbitrarily periodic orbits. The results here, in particular the finding that the set of m sequences is a subset of the U sequences of one-dimensional chaotic unimodal maps, can be useful to the understanding, control, and application of chaos in various fields.[2,8–14,21–25]

II. *U* AND *m* SEQUENCES

It is not easy to find a long UPO in the chaotic sea even for relatively simple dynamical systems. Usually much numerical effort is required and there are also methodical problems.[26–28] An important step was made by Metropolis, Stein, and Stein (MSS),[20] who proved that an infinite sequence of finite limit sets exists in any 1D unimodal map. This sequence constitutes the most interesting family of finite limit sets in virtue of the universality of their structures within the class of unimodal maps. To uncover these kneading sequences that form a MSS table, one can either employ the auxiliary construction approach[20] or a technique based on the periodic window theorem.[21] Since a SPO eventually becomes an UPO when the system becomes chaotic, the MSS table helps us to enumerate the set of UPOs up to a certain length. However, the family of U sequences can be very large, especially when the period is long. As shown in Table I, for example, there are 16 and 2048 distinct U sequences of periods 7 and 15, respectively. This is to be compared with 2 and 2 distinct m sequences with the same periods. When the period $L=2^m-1$, the number of U sequences is given by $N_U=2^L/(L+1)$.[21] Obviously, it is impractical to list a complete set of U sequences, especially for long (say, $L=2^{20}-1$) periods.[26,27]

TABLE I. The number N_m of m sequences and their preferred pairs for $3\leqslant m\leqslant 10$. This table also gives a comparison of m sequences with U sequences. The symbol "-" indicates that the number of U sequences for the case of period -2^m may be calculated by using $N_U=2^L/(L+1)$, where $L=2^m-1$.

m	L	N_m	N_U
3	7	2	16
4	15	2	2048
5	31	6	67108864
6	63	6	144115188075855872
7	127	18	1329227995784915872903807060280344576
8	255	16	226156424291633194186662080095093570025917938800079226639565593765455331328
9	511	48	⋯
10	1023	60	⋯

We start with a brief review of the LFSR, which is a typical finite-state machine and has been extensively studied.[11] Mathematically, a LFSR can be expressed as a feedback function $f(x)=\Sigma_{i=1}^{\oplus m}c_i x^i\oplus 1$, where $\oplus$ and $\Sigma^{\oplus}$ denote modulo-2 additions, and c_i and $x\in\{0,1\}$ are the feedback connections and states, respectively. With appropriate choice of the line feedback c in this function, one can obtain an m sequence. For example, using the line feedback $c=[1001]$ with the initial state $x=[0001]$, it is easy to get a period-15 m sequence 100011110101100. Surprisingly, this PN code corresponds to one of the U sequences of 1D unimodal maps.[20] This finding stimulates us to consider the question: does an arbitrary m sequence correspond to a U sequence? In principle, the answer should be direct if we have the completed sets of these two sequences.

Owing to the problem mentioned above concerning the U sequences,[28] we shall concentrate on m sequences. Using suitable linear combinations of the feedback taps c, it is easy to generate the maximal-length binary periodic sequence. An m sequence generated by an m-stage shift register with linear feedback has a period of $L=2^m-1$ bits. Each period has a sequence of 2^{m-1} ones and $2^{m-1}-1$ zeros. On the other hand, m sequences satisfy the minimum autocorrelation $C_{au}(\tau)=L$ if $\tau=0$, and -1 if $1\leqslant\tau\leqslant L-1$, where τ is time delay. To construct a LFSR for generating a useful code sequence, various analytical, experimental, and computational methods have been introduced.[12] Concerning the nature of the polynomials, there are two essentially different viewpoints that one may adopt. When 2^m-1 is a prime, one can directly compute the number of m sequences through a table of irreducible polynomials based on the sieve method. When 2^m-1 is not a prime, the sieve method must be supplemented by a deletion of certain extra polynomials. This deletion can be accomplished by applying synthetic techniques.[11]

Based on these properties of m sequences, we here present a simple method to compute the limit set of m sequences of length $L=2^m-1$ using the following steps: (1) without loss of generality, set initial states as $x_{i=m,m-1,\ldots,1}=[00\cdots01]$; (2) check the balance of the sequence generated by a function $f(x)$ with all possible feedback taps to ensure that the output sequence contains 2^{m-1} ones and $2^{m-1}-1$ zeros; (3) check autocorrelation of the output sequence to ensure that a new set of sequences satisfies the minimum correlation $C_{au}(\tau)$ given above. The sequences thus obtained then constitute a limit set of infinite m sequences. As an example, we consider the case $m=3$. First, we let the initial state be $x=[001]$. Four possible connections are $c=\{[100],[110],[101],[111]\}$. That is, $f=\{x^3+1, x^3+x^2+1, x^3+x+1, x^3+x^2+x+1\}$. We then get four sequences {1001001, 1001011, 1001110, 1001100}. The corresponding autocorrelations are $C_{au}=\{7\ -1\ -5\ 3\ 3\ -5\ -1,\ 7\ -1\ -1\ -1\ -1\ -1\ -1,\ 7\ -1\ -1\ -1\ -1\ -1\ -1,\ 7\ -1\ -5\ 3\ 3\ -5\ -1\}$. Therefore, these two sequences {1001011, 1001110}, corresponding to the two connections $c=[110]$ and $[101]$, respectively, form the set of m sequences with $m=3$. Using the above algorithm, it is straightforward to compute the set of m sequences.

Once the limit set of m sequences is obtained, we still need a method to check if these sequences belong to U sequences of chaotic systems. Recall that different SPO may appear when a bifurcation parameter is varied. For the logistic map $x_{n+1}=1-ax_n^2$, where x_n is the system state and $a \in \mathfrak{R}$ is the system parameter, for example, an arbitrarily given SPO must correspond to a real value of the system parameter a. According to the standard word-lifting technique of symbolic dynamics, this parameter value a for the logistic map may be calculated by using the iteration form:[21,29] $a_{n+1}=\sqrt{a_n+\sqrt{a_n \pm \cdots \pm \sqrt{a_n}}}$, where the signs "+" and "−" are associated with two symbols "1" and "0," respectively.

Because the bifurcation parameter $a \in (0,2]$ of the logistic map must be real, for an arbitrary U sequence its corresponding iteration solution $a=\lim_{n\to\infty} a_n$ will converge to a real value. In other words, a binary sequence is an admissible U sequence *if and only if* a real convergence value a of the corresponding orbit of the logistic map exists. If the iteration value a gives a complex value, the corresponding binary sequence does *not* belong to an admissible U sequence. To verify this, we first examine all possible sequences generated by the LFSR for $m \leqslant 10$. Numerical results show that four sequences have complex values, for example[30] $a \approx 1.9933694089997989 - i3.80024 \times 10^{-10}$ for the sequence 1000011111010100110001000011111 with the connection. [10001] Therefore, not all sequences generated by the LFSR belong to a subset of U sequences.

We now explore possible links between the m sequences and U sequences. When the period is relatively short (say, $L \leqslant 30$), we can calculate the U sequences by using the periodic window theorem.[21] Table II is a list of all the U sequences for $L=7$. Clearly, the sequences 1001011 and 1001110 are two m sequences. In fact, the two double-underlined sequences in Table II are a mirror transformation of these two m sequences. The mirror transformation P is defined as transforming zero to one and one to zero, that is, $P\cdot 0=1$ and $P\cdot 0=1$. We may then write a mirror set $S_{\bar{m}}=P\cdot S_m$ of m sequences (S_m). For example, we change $1001011 \to 0110100$, then shift the latter to get 1000110. This is a U sequence. Obviously, the feedback function of a mirror sequence of m sequences is $f(x)=(\Sigma_{i=1}^{\oplus m} c_i x^i \oplus 1) \oplus 1$. Thus, any sequence taken from the set $S_{\bar{m}}$ contains $2^{m-1}-1$ ones and 2^{m-1} zeros, and its autocorrelation is the same as

TABLE II. A list of all U sequences with period $L=7$. The iteration value a of the logistic map corresponds to the superstable parameter.

Index	Sequence	a_{real}	a_{imag}
1	1011101	1.3815474844320617	0
2	1011111	1.5218172316712510	0
3	1011011	1.7110794700131522	0
4	1001011	1.8100013857280119	0
5	1001010	1.8517300494106415	0
6	1001110	1.8700038808287653	0
7	1001111	1.8969179946788330	0
8	1001101	1.9170982771134220	0
9	1000100	1.9417820903181982	0
10	1000101	1.9607589871974289	0
11	1000111	1.9721998383668757	0
12	1000110	1.9816557862760444	0
13	1000010	1.9887932743094712	0
14	1000011	1.9943329667155349	0
15	1000001	1.9979629155977143	0
16	1000000	1.9997740486937274	0

the corresponding m sequence. With the use of such a mirror transformation, we can directly generate a U sequence by taking the initial state $x=[1\cdots 10]$. As an example, in Fig. 1 we show how two U sequences are generated from a three-stage LFSR and its mirror version.

As shown in Table I, the number (N_U) of U sequences is much larger than the number (N_m) of m sequences, that is, $N_U \ll N_m$. The next task is to check the iteration values (a in the logistic map) of all the m sequences with a given periodic length. We have examined all m sequences and its mirror-shifted version for $2 \leqslant m \leqslant 14$, as well as some longer sequences listed in Table 3.7 of Dixon[12] for $15 \leqslant m \leqslant 20$. Through extensive computation, we found that all these m sequences (examples are given in Table III) indeed give real iteration values. We can thus make the conclusion: *any set of m sequences and its mirror-shifted version is a subset of the U sequences*. Due to the fact that a SPO turns into an UPO when the system enters the chaotic state, we can apparently reach the conclusion that a subset of UPOs of chaotic 1D unimodal maps also corresponds to the set of m sequences. In addition, we should mention that Gold codes with certain

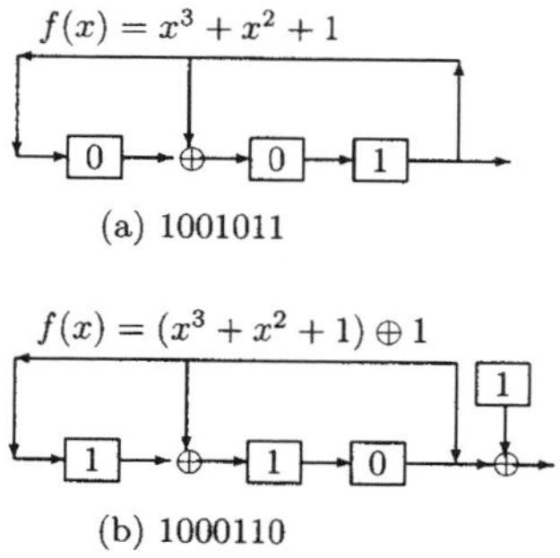

FIG. 1. (a) Generating two U sequences by employing three-stage shift register with linear feedback connection $c=[101]$, and (b) its corresponding mirror sequence generator.

TABLE III. A list of part m sequences for $m=20$. [12]. The initial condition [00...01] is used. Here, the iteration value a and ϵ are calculated by using word-lifting technique of symbolic dynamics of the logistic map, a the system parameter and ϵ the control parameter of the limiter control.

Index	Maximal taps	a [real, imag]	ϵ
1	10001001100101101011	[1.9446134990754280,0]	0.0912607988197669
2	10001001010011001111	[1.9497349974437252,0]	0.0886468944110794
3	10001001001110011110	[1.9525318980897151,0]	0.0869129052869011
4	10001011110011100100	[1.9575937983215543,0]	0.0824648228937067
5	10001010010110011011	[1.9644369800640302,0]	0.0762138551329609
6	10001110110011010010	[1.9696043203037448,0]	0.0702707134721874
7	10001110101101001100	[1.9713967053697421,0]	0.0683877400079509
8	10001111100110010100	[1.9724817187035448,0]	0.0670246734964305
9	10001100010010111110	[1.9840372769290706,0]	0.0518138039309020
10	10001100010111110100	[1.9841946547459823,0]	0.0516163288196166
11	10000101100011101110	[1.9889085303238492,0]	0.0426681106710850
12	10000101110111000110	[1.9900937834830228,0]	0.0404409706866883
13	10000101110110110001	[1.9901642546460481,0]	0.0403062580837193
14	10000101011111001001	[1.9908542276025702,0]	0.0388932288815860
15	10000111110101001001	[1.9933732540088001,0]	0.0331445415803064
16	10000110010011111010	[1.9955038991773151,0]	0.0273498335105652
17	10000110010010101111	[1.9955192056809790,0]	0.0273049094508865
18	10000110001101101110	[1.9961394159360080,0]	0.0254210453543694
19	10000010110111001110	[1.9973664089105774,0]	0.0209161901643347
20	10000010111011100101	[1.9975406242027876,0]	0.0202182567412865

positional shift are also found to belong to the subset of UPOs (see Table IV). This has the important implication that even periodic sequences (at least for the m sequence and Gold codes) of chaotic systems can produce PN codes with good randomness and correlation properties.

TABLE IV. Period-15 Gold sequences generated by using pair of connections {[1 1 0 0], [1 0 0 1]}. Here, the symbol "→" stands for the position shift toward the right (four sequences labeled with shift "0" are four m sequence), and "order" stands for the order of the parameter a in the logistic map (also the order in the MSS table). The sequence whose order is indicated by "UPO" cannot be found from the MSS table for any position shift. However, it does correspond to an UPO at a certain positional shift.

Index	Gold sequences	←	U sequences	Order
1	100010011010111	0	100010011010111	554
2	100011110101100	0	100011110101100	920
3	000001101111011	14	100000110111101	1756
4	100101110001110	7	100011101001011	857
5	101101001100101	5	100110010110110	482
6	111100110110011	3	100110110011111	466
7	011111000011111	5	100001111101111	1396
8	011000101000110	8	100011001100010	1107
9	010111111110100	12	100010111111110	674
10	001001010010000	10	100000010010100	1818
11	110100001011000	3	100001011000110	1222
12	001110111001001	14	100111011100100	UPO
13	111011011101010	2	101101110101011	42
14	010000010101101	1	100000101011010	1655
15	000110000100010	13	100001100001000	1549
16	101010100111100	6	100111100101010	347
17	110011100000001	6	100000001110011	1955

III. GENERATING *m* SEQUENCES VIA UNIMODAL MAPS

With the knowledge of UPOs of 1D unimodal maps that possess good randomness and correlation properties, the next question is whether we can extract them from the chaotic system. To stabilize UPO in 1D unimodal maps, a simple method is to transform the UPO into its corresponding SPO.[18,19,31] By using the word-lifting technique of symbolic dynamics, we can then estimate the control parameter of a superstable periodic window.[18,19] To illustrate the approach, we shall still use the logistic map. Its corresponding limiter control system may be written into[31–34]

$$x_{n+1}=f(x_n)=\begin{cases}1-ax_n^2 & \text{if } f(x_n)<h,\\ h & \text{if } f(x_n)\geqslant h,\end{cases} \tag{1}$$

where h is a limited height. Generalizing the standard symbolic dynamics to this system, we obtain the control parameter by an iteration form,[18,19]

$$\epsilon_c=\sqrt{\frac{1}{a}\left[1-\frac{1}{a}g(a,x_0)\right]}, \tag{2}$$

where

$$g(a,x_0)=\sqrt{a+\sqrt{a\cdots\pm\sqrt{a(1-x_0)}}} \tag{3}$$

and x_0 is defined by

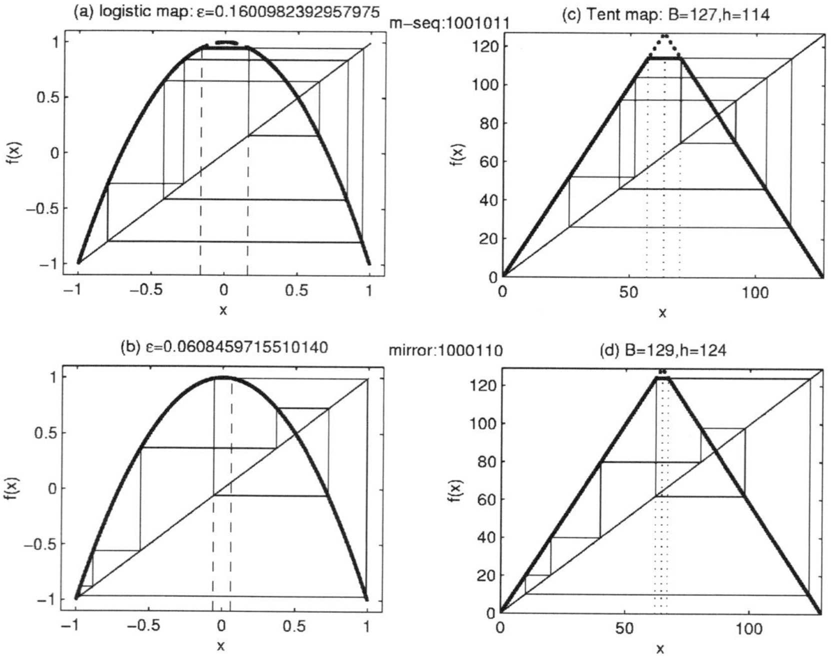

FIG. 2. Generating a period-7 m sequence and its mirror sequence using 1D unimodal maps with the limiter control scheme; (a) and (b) for the logistic map; (c) and (d) for the digital tent map. In order to transform the desired m sequence into the same periodic SPO of the system, we implement here a period-7 binary sequence by replacing the last bit 1 (or 0) with the symbol C_R (or C_L), which are associated with the right and left boundaries of the control window shown in the four subfigures.

$$x_0 = \begin{cases} -\epsilon_c & \text{if } C_s = C_L, \\ 0 & \text{if } C_s = C_0, \\ \epsilon_c & \text{if } C_s = C_R. \end{cases} \tag{4}$$

In Table III, we give also the control parameter values ϵ. However, it should be mentioned that using these values cannot perhaps generate the very long periodic sequences robustly, since any rounding error of the iteration map can cause the calculated value to go out of the accuracy range. To stabilize a long UPO on its nearest SPO of the controlled logistic map, a more accurate iteration value ϵ (thus requiring a high-bit computer) is necessary. On the other hand, one may employ an integer system, such as the digital tent map[36,37] to solve the rounding error problem. The controlled digital tent map may be written as

$$x_{n+1} = \begin{cases} 2x_n & \text{if } 0 \leqslant f(x_n) < h/2, \\ h & \text{if } h/2 \leqslant f(x_n) \leqslant B - h/2, \\ 2(B - x_n) & \text{if } B - h/2 < f(x_n) \leqslant B, \end{cases} \tag{5}$$

where the state variable x_n and the base B are integers, and the function f is defined as $x_{n+1}=f(x_n,B,h)$. We also define a symbol rule

$$s = \begin{cases} R & \text{if } B - h/2 < x_n \leqslant B, \\ C_R & \text{if } B/2 < x_n \leqslant B - h/2, \\ C_0 & \text{if } x_n = B/2, \\ C_L & \text{if } h/2 \leqslant x_n < B/2, \\ L & \text{if } 0 \leqslant x_n < h/2, \end{cases} \tag{6}$$

and set $f_R^{-1}=B-x_0/2$ and $f_L^{-1}=x_0/2$, and

$$B - \underbrace{\tfrac{1}{2}\cdot\tfrac{1}{2}}_{L}\cdots\Big[\underbrace{B - \tfrac{1}{2}(\cdots x_0)}_{R}\Big] = h, \tag{7}$$

where

$$x_0 = \begin{cases} h/2 & \text{if } C_s = C_L, \\ B/2 & \text{if } C_s = C_0, \\ B - h/2 & \text{if } C_s = C_R. \end{cases} \tag{8}$$

Here, the state variable x_n and the base B are integers. For an arbitrary UPO sequence, the expression (8) can give accurate analytical values $[B,h]$.[19] Using these values, we can then easily generate the corresponding binary sequences.

As examples, in Fig. 2 we show how the controlled logistic map and digital tent map generate a period-7 m sequence and its corresponding mirror sequence. Figure 3 shows that a period-31 m sequence 1000011001001111101110001010110 (Ref. 11) has been successfully generated by using the digital tent map, In principle, the approach is capable of generating an arbitrarily long periodic sequence if we have an appropriately high-bit machine. To stabilize a period-L m sequence via the limiter controller, one needs possibly a 2^{2^L}-bit machine because the m sequence is derived from the class of irreducible polynomials and it has anomalous superexponential scaling property in the limiter control approach.[18,19,34]

IV. DISCUSSION AND CONCLUSION

We have shown through extensive computation that chaotic trajectories bear codes with good randomness and correlation properties. Remarkably, it is found that the codes from

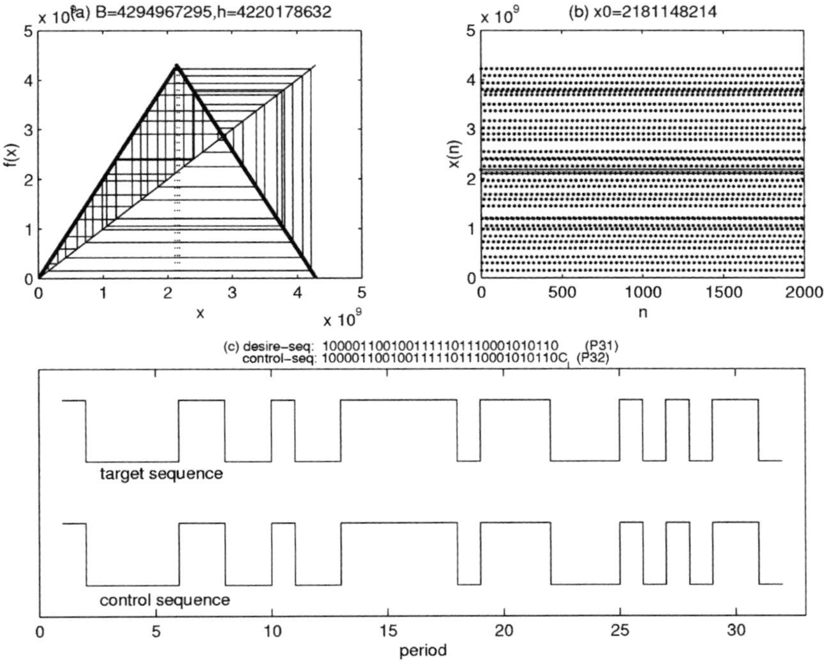

FIG. 3. A 31-period m sequence is generated by the digital tent map. (a) The stabilization procedure of the sequence; (b) the time series; and (c) the binary representation.

a subset of SPOs (also UPOs) of the class of 1D chaotic unimodal maps correspond to the set of m sequences (and at certain position shift the Gold code as well). As the set of U sequences of an 1D unimodal map constitutes a large set of possible codes (for given length L), it has a size of about $2^L/(L+1)$ when $L=2^m-1$). It is thus not surprising that it contains the set of random sequences. This is somewhat analogous to the fact that the set of all possible binary sequences of length L (of size 2^L) contains the set of random sequences. However, there is a major difference: if we can determine the set of UPOs with good randomness and correlation properties, we should be able to generate these sequence in the set by employing the corresponding chaotic system containing them. The same is clearly not true for the set of binary sequences of length L. In fact, in this work we have developed a technique based on a simple limiter control of unimodal maps to transfer UPO sequences into the corresponding SPO sequences. In particular, we have shown that it is possible to generate the set of m sequences by controlling the chaotic dynamics of unimodal maps. Finally, it should be mentioned that the results here may also have applications in other branches where chaos is involved.[2–7,22–25,31,32,35]

ACKNOWLEDGMENTS

The project was sponsored by SRF for ROCS, SEM. One of us (C.T.Z.) is also supported by the National Natural Science Foundation of China, Grant No. 10575013.

[1]E. N. Lorenz, J. Atmos. Sci. **20**, 130 (1963); *The Essence of Chaos* (University of Washington Press, Washington, D.C., 1993).
[2]E. Ott, *Chaos in Dynamical Systems* (Cambridge University Press, Cambridge, 1993), and references therein.
[3]E. Bollt, Y. C. Lai, and C. Grebogi, Phys. Rev. Lett. **79**, 3787 (1997).
[4]E. M. Bollt, T. Stanford, Y. C. Lai, and K. Życzkowski, Phys. Rev. Lett. **85**, 3524 (2000), and references therein.
[5]See, for example, S. Boccalett, C. Grebogi, Y. C. Lai, H. Mancini, and D. Maza, Phys. Rep. **329**, 103 (2000).
[6]S. D. Pethel, N. J. Corron, Q. R. Underwood, and K. Myneni, Phys. Rev. Lett. **90**, 254101 (2003).
[7]S. D. Pethel, N. J. Corron, and E. M. Bollt, Phys. Rev. Lett. **96**, 034105 (2006).
[8]B. Sklar, *Digital Communications, Fundamentals and Applications* (Prentice-Hall, Upper Saddle River, NJ, 2001).
[9]B. Schneier, *Applied Cryptography* (Wiley, New York, 1996).
[10]J. Daemen and V. Dijmen, *The Design of Rijndael* (Springer, Berlin, 2002).
[11]S. W. Golomb, *Shift Register Sequences* (Aegean Park Press, Orlando, 1982).
[12]R. C. Dixon, *Spread Spectrum Systems with Commercial Applications* (Wiley, New York, 1994).
[13]T. Kohda and A. Tsuneda, in *Chaotic Electronics in Telecommunications*, edited by M. P. Kennedy, R. Rovatti, and G. Setti (CRC Press, Columbus, OH, 2000).
[14]T. L. Carroll, IEEE Trans. Comput.-Aided Des. **47**, 443 (2000).
[15]L. Cong and X. F. Wu, IEEE Trans. Comput.-Aided Des. **48**, 521 (2001).
[16]Chaotic systems with good randomness (ergodic) and correlation ("Lebesgue spectrum") properties are commonly assumed to be suitable for spread-spectrum communications.
[17]D. S. Broomhead, J. P. Huke, and M. R. Muldoon, Dyn. Stab. Syst. **14**, 95 (1999).
[18]C. T. Zhou and M. Y. Yu, Phys. Rev. E **71**, 016204 (2005).
[19]C. T. Zhou, Chaos **16**, 013109 (2006).
[20]N. Metropolis, M. L. Stein, and P. R. Stein, J. Comb. Theory, Ser. A **15**, 25 (1973).
[21]B. L. Hao and W. M. Zheng, *Applied Symbolic Dynamics and Chaos* (World Scientific, Singapore, 1998).
[22]C. T. Zhou, C. H. Lai, and M. Y. Yu, J. Math. Phys. **38**, 5225 (1997).
[23]C. H. Lai, C. T. Zhou, and M. Y. Yu, Phys. Scr. **59**, 198 (1999).
[24]M. Y. Yu, Phys. Scr. **T82**, 10 (1999).
[25]T. F. Cai, T. X. Cai, C. T. Zhou, and M. Y. Yu, Phys. Scr. **66**, 187 (2002).
[26]P. Schmelcher and F. K. Diakonos, Phys. Rev. Lett. **78**, 4733 (1997).
[27]P. Schmelcher and F. K. Diakonos, Phys. Rev. E **57**, 2739 (1998).
[28]R. L. Davidchack and Y. C. Lai, Phys. Rev. E **60**, 6172 (1999).

[29]H. Kaplan, Phys. Lett. **97**, 365 (1983).
[30]The other three sequences are generated by using the connections {1111010, 1010111, 1100001011}, respectively. The corresponding values a of their iterative solutions are given approximately by: $1.9994560709127616+i1.69318\times10^{-25}$, $1.9997173997556472-i7.70719\times10^{-29}$, $1.9999945826320733-i3.43094\times10^{-148}$.
[31]N. J. Corron, S. D. Pethel, and B. A. Hopper, Phys. Rev. Lett. **84**, 3835 (2000).
[32]N. J. Corron and S. Pethel, Chaos **12**, 1 (2002).
[33]C. Wagner and R. Stoop, Phys. Rev. E **63**, 017201 (2000).
[34]R. Stoop and C. Wagner, Phys. Rev. Lett. **90**, 154101 (2003).
[35]S. Sinha, Phys. Rev. E **49**, 4832 (1994); **63**, 036212 (2001); S. Sinha and W. L. Ditto, Phys. Rev. Lett. **81**, 2156 (1998); **60**, 363 (1999).
[36]P. H. Borcherds, and G. P. McCauley, Chaos, Solitons Fractals **3**, 451 (1993).
[37]A. Dornbusch and J. P. de Gyvez, Proc. IEEE International Symp. Circuits and Systems 5, 454 (1999).

PHYSICAL REVIEW E **73**, 026209 (2006)

X-wave solutions of complex Ginzburg-Landau equations

C. T. Zhou,[1] M. Y. Yu,[2] and X. T. He[1]
[1]*Institute of Applied Physics and Computational Mathematics, P.O. Box 8009, Beijing 100088, People's Republic of China*
[2]*Institut für Theoretische Physik I, Ruhr-Universität Bochum, D-44780 Bochum, Germany*
(Received 29 August 2005; published 14 February 2006; corrected 22 February 2006)

A solution in the form of X-wave patterns of the complex Ginzburg-Landau equation with a harmonic background inhomogeneity is obtained. The pattern can be attributed to the effects of the harmonic potential and the boundary configuration and size. By varying the harmonic of the background potential, the competition among three types of wave patterns: spiral, X, and target, is investigated by following the evolution of the Fourier modes.

DOI: 10.1103/PhysRevE.73.026209 PACS number(s): 89.75.Kd, 05.45.Jn, 05.45.Yv, 82.40.Np

I. INTRODUCTION

Spiral and target wave patterns have often been observed in reaction-diffusion systems such as the Belousov-Zhabotinsky chemical reaction [1,2]. They have also been found in biological systems such as the cardiac muscle tissue and slime mold colonies of *Dictyostelium* [3], as well as physical systems such as the planar dc semiconductor-gas discharge [4] and large-aspect-ratio lasers [5]. Spiral and target waves are usually investigated theoretically by means of the complex Ginzburg-Landau equation (CGLE), which is a universal model for describing the evolution of nearly coherent waves. Depending on the problem involved, it can appear in various forms involving complex coefficients and nonlinearities [6–9]. The CGLE has been extensively applied to different physical, chemical, and biological systems, such as transversely extended laser, electrohydrodynamic convection in liquid crystal, Bose-Einstein condensate, fluid and chemical turbulence, planar gas discharge, plasma surface-wave oscillation, bluff body motion, etc. [1–12], and is closely related to the nonlinear Schrödinger [13,14] or Gross-Pitaevsky equation. Waves as well as pattern solutions of various forms and limits of the CGLE have been theoretically investigated by several authors [4,15–17], and it has been found that the actual solution, in particular, the target pattern, strongly depends on the medium inhomogeneity and the boundary condition.

Existing results indicate that singly charged spiral solutions of the two-dimensional (2D) CGLE are dynamically stable in some parameter regimes [9], but spirals with zero topological charge are always unstable. Target waves arising from boundary effects have also been observed [15]. By introducing localized inhomogeneity in the CGLE, Hendrey and co-workers [16,17] found that stationary and breathing target waves can also occur. Stationary target patterns have been observed in the light-sensitive Belousov-Zhabotinsky reaction [18].

On the other hand, X- or cross-shaped wave patterns have also been observed in various physical systems. They are well known in the context of linear propagation of acoustic and electromagnetic waves [19–23]. By using beam-shaping techniques, one can also obtain X waves in optical [24] and microwave radiation [25]. Recently, in connection with self-trapped optical wave packets the investigation of X waves has been extended to the nonlinear regime [26–28]. X waves can be spontaneously generated because of nonlinearities in the source or medium through wave-matter interaction. They have been found to be rather robust and can propagate or survive for long times or distances. Such waves can thus be expected in many fields. In this paper, we show numerically that if the medium is quadratic, or harmonic, the CGLE can admit X-wave solutions. The formation and stability behavior of the latter are explored. It is found that the X-shaped patterns depend on the harmonic number of the inhomogeneity as well as the boundary condition, and that they are robust.

The paper is organized as follows: In Sec. II, the problem is formulated. In Sec. III, we numerically investigate the generation of spiral-, X-, and target-wave patterns. In Sec. IV, the competition among the three types of wave patterns is analyzed. The formation of X waves is investigated in terms of the system size and the background inhomogeneity. In Sec. V, the effects of the boundary shape and type as well as the stability of the X wave is studied. The results are summarized and discussed in the final section.

II. FORMULATION

The 2D CGLE of interest here is given by

$$\partial_t \Psi = \mu(\mathbf{x})\Psi + (1+i\alpha)\nabla_\perp^2 \Psi - (1+i\beta)|\Psi|^2\Psi, \qquad (1)$$

where $\Psi(x,y,t)$ is a complex function of time t and space $\mathbf{x}$, and $\nabla_\perp^2 = \partial_x^2 + \partial_y^2$. Equation (1) is a universal model for many phenomena near the threshold of a long-wavelength oscillatory instability. In Eq. (1), $\mu(\mathbf{x})$ characterizes the property of the medium and gives the linear frequency and/or the growth and/or damping rate of the normal oscillations, the second term corresponds to spatial diffusion (real part) and dissipation (imaginary part), and the last term corresponds to the nonlinear effects (with an imaginary part describing the nonlinear frequency shift). In the limit $(\alpha,\beta)\to\infty$, Eq. (1) reduces to the widely investigated cubic nonlinear Schrödinger or Gross-Pitaevsky equation. Depending on the parameters α and β, defect or phase turbulence of the complex quantity Ψ may occur, resulting in spatiotemporal

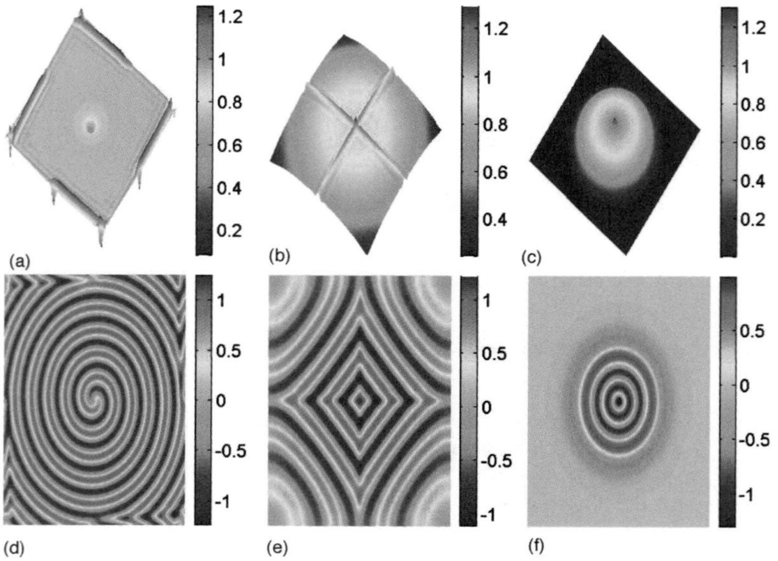

FIG. 1. (Color online) Examples of the three stationary wave patterns at $t=200$. The labels (a), (b), and (c) correspond to three subplots of the first row (from the left to right) and (d), (e), and (f) correspond to three subplots of the second row. (a), (b), and (c) give the amplitude $|\Psi(x,y,t)|$ and (d), (e), and (f) give the corresponding real part $\mathrm{Re}\,\Psi(x,y,t)$. The parameters of the CGLE are $[\alpha,\beta]=[3.5,-0.34]$, and the size of the system is $L=200\times 200$. The corresponding external frequencies are $\Omega=0$ in (a) and (d), $\Omega=0.01$ in (b) and (e), and $\Omega=0.02$ in (c) and (f).

chaotic solitons [9,29].

In the limit $\mu=1$, analysis of the local bifurcation properties indicates that the plane-wave solution [29], $\Psi=\sqrt{1-k^2}e^{i(\mathbf{k}\cdot\mathbf{x}-\omega t)}$ with $|k|\leqslant 1$ and frequency $\omega=\beta+(\alpha-\beta)k^2$, is linearly stable in the range $|k|\leqslant k_c$ with $k_c=\sqrt{(1+\alpha\beta)/(3+\alpha\beta+2\beta^2)}$. The stability range vanishes at the Benjamin-Feir-Newell line $\alpha\beta=-1$, and no stable plane-wave solution exists for $\alpha\beta<-1$. For a wide range of α and β, the 2D CGLE (1) possesses rotating spiral solutions in the form [30]

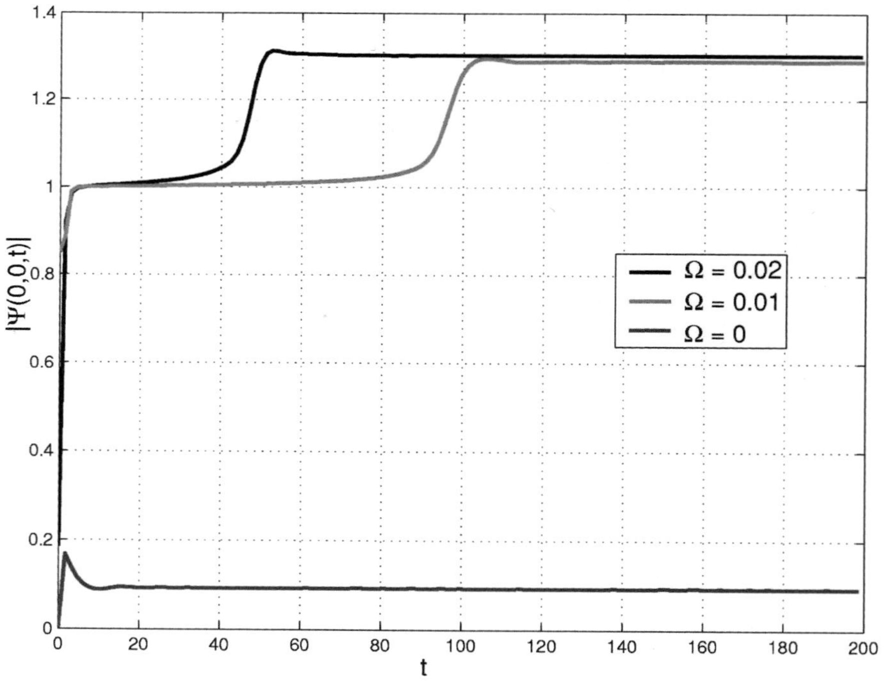

FIG. 2. (Color online) The evolution of the amplitude $|\Psi(0,0,t)|$ at the center of the simulated system for $\Omega=0$, 0.01, and 0.02, corresponding to the spiral, X, and target patterns, respectively.

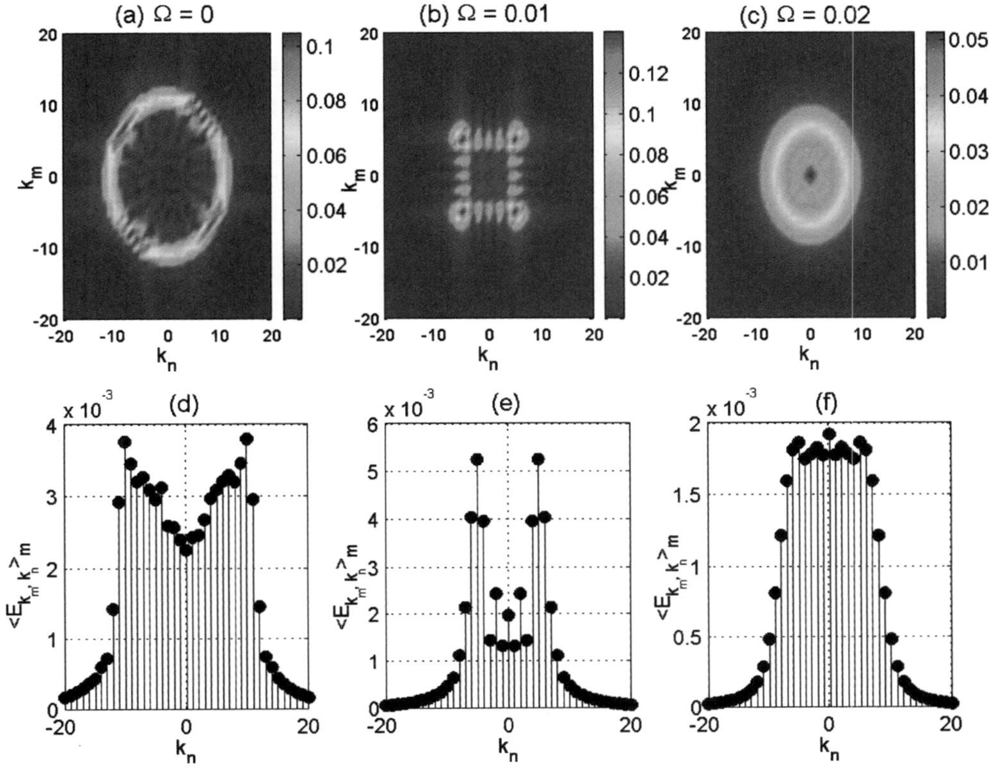

FIG. 3. (Color online) Spectral structures corresponding to Fig. 1. (a), (b), and (c) give the spectral amplitude $|\Psi(k_m,k_n,t)|$ and (d), (e), and (f) give the average spectrum $\langle|\Psi(k_m,k_n,t)|\rangle_m$. The external frequencies are (a) and (d): $\Omega=0$, (b) and (e): $\Omega=0.01$, and (c) and (f): $\Omega=0.02$, corresponding to the spiral, X, and target patterns, respectively.

$$\Psi(r,\theta,t)=F(r)e^{i[-\omega t+m\theta+\phi(r)]}$$

with $\Psi=0$ at the spiral core. For a target wave (i.e., when $\theta=0$), concentric oscillations propagate from an oscillatory centrum.

In many physical applications of the CGLE, the medium is inhomogeneous, or $\mu=\mu(\mathbf{x})$. In this case stable superspiral [31] as well as target waves can appear if there is exponential spatial dependence in both the real and imaginary parts of $\mu(r)$ [16,17]. It has been found that the lowest-order effect of the inhomogeneity is a strong dependence of the local frequency and growth rate of the excitation in space [16,17]. Here we shall use the often-invoked harmonic- or quadratic-potential model [7,8] for the inhomogeneity: $\mu(r)=1-\frac{1}{2}\Omega^2r^2$. Physically, this dependence approximates the bottom of any trapping potential. It allows trapping of waves or particles in a simple potential well, and the corresponding Einstein frequency is Ω [32]. Such a potential has often been associated with classical systems describing crystal structure, as well as with the more recent vortex-dynamics systems such as that related to Bose-Einstein condensates [33–36]. It is also used in more practical systems such as optical-fiber and plasma guides and traps [7,8]. The potential can appear naturally (as in crystals), manufactured (as in optical fibers), or externally introduced (such as by electric and magnetic fields in plasmas and rotation in fluids), or by other means (such as by shining laser light on the light-sensitive Belousov-Zhabotinsky reaction [18]).

By means of numerical experiments, we look for wave-pattern solutions of Eq. (1) for different Ω. For definitiveness, we shall assume $\alpha=3.5$ and $\beta=-0.34$, that are typical for studying spiral waves [29]. The numerical solution is performed using fast Fourier transform for the space variables x and y, and a variable-step fourth-order Runge-Kutta scheme for the time variable. Such a periodic boundary condition approach is applicable to many physical systems and is frequently used in numerical simulations. Unless otherwise stated, the system size is $L=200\times200$, and the number of Fourier modes is taken to be between $2^7\times2^7$ and $2^9\times2^9$. As an initial condition, in order not to favor any particular type of solution, we use a random distribution of small-amplitude fluctuations. For $\Omega=0$, it is found that *any* initial condition consisting of small-amplitude waves with a topological charge will rapidly evolve into spiral waves.

III. SPIRAL-, X-, AND TARGET WAVES

Three types of wave patterns corresponding to $\Omega=0$, 0.01, and 0.02 are shown in Figs. 1(a)–1(c), respectively. The

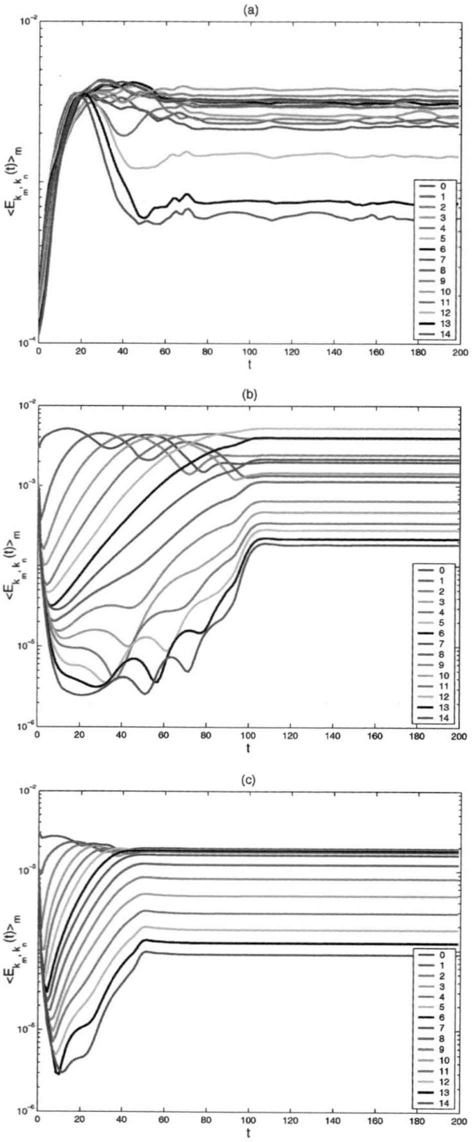

FIG. 4. (Color online) Evolution of the first 15 averaged spectral amplitudes (inset) $\langle|\Psi(k_m,k_n,t)|\rangle_m$ (for $n=0,\ldots,14$) for (a) $\Omega=0$, (b) $\Omega=0.01$, and (c) $\Omega=0.02$, corresponding to the spiral, X, and target patterns, respectively. The very distinct evolution characteristics (see text) of the three wave patterns can be observed.

snapshots are taken at $t=200$. The upper row shows $|\Psi(x,y,t)|$ and the lower row shows $\mathrm{Re}\,\Psi(x,y,t)$. Figure 1(a) shows that when $\Omega=0$, the amplitude represents a point source in a spiral core, corresponding to a typical spiral pattern for the real part $\mathrm{Re}\,\Psi$, given in Fig. 1(d), of the complex field $\Psi(x,y,t)$. Recall that the real part and the phase of the latter are, in general, of the same topological structure. The case $\Omega=0.02$ corresponds to the stationary target-wave pattern and has been investigated for homogeneous as well as inhomogeneous CGLE [16,17]. The spirals rotate in time at the Hopf frequency ω given by the imaginary part of the eigenvalue found in the linear analysis. The homogeneous case yields spirals with the same frequency ω. The spiral centers can be identified by the dark spots in the amplitude plot given by Fig. 1(a) or as the center of the spiral waves in the $\mathrm{Re}\,\Psi$ phase plot, given by Fig. 1(d). However, when the homogeneity is destroyed by an external harmonic potential, the wave energy propagates from the outer domain to the center within a short time, leading to the disappearance of the point singularity at the center of the spiral waves. As long as Ω is sufficiently larger than the characteristic oscillation frequency ω of the system, most of the energy is concentrated in the center region of the system and the phase becomes concentric oscillations propagating from an oscillatory centrum. When the equilibrium corresponding to these two frequencies is reached before the waves from the center can arrive at the boundary, one easily gets a stationary target-wave pattern. The corresponding patterns of $|\Psi|$ and $\mathrm{Re}\,\Psi$ are shown in Figs. 1(c) and 1(f).

However, if the interaction of the outward wave with the boundary occurs before equilibrium is reached, the boundary effect would lead to a new spatially local wave pattern, whose amplitude maxima are at the centers of the four boundary walls. The role of the real part [Fig. 1(d)] of the complex field (or the phase) is enhanced by the boundary effect. The balance between dispersion and trapping by the harmonic background potential as well as the walls results in an interesting X-wave structure, as shown in Fig. 1(b). It is found that the very-well-defined coherent X pattern can occur in the parameter regime $\Omega\in[0.005,0.015]$. Unlike the spiral waves, the existence of X waves depends sensitively on the size of the system (with fixed Ω), a property which will be further discussed below. Finally, it is worth noting that the three wave patterns in Figs. 1(d)–1(f) are of similar topological structure as that of low-dimensional phase trajectories (with respect to the node-, saddle- and center-point behavior) in dynamical systems. This similarity leads to the still unanswered question of whether there is topological relationship between low- and infinite-dimensional systems, a problem that is beyond the scope of the present work.

To see how the external harmonic potential affects the formation of the wave patterns, we have also analyzed the time evolution of the wave amplitude at the system center. Figure 2 shows that the defect in the spiral waves can occur in the presence of the external potential. An initial core with low energy can absorb traveling waves and rapidly become a high-energy center because of the harmonic potential. On the other hand, dispersion can cause an outflow of wave energy. A stationary state occurs when a balance among dispersion and/or dissipation, harmonic-potential trapping, and nonlinear effects is reached. Figure 2 shows that with different Ω values, stationary wave patterns are eventually formed, and the time needed to reach the asymptotic stationary state becomes shorter with increasing Ω. We note that the formation of the X wave is associated with a long ($5\leq t\leq 110$) evolution time, while the target wave reaches its final stationary state already at $t\sim 50$.

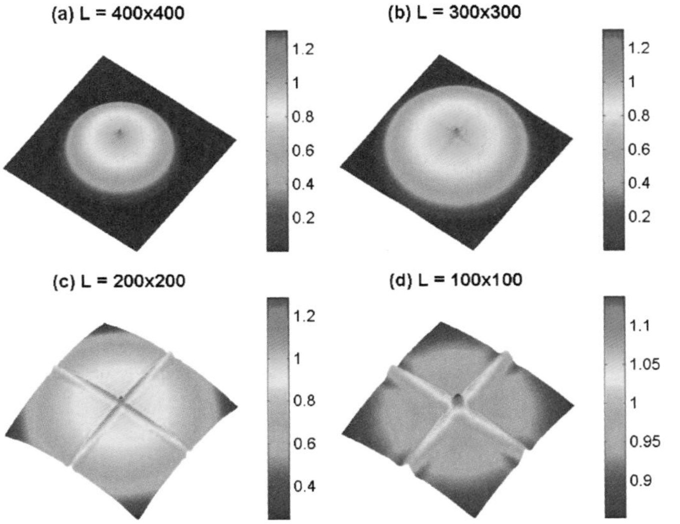

FIG. 5. (Color online) Dependence of the amplitude $|\Psi(x,y,t=200)|$ on the size of the system. (a) $L=400\times400$, (b) $L=300\times300$, (c) $L=200\times200$, and (d) $L=100\times100$. The other parameters are the same as in Fig. 1.

IV. FORMATION OF X WAVES

In order to better understand the X waves, it is useful to analyze the energy distribution associated with the coherent patterns and their formation. The 2D Fourier spectrum $|\Psi(k_m,k_n)|$, where k_m and k_n are the wave numbers in the x and y directions, are given in Fig. 3. The parameters are the same as that for Fig. 1. One can see that each wave pattern has a distinct spectral distribution. For the spiral wave [Fig. 3(a)], much of the energy is nonuniformly distributed in a ring structure and there is a dip at $\mathbf{k}=\mathbf{0}$. For the X wave [Fig. 3(b)], the energy is distributed in the form of concentrate squares, with the energy maxima in each square at the four corners. The absence of the isotropic ring structure here implies a dominance of the wall effects. For the target wave [Fig. 3(c)], we again have an isotropic, but now table-top structure. Since the energy spectra for the three patterns exhibit x,y symmetry, for analyzing the energy distribution more quantitatively one can take the mode average over one of the two wave-number spaces. In Figs. 3(d)–3(f), we show $\langle|\Psi(k_m,k_n,t)|\rangle_m$, where $\langle\cdots\rangle_m$ indicates that the average has been taken over the Fourier modes in the x direction. For the spiral wave [Figs. 3(a) and 3(d)], the spectrum is peaked at $k_n=\pm10$ and its decrease as $|k_n|\to0$ is almost monotonic. For the X wave [Figs. 3(b) and 3(e)], the spectrum is sharply

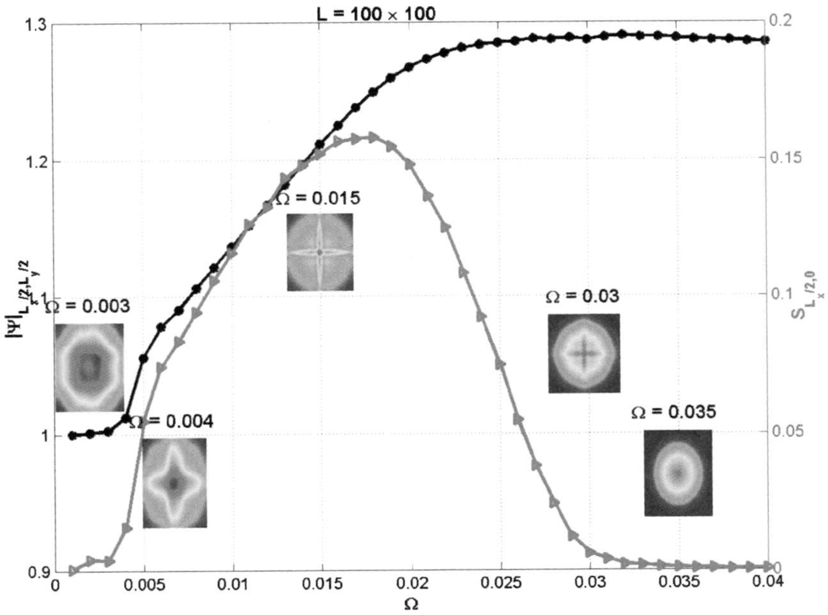

FIG. 6. (Color online) Dependence of the amplitude at the system center (triangles) and the boundary point $(L_x/2,0)$ (circles) on Ω at $t=200$ for $L=100\times100$, $\alpha=3.5$, and $\beta=-0.34$. The transition from the X to the target pattern as Ω increases can be seen from the insets.

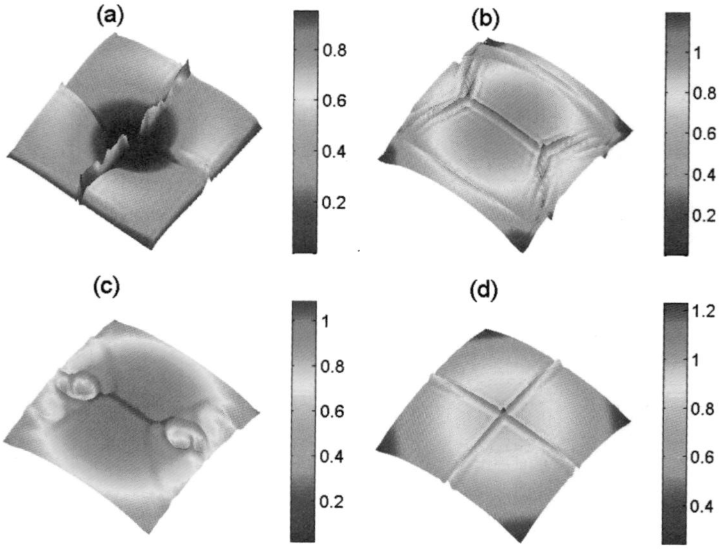

FIG. 7. (Color online) Evolution of a perturbed X pattern. The field at $t=200$ in Fig. 1(b) is perturbed by adding a phase $e^{im\theta}$ (with charge $m=3$) in the period $t=200-205$. (a), (b), and (c) show the transient states of the amplitude $|\Psi(x,y,t)|$ at $t=201.109$, 227.641, and 302.503. (d) shows the final asymptotic or stationary state.

peaked at the two modes $k_n=\pm 5$, and the energy decrease as $|k_n|\rightarrow 0$ is in an oscillatory manner. Most of the energy is in the modes $k_n=\pm 5$ and their neighbors $k_n=\pm 4, \pm 6$. For the target waves [Figs. 3(a) and 3(d)], on the other hand, most of the energy is evenly distributed among the first 13 modes $k_n=0, \pm 1, \ldots, \pm 6$.

In the formation process of the coherent patterns, energy exchange or cascade among the different Fourier modes plays an important role. To investigate this, we follow the evolution of the average energy of the first 50 modes starting from the initial random state. The results are given in Fig. 4. Distinct evolution behavior of the three patterns can be observed. To which pattern the system eventually evolves depends on how energy is partitioned. For the spiral wave, the nonlinear mode-mode interaction is quite complicated. Figure 4(a) shows that an energy exchange among the modes between $k_n=1$ and $k_n=9$ always occurs for $t>50$. Thus, the spiral-wave solution found here appears to be a stationary solution of the 2D CGLE. Figures 4(b) and 4(c) show that for both the X and target waves, no mode-mode energy exchange takes place after a certain time, indicating that these patterns are stationary or quasistationary. The times needed to form stationary coherent states are approximately 110 and 50 for the X and target waves, respectively. Moreover, the energy cascade or self-organization process in the Fourier space for these two patterns is completely different. For tar-

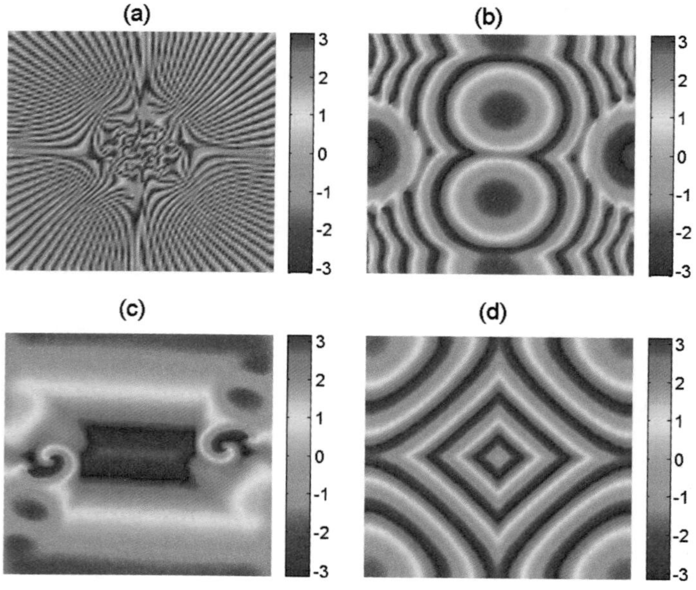

FIG. 8. (Color online) The phase [Im $\Psi(x,y,t)$] states corresponding to the evolution of the perturbation in Fig. 7. The phase scale is in arbitrary units.

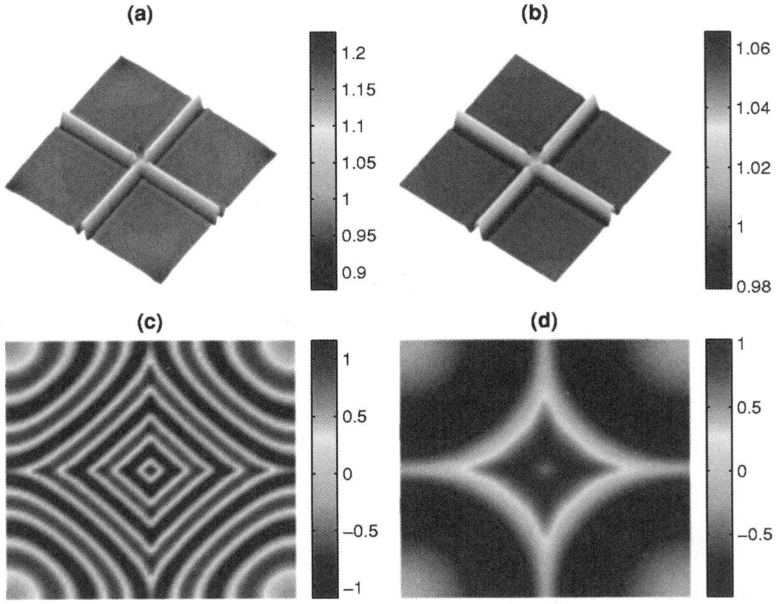

FIG. 9. Evolution of the X wave appearing in 1(b) with the inhomogeneity turned off (Ω set to zero) at $t=200$. (a) and (c) show the amplitude ($|\Psi(x,y,t)|$) and phase [Im $\Psi(x,y,t)$] of the transient state at $t=202.48$, respectively. (b) and (d) show the same parameters for the asymptotic state at $t=400$.

get waves, the energy in the higher-order modes ($k_n>7$) is first partitioned, then the energy among the low-order modes ($k=0,\ldots,7$) is reorganized during $20\leqslant t\leqslant 50$ until the process stops. Figure 4(b) shows that for the X wave there exist two independently interacting groups of modes. The two groups are separated by the modes $k_n=8,9,10$, that do not exchange their energy among themselves nor with the other modes. On the other hand, during $20\leqslant t\leqslant 110$ strong energy exchange is observed in each of the two ($0\leqslant k_n\leqslant 7$ and $11\leqslant k_n\leqslant 14$) interacting groups. Thus, the formation of the patterns can be distinguished by the energy partition process among the Fourier modes.

Having discussed the energy partition process of the three coherent wave patterns, we now consider the mechanism of forming them. The dynamics of the spiral and target waves has already been discussed in the existing literature [9,16,17]. It is believed that the formation of target waves is due to boundary [5,15] or local-inhomogeneity [16,17] effects. With the latter mechanism, Hendrey and co-workers [16,17] found that the lowest-order inhomogeneity effect is sufficient to form a target wave. On the other hand, as pointed out by Aranson *et al.* [5] in connection with a modified complex Swift-Hohenberg equation, boundary effects are important in selecting the different patterns in large-aperture lasers. They showed that traveling waves with well-defined wave numbers emitted by the boundary can collide in the interior of the domain, forming a sink with shocklike structure. It is thus of interest to see if a similar mechanism can lead to the X-shaped coherent wave pattern. To investigate boundary-driven selection processes for the different patterns, we fix $\Omega=0.01$ and vary the size of the system. As shown in Fig. 1(b), the parameter values $\Omega=0.01$ and $L=200\times 200$ result in an X wave. In order to see the boundary dependence, we consider three cases: $L=400\times 400$, 300×300, and 100×100. Figure 5 shows the amplitude of $\Psi(x,y,t=200)$ for different system sizes. Figure 5(a) indicates that for a large system the final asymptotic state is a target pattern. In Fig. 5(b), where $L=300\times 300$, one can see that a weak X wave is seeded within a target pattern. With decreasing system size, the X-wave pattern becomes dominant, as shown in Figs. 5(c) and 5(d). It is thus clear that, besides the external harmonic potential, size and boundary effects are important in determining if X waves can occur.

V. EFFECT OF SIZE, BOUNDARY, AND INHOMOGENEITY

To illustrate the effects of size, boundary, and the harmonic parameter Ω on the X wave, we first consider the shocklike structure at the boundary. Figure 6 shows the amplitude $|\Psi(L_x/2,L_y/2)|$ at the center of the system and the structural parameter $S_{L_x/2,0}\equiv \partial_x\Psi|_{L_x/2,0}$ for different Ω values. With increasing Ω, the amplitude becomes larger until the energy at the center saturates. Figure 6 shows that an X wave appears when $\Omega=0.004$. When $\Omega=0.018$, the X wave becomes clearly defined. When Ω is further increased, a target wave with an embedded X wave at the center starts to appear. When $\Omega\geqslant 0.035$, only the target wave remains.

The stability of the X pattern with respect to topological-charge perturbations should be considered. For convenience, we start with the X wave at $\Psi(t=200)$ as given in Fig. 1(b). To simulate the perturbation, we add topological charges to the complex field by letting $\Psi(x,y,t)=\Psi(x,y,t-\delta t)e^{im\theta}$, where the charge is $m=3$ and δt is the time step in the calculation, $\theta=\tan^{-1}(x/y)$. The perturbation persists for $t=200-205$. That is, the perturbation is represented by the azimuthal phase $e^{im\theta}$ and it is not of small amplitude. Since we have shown above that when the initial condition consists of small amplitude waves with topological charge the system

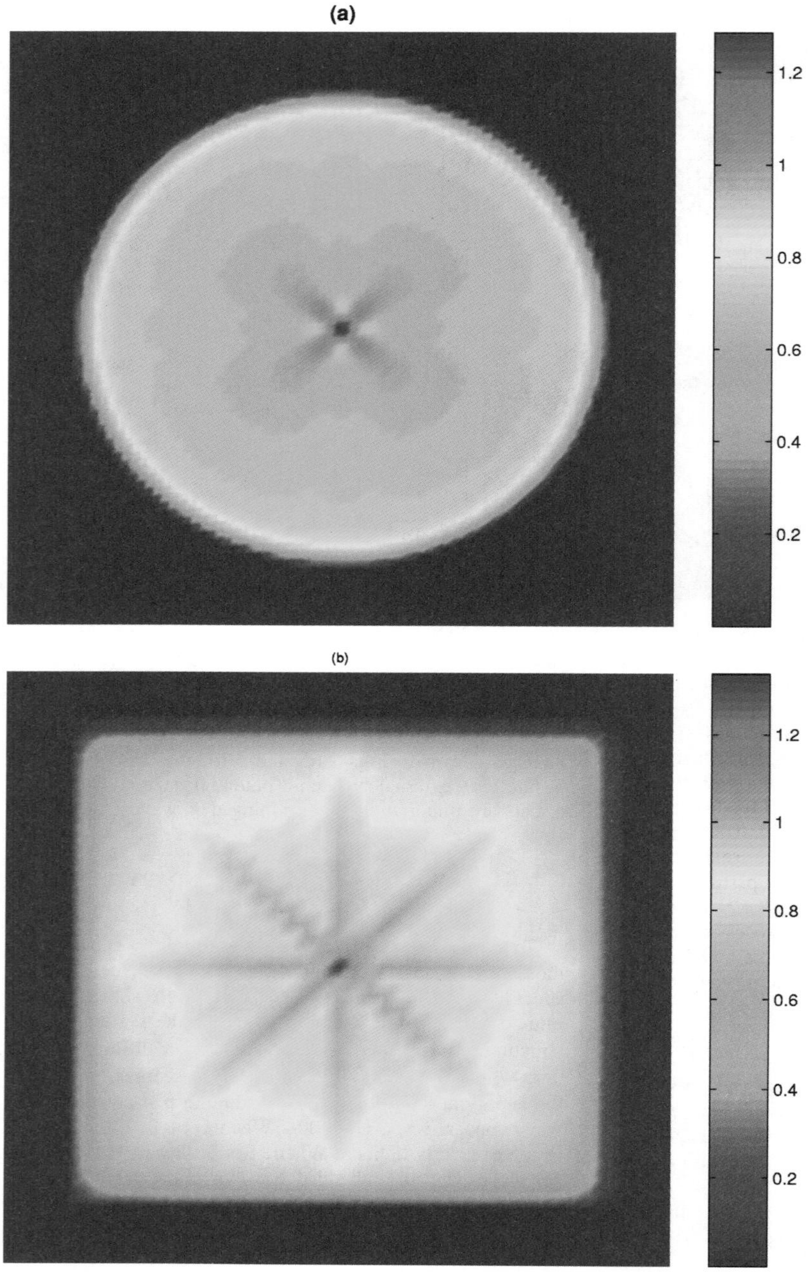

FIG. 10. (Color online) Stationary patterns for circular (a) and square (b) absorbing boundaries. The radius of the circular domain is $R=80$, and the side of the square domain is $L=160$. The parameters of the CGLE are $\alpha=3.5$, $\beta=-0.34$, and $\Omega=0.01$.

can rapidly evolve into a spiral wave, one might expect that the X pattern might be destroyed by the present topologically charged nonlinear perturbation. This turns out to be not the case. Figure 7(a) is a snapshot taken at $t=201.109$. It shows that as soon as the perturbation is introduced at $t=200$, the symmetric X pattern is rapidly disrupted. The corresponding phase, shown in Fig. 8(a), also reveals fine structuring near the system center. Later in the evolution, target-like and spiral-like patterns appear, but in a piecewise manner, as shown in Figs. 8(b) and 8(c). However, eventually the original X pattern is recovered, as shown in Figs. 7(d) and 8(d). Thus, the X wave appears to be stable to topological-charge perturbations. Since the finite-amplitude perturbed state can also be considered as a new initial state, this analysis also implies that the X wave does not depend on the initial conditions. It is only governed by the harmonic-potential parameter Ω, the boundary configuration, and the system size.

We have shown that with a random initial distribution of energy, the X-wave solution can only be obtained in the presence of the harmonic inhomogeneity. An unsolved question is whether the X wave can also be a stationary solution of the *homogeneous* CGLE. To answer this question, we start with the X wave at $\Psi(t=200)$ as given in Fig. 1(b), and turn off the harmonic inhomogeneity at $t=200$ by setting $\Omega=0$. In Fig. 9 we see that in the asymptotic state almost all of the energy becomes concentrated in the cross region of the system. Except at the center point, the energy is also more uniformly spread inside the cross. This behavior can be expected since in the absence of the trapping potential the background is uniform. By comparing Figs. 9(a) and 9(b), we see that the cross-wave pattern evolves very slowly, with more and more localization of energy in the cross. Comparing Figs. 9(c) and 9(d) we also note that the oscillation period becomes much longer as the asymptotic state is reached. This further indicates that the stationary X-wave solution may be associated with an intrinsic state of CGLE. However, in order to obtain such a stationary solution dynamically for a homogeneous system, appropriated initial conditions must be used.

Finally, we consider two additional types of boundaries. Figures 10(a) and 10(b) show $|\Psi(t=200)|$ for *absorbing* (in the sense that there is an infinitely deep potential well there [5]) circular and square boundaries, respectively. In Fig. 10(a), we see that although the system is azimuthally symmetric, one still finds an X wave at the center, with a fairly uniform energy distribution. However, the well-defined intense X-wave pattern in Fig. 1(b) for the periodic-boundary case is not recovered. We can better preserve (trap) the energy by increasing Ω, but then the asymptotic state becomes a target wave (not shown), trapped at the center by the strong harmonic potential. Figure 10(b) shows the stationary state for an *absorbing* square boundary. The pattern consists of a double-cross structure. In these figures the cross patterns are of much lower energy (compared with that for periodic boundary) because the energy is more uniformly spread across the domain as well as absorbed by the boundary. Thus, we can conclude that the well-defined single-X pattern appearing in Fig. 1 is dependent on the boundary as well as the harmonic potential, but for different boundary and harmonic conditions it can appear and behave quite differently.

VI. CONCLUSION

In an infinite homogeneous system, the waves form a continuum in the wave-number space. The existence of boundary and inhomogeneity severely limits the selection of wave numbers. As a result, the possible solutions necessarily depend on the boundary and inhomogeneity conditions. By varying the degree of the harmonic inhomogeneity, we have shown that, depending on the latter, three different types of wave patterns can appear as stationary solutions of the CGLE. In particular, we have shown that an inhomogeneous 2D CGLE in the presence of a simple harmonic potential admits an X-wave solution. The latter can be attributed to collision of outward traveling waves with the system boundary. Depending on the harmonics of the inhomogeneity, one can obtain spiral, target, as well as X patterns. By analyzing the energy spectra as well as by following the evolution of the nonlinear phase perturbation, we have shown that the X waves are nonspreading and quite robust, as have been observed experimentally. We have also introduced a simple method to control the formation of different coherent patterns through the background harmonic potential in the 2D CGLE.

It should be pointed out that the X waves are fundamentally different from the spiral and target waves. Their properties as well as their relation to other localized phenomena, such as solitons and oscillons, still remain to be investigated. On the other hand, since the CGLE has been invoked as a paradigm in a large number of physical, chemical, biological, and other systems [1–7,9–12,15–28], the present results on the robust nonspreading X-wave pattern indicate that the latter can occur in many different systems, and can thus underline future applications. On the other hand, the general properties and exact existence conditions of the X (or double-X) waves are still unknown, and should be of much interest for further analytical and experimental investigations.

ACKNOWLEDGMENTS

This work is supported by the National Natural Science Foundation of China (Grant No. 10575013), National Basic Research Project "Nonlinear Science in China", and SRF for ROCS, SEM.

[1] Q. Ouyang and J. M. Flesselles, Nature (London) **379**, 143 (1996).
[2] L. Q. Zhou and Q. Ouyang, Phys. Rev. Lett. **85**, 1650 (2000).
[3] K. J. Lee, E. C. Cox, and R. E. Goldstein, Phys. Rev. Lett. **76**, 1174 (1996).
[4] Y. A. Astrov, I. Müller, E. Ammelt, and H.-G. Purwins, Phys. Rev. Lett. **80**, 5341 (1998).
[5] I. Aranson, D. Hochheiser, and J. V. Moloney, Phys. Rev. A **55**, 3173 (1997).
[6] O. M. Gradov, L. Stenflo, and M. Y. Yu, Phys. Fluids B **5**, 1922 (1993).
[7] M. C. Cross and P. C. Hohenberg, Rev. Mod. Phys. **65**, 851 (1993).
[8] D. H. L. Dubin and T. M. O'Neil, Rev. Mod. Phys. **71**, 87 (1999).
[9] See, for example, I. S. Aranson and L. Kramer, Rev. Mod. Phys. **74**, 99 (2002), and the references therein.
[10] N. G. Berloff, Phys. Rev. Lett. **94**, 010403 (2005).
[11] L. Gurevich, A. S. Moskalenko, A. L. Zanin, Yu. A. Astrov, and H.-G. Purwins, Phys. Lett. A **307**, 299 (2003).
[12] L. Stenflo and M. Y. Yu, Nature (London) **384**, 224 (1996).
[13] C. T. Zhou, X. T. He, and S. G. Chen, Phys. Rev. A **46**, 2277

(1992).
[14] C. Zhou, X. T. He, and T. Cai, Phys. Rev. E **50**, 4136 (1994).
[15] V. M. Eguiluz, E. Hernandez-Garcia, and O. Piro, Int. J. Bifurcation Chaos Appl. Sci. Eng. **9**, 2209 (1999).
[16] M. Hendrey, E. Ott, and T. M. Antonsen, Phys. Rev. Lett. **82**, 859 (1999).
[17] M. Hendrey, K. Nam, P. Guzdar, and E. Ott, Phys. Rev. E **62**, 7627 (2000).
[18] V. Petrov, Q. Ouyang, G. Li, and H. L. Swinney, J. Phys. Chem. **100**, 18992 (1996).
[19] J. Lu and J. F. Greenleaf, IEEE Trans. Ultrason. Ferroelectr. Freq. Control **39**, 441 (1992).
[20] P. R. Stepanishen and J. Sun, J. Acoust. Soc. Am. **102**, 3308 (1997).
[21] P. Saari and K. Reivelt, Phys. Rev. Lett. **79**, 4135 (1997).
[22] J. Salo, J. Fagerholm, A. T. Friberg, and M. M. Salomaa, Phys. Rev. Lett. **83**, 1171 (1999).
[23] J. Salo, J. Fagerholm, A. T. Friberg, and M. M. Salomaa, Phys. Rev. E **62**, 4261 (2000).
[24] P. Saari and K. Reivelt, Phys. Rev. Lett. **79**, 4135 (1997).
[25] D. Mugnai, A. Ranfagni, and R. Ruggeri, Phys. Rev. Lett. **84**, 4830 (2000).
[26] C. Conti, S. Trillo, P. Di Trapani, G. Valiulis, A. Piskarskas, O. Jedrkiewicz, and J. Trull, Phys. Rev. Lett. **90**, 170406 (2003).
[27] P. Di Trapani, G. Valiulis, A. Piskarskas, J. Trull, C. Conti, and S. Trillo, Phys. Rev. Lett. **91**, 093904 (2003).
[28] S. Droulias, K. Hizanidis, J. Meier, and D. N. Christodoulides, Opt. Express **13**, 1827 (2005).
[29] H. Chaté, Nonlinearity **7**, 185 (1994).
[30] Here r and θ are the polar coordinates. The integer $m = 0, \pm 1, \ldots$, denotes the topological charge, ω is the rigid rotation frequency of spirals, $F(r) > 0$ is the amplitude, $\psi(r)$ is the phase of the spiral. The target waves do not have a phase singularity at the center. A target wave can be thought of as having no topological charge, or $m = 0$.
[31] L. Brusch, A. Torcini, and M. Bär, Phys. Rev. Lett. **91**, 108302 (2003).
[32] S. Ichimaru, *Statistical Plasma Physics Vol. II: Condensed Plasmas* (Addison-Wesley, Reading, Massachusetts, 1994).
[33] L. Khaykovich F. Schreck, G. Ferrari, T. Bourdel, J. Cubizolles, L. D. Carr, Y. Castin, and C. Salomon, Science **296**, 1290 (2002).
[34] G. Theocharis, D. J. Frantzeskakis, P. G. Kevrekidis, B. A. Malomed, and Y. S. Kivshar, Phys. Rev. Lett. **90**, 120403 (2003).
[35] M. Soljačić and M. Segev, Phys. Rev. Lett. **86**, 420 (2001).
[36] V. E. Zakharov and S. V. Nazarenkoc, Physica D **201**, 203 (2005).

APPLIED PHYSICS LETTERS **91**, 221114 (2007)

Complex dynamics of femtosecond terawatt laser pulses in air

Bin Qiao[a)] and C. H. Lai
Department of Physics, National University of Singapore, Singapore 117542, Singapore
C. T. Zhou and X. T. He
Institute of Applied Physics and Computational Mathematics, Beijing 100088, People's Republic of China
Xingang Wang
Temasek Laboratories, National University of Singapore, Singapore 117508, Singapore

(Received 3 October 2007; accepted 9 November 2007; published online 30 November 2007)

Complex dynamics of femtosecond terawatt laser pulses in air is investigated theoretically and numerically by considering both $\chi^{(5)}$ susceptibility and multiphoton ionization. Our investigation shows that these two high-order nonlinearities, acting as a Hamiltonian perturbation, can destroy homoclinic orbit connection of phase trajectory and result in spatial chaos and complex pattern of laser wave field. Due to the chaotic behavior of laser field, laser filamentation appears when laser intensity increases close to the ionization threshold. © *2007 American Institute of Physics*. [DOI: 10.1063/1.2819611]

Propagation of femtosecond terawatt laser pulses[1] in air has attracted wide interest due to its applications in lightning discharge control and atmospheric remote sensing.[2–4] Self-channelling of initial laser pulse experiences an early self-focusing (SF) stage caused by the Kerr response of air, which leads to a sharp increase of laser intensity. When laser intensity increases close to the ionization threshold, $\chi^{(5)}$ susceptibility originated from the expansion of the nonlinear polarization vector and multiphoton ionization (MPI) of air molecules play important roles in the propagation dynamics.[5,6]

SF, filamentation, and soliton of laser pulse in air have been investigated.[6–10] In nature, these phenomena are the results of nonlinear development of modulational instabilities and are associated with the nonlinear complex dynamics. However, the latter has not been studied in detail. Thus, the study of complex dynamics including chaos and patterns for terawatt laser pulse in air is of interest to the understanding of its nonlinear propagation behavior. In this letter, we investigate the chaotic behavior of laser pulse in air by considering a one-dimensional extended nonlinear Schrödinger (NLS) equation.

According to the standard propagation equation for femtosecond laser field in air, we can model its envelope evolution by an extended NLS equation[6,8] in one dimension as

$$i\partial_z E + \partial_x^2 E + f(|E|^2)E = 0, \tag{1}$$

where the nonlinear term $f(|E|^2)=\alpha|E|^2-\epsilon|E|^4-\gamma|E|^{2K}$ includes Kerr focusing of air, $\chi^{(5)}$ susceptibility originated from the expansion of nonlinear polarization, and plasma generation[11] by MPI of air molecules. For pulse duration $t_p=250$ fs and wavelength $\lambda_0=800$ nm, their coefficients take values $\alpha=0.446$, $\epsilon=7.3\times10^{-7}\ \mathrm{cm}^2/w_0^2$, and $\gamma=8.4\times10^{-40}\ \mathrm{cm}^{2(K-2)}/w_0^{2k-2}$ with w_0 laser beam transverse width in unit of centimeter. The variables z, x, and E are normalized by the units of $4z_f$, w_0, and $\sqrt{P_{\mathrm{cr}}/4\pi w_0^2}$, respectively, where $z_f=\pi w_0^2/\lambda_0$ is the Rayleigh length and $P_{\mathrm{cr}}=2.55$ GW the critical power for SF.

The Hamiltonian of NLS equation (1) can be written as $H=H_0+H_1$, where $H_0=\int(|E_x|^2-\frac{1}{2}\alpha|E|^4)dx$ and $H_1=\int[\frac{1}{3}\epsilon|E|^6+(1/K+1)\gamma|E|^{2K+2}]dx$. If only Kerr effect is considered, $H=H_0$ is integrable corresponding to a cubic NLS (CNLS) equation. $\chi^{(5)}$ susceptibility and MPI effect act as a Hamiltonian perturbation H_1 to H_0. Figure 1(a) gives comparisons of these three effects for different laser intensities I_0. For $I_0<10^{12}\ \mathrm{W/cm^2}$, purely Kerr nonlinearity plays a major role with $H\approx H_0$ near integrable. For $10^{12}<I_0<2\times10^{13}\ \mathrm{W/cm^2}$, $\chi^{(5)}$ susceptibility starts to act as a small perturbation H_1. For $I_0>2\times10^{13}\ \mathrm{W/cm^2}$, MPI appears, laser propagation is governed by all three effects with H nonintegrable. When I_0 further increases to larger than $5\times10^{13}\ \mathrm{W/cm^2}$, MPI dominates $\chi^{(5)}$ susceptibility and laser beams reveal complex dynamics, including SF and filamentation.

We now analyze some general nonlinear behavior of NLS equation (1). For definitiveness, we consider nonlinear modulation of a plane wave solution $E_s(z)=E_0e^{i\lambda z}$, which satisfies $\lambda=f(|E_0|^2)$. In the presence of small perturbations, we can write $E(x,z)=E_s(z)+\delta E(x,z)$, where $\delta E(x,z)\ll E_s(z)$. Linearizing Eq. (1), we obtain

$$\begin{pmatrix} i\partial_z+\partial_x^2+L & L \\ L & -i\partial_z+\partial_x^2+L \end{pmatrix}\begin{pmatrix}\delta E\\ \delta E^*\end{pmatrix}=0, \tag{2}$$

where $L=f'(|E_0|^2)|E_0|^2$ and $f'(u)=\partial_u f$.

For periodic boundary conditions, the eigenfunction $\delta E(x,z)$ of Eq. (2) can be defined as[12]

$$\begin{pmatrix}\delta E\\ \delta E^*\end{pmatrix}=\begin{pmatrix}\eta e^{i\mu z}\\ \eta^* e^{-i\mu^* z}\end{pmatrix}\cos(k_x x), \tag{3}$$

where η and η^* are small parameters and k_x is the wave number of perturbations. Some algebra yields the eigenvalue

$$\mu=2[f'(|E_0|^2)|E_0|^2-k_x^2]\pm i\sqrt{k_x^2[2f'(|E_0|^2)|E_0|^2-k_x^2]} \tag{4}$$

and the growth rate $\Gamma=\sqrt{k_x^2[2f'(|E_0|^2)|E_0|^2-k_x^2]}$. The wave number of the most unstable mode is then $k_{x,\max}=\sqrt{f'(|E_0|^2)|E_0|^2}$. Inserting Eqs. (3) and (4) into Eq. (2), we easily obtain that $\delta E/\delta E^*=\pm i$.

a)Electronic mail: qiaobin3@gmail.com

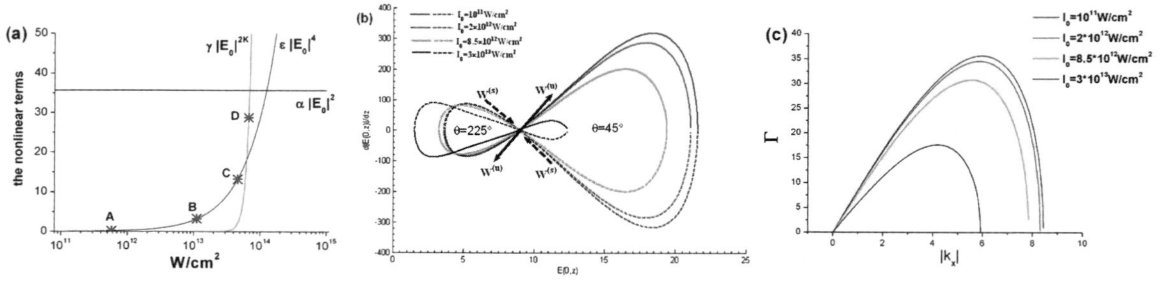

FIG. 1. (Color online) (a) Comparison of three nonlinear effects for different I_0. Asterisk symbols mark the points with I_0 of 5.82×10^{11}, 1.13×10^{13}, 4.6×10^{13}, and 6×10^{13} W/cm², respectively. (b) Stable $W^{(s)}$ and unstable $W^{(u)}$ manifolds for the saddle point $(E_0,0)$ with $\theta=45°$ and 225°, where the solid curve is calculated by NLS equation (1) with $t>0$ and the dashed curve with $t<0$. (c) The growth rate Γ with the corresponding I_0.

To analyze the nonlinear behavior of the system (1), we construct phase space $(|E|,d_z|E|)$ at $x=0$. As the eigenvalue μ has a pair of conjugated complex roots, we can determine that $(E_0,0)$ is associated with a saddle (or hyperbolic fixed) point in phase space. Furthermore, from Eq. (3) the initial condition can be chosen as

$$E(x,0)=E_0+\eta\exp(i\theta)\cos(k_x x). \tag{5}$$

We, therefore, find that the unstable manifolds $W^{(u)}$ possessing the hyperbolic fixed point correspond to $\theta=45°$ and 225°, and the stable manifolds $W^{(s)}$ correspond to $\theta=135°$ and 315°.[12] This analysis is completely consistent with our numerical results given in Fig. 1(b).

Figure 1(c) plots the unstable growth rate Γ for different I_0. From Figs. 1(c) and 1(a), we can understand that laser pulse with lower intensity ($I_0<10^{12}$ W/cm²) follows the early SF caused by purely Kerr response of air, which compresses laser beam in the transverse plane and increases the intensity sharply [see Fig. 1(c), Γ is very large for $I_0=10^{11}$ W/cm²]. With increasing I_0, both $\chi^{(5)}$ susceptibility and MPI become important mechanisms, the growth of laser field amplitude is quickly arrested by such two effects [see Fig. 1(c), Γ much decreases when I_0 increases, in particular, for $I_0=3\times10^{13}$ W/cm²].

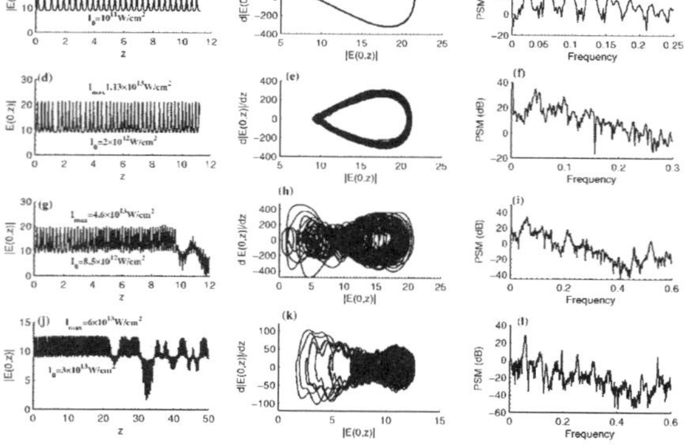

FIG. 2. Complex dynamics of laser fields with I_0 of 10^{11} [(a)–(c)], 2×10^{12} [(d)–(f)], 8.5×10^{12} [(g)–(i)], and 3×10^{13} W/cm² [(j)–(l)], respectively. (a), (d), (g), and (j) are evolutions of laser field envelope; (b), (e), (h), and (k) are the phase-space trajectories; and (c), (f), (i), and (l) are the power spectra.

For the Hamiltonian system (1), the Kolmogorov-Arnold-Moser (KAM) torus is determined by its initial energy.[13] If we use an arbitrary initial condition, it is generally difficult to give a clear description of its complex dynamics. However, our analyses shown in Fig. 1(b) clearly illustrate that the initial condition (5) is suitable. Using such initial condition, we now investigate its complex dynamics including spatial chaos and patterns by numerically solving NLS equation (1). The standard splitting-step spectral method[12,14] is improved. We choose input laser power $P_{in}=10P_{cr}$ ($E_0=8.94$) with noise level of 0.01 ($\eta=0.089$), initial phase $\theta=45°$, and consider the most unstable mode with $k_x=k_{x,max}$. The spatial period of the system is taken to be $L=2\pi/k_{x,max}$. Initial laser intensities $I_0=10^{11}$, 3×10^{12}, 8.5×10^{12}, and 3×10^{13} W/cm² are considered.

Figures 2 and 3 give solutions of NLS equation (1). For $I_0=10^{11}$ W/cm², as analyzed above, laser propagation obeys an integrable CNLS equation with only Kerr effect. Laser fields reveal periodic oscillation with the maximum intensity $I_{max}=5.82\times10^{11}$ W/cm² [see Fig. 2(a) and point A in Fig. 1(a)]. The phase trajectory is a homoclinic orbit (HMO) connection, where the stable manifold smoothly joins the unstable manifold at the saddle point $(E_0,0)$ [see Fig. 2(b)]. Figure 3(a) shows its spatially coherent pattern structure of laser fields.

When I_0 increases to greater than 10^{12} W/cm², $\chi^{(5)}$ susceptibility and MPI play role as a Hamiltonian perturbation H_1, we see from Figs. 2(d)–2(l) that HMO connection is

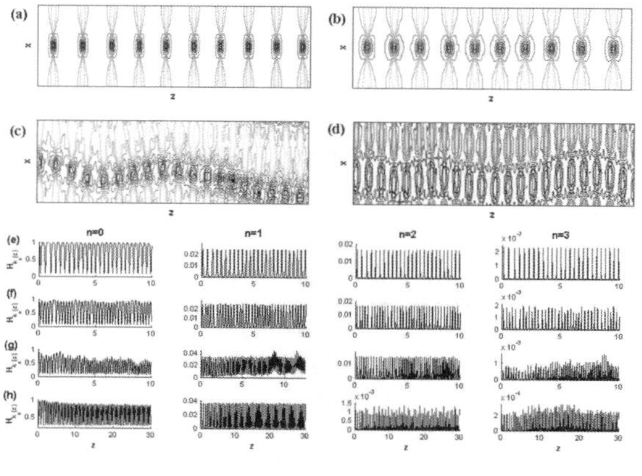

FIG. 3. The corresponding complex patterns of laser fields [(a)–(d)], and the energy evolution in the first four Fourier modes [(e)–(h)].

broken down, laser fields oscillate aperiodically, and irregular motions appear. For $I_0=2\times10^{12}$ W/cm^2, we see in Fig. 2(d) that laser fields oscillate quasiperiodically with $I_{\max}=1.13\times10^{13}$ W/cm^2, where $\chi^{(5)}$ susceptibility starts to act [see point B in Fig. 1(a)]. Figure 2(e) shows the typical weak chaotic behavior:[13] the phase trajectory is a band, the KAM torus has a small thickness but is not completely broken down. Figure 3(b) shows that laser field pattern structure still remains fairly coherent, but irregular substructures appear.

For $I_0=8.5\times10^{12}$ and 3×10^{13} W/cm^2, laser fields oscillate aperiodically with $I_{\max}=4.6\times10^{13}$ and 6×10^{13} W/cm^2, respectively [see Figs. 2(g) and 2(j)], MPI effect appears [see point C in Fig. 1(a)] and dominate [see point D] it combined with $\chi^{(5)}$ susceptibility completely breaks down KAM tori and strong chaos occurs. Irregular HMO crossings exist in phase space [see Fig. 2(h) and 2(k)], demonstrating the presence of stochastic motions for complex dynamics. The power spectra [Figs. 2(i) and 2(l)] indicate broadband structures and noiselike spectra being typical of strong chaos. We also see from Fig. 2(k) that when MPI dominates, the growth of laser field is heavily arrested with $d_z|E(0,z)|$ much smaller than the other cases. Such strong chaos corresponds to complex patterns with coherence completely broken down, shown in Fig. 3(c) and 3(d). Figs. 3(e)–3(h) show the evolution of the energy in the first four Fourier modes, in which the system's energy is defined as $\mathcal{H}=\Sigma_n H_{K_n}=\Sigma_n|E_{K_n}|^2$. It can be seen that in all cases large part of $\mathcal{H}$ lies in the first and second modes, leading to formation of solitarylike structures. Moreover, Figs. 3(e)–3(h) also show that coherent, near-coherent, and complex patterns are relevant to the stochastic evolution of the energy in the third and fourth modes.

These complicated dynamical phenomena are the characteristic of laser propagation in a real medium. In laser fusion studies, it is well known that the focusing becomes a chaotic oscillation rather than a catastrophic process due to relativistically ponderomotive effects.[1] For our problems, the collapse process is arrested and a subsequent propagation regime takes place in the form of a dynamical equilibrium among $\chi^{(5)}$ susceptibility, MPI effects, and SF.[4–6] When plasma is generated by MPI, its defocusing acts not only as a higher-order nonlinearity which saturates SF locally but also as a global effect resulting in complex propagation dynamics. In particular, since filamentation is associated with a small-scale transient SF behavior of laser beams, it is obvious that the pattern of filamentation beams would be quite complex due to the appearance of homoclinic chaos, as shown in Figs. 3(c) and 3(d). The stochastic partition of the system energy is the source of complexity in forming multiple filamentation beams. Thus, it is technically difficult to control a stable propagation process of multiple filamentation beams when the laser intensity becomes high.

In conclusion, we have studied complex dynamics of femtosecond terawatt laser pulse in air. Our theoretical and numerical results clearly illustrate that laser filamentation phenomena are associated with the homoclinic chaotic behaviors of laser fields. Furthermore, we show that the nonlinear dynamical behavior of laser fields depends on the intensity of laser pulses. Physically, the complex pattern of laser filamentation is attributed to the stochastic partition of the system energy in Fourier modes. We believe that these findings are general and will give useful implications to laser propagation dynamics, especially the filamentation process, in various nonlinear media.

C.T.Z. thanks the great hospitality of NUS during his visit. This work is partially supported by NSFC (Grant Nos. 10575013, 10576007, and 10335020) and National 973 Project "Nonlinear Science" in China.

[1] P. Gibbon, *Short Pulse Laser Interactions with Matter—An Introduction* (Imperial College Press, London, 2005), pp. 243–261.
[2] R. Ackermann, E. Salmon, N. Lascoux, J. Kasparian, P. Rohwetter, K. Stelmaszczyk, S. Li, A. Lindinger, L. Wöste, P. Béjot, L. Bonacina, and J. P. Wolf, Appl. Phys. Lett. **89**, 171117 (2006).
[3] K. Stelmaszczyk, P. Rohwetter, G. Méjean, J. Yu, E. Salmon, J. Kasparian, R. Ackermann, J. P. Wolf, and L. Wöste, Appl. Phys. Lett. **85**, 3977 (2004).
[4] L. Wöste, C. Wedekind, H. Wille, P. Rairoux, B. Stein, S. Nikolov, Ch. Werner, S. Niedermeier, H. Schillinger, and R. Sauerbrey, Laser Optoelektron. **29**, 51 (1997).
[5] W. Yu, M. Y. Yu, J. Zhang, L. J. Qian, X. Yuan, P. X. Lu, R. X. Li, Z. M. Sheng, J. R. Liu, and Z. Z. Xu, Phys. Plasmas **11**, 5360 (2004).
[6] L. Bergé, S. Skupin, F. Lederer, G. Méjean, J. Yu, J. Kasparian, E. Salmon, J. P. Wolf, M. Rodriguez, L. Wöste, R. Bourayou, and R. Sauerbrey, Phys. Rev. Lett. **92**, 225002 (2004), and references therein.
[7] H. Yang, J. Zhang, W. Yu, Y. J. Li, and Z. Y. Wei, Phys. Rev. E **65**, 016406 (2001).
[8] A. Vincotte and L. Bergé, Phys. Rev. Lett. **95**, 193901 (2005).
[9] B. F. Shen, M. Y. Yu, and R. X. Li, Phys. Rev. E **70**, 036403 (2004).
[10] L. Stenflo, M. Y. Yu, and P. K. Shukla, Phys. Scr. **40**, 257 (1989).
[11] H. Hora, *Plasmas at High Temperature and Density: Applications and Implications of Laser-Plasma Interaction* (Springer, New York, 1991), pp. 1–26.
[12] C. T. Zhou, X. T. He, and S. G. Chen, Phys. Rev. A **46**, 2277 (1992), and references therein.
[13] A. J. Lichtenberg and M. A. Lieberman, *Regular and Chaotic Dynamics* (Springer, New York, 1992), pp. 1–20.
[14] C. T. Zhou, M. Y. Yu, and X. T. He, Phys. Rev. E **73**, 026209 (2005).

COMMUNICATIONS IN COMPUTATIONAL PHYSICS
Vol. **4**, No. 5, pp. 1129-1150

Commun. Comput. Phys.
November 2008

Progress of Pattern Dynamics in Plasma Waves

B. Qiao[1,2,3], C. T. Zhou[3,4,5,*], X. T. He[3,4,5] and C. H. Lai[2,6]

[1] *Center for Plasma Physics, Department of Physics and Astronomy, Queen's University Belfast, Belfast BT7 1NN, UK.*
[2] *Department of Physics, National University of Singapore, Singapore 117542.*
[3] *Institute of Applied Physics and Computational Mathematics, P. O. Box 8009, Beijing 100088, China.*
[4] *Center for Applied Physics and Technology, Peking University, Beijing 100871, China.*
[5] *Institute for Fusion Theory and Simulation, Zhejiang University, Hangzhou 310027, China.*
[6] *Beijing-Hong Kong-Singapore Joint Centre for Nonlinear & Complex Systems (Singapore), National University of Singapore, Singapore 117542.*

Received 29 February 2008; Accepted (in revised version) 8 July 2008

Available online 9 September 2008

Abstract. This paper is concerned with the pattern dynamics of the generalized nonlinear Schrödinger equations (NSEs) related with various nonlinear physical problems in plasmas. Our theoretical and numerical results show that the higher-order nonlinear effects, acting as a Hamiltonian perturbation, break down the NSE integrability and lead to chaotic behaviors. Correspondingly, coherent structures are destroyed and replaced by complex patterns. Homoclinic orbit crossings in the phase space and stochastic partition of energy in Fourier modes show typical characteristics of the stochastic motion. Our investigations show that nonlinear phenomena, such as wave turbulence and laser filamentation, are associated with the homoclinic chaos. In particular, we found that the unstable manifolds $W^{(u)}$ possessing the hyperbolic fixed point correspond to an initial phase $\theta = 45^\circ$ and 225°, and the stable manifolds $W^{(s)}$ correspond to $\theta = 135^\circ$ and 315°.

PACS: 47.54.-r, 05.45.Yv, 52.25.Gj, 52.35.Mw
Key words: Pattern dynamics, homoclinic chaos, nonlinear Schrödinger equations, plasma waves.

1 Introduction

It is well known that the generalized nonlinear Schrödinger equation (NSE) [1–25] is of the form

$$iE_t + \partial_x^2 E + F(|E|^2)E = 0, \tag{1.1}$$

*Corresponding author. *Email addresses:* b.qiao@qub.ac.uk (B. Qiao), zcangtao@iapcm.ac.cn (C. T. Zhou), xthe@iapcm.ac.cn (X. T. He), phylaich@nus.edu.sg (C. H. Lai) where $E(x,t)$ is the complex amplitude of waves, and t and x are time and space variables, respectively. The function F is used to describe different physical processes [20–23], such as plasma physics, nonlinear optics, fluid dynamics, superconductivity theory and Bose-Einstein condensates (BEC) etc. The generalized NSE is one of the basic evolution models for nonlinear process in various branches of the conservative systems. Early applications of the NSE were in the context of nonlinear optics where it described the propagation of light beams in nonlinear media [24]. Also it has been applied to gravity waves on deep water, for which the predicted modulational instability and envelope soliton formation have been clearly demonstrated experimentally [25]. In plasma physics, a large number of nonlinear processes, such as nonlinear hydromagnetic waves [26], small-K condensation of weak turbulence in nonlinear plasmas, Langmuir waves in electrostatic plasmas, femtosecond (fs) laser pulse in air, medium-intensity laser in underdense plasma and intense laser pulse in relativistic plasmas, can all be effectively modeled by the generalized NSE (1.1) with different potential function F.

For the generalized NSE (1.1), the Lagrangian density is

$$L=\frac{i}{2}(E^*E_t-EE_t^*)-|E_x|^2+f(|E|^2), \tag{1.2}$$

where $f(|E|^2)=\int_0^{|E|^2}F(s)ds$. The system described by it is a conservative system. According to the Norther theorem, we can obtain the following invariants: the quasiparticle number

$$N=\int|E|^2dx, \tag{1.3}$$

the momentum

$$P=i\int E^*E_xdx, \tag{1.4}$$

and the Hamiltonian quantity

$$H=\int[|E_x|^2-f(|E|^2)]dx. \tag{1.5}$$

For such a conservative Hamiltonian system (1.1) or (1.2), the nonlinear dynamics including solitary waves and patterns are very important. In nature, most nonlinear phenomena, such as Langmuir wave collapse, laser self-focusing and filamentation, are all associated with the basic nonlinear dynamics of the system and are the results of nonlinear development of modulational instability. However, the latter has not been systematically studied, while previous work has concentrated on the solitary wave solutions and the singular solutions. The study of complex dynamics including chaos and patterns for the generalized NSE (1.1) associated with different physical problems is of interest to the understanding of various nonlinear phenomena in plasmas.

In 1992, we found for the first time that an one-dimensional cubic-quintic nonlinear Schrödinger equation [1,2] is non-integrable [3] due to high-order Hamiltonian perturbation. Physically, our results showed that the coherent pattern of plasma electron (Langmuir) waves can be broken down due to high-order field interaction of Langmuir waves and ion-acoustic waves. Then we completed some systematic investigations on pattern dynamics of the generalized nonlinear Schrödinger equations [4–12], Zakharov equations [13–15] and complex Ginzburg-Landau equations [16, 17] in one- and two-dimensional spaces, respectively. Our investigations showed that nonlinear phenomena, such as wave turbulence and laser filamentation, etc., are associated with the homoclinic chaos. In particular, we found that the unstable manifolds $W^{(u)}$ possessing the hyperbolic fixed point correspond to an initial phase $\theta=45°$ and $225°$, and the stable manifolds $W^{(s)}$ correspond to $\theta=135°$ and $315°$. Recently, we have applied our method to investigate complex dynamics of femtosecond terawatt laser pulses in air [18] and relativistic lasers in plasmas [19]. Our investigations further show that our previously theoretical methods and numerical strategies are applicable for the study of different physical systems. Pattern dynamics in nonequilibrium system [27], in nonlinear optics [28], and in chemical systems [29] have been all reviewed. This paper mainly covers pattern dynamics of the generalized nonlinear Schrödinger equations related to various nonlinear physical problems in plasmas.

The paper is organized as follows. In Section 2, we give a qualitative nonlinear analysis of NSE (1.1) and constitute an available phase space and initial condition to describe its nonlinear behavior. In Sections 3-7, we analyze the dynamics in detail respectively of cubic NSE (CNSE) for weak turbulence in plasmas, quintic NSE for langmuir waves in plasmas, higher-order NSE for fs laser pulse in air, exponential NSE for medium-intensity laser in underdense plasmas and relativistic NSE (RNSE) for intense laser in relativistic plasmas. Some conclusion and discussion are given in Section 8.

2 Qualitative analysis

For the continuum Hamiltonian system (1.1) or (1.2), the dynamic description is dependent on the choice of the initial condition. In particular, we cannot give a reasonable explanation if we take an arbitrary initial condition. Therefore, it is necessary that we choose an available initial condition and constitute a reasonable phase space in order to discuss the nonlinear evolution properties of this system. For definitiveness, we only deal with the developing behavior of an initial homogeneous state due to the modulational instability. In other words, we consider the nonlinear modulation of the plane wave solution as in [3]

$$E_s(t)=E_0\exp(i\lambda t), \tag{2.1}$$

where $\lambda=F(|E_0|^2)$ for $E_0\neq 0$.

In the presence of small perturbations, we can write

$$E(x,t)=E_s(t)+\delta E(x,t), \tag{2.2}$$

where $\delta E(x,t) \ll E_s(t)$. Linearizing the NSE (1.1), we obtain the eigen equation for the perturbation field $\delta E(x,t)$ as

$$\begin{pmatrix} i\partial_t+\partial_x^2+L & L \\ L & -i\partial_t+\partial_x^2+L \end{pmatrix}\begin{pmatrix} \delta E \\ \delta E^* \end{pmatrix}=0, \tag{2.3}$$

where $L=F'(|E_0|^2)|E_0|^2$ and $F'(u)=\partial_u F$.

For periodic boundary conditions, the eigenfunction $\delta E(x,t)$ can be defined [3–12] as

$$\begin{pmatrix} \delta E \\ \delta E^* \end{pmatrix}=\begin{pmatrix} \epsilon\exp(i\mu t) \\ \epsilon^*\exp(-i\mu^* t) \end{pmatrix}\cos(kx), \tag{2.4}$$

where ϵ and ϵ^* are small parameters, and k is the wave number of the perturbation. Some algebra yields the eigenvalue [5]

$$\mu=2\left[F'(|E_0|^2)|E_0|^2-k^2\right]\pm i\sqrt{k^2\left[2F'(|E_0|^2)|E_0|^2-k^2\right]}, \tag{2.5}$$

and the growth rate

$$\Gamma=\mathrm{Im}(\mu)=\sqrt{k^2\left[2F'(|E_0|^2)|E_0|^2-k^2\right]}. \tag{2.6}$$

The wave number of the most unstable mode is then $K_{max}=\sqrt{f'(|E_0|^2)|E_0|^2}$. Inserting Eqs. (2.4) and (2.5) into Eq. (2.3), we easily obtain that [3]

$$\delta E/\delta E^*=\pm i. \tag{2.7}$$

To analyze the nonlinear behavior of the system (1.1), we construct phase space ($|E|$, $d_t|E|$) at $x=0$ [10, 11, 13, 30]. From Eq. (2.5), we see that the eigenvalue μ has a pair of conjugated complex roots, thus we can determine that $(E_0,0)$ is associated with a saddle (or hyperbolic fixed) point in phase space [3]. Furthermore, from Eq. (2.4) the initial condition can be chosen as

$$E(x,0)=E_0+\eta\exp(i\theta)\cos(k_x x). \tag{2.8}$$

We therefore find that the unstable manifolds $W^{(u)}$ possessing the hyperbolic fixed point correspond to $\theta=45°$ and $225°$, and the stable manifolds $W^{(s)}$ correspond to $\theta=135°$ and $315°$ [3–12].

3 Cubic NSE for coherent solitons in plasmas

The dynamical structure of Hamiltonian systems is sensitive to initial conditions [31–33]. The analysis above gives us a practical initial condition as well as a suitable phase space for describing the nonlinear behavior of the system that can be described by the generalized NSE (1.1). Using such an initial condition and phase space, we can study various physical systems, once its potential function F in NSE (1.1) is determined.

First, for the cubic NSE (CNSE), the potential function is taken as

$$\text{(i)} \qquad F(|E|^2) = \alpha |E|^2, \tag{3.1}$$

where α is a parameter and the Hamiltonian can be written as

$$H_0 = \int \left(|E_x|^2 - \frac{1}{2}\alpha |E|^4 \right) dx, \tag{3.2}$$

which is integrable. The cubic NSE has been widely studied in nonlinear optics and fibers etc. [34,35]. Recently, the stabilization of high-order solutions of the cubic NSE has been discussed [36]. In nonlinear plasmas, it describes the nonlinear interaction between Langmuir wave and ion-acoustic wave in the subsonic regime, where the interaction of the second-order fields is only considered. In the initial stages of field evolutions, many physical phenomena, such as solitons developed by modulational instabilities, can be explained in terms of this model.

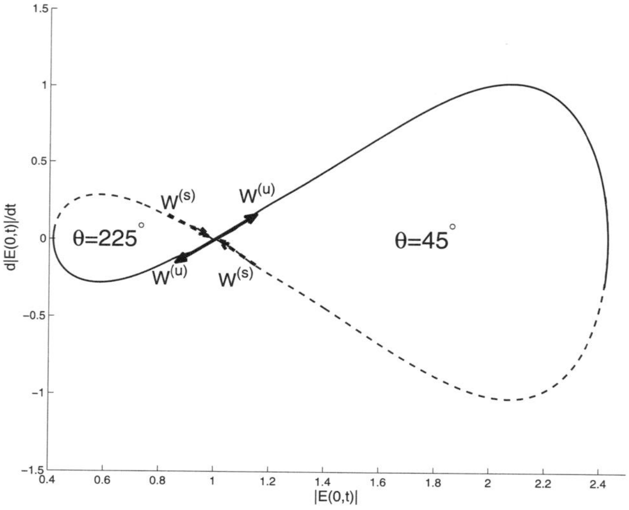

Figure 1: Stable $W^{(s)}$ and unstable $W^{(u)}$ manifolds of the hyperbolic fixed point $(E_0,0)$ with $\theta=45°$ and $225°$ for the CNSE [Eq. (1.1) combined with Eq. (3.1)], where the solid curve is computed with $t>0$ and the dashed curve is computed with $t<0$.

In [10], we gave the solution of the CNSE [Eqs. (1.1) and (3.1)] by taking the initial condition (2.8) and choosing $E_0=1$, $\alpha=1$, $\eta=0.1$, $\theta=45°$ and $225°$ respectively. The standard splitting-step spectral method [3,10,11] is improved [12] in order to better preserve the conserved quantities (1.3)-(1.5). The spatial period of the system is taken to be $L=2\pi/K_{max}$. From Fig. 1, we see the theoretical analysis in Section 2 is completely verified. The CNSE is fully integrable due to the existence of Lax pair, i.e., H_0 is integrable. One solution corresponding to the periodic oscillations is shown in Fig. 2. The corresponding phase-space trajectory is a homoclinic orbit (HMO). The stable manifold smoothly joins the unstable manifold at the saddle point (E_0,0), i.e., HMO connection, and the well-known Fermi-Pasta-Ulam (FPU) recurrence exists, as shown in Fig. 2(b). The fundamental frequency of the periodic solution is shown in Fig. 2(c). Because of the exact FPU recurrence, a spatially coherent pattern appears, shown in Fig. 2(d).

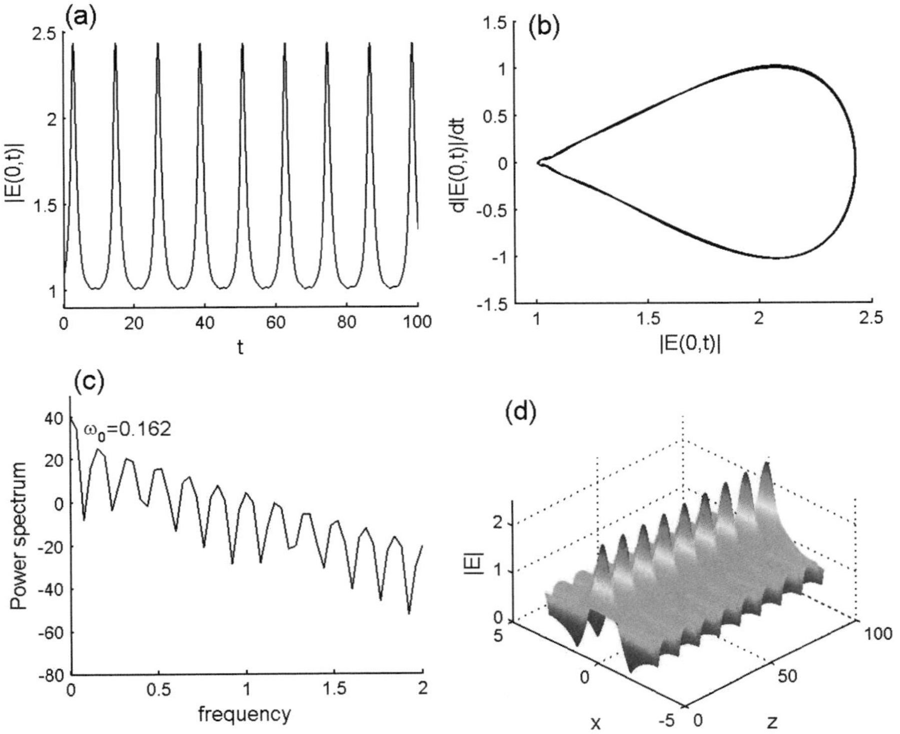

Figure 2: Dynamics of the CNSE [Eqs. (1.1) and (3.1)] for $E_0=1$, $\eta=0.1$ and $\theta=45°$. (a) the evolution of laser field envelope at $x=0$; (b) the trajectories of the phase space ($E(0,t)$, $d|E(0,t)|/dt$); (c) the corresponding power spectra; and (d) the corresponding spatial patterns.

To display this recurrence, the evolution of the energy contained in Fourier modes was measured. In Fourier space, the energy of the system is defined as

$$\mathcal{H}=\sum_n \mathcal{H}_{K_n}=\sum_n |E_{K_n}|^2, \tag{3.3}$$

and the initial energy is added to the mode K_{max}. From Fig. 3, we observe that a large part of the energy in the system lies in the first and second modes. Furthermore, the evolution of the energy in all modes is periodic, which is consistent with the periodic recurrence. This means that the unstable modulation first grows exponentially as predicted by Benjamin and Feir [37] but eventually the solution would demodulate and return to a near-uniform state. Thus, solitons developed by modulational instability keep their spatially coherent structures and temporally periodic evolutions.

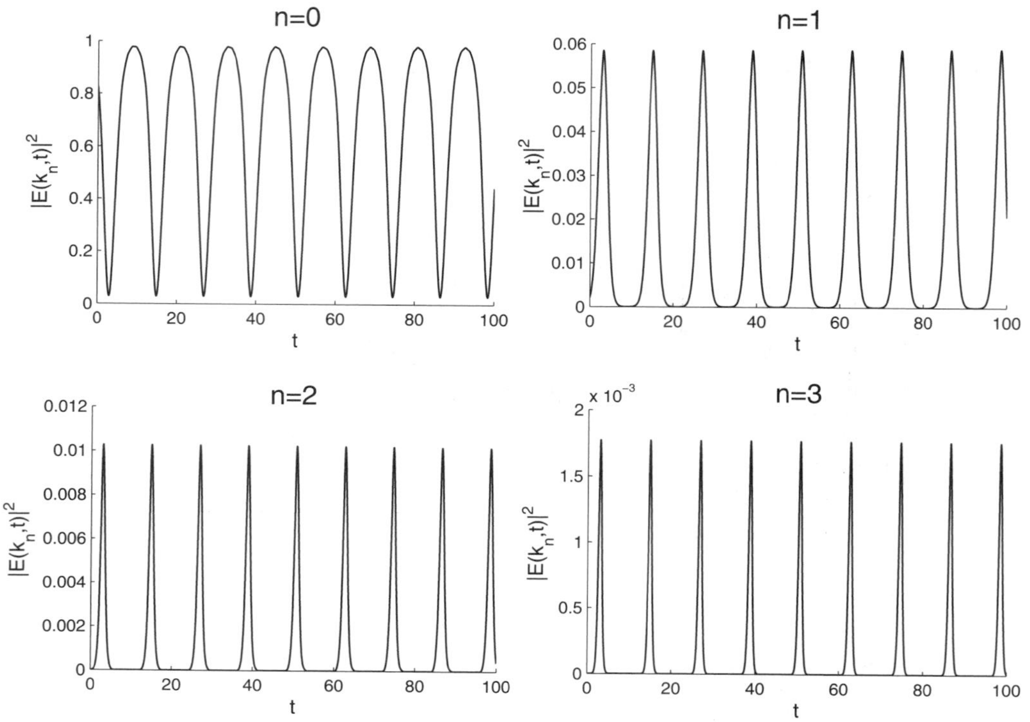

Figure 3: The energy evolution in the first four Fourier modes for the CNSE [Eqs. (1.1) and (3.1)], where the parameters are the same as Fig. 2.

4 Quintic NSE for Langmuir waves in electrostatic plasmas

For Langmuir waves in electrostatic plasmas, the beat-frequency interaction between large amplitude parts of high frequency fields and particles can occur in strong turbulent stage of plasma instability. The other physical effects, for example, damping and dissipation, have to be considered and the CNSE is no longer valid. Under the static approximation, from Vlasov-Maxwell equations, He [1] has obtained a quintic NSE describing the nonlinear interaction between Langmuir waves and electrons in plasmas, where the potential $F(|E|^2)$ can be expressed as

$$\text{(ii)} \qquad F(|E|^2)=|E|^2-g|E|^4, \tag{4.1}$$

where $E(x,t)$ is the slowly-varying complex amplitude of Langmuir wave fields, and g is the coupling constant of Langmuir wave fields with electrons, which depends on electron temperature and density, etc. The Hamiltonian of the quintic NSE [Eq. (1.1) combined with the potential (4.1)] can be written as

$$H=H_0+H_1, \tag{4.2}$$

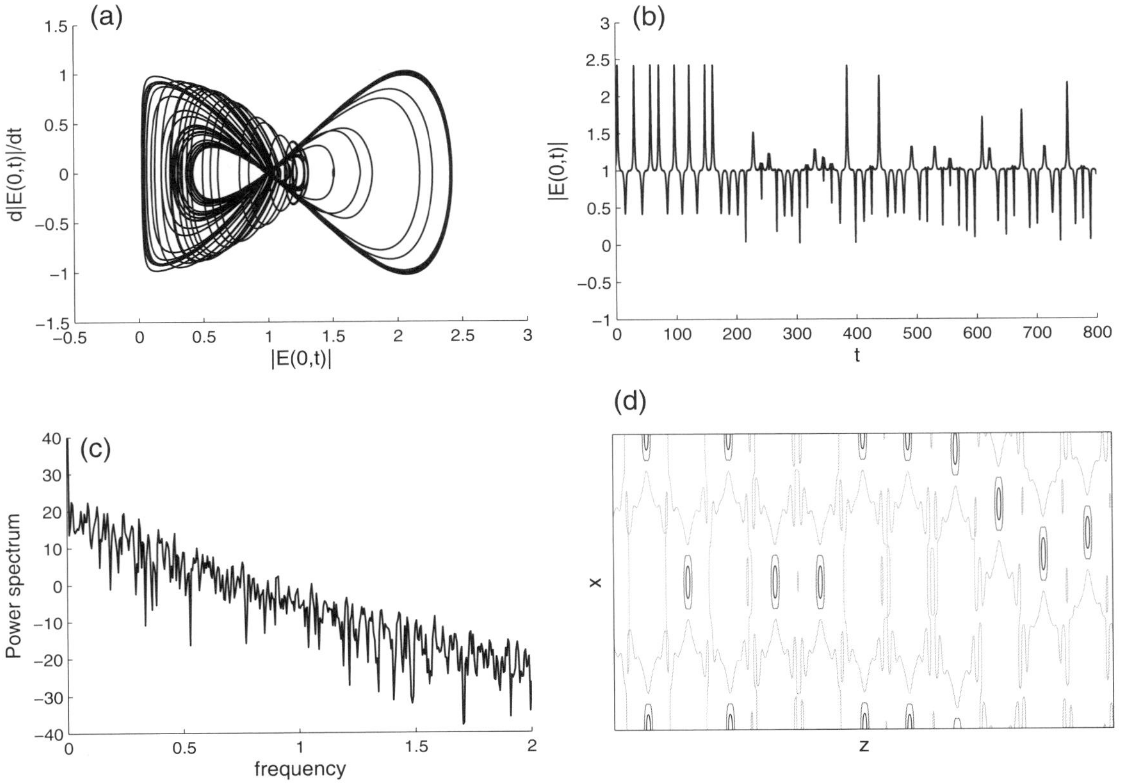

Figure 4: Dynamics of the quintic NSE [Eqs. (1.1) and (4.1)] for $E_0=1$, $\eta=0.1$ and $\theta=45°$. (a) the trajectories of the phase space $(E(0,t), d|E(0,t)|/dt)$; (b) the evolution of laser field envelope at $x=0$; (c) the corresponding power spectra; and (d) the corresponding spatial patterns.

where H_0 is the one for the integrable CNSE [Eq. (3.2) with $\alpha=1$] and

$$H_1=\int \frac{1}{3}g|E|^6 dx \tag{4.3}$$

acts as a quintic Hamiltonian perturbation to H_0.

Much attention has been focused on analyzing the behavior of high-order NSEs involving quintic terms [38,39]. In 1992, we [3] systematically studied the dynamic properties of such quintic NSE for Langmuir waves in plasmas. We found that due to H_1, the NSE integrability is broken down and chaos and complex patterns of wave fields are formed. Fig. 4(a) clearly illustrates the chaotic behavior of the wave fields. Fig. 4(b) shows the presence of irregular HMO crossings, where the irregular motions appear nearby HMO. These demonstrate the presence of stochastic motions for complex dynamics. The power spectra [Fig. 4(c)] also clearly indicate the broadband structures and noiselike spectra being typical of chaotic evolutions. Fig. 4(d) shows that the Langmuir wave field patterns are no longer coherent but still keep their spatially localized structures. Physically, this means that the high-order nonlinear effects occurring during the long-term evolution of Langmuir waves in plasmas result in the conversion of wave field patterns from coherence to turbulence. Furthermore, the conversion is chaotic but keeps their spatially localized structures.

5 Higher-order NSE for femtosecond laser pulse in air

Propagation of femtosecond (fs) Terawatt (TW) laser pulses [23] in air has attracted wide interest due to its applications in lightning discharge control and atmospheric remote sensing [40–42]. Self-channeling of initial laser pulse experiences an early self-focusing (SF) stage caused by the Kerr response of air, which leads to a sharp increase of laser intensity. When laser intensity increases close to the ionization threshold, $\chi^{(5)}$ susceptibility originated from expansion of the nonlinear polarization vector and multiphoton ionization (MPI) of air molecules play important roles in the propagation dynamics [43].

Recently, we [18] applied the above method to investigate the complex dynamics of such laser pulses in air by considering the NSE (1.1) with the potential function

$$\text{(iii)} \qquad F(|E|^2)=\alpha|E|^2-\epsilon|E|^4-\gamma|E|^{2K}, \tag{5.1}$$

which includes Kerr focusing of air, $\chi^{(5)}$ susceptibility originated from expansion of nonlinear polarization, and plasma generation by MPI of air molecules. It is noted that the time variable t in Eq. (1.1) has been replaced with the propagation distance z for the following problems. For pulse duration $t_p=250\text{fs}$ and wavelength $\lambda_0=800\text{nm}$, their coefficients take values $\alpha=0.446$, $\epsilon=7.3\times10^{-7}[\text{cm}^2/w_0^2]$ and $\gamma=8.4\times10^{-40}[\text{cm}^{2(K-2)}]/w_0^{2k-2}$ with w_0 laser beam transverse width in unit of cm. The variables z, x and E are normalized by the units of $4z_f$, w_0, and $\sqrt{P_{cr}/4\pi w_0^2}$, respectively, where $z_f=\pi w_0^2/\lambda_0$ is the Rayleigh length and $P_{cr}=2.55\text{GW}$ the critical power for SF.

The Hamiltonian of such a NSE [Eqs. (1.1) and (5.1)] can also be written as Eq. (4.2), where

$$H_1=\int\left(\frac{1}{3}\epsilon|E|^6+\frac{1}{K+1}\gamma|E|^{2K+2}\right)dx. \tag{5.2}$$

If only the Kerr effect is considered, $H=H_0$ [Eq. (3.2)] is integrable corresponding to a CNSE. $\chi^{(5)}$ susceptibility and MPI effect act as a Hamiltonian perturbation H_1 to H_0. Fig. 5(a) gives comparisons of these three effects for different laser intensities I_0. For $I_0<10^{12}\text{W/cm}^2$, purely Kerr nonlinearity plays a major role with $H\approx H_0$ integrable. For $10^{12}<I_0<2\times10^{13}\text{W/cm}^2$, $\chi^{(5)}$ susceptibility starts to act as a small perturbation H_1. For $I_0>2\times10^{13}\text{W/cm}^2$, MPI appears, laser propagation is governed by all three effects with H nonintegrable. When I_0 further increases to larger than $5\times10^{13}\text{W/cm}^2$, MPI dominates $\chi^{(5)}$ susceptibility and laser beams reveal complex dynamics, including SF and filamentation.

Our numerical results given in Fig. 5(b) completely verify the theoretical analysis obtained in Section 2, where $W^{(u)}$ possessing the hyperbolic fixed point correspond to $\theta = 45°$ and $225°$, and $W^{(s)}$ correspond to $\theta = 135°$ and $315°$. Fig. 5(c) plots the unstable growth rate Γ for different I_0. From Figs. 5(c) and (a), we can understand that laser pulse with lower intensity ($I_0 < 10^{12}\mathrm{W/cm^2}$) follows the early SF caused by purely Kerr response of air, which compresses laser beam in the transverse plane and increases the intensity sharply [see Fig. 5(c), Γ is very large for $I_0 = 10^{11}\mathrm{W/cm^2}$]. With increasing I_0, both $\chi^{(5)}$ susceptibility and MPI become important mechanisms, the growth of laser field amplitude is quickly arrested by such two effects [see Fig. 5(c), Γ much decreases when I_0 increases, in particular, for $I_0 = 3\times10^{13}\mathrm{W/cm^2}$].

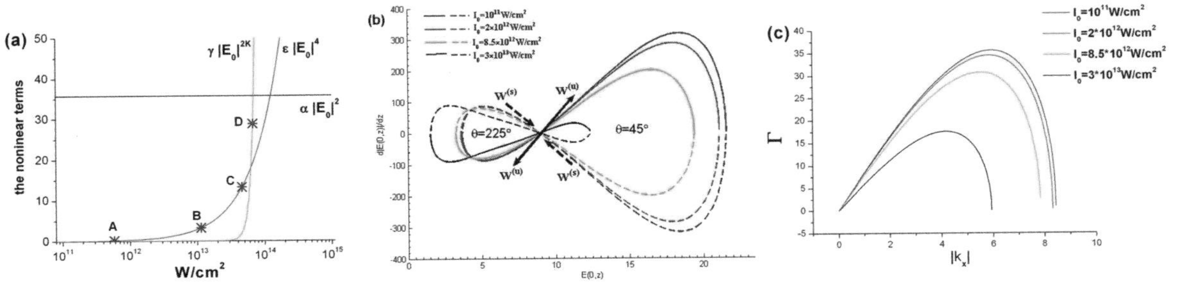

Figure 5: (a) Comparison of three nonlinear effects for different I_0 in Eq. (5.1). Asterisk symbols mark the points with I_0 respectively of 5.82×10^{11}, 1.13×10^{13}, 4.6×10^{13} and $6\times10^{13}\mathrm{W/cm^2}$. (b) Stable $W^{(s)}$ and unstable $W^{(u)}$ manifolds for the saddle point $(E_0, 0)$ with $\theta = 45°$ and $225°$, where the solid curve is calculated by NSE's (1.1) and (5.1) with $z > 0$ and the dashed curve with $z < 0$. (c) The growth rate Γ with the corresponding I_0.

We choose input laser power $P_{in} = 10P_{cr}$ ($E_0 = 8.94$) with noise level 0.01 ($\eta = 0.089$), initial phase $\theta = 45°$, and consider the most unstable mode with $k = K_{max}$. Initial laser intensities $I_0 = 10^{11}$, 3×10^{12}, 8.5×10^{12} and $3\times10^{13}\mathrm{W/cm^2}$ are considered, respectively. Figs. 6 and 7 plot the solutions. For $I_0 = 10^{11}\mathrm{W/cm^2}$, as analyzed above, the laser propagation obeys an integrable CNSE with only Kerr effect. Laser fields reveal periodic oscillation with the maximum intensity $I_{max} = 5.82\times10^{11}\mathrm{W/cm^2}$ [see Fig. 6(a) and point "A" in Fig. 5(a)]. The phase trajectory is a HMO connection [see Fig. 6(b)]. Fig. 7(a) shows its spatially coherent pattern structure of the laser fields.

When I_0 increases to greater than $10^{12}\mathrm{W/cm^2}$, $\chi^{(5)}$ susceptibility and MPI play roles as a Hamiltonian perturbation H_1, we see from Figs. 6(d)-(l) that HMO connection is broken down, laser fields oscillate aperiodically and irregular motions appear. For $I_0 = 2\times10^{12}\mathrm{W/cm^2}$, we see in Fig. 6(d) that laser fields oscillate quasi-periodically with $I_{max} = 1.13\times10^{13}\mathrm{W/cm^2}$, where $\chi^{(5)}$ susceptibility starts to act [see point "B" in Fig. 5(a)]. Fig. 6(e) shows the typical weak chaotic behavior [31]: the phase trajectory is a band, the KAM torus has a small thickness but is not completely broken down. Fig. 7(b) shows that laser field pattern structure still remains fairly coherent, but irregular sub-structures appear.

For $I_0 = 8.5\times10^{12}$ and $3\times10^{13}\mathrm{W/cm^2}$, laser fields oscillate aperiodically with respectively $I_{max} = 4.6\times10^{13}$ and $6\times10^{13}\mathrm{W/cm^2}$ [see Figs. 6(g) and (j)], MPI effect appears [see point "C" in Fig. 5(a)] and dominates [see point "D"], it combined with $\chi^{(5)}$ susceptibil-

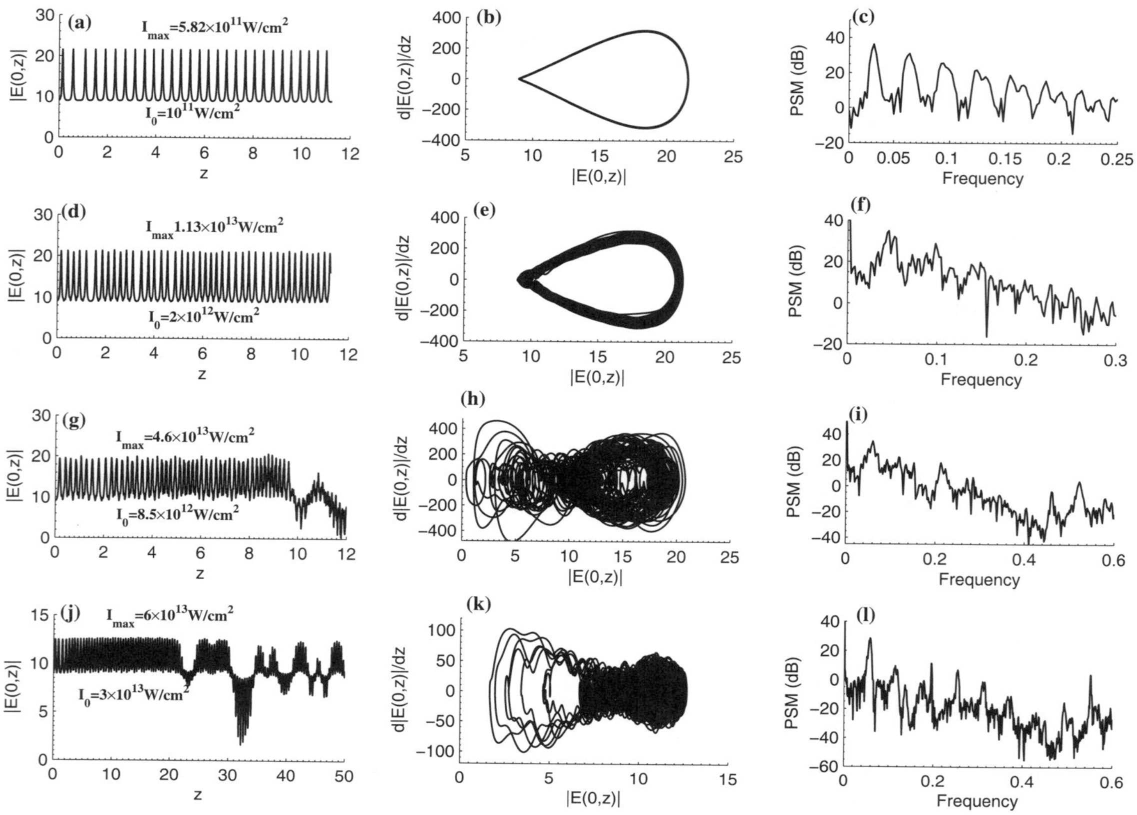

Figure 6: Complex dynamics of laser fields in air with I_0 respectively of 10^{11} [(a)-(c)], 2×10^{12} [(d)-(f)], 8.5×10^{12} [(g)-(i)] and $3\times10^{13}\text{W/cm}^2$ [(j)-(l)]. (a), (d), (g) and (j) are evolutions of laser field envelope; (b), (e), (h) and (k) are the phase-space trajectories; (c), (f), (i) and (l) are the power spectra.

ity completely breaks down KAM tori and strong chaos occurs. Irregular HMO crossings exist in phase space [see Figs. 6(h) and (k)], demonstrating the presence of stochastic motions for complex dynamics. The power spectra [Figs. 6(i) and (l)] indicate broadband structures and noiselike spectra being typical of strong chaos. Such strong chaos corresponds to complex patterns with coherence completely broken down, shown in Figs. 7(c) and (d). Figs. 7 (e)-(h) show the evolution of the energy in the first four Fourier modes, in which the system's energy is defined as Eq. (3.3). It shows that coherent, near-coherent and complex patterns are relevant to the stochastic evolution of the energy in the third and fourth modes.

These complicated dynamical phenomena are the characteristic of laser propagation in air. It clearly illustrates that laser filamentation phenomena are associated with the homoclinic chaotic behaviors of laser fields. Furthermore, it shows that the nonlinear dynamical behavior of laser fields depends on the intensity of laser pulses. Physically, the complex pattern of laser filamentation is attributed to the stochastic partition of the system energy in Fourier modes.

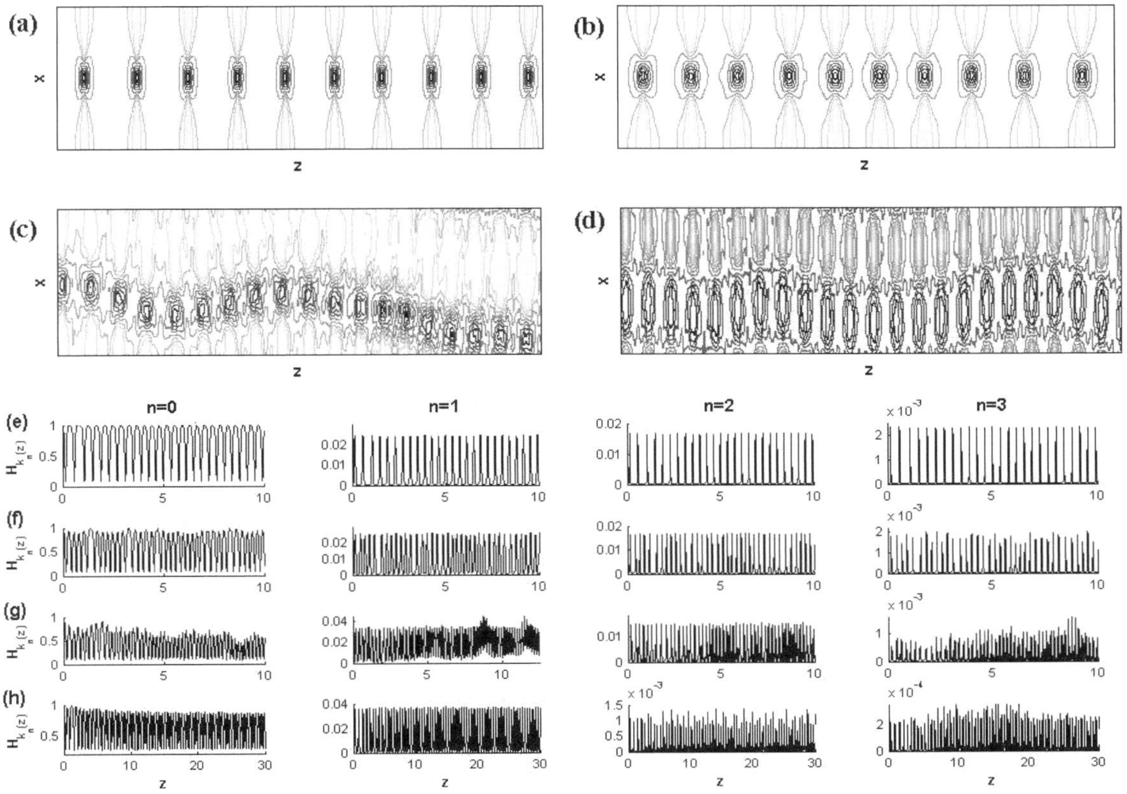

Figure 7: The corresponding complex patterns of laser fields in air [(a)-(d)], and the energy evolution in the first four Fourier modes [(e)-(h)] for the same parameters as Fig. 6.

6 Exponential NSE for medium-intensity laser in underdense plasmas

For a laser pulse with intensity from 10^{15} to $10^{17}\mathrm{W/cm^2}$ interacting with underdense plasmas, considering the Maxwell equations and the electron motion equations, one can also obtain the dimensionless NSE (1.1) in static approximation to describe its dynamics, where the potential function is [10, 11]

$$\text{(iv)} \qquad F(|E|^2)=1-N_e=\frac{1}{2g}(1-e^{-2g|E|^2}). \tag{6.1}$$

Here g is a parameter. When $g=0$, the exponential NSE (ENSE) [Eqs. (1.1) and (6.1)] becomes the above well-known integrable CNSE. The Hamiltonian perturbation here can be written as

$$H_1=\int\left\{\frac{1}{2}|E|^4-\frac{1}{2g}\left[|E|^2+\frac{1}{2g}(e^{-2g|E|^2}-1)\right]\right\}dx. \tag{6.2}$$

where H_0 is also the one (3.2) of CNSE with $\alpha=1$.

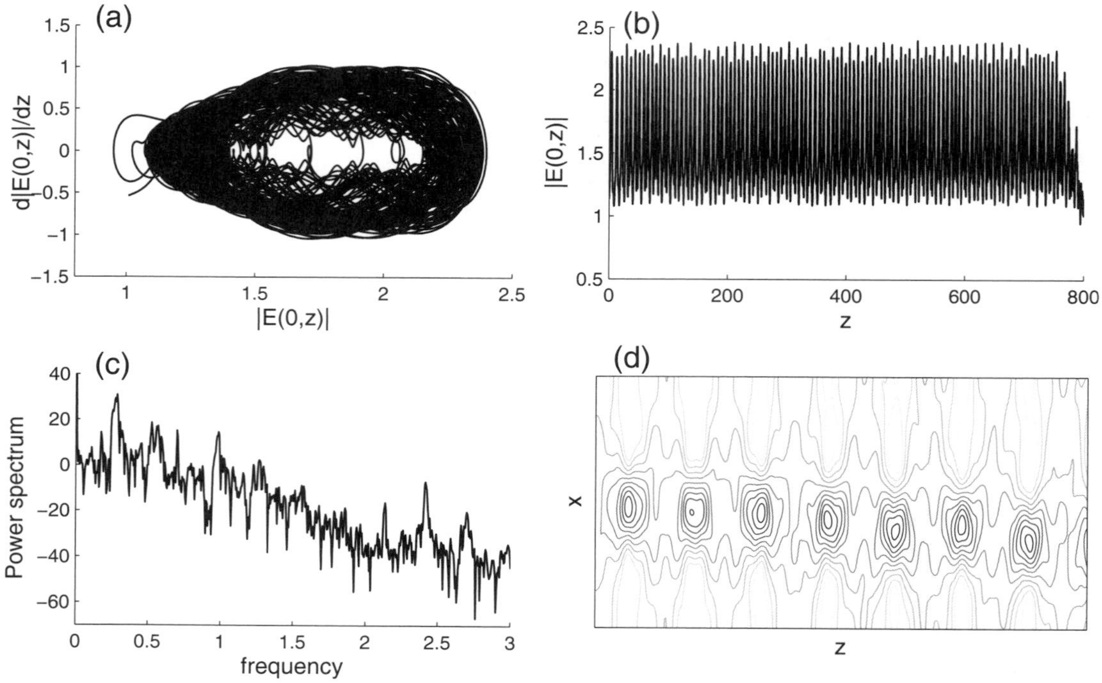

Figure 8: Solutions of Eqs. (1.1) and (6.1) with $\eta=0.1$, $\theta=45^\circ$ and $g=0.1$. (a) phase trajectories illustrate the irregular HMO crossings; (b) propagation of the field amplitude; (c) continuous power spectrum shows the chaotic characteristic; (d) complex patterns with coherence broken down.

In 1994 [11], we also studied the complex dynamics of such laser pulse propagation in plasmas by numerically solving the ENSE [Eqs. (1.1) and (6.1)]. According to the theoretical results obtained in Section 2, the initial condition here should be

$$E_0=\sqrt{(1/2g)l_n(1-2g)^{-1}} \quad \text{for } 0\le g<\frac{1}{2}$$

and $[\sqrt{(1/2g)l_n(1-2g)^{-1}},0]$ corresponds to a saddle point in phase space. We choose an initial position that lies in the nearby saddle point $(E_0,0)$, that is, $\eta=0.1$, $\theta=45.225^\circ$. For $g=0$, The ENSE [Eqs. (1.1) and (6.1)] becomes a CNSE, the HMO connection is shown in Fig. 1 and the coherent pattern is shown in Fig. 2(d) in Section 3. As $g\neq 0$, however, we find that the stable and unstable manifolds in phase space do not smoothly join together and the irregular HMO crossings exist as shown in Fig. 8(a), which illustrates that the current system is nonintegrable. $|E(0,z)|$ experiences stochastic oscillations [see Fig. 8(b)] with a continuous nonperiodic spectrum [see Fig. 8(c)]. Fig. 8(d) shows the irregular patterns with coherence broken down. Similarly to the above, the irregular patterns are associated with the partition of energy in Fourier modes. As shown in Fig. 9, the energy in the system, which is initially confined to the master mode, would spread to many slaved harmonic modes because of the nonlinear interaction, but would not regroup into the original lowest mode.

To illustrate the route to spatial chaos by the Hamiltonian perturbation H_1 (6.2), we further fix parameters $\theta = 45.225°$, $\eta = 0.1$ and vary g. From Fig. 10, we find the foundational frequency $\omega_0 = 0.2513$ for $g = 0$. When $g = 0.0002$, the wave field still seems to propagate with periodic behavior, but some small peaks in the power spectrum appear and the base frequencies are not uniquely defined. When $g = 0.0008$, in particular, more peaks are produced (but still countable), although the solution does not recur within the finite distance. In a sense, these solutions are called quasiperiodic solutions. With the increase of parameter g, the continuous power spectrum reveals the chaotic behavior (see Fig. 10, in the case of g=0.01, and Fig. 8). From the standpoint of nonlinear dynamics, the base frequency of the cubic NSE is unique (ω_0) for our parameter due to the integrability of the system. However, the nonintegrable perturbation H_1 (6.2) will make the frequency shift, that is, $\omega_i = \omega_0 + \triangle\omega_i$. When the parameter g is quite small, there exist finite countable frequencies. With the increase of g, the oscillatory overlapping of the modes could occur [31], which leads to the formation of the stochastic layer and appearance of the continuous power spectrum.

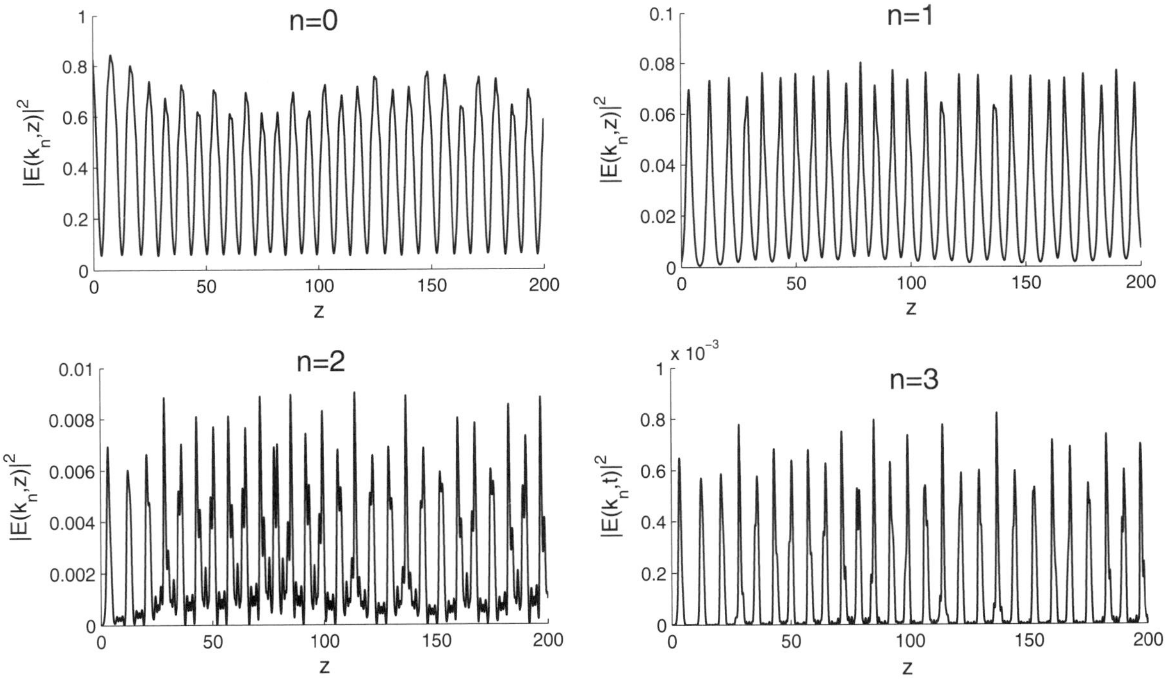

Figure 9: Propagation of energy in the first four Fourier modes, where the parameters are the same as Fig. 8.

In fact, for medium-intensity laser propagation in plasma, the response of the plasma density to the laser fields is only exhibited by the nonlinear ponderomotive force, which corresponds to the exponential nonlinear term in Eq. (6.1). From the investigation, we understand that such a response would act as a nonintegrable perturbation H_1 (6.2) and lead to the stochastic propagation of laser beams.

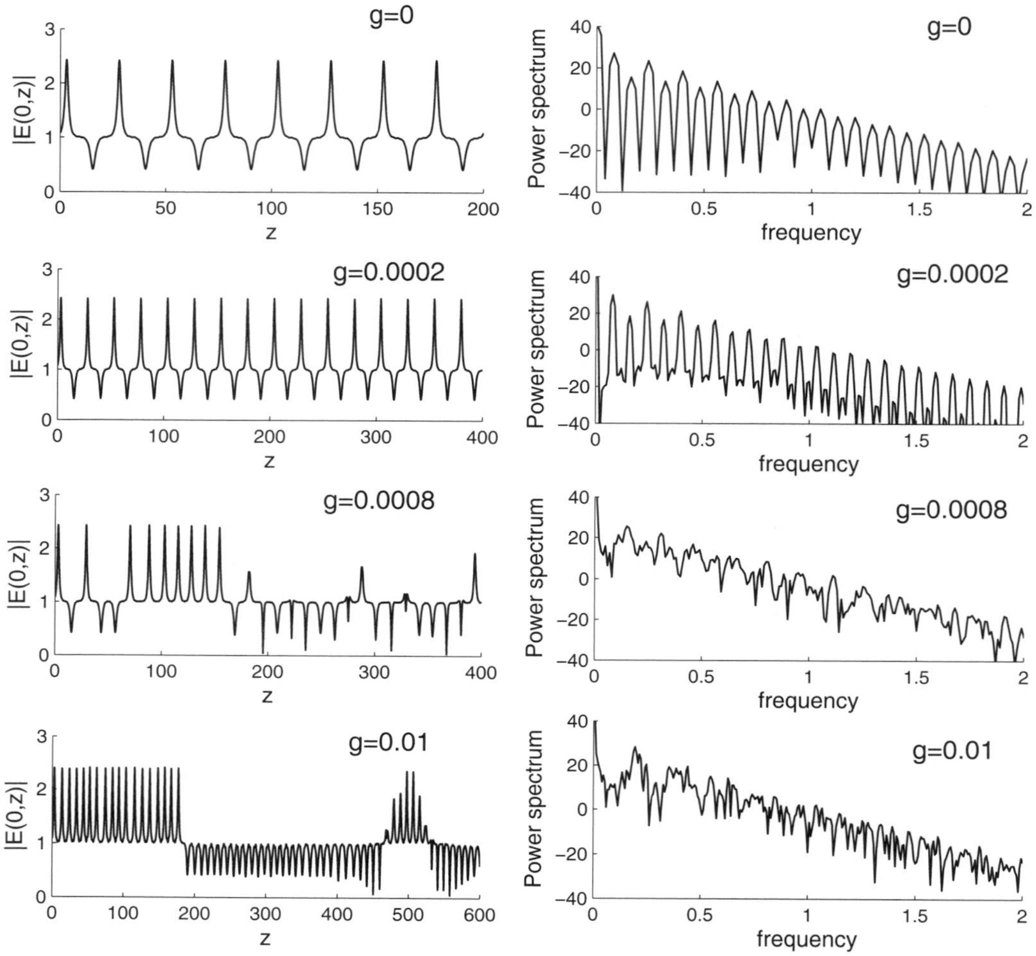

Figure 10: The periodic solution, quasiperiodic solutions, and chaotic solutions for Eqs. (1.1) and (6.1) with $\eta=0.1$ and $\theta=45.225°$. (a) Propagation of the field amplitude; (b) power spectrum.

7 Relativistic NSE for intense laser pulse in plasmas

Interest in the interaction of relativistically intense laser (RIL, at intensities 10^{18} W/cm^2 or above) with plasma has been stimulated by the rapid development of high-power lasers and their applications to fast ignition of fusion fuel, compact X-ray sources and particle accelerators, etc. [23,44,45]. In the RIL the field, electron quiver velocity is highly relativistic and the ponderomotive pressure is much stronger than the plasma pressure. The propagation dynamics of RIL in plasmas are still unclear.

Recently, we [19] also studied the dynamics of RIL in plasmas by using the above method. Considering a circularly-polarized RIL in an initially-homogeneous underdense plasma and introducing the normalizations $E\rightarrow m_{e0}\omega cE/e$, $x\rightarrow k_p^{-1}x$, and $z\rightarrow 2kk_p^{-2}z$, we obtain the RNSE describing the evolution of laser-field envelope [22,46–48], i.e., NSE (1.1) combined with the potential function

$$\text{(v)} \qquad F(|E|^2) = 1 - \frac{N_e}{\gamma} = 1 - \frac{\max[0, 1 + \partial_x^2 \sqrt{1+|E|^2}]}{\sqrt{1+|E|^2}}, \tag{7.1}$$

where the laser intensity is given by $I = (|E|^2 \times 1.37 \times 10^{18} / \lambda^2)$ W/cm^2. The system (1.1) and (7.1) include the effects of relativistic mass variation and ponderomotive force, as well as diffraction caused by the finite pulse aperture. The Hamiltonian perturbation is

$$H_1 = \int \left[\frac{1}{4}|E|^4 + 2(\sqrt{1+|E|^2} - 1) - |E|^2 - \left| \partial_x \sqrt{1+|E|^2} \right|^2 \right] dx, \tag{7.2}$$

where H_0 expressed as Eq. (3.2) with $\alpha = 1/2$ is the integrable Hamiltonian for CNSE. Fig. 11 of $W^{(u)}$ and $W^{(s)}$ manifolds is also completely consistent with the theoretical results in Section 2.

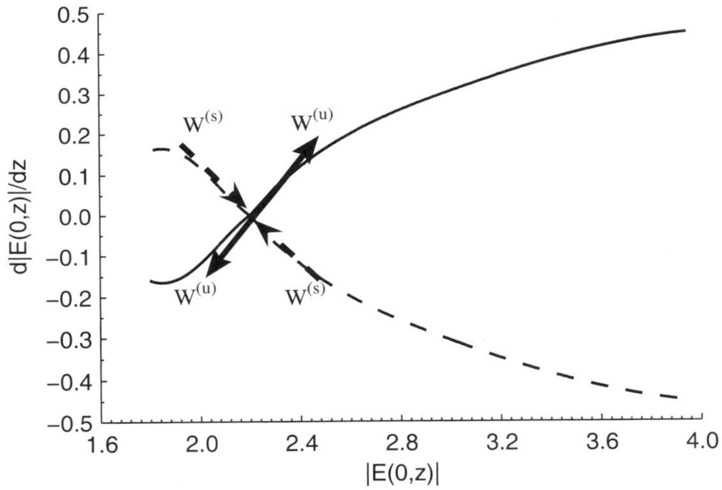

Figure 11: Stable $W^{(s)}$ and unstable $W^{(u)}$ manifolds of the hyperbolic fixed point $(E_0, 0)$ with $\theta = 45°$ and $225°$ for the RNSE [Eqs. (1.1) and (7.1)]. The solid curve is for $z > 0$ and the dashed curve is for $z < 0$.

Physically, the relativistic mass variation and ponderomotive nonlinearities cannot be separated, since they are from the same electron dynamics. However, in order to see their individual contributions to the complexity of the laser-plasma interaction and formation of spatial chaos and patterns, we numerically solve the RNSE [Eqs. (1.1) and (7.1)] with the two cases: (i) the purely relativistic mass variation nonlinearity, and (ii) the combined relativistic mass variation and ponderomotive nonlinearities.

The RNSE [Eqs. (1.1) and (7.1)] with only the relativistic mass variation nonlinearity becomes

$$iE_z + E_{xx} + \left(1 - \frac{1}{\sqrt{1+|E|^2}}\right) E = 0 \tag{7.3}$$

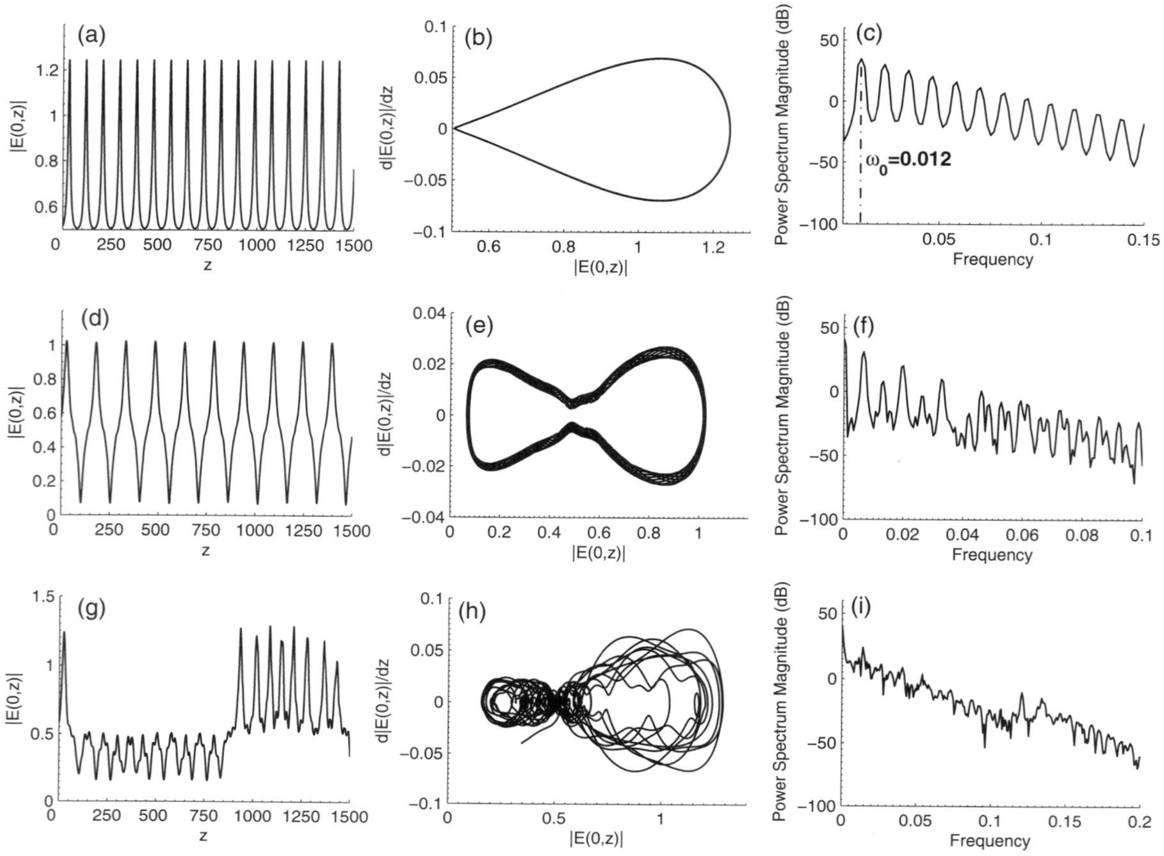

Figure 12: The solutions of the CNSE [(a)-(c)], Eq. (7.3) [(d)-(f)], RNSE [Eqs. (1.1) and (7.1)] [(g)-(i)], for $E_0=0.5$ and $\theta=45°$, at $x=0$. (a), (d), and (g) show the evolution of laser field envelope. (b), (e), and (h) show the structures of the phase space $(E(0,z),\ d|E(0,z)|/dz)$. (c), (f), and (i) show the corresponding power spectra. (h) indicates the irregular homoclinic orbit crossings.

by ignoring the ponderomotive nonlinearity. Eq. (7.3) is physically valid if the condition $k_p^{-2}\nabla^2\gamma \ll 1$ is satisfied [49], where the relativistic nonlinearity prevails over the ponderomotive nonlinearity.

In the weakly relativistic regime with laser intensity $I \leq 1.37\times10^{18}$ W/cm^2, *i.e.*, $|E|\leq 1$, we can expand Eq. (7.3) to obtain the CNSE [Eqs. (1.1) and (3.1)]. One can then treat the higher-order relativistic mass variation and ponderomotive nonlinearities as Hamiltonian perturbations H_1. We choose parameters $E_0=0.5$ (corresponding to $I_0=6.85\times10^{17}$ W/cm^2), $\eta=0.1$, and $\theta=45°$. As analyzed in Section 3, the CNSE is fully integrable, one solution corresponding to the periodic oscillations is shown in Figs. 12(a) and (b). However, the integrability breaks down if high-order relativistic mass variation and ponderomotive nonlinearities are taken into account. With the full but only purely relativistic mass variation nonlinearity [Eq. (7.3)], the phase-space trajectory is in the form of a band, the KAM tori becomes thicker but are not completely destroyed [as shown in Fig. 12(e)], since the solution now contains more subharmonics [see Fig. 12(f)]. This represents a non-chaotic "quasiperiodic" behavior. Under the combined action of the relativistic mass variation and ponderomotive nonlinearities [RNSE's (1.1) and (7.1)], the periodic oscillations disappear and stochastic behavior occurs, as can be seen in Fig. 12(g). On can see from Fig. 12(h) that irregular HMO crossings appear in the phase space. The corresponding dynamical system then is a chaotic one. The power spectra in Fig. 12(i) also shows typical noisy behavior of chaotic motion.

If the laser intensity I is greater than 1.37×10^{18} W/cm^2, corresponding to $|E_0|>1$, both the relativistic mass variation and ponderomotive nonlinearities are important. We use the parameters $E_0=2$ (or $I_0=2.75\times10^{18}$ W/cm^2), $\eta=0.1$, and $\theta=45^\circ$ for numerical calculation. For Eq. (7.3), an interesting phenomenon can be seen in Figs. 13(a) and (b), pseudo-recurrence appears with the laser field exhibiting almost periodic behavior and the phase trajectories are nearly regular with the motion being smooth in small regions. This means that the KAM tori are not completely destroyed by only the relativistic mass variation. However, with both nonlinear relativistic mass variation and ponderomotive effects taken into account [Eqs. (1.1) and (7.1)], the KAM tori are completely destroyed. The laser field envelope displays typical chaotic behavior, as can be seen in Fig. 13(d). Fig. 13(e) shows the irregular HMO crossings in the phase space. Fig. 13(f) shows the broadband structure and the noisy spectra that are typical of chaotic motion.

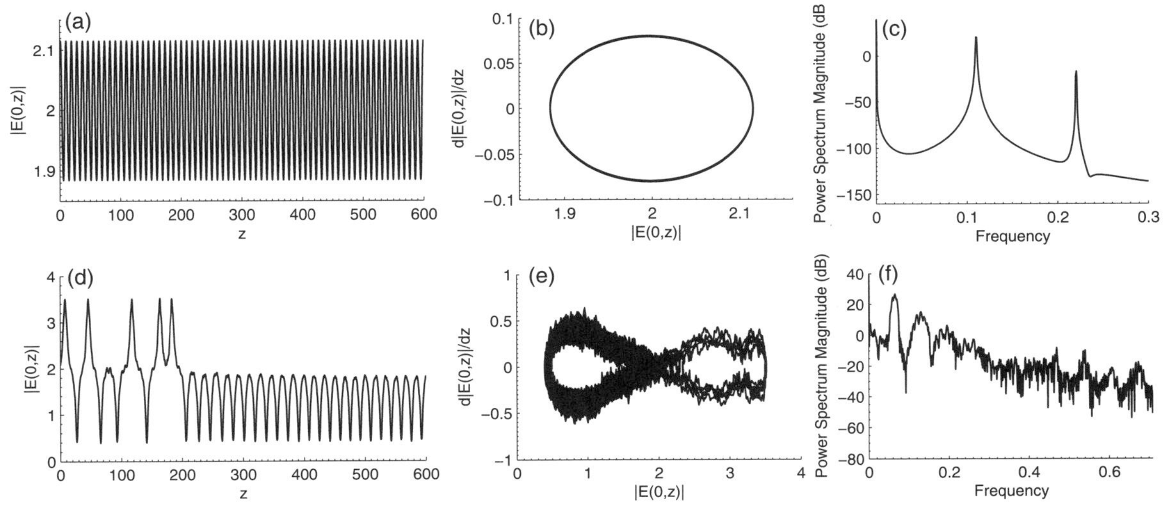

Figure 13: The solutions of Eq. (7.3) [(a)-(c)] and RNSE [Eqs. (1.1) and (7.1)] [(d)-(f)] for $E_0=2.0$ and $\theta=45^\circ$ at $x=0$. (a) and (d) show the evolution of laser field envelope, (b) and (e) show the structures of phase space, (c) and (f) show the power spectra corresponding to (a) and (d), respectively.

The chaotic behavior of the RNSE leads to the appearance of complex patterns for RIL wave field in plasmas. Fig. 14 shows the pattern structures of the CNSE, Eq. (7.3) and the RNSE [Eqs. (1.1) and (7.1)], respectively. With only the fully relativistic mass variation effect, because the KAM tori is not completely destroyed and laser field evolution is only weakly chaotic, its pattern structure in general remains fairly coherent, but irregular sub-

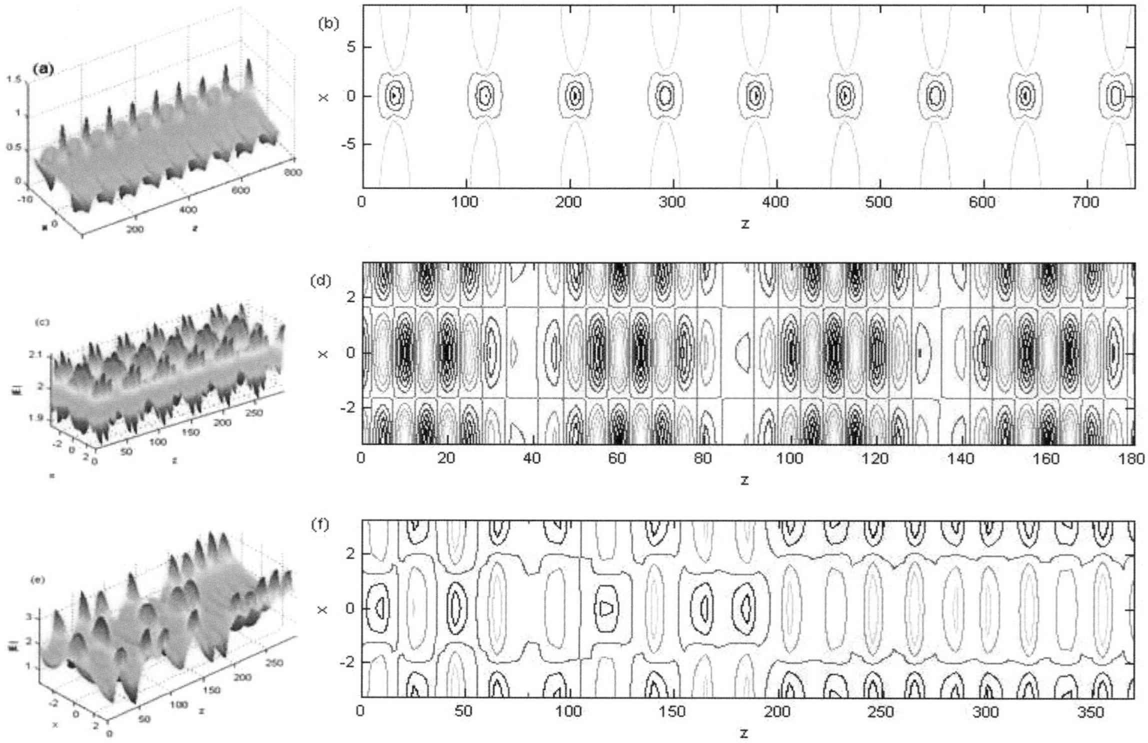

Figure 14: Spatial patterns: (a) and (b) show the coherent patterns of the CNSE for $E_0=0.5$ and $\theta=45^\circ$, (c) and (d) show the nearly coherent patterns of Eq. (7.3) with only the relativistic mass variation effect, and (e) and (f) show the complex patterns of the RNSE [Eqs. (1.1) and (7.1)] with both the relativistic mass variation and ponderomotive nonlinearities, for $E_0=2.0$ and $\theta=45^\circ$.

structures appear, as shown in Figs. 14(c) and (d). This near-coherent behavior can also be seen in Figs. 15(b): the energy evolution is periodic in the first two Fourier modes, but aperiodic in the third and fourth modes. With both the fully relativistic mass variation and ponderomotive nonlinearities, the KAM tori are completely destroyed and strong chaos occurs. The corresponding complex patterns can be seen in Figs. 14(e) and (f). The localized structures remain but they are rather irregular. Fig. 15(c) shows that the evolution of the energy in the first mode is quasi-periodic, so that spatially localized structures remain. However, the evolution of the energy in the second and higher modes exhibits stochastic, resulting in irregular complex patterns.

All these show that the relativistic mass variation and ponderomotive nonlinearities in the RNSE (7.1) lead to the chaotic behaviors of laser fields. Coherent structures are replaced by complex patterns, HMO crossings exist in the phase space and stochastic partition of energy in Fourier modes occurs. Furthermore, the ponderomotive nonlinearity plays a more important role in completely destroying the periodicity of laser field and forming chaos and complex patterns. With only the relativistic mass variation nonlinearity, laser field tends to remain pseudo-periodic with pseudo-recurrent phase orbit, and the patterns remain fairly coherent.

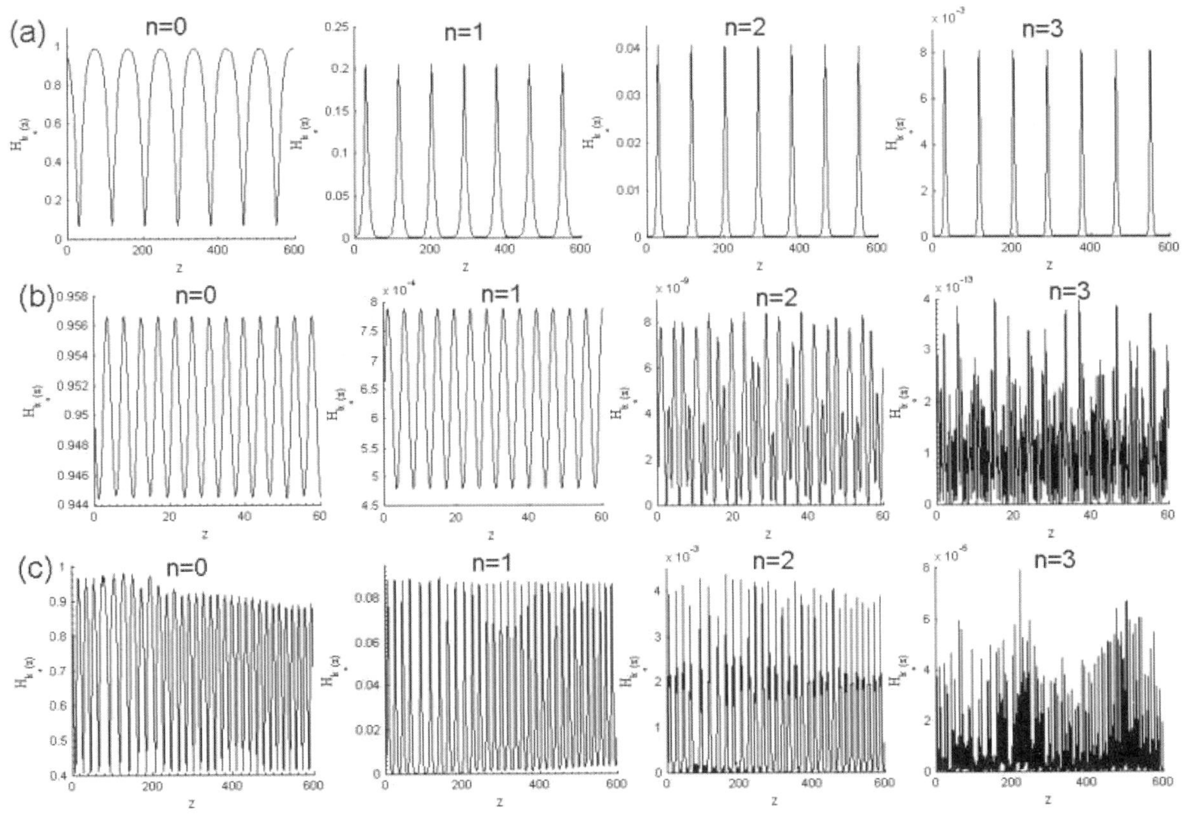

Figure 15: Energy evolution of the first four Fourier modes for the CNSE, Eq. (7.3), and RNSE [Eqs. (1.1) and (7.1)] The parameters are the same as in Fig. 14.

8 Conclusion and discussion

We have given a systematic review of the complex dynamics including chaos and patterns of a series of generalized NSEs with different potential functions for various nonlinear physical problems in plasmas. Our studies from the early Langmuir wave turbulence to the recent laser filaments in plasmas all show that the high-order nonlinear effects, acting as the Hamiltonian perturbation, break down the NSE integrability and lead to the chaotic behaviors of the system. The corresponding nonlinear phenomena are just associated with this homoclinic chaos. In particular, we found that the unstable manifolds $W^{(u)}$ possessing the hyperbolic fixed point correspond to an initial phase $\theta = 45^\circ$ and 225°, and the stable manifolds $W^{(s)}$ correspond to $\theta = 135^\circ$ and 315°. We believe these findings are general and will give useful implication to the understanding of nonlinear wave and laser propagation dynamics in plasmas.

As far as various branches of physics are concerned, the presence of the stochastic wave fields (Langmuir wave field or laser wave field) depends on the evolution processes of systems. In other words, the lowest order nonlinear term $|E|^2E$ is the predominant nonlinear mechanism in the evolving initial stages. Some important phenomena, such as coherent structures and Langmuir wave collapse, etc., can be reasonably explained by making use of the cubic NSE. In the strongly nonlinear stages, other physical effects could play an important role in the process of plasma instabilities. For example, Landau damping of high-frequency waves, $\chi^{(5)}$ susceptibility, multiphoton ionization, relativistic ponderomotive force and so on become quite significant in different problems. These higher-order nonlinear effects make complex dynamic behaviors occur. In a sense, these complex dynamic phenomena are the characteristic of wave propagation. It is shown that the focusing (or collapse) of wave becomes a chaotic oscillation, rather than a catastrophic process due to the higher-order nonlinear effects. Moreover, the latter acts not only as a locally saturating nonlinearity, but also as a global Hamiltonian perturbation resulting in complex propagation dynamics.

Acknowledgments

B.Q. acknowledges the financial support of Department of Physics, National University of Singapore. CTZ would like to thank the great hospitality of the Temasek Laboratories of National University of Singapore, where part of work was performed during his visits. This work is also supported by the National Natural Science Foundation of China grant Nos. 10575013 and 10576007, and partially by the National Basic Research Program of China (973) (2007CB814802 and 2007CB815101).

References

[1] X. T. He, Acta Phys. Sin. 30, 1415 (1981).
[2] X. T. He, Acta Phys. Sin. 31, 1317 (1982).
[3] C. T. Zhou, X. T. He, S. G. Chen, Phys. Rev. A 46, 2277 (1992), also the references therein.
[4] C. T. Zhou, S. P.Zhu, X. T. He, Phys. Rev. A 46, 3486 (1992).
[5] X. T. He, C. T. Zhou, J. Phys. A 26, 4123 (1993).
[6] C. T. Zhou, X. T. He, Comm. Theor. Phys. 19, 231 (1993).
[7] C. T. Zhou, X. T. He, Chinese Phys. Lett. 10, 290 (1993).
[8] C. T. Zhou, X. T. He, Mod. Phys. Lett. B 8 1503 (1994).
[9] C. T. Zhou, X. T. He, Phys. Scr. 50, 415 (1994).
[10] C. T. Zhou, X. T. He, Phys. Rev. E 49, 4417 (1994).
[11] C. T. Zhou, X. T. He, T. X. Cai, Phys. Rev. E 50, 4136 (1994).
[12] C. T. Zhou, C. H. Lai, Int. J. Mod. Phys. C 7, 775 (1996).
[13] X. T. He, C. Y. Zheng, Phys. Rev. Lett. 74, 78 (1995).
[14] Y. Tan, X. T. He, S. G. Chen, Phys. Rev. A 45, 6109 (1992).
[15] C. Y. Zheng, X. T. He, Y. Tan, Modern Phys. Lett. B 8, 833 (1994).
[16] C. T. Zhou, M. Y. Yu, X. T. He, Phys. Rev. E 73, 026209 (2006).
[17] C. T. Zhou, Chaos 16, 013124 (2006).
[18] B. Qiao, C. H. Lai, C. T. Zhou, X. T. He, X. G. Wang, Appl. Phys. Lett. 91, 221114 (2007).
[19] B. Qiao, C. H. Lai, C. T. Zhou, X. T. He, X. G. Wang, Phys. Plasmas 14, 112301 (2007).

[20] See for example, Y. S. Kivshar, G. P. Agrawal, Optical Solitons: From Fibers to Photonic Cryptals, Academic Press, London, 2003.
[21] K. Okamota, Fundamentals of Optical Waveguides, Academic Press, London, 2000.
[22] A. V. Borovsky, A. L. Galkin, A. B. Shiryaev, T. Auguste, Laser Physics at Relativistic Intensities, Springer, Berlin, 2003.
[23] P. Gibbon, Short Pulse Laser Interactions with Matter - An Introduction, Imperial College Press, London, 2005.
[24] P. L. Kelley, Phys. Rev. Lett. 15, 1005 (1965).
[25] M. J. Alblowitz, H. Segur, J. Fluid Mech. 93, 691 (1979).
[26] T. Taninuti, H. Washimi, Phys. Rev. Lett. 21, 209 (1968).
[27] A. I. Olemskoi, V. F. Klepikov, Phys. Reports 338, 571 (2000).
[28] F. T. Arecchi, S. Boccaletti, P. L. Ramazza, Phys. Reports 318, 1 (1999).
[29] A. S. Mikhailova and K. Showalterb, Phys. Reports 425, 79 (2006).
[30] T. H. Moon, Phys. Rev. Lett. 64, 412 (1990).
[31] A. J. Lichtenberg, M. A. Lieberman, Regular and Chaotic Dynamics, Springer, New York, 1992, pp. 1-20.
[32] C. T. Zhou, C. H. Lai, M. Y. Yu, J. Math. Phys. 38, 5225 (1997); Phys. Scr. 55, 394 (1997).
[33] G. M. Zaslavsky, R. Z. Sagdeev, D. A. Usikov, A. A. Chernikov, translated by A. R. Sagdeeva, Weak Chaos and Quasi-Regular Patterns, Cambridge University Press, Cambridge, 1991, pp. 21-35 and pp. 86-108.
[34] C. Sulem, P. Sulem, The Nonlinear Schröinger Equation: Self-focusing and Wave Collapse, Springer, Berlin, 2000.
[35] A. Hasegawa, Optical Solitons in Fibers, Springer-Verlag, Berlin, 1989.
[36] A. Alexandrescu, G. D. Montesinos, V. M. Pérez-García, Phys. Rev. E 75, 046609 (2007).
[37] T. B. Benjamin, J. E. Feir, J. Fluid Mech. 27, 417 (1967).
[38] A. Cloot, B. M. Herbst, J. A. C. Weideman, J. Comput. Phys. 86, 127 (1990)
[39] B. J. Lemesnrier, G. Papanicollaon, C. Sulem, P. L. Sulem, Physica D 31, 79 (1988).
[40] R. Ackermann, E. Salmon, N. Lascoux, J. Kasparian, P. Rohwetter, K. Stelmaszczyk, S. Li, A. Lindinger, L. Wöste, P. Béjot, L. Bonacina, J. P. Wolf, Appl. Phys. Lett. 89, 171117 (2006).
[41] K. Stelmaszczyk, P. Rohwetter, G. Méjean, J. Yu, E. Salmon, J. Kasparian, R. Ackermann, J. P. Wolf, L. Wöste, Appl. Phys. Lett. 85, 3977 (2004).
[42] L. Wöste C. Wedekind, H. Wille, P. Rairoux, B. Stein, S. Nikolov, Ch. Werner, S. Niedermeier, H. Schillinger, R. Sauerbrey, Laser Optoelektron 29, 51 (1997).
[43] L. Bergé, S. Skupin, F. Lederer, G. Méjean, J. Yu, J. Kasparian, E. Salmon, J. P. Wolf, M. Rodriguez, L. Wöste, R. Bourayou, R. Sauerbrey, Phys. Rev. Lett. 92, 225002 (2004), also the references therein.
[44] C. T. Zhou and X. T. He, Appl. Phys. Lett. 90, 031503 (2007); ibid. 92, 071502 (2008); ibid. 92, 151502 (2008); Opt. Lett. 32, 2444 (2007).
[45] C. T. Zhou, M. Y. Yu, X. T. He, J. Appl. Phys. 101, 103302 (2007); Europhys. Lett. 79, 35001(2007); Laser Part. Beam 25, 313 (2007); Phys. Plasmas 13 092109 (2006).
[46] G. Z. Sun, E. Ott, Y. C. Lee, P. Guzdar, Phys. Fluids 30, 526 (1987).
[47] A. B. Borisov, O. B. Shiryaev, A. McPherson, K. Boyer, C. K. Rhodes, Plasma Phys. Control. Fusion 37, 569 (1995).
[48] M. Y. Yu, P. K. Shukla, K. H. Spatschek, Phys. Rev. A 18, 1591 (1978).
[49] P. Sprangle, A. Zigler, E. Esarey, Appl. Phys. Lett. 58, 346 (1991).

PHYSICAL REVIEW E **87**, 053103 (2013)

Pattern dynamics and filamentation of femtosecond terawatt laser pulses in air including the higher-order Kerr effects

T. W. Huang,[1] C. T. Zhou,[1,2,3,*] and X. T. He[1,2]
[1]*HEDPS, Center for Applied Physics and Technology, Peking University, Beijing 100871, People's Republic of China*
[2]*Institute of Applied Physics and Computational Mathematics, Beijing 100094, People's Republic of China*
[3]*Science College, National University of Defense Technology, Changsha 410073, People's Republic of China*
(Received 9 January 2013; published 9 May 2013)

Plasma defocusing and higher-order Kerr effects on multiple filamentation and pattern formation of ultrashort laser pulse propagation in air are investigated. Linear analyses and numerical results show that these two saturable nonlinear effects can destroy the coherent evolution of the laser field, and small-scale spatial turbulent structures rapidly appear. For the two-dimensional case, numerical simulations show that blow-up-like solutions, spatial chaos, and pseudorecurrence can appear at higher laser intensities if only plasma defocusing is included. These complex patterns result from the stochastic evolution of the higher- or shorter-wavelength modes of the laser light spectrum. From the viewpoint of nonlinear dynamics, filamentation can be attributed to the modulational instability of these spatial incoherent localized structures. Furthermore, filament patterns associated with multiphoton ionization of the air molecules with and without higher-order Kerr effects are compared.

DOI: 10.1103/PhysRevE.87.053103 PACS number(s): 52.38.Hb, 52.35.Mw, 42.65.Sf

I. INTRODUCTION

In recent years, considerable interest in the self-guided propagation of femtosecond terawatt laser pulses in air has been stimulated because of the many potential applications [1–14] such as in lighting discharge control, lidar remote sensing, generation of few-cycle light bullets, and so on. The propagation of ultrashort laser pulses in air is also rich in fundamental nonlinear physics [12–20], including generation of white light, third-order harmonic, terahertz radiation, and conical emission. It is commonly believed that dynamic equilibrium between Kerr compression and plasma diffraction allows femtosecond terawatt laser pulses to self-channel and propagate over a long distance in air. The second-order Kerr self-focusing effect arising from the third-order susceptibility can induce a positive nonlinear refractive index of the air. In principle, for input laser beam having a power above the threshold power (P_{cr}) for self-focusing in air, one can expect that the beam first undergoes a compression process in the diffraction plane and narrower beams can be generated by self-induced waveguiding. However, there is an upper limit to the optical intensity of the light beams. The limit is due to ionization of air molecules when the laser intensity becomes larger than a few tens of $\mathrm{TW/cm^2}$. In a parameter range with the input laser intensity less than $10^{14}\ \mathrm{W/cm^2}$, the optically induced ionization can be modeled within the framework of lowest-order perturbation theory and multiphoton ionization processes would be dominant [21], where multiple photons are absorbed by the air molecules and an electron plasma can be generated. The resulting electron plasma has a diverging effect on the light pulse, since it has a negative contribution to the refractive index of the air. Dynamic balance between the Kerr self-focusing and the diffraction of the plasma results in filamentation of the light beam that can last for a distance much longer than the Rayleigh length [10,11]. The first experimental observation of long-distance self-guided laser propagation in air was shown in the mid-1990s, where a plasma channel as long as 20 m was observed [12]. Later, it was shown that an intense femtosecond pulse laser beam launched vertically into the sky can generate white light at a distance of 10 km [13,14]. The optical intensities inside a filament can reach a few tens of $\mathrm{TW/cm^2}$. Such laser-induced supercontinuum light can be useful for detecting pollutants in the atmosphere [13].

There exist many theories, simulations, and experiments on self-channeling of intense femtosecond laser pulses in air [1–62]. The critical power for self-focusing is one of the key parameters in laser filamentation in air. As is well known, if the pulse power is less than a few critical powers, only one filament is formed. At much higher powers, a single laser pulse can form several filaments—a process known as multiple filamentation. The generation of multiple filaments is usually explained in terms of modulation instability of the laser beam [22] and an optically turbulent light guide model [23]. It has been proposed that the multiple filaments can be regarded as the propagation of a group of interacting light filaments [24]. In the proposed models, plasma diffraction due to multiphoton ionization of the air molecules is invoked for the formation of the self-guided light channel. However, for balancing the second-order nonlinear Kerr focusing, the saturation of the Kerr effect or the so-called higher-order Kerr effects (HOKE) have also been suggested to play an important role [25–46]. It is well known that the high-order saturation effect in a nonlinear Schrödinger equation (NLSE) can lead to spatialtemporal chaos and complicated patterns of light waves propagating in a medium [47]. Recently, some experiments have measured the coefficients of the higher-order terms up to n_8 in nitrogen and oxygen and n_{10} for argon [25]. The data showed that the different-ordered nonlinear Kerr terms in the refractive index have different signs, and the higher-order Kerr effects can provide a defocusing effect. Filaments can then be formed when the nonlinear refractive index change signs from positive to negative with the increase of laser intensity [26–28]. However, it is mentioned that subsequent experimental results do not support such a higher-order Kerr

*zcangtao@iapcm.ac.cn

model [29–33]. Recently, some attention has been paid to these two mechanisms of laser filamentation in air [25–46]. It is found that HOKE can improve the quantitative modeling of laser filamentation with respect to the peak intensity and electron density in the laser filaments [34,35]. However, these results cannot be reproduced by other experiments and simulations [29–33,36–38]. On the other hand, the HOKE model cannot fully produce the experimental conical emissions although the classical model can [39]. A more detailed understanding of laser filamentation in air is, therefore, still needed.

Both plasma diffraction and HOKE can balance nonlinear Kerr focusing in laser propagation in air. It should be noted that these two models balance Kerr focusing for different parameter ranges. For instance, for infrared laser pulses with duration of about 100 fs, the nonlinear refractive index saturates and goes to negative at a lower intensity in HOKE model than that in classical case [25,34,35,48], where the plasma defocusing is considered a saturating nonlinearity. For such generalized saturable NLSEs, our previous investigations [47] illustrated that spatial chaos and complicated patters can be formed due to high-order saturable effects. In particular, it is found that the nonlinear propagation of ultrashort laser pulses in air can become incoherent and chaotic by invoking a fifth-order susceptibility [49]. We note that filamentation instability, solitary waves, and spatial chaotic patterns are the result of nonlinear development of the modulational instability for unstable wave modes [63–66]. It is, therefore, of interest to investigate problems of pattern dynamics and filamentation of femtosecond terawatt laser pulses in air including higher-order Kerr nonlinearity and plasma diffraction. Such an investigation may throw a new light onto the better understanding of the filamentation process in air.

In the present paper, we theoretically and numerically investigate spatial chaos and multiple filamentation of ultrashort laser pulses in air by taking into account multiphoton ionization of the air molecules without and with higher-order Kerr nonlinear effects. The numerical simulations are carried out based on a NLSE including HOKE [25] and a high-order plasma saturation term that describes the multiphoton ionization of the air molecules [48]. We aim to provide a deeper understanding of the influence of the higher-order nonlinear effects (including both HOKE and plasma saturable effects) on the nonlinear propagation of ultrashort laser pulses in air. In particular, we hope to establish a link between nonlinear dynamical theory and filamentation processes. The paper is organized as follows. In Sec. II, we give a brief introduction of the reduced physical model describing the ultrashort laser pulses propagating in air, and qualitative analysis of the physical model is also given. The spatial chaos and pattern dynamics in one and two dimensions are discussed in Sec. III and Sec. IV, respectively. The effect of HOKE and plasma diffraction on multiple filamentation of femtosecond terawatt laser pulses in air is investigated in Sec. V. A conclusion is given in the last section.

II. PHYSICAL MODEL AND QUALITATIVE ANALYSIS

The generalized NLSE is often considered a universal model describing wave collapse and strong turbulence [66,67]. It has widely been applied in continuum mechanics, nonlinear optics, plasma physics, Bose-Einstein condensates [67], and so on. Recently, a full $3D+1$ dimensional NLSE model describing a linearly polarized paraxial laser beam propagating in air was developed [50,51]. In this model, the vectorial effects [68] due to tight focusing and nonparaxial propagation of the laser are not considered. Moreover, numerical integration of the full $3D+1$ dimensional model over long distances is often limited by the computer capacity. If we only focus on the spatial dynamics of the laser pulse, a time-averaged $2D+1$ dimensional model that can well reproduce the qualitative features of the experimentally observed patterns can be employed [48]. In addition, such a reduced model can provide a reasonable approximation of fluence patterns developed by the full $(3D+1)$-dimensional model, which makes it be widely used to capture the transverse dynamics of the filamentation phenomenon [48,49,69]. For multiple filamentation problem, such a reduced model can be employed to capture the main characteristics of the multiple-filament patterns. Including HOKE, this model for the evolution of the scalar envelope $E(x,y,z)$ in the frame traveling at the pulse velocity is given by the equation [34,50]

$$\partial_z E = i\frac{1}{2k_0 n_0}\nabla_\perp^2 E + ik_0\Delta n E - i\gamma|E|^{2K}E, \qquad (1)$$

where $\nabla_\perp^2 = \partial_{xx}^2 + \partial_{yy}^2$ is the transverse Laplace operator, $k_0 = 2\pi n_0/\lambda_0$ is the wave number in vacuum, n_0 is the linear index of refraction of the medium, λ_0 $(= 800$ nm) is the central wavelength of the laser, $\Delta n = \frac{n_2}{\sqrt{3}}|E|^2 + \frac{n_4}{\sqrt{5}}|E|^4 + \frac{n_6}{\sqrt{7}}|E|^6 + \frac{n_8}{3}|E|^8$ is the nonlinear refractive index of the medium, the scalar envelop $E(x,y,z)$ is defined as $|E|^2 = I$, and I is the incident laser intensity. The last term on the right-hand side of Eq. (1) represents the plasma defocusing effect. Here $\gamma = k_0\sigma_K\rho_{nt}\sqrt{\pi/(8K)}T/(2n_c)$, where $\sigma_K \approx 2.88\times10^{-99}$ s^{-1} cm^{2K}/W^K is the coefficient of multiphoton ionization, K is the number of photons, say $K=8$ for O_2 gas, $\rho_{\rm nt}\approx 5.4\times10^{18}$ cm^{-3} is the neutral density, $n_c \approx 1.8\times10^{21}$ cm^{-3} is the critical plasma density, and $T\approx t_p/10$ with $t_p =$ 85 fs is the half-width of the laser pulse duration. It is noted here that multiphoton absorption (MPA) is not considered in our system in order to obtain a better understanding of the difference between the two different saturable effects, as well for the fact that in the presence of HOKE the MPA does not play an important role [34,42,43] in filamentation. To be more concrete, in most of the simulations we assume the propagation medium is O_2 gas. For comparison, in Sec. III we also discuss several simulations for N_2. The corresponding experimental parameters [25] are listed in Table I.

The system of Eq. (1) is nondissipative and can be expressed in a dimensionless form as

$$i\partial_z E(\mathbf{x},z) + \nabla_\perp^2 E(\mathbf{x},z) + F(|E|^2)E(\mathbf{x},z) = 0, \qquad (2)$$

where $F(|E|^2) = \alpha|E|^2 + \beta(\beta_4|E|^4 + \beta_6|E|^6 + \beta_8|E|^8) - \gamma_K|E|^{2K}$ and $\mathbf{x}$, E, z are normalized by w_0, $\sqrt{P_{\rm cr}/4\pi w_0^2}$, and $4z_f$, respectively. Here $z_f = \pi w_0^2/\lambda_0$ is the Rayleigh length and $P_{\rm cr}\approx 3.77\lambda_0^2/8\pi n_0 n_2$ is the critical power for self-focusing. For the O_2 case with $\lambda_0 = 800$ nm, we have $\alpha = 0.544$, $\beta = -1$, $\beta_4 = 6.541\times10^{-6}$ $/(w_0/\mathrm{cm})^2$, $\beta_6 = -2.4364\times10^{-10}$ $/(w_0/\mathrm{cm})^4$, $\beta_8 = 4.489\times10^{-14}$ $/(w_0/\mathrm{cm})^6$,

TABLE I. Indices of the higher-order refraction terms in both N_2 and O_2.

Mediums	$n_2(10^{-19}\text{cm}^2/\text{W})$	$n_4(10^{-33}\text{cm}^4/\text{W}^2)$	$n_6(10^{-46}\text{cm}^6/\text{W}^3)$	$n_8(10^{-59}\text{cm}^8/\text{W}^4)$
N_2	1.1	−0.53	1.4	−0.44
O_2	1.6	−5.16	4.75	−2.1

$\gamma_K = 2.7115 \times 10^{-37}\ /(w_0/\text{cm})^{2K-2}$, and $w_0 = 0.3$ cm is the transverse width of the laser beam. These values will be used in Sec. IV.

We now consider the invariants of our system. Equation (2) can be derived from the Lagrangian density $L = \frac{i}{2}(E^*E_z - EE_z^*) - |\nabla_\perp E|^2 + \frac{1}{2}\alpha|E|^4 + \beta(\frac{1}{3}\beta_4|E|^6 + \frac{1}{4}\beta_6|E|^8 + \frac{1}{5}\beta_8|E|^{10}) - \frac{1}{K+1}\gamma_K|E|^{2K+2}$. According to the Noether's theorem, we can obtain three invariants [66]: the wave power $P = \int |E|^2 d\mathbf{x}$, the momentum $M = \frac{i}{2}\int (E_\mathbf{x}E^* - EE_\mathbf{x}^*)d\mathbf{x}$, and the Hamiltonian

$$H = \int [|\nabla_\perp E|^2 - \frac{1}{2}\alpha|E|^4 - \beta\left(\frac{1}{3}\beta_4|E|^6 + \frac{1}{4}\beta_6|E|^8 + \frac{1}{5}\beta_8|E|^{10}\right) + \frac{1}{K+1}\gamma_K|E|^{2K+2}]d\mathbf{x}. \quad (3)$$

From Eq. (3), the Hamiltonian of the system (2) can be separated into two parts: $H = H_0 + H_1$, where $H_0 = \int (|\nabla_\perp E|^2 - \frac{\alpha}{2}|E|^4)d\mathbf{x}$ is the Hamiltonian of cubic NLSE and $H_1 = \int [-\beta(\frac{1}{3}\beta_4|E|^6 + \frac{1}{4}\beta_6|E|^8 + \frac{1}{5}\beta_8|E|^{10}) + \frac{1}{K+1}\gamma_K|E|^{2K+2}]d\mathbf{x}$ is the Hamiltonian perturbation of system (2) induced by HOKE and plasma diffraction.

For Eq. (2), a simple plane-wave solution can be written as $E_s = E_0 e^{iF(E_0^2)z}$. Now we give a qualitative analysis of the dynamic behavior of this homogeneous solution exposed to the modulational instability. Let $\delta E(\mathbf{x},z)$ be a small perturbation from E_s, so $E(\mathbf{x},z) = E_s(z) + \delta E(\mathbf{x},z)$, where $|\delta E| \ll |E_s|$. Linearizing Eq. (2) around E_s, we obtain the equation for the eigenfunction $\delta E(\mathbf{x},z)$,

$$\begin{pmatrix} i\partial_z + \nabla_\perp^2 + \text{Re} & h(z) \\ h^*(z) & -i\partial_z + \nabla_\perp^2 + \text{Re} \end{pmatrix} \begin{pmatrix} \delta E \\ \delta E^* \end{pmatrix} = 0, \quad (4)$$

where $\text{Re} = F(E_0^2) + F'(E_0^2)E_0^2$, $h(z) = F'(E_0^2)E_s^2$, and $F'(u) = \partial_u F$.

Assuming periodic boundary conditions, $\delta E(x,z)$ can be expressed as [70]

(i) a one-dimensional case,

$$\begin{pmatrix} \delta E \\ \delta E^* \end{pmatrix} = \begin{pmatrix} \epsilon e^{i\lambda z} \\ \epsilon^* e^{-i\lambda^* z} \end{pmatrix} \cos(k_x x), \quad (5)$$

(ii) a two-dimensional case,

$$\begin{pmatrix} \delta E \\ \delta E^* \end{pmatrix} = \begin{pmatrix} \epsilon e^{i\lambda z} \\ \epsilon^* e^{-i\lambda^* z} \end{pmatrix} \cos(k_x x)\cos(k_y y), \quad (6)$$

where ϵ and ϵ^* are small parameters and k is the wave number of the perturbation. Inserting Eqs. (5) and (6) into (4), we obtain the eigenvalue

$$\lambda = F(E_0^2) \pm ik\sqrt{2F'(E_0^2)E_0^2 - k^2}, \quad (7)$$

and the linear growth rate $\Gamma = k\sqrt{2F'(E_0^2)E_0^2 - k^2}$ of the modulational instability for $k^2 < 2F'(E_0^2)E_0^2$, where $k = k_x$ for the one-dimensional case and $k = \sqrt{k_x^2 + k_y^2}$ for the two-dimensional case. The wave number of the most unstable mode is then $k_{\max} = E_0\sqrt{F'(E_0^2)}$ for the one-dimensional case and $k_{x,\max} = k_{y,\max} = E_0\sqrt{F'(E_0^2)/2}$ for the two-dimensional case, respectively. Inserting Eqs. (5) and (7) into Eq. (4), we obtain $\frac{\delta E}{\delta E^*}|_{z=0} = \pm i$, which is very important for analyzing the nonlinear behavior of Eq. (2). In order to illustrate

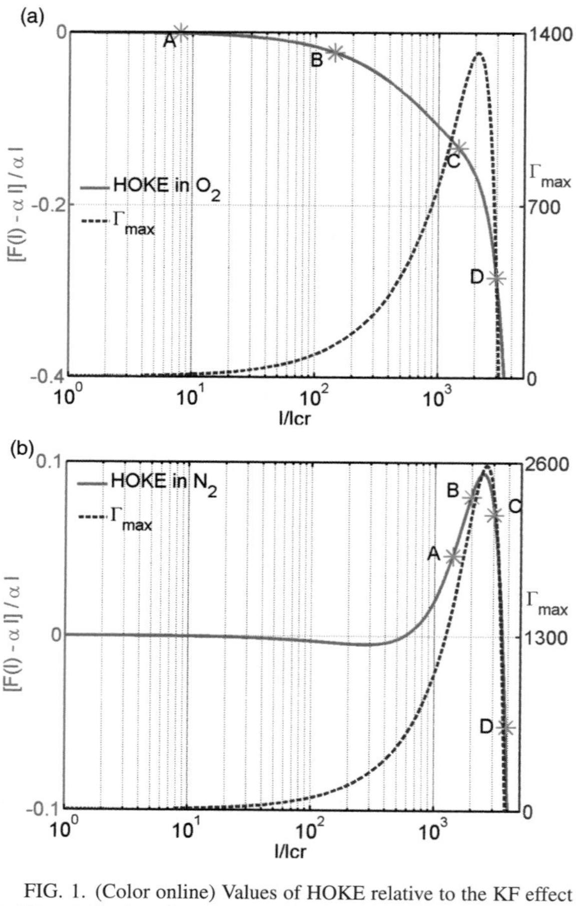

FIG. 1. (Color online) Values of HOKE relative to the KF effect (solid red line) and most unstable growth rate of modulational instability (dashed blue line) at different intensity of input laser (I_0) for the O_2 case (a) and the N_2 case (b). Asterisk symbols in (a) mark the points with I_0 of $8I_{cr}$, $150I_{cr}$, $1500I_{cr}$, $3000I_{cr}$, respectively, and in (b) the asterisk symbols correspond to $1500I_{cr}$, $2000I_{cr}$, $3000I_{cr}$, $3700I_{cr}$, respectively, where $I_{cr} = P_{cr}/4\pi w_0^2 \approx 5.305 \times 10^9\text{W/cm}^2$ for O_2 and $7.717 \times 10^9\text{W/cm}^2$ for N_2.

its importance, we construct the phase space $(|E|,d|E|/dz)$ at $\mathbf{x}=0$ and assume the initial condition $E(x,0)=E_0+\epsilon e^{i\theta}\cos(k_x x)$ for the one-dimensional case and $E(x,y,0)=E_0+\epsilon e^{i\theta}\cos(k_x x)\cos(k_y y)$ for the two-dimensional case, where ϵ is a small real parameter. Although such a low-dimensional phase space is only a projection of high-dimensional space, it can reasonably describe homoclinic crossing of the one-dimensional cubic integrable NLSE equation (see Fig. 2(b); also see Fig. 3(a) in Ref. [70]). Obviously, the unstable manifolds of the hyperbolic fixed point $(E_0,0)$ in this phase space correspond to $\theta=45^\circ$ and 225°, and the stable manifolds correspond to $\theta=135^\circ$ and 315°. To obtain both stable and unstable manifolds (corresponding to a homoclinic orbit (HMO) [49,63–65,71]) for the hyperbolic fixed point $(E_0,0)$, one can directly solve Eq. (2) [47,49,70] by choosing $\pm h$ (h is the numerical step).

III. SPATIAL CHAOS IN ONE-DIMENSIONAL TRANSVERSE SPACE

We now turn to numerical simulations of the nonlinear system (2). First, we would discuss the influence of HOKE on the pattern dynamics of ultrashort laser pulses in air for the one-dimensional case. The two-dimensional problem will be discussed in the next section. When only HOKE is considered, Eq. (2) in the one-dimensional case can be written as

TABLE II. Coefficients of the Kerr refraction terms in Eq. (8).

Mediums	α	$\beta_4(10^{-4})$	$\beta_6(10^{-8})$	$\beta_8(10^{-11})$
N_2	3.77/4	0.3504	−7.1431	1.7324
O_2	3.77/4	1.6251	−7.9583	1.8472

$$i\partial_z E(x,z)+\partial_x^2 E(x,z)+F(|E|^2)E(x,z)=0, \quad (8)$$

where $F(|E|^2)=\alpha|E|^2-(\beta_4|E|^4+\beta_6|E|^6+\beta_8|E|^8)$. In this case, we consider the different propagation mediums such as N_2 and O_2. In particular, for N_2, HOKE has a positive contribution to the refractive index of air when the intensity of the laser pulse is in the region $4.49\times10^{12}<I_0$ (W/cm^2) $<2.74\times10^{13}$, which could induce a quite different behavior on the pattern dynamics compared with the case of O_2 [69]. The coefficients in $F(|E|^2)$ for both N_2 and O_2 are listed in Table II, where $w_0=0.3$ cm and $\lambda_0=800$ nm are taken.

According to the analysis in Sec. II, we have given a reasonable initial condition and a suitable phase space to analyze nonlinear dynamic properties of Eq. (8). In our numerical simulations, we assume different input laser

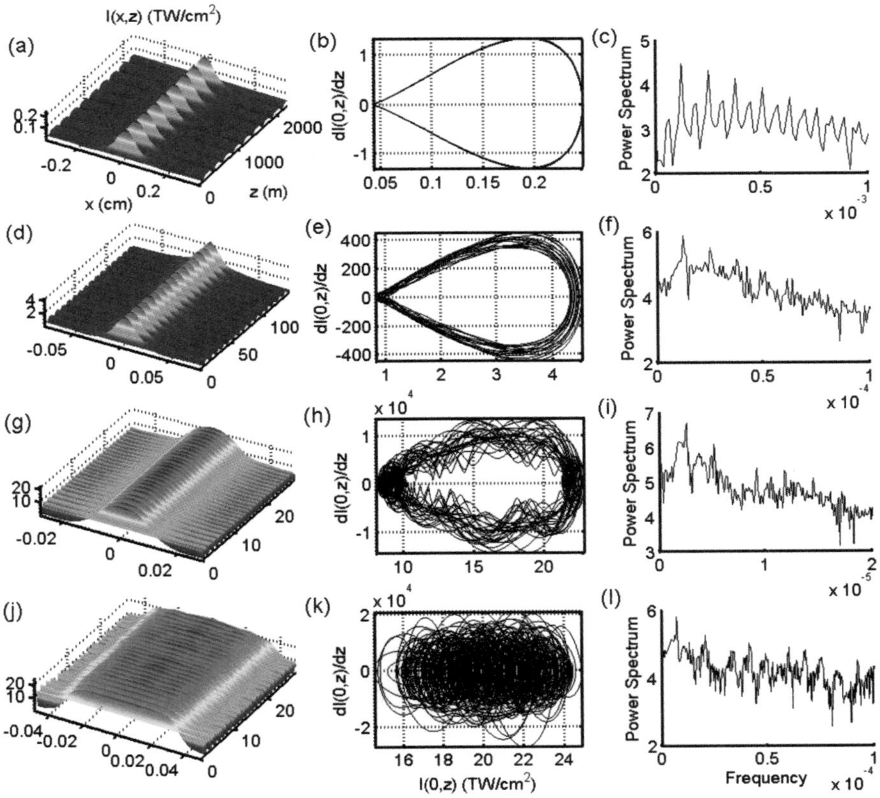

FIG. 2. (Color online) Pattern structures of laser fields [(a), (d), (g), and (j)], phase space computed from the series $(|E(0,z)|\sim z)$ [(b), (e), (h), and (k)] and the corresponding power spectrum (arbitrary unit) [(c), (f), (i), and (l)] with I_0, respectively, of $8I_{cr}$ [(a)–(c)], $150I_{cr}$ [(d)–(f)], $1500I_{cr}$ [(g)–(i)], and $3000I_{cr}$ [(j)–(l)] for the O_2 case, corresponding to the points in Fig. 1(a).

intensities (different E_0), with a perturbation level of $0.01E_0$ and initial phase $\theta = 45°$. We also consider the most unstable mode with $k = k_{max}$. In addition, the standard spectral method has been employed for transverse space (x direction) integration with a periodic length $L = 2\pi/k_{max}$ and the fourth-order Runge-Kutta method with variable step for propagation along the z direction.

If only the Kerr focusing (KF) is considered, the Hamiltonian of the system (8) becomes $H = H_0$. Equation (8) reduces to the well-known cubic NLSE that is fully integrable in the one-dimension case due to the existence of a Lax pair [72]. For the cubic NLSE, a class of periodic solutions and solitons can be obtained by use of an inverse scattering transform, and the solitons developed by modulational instabilities keep their spatially coherent structures and temporally periodic evolutions [70–73]. In terms of our dynamical system (8), HOKE acts as a Hamiltonian perturbation H_1 to H_0. In Fig. 1, we give the values of HOKE relative to the KF term and most unstable growth rate $\Gamma_{max} = F'(E_0^2)E_0^2$ at different I_0 for both the N_2 and O_2 cases. In the context of the specific coefficients provided in the experiments [25], it is shown that for O_2 HOKE can always provide a negative contribution to the refractive index; however, for N_2, HOKE cannot. Figure 1(b) shows that when I_0 lies in the region $4.49 \times 10^{12} < I_0$ (W/cm^2) $< 2.74 \times 10^{13}$, HOKE becomes a positive perturbation due to rapid increase of the high-order focusing term $\beta_6|E|^6$ in HOKE, and, correspondingly, Γ_{max} increases quickly. In addition, the weaker defocusing term in HOKE for N_2 leads to a much larger Γ_{max} than that for O_2, which would result in a quite different behavior of the evolutions of laser field in the two mediums.

Figure 2 shows the solutions of Eq. (8) at different I_0 for O_2 case. When I_0 is small enough [see point A in Fig. 1(a)], HOKE relative to the KF effect is very small and the perturbation H_1 from HOKE can be neglected. As illustrated above, laser propagation obeys an integrable cubic NLSE with only the KF term. A typical periodic recurrent solution is obtained in Fig. 2(a): spatially coherent pattern structures appear and propagate with an invariable velocity. In a sense, the periodic motion consists of two stages as predicted by Benjamin and Fair [74]. The unstable modulation of the uniform solution would first grow at an exponential rate, but eventually the solution would demodulate and return to a near-uniform state. The corresponding phase-space trajectory is a HMO or the Kolmogorov-Arnold-Moser (KAM) tori, where the unstable manifold smoothly joins with the stable manifold at the saddle point (I_0,0) [70]. Such a soliton-like structure has also been obtained for relativistically intense

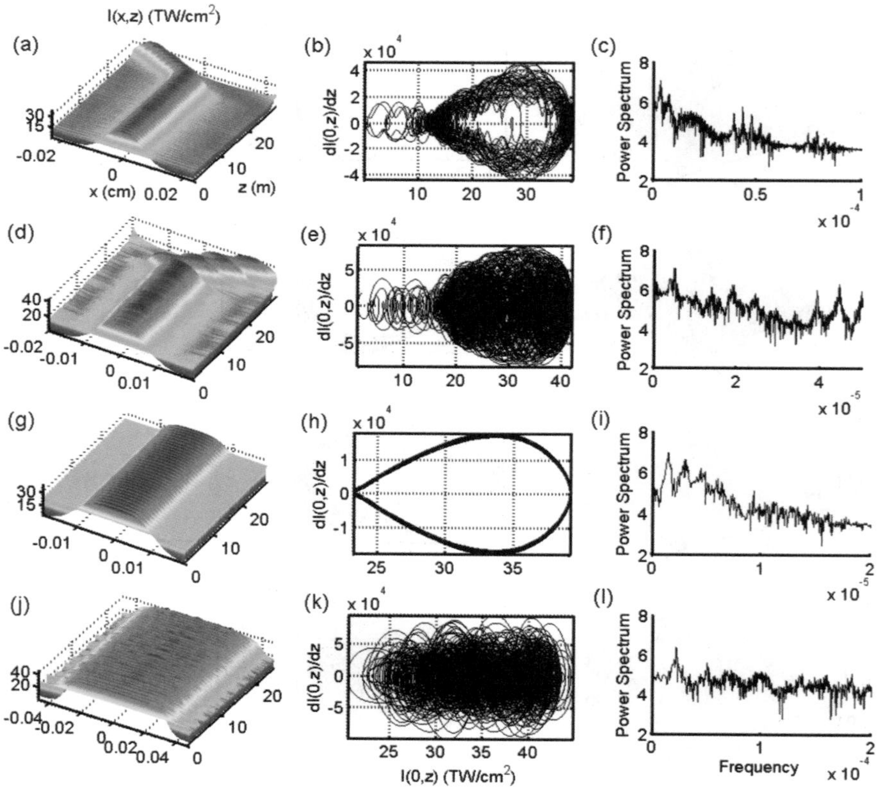

FIG. 3. (Color online) Pattern structures of laser fields [(a), (d), (g), and (j)], phase space computed from the series ($|E(0,z)| \sim z$) [(b), (e), (h), and (k)] and corresponding power spectrum (arbitrary unit) [(c), (f), (i), and (l)] with I_0, respectively, of $1500I_{cr}$ [(a)–(c)], $2000I_{cr}$ [(d)–(f)], $3000I_{cr}$ [(g)–(i)], and $3700I_{cr}$ [(j)–(l)] for the N_2 case, corresponding to the points in Fig. 1(b).

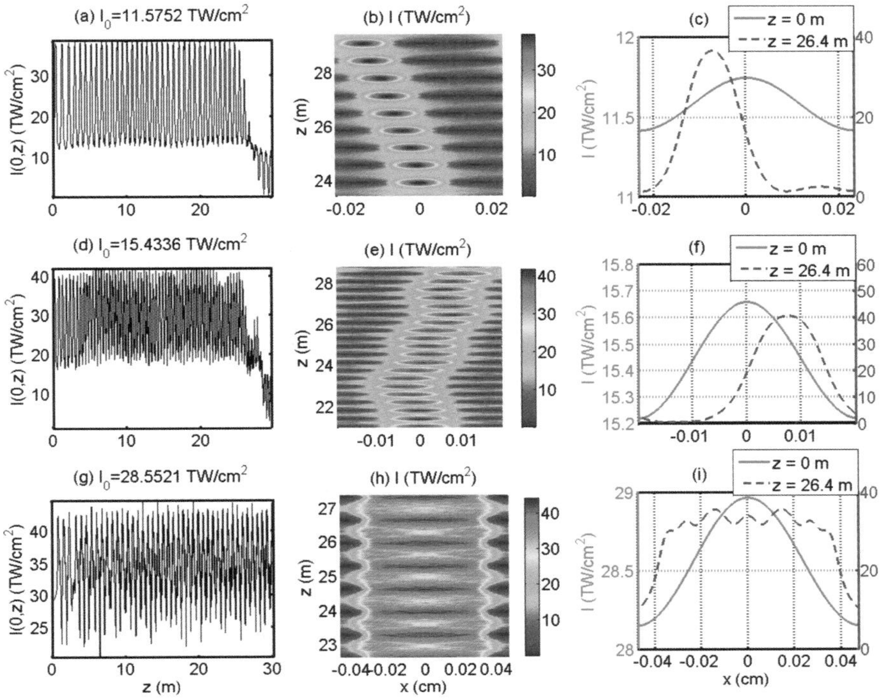

FIG. 4. (Color online) Comparison of the two different complicated spatial patterns with I_0 respectively of $1500 I_{cr}$ [(a)–(c)], $2000 I_{cr}$ [(d)–(f)], and $3700 I_{cr}$ [(g)–(i)] for N_2 case. [(a), (d), and (g)] Evolutions of laser field at $x = 0$. [(b), (e), and (h)] The contours of laser intensity $I(x,z)$. [(c), (f), and (i)] The profile of laser intensity at $z = 0$ (solid red line) and 26.4 m (dashed blue line).

laser in underdense plasmas [75,76], where the relativistic effect and the ponderomotive force dominate the propagation. For $I_0 = 150 I_{cr}$ [see point B in Fig. 1(a)], the first defocusing term in HOKE ($\beta_4|E|^4$) cannot be neglected and HOKE begins to play an important role. As a result, the laser fields exhibit a weak chaotic behavior [63,64]: The pattern structure does not recur within the finite distance and KAM tori in phase space become much thicker, as shown in Fig. 2(e). In a sense, this kind of solution is also called a quasiperiodic solution [70]. The corresponding power spectra shown in Fig. 2(f) indicate that many subharmonics appear compared with that in Fig. 2(c).

When laser intensity increases to $1500 I_{cr}$ (or 7.96×10^{12} W/cm^2) [see point C in Fig. 1(a)], the second focusing term in HOKE ($\beta_6|E|^6$) becomes a large perturbation but is dominated by the stronger defocusing effect in HOKE. At this point, HOKE can completely break down the KAM tori in phase space and the noiselike power spectra shown in Fig. 2(i) demonstrate that strong chaos occurs. However, for N_2, as shown in Fig. 1(b), when the laser intensity I_0 increases to 4.49×10^{12} W/cm^2, the second focusing term in HOKE ($\beta_6|E|^6$) dominates $\beta_4|E|^4$ and becomes an important mechanism. As a result, the HOKE turns from a negative perturbation to a positive perturbation, leading to the sharp increase of the maximum growth rate of modulational instability and quickly breakdown of the system. Complicated pattern structures with a continuous phase shift and off-axis evolution are observed for $4.49 \times 10^{12} < I_0$ (W/cm^2) $< 1.91 \times 10^{13}$, in which the focusing term $\beta_6|E|^6$ becomes a large perturbation [69]. Such complex patterns differ substantially from that given in Fig. 2(g). Particularly, for $I_0 = 1500 I_{cr}$ ($\approx 1.16 \times 10^{13}$ W/cm^2) [see point A in Fig. 1(b)], the laser intensity $I(x,z)$ experiences a stochastic oscillation at sufficiently far distance and the pattern drifting phenomenon [77] is shown in Fig. 3(a) [also Fig. 4(b)]. For $I_0 = 2000 I_{cr}$ ($\approx 1.54 \times 10^{13}$ W/cm^2) [see point B in Fig. 1(b)], more complicated pattern structures with oscillated centroids that differ from the pattern drifting are shown in Fig. 3(d) [also Fig. 4(e)]. It is noted that the found off-axis evolution, as shown in Figs. 4(b) and 4(e), corresponds to the most unstable stage of the system. When the modulational instability is strong enough (a very large Γ_{max}), the laser fields with quite small noise can easily develop into such complex structures. In addition, the off-axis propagation of high-intensity laser pulses in underdense plasmas can lead to laser hosing instability [78]. Thus, the off-axis behavior should be a general characteristic of laser pulse propagation in the real medium, where the asymmetric factors of the laser field would be quickly enlarged by the modulational instability. Furthermore, the corresponding phase-space trajectories shown in Figs. 3(b) and 3(e) demonstrate that the KAM tori are completely broken down and irregular HMO crossings gradually appear with the off-axis evolution of the laser field. As we have known, irregular HMO crossings are associated with spatiotemporal chaos [70,71] and pattern competition [77]. Differing from pattern competition that experiences a phase jump from $\theta = 45°$ to $\theta = 225°$, these two complicated patterns shown in

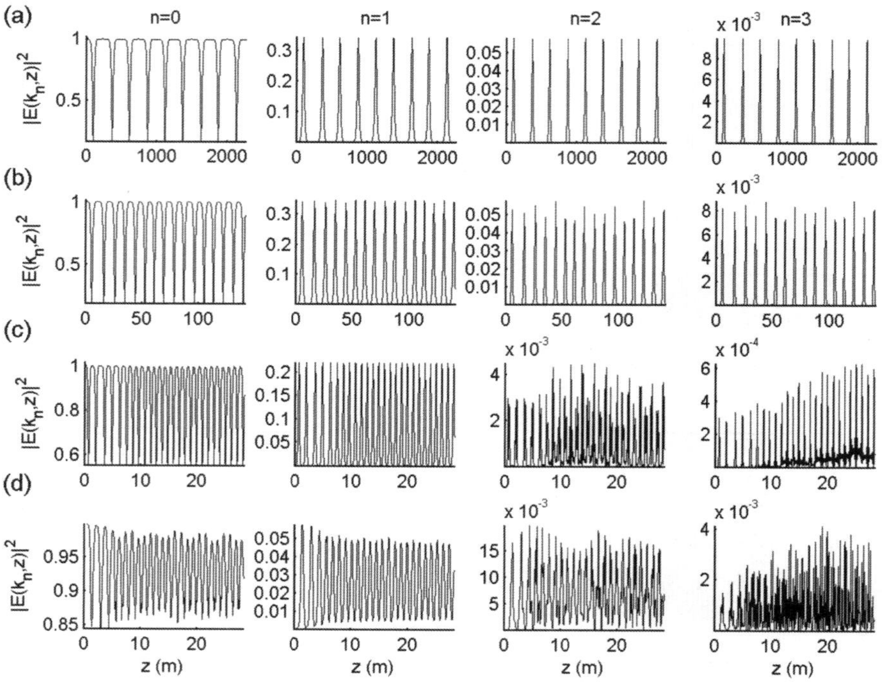

FIG. 5. The energy evolution in the Fourier modes for the O_2 case with I_0, respectively, of $8I_{cr}$ (a), $150I_{cr}$ (b), $1500I_{cr}$ (c), and $3000I_{cr}$ (d) as shown in Fig. 1(a).

Figs. 3(a) and 3(d) experience a continuous phase shift from $\theta = 45^\circ$ to $\theta = 225^\circ$. On the other hand, the trajectories of off-axis evolution are extremely sensitive to the initial laser intensity, which is typical for a chaotic system [63,64], further demonstrating that strong chaos occur in this case.

For the N_2 case, when laser intensity increases further to 2.74×10^{13} W/cm^2, the third defocusing term in HOKE ($\beta_8|E|^8$) dominates the focusing term $\beta_6|E|^6$ and results in another transition for HOKE from a positive perturbation to a negative perturbation again. In addition, the negative HOKE experiences a dramatic increase with the increase of laser intensity [see Fig. 1(b)]. For I_0 (W/cm^2) $> 2.74 \times 10^{13}$, another complicated pattern which differs completely from the ones discussed above is observed. Particularly, for $I_0 \approx 2.86 \times 10^{13}$ W/cm^2 [see point D in Fig. 1(b)], the complicated pattern and relevant stochastic phase-space trajectory are shown in Figs. 3(j) and 3(k). In Fig. 4, we give a comparison between the off-axis patterns and this complicated pattern. It is shown that at the same propagation distance z, the former exhibits off-axis behavior, while the latter first experiences a quick spatial diffraction and then develops into many small-scale spatial structures. Because of the interaction of these small-scale waves, their spatial structures experience a stochastic change with the propagation of laser field, leading to the chaotic behavior of laser field shown in Fig. 4(g). The interaction of different short waves can lead to more complicated spatial structures. It is also important to point out that although the corresponding laser field propagates with a quite complicated oscillation behavior, Fig. 4(h) shows that their patterns are still of on-axis structures. In addition, for $I_0 \approx 2.3 \times 10^{13}$ W/cm^2 [see point C in Fig. 1(b)], Fig. 3(g) shows that the laser propagation experiences a quasiperiodic process. Therefore, our simulations demonstrate that the off-axis pattern [in Figs. 3(a) and 3(d)] can again become the on-axis pattern [in Fig. 3(j)] by a quasiperiodic pattern [as shown in Fig. 3(g)]. Such complicated patterns have also been shown in Fig. 2(j) when laser intensity increases to 1.06×10^{13} W/cm^2. Their corresponding continuous power spectra shown in Fig. 2l and Fig. 3l reveal the chaotic behavior.

In order to analyze the mechanism that leads to these chaotic behaviors, we further investigate the evolution of energy contained in Fourier modes. In Fourier space, the energy of the system (8) can be defined as

$$H = \sum_n H_{K_n} = \sum_n |E_{K_n}|^2, \tag{9}$$

where the initial energy is in the mode k_{max}. Figure 5 shows the evolution of the energy in the first four Fourier modes corresponding to the solutions shown in Fig. 2. Obviously, a large part of the energy lies in the first and second Fourier modes. For $I_0 = 8I_{cr}$, the evolution of energy in all modes is periodic [see Fig. 5(a)], which is consistent with periodic recurrent patterns shown in Fig. 2(a). For $I_0 = 150I_{cr}$, the quasiperiodic evolution of energy in Fourier modes shown in Fig. 5(b) also agrees with the quasiperiodic solution shown in Fig. 2(d), whereas for $I_0 = 1500I_{cr}$ and $3000I_{cr}$, Figs. 5(c) and 5(d) show that energy in the system would spread from the lowest mode to many higher modes due to HOKE and would not regroup into the original lowest mode. We can also see from Figs. 5(c) and 5(d) that the quasiperiodic evolution

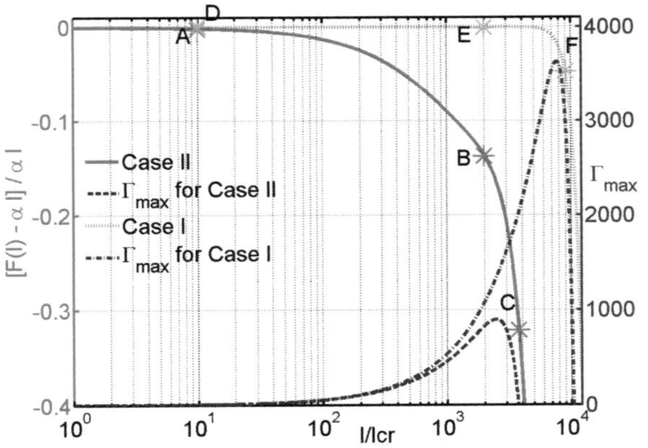

FIG. 6. (Color online) Comparison of the values of nonlinear terms (solid red line for Case II and dotted red line for Case I) and the corresponding most unstable growth rate $\Gamma_{\max}$ of modulational instability (dashed blue line for Case II and dash-dotted blue line for Case I) in two cases, where the asterisk symbols in Case I mark the points with I_0 of $10I_{cr}$, $2000I_{cr}$, $9000I_{cr}$, and in Case II the asterisk symbols correspond to $10I_{cr}$, $2000I_{cr}$, $3600I_{cr}$, respectively. Here, $I_{cr} = P_{cr}/4\pi w_0^2 \approx 5.305 \times 10^9 \text{W/cm}^2$.

of the energy in the first two modes dominates the spatially localized structures being kept. However, the evolution of energy in higher modes clearly exhibits stochastic behavior due to the strong mode-mode interactions, which would lead to the presence of complicated spatial patterns, as shown in Figs. 2(g) and 2(j).

IV. SPATIOTEMPORAL PATTERNS IN TWO-DIMENSIONAL TRANSVERSE SPACE

In this section, we turn to discussing the dynamic characteristics of the system (2) in two-dimensional transverse space. In this case, both the HOKE and plasma diffraction are considered. As we have known, the filamentation processes in air are mainly investigated in high-dimensional space [58–62]; however, nonlinear properties of ultrashort laser propagating in air, including the spatial chaos and pattern formation, have not been investigated in high-dimensional spaces. For two-dimensional NLSE, the singular focusing solution exists [79–81], pseudorecurrence can also appear when the saturable nonlinear effects are strong enough [82], and even a solitary solution can be produced with an appropriate positive potential [83]. We here investigated the nonlinear pattern dynamics of the system (2) in two-dimensional transverse spaces by taking into account the high-order plasma saturation effect without and with higher-order Kerr nonlinear effects. In the numerical simulation, we employ the same initial condition and numerical method as discussed in Sec. III. The most unstable mode with $k_x = k_{x,\max}$ and $k_y = k_{y,\max}$ is considered and we assume the medium is O_2.

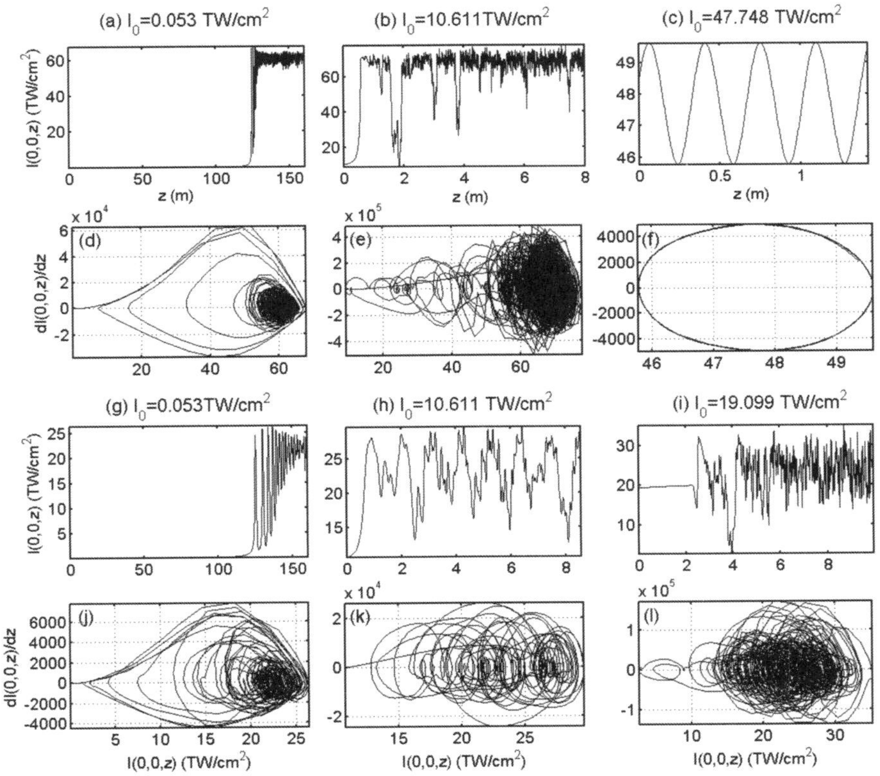

FIG. 7. Evolutions of laser fields at $(x,y) = 0$ [(a)–(c) and (g)–(i)] in two cases and corresponding phase space [(d)–(f) and (j)–(l)] with the same I_0 as given in Fig. 6. The upper two rows are for Case I and the bottom two rows are for Case II.

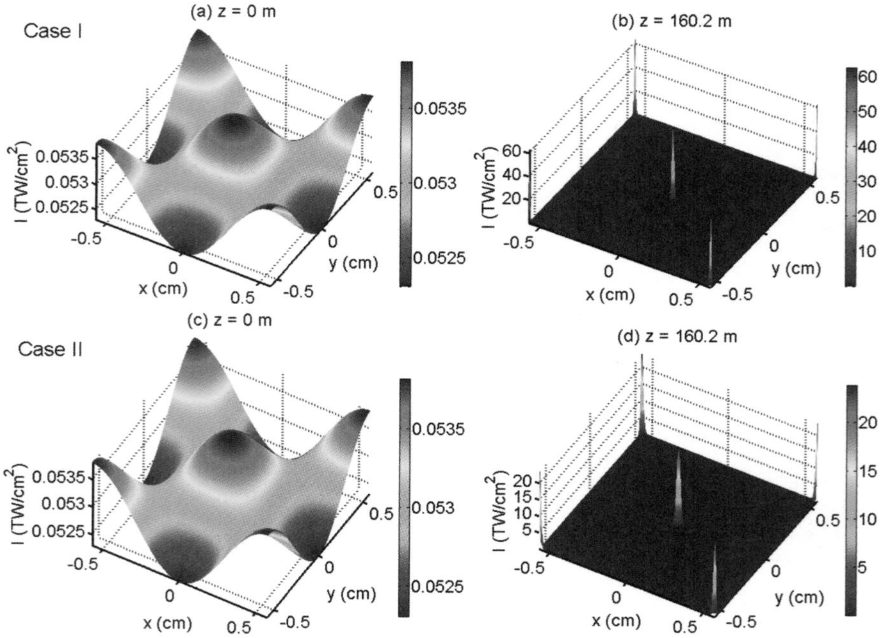

FIG. 8. (Color online) The transverse intensity patterns at different propagation distance in Case I [(a)–(b)] and Case II [(c)–(d)] with an initial laser intensity of $I_0 = 10I_{cr}$, which are corresponding to the points A and D shown in Fig. 6.

To have a clear comparison between the two different saturable effects, i.e, the plasma saturation effect and the higher-order Kerr nonlinear effects, in the following we consider two cases and refer to them as Case I and Case II. Case I corresponds to $\beta = 0$, and Case II corresponds to $\beta = -1$ in the nonlinear term $F(|E|^2)$ of Eq. (2). Thus, in Case I the Kerr terms are truncated to n_2 and the plasma saturation term is included; however, in Case II both the higher-order Kerr nonlinear effects and the plasma saturation effect are considered. Figure 6 shows the values of higher-order nonlinear terms relative to KF term and the most unstable growth rate of modulational instability Γ_{max} in Case I and Case II. It is shown that in the presence of HOKE for our parameter range, the high-order saturation term in Case II is much stronger than that in Case I for a wide range of intensities. In addition, Γ_{max} in Case I is about 4 times higher than that in Case II.

Figures 7–10 give the solutions of Eq. (2) in two different cases. At relatively low input intensity, say $10I_{cr}$ (see points A and D in Fig. 6), the high-order saturable effects are initially negligible. Equation (2) then reduces to the two-dimensional cubic NLSE that is not integrable. For the two-dimensional cubic NLSE, singular solutions would occur at the points of self-focusing, where the time of blowup is finite and the field intensities exhibit catastrophic processes [47,79–81]. As shown in Figs. 7(a) and 7(g), in both cases, the laser first experiences a focusing process and then energy in the laser fields would quickly converge into the filament structures shown in Figs. 8(b) and 8(d). However, the catastrophic process would be arrested by the defocusing effects arising from the plasma or higher-order Kerr nonlinearity with the increase of laser intensity. As a result, energy in the laser fields flows from the filament regions to the peripheral regions, and, correspondingly, the peak laser intensity decreases. However, the laser power is still high and the focusing process occurs again. Such a process is the well-known multiple-refocusing phenomenon [50,84,85], which is clearly shown in Figs. 7(a) and 7(g). Their corresponding phase-space trajectories in Figs. 7(d) and 7(j) reveal an attractor-like structure. However, their irregular phase-space structures and the irregular distribution of energy in Fourier modes shown in Fig. 11(a) indicate a chaotic behavior. In addition, due to lower peak intensity in Case II, the filament size shown in Fig. 8(d) is larger than that in Case I, shown in Fig. 8(b).

When the initial laser intensity increases to $2000I_{cr}$ (see points B and E in Fig. 6), the high-order saturable effects cannot be neglected and the maximum growth rate of modulational instability quickly increases, as shown in Fig. 6. Figures 7(b) and 7(h) show that the laser field oscillates stochastically and the relevant phase-space trajectories also demonstrate stochastic behavior. On the other hand, we note from Fig. 9 that the characteristic scale length of the patterns in Case II is about 2 times larger than that in Case I. This can be understood from the relation $\lambda_{max} = 2\pi/\sqrt{\Gamma_{max}}$, where λ_{max} refers to the wavelength of the most unstable mode. In view of the curve for Γ_{max} in Fig. 6, one finds that the wavelength of modulational instability in Case II is approximately 2 times longer than that in Case I, leading to a large- or small-scale pattern structure in Cases II and I, respectively, as shown in Fig. 9.

With an increase of the initial laser intensity, in Case I a very interesting phenomenon is shown in Figs. 7(c) and 7(f). It is seen that when the laser intensity increases further

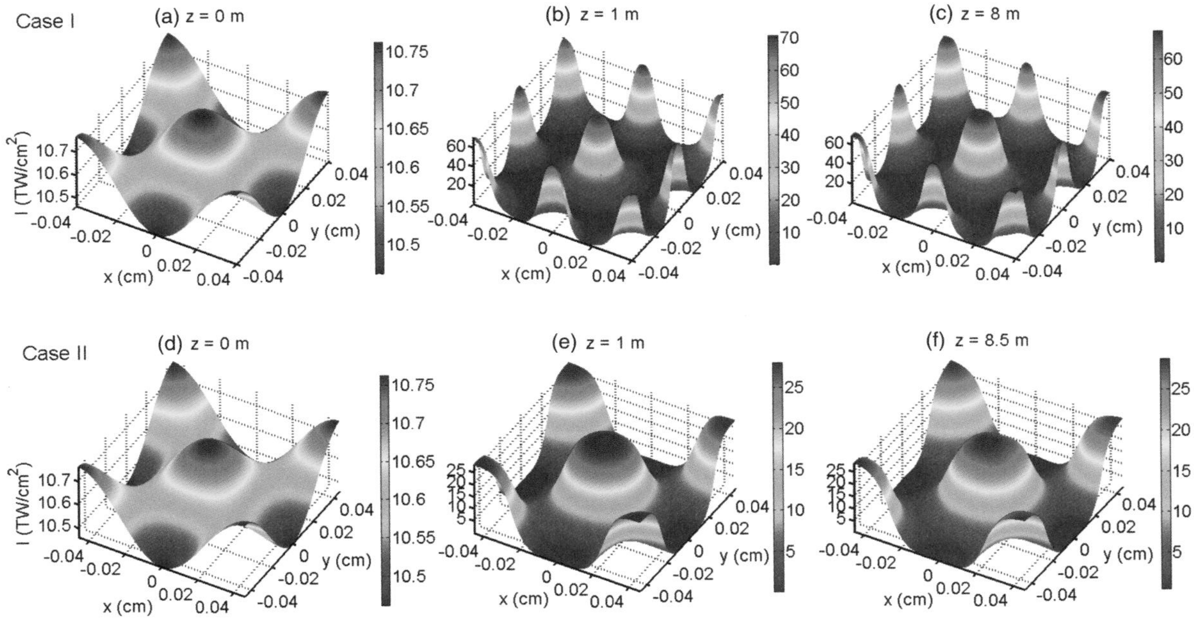

FIG. 9. (Color online) The transverse intensity patterns at different propagation distance in Case I [(a)–(c)] and Case II [(d)–(f)] with an initial laser intensity of $I_0 = 2000I_{cr}$, which are corresponding to the points B and E shown in Fig. 6.

to $9000I_{cr}$ (see point F in Fig. 6), pseudorecurrence [82] can appear. Actually, in this case nonlinear perturbation from the high-order saturation term becomes larger. However, the maximum growth rate of the modulational instability decreases sharply to zero in this region (see Fig. 6), which suggests that the instability cannot develop, eventually leading to the pseudorecurrent solution. Physically, we can consider that in this region plasma defocusing is strong enough to prevent the Kerr focusing singularity. In earlier works [47,82], pseudorecurrence was also observed when the saturation effect

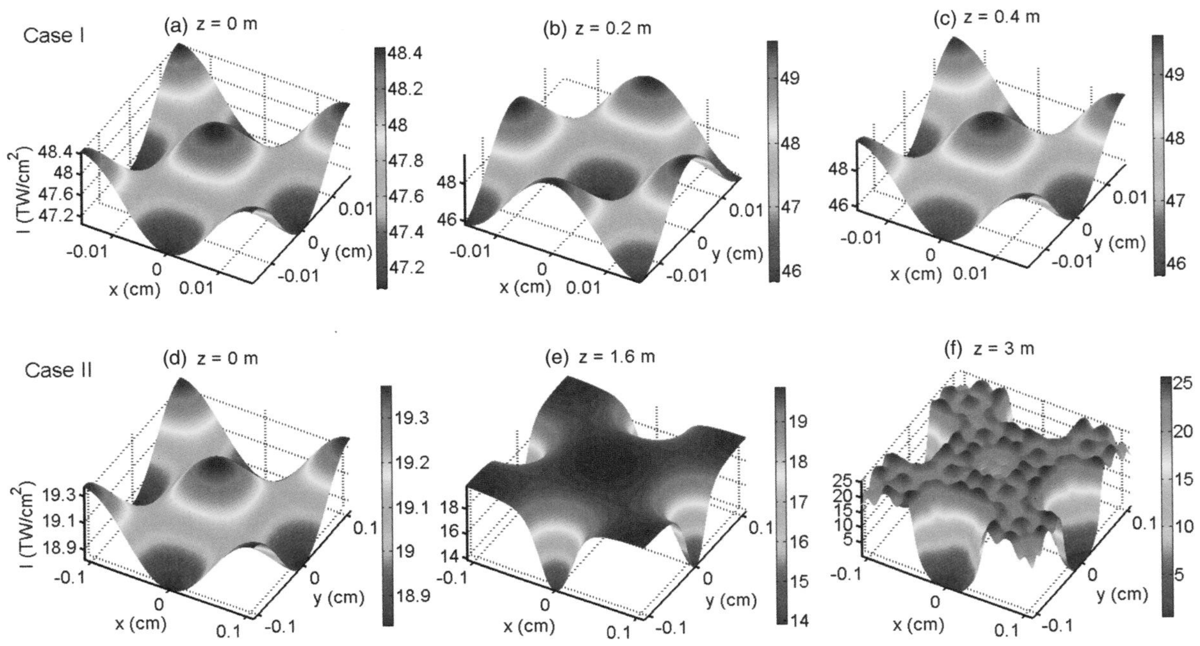

FIG. 10. (Color online) The transverse intensity patterns at different propagation distance in Case I [(a)–(c)] and Case II [(d)–(f)] with an initial laser intensity of $I_0 = 9000I_{cr}$ and $I_0 = 3600I_{cr}$, respectively, which correspond to the points C and F shown in Fig. 6.

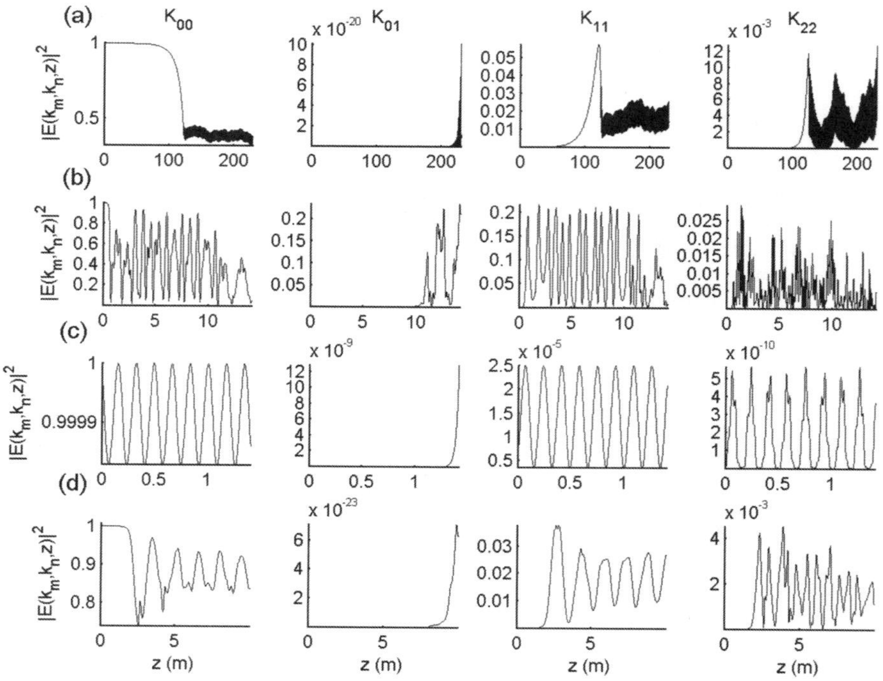

FIG. 11. The energy evolution in the Fourier modes with I_0, respectively, of $10I_{cr}$ (a), $2000I_{cr}$ (b), $9000I_{cr}$ (c) in Case I and $3600I_{cr}$ (d) in Case II.

was very strong and the modulational instability was weak. In contrast, for Case II, the strong HOKEs would induce a quite different behavior on the pattern dynamics, as shown in Figs. 7(i) and 7(l). Figure 10 shows the transverse intensity patterns at three appropriate propagation distances. We can see that in Case II, the laser fields first develop into a uniform X-type structure due to the modulational instability. Nevertheless, such uniform structures are quite unstable and small-scale turbulent structures quickly appear on this X-type pattern [86] as the laser propagates, which is in good agreement with the solution in the one-dimensional case, shown in Fig. 4(h). In addition, the complex interactions of these small-scale structures lead to the stochastic behavior of the laser evolution and phase-space trajectory shown in Figs. 7(i) and 7(l).

To clearly illustrate the mechanism that leads to the formation of these complicated patterns, we again investigate the evolution of energy contained in Fourier modes. In Fourier space, we can define the energy of system (2) as [70]

$$H = \sum_{K_{n,m}} H_{K_n,K_m} = \sum_{K_{n,m}} |E(K_n,K_m,z)|^2, \qquad (10)$$

where $K_{n,m} = (K_n, K_m)$ represents the n-th Fourier mode in x space and the m-th mode in y space. The evolution of energy in the Fourier modes is shown in Fig. 11. Figure 11(a) corresponds to the condition of point D in Fig. 6 in Case I. It can be seen that as the laser propagates, the energy in the initial mode would slowly spread to higher diagonal modes at first, say K_{11} and K_{22}, and, correspondingly, the energy in the K_{11} and K_{22} modes increases. However, at the distance ($z \approx 120$ m) where the peak laser intensity experiences a sharp increase [see Fig. 7(a)], the energy in the first three diagonal modes all decreases in a short propagation distance [see Fig. 11(a)], which implies that the energy in the system transfers from the high-order diagonal modes (K_{11} and K_{22}) to much higher-order diagonal modes. Physically, the energy in the laser fields would be concentrated on the higher-order or shorter-wavelength Fourier modes during the focusing process. Such process may be the intrinsic scheme of the formation of laser filament in air.

Figure 11(b) shows the case of point E in Fig. 6 in Case I. The stochastic evolution of energy in the high-order diagonal modes leads to the chaotic behavior of laser fields shown in Fig. 7(b). In particular, it is noted that for $z > 10$ m the energy in the diagonal modes decreases but in the nondiagonal modes increases. As a result, a considerable part of the energy lies in the nondiagonal Fourier modes. The abnormal partition of energy in Fourier modes would result in the formation of off-axis patterns [47]. Here, we do not show the off-axis patterns. It seems that such off-axis patterns are quite sensitive to the maximum growth rate of modulational instability. When the modulational instability is strong enough (a large Γ_{max}), the laser field would easily develop into these complicated patterns. The results in the one-dimensional case also verify this point.

We now discuss the pseudorecurrent case shown in Fig. 7(c). We see from Fig. 11(c) that the energy in the first two diagonal modes exhibits periodic motion but in high-order mode K_{22} is not exactly periodic, which indicates that the solution is pseudorecurrent rather than exactly recurrent as shown in Fig. 2(a) for the one-dimensional case. Figure 11(d)

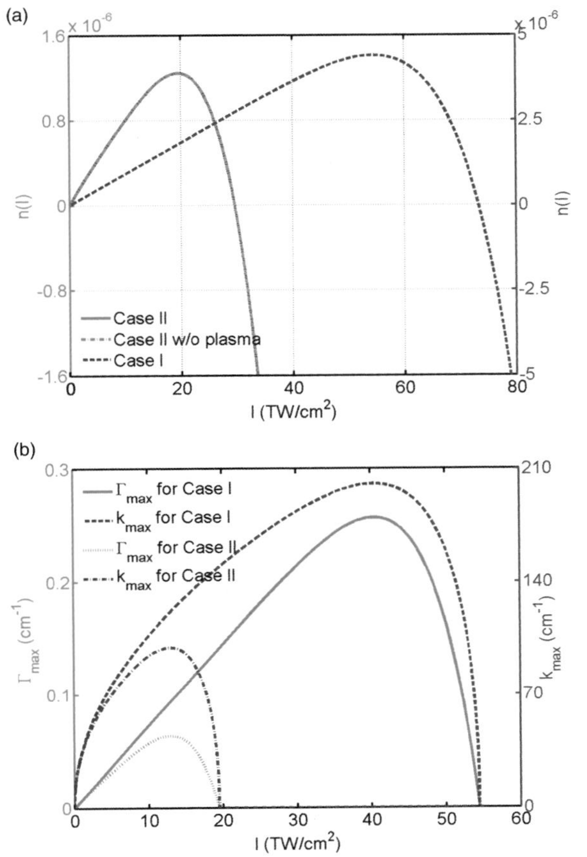

FIG. 12. (Color online) (a) Comparison of the nonlinear refractive index $n(I)$, where $n(I) = \frac{n_2}{\sqrt{3}}I - \frac{\gamma}{k_0}I^K$ in Case I and $n(I) = \Delta n - \frac{\gamma}{k_0}I^K$ in Case II. Here $\Delta n = \frac{n_2}{\sqrt{3}}I + \frac{n_4}{\sqrt{5}}I^2 + \frac{n_6}{\sqrt{7}}I^3 + \frac{n_8}{3}I^4$ is the nonlinear refractive index in Case II without the plasma. (b) Comparison of the most unstable growth rate of modulational instability $\Gamma_{\max}$ and the wave number of the most unstable modes in two cases, where $\Gamma_{\max} = k_0\frac{n_2}{\sqrt{3}}I - \gamma K I^K$ in Case I and $\Gamma_{\max} = k_0(\frac{n_2}{\sqrt{3}}I + \frac{2n_4}{\sqrt{5}}I^2 + \frac{3n_6}{\sqrt{7}}I^3 + \frac{4n_8}{3}I^4) - \gamma K I^K$ in Case II, and the corresponding wave number of the most unstable modes satisfy the relation $k_{\max} = \sqrt{2k_0\Gamma_{\max}}$.

also shows the case of point C in Fig. 6 in Case II. It is found that the regular partition of energy in the first two diagonal modes dominates the main X-type structure shown in Fig. 10(e). However, the irregular behavior of energy in the high-order mode (K_{22}) leads to the appearance of the small-scale turbulent structures, as shown in Fig. 10(f). In addition, it can be seen from Figs. 11(c) and 11(d) that a small part of the energy is distributed in the nondiagonal mode (K_{01}); however, the ratio is still negligible compared with that in Fig. 11(b), which does not severely affect the symmetry of the intensity patterns, as seen in Fig. 10.

V. MULTIPLE FILAMENTATION SIMULATIONS

In the above sections, we have given a detailed comparison of nonlinear pattern dynamics in two different cases. Now we turn to the simulation of multiple filamentation of ultrashort laser. Filamentation has extensively been found in laser or particle beam propagations in air or plasmas [1–62,87]. As far as multiple filamentation of laser beams propagating in air is concerned, most of the numerical simulations are based on the so-called classical model where the plasma defocusing is considered as the saturating nonlinearity [48,54–58]. Here we analyze multiple filamentation of ultrashort laser pulses in air by comparing both plasma saturation effect and higher-order Kerr nonlinearities. Equation (1) in two-dimensional spaces is also solved using the standard spectral split-step scheme [47,70]. In the transverse directions, a fixed grid of high resolution ($\Delta x = \Delta y \leqslant 20\,\mu$m) is taken. A sufficiently large simulation box $[(L_x, L_y) \geqslant 6w_0]$ guarantees free propagation of the laser pulse [48]. The input laser beam is modeled by a super-Gaussian profile

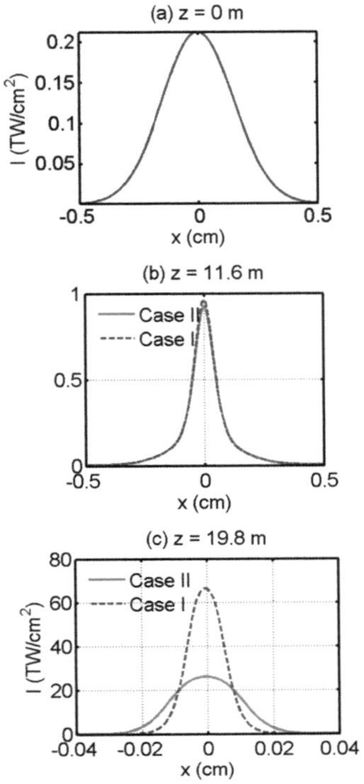

FIG. 13. (Color online) The intensity profile $I(x,0)$ at different propagation distances in two cases with an initial laser power of $P_{\rm in} = 5P_{\rm cr}$.

$$E(x,y,z=0) = \sqrt{\frac{2P_{\rm in}}{\pi w_0^2}}\exp\left[-\frac{x^2+y^2}{w_0^2}\right]^m, \qquad (11)$$

where $P_{\rm in}$ is the input laser power, w_0 is the initial beam waist, and m is the index to define the super-Gaussian beam with $m = 1$ corresponding to a Gaussian beam. In our simulations, we consider the initial beam waist as 0.3 cm and the input laser power on a level of low ($P_{\rm in} = 5P_{\rm cr}$), moderate ($P_{\rm in} = 80P_{\rm cr}$), and high ($P_{\rm in} = 1000P_{\rm cr}$) powers, respectively, and $m = 1$ is assumed in the simulations. In addition, the propagation medium is assumed to be O_2.

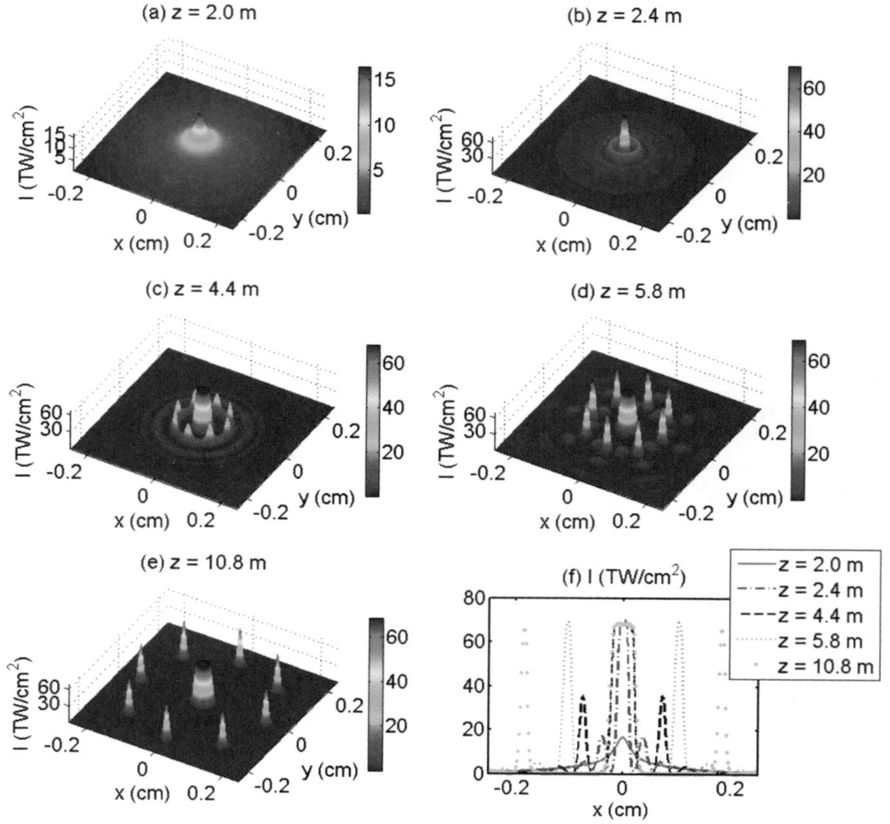

FIG. 14. (Color online) The transverse intensity patterns at different propagation distance in Case I [(a)–(e)] with a laser power of $P_{in} = 80P_{cr}$ and the corresponding intensity profiles (f) at $y = 0$, where the displayed space scale is 0.5×0.5 cm (scale of the simulation box is 2×2 cm).

Figure 12 shows the nonlinear refractive index $n(I)$ and the most unstable growth rate Γ_{max} of the modulational instability of a plane wave [51,85,88] in Case I and Case II as defined in Sec. IV. With the increase of laser intensity, it can be seen that the total nonlinear refractive index decreases from positive to negative values. Particularly, intensity clamping occurs when the total nonlinear refractive index equals zero [89–91]. However, it is necessary to mention that intensity clamping depends much on the focusing condition of the laser beam, which may result in a champing intensity that does not correspond to the zero value of the total nonlinear refractive index [92–95]. It can be seen from Fig. 12 that Case II yields a lower intensity for $n(I) = 0$, which directly determines a lower peak intensity in laser filaments in Case II and is in excellent agreement with the simulation results (28.6 TW/cm^2 vs 72.9 TW/cm^2) shown in Figs. 14 and 15. In addition, Γ_{max} in Case I is about 4 times higher than that in Case II, leading to faster formation of filaments in Case I (see Figs. 14 and 15). Figure 12(b) also shows that the wave number of the most unstable modes in Case I is much larger than that in Case II, demonstrating that the propagation of ultrashort laser in air in Cases I and II is dominated by a short- and long-wavelength modulation, respectively.

Figure 13 shows the process of single filament formation in the two cases for a low laser power of $5P_{cr}$. In both cases, the laser profiles collapse towards a well-known "Towns profile" [see Fig. 13(b)] during the first self-focusing stage [96,97]. At this time, the saturable defocusing effects do not take effect yet. Therefore, "Towns Profiles" in the two cases are almost the same, as shown in Fig. 13(b). With the increase of laser intensity, the plasmas are generated or HOKEs come into play. Then the triggered defocusing effects will counteract the focusing effect and arrest the spatial collapse of the laser beam. Finally, the balance between the focusing effects and defocusing effects sets a limit of the minimum beam diameter and the highest intensity. The final equilibrium state of the laser beam corresponds to the formation of single laser filament. Figure 13(c) shows the intensity profile of the laser filament in the two cases. It can be seen that the filament in Case II has a wider diameter (254 μm vs 138 μm) but a lower peak intensity (26.07 TW/cm^2 vs 66.47 TW/cm^2) than that in Case I, which is basically consistent with the points of $n(I) = 0$.

We now consider the case of multiple filament formation with the initial laser power of $80P_{cr}$. Figures 14 and 15 show the phase of multiple filament formation and propagation in the two cases. As illustrated above, the laser would first evolve

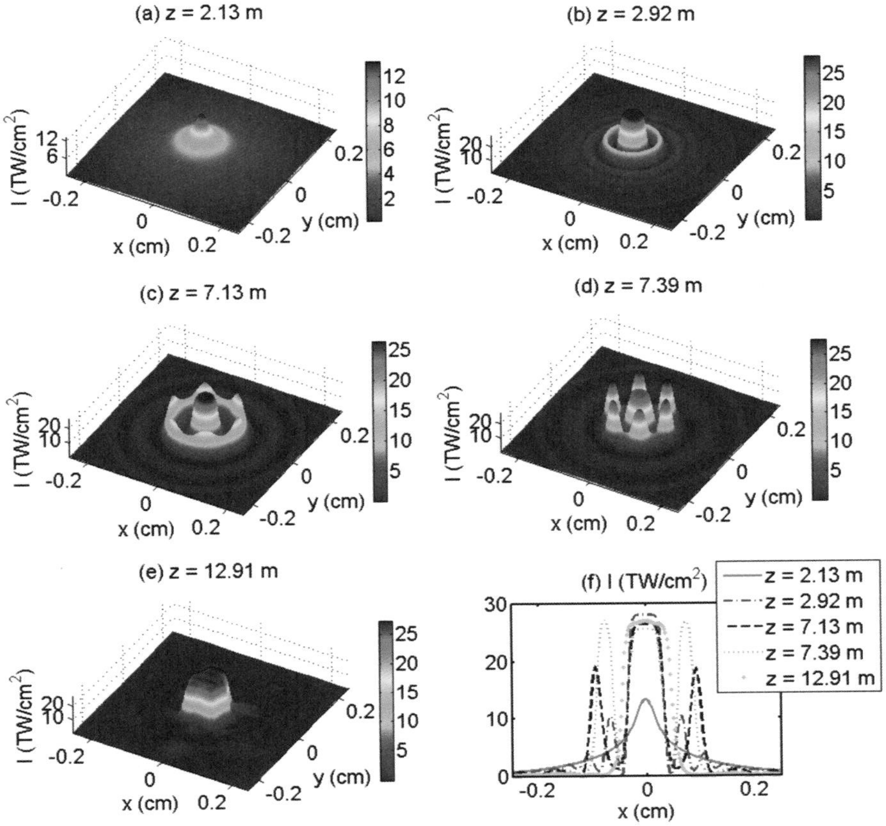

FIG. 15. (Color online) The transverse intensity patterns at different propagation distance in Case II [(a)–(e)] with a laser power of $P_{\text{in}} = 80P_{\text{cr}}$ and the corresponding intensity profiles (f) at $y = 0$, where the displayed space scale is 0.5×0.5 cm (scale of the simulation box is 2×2 cm).

to the "Towns profile" [see Figs. 14(a) and 15(a)], which is the typical characteristic for a Gaussian beam [96,97]. Then the defocusing effects arising from the plasma or higher-order Kerr nonlinear effects counteract the optical self-focusing by diverging into a series of concentric ring structures [see Figs. 14(b) and 15(b)], as have been observed in existing experiments and numerical simulations [98–102]. The ring structures may result from the interference of the convergent background field and the divergent field due to the reflection of the front part of the pulse from the surface [98,99]. One the other hand, we note from Fig. 18 that before the occurrence of filaments, there exists an oscillation between the central focal spot and the ring structures. Such an interesting phenomenon has also been observed in previous work [100], which could be due to the competition between the convergent field and divergent field. Finally, only the interior ring with higher refractive index can quickly break up into narrow filaments by the modulational instability [see Figs. 14(c) and 15(c)]. The formation and breakup of the ring structures have also been observed for intense lasers in near-critical plasmas by use of three-dimensional particle-in-cell simulations [103]. Thus, the appearance and breakup of the ring structures should be a typical characteristic to multiple filamentation of lasers propagating in the medium. Furthermore, Figs. 14(d) and 15(d) show that the interactions among the filaments differ markedly in the two cases. In Case I, propagation of the external filaments exhibits transverse deflection [see Figs. 14(e) and 18(a)], as observed in some previous experiments [48]. On the contrary, in Case II, the filaments rapidly coalesce to a central core, resulting in a thick light bullet [see Figs. 15(e) and 18(b)]. The mechanism of the two different propagation modes can be understood in terms of the evolution of the overall mean-square radius $\langle r_\perp^2 \rangle$ of the beam. For our conservative system (1), we can rewrite the evolution equation for $\langle r_\perp^2 \rangle$ as [85]

$$P d_z^2 \langle r_\perp^2 \rangle = 8\left\{ H + \int [2F(|E|^2) - f(|E|^2)|E|^2] d\vec{r} \right\}, \quad (12)$$

where $P = \int |E|^2 d\vec{r}$ is the laser power, $\langle r_\perp^2 \rangle$ is defined as $\int r_\perp^2 |E|^2 d\vec{r}/P$, $H = \int [\frac{1}{2k_0 n_0}(|\partial_x E|^2 + |\partial_y E|^2) - F(|E|^2)] d\vec{r}$ is the Hamiltonian of the system (1), $F(|E|^2) \equiv \int_0^{|E|^2} f(s) ds$, $f(s) = k_0 \Delta n(s) - \gamma s^K$, and $s = |E|^2$. For two or more in-phase filaments with a mean separation distance δ, the Hamiltonian H can be expanded as $H = H_{\text{free}} + H_{\text{int}}(\delta)$, where

H_{free} refers to the contribution of each individual filament and $H_{\text{int}}(\delta)$, which decreases exponentially with δ^2 (see Ref. [85] for more details), describes the interaction between the filaments. In addition, the critical separation distance δ_c, below which the filaments can coalesce, is given by the zeros of the right-hand side of Eq. (12). When the separation distance is above δ_c, no mutual attraction between the filaments is possible and the filamentation would develop independently [85,104]. In the two cases, it is found that in Case II δ_c is larger than that in Case I. Moreover, due to a larger waist (ρ) of filaments in Case II, the relative separation distance (δ/ρ) of the filaments in Case II is smaller than that in Case I, which suggests that the filaments in Case II are easier to attract each other. Accordingly, in Case I no mutual attraction is observed for filaments at wider separation distances. Instead, the filaments diverge along the divergent tide from the reflection part of the pulse. However, in Case II, filaments with small ($\delta < \delta_c$) separation distances can provide strong attraction force to seize the divergent tide and turn to a fusion process, which enhances the channel stability and leads to prolongation of the propagating distance [24,48]. In order to illustrate the robustness of the two different propagation modes, we further do the simulation by choosing some neighboring parameters. It is found that for $90P_{\text{cr}}$ and $100P_{\text{cr}}$, the similar characteristics analyzed above can also occur. In addition, it is interesting to note that the propagation dynamics of laser beams shown in Figs. 14 and 15 are very similar to the behavior of spatial solitons, including the breakup of high-order solitons and the interaction dynamics of different solitons [105]. In fact, such solitonlike dynamics are the typical characteristics of our reduced model [48], especially for the conservative case.

Figure 17, which is for a much higher laser power, namely $P_{\text{in}} = 1000P_{\text{cr}}$, shows the intensity patterns for the two cases at appropriate propagation distances. From Fig. 16(a) we see that the initial refractive index in the central region satisfies $n(I) > 0$ in Case I but $n(I) < 0$ in Case II. Accordingly, in Case II the laser first experiences defocusing, especially when $n(I)$ is near the minimum, and energy of the laser fields would flow into the region with maximum refractive index that corresponds to $dn/dr = 0$ [see Fig. 17(e)]. As a result, higher-order Kerr nonlinear effects produce a deep hole in the transverse intensity distribution and most of the light energy is accumulated in a cylindrical optical channel, shown as the thick (red) ring in Fig. 17(f). Differing from the unstable rings shown in Figs. 14 and 15, such a channel, which can trap the intense laser light for several meters, as can be seen in Fig. 18(d), corresponds to $n(I) \approx 0$ and can make the system become more compact. From Fig. 17(f) we can also see that rings are formed outside as well as inside the channel due to the defocusing effects acting on this channel. Figures 17(g) and 17(h) show that only the inside rings would break up into filaments and then merge with the optical channel. It can also be seen that small-scale turbulent structures appear in the optical channel [see Figs. 17(g) and 17(h)], but they do not affect the robustness of the channel. Such local complex structures may be characteristic of Case II, also discussed in Secs. III and IV, and can be attributed to energy distribution among the small-scale Fourier modes typical for localized turbulence and chaos [66,70,106]. On the other hand, in Case I filaments emerge from the rings outside

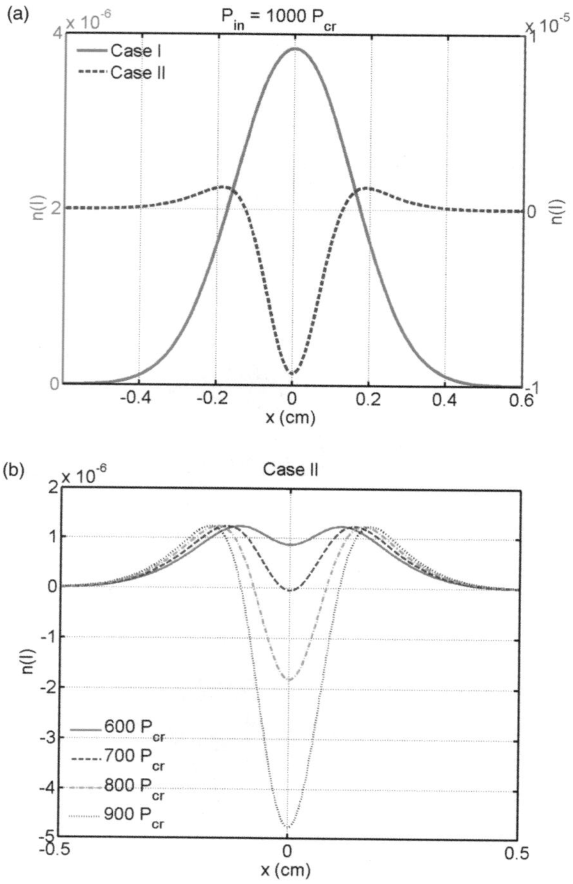

FIG. 16. (Color online) The initial refractive index distribution along the axis of $y = 0$ for the two cases with a laser power of $P_{\text{in}} = 1000P_{\text{cr}}$ (a) and the initial refractive index distribution in Case II at different input powers (b).

a central optical pillar [see Figs. 17(a) and 17(b)]. As the laser propagates, the interior filaments would redistribute due to mutual attraction. However, the exterior filaments would move outwards along with the divergent tide, as shown in Figs. 17(c) and 17(d). Figure 16(b) shows the initial refractive index at different input powers in Case II. We can see that the location of maximum refractive index deviate from the central spot and the deviation increases with the the input power, which could result in the formation of optical channel with variable radius. Thus, in our parameter range the channel structure appears to be typical in Case II at very high input powers ($\geqslant 600P_{\text{cr}}$). Since the characteristic scale length of the system differs significantly when different mechanisms, such as the ones considered here for our parameter range, of filamentation are in operation, experimental records of the spatial filament patterns would possibly be a new basis for identifying the regime of filamentation in terawatt laser propagation in air. However, we should mention that our simulation results as given in Figs. 14–18 are mainly based on the measurement

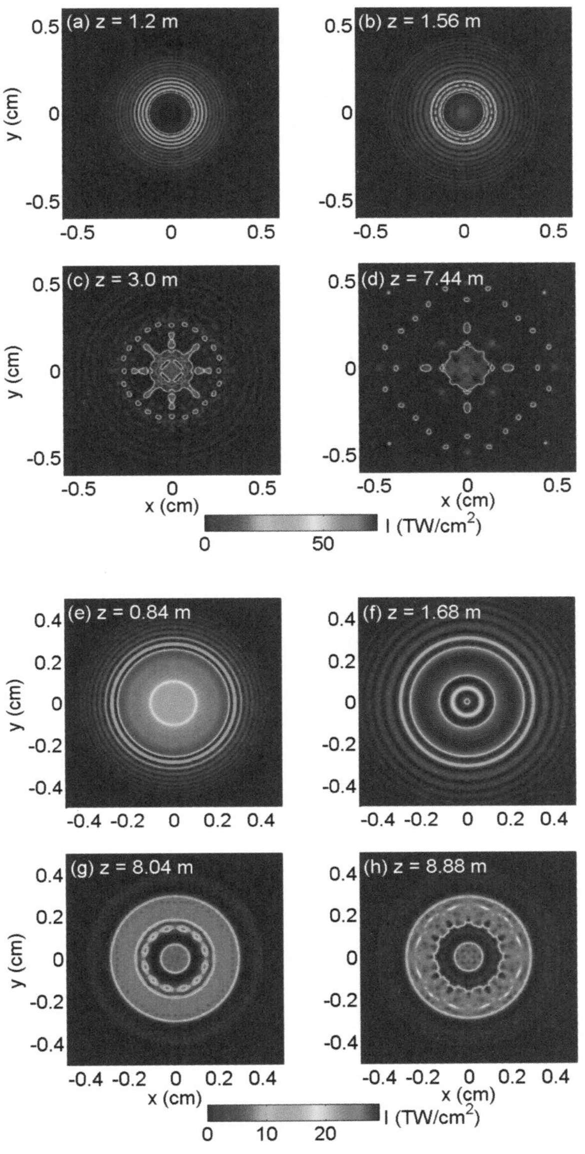

FIG. 17. (Color online) The transverse laser intensity patterns at different propagation distance in Case I [(a)–(d)] and Case II [(e)–(h)] with a laser power of $P_{\rm in} = 1000P_{\rm cr}$, where the displayed space scale is 1.2 × 1.2 cm in Case I and 1.0 × 1.0 cm in Case II (scale of the simulation box is 3 × 3 cm).

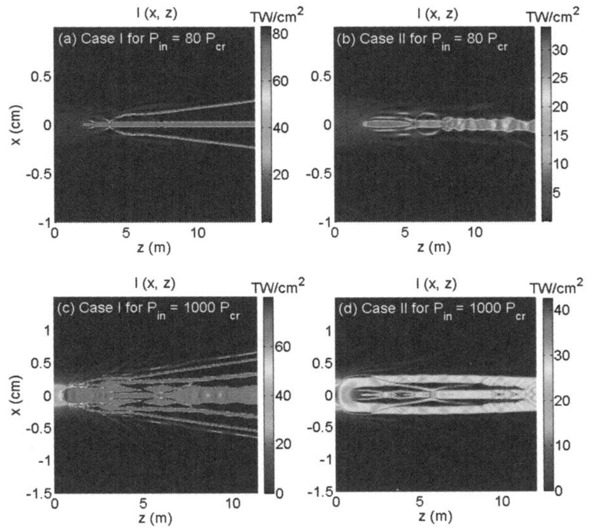

FIG. 18. (Color online) The laser intensity profile $I(x,0,z)$ along the propagation direction at two cases; the top row is for $P_{\rm in} = 80P_{\rm cr}$ and the bottom row is for $P_{\rm in} = 1000P_{\rm cr}$.

data (as shown in Tables I and II). To observe these different filamentation patterns experimentally, they possibly have to follow the specific values of parameters and experimental conditions as discussed in Ref. [25].

The difference in the results for the two cases considered here originates from the characteristics of the modulational instability. The modulational instability growth rate of a plane wave given by Eq. (1) can be expressed as $\Gamma = k\sqrt{f'(I_0)I_0 - k^2/(4k_0^2)}$, where $f(I_0) = k_0\Delta n(I_0) - \gamma I_0^K$ and I_0 is the incident laser intensity. The corresponding wave number of the most unstable mode is $k_{\max} = k_0\sqrt{2f'(I_0)I_0}$, and it satisfies the relation $k_{\max} = \sqrt{2k_0\Gamma_{\max}}$. In view of the curve for $k_{\max}$ in Fig. 12(b), it can be seen that the wave number of modulational instability in Case I is much larger than that in Case II. That is, in Cases I and II the laser propagation suffers mainly from short- and long-wavelength modulation, which leads to a smaller and larger filament size, respectively, as shown in Figs. 14 and 15. As a result, the short-wavelength modulation in Case I leads to formation of small-scale filaments, and the long-wavelength modulation in Case II leads to a thick optical bullet. Figure 12(a) also shows that neglect of the high-order plasma saturation term in Case II does not significantly affect the beam dynamics, demonstrating that the higher-order Kerr nonlinear effect would become important in the context of our specific parameters. This is expected, since in our parameter range the peak laser intensity always remains below 4×10^{13} W/cm^2 in Case II, but in Case I plasma defocusing can counteract Kerr focusing [101] for $I \geqslant (\frac{n_2k_0}{\sqrt{3K\gamma}})^{1/(K-1)} \approx 5.4 \times 10^{13}$ W/cm^2. In addition, our estimations here is qualitatively consistent with the parameter range demonstrated in Refs. [40,41,45,46], where it was concluded that the filamentation process is dominated by plasma for long pulses at short wavelengths and by HOKEs for short pulses at long wavelengths. Finally, we should give a brief discussion on the effect of noise. In our further simulations, we have added a small noise level in Eq. (11). It is found that the filament patterns, as shown in Figs. (14), (15), and 17, can be kept when the noise level is very small. In particular, for very high powers ($\geqslant 600P_{\rm cr}$), the filament patterns shown in Fig. 17 are quite robust, even when subjected to a high noise level.

VI. SUMMARY AND DISCUSSION

In summary, we have systematically investigated the nonlinear pattern dynamics and multiple filamentation of terawatt laser pulses in air by using a generalized saturable nonlinear Schrödinger equation including a high-order plasma saturation term without and with HOKEs. In the one-dimensional case, the latter is investigated in detail for O_2 and N_2. Our theoretical and numerical results show that the Hamiltonian perturbation induced by the plasma and/or higher-order Kerr nonlinearity can destroy the coherent propagation of the laser and result in spatially complex incoherent structures. In the context of the specific parameters used here, it is found that higher-order Kerr nonlinearity can lead to quite different behaviors of the pattern dynamics in both one- and two-dimensional spaces, where turbulent small-scale spatial structures rapidly appear as the laser propagates in air. In particular, in the one-dimensional case, for N_2 with our parameter choice, the higher-order Kerr nonlinearity cannot always provide the negative contribution to the refractive index in the regime $4.49 \times 10^{12} < I_0(\mathrm{W/cm^2}) < 2.74 \times 10^{13}$ that is needed to balance the KF. As a result, in the regime $4.49 \times 10^{12} < I_0(\mathrm{W/cm^2}) < 1.91 \times 10^{13}$, where the focusing term $\beta_6|E|^6$ in HOKE gives a large perturbation term and the system becomes rather unstable (a large $\Gamma_{\max}$), a typical chaotic behavior that is associated with a continuous phase shift and complicated spatial structures is shown and its phase-space trajectory corresponds to the formation of irregular HMO crossings. Our Fourier mode analyses further confirm that the spatially complicated structures are associated with the stochastic evolution of energy in higher-order, or shorter-wavelength, modes. For the two-dimensional case, our numerical results demonstrate that the oscillation filamentation, chaotic evolution, spatial complicated patterns, and pseudorecurrence can appear with increase of the saturable nonlinear effects in Case I where the Kerr terms are truncated to n_2. We find that the filamentation corresponds to concentrations of energy in shorter-wavelength diagonal modes, and the pseudorecurrence phenomenon occurs only when the saturation effects are strong enough.

Furthermore, our results on the multiple filamentation shows that in our parameter range for laser pulses at moderate power levels, plasma saturation effects can lead to production of more small-scale filament structures, but higher-order Kerr nonlinearity can lead to different filamentation patterns and produce a thick optical bullet. At much higher powers, a typical optical channel, capable of conveying a high-power laser for a long distance, can be generated if the higher-order Kerr nonlinearity is included. Such a different behavior induced by the latter can enhance the long-distance propagation of the laser. Because a wave beam with a small perturbation can be developed into very rich localized pattern structures by the modulational instability, such local structures may be the early form of filament formation. The formation of different nonlinear structures of laser fields propagating in air would eventually evolve into complex filamentation patterns, as shown in our results. Their corresponding nonuniformities in the beam profile can be attributed to the stochastic evolution of the shorter-wavelength modes (or higher-order Fourier modes), where strong mode-mode interactions triggered by the nonlinear saturable effects may enlarge the local inhomogeneities on the beam profile and modulational instability can broaden the frequency spectrum of the incident laser. In particular, for self-focusing, the energy spectrum of the laser should experience a transition from the long-wavelength modes to the shorter-wavelength ones. For different nonlinear effects, different pattern structures and filamentation phenomena of femtosecond terawatt laser pulses in air therefore can appear due to the nonlinear evolution of these localized structures by the modulational instability.

ACKNOWLEDGMENTS

This work is supported by the National Natural Science Foundation of China (Grants No. 91230205, No. 10974022, and No. 10835003), the National High-Tech 863 Project, and the National Basic Research 973 Project (Grant No. 2013CB834100). T.W.H. thanks S. D. Yao, C. Z. Xiao, Q. Jia, D. Wu, B. Liu, S. Zhao, W. W. Wang, H. Z. Fu, S. Z. Wu, and F. L. Zheng for useful discussions.

[1] B. L. Fontaine, F. Vidal, Z. Jiang, C. Y. Chien, D. Comtois, A. Desparois, T. W. Johnston, J. C. Kieffer, H. Pépin, and H. P. Mercure, Phys. Plasmas **6**, 1615 (1999).
[2] P. Rairoux, H. Schillinger, S. Niedermeier, M. Rodriguez, F. Ronneberger, R. Sauerbrey, B. Stein, D. Waite, C. Wedekind, H. Wille, L. Wöste, and C. Ziener, Appl. Phys. B **71**, 573 (2000).
[3] A. Biswas, Appl. Maths. Comput. **153**, 369 (2004); Chaos Solitons Fractals **14**, 673 (2002).
[4] L. Bergé and S. Skupin, Phys. Rev. Lett. **100**, 113902 (2008).
[5] X. M. Zhao, J.-C. Diels, C. Y. Wang, and J. M. Elizondo, IEEE J. Quantum Electron. **31**, 599 (1995).
[6] L. Wöste, C. Wedekind, H. Wille, P. Rairoux, B. Stein, S. Nikolov, C. Werner, S. Niedermeier, F. Ronneberger, H. Schillinger, and R. Sauerbrey, Laser Optoelektron. **29**, 51 (1997).
[7] A. Couairon, Eur. Phys. J. D **27**, 159 (2003).
[8] A. Couairon, J. Biegert, C. P. Haurt, W. Kornells, F. W. Helbing, U. Keller, and A. Mysyrowicz, J. Mod. Opt. **53**, 75 (2006).
[9] C. P. Hauri, W. Kornelis, F. W. Helbing, A. Heinrich, A. Couairon, A. Mysyrowicz, J. Biegert, and U. Keller, Appl. Phys. B **79**, 673 (2004).
[10] J. R. Peñano, P. Sprangle, B. Hafizi, A. Ting, D. F. Gordon, and C. A. Kapetanakos, Phys. Plasmas **11**, 2865 (2004).
[11] I. Alexeev, A. Ting, D. F. Gordon, E. Briscoe, J. R. Penano, R. F. Hubbard, and P. SprangleAppl. Phys. Lett. **84**, 4080 (2004).
[12] A. Braun, G. Korn, X. Liu, D. Du, J. Squier, and G. Mourou, Opt. Lett. **20**, 73 (1995).
[13] J. Kasparian, J. Wolf, Y. B. André, G. Méchain, G. Méjean, B. Prade, P. Rohwetter, E. Salmon, K. Stelmaszczyk, J. Yu, A. Mysyrowicz, R. Sauerbrey, L. Woeste, and J. P. Wolf, Opt. Express **16**, 466 (2008).
[14] J. Kasparian, M. Rodriguez, G. Mejean, J. Yu, E. Salmon, H. Wille, R. Bourayou, S. Frey, Y.-B. André, A. Mysyrowicz,

R. Sauerbrey, J. P. Wolf, and L. Wöste, Science **301**, 61 (2003).
[15] E. T. J. Nibbering, P. F. Curley, G. Grillon, B. S. Prade, M. A. Franco, F. Salin, and A. Mysyrowicz, Opt. Lett. **21**, 62 (1996).
[16] H. Yang, J. Zhang, J. Zhang, L. Z. Zhao, Y. J. Li, H. Teng, Y. T. Li, Z. H. Wang, Z. L. Chen, Z. Y. Wei, J. X. Ma, W. Yu, and Z. M. Sheng, Phys. Rev. E **67**, 015401 (2003).
[17] J. M. Dai, X. Xie, and X. C. Zhang, Phys. Rev. Lett. **97**, 103903 (2006).
[18] H. Zhong, N. Karpowicz, and X. C. Zhang, Appl. Phys. Lett. **88**, 261103 (2006).
[19] C. D'Amico, A. Houard, M. Franco, B. Prade, A. Mysyrowicz, A. Couairon, and V. T. Tikhonchuk, Phys. Rev. Lett. **98**, 235002 (2007).
[20] C. D'Amico, A. Houard, S. Akturk, Y. Liu, J. L. Bloas, M. Franco, B. Prade, A. Couairon, V. T. Tikhonchuk, and A. Mysyrowicz, New J. Phys. **10**, 013015 (2008).
[21] L. V. Keldysh, Zh. Eksp. Teor. Fiz. **47**, 1945 (1964) [Sov. Phys. JETP **20**, 1307 (1965)].
[22] V. I. Bespalov and V. I. Talanov, Zh. Eksp. Teor. Fiz. Pis'ma Red. **3**, 471 (1966) [JETP Lett. **3**, 307 (1966)].
[23] M. Mlejnek, M. Kolesik, J. V. Moloney, and E. M. Wright, Phys. Rev. Lett. **83**, 2938 (1999).
[24] T. T. Xi, X. Lu, and J. Zhang, Phys. Rev. Lett. **96**, 025003 (2006).
[25] V. Loriot, E. Hertz, O. Faucher, and B. Lavorel, Opt. Express **17**, 13429 (2009); **18**, 3011 (2010).
[26] A. Couairon, Phys. Rev. A **68**, 015801 (2003).
[27] N. Aközbek, C. M. Bowden, A. Talebpour, and S. L. Chin, Phys. Rev. E **61**, 4540 (2000).
[28] N. Aközbek, M. Scalora, C. M. Bowden, and S. L. Chin, Opt. Commun. **191**, 353 (2001).
[29] M. Kolesik, D. Mirell, J.-C. Diels, and J. V. Moloney, Opt. Lett. **36**, 3685 (2010).
[30] Y.-H. Chen, S. Varma, T. M. Antonsen, and H. M. Milchberg, Phys. Rev. Lett. **105**, 215005 (2010).
[31] J. K. Wahlstrand, Y.-H. Cheng, Y.-H. Chen, and H. M. Milchberg, Phys. Rev. Lett. **107**, 103901 (2011).
[32] P. Polynkin, M. Kolesik, E. M. Wright, and J. V. Moloney, Phys. Rev. Lett. **106**, 153902 (2011).
[33] J. M. Brown, E. M. Wright, J. V. Moloney, and M. Kolesik, Opt. Lett. **37**, 1604 (2012).
[34] P. Béjot, J. Kasparian, S. Henin, V. Loriot, T. Vieillard, E. Hertz, O. Faucher, B. Lavorel, and J. P. Wolf, Phys. Rev. Lett. **104**, 103903 (2010).
[35] M. Petrarca, Y. Petit, S. Henin, R. Delagrange, P. Béjot, and J. Kasparian, Opt. Lett. **37**, 4347 (2012).
[36] A. Couairon and A. Mysyrowicz, Phys. Rep. **441**, 47 (2007).
[37] S. L. Chin, S. A. Hosseini, W. Liu, Q. Luo, F. Théberge, N. Aközbek, A. Becker, V. P. Kandidov, O. G. Kosareva, and H. Schroeder, Can. J. Phys. **83**, 863 (2005).
[38] S. Tzortzakis, M. A. Franco, Y.-B. Andre, A. Chiron, B. Lamouroux, B. S. Prade, and A. Mysyrowicz, Phys. Rev. E **60**, 3505(R) (1999).
[39] O. Kosareva, J. Daigle, N. Panov, T. J. Wang, S. Hosseini, S. Yuan, G. Roy, V. Makarov, and S. L. Chin, Opt. Lett. **36**, 1035 (2011).
[40] V. Loriot, P. Béjot, W. Ettoumi, Y. Petit, J. Kasparian, S. Henin, E. Hertz, B. Lavorel, O. Faucher, and J. Wolf, Laser Phys. **21**, 1319 (2011).
[41] P. Béjot, E. Hertz, J. Kasparian, B. Lavorel, J. P. Wolf, and O. Faucher, Phys. Rev. Lett. **106**, 243902 (2011).
[42] H. T. Wang, C. Y. Fan, P. F. Zhang, C. H. Qiao, J. H. Zhang, and H. M. Ma, Opt. Express **18**, 24301 (2010).
[43] H. T. Wang, C. Y. Fan, P. F. Zhang, C. H. Qiao, J. H. Zhang, and H. M. Ma, J. Opt. Soc. Am. B **28**, 2081 (2011).
[44] W. Ettoumi, P. Béjot, Y. Petit, V. Loriot, E. Hertz, O. Faucher, B. Lavorel, J. Kasparian, and J. P. Wolf, Phys. Rev. A **82**, 033826 (2010).
[45] H. T. Wang, C. Y. Fan, H. Shen, P. F. Zhang, and C. H. Qiao, Opt. Commun. **293**, 113 (2012).
[46] C. Brée, A. Demircan, and G. Steinmeyer, Phys. Rev. A **85**, 033806 (2012).
[47] C. T. Zhou, X. T. He, and T. X. Cai, Phys. Rev. E **50**, 4136 (1994); C. T. Zhou and X. T. He, *ibid.* **49**, 4417 (1994); X. T. He and C. T. Zhou, J. Phys. A **26**, 4123 (1993).
[48] S. Skupin, L. Bergé, U. Peschel, F. Lederer, G. Méjean, J. Yu, J. Kasparian, E. Salmon, J. P. Wolf, M. Rodriguez, L. Wöste, R. Bourayou, and R. Sauerbrey, Phys. Rev. E **70**, 046602 (2004).
[49] B. Qiao, C. T. Zhou, and C. H. Lai, Appl. Phys. Lett. **91**, 221114 (2007).
[50] M. Mlejnek, E. M. Wright, and J. V. Moloney, Opt. Lett. **23**, 382 (1998).
[51] A. Couairon, and L. Bergé, Phys. Plasmas **7**, 193 (2000).
[52] D. Novoa, H. Michinel, and D. Tommasini, Phys. Rev. Lett. **105**, 203904 (2010).
[53] A. Couairon, S. Tzortzakis, L. Bergé, M. Franco, B. Prade, and A. Mysyrowicz, J. Opt. Soc. Am. B **19**, 1117 (2002).
[54] G. Fibich, S. Eisenmann, B. Ilan, and A. Zigler, Opt. Lett. **29**, 1772 (2004).
[55] A. Dubietis, G. Tamošauskas, G. Fibich, and B. Ilan, Opt. Lett. **29**, 1126 (2004).
[56] M. Centurion, Y. Pu, M. Tsang, and D. Psaltis, Phys. Rev. A **71**, 063811 (2005).
[57] A. Houard, M. Franco, B. Prade, A. Durécu, L. Lombard, P. Bourdon, O. Vasseur, B. Fleury, C. Robert, V. Michau, A. Couairon, and A. Mysyrowicz, Phys. Rev. A **78**, 033804 (2008).
[58] G. Méchain, A. Couairon, M. Franco, B. Prade, and A. Mysyrowicz, Phys. Rev. Lett. **93**, 035003 (2004).
[59] W. Liu, S. A. Hosseini, Q. Luo, B. Ferland, S. L. Chin, O. G. Kosareva, N. A. Panov, and V. P. Kandidov, New J. Phys. **6**, 6 (2004).
[60] P. P. Kiran, S. Bagchi, S. R. Krishnan, C. L. Arnold, G. R. Kumar, and A. Couairon, Phys. Rev. A **82**, 013805 (2010).
[61] S. A. Hosseini, Q. Luo, B. Ferland, W. Liu, S. L. Chin, O. G. Kosareva, N. A. Panov, N. Aközbek, and V. P. Kandidov, Phys. Rev. A **70**, 033802 (2004).
[62] N. A. Panov, O. G. Kosareva, V. P. Kandidov, N. Aköbek, M. Scalora, and S. L. Chin, Quantum Electron. **37**, 1153 (2007).
[63] A. J. Lichtenberg and M. A. Lieberman, *Regular and Stochastic Motion* (Springer-Verlag, New York, 1983).
[64] E. Ott, *Chaos in Dynamical Systems* (Cambridge University Press, Cambridge, 1993).
[65] C. T. Zhou, C. H. Lai, and M. Y. Yu, Phys. Scr. **55**, 394 (1997); J. Maths. Phys. **38**, 5225 (1997); C. T. Zhou and X. T. He, Phys. Scr. **50**, 415 (1994).
[66] C. Sulem and P. L. Sulem, *The Nonliear Schrödinger Equation: Self-Focusing and Wave Collapse* (Springer, New York, 1999).

[67] P. A. Robinson, Rev. Mod. Phys. **69**, 507 (1997).
[68] G. Fibich and B. Ilan, Opt. Lett. **26**, 840 (2001).
[69] T. W. Huang, C. T. Zhou, and X. T. He, AIP Adv. **2**, 042190 (2012).
[70] C. T. Zhou, X. T. He, and S. G. Chen, Phys. Rev. A **46**, 2277 (1992).
[71] H. T. Moon, Phys. Rev. Lett. **64**, 412 (1990).
[72] V. E. Zakharov and A. B. Shabat, Zh. Eksp. Teor. Fiz. **61**, 118 (1971) [Sov. Phys. JETP **34**, 62 (1972)].
[73] R. K. Bullough and P. J. Caudrey, *Soliton* (Springer-Verlag, Berlin, 1980).
[74] T. B. Benjamin and J. E. Feir, J. Fluid Mech. **27**, 417 (1967).
[75] B. Qiao, C. H. Lai, C. T. Zhou, X. T. He, X. G. Wang, and M. Y. Yu, Phys. Plasmas **14**, 112301 (2007).
[76] T. Höuser, W. Scheid, and H. Hora, J. Opt. Soc. Am. B **5**, 2029 (1988).
[77] Y. Tan and J. M. Mao, Phys. Rev. E **57**, 381 (1998); J. Phys. A **33**, 9119 (2000).
[78] C. Ren and W. B. Mori, Phys. Plasmas **8**, 3118 (2001).
[79] J. J. Rasmussen and K. Rypdal, Phys. Scr. **33**, 481 (1986).
[80] D. W. McLaughlin, G. C. Papanicolaou, C. Sulem, and P. L. Sulem, Phys. Rev. A **34**, 1200 (1986).
[81] S. N. Vlasov, L. V. Piskunova, and V. I. Talanov, Zh. Eksp. Teor. Fiz. **95**, 1945 (1989) [Sov. Phys. JETP **68**, 1125 (1989)].
[82] N. N. Akhmediev, D. R. Heatley, G. I. Stegeman, and E. M. Wright, Phys. Rev. Lett. **65**, 1423 (1990).
[83] L. Bergé, Phys. Rev. E **62**, R3071 (2000).
[84] W. Liu, S. L. Chin, O. Kosareva, I. S. Golubtsov, and V. P. Kandidov, Opt. Commun. **225**, 193 (2003)
[85] L. Bergé, C. Gouédard, J. S. Eriksen, and H. Ward, Physica D **176**, 181 (2003).
[86] C. T. Zhou, M. Y. Yu, and X. T. He, Phys. Rev. E **73**, 026209 (2006).
[87] C. T. Zhou, L. Y. Chew, and X. T. He, Appl. Phys. Lett. **97**, 051502 (2010); C. T. Zhou, X. G. Wang, S. Z. Wu, H. B. Cai, F. Wang, and X. T. He, *ibid.* **97**, 201502 (2010); C. T. Zhou, X. G. Wang, and X. T. He, Phys. Plasmas **17**, 083103 (2010).
[88] P. Béjot, B. Kibler, E. Hertz, B. Lavorel, and O. Faucher, Phys. Rev. A **83**, 013830 (2011).
[89] J. Kasparian, R. Sauerbrey, and S. L. Chin, Appl. Phys. B **71**, 877 (2000).
[90] A. Becker, N. Aközbek, K. Vijayalakshmi, C. M. Bowden, and S. L. Chin, Appl. Phys. B **73**, 287 (2001).
[91] W. Liu, S. Petit, A. Becker, N. Aközbek, C. M. Bowden, and S. L. Chin, Opt. Commun. **202**, 189 (2002).
[92] F. Théberge, W. W. Liu, P. Tr. Simard, A. Becker, and S. L. Chin, Phys. Rev. E **74**, 036406 (2006).
[93] M. B. Gaarde and A. Couairon, Phys. Rev. Lett. **103**, 043901 (2009).
[94] D. S. Steingrube, E. Schulz, T. Binhammer, M. B. Gaarde, A. Couairon, U. Morgner and M. Kovačev, New. J. Phys. **13**, 043022 (2011).
[95] P. P. Kiran, S. Bagchi, C. L. Arnold, S. R. Krishnan, G. R. Kumar, and A. Couairon, Opt. Express **18**, 21504 (2010).
[96] K. D. Moll, A. L. Gaeta, and G. Fibich, Phys. Rev. Lett. **90**, 203902 (2003).
[97] T. D. Grow, A. A. Ishaaya, L. T. Vuong, A. L. Gaeta, N. Gavish, and G. Fibich, Opt. Express **14**, 5468 (2006).
[98] S. L. Chin, N. Aközbek, A. Proulx, S. Petit, and C. M. Bowden, Opt. Commun. **188**, 181 (2001).
[99] S. L. Chin, S. Petit, W. Liu, A. Iwasaki, M. C. Nadeau, V. P. Kandidov, O. G. Kosareva, and K. Y. Andrianov, Opt. Commun. **210**, 329 (2002).
[100] K. Konno and H. Suzuki, Phys. Scr. **20**, 382 (1979).
[101] S. Tzortzakis, L. Bergé, A. Couairon, M. Franco, B. Prade, and A. Mysyrowicz, Phys. Rev. Lett. **86**, 5470 (2001).
[102] G. Méchain, A. Couairon, Y.-B. André, C. D'Amico, M. Franco, F. B. Prade, S. Tzortzakis, A. Mysyrowicz, and R. Sauerbrey, Appl. Phys. B **79**, 379 (2004)
[103] A. Pukhov and J. Meyer-ter-Vehn, Phys. Rev. Lett. **76**, 3975 (1996).
[104] Y. Y. Ma, X. Lu, T. T. Xi, Q. H. Gong, and J. Zhang, Appl. Phys. B **93**, 463 (2008).
[105] M. Segev, Opt. Quantum Electron. **30**, 503 (1998).
[106] Y. H. Liu, L. Y. Chew, and M. Y. Yu, Phys. Rev. E **78**, 066405 (2008).

等离子体自生磁场

自生磁场是物质内在因素相互作用、自发产生电流驱动的准静态磁场，对物质演化和天体自然现象的发生有着重要作用。早在1983年，贺贤土院士首次提出，在强激光与物质作用时，激光通道内有质动力驱动强非线性电流，会产生较强的准静态磁场[quasi-static magnetic（QSM）field]。他从非相对论动理论Vlasov-Maxwell方程导得了圆偏振（circularly polarized，CP）激光作用下的轴向自生磁场，并解释了激光通道内电子加速过程。

2000年前后，贺贤土院士又将研究拓展到相对论超强激光与等离子体作用通道中产生的上亿高斯QSM，并指导学生由相对论动理论分布函数方程自洽和统一地导得了圆偏振和线偏振（linearly polarized，LP）激光与等离子体作用下的轴向和角向QSM方程，并给出了其与激光强度的定标关系。数值模拟和国际同行的实验都证实了这样的强自生磁场的存在和该理论模型的正确性。与此同时，贺贤土院士与其团队成员将自生磁场理论应用到相对论电子加速和高能量密度等离子体磁重联及相对论磁重联现象的研究中，为解释一些极端天体物理现象做出了探索性研究。

贺贤土院士发展的一套系统的动理学QSM理论，对了解激光等离子体非线性作用下准静态磁场产生及其应用具有重要意义，也为解释天体物理现象提供了启发，相关研究结果在国际上被大量引用。

第32卷 第3期 物理学报 Vol. 32, No. 3
1983年3月 ACTA PHYSICA SINICA Mar., 1983

等离子体波-粒子非线性作用的有质动力和自生磁场效应

贺贤土
(中国科学院原子能研究所)

1982年2月13日收到

提　　要

本文从 Vlasov-Maxwell 方程组出发，用自洽场方法首先建立了低频振荡粒子分布函数、密度与高频电场、低频磁场在 Fourier 表象中的耦合关系，然后对低频线性介电函数作各种近似展开，在时空表象中得到了包括磁场效应、有质动力和 Landau 阻尼的一套非线性作用方程组最后还给出了 Lagrangian 密度和守恒量，并简单地讨论了磁场能否促进三维 Soliton 形成问题.

一、引　　言

本文是文献[1]的继续. 在文献[1]中，我们只考虑了低频场的离子声波效应，从而高频场方程中的非线性高频电流内只包含了高频场力与粒子低频振荡的作用部分；至于低频场力与粒子高频振荡的相互作用项，由于因子 $\boldsymbol{k}\cdot\boldsymbol{v}/\omega_{pe}\ll 1$ 被略去，因而高频场的变化只取决于通常的有质动力效应. 但是当低频场力包含横波时，正如本文所表明的那样，其所激发的非线性电流的有旋分量可以导致等离子体中磁场的产生. 这种自生磁场的作用虽然包含有 β_e^2 因子，但在电子温度较高的系统中，它的效应与有质动力相比是不能忽略的.

早在1971年时，激光等离子体中自生磁场效应就被发现[2]，Chase 等人[3]和 Stamper, Tidman[4] 从宏观磁流体力学方程出发研究了这个问题，他们指出：当粒子流体密度和温度的梯度方向不平行时，或者有较大的辐射压时都可产生强磁场. 但是，实际上等离子体中自生强磁场还可由我们这里讨论的两个或两个以上的强高频电场不平行传播的非线性无碰撞作用来产生. 这样的高频场非线性拍频成为低频时就可导致低频横电场的激发，从而由感应定律就导致低频磁场的形成. 近几年来，由于激光等离子体、天体等离子体工作的发展以及与它们有关的强湍流理论的研究，这种自生磁场效应开始引起注意，Kono 等人[5]就计算过这种磁场的形式. 最近 Haar 和 Tsytovich[6] 又进一步在双流体力学理论框架下把它应用于天体等离子体由调制不稳定性激发的强湍流研究中. 本文的目的便是从 Vlasov 方程和电磁场的 Maxwell 方程组出发，利用我们在文献[1]中发展的方法，在二级场近似下，仔细地导得包含有质动力和上述自生磁场效应的波和粒子相互作用方程组. 我们在文献[1]中也已提到，从粒子速度分布函数耦合电磁场的自洽场方法去导得上述方程组是十分必要的，这不仅是在推导中考虑了细致的微观过程，所作的近似要比流体力学方法明确得多，而且可以确切得到各种近似下包括 Landau 阻尼效应的方程组形式．最后我们还将给出包含自生磁场效应的无阻尼方程组的拉氏密度和守恒量．

二、Fourier 表象中粒子与波的非线性耦合关系

把粒子分布函数、电磁场量和非线性电流分成高频和低频振荡两部分，引入 F 变换：$(k)=\int(\boldsymbol{x},t)e^{i\omega t-i\boldsymbol{k}\cdot\boldsymbol{x}}d^2x\,dt$，这里 $k\equiv\{\boldsymbol{k},\omega\}$，从 Vlasov-Maxwell 方程组就得到 F 表象中诸量表达式如下：

粒子高频分布：

$$
\begin{aligned}
f_\alpha(k)\equiv f_\alpha(k,\boldsymbol{v})=\frac{e_\alpha}{im_\alpha(\omega-\boldsymbol{k}\cdot\boldsymbol{v})}\Big\{&\boldsymbol{G}_h(k)\cdot\frac{\partial F'_{0\alpha}}{\partial\boldsymbol{v}}\\
&+\int\boldsymbol{G}_h(k_1)\cdot\frac{\partial\delta F_\alpha(q)}{\partial\boldsymbol{v}}\delta^4(k-k_1-q)\frac{dk_1dq}{(2\pi)^4}\\
&+\int\boldsymbol{G}_d(q)\cdot\frac{\partial f_\alpha(k_1)}{\partial\boldsymbol{v}}\delta^4(k-k_1-q)\frac{dk_1dq}{(2\pi)^4}\Big\}.
\end{aligned}\tag{1}
$$

与文献[1]不一样，在(1)式中我们保留了低频场力与粒子高频振荡作用的第三项，以后将会看到它就是我们现在感兴趣的高频电流激发低频磁场有关的项．

粒子低频分布扰动部分：

$$
\begin{aligned}
\delta F_\alpha(q)\equiv\delta F_\alpha(q,\boldsymbol{v})=\frac{e_\alpha}{im_\alpha(\Omega-\boldsymbol{q}\cdot\boldsymbol{v})}\Big\{&\boldsymbol{G}_d(q)\cdot\frac{\partial F'_{0\alpha}}{\partial\boldsymbol{v}}\\
&+\int_{(d)}\boldsymbol{G}_h(k_1)\cdot\frac{\partial f_\alpha(k_2)}{\partial\boldsymbol{v}}\delta^4(q-k_1-k_2)\frac{dk_1dk_2}{(2\pi)^4}\Big\}.
\end{aligned}\tag{2}
$$

我们只感兴趣频率接近等离子体频率的高频场拍频作用和二级场拍频效应，所以(1)式中略去了高频场向更高频的拍频作用项，(2)式中略去了两个低频场向更低频拍频作用项．(1)，(2)式中积分号 $\int_{(d)}$ 表示积分只作用于两个高频场向低频波拍频过程，脚标"h"和"d"分别为高频、低频记号．其余表示：$dk\equiv d^3k\,d\omega$，$\delta^4(k)=\delta(\omega)\delta^3(\boldsymbol{k})$，$F'_{0\alpha}=n_0F_{0\alpha}(\boldsymbol{v})$，以后 $F_{0\alpha}$ 取为 Maxwell 分布，满足 $\int F_{0\alpha}(\boldsymbol{v})d^3v=1$，$n_0=n_{0\alpha}$，$\alpha=i,e$．(1)，(2)式中低频场力

$$
\boldsymbol{G}_d(q)=\sum_{\sigma=l,t}\boldsymbol{e}_{\boldsymbol{q}}^{\sigma\prime}E_d^\sigma(q).\tag{3}
$$

这里 $\sigma=l$（纵波），$\sigma=t$（横波）；$E_d^\sigma(q)=\boldsymbol{e}_{\boldsymbol{q}}^\sigma\cdot\boldsymbol{E}_d^\sigma(q)$ 为 σ 波低频电场标量强度，对横波 $\boldsymbol{E}_d^t$ 已作了两个偏振方向求和；$\boldsymbol{e}_{\boldsymbol{q}}^\sigma$ 为单位偏振矢量，$\boldsymbol{e}_{\boldsymbol{q}}^\sigma\cdot\boldsymbol{e}_{\boldsymbol{q}}^\sigma=1$，我们约定 $\boldsymbol{e}_{\boldsymbol{q}}^\sigma=-\boldsymbol{e}_{-\boldsymbol{q}}^\sigma$，$E_d^\sigma(-q)=-E_d^\sigma(q)^*$，"$*$"表示复数共轭．(3) 式中

$$
\boldsymbol{e}_{\boldsymbol{q}}^{\sigma\prime}=\left(1-\frac{\boldsymbol{q}\cdot\boldsymbol{v}}{\Omega}\right)\boldsymbol{e}_{\boldsymbol{q}}^\sigma+\frac{\boldsymbol{e}_{\boldsymbol{q}}^\sigma\cdot\boldsymbol{v}}{\Omega}\boldsymbol{q}.\tag{4}
$$

显然 $\boldsymbol{e}_{\boldsymbol{q}}^{\sigma\prime}\cdot\boldsymbol{v}=\boldsymbol{e}_{\boldsymbol{q}}^\sigma\cdot\boldsymbol{v}$．当磁场不存在时，$\boldsymbol{G}_d$ 就化为通常的 Poisson 场力．关于高频场力

$\boldsymbol{G}_h(k)$ 有类似于 (3), (4) 式的形式，这里不再说明.

高频和低频标量场 $E_h(k)$ 和 $E_d(q)$ 分别满足方程

$$\epsilon_h^\sigma(k)E_h^\sigma(k_1) = -\frac{4\pi}{\omega}\sum_\alpha \frac{e_\alpha^2}{m_\alpha}\int \frac{\boldsymbol{e_k}\cdot\boldsymbol{v}}{\omega-\boldsymbol{k}\cdot\boldsymbol{v}}\left[\sum_{\sigma_1}\boldsymbol{e}_1'\cdot\frac{\partial\delta F_\alpha(q)}{\partial\boldsymbol{v}}E_h^{\sigma_1}(k_1)\right.$$
$$\left.+\sum_{\sigma_d}\boldsymbol{e}_d'\cdot\frac{\partial f_\alpha(k_1)}{\partial\boldsymbol{v}}E_d^{\sigma_d}(q)\right]\delta^4(k-k_1-q)\frac{dk_1dq}{(2\pi)^4}d^3v. \tag{5}$$

$$\epsilon_d^\sigma(q)E_d^\sigma(q) = -\frac{4\pi}{\Omega}\sum_{\alpha\sigma_1}\frac{e_\alpha^2}{m_\alpha}\int_{(d)}\frac{\boldsymbol{e}_d\cdot\boldsymbol{v}}{\Omega-\boldsymbol{q}\cdot\boldsymbol{v}}\boldsymbol{e}_1'\cdot\frac{\partial f_\alpha(k_2)}{\partial\boldsymbol{v}}$$
$$\times E_h^{\sigma_1}(k_1)\delta^4(q-k_1-k_2)\frac{dk_1dk_2}{(2\pi)^4}d^3v, \tag{6}$$

其中 $\epsilon_h^\sigma(k)$, $\epsilon_d^\sigma(q)$ 分别为高频和低频线性介电函数；因子 $(\omega-\boldsymbol{k}\cdot\boldsymbol{v})^{-1}$ 在速度积分时一律指 $(\omega-\boldsymbol{k}\cdot\boldsymbol{v}+i0_+)^{-1}$ 形式，积分线路沿实轴上岸向下绕过极点；对于因子 $(\Omega-\boldsymbol{q}\cdot\boldsymbol{v})^{-1}$ 也同样. 为了简便，(5)，(6)式还应用了简化表示 $\boldsymbol{e}_1\equiv\boldsymbol{e}_{k_1}^{\sigma_1}$, $\boldsymbol{e}_d\equiv\boldsymbol{e}_q^\sigma$，以下亦同样.

现在以 (1) 式中 $f_\alpha(k_2)$ 代入 (6) 式，取到二级场并略去离子对非线性电流的贡献，得

$$E_d^{\sigma_d}(q) = \frac{\omega_{pe}^2e}{im_e}\sum_{\sigma_1\sigma_2}\int_{(d)}\frac{H_e^{\sigma_d\sigma_1\sigma_2}(q,k_1,k_2)}{\Omega\epsilon_d^{\sigma_d}(q)}E_h^{\sigma_1}(k_1)E_h^{\sigma_2}(k_2)\delta^4(q-k_1-k_2)$$
$$\times\frac{dk_1dk_2}{(2\pi)^4}. \tag{7}$$

这里 $\omega_{pe}=\left(\frac{4\pi n_0e^2}{m_e}\right)^{1/2}$ 为电子等离子体频率而 (7) 式中激发非线性电流跃迁矩阵元

$$H_\alpha^{\sigma_d\sigma_1\sigma_2}(q,k_1,k_2)\equiv\frac{1}{2}\int\frac{\boldsymbol{e}_d\cdot\boldsymbol{v}}{\Omega-\boldsymbol{q}\cdot\boldsymbol{v}}\left[\boldsymbol{e}_1'\cdot\frac{\partial}{\partial\boldsymbol{v}}\frac{\boldsymbol{e}_2'\cdot\frac{\partial F_{0\alpha}}{\partial\boldsymbol{v}}}{\omega_2-\boldsymbol{k}_2\cdot\boldsymbol{v}}+1\longleftrightarrow2\right]d^3v. \tag{8}$$

上式对脚标 1, 2 对称，在 $(\omega-\boldsymbol{k}\cdot\boldsymbol{v})_{1,2}=0$ 无奇点条件下，无论 σ_d, σ_1, σ_2 为纵波和横波均可证明有下列性质:

$$\frac{1}{\Omega}H_\alpha^{\sigma_d\sigma_1\sigma_2}(q,k_1,k_2)=\frac{1}{\omega_2}H_\alpha^{\sigma_2\sigma_d\sigma_1}(k_2,q,-k_1)\quad \alpha=i,e. \tag{9}$$

从 (2) 和 (7) 式可见，低频场和粒子低频振荡分布函数最低阶为二级高频场效应.

注意到我们只感兴趣非线性拍频效应，高频波的线性 Landau 阻尼自然可略去，于是这部分迴路积分只保留了主值，所以无奇点的关系式 (9) 在以下讨论中总是能很好适用. 同时也由于我们研究的是非线性效应，高频波主值中 Doppler 频移部分可以取到 $o(\boldsymbol{k}\cdot\boldsymbol{v}/\omega)$ 项. 经这样考虑后，上述关系式可以明显地简化.

对 $\sigma=t$,

$$2H_e^{\sigma_2 t\sigma_1}(k_2,q,-k_1)=\int\frac{\boldsymbol{e}_2\cdot\boldsymbol{v}}{\omega_2-\boldsymbol{k}_2\cdot\boldsymbol{v}}\left[\boldsymbol{e}_d^{t\prime}\cdot\frac{\partial}{\partial\boldsymbol{v}}\frac{\boldsymbol{e}_1'\cdot\frac{\partial F_{0e}}{\partial\boldsymbol{v}}}{\omega_1-\boldsymbol{k}_1\cdot\boldsymbol{v}}-\boldsymbol{e}_1'\cdot\frac{\partial}{\partial\boldsymbol{v}}\frac{\boldsymbol{e}_d^{t\prime}\cdot\frac{\partial F_{0e}}{\partial\boldsymbol{v}}}{\Omega-\boldsymbol{q}\cdot\boldsymbol{v}}\right]d^3v$$
$$=\frac{1}{\omega_1\omega_2\Omega}\int\boldsymbol{e}_2\cdot\boldsymbol{v}\,[\boldsymbol{e}_d^t(\Omega-\boldsymbol{q}\cdot\boldsymbol{v})+\boldsymbol{e}_d^t\cdot\boldsymbol{v}\,\boldsymbol{q}]\cdot\frac{\partial}{\partial\boldsymbol{v}}\boldsymbol{e}_1\cdot\frac{\partial F_{0e}}{\partial\boldsymbol{v}}d^3v$$
$$+\frac{\boldsymbol{e}_1\cdot\boldsymbol{e}_2}{\omega_2\Omega}\int\left[(\Omega-\boldsymbol{q}\cdot\boldsymbol{v})\boldsymbol{e}_d^t\cdot\frac{\partial F_{0e}}{\partial\boldsymbol{v}}+\boldsymbol{e}_d^t\cdot\boldsymbol{v}\boldsymbol{q}\cdot\frac{\partial F_{0e}}{\partial\boldsymbol{v}}\right]$$

$$\times(\Omega-\boldsymbol{q}\cdot\boldsymbol{v})^{-1}d^3v+o\left(\frac{\boldsymbol{k}\cdot\boldsymbol{v}}{\omega}\right)_{1,2}. \tag{10}$$

因为

$$\int\frac{\boldsymbol{e}_d^t\cdot\boldsymbol{v}\boldsymbol{q}\cdot\dfrac{\partial F_{0e}}{\partial\boldsymbol{v}}}{\Omega-\boldsymbol{q}\cdot\boldsymbol{v}}d^3v=\int\frac{\boldsymbol{q}\cdot\boldsymbol{v}\boldsymbol{e}_d^t\cdot\dfrac{\partial F_{0e}}{\partial\boldsymbol{v}}}{\Omega-\boldsymbol{q}\cdot\boldsymbol{v}}d^3v=\Omega\int\frac{\boldsymbol{e}_d^t\cdot\dfrac{\partial F_{0e}}{\partial\boldsymbol{v}}}{\Omega-\boldsymbol{q}\cdot\boldsymbol{v}}d^3v=0.$$

这样，(10)式中第二个等号右端只剩下第一项，即有

$$2H_e^{\sigma_2 t\sigma_1}(k_2,q,-k_1)=\frac{1}{\omega_1\omega_2\Omega}\boldsymbol{e}_d^t\cdot(\boldsymbol{q}\times(\boldsymbol{e}_1\times\boldsymbol{e}_2))+o\left(\frac{\boldsymbol{k}\cdot\boldsymbol{v}}{\omega}\right)_{1,2}. \tag{11}$$

对 $\sigma=l$，文献[1]给出了

$$2H_e^{\sigma_2 l\sigma_1}(k_2,q,-k_1)=\frac{|\boldsymbol{q}|}{\omega_{pe}^2\omega_2}(\boldsymbol{e}_1\cdot\boldsymbol{e}_2)[\epsilon_e^l(q)-1]+o\left(\frac{\boldsymbol{k}\cdot\boldsymbol{v}}{\omega}\right)_{1,2}. \tag{12}$$

利用(11),(12)式可以得到低频场的进一步表达式. 注意到我们只研究两个高频场向低频场的拍频过程，所以求和号 $\sum\limits_{\sigma_1\sigma_2}$ 只有两种组合，于是(6)式两边乘上 $\boldsymbol{e}_d\cdot\boldsymbol{e}_d$ 后，对 $\sigma_d=t$，

$$\boldsymbol{E}_d^t(q)=\frac{e}{im_e\omega_{pe}\Omega\epsilon_d^t(q)}\sum_{\sigma_1^+\sigma_2^-}\int[\boldsymbol{q}\times(\boldsymbol{E}_h^{\sigma_1^+}(k_1)\times\boldsymbol{E}_h^{\sigma_2^-}(k_2)^*)]$$
$$\times\delta^4(q-k_1+k_2)\frac{dk_1dk_2}{(2\pi)^4}, \tag{13}$$

其中 $\sigma^\pm$ 分别表示四度动量为 $\pm k$ 的波，

$$\epsilon_d^t(q)=1-\frac{c^2q^2}{\Omega^2}+\sum_\alpha\frac{\omega_{p\alpha}^2}{\Omega}\int\frac{\boldsymbol{e}_d^t\cdot\boldsymbol{v}\boldsymbol{e}_d^{t'}\cdot\dfrac{\partial F_{0\alpha}}{\partial\boldsymbol{v}}}{\Omega-\boldsymbol{q}\cdot\boldsymbol{v}}d^3v$$
$$=1-\frac{c^2q^2}{\Omega^2}+\sum_\alpha\frac{\omega_{p\alpha}^2}{\Omega^2}\zeta_\alpha Z(\zeta_\alpha). \tag{14}$$

这里 $\zeta_\alpha=\dfrac{\Omega}{\sqrt{2}qv_{T_\alpha}}$，$v_{T_\alpha}$ 为粒子热速度，c 为光速，

$$Z(\zeta_\alpha)^{[7]}=\begin{cases}-2\zeta_\alpha+\dfrac{4}{3}\zeta_\alpha^3\cdots+i\sqrt{\pi}e^{-\zeta_\alpha^2} & \text{对 } |\zeta_\alpha|\ll1,\\ -\dfrac{1}{\zeta_\alpha}-\dfrac{1}{2\zeta_\alpha^3}-\cdots & \text{对 } |\zeta_\alpha|\gg1.\end{cases} \tag{15}$$

对 $\sigma_d=l$，

$$\boldsymbol{E}_d^l(q)=\frac{(\epsilon_e^l(q)-1)}{\epsilon_d^l(q)}\frac{e}{im_e\omega_{pe}^2}\boldsymbol{q}\sum_{\sigma_1^+\sigma_2^-}\int\boldsymbol{E}_h^{\sigma_1^+}(k_1)\cdot\boldsymbol{E}_h^{\sigma_2^-}(k_2)^*$$
$$\times\delta^4(q-k_1+k_2)\frac{dk_1dk_2}{(2\pi)^4}, \tag{16}$$

其中

$$\epsilon_\alpha^l(q) = 1 + \frac{\omega_{p\alpha}^2}{|\boldsymbol{q}|}\int \frac{\boldsymbol{e}_d^l \cdot \dfrac{\partial F_{0\alpha}}{\partial \boldsymbol{v}}}{\Omega - \boldsymbol{q}\cdot\boldsymbol{v}} d^3v, \tag{17}$$

$$\epsilon_d^l(q) = \sum_{\alpha=i,e} \epsilon_\alpha^l(q) - 1. \tag{18}$$

从(13)，(16)式可以清楚地看到，低频场在纵波情况下，就是高频纵场或横场拍频形成的有质动力源；在横波情况下，则是两个互不平行的高频场拍频作用激发非线性电流的结果，它是低频磁场源．应用电磁感应定律，立刻可得 Vlasov-Maxwell 体系中自生低频磁场为

$$\boldsymbol{B}_d^t(q) = \frac{c}{\Omega}\boldsymbol{q} \times \boldsymbol{E}_d^t(q). \tag{19}$$

由于强自生磁场的出现，低频振荡带电粒子绕磁力线"螺旋"运动，有可能阻滞高频随机场导致的无碰撞扩散过程．

为了进一步给出 $\boldsymbol{B}_d(q)$，需要给出高频场方程的形式，以 (2) 式代入 (5) 式后得

$$\epsilon_h^\sigma(k) E_h^\sigma(k) = \mathrm{I}^\sigma(k) + \mathrm{II}^\sigma(k), \tag{20}$$

其中

$$\begin{aligned}\mathrm{I}^\sigma(k) &= -\frac{4\pi}{\omega}\sum_{\sigma_1}\frac{e^2}{m_e}\int \frac{\boldsymbol{e}_k\cdot\boldsymbol{v}}{\omega - \boldsymbol{k}\cdot\boldsymbol{v}}\boldsymbol{e}_1' \cdot \frac{\partial \delta F_e(q)}{\partial \boldsymbol{v}} E_h^{\sigma_1}(k_1)\delta^4(k-k_1-q)\frac{dk_1dq}{(2\pi)^4} \\ &= \sum_{\sigma_1}\frac{\omega_{pe}^2}{\omega^2}\int (\boldsymbol{e}_k\cdot\boldsymbol{e}_1)E_h^{\sigma_1}(k_1)\frac{n_e(q)}{n_0}\delta^4(k-k_1-q)\frac{dk_1dq}{(2\pi)^4} + o\left(\frac{\boldsymbol{k}\cdot\boldsymbol{v}}{\omega}\right).\end{aligned} \tag{21}$$

文献 [1] 中已给出了(21)式的详细形式，它是低频波引起的低频振荡电子密度扰动 n_e 与高频场作用项，

$$n_e(q) = \int \delta F_e(q) d^3v. \tag{22}$$

$$\begin{aligned}\mathrm{II}^\sigma(k) &= -\frac{4\pi}{\omega}\sum_{\sigma_d=l,t}\frac{e^2}{m_e}\int\frac{\boldsymbol{e}_k\cdot\boldsymbol{v}}{\omega-\boldsymbol{k}\cdot\boldsymbol{v}}\boldsymbol{e}_d' \cdot \frac{\partial f_e(k_1)}{\partial \boldsymbol{v}} E_d^{\sigma_d}(q)\delta^4(k-k_1-q)\frac{dk_1dq}{(2\pi)^4}d^3v \\ &= \frac{1}{i\omega}\frac{\omega_{pe}^2 e}{m_e\Omega}\sum_{\sigma_d\sigma_1}\int\frac{\boldsymbol{e}_k\cdot\boldsymbol{v}}{\omega-\boldsymbol{k}\cdot\boldsymbol{v}}[(\Omega-\boldsymbol{q}\cdot\boldsymbol{v})\boldsymbol{e}_d + \boldsymbol{e}_d\cdot\boldsymbol{v}\boldsymbol{q}]\cdot\frac{\partial}{\partial\boldsymbol{v}}\frac{\boldsymbol{e}_1'\cdot\dfrac{\partial F_{0e}}{\partial\boldsymbol{v}}}{\omega_1-\boldsymbol{k}_1\cdot\boldsymbol{v}} \\ &\quad\times E_h^{\sigma_1}(k_1)E_d^{\sigma_d}(q)\delta^4(k-k_1-q)\frac{dk_1dq}{(2\pi)^4}d^3v \\ &= +\frac{\omega_{pe}e}{i\omega^2 m_e\Omega}\sum_{\sigma_1,\sigma_d}\int \boldsymbol{e}_k\cdot[\boldsymbol{q}\boldsymbol{e}_d\cdot\boldsymbol{e}_1 - \boldsymbol{e}_d\boldsymbol{q}\cdot\boldsymbol{e}_1]E_h^{\sigma_1}(k_1)E_d^{\sigma_d}(q) \\ &\quad\times\delta^4(k-k_1-q)\frac{dk_1dq}{(2\pi)^4} + o\left(\frac{\boldsymbol{k}\cdot\boldsymbol{v}}{\omega}\right)_{1,2}.\end{aligned} \tag{23}$$

上式对 $\sum_{\sigma_d=l,t}$ 求和时，准确到 $o(\boldsymbol{k}\cdot\boldsymbol{v}/\omega)_{1,2}$，只留下低频横波项，这也就是我们为什么在文献 [1] 的粒子高频分布函数中不包括第三项的原因．现在 (20) 式为

$$\epsilon_h^\sigma(k)E_h^\sigma(k)=\frac{\omega_{pe}^2}{\omega^2}\sum_{\sigma_1}\int\left\{\frac{n_e(q)}{n_0}(\boldsymbol{e_k}\cdot\boldsymbol{e}_1)E_h^{\sigma_1}(k_1)-i\frac{e}{m_e\omega_{pe}\Omega}\boldsymbol{e_k}\cdot[\boldsymbol{e}_1\times(\boldsymbol{q}\times\boldsymbol{e}_d)]\right.$$
$$\left.\times E_h^{\sigma_1}(k_1)E_d^t(q)\right\}\delta^4(k-k_1-q)\frac{dk_1dq}{(2\pi)^4}. \tag{24}$$

再一次应用电磁感应定律,得高频场与低频磁场耦合关系为

$$\epsilon_h^\sigma(k)E_h^\sigma(k)=\frac{\omega_{pe}^2}{\omega^2}\sum_{\sigma_1}\int\left\{\boldsymbol{e_k}\cdot\boldsymbol{E}_h^{\sigma_1}(k_1)\frac{n_e(q)}{n_0}-\frac{ie}{m_ec\omega_{pe}}\boldsymbol{e_k}\cdot[\boldsymbol{E}_h^{\sigma_1}(k_1)\times\boldsymbol{B}_d(q)]\right\}$$
$$\times\delta^4(k-k_1-q)\frac{dk_1dq}{(2\pi)^4}. \tag{25}$$

最后,写出低频振荡电子密度扰动的表达式,取到二级场得

$$n_e(q)=\sum_{\sigma_1\sigma_2\sigma_d}\frac{\omega_{pe}^2}{4\pi m_e}\int_{(d)}\left\{\omega_{pe}^2\frac{H_e^{\sigma_d\sigma_1\sigma_2}(q,k_1,k_2)}{\epsilon_d^{\sigma_d}(q)\Omega}\int\frac{\boldsymbol{e}_d'\cdot\frac{\partial F_{0e}}{\partial\boldsymbol{v}}}{\Omega-\boldsymbol{q}\cdot\boldsymbol{v}}d^3v-\mathrm{I}_e^{\sigma_1\sigma_2}(q,k_1,k_2)\right\}$$
$$\times E_h^{\sigma_1}(k_1)E_h^{\sigma_2}(k_2)\delta^4(q-k_1-k_2)\frac{dk_1dk_2}{(2\pi)^4}, \tag{26}$$

其中

$$\mathrm{I}_\alpha^{\sigma_1\sigma_2}(q,k_1,k_2)=\frac{1}{2}\int\frac{1}{\Omega-\boldsymbol{q}\cdot\boldsymbol{v}}\left[\boldsymbol{e}_1'\cdot\frac{\partial}{\partial\boldsymbol{v}}\frac{\boldsymbol{e}_2'\cdot\frac{\partial F_{0\alpha}}{\partial\boldsymbol{v}}}{\omega_2-\boldsymbol{k}_2\cdot\boldsymbol{v}}+1\longleftrightarrow2\right]d^3v,$$
$$\alpha=i,\ e. \tag{27}$$

$\mathrm{I}_\alpha^{\sigma_1\sigma_2}$ 对脚标 1, 2 对称,不难证明

$$\mathrm{I}_\alpha^{\sigma_1\sigma_2}(q,k_1,k_2)=\frac{|\boldsymbol{q}|}{\Omega}H_\alpha^{l\sigma_1\sigma_2}(q,k_1,k_2)\approx\frac{|\boldsymbol{q}|^2}{2\omega_2^2\omega_{pe}^2}(\boldsymbol{e}_1\cdot\boldsymbol{e}_2)[\epsilon_e^l(q)-1]. \tag{28}$$

注意到横波 $\int\frac{\boldsymbol{e}_d^{t'}\cdot\frac{\partial F_{0e}}{\partial\boldsymbol{v}}}{\Omega-\boldsymbol{q}\cdot\boldsymbol{v}}d^3v=\int\frac{\boldsymbol{e}_d^t\cdot\boldsymbol{v}\boldsymbol{q}\cdot\frac{\partial F_{0e}}{\partial\boldsymbol{v}}}{\Omega-\boldsymbol{q}\cdot\boldsymbol{v}}=0$, 最后 (26) 式变为

$$n_e(q)=\frac{\omega_{pe}^2}{2\pi m_e}\frac{|\boldsymbol{q}|\epsilon_i^l(q)}{\epsilon_d^l(q)\Omega}\sum_{\sigma_1^+\sigma_2^-}\int H_e^{l\sigma_1^+\sigma_2^-}(q,k_1,-k_2)E_h^{\sigma_1^+}(k_1)E_h^{\sigma_2^-}(k_2)^*$$
$$\times\delta^4(q-k_1+k_2)\frac{dk_1dk_2}{(2\pi)^4}$$
$$=-\frac{1}{4\pi m_e}\frac{\epsilon_i^l(q)(\epsilon_e^l(q)-1)}{\epsilon_d^l(q)}\frac{|\boldsymbol{q}|^2}{\omega_{pe}^2}$$
$$\times\sum_{\sigma_1^+\sigma_2^-}\int\boldsymbol{E}_h^{\sigma_1^+}(k_1)\cdot\boldsymbol{E}_h^{\sigma_2^-}(k_2)^*\delta^4(q-k_1+k_2)\frac{dk_1dk_2}{(2\pi)^4}. \tag{29}$$

这里,我们看到,准确到 $o(\boldsymbol{k}\cdot\boldsymbol{v}/\omega)_{1,2}$ 阶, 低频振荡粒子密度扰动完全由低频纵场 (离子声) 驱动所产生,自发磁场不发生明显的作用,它只通过非线性有旋电流与高频场的耦合来间接发生对低频密度扰动的影响.

方程 (19), (25) 和 (29) 构成了 F 表象中波与粒子非线性作用的耦合关系.

三、时空表象中粒子与波非线性作用方程组

为了得到粒子与场动力学方程组在时空坐标中的表达式，进行 Fourier 逆变换后需要对低频波的线性介电函数部分作各种近似展开，因为我们讨论的是非线性色散关系支配下有限幅波的相互作用，线性介电函数部分不可能有零值，故有多种情况出现，现在分别讨论于下：

1. 当低频波频率 $\boldsymbol{\Omega}$ 处在 $|\boldsymbol{q}|v_{T_i} \ll \Omega \ll |\boldsymbol{q}|v_{T_e}$ 之间时，低频波的电子、离子 Landau 阻尼可忽略，于是

$$\epsilon_i^l(q)(\epsilon_e^l(q)-1)/\epsilon_d^l(q) \approx -\omega_{pi}^2(\Omega^2 - q^2v_s^2)^{-1} \quad (\text{离子声}), \tag{30}$$

$$\epsilon_d^t(q) \approx 1 - \frac{c^2q^2}{\Omega^2} - \frac{1}{q^2\lambda_e^2}\left(1 + \frac{q^2v_s^2}{\Omega^2}\right). \tag{31}$$

这里 $v_s = \sqrt{T_e/M}$，T_e 为电子温度，M 为离子质量；ω_{pi} 为离子等离子体频率，λ_e 为电子 Debye 长度．因为 $c|\boldsymbol{q}| \gg \Omega$，$|\boldsymbol{q}|\lambda_e \ll 1$，所以

$$\epsilon_d^t(q) \approx -\frac{1}{q^2\lambda_e^2}\left(\frac{c^2\lambda_e^2q^4}{\Omega^2} + \frac{q^2v_s^2}{\Omega^2} + 1\right). \tag{32}$$

利用 $-i\Omega \to \frac{\partial}{\partial t}$，$i\boldsymbol{q} \to \nabla$，$\boldsymbol{E}^\sigma = \boldsymbol{E}^{\sigma+} + \boldsymbol{E}^{\sigma-}$，$\boldsymbol{E} = \sum\limits_\sigma \boldsymbol{E}^\sigma$，对 (29) 式逆变换后得密度扰动方程(这里及以下已略去高频场脚标 h 记号)

$$\left(\frac{\partial^2}{\partial t^2} - v_s^2\nabla^2\right)n_e(\boldsymbol{x}, t) = \nabla^2\frac{|\boldsymbol{E}(\boldsymbol{x}, t)|^2}{8\pi M}. \tag{33}$$

这里 $\boldsymbol{E}$ 包含纵波和横波．

注意到横场的旋量性质，纵场的 $\boldsymbol{E}^l(k) = \boldsymbol{kk}\cdot\boldsymbol{E}^l(k)/k^2$ 关系以及 $\sum\limits_{\sigma=l,t}\boldsymbol{e}_{\boldsymbol{k}}^\sigma(\boldsymbol{e}_{\boldsymbol{k}}^\sigma\cdot\boldsymbol{e}_1^{\sigma_1}) = \boldsymbol{\sigma}_1^{\sigma_1}$，对 (25) 式两边乘 $\boldsymbol{e}_{\boldsymbol{k}}$ 并进行 σ 求和，最后得逆变换后高频场方程为

$$\frac{\partial^2}{\partial t^2}\boldsymbol{E} + c^2\nabla\times(\nabla\times\boldsymbol{E}) - 3v_{T_e}^2\nabla\nabla\cdot\boldsymbol{E}$$
$$= -\omega_{pe}^2\left(1 + \frac{n_e}{n_0}\right)\boldsymbol{E} + i\,\frac{\omega_{pe}e}{m_ec}\boldsymbol{E}\times\boldsymbol{B}. \tag{34}$$

(34) 式中的 $\boldsymbol{B}$ 可由 (13) 和 (19) 式并利用 (32) 式作逆变换后得到，即

$$\left[\frac{\partial^2}{\partial t^2} - (\lambda_e^2c^2\nabla^2 - v_s^2)\nabla^2\right]\boldsymbol{B} = \frac{\lambda_e^2ce}{i2m_e\omega_{pe}}\nabla^2\nabla\times[\nabla\times(\boldsymbol{E}\times\boldsymbol{E}^*)]. \tag{35}$$

在条件

$$|\boldsymbol{q}|\lambda_e \gg \frac{v_s}{c} \quad 及 \quad |\boldsymbol{q}|\lambda_e \gg \frac{\Omega}{|\boldsymbol{q}|v_{T_e}}\beta_e \quad \left(\beta_e = \frac{v_{T_e}}{c}\right), \tag{36}$$

下，由 (13)，(19)，(32) 式，则 (35) 式退化为

$$\boldsymbol{B} = \frac{e}{i2m_e\omega_{pe}c}\boldsymbol{E}\times\boldsymbol{E}^*. \tag{37}$$

这时,方程(34)为

$$\frac{\partial^2}{\partial t^2}\boldsymbol{E}+c^2\nabla\times(\nabla\times\boldsymbol{E})-3v_{T_e}^2\nabla\nabla\cdot\boldsymbol{E}=-\omega_{pe}^2\left(1+\frac{n_e}{n_0}\right)\boldsymbol{E}$$
$$+\frac{\omega_{pe}^2\beta_e^2}{8\pi n_0T_e}\boldsymbol{E}\times(\boldsymbol{E}\times\boldsymbol{E}^*). \tag{38}$$

方程(33),(34),(35)或(33),(38)便是无阻尼时的方程组.

2. 当 $\Omega<|\boldsymbol{q}|v_{T_e}$ 时,电子的 Landau 阻尼不能忽略,取到一级项便有

$$\frac{\epsilon_i^l(q)[\epsilon_e^l(q)-1]}{\epsilon_d^e(q)}\approx-\omega_{pi}^2\left(1+i\sqrt{\frac{\pi}{2}}\frac{\Omega}{|\boldsymbol{q}|v_{T_e}}e^{-\frac{\Omega^2}{2q^2v_{Te}^2}}\right)$$
$$\times\left(\Omega^2-q^2v_s^2+i\sqrt{\frac{\pi}{2}}\frac{\Omega^3}{|\boldsymbol{q}|v_{T_e}}e^{-\frac{\Omega^2}{2q^2v_{Te}^2}}\right)^{-1}$$
$$\approx-\omega_{pi}^2(\Omega^2-q^2v_s^2)^{-1}\left(1+i\sqrt{\frac{\pi}{2}}\frac{q^2v_s^2}{(\Omega^2-q^2v_s^2)}\right.$$
$$\left.\times\frac{\Omega}{|\boldsymbol{q}|v_{T_e}}e^{-\frac{\Omega^2}{2q^2v_{Te}^2}}\right)^{-1}, \tag{39}$$

$$\epsilon_d^e(q)\approx-\frac{1}{q^2\lambda_e^2\Omega^2}\left\{c^2q^4\lambda_e^2+q^2v_s^2+\Omega^2-i\sqrt{\frac{\pi}{2}}\Omega|\boldsymbol{q}|v_{T_e}e^{-\frac{\Omega^2}{2q^2v_{Te}^2}}\right\}. \tag{40}$$

这时密度扰动方程变为

$$\left(\frac{\partial^2}{\partial t^2}-v_s^2\nabla^2\right)n_e+\Gamma_e\frac{\partial n_e}{\partial t}=\nabla^2\frac{|\boldsymbol{E}|^2}{8\pi M}, \tag{41}$$

其中

$$\Gamma_e\frac{\partial n_e}{\partial t}=\sqrt{\frac{\pi m_e}{2M}}v_s\frac{\partial}{\partial t}\int|\boldsymbol{q}|e^{i\boldsymbol{q}\cdot\boldsymbol{x}}n_e(\boldsymbol{q},t)\frac{d^3q}{(2\pi)^3}$$
$$=-(2\pi)^{-3/2}\left(\frac{m_e}{M}\right)^{1/2}v_s\nabla^2\frac{\partial}{\partial t}\int\frac{n_e(\boldsymbol{x}',t)}{|\boldsymbol{x}-\boldsymbol{x}'|^2}d^3\boldsymbol{x}'. \tag{42}$$

磁场方程(35)变为

$$\left[\frac{\partial^2}{\partial t^2}+\gamma_e\frac{\partial}{\partial t}-(\lambda_e^2c^2\nabla^2-v_s^2)\nabla^2\right]\boldsymbol{B}=(35)\text{式右边}, \tag{43}$$

其中

$$\gamma_e\frac{\partial}{\partial t}\boldsymbol{B}=-\sqrt{\frac{\pi}{2}}v_{T_e}\frac{\partial}{\partial t}\int|\boldsymbol{q}|e^{i\boldsymbol{q}\cdot\boldsymbol{x}}\boldsymbol{B}(\boldsymbol{q},t)\frac{d^3q}{(2\pi)^3}$$
$$=(2\pi)^{-3/2}v_{T_e}\nabla^2\frac{\partial}{\partial t}\int\frac{\boldsymbol{B}(\boldsymbol{x}',t)}{|\boldsymbol{x}-\boldsymbol{x}'|^2}d^3\boldsymbol{x}'. \tag{44}$$

一般情况下条件(36)式极易满足,这时(43)式便为

$$\nabla\times\nabla\times\boldsymbol{B}=\frac{e}{i2m_e\omega_{pe}c}\nabla\times[\nabla\times(\boldsymbol{E}\times\boldsymbol{E}^*)]$$
$$-(2\pi)^{-3/2}\frac{\omega_{pe}^2}{c^2v_{T_e}}\frac{\partial}{\partial t}\int\frac{\boldsymbol{B}(\boldsymbol{x}',t)}{|\boldsymbol{x}-\boldsymbol{x}'|^2}d^3\boldsymbol{x}'. \tag{45}$$

上述形式磁场方程曾被 Kono 等人[5]得到过，右端第二项就是非线性 Landau 阻尼项．应当指出，文献[5]中这一项的前面是个加号，对空间阻尼项来说，这是不对的．

3. 当 $\Omega \gtrsim |\boldsymbol{q}| v_{T_i}$ 时，离子的 Landau 阻尼起作用，这时

$$\frac{\epsilon_i^l(q)[\epsilon_e^l(q)-1]}{\epsilon_d^l(q)} = -q^2v_s^2\left(1 - i\sqrt{\frac{\pi}{2}}\left(\frac{T_e}{T_i}\right)^{3/2}\frac{\Omega^3}{(qv_s)^3}e^{-\frac{\Omega^2}{2q^2v_{T_i}^2}}\right)$$

$$\times\left(\Omega^2 - q^2v_s^2 + i\sqrt{\frac{\pi}{2}}\left(\frac{T_e}{T_i}\right)^{3/2}\frac{\Omega^3}{qv_s}e^{-\frac{\Omega^2}{2q^2v_{T_i}^2}}\right)^{-1}$$

$$\approx -q^2v_s^2\left(\Omega^2 - q^2v_s^2 + i\sqrt{\frac{\pi}{2}}\Omega|\boldsymbol{q}|v_{T_i}e^{-\frac{1}{2}}\right)^{-1}, \tag{46}$$

$$\epsilon_d^t(q) = -\frac{1}{q^2\lambda_e^2\Omega^2}\left\{c^2\lambda_e^2q^4 + q^2v_s^2 + \Omega^2 - i\sqrt{\frac{\pi}{2}}\left(\frac{m_e}{M}\frac{T_e}{T_i}\right)^{1/2}\Omega q v_{T_e}e^{-\frac{1}{2}}\right\}. \tag{47}$$

于是(33)和(35)式便分别为

$$\left(\frac{\partial^2}{\partial t^2} - v_s^2\nabla^2\right)n_e + \Gamma_i\frac{\partial}{\partial t}n_e = \nabla^2\frac{|\boldsymbol{E}|^2}{8\pi M}, \tag{48}$$

$$\left[\frac{\partial^2}{\partial t^2} + \gamma_i\frac{\partial}{\partial t} - (\lambda_e^2c^2\nabla^2 - v_s^2)\nabla^2\right]\boldsymbol{B} = (35)\text{ 式右边}, \tag{49}$$

其中

$$\Gamma_i\frac{\partial n_e}{\partial t} = -(2\pi)^{-3/2}e^{-\frac{1}{2}}v_{T_i}\nabla^2\frac{\partial}{\partial t}\int\frac{n_e(\boldsymbol{x}',t)}{|\boldsymbol{x}-\boldsymbol{x}'|^2}d^3x', \tag{50}$$

$$\gamma_i\frac{\partial\boldsymbol{B}}{\partial t} = (2\pi)^{-3/2}e^{-\frac{1}{2}}\left(\frac{m_eT_e}{MT_i}\right)^{1/2}v_{T_e}\nabla^2\frac{\partial}{\partial t}\int\frac{\boldsymbol{B}(\boldsymbol{x}',t)}{|\boldsymbol{x}-\boldsymbol{x}'|^2}d^3x' \tag{51}$$

分别为对密度扰动和自生磁场的 Landau 阻尼项．

当取条件(36)式时，由(49)与(50)式便得

$$\nabla\times\nabla\times\boldsymbol{B} = \frac{e}{i2m_e\omega_{pe}c}\nabla\times\nabla\times(\boldsymbol{E}\times\boldsymbol{E}^*)$$

$$-\frac{\omega_{pe}^2}{(2\pi)^{3/2}e^{1/2}v_{T_e}c^2}\xi_0\frac{\partial}{\partial t}\int\frac{\boldsymbol{B}(\boldsymbol{x}',t)}{|\boldsymbol{x}-\boldsymbol{x}'|^2}d^3x', \tag{52}$$

式中 $\xi_0 = \left(\frac{m_eT_e}{MT_i}\right)^{1/2} \ll 1$．(50)，(51)式中的 $e^{-\frac{1}{2}}$ 及(52)式中的 $e^{\frac{1}{2}}$ 均为自然对数底开方．

4. 当 $\Omega \ll |\boldsymbol{q}| v_{T_i}$ 时(即所谓静态近似)

$$\frac{\epsilon_i^l(q)[\epsilon_e^l(q)-1]}{\epsilon_d^l(q)} = \frac{1}{q^2\lambda_e^2}\frac{T_e}{(T_i+T_e)}, \tag{53}$$

$$\epsilon_d^t(q) = -\frac{1}{q^2\lambda_e^2}\left(\frac{c^2q^4\lambda_e^2}{\Omega^2} + \frac{T_e+T_i}{T_i}\right). \tag{54}$$

于是(33)和(35)式变为

$$n_e = -|\boldsymbol{E}|^2/8\pi(T_i+T_e), \tag{55}$$

$$\left(\left(1+\frac{T_e}{T_i}\right)\frac{\partial^2}{\partial t^2} - \lambda_e^2c^2\nabla^4\right)\boldsymbol{B} = \frac{ec\lambda_e^2}{i2m_e\omega_{pe}}\nabla^2\nabla\times\nabla\times(\boldsymbol{E}\times\boldsymbol{E}^*). \tag{56}$$

当 $|\boldsymbol{q}|\lambda_e \gg \left(1+\frac{T_e}{T_i}\right)^{1/2}\frac{\Omega}{c|\boldsymbol{q}|}$ 时，与得到方程(37)式时类似讨论，有

$$\boldsymbol{B}=\frac{e}{i2m_e\omega_{pe}c}\boldsymbol{E}\times\boldsymbol{E}^*. \tag{57}$$

(55)和(57)式代入(34)式就得到一个单一 $\boldsymbol{E}$ 的方程.

5. 当 $\Omega \gg |\boldsymbol{q}|v_{T_e}$ 时，即快振荡过程. 这时没有低频粒子密度扰动，只有电子 Langmuir 振荡. 但是如果有高频 Pump 场入射，拍频后可以产生低频磁场，因为这时

$$\varepsilon_d^t(q)\approx-\frac{1}{\Omega^2}(c^2q^2+\omega_{pe}^2), \tag{58}$$

自生磁场方程便为

$$\nabla\times\nabla\times\boldsymbol{B}=\frac{e}{i2m_e\omega_{pe}c}\nabla\times\nabla\times(\boldsymbol{E}\times\boldsymbol{E}^*)-\frac{\omega_{pe}^2}{c^2}\boldsymbol{B}. \tag{59}$$

可见自生磁场的空间阻尼长度 $\sim c/\omega_{pe}=\lambda_e(c/v_{T_e})\gg\lambda_e$. 方程(59)与文献[5]所得的一样.

至此，我们在各种近似下导得了缓变粒子密度扰动 n_e，快变电场 $\boldsymbol{E}$ 和缓变磁场 $\boldsymbol{B}$ 之间耦合关系的动力学方程组. 我们看到除了不感兴趣的 $\Omega\gg|\boldsymbol{q}|v_{T_e}$ 近似以外，如果 Landau 阻尼可忽略，在极易满足(36)式条件下，磁场方程都可用形如(37)式的静态形式表示出来，这时方程组简化为 n_e 和 $\boldsymbol{E}$ 的两个方程，它们最近已被文献[6]用于关于天体等离子体的强湍流研究中.

为了使方程便于求解，我们需要把快变电场方程变成缓变振幅 $\boldsymbol{\psi}$ 的方程. 令

$$\boldsymbol{E}=\frac{1}{2}(\boldsymbol{\psi}e^{-i\omega_{pe}t}+\text{c. c.}) \quad 及取 \quad \omega_{pe}\boldsymbol{\psi}\gg\frac{\partial}{\partial t}\boldsymbol{\psi},$$

便使(34)式变为

$$i2\omega_{pe}\frac{\partial}{\partial t}\boldsymbol{\psi}-c^2\nabla\times(\nabla\times\boldsymbol{\psi})+3v_{T_e}^2\nabla\nabla\cdot\boldsymbol{\psi}=\omega_{pe}^2\frac{n_e}{n_0}\boldsymbol{\psi}-\frac{i\omega_{pe}e}{m_ec}\boldsymbol{\psi}\times\boldsymbol{B}. \tag{60}$$

在磁场和密度扰动均为静态近似下，即如 $\Omega\ll qv_{T_i}$ 时，便得包含磁场效应的非线性 Schrödinger 方程

$$i2\omega_{pe}\frac{\partial}{\partial t}\boldsymbol{\psi}-c^2\nabla\times(\nabla\times\boldsymbol{\psi})+3v_{T_e}^2\nabla\nabla\cdot\boldsymbol{\psi}$$
$$=\omega_{pe}^2\left[-\frac{|\boldsymbol{\psi}|^2}{16\pi n_0(T_e+T_i)}\boldsymbol{\psi}-\frac{\beta_e^2}{16\pi n_0T_e}\boldsymbol{\psi}\times(\boldsymbol{\psi}\times\boldsymbol{\psi}^*)\right]. \tag{61}$$

四、Lagrangian 密度和守恒量

现在讨论无阻尼方程(33)和(60)的运动积分，其中取磁场 $\boldsymbol{B}=\frac{e}{i4m_e\omega_{pe}c}\boldsymbol{\psi}\times\boldsymbol{\psi}^*$. 首先对它们进行无量纲化，作如下替换：

$$\frac{2}{3}\omega_{pe}\mu t\to t,\quad \frac{2}{3}\frac{\sqrt{\mu}}{\lambda_e}\boldsymbol{x}\to\boldsymbol{x},\quad \frac{3n_e}{4\mu n_0}\to n,$$

$$\frac{\sqrt{3}\,\boldsymbol{\psi}}{(64\pi\mu n_0 T_e)^{1/2}} \rightarrow \boldsymbol{\varepsilon}, \quad \mu = \frac{m_e}{M}, \quad \alpha_e^2 = \frac{c^2}{3v_{T_e}^2}. \tag{62}$$

这样，方程组 (33)，(60) 式便变为

$$\left(\frac{\partial^2}{\partial t^2} - \nabla^2\right) n = \nabla^2 |\boldsymbol{\varepsilon}|^2, \tag{63}$$

$$i\frac{\partial}{\partial t}\boldsymbol{\varepsilon} - \alpha_e^2 \nabla \times (\nabla \times \boldsymbol{\varepsilon}) + \nabla\nabla \cdot \boldsymbol{\varepsilon} = n\boldsymbol{\varepsilon} - \beta_e^2 \boldsymbol{\varepsilon} \times (\boldsymbol{\varepsilon} \times \boldsymbol{\varepsilon}^*). \tag{64}$$

从 (63)，(64) 式可以得到拉氏密度函数

$$\begin{aligned}\mathscr{L} = &\frac{1}{2} i(\boldsymbol{\varepsilon}^* \cdot \boldsymbol{\varepsilon}_t - \boldsymbol{\varepsilon}_t^* \cdot \boldsymbol{\varepsilon}) - \alpha_e^2 (\nabla \times \boldsymbol{\varepsilon}) \cdot (\nabla \times \boldsymbol{\varepsilon}^*) - (\nabla \cdot \boldsymbol{\varepsilon})(\nabla \cdot \boldsymbol{\varepsilon}^*) - \varphi_t (\boldsymbol{\varepsilon} \cdot \boldsymbol{\varepsilon}^*) \\ &+ \frac{1}{2}[(1 + \beta_e^2)(\boldsymbol{\varepsilon} \cdot \boldsymbol{\varepsilon}^*)^2 - \beta_e^2 (\boldsymbol{\varepsilon}^* \cdot \boldsymbol{\varepsilon}^*)(\boldsymbol{\varepsilon} \cdot \boldsymbol{\varepsilon}) + \varphi_t^2 - (\nabla\varphi)^2],\end{aligned} \tag{65}$$

其中 φ 相当于流体力学流势，即流 $\boldsymbol{u} = -\nabla\varphi$；$\varphi_t = n + |\boldsymbol{\varepsilon}|^2$. 利用变分方程很容易证明从 (65) 可导得 (63) 和 (64). 现在从 (65) 式可得“能量-动量”张量密度

$$T_{\mu\nu} = \mathscr{L}\delta_{\mu\nu} - \Phi^r_{,\mu} \frac{\partial \mathscr{L}}{\partial \Phi^r_{,\nu}}. \tag{66}$$

这里

$$\Phi^r = (\boldsymbol{\varepsilon}, \boldsymbol{\varepsilon}^*; \boldsymbol{\varepsilon}_t, \boldsymbol{\varepsilon}_t^*; \cdots; \varphi_t; \nabla\varphi), \tag{67}$$

$$\Phi^r_{,\mu} = \left(\frac{\partial \Phi^r}{\partial x}, \cdots\right). \tag{68}$$

于是由 (66) 式可得线动量守恒量

$$P_\mu = \int T_{\mu 0} d^3x = \int \left\{ \sum_j \frac{1}{2} i \left(\varepsilon_j \frac{\partial}{\partial x_\mu} \varepsilon_j^* - \varepsilon_j^* \frac{\partial}{\partial x_\mu} \varepsilon_j \right) + n u_\mu \right\} d^3\boldsymbol{x}, \tag{69}$$

能量守恒量

$$\begin{aligned} H &= -\int T_{00} d^3\boldsymbol{x} \\ &= \int \left\{ \alpha_e^2 |\nabla \times \boldsymbol{\varepsilon}|^2 + |\nabla \cdot \boldsymbol{\varepsilon}|^2 + n|\boldsymbol{\varepsilon}|^2 - \frac{1}{2}[\beta_e (|\boldsymbol{\varepsilon}|^4 - \boldsymbol{\varepsilon}^{*2}\boldsymbol{\varepsilon}^2) - n^2 - u^2] \right\} d^3\boldsymbol{x}. \end{aligned} \tag{70}$$

去掉磁场效应后 (70) 式退化为 (9) 式中给出的形式.

最后，由规范不变，得四度流守恒量

$$j_\mu = -i\left(\frac{\partial \mathscr{L}}{\partial \Phi_{,\mu}} \Phi - \frac{\partial \mathscr{L}}{\partial \Phi^*_{,\mu}} \Phi^*\right). \tag{71}$$

由此得等离子体中准粒子数守恒量

$$N = \int j_0 d^3x = \int |\boldsymbol{\varepsilon}|^2 d^3x. \tag{72}$$

如果对 (61) 式无量纲化，取 $\sqrt{3}\,\boldsymbol{\psi}/(64\pi n_0 \mu(T_i + T_e))^{1/2} \rightarrow \boldsymbol{\varepsilon}, \beta = \left(\frac{T_e + T_i}{m_e c^2}\right)^{1/2}$，而其余量同 (62) 式就得非线性 Schrödinger 方程（无量纲化）

$$i\frac{\partial}{\partial t}\boldsymbol{\varepsilon} - \alpha_e^2 \nabla \times (\nabla \times \boldsymbol{\varepsilon}) + \nabla\nabla \cdot \boldsymbol{\varepsilon} + |\boldsymbol{\varepsilon}|^2 \boldsymbol{\varepsilon} + \beta^2 \boldsymbol{\varepsilon} \times (\boldsymbol{\varepsilon} \times \boldsymbol{\varepsilon}^*) = 0. \tag{73}$$

(73) 式的拉氏密度函数为

$$\mathscr{L}_s = \frac{1}{2} i(\boldsymbol{\varepsilon}^* \cdot \boldsymbol{\varepsilon}_t - \boldsymbol{\varepsilon}_t^* \cdot \boldsymbol{\varepsilon}) - \alpha_c^2 |\nabla \times \boldsymbol{\varepsilon}|^2 - |\nabla \cdot \boldsymbol{\varepsilon}|^2 + \frac{1}{2} |\boldsymbol{\varepsilon}|^4 + \frac{\beta^2}{2}(|\boldsymbol{\varepsilon}|^4 - \boldsymbol{\varepsilon}^{*2}\boldsymbol{\varepsilon}^2). \quad (74)$$

在一维空间情况下 (65) 和 (74) 式中含 β_c 或 β 因子项抵消. 它们的拉氏密度函数分别为熟知的

$$\mathscr{L}^{(1)} = \frac{1}{2} i(\varepsilon^* \varepsilon_t - \varepsilon_t^* \varepsilon) - |\varepsilon_x|^2 - \varphi_t |\varepsilon|^2 + \frac{1}{2} |\varepsilon|^4 + \frac{1}{2}(\varphi_t^2 - \varphi_x^2), \quad (75)$$

$$\mathscr{L}_s^{(1)} = \frac{1}{2} i(\varepsilon^* \varepsilon_t - \varepsilon_t^* \varepsilon) - |\varepsilon_x|^2 + \frac{1}{2} |\varepsilon|^4. \quad (76)$$

从它们很快可以得到一维守恒量形式.

五、结果与讨论

1. 本文全部结果对高频波取到 $o(\boldsymbol{k} \cdot \boldsymbol{v}/\omega)$ 展开. 这意味着高频磁场的 Lorentz 力已被忽略. 这时取低频场为纵波和横波,我们便扩展了文献 [1] 的结果,成为包括自生磁场效应的方程组. 这样,除了有质动力驱动着粒子密度变化外,自生磁场影响着高频波的发展从而影响低频粒子密度的变化. 还应当提到的是,这里只考虑了两个高频波拍频产生低频波效应. 当调制不稳定性使场强足够大时,四级场以上高频波向低频波的拍频作用不能忽略,在低频波为纵波情况下我们已在文献 [8] 中作了讨论.

2. 从 (38) 或 (61) 式可以看到,当系统电子温度 T_e 很高时,调制不稳定性发展后期,只要两个高频波(例 $\boldsymbol{k}_1$, $\boldsymbol{k}_2$) 传播方向不平行,非线性有旋电流的贡献就不能忽略,磁场项 $\beta_c/8\pi n_0 T_e |\boldsymbol{\varepsilon} \times \boldsymbol{\varepsilon}^*|$ 就会起到一定作用. 但在通常情况下, 电子的 n_e/n_0 贡献仍然是主要的.

3. 从方程可以直接看到自生磁场只在大于一维情况下发生作用. 因此, 一维 Soliton 的形成仍然是有质动力与色散力平衡的结果. 在大于一维情况下文献 [10] 已指出,无磁场的三维 Soliton 不可能存在. 在有磁场时可作类似直观地讨论. 如引进波的宽度为 δ, 空间维数为 d, 显然与有质动力有关的非线性凝聚项中场能 $|\boldsymbol{\varepsilon}|^2$ 和磁场项 $|\boldsymbol{\varepsilon} \times \boldsymbol{\varepsilon}^*|$ 都随 δ^{-d} 变化,而色散项算子部分以 δ^{-2} 变化. 因此,当增大或缩小 δ 时与无磁场时有相同的效果: $d = 1$ 时,总是稳定,即可形成一维 Soliton. 但是, $d = 3$ 时,磁场项的存在仍然不可能维持三维 Soliton 的形成,不过 Caviton 形式的收缩是不可避免的.

附注 本文寄出以后,我们见到了 Kono 等人[11] 1981 年发表的文章[11], 其中也讨论了包括自生磁场在内的动力学方程组问题. 但是,他们只有自生磁场部分从 Vlasov-Maxwell 方程组导得,方法与我们不一样;至于动力学方程组的其余部分均采用了流体力学近似. 因此,非线性 Landau 阻尼效应只出现在磁场方程中而不出现在密度方程中. 这些, 我们已在文献[8]中作了评论.

参　考　文　献

[1] 贺贤土，核聚变，**1** (1980)，245.
[2] J. A. Stamper, *et al.*, *Phys. Rev. Lett.*, **26**(1971), 1012.
[3] J. B. Chase, *et al.*, *Phys. Fluids*, **16**(1973), 1142.
[4] J. A. Stamper and Tidman, *Phys. Fluids*, **16**(1973), 2024.
[5] M. Kono. *et al.*, *Phys. Lett. A*, **77**(1980), 27.
[6] D. ter Haar and V. N. Tsytovich, *Phys. Report*, **73**(1981), 177.
[7] B. D. Fried and S. D. Conte, The Plasma Dispersion Function, (1961).
[8] 贺贤土，物理学报，**31** (1982) 1317.
[9] J. Gibbons, *et al.*, *J. Plasma Physics*, **17**(1977), 153.
[10] S. G. Thornhill and D. ter Haar, *Phys. Report* **43C**(1978), 45.
[11] M. Kono, *et al.*, *J. Plasma Phys.*, **26** part 1(1981), 123.

THE PONDERMOTIVE FORCE AND MAGNETIC FIELD GENERATION EFFECTS RESULTING FROM THE NON-LINEAR INTERACTION BETWEEN PLASMA-WAVE AND PARTICLES

HE XIAN-TU

(*Institute of Atomic Energy, Academia Sinica*)

ABSTRACT

Starting from the Vlasov-Maxwell equations by using the method of Self-Consistent field theory, we set up first the coupling relation between distribution function and density of low-frequency oscillating particles and high-frequency EF, low-frequency MF in fourier representation. Then, in the time-space representation, a set of equations for the nonlinear interaction including the effects of magnetic field generation, pondermotive force and Landau damping were derived through the expansion of the low-frequency linear dielectric function under various approximations. Finally, we also gave the Lagrangian density and conserved quantities and discussed briefly whether the magnetic field can promote the formation of soliton in three dimensions.

Plasma Phys. Control. Fusion 35 (1993) 291–299. Printed in the UK

Ponderomotive force in unmagnetized Vlasov plasma

Shao-ping Zhu and X T He
Centre for Nonlinear Studies, Institute of Applied Physics and Computational Mathematics, PO Box 8009, Beijing 100088, People's Republic of China

Received 24 June 1992, in final form 7 September 1992

Abstract. An expression for ponderomotive force caused by electromagnetic fields is derived from Valsov–Maxwell equations in a non-stationary, unmagnetized plasma. Comparison between our result and earlier ones is given.

1. Introduction

The problem of finding a time-averaged ponderomotive force caused by electromagnetic fields in plasma has been treated in many papers [1–6]. Various methods such as kinetic [1], single-particle [2], hydrodynamic [3–5] and stress tensor methods [3, 6, 7], have been used in the past. But different expressions for pondermotive force appeared. Statham *et al* [3] reviewed various methods which have been used in earlier literature for calculating this force and discussed the expressions for it. The hydrodynamic and stress tensor calculation give the expression for ponderomotive force in unmagnetized non-stationary Langmuir plasma as follows [3]:

$$\boldsymbol{F} = -\frac{1}{8\pi}\frac{\omega_{pe}^2}{\omega^2}\nabla\langle E_f^2\rangle + \frac{i\omega_{pe}^2}{16\pi\omega^3}\frac{\partial}{\partial t}[\nabla\times(\boldsymbol{E}\times\boldsymbol{E}^*)] + \frac{i\omega_{pe}^2}{8\pi\omega^3}\left[\left(\frac{\partial\boldsymbol{E}}{\partial t}\cdot\nabla\right)\boldsymbol{E}^* - \left(\frac{\partial\boldsymbol{E}^*}{\partial t}\cdot\nabla\right)\boldsymbol{E}\right] \tag{1.1}$$

where

$$\boldsymbol{E}_f(\boldsymbol{x}, t) = \tfrac{1}{2}(\boldsymbol{E}(\boldsymbol{x}, t)e^{-i\omega t} + \text{cc}). \tag{1.2}$$

The first term on the right-hand side of equation (1.1) is well known, the non-potential part of equation (1.1) is due to slow-timescale current in non-stationary plasma where the fast-timescale field amplitude has a dependence on time as well as on space. On the other hand, the kinetic [1] and single-particle [2] calculations fail to produce the third term on the right-hand side of equation (1.1). This is an interesting problem. An interpretation has been given to this point, but 'the reason for this apparent anomaly in the kinetic calculation is not properly understood [3]. So, in our opinion, it is very necessary to reinvestigate ponderomotive force in detail on the basis of the kinetic equation.

The purpose of this paper is to derive the ponderomotive force based on Vlasov–Maxwell equations in unmagnetized plasma. We derive an expression for ponderomotive force caused by electromagnetic fields from Vlasov–Maxwell equations in unmagnetized plasma. Our result contains all the terms of equation (1.1).

We also show that the reason why the third term on the right-hand side of equation (1.1) was not produced in earlier kinetic calculations is because the slow-timescale nonlinear electric current transition matrix was not calculated correctly.

2. Slow-timescale distribution function

Consider the Vlasov–Maxwell system of equations:

$$\left\{\frac{\partial}{\partial t}+\boldsymbol{v}\cdot\frac{\partial}{\partial \boldsymbol{x}}+\frac{e_\alpha}{m_\alpha}\boldsymbol{G}(\boldsymbol{x},t)\cdot\frac{\partial}{\partial \boldsymbol{v}}\right\}f_\alpha(\boldsymbol{x},\boldsymbol{v},t)=0 \tag{2.1}$$

$$\nabla\times\boldsymbol{E}=-\frac{1}{c}\frac{\partial \boldsymbol{B}}{\partial t} \tag{2.2}$$

$$\nabla\times\boldsymbol{B}=\frac{1}{c}\frac{\partial \boldsymbol{E}}{\partial t}+\frac{4\pi}{c}\boldsymbol{j} \tag{2.3}$$

$$\boldsymbol{j}=\sum_\alpha\int \boldsymbol{v}f_\alpha(\boldsymbol{x},\boldsymbol{v},t)\,\mathrm{d}\boldsymbol{v} \tag{2.4}$$

$$\boldsymbol{G}(\boldsymbol{x},t)=\boldsymbol{E}(\boldsymbol{x},t)+\frac{1}{c}\boldsymbol{v}\times\boldsymbol{B} \tag{2.5}$$

where the index α ($\alpha=\mathrm{i,e}$) represent ion and electron respectively and other notations are standard.

We split the particle distribution functions up into their background, slow- and fast-timescale parts as follows

$$f_\alpha=f_{\alpha 0}+f_{\alpha \mathrm{s}}+f_{\alpha \mathrm{f}}. \tag{2.6}$$

Also, we divide the electromagnetic field into slow- and fast-timescale parts. Then, the slow-timescale part of equation (2.1) is

$$\left\{\frac{\partial}{\partial t}+\boldsymbol{v}\cdot\frac{\partial}{\partial \boldsymbol{x}}\right\}f_{\alpha\mathrm{s}}=-\frac{e_\alpha}{m_\alpha}\boldsymbol{G}_\mathrm{s}\cdot\frac{\partial}{\partial \boldsymbol{v}}f_{\alpha0}-\frac{e_\alpha}{m_\alpha}\left(\boldsymbol{G}_\mathrm{f}\cdot\frac{\partial}{\partial \boldsymbol{v}}f_{\alpha\mathrm{f}}\right)_\mathrm{s} \tag{2.7}$$

and the fast-timescale part

$$\left\{\frac{\partial}{\partial t}+\boldsymbol{v}\cdot\frac{\partial}{\partial \boldsymbol{x}}\right\}f_{\alpha\mathrm{f}}=-\frac{e_\alpha}{m_\alpha}\boldsymbol{G}_\mathrm{f}\cdot\frac{\partial}{\partial \boldsymbol{v}}f_{\alpha0}. \tag{2.8}$$

Where we only keep the second-order nonlinearity in equation (2.7).

Taking the Fourier transform of equations (2.7) and (2.8) with respect to both time and space

$$f_{\alpha\mathrm{s}}(q)=\frac{e_\alpha}{im_\alpha(\Omega-\boldsymbol{q}\cdot\boldsymbol{v})}\left(\boldsymbol{G}_\mathrm{s}(q)\cdot\frac{\partial}{\partial \boldsymbol{v}}f_{\alpha0}+\int_{(\mathrm{s})}\boldsymbol{G}_\mathrm{f}(k_1)\cdot\frac{\partial}{\partial \boldsymbol{v}}f_{\alpha\mathrm{f}}(k_2)\delta^4(q-k_1-k_2)\frac{\mathrm{d}^4k_1\,\mathrm{d}^4k_2}{(2\pi)^4}\right) \tag{2.9}$$

$$f_{\alpha\mathrm{f}}(q)=\frac{e_\alpha}{im_\alpha(\omega-\boldsymbol{k}\cdot\boldsymbol{v})}\boldsymbol{G}_\mathrm{f}(k)\cdot\frac{\partial}{\partial \boldsymbol{v}}f_{\alpha0} \tag{2.10}$$

where $q=(\boldsymbol{q},\Omega)$, $k=(\boldsymbol{k},\omega)$; the symbol $\int_{(s)}$ means the process of two fast-timescale motions producing one slow-timescale motion.

The slow-timescale distribution function for electrons reads

$$f_{es}(q)=-\frac{e}{im_e(\Omega-\boldsymbol{q}\cdot\boldsymbol{v})}\sum_{\sigma_s}E_s^{\sigma_s}(q)\boldsymbol{e}_s'^{\sigma_s}(q)\cdot\frac{\partial}{\partial\boldsymbol{v}}f_{e0}$$
$$-\frac{\omega_{pe}^2}{4\pi m_e n_0(\Omega-\boldsymbol{q}\cdot\boldsymbol{v})}\int_{(s)}\sum_{\sigma_1\sigma_2}\boldsymbol{e}_f'^{\sigma_1}(k_1)\cdot\frac{\partial}{\partial\boldsymbol{v}}\left[\frac{\boldsymbol{e}_f'^{\sigma_2}\cdot(\partial/\partial\boldsymbol{v})f_{e0}}{\omega_2-\boldsymbol{k}_2\cdot\boldsymbol{v}}\right]$$
$$\times E_f^{\sigma_1}(k_1)E_f^{\sigma_2}(k_2)\delta^4(q-k_1-k_2)\frac{d^4k_1\,d^4k_2}{(2\pi)^4} \tag{2.11}$$

where $E_s^{\sigma_s}(q)$ and $\boldsymbol{e}_s'^{\sigma_s}(q)$ are defined as follows [8]

$$\boldsymbol{G}_s(q)=\sum_{\sigma_s=t,l}E_s^{\sigma_s}(q)\boldsymbol{e}_s'^{\sigma_s}(q) \tag{2.12}$$

$$\boldsymbol{e}_s'^{\sigma_s}(q)=\left(1-\frac{\boldsymbol{q}\cdot\boldsymbol{v}}{\Omega}\right)\boldsymbol{e}_s^{\sigma_s}(q)+\frac{\boldsymbol{e}_s^{\sigma_s}(q)\cdot\boldsymbol{v}}{\Omega}\boldsymbol{q}. \tag{2.13}$$

We have a convention: $E_s^{\sigma_s *}(q)=-E_s^{\sigma_s}(-q)$ and $\boldsymbol{e}_s^{\sigma_s *}(q)=-\boldsymbol{e}_s^{\sigma_s}(-q)$, where * denotes the complex conjugate. Similarly, for fast-timescale field E_f^{σ} and $\boldsymbol{e}_f^{\sigma}(k)$, we have the same expression. For simplicity, we replace $\boldsymbol{e}_s^{\sigma_s}(q)$ by $\boldsymbol{e}_s^{\sigma_s}$ in the following discussion, also for the fast-timescale field.

3. Slow-timescale electric current

Ignoring the ion contribution to electric current due to much larger ions mass, we have the slow-timescale electric current

$$\boldsymbol{j}_s(q)=-e\int\boldsymbol{v}f_{es}(q)\,d^3\boldsymbol{v}=\boldsymbol{j}_{1s}(q)+\boldsymbol{j}_{2s}(q) \tag{3.1}$$

where

$$\boldsymbol{j}_{1s}(q)=\int\frac{e^2\boldsymbol{v}}{im_e(\Omega-\boldsymbol{q}\cdot\boldsymbol{v})}\sum_{\sigma_s}E_s^{\sigma_s}(q)\boldsymbol{e}_s'^{\sigma_s}(q)\cdot\frac{\partial}{\partial\boldsymbol{v}}f_{e0}\,d^3\boldsymbol{v} \tag{3.2}$$

$$\boldsymbol{j}_{2s}(q)=j_{2s}^{t}(q)\boldsymbol{e}_s^{t}+j_{2s}^{l}(q)\boldsymbol{e}_s^{l} \tag{3.3}$$

$$j_{2s}^{\sigma_s}(q)=\frac{\omega_{pe}^2 e}{4\pi n_0 m_e}\sum_{\sigma_1\sigma_2}\int_{(s)}H_e^{\sigma_s\sigma_1\sigma_2}(q,k_1,k_2)E_f^{\sigma_1}(k_1)E_f^{\sigma_2}(k_2)\delta^4(q-k_1-k_2)\frac{d^4k_1\,d^4k_2}{(2\pi)^4}. \tag{3.4}$$

Here, $H_s^{\sigma_s\sigma_1\sigma_2}(q,k_1,k_2)$ is the nonlinear electric current transition matrix,

$$H_e^{\sigma_s\sigma_1\sigma_2}(q,k_1,k_2)=\frac{1}{2}\int\frac{\boldsymbol{e}_s^{\sigma_s}\cdot\boldsymbol{v}}{\Omega-\boldsymbol{q}\cdot\boldsymbol{v}}\left\{\boldsymbol{e}_f'^{\sigma_1}\cdot\frac{\partial}{\partial\boldsymbol{v}}\left(\frac{\boldsymbol{e}_f'^{\sigma_2}\cdot(\partial/\partial\boldsymbol{v})f_{e0}}{\omega_2-\boldsymbol{k}_2\cdot\boldsymbol{v}}\right)+(1\leftrightarrow 2)\right\}d^3\boldsymbol{v} \tag{3.5}$$

where $(1\leftrightarrow 2)$ represents the preceding term with the indices 1 and 2 interchanged.

According to the definition of nonlinear current transition matrix, we have

$$H_\alpha^{\sigma_s\sigma_1\sigma_2}(q,k_1,k_2)=H_\alpha^{\sigma_s\sigma_1\sigma_2}(q,k_2,k_1). \tag{3.6}$$

When $\Omega=\omega_1+\omega_2$, $\boldsymbol{q}=\boldsymbol{k}_1+\boldsymbol{k}_2$ and $(\omega-\boldsymbol{k}\cdot\boldsymbol{v})_{1,2}\neq 0$, it can be shown that the following relational expression holds [9] (see the appendix)

$$\frac{1}{\Omega}H_\alpha^{\sigma_s\sigma_1\sigma_2}(q,k_1,k_2)=\frac{1}{\omega_2}H_\alpha^{\sigma_2\sigma_s\sigma_1}(k_2,q,-k_1). \tag{3.7}$$

Because we are only interested in the process of two fast-timescale motions producing one slow-timescale motion, the linear Landau damping effect for the fast-timescale wave can be neglected naturally, and the Doppler frequency shift in the principal value may be taken to the term of first order of $(\boldsymbol{k}\cdot\boldsymbol{v})/\omega$. Using relation (3.7), after a lengthy calculation, we find $H_c^{\sigma_s\sigma_1\sigma_2}(q,k_1,k_2)$ approximately

$$\begin{aligned}H_\alpha^{t\sigma_1\sigma_2}(q,k_1,k_2)\approx{}&\frac{n_0}{2\omega_1\omega_2^2}\boldsymbol{e}_s^t\cdot[\boldsymbol{q}\times[\boldsymbol{e}_f^{\sigma_1}\times\boldsymbol{e}_f^{\sigma_2}]]\\&-\frac{\Omega I_1(q)}{2}\boldsymbol{e}_s^t\cdot\left(\frac{(\boldsymbol{e}_f^{\sigma_2}\cdot\boldsymbol{k}_1)\boldsymbol{e}_f^{\sigma_1}}{\omega_1\omega_2^2}+\frac{(\boldsymbol{e}_f^{\sigma_1}\cdot\boldsymbol{k}_2)\boldsymbol{e}_f^{\sigma_2}}{\omega_2^3}\right)\end{aligned} \tag{3.8}$$

$$\begin{aligned}H_\alpha^{l\sigma_1\sigma_2}(q,k_1,k_2)\approx{}&\frac{\boldsymbol{e}_f^{\sigma_1}(k_1)\cdot\boldsymbol{e}_f^{\sigma_2}(k_2)\,|\boldsymbol{q}|\,\Omega}{2\omega_2^2}I_0(q)\\&-\frac{\Omega I_1(q)}{2}\boldsymbol{e}_s^l\cdot\left(\frac{(\boldsymbol{e}_f^{\sigma_2}\cdot\boldsymbol{k}_1)\boldsymbol{e}_f^{\sigma_1}}{\omega_1\omega_2^2}+\frac{(\boldsymbol{e}_f^{\sigma_1}\cdot\boldsymbol{k}_2)\boldsymbol{e}_f^{\sigma_2}}{\omega_2^3}\right)\end{aligned} \tag{3.9}$$

where

$$I_0(q)=-\int\frac{f_{e0}\,d^3\boldsymbol{v}}{(\Omega-\boldsymbol{q}\cdot\boldsymbol{v})^2} \tag{3.10}$$

$$I_1(q)=\int\frac{f_{e0}\,d^3\boldsymbol{v}}{(\Omega-\boldsymbol{q}\cdot\boldsymbol{v})}. \tag{3.11}$$

Noting the meaning of $\int_{(s)}$, we can replace the summation of $\sum_{\sigma_1\sigma_2}$ by two kinds of combination $\sum_{\sigma_1^-\sigma_2^+}$ and $\sum_{\sigma_1^+\sigma_2^-}$, where σ_1^+ and σ_1^- represent the waves with $(\boldsymbol{k}_1,\omega_1)$ and $(-\boldsymbol{k}_2,-\omega_2)$ respectively.

$$\begin{aligned}\boldsymbol{j}_{2s}(q)={}&\frac{\omega_{pe}^2e}{4\pi m_e\omega_1\omega_2^2}\int\sum_{\sigma_1^+\sigma_2^-}\boldsymbol{q}\times[\boldsymbol{E}_f^{\sigma_1^+}(k_1)\times\boldsymbol{E}_f^{\sigma_2^-*}(k_2)]\delta^4(q-k_1+k_2)\frac{d^4k_1\,d^4k_2}{(2\pi)^4}\\&+\frac{eI_0(q)\omega_{pe}^2\boldsymbol{q}\Omega}{4\pi n_0m_e\omega_2\omega_1}\int\sum_{\sigma_1^+\sigma_2^-}[\boldsymbol{E}_f^{\sigma_1^+}(k_1)\cdot\boldsymbol{E}_f^{\sigma_2^-*}(k_2)]\delta^4(q-k_1+k_2)\frac{d^4k_1\,d^4k_2}{(2\pi)^4}\\&-\frac{\Omega\omega_{pe}^2eI_1(q)}{4\pi n_0m_e\omega_2^3}\int\sum_{\sigma_1^+\sigma_2^-}[\boldsymbol{E}_f^{\sigma_1^+}(k_1)\cdot\boldsymbol{k}_2]\boldsymbol{E}_f^{\sigma_2^-*}(k_2)\delta^4(q-k_1+k_2)\frac{d^4k_1\,d^4k_2}{(2\pi)^4}\\&-\frac{\Omega\omega_{pe}^2eI_1(q)}{4\pi n_0m_e\omega_1\omega_2^2}\int\sum_{\sigma_1^+\sigma_2^-}[\boldsymbol{E}_f^{\sigma_2^-*}(k_2)\cdot\boldsymbol{k}_1]\boldsymbol{E}_f^{\sigma_1^+}(k_1)\delta^4(q-k_1+k_2)\frac{d^4k_1\,d^4k_2}{(2\pi)^4}.\end{aligned} \tag{3.12}$$

4. The expression for ponderomotive force

According to the slow-timescale electric current, we define a nonlinear force

$$\boldsymbol{F}_N(q)=\frac{i\Omega m_e\boldsymbol{j}_{2s}(q)}{e}. \tag{4.1}$$

When thermal pressure can be neglected, that is when $\Omega \gg \boldsymbol{q} \cdot \boldsymbol{v}$, $\boldsymbol{F}_{\mathrm{N}}(q)$ gives the ponderomotive force in the Fourier representation of both time and space. In this case, we have

$$I_0(q) = -\frac{n_0}{\Omega^2} \tag{4.2}$$

$$I_1(q) = \frac{n_0}{\Omega}. \tag{4.3}$$

Then, the ponderomotive force caused by electromagnetic fields is given by the following formula

$$\begin{aligned}\boldsymbol{F}_{\mathrm{P}}(q) = & \frac{\mathrm{i}\Omega\omega_{\mathrm{pe}}^2}{4\pi\omega_1\omega_2^2}\int\sum_{\sigma\dagger\sigma\bar{2}} \boldsymbol{q}\times(\boldsymbol{E}_{\mathrm{f}}^{\sigma\dagger}(k_1)\times\boldsymbol{E}_{\mathrm{f}}^{\sigma\bar{2}*}(k_2))\delta^4(q-k_1+k_2)\frac{\mathrm{d}^4k_1\,\mathrm{d}^4k_2}{(2\pi)^4} \\ & -\frac{\mathrm{i}\omega_{\mathrm{pe}}^2\boldsymbol{q}}{4\pi\omega_2\omega_1}\int\sum_{\sigma\dagger\sigma\bar{2}}(\boldsymbol{E}_{\mathrm{f}}^{\sigma\dagger}(k_1)\cdot\boldsymbol{E}_{\mathrm{f}}^{\sigma\bar{2}*}(k_2))\delta^4(q-k_1+k_2)\frac{\mathrm{d}^4k_1\,\mathrm{d}^4k_2}{(2\pi)^4} \\ & -\frac{\mathrm{i}\Omega\omega_{\mathrm{pe}}^2}{4\pi\omega_2^3}\int\sum_{\sigma\dagger\sigma\bar{2}}(\boldsymbol{E}_{\mathrm{f}}^{\sigma\dagger}(k_1)\cdot\boldsymbol{k}_2)\boldsymbol{E}_{\mathrm{f}}^{\sigma\bar{2}*}(k_2)\delta^4(q-k_1+k_2)\frac{\mathrm{d}^4k_1\,\mathrm{d}^4k_2}{(2\pi)^4} \\ & -\frac{\mathrm{i}\Omega\omega_{\mathrm{pe}}^2}{4\pi\omega_1\omega_2^2}\int\sum_{\sigma\dagger\sigma\bar{2}}(\boldsymbol{E}_{\mathrm{f}}^{\sigma\bar{2}*}(k_2)\cdot\boldsymbol{k}_1)\boldsymbol{E}_{\mathrm{f}}^{\sigma\dagger}(k_1)\delta^4(q-k_1+k_2)\frac{\mathrm{d}^4k_1\,\mathrm{d}^4k_2}{(2\pi)^4}.\end{aligned} \tag{4.4}$$

Taking the inverse Fourier transform of above equation with respect to both time and space

$$\begin{aligned}\boldsymbol{F}_{\mathrm{P}}(\boldsymbol{x},t) = & -\frac{1}{8\pi}\frac{\omega_{\mathrm{pe}}^2}{\omega^2}\nabla|\boldsymbol{E}_{\mathrm{f}}|^2+\frac{\mathrm{i}\omega_{\mathrm{pe}}^2}{8\pi\omega^3}\frac{\partial}{\partial t}[\nabla\times(\boldsymbol{E}_{\mathrm{f}}\times\boldsymbol{E}_{\mathrm{f}}^*)] \\ & +\frac{\mathrm{i}\omega_{\mathrm{pe}}^2}{8\pi\omega^3}\frac{\partial}{\partial t}[(\boldsymbol{E}_{\mathrm{f}}\cdot\nabla)\boldsymbol{E}_{\mathrm{f}}^*]-\frac{\mathrm{i}\omega_{\mathrm{pe}}^2}{8\pi\omega^3}\frac{\partial}{\partial t}[(\boldsymbol{E}_{\mathrm{f}}^*\cdot\nabla)\boldsymbol{E}_{\mathrm{f}}]\end{aligned} \tag{4.5}$$

where we make an approximation of $\omega_1 \approx \omega_2 \approx \omega$; ω is a characteristic frequency, and

$$|\boldsymbol{E}_{\mathrm{f}}|^2 = 2\int\sum_{\sigma\dagger\sigma\bar{2}}[\boldsymbol{E}_{\mathrm{f}}^{\sigma\dagger}(k_1)\cdot\boldsymbol{E}_{\mathrm{f}}^{\sigma\bar{2}*}(k_2)]\exp[(\boldsymbol{k}_1-\boldsymbol{k}_2)\cdot\boldsymbol{x}-(\omega_1-\omega_2)t]\cdot\frac{\mathrm{d}^4k_1\,\mathrm{d}^4k_2}{(2\pi)^8} \tag{4.6}$$

$$\boldsymbol{E}_{\mathrm{f}}\times\boldsymbol{E}_{\mathrm{f}}^* = 2\int\sum_{\sigma\dagger\sigma\bar{2}}[\boldsymbol{E}_{\mathrm{f}}^{\sigma\dagger}(k_1)\times\boldsymbol{E}_{\mathrm{f}}^{\sigma\bar{2}*}(k_2)]\exp[(\boldsymbol{k}_1-\boldsymbol{k}_2)\cdot\boldsymbol{x}-(\omega_1-\omega_2)t]\cdot\frac{\mathrm{d}^4k_1\,\mathrm{d}^4k_2}{(2\pi)^8}. \tag{4.7}$$

Substituting equation (1.2) into equation (4.5), we obtain

$$\begin{aligned}\boldsymbol{F}_{\mathrm{P}}(\boldsymbol{x},t) = & -\frac{1}{8\pi}\frac{\omega_{\mathrm{pe}}^2}{\omega^2}\nabla|\boldsymbol{E}_{\mathrm{f}}|^2+\frac{\mathrm{i}\omega_{\mathrm{pe}}^2}{16\pi\omega^3}\frac{\partial}{\partial t}[\nabla\times(\boldsymbol{E}\times\boldsymbol{E}^*)] \\ & +\frac{\mathrm{i}\omega_{\mathrm{pe}}^2}{16\pi\omega^3}\frac{\partial}{\partial t}[(\boldsymbol{E}\cdot\nabla)\boldsymbol{E}^*-(\boldsymbol{E}^*\cdot\nabla)\boldsymbol{E}].\end{aligned} \tag{4.8}$$

After simplification

$$\begin{aligned}\boldsymbol{F}_{\mathrm{P}}(\boldsymbol{x},t) & = -\frac{1}{8\pi}\frac{\omega_{\mathrm{pe}}^2}{\omega^2}\nabla|\boldsymbol{E}_{\mathrm{f}}|^2+\frac{\mathrm{i}\omega_{\mathrm{pe}}^2}{16\pi\omega^3}\frac{\partial}{\partial t}[\boldsymbol{E}(\nabla\cdot\boldsymbol{E}^*)-\boldsymbol{E}^*(\nabla\cdot\boldsymbol{E})] \\ & = -\frac{1}{8\pi}\frac{\omega_{\mathrm{pe}}^2}{\omega^2}\nabla|\boldsymbol{E}_{\mathrm{f}}|^2+\frac{\mathrm{i}\omega_{\mathrm{pe}}^2}{4\omega^3}\frac{\partial}{\partial t}[\boldsymbol{E}\rho^*-\boldsymbol{E}^*\rho]\end{aligned} \tag{4.9}$$

where ρ is the fast-timescale charge density perturbation. Equation (4.9) is our central result. The physics in equation (4.9) is clear, the first term on the right-hand side comes from the wave–wave interaction, and the second term is due to the interaction between wave and fast-timescale density perturbation which is driven by the fast-timescale electric field.

5. Discussions

Equation (4.8) can be read

$$F_P = -\frac{1}{8\pi}\frac{\omega_{pe}^2}{\omega^2}\nabla\langle E_f^2\rangle + \frac{i\omega_{pe}^2}{16\pi\omega^3}\frac{\partial}{\partial t}[\nabla\times(E\times E^*)]$$
$$+\frac{i\omega_{pe}^2}{8\pi\omega^3}\left[\left(\frac{\partial E}{\partial t}\cdot\nabla\right)E^* - \left(\frac{\partial E^*}{\partial t}\cdot\nabla\right)E\right]$$
$$+\frac{i\omega_{pe}^2}{16\pi\omega^3}\left[\frac{\partial E}{\partial t}\times(\nabla\times E^*) - \frac{\partial E^*}{\partial t}\times(\nabla\times E) + E^*\times\left(\nabla\times\frac{\partial E}{\partial t}\right)\right.$$
$$\left.-E\times\left(\nabla\times\frac{\partial E^*}{\partial t}\right)\right] - \frac{i\omega_{pe}^2}{16\pi\omega^3}\nabla\left(\frac{\partial E}{\partial t}\cdot E^* - \frac{\partial E^*}{\partial t}\cdot E\right). \tag{5.1}$$

Equation (5.1) is given by [3], apart from the last term on right-hand side. In a recent paper, Lee and Parks [5] calculated the ponderomotive force in a warm two-fluid plasma in Fourier representation. It is easy to prove that the right-hand side of their equation (35) is just the last three terms on the right-hand side of equation (4.8) if a stationary and cold medium is assumed in [5]. Equation (5.1) may also be compared with equation (1.1). We can now see that the earlier kinetic [1] calculation not only failed to produce the third term on the right-hand side of equation (5.1) (and equation (1.1)), but also several additional terms. These additional terms come from a proper calculation of the nonlinear electric current transition matrix, as indicated by equations (3.8) and (3.9).

We discuss the ponderomotive force caused by electromagnetic fields, but the fast-timescale magnetic field obviously does not appear in equation (4.8). The reason is that we only keep the terms of first-order of $(k\cdot v)/\omega$ while the contribution of the magnetic field to nonlinear electric current is of a higher order. It should be pointed out that a time derivative of the fast-timescale magnetic field is not ignored though the magnetic field itself can be neglected. So the terms containing $\nabla\times E$ in equation (5.1) are not contradictory. In the above calculation, we neglect the effect of thermal motion. But, in many applications this effect is important. Because of the purpose of this paper we leave this problem for the future.

Acknowledgment

We would like to thank the referees for their comments and good suggestions.

Appendix. Proof of equation (3.7)

According to the definition of nonlinear electric current matrix $H_e^{\sigma_s\sigma_1\sigma_2}(q, k_1, k_2)$ we now prove that

$$\frac{1}{\Omega}\int \frac{e_s^{\sigma_s}\cdot v}{\Omega-q\cdot v}\left[e_f^{\prime\sigma_1}\cdot\frac{\partial}{\partial v}\left(\frac{e_f^{\prime\sigma_2}\cdot(\partial/\partial v)f_{e0}}{\omega_2-k_2\cdot v}\right)+(1\leftrightarrow 2)\right]d^3v$$
$$-\frac{1}{\omega_2}\int\frac{e_f^{\sigma_2}\cdot v}{\omega_2-k_2\cdot v}\left[e_s^{\prime\sigma_s}\cdot\frac{\partial}{\partial v}\left(\frac{e_f^{\prime\sigma_1}\cdot(\partial\partial v)f_{e0}}{\omega_1-k_1\cdot v}\right)\right.$$
$$\left.-e_f^{\prime\sigma_1}\cdot\frac{\partial}{\partial v}\left(\frac{e_s^{\prime\sigma_s}\cdot(\partial\partial v)f_{e0}}{\Omega-q\cdot v}\right)\right]d^3v=0. \tag{A.1}$$

Making partial integration, we have

$$(A.1)=-\int\frac{e_f^{\prime\sigma_1}\cdot(\partial/\partial v)f_{e0}}{\omega_1-k_1\cdot v}\left(e_f^{\prime\sigma_2}\cdot\frac{\partial}{\partial v}\frac{e_s^{\sigma_s}\cdot v}{\Omega(\Omega-q\cdot v)}+e_s^{\prime\sigma_s}\cdot\frac{\partial}{\partial v}\frac{e_f^{\sigma_2}\cdot v}{\omega(\omega_2-k_2\cdot v)}\right)d^3v$$
$$-\int\left(\frac{e_f^{\prime\sigma_2}\cdot(\partial/\partial v)f_{e0}}{\omega_2-k_2\cdot v}e_f^{\prime\sigma_1}\cdot\frac{\partial}{\partial v}\frac{e_s^{\sigma_s}\cdot v}{\Omega(\Omega-q\cdot v)}\right.$$
$$\left.+\frac{e_s^{\prime\sigma_s}\cdot(\partial/\partial v)f_{e0}}{\Omega-q\cdot v}e_f^{\prime\sigma_1}\cdot\frac{\partial}{\partial v}\frac{e_f^{\sigma_2}\cdot v}{\omega_2(\omega_2-k_2\cdot v)}\right)d^3v=\int Y\,d^3v \tag{A.2}$$

$$Y=\frac{\partial}{\partial v}f_{e0}\cdot\left(\frac{e_f^{\prime\sigma_1}e_f^{\prime\sigma_2}\cdot e_s^{\prime\sigma_s}-e_f^{\prime\sigma_2}e_f^{\prime\sigma_1}\cdot e_s^{\prime\sigma_s}}{(\Omega-q\cdot v)^2(\omega_2-k_2\cdot v)}+\frac{e_f^{\prime\sigma_1}e_f^{\prime\sigma_2}\cdot e_s^{\prime\sigma_s}-e_s^{\prime\sigma_s}e_f^{\prime\sigma_1}\cdot e_f^{\prime\sigma_2}}{(\Omega-q\cdot v)(\omega_2-k_2\cdot v)^2}\right). \tag{A.3}$$

Substituting equation (2.13) into equation (A.3) and using the relations $\Omega=\omega_1+\omega_2$, $q=k_1+k_2$, we have

$$\Omega\omega_1\omega_2Y=\frac{\partial}{\partial v}f_{eo}\cdot\left(e_s^{\sigma_s}e_f^{\sigma_1}\cdot e_f^{\sigma_2}-e_f^{\sigma_2}e_f^{\sigma_1}\cdot e_s^{\sigma_s}\right.$$
$$+\frac{1}{\Omega-q\cdot v}(k_1e_f^{\sigma_1}\cdot ve_f^{\sigma_2}\cdot e_s^{\sigma_s}-e_f^{\sigma_1}e_s^{\sigma_s}\cdot k_2e_f^{\sigma_2}\cdot v$$
$$+k_2e_f^{\sigma_2}\cdot ve_f^{\sigma_1}\cdot e_s^{\sigma_s}-e_f^{\sigma_2}e_f^{\sigma_1}\cdot qe_s^{\sigma_s}\cdot v-e_f^{\sigma_2}k_1\cdot e_s^{\sigma_s}e_f^{\sigma_1}\cdot v$$
$$+qe_s^{\sigma_s}\cdot ve_f^{\sigma_1}\cdot e_f^{\sigma_2})+\frac{\omega_2-k_2\cdot v}{\Omega-q\cdot v}(e_f^{\sigma_2}e_f^{\sigma_1}\cdot e_s^{\sigma_s}-e_f^{\sigma_1}e_f^{\sigma_2}\cdot e_s^{\sigma_s})$$
$$+\frac{\omega_2-k_2\cdot v}{(\Omega-q\cdot v)^2}e_s^{\sigma_s}\cdot v(e_f^{\sigma_2}e_f^{\sigma_1}\cdot q-e_f^{\sigma_1}e_f^{\sigma_2}\cdot q)$$
$$+\frac{e_s^{\sigma_s}\cdot v}{(\Omega-q\cdot v)^2}(k_1e_f^{\sigma_1}\cdot ve_f^{\sigma_2}\cdot q-e_f^{\sigma_1}q\cdot k_2e_f^{\sigma_2}\cdot v$$
$$+k_2e_f^{\sigma_2}\cdot ve_f^{\sigma_1}\cdot q-e_f^{\sigma_2}q\cdot k_1e_f^{\sigma_1}\cdot v)$$
$$+\frac{1}{(\Omega-q\cdot v)(\omega_2-k_2\cdot v)}(qe_f^{\sigma_1}\cdot ve_s^{\sigma_s}\cdot k_2e_f^{\sigma_2}\cdot v$$
$$-k_2q\cdot e_s^{\sigma_s}e_f^{\sigma_1}\cdot ve_f^{\sigma_2}\cdot v-k_2q\cdot e_f^{\sigma_1}e_f^{\sigma_2}\cdot ve_s^{\sigma_s}\cdot v$$
$$-k_2q\cdot e_f^{\sigma_2}e_f^{\sigma_1}\cdot ve_s^{\sigma_s}\cdot v+$$

$$
\begin{aligned}
&+\boldsymbol{q}\boldsymbol{e}_s^{\sigma_s}\cdot\boldsymbol{v}\boldsymbol{k}_2\cdot\boldsymbol{e}_f^{\sigma_2}\boldsymbol{e}_f^{\sigma_1}\cdot\boldsymbol{v}+\boldsymbol{q}\boldsymbol{e}_s^{\sigma_s}\cdot\boldsymbol{v}\boldsymbol{k}_2\cdot\boldsymbol{e}_f^{\sigma_1}\boldsymbol{e}_f^{\sigma_2}\cdot\boldsymbol{v}) \\
&+\frac{\boldsymbol{e}_s^{\sigma_s}\cdot\boldsymbol{v}\boldsymbol{e}_f^{\sigma_1}\cdot\boldsymbol{v}\boldsymbol{e}_f^{\sigma_2}\cdot\boldsymbol{v}}{(\Omega-\boldsymbol{q}\cdot\boldsymbol{v})^2(\omega_2-\boldsymbol{k}_2\cdot\boldsymbol{v})}(\boldsymbol{q}\boldsymbol{q}\cdot\boldsymbol{k}_2-\boldsymbol{k}_2\boldsymbol{q}\cdot\boldsymbol{q}) \\
&+\frac{\Omega-\boldsymbol{q}\cdot\boldsymbol{v}}{\omega_2-\boldsymbol{k}_2\cdot\boldsymbol{v}}(\boldsymbol{e}_f^{\sigma_1}\boldsymbol{e}_s^{\sigma_s}\cdot\boldsymbol{e}_f^{\sigma_2}-\boldsymbol{e}_s^{\sigma_s}\boldsymbol{e}_f^{\sigma_1}\cdot\boldsymbol{e}_f^{\sigma_2}) \\
&+\frac{1}{\omega_2-\boldsymbol{k}_2\cdot\boldsymbol{v}}(\boldsymbol{e}_f^{\sigma_1}\boldsymbol{e}_f^{\sigma_2}\cdot\boldsymbol{q}\boldsymbol{e}_s^{\sigma_s}\cdot\boldsymbol{v}+\boldsymbol{k}_1\boldsymbol{e}_f^{\sigma_1}\cdot\boldsymbol{v}\boldsymbol{e}_f^{\sigma_2}\cdot\boldsymbol{e}_s^{\sigma_s} \\
&+\boldsymbol{e}_s^{\sigma_s}\boldsymbol{e}_f^{\sigma_1}\cdot\boldsymbol{k}_2\boldsymbol{e}_f^{\sigma_2}\cdot\boldsymbol{v}-\boldsymbol{e}_s^{\sigma_s}\boldsymbol{k}_1\cdot\boldsymbol{e}_f^{\sigma_2}\boldsymbol{e}_f^{\sigma_1}\cdot\boldsymbol{v}-\boldsymbol{q}\boldsymbol{e}_s^{\sigma_s}\cdot\boldsymbol{v}\boldsymbol{e}_f^{\sigma_1}\cdot\boldsymbol{e}_f^{\sigma_2} \\
&-\boldsymbol{k}_2\boldsymbol{e}_f^{\sigma_2}\cdot\boldsymbol{v}\boldsymbol{e}_f^{\sigma_1}\cdot\boldsymbol{e}_s^{\sigma_s}) \\
&+\frac{\Omega-\boldsymbol{q}\cdot\boldsymbol{v}}{(\omega_2-\boldsymbol{k}_2\cdot\boldsymbol{v})^2}\boldsymbol{e}_f^{\sigma_2}\cdot\boldsymbol{v}(\boldsymbol{e}_f^{\sigma_1}\boldsymbol{e}_s^{\sigma_s}\cdot\boldsymbol{k}_2-\boldsymbol{e}_s^{\sigma_s}\boldsymbol{e}_f^{\sigma_1}\cdot\boldsymbol{k}_2) \\
&+\frac{\boldsymbol{e}_f^{\sigma_2}\cdot\boldsymbol{v}}{(\omega_2-\boldsymbol{k}_2\cdot\boldsymbol{v})^2}(\boldsymbol{e}_f^{\sigma_1}\boldsymbol{k}_2\cdot\boldsymbol{q}\boldsymbol{e}_s^{\sigma_s}\cdot\boldsymbol{v}+\boldsymbol{k}_1\boldsymbol{e}_s^{\sigma_s}\cdot\boldsymbol{k}_2\boldsymbol{e}_f^{\sigma_1}\cdot\boldsymbol{v} \\
&-\boldsymbol{e}_s^{\sigma_s}\boldsymbol{k}_1\cdot\boldsymbol{k}_2\boldsymbol{e}_f^{\sigma_1}\cdot\boldsymbol{v}-\boldsymbol{q}\boldsymbol{e}_s^{\sigma_s}\cdot\boldsymbol{v}\boldsymbol{e}_f^{\sigma_1}\cdot\boldsymbol{k}_2) \\
&+\frac{\boldsymbol{e}_s^{\sigma_s}\cdot\boldsymbol{v}\boldsymbol{e}_f^{\sigma_1}\cdot\boldsymbol{v}\boldsymbol{e}_f^{\sigma_2}\cdot\boldsymbol{v}}{(\Omega-\boldsymbol{q}\cdot\boldsymbol{v})(\omega_2-\boldsymbol{k}_2\cdot\boldsymbol{v})^2}(\boldsymbol{q}\boldsymbol{k}_2\cdot\boldsymbol{k}_2-\boldsymbol{k}_2\boldsymbol{q}\cdot\boldsymbol{k}_2)\Big) \\
=&\frac{\partial f_{0\alpha}}{\partial\boldsymbol{v}}\cdot(\boldsymbol{e}_s^{\sigma_s}\boldsymbol{e}_f^{\sigma_1}\cdot\boldsymbol{e}_f^{\sigma_2}-\boldsymbol{e}_f^{\sigma_2}\boldsymbol{e}_f^{\sigma_1}\cdot\boldsymbol{e}_s^{\sigma_s}) \\
&+\frac{\partial}{\partial\boldsymbol{v}}\cdot\Big((\boldsymbol{e}_f^{\sigma_1}\boldsymbol{e}_f^{\sigma_2}-\boldsymbol{e}_f^{\sigma_2}\boldsymbol{e}_f^{\sigma_1})\cdot\frac{\partial f_{0\alpha}}{\partial\boldsymbol{v}}\frac{\boldsymbol{e}_s^{\sigma_s}\cdot\boldsymbol{v}}{\Omega-\boldsymbol{q}\cdot\boldsymbol{v}}(\omega_2-\boldsymbol{k}_2\cdot\boldsymbol{v}) \\
&+(\boldsymbol{e}_s^{\sigma_s}\boldsymbol{e}_f^{\sigma_1}-\boldsymbol{e}_f^{\sigma_1}\boldsymbol{e}_s^{\sigma_s})\cdot\frac{\partial f_{0\alpha}}{\partial\boldsymbol{v}}\frac{\boldsymbol{e}_f^{\sigma_2}\cdot\boldsymbol{v}}{\omega_2-\boldsymbol{k}_2\cdot\boldsymbol{v}}(\Omega-\boldsymbol{q}\cdot\boldsymbol{v}) \\
&+(\boldsymbol{e}_f^{\sigma_2}\boldsymbol{k}_1-\boldsymbol{k}_1\boldsymbol{e}_f^{\sigma_2})\cdot\frac{\partial f_{0\alpha}}{\partial\boldsymbol{v}}\frac{\boldsymbol{e}_s^{\sigma_s}\cdot\boldsymbol{v}\boldsymbol{e}_f^{\sigma_1}\cdot\boldsymbol{v}}{\Omega-\boldsymbol{q}\cdot\boldsymbol{v}} \\
&+(\boldsymbol{e}_f^{\sigma_1}\boldsymbol{k}_2-\boldsymbol{k}_2\boldsymbol{e}_f^{\sigma_1})\cdot\frac{\partial f_{0\alpha}}{\partial\boldsymbol{v}}\frac{\boldsymbol{e}_s^{\sigma_s}\cdot\boldsymbol{v}\boldsymbol{e}_f^{\sigma_2}\cdot\boldsymbol{v}}{\Omega-\boldsymbol{q}\cdot\boldsymbol{v}} \\
&+(\boldsymbol{e}_s^{\sigma_s}\boldsymbol{k}_1-\boldsymbol{k}_1\boldsymbol{e}_s^{\sigma_s})\cdot\frac{\partial f_{0\alpha}}{\partial\boldsymbol{v}}\frac{\boldsymbol{e}_f^{\sigma_1}\cdot\boldsymbol{v}\boldsymbol{e}_f^{\sigma_2}\cdot\boldsymbol{v}}{\omega_2-\boldsymbol{k}_2\cdot\boldsymbol{v}} \\
&+(\boldsymbol{q}\boldsymbol{e}_f^{\sigma_1}-\boldsymbol{e}_f^{\sigma_1}\boldsymbol{q})\cdot\frac{\partial f_{0\alpha}}{\partial\boldsymbol{v}}\frac{\boldsymbol{e}_f^{\sigma_2}\cdot\boldsymbol{v}\boldsymbol{e}_s^{\sigma_s}\cdot\boldsymbol{v}}{\omega_2-\boldsymbol{k}_2\cdot\boldsymbol{v}} \\
&+(\boldsymbol{k}_2\boldsymbol{q}-\boldsymbol{q}\boldsymbol{k}_2)\cdot\frac{\partial f_{0\alpha}}{\partial\boldsymbol{v}}\frac{\boldsymbol{e}_f^{\sigma_1}\cdot\boldsymbol{v}\boldsymbol{e}_f^{\sigma_2}\cdot\boldsymbol{v}\boldsymbol{e}_s^{\sigma_s}\cdot\boldsymbol{v}}{(\Omega-\boldsymbol{q}\cdot\boldsymbol{v})(\omega_2-\boldsymbol{k}_2\cdot\boldsymbol{v})}\Big).
\end{aligned}
$$

Then, equation (3.7) has been proved.

References

[1] Kono M, Skoric M M and Ter Haar D 1981 *J. Plasma Phys.* **26** 123
[2] Cary J and Kaufman A N 1981 *Phys. Fluids* **24** 1238

[3] Statham T and Ter Haar D 1983 *Plasma Phys.* **25** 681
[4] Karpman V I and Shagalov A G 1982 *J. Plasma Phys.* **27** 215
[5] Lee N C and Parks G K 1983 *Phys. Fluids* **26** 724
[6] Washimi H and Karpman V I 1976 *Sov. Phys.–JETP* **44** 528
[7] Landau L D and Lifshitz E M 1959 *Electrodynamics of Continuous Media* (Oxford: Pergamon)
[8] He X T 1983 *Acta. Physica Sinica* **32** 325 (in Chinese)
[9] Yu M 1984 private communication

PHYSICS OF PLASMAS VOLUME 6, NUMBER 7 JULY 1999

等离子体自生磁场 477

Magnetic field generated by ionization front produced by intense laser radiating gas

Guihua Zeng
National Key Laboratory on ISDN of XiDian University, Xi'an, 710071, China
Xiantu He
Institute of Applied Physics and Computational Mathematics, Beijing 100088, China

(Received 27 July 1998; accepted 10 March 1999)

A general evolution equation for the quasistatic magnetic field (QMF) in a relativistic regime is derived, and an ionization front mechanism for QMF generation is proposed. The properties of the QMF produced by the ionization front mechanism are investigated numerically: The direction of the QMF changes periodically along the direction of the laser propagation, and the generated magnetic field depends on the temperature of the electron plasma and on the effective radial width of the ionization front. © *1999 American Institute of Physics.* [S1070-664X(99)02806-2]

The new generation of laser systems that produces intense short laser pulses[1] makes possible the study of the generation of the quasistatic magnetic field (QMF) and its properties in a new regime, in which the relativistic effect and the ponderomotive effect are significant.[2,3] In this relativistic regime, the proposed mechanisms for QMF generation include the ponderomotive mechanism,[4–7] inverse Faraday effect,[8–10] and vortex dynamics mechanism.[11] The QMF has relevant consequences: it can influence the propagation of the laser pulse,[8] modify the plasma density profile,[12] and reduce the growth rate of the Raman backscattering instability.[13]

The main objective of this work is to derive a general evolution equation for the QMF in a relativistic regime, and use the derived equation to investigate the magnetic field driven by the ionization front mechanism. Previously, a nonrelativistic equation was derived in Ref. 11, and a similar equation in a uniform rarefied plasma was obtained in Ref. 6 by the perturbation theory. However, the relativistic effect is so significant that it cannot be neglected. With this consideration, we develop the previous theory and derive a relativistic evolution equation. Then a new mechanism for QMF generation, in which the magnetic field is driven by the ionization front, is proposed. By using the model of ionization front presented in Ref. 14, the magnetic field and its properties are investigated.

One obtains the following equation by Maxwell's equations:

$$\nabla^2\mathbf{B}-\frac{1}{c^2}\frac{\partial^2\mathbf{B}}{\partial t^2}=\frac{4\pi}{c}\nabla\times\mathbf{J}, \tag{1}$$

where $\mathbf{J}=-en\mathbf{V}$, n and $\mathbf{V}$ are the electron density and velocity in plasma, respectively. $\mathbf{B}$ is the total magnetic field, $\mathbf{B}=\mathbf{B}_L+\mathbf{B}_s$, $\mathbf{B}_L$ represents the magnetic field component of the laser, which is oscillating at the fast laser frequency ω, and $\mathbf{B}_s$ is the self-generated magnetic field. One can easily find that the right-hand side of Eq. (1) equals $-4\pi e/c(\nabla n\times\mathbf{V}+n\nabla\times\mathbf{V})$. Inserting $\mathbf{P}=m\gamma\mathbf{V}$ into the above expression and using the relationship $e\mathbf{B}/c=\nabla\times\mathbf{P}$ presented in Ref. 5 ($\mathbf{P}$ denotes the canonical momentum), the right-hand side of Eq. (1) becomes $(k_p^2/mc)\nabla(n/\gamma n_0)\times\mathbf{P}-(nk_p^2/\gamma n_0)\mathbf{B}$. Moving the term $(nk_p^2/\gamma n_0)\mathbf{B}$ to the left-hand side of Eq. (1) and averaging over the fast laser frequency time scale, but not over the laser pulse time scale, one obtains

$$\nabla^2\mathbf{b}_s-\frac{\partial^2\mathbf{b}_s}{\partial\tau^2}+\frac{1}{\gamma}\mathbf{b}_s=\left\langle\nabla\left(\frac{n}{n_0\gamma}\right)\times\mathbf{p}\right\rangle, \tag{2}$$

where $\mathbf{b}_s=(e/m\omega_p c)\mathbf{B}_s$. $\langle\cdots\rangle$ denotes the averaging over the fast laser frequency. ω_p is the plasma wave frequency. $\mathbf{p}=\mathbf{P}/mc$ is the normalized momentum of the electron plasma. n_0 is the ambient electron plasma density. $\gamma=\sqrt{1+p^2}$ is the relativistic factor. For concision and convenience, the normalized QMF $\mathbf{b}_s$, momentum $\mathbf{p}$, and the variables $\rho=k_p r$, $\theta=k_p\varphi$, $\eta=k_p z$, $\tau=\omega_p t$ are used in Eq. (2), and cylindrical coordinates are chosen. The Laplace operator takes the form

$$\nabla^2=\frac{1}{\rho}\frac{\partial}{\partial\rho}\left(\rho\frac{\partial}{\partial\rho}\right)+\frac{1}{\rho^2}\frac{\partial^2}{\partial\theta^2}+\frac{\partial^2}{\partial\eta^2}.$$

Equation (2) is valid for the ponderomotive mechanism and inverse Faraday mechanism. It does not contain restrictions for the laser parameters (i.e., polarization and intensities) and plasma characteristics (i.e., underdense, overdense plasma, and ionization front).

The relativistic effect plays an important role in the QMF generation. Let

$$\mathbf{S}=\left\langle\nabla\left(\frac{n}{n_0\gamma}\right)\times\mathbf{p}\right\rangle.$$

$\mathbf{S}$ can be viewed as the source of QMF generation, it depends on the relativistic factor γ. In addition, the linear term $\mathbf{b}_s/\gamma$ of Eq. (2) also depends on γ. This implies that the relativistic effect significantly influences the magnitude, growth, and saturation of the QMF. In the weakly relativistic regime $[(v/c)^2\ll 1]$, Eq. (2) has a form analogous to that presented in Refs. 6 and 8.

贺贤土论文选集
478

The QMF can be generated by several mechanisms; in the following we propose a new mechanism for the QMF generation. As an intense short laser pulse or an electron beam propagates in a gas, it quickly ionizes the gas at or near the front of the pulse, and thereby sets up an ionization front. This ionization of the gas produces a density gradient in the ionization front. This density gradient will induce a movement of the electron plasma in the region of the ionization front and of its wake, that will produce a current. The established electron current density gives rise to a QMF directed along an azimuthal direction. This induced magnetic field moves with the ionization front. It is appropriate to emphasize differences between the ponderomotive and ionization front mechanism for QMF generation. The former[6] is generated by the ponderomotive force of laser pulse and the latter is generated by the density gradient of ionization front. So, the corresponding generated magnetic fields have different properties.

For convenience, we use the ionization front model presented in Ref. 14 to investigate the properties of the generated magnetic field. In this model, the ionization front, produced by intense short laser pulse radiating gas, is modeled as a pressure force due to the density gradient, and the ionizing agent is assumed to be constant in time. This model is valid if the ionizing agent causing the force does not change significantly over the time of interest. When the velocities of electron plasma induced by the ionization front are weakly relativistic ($|v/c|\ll 1$), we employ a linear two-dimensional fluid model with axial symmetry, and we assume a Maxwellian distribution for the thermal velocities. The motion of the plasma electrons induced by the pressure force can be described by the momentum equation, $\partial \mathbf{P}/\partial t=\nabla(e\Phi+(T_e/n_0)n_{iz}(r,z,t))$, where T_e is the electron temperature, assumed to be a constant in the calculation. n_{iz} is the ionization front density profile, n_0 is the density of background plasma, and Φ is the Wakefield potential produced by the ionization front, which satisfies the Possion's equation.[14] Φ is given by $\Phi=k_p\int_{-\infty}^{\xi}d\xi' \sin k_p(\xi-\xi')(T_e n_{iz}/en_0)$, where $\xi=z-v_g t$, v_g is the group velocity. Combining the linearized continuity equation and Poisson's equation, we obtain

$$\frac{n}{n_0\gamma}=1+\frac{T_e}{e}\nabla^2(\sin(k_p\xi)e^{-(r^2/r_0^2)}), \tag{3}$$

$$\mathbf{P}=\frac{T_e}{\omega_p}\nabla[(1-\cos(k_p\xi))e^{-(r^2/r_0^2)}], \tag{4}$$

where we use the ionization front profile presented in Ref. 14,

$$n_{iz}=\frac{n_0}{\omega_p}e^{-(r^2/r_0^2)}\delta\left(\frac{\xi}{c}\right), \tag{5}$$

with $\delta(\xi/c)$ being the Dirac function. This form of ionization profile is valid when the ionization profile time scale τ_i is smaller than the plasma time scale τ_p, i.e., $\tau_i\ll\tau_p$.

We may obtain the azimuthal component of the quasi-static magnetic field produced by the ionization front by Eqs. (2)–(4), but the analytic solution is very difficult. Because of the symmetry, the magnetic field is independent of the θ coordinate. Assuming $\partial^2 b_s/\partial\tau^2\simeq 0$ in Eq. (2), and using the Hankel transform, one gets the solution for the generated magnetic field in the form,

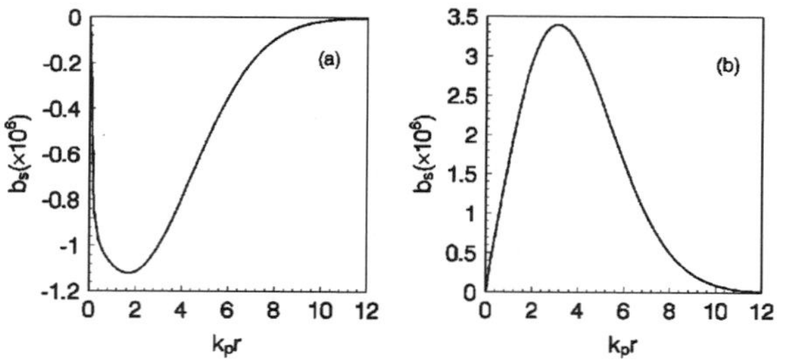

FIG. 1. The variation of the normalized QMF with radial coordinates. The parameters are $T_e=0.03$, $k_pr_0=6.28$, and (a) $k_p\xi=11.4$, (b) $k_p\xi=16.2$.

$$b_s=\int_0^{\infty}\lambda J_0(\lambda r)S(\lambda)d\lambda, \tag{6}$$

where $J_0(\lambda r)$ is the zero-order Bessel function and $S(\lambda)$ can be obtained by

$$S(\lambda)=\frac{r_g}{2\lambda}\int_0^{\xi}d\xi(e^{r_g\lambda(\xi-t)}-e^{-r_g\lambda(\xi-t)})\left\langle\nabla\left(\frac{n}{n_0\gamma}\right)\times\mathbf{p}\right\rangle,$$

where $r_g^2=1/(1-\beta_g^2)$, $\beta_g=v_g/c$.

Some properties of the QMF can be derived by the numerical solutions of Eq. (2). First, we consider the radial distribution of the generated magnetic field. The variation of magnetic field with radial coordinate is shown in Fig. 1. In Fig. 1(a) ($k_p\xi=11.4$) one sees that the distribution of QMF in the radial direction has a similar structure to the QMF produced by the ponderomotive mechanism.[6] However, the magnetic field does not always keep the distribution shown in Fig. 1(a), another radial distribution appears at a different position in the ionization front mechanism, as plotted in Fig. 1(b) ($k_p\xi=16.2$). The two distributions of QMF have similar radial structure but the generated magnetic fields have opposite direction. Positions corresponding to the maximum of the induced magnetic field are different between Figs. 1(a) and 1(b): in Fig. 1(a) the maximum is located at $k_pr=2.4$; however, in Fig. 1(b), it is located at $k_pr=3.0$.

Second, we consider the longitudinal distribution of the QMF. Figure 2 shows the normalized magnetic field vs the

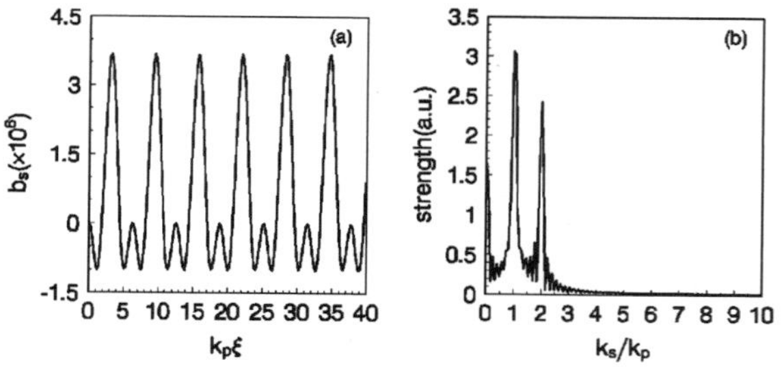

FIG. 2. (a) The normalized QMF vs the dimensionless longitudinal variable taken at the distance $k_pr=3.0$. (b) The forward Fourier transform of (a). The parameters are $T_e=0.03$, $k_pr_0=6.28$.

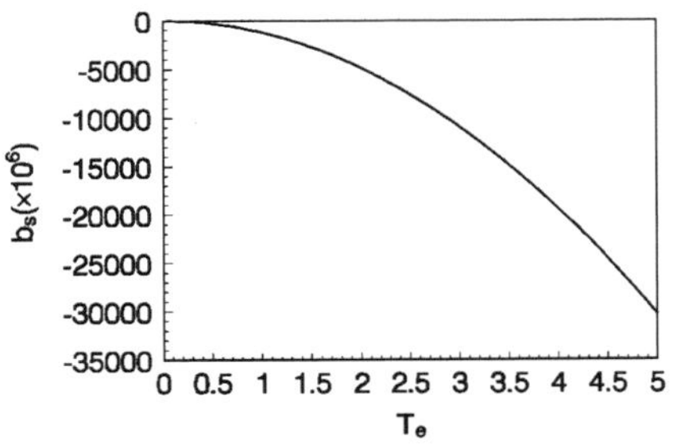

FIG. 3. The normalized QMF vs the normalized temperature at $k_p r = 2.41$, $k_p \xi = 11.4$, $k_p r_0 = 6.28$.

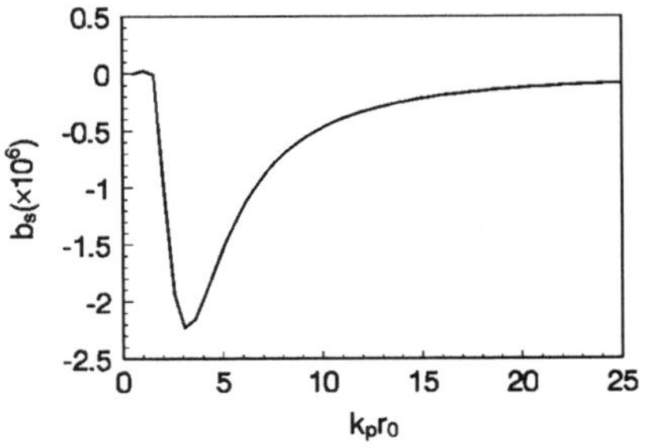

FIG. 4. The normalized QMF vs the radial width $k_p r_0$ of the ionization front profile. The normalization temperature is $T_e = 0.03$, and $k_p r = 2.41$, $k_p \xi = 11.4$.

dimensionless longitudinal variable taken at the distance $k_p r = 3.0$. The generated magnetic field periodically varies with positive and negative values along the longitudinal direction. This means that the direction of the QMF changes periodically. In addition, the magnitude of the QMF is larger where the QMF assumes positive values. This may also be found in Fig. 1. By applying the forward Fourier transform, we find that the oscillation of the magnetic field contains three harmonic components in the longitudinal direction [see Fig. 2(b)]. One of these is the zero-frequency component, implying that a homogeneous magnetic field is generated. The remainders are $k_s = k_p$ and $k_s = 2k_p$, where k_s is the wave number of the quasistatic magnetic field.

Finally, we investigate the dependence of QMF on the temperature of the electron plasma and the effective radial width of the ionization front. Figure 3 illustrates the variation of the magnetic field with the temperature. For convenience, the temperature of the electron plasma is normalized as $T_e' = (T_e / mc^2)$. We find that the higher temperature is more effective for the generation of quasistatic magnetic fields. The QMF is also sensitive to the effective radial width $k_p r_0$ of the ionization front. The generated magnetic field reaches its maximum at some specific effective radial width. Figure 4 shows that the magnetic field tends to saturate with a large $k_p r_0$.

In conclusion, a general equation for the QMF in the relativistic regime is derived and a new mechanism for QMF generated by the ionization front is proposed. By using the model of the ionization front presented in Ref. 14, the properties of the QMF generated by the new mechanism have been investigated. In the longitudinal direction the magnetic field varies periodically between positive and negative values with periodicity equal to π / ω_p. The higher electron temperature is more effective for the QMF. In addition, for effective generation of a magnetic field, a proper radial width $k_p r_0$ of the ionization front is needed. The main potential application of the magnetic field produced by the ionization front might be found in problems like fast ignitor, plasma channels formed by shock wave, and the frequency shift produced by the ionization front.

[1] B. Luther-Davies, E. G. Gamalii, Y. Wang *et al.*, Sov. J. Quantum Electron. **22**, 289 (1992); G. Mourou and D. Umstadter, Phys. Fluids B **4**, 2315 (1992).
[2] P. Sprangle and E. Esarey, Phys. Fluids B **4**, 2241 (1992); S. V. Bulanov and N. M. Naumova, Phys. Plasmas **1**, 745 (1994); A. V. Borovskii, V. V. Korobkin, and A. M. Prokhorov, JETP **79**, 81 (1994).
[3] A. B. Borisov, A. V. Borovskiy, V. V. Korobkin *et al.*, Phys. Rev. Lett. **68**, 2309 (1992).
[4] S. C. Wilks, W. L. Kruer, M. Tabak, and A. B. Langdon, Phys. Rev. Lett. **69**, 1383 (1992).
[5] R. N. Sudan, Phys. Rev. Lett. **70**, 3075 (1993).
[6] L. Gorbunov, P. Mora, and T. M. Antonsen, Jr., Phys. Rev. Lett. **76**, 2495 (1996).
[7] V. K. Tripathi and C. S. Liu, Phys. Plasmas **1**, 990 (1994).
[8] A. D. Steiger and C. H. Woods, Phys. Rev. A **5**, 1467 (1972).
[9] G. A. Askar'yan, S. V. Bulanov, F. Pegoraro, and A. M. Pukhov, JETP Lett. **60**, 251 (1994); A. Pukhov and J. Meyer-ter-vehn, Phys. Rev. Lett. **76**, 3975 (1996).
[10] V. Yu. Bychenkov and V. T. Tikhonchuk, Laser Part. Beams **14**, 55 (1996).
[11] S. V. Bulanov, M. Lontano, T. Zh. Esirkepov, F. Pegoraro, and A. M. Pukhov, Phys. Rev. Lett. **76**, 3562 (1996).
[12] P. Vyas and M. P. Srivastava, Phys. Plasmas **2**, 2835 (1995).
[13] N. M. Laham, Phys. Scr. **52**, 688 (1995).
[14] D. L. Fisher and T. Tajima, Phys. Rev. E **53**, 1844 (1996).

CHIN.PHYS.LETT. Vol. 17, No. 7 (2000) 517

Dipole AlfvenVortex with Finite Ion Larmor Radius in a Low-Beta Plasma *

WANG Xu-Yu(王旭宇)[1,2], HE Xian-Tu(贺贤土)[1], LIU Zhen-Xing(刘振兴)[2], CAO Jin-Bin(曹晋滨)[2]
[1]*Institute of Applied Physics & Computational Mathematics, Beijing 100088*
[2]*Laboratory of Space Weather, Center for Space Science & Applied Research, Chinese Academy of Sciences, Beijing 100080*

(Received and accepted by WANG Long 17 March 2000)

A set of nonlinear fluid equations which include the effects of ion gyroradius is derived to describe Alfven vortex. The correction of finite ion gyroradius to the Alfven vortex in the inertial region is much more significant than that in the kinetic region. The amplitude of the vortex is enhanced in both regions. The scale of the vortex in the kinetic region becomes larger whereas it becomes smaller in the inertial region.

PACS: 52.35.Bj, 52.25.Dg, 94.10.Rk

Kinetic Alfven waves have been invoked in association with auroral current and particle acceleration since the pioneer work of Hasegawa *et al.*[1] and Gortz *et al.*[2] Observations from the Freja or IC-B-1300 satellite indicate that the auroral plasma can evolve to highly coherent electromagnetic structures.[3,4] They are interpreted as solitary kinetic Alfven waves (SKAW's) or Alfven vortex. Since the auroral plasma has both cold and hot components from the ionosphere or from the plasma sheet, the hot ions may have a significant effect on these coherent structures. By using the gyrokinetic theory, one-dimensional SKAW's with a finite ion gyroradius correction has been considered by Wang *et al.*[5] When the same method was used to investigate the Alfven vortex, the resultant mode coupling equations are too complex to be useful. In this respect the fluid description would be more convenient. The purpose of our paper is to derive a set of two-fluid equations with finite ion gyroradius that can be used to study the Alfven vortex. There are two possible approaches to solve this problem. The rigorous method is to reduce the magnetohydrodynamic (MHD) equations from moments of Vlasov equation, truncate the resulting set of coupled equations by choosing a closure condition, and solve the MHD equations obtained by expanding them in a series of the ω/Ω_i powers (ω is the wave frequency, Ω_i the ion gyrofrequency).[6] But the resultant equations are not quite manageable and the vortex solution can not be obtained from them. So we take an alternative method: adding the ion diamagnetic drift to the convection derivative of the ion equation of motion.[7]

In view of the low-β (β is the ratio of plasma thermal pressure to the magnetic pressure) plasma approximation, compressional magnetic field effects are neglected. The disturbed electromagnetic fields can be expressed as

$$\boldsymbol{E} = -\nabla\varphi - \hat{e}_z\partial_t A_z, \boldsymbol{B} = \nabla A_z \times \hat{e}_z. \tag{1}$$

Then the perpendicular components of the electron and ion drift velocity are given, respectively, by

$$\boldsymbol{v}_{e\perp} = \frac{\hat{e}_z \times \nabla\varphi}{B_0} + \frac{v_{ez}\boldsymbol{B}}{B_0}, \tag{2}$$

$$\boldsymbol{v}_{i\perp} = \frac{\hat{e}_z \times \nabla\varphi}{B_0} - \frac{1}{B_0\Omega_i}\Big[\partial_t + \Big(\frac{\hat{e}_z \times \nabla\varphi}{B_0} + \frac{T_i\hat{e}_z \times \nabla \ln n_i}{eB_0}\Big)\cdot\nabla\Big]\nabla_\perp\varphi + \frac{v_{iz}\boldsymbol{B}}{B_0}, \tag{3}$$

where φ is the scale potential, A_z the parallel component of the vector potential, n_e the electron density, T_i the ion temperature and B_0 the external magnetic field. When the parallel and perpendicular current caused by the difference of electron and ion motion are introduced into the current closure condition $\nabla\cdot\boldsymbol{j} = 0$, we have

$$\frac{d_0}{dt}\nabla_\perp^2\varphi + v_A^2\frac{d}{dz}\nabla_\perp^2 A_z = 0, \tag{4}$$

where, $\frac{d_0}{dt} = \partial_t + \Big(\frac{\hat{e}_z \times \nabla\varphi}{B_0} + \frac{T_i\hat{e}_z \times \nabla \ln n_i}{eB_0}\Big)\cdot\nabla,$

$$\frac{d}{dz} = \partial_z + \frac{\nabla A_z \times \hat{e}_z\cdot\nabla}{B_0}.$$

By using Eq. (5) and the current closure condition, we get the electron continuity equation as

$$\frac{dn}{dt} + n_0\frac{d}{dz}v_{iz} + \frac{1}{\mu_0 e}\frac{d}{dz}\nabla_\perp^2 A_z = 0, \tag{5}$$

where, $\frac{d}{dz} = \partial_t + \boldsymbol{\nu}_E\cdot\nabla$.

The parallel motion equations for ions and electrons are, respectively, written as:

$$m_i\frac{dv_{iz}}{dt} + e\Big(\frac{d\varphi}{dz} + \frac{\partial A_z}{\partial t}\Big) = 0, \tag{6}$$

$$m_e\Big(\frac{dv_{ez}}{dt} + \frac{1}{\mu_0 e n_0}\frac{d}{dt}\nabla_\perp^2 A_z\Big) = e\Big(\frac{d\varphi}{dz} + \frac{\partial A_z}{\partial t}\Big) - \frac{T_e}{n_0}\frac{dn}{dz}. \tag{7}$$

The coupling Eqs. (4)−(7) are for two-fluid equations with finite ion gyroradius, which may be used to study the Alfven vortex. In the one-dimensional case, all variables are functions of only one variable. This implies that the convection term

$$\frac{\hat{e}_z \times \nabla\varphi}{B_0}\cdot\nabla = 0, \frac{d_0}{dt}, \frac{d}{dt} \to \partial_t, \frac{d}{dz} \to \partial_z,$$

*Supported by the National Natural Science Foundation of China under Grant No. 49784003, and China Postdoctoral Science Foundation.

and thus the equations reduce to those of the one-dimensional model.[8] The left-hand side of Eq. (7) can be omitted in the kinetic limit region ($m_e/m_i \ll \beta \ll 1$) as the dispersion of kinetic Alfven waves is controlled by the effects of the electron pressure and ion gyroradius. While in the inertial limit region ($\beta \ll m_e/m_i$), this term can not be omitted.

The dipole vortex solution of the above coupling equations had been studied by Liu and Horton.[9] Since the procedure is very tedious, we only give the results in the kinetic region. The solution in the inertial region has a similar form and is thus omitted here. We represent the vortex disturbed field in formula (1) by the scalar potential φ and the parallel component of the vector potential A_z. In addition, instead of A_z it is convenient to use the longitudinal Ψ. Then instead of Eq. (1), the disturbed field can be written as

$$E_\perp = -\nabla_\perp \varphi, E_\| = -\nabla_\| \Psi, \partial_t \boldsymbol{B} = -\partial_z[\nabla(\varphi - \Psi) \times \hat{e}_z], \tag{8}$$

Thus the dipole vortex solution can be represented as $\varphi = \varphi(r)\cos\theta$, $\Psi = \Psi(r)\cos\theta$, where

$$\varphi(r) = \begin{cases} \varphi(a)\left[e_1 \dfrac{K_1(\chi r/a)}{K_1(\chi)} + e_2 \dfrac{a}{r}\right] & (r > a) \\ \varphi(a)\left[\mathrm{d}_1 \dfrac{J_1(\gamma r/a)}{J_1(\gamma)} + \mathrm{d}_2 \dfrac{I_1(\lambda r/a)}{I_1(\lambda)} + \dfrac{r}{a}\right], & (r < a) \end{cases} \tag{9}$$

$$\Psi(r) = \begin{cases} \dfrac{\varphi(a)\rho_s^2}{1 - (\alpha c_s/u)^2}(\dfrac{\chi}{a})^2 e_1 \dfrac{K_1(\chi r/a)}{K_1(\chi)}, \quad (r > a) \\ \dfrac{\varphi(a)\rho_s^2}{[1 - (\alpha c_s/u)^2](1 + c_1\rho_i^2)}\left[\xi_1 d_1 \dfrac{J_1(\gamma r/a)}{J_1(\gamma)} \right. \\ \left. +\xi_2 \mathrm{d}_2 \dfrac{I_1(\lambda r/a)}{I_1(\lambda)}\right], \quad (r < a), \end{cases} \tag{10}$$

$r > a$ defines the outer region of the vortex, $r < a$ the inner region. J_1, I_1 and K_1 are the first-order Bessel function, modified Bessel function, and McDonald function, respectively. The constants e_1, e_2, d_1 and d_2 can be determined from the condition that φ, Ψ, $\partial_r\varphi$, $\partial_r\Psi$, $\nabla_\perp^2\varphi$, and $\partial_\perp^2\Psi$ must be continuous at $r = a$.

$$\xi_1 = -(\frac{\gamma}{a})^2 - c_1(1 - \frac{\alpha^2 c_s^2}{u^2}), \xi_2 = (\frac{\lambda}{a})^2 - c_1(1 - \frac{\alpha^2 c_s^2}{u^2}),$$

$$C_n^2 = \frac{a^2(1 - \alpha^2 c_s^2/u^2)}{\rho_s^2}, p = 1 - \frac{u^2}{\alpha^2 v_A^2} + c_1\rho_s^2(1 + \frac{T_i}{T_e}),$$

$$\mu = c_1\rho_s^2 \frac{u^2}{\alpha^2 v_A^2}, \lambda^2 = \frac{\sqrt{p^2 + 4\mu} - p}{2} C_n^2$$

$$\gamma^2 = \frac{\sqrt{p^2 + 4\mu} + p}{2} C_n^2$$

$$\chi^2 = [1 - (u/\alpha v_A)^2] C_n^2$$

$$(k_\|/k_\perp = \alpha, \omega/k_\perp = u, c_1 = \text{const}).$$

The radial dependences of φ, Ψ, and $(\varphi - \Psi)$ in the kinetic and inertial limit region are plotted at the different temperatures, which are shown in Figs. 1 and 2, respectively. Due to the finite ion gyroradius effects,

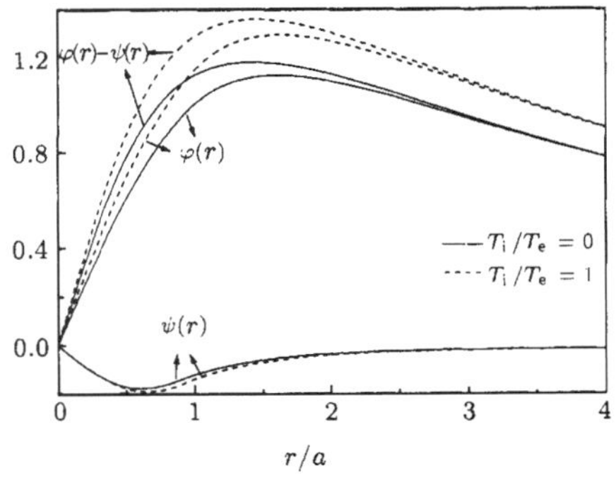

Fig. 1. Radial dependence of $\varphi(r)$, $\Psi(r)$ and $\varphi(r) - \Psi(r)$, normalized by $\varphi(a) = uB_0 a$ (a is the scale of the vortex) in the kinetic limit. The parameters we choose are: $\beta = 0.1$, $m = 0.95$ $C_n = 4.09$.

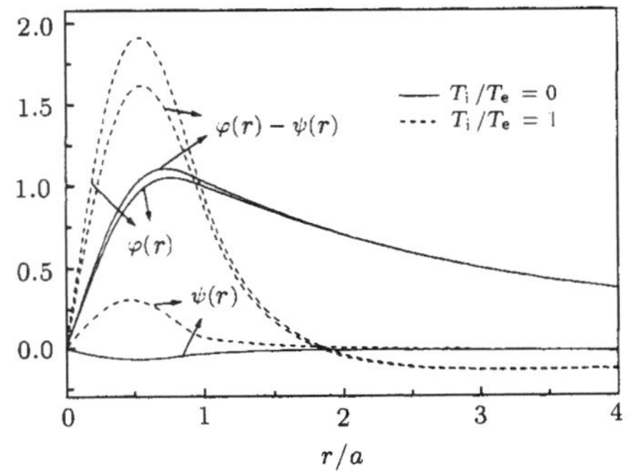

Fig. 2. Radial dependence of $\varphi(r)$, $\Psi(r)$ and $\varphi(r) - \Psi(r)$, normalized by $\varphi(a) = uB_0 a$ (a is the scale of the vortex) in the inertial limit. The parameters we choose are: $\beta = 1/18370$, $m = 0.95$, $C_n = 6.65$.

(1) the vortex scale is enlarged in the kinetic region and reduced in the inertial region; (2) the amplitude of the vortex is enhanced in both regions.

In summary, we have generalized the previous nonlinear Alfven wave equations in plasma with cold ions to plasmas with hot ions. The results will be of fundamental importance in understanding the fine Alfvenic structures in the auroral plasma.

REFERENCES

1 A. Hasegawa, J. Geophys. Res. 81 (1976) 5083.
2 C. K. Goertz and R. W. Boswell, J. Geophys. Res. 84 (1979) 7239.
3 J. E. Wanlund, P. Louarn, T. Chust, H. de Feraudy, A. Roux, B. Holback, P. O. Dovner and G. Holmgren, Geophys. Res. Lett. 21 (1994) 1831.
4 V. M. Chmyrev, S. V. Bilichenko, O. A. Pokhotelov, V. A. Marchenko, V. I. Lazarev, A. V. Streltsov and L. Stenflo, Physica Scripta, 38 (1988) 841.
5 Xu-Yu Wang, Wang Xue-yi, Z. X. Liu and Z. Y. Li, Phys. Plasmas, 5 (1998) 3447.
6 V. A. Marchenko, R. E. Denton and M. K. Hudson, Phys. Plasmas, 3 (1996) 3861.
7 A. Hasegawa and M. Wakatani, Phys. Fluids, 26 (1983) 2770.
8 D. J. Wu, G. L. Huang, D. Y. Wang and C. G. Falthammar, Phys. Plasmas, 3 (1996) 2879.
9 J. Liu and W. Horton, J. Plasma Phys. 36 (1986) 1.

CHIN.PHYS.LETT. Vol. 17, No. 10 (2000) 746

Quasistatic Magnetic Field Generation by an Intense Ultrashort Laser Pulse in Underdense Plasma *

ZHENG Chun-Yang(郑春阳), ZHU Shao-Ping (朱少平), HE Xian-Tu(贺贤土)
Institute of Applied Physics and Computational Mathematics, P.O.Box 8009, Beijing 100088

(Received 21 December 1999)

The quasistatic magnetic field created in the interaction of intense ultrashort laser pulses with underdense plasmas has been investigated by two-dimensional particle simulation. The relativistic ponderomotive force and plasma wave excited in self-modulation processes can drive intense electron current mainly in the propagation direction. As a result, an azimuthal, multi-mega Gauss order quasi-static magnetic field can be generated around the laser beam.

PACS: 52.40.Nk, 52.35.Mw, 52.25.Rv

The ultra-intense (over $10^{19}\,\mathrm{W/cm^2}$) and ultra-short (subpicosecond) fields attained in current laser pulses produce novel effects in plasmas, which are of fundamental and practical interest.[1-4] Among them the generation of quasistatic magnetic field[3,4] is a hot topic. The presence of this field could considerably affect the nonlinear plasma dynamics, especially for the laser pulse propagation in a plasma and the generation of relativistic electron currents, which are crucial to the fast ignitor scheme for inertial confinement fusion.[1,2] Much has been written[3-7] on the generation mechanism and condition of the field as well as the consequences of such generation. But many problems, in our opinion, are still open because of the absence of describing the evolution of laser plasma as a whole.

In this paper, we first give a simple theoretical analysis for the quasistic magnetic field generation in the weak relativistic approximation. Then we show the simulation results using an electromagnetic, relativistic two-dimensional particle code. The simulation shows that the fundamental spatiotemporal features of the magnetic field are consistent with the theoretical expectation. We propose that the relativistic poderomotive force and laser wake field can coexist and act jointly to drive intense electron current in the laser propagation direction. As a result, an azimuthal quasi-static magnetic field about 80 MG is generated in the edge of laser beam with intensity about $2\times10^{19}\,\mathrm{W/cm^2}$.

Considering Maxwells equations and cold fluid equations, splitting the physical variables into their slow-time-scale and fast-time-scale parts, we can easily get the equation for quasistatic magnetic field $\boldsymbol{B}$:

$$\nabla^2\boldsymbol{B}-\frac{1}{c^2}\frac{\partial^2\boldsymbol{B}}{\partial t^2}=-\frac{4\pi}{c}\nabla\times\boldsymbol{J},\tag{1}$$

where $\boldsymbol{J}=en\boldsymbol{v}$, n and $\boldsymbol{v}$ are the electron density and velocity.

Using conservation law for the generalized vortex $(e/c)\boldsymbol{B}=-\nabla\times\boldsymbol{P}$ presented in Ref. 5 and the relation $\nabla\times(n\boldsymbol{v})=\nabla n\times\boldsymbol{v}+n\nabla\times\boldsymbol{v}$, Eq. (1) can be rewritten as

$$\nabla^2\boldsymbol{B}-\frac{1}{c^2}\frac{\partial^2\boldsymbol{B}}{\partial t^2}-\frac{\omega_p^2 n}{c^2\gamma n_0}\boldsymbol{B}=\frac{\omega_p^2}{ec}\nabla\left(\frac{n}{\gamma n_0}\right)\times\boldsymbol{P},\tag{2}$$

$\boldsymbol{P}=\gamma m\boldsymbol{v}$ is the slow-time-scale momentum and satisfies the following equation:

$$\frac{\partial\boldsymbol{P}}{\partial t}=e\boldsymbol{E}-mc^2\nabla\gamma,\tag{3}$$

where $\gamma=(1-v^2/c^2)^{-1/2}=(1+P^2/m^2c^2)^{1/2}$ is the relativistic factor.

In linear polarized case, the slow component of electron momentum is mainly in the direction of the laser propagation and comes from both the longitudinal wake field and ponderomotive force of the laser pulses. Hence an azimuthal magnetic field can be generated if there is significant transverse plasma density variation.

In the weak-relativistic regime, the density variation and wake field can be obtained by solving the linear fluid equations and Poisson's equation. The electron density oscillation produced in the wake of the laser pulse can be given by[8]

$$\begin{aligned}\frac{\delta n}{n}&=A\left[1+\left(\frac{2c}{\omega_p a_0}\right)^2\left(1-\frac{r^2}{a_0^2}\right)\right]\exp\left(-\frac{r^2}{a_0^2}\right)\\&\quad\cdot\sin\left[\omega_p\left(t-\frac{y}{c}\right)\right],\\A&=\frac{m\langle v_{osc}^2/c^2\rangle}{2e}\left(\frac{\omega_p\tau_0}{2}\right)\exp\left[-\left(\frac{\omega_p\tau_0}{2}\right)^2\right],\end{aligned}\tag{4}$$

where τ_0 is the pulse length and a_0 the pulse width.

The quasistatic magnetic field can be estimated by using Eqs. (2)–(4). A typical result is given by assuming $\lambda_p\sim\tau_0\ll a_0$, where $\lambda_p=\omega_p/c$ is the plasma wavelength, and it reads

$$\begin{aligned}\frac{eB}{mc\omega_p}&\sim C\frac{\langle v_{osc}^2/c^2\rangle^2}{(\omega_p\tau_0)^2}\exp\left(-\frac{2r^2}{a_0^2}\right)\\&\quad\cdot\sin^2(\omega_p t-k_p y),\\C&=\omega_p\frac{4\sqrt{\pi}r}{c}\left[\frac{1}{2}+\left(\frac{2c}{\omega_p a_0}\right)^2\left(1-\frac{r^2}{2a_0^2}\right)\right].\end{aligned}\tag{5}$$

It should be noted that the magnetic field only has an azimuthal component and vanishes at the beam center with the maximum towards the beam edge. In the

*Supported by the National High-Tech. ICF Committee in China, the National Natural Science Foundation of China under Grant No. 19735620, and National Basic Research Project "Nonlinear Science" in China.

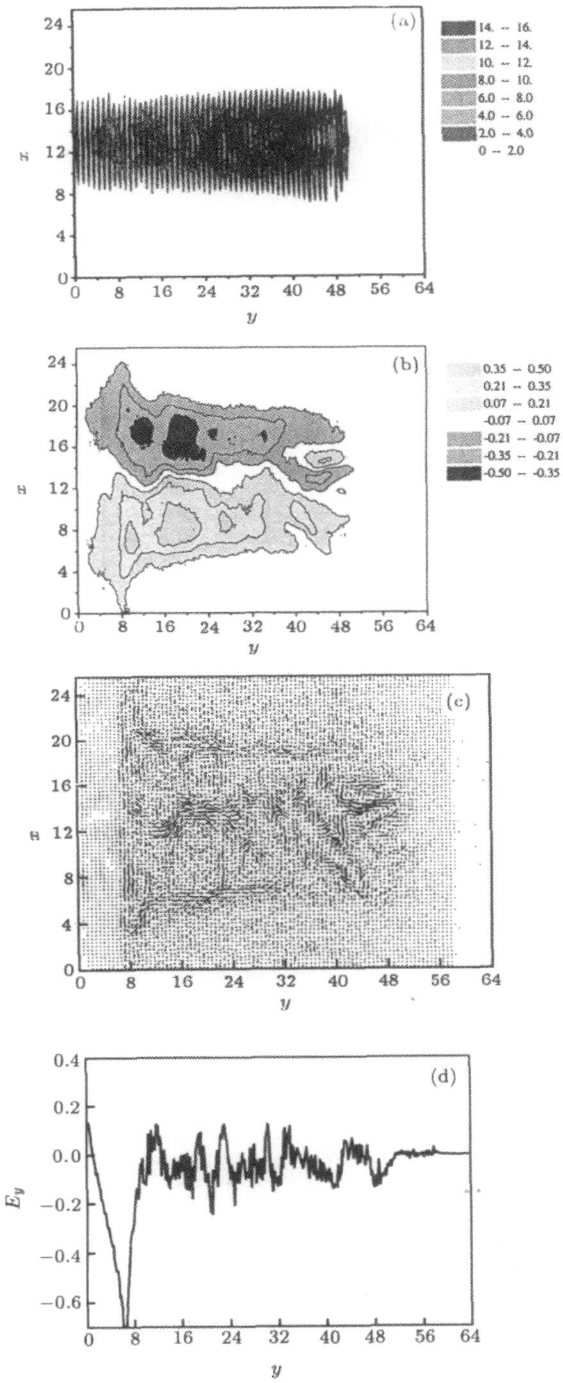

Fig. 1. Laser interaction with plasma and the generation of quasistatic magnetic field at $t = 60(2\pi/\omega_0)$: (a) contour plots of the electromagnetic energy density; (b) contour plots of the quasistatic magnetic field B_z, in unit of $mc\omega_0/e \simeq 10^8$ G for $\lambda_0 = 1.06\,\mu$m; (c) distribution of the vector electron current density; (d) the longitudinal wake field at the channel axis(in unit of $mc\omega_0/e$), x, y are in unit of λ_0.

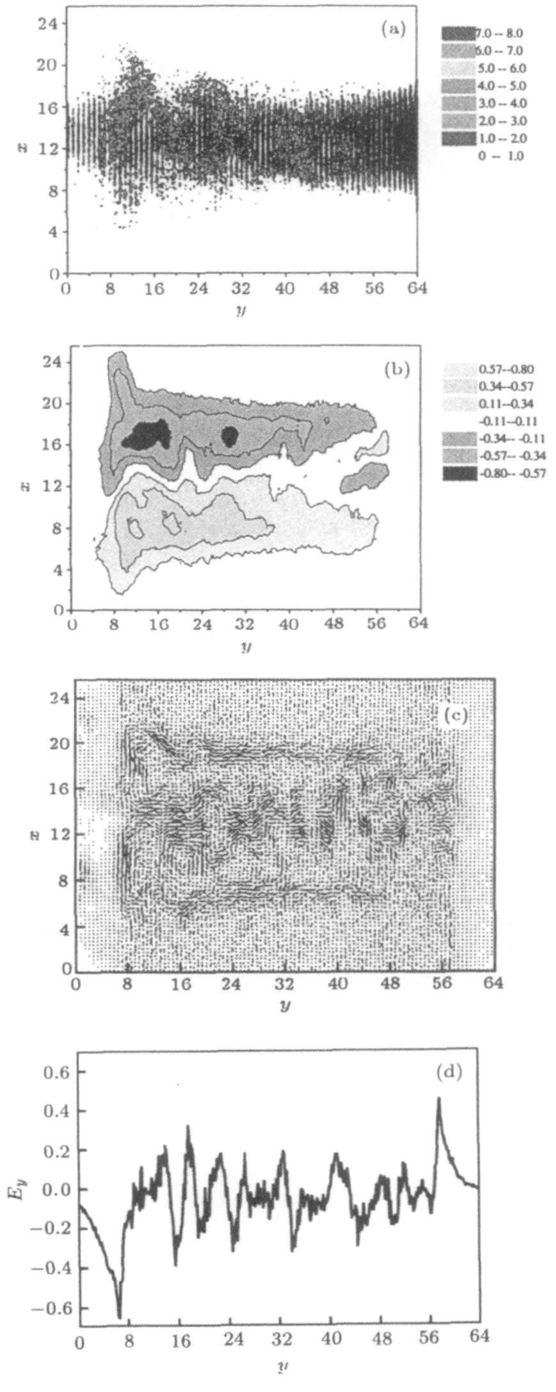

Fig. 2. Laser interaction with plasma and the generation of quasistatic magnetic field at $t = 80(2\pi/\omega_0)$: (a) contour plots of the electromagnetic energy density;(b) contour plots of the quasistatic magnetic field B_z; (c) distribution of the vector electron current density; (d) the longitudinal wake field at the channel axis.

laser propagation direction y, the field has a uniform part and a $\lambda_p/2$ modulation.

In order to investigate origination of quasistatic magnetic field generally, a two-dimensional particle code is written. Maxwell's equations and relativistic particle motion equations are solved. The numerical procedure used in the code is similar to that in Ref. 9. We consider a 2-dimensional homogeneous plasma slab in the $x-y$ plane $(0-L_x, 0-L_y)$ described in Cartesian geometry. A linearly S-polarized electromagnetic wave of frequency ω_0 is normally incident on the plasma slab at $y=0$ and propagates along the y direction. The boundary conditions are open-sided for fields and mirror reflection for particles in the y-direction. In the x-direction, periodic boundary conditions are selected. The spatial length is normalized by laser wave length λ_0 and the time is normalized by the reciprocal of laser frequency $1/\omega_0$. The plasma-vacuum interfaces are localized at $L_y/10$ and $9L_y/10$. The initial temperatures are 0.1 keV for particles. The number of spatial grids are 320×800, the size of the spatial mesh is $0.08\lambda_0$ and number of particles is 6×10^6, respectively.

The incident laser electromagnetic field vectors are given as follows:

$$(E_z, B_x) = (E_0, B_0)e^{-(t-t_0)^2/\tau_0^2}e^{-(x-L_x/2)^2/a_0^2}, \quad (6)$$

where $E_0(B_0)$ is the electric(magnetic) field of the incident wave and $a = eE_0/(m\omega_0 c)$ the normalized amplitude. The laser pulse is initialized outside the plasma in the region $y<0, t_0=2$.

Several different simulation runs were carried out by varying the laser pulse and plasma parameters. We here introduce simulation results for one typical run. The simulation parameters are that: pulse duration $\tau_0 = 275$ fs, transverse width $a_0 = 5\lambda_0$ and electron density $n = 0.16n_{cr}$, where n_{cr} is the critical density. The laser intensity is taken to be 2×10^{19} W/cm^2

Results of the simulation at two different time $t = 60(2\pi/\omega_0)$ and $t = 80(2\pi/\omega_0)$ are presented in Figs. 1 and 2, the quantities plotted are (a)the contour plots of the electromagnetic energy density; (b)contour plots of the z component of the quasistatic magnetic field (average over one laser period); (c)distribution of the vector electron current density and (d) the longitudinal wake field at the channel axis.

It can be seen from Figs. 1(a) and 2(a) that the laser beam is modulated along the propagation direction due to relativistic self-modulational instabilities,[10] and the modulation persists and becomes more obvious at $t=80$ (Fig. 2(a)), focusing and defocusing regions are formed on the back of the pulse. The magnetic field exhibits the structure with two ribbons of opposite direction behind the laser pulse. The light regions indicate upward(positive) magnetic fields. The dark regions are for downward (negative) fields (see Figs. 1(b) and 2(b)). The quasistatic magnetic field is generated by the average relativistic electron currents, which flow mainly in the propagation direction near the channel axis, and flow in the opposite direction in the surrounding regions as can be seen in Figs. 1(c) and 2(c). Figures 1(d) and 2(d) give the longitudinal wake field at the channel axis. The wake field was excited initially by the ponderomotive force associated with the sharp front of pulse, and can strongly affect the focusing properties of the pulse body by introducing modulation in the uniform background plasma density. In the meanwhile, the wake field amplitude increases. Making a comparison between Fig. 1 and Fig. 2, we find that these processes are growing as a function of both the propagation distance and the distance behind the pulse front, and proceed in a highly nonlinear manner. The maximum value of the quasistatic magnetic field rises from 50 MG up to 80 MG.

So, the generation of magnetic field can be explained shortly. The electrons are initially pushed forward by the ponderomotive force due to the sharp front of pulse, then along with the formation of modulation regions on the back of pulse caused by the density depression of the wake wave, the ponderomotive force and wake field (the maximum of the latter always lags that of the former) have both forward and backward forces to electrons, therefore the modulation of electron current is performed. The electron current is very strong in tight focusing regions, such as around $y = 16, 28(\lambda_0)\cdots$ at $t=80$ (see Fig.2(c)). As a result, the maxima of the quisistatic magnetic field are also generated in these regions.

In summary, it was clarified that the coexistence and joint action of relativistic poderomotive force and plasma wake wave are responsible for the generation of quasistatic magnetic field. In the weak relativistic approximation, the relation of wake field generation and ponderomotive fore is simple, so a regular periodically $\lambda_p/2$ modulation to the magnetic field in the propagation direction exists. In the strong relativistic regime, simulation results show that the fundamental spatiotemporal features of the magnetic field and generation mechanism are consistent with the theoretical expectation. From the theoretical model, it is also shown that the magnitude of the magnetic field increases as the plasma density increasing. It implies that more strong magnetic field will be generated in higher density (critical to overdense), which is left for future studies.

REFERENCES

1. M. Tabak *et al.*, Phys. Plasmas, 1 (1994) 1626.
2. A. Pukhov and J. Meyer-ter Vehn, Phys. Rev. Lett. 76 (1996) 3975.
3. S. C. Wilks, W. L. Kruer, M. Tabak and A. B. Langdon, Phys. Rev. Lett. 69 (1992) 1383.
4. J. A. Stamper, Laser Part. Beams, 9 (1991) 841; J. A. Stamper *et al.*, Phys. Rev . Lett. 34 (1975) 138.
5. R. N. Sudan, Phys. Rev. Lett. 70 (1993) 3075; V. K. Tripathi and C. S. Liu, Phys. Plasmas, 1 (1994) 990.
6. V. I. Berezhiani *et al.*, Phys. Rev. E 55 (1997) 995; V. Y. Bychenkov *et al.*, Zh. Eksp. Teor. Fiz. 105 (1994) 118 [JETP, 78 (1994) 62].
7. J. Mason *et al.*, Phys. Rev. Lett. 80 (1998) 524; M. Borghesi *et al.*, Phys. Rev. Lett. 81 (1998) 112.
8. J. R. Marques *et al.*, Phys. Plasmas, 5 (1998) 1162.
9. C. K. Birdsall and A. B. Langdon, *Plasma physics via Computer simulation* (McGraw-Hill, New York, 1985).
10. E. Esarey *et al.*, Comments Plasma Phys. Controlled Fusion, 12 (1989) 191.

等离子体自生磁场

485

Slow-time-scale magnetic fields driven by fast-time-scale waves in an underdense relativistic Vlasov plasma

Shao-ping Zhu, X. T. He, and C. Y. Zheng
Institute of Applied Physics and Computational Mathematics, P.O. Box 8009, Beijing 100088, People's Republic of China

(Received 5 September 2000; accepted 27 September 2000)

Slow-time-scale magnetic fields driven by fast-time-scale electromagnetic waves or plasma waves are examined from the perspective of the Vlasov–Maxwell equations for a relativistic Vlasov plasma. An equation for slow-time-scale magnetic field is obtained. The field proposed in the present paper is a result of wave–wave beating which drives a solenoidal current. The magnitude of the slow-time-scale magnetic field proposed here can be as high as 20 MG at the critical surface for a laser intensity $I=10^{18}\,\mathrm{W/cm^2}$ at wavelength $\lambda_0=1.05\,\mu\mathrm{m}$. The predicted magnetic field is observed in two-dimensional particle simulations presented here. © *2001 American Institute of Physics.* [DOI: 10.1063/1.1330200]

I. INTRODUCTION

The first experiment of the slow-time-scale magnetic field created in laser produced plasmas was observed by Stamper *et al.*[1] in 1971. They explained the generating mechanism in terms of a thermoelectric source of $\nabla n_e \times \nabla T_e$, where T_e and n_e are electron temperature and density, respectively. The development of laser chirped pulse amplification (CPA) technique applications, such as the fast ignition scheme for inertial confinement fusion,[2] have generated worldwide interest in the study of the propagation of an intense short-pulse laser through relativistic plasma. This has been investigated by experiments[3] and numerical simulations.[4] An extremely intense slow-time-scale magnetic field was observed[4] in the interaction of an ultraintense short laser pulse with an overdense plasma target. It is expected that a slow-time-scale magnetic field can play an important role in the complex interaction between the ultraintense short-pulse laser waves and a relativistic background plasma. In order to explain the simulation results by Wilks *et al.*,[4] Sudan[5] proposed a collisionless mechanism for slow-time-scale magnetic field generation in the overdense plasma in which the slow-time-scale magnetic field is the result of a slow-time-scale current driven by the spatial gradient and the temporal variation of the ponderomotive force exerted by the laser on the plasma electrons. Tripathi and Liu[6] proposed a collisionless mechanism for the slow-time-scale magnetic fields generation near the critical layer in which the magnetic field is driven by the large gradient in plasma density. And recently Mason and Tabak[7] numerically simulated a magnetic field generation that is traced to the ponderomotive push on background electrons. It is also known that a strong slow-time-scale magnetic field is generated in the underdense plasma region.[8] The generation mechansim for the slow-time-scale magnetic field in underdense plasma due to inverse Faraday effect for a circularly polarized laser pulse was studied by several authors.[9–12] In fact, about two decades ago, a collisionless mechanism for the slow-time-scale magnetic field in the nonrelativistic underdense plasma was proposed by Kono *et al.*,[13] which is generated by the beat interaction between two longitudinal Langmuir waves which drive a nonlinear solenoidal current. A generation mechanism in nonrelativistic plasma for the slow-time-scale magnetic field driven by the beat interaction between two fast-time-scale electromagnetic field, was proposed by one of the present authors[14] in 1983. As we showed above, there are a lot of theoretical explanations for the generation mechanism of the field. However, the generation mechanism of the slow-time-scale magnetic field, in our opinion, is still an open problem. In this paper, we propose a new collisionless mechanism for the generation of the slow-time-scale magnetic fields that are driven by relativistic solenoidal current of ultraintense short-pulse laser waves, or by Langmuir waves generated from the parametric excitation, such as, relativistic stimulated Raman scattering (RSRS) in a relativistic channel plasma. The slow-time-scale magnetic field is the result of the nonlinear interaction of two fast-time-scale fields. In an analytical derivation the key point is to calculate consistently and correctly the nonlinear electric current matrix, which depends on the compex vector operation (especially for transverse waves) and the integral of the particle velocity distribution, which certainly requires kinetic theory. In order to check our analytical results, we here run two-dimensional particle simulations. The simulations clearly show the generation of the proposed slow-time-scale magnetic field, which approaches the magnitude of one driven by ponderomotive force.

II. SLOW-TIME-SCALE MAGNETIC FIELDS DERIVED FROM KINETIC THEORY

We now start our kinetic discussion of a mechanism for slow-time-scale magnetic field generation driven by the fast-time-scale waves in an underdense plasma. Consider the relativistic Vlasov–Maxwell system of equations,

$$\left(\frac{\partial}{\partial t}+\frac{c\mathbf{p}_\alpha}{p_{0\alpha}}\cdot\frac{\partial}{\partial \mathbf{x}}+e_\alpha \mathbf{G}(\mathbf{x},t)\cdot\frac{\partial}{\partial \mathbf{p}_\alpha}\right)f_\alpha(\mathbf{x},\mathbf{p}_\alpha,t)=0, \tag{1}$$

$$\nabla\times\mathbf{E}=-\frac{1}{c}\frac{\partial\mathbf{B}}{\partial t}, \tag{2}$$

$$\nabla\times\mathbf{B}=\frac{1}{c}\frac{\partial\mathbf{E}}{\partial t}+\frac{4\pi}{c}\mathbf{j}, \tag{3}$$

$$\mathbf{j}=\sum_{\alpha}\int e_\alpha\frac{c\mathbf{p}_\alpha}{p_{0\alpha}}f_\alpha(\mathbf{x},\mathbf{p}_\alpha,t)d^3\mathbf{p}_\alpha, \tag{4}$$

$$\mathbf{G}(\mathbf{x},t)=\mathbf{E}(\mathbf{x},t)+\frac{\mathbf{p}_\alpha\times\mathbf{B}}{p_{0\alpha}}, \tag{5}$$

where the index $\alpha(\alpha=i,e)$ represents ions and electrons, respectively, $p_{0\alpha}=\sqrt{\mathbf{p}_\alpha^2+m_\alpha^2c^2}$, and the other notations are standard.

We split the particle distribution functions up into their background, slow-time-scale and fast-time-scale parts,

$$f_\alpha=n_{\alpha 0}f_{\alpha 0}+f_{\alpha s}+f_{\alpha f}, \tag{6}$$

where $f_{\alpha 0}$ is the background distribution function, and the isotrope for $f_{\alpha 0}$ with $\int f_{\alpha 0}d^3\mathbf{p}_\alpha=1$ is assumed in the following discussion. Also, we divide electric fields into slow-time-scale and fast-time-scale parts. Making Fourier-transform with respect to both space and time, we obtain the slow-time-scale distribution function in the Fourier representation

$$\begin{aligned}f_{\alpha s}(q,\mathbf{p}_\alpha)=&\frac{e_\alpha n_{\alpha 0}}{i\left(\Omega-\dfrac{c\mathbf{q}\cdot\mathbf{p}_\alpha}{p_{0\alpha}}\right)}\sum_{\sigma_s}E_s^{\sigma_s}(q)\mathbf{e}_s'^{\sigma_s}(q)\cdot\frac{\partial f_{\alpha 0}}{\partial\mathbf{p}_\alpha}\\&-\frac{e_\alpha^2 n_{\alpha 0}}{\left(\Omega-\dfrac{c\mathbf{q}\cdot\mathbf{p}_\alpha}{p_{0\alpha}}\right)}\int_{(s)}\sum_{\sigma_1,\sigma_2}\mathbf{e}_f'^{\sigma_1}(k_1)\\&\cdot\frac{\partial}{\partial\mathbf{p}_\alpha}\frac{\mathbf{e}_f'^{\sigma_2}(k_2)\cdot\dfrac{\partial f_{\alpha 0}}{\partial\mathbf{p}_\alpha}}{\left(\omega_2-\dfrac{c\mathbf{k}_2\cdot\mathbf{p}_\alpha}{p_{0\alpha}}\right)}E_f^{\sigma_1}(k_1)E_f^{\sigma_2}(k_2)\\&\times\delta^4(q-k_1-k_2)\frac{d^4k_1d^4k_2}{(2\pi)^4},\end{aligned} \tag{7}$$

where $E_f^{\sigma_i}(k_i)$ $(i=1,2)$ is the amplitude of the fast-time-scale electric field (such as laser electric and Langmuir fields in a laser produced plasma). In Eq. (7) we only keep the lowest nonlinear terms, and the symbol $\int_{(s)}$ refers to a slow-time-scale motion caused by the difference beat of two fast-time-scale motions. Also $k=(\mathbf{k},\omega)$ and $q=(\mathbf{q},\Omega)$, $E_s^{\sigma_s}(q)$, and $\mathbf{e}_s'^{\sigma_s}(q)$ are defined as follows:[14,15]

$$\mathbf{G}_s(q)=\sum_{\sigma_s=t,l}E_s^{\sigma_s}(q)\mathbf{e}_s'^{\sigma_s}(q), \tag{8}$$

$$\mathbf{e}_s'^{\sigma_s}(q)=\left[1-\frac{c\mathbf{q}\cdot\mathbf{p}_\alpha}{\Omega p_{0\alpha}}\right]\mathbf{e}_s^{\sigma_s}(q)+\frac{c\mathbf{e}_s^{\sigma_s}(q)\cdot\mathbf{p}_\alpha}{\Omega p_{0\alpha}}\mathbf{q}, \tag{9}$$

where t indicates the transverse field and l is the longitudinal field. We use the convention $E_s^{\sigma_s^*}(q)=-E_s^{\sigma_s}(-q)$ and $\mathbf{e}_s^{\sigma_s^*}(q)=-\mathbf{e}_s^{\sigma_s}(-q)$, where $*$ notes the complex conjugate. Similarly, we have the same expression for the fast-time-scale field.[16]

Having the slow-time-scale particle distribution functions and ignoring the ion contribution to electric current due to much larger mass, we get slow-time-scale electric fields,

$$\begin{aligned}\Theta_s^{\sigma_s}(q)E_s^{\sigma_s}(q)&=-\frac{i4\pi n_0e^3}{\Omega}\sum_{\sigma_1,\sigma_2}\int_{(s)}\Pi_e^{\sigma_s,\sigma_1,\sigma_2}(q,k_1,k_2)\\&\times E_f^{\sigma_1}(k_1)E_f^{\sigma_2}(k_2)\delta^4(q-k_1-k_2)\frac{d^4k_1d^4k_2}{(2\pi)^4},\end{aligned} \tag{10}$$

where

$$\begin{aligned}\Theta_s^l(q)=&1+\sum_\alpha\sum_{\sigma_s'}\frac{\omega_{p\alpha}^2m_\alpha}{\Omega|\mathbf{q}|}\\&\cdot\int\frac{c(\mathbf{q}\cdot\mathbf{p}_\alpha)\mathbf{e}_s^{\sigma_s'}(q)\cdot\dfrac{\partial f_{\alpha 0}}{\partial\mathbf{p}_\alpha}}{p_{0\alpha}\left(\Omega-\dfrac{c\mathbf{q}\cdot\mathbf{p}_\alpha}{p_{0\alpha}}\right)}d^3\mathbf{p}_\alpha,\quad\text{for }\sigma_s=l,\end{aligned} \tag{11}$$

$$\begin{aligned}\Theta_s^t(q)=&1-\frac{c^2|\mathbf{q}|^2}{\Omega^2}+\sum_\alpha\sum_{\sigma_s'}\frac{\omega_{p\alpha}^2m_\alpha}{\Omega}\\&\times\int\frac{c\mathbf{e}_s^t(q)\cdot\mathbf{p}_\alpha\mathbf{e}_s^{\sigma_s'}(q)\cdot\dfrac{\partial f_{\alpha 0}}{\partial\mathbf{p}_\alpha}}{p_{0\alpha}\left(\Omega-\dfrac{c\mathbf{q}\cdot\mathbf{p}_\alpha}{p_{0\alpha}}\right)}d^3\mathbf{p}_\alpha\quad\text{for }\sigma_s=t,\end{aligned} \tag{12}$$

$$\begin{aligned}\Pi_\alpha^{\sigma_s\sigma_1\sigma_2}&(q,k_1,k_2)\\&=\frac{1}{2}\int\left[\frac{c\mathbf{e}_s^{\sigma_s}(q)\cdot\mathbf{p}_\alpha}{p_{0\alpha}\left(\Omega-\dfrac{c\mathbf{q}\cdot\mathbf{p}_\alpha}{p_{0\alpha}}\right)}\mathbf{e}_f'^{\sigma_1}(k_1)\right.\\&\left.\cdot\frac{\partial}{\partial\mathbf{p}_\alpha}\frac{\mathbf{e}_f'^{\sigma_2}(k_2)\cdot\dfrac{\partial f_{\alpha 0}}{\partial\mathbf{p}_\alpha}}{\left(\omega_2-\dfrac{c\mathbf{k}_2\cdot\mathbf{p}_\alpha}{p_{0\alpha}}\right)}+(1\leftrightarrow 2)\right]d^3\mathbf{p}_\alpha,\end{aligned} \tag{13}$$

where $(1\leftrightarrow 2)$ represents the preceding term with the indices 1 and 2 interchanged. $\Pi_\alpha^{\sigma_s\sigma_1\sigma_2}(q,k_1,k_2)$ is a nonlinear electric current matrix. Here and below $\omega_{p\alpha}$ is the plasma frequency defined by the rest mass m_α.

With $\Omega=\omega_1+\omega_2$, $\mathbf{q}=\mathbf{k}_1+\mathbf{k}_2$, and $(\omega-c\mathbf{k}\cdot\mathbf{p}_\alpha/p_{0\alpha})_{1,2}\neq 0$, after tedious calculations, one can show that the following relation holds for relativistic plasma (see Appendix):

$$\frac{1}{\Omega}\Pi_\alpha^{\sigma_s\sigma_1\sigma_2}(q,k_1,k_2)=\frac{1}{\omega_2}\Pi_\alpha^{\sigma_2\sigma_s\sigma_1}(k_2,q,-k_1). \tag{14}$$

Because we are only interested in the process of two fast-time-scale motions producing one slow-time-scale response in Eq. (14), the linear Landau damping effect for the fast-time-scale wave can be neglected. That is, the Doppler frequency shift in the principal value can be ignored. Using relation (14) and considering the condition $\Omega \ll c|\mathbf{q}|$, we find

$$\Pi_{\alpha}^{t\sigma_1\sigma_2}(q,k_1,k_2) = \frac{\mathbf{e}_s^t\cdot[\mathbf{q}\times(\mathbf{e}_1\times\mathbf{e}_2)]}{m_e^2\omega_2^2\omega_1}\beta_3 - \frac{\Omega\beta_4}{m_e^2c^2\omega_2^2|\mathbf{q}|^2}\mathbf{e}_s^t \cdot[\mathbf{q}(\mathbf{e}_1\cdot\mathbf{e}_2)+\mathbf{e}_1(\mathbf{k}_2\cdot\mathbf{e}_2)+\mathbf{e}_2(\mathbf{k}_1\cdot\mathbf{e}_1)-\mathbf{e}_2\times(\mathbf{k}_1\times\mathbf{e}_1) - \mathbf{e}_1(\mathbf{k}_2\times\mathbf{e}_2)]. \tag{15}$$

Under the condition $v_i|\mathbf{q}|\ll\Omega\ll v_e|\mathbf{q}|$, we have

$$\Theta_s^t(q) = -\frac{c^2|\mathbf{q}|^2}{\Omega^2} - \frac{\omega_{pe}^2}{c^2|\mathbf{q}|^2}\beta_1 - \frac{\omega_{pi}^2}{\Omega^2}\beta_2. \tag{16}$$

In Eqs. (15) and (16) the coefficients related to relativistic effect are defined as follows:

$$\beta_1 = m_e c\int\left(\frac{p_{0e}}{|\mathbf{p}_e|^2} - \frac{3}{p_{0e}}\right)f_{e0}d^3\mathbf{p}_e, \quad \beta_2 = m_i c\int\left(\frac{1}{p_{0i}} - \frac{|\mathbf{p}_i|^2}{3p_0^3}\right)f_{i0}d^3\mathbf{p}_i, \tag{17}$$

and

$$\beta_3 = m_e^2c^2\int\frac{1}{p_{0e}^2}\left(\frac{1}{2} - \frac{|\mathbf{p}_e|^2}{3p_{0e}^2}\right)f_{e0}d^3\mathbf{p}_e, \quad \beta_4 = m_e^2c^2\int\left(\frac{1}{p_{0e}^2} - \frac{|\mathbf{p}_e|^2}{6p_{0e}^4}\right)f_{e0}d^3\mathbf{p}_e. \tag{18}$$

For the ions, the relativistic effect can be ignored, so that $\beta_2=1$. Figure 1 shows the dependence of β_3 and β_4 on the background electron temperature T_e. For the nonrelativistic limitation, $\beta_3=1/2$ and $\beta_4=1$. For the condition $T_e=50\,\text{keV}$, we get that $\beta_3=0.33$ and $\beta_4=0.73$. The dependence of β_1 on the background electron temperature T_e is given in Fig. 2.

Substituting Eqs. (15) and (16) into electric field (10) and using Faraday's Law $\mathbf{B}_s(q)=(c/\Omega)\mathbf{q}\times\mathbf{E}_s^t(q)$, and transforming the magnetic field equation from Fourier representation into time-space representation, we obtain the equation for the slow-time-scale magnetic fields driven by transverse or longitudinal fields as follows:

$$\left(-c^2\nabla^4 + \frac{\omega_{pe}^2\beta_1}{c^2}\frac{\partial^2}{\partial t^2} + \omega_{pi}^2\beta_2\nabla^2\right)\mathbf{B}(\mathbf{x},t) = -\frac{i4\pi c n_0 e^3\beta_3}{m_e^2\omega_0^3}\nabla^2\nabla\times[\nabla\times(\mathbf{E}_f(\mathbf{x},t)\times\mathbf{E}_f^*(\mathbf{x},t))] + \frac{2\pi n_0 e^3\beta_4}{m_e^2\omega_0^2 c}\frac{\partial}{\partial t}\nabla\times\{\mathbf{E}_f(\mathbf{x},t)\nabla\cdot\mathbf{E}_f^*(\mathbf{x},t)+\mathbf{E}_f^*(\mathbf{x},t)\nabla \cdot\mathbf{E}_f(\mathbf{x},t) - \mathbf{E}_f(\mathbf{x},t)\times[\nabla\times\mathbf{E}_f^*(\mathbf{x},t)] - \mathbf{E}_f^*(\mathbf{x},t) \times[\nabla\times\mathbf{E}_f(\mathbf{x},t)]\}. \tag{19}$$

We can see from Eq. (19) that the slow-time-scale magnetic field comes from three terms. The first driven source is the nonlinear solenoidal current, the second one is the interaction

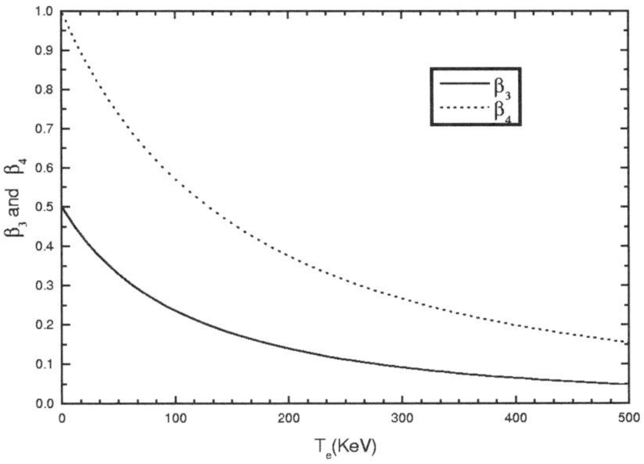

FIG. 1. The dependence of β_3 and β_4 on the electron temperature T_e. For the unrelativistic limitation, $\beta_3=1/2$ and $\beta_4=1$. For the condition $T_e=50\,\text{keV}$, $\beta_3=0.33$, and $\beta_4=0.73$.

of the fast-time-scale wave and the electron density perturbation excited by the fast-time-scale electric fields, and the third one is the coupling between fast-time-scale electric fields and the magnetic field.

If the fast-time-scale wave is the laser field, we have

$$\left(-c^2\nabla^4 + \frac{\omega_{pe}^2\beta_1}{c^2}\frac{\partial^2}{\partial t^2} + \omega_{pi}^2\beta_2\nabla^2\right)\mathbf{B}(\mathbf{x},t) = -\frac{i4\pi n_0 c e^3\beta_3}{m_e^2\omega_0^3}\nabla^2\nabla\times[\nabla\times(\mathbf{E}_f(\mathbf{x},t)\times\mathbf{E}_f^*(\mathbf{x},t))] - \frac{2\pi e^3\beta_4}{m_e^2\omega_0^2c^2}\frac{\partial}{\partial t}\nabla\times\left\{\mathbf{E}_f(\mathbf{x},t)\times\frac{\partial\mathbf{B}_f^*(\mathbf{x},t)}{\partial t} + \mathbf{E}_f^*(\mathbf{x},t)\times\frac{\partial\mathbf{B}_f(\mathbf{x},t)}{\partial t}\right\}. \tag{20}$$

When the condition $|\mathbf{q}|\lambda_e \gg (v_{te}\sqrt{\beta_1}/c^2)(\Omega/|\mathbf{q}|)$ holds and ignoring the ion effect, we obtain the slow-time-scale magnetic fields in the static approximation,

$$\mathbf{B}(\mathbf{x},t) = -\frac{i\omega_{pe}^2 e\beta_3}{m_e c\omega_0^3}\mathbf{E}_f(\mathbf{x},t)\times\mathbf{E}_f^*(\mathbf{x},t), \tag{21}$$

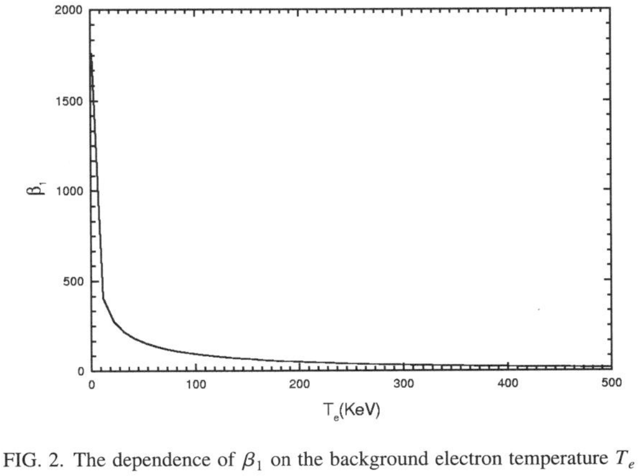

FIG. 2. The dependence of β_1 on the background electron temperature T_e.

where ω_{pe} is the electron plasma frequency defined by the rest mass m_e. The magnitude of the slow-time-scale magnetic field can be written in the form,

$$|\mathbf{B}(\mathbf{x},t)| \sim \frac{4\times 10^7 \times \sqrt{\pi}\omega_{pe}^3 \beta_3}{c^2\sqrt{n_0 m_e}\omega_0^3} I\langle \sin\theta\rangle$$
$$= 7.15\times 10^6 \beta_3 \lambda_0^3 \left(\frac{n_0}{10^{20}}\right)\left(\frac{I}{10^{18}}\right)\langle \sin\theta\rangle, \quad (22)$$

where I is the laser intensity in the unit of W/cm^2, λ_0 laser wavelength in micron, n_0 is electron number density, and β_3 depends on the background plasma temperature T_e and tends to 1/2 with $T_e \to 0$. We can see in Eq. (21) that the cross fields from laser waves or plasma waves provides a difference beat force to drive the background electrons and hot electrons to form the solenoidal current $\sim \beta_3 \nabla \times \{n_0[\mathbf{E}_f(\mathbf{x},t)\times \mathbf{E}_f^*(\mathbf{x},t)]\}$. Thus, the slow-time-scale magnetic field occurs. In practice, the relativistic effect reduces the slow-time-scale magnetic field. However, the ultraintense laser enhances it to be proportional to laser intensity I, so that the magnetic field in the relativistic plasma can reach tens of MG, if the laser intensity $I \simeq 10^{18}$ W/cm^2 with wavelength $\lambda_0 = 1.05\ \mu$m and $\omega_0 \simeq \omega_{pe}$. Indeed, it approaches to the magnitude of magnetic field of the laser itself.

Figure 3 shows the magnitude of the slow-time-scale magnetic field for $I = 10^{18}$ W/cm^2 and $T_e = 20$ keV in the underdense plasma region for the two cases of the wavelength $\lambda_0 = 1.05\ \mu$m and $\lambda_0 = 0.35\ \mu$m. For $I = 10^{18}$ W/cm^2 and $\lambda_0 = 1.05\ \mu$m, the slow-time-scale magnetic field at the critical surface $|\mathbf{B}(\mathbf{x},t)| \sim 17$ MG, and the magnetic field at the one-fourth critical surface $|\mathbf{B}(\mathbf{x},t)| \sim 4.3$ MG, where we have set $\langle \sin\theta\rangle = 0.5$. For $I = 10^{18}$ W/cm^2 and $\lambda_0 = 0.35\ \mu$m, the slow-time-scale magnetic field at the critical surface $|\mathbf{B}(\mathbf{x},t)| \sim 5.7$ MG and the magnetic field on the one-fourth critical surface $|\mathbf{B}(\mathbf{x},t)| \sim 1.4$ MG, where we have also set $\langle \sin\theta\rangle = 0.5$. Figure 4 shows the dependence of the magnitude of the slow-time-scale magnetic field on the background electron temperature T_e on the critical surface for $I = 10^{18}$ W/cm^2 and $\lambda_0 = 1.05\ \mu$m, where $\langle \sin\theta\rangle = 0.5$. The slow-time-scale magnetic field decreases from 20 MG to 2 MG as the temperature increases from 1 keV to 500 keV.

III. PARTICLE SIMULATION

In order to check our analytical model, a two-dimensional code was constructed. The laser beam propagates along the y-direction. The system is periodic in the x-direction, which is perpendicular to the laser propagation direction. The simulation box is bounded by $(0, L_x)$ and $(0, L_y)$. A circularly polarized laser of frequency ω_0 is normally incident on the plasma slab at $y = 0$, and propagates along the y direction. The spatial length is normalized by the laser length λ_0, and the time is normalized by the reciprocal of laser frequency $1/\omega_0$. The plasma–vacuum interfaces are localized at $L_y/10$ and $9L_y/10$. The number of spatial grids is 1280×800, the size of the spatial mesh is $0.08\lambda_0$, and the number of particles is 7×10^6, respectively. The temporal pulse shape and the radial intensity profile of the incident

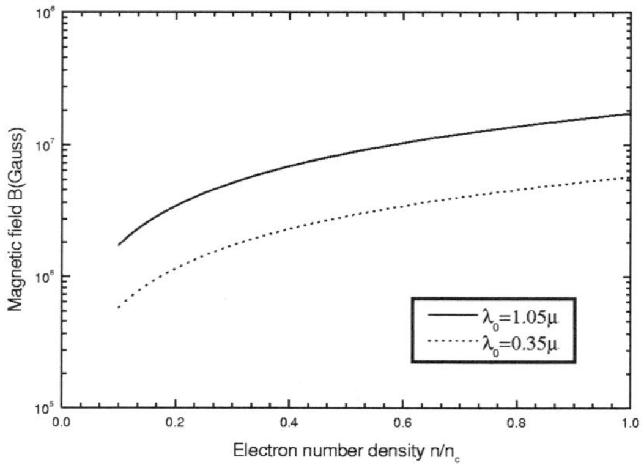

FIG. 3. The magnitude of the slow-time-scale magnetic field in the underdense plasma region for the two cases of the wavelength $\lambda_0 = 1.05\ \mu$ and $\lambda_0 = 0.35\ \mu$, where the laser intensity $I = 10^{18}$ W/cm^2 and the background electron temperature $T_e = 20$ keV.

laser field vectors $E_z + iE_x$ are both Gaussian. The laser pulse is initialized outside the plasma in the region $y < 0$.

Several simulation runs have been performed with different parameters for initial electron density n_e from $0.01n_c$ to $0.45n_c$ and laser intensity from 10^{17} W/cm^2 to 10^{20} W/cm^2. Figure 5 shows some typical simulation results for pulse duration 275 fs, transverse width $5\lambda_0$, electron density $n_e = 0.16n_c$, laser intensity $I = 10^{19}$ W/cm^2 and laser wave length $\lambda_0 = 1.06\ \mu$m. As predicted in the analytical model, along the laser propagation direction the slow-time-scale magnetic field B_y, proportional to $E_f \times E_f^*$ and driven by the solenoidal current $J_z = [\nabla \times (E_f \times E_f^*)]_z$, is clearly observed in Figs. 5(a) and 5(c), respectively. J_z, as shown in Fig. 5(c), consists of two kinds of electrons that are in solenoidal motion and in opposite directions. Correspondingly, B_y, as shown in Fig. 5(a), has two parts, the negative and the positive direction along the y axis. It is seen from Fig. 5 that the maximum amplitude of the magnetic field $|B_y|$ is the order of 30 MG at time $t = 40(2\pi/\omega_0)$. With the present

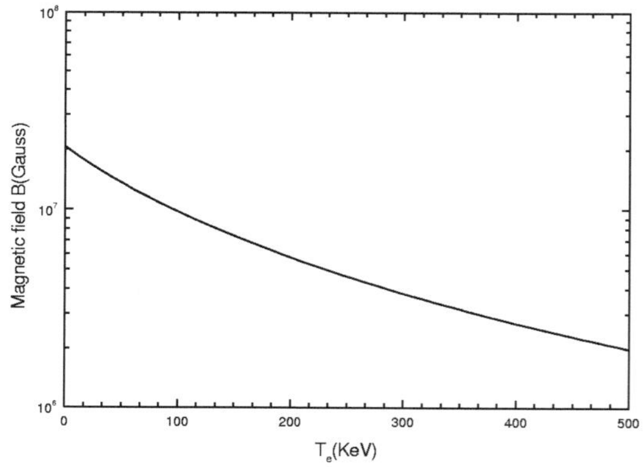

FIG. 4. The dependence of the magnitude of the slow-time-scale magnetic field on the background electron temperature T_e on the critical surface for the laser with the wavelength $\lambda_0 = 1.05\ \mu$ and the intensity $I = 10^{18}$ W/cm^2, where $\langle \sin\theta\rangle = 0.5$.

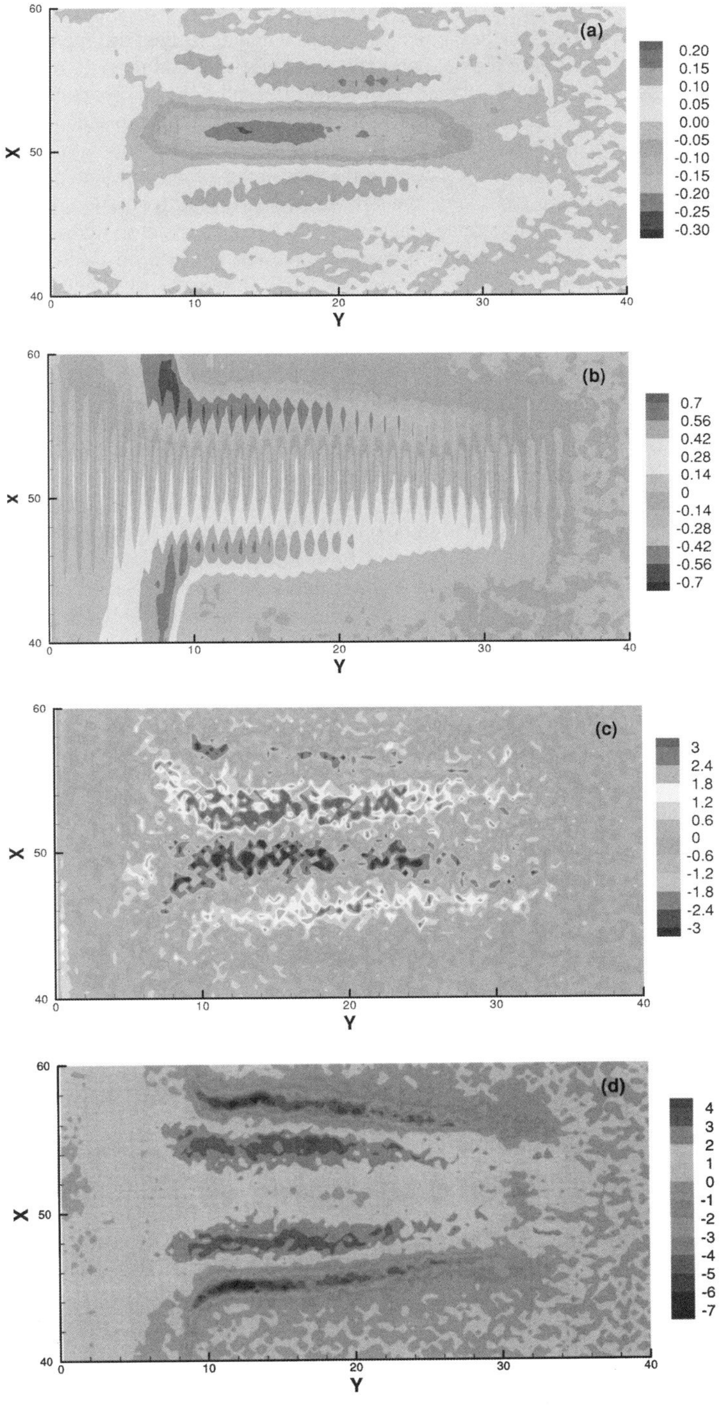

FIG. 5. The contour plots of the slow-time-scale magnetic field $B_y(a)$, the slow-time-scale magnetic field $B_z(b)$, the slow-time-scale electric current $\langle J_z\rangle(c)$ and the slow-time-scale electric current $\langle J_y\rangle(d)$ at the time $t=40(2\pi/\omega_0)$, where magnetic field is in units of 10^8 G and x, y are in units λ_0.

simulation parameters, $|B_y|$ from Eq. (22) is approximately 28 MG. According to Eq. (22), the slow-time-scale magnetic field is dependent on the angle θ. In order to check this point, we performed simulation runs varying the angle θ. The simulation results listed in Table I are qualitatively consistent with the theoretical prediction, where laser intensity $I=2\times10^{17}\,\mathrm{W/cm}^2$ and initial electron density $n_e=0.01n_c$.

From such numerical simulations, we have found that the slow-time-scale magnetic field B_y coexists with the slow-

贺贤土论文选集 490

TABLE I. The maximum value of magnetic field vs θ.

θ	90	86.4	72	64.8	45	36	30
B_y	1.8	1.6	1.4	1.2	0.8	0.6	0.5

time-scale magnetic field B_z [Fig. 5(b)] that is driven by the ponderomotive force $\nabla|E_f|^2$. Figure 5(d) shows that J_y, the electric current for B_z, includes the return current. As seen in Fig. 5, $B_z \simeq 2.5 B_y$. As a result, B_y is comparable to B_z. Obviously, both of them could affect the motion of electrons in the plasma.

IV. CONCLUSIONS

We have discussed the mechansim of the slow-time-scale magnetic field generation in the underdense plasma driven by ultraintense laser. The slow-time-scale magnetic fields comes from three source terms. The first driven source is the nonlinear solenoidal current, the second is the interaction of the fast-time-scale wave and the electron density perturbation excited by the fast-time-scale electric fields, and the third one is the coupling between fast-time-scale electric fields and the magnetic fields. On the critical surface, the magnitude of the slow-time-scale magnetic field proposed by us can be as high as 20 MG for the laser intensity $I = 10^{18}\,\mathrm{W/cm^2}$ at $\lambda_0 = 1.05\,\mu\mathrm{m}$. For $I = 10^{18}\,\mathrm{W/cm^2}$ at $\lambda_0 = 0.35\,\mu\mathrm{m}$, the slow-time-scale magnetic field at the one-fourth critical surface is about 1 MG. Via particle simulation, we have investigated the generation of slow-time-scale magnetic fields. Our theoretical model is supported by simulation results. For the purposes of laser–plasma acceleration and fast ignitor, the ultraintense laser with intensities in $10^{20}-10^{21}\,\mathrm{W/cm^2}$ is propagating in the underdense plasma, so that a slow-time-scale magnetic fields with hundreds of MG can be generated. Such a magnetic field is important because the field will affect the electron motion. As a result, it will affect the electron heat transport and the propagation of laser waves in plasma, which is still an open problem and under further investigation.

As we showed in the beginning of the present paper, there are several theoretical models for the generation of the slow-time-scale magnetic fields in an underdense plasma region. Here, we compare the present model with the earlier ones. The direction of the slow-time-scale magnetic field given in Ref. 6 is parallel to the pumping laser field, but the magnetic field in the present paper is perpendicular to it. The slow-time-scale magnetic field proposed in Refs. 11 and 12 is generated due to a nonlinear electric current $\mathbf{j}_{\mathrm{NL}} \sim \mathbf{E}_0 \nabla \cdot \mathbf{E}_0^* - \mathrm{c.c.}$ (where $\mathbf{E}_0$ is the slow-time-scale amplitude of laser pulse), which is different from the present one. We have shown that the magnitude of the slow-time-scale magnetic field given by the present model is qualitatively consistent with particle simulations. We think that the present mechanism is a possible cause for the generation of slow-time-scale magnetic fields in an underdense plasma.

ACKNOWLEDGMENTS

This work was supported by National Hi-Tech Inertial Confinement Fusion Committee of China, National Natural Science Foundation of China, Grant No. 19735002, and National Basic Research Project ''Nonlinear Science'' in China.

APPENDIX: PROOF OF EQ. (14)

Let $\mathbf{e}_s' = \mathbf{e}_s'^{\sigma_s}(q)$, $\mathbf{e}_1' = \mathbf{e}_f'^{\sigma_1}(k_1)$, $\mathbf{e}_2' = \mathbf{e}_f'^{\sigma_2}(k_2)$, $\mathbf{e}_s = \mathbf{e}_s^{\sigma_s}(q)$, $\mathbf{e}_1 = \mathbf{e}_f^{\sigma_1}(k_1)$, $\mathbf{e}_2 = \mathbf{e}_f^{\sigma_2}(k_2)$, Then

$$\Pi_\alpha^{\sigma_s\sigma_1\sigma_2}(q,k_1,k_2) = \frac{1}{2}\int \left[\frac{\dfrac{c\mathbf{e}_s\cdot\mathbf{p}_\alpha}{p_{0\alpha}}}{\Omega - \dfrac{c\mathbf{q}\cdot\mathbf{p}_\alpha}{p_{0\alpha}}} \mathbf{e}_1'\cdot\frac{\partial}{\partial\mathbf{p}_\alpha} \frac{\mathbf{e}_2'\cdot\dfrac{\partial f_{\alpha 0}}{\partial\mathbf{p}_\alpha}}{\omega_2 - \dfrac{c\mathbf{k}_2\cdot\mathbf{p}_\alpha}{p_{0\alpha}}} + (1\leftrightarrow 2) \right] d^3\mathbf{p}_\alpha. \tag{A1}$$

We now prove that

$$\begin{aligned} &\frac{1}{\Omega}\int \left[\frac{\dfrac{c\mathbf{e}_s\cdot\mathbf{p}_\alpha}{p_{0\alpha}}}{\Omega - \dfrac{c\mathbf{q}\cdot\mathbf{p}_\alpha}{p_{0\alpha}}} \mathbf{e}_1'\cdot\frac{\partial}{\partial\mathbf{p}_\alpha} \frac{\mathbf{e}_2'\cdot\dfrac{\partial f_{\alpha 0}}{\partial\mathbf{p}_\alpha}}{\omega_2 - \dfrac{c\mathbf{k}_2\cdot\mathbf{p}_\alpha}{p_{0\alpha}}} + (1\leftrightarrow 2) \right] d^3\mathbf{p}_\alpha, \\ &-\frac{1}{\omega_2}\int \frac{\dfrac{c\mathbf{e}_2\cdot\mathbf{p}_\alpha}{p_{0\alpha}}}{\omega_2 - \dfrac{c\mathbf{k}_2\cdot\mathbf{p}_\alpha}{p_{0\alpha}}} \left\{ \mathbf{e}_s'\cdot\frac{\partial}{\partial\mathbf{p}_\alpha} \frac{\mathbf{e}_1'\cdot\dfrac{\partial f_{\alpha 0}}{\partial\mathbf{p}_\alpha}}{\omega_1 - \dfrac{c\mathbf{k}_1\cdot\mathbf{p}_\alpha}{p_{0\alpha}}} - \mathbf{e}_1'\cdot\frac{\partial}{\partial\mathbf{p}_\alpha} \frac{\mathbf{e}_s'\cdot\dfrac{\partial f_{\alpha 0}}{\partial\mathbf{p}_\alpha}}{\Omega - \dfrac{c\mathbf{q}\cdot\mathbf{p}_\alpha}{p_{0\alpha}}} \right\} d^3\mathbf{p}_\alpha = 0. \end{aligned} \tag{A2}$$

Making partial integration, we have

$$(A2)=\int\left\{-\frac{1}{\Omega}\frac{\mathbf{e}_2'\cdot\dfrac{\partial f_{\alpha0}}{\partial\mathbf{p}_\alpha}}{\omega_2-\dfrac{c\mathbf{k}_2\cdot\mathbf{p}_\alpha}{p_{0\alpha}}}\mathbf{e}_1'\cdot\left[\frac{c\mathbf{e}_s}{\left(\Omega-\dfrac{c\mathbf{q}\cdot\mathbf{p}_\alpha}{p_{0\alpha}}\right)p_{0\alpha}}-\frac{c(\mathbf{e}_s\cdot\mathbf{p}_\alpha)\mathbf{p}_\alpha}{\left(\Omega-\dfrac{c\mathbf{q}\cdot\mathbf{p}_\alpha}{p^0}\right)p_0^3}+\frac{c(\mathbf{e}_s\cdot\mathbf{p}_\alpha)}{p_{0\alpha}\left(\Omega-\dfrac{c\mathbf{q}\cdot\mathbf{p}_\alpha}{p^0}\right)^2}\left[\frac{c\mathbf{q}}{p_{0\alpha}}-\frac{c(\mathbf{q}\cdot\mathbf{p}_\alpha)\mathbf{p}_\alpha}{p_0^3}\right]\right]\right.$$

$$-\frac{1}{\Omega}\frac{\mathbf{e}_1'\cdot\dfrac{\partial f_{\alpha0}}{\partial\mathbf{p}_\alpha}}{\omega_1-\dfrac{c\mathbf{k}_1\cdot\mathbf{p}_\alpha}{p_{0\alpha}}}\mathbf{e}_2'\cdot\left[\frac{c\mathbf{e}_s}{\left(\Omega-\dfrac{c\mathbf{q}\cdot\mathbf{p}_\alpha}{p_{0\alpha}}\right)p_{0\alpha}}-\frac{c(\mathbf{e}_s\cdot\mathbf{p}_\alpha)\mathbf{p}_\alpha}{\left(\Omega-\dfrac{c\mathbf{q}\cdot\mathbf{p}_\alpha}{p^0}\right)p_0^3}+\frac{c(\mathbf{e}_s\cdot\mathbf{p}_\alpha)}{p_{0\alpha}\left(\Omega-\dfrac{c\mathbf{q}\cdot\mathbf{p}_\alpha}{p^0}\right)^2}\left[\frac{c\mathbf{q}}{p_{0\alpha}}-\frac{c(\mathbf{q}\cdot\mathbf{p}_\alpha)\mathbf{p}_\alpha}{p_0^3}\right]\right]+\frac{1}{\omega_2}$$

$$\times\frac{\mathbf{e}_1'\cdot\dfrac{\partial f_{\alpha0}}{\partial\mathbf{p}_\alpha}}{\omega_1-\dfrac{c\mathbf{k}_1\cdot\mathbf{p}_\alpha}{p_{0\alpha}}}\mathbf{e}_s'\cdot\left[\frac{c\mathbf{e}_2}{\left(\omega_2-\dfrac{c\mathbf{k}_2\cdot\mathbf{p}_\alpha}{p_{0\alpha}}\right)p_{0\alpha}}-\frac{c(\mathbf{e}_2\cdot\mathbf{p}_\alpha)\mathbf{p}_\alpha}{\left(\omega_2-\dfrac{c\mathbf{k}_2\cdot\mathbf{p}_\alpha}{p^0}\right)p_0^3}+\frac{c(\mathbf{e}_2\cdot\mathbf{p}_\alpha)}{p_{0\alpha}\left(\omega_2-\dfrac{c\mathbf{k}_2\cdot\mathbf{p}_\alpha}{p^0}\right)^2}\left[\frac{c\mathbf{k}_2}{p_{0\alpha}}-\frac{c(\mathbf{k}_2\cdot\mathbf{p}_\alpha)\mathbf{p}_\alpha}{p_0^3}\right]\right]$$

$$-\frac{1}{\omega_2}\frac{\mathbf{e}_s'\cdot\dfrac{\partial f_{\alpha0}}{\partial\mathbf{p}_\alpha}}{\Omega-\dfrac{c\mathbf{q}\cdot\mathbf{p}_\alpha}{p_{0\alpha}}}\mathbf{e}_1'\cdot\left[\frac{c\mathbf{e}_2}{\left(\omega_2-\dfrac{c\mathbf{k}_2\cdot\mathbf{p}_\alpha}{p_{0\alpha}}\right)p_{0\alpha}}-\frac{c(\mathbf{e}_2\cdot\mathbf{p}_\alpha)\mathbf{p}_\alpha}{\left(\omega_2-\dfrac{c\mathbf{k}_2\cdot\mathbf{p}_\alpha}{p^0}\right)p_0^3}\right.$$

$$\left.\left.+\frac{c(\mathbf{e}_2\cdot\mathbf{p}_\alpha)}{p_{0\alpha}\left(\omega_2-\dfrac{c\mathbf{k}_2\cdot\mathbf{p}_\alpha}{p^0}\right)^2}\left[\frac{c\mathbf{k}_2}{p_{0\alpha}}-\frac{c(\mathbf{k}_2\cdot\mathbf{p}_\alpha)\mathbf{p}_\alpha}{p_0^3}\right]\right]\right\}d^3\mathbf{p}_\alpha$$

$$=\frac{1}{\Omega\omega_1\omega_2}\int\frac{c}{p_{0\alpha}}\frac{\partial f_{\alpha0}}{\partial\mathbf{p}_\alpha}\cdot\left\{\frac{\mathbf{e}_1'[\mathbf{e}_2'\cdot\mathbf{e}_s']-\mathbf{e}_2'[\mathbf{e}_1'\cdot\mathbf{e}_s']}{\left(\Omega-\dfrac{c\mathbf{q}\cdot\mathbf{p}_\alpha}{p_{0\alpha}}\right)^2\left(\omega_2-\dfrac{c\mathbf{k}_2\cdot\mathbf{p}_\alpha}{p_{0\alpha}}\right)}+\frac{\mathbf{e}_1'[\mathbf{e}_2'\cdot\mathbf{e}_s']-\mathbf{e}_s'[\mathbf{e}_1'\cdot\mathbf{e}_2']}{\left(\Omega-\dfrac{c\mathbf{q}\cdot\mathbf{p}_\alpha}{p_{0\alpha}}\right)\left(\omega_2-\dfrac{c\mathbf{k}_2\cdot\mathbf{p}_\alpha}{p_{0\alpha}}\right)^2}\right\}d^3\mathbf{p}_\alpha-\int\frac{c}{p_0^3}\frac{\partial f_{\alpha0}}{\partial\mathbf{p}_\alpha}$$

$$\cdot\left\{\frac{\mathbf{e}_1'[\mathbf{p}_\alpha\cdot\mathbf{e}_2'][\mathbf{p}_\alpha\cdot\mathbf{e}_s']-\mathbf{e}_2'[\mathbf{p}_\alpha\cdot\mathbf{e}_1'][\mathbf{p}_\alpha\cdot\mathbf{e}_s']}{\left(\Omega-\dfrac{c\mathbf{q}\cdot\mathbf{p}_\alpha}{p_{0\alpha}}\right)^2\left(\omega_2-\dfrac{c\mathbf{k}_2\cdot\mathbf{p}_\alpha}{p_{0\alpha}}\right)}+\frac{\mathbf{e}_1'[\mathbf{p}_\alpha\cdot\mathbf{e}_2'][\mathbf{p}_\alpha\cdot\mathbf{e}_s']-\mathbf{e}_s'[\mathbf{p}_\alpha\cdot\mathbf{e}_1'][\mathbf{p}_\alpha\cdot\mathbf{e}_2']}{\left(\Omega-\dfrac{c\mathbf{q}\cdot\mathbf{p}_\alpha}{p_{0\alpha}}\right)\left(\omega_2-\dfrac{c\mathbf{k}_2\cdot\mathbf{p}_\alpha}{p_{0\alpha}}\right)^2}\right\}d^3\mathbf{p}_\alpha. \tag{A3}$$

After tedious calculation, we can obtain

$$(A2)=\frac{1}{\Omega\omega_1\omega_2}\int\frac{c}{p_{0\alpha}}\frac{\partial f_{\alpha0}}{\partial\mathbf{p}_\alpha}\cdot\left\{\frac{\mathbf{e}_1'[\mathbf{e}_2'\cdot\mathbf{e}_s']-\mathbf{e}_2'[\mathbf{e}_1'\cdot\mathbf{e}_s']}{\left(\Omega-\dfrac{c\mathbf{q}\cdot\mathbf{p}_\alpha}{p_{0\alpha}}\right)^2\left(\omega_2-\dfrac{c\mathbf{k}_2\cdot\mathbf{p}_\alpha}{p_{0\alpha}}\right)}+\frac{\mathbf{e}_1'[\mathbf{e}_2'\cdot\mathbf{e}_s']-\mathbf{e}_s'[\mathbf{e}_1'\cdot\mathbf{e}_2']}{\left(\Omega-\dfrac{c\mathbf{q}\cdot\mathbf{p}_\alpha}{p_{0\alpha}}\right)\left(\omega_2-\dfrac{c\mathbf{k}_2\cdot\mathbf{p}_\alpha}{p_{0\alpha}}\right)^2}\right\}d^3\mathbf{p}_\alpha$$

$$-\frac{1}{\Omega\omega_1\omega_2}\int\frac{c}{p_0^3}\frac{\partial f_{\alpha0}}{\partial\mathbf{p}_\alpha}\cdot\left\{\frac{\mathbf{e}_1'[\mathbf{p}_\alpha\cdot\mathbf{e}_2'][\mathbf{p}_\alpha\cdot\mathbf{e}_s']-\mathbf{e}_2'[\mathbf{p}_\alpha\cdot\mathbf{e}_1'][\mathbf{p}_\alpha\cdot\mathbf{e}_s']}{\left(\Omega-\dfrac{c\mathbf{q}\cdot\mathbf{p}_\alpha}{p_{0\alpha}}\right)^2\left(\omega_2-\dfrac{c\mathbf{k}_2\cdot\mathbf{p}_\alpha}{p_{0\alpha}}\right)}\right.$$

$$\left.+\frac{\mathbf{e}_1'[\mathbf{p}_\alpha\cdot\mathbf{e}_2'][\mathbf{p}_\alpha\cdot\mathbf{e}_s']-\mathbf{e}_s'[\mathbf{p}_\alpha\cdot\mathbf{e}_1'][\mathbf{p}_\alpha\cdot\mathbf{e}_2']}{\left(\Omega-\dfrac{c\mathbf{q}\cdot\mathbf{p}_\alpha}{p_{0\alpha}}\right)\left(\omega_2-\dfrac{c\mathbf{k}_2\cdot\mathbf{p}_\alpha}{p_{0\alpha}}\right)^2}\right\}d^3\mathbf{p}_\alpha$$

$$=\frac{1}{\Omega^2\omega_1^2\omega_2^2}\int\left\{\frac{\partial}{\partial\mathbf{p}_\alpha}\cdot\left[\frac{c\mathbf{e}_1\cdot\mathbf{p}_\alpha}{p_{0\alpha}}(\mathbf{e}_2\mathbf{e}_s-\mathbf{e}_s\mathbf{e}_2)\cdot\frac{\partial}{\partial\mathbf{p}_\alpha}\right]\right.$$

$$+\frac{\partial}{\partial\mathbf{p}_\alpha}\cdot\left[\frac{1}{\left(\Omega-\dfrac{c\mathbf{q}\cdot\mathbf{p}_\alpha}{p_{0\alpha}}\right)\left(\omega_2-\dfrac{c\mathbf{k}_2\cdot\mathbf{p}_\alpha}{p_{0\alpha}}\right)}\frac{c^3(\mathbf{e}_1\cdot\mathbf{p}_\alpha)(\mathbf{e}_2\cdot\mathbf{p}_\alpha)(\mathbf{e}_s\cdot\mathbf{p}_\alpha)}{p_0^3}(\mathbf{k}_2\mathbf{q}-\mathbf{q}\mathbf{k}_2)\cdot\frac{\partial}{\partial\mathbf{p}_\alpha}\right]$$

$$+\frac{\partial}{\partial\mathbf{p}_\alpha}\cdot\left[\frac{\Omega-\dfrac{c\mathbf{q}\cdot\mathbf{p}_\alpha}{p_{0\alpha}}}{\omega_2-\dfrac{c\mathbf{k}_2\cdot\mathbf{p}_\alpha}{p_{0\alpha}}}\frac{c(\mathbf{e}_2\cdot\mathbf{p}_\alpha)}{p_0^3}(\mathbf{e}_s\mathbf{e}_1-\mathbf{e}_1\mathbf{e}_2)\cdot\frac{\partial}{\partial\mathbf{p}_\alpha}\right]+\frac{\partial}{\partial\mathbf{p}_\alpha}\cdot\left[\frac{\omega_2-\dfrac{c\mathbf{k}_2\cdot\mathbf{p}_\alpha}{p_{0\alpha}}}{\Omega-\dfrac{c\mathbf{q}\cdot\mathbf{p}_\alpha}{p_{0\alpha}}}\frac{c(\mathbf{e}_2\cdot\mathbf{p}_\alpha)}{p_0^3}(\mathbf{e}_1\mathbf{e}_2-\mathbf{e}_2\mathbf{e}_1)\cdot\frac{\partial}{\partial\mathbf{p}_\alpha}\right]$$

$$+\frac{\partial}{\partial\mathbf{p}_\alpha}\cdot\left[\frac{1}{\Omega-\dfrac{c\mathbf{q}\cdot\mathbf{p}_\alpha}{p_{0\alpha}}}\frac{c^2(\mathbf{e}_s\cdot\mathbf{p}_\alpha)(\mathbf{e}_1\cdot\mathbf{p}_\alpha)}{p_0^2}(\mathbf{e}_2\mathbf{k}_1-\mathbf{k}_1\mathbf{e}_2)\cdot\frac{\partial}{\partial\mathbf{p}_\alpha}\right]+\frac{\partial}{\partial\mathbf{p}_\alpha}$$

$$\cdot\left[\frac{1}{\Omega-\dfrac{c\mathbf{q}\cdot\mathbf{p}_\alpha}{p_{0\alpha}}}\frac{c^2(\mathbf{e}_s\cdot\mathbf{p}_\alpha)(\mathbf{e}_2\cdot\mathbf{p}_\alpha)}{p_0^2}(\mathbf{e}_1\mathbf{k}_2-\mathbf{k}_2\mathbf{e}_1)\cdot\frac{\partial}{\partial\mathbf{p}_\alpha}\right]+\frac{\partial}{\partial\mathbf{p}_\alpha}\cdot\left[\frac{1}{\omega_2-\dfrac{c\mathbf{k}_2\cdot\mathbf{p}_\alpha}{p_{0\alpha}}}\frac{c^2(\mathbf{e}_1\cdot\mathbf{p}_\alpha)(\mathbf{e}_2\cdot\mathbf{p}_\alpha)}{p_0^2}(\mathbf{e}_s\mathbf{k}_1-\mathbf{k}_1\mathbf{e}_s)\cdot\frac{\partial}{\partial\mathbf{p}_\alpha}\right]$$

$$\left.+\frac{\partial}{\partial\mathbf{p}_\alpha}\cdot\left[\frac{1}{\omega_2-\dfrac{c\mathbf{k}_2\cdot\mathbf{p}_\alpha}{p_{0\alpha}}}\frac{c^2(\mathbf{e}_s\cdot\mathbf{p}_\alpha)(\mathbf{e}_2\cdot\mathbf{p}_\alpha)}{p_0^2}(\mathbf{e}_s\mathbf{q}-\mathbf{q}\mathbf{e}_1)\cdot\frac{\partial}{\partial\mathbf{p}_\alpha}\right]\right\}f_{\alpha0}d^3\mathbf{p}_\alpha=0. \quad \text{(A4)}$$

Then Eq. (14) has been proved. It should be noted that the isotropy of distribution function $f_{\alpha0}$ has been used in the above proof.

[1]J. A. Stamper, K. Papadopoulos, R. N. Sudan, S. O. Dean, E. A. Mclean, and J. M. Dawson, Phys. Rev. Lett. **26**, 1012 (1971).
[2]M. Tabak, J. Hammer, M. E. Glinsky, W. L. Krue, S. C. Wilks, J. Wodwirth, E. M. Campbell, and M. D. Perry, Phys. Plasmas **1**, 1626 (1994).
[3]A. B. Borisov, A. V. Borovskiy, V. V. Korobkin *et al.*, Phys. Rev. Lett. **68**, 2309 (1992).
[4]S. C. Wilks, W. L. Kruer, M. Tabak, and A. B. Langdon, Phys. Rev. Lett. **69**, 1383 (1992).
[5]R. N. Sudan, Phys. Rev. Lett. **70**, 3075 (1993).
[6]V. K. Tripathi and C. S. Liu, Phys. Plasmas **1**, 990 (1994).
[7]R. J. Mason and M. Tabak, Phys. Rev. Lett. **80**, 524 (1998).
[8]J. Fuchs, G. Malka, J. C. Adam *et al.*, Phys. Rev. Lett. **80**, 1658 (1998).
[9]A. D. Steiger and C. H. Woods, Phys. Rev. A **5**, 1467 (1972).
[10]T. Lehner, Phys. Scr. **49**, 704 (1994).
[11]V. Yu Bychenkov, V. I. Demin, and V. T. Tikhonchuk, JETP **78**, 62 (1994).
[12]V. Yu Bychenkov and V. T. Tikhonchuk, Laser Part. Beams **14**, 55 (1996).
[13]M. Kono, M. M. Skoric, and D. ter. Haar, J. Plasma Phys. **26**, 123 (1981).
[14]X. T. He, Acta Sci. Sin. **32**, 325 (1983).
[15]S. P. Zhu, Phys. Fluids B **5**, 1024 (1993).
[16]S. P. Zhu and X. T. He, Plasma Phys. Controlled Fusion **36**, 291 (1993).

Generation of Helical and Axial Magnetic Fields by the Relativistic Laser Pulses in Under-dense Plasma: Three-Dimensional Particle-in-Cell Simulation*

ZHENG Chun-Yang(郑春阳), ZHU Shao-Ping(朱少平), HE Xian-Tu(贺贤土)
Institute of Applied Physics and Computational Mathematics, PO Box 8009, Beijing 100088

(Received 4 April 2002)

The quasi-static magnetic fields created in the interaction of relativistic laser pulses with under-dense plasmas have been investigated by three-dimensional particle-in-cell simulation. The relativistic ponderomotive force can drive an intense electron current in the laser propagation direction, which is responsible for the generation of a helical magnetic field. The axial magnetic field results from a difference beat of wave–wave, which drives a solenoidal current. In particular, the physical significance of the kinetic model for the generation of the axial magnetic field is discussed.

PACS: 52.40.Nk, 52.35.Mw, 52.25.Rv

Experiments of ultra-intense short laser pulse interaction with plasmas have stimulated great interest in the studies of quasi-static magnetic field generation. The presence of this field could considerably affect the nonlinear plasma dynamics, especially the laser pulse propagation in a plasma and the motion of relativistic electrons, which are crucial for realization of the "fast ignitor" project.[1,2] Many researchers have discussed[3−7] the generation mechanism and effects of this field. However, many problems have not been clarified because of the intrinsically complex nature, as detailed information is needed on the space-time behaviour of involving fields, the electron distribution and the very nonlinear stage of this process. In this Letter, we present, for the first time to our knowledge, the results of a numerical simulation of three-dimensional regimes of the propagation of a laser pulse through a plasma. A highly efficient, completely relativistic, three-dimensional electromagnetic code that is a realization of the particle-in-cell (PIC) method[8] is used to study the generation of slow-timescale magnetic fields.

The axial slow-timescale magnetic field (in the direction of laser propagation) driven by fast-timescale electromagnetic waves or plasma waves are examined from the perspective of the Vlasov–Maxwell equations for a relativistic Vlasov plasma. An equation for a slow-timescale magnetic field is obtained.[10] A nonuniformity or time variation of the ponderomotive force can serve as a source of the helical magnetic field (along the periphery of the pulse). A theory of this mechanism was derived in Ref. [5] near the boundary of a plasma of supercritical density. For the generation of a helical magnetic field in uniform under-dense plasma, we have also given the theoretical model and two-dimensional PIC simulation. It was confirmed that the relativistic ponderomotive force and the laser wake field concur to drive the intense electron current in the laser propagation direction, which is responsible for the generation of this helical magnetic field.[9] The present work treats the generation mechanism of both magnetic fields by PIC simulation in three-dimensional geometry simultaneously. The spontaneous magnetic field driven by a relativistic solenoidal current is comparable to that driven by a relativistic ponderomotive current. A richer and clearer physical picture of the relativistic laser plasma interaction is displayed.

We consider a three-dimensional homogeneous plasma slab described in Cartesian geometry $xyz(0-L_x, 0-L_y, 0-L_z)$. A circularly polarized laser of frequency ω_0 is normally incident on the plasma slab at $y=0$ and propagates along the y-direction. The boundary conditions are open-sided for fields and mirror reflection for particles in the y-direction. In the x- and z-directions, the periodic boundary conditions are chosen. The spatial length is normalized by the laser length λ_0, and the time is normalized by the reciprocal of laser frequency $1/\omega_0$, a spatial mesh of $110\times160\times110$ cells $(8.5\times12.5\times8.5\lambda_0)$ and up to 5×10^6 particles (ion mass is $1836\,m_e$) are used. The initial plasma temperature is 20 keV and $n/n_c=0.16$ is chosen. The laser pulse is initialized outside the plasma in the region $y<0$. The transverse intensity profile of incident laser field vectors $E_z+\mathrm{i}E_x$ is Gaussian with width $l_\perp\approx3\lambda_0$. The intensity in one temporal pulse at first increases sharply in the initial duration of $3\lambda_0/c$ (called the pulse front) and is then kept constant. The peak intensity is varied from $I=10^{19}$ to $10^{20}\,\mathrm{W/cm^2}$. In this regime, the directed motion of electrons is much larger than the thermal motion, so the electron temperature effect is negligible.

The energy conservation relation

$$\frac{\partial}{2\partial t}\int(E^2+B^2)\mathrm{d}V=-\int(\boldsymbol{E}\times\boldsymbol{B})\cdot\mathrm{d}s-\frac{\omega_{p0}^2}{\omega_0^2}\int\boldsymbol{j}\cdot\boldsymbol{E}\mathrm{d}V$$

* Supported by the National High-Tech ICF Committee in China, the National Basic Science Project and Foundation of China Academy of Engineering Physics (2000 Z0206) and the National Key Laboratory Foundation of China (51480030101ZW0902).

is first checked to ensure that the simulation results are reliable.

The results for one typical run $I = 2\times 10^{19}\ \mathrm{W/cm^2}$ are presented in longitudinal (Fig. 1) and transverse (Fig. 2) cuts through the plasma volume. Above this laser intensity, the amplitude of self-generated magnetic fields does not increase significantly with the increasing laser intensity due to nonlinear saturation (such as the electronic cavity effect). From Fig. 1(a), we can see the depletion of the pulse just behind the pulse front, which may result in the downshift of frequency of this portion of the pulse and the excitation of the intense electrostatic field, which lags behind the pulse front.[9] This process is developed in a nonlinear recurrent manner (involving the reversal of energy between the electric wake field and pulse field) and can result in acceleration and heating of plasma electrons. It can been seen that a more concentrated intense current is generated in the laser propagation direction, as shown in Fig. 1(c). A detailed investigation of the above process and the related phenomena is of considerable interest, but it is out of the scope of this Letter. Here we only concentrate on the generation of the spontaneous magnetic field.

The helical magnetic field is driven by a relativistic ponderomotive current and can be simply estimated from the longitudinal current (Fig. 2(c)) by using

$$(\nabla \times \langle \boldsymbol{B} \rangle)_y = \frac{4\pi}{c}\langle J_y \rangle.$$

The maximum cycle-averaged intensity is found to be 40 MG from Fig. 1(b). It vanishes at the pulse axis and reaches a maximum at the edges (Fig. 2(c)). It is also noted that the azimuthal magnetic field is always negative. Therefore, it has a focusing effect on relativistic electrons, which can overtake the leading edge of the pulse, as seen in Fig. 1(c).

As predicted in the analytical model,[10] along the laser propagation direction the axial magnetic field B_y, which is driven by the solenoidal current $\boldsymbol{J}_\perp = \beta\nabla \times (n_0 \boldsymbol{E}_f \times \boldsymbol{E}_f^*)$ and is proportional to

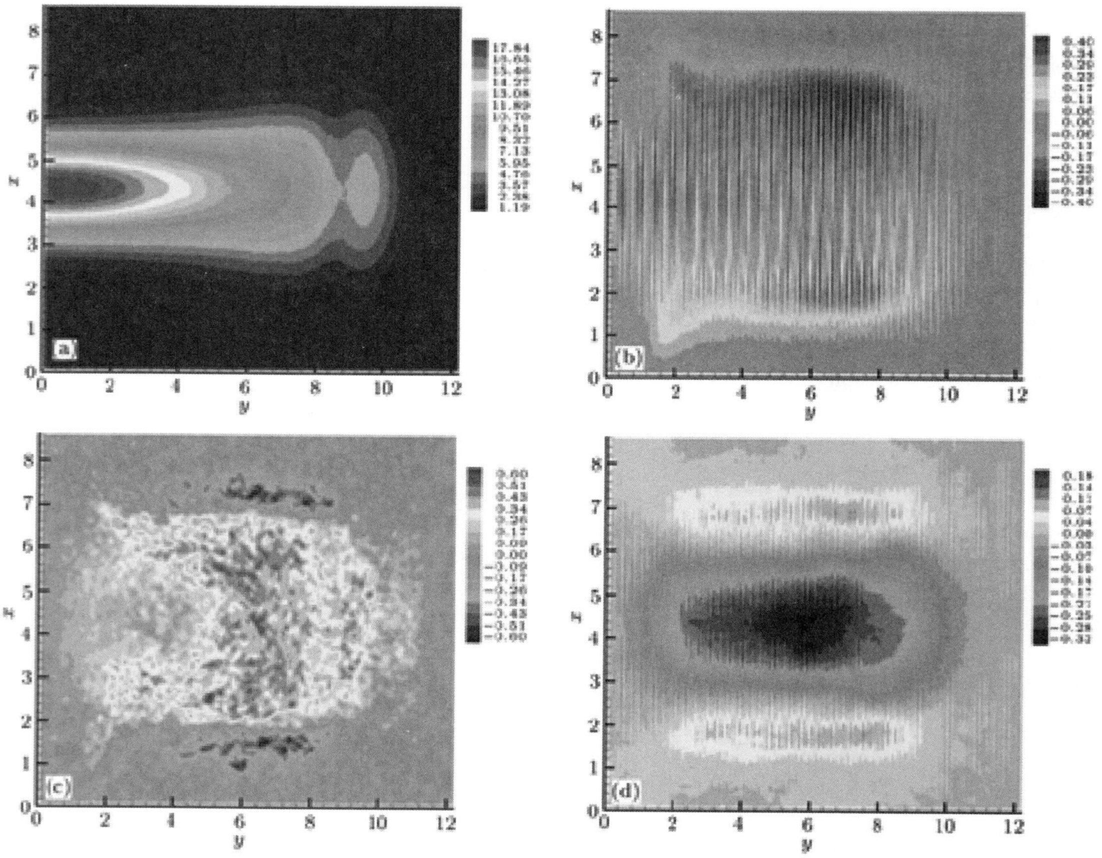

Fig. 1. Interaction of laser light with plasma and generation of quasi-static magnetic field (averaged over one laser period) at $t = 16(\pi/\omega_0)$, $I = 2\times 10^{19}\ \mathrm{W/cm^2}$, along the pulse axis in the $x-y$ plane at $z = L_z/2$. (a) Contour of the electromagnetic energy density in units of $I\lambda_0^2$, i.e., $\mu\mathrm{m}\cdot\mathrm{cm}^2/1.368\times 10^{18}$ W. (b) Contour of the quasi-static magnetic field B_z in units of $mc\omega_0/e \cong 10^8$ G for $\lambda_0 = 1.06\mu$m. The fast oscillatory part is from the laser magnetic field. (c) Contour plots of the longitudinal current $\langle J_y \rangle$ in units of en_0c. (d) Contour plots of the axial magnetic field $\langle B_y \rangle$.

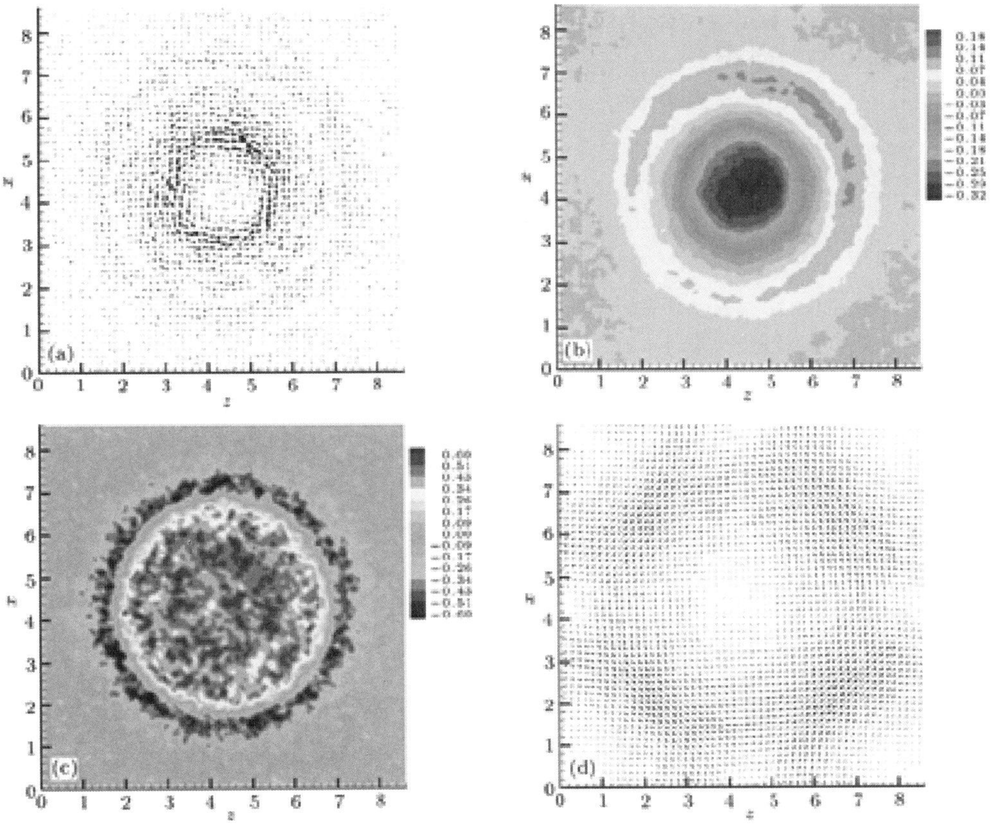

Fig. 2. Transverse cuts (x, z) at $y = L_y/2$. (a) Distribution of the vector of transverse electron current density. (b) Contour of the axial magnetic field $\langle B_y \rangle$. (c) Contour of the longitudinal current $\langle J_y \rangle$. (d) Distribution of the vector helical magnetic field.

$$\frac{\mathrm{i}\omega_{\mathrm{pe}}^2 e\beta}{m_e c\omega_0^3}\boldsymbol{E}_f\times\boldsymbol{E}_f^* \approx 7.15\times10^6\beta\lambda_0^3\Big(\frac{n_0}{10^{20}}\Big)\Big(\frac{I}{10^{18}}\Big)\langle\sin\theta\rangle,$$

is clearly observed in Fig. 2, where β is about 0.3 for $T_e = 20\,\mathrm{keV}$. Here, as shown in Fig. 2(a), $J_\perp$ consists of two kinds of electrons that are in the solenoidal motion and in the opposite direction. Correspondingly, as shown in Fig. 2(b), B_y has two parts, i.e. the negative and positive directions along the y-axis. The maximum amplitude of the magnetic field $|B_y|$ is of the order of 30 MG at time $t = 16(2\pi/\omega_0)$. With the present simulation parameters, from the theoretical model,[10] $|B_y|$ is approximately 35 MG. The simulation agrees well with the theoretical prediction.

We can see that the axial magnetic field is comparable to the helical field. Obviously, both of these could affect the motion of hot electrons (pinch and gyro-motion) in fast ignition.

In conclusion, we have presented the first three-dimensional PIC simulations. The generation of the helical magnetic field is due to the relativistic ponderomotive force, which can drive an intense electron current in the laser propagation direction. The axial magnetic field results from a difference beat of wave–wave, which drives a solenoidal current. Both the fields are important to the motion of relativistic electrons. In particular, the physical significance of the kinetic model for the generation of the axial magnetic field is discussed.

References

[1] Tabak M et al 1994 *Phys. Plasmas* **1** 1626
[2] Pukhov A and Meyer-ter Vehn J 1996 *Phys. Rev. Lett.* **76** 3975
[3] Wilks S C et al 1992 *Phys. Rev. Lett.* **69** 1383
[4] Stamper J A 1991 *Laser Part. Beams* 9841
Stamper J A et al 1975 *Phys. Rev. Lett.* **4** 138
[5] Sudan R N 1997 *Phys. Rev. Lett.* **70** 3075
Tripathi V K and Liu C S 1994 *Phys. Plasmas* **1** 990
[6] Berezhiani V I et al 1997 *Phys. Rev.* E **55** 995
Bychenkov V Y et al 1994 *Zh. Eksp. Teor. Fiz.* **105** 118
[7] Mason J et al 1998 *Phys. Rev. Lett.* **80** 524
Borghesi M et al 1998 *Phys. Rev. Lett.* **81** 112
[8] Birdsall C K and Langdon A B 1985 *Plasma Physics via Computer Simulation* (New York: McGraw-Hill)
[9] Zheng C Y, Zhu S P and He X T 2000 *Chin. Phys. Lett.* **17** 746
[10] Zhu S P, He X T and Zheng C Y 2001 *Phys. Plasmas* **8** 321

PHYSICS OF PLASMAS VOLUME 10, NUMBER 10 OCTOBER 2003

A theoretical model for a spontaneous magnetic field in intense laser plasma interaction

Shao-ping Zhu, C. Y. Zheng, and X. T. He
Institute of Applied Physics and Computational Mathematics, P.O. Box 8009, Beijing 100088, People's Republic of China

(Received 8 April 2003; accepted 18 July 2003)

The experiment results on spontaneous magnetic fields given by Najmudin *et al.* [Phys. Rev. Lett. **87**, 215004 (2001)] can be explained by the kinetic model proposed by the present authors [Phys. Plasmas **8**, 312 (2001)]. The dependence of the peak spontaneous magnetic field on the laser intensity is discussed. It is found the ratio of the number of thermal electrons to the total number of electrons is an important parameter. © *2003 American Institute of Physics.*
[DOI: 10.1063/1.1608936]

The fast ignition scheme for inertial confinement fusion[1] has generated worldwide interest in the interaction of an intense short-pulse laser with plasma.[2,3] Among the various nonlinear effects which may occur in plasma interacting with intense short-pulse laser, the generation of spontaneous magnetic fields is one of the most interesting and significant problems because the fields could have considerable influence on nonlinear plasma dynamics. An extremely intense spontaneous magnetic field has been observed by both particle simulation[3,4] and experiment measurement[5,6] in the interaction of ultra-intense short laser pulses with a plasma target. The generation mechanism for the spontaneous magnetic field has been studied by a number of authors.[7–15] A kinetic generation mechanism for the spontaneous magnetic field driven by the beat interaction between two electromagnetic fields was proposed by one of the present authors[16] in 1983 and an extension of Ref. 16 was given in Ref. 17. In Ref. 18, we discussed the spontaneous magnetic field using the relativistic Vlasov-Maxwell equations. Recently, a measurement of the spontaneous magnetic field was reported.[6] In the present paper, we show that the experimental results given in Ref. 6 can be explained by our model proposed in Ref. 18.

Starting from the relativistic Vlasov-Maxwell system of equations, we can obtain the spontaneous fields (the detailed treatment can be found in Ref. 18)

$$c^2\nabla^2\mathbf{B}_s(\mathbf{x},t)=\frac{ice\omega_{pe}^2\beta}{m_e\omega_0^3}\nabla\times\nabla\times(\mathbf{E}^*\times\mathbf{E}), \tag{1}$$

$$\beta=m_e^2c^2\int\frac{1}{p_{0e}^2}\left(\frac{1}{2}-\frac{|\mathbf{p}_e|^2}{3p_{0e}^2}\right)f_{0e}\,d^3\mathbf{p}_e, \tag{2}$$

where $p_{0e}=\sqrt{m_e^2c^2+|\mathbf{p}_e|^2}$, f_{0e} is the relativistic electron distribution function. The dependence of β on the electron temperature is plotted in Fig. 1. The curve in Fig. 1 can be fitted by function $\beta=0.32e^{-T_e/88}+0.18e^{-T_e/400}$, where T_e is the electron temperature in the unit of KeV. For a laser field $\mathbf{E}_L(\mathbf{x},t)$, $\mathbf{E}(\mathbf{x},t)$ in Eq. (1) is defined as

$$\mathbf{E}_L(\mathbf{x},t)=\mathbf{E}(x,y,t)e^{ik_0z-i\omega_0t}+\text{c.c.} \tag{3}$$

It is should be noted that the spontaneous fields given by Eq. (1) satisfy the necessary condition $\nabla\cdot\mathbf{B}_s=0$.

For a laser beam with wave number $k_0=2\pi/\lambda_0$ and intensity $I=I_0[g(x,y)f(t)]^2$, where $g(x,y)$ is the spatial profile of laser beam and $f(t)$ is temporal profile of laser beam, the electromagnetic fields are usually expressed as

$$\mathbf{E}_L(\mathbf{x},t)=E_0(x,y,t)(\mathbf{e}_x+i\alpha\mathbf{e}_y)e^{ik_0z-i\omega_0t}+\text{c.c.}, \tag{4}$$

$$E_0(x,y,t)=\sqrt{\frac{I_0}{2(1+\alpha^2)}}g(x,y)f(t), \tag{5}$$

where $\alpha=0,\pm1$ for a linearly polarized (LP) and circularly polarized (CP) laser, respectively. As noted by Haines,[15] $\mathbf{E}_L(\mathbf{x},t)$ given by Eq. (4) is not correct because it does not satisfy the Poisson equation $\nabla\cdot\mathbf{E}=0$. Here we write the field in the form

$$\mathbf{E}_L(\mathbf{x},t)=E_0(x,y,t)(\mathbf{e}_x+i\alpha\mathbf{e}_y)e^{ik_0z-i\omega_0t}+E_z(\mathbf{x},t)\mathbf{e}_z+\text{c.c.} \tag{6}$$

By Poisson equation, we have

$$\mathbf{E}_z(\mathbf{x},t)=\frac{i}{k_0}\left(\frac{\partial E_0}{\partial x}+i\alpha\frac{\partial E_0}{\partial y}\right)\mathbf{e}_ze^{ik_0z-i\omega_0t}+\text{c.c.} \tag{7}$$

Introducing the normalized transform, $t\rightarrow t/\omega_0$, $\mathbf{x}\rightarrow c\mathbf{x}/\omega_0$, $\mathbf{E}\rightarrow\mathbf{E}(m_ec\omega_0/e)$, $\mathbf{B}\rightarrow\mathbf{B}(m_ec\omega_0/e)$, we have equations for the spontaneous fields,

$$B_{sx}=n_eI_Lf^2\alpha^2\beta\frac{\partial g^2(x,y)}{\partial y}, \tag{8}$$

$$B_{sy}=-n_eI_Lf^2\beta\frac{\partial g^2(x,y)}{\partial x}, \tag{9}$$

$$B_{sz}=2\alpha n_e\beta I_Lf^2g^2, \tag{10}$$

where n_e is the electron density in the unit of critical density and $I_L=I_0\lambda_0^2/1.368\times10^{18}/2(1+\alpha^2)$. In the cylindrical coordinates, the above equations become

$$B_{sr}=n_eI_Lf^2\beta\frac{dg^2(r)}{dr}(\alpha^2-1)\sin\theta\cos\theta, \tag{11}$$

1070-664X/2003/10(10)/4166/3/$20.00

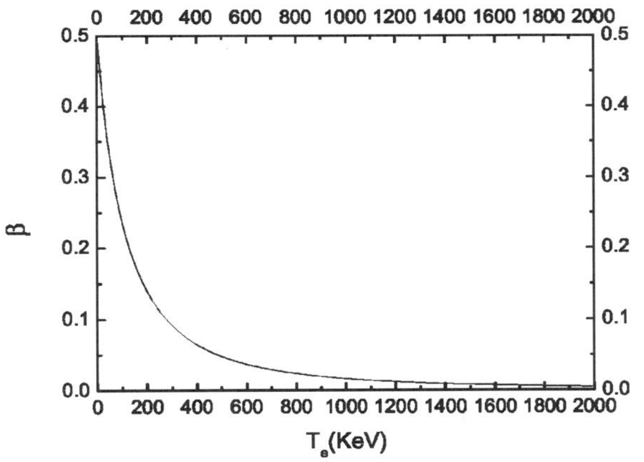

FIG. 1. The dependence of β on the electron temperature, $\beta=0.5$ when $T_e=0$.

$$B_{s\theta}=-n_e I_L f^2 \beta \frac{dg^2(r)}{dr}(\cos^2\theta+\alpha^2\sin^2\theta), \tag{12}$$

$$B_{sz}=2\alpha n_e I_L f^2 g^2 \beta. \tag{13}$$

$B_{s\theta}$ and B_{sr} are dependent of azimuthal angle θ for the LP case, but independent of θ for the CP laser. From Eqs. (11)–(13), we can see that there exists B_{sz} only for the CP laser beam. Equations (11)–(13) show that B_{sz} is proportional to the intensity of the laser beam, but $B_{s\theta}$ and B_{sr} are determined by both the intensity and profile of the laser beam.

Letting $g(r)=e^{-(r/a)^n}$ and $f(t)=e^{-b[(t-t_0)/t_0]^m}$, Fig. 2 shows B_{sz} for the case of CP laser beam with wavelength $\lambda_0=1.05$ and two different intensities $I_0=6.7\times10^{18}$ W cm^{-2} [Fig. 2(a)] and $I_0=1.34\times10^{19}$ W cm^{-2} [Fig. 2(b)], where $n=2$, $a=1/(10\pi)^2$ (laser focal spot radius is $5\lambda_0$), $f(t)=1$ and plasma density $n_e=0.028$ are adopted. The parameters used here are consistent with those given in the measurement of the spontaneous magnetic fields.[6] The electron temperature T_e is assumed to be 300 KeV for Fig. 2. It can be seen that the profile of B_{sz} is the same as in Fig. 2 of Ref. 6. The peak spontaneous magnetic field is about 0.8 megagauss for $I_0=6.7\times10^{18}$ W cm^{-2}, which is less than the experiment result 2 megagauss. As mentioned in Ref. 6, $I_0=6.7\times10^{18}$ W cm^{-2} is the "vacuum" intensity, the actual intensity in localized regions may be higher than this due to self-focusing. Figure 2(b) shows that the peak magnetic field is 1.6 megagauss for $I_0=1.34\times10^{19}$ W cm^{-2}, which is consistent with the experiment result of 2 megagauss. We plot $B_{s\theta}$ in Fig. 3, where $I_0=1.34\times10^{19}$ W cm^{-2} and other parameters are same as Fig. 2. It is found that $B_{s\theta}$ is very weak, which is inconsistent with the result obtained by the inverse Faraday effect. Unfortunately, no result is given for $B_{s\theta}$ in Ref. 6. One point should be noted which is that $\nabla\times B_s=\mathbf{j}$ is used in many early works and the electric current $\mathbf{j}$ is driven by ponderomotive force $\nabla\gamma$, where $\gamma=\sqrt{1+(1+\alpha^2)I\lambda_0^2/2\times1.38\times10^{18}}$ is a relativistic factor. It is likely that $\nabla\cdot\mathbf{j}\neq0$ or $\nabla\cdot\mathbf{B}_s\neq0$. On the other hand, the modification of the laser beam by the plasma is not included in our model.[18] This is an important problem which is under consideration.

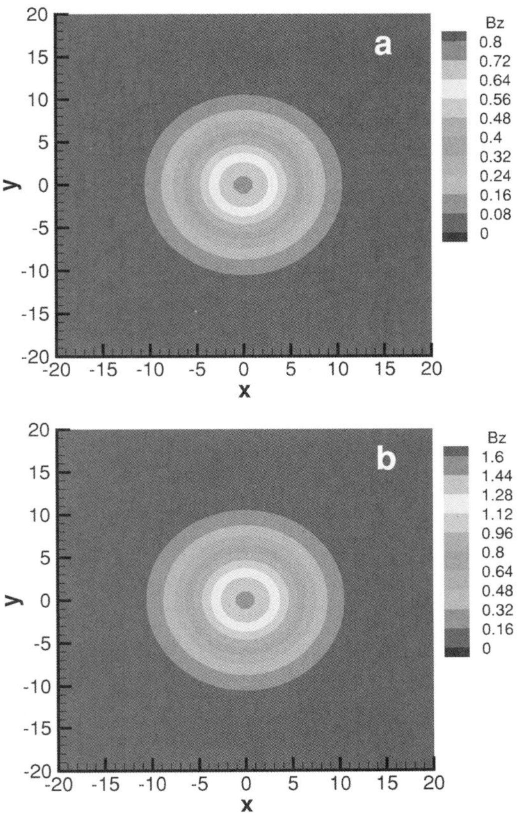

FIG. 2. B_{sz} for the case of CP laser beam with wavelength $\lambda_0=1.05$ and two different intensities $I_0=6.7\times10^{18}$ W cm^{-2} (a), and $I_0=1.34\times10^{19}$ W cm^{-2} (b), where $n=2$, $a=1/(10\pi)^2$ (laser focal spot radius is $5\lambda_0$), $f(t)=1$, $n_e=0.028$ and $T_e=300$ KeV. (x,y) is in the unit of laser wavelength.

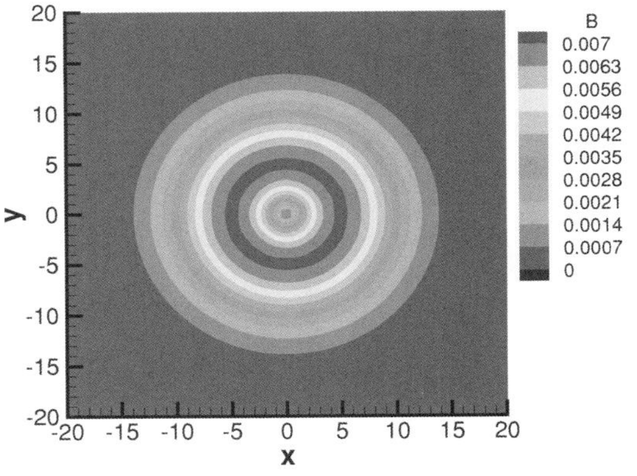

FIG. 3. $B_{s\theta}$ given by the present model for the case of $I_0=1.34\times10^{19}$ W cm^{-2}, other parameters are same as Fig. 2. (x,y) is in the unit of laser wavelength.

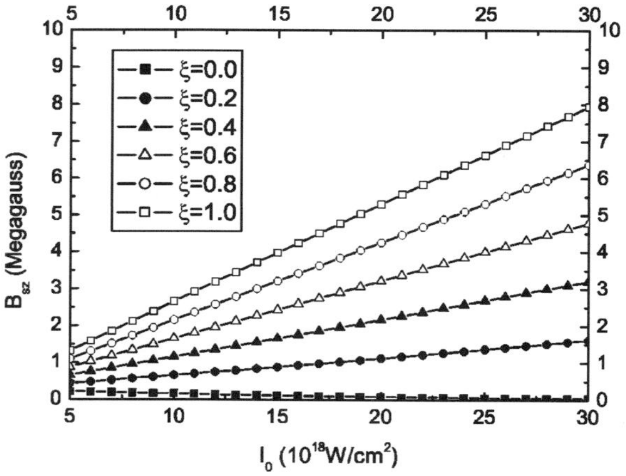

FIG. 4. The dependence of the peak B_{sz} on the laser intensity I_0 for different ξ, where Wilk's scaling law $T_{re}=511(\gamma-1)$ is used and the thermal electron temperature $T_{te}=10$ KeV.

In real intense laser plasma interaction, some of electrons are accelerated into the relativistic regime. It is reasonable to assume that there are two groups of the electrons in the system, thermal ones and relativistic ones. So the spontaneous magnetic field can be written as

$$B_{sz}=2\alpha I_L n_e f^2 g^2(\xi\beta(T_{te})+(1-\xi)\beta(T_{re})), \quad (14)$$

where ξ is the ratio of the number of thermal electrons to the total number of electrons, T_{te} is the temperature for thermal electrons and T_{re} is the temperature for relativistic electrons. From Eq. (14), we can find that the higher the relativistic electron temperature T_{re} the smaller the spontaneous magnetic fields for a given ξ, laser intensity I_0 and electron density n_e. If I_0, T_{re}, and n_e are fixed, then the smaller ξ (the higher the number of relativistic electrons), the smaller the spontaneous magnetic fields. If all of electrons are relativistic, that is, $\xi=0$, then we can conclude that the higher the number of relativistic electrons the higher the spontaneous magnetic fields.

Figure 4 shows the dependence of the peak B_{sz} on the laser intensity I_0 for different ξ, where Wilk's scaling law $T_{re}=511(\gamma-1)$ is used and the thermal electron temperature $T_{te}=10$ KeV. For the case of $\xi=0$, the peak spontaneous magnetic field decreases as the laser intensity increases, which is suggested by Sheng's model[13] (See the solid line in Fig. 3 of Ref. 6). But experiment results[6] show that the peak spontaneous magnetic field increases with laser intensity. We can see from Fig. 4 that the peak B_{sz} grows up as intensity I_0 increases when $\xi>0.2$.

In conclusion, the experiment results on spontaneous magnetic fields given in Ref. 6 can be explained by the kinetic model proposed by the present authors.[18] We also discussed the dependence of the peak magnetic field B_{sz} on the laser intensity. It is found the ratio of the number of thermal electrons to the total number of electrons is an important parameter.

ACKNOWLEDGMENTS

This work was supported by National Hi-Tech Inertial Confinement Fusion Committee of China, National Natural Science Foundation of China No. 10135010, Special Funds for Major State Basic Research Projects in China, and Science Foundation of CAEP.

[1] M. Tabak, J. Hammer, M. E. Glinsky, W. L. Kruer, S. C. Wilks, J. Woodworth, E. M. Campbell, M. D. Perry, and R. J. Mason, Phys. Plasmas **1**, 1626 (1994).
[2] A. B. Borisov, A. V. Borovskiy, V. V. Korobkin *et al.*, Phys. Rev. Lett. **68**, 2309 (1992).
[3] S. C. Wilks, W. L. Kruer, M. Tabak, and A. B. Langdon, Phys. Rev. Lett. **69**, 1383 (1992).
[4] A. Pukhov and J. Meyer-ter-Vehn, Phys. Rev. Lett. **76**, 3975 (1996).
[5] J. Fuchs, G. Malka, J. C. Adam *et al.*, Phys. Rev. Lett. **80**, 1658 (1998).
[6] Z. Najmudin, M. Tatarakis, A. Pukhov *et al.*, Phys. Rev. Lett. **87**, 215004 (2001).
[7] R. N. Sudan, Phys. Rev. Lett. **70**, 3075 (1993).
[8] V. K. Tripathi and C. S. Liu, Phys. Plasmas **1**, 990 (1994).
[9] R. J. Mason and M. Tabak, Phys. Rev. Lett. **80**, 524 (1998).
[10] A. D. Steiger and C. H. Woods, Phys. Rev. A **5**, 1467 (1972).
[11] T. Lehner, Phys. Scr. **49**, 704 (1994).
[12] V. Yu Bychenkov and V. T. Tikhonchuk, Laser Part. Beams **14**, 55 (1996).
[13] Z. M. Sheng and J. Meyer-ter-Vehn, Phys. Rev. E **54**, 1833 (1996).
[14] V. I. Berezhiani, S. M. Mahajan, and N. L. Shatashvili, Phys. Rev. E **55**, 995 (1997).
[15] M. G. Haines, Phys. Rev. Lett. **87**, 135005 (2001).
[16] X. T. He, Acta Phys. Sin. **32**, 325 (1983).
[17] S. P. Zhu, Phys. Fluids B **5**, 1024 (1993).
[18] S. P. Zhu, X. T. He, and C. Y. Zheng, Phys. Plasmas **8**, 321 (2001).

EUROPHYSICS LETTERS 15 December 2005

Europhys. Lett., **72** (6), pp. 955–961 (2005)
DOI: 10.1209/epl/i2005-10329-2

Fluid theory of magnetic-field generation in intense laser-plasma interaction

BIN QIAO[1], X. T. HE[2,3] and SHAO-PING ZHU[2]

[1] *Graduate School of China Academy of Engineering Physics*
P. O. Box 2101, Beijing 100088, PRC
[2] *Institute of Applied Physics and Computational Mathematics*
P. O. Box 8009, Beijing 100088, PRC
[3] *Department of Physics, Zhejiang University - Hangzhou 310027, PRC*

received 11 July 2005; accepted in final form 14 October 2005
published online 16 November 2005

PACS. 52.38.Fz – Laser-induced magnetic fields in plasmas.
PACS. 52.35.Mw – Nonlinear phenomena: waves, wave propagation, and other interactions (including parametric effects, mode coupling, ponderomotive effects etc.).
PACS. 52.57.Kk – Fast ignition of compressed fusion fuels.

Abstract. – Using the ten-moment Grad system of hydrodynamic equations, a self-consistent fluid model is presented for the generation of spontaneous magnetic fields in intense laser-plasma interaction. The generalized vorticity is not conserved as opposed to previous studies, for the nondiagonal stress force is considered. Equations for both the axial magnetic field B_z and the azimuthal one B_θ are simultaneously derived from a fluid scheme for the first time in the quasi-static approximation, where the low-frequency phase speed v_p is much smaller than the electron thermal speed v_{te}. It is found that the condition $v_p \gg v_{te}$, widely used as cold-fluid approximation, where B_z is incomplete, is improper. The profiles of B_z and B_θ as well as the plasma density cavitation are analyzed. Their dependences on the laser intensity are also discussed.

In the study of the interaction of an intense short-pulse laser with relativistic plasma [1], the generation of spontaneous magnetic fields is most significant because these fields have considerable influence on the whole nonlinear plasma dynamics. The study of this problem has wide applications in the fast-ignition scheme [2] and particle acceleration. Although much effort [3–7] has been done, there still exists a great deal of controversy concerning the proper derivation for the spontaneous magnetic fields. It has been argued in refs. [8, 9] that only a kinetic treatment can give the correct expression and that all fluid attempts yield unsatisfactory results. So far, no fluid model can give complete spontaneous magnetic fields including the axial component B_z and the azimuthal one B_θ, as in the kinetic theory [9–12].

We believe some facts should be noted during the self-consistent derivation of spontaneous magnetic fields. 1) In intense laser-plasma interaction, there exist strong rotational motions (currents) and large vortexes are generated, thus the fluid viscosity force (the nondiagonal stress tensor) should be considered. Ideal-fluid assumption in most studies is improper. 2) In this case, the generalized vorticity is not conserved as opposed to previous ideal-fluid models [5, 13]. 3) With intense laser, the plasma is warmed and the condition $v_p \gg v_{te}$ (v_p is the low-frequency phase speed and v_{te} the electron thermal speed) widely used as cold-fluid approximation [5, 13] is improper. The condition $v_p \ll v_{te}$ in the quasi-static approximation should be satisfied.

In this paper, we consider an intense laser propagating in an initially uniform plasma. Ion motion and the excitation of Langmuir waves are ignored. The electron motion can be well described by the ten-moment Grad system of hydrodynamic equations [14, 15], which is a closed system of equations providing an approximate solution to the Boltzmann equation and is composed of ten independent components of variables including density n, momentum $\boldsymbol{p}_i$ and stress tensor $\boldsymbol{\Pi}_{ij}$. We give the system of equations as

$$\frac{\partial \boldsymbol{p}_i}{\partial t} + \boldsymbol{v}\cdot\nabla\boldsymbol{p}_i = -e\left[\boldsymbol{E}+\frac{1}{c}(\boldsymbol{v}\times\boldsymbol{B})\right]_i + \frac{\partial\boldsymbol{\Pi}_{ij}}{\partial r_j}, \tag{1}$$

$$\begin{aligned}\frac{\partial\boldsymbol{\Pi}_{ij}}{\partial t} &+ \frac{\partial}{\partial r_s}(\boldsymbol{u}_s\boldsymbol{\Pi}_{ij}) + \boldsymbol{\Pi}_{is}\frac{\partial \boldsymbol{u}_j}{\partial r_s} + \boldsymbol{\Pi}_{js}\frac{\partial \boldsymbol{u}_i}{\partial r_s} - \frac{2}{3}\delta_{ij}\boldsymbol{\Pi}_{sl}\frac{\partial \boldsymbol{u}_l}{\partial r_s} + \\ &+\frac{e}{mc}\boldsymbol{B}_l(e_{isl}\boldsymbol{\Pi}_{js}+e_{jsl}\boldsymbol{\Pi}_{is}) = v_{te}^2\left(\frac{\partial\boldsymbol{p}_i}{\partial r_j}+\frac{\partial\boldsymbol{p}_j}{\partial r_i}-\frac{2}{3}\delta_{ij}\nabla\cdot\boldsymbol{p}\right),\end{aligned} \tag{2}$$

where the electric field $\boldsymbol{E}$ and magnetic field $\boldsymbol{B}$ are governed by Maxwell equations as

$$\nabla\times\boldsymbol{E} = -\frac{1}{c}\frac{\partial\boldsymbol{B}}{\partial t}, \tag{3}$$

$$\nabla\times\boldsymbol{B} = \frac{1}{c}\frac{\partial\boldsymbol{E}}{\partial t}+\frac{4\pi}{c}\boldsymbol{j} = \frac{1}{c}\frac{\partial\boldsymbol{E}}{\partial t}-\frac{4\pi e}{c}n\boldsymbol{v}, \tag{4}$$

$$\nabla\cdot\boldsymbol{E} = 4\pi(n_0-n)e. \tag{5}$$

Here $-e$, m, n and $v_{te}=\sqrt{T_e/m}$ represent the charge, mass, density and thermal velocity of electrons. n_0 is the initial density. $\boldsymbol{p}_i=\gamma m\boldsymbol{v}_i$ is the electron momentum and $\gamma_i=(1+\boldsymbol{p}_i^2/m^2c^2)^{1/2}$ is the relativistic factor. The stress tensor is written as $\boldsymbol{\Pi}_{ij}=\frac{1}{n_0}m\sigma_{ij}$, where the diagonal components represent the normal pressure $P_{ij}=\delta_{ij}mn_0T_e$ and the non-diagonal components represent the viscosity force. For the incompressible plasma electron fluid, defining the stress force as $\boldsymbol{F}_i=\frac{\partial\boldsymbol{\Pi}_{ij}}{\partial r_j}$, eq. (2) can be much simplified as $\frac{\partial \boldsymbol{F}_i}{\partial t}=\frac{1}{\gamma}v_t^2\nabla^2\boldsymbol{p}_i-\frac{iv_t^2}{4\omega_0 m\gamma^2}\nabla^2\nabla\times(\boldsymbol{p}_i\times\boldsymbol{p}_i)$.

Now we divide both electron motion and fields into slow- and fast-time-scale parts as $\boldsymbol{p}=\langle\boldsymbol{p}\rangle+\frac{1}{2}\tilde{\boldsymbol{p}}e^{-i\omega_0 t}+\frac{1}{2}\tilde{\boldsymbol{p}}^*e^{i\omega_0 t}$, $\boldsymbol{E}=\langle\boldsymbol{E}\rangle+\frac{1}{2}\tilde{\boldsymbol{E}}e^{-i\omega_0 t}+\frac{1}{2}\tilde{\boldsymbol{E}}^*e^{i\omega_0 t}$ and $\boldsymbol{B}=\langle\boldsymbol{B}\rangle+\frac{1}{2}\tilde{\boldsymbol{B}}e^{-i\omega_0 t}+\frac{1}{2}\tilde{\boldsymbol{B}}^*e^{i\omega_0 t}$, where the angular bracket $\langle\ \rangle$ denotes averaging over the fast time period $(2\pi/\omega_0)$ and $*$ denotes the complex conjugate for the fast-time-scale parts. Here and below, the subscript i is and will be not written for simplification. Note that here $\langle\boldsymbol{B}\rangle$ just represents the spontaneous magnetic field, and $\tilde{\boldsymbol{E}}$ represents the laser electric field. Introducing the generalized vorticity $\Omega\equiv\nabla\times\boldsymbol{p}-\frac{e}{c}\boldsymbol{B}$ and splitting Ω similarly, the slow-time-scale parts of eqs. (1), (2), (4) can be simplified as

$$\frac{\partial\langle\boldsymbol{p}\rangle}{\partial t} = -e\langle\boldsymbol{E}\rangle-\frac{1}{4m\gamma}\nabla\langle\tilde{\boldsymbol{p}}\cdot\tilde{\boldsymbol{p}}^*\rangle+\frac{1}{4m\gamma}\langle\tilde{\boldsymbol{p}}\times\tilde{\boldsymbol{\Omega}}^*+\tilde{\boldsymbol{p}}^*\times\tilde{\boldsymbol{\Omega}}\rangle+\langle\boldsymbol{F}\rangle, \tag{6}$$

$$\frac{\partial\langle\boldsymbol{F}\rangle}{\partial t} = \frac{1}{\gamma}v_t^2\nabla^2\langle\boldsymbol{p}\rangle-\frac{iv_t^2}{4\omega_0 m\gamma^2}\nabla^2\nabla\times\langle\tilde{\boldsymbol{p}}\times\tilde{\boldsymbol{p}}^*\rangle, \tag{7}$$

$$\nabla\times\langle\boldsymbol{B}\rangle = \frac{1}{c}\frac{\partial\langle\boldsymbol{E}\rangle}{\partial t}-\frac{4\pi e}{m\gamma c}\langle n\rangle\langle\boldsymbol{p}\rangle-\frac{\pi e}{m\gamma c}\langle\tilde{n}\tilde{\boldsymbol{p}}^*+\tilde{n}^*\tilde{\boldsymbol{p}}\rangle, \tag{8}$$

where $\langle\gamma\rangle=\gamma$ is assumed. Applying the curl operator on eq. (6) and using eq. (3), we get

$$\frac{\partial\langle\boldsymbol{\Omega}\rangle}{\partial t} = \frac{1}{4m\gamma}\nabla\times\langle\tilde{\boldsymbol{p}}\times\tilde{\boldsymbol{\Omega}}^*+\tilde{\boldsymbol{p}}^*\times\tilde{\boldsymbol{\Omega}}\rangle+\nabla\times\langle\boldsymbol{F}\rangle. \tag{9}$$

From eq. (9), we can see the generalized vorticity $\langle\boldsymbol{\Omega}\rangle$ is not conserved and is generated by the nondiagonal stress force due to nonzero electron thermal motion in plasmas. From eqs. (6)-(9), after some calculation, we obtain the equation for $\langle\boldsymbol{p}\rangle$ in the lowest order as

$$c^2\nabla\times\nabla\times\langle\boldsymbol{p}\rangle+\frac{\partial^2\langle\boldsymbol{p}\rangle}{\partial t^2}+\frac{1}{4m\gamma}\frac{\partial}{\partial t}\nabla\langle\boldsymbol{p}\cdot\boldsymbol{p}^*\rangle+\omega_{pe}^2\frac{\langle n\rangle}{n_0}\frac{\langle\boldsymbol{p}\rangle}{\gamma}-\frac{v_{te}^2}{\gamma}\nabla^2\langle\boldsymbol{p}\rangle-c^2\nabla\times\langle\boldsymbol{\Omega}\rangle=$$
$$=-\frac{\omega_{pe}^2}{4n_0\gamma}\langle\tilde{n}\tilde{\boldsymbol{p}}^*+\tilde{n}^*\tilde{\boldsymbol{p}}\rangle-i\frac{v_{te}^2}{4m\omega_0\gamma^2}\nabla^2\nabla\times\langle\tilde{\boldsymbol{p}}\times\tilde{\boldsymbol{p}}^*\rangle,\quad(10)$$

where $\omega_{pe}=\sqrt{4\pi n_0e^2/m}$ is defined. Note eq. (10) includes the generalized vorticity $\langle\boldsymbol{\Omega}\rangle$.

Similarly, to the lowest order, for $\omega_0\gg v_{te}\nabla$, we get fast-time-scale quantities, respectively, as $\tilde{\boldsymbol{p}}=-\frac{ie}{\omega_0}\tilde{\boldsymbol{E}}$, $\tilde{\boldsymbol{F}}=\frac{ev_{te}^2}{\gamma\omega_0^2}\nabla^2\tilde{\boldsymbol{E}}$, $\tilde{\boldsymbol{\Omega}}=0$. Using the continuity equation $\frac{\partial n}{\partial t}+\nabla\cdot(n\frac{\boldsymbol{p}}{m\gamma})=0$, we get

$$\langle\tilde{n}\tilde{\boldsymbol{p}}^*+\tilde{n}^*\tilde{\boldsymbol{p}}\rangle=i\frac{e^2}{m\omega_0^3}\tilde{\boldsymbol{E}}\nabla\cdot\frac{\langle n\rangle\tilde{\boldsymbol{E}}^*}{\gamma}+\text{c.c.}\quad(11)$$

Making a second differential for eq. (10) related to t, getting $\frac{\partial^2\langle\boldsymbol{\Omega}\rangle}{\partial t^2}$ from eqs. (7), (9) and then substituting $\frac{\partial^2\langle\boldsymbol{\Omega}\rangle}{\partial t^2}$ and eq. (11) into it, we obtain the coupling equation for $\langle\boldsymbol{p}\rangle$ as

$$\frac{\partial^2}{\partial t^2}\left[c^2\nabla\times\nabla\times\langle\boldsymbol{p}\rangle+\frac{\partial^2\langle\boldsymbol{p}\rangle}{\partial t^2}+\omega_{pe}^2\frac{\langle n\rangle}{n_0}\frac{\langle\boldsymbol{p}\rangle}{\gamma}+\frac{e^2}{4m\omega_0^2\gamma}\frac{\partial}{\partial t}\nabla\langle\boldsymbol{E}\cdot\boldsymbol{E}^*\rangle+\right.$$
$$\left.+\frac{\omega_{pe}^2e^2}{4mn_0\omega_0^3}\left(\frac{i\tilde{\boldsymbol{E}}}{\gamma}\nabla\cdot\frac{\langle n\rangle\tilde{\boldsymbol{E}}^*}{\gamma}+\text{c.c.}\right)\right]-\frac{\partial^2}{\partial t^2}\left[\frac{v_{te}^2}{\gamma}\nabla^2\langle\boldsymbol{p}\rangle-i\frac{v_{te}^2e^2}{4m\omega_0^3\gamma^2}\nabla^2\nabla\times\langle\tilde{\boldsymbol{E}}\times\tilde{\boldsymbol{E}}^*\rangle\right]=$$
$$=c^2\frac{v_{te}^2}{\gamma}\nabla^2\nabla\times\left[\langle\boldsymbol{p}\rangle-\frac{ie^2}{4m\omega_0^3\gamma}\nabla\times\langle\tilde{\boldsymbol{E}}\times\tilde{\boldsymbol{E}}^*\rangle\right].\quad(12)$$

From eq. (12), we can see that the relation between v_{te} and the low-frequency phase speed $v_p=\frac{\omega_s}{q}$ (ω_s is the frequency, and q the wave number) is a key condition, for $\frac{\partial}{\partial t}=\omega_s$ and $v_{te}\nabla=v_{te}q$ here.

In intense laser-plasma interaction, the plasma electron is warmed and v_{te} cannot be approximately regarded as zero anylonger. Thus, the previous widely used condition $v_p\gg v_{te}$ for the cold-fluid approximation [5, 13] is not satisfied anylonger. Note here that the low frequency ω_s is indeed a beat frequency of two high frequencies (ω_1, ω_2), as $\omega_s=\omega_1-\omega_2$. In the quasi-static approximation, it is reasonable to assume that ω_2 is very close to ω_1 and both of them are close to ω_0. Thus, ω_s is much smaller than the characteristic oscillating frequency of plasma electrons ω_{pe}, which is determined by v_{te} as $\omega_{pe}=v_{te}/\lambda_{De}$ (λ_{De} is the electron Debye length) and has the same scale as ω_0. When the spatial scale $\frac{1}{q}$ of the low-frequency field is comparable to or smaller than λ_{De}, the condition $v_p\ll v_{te}$ is always satisfied. In fact, the beat frequency ω_s is approximately equivalent to the ion-acoustic frequency, *i.e.*, $\omega_s=qv_s=q\sqrt{\frac{T_e}{m_i}}$ (m_i is the ion mass) and thus $v_p=v_s\ll v_{te}$ is always satisfied.

Under the condition $v_p\ll v_{te}$ for quasi-static approximation, the right-hand side of eq. (12) is dominant, the left-hand side can be ignored as zero, thus from eq. (12), we get

$$\langle\boldsymbol{p}\rangle=\frac{ie^2}{4m\omega_0^3\gamma}\nabla\times\langle\tilde{\boldsymbol{E}}\times\tilde{\boldsymbol{E}}^*\rangle.\quad(13)$$

Substituting eqs. (11), (13) into eq. (8), ignoring $\frac{\partial\langle E\rangle}{\partial t}$ due to $\frac{\partial}{\partial t}\ll v_{te}\nabla$, we get $\langle\boldsymbol{B}\rangle$ as

$$\nabla\times\langle\boldsymbol{B}\rangle=-i\frac{\omega_{pe}^2e}{4m\omega_0^3c\gamma^2}\frac{\langle n\rangle}{n_0}\nabla\times\langle\tilde{\boldsymbol{E}}\times\tilde{\boldsymbol{E}}^*\rangle-\frac{\omega_{pe}^2e}{4m\omega_0^3c}\left(\frac{i\tilde{\boldsymbol{E}}}{\gamma}\nabla\cdot\frac{\langle n\rangle\tilde{\boldsymbol{E}}^*}{n_0\gamma}+\text{c.c.}\right).\quad(14)$$

Substituting eq. (13) into eq. (7) and making its divergence, we get $\frac{\partial\nabla\cdot\boldsymbol{F}}{\partial t}=0$, *i.e.*, $\nabla\cdot\boldsymbol{F}=0$. Thus, from eqs. (5), (6), we get the slow-time-scale plasma density $\frac{\langle n\rangle}{n_0}$ as

$$\frac{\langle n\rangle}{n_0} = 1 + \frac{e^2}{4m^2\omega_{pe}^2\omega_0^2\gamma}\nabla^2\langle\tilde{\boldsymbol{E}}\cdot\tilde{\boldsymbol{E}}^*\rangle. \tag{15}$$

Equation (14) combined with eq. (15) together gives the axial spontaneous magnetic fields and their two driven sources. The first source is the rotational current as the result of the beating interaction high-frequency electromagnetic waves (laser wave fields) $\langle\tilde{\boldsymbol{E}}\times\tilde{\boldsymbol{E}}^*\rangle$, the second one is the rotational current as the result of the coupling interaction between the high-frequency electron quiver velocity and the high-frequency electron density perturbation (see eq. (14)). Note such two sources are unitedly and simultaneously given for the first time. In most of the previous fluid models [5, 13], the viscosity force is not considered and the cold-fluid approximation $v_p \gg v_{te}$ is improperly used, which leads to the absence of the first source and the incomplete axial magnetic field. From eqs. (13), (14), we can also prove that the generalized vorticity is not conserved when the viscosity force is considered, as opposed to previous ideal-fluid models [5, 13]. The generated vorticity induced by high-frequency fields is $\nabla\times\langle\boldsymbol{\Omega}\rangle = \frac{e^2}{4m\omega_0^3\gamma}[i\frac{\omega_{pe}^2}{c^2\gamma}\frac{\langle n\rangle}{n_0}\nabla\times\langle\tilde{\boldsymbol{E}}\times\tilde{\boldsymbol{E}}^*\rangle + \frac{\omega_{pe}^2}{c^2}(i\tilde{\boldsymbol{E}}\nabla\cdot\frac{\langle n\rangle\tilde{\boldsymbol{E}}^*}{n_0\gamma} + \text{c.c.}) - i\nabla^2\nabla\times\langle\tilde{\boldsymbol{E}}\times\tilde{\boldsymbol{E}}^*\rangle]$. Another fact that should be noted is that the inverse Faraday effect is included in our model. Let us apply the curl operator on eq. (6), noticing that $\tilde{\boldsymbol{\Omega}} = 0$ and $\nabla\times\langle\boldsymbol{p}\rangle = \langle\boldsymbol{\Omega}\rangle + \frac{e}{c}\langle\boldsymbol{B}\rangle$ and considering eq. (9), we easily get $\frac{\partial\langle\boldsymbol{B}\rangle}{\partial t} = -c\nabla\times\langle\boldsymbol{E}\rangle$, which is just the equation describing the inverse Faraday effect.

For simplicity, now we consider the interaction of an underdense plasma with an intense laser. We assume that the laser beam propagates along the z-axis direction with wave number $k_0 = \frac{2\pi}{\lambda_0}$, under the Coulomb gauge, the electric fields are usually expressed as $\boldsymbol{E}_L(\boldsymbol{x},t) = \frac{1}{2}E_0(r,t)(\boldsymbol{e}_r + i\alpha\boldsymbol{e}_\theta)e^{i\alpha\theta}e^{ik_0z-i\omega_0t} + \text{c.c.}$, where $\alpha = 0, \pm1$ for linearly polarized (LP) and circularly polarized (CP) laser, respectively. The laser intensity is assumed to have Gaussian distribution $I = I_0e^{-r^2/r_0^2-(z-ct)^2/L^2}$ and $E_0 = \sqrt{I}$, where L is the pulse longitudinal width, and r_0 the transverse radius. Thus the above high-frequency electric field is expressed as $\tilde{\boldsymbol{E}} = E_0(r,t)(\boldsymbol{e}_r + i\alpha\boldsymbol{e}_\theta)e^{i\alpha\theta}e^{ik_0z}$ and $\gamma = (1+\frac{\boldsymbol{P}^2}{m^2c^2})^{1/2} = [1+\frac{(1+\alpha^2)e^2E_0^2}{2m^2c^2\omega^2}]^{1/2} = [1+\frac{(1+\alpha^2)}{2.76}I_0\lambda_{\mu m}^2e^{-r^2/r_0^2-(z-ct)^2/L^2}]^{1/2}$ is obtained. Introducing the normalized transform $t\to\frac{t}{\omega_0}, \boldsymbol{x}\to\frac{c\boldsymbol{x}}{\omega_0}, E\to E\frac{m_ec\omega_0}{e}, B\to B\frac{m_ec\omega_0}{e}$, using B_z to represent $\langle B_z\rangle$ and noticing that $\nabla\cdot\tilde{\boldsymbol{E}} \simeq 0$ is satisfied for a laser beam propagating in an underdense plasma (see ref. [12]), we get the normalized B_z from eqs. (14),(15),

$$\frac{\partial B_z}{\partial r} = -\frac{\alpha}{1+\alpha^2}n_0\frac{1}{\gamma}\frac{\partial}{\partial r}\left(\frac{\langle n\rangle}{n_0}\frac{\gamma^2-1}{\gamma}\right), \tag{16}$$

where

$$\frac{\langle n\rangle}{n_0} = 1 - \frac{1}{n_0r_0^2}\frac{\gamma^2-1}{\gamma}\left(2-\frac{r^2}{r_0^2}\frac{\gamma^2+1}{\gamma^2}\right). \tag{17}$$

Equation (16) shows that B_z is only generated by CP laser ($\alpha=\pm1$). For LP laser ($\alpha=0$), B_z is zero.

The transverse profile of $|B_z|$ calculated by eqs. (16), (17) is plotted in fig. 1(a), where the laser intensity is $I_0 = 10^{19}\,\text{W/cm}^2$ with wavelength $\lambda = 1.05\,\mu\text{m}$, the transverse focused radii are, respectively, $r_0 = 1\lambda,\ 2\lambda,\ 3\lambda$. The plasma density is $n_0 = 0.36n_{cr}$. The parameters are consistent with those after focusing on the 3D particle simulation of ref. [12]. It can be seen that B_z dominates the "inner" zone near the laser axis and the peak $|B_z|$ is about 20–26 MG at the axis for $r_0 = 1$–3λ. Either the profile or the peak value of $|B_z|$ is very close to the simulation results [12]. Recent experiment [16] reports that a B_z of about 2 MG is generated by CP laser with "vacuum" $I_0 = 6.7\times10^{18}\,\text{W/cm}^2$ and "vacuum" $r_0 = 10\lambda$ ($\lambda = 1.05\,\mu\text{m}$), where $n_0 = 2.8\times10^{19}\,\text{cm}^{-3}$. Using equations (16), (17), $|B_z| \sim 1.93\,\text{MG}$ is estimated under the same

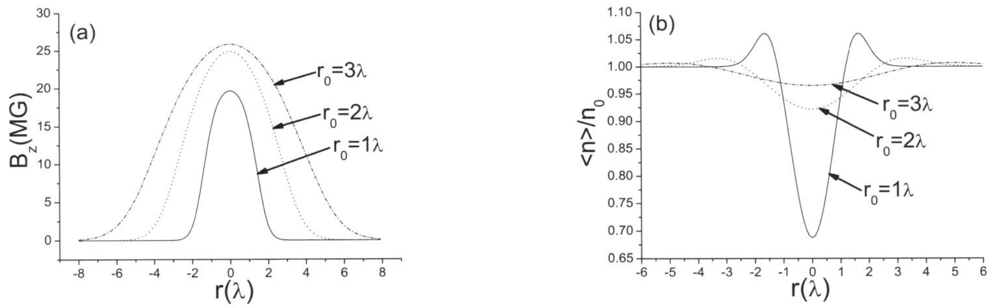

Fig. 1 – The transverse profiles of (a) the axial magnetic field B_z and (b) the plasma density cavitation $\langle n\rangle/n_0$ as a function of the radius r calculated by eqs. (16), (17), for a Gaussian laser beam with intensity $I_0 = 10^{19}\,\mathrm{W/cm^2}$, wavelength $\lambda = 1.05\,\mu\mathrm{m}$, and transverse radii $r_0 = 1\lambda,\ 2\lambda,\ 3\lambda$. The plasma density is $n_0 = 0.36n_{cr}$.

parameters, where we assume I_0 be two times the "vacuum" laser intensity and r_0 decreased down to 5λ, due to self-focusing. The result is also very close to that in experiment [16].

Figure 1(b) plots the transverse profile of plasma density $\langle n\rangle/n_0$ by eq. (17) for the same parameters as fig. 1(a). We can evidently see that electrons in the region $r < r_0$ are expelled by the ponderomotive force so that $\langle n\rangle/n_0 < 1$, and the expelled electrons will be deposited at a cavity edge with $r \succeq r_0$, where $\langle n\rangle/n_0 > 1$; then when $r \gg r_0$, $\langle n\rangle/n_0 = 1$ is taken again.

The experimental results of [16] show that B_z increases with laser intensity I_0. Figure 2 plots the dependences of $|B_z|$ and $\frac{\langle n\rangle}{n_0}$ on I_0, where other parameters are consistent with those in fig. 1. It can be seen that the peak $|B_z|$ increases with I_0, while the plasma density near the axis $(\frac{\langle n\rangle}{n_0})_{r=0}$ decreases as I_0 increases. When I_0 rises high enough, the peak B_z does not increase anymore. However, in a dense plasma, the fields can be quite strong for high-intensity laser. Indeed, for $I_0 = 4\times10^{20}\,\mathrm{W/cm^2}$ ($\lambda = 1.05\,\mu\mathrm{m}$), and $n_0 = 0.9\times10^{21}\,\mathrm{cm^{-3}}$, B_z turns out to be 150 MG.

The above analysis is only for the azimuthal momentum $\langle P_\theta\rangle$, which generates B_z. For the azimuthal magnetic field B_θ, the longitudinal momentum $\langle p_z\rangle$ should be studied. For the same Gaussian laser beam propagating in an underdense plasma as above, from eq. (12), assuming $\frac{\partial\langle p_r\rangle}{\partial z} \ll \frac{\partial\langle p_z\rangle}{\partial r}$, we get the normalized coupling equation of $\langle p_z\rangle$ as

$$\frac{\partial^2\langle p_z\rangle}{\partial\xi^2} - \frac{\partial^2\langle p_z\rangle}{\partial r^2} - \frac{1}{r}\frac{\partial\langle p_z\rangle}{\partial r} + n_0\frac{\langle n\rangle}{n_0\gamma}\langle p_z\rangle = \frac{1}{4\gamma}\frac{\partial^2\langle\tilde{\boldsymbol{E}}\cdot\tilde{\boldsymbol{E}}^*\rangle}{\partial\xi^2} = \frac{\partial^2\gamma}{\partial\xi^2}, \tag{18}$$

where $\xi = z - t$ is taken. Notice that B_θ is mainly produced by the longitudinal current,

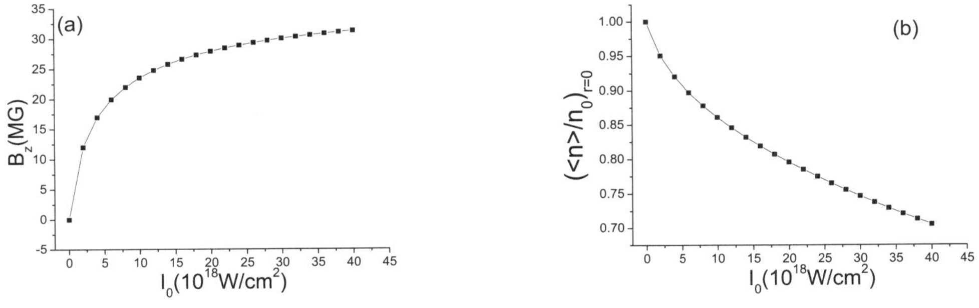

Fig. 2 – The dependence of the peak B_z (a) and $\langle n\rangle/n_0$ at the axis ($r = 0$) (b) on the laser intensity I_0, where the transverse radius r_0 is taken equal to 1.5λ and the other parameters are consistent with fig. 1.

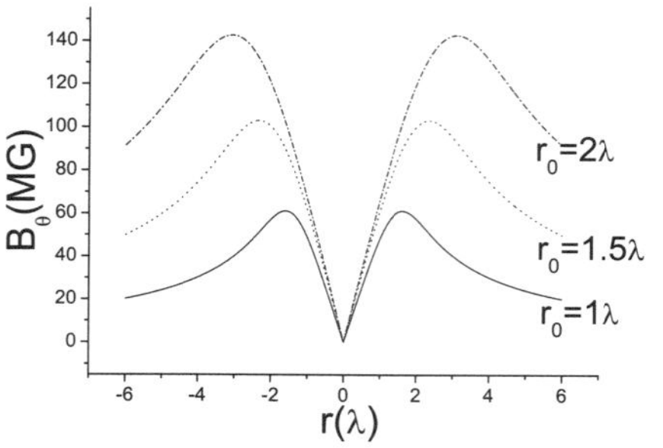

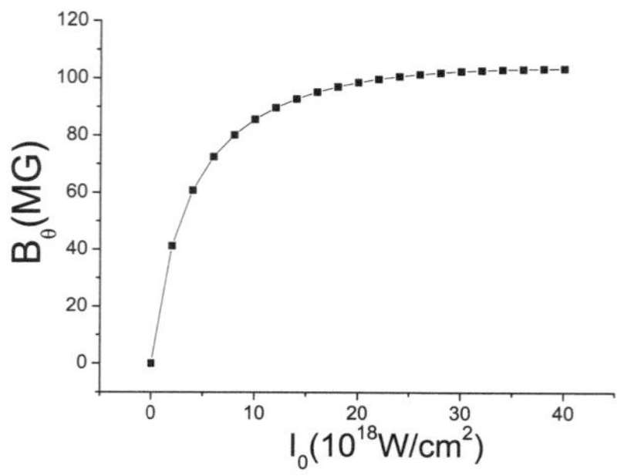

Fig. 3 – The transverse profiles of the azimuthal magnetic field B_θ calculated by eq. (19) with the same parameters as in fig. 1, except for r_0 which is taken as 1λ, 1.5λ, and 2λ, respectively.

Fig. 4 – The dependence of the peak B_θ on the laser intensity I_0, where the laser beam transverse radius r_0 is taken equal to 1.5λ and the other parameters are consistent with fig. 1.

which dominates near the axis. That is to say, B_θ is mainly decided by p_z near the axis. Thus, it is reasonable to ignore $\frac{\partial\langle p_z\rangle}{\partial r}$ and $n_0\frac{\langle n\rangle}{n_0\gamma}\langle p_z\rangle$ in eq. (18) and then eq. (18) is simplified as $\frac{\partial^2\langle p_z\rangle}{\partial\xi^2}=\frac{\partial^2\gamma}{\partial\xi^2}$. Assuming initially $\gamma=1$ when $\langle p_z\rangle=0$, then we obtain that $\langle p_z\rangle=\gamma-1$.

Ignoring $\frac{\partial\langle \boldsymbol{E}\rangle}{\partial t}$, using the symbol B_θ to represent $\langle B_\theta\rangle$, from eq. (8), we get the normalized B_θ as

$$\frac{1}{r}\frac{\partial}{\partial r}(rB_\theta)=-n_0\frac{\langle n\rangle}{n_0}\frac{\gamma-1}{\gamma}. \tag{19}$$

Equation (19) shows that both the CP laser ($\alpha=\pm1$) and LP laser ($\alpha=0$) can generate B_θ.

The transverse profile of $|B_\theta|$ by eq. (19) is plotted in fig. 3, where the parameters are consistent with those of fig. 1, except for r_0, which is taken as 1λ, 1.5λ, and 2λ, respectively. It can be seen that B_θ dominates the "outer" zone away from the laser axis, is close to zero near the axis, increases gradually with r until it reaches its peak value at about $r=r_0$, then decreases due to return currents. The peak B_θ at $r=r_0$ is about 60–140 MG for $r_0=1$–2λ. The results and the profile are both close to the particle simulation [12]. Experiment [17] reports that $B_\theta\sim 35$–70 MG is produced by LP laser with "vacuum" $I_0=4.7\times10^{18}\,\mathrm{W/cm^2}$ and $r_0=4\lambda$ ($\lambda=1.05\,\mu$m), where $n_0=2\times10^{20}\,\mathrm{cm^{-3}}$. We also assume that the focused I_0 is two times its "vacuum" value as $I_0=9.4\times10^{18}\,\mathrm{W/cm^2}$ and the focused $r_0=1.5\lambda$ due to self-focusing. Using eq. (19), the peak B_θ is estimated to be about 52 MG, which agrees well with the experimental result of [17]. Figure 4 plots the dependence of B_θ on the laser intensity I_0, where $r_0=1.5\lambda$ and the other parameters are the same as in fig. 1. It shows that the peak $|B_\theta|$ also increases with laser intensity I_0.

Recently, observations of nearly Giga-Gauss (GG) B_θ generated in ultra-intense laser over-dense plasma interaction are reported by Tatarakis *et al.* [18] and Wagner *et al.* [19], which can be explained by the scaling law of Sudan [3]. As mentioned in refs. [18, 19], the minimum B_θ should be generated at the nonrelativistic critical density surface ($n_0\simeq n_{cr}$), the higher B_θ should be generated in the higher density region ($n_{cr}<n_0<\gamma n_{cr}$) due to laser hole-boring. For the same parameters as those of the experiments, *i.e.*, $I_0=9\times10^{19}\,\mathrm{W/cm^2}$ and $\lambda=1.05\,\mu$m, assuming that the Gaussian distribution of laser intensity is still valid for $n_0<\gamma n_{cr}=6n_{cr}$, we can calculate B_θ by our model too. Using eq. (19), peaks $B_\theta=360$ MG, 466 MG, 795 MG (0.36 GG, 0.47 GG, 0.79 GG) are estimated, respectively, at densities $n_0=0.95n_{cr}, 1.2n_{cr}, 2.4n_{cr}$, which are close to the experimental results of [18, 19]. Certainly, this is a rough estimate, for the laser hole-boring effect is a very complex topic.

In conclusion, a self-consistent fluid model is presented for the spontaneous magnetic fields including the axial part B_z and the azimuthal one B_θ generated by an intense CP and LP laser propagating in an initially uniform plasma. The ten-moment Grad system of hydrodynamic equations is used, where the nondiagonal stress tensor (the fluid viscosity force) is taken into account, for there exist strong rotational motions (vortexes) in intense laser warm plasma interaction. The generalized vorticity is proved to be not conserved, as opposed to previous studies with ideal-fluid assumption. This hydrodynamic derivation yields exact and complete expressions for both B_z and B_θ for the first time. In the quasi-static approximation $v_p \ll v_{te}$, B_z is driven by two rotational currents generated only by CP laser, one is the result of the beating interaction of the high-frequency electric fields of CP laser, and the other is the result of the coupling interaction between the high-frequency electron quiver velocity and the high-frequency electron density perturbation. These two sources are simultaneously given for the first time. B_θ is driven by the irrotational current of ponderomotive force. It is shown that the condition $v_p \gg v_{te}$, widely used as cold-fluid approximation in previous studies, where the first source of B_z is not given and B_z is incomplete, is improper. The results calculated by our model are compared with those of the particle simulation of ref. [12] and with the experiments [16–19]. The transverse profiles of both B_z and B_θ as well as the plasma density cavitation $\langle n\rangle/n_0$ are analyzed. Their dependences on the laser intensity are also discussed.

* * *

This work was supported by National Hi-Tech ICF Committee of China, NSFC Grant No. 10335020, 10375011 and 10135010, National Basic Research Project "nonlinear Science".

REFERENCES

[1] BORISOV A. B. *et al.*, *Phys. Rev. Lett.*, **68** (1992) 2309; WILKS S. C. *et al.*, *Phys. Rev. Lett.*, **69** (1992) 1383; HONDA M. *et al.*, *Phys. Plasmas*, **7** (2000) 1302.
[2] TABAK M. *et al.*, *Phys. Plasmas*, **1** (1994) 1626.
[3] SUDAN R. N., *Phys. Rev. Lett.*, **70** (1993) 3075; LEHNER T., *Phys. Scr.*, **49** (1994) 704.
[4] SHENG Z. M. and MEYER-TER-VEHN J., *Phys. Rev. E.*, **54** (1996) 1833.
[5] BEREZHIANI V. I., MAHAJAN S. M. and SHATASHVILI N. L., *Phys. Rev. E.*, **55** (1997) 995.
[6] HAINES G., *Phys. Rev. Lett.*, **87** (2001) 135005.
[7] KIM A., TUSHENTSOV M., ANDERSON D. and LISAK M., *Phys. Rev. Lett.*, **89** (2002) 095003.
[8] ŜKORIĆ M. M., *Laser Part. Beams*, **5** (1987) 83.
[9] BEREZHIANI V. I., TSKHAKAYA D. D. and AUER G., *J. Plasma Phys.*, **38** (1987) 139.
[10] BEZZERIDES B. *et al.*, *Phys. Rev. Lett.*, **38** (1977) 495; MORA P. and PELLAT R., *Phys. Fluids*, **22** (1979) 2408; KONO M., *et al.*, *Phys. Lett. A*, **77** (1980) 27.
[11] ZHU SHAO-PING, HE X. T. and ZHENG C. Y., *Phys. Plasmas*, **8** (2001) 312.
[12] QIAO BIN, ZHU SHAO-PING, ZHENG C. Y. and HE X. T., *Phys. Plasmas*, **12** (2005) 053104.
[13] GORBUNOV L. M., *Usp. Fiz. Nauk*, **109** (631) 1973 (*Sov. Phys. Usp.*, **16** (1973) 217); TSINTSADZE N. L. *et al.*, *Zh. Eksp. Teor. Fiz.*, **72** (1977) 48016 (*Sov. Phys. JETP*, **45** (1977) 252).
[14] GRAD H., *Commun. Pure Appl. Math.*, **2** (1949) 331; KIRII A. YU. *et al.*, *Zh. Tekh. Fiz.*, **39** (1969) 773 (*Sov. Phys. Tech. Phys.*, **14** (1969) 583); ALEXANDER N. GORBAN *et al.*, *Phys. Rev. E*, **54** (1996) R3109.
[15] ALIEV YU. M., FROLOV A. A. and STENFLO L. *et al.*, *Phys. Fluids B*, **2** (1990) 34.
[16] NAJMUDIN Z., TATARAKIS M. *et al.*, *Phys. Rev. Lett.*, **87** (2001) 215004-1.
[17] FUCHS J., MALKA G. *et al.*, *Phys. Rev. Lett.*, **80** (1998) 1658.
[18] TATARAKIS M. *et al.*, *Nature*, **415** (2002) 280; *Phys. Plasmas*, **9** (2002) 2244.
[19] U. WAGNER, *Phys. Rev. E*, **70** (2004) 026401.

PHYSICS OF PLASMAS **12**, 044505 (2005)

Magnetic field generation and relativistic electron dynamics in circularly polarized intense laser interaction with dense plasma

C. Y. Zheng[a)]
Institute of Applied Physics and Computational Mathematics, P. O. Box 8009, Beijing 100088, China
X. T. He
Institute of Applied Physics and Computational Mathematics, P. O. Box 8009, Beijing 100088, China and Department of Physics, Zhejiang University, Hang Zhou 310027, China
S. P. Zhu
Institute of Applied Physics and Computational Mathematics, P. O. Box 8009, Beijing 100088, China

(Received 17 August 2004; accepted 24 January 2005; published online 25 March 2005)

The generation of both axial and azimuthal magnetic field components by the intense laser-dense plasma interaction and relativistic electron dynamics are studied in theoretical analysis and three-dimensional particle-in-cell simulations. For a circularly polarized laser, the cylindrical symmetry can be applied. Interacting with this laser, both axial and the azimuthally magnetic fields are generated and play an essential role in hot electron beam collimation, stabilization, and shaping. It is significantly different from the linearly polarized case, where the azimuthal current and the axial magnetic field are absent, therefore the axial current density gets filamentary early in the ramp plasma and is focused into a single channel in the dense (with a density larger than the critical density n_c) plasma later, and the electron and ion density distributions as well as the azimuthally magnetic field profile are broadened in the polarized direction to form an ellipse-like structure in the transverse plane. Instead, the results for the circularly polarized laser analytically and numerically show a much better collimation of the relativistic electron beams than in the case of the linearly polarized laser, due to the axial field component generated by azimuthal electron dynamics. © *2005 American Institute of Physics.* [DOI: 10.1063/1.1875612]

Fast ignition (FI), as a novel ignition concept of the inertial confinement fusion, is extensively studied in recent years.[1–8] Two of the critical issues for FI are the decay of the intense laser and the collimation of relativistic electron beam (REB) in a relatively long plasma channel. So far, many studies were on the above issues and most of the previous work was focused on the case of linear polarized (LP) lasers.[5–10] For a LP laser, electrons undergo a quiver motion in polarized direction, which leads to diverging the axial current and the quasistatic azimuthal magnetic field B_θ from the axial symmetry. As a result, the collimation of REB is distorted by instabilities due to the asymmetry. On the other hand however, for a circularly polarized (CP) laser, an azimuthal current J_θ, which is caused by the azimuthal motion of the electrons driven by the rotating CP laser fields, generates an axial quasistatic magnetic field B_y in addition to the axial current J_y and the azimuthal quasistatic magnetic field B_θ.[11–13] The axial field B_y forces the electrons to move helically around the axis while the azimuthal field B_θ produces a centripetal force to pinch electrons. Therefore the CP laser is more favorable for the collimation of REB than the LP lasers. In this paper, we revisit the quasistatic magnetic field generation and propose a model for it. In theoretical analysis and a numerical simulation by using three-dimensional particle-in-cell code, we investigate the process that generates the quasistatic magnetic fields and drives the electron dynamics. Simulation results then are shown and compared with analytic predictions. It is found that the CP laser significantly affects the collimation of the REB.

The momentum equation for relativistic electrons can be divided into a parallel and a transverse part, where "parallel" ("transverse") stands for parallel (perpendicular) to the laser propagation (axial) direction. Then, with the generalized vorticity law, the parallel momentum $P_\parallel$ can be calculated from

$$\partial P_\parallel/\partial t = e\nabla_\parallel\varphi - mc^2\nabla_\parallel(\gamma-1), \tag{1}$$

where the parallel momentum $P_\parallel$ is related to the parallel velocity as $v_\parallel = P_\parallel/\gamma m$, $mc^2\nabla_\parallel(\gamma-1)$ is the axial ponderomotive force, and φ is the electrostatic potential. Correspondingly, the transverse momentum $P_\perp$ is given by

$$\partial P_\perp/\partial t = -e(E_L+E_S) - mc^2\nabla_\perp(\gamma-1), \tag{2}$$

where $mc^2\nabla_\perp(\gamma-1)$ is again the ponderomotive force but perpendicular to the axis, E_L is the laser field, E_s is a transverse electric field excited by the azimuthal current due to the electron rotation driven only by the CP laser field and vanishes in the LP laser case.[12] The laser field is expressed as $E_L = E_0(e_r + i\alpha e_\theta)e^{i\alpha\theta+i\psi}/2 + \mathrm{c.c.}$ in the cylindrical coordinates, where $\alpha=0$ for the LP laser and $\alpha=\pm1$ for the CP laser. In Eqs. (1) and (2), the relativistic factor γ is expressed by $\gamma=(1+P^2/m^2c^2)^{1/2}=[1+(1+\alpha^2)I\lambda^2/2.76]^{1/2}$, where the laser wavelength λ is in microns and the laser intensity $I=cE_0^2/(8\pi)$ is measured in 10^{18} W/cm^2.

Electric current is proportional to nv, i.e., the product of electron density n and velocity v. Dividing both electron

a)Author to whom correspondence should be addressed. Electronic mail: zheng_chunyang@iapcm.ac.cn density and velocity into a "fast" (in the laser field oscillation time scale) and a "slow" (in the quasistatic magnetic field time scale) parts, we then write the current density as follows:

$$\mathbf{J} = -e[\langle n_f \mathbf{v}_f\rangle_S + n_0 \mathbf{v}_S], \tag{3}$$

where the subscripts "f" and "s" stand for the fast and slow time scales, respectively, n_0 is the plasma background density, and $\langle n_f \mathbf{v}_f\rangle_S$ represents the slow motion average of the coupling interaction of two fast-time-scale motions.

First, let us discuss the transverse current $\mathbf{J}_\perp$. Noticing $\nabla\cdot\mathbf{E}_L=0$ and considering a uniform plasma, we obtain the coupling contribution $-e\langle n_f \mathbf{v}_f\rangle_s$ due to the "fast" laser field E_L, in the case with a cylindrical symmetry

$$\mathbf{J}_{\perp 1} = -\frac{\alpha}{2\pi(1+\alpha^2)} n_0 c e \lambda \frac{\partial}{\partial r}\left(\ln\gamma + \frac{1}{2\gamma^2}\right)\mathbf{e}_\theta, \tag{4}$$

where $\mathbf{e}_\theta$ is the azimuthal unit vector. We can see that this contribution to the current density is clearly a CP laser effect. The corresponding "slow" magnetic field is then axial, along with the laser propagation direction e_y as

$$B_{1y} = \frac{2\alpha}{1+\alpha^2} n_0 e \lambda\left(\ln\gamma + \frac{1}{2\gamma^2} - \frac{1}{2}\right), \tag{5}$$

where the boundary condition $B_{1y}=0$ at $r\to\infty$ is used. Equation (5) shows that $B_{1y}=0$ if $\gamma=1$ or $\alpha=0$ (for LP lasers).

The other "slow" axial magnetic field B_{2y} is generated by the current density $\mathbf{J}_{\perp 2}=-en_0\mathbf{v}_s$ driven by the "slow" transverse electric field E_s as shown in Eq. (2). Again under the uniform plasma approximation as in Ref. 12, we can write

$$B_{2y} = -4[\alpha/(1+\alpha^2)] n_0 e \lambda \beta(\gamma^2 - 1), \tag{6}$$

where

$$\beta = \int d^3 p f(p)\left(1 + \frac{p^2}{m^2c^2}\right)^{-1}\left[\frac{1}{2} - \frac{p^2}{3(m^2c^2+p^2)}\right], \tag{7}$$

$f(p)$ is a relativistic electron distribution. Again, we have $B_{2y}=0$ for the LP laser with $\alpha=0$. For a beam distribution $f(p)=\delta(p-p_0)$, where $p_0^2/m^2c^2=\gamma^2-1$ is determined by a general relationship from a ponderomotive potential $(\gamma-1)\times 0.511 = T_h(\mathrm{MeV})$ as in Ref. 10 with T_h the hot electron temperature, we can easily calculate that $\beta=0.5\gamma^{-2}[1-2(\gamma^2-1)/(3\gamma^2)]$. Then Eq. (6) has a form of

$$B_{2y} = \mp n_0 e\lambda(\gamma^2-1)[1-2(\gamma^2-1)/(3\gamma^2)]/\gamma^2, \tag{8}$$

which up to the factor of $1-2(\gamma^2-1)/(3\gamma^2)$ was shown in Ref. 11. It can also be seen that for cold plasmas, $\beta=1/2$; and β decreases as T_h increases.

Now we discuss the azimuthal magnetic field B_θ driven by the axial current density $\mathbf{J}_\parallel=-n_0 e\mathbf{v}_s$. Assume a Gaussian laser intensity $I(r,y,t)\propto \exp(-\eta^2/L^2 - r^2/R^2)$, where $\eta=y-ct$, L and R are the pulse width and the spot radius, respectively, then we get from Eq. (1)

$$\mathbf{J}_\parallel = -n_0 ec[(\gamma-1)/\gamma]\mathbf{e}_y. \tag{9}$$

Then the azimuthal magnetic field for both LP and CP lasers reads

$$B_\theta = -2\pi n_0 e r\left[1 - \frac{R^2}{r^2}\left(\ln\frac{\gamma_0-1}{\gamma_0+1} - \ln\frac{\gamma-1}{\gamma+1}\right)\right], \tag{10}$$

with $\gamma_0=\gamma(r=0)$. Equation (10) shows that $B_\theta(r,y,t)=0$ at $r=0$ and increases gradually with r. Physically, B_θ should have a maximum as r increases if the return current is considered. We discuss the following heuristic examples. Taking the parameters: $I_0=1\times 10^{19}$ W/cm^2, $n_0=10^{21}$ cm^{-3}, $\lambda=1.06\ \mu$m, $\beta=0.1$, and a focused spot $R=2.5\lambda$, we can obtain B_{2y}(MG)$=-72,-57$ for $r=0,R/2$, respectively, and B_θ(MG)$=0,-52,-160$ for $r=0,R/10,R/3$, respectively. Equations (5), (6), and (10) show that $B_y=B_{1y}+B_{2y}$ dominates the "inner" zone near the axis and B_θ dominates the "outer" zone from the axis.

To compute the "slow" axial B_y field, we have used a uniform plasma approximation. On the other hand, a cold electron return current was not considered to obtain the azimuthal B_θ field. Nevertheless, the model still provides a qualitative description of B_θ-field generation. Furthermore, in order to more precisely describe the laser-electron interaction process we numerically simulate the laser-plasma interaction process by using three-dimensional PIC code LARED-P. The numerical procedure used in the code is similar to that in Ref. 14. Also, to check the applicability of the analytic model, we then compare the simulation output with the B_y field argument in the analytic model.

The simulation box is a cuboid. The laser beam is propagating from left to right along the y direction with an absorbing boundary for laser fields and mirror reflection boundary for particles. Periodic boundary conditions are applied in the other two directions (x,z), i.e., the transverse plane, and z is the laser polarization direction for LP lasers. The number of spatial grid points in the simulation box is $320\times 640\times 320(12.7\lambda\times 25.5\lambda\times 12.7\lambda)$, and 7.8×10^8 of particles are loaded. The initial plasma density is homogeneous in the transverse plane, but after a vacuum region of $0\leqslant y<5.1\lambda$, linearly rising along the y direction from $0.5n_c$ at $y=5.1\lambda$ to $4n_c$ at $y=12.2\lambda$, and is then keeping at the level of $4n_c$ over an interval $\Delta y=10.8\lambda$, where n_c is the non-relativistic critical density. And there is again a vacuum region of $23\lambda<y\leqslant 25.5\lambda$ at the back of the plasma. The rest mass ratio of ion to electron is $m_i/m_e=4500$. The initial plasma temperatures are $T_e=T_i=16$ keV. The incident laser pulse with a wavelength of $\lambda=1.06\ \mu$m has a Gaussian spatial profile with a width of 6λ. And temporally, the laser pulse ramps up in the first three laser cycles in a Gaussian profile, and then remains at the intensity level that is $I=10^{19}$ W/cm^2 for the CP laser and $I=2\times 10^{19}$ W/cm^2 for the LP laser.

In the simulation, the laser beam enters the density ramp at $y=5.1\lambda$ from vacuum. It is found that the laser beam, either the LP or the CP, is gradually focused into a spot with a spatial scale of $O(\lambda)$ and the relativistic factor γ reaches its maximum $\gamma_m\cong 6.4$ near the interface between the ramp and dense regions. In the beam front, γ is about 3.6 and the hole boring velocity is close to $0.03c$.

For the LP laser, electrons in the ramp plasma oscillate in the z direction driven by the linearly polarized electric field E_z and are accelerated in the y direction by the axial ponderomotive force to form an axial current in the early

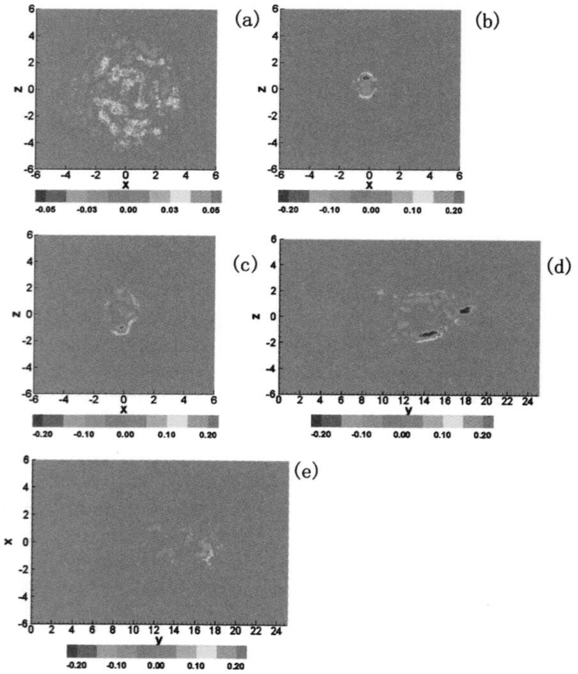

FIG. 1. (Color online). For the LP laser, the transverse (x,z) cuts of the axial current density J_y, scaled by $4n_c ec$ at (a) $y=6.4\lambda$, $t=48\tau$, with filaments emerging; and (b) at $y=15.36\lambda$, $t=112\tau$, and (c) $t=192\tau$, in the dense plasma. Also the longitudinal cuts of the axial current J_y along the pulse axis at $t=192\tau$, (d) in the (y,z) plane, and (e) in the (x,y) plane.

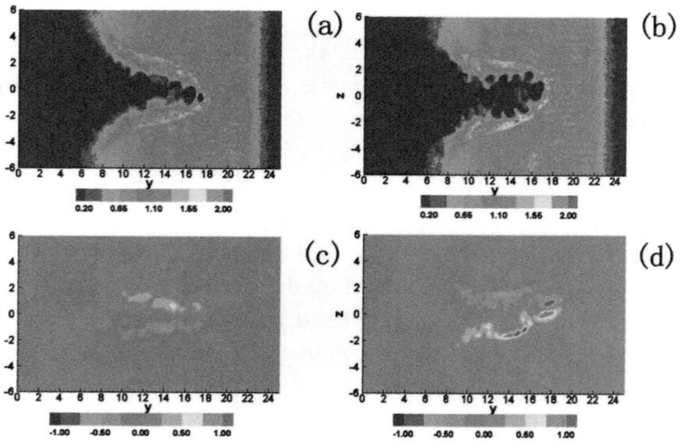

FIG. 2. (Color online). For the LP laser, the longitudinal cuts of electron density, along the pulse axis at 192τ, (a) in the (x,y) plane, and (b) in the (y,z) plane, as well as the longitudinal cuts of the magnetic field B_θ at 192τ, (c) in the (x,y) plane, and (d) in the (y,z) plane.

stage. This axial current density J_y is counteracted significantly by a return current in the opposite direction and torn into filaments due to joint interaction of Weibel-like and filamentary instabilities as shown in Fig. 1(a), at $y=6.4\lambda$ and $t=48\tau(\tau=2\pi/\omega\approx 3.3$ fs), just after the laser arrives. Then, the thin current filaments merge into one that self-focuses and propagates in the dense plasma following laser beam, as seen in Fig. 1(b), at $y=15.36\lambda$ and $t=112\tau$. Figures 1(c)–1(e) display current evolution at a later stage of $t=192\tau$, showing that the deeper the current penetrates into the dense plasma, the wider the current channel spreads in the polarized z direction and an ellipse-like current structure is formed in the transverse (x,z) plane with a ratio $\Delta\approx 2$ (of the major axis in z direction to the minor axis in x direction). This phenomenon was also observed in the previous PIC simulation.[9] Cavitation is found in the electron density plotted in Figs. 2(a) and 2(b) at $t=192\tau$. The electron expulsion from the high beam intensity region generates a strong electrostatic field, dragging ions moving radially with electrons to form a plasma density cavity. Thus, the electron density n_e and the ion density n_i have almost the same structure and are likewise broadened in the polarized direction, where the width of n_e in the x direction is clearly narrower than that in the polarized direction. Figures 2(c) and 2(d) show that the magnitude of the azimuthal magnetic field $|B_\theta|$ vanishes in the vicinity of the axis, in good agreement with the analytic relation (10). Figures 2(c) and 2(d) show that magnetic field has a similar elongated structure to the charged particle densities in the polarization direction. We then can write

$$B/B_c\sim(2\lambda/L)(I/I_0),$$

where $B_c=mc\omega/e$, $I=\int JdS$ is the total current through the ellipse, $I_0\equiv mc^3/e=17$ kA, and L is the girth of the ellipse. For light of wavelength $\lambda=1.06\ \mu$m, one obtains $B_c\approx 100$ MG. Figure 1(c) shows that the major and the minor radii for the ellipse are about 1.75λ and 0.9λ, respectively. Correspondingly, the girth of the ellipse $L\cong 8.7\lambda$, and the averaged current density inside the channel in the dense plasma is approximately about $0.07(4n_c ec)$, then $I=\int JdS\approx 77$ kA. Then we can calculate that $|B_\theta|\sim 1.04B_c\approx 104$ MG. The profile of $|B_\theta|$ given by simulation is then in certain aspects consistent with the analytic result Eq. (10). However due to the fact that fast electron flow can be neutralized by a cold return flow, in the outer zone far from the axis $|B_\theta|$ can be greatly reduced.[9]

For the CP laser, J_y and J_θ form a helical current density distribution with its axis along the y direction. Figures 3(a)–3(c) show results of the simulation in transverse cuts for the longitudinal current J_y through the plasma volume. In comparison with the LP case plotted in Fig. 1(a), the current filaments shown in Fig. 3(a) at $y=6.4\lambda$ and $t=48\tau$ in the ramp plasma are not obvious. The electrons are driven by the electric fields E_x and E_z of the laser, and the electron kinetic energy in either polarized direction is proportional to $|E_x|^2_{\rm CP}$ or $|E_z|^2_{\rm CP}$. If the laser power is the same for the LP and the CP lasers, we have a relation $|E_x|^2_{\rm CP}=|E_z|^2_{\rm CP}=\frac{1}{2}|E_z|^2_{\rm LP}$. Therefore, the Weibel-like instability may be reduced or even stabilized in the CP laser in comparison with the LP laser. Such a current density following the CP laser is focused into a cylindrically symmetrical spot in the dense plasma as seen in Figs. 3(b) and 3(c) at $y=15.36\lambda$, $t=112\tau, 192\tau$. The longitudinal (x,y) cuts along the laser pulse axis are shown in Fig. 3(d) for the electron density $n_e/(4n_c)$ and in Fig. 3(e) for the axial magnetic field B_y, while the corresponding profiles in the (y,z) cuts are almost the same as in (x,y) cuts. The density n_e and n_i are depleted by the radial ponderomotive force to form a symmetric bell-shape structure with a high

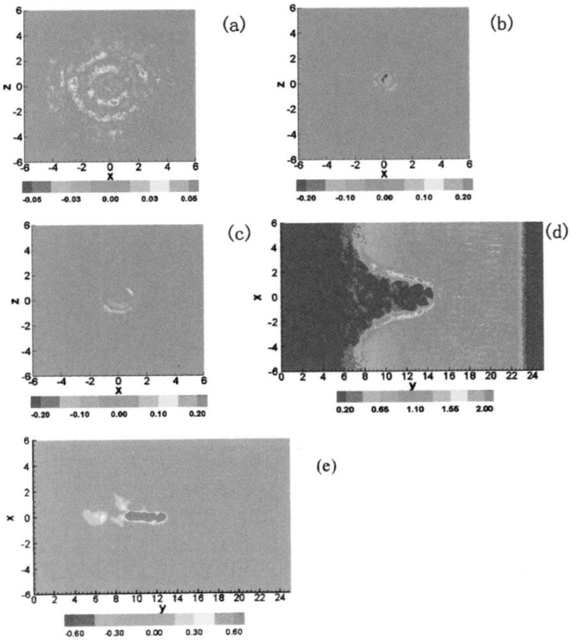

FIG. 3. (Color online). For the CP laser, the transverse cuts of the axial current density J_y scaled by $4n_c ec$, (a) at $y=6.4\lambda$, $t=48\tau$, and (b) at $y=15.36\lambda$, $t=112\tau$ and (c) $t=192\tau$. Also the longitudinal (x,y) cut along the pulse axis for (d) the electron density n_e at $t=112\tau$ and (e) the axial field B_y at $t=48\tau$.

density "shell" of $n_e/4n_c \cong 1.5$ at $t=112\tau$. The reason why electrons have such a distribution is that they are expelled or accelerated at the center of the spot. The axial magnetic field B_y with the cylindrical symmetry reaches its maximum of 60 MG at $t=48\tau$ as seen in Fig. 3(e), then gradually drops down to 40 MG, except for the channel front where the maximal value is maintained. Making use of Eq. (8), the peak magnetic field $B_y=73$ MG is estimated, very closed to simulation result. Figure 3(b) shows that B_y traps the relativistic electrons inside the channel with a Larmor radius about 1.7λ (for $B_y=40$ MG and $\gamma=4$), and helically to go along the y axis and as a result, it collimates the relativistic electrons in this channel to avoid a large divergent angle as occurred in the LP laser case.[2,6–9] For the other component, B_θ reaches its maximal value of ±150 MG initially, then falls to ±100 MG at a later time, and pinches the current density toward the axis, similar to the LP laser. However, with a cylindrically symmetric structure in the CP case, different from the anisotropic LP laser, the pinch effect of B_θ now is isotropic. Thus, both magnetic fields generated by the CP laser-plasma interaction stabilize and collimate the relativistic electron motion in cylindrical symmetry.

In conclusion, the magnetic field generation mechanism and electron dynamics in the intense laser-dense relativistic plasma interaction are studied analytically and numerically. The analytic model suggests a mechanism to generate an axial magnetic field B_y by the CP laser electrical field. This B_y field can serve as a guide-center to confine the electron moving helically along the axis. Also the cylindrical symmetry of the CP laser makes the azimuthal field B_θ isotropic, different from the ellipse geometry in the LP laser case. These predictions are in agreements with the particle simulations carried out in this paper. It is observed in the simulation that for CP lasers relativistic electrons with energy higher than MeV are unable to escape from the highly compressed fuel region with a size of tens of microns, and thus have enough time in overdense plasma to exchange energy to ions through collision. For the LP laser, it is observed that relativistic electrons oscillate up and down in the polarized direction, and the current propagating in overdense plasma becomes filamentary and severely distorted. These results show that the circularly polarized laser is favorable to the collimation of the relativistic electron beams in comparison with the linearly polarized laser.

ACKNOWLEDGMENTS

The authors would like to thank Dr. X. G. Wang for fruitful discussions.

This work was supported in part by the National Hi-Tec ICF Committee of China, NSFC Grant Nos. 10335020 and 10375011, and National Basic Research Project "Nonlinear Science" in China.

[1]M. Tabak, J. Hammer, M. E. Glinsky, W. L. Kruer, S. C. Wilks, J. Woodworth, E. M. Campbell, M. D. Perry, and R. J. Mason, Phys. Plasmas **1**, 1626 (1994).
[2]R. Kodama, P. A. Norreys, K. Mima *et al.*, Nature (London) **412**, 798 (2001).
[3]E. S. Weibel, Phys. Rev. Lett. **2**, 83 (1958).
[4]L. O. Silva, R. A. Fonseca, J. W. Tonge, J. M. Dawson, and W. B. Mori, Phys. Plasmas **9**, 2458 (2002).
[5]Y. Sentoku, K. Mima, P. Kaw, and K. Nishikawa, Phys. Rev. Lett. **90**, 155001 (2003); Y. Sentoku, K. Mima, Z-M. Sheen, P. Kew, and K. Nishikawa, Phys. Rev. E **65**, 046408 (2002).
[6]S. C. Wilks, W. L. Kruer, M. Tabak, and A. B. Langdon, Phys. Rev. Lett. **69**, 1383 (1992); V. K. Tripathi and C. S. Liu, Phys. Plasmas **1**, 990 (1994).
[7]R. J. Mason and M. Tabak, Phys. Rev. Lett. **80**, 524 (1998); J. Fuchs, G. Malka, J. C. Adam *et al.*, *ibid.* **80**, 1658 (1998); M. Honda, J. Meyer-ter-Vehn, and A. Pukhov, Phys. Plasmas **7**, 1302 (2000).
[8]A. S. Sandhu, A. K. Dharmadhikari, P. P. Rajeev, G. R. Kumar, S. Sengupta, A. Das, and P. K. Kaw, Phys. Rev. Lett. **89**, 225002 (2002); K. Nishihara, T. Honda, S. V. Bulanov, and Z-M. Sheng, Proc. SPIE **3886**, 90 (1999).
[9]A. Pukhov and J. Meyer-ter-Vehn, Phys. Rev. Lett. **76**, 3975 (1996); **79**, 2686 (1997).
[10]S. C. Walks, Phys. Fluids B **5**, 2603 (1993).
[11]Z. M. Sheng and J. Meyer-ter-Vehn, Phys. Rev. E **54**, 1833 (1996).
[12]S. P. Zhu, X. T. He, and C. Y. Zheng, Phys. Plasmas **8**, 321 (2001).
[13]Z. Najmudin, M. Tamaracks, A. Pushover *et al.*, Phys. Rev. Lett. **87**, 215004 (2001).
[14]C. K. Birdsall and A. B. Langdon, *Plasma physics via Computer simulation* (McGraw-Hill, New York, 1985).

PHYSICS OF PLASMAS **12**, 053104 (2005)

Quasistatic magnetic and electric fields generated in intense laser plasma interaction

Bin Qiao[a)]
Graduate School of China Academy of Engineering Physics, P.O. Box 2101, Beijing 100088, People's Republic of China

Shao-ping Zhu and C. Y. Zheng
Institute of Applied Physics and Computational Mathematics, P.O. Box 8009, Beijing 100088, People's Republic of China

X. T. He
Institute of Applied Physics and Computational Mathematics, P.O. Box 8009, Beijing 100088, People's Republic of China and Department of Physics, Zhejiang University, Hangzhou 310027, People's Republic of China

(Received 30 November 2004; accepted 8 February 2005; published online 20 April 2005)

A self-consistent kinetic model based on relativistic Vlasov–Maxwell equations is presented for the generation of quasistatic spontaneous fields, i.e., both the quasistatic magnetic (QSM) field and the quasistatic electric (QSE) field, in intense laser plasma interaction. For the circularly polarized laser, QSM field includes two parts, the axial part B_z as well as the azimuthal B_θ; the QSE field $\mathbf{E}_s$, corresponding to the space-charge potential, forms a plasma density channel. For the linearly polarized laser, B_z is absent. Equations for B_z, B_θ, and $\mathbf{E}_s$ are uniformly derived from one self-consistent model under the static-state approximation, which satisfies the conservation law of charge. The profile of the plasma density channel and the dependence of the peak QSM fields on the laser intensity are discussed. The experiment and simulation results are explained by the model. The predicted QSM and QSE fields are also observed in the three-dimensional particle simulation.
 [DOI: 10.1063/1.1889090]

I. INTRODUCTION

Since the fast ignition scheme[1] for inertial confinement fusion was put forward, much attention has been given to the interaction of an intense short-pulse laser with relativistic plasma.[2–4] Studies show that various nonlinear effects[5–8] can lead to the generation of many nonlinear spontaneous fields. Among them, quasistatic spontaneous fields, including the quasistatic magnetic (QSM) field $\mathbf{B}_s$ and quasistatic electric (QSE) field $\mathbf{E}_s$, have been considered to be most critical. The former has a very important influence on the collimation of hot electrons that deposit energy at the edge of the high-compressed density fuel to form an ignition spark with a size of about 10 μm. The latter affects the formation of the plasma density channel, through which an intense laser can propagate a long distance. The study of this issue also has wide applications in analyzing the mechanism of particle acceleration and x-ray generation, as well as some phenomena of astrophysics, such as the radiation of the pulsar.[9,10] Some efforts[11–17] have been devoted to the study of the axial magnetic field B_z from a fluid scheme. However, these fluid models are not perfect. So far, to our knowledge, no work gives a complete QSM field $\mathbf{B}_s$, which should include both the axial part B_z and the azimuthal part B_θ, in particular, B_θ and $\mathbf{E}_s$ have not been given. Some of the present authors[18–20] proposed a kinetic model for B_z. However, unfortunately, except for B_z, B_θ and $\mathbf{E}_s$ are also not given in Refs. 18–20, for there the nonlinear currents are incomplete, dissatisfying the conservation law of charge, and the direction of the slow-time-scale wave-vector q is improperly assumed to be parallel to that of the laser wave vector k_0. In principle, a self-consistent kinetic model satisfying the conservation law of charge $(\partial\rho/\partial t)+\nabla\cdot\mathbf{j}=0$ (ρ and $\mathbf{j}$ are charge and current, respectively) should give a reasonable expectation for all quasistatic fields, including the complete $\mathbf{B}_s$ (B_z and B_θ) and $\mathbf{E}_s$.

In this paper, we establish such a self-consistent kinetic model to uniformly discuss the generation mechanism of both the complete QSM field (including B_z and B_θ) and the QSE field $\mathbf{E}_s$ from relativistic Vlasov–Maxwell equations. It is found that the nonlinear beat interaction of laser fields with relativistic plasma causes a rotational (azimuthal) current in second order to generate B_z, while the axial irrotational current driven by the ponderomotive force generates B_θ. The ponderomotive force also expels electrons (self-channeling) to form an electron density channel, and accordingly the QSE field $\mathbf{E}_s$ is generated, corresponding to the space-charge potential $\nabla\phi_s$. We deduce the equations for $\mathbf{E}_s$, B_z and B_θ, among which the equations of B_θ and $\mathbf{E}_s$ are explicitly given for the first time. For a given laser beam, we discuss the approximate solutions, analyze the formation of the plasma density channel from $\mathbf{E}_s$, and explain the experiment and simulation results for B_z (Ref. 21) and B_θ.[22,23] We also analyze the dependence of the peak magnetic field on the laser intensity. In order to check our analytical results, we run a three-dimensional (3D) particle-in-cell (PIC) simulation to make further study and analysis.

In Sec. II, the kinetic model for the QSM and QSE fields

[a)]Electronic mail: bin_qiao1979@yahoo.com

1070-664X/2005/12(5)/053104/12/$22.50

is given. The QSE field $\mathbf{E}_s$ and the profile of the plasma density channel for a given laser beam are studied in Sec. III. The complete QSM fields, including B_z and B_θ for a given laser beam, are given in Sec. IV. In Sec. V, we compare our solutions with the results of experimental measurements.[21,22] In Sec. VI, we explain the particle simulation results[23] and make further study using our 3D PIC simulation code "LARAD-P." And, the conclusion is given in Sec. VII.

II. KINETIC MODEL FOR QSM AND QSE FIELDS

In this section, we present our kinetic model and give equations for the QSM and QSE fields generated by laser fields interacting with an underdense plasma. Consider the relativistic Vlasov–Maxwell equations

$$\left[\frac{\partial}{\partial t}+\frac{c\mathbf{p}_\alpha}{p_{0\alpha}}\cdot\frac{\partial}{\partial \mathbf{x}}+e_\alpha\mathbf{G}(\mathbf{x},t)\cdot\frac{\partial}{\partial \mathbf{p}_\alpha}\right]f_\alpha(\mathbf{x},\mathbf{p}_\alpha,t)=0, \tag{1}$$

$$\nabla\times\mathbf{E}=-\frac{1}{c}\frac{\partial\mathbf{B}}{\partial t},\quad \nabla\times\mathbf{B}=\frac{1}{c}\frac{\partial\mathbf{E}}{\partial t}+\frac{4\pi}{c}\mathbf{j}, \tag{2}$$

$$\mathbf{G}(\mathbf{x},t)=\mathbf{E}(\mathbf{x},t)+\frac{\mathbf{p}_\alpha\times\mathbf{B}}{p_{0\alpha}}, \tag{3}$$

$$\mathbf{j}=\sum_\alpha\int e_\alpha\frac{c\mathbf{p}_\alpha}{p_{0\alpha}}f_\alpha(\mathbf{x},\mathbf{p}_\alpha,t)d^3\mathbf{p}_\alpha, \tag{4}$$

where the index α $(\alpha=i,e)$ represents ions and electrons, respectively; $p_{0\alpha}=\dot{\gamma}_\alpha m_\alpha c$, $\mathbf{p}_\alpha=\gamma_\alpha m_\alpha\mathbf{v}$, $\gamma_\alpha=\sqrt{1+\mathbf{p}_\alpha^2/m_\alpha^2c^2}$, namely, $p_{0\alpha}=\sqrt{p_\alpha^2+m_\alpha^2c^2}$, are taken; and the other notations are standard. We split the particle distribution functions up into their background, slow-time-scale and fast-time-scale parts,

$$f_\alpha=n_{\alpha0}f_{\alpha0}+f_{\alpha s}+f_{\alpha f}, \tag{5}$$

where $n_{\alpha0}$ represents the initial plasma density and the isotrope for $f_{\alpha0}$ with $\int f_{\alpha0}d^3\mathbf{p}_\alpha=1$ is assumed. Also, we divide fields and currents into slow-time-scale and fast-time-scale parts. Making Fourier transforms with respect to both space and time for Eqs. (1)–(4) and defining $k=(\mathbf{k},\omega)$ as wave vector and frequency of fast time scale, $q=(\mathbf{q},\Omega)$ as those of slow time scale, we can obtain the slow-time-scale distribution function in the Fourier representation from Eqs. (1)–(4) as

$$\begin{aligned}f_{\alpha s}(q,\mathbf{p}_\alpha)=&\frac{e_\alpha n_{\alpha0}}{i\left(\Omega-\dfrac{c\mathbf{q}\cdot\mathbf{p}_\alpha}{\mathbf{p}_{0\alpha}}\right)}\sum_s E_s^{\sigma_s}\mathbf{e}_s^{\sigma_s'}(q)\cdot\frac{\partial f_{\alpha0}}{\partial\mathbf{p}_\alpha}\\&-\frac{e_\alpha^2n_{\alpha0}}{\left(\Omega-\dfrac{c\mathbf{q}\cdot\mathbf{p}_\alpha}{\mathbf{p}_{0\alpha}}\right)}\int_{(s)}\sum_{\sigma_1\sigma_2}\mathbf{e}_{\mathbf{f}}'^{\sigma_1}(k_1)\\&\cdot\frac{\partial}{\partial\mathbf{p}_\alpha}\frac{\mathbf{e}_{\mathbf{f}}'^{\sigma_2}(k_2)\cdot\dfrac{\partial f_{\alpha0}}{\partial\mathbf{p}_\alpha}}{\left(\omega_2-\dfrac{c\mathbf{k}_2\cdot\mathbf{p}_\alpha}{\mathbf{p}_{0\alpha}}\right)}\\&\times E_f^{\sigma_1}(k_1)E_f^{\sigma_2}(k_2)\delta^4(q-k_1-k_2)\frac{d^4k_1d^4k_2}{(2\pi)^4},\end{aligned} \tag{6}$$

where $E_f^{\sigma_i}(k_i)(i=1,2)$ is the amplitude of the fast-time-scale electric field (such as laser electric field, Langmuir field, and wakefield in laser produced plasma). In Eq. (6), we keep only the lowest nonlinear terms, and the symbol $\int_{(s)}$ refers to a slow-time-scale motion caused by the different beat of two fast-time-scale motions. Also, $E_s^{\sigma_s}(q)$, $\mathbf{G}_s(q)$, $\mathbf{e}_s'^{\sigma_s}(q)$, and $j_s^{\sigma_s}(q)$ are defined as[18–20] $\mathbf{E}_s(q)=\Sigma_{\sigma_s=t,l}E_s^{\sigma_s}\mathbf{e}^{\sigma_s}$, $\mathbf{G}_s(q)=\Sigma_{\sigma_s=t,l}E_s^{\sigma_s}\mathbf{e}'^{\sigma_s}$, $\mathbf{e}_s'^{\sigma_s}(q)=[1-(c\mathbf{q}\cdot\mathbf{p}_\alpha/\Omega p_{0\alpha})]\mathbf{e}_s^{\sigma_s}(q)+(c\mathbf{e}_s^{\sigma_s}(q)\cdot\mathbf{p}_\alpha/\Omega p_{0\alpha})\mathbf{q}$, $\mathbf{j}_s(q)=\Sigma_{\sigma_s=l,t}j_s^{\sigma_s}(q)\mathbf{e}_s^{\sigma_s}$, where t indicates the transverse part and l the longitudinal part, respectively, perpendicular and parallel to the slow-time-scale wave vector $\mathbf{q}$.

We use the convention $E_s^{\sigma_s *}(q)=-E_s^{\sigma_s}(-q)$ and $e_s^{\sigma_s *}(q)=-e_s^{\sigma_s}(-q)$, where $*$ notes the complex conjugate. Similarly, we have the same convention for the fast-time-scale field. Substituting Eq. (6) into Eq. (4) and ignoring the ion contribution to electric current due to much larger mass, we get the slow-time-scale electric currents and electric fields, respectively, as

$$\begin{aligned}j_s^{\sigma_s}=&-in_{e0}e^2\mathbf{e}_s^{\sigma_s}\cdot\int\frac{c\mathbf{p}_e}{p_{0e}\left(\Omega-\dfrac{c\mathbf{q}\cdot\mathbf{p}_e}{p_{0e}}\right)}E_s^{\sigma_s}(q)\mathbf{e}_s^{\sigma_s1}\\&\cdot\frac{\partial f_{e0}}{\partial\mathbf{p}_e}d^3\mathbf{p}_e+n_0e^3\sum_{\sigma_1\sigma_2}\int_{(s)}\Pi_e^{\sigma_s\sigma_1\sigma_2}(q,k_1,k_2)\\&\times E_f^{\sigma_1}(k_1)E_f^{\sigma_2}(k_2)\delta^4(q-k_1-k_2)\frac{d^4k_1d^4k_2}{(2\pi)^4},\end{aligned} \tag{7}$$

$$\begin{aligned}\Theta_s^{\sigma_s}(q)E_s^{\sigma_s}(q)=&-\frac{i4\pi n_{e0}e^3}{\Omega}\sum_{\sigma_1\sigma_2}\int_{(s)}\Pi_e^{\sigma_s\sigma_1\sigma_2}(q,k_1,k_2)\\&\times E_f^{\sigma_1}(k_1)E_f^{\sigma_2}(k_2)\delta^4(q-k_1-k_2)\frac{d^4k_1d^4k_2}{(2\pi)^4},\end{aligned} \tag{8}$$

where

$$\Theta_s^{\sigma_s}(q) = 1 - \frac{c^2|\mathbf{q}|^2}{\Omega^2} + \frac{c^2(\mathbf{q}\cdot\mathbf{e}_s^{\sigma_s})^2}{\Omega^2} + \sum_{\alpha,\sigma_s'} \frac{\omega_{p\alpha}^2 m_\alpha}{\Omega}\mathbf{e}_s^{\sigma_s}\cdot \times \int \frac{c\mathbf{p}_\alpha \mathbf{e}_s^{\sigma_s'}\cdot \dfrac{\partial f_{\alpha 0}}{\partial \mathbf{p}_\alpha}}{p_{0\alpha}\left(\Omega - \dfrac{c\mathbf{q}\cdot\mathbf{p}_\alpha}{p_{0\alpha}}\right)} d^3\mathbf{p}_\alpha, \quad (9)$$

$$\Pi_e^{\sigma_s\sigma_1\sigma_2}(q,k_1,k_2) = \frac{1}{2}\mathbf{e}_s^{\sigma_s}\cdot\int\left[\frac{c\mathbf{p}_e}{p_{0e}\left(\Omega - \dfrac{c\mathbf{q}\cdot\mathbf{p}_e}{p_{0e}}\right)}\mathbf{e}_f^{\prime\sigma_1}(k_1)\cdot\frac{\partial}{\partial\mathbf{p_e}}\frac{\mathbf{e}_f^{\prime\sigma_2}(k_2)\cdot\dfrac{\partial f_{e0}}{\partial\mathbf{p_e}}}{\left(\omega_2 - \dfrac{c\mathbf{k}_2\cdot\mathbf{p_e}}{p_{0e}}\right)} + (1\leftrightarrow 2)\right]d^3\mathbf{p}_e \quad (10)$$

are defined and $(1\leftrightarrow 2)$ represents the preceding term with indices 1 and 2 interchanged. $\Pi_e^{\sigma_s\sigma_1\sigma_2}(q,k_1,k_2)$ is a nonlinear electric current matrix. $\omega_{p\alpha}$ is the plasma frequency defined by the rest mass m_α as $\omega_{p\alpha}=\sqrt{4\pi n_{\alpha 0}e^2/m_e}$. Here it should be noted that there are two important points of differences between the kinetic and fluid theories. On the one hand, in kinetic theory, the slow-time-scale electric current is determined only by the slow-time-scale distribution function f_{es} as $\mathbf{j}_s=-\int e(c\mathbf{p}_e/p_{0e})f_{es}d^3\mathbf{p}_e$, which corresponds to the whole of the three terms $n_0\mathbf{v}_s+n_s\mathbf{v}_s+n_f\mathbf{v}_f$ ($\mathbf{v}_s$ and $\mathbf{v}_f$ are, respectively, the slow-time-scale and fast-time-scale velocities) in fluid theory. On the other hand, the integration of f_{es} over momentum, $n_s=\int f_{es}d^3\mathbf{p}_e$, is the plasma density variation, which is usually not equal to zero if $f_{es}\neq 0$. Accordingly, the conclusion can be obtained that the QSM field $\mathbf{B}_s$ is only generated in the existence of plasma density variation n_s, for the slow-time-scale currents are the basic source for the generation of $\mathbf{B}_s$. If $n_s=0$, no $\mathbf{B}_s$ will be generated. This conclusion is consistent with that of the usual fluid model. However, in the fluid model, n_s is usually inconsistently assumed as in Refs. 15 and 16, rather than self-consistently included as in our model.

As mentioned above, the slow-time-scale motion is caused by the different beat of two fast-time-scale motions, thus it is reasonable to assume $\Omega=\omega_1+\omega_2$, $\mathbf{q}=\mathbf{k}_1+\mathbf{k}_2$, and $\omega-c\mathbf{k}\cdot\mathbf{p}_\alpha/p_{0\alpha}\neq 0$. After tedious calculations, one can show that the following relation holds for relativistic plasma (see the Appendix):

$$\frac{1}{\Omega}\Pi_e^{\sigma_s\sigma_1\sigma_2}(q,k_1,k_2) = \frac{1}{\omega_2}\Pi_e^{\sigma_2\sigma_s\sigma_1}(k_2,q,-k_1). \quad (11)$$

As the conditions $v_{\text{ti}}|\mathbf{q}|\ll\Omega\ll v_{\text{te}}|\mathbf{q}|$, $\Omega\ll c|\mathbf{q}|$, and $\omega\gg|\mathbf{k}|v_{\text{te}}$ are usually satisfied under the static-state approximation, using relation (11), we can obtain the approximate expressions of Eqs. (9) and (10) as

$$\Pi_e^{t\sigma_1\sigma_2}(q,k_1,k_2) = \mathbf{e}_s^t\cdot\frac{[\mathbf{q}\times\beta_1(\mathbf{e}_1\times\mathbf{e}_2)]}{m_e^2\omega_2^2\omega_1} + \mathbf{e}_s^t\cdot\frac{[\mathbf{e}_1(\mathbf{q}\cdot\beta_2\mathbf{e}_2)]}{m_e^2\omega_2^2\omega_1} + \frac{\pi(\mathbf{q}\cdot\beta_3\mathbf{e}_2)(\mathbf{q}\cdot\mathbf{e}_1)}{16m_e^2\omega_2^2c|\mathbf{q}|^2} + \frac{\pi\beta_4(\mathbf{e}_1\cdot\mathbf{e}_2)}{4m_e^2\omega_2^2c}, \quad (12)$$

$$\Pi_e^{l\sigma_1\sigma_2}(q,k_1,k_2) = \frac{\Omega}{m_e^2\omega_2^2c^2|\mathbf{q}|}\beta_5(\mathbf{e}_1\cdot\mathbf{e}_2), \quad (13)$$

$$\Theta_s^t(q) = -\frac{c^2|\mathbf{q}|^2}{\Omega^2} - \frac{\omega_{pe}^2}{c^2|q|^2}\beta_6 - \frac{\omega_{pi}^2}{\Omega^2}\beta_7 \simeq -\frac{c^2|\mathbf{q}|^2}{\Omega^2},$$
$$\Theta_s^l = 1 + \frac{\omega_{pe}^2}{c^2|q|^2}\beta_6 - \frac{\omega_{pi}^2}{\Omega^2}\beta_7 \simeq \frac{\omega_{pe}^2}{c^2|q|^2}\beta_6. \quad (14)$$

In Eqs. (12)–(14), the coefficients related to the relativistic effect are defined as $\beta_1=m_e^2c^2\int 1/p_{0e}^2[(1/2)-(|\mathbf{p}_e|^2/3p_{0e}^2)]f_{e0}d^3\mathbf{p}_e$, $\beta_2=m_e^2c^2\int(|\mathbf{p}_e|^2/3p_{0e}^4)f_{e0}d^3\mathbf{p}_e$, $\beta_3=m_e^2c^2\int(|\mathbf{p}_e|/p_{0e}^3)f_{e0}d^3\mathbf{p}_e$, $\beta_4=m_e^2c^2\int[(1/p_{0e}|\mathbf{p}_e|)-(3|\mathbf{p}_e|/4p_{0e}^3)]f_{e0}d^3\mathbf{p}_e$, $\beta_5=m_e^2c^2\int[(1/2|\mathbf{p}_e|^2)+(|\mathbf{p}_e|^2/3p_{0e}^4)-(1/2p_{0e}^2)]f_{e0}d^3\mathbf{p}_e$, $\beta_6=m_ec\int[(1/p_{0e})+(p_{0e}/|p_e|^2)]f_{e0}d^3\mathbf{p}_e$, and $\beta_7=m_ic\int[(1/p_{0i})-(p_i^2/3|p_{0i}|^3)]f_{i0}d^3\mathbf{p}_i$. The dependence of the coefficients $\beta_1-\beta_5$ on the electron temperature is plotted in Fig. 1(a), and β_6, β_7 in Fig. 1(b). In Eqs. (14), the condition $1\gg|\mathbf{q}|\lambda_e\gg(v_{\text{te}}\sqrt{\beta_6/c^2})(\Omega/|\mathbf{q}|)$ holds and the ion effect is ignored. From Eq. (12), we can see that the nonlinear current matrix calculated in Refs. 18–20 is incomplete, missing the latter three terms of Eq. (12). We find there are two main reasons for this slip. One is that in Ref. 19 the direction of $\mathbf{q}$ is improperly taken for the direction of $\mathbf{k}_1$ or $\mathbf{k}_2$ and accordingly $\mathbf{e}_s^t\cdot\mathbf{k}_i=0(i=1,2)$ is improperly taken. The other is that $\mathbf{e}_s^t$ has been incorrectly regarded as a fixed direction perpendicular to the direction of $\mathbf{q}$, rather than a plane including two possible directions that are both perpendicular to $\mathbf{q}$, and thus many terms of the currents $\mathbf{j}_s^t$ in this plane have been incorrectly averaged to zero as $\mathbf{j}_s^t$ is projected on a line other than a plane, when $\Pi_e^{t\sigma_1\sigma_2}(q,k_1,k_2)$ is calculated.

From Eqs. (12)–(14) and (7)–(10), transforming into space-time coordinates, after tedious calculation, we obtain the slow-time-scale electric fields and the nonlinear currents, respectively, perpendicular and parallel to $\mathbf{q}$ as

$$\nabla^2\nabla^2\mathbf{E}_s^t(\mathbf{x},t) = -i\frac{\omega_{\text{pe}}^2e}{m_e\omega_0^3c^2}\frac{\partial}{\partial t}\nabla^2\boldsymbol{\nabla}\times[\beta_1\mathbf{E}_f(\mathbf{x},t)\times\mathbf{E}_f^*(\mathbf{x},t)] - i\frac{\omega_{\text{pe}}^2e}{m_e\omega_0^3c^2}\frac{\partial}{\partial t}\nabla^2[\mathbf{E}_f(\mathbf{x},t)\boldsymbol{\nabla}\cdot\beta_2\mathbf{E}_f^*(\mathbf{x},t)] + \frac{\pi\omega_{\text{pe}}^2e}{16m_e\omega_0^2c^3}\mathbf{e}_s^t\frac{\partial}{\partial t}\boldsymbol{\nabla}\cdot[\beta_3\mathbf{E}_f^*(\mathbf{x},t)\boldsymbol{\nabla}\cdot\mathbf{E}_f(\mathbf{x},t)] + \frac{\pi\omega_{\text{pe}}^2e}{4m_e\omega_0^2c^3}\mathbf{e}_s^t\frac{\partial}{\partial t}\nabla^2\beta_4|\mathbf{E}_f(\mathbf{x},t)|^2, \quad (15)$$

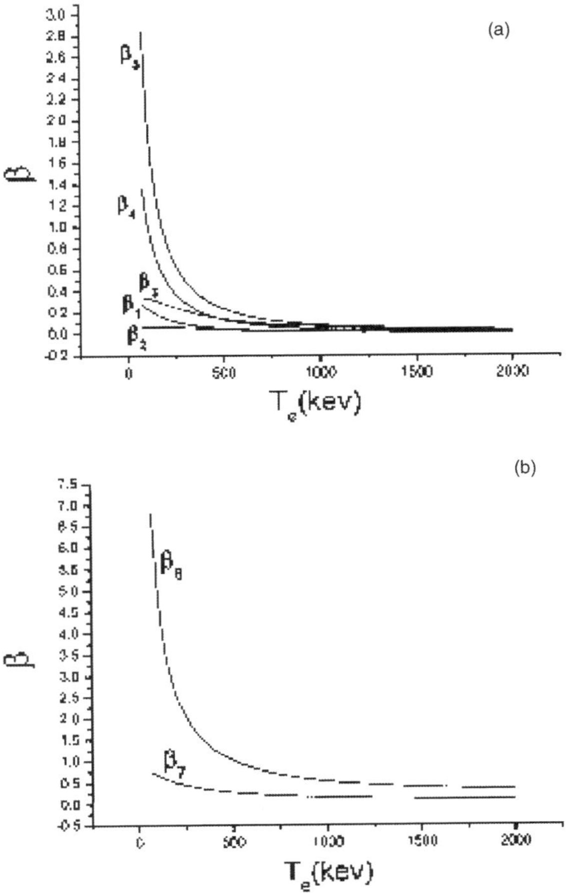

FIG. 1. The dependence of the relativistic coefficients β_i ($i=1-7$) on the temperature of the electrons, where (a) plots of $\beta_1-\beta_5$ and (b) plots of β_6 and β_7.

$$\begin{aligned}\nabla^2 \mathbf{j}_s^t(\mathbf{x},t) = &-i\frac{\omega_{pe}^2 e}{4\pi m_e\omega_0^3}\nabla^2 \nabla\times[\beta_1 \mathbf{E}_f(\mathbf{x},t)\times\mathbf{E}_f^*(\mathbf{x},t)] \\ &- i\frac{\omega_{pe}^2 e}{4\pi m_e \omega_0^3}\nabla^2[\mathbf{E}_f(\mathbf{x},t)\,\nabla\cdot\beta_2\mathbf{E}_f^*(\mathbf{x},t)] \\ &+ \frac{\omega_{pe}^2 e}{64 m_e\omega_0^2 c}\mathbf{e}_s^t\,\nabla\cdot[\beta_3\mathbf{E}_f^*(\mathbf{x},t)\,\nabla\cdot\mathbf{E}_f(\mathbf{x},t)] \\ &+ \frac{\omega_{pe}^2 e}{16 m_e\omega_0^2 c}\mathbf{e}_s^t\nabla^2\beta_4|\mathbf{E}_f(\mathbf{x},t)|^2,\end{aligned} \tag{16}$$

$$\mathbf{E}_s^l = -\frac{e\beta_5}{m_e\omega_0^2\beta_6}\nabla|\mathbf{E}_f|^2, \tag{17}$$

$$\mathbf{j}_s^l = \frac{e\beta_5}{4\pi m_e\omega_0^2\beta_6}\frac{\partial}{\partial t}\nabla|\mathbf{E}_f|^2. \tag{18}$$

It can be seen from Eqs. (15) and (17) that the transverse electric field $\mathbf{E}_s^t$ is slowly varying, and its effects on electron motion and other nonlinear plasma dynamics are embodied by the QSM field $\mathbf{B}_s$ as $\nabla\times\mathbf{E}_s^t=-1/c(\partial\mathbf{B}_s/\partial t)$; the longitudinal electric field $\mathbf{E}_s^l$ is just the QSE field, which corresponds to space-charge potential as $\mathbf{E}_s^l=-\nabla\phi_s$. It can also be seen from Eq. (16) that the transverse currents $\mathbf{j}_s^t$ are composed of four kinds of currents. The first is the nonlinear helical (rotational) current driven by different beats of fast-time-scale fields (laser fields), which will generate B_z. The second is the nonlinear azimuthal current driven by interactions of fast-time-scale velocity v_f and fast-time-scale electron density perturbation n_f, which corresponds to the current $n_f v_f$ in the fluid model and will generate B_z as well. However, in our kinetic model, the gradient of plasma temperature is assumed to be small, where T_e represents an averaged temperature of plasma electrons in the main region of laser plasma interaction, i.e., $\nabla\beta=0$; and for an intense laser propagating in underdense plasma, $\nabla\cdot\mathbf{E}_f\simeq 0$ can be approximately satisfied because n_f is much smaller, which will be explained in the following paragraph. Thus this current vanishes. The third term always equals zero for $\nabla\cdot\mathbf{E}_f\simeq 0$. The fourth is the nonlinear irrotational current driven by the ponderomotive force, which will generate B_θ. It should be noted that these transverse currents $\mathbf{j}_s^t$ of Eq. (17) are the basic sources for generation of QSM fields as $\nabla\times\mathbf{B}_s=(4\pi/c)\mathbf{j}_s^t$, and the longitudinal currents $\mathbf{j}_s^l$ of Eq. (18) are to keep the charge conservation as $(\partial\rho_s/\partial t)+\nabla\cdot\mathbf{j}_s^l=0$. In other words, our model satisfies the conservation law of charge through the longitudinal currents.

When the laser beam propagates in plasma, a longitudinal fast-time-scale electric field E_z will be generated by the interaction of laser field E_L with the fast-time-scale plasma density variation n_f as $\nabla\cdot\mathbf{E}_f=-4\pi e n_f$; and a plasma wave field (wakefield) $\mathbf{E}_W$ of frequency ω_{pe} may also be generated via a self-modulation or stimulated Raman scattering instability, which always exists behind the laser pulse. That is, the fast-time-scale electric field $\mathbf{E}_f$ should include $\mathbf{E}_L$, E_z, and $\mathbf{E}_W$. As mentioned above, we only keep the second-order terms representing nonlinear interactions of laser fields and plasmas. The wakefield $\mathbf{E}_W$, as a subfield to laser field, here can be ignored, because the amplitude of the wakefield is much smaller than the laser field for underdense plasmas as $|\mathbf{E}_w|\sim 0.1-0.3(mc\omega_p/e)=0.1-0.3(n_0/n_{cr})^{1/2}|\mathbf{E}_L|\ll|\mathbf{E}_L|$, which is shown in Ref. 24. For a laser beam with wave number $k_0=2\pi/\lambda_0$, assuming it propagates along the z-axis direction, under the Coulomb gauge, the electric fields are usually expressed as $\mathbf{E}_L(\mathbf{x},t)=\frac{1}{2}E_0(r,\theta,t)(\mathbf{e}_r+i\alpha\mathbf{e}_\theta)e^{i\alpha\theta}e^{ik_0z-i\omega_0t}+\text{c.c.}$, where $\alpha=0,\pm1$ for the linearly polarized (LP) and the circularly polarized (CP) laser, respectively. We assume the laser intensity is of Gaussian distribution, as $I=I_0e^{-r^2/r_0^2-(z-ct)^2/L^2}$ and $E_0=\sqrt{I}$, where L is the pulse longitudinal width and r_0 the transverse radius. We can approximately estimate the value of E_z through the fluid equations. From the continuity equation $(\partial n_f/\partial t)+\nabla\cdot(n_{e0}\mathbf{v}_f)=0$ and the motion equation for plasma electrons quivering in the laser field $\partial\mathbf{p}_f/\partial t=\partial\gamma m\mathbf{v}_f/\partial t=-e\mathbf{E}_L$ $[\gamma=[1+(\mathbf{p}_f^2/m_e^2c^2)]^{1/2}\simeq[1+(1+\alpha^2)I\lambda_{\mu m}^2/2.76]^{1/2}]$, we get $n_f=-(en_{e0}/\omega_0^2m_e)\nabla\cdot(\mathbf{E}_L/\gamma)$. Then, from Poisson's equation, $\nabla\cdot\mathbf{E}_f=ik_0E_z+\nabla_\perp\cdot\mathbf{E}_L=-4\pi e n_f$, we can obtain E_z as $E_z=i/k_0[1-(\omega_{pe}^2/\omega_0^2\gamma)]\nabla_\perp\cdot\mathbf{E}_L-(i/k_0)$

$\times(\omega_{pe}^2/\omega_0^2)\mathbf{E}_L\cdot\nabla_\perp(1/\gamma)$+c.c. After substituting the detailed Gaussian intensity distribution into them, we get $n_f=(en_{e0}/2\omega_0^2m_e)(1/\gamma^3)E_0e^{i\alpha\theta}e^{ik_0z-i\omega_0t}(r/r_0^2)$+c.c. and $E_z=-(i/2k_0)(r/r_0^2)E_0e^{i\alpha\theta}e^{ik_0z-i\omega_0t}+(i/2k_0)(r/r_0^2)(\omega_{pe}^2/\omega_0^2)(1/\gamma^3)\times E_0e^{i\alpha\theta}e^{ik_0z-i\omega_0t}$+c.c. For intense laser propagating in underdense plasma, $\gamma>1$ and $\omega_{pe}^2/\omega_0^2<1$ should be taken, assuming the laser beam transverse radius r_0 to be about 2λ and intensity to be $I_0=10^{19}$ W/cm^2 ($\gamma=3$) after focus, $|E_z|<(r/k_0r_0^2)E_0<(1/4\pi)E_0\ll E_0$ ($|E_z|\to 0$ when $r\to 0$) and $|n_f|<(r/k_0r_0^2)(\sqrt{\gamma^2-1}/\gamma^3)n_{e0}<(1/k_0r_0\gamma^2)n_{e0}<(1/36\pi)n_{e0}\ll n_{e0}$ are satisfied. That is, for intense laser propagating in underdense plasma, $|E_z|\ll|E_L|$ and $|n_f|\ll n_{e0}$ are both satisfied. Thus, it is reasonable to ignore E_z itself in the detailed calculation and to approximately take the relation $\nabla\cdot\mathbf{E}_f\simeq 0$. But note that the effect of ik_0E_z is not ignored in our model as the relation $\nabla\cdot\mathbf{E}_f\simeq 0$ is used above, for the value of ik_0E_z is comparable with that of $\nabla_\perp\cdot\mathbf{E}_L$.

Ignoring $\partial\mathbf{E}_L/\partial t$ and using the Ampere law $\nabla\times\mathbf{B}_s=(4\pi/c)\mathbf{j}_s^t$, we obtain the QSE field $\mathbf{E}_s$ and the QSM field $\mathbf{B}_s$ as,

$$\mathbf{E}_s=-\frac{e\beta_5}{m_e\omega_0^2\beta_6}\nabla|\mathbf{E}_L|^2, \tag{19}$$

$$\nabla\times\mathbf{B}_s(\mathbf{x},t)=-i\frac{\omega_{pe}^2e}{m_e\omega_0^3c}\nabla\times[\beta_1\mathbf{E}_L(\mathbf{x},t)\times\mathbf{E}_L^*(\mathbf{x},t)]+\frac{\pi\omega_{pe}^2e}{4m_e\omega_0^2c^2}\mathbf{e}_s^t\beta_4|\mathbf{E}_L(\mathbf{x},t)|^2. \tag{20}$$

Note that $\mathbf{E}_s$ and $\mathbf{B}_s$ in Eqs. (19) and (20) are quasistatic, assuming $\Omega\ll|\mathbf{q}|v_{te}$. That is, they do have important average effects (time-averaged over one laser period $T=2\pi/\omega_0$) on plasma dynamics, especially on electron motion, such as the ponderomotive force, while the effects of all fast-time-scale fields are averaged to zero.

III. QUASISTATIC ELECTRIC FIELD AND PLASMA DENSITY CHANNEL

In this section, we deduce $\mathbf{E}_s$ and analyze the profile of the plasma density channel for the given focused laser beam as that in Sec. II. Assuming the laser beam propagates along the z-axis direction with wave number $k_0=2\pi/\lambda_0$, under the Coulomb gauge, the electric fields are usually expressed as $\mathbf{E}_L(\mathbf{x},t)=\frac{1}{2}E_0(r,\theta,t)(\mathbf{e}_r+i\alpha\mathbf{e}_\theta)e^{i\alpha\theta}e^{ik_0z-i\omega_0t}$+c.c, where $\alpha=0,\pm1$ for the linearly polarized (LP) and the circularly polarized (CP) laser, respectively. We assume laser intensity is of Gaussian distribution as $I=I_0e^{-r^2/r_0^2-(z-ct)^2/L^2}$ and $E_0=\sqrt{I}$, where L is the pulse longitudinal width, and r_0 the transverse radius. All these descriptions of the laser beam are the same as those in Sec. II. Introducing the normalized transform $t\to t/\omega_0$, $\mathbf{x}\to c\mathbf{x}/\omega_0$, $E\to E(m_ec\omega_0/e)$, $B\to B(m_ec\omega_0/e)$, we can obtain $\mathbf{E}_s$, plasma density variation n_s through Poisson equation $\nabla\cdot\mathbf{E}_s^l=-4\pi en_s$ and plasma density divided by its initial value $n_e/n_{e0}=(n_{e0}+n_s)/n_{e0}$ as

$$\mathbf{E}_s=\frac{(1+\alpha^2)}{2}\frac{\beta_5}{\beta_6}I_0e^{-(z-t)^2/L^2}e^{-r^2/r_0^2}\left(\frac{r}{r_0^2}\mathbf{e}_r+\frac{z-t}{L^2}\mathbf{e}_z\right), \tag{21}$$

$$n_s=-(1+\alpha^2)\frac{\beta_5}{\beta_6}I_0e^{-(z-t)^2/L^2}e^{-r^2/r_0^2}\frac{1}{r_0^2}\left(1-\frac{r^2}{r_0^2}\right), \tag{22}$$

$$\frac{n_e}{n_{e0}}=1-\frac{(1+\alpha^2)}{n_{e0}}\frac{\beta_5}{\beta_6}I_0e^{-(z-t)^2/L^2}e^{-r^2/r_0^2}\frac{1}{r_0^2}\left(1-\frac{r^2}{r_0^2}\right). \tag{23}$$

Equations (22) and (23) just describe the transverse profile of plasma density channel, ignoring the longitudinal density variation due to $L\gg r_0$. It can be seen that n_en_{e0} is smaller than 1 in the region of $r<r_0$, that is, a density channel is formed; when $r\gtrsim r_0$, then $n_e/n_{e0}>1$ is taken; and when $r\to\infty$, then $n_e/n_{e0}=1$ is satisfied.

We can see that the expression of the right-hand side of Eq. (19) or Eq. (21) is similar to that of the ponderomotive force except that the sign is opposite. This can be easily understood, as it is just the ponderomotive force that expels some electrons from the axis so that a space-charge potential is formed to generate $\mathbf{E}_s$. This is also just the mechanism for both the generation of QSE field and the formation of plasma density channel. In the meantime, it should be noted that the hot electrons with temperature T_{he} are mainly originated from this part of electrons, which are accelerated by ponderomotive force in the process of being expelled.

IV. QUASISTATIC MAGNETIC FIELD

In this section, we give the solutions of the QSM fields B_z and B_θ for the given focused laser beam described in Secs. II and III. For real intense laser plasma interaction, in the main interacting region, some of electrons are warmed by laser fields due to heat exchange, others are accelerated into the relativistic regime by the ponderomotive force, as mentioned at the end of Sec. III. It is reasonable to assume that there are two groups of electrons in the interacting region, thermal ones and hot ones. Certainly, there are still a large number of cold electrons in the noninteracting region. But they do not evidently affect the generation of the QSM fields and thus we ignore their contribution. If we take the parameter ξ to represent the ratio of the number of thermal electrons to that of total electrons, we get the normalized equations, respectively, of B_z and B_θ from Eq. (20) as follows:

$$B_z=-\frac{1}{2}\alpha n_{e0}[\xi\beta_1(T_{te})+(1-\xi)\beta_1(T_{he})]I_0e^{-r^2/r_0^2-(z-ct)^2/L^2}, \tag{24}$$

$$B_\theta=\frac{\pi(1+\alpha^2)}{64}n_{e0}[\xi\beta_4(T_{te})+(1-\xi)\beta_4(T_{he})]I_0e^{-(z-ct)^2/L^2}\frac{r_0^2}{r}(1-e^{-r^2/r_0^2}), \tag{25}$$

where T_{te} is the averaged temperature for thermal electrons and T_{he} is the averaged temperature for hot electrons. In Eq. (25), we also assume $\mathbf{q}$ makes an angle θ_q with the axis z and $\langle\mathbf{e}_z\cdot\mathbf{e}_s^t\rangle=\langle\sin\theta_q\rangle=1/2$.

From Eqs. (20), (24), and (25), the mechanism for QSM field generation can be obtained where the beat interactions of CP laser fields drive a rotational azimuthal current [the first term of Eq. (20)] to generate B_z of Eq. (24), in which electrons following the CP laser ($\alpha=\pm 1$) move spirally around the laser propagation direction; the ponderomotive force drives a irrotational axial current [the second term of Eq. (20)] for the generation of B_θ of Eq. (25). Yet, for LP laser fields ($\alpha=0$), which are only along a fixed linearly polarized direction, the rotational current should vanish and B_z is nothing, as seen in Eq. (24).

V. COMPARING WITH EXPERIMENTAL RESULTS

Although the QSE fields can not be easily measured, there are some experiments involving the QSM fields. Recently, measurements of the QSM fields were reported by Najmudin *et al.*[21] and Fuchs *et al.*[22] In Ref. 21, Najmudin *et al.* paid attention to the QSM fields parallel to the laser beam propagation direction, that is, B_z in the present paper. But Fuchs *et al.*[22] measured the magnetic fields perpendicular to the laser beam propagation direction, that is, B_θ in the present paper. We compare our theoretical expectation with the experiments in this section.

Before the detailed comparisons, we have to give some reasonable estimations of three parameters ξ, T_{te}, and T_{he} due to their complexity. The values of these parameters may have some little effects on the numerical value of B_z and B_θ, but do not essentially affect the physical character of B_z and B_θ and the comparisons between our model and the experiments, as well as the following numerical simulation. First, as mentioned above and in the end of Sec. III, if we approximately consider that hot electrons are mainly produced by the acceleration of ponderomotive force ignoring fast-varying density variation, we can get the approximate ξ from Eqs. (22) and (23) as

$$\xi \simeq \frac{n_e}{n_{e0}} = 1 - \frac{(1+\alpha^2)}{n_{e0}}\frac{\beta_5}{\beta_6} I_0 e^{-(z-t)^2/L^2} e^{-r^2/r_0^2}\frac{1}{r_0^2}\left(1-\frac{r^2}{r_0^2}\right). \tag{26}$$

Second, it is widely proved that the Wilk's scaling law can be used to determine the averaged hot electron temperature as $T_{he}=511(\gamma-1)$ $[\gamma=(1+[(1+\alpha^2)I_0\lambda_{\mu m}^2/2\times 1.38\times 10^{18}])^{1/2}]$. Lastly, the value estimation of the averaged thermal electron temperature T_{te} is the most difficult. So far, to our knowledge, no paper has give the exact value of T_{te}. Through a large number of numerical calculations and simulation, it is found that T_{te} seems to be several tens of keV under the above estimation of ξ and T_{he}. Here we simply assume all following T_{te} to be 75 keV. It is also found that if ξ is chosen to be smaller, T_{te} may be chosen to be lower; and in contrast, ξ greater, T_{te} higher.

A CP laser beam with "vacuum" intensity $I_0=6.7\times 10^{18}$ W/cm^2 is used in the experiment given in Ref. 21, the main parameters are laser wavelength $\lambda=1.05$ μm, vacuum laser transverse radius $r_0=10\lambda$, laser pulse duration $\tau=1$ ps and plasma density $n_{e0}=2.8\times 10^{19}$ cm^{-3}. As mentioned in Ref. 21, the laser beam may be self-focused in plasma, the intensity may be enhanced, while the transverse radius decreased in some regions. Here we assume I_0 can be two times of vacuum laser intensity and r_0 be decreased to 5λ. Figures 2(a) and 2(b) show B_z calculated by Eq. (24) for the CP laser beam with wavelength $\lambda=1.05$, $z-ct=0$, and $n_{e0}=2.8\times 10^{19}$ cm^{-3} for (a) intensity $I_0=6.7\times 10^{18}$ W/cm^2, transverse pulse radius $r_0=10\lambda$ and (b) $I_0=1.34\times 10^{19}$ W/cm^2, $r_0=5\lambda$, respectively. As mentioned above, $T_{he}=511(\gamma-1)$ and $T_{te}=75$ kev are taken and ξ is chosen as Eq. (26). Other parameters accord with those given in Ref. 21. It can be seen B_z dominates in the region near the axis and the profile of B_z is consistent with Fig. 2 of Ref. 21. The peak magnetic field B_z is about (a) 2.13 megagauss (MG) and (b) 3.07 MG at the laser axis $r=0$, both of which are very close to the experiment result of 1.6–3.2 MG.

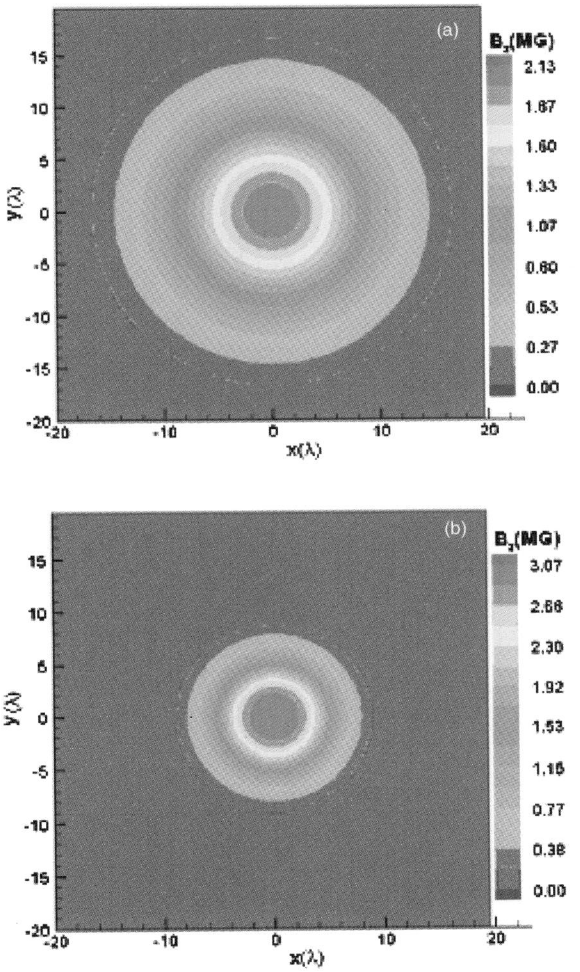

FIG. 2. (Color online). B_z calculated by Eq. (24) for the CP laser beam with $\lambda=1.05$ μm and (a) $I_0=6.7\times 10^{18}$ W/cm^2, $r_0=10\lambda$; (b) $I_0=1.34\times 10^{19}$ W/cm^2, $r_0=5\lambda$, where $z-ct=0$, $n_e=2.8\times 10^{19}$ cm^{-3}, $T_{te}=75$ keV, and $T_{re}=511(\gamma-1)$ are adopted.

The experiment measurement on B_θ generated by a LP laser is reported in Ref. 22. The main parameters are vacuum laser intensity $I_0=4.7\times 10^{18}$ W/cm^2, laser wavelength λ

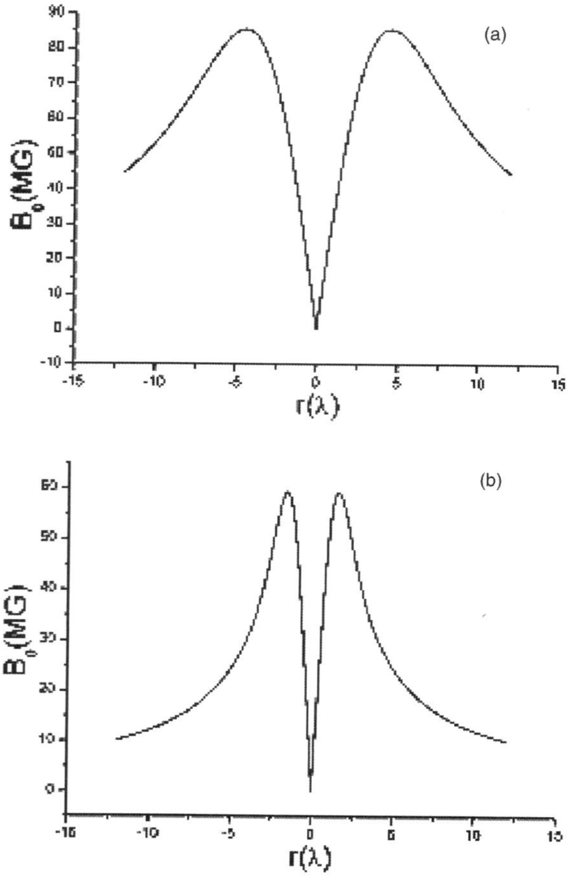

FIG. 3. B_θ calculated by Eq. (25) for the LP laser beam with $\lambda=1.05\ \mu$m, where (a) $I_0=4.7\times10^{18}\ \mathrm{W/cm^2}$, $r_0=4\lambda$; (b) $I_0=9.4\times10^{18}\ \mathrm{W/cm^2}$, $r_0=1.33\lambda$, where $n_e=2\times10^{20}\ \mathrm{cm^{-3}}$, $T_{\mathrm{te}}=75$ keV, $T_{\mathrm{re}}=511(\gamma-1)$, $z-ct=0$ are adopted.

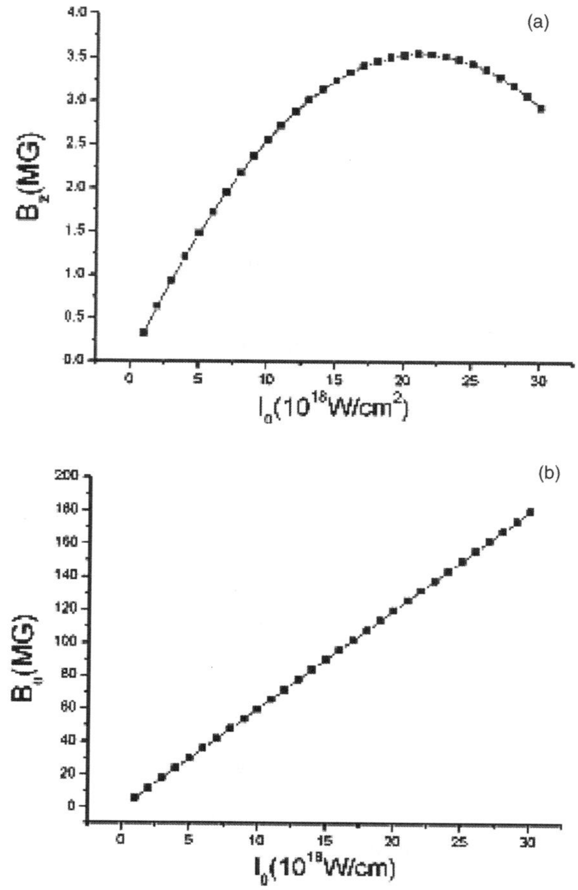

FIG. 4. The dependence of the peak B_z (a) and B_θ (b) on the laser intensity I_0, where (a) $r_0=5\lambda$ and (b) $r_0=4/3\lambda$ are taken after self-focusing, other parameters are consistent with Figs. 2 and 3, respectively.

$=1.058\ \mu$m, vacuum laser transverse radius $r_0=4\lambda$, laser pulse duration $\tau=600$ fs, and plasma density $n_{e0}=2\times10^{20}\ \mathrm{cm^{-3}}$. Figures 3(a) and 3(b) show B_θ calculated by Eq. (25) for the LP laser beam with $\lambda=1.05\ \mu$m, where (a) $I_0=4.7\times10^{18}\ \mathrm{W/cm^2}$, $r_0=4\lambda$ and (b) $I_0=9.4\times10^{18}\ \mathrm{W/cm^2}$, $r_0=\frac{4}{3}\lambda$, as two times of vacuum laser intensity and one-third of the initial transverse radius due to self-focusing, and $n_{e0}=2\times10^{20}\ \mathrm{cm^{-3}}$ is adopted. The parameters used here are in accordance with those in Ref. 22. The averaged thermal electron temperature T_{te} is also assumed to be 75 keV, ξ is also chosen as Eq. (26) and T_{he} is determined by Wilk's scaling law $T_{\mathrm{he}}=511(\gamma-1)$, as well. It can be seen from Figs. 3(a) and 3(b) that the peak B_θ is about (a) 85 MG and (b) 59 MG, respectively, which is either very close to the results 35–70 MG in Ref. 22. We can see that B_θ is close to zero near the axis, increases gradually with r until reaches its maximum value at about $r=r_0$, and then decreases due to return currents.

Notice that the experiment results[21,22] both show that the QSM fields increase with laser intensity. But few models can give the explanation of the dependence of the peak QSM fields B_z and B_θ on the laser intensity. Sheng's model[12] suggests that the peak B_z decreases as laser intensity increases (see theoretical curve in Fig. 3 of Ref. 21), which is in contradiction with the experimental result given in Ref. 21. The reason is that in Sheng's model[12] all electrons are assumed to be hot (relativistic) accelerated by ponderomotive force, that is, $\xi=0$. However in fact, as mentioned in our model, not all electrons had been accelerated into the relativistic regime, there are some thermal electrons, that is, $\xi\neq0$. Figures 4(a) and 4(b) plot the dependence of (a) B_z and (b) B_θ on the laser intensity I_0, where (a) $r_0=5\lambda$ and (b) $r_0=4/3\lambda$ are taken after self-focusing, other parameters are consistent with the above experiments given in Refs. 21 and 22, respectively. It can be seen from Fig. 4(a) that B_z increases with laser intensity I_0, which accords with the experimental results. It also should be noted that while I_0 rises to high enough that all electrons at $r=0$ were expelled and accelerated to be hot, i.e., $n_e\to0$ and $\xi\to0$ here, the peak B_z then decreases if I_0 still increases, as seen in Fig. 4(a), because the peak B_z is mainly determined by electrons at the axis $r=0$. However, the peak B_θ is determined mainly by electrons at $r=r_0$, where all elec-

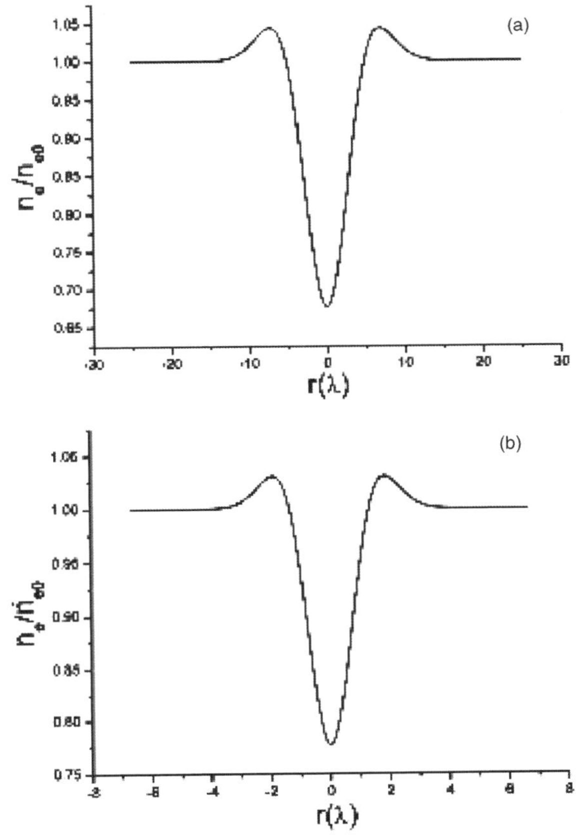

FIG. 5. The plasma density divided by its initial value, that is, the transverse profile of the corresponding plasma density channel, under the same parameters as Figs. 2(b) and 3(b), respectively.

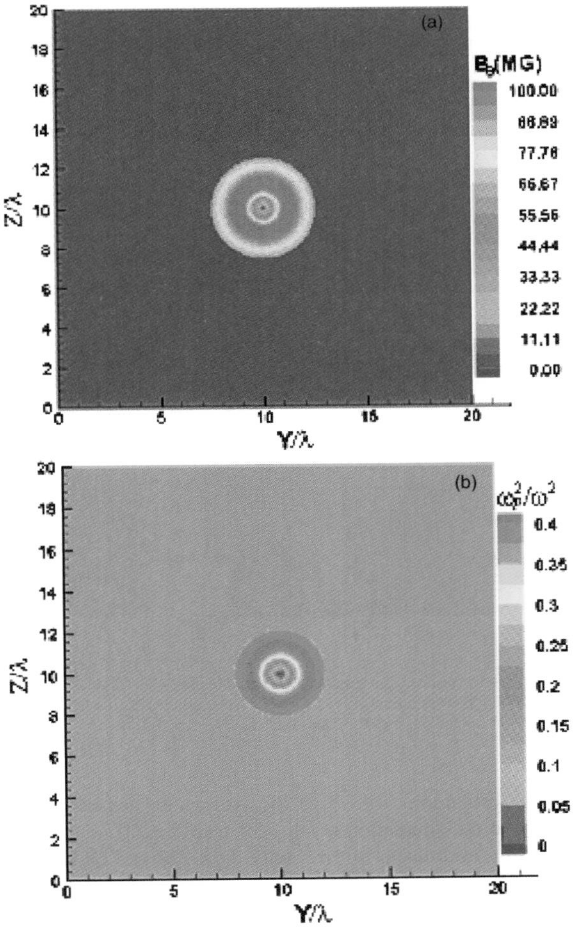

FIG. 6. (Color online). (a) Plots B_θ calculated by Eq. (24) for LP laser, where $I_0=2.48\times 10^{19}$ W/cm^2 as two times of "vacuum" intensity, $r_0=1\lambda$ after self-focusing, and $n_{e0}=0.36n_{cr}$ are taken; (b) transverse (Y,Z) plots $\omega_p^2/\omega^2=n_e/n_{cr}$ using Eq. (23), where $I_0=4.5\times 10^{20}$ W/cm^2, $r_0=1\lambda$, and $n_{e0}=0.36n_{cr}$ are adopted.

trons are thermal and $\xi=1$ is always satisfied, so B_θ will increase all along with I_0, as seen in Fig. 4(b). This explanation can be more explicitly seen from Figs. 5(a) and 5(b), which plots the corresponding plasma density divided by its initial value calculated by Eq. (23), that is, just describing the transverse profile of the corresponding plasma density channel, under the same parameters as Figs. 2(b) and 3(b), respectively. From Figs. 5(a) and 5(b) we can evidently see that electrons in the region $r<r_0$ are expelled by ponderomotive force so that $n_e/n_{e0}<1$, i.e., $\xi<1$; at the place $r=r_0$, $n_e/n_{e0}=1$, i.e., $\xi=1$; and the expelled electrons will deposited at a cavity edge of $r\gtrsim r_0$, where $n_e/n_{e0}>1$ should be satisfied; then when $r\gg r_0$, $n_e/n_{e0}=1$ should be taken again.

VI. PARTICLE SIMULATION

In this section, we check our analytical model through the particle simulations. The values of parameters ξ, T_{te}, and T_{he} are approximately estimated in the same way as Sec. V. First, we compare our theoretical expectation with the simulation of Pukhov and Meyer-ter-Vehn.[23] A LP laser beam with vacuum intensity $I_0=1.24\times 10^{19}$ W/cm^2 and vacuum transverse radius $r_0=6\lambda$ was used in the simulation given in Ref. 23. The main parameters are laser wavelength $\lambda=1$ μm and plasma density $n_{e0}=0.36n_{cr}$. From Figs. (1)–(4) of Ref. 23, we can see the results that (i) the laser beam will be self-focused when it enters plasmas, the intensity can be enhanced to be $1.24\times 10^{19}-6.0\times 10^{20}$ W/cm^2, transverse radius can be collapsed to be $1-2\lambda$; (ii) about 50–100 MG azimuthal QSM field is generated; (iii) a density cavitation is formed and guides the laser beam. Using our model, we plot the azimuthal QSM field B_θ calculated by Eq. (24) in Fig. 6(a), where $I_0=2.48\times 10^{19}$ W/cm^2 as two times of vacuum intensity, $r_0=1\lambda$ after self-focusing and $n_{e0}=0.36n_{cr}$ are taken. The parameters accord with that given in Ref. 23. It can be seen that the profile of B_θ in Fig. 6(a) is very similar to that of Fig. 2 in Ref. 23. The peak B_θ is 100 MG, also very close to the simulation results 50–100 MG. Figure 6(b) shows transverse (Y,Z) plots $\omega_p^2/\omega^2=n_e/n_{cr}$ using Eq. (23). The parameters are $I_0=4.5\times 10^{20}$ W/cm^2, $r_0=1\lambda$, and $n_{e0}=0.36n_{cr}$, which are consistent with that of Fig. 4 in Ref. 23. As can be seen from the figure, our model can give well the

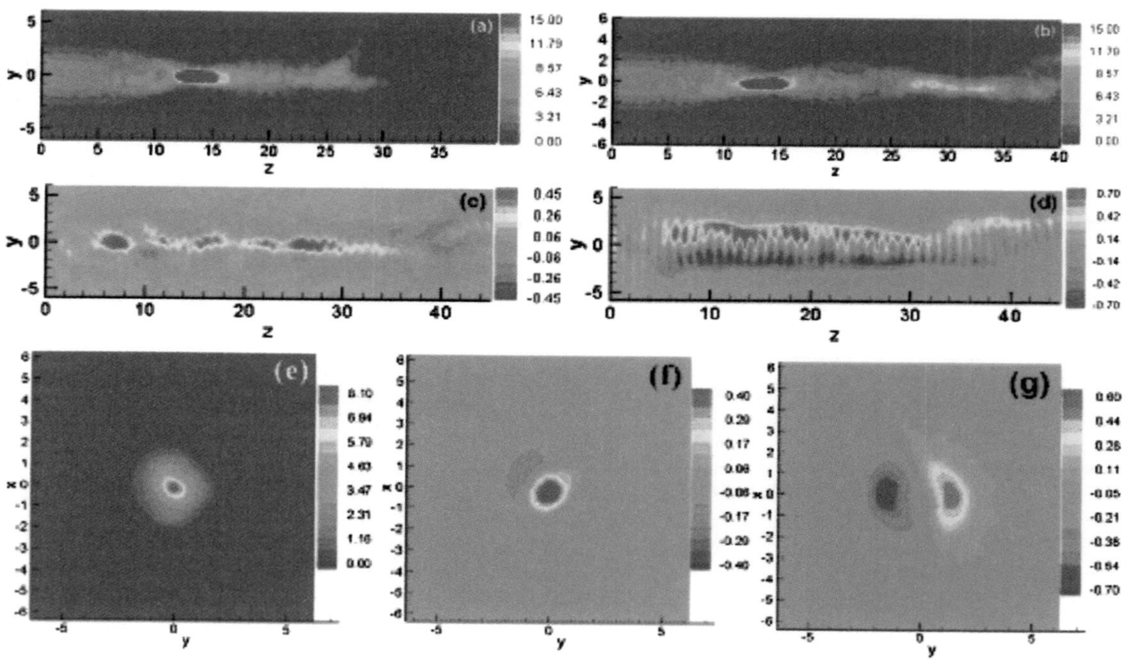

FIG. 7. (Color online). Instantaneous laser intensity (γ^2-1) longitudinal (z,y) for the CP laser cuts at $t=28\tau$ (a) and $t=48\tau$ (b), respectively. At $t=48\tau$ magnetic fields longitudinal (z,y) for the CP laser cut, (c) B_z and (d) B_x (azimuthal), respectively. The profiles of (e) I_0, (f) B_z, and (g) B_x (azimuthal) transverse (y,x) cut at $z=26.25\lambda$, $t=48\tau$. The units of x,y,z are λ.

density cavitation, whose profile and value are both very similar to the simulation results given in Ref. 23.

Then, in order to further study the QSM and QSE fields for the CP laser beam, we commence with the numerical simulation using our 3D PIC code LARAD-P. The CP laser beam ($\lambda=1.054$ μm) with an initial transverse Gaussian width $d_0=6\lambda$ is incident from the left $z=0$ in the intensity $I=5\times10^{18}$ W/cm^2 to the plasmas. The initial plasma density is $n_e=0.36n_{cr}$, where n_{cr} is the nonrelativistic critical density; the rest mass ratio for ion and electron is $m_i/m_e=4500$; $T_e=16$ keV; 3×10^8 electrons and ions are uniformly distributed in $160\times660\times160(12.7\lambda\times52.5\lambda\times12.7\lambda)$ spatial meshes.

The numerical simulation results show that the laser beam is strongly focused when it enters plasmas, the laser intensity increases from $I_0=5\times10^{18}$ W/cm$^2(\gamma^2-1=4)$ to $I_0=1.86\times10^{19}$ W/cm$^2(\gamma^2-1=15)$, while the transverse width decreases to $d_0\approx2\lambda-3\lambda$, as seen in Figs. 7(a) and 7(b). For the CP laser, the longitudinal profiles of the axial magnetic field B_z and the azimuthal one B_x at $t=48\tau$ are illustrated in Figs. 7(c) and 7(d). At $t=48\tau$, $|B_z|=20-45$ MG is distributed in $z=14\lambda-30\lambda$, approximately uniform in the transverse directions with the transverse diameter $d=1\lambda$ and dominates in the region near the axis; $|B_x|=70$ MG is distributed in $z=8\lambda-28\lambda$ and dominates at the outlying zone where the transverse diameter is $d_0\approx2\lambda-3\lambda$. Figures 7(e)–7(g), respectively, plot the profiles of I_0, B_z, and B_x, which transverse cuts at $z=26.25\lambda$, $t=48\tau$. It can be seen that the shapes of the transverse profiles for B_z and B_x are both consistent with our theoretical expectation [just as B_z in Fig. 2 and B_θ in Fig. 3 or Fig. 6(a)]. And it can be seen that the peaks $|B_z|$ and $|B_x|$ are, respectively, about 40 MG and 70 MG, where $I_0=10^{19}$ W/cm$^2(\gamma^2-1=8)$ and $d_0\approx2\lambda-3\lambda$ are satisfied. Adopting the same parameters and using Eqs. (24) and (25), the peak $|B_z|=24.6$ MG and $|B_x=\frac{1}{\sqrt{2}}B_\theta|=98.4$ MG are estimated by our model, which are both very close to the simulation results. Figure 8 plots the corresponding QSE fields and plasma density channel, (a) and (b), respectively, illustrate the longitudinal profiles of the radial QSE field $E_{sx}=\mathbf{e}_x\cdot\mathbf{E}_s$ (E_{sr}) and axial QSE field $E_{sz}=\mathbf{e}_z\cdot\mathbf{E}_s$ at $t=92\tau$; (c)–(e), respectively, plot the transverse profiles of I_0, E_{sx}, and E_{sz} transverse cutting at $z=26.25\lambda$, $t=92\tau$; (f) and (g), respectively, plot the longitudinal and transverse profiles of plasma density channel n_e/n_{e0} cutting, respectively, as (a) and (d). As seen from Figs. 8(a) and 8(b), $|E_{sx}|=30$ CGSE (E_{sr}) is distributed in $z=10\lambda-45\lambda$ and dominates at the outlying zone about $d_0=2-3\lambda$, $|E_{sz}|=20$ CGSE is distributed in $z=8\lambda-28\lambda$ dominating in the region near the axis. These profiles of E_{sx} and E_{sz} are both consistent with our theoretical expectation [can be seen from Eq. (21)], which can be more explicitly seen from the transverse profiles of them in Figs. 8(d) and 8(e). From Figs. 8(c)–8(e), we can also see that the peak $|E_{sx}|$ and $|E_{sz}|$ are, respectively, about 30 CGSE and 20 CGSE for $I_0=1.29\times10^{19}$ W/cm$^2(\gamma^2-1=10.4)$ and $d_0=2\lambda$. Adopting the same parameters and using Eq. (21), the peak $|E_{sx}|=\frac{1}{\sqrt{2}}E_{sr}=29.3$ CGSE and $|E_{sz}|=14.7$ CGSE (assuming laser beam longitudinal width $L=2r_0=2\lambda$ after focused) are estimated by our model, which are both very close to the simulation results. From Figs. 8(f) and 8(g), we can see that

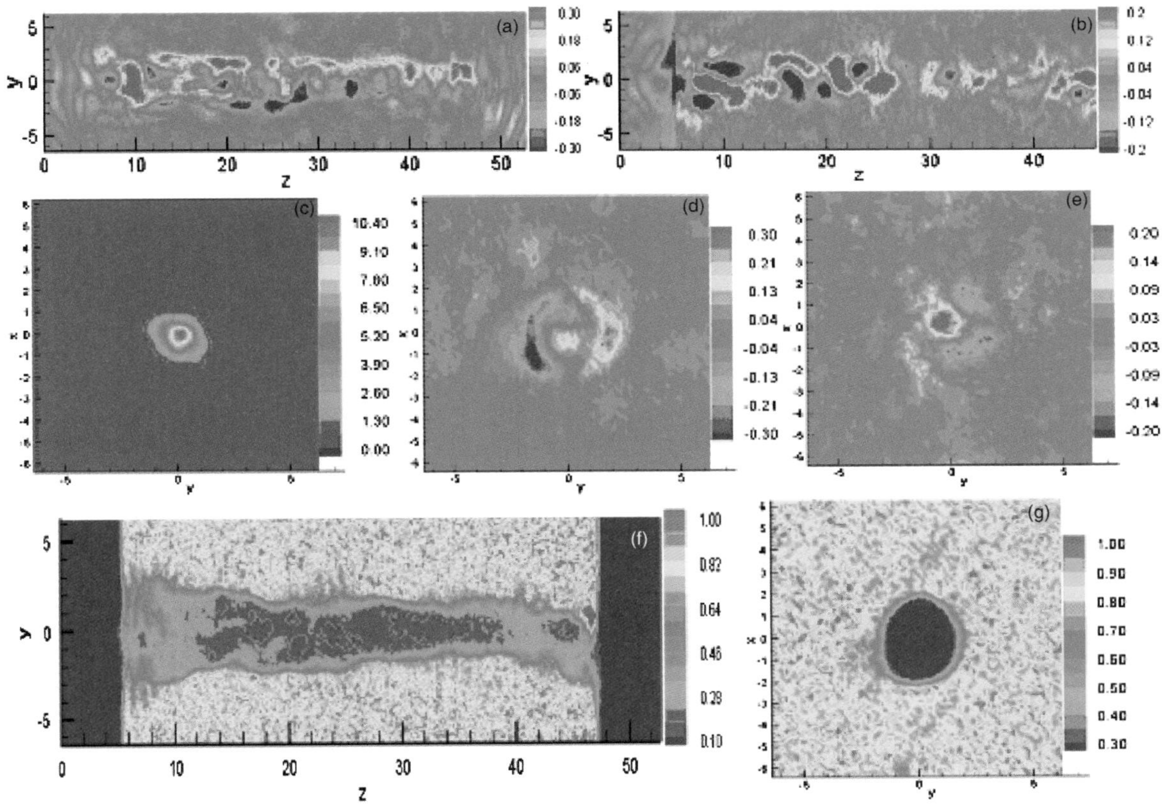

FIG. 8. (Color online). The longitudinal profiles of (a) radial QSE field $E_{sx}=(1/\sqrt{2})E_{sr}$ and (b) axial QSE field E_{sz} cut at $t=92\tau$. The transverse profiles of I_0 (c), E_{sx} (d), and E_{sz} (e) cut at $z=26.25\lambda$, $t=92\tau$; the longitudinal and transverse profiles of plasma density channel n_e/n_{e0} cut at $t=92\tau$, $z=26.25\lambda$, respectively, of (f) and (g).

a plasma density channel $[n_e/n_{e0}=0.3$ in Fig. 8(g)] can be formed and guide the laser propagation, which also basically accords with our theoretical prediction. $n_e/n_{e0}=0.4$ is estimated using Eq. (22) with the same parameters as Fig. 8(g).

VII. CONCLUSION

In conclusion, we have studied the generation of both the QSM field and the QSE field in intense laser plasma interaction analytically and numerically, which includes the axial QSM field B_z as well as the azimuthal B_θ, and the QSE field $\mathbf{E}_s$. A self-consistent kinetic model satisfying the conservation law of charge is set up, which can give a good uniform explanation of the generation mechanism of all these fields and even the formation of the plasma density channel. We compare our theoretical expectations with the results of the experiments[21,22] and Pukhov's simulation.[23] Through our own particle simulation, we further investigate and verify our model. It is found that B_z as high as about 40 MG and B_θ about 70 MG can both be generated by the CP laser in a relativistic plasma with laser intensity $I_0=5\times10^{18}$ W/cm^2; but for the LP laser B_z is absent. In the meantime, E_{sr} about 30 CGSE and E_{sz} about 20 CGSE can both be generated by LP or CP laser. It is also found that the peak magnetic fields depend heavily on the laser intensity and the ratio of the number of thermal electrons to that of total electrons. As, actually, not all electrons have been accelerated into the relativistic regime by intense laser beam, the peak magnetic fields B_z and B_θ both increase with the laser intensity. For the purposes of laser plasma acceleration and fast ignition, such quasistatic fields are important because the fields will affect the electron motion and the laser propagation, which are still unclear and under further investigation.

ACKNOWLEDGMENTS

This work was supported by National Hi-Tech Inertial Confinement Fusion Committee of China, National Natural Science Foundation of China Under Grant No. 10135010, Special Funds for Major State Basic Research Projects in China, National Basic Research Project "nonlinear Science" in China, and Science Foundation of CAEP.

APPENDIX: PROOF OF EQ. (11)

Let $\mathbf{e}'_s=\mathbf{e}'^{\sigma_s}_s(q)$, $\mathbf{e}'_1=\mathbf{e}'^{\sigma_1}_f(k_1)$, $\mathbf{e}'_2=\mathbf{e}'^{\sigma_2}_f(k_2)$, $\mathbf{e}_s=\mathbf{e}^{\sigma_s}_s(q)$, $\mathbf{e}_1=\mathbf{e}^{\sigma_1}_f(k_1)$, $\mathbf{e}_2=\mathbf{e}^{\sigma_2}_f(k_2)$, then

$$\Pi_e^{\sigma_s\sigma_1\sigma_2}(q,k_1,k_2)=\frac{1}{2}\int\left[\frac{c\mathbf{e}_s\cdot\mathbf{p}_e}{p_{0e}\left(\Omega-\frac{c\mathbf{q}\cdot\mathbf{p}_e}{p_{0e}}\right)}\mathbf{e}_1'\frac{\partial}{\partial\mathbf{p_e}}\frac{\mathbf{e}_2'\cdot\frac{\partial f_{e0}}{\partial\mathbf{p_e}}}{\left(\omega_2-\frac{c\mathbf{k}_2\cdot\mathbf{p_e}}{p_{0e}}\right)}+(1\leftrightarrow 2)\right]d^3\mathbf{p}_e. \tag{A1}$$

We now prove that

$$\frac{1}{\Omega}\int\left[\frac{c\mathbf{e}_s\cdot\mathbf{p}_e}{p_{0e}\left(\Omega-\frac{c\mathbf{q}\cdot\mathbf{p}_e}{p_{0e}}\right)}\mathbf{e}_1'\cdot\frac{\partial}{\partial\mathbf{p_e}}\frac{\mathbf{e}_2'\cdot\frac{\partial f_{e0}}{\partial\mathbf{p_e}}}{\left(\omega_2-\frac{c\mathbf{k}_2\cdot\mathbf{p_e}}{p_{0e}}\right)}+(1\leftrightarrow 2)\right]d^3\mathbf{p}_e-\frac{1}{\omega_2}\int\frac{c\mathbf{e}_2\cdot\mathbf{p}_e}{p_{0e}\left(\omega_2-\frac{c\mathbf{k}_2\cdot\mathbf{p}_e}{p_{0e}}\right)}$$
$$\times\left[\mathbf{e}_s'\cdot\frac{\partial}{\partial\mathbf{p_e}}\frac{\mathbf{e}_1'\cdot\frac{\partial f_{e0}}{\partial\mathbf{p_e}}}{\left(\omega_1-\frac{c\mathbf{k}_1\cdot\mathbf{p_e}}{p_{0e}}\right)}-\mathbf{e}_1'\cdot\frac{\partial}{\partial\mathbf{p_e}}\frac{\mathbf{e}_s'\cdot\frac{\partial f_{e0}}{\partial\mathbf{p_e}}}{\left(\Omega-\frac{c\mathbf{q}\cdot\mathbf{p_e}}{p_{0e}}\right)}\right]d^3\mathbf{p}_e=0. \tag{A2}$$

$$\text{Equation (A2)}=\int\Bigg\{-\frac{1}{\Omega}\frac{\mathbf{e}_2'\cdot\frac{\partial f_{e0}}{\partial\mathbf{p_e}}}{\left(\omega_2-\frac{c\mathbf{k}_2\cdot\mathbf{p_e}}{p_{0e}}\right)}\mathbf{e}_1'\cdot\left[\frac{c\mathbf{e}_s}{\left(\Omega-\frac{c\mathbf{q}\cdot\mathbf{p_e}}{p_{0e}}\right)p_{0e}}-\frac{c(\mathbf{e}_s\cdot\mathbf{p}_e)\mathbf{p}_e}{\left(\Omega-\frac{c\mathbf{q}\cdot\mathbf{p_e}}{p_{0e}}\right)p_{0e}^3}+\frac{c(\mathbf{e}_s\cdot\mathbf{p_e})}{p_{0e}\left(\Omega-\frac{c\mathbf{q}\cdot\mathbf{p_e}}{p_{0e}}\right)^2}\right.$$
$$\left.\times\left(\frac{c\mathbf{q}}{p_{0e}}-\frac{c(\mathbf{q}\cdot\mathbf{p}_e)\mathbf{p}_e}{p_{0e}^3}\right)\right]-\frac{1}{\Omega}\frac{\mathbf{e}_1'\cdot\frac{\partial f_{e0}}{\partial\mathbf{p_e}}}{\left(\omega_1-\frac{c\mathbf{k}_1\cdot\mathbf{p_e}}{p_{0e}}\right)}\mathbf{e}_2'\cdot\left[\frac{c\mathbf{e}_s}{\left(\Omega-\frac{c\mathbf{q}\cdot\mathbf{p_e}}{p_{0e}}\right)p_{0e}}-\frac{c(\mathbf{e}_s\cdot\mathbf{p}_e)\mathbf{p}_e}{\left(\Omega-\frac{c\mathbf{q}\cdot\mathbf{p_e}}{p_{0e}}\right)p_{0e}^3}\right.$$
$$\left.+\frac{c(\mathbf{e}_s\cdot\mathbf{p_e})}{p_{0e}\left(\Omega-\frac{c\mathbf{q}\cdot\mathbf{p_e}}{p_{0e}}\right)^2}\left(\frac{c\mathbf{q}}{p_{0e}}-\frac{c(\mathbf{q}\cdot\mathbf{p}_e)\mathbf{p}_e}{p_{0e}^3}\right)\right]+\frac{1}{\omega_2}\frac{\mathbf{e}_1'\cdot\frac{\partial f_{e0}}{\partial\mathbf{p_e}}}{\left(\omega_1-\frac{c\mathbf{k}_1\cdot\mathbf{p_e}}{p_{0e}}\right)}\mathbf{e}_s'\cdot\left[\frac{c\mathbf{e}_2}{\left(\omega_2-\frac{c\mathbf{k}_2\cdot\mathbf{p_e}}{p_{0e}}\right)p_{0e}}\right.$$
$$\left.-\frac{c(\mathbf{e}_2\cdot\mathbf{p}_e)\mathbf{p}_e}{\left(\omega_2-\frac{c\mathbf{k}_2\cdot\mathbf{p_e}}{p_{0e}}\right)p_{0e}^3}+\frac{c(\mathbf{e}_2\cdot\mathbf{p_e})}{p_{0e}\left(\omega_2-\frac{c\mathbf{k}_2\cdot\mathbf{p_e}}{p_{0e}}\right)^2}\left(\frac{c\mathbf{k}_2}{p_{0e}}-\frac{c(\mathbf{k}_2\cdot\mathbf{p}_e)\mathbf{p}_e}{p_{0e}^3}\right)\right]$$
$$-\frac{1}{\omega_2}\frac{\mathbf{e}_s'\cdot\frac{\partial f_{e0}}{\partial\mathbf{p_e}}}{\left(\Omega-\frac{c\mathbf{q}\cdot\mathbf{p_e}}{p_{0e}}\right)}\mathbf{e}_1'\cdot\left[\frac{c\mathbf{e}_2}{\left(\omega_2-\frac{c\mathbf{k}_2\cdot\mathbf{p_e}}{p_{0e}}\right)p_{0e}}-\frac{c(\mathbf{e}_2\cdot\mathbf{p}_e)\mathbf{p}_e}{\left(\omega_2-\frac{c\mathbf{k}_2\cdot\mathbf{p_e}}{p_{0e}}\right)p_{0e}^3}+\frac{c(\mathbf{e}_2\cdot\mathbf{p_e})}{p_{0e}\left(\omega_2-\frac{c\mathbf{k}_2\cdot\mathbf{p_e}}{p_{0e}}\right)^2}\left(\frac{c\mathbf{k}_2}{p_{0e}}\right.\right.$$
$$\left.\left.-\frac{c(\mathbf{k}_2\cdot\mathbf{p}_e)\mathbf{p}_e}{p_{0e}^3}\right)\right]\Bigg\}d^3\mathbf{p}_e=\frac{1}{\Omega\omega_1\omega_2}\int\frac{c}{p_{0e}}\frac{\partial f_{e0}}{\partial\mathbf{p}_e}\cdot\left\{\frac{\mathbf{e}_1'[\mathbf{e}_2'\cdot\mathbf{e}_s']-\mathbf{e}_2'[\mathbf{e}_1'\cdot\mathbf{e}_s']}{\left(\Omega-\frac{c\mathbf{q}\cdot\mathbf{p_e}}{p_{0e}}\right)^2\left(\omega_2-\frac{c\mathbf{k}_2\cdot\mathbf{p_e}}{p_{0e}}\right)}\right.$$
$$\left.+\frac{\mathbf{e}_1'[\mathbf{e}_2'\cdot\mathbf{e}_s']-\mathbf{e}_s'[\mathbf{e}_1'\cdot\mathbf{e}_2']}{\left(\Omega-\frac{c\mathbf{q}\cdot\mathbf{p_e}}{p_{0e}}\right)\left(\omega_2-\frac{c\mathbf{k}_2\cdot\mathbf{p_e}}{p_{0e}}\right)^2}\right\}d^3\mathbf{p}_e-\frac{1}{\Omega\omega_1\omega_2}\int\frac{c}{p_{0e}^3}\frac{\partial f_{e0}}{\partial\mathbf{p}_e}\cdot\left\{\frac{\mathbf{e}_1'[\mathbf{p}_e\cdot\mathbf{e}_2'][\mathbf{p}_e\cdot\mathbf{e}_s']-\mathbf{e}_2'[\mathbf{p}_e\cdot\mathbf{e}_1'][\mathbf{p}_e\cdot\mathbf{e}_s']}{\left(\Omega-\frac{c\mathbf{q}\cdot\mathbf{p_e}}{p_{0e}}\right)^2\left(\omega_2-\frac{c\mathbf{k}_2\cdot\mathbf{p_e}}{p_{0e}}\right)}\right.$$
$$\left.+\frac{\mathbf{e}_1'[\mathbf{p}_e\cdot\mathbf{e}_2'][\mathbf{p}_e\cdot\mathbf{e}_s']-\mathbf{e}_s'[\mathbf{p}_e\cdot\mathbf{e}_1'][\mathbf{p}_e\cdot\mathbf{e}_2']}{\left(\Omega-\frac{c\mathbf{q}\cdot\mathbf{p_e}}{p_{0e}}\right)\left(\omega_2-\frac{c\mathbf{k}_2\cdot\mathbf{p_e}}{p_{0e}}\right)^2}\right\}d^3\mathbf{p}_e(A3). \tag{A3}$$

After tedious calculation, we can further deduce Eq. (A2) from Eq. (A3) as

$$
\begin{aligned}
\text{Eq. (A2)} = \frac{1}{\Omega^2\omega_1^2\omega_2^2}\int \Bigg\{ & \frac{\partial}{\partial \mathbf{p}_e}\cdot\left[\frac{c\mathbf{e}_1\cdot\mathbf{p}_e}{p_{0e}}(\mathbf{e}_2\mathbf{e}_s-\mathbf{e}_s\mathbf{e}_2)\cdot\frac{\partial}{\partial \mathbf{p}_e}\right] \\
& + \frac{\partial}{\partial \mathbf{p}_e}\left[\frac{1}{\left(\Omega-\frac{c\mathbf{q}\cdot\mathbf{p}_e}{p_{0e}}\right)\left(\omega_2-\frac{c\mathbf{k}_2\cdot\mathbf{p}_e}{p_{0e}}\right)}\frac{c^3(\mathbf{e}_1\cdot\mathbf{p}_e)(\mathbf{e}_2\cdot\mathbf{p}_e)(\mathbf{e}_s\cdot\mathbf{p}_e)}{p_{0e}^3}(\mathbf{k}_2\mathbf{q}-\mathbf{q}\mathbf{k}_2)\cdot\frac{\partial}{\partial \mathbf{p}_e}\right] \\
& + \frac{\partial}{\partial \mathbf{p}_e}\cdot\left[\frac{\Omega-\frac{c\mathbf{q}\cdot\mathbf{p}_\mathbf{e}}{p_{0e}}}{\omega_2-\frac{c\mathbf{k}_2\cdot\mathbf{p}_\mathbf{e}}{p_{0e}}}\frac{c\mathbf{e}_2\cdot\mathbf{p}_e}{p_{0e}^3}(\mathbf{e}_s\mathbf{e}_1-\mathbf{e}_1\mathbf{e}_s)\cdot\frac{\partial}{\partial \mathbf{p}_e}\right] + \frac{\partial}{\partial \mathbf{p}_e}\cdot\left[\frac{\omega_2-\frac{c\mathbf{k}_2\cdot\mathbf{p}_\mathbf{e}}{p_{0e}}}{\Omega-\frac{c\mathbf{q}\cdot\mathbf{p}_\mathbf{e}}{p_{0e}}}\frac{c\mathbf{e}_2\cdot\mathbf{p}_e}{p_{0e}^3}(\mathbf{e}_1\mathbf{e}_2-\mathbf{e}_2\mathbf{e}_1)\cdot\frac{\partial}{\partial \mathbf{p}_e}\right] \\
& + \frac{\partial}{\partial \mathbf{p}_e}\cdot\left[\frac{1}{\Omega-\frac{c\mathbf{q}\cdot\mathbf{p}_\mathbf{e}}{p_{0e}}}\frac{c^2(\mathbf{e}_s\cdot\mathbf{p}_e)(\mathbf{e}_2\cdot\mathbf{p}_e)}{p_{0e}^2}(\mathbf{e}_2\mathbf{k}_1-\mathbf{k}_1\mathbf{e}_2)\cdot\frac{\partial}{\partial \mathbf{p}_e}\right] + \frac{\partial}{\partial \mathbf{p}_e}\cdot\left[\frac{1}{\Omega-\frac{c\mathbf{q}\cdot\mathbf{p}_\mathbf{e}}{p_{0e}}}\frac{c^2(\mathbf{e}_s\cdot\mathbf{p}_e)(\mathbf{e}_2\cdot\mathbf{p}_e)}{p_{0e}^2}(\mathbf{e}_1\mathbf{k}_2 \right. \\
& \left. -\mathbf{k}_2\mathbf{e}_1)\cdot\frac{\partial}{\partial \mathbf{p}_e}\right] + \frac{\partial}{\partial \mathbf{p}_e}\left[\frac{1}{\omega_2-\frac{c\mathbf{k}_2\cdot\mathbf{p}_\mathbf{e}}{p_{0e}}}\frac{c^2(\mathbf{e}_1\cdot\mathbf{p}_e)(\mathbf{e}_2\cdot\mathbf{p}_e)}{p_{0e}^2}(\mathbf{e}_s\mathbf{k}_1-\mathbf{k}_1\mathbf{e}_s)\cdot\frac{\partial}{\partial \mathbf{p}_e}\right] \\
& + \frac{\partial}{\partial \mathbf{p}_e}\left[\frac{1}{\omega_2-\frac{c\mathbf{k}_2\cdot\mathbf{p}_\mathbf{e}}{p_{0e}}}\frac{c^2(\mathbf{e}_s\cdot\mathbf{p}_e)(\mathbf{e}_2\cdot\mathbf{p}_e)}{p_{0e}^2}(\mathbf{e}_s\mathbf{q}-\mathbf{q}\mathbf{e}_s)\cdot\frac{\partial}{\partial \mathbf{p}_e}\right]\Bigg\} d^3\mathbf{p}_e = 0. \qquad \text{(A4)}
\end{aligned}
$$

Then Eq. (11) has been proved. It should be noted that the isotropy of distribution function f_{e0} has been used in the above proof.

[1] M. Tabak, J. Hammer, M. E. Glinsky, W. L. Kruer, S. C. Wilks, J. Woodworth, E. M. Campbell, M. D. Perry, and R. J. Mason, Phys. Plasmas **1**, 1626 (1994).
[2] A. B. Borisov, A. V. Borovskiy, V. V. Korobkin *et al.*, Phys. Rev. Lett. **68**, 2309 (1992).
[3] S. C. Wilks, W. L. Kruer, M. Tabak, and A. B. Langdon, Phys. Rev. Lett. **69**, 1383 (1992).
[4] M. Honda, J. Meyer-ter-Vehn, and A. Pukhov, Phys. Plasmas **7**, 1302 (2000).
[5] R. Kodama, P. A. Norreys, K. Mima *et al.*, Nature (London) **412**, 798 (2001).
[6] M. H. Key, T. E. Cowan, B. A. Hammel *et al.*, *Proceedings of the 17th IAEA Fusion Energy Conference, Yokohama, 1998* (International Atomic Energy Agency, Vienna, to be published).
[7] A. Pukhov and J. Meyer-ter-Vehn, Phys. Plasmas **5**, 1880 (1998).
[8] M. Tanimoto, S. Kato, E. Miura, N. Saito, and K. Koyama, Phys. Rev. E **68**, 026401 (2003).
[9] P. Goldreich and W. H. Julian, Astrophys. J. **157**, 869 (1969).
[10] D. B. Melrose and M. E. Gedalin, Astrophys. J. **521**, 351 (1999).
[11] S. C. Wilks, Phys. Fluids B **B5**, 2603 (1993).
[12] Z. M. Sheng and J. Meyer-ter-Vehn, Phys. Rev. E **54**, 1833 (1996).
[13] L. Gorbunov, P. Mora, and T. M. Antonsen, Jr., Phys. Rev. Lett. **76**, 2495 (1996).
[14] V. I. Berezhiani, S. M. Mahajan, and N. L. Shatashvili, Phys. Rev. E **55**, 995 (1997).
[15] I. Yu. Kostyukov, G. Shvets, N. J. Fisch, and J. M. Rax, Phys. Plasmas **9**, 636 (2002).
[16] A. Kim, M. Tushentsov, D. Anderson, and M. Lisak, Phys. Rev. Lett. **89**, 095003 (2002).
[17] A. Pukhov, Z. M. Sheng, and J. Meyer-ter-Vehn, Phys. Plasmas **6**, 2847 (1999).
[18] X. T. He, Acta Phys. Sin. **32**, 325 (1983).
[19] S.-p. Zhu, X. T. He, and C. Y. Zheng, Phys. Plasmas **8**, 312 (2001).
[20] S.-p. Zhu, C. Y. Zheng, and X. T. He, Phys. Plasmas **10**, 4166 (2003).
[21] Z. Najmudin, M. Tatarakis, A. Pukhov *et al.*, Phys. Rev. Lett. **87**, 215004-1 (2001).
[22] J. Fuchs, G. Malka, J. C. Adam *et al.*, Phys. Rev. Lett. **80**, 1658 (1998).
[23] A. Pukhov and J. Meyer-ter-Vehn, Phys. Rev. Lett. **76**, 3975 (1996).
[24] A. Ting, K. Krushelnick, C. I. Moore, H. R. Burris, E. Esarey, J. Krall, and P. Sprangle, Phys. Rev. Lett. **77**, 5377 (1996); S. P. Le Blanc, M. C. Downer, R. Wagner, S.-Y. Chen, A. Maksimchuk, G. Mourou, and D. Umstadter, *ibid.* **77**, 5381 (1996); E. Esarey, B. Hafizi, R. Hubbard, and A. Ting, *ibid.* **80**, 5552 (1998).

PHYSICS OF PLASMAS **13**, 053106 (2006)

Fluid theory for quasistatic magnetic field generation in intense laser plasma interaction

Bin Qiao[a)]
Graduate School, China Academy of Engineering Physics, P.O. Box 2101, Beijing 100088, People's Republic of China

X. T. He
Institute of Applied Physics and Computational Mathematics, P.O. Box 8009, Beijing 100088, People's Republic of China and Department of Physics, Zhejiang University, Hangzhou 310027, People's Republic of China

Shao-ping Zhu
Institute of Applied Physics and Computational Mathematics, P. O. Box 8009, Beijing 100088, People's Republic of China

(Received 4 January 2006; accepted 6 April 2006; published online 12 May 2006)

Based on the ten-moment Grad system of hydrodynamic equations, a self-consistent fluid model is presented for the generation of quasistatic magnetic fields in relativistic intense laser plasma interaction. In this model, the nondiagonal stress tensor is taken into account and the generalized vorticity is proved to be not conserved, which are different from previous ideal fluid models. In the quasistatic approximation, where the low-frequency phase speed v_p is much smaller than the electron thermal speed v_{te}, the axial magnetic field B_z and the azimuthal one B_θ are derived. It is found that the condition $v_p \gg v_{te}$ used as the cold fluid approximation by previous papers is improper, where the derived B_z is incomplete and one magnetization current for B_z associated with the electron thermal motion does not appear. The profiles of both B_z and B_θ are analyzed. Their dependence on the laser intensity is discussed. © *2006 American Institute of Physics.* [DOI: 10.1063/1.2200298]

I. INTRODUCTION

Recent remarkable progress in the development of intense ($I\lambda^2 > 10^{18}$ W μm^2/cm^2) short pulse (<1 ps) lasers has stimulated worldwide interest in relativistic intense laser plasma interaction.[1] In this interaction, the generation of huge quasistatic magnetic fields is one of the most significant phenomena which has considerable influences on the whole nonlinear plasma dynamics. The study of this problem has wide applications in fast ignition scheme[2] for inertial confinement fusion, particle acceleration,[3] and laboratory astrophysics.[4] The experiments[5,6] report that magnetic fields about tens of mega-Gauss (MG), including the axial component B_z and the azimuthal one B_θ, are generated in underdense plasma. Although much effort[7–13] has been devoted to the quasistatic magnetic field generation, a self-consistent fluid theory still does not exist. It has been argued in Refs. 14 and 15 that only a kinetic treatment can give the correct expression for B_z and B_θ, and that all fluid attempts yield inconsistent and unsatisfactory results. So far, to our knowledge, most fluid models adopt the cold ideal fluid approximation and have not given the complete B_z and B_θ as the kinetic theory.[15–18] Especially, one of the magnetization currents associated with the electron thermal motion for B_z[15–18] does not appear in all the cold fluid models.

We believe that some of the following problems exist in the early publications with the fluid models. (1) In relativistic intense laser plasma interaction, there exist strong rotational motions (currents); momentum (velocity) exchanges between different fluid particles are important due to nonlinear effects and thus the nondiagonal stress tensor should be taken into account. Ideal fluid assumption in most studies is improper. (2) In such case, the generalized vorticity is not conserved, in contrast to previous ideal fluid models.[11,19] (3) For intense laser plasma interaction, the plasma is heated and the thermal motion of electrons is not negligible; thus, the condition $v_p \gg v_{te}$ (where v_p is the low-frequency phase speed and v_{te} the electron thermal speed) widely used as cold fluid approximation[11,19] is improper. The condition $v_p \ll v_{te}$ in the quasistatic approximation should be satisfied.

In this paper, we present a self-consistent fluid model for the generation of quasistatic magnetic fields in the interaction of intense laser with an initially uniform plasma, where the electron motions are described by the ten-moment Grad system of hydrodynamic equations.[20,21] Equations for both B_z and B_θ are derived simultaneously from such a self-consistent model in the quasistatic approximation $v_p \ll v_{te}$. It is shown that the condition $v_p \gg v_{te}$ widely used as cold fluid approximation is improper, under which the derived B_z is incomplete. We prove the generalized vorticity is not conserved, which is different from previous ideal fluid models.

II. DERIVATION OF QUASISTATIC MAGNETIC FIELDS

We consider the interaction of a short intense laser pulse with an initially uniform plasma. The laser pulse is assumed to be short with a time duration (T_l) less than the character-

[a)]Electronic mail: bin_qiao1979@yahoo.com

istic time for the ion response ω_i^{-1} (ω_i^{-1} is the ion Langmuir frequency), so that the ion motion can be neglected. Meanwhile, the pulse is also assumed sufficiently long as $T_l \gg \omega_{pe}^{-1}$ (ω_{pe} is the electron Langmuir frequency), here usually satisfying $0.1\ \mathrm{ps} < T_l < 1\ \mathrm{ps}$, so that the wake-field effects (the excitation of Langmuir waves) are negligible and the quasistatic magnetic field is mainly generated within the laser pulse. The plasma is assumed collisionless and the electron temperature gradient is assumed negligible, that is, the electron heat conduction is negligible. In this case, the electron motion obeys the ten-moment Grad system of hydrodynamic equations,[20,21]

$$\frac{\partial \mathbf{p_i}}{\partial t} + \mathbf{v}\cdot\nabla\mathbf{p_i} = -e\left[\mathbf{E} + \frac{1}{c}(\mathbf{v}\times\mathbf{B})\right]_i + \frac{\partial \mathbf{\Pi}_{ij}}{\partial r_j}, \qquad (1)$$

$$\frac{\partial \mathbf{\Pi}_{ij}}{\partial t} + \frac{\partial}{\partial r_s}(\mathbf{v}_s\mathbf{\Pi}_{ij}) + \mathbf{\Pi}_{is}\frac{\partial \mathbf{v}_j}{\partial r_s} + \mathbf{\Pi}_{js}\frac{\partial \mathbf{v}_i}{\partial r_s} - \frac{2}{3}\delta_{ij}\mathbf{\Pi}_{sl}\frac{\partial \mathbf{v}_l}{\partial r_s} + \frac{e}{mc}\mathbf{B}_l(e_{isl}\mathbf{\Pi}_{js} + e_{jsl}\mathbf{\Pi}_{is}) = v_{te}^2\left(\frac{\partial \mathbf{p}_i}{\partial r_j} + \frac{\partial \mathbf{p}_j}{\partial r_i} - \frac{2}{3}\delta_{ij}\nabla\cdot\mathbf{p}\right), \qquad (2)$$

which is a closed system of equations derived following the Boltzmann equation. The ten moments are composed of five dynamical moments including density n, momentum $\mathbf{p}_i$, electron temperature T_e, and five independent components of the stress tensor including the diagonal part P_{ij} and the nondiagonal part $n_0\mathbf{\Pi}_{ij}$. The electric field $\mathbf{E}$ and magnetic field $\mathbf{B}$ are governed by the Maxwell equations as

$$\nabla\times\mathbf{E} = -\frac{1}{c}\frac{\partial \mathbf{B}}{\partial t}, \qquad (3)$$

$$\nabla\times\mathbf{B} = \frac{1}{c}\frac{\partial \mathbf{E}}{\partial t} + \frac{4\pi}{c}\mathbf{j} = \frac{1}{c}\frac{\partial \mathbf{E}}{\partial t} - \frac{4\pi e}{c}n\mathbf{v}, \qquad (4)$$

$$\nabla\cdot\mathbf{E} = 4\pi(n_0 - n)e. \qquad (5)$$

Here, $-e$, m, n, and T_e represent the charge, the mass, the density, and the temperature of the electrons, respectively, n_0 is initial plasma density and $v_{te}=\sqrt{T_e/m}$ is the electron thermal velocity, $\mathbf{p}=\gamma m\mathbf{v}$ is the electron momentum, and $\gamma=(1+\mathbf{p}^2/m^2c^2)^{1/2}$ is the relativistic factor, c is the speed of light, and e_{ist} is the unit antisymmetric tensor. The nondiagonal stress tensor density in Eq. (1) is written as $\mathbf{\Pi}_{ij}=\sigma_{ij}/n_0$, whose evolution is governed by Eq. (2), where the ten-moment approximations have been adopted. It is worth noticing that Eqs. (1) and (2) can be derived from the standard relativistic hydrodynamics of Landau-Lifshitz[22] or Pomraning,[23] and that Eq. (2) is indeed the ten-moment approximate equilibrium equation for the stress tensor of Balescu[24] in the relativistic regime. The right-hand side of Eq. (2) is indeed the linear viscous (nondiagonal) stress tensor of the lowest order in the relativistic regime as $\sigma_{ij}=v_{te}^2(\partial\mathbf{p}_i/\partial r_j+\partial\mathbf{p}_j/\partial r_i-2\delta_{ij}\nabla\cdot\mathbf{p}/3)$. Comparing with the theory of Landau-Lifshitz,[22] the viscosity coefficient η here should be approximately equal to γv_{te}^2, i.e., $\eta=\gamma v_{te}^2 \propto \gamma T_e$. For intense laser, the heat pressure is much smaller than the light pressure of laser; thus, the diagonal (heat) stress tensor $P_{ij}=\delta_{ij}\gamma n_0 T_e$ is also ignored in Eq. (1). Besides, we assume the temperature to be homogeneous, because the electromagnetic fields are transverse and the plasma is supposed to be initially homogeneous. This implies that the electron fluid is incompressible, and that the general equation (2) thus can be much simplified due to $\nabla\cdot\mathbf{p}=0$.

Introducing the generalized vorticity $\mathbf{\Omega}\equiv\nabla\times\mathbf{p}-e\mathbf{B}/c$ and defining the stress force as $\mathbf{F}_i=\partial\mathbf{\Pi}_{ij}/\partial r_j$, applying the curl operator on Eq. (1) and using Eq. (3), we obtain that

$$\frac{\partial \mathbf{\Omega}}{\partial t} = \frac{1}{m\gamma}\nabla\times(\mathbf{p}\times\mathbf{\Omega}) + \nabla\times\mathbf{F}. \qquad (6)$$

Here and below, the subscript i is not and will be not written for simplification.

Now, we divide both electron motion and fields into slow- and fast-time-scale parts as $\mathbf{p}=\langle\mathbf{p}\rangle+\tilde{\mathbf{p}}\exp(-i\omega_0 t)/2+\tilde{\mathbf{p}}^*\exp(i\omega_0 t)/2$, $\mathbf{E}=\langle\mathbf{E}\rangle+\tilde{\mathbf{E}}\exp(-i\omega_0 t)/2+\tilde{\mathbf{E}}^*\exp(i\omega_0 t)/2$, and $\mathbf{B}=\langle\mathbf{B}\rangle+\tilde{\mathbf{B}}\exp(-i\omega_0 t)/2+\tilde{\mathbf{B}}^*\exp(i\omega_0 t)/2$, where the angular bracket $\langle\rangle$ denotes averaging over the fast time period $(2\pi/\omega_0)$ and $*$ denotes the complex conjugate for the fast-time-scale parts. Note here $\langle\mathbf{B}\rangle$ just represents the quasistatic magnetic field, and $\tilde{\mathbf{E}}$ represents the laser electric field. Similarly, we split Ω, $\mathbf{\Pi}_{ij}$, and $\mathbf{F}$. Using the relation $\nabla\cdot\mathbf{p}=0$ for the incompressible fluid as analyzed above, from Eq. (2), we easily get the fast-time-scale $\tilde{\mathbf{F}}$ in the lowest order as $\partial\tilde{\mathbf{F}}/\partial t=v_{te}^2\nabla^2\tilde{\mathbf{p}}$ and $\tilde{\mathbf{F}}=iv_{te}^2\nabla^2\tilde{\mathbf{p}}/\omega_0$. Then, from Eqs. (1)–(6), using $\nabla\cdot\mathbf{p}=0$ and $\tilde{\mathbf{F}}=iv_{te}^2\nabla^2\tilde{\mathbf{p}}/\omega_0$, after tedious algebra, a set of equations for the slow-time-scale physical quantities is obtained as

$$\frac{\partial\langle\mathbf{p}\rangle}{\partial t} = -e\langle\mathbf{E}\rangle - \frac{1}{4m\gamma}\nabla\langle\tilde{\mathbf{p}}\cdot\tilde{\mathbf{p}}^*\rangle + \frac{1}{4m\gamma}\langle\tilde{\mathbf{p}}\times\tilde{\mathbf{\Omega}}^* + \tilde{\mathbf{p}}^*\times\tilde{\mathbf{\Omega}}\rangle + \langle\mathbf{F}\rangle, \qquad (7)$$

$$\frac{\partial\langle\mathbf{F}\rangle}{\partial t} = v_{te}^2\nabla^2\langle\mathbf{p}\rangle - \frac{iv_{te}^2}{4\omega_0 m\gamma}\nabla^2\nabla\times\langle\tilde{\mathbf{p}}\times\tilde{\mathbf{p}}^*\rangle, \qquad (8)$$

$$\nabla\times\langle\mathbf{B}\rangle = \frac{1}{c}\frac{\partial\langle\mathbf{E}\rangle}{\partial t} - \frac{4\pi e}{m\gamma c}\langle n\rangle\langle\mathbf{p}\rangle - \frac{\pi e}{m\gamma c}\langle\tilde{n}\tilde{\mathbf{p}}^* + \tilde{n}^*\tilde{\mathbf{p}}\rangle, \qquad (9)$$

$$\frac{\partial\langle\mathbf{\Omega}\rangle}{\partial t} = \frac{1}{4m\gamma}\nabla\times\langle\tilde{\mathbf{p}}\times\tilde{\mathbf{\Omega}}^* + \tilde{\mathbf{p}}^*\times\tilde{\mathbf{\Omega}}\rangle + \nabla\times\langle\mathbf{F}\rangle, \qquad (10)$$

where $\langle\gamma\rangle=\gamma$ is assumed. From Eq. (10), we can see the generalized vorticity $\langle\mathbf{\Omega}\rangle$ is not conserved and is generated by the nondiagonal stress force due to nonzero electron thermal motion in plasmas. For the previous ideal cold fluid model $v_{te}\nabla \ll \partial/\partial t$, the right-hand side of Eq. (8) can be ignored as zero; thus, $\mathbf{F}=0$, and the vorticity $\langle\mathbf{\Omega}\rangle$ is conserved.

Substituting $\langle\mathbf{E}\rangle$ from Eq. (7), $\partial\langle\mathbf{F}\rangle/\partial t$ from Eq. (8), and $\mathbf{B}=c(\nabla\times\mathbf{p}-\mathbf{\Omega})/e$ into Eq. (9), by use of Eq. (10) and keeping only the lowest nonlinear terms, we obtain the coupling equation for $\langle\mathbf{p}\rangle$ in second order as

$$c^2 \nabla \times \nabla \times \langle \mathbf{p} \rangle + \frac{\partial^2 \langle \mathbf{p} \rangle}{\partial t^2} + \frac{1}{4m\gamma}\frac{\partial}{\partial t} \nabla \langle \mathbf{p} \cdot \mathbf{p}^* \rangle + \omega_{pe}^2 \frac{\langle n \rangle}{n_0}\frac{\langle \mathbf{p} \rangle}{\gamma} - v_{te}^2 \nabla^2 \langle \mathbf{p} \rangle - c^2 \nabla \times \langle \mathbf{\Omega} \rangle = -\frac{\omega_{pe}^2}{4n_0\gamma}\langle \tilde{n}\tilde{\mathbf{p}}^* + \tilde{n}^*\tilde{\mathbf{p}} \rangle - i\frac{v_{te}^2}{4m\omega_0\gamma}\nabla^2 \nabla \times \langle \tilde{\mathbf{p}} \times \tilde{\mathbf{p}}^* \rangle, \tag{11}$$

where $\omega_{pe}=\sqrt{4\pi n_0 e^2/m}$. Note that Eq. (11) includes the averaged generalized vorticity $\langle \mathbf{\Omega} \rangle$.

Similarly, to the lowest order, for $\omega_0 \gg v_{te}\nabla$ we easily get the fast-time-scale quantities, respectively, as $\tilde{\mathbf{p}}=-ie\tilde{\mathbf{E}}/\omega_0$, $\tilde{\mathbf{F}}=iv_{te}^2\nabla^2\tilde{\mathbf{p}}/\omega_0=ev_{te}^2\nabla^2\tilde{\mathbf{E}}/\omega_0^2$, and $\tilde{\mathbf{\Omega}}=\nabla\times\tilde{\mathbf{p}}-e\tilde{\mathbf{B}}/c=-ie\nabla\times\tilde{\mathbf{E}}/\omega_0+ie\nabla\times\tilde{\mathbf{E}}/\omega_0=0$. Using the continuity equation $\partial\tilde{n}/\partial t+\nabla(\langle n\rangle\tilde{\mathbf{p}}/m/\gamma)=0$, we get that

$$\langle \tilde{n}\tilde{\mathbf{p}}^* + \tilde{n}^*\tilde{\mathbf{p}} \rangle = i\frac{e^2}{m\omega_0^3}\tilde{\mathbf{E}}\,\nabla \cdot \frac{\langle n \rangle \tilde{\mathbf{E}}^*}{\gamma} + \text{c.c.} \tag{12}$$

Making the second differential for Eq. (11) related to t, getting $\partial^2\langle\mathbf{\Omega}\rangle/\partial t^2$ from Eqs. (8) and (10) and then substituting $\partial^2\langle\mathbf{\Omega}\rangle/\partial t^2$ and Eq. (12) into it, we obtain the coupling equation for $\langle \mathbf{p} \rangle$ as

$$\frac{\partial^2}{\partial t^2}\left[c^2 \nabla \times \nabla \times \langle \mathbf{p} \rangle + \frac{\partial^2 \langle \mathbf{p} \rangle}{\partial t^2} + \omega_{pe}^2 \frac{\langle n \rangle}{n_0}\frac{\langle \mathbf{p} \rangle}{\gamma} + \frac{e^2}{4m\omega_0^2\gamma} \times \frac{\partial}{\partial t}\nabla \langle \mathbf{E}\cdot\mathbf{E}^* \rangle + \frac{\omega_{pe}^2 e^2}{4mn_0\omega_0^3}\left(\frac{i\tilde{\mathbf{E}}}{\gamma}\nabla \cdot \frac{\langle n \rangle \tilde{\mathbf{E}}^*}{\gamma} + \text{c.c.}\right)\right] - \frac{\partial^2}{\partial t^2}\left[v_{te}^2\nabla^2\langle \mathbf{p} \rangle - i\frac{v_{te}^2 e^2}{4m\omega_0^3\gamma}\nabla^2 \nabla \times \langle \tilde{\mathbf{E}} \times \tilde{\mathbf{E}}^* \rangle\right] = v_{te}^2\nabla^2\left[c^2 \nabla \times \nabla \times \left(\langle \mathbf{p} \rangle - \frac{ie^2}{4m\omega_0^3\gamma}\nabla \times \langle \tilde{\mathbf{E}} \times \tilde{\mathbf{E}}^* \rangle\right)\right]. \tag{13}$$

One can see that the relation between the electron thermal speed v_{te} and the characteristic low-frequency phase speed $v_p=\omega_s/q$ is a key factor to decide which terms are dominant in Eq. (13), where $\partial/\partial t \sim \omega_s$ and $v_{te}\nabla \sim v_{te}q$, i.e., ω_s and q are the characteristic frequency and wave number of the slowly varying variables, respectively.

In intense laser plasma interaction, plasma electrons are heated and the thermal motion of electrons is not negligible, that is, $v_{te}=\sqrt{T_e/m}$ cannot be regarded as zero. Thus, the previous widely used condition $v_p \gg v_{te}$ for cold fluid approximation[11,19] is not satisfied anymore. The quasistatic approximation should be considered. Note here the low-frequency motions and fields are generated by the nonlinear beating interaction of two high-frequency fields or motions. The low-frequency ω_s is indeed a difference of two high-frequency (ω_1, ω_2) as $\omega_s=\omega_1-\omega_2$. In quasistatic approximation, it is reasonable to assume ω_2 is very close to ω_1 and both of them are close to ω_0. Thus, ω_s is much smaller than the characteristic oscillating frequency of plasma electrons ω_{pe}, which is determined by v_{te} as $\omega_{pe}=v_{te}/\lambda_{\text{De}}$ (where λ_{De} is electron Debye length) and is in the same scale of ω_0. When the spatial scale q^{-1} of low-frequency field is comparable to λ_{De}, the quasistatic approximate condition $v_p \ll v_{te}$ is always satisfied in warm plasma interacting with intense laser. Usually, the quasistatic approximation is in an ion-acoustic oscillation time scale such that $\omega_s=qv_s=q\sqrt{T_e/m_i}$ (where m_i is ion mass), and thus $v_p=v_s \ll v_{te}$ is always satisfied.

Under the condition $v_p \ll v_{te}$ for the quasistatic approximation, i.e., $\partial/\partial t \ll v_{te}\nabla$, the right-hand side of Eq. (13) is dominant; the left-hand side can be ignored as zero. Thus, Eq. (13) can be simplified and we get $\langle \mathbf{p} \rangle$ as

$$\langle \mathbf{p} \rangle = \frac{ie^2}{4m\omega_0^3\gamma}\nabla \times \langle \tilde{\mathbf{E}} \times \tilde{\mathbf{E}}^* \rangle. \tag{14}$$

Substituting Eqs. (12) and (14) into Eq. (9), ignoring $\partial\langle\mathbf{E}\rangle/\partial t$ due to $\partial/\partial t \ll c\nabla$, we get the quasistatic magnetic field as

$$\nabla \times \langle \mathbf{B} \rangle = -i\frac{\omega_{pe}^2 e}{4m\omega_0^3 c\gamma^2}\frac{\langle n \rangle}{n_0}\nabla \times \langle \tilde{\mathbf{E}} \times \tilde{\mathbf{E}}^* \rangle - \frac{\omega_{pe}^2 e}{4m\omega_0^3 c}\left(\frac{i\tilde{\mathbf{E}}}{\gamma}\nabla \cdot \frac{\langle n \rangle \tilde{\mathbf{E}}^*}{n_0\gamma} + \text{c.c.}\right) \tag{15}$$

Substituting Eq. (14) into Eq. (8) and making it divergent, we get $\partial\nabla\cdot\mathbf{F}/\partial t=0$, i.e., $\nabla\cdot\mathbf{F}=0$. Thus, From Eqs. (5) and (7), we get the slow-time-scale plasma density $\langle n \rangle/n_0$ as

$$\frac{\langle n \rangle}{n_0} \simeq 1 + \frac{e^2}{4m^2\omega_{pe}^2\omega_0^2\gamma}\nabla^2\langle \tilde{\mathbf{E}} \cdot \tilde{\mathbf{E}}^* \rangle. \tag{16}$$

Equations (15) and (16) together give the axial quasistatic magnetic field and its two driven sources. One is the rotational current as the result of the beating interaction of high-frequency electromagnetic waves (laser wave fields) $\langle \tilde{\mathbf{E}} \times \tilde{\mathbf{E}}^* \rangle$, another is the rotational current as the result of the coupling interaction between the high-frequency electron quiver velocity and the high-frequency electron density perturbation. Such two sources are simultaneously given with a fluid model. Note that, although it seems $\langle \mathbf{B} \rangle$ is independent of the temperature if only seen from the expressions [Eq. (15)], in fact $\langle \mathbf{B} \rangle$, especially the first driven source of it $\langle \mathbf{E} \times \mathbf{E}^* \rangle$, is just associated with the thermal motion of electrons and is generated due to the thermal effect v_{te}. In previous ideal fluid models,[11,19] the nondiagonal stress tensor is not considered and the cold fluid approximation $v_p \gg v_{te}$ is improperly used, so the derived axial magnetic field is incomplete in the absence of the first source. Not having taken the thermal effect into account is exactly the reason why previous cold fluid models cannot give the complete expression of $\langle \mathbf{B} \rangle$.

The generalized vorticity can be obtained from Eqs. (14) and (15) as

$$\nabla\times\langle\boldsymbol{\Omega}\rangle=\frac{e^2}{4m\omega_0^3\gamma}\Bigg[i\frac{\omega_{pe}^2}{c^2\gamma}\frac{\langle n\rangle}{n_0}\nabla\times\langle\tilde{\mathbf{E}}\times\tilde{\mathbf{E}}^*\rangle+\frac{\omega_{pe}^2}{c^2}\left(i\tilde{\mathbf{E}}\nabla\cdot\frac{\langle n\rangle\tilde{\mathbf{E}}^*}{n_0\gamma}+\text{c.c.}\right)-i\nabla^2\nabla\times\langle\tilde{\mathbf{E}}\times\tilde{\mathbf{E}}^*\rangle\Bigg]. \tag{17}$$

Equation (17) gives the generated vorticity $\langle\boldsymbol{\Omega}\rangle$ induced by high-frequency fields. This proves $\langle\boldsymbol{\Omega}\rangle$ is not conserved when the nondiagonal stress tensor is considered, which is different from previous ideal fluid models.[11,19]

In previous fluid studies, the condition $v_p \gg v_{te}$ is used as a cold fluid approximation. For $v_p \gg v_{te}$, i.e., $\partial/\partial t \gg v_{te}\nabla$, we get similarly from Eq. (13) that $\partial^2\langle\mathbf{p}\rangle/\partial t^2+c^2\nabla\times\nabla\times\langle\mathbf{p}\rangle+\omega_{pe}^2\langle n\rangle\langle\mathbf{p}\rangle/n_0/\gamma=-\omega_{pe}^2e^2[(i\tilde{\mathbf{E}}/\gamma)\nabla(\langle n\rangle\tilde{\mathbf{E}}^*/n_0/\gamma)+\text{c.c.}]/(4m\omega_0^3)-e^2/4m\omega_0^2\gamma\partial/\partial t\nabla\langle\mathbf{E}\cdot\mathbf{E}^*\rangle$ and $\nabla\times\langle\mathbf{B}\rangle=c\nabla\times\nabla\times\langle\mathbf{p}\rangle/e+(e/4m\omega_0^2c\gamma)\partial/\partial t\nabla\langle\mathbf{E}\cdot\mathbf{E}^*\rangle$. The axial part of $\langle\mathbf{B}\rangle$ here is consistent with that obtained in Ref. 11 [see Eqs. (11) and (12) in Ref. 11]. However, as can be seen, the axial component of $\langle\mathbf{B}\rangle$ given here and in Ref. 11 is incomplete, lacking the term driven by the first source $\langle\tilde{\mathbf{E}}\times\tilde{\mathbf{E}}^*\rangle$ of Eq. (15). This proves that this term is associated with the thermal motion of electrons, and that the condition $v_p \gg v_{te}$ used as the cold fluid approximation in previous models is improper.

III. AXIAL MAGNETIC FIELD FOR A FOCUSED LASER BEAM

Now, we consider the interaction of an intense focused laser beam with underdense plasma for simplicity. As mentioned in the Introduction, the laser pulse is assumed sufficiently long with time duration $T_l \gg \omega_{pe}^{-1}$, usually satisfying $0.1\ \text{ps} < T_l < 1\ \text{ps}$, so that the wake-field effects (the excitation of Langmuir waves) are negligible and the quasistatic magnetic field is mainly generated within the laser pulse. Assuming laser beam propagates along the z-axis direction with wave number $k_0=2\pi/\lambda_0$, under the Coulomb gauge, the electric fields are usually expressed as $\mathbf{E}_L(\mathbf{x},t)=1/2E_0(r,t)\times(\mathbf{e}_r+i\alpha\mathbf{e}_\theta)e^{i\alpha\theta}e^{ik_0z-i\omega_0t}+\text{c.c.}$, where $\alpha=0,\pm1$ for the linearly polarized (LP) and the circularly polarized (CP) laser, respectively. We assume the laser intensity has a transverse Gaussian distribution as $I=I_0e^{-r^2/r_0^2}$ and $E_0=\sqrt{I}$, where r_0 the transverse radius. Thus ,the above high-frequency electric field can be expressed as $\tilde{\mathbf{E}}=E_0(r,t)(\mathbf{e}_r+i\alpha\mathbf{e}_\theta)e^{i\alpha\theta}e^{ik_0z}$ and $\tilde{\mathbf{E}}^*=E_0(r,t)(\mathbf{e}_r-i\alpha\mathbf{e}_\theta)e^{-i\alpha\theta}e^{-ik_0z}$. Then, we approximately get the relativistic coefficient $\gamma=[1+(1+\alpha^2)e^2E_0^2/(2m^2c^2\omega^2)]^{1/2}=[1+(1+\alpha^2)I_0\lambda_{\mu m}^2e^{-r^2/r_0^2}/2.76]^{1/2}$. Introducing the normalized transform $t\rightarrow t/\omega_0, \mathbf{x}\rightarrow c\mathbf{x}/\omega_0, E\rightarrow m_ec\omega_0E/e, B\rightarrow m_ec\omega_0B/e$, and using the symbol B_z to represent $\langle B_z\rangle$ for simplicity, we can get the normalized axial magnetic field B_z from Eqs. (15) and (16) as

$$\frac{\partial B_z}{\partial r}=-\frac{\alpha}{1+\alpha^2}n_0\frac{1}{\gamma}\frac{\partial}{\partial r}\left(\frac{\langle n\rangle}{n_0}\frac{\gamma^2-1}{\gamma}\right), \tag{18}$$

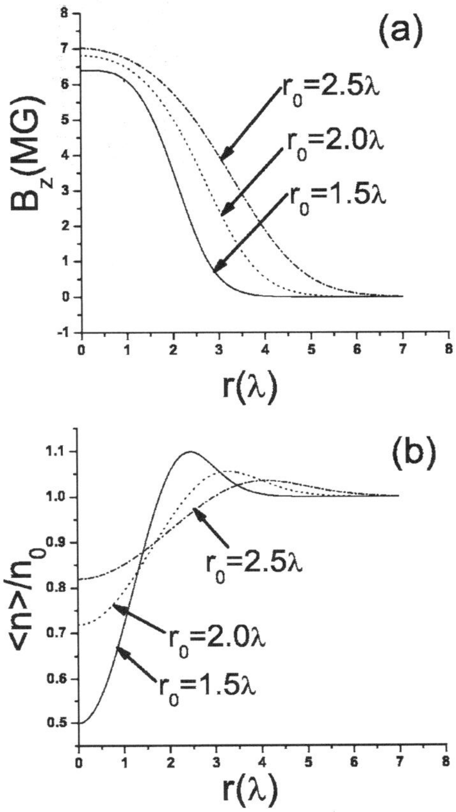

FIG. 1. The transverse profiles of (a) axial magnetic field B_z and (b) plasma density cavitation $\langle n\rangle/n_0$ calculated by Eqs. (18) and (19), for a Gaussian laser beam with intensity $I_0=10^{19}\ \text{W/cm}^2$, wavelength $\lambda=1.05\ \mu\text{m}$, transverse radius $r_0=1.5\lambda, 2\lambda, 2.5\lambda$, and pulse duration 0.6 ps. The initial plasma density is $n_0=0.1n_{cr}$.

$$\frac{\langle n\rangle}{n_0}=1-\frac{1}{n_0r_0^2}\frac{\gamma^2-1}{\gamma}\left(2-\frac{r^2}{r_0^2}\frac{\gamma^2+1}{\gamma^2}\right). \tag{19}$$

Equation (18) shows B_z is only generated by CP laser ($\alpha=\pm1$). For LP laser ($\alpha=0$), B_z is zero.

Figure 1(a) plots the transverse profile of $|B_z|$ calculated by Eqs. (18) and (19), where laser intensity is $I_0=10^{19}\ \text{W/cm}^2$ with wavelength $\lambda=1.05\ \mu\text{m}$; transverse focused radius are, respectively, $r_0=1.5\lambda, 2\lambda, 2.5\lambda$. The laser pulse duration is assumed to be 0.6 ps. Plasma density is $n_0=0.1n_{cr}$. It can be seen that B_z dominates the "inner" zone near the laser axis and the peak $|B_z|$ is about 6.41–7.02 MG at $r=0$ for $r_0=1.5$–2.5λ. The B_z profile is consistent with the experiment result[5] and the simulation result.[18] A recent experiment[5] reported that about 2 MG B_z is generated by CP laser with incident intensity $I_0=6.7\times10^{18}\ \text{W/cm}^2$, transverse radius $r_0=10\lambda$ ($\lambda=1.05\ \mu\text{m}$), and laser pulse duration 0.9–1.2 ps, where plasma density $n_0=2.8\times10^{19}\ \text{cm}^{-3}$. Using Eqs. (18) and (19), $|B_z|$ of about 2.17 MG is estimated for the same parameters, where we choose I_0 to be two times the incident laser intensity and r_0 to be 5λ due to self-focusing, according to the experimental results. The result is very close to that in the experiment.[5] We also calculate the $|B_z|$ given in

Ref. 11 with the same parameters as above. The results show the $|B_z|$ of Ref. 11 is quite underestimated as only 0.75–1.98 MG for the parameters of Fig. 1(a) and 0.25 MG for those of the experiment.[5] This further proves the cold fluid approximation $v_p \gg v_{te}$ is improper, where B_z is incomplete and quite underestimated.

The transverse profile of plasma density $\langle n\rangle / n_0$ by Eq. (19) is plotted in Fig. 1(b) for the same parameters as Fig. 1(a). One can see clearly that electrons in the region $r < r_0$ are expelled by the ponderomotive force so that $\langle n\rangle / n_0 < 1$; the expelled electrons are deposited at a cavity edgy of $r \geqslant r_0$, where $\langle n\rangle / n_0 > 1$ is satisfied; then, when $r \gg r_0$, $\langle n\rangle / n_0 = 1$ is taken again. It can also be seen that when the laser beam transverse radius r_0 is decreased, more electrons in the region $r < r_0$ are expelled; thus, $\langle n\rangle / n_0$ in this region is decreased and B_z is decreased, too. It illustrates that electrons near the axis are easier to be expelled for the smaller transverse radius of laser beam. That is, the laser beam with smaller transverse width is more beneficial for the formation of density cavitation channel.

The experiment results[5] show that B_z increases with laser intensity I_0. But, few fluid models can give the analysis for the dependence of the peak B_z on the laser intensity. We calculate and plot the dependence of the peak $|B_z|$ and $\langle n\rangle / n_0$ on the laser intensity I_0 in Fig. 2(a) and 2(b), where the parameters are the same as those in Fig. 1(a). It can be seen that the peak $|B_z|$ increases with the laser intensity I_0, while the plasma density near the axis $(\langle n\rangle / n_0)_{r=0}$ decreases as I_0 increases.

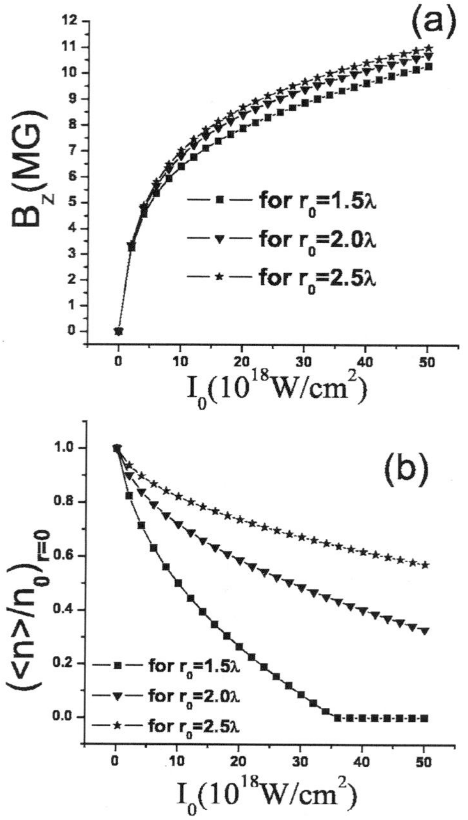

FIG. 2. The dependence of the peak (a) B_z and (b) $\langle n\rangle / n_0$ at the axis ($r=0$) on the laser intensity I_0 for different transverse radii $r_0 = 1.5\lambda, 2\lambda, 2.5\lambda$. Other parameters are the same as Fig. 1.

IV. AZIMUTHAL MAGNETIC FIELD FOR A FOCUSED LASER BEAM

For the azimuthal magnetic field B_θ, the coupling equation of the longitudinal momentum $\langle p_z\rangle$ should be studied. Here, we consider the same Gaussian laser beam propagating in an underdense plasma as Sec. III. The laser pulse is also assumed sufficiently long with duration $T_l \gg \omega_{pe}^{-1}$, here usually satisfying 0.1 ps $< T_l <$ 1 ps, so that the wake-field effects are negligible and the quasistatic magnetic field is generated dominantly within the laser pulse, which is consistent with Sec. III.

From Eq. (13), under the conditions either $v_p \ll v_{te}$ or $v_p \gg v_{te}$, and assuming $\partial\langle p_r\rangle / \partial z \ll \partial\langle p_z\rangle / \partial r$, we have the normalized coupling equation for $\langle p_z\rangle$,

$$\frac{\partial^2\langle p_z\rangle}{\partial\xi^2} - \frac{\partial^2\langle p_z\rangle}{\partial r^2} - \frac{1}{r}\frac{\partial\langle p_z\rangle}{\partial r} + n_0\frac{\langle n\rangle}{n_0\gamma}\langle p_z\rangle = \frac{1}{4\gamma}\frac{\partial^2\langle \tilde{\mathbf{E}}\cdot\tilde{\mathbf{E}}^*\rangle}{\partial\xi^2} = \frac{\partial^2\gamma}{\partial\xi^2}, \tag{20}$$

where $\xi = z - v_g t$ is approximately taken for underdense plasma. Notice the azimuthal magnetic field B_θ is mainly produced by the longitudinal current, which dominates near the axis. That is to say, B_θ is mainly decided by the longitudinal momentum p_z near the axis. Thus, it is reasonable to ignore $\partial\langle p_z\rangle / \partial r$ and $n_0\langle n\rangle\langle p_z\rangle / n_0\gamma$ in Eq. (20), both of which are close to zero near the axis. Then, Eq. (20) can be simplified as $\partial^2\langle p_z\rangle / \partial\xi^2 = \partial^2\gamma / \partial\xi^2$. Assuming initially $\gamma = 1$ when $\langle p_z\rangle = 0$, we thus obtain that $\langle p_z\rangle = \gamma - 1$.

Here, we mainly consider B_θ generated within the laser pulse ignoring the wake-field effect; thus, $\partial\langle \mathbf{E}\rangle / \partial t$ can be ignored. Using B_θ to represent $\langle B_\theta\rangle$ for simplicity, from Eq. (9), we easily get the normalized azimuthal magnetic field B_θ as

$$\frac{1}{r}\frac{\partial}{\partial r}(rB_\theta) = -n_0\frac{\langle n\rangle}{n_0}\frac{\gamma - 1}{\gamma}. \tag{21}$$

Equation (21) shows that both CP laser ($\alpha = \pm 1$) and LP laser ($\alpha = 0$) can generate the azimuthal magnetic field B_θ. B_θ is generated by the longitudinal current driven by the ponderomotive force.

Note here B_θ is proportional to the laser intensity I_0, while in Refs. 25 and 26, B_θ is proportional to I_0^2. This discrepancy can be explained. In Refs. 25 and 26, B_θ is generated by the wake field ϕ [see Eq. (6) in Ref. 25 and Eqs. (12), (26), and (27) in Ref. 26], which is already a subfield excited by laser field E_L as $\phi \sim |E_L|^2 \sim I_0$ and always exists behind the laser pulse. Accordingly, B_θ in Refs. 25 and 26 is in fourth-order nonlinearity and is proportional to the square of laser intensity as $B_\theta \sim |\phi|^2 \sim |E_L|^4 \sim I_0^2$ or $B_\theta \sim |E_L|^2 |\phi|$

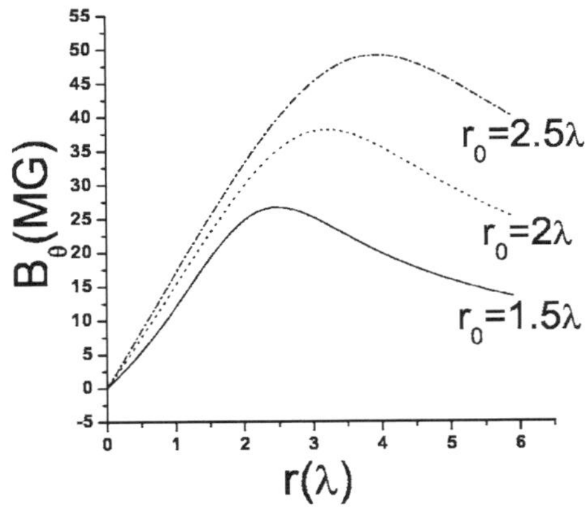

FIG. 3. The transverse profiles of azimuthal magnetic field B_θ calculated by Eq. (21), where the parameters are the same as Fig. 1.

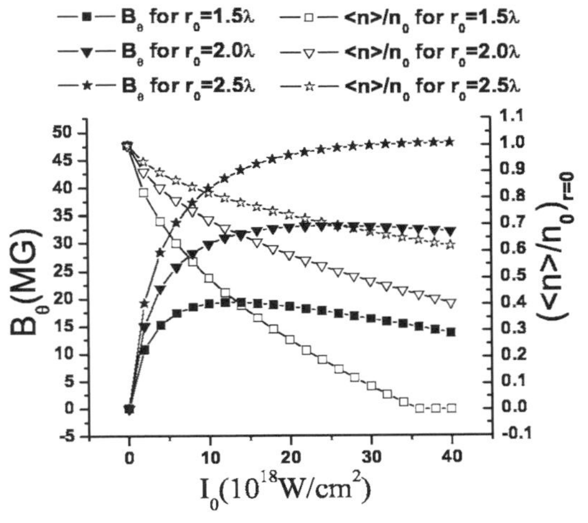

FIG. 4. The dependence of the peak B_θ at $r=r_0$ (the line with the solid symbol "■, ▼, ★") and $\langle n\rangle/n_0$ at the axis ($r=0$) (the line with the corresponding hollow symbol up) on the laser intensity I_0, where the laser beam transverse radii are $r_0=1.5\lambda, 2\lambda, 2.5\lambda$, respectively, and other parameters are the same as Fig. 1.

$\sim|E_L|^4\sim I_0^2$. However, B_θ here is generated directly by the laser field within the laser pulse; thus, it is in second order and proportional to I_0 as $B_\theta\sim|E_L|^2\sim I_0$. We only keep up to second order, ignoring the fourth-order terms for the effects of wake field. As a subfield to the laser field, the wake field E_w is much smaller than the laser field for underdense plasma as $E_w\sim 0.1-0.3mc\omega_p/e=0.1-0.3(n_0/n_{cr})^{1/2}E_L\ll E_L$ within the laser pulse, which has been explained in Ref. 18.

The transverse profile of $|B_\theta|$ calculated by Eq. (21) is plotted in Fig. 3, where the parameters are consistent with Fig. 1. It can be seen that $|B_\theta|$ dominates the "outer" zone away from the laser axis, is close to zero near the axis, increases gradually with r until reaching its peak value at about $r=r_0$, then decreases due to the effects of return currents. The peak $|B_\theta|$ at $r=r_0$ is about 26.7–49.1 MG for $r_0=1.5-2.5\lambda$. The profiles are close to that of the particle simulation,[18] where the pulse duration is about 0.46 ps. An earlier experiment[6] reported a measurement of B_θ about 35–70 MG produced in the plasma with $n_0=2\times10^{20}$ cm^{-3}, where the LP laser is incident with $I_0=4.7\times10^{18}$ W/cm^2, $r_0=4\lambda$ ($\lambda=1.05$ μm), and pulse duration 0.6 ps. Taking the focused intensity to be two times the incident value as $I_0=9.4\times10^{18}$ W/cm^2 and focused $r_0=2\lambda$ due to self-focusing, according to the experimental results, using Eq. (21), the peak $|B_\theta|$ is calculated to be about 55.4 MG, which accords well with the experiment result.[6]

Figure 4 plots the dependence of B_θ on the laser intensity I_0, where the parameters are consistent with Fig. 1. It shows that the peak B_θ first increases with the laser intensity I_0 in the same way as B_z. When intensity I_0 rises so high that about half of the electrons near the axis are expelled, i.e. $(\langle n\rangle/n_0)_{r=0}<1/2$, the peak B_θ does not increase anymore and starts to decrease as I_0 still continues to rise [see the line with the symbol "■" for B_θ and the line with the symbol "□" for $(\langle n\rangle/n_0)_{r=0}$ in Fig. 4]. This can be explained from the effect of the return current. As is known, B_θ is generated by the axial forward current $\mathbf{j}_{\text{forward}}$ and the return current $\mathbf{j}_{\text{return}}$ compensates the forward current. Usually, $\mathbf{j}_{\text{forward}}$ dominates near the axis and $\mathbf{j}_{\text{return}}$ dominates the outer zone far away from the axis. When the laser intensity I_0 increases, at first, $\mathbf{j}_{\text{forward}}$ near the axis is larger than $\mathbf{j}_{\text{return}}$; thus, B_θ increases with I_0. However, as the electron density near the axis $(\langle n\rangle/n_0)_{r=0}$ decreases as I_0 increases, more electrons near the axis are expelled; the current $\mathbf{j}_{\text{forward}}$ becomes smaller while $\mathbf{j}_{\text{return}}$ becomes larger. When about half of the electrons near the axis are expelled $(\langle n\rangle/n_0)_{r=0}<1/2$, the growth of $\mathbf{j}_{\text{return}}$ with laser intensity gets larger than that of $\mathbf{j}_{\text{forward}}$. Thus, B_θ does not increase anymore and starts to decrease with laser intensity.

V. CONCLUSION

In conclusion, the quasistatic magnetic fields including the axial part B_z and the azimuthal one B_θ generated by intense CP and LP laser pulses propagating in an initially uniform plasma are studied theoretically by a self-consistent fluid model. The model adopts the ten-moment Grad system of hydrodynamic equations in combination with the Maxwell equations, where the nondiagonal stress tensor is taken into account. The generalized vorticity is proved to be not conserved, which is different from previous ideal fluid models. The hydrodynamic derivation yields the complete expressions for both B_z and B_θ. In the quasistatic approximation $v_p\ll v_{te}$, B_z is driven by two rotational currents existing only for CP laser; one is the result of the beating interaction of high-frequency laser electric fields, and the other is the result of the coupling interaction between the electron high-frequency quiver velocity and the high-frequency density perturbation. It is found that the cold fluid approximate condition $v_p\gg v_{te}$ used in previous studies is improper, which causes the incomplete B_z lacking the first driven source associated with the thermal motion of electrons. B_θ is driven by the irrotational current of the ponderomotive force. The results of B_z and B_θ for underdense plasma calculated by our model are in agreement with earlier experimental results.[5,6]

The transverse profiles of both B_z and B_θ as well as plasma density $\langle n\rangle/n_0$ are analyzed. Their dependence on the laser intensity is discussed in detail.

ACKNOWLEDGMENTS

This work was supported in part by the National Hi-Tech Inertial Confinement Fusion (ICF) Committee of China, National Science Foundation of China (NSFC) Grants No. 10335020, 10576007, 10375011, and 10135010, National Basic Research Project "Nonlinear Science" in China, and Science Foundation of CAEP.

[1] A. B. Borisov, A. V. Borovskiy, V. V. Korobkin *et al.*, Phys. Rev. Lett. **68**, 2309 (1992); S. C. Wilks, W. L. Kruer, M. Tabak, and A. B. Langdon, Phys. Rev. Lett. **69**, 1383 (1992); M. Honda, J. Meyer-ter-Vehn, and A. Pukhov, Phys. Plasmas **7**, 1302 (2000).
[2] M. Tabak, J. Hammer, M. E. Glinsky, W. L. Kruer, S. C. Wilks, J. Woodworth, E. M. Campbell, M. D. Perry, and R. J. Mason, Phys. Plasmas **1**, 1626 (1994).
[3] M. Schmitz and H.-J. Kull, Laser Phys. **12**, 443 (2002); M. Y. Yu, W. Yu, Z. Y. Chen, J. Zhang, Y. Yin, L. H. Cao, P. X. Lu, and Z. Z. Xu, Phys. Plasmas **10**, 2468 (2003); H. Liu, X. T. He, and S. G. Chen, Phys. Rev. E **66**, 066409 (2004); B. Qiao, X. T. He, S. Zhu, and C. Y. Zheng, Phys. Plasmas **12**, 083102 (2005).
[4] B. H. Ripin, C. K. Manka, T. A. Peyser, E. A. McLean, J. A. Stamper, A. N. Mostovych, J. Grun, K. Kearney, J. R. Crawford, and J. D. Huba, Laser Part. Beams **8**, 183 (1990).
[5] Z. Najmudin, M. Tatarakis, A. Pukhov *et al.*, Phys. Rev. Lett. **87**, 215004 (2001).
[6] J. Fuchs, G. Malka, J. C. Adam *et al.*, Phys. Rev. Lett. **80**, 1658 (1998).
[7] M. Tatarakis, I. Watts, F. N. Beg *et al.*, Nature (London) **415**, 280 (2002); Phys. Plasmas **9**, 2244 (2002).
[8] U. Wagner, M. Tatarakis, and A. Gopal *et al.*, Phys. Rev. E **70**, 026401 (2004).
[9] R. N. Sudan, Phys. Rev. Lett. **70**, 3075 (1993); V. K. Tripathi and C. S. Liu, Phys. Plasmas **1**, 990 (1994); R. J. Mason and M. Tabak, Phys. Rev. Lett. **80**, 524 (1998); T. Lehner, Phys. Scr. **49**, 704 (1994).
[10] Z. M. Sheng and J. Meyer-ter-Vehn, Phys. Rev. E **54**, 1833 (1996).
[11] V. I. Berezhiani, S. M. Mahajan, and N. L. Shatashvili, Phys. Rev. E **55**, 995 (1997).
[12] M. G. Haines, Phys. Rev. Lett. **87**, 135005 (2001).
[13] A. Kim, M. Tushentsov, D. Anderson, and M. Lisak, Phys. Rev. Lett. **89**, 095003 (2002).
[14] M. M. Korić, Laser Part. Beams **5**, 83 (1987).
[15] V. I. Berezhiani, D. D. Tskhakaya, and G. Auer, J. Plasma Phys. **38**, 139 (1987).
[16] B. Bezzerides, D. F. Forslund, and E. L. Lindman, Phys. Rev. Lett. **38**, 495 (1977); P. Mora and R. Pellat, Phys. Fluids **22**, 2408 (1979); M. Kono, M. M. ŜKorić, and D. ter Haar, Phys. Lett. **77**, 27 (1980).
[17] S. Zhu, X. T. He, and C. Y. Zheng, Phys. Plasmas **8**, 312 (2001).
[18] B. Qiao, S. Zhu, C. Y. Zheng, and X. T. He, Phys. Plasmas **12**, 053104 (2005).
[19] L. M. Gorbunov, Usp. Fiz. Nauk **109**, 631 (1973) [Sov. Phys. Usp. **16**, 217 (1973)]; N. L. Tsintsadze and D. D. Tskhakaya, Zh. Eksp. Teor. Fiz. **72**, 480 (1977) [Sov. Phys. JETP **45**, 252 (1977)]; A. Bourdier and X. Fortin, Phys. Rev. A **20**, 2154 (1979).
[20] H. Grad, Commun. Pure Appl. Math. **2**, 331 (1949); A. Yu. Kirii and V. P. Sillin, Zh. Tekh. Fiz. **39**, 773 (1969) [Sov. Phys. Tech. Phys. **14**, 583 (1969)].
[21] Yu. M. Aliev, A. A. Frolov, L. Stenflo, and P. K. Shukla, Phys. Fluids B **2**, 34 (1990).
[22] L. D. Landau and E. M. Lifshitz, *Fluid Mechanics*, 2nd ed.(Pergamon Press, Oxford, 1987).
[23] G. C. Pomraning, *The Equations of Radiation Hydrodynamics*, 1st ed. (Pergamon Press, Oxford, 1973).
[24] R. Balescu, *Transport Processes in Plasmas*, Vol. 1, Classical Transport (North-Holland, Amsterdam, 1988).
[25] L. Gorbunov, P. Mora, and T. M. Antonsen, Jr., Phys. Rev. Lett. **76**, 2495 (1996); Phys. Plasmas **4**, 3764 (1998).
[26] Z. M. Sheng, J. Meyer-ter-Vehn, and A. Pukhov, Phys. Plasmas **5**, 3764 (1998).

Laser and Particle Beams (2009), *27*, 471–474. Printed in the USA.
Copyright © 2009 Cambridge University Press 0263-0346/09 $20.00
doi:10.1017/S026303460999019X

Generation of strong magnetic fields from laser interaction with two-layer targets

S.Z. WU,[1] C.T. ZHOU,[2,3] X.T. HE,[2,3] AND S.-P. ZHU[2]
[1]Graduate School of the Chinese Academy of Engineering Physics, Beijing, People's Republic of China
[2]Institute of Applied Physics and Computational Mathematics, Beijing, People's Republic of China
[3]Center for Applied Physics and Technology, Peking University, Beijing, People's Republic of China

(RECEIVED 30 March 2009; ACCEPTED 10 May 2009)

Abstract

A two-layer target irradiated by an intense laser to generate strong interface magnetic field is proposed. The mechanism is analyzed through a simply physical model and investigated by two-dimensional particle-in-cell simulation. The effect of laser intensity on the resulting magnetic field strength is also studied. It is found that the magnetic field can reach up to several ten megagauss for laser intensity at 10^{19} Wcm^{-2}.

Keywords: Electron beams; Intense laser-plasma interaction; PIC simulation; Strong magnetic field

1. INTRODUCTION

The self-generated magnetic field (Stamper *et al.*, 1971; Sudan, 1993; Tatarakis *et al.*, 2002; Wu *et al.*, 2009; Zhou & He, 2008*a*, 2008*b*) during laser-plasma interaction has been widely investigated theoretically and experimentally over the past decades, because of many potential applications (Ghoranneviss *et al.*, 2008; Hora *et al.*, 1984, 2002, 2008*b*; Hora & Hoffmann, 2008*a*; Tan & Min, 1985; Zhou & He, 2007*a*, 2007*b*, 2007*c*), such as in fusion science, beam collimation, and electron acceleration, etc. When ultra-intense ($>10^{18}$ Wcm^{-2} μm^{-2}) lasers interact with overdense or solid targets, relativistic electrons are generated at the relativistic critical density. These fast electron currents will penetrate into the target and become a source of strong magnetic field, which can be up to 100 megagauss (MG) for laser intensities of 10^{20} Wcm^{-2}.

Recently, there have been works on the generation and application of strong interface magnetic fields. With specially designed target structures, both hybrid-Vlasov-Fokker-Planck (HVFP) (Robinson *et al.*, 2007) and hybrid-fluid-particle-in-cell (HFPIC) (Zhou *et al.*, 2008*a*, 2008*b*) simulations show that strong interface magnetic fields are generated because of the gradients in the resistivity and density at the material interface. However, in realistic experiments and applications, the fast electrons are generated during the laser-plasma interaction (LPI). The earlier HVFP or HFPIC simulations cannot consider this process directly. Instead, in these simulations, the electron beams are injected and are assumed to satisfy certain scaling law (Wilks *et al.*, 1992). By making use of the explicit two-dimensional particle-in-cell (PIC) simulation and including the LPI directly, here we investigate the generation of strong interface magnetic fields in a two-layer target irradiated by ultra-intense lasers. The effect of the laser intensity on the field strength is also studied. It is found that the interface magnetic field can reach up to tens MG for laser intensity at 10^{19} Wcm^{-2}, and even hundreds of MG for 10^{20} Wcm^{-2}.

2. PHYSICAL MODEL

The self-generated magnetic field from the intense laser-produced fast electron flux is given by (Bell *et al.*, 1998): $\partial \mathbf{B}/\partial t = \eta \nabla \times \mathbf{j}_f + \nabla \eta \times \mathbf{j}_f$, where $\mathbf{j}_f$ is the fast electron current density, and η is the space-dependent resistivity. Since the standard explicit PIC simulation cannot include the effect of resistivity, the magnetic field is due to the density gradient alone.

We now consider the target shown in Figure 1. It consists of two plasma layers of different densities (layers 1 and 2 with densities n_1 and n_2, respectively). For laser-driven electron beam with currents larger than the Alfvén limit, a return current $\mathbf{j}_r$ moving in the opposite direction will be induced to establish charge and current equilibrium. The local net current density is thus $\mathbf{j}_{\text{net}} \equiv \mathbf{j}_r + \mathbf{j}_f \sim \mu_0^{-1} \nabla \times \mathbf{B}$,

Address correspondence and reprint requests to: C.T. Zhou, Institute of Applied Physics and Computational Mathematics, P. O. Box 8009, Beijing 100088, People's Republic of China. E-mail: zcangtao@iapcm.ac.cn

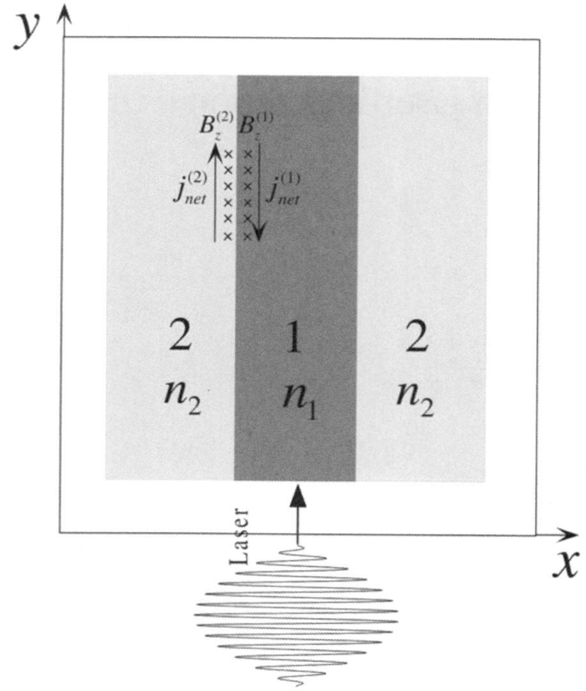

Fig. 1. A laser pulse incident on a two-layer target (layers 1 and 2 with plasma densities n_1 and n_2, respectively), and the self-generated interface magnetic field.

corresponding to $\mathbf{I}_A \approx 17.1\ \beta\gamma\text{kA}$ (Batani *et al.*, 2002) in a width on the order of 0.1 μm.

The net currents on the two sides of the interface are different due to the density difference, but as a whole, there is a current balance: $j_{\text{net}}^{(1)} \sim -j_{\text{net}}^{(2)}$ (see Fig. 1). Therefore, a magnetic field $B_z = B_z^{(1)} + B_z^{(2)}$ will be generated near the interface according to Ampere's Law. We can write

$$B_z \sim \mu_0 j_{\text{net}}^{(1)} L_1 + \mu_0 j_{\text{net}}^{(2)} L_2, \tag{1}$$

where L_i ($i = 1, 2$) is the transverse width of the current. The fast electron current j_f can be estimated from the scaling law (Wilks *et al.*, 1992):

$$j_f = \frac{\eta_a I_{L,18}}{0.511\left(\sqrt{1 + I_{L,18}\lambda_0^2/1.38 - 1}\right)}, \tag{2}$$

where $I_{L,18}$, λ_0, j_f represent the laser intensity, wavelength, and current density, in units of 10^{18} Wcm^{-2}, μm, and 10^{12} Acm^{-2}, respectively, and η_a is the laser absorption rate. For a typical intensity of $I_L = 10^{19}$ Wcm^{-2} and 30% absorption rate, the magnetic field strength can reach more than 30 MG according to Eqs. (1) and (2). It is also expected that the field strength will increase with increasing incident laser intensity.

3. SIMULATION RESULTS

Since we are mainly interested in the generation of the interface magnetic field by the density effect, the standard two-dimensional PIC technique is sufficient (Birdsall & Langdon, 1985). Our target configuration is shown in Figure 1. The two-dimensional (x, y) simulation box is 20 μm × 20 μm. The mesh resolution is 1200 × 1200 cells with 50 electrons and 50 ions in each cell. The initial temperature of the plasma electrons and ions is 1 keV. As shown in Figure 1, there are vacuum regions on the four sides of the target, 1 μm in the left and the right sides of the plasma (in the x direction), 2 μm in the front, and 1 μm behind the target (in the y direction). The periodic boundary condition is used in the x direction and the outgoing boundary condition in the y direction. The width of the inner layer (layer 1) is 4 μm in the x direction. The laser is incident from the bottom along the y direction with a rise time of about 4 fs, after which it remains at the constant peak intensity. The laser transverse profile (along the x direction) is super-gaussian with a 2.4 μm spot size. The laser wavelength is chosen to be $\lambda = 1.05$ μm.

We now consider two cases. The peak intensity of the incident laser is $I_L = 10^{19}$ Wcm^{-2} in case I, and 6×10^{19} Wcm^{-2} in case II. The target density profiles for the two cases are the same: $n_1 = 20n_c$, $n_2 = 50n_c$, where n_c is the critical density.

Figures 2a and 2c show the net current density and the corresponding magnetic field at $t = 160$ fs for case I. It is clearly seen that the net currents on the two sides of the interface have opposite signs, and the values are up to the order of 10^{12} Acm^{-2}. Such strong opposite currents generate a strong magnetic field along the interface (see Fig. 2c). In case I, the interface magnetic field can be more than 30 MG. The similar net currents and interface magnetic fields also appear in case II, as shown in Figures 2b and 2d. It is noted that the values for the current and magnetic field in case II are much larger than that in case I, as shown in Figure 2. According to Eqs. (1) and (2), higher laser intensity will generate higher current, and thus higher magnetic field.

In order to show the effect of laser intensity on the interface magnetic generation more quantitatively, several other cases have also been considered: 2×10^{19} Wcm^{-2}, 4×10^{19} Wcm^{-2}, 8×10^{19} Wcm^{-2}, and 10^{20} Wcm^{-2}. The other simulation parameters are identical to these in cases I and II. The strength of the resulting magnetic fields as a function of the incident laser intensity is shown in Figure 3. The line with the error bars is from the simulation, confirming that the field strength increases with the laser intensity. The analytical estimate based on Eqs. (1) and (2) is given in Figure 3 by the line with triangles. The good agreement demonstrates that the simple physical model is useful for estimating the interface magnetic field strength.

4. SUMMARY

To summarize, we have proposed a two-layer target for generating strong interface magnetic fields. The scheme is

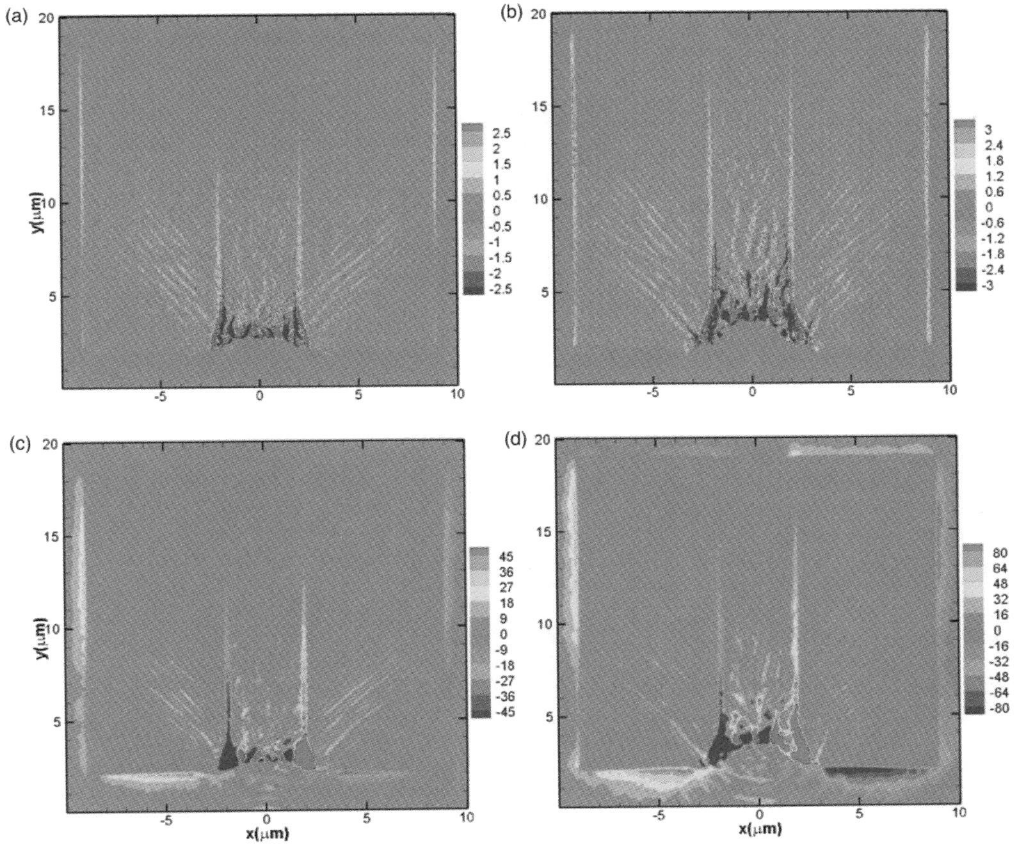

Fig. 2. (Color online) The net current density in the y direction j_{net} (in 10^{12} Acm^{-2}) at 160 fs: (**a**) for case I and (**b**) case II. The corresponding magnetic fields B_z (in MG): (**c**) case I and (**d**) case II.

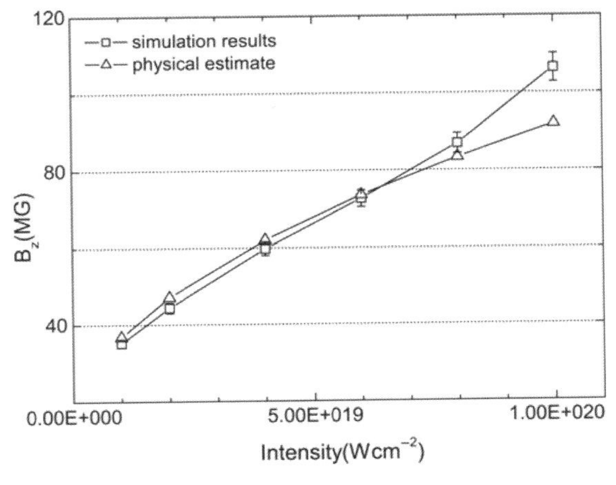

Fig. 3. The interface magnetic field B_z (in MG) against the incident laser intensity: simulation results (squares with the error bars) and physical estimate from Eqs. (1) and (2) (the line with triangles).

analyzed with a simple analytical model and the process is investigated using two-dimensional PIC simulation. It is found that the interface field can reach several tens MG at laser intensities of 10^{19} Wcm^{-2}. Furthermore, we have also studied the effect of laser intensity on the magnetic field strength. Our simulation results agree very well with that from the analytical model. Thus, the proposed scheme as well as the physical model should be useful in the applications such as in fast electron collimation and high energy density physics, where strong magnetic fields are required.

ACKNOWLEDGMENTS

Wu would like to thank Z.J. Liu for his helpful discussions on PIC simulation. This work is supported by the National Natural Science Foundation of China Grant Nos. 10835003 and 10575013 and the National High-Tech ICF Committee and partially by the National Basic Research Program of China (973)(2007CB815101).

REFERENCES

Batani, D. (2002). Transport in dense matter of relativistic electrons produced in ultra-high-intensity laser interactions. *Laser Part. Beams* **20**, 321–336.

Bell, A.R., Davies, J.R. & Guérin, S.M. (1998). Magnetic field in short-pulse high-intensity laser-solid experiments. *Phys. Rev. E* **58**, 2471–2473.

Birdsall, C.K. & Langdon, A.B. (1985). *Plasma Physics via Computer Simulation*. New York: McGraw-Hill Inc.

Ghoranneviss, M., Malekynia, B., Hora, H., Miley, G.H. & He, X.T. (2008). Inhibition factor reduces fast ignition threshold of laser fusion using nonlinear force driven block ignition. *Laser Part. Beams* **26**, 105–111.

Hora, H., Lalousis, P. & Eliezer, S. (1984). Analysis of the inverted double-layers produced by nonlinear forces in laser-produced plasmas. *Phys. Rev. Lett.* **53**, 1650–1652.

Hora, H., Hoelss, M., Scheid, W., Wang, J.X., HO, Y.K., Osman, F. & Castillo, R. (2000). Principle of high accuracy of the nonlinear theory for electron acceleration in vacuum by lasers at relativistic intensities. *Laser Part. Beams* **18**, 135–144.

Hora, H. & Hoffmann, D.H.H. (2008*a*). Using petawatt laser pulses of picosecond duration for detailed diagnostics of creation and decay processes of B-mesons in the LHC. *Laser Part. Beams* **26**, 503–505.

Hora, H., Malekynia, B., Ghoranneviss, M., Miley, G.H. & He, X.T. (2008*b*). Twenty times lower ignition threshold for laser driven fusion using collective effects and the inhibition factor. *Appl. Phys. Lett.* **93**, 0111011-3.

Robinson, A.P.L. & Sherlock, M. (2007). Magnetic collimation of fast electrons produced by ultraintense laser irradiation by structuring the target composition. *Phys. Plasams* **14**, 083105.

Stamper, J.A., Papadopoulos, K., Sudan, R.N., Dean, S.O., McLean, E.A. & Dawson, J.M. (1971). Electron acceleration by high current-density relativistic electron bunch in plasmas. *Laser Part. Beams* **26**, 1012–1015.

Sudan, R. (1993). Mechanism for the generation of 109 G magnetic fields in the interaction of ultraintense short laser pulse with an overdense plasma target. *Phys. Rev. Lett.* **70**, 3075–3078.

Tan, W. & Min, G. (1985). Thermal flux limitation and thermal conduction inhibition in laser plasmas. *Laser Part. Beams* **3**, 243–250.

Tatarakis, M., Watts, I., Beg, F.N., Dangor, A.E., Krushelnick, K., Wagner, U., Norreys, P.A. & Clark, E.L. (2002). Measurements of ultrastrong magnetic fields during relativistic laserplasma interactions. *Phys. Plasmas* **9**, 2244.

Wilks, S.C., Kruer, W.L., Tabak, M. & Langdon, A.B. (1992). Absorption of ultra-intense laser pulses, *Phys. Rev. Lett.* **69**, 1383–1386.

Wu, S.Z., Liu, Z.J., Zhou, C.T. & Zhu, S.P. (2009). Density effects on collimation of energetic electron beams driven by two intense laser pulses. *Phys. Plasmas* **16**, 043106.

Yu, M.Y. & Luo, H.Q. (2008). A note on the multispecies model for identical particles. *Phys. Plasmas* **15**, 024504.

Zhou, C.T. & He, X.T. (2007*a*). Influence of a large oblique incident angle on energetic protons accelerated from solid-density plasmas by ultraintense laser pulses. *Appl. Phys. Lett.* **90**, 031503.

Zhou, C.T., Yu, M.Y. & He, X.T. (2007*b*). Electron acceleration by high current-density relativistic electron bunch in plasmas. *Laser Part. Beams* **25**, 313–319.

Zhou, C.T. & He, X.T. (2007*c*). Intense laser-driven energetic proton beams from solid density targets. *Opt. Lett.* **32** 2444–2446.

Zhou, C.T. & He, X.T. (2008*a*). Intense-laser generated relativistic electron transport in coaxial two-layer targets., *Appl. Phys. Lett.* **92**, 0715021-3.

Zhou, C.T. & He, X.T. (2008*b*). Laser-produced energetic electron transport in overdense plasmas by wire guiding. *Appl. Phys. Lett.* **92**, 1515021-3.

PHYSICS OF PLASMAS **18**, 023106 (2011)

Magnetic collimation of fast electrons in specially engineered targets irradiated by ultraintense laser pulses

Hong-bo Cai,[1,2,a)] Shao-ping Zhu,[1] X. T. He,[1,2,3,b)] Si-zhong Wu,[1] Mo Chen,[1] Cangtao Zhou,[1] Wei Yu,[4] and Hideo Nagatomo[5]
[1]*Institute of Applied Physics and Computational Mathematics, Beijing 100094, People's Republic of China*
[2]*Center for Applied Physics and Technology, Peking University, Beijing 100871, People's Republic of China*
[3]*Institute for Fusion Theory and Simulation, Zhejiang University, Hangzhou 310027, People's Republic of China*
[4]*Shanghai Institute of Optics and Fine Mechanics, Shanghai 201800, People's Republic of China*
[5]*Institute of Laser Engineering, Osaka University, 2-6 Yamadaoka, Suita, Osaka 565-0871, Japan*

(Received 6 December 2010; accepted 17 January 2011; published online 9 February 2011)

The efficient magnetic collimation of fast electron flow transporting in overdense plasmas is investigated with two-dimensional collisional particle-in-cell numerical simulations. It is found that the specially engineered targets exhibiting either high-resistivity-core-low-resistivity-cladding structure or low-density-core-high-density-cladding structure can collimate fast electrons. Two main mechanisms to generate collimating magnetic fields are found. In high-resistivity-core-low-resistivity-cladding structure targets, the magnetic field at the interfaces is generated by the gradients of the resistivity and fast electron current, while in low-density-core-high-density-cladding structure targets, the magnetic field is generated by the rapid changing of the flow velocity of the background electrons in transverse direction (perpendicular to the flow velocity) caused by the density jump. The dependences of the maximal magnetic field on the incident laser intensity and plasma density, which are studied by numerical simulations, are supported by our analytical calculations. © *2011 American Institute of Physics*. [doi:10.1063/1.3553453]

I. INTRODUCTION

Fast ignition[1,2] relies on efficient energy coupling of a petawatt-picosecond (10^{15} W, 10^{-12} s) laser to heat the hot-spot region in the dense deuterium-tritium fuel to temperatures ~ 10 keV. The laser energy is first deposited into fast electrons in the interaction region, which is usually 50–100 μm away from the dense core, and the resulting energy flux transports into the dense core. However, studies have shown that electron divergence increases with laser intensity.[3,4] Shadowgraphy and K_α imaging measurements revealed a large divergence angle of $\sim 50°$ (Refs. 5 and 6) as the intensity reaches $I_0 \sim 10^{20}$ W/cm^2. Therefore, propagation and collimation of fast electron beam in overdense plasmas are of particular importance since they will affect the laser coupling efficiency.

In recent experiments at Rutherford Appleton Laboratory employing VULCAN petawatt laser system, sandwich targets[7] [a thin slab of tin (Sn) with aluminum (Al) at either side] or high-resistivity-core-low-cladding structure targets[8] [iron (Fe) wires surrounded by aluminum (Al)] are found to take advantages in collimating fast electrons. The mechanism proposed to transversely confine fast electrons relies on the fact that in the structure targets, steep resistivity gradient along the boundary between different materials will lead to rapid magnetic field growth and consequently guiding of fast electrons. The most familiar example of this effect is introduced by Robinson and Sherlock,[9] who proposed a radically different route to artificially forcing collimated beam transport to occur. In his study, a structured target in which the resistivity varies in the direction transverse to the electron beam propagation was studied. The resistivity jump then leads to the growth of the magnetic field.

Furthermore, numerical simulations by Zhou *et al.*[10,11] and Wu *et al.*[12] showed that low-density-core-high-density-cladding structure target can also generate a mega-Gauss (MG) interfaces magnetic field, which collimates fast electrons even better. However, in such targets, the magnetic field generated by the gradient of resistivity and current density should play a role in scattering electrons, instead of collimating them. Therefore, their results are a clear indication of that in a low-density-core-high-density-cladding structure target, other mechanisms tend to be dominant for the generation of magnetic field. Such spontaneous magnetic field can be explained by several mechanisms, including anomalous resistivity.[12] However, there still does not exist a well-established and satisfactory theory, in particular, for quantitatively predicting the magnetic field and modeling its dependence on plasma parameters.

In this paper, collimation effect of quasistatic magnetic field on transport of fast electron beam in specially engineered structure targets is investigated by using collisional particle-in-cell (PIC) simulations. The mechanisms governing the generation of magnetic field are studied in details. Two different generation mechanisms are discussed. One is the large resistivity gradient along the interfaces between different materials, the other is nonparallel density gradient and fast electron current at the interfaces. The former is consis-

a)Electronic mail: caihonb@yahoo.com.cn.
b)Electronic mail: xthe@iapcm.ac.cn.

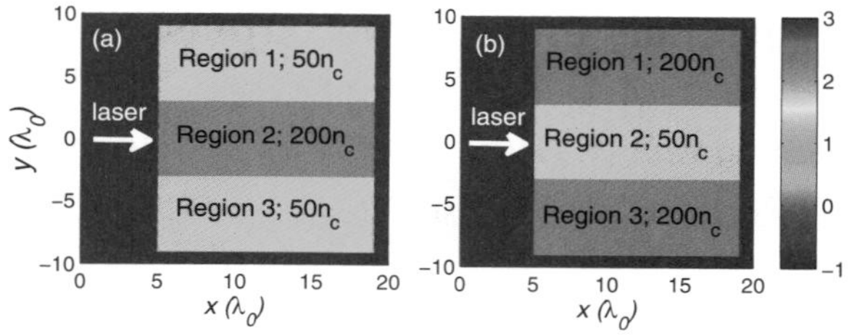

FIG. 1. (Color online) A view of initial plasma density in logarithm scale. (a) is for the high-density-core target and (b) is for the low-density-core target. Here the density is normalized by the critical density n_c.

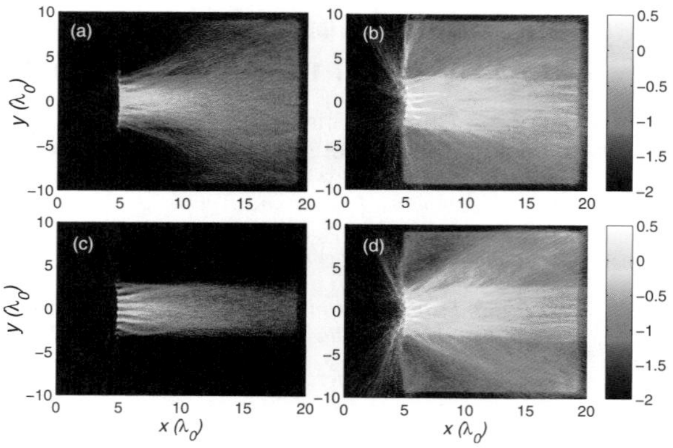

FIG. 2. (Color online) Electron energy density in logarithm scale for [(a) and (c)] high-density-core target and [(b) and (d)] low-density-core target at time $t=264$ fs. (a) and (b) are for the collisionless cases and (c) and (d) are for the collisional cases. Here, the electron energy density is normalized by $m_ec^2n_c$.

tent with the hybrid-Vlasov–Fokker–Planck's simulation results[9] and experimental results.[7,8] The latter is consistent with our analytical model. The paper is organized as follows. In Sec. II, we present the excitation of quasistatic magnetic fields by the gradient of resistivity and the electron collimation at the density/resistivity jump. In Sec. III, we discuss the excitation of quasi-static magnetic fields by the rapid changing of the flow velocity of the background electrons. In Sec. IV, a summary of the results is given.

II. EXCITATION OF QUASI-STATIC MAGNETIC FIELDS AT THE DENSITY/RESISTIVITY JUMP

A. Description of the physical modeling

In order to describe collisional effect on the generation of magnetic fields and to confirm the hybrid-Vlasov–Fokker–Planck's simulations,[9] first we study two specially engineered structure targets with a 2D3V collisional PIC simulation code ASCENT.[13] Figure 1 is a sketch of the geometry of the simulations. Here, we studied two different targets: (a) high-density-core target and (b) low-density-core target. In the high density regions [region 1 and region 3 in Fig. 1(a) and region 2 in Fig. 1(b)], the plasma consists of two species: electrons and fully ionized carbon (charge state $Z_i=6$) with $m_i/m_eZ_i=1836\times 12/6$. In the lower density region [region 2 in Fig. 1(a) and region 1 and region 3 in Fig. 1(b)], the plasma consists of two species: electrons and protons. The widths of the three regions are: $6\lambda_0$, $6\lambda_0$, and $6\lambda_0$, respectively. The densities of three regions are $200n_c$, $50n_c$, and $200n_c$ for Fig. 1(a) and $50n_c$, $200n_c$, and $50n_c$ for Fig. 1(b). The p-polarized laser pulse at $\lambda_0=1.06$ μm wavelength and 1.5×10^{19} W/cm^2 intensity irradiates the target from the left boundary. The intensity profile is Gaussian in the y direction with a spot size of 5.0 μm [full width at half maximum (FWHM)]. The laser rises in $15T_0$, where T_0 is the laser period, after which the laser amplitude is kept constant.

The size of the simulation box is $30\lambda_0\times 24\lambda_0$. Here, we use a grid size of $\Delta x=\Delta y=\lambda_0/128$ with 3840×3072 grid cells. The time step used is $0.005T_0$. During the simulation, we set 49 electrons and 49 ions per cell. The total number of particles is about 7.5×10^8. For both the fields and particles, we use absorbing boundary condition in both x and y directions. Furthermore, in order to reduce the restrictions on the grid size compared with the Debye length, we used a fourth order interpolation scheme to evaluate fields and current.[13] In our code ASCENT, the binary collision module is based on Takizuka and Abe's model,[14,15] and is extended to treat collisions between weighted macroparticles and among different species of macroparticles relativistically.

B. Collisional effect on magnetic field generation and electron collimation

The preceding geometry is very close to the simulations performed in Ref. 12, but our work was performed with several differences: (1) binary Monte-Carlo collision[15] is included, (2) two materials with different Z are considered, and (3) the density is much higher. These three points enable us to study the magnetic fields caused by the resistivity jump since the Spitzer resistivity (ideal gas specific heat capacity) relies only on the charge state Z and temperature T_c (Ref. 9), $\eta=10^{-4}Z\ln\Lambda/T_c^{3/2}$.

To determine the collisional effect on the generation of magnetic fields and the collimation of fast electrons, we studied the cases with and without collision for both the high-density-core target and low-density-core target. In Fig. 2, the energy density distributions of electrons with energies between $0.5\leq E_e[\mathrm{MeV}]\leq 5.0$ are plotted. It is clearly shown that only the high-density-core target with collision [Fig. 2(c)] can collimate fast electrons. For collisionless case [Fig. 2(a)] the fast electron beam has a large divergence angle even for the high-density-core target. By comparing the electron energy density in Figs. 2(a) and 2(c), we may therefore assert that the resistivity variation at the interfaces plays an important role in guiding electron beam transport. Alternatively, for the low-density-core target as shown in Figs. 2(b) and 2(d), the fast electrons are not collimated for

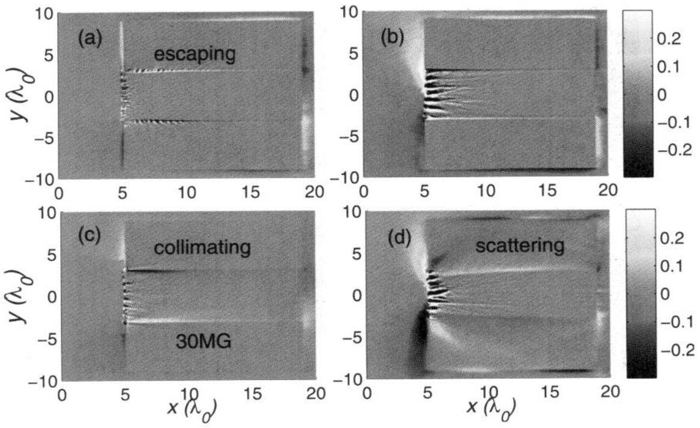

FIG. 3. (Color online) Contours of spontaneous magnetic fields for [(a) and (c)] high-density-core target and [(b) and (d)] low-density-core target at time t=264 fs. (a) and (b) are for the collisionless cases and (c) and (d) are for the collisional cases. The unit of the magnetic field is $m_e\omega_0 c/e$ (1 unit $\simeq$100 MG).

both the collisional and collisionless cases. We notice that in the collisional case, the fast electrons are even scattered at the interfaces between different density regions [see Fig. 2(d)].

C. Generation mechanism of magnetic fields

In such overdense plasmas, the reason to collimate fast electrons is the spontaneously generated magnetic fields. The generation of magnetic fields during the propagation of fast electrons through a solid target can be described by combining a simple Ohm's law ($\mathbf{E}=-\eta\mathbf{j}_f$) with the Faraday's law to yield[16]

$$\frac{\partial \mathbf{B}}{\partial t}=c[\eta\nabla\times\mathbf{j}_f+\nabla\eta\times\mathbf{j}_f], \qquad (1)$$

where η is the resistivity, and $\mathbf{j}_f$ is the fast electron current density. In solid density region, the generation of the collimating magnetic field is mainly due to the interplay of two effects. It has been explained in Refs. 17 and 18. The first term on the right-hand side of Eq. (1) generates a magnetic field that pushes fast electrons toward regions of higher fast electron current density, while the second term pushes fast electrons toward regions of higher resistivity. It is evident in Fig. 3, for the high-density-core target with collision, the generated magnetic fields at the interfaces play a role in collimating fast electrons [Fig. 3(c)]. In this case, the resistivity jump, which is caused by the difference of the charge state according to the Spitzer resistivity, dominates the generation of the magnetic field. Here, the spontaneously generated magnetic field can be as high as 30 MG. While in the collisionless case [Fig. 3(a)], the generated magnetic field at the interfaces where alternate yellow and dark short scale filaments suggests the excitation of Weibel-like instability.[19] However, it has no contribution to the collimation of fast electrons. For the low-density-core target with collision [shown in Fig. 3(d)], the generated magnetic field tends to scatter the fast electrons. We note that for the collisionless cases[Figs. 3(a) and 3(b)], the magnetic field cannot be generated from the resistivity jump since there is no resistivity.

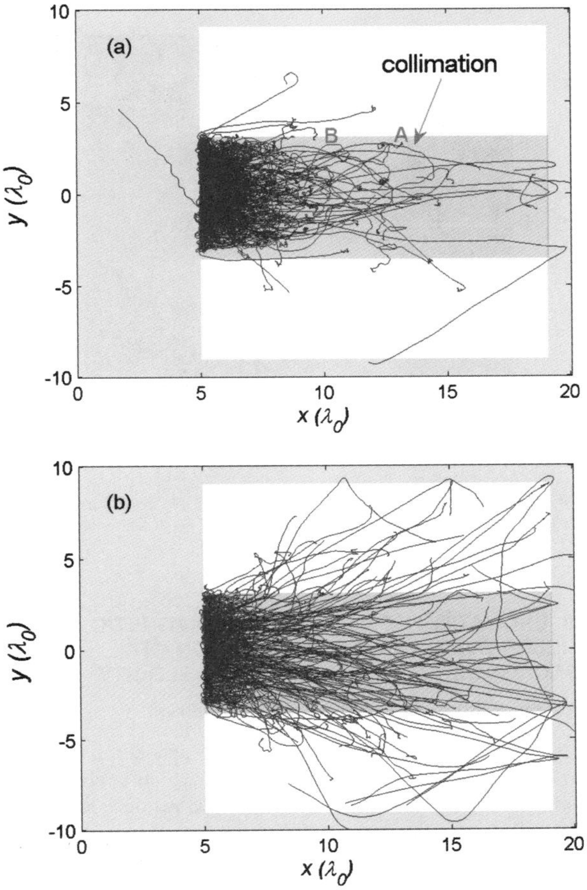

FIG. 4. (Color online) Typical trajectories of the fast electrons for high-density-core target (a) with collision and (b) without collision, showing the guiding and collimation of the electrons. Initially, the fast electrons are sampled randomly in the overdense plasma region near the laser plasma interfaces.

These results are consistent with the experimental and simulation results.[7–9]

The electron trajectories in Fig. 4 show more clearly that the outgoing fast electrons [for example, Marks A and B in Fig. 4(a)] are reflected by the magnetic field at the interfaces between different regions for the case of high-density-core target with collision [see Fig. 4(a)]. Here, the tracked electrons are sampled randomly at the beginning near the laser plasma interface. By comparing the trajectories in Figs. 4(a) and 4(b), one can see that the energy loss of fast electrons in the collisional case [Fig. 4(a)] is more serious comparing with that in the collisionless case [Fig. 4(b)] due to the high resistivity. This will result in a drop in coupling efficiency from laser to compressed core in fast ignition. Therefore, it is better to reduce the energy loss during the transport of fast electron beam. In the next section, we will show another kind of specially engineered target, which is expected to collimate fast electrons as well as to reduce their energy loss.

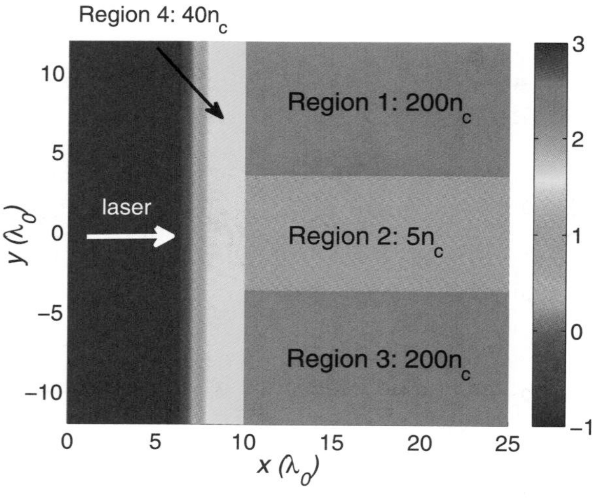

FIG. 5. (Color online) The initial profile of the plasma density in logarithm scale of the low-density-core-high-density-cladding structure target. Here the density is normalized by the critical density n_c.

III. EXCITATION OF QUASI-STATIC MAGNETIC FIELDS BY TRANSVERSE VARIATION OF BACKGROUND ELECTRON FLOW VELOCITY

A. Description of the physical modeling

In Sec. II, We have reproduced the generation of magnetic fields by the resistivity gradient, which confirms the hybrid-Vlasov–Fokker–Planck simulation results.[9] However, we still need to answer the following question: what is the reason for the discrepancy between hybrid-Vlasov–Fokker–Planck simulations[9] and PIC simulations?[12] In this section, we study another generation mechanism of magnetic fields in specially engineered low-density-core-high-density-cladding structure target. Figure 5 is a sketch of the geometry of the simulations. In this case, the middle region has been replaced by a lower density hydrogen plasma than that in Fig. 1(b). The widths of the three regions are $8.4\lambda_0$, $7.2\lambda_0$, and $8.4\lambda_0$, respectively. The densities of the three regions are $200n_c$, $5n_c$, and $200n_c$, respectively. In order to avoid the ruin of the hydrogen plasma by laser pressure, we placed another high density target ($40n_c$, region 4) in front of the low-density-core-high-density-cladding structure target. Furthermore, in order to avoid the reflux of fast electrons, we set cooling buffers on the boundaries in our simulations. Other parameters are the same with those in Sec. II.

B. Collisional effect on magnetic field generation and electron collimation

In Fig. 6, the electron energy density in logarithm scale with electron energies between $0.5 \leq E[\text{MeV}] \leq 5.0$ is plotted. It is clearly seen that the fast electrons are collimated well at later time ($t \geq 500$ fs) for both cases without and with collision [Figs. 6(b) and 6(d)]. In the collisional case, the electron beam has a larger divergence angle [Fig. 6(c)] than that in the collisionless case [Fig. 6(a)] at early time ($t=264$ fs). However, since the magnetic field develops further into the target as time goes on, one can expect that fast electrons will be also collimated well even for collisional case at later time. In comparison with the collisional case in Figs. 6(c) and 6(d), in the collisionless case in Figs. 6(a) and 6(b), the electrons are collimated much faster and better. Thus, we may conclude that for the present target the resistivity plays a negative role in collimating fast electrons and other mechanisms dominate the generation of magnetic fields. Actually, the physics of the role of the resistivity is straightforward. In the case of a target where the inner-core is less resistive than the cladding, Eq. (1) leads to a conclusion that a scattering magnetic field should be generated [also seen in Fig. 3(d)], which partly cancels out the collimating magnetic field.

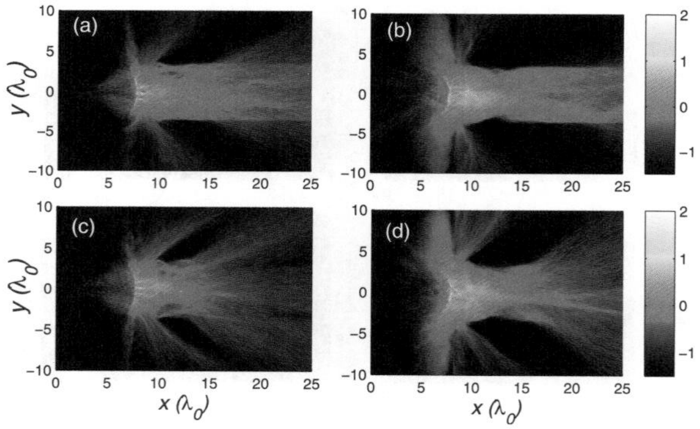

FIG. 6. (Color online) Electron energy density in logarithm scale at two different times [(a) and (c)] $t=264$ fs and [(b) and (d)] $t=500$ fs. (a) and (b) are for the collisionless cases and (c) and (d) are for collisional cases. Here the electron energy density is normalized by $m_e c^2 n_c$.

Figure 7 shows the quasi-static magnetic fields at two different times: [Figs. 7(a) and 7(c)] $t=264$ fs and [Figs. 7(b) and 7(d)] $t=500$ fs. One can see that both the collisionless [Figs. 7(a) and 7(b)] and collisional [Figs. 7(c) and 7(d)] cases can generate a magnetic field at the interfaces of the density jump. The magnetic fields at the interfaces first ap-

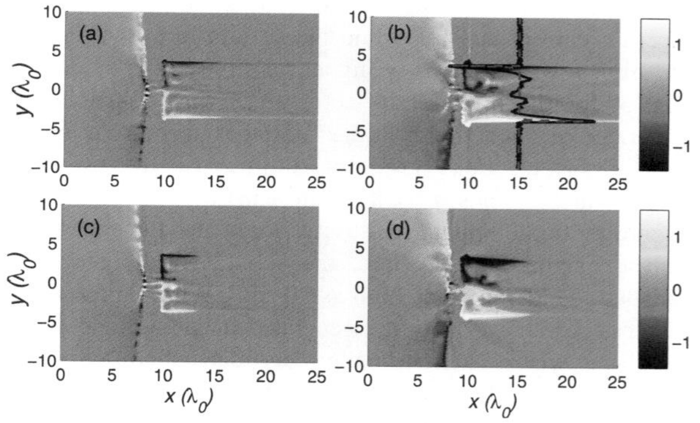

FIG. 7. (Color online) The spontaneous magnetic fields at times [(a) and (c)] $t=264$ fs and [(b) and (d)] $t=500$ fs. (a) and (b) are for the collisionless cases and (c) and (d) are for the collisional cases. The solid line on (b) is a slice of spontaneous magnetic fields in simulation at $x=15\lambda_0$, the dashed line for the analytical result. Here, the magnetic fields are averaged over one laser period. The unit of the magnetic field is $m_e\omega_0 c/e$ (1 unit $\simeq$ 100 MG).

pear at the region close to the laser plasma interaction region. As time goes on, the magnetic fields grow to the rear side of the target because of the transport of fast electrons. It should be pointed out that the magnetic fields in Figs. 7(c) and 7(d) develop much slower than those in Figs. 7(a) and 7(b) because of the resistivity. However, their magnitudes almost maintain the same level as those in Figs. 7(a) and 7(b). We note that for all cases the magnetic fields can reach as high as 150 MG, which is much higher than those in the cases presented in Sec. II.

C. Generation mechanism of magnetic field

Several mechanisms can produce such mega-Gauss spontaneous magnetic fields in overdense plasma region, including resistivity jump,[9] anomalous resistivity,[12] and so on. However, in the present case, the rapid changing of the flow velocity ($\mathbf{e}_x$) of the background return electrons in transverse direction ($\mathbf{e}_y$) caused by the density jump dominates the generation of spontaneous interface field. The detailed analytical description has been presented in Ref. 20. Here, to simplify the analysis, we assume that a uniform fast electron beam of density n_h propagates with an average velocity v_h along the x-axis. The fast electron beam passing through the plasma quickly generates a return current of background electrons v_{e0}. This plasma flow in the return current, whose velocity is much faster than the electron thermal velocity, is responsible for the spontaneous generated magnetic fields inside the beam pulse.[21,22] For purpose of the interface magnetic field excitation study, we assume that the fast electron beam motion is unperturbed. Since the magnetic field can develops in a very short time (of the order of hundred femtoseconds), the combined system of the fast electron beam and the response of the background plasma can be suitably treated by single fluid electron magnetohydrodynamic description. From the electron fluid equations (the continuity equation and force balance equation), we can get the relationship between the self-generated magnetic field and electron flow velocity,[20–23]

$$\mathbf{B} = \frac{c}{e}\nabla \times \mathbf{p_e}. \tag{2}$$

where $\mathbf{p_e}=m_e\gamma_e\mathbf{v}_{e0}$ is the momentum of the background electron flow and $\gamma_e=1/\sqrt{1-v_{e0}^2/c^2}$ is the relativistic factor. Maxwell's equations for the self-generated electric and magnetic fields, $\mathbf{E}$, and $\mathbf{B}$, are given by

$$\nabla \times \mathbf{B} = -\frac{4\pi e}{c}(n_{e0}\mathbf{p}_e/m_e\gamma_e + n_h\mathbf{v}_h) + \frac{1}{c}\frac{\partial \mathbf{E}}{\partial t}. \tag{3}$$

Here, for a long beam with pulse length $l_b \gg v_h/\omega_p$, where v_h and ω_p are the fast electron beam velocity and background plasma frequency, the displacement current $(1/c)(\partial\mathbf{E}/\partial t)$ is of order $(v_h/\omega_p l_b)^2 \ll 1$ compared to the electron current. In the following discussion, the displacement current $(1/c)(\partial\mathbf{E}/\partial t)$ is neglected.[17,18,24] Equation (3) gives

$$\mathbf{p}_e = -\frac{mc\gamma_e}{4\pi e n_{e0}}\nabla \times \mathbf{B} - m\gamma_e n_h \mathbf{v}_h/n_{e0}. \tag{4}$$

Substituting Eq. (4) into Eq. (2), we obtain the equations for self-generated magnetic field,

$$\frac{mc^2}{4\pi e^2}\nabla \times \left(\frac{\gamma_e \nabla \times \mathbf{B}}{n_{e0}}\right) + \mathbf{B} = -\frac{mc}{e}\nabla \times \left(\frac{\gamma_e n_h \mathbf{v}_h}{n_{e0}}\right). \tag{5}$$

When fast electron beam velocity and background electron density is finite, then the magnetic field $\mathbf{B}=\nabla\times\mathbf{a}$ is finite. The source is $\nabla\times(\gamma_e n_h \mathbf{v}_h/n_{e0}) \approx -(\gamma_e/e n_{e0}^2)\nabla n_{e0}\times\mathbf{j}_h$, which means the generation of self-generated magnetic field is indeed due to the nonparallel density gradient and fast electron current. For the nonrelativistic case, $\gamma_e \sim 1$. It is not easy to solve Eq. (5) directly even though Eq. (5) shows the physics clearly. However, we can still derived the magnetic field from Eqs. (2) and (4). Operating on Eq. (2) with $\nabla\times$, and substituting it into Eq. (4), we obtain the equation for the flow velocity of the background electrons[20,25]

$$\frac{m_e c^2}{4\pi e^2}\frac{d^2 v_{e0}(y)}{dy^2} = n_{e0}v_{e0}(y) + n_h v_h. \tag{6}$$

In the present case, the low-density-core-high-density-cladding structure background plasma has three different regions with ion densities n_{i1} for region (1) $y<-d$, n_{i2} for region (2) $-d\le y\le d$, and n_{i3} for region (3) $y>d$, as shown in Fig. 5. Since the fast electron beam may be affected by this background plasma at later time, we assume the densities of the fast electron beams are (1) n_{h1}, (2) n_{h2}, and (3) n_{h1}, respectively. From Eq. (6), we know that the fast electron beam is neutralized by the plasma return current everywhere except at the interfaces over a characteristic transverse distance. Together with the continuous boundary conditions at the interfaces, the spontaneous magnetic field B_0 can be solved analytically from Eq. (6),

$$\frac{eB_0}{m_e c} = -\frac{dv_{e0}}{dy} = \begin{cases} \dfrac{-1}{(\delta_{pe1}+\delta_{pe2})}\left(\dfrac{n_{h2}v_{h2}}{n_2} - \dfrac{n_{h1}v_{h1}}{n_1}\right)\exp\left[\dfrac{y+d}{\delta_{pe1}}\right], & y<-d \\ \dfrac{1}{(\delta_{pe1}+\delta_{pe2})}\left(\dfrac{n_{h2}v_{h2}}{n_2} - \dfrac{n_{h1}v_{h1}}{n_1}\right)\left\{\exp\left[-\dfrac{y+d}{\delta_{pe2}}\right] - \exp\left[\dfrac{y-d}{\delta_{pe2}}\right]\right\}, & -d\le y\le d \\ \dfrac{1}{(\delta_{pe1}+\delta_{pe2})}\left(\dfrac{n_{h2}v_{h2}}{n_2} - \dfrac{n_{h1}v_{h1}}{n_1}\right)\exp\left[-\dfrac{y-d}{\delta_{pe1}}\right], & y>d \end{cases}, \tag{7}$$

where $\delta_{pej}=c/\omega_{pej}(j=1,2,3)$ is the electron skin depth and $\omega_{pej}=\sqrt{4\pi n_j e^2/m_e}(j=1,2,3)$ is the electron plasma frequency for the background plasma in the jth region. n_j and v_{hj} denote the electron density of the background plasmas and the velocity of the fast electrons, respectively. Here, $n_3=n_1$, $n_{h3}=n_{h1}$, and $v_{h3}=v_{h1}$ have been used because of the symmetry. From Eq. (7), one can see that the spontaneous magnetic field B_0 exists within a layer of width δ_{pe} near the interfaces of different regions, $y=\pm d$. We notice that B_0 peaks at the interfaces and evanesces exponentially to both sides. In the lower density plasma region, the magnetic field tunnels much deeper into the target since the skin depth $\delta_{pe}\propto n_e^{-1}$. It should emphasize that the velocity variation and magnetic field occur even when $n_h(y)$ is laterally constant, i.e., $n_{h1}=n_{h2}$. The analytical result is also drawn on Fig. 7(b) (the green dashed line). It is found that the analytical result is consistent with the simulation result (the dark solid line) very well.

It should be pointed out that when the magnetic field develops, it will disturb both the incident fast electron beam and return electron flow in a real system. This is certainly the reason why we develop this magnetic field—to collimate fast electrons. In a real case, when the magnetic field develops, the fast electrons are collimated in the inner region (region 2), some electrons are even recirculated inside the inner region by the ambient magnetic field. Thus, a higher fast electron beam density is expected. From Eq. (7), we know that the magnetic field will develop to higher level in a wider region. It also can be seen in Fig. 7. Indeed, the interface magnetic field will develop when there is a forward fast electron beam pass the interfaces [Eq. (5)].

In the fast ignition scenario, the laser-generated incident fast electrons will likely enter with a conical spread in angle and a distribution of velocities. Such fast electrons constitute a large forward current density under the pressure gradient of fast electrons. The fast electron current travels into the plasma and sets up a space-charge electric field (thermoelectric effects) that acts to decelerate the fast electrons. This current exceeds the Alfven current limit, it would produce intense self-consistent magnetic fields bending the electron trajectories backward, and preventing the forward movement if it happens in vacuum.[26] However, in overdense plasmas, the generated electric field also accelerates the background plasma electrons in the opposite direction, providing a return current, which largely cancels the forward fast electron current.[27] This allows a larger fast electron current to propagate into the overdense plasma. However, it is unstable to relativistic electromagnetic two stream instabilities which break the streams into the magnetic channels. Some fast electrons are affected by these instabilities and are accumulated near the laser plasma interaction region, as shown in Fig. 7. To actually solve the two stream and Weibel instabilities, the suggested model is not enough; yet it is not the purpose of our study. Indeed, we focus that the interface magnetic field will develop when there is a forward fast electron beam pass the interfaces. Remarkably though, our explanation on why the 100 MG interface magnetic field at the interfaces is generated applies to the self-consistent problem just as well.

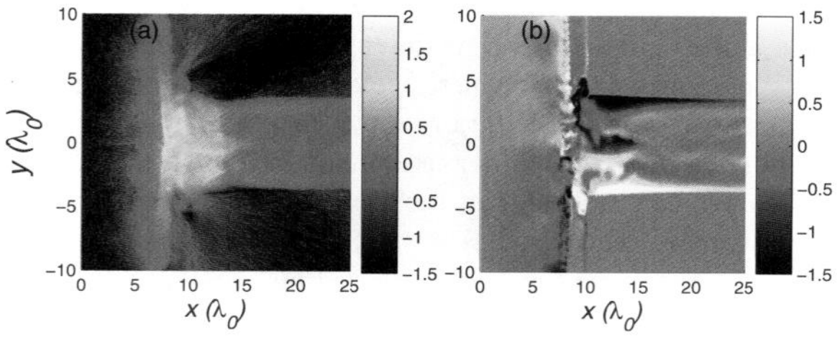

FIG. 8. (Color online) (a) The electron energy density in logarithm scale and (b) spontaneous magnetic field for the case irradiated by a larger spot size laser with $r_0=8$ μm (FWHM) at time $t=500$ fs. Here the electron energy density is normalized by $m_e c^2 n_c$.

D. Dependence of magnetic field on laser intensity and plasma density

It is worth stressing that no significant contribution of the fast electron beam in the cladding region (region 1 and region 3) to the magnetic field is observed. In another word, there is no much difference for the case irradiates with a larger spot size laser or a finite spot size laser. In this case, fast electron flows in region 1 and region 3 are very weak, $n_{b1}=n_{b3}\approx 0$. From Eq. (7), we can see that the magnetic field is only slightly influenced, which is also clearly shown in Fig. 8. For the case $Z_1 n_{i1}\gg Z_2 n_{i2}\gg n_{b2}$, Eq. (7) gives the maximum spontaneous magnetic field

$$B_{\max}\approx\frac{m_e\omega_0}{en_c}\frac{n_{h2}v_{h2}}{\sqrt{n_2/n_c}}$$

$$\simeq 200\frac{\eta}{f}\frac{I_{18}^{1/2}\lambda_{\mu m}^{-1/2}}{[(Z_2 n_{i2}-n_{h2})/n_c]^{1/2}}\frac{v_{h2}}{c}\quad(\text{MG}).\qquad(8)$$

Here, the density of fast electrons required to carry the energy flux has been approximated by $n_{h2}\simeq(2\eta/f)I_{18}(I_{18}\lambda_{\mu m}^2)^{-1/2}$.[18] At these high intensities we have approximated $2\eta/f\simeq 1/\sqrt{2}$. Equation (8) means that the magnitude of the magnetic field is strongly related to the incident laser intensity and the background electron density of the low-density-core. This provides an interesting study when experiments begin to enter this regime. Figure 9(a) shows how the peak spontaneous magnetic field $B_{\max}$ de-

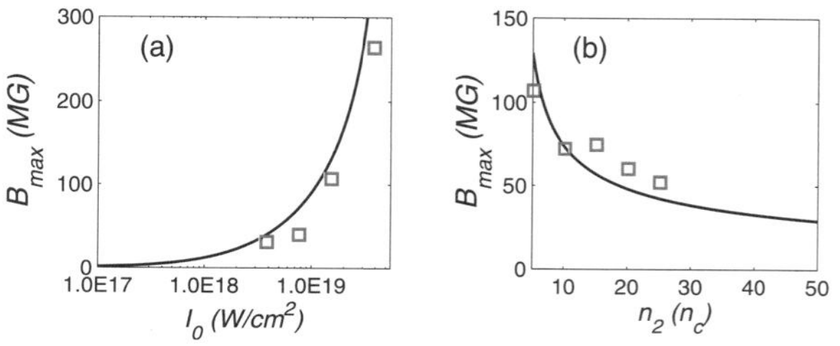

FIG. 9. (Color online) (a) Peak spontaneous magnetic field as a function of the incident laser intensity. Here, the plasma density is fixed with $Z_2 n_{i2}=5n_c$. (b) Peak spontaneous magnetic field as a function of the plasma density. Here the laser intensity is fixed with $I_L\lambda_0^2=1.5\times10^{19}$ W cm^{-2} μm^2. Solid lines are derived from Eq. (3), and squares denote the peak magnetic field obtained from PIC simulation. The simulation data are the average of the maximum values of all slices between $x=(15\lambda_0,17\lambda_0)$.

pends upon the incident laser intensity for fixed initial plasma density. One can see that $B_{\max}$ increases for fixed plasma density and growing intensity. Figure 9(b) shows that the peak spontaneous magnetic field $B_{\max}$ decreases with increasing initial plasma density for fixed laser intensity. One can see that our analytical results (solid curves) are consistent well with the simulation results (squares). However, it should be pointed out that in this section we focus on the magnetic field generated by the rapid changing of the flow velocity of the background electrons. Therefore, both the analytical results described in Eqs. (7) and (8) and the simulations in Fig. 9 take no account of the resistivity.

Up to now, we have shown that the low-density-core target whose core region is moderately overdense can collimate fast electrons well. But the pending questions concerning electron collimation is the discrepancy between hybrid-Vlasov–Fokker–Planck simulations[9] and standard PIC simulations.[12] The former has shown that the low-density-core target cannot collimate fast electrons. The answer is quite straightforward when we compare the low-density-core targets in Sec. II and this section. As discussed above, the magnetic field can be generated by two different mechanisms: (a) the gradient of the resistivity and fast electron current and (b) the rapid changing of the flow velocity of the background electrons. The latter is shown to dominate for the low-density-core target whose core region is moderately overdensed. Since it is inversely proportional to the square root of the density of the core, it is therefore recognized that the former tends to be dominant as the increasing of the core density. Remember that in the low-density-core target case, magnetic fields generated by the former play a role in scattering electrons. Therefore, the PIC simulation results are consistent with those of the hybrid-Vlasov–Fokker–Planck simulations. Actually, it has already been shown in Fig. 3(d).

IV. SUMMARY AND DISCUSSION

In summary, we have performed 2D3V collisional PIC simulations to study the magnetic collimation of fast electrons in different specially engineered multilayer targets irradiated by ultraintense laser pulses. The electron collimation, observed in many experiments, is due to the generation of magnetic fields by two different mechanisms: (1) the gradients of the resistivity for high-density-core targets and (2) nonparallel density gradient and fast electron current. These two generation mechanisms of magnetic fields have been analyzed in detail. In particular, it is found that for low-density-core targets, the collimating magnetic field can even reach as high as hundreds of MG. Furthermore, theoretical and numerical calculations revealed this magnetic field is proportional to the fast electron current and inversely proportional to the square root of the electron density of the inner core. Though it is not yet possible to draw final conclusions for how efficient is the electron transport from interaction region to the compressed core in fast ignition, our results indicate that such kind of structure targets have a potential to collimate and transport fast electrons with much less energy loss because of the low resistivity.

ACKNOWLEDGMENTS

We gratefully acknowledge C. Y. Zheng, L. H. Cao, Z. J. Liu, T. Johzaki, A. Sunahara, and K. Mima for fruitful discussions. This work was supported by the National Natural Science Foundation of China (Grant Nos. 11045001, 11005011, 10835003, and 10935003), National Basic Research Program of China (973 Program Nos. 2011CB808104, 2007CB814802, and 2007CB815101), and the National High-Tech ICF Committee of China.

[1] M. Tabak, J. Hammer, M. E. Glinsky, W. L. Kruer, S. C. Wilks, J. Woodworth, E. M. Campbell, M. D. Perry, and R. J. Mason, Phys. Plasmas **1**, 1626 (1994).
[2] S. Atzeni, A. Schiavi, J. J. Honrubia, X. Ribeyre, G. Schurtz, Ph. Nicolai, M. Olazabal-Loume, C. Bellei, R. G. Evans, and J. R. Davies, Phys. Plasmas **15**, 056311 (2008).
[3] A. Debayle, J. J. Honrubia, E. d'Humieres, and V. T. Tikhonchuk, Phys. Rev. E **82**, 036405 (2010).
[4] K. L. Lancaster, J. S. Green, D. S. Hey, K. U. Akli, J. R. Davies, R. J. Clarke, R. R. Freeman, H. Habara, M. H. Key, R. Kodama, K. Krushelnick, C. D. Murphy, M. Nakatsutsumi, P. Simpson, R. Stephens, C. Stoeckl, T. Yabuuchi, M. Zepf, and P. A. Norreys, Phys. Rev. Lett. **98**, 125002 (2007).
[5] J. S. Green, V. M. Ovchinnikov, R. G. Evans, K. U. Akli, H. Azechi, F. N. Beg, C. Bellei, R. R. Freeman, H. Habara, R. Heathcote, M. H. Key, J. A. King, K. L. Lancaster, N. C. Lopes, T. Ma, A. J. Mackinnon, K. Markey, A. McPhee, Z. Najmudin, P. Nilson, R. Onofrei, R. Stephens, K. Takeda, K. A. Tanaka, W. Theobald, T. Tanimoto, J. Waugh, L. Van Woerkom, N. C. Woolsey, M. Zepf, J. R. Davies, and P. A. Norreys, Phys. Rev. Lett. **100**, 015003 (2008).
[6] C. Ren, M. Tzoufras, F. S. Tsung, W. B. Mori, S. Amorini, R. A. Fonseca, L. O. Silva, J. C. Adam, and A. Heron, Phys. Rev. Lett. **93**, 185004 (2004).
[7] S. Kar, A. P. L. Robinson, D. C. Carroll, O. Lundh, K. Markey, P. McKenna, P. Norreys, and M. Zepf, Phys. Rev. Lett. **102**, 055001 (2009).
[8] B. Ramakrishna, S. Kar, A. P. L. Robinson, D. J. Adams, K. Markey, M. N. Quinn, X. H. Yuan, P. McKenna, K. L. Lancaster, J. S. Green, R. H. H. Scott, P. A. Norreys, J. Schreiber, and M. Zepf, Phys. Rev. Lett. **105**, 135001 (2010).
[9] A. P. L. Robinson, M. Sherlock, and P. A. Norreys, Phys. Rev. Lett. **100**, 025002 (2008); A. P. L. Robinson and M. Sherlock, Phys. Plasmas **14**, 083105 (2007).
[10] C. T. Zhou, X. G. Wang, S. Z. Wu, H. B. Cai, F. Wang, and X. T. He, Appl. Phys. Lett. **97**, 201502 (2010).
[11] C. T. Zhou, X. G. Wang, S. C. Ruan, S. Z. Wu, L. Y. Chew, M. Y. Yu, and X. T. He, Phys. Plasmas **17**, 083103 (2010); C. T. Zhou and X. T. He, *ibid.* **15**, 123105 (2008); C. T. Zhou, X. T. He, and M. Y. Yu, Appl. Phys. Lett. **92**, 071502 (2008); **92**, 151502 (2008).
[12] S. Z. Wu, C. T. Zhou, and S. P. Zhu, Phys. Plasmas **17**, 063103 (2010).
[13] H. B. Cai, K. Mima, W. M. Zhou, T. Jozaki, H. Nagatomo, A. Sunahara, and R. J. Mason, Phys. Rev. Lett. **102**, 245001 (2009).
[14] T. Takizuka and H. Abe, J. Comput. Phys. **25**, 205 (1977).
[15] Y. Sentoku and A. J. Kemp, J. Comput. Phys. **227**, 6846 (2008).
[16] A. R. Bell, J. R. Davies, and S. M. Guerin, Phys. Rev. E **58**, 2471 (1998).
[17] J. R. Davies, Phys. Rev. E **68**, 056404 (2003).
[18] A. B. Bell, A. P. L. Robinson, M. Sherlock, R. J. Kingham, and W. Rozmus, Plasma Phys. Controlled Fusion **48**, R37 (2006).
[19] F. Califano, T. Cechi, and C. Chiuderi, Phys. Plasmas **9**, 451 (2002).
[20] H. B. Cai, S. P. Zhu, M. Chen, S. Z. Wu, X. T. He, and K. Mima, "Magnetic field generation and electron collimation analysis for propagating fast electron beams in overdense plasmas," Phys. Rev. E (in press).
[21] I. D. Kaganovich, E. A. Startsev, and R. C. Davidson, Phys. Plasmas **11**, 3546 (2004).
[22] I. D. Kaganovich, R. C. Davidson, M. A. Dorf, E. A. Startsev, A. B. Sefkow, E. P. Lee, and A. Friedman, Phys. Plasmas **17**, 056703 (2010).
[23] I. D. Kaganovich, G. Shvets, E. A. Startsev, and R. C. Davidson, Phys. Plasmas **8**, 4180 (2001); I. D. Kaganovich, E. A. Startsev, A. B. Sefkow, and R. C. Davidson, Phys. Rev. Lett. **99**, 235002 (2007).

[24]I. D. Kaganovich, E. A. Startsev, A. B. Sefkow, and R. C. Davidson, Phys. Plasmas **15**, 103108 (2008).
[25]E. A. Startsev, R. C. Davidson, and M. Dorf, Phys. Plasmas **16**, 092101 (2009).
[26]Y. Sentoku, K. Mima, P. Kaw, and K. Nishikawa, Phys. Rev. Lett. **90**, 155001 (2003).
[27]M. Sherlock, A. R. Bell, R. J. Kingham, A. P. L. Robinson, and R. Bingham, Phys. Plasmas **14**, 102708 (2007).

PHYSICAL REVIEW E **83**, 036408 (2011)

Magnetic-field generation and electron-collimation analysis for propagating fast electron beams in overdense plasmas

Hong-bo Cai,[1,2,*] Shao-ping Zhu,[1] Mo Chen,[1] Si-zhong Wu,[1] X. T. He,[1,2,3] and Kunioki Mima[4,5]
[1]*Institute of Applied Physics and Computational Mathematics, Beijing 100094, People's Republic of China*
[2]*Center for Applied Physics and Technology, Peking University, Beijing 100871, People's Republic of China*
[3]*Institute for Fusion Theory and Simulation, Zhejiang University, Hangzhou 310027, People's Republic of China*
[4]*Institute of Laser Engineering, Osaka University, 2-6 Yamadaoka, Suita, Osaka 565-0871, Japan*
[5]*The Graduate School for the Creation of New Photonics Industries, 1955-1 Kurematsu, Nishiku, Hamamatsu, Sizuoka 431-1202, Japan*

(Received 19 November 2010; revised manuscript received 25 January 2011; published 17 March 2011)

An analytical fluid model is proposed for artificially collimating fast electron beams produced in the interaction of ultraintense laser pulses with specially engineered low-density-core–high-density-cladding structure targets. Since this theory clearly predicts the characteristics of the spontaneously generated magnetic field and its dependence on the plasma parameters of the targets transporting fast electrons, it is of substantial relevance to the target design for fast ignition. The theory also reveals that the rapid changing of the flow velocity of the background electrons in a transverse direction (perpendicular to the flow velocity) caused by the density jump dominates the generation of a spontaneous interface magnetic field for these kinds of targets. It is found that the spontaneously generated magnetic field reaches as high as 100 MG, which is large enough to collimate fast electron transport in overdense plasmas. This theory is also supported by numerical simulations performed using a two-dimensional particle-in-cell code. It is found that the simulation results agree well with the theoretical analysis.

DOI: 10.1103/PhysRevE.83.036408   PACS number(s): 52.57.Kk, 52.38.Kd, 52.65.Rr

I. INTRODUCTION

In the concept of fast ignition (FI) [1], a highly compressed deuterium-tritium pellet, with a density of $\sim$300 g/cc and a radius of $\sim$10–20 μm at its core, is ignited by a $\sim$10 ps, 10 kJ intense flux of MeV electrons (or ions). These high-energy particles are generated by the absorption of an intense petawatt (10^{15} W) laser, at the edge of the pellet, which is usually $\sim$50 μm away from the dense core [2]. In order to deliver enough energy ($\sim$10 kJ) into the pellet in $\sim$10 ps, the ignition laser beam should reach the intensity of $I_0 \sim 10^{20}$ W/cm^2 if the coupling from the laser to the core plasma is about 20% for a well-collimated beam [3]. However, studies have shown that the electron divergence increases with the laser intensity [4,5]. Shadowgraphy and K_α imaging measurements revealed an electron beam divergence of half angle of half maximum larger than 50° [6]. Therefore, of particular importance is the possibility of collimating the fast electron beams with the size of the compressed core.

To date, several works have been done regarding the control of the divergence of fast electron beams [7–13]. Robinson *et al.*'s hybrid-Vlasov-Fokker-Planck simulations [8] have shown some promising results, where the fast electrons are collimated in targets exhibiting a high-resistivity-core–low-resistivity-cladding structure analogous to optical waveguides. Recent experiments have also demonstrated this electron collimation in targets with a resistivity boundary [9,13]. In these cases, the magnetic-field growth can be derived from Faraday's law and Ohm's law as [14,15]

$$\frac{\partial \mathbf{B}}{\partial t} = c[\eta \nabla \times \mathbf{j}_h + (\nabla \eta) \times \mathbf{j}_h],$$

where η is the resistivity and $\mathbf{j}_h$ is the fast electron current density. Furthermore, numerical simulations by Zhou *et al.* [11] and Wu *et al.* [12] showed that a low-density-core–high-density-cladding structure target can also generate a megagauss (MG) interface magnetic field, which collimates fast electrons even better. However, according to the above equation, the magnetic field generated from the gradient of resistivity and current density should cause divergence of an electron beam, instead of collimating a beam. Therefore, their results clearly indicate that in a low-density-core–high-density-cladding structure target, other mechanisms tend to be dominant for the generation of a magnetic field. Such a spontaneous magnetic field has been explained by several mechanisms, such as anomalous resistivity [12]. However, there still does not exist a well-established and satisfactory theory, in particular, for quantitatively predicting the magnetic field and modeling its dependence on the plasma parameters.

In this paper, an analytical model describing this scenario is presented. The model, describing a uniform fast electron beam propagating in a low-density-core–high-density-cladding structure target, shows clearly the formation of the background return current flow and the generation of the spontaneous magnetic field. The mechanisms governing the generation of the magnetic field are studied in detail. It is found that the spontaneous magnetic field peaks at the interface and evanesces exponentially into the inner target over a characteristic skin depth. It is also found that the maximum magnetic field at the interface is proportional to the fast electron current and is inversely proportional to the square root of the background electron density of the inner target. Two-dimensional particle-in-cell (PIC) simulations have been run to verify this model. It is shown that the simulation results are very consistent with the prediction of our analytical model.

*caihonb@yahoo.com.cn

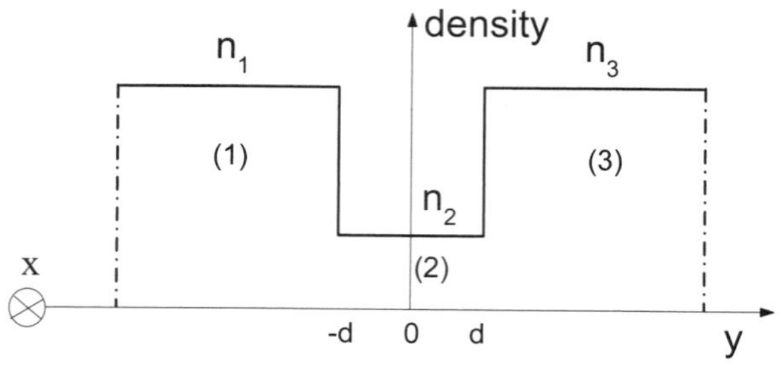

FIG. 1. Schematic of initial plasma density.

The paper is organized as follows. In Sec. II, the analytical model describing this scenario is studied in detail. In Sec. III, we present the PIC simulation for the generation of the magnetic field and the collimation of fast electron beams. In Sec. IV, a summary of the results is given.

II. ANALYTICAL DESCRIPTION OF THE SPONTANEOUSLY GENERATED MAGNETIC FIELD

We consider an ultraintense laser pulse with a large spot size normally irradiating on an unmagnetized low-density-core–high-density-cladding structure plasma with density profile as shown in Fig. 1, where ions form a fixed and neutralized background. A fast electron beam is generated at the laser plasma interface. If the spot size is large enough, we can simply assume that an equally infinite and uniform fast electron beam of density n_h propagates with an average velocity v_h along the x axis (longitudinal direction). The fast electron beam passing through the plasma quickly generates a return current of background thermal electrons. Suppose for a moment that the beam head is not neutralized by a balancing return current as the fast electron beam passes through the target. From the Maxwell equations, we get $\partial\mathbf{E}/\partial t = -4\pi\mathbf{j}_h$ [16]. The induced electric field is sufficiently strong to accelerate the background thermal electrons flowing in x direction with average velocity $v_{e0}(y)$.

For purpose of the interface magnetic-field excitation study, we assume that the fast electron beam motion is unperturbed. Since the magnetic field can develop in a very short time (of the order of a hundred femtoseconds), the combined system of the fast electron beam and the response of the background plasma can be suitably treated by a single fluid electron magnetohydrodynamic (EMHD) description. The EMHD equations, including electron-fluid equations and Maxwell equations, comprise a complete system of equations describing the electron response to the propagating fast beam pulse. The electron-fluid equations consist of the continuity equation

$$\frac{\partial n_e}{\partial t} + \nabla\cdot(n_e\mathbf{v}_{e0}) = 0, \tag{1}$$

and force balance equation

$$\frac{\partial\mathbf{p}_e}{\partial t} + (\mathbf{v}_{e0}\cdot\nabla)\mathbf{p}_e = -e\left(\mathbf{E} + \frac{1}{c}\mathbf{v}_{e0}\times\mathbf{B}\right) - \nabla P/n_e - \nu_{ei}\mathbf{p}_e, \tag{2}$$

where $-e$ is the electron charge, $\mathbf{v}_{e0}$ is the flow velocity of the background electrons, $\mathbf{p}_e = m_e\gamma_e\mathbf{v}_{e0}$ is the momentum of the background electrons, m_e is the electron rest mass, $\gamma_e = 1/\sqrt{1 - v_{e0}^2/c^2}$ is the relativistic factor, ∇P is the pressure gradient, and ν_{ei} is the e-i collision frequency. Eq. (2) can be simplified to

$$\frac{\partial(\mathbf{p}_e - \mathbf{a})}{\partial t} + \mathbf{v}_{e0}\times\nabla\times(\mathbf{p}_e - \mathbf{a}) = -\nabla\gamma_e + \nabla\phi - \nabla P/n_e - \nu_{ei}\mathbf{p}_e, \tag{3}$$

where $\mathbf{a}$ and ϕ are the vector and scalar potentials satisfying Coulomb gauge $\nabla\cdot\mathbf{a} = 0$. Operating on the electron momentum given by Eq. (3) with $\nabla\times$, we obtain the equation describing the transverse motion,

$$\frac{\partial\mathbf{\Omega}}{\partial t} + \mathbf{v}_{e0}\times\nabla\times\mathbf{\Omega} = \nabla n_e\times\nabla P/n_e^2 - \nabla\times(\nu_{ei}\mathbf{p}_e). \tag{4}$$

Here, $\mathbf{\Omega} = \nabla\times\mathbf{p_e} - e\mathbf{B}/c$ is the generalized vorticity. In the present study, due to the fast motion of the fast electron beam through the plasma, a flow in the return current is generated in the plasma with the flow velocity much faster than the electron thermal velocity. In such cases the electron pressure term can be neglected, in contrast to the case of slow beam pulses. Furthermore, there is no resistivity in the collisionless case ($\nu_{ei} = 0$). Therefore, vorticity is conserved [17–19]. If, before the arrival of the beam, we have $\nabla\times\mathbf{p_e} = 0$ and $\mathbf{B} = 0$, then $\mathbf{\Omega}(t = 0) = 0$. Eq. (4) then tells us that $\mathbf{\Omega} = 0$ for all time, so we can immediately write [20–23]

$$\mathbf{B} = \frac{c}{e}\nabla\times\mathbf{p_e}. \tag{5}$$

Similar processes have been discussed in Refs. [21–23]. This results in the relation $\frac{dv_{e0}}{dy} = -\frac{eB_0(y)}{m_e c}$ for the nonrelativistic case in the present study. Maxwell equations for the self-generated electric and magnetic fields, $\mathbf{E}$ and $\mathbf{B}$, are given by

$$\nabla\times\mathbf{B} = \frac{4\pi}{c}(-en_e\mathbf{p}_e/m_e\gamma_e + \mathbf{j}_h) + \frac{1}{c}\frac{\partial\mathbf{E}}{\partial t}. \tag{6}$$

Here, for a long beam with pulse length $l_b \gg v_h/\omega_p$, where v_h and ω_p are the fast electron beam velocity and background plasma frequency, the displacement current $\frac{1}{c}\frac{\partial\mathbf{E}}{\partial t}$ is of the order of $(v_h/\omega_p l_b)^2 \ll 1$ compared to the electron current. Therefore, in the following discussion, the displacement current is neglected [15,16,23]. Eq. (6) gives

$$\mathbf{p}_e = -\frac{m_e c\gamma_e}{4\pi e n_e}\nabla\times\mathbf{B} + m_e\gamma_e\mathbf{j}_h/en_e. \tag{7}$$

Substituting Eq. (7) into Eq. (5), we obtain the equation for the self-generated magnetic field,

$$\frac{m_e c^2}{4\pi e^2}\nabla\times\left(\frac{\gamma_e\nabla\times\mathbf{B}}{n_e}\right) + \mathbf{B} = \frac{m_e c\gamma_e}{e^2}\left(\frac{1}{n_e}\nabla\times\mathbf{j}_h - \frac{1}{n_e^2}\nabla n_e\times\mathbf{j}_h\right). \tag{8}$$

Equation (8) shows clearly that, in a collisionless case, the source of the self-generated magnetic field is $\frac{m_e c\gamma_e}{e^2}(\frac{1}{n_e}\nabla\times\mathbf{j}_h - \frac{1}{n_e^2}\nabla n_e\times\mathbf{j}_h)$, which means a target with sharp radial density boundary should build up strong magnetic fields at the boundary between a low-density core and high-density cladding, which act to collimate the flow of fast electrons. The mechanism for magnetic-field generation described here is distinct from magnetic-field generation arising from the baroclinic source ($\partial\mathbf{B}/\partial t \propto \nabla n \times \nabla T$), which requires a temperature gradient [24], and from the resistivity gradient($\frac{\partial\mathbf{B}}{\partial t} = c[\eta\nabla \times \mathbf{j_h} + (\nabla\eta) \times \mathbf{j_h}]$), which requires the collision [16]. The mechanism considered here needs the density gradient instead of the resistivity gradient. The first term on the right-hand side of Eq. (8) generates a magnetic field that pushes fast electrons toward regions of higher fast electron current density, while the second term pushes fast electrons toward regions of lower density. It is not easy to solve Eq. (8) directly, even though Eq. (8) shows the physics clearly. However, we can still derive the magnetic field from Eqs. (5) and (7). Operating on Eq. (5) with $\nabla\times$, and substituting it into Eq. (7), we obtain

$$\frac{m_e c^2}{4\pi e^2}\frac{d^2 v_{e0}(y)}{dy^2} = n_e v_{e0}(y) + n_h v_h. \tag{9}$$

In the present case, the low-density-core–high-density-cladding structure background plasma has three different regions with ion densities as follows: n_{i1} for region 1 with $y < -d$, n_{i2} for region 2 with $-d \leqslant y \leqslant d$, and n_{i3} for region 3 with $y > d$, as shown in Fig. 1. Since the fast electron beam may be affected by this background plasma, we assume the densities of the fast electron beams for regions 1, 2, and 3 are n_{h1}, n_{h2}, and n_{h3}, respectively. Therefore, we can rewrite Eq. (9) as

$$\begin{aligned}
&\left[\delta_{pe1}^2\frac{d^2}{dy^2} - 1\right]v_{e0} = \frac{n_{h1}}{n_1}v_{h1}, \quad y < -d,\\
&\left[\delta_{pe2}^2\frac{d^2}{dy^2} - 1\right]v_{e0} = \frac{n_{h2}}{n_2}v_{h2}, \quad -d \leqslant y \leqslant d\\
&\left[\delta_{pe3}^2\frac{d^2}{dy^2} - 1\right]v_{e0} = \frac{n_{h3}}{n_3}v_{h3}, \quad y > d,
\end{aligned} \tag{10}$$

where $\delta_{pej} = c/\omega_{pej}$ $(j = 1, 2, 3)$ is the collisionless electron skin depth and $\omega_{pej} = \sqrt{4\pi n_j e^2/m_e}$ $(j = 1, 2, 3)$ is the electron plasma frequency for the background plasma in the jth region. Charge neutrality requires that $n_j = Z_j n_{ij} - n_{hj}$ $(j = 1, 2, 3)$, where n_j is the background thermal electron density and Z_j is the charge state for the jth region. That is to say, some of the thermal electrons will be expelled out of the regions where fast electrons arrive. We will discuss this physical process in detail in the next section. From Eq. (10), the flow velocity of the background thermal electrons can be solved analytically as follows:

$$v_{e0}(y) = \begin{cases} c_1\exp[(y+d)/\delta_{pe1}] - n_{h1}v_{h1}/n_1, & y < -d,\\ c_2\exp(-y/\delta_{pe2}) + c_3\exp(y/\delta_{pe2}) - n_{h2}v_{h2}/n_2, & -d \leqslant y \leqslant d\\ c_4\exp[-(y-d)/\delta_{pe3}] - n_{h3}v_{h3}/n_3, & y > d \end{cases} \tag{11}$$

The constants c_1, c_2, c_3,and c_4 can be determined from the boundary conditions:

$$c_1 = \frac{\delta_{pe1}}{\delta_{pe2}}\frac{\alpha_2\beta_1 + \alpha_1\beta_2\exp(2d/\delta_{pe2}) - \alpha_1\exp(-2d/\delta_{pe2}) - \alpha_2}{\beta_1\exp(-2d/\delta_{pe2}) - \beta_2\exp(2d/\delta_{pe2})},$$

$$c_2 = -\frac{\alpha_2\beta_1\exp(-d/\delta_{pe2}) + \alpha_1\beta_2\exp(d/\delta_{pe2})}{\beta_1\exp(-2d/\delta_{pe2}) - \beta_2\exp(2d/\delta_{pe2})},$$

$$c_3 = -\frac{\alpha_1\exp(-d/\delta_{pe2}) + \alpha_2\exp(d/\delta_{pe2})}{\beta_1\exp(-2d/\delta_{pe2}) - \beta_2\exp(2d/\delta_{pe2})},$$

$$c_4 = \frac{\delta_{pe3}}{\delta_{pe2}}\frac{-\alpha_1\beta_2 - \alpha_2\beta_1\exp(-2d/\delta_{pe2}) + \alpha_1 + \alpha_2\exp(2d/\delta_{pe2})}{\beta_1\exp(-2d/\delta_{pe2}) - \beta_2\exp(2d/\delta_{pe2})},$$

where

$$\alpha_1 = \frac{\frac{n_{h2}v_{h2}}{n_2} - \frac{n_{h1}v_{h1}}{n_1}}{\frac{\delta_{pe1}}{\delta_{pe2}} + 1}, \quad \alpha_2 = \frac{\frac{n_{h2}v_{h2}}{n_2} - \frac{n_{h3}v_{h3}}{n_3}}{\frac{\delta_{pe3}}{\delta_{pe2}} - 1}, \quad \beta_1 = \frac{\delta_{pe1} - \delta_{pe2}}{\delta_{pe1} + \delta_{pe2}}, \quad \text{and} \quad \beta_2 = \frac{\delta_{pe3} + \delta_{pe2}}{\delta_{pe3} - \delta_{pe2}}.$$

Substituting Eq. (11) into Eq. (5) gives

$$\frac{eB_0}{m_e c} = -\frac{dv_{e0}(y)}{dy} = \begin{cases} -\frac{c_1}{\delta_{pe1}}\exp[(y+d)/\delta_{pe1}], & y < -d\\ \frac{1}{\delta_{pe2}}[c_2\exp(-y/\delta_{pe2}) - c_3\exp(y/\delta_{pe2})], & -d \leqslant y \leqslant d,\\ \frac{c_4}{\delta_{pe3}}\exp[-(y-d)/\delta_{pe3}], & y > d. \end{cases} \tag{12}$$

Equations (11) and (12) describe the background electron flow velocity and the spontaneous magnetic field. For simplicity, we assume a symmetric low-density-core–high-density-cladding structure target, which gives $n_1 = n_3$, $n_{h1} = n_{h3}$, and $v_{h1} = v_{h3}$. Furthermore, since the lower density region is much thicker than the skin depth $d \gg \delta_{pe2}$, it allows us to neglect the terms with $\exp(-d/\delta_{pe2})$. Therefore, we can simplify the constants as follows: $c_1 = c_4 = \frac{\delta_{pe1}}{\delta_{pe1}+\delta_{pe2}}(\frac{n_{h2}v_{h2}}{n_2} - \frac{n_{h1}v_{h1}}{n_1})$ and $c_2 = c_3 = \frac{\delta_{pe2}}{\delta_{pe1}+\delta_{pe2}}(\frac{n_{h2}v_{h2}}{n_2} - \frac{n_{h1}v_{h1}}{n_1})\exp(-d/\delta_{pe2})$. Using Eq. (12), we obtain

$$\frac{eB_0}{m_e c} = \begin{cases} \frac{-1}{(\delta_{pe1}+\delta_{pe2})}\left(\frac{n_{h2}v_{h2}}{n_2} - \frac{n_{h1}v_{h1}}{n_1}\right)\exp\left[\frac{y+d}{\delta_{pe1}}\right], & y < -d, \\ \frac{1}{(\delta_{pe1}+\delta_{pe2})}\left(\frac{n_{h2}v_{h2}}{n_2} - \frac{n_{h1}v_{h1}}{n_1}\right)\{\exp[-\frac{y+d}{\delta_{pe2}}] - \exp[\frac{y-d}{\delta_{pe2}}]\}, & -d \leqslant y \leqslant d, \\ \frac{1}{(\delta_{pe1}+\delta_{pe2})}\left(\frac{n_{h2}v_{h2}}{n_2} - \frac{n_{h1}v_{h1}}{n_1}\right)\exp\left[-\frac{y-d}{\delta_{pe1}}\right], & y > d. \end{cases} \quad (13)$$

Figure 2 shows the schematic of the background electron flow velocity and the spontaneous magnetic field. It is evident in Fig. 2 that, for this kind of low-density-core–high-density-cladding structure target, the fast electron current is neutralized by the return current of thermal electrons everywhere except at the interfaces of different density regions over a characteristic transverse distance $\Delta y_\perp = \delta_{pe}$ [20]. Hence, the spontaneous magnetic field B_0 exists within a layer of width δ_{pe} near the interfaces of different regions, $y = \pm d$. We notice that B_0 peaks at the interface and evanesces exponentially to both sides over the skin depth. Since the skin depth is $\delta_{pe} \propto n_e^{-1}$ in the lower density plasma region, the magnetic field tunnels much deeper into the region, as shown in Fig. 2. For the case $n_1 \gg n_2$, Eq. (13) gives the maximum spontaneous magnetic field

$$B_{\max} \approx \frac{m_e\omega_0}{en_c}\frac{n_{h2}v_{h2}}{\sqrt{n_2/n_c}} \simeq 200\frac{\eta}{f}\frac{I_{18}^{1/2}\lambda_{\mu m}}{[(Z_2n_{i2}-n_{h2})/n_c]^{1/2}}\frac{v_{h2}}{c}\ \text{MG}. \quad (14)$$

Here, the density of fast electrons required to carry the energy flux has been approximated by $n_{h2} \simeq \frac{2\eta}{f}(I_{18}\lambda_{\mu m}^2)^{1/2}n_c$ [16]. At these high intensities, we have approximated $2\eta/f \simeq 1/\sqrt{2}$. Equation (14) means that the magnitude of the magnetic field is strongly related to the incident laser intensity and the background electron density of the low-density core. This provides an interesting study when experiments begin to enter this regime. As a numerical example, we evaluate Eq. (14) for the case of the low-density-core–high-density-cladding structure target with $Z_1n_{i1} = Z_3n_{i3} = 200n_c$ and $Z_2n_{i2} = 10n_c$, irradiated by a laser light with $I\lambda_0^2 = 1.5 \times 10^{19}$ W/cm^{-2} μm^2, and assume $n_{h2} \approx 3n_c$ and $v_{h2} \sim c$ to obtain the qualitative estimate of the maximum magnetic field at the interface, $B_{\max} \approx$ 110 MG.

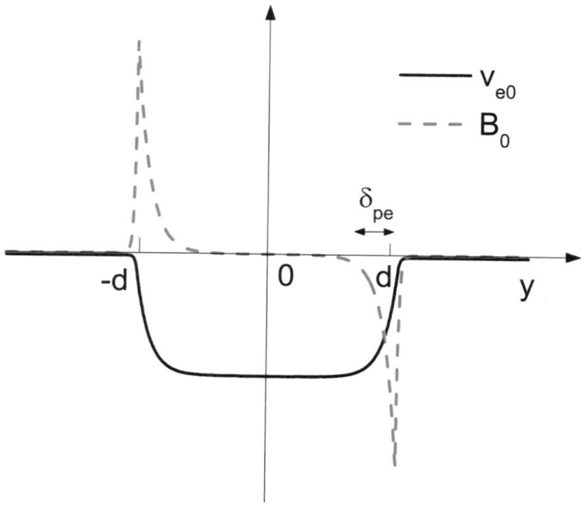

FIG. 2. (Color online) Schematic of the profile of the background electron flow velocity $v_{e0}(y)$ and the spontaneous magnetic field $B_0(y)$.

It is worth stressing that no significant contribution of the fast electron beam in the cladding regions (regions 1 and 3) to the magnetic field has been found. In other words, there is not much difference between the cases irradiating with a very large spot size laser and a finite spot size laser. In the case with a finite laser spot size, fast electron flow in region 1 and region 3 may be very weak, $n_{h1} = n_{h3} \approx 0$. From Eqs. (13) and (14), we can see that the magnetic field is only slightly influenced.

This simple analytical model is helpful for gaining insight into certain aspects of the generation of a spontaneous magnetic field and collimation of electron beams when fast electron beams propagate in a well-engineered low-density-core–high-density-cladding structure target. Unfortunately, the vast majority of experiments are inherently more complex, and these complications must be accounted for if a proper analysis of the physical process is to be done. The first, and most important, complexity we will discuss is that of the potential of beam-plasma instabilities. For this case, PIC simulation is a powerful tool to study it further.

III. NUMERICAL STUDY OF THE GENERATION OF A MAGNETIC FIELD

In order to describe the generation of magnetic fields in more detail, first we study low-density-core–high-density-cladding structure targets with the 2D3V (two spatial and three velocity components) PIC code ASCENT [25]. Figure 3 is a sketch of the geometry of the simulations. Regions 1 and 3 are fully ionized carbon (charge state $Z_i = 6$), with $m_i/m_eZ_i = 1836 \times 12/6$. Region 2 is the low-density hydrogen plasma. The widths of the three regions are $8.4\lambda_0$, $7.2\lambda_0$, and $8.4\lambda_0$, respectively. The densities of the three regions are $200n_c$, $5n_c$, and $200n_c$, respectively. In order to avoid ruining the hydrogen plasma by the laser pressure, we place another high-density target ($40n_c$) in front of the low-density-core–high-density-cladding structure target. The p-polarized laser pulse at $\lambda_0 = 1.06$ μm wavelength and 1.5×10^{19} W/cm^2 intensity irradiates the target from the left boundary. The

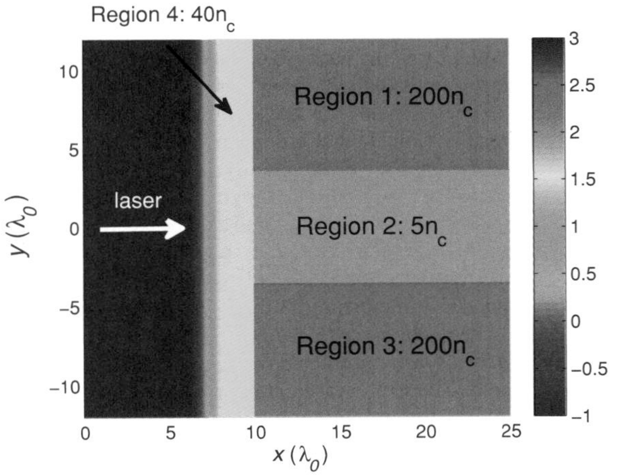

FIG. 3. (Color online) The initial density profile of the low-density-core–high-density-cladding structure target on a logarithmic scale.

intensity profile is Gaussian in the y direction with a spot size of 5.0 μm (full width at half maximum). The laser rises in $15T_0$, where T_0 is the laser period, after which the laser amplitude is kept constant.

The size of the simulation box is $30\lambda_0 \times 24\lambda_0$. Here, we use 3840×3072 grid cells with a grid size of $\Delta x = \Delta y = \lambda_0/128$. The time step used is $0.005T_0$, where T_0 denotes the laser period and its value is 3.3 fs for a 1.05-μm laser wave. During the simulation, we set 49 electrons and 49 ions per cell. The total number of particles is about 7.5×10^8. For both the fields and particles, we use the absorbing boundary condition in both x and y directions. Furthermore, in order to reduce the restrictions on the grid size compared with the Debye length, we use a fourth-order interpolation scheme to evaluate the fields and current [25]. In order to avoid the reflux of fast electrons, we set cooling buffers at the boundaries in our simulation.

Figure 4(a) shows the spontaneous magnetic field at time $t = 500$ fs. Clearly, a very large magnetic field has been generated at the interfaces, which plays a role in collimating fast electrons. Moreover, we observe that there are several

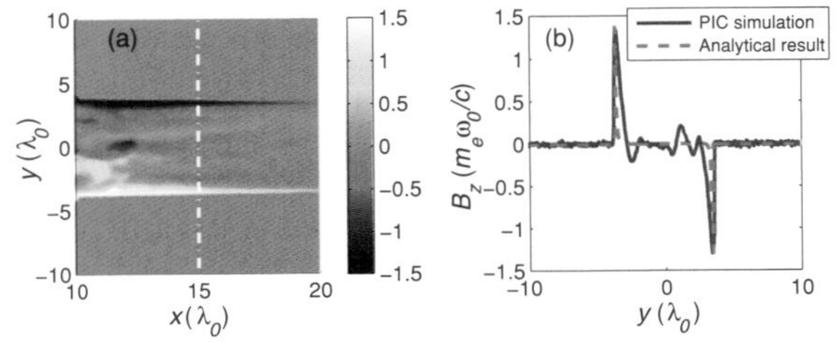

FIG. 4. (Color online) (a) The spontaneous magnetic fields at time $t = 500$ fs. (b) A slice of spontaneous magnetic fields at $x = 15\lambda_0$; the solid line is for the simulation result and the dash-dotted line is for the analytical result. Here, the magnetic fields are averaged over one laser period. The units of the magnetic field are $m_e\omega_0 c/e$ (1 unit = 100 MG).

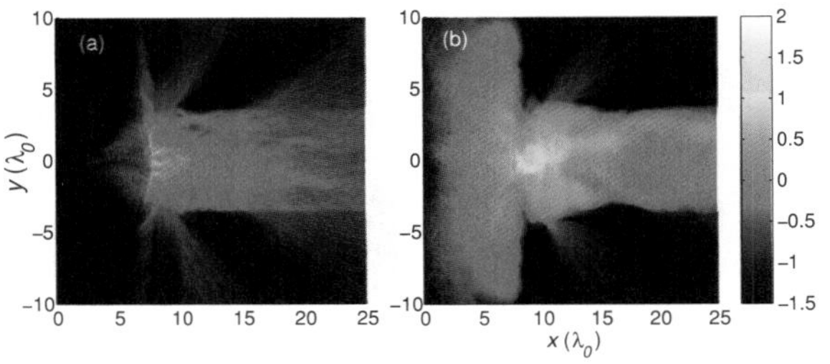

FIG. 5. (Color online) The electron energy density on a logarithmic scale at two different times, (a) $t = 264$ fs, and (b) $t = 1000$ fs. Here the electron energy density is normalized by $m_e c^2 n_c$.

stronger magnetic peaks in the inner target. These peaks are due to the formation of filament due to the Weibel instability [26]. Figure 4(b) shows a slice of spontaneous magnetic fields at $x = 15\lambda_0$; the solid line is for the simulation result and the dashed line is for the analytical result. We can see that the analytical result is consistent with the simulation result. However, the magnetic field is larger at the lower x position [as shown in Fig. 4(a)]. This is due to the complex beam-plasma instabilities (not only due to the present mechanism). At the position $x = 15\lambda_0$, the instabilities are still not developed, while the magnetic field is well developed. That is the reason why we choose to compare the analytical results with the simulated results at $x = 15\lambda_0$ or nearby. It should be noted that the suggested model is not enough to actually solve the two stream and Weibel instabilities, yet that is not the purpose of our study. Indeed, we focus on whether the interface magnetic field will develop when there is a forward fast electron beam passing the interfaces. Remarkably though, our explanation of why the 100 MG interface magnetic fields at the interfaces are generated applies to the self-consistent problem just as well.

We next examine how the spontaneous magnetic fields affect the transport of fast electron beams. In Fig. 5 the energy density distributions of electrons with energy between $0.5 \leqslant E_e \leqslant 5.0$ MeV are plotted. It is clearly seen that the fast electrons generated at the laser plasma interface have a large divergence angle. As time goes on, the electrons are highly collimated after the generation of the magnetic field, and few electrons can "leak" out into the high-density cladding.

For the fast electron beam with currents greater than the Alfvén limit, a return current moving in the opposite direction establishes approximately a charge- and current-neutral equilibrium [11,27]. Our theoretical model in Sec. II is quasineutral in both charge density and current density at the regions, except for the interfaces. Now we show how the neutralization is established. Figure 6 shows that, for electrons with energy between 1 and 50 keV [subplot (a)], and 100 and 5000 keV [subplot (b)], the mass current density extends from the layer of interaction is $\sim 7\lambda_0$ up to $25\lambda_0$ inside the plasma. This mass current density is superimposed on the structure of the spontaneous magnetic field that is mostly concentrated in the interaction region and low-density-core region. The peak value of the magnetic field is over 100 MG, which acts to collimate forward fast electrons and to scatter the returned

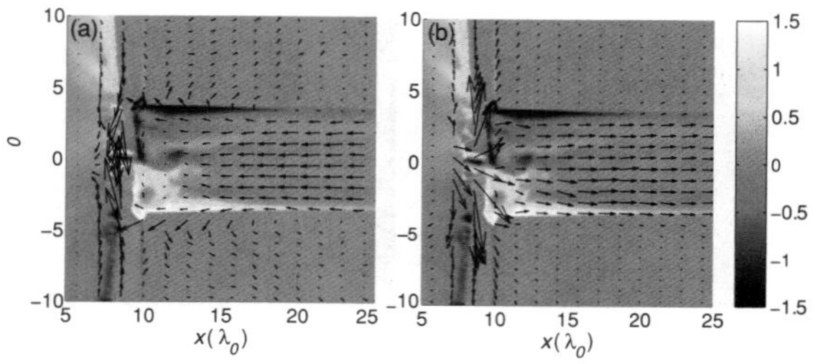

FIG. 6. (Color online) The mass flow of electrons at time $t = 500$ fs with different energies (a) $E = (1,50)$ keV and (b) $E = (100,5000)$ keV. This mass flow is superimposed on the spontaneous magnetic field at $t = 500$ fs. The units of the magnetic field are $m_e\omega_0 c/e$ (1 unit = 100 MG).

background electrons. Figure 6(a) shows clearly that some return electrons are scattered out of the low-density-core region to keep the charge neutralization. Meanwhile, the returned background electrons are accelerated to higher velocity to maintain the current neutralization. Figure 6(b) shows that the fast electron beams are guided well by the magnetic field along the interface of the low-density-core region. It is also found in Fig. 6, when the fast electron pressure is developed, the inner interfaces of the plasmas of regions 1 and 3 are eroding, resulting in the outside expansion of the magnetic fields at the interfaces of the lower x position. However, such considerations are ignored in our simple model since the theory is developed in the framework of EMHDs.

IV. SUMMARY AND DISCUSSION

In summary, we have developed a self-consistent model to explore the generation mechanisms of a spontaneous magnetic field at the interfaces of a low-density-core–high-density-cladding structure target when a relativistic electron beam flows inside. The model revealed that the rapid changing of the flow velocity of the background electrons caused by the density jump dominates the generation of the spontaneous interface magnetic field for these kinds of targets. Our analytical results indicate that the magnetic field peaks at the interface, and evanesces exponentially into the inner region over a characteristic skin depth. It is found that the maximal magnetic field, which is proportional to the fast electron current and inverse proportional to the square root of the density of the inner region, can be as high as hundreds of megagauss. This should provide an interesting study for target design when experiments begin to enter this regime. It is also found that the fast electrons are collimated very well because of this magnetic field. These processes are verified with 2D3V PIC simulations.

While the generation of a spontaneous magnetic field at the interface of a low-density-core–high-density-cladding structure target and the collimation of fast electrons have been studied here in detail, the lifetime of this kind of target during the implosion compression in a fast ignition scenario still needs further study. In this scenario, radiation hydrodynamic code is needed to further study the implosion physics.

ACKNOWLEDGMENTS

We gratefully acknowledge C. Y. Zheng, L. H. Cao, Z. J. Liu, H. Nagatomo, T. Johzaki, and A. Sunahara for fruitful discussions. This work was supported by the National Natural Science Foundation of China (Grants No. 11045001, No. 11005011, No. 10835003, and No. 10935003), National Basic Research Program of China (973 Programs No. 2011CB808104, No. 2007CB814802) and the National High-Tech ICF Committee of China.

[1] M. Tabak *et al.*, Phys. Plasmas **1**, 1626 (1994).
[2] C. Ren, M. Tzoufras, F. S. Tsung, W. B. Mori, S. Amorini, R. A. Fonseca, L. O. Silva, J. C. Adam, and A. Heron, Phys. Rev. Lett. **93**, 185004 (2004).
[3] S. Atzeni, A. Schiavi, J. J. Honrubia, X. Ribeyre, G. Schurtz, Ph. Nicolai, M. Olazabal-Loume, C. Bellei, R. G. Evans, and J. R. Davies, Phys. Plasmas **15**, 056311 (2008).
[4] A. Debayle, J. J. Honrubia, E. d'Humieres, and V. T. Tikhonchuk, Phys. Rev. E **82**, 036405 (2010).
[5] K. L. Lancaster, J. S. Green, D. S. Hey, K. U. Akli, J. R. Davies, R. J. Clarke, R. R. Freeman, H. Habara, M. H. Key, R. Kodama, K. Krushelnick, C. D. Murphy, M. Nakatsutsumi, P. Simpson, R. Stephens, C. Stoeckl, T. Yabuuchi, M. Zepf, and P. A. Norreys, Phys. Rev. Lett. **98**, 125002 (2007).
[6] J. S. Green, V. M. Ovchinnikov, R. G. Evans, K. U. Akli, H. Azechi, F. N. Beg, C. Bellei, R. R. Freeman, H. Habara, R. Heathcote, M. H. Key, J. A. King, K. L. Lancaster, N. C. Lopes, T. Ma, A. J. Mackinnon, K. Markey, A. McPhee, Z. Najmudin, P. Nilson, R. Onofrei, R. Stephens, K. Takeda, K. A. Tanaka, W. Theobald, T. Tanimoto, J. Waugh, L. Van Woerkom, N. C. Woolsey, M. Zepf, J. R. Davies, and P. A. Norreys, Phys. Rev. Lett. **100**, 015003 (2008).
[7] J. A. King, K. U. Akli, R. R. Freeman, J. Green, S. P. Hatchett, D. Hey, P. Jamangi, M. H. Key, J. Koch, K. L. Lancaster, T. Ma, A. J. MacKinnon, A. MacPhee, P. A. Norreys, P. K. Patel, T. Phillips, R. B. Stephens, W. Theobald, R. P. J. Town, L. Van Woerkom, B. Zhang, and F. N. Beg, Phys. Plasmas **16**, 020701 (2009).
[8] A. P. L. Robinson, M. Sherlock, and P. A. Norreys, Phys. Rev. Lett. **100**, 025002 (2008); A. P. L. Robinson and M. Sherlock, Phys. Plasmas **14**, 083105 (2007).
[9] S. Kar, A. P. L. Robinson, D. C. Carroll, O. Lundh, K. Markey, P. McKenna, P. Norreys, and M. Zepf, Phys. Rev. Lett. **102**, 055001 (2009).
[10] M. Borghesi, A. J. Mackinnon, A. R. Bell, G. Malka, C. Vickers, O. Willi, J. R. Davies, A. Pukhov, and J. Meyer-ter-Vehn, Phys. Rev. Lett. **83**, 4309 (1999).
[11] C. T. Zhou *et al.*, Phys. Plasmas **17**, 083103 (2010); **15**, 123105 (2008); **92**, 071502 (2008); **92**, 151502 (2008); **97**, 201502 (2010).
[12] S. Z. Wu, C. T. Zhou, and S. P. Zhu, Phys. Plasmas **17**, 063103 (2010).
[13] B. Ramakrishna, S. Kar, A. P. L. Robinson, D. J. Adams, K. Markey, M. N. Quinn, X. H. Yuan, P. McKenna, K. L.

Lancaster, J. S. Green, R. H. H. Scott, P. A. Norreys, J. Schreiber, and M. Zepf, Phys. Rev. Lett. **105**, 135001 (2010).
[14] A. R. Bell, J. R. Davies, and S. M. Guerin, Phys. Rev. E **58**, 2471 (1998).
[15] J. R. Davies, Phys. Rev. E **68**, 056404 (2003).
[16] A. B. Bell, A. P. L. Robinson, M. Sherlock, R. J. Kingham, and W. Rozmus, Plasma Phys. **48**, R37 (2006).
[17] L. M. Gorbunov, P. Mora, and T. M. Antonsen Jr., Phys. Plasmas **4**, 4358 (1997).
[18] P. Gibbon, *Short Pulse Laser Interactions with Matter* (Imperial College Press, London, 2005).
[19] B. Qiao, X. T. He, and S. P. Zhu, Europhys. Lett. **72**, 955 (2005).
[20] E. A. Startsev, R. C. Davidson, and M. Dorf, Phys. Plasmas **16**, 092101 (2009).
[21] I. D. Kaganovich, E. A. Startsev, A. B. Sefkow, and R. C. Davidson, Phys. Rev. Lett. **99**, 235002 (2007); I. D. Kaganovich, G. Shvets, E. A. Startsev, and R. C. Davidson, Phys. Plasmas **8**, 4180 (2001).
[22] I. D. Kaganovich, E. A. Startsev, and R. C. Davidson, Phys. Plasmas **11**, 3546 (2004); I. D. Kaganovich *et al.*, *ibid.* **17**, 056703 (2010).
[23] I. D. Kaganovich, E. A. Startsev, A. B. Sefkow, and R. C. Davidson, Phys. Plasmas **15**, 103108 (2008).
[24] J. A. Stamper, K. Papadopoulos, R. N. Sudan, S. O. Dean, E. A. McLean, and J. M. Dawson, Phys. Rev. Lett. **26**, 1012 (1971).
[25] H. B. Cai, K. Mima, W. M. Zhou, T. Jozaki, H. Nagatomo, A. Sunahara, and R. J. Mason, Phys. Rev. Lett. **102**, 245001 (2009).
[26] F. Califano, D. Del Sarto, and F. Pegoraro, Phys. Rev. Lett. **96**, 105008 (2006).
[27] R. J. Mason, Phys. Rev. Lett. **96**, 035001 (2006).

EPJ Web of Conferences **59**, 17017 (2013)
DOI: 10.1051/epjconf/20135917017

Study on magnetic field generation and electron collimation in overdense plasmas

Hongbo Cai[1,2,a], Shaoping Zhu[1], X.T. He[1,2] and K. Mima[3]

[1] *Institute of Applied Physics and Computational Mathematics, Beijing 100094, China*
[2] *Center for Applied Physics and Technology, Peking University, Beijing 100871, China*
[3] *Institute of Laser Engineering, Osaka University, 2-6 Yamadaoka, Suita, Osaka 565-0871, Japan*

Abstract. An analytical fluid model is proposed for artificially collimating fast electron beams produced in interaction of ultraintense laser pulses with specially engineered sandwich structure targets. The theory reveals that in low-density-core structure targets, the magnetic field is generated by the rapid change of the flow velocity of the background electrons in transverse direction (perpendicular to the flow velocity) caused by the density jump. It is found that the spontaneously generated magnetic field reaches as high as 100 MG, which is large enough to collimate fast electron transport in overdense plasmas. This theory is also supported by numerical simulations performed using a two-dimensional particle-in-cell code. It is found that the simulation results agree well with the theoretical analysis.

1. INTRODUCTION

In the fast ignition scheme (FI) [1], a high-efficiency coupling of an ultraintense beam to the compressed core is required to save igniter energy. Crucial issues for the scheme are the efficient generation of well-collimated fast electron beams and their transport from the critical density layer to the compressed fuel. To date, several works have been done regarding the control of the divergence of fast electron beams [2–5]. Robinson et al.'s hybrid-Vlasov-Fokker-Planck simulations [2] have shown some promising results, the fast electrons are collimated in targets exhibiting a high-resistivity-core-low-resistivity-cladding structure analogous to optical waveguides. Recent experiments have also demonstrated this electron collimation in targets with resistivity boundary [3, 4]. In these cases, the magnetic field growth can be derived from the Faraday's law and the Ohm's law as [2]

$$\frac{\partial \mathbf{B}}{\partial t} = c[\eta \nabla \times \mathbf{j_h} + (\nabla \eta) \times \mathbf{j_h}], \tag{1}$$

where η is the resistivity and $\mathbf{j}_h$ is the fast electron current density. Furthermore, numerical simulations by Zhou et al. [5] showed that low-density-core-high-density-cladding structure target can also generate a mega-Gauss interface magnetic field, which collimates fast electrons even better. However, according to the above equation, the magnetic field generated from the gradient of resistivity and current density should cause divergence of an electron beam, instead of collimating a beam. Therefore, their results clearly indicate that in a low-density-core-high-density-cladding structure target, other mechanisms tend to be dominant for the generation of magnetic field. To date, there still does not exist a

[a] e-mail: cai_hongbo@iapcm.ac.cn

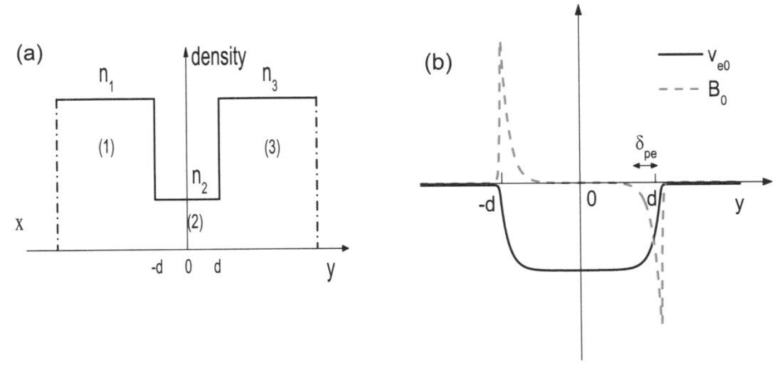

Figure 1. (a) A view of initial plasma density; (b) schematic of the profile of the background electron flow velocity $v_{e0}(y)$ and the spontaneous magnetic field $B_0(y)$.

well-established and satisfactory theory, in particular, to quantitatively predict the magnetic field and to model its dependence on the plasma parameters.

In this paper, an analytical model describing this scenario is presented. The model, describing a uniform fast electron beam propagating in a low-density-core-high-density-cladding structure target, shows clearly the formation of the background return current flow and the generation of the spontaneous magnetic field. The mechanisms governing the generation of magnetic field are studied in detail. Two-dimensional particle-in-cell (PIC) simulations have been run to verify this model. It is shown that the simulation results are consistent with the predictions of our analytical model.

2. ANALYTICAL DESCRIPTION OF THE SPONTANEOUSLY GENERATED MAGNETIC FIELD

We consider an ultraintense laser pulse with a large spot size normally irradiating an unmagnetized low-density-core-high-density-cladding structure plasma with density profile as shown in Fig. 1(a), where ions form a fixed and neutralizing background. A fast electron beam is generated at the laser plasma interface. If the spot size is large enough, we can simply assume that an equally infinite and uniform fast electron beam of density n_h propagates with an average velocity v_h along the x-axis (longitudinal direction). For purpose of the interface magnetic field excitation study, we assume that the fast electron beam motion is unperturbed. Since the magnetic field can develop in a very short time (of the order of hundred femtoseconds), the combined system of the fast electron beam and the response of the background plasma can be suitably treated by single fluid electron magnetohydrodynamic (EMHD) description. From the electron fluid equations (the continuity equation and force balance equation), we can get the relationship between the self-generated magnetic field and electron flow velocity [6, 7],

$$\mathbf{B} = \frac{c}{e}\nabla \times \mathbf{p_e}, \tag{2}$$

where $\mathbf{p_e} = m_e\gamma_e\mathbf{v}_{e0}$ is the momentum of the background electron flow and $\gamma_e = 1/\sqrt{1 - v_{e0}^2/c^2}$ is the relativistic factor. Maxwell's equations for the self-generated electric and magnetic fields, $\mathbf{E}$, and $\mathbf{B}$, are given by

$$\nabla \times \mathbf{B} = -\frac{4\pi e}{c}(n_{e0}\mathbf{p}_e/m_e\gamma_e + n_h\mathbf{v}_h) + \frac{1}{c}\frac{\partial \mathbf{E}}{\partial t}\,. \tag{3}$$

Here, for a long beam with pulse length $l_b \gg v_h/\omega_p$, where v_h and ω_p are the fast electron beam velocity and background plasma frequency, the displacement current $\frac{1}{c}\frac{\partial \mathbf{E}}{\partial t}$ is of order $(v_h/\omega_p l_b)^2 \ll 1$ compared to the electron current. In the following discussion, the displacement current $\frac{1}{c}\frac{\partial \mathbf{E}}{\partial t}$ is neglected [6]. Eq. (3) gives

$$\mathbf{p}_e = -\frac{mc\gamma_e}{4\pi e n_{e0}}\nabla\times\mathbf{B} - m\gamma_e n_h \mathbf{v}_h/n_{e0}. \tag{4}$$

Substituting Eq. (4) into Eq. (2), we obtain the equations for self-generated magnetic field,

$$\frac{mc^2}{4\pi e^2}\nabla\times\left(\frac{\gamma_e\nabla\times\mathbf{B}}{n_{e0}}\right)+\mathbf{B} = -\frac{mc}{e}\nabla\times\left(\frac{\gamma_e n_h\mathbf{v}_h}{n_{e0}}\right). \tag{5}$$

When fast electron beam velocity and background electron density is finite, then the magnetic field $\mathbf{B}=\nabla\times\mathbf{a}$ is finite. The source is $\nabla\times\left(\frac{\gamma_e n_h\mathbf{v}_h}{n_{e0}}\right)\approx-\frac{\gamma_e}{n_{e0}^2}\nabla n_{e0}\times\mathbf{j}_h$, which means the generation of self-generated magnetic field is due to the nonparallel density gradient and fast electron current. For the nonrelativistic case, $\gamma_e\sim 1$. It is not easy to solve Eq. (5) directly even though Eq. (5) shows the physics clearly. However, we can still derive the magnetic field from Eqs. (2) and (4). Operating on equation (2) with $\nabla\times$, and substituting it into Eq. (4), we obtain the equation for the flow velocity of the background electrons [6, 7]

$$\frac{m_e c^2}{4\pi e^2}\frac{d^2 v_{e0}(y)}{dy^2} = n_{e0}v_{e0}(y)+n_h v_h. \tag{6}$$

In the present case, the sandwich background plasma has three different regions with ion densities n_{i1} for region (1) $y<-d$, n_{i2} for region (2) $-d\le y\le d$, and n_{i3} for region (3) $y>d$. Since the fast electron beam may be affected by this background plasma at later time, we assume the densities of the fast electron beams are (1) n_{h1}, (2) n_{h2}, and (3) n_{h1}, respectively. From Eq. (6), we know that the fast electron beam is neutralized by the plasma return current everywhere except at the interfaces over a characteristic transverse distance. Together with the continuous boundary conditions at the interfaces, the spontaneous magnetic field B_0 can be solved analytically from Eq. (6),

$$\frac{eB_0}{m_e c} = \frac{dv_{e0}}{dy} = \begin{cases} \frac{1}{(\delta_{pe1}+\delta_{pe2})}\left(\frac{n_{h2}v_{h2}}{n_2}-\frac{n_{h1}v_{h1}}{n_1}\right)\exp\left[\frac{y+d}{\delta_{pe1}}\right], y<-d \\ \frac{1}{(\delta_{pe1}+\delta_{pe2})}\left(\frac{n_{h2}v_{h2}}{n_2}-\frac{n_{h1}v_{h1}}{n_1}\right)\left\{\exp\left[-\frac{y+d}{\delta_{pe2}}\right]-\exp\left[\frac{y-d}{\delta_{pe2}}\right]\right\}, -d\le y\le d, \\ \frac{-1}{(\delta_{pe1}+\delta_{pe2})}\left(\frac{n_{h2}v_{h2}}{n_2}-\frac{n_{h1}v_{h1}}{n_1}\right)\exp\left[-\frac{y-d}{\delta_{pe1}}\right], y>d \end{cases} \tag{7}$$

where $\delta_{pej}=c/\omega_{pej}(j=1,2,3)$ is the electron skin depth and $\omega_{pej}=\sqrt{4\pi n_j e^2/m_e}(j=1,2,3)$ is the electron plasma frequency for the background plasma in the jth region. n_j and v_{hj} denote the electron density of the background plasmas and the velocity of the fast electrons, respectively. Here, $n_3=n_1$, $n_{h3}=n_{h1}$, and $v_{h3}=v_{h1}$ have been used because of the symmetry. From Eq. (7), one can see that the spontaneous magnetic field B_0 exists within a layer of width δ_{pe} near the interfaces of different regions, $y=\pm d$. We notice that B_0 peaks at the interfaces and is exponentially evanescent to both sides. In the lower density plasma region, the magnetic field tunnels much deeper into the target since the skin depth $\delta_{pe}\propto n_e^{-1}$. It should be emphasized that the velocity variation and magnetic field occur even when $n_h(y)$ is laterally constant, i.e., $n_{h1}=n_{h2}$. The analytical results of background electron flow velocity and the spontaneous magnetic field are drawn on Fig. 1(b).

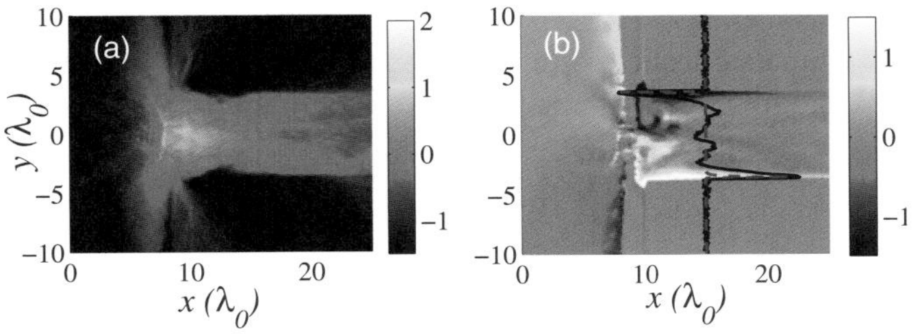

Figure 2. (a) Electron energy density in logarithm scale and (b) spontaneous magnetic fields at t = 500 fs (1 unit $\simeq$ 100 MG).

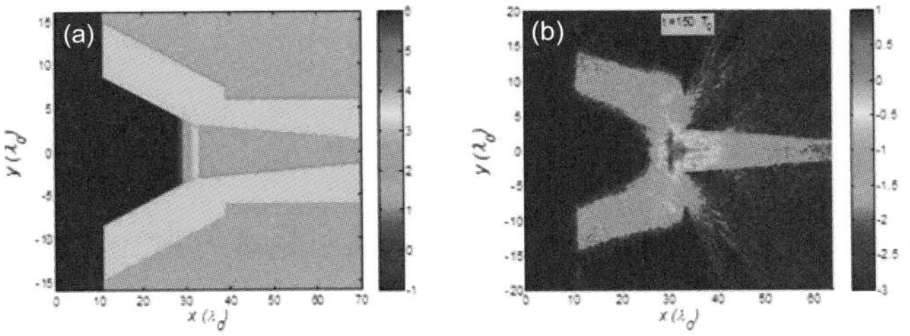

Figure 3. (a) Initial density profile of cone target with a funnel structure; (b) Electron energy density in logarithm scale at time 500 fs.

3. NUMERICAL STUDY OF THE GENERATION OF THE MAGNETIC FIELD

In order to describe the generation of magnetic fields in more details, firstly we study low-density-core-high-density-cladding structure targets with the 2D3V PIC code ASCENT [7]. The cladding regions are fully ionized carbon (charge state $Z_i = 6$) with $m_i/m_e Z_i = 1836 * 12/6$. The core region is the low density hydrogen plasma. The widths of three regions are: $8.4\lambda_0$, $7.2\lambda_0$, $8.4\lambda_0$, respectively. The densities of three regions are: $200n_c$, $5n_c$, $200n_c$, respectively. In order to avoid the ruin of the hydrogen plasma by the laser pressure, we place another high density target ($40n_c$) in front of the sandwich target. The p-polarized laser pulse at $\lambda_0 = 1.06\,\mu$m wavelength and 1.5×10^{19} W/c m^2 intensity irradiates the target from the left boundary. The intensity profile is Gaussian in the y direction with a spot size of 5.0 μm (FWHM). The laser rises in $15T_0$, after which the laser amplitude is kept constant.

In Fig. 2 (a) the energy density distributions of electrons with energy between $0.5 \leq E_e[MeV] \leq 5.0$ at $t = 500$ fs are plotted. It is clearly seen that the fast electrons generated at the laser plasma interface have a large divergence angle. As time goes on, the electrons are highly collimated after the generation of the magnetic field, few electrons can "leak" out into the high-density-cladding. We also plot the spontaneous magnetic field at time $t = 500$ fs on Fig. 2(b). Clearly, magnetic field of the order of 100 MG has been generated at the interfaces, which plays a role in collimating fast electrons. Note that the green dashed line is the analytical results from Eq. (7). It is found that the analytical result is consistent with the simulation result (the dark solid line) very well.

In cone-guided fast ignition, the laser energy is firstly deposited into fast electrons in the interaction region at the cone tip [8], which is usually 50 to 100 microns away from the dense core, and the resulting energy flux transports into the dense core. In order to collimate the fast electrons and enhance the laser coupling efficiency, we designed a cone target with a funnel structure, as shown in Fig. 3(a). From the above analysis, we expect that huge magnetic field will be generated at the inner surface of the funnel target, which can collimate fast electrons produced at the cone tip. Fig. 3(b) shows the energy density distributions of electrons with energy between $0.5 \leq E_e[MeV] \leq 5.0$ at $t = 500$ fs. We can see that the fast electrons are well collimated inside the funnel structure. Detailed computations and analysis in this last case are currently in progress, and will be the object of a future publication.

This work was supported by the National Natural Science Foundation of China (Grant Nos. 11005011, 11275028, and 10935003), and key Laboratory of Sci. & Tech. for National Defense of CAEP (Grant No. 9140C690401110C6908).

References

[1] M. Tabak *et al.*, Phys. Plasmas **1**, 1626 (1994)
[2] A.P.L. Robinson *et al.*, Phys. Plasmas **14**, 083105 (2007)
[3] S. Kar *et al.*, Phys. Rev. Lett. **102**, 055001 (2009)
[4] B. Ramakrishna *et al.*, Phys. Rev. Lett. **105**, 135001 (2010)
[5] C. T. Zhou et al. Phys. Plasmas **17**, 083103 (2010)
[6] E. A. Startsev, R. C. Davidson, and M. Dorf, Phys. Plasmas **16**, 092101 (2009)
[7] H.B. Cai *et al.*, Phys. Rev. E **83**, 036408 (2011)
[8] H. B. Cai *et al.*, Phys. Rev. Lett. **102**, 245001 (2009)

PHYSICAL REVIEW E **89**, 031101(R) (2014)

Three-dimensional fast magnetic reconnection driven by relativistic ultraintense femtosecond lasers

Y. L. Ping,[1,*] J. Y. Zhong,[1] Z. M. Sheng,[2,5] X. G. Wang,[3] B. Liu,[4] Y. T. Li,[5] X. Q. Yan,[3,4] X. T. He,[4] J. Zhang,[2,5] and G. Zhao[1,†]

[1]*Key Laboratory of Optical Astronomy, National Astronomical Observatories, Chinese Academy of Sciences, Beijing 100012, China*

[2]*Key Laboratory for Laser Plasmas (MOE) and Department of Physics and Astronomy, Shanghai Jiao Tong University, Shanghai 200240, China*

[3]*State Key Laboratory of Nuclear Physics and Technology, School of Physics, Peking University, Beijing 100871, China*

[4]*Center for Applied Physics and Technology, Peking University, Beijing 100084, China*

[5]*Beijing National Laboratory of Condensed Matter Physics, Institute of Physics, Chinese Academy of Sciences, Beijing 100080, China*

(Received 19 July 2012; revised manuscript received 12 June 2013; published 7 March 2014)

Three-dimensional fast magnetic reconnection driven by two ultraintense femtosecond laser pulses is investigated by relativistic particle-in-cell simulation, where the two paralleled incident laser beams are shot into a near-critical plasma layer to form a magnetic reconnection configuration in self-generated magnetic fields. A reconnection X point and out-of-plane quadrupole field structures associated with magnetic reconnection are formed. The reconnection rate is found to be faster than that found in previous two-dimensional Hall magnetohydrodynamic simulations and electrostatic turbulence contribution to the reconnection electric field plays an essential role. Both in-plane and out-of-plane electron and ion accelerations up to a few MeV due to the magnetic reconnection process are also obtained.

DOI: 10.1103/PhysRevE.89.031101 PACS number(s): 52.35.Vd, 52.30.−q, 52.50.Jm, 52.65.Rr

Magnetic reconnection (MR) [1,2] occurs widely in fast energy release processes such as in solar flares [3,4], coronal mass ejections, interaction of solar and magnetosphere [5], certain explosive astrophysical events [6], and fusion plasma instabilities [7]. Moreover, the relativistic magnetic reconnection may play an important role for energy conversion in relativistic objects, for example, hard x-ray and higher energy spectrum bursts in solar flares [8], pulsars [9], gamma-ray bursts [10], and active galactic nuclei [11]. With the development of high energy and high power long lasers, there is an increasing interest to investigate this kind of process in the laboratory [12–15]. Related two-dimensional particle-in-cell (2D PIC) simulation has been carried out with preassumed laser-produced plasma bubble structures [16].

These laser-driven MR configurations were realized with magnetic fields typically at the mega-gauss (MG) level produced by nonparallel temperature and density gradients [12–15], which are located near the target surface. On the other hand, it is well known that the intensive quasistatic magnetic fields over 100 MG can be generated by the relativistic intense lasers interacting with plasma, as shown numerically [17–19] and theoretically [20–22]. Near giga-Gauss (GG) magnetic fields have been measured in intense laser interaction with solid targets, where the region of strong magnetic fields is located both at surface and inside the target [23]. Due to dynamic evolution of the magnetic fields, magnetic interaction [24] and reconnection [25] have been foreseen in such laser-produced plasmas.

In this Rapid Communication, we report a relativistic PIC simulation using KLAP code [26,27] for a three-dimensional (3D) MR process occurring in intense femtosecond laser pulse produced plasmas. Two femtosecond laser pulses, parallel to each other, are shot into a target of near critically dense plasma layer. Fast reconnection of the GG level magnetic field inside the target is then demonstrated. Such a magnetic reconnection process can be possibly realized experimentally with subpicosecond subpetawatt intense laser systems such as LFEX [28] and OMEGA-EP [29].

The simulation is performed in a box of $L_{x0} = L_{y0} = 24\ \mu$m and $L_{z0} = 40\ \mu$m, with $480 \times 480 \times 800$ cells and 10^9 quasiparticles. Two identical circularly polarized laser beams shoot at the target and then propagate in parallel in the box along the z axis. The lasers have a peak intensity of 5×10^{20} W/cm^2 corresponding to a normalized laser vector potential $a_0 = 13.5$, a spot diameter of 3 μm, the wavelength $\lambda_0 = 1\ \mu$m, and the laser period $T_0 = \lambda_0/c$. The incident laser pulse has a rising front of two laser cycles followed by a flat top in the z direction and is a Gaussian form in the x,y directions. The distance between two laser centroids is 8 μm. A near-critical density plasma target is located in $5\ \mu\text{m} < z < 35\ \mu$m with nonuniform density in the z direction and uniform density in the x,y directions. The density linearly increases to a critical density from 5 to 10 μm, and then remains constant for $10\ \mu\text{m} < z < 15\ \mu$m, then linearly decreases as $n = [1 - (z - z0)/L_0]n_c$, $L_0 = 20\ \mu$m for $15\ \mu\text{m} < z < 35\ \mu$m, where $z_0 = 15\ \mu$m. In the transverse directions, periodic boundary conditions are applied for particles and fields. Particles are reflected and fields are absorbed at the boundaries in the longitudinal direction. The temperatures of the initial ion and electron are 0.01 and 10 keV, respectively. The ions are simply protons with the mass ratio $m_p/m_e = 1836$. The critical density is $n_c = m_e\omega_0^2/4\pi e^2 = 1.15 \times 10^{21}$ cm^{-3}, where m_e is the electron rest mass, ω_0 represents the laser angular frequency, and e is the element charge. It has been shown by Nakamura *et al.* [30] and Bulanov *et al.* [31] that the produced magnetic fields can expand laterally when intense laser beams propagate in inhomogeneous plasmas from the higher density to the lower density or vacuum.

When an ultraintense laser propagates in underdense plasmas, both electron and ion density channels are formed. For the circularly polarized laser pulses, an axial field B_z can be generated through the inverse Faraday effect [20–22],

*ylping@bao.ac.cn

†gzhao@bao.ac.cn

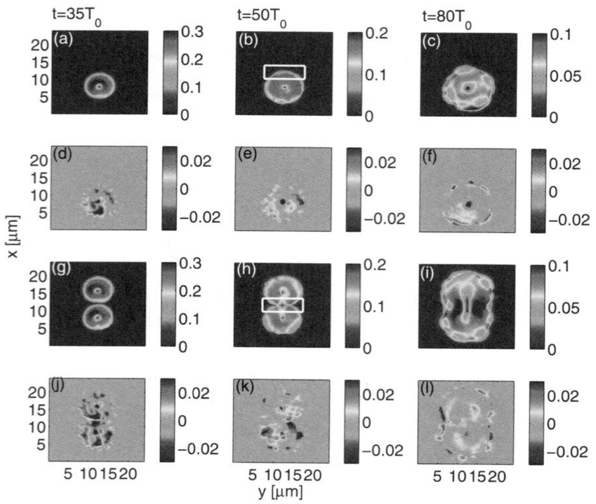

FIG. 1. (Color online) Snapshots (at $z = 20.7\ \mu$m from $35T_0$ to $80T_0$) of magnetic fields **B**. (a)–(c) Azimuthal magnetic fields B_θ and (d)–(f) out-of-plane magnetic fields B_z produced by a single incident laser. (g)–(i) B_θ and (j)–(l) B_z by two incident lasers. Here, the magnetic fields are normalized by the initial laser field $B_0 = \sqrt{I/\epsilon c}/c = 1.45$ GG. The region in the white line surrounded box contains a reconnection region and an outflow region.

and an azimuthal field B_θ is produced from the longitudinal current driven by the laser acceleration [32,33]. The evolution of the magnetic fields, normalized to the initial laser field $B_0 = \sqrt{I/\epsilon c}/c = 1.45$ GG, from $35T_0$ to $80T_0$ in plasma at $z = 20.7\ \mu$m is shown in Figs. 1(g)–1(l). As a reference, the field generated by a single incident laser pulse is also plotted in Figs. 1(a)–1(f). Clearly the azimuthal magnetic field expands laterally due to the force acting on the vortex in the $\nabla n \times \mathbf{\Omega}$ direction [34] with its topological structure kept unchanged, as an expanding bubble with an O-point-type null at the center. The evolution of the azimuthal magnetic field produced by the two lasers is shown in Figs. 1(g)–1(i). Irradiated by the lasers, the azimuthal fields expand toward each other at $t = 35T_0$. At this moment, the magnetic fields of two bubbles do not yet meet [Fig. 1(g)]. At $t = 50T_0$, however, much of the magnetic flux is already reconnected, as shown in Fig. 1(h). The X-point-type magnetic null is then found between the two bubbles. It is the place where the magnetic field vanishes as the magnetic bubbles with different polarities meet one another, that MR occurs. As the process goes on, the two bubbles start to merge together into a single big bubble as shown in Fig. 1(i). The area with heavily diminished field strength also increases, where the magnetic energy is converted to kinetic energy of electrons and ions, forming outflow jets along the y direction and out-of-plane acceleration along the z direction, as shown in Fig. 4 later. At $t = 80T_0$, the two bubbles merge with each other completely and the new magnetic topological structure is formed.

The out-of-plane magnetic fields B_z generated by a single laser pulse and two lasers are also shown in Figs. 1(d)–1(f) and 1(j)–1(l), respectively. In the case of one laser [Fig. 1(d)–1(f)], it is found that the magnetic field structure is always nearly centrosymmetric. In the case of two lasers [Figs. 1(j)–1(l)], however, a quadrupole structure appears around the reconnection site, which is a typical signature of the Hall effect [1,35]. The reconnection process is clearly collisionless since the width of the diffusion region [$<5\ \mu$m as shown in Fig. 3(b)] is much less than the ion skin depth. The ion skin depth is calculated as $d_i = c/\omega_{pi} \approx 8.3\ \mu$m at $z = 20.7\ \mu$m, where $\omega_{pi} = (4\pi e^2 n_0/m_i)^{1/2}$ is the ion plasma frequency. However, the electron skin depth is $d_e = c/\omega_{pe} \approx 0.71\ \mu$m, where the electron plasma frequency is $\omega_{pe} = (4\pi e^2 n_0/\gamma_0 m_e)^{1/2}$ with

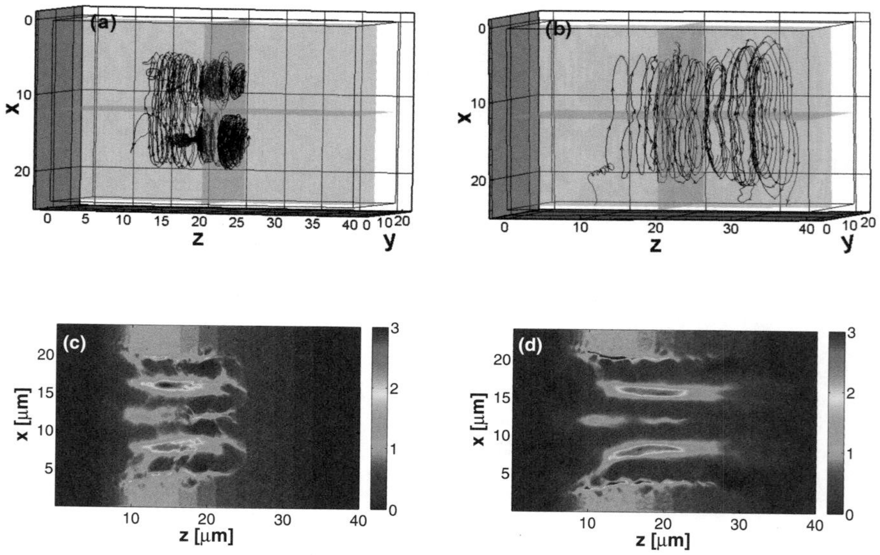

FIG. 2. (Color online) (a), (b) Snapshots of three-dimensional magnetic field lines (a) at $35T_0$ and (b) at $50T_0$. (c), (d) Corresponding electron density normalized by n_c in the x-z plane at $y = 12\ \mu$m.

the relativistic factor $\gamma_0 \approx \sqrt{1+a_0^2}$. Since the reconnection region size is less than d_i, the Hall magnetohydrodynamics (MHD) model may not be applied and the electron MHD (EMHD) model is used in the fluid approach [36]. In EMHD, the generalized vorticity should be frozen in the electron component if the plasma density is initially uniform. In the region where the displacement current effect is significant, the general vorticity of $\mathbf{\Omega}_e = -e(1 - d_e^2\nabla^2)\mathbf{B}/c$ should be rewritten as $\mathbf{\Omega}_e = -\frac{e}{c}[1 - d_e^2(\nabla^2 - \frac{1}{c^2}\frac{\partial^2}{\partial t^2})]\mathbf{B}$. Then the optical propagating operator of $\nabla^2 - \partial^2/c^2\partial t^2$ can then spread the electromagnetic fluctuations, in the small scale of and below d_e, out of the reconnection region. Additionally, the nonuniformity of the plasma induces the Biermann battery effect, also called the baroclinic effect in fluid theory, of $\nabla n_e \times \nabla T_e$ to break the frozen-in condition and meanwhile to generate the magnetic field.

The 3D magnetic field configurations and corresponding electron density are plotted in Fig. 2. At $t = 35T_0$ when the laser pulses have not yet reached most of the right half of the simulation box, two magnetic flux tubes have already clearly shaped [shown in Fig. 2(a)] and the lasers transport in the plasmas in Fig. 2(c). Then, a fully developed 3D magnetic reconnection configuration is formed at $t = 50T_0$ [Fig. 2(b)] with the two flux tubes reconnecting into one in the outer region.

Figures 3(a)–3(d) show the reconnection electric field at two characteristic moments ($35T_0$ and $50T_0$). At $t = 35T_0$ reconnection has not yet fully developed and the electric field $\langle E_z\rangle$ between the bubbles is very low. As time increases the reconnection electric field component $\langle E_z\rangle$ at the X point starts to increase and reaches its maximum at $t = 50T_0$.

Another important parameter for MR is the reconnection rate $c\langle E_z\rangle/\langle B_A V_A\rangle$, where the mean field of a variable A, $\langle A\rangle$, is smoothened over the fast oscillation and then averaged over the z direction, and its fluctuation $\delta A = A - \langle A\rangle$, $\langle B_A\rangle$ is asymptotic magnetic field strength and $\langle V_A\rangle$ is the corresponding Alfvén speed, as $\langle V_A\rangle = \langle B_A\rangle/\sqrt{4\pi\rho}$, where $\rho = m_p n_i + m_e n_e$ is the mass density of the plasma with n_i and n_e being the ion and electron number densities, respectively. The reconnection electric field $\langle E_z\rangle$ on the line of $y = 12\ \mu$m at $35T_0$ and $50T_0$ is then plotted in Figs. 3(c) and 3(d). As shown above, at $t = 35T_0$, flux tubes have not fully formed in the simulation domain and the corresponding reconnection field near the X point $x = 12\ \mu$m are very small. At $t = 50T_0$ reconnection is proceeding throughout the box and the reconnection electric field reaches its maximum. The *asymptotic* Alfvén speed is $\langle V_A\rangle = 1.483 \times 10^9$ cm/s (nearly 1/20 of the speed of light in vacuum) with $\langle B_A\rangle = 0.1$ at $x = 9.75\ \mu$m. The reconnection rate is then calculated to be $c\langle E_z\rangle/\langle B_A V_A\rangle = 6.72$, almost two orders of magnitude higher than the typical value 0.1 obtained by 2D MHD simulations for Hall magnetic reconnection [37]. In fact, in the short pulse laser-driven duration, the ion is approximately at rest in comparison with the electron, as shown in Fig. 4. Then the reconnection electrical field should be scaled by the electron Alfvén velocity $V_{Ae} \sim c/\sqrt{\epsilon} \approx 0.9992c$, where $\epsilon = 1 + 2(\omega_{pe}/\Omega_c)^2$ is the relative permittivity with the electron cyclotron frequency $\Omega_c = eB/m_e c$, as $c\langle E_z\rangle/\langle B_A V_{Ae}\rangle \approx 0.44$. It is still much faster than that of 2D MHD simulations, a significant feature of strong driven fast reconnection.

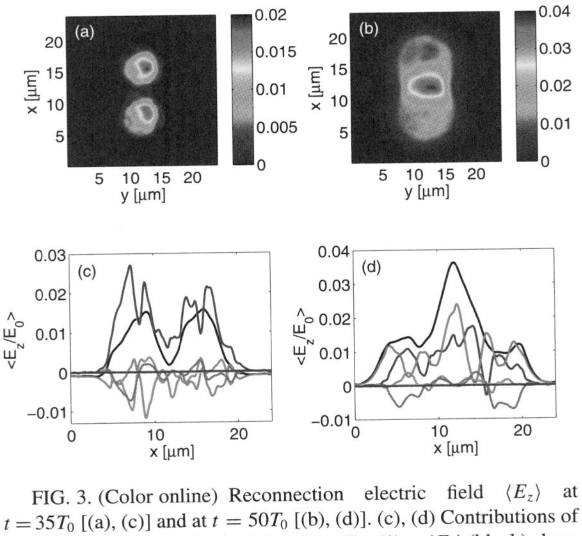

FIG. 3. (Color online) Reconnection electric field $\langle E_z\rangle$ at $t = 35T_0$ [(a), (c)] and at $t = 50T_0$ [(b), (d)]. (c), (d) Contributions of terms in the generalized Ohm's law from Eq. (1) to $\langle E_z\rangle$ (black) along the x axis at (c) and (d) for $y = 12\ \mu$m, as $\frac{1}{e\langle n_e\rangle}(\langle\mathbf{j}\rangle \times \langle\mathbf{B}\rangle)_z$ (purple), $-\frac{1}{e\langle n_e\rangle}\langle\nabla\cdot\mathbf{P}_e\rangle_z$ (blue), $\frac{m_e}{e^3\langle n_e\rangle}(\langle\mathbf{j}\rangle\cdot\nabla\langle\frac{\mathbf{j_z}}{\mathbf{n_e}}\rangle)$ (brown), $\frac{m_e}{\langle n_e\rangle e^2}\frac{\partial\langle\mathbf{j_z}\rangle}{\partial t}$ (yellow), $-\frac{1}{\langle n_e\rangle}\langle\delta n_e\delta E_z\rangle$ (red), and $\frac{1}{e\langle n_e\rangle}\langle\delta\mathbf{j}\times\delta\mathbf{B}\rangle_z$ (green). $\langle A\rangle$ means that variable A is smoothed over the fast oscillation and then averaged over the z direction from 5 to 35 μm. Here, the electric fields is normalized by the initial laser field $E_0 = cB_0 = 4.34 \times 10^{13}$ V/m.

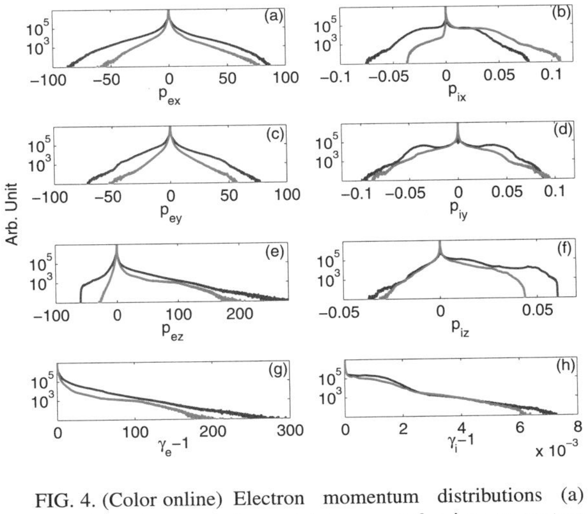

FIG. 4. (Color online) Electron momentum distributions (a) $p_{ex} = \gamma_e\beta_{ex}$, (c) $p_{ey} = \gamma_e\beta_{ey}$, and (e) $p_{ez} = \gamma_e\beta_{ez}$; ion momentum distributions (b) $p_{ix} = \gamma_i\beta_{ix}$, (d) $p_{iy} = \gamma_y\beta_{iy}$, and (f) $p_{iz} = \gamma_i\beta_{iz}$, as well as their energy distributions (g) for electrons normalized by $m_e c^2$ and (h) for ions normalized by $m_i c^2$, in the white line surrounded box ($10\ \mu\text{m} < x < 14\ \mu\text{m}, 7\ \mu\text{m} < y < 17\ \mu\text{m}$) shown in Figs. 1(b) and 1(h) with $11\ \mu\text{m} < z < 35\ \mu\text{m}$ at $t = 50T_0$. The red lines are for a laser and the blue lines are for two lasers.

From the generalized Ohm's law [38,39] of the mean field, $\langle E_z \rangle$ can be written as

$$\langle E_z\rangle = \frac{1}{e\langle n_e\rangle}(\langle \mathbf{j}\rangle \times \langle \mathbf{B}\rangle)_z + \frac{-1}{e\langle n_e\rangle}\langle \nabla \cdot \mathbf{P}_e\rangle_z + \frac{m_e}{e^3\langle n_e\rangle}\left(\langle \mathbf{j}\rangle \cdot \nabla\left\langle \frac{\mathbf{j_z}}{\mathbf{n_e}}\right\rangle\right) + \frac{\mathbf{m_e}}{\langle \mathbf{n_e}\rangle \mathbf{e}^2}\frac{\partial \langle \mathbf{j_z}\rangle}{\partial \mathbf{t}} + \frac{-1}{\langle n_e\rangle}\langle \delta n_e \delta E_z\rangle + \frac{1}{e\langle n_e\rangle}\langle \delta \mathbf{j} \times \delta \mathbf{B}\rangle_z. \quad (1)$$

In Eq. (1), $\frac{-1}{\langle n_e\rangle}\langle \delta n_e \delta E_z\rangle$ is the electrostatic (ES) turbulence contribution and $\frac{1}{e\langle n_e\rangle}\langle \delta \mathbf{j} \times \delta \mathbf{B}\rangle_z$ is the electromagnetic (EM) turbulence contribution to the reconnection field, while the first term is the Hall field, the second is the pressure gradient effect, and the third and the fourth are the electron inertial terms. In Figs. 3(c) and 3(d), the distributions of each of those terms and the reconnection electric field along the x direction are given at $t = 35T_0$ and $t = 50T_0$, respectively. It can be seen that both the electron pressure tensor gradient and ES turbulence dominate the reconnection process near the X point at $t = 50T_0$, different from previous studies in 3D reconnection where the EM turbulence plays an important role at the X point, while the ES turbulence contribution is very small [38,39]. This is due to the fact that the ultraintense lasers are injected continuously into the plasmas to generate intensive ES fluctuations propagating in the z direction, and the EM propagating operator in the general vorticity spreads the EM fluctuations out of the reconnection region.

A major consequence of reconnection is that the magnetic energy can be partially converted to particle kinetic energy. We then focus on the particle acceleration in the volume around the diffusion area, $10\ \mu\text{m} < x < 14\ \mu\text{m}$, $7\ \mu\text{m} < y < 17\ \mu\text{m}$, and $11\ \mu\text{m} < z < 35\ \mu\text{m}$ for MR at $t = 50T_0$. The distribution functions for the transverse and longitudinal momentum of the electrons and ions in this volume are then shown in Figs. 4(a)–4(f), respectively. In comparison with the single laser case (illustrated by red lines, without MR), clearly both electrons and ions are significantly accelerated in the MR process. For the single laser case the expansion of the bubble in the x direction can be seen in Fig. 4(b) for p_{ix}. Clearly, for the two laser case, ions are decelerated in the x direction in the selected box due to reconnection. On the other hand, electrons are still accelerated in the x direction though there should be a stagnation point of electron flow at the X point. This is due to the compressibility and z-direction acceleration of the electrons, and might be a reason why the reconnection rate is much higher than that of 2D MHD cases. For the longitudinal momentum p_z, more electrons are accelerated towards the $-z$ direction than that in the single laser case, while protons are accelerated in the $+z$ direction, due to the induced E_z field, as shown in Fig. 3. In the outflow directions, however, from Figs. 4(c) and 4(d), it is found that both ions and electrons are accelerated substantially. It is shown clearly that a great portion of high energy protons are ejected to both sides together with peaks at $\pm V_A$ [Fig. 4(d)], with $p_{iy} \approx \pm 0.04$ [Fig. 4(d)].

In Fig. 4(g), the energy spectra of out-of-plane electrons and ions are shown at $t = 50T_0$ in the diffusion region around the X-point region. With reconnection, the electron spectrum clearly shows a collective acceleration by a $\Delta\gamma \sim 20$–50 almost throughout the spectrum, corresponding to an energy gain of MeV level. It can also be seen in Fig. 4(c) that the electron momentum p_{ey} in the reconnection outflow direction approximately peaks at ± 50. Therefore, our results provide evidence of particle acceleration and ultrahigh energy electron generation by magnetic reconnection in the X-point region in intense laser generated plasmas.

In summary, fast magnetic reconnection driven by two intense femtosecond lasers is studied by a self-consistent 3D relativistic PIC simulation. Magnetic reconnection geometry is built up while two identical intense laser pulses propagate in a near-critical density plasma. Evolution of the topological structure of magnetic reconnection is plotted to reveal the magnetic X- point in plane (of reconnection) and the out-of-plane quadrupole magnetic field component. The reconnection rate is calculated as $c\langle E_z\rangle/\langle B_A V_{Ae}\rangle = 0.44$ scaled by the electron Alfvén velocity instead of the Alfvén velocity. By the initial driven laser field E_0, the reconnection rate can be scaled as $\langle E_z\rangle = 0.036$, larger than that previously obtained in two-fluid models. It shows clearly the strong driven and 3D reconnection features. Investigating the generalized Ohm's law, we find that the electron pressure tensor gradient and electrostatic turbulence terms dominate the reconnection process. By analyzing momentum distributions for electrons and ions in the reconnection area, both of them are accelerated in the reconnection outflow directions. For the longitudinal direction, electrons are accelerated backward while ions are accelerated forward by the induced reconnection electric field. Moreover, the energy spectra of electrons show ultrahighly energetic electron acceleration by an energy gain of MeV with $\Delta\gamma \sim 20$–50. It may provide new mechanisms for some explosive astrophysical phenomena.

J.Y.Z. and Y.L.P. acknowledge the support of the K. C. Wong Education Foundation. The authors are grateful to Dr. H. Ji of PPPL for helpful discussions. This work was supported by the National Basic Research Program of China (Grants No. 2013CBA01502, No. 2013CBA01503, and No. 2013CBA01504) and by the National Natural Science Foundation of China (Grants No. 10905004, No. 11135012, No. 11121504, No. 11273033, No. 11220101002, No. 11261140326, and No. 10925421). Computer runs were performed on the Laohu high performance computer cluster of the National Astronomical Observatories, Chinese Academy of Sciences (NAOC).

[1] M. Yamada, R. Kulsrud, and H. Ji, Rev. Mod. Phys. **82**, 603 (2010).
[2] D. Biskamp, *Magnetic Reconnection in Plasmas* (Cambridge University Press, Cambridge, England, 2000).
[3] P. A. Sweet, in *The Neutral Point Theory of Solar Flares*, edited by B. Lehnert, IAU Symposia 6 (Cambridge University Press, Cambridge, England, 1958), p. 123.
[4] E. N. Parker, Geophys. Res. **62**, 509 (1957).

[5] P. Brady, T. Ditmire, W. Horton, M. L. Mays, and Y. Zakharov, Phys. Plasmas **16**, 043112 (2009).
[6] J. Goodman and D. Uzdensky, Astrophys. J. **688**, 555 (2008).
[7] J. B. Taylor, Rev. Mod. Phys. **58**, 741 (1986).
[8] R. P. Lin *et al.*, Astrophys. J. **595**, L69 (2003).
[9] Yu. E. Lyubarskii, Astron. Astrophys. **308**, 809 (1996); Y. Lyubarsky and J. G. Kirk, Astrophys. J. **547**, 437 (2001); J. G. Kirk and O. Skjraasen, *ibid.* **591**, 366 (2003).
[10] G. Drenkhahn, Astron. Astrophys. **387**, 714 (2002); B. Zhang, and H. Yan, Astrophys. J. **726**, 90 (2011); J. C. McKinney and D. A. Uzdensky, Mon. Not. R. Astron. Soc. **419**, 573 (2012).
[11] T. Di Matteo, Mon. Not. R. Astron. Soc. **299**, L15 (1998); G. T. Birk, A. R. Crusius-Waetzel, and H. Lesch, Astrophys. J. **559**, 96 (2001).
[12] P. M. Nilson *et al.*, Phys. Rev. Lett. **97**, 255001 (2006).
[13] P. M. Nilson *et al.*, Phys. Plasmas **15**, 092701 (2008).
[14] C. K. Li, C. K. Li, F. H. Séguin, J. A. Frenje, J. R. Rygg, R. D. Petrasso, R. P. J. Town, O. L. Landen, J. P. Knauer, and V. A. Smalyuk, Phys. Rev. Lett. **99**, 055001 (2007).
[15] J. Zhong *et al.*, Nat. Phys. **6**, 984 (2010).
[16] W. Fox, A. Bhattacharjee, and K. Germaschewski, Phys. Rev. Lett. **106**, 215003 (2011).
[17] A. Pukhov and J. Meyer-ter-Vehn, Phys. Rev. Lett. **76**, 3975 (1996).
[18] J. Fuchs, G. Malka, J. C. Adam *et al.*, Phys. Rev. Lett. **80**, 1658 (1998).
[19] M. Tanimoto, S. Kato, E. Miura, N. Saito, K. Koyama, and J. K. Koga, Phys. Rev. E **68**, 026401 (2003).
[20] Xian-Tu He, Acta Phys. Sin. **32**, 325 (1983) (in Chinese).
[21] Z. M. Sheng and J. Meyer-ter-Vehn, Phys. Rev. E **54**, 1833 (1996).
[22] B. Qiao, S.-p. Zhu, C. Y. Zheng, and X. T. He, Phys. Plasmas **12**, 053104 (2005).
[23] U. Wagner *et al.*, Phys. Rev. E **70**, 026401 (2004).
[24] G. A. Askar'yan *et al.*, JETP Lett. **60**, 251 (1994).
[25] G. A. Askar'yan *et al.*, Comments Plasma Phys. Controlled Fusion **17**, 35 (1995).
[26] M. Chen *et al.*, Chin. J. Comput. Phys. **25**, 43 (2008); X. Q. Yan, C. Lin, Z. M. Sheng, Z. Y. Guo, B. C. Liu, Y. R. Lu, J. X. Fang, and J. E. Chen, Phys. Rev. Lett. **100**, 135003 (2008).
[27] B. Liu, H. Y. Wang, J. Liu, L. B. Fu, Y. J. Xu, X. Q. Yan, and X. T. He, Phys. Rev. Lett. **110**, 045002 (2013).
[28] H. Shiraga *et al.*, Plasma Phys. Control. Fusion **53**, 124029 (2011).
[29] J.-C. Gauthier, B. Hammel, H. Azechi, and C. Labaune, J. Phys. IV France **133**, V-6 (2006).
[30] T. Nakamura, S. V. Bulanov, T. Z. Esirkepov, and M. Kando, Phys. Rev. Lett. **105**, 135002 (2010).
[31] S. S. Bulanov *et al.*, Phys. Plasmas **17**, 043105 (2010).
[32] A. Pukhov, Z. M. Sheng, and J. Meyer-ter-Vehn, Phys. Plasmas **6**, 2847 (1999).
[33] Z. M. Sheng, K. Mima, Y. Sentoku, M. S. Jovanovic, T. Taguchi, J. Zhang, and J. Meyer-ter-Vehn, Phys. Rev. Lett. **88**, 055004 (2002).
[34] J. Nycander and M. B. Isichenko, Phys. Fluids B **2**, 2042 (1990); S. K. Yadav, A. Das, and P. Kaw, Phys. Plasmas **15**, 062308 (2008).
[35] M. Yamada, Phys. Plasmas **14**, 058102 (2007).
[36] S. V. Bulanov, F. Pegoraro, and A. S. Sakharov, Phys. Fluids B **4**, 2499 (1992); L. C. Steinhauer and A. Ishida, Phys. Plasmas **5**, 2609 (1998); S. V. Bulanov, T. Zh. Esirkepov, D. Habs, F. Pegoraro, and T. Tajima, Eur. Phys. J. D **55**, 483 (2009).
[37] J. D. Huba and L. I. Rudakov, Phys. Rev. Lett. **93**, 175003 (2004).
[38] H. Che, J. F. Drake, and M. Swisdak, Nature (London) **474**, 184 (2011).
[39] Keizo Fujimoto and Richard D. Sydora, Phys. Rev. Lett. **109**, 265004 (2012).

20pD2-2

Self-Generated Magnetic Dipoles in Weakly Magnetized Beam-Plasma System

Qing Jia[1,2], Kunioki Mima[2,3], Hong-bo Cai[4], Toshihiro Taguchi[5], Hideo Nagatomo[2], and X. T. He[1,4]

[1]*Center for Applied Physics and Technology, Peking University, Beijing, China*
[2]*Institute of Laser Engineering, Osaka University, Osaka, Japan*
[3]*Graduate School for the Creation of New Photonics Industries, Hamamatsu, Japan*
[4]*Institute of Applied Physics and Computational Mathematics, Beijing, China*
[5]*Department of Electric and Electronic Engineering, Setsunan University, Osaka, Japan*

A new self-generation mechanism of magnetic dipoles and the anomalous energy dissipation of fast electrons in the magnetized beam-plasma system are presented. Our particle-in-cell simulations show that the magnetic dipoles are self-organized and play important role for the beam energy dissipation. These dipoles drift slowly with a quasi-steady velocity depending on the magnetic gradient of the dipole and the imposed external magnetic field. The formation of dipoles provides a new anomalous energy dissipation mechanism of the relativistic electron beam, and this mechanism plays important roles in the fast ignition and the collisionless shock.

1. Introduction

The resulting two-stream instability (TSI) and the Weibel instability (WI) in the beam-plasma system have been widely investigated in the fast ignition and astrophysics. Recently, an external magnetic field is proposed to collimate the fast electrons in fast ignition [1], and strong external magnetic field of the order of kilo Tesla has already been produced in laser driven coil [2]. A variety of theoretical studies [3] indicate the important role of the inverse cascade in the nonlinear evolution of the beam-plasma system without external magnetic field. Here, we investigate the role of self-generated magnetic dipoles (MDs) in the magnetized plasma in relation with the anomalous energy dissipation [4] in the magnetized fast ignition, and suggest implications to the formation of collisionless shocks.

2. Simulations

We carry out the 2D3V PIC simulation code Ascent to study the transport of the fast electron beam in the magnetized dense plasma. The simulation plane is parallel to the external magnetic field $\mathbf{B}_0$. The simulation region is $25.6\lambda_0$ in the $\mathbf{x}$ direction and $12.8\lambda_0$ in the $\mathbf{y}$ direction. Periodic boundary conditions are applied for particles and fields in both directions. The density ratio of the beam and background electrons is $n_b/n_p=1/6$ with the drift velocities of $\mathbf{v}_b=0.978c$ for the beam and $\mathbf{v}_p=-0.163c$ for the background electrons. The total simulation time is $60T_0$. We have performed a number of simulations without and with external magnetic field of different amplitudes. (The fields are normalized to the characteristic field $mc\omega_p/e$.)

3. Results

Figure 1(a) shows the temporal evolution of the total energy of the out-of-plane magnetic field B_z and the electric field E_y and E_x, which are normalized to the initial total kinetic energy in the system for simulations with different amplitude of external magnetic field. In the linear stage of the WI and TSI, the field energies grow up exponentially at nearly the same growth rate for $\mathbf{B}_0$=0.15 and $\mathbf{B}_0$=0, as shown in Fig.1(b). In the case with strong $\mathbf{B}_0$=1.0, the growth rates are smaller. The growth rates in simulations approximately agree with the analytic analysis. Then, after the saturation and nonlinear damping due to the inverse cascade, the energy $\langle B_z^2\rangle$ reaches a quasi-steady level in the $\mathbf{B}_0$=0 and $\mathbf{B}_0$=1.0 cases. However, in the weak external magnetic field case, the energy $\langle B_z^2\rangle$ starts to grow up anomalously. The final energy in $\mathbf{B}_0$=0.15 case is nearly one order of magnitude higher than the other cases as shown in Fig.1(a). Furthermore, compare the momentum spread of the beam and background electrons in the unmagnetized and weakly magnetized cases, it is found that the beam

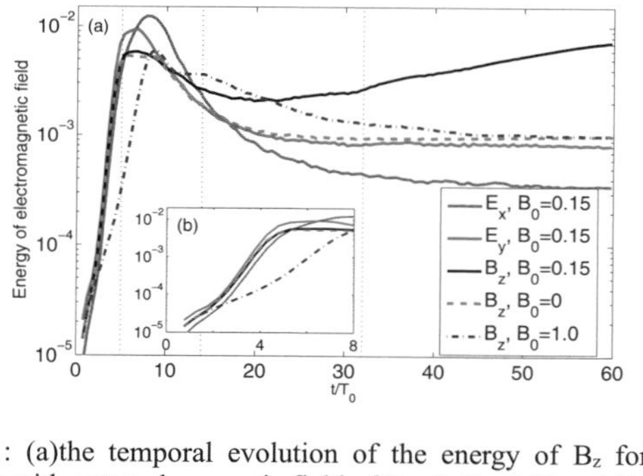

Fig.1: (a)the temporal evolution of the energy of B_z for the cases with external magnetic field of B_0=0, B_0=0.15 and B_0=1.0, and the energy of E_x and E_y for the B_0=0.15 case. The inserted figure (b) is the temporal evolution in the linear stage.

electrons are significantly decelerated together with the background electrons strongly heated up in the weakly magnetized case, and about 5.4% more total beam energy is lost in the weakly magnetized case than the unmagnetized case.

4. Analysis

Look into the dynamics of the magnetic field fluctuations in $\mathbf{B_0}$=0.15 case, we can observe that during the initial nonlinear stage, some small MD-like structures appear due to the possible merging and crossing of current filaments (the magnetic fluctuations correspond to the current fluctuations.) These MD seeds can survive a long enough time to grow up and finally self-organize the coherent MDs shown in Fig.2(a). The grow-up of MD seeds and the formation of MDs are coincident with the slow nonlinear increase of energy $\langle B_z^2 \rangle$. In addition, these dipoles drift in the negative x direction at a quasi-steady velocity and the amplitude of the dipole gradually increases during the drift.

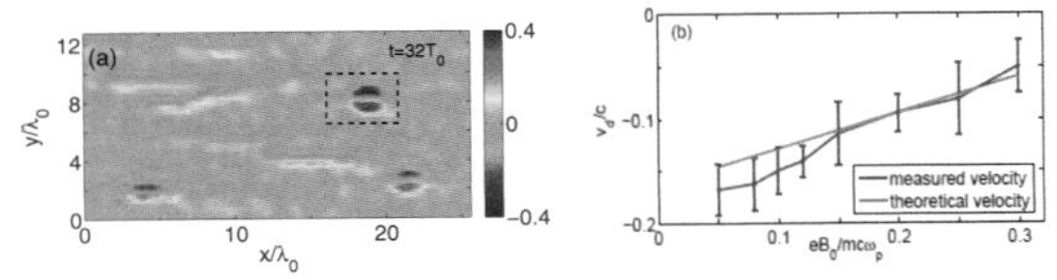

Fig.2: (a) Snapshot of the magnetic field B_z for B_0=0.15 case at t=32T_0. (b)Measured average drift velocities and the theoretical velocities of the dipoles for different amplitudes of external magnetic field.

By resorting to 2D EMHD theory, we deduce that the relationship between the drift velocity of the MDs and the external magnetic field as

$$v_d = v_b - (\frac{\partial b}{\partial y})_{(x_0, y_0)} + \frac{k}{1+k^2} B_0 \qquad (1)$$

Here the first term is the background flow velocity, the second term denotes the magnetic gradient at the center of one vortex of the dipole, k is the curvature of the B_z field at the vortex center. It is found that the off-set linear dependence of the measured drift velocity on the external magnetic field amplitude agrees approximately with the EMHD analysis as shown in Fig.2(b).

5. Discussions

Next, we investigate the dynamics of electrons around the MDs to look for the reason of the anomalous energy dissipation. The trajectories of the beam electrons shows that due to the large B_z of the dipole, the beam electrons moving forward are deflected and focused by the Lorentz force. Similarly, the background electrons moving backward are scattered out of the MD, and the overall electron density is reduced in the MD. Thus a positive electrostatic potential well is excited

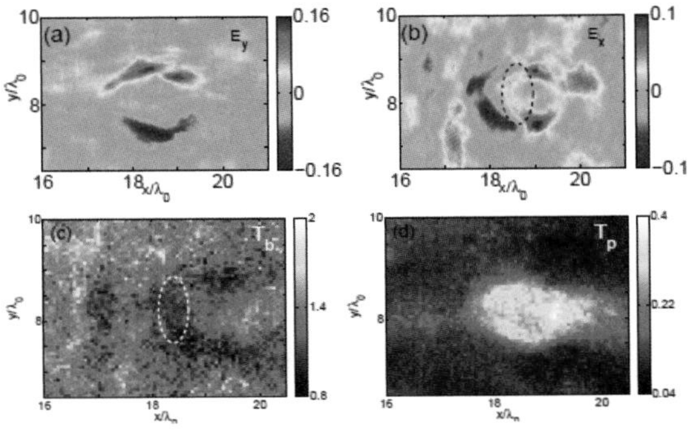

Fig.3: (a) and (b) respectively shows the detailed structure of the electric field E_y and E_x at t=32T_0 of the dipole marked in Fig.2(a). (c) and (d) represent the average energy (unit: MeV) per electron for the beam and background electrons in the same region respectively.

accordingly, and the detailed structures of the E_y and E_x are shown in Fig.3(a) and (b). It is worth noting that there is a positive E_x component in the center of the MD as marked in Fig.3(b). This field is the inductive electromagnetic component when the MD grows, and this field contributes to decelerate the beam electrons. Look at the average energy distribution of the beam electrons shown in Fig.4(c), the average energy is lower for the beam electrons in the MD, and the deceleration region corresponds to the electromagnetic Ex component. In addition, the average energy of the background electrons is dramatically high in the MD as shown in Fig.4(d), that is because some extremely low energetic background electrons trapped by the positive potential well, and gradually get accelerated during the oscillations inside the well.

In summary, we have studied the microphysics of the self-organized MDs in the nonlinear evolution stage of the instability in a weakly magnetized beam-plasma. It is found that the MDs serve as medium for the energy transfer between the beam and background electrons. Thus, the formation of MDs not only contributes to the anomalous energy dissipation in the fast electron beam transport in fast ignition, but also provides new insights that the formation of the collisionless shocks could be stronger in the weakly magnetized system.

References

[1] D. J. Strozzi et al.,Phys. Plasmas **19** (2012) 072711.
[2] S. Fujioka et al., Sci. Rep. **3** (2013) 1170.
[3] F. Califano et al., Phys. Rev. Lett. 84 (2000) 3602; T. Nakamura and K. Mima, Phys. Rev. Lett. 100 (2008) 205006.
[4] Y. Sentoku et al., Phys. Rev. Lett. 90 (2003) 155001; T. Taguchi et al., Proceedings of the 8th IFSA, Nara, Japan.

PHYSICAL REVIEW E **93**, 033206 (2016)

Characterization of magnetic reconnection in the high-energy-density regime

Z. Xu,[1] B. Qiao,[1,2,*] H. X. Chang,[1] W. P. Yao,[1] S. Z. Wu,[3] X. Q. Yan,[1] C. T. Zhou,[1,3] X. G. Wang,[1] and X. T. He[1,3]
[1]*Center for Applied Physics and Technology, HEDPS, and State Key Laboratory of Nuclear Physics and Technology, School of Physics, Peking University, Beijing 100871, People's Republic of China*
[2]*Collaborative Innovation Center of Extreme Optics, Shanxi University, Taiyuan, Shanxi 030006, People's Republic of China*
[3]*Institute of Applied Physics and Computational Mathematics, Beijing 100094, People's Republic of China*

(Received 29 August 2015; published 24 March 2016)

The dynamics of magnetic reconnection (MR) in the high-energy-density (HED) regime, where the plasma inflow is strongly driven and the thermal pressure is larger than the magnetic pressure ($\beta > 1$), is reexamined theoretically and by particle-in-cell simulations. Interactions of two colliding laser-produced plasma bubbles with self-generated poloidal magnetic fields of, respectively, antiparallel and parallel field lines are considered. Through comparison, it is found that the quadrupole magnetic field, bipolar poloidal electric field, plasma heating, and even the out-of-plane electric field can appear in both cases due to the mere plasma bubble collision, which may not be individually recognized as evidences of MR in the HED regime separately. The Lorentz-invariant scalar quantity $D_e \simeq \gamma_e \boldsymbol{j} \cdot (\boldsymbol{E} + \boldsymbol{v}_e \times \boldsymbol{B})$ $\{\gamma_e = [1 - (v_e/c)^2]^{-1/2}\}$in the electron dissipation region is proposed as the key sign of MR occurrence in this regime.

DOI: 10.1103/PhysRevE.93.033206

Magnetic reconnection (MR) [1], breaking and reorganizing the topology of the magnetic field dramatically, is a fundamental process observed in space, laboratory, and astrophysics [2–5]. MR provides a mechanism for fast release of magnetic energy to plasma kinetic and thermal energy with heat, flows, and jets. Investigations of the dynamics of such explosive events have aroused broad interest for half a century, which, however, depend critically on the plasma and inflow conditions. MR in tenuous, quasisteady, cold plasmas has been widely studied [6–9], where the thermal pressure in plasmas is smaller than the magnetic pressure, i.e., $\beta = 2n_0\mu_0 T_{e0}/B_0^2 < 1$. However, MR in strongly driven $\beta > 1$ plasmas has not been previously accessible in laboratory, which occurs frequently in astrophysics, such as those at the Earth's magnetopause [10] (where the driver is the super Alfvénic solar wind), the solar photosphere [11], the heliopause [12], and the supernova remnants.

Thanks to the rapid progresses of laser technology, recent high-energy-density (HED) experiments using a pair of colliding plasma bubbles produced from laser-irradiated solids [13–15] are capable of providing such an extreme condition of MR, where the magnetic field is self-generated through the Biermann battery effect or externally applied. In traditional experiments, various observed phenomena, such as the out-of-plane quadrupole magnetic field, bipolar poloidal electric field, and plasma heating, were recognized as evidences for MR, and in their simulations it was also generally admitted that they are consequences of MR [16]. However, in this strongly driven, high-β regime, the hydrodynamic plasma inflow is a significant source of energy and the compression and amplification of magnetic fields are expected to play significant roles in the reconnection dynamics, which are inherently distinct from the traditional MR scheme. Many of the experimentally observed phenomena can be induced merely by the plasma bubble collision instead of MR, where electron and ion species show different behaviors. It is necessary to distinguish the consequences caused by MR from those by hydrodynamic colliding effects and determine a key sign of MR occurrence in the HED regime.

In this paper, we reexamine the MR dynamics in the HED regime with the help of theory and particle-in-cell (PIC) simulations. Inspired from laser-solid interaction physics, we consider the interactions of two colliding laser-produced plasma bubbles with self-generated poloidal magnetic fields of, respectively, antiparallel and parallel field lines, where no MR should occur to the latter. Through comparing the simulation results of two cases, we address the unique characteristics of MR in the HED regime prior, during, and after the reconnection process so that the causes and consequences of MR can be clearly distinguished from those by the mere plasma bubble collision. A key sign of MR occurrence in the HED regime is proposed, which helps understanding of the extreme high-β MR phenomena in astrophysics and guiding future HED experiments of laboratory astrophysics studies.

The two-dimensional (2D) relativistic PIC simulations are carried out in the (x,y) coordinates with a rectangular domain $[-L_x,L_x]$ and $[-L_y,L_y]$. Similar to those previously presented in Refs. [17–19], the simulations directly track the interaction of a pair of laser-produced colliding plasma bubbles, which are driven by the substantial energy stored in the plasma flows, while the formation process of them is neglected. The initial condition corresponds to a time about halfway after the bubbles have been created, have expanded, and have generated their magnetic ribbons, but before the pair of bubbles has begun to interact, shown in Fig. 1. Two cases of magnetic ribbons with the antiparallel (named as the AP case) and parallel (P case) field lines are, respectively, considered, with both achievable in experiments [20,21]. Therefore, two half bubbles are taken and centered at $(0,L_y)$ and $(0,-L_y)$ with radius vectors of $\boldsymbol{r}^{(1)} = (x,y+L_y)$ and $\boldsymbol{r}^{(2)} = (x,y-L_y)$, respectively. As shown in Fig. 1, the AP case (P case similarly) is set up with the initial density $n = n^{(1)} + n^{(2)} + n_b$, expansion velocity $\boldsymbol{V} = \boldsymbol{V}^{(1)} + \boldsymbol{V}^{(2)}$, and magnetic field $\boldsymbol{B} = \boldsymbol{B}^{(1)} + \boldsymbol{B}^{(2)}$, where

*Author to whom correspondence should be addressed: bqiao@pku.edu.cn

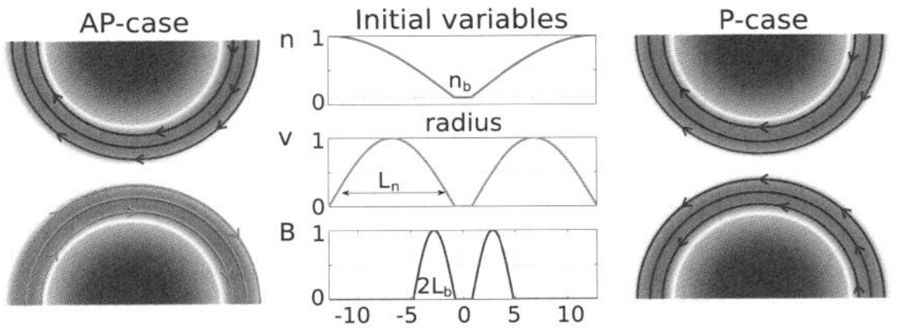

FIG. 1. PIC simulation setup: modeling for the collision of a pair of laser-produced expanding plasma bubbles with self-generated antiparallel (left) and parallel (right) magnetic field ribbons, named, respectively, as the AP and P cases. The middle plots show the initial variables of density n, radial expanding velocity $|\boldsymbol{V}|$, and magnetic field $|\boldsymbol{B}|$ vs radius, normalized by n_0, V_0, B_0, and d_i, respectively. Detailed simulation parameters are shown in the text.

n_b is the background density and $n^{(i)}$, $\boldsymbol{V}^{(i)}$, and $\boldsymbol{B}^{(i)}$ are $n^{(i)} = (n_0 - n_b)\cos^2(\pi r^{(i)}/2L_n)H_1$, $\boldsymbol{V}^{(i)} = V_0 \sin(\pi r^{(i)}/L_n)\hat{\boldsymbol{r}}^{(i)} H_1$, and $\boldsymbol{B}^{(i)} = B_0 \sin[\pi(L_n - r^{(i)})/2L_b]\hat{\boldsymbol{r}}^{(i)} \times \hat{\boldsymbol{z}} H_2$, respectively. n_0, V_0, and B_0 are initial parameters. H_1 and H_2 are the Heaviside step functions $H(x)$ as $H_1 = H(L_n - r^{(i)})$, $H_2 = H\{[r^{(i)} - (L_n - 2L_b)](L_n - r^{(i)})\}$, where L_n is the bubble radius and L_b is the half width of the ribbons.

For simplicity, we restrict the parameters to obtain only one X point of MR so that no plasmoid is formed [18,22]. The ion-electron mass ratio m_i/m_e is taken to be 100, and their temperatures are initially uniform as $T_e = T_i = T_0 = 0.025 m_e c^2$. The above lengths are taken as $L_x = 25.6 d_i$, $L_y = 12.8 d_i$, $L_n = 12 d_i$, and $L_b = 2 d_i$, where $d_i = c/\omega_{pi}$ is the ion skin depth and c is the light speed. The background density is $n_b = 0.1 n_0$, and B_0 is chosen to make the plasma beta $\beta_e = 2 n_0 \mu_0 T_{e0}/B_0^2 = 5$ and the Alfvén speed $V_A = c/100$. The plasma expansion velocity is taken as $V_0 = 2.0 C_s$, $0.8 C_s$, and $0.2 C_s$, respectively, corresponding to the ram pressure $\beta_{\mathrm{ram}} = \frac{1}{2}\rho V_0^2/(B^2/2\mu_0) \approx 16.5$, 2.6, and 0.2. Spatially 1024×512 cells and 144 particles per cell are employed in the simulations. Reflecting boundary conditions for particles and conducting boundary conditions for fields are employed. For convenience, simulation results in the following are presented in the normalized SI units: density to n_0, velocity to c, magnetic field to B_0, length to d_i, time to the ion cyclotron frequency Ω_{ci}^{-1}, temperature to T_0, electric current to $n_0 e c$, and electric field to $B_0 c$. Comparing with the experiments, e.g., for SG-II reconnection experimental parameters in [19] as $n_e \sim 3 \times 10^{19}\,\mathrm{cm}^{-3}$, $d_i \sim 50\,\mu\mathrm{m}$, and $B_0 \sim 3\mathrm{MG}$, the space and time scales of our simulation are, respectively, about $10 d_i \sim 500\,\mu\mathrm{m}$ and $4\Omega_{ci}^{-1} \sim 4\,\mathrm{ns}$. Note that the modeling capability of such large space and nanosecond time scales by PIC simulations is due to the low ion-electron mass ratio and high initial temperature we take as above. This approach has been popularly used in the magnetic reconnection simulations such as Refs. [8,17–19,21,22].

First, let us look at the basic physical picture of the interaction of colliding plasma bubbles, i.e., the evolutions of magnetic field topologies and plasma density distributions for both cases. Figure 2(a) plots the MR reconnection flux $\Psi/B_0 d_i = (\int B \times dl)/B_0 d_i$ vs time t. We can see clearly from the red (dark gray) circle line that the MR in the AP case can be roughly divided into three stages as expansion ($\tau = 0 \sim 0.8$), stagnation ($\tau = 0.8 \sim 2.1$), and poststagnation ($\tau > 2.1$) stages. However, for the P case, there is no reconnection flux increase for the whole interaction, shown in the black line, because no MR occurs.

During the bubble expansion stage, the out-of-plane current j_z is generated by $\boldsymbol{j} = \nabla \times \boldsymbol{B}/\mu_0$, which together with the poloidal magnetic field establishes a $\boldsymbol{j} \times \boldsymbol{B}$ force, pushing plasmas radially outward, similarly to those described in Ref. [18]. The magnetic fields $\boldsymbol{B}$ at $t = 1.2$ are shown in Figs. 2(b) and 2(e), respectively. For both cases, two expanding plasma bubbles squeeze each other near the interaction center ($x = 0$ and $y = 0$), enhancing $\boldsymbol{B}$ up to 2.5 times of the initial B_0. Meanwhile, plasmas are piled up with the maximum densities both increased from 0.1 to 0.6, which, however, is confined inside the respective bubbles by $\boldsymbol{B}$ at this stage, as shown in Figs. 2(h) and 2(k).

At later time, for the AP case, due to the antiparallel magnetic field lines, the currents in the center of two plasma bubbles add up, and a narrow current layer forms there due to bubble collision and squeezing, which is shown in the inset

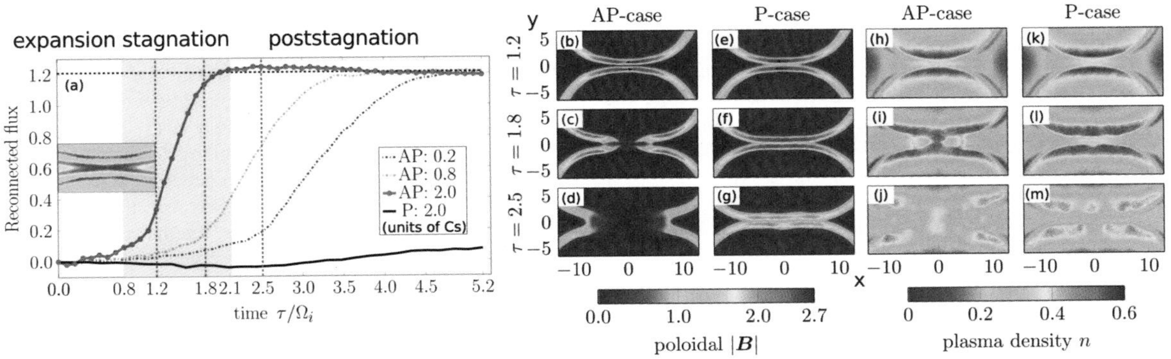

FIG. 2. Simulation results of Fig. 1: (a) The reconnected flux $\Psi/B_0 d_i$ vs time τ for both AP and P cases, where the dashed green (light gray) and blue (gray) lines show those for the AP case when the driven inflow velocity V_0 decreases to 0.8 and 0.2 (units of C_s). The inset in (a) shows that the narrow current layer (current density j_z) forms at time $t = 1.2$. (b–g) The poloidal magnetic fields B at $\tau = 1.2$, 1.8, and 2.5 for, respectively, AP (b–d) and P (e–g) cases. (h–m) The corresponding plasma density distributions n_e for, respectively, AP (h–j) and P (k–m) cases.

(current density map j_z) of Fig. 2(a) at time $t = 1.2$. The magnitude of this narrow current layer increases rapidly because of magnetic field piling up. When the current layer becomes too intense, it breaks up through self-driven instabilities or turbulences (such as Weibel/or beam instabilities), which leads to the onset of MR. However, for the P case, since the currents in the center between two plasma bubbles cancel out each other, no MR can occur although the bubble colliding and magnetic field piling up still exist. Therefore, considerable differences of the magnetic field configuration between two cases can be seen in Figs. 2(c) and 2(f) at $\tau = 1.8$. For the AP case [Fig. 2(c)], the magnitude of the field drops to almost zero at the interaction center due to the magnetic annihilation, which allows more background plasmas entering in, as shown in Fig. 2(i). However, no significant changes take place in the P case [Fig. 2(f)] from $\tau = 1.2$ to 1.8 and even later 2.5, where the enhanced magnetic field just becomes wider along the x direction because of bubble squeezing between each other. Because of the persistence of magnetic field confinement, less background plasmas enter into the interaction center, as shown in Fig. 2(l).

At $t = 2.5$ during the poststagnation stage, for the AP case, the magnetic field line is totally reconnected, with zero magnitude at the center, where plasmas in one bubble can flow freely into the other one [see Figs. 2(d) and 2(j)]. This clearly shows that MR results in breaking and reorganizing of magnetic field topology, and relaxing of plasma thermal and magnetic pressure at the center due to plasma bubble coalescence.

Note that the dashed green (light gray) and blue (gray) lines in Fig. 2(a) plot the MR flux evolving with time for the weaker-drive cases of $V_0 = 0.2C_s$ and $0.8C_s$, respectively, which show that the MR rate significantly decreases when the drive inflow becomes weaker. In other words, the fast reconnection rate of MR in the HED regime is strongly influenced by the drive inflow speed.

Second, let us distinguish the consequences of MR from those merely of the plasma bubble collision in the HED regime, where the two-fluid effect [18] may play roles. In this regime, due to the strongly driven inflow and the large thermal pressure ($\beta > 1$), ions are unmagnetized no matter whether they are in the magnetic ribbons, while electrons are easily magnetized in the ribbons, moving along the magnetic field. When two plasma bubbles collide and squeeze each other, the magnetized electrons are well frozen along the magnetic field line, being pushed deeper than ions towards the interaction center. Because of the poloidal magnetic field ribbon configuration, electrons move from the upper bubble (upstream) down towards the center, and then shift along the magnetic field line going backwards. However, ions behave differently. Therefore, the two-fluid effect may show up, where different species in the system satisfy different constraint conditions.

As we know, the MR process generally has many observable consequences such as generation of various electromagnetic fields and plasma heating [1,4,23,24], which are previously taken as the evidences of MR occurrence in experiments [14,25,26]. However, we show here that in the HED regime many of them can be actually caused by the mere plasma bubble collision instead of MR, in contrast to previous thought

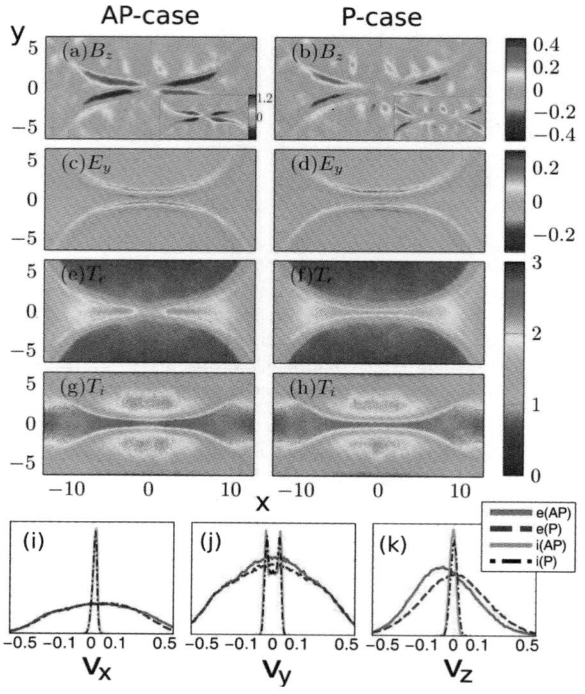

FIG. 3. Simulation results of Fig. 1 at time $\tau = 1.2$: (a,b) out-of-plane magnetic field B_z, (c,d) bipolar poloidal electric field E_y at time $\tau = 1.4$, (e,f) electron temperature T_e, (g,h) ion temperature T_i, (i–k) and velocity distributions (v_x, v_y, v_z) of electrons and ions at location $x = 4$–7 (units of d_i), $y = -0.5$–0.5 (units of d_i) for the AP and P cases, respectively.

[14,18]. Figure 3 shows the out-of-plane magnetic field B_z [(a) and (b)] and the in-plane electric field E_y [(c) and (d)] at $\tau = 1.2$. It can be clearly seen that B_z shows the expected quadrupole configuration, but it appears in both cases having similar magnitude. The reason is that both B_z here come from different behaviors of magnetized electrons and unmagnetized ions, i.e., two-fluid effects. We can also see that the bipolar electric field E_y near the center is generated for both cases. The physical origin of E_y is also from the decoupling of electrons and ions near the magnetic ribbon when two plasma bubbles are decelerated by each other, where the expanding, unmagnetized electrons are obstructed by the magnetic field in the ribbon, while the piston ions can penetrate into the ribbon with greater inertia, increasing the electric potential of the layer. Therefore, in the HED regime, the mere observation of the out-of-plane quadrupole magnetic field and/or the bipolar electric field does not immediately indicate the occurrence of MR, although indeed the hall fields play significant roles in MR [8,25,27,28]. Actually, the hall fields characterize the electron inflow speed at later time in the AP case when the MR is ongoing. Strong outflow is established near the center, breaking the bottleneck of inflow in the upstream (and downstream), where B_z increases significantly by $\sim$3 times stronger, as shown in the inset of Fig. 3(a). Meanwhile, in the P case, the squeezing effect decreases, leading to weakening of

the B_z fields in the center, shown in the inset of Fig. 3(b). Note that a couple of extra laminar electric fields also appear near the inner bubble ribbons, shown in Figs. 3(c) and 3(d), which is a new feature in the colliding bubbles due to the two-fluid effects.

Observation of ion and electron heating was also regarded as evidence for general MR [14,23,29,30]. However, here we do not see any considerable differences between two cases, as shown in Figs. 3(e)–3(h). This indicates that in the HED regime electrons and ions are mainly heated when they are bumping between the piled up magnetic fields, which can be caused by the mere plasma bubble collision. Note that here the temperature is calculated by the formula $T = m[(v_x - \overline{v}_x)^2 + (v_y - \overline{v}_y)^2 + (v_z - \overline{v}_z)^2]/3$, where $\overline{v}$ indicates the averaged velocity. The velocity (v_x, v_y, v_z) distribution functions of electrons and ions located in the region with x of $[4d_i \sim 7d_i]$ and y of $[-0.5d_i \sim 0.5d_i]$ for two cases are shown in, respectively, Figs. 3(i)–3(k). We can see that the electron distributions are mostly Maxwellian, while some of ions are additionally heated in the y direction by the bubble collision because ions having the greater mass are not all confined by the magnetic field, as shown in Figs. 3(g), 3(h), and 3(j). Moreover, the reconnection electric field E_z in the AP case contributes to the acceleration of electrons in the z direction when the electrons flow out, shown in Fig. 3(k).

Lastly, let us determine the potential key signs of MR occurrence in the HED regime. A popular criterion in 2D geometry of MR is to identify the out-of-plane electric field E_z at the central "electron diffusion region" (EDR), as shown in Fig. 4(a). And the value of E_z normalized to B_0c (the usual normalized parameters $BV_A/B_0c \sim 0.12$) is generally used to measure the reconnection rate of MR. The dynamics of E_z generation is similar to the well-studied Harris current reconnection scenario, obeying the generalized Ohm's law $\boldsymbol{E} = -\boldsymbol{v}_e \times \boldsymbol{B} - \nabla \cdot \boldsymbol{P}_e/n_e e - m_e/e(\partial \boldsymbol{v}_e/\partial t + \boldsymbol{v}_e \cdot \nabla \boldsymbol{v}_e)$. For the AP case here, E_z near the center is mainly supported by $\boldsymbol{v}_e \times \boldsymbol{B}$, and in the center is mainly supported by the off-diagonal electron pressure tensor $E_z \approx -1/ne(\partial_x P_{exz} + \partial_y P_{eyz})$. We find that more than half contribution comes from the component $\partial_x P_{exz}$, plotted in Fig. 4(e), consistent with Ref. [17]. Assuming $P_{exz} \sim nm_e < v_{ex} >_d < v_{ez} >_a$ and the term $< v_{ez} >_a \sim \sqrt{T_e/m_e}$, where d means the drifting out particles and a means the accelerated meandering particles [31], E_z can be estimated as $E_z \sim \partial_x < v_{ex} >_d \sqrt{2T_e m_e}/e \sim v_{ex}\sqrt{2T_e m_e}/(\delta e)$. In the simulations, $\delta \approx 1.0d_i$, $T_e \approx 0.04 m_e c^2$, and $v_{ex} \approx 0.1c$, so we estimate the reconnection rate as $E_z/c \sim 0.03B_0$. By comparing the results of $V_0 = 2$, 0.8, and $0.2C_s$, respectively, we find different driven inflow velocity results in different electron temperature in the EDR and therefore different reconnection rate of MR. For the P case, since two plasma bubbles have the opposite drifting velocities $\boldsymbol{v_e}$ but the same direction of magnetic fields $\boldsymbol{B}$, the induced electric fields E_z have opposite directions canceling out each other, resulting in zero E_z in the EDR, shown in Fig. 4(b). However, if two plasma bubbles are asymmetric, E_z cannot be completely offset, which can still be observed even in the P case, as shown in Fig. 4(i). This suggests that the mere E_z is not a good sign of MR occurrence in the HED regime, because the null lines of the magnetic field are not easy to identify in a tiny region in such experiments. The out-of-plane electric field resulted from the

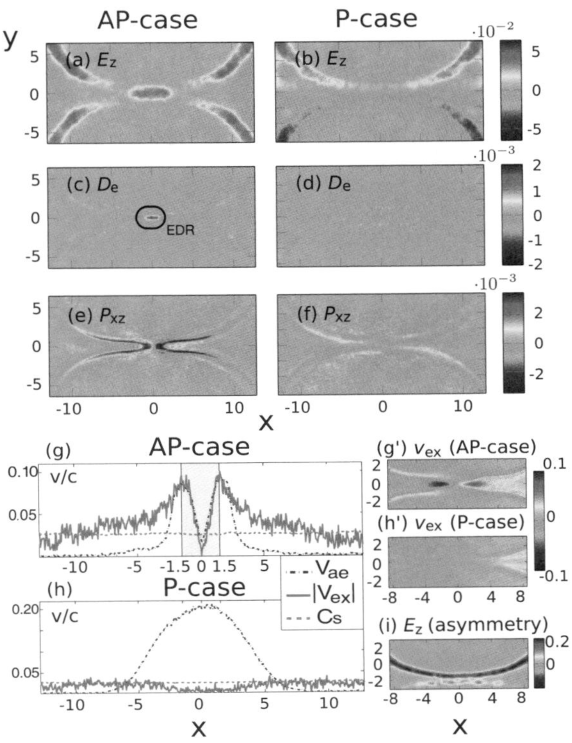

FIG. 4. Snapshots at $\tau = 1.4$ in the simulation of Fig. 1 for the AP and P case, respectively: (a,b) the averaged out-of-plane electric field E_z/cB_0, (c,d) the electron-frame dissipation measure D_e, (e,f) the component of the electron pressure tensor $P_{exz}/n_0 m_e c^2$, (g,h) electron outflow velocity v_{ex} at line $y = 0$, and (g',h') contours of electron outflow velocity v_{ex}. (i) E_z/cB_0 for the asymmetric P case.

MR should be carefully checked, where the contribution from the term $\boldsymbol{v} \times \boldsymbol{B}$ needs to be excluded.

Electron jets have been previously recognized as another important consequence of MR [14,19,20], which are embedded in the electron outflows along the x direction. For the AP case, from Figs. 4(g) and 4(g′), we can see clearly the jet configuration forms, where the outflow velocity V_{ex} is close to the local electron Alfvén velocity $V_{ae} = |\boldsymbol{B}|/\sqrt{\mu_0 m_e n}$ in the region $-1.5 < x < 1.5$ and drops to the ion sound speed C_s going outwards. If we define $C_s = \sqrt{\Gamma T/m} \sim 0.025$ as a lower limit for jet velocity, where $\Gamma = 5/3$ is the adiabatic coefficient, the length of the MR outflow electron jet can be estimated as $8d_i$ (the width of the red line above C_s) [see Fig. 4(g)]. However, for the P case [Figs. 4(h) and 4(h′)], the purely bubble collision cannot support such strong outflow, where the maximum outflow velocity V_{ex} is four times slower, close to C_s. Therefore, to distinguish the electron jets as an evidence of MR in the HED regime, the ratio of the jet velocity to the local sound speed, $R_v = V_{ae}/C_s \sim \sqrt{2m_i/\beta m_e} > 1$, should be carefully checked.

In view of the above, the remaining question is what should be a key sign of MR in the HED regime. We propose to measure the Lorentz-invariant scalar quantity [32] D_e as

$$D_e = \gamma_e[\boldsymbol{j} \cdot (\boldsymbol{E} + \boldsymbol{v}_e \times \boldsymbol{B}) - \rho_e(\boldsymbol{v_e} \cdot \boldsymbol{E})], \tag{1}$$

where $\gamma_e = [1 - (v_e/c)^2]^{-1/2}$ is the Lorentz factor and $\rho_e = (n_i - n_e)e$ is the charge density. For the AP case [Fig. 4(c)], we can clearly see that a remarkable D_e appears in the EDR center, where the nonideal Joule dissipation component $j_z \cdot (\boldsymbol{E} + \boldsymbol{v_e} \times \boldsymbol{B})_z$ approximately contributes 90% of the measured D_e and the charge term $\rho_c \boldsymbol{v_e} \cdot \boldsymbol{E}$ contributes as little as $< 7\%$, similar to that in Ref. [33]. After a more detailed analysis, we find that $j_z \cdot E_z$ dominates over other terms in D_e, where j_z and E_z may potentially be measurable in future experiments. However, for the P case, no matter whether the bubbles are symmetric or not, D_e always equals to zero as there is no dissipation region, shown in Fig. 4(d). As a result, we propose D_e in the EDR as a key evidence of MR occurrence.

In summary, we have reexamined the magnetic reconnection dynamics in the HED regime theoretically and by PIC simulations. The interaction of colliding laser-produced plasma bubbles with self-generated poloidal magnetic fields of, respectively, antiparallel and parallel field lines are considered. It is found that many of the previously observed characteristics of MR [14,18–20], such as the out-of-plane quadrupole magnetic field, bipolar poloidal electric field, plasma heating, and even the out-of-plane electric field, can be induced by the mere plasma bubble collision, which do not necessarily indicate MR occurrence in the HED regime. We propose the Lorentz-invariant scalar quantity $D_e \simeq \gamma_e \boldsymbol{j} \cdot (\boldsymbol{E} + \boldsymbol{v}_e \times \boldsymbol{B})$ in the electron dissipation region as the key sign of MR occurrence in the HED regime. These findings will significantly help understanding of the extreme high-β MR phenomena in astrophysics and guiding of future HED experiments of laboratory astrophysics studies.

This work is supported by the National Natural Science Foundation of China (NSFC), Grants No. 11575298, No. 91230205, No. 11575031, and No. 11175026; the National Basic Research 973 Project, Grants No. 2013CBA01500 and No. 2013CB834100; and the National High-Tech 863 Project. B.Q. acknowledges support from the Thousand Young Talents Program of China. The computational resources are supported by the Special Program for Applied Research on Super Computation of the NSFC-Guangdong Joint Fund (the second phase).

---

[1] *Reconnection of Magnetic Fields: Magnetohydrodynamics and Collisionless Theory and Observations*, edited by J. Birn and E. Priest (Cambridge University Press, Cambridge, 2007).
[2] S. Masuda, T. Kosugi, H. Hara, S. Tsuneta, and Y. Ogawara, Nature (London) **371**, 495 (1994).
[3] T. D. Phan, J. T. Gosling, M. S. Davis, R. M. Skoug, M. Øieroset, R. P. Lin, R. P. Lepping, D. J. McComas, C. W. Smith, H. Reme, and A. Balogh, Nature (London) **439**, 175 (2006).
[4] E. G. Zweibel and M. Yamada, Annu. Rev. Astron. Astrophys. **47**, 291 (2009).
[5] R. A. Treumann and W. Baumjohann, Front. Phys. **1**, 31 (2013).
[6] P. A. Sweet, in *Electromagnetic Phenomena in Cosmical Physics*, edited by B. Lehnert (Cambridge University Press, New York, 1958), p. 123.
[7] E. N. Parker, J. Geophys. Res. **62**, 509 (1957).
[8] J. Birn, J. F. Drake, M. A. Shay, B. N. Rogers, R. E. Denton, M. Hesse, M. Kuznetsova, Z. W. Ma, A. Bhattacharjee, A. Otto, and P. L. Pritchett, J. Geophys. Res. **106**, 3715 (2001).
[9] Y. Ren, M. Yamada, H. Ji, S. P. Gerhardt, and R. Kulsrud, Phys. Rev. Lett. **101**, 085003 (2008).
[10] H. Hasegawa, M. Fujimoto, T.-D. Phan, H. Rème, A. Balogh, M. W. Dunlop, C. Hashimoto, and R. TanDokoro, Nature (London) **430**, 755 (2004).
[11] Y. E. Litvinenko, Astrophys. J. **456**, 840 (1996).
[12] M. Opher, J. F. Drake, M. Swisdak, K. M. Schoeffler, J. D. Richardson, R. B. Decker, and G. Toth, Astrophys. J. **734**, 71 (2011).
[13] B. A. Remington, R. P. Drake, and D. D. Ryutov, Rev. Mod. Phys. **78**, 755 (2006).
[14] P. M. Nilson, L. Willingale, M. C. Kaluza, C. Kamperidis, S. Minardi, M. S. Wei, P. Fernandes, M. Notley, S. Bandyopadhyay, M. Sherlock, R. J. Kingham, M. Tatarakis, Z. Najmudin, W. Rozmus, R. G. Evans, M. G. Haines, A. E. Dangor, and K. Krushelnick, Phys. Rev. Lett. **97**, 255001 (2006).
[15] C. K. Li, F. H. Séguin, J. A. Frenje, J. R. Rygg, R. D. Petrasso, R. P. J. Town, O. L. Landen, J. P. Knauer, and V. A. Smalyuk, Phys. Rev. Lett. **99**, 055001 (2007).
[16] M. Yamada, R. Kulsrud, and H. Ji, Rev. Mod. Phys. **82**, 603 (2010).
[17] W. Fox, A. Bhattacharjee, and K. Germaschewski, Phys. Rev. Lett. **106**, 215003 (2011).
[18] W. Fox, A. Bhattacharjee, and K. Germaschewski, Phys. Plasmas **19**, 056309 (2012).
[19] Q.-L. Dong, S.-J. Wang, Q.-M. Lu, C. Huang, D.-W. Yuan, X. Liu, X.-X. Lin, Y.-T. Li, H.-G. Wei, J.-Y. Zhong, J.-R. Shi, S.-E. Jiang, Y.-K. Ding, B.-B. Jiang, K. Du, X.-T. He, M. Y. Yu, C. S. Liu, S. Wang, Y.-J. Tang, J.-Q. Zhu, G. Zhao, Z.-M. Sheng, and J. Zhang, Phys. Rev. Lett. **108**, 215001 (2012).
[20] J. Zhong, Y. T. Li, X. G. Wang, J. Q. Wang, Q. L. Dong, C. J. Xiao, S. J. Wang, X. Liu, L. Zhang, L. An, F. L. Wang, J. Q. Zhu, Y. A. Gu, X. T. He, G. Zhao, and J. Zhang, Nat. Phys. **6**, 984 (2010).
[21] M. J. Rosenberg, C. K. Li, W. Fox, I. Igumenshchev, F. H. Séguin, R. P. J. Town, J. A. Frenje, C. Stoeckl, V. Glebov, and R. D. Petrasso, Phys. Plasmas **22**, 042703 (2015).
[22] S. Lu, Q. M. Lu, Q. L. Dong, C. Huang, S. Wang, J. Q. Zhu, Z. M. Sheng, and J. Zhang, Phys. Plasmas **20**, 112110 (2013).

[23] T. D. Phan, M. A. Shay, J. T. Gosling, M. Fujimoto, J. F. Drake, G. Paschmann, M. Oieroset, J. P. Eastwood, and V. Angelopoulos, Geophys. Res. Lett. **40**, 4475 (2013).

[24] M. Øieroset, T. D. Phan, M. Fujimoto, R. P. Lin, and R. P. Lepping, Nature (London) **412**, 414 (2001).

[25] Y. Ren, M. Yamada, S. Gerhardt, H. Ji, R. Kulsrud, and A. Kuritsyn, Phys. Rev. Lett. **95**, 055003 (2005).

[26] M. Yamada, J. Yoo, J. Jara-Almonte, H. T. Ji, R. M. Kulsrud, and C. E. Myers, Nat. Commun. **5**, 4774 (2014).

[27] M. Hesse and S. Zenitani, Phys. Plasmas **14**, 112102 (2007).

[28] R. Smets, N. Aunai, G. Belmont, C. Boniface, and J. Fuchs, Phys. Plasmas **21**, 062111 (2014).

[29] J. F. Drake, M. Swisdak, T. D. Phan, P. A. Cassak, M. A. Shay, S. T. Lepri, R. P. Lin, E. Quataert, and T. H. Zurbuchen, J. Geophys. Res. [Space Phys.] **114**, 14 (2009).

[30] J. Yoo, M. Yamada, H. Ji, and C. E. Myers, Phys. Rev. Lett. **110**, 215007 (2013).

[31] A. Divin, S. Markidis, G. Lapenta, V. S. Semenov, N. V. Erkaev, and H. K. Biernat, Phys. Plasmas **17**, 122102 (2010).

[32] S. Zenitani, M. Hesse, A. Klimas, and M. Kuznetsova, Phys. Rev. Lett. **106**, 195003 (2011).

[33] S. Zenitani, M. Hesse, A. Klimas, C. Black, and M. Kuznetsova, Phys. Plasmas **18**, 122108 (2011).

激光等离子体相互作用和粒子加速

激光与等离子体相互作用物理是高能量密度物理研究的核心内容之一，是一些重要高技术应用的科学基础。目前，激光技术正向着纳秒级脉冲大能量和超短脉冲高聚焦强度两个方向发展。关于前者，重要的激光装置包括美国的NIF和我国的神光系列装置等，NIF输出蓝波激光能量最高可达2MJ，用来研究ICF、强冲击波下的物态方程、不透明度等重要物理问题。关于后者，目前激光峰值聚焦强度已突破10^{22}W/cm^2，这种强激光与等离子体相互作用会产生大量全新的物理现象，它是超强激光驱动高能粒子加速器、诊断源、“快点火”源、正负电子等离子体、带电粒子束输运等应用的基础。

在针对大能量激光等离子体非相对论相互作用研究方面，贺贤土院士领导团队进行了大量有关激光聚变的研究，例如受激Raman散射（stimulated Raman scattering，SRS）、受激Brillouin散射（stimulated Brillouin scattering，SBS）、双等离子体衰变（two-plasmon decay，TPD），以及交叉束能量转移（crossed-beam energy transfer，CBET）等，取得了很多对ICF研究有重要价值的结果。

在相对论激光等离子体相互作用产生和带电粒子加速方面，贺贤土院士提出了利用CP激光产生的轴向和径向电磁场加速相对论电子束的共振加速模型，加速电子电荷密度远大于通常的尾场准静态自生加速，而且准直性好，具有重要的实际应用价值。这一模型被国际同行大量应用，有时也被称为“磁光加速模型”。

在离子加速方面，贺贤土院士提出了一种新的相对论质子TeV能量加速模型。研究表明，质子在由激光与薄靶作用形成的电荷分离的双壳层中预加速，当双壳层中电子先进入组合的气体靶中产生“雪耙”效应并形成强空间电荷场时，质子可以进一步被加速，达到TeV能量。他还与合作者及学生研究了多种离子加速机制，可为研制超高能加速器代替高能射频加速器［例如大型强子对撞机(Large Hadron Collider,LHC)］提供物理参考。

在激光聚变“快点火”研究中，贺贤土院士提出用CP激光与等离子体作用产生的轴向强磁场和外加轴向磁场分别控制入射和传输电子的发散角，使电子能量有效沉积和形成点火热斑。他领导团队进行的理论和实验研究表明“快点火”具有很好的效果。

PHYSICAL REVIEW E **69**, 066409 (2004)

Resonance acceleration of electrons in combined strong magnetic fields and intense laser fields

Hong Liu,[1,4,*] X. T. He,[2,3] and S. G. Chen[2]
[1]*Graduate School, China Academy of Engineering Physics, Beijing P. O. Box 2101, Beijing 100088, People's Republic of China*
[2]*Institute of Applied Physics and Computational Mathematics, Beijing P. O. Box 8009, Beijing 100088, People's Republic of China*
[3]*Department of Physics, Zhejiang University, Hangzhou 310027, People's Republic of China*
[4]*Basic Department, Beijing Material College, Beijing 101149, People's Republic of China*
(Received 25 June 2003; revised manuscript received 17 October 2003; published 11 June 2004)

The acceleration mechanism of electrons in combined strong axial magnetic fields and circularly polarized laser pulse fields is investigated by solving the dynamical equations for relativistic electrons both numerically and analytically. We find that the electron acceleration depends not only on the laser intensity, but also on the ratio between electron Larmor frequency and laser frequency. As the ratio approaches unity, a clear resonance peak is observed, corresponding to the laser-magnetic resonance acceleration. Away from the resonance regime, the strong magnetic fields still affect the electron acceleration dramatically. We derive an approximate analytical solution of the relativistic electron energy in adiabatic limit, which provides a full understanding of this phenomenon. Application of our theory to fast ignition of inertial confinement fusion is discussed.

DOI: 10.1103/PhysRevE.69.066409 PACS number(s): 52.38.Kd, 41.75.Lx

I. INTRODUCTION

Recently, strong magnetic fields have achieved much attention in laser-matter interaction [1] and other physical fields [2]. An interesting question of how strong quasistatic magnetic fields influence electron acceleration has been discussed extensively, e.g., B-loop acceleration of Pukhov and Meyer-ter-Vehn [3], direct electron acceleration of Tanimoto *et al.* [4]. Here, the strong magnetic fields are generated by high electron current during short-pulse high-intense laser-plasma interactions [5], namely, spontaneous magnetic fields [6,7]. It is observed that a linearly polarized (LP) laser pulse can only generate quasistatic toroidal magnetic fields [4,8], whereas a circularly polarized (CP) laser pulse can generate toroidal as well as axial magnetic fields [7,9]. A recent experiment [10] has measured dc 3.4×10^8 G magnetic field on the surface of solid target in the LP laser-plasma interaction. Much higher magnetic field of strength up to 10^{12} G has been found on the surface of neutron stars [2].

On the other hand, in the fast ignition scheme [11] of inertial confinement fusion (ICF), the production of collimating electrons with moderate energy (<2 MeV) is a key point. Therefore, how the spontaneous magnetic fields affect election acceleration and electron collimation is a problem of great interest. Although experiments [12,13] and three-dimensional (3D) particle-in-cell (PIC) simulations [14] clearly demonstrate that the fact of strong currents of energetic 10–100 MeV electrons manifest themselves in a giant quasistatic magnetic field with up to 100 MG amplitude, the electrons with such high energy do not easily stop in highly compressed deuterium tritium (DT) fuel core for fast ignition. To our knowledge, on these topics linearly polarized short-pulse and high-intense lasers have been intensely considered in theoretical and experimental studies.

In studying the acceleration of electrons, the single test electron model is a simple but effective one. It has been used to analyze the direct laser acceleration of relativistic electrons in plasma channels. For example, Schmitz and Kull [15] have analyzed the LP laser system with self-generated static electric field and discussed the electron resonant acceleration mechanism. Even though the simple model is not fully self-consistent, it can help us achieve essential insight into the physical process of electron acceleration involved in the laser-plasma interactions. In this paper, we extend this study to consider combined CP laser fields and strong magnetic fields, exploiting the single test electron model to investigate the electron acceleration. We also discuss laser-magnetic resonance acceleration (LMRA) mechanism of electrons in strong laser and axial magnetic fields. In our simulation, the laser field is a Gaussian profile. The axial magnetic field is considered as a constant field and a quassistatic field with Gaussian profile, respectively. A fully relativistic single particle code is developed to investigate the dynamical properties of the energetic electrons. We find a big difference between the CP laser case and the LP laser case. In the CP laser system there is only one resonance peak, but in the LP laser system there are two resonance peaks. More importantly, at the magnetic field corresponding to the disappeared resonance peak, we find that energetic electrons can be successively collimated and accelerated to a moderate energy <2 MeV. This kind of electrons have potential application in the fast-ignitor scheme of ICF.

Our paper is organized as follows. In Sec. II, we derive the dynamical equations describing relativistic electrons in combined strong axial magnetic fields and the CP laser fields. These equations will be solved both numerically and analytically. We describe LMRA in a Gaussian CP beam with static axial magnetic field. An approximately analytical solution of relativistic electron energy is obtained, which gives a good explanation for our numerical simulations. This equation also suits a Gaussian profile quasistatic axial magnetic field which is considered as a spontaneous axial magnetic field. Our discussion and conclusion are given in Sec. III. The potential applications in fast ignition scheme of ICF are also discussed.

*Email address: Liuhong_cc@yahoo.com.cn

II. GAUSSIAN CP LASER PULSE MODEL

In this section, we use 3D test electron model to simulate electron acceleration in plasma channel, i.e., a cylindrically symmetric plasma with a uniform charge density n and a uniform current density $\mathbf{j}$ along the z axis. Here, we assume that the electrons have been partially expelled from the channel by the ponderomotive force. The focused laser pulse with frequency ω propagates in positive $\hat{\mathbf{z}}$ direction along the channel with a phase velocity v_{ph}. The corresponding electromagnetic fields $\mathbf{E}=-\nabla\Phi-\partial_t\mathbf{A}/c$ and $\mathbf{B}=\nabla\times\mathbf{A}$ are derived from potentials Φ and $\mathbf{A}$ by

$$\frac{e\Phi}{mc^2}=-\tfrac{1}{2}k_E(kr)^2, \tag{1}$$

$$\frac{e\mathbf{A}}{mc^2}=\mathbf{a}-\tfrac{1}{2}k_B(\hat{\mathbf{x}}y-\hat{\mathbf{y}}x), \tag{2}$$

where m denotes the electron mass, $-e$ the electron charge, c the light velocity, and $k=\omega/c$, $r=\sqrt{x^2+y^2}$. The model contains three dimensionless parameters,

$$k_E=\frac{n}{2en_c},\quad k_B=\frac{|\mathbf{j}|}{2en_cc}, \tag{3}$$

$$\begin{aligned}\mathbf{a}&=a_0e^{-(x^2+y^2)/R_0^2}e^{-(kz-\omega t)^2/k^2L^2}[\cos(\omega t-kz)\hat{\mathbf{x}}\\&\quad+\delta\sin(\omega t-kz)\hat{\mathbf{y}}]\\&=a_x+\delta a_y,\end{aligned} \tag{4}$$

where the critical density $n_c=m\omega^2/(4\pi e^2)$, the plasma frequency equals the light frequency, and L and R_0 are the pulse width and minimum spot size, respectively. The two components of the laser amplitude $\mathbf{a}$ take the form $a_x=a_0e^{-(x^2+y^2)/R_0^2}e^{-(kz-\omega t)^2/k^2L^2}[\cos(\omega t-kz)]$ and $a_y=a_0e^{-(x^2+y^2)/R_0^2}e^{-(kz-\omega t)^2/k^2L^2}[\sin(\omega t-kz)]$, respectively.

For the irradiance of the femtosecond laser pulses, the plasma ions have no time to respond to the laser and therefore can be assumed to be immobile. δ equals 0, 1, and -1, corresponding to linear, right-hand, and left-hand circular polarization, respectively. For simplicity, in the following discussions we assume that the phase velocity of the laser pulse equals the light velocity, i.e., $v_{ph}=c$. The main results obtained can be readily extended to the case of $v_{ph}\neq c$.

To describe an axial spontaneous quasistatic magnetic field that exists in the relativistic CP laser-plasma interaction systems [7,9], we use following approximate expression:

$$\mathbf{b}_z=-\delta b_{z0}|a|^2\hat{\mathbf{z}}, \tag{5}$$

where b_{z0} is a parameter, its value can be fitted by numerical simulations or practical experiments.

We first consider a static axial magnetic field and then extend our conclusion to a Gaussian profile quasistatic axial magnetic field. The above explicit expression clearly indicates that the self-generated magnetic field $\mathbf{b}_z$ is oriented along the negative laser propagation direction for the right-hand circular polarization ($\delta=1$). For left-hand circular polarization ($\delta=-1$), the self-generated magnetic field $\mathbf{b}_z$ is oriented along the laser propagation direction. Obviously, for the LP laser beam ($\delta=0$) there is no spontaneous magnetic field along the axial direction. Horovitz *et al.* [16] and Najmudin *et al.* [17] have measured the axial magnetic field $\mathbf{b}_z$ for the intense CP lasers. They found that, in underdense helium plasma the maximum strength of the magnetic field is of the order 10 kG for the laser intensity $10^{13}-10^{14}$ W/cm^2, and of the order 7 MG for the laser intensity 10^{19} W/cm^2, respectively. For near-overdense plasma, the peak value of the self-generated magnetic field is around hundreds of MG from the theoretical analysis [7]. The spatial distribution of the axial magnetic field $\mathbf{b}_z$ takes the same Gaussian shape as a^2, as plotted in Fig. 1. For convenience, we first consider the simplest case of the constant magnetic field, i.e., $b_z=k_B=const$. The case with Gaussian profile of $\mathbf{b}_z$ will be discussed later for comparison.

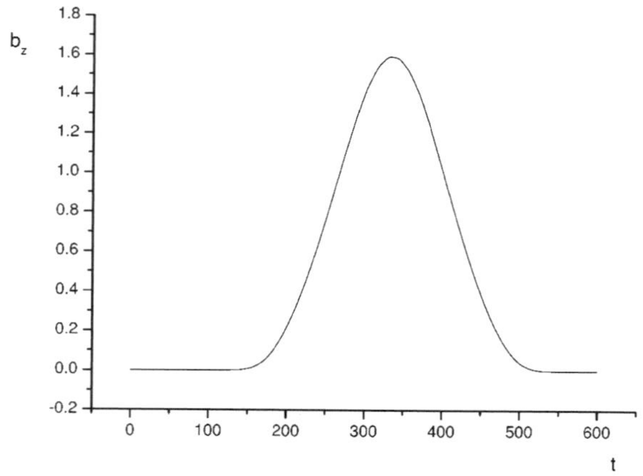

FIG. 1. The profile of a rough spontaneous quasistatic magnetic field (units=100 MG) as a function of time (in the units of ω^{-1}) generated by a left-hand Gaussian profile CP laser (propagation is in positive $\hat{\mathbf{z}}$ direction).

To discuss the acceleration of electrons in the channel, we consider the dynamical equations,

$$\frac{d\mathbf{p}}{dt}=-e\left(\mathbf{E}+\frac{1}{c}\mathbf{v}\times\mathbf{B}\right), \tag{6}$$

with the relativistic momentum,

$$\mathbf{p}=\gamma m\mathbf{v},\quad \gamma=\frac{1}{\sqrt{1-v^2/c^2}}. \tag{7}$$

We assume that the trajectory of a test electron starts at $\mathbf{v}_0=0$. Equation (1) and (2) yield

$$\frac{dp_x}{dt}=(v_z-1)\frac{\partial a_x}{\partial z}+v_y\left(\frac{\partial a_x}{\partial y}-\delta\frac{\partial a_y}{\partial x}\right)-v_yb_z-k_Ex, \tag{8}$$

$$\frac{dp_y}{dt}=(v_z-1)\delta\frac{\partial a_y}{\partial z}+v_x\left(\delta\frac{\partial a_y}{\partial x}-\frac{\partial a_x}{\partial y}\right)+v_xb_z-k_Ey, \tag{9}$$

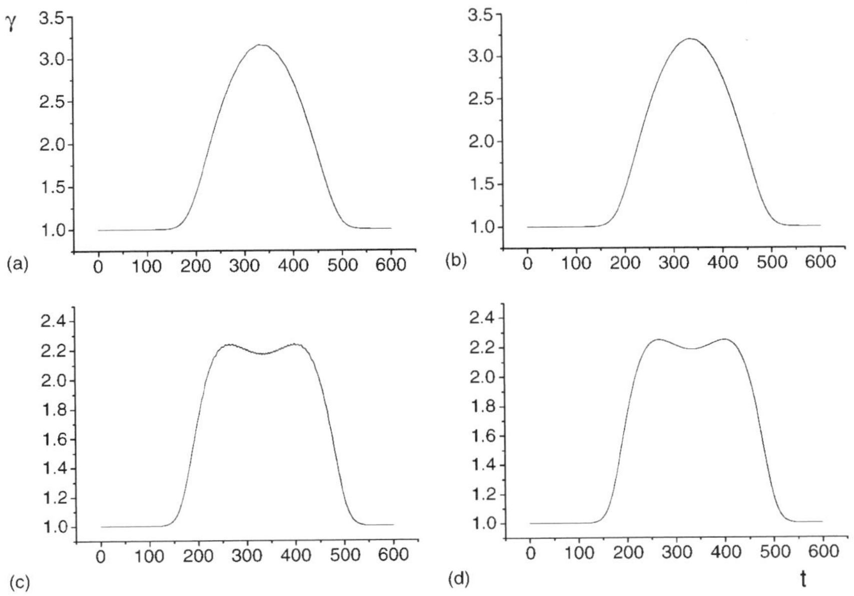

FIG. 2. The electron energy γ in units of mc^2 as a function of time in the units of ω^{-1} of the CP laser. (a) and (c) The numerical solutions by the Eqs. (8)–(10). (b) and (d) The analytical solutions by Eq. (19). The parameters $\delta=-1$, $a_0=4$, $k_E=0$, $\mathbf{r}_0=0.1$ (in the units of k^{-1}), $\mathbf{v}_0=0$, for the static case $b_z=k_B=0.9$ (corresponding to $B_z=90$ MG) shown in (a) and (b), for the Gaussian profile case B_z ($|b_{z0\ \max}|=1.6$ corresponding to $B_{z\ \max}=160$ MG) shown in (c) and (d).

$$\frac{dp_z}{dt}=-v_x\frac{\partial a_x}{\partial z}-v_y\delta\frac{\partial a_y}{\partial z}, \tag{10}$$

$$\frac{d\gamma}{dt}=-v_x\frac{\partial a_x}{\partial z}-v_y\delta\frac{\partial a_y}{\partial z}-k_Exv_x-k_Eyv_y, \tag{11}$$

where we have used the dimensionless variables

$$\mathbf{a}=\frac{e\mathbf{A}}{m_ec^2},\quad \Phi=\frac{e\mathbf{\Phi}}{m_ec^2},\quad \mathbf{b}=\frac{e\mathbf{B}}{m_ec\omega},\quad \mathbf{v}=\frac{\mathbf{u}}{c},$$

$$\mathbf{p}=\frac{\mathbf{p}}{m_ec}=\gamma\mathbf{v},\quad t=\omega t,\quad r=kr. \tag{12}$$

Using Eqs. (8)–(10), we choose different initial positions to investigate the electron dynamics for a Gaussian profile laser pulse. We first neglect the static electric field for simplicity. Because initial velocity can be transformed to initial position in our single test electron case, we keep the initial velocity at rest and change the initial positions of the test electrons. We assume that the trajectory of a test electron starts from $\mathbf{v}_0=0$ and $z_0=4$ L at $t=0$, while the center of laser pulse locates at $z=0$, then the classical trajectory is fully determined by Eqs. (8)–(10). Now we choose following parameters that are available in present experiments, i.e., $L=10\lambda$, $R_0=5\lambda$ ($\lambda=1.06\ \mu$m), $\delta=-1$, $a_0=4$ (corresponding to $I=2\times10^{19}$ W/cm^2), $k_E=0$, $\mathbf{r}_0=0.1$. We then trace the temporal evolution of electron energy, i.e., $\gamma\sim t$, and plot the results in Fig. 2.

Obtaining an exactly analytical solution of Eqs. (8)–(11) is impossible because of their nonlinearity. However, we notice that the second term on the left side of Eqs. (8) and (9) possesses symmetric form, which is found to be a negligibly small quantity after careful analysis. Then, approximately analytical solutions of Eqs. (8)–(11) in adiabatic limit can be obtained after setting $k_E=0$.

We first consider the right-hand CP laser. From the phase of the laser pulse, we have the equation,

$$\frac{d\eta}{dt}=\omega(1-v_z), \tag{13}$$

where $\eta=\omega t-kz$. Then, from Eqs. (10) and (11), we can easily arrive at the second useful relation under the initial condition $\mathbf{v}_0=0$ at $t=0$,

$$\gamma v_z=\gamma-1. \tag{14}$$

Finally, the energy-momentum equation yields

$$\gamma^2=1+(\gamma v_x)^2+(\gamma v_y)^2+(\gamma v_z)^2. \tag{15}$$

The solution of the electron momentum has symmetry. In adiabatic limit and under the initial condition of zero velocity electron, we find the solutions of Eqs. (8) and (9) taking the form,

$$\omega p_x=-b_z\cos(\omega t-kz), \tag{16}$$

$$\omega p_y=b_z\sin(\omega t-kz). \tag{17}$$

Substituting Eqs. (16) and (17) into Eq. (8), and using Eqs. (13)–(15), we obtain an equation having a resonance point (singularity) at a positive $b_z(=\omega)$,

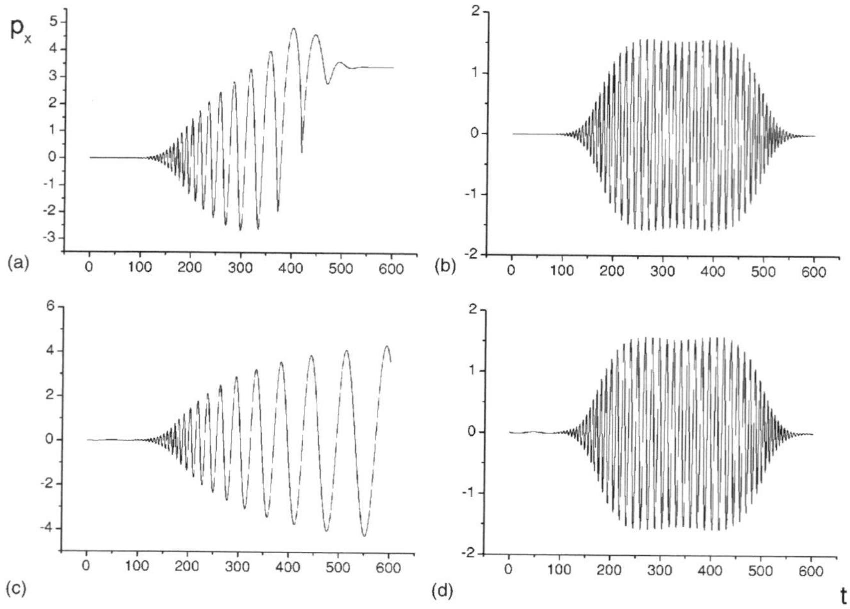

FIG. 3. The transverse momentum p_x in units of mc as a function of time in the units of ω^{-1} of the CP laser. The initial dimensionless parameters $\mathbf{v}_0=0$, $x_0=0.1$, $y_0=0$, and the dimensionless field parameters (a) $a_0=4$, $b_z=k_E=0$, (b) $a_0=4$, $b_z=1.6$, $k_E=0$, (c) $a_0=4$, $b_z=0$, $k_E=0.01$, (d) $a_0=4$, $b_z=1.6$, $k_E=0.01$.

$$\gamma=1+\frac{1}{2}\frac{a^2}{\left(1-\frac{b_z}{\omega}\right)^2}. \tag{18}$$

The same steps can be repeated for the left-hand CP laser, the solution with a resonance point in the negative b_z is then obtained,

$$\gamma=1+\frac{1}{2}\frac{a^2}{\left(1+\frac{b_z}{\omega}\right)^2}. \tag{19}$$

Equations (18) and (19) are approximately analytical energy solutions of Eqs. (8)–(11). For the left-hand CP laser, the analytic results of $\gamma\sim t$ from Eq. (19) are shown in Fig. 2(b) for $b_z=k_B=0.9$, and in Fig. 2(d) for $|b_{z0\ \max}|=1.6$ (Gaussian profile). From the above analytic solutions and our numerical calculations, we find that the strong self-generated magnetic fields reduce the maximum energy gain and produce a local minimum in $\gamma\sim t$ curves, which can be clearly seen by comparing Figs. 2(a) and 2(b) with Figs. 2(c) and 2(d).

In order to get an analytical expression of electron energy γ at the exact resonance point ($b_z=\omega$), we plug following the approximate solutions into the dynamical equations:

$$\omega p_x=c(t)\sin(\omega t-kz), \tag{20}$$

$$\omega p_y=c(t)\cos(\omega t-kz), \tag{21}$$

where $c(t)$ is a coefficient to be fitted. Careful analysis gives the solution at $t\rightarrow\infty$ in the following approximate expression

$$\gamma\approx\left(\frac{3}{\sqrt{2}}a\omega t\right)^{2/3}. \tag{22}$$

It indicates that the resonance between the laser fields and magnetic fields will drive the energy of electrons to infinity with a $2/3$ power law in time. But in actual experimental conditions, Eq. (22) is hard to occur.

When the self-generated static electric field is considered, the resonant acceleration of electron can occur as well. This has been analyzed by Schmitz and Kull [15]. In order to compare it with our LMRA mechanism, we plot the numerical solutions in Figs. 3–5 for (a) CP laser field, (b) CP laser field and quasistatic $\mathbf{b}_z$, (c) CP laser field and static electric field, (d) CP laser field, quasistatic $\mathbf{b}_z$, and static electric field. Figure 3 shows the temporal evolution of the transverse momentum $p_x(t)$. The initial conditions $\mathbf{v}_0=0$, $x_0=0.1$, $y_0=0$, and the field parameters are (a) $a_0=4$, $b_z=k_E=0$, (b) $a_0=4$, $b_z=1.6$, $k_E=0$, (c) $a_0=4$, $b_z=0$, $k_E=0.01$, (d) $a_0=4$, $b_z=1.6$, $k_E=0.01$. Figure 4 shows the evolution of the energy $\gamma(t)$. Figure 5 shows the 3D trajectories of the electrons.

It is well known, in off-resonance regime, e.g., around $b_z=0$, a Gaussian-wave pulse can accelerate electrons and the accelerated electron has a large scattering angle [18]. This is the so-called ponderomotive scattering, as shown in Figs. 3–5(a). We like to emphasize that at relativistic intensities laser ($I>10^{18}$ W/cm^2) the electron drift velocity is very slow but not slow enough. In fact, for $a_0=2$ the drift velocity is the same order as the quiver velocity. When $a_0=4$, $\mathbf{p}_\perp=\mathbf{a}$, $p_z=a^2/2$ from $\gamma=\sqrt{1+p_\perp^2+p_z^2}$, then $\gamma_{\max}=9$. The nonlinear ponderomotive scattering angle (in vacuum) $\theta=\arctan\sqrt{2/(\gamma-1)}\approx 28°$. Such large scattering angles will be unfavorable to the fast ignition of the high compressed fuel.

When a strong self-generated axial magnetic field exists, from analytic solutions Eqs. (18) and (19), the magnetic field

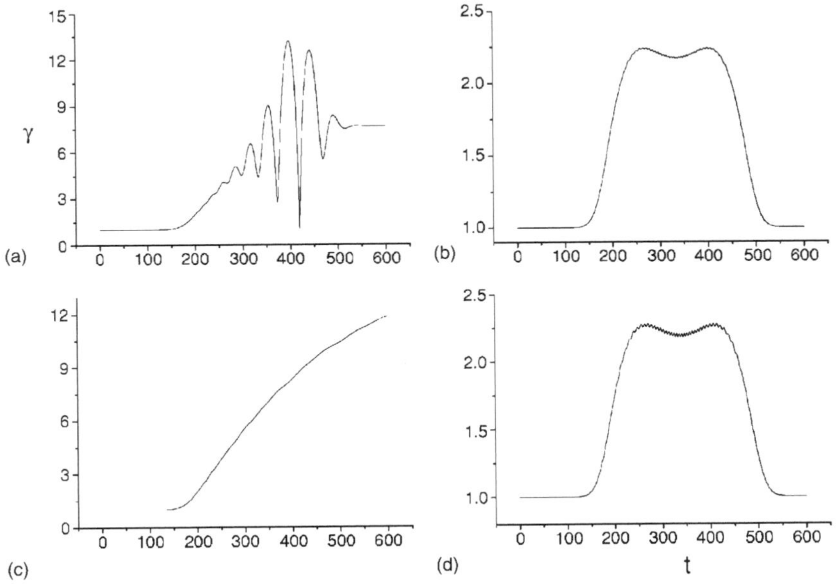

FIG. 4. The electron energy γ in units of mc^2 as a function of time in the units of ω^{-1} with the same parameters of Figs. 2(a)–2(d), respectively.

will reduce the longitudinal and transverse momentum of the electrons, as shown in Figs. 3–5(b). The electrons that initially locate on the center axis of the laser beam will move around z axis. The maximum electron energy (<2 MeV) can be calculated from Eqs. (18) and (19) and can also be found from $\gamma \sim t$ curve in Fig. 4(b). From the above discussions we find that strong axial magnetic fields will influence the electron motion and change the distribution in the three velocity components of the electrons. The existence of dc axial magnetic field changes the scattering angle of accelerated electron dramatically. It provides a guiding center for the accelerated electrons, and finally can collimate the electrons.

Figures 3–5(c) show the resonant acceleration in presence of a static electric field ($k_E=0.01$). The electron energy increases and its trajectory keeps around the z axis. This can be explained by the fact that the electron rides on the laser field with the help of static electric field in the proper phase and gains substantial energy, traveling with the driving laser pulse.

Figures 3–5(d) show the dominating effects of LMRA where the static electric field and self-generated axial magnetic field coexist. Here, because the dimensionless self-generated axial magnetic field ($b_{z\ \max}=1.6$) is much larger than the static electric field ($k_E=0.01$), the effects of the static electric field can be ignored.

To give a clear picture of the difference in performance between the CP laser and the LP laser, we plot $\gamma \sim b_z$, i.e., the normalized energy (units of mc^2) with respect to the normalized magnetic field (unit=100 MG) with different laser intensity, for the case of the plane wave lasers and a constant axial magnetic field, in Fig. 6, where the analytic work is given by Salamin and Faisal [19]. Figures 6(a) and 6(b) are for the CP laser case. If we assume the constant magnetic field is parallel to the laser propagation direction, then $\gamma \sim b_z$ is shown in Fig. 6(a) for the right-hand CP, in Fig. 6(b) for the left-hand CP, and in Fig. 6(c) for the LP laser. An obvious resonance behavior can be observed near the resonance point, i.e., the dimensionless magnetic field

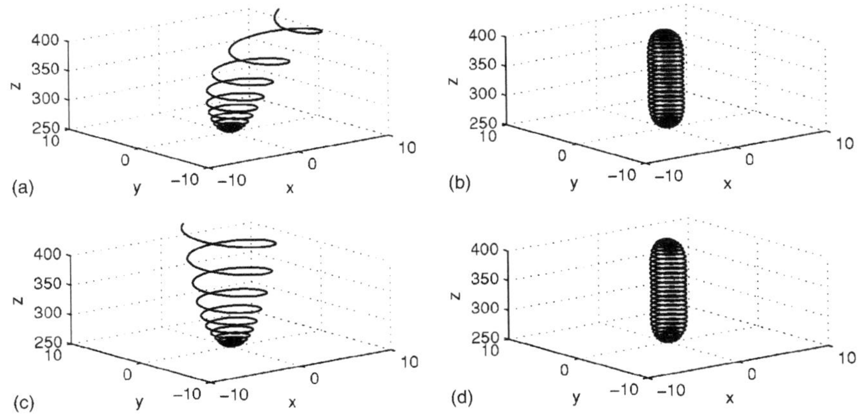

FIG. 5. The 3D electron trajectories (in the units of k^{-1}) with the same parameters of Figs. 2(a)–2(d), respectively.

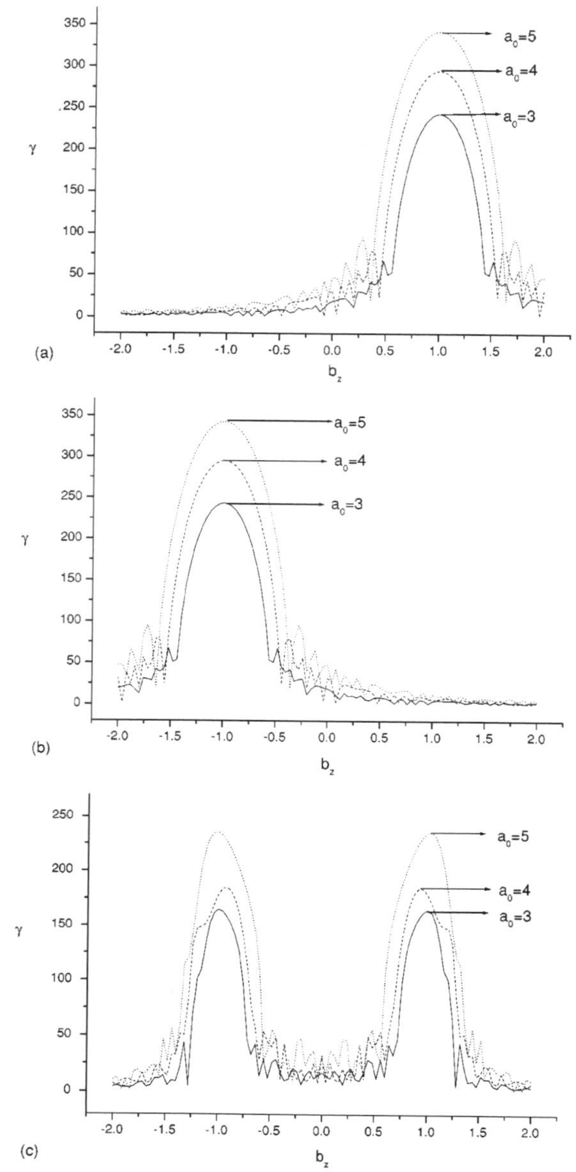

FIG. 6. $\gamma \sim b_z$ line of normalized energy along with normalized static magnetic field (unit=100 MG) with the different intensity a of plane wave laser. The dot line $a=5$, the dash line $a=4$, and the solid line $a=3$. The initial position starts from $\mathbf{r}=\mathbf{v}=0$ when $t=0$. Each achievable value γ is obtained at the same time $t=600$ in the units of ω^{-1}. (a) Right-hand CP laser. (b) Left-hand CP laser. (c) LP laser.

$b_z(=eB_z/m_e c\omega)$ equals the classical Larmor frequency Ω $(=eB_z/m_e c)$ of electron. The electron gains energy efficiently near the resonance point $(\Omega \approx b_z)$. The resonance regimes can be typically classified into three regimes: (a) exact resonance, (b) off-resonance $(b_z=0)$, (c) far away from resonance (only for CP laser). In the regime (c), at the magnetic field corresponding to the disappeared resonance peak of the LP laser case, we find that energetic electrons can be successively collimated and accelerated to a moderate energy <2 MeV. This kind of energetic electrons are suitable for the fast-ignitor ICF scheme.

III. DISCUSSIONS AND CONCLUSIONS

Using a single test electron model, we investigate the acceleration mechanism of energetic electrons in combined strong axial magnetic fields and circularly polarized laser fields. The axial magnetic field is considered as a constant field and a quasistatic axial magnetic field with Gaussian profile, respectively. An analytic solution of electron energy is obtained, showing a good agreement with numerical simulations for the physical parameters available for laboratory experiments. We find that the electron acceleration depends not only on the laser intensity, known as the pondermotive acceleration, but also on the ratio between electron classical Larmor frequency and the laser frequency. As the ratio approaches unity, a clear resonance peak is observed, that is, the LMRA. Away from the resonance regime, the strong magnetic fields still affect electron acceleration dramatically.

In contrast to the resonance acceleration described in Ref. [15] which dominates static electric field, our LMRA mechanism dominates in the case of strong quasistatic magnetic fields. In fast-ignitor scheme, there exists a high self-generated axial magnetic field if the CP laser is used. This fact provides an opportunity to acquire electrons with suitable energy.

To illustrate the difference between the CP laser and the LP laser, we plot the plane wave LP and CP laser of normalized electron energy with respect to the normalized static magnetic field, respectively. In the CP laser system there is only one resonance peak, whereas in the LP laser case there are two apparent resonance peaks. Obviously the electron energy gain increases with the intensity of the laser. In the CP laser system, the maximum energy will be higher than in LP case by about 20%. So the efficiency of energy transfer will be higher than in LP case.

In summary, the resonance acceleration mechanism of electrons in combined intense laser fields and strong magnetic fields (LMRA) is discussed in this paper. In absence of magnetic fields $(b_z=0)$, the electrons gain energy through usual ponderomotive acceleration that has been discussed extensively. In this case the laser polarization is not important, because the ponderomotive potential makes transverse momentum of energetic electrons equal in x-axis and y-axis direction. For this reason, the LP laser has been discussed thoroughly in recent years whereas the CP laser is relatively seldom mentioned. However, the fast ignition scheme requires the electrons with energy less than 2 MeV to match the stopping distance in compressed DT fuel core, which seems not easily satisfied by the LP laser system. In the present paper we turn to investigate the CP laser combined with strong quasistatic magnetic fields. We find that in the CP laser system there is only one resonance peak, while in the LP laser system there are two resonance peaks. More importantly, at the magnetic field corresponding to the disappeared resonance peak, the energetic electrons can be successively collimated and accelerated to a moderate energy (<2 MeV), with an intense CP laser ($I > 10^{18}$ W/cm^2) and axial self-generated quasistatic magnetic field ($Bs \approx 10^8$ G). This kind of energetic electrons have potential applications in the fast-ignitor scheme of ICF.

ACKNOWLEDGMENTS

H.L. thanks J. Liu for revising the manuscript. This work was supported by National Hi-Tech Inertial Confinement Fusion Committee of China, National Natural Science Foundation of China, National Basic Research Project "nonlinear Science" in China, and National Key Basic Research Special Foundation.

[1] A. Pukhov, Rep. Prog. Phys. **66**, 47 (2003).
[2] Dong Lai, Rev. Mod. Phys. **73**, 629 (2001).
[3] A. Pukhov and J. Meyer-ter-Vehn, Phys. Plasmas **5**, 1880 (1998).
[4] Mitsumori Tanimoto *et al.*, Phys. Rev. E **68**, 026401 (2003).
[5] Shi-Bing Liu, Shi-Gang Chen, and Jie Liu, Phys. Rev. E **55**, 3373 (1997).
[6] L. Gorbunov, P. Mora, and T. M. Antonsen, Phys. Rev. Lett. **76**, 2495 (1996).
[7] Z. M. Sheng and J. Meyer-ter-Vehn, Phys. Rev. E **54**, 1833 (1996).
[8] R. N. Sudan, Phys. Rev. Lett. **70**, 3075 (1993).
[9] Shao-ping Zhu, X. T. He, and C. Y. Zheng, Phys. Plasmas **8**, 321 (2001).
[10] M. Tatarakis *et al.*, Nature (London) **415**, 280 (2002).
[11] M. Tabak *et al.*, Phys. Plasmas **1**, 1626 (1994).
[12] M. Tatarakis *et al.*, Phys. Rev. Lett. **81**, 999 (1998).
[13] G. Malka *et al.*, Phys. Rev. E **66**, 066402 (2002).
[14] A. Pukhov and J. Meyer-ter-Vehn, Phys. Rev. Lett. **76**, 3975 (1996).
[15] M. Schmitz and H.-J. Kull, Laser Phys. **12**, 443 (2002).
[16] Y. Horovitz *et al.*, Phys. Rev. Lett. **78**, 1707 (1997).
[17] Z. Najmudin *et al.*, Phys. Rev. Lett. **87**, 215004 (2001).
[18] F. V. Hartemann *et al.*, Phys. Rev. E **51**, 4833 (1995).
[19] Yousef I. Salamin and Farhad H. M. Faisal, Phys. Rev. A **58**, 3221 (1998).

PHYSICS OF PLASMAS **12**, 083102 (2005)

Electron acceleration in combined intense laser fields and self-consistent quasistatic fields in plasma

Bin Qiao[a)]
Graduate School of China Academy of Engineering Physics, P. O. Box 2101, Beijing 100088, People's Republic of China

X. T. He
Institute of Applied Physics and Computational Mathematics, P. O. Box 8009, Beijing 100088, People's Republic of China and Department of Physics, Zhejiang University, Hangzhou 310027, People's Republic of China

Shao-ping Zhu and C. Y. Zheng
Institute of Applied Physics and Computational Mathematics, P. O. Box 8009, Beijing 100088, People's Republic of China

(Received 4 April 2005; accepted 11 July 2005; published online 19 August 2005)

The acceleration of plasma electron in intense laser-plasma interaction is investigated analytically and numerically, where the conjunct effect of laser fields and self-consistent spontaneous fields (including quasistatic electric field $\mathbf{E}_s^l$, azimuthal quasistatic magnetic field $B_{s\theta}$ and the axial one B_{sz}) is completely considered for the first time. An analytical relativistic electron fluid model using test-particle method has been developed to give an explicit analysis about the effects of each quasistatic fields. The ponderomotive accelerating and scattering effects on electrons are partly offset by $\mathbf{E}_s^l$, furthermore, $B_{s\theta}$ pinches and B_{sz} collimates electrons along the laser axis. The dependences of energy gain and scattering angle of electron on its initial radial position, plasma density, and laser intensity are, respectively, studied. The qualities of the relativistic electron beam (REB), such as energy spread, beam divergence, and emitting (scattering) angle, generated by both circularly polarized (CP) and linearly polarized (LP) lasers are studied. Results show CP laser is of clear advantage comparing to LP laser for it can generate a better REB in collimation and stabilization. © *2005 American Institute of Physics.* [DOI: 10.1063/1.2010247]

I. INTRODUCTION

Recent remarkable progress in the development of laser chirped pulse amplification (CPA) technique[1] has made ultraintense short-pulse lasers with intensity of $10^{19}-10^{21}$ W/cm^2 available in laboratories. Such ultraintense lasers provide a strong field gradient of 10^4-10^6 MV/m, which is much higher than that of conventional accelerators (<100 MV/m). Thus laser-plasma accelerators have been proposed as a next generation of compact accelerators to produce the energetic particle beams, especially relativistic electron beams (REB).[2–14] In the present laser-plasma accelerators, relativistic electrons are generated through two approaches. One is wakefield acceleration,[2–10] through the breaking of large-amplitude relativistic plasma waves created in the wake of laser pulse, which only occurs in the small broken-wave regime always behind the laser pulse. The other is direct acceleration[11–14] through the direct interaction between the laser fields and the electrons in plasma within the laser pulse. However, in real wide applications from medicine and biology to high-energy physics, the required REB should be well collimated, quasimonoenergetic and with small beam divergence. As well known, when an intense laser propagates in plasma, strong spontaneous quasistatic fields including quasistatic magnetic (QSM) and quasistatic electric (QSE) fields are self-generated. These fields may significantly affect the acceleration of plasma electrons and may be benefit for the production of a REB with good quality, such as the collimating effect of axial QSM field and so on.

In this paper, we just want to study the conjunct effect of laser fields and the complete self-consistent QSE and QSM fields on the plasma electron acceleration as well as their synthetic effect on the quality of the produced REB. Certainly, first of all, the key points are self-consistently deriving all quasistatic fields and completely considering their synthetical effect on electron acceleration. However, so far, to our knowledge, no work[11–14] has completely taken all the self-consistent spontaneous quasistatic fields into account, which should include the QSE fields E_s^l and QSM fields (both axial B_{sz} and azimuthal $B_{s\theta}$) self-generated in laser-plasma interaction (see Ref. 15). Recently, a self-consistent kinetic model has been proposed by the present authors[15] for the complete quasistatic fields, including both B_{sz} and $B_{s\theta}$ as well as $\mathbf{E}_s^l$. Analytical expressions of $B_{s\theta}$ and $\mathbf{E}_s^l$ were given for the first time. All the predicted quasistatic fields accord well with the experimental[16] and simulation[23] results. We use the results of QSE and QSM fields given in Ref. 15 here. An analytical electron fluid model using test-particle method other than the usual single-particle method[2–14] has been developed, which embodies the fluid property of plasma electrons. Using this model, we make an explicit analysis of various effects of each spontaneous quasistatic fields as well

a)Electronic mail: bin_qiao1979@yahoo.com

as their synthetical effect on electron acceleration, and thus give the basic physical picture of electron acceleration in laser-plasma interaction. A fully relativistic test-particle fluid code is also developed to solve the series of the corresponding equations given by the model. The dependences of final energy gain and scattering angle of electron on its initial position, plasma density, and laser intensity are, respectively, studied for both circularly polarized (CP) laser and linearly polarized (LP) laser. We further approximately analyze and compare the quality of the REB generated, respectively, by CP laser and LP laser, such as energy spread, beam divergence, and emitting (scattering) angle through choosing one test electron to represent the large number of electrons at the same place.

The paper is organized as follows. In Sec. II, we derive the fluid equations describing the relativistic electron motion in combined laser fields and spontaneous fields. These equations are analytically simplified using the generalized vorticity conservation law. Laser fields, spontaneous quasistatic fields, and ponderomotive force for a focused laser beam are given in Sec. III. In Sec. IV, we numerically solve the fluid equations using three-dimensional (3D) test-particle code and give the analysis about the acceleration of plasma electrons as well as the quality of REB. Discussion and conclusion are given in Sec. V.

II. THE FLUID MOTION EQUATIONS FOR RELATIVISTIC ELECTRONS

The motion of relativistic electrons in a plasma interacted with intense laser beam is described by the Euler fluid equation,

$$\frac{\partial \mathbf{p}}{\partial t}+\mathbf{v}\cdot\nabla\mathbf{p}=-e\left[\mathbf{E}+\frac{1}{c}(\mathbf{v}\times\mathbf{B})\right], \tag{1}$$

where the electric and magnetic fields $\mathbf{E}$ and $\mathbf{B}$ include the laser ones $\mathbf{E}_L$, $\mathbf{B}_L$, the spontaneous slow-time-scale (including slowly varying and quasistatic) ones $\mathbf{E}_s$, $\mathbf{B}_s$, and the spontaneous fast-time-scale ones $\mathbf{E}_f$, $\mathbf{B}_f$. Under the condition of the Coulomb gauge $\nabla\cdot\mathbf{A}=0$, the laser electric field has only transverse part as $\mathbf{E}_L=-(1/c)(\partial\mathbf{A}_L/\partial t)$, while the spontaneous fields have both the transverse part $\mathbf{E}^t_{s,f}=-(1/c)\times(\partial\mathbf{A}_{s,f}/\partial t)$ and longitudinal part $\mathbf{E}^l_{s,f}=-\nabla\phi_{s,f}$. Thus $\mathbf{E}$ and $\mathbf{B}$ are expressed using potentials as follows:

$$\mathbf{E}=\mathbf{E}_L+\mathbf{E}_s+\mathbf{E}_f=-\nabla\phi-\frac{1}{c}\frac{\partial\mathbf{A}}{\partial t}$$

$$=-\nabla\Phi_s-\nabla\Phi_f-\frac{1}{c}\frac{\partial\mathbf{A}_L+\mathbf{A}_s+\mathbf{A}_f}{\partial t}, \tag{2}$$

$$\mathbf{B}=\mathbf{B}_L+\mathbf{B}_s+\mathbf{B}_f=\nabla\times\mathbf{A}=\nabla\times(\mathbf{A}_L+\mathbf{A}_s+\mathbf{A}_f). \tag{3}$$

Substituting Eqs. (2) and (3) into Eq. (1), introducing the normalized transform $\mathbf{v}\to\mathbf{u}c$, $\mathbf{p}\to\mathbf{p}m_0c=\gamma\mathbf{u}m_0c$, $\phi\to(m_0c^2/e)\phi$, $\mathbf{A}\to(m_0c^2/e)\mathbf{a}$, $\mathbf{x}\to c\mathbf{x}/\omega_0$, $t\to t/\omega_0$, and considering the averaged ponderomotive force of the electron fluid to be written as $\nabla\Gamma=\nabla\sqrt{1+\mathbf{p}^2}=\mathbf{u}\times(\nabla\times\mathbf{p})+(\mathbf{u}\cdot\nabla)\mathbf{p}$, we get that

$$\frac{\partial(\mathbf{p}-\mathbf{a})}{\partial t}=\nabla\phi-\nabla\Gamma+\mathbf{u}\times\nabla\times(\mathbf{p}-\mathbf{a}). \tag{4}$$

Additional simplification is achieved by applying the conservation law of generalized vorticity $\mathbf{Q}$, where $\mathbf{Q}=\nabla\times(\mathbf{p}-\mathbf{a})$ is defined. In fluid mechanics, for incompressible flow, the vorticity $\nabla\times\mathbf{Q}=0$ is conserved along the path of a fluid element. Moreover, if all the streamlines originate from the region where the vorticity is equal to zero, then the vorticity is equal to zero everywhere and the flow is eddy-free. In the background plasma without the laser beam pulse ($\mathbf{a}=0$), $\nabla\times\mathbf{p}=0$ and $\mathbf{Q}=0$ should be taken. Therefore, it follows from the conservation of generalized vorticity, similar to fluid mechanics, that $\mathbf{Q}$ is equal to zero in the laser beam pulse as well, i.e.,

$$\nabla\times(\mathbf{p}-\mathbf{a})=0. \tag{5}$$

Note that Eq. (5) is a consequence of the fluid description other than *a priori* assumption. Such a simplification is reasonable and also done in many papers.[17] Substituting Eq. (5) into Eq. (4) yields

$$\frac{\partial\mathbf{p}}{\partial t}=\frac{\partial\mathbf{a}}{\partial t}+\nabla\phi-\nabla\Gamma. \tag{6}$$

As mentioned above, $\mathbf{a}=\mathbf{a}_L+\mathbf{a}_s+\mathbf{a}_f$ and $\phi=\phi_s+\phi_f$ are taken, we get that

$$\frac{\partial\mathbf{p}}{\partial t}=-\mathbf{E}_L-\mathbf{E}^t_s-\mathbf{E}^l_s-\mathbf{E}^t_f-\mathbf{E}^l_f-\nabla\Gamma. \tag{7}$$

Equations (6) and/or (7) are the fluid motion equations for relativistic plasma electron. In Eq. (7), $\mathbf{E}_f$ always refers to the amplitude of plasma wave field (wake field) generated via a self-modulation[18] or stimulated Raman scattering instability,[19] which always exists behind the laser pulse. As mentioned in Refs. 10 and 15, the amplitude of the wake field is much smaller than the laser field for underdense plasma as $E_w\sim0.1-0.3(mc\omega_p/e)=0.1-0.3(n_0/n_{\rm cr})^{1/2}E_L\ll E_L$. Thus within the laser pulse, the effect of the plasma wave field (wake field) $\mathbf{E}_f$ is smaller enough to be ignored comparing to that of the laser field $\mathbf{E}_L$. In our model, we ignore the effect of the plasma wave (wake field), that is, ignoring the terms of $\mathbf{E}^l_f$ and $\mathbf{E}^t_f$ in Eq. (7). Note that in Eq. (7) under the static-state approximation, $\mathbf{E}^t_s$ is a slowly varying electric field which here embodies the effect of the QSM field B_s as $\nabla\times\mathbf{E}^t_s=-(1/c)(\partial\mathbf{B}_s/\partial t)$, while $\mathbf{E}^l_s$ is just the QSE field which corresponds to the space-charge potential as $\mathbf{E}^l_s=-\nabla\phi_s$ and partly compensates the effect of the ponderomotive force.

Assuming that the laser propagates along the z-axis direction, expressing Eq. (7) into three components of $\mathbf{e}_x$, $\mathbf{e}_y$, and $\mathbf{e}_z$, directions together with three differential equations about the spatial position (x,y,z) of electrons, we get a series of equations describing the electron momentum and trajectory as follows:

$$\frac{\partial p_x}{\partial t} = -E_{Lx} - E_{sx}^t - E_{sx}^l - \frac{\partial \Gamma}{\partial x},$$

$$\frac{\partial p_y}{\partial t} = -E_{Ly} - E_{sy}^t - E_{sy}^l - \frac{\partial \Gamma}{\partial y},$$

$$\frac{\partial p_z}{\partial t} = -E_{sz}^t - E_{sz}^l - \frac{\partial \Gamma}{\partial z},$$

$$\frac{dx}{dt} = \frac{p_x}{\sqrt{1+p_x^2+p_y^2+p_z^2}}, \quad \frac{dy}{dt} = \frac{p_y}{\sqrt{1+p_x^2+p_y^2+p_z^2}},$$

$$\frac{dz}{dt} = \frac{p_z}{\sqrt{1+p_x^2+p_y^2+p_z^2}}. \tag{8}$$

Equations (8) are the basic equations for compiling our three-dimensional (3D) test-particle code.

III. QUASISTATIC FIELDS AND PONDEROMOTIVE FORCE FOR A FOCUSED LASER BEAM

In this section, we derive the laser fields, the spontaneous quasistatic fields, and the averaged ponderomotive force of fluid in Eqs. (8) for a focused laser beam with intensity of a Gaussian distribution. Yariv[20] and Feng *et al.*[21] have expressed the focused laser beam well as

$$I = I_0 \exp\left(-\frac{2\eta^2}{L^2} - \frac{2r^2}{b^2}\right)\frac{b_0^2}{b^2},$$

$$\mathbf{a}_L = \sqrt{I_0}\exp\left(-\frac{\eta^2}{L^2} - \frac{r^2}{b^2}\right)\frac{b_0}{b}\hat{a}, \tag{9}$$

where $\hat{a}=\cos\phi\mathbf{e}_x+\alpha\sin\phi\mathbf{e}_y$, $r^2=x^2+y^2$ are chosen, and $\alpha=0,\pm 1$ for LP and CP lasers, respectively. I is the laser intensity, I_0 represents its amplitude and $\mathbf{a}_L$ the vector potential of laser field. L and b are the pulse longitudinal width and the beam transverse radius, and $b=b_0(1+z^2/z_f^2)^{1/2}$, where b_0 is the minimum beam transverse radius at the focus point and $z_f=b_0^2/2$ is the corresponding Rayleigh length. $\phi=\phi_p-\phi_G-\phi_0+\phi_R$, where $\phi_p=z-t=\eta$, $\phi_G=\arctan(z/z_f)$, $\phi_R=(x^2+y^2)/2R(z)$, and $R(z)=z(1+z_f^2/z^2)$. Note that ϕ_0 is a constant, ϕ_p is the plane-wave phase, ϕ_G is the Guoy phase associated with the fact that a Gaussian beam undergoes a total phase change of π as z changes from $-\infty$ to $+\infty$, ϕ_R is the phase associated with the curvature of the wave front intersecting the beam axis at the coordinate z. Note that when a laser beam propagates in plasma, a fast-time-scale electric field E_z will be also generated by the interaction of the laser field $\mathbf{E}_L$ with the fast-time-scale plasma density variation n_f as $\nabla\cdot\mathbf{E}_f=ik_0E_z+\nabla_\perp\cdot\mathbf{E}_L=-4\pi e n_f$. In Ref. 15, it has been proven that for an intense laser propagating in underdense plasma, the field is predominantly transverse $(E_\perp)$, i.e., $E_z \ll E_L$. Thus, it is reasonable to ignore E_z itself in the detailed calculation. Assuming $\phi_0=0$, we can easily get the laser electric fields as

$$E_{Lx} = \sqrt{I}\cos\left(z - t - \arctan\left(\frac{z}{z_f}\right) + \frac{x^2+y^2}{2R(z)}\right),$$

$$E_{Ly} = -\alpha\sqrt{I}\sin\left(z - t - \arctan\left(\frac{z}{z_f}\right) + \frac{x^2+y^2}{2R(z)}\right). \tag{10}$$

A. Spontaneous quasistatic fields

As mentioned in the Introduction, the self-consistent derivation of quasistatic fields generated in laser-plasma interaction is very important, which directly determines whether we can get the real physical picture of plasma electron acceleration in intense laser-plasma interaction or not. Recently, a self-consistent kinetic model for such quasistatic fields has been proposed by the present author.[15] In Ref. 15, The expressions of the QSE field $\mathbf{E}_s^l$ and the QSM field $\mathbf{B}_s$ as well as the corresponding slowly varying electric filed $\mathbf{E}_s^t$ are given as [see Eqs. (19) and (20) in Ref. 15],

$$\mathbf{E}_s^l = -\frac{e\beta_5}{m_e\omega_0^2\beta_6}\nabla|\mathbf{E}_L|^2, \tag{11}$$

$$\nabla\times\mathbf{B}_s(\mathbf{x},t) = -i\frac{\omega_{pe}^2 e}{m_e\omega_0^3 c}\nabla\times[\beta_1\mathbf{E}_L(\mathbf{x},t)\times\mathbf{E}_L^*(\mathbf{x},t)] + \frac{\pi\omega_{pe}^2 e}{4m_e\omega_0^2c^2}\mathbf{e}_s^t\beta_4|\mathbf{E}_L(\mathbf{x},t)|^2, \tag{12}$$

$$-\nabla^2\mathbf{E}_s^t(\mathbf{x},t) = -i\frac{\omega_{pe}^2 e}{m_e\omega_0^3c^2}\frac{\partial}{\partial t}\nabla\times(\beta_1\mathbf{E}_L\times\mathbf{E}_L) - \frac{\pi\omega_{pe}^2 e}{4m_e\omega_0^2c^3}\beta_4\frac{\partial}{\partial t}(\mathbf{e}_s^t\cdot\mathbf{e}_z)|\mathbf{E}_L|^2\mathbf{e}_z, \tag{13}$$

where the coefficients related to the relativistic effect are defined as $\beta_1=m_e^2c^2\int(1/p_{0e}^2)[(1/2)-(|\mathbf{p}_e|^2/3p_{0e}^2)]f_{e0}d^3\mathbf{p}_e$, $\beta_4=m_e^2c^2\int[(1/p_{0e}|\mathbf{p}_e|)-(3|\mathbf{p}_e|/4p_{0e}^3)]f_{e0}d^3\mathbf{p}_e$, $\beta_5=m_e^2c^2\int[(1/2|\mathbf{p}_e|^2)+(|\mathbf{p}_e|^2/3p_{0e}^4)-(1/2p_{0e}^2)]f_{e0}d^3\mathbf{p}_e$, and $\beta_6=m_ec\int[(1/p_{0e})+(p_{0e}/|\mathbf{p}_e|^2)]f_{e0}d^3\mathbf{p}_e$. The first term of Eq. (12) represents the expression of axial quasistatic magnetic (QSM) field B_{sz}, while the second term of Eq. (12) represents the azimuthal QSM field $B_{s\theta}$.

Note that $B_{s\theta}$ here of the second term in Eq. (12) is in second order and proportional to the square of the laser amplitude, while in Ref. 22, $B_{s\theta}$ is in fourth order and proportional to the fourth power of the laser amplitude, where the second-order $B_{s\theta}$ is zero. This discrepancy can be explained. In the paper of Gorbunov *et al.* (Ref. 22), the cold fluid description is adopted, i.e., the plasma is assumed cold as electron temperature $T_e\simeq 0$, thus no second-order QSM field exists. Here in kinetic theory, we consider that the plasma is warm, thus the second-order QSM field including $B_{s\theta}$ does exist as a finite-temperature effect. If we assume the temperature of plasma electrons in our manuscript to be zero too, equivalent to the cold fluid description in Ref. 22, that is, we take the initial electron distribution function as $f_{e0}=\delta(p_x)\delta(p_y)\delta(p_z)$, then the relativistic coefficient β_4 in

Eq. (13) is zero and no second-order QSM field is recovered, which accords with the results of Ref. 22. Thus, the discrepancy between the results of our manuscript and those of Ref. 22 actually originates in the two different approaches, the cold fluid description in Ref. 22 and the kinetic one here and in Ref. 15.

Now, we give the detailed expression of these QSE and QSM fields for the above given laser beam. As also mentioned in Ref. 15, for real intense laser-plasma interaction, some of the electrons are accelerated into the relativistic regime. It is reasonable to assume that there are two groups of electrons in the system, thermal ones and relativistic ones. If we take the parameter ξ to represent the ratio of the number of thermal electrons to that of total electrons, substituting Eqs. (9) and (10) into Eqs. (11)–(13), noticing that $\partial/\partial t=(v_z-1)(\partial/\partial\eta)$ should be taken, introducing the normalized transform $t\rightarrow t/\omega_0$, $\mathbf{x}\rightarrow c\mathbf{x}/\omega_0$, $E\rightarrow E(m_e c\omega_0/e)$, and $B\rightarrow B(m_e c\omega_0/e)$, we get B_{sz}, $B_{s\theta}$, and three components of the corresponding $E^t_s|_{x,y,z}$, as well as $E^l_s|_{x,y,z}$ as

$$B_{sz}=-\frac{1}{2}\alpha n_{e0}[\xi\beta_1(T_{\mathrm{te}})+(1-\xi)\beta_1(T_{\mathrm{re}})]I_0 e^{-2r^2/b_0^2-2\eta^2/L^2}\frac{b_0^2}{b^2},$$

$$B_{s\theta}=\varepsilon\frac{\pi(1+\alpha^2)}{128}n_{e0}[\xi\beta_4(T_{\mathrm{te}})+(1-\xi)\beta_4(T_{\mathrm{re}})]I_0 e^{-2\eta^2/L^2}\frac{b^2}{r^2}(1-e^{-2r^2/r_0^2})\frac{b_0^2}{b^2}, \qquad (14)$$

$$\left(\frac{\partial^2}{\partial x^2}+\frac{\partial^2}{\partial y^2}+\frac{\partial^2}{\partial\eta^2}\right)E^t_{sx}=8\alpha n_{e0}[\xi\beta_1(T_{\mathrm{te}})+(1-\xi)\beta_1(T_{\mathrm{re}})]\times(1-v_z)I\left[\frac{\eta}{L^2}+\frac{1}{2}\frac{z/z_f^2}{(1+z^2/z_f^2)}-\frac{r^2}{b^2}\frac{z/z_f^2}{(1+z^2/z_f^2)}\right]\frac{y}{b^2},$$

$$\left(\frac{\partial^2}{\partial x^2}+\frac{\partial^2}{\partial y^2}+\frac{\partial^2}{\partial\eta^2}\right)E^t_{sy}=-8\alpha n_{e0}[\xi\beta_1(T_{\mathrm{te}})+(1-\xi)\beta_1(T_{\mathrm{re}})](1-v_z)I\left[\frac{\eta}{L^2}+\frac{1}{2}\frac{z/z_f^2}{(1+z^2/z_f^2)}-\frac{r^2}{b^2}\frac{z/z_f^2}{(1+z^2/z_f^2)}\right]\frac{x}{b^2},$$

$$\left(\frac{\partial^2}{\partial x^2}+\frac{\partial^2}{\partial y^2}+\frac{\partial^2}{\partial\eta^2}\right)E^t_{sz}=\varepsilon\frac{\pi(1+\alpha^2)}{8}n_{e0}[\xi\beta_4(T_{\mathrm{te}})+(1-\xi)\beta_4(T_{\mathrm{re}})](1-v_z)I\left[\frac{\eta}{L^2}+\frac{1}{2}\frac{z/z_f^2}{(1+z^2/z_f^2)}-\frac{r^2}{b^2}\frac{z/z_f^2}{(1+z^2/z_f^2)}\right], \qquad (15)$$

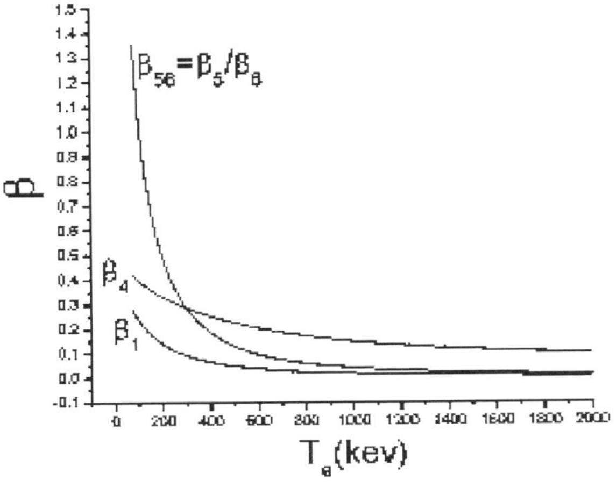

FIG. 1. The dependence of the relativistic coefficients β_1, β_4, and $\beta_{56}=\beta_5/\beta_6$ on the plasma electron temperature T_e.

$$E^l_{sx}=(1+\alpha^2)[\xi\beta_{56}(T_{\mathrm{te}})+(1-\xi)\beta_{56}(T_{\mathrm{re}})]\frac{x}{b^2}I,$$

$$E^l_{sy}=(1+\alpha^2)[\xi\beta_{56}(T_{\mathrm{te}})+(1-\xi)\beta_{56}(T_{\mathrm{re}})]\frac{y}{b^2}I,$$

$$E^l_{sz}=(1+\alpha^2)[\xi\beta_{56}(T_{\mathrm{te}})+(1-\xi)\beta_{56}(T_{\mathrm{re}})]\times I\left[\frac{\eta}{L^2}+\frac{1}{2}\frac{z/z_f^2}{(1+z^2/z_f^2)}-\frac{r^2}{b^2}\frac{z/z_f^2}{(1+z^2/z_f^2)}\right], \qquad (16)$$

where T_{te} is the averaged temperature for thermal electrons, T_{re} is the averaged temperature for relativistic electrons and $\beta_{56}=\beta_5/\beta_6$ is defined. The dependence of the relativistic coefficients β_1, β_4, and β_{56} on the electron temperature T_e is plotted in Fig. 1. The curves in Fig. 1 can be, respectively, fitted by the functions $\beta_1=0.36e^{-T_e/142}+0.07e^{-T_e/767}$, $\beta_4=2.90e^{-T_e/66}+0.56e^{-T_e/363}$, and $\beta_{56}=0.29e^{-T_e/346}+0.18e^{-T_e/3145}$. In Eqs. (14) and (15), we also assume $\mathbf{q}$ makes an angle of θ_q with the axis z, and $\langle\mathbf{e}_z\cdot\mathbf{e}^t_s\rangle=\langle\sin\theta_q\rangle=1/2$ is assumed. It should be noted that the effect of the axial QSM field B_{sz} is embodied by the slowly varying electric field $E^t_{sx,y}$ in Eq. (8) and the azimuthal one $B_{s\theta}$ is embodied by E^t_{sz}. In the second equation of Eqs. (14) and the third equation of Eqs. (15), ε is a coefficient less than 1.0, which indicates that $B_{s\theta}$, $\mathbf{E}^t_{sz}$ (the nonlinear transverse currents $\mathbf{j}^t_s$) are strongly compensated by the return currents.[23–25] In Refs. 23 and 25, simulations demonstrated that ε is approximately equal to 0.1. In this paper, we also choose $\varepsilon=0.1$ in the following simulation.

B. Ponderomotive force

In Eqs. (8), the ponderomotive force $\nabla\Gamma$ refers to the averaged one for the whole electron fluid. Firstly, we derive the approximate averaged energy Γ for the whole electron fluid in which electrons oscillating and moving forward along the laser propagation direction within laser fields ignoring the change of the laser beam transverse radius b. That is, b is assumed to be a constant having no relation to z. Under this assumption, the potential of spontaneous fields a_{sz}^t and ϕ_s, corresponding to the third equation of Eqs. (15) and (16), respectively, can be assumed to depend on z and t only through η as $\partial a_{sz}^t/\partial t=-\partial a_{sz}^t/\partial \eta$ and $\partial \phi_s/\partial z=\partial \phi_s/\partial \eta$. Thus, from Eq. (6) or the third equation of Eqs. (8), ignoring the fast-time-scale spontaneous fields, we get that $-\partial p_z/\partial \eta = -\partial a_{sz}^t/\partial \eta+\partial \phi_s/\partial \eta-\partial \Gamma/\partial \eta$, that is,

$$p_z - a_{sz}^t + \phi_s - \Gamma = C(\text{constant}). \tag{17}$$

As initially electrons are always assumed to be far away from the laser, $p_x=p_y=p_z=0$ should be chosen, that is, $\Gamma=1$ should be taken. Substituting this initial condition into Eq. (17), we get $C=\phi_s-a_{sz}^t-1$. Then we get that $p_z=\Gamma-1$ from Eq. (17). As well known, $\mathbf{p}_\perp^2=p_x^2+p_y^2=\mathbf{a}_L^2$, so we can easily get the averaged fluid energy Γ and z-axial velocity p_z as

$$\Gamma=\sqrt{1+\mathbf{p}_\perp^2+\mathbf{p}_z^2}=1+\frac{\mathbf{a}_L^2}{2}=1+\frac{1+\alpha^2}{4}I,$$

$$p_z=\frac{\mathbf{a}_L^2}{2}=\frac{1+\alpha^2}{4}I. \tag{18}$$

Substituting Eqs. (9) into Eq. (18), then we get the averaged ponderomotive force of electron fluid approximately as

$$\nabla\Gamma=-(1+\alpha^2)I\frac{x}{b^2}\mathbf{e}_x-(1+\alpha^2)I\frac{y}{b^2}\mathbf{e}_y-(1+\alpha^2)\times I\left[\frac{z-t}{L^2}+\frac{1}{2}\frac{z/z_f^2}{(1+z^2/z_f^2)}-\frac{r^2}{b^2}\frac{z/z_f^2}{(1+z^2/z_f^2)}\right]\mathbf{e}_z. \tag{19}$$

It should be noted here that Γ of Eq. (18) and $\nabla\Gamma$ of Eq. (19) are all approximate expressions that the change of the laser beam transverse radius is ignored. Because the exact expression for the ponderomotive force in intense focused laser beam interacted with relativistic plasma is still not clear so far, we have to use the approximate expression (19) for the averaged ponderomotive force of fluid to substitute $\nabla\Gamma$ in Eqs. (7) and (8) in our following numerical calculation. In general, this approximation is reasonable and has been always done in other earlier papers,[26,27] because the most important factor p_z of the ponderomotive force for fluid has already been considered. It should be demonstrated that, except for this approximation for $\nabla\Gamma$, all other physical variants are exactly derived and calculated in our following numerical calculation considering the change of the beam transverse radius, such as $\mathbf{E}_s^t$ and $\mathbf{E}_s^l$ of Eqs. (15) and (16). The energy of test electron γ is also calculated exactly as $\gamma=\sqrt{1+p_x^2+p_y^2+p_z^2}$ in the fourth equations of Eqs. (8).

IV. NUMERICAL SOLUTION AND ANALYSIS

Substituting the corresponding quasistatic fields and the averaged ponderomotive force discussed in Sec. III into Eqs. (8), we develop a three-dimensional (3D) test-particle code using Runge-Kutta method to solve the motion equations (8) of relativistic electrons. Using this code, we can numerically simulate the acceleration of plasma electron. In this section, we give the basic physical picture of plasma electron acceleration for, respectively, circularly polarized (CP) and linearly polarized (LP) lasers, analyze its dependences on initial radial position, plasma density, and laser intensity, respectively, and approximately analyze the quality of the formed relativistic electron beam (REB). We choose the parameters of the laser beam as intensities $I_0=6.2\times10^{18}$ W/cm^2 for CP laser ($\alpha=\pm1$) and/or $I_0=1.24\times10^{19}$ W/cm^2 for LP laser ($\alpha=0$) [keeping the averaged fluid ponderomotive force $\nabla\Gamma$ of Eq. (19) equal], wavelength $\lambda=1.054$ μm, pulse width $L=10\lambda$, and beam transverse radius at the focus point $b_0=5\lambda$. The initial plasma density is chosen as $n_{e0}=0.36n_{cr}$, where n_{cr} is the nonrelativistic critical density; plasma electrons are uniformly distributed and initially assumed to be far away from the laser beam. Here we assume test electrons initially locate at $z_0=4L$ and keep the their initial velocity at rest, i.e., $p_{x0}=p_{y0}=p_{z0}=0$, while the center of the laser pulse locates at $z=0$. As mentioned above, there are two groups of electrons in laser-plasma interaction, thermal electrons and hot electrons. The thermal electron temperature T_{te} is always from several keV to tens of keV, here we simply assumed the averaged T_{te} to be 10 keV. The relativistic electron temperature T_{re} is always determined by Wilk's scaling law as $T_{re}=511(\gamma_\perp-1)=511(\sqrt{1+[(1+\alpha^2)I_0\lambda_{\mu m}^2/2.76]}-1)$ keV. The ratio of the number of thermal electrons to that of total electrons ξ is assumed to be 0.3. Figure 2 plots the profiles of the corresponding spontaneous quasistatic fields under such parameters of laser and plasma; (a) transverse profiles of axial QSM field $|B_{sz}|$ (only exists for CP laser) and azimuthal one $|B_{s\theta}|$ calculated by Eqs. (14) cut at $z-t=0$, (b) transverse profiles of transverse (radial) QSE field E_{sr}^l calculated by Eqs. (16) and the corresponding equivalent "field" of radial ponderomotive force $m_0c^2\nabla_r\Gamma/e$ calculated by Eq. (19) cut at $z-t=0$, and (c) longitudinal profiles of longitudinal QSE field E_{sz}^l and the corresponding equivalent field of longitudinal ponderomotive force $m_0c^2\Delta_z\Gamma/e$ cut at $r=0$. From Fig. 2(a), we can see that B_{sz} gets to its peak value about 13.5 MG at the z axis ($r=0$) and decreases while r is increased, $B_{s\theta}$ is close to zero at the z axis ($r=0$), increases gradually with r until reaching its maximum value of 25.6 MG at about $r=b_0$, then decreases due to the return currents. From Figs. 2(b) and 2(c), we can also see that in laser-plasma interaction, the ponderomotive force $m_0c^2\Delta\Gamma$ is dominant and $\mathbf{E}_s^l$ can only partly offset the effect of the ponderomotive force.

A. Basic physical picture of electron acceleration

We trace the temporal evolution of momentum $\mathbf{p}$, normalized energy $\gamma=\sqrt{1+p_x^2+p_y^2+p_z^2}$, scattering angle with relation to the laser propagation axis (z axis), $\theta=\tan^{-1}(\sqrt{p_x^2+p_y^2}/p_z)$, and 3D trajectory (x,y,z) for test electron initially at $r=\sqrt{x_0^2+y_0^2}=0.4b_0$, $z_0=4L$ and plot the results in Figs. 3–6, respectively. Parts (a) and (b) of each figures are under the case neglecting the effects of spontaneous quasistatic fields, which are, respectively, for LP laser and CP laser; (c) and (d), which are also, respectively, for LP laser and CP laser, are under the case considering the effects of spontaneous quasistatic fields including QSE fields $\mathbf{E}_s^l$ and

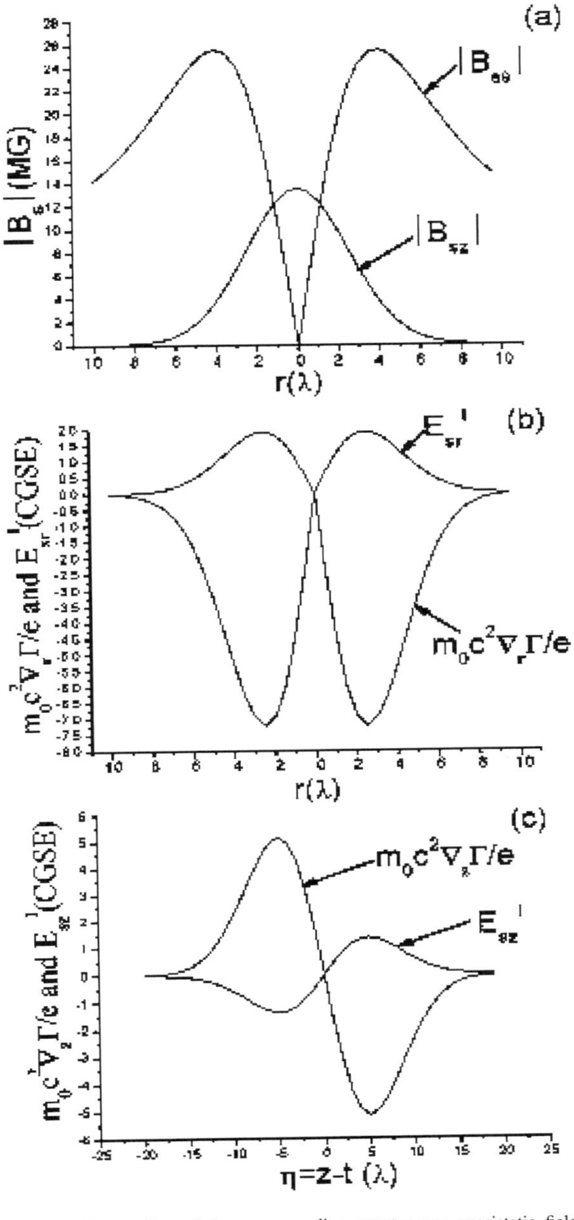

FIG. 2. The profiles of the corresponding spontaneous quasistatic fields under parameters of laser and plasma as $I_0=6.2\times10^{18}$ W/cm^2 for CP laser and/or $I_0=1.24\times10^{19}$ W/cm^2 for LP laser, $b_0=5\lambda$, $L=10\lambda$, and $n_{e0}=0.36n_{\rm cr}$. (a) Transverse profiles of $|B_{sz}|$ (only exists for CP laser) and $|B_{s\theta}|$ calculated by Eqs. (14) cut at $z-t=0$, (b) transverse profiles of transverse (radial) QSE field E^l_{sr} calculated by Eqs. (16) and the corresponding equivalent field of radial ponderomotive force $m_0c^2\nabla_r\Gamma/e$ calculated by Eq. (19) cut at $z-t=0$, (c) longitudinal profiles of longitudinal QSE field E^l_{sz} and the corresponding equivalent field of longitudinal ponderomotive force $m_0c^2\nabla_z\Gamma/e$ cut at $r=0$.

QSM fields $\mathbf{B}_s$ embodied in Eqs. (8) by $\mathbf{E}^l_s$. We give an explicit analysis about the effects of the ponderomotive force and each quasistatic fields on electron acceleration from the

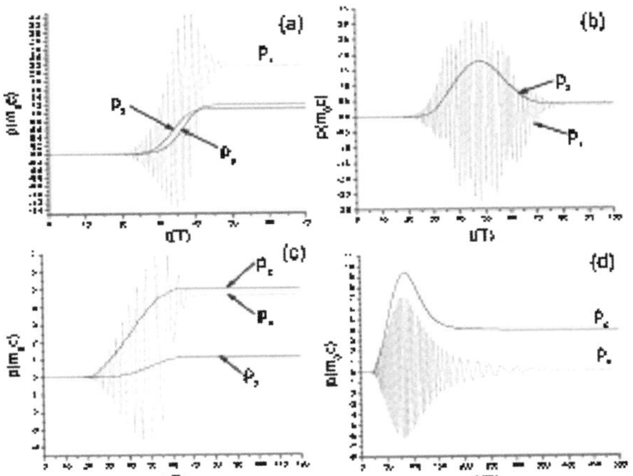

FIG. 3. The momentum **p** in units of m_0c as a function of time t in units of $T=2\pi c/\omega$. The initial test electron parameters are $p_{x0}=p_{y0}=p_{z0}=0$, $r=0.4b_0$, and $z_0=4L$; the center of the laser pulse locates at $z=0$ and the laser beam parameters are $I_0=6.2\times10^{18}$ W/cm^2 for CP laser ($\alpha=\pm1$) and/or $I_0=1.24\times10^{19}$ W/cm^2 for LP laser ($\alpha=0$) [keeping $\nabla\Gamma$ of them in Eqs. (8) and (19) equal], $\lambda=1.054$ μm, $L=10\lambda$, and $b_0=5\lambda$. (a) is for LP laser in the absence of spontaneous quasistatic fields, (b) is for CP laser in the absence of spontaneous quasistatic fields, (c) is for LP laser fields combined with spontaneous quasistatic fields including $\mathbf{E}^l_s$ and $B_{s\theta}$, and (d) is for CP laser fields combined with spontaneous quasistatic fields including $\mathbf{E}^l_s$, B_{sz}, and $B_{s\theta}$. For CP laser, because p_x is antisymmetric with p_y, here only p_x is plotted.

above figures and thus give the basic physical picture of electron acceleration, respectively, for LP and CP lasers.

First, for LP laser, due to the asymmetric acceleration of laser field, electron is accelerated and scattered by the ponderomotive force especially by the radial (transverse) ponderomotive force in the ascending (leading) front of laser field, then soon leaves the interaction region before being decelerated, thus finally satisfies $p_x>p_z$ as seen in part (a) of Fig. 3, and as a result, acquires a net energy gain as $\gamma-1$

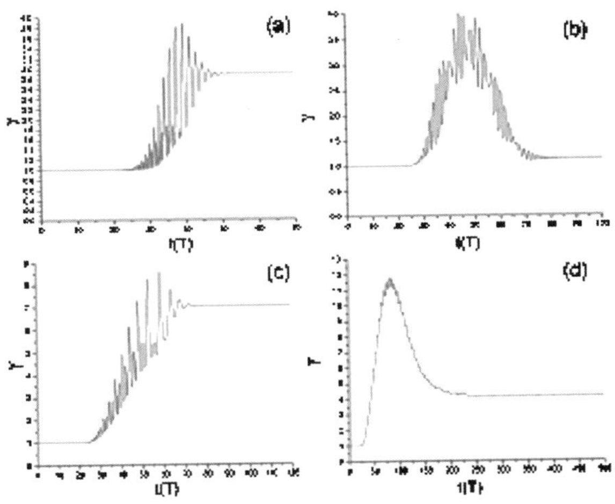

FIG. 4. The normalized test electron energy $\gamma=\sqrt{1+p_x^2+p_y^2+p_z^2}$ in units of m_0c^2 as a function of time t in units of $T=2\pi c/\omega$. The parameters and the conditions of (a)–(d) are the same as Fig. 3.

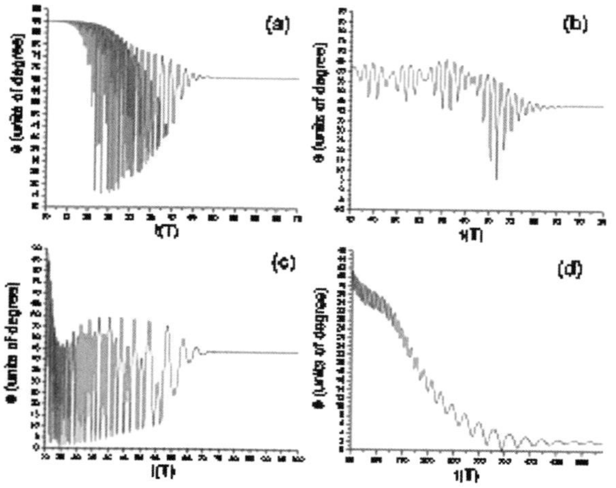

FIG. 5. The scattering angle $\theta=\tan^{-1}(\sqrt{p_x^2+p_y^2}/p_z)$ as a function of time t in units of $T=2\pi c/\omega$. The parameters and the conditions of (a)–(d) are the same as Fig. 3.

$=1.9$ and scattering angle of $\theta=66.3°$, as seen in part (a) of Figs. 4 and 5. Considering the effects of QSE field $\mathbf{E}_s^l$ and azimuthal QSM field $B_{s\theta}$, on the one hand, in order to keep the conservation of charge, E_s^l corresponding to space-charge potential will partly offset the effects of ponderomotive force including accelerating and scattering effects; on the other hand, $B_{s\theta}$ not only pinches electrons but also helps to accelerate electrons (in Eqs. (8), this effect is embodied through E_{sz}^t as mentioned above), so that the acceleration in the z-axis direction of electron has been enhanced. Thus, after considering the effects of quasistatic fields, finally electron satisfies $p_x \sim p_z$ as seen in part (c) of Fig. 3, gets much more energy gain of $\gamma-1=6.06$ and smaller scattering angle of $\theta=43.7°$ as seen in part (c) of Figs. 4 and 5.

Secondly, for the CP laser, because electron circularly (spirally) moves following the axial-symmetric laser electric fields, the scattering effect of the radial ponderomotive force is not obvious, electron is mainly accelerated in the ascending (leading) front, decelerated in the descending (trailing) part of the laser pulse by the longitudinal ponderomotive force, finally satisfying $p_x \approx p_z \to 0$ as seen in part (b) of Fig. 3, and as a result, does not acquire a net energy $\gamma-1\simeq 0$ but gets a scattering angle of $\theta=43°$ by the radial ponderomotive force, as seen in part (b) of Figs. 4 and 5. Considering the effect of spontaneous quasistatic fields, except the effects of E_s^l and $B_{s\theta}$ mentioned above, the axial QSM field B_{sz} generated only by CP laser ($E_{sx,y}^t$) will collimate the electron so that finally electron satisfies $p_z > p_x$ as seen in part (d) of Fig. 3, thus lastly gets a net energy of $\gamma-1\simeq 3.12(T_e=0.511(\gamma-1)=1.6\text{ MeV})$ and almost zero-degree scattering angle of $\theta=2.3°$, as seen in part (d) of Figs. 4 and 5.

Figure 6 plots the trajectories of test electron, respectively, under each cases, which cut at either $x=10\lambda$ or $y=10\lambda$. As can be seen, electron only moves to (a) $z=47\lambda$, (c) $z=70\lambda$ being scattered with 10λ transverse distance for LP laser, while for CP laser, electron can move along to (b) $z=64\lambda$, and considering more existence of B_{sz}, electron can move spirally in collimation to (d) $z=300\lambda$. That is, electron can move a longer distance along the z axis for a fixed transverse scattering region considering the effects of quasistatic fields, especially the pinching effect of $B_{s\theta}$ and collimating effect of B_{sz}. And electrons can move even much longer in the case of CP laser than that in the case of LP laser due to more existence of B_{sz}. Another important point that should be noted is that in the case of (d), when test electron gets to $z=300\lambda$, the laser beam has already passed electron far away with $z-t\simeq-20\lambda=-2L$ and the velocity of electron is about $0.97c$. That is, to say, the generated relativistic electrons can move with a velocity about c for a long distance, where the laser field has been absent. This is a promising for application to generate the high-energy REB, which can transport divorced from the laser.

All figures above demonstrate the following conclusions. (1) Spontaneous fields including QSM and QSE fields in laser-plasma interaction are favorable and prerequisite for the generation of collimated electrons with a moderate energy. (2) Due to more existence of B_{sz}, CP laser has more advantage to collimate relativistic electrons along the laser propagation axis (z axis) with almost zero scattering angle comparing to LP laser. (3) The generated relativistic electrons can move with a velocity about c divorced from laser, which is promising for application to generating the REB. (4) The resonance proposed in Ref. 14 does not happen here when we completely consider the synthetical effect of all self-consistent quasistatic fields in real intense laser-plasma interaction.

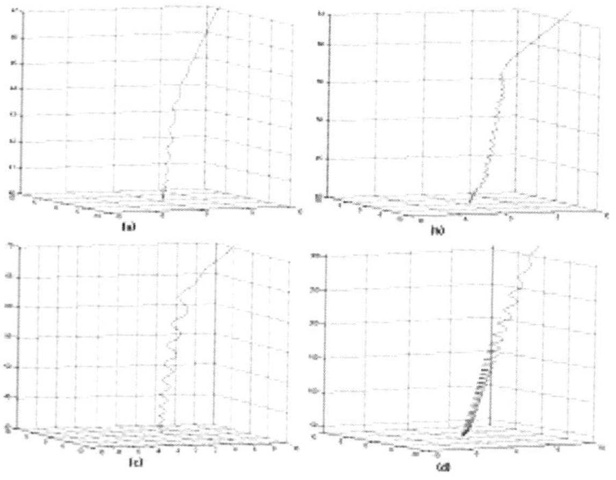

FIG. 6. The 3D electron trajectories (x, y, z in units of λ) with the same parameters of Figs. 3(a)–3(d), respectively. The parameters and the conditions of (a)–(d) are the same as Fig. 3.

B. Electron acceleration with different initial radial position and quality of relativistic electron-beam

When the test electrons are put on different initial positions in plasmas, here especially referring to different radial initial positions away from the laser propagation axis (z axis), their final energy and the scattering angle are different

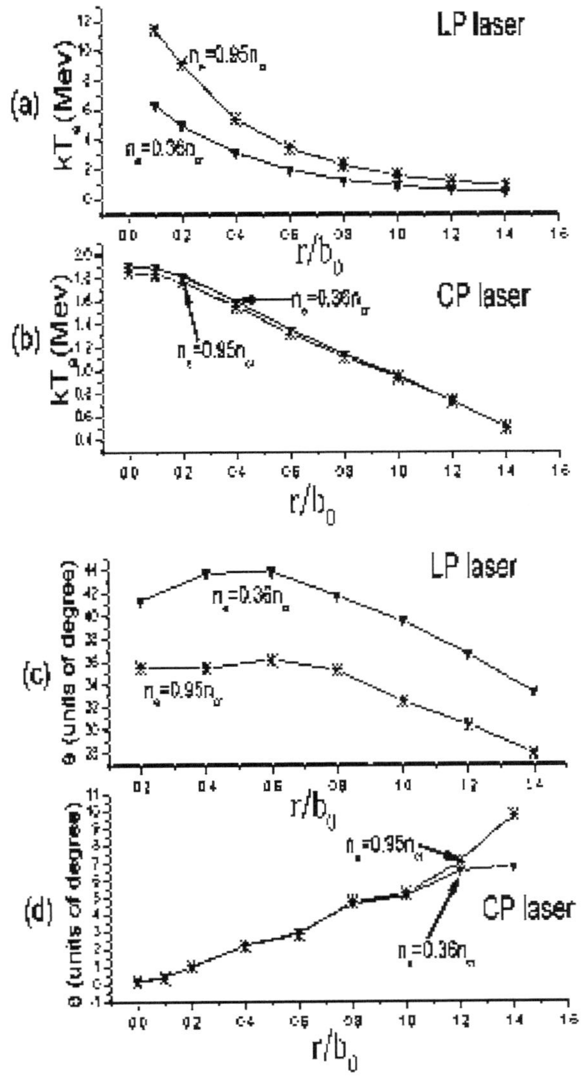

FIG. 7. The dependence of test electron final energy $kT_e=0.511(\gamma-1)$ MeV (a) and (b), scattering angle $\theta=\tan^{-1}(\sqrt{p_x^2+p_y^2}/p_z)$ (c) and (d) on its initial radial position r/b_0 in laser-produced plasma. Parts (a) and (c) are for LP laser, (b) and (d) for CP laser. The line with the symbol ▼ is under the case $n_e=0.36n_{\rm cr}$ and the line with the symbol $*$ is under the case $n_e=0.95n_{\rm cr}$. Other parameters are the same as Fig. 3.

due to acceleration. Figure 7 shows the final energy kT_e, scattering angle θ of test electron with different initial radial position in plasma (see the lines with the symbols "▼" and "$*$" which are, respectively, under $n_e=0.36n_{\rm cr}$ and $n_e=0.95n_{\rm cr}$), where test electron is under the effects of combined LP/CP laser field and spontaneous fields including QSE and QSM fields. Other parameters are consistent with Fig. 3. It can be seen from Fig. 7 that the nearer the electron initially is to the z axis, for LP laser, the more energy gain, the bigger scattering angle electron will obtain; while for CP laser, the more energy gain, the smaller scattering angle, electron will obtain. These results can be easily explained. For LP laser, the nearer the electron initially is to the z axis, the longer the time of electron acceleration by ponderomotive force before being expelled and the smaller the azimuthal QSM fields $B_{s\theta}$ (as can be seen from Fig. 2) are, which play a pinching effect on electrons, so that electron can obtain more energy and bigger scattering angle; while for CP laser, due to the more existence of axial QSM field B_{sz}, which dominates near the axis and collimates electrons along the z axis, the smaller scattering angle electron will obtain.

If we approximately take each test electron to represent a large number of plasma electrons at/near the same position, we can also analyze some qualities of the relativistic electron beam (REB) in laser-plasma interactions from the above Fig. 7, which is formed by the acceleration of large number of plasma electrons with different positions. This REB may have better qualities than that produced by other conventional accelerator, due to the above-analyzed advantageous effects of QSM and QSE fields. For the LP laser ($I_0=1.24\times10^{19}$ W/cm^2) with plasma density of $n_e=0.36n_{\rm cr}$, it can be seen from Figs. 7(a) and 7(c) that all electrons in the region of $r<1.4b_0$, $r\neq0$ are accelerated to be relativistic with energy from 450 keV($\gamma-1=0.88$) to 6.34 MeV($\gamma-1=12.4$) and scattering angle from 33° to 44° with relation to the z axis. That is, to say, a REB with beam divergence of 44° $-33°=11°$ is emitted along the direction at an angle of about 45° with the laser propagation axis. Due to zero ponderomotive force at $r=0$, we find that electrons on the z axis ($r=0$) are accelerated to have much smaller energy 1.75 MeV and smaller scattering angle of 11°. For the CP laser ($I_0=6.2\times10^{18}$ W/cm^2) with $n_e=0.36n_{\rm cr}$, the axial QSM field B_{sz} will collimate electrons, one should see from Figs. 7(b) and 7(d) that most electrons in a very large region $r<1.4b_0$ near the axis are accelerated to the required moderate energy 500 keV$<kT_e<$1.9 MeV($1.0<\gamma-1<3.7$) and move along the z axis in collimation with a very small scattering angle θ from 0.2° to 7°. This demonstrates that a much better REB source with beam divergence of about 7° and moderate energy <2 MeV is emitted along the laser propagation direction in CP-laser-produced plasma. Thus the conclusion can be obtained that both CP laser and LP laser can generate a well-directed REB, however, CP laser has more advantage comparing to LP laser because the it can generate a REB with smaller beam divergence and smaller energy spread and almost zero scattering (emitting) angle. This conclusion suggests that if CP laser is used in the future laser-plasma accelerator applications, a much better REB in collimation and stabilization can be obtained. The quality of REB generated by laser-plasma accelerator may be better than that obtained by conventional particle accelerators due to the effects of strong spontaneous quasistatic fields self-generated in laser-plasma interaction.

C. Electron acceleration with different plasma density and laser intensity

We increase the plasma density up to near critical to see the plasma electron acceleration by intense laser. The lines with the symbols "$*$" in Fig. 7 plot the final energy kT_e, scattering angle θ of test electron under the near-critical plasma with density $n_e=0.95n_{cr}$, other parameters are the same. It can be seen from Fig. 7 that the plasma density has little effect on the single test electron acceleration behavior such as final energy and scattering angle. For CP laser, the final energy and scattering angle vary little corresponding to those under $n_e=0.36n_{cr}$; while for LP laser, the final energy increases a little higher and the scattering angle becomes a little smaller. Thus, the conclusion can be given that as far as single test electron is concerned, whether in underdense plasma or in near-critical plasma, it has the same acceleration process/mechanism; the above basic physical picture of plasma electron acceleration do not change. The reason can be easily explained from the expressions of all fields in Sec. III. As can be seen there, plasma density has only effects on QSM fields [see Eqs. (14) and (15)], which only play roles in collimating and pinching electrons; while the final energy of test particle (single electron) is mainly decided by the ponderomotive force and QSE fields, which both have no relation to plasma density [see Eqs. (16) and (19)]. For CP laser, electron is collimated very near the laser axis by the strong axial QSM field B_{sz}, the effect of the azimuthal one $B_{s\theta}$ is almost none [see $B_{s\theta}$ in Fig. 2(a)], and thus the energy and scattering angle of electron almost do not change with respect those of $n_e=0.36n_{cr}$. For LP laser, in the absence of B_{sz}, the strengthened $B_{s\theta}$ helps to accelerate and pinch electron, and thus electron gets a little more energy and smaller scattering angle. We should demonstrate that the number of the relativistic electrons produced in near-critical plasma (such as $n_e=0.95n_{cr}$) is much larger than that in underdense plasma (such as $n_e=0.36n_{cr}$). The study of this problem is beyond the scope of test-particle method in our paper.

If the laser intensity I_0 changes, the acceleration of test electron varies. Figure 8 gives the changes of normalized final energy γ, final scattering angle θ with different I_0. The initial position of test electron is $r=0.2b_0$, $z_0=4L$, and other parameters are consistent with the above. In order to keep the approximate ponderomotive force of Eq. (19) for LP laser equal to that of CP laser, we take intensity $I_{LP}=2I_{CP}=2I_0$ for I_0 from 1×10^{18} to 2×10^{19} W/cm^2. As can be seen from Eqs. (14)–(16) and (19), with I_0 is increased, $B_{s\theta}$, B_{sz}, $\mathbf{E}_s^l$, and ponderomotive force $\nabla\Gamma$ are all strengthened. As analyzed above, the acceleration considering spontaneous quasistatic fields of electrons depends mainly on the ponderomotive force $\nabla\Gamma$, thus the electron energy γ is increased with I whether for LP laser or CP laser, as seen in Fig. 8(a). And also as analyzed above, $B_{s\theta}$ helps to pinch electron, B_{sz} collimates electron, so the scattering angle θ of electron is decreased while I_0 is strengthened as seen in Fig. 8(b). As can be seen from Fig. 8, for CP laser, when I_0 increase to 2×10^{19} W/cm^2, the final electron energy can reach about $\gamma=18$, i.e., energy $kT_e=8.6$ MeV and the scattering angle is still about zero. Our calculation results show that when $I_0=10^{20}$ W/cm^2, the electron final energy can reach very high about $kT_e=70$ MeV and scattering angle about 3° for CP laser. Such high-energy electrons with almost zero scattering angle can form the REB with good quality for application.

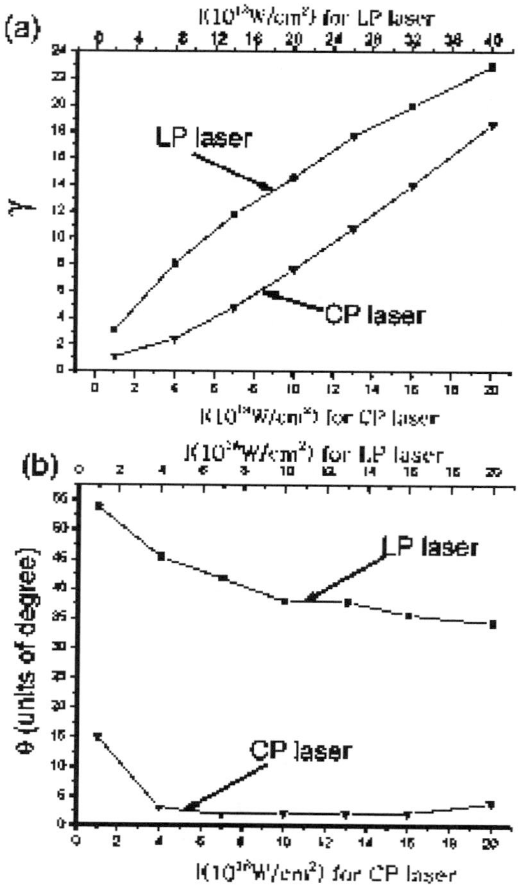

FIG. 8. The dependence of test electron normalized final energy γ (a), scattering angle θ (b) on the laser intensity I_0 in units of 10^{18} W/cm^2 for initial $r=0.2b_0$, $z_0=4L$. Other parameters are the same as Fig. 3.

V. DISCUSSION AND CONCLUSION

Using test-particle method, we have established an analytical electron fluid model to investigate the acceleration mechanism of plasma electrons in combined intense focused laser fields and self-consistent spontaneous quasistatic fields. The spontaneous quasistatic fields including quasistatic electric (QSE) fields $\mathbf{E}_s^l$ and quasistatic magnetic (QSM) fields (both axial B_{sz} and azimuthal $B_{s\theta}$) are not only self-consistently derived from kinetic theory (Ref. 15) but also completely considered for the first time in the model, other than those assumed as Refs. 11–14. We develop a fully relativistic 3D test-particle fluid code using Runge-Kutta method based on this model. Through numerical calculation, we find that the spontaneous quasistatic fields play important roles in accelerating plasma electrons and form the relativistic electron beam (REB), such as $\mathbf{E}_s^l$ partially offsets the ponderomotive acceleration and scattering effects on electrons, $B_{s\theta}$ helps to accelerate and pinch electrons, and B_{sz} makes collimating electrons to move along near the laser propagation axis. The dependences of final energy gain and scattering angle of test electron on its initial radial position, plasma density, and laser intensity are, respectively, studied. We find that plasma electrons can be accelerated to form a well-directed REB under the combined laser fields and spontaneous quasistatic fields, which would transport divorced from laser. For CP laser (intensity $I_0=6.2\times10^{18}$ W/cm^2) combined with B_{sz} about 13.5 MG and $B_{s\theta}$ about 25.6 MG (plasma density $n_e=0.36n_{cr}$), plasma electrons in a radial region $r<1.4b_0$ are accelerated to form a REB source with mean energy of (kT_e) 500 keV−1.9 MeV and beam divergence $<7°$, which is emitted along the laser propagation axis (z axis). For LP laser ($I_0=1.24\times10^{19}$ W/cm^2), B_{sz} is nothing, a REB source with kT_e of 450 keV–6.5 MeV and beam divergence of 11° are formed by acceleration of plasma electrons in the range of $r<1.4b_0$, $r\neq0$ and emitted along the direction at the angle of about 45° with the z axis. When the laser intensity I_0 increases to 10^{20} W/cm^2, the energy of the produced REB can reach very high about 70 MeV. Thus, it is demonstrated that intense laser directly interacting with plasma can generate the REB with very good qualities and high energy, where the conjunct effect of laser fields and self-consistent quasistatic fields is considered. The results also show that the circularly polarized laser is favorable for generating the better relativistic electron beams in collimation and stabilization in comparison with the linearly polarized laser.

ACKNOWLEDGMENTS

This work was supported in part by the National Hi-Tech Inertial Confinement Fusion (ICF) Committee of China, National Science Foundation of China (NSFC) Grant Nos. 10335020, 10375011, and 10135010, National Basic Research Project "Nonlinear Science" in China, and Science Foundation of CAEP.

[1] D. Strickland and G. Mourou, Opt. Commun. **56**, 219 (1985); G. Mourou, C. P. J. Barty, and M. D. Perry, Phys. Today **51**(1), 22 (1998).
[2] T. Tajima and J. M. Dawson, Phys. Rev. Lett. **43**, 267 (1979).
[3] C. Joshi and T. Katsouleas, Phys. Today **56**, 47 (2003).
[4] E. Esarey, P. Sprangle, J. Krall, and A. Ting, IEEE Trans. Plasma Sci. **24**, 252 (1996).
[5] R. Wagner, S. Y. Chen, A. Maksimchuk, and D. Umstadter, Phys. Rev. Lett. **78**, 3125 (1997).
[6] A. Ting, C. I. Moore, K. Krushelnick, C. Manka, E. Esarey, P. Sprangle, R. Hubbard, H. R. Burris, R. Fischer, and M. Baine, Phys. Plasmas **4**, 1889 (1997).
[7] K. C. Tzeng and W. B. Mori, Phys. Rev. Lett. **81**, 104 (1998).
[8] A. Modena, Z. Najmudin, A. E. Dangor, C. E. Clayton, K. A. Marsh, C. Joshi, V. Malka, C. B. Darrow, C. Danson, D. Neely, and F. N. Walsh, Nature (London) **377**, 606 (1995).
[9] M. I. K. Santala, Z. Najmudin, E. L. Clark, M. Tatarakis, K. Krushelnick, A. E. Dangor, V. Malka, J. Faure, R. Allott, and R. J. Clarke, Phys. Rev. Lett. **86**, 1227 (2001).
[10] A. Pukhov and J. Meyer-ter-Vehn, Appl. Phys. B: Lasers Opt. **74**, 355 (2002).
[11] A. Pukhov and J. Meyer-ter-Vehn, Phys. Plasmas **5**, 1880 (1998); A. Pukhov, Z. M. Sheng, and J. Meyer-ter-Vehn, Phys. Plasmas **6**, 2847 (1999).
[12] M. Schmitz and H.-J. Kull, Laser Phys. **12**, 443 (2002); M. Tanimoto, S. Kato, E. Miura, N. Saito, K. Koyama, and J. K. Koga, Phys. Rev. E **68**, 026401 (2003).
[13] M. Y. Yu, Wei Yu, Z. Y. Chen, J. Zhang, Y. Yin, L. H. Cao, P. X. Lu, and Z. Z. Xu, Phys. Plasmas **10**, 2468 (2003).
[14] Hong Liu, X. T. He, and S. G. Chen, Phys. Rev. E **66**, 066409 (2004).
[15] B. Qiao, S.-P. Zhu, C. Y. Zheng, and X. T. He, Phys. Plasmas (to be published).
[16] Z. Najmudin, M. Tatarakis, A. Pukhov *et al.*, Phys. Rev. Lett. **87**, 215004–1 (2001); J. Fuchs, G. Malka, J. C. Adam *et al.*, Phys. Rev. Lett. **80**, 1658 (1998).
[17] V. I. Berezhiani, S. M. Mahajan, and N. L. Shatashvili, Phys. Rev. E **55**, 995 (1997); L. M. Gorbunov, Usp. Fiz. Nauk **109**, 631 (1973) [Sov. Phys. Usp. **16**, 217 (1973)]; N. L. Tsintsadze and D. D. Tskhakaya, Zh. Eksp. Teor. Fiz. **72**, 480 (1977) [Sov. Phys. JETP **45**, 252 (1977)]; A. Bourdier and X. Fortin, Phys. Rev. A **20**, 2154 (1979).
[18] X. F. Wang, R. Fedosejevs, and G. D. Tsakiris, Opt. Commun. **146**, 363 (1998).
[19] D. Gordon, K. C. Tzeng, C. E. Clayton *et al.*, Phys. Rev. Lett. **80**, 2133 (1998).
[20] A. Yariv, *Quantum Electronics*, 2nd ed. (Wiley, New York, 1975).
[21] F. He, W. Yu, P. Lu, H. X. L. Qian, B. Shen, X. Yuan, R. Li, and Z. Xu, Phys. Rev. E **68**, 046407 (2003).
[22] L. Gorbunov, P. Mora, and T. M. Antonsen, Jr., Phys. Rev. Lett. **76**, 2495 (1996).
[23] A. Pukhov and J. Meyer-ter-Vehn, Phys. Rev. Lett. **76**, 3975 (1996).
[24] C. Y. Zheng, X. T. He, and S. P. Zhu, Phys. Plasmas **12**, 044505 (2005).
[25] A. Pukhov and J. Meyer-ter-Vehn, Phys. Rev. Lett. **79**, 2686 (1997).
[26] P. Mora and B. Quesnel, Phys. Rev. Lett. **80**, 1351 (1998); B. Quesnel and P. Mora, Phys. Rev. E **58**, 3719 (1998).
[27] W. Yu, M. Y. Yu, J. X. Ma, Z. M. Sheng, J. Zhang, H. Daido, S. B. Liu, Z. Z. Xu, and R. X. Li, Phys. Rev. E **61**, 2220 (2000).

Appl. Phys. B 82, 93–97 (2006)
DOI: 10.1007/s00340-005-1974-3

Applied Physics B
Lasers and Optics

H. LIU[1,4,✉]
X.T. HE[2,3]
H. HORA[5]

Additional acceleration and collimation of relativistic electron beams by magnetic field resonance at very high intensity laser interaction

[1] Basic Department, Beijing Materials Institute, Beijing 101149, P.R. China
[2] Institute of Applied Physics and Computational Mathematics, Beijing P.O. Box 8009, Beijing 100088, P.R. China
[3] Department of Physics, Zhejiang University, Hangzhou 310027, P.R. China
[4] Laboratory of Optical Physics, Institute of Physics, Chinese Academy of Science, Beijing 100080, P.R. China
[5] Department of Theoretical Physics, University of New South Wales, Sydney 2052, Australia

Received: 21 January 2005/Revised version: 18 July 2005
Published online: 26 November 2005 •

ABSTRACT For the interpretation of experiments for acceleration of electrons at interaction up to nearly GeV energy in laser produced plasmas, we present a new model using interaction magnetic fields. In addition to the ponderomotive acceleration of highly relativistic electrons at the interaction of very short and very intense laser pulses, a further acceleration is derived from the interaction of these electron beams with the spontaneous magnetic fields of about 100 MG. This additional acceleration is the result of a laser-magnetic resonance acceleration (LMRA) around the peak of the azimuthal magnetic field. This causes the electrons to gain energy within a laser period. Using a Gaussian laser pulse, the LMRA acceleration of the electrons depends on the laser polarization. Since this is in the resonance regime, the strong magnetic fields affect the electron acceleration considerably. The mechanism results in good collimated high energetic electrons propagating along the center axis of the laser beam as has been observed by experiments and is reproduced by our numerical simulations.

PACS 41.75.Jv; 52.38.Kd; 52.65.Cc

1 Introduction

The use of very short focused laser pulses of picosecond or less duration with powers of TW (Terawatt = 10^{12} W) and up of PW (Petawatt = 10^{15} W) and beyond arrived at a new category of interactions. The phenomena to be discussed now are polarization dependence effects [1–10], deviations of the generated plasmas from nearly space charge neutralization as in the earlier cases, and in relativistic effects. The earlier observed cases could mostly be described by space neutralized plasma hydrodynamics even including relativistic self focusing [11–15] generated ions of several 100 MeV energy because the Debye lengths involved were sufficiently short and the internal electric fields [16] were not yet of dominating influence. The acceleration of free electrons by laser fields was well discussed separately [15, 17, 18] and resulted in some agreement with ps–TW measurements [19]. Nevertheless very high density relativstic electron beams were measured recently, whichresulted in a new situation where relativistic self-focusing, plasma motion, and the beam generation described by particle-in-cell (PIC) methods [10] were covering the phenomena not completely depending on each situation.

The generation of extremely high magnetic fields in laser-produced plasmas [20] was known since a long time but there is a basically new situation with the very intense relativistic electron beams and their mutual interaction with the very high magnetic fields. We present here studies of these interactions of the electron-beams with the magnetic fields as a basic new laser-magnetic resonance acceleration (LMRA) mechanism which results in a kind of pinch effect with a high degree of collimation of the electron beams. As the fast-ignitor (FI) concept [21] for inertial confinement fusion relies on many new phenomena such as explosive channel formation [22] and self-generated huge magnetic field [20], much interest in ultraintense laser-plasma interaction studies appeared. More interestingly, the gigagauss self-generated azimuthal magnetic field has been observed recently [9].

We found that two different fast electrons exist in the presence of self-generated azimuthal magnetic field. Two sources of multi-MeV electrons are distinguished from the relativistic laser–plasma interaction. The first is that the electron acceleration depends on the laser intensity, known as the ponder-motive acceleration [16–19]. The second is that, around the peak of azimuthal magnetic field, LMRA partly occurs which causes the electron to gain energy from the ratio between electron Larmor frequency and laser frequency within one laser period [23]. If we consider a linearly polarized (LP) laser pulse, the LMRA results in a dependence of the laser accelerated electrons on the laser polarization. Because in the resonance regime, the strong magnetic field affects the electron acceleration dramatically. Only just from the second source, polarization dependence of electron is appeared. In our knowledge, these different sources of fast electrons are mentioned for the first time. This clears up many experiments and PIC simulations which are related with polarization dependence phenomena.

A fully relativistic single particle code is developed to investigate the dynamical properties of the energetic electrons.

✉ Fax: +86-10-58800567, E-mail: liuhong_cc@yahoo.com.cn

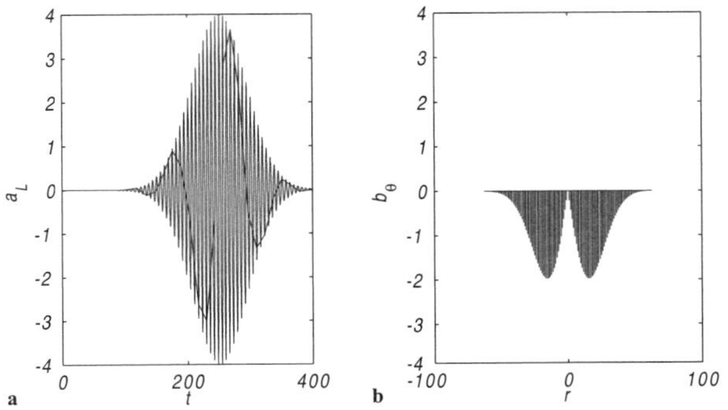

FIGURE 1 (**a**) The profile of laser intensity a as a function of time in the units of ω^{-1}, (**b**) the profile of quasistatic magnetic field b_θ (units 1 corresponding to 100 MG) as a function of circle radius in the units of k^{-1}

The single test electron model is a simple but effective one. It has been used to analyze the direct laser acceleration of relativistic electrons in plasma channels, e.g., M. Schmitz et al. [24] have analyzed the LP laser system with self-generated static electric field and discussed the electron resonant acceleration mechanism. K.P. Singh [25] found that resonance occurs between the electrons and electric field of the laser pulse. We find a big difference between the pondermotive acceleration and LMRA mechanism in this paper. We discuss LMRA mechanism of electrons in strong LP laser pulses and self-generated azimuthal magnetic fields. In our simulation, the laser field is a Gaussian profile and the quasistatic magnetic field is in a circle profile which lies on the laser intensity and as a function of circle radius. Our paper is organized as follows. In Sect. 2 we describe the dynamical behavior of relativistic electrons in a combined LP laser and quasistatic azimuthal magnetic field numerically. Two regimes (the peak of laser and the peak of quasistatic magnetic field) and two typical directions (in polarization direction and 90° turned from the polarization direction) are discussed. We also show an approximate analytical equation which provides a full understanding of the LMRA mechanism for comparison. The numerical results clearly demonstrate that LMRA partly occurs within one laser period. Our discussion and conclusion are given in Sect. 3.

2 Gaussian LP laser pulse model

The relativistic Lorentz force equations with quasistatic magnetic-field which is perpendicular to the laser propagation direction are

$$\frac{d\boldsymbol{p}}{dt} = \frac{\partial \boldsymbol{a}}{\partial t} - \boldsymbol{v} \times (\nabla \times \boldsymbol{a} + \boldsymbol{b}_\theta)\,, \tag{1}$$

$$\frac{d\gamma}{dt} = \boldsymbol{v} \cdot \frac{\partial \boldsymbol{a}}{\partial t}\,, \tag{2}$$

where $\boldsymbol{a}$ is the normalized vector potential. $\boldsymbol{b}_\theta$ is the normalized azimuthal magnetic field, $\boldsymbol{v}$ is the normalized velocity of electron, $\boldsymbol{p}$ is the normalized relativistic momentum, $\gamma = (1-v^2)^{-1/2}$ is the relativistic factor or normalized energy. Their dimensionless forms are $\boldsymbol{a} = \frac{e\boldsymbol{A}}{m_e c^2}$, $\boldsymbol{b} = \frac{e\boldsymbol{B}}{m_e c\omega}$, $\boldsymbol{v} = \frac{\boldsymbol{u}}{c}$, $\boldsymbol{p} = \frac{\boldsymbol{p}}{m_e c} = \gamma \boldsymbol{v}$, $t = \omega t$, $r = kr \cdot m_e$ and e are the electric mass and charge, respectively, c is the light velocity. $k = 2\pi/\lambda$ is the wave number, λ is the wave length. We assume that the laser propagation is in positive $\hat{\boldsymbol{z}}$ direction along the plasma channel with a phase velocity v_{ph}. For simplicity, in the following discussions we assume that the phase velocity of the laser pulse equals to the light velocity, i.e., $v_{\mathrm{ph}} = c$. The main results obtained can be readily extended to the case of $v_{\mathrm{ph}} \neq c$. For the irradiance of the femtosecond laser pulses, the plasma ions have no time to respond to the laser and therefore can be assumed to be immobile. We have used the Coulomb gauge. Here, because the dimensionless self-generated azimuthal magnetic field e.g., 2 is much larger than the static electric field e.g., 0.01, the effects of the static electric field can be ignored.

For a linearly focused Gaussian profile laser with frequency ω along in plasma channel can be modeled as

$$\boldsymbol{a} = a_0 \mathrm{e}^{-\frac{x^2+y^2}{R_0^2}} \cdot \mathrm{e}^{-\frac{(kz-\omega t)^2}{k^2 L^2}} \cdot \cos(kz - \omega t)\hat{\boldsymbol{x}} \tag{3}$$

where the critical density $n_c = m\omega^2/(4\pi e^2)$, the plasma frequency is equal to the light frequency, L and R_0 are the pulse width and minimum spot size, respectively. The laser pulse is a transverse wave satisfying $\boldsymbol{k} \cdot \boldsymbol{a} = 0$. Its profile shows in Fig. 1a.

For the background field, the generation of an azimuthal quasistatic magnetic field (b_θ) has been discussed by many authors and observed in experiments. The typical work related with ultraintense short laser pulseswith overdense plasma interaction has been done by R.N. Sudan [20] in 1993. He proposed that the mechanism for magnetic field generation is a result of dc currents driven by the spatial gradients and temporal variations of the ponderomative force exerted by the laser on the plasma electrons. In a recent experiment, M. Tatarkis et al. observed the peak of $\boldsymbol{b}_\theta$ at least hundreds of MG as given in [9]. Our model uses $\boldsymbol{b}_\theta$ in the form:

$$b_\theta = -b_{\theta 0} \cdot r \cdot \langle a^2 \rangle \hat{\boldsymbol{\theta}} \tag{4}$$

The above explicit expression clearly indicates that the self-generated magnetic field $\boldsymbol{b}_\theta$ is an oriented circle. It is caused by the longitudinal electron currents motion. We have assumed that ion immobile. $b_{\theta 0}$ is an approximately coefficient, including slow-time changed plasma parameters. $r = \sqrt{x^2+y^2}$ is the distance from the axis. $\langle\,\rangle$ denotes the time average over one laser period. $r \cdot \langle a^2 \rangle$ decide the structure of $\boldsymbol{b}_\theta$. The peak of $\boldsymbol{b}_\theta$ is located at the $R_0/2$ laser spot. Although the reality is far more complex and the form will be significantly different. We use the rough profile to investigate the dynamics of the fast electron. Its profile shows in Fig. 1b.

We assume that the trajectory of a test electron starts at $\boldsymbol{v}_0 = 0$. Equations (1) and (2) yield

$$\frac{dp_x}{dt} = (v_z - 1)\frac{\partial a}{\partial z} + v_y \frac{\partial a}{\partial y} + v_z b_\theta \cos\theta \tag{5}$$

$$\frac{dp_y}{dt} = -v_x \frac{\partial a}{\partial y} + v_z b_\theta \sin\theta \tag{6}$$

$$\frac{dp_z}{dt} = -v_x \frac{\partial a}{\partial z} - v_x b_\theta \cos\theta - v_y b_\theta \sin\theta \tag{7}$$

$$\frac{d\gamma}{dt} = -v_x \frac{\partial a}{\partial z} \tag{8}$$

where $\boldsymbol{b}_\theta = b_\theta(-\sin\theta\hat{\boldsymbol{x}} + \cos\theta\hat{\boldsymbol{y}})$, $\theta = \mathrm{arctg}(\frac{y}{x})$. An exact analytical solution of (5)–(8) is impossible because of their

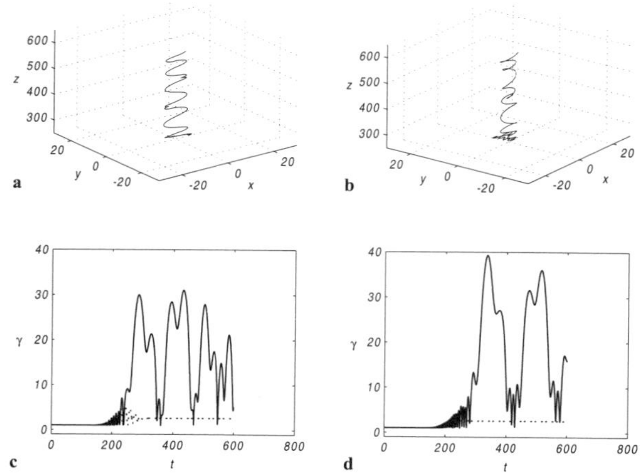

FIGURE 2 Electron in combined $\boldsymbol{a}$ and $\boldsymbol{b}_\theta$ fields. (**a**) The trajectory of electron start at $x_0=\lambda$, $y_0=0$ (in polarization direction $\hat{\boldsymbol{x}}$), (**b**) The trajectory of electron start at $x_0=0$, $y_0=\lambda$ (90° turned from the direction of polarization). (**c**) and (**d**) Electron energy γ in units of mc^2 as a function of time in the units of ω^{-1}. Other parameters is corresponding to (**a**) and (**b**), respectively, with $\boldsymbol{b}_\theta$ (*solid line*) and without $\boldsymbol{b}_\theta$ (*dashed line*)

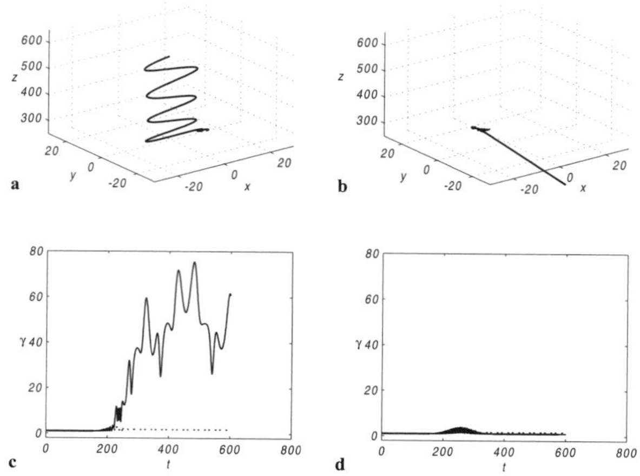

FIGURE 3 The parameters are same with Fig. 2 but only (**a**) $x_0=R_0/2$, $y_0=0$ (in the peak of $\boldsymbol{b}_\theta$) (**b**) $x_0=0$, $y_0=R_0/2$ (90° turned from the direction of polarization)

nonlinearity. Nevertheless, these equations reveal the mechanism of acceleration and collimation and will be solved numerically.

Using (5)–(8), we choose a different initial position to investigate the electron dynamics of a LP Gaussian profile laser pulse. Because the initial velocity can be transformed to initial the position in our single test electron case, we keep the initial velocity at rest and change the initial positions of the test electrons. We assume that the trajectory of a test electron starts from $\boldsymbol{v}_0=0$ and $z_0=4L$ at $t=0$, while the center of laser pulse locates at $z=0$, then the classical trajectory is then fully determined by (5)–(8). Now we choose following parameters that are available in present experiments, i.e., $L=10\lambda$, $R_0=5\lambda(\lambda=1.06\,\mu\text{m})$, $a_0=4$ (corresponding to $I\approx 2\times10^{19}\,\text{W/cm}^2$), $b_{\theta 0}=2$ (corresponding to $B_\theta\approx 200\,\text{MG}$). Then, we trace the temporal evolution of electron energy and trajectory and plot the results in Figs. 2–3.

In order to explain the simulation results, we excerpt an analytical equation which has been obtained in [23]

$$\gamma = 1+\frac{1}{2}\frac{a^2}{\left(1+\frac{b}{\omega}\right)^2}\,. \tag{9}$$

where a and b are the local magnitude of the laser and the quasistatic magnetic field of the energetic electron. Although the equation is derived from a model which contains a circularly polarized laser and an axial static magnetic field, it indicates the resonance between the laser field and the magnetic field. Because the energy (9) is independent of time, we use it to explain what drives the energy of electron to high energy along the strong magnetic field presence. That is when the LMRA occurs or partly occurs within laser period, the electron will gain energy from the near resonance point (singularity) at a negative b_θ ($\approx\omega$). The electron acceleration depends not only on the laser intensity, but also on the ratio between electron Larmor frequency and laser frequency.

3 Numerical results

Figures 2 and 3 show the track of a test electron and its correspondent net energy gain in the combined $\boldsymbol{a}$ and $\boldsymbol{b}_\theta$ fields from different regimes, e.g., the peak of laser and the peak of quasistatic magnetic field, respectively. (a)-(c) The trajectory of electron start at $x_0=\lambda$, $y_0=0$ (in polarization direction $\hat{\boldsymbol{x}}$), (b)-(d) The trajectory of electron start at $x_0=0$, $y_0=\lambda$ (90° turned from the direction of polarization). The difference trajectory of the test electron and its correspondent energy gain from the different initial positions can be compared. We also show the electron energy γ (in the units of mc^2) as a function of time (in the units of ω^{-1}) in dashed line for the case of without $\boldsymbol{b}_\theta$ in Figs. 2–3c and d. We like to emphasize that at relativistic intensities laser ($I>10^{18}\,\text{W/cm}^2$) the electron drift velocity is very slow but not slow enough. In fact, for $a_0=2$ the drift velocity is the same order of the quiver velocity. When $a_0=4$, $\boldsymbol{p}_\perp=\boldsymbol{a}$, $p_z=\frac{a^2}{2}$ from $\gamma=\sqrt{1+p_\perp^2+p_z^2}$, then $\gamma_{\max}=9$. The nonlinear ponderomotive scattering angle [19] (in vacuum) $\theta=\text{arctg}\sqrt{\frac{2}{\gamma-1}}\approx 28°$. The electron momentums of two transverse directions are independent on the laser polarization. But such large scattering angles will be unfavorable to the fast ignition of the high compressed fuel. When a strong self-generated azimuthal magnetic field presence, things will be changed. When $\boldsymbol{a}$ and $\boldsymbol{b}_\theta$ coexist, e.g., $a=4$, $b_\theta=-0.6$, using our derived (9) we can estimate that $\gamma_{\max}\approx 51$ which is very close to our simulation results. Within one laser period, the LMRA mechanism can partly occur and give chance to let electron rest in one phase of the laser for a while. This relatively rest makes the electron in a slowly motion and gain energy from the laser field. The efficiency of energy transfer will be high. This is the laser-magnetic resonance acceleration. It's very different from the pondermotive acceleration which does not concern the laser period. In our profile of $\boldsymbol{b}_\theta$, a lower component quassistatic magnetic field exists in the center regime. So the electron energy gain is high than the ponderomotive acceleration energy. Anyway in this regime

($r \leq \lambda$) ponderomotive acceleration is in dominant and polarization independence still remain.

In Fig. 3 the parameters of initial position were changed to (a) $x_0 = R_0/2$, $y_0 = 0$, (b) $x_0 = 0$, $y_0 = R_0/2$ (around the peak of $\boldsymbol{b}_\theta$). One can find that evidence deform appears in solid line between Fig. 3a–c and b–d. Polarization dependence is a main feature in this regime. If the initial position is in the polarization direction, the electron has quiver energy to let the LMRA occurs, otherwise when the initial position is not in polarization direction, the electron has no quiver energy to utilize. This is the one reason which makes the low efficiency of energy transfer than circularly polarized (CP) laser case. When $\boldsymbol{a}$ and $\boldsymbol{b}_\theta$ coexist, e.g., $a = 2$, $b_\theta = -0.85$, using our derived (9), we can estimate that $\gamma_{\max} \approx 89$ which is very close to our simulation results. Evident pondermotive acceleration still be shown in dot line in Fig. 3c and d, its value relatively smaller than that in the center regime. The electron polarization dependence is controlled by the competition of the amplitude of $\boldsymbol{a}$ and $\boldsymbol{b}_\theta$. If the value of $\boldsymbol{a}$ is in dominant, e.g., in the center regime, the polarization dependence is not evident, shows in Fig. 2. If the value of $\boldsymbol{b}_\theta$ is larger than $\boldsymbol{a}$, e.g., in the second regime, polarization dependent appears which shows in Fig. 3. Because of the different initial position in or not in the polarization direction, the electron has a different chance to utilize quiver energy and make LMRA to occur.

Our simulations satisfy with the phenomena which have been reported by experiments and numerical simulations e.g., [4–7, 10]. For example in [4], the authors pointed out that a narrow plasma jet is formed at the rear surface which is consistent with a beam of fast electrons traveling through the target, collimated by a magnetic field in the target. In [5] L. Gremillet et al. observed two narrow long jets originating from the focal spot. These may be caused by the $\boldsymbol{b}_\theta$ in the second region. Even the $\boldsymbol{b}_\theta$ has more than one peak, more electron jets can be produced. As given in [6] the snake like electron orbit is very similar to our Fig. 2a and b and Fig. 3a. If the amplitude of $\boldsymbol{a}$ and $\boldsymbol{b}_\theta$ can be comparable, an elliptical heating area appears, such as pointed out by Kodama et al. in experiment [7]. In [10] A. Pukhov et al. pointed out that distribution of electron current $\boldsymbol{J}$ and quasistatic magnetic field $\boldsymbol{B}$ at the positions of tight focusing is elongated in the direction of polarization and heavy relativistic electrons sprayed in the direction of polarization. From above analysis, the polarization dependence of LP system is a typical different feature with a circularly polarized (CP) system when the self-generated azimuthal magnetic fields are present.

4 Discussions and conclusions

Using a single test electron model, we study the energetic electrons in a combined strong azimuthal magnetic field and a Gaussian profile linearly polarized laser field numerically. Two different sources of fast electron are distinguished. In the presence of a magnetic field in the LP system, polarization independence is being modified by the increasing value of magnetic field. If the laser intensity is dominant, the polarization dependence is not evident. If the value of the magnetic field becomes comparable with the laser intensity, the polarization dependence appears. Comparing with an energy analytic solution of electron which contains the laser-magnetic resonance acceleration mechanism, we point out that a strong quasistatic magnetic field affects electron acceleration dramatically from the ratio between the Larmor frequency and the laser frequency. As the ratio approaches unity, clear resonance peaks are observed. From the physical parameters available for laboratory experiments, we find that the electron acceleration depends not only on the laser intensity, but also on the ratio between electron Larmor frequency and the laser frequency. The different fast electrons which are produced by LMRA and pondermotive acceleration mechanism give a clear explanation of the polarization dependent phenomena. Because the LMRA relates to the laser period, an averaged calculation over one laser period will lost the effect of $\boldsymbol{b}_\theta$. This is different from the pondermotive acceleration mechanism.

The study of relativistic strong laser pulses along with hundreds of MG azimuthal quasistatic magnetic fields is a complex process, related with several mechanisms. In this paper we simply treat the laser pulse in channel and the quasistatic magnetic fields, ignoring that the energetic electrons interact with and are deflected by background particles. Our purpose is to make clear how the fast electron behaves in the presence of a magnetic field. The polarization dependent phenomena only appears in the LP laser case, whereas the CP laser case does not. Thus, the efficiency of energy transfer will be different. The value in the CP case is higher than in the LP case. For the quasistatic magnetic field modifying the polarization independence is very important in laser–plasma interactions; it will have good application in the fast-ignitor scheme and particle accelerators.

An experimental clarification of the here described acceleration against the alternative schemes may have to focus on the use of varying polarization of the interacting laser beam because a manupulation of the 100 MGauss magnetic self generated fields may be not easy. We have to be aware to separate the different mechanisms. From the first publication using the nonlinearities for the direct laser interaction with free electrons [17] by clarifying the absolute maximum electron energy, which can be gained by fully rectified laser pulses [13], realistic examples including the longitudinal components of the laser field [1] resulted in electron energies always a little lower than the maximum values [13, 19]. Even the mechanisms of trapping the electrons in the laser beam [26] arrived only close to the mentioned [13] maximum value. However, there are experiments [27, 28] showing that at least a small group of electrons with higher energies were detected. The question is whether this is due to a repetitive free-wave acceleration process [13, 17, 18], or by a new kind of trapping mechanisms calculated by Pukhov et al. [10, 30] or whether this is due to the magnetic field effect which we described here, or by a mixture of all these processes and of perhaps not yet known mechanisms. The advantage to distinguish the here described magnetic field process will be in the appropriate use of the polarization of the laser field.

ACKNOWLEDGEMENTS This work was supported by National Natural Science Foundation of China (Grant No. 10476033), National

Hi-Tech Inertial Confinement Fusion Committee of China, National Basic Research Project Nonlinear Science in China, and National Key Basic Research Special Foundation.

REFERENCES

1 L. Cicchitelli, H. Hora, R. Postle, Phys. Rev. A **41**, 3727 (1990)
2 B. Quesnel, P. Mora, Phys. Rev. E **58**, 3719 (1998)
3 D. Giulietti, M. Galimberti, A. Giulietti , D. Giulietti, M. Galimberti, A. Giulietti, L.A. Gizzi, R. Numico, P. Tomassini, M. Borghesi, V. Malka, S. Fritzler, M. Pittman, K. Ta Phouc, A. Pukhov, Phys. Plasmas **9**, 3655 (2002)
4 M. Tatarakis, J.R. Davies, P. Lee, P.A. Norreys, N.G. Kassapakis, F.N. Beg, A.R. Bell, M.G. Haines, A.E. Dangor, Phys. Rev. Lett. **81**, 999 (1998)
5 L. Gremillet, F. Amiranoff, S.D. Baton, J.-C. Gauthier, M. Koenig, E. Martinolli, F. Pisani, G. Bonnaud, C. Lebourg, C. Rousseaux, C. Toupin, A. Antonicci, D. Batani, A. Bernardinello, T. Hall, D. Scott, P. Norreys, H. Bandulet, H. Pépin, Phys. Rev. Lett. **83**, 5015 (1999)
6 B.F. Lasinski, A.B. Langdon, S.P. Hatchett, M.H. Key, M. Tabak, Phys. Plasmas **6**, 2041 (1999)
7 R. Kodama, P.A. Norreys, K. Mima, A.E. Dangor, R.G. Evans, H. Fujita, Y. Kitagawa, K. Krushelnick, T. Miyakoshi, N. Miyanaga, T. Norimatsu, S.J. Rose, T. Shozaki, K. Shigemori, A. Sunahara, M. Tampo, K.A. Tanaka, Y. Toyama, T. Yamanaka, M. Zepf, Nature **412**, 798 (2001)
8 C. Gahn, G.D. Tsakiris, A. Pukhov, J. Meyer-ter-Vehn, G. Pretzler, P. Thirolf, D. Habs, K.J. Witte, Phys. Rev. Lett. **83**, 4772 (1999)
9 M. Tatarakis, I. Watts, F.N. Beg, E.L. Clark, A.E. Dangor, A. Gopal, M.G. Haines, P.A. Norreys, U. Wagner, M.-S. Wei, M. Zepf, K. Krushelnick, Nature **415**, 280 (2002); M. Tatarakis, A. Gopal, I. Watts, F.N. Beg, A.E. Dangor, K. Krushelnick, U. Wagner, P.A. Norreys, E.L. Clark, M. Zepf, R.G. Evans, Phys. Plasmas **9**, 2244 (2002)
10 A. Pukhov, J. Meyer-ter-Vehn, Phys. Rev. Lett. **76**, 3975 (1996); A. Pukhov, J. Meyer-ter-Vehn, Phys. Plasmas **5**, 1880 (1998)
11 H. Hora, J. Opt. Soc. Am. **65**, 882 (1975)
12 D.A. Jones, E.L. Kane, P. Lalousis, P. Wiles, H. Hora, Phys. Fluids **25**, 2295 (1982)
13 T. Häuser, W. Scheid, H. Hora, Phys. Rev. A **45**, 1278 (1992)
14 H. Haseroth, H. Hora, Laser Part. Beams **14**, 393 (1996)
15 E. Esarey, P. Sprangle, J. Krall, A. Ting, IEEE J. Quantum Electron. **QE-33**, 1879 (1997)
16 S. Eliezer, H. Hora, Phys. Rep. **172**, 339 (1989)
17 H. Hora, Nature **333**, 337 (1988)
18 F.V. Hartemann, J.R. Van Meter, A.L. Troha, E.C. Landahl, N.C. Luhmann Jr., H.A. Baldis, A. Gupta, A.K. Kerman, Phys. Rev. E **58**, 5001 (1998)
19 H. Hora, M. Hoelss, W. Scheid, J.W. Wang, Y.K. Ho, F. Osman, R. Castillo, Laser Part. Beams **18**, 135 (2000)
20 R.N. Sudan, Phys. Rev. Lett. **70**, 3075 (1993)
21 M. Tabak, J. Hammer, M.E. Glinsky, W.L. Kruer, S.C. Wilks, J. Woodworth, E.M. Campbell, M.D. Perry, R.J. Mason, Phys. Plasmas **1**, 1626 (1994)
22 S.-Y. Chen, G.S. Sarkisov, A. Maksimchuk, R. Wagner, D. Umstadter, Phys. Rev. Lett. **80**, 2610 (1998)
23 H. Liu, X.T. He, S.G. Chen, Phys. Rev. E **69**, 066409 (2004)
24 M. Schmitz, H.-J. Kull, Laser Phys. **12**, 443 (2002)
25 K.P. Singh, Phys. Plasmas **11**, 3992 (2004)
26 J.X. Wang, Y.K. Ho, Q. Kong, L.J. Zhu, L. Feng, W. Scheid, H. Hora, Phys. Rev. E, **58**, 6576 (1998)
27 S. Weber, G. Riazuelo, P. Michel, R. Loubere, F. Walraet, V.T. Tikhonchuk, V. Malka, J. Ovadia, G. Bonnard, Laser Part. Beams **22**, 189 (2004)
28 V. Malka, S. Fritzler, Laser Part. Beams **22**, 399 (2004)
29 Y. Glinec, J. Faure, A. Pukhov, S. Kiselev, S. Gordienko, B. Mercier, V. Malka, Laser Part. Beams **23**, 161 (2005)
30 O. Shorokov, A. Pukhov, Laser Part. Beams **22**, 175 (2004)

Laser and Particle Beams (2008), *26*, 51–59. Printed in the USA.
Copyright © 2008 Cambridge University Press 0263-0346/08 $20.00
doi: 10.1017/S0263034608000086

Resonant acceleration of electrons by intense circularly polarized Gaussian laser pulses

H.Y. NIU,[1] X.T. HE,[2,3,4] B. QIAO,[2,5] AND C.T. ZHOU[2,3,4]
[1]Graduate School of China Academy of Engineering Physics, Beijing, People's Republic of China
[2]Institute of Applied Physics and Computational Mathematics, Beijing, People's Republic of China
[3]Center for Applied Physics and Technology, Peking University, Beijing, People's Republic of China
[4]Institute for Fusion Theory and Simulation, Zhejiang University, Hangzhou, People's Republic of China
[5]Department of Physics, National University of Singapore, Singapore

(RECEIVED 20 January 2008; ACCEPTED 3 February 2008)

Abstract

Resonant acceleration of plasma electrons in combined circularly polarized Gaussian laser fields and self-generated quasistatic fields has been investigated theoretically and numerically. The latter includes the radial quasistatic electric field, the azimuthal quasistatic magnetic field and the axial one. The resonant condition is theoretically given and numerically testified. The results show some of the resonant electrons are accelerated to velocities larger than the laser group velocity and thus gain high energy. For peak laser intensity $I_0 = 1 \times 10^{20}$ W cm^{-2} and plasma density $n_0 = 0.1n_{cr}$, the relativistic electron beam with energies increased from 207 MeV to 262 MeV with a relative energy width around 24% and extreme low beam divergence less than 1° has been obtained. The effect of laser intensity and plasma density on the final energy gain of resonant electrons is also investigated.

1. INTRODUCTION

Acceleration of electrons from short-pulse and high-intense laser ($\geq 10^{18}$ W/cm^2) interacting with plasmas is an important issue for electron accelerators (Hora *et al.*, 2000; Joshi & Katsouleas, 2003; Faure *et al.*, 2004; Geddes *et al.*, 2004; Borghesi *et al.*, 2007; Esarey *et al.*, 2007; Gupta & Suk, 2007; Karmakar & Pukhov, 2007; Xu *et al.*, 2007; Zhou *et al.*, 2007) and the fast ignition scheme (Tabak *et al.*, 1994; Kodama *et al.*, 2001; Nuckolls *et al.*, 2002; Hora, 2007*a*, 2007*b*) or beam fusion (Hoffmann *et al.*, 2005). Recently, different acceleration mechanisms have been investigated for laser interaction with both underdense and overdense plasmas (Gibbon, 2005; Salamin *et al.*, 2006; Stait-Gardner & Castillo, 2006; Zhou *et al.*, 2006). The idea of using the plasma wave excited by an intense laser to accelerate electrons was first suggested by Tajima and Dawson (1979) more than 20 years ago. In the so-called standard laser wakefield acceleration scheme, the wake plasma wave is most efficiently generated by the ponderomotive force of the laser pulse when its duration τ_L is close to half of the plasma wave period (Hamster *et al.*, 1993; Siders *et al.*, 1996). This is the same as acceleration of electrons in vacuum by the electromagntic laser field (Evans, 1988; Hora, 1988) where it was confirmed that the highest acceleration is achieved with half optical waves or "rectified waves" (Scheid & Hora, 1989, Hora *et al.*, 2002).

On the other hand, the nonlinear ponderomotive force also plays an important role in direct laser acceleration (Mckinstrie & Startsev, 1996; Quesnel & Mora, 1998; Stupakov & Zolotorev, 2001; Schmitz & Kull, 2002; Kong *et al.*, 2003). At high laser intensities, the relativistic motion starts to operate, causing the electron to be directed forward as well as sideways, and forming laser-plasma channels. It is known that ponderomotive expulsion of background plasma electrons from such plasma channels creates a radial quasistatic electric (QSE) field $\mathbf{E}_s$, and the current of electrons accelerated by axial ponderomotive force generates the azimuthal quasistatic magnetic (QSM) field $\mathbf{B}_{s\theta}$. For a linearly polarized (LP) laser, both fields trap a relativistic electron in the channel and result in its betatron oscillation along the laser polarization. Therefore, an electron undergoes a strong downshifted optical frequency when it is accelerated in the laser propagation direction. As a result, the transverse betatron oscillation may be in resonance with the LP laser if the downshift is strong enough (Pukhov *et al.*, 1999; Tsakiris

Address correspondence and reprint requests to: H.Y. Niu, Graduate School of China Academy of Engineering Physics, PO Box 2101 Beijing 100088, People's Republic of China. E-mail: niuhaiyan221@126.com et al., 2000; Yu et al., 2003; Liu & Tripathi, 2005, 2006), leading to a considerable acceleration of electrons.

However, the interaction of a circularly polarized (CP) laser with plasma can additionally generate an axial QSM field $\mathbf{B}_{sz}$ due to the rotational motion of relativistic electrons, besides the radial QSE field $\mathbf{E}_s$ and the azimuthal QSM field $\mathbf{B}_{s\theta}$. The field $\mathbf{B}_{sz}$ may effectively trap and collimate the accelerated electrons. Consequently, the accelerated electrons driven by CP laser may have a smaller divergent angle than using LP laser (Sheng & Meyer-ter-vehn, 1996; Qiao et al., 2005*a*). Furthermore, a sharp resonance peak is found in the interaction of a planar CP laser with a strong axial magnetic field (Liu et al., 2004). To the best of our knowledge, no work has been conducted to investigate the resonant acceleration of plasma electrons in combined CP Gaussian laser fields and complete self-generated qausistatic fields so far, but it may be significant in practice. In this paper, we investigate the resonant acceleration of plasma electrons under the action of the combined laser fields and self-generated quasistatic fields. A Gaussian CP laser dependent on both space and time is introduced and the self-generated fields, including radial QSE field $\mathbf{E}_s$, azimuthal QSM field $\mathbf{B}_{s\theta}$, and axial QSM field $\mathbf{B}_{sz}$ are all taken into consideration. The resonant condition is analytically derived and numerically testified. The qualities of relativistic electron beam (REB) produced by resonant acceleration, including high energy, quasi-monoenergy, and small divergent angles, are discussed due to potential applications. The dependence of the final energy gain of accelerated electrons on laser intensity and plasma density is also investigated.

2. PHYSICAL MODEL AND RESONANT CONDITION

Consider a CP laser pulse with pulse duration τ_L and spot radius r_0 propagating through a unmagnetized underdense plasma with uniform density n_0. The CP laser electric fields with Gaussian profiles both in space and time are

$$E_{Lx} = E_L \cos\phi, \tag{1}$$

$$E_{Ly} = \alpha E_L \sin\phi, \tag{2}$$

$$E_{Lz} = E_L\left(-\frac{x}{kr_0^2}sin\phi + \alpha\frac{y}{kr_0^2}cos\phi\right), \tag{3}$$

where $E_L(r,z,t) = E_0\exp\{-(r^2/2r_0^2) - (t-(z-z_0)/v_g)^2/2\tau_L^2)\}$, $\phi = \omega t - kz$, $k = \omega\eta/c$, $\eta \approx [1-\omega_{pe}^2/\omega^2\ (1+a_0^2)^{1/2}]^{1/2}$, $a_0 = eE_0/m\omega c$, $\omega_{pe} = \sqrt{4\pi n_0 e^2/m_e}$, $-e$ and m are electron charge and rest mass, c is the velocity of light in vacuum, v_g is the laser group velocity in plasma, z_0 the initial position of the laser pulse center, and $\alpha = \pm 1$, 0 represents the circular and linear polarization, respectively. In Eqs. (1)–(3), we have ensured that $\nabla\cdot\mathbf{E_L} = 0$. The magnetic field related to the laser pulse is given by $\nabla\times\mathbf{E_L} = -\partial\mathbf{B_L}/\partial t$ and can be formulated as follows:

$$B_{Lx} = -\alpha\eta E_L\left(1+\frac{1}{k^2r_0^2}\right)\sin\phi, \tag{4}$$

$$B_{Ly} = \eta E_L\left(1+\frac{1}{k^2r_0^2}\right)\cos\phi, \tag{5}$$

$$B_{Lz} = -\eta E_L\left(\alpha\frac{x}{kr_0^2}cos\phi + \frac{y}{kr_0^2}sin\phi\right). \tag{6}$$

As is well-known, when an intense laser propagates in plasma, strong QSE fields, and QSM fields are self-generated by various nonlinear effects and relativistic effects (Pukhov & Meyer-ter-vehn, 1996; Qiao et al., 2005*b*). These quasistatic fields will heavily affect the electron motion and acceleration. First, the ponderomotive force $\mathbf{F}_p$ produced by laser pulse expels some electrons away from the laser axis so that a space-charge potential $\mathbf{\Phi}$ is formed, which corresponds to the QSE field $\mathbf{E}_s = -\nabla\Phi$. The longitudinal relativistic current driven by the ponderomotive force generates the azimuthal QSM field $\mathbf{B}_{s\theta}$, while the rotational current caused by the nonlinear beat interaction of the CP laser fields with relativistic electrons produces the axial QSM field $\mathbf{B}_{sz}$. Particle-in-cell (PIC) simulations (Zheng et al., 2005) have shown that strong azimuthal and axial QSM fields exist in the interaction of the CP laser with dense plasma. Some of the present authors (Qiao et al., 2005*b*) have put forword a self-consistent quasistatic fields model based on the relativistic Vlasov-Maxwell equations. Introducing the normalized transform $t\to\omega t$, $\mathbf{x}\to\omega\mathbf{x}/c$, $E\to eE/m_e\omega c$, $B\to eB/m_e\omega c$, the model mentioned above is given by

$$\mathbf{E}_s = k_E\frac{r}{r_0}\exp\left\{-\frac{r^2}{r_0^2} - \frac{(t-(z-z_0)/v_g)^2}{\tau_L^2}\right\}\mathbf{e}_r, \tag{7}$$

$$\mathbf{B}_{s\theta} = -k_B\frac{r_0}{r}(1-e^{-r^2/r_0^2})\exp\left\{-\frac{(t-(z-z_0)/v_g)^2}{\tau_L^2}\right\}\mathbf{e}_\theta, \tag{8}$$

$$\mathbf{B}_{sz} = -b_Z\exp\left\{-\frac{r^2}{r_0^2} - \frac{(t-(z-z_0)/v_g)^2}{\tau_L^2}\right\}\mathbf{e}_z, \tag{9}$$

where the coefficients $k_E = (1+\alpha^2)\beta_{56}I_0/r_0$, $k_B = \epsilon(\pi(1+\alpha^2)/64)n_0r_0I_0[\xi\beta_4(T_{te}) + (1-\xi)\beta_4(T_{re})]$, and $b_Z = (1/2)\alpha n_0I_0[\xi\beta_1(T_{te}) + (1-\xi)\beta_1(T_{re})]$ are related to laser and plasma parameters such as n_0, I_0, etc, and among which other quantities β_1, β_4, β_{56}, ξ, T_{te}, T_{re}, and ϵ are defined (Qiao et al., 2005*a*, 2005*b*).

Based on the above discussion, a test electron model is investigated. The motion of the test electron in the presence of electromagnetic fields $\mathbf{E}$ and $\mathbf{B}$ is described by the Lorentz equation

$$\frac{d\mathbf{p}}{dt} = -(\mathbf{E}+\mathbf{v}\times\mathbf{B}), \tag{10}$$

together with an energy equation

$$\frac{d\gamma}{dt} = -\mathbf{E} \cdot \mathbf{v}, \tag{11}$$

where $\mathbf{p} = \gamma\mathbf{v}$ and $\gamma = (1 + p^2)^{1/2}$. Note that here the electric field $\mathbf{E}$ and magnetic field $\mathbf{B}$ include the laser fields $\mathbf{E}_L$, $\mathbf{B}_L$, and the self-generated quasistatic ones $\mathbf{E}_s$, $\mathbf{B}_s$.

As pointed out (Pukhov & Meyer-ter-vehn, 1999), the electron motion in the presence of both self-generated fields $\mathbf{E}_s$, $\mathbf{B}_s$, and laser fields $\mathbf{E}_L$, $\mathbf{B}_L$ can be seen as a driven oscillation. The electron makes betatron oscillations in the self-generated electric and magnetic fields and, the laser pulse acts as a driving source. Obviously it is difficult to obtain an exact solution of Eqs. (10)–(11) because of their nonlinearity. Nevertheless, we are interested in the condition of resonance between the electron and the laser pulse. The transverse betatron oscillation of electrons can be derived by Eqs. (10)–(11). If the magnitude of transverse displacement of an electron is much smaller than the laser spot size, i.e., $r^2 \ll r_0^2$, then E_{Lz} and B_{Lz} may have a negligible influence on the motion of electrons. Note the fact that the self-generated fields are slowly variant in laser-plasma interactions, it will be assumed to be static in the analytical discussion. The effect of time-dependence is included in the numerical calculations presented in Section 3. Under these assumptions, the equations governing p_x and p_y are

$$\frac{d^2p_x}{dt^2} + \omega_b^2 p_x = (1 - \eta v_z)E_L \sin(t - z/v_{ph}), \tag{12}$$

$$\frac{d^2p_y}{dt^2} + \omega_b^2 p_y = -\alpha(1 - \eta v_z)E_L \cos(t - z/v_{ph}), \tag{13}$$

where

$$\omega_b = \frac{1}{\sqrt{\gamma}}\left[\frac{k_E}{r_0}\left(1 - \frac{r^2}{r_0^2}\right) + \frac{k_B v_z}{r_0} + \frac{\alpha^2 b_Z^2}{\gamma}\left(1 - \frac{2r^2}{r_0^2}\right)\right]^{1/2}, \tag{14}$$

v_{ph} is the laser phase velocity. On the right-hand side of Eqs. (12) and (13), $\gamma \gg 1$ is assumed and consequently the terms E_L/γ, E_L/γ^2 are ignored in comparison with the term of E_L. The derivative of longitudinal velocity v_z is also neglected in the above derivation due to the slow variation of v_z when the electron has been pre-accelerated. In Eq. (14), we have expanded $e^{-r^2/r_0^2} \approx 1 - r^2/r_0^2$. Note that ω_b is the same for both the left-hand and the right-hand CP lasers and the b_Z-term is absent for LP lasers. It can be seen that the electron makes betatron oscillations driven by CP laser with betatron frequency $\simeq \omega_b$. When the driver frequency $\omega - kv_z$ equals the betatron frequency ω_b, i.e.,

$$\frac{1}{\sqrt{\gamma}}\left[\frac{k_E}{r_0}\left(1 - \frac{r^2}{r_0^2}\right) + \frac{k_B v_z}{r_0} + \frac{\alpha^2 b_Z^2}{\gamma}\left(1 - \frac{2r^2}{r_0^2}\right)\right]^{1/2} = 1 - \frac{v_z}{v_{ph}}, \tag{15}$$

a resonance occurs, leading to an effective energy exchange between the pulse and electrons. In fact, after the electrons are pre-accelerated, ω_b changes slowly on the time scale of one betatron oscillation. In this case, the transverse electron motion may be written

$$p_x = c(t)\cos\theta_b,\ p_y = \alpha c(t)\sin\theta_b,\ \frac{d\theta_b}{dt} = \omega_b, \tag{16}$$

where $c(t)$ is the magnitude of transverse momentum. Using this value of p_x, p_y in Eq. (11) and neglecting the rv_r-term as it averages out to be zero over one oscillation period, the electron energy equation is

$$\frac{d\gamma}{dt} = -\frac{(1 + \alpha^2)c(t)}{2\gamma}E_L[\cos\psi^- + (1 - \alpha^2)\cos\psi^+], \tag{17}$$

where $\psi^- = \theta_b - (t - z/v_{ph})$ and $\psi^+ = \theta_b + (t - z/v_{ph})$ are the slow ponderomotive phase and the fast ponderomotive phase, respectively. It can be seen from Eq. (17) that electrons are accelerated when their ponderomotive phase satisfies $\pi/2 < \psi^- < 3\pi/2$, and are decelerated otherwise. The maximum acceleration is attained if the betatron oscillations are exactly in counterphase with the laser electric field. In particular, if the ponderomotive phase ψ^- approximately maintains an invariable value, i.e., $d\psi^-/dt = \omega_b - (1 - v_z/v_{ph}) \simeq 0$, a resonance takes place. In this case, as displayed by the last term of Eq. (17), there exists a fast oscillation $\cos\psi^+ \simeq \cos 2\omega_b t$ in γ or p_z for the LP lasers while it is absent in the case of CP lasers.

It can also be implied from Eq. (17) that the main accelerating term in the z-direction is $-\mathbf{v} \times \mathbf{B}_L$. The final energy of accelerated electrons can be approximately expressed by the combination of variation of p_z and γ. Multiplying the z-product of Eq. (10) by η and subtracting from Eq. (11), we get

$$\frac{dp_z}{dt} - \eta\frac{d\gamma}{dt} = \frac{\eta k_E}{r_0}e^{-r^2/r_0^2}rv_r + \frac{k_B r_0}{r^2}(1 - e^{-r^2/r_0^2}))rv_r - \frac{\eta}{k^2 r_0^2}E_L(v_x\cos\phi + \alpha v_y \sin\phi), \tag{18}$$

where $v_r = dr/dt$. If the last term of the right-hand side of Eq. (18) (induced by the second terms of B_{Lx} and B_{Ly}) were negligible (valid for $a_0 \ll (2\pi r_0/\lambda)^2$), then Eq. (18) can be integrated

$$\gamma = \frac{\gamma_0(v_{0z}-\eta) + (1/2r_0)(k_E\eta + 2k_B)(r^2 - r_{in}^2) - (k_E\eta/4r_0^3)(r^4 - r_{in}^4)}{v_z - \eta}, \quad (19)$$

where γ_0, r_{in} and p_{0z} are the initial values of energy, radial coordinate, and z-directional momentum of electrons, respectively. This condition indicates that the final value of γ is a complicated function that relies not only on the initial and final longitudinal velocities, but also on the radial coordinates of the electron at the beginning and at the end of the interaction besides the rest of the plasma and laser parameters. However, the equation implies that a sudden increase in γ may happen when v_z approaches v_g. Furthermore, it can also be seen that if the radial QSE field and the azimuthal QSM field are absent, i.e., $k_E = k_B = 0$, then v_z is always less than v_g for an electron initially having $v_{0z} < v_g$. This demonstrates that the electron initially located in front of the pulse center is overtaken first and then is overrun by the pulse. On the contrary, if the QSE field and the azimuthal QSM field are so large as to satisfy $k_E\eta + 2k_B > 2r_0\gamma_0(\eta - v_{0z})/(r^2 - r_{in}^2)$ (here the smaller term $(k_E v_g/4r_0^3)$ $(r^4 - r_{in}^4)$ is ignored), v_z will gradually approach and may surpass v_g, thus the electron overtaken by the pulse first may always keep ahead of the pulse center until the end of the interaction, so gains more energy.

3. NUMERICAL ANALYSIS ON ELECTRON ACCELERATION

To further analyze the electron acceleration process, Eqs. (10)–(11) are numerically solved. We choose the parameters of laser beam as wavelength $\lambda = 1.06$ μm, the spot radius $r_0 = 4\lambda$ and the pulse duration $\tau_L = 100$ fs. We assume that the test electron initially locates at $z = 0$ while the laser pulse center is at $z_0 = -4c\tau_L$. We trace the motion of electron accelerated by the laser field combined with the self-generated QSE and QSM fields. For the present self-generated fields model, we have assumed the parameters are $T_{te} = 5$ KeV and $\xi = 0.5$ and consequently the coefficients k_E, k_B, and b_Z can be calculated based on the definitions in Section 2. For example, if the peak laser intensity $I_0 = 1 \times 10^{20}$ W cm^{-2} and the plasma density $n_0 = 0.1n_{cr}$ are assumed, the peak azimuthal QSM field is about 255 megagauss (MG) near r_0; The axial QSM field gets to its peak value of 90 MG at the z axis ($r = 0$) and decreases as r increases; And the self-generated QSE field reaches its maximum value of 70 CGSE at $r \simeq 0.7r_0$.

Figure 1 shows electron trajectories in three-dimensional (3D) plane and the variation of the relative distance $\Delta z = z - z_L$ and of the total energy with z. The electron rotates around the direction of the laser propagation and the direction of the rotation is such that it should generate an axial magnetic field in the negative z direction by the azimuthal current. The relative distance Δz decreases first, gets to its minimum value at $z \approx 200\lambda$. Note that the minimum distance just corresponds to $v_z = v_g$. The following increase of Δz indicates the electron has been accelerated to velocity larger than v_g. A clear sharp increase in γmc^2 occurs due to the match between the electron betatron frequency ω_b and the Doppler shifted laser frequency $\omega - kv_z$. That is to say, a resonance takes place. This can be explained in Figure 2, where a phase comparison of p_x and B_{Ly} is shown. One can see that an almost locked phase with $\psi^- \simeq \pi$ between p_x and B_{Ly} occurs from $z \approx 180\lambda$ to $z \approx 700\lambda$, i.e., $d\psi^-/\mathrm{dt} \approx 0$, which indicates the resonance condition $\omega_b \simeq \omega - kv_z$ holds. After several synchrotron rotations in the ponderomotive bucket, the phase space is mixed (about $z > 700\lambda$) and deceleration occurs. When the resonance takes place, as predicted by Eq. (17) in Section 2, the effect of polarization on the evolution of γ or p_z is apparent. In the case of LP laser, the fast oscillations at $2\omega_b$ accompanying the increase of γ or p_z are clearly seen in the p_z plot of Figure 3b. These fast oscillations are absent for CP laser because of the rotation of the CP laser electric field. In Pukhov *et al.* (1999), two-dimensional PIC simulation shows that there exists electron density modulation with two times per laser wavelength in LP laser-plasma interactions. The reason is that the transverse velocity of the resonant electrons oscillates with the laser period while the longitudinal velocity oscillating twice per laser period, leading to the electron bunching in space two times per laser wavelength. It can be seen from the resonant condition (Eq. (15)) that besides the radial QSE field $\mathbf{E}_s$ and the azimuthal QSM field $\mathbf{B}_{s\theta}$, the axial QSM field $\mathbf{B}_{sz}$ also makes partial contribution towards betatron oscillations. Furthermore, $\mathbf{B}_{s\theta}$ and $\mathbf{B}_{sz}$ play an pinch and collimation role in the acceleration process while $\mathbf{E}_s$ partly offsets the radial ponderomotive force of the pulse.

The continuous increase of γ leads to the decrease of ω_b and the resonant condition is violated. The electron dephases and acceleration ceases. However, since the electron has been accelerated to velocity larger than v_g, the laser electric field (seen by the electron) becomes more and more smaller. When the electron leaves the pulse, its energy gain still maintains a high level. It should be mentioned that the coincidence of ω_b with $\omega - kv_z$ just $v_z < v_g$ is a determinative condition that electron attain high energy. Some electrons make betatron oscillations but dephase just $v_z < v_g$ and consequently are run over by the pulse and only gain small energy.

The energetic electrons that attain resonance and have $v_z > v_g$ always move ahead of the pulse and emit from the plasma channel first. The final energy gain and the scattering angles of these electrons are of our concern. The numerical calculations demonstrate that the electrons originated from the vicinity of the axis are more likely to be trapped by the ponderomotive bucket and attain high energies. The final scattering angles and spatial distribution of these energetic electrons are displayed in Figure 4, where total 180 electrons are uniformly distributed within a disc with radius $r = r_0/4$

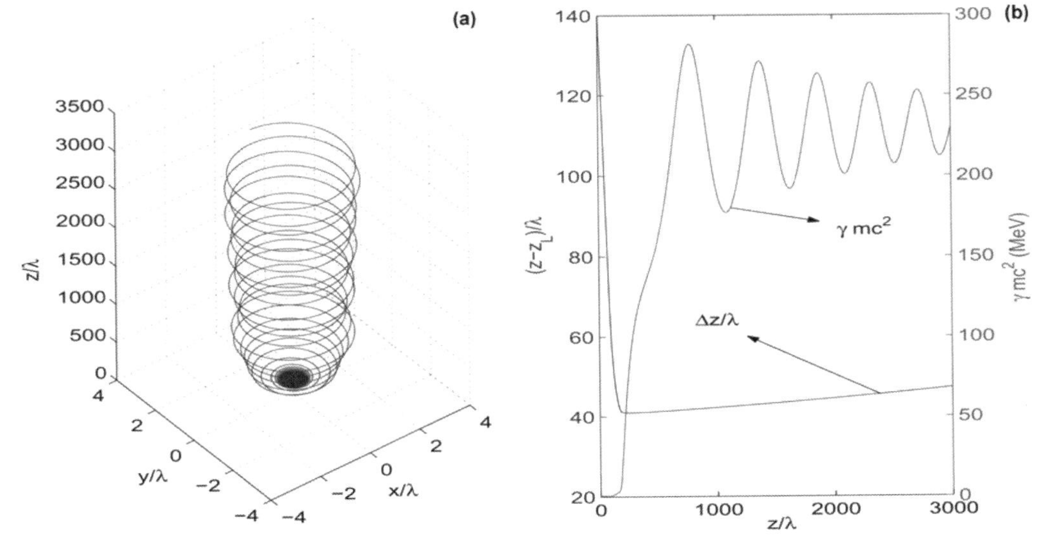

Fig. 1. (**a**) Electron trajectories in three-dimensional plane and (**b**) the electron position relative to the pulse center and the energy gain of the electron as a function of z in the interaction of an electron with a propagating laser pulse. The parameters of laser and plasma are $I_0 = 1 \times 10^{20}\ \mathrm{W/cm^2}$ and $n_0 = 0.1n_{cr}$. The initial conditions are $p_{x0} = p_{y0} = 0$, $\gamma_0 = 1.01$, and $r_{in} = 0$. Here z_L is the position of the propagating laser pulse center.

placed at $z = 0$ initially, and their velocities approximately satisfy Maxwellian distribution with averaged temperature of 5 KeV. At the end of the interaction, we record the final energy, its radial position and its final momentum. Only electrons having energies larger than 10 MeV are shown. The scattering angle of electrons relative to the z-axis is $\theta_r = \arctan(p_\perp/p_\parallel)$. In fact, we can derive $p_\perp/p_\parallel = \sqrt{(\gamma^2-1)/(\gamma^2 v_\parallel^2)-1}$ from the equation $\gamma^2 = 1 + p_\perp^2 + p_\parallel^2$. The scattering angle of the electrons obtaining

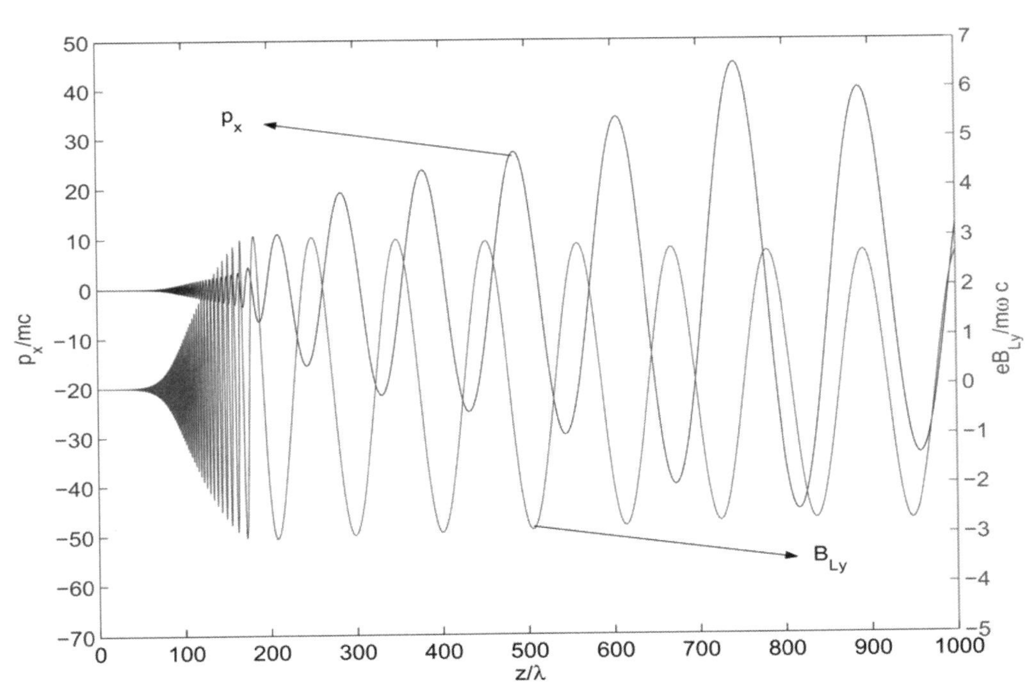

Fig. 2. Variation of p_x and B_{Ly} as a function of z for the same interaction as Fig. 1.

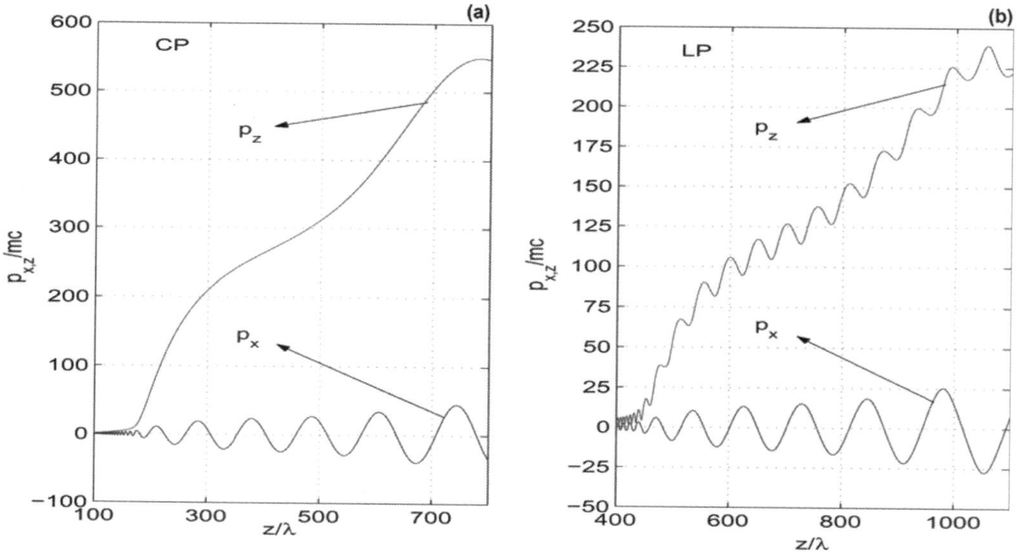

Fig. 3. The x- and z-momentum as a function of z in the interaction of an electron with a propagating laser pulse for different polarization. (**a**) CP, $\gamma_0 = 1.01$ and (**b**) LP, $\gamma_0 = 1.5$. Other parameters are the same as Fig. 1.

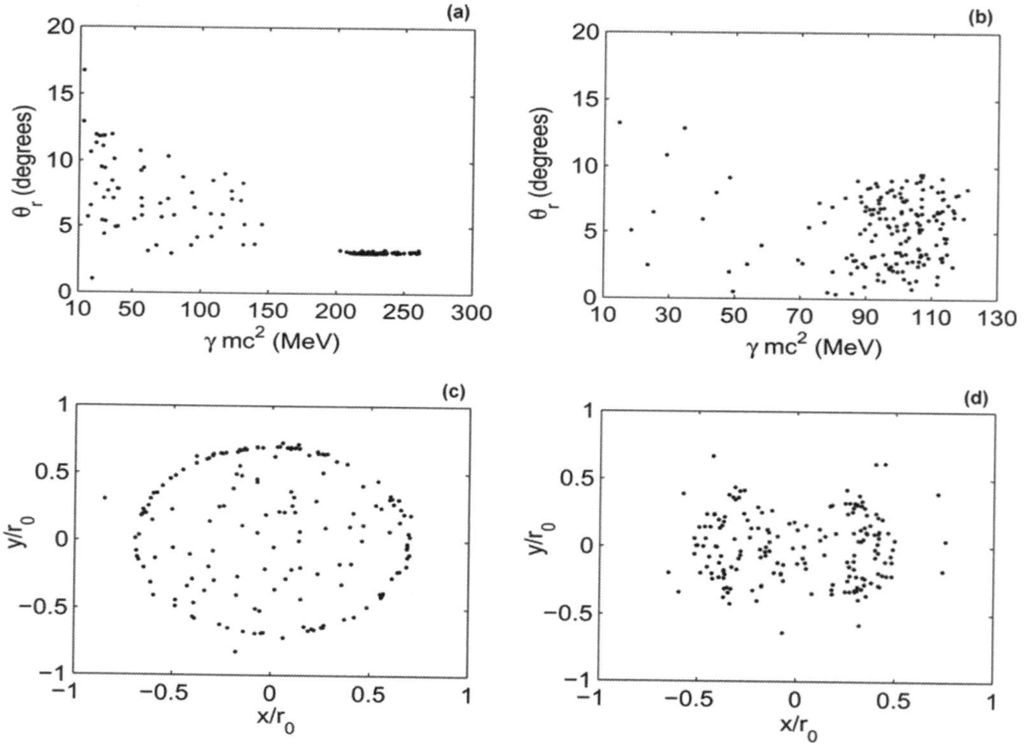

Fig. 4. (**a**) and (**b**) The distribution in $\gamma - \theta_r$ space, and (**c**) and (**d**) the spatial distribution of the energetic electrons in the $x - y$ plane for different polarization when the interaction of electrons with the laser pulse is over. The total 180 electrons are initially ($t = 0$) located uniformly within a disc with radius $r = r_0/4$ placed at $z = 0$. Their initial velocities approximately yield Maxwellian distribution with averaged temperature 5 KeV. Here the plasma density $n_0 = 0.1n_{cr}$ is assumed. The peak laser intensity $I_0 = 1 \times 10^{20}$ W/cm^2 and $I_0 = 6 \times 10^{19}$ W/cm^2 are chosen for CP and LP, respectively.

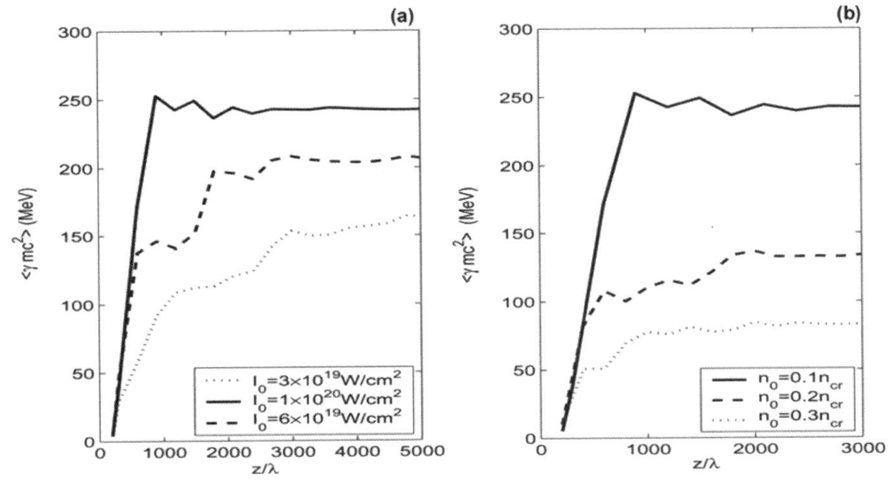

Fig. 5. Variation of the averaged energy gain $\langle \gamma mc^2 \rangle$ over several betatron oscillations as a function of z for (**a**) different laser intensities, $n_0 = 0.1 n_{cr}$ and (**b**) different plasma densities, $I_0 = 1 \times 10^{20}$ W/cm^2. The initial conditions of the electron are $r_{in} = 0$ and $\gamma_0 = 1.0$.

$\gamma \gg 1$ and $v_z > v_g$ is doomed to be small. The numerical results show that most of the electrons attain resonance and the energy gain spreads from 207 MeV to 262 MeV with an relative energy width $\Delta E/\langle E \rangle \simeq 24\%$. These resonant electrons separate from the pulse with scattering angle around 3° with respect to z axis (see Fig. 4a). Moreover, the beam divergence of REB is less than 1° after the interaction is over. For comparison, The corresponding results in the case of LP laser are shown in Figure 4b. It can be seen that the REB driven by the LP laser has a larger beam divergence around $10° - 1° = 9°$. The attainment of such REB is relevant to the motion of electrons. The electrons driven by the CP laser helically move within a Larmor radius induced by the axial QSM field until leaving the pulse. The spatial distribution of these electrons is almost circular (see Fig. 4c). While the motion of electrons driven by the LP pulse is about elliptical and predominantly along the direction of laser polarization (see Fig. 4d). By comparison, the REB produced by CP laser-plasma interaction with good qualities of high energy, quasi-monoenergy, and extreme low beam divergence may be a better candidate in future laser-plasma accelerator.

We turn to the dependence of the energy gain of the energetic electrons on laser intensity and plasma density. The different laser intensities $I_0 = 30, 60, 100$ ($\times 10^{18}$ W cm^{-2}) and different plasma densities $n_0 = 0.1 n_{cr}$, $0.2 n_{cr}$, $0.3 n_{cr}$ are chosen. The self-generated fields are correspondingly changed by Eqs. (7)–(9). Figure 5 shows that the final energy gain of accelerated electrons increases with laser intensity but decreases with plasma density. In fact, in terms of Eq. (17), the electron energy gain is proportional to the laser electric field E_0. At higher plasma density, the laser pulse travels with a lower group velocity v_g and the electrons are earlier to be accelerated to reach v_g and then are pushed away from the pulse. As a result, the duration of interaction between the electron and the laser pulse decreases, leading to the reduction of the energy gain.

A quantity that is often quoted in discussions of particle accelerators is the energy gradient, defined as the energy gained by the particle per unit of distance along its trajectory. It is usually expressed in units of MeV m^{-1}. For example, the maximum energy gradient achievable in radio-frequency (rf) conventional accelerators is 100 MeV m^{-1}. In our case, the energy gain in general is about tens or hundreds of MeV with acceleration length around several millimeters. So we can attain the energy gradient about tens or hundreds of GeV m^{-1}, which is 2~3 orders larger than the rf accelerators.

4. CONCLUSION

In conclusion, the resonance acceleration of plasma electrons in interaction of an intense short laser with underdense plasma is theoretically and numerically studied by using a test particle model, where the joint effect of the CP Gaussian laser fields and the self-generated quasistatic fields, including the radial QSE field, the azimuthal QSM field, and the axial QSM field, is included. It is found that the resonance takes place when the betatron oscillation frequency of electrons in self-generated quasistatic fields coincides with the Doppler shifted laser frequency. Due to resonant acceleration, some electrons attain energy directly from the laser pulse and thus form a well qualified REB with high energy, quasi-monoenergy and extreme low divergent angle. It is also found that the final energy of the REB increases with laser intensity but decreases with plasma density. Our results show that the REB produced by CP laser-plasma interaction has better collimation comparing to the LP case for the resonant electrons move helically within a

Larmor radius induced by the axial QSM field. We believe that our results can give some enlightenment on the future laser-plasma based accelerators.

ACKNOWLEDGMENTS

We acknowledge helpful discussions with H. B. Cai on this work. This work is supported by the National Natural Science Foundation of China, Grant Nos. 10576007, 10335020, and 10575013, and the National High-Tech ICF Committee in China, and partially by the National Basic Research Program of China (973) (2007CB814802 and 2007CB815101).

REFERENCES

Borghesi, M., Kar, S., Romagnani, L., Toncian, T., Antici, P., Audebert, P., Brambrink, E., Ceccheerini, F., Cecchetti, E.F., Fuchs, J., Galimberti, M., Gizzi, L.A., Grismayer, T., Lyseikina, T., Jung, R., Macchi, A., Mora, P., Osterholz, J., Schiavi, A. & Willi, O. (2007). Stochastic heating in ultra high intensity laser-plasma interaction. *Laser Part. Beams* **25**, 169–168.

Esarey, E., Schroeder, C.B., Cormier-Michel, E., Shadwick, B.A., Geddes, C.G.R. & Leemans, W.P. (2007). Thermal effects in plasma-based accelerators. *Phys. Plasmas* **14**, 056707.

Evans, R.G. (1988) Particle accelerators-the light that never was. *Nature* **333**, 296–297.

Faure, J., Glinec, Y., Pukhov, A., Kiselev, S., Gordienko, S., Lefebvre, E., Rousseau, J.-P., Burgy, F. & Malka, V. (2004). A laser-plasma accelerator producing monoenergetic electron beams. *Nature* **431**, 541–544.

Geddes, C.G.R., Toth, C., Tilborg, J.V., Esarey, E., Schroeder, C.B., Bruhwiler, D., Nieter, C., Cary, J. & Leemans, W.P. (2004). High-quality electron beams from a laser wakefield accelerator using plasma-channel guiding. *Nature* **431**, 538–541.

Gibbon, P. (2005). *Short Pulse Laser Interaction with Matter-An Introduction*. London: Imperial College Press.

Gupta, D.N. & Suk, H. (2007) Electron acceleration to high energy by using two chirped lasers. *Laser Part. Beams* **25**, 31–36.

Hamster, H., Sullivan, A., Gordon, S., White, W. & Falcone, R.W. (1993). Subpicosecond, electromagnetic pulses from intense laser-plasma interaction. *Phys. Rev. Lett.* **71**, 2725–2728.

Hirshfield, J.L. & Wang, C.B. (2000). Laser-driven electron cyclotron autoresonance accelerator with production of an optically chopped electron beam. *Phys. Rev. E* **61**, 7252–7255.

Hoffmann, D.H.H., Blazevic, A., Ni, P., Rosemej, P., Roth, M., Tahir, N.A., Tauschwitz, A., Udrea, S., Varentsov, D., Weyrich, K., Maron, Y. (2005). Present and future perspectives for high energy density physics with intense heavy ion and laser beams. *Laser Part. Beams* **23**, 47–53.

Hora, H. (1988). Particle acceleration by superposition of frequency-controlled laser pulses. *Nature* **333**, 337–338.

Hora, H., Hoelss, M., Scheid, W., Wang, J.X., Ho, Y.K., Osman, F. & Castillo, R. (2000). Principle of high accuracy of the nonlinear theory for electron acceleration in vacuum by lasers at relativistic intensities. *Laser Part. Beams* **18**, 135–144.

Hora, H., Osman, F., Castillo, R., Collins, M., Stait-Gardener, T., Chan, W.K., Hölss, M., Scheid, W., Wang, J.J. & Ho, Y.K. (2002). Laser-generated pair production and Hawking-Unruh radiation. *Laser Part. Beams* **20**, 79–86.

Hora, H. (2007*a*). New aspects for fusion energy using inertial confinement. *Laser Part. Beams* **25**, 37–46.

Hora, H., Badziak, J., Read, M.N., Li, Yu-Tong, Liang, Tian-Jiao, C Ang, Yu, Liu Hong, Sheng, Zheng-Ming, Zhang, J., Osmanm, F., Mi-Ley, G.H., Zhang, Wei-Yan., He, Xian-Tu, Peng, Han-Sheng, Glowacz, S., Jablonski, G., Wolowski, J., Skladanovski, Z., Jungwirth, K., Rohlena, K. & Ulschmied, J. (2007*b*). Fast ignition by laser driven particle beams of very high intensity. *Phys., Plasmas* **14**, 072701–072717.

Joshi, C. & Katsouleas, T. (2003). Plasma Accelerators at the Energy Frontier and on Tabletops. *Phys. Today* **56**(6), 47–53.

Karmakar, A. & Pukhov, A. (2007). Collimated attosecond GeV electron bunches from ionized high-Z material by radially polarized ultr-relativistic laser pulses. *Laser Part. Beams* **25**, 371–378.

Kodama, R., Norreys, P.A., Mima, K., Dangor, A.E., Evans, R.G., Fujita, H., Kitagawa, Y., Krushelnick, K., Miyakoshi, T., Miyanaga, N., Norimatsu, T., Rose, S.J., Shozaki, T., Shigemori, K., Sunahara, A., Tampo, M., Tamaka, K.A., Toyama, Y., Yamanaka, T. & Zepf, M. (2001). Fast heating of ultrahigh-density plasma as a step towards laser fusion ignition. *Nature* **412**, 798–802.

Kong, Q., Miyazaki, S., Kawata, S., Miyauchi, K., Nakajima, K., Masuda, S., Miyanaga, N. & Ho, Y.K. (2003). Electron bunch acceleration and trapping by the ponderomotive force of an intense short-pulse laser. *Phys. Plasmas* **10**, 4605–4608.

Liu, C.S. & Tripathi, V.K. (2005). Ponderomotive effect on electron acceleration by plasma wave and betatron resonance in short pulse laser. *Phys. Plasmas* **12**, 043103.

Liu, H., He, X.T. & Chen, S.G. (2004). Resonance acceleration of electrons in combined strong magnetic fields and intense laser fields. *Phys. Rev. E* **66**, 066409.

Liu, H., He, X.T. & Hora, H. (2006). Additional acceleration and colliation of relativistic electron beams by magnetic field resonance at very high intensity laser interaction. *Appl. Phys. B: Lasers Opt.* **82**, 93–97.

Mckinstrie, C.J. & Startsev, E.A. (1996). Electron acceleration by a laser pulse in a plasma. *Phys. Rev. E* **54**, R1070–R1073.

Nuckolls, J.L. & Wood, L. (2002) Future of Inertial Fusion Energy, LLNL Preprint UCRL-JC-149860, Sept. 2002, (available to the public www.ntis.gov/).

Nuckolls, J.L. & Woods, L. (2002) Future of Inertial Fusion Energy. Proceedings International Conference on Nuclear Energy Systems ICNES Albuquerque, NM. 2002, edited by T.A. Mehlhorn (Sandia National Labs., Albuquerque, NM) p.171–176.

Pukhov, A. & Meyer-ter-vehn, J. (1996). Relativistic Magnetic Self-Channeling of Light in Near-Critical Plasma: Three-Dimensional Particle-in-Cell Simulation. *Phys. Rev. Lett.* **76**, 3975–3978.

Pukhov, A., Sheng, Z.M. & Meyer-terr-vehn, J. (1999). Particle acceleration in relativistic laser channels. *Phys. Plasmas* **6**, 2847–2854.

Qiao, B., He, X.T., Zhu, S.P. & Zheng, C.Y. (2005*a*). Electron acceleration in combined intense laser fields and self-consistent quasistatic fields in plasma. *Phys. Plasmas* **12**, 083102.

Qiao, B., Zhu, S.P., He, X.T. & Zheng, C.Y. (2005*b*). Quasistatic magnetic and electric fields generated in intense laser plasma interaction. *Phys. Plasmas* **12**, 053104.

Quesnel, B. & Mora, P. (1998). Theory and simulation of the interaction of ultraintense laser pulses with electrons in vacuum. *Phys. Rev. E* **58**, 3719–3723.

Salamin, Y.I., Hu, S.X., Hatsagortsyan, K.Z. & Keitel, C.H. (2006). Relativistic high-power laser-matter interactions. *Phys. Reports* **427**, 41–155.

Scheid, W., & Hora, H. (1989) On electromagnetic acceleration by plane transverse electromagnetic pulses in vacuum. *Laser Part. Beams* **7**, 315–332.

Schmitz, M. & Kull, H.J. (2002). Single-Electron Model of Direct Laser Acceleration in Plasma Channels. *Laser Phys.* **12**, 443–448.

Sheng, Z.M. & Meyer-ter-vehn, J. (1996). Inverse Faraday effect and propagation of circularly polarized intense laser beams in plasmas. *Phys. Rev. E* **54**, 1833–1842.

Stait-Gardner, T., & Castillo, R. (2006) Difference between Hawking-Unruh radiation derived from studies about pair production by lasers in vacuum. *Laser Part. Beams* **24**, 579–604.

Siders, C.W., Leblanc, S.P., Fisher, D., Tajima, T., Downer, M.C., Babine, A., Stepanov, A. & Sergeev, A. (1996). Laser Wakefield Excitation and Measurement by Femtosecond Longitudinal Interferometry. *Phys. Rev. Lett.* **76**, 3570–3573.

Stupakov, G.V. & Zolotorev, M.S. (2001). Ponderomotive Laser Acceleration and Focusing in Vacuum for Generation of Attosecond Electron Bunches. *Phys. Rev. Lett.* **86**, 5274–5277.

Tabak, M., Ammer, J.H., Glinsky, M.E., Kruer, W.L., Wilke, S.C., Woodworth, J., Campbell, E.M. & Perry, M.D. (1994). Ignition and high gain with ultrapowerful lasers. *Phys. Plasmas* **1**, 1626–1634.

Tajima, T.T. & Dawson, J.M. (1979). Laser Electron Accelerator. *Phys. Rev. Lett.* **43**, 267–270.

Tsakiris, G.D., Gahn, C. & Tripathi, V.K. (2000). Laser induced electron acceleration in the presence of static electric and magnetic fields in a plasma. *Phys. Plasmas* **7**, 3017–3030.

Xu, J.J., Kong, Q., Chen, Z., Wang, P.X., Wang., W., Lin, D. & Ho, Y.K. (2007) Polarization effect on vacuum laser acceleration. *Laser Part. Beams* **25**, 253–258.

Yu, M.Y., Yu, W., Chen, Z.Y., Zhang, J., Yin, Y., Cao, L.H., Lu, P.X. & Xu, Z.Z. (2003). Electron acceleration by an intense short-pulse laser in underdense plasma. *Phys. Plasmas* **10**, 2468–2474.

Zheng, C.Y., He, X.T. & Zhu, S.P. (2005). Magnetic field generation and relativistic electron dynamics in circularly polarized intense laser interaction with dense plasma. *Phys. Plasmas* **12**, 044505.

Zhou, C.T., He, X.T. & Yu, M.Y. (2006). A comparison of ultrarelativsic electron- and positron-bunch propagation in plasmas. *Phys. Plasmas* **13**, 092109.

Zhou, C.T., Yu, M.Y. & He, X.T. (2007). Electron acceleration by high current-density relativistic electron bunch in plasmas. *Laser Part. Beams* **25**, 313–319.

PHYSICS OF PLASMAS **16**, 013104 (2009)

Enhanced resonant acceleration of electrons from intense laser interaction with density-attenuating plasma

H. Y. Niu,[1,a)] X. T. He,[2] C. T. Zhou,[2] and Bin Qiao[3]
[1]*Graduate School of China Academy of Engineering Physics, P.O. Box 2101, Beijing 100088, People's Republic of China*
[2]*Institute of Applied Physics and Computational Mathematics, P.O. Box 8009, Beijing 100088, People's Republic of China; Center for Applied Physics and Technology, Peking University, Beijing 100871, People's Republic of China; and Institute for Fusion Theory and Simulation, Zhejiang University, Hangzhou 310027, People's Republic of China*
[3]*Department of Physics and Astronomy, Center for Plasma Physics, Queen's University Belfast, Belfast BT7 1NN, United Kingdom*

(Received 10 October 2008; accepted 5 December 2008; published online 13 January 2009)

An enhanced resonant acceleration scheme for electrons by intense circularly polarized laser pulse in a plasma with slowly attenuating density is proposed. As it propagates, the phase velocity and Doppler-shifted frequency of the laser both gradually decrease, so that the electrons moving in the combined laser and spontaneous fields can retain betatron resonance for a rather long time and effectively acquire energy from the laser. The theoretical analysis is verified by test-particle numerical calculations. It is shown that well-collimated GeV electron beams with very low beam divergence can be produced. © *2009 American Institute of Physics*. [DOI: 10.1063/1.3057445]

I. INTRODUCTION

With the rapid development of laser technology, relativistic electrons generated in intense laser–plasma interaction have received much interest. Relativistic electron beams have wide potential applications in laser-based particle accelerators, fast ignition for inertial confinement fusion, etc.[1–5] The wakefield acceleration mechanism based on the trapping of electrons in the plasma waves has been studied and experimentally verified.[6–10] However, it relies on complex laser modulation and electron injection, and the density of the generated electrons as well as the energy conversion efficiency are both very low.

Recently, direct resonant acceleration of electrons in intense laser–plasma interaction has been proposed.[11–18] In this scheme, plasma electrons oscillate in the spontaneous quasistatic electric (QSE) and quasistatic magnetic (QSM) fields[19] like a betatron oscillator and those with their betatron frequency coinciding with the Doppler-shifted laser frequency are resonantly trapped in the laser pulse. Therefore, they can efficiently acquire energy from the latter. Due to the existence of a quasistatic axial magnetic field, relativistic electron beams generated by resonant acceleration in the interaction of circularly polarized (CP) lasers with plasmas have much smaller beam divergence and better collimation than those with linearly polarized lasers.[20] However, as a resonant electron moves faster, its mass increases and the betatron frequency decreases, and it becomes rapidly off-resonant. Thus, the maximum energy obtained in this scheme is limited.

In this paper, we propose a new method for enhancing the resonant acceleration in intense laser–plasma interaction by using a plasma with slowly attenuating density in the laser propagation direction. In such a plasma, the laser phase velocity $v_{\rm ph}$ and Doppler-shifted laser frequency $\omega_0(1-v_z/v_{\rm ph})$ both decrease gradually during the propagation, which can effectively offset the reduction of ω_b due to the relativistic effect. As a result, the electrons can be resonantly accelerated for a longer time and attain much higher energy. The numerical results show that relativistic electron beams with GeV energy and low beam divergence are generated in the present enhanced resonant acceleration scheme.

This paper is organized as follows. In Sec. II, the theoretical model for resonant acceleration from CP laser interaction with density-attenuating plasma is presented. In Sec. III, the results from the test-particle method are given. A conclusion is given in Sec. IV.

II. THEORETICAL MODEL AND RESONANT CONDITION

We consider a CP laser pulse of duration τ_L and spot radius r_0 propagating in an underdense plasma with its density attenuating along the laser axis z. The electric fields of the Gaussian laser pulse can be expressed as

$$E_{Lx}=E_L\cos\phi, \quad (1)$$

$$E_{Ly}=\alpha E_L\sin\phi, \quad (2)$$

$$E_{Lz}=E_L\left(-\frac{x}{kr_0^2}\sin\phi+\alpha\frac{y}{kr_0^2}\cos\phi\right), \quad (3)$$

where $E_L(r,z,t)=E_0\exp\{-r^2/2r_0^2-(t-(z-z_0)/v_g)^2/2\tau_L^2\}$, $\phi=\omega_0 t-kz$, $k=\eta\omega_0/c$, $\eta\approx[1-\omega_{pe}^2/\omega_0^2(1+a_0^2)^{1/2}]^{1/2}$, $a_0=eE_0/m\omega_0 c$, $\omega_{pe}=\sqrt{4\pi n_p(z)e^2/m}$, $-e$ and m are electron charge and rest mass, c is the velocity of light in vacuum, $n_p(z)$ is the axially attenuating plasma density, v_g is the laser group velocity in the plasma, z_0 is the initial position of the laser pulse center, and $\alpha=\pm1$ represent the left- and right-

[a)]Electronic mail: niuhaiyan221@126.com.

hand circular polarizations, respectively. We note that $\nabla\cdot\mathbf{E_L}=0$. The magnetic field related to the laser pulse can be obtained from $\nabla\times\mathbf{E_L}=-\partial\mathbf{B_L}/\partial t$.

As is well known, when such an intense CP laser propagates in a plasma, strong QSE and QSM fields are spontaneously generated by various nonlinear and relativistic effects.[19,21] These fields can strongly affect the electron motion. First, the ponderomotive force $\mathbf{F}_p$ produced by laser pulse expels some electrons from the laser axis so that a space-charge potential Φ is formed, which corresponds to the QSE field $\mathbf{E}_s=-\nabla\Phi$. The longitudinal relativistic current driven by the ponderomotive force generates the azimuthal QSM field $\mathbf{B}_{s\theta}$, while the rotational current caused by the nonlinear beat interaction of the CP laser fields with relativistic electrons produces the axial QSM field $\mathbf{B}_{sz}$. Particle-in-cell simulations[22] have shown that strong azimuthal and axial QSM fields exist in the interaction of the CP laser with dense plasmas. The spontaneous fields can be obtained from an analytical model based on the relativistic Vlasov–Maxwell equations.[19] Introducing $t\rightarrow\omega_0 t$, $\mathbf{x}\rightarrow\omega\mathbf{x}/c$, $E\rightarrow eE/m\omega_0 c$, and $B\rightarrow eB/m\omega_0 c$, we have

$$\mathbf{E}_s=k_E\frac{r}{r_0}\exp\left\{-\frac{r^2}{r_0^2}-\frac{[t-(z-z_0)/v_g]^2}{\tau_L^2}\right\}\mathbf{e}_r, \tag{4}$$

$$\mathbf{B}_{s\theta}=-k_B\frac{r_0}{r}\left(1-e^{-r^2/r_0^2}\right)\exp\left\{-\frac{[t-(z-z_0)/v_g]^2}{\tau_L^2}\right\}\mathbf{e}_\theta, \tag{5}$$

$$\mathbf{B}_{sz}=-b_Z\exp\left\{-\frac{r^2}{r_0^2}-\frac{[t-(z-z_0)/v_g]^2}{\tau_L^2}\right\}\mathbf{e}_z. \tag{6}$$

In a slowly varying plasma, the coefficients are chosen as $k_E=(1+\alpha^2)\beta_{56}I_0/r_0$, $k_B=\epsilon[\pi(1+\alpha^2)/64]n_p(z)r_0I_0[\xi\beta_4(T_{\mathrm{te}})+(1-\xi)\beta_4(T_{\mathrm{re}})]$, and $b_Z=\frac{1}{2}\alpha n_p(z)I_0[\xi\beta_1(T_{\mathrm{te}})+(1-\xi)\beta_1(T_{\mathrm{re}})]$, which are related to the laser and plasma parameters such as $n_p(z)$, I_0, etc. In these expressions, the other quantities such as T_{te}, T_{re}, ξ, ϵ, β_1, β_4, and β_{56} are defined as follows.[15,19] T_{te} and T_{re} are the averaged temperature for thermal and relativistic electrons, respectively. ξ represents the ratio of the number of thermal electrons to that of total electrons. ϵ is a coefficient less than 1.0, which indicates that $B_{s\theta}$ is strongly compensated by the return currents. The coefficients related to the relativistic effect are defined as $\beta_1=\int(1/p_{0e}^2)[(1/2)-(|\mathbf{p}_e|^2/3p_{0e}^2)]f_{e0}d^3\mathbf{p}_e$, $\beta_4=\int[(1/p_{0e}|\mathbf{p}_e|)-(3|\mathbf{p}_e|/4p_{0e}^3)]\times f_{e0}d^3\mathbf{p}_e$, $\beta_5=\int[(1/2|\mathbf{p}_e|^2)+(|\mathbf{p}_e|^2/3p_{0e}^4)-(1/2p_{0e}^2)]f_{e0}d^3\mathbf{p}_e$, $\beta_6=\int[(1/p_{0e})+(p_{0e}/|p_e|^2)]f_{e0}d^3\mathbf{p}_e$, and $\beta_{56}=\beta_5/\beta_6$, where $p_{0e}=\sqrt{1+p_e^2}$ and f_{e0} is the distribution function of background electrons. We have assumed that the plasma density $n_p(z)$ is slowly attenuating so that the model of Ref. 19 remains valid.

The motion of a test particle is described by the Lorentz equation

$$\frac{d\mathbf{p}}{dt}=-(\mathbf{E}+\mathbf{v}\times\mathbf{B}), \tag{7}$$

where $\mathbf{p}=\gamma\mathbf{v}$, $\gamma=(1+p^2)^{1/2}$, $\mathbf{E}=\mathbf{E}_L+\mathbf{E}_s$, and $\mathbf{B}=\mathbf{B}_L+\mathbf{B}_s$. As pointed out in Ref. 11, the electron motion in the presence of both spontaneous fields $\mathbf{E}_s$, $\mathbf{B}_s$ and the laser fields $\mathbf{E}_L$, $\mathbf{B}_L$ can be seen as a driven oscillation. The electron makes betatron oscillations in the spontaneous electric and magnetic fields and the laser pulse acts as a driving source. When the driver frequency ω_0-kv_z coincides with the betatron frequency ω_b, a resonance occurs, leading to an effective energy exchange between the laser and electrons. It is difficult to obtain an exact solution of Eq. (7) because of the high nonlinearity. However, we are mainly interested in the resonant condition. If the magnitude of transverse displacement of an electron is much smaller than the laser spot size, i.e., $r^2\ll r_0^2$, then E_{Lz} and B_{Lz} will have a negligible influence on the electron motion. Since in the laser–plasma interaction the spontaneous fields are very slowly varying, in the analysis they can be assumed to be static. The time dependence is, however, included in the numerical calculations presented in Sec. III. Accordingly, we can approximately obtain the betatron oscillation frequency from the transverse motion of the electron.[20] Thus, the normalized resonance condition $\omega_b=1-v_z/v_{\mathrm{ph}}$ can be written as

$$\frac{1}{\sqrt{\gamma}}\left[\frac{k_E}{r_0}\left(1-\frac{r^2}{r_0^2}\right)+\frac{k_Bv_z}{r_0}+\frac{\alpha^2b_Z^2}{\gamma}\left(1-\frac{2r^2}{r_0^2}\right)\right]^{1/2}=1-\frac{v_z}{v_{\mathrm{ph}}}. \tag{8}$$

As discussed in Ref. 20, the match between ω_b and $1-v_z/v_{\mathrm{ph}}$ will lead to a steep increase in γ. However, it can also be seen from the left-hand side of Eq. (8) that ω_b decreases with increase in γ. The increase in v_z on the right-hand side of Eq. (8) is relatively small after the electron has been preaccelerated. Thus, the resonant condition is violated, leading to termination of the electron acceleration. However, in a density-attenuating plasma, we can expand $1-v_z/v_{\mathrm{ph}}\simeq 1-v_z[1-(n_p(z)/2)(1/\sqrt{1+a_0^2})]$. The decrease in $1-v_z/v_{\mathrm{ph}}$ caused by the decrease in the plasma density can effectively offset the reduction in ω_b owing to the increase in γ. As a result, the resonance duration between the pulse and electron can be greatly extended, leading to strongly high energy gain.

The electron acceleration induced by the match between ω_b and $1-v_z/v_{\mathrm{ph}}$ can be attributed to a phase locking between the electron and the pulse. In fact, after the electrons are preaccelerated, ω_b changes slowly on the time scale of one betatron oscillation, and the rotational motion of the electron driven by the CP pulse can be written

$$p_x=b(t)\cos\theta_b,\quad p_y=\alpha b(t)\sin\theta_b,\quad \frac{d\theta_b}{dt}=\omega_b, \tag{9}$$

where $b(t)$ is the magnitude of transverse momentum. Inserting this value of p_x, p_y into $d\gamma/dt=-\mathbf{E}\cdot\mathbf{v}$, and neglecting the rv_r-term since it averages to zero over an oscillation period, the energy equation becomes

$$\frac{d\gamma}{dt}=-\frac{b(t)}{\gamma}E_L\cos\Psi, \tag{10}$$

where $\Psi=\theta_b-(t-z/v_{\mathrm{ph}})$ is the phase shift between the electron and the laser, or the so-called slow ponderomotive phase.[11] We note that a steady phase shift ($d\Psi/dt\simeq 0$) corresponds to $\omega_b=1-v_z/v_{\mathrm{ph}}$ [for slowly varying plasma den-

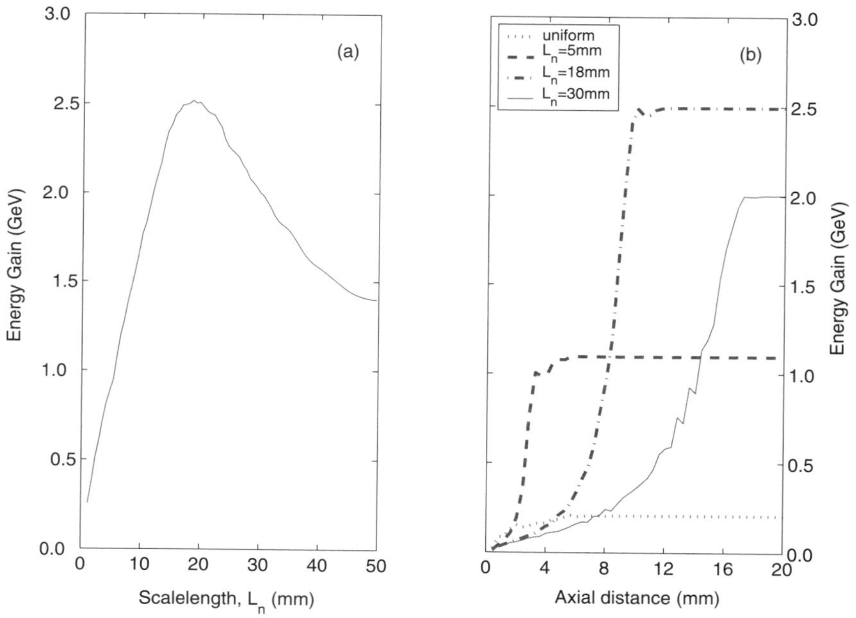

FIG. 1. (Color online) (a) The energy gain of the electron as a function of density scalelengths varying from 1 to 50 mm. (b) Variation of the averaged energy gain over several betatron oscillations with the axial distance for a uniform plasma (dotted line) and an axially decreasing plasma with the initial plasma density $n_0=0.3n_{cr}$. In the latter, three representative scalelengths are chosen such that $L_n=5$ mm (dashed line), $L_n=18$ mm (dash-dotted line), and $L_n=30$ mm (solid line). The electron is initially located at the axis.

sity, or $z(dk/dz) \ll k$]. The electrons in the acceleration phase can continuously obtain energy from the pulse if Ψ is approximately constant. Maximum acceleration is attained if the betatron oscillation is exactly out of phase with the laser electric field, i.e., the phase shift is $\Psi=\pi$.

III. NUMERICAL RESULTS ON ELECTRON ACCELERATION

To further analyze the electron acceleration process, the equation of motion of the electron is numerically solved. The laser parameters are wavelength $\lambda=1.06$ μm, spot radius $r_0=4\lambda$, pulse duration $\tau_L=100$ fs, and peak intensity $I_0=1\times10^{20}$ W cm^{-2}. For simplicity, we choose a linearly attenuating density profile $n_p(z)=n_0(1-z/L_n)$, where n_0 is the initial plasma density, $L_n=-[(1/n_0)(dn_p/dz)]^{-1}$ is the density scalelength. For the present spontaneous field model, we have assumed $T_{te}=5$ keV, $\xi=0.3$ and $\epsilon=0.1$. T_{re} is always determined by Wilk's scaling law as $T_{re}=511\{\sqrt{1+[(1+\alpha^2)I_0\lambda^2/2.76]}-1\}$ keV. Thus, the coefficients k_E, k_B, and b_Z can be calculated from the definitions in Sec. II. We further assume that the test electron is initially at rest and located at $z=0$, and the laser pulse center is at $z_0=-4c\tau_L$. We then follow the motion of the electron in the CP laser and the spontaneous QSE and QSM fields.

Figure 1(a) shows the variation of the gained energy as a function of density scalelengths. It can be seen that there exists an optimum density scalelength about $L_n=18$ mm at which the electron attains the maximum energy of 2.5 GeV. In fact, the optimum value of density scalelength leads to the optimum phase match of the electron with the laser pulse. For other density profiles, the slightly weak phase-locked efficiency results in energy gain lower than that for the optimum profile. However, such an energy gain is still larger than that gained in an uniform plasma. This comparison can be seen from Fig. 1(b), where the evolution of the energy gains for three representative scalelengths and that in a uniform plasma for the same laser conditions are simultaneously shown. For the latter, the electron is accelerated within a distance of about 4 mm and consequently maintains an energy level of 200 MeV. The other three lines demonstrate that the electron experiences comparatively long acceleration distance and attains sharper energy increase. Among these, the most efficient acceleration is displayed by the dash-dotted line, which corresponds to $L_n=18$ mm. Such acceleration efficiency can be confirmed by the phase comparison between p_x and B_{Ly} (see Fig. 2). It appears that the electron can almost keep in phase with the pulse. Especially for $z\in(8.5,\ 9.5$ mm), an almost locked phase shift $\Psi\simeq\pi$ between p_x and B_{Ly} takes place, which indicates the resonance

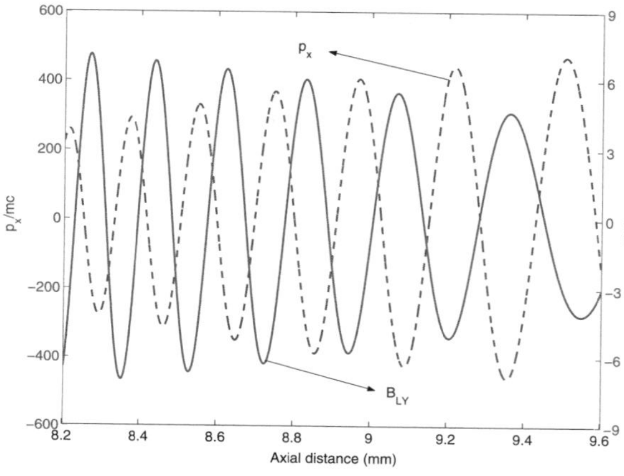

FIG. 2. (Color online) Variation of p_x and B_{Ly} as a function of axial distance for the interaction shown in Fig. 1(b) when $L_n=18$ mm.

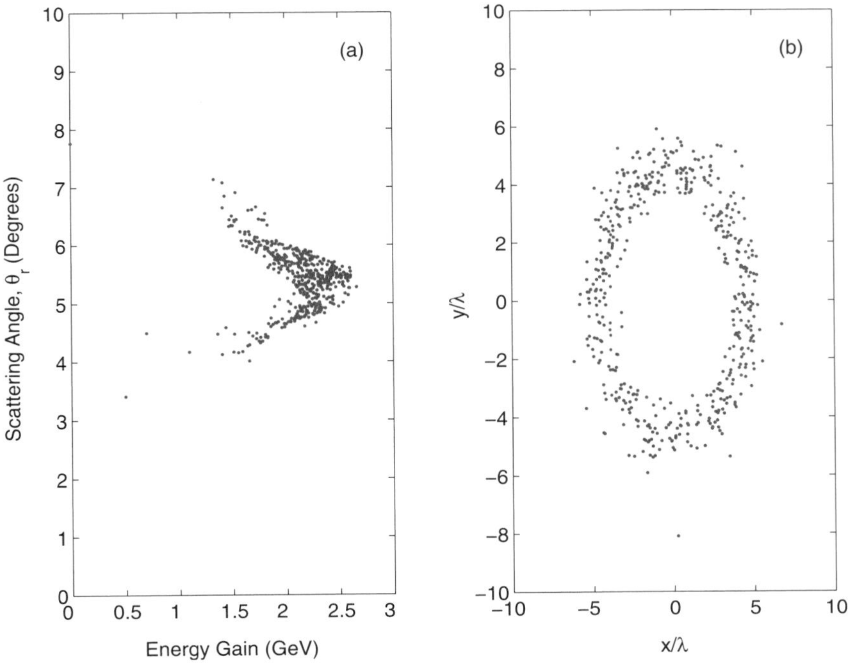

FIG. 3. (Color online) (a) The scattering angles vs the energy gain of electrons and (b) the resultant spatial distribution of the electrons in the $x-y$ plane when the interaction between the electrons and the laser pulse is over. A total of 500 electrons are initially ($t=0$) located uniformly within a disk with radius $r=r_0/4$ placed at $z=0$. Their initial velocities approximately yield Maxwellian distribution with averaged temperature 5 keV. The density scalelengths are taken to be optimal, i.e., $L_n=18$ mm, and $n_0=0.3n_{cr}$.

condition $\omega_b \simeq \omega - kv_z$ holds and consequently the electron attains the most efficient acceleration. This is consistent with the prediction by Eq. (10). We note that the increase of γ is mainly from the longitudinal momentum p_z. It is the ponderomotive force $-(e/c)\mathbf{v}\times\mathbf{B}_L$ that converts the laser energy to the longitudinal momentum of the electron. The acceleration ceases when the dephasing occurs at $z>9.6$ mm.

We now consider the scattering angle, the energy spread, and the spatial distribution of the resonant electrons. The results are displayed in Fig. 3, where the density scalelength is chosen as the optical value. Initially, 500 electrons are uniformly distributed in a disk of radius $r=r_0/4$ at $z=0$. Their velocities satisfy approximately a Maxwellian distribution with average temperature of 5 keV. At the end of the interaction we record the final energy, the radial position, and final momentum of each electron. The scattering angle of the electrons relative to the z axis is $\theta_r=\arctan(p_\perp/p_\parallel)$. In fact, from the equation $\gamma^2=1+p_\perp^2+p_\parallel^2$ we obtain $p_\perp/p_\parallel=\sqrt{(\gamma^2-1)/(\gamma^2 v_\parallel^2)-1}$. For electrons satisfying $\gamma \gg 1$, the angle θ_r is always small. The relative energy spread is calculated by $\Delta E/\langle E\rangle$ (here $\langle E\rangle$ is the mean energy of electron beam and ΔE is the absolute energy width). The numerical results show that the energy gain of most electrons is from 1.6 to 2.6 GeV with a spread $\Delta E/\langle E\rangle \simeq 45\%$. Most resonant electrons are emitted with scattering angle around 5.5° with respect to the z axis [see Fig. 3(a)]. The divergence angle of the electron beam is about $6°-4.5°=1.5°$ when the interaction is over. This result can be attributed to the fact that the electrons driven by the CP laser move helically within a Larmor radius induced by the axial QSM field. The resultant spatial distribution of these electrons is almost circular [see Fig. 3(b)] due to the rotation of laser electric field. Thus, the relativistic electron beam produced by CP laser interacting with a plasma having a negative density gradient has very high energy, relatively small energy spread, and extremely low beam divergence, and can serve as a high-quality laser-plasma accelerator.

IV. CONCLUSION

In this paper, we have investigated an enhanced betatron resonance scheme for electron acceleration by an intense CP laser pulse in a plasma with decreasing density. It is found that the electrons can be phase-stably trapped in the laser pulse and the spontaneous quasistatic fields for a long time. Monoenergetic relativistic electron beams with GeV-level energy and very low divergence can be generated. The theoretical model is verified by relativistic test-particle numerical calculations. Although our approach is not fully self-consistent, it can still yield insight to the physical processes involved in the proposed acceleration scheme, and the results should be useful for future plasma-based particle accelerators.

ACKNOWLEDGMENTS

This work was supported by the National Natural Science Foundation of China, Grant Nos. 10576007, 10575013, and 10835003, the National Basic Research Program of China (973) (Grant Nos. 2007CB814802 and 2007CB815101), and the National High-Tech ICF Committee in China.

[1] M. Tabak, J. Hammer, M. E. Glinsky, W. L. Kruer, S. C. Wilks, J. Woodworth, E. M. Campbell, M. D. Perry, and R. J. Mason, Phys. Plasmas **1**, 1626 (1994).

[2]C. Joshi and T. Katsouleas, Phys. Today **56** (6), 47 (2003); K. Nakajima, D. Fisher, and T. Kawakubo, Phys. Rev. Lett. **74**, 4428 (1995).
[3]J. Faure, Y. Glinec, A. Pukhov, S. Kiselev, S. Gordienko, E. Lefebvre, J. P. Rousseau, F. Burgy, and V. Malka, Nature (London) **431**, 541 (2004); S. P. D. Mangles, C. D. Murphy, Z. Najmudin, A. G. R. Thomas, J. G. Collier, A. E. Dangor, E. J. Divall, P. S. Foster, J. G. Gallacher, C. J. Hooker, D. A. Jaroszynski, A. J. Langley, W. B. Mori, P. A. Norreys, F. S. Tsung, R. Viskup, B. R. Walton, K. Krushelnick, *ibid.* **431**, 535 (2004); C. G. R. Geddes, C. Toth, J. V. Tilborg, E. Esarey, C. B. Schroeder, D. Bruhwiler, C. Nieter, J. Cary, and W. P. Leemans, *ibid.* **431**, 538 (2004).
[4]E. Esarey, C. B. Schroeder, E. Cormier-Michel, B. A. Shadwick, C. G. R. Geddes, and W. P. Leemans, Phys. Plasmas **14**, 056707 (2007).
[5]C. T. Zhou, X. T. He, and M. Y. Yu, Phys. Plasmas **13**, 092109 (2006); Laser Part. Beams **25**, 313 (2007).
[6]T. T. Tajima and J. M. Dawson, Phys. Rev. Lett. **43**, 267 (1979).
[7]C. W. Siders, S. P. Leblanc, D. Fisher, T. Tajima, M. C. Downer, A. Babine, A. Stepanov, and A. Sergeev, Phys. Rev. Lett. **76**, 3570 (1996).
[8]P. Sprangle, B. Hafizi, J. R. Penano, R. F. Hubbard, A. Ting, C. I. Moore, D. F. Gordon, A. Zigler, D. Kaganovich, and T. M. Antonsen, Jr., Phys. Rev. E **63**, 056405 (2001).
[9]G. Fubiani, E. Esarey, C. B. Schroeder, and W. P. Leemans, Phys. Rev. E **73**, 026402 (2006).
[10]P. Gibbon, *Short Pulse Laser Interaction with Matter—An Introduction* (Imperial College Press, London, 2005), p. 64.
[11]A. Pukhov, Z. M. Sheng, and J. Meyer-ter-Vehn, Phys. Plasmas **6**, 2847 (1999).
[12]Q. Kong, S. Miyazaki, S. Kawata, K. Miyauchi, K. Nakajima, S. Masuda, and N. Miyanaga, Phys. Plasmas **10**, 4605 (2003).
[13]H. Liu, X. T. He, and S. G. Chen, Phys. Rev. E **69**, 066409 (2004); H. Liu, X. T. He, and H. Hora, Appl. Phys. B: Lasers Opt. **82**, 93 (2006).
[14]G. D. Tsakiris, C. Gahn, and V. K. Tripathi, Phys. Plasmas **7**, 3017 (2000).
[15]B. Qiao, X. T. He, S. P. Zhu, and C. Y. Zheng, Phys. Plasmas **12**, 083102 (2005).
[16]M. Y. Yu, W. Yu, Z. Y. Chen, J. Zhang, Y. Yin, L. H. Cao, P. X. Lu, and Z. Z. Xu, Phys. Plasmas **10**, 2468 (2003).
[17]H. Hora, M. Hoelss, W. Scheid, J. X. Wang, Y. K. Ho, F. Osman, and R. Castillo, Laser Part. Beams **18**, 135 (2000).
[18]W. Yu, V. Bychenkov, Y. Sentoku, M. Y. Yu, Z. M. Sheng, and K. Mima, Phys. Rev. Lett. **85**, 570 (2000).
[19]B. Qiao, S. P. Zhu, X. T. He, and C. Y. Zheng, Phys. Plasmas **12**, 053104 (2005).
[20]H. Y. Niu, X. T. He, B. Qiao, and C. T. Zhou, Laser Part. Beams **26**, 51 (2008).
[21]A. Pukhov and J. Meyer-ter-Vehn, Phys. Rev. Lett. **76**, 3975 (1996); **79**, 2686 (1997).
[22]C. Y. Zheng, X. T. He, and S. P. Zhu, Phys. Plasmas **12**, 044505 (2005); S. P. Zhu, X. T. He, and C. Y. Zheng, *ibid.* **8**, 321 (2001).

Generating Overcritical Dense Relativistic Electron Beams via Self-Matching Resonance Acceleration

B. Liu,[1] H. Y. Wang,[1] J. Liu,[2] L. B. Fu,[2] Y. J. Xu,[1] X. Q. Yan,[1,*] and X. T. He[1,2,†]

[1]*Key Laboratory of HEDP of the Ministry of Education, CAPT, and State Key Laboratory of Nuclear Physics and Technology, Peking University, Beijing 100871, China*

[2]*Institute of Applied Physics and Computational Mathematics, Beijing 100088, China*

(Received 19 July 2012; published 22 January 2013)

We show a novel self-matching resonance acceleration regime for generating dense relativistic electron beams by using ultraintense circularly polarized laser pulses in near-critical density plasmas. When the self-generated quasistatic axial magnetic field is strong enough to pinch and trap thermal relativistic electrons, an overdense electron bunch is formed in the center of the laser channel. In the trapping process, the electron betatron frequencies and phases can be adjusted automatically to match the resonance condition. The matched electrons are accelerated continuously and a collimated electron beam with overcritical density, helical structure, and plateau profile energy spectrum is hence generated.

DOI: 10.1103/PhysRevLett.110.045002 PACS numbers: 52.38.Kd, 52.27.Ny, 52.38.Fz, 52.59.−f

The rapid development of ultraintense laser technology has opened new and active research fields in laser plasma interactions [1] ranging from fast ignition of inertial confinement fusion [2], laboratory astrophysics [3], to development of compact sources of high-energy particles, such as electrons, ions, and gamma photons [4,5].

In laser plasma interactions, the resonance between betatron motion of electrons and ultraintense laser pulses is an interesting phenomenon and attracts much attention in both electron acceleration [6,7] and gamma photon production [5]. A betatron oscillating relativistic electron, confined by the self-generated quasistatic transverse fields, can be resonant with the laser pulse and gain energy efficiently from the laser fields. This is an inverse process of the ion-channel laser [8]. In order to accelerate electron beam in such a way, the resonance condition must be realized: the frequency and phase of the electron betatron motion should be matched with the laser pulse in the electron comoving frame. However, by using linearly polarized (LP) laser pulses, the preacceleration process necessary to reach the resonant behavior is stochastic in nature, and only few electrons experience the betatron resonance acceleration [9], which severely limits the density and number of energetic electrons, and lacks practical applications.

In this Letter, by using ultraintense circularly polarized (CP) laser pulses in near-critical density plasmas, we report on a novel self-matching resonance acceleration (SMRA) regime, where the electron betatron frequencies and phases are adjusted automatically to match the resonance condition. In such ultraintense and near-critical conditions, the self-generated quasistatic magnetic fields play an important role in many phenomena, including the relativistic magnetic self-channeling of light [10], magnetic-dipole vortex generation [11], ion acceleration by magnetic field annihilation [12], and laser shaping by plasma lens [13]. In particular, quasistatic axial magnetic fields can be generated by CP laser pulses, which have been widely investigated both theoretically and experimentally [14]. In the presence of axial magnetic field, electron acceleration by CP laser pulses shows interesting and complex properties [15,16]. With the aid of analytical modeling and 3D simulations, we found that when the self-generated quasistatic axial magnetic field is strong enough to pinch and trap the thermal relativistic electrons, an overdense electron bunch can appear in the center of the plasma channel produced by the laser pulse. In this process, the frequencies and phases of the trapped electrons can be self-matched with that of the laser pulse in the electron comoving frame, and the resonance conditions are satisfied. The matched electrons exhibit similar behavior: executing betatron rotation around the axial magnetic field and being accelerated continuously along the laser propagation direction. Hence a collimated relativistic electron beam with overcritical density and helical structure is generated.

In the CP laser evacuated channel, strong quasistatic magnetic fields, both in the axial and azimuthal direction are generated. The self-generated axial magnetic field B_{Sz} provides a "trapping effect" on relativistic electrons. An upper bound of perpendicular momentum for the trapped electrons is defined, when the radius of the Larmor motion is equal to the spot size of the axial magnetic field, as

$$p_{\mathrm{tr}} = 2eRB_{Sz}, \quad (1)$$

where p_{tr} is called trapping momentum, R is the radius of the spot size of the axial magnetic field, and e is the elementary charge. On the other hand, the quasistatic azimuthal magnetic field produces a "resonance effect." In order to understand the mechanism of the resonance, we employ the single-particle dynamics model, which has been widely investigated in many similar problems [17]. It is noticed that, in near critical density plasmas, the collective behavior of the electrons may begin to modify the dynamics away from what one obtains from single-particle orbits alone [18]. In the LP laser field, the collective behavior may cause density asymmetry in the perpendicular direction, while in our CP laser case, since the laser field is symmetrical, such collective effects are negligible. Considering a right-hand CP plane wave with frequency ω_0, the transverse electromagnetic fields are $E_{Lx} = E_L \cos\phi$, $E_{Ly} = E_L \sin\phi$, $B_{Lx} = -E_{Ly}/v_{\rm ph}$, $B_{Ly} = E_{Lx}/v_{\rm ph}$, where $\phi = kz - \omega_0 t$, $v_{\rm ph} = \omega_0/k$ is the phase velocity, and k is the wave number. The self-generated quasistatic azimuthal magnetic field can be expressed as $\boldsymbol{B}_{S\theta} = -B_{Sy}\boldsymbol{i} + B_{Sx}\boldsymbol{j}$. For an electron moving in combination with the self-generated quasistatic magnetic field (both in axial and azimuthal direction) and the laser pulse, the equations of the transverse motion are $dp_x/dt = -e\kappa E_L \cos\phi + ev_z B_{Sy} - ev_y B_{Sz}$, $dp_y/dt = -e\kappa E_L \sin\phi - ev_z B_{Sx} + ev_x B_{Sz}$, where $\kappa = 1 - v_z/v_{\rm ph}$. It is difficult to solve the equations analytically without any approximation. We focus on the resonance electrons, for which the longitudinal velocity v_z and the Lorentz factor γ are slow variables compared to the fast variables p_x and p_y. It is resonable to assume that $\dot{v}_z \to 0$ and $\dot{\gamma} \to 0$. According to the circular symmetry of the azimuthal magnetic field, we have $\partial B_{S\theta}/\partial r = -\partial B_{Sy}/\partial x = \partial B_{Sx}/\partial y$. Then the derivative of the equations becomes

$$\frac{d^2 p_x}{dt^2} + \Omega_\theta^2 p_x + \Omega_z \frac{dp_y}{dt} = m_e c a_L \omega_L^2 \sin\omega_L t, \quad (2)$$

$$\frac{d^2 p_y}{dt^2} + \Omega_\theta^2 p_y - \Omega_z \frac{dp_x}{dt} = m_e c a_L \omega_L^2 \cos\omega_L t, \quad (3)$$

where $a_L = eE_L/m_e c\omega_0$ is the normalized laser amplitude, $\omega_L = \kappa\omega_0$ is the frequency of the laser pulse in the electron comoving frame, $\Omega_\theta = \sqrt{\frac{ev_z}{\gamma m_e}\frac{\partial}{\partial r}B_{S\theta}}$, $\Omega_z = eB_{Sz}/\gamma m_e$, and m_e, c are the electron mass and the speed of light in vacuum, respectively. By ignoring small oscillating terms, the transverse momenta can be solved as $p_x(t) = -p_\perp(t)\cos\omega_+ t$, $p_y(t) = p_\perp(t)\sin\omega_+ t$, where the perpendicular momentum is

$$p_\perp(t) = \frac{2m_e c a_L \omega_L^2}{(\omega_B - \omega_L)(\omega_B + \omega_L + \Omega_z)} \sin\omega_- t \quad (4)$$

and the frequency of the betatron motion $\omega_B = \sqrt{\Omega_\theta^2 + (\Omega_z/2)^2} - \Omega_z/2$, here $\omega_+ = (\omega_B + \omega_L)/2$, $\omega_- = (\omega_B - \omega_L)/2$. It is clearly seen that resonance occurs when $\omega_B \approx \omega_L$.

The resonance acceleration of electrons in CP laser pulses is usually investigated under the condition of $\gamma \gg 1$ (e.g., in Ref. [16]). However, the axial magnetic field B_{Sz} contributes only a small quantitative modification to the resonance effect, and is negligible when $\gamma \gg 1$. In this sense, we can say that the resonance effect is dominated by the azimuthal magnetic field $B_{S\theta}$. On the other hand, in the trapping effect, the axial magnetic field B_{Sz} plays a crucial role. In order to understand the electron dynamics comprehensively, we must investigate the transition process from the trapping regime to the resonance regime. If an electron is initially trapped in the self-generated quasistatic axial magnetic field B_{Sz}, the only way to escape is to make the amplitude of the perpendicular momentum described in Eq. (4) exceed the trapping momentum $p_{\rm tr}$, i.e.,

$$\left|\frac{2m_e c a_L \omega_L^2}{(\omega_B - \omega_L)(\omega_B + \omega_L + \Omega_z)}\right| > 2eRB_{Sz}, \quad (5)$$

or

$$\frac{\Delta\omega}{\omega_L} < \left(1 + \frac{\Omega_z}{2\omega_L}\right)\left(\sqrt{1 + \frac{m_e c a_L}{eRB_{Sz}} \Big/ \left(1 + \frac{\Omega_z}{2\omega_L}\right)^2} - 1\right), \quad (6)$$

where $\Delta\omega = |\omega_B - \omega_L|$. When the axial magnetic field is strong enough to make

$$\frac{m_e c a_L}{eRB_{Sz}} \Big/ \left(1 + \frac{\Omega_z}{2\omega_L}\right)^2 \ll 1, \quad (7)$$

we have

$$\frac{\Delta\omega}{\omega_L} < \frac{a_L}{2\pi(R/\lambda)[(B_{Sz}/B_0)^2 + 2(B_{Sz}/B_0)]}, \quad (8)$$

where $B_0 = m_e\omega_0/e$, $\lambda = 2\pi c/\omega_0$, and $\gamma\kappa \sim 1$ is taken for the trapped electrons. The stronger the axial magnetic field B_{Sz}, the closer the right-hand side of Eq. (8) is to zero, i.e., $\Delta\omega/\omega_L \to 0$, which means that only exactly matched electrons can escape. Since the trapped electrons are driven by the rapidly varying laser fields, they have enough opportunity to adjust their frequencies and phases to match the resonance conditions. The trapping effect of the axial magnetic field and the resonance effect of the azimuthal magnetic field together develop a self-matching mechanism. In this situation, the trapped electrons form a "reservoir", which provides resonance electrons continually.

To investigate the features of SMRA, we carried out simulations using a fully relativistic three-dimensional (3D) particle-in-cell (PIC) code (KLAP) [19]. A right-hand CP laser beam with central wavelength $\lambda = 1\ \mu$m, wave period $T = \lambda/c$, rise time $5T$, peak intensity $I_0 = 5 \times 10^{20}$ W/cm^2, and spot diameter 4 μm (FWHM), is normally incident from the left boundary ($z = 0$) of a $80 \times 16 \times 16\ \mu$m^3 simulation box with a grid of $960 \times 192 \times 192$ cells. A near-critical density plasma target consisting of electrons and protons is located in $6\ \mu$m $< z < 77\ \mu$m. In the laser propagation direction, the plasma density rises linearly from 0 to $n_0 = n_c$ in a distance of 5 μm, and then remains constant, where $n_c = m_e\omega_0^2\epsilon_0/e^2$ is the critical plasma density, and ϵ_0 is the vacuum permittivity. In the radial direction, the density is uniform. The number of particles used in the simulation is 5×10^8 for each species (16 particles per cell for each species corresponds to n_0). An initial electron temperature T_e of 150 keV is used to resolve the initial Debye length ($T_i = 10$ eV initially).

Simulation results at $t = 80T$ are plotted in Fig. 1. The incident laser beam first propagates through an unstable filamentary stage and then collapses into a single channel, which is known as relativistic magnetic self-channeling [10], as is shown in Fig. 1(a) in the longitudinal cut. Both electrons and ions are expelled by the self-focused laser pulse. The unique feature here is that a high density electron beam is generated in the center of the laser channel, as is shown in Fig. 1(b), where the beam is labeled by a black dashed box. The radius of the beam is less than 1 μm, while the radius of the channel is about 3 μm. The maximum density of the beam is up to $3n_c$, which is far greater than the initial density of the plasma. Most of the energy of the electrons is localized in the beam, as is shown by the energy density distribution in Fig. 1(c). The points with the maximum energy density are distributed at the edge of the beam. The self-generated quasistatic azimuthal and axial magnetic fields are shown in Figs. 1(d) and 1(e). In the simulation, the self-focused laser intensity is close to its initial value; then we can use $a_L \sim 13.4$. The axial magnetic field $B_{Sz} \sim 1.4B_0$, with the radius of the spot size $R \sim 0.8$ μm, is strong enough to make the frequency of the betatron motion satisfy $\Delta\omega/\omega_L < 0.6$. Figure 1(f) plots the isosurface of the energy density with isosurface value $100n_c m_e c^2$ in 3D, which shows a helical structure with the same period as the laser pulse.

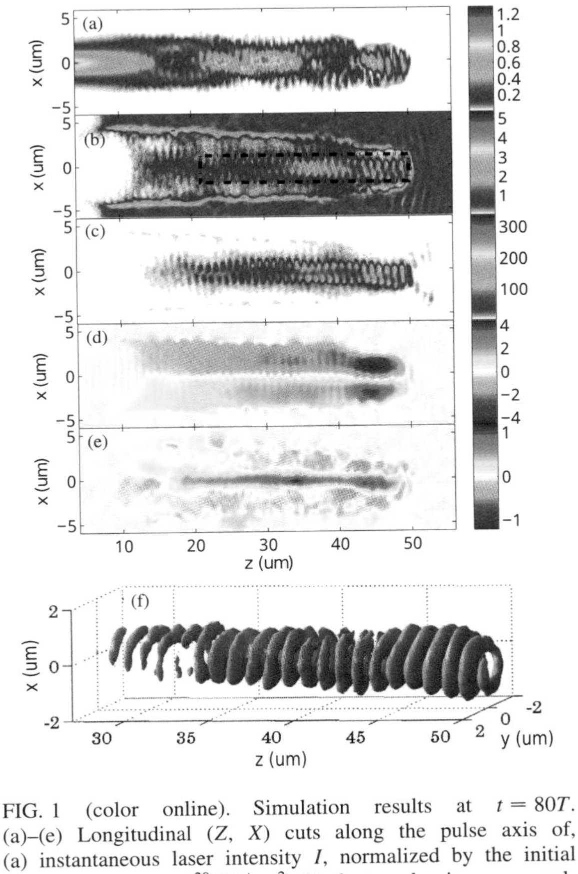

FIG. 1 (color online). Simulation results at $t = 80T$. (a)–(e) Longitudinal (Z, X) cuts along the pulse axis of, (a) instantaneous laser intensity I, normalized by the initial intensity $I_0 = 5 \times 10^{20}$ W/cm^2; (b) electron density n_e, normalized by the critical density n_c; (c) electron energy density normalized by $n_c m_e c^2$; (d), (e) self-generated azimuthal and axial magnetic fields $B_{S\theta}$ and B_{Sz}, averaged over laser period, normalized by $m_e\omega_0/e$. (f) 3D isosurface distribution of electron energy density with isosurface value $100n_c m_e c^2$.

In order to illustrate SMRA in detail the evolutions for a resonance electron of the perpendicular momentum $p_\perp$, the radius of motion r, the Lorentz factor γ, the perpendicular velocity $v_\perp$, and the frequencies ω_B and ω_L, are plotted in Figs. 2(a)–2(e). The electron is first dragged into the center of the channel from the channel boundary by the self-generated fields. In the center of the channel, the evolution can be divided into two processes, the trapping process (I), and the resonance process (II). In the trapping process, the electron, with a lower perpendicular momentum $p_\perp < p_{\rm tr}$, is trapped by the quasistatic axial magnetic field. There is no obvious energy gain when the electron is trapped. The frequency of the betatron motion ω_B and the frequency of the laser pulse in the electron comoving frame ω_L both vary rapidly. When the resonance condition is achieved, i.e., the frequencies and phases are matched, the evolution goes into the resonance process. In the resonance process, with the perpendicular momentum exceeding the trapping momentum, the electron is out of the axial magnetic field area, and into the azimuthal magnetic field dominant regime. The Lorentz factor increases dramatically until the electron catches up to the electrostatic sheath field in the front of the laser pulse at $90T$. The perpendicular velocity $v_\perp$ is suppressed to about $0.2c$. The matching of frequencies ω_B and ω_L, and the matching of the corresponding phases (the time integral of the frequency curve), are both maintained. After $100T$, the electron exceeds the front of the laser pulse in the propagation direction and moves freely. The corresponding transverse trajectory of the electron projected in the quasistatic axial magnetic field B_{Sz} is shown in Fig. 2(f). In the trapping process [marked by green line (bright)], the electron is trapped in the axial magnetic field. In the resonance process [marked by blue line (dark)], the electron is thrown out of the axial magnetic field, and then rotates around the axial magnetic field. All resonance electrons show similar dynamic properties, which can be explained by the SMRA model. In the resonance process, $\omega_B \approx \omega_L$, we have $p_\perp = m_e c a_L \omega_L t/2$, $p_x = -p_\perp \cos\omega_L t$, and $p_y = p_\perp \sin\omega_L t$. Note that $\omega_L t \approx -\phi$; i.e., the betatron motion is synchronous with the laser pulse, which ensures that the electron is always in the acceleration phase.

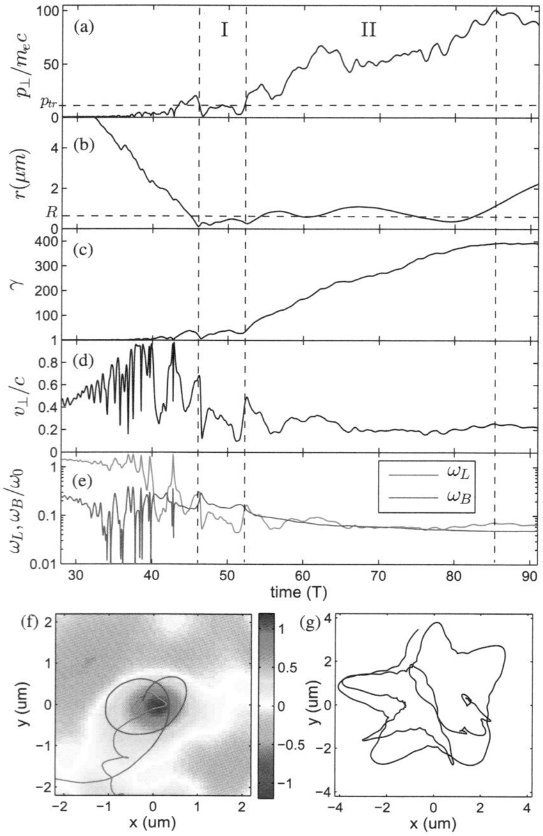

FIG. 2 (color online). Simulation results of sample electrons. (a)–(e) The evolutions, for a resonance electron, are shown of, (a) the perpendicular momentum $p_\perp$, normalized by m_ec; (b) the radius of motion r; (c) the Lorentz factor γ; (d) the perpendicular velocity $v_\perp$, normalized by c; (e) the frequencies ω_B and ω_L, normalized by ω_0. The three vertical dashed lines indicate the two processes: I, the trapping process, and II, the resonance process. (f) The corresponding transverse trajectory of the resonance electron projected in the quasistatic axial magnetic field B_{Sz}. The quasistatic axial magnetic field is cut at the position of the selected electron at $t = 50T$, normalized by $m_e\omega_0/e$. The two processes are marked by the green line (bright) and the blue line (dark), respectively. (g) The transverse trajectory of a nonresonance electron.

By taking p_x and p_y into the energy equation $m_ec^2d\gamma/dt = -eE_L(p_x\cos\phi + p_y\sin\phi)/\gamma m_e$, the Lorentz factor is determined as $\gamma = a_L\omega_0t\sqrt{\kappa/2} + 1$. At the beginning of the acceleration, $\gamma \approx 1$, the transverse motion is an Archimedes' spiral. The electron is thrown out of the axial magnetic field along a spiral trajectory. When $\gamma \gg 1$, the perpendicular velocity $v_\perp = p_\perp/\gamma m_e \to c\sqrt{\kappa/2}$, is approximately a constant, which means that the transverse motion is a circle. The acceleration mechanism of the resonance electron in the self-generated azimuthal magnetic field is very similar to that of the original inverse free-electron-laser accelerator by means of circularly polarized electromagnetic waves in static helical magnet field [20]. Since the axial magnetic field is generated by the transverse rotating electrons, the radius of the circle is equal to the radius of the spot size of the axial magnetic field, thus $R = v_\perp/\omega_B \to \lambda/2\pi\sqrt{2\kappa}$, and the self-generated axial magnetic field is estimated as $B_z = R\mu_0en_ev_\perp = n_eB_0/2n_c$, where n_e is the density of the electron beam. The transverse trajectory of a nonresonance electron is shown in Fig. 2(g). For the nonresonance electron, when it passes through the center of the channel, the perpendicular momentum is too large to allow the electron to be trapped by the axial magnetic field. The amplitude of the transverse motion of the nonresonance electron is about 3 μm, which is almost the radius of the channel.

Figure 3(a) shows the time evolution of the maximum electron energy, which increases dramatically soon after the laser irradiates on the plasma and reaches a saturation value 240 MeV after $100T$. The energy spectrum of the electrons at $t = 100T$ shows a rather plateau profile energy spectrum, as is shown in Fig. 3(b). The angular distribution of electron energy is plotted in Fig. 3(c), which shows that the divergence angle (full angle) of the high energy electrons is about 0.15 rad and the peak of the angle distribution is at 0.2 rad, suggesting that the output electron beam is highly collimated and the accelerated electrons are executing collective betatron motion with perpendicular velocity $\langle v_\perp\rangle \sim 0.2c$. The SMRA model predicts a divergence angle (full angle) of the accelerated electrons, $\Delta\theta = (\langle v_\perp\rangle/c)(\Delta\omega/\omega_L) \sim 0.12$ rad, which agrees with the

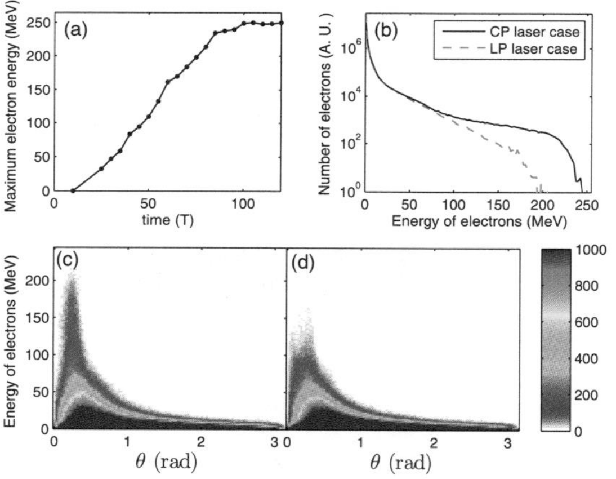

FIG. 3 (color online). (a) Time evolution of the maximum electron energy. (b) Energy spectrum of electrons at $t = 100T$ for the CP laser pulse (black solid line) and the LP laser pulse (red dashed line) with the same laser intensity. (c), (d) Angular distribution of electron energy at $t = 100T$ for the CP laser pulse and the LP laser pulse with the same laser intensity, respectively.

simulation results. In order to highlight the advantages of the SMRA regime, we compared the case by using a LP laser pulse with the same laser intensity where the quasi-static axial magnetic field is absent. The energy spectrum of electrons for the LP laser case is shown in Fig. 3(b) by a red dashed line, which shows a thermal-like distribution, as is mentioned in Ref. [6]. The angular distribution of electron energy for the LP laser case is shown in Fig. 3(d), which shows a larger divergence angle and a lower energy peak.

In summary, the SMRA regime has been proposed with the aid of analytical modeling and 3D-PIC simulations. In combination of the self-generated magnetic fields (both in axial and azimuthal direction) and the laser pulse, relativistic electrons can experience two processes: in the trapping process, the electron betatron frequencies and phases are adjusted automatically to match the resonance conditions; in the resonance process, the matched electrons are accelerated continuously along the laser propagation direction. All resonance electrons show similar dynamic behavior, which can be explained by the SMRA model. This SMRA regime, in which a collimated relativistic electron beam with overcritical density, helical structure, and plateau profile energy spectrum, can be generated, might be very promising for x-ray production or ion acceleration.

This work was supported by National Natural Science Foundation of China (Grants No. 11025523, No. 10974022, No. 10935002, No. 11105009, No. 11205010, No. J1103206, No. 10905004, and No. 10835003) and MOST Grand Instrument project (2012YQ030142). The authors would like to acknowledge fruitful discussions with B. Qiao.

*X.Yan@pku.edu.cn
†xthe@iapcm.ac.cn

[1] G. A. Mourou, T. Tajima, and S. Bulanov, Rev. Mod. Phys. **78**, 309 (2006).

[2] M. Tabak, J. Hammer, M.E. Glinsky, W.L. Kruer, S.C. Wilks, J. Woodworth, E.M. Campbell, M.D. Perry, and R.J. Mason, Phys. Plasmas **1**, 1626 (1994); M. Roth *et al.*, Phys. Rev. Lett. **86**, 436 (2001); V. Bychenkov, W. Rozmus, A. Maksimchuk, D. Umstadter, and C.E. Capjack, Plasma Phys. Rep. **27**, 1017 (2001); S. Atzeni, M. Temporal, and J.J. Honrubia, Nucl. Fusion **42**, L1 (2002).

[3] B.A. Remington, D. Arnett, R. Paul Drake, and H. Takabe, Science **284**, 1488 (1999); S.V. Bulanov, T.Zh. Esirkepov, D. Habs, F. Pegoraro, and T. Tajima, Eur. Phys. J. D **55**, 483 (2009).

[4] E. Esarey, C. Schroeder, and W. Leemans, Rev. Mod. Phys. **81**, 1229 (2009); T. Tajima and J.M. Dawson, Phys. Rev. Lett. **43**, 267 (1979); E.L. Clark *et al.*, *ibid.* **84**, 670 (2000); A. Maksimchuk, S. Gu, K. Flippo, D. Umstadter, and V.Y. Bychenkov, *ibid.* **84**, 4108 (2000); A. McPherson, B.D. Thompson, A.B. Borisov, K. Boyer, and C.K. Rhodes, Nature (London) **370**, 631 (1994); S.P. Hatchett *et al.*, Phys. Plasmas **7**, 2076 (2000).

[5] S. Cipiccia *et al.*, Nat. Phys. **7**, 867 (2011).

[6] A. Pukhov and J. Meyer-ter-Vehn, Phys. Plasmas **5**, 1880 (1998); A. Pukhov, Z.-M. Sheng, and J. Meyer-ter-Vehn, *ibid.* **6**, 2847 (1999); C. Gahn, G. Tsakiris, A. Pukhov, J. Meyer-ter-Vehn, G. Pretzler, P. Thirolf, D. Habs, and K. Witte, Phys. Rev. Lett. **83**, 4772 (1999).

[7] G.D. Tsakiris, C. Gahn, and V.K. Tripathi, Phys. Plasmas **7**, 3017 (2000); D.N. Gupta and C.-M. Ryu, *ibid.* **12**, 053103 (2005); C.S. Liu and V.K. Tripathi, *ibid.* **12**, 043103 (2005); D.N. Gupta and H. Suk, *ibid.* **13**, 013105 (2006).

[8] D.H. Whittum, A.M. Sessler, and J.M. Dawson, Phys. Rev. Lett. **64**, 2511 (1990).

[9] See, e.g., S.P.D. Mangles *et al.*, Phys. Rev. Lett. **94**, 245001 (2005).

[10] A. Pukhov and J. Meyer-ter-Vehn, Phys. Rev. Lett. **76**, 3975 (1996).

[11] T. Nakamura and K. Mima, Phys. Rev. Lett. **100**, 205006 (2008).

[12] Y. Fukuda *et al.*, Phys. Rev. Lett. **103**, 165002 (2009); T. Nakamura, S.V. Bulanov, T.Z. Esirkepov, and M. Kando, *ibid.* **105**, 135002 (2010); S.S. Bulanov *et al.*, Phys. Plasmas **17**, 043105 (2010).

[13] H.Y. Wang, C. Lin, Z. Sheng, B. Liu, S. Zhao, Z. Guo, Y. Lu, X. He, J. Chen, and X. Yan, Phys. Rev. Lett. **107**, 265002 (2011).

[14] Z. Najmudin *et al.*, Phys. Rev. Lett. **87**, 215004 (2001); A. Kim, M. Tushentsov, D. Anderson, and M. Lisak, *ibid.* **89**, 095003 (2002); A. Frolov, Plasma Phys. Rep. **30**, 698 (2004); I.Y. Kostyukov, G. Shvets, N.J. Fisch, and J.M. Rax, Phys. Plasmas **9**, 636 (2002); S. Zhu, C.Y. Zheng, and X.T. He, *ibid.* **10**, 4166 (2003); B. Qiao, S.-p. Zhu, C.Y. Zheng, and X.T. He, *ibid.* **12**, 053104 (2005); C.Y. Zheng, X.T. He, and S.P. Zhu, *ibid.* **12**, 044505 (2005); N. Naseri, V.Y. Bychenkov, and W. Rozmus, *ibid.* **17**, 083109 (2010).

[15] H. Liu, X.T. He, and S.G. Chen, Phys. Rev. E **69**, 066409 (2004); A. Sharma and V.K. Tripathi, Phys. Plasmas **12**, 093109 (2005); B. Qiao, X.T. He, S.-p. Zhu, and C.Y. Zheng, *ibid.* **12**, 083102 (2005); H.Y. Niu, X.T. He, C.T. Zhou, and B. Qiao, Phys. Plasmas **16**, 013104 (2009).

[16] H.Y. Niu, X.T. He, B. Qiao, and C.T. Zhou, Laser Part. Beams **26**, 51 (2008).

[17] See, e.g., K. Németh, B. Shen, Y. Li, H. Shang, R. Crowell, K. Harkay, and J. Cary, Phys. Rev. Lett. **100**, 095002 (2008); A.V. Arefiev, B.N. Breizman, M. Schollmeier, and V.N. Khudik, *ibid.* **108**, 145004 (2012).

[18] L. Yin, B. Albright, K. Bowers, D. Jung, J. Fernández, and B. Hegelich, Phys. Rev. Lett. **107**, 045003 (2011); L. Yin, B.J. Albright, D. Jung, R.C. Shah, S. Palaniyappan, K.J. Bowers, A. Henig, J.C. Fernndez, and B.M. Hegelich, Phys. Plasmas **18**, 063103 (2011).

[19] X.Q. Yan, C. Lin, Z. Sheng, Z. Guo, B. Liu, Y. Lu, J. Fang, and J. Chen, Phys. Rev. Lett. **100**, 135003 (2008); Z. Sheng, K. Mima, J. Zhang, and H. Sanuki, *ibid.* **94**, 095003 (2005).

[20] R.B. Palmer, J. Appl. Phys. **43**, 3014 (1972).

www.nature.com/scientificreports

SCIENTIFIC REPORTS

OPEN

Received: 20 January 2015
Accepted: 21 September 2015
Published: 27 October 2015

Dense Helical Electron Bunch Generation in Near-Critical Density Plasmas with Ultrarelativistic Laser Intensities

Ronghao Hu[1,*], Bin Liu[1,2,*], Haiyang Lu[1], Meilin Zhou[1], Chen Lin[1], Zhengming Sheng[3], Chia-erh Chen[1], Xiantu He[1,2] & Xueqing Yan[1]

The mechanism for emergence of helical electron bunch(HEB) from an ultrarelativistic circularly polarized laser pulse propagating in near-critical density(NCD) plasma is investigated. Self-consistent three-dimensional(3D) Particle-in-Cell(PIC) simulations are performed to model all aspects of the laser plasma interaction including laser pulse evolution, electron and ion motions. At a laser intensity of 10^{22} W/cm^2, the accelerated electrons have a broadband spectrum ranging from 300 MeV to 1.3 GeV, with the charge of 22 nano-Coulombs(nC) within a solid-angle of 0.14 Sr. Based on the simulation results, a phase-space dynamics model is developed to explain the helical density structure and the broadband energy spectrum.

High energy electron beams have broad applications in research fields like high energy physics, Inertial Confined Fusion Fast Ignition scheme[1], radiography[2], synchrotron radiation[3–7], etc. Laser plasma accelerators[8–10], first proposed by T. Tajima and J. M. Dawson[11], have draw much attension due to its huge acceleration gradients, typically on the order of tens to hundreds of GV/m. Remarkable advances have been made both theoretically and experimentally. Two significant acceleration regimes have been proposed, known as the laser wake-field acceleration(LWFA)[11–17] and the direct laser acceleration(DLA)[18,19].

LWFA regime has been studied thoroughly in theories[8,11,13,14], simulations[12,14,15,20] and experiments[16,17,21–24]. It is promising for generating high energy monoenergetic electron beams with the energies upto several GeVs[16,17,21–24]. However, dephasing, or the phase slipping between the ultrarelativistic electrons and the subluminal acceleration fields, limits the energy gain of LWFA. To achieve a longer acceleration length before dephasing[22], low density plasmas, typically with $n_e < 10^{20}\,\mathrm{cm}^{-3}$, were used in LWFA. As a consequence, the total charges of accelerated bunches are limited under 10 nC[20,25–27].

At higher plasma densities, the DLA mechanism makes a significant contribution or even become the dominant to accelerate electrons[20,28]. In DLA regime, only a small fraction of electrons can be accelerated to relatively high energy[2,18,19,29], which strongly limits its potential applications. To obtain more energetic electrons within a single laser shot, higher laser intensity and higher plasma density should be employed. Higher laser intensity can be expected as more powerful femtosecond laser could be available in the near future[30–32]. Higher density plasmas, like near-critical density(NCD) plasmas, can be deployed for electron acceleration[33].

The density modulation of electrons can be seen as a sign of DLA[18]. HEB was observed with PIC simulations in ref. 33 but the mechanism of its emergence lacks proper physical explanation. Besides, the performance of DLA with laser intensity over 10^{21} W/cm^2 has not been studied systematically. At

[1]State Key Laboratory of Nuclear Physics and Technology, and Key Laboratory of HEDP of the Ministry of Education, CAPT, Peking University, Beijing, 100871, China. [2]Institute of Applied Physics and Computational Mathematics, Beijing, 100088, China. [3]Department of Physics, Shanghai Jiao Tong University, Shanghai, 200240, China. [*]These authors contributed equally to this work. Correspondence and requests for materials should be addressed to H.L. (email: hylu@pku.edu.cn) or X.Y. (email: x.yan@pku.edu.cn)

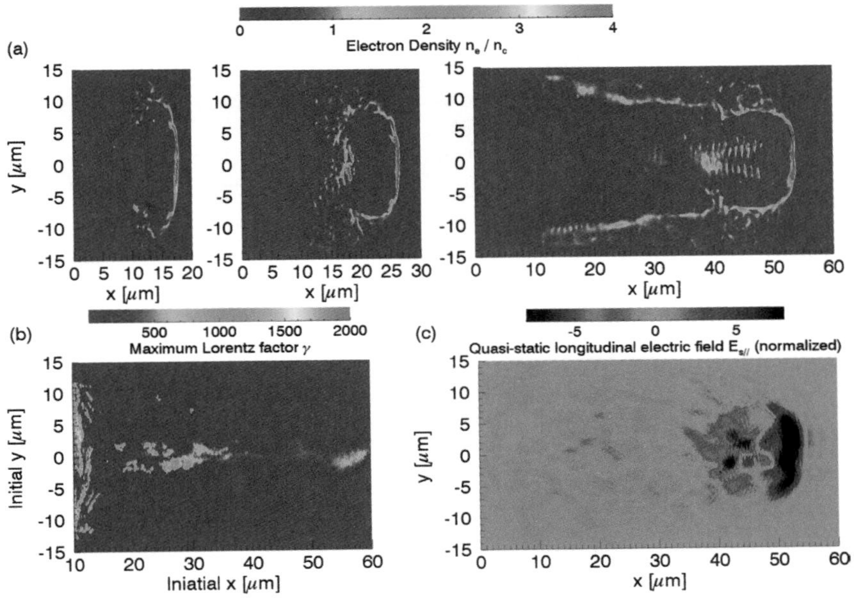

Figure 1. (**a**) Electron density at $t = 20\,T_L$, $30\,T_L$ and $60\,T_L$. (**b**) The maximum Lorentz factor γ for electrons from different intial positions. (**c**) The longitudinal electric field $E_{S\|}$ at $t = 60\,T_L$.

relatively low laser intensities, the energy spectrums of electrons were Maxwellian-like with the "effective temperature" grows as the square root of the intensity[18,19]. When only the DLA electrons were selected from the heated background plasma in the PIC simulations, the energy spectrum is broadband which contains more information of the DLA performance[34]. In this paper, a theoretical model of electron phase-space dynamics was presented to explain the mechanism for emergence of HEB and PIC simulations were performed in a slab of NCD plasma interacting with a circularly polarized laser pulse, which was focused to a peak intensity of $10^{22}\,\mathrm{W/cm^2}$ (ref. 44).

Simulation and Results

The PIC simulation was performed with 3D KLAP codes[35–37]. The circularly polarized laser pulse is incident from the left boundary of the simulation box. The simulation box is sampled by 1000 cells in light propagation direction and 240 cells in each transverse direction, corresponding to a volume of $80\,\mu\mathrm{m} \times 30\,\mu\mathrm{m} \times 30\,\mu\mathrm{m}$ in real space. The incident laser pulse, with a wavelength of $\lambda_L = 1.0\,\mu\mathrm{m}$, is focused to a spot diameter of $3.6\,\mu\mathrm{m}$ in full width at half maximum(FWHM) with a Gaussian transverse field profile, which results in a peak intensity of about $10^{22}\,\mathrm{W/cm^2}$. The focus is inside the simulation volume $40\,\mu\mathrm{m}$ away from the left boundary. The laser pulse has a trapezoidal envelope in time domain, with linear rising and falling edges of $5\,T_L$, where T_L is the light period, and a total duration of $45\,T_L$ in FWHM. The target is located between $10\,\mu\mathrm{m}$ to $60\,\mu\mathrm{m}$ from the left boundary, with an uniform electron density of $5.5 \times 10^{20}\,\mathrm{cm^{-3}}$, about half of the critical density for light with a wavelength of $1\,\mu\mathrm{m}$. Two species of particles are included in the simulations, electrons and protons, with about 2.88×10^8 macro-particles for each specie. The simulation can be divided into injection and acceleration stages. The injection stage is before $30\,T_L$. A bubble-like structure was formed near the target surface, shown in Fig. 1(a), after the pulse front hits the target surface. Electrons were expeled from light axis by the transverse ponderomotive force, but the ions were barely moved, thus the transverse quasistatic electric field was formed due to the charge separation. The injection process, similar to the self-injection process of the LWFA[12,13,38,39], is not continuous as most of the accelerated electrons are from the front surface of the target, shown in Fig. 1(b). An over-critical density electron bunch was formed at the rear of the bubble, shown in Fig. 1(a), and the rest of electrons were prevented from injection by the strong radial electric field generated by the bunch. The bubble-like structure evolves into a plasma channel in the following acceleration stage. The electrons gain energy through the strong light field and longitudinal quasi-static electric field inside the channel, shown in Fig. 1(c). Along with the acceleration, the electron bunch is modulated to a helix by the light field with the thread pitch roughly equal to the light wavelength, as shown in Fig. 2(a). After exiting the target, the helical electron bunch will spread with a cone angle of $14° \pm 1.7°$, shown in Fig. 2(b). The output energy spectrum, which is broadband and roughly from 300 MeV to 1.3 GeV, is very different from the Maxwellian-like spectrums of previous work of DLA[2,18,19,29]. Particle movement tracking were performed to have an insight view of the behavior of the high energy electrons in the acceleration stage. It could be found that the radii of the electrons are varying slowly in the acceleration stage and finally close to each other when exit from the

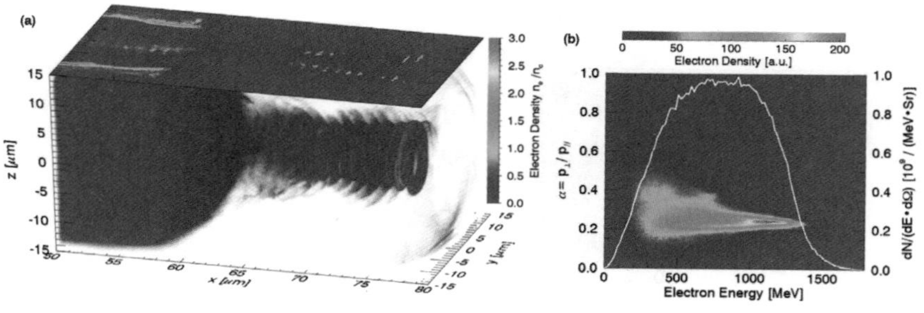

Figure 2. (**a**) The volume plot and contour plot on the $z=0$ plane of the electron density at $t=90\,T_L$. (**b**) The electon density distribution in phase-space ($\gamma, \alpha = p_\perp/p_\parallel$) at the same time. The electrons are selected from the the cylindrical volume from $x=62\,\mu m$ to $80\,\mu m$, with a radius of $6\,\mu m$. The white line is the energy spectrum with $0.2<\alpha<0.3$. The spectrum should be similar to that of a detector placed at an angle of $14° \pm 1.7°$ to the laser propagation direction.

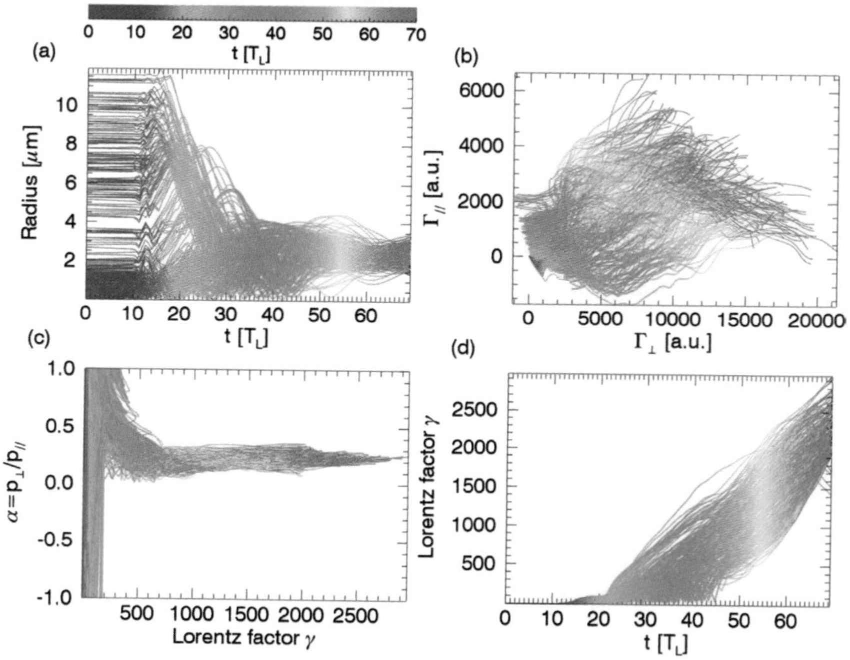

Figure 3. The trajectories of the PIC sample electrons in phase-space (a) (t, r), (b) (γ, t), (c) $(\alpha = p_\perp/p_\parallel, \gamma)$, (d) $(\Gamma_\perp, \Gamma_\parallel)$ from $t=0\,T_L$ to $70\,T_L$.

acceleration stage even they are initially different as shown in Fig. 3(a). The trajectories of electrons in phase-space ($\Gamma_\parallel = -e\int \mathbf{v}_\parallel \cdot \mathbf{E}_\parallel dt, \Gamma_\perp = -e\int \mathbf{v}_\perp \cdot \mathbf{E}_\perp dt$) ($\parallel$ indicates the direction of light propagation, $\perp$ indicates the directions perpendicular to the light propagation, **v** is the velocity of electron and **E** is the electric field), shown in Fig. 3(b), indicate that electrons gain energies in the longitudinal and the transverse directions, but the major part is in the transverse directions, which indicates that DLA is dominant mechanism during the acceleration[19,29]. Figure 3(c) shows that the ratios of momentums of the transverse against the longitudinal are almost contants in the acceleration stage, which is coincident with Fig. 2(b). Figure 3(d) shows that the energies of electrons increase almost linearly with time and the maximum acceleration gradient is over 20 TeV/m.

Electron Phase-space Dynamics

The behavior of electrons in the phase-space (ψ, γ) is the key to explain the HEB generation and acceleration mechanism. Here ψ is the ponderomotive phase and γ is the Lorentz factor of electrons. ψ can be seen as the relative angle of the light electric field vector and the electron transverse momentum, i.e. $\psi = \theta - \phi$, where θ and ϕ are the polar angles of the electron transverse momentum and light electric vector respectively. γ is the Lorentz factor of electrons. Electron motions inside the ion channel are already well-understood from the previous work[15,18,29,33]. Electrons inside the ion channel are trapped by the strong self-generated quasi-static electric and magnetic fields[18,33,40], i. e., the radial electric field $\mathbf{E}_{Sr}$, the azimuthal magnetic field $\mathbf{B}_{S\theta}$ and the longitudinal magnetic field $\mathbf{B}_{Sz}$. The trapped electrons undergo betatron-like oscillations in the self-generated fields, the transverse Lorentz equation can be written as

$$\frac{d\gamma m_e \mathbf{v}_\perp}{dt} = -e[\mathbf{E}_L + \mathbf{E}_{Sr} + \mathbf{v}_\parallel \times (\mathbf{B}_L + \mathbf{B}_{Sr}) + \mathbf{v}_\perp \times \mathbf{B}_{Sz}], \tag{1}$$

where m_e is the electron mass at rest, $\mathbf{E}_L$ and $\mathbf{B}_L$ are the light electromagnetic fields. In the SI units, $E_L = v_{ph} \cdot B_L$, where v_{ph} is the phase velocity of light in plasma and is larger than the light velocity in vacuum c[18]. As the electrons co-propagate with the laser pulse, the $\mathbf{v}_\parallel \times \mathbf{B}_L$ force is antiparallel to the electric force $\mathbf{E}_L$ and the total Lorentz force of the light field, $\mathbf{F}_L = -e(\mathbf{E}_L + \mathbf{v}_L \times \mathbf{B}_L)$, is small compared to the Lorentz force of the quasi-static fields. The PIC simulation results show that the term $\mathbf{v}_\perp \times \mathbf{B}_{Sz}$ can be neglected comparing to other terms. Based on simulations and previous work[18,33,40], the radial electric field and the azimuthal magnetic field profile near the light propagation axis can be written as $\mathbf{E}_{Sr} = k_E r \hat{e}_r$, $\mathbf{B}_{S\theta} = -k_B r \hat{e}_\theta$, where k_E and k_B are constants, $\hat{e}_r$ and $\hat{e}_\theta$ are the unit vectors of the radial and azimuthal directions respectively. Although the projected transverse motion is elliptical, as shown in Fig. 3(a), one can always use a simple circular motion to approximate. Assuming the electrons undergo circular motions with fixed radii, the angular frequency of the circular motion, i.e. the betatron frequency then can be derived from Eq. (1) as[18,29,41]

$$\frac{d\theta}{dt} = \omega_\beta = \sqrt{\frac{e}{\gamma m_e}(v_\parallel k_B + k_E)}\,. \tag{2}$$

The light frequency witnessed by the electrons is $\frac{d\phi}{dt} = \omega_L = (1 - v_\parallel / v_{ph})\omega_0$, where ω_0 is the light frequency in laboratory frame. The time derivative of the ponderomotive phase can be written as

$$\frac{d\psi}{dt} = \omega_\beta - \omega_L = \sqrt{\frac{e}{\gamma m_e}(v_\parallel k_B + k_E)} - (1 - v_\parallel / v_{ph})\omega_0, \tag{3}$$

The time derivative of the electron energy is

$$m_e c^2 \frac{d\gamma}{dt} = -e\mathbf{v} \cdot \mathbf{E} = -ev_\perp E_L \cos\psi - ev_\parallel E_{S\parallel}. \tag{4}$$

Here E_{Sr} is ignored as the radii of electrons vary slowly in the acceleration stage as shown in Fig. 3(a). Also according to PIC simulations, in the acceleration stage, $v_\perp$, $v_\parallel$, k_B, k_E, and $E_{S\parallel}$ can be treated as constants as all of them vary slowly enough. A Hamiltonian can be obtained from Eqs. (3) and (4) as the phase-space spanned by (ψ, γ) is conserved.

$$\mathscr{H} = \frac{e}{m_e c^2} v_\perp E_L \sin\psi + \frac{e}{m_e c^2} v_\parallel E_{S\parallel}\psi + 2\sqrt{\frac{e}{m_e}(v_\parallel k_B + k_E)\gamma} - (1 - v_\parallel / v_{ph})\omega_0 \gamma \tag{5}$$

For convenience, ψ is redefined as the remainder of ψ divided by 2π, noted as $\left(\psi + \frac{\pi}{2}\right) mod\,(2\pi)$ in the following discussion. The ponderomotive phase indicates the direction of energy exchange between electrons and the light field. The electrons decelerate when $-\frac{\pi}{2} < \psi < \frac{\pi}{2}$, and accelerate when $\frac{\pi}{2} < \psi < \frac{3\pi}{2}$. When the longitudinal electric field is ignored in Eq. (5), the Hamiltonian $\mathscr{H}$ is symmetrical and periodic, and there is a separatrix and a fixed point at phase $\psi_c = \frac{\pi}{2}$ and Lorentz factor $\gamma_c = \frac{e(v_\parallel k_B + k_E)}{m_e(1 - v_\parallel / v_{ph})^2 \omega_0^2}$ in the phase-space, as shown in Fig. 4(a). The orbits of electrons, initially located between the separatrix and the fixed point, are closed and the ponderomotive phases of these electrons grow and then reduce slowly, so they can stay in the acceleration phase longer enough and get effectively accelerated. The electrons below the separatrix can never be accelerated because their orbits are open and their phases grow so rapidly that they can not stay in the acceleration phase long enough. With the longitudinal electric field included, the electrons below the separatrix are first accelerated by the longitudinal field, and when they are above the separatrix, they experience slower phase movements and gain energy in the acceleration phase, shown in Figs. 3(d), 4(b) and 4(c). In the theoretical model, the fixed point mentioned above is a center node, which means electrons can not get close to it. But in PIC simulations, some electrons move around the center similar as in the theoretical model while some others get close to this fixed point and are trapped nearby, causing the electron density around this fixed point

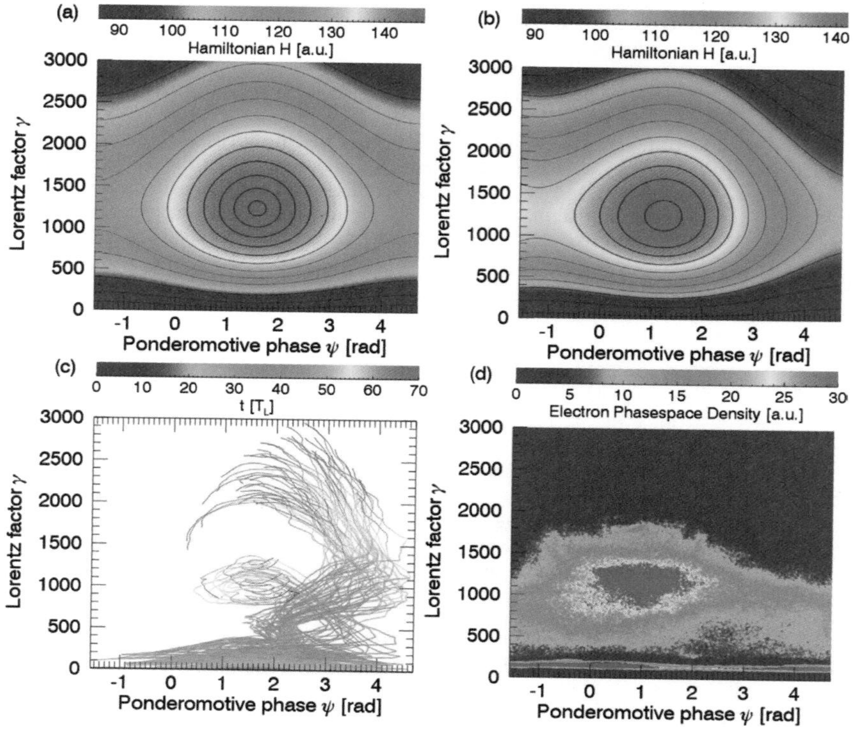

Figure 4. (**a**) Hamiltonian and trajectories in phase-space (ψ, γ), with longitudinal electric field ignored. (**b**) Same as (**a**), but with longitudinal electric field included. Although the phase-space is not periodic, the trajectries are mannually arranged into one period as ψ is redefined as the remainder $\left(\psi + \frac{\pi}{2}\right) mod\,(2\pi)$. (**c**) Trajectories of the PIC sample electrons. (**d**) Electron density distribution in phase-space, selected from the volume from $x = 44\,\mu m$ to $49\,\mu m$, within the radius of $5\,\mu m$ at $t = 60\,T_L$.

to be larger than that elsewhere, as shown in Figs. 4(c) and 4(d). This is because the parameters, which are assumed to be constants in the theoretical model, may be varying slowly in time and space, which leads to the crossing of phase trajectories and converting of the fixed point property. Moreover, the special phase-space distribution, shown in Fig. 4(d), is revealed by the helical density structure of the electron beam in real space, shown in Fig. 1(a). As most of electrons are located near the fixed point, their ponderomotive phases and radii are close to each other. The electric vector of circularly polarized light is rotating in the propagation direction, so the electrons, with similar relative polar angles to the electric vector and similar radii, form a helix with the thread pitch roughly equal to the light wavelength. Figure 4(d) also indicates that the central energy of accelerated electron beam is given by the fixed point, which is roughly $\gamma_c m_e c^2$. The constants k_E and k_B is relative to the plasma density and can be expressed as $k_E = ck_B = \frac{en_e}{4\varepsilon_0}$[39], where ε_0 is the vacuum permittivity. The phase velocity of circularly polarized long duration laser pulse can be approximated as $v_{ph} = c/\sqrt{1 - \omega_p^2/(\omega_0^2 \gamma_\perp)} = c/\sqrt{1 - n_e/(n_c\sqrt{1 + a_0^2})} \simeq c/\sqrt{1 - n_e/(n_c a_0)}$[42], where ω_p is the plasma frequency, $\gamma_\perp = \sqrt{1 + a_0^2}$ is the average transverse Lorentz factor for circularly polarized laser, n_c is the critical density, a_0 is the normalized laser amplitude. γ_c can be expressed as

$$\gamma_c = \frac{(v_\parallel/c) \cdot (n_e/n_c)}{4[1 - (v_\parallel/c)\sqrt{1 - n_e/(n_c a_0)}\,]^2} \tag{6}$$

From Eq. (6), with the increase of laser intensity and plasma density, the corresponding energy of the fixed point is increasing. In the previous work[18,29], the laser intensity and plasma density are small, the low energy fixed point is covered by the Maxwellian-like spectrum of the heated plasma. With the increasing of the plasma density and laser intensity, γ_c is also increasing and the broadband spectrum like in Fig. 2(b) can be observed.

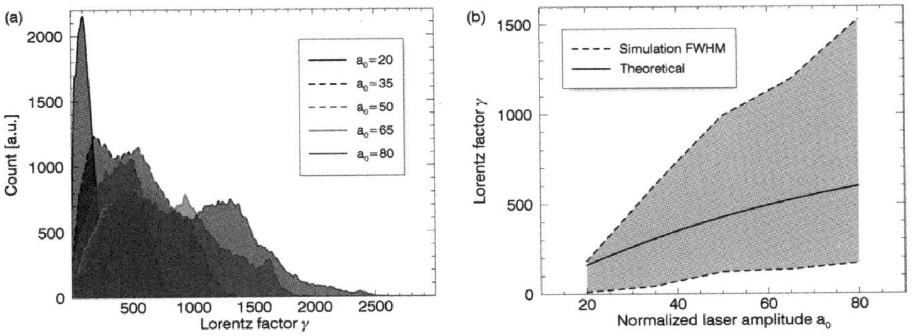

Figure 5. (**a**) Energy spectrums of DLA electrons with different laser amplitudes. (**b**) The FWHM of the energy spectrums in (**a**) and the Lorentz factor of the theoretical fixed point obtained with Eq. (6).

To verify the acceleration mechanism, a set of 2D PIC simulations were performed with linearly polarized laser pulses. The results are in agreement with the circularly polarized case except the electron bunches are planar rather than helical. The spectrums for different laser amplitudes, shown in Fig. 5(a), are also broadband and the bandwidths grow almost linearly with the increasing of the laser amplitudes as shown in Fig. 5(b). More energetic electrons will be obtained with higher laser intensities.

Discussion

In this paper the mechanism for emergence of HEB from circularly polarized laser pulse propagating in NCD plasma is explained with the electron phase-space dynamics model. At a laser intensity of $10^{22}\,\mathrm{W/cm^2}$, the generated HEB has a broadband spectrum ranging from 300 MeV to 1.3 GeV with a charge of 21.6 nC. With the increase of laser peak intensity, the bandwidths of the energy spectrum would grow and more energetic electrons would be obtained. Electron bunches with helical density structures are promising for generating synchrotron radiations with angular momentums[43].

Methods

The 2D PIC simulations are performed with fixed plasma density of $0.5\,n_c$, and target length of $60\,\mu$m. The laser pulses have the same duration of about 300 femtoseconds and waist radius of $3\,\mu$m. The energy spectrums are for electrons selected from the area from $10\,\mu$m to $20\,\mu$m behind the target within a radius of $5\,\mu$m and a spreading angle of $\pm 25°$.

Figures 4(a,b) are obtained with the parameters as follows: $E_L = 60.0$, $E_{S\parallel} = 0.0$ for (*a*) and -4.0 for (*b*), $v_\parallel = 0.9806$, $v_\perp = 0.1961$, $v_{ph} = 1.101$, $k_E = 5.0$, $k_B = 10.0$. All of the parameters are normalized as in ref. 18.

The theoretical curve in Fig. 5(b) are obtained with $v_\parallel = 0.99c$ and $n_e = 0.5n_c$. With $\gamma_\perp = \sqrt{1 + \frac{a_0^2}{2}}$ for linearly polarized laser pulses, Eq. (6) is modified as $\gamma_c = \frac{(v_\parallel / c) \cdot (n_e / n_c)}{4[1 - (v_\parallel / c)\sqrt{1 - (\sqrt{2}\,n_e) / (n_c a_0)}\,]^2}$.

References

1. Pukhov, A. & Meyer-ter Vehn, J. Laser hole boring into overdense plasma and relativistic electron currents for fast ignition of icf targets. *Phys. Rev. Lett.* **79,** 2686–2689 (1997).
2. Mangles, S. P. D. *et al.* Table-top laser-plasma acceleration as an electron radiography source. *Laser Part. Beams* **24,** 185–190 (2006).
3. Schlenvoigt, H. P. *et al.* A compact synchrotron radiation source driven by a laser-plasma wakefield accelerator. *Nat. Phys.* **4,** 130–133 (2007).
4. Fuchs, M. *et al.* Laser-driven soft-x-ray undulator source. *Nat. Phys.* **5,** 826–829 (2009).
5. Kneip, S. *et al.* Bright spatially coherent synchrotron x-rays from a table-top source. *Nat. Phys.* **6,** 980–983 (2010).
6. Cipiccia, S. *et al.* Gamma-rays from harmonically resonant betatron oscillations in a plasma wake. *Nat. Phys.* **7,** 867–871 (2011).
7. Phuoc, K. T. *et al.* All-optical compton gamma-ray source. *Nat. Photon.* **6,** 308–311 (2012).
8. Esarey, E., Schroeder, C. B. & Leemans, W. P. Physics of laser-driven plasma-based electron accelerators. *Rev. Mod. Phys.* **81,** 1229–1285 (2009).
9. Malka, V. *et al.* Principles and applications of compact laser–plasma accelerators. *Nat. Phys.* **4,** 447–453 (2008).
10. Hooker, S. M. Developments in laser-driven plasma accelerators. *Nat. Photon.* **7,** 775–782 (2013).
11. Tajima, T. & Dawson, J. M. Laser electron accelerator. *Phys. Rev. Lett.* **43,** 267–270 (1979).
12. Kalmykov, S. Y. *et al.* Electron self-injection into an evolving plasma bubble: quasi-monoenergetic laser-plasma acceleration in the blowout regime. *Phys. Plasmas* **18,** 056704-1–056704-9 (2011).
13. Esirkepov, T., Bulanov, S. V., Yamagiwa, M. & Tajima, T. Electron, positron, and photon wakefield acceleration: trapping, wake overtaking, and ponderomotive acceleration. *Phys. Rev. Lett.* **96,** 014803-1–014803-4 (2006).
14. Lu, W., Huang, C., Zhou, M., Mori, W. B. & Katsouleas, T. Nonlinear theory for relativistic plasma wakefields in the blowout regime. *Phys. Rev. Lett.* **96,** 165002-1–165002-4 (2006).
15. Németh, K. *et al.* Laser-driven coherent betatron oscillation in a laser-wakefield cavity. *Phys. Rev. Lett.* **100,** 095002-1–095002-4 (2008).
16. Leemans, W. P. *et al.* GeV electron beams from a centimetre-scale accelerator. *Nat. Phys.* **2,** 696–699 (2006).

17. Lu, W. *et al.* Generating multi-GeV electron bunches using single stage laser wakefield acceleration in a 3d nonlinear regime. *Phys. Rev. ST Accel. Beams* **10,** 061301-1–061301-12 (2007).
18. Pukhov, A., Sheng, Z. M. & Meyer-ter Vehn, J. Particle acceleration in relativistic laser channels. *Phys. Plasmas* **6,** 2847–2854 (1999).
19. Gahn, C. *et al.* Multi-MeV electron beam generation by direct laser acceleration in high-density plasma channels. *Phys. Rev. Lett.* **83,** 4772–4775 (1999).
20. Pukhov, A. & Meyer-ter Vehn, J. Laser wake field acceleration: the highly non-linear broken-wave regime. *Appl. Phys. B* **74,** 355–361 (2002).
21. Leemans, W. *et al.* Multi-GeV electron beams from capillary-discharge-guided subpetawatt laser pulses in the self-trapping regime. *Phys. Rev. lett.* **113,** 245002-1–245002-5 (2014).
22. Hafz, N. A. *et al.* Stable generation of GeV-class electron beams from self-guided laser–plasma channels. *Nat. Photon.* **2,** 571–577 (2008).
23. Kim, H. T. *et al.* Enhancement of electron energy to the multi-GeV regime by a dual-stage laser-wakefield accelerator pumped by petawatt laser pulses. *Phys. Rev. Lett.* **111,** 165002-1–165002-5 (2013).
24. Wang, X. *et al.* Quasi-monoenergetic laser-plasma acceleration of electrons to 2 GeV. *Nat. Commun.* **4,** 1–9 (2013).
25. Wen, M. *et al.* Generation of high charged energetic electrons by using multiparallel laser pulses. *Phys. Plasmas* **17,** 103113-1–103113-5 (2010).
26. Faure, J. *et al.* A laser–plasma accelerator producing monoenergetic electron beams. *Nature* **431,** 541–544 (2004).
27. Malka, V. *et al.* Characterization of electron beams produced by ultrashort (30 fs) laser pulses. *Phys. Plasmas* **8,** 2605–2608 (2001).
28. Shaw, L. J. *et al.* Role of direct laser acceleration in energy gained by electrons in a laser wakefield accelerator with ionization injection. *Plasma Phys. Control. Fusion* **56,** 084006-1–084006-7 (2014).
29. Pukhov, A. Strong field interaction of laser radiation. *Rep. Prog. Phys.* **66,** 47–101 (2003).
30. Mourou, G. & Tajima, T. The extreme light infrastructure: optics' next horizon. *Opt. Photonics News* **22,** 47–51 (2011).
31. Mourou, G. *et al.* Exawatt-zettawatt pulse generation and applications. *Opt. Commun.* **285,** 720–724 (2012).
32. Brocklesby, W. *et al.* Ican as a new laser paradigm for high energy, high average power femtosecond pulses. *Eur. Phys. J. ST* **223,** 1189–1195 (2014).
33. Liu, B. *et al.* Generating overcritical density relativistic electron beams via self-matching resonance acceleration. *Phys. Rev. Lett.* **110,** 045002-1–045002-5 (2013).
34. Wang, H. Y. *et al.* Efficient and stable proton acceleration by irradiating a two-layer target with a linearly polarized laser pulse. *Phys. Plasmas* **20,** 013101-1–013101-6 (2013).
35. Chen, M., Sheng, Z., Zheng, J. & Ma, Y. Development and application of multi-dimensional particle-in-cell codes for of laser plasma interactions investigation. *Chin. J. Comput. Phys.* **25,** 43–50 (2008).
36. Yan, X. Q. *et al.* Generating high-current monoenergetic proton beams by a circularly polarized laser pulse in the phase-stable acceleration regime. *Phys. Rev. Lett.* **100,** 135003-1–135003-4 (2008).
37. Sheng, Z., Mima, K., Zhang, J. & Sanuki, H. Emission of electromagnetic pulses from laser wakefields through linear mode conversion. *Phys. Rev. Lett.* **94,** 095003-1–095003-4 (2005).
38. Kalmykov, S., Yi, S. A., Khudik, V. & Shvets, G. Electron self-injection and trapping into an evolving plasma bubble. *Phys. Rev. Lett.* **103,** 135004-1–135004-4 (2009).
39. Kostyukov, I., Pukhov, A. & Kiselev, S. Phenomenological theory of laser-plasma interaction in "bubble" regime. *Phys. Plasmas* **11,** 5256–5264 (2004).
40. Qiao, B., Zhu, S. P., Zheng, C. Y. & He, X. T. Quasistatic magnetic and electric fields generated in intense laser plasma interaction. *Phys. Plasmas* **12,** 053104-1–053104-12 (2005).
41. Niu, H. Y., He, X. T., Qiao, B. & Zhou, C. T. Resonant acceleration of electrons by intense circularly polarized gaussian laser pulses. *Laser Part. Beams* **26,** 51–60 (2008).
42. Decker, C. & Mori, W. Group velocity of large-amplitude electromagnetic waves in a plasma. *Phys. Rev. E* **51,** 1364–1375 (1995).
43. Hemsing, E. Dunning, M., Hast, C., Raubenheimer, T. & Xiang, D. First characterization of coherent optical vortices from harmonic undulator radiation. *Phys. Rev. Lett.* **113,** 134803-1–134803-5 (2014).
44. Bahk, S.-W. *et al.* Generation and characterization of the highest laser intensities (10^{22} w/cm^2). *Opt. lett.* **29,** 2837–2839 (2004).

Acknowledgements

The 3D PIC simulatons were carried out in Shanghai Super Computation Center and Super Computation Center in Max Planck Institute. This work was supported by National Basic Research Program of China (Grant No. 2013CBA01502), National Natural Science Foundation of China (Grant Nos. 11025523, J1103206, 11575011) and National Grand Instrument Project (2012YQ030142).

Author Contributions

H.L., X.Y., C.C. and X.H. conducted the work. B.L., R.H., M.Z. and X.Y. developed the basic theory. R.H. and B.L. carried out all simulations. Some details of the physics are clarified by C.L., H.L. and Z.S. The manuscript is written by R.H. and H.L. All authors reviewed the manuscript. R.H. and B.L. contributed equally to this work.

Additional Information

Competing financial interests: The authors declare no competing financial interests.

How to cite this article: Hu, R. *et al.* Dense Helical Electron Bunch Generation in Near-Critical Density Plasmas with Ultrarelativistic Laser Intensities. *Sci. Rep.* **5**, 15499; doi: 10.1038/srep15499 (2015).

www.nature.com/scientificreports

SCIENTIFIC REPORTS

OPEN

Received: 05 October 2016
Accepted: 20 February 2017
Published: 24 March 2017

Brilliant petawatt gamma-ray pulse generation in quantum electrodynamic laser-plasma interaction

H. X. Chang[1], B. Qiao[1,2], T. W. Huang[1], Z. Xu[1], C. T. Zhou[1,3], Y. Q. Gu[4], X. Q. Yan[1], M. Zepf[5] & X. T. He[1,2,3]

We show a new resonance acceleration scheme for generating ultradense relativistic electron bunches in helical motions and hence emitting brilliant vortical γ-ray pulses in the quantum electrodynamic (QED) regime of circularly-polarized (CP) laser-plasma interactions. Here the combined effects of the radiation reaction recoil force and the self-generated magnetic fields result in not only trapping of a great amount of electrons in laser-produced plasma channel, but also significant broadening of the resonance bandwidth between laser frequency and that of electron betatron oscillation in the channel, which eventually leads to formation of the ultradense electron bunch under resonant helical motion in CP laser fields. Three-dimensional PIC simulations show that a brilliant γ-ray pulse with unprecedented power of 6.7 PW and peak brightness of 10^{25} photons/s/mm^2/mrad^2/0.1% BW (at 15 MeV) is emitted at laser intensity of 1.9×10^{23} W/cm^2.

γ-ray is an electromagnetic radiation with extremely high frequency and high photon energy. As a promising radiation source, it has a broad range of applications in material science, nuclear physics, antimatter physics[1–3], logistics for providing shipment security, medicine[4] for sterilizing medical equipment and for treating some forms of cancer, e.g. gamma-knife surgery[5]. γ-ray from distant space can also provide insights into many astrophysical[6,7] phenomena, including γ-ray bursts, cosmic ray acceleration at shock wave front, and emission from pulsar.

Generating intense bursts of high-energy radiation usually requires the construction of large and expensive particle accelerators[8,9]. Laser-driven accelerators offer a cheaper and smaller alternative, and they are now capable of generating bursts of γ-rays[10]. γ-ray generation has been demonstrated in a number of experiments on laser interactions with solid and gas targets, where the main mechanism is the Bremsstrahlung radiation of fast electrons interacting with high-Z material targets[11–17]. However, due to the small bremsstrahlung cross-section, the conversion efficiency of this scheme is rather low. Further, the broad divergence and large size of fast electron source also limit the achievable brightness of the generated γ-rays. γ-ray can also be produced by the nonlinear Compton backscattering, in which an electron beam accelerated by laser wakefields interacts with a counterpropagating laser pulse[18–22]. However, the number of electrons accelerated by laser wakefields in underdense plasmas is small, which also leads to rather low peak brightness of the produced γ-rays. Recent experiment[17] demonstrates that the peak brightness of the γ-ray pulse at 15 MeV can reach only the order of 10^{20} photons/s/mm^2/mrad^2/0.1% BW.

With the progress of laser technology, laser intensities of 5×10^{22} W/cm^2 are now available[23] and are expected to reach the order of 10^{23}–10^{24} W/cm^2 in the next few years[24], where the quantum electrodynamic (QED) effects play role in their interaction with plasmas. In the QED laser-plasma interaction regime, a promising mechanism for production of γ-ray photons is the nonlinear synchrotron radiation[25–29,31] of ultrarelativistic electrons in the laser fields, i.e., $e + n\gamma_l \rightarrow e' + \gamma^{30}$. It is shown that γ-ray photons with the maximum energy extending to

[1]Center for Applied Physics and Technology, HEDPS, State Key Laboratory of Nuclear Physics and Technology, and School of Physics, Peking University, Beijing, 100871, China. [2]Collaborative Innovation Center of IFSA (CICIFSA), Shanghai Jiao Tong University, Shanghai 200240, China. [3]Institute of Applied Physics and Computational Mathematics, Beijing 100094, China. [4]Science and Technology on Plasma Physics Laboratory, Mianyang 621900, China. [5]Department of Physics and Astronomy, Queen's University Belfast, Belfast BT7 1NN, United Kingdom. Correspondence and requests for materials should be addressed to B.Q. (email: bqiao@pku.edu.cn)

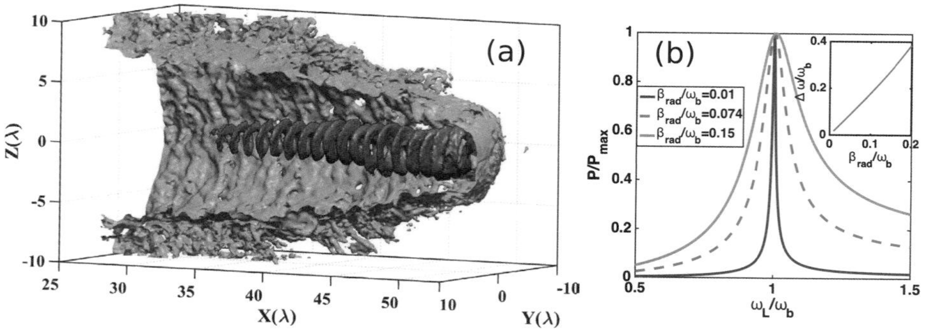

Figure 1. (**a**) 3D isosurface distributions for electron densities of the plasma channel (blue) and the helical electron bunch (red), where the isosurface values are respectively 30 n_c and 50n_c; part (**b**) shows the resonance curve of the transverse momentum with $\beta_{rad}/\omega_b = 0.01$ (blue solid), 0.074 (red dash) and 0.15 (green solid); Inset of (**b**) shows the resonance bandwidth $\Delta\omega$ with the radiation reaction factor β_{rad}.

100 s MeV can be generated by irradiating a solid target with an ultraintense laser[26–28]. However, for laser interaction with steep solid density targets, where no preplasmas exist, the γ-ray emission occurs only in the small skin-depth region[2,26–28], therefore, the conversion efficiency from laser to γ-rays is still low, and the peak brightness of γ-rays is also limited. Recently, a self-matching resonance acceleration scheme[32,33] in near-critical plasmas by circularly polarized laser pulses has been explored, which can generate much denser relativistic electron beams than the case of direct laser acceleration with linearly polarized lasers[34,35]. However, the laser intensity used there is comparatively low in the non-QED regime, electron resonance acceleration is dominantly governed by only the self-generated electromagnetic fields in the plasma, which limits both the energy and the density of the electron bunch for synchrotron radiation.

In this paper, by using a near-critical plasma interaction with ultraintense circularly polarized (CP) laser pulses, we report on a new resonance acceleration scheme in the QED regime for generating ultradense ultrarelativistic electron bunches in helical motions [see Fig. 1(a)] and therefore emitting brilliant vortical γ-ray pulses. In this QED scheme, on the one hand, because of the quantum radiation losses, the transverse phase space of electrons is confined[36,31], and the electrons are easily trapped in the center of laser-produced plasma channel; on the other hand, due to the additional contribution of radiation reaction recoil force, the resonance bandwidth[37] between laser frequency (in the electron rest frame) and that of electron betatron oscillation under quasistatic electromagnetic fields in the channel is significantly broadened, that is, the resonance condition is much relaxed. Both of these effects result in formation of an ultradense electron bunch under resonant helical acceleration in CP laser fields, where both the particle number and energy are much larger than those under only direct laser acceleration (DLA) by linearly polarized lasers[29,31]. Furthermore, the synchrotron radiation efficiency is much enhanced by the resonant electrons' helical motion feature in the self-generated axial and azimuthal magnetic fields due to the use of CP lasers, comparing with that by linearly polarized lasers[29,31], eventually leading to production of brilliant petawatt vortical γ-ray pulses. Three dimensional (3D) particle-in-cell (PIC) simulations show that brilliant γ-ray pulses with unprecedented peak brightness of 10^{25} photons/s/mm/mrad^2/0.1% BW at 15 MeV and power of 6.7 PW are produced at laser intensity of 1.9×10^{23} W/cm^2.

Theoretical Analysis

The properties of γ-ray radiation depend strongly on electron dynamics. Let' s start with the dynamics of a single electron interacting with the laser and self-generated electromagnetic fields in laser-produced plasma channel by taking into account of the QED effects. In the ultrarelativistic limit $\gamma_e \gg 1$, the radiation reaction force can be approximately written as[38–40]

$$\mathbf{F}_{rad} \approx G_e \mathbf{F}_{LL} = -G_e \varepsilon_{rad} m_e c \omega_0 \beta a_S^2 \chi_e^2, \tag{1}$$

where $\mathbf{F}_{LL}$ is the classical radiation reaction force in the Landau-Lifshitz form[38]. In order to take the quantum effect of radiation reaction force into account, we use a quantum-mechanically corrected factor G_e[39,40], which reduces the amount of electrons' unphysical energy loss due to the overestimation of the emitted photon energy in classical calculations. $\varepsilon_{rad} = 4\pi r_e/3\lambda$, $r_e = e^2/m_e c^2$ is the electron radius. $\boldsymbol{\beta} = \mathbf{v}/c$, $a_S = eE_S/m_e\omega c$ corresponds to the QED critical field E_S ($E_S = \alpha e/r_e^2$), and $\alpha = e^2/\hbar c$ is the fine-structure constant. e, m_e, $\mathbf{v}$ are electron charge, mass, and velocity, respectively, ω_0 and λ refer to laser frequency and wavelength, and c is light speed. The probability of γ-photon emission by an electron is characterized by the relativistic gauge-invariant parameter $\chi_e = (\gamma_e/E_S)[(\mathbf{E} + \beta \times \mathbf{B})^2 - (\beta \cdot \mathbf{E})^2]^{1/2}$, where γ_e is the Lorentz factor. The QED effects are negligible for $\chi_e \ll 1$ but play an important role for $\chi_e \gtrsim 1$.

Assuming a CP plane laser propagating along x-direction, the laser fields are $E_{Ly} = E_L \cos\phi$, $E_{Lz} = E_L \sin\phi$, $B_{Ly} = -E_{Lz}/\nu_{ph}$, $B_{Lz} = E_{Ly}/\nu_{ph}$, where E_L refers to their amplitude. The phase is $\phi = kx - \omega_0 t$ and the phase velocity is $\nu_{ph} = \omega_0/k$, where ω_0 and k are laser frequency and wave number. The self-generated electromagnetic fields in the plasma channel are assumed to be E_{Sy}, E_{Sz}, B_{Sy}, B_{Sz} transversely and B_{Sx}

longitudinally. Considering the quantum radiation reaction force Eq. (1), the electron's transverse motion in channel can be described as $dp_y/dt = -e\kappa E_L \cos\phi + ev_x B_{Sz} - ev_z B_{Sx} - eE_{Sy} - \Lambda v_y$, and $dp_z/dt = -e\kappa E_L \cos\phi + ev_x B_{Sy} - ev_y B_{Sx} - eE_{Sz} - \Lambda v_z$, where $\kappa = 1 - \nu_x/\nu_{ph}$ and $\Lambda = \varepsilon_{rad} G_e m_e \omega_0 a_S^2 \chi_e^2$. For high-energy electrons, it is reasonable to assume that $\dot{v}_x \approx 0, \dot{\gamma}_e \approx 0$ and $\dot{\chi}_e \approx 0$, as they are slowly-varying comparing with the fast-varying p_y and p_z. Further assuming $\kappa_E = \delta E_{Sy}/\delta y = \delta E_{Sz}/\delta z, \kappa_B = \delta B_{Sy}/\delta z = -\delta B_{Sz}/\delta y$, we obtain that

$$\frac{d^2 p_y}{dt^2} + \Omega_\theta^2 p_y + \Omega_x \frac{dp_z}{dt} + \beta_{rad} \frac{dp_y}{dt} = m_e c a_0 \omega_L{}^2 \sin\omega_L t, \tag{2}$$

$$\frac{d^2 p_z}{dt^2} + \Omega_\theta^2 p_z - \Omega_x \frac{dp_y}{dt} + \beta_{rad} \frac{dp_z}{dt} = m_e c a_0 \omega_L{}^2 \cos\omega_L t. \tag{3}$$

Here $\Omega_\theta = \sqrt{\frac{e(\kappa_B v_z + \kappa_E)}{\gamma_e m_e}}$, $\Omega_x = \frac{eB_{Sx}}{\gamma_e m_e}$ and $\beta_{rad} = \frac{\Lambda}{\gamma_e m_e}$. $\omega_L = \kappa\omega_0$ refers to laser frequency in the electron rest frame, and $a_0 = eE_L/m_e\omega_0 c$ is the normalized laser amplitude. In Eqs (2) and (3), the third term Ω_x is mainly distributed on the laser axis and can be neglected for the ultrarelativistic electrons in the electron bunch. Therefore, the betatron oscillation frequency can be estimated as $\omega_b \simeq \Omega_\theta$. For electrons under resonance acceleration in laser fields, one can assume $p_{y,z} \simeq P_{rad}\cos(\omega_L t + \varphi)$, where P_{rad} is the momentum amplitude and φ is its initial phase. Substituting $p_{y,z}$ into Eqs (2) and (3), one has

$$P_{rad} = \frac{m_e c a_0 \omega_L^2}{\sqrt{(\beta_{rad}\omega_L)^2 + (\omega_L^2 - \omega_b^2)^2}} \tag{4}$$

From Eq. (4), the amplitude of electron transverse oscillation can be obtained as

$$R_{rad} = \frac{c a_0 \omega_L}{\gamma_e \sqrt{(\beta_{rad}\omega_L)^2 + (\omega_L^2 - \omega_b^2)^2}} \tag{5}$$

On the one hand, Eqs (4) and (5) show that P_{rad} and R_{rad} decrease when the radiation reaction factor β_{rad} increases, that is, the transverse phase space of the electrons is confined by the radiation reaction force[36]. This helps trapping of a great amount of electrons in the plasma channel center. On the other hand, by taking $dP_{rad}/d\omega_L = 0$ from Eq. (4), one can get the resonance condition between laser frequency in the electron instantaneous rest frame and that of electron betatron oscillation ω_r in the plasma channel as

$$\omega_r = \frac{\omega_b^2}{\sqrt{\omega_b^2 - \beta_{rad}^2/2}} = \omega_L \tag{6}$$

Here, one can treat Eq. (4) as a function of $P_{rad}(\omega_L)$, and the resonance curves for different radiation reaction factors β_{rad} are plotted in Fig. 1(b). It shows that the resonance bandwidth $\Delta\omega$, i.e., the full-width-at-half-maximum (FWHM) value of the resonance curve[37], is significantly broadened when the radiation reaction factor β_{rad} increases [See inset figure of Fig. 1(b)]. Results indicate that the resonance condition of the accelerated electrons is much relaxed. Both of these effects eventually result in formation of an ultradense electron bunch under resonant helical acceleration by laser, and hence emission of unprecedented brilliant vortical γ-ray pulses.

Simulation and Results

To verify our scheme, 3D PIC simulations are carried out using the QED-PIC code EPOCH[41], which takes into account of the QED effects in the synchrotron radiation of γ-rays by using a Monte Carlo algorithm[42]. From Fig. 1(a), we clearly see that an ultradense helical electron bunch is formed in laser-produced plasma channel, which undergoes resonance acceleration by the laser pulses. For comparison, simulations with the QED calculation switched off are also carried out. Figure 2 plots electron density maps in plane $z=0$ at different times for the cases with (upper row) and without (lower row) the QED effects taken into account. At early time $t=30T_0$ [2(a) and 2(e)], both of the cases show similar characters that a number of electrons are firstly injected into the center of the plasma channel. However, at later time $t=60T_0$, they show completely different physics. For the case without the QED effects, most electrons in the channel do not satisfy the narrow resonance condition, and the focusing force provided by the self-generated electromagnetic fields is not strong enough to offset the laser radial ponderomotive force[31], so that they are expelled from the plasma channel, as shown in Fig. 2(f). For the case with the QED effects, as predicted by our theory, the radiation reaction force provides not only "trapping" but also"resonant" effects on electrons in the plasma channel, where a great amount of electrons are trapped and undergo direct resonance acceleration by intense lasers, shown in Fig. 2(b). At $t=80T_0$, in the case with QED effects, the transverse phase space of electrons is adequately confined by the radiation reaction recoil force, shown in Fig. 2(d), compared with that in Fig. 2(h). We see from Fig. 2(c) that an ultradense relativistic electron bunch with density above $100n_c$ is formed in the channel under helical resonant motion in CP laser fields. Such an ultradense helical electron bunch leads to generation of strong axial magnetic field B_{Sx} up to 5.0×10^5T and azimuthal B_{Sz} up to 8.0×10^5T [See Fig. 3(a)]. The axial magnetic field helps to trap the background electrons near the laser axis undergoing pre-acceleration to hit the resonance condition[32]. The azimuthal magnetic field in turn

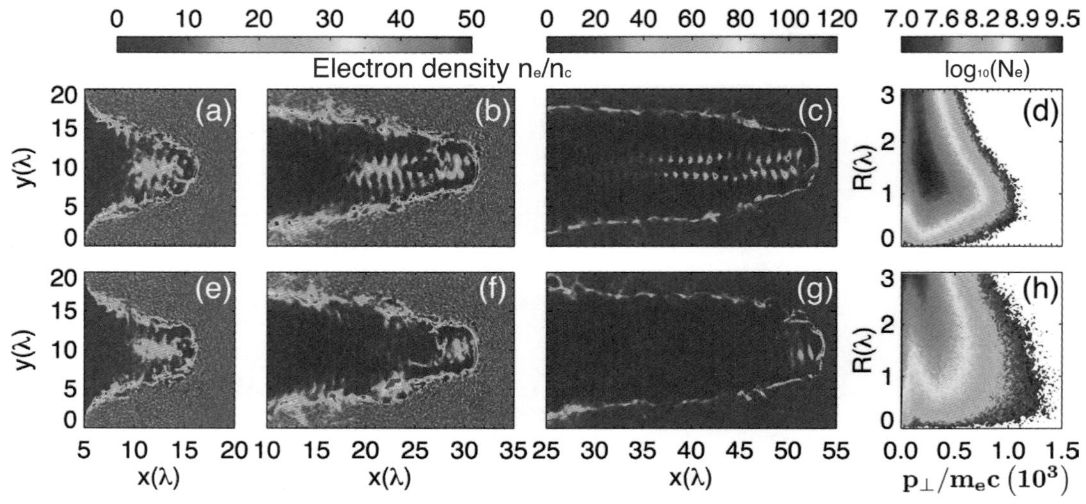

Figure 2. Density maps (in units of n_c) of electrons in the plane z = 0 at $t = 30T_0$ [(**a**) and (**e**)], $60T_0$ [(**b**) and (**f**)], $80T_0$ [(**c**) and (**g**)] for a CP laser pulse at intensity 1.9×10^{23} W/cm^2 interaction with plasmas at densities $10n_c$ in the cases with (upper row) and without (lower row) the QED effects taken into account, respectively. (**d**) and (**h**) show the corresponding transverse phase space distributions of electrons at $80T_0$.

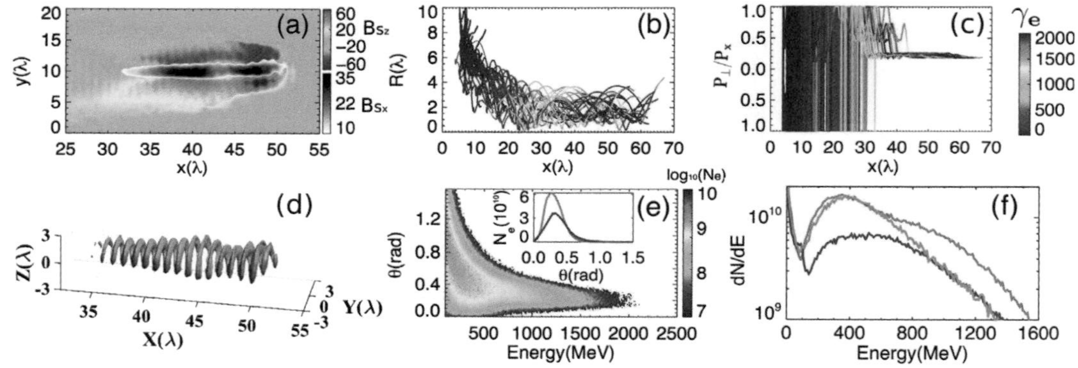

Figure 3. Properties of resonant electrons in the channel of Fig. 2: (**a**) self-generated magnetic field B_{Sz} and B_{Sx} (normalised by $m_e\omega_0/e$) in the plane z = 0 at $t = 80T_0$; (**b**) and (**c**) typical electron motion trajectories ($R = \sqrt{y^2 + z^2}$, x) and ($p_\perp/p\|$, x), where the color bar shows increase of γ_e with time; (**d**) 3D isosurface distributions for electron energy density with the isosurface value at $1.2 \times 10^4 n_c m_e c^2$; (**e**) the angular distribution of electrons; (**f**) the energy spectra of electrons in cases of with (red) and without (blue) the QED effects for CP laser and in the case of with QED effects for LP laser (green); Inset of (**e**) shows the distribution of the number of the electrons (energy above 500 MeV) with the polar angle.

not only provides additional confined forces to help trapping and achieving resonance acceleration of electrons, but also significantly enhances the probability of γ-photon emission[29].

Figure 3(b) shows the typical electron motion trajectories under resonance acceleration. It can be seen that the electrons are injected from the front interaction surface and the wall of the plasma channel. As the role of quantum radiation losses increases, the transverse phase space of these electrons is confined[36], and they are easily trapped in the plasma channel. Further with the aid of the radiation reaction force, as expected, a great amount of these electrons undergo resonance acceleration in CP laser fields, whose energies increase dramatically (see the color evolutions of the lines). For electrons undergoing resonance acceleration, their energies are gradually transferred from the transverse component into the longitudinal one by the $\mathbf{v} \times \mathbf{B}$ force, whose $p_\perp/p_\|$ eventually decreases to a small value of 0.2, shown in Fig. 3(c). Figure 3(d) plots the energy density distribution of electrons at an isosurface value of $1.2 \times 10^4 n_c m_e c^2$. It can be seen that an ultradense, helical relativistic electron bunch is formed, in which the electron maximum energy can reach 2 GeV [see 3(e) and 3(f)] and the total charge of

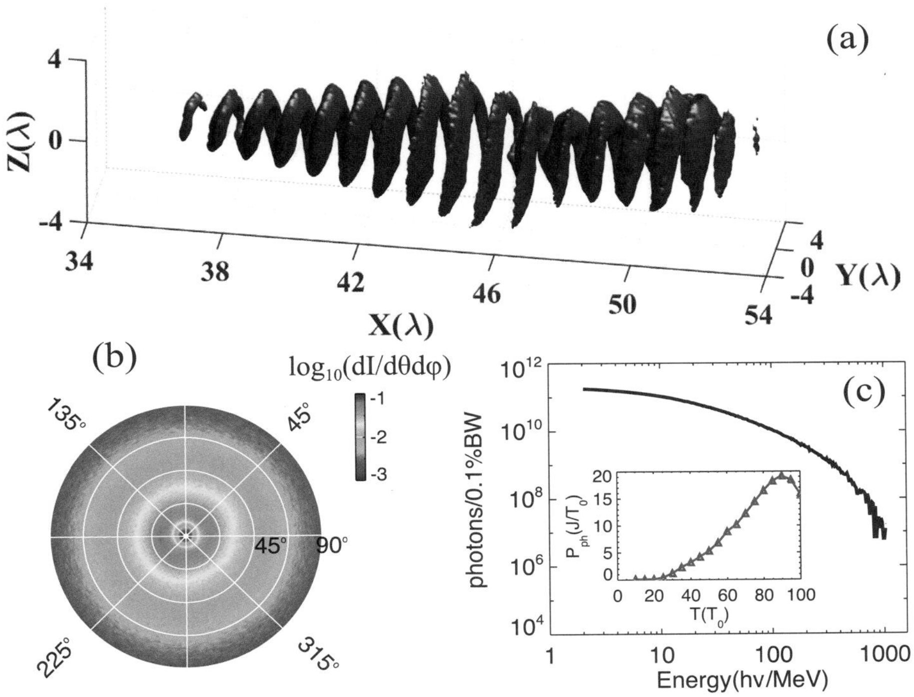

Figure 4. γ-ray pulse emitted in simulations: (**a**) 3D isosurface distribution of the γ-ray's energy density at $5.0 \times 10^3 n_c m_e c^2$; (**b**) the angular distribution of γ-ray energy for photons with energy above 2.0 MeV, where the polar and azimuthal angles are θ and φ, respectively; (**c**) the energy spectrum of photons, where the photon number is calculated in 0.1% bandwidth (BW); The inset in (**c**) shows the total radiation power $P_{ph} = W_{ph}/T_0$, which is defined as the emitted photon energy per laser period.

electrons with energy above 500 MeV is about 200 nanoCoulomb [see the red line in 3(f)]. By comparing these high qualities of the beam with those without the QED effects, shown by the blue lines in Fig. 3(e) inset and 3(f), we conclude that the QED effects lead to great increase in the particle number and decrease in the divergence angle of high-energy electrons, in consistence with our theoretical expectations. The energy spectrum of electrons by using a LP laser with the same other parameters is shown by the green line in Fig. 3(f), which shows much lower cutoff energy and smaller number of high energy electrons, eventually leading to much weaker γ-ray emission.

When the ultradense electron bunch undergoes resonance acceleration in CP laser fields, high-energy γ photons can be synchronously emitted. Figure 4(a) plots the 3D isosurface distribution of the γ-ray photon energy density at $t = 80T_0$. It shows that a brilliant, vortical γ-ray pulse with energy density above $5.0 \times 10^3\, n_c m_e c^2$, transverse size of 2 μm and duration of 40 fs is generated. The radiation power can increase to 18 J/T_0 (6.7 PW) and then decrease with the dissipation of the laser pulse [See the inset figure of Fig. 4(c)], more γ photons and higher energy conversion efficiency from laser to γ-ray can be obtained when the laser pulse is fully dissipated. At $t = 100T_0$, the total number of γ-ray photons with energy above 2.0 MeV is about 3.08×10^{14} with a 15.9 MeV mean energy, and the total energy of γ-ray photons is about 780 J, which corresponds to 22.91% from the laser energy, significantly higher than that in the case using LP laser (16.15%). The energy spectrum of the γ photons at $80T_0$ is plotted in Fig. 4(c). The maximum photon energy exceeds 1.0 GeV, and the peak brightness of the γ-ray pulse at 15 MeV is 1.4×10^{25} photons/s/mm^2/mrad^2/0.1% BW. At $t = 100T_0$, it can arrive at 3.5×10^{25} photons/s/mm^2/mrad^2/0.1% BW. To the best of our knowledge, this is the γ-ray source with the highest peak brightness in tens-MeV regime ever reported in the literature. From Fig. 4(b), we can also see that the high-energy photons are highly collimated.

Discussion

The scaling properties of the emitted γ-rays by the proposed scheme have also been investigated. For a fixed self-similar parameter $S = n_e/a_0 n_c = 1/30$, which strongly determines the electron dynamics in near-critical plasmas as discussed in ref. 43, a series of simulations are carried out with different laser intensities, i.e., a_0. As shown in Fig. 5(c), our scheme still works at lower $a_0 = 150$, though the density of the electron bunch drops, which leads to lower energy conversion efficiency of 8.0% from laser pulse to γ photons [Fig. 5(a)]. When the laser intensity increases, the energy conversion efficiency from laser to electrons drops and that from laser to photons grows up to 27% and then becomes saturate, shown by the blue lines in Fig. 5(a). And due to the enhanced resonance

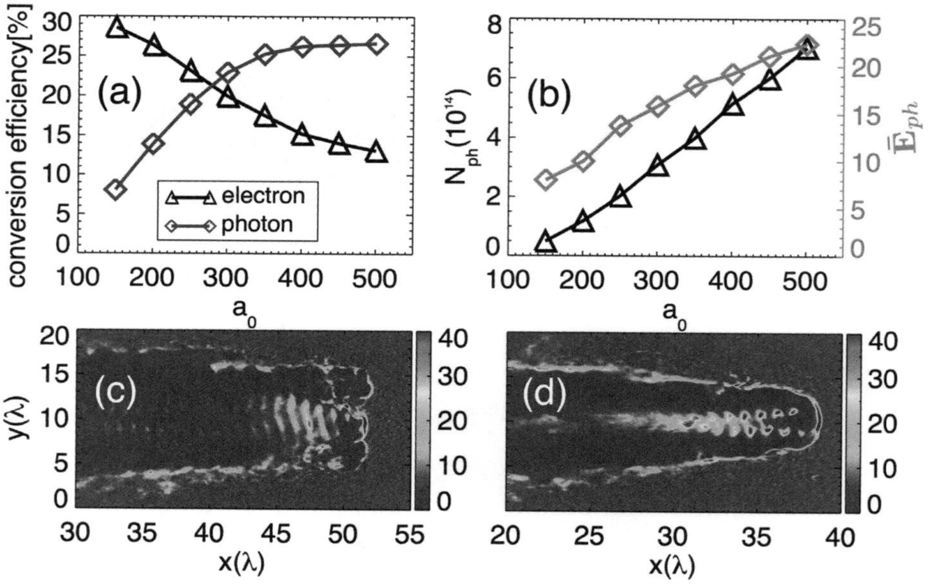

Figure 5. The scaling properties of the emitted γ-ray: (**a**) the conversion efficiencies from laser to electrons (black) and photons (blue) varying with different laser amplitude a_0; (**b**) the photon number (black) and mean energy (red) varying with a_0. (**c**) and (**d**) Respectively show the density maps (in units of n_c) of electrons in the plane z = 0 for the case at normalized laser intensity of $a_0 = 150$ and the case with a reasonable temporal profile of $a = 250\sin^2(\pi t/22T_0)$.

acceleration under stronger radiation reaction recoil force and self-generated electromagnetic fields, both the number and mean energy of the emitted γ-photons increase as well with the laser intensity, shown in Fig. 5(b). Different from the previous theoretical and numerical results [43], which is based on the betatron radiation properties in the non-QED regime with linearly polarized lasers ($a_0 \le 80$), here the scaling shows to be as linear as $N_{ph} \propto a_0$ and $\overline{E}_{ph} \propto a_0$ respectively.

To verify that our scheme still works for a more reasonable laser pulse, an additional simulation is performed. The laser pulse has a exact temporal profile of $a = 250\sin^2(\pi t/22T_0)$ with a duration of $11T_0$ (29.3 fs) and total energy of 266 J, which can be achieved, for example, with the ELI laser[24] under development and Vulcan[44] (planned updating) in the near future. The electron density map in plane $z = 0$ is plotted in Fig. 5(d). It can be clearly seen that an ultradense electron bunch is still formed undergoing stable resonance acceleration in the plasma channel and then brilliant γ-ray pulse is generated. As a result, 4.1×10^{13} γ photons are emitted and the energy conversion efficiency from laser pulse to γ photons can reach as high as 33%. Therefore, this robust electron acceleration and γ-ray emission scheme still works if a more reasonable laser pulse is used.

In this paper, we have reported a novel electron resonance acceleration scheme in the QED regime of CP laser-plasma interactions, where the quantum radiation loss helps trapping of electrons and the radiation reaction recoil force significantly relaxes the resonance condition between electrons and lasers. A great amount of electrons gather around the center of the plasma channel and undergo resonance acceleration, forming an ultradense, vortical relativistic electron bunch. As a result, unprecedentedly brilliant petawatt γ-ray pulses can be obtained.

Methods

The 3D PIC simulations are carried out using the QED-PIC code EPOCH. In the simulations, 900 cells longitudinally along the x axis and 240^2 cells transversely along y and z axes constitute a $75 \times 20 \times 20\lambda^3$ simulation box. A fully-ionized hydrogen plasma target with an uniform electron density of $1.7 \times 10^{22}\,\mathrm{cm}^{-3}$ ($10n_c$) is located from $x = 5$ to $75\,\lambda$. Each cell of plasma is filled with 12 pseudoelectrons and 12 pseudoprotons. A CP laser pulse with peak intensity $1.9 \times 10^{23}\,\mathrm{W/cm^2}$ and wavelength $\lambda = 0.8\,\mu\mathrm{m}$ propagates from the left boundary into target. The laser pulse has a transverse Gaussian profile of FWHM radius $r_0 = 4\lambda$ and a square temporal profile of durations $\tau = 60T_0$ ($T_0 = 2\pi/\omega_0$), which is composed of 30 T_0 sinusoidal rising and 30 T_0 constant parts.

The resonance curve of the transverse momentum in Fig. 1(b) is obtained by Eq. (4). The value $\beta_{rad}/\omega_b = 0.074$ is estimated from the parameters gotten in our simulation. For comparison in Fig. 3(f), under the premise of ensuring the same laser intensity, we have also carried out simulations that the incident laser is linearly polarized. The spectrums here are for electrons within a radius of 3 λ and a spreading angle of 0.38 rad. To investigate the scaling properties of the emitted γ-ray in Fig. 5, we fix the self-similar parameter $S = 1/30$ and the other parameters in the simulations are same. In order to check the accuracy of our simulation results, we have also carried out the simulation at a higher resolution with spatially half of the current grid size, which shows almost the same results as here.

References

1. Nerush, E. N. *et al.* Laser field absorption in self-generated electron-positron pair plasma. *Phys. Rev. Lett.* **106,** 035001 (2011).
2. Ridgers, C. P. *et al.* Dense electron-positron plasmas and ultraintense γ-rays from laser-irradiated solids. *Phys. Rev. Lett.* **108,** 165006 (2012).
3. Chang, H. X. *et al.* Generation of overdense and high-energy electron-positron-pair plasmas by irradiation of a thin foil with two ultraintense lasers. *Phys. Rev. E* **92,** 053107 (2015).
4. Hegelich, B. M. *et al.* Laser acceleration of quasi-monoenergetic MeV ion beams. *Nature* **439,** 441–444 (2006).
5. Ganz, J. *Gamma Knife Neurosurgery* (Springer-Verlag, Wien, 2011).
6. Wardle, J. F. C., Homan, D. C., Ojha, R. & Roberts, D. H. Electron-positron jets associated with the quasar 3C279. *Nature* **395,** 457–461 (1998).
7. Aharonian, F. A. *et al.* High-energy particle acceleration in the shell of a supernova remnant. *Nature* **432,** 75–77 (2004).
8. Ballam, J. *et al.* Total and partial photoproduction cross sections at 1.44, 2.8, and 4.7 GeV. *Phys. Rev. Lett.* **23,** 498–501 (1969).
9. Schoenlein, R. W. *et al.* Femtosecond x-ray pulses at 0.4 Å generated by 90° Thomson Scattering: A tool for probing the structural dynamics of materials. *Science* **274,** 236–238 (1996).
10. Corde, S. *et al.* Femtosecond x rays from laser-plasma accelerators. *Rev. Mod. Phys.* **85,** 1–48 (2013).
11. Giulietti, A. *et al.* Intense γ-ray source in the giant-dipole-resonance range driven by 10-TW laser pulses. *Phys. Rev. Lett.* **101,** 105002 (2008).
12. Courtois, C. *et al.* Effect of plasma density scale length on the properties of bremsstrahlung x-ray sources created by picosecond laser pulses. *Physics of Plasmas* **16,** 013105 (2009).
13. Galy, J., Maučec, M., Hamilton, D. J., Edwards, R. & Magill, J. Bremsstrahlung production with high-intensity laser matter interactions and applications. *New Journal of Physics* **9,** 23 (2007).
14. Ledingham, K. W. D. & Galster, W. Laser-driven particle and photon beams and some applications. *New Journal of Physics* **12,** 045005 (2010).
15. Chen, H. *et al.* Relativistic positron creation using ultraintense short pulse lasers. *Phys. Rev. Lett.* **102,** 105001 (2009).
16. Chen, H. *et al.* Relativistic quasimonoenergetic positron jets from intense laser-solid interactions. *Phys. Rev. Lett.* **105,** 015003 (2010).
17. Sarri, G. *et al.* Ultrahigh brilliance multi-MeV γ-ray beams from nonlinear relativistic Thomson Scattering. *Phys. Rev. Lett.* **113,** 224801 (2014).
18. PhuocK., T. *et al.* All-optical compton gamma-ray source. *Nat Photon* **6,** 308–311 (2012).
19. Thomas, A. G. R., Ridgers, C. P., Bulanov, S. S., Griffin, B. J. & Mangles, S. P. D. Strong radiation-damping effects in a gamma-ray source generated by the interaction of a high-intensity laser with a wakefield-accelerated electron beam. *Phys. Rev. X* **2,** 041004 (2012).
20. Powers, N. D. *et al.* Quasi-monoenergetic and tunable x-rays from a laser-driven compton light source. *Nat Photon* **8,** 28–31 (2014).
21. Liu, C. *et al.* Generation of 9 MeV γ-rays by all-laser-driven compton scattering with second-harmonic laser light. *Opt. Lett.* **39,** 4132–4135 (2014).
22. Sarri, G. *et al.* Generation of neutral and high-density electron-positron pair plasmas in the laboratory. *Nat Commun* **6,** 6747 (2015).
23. Martinez, M. *et al.* In *Boulder Damage Symposium XXXVII: Annual Symposium on Optical Materials for High Power Lasers*, (SPIE-International Society for Optical Engineer- ing, Bellingham, WA), p. 59911N (2005).
24. The extreme light infrastructure: http://www.eli-laser.eu.
25. Nakamura, T. *et al.* High-power γ-ray flash generation in ultraintense laser-plasma interactions. *Phys. Rev. Lett.* **108,** 195001 (2012).
26. Brady, C. S., Ridgers, C. P., Arber, T. D., Bell, A. R. & Kirk, J. G. Laser absorption in relativistically underdense plasmas by synchrotron radiation. *Phys. Rev. Lett.* **109,** 245006 (2012).
27. Brady, C. S., Ridgers, C. P., Arber, T. D. & Bell, A. R. Synchrotron radiation, pair production, and longitudinal electron motion during 10–100 PW laser solid interactions. *Physics of Plasmas* **21,** 033108 (2014).
28. Brady, C. S., Ridgers, C. P., Arber, T. D. & Bell, A. R. Gamma-ray emission in near critical density plasmas. *Plasma Physics and Controlled Fusion* **55,** 124016 (2013).
29. Stark, D. J., Toncian, T. & Arefiev, A. V. Enhanced multi-MeV photon emission by a laser-driven electron beam in a self-generated magnetic field. *Phys. Rev. Lett.* **116,** 185003 (2016).
30. Breit, G. & Wheeler, J. A. Collision of two light quanta. *Phys. Rev.* **46,** 1087–1091 (1934).
31. Ji, L. L., Pukhov, A., Kostyukov, I. Y., Shen, B. F. & Akli, K. Radiation-reaction trapping of electrons in extreme laser fields. *Phys. Rev. Lett.* **112,** 145003 (2014).
32. Liu, B. *et al.* Generating overcritical dense relativistic electron beams via self-matching resonance acceleration. *Phys. Rev. Lett.* **110,** 045002 (2013).
33. Hu, R. *et al.* Dense helical electron bunch generation in near-critical density plasmas with ultrarelativistic laser intensities. *Scientific Reports* **5,** 15499 (2015).
34. Pukhov, A., Sheng, Z. M. & Meyer-ter-Vehn, J. Particle acceleration in relativistic laser channels. *Physics of Plasmas* **6,** 2847–2854 (1999).
35. Gahn, C. *et al.* Multi-MeV electron beam generation by direct laser acceleration in high-density plasma channels. *Phys. Rev. Lett.* **83,** 4772 (1999).
36. Green, D. G. & Harvey, C. N. Transverse spreading of electrons in high-intensity laser fields. *Phys. Rev. Lett.* **112,** 164801 (2014).
37. Landau, L. D. & Lifshitz, E. M., *Mechanics: Volume 1 of Course of Theoretical Physics* (Elsevier, Oxford, 1976).
38. Landau, L. D. & Lifshitz, E. M., *The Classical Theory of Fields*, (Pergamon, Oxford, 1975).
39. Bulanov, S. S., Schroeder, C. B., Esarey, E. & Leemans, W. P. Electromagnetic cascade in high-energy electron, positron, and photon interactions with intense laser pulses. *Phys. Rev. A* **87,** 062110 (2013).
40. Esirkepov, T. Z. *et al.* Attractors and chaos of electron dynamics in electromagnetic standing waves. *Physics Letters A* **379,** 2044–2054 (2015).
41. Arber, T. D. *et al.* Contemporary particle-in-cell approach to laser-plasma modelling. *Plasma Physics and Controlled Fusion* **57,** 113001 (2015).
42. Duclous, R., Kirk, J. G. & Bell, A. R. Monte carlo calculations of pair production in high-intensity laser-plasma interactions. *Plasma Physics and Controlled Fusion* **53,** 015009 (2011).
43. Huang, T. W. *et al.* Characteristics of betatron radiation from direct-laser-accelerated electrons. *Phys. Rev. E* **93,** 063203 (2016).
44. The vulcan 10 petawatt project: http://www.clf.stfc.ac.uk/CLF/Facilities/Vulcan/The+Vulcan+10+Petawatt+Project/14684.aspx

Acknowledgements

This work is supported by the National Natural Science Foundation of China, Grants No. 11575298, No. 91230205, No. 11575031, and No. 11175026; the NSAF, Grant No. U1630246; the National key research and development program No. 2016YFA0401100; the National Science Challenging Program; the National Basic Research 973 Projects No. 2013CBA01500 and No. 2013CB834100, and the National High-Tech 863 Project. B.Q. acknowledges the support from the Thousand Young Talents Program of China. M.Z. acknowledge support from the Engineering and Physical Sciences Research Council (EPSRC), Grant No. EP/1029206/1. The computational resources are supported by the Special Program for Applied Research on Super Computation of the NSFC-Guangdong Joint Fund (the second phase). H.X.C. acknowledges the comments and suggestions from the anonymous referees.

Author Contributions

B.Q., C.T.Z. and X.T.H. conducted the work. B.Q., H.X.C.,T.W.H. and Z.X. developed the basic theory. H.X.C. carried out all simulations. Some detail of the physics are clarified by Y.Q.G., X.Q.Y. and M.Z. The manuscript is written by H.X.C. and B.Q. All authors reviewed the manuscript.

Additional Information

Competing Interests: The authors declare no competing financial interests.

How to cite this article: Chang, H. X. *et al.* Brilliant petawatt gamma-ray pulse generation in quantum electrodynamic laser-plasma interaction. *Sci. Rep.* 7, 45031; doi: 10.1038/srep45031 (2017).

Publisher's note: Springer Nature remains neutral with regard to jurisdictional claims in published maps and institutional affiliations.

APPLIED PHYSICS LETTERS **110**, 021102 (2017)

Collimated gamma photon emission driven by PW laser pulse in a plasma density channel

T. W. Huang,[1] C. T. Zhou,[1,2,3,a)] H. Zhang,[3] S. Z. Wu,[3] B. Qiao,[2,b)] X. T. He,[2,3] and S. C. Ruan[1,c)]
[1]*College of Optoelectronic Engineering, Shenzhen University, Shenzhen 518060, People's Republic of China*
[2]*HEDPS, Center for Applied Physics and Technology and School of Physics, Peking University, Beijing 100871, People's Republic of China*
[3]*Institute of Applied Physics and Computational Mathematics, Beijing 100094, People's Republic of China*

(Received 26 November 2016; accepted 30 December 2016; published online 9 January 2017)

We use three-dimensional particle-in-cell simulations to demonstrate that a plasma density channel can stably guide the petawatt laser pulse in near critical plasmas. In this regime, a directed, collimated, and micro-sized gamma photon beam is emitted by the direct-laser accelerated electrons along the channel axis. While in the case without the plasma density channel, the laser tilting behavior leads to the generation of randomly deflected gamma photon beams with a large divergence angle and transverse source size. In addition, in the plasma density channels, the divergence angle of the gamma photon beams can be much reduced by using a smaller value of n_0/a_0n_c. The energy conversion efficiency can also be improved by increasing the laser power or the plasma density. This regime provides an efficient and compact approach for the production of high quality gamma photon beams. *Published by AIP Publishing.* [http://dx.doi.org/10.1063/1.4973972]

In the past several decades, high power laser pulses have led to the development of many research areas,[1–7] such as inertial confinement fusion, laboratory astrophysics, nuclear physics, and plasma accelerators. Particularly, with the rapid improvement of laser intensity in recent years, the generation of high quality gamma photon beams in an all-optical approach has simulated great interest due to the critical role they play in many significant applications, including the cancer radiation therapy,[8] the probing of hot dense matter,[9] inspection of nuclear waste packages,[10] and the phototransmutation of long-lived radiative isotopes.[11]

Several schemes have been proposed to produce intense gamma photon beams in laser plasma interactions, including the laser driven bremsstrahlung emission[12–16] and the all-optical Compton scattering source.[17–22] Laser based gamma photon sources could potentially offer a cheaper, smaller, shorter, and brighter alternative compared with the conventional γ–ray sources. This provides the opportunity to achieve the table-top γ–ray sources with much higher spatial and temporal resolution than the conventional ones. In addition, the most recent Compton scattering experiment[22] using laser wakefield accelerated electron beams has demonstrated that the brilliance of the gamma photon source is about six orders of magnitude higher than that based on the conventional accelerators. Currently, high power lasers with the intensity on the order of $10^{22}\mathrm{W/cm^2}$ are already achievable[23] and the laser intensity could reach $10^{23}\mathrm{W/cm^2}$ in the next few years.[24] Such high intensities would give rise to a novel regime of multi-MeV gamma photon emission through the nonlinear synchrotron radiation of direct laser accelerated electrons in near critical plasmas. Recently, the prospects of realizing this regime have attracted much attention and simulated numerous studies.[25–31] This regime has shown its advantages on the production of high quality gamma photon beams in many aspects, such as the enhancement of energy conversion efficiency, photon flux, and peak brilliance.[31] However, recent experimental results[32] demonstrate that the divergence angle and transverse size of the radiation source generated in this regime are very large, which severely limits the observed brilliance of the radiation source. In previous works, little attention has been paid to the reduction of the divergence angle of the gamma photon beams. Especially for a relatively long time propagation, a large divergence angle would lead to the rapid broadening of the source size and reduction of the peak brilliance, which greatly restrict the usefulness of the gamma photon beams generated in this regime.

In this Letter, we propose that the employment of a plasma density channel can greatly reduce the divergence angle and transverse size of the gamma photon beams that are generated in the direct laser acceleration (DLA) regime. In this case, the petawatt (PW) laser pulse is well guided and electrons are adequately accelerated along the channel axis; as a result, a directed and highly collimated gamma photon beam is generated. In contrast, in the case without a plasma density channel, the laser pulse experiences the transverse tilting, leading to the production of a randomly deflected gamma photon beam that has a larger divergence angle and transverse size. In addition, the scaling study indicates that in the plasma density channels, the divergence angle of the gamma photon beam is determined by the parameter n_0/a_0n_c and the energy conversion efficiency can also be improved by increasing the input laser power or the plasma density in a suitable parameter region.

The schematic of gamma photon emission driven by the PW laser pulse in near critical plasmas is shown in Fig. 1(a). The high power laser pulse will first propagate into the near

a)Electronic mail: zcangtao@iapcm.ac.cn
b)Electronic mail: bqiao@pku.edu.cn
c)Electronic mail: scruan@szu.edu.cn

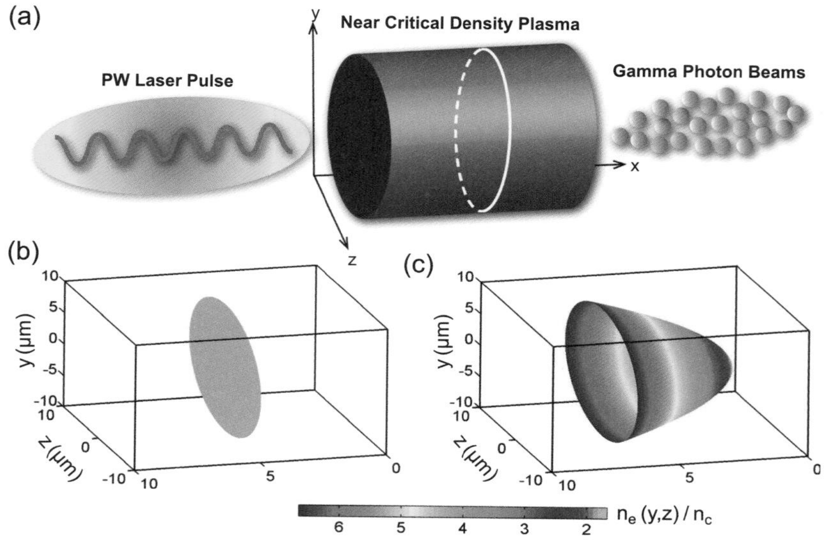

FIG. 1. The schematic of gamma photon emission driven by PW laser pulses in near critical plasmas. (b) and (c) The transverse density profiles of the near critical plasmas, which, respectively, correspond to the case without a preformed plasma channel (b) and the case with a preformed plasma channel (c).

critical plasmas and its ponderomotive force expels the electrons from the central region to form a plasma channel. In this case, electrons would be confined by the channel fields and execute the betatron oscillations. They would be directly accelerated by the laser field when the resonance between the betatron oscillation and the oscillating laser field is excited. This is the so-called direct laser acceleration regime.[33] In this acceleration process, high energy and high flux gamma photons are emitted through the transverse betatron oscillations of electrons in the channel fields.[31] However, when the high power laser pulse propagates into the plasmas, it will suffer from many instabilities.[34–38] Especially, the transverse asymmetric perturbations will lead to the tilting of the laser propagation direction.[37,38] As a result, the accelerated electron beam along with the emitted gamma photons will veer off the axis.[30] In our study, a preformed plasma density channel, with a radially symmetric density profile that has a minimum on its central axis, as shown in Fig. 1(c), is suggested to optimize the production of gamma photon beams in this regime. As a comparison, the case without the plasma density channel, in which the plasma is uniformly distributed in a transverse place, as shown in Fig. 1(b), is also considered.

In the following, three-dimensional (3D) particle-in-cell (PIC) simulations were carried out using the EPOCH code.[39] The radiation reaction effect on the electron dynamics is calculated self-consistently. In the simulations, a linearly polarized Gaussian laser beam is irradiated from the left boundary. The central laser wavelength is $\lambda = 1\mu$m. The amplitude of the normalized laser electric field is set as $a_0 = 100$. The laser has a flat-topped temporal profile with a duration of $60T_L$, where $T_L = \lambda/c$ refers to the laser cycle and c is the light speed in vacuum. The laser beam is focused into a radius of 2μm. The corresponding peak laser power (P_L) is 0.86PW and the laser energy (E_L) is 86J. The near critical plasmas with different transverse profiles, as shown in Figs. 1(b) and 1(c), are considered. The plasma density profile is defined as $n_e(y,z) = n_0(1 + \alpha\frac{y^2+z^2}{L^2})$, where the transverse size of the plasma channel is set as $L = 10\lambda$. In the case without the plasma density channel, the plasma parameters are set as $n_0 = 5n_c$ and $\alpha = 0$, where n_c refers to the critical plasma density. This case is labelled as case I in the following discussions. In the case with the preformed plasma channel, the parameters correspond to $n_0 = 5/3n_c$ and $\alpha = 3$, which is labelled as case II in the following discussions. In these two cases, the total number of electrons and ions is fixed, which means that the integration $\int\int\int n_e(y,z)dydzdx = n_0(1 + \frac{2}{3}\alpha)V$ is fixed as constant; here, V refers to the volume of the simulation box. The simulation box is featured as $120 \times 20 \times 20\mu$m with a grid of $1440 \times 240 \times 240$ cells and eight particles of each species per cell. The initial electron temperature is assumed to be 1 keV.

Fig. 2 shows the corresponding 3D PIC simulation result at $T = 110T_L$. At this moment, one can clearly observe the differences in the above two cases. It is shown that in case I, the laser pulse will experience the serious transverse tilting for a relatively long time propagation at this moment, so will the accelerated electron beam, as indicated in Figs. 2(a) and 2(c).

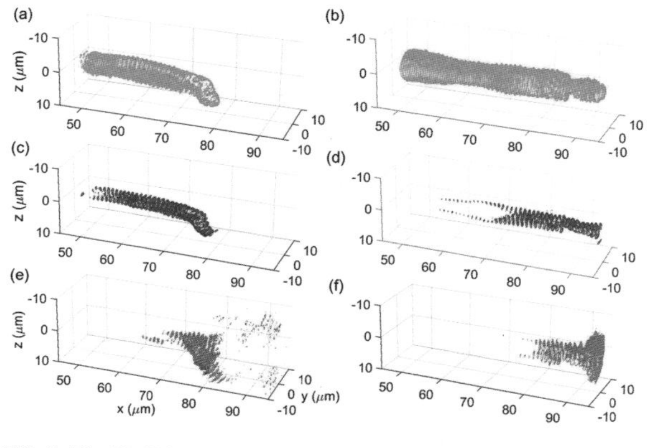

FIG. 2. The 3D PIC simulation results at $T = 110T_L$ for different cases. (a) and (b) The isosurface distribution of the laser intensity, where the isovalue is set as 2×10^{20}W/cm^2. (c) and (d) The isosurface distribution of the electron energy density, where the isovalue is set as $1000n_cm_ec^2$. (e) and (f) The isosurface distribution of the photon energy density, where the isovalue is set as $16n_cm_ec^2$. (a), (c), and (e) The simulation results for case I. (b), (d), and (f) The simulation results for case II.

For the synchrotron radiation of ultra-relativistic electrons, the photons are emitted almost parallel with the electron momentum. As a result, the high energy photon beam is also deflected along with the electron beam, as shown in Fig. 2(e). In addition, this kind of tilting has the characteristic of randomness, which implies that this behavior is unpredictable, while it is shown from Fig. 2(b) that the tilting behavior is suppressed by employing a plasma density channel. Once the laser wavefront is perturbed and tilted in the plasmas, then the preformed plasma density channel will induce a transverse density asymmetry on the laser pulse. Particularly, its inner side closest to the channel axis will experience a lower plasma density than the outer side, which would lead to a phase velocity gradient with a lower phase velocity in the inner side than that in the outer side. Therefore, the plasma density profile will correct the titled wavefront. As a result, the PW laser pulse is well guided and the electrons are directed along the channel axis, as shown in Fig. 2(d), and they can be adequately accelerated by the laser field in the plasma channel. Accordingly, a directed gamma photon beam is produced along the channel axis, as shown in Fig. 2(f).

Fig. 3 shows the angular distribution of the emitted gamma photons. It is shown that the high energy photons are distributed asymmetrically in case I, as demonstrated in Fig. 3(a). This indicates the deflection of the photon beam in a certain direction in transverse space, as shown in Fig. 2(e). While in case II, the gamma photons are symmetrically distributed and they are mainly emitted along the polarization plane that crosses the central channel axis, as indicated in Fig. 3(b). In addition, it is noted that in this case the gamma photons are well collimated within an emission angle less than 20°. It is much smaller than that in case I, in which the gamma photons are divergently distributed within an emission angle about 40°. Correspondingly, in case I, the gamma photons are off-axis distributed with a large transverse size, as shown in Fig. 3(c). In contrast, the emitted gamma photons are directed along the central axis with a smaller transverse size in case II. In spite of the optimization, it is noted that the total number of high energy gamma photons ($\epsilon > 1$MeV) in case II (3.7×10^{12}) is lower than that in case I (9.6×10^{12}); here, ϵ refers to the photon energy. In addition, the total energy of the gamma photons in case I (8.5J) is also larger than that in case II (3.9J). This indicates that in case I the laser pulse is more efficiently coupled with the plasmas because of a larger plasma density in the central region than that in case II. In case I, the high energy (≥ 100MeV) electron beam has a charge number of 149nC, while in case II the corresponding charge number is about 109nC. Thus, one can conclude that the employment of a plasma density channel can greatly reduce the divergence angle and transverse size of the gamma photon beams, but at the cost of decreasing the energy conversion efficiency.

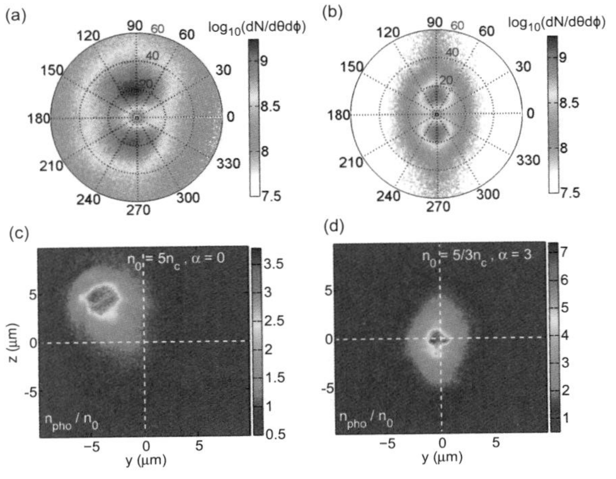

FIG. 3. (a) and (b) The corresponding angular distribution of the emitted high energy gamma photons (≥ 1MeV), where $\theta = \arctan(p_\perp/p_x)$, $\phi = \arctan(p_y/p_z)$, and $p_\perp = \sqrt{p_y^2 + p_z^2}$. (c) and (d) The normalized photon density (n_{pho}/n_0) distribution in transverse place at the laser wavefront positions. (a) and (c) The simulation results for case I. (b) and (d) The simulation results for case II.

In order to understand the scaling properties of gamma photon emission in the plasma density channels, a physical model can be established. To begin, the laser power (P_L) is assumed to be constant for a given laser pulse, then one has $P_L \propto I_0 R_0^2 = I_f R_f^2$, where R_0 denotes the initial beam radius, $I_0 = a_0^2 n_c m_e c^3/2$ is the initial laser intensity, m_e is the rest electron mass, R_f refers to the self-shaped beam radius, $I_f = a_f^2 n_c m_e c^3/2$ and a_f refers to the normalized amplitude of the self-shaped laser beam. It is known that for ultra-intense laser pulse propagating in the plasma channel, it will self-organize into a laser beam with a radius of $R_f = 2\sqrt{a_f}c/\omega_p$,[34,35] where ω_p is the plasma frequency. Thus, one has $a_f = a_0\left(\frac{n_0}{a_0 n_c}\right)^{1/3}\left(\frac{\pi R_0}{\lambda}\right)^{2/3}$. Then according to the radiation properties of the direct laser accelerated electrons,[31] one can easily get the scaling as $E_{ph} \propto \hbar\omega_0 a_f^5\left(\frac{a_f n_c}{n_0}\right)^2 n_c \lambda^2 c\tau$ and $\theta_d \propto \sqrt{\frac{n_0}{a_f n_c}}$, where $\hbar$ is the Planck constant, ω_0 is the laser frequency, τ is the pulse duration, E_{ph} refers to the total photon energy, and θ_d refers to the divergence angle of the emitted photons. The energy conversion efficiency of the emitted photons can be expressed as $\frac{E_{ph}}{E_L} \propto a_f^3\left(\frac{a_f n_c}{n_0}\right)$, where the input laser energy scales as $E_L \propto a_0^2 R_0^2 n_c m_e c^3 \tau/2$. By substituting a_f into the above expressions, one can get the energy conversion efficiency as

$$\frac{E_{ph}}{E_L} \propto a_0^3\left(\frac{n_0}{a_0 n_c}\right)^{1/3} \propto \left(\frac{P_L}{n_c m_e c^3 \lambda^2}\right)^{4/3}\left(\frac{n_0}{n_c}\right)^{1/3}, \quad (1)$$

and the divergence angle of the emitted photons scales as

$$\theta_d \propto \left(\frac{n_0}{a_0 n_c}\right)^{1/3}. \quad (2)$$

It is noted that these scalings are different from that in previous works,[31] in which a self-matched propagation regime was considered. Eq. (1) indicates that the energy conversion efficiency of the gamma photons would increase with the plasma density, but correspondingly, the divergence angle is also broadened, as indicated by Eq. (2). In addition, with the increase of the input laser power, the energy conversion efficiency would also be increased. The divergence angle of the gamma photons can be controlled by adjusting the single parameter n_0/a_0n_c.

To verify the above scalings, a series of two-dimensional PIC simulations are conducted. The effects of different density profiles are also investigated. It is shown that for $\alpha \geq 1$, the tilting behavior of the PW laser pulse in near critical plasmas can be effectively suppressed. Here, $\alpha = 3$ is employed to explore the underlying dependency of gamma photon emissions on the plasma density (n_0) and the input laser power (P_L). The plasma density channels lead to the generation of directed and collimated gamma photon beams. The laser pulse duration and beam radius are fixed as constants, which are the same as those in Fig. 2. The corresponding simulation results are shown in Fig. 4. It can be seen from Fig. 4(a) that the energy conversion efficiency of the gamma photons increases with the plasma density but at the expense of broadening the divergence angle, as shown in Fig. 4(b). This conclusion agrees well with the theoretical predications. But it is noted that when the plasma density is relatively high ($n_0/a_0n_c > 0.15$), the energy conversion efficiency is decreased as in this case the DLA regime gradually loses its efficacy[31] and the laser hole boring process[40] plays the dominant role. Fig. 4(c) shows that the energy conversion efficiency of the gamma photons can be much enhanced by improving the laser power, about seven percents of the laser energy can be converted into the gamma photon beam by using a 2PW laser beam. Fig. 4(d) shows that the gamma photons are highly collimated along the channel axis and the divergence angles are invariable for a fixed value of n_0/a_0n_c, which is in good agreement with the theoretical predications.

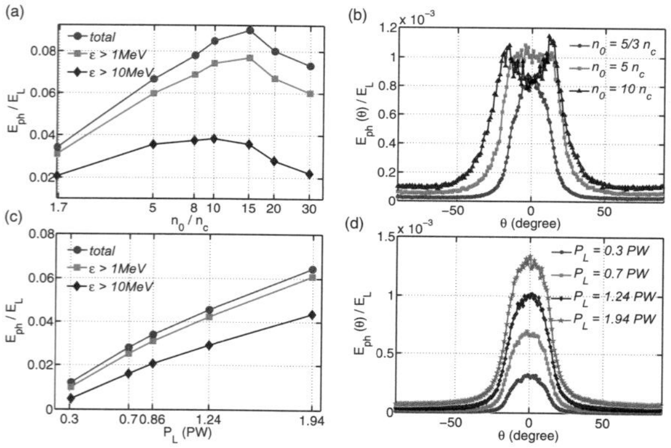

FIG. 4. The parameter study on the gamma photon emissions in plasma density channels. (a) The calculated energy conversion efficiency of gamma photons with different energy levels for different plasma densities n_0. (b) The corresponding angularly resolved energy distribution of high energy gamma photons ($\geq$ 1MeV) for different plasma densities n_0. In these simulations (for (a) and (b)), the laser parameters are kept unchanged, which are the same as those in Fig. 2. (c) The energy conversion efficiency of gamma photons with different energy levels for different input laser powers P_L. (d) The corresponding angularly resolved energy distribution of high energy gamma photons ($\geq$ 1MeV) for different input laser powers P_L. In these simulations (for (c) and (d)), the parameter n_0/a_0n_c is fixed as 1/60. The laser pulse duration and radius are all kept as constants, which are the same as those in Fig. 2. In all the simulations (for (a) and (d)), the steepness of the plasma channel is fixed as $\alpha = 3$ and only the forward-going photons with $p_x > 0$ are calculated. The total photon energy is normalized by the input laser energy (E_L). The θ is calculated as $\theta = \arctan(p_y/p_x)$.

In summary, the gamma photon emission process driven by PW laser pulses in a plasma density channel is investigated. It is demonstrated that the plasma density channel can stably guide the PW laser pulse in near critical plasmas, leading to the generation of directed and highly collimated gamma photon beams. The scaling properties of gamma photon emission in plasma density channels are also explored, indicating that the energy conversion efficiency increases with the input laser power and the plasma density. The divergence angle can also be much decreased as the value of n_0/a_0n_c is decreased. The employment of a proper plasma density channel can lead to the generation of high quality gamma photon beams, which have considerable energies and are highly collimated, directed, and micro-sized. Such high quality gamma photon sources would simulate great interest in a broad range of applications, including the imaging of weakly absorbing and ultra-dense matter, time-resolved probing of the nucleus, inspection of nuclear waste packages, and the photo-transmutation of long-lived radiative isotopes.

This work was supported by the National Key Programme for S&T Research and Development, Grant No. 2016YFA0401100, the National Natural Science Foundation of China (NSFC), Grant Nos. 11575298, 91230205, 11575031, and 11175026, the National Basic Research 973 Project, No. 2013CB834100, the NSAF, Grant No. U1630246, and the National High-Tech 863 Project. The computational resources are supported by the Special Program for Applied Research on Super Computation of the NSFC-Guangdong Joint Fund (the second phase). The EPOCH code was developed under the UK EPSRC Grant Nos. EP/G054940/1, EP/G055165/1, and EP/G056803/1. B.Q. acknowledges the support from Thousand Young Talents Program of China. T.W.H. would like to thank Q. Wang, K. D. Xiao, L. B. Ju, H. X. Chang, Y. X. Zhang, W. P. Yao, and Z. Xu at Peking University for their help and useful discussions in this work.

[1]T. Tajima and J. M. Dawson, Phys. Rev. Lett. **43**, 267 (1979).
[2]C. K. Li, D. D. Ryutov, S. X. Hu, M. J. Rosenberg, A. B. Zylstra, F. H. Seguin, J. A. Frenje, D. T. Casey, M. G. Johnson, M. J.-E. Manuel, H. G. Rinderknecht, R. D. Petrasso, P. A. Amendt, H. S. Park, B. A. Remington, S. C. Wilks, R. Betti, D. H. Froula, J. P. Knauer, D. D. Meyerhofer, R. P. Drake, C. C. Kuranz, R. Young, and M. Koenig, Phys. Rev. Lett. **111**, 235003 (2013).
[3]C. Gahn, G. D. Tsakiris, G. Pretzler, K. J. Witte, C. Delfin, C.-G. Wahlstr?m, and D. Habs, Appl. Phys. Lett. **77**, 2662 (2000).
[4]C. T. Zhou, L. Y. Chew, and X. T. He, Appl. Phys. Lett. **97**, 051502 (2010).
[5]X. T. He, J. W. Li, Z. F. Fan, L. F. Wang, J. Liu, K. Lan, J. F. Wu, and W. H. Ye, Phys. Plasmas **23**, 082706 (2016).
[6]T. P. Yu, L. X. Hu, Y. Yin, F. Q. Shao, H. B. Zhuo, Y. Y. Ma, X. H. Yang, W. Luo, and A. Pukhov, Appl. Phys. Lett. **105**, 114101 (2014).
[7]S. Corde, K. Ta Phuoc, G. Lambert, R. Fitour, V. Malka, and A. Rousse, Rev. Mod. Phys. **85**, 1 (2013).
[8]K. J. Weeks, V. N. Litvinenko, and J. M. J. Madey, Med. Phys. **24**, 417 (1997).
[9]A. Ben-Ismaïl, O. Lundh, C. Rechatin, J. K. Lim, J. Faure, S. Corde, and V. Malka, Appl. Phys. Lett. **98**, 264101 (2011).
[10]C. P. Jones, C. M. Brenner, C. A. Stitt, C. Armstrong, D. R. Rusby, S. R. Mirfayzi, L. A. Wilson, A. Alejo, H. Ahmed, R. Allott, N. M. H. Butler, R. J. Clarke, D. Haddock, C. Hernandez-Gomez, A. Higginson, C. Murphy, M. Notley, C. Paraskevoulakos, J. Jowsey, P. McKenna, D. Neely, S. Kar, and T. B. Scott, J. Hazard. Mater. **318**, 694 (2016).
[11]K. W. D. Ledingham, J. Magill, P. McKenna, J. Yang, J. Galy, R. Schenkel, J. Rebizant, T. McCanny, S. Shimizu, L. Robson, R. P. Singhal,

M. S. Wei, S. P. D. Mangles, P. Nilson, K. Krushelnick, R. J. Clarke, and P. A. Norreys, J. Phys. D: Appl. Phys. **36**, L79 (2003).
[12]J. D. Kmetec, C. L. Gordon III, J. J. Macklin, B. E. Lemoff, G. S. Brown, and S. E. Harris, Phys. Rev. Lett. **68**, 1527 (1992).
[13]R. D. Edwards, M. A. Sinclair, T. J. Goldsack, K. Krushelnick, F. N. Beg, E. L. Clark, A. E. Dangor, Z. Najmudin, M. Tatarakis, B. Walton, M. Zepf, K. W. D. Ledingham, I. Spencer, P. A. Norreys, R. J. Clarke, R. Kodama, Y. Toyama, and M. Tampo, Appl. Phys. Lett. **80**, 2129 (2002).
[14]Y. Glinec, J. Faure, L. Le Dain, S. Darbon, T. Hosokai, J. J. Santos, E. Lefebvre, J. P. Rousseau, F. Burgy, B. Mercier, and V. Malka, Phys. Rev. Lett. **94**, 025003 (2005).
[15]A. Giulietti, N. Bourgeois, T. Ceccotti, X. Davoine, S. Dobosz, P. DOliveira, M. Galimberti, J. Galy, A. Gamucci, D. Giulietti, L. A. Gizzi, D. J. Hamilton, E. Lefebvre, L. Labate, J. R. Marquès, P. Monot, H. Popescu, F. Réau, G. Sarri, P. Tomassini, and P. Martin, Phys. Rev. Lett. **101**, 105002 (2008).
[16]S. Cipiccia, S. M. Wiggins, R. P. Shanks, M. R. Islam, G. Vieux, R. C. Issac, E. Brunetti, B. Ersfeld, G. H. Welsh, M. P. Anania, D. Maneuski, N. R. C. Lemos, R. A. Bendoyro, P. P. Rajeev, P. Foster, N. Bourgeois, T. P. A. Ibbotson, P. A. Walker, V. O. Shea, J. M. Dias, and D. A. Jaroszynski, J. Appl. Phys. **111**, 063302 (2012).
[17]K. Ta Phuoc, S. Corde, C. Thaury, V. Malka, A. Tafzi, J. P. Goddet, R. C. Shah, S. Sebban, and A. Rousse, Nat. Photonics **6**, 308 (2012).
[18]A. G. R. Thomas, C. P. Ridgers, S. S. Bulanov, B. J. Griffin, and S. P. D. Mangles, Phys. Rev. X **2**, 041004 (2012).
[19]N. D. Powers, I. Ghebregziabher, G. Golovin, C. Liu, S. Chen, S. Banerjee, J. Zhang, and D. P. Umstadter, Nat. Photonics **8**, 28 (2014).
[20]C. Liu, G. Golovin, S. Chen, J. Zhang, B. Zhao, D. Haden, S. Banerjee, J. Silano, H. Karwowski, and D. Umstadter, Opt. Lett. **39**, 4132 (2014).
[21]G. Sarri, D. J. Corvan, W. Schumaker, J. M. Cole, A. Di Piazza, H. Ahmed, C. Harvey, C. H. Keitel, K. Krushelnick, S. P. D. Mangles, Z. Najmudin, D. Symes, A. G. R. Thomas, M. Yeung, Z. Zhao, and M. Zepf, Phys. Rev. Lett. **113**, 224801 (2014).
[22]C. H. Yu, R. Qi, W. T. Wang, J. S. Liu, W. T. Li, C. Wang, Z. J. Zhang, J. Q. Liu, Z. Y. Qin, M. Fang, K. Feng, Y. Wu, Y. Tian, Y. Xu, F. X. Wu, Y. X. Leng, X. F. Weng, J. H. Wang, F. L. Wei, Y. C. Yi, Z. H. Song, R. X. Li, and Z. Z. Xu, Sci. Rep. **6**, 29518 (2016).
[23]C. Danson, D. Hillier, N. Hopps, and D. Neely, High Power Laser Sci. Eng. **3**, e3 (2015).
[24]D. Powell, Nature **500**, 264 (2013).
[25]C. S. Brady, C. P. Ridgers, T. D. Arber, and A. R. Bell, Plasma Phys. Controlled Fusion **55**, 124016 (2013).
[26]L. L. Ji, A. Pukhov, I. Y. Kostyukov, B. F. Shen, and K. Akli, Phys. Rev. Lett. **112**, 145003 (2014).
[27]B. Liu, R. H. Hu, H. Y. Wang, D. Wu, J. Liu, J. E. Chen, J. Meyer-ter-Vehn, X. Q. Yan, and X. T. He, Phys. Plasmas **22**, 080704 (2015).
[28]H. Y. Wang, B. Liu, X. Q. Yan, and M. Zepf, Phys. Plasmas **22**, 033102 (2015).
[29]X. L. Zhu, Y. Yin, T. P. Yu, F. Q. Shao, Z. Y. Ge, W. Q. Wang, and J. J. Liu, New J. Phys. **17**, 053039 (2015).
[30]D. J. Stark, T. Toncian, and A. V. Arefiev, Phys. Rev. Lett. **116**, 185003 (2016).
[31]T. W. Huang, A. P. L. Robinson, C. T. Zhou, B. Qiao, B. Liu, S. C. Ruan, X. T. He, and P. A. Norreys, Phys. Rev. E **93**, 063203 (2016).
[32]S. Kneip, S. R. Nagel, C. Bellei, N. Bourgeois, A. E. Dangor, A. Gopal, R. Heathcote, S. P. D. Mangles, J. R. Marques, A. Maksimchuk, P. M. Nilson, K. T. Phuoc, S. Reed, M. Tzoufras, F. S. Tsung, L. Willingale, W. B. Mori, A. Rousse, K. Krushelnick, and Z. Najmudin, Phys. Rev. Lett. **100**, 105006 (2008).
[33]A. Pukhov, Z. M. Sheng, and J. Meyer-ter-Vehn, Phys. Plasmas **6**, 2847 (1999).
[34]A. B. Borisov, A. V. Borovskiy, A. Mcpherson, K. Boyer, and C. K. Rhodes, Plasma Phys. Controlled Fusion **37**, 569 (1995).
[35]T. W. Huang, C. T. Zhou, A. P. L. Robinson, B. Qiao, H. Zhang, S. Z. Wu, H. B. Zhuo, P. A. Norreys, and X. T. He, Phys. Rev. E **92**, 053106 (2015); T. W. Huang, C. T. Zhou, and X. T. He, Laser Part. Beams **33**, 347 (2015).
[36]L. B. Ju, T. W. Huang, K. D. Xiao, G. Z. Wu, S. L. Yang, R. Li, Y. C. Yang, T. Y. Long, H. Zhang, S. Z. Wu, B. Qiao, S. C. Ruan, and C. T. Zhou, Phys. Rev. E **94**, 033202 (2016).
[37]C. Ren and W. B. Mori, Phys. Plasmas **8**, 3118 (2001).
[38]L. Ceurvorst, N. Ratan, M. C. Levy, M. F. Kasim, J. Sadler, R. H. H. Scott, R. M. G. M. Trines, T. W. Huang, M. Skramic, M. Vranic, L. O. Silva, and P. A. Norreys, New J. Phys. **18**, 053023 (2016).
[39]T. D. Arber, K. Bennett, C. S. Brady, A. Lawrence-Douglas, M. G. Ramsay, N. J. Sircobe, P. Gillies, R. G. Evans, H. Schmitz, A. R. Bell, and C. P. Ridgers, Plasma Phys. Controlled Fusion **57**, 113001 (2015).
[40]A. P. L. Robinson, P. Gibbon, M. Zepf, S. Kar, R. G. Evans, and C. Bellei, Plasma Phys. Controlled Fusion **51**, 024004 (2009).

JOURNAL OF APPLIED PHYSICS **101**, 103302 (2007)

Density effect on proton acceleration from carbon-containing high-density thin foils irradiated by high-intensity laser pulses

C. T. Zhou[a)]
Institute of Applied Physics and Computational Mathematics, P. O. Box 8009, Beijing 100088, People's Republic of China and Institute for Fusion Theory and Simulation, Zhejiang University, Hangzhou 310027, People's Republic of China

M. Y. Yu
Institut für Theoretische Physik I, Ruhr-Universität Bochum, D-44780 Bochum, Germany and Institute for Fusion Theory and Simulation, Zhejiang University, Hangzhou 310027, People's Republic of China

X. T. He
Institute of Applied Physics and Computational Mathematics, P. O. Box 8009, Beijing 100088, People's Republic of China and Institute for Fusion Theory and Simulation, Zhejiang University, Hangzhou 310027, People's Republic of China

(Received 16 February 2007; accepted 10 March 2007; published online 22 May 2007)

The acceleration of protons in dense plastic foils irradiated by ultrahigh intensity laser pulses is simulated using a two-dimensional hybrid particle-in-cell scheme. For the chosen parameters of the overdense foils of densities $\rho=0.2$, 1, and $3\ \mathrm{g/cm^3}$ and of an ultrahigh intensity ($2\times10^{20}\ \mathrm{W/cm^2}$) laser pulse, our simulations illustrate that a high-density target is favorable to high collimation of the target-normal-sheath acceleration protons but less energy for a short acceleration time (<100 fs). In particular, the difference of strong local heating of the carbon ion for different plasma densities is clearly observed at both the front and rear surfaces of thin solid targets, suggesting that the effect of the density and composition of the targets are also important for correctly simulating energetic ion generation in ultraintense laser-solid interactions. © *2007 American Institute of Physics*. [DOI: 10.1063/1.2730565]

I. INTRODUCTION

Since the observation of energetic protons from the rear surface of a thin target irradiated by an ultraintense laser pulse[1] there has been much interest in the production and acceleration of protons and heavier ions.[2–21] Laser-driven ion acceleration has many applications, especially in the medical, high-energy, and fusion physics. Novel fast-ignition schemes based on energetic ions have, in particular, received much attention.[22–26]

Protons from contaminant water vapor and hydrocarbons at both the front and rear surfaces of thin solid targets, or foils, can be accelerated[19,27] by exawatt or above laser pulses. In the experiments, plastic targets such as C_8H_8 and $H_8C_{10}O_4$ Mylar foils are often used because of their considerably higher hydrogen content compared to that (due to contamination) of metal targets. For laser intensities in the moderately relativistic regime, relevant to many fast-ignition schemes, the most energetic and collimated protons are generated by sheath acceleration at the rear foil surface.[28] At higher laser intensities, a shock front propagating into the plasma is created by the laser pulse. The electrostatic field of the shock front accelerates the plasma protons. The latter can attain very high energies since upon exiting the foil they are further accelerated by the sheath field.

Since the protons are accelerated by the electron sheath on the target surface, their spatial and angular characteristics are determined by the electron sheath spatial distribution. Therefore, the emission characteristic of proton beams will depend on the target material, the target surface roughness, the target shape, and the laser focal distribution on the target.[19] Although laser-driven acceleration of protons in low-density foils has been intensively investigated,[2–21] it is still not clear if the acceleration mechanisms and results would remain applicable for high-density foils. In this article we consider the dependence of proton acceleration on the foil density.

II. DESCRIPTION OF SIMULATION

For investigating proton acceleration, several numerical techniques, such as the gridless tree method,[29,30] various fluid models,[31–33] and particle-in-cell (PIC) simulations,[34] have been employed. Furthermore, a purely collisionless acceleration in plasma foils by the ponderomotive force could be treated analytically.[35] The standard PIC simulation still remains the major quantitative tool for studying laser-plasma interaction for plasma densities less than, say $2-5\gamma n_{cr}$, where γ is the relativistic factor and n_{cr} is the critical density. However, the standard PIC technique is not practical for solid-density targets where the electron density is much higher than critical.

One method for simulating such high-density plasmas is to use the hybrid technique.[36] Recently there have been several hybrid approaches to deal with overdense plasma problems. They are the inertialess electron model,[37] the implicit fluid moment model,[38,39] and the direct-implicit fully kinetic PIC model.[40,41] More recently, some improvements, based on

a)Electronic mail: zcangtao@iapcm.ac.cn

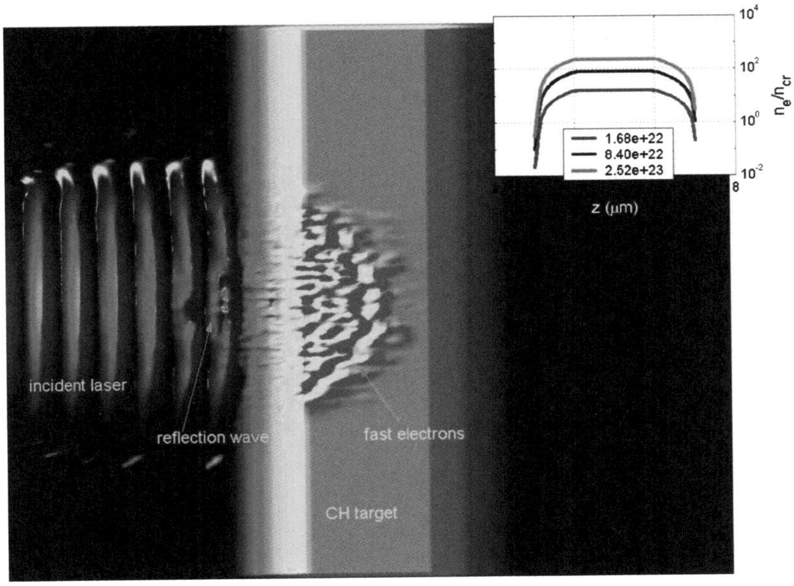

FIG. 1. (Color online) Artistic representation of the laser-slab interaction for $I_0=2\times 10^{20}$ W/cm^2 and $\sigma_z=3$ μm. The inset shows the three electron-density profiles investigated. The figure shows that the energized electrons are moving across the slab and the laser is being reflected from the target surface.

the first two approaches for treating beam electrons by PIC and the background plasma currents in a resistive fluid description, have also been discussed.[42,43] Although fluid treatment requires less computation than particle and may easily solve the problem of large density variation, the solution with magnetic fields and multiple dimensions in fluid moment method can be complicated and require multiple iterations. In particular, higher order moments can be important in this scheme. On the other hand, the inertialess electron model makes use of Ohm's law for electron equations of motion. The Ohm's model has been successfully used for transport[44] in collisional matter. The net current and resistive instabilities are also well modeled by introducing inertial effects in plasma dielectric function. Moreover, the numerical solutions do not have to be implicit and are thus fast, but the details of laser plasma interaction cannot be adequately modeled due to the difficulty to conserve charge.[40] Our present work relies on the fully electromagnetic implicit hybrid PIC code. Our numerical algorithms are somewhat similar to those developed by Welch *et al.*,[40,41] in which both PIC kinetic and fluid electrons are included. In our code, Spitzer resistivity[45] for the background electrons is also assumed, which is valid for high temperature plasmas.[46] Our code can now treat a two-dimensionally (2D) Cartesian or cylindrical geometrices.[47–49] The 2D treatment gives a reasonable approximation to reality as far as effects we want to show.

In order to enhance the density effect, we shall consider a plastic target foil (Fig. 1), represented by a fully dissociated $C^+H_2^+$ plasma at 5 eV. The plasma thus contains electrons, protons, and carbon ions. The plasma densities considered are (I) $\rho=0.2$, (II) 1, and (III) 3 g/cm^3, corresponding to the electron, proton, and carbon-ion number densities $n_{0,e}$ [cm^{-3}]=[1.68×10^{22}, 8.4×10^{22}, 2.52×10^{23}], n_{0,H^+}[cm^{-3}] =[0.24×10^{22}, 1.2×10^{22}, 3.6×10^{22}], and n_{0,C^+}[cm^{-3}] =[1.2×10^{22}, 6×10^{22}, 1.8×10^{23}], respectively. The two-dimensional (x, z) simulation box is 20 μm$\times$20 μm. The thin plasma slab, initially uniform in the transverse direction, is located in $3<z[\mu\mathrm{m}]<7$ and bounded by two vacuum regions. On both sides of the slab there is a 1 μm linear rise to the uniform density (n_0) region at the center. The initial density at $z=3$ μm is taken to be $n_0/100$, so that only in case I the critical density is on the front ramp. There are 400 uniform grids in the laser-polarization direction (x). The grid in the laser-propagation direction (z) is variable according to need. The grid step δz in the two vacuum regions is 0.05 μm, enough to resolve the laser wavelength. To include the details of the laser-plasma interaction at the relativistic critical density, we use much smaller grid steps in the plasma slab, namely 16 particles of each species per cell. The temporal resolution is $\delta t=0.02$ fs. Outgoing-particle boundary conditions are used. The electromagnetic fields are periodic in the transverse direction.

The incoming laser is launched from the left boundary of the simulation box with its electric field in the simulation plane. The Gaussian p-polarized laser pulse has a peak intensity of $I_0=2\times 10^{20}$ W/cm^2 and a wavelength of $\lambda_0=1$ μm, with a rise time of 10 fs, corresponding to the time for the light to cross the 3 μm of vacuum to the front surface of the plasma foil. It is focused on the target plasma with the full width at half maximum spot size $\sigma_x=3$ μm at $z=3$ μm.

Even for the lowest-density case ($n_{e0}<\gamma n_{\mathrm{cr}}$), the foil is not transparent to the laser light. Figure 1 illustrates the interaction process. One can see reflection of the laser light and deformation of the front electron surface by the ponderomotive pressure, which energizes and pushes the electrons into the slab.

III. DENSITY EFFECT ON PROTON ACCELERATION

At the very beginning of the interaction, the accelerated front-side electrons traverse the plasma slab and exit into the rear vacuum region. Huge (several times 10^{10} V/cm) space-

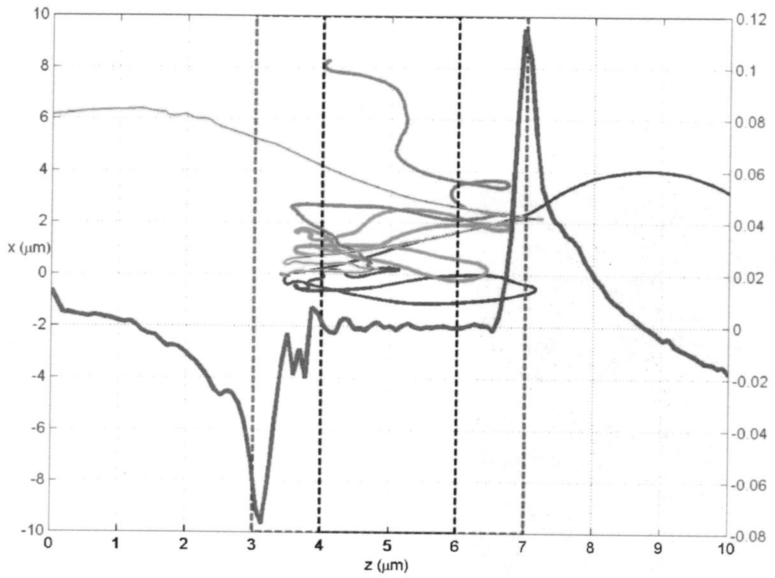

FIG. 2. (Color online) Trajectories of several typical electrons (for $t=5-100$ fs) initially at $(x, z)=(0, 3.5)$ μm for case II show the effects of electron reflection, stochastic heating, and collisions. The thick green solid line is the spatial-averaged longitudinal electric field E_z in units of the transverse E_0 field of the laser. The red dashed lines indicate the initial target surfaces in the simulation box and the black dashed lines indicate the initial surfaces of the 2-μm-thick homogeneous section of the plasma at density n_0.

charge electrostatic fields at the front and backsides of the slab are generated, as shown in Fig. 2. These fields create across the plasma slab an electrostatic potential well. Thus, before their eventual exit into the front or back vacuum regions, the laser-driven electrons can experience multiple reflections, since their mean free path is much larger than the slab thickness. Furthermore, in conjunction with the large self-generated magnetic field, the energized electrons spread along and heat the slab surfaces. These phenomena are illustrated in Fig. 2 by several trajectories of electrons initially at rest at $(x, z)=(0, 3.5)$ μm.

Figure 3 shows the density profiles of the electrons, protons, and C^+ ions at $t=75$ fs. The top, center, and lower rows correspond to cases I, II, and III, respectively, and the left, center, and right columns correspond to the electrons, protons, and C^+ ions, respectively. Since the laser pulse is reflected after the impact with the plasma slab, the recirculating electrons prevent the formation of a *bona fide* electron hole in or at the front side of the slab in all three cases (Fig. 3, left column). For the lowest-density case I [panel (a)], there are many more fast electrons at both the front and rear vacuum regions than the higher-density cases II and III [panels (d) and (g)]. This is because for that case there is a small underdense region in the front density ramp. The electrons thus also experience resonant absorption, $j\times B$ heating, betatron acceleration, laser self-focusing, etc.[50–52] The higher-density cases II and III have much steeper density gradients at the surface. The space-charge fields set up by the expelled electrons at the front and backslab surfaces accelerate the surface ions in both the backward and forward directions. At

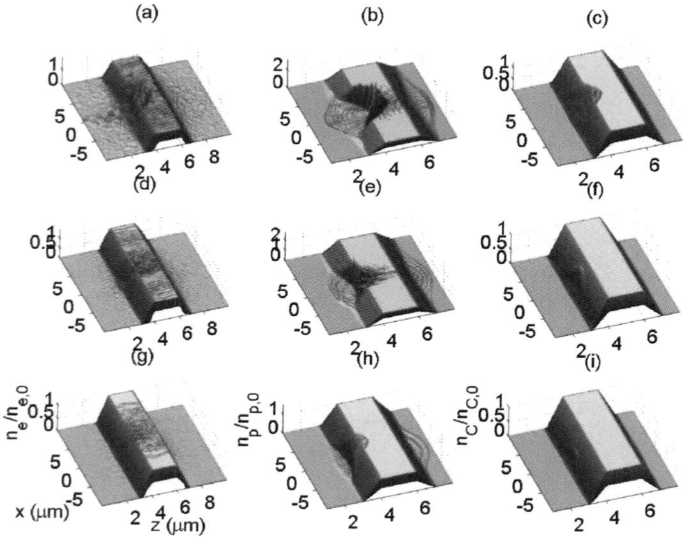

FIG. 3. (Color online) Snapshots of the density profile (in units of the initial maximum particle density n_0) of plasma electrons, protons, and C^+ ions at the simulation time $t=75$ fs. Left column: the electron density, middle column: the proton density, and right column: the C^+ ion density. The top row: case I ($\rho=0.2$ g/cm^3), the second row: case II ($\rho=1$ g/cm^3), and the bottom row: case III ($\rho=3$ g/cm^3).

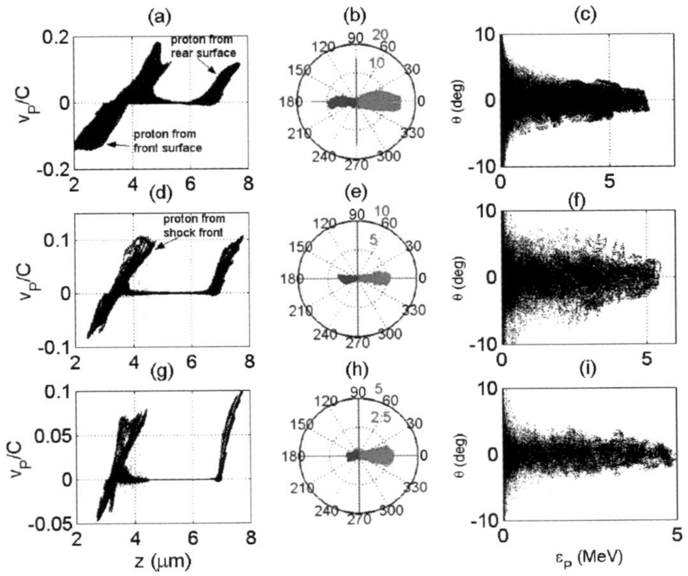

FIG. 4. (Color online) Proton velocities and angular distribution at t=75 fs. Left column: the velocities (or momenta), center column: the angular distribution of protons from the front surface, and right column: the angular distribution of proton energy. The energy values of the radial coordinate are given in red. The blue and red dots represent the backward ($v_z<0$) and forward ($v_z>0$) propagating energetic protons, respectively. Top row: case I, center row: case II, and bottom row: case III.

the front slab surface, the ions are driven backward into the left vacuum region by laser ablation and/or ponderomotive force, explosion, and forward into the target by the laser-generated collisionless shock. Accordingly, an ion-deficient region, or ion hole, is formed on the front surface. At the rear of the target, the surface ions are driven into the right vacuum region by the target-normal-sheath acceleration (TNSA) mechanism.

Depending on the target thickness and the laser intensity, shock-accelerated ions can even have higher energies than TNSA-accelerated ions. On the other hand, ion hole boring by the laser ponderomotive force depends on the densities, masses, and charge numbers of the plasma particles, as well as the laser intensity and wavelength. If we neglect the C^+ ions, balance between the photon and plasma pressures gives a hole-boring speed[14,50,53,54] $u=\{[Zm_en_{cr}/(2m_pn_e)I_0\lambda_0^2/(1.37\times10^{18})]\}^{1/2}c$, where Z is the charge, m_e is the electron mass, m_p is the proton mass, and c is the speed of light in vacuum. For the plasma densities considered, the hole-boring speeds are $0.051c$, $0.0228c$, and $0.0132c$. The corresponding hole depths are then 1.148, 0.513, and 0.296 μm, respectively. The latter agree well with the simulation results (1.53, 0.68, 0.35) μm, respectively, as given in Fig. 3, center column. The small discrepancy can be attributed to the plasma density ramp at the front of the slab and the C^+ ions that are present in the simulation. The effect of the latter can be seen from the relative modifications of the C^+ density profiles for the three cases in Fig. 3, right column. Thus, for a short duration (<100 fs) the simple estimate is still applicable for the proton hole-boring speed. However, for longer durations (for example, the order of a picosecond), the change of the estimate should include the effect of the C^+ ions, which will be investigated in our future work.

Shock-induced strong density compression at the edge of the proton and C^+ holes can also be observed in Fig. 3, center and right columns, respectively. In particular, the proton density at the front of the hole can reach about twice the initial density [Figs. 3(b) and 3(e)]. As expected, the proton and C^+ shock fronts do not overlap. This contributes to the complexity of the electric field structure at the front target surface and the electron trajectory there, as can be seen in Fig. 3.

For completeness, Fig. 4 shows the distributions of the longitudinal proton velocities at t=75 fs. The effects of the acceleration mechanisms operating at the front and back target surfaces can be observed. In particular, the left column shows that at lower density shock acceleration is more effective than TNSA, but at higher density the latter becomes more effective, although the efficiency of both mechanisms decreases with the plasma density. In case I, where several acceleration mechanisms are operative, there is considerable backward acceleration of the front-surface protons. In all three cases, at the backside of the target the surface protons experiencing TNSA can reach $v_p\gtrsim0.1c$. However, their number decreases with the increase of the plasma density. The velocity as well as the number of the shock-driven protons both decrease with the increase of the plasma density. The decrease of the velocity of the shock-driven forward protons is consistent with the earlier discussion on proton hole boring.

The center column of Fig. 4 shows the angular distribution of the proton energy at the front target surface. We see that the forward moving protons are higher in number and energy than the backward moving ones. The peak energies for the backward-accelerated protons are about 10, 3, and 1 MeV for cases I, II, and III, respectively, and that for the forward-accelerated protons are about 17, 5, and 2.5 MeV, respectively. The right column of Fig. 4 shows the angular distribution of proton energy distribution at the rear surface of the plasma slab. The peak proton energies are approximately 7, 5.5, and 5 MeV, for cases I, II, and III, respectively. It is mentioned that electrons on the target front side are expelled from high-intensity regions by the ponderomotive potential of the laser until it is balanced by the electrostatic potential arising from the charge separation. When the laser

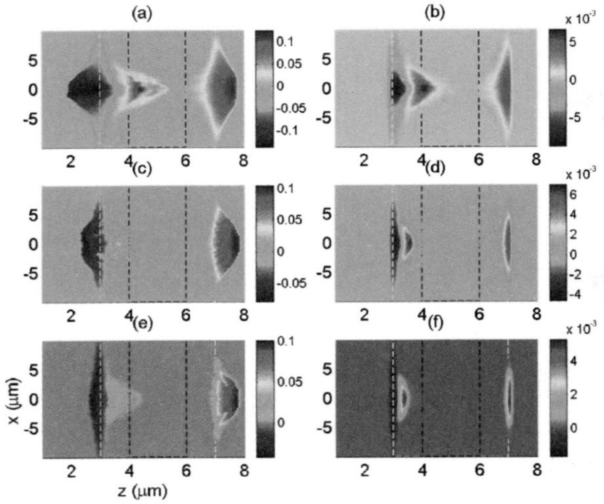

FIG. 5. (Color online) The intensity distribution of v_z (in units of c) of the protons and C^+ ions at $t=75$ fs. Left column: the proton velocity distribution and right column: the C^+ ion velocity distribution. Top row: case I, center row: case II, and bottom row: case III. The white dashed lines indicate the initial target surfaces in the simulation box and the black dashed lines indicate the initial surfaces of the 2-μm-thick homogeneous section of the plasma at density n_0.

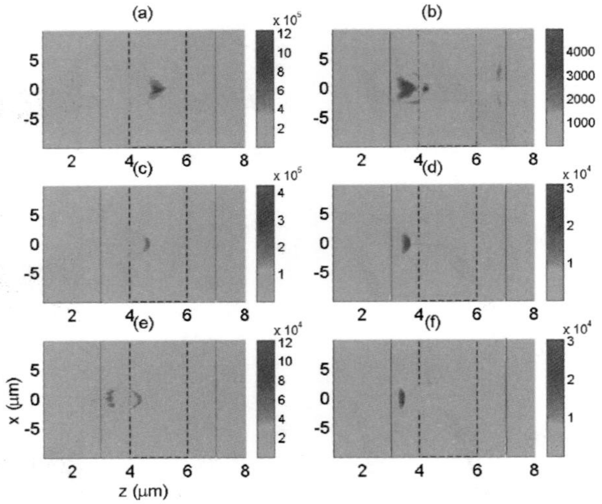

FIG. 6. (Color online) Snapshots of the thermal energy (in units of eV) of the protons and C^+ ions at $t=75$ fs. The descriptions are the same as in Fig. 5.

pulse is longer than the acceleration time $\tau=\lambda_0/c \times \sqrt{m_p/m_e\gamma}$, it was shown[55] that protons can initially gain kinetic energies approaching the ponderomotive potential, $\Phi_p=m_ec^2(\gamma-1)=m_ec^2\sqrt{(1+I_0\lambda_0^2/1.37\times 10^{18})^{1/2}-1}$. For our laser parameters, we have $\tau\approx 40$ fs, which is two times shorter than our pulse duration. Thus we can estimate[56] the cutoff energies of $1.5^2\times\Phi_p\approx 12.8$ MeV, which approximately agree with the numerical result for case I. However, it is noted that the estimation is very inappropriate for cases II and III. This is because the acceleration time should also be relevant to the density of thin solid target, as seen in Figs. 3 and 4.

Figure 4 shows that the most energetic protons are from shock acceleration in the lowest-density target. However, the TNSA protons at the backside of the slab have a much smaller emission cone than those accelerated from the front surface. The bright center spots ($<5°$) shown in the right column in Fig. 4 are well collimated. In fact, the highest-density case III [Fig. 4(i)] yields the most collimated proton beam. But the latter also has a lower maximum proton energy.

As a final point, we consider the effect of the C^+ ions. Since the C^+ ion has a longer response time due to its heavier mass, one generally thinks that the ion is immobile in 100 fs. However, Figs. 5 and 6 clearly show that both the proton and the C^+ ion strongly respond to both the laser field and space-charge electrostatic fields at the front and backsides of the plasma slab, although the accelerated C^+ ions indeed have much smaller velocity values than the protons, as compared to Figs. 5(b), 5(d), and 5(f) with Figs. 5(a), 5(c), and 5(e). The thermal energy of the accelerated protons and C^+ ions are also given in Fig. 6. Clearly, Figs. 6(a), 6(c), and 6(e) illustrate that the thermal energy of the protons accelerated from the shock front decreases with the increase of the plasma density. In particular, the thermal energy of the protons at the rear surface is quite small, which indicates that the divergent angle of TNSA-accelerated protons also decreases with increasing the plasma density. Furthermore, it is shown from Figs. 6(b), 6(d), and 6(f) that strong local heating of the carbon ions is clearly distinguishable for different plasma densities. In particular, shock-acceleration C^+ ions can have a higher thermal energy than TNSA-accelerated C^+ ions in a short time. We must emphasize that the peak energy of the accelerated protons and C^+ ions at both the front and rear surfaces of thin solid targets is possibly different with the simulation of petawatt experiments in proton fast ignition. To have a better description of the accelerated energy, a benchmark has to be set for future numerical descriptions of the complex process of ion accelerating in large-size laser-target interactions.

IV. CONCLUSION

In conclusion, we have used a two-dimensional hybrid particle-in-cell-fluid code to investigate the effect of the target density on energetic proton production in the interaction of a thin solid-density plastic slab with an ultraintense laser pulse. In our simulations, overdense foils of densities $\rho=0.2$, 1, and 3 g/cm^3 are considered. Traditional PIC techniques are impractical for studying such overdense plasmas. Our simulations show that at higher plasma densities the accelerated protons exiting at the rear foil surface are as expected of less energy but are more collimated. Furthermore, we show that TNSA-accelerated C^+ ions indeed have quite small forward-velocity values in a short time (<100 fs), but the difference of thermal energy of the protons and C^+ ions can be clearly distinguishable for different plasma densities. Our results also indicate that further investigations on the effect of the density and composition of the targets in laser interaction with larger-size solid-targets are warranted.

ACKNOWLEDGMENTS

This work is supported by the National Natural Science Foundation of China, Grant Nos. 10575013, 10576007, and 10335020, and the National High-Tech ICF Committee in China.

[1]R. A. Snavely *et al.*, Phys. Rev. Lett. **85**, 2945 (2000).
[2]S. C. Wilks *et al.*, Phys. Plasmas **8**, 542 (2001), and references therein.
[3]M. Borghesi *et al.*, Phys. Rev. Lett. **92**, 055003 (2004).
[4]I. Spencer *et al.*, Nucl. Instrum. Methods Phys. Res. B **183**, 449 (2001).
[5]U. Amaldi, Nucl. Phys. A **654**, C375 (1999).
[6]Y. Sentoku, V. Yu. Bychenkov, K. Flippo, A. Maksimchuk, K. Mima, G. Mourou, Z. M. Sheng, and D. Umstadter, Appl. Phys. B: Lasers Opt. **74**, 207 (2002).
[7]J. A. Cobble, R. P. Johnson, T. E. Cowan, N. Renard-Le Galloudec, and M. Allen, J. Appl. Phys. **92**, 1775 (2002).
[8]E. Lefebvre, F. d'Humires, S. Fritzler, and V. Malka, J. Appl. Phys. **100**, 113308 (2006).
[9]F. N. Beg *et al.*, Appl. Phys. Lett. **84**, 2766 (2004).
[10]J. Badziak, S. Jablonski, and S. Glowacz, Appl. Phys. Lett. **89**, 061504 (2006).
[11]J. Badziak, S. Glowacz, S. Jablonski, P. Parys, J. Wolowski, and H. Hora, Laser Part. Beams **23**, 401 (2005).
[12]B. M. Hegelich *et al.*, Nature (London) **439**, 441 (2006).
[13]M. Schnurer, S. Ter-Avetisyan, S. Busch, M. P. Kalachnikov, E. Risse, W. Sandner, and P. V. Nickles, Appl. Phys. B: Lasers Opt. **78**, 895 (2004).
[14]W. Yu, H. Xu, F. He, M. Y. Yu, S. Ishiguro, J. Zhang, and A. Y. Wong, Phys. Rev. E **72**, 046401 (2005).
[15]O. Klimo and J. Limpouch, Laser Part. Beams **24**, 107 (2006).
[16]S. Miyazaki, N. Ozaki, R. Sonobe, Q. Kong, S. Kawata, A. A. Andreev, and J. Limpouch, Laser Part. Beams **24**, 157 (2006).
[17]S. Kawata *et al.*, Laser Part. Beams **23**, 61 (2005).
[18]S. Ter-Avetisyan *et al.*, Phys. Rev. Lett. **96**, 145006 (2006).
[19]M. Borghesi, J. Fuchs, S. V. Bulanov, A. J. Mackinnon, P. K. Patel, and M. Roth, Fusion Sci. Technol. **49**, 412 (2006), and references therein.
[20]E. d'Humières, E. Lefebvre, L. Gremillet, and V. Malka, Phys. Plasmas **12**, 062704 (2005).
[21]H. Hora *et al.*, Opt. Commun. **207**, 333 (2002).
[22]M. Roth *et al.*, Phys. Rev. Lett. **86**, 436 (2001).
[23]M. H. Key, R. R. Greeman, S. P. Hatchett, A. J. Mackinnon, P. K. Patel, R. A. Snavely, and R. B. Stephens, Fusion Sci. Technol. **49**, 440 (2006), and references therein.
[24]J. Badziak, S. Glowacz, H. Hora, S. Jablonski, and J. Wolowski, Laser Part. Beams **24**, 249 (2006).
[25]H. Hora *et al.*, Laser Part. Beams **23**, 423 (2005).
[26]P. K. Patel *et al.*, Phys. Rev. Lett. **91**, 125004 (2003).
[27]T. Z. Esirkepov *et al.*, Phys. Rev. Lett. **89**, 175003 (2002).
[28]J. Fuchs *et al.*, Phys. Rev. Lett. **94**, 045004 (2005).
[29]P. Gibbon, E. L. Clark, R. G. Evans, and M. Zepf, Phys. Plasmas **11**, 4032 (2004).
[30]P. Gibbon, Phys. Rev. E **72**, 026411 (2005).
[31]P. Mora, Phys. Rev. E **72**, 056401 (2005).
[32]A. F. Lifshitz, J. Faure, Y. Glinec, V. Malka, and P. More, Laser Part. Beams **24**, 255 (2006).
[33]S. P. D. Mangels *et al.*, Laser Part. Beams **24**, 185 (2006).
[34]C. K. Birdsall and A. B. Langdon, *Plasma Physics via Computer Simulation* (McGraw-Hill, New York, 1985).
[35]M. T. Batchelor and R. T. Stening, Laser Part. Beams **3**, 189 (1985).
[36]A. S. Lipatov, *The Hybrid Multiscale Simulation Technology* (Springer, Berlin, 2002).
[37]L. Gremillet, G. Bonnaud, and F. Amiranoff, Phys. Plasmas **9**, 941 (2002).
[38]R. J. Mason, E. S. Doaa, and B. J. Albright, Phys. Rev. E **72**, 015401 (2005).
[39]J. R. Davies, Phys. Rev. E **68**, 056404 (2003).
[40]D. R. Welch, D. V. Rose, B. V. Oliver, and R. E. Clark, Nucl. Instrum. Methods Phys. Res. A **464**, 134 (2001).
[41]D. R. Welch, D. V. Rose, M. E. Cuneo, R. B. Campbell, and R. A. Mehlhorn, Phys. Plasmas **13**, 063105 (2006).
[42]R. J. Mason, Phys. Rev. Lett. **96**, 035001 (2006).
[43]J. J. Honrubia, M. Kaluza, J. Schreiber, G. D. Tsakiris, and J. Meyer-ter-Vehn, Phys. Plasmas **12**, 052708 (2005).
[44]R. R. Freeman, D. Batani, S. Baton, M. Key, and R. Stephens, Fusion Sci. Technol. **49**, 297 (2006).
[45]Y. T. Lee and R. M. More, Phys. Fluids **25**, 1237 (1984).
[46]M. Tabak *et al.*, Phys. Plasmas **12**, 057305 (2005).
[47]C. T. Zhou, X. T. He, and M. Y. Yu, Phys. Plasmas **13**, 092109 (2006).
[48]C. T. Zhou, M. Y. Yu, and X. T. He, Laser Part. Beams **52** (No. 2) (2007).
[49]C. T. Zhou and X. T. He, Appl. Phys. Lett. **90**, 031503 (2007).
[50]See, for example, P. Gibbon, *Short Pulse Laser Interactions with Matter—An Introduction* (Imperial College Press, London, 2005).
[51]M. Y. Yu, W. Yu, Z. Y. Chen, J. Zhang, Y. Yin, L. H. Cao, P. X. Lu, and Z. Z. Xu, Phys. Plasmas **10**, 2468 (2003).
[52]H. Hora, M. Hoelss, W. Scheid, J. X. Wang, Y. K. Ho, F. Osman, and R. Castillo, Laser Part. Beams **18**, 135 (2000).
[53]S. C. Wilks, W. L. Kruer, M. Tabak, and A. B. Langdon, Phys. Rev. Lett. **69**, 1383 (1992).
[54]A. Pukhov and J. Mayer-ter-Vehn, Phys. Rev. Lett. **79**, 2686 (1997).
[55]Y. Sentoku, T. E. Cowan, A. Kemp, and H. Ruhl, Phys. Plasmas **10**, 2009 (2003).
[56]M. Kaluza, J. Schreiber, M. I. K. Santala, G. D. Tsakiris, K. Eidmann, J. Meyer-ter-Vehn, and K. J. Witte, Phys. Rev. Lett. **93**, 045003 (2004).

APPLIED PHYSICS LETTERS **90**, 031503 (2007)

Influence of a large oblique incident angle on energetic protons accelerated from solid-density plasmas by ultraintense laser pulses

C. T. Zhou[a] and X. T. He
Institute of Applied Physics and Computational Mathematics, P.O. Box 8009, Beijing 100088, People's Republic of China and Institute for Fusion Theory and Simulation, Zhejiang University, Hangzhou 310027, People's Republic of China

(Received 10 November 2006; accepted 13 December 2006; published online 18 January 2007)

The acceleration of energetic electron, proton, and heavy ion beams produced by ultrahigh-intensity laser pulses through thin plastic targets is studied using two-dimensional hybrid particle-in-cell simulation. When the laser is incident at a large angle, the proton beams accelerated from the front and rear surfaces of the target deviate from the normal direction because of the formation of non-Gaussian asymmetric sheath field at the target surfaces. In particular, the simulations clearly show that the proton beam in the backward direction can have higher Bragg peak energy than that of the forward direction if the incident angle is sufficiently large. © *2007 American Institute of Physics*. [DOI: 10.1063/1.2432242]

Although ion acceleration driven by laser-plasma interaction has been investigated for more than two decades,[1] only recently has multi-MeV proton acceleration in thin foils by ultraintense laser pulses received great attention due to its extensive applications.[2–7] When the laser pulse hits the target, electrons driven out of the target front (laser) side by the ponderomotive force set up electrostatic fields that accelerate protons backward against the laser direction. On the other hand, the electrons in the target front side can also be accelerated forward by the ponderomotive force and propagate through the target, forming a thin Debye sheath at the target rear side. When the laser is along the target normal direction or at a small angle,[8] the emission cone of the protons is still aligned along the target *normal* direction, since the field gradients on both target surfaces responsible for the acceleration of the protons are parallel to the surface and, in particular, perpendicular to the rear surface. Furthermore the electron sheath has a Gaussian profile, and the central region as well as the edges of the sheath will expel protons normal to the surface. The Bragg peak proton energy is at the center of the resulting Gaussian proton beam.

Over the past years, research is mainly focused on the forward-accelerated proton beams as these should exhibit higher beam quality and higher energies. Experimentally, different laser systems with different solid targets have been considered. The laser intensities range from a few times $I_0 = 10^{18}$ to 10^{20} W/cm^2. In particular, different target materials,[3,4] involving monolayer metallic or plastic targets or double-layer foils, have been explored. A common conclusion[5,9] is that proton beams accelerated by ultraintense laser pulses exhibit a remarkable degree of beam collimation and laminarity and high cutoff energy, and the emission is along the normal to the rear target surface. In addition to proton acceleration in the forward direction, electron expansion at the front surface of the target can also lead to proton acceleration in the backward direction.[2,5] The Bragg peak proton energy, beam emission direction, and other properties of the accelerated protons depend strongly on the laser intensity and pulse length, as well as the target thickness.

Since the protons are accelerated by the electron sheath on the target surface, their spatial and angular characteristics are determined by the electron sheath spatial distribution.[3–5,7,9,10] Therefore the acceleration and transport processes of the fast electrons through the target have a crucial effect on the ions accelerated from the target surface. Surface acceleration of fast electrons from a foil target irradiated by ultraintense laser pulses has recently been discussed,[11] but the effect of large oblique incident angle on the energetic ions accelerated from solid-density plasmas has never been studied in detail. In this letter, we investigate this problem by two-dimensional simulation of solid-density plastic slab plasmas irradiated by the laser. It should be mentioned that the standard particle-in-cell (PIC) technique is not applicable for simulating a real solid target because of present-day limitations in computation resources. Recently, a hybrid fluid PIC code[12] has been employed to simulate electron and proton acceleration by laser-solid interaction. We have developed a two-dimensional electromagnetic hybrid PIC code including both PIC kinetic and fluid electrons to study in detail energetic-particle generation and acceleration from solid-density plastic foils.

We start with a brief illustration of our numerical model and physical parameters. The plasma is initialized at 5 eV with fixed charge-state ions. The plastic slab was initialized as a fully dissociated $C^+H_2^+$ plasma with solid density, corresponding to the number densities $[n_{0,C^+}, n_{0,H_2^+}, n_{0,e}](\mathrm{cm}^{-3}) = [6\times10^{22}, 1.2\times10^{22}, 8.4\times10^{22}]$ of carbon ion, proton, and electron, respectively. The actual simulation box (x,z) is $50\times50\ \mu\mathrm{m}^2$. There are 800 uniform grids in the laser polarization (x) direction. However, the grid in the laser propagation (z) direction is nonuniform. The grid step δz within the vacuum regions is 0.05 μm, enough to resolve the laser wavelength. To solve the laser-plasma interaction problem under relativistic critical density, we have chosen a much smaller grid step for the plasma slab. In the simulations, there are 16 particles of each species (C^+ ion, H^+ proton, PIC, and fluid electrons) per cell. The temporal resolution is $\delta t = 0.02$ fs. Outgoing-particle boundary conditions are used.

The electromagnetic fields are periodic in the transverse direction. The incoming laser light is launched from the left

[a]Electronic mail: zcangtao@iapcm.ac.cn

0003-6951/2007/90(3)/031503/3/$23.00

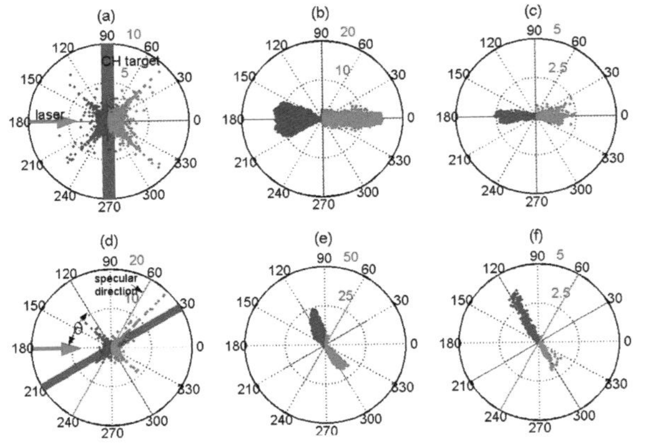

FIG. 1. (Color online) Angular distributions of electron, proton, and C^+ ion energies at $t=500$ fs. The frames of the first and second rows correspond to $\theta=0°$ and 60°, and the first, second, and third columns show the electrons, protons, and C^+ ions, respectively. The CH target (green), the incoming laser, and the angle between the laser and the target normal are shown in (a) and (d). The blue and red dots represent the backward ($v_z<0$) and forward ($v_z>0$) particles. The energy levels (in units of MeV, radial circles) are labeled in red.

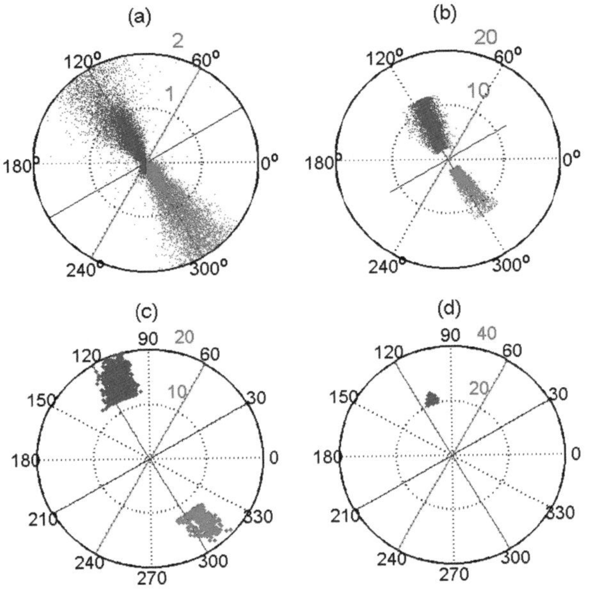

FIG. 2. (Color online) Angular distribution of proton energy, corresponding to Fig. 1(e), at four different energy levels for $\theta=60°$; (a) $\epsilon_p<2$ MeV, (b) $2\leqslant\epsilon_p<12$ MeV, (c) $12\leqslant\epsilon_p<20$ MeV, and $\epsilon_p\geqslant 20$ MeV.

boundary with the peak intensity $I_0=3\times10^{20}$ W/cm^2 for $\lambda_0=1$ μm and rise time of 60 fs. The electric field of the laser is in the simulation plane (p polarization) and has a transverse spot size (Gaussian radial profile) on the target of 6 μm full width at half maximum. The laser beam is focused at the edge of the high-density region. In the numerical simulations, we consider two cases: $\theta=0°$ and $\theta\neq0°$, where θ is the angle between the laser axis and the target normal direction. For $\theta=0°$, the plasma slab is located in $19<z(\mu\text{m})<26$ and initially uniform in density in the transverse direction. In the propagation direction there is a 1 μm linear rise to the specified uniform density n_0 on both sides, but the initial lowest density at $z=19$ μm is taken to be $n_0/100$. For $\theta\neq0°$, the plasma density and the incident laser are the same as in $\theta=0°$, but the target is rotated with respect to the laser axis by an angle θ, and the target thickness is reduced to 2 μm with a 0.1 μm linear rise in order to rule out thickness effects.

Figure 1(a) shows that emission of the accelerated electrons for $\theta=0°$ is of X-like angular structure. This is because the half-angle of the divergence of the very high-energy electron beam at the target surface may be approximately given by $\alpha=\pm\text{tg}^{-1}P_\perp/P_\parallel\approx\pm45°$ (since $P_\perp\approx P_\parallel$), where $P_\perp$ and $P_\parallel$ are the transverse and parallel momenta, respectively. For $\theta=60°$, however, a fast electron beam along the target surface is generated by the quasistatic surface, electromagnetic fields.[11] Fig. 1(d) indeed indicates that the maximum electron energy (~20 MeV) of the surface electrons is close to the laser-side specular direction, but the maximum energy of the surface electrons at the rear-side specular direction is only 5 MeV. On the other hand, the electrons that are electrostatically confined on the target surface can contribute to the acceleration of protons and heavy ions. For $\theta=0°$, the charge-separation field follows a Gaussian profile.[3,4,9] However, with a large oblique incident angle, the symmetrical spatial structure is not seen. The electron recirculating process, further increases the effective electron density on the front and rear surfaces. Therefore, the energy, spatial and angular characteristics of the protons will also depend on the incident angle.

The angular distributions of the protons and heavy ions for normal and oblique incidence are shown in Figs. 1(b) and 1(c) and Figs. 1(e) and 1(f). As expected, the proton and carbon ion beams for normal incidence are of good symmetry with respect to the normal direction. Furthermore, a smaller emission angle of the forward proton beam, as well as a higher cutoff proton energy, is observed. For $\theta=60°$, Fig. 1(f) shows that the carbon ion beams at the target front and rear sides are accelerated mainly along target normal, but Fig. 1(e) shows that the proton beams at the two sides deviate from the normal.

To highlight the details of the accelerated proton energy and angular spread, Fig. 2 shows the same angular distribution of protons as in Fig. 1(e) by distinguishing four different energy levels. For $\epsilon_p<2$ MeV, Fig. 2(a) shows that the protons are accelerated along the target normal on both the front and rear surfaces of the target. For $2\leqslant\epsilon_p<12$ MeV, Fig. 2(b) indicates that the emission cone of the protons is still almost along the target normal direction. However, Fig. 2(c), for $12\leqslant\epsilon_p<20$ MeV, clearly illustrates that proton beams are no longer perpendicular to the target normal. Figure 2(d) further shows that protons with $\epsilon_p\geqslant20$ MeV in the backward direction are quite off the target normal. The deviation angle is about 12°, which differs from the analytic result[13] for a plane wave. Moreover, the backward proton beam has a higher cutoff energy than the forward beam.

Figure 3(a) shows the proton density distribution. We can see that the protons are indeed not accelerated normal to the target surface. Figures 3(b) and 3(c) also illustrate that the energy distribution at the backward and forward directions is no longer of a symmetrical Gaussian-like structure. In particular, two Bragg energy peaks are observed in the backward-accelerated proton beam. These results further confirm that the front-surface electrons[11] confined by the

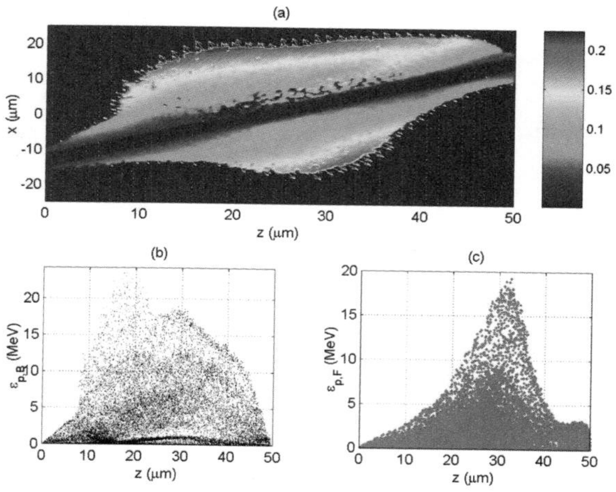

FIG. 3. (Color online) (a) Proton velocity **v** (in units of C) and its intensity distribution at t=500 fs. (b) and (c) show the corresponding energy distribution of both backward and forward directions, respectively.

quasistatic surface electromagnetic fields can seriously influence on the emission of the backward-accelerated protons.

The conversion efficiency of laser energy into particle energy has also been analyzed. For $\theta=0°$, about 35% laser is converted into a fast electron population with temperature larger than 1 MeV. This conversion efficiency is somewhat lower than the known scaling law $\eta \approx 0.000\,175 \times I^{0.2661}$, where I is in W/cm^2. For $\theta=60°$, however, a fraction of the laser electric field is reflected in the specular direction, leading to only 18% conversion efficiency. Moreover, the rate of electron energy transfer to the protons and C^+ ions is found to increase initially from zero and peak at the level associated with self-similar expansion.[4] The corresponding rates of electron energy transfer to the high-energy (>2 MeV) proton beams for $\theta=0°$ and 60° are about 14% and 25%, respectively. The high conversion efficiency for $\theta=60°$ is consistent with the angular distribution of proton energy, as given in Fig. 1(e).

In conclusion, we have used a two-dimensional hybrid particle-in-cell code to investigate the influence of a large oblique incident angle on the energetic protons accelerated from a thin plastic target by an ultraintense laser pulse. Our integrated simulation scheme clearly illustrates for the first time that the proton beams accelerated from the front and rear target surfaces are *not* completely aligned along the target normal. This is because the electron sheath on the surface does not follow a symmetric Gaussian spatial profile if the incident angle is sufficiently large. Moreover, the proton peak energy at the target front side can be higher than that at the rear. The results here are clearly essential in realizing many potential applications of plasma proton accelerators in, for example,[3–7] proton radiography, medical isotope preparation, cancer therapy, isochoric heating, fast ignition of fusion fuel, and in other high-energy density physics. In particular, energetic protons at the front of the target can be useful for compact neutron sources.

This work is supported by the National Natural Science Foundation of China, Grant Nos. 10575013, 10576007, and 10335020, and the National High-Tech ICF Committee in China.

[1]S. J. Gitomer, R. D. Jones, F. Begay, A. W. Ehler, J. F. Kephart, and R. Kristal, Phys. Fluids **29**, 2679 (1986).
[2]J. Fuchs, P. Antici, E. D'humiéres, E. Lefebvre, M. Borghesi, E. Brambrink, C. A. Cecchetti, M. Kaluza, V. Malka, M. Manclossi, S. Meyroneinc, P. Mora, J. Schreiber, T. Toncian, H. Pépin, and P. Audebert, **2**, 48 (2006).
[3]M. Borghesi, J. Fuchs, S. V. Bulanov, A. J. Mackinnon, P. K. Patel, and M. Roth, Adv. Stud. Contemp. Math. **49**, 412 (2006), and references therein.
[4]M. H. Key, R. R. Greeman, S. P. Hatchett, A. J. Mackinnon, P. K. Patel, R. A. Snavely, and R. B. Stephens, Fusion Sci. Technol. **49**, 440 (2006), and references therein.
[5]J. Fuchs, Y. Sentoku, S. Karsch, J. Cobble, P. Audebert, A. Kemp, A. Nikroo, P. Antici, E. Brambrink, A. Blazevic, E. M. Campbell, J. C. Fernández, J.-C. Gauthier, M. Geissel, M. Hegelich, H. Ppin, H. Popescu, N. Renard-LeGalloudec, M. Roth, J. Schreiber, R. Stephens, and T. E. Cowan, Phys. Rev. Lett. **94**, 045004 (2005).
[6]F. N. Beg, M. S. Wei, A. E. Dangor, A. Gopal, M. Tatarakis, K. Krushelnick, P. Gibbon, E. L. Clark, R. G. Evans, K. L. Lancaster, P. A. Norreys, K. W. D. Ledingham, P. McKenna, and M. Zepf, Appl. Phys. Lett. **84**, 2766 (2004).
[7]J. Badziak, S. Jablonski, and S. Glowacz, Appl. Phys. Lett. **89**, 061504 (2006).
[8]H. Habara, R. Kodama, Y. Sentoku, N. Izumi, Y. Kitagwa, K. A. Tanaka, K. Mima, and T. Yamanaka, Phys. Rev. E **70**, 046414 (2004).
[9]S. C. Wilks, A. B. Langdon, T. E. Cowan, M. Roth, M. Singh, S. Hatchett, M. H. Key, D. Pennington, A. MacKinnon, and R. A. Snavely, Phys. Plasmas **8**, 542 (2001), and references therein.
[10]W. Yu, V. Bychenkov, Y. Sentoku, M. Y. Yu, Z. M. Sheng, and K. Mima, Phys. Rev. Lett. **85**, 570 (2000).
[11]Y. T. Li, X. H. Yuan, M. H. Xu, Z. Y. Zheng, Z. M. Sheng, M. Chen, Y. Y. Ma, W. X. Liang, Q. Z. Yu, Y. Zhang, F. Liu, Z. H. Wang, Z. Y. Wei, W. Zhao, Z. Jin, and J. Zhang, Phys. Rev. Lett. **96**, 165003 (2006).
[12]D. R. Welch, D. V. Rose, M. E. Cuneo, R. B. Campbell, and R. A. Methlhorn, Phys. Plasmas **13**, 063105 (2006).
[13]Z. M. Sheng, Y. Sentoku, K. Mima, J. Zhang, W. Yu, and J. Mayer-ter-Vehn, Phys. Rev. Lett. **85**, 5340 (2000).

epl A LETTERS JOURNAL EXPLORING THE FRONTIERS OF PHYSICS

September 2011

EPL, **95** (2011) 55005
doi: 10.1209/0295-5075/95/55005

www.epljournal.org

An ultra-short and TeV quasi-monoenergetic ion beam generation by laser wakefield accelerator in the snowplow regime

F. L. ZHENG[1,2(a)], S. Z. WU[2,4], C. T. ZHOU[2,4], H. Y. WANG[2,3], X. Q. YAN[2,3(b)] and X. T. HE[2,4(c)]

[1] *Graduate School, China Academy of Engineering Physics - P.O. Box 8009, Beijing 100088, China*
[2] *Center for Applied Physics and Technology, Peking University - Beijing 100871, China*
[3] *State Key Laboratory of Nuclear Physics and Technology, Peking University - Beijing 100871, China*
[4] *Institute of Applied Physics and Computational Mathematics - P.O. Box 8009, Beijing 100088, China*

received 8 April 2011; accepted in final form 20 July 2011
published online 25 August 2011

PACS 52.38.Kd – Laser-plasma acceleration of electrons and ions
PACS 41.75.Jv – Laser-driven acceleration
PACS 52.35.Mw – Nonlinear phenomena: waves, wave propagation, and other interactions (including parametric effects, mode coupling, ponderomotive effects, etc.)

Abstract – A new laser-plasma ion acceleration mechanism, snowplow ion acceleration, is proposed using an ultra-relativistically intense laser pulse irradiating on a combination target. When the thickness of the foil D is less than the length where the double-layer consisting of electron and ion layers is formed, the relativistic ion beam pre-accelerated by radiation pressure acceleration can be trapped and accelerated to the TeV level by the laser plasma wakefield over a long distance in the snowplow regime. Based on the classic wakefield theory, the snowplow structure can control the beam quality in terms of pulse duration, leading to the fact that the heavy-ion beam is theoretically shorter than half of the total wakefield structure, the size of the acceleration region. An analytical model is established, suggesting that ultra-short (70 μm) and ultra-highly energetic (3.2 TeV) carbon ion bunches generated by a centimeter-scale laser wakefield acceleration are expected, driven by a circularly polarized (CP) laser pulse with intensity of 10^{23} W/cm^2 and duration of 66 fs. The particle-in-cell simulations agree well with theoretical results.

Introduction. – Laser plasma accelerators have produced high-quality ion beams that compare favourably with conventional acceleration techniques in terms of emittance, brightness and pulse duration [1]. A long-term difficulty associated with laser-plasma acceleration —the very broad, exponential energy spectrum of the emitted particles— has been overcome recently for electron beams [2]. Quasi-monoenergetic GeV electron beams can be generated by a centimeter-scale laser wakefield accelerator [3]. Here we report an acceleration scheme for ions, especially the generation of TeV quasi-monoenergetic C^{6+} carbon ion beams using the ultra-relativistically intense laser-plasma accelerator in the snowplow regime. Such laser-driven, quasi-monoenergetic TeV ion source may enable significant advances in the development of compact ultra-high energy ion accelerator [4] and laboratory astrophysics [5]. All these applications may be enabled by the near-term high-intensity lasers, which will be capable to produce pulses with intensity of 10^{22}–10^{24} W/cm^2 [6].

There are several approaches in producing high-energy carbon ion beams. One is the target normal sheath acceleration (TNSA) [7] using linearly polarized laser pulse irradiating on a thick target, yielding carbon C^{5+} peak energy of 3.5 MeV/u [8]. Varying the target thickness to nanometer scale, an unprecedented maximum energy of 185 MeV (15 MeV/u) for C^{6+} beams is achieved experimentally [9], which is attributed to the relativistic transparency regime. In the version of radiation pressure acceleration [10], a quasi-monoenergetic C^{6+} beam with peak centered at 30 MeV is obtained by the circularly polarized laser pulse irradiating on a nanometer-thin DLC foil [11]. Another main scheme, dubbed "laser break-out afterburner" (BOA), enables the acceleration of carbon ions to more than 2 GeV theoretically [12]. The limitation with the above laser-driven ion acceleration mechanism has been an acceleration length shorter than hundreds of micrometers, although the accelerating field is over tens of TeV/m. Therefore it is difficult to further increase the

(a) E-mail: zhengfl_ccnu@yahoo.com.cn
(b) E-mail: X.Yan@pku.edu.cn
(c) E-mail: xthe@iapcm.ac.cn

ion kinetic energy to very high energy (*i.e.*, TeV). Lately Shen [13] and Yu [14] reported that a laser pulse with an ultra-relativistic intensity can excite a high-amplitude electric field in the underdense plasma, which can accelerate protons to tens of GeV. Recently we have reported for the first time that sub-TeV quasi-monoenergetic proton bunches can be generated by an ultra-relativistically intense laser wakefield accelerator, which is attributed to the snowplow ion acceleration (SPA) [15].

Acceleration scheme in generating ultra-short and TeV quasi-monoenergetic heavy-ion beam. – In this letter, we present the first theoretical study indicating that besides protons also high-quality, ultra-high energy ion beams can be generated by the centimeter-scale laser wakefield accelerator in the snowplow regime. An analytic model is established for heavy-ion acceleration based on the relativistic dynamical equation. Our model shows that the maximum energy of heavy-ion scales with the charge number Z in comparison with protons by the laser wakefield accelerator and the theoretically maximum energy of about 3.2 TeV for carbon ion beam has been verified by PIC simulations. A combined target shows that a planar carbon target is followed by underdense background gas. The whole stable ion acceleration regime is composed of two connected stages: firstly, relativistic heavy-ion beams generated by radiation pressure acceleration (RPA) in the laser-foil interaction and then snowplow ion acceleration (SPA) over a long distance for ion beams in the gas plasma. At the beginning of the laser-foil interaction, the electrons are pushed at first by the steady ponderomotive force and quickly piled up in a compressed layer in front of the laser pulse. Since the thickness of the foil D is less than the length where the double layer consisting of electron and ion layers is formed and $D > l_s$, the plasma skin depth, resulting in that light pressure exerted on the electrons, is larger than the charge-separated force. Hence the electron layer runs faster than the carbon ion layer and the double layers cannot be formed in the laser-foil stage. The C^{6+} carbon ion beams are rapidly accelerated to the relativistic level by the strong charge-separated field. Then the CP pulse behind the highly compressed electron layer traverses the foil and drives the nonlinear electric field in the gas plasma, in the wake of the laser pulse which can capture and continuously accelerate the relativistic C^{6+} ions over a long distance. Carbon ions are mainly accelerated by the laser-plasma wakefield in the snowplow regime in the underdense gas region, which is totally different from the laser-foil interaction where the Rayleigh-Taylor instability occurs. This disadvantageous effect is thus ignored in the present scheme and a very stable acceleration structure over a long distance is available. This acceleration scenario is referred to as the so-called snowplow ion acceleration which is connected closely with ref. [16]. In order to explore the main physics mechanism and construct an analytical model, in this letter the one-dimensional theory and PIC simulations are considered.

Dynamical process for the ion beam in the snowplow regime. – For a species of heavy ions in the snowplow regime, the dynamical equation of the relativistic heavy ion can be given as

$$\frac{\mathrm{d}P}{q_i \mathrm{d}t} = E_x, \tag{1}$$

where the momentum of the heavy ion is equal to $P = \gamma m_i c\beta$, and E_x is the longitudinal electrostatic field. Here γ is the relativistic factor, m_i the heavy-ion mass, q_i the charge quantity of the heavy ion, c the light speed in the vacuum space, $\beta = v/c$, v the velocity of the ion, respectively. Substituting the above physical quantities back into eq. (1), we can get the relativistic factor for the heavy ion in the snowplow regime in integration form,

$$\gamma \simeq \int \frac{q_i}{m_i}\frac{E_x}{c}\mathrm{d}t \simeq \int \frac{q_i}{m_i}\frac{E_x}{c^2}\mathrm{d}x. \tag{2}$$

Adopting the scaling law in the snowplow regime [15], the maximum longitudinal electrostatic field that the heavy ion experiences is $E_x = 3\pi(a_0 n_e/n_c)^{1/2}E_0$ with dephasing length $L_{dp} \simeq \frac{1}{36\pi^2}(a_0 n_c/n_e)^{3/2}\lambda_l$, where the electric-field strength is normalized by $E_0 = m_e\omega c/e$, a_0 is the dimensionless laser pulse amplitude, n_e the underdense plasma density, $n_c = \pi m_e c^2/e^2\lambda_l^2$ the critical plasma density, m_e the electron mass, e the electron charge, λ_l the laser wavelength in the vacuum, respectively. Therefore, the maximum energy of the heavy ion with charge mass ratio $\frac{Z}{A}$ can be given as

$$W_{max} = (\gamma_{max} - 1)m_i c^2 \simeq \frac{Z}{6}\frac{a_0^2 n_c}{n_e} m_e c^2. \tag{3}$$

This implies that in the snowplow regime the ion energy scales with the charge number Z in comparison with the proton. Based on eq. (3), the maximum energy for C^{6+} carbon ion beams theoretically is 6 times higher than that of the proton beams by the laser wakefield accelerator. In addition, we define η as the dephasing length ratio between heavy ion and proton in the snowplow regime. Considering that the initial velocity of the heavy ion is less than that of the proton due to the lower charge mass ratio in the initial injection into the wakefield stage, the heavy ion will undergo a longer acceleration distance than the proton with dephasing length ratio $\eta \geqslant 1$. According to the classic wakefield theory [17], the snowplow structure has length $l_0 \simeq (\frac{a_0 n_c}{n_e})^{1/2}\lambda_l$, which can control the ion beam quality in terms of pulse duration, corresponding to the length of the highly energetic ion beam which is less than $\frac{1}{2}l_0$, the size of acceleration region in the snowplow structure.

Direct numerical simulation by particle-in-cell code. – Accordingly, we carried out a simulation using a fully relativistic particle-in-cell code (KLAP) [14,18] to the show dynamical acceleration process. In our simulation, we

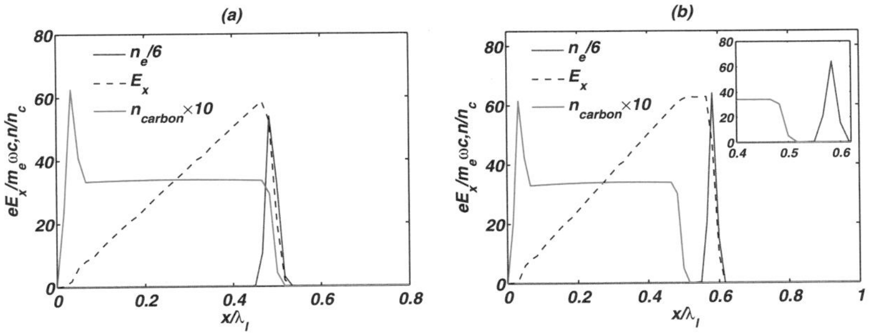

Fig. 1: (Color online) Snapshots of the laser pulse interaction with a combined target electron density n_e/n_c (blue solid line), C^{6+} ion density n_{carbon}/n_c (the red solid line) and longitudinal electric field $eE_x/m_e\omega c$ (black dashed line) at $t=0.5T_l$ (a) and the distance between the electron layer and the carbon ion layer (in the inset) at $t=0.6T_l$ (b). The field strength is normalized by $E_0=m_e\omega c/e$.

consider a CP laser pulse with wavelength $\lambda_l=1\,\mu\text{m}$ and peak intensity $I_0=1.7125\times10^{23}\,\text{W/cm}^2$, corresponding to the peak dimensionless laser amplitude $a_0=eA/m_ec^2=250$, where A is the laser vector potential. The laser pulse has a longitudinal flap-topped shape for $20\lambda_l$. At time $t=0$ it is normally incident from the left on a uniform, fully ionized carbon foil of thickness $D=0.5\,\mu\text{m}$ and normalized density $N=n_0/n_c=20$, where the electron density is given in units of the critical density n_c. Behind the carbon foil is the tenuous underdense gas with a charge-to-mass ratio of 1/3 and plasma density $n_e=0.01n_c$. The length of the gas plasma is $15000\lambda_l$ (1.5 cm).

Non–double-layer regime in the laser-foil acceleration process for the heavy ion. – Initially, for CP pulses, the ponderomotive force without no oscillating component quickly pushes electrons to the rear of the foil to form a highly compressed electron layer, when the laser pulse impinges on the carbon target surface. The electrons that piled up there thus reach a density far greater than γn_c, which prevents the laser pulse passing through the foil. The dynamic evolution processes of laser-foil interaction are simulated as shown in fig. 1(a); $\gamma=(1+I\lambda_l^2/2.74\times10^{18})^{1/2}$, where I is the laser intensity, λ_l is in units of micrometer. Meanwhile the C^{6+} ions in the foil are accelerated by the separated charge field. Since the thickness of the foil satisfies

$$l_s<D<\frac{1}{2\pi}\frac{n_c}{n_0}a_0\lambda_l, \tag{4}$$

indicating that the charge-separated force is smaller than the light pressure exerted on the electrons, the balance condition to form double layers (consisting of electrons and carbon ions) loses effect. The electron layer in the rear of the foil is pushed out of the foil by the laser pulse soon before the double layer is formed. At the laser-foil interaction stage, because of smaller charge mass ratio $Z/A=1/2$, carbon ions respond more slowly than protons. A flat-top–shaped laser pulse at ultra-relativistic intensity is thus used to induce an electrostatic field higher than that induced by the laser pulse with slowly rising ramp, where heavy ion can be pre-accelerated to reach the relativistic level. Otherwise, the non-relativistic heavy ion would thus be slipping out of the acceleration phase at the initial injection stage into wakefield, due to a velocity much smaller than the phase velocity of the wakefield, which is approximately equal to the group velocity of the laser pulse. This acceleration process of the ion in the laser-foil interaction stage can be regarded as the non–double-layer regime.

Main acceleration for the relativistic heavy ion by the wakefield to the TeV level in the snowplow regime. – Then the electron layer runs in the low-density gas and is further pushed by the laser pulse like light-sail to generate "snowplow" effects on the enhancement of the electron number. The electron layer can continuously be compensated by the background plasma in the snowplow process, although the fraction of electrons in the snowplow layer is lost due to the drag of the electric field. Therefore, the snowplow layer ahead of the laser pulse in the gas plasma drives a large electrostatic field and a stable acceleration structure is thus formed. Then the C^{6+} carbon ion beams that have been accelerated to the relativistic level run out of the foil and are injected into the wakefield as depicted in fig. 1(b). The ion bunches start a long march in gas by the wakefield in the snowplow regime and thereby TeV level monoenergetic C^{6+} ion beams are generated. Here we refer to this acceleration scheme as snowplow regime.

Since the velocity of the injected relativistic C^{6+} carbon ions is still smaller than that of the laser pulse at the beginning of the snowplow regime, the distance between the ion beam and the laser front increases to reach the maximum $\frac{1}{4}l_0\simeq20\,\mu\text{m}$. C^{6+} carbon ion beams are gradually accelerated and finally overrun the snowplow region due to the continuous erosion of the laser pulse. Once the C^{6+} carbon ion beam is injected into the wakefield, its maximum energy depends on the laser pump depletion

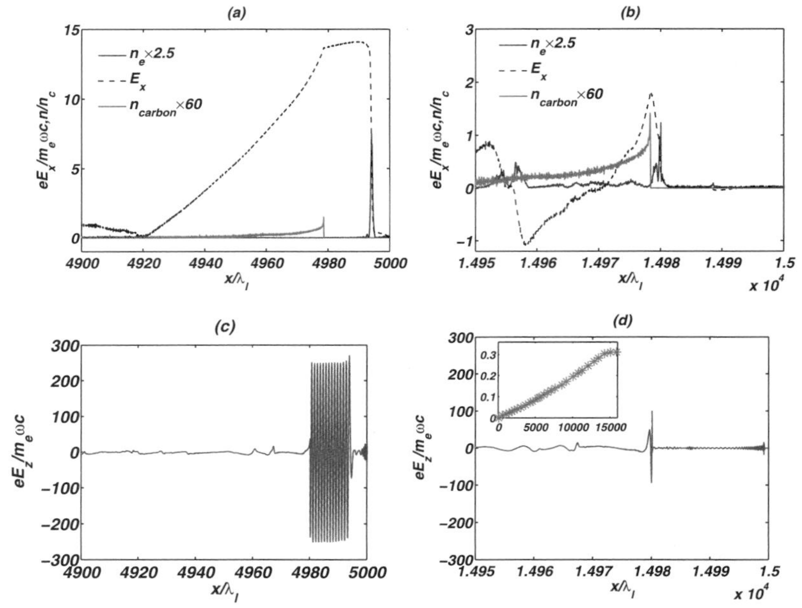

Fig. 2: (Color online) Snapshots of the snowplow structure (a) electron density n_e/n_c (blue solid line), C^{6+} ion density n_{carbon}/n_c (red solid line) and longitudinal electric field $eE_x/m_e\omega c$ at $t = 5000T_l$ (black dashed line) and $t = 15000T_l$ (b); laser field $eE_z/m_e\omega c$ at $t = 5000T_l$ (c); laser field $eE_z/m_e\omega c$ at $t = 15000T_l$ (purple solid line) and the laser pulse energy transfer efficiency to carbon ions *vs.* time in units of laser cycles (in the inset) (d).

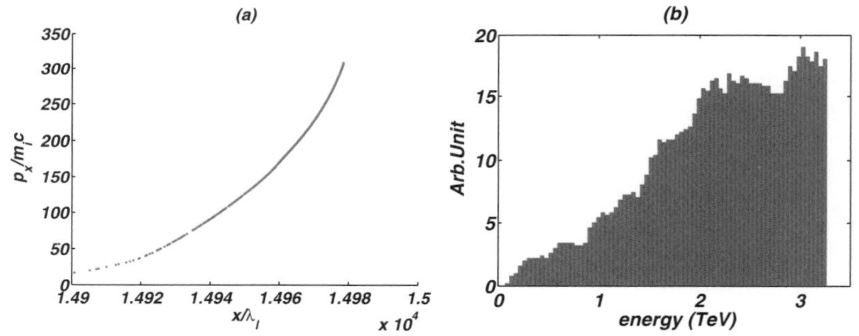

Fig. 3: (Color online) Longitudinal phase space for carbon ions p_x/m_ic (a) and energy spectrum of trapped carbon ions (b) from simulations at $t = 15000T_l$.

length and ion dephasing length [19]. The simulation snapshots of the electron density, the carbon ion density, the laser pulse and the electrostatic field are plotted in fig. 2. The heavy-ion beam has been continuously accelerated and bunched until the laser pulse is completely depleted at $t = 15000T_l$ as described in fig. 2(c) and (d), meaning that the pump depletion length equals the dephasing length. When the carbon ion beam approaches the laser front, the beam loading effect causes a substantial reduction of the electrostatic field as shown in fig. 2(a) and (b). On the other hand, fig. 4(a) indicates that the maximum energy of carbon ion increases rapidly *vs.* time in units of laser cycles by the wakefield and then it is saturated at $t = 15000T_l$. According to our analytical model in characterizing the snowplow regime [15], the theoretically maximum values of the longitudinal electrostatic field is $E_x = 14.9$ in units of E_0, the maximum energy of the carbon ion beam is $W_{max} = 3.2$ TeV with the corresponding relativistic factor $\gamma \simeq 300$ by eq. (2), respectively. They are consistent with the simulation results as shown in fig. 3(a) and fig. 3(b). The simulation shows that 5.01% of carbon ions from the thin foil have been trapped and the maximum ion energy is 3.3 TeV. It is found that the energy transfer efficiency from the laser pulse converted to carbon ions is greater than 30% as shown in the inset of fig. 2(d), which is extremely close

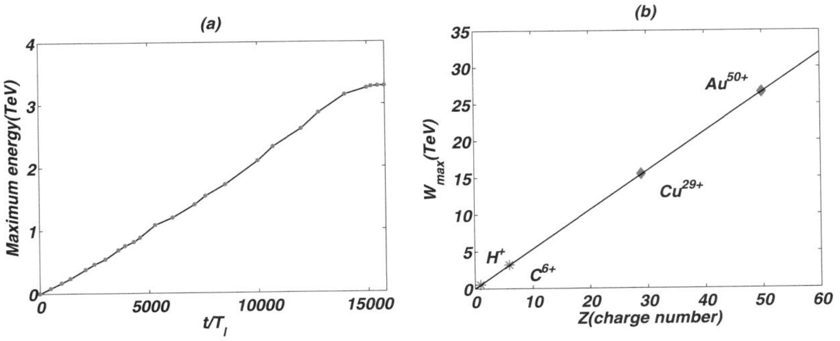

Fig. 4: (Color online) Maximum energy of the carbon ion by the wakefield acceleration in the snowplow regime *vs.* time in units of laser cycles (a); maximum energy for different ions with charge number Z in the snowplow regime by the ultra-relativistically intense laser wakefield accelerator (b).

to that for electron acceleration in the bubble regime [3]. Compared with the proton in the snowplow regime, the dephasing length of carbon ion beam is $15000\,\mu$m, corresponding to dephasing length ratio between carbon ion and proton, $\eta = 1.34$. These agree well with our theoretical estimates.

Wakefield structure in modulating carbon ion duration. – Besides the ultra-high energy for the ion, the beam is compressed in the phase space by the snowplow electrostatic field, which can control the beam quality in terms of pulse duration. Since the acceleration region in the snowplow structure is approximately $\frac{1}{2}(a_0 n_c/n_e)^{1/2}\lambda_l$, half of the total wakefield length, it leads to the fact that the C^{6+} carbon ion beam is theoretically less than $80\,\mu$m. The final quasi-monoenergetic ion beam is about $70\,\mu$m in our simulation, which agrees well with our theoretical estimate. Therefore, the ultra-short and TeV level C^{6+} carbon ion beam can be generated by our acceleration scheme.

Conclusions and discussions. – Based on the scaling law of eq. (3), the maximum energy of the ion beam increases linearly with the charge number Z as shown in fig. 4(b). Therefore, Cu^{29+} and Au^{50+} ions with maximum energy of 16 TeV and 25 TeV can be generated in principle according to our acceleration scheme. It is also easy to expect a second-stage acceleration in which we take only the second part of the present method (*i.e.*, the lower-density wakefield acceleration). This implementation of two stages allows us i) to discriminate particles in the lower-energy tail and to decrease the energy spread of the proton beam; ii) to further increase the ion energy. We can repeat this over many stages for accelerated heavy ions reaching as high as PeV, which can be used to probe the texture of vacuum and study astrophysical data of the primordial gamma-ray bursts (GRBs) [20].

In conclusion, the ultra-short and TeV quasi-monoenergetic ion beam can be generated by the centimeter-scale laser wakefield accelerator, excited by an ultra-relativistic laser pulse. An analytic model is developed for heavy-ion acceleration based on the relativistic dynamical equation. Our model suggests that the maximum energy of the heavy-ion scales with the charge number Z in comparison with the proton accelerated by the laser wakefield accelerator in the snowplow regime. The theoretically maximum energy of about 3.2 TeV for the carbon C^{6+} beam is 6 times higher than that of the proton, which has been verified by a one-dimensional particle-in-cell simulation. The length of the final ion beam is about $70\,\mu$m. Such a short bunch of ions may excite the plasma wakefield to accelerate electrons to several TeV, as Caldwell proposed [21]. The TeV ion beam can be also applied to high-energy ion collisions for exploring the primordial state of hadronic matter called quark-gluon plasma [22].

* * *

This work was supported by National Nature Science Foundation of China (Grant Nos. 10935002, 10835003, 11025523, 10975023) and National Basic Research Program of China (Grant No. 2011CB808104). XQY would like to acknowledge the support from the Alexander von Humboldt Foundation.

REFERENCES

[1] Maksimchuk A. *et al.*, *Phys. Rev. Lett.*, **84** (2000) 4108; Clark E. L. *et al.*, *Phys. Rev. Lett.*, **85** (2000) 1654; Snavely R. A. *et al.*, *Phys. Rev. Lett.*, **85** (2000) 2945; Cowan T. E. *et al.*, *Phys. Rev. Lett.*, **92** (2004) 204801.
[2] Mangles S. *et al.*, *Nature*, **431** (2004) 535; Geddes C. *et al.*, *Nature*, **431** (2004) 538; Faure J. *et al.*, *Nature*, **431** (2004) 541.
[3] Tajima T. and Dawson J. M., *Phys. Rev. Lett.*, **43** (1979) 267; Pukhov A. and Meyer-ter-Vehn J., *Appl. Phys. B*, **74** (2002) 355; Leemans W. P. *et al.*, *Nat. Phys.*, **2** (2006) 696; Lu W. *et al.*, *Phys. Rev. ST Accel. Beams*, **10** (2007) 061301.
[4] Shearer J. W. *et al.*, *Phys. Rev. A*, **8** (1973) 1582.

[5] WILKS S. C. *et al.*, *Phys. Plasma*, **8** (2001) 542.
[6] http://www.extreme-light-infrastructure.eu.
[7] HATCHETT S. *et al.*, *Phys. Plasma*, **7** (2000) 2076; WILKS S. *et al.*, *Phys. Plasma*, **8** (2001) 542.
[8] SCHWOERER H. *et al.*, *Nature*, **439** (2006) 26; ESIRKEPOV T. *et al.*, *Phys. Rev. Lett.*, **89** (2002) 17; HEGLICH B. M. *et al.*, *Nature (London)*, **439** (2006) 441.
[9] HENIG A. *et al.*, *Phys. Rev. Lett.*, **103** (2009) 045002.
[10] MACCHI A. *et al.*, *Phys. Rev. Lett.*, **94** (2005) 165003; ZHANG X. *et al.*, *Phys. Plasmas*, **14** (2007) 123108; YAN X. Q. *et al.*, *Phys. Rev. Lett.*, **100** (2008) 135003; KLIMO O. *et al.*, *Phys. Rev. ST Accel. Beams*, **11** (2008) 031301; ROBINSON A. P. L. *et al.*, *New J. Phys.*, **10** (2008) 013021; RYKOVANOV S. G. *et al.*, *New J. Phys.*, **10** (2008) 113005; LIU C. S. *et al.*, *AIP Conf. Proc.*, **1061** (2008) 246; HENIG A. *et al.*, *Phys. Rev. Lett.*, **103** (2010) 245003.
[11] HENIG A. *et al.*, *Phys. Rev. Lett.*, **103** (2009) 245003.
[12] YIN L. *et al.*, *Laser Part. Beams*, **24** (2006) 291; YIN L. *et al.*, *Phys. Plasmas*, **14** (2007) 056706.
[13] SHEN B. F. *et al.*, *Phys. Rev. ST Accel. Beams*, **12** (2009) 121301.
[14] YU L. L. *et al.*, *New J. Phys.*, **12** (2010) 045021.
[15] ZHENG F. L. *et al.*, in preparation, arXiv:1101.2350v2 [physics.plasm-ph].
[16] ASHOUR-ABDALLA M. *et al.*, *Phys. Rev. A*, **23** (1981) 1906.
[17] ESAREY E. and SPRANGLE P., *IEEE Trans. Plasma Sci.*, **24** (1996) 252; ESAREY E. and PILLOFF M., *Phys. Plasmas*, **2** (1995) 1432.
[18] YAN X. Q. *et al.*, *Phys. Rev. Lett.*, **103** (2009) 135001.
[19] LU W. *et al.*, *Phys. Rev. Lett.*, **96** (2006) 165002.
[20] TAJIMA T., KANDO M. and TESHIMA M., *Prog. Theor. Phys.*, **125** (2011) 3.
[21] CALDWELL A. *et al.*, *Nat. Phys.*, **5** (2009) 363.
[22] HWA RUDOLPH C. and WANG XIN-NIAN, *Quark-Gluon Plasma*, Vol. **3**, 1st edition (World Scientific Publishing Company) 2004.

Laser Shaping of a Relativistic Intense, Short Gaussian Pulse by a Plasma Lens

H. Y. Wang,[1] C. Lin,[1,*] Z. M. Sheng,[2] B. Liu,[1] S. Zhao,[1] Z. Y. Guo,[1] Y. R. Lu,[1] X. T. He,[1] J. E. Chen,[1] and X. Q. Yan[1,†]

[1]*State Key Laboratory of Nuclear Physics and Technology, Peking University, Beijing 100871, China, and Key Lab of High Energy Density Physics Simulation, CAPT, Peking University, Beijing 100871, China*

[2]*Key Laboratory for Laser Plasmas (MoE) and Department of Physics, Shanghai Jiao Tong University, Shanghai 200240, China and Beijing National Laboratory of Condensed Matter Physics, Institute of Physics, CAS, Beijing 100190, China*

(Received 17 September 2011; published 22 December 2011)

By 3D particle-in-cell simulation and analysis, we propose a plasma lens to make high intensity, high contrast laser pulses with a steep front. When an intense, short Gaussian laser pulse of circular polarization propagates in near-critical plasma, it drives strong currents of relativistic electrons which magnetize the plasma. Three pulse shaping effects are synchronously observed when the laser passes through the plasma lens. The laser intensity is increased by more than 1 order of magnitude while the initial Gaussian profile undergoes self-modulation longitudinally and develops a steep front. Meanwhile, a nonrelativistic prepulse can be absorbed by the overcritical plasma lens, which can improve the laser contrast without affecting laser shaping of the main pulse. If the plasma skin length is properly chosen and kept fixed, the plasma lens can be used for varied laser intensity above 10^{19} W/cm^2.

DOI: 10.1103/PhysRevLett.107.265002 PACS numbers: 52.38.Kd, 41.75.Jv, 52.35.Mw, 52.59.−f

The recent development of ultrashort-pulse high peak power laser systems enables us to investigate high field science under extreme conditions [1], which opened new and active research fields such as fast ignition of inertial confinement fusion [2], laboratory astrophysics [3], and development of a compact source of high-energy electrons and ions [4].

Generation of high-energy ions by ultraintense laser pulses has been intensively studied due to its wide range of applications [5,6]. Radiation pressure acceleration has been proposed as a promising route to obtain high-quality ion beams in a much more efficient way [7–13], compared to the target normal sheath acceleration [14–16]. In order to accelerate ions to a relativistic velocity in the radiation pressure acceleration regime, hole-boring effects [10] and transverse instabilities [17] should be restrained, which normally require extremely high laser intensity ($> 10^{21}$ W/cm^2), a sharp rising front, and high temporal laser contrast ($> 10^{10}$). Such a laser pulse is also promising for attosecond UV and x-ray sources from plasma surfaces of solid foil [18].

In this Letter we report on a laser-driven plasma lens that can transversely focus the laser beam to the sublaser wavelength in the radius and enhance the laser intensity by more than 1 order of magnitude, while temporally steepening the Gaussian pulse and improving the laser contrast. One fundamental nonlinear effect of laser-plasma interaction arises from the relativistic motion of the electrons in the intense laser field [19]. In particular, the transverse and axial profiles of the laser pulse can be changed by relativistic self-focusing (RSF) [20] and relativistic self-phase modulation (RSPM) [21]. Laser shaping by nonlinear interactions of an intense pulse in underdense plasma is discussed by weekly nonlinear theory and simulations [22]. The RSF of a linearly polarized laser beam in a near-critical plasma has been investigated by Pukhov and Meyer-ter-Vehn, which shows that the magnetic interaction plays an important role during RSF in the 3D simulations [23]. For near-critical plasma, the Raman instability that otherwise destroys the pulse is prohibited. RSPM can be very effective in this density region and results in a smooth pulse self-compression [24]. For relativistically strong ($|a| \geq 1$) laser pulses, however, the weak nonlinearity approximation is not valid anymore, because the nonlinear interactions can be rather complex and must be understood in conjunction with generation of a strongly nonlinear wakefield, erosion of the leading edge, and generation of the quasistatic magnetic field, etc. [1]. By means of 3D simulations we found that RSF, RSPM, and relativistic transparency effects are dynamically related together in laser interaction with the near-critical plasma. The laser pulse is focused to a smallest spot size and the focused intensity can be increased by a factor of 25. At the same time the initial Gaussian pulse becomes steepened, meanwhile a nonrelativistic prepulse can be absorbed by the overcritical plasma and the laser contrast is improved. For the proper chosen plasma lens parameters, the three lens effects above are effective for varied laser intensity above 10^{19} W/cm^2. Therefore the plasma slab can be used as a plasma lens to generate high intensity, high contrast, and steepened laser pulses, which is quite challenging for state of the art laser technology.

In the relativistic regime magnetic interaction appears to be due to the fact that electrons accelerated inside a self-focused laser pulse produce electric currents in the plasma and an associated quasistatic magnetic field. The electron velocity is limited by the velocity of light in vacuum, so the electron current density is approximately equal to en_ec, where e,n_e, and c is the charge of electron, electron density, and speed of light in vacuum, respectively. Omitting the numerical factors, the quasistatic magnetic field at the distance r from the axis reads:

$$B_s = (en_e)2\pi r. \quad (1)$$

For a uniform plasma, the self-generated magnetic field vanishes at the channel axis and reaches a maximum at the channel edges. The quasistatic magnetic field pinches the relativistic electrons into a channel with a radius r assumed to equal the diameter of Larmor motion in quasimagnetic field B_s, which has been testified in 3D simulation later.

$$r = 2\gamma mc^2/eB_s, \quad (2)$$

where $\gamma = \sqrt{a^2+1} \sim a$ for $a \gg 1$. Here $a = eE_L/m_e\omega c$ is the normalized vector potential for a laser with electric field E_L and laser frequency ω. Combining Eqs. (1) and (2) we have:

$$r = \frac{1}{\pi}\sqrt{an_c/n_e}\lambda, \quad (3)$$

and then we obtain the magnetic field:

$$B_s = \sqrt{an_e/n_c}B_0, \quad (4)$$

where $B_0 = m_e c\omega/e$. For the light of wavelength $\lambda = 2\pi c/\omega = 1\ \mu$m, one obtains $B_0 = 107.1$ MG. The electric currents leads to a redistribution of fast electrons, which in turn changes the refractive index $\eta = [1 - \omega_p^2/(\gamma\omega^2)]^{1/2}$. As the laser has a Gaussian profile, the index of refraction is peaked along the propagation axis. The phase velocity $v_p = c/\eta$ is smaller on axis than off axis and RSF occurs. As $v_p v_g = c^2$, the group velocity of the laser pulse vertex is larger than the pulse front, which causes the RSPM process. These two processes and pinch effects of the magnetic field B_s exist in the laser-plasma interactions, which can reshape the laser profile in both transverse and longitudinal directions.

To illustrate specific features of the laser shaping in three-dimensional regimes, we carried out 3D simulations using a fully relativistic particle-in-cell code (KLAP3D) [9,25]. The simulation box is $60\lambda \times 20\lambda \times 20\lambda$ and contains $600 \times 200 \times 200$ cells. A circularly polarized laser pulse with a Gaussian envelope $a = a_0\exp[-(y-y_0)^2/r_0^2]\exp[-(x-x_0)^2/r_0^2]\exp[-(t-t_0)^2/\tau^2]$ in both the longitudinal (z) and transverse (x, y) directions is normally incident from the left side ($z = 0\ \mu$m). We shall set $a_0 = 16.5$, corresponding to a peak laser intensity of 7.46×10^{20} W cm^{-2}, $t_0 = 50$ T, $\tau = 25$ T, $r_0 = 6\lambda$,

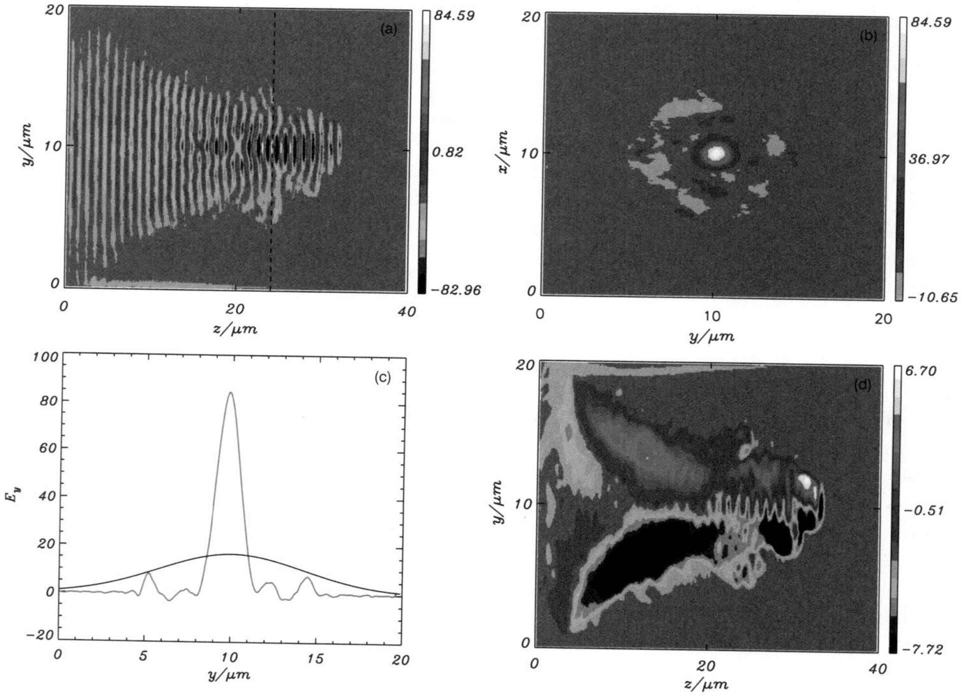

FIG. 1 (color online). Transverse pulse self-focusing at $t = 72$ T: (a) (Z, Y) section of E_y at $x = 10\ \mu$m (in units of $m_e c\omega/e$), the dashed line marks transverse sections shown in (b),(c); (b) (X, Y) section of E_y at $z = 24.1\ \mu$m; (c) E_y along y axis at $z = 24.1\ \mu$m, $x = 10\ \mu$m, the black line shows the initial laser transverse profile. (d) (Z, Y) section of quasistatic magnetic field B_x (in units of $m_e\omega/e$) at $t = 72$ T, averaged over 2 laser periods.

$y_0 = 10\lambda$ and $x_0 = 10\lambda$, where $T = 3.3$ fs is the laser period. The uniform carbon plasma of density $n_0 = 2.4n_c$ is placed between $3\lambda \leq z \leq 60\lambda$, where n_c is the critical plasma density. Each cell is filled with 8 quasiparticles. The initial temperature of the electrons and carbon ions is 5 eV.

The transverse pulse self-focusing process at $t = 72$ T is shown in Fig. 1. The incident beam first propagates through an unstable filamentary stage and then collapses into a single channel as shown in Fig. 1(a). The laser pulse is focused to a smallest spot size at $z = 24.1$ μm, corresponding to a laser self-focusing length of about 21.1 μm. The normalized electric field E_y is 84.59 which is increased by a factor of 5 and the laser intensity is 25 times higher than the initial one. The (X, Y) section of E_y in Fig. 1(b) shows that the pulse self-focusing process is symmetrical. The laser retains its Gaussian radial profile within the channel with a radius of about 0.9λ as seen in Fig. 1(c), which is in good agreement with Equation. (3) (theory predicts 0.83λ). The coalescence of the filaments by magnetic interaction are well explained on the basis of the magnetic interaction by Askaryan *et al.* [26]. In Fig. 1(d), the (Z, Y) section of the quasistatic magnetic field B_x averaged over 2 laser periods is plotted. This field vanishes at the channel axis and reaches a maximum at the edges. The maximum magnetic field is about 6.7 at the edges, which agrees well with the theoretical prediction of 6.29 by Eq. (4).

Figure 2 shows how the pulse profile changes in the longitudinal direction by laser compression during its propagation through different plasma planes. It shows an efficient laser compression at $z = 24.1$ μm, where the laser propagates a plasma slab length that is the same as the self-focusing length. At the same time the laser pulse developed a steep front of about 8 T. This implies that a initial Gaussian profile is nearly transformed into a quasi-step function pulse. For a plasma length longer or shorter than the self-focusing length, the laser compression is less efficient.

We now study the optimistic parameters for the plasma lens. Considering laser shaping of an intense laser pulse, three typical parameters are critical for the process: (i) increase of the normalized vector potential on axis $f_0 = a_{\max}/a_0$; (ii) laser rise time r_T; (iii) laser transmission efficiency t_E. Figure 3(a) shows the influence of plasma density on these three parameters with $a_0 = 16.5$. The normalized vector potential on axis can be enhanced by a factor of 5 for plasma density varied from $1n_c$ to $5n_c$. We should note that the laser transmission efficiency t_E decreases with plasma density increasing, while the laser rise time r_T decreases rapidly to about 8 T at $n_e = 2.4n_c$ and becomes saturated. The laser transmission efficiency is about 30% at this point while it can be as high as 60% with plasma density of $n_e = 1.5n_c$ if the pulse rise time can be compromised. The quasistatic magnetic field B_s and the channel radius r observed in simulation are consistent with the theoretical value from Eqs. (3) and (4) as shown in Figs. 3(b) and 3(d). For a fixed plasma skin length $l_s/\lambda = \sqrt{an_c/n_e} = 2.6$, the enhancement of laser amplitude, laser rise time, transmission, and the channel radius r are nearly the same for varied laser intensity ranging from 10^{19} W/cm^2 to 10^{21} W/cm^2 as shown in Figs. 3(c) and 3(d).

This means that the plasma lens works for different laser intensities when keeping the plasma skin length fixed. As a result, if we choose a proper plasma density (satisfy $l_s/\lambda \sim 2.6$) and a proper plasma slab length (equal the self-focusing length), the laser focusing and compressing can be realized at the same time and same position. This provides an efficient way to generate a high intensity laser pulse with a sharp rising front by laser shaping in both the transverse and longitudinal directions.

The improvement of the laser pulse contrast (the intensity ratio between the main pulse and the prepulse) is

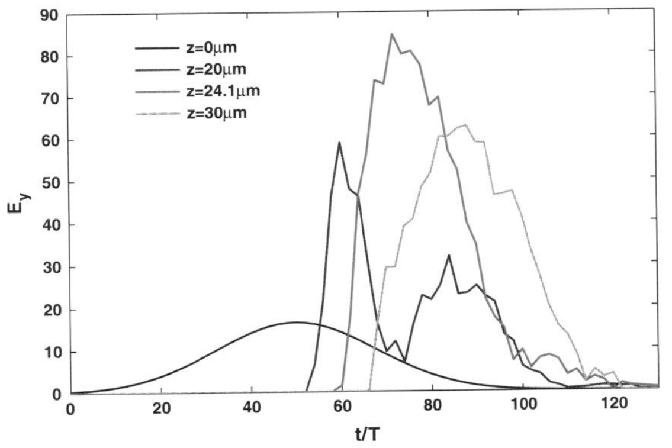

FIG. 2 (color online). Longitudinal pulse compression: On-axis envelope profile E_y at $z = 0$ μm, $z = 20$ μm, $z = 24.1$ μm, and $z = 30$ μm.

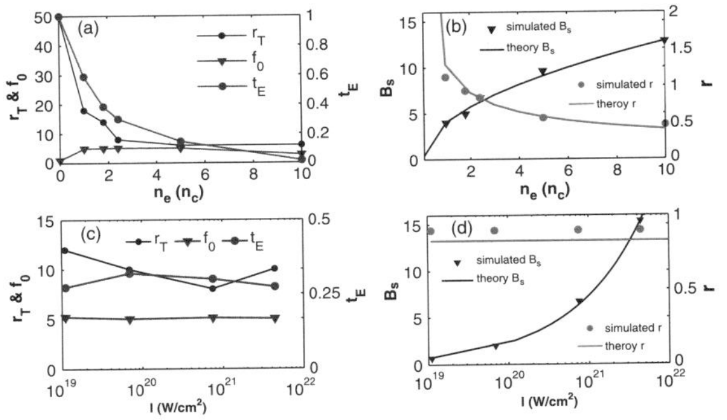

FIG. 3 (color online). Pulse shaping for varied plasma density and laser intensity (a) f_0, r_T and t_E for varied plasma density at $a_0 = 16.5$; (b) B_s and r for varied plasma density at $a_0 = 16.5$; (c) f_0, r_T and t_E for varied laser intensity with fixed plasma skin length $l_s/\lambda = \sqrt{an_c/n_e} = 2.6$; (d) B_s and r for varied laser intensity with fixed plasma skin length $l_s/\lambda = \sqrt{an_c/n_e} = 2.6$.

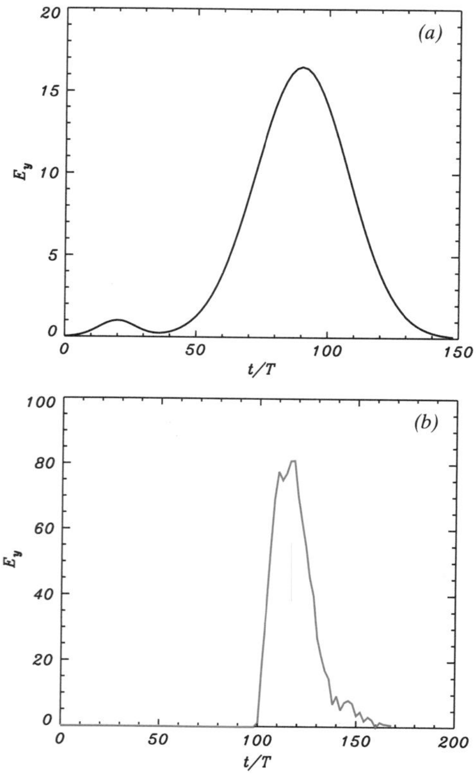

FIG. 4 (color online). Absorbing of laser prepulse: (a) On-axis envelope profile E_y at $z = 0\ \mu$m; (b) On-axis envelope profile E_y at $z = 24.1\ \mu$m.

promising for various applications. Relativistic transparency provides a way to achieve laser pulse shaping and generate a high contrast laser pulse [27]. In particle-in-cell (PIC) stimulations here, we study the prepulse absorbing process. We set a prepulse of $a = 1$, $\tau = 10$ T and $r_0 = 6\lambda$, which is 40 T before the main pulse as shown in Fig. 4(a). It is observed that the prepulse can be absorbed in the plasma without affecting the shaping of the main pulse as shown in Fig. 4(b). This shows the plasma lens can also improve the contrast of laser pulse.

In summary we propose a scheme to make a high intensity laser pulse with a high contrast and a steep rise front using a plasma lens. As relativistic self-focusing (RSF), relativistic self-phase modulation (RSPM) and relativistic transparency are dynamically related together in laser interaction with the near-critical plasma, the lens has three shaping effects: (I) pulse focusing that results in laser intensity enhancement; (II) laser profile steepening that can transform a Gaussian pulse into a quasistep function pulse; (III) absorption of nonrelativistic prepulse. The transmission efficiency of the lens can be as high as 60%. The generated laser pulse should be useful for many applications such as generation of high-energy ions and electrons. The optimistic lens parameter (plasma skin length) is determined in 3D simulations. This scheme has no limitation for laser intensity above 10^{19} W/cm^2, since it involves only laser-plasma interaction.

This work was supported by the National Natural Science Foundation of China (Grants No. 10935002, No. 10835003, and No. 11025523) and the National Basic Research Program of China (Grant No. 2011CB808104).

*linchen_0812@pku.edu.cn
†x.yan@pku.edu.cn

[1] G. A. Mourou *et al.*, Rev. Mod. Phys. **78**, 309 (2006).
[2] M. Roth *et al.*, Phys. Rev. Lett. **86**, 436 (2001); V. Yu. Bychenkov *et al.*, Plasma Phys. Rep. **27**, 1017 (2001); S. Atzeni *et al.*, Nucl. Fusion **42**, L1 (2002).
[3] B. A. Remington *et al.*, Science **284**, 1488 (1999); S. V. Bulanov *et al.*, Eur. Phys. J. D **55**, 483 (2009).
[4] T. Tajima and J. M. Dawson, Phys. Rev. Lett. **43**, 267 (1979); E. Esarey, C. B. Schroeder, and W. P. Leemans, Rev. Mod. Phys. **81**, 1229 (2009); S. P. Hatchett *et al.*, Phys. Plasmas **7**, 2076 (2000); E. L. Clark *et al.*, Phys. Rev. Lett. **84**, 670 (2000); A. Maksimchuk *et al.*, *ibid.* **84**, 4108 (2000); A. McPherson *et al.*, Nature (London) **370**, 631 (1994).
[5] M. Borghesi *et al.*, Phys. Plasmas **9**, 2214 (2002).
[6] S. V. Bulanov *et al.*, Phys. Lett. A **299**, 240 (2002); T. Zh. Esirkepov *et al.*, Phys. Rev. Lett. **89**, 175003 (2002).
[7] S. S. Bulanov *et al.*, Med. Phys. **35**, 1770 (2008); B. Eliasson *et al.*, New J. Phys. **11**, 073006 (2009); A. Macchi, S. Veghini, and F. Pegoraro, Phys. Rev. Lett. **103**, 085003 (2009); Z. M. Zhang *et al.*, Phys. Plasmas **17**, 043110 (2010); T. Esirkepov *et al.*, Phys. Rev. Lett. **92**, 175003 (2004); A. Henig *et al.*, Phys. Rev. Lett. **103**, 045002 (2009); S. G. Rykovanov *et al.*, New J. Phys. **10**, 113005 (2008); C. S. Liu *et al.*, AIP Conf. Proc. **1061**, 246 (2008).
[8] X. Zhang *et al.*, Phys. Plasmas **14**, 123108 (2007).
[9] X. Q. Yan *et al.*, Phys. Rev. Lett. **100**, 135003 (2008).
[10] A. Macchi *et al.*, Phys. Rev. Lett. **94**, 165003 (2005).
[11] O. Klimo *et al.*, Phys. Rev. ST Accel. Beams **11**, 031301 (2008).
[12] A. P. L. Robinson *et al.*, New J. Phys. **10**, 013021 (2008).
[13] M. Chen *et al.*, Phys. Rev. Lett. **103**, 024801 (2009); X. Q. Yan *et al.*, Phys. Rev. Lett. **103**, 135001 (2009); S. V. Bulanov *et al.*, Phys. Rev. Lett. **104**, 135003 (2010); B. Qiao *et al.*, Phys. Rev. Lett. **102**, 145002 (2009).
[14] R. A. Snavely *et al.*, Phys. Rev. Lett. **85**, 2945 (2000).
[15] H. Schwoerer *et al.*, Nature (London) **439**, 445 (2006); B. M. Hegelich *et al.*, Nature (London) **439**, 441 (2006); A. J. Mackinnon *et al.*, Phys. Rev. Lett. **88**, 215006 (2002); S. P. Hatchett *et al.*, Phys. Plasmas **7**, 2076 (2000); L. Willingale *et al.*, Phys. Rev. Lett. **96**, 245002 (2006); L. Robson *et al.*, Nature Phys. **3**, 58 (2006); A. Yogo *et al.*, Phys. Rev. E **77**, 016401 (2008).

[16] J. Fuchs *et al.*, Nature Phys. **2**, 48 (2005).
[17] F. Pegoraro and S. V. Bulanov, Phys. Rev. Lett. **99**, 065002 (2007).
[18] C. D. Tsakiris *et al.*, New J. Phys. **8**, 19 (2006); Y. Nomura *et al.*, Nature Phys., **5**, 124 (2008).
[19] W. B. Mori, IEEE J. Quantum Electron. **33**, 1942 (1997), and references therein.
[20] A. G. Litvak, Zh. Eksp. Teor. Fiz. **57**, 629 (1969) [Sov. Phys. JETP **30**, 344 (1970)]; P. Sprangle, C. M. Tang, and E. Esarey, IEEE Trans. Plasma Sci. **15**, 145 (1987); G.-Z. Sun *et al.*, Phys. Fluids **30**, 526 (1987); X. L. Chen and R. N. Sudan, Phys. Rev. Lett. **70**, 2082 (1993); E. Esarey *et al.*, IEEE J. Quantum Electron. **33**, 1879 (1997), and references therein.
[21] C. Max, J. Arons, and A. B. Langdon, Phys. Rev. Lett. **33**, 209 (1974); C. J. Mackinstrie and R. Bingham, Phys. Fluids B **4**, 2626 (1992).
[22] C. Ren *et al.*, Phys. Rev. E **63**, 026411 (2001); S. S. Bulanov *et al.*, Phys. Plasmas **17**, 043105 (2010).
[23] A. Pukhov and J. Meyer-ter-Vehn, Phys. Rev. Lett. **76**, 3975 (1996).
[24] O. Shorokhov, A. Pukhov, and I. Kostyukov, Phys. Rev. Lett. **91**, 265002 (2003).
[25] Z. M. Sheng *et al.*, Phys. Rev. Lett. **94**, 095003 (2005).
[26] G. A. Askar'yan *et al.*, JETP Lett. **60**, 251 (1994).
[27] A. V. Vshivkov *et al.*, Nucl. Instrum. Methods Phys. Res., Sect. A **410**, 493 (1998); A. V. Vshivkov *et al.*, Phys. Plasmas **5**, 2727 (1998).

PHYSICS OF PLASMAS **19**, 023111 (2012)

Sub-TeV proton beam generation by ultra-intense laser irradiation of foil-and-gas target

F. L. Zheng,[1,2] H. Y. Wang,[1] X. Q. Yan,[1,a)] T. Tajima,[3] M. Y. Yu,[4,5] and X. T. He[1,6,a)]
[1]*Key Laboratory of HEDP of Ministry of Education, CAPT, State Key Laboratory of Nuclear Physics and Technology, Peking University, Beijing 100871, China*
[2]*Graduate School, China Academy of Engineering Physics, P.O. Box 8009, Beijing 100088, People's Republic of China*
[3]*Fakultät für Physik, LMU München, D-85748 Garching, Germany*
[4]*Institute of Fusion Theory and Simulation, Zhejiang University, Hangzhou 310027, China*
[5]*Institut für Theoretische Physik I, Ruhr-Universität Bochum, D-44780 Bochum, Germany*
[6]*Institute of Applied Physics and Computational Mathematics, P. O. Box 8009, Beijing 100088, China*

(Received 5 September 2011; accepted 11 January 2012; published online 17 February 2012)

A two-phase proton acceleration scheme using an ultra-intense laser pulse irradiating a proton foil with a tenuous heavier-ion plasma behind it is presented. The foil electrons are compressed and pushed out as a thin dense layer by the radiation pressure and propagate in the plasma behind at near the light speed. The protons are in turn accelerated by the resulting space-charge field and also enter the backside plasma, but without the formation of a quasistationary double layer. The electron layer is rapidly weakened by the space-charge field. However, the laser pulse originally behind it now snowplows the backside-plasma electrons and creates an intense electrostatic wakefield. The latter can stably trap and accelerate the pre-accelerated proton layer there for a very long distance and thus to very high energies. The two-phase scheme is verified by particle-in-cell simulations and analytical modeling, which also suggests that a 0.54 TeV proton beam can be obtained with a 10^{23} W/cm^2 laser pulse. © *2012 American Institute of Physics*. [doi:10.1063/1.3684658]

I. INTRODUCTION

Proton beams with tens to hundreds MeV energies are useful for medical imaging,[1] cancer therapy,[2] fast ignition in inertial fusion,[3] conversion of radioactive wastes,[4] and protons with TeV energies are relevant to high-energy physics[5] and astrophysics.[6] Existing studies have shown that proton beams at the GeV level can be obtained by radiation pressure acceleration (RPA) using $>10^{22}$ W/cm^2 lasers.[7–12] In that scheme the light pressure in the plasma is balanced by the charge-separation field, so that the electron and proton layers co-move as a double-layer.[13–20] However, because of instability the acceleration length is limited to less than hundreds of micrometers and it is difficult to increase the ion energy. On the other hand, quasi-monoenergetic GeV electrons can be generated by the laser wakefield acceleration (LWFA) scheme.[21–24] Recently, Shen *et al.*[25] found that a laser pulse at ultra-relativistic intensity can excite a wake field that can capture protons from the underdense plasma and accelerate them to tens of GeV. Furthermore, Yu *et al.*[26] found that by placing an underdense plasma behind a thin-foil target, 60 GeV protons can be obtained by the fast-moving double-layer. As a laser pulse propagates in the underdense plasma region, it excites a plasma wakefield with a linear profile which has a negative back part and positive front part. The latter can accelerate protons. Motivated by this, we anticipate that the pre-accelerated protons by RPA can be further accelerated over a long distance by the front positive wakefield as long as they are trapped in the acceleration field.

In this paper, we show by particle-in-cell simulation and analytical modeling that with a suitable combination of RPA and LWFA, TeV protons can be generated using a laser of the same intensity. In the proposed two-phase acceleration scheme, an ultra-intense laser irradiates a foil target with a backside plasma, as shown in Fig. 1(a). The foil thickness d satisfies,

$$l_0 < d < D, \tag{1}$$

where l_0, n_e, n_c, and λ_l are the plasma skin depth, plasma density, critical plasma density, and laser wavelength, respectively, $D = \frac{1}{2\pi}\frac{n_c}{n_e}a_0\lambda_l$ is the initial thickness of the foil,[13–20] $a_0 = eA/m_e\omega_l c$, A is the laser vector potential, and c is the speed of light. The condition means that the light pressure exerted on the electrons is larger than the space-charge force, i.e., the condition for double-layer formation is not satisfied. The RP quickly compresses and pushes the electrons to the rear of the foil, as shown in Fig. 2(a). The electron layer can reach a density far greater than γn_c, where $\gamma = (1+a_0^2)^{1/2}$. Meanwhile, the protons in the foil are accelerated by the space-charge field. The electron-layer in rear of the foil is pushed out of the thin foil by the RP just before quasisteady electron-ion double layers can form. The electron layer thus propagates in the underdense gas while being continuously accelerated by the laser pulse like a light-sail. Due to the snowplow effect, the layer is enhanced by the electrons from the backside plasma[27] and a wake field is formed. The large electric field of the latter in turn accelerates the pre-accelerated (by RPA) proton layer (PL) from the foil, starting the stable

[a)]Authors to whom correspondence should be addressed. Electronic addresses: X.Yan@pku.edu.cn and xthe@iapcm.ac.cn.

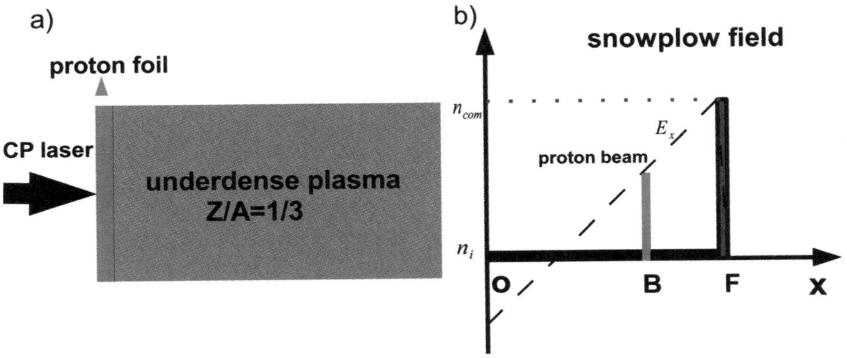

FIG. 1. (Color online) (a) The two-stage acceleration scheme. The initial density of the hydrogen foil is $n_0/n_c = 20$, the thickness is $d = 0.5\lambda_l$, the density of the tenuous backside plasma is $n_e/n_c = 0.01$, its length is 12 000λ_l. (b) Sketch of the snowplow process, the dynamic density of the ions is n_i and the electron density in the snowplow layer is n_{com}. The position B indicates the laser pulse front and A is an arbitrary point in the snowplow region.

long-distance acceleration in the backside plasma, as shown in Fig. 1(b). This is possible because the heavy ions of the backside plasma remain almost unaffected by the wake electric field. The two-phase scheme therefore differs significantly from the existing acceleration schemes invoking the double-layer, where $d \sim D$ and the accelerated protons can quickly overtake the double layers, so that their energy gain is limited because of decoupling of the protons from the accelerating field.[26]

II. THE TWO-PHASE ACCELERATION SCHEME

It is instructive to first discuss our one-dimensional analytical model, which extends that of Esarey[29] for ultrarelativistic-laser driven plasma wakefields. The evolution of such a wakefield is characterized by (i) lengthening of the wakefield size $l_s = (\gamma n_c/n_e)^{1/2}\lambda_l$, and (ii) enhancement of the wavebreaking field $E_B = 3\pi(\gamma n_e/n_c)^{1/2}E_0$, where $E_0 = m_e\omega_l c/e$. That is, both l_s and E_B scale with $\gamma^{1/2}$. When d satisfies Eq. (1), the pre-accelerated PL can be captured by the wakefield in the underdense plasma and be stably accelerated in the so-called snowplow regime of LWFA over a long distance. The wakefield is composed of two components, which are generated by the compressed dense electron layer and back-side plasma ions. Accordingly, using Poisson equation we can estimate the electrostatic field E_A at an point x_A in the snowplow region, (see Fig. 1(b)). The contribution of the compressed dense electron layer at position x_A is close to $E'_A = 4\pi e n_e l_s/2$ and the contribution of backside plasma ions is $E''_A = 4\pi e n_i x_A - 4\pi e n_i(l_s - x_A)$. The longitudinal electrostatic field at x_A is thus $E_A = E'_A + E''_A = 4\pi e n_e (2x_A - l_s/2)$. Thus the maximum snowplow field at the laser front x_B is

$$E_B = 3\pi(\gamma n_e/n_c)^{1/2}E_0. \tag{2}$$

The captured pre-accelerated foil PL is at first slower than the laser pulse as well as the electron layer because of the

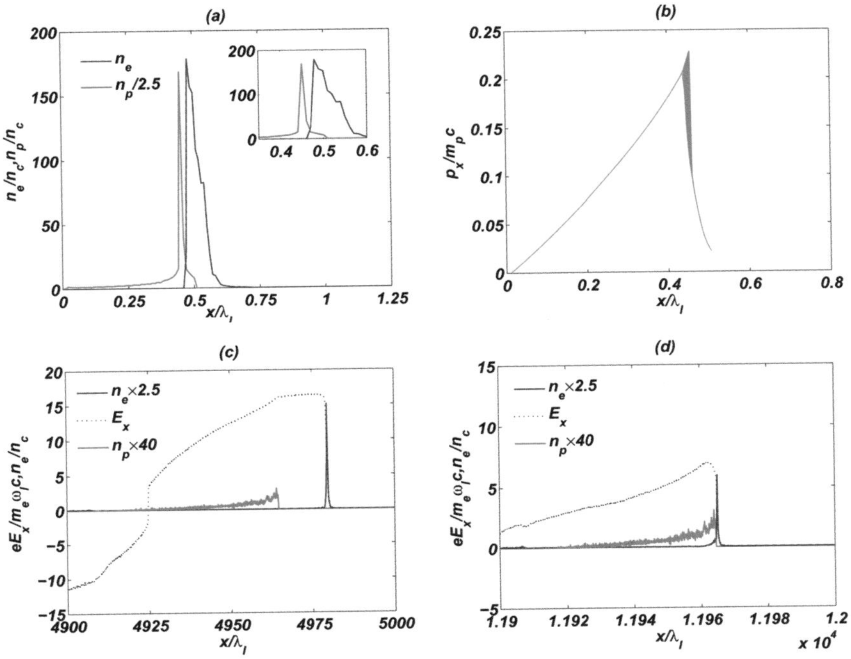

FIG. 2. (Color online) (a) Electron density n_e/n_c, proton density n_p/n_c, and the distance between the electron and PLs (see inset), and (b) longitudinal proton phase space p_x/m_pc at $t = 5.8T_l$, where m_p is proton mass. The longitudinal electrostatic field $eE_x/m_e\omega_l c$ (dashed line) in the snowplow regime at (c) $t = 5000T_l$ and (d) $t = 12\ 000T_l$. The initial plasma parameters are given in Fig. 1.

condition (1), which, however, weakens rapidly as the electrons there are pulled back by the space-charge force.[28] On the other hand, the PL is stably accelerated in the wake field and can achieve very high energy after some distance. It is convenient to introduce a length, namely $l_{\rm inj}$, between the PL and the laser-pulse front. Once the PL enters into the wakefield, its maximum achievable energy depends on the laser depletion length and proton decoupling (from the accelerating field) length. In the present snowplow regime, the laser strongly interacts with only the compressed electron layer because its rear part is in a region void of electrons. As a result, the energy loss is localized near the laser front, causing the front part of the laser to etch backward as it propagates. Pump depletion is thus the limiting mechanism for the survival of the wakefield. The laser depletion length L_{pd} is given by[29]: $E_x^2 L_{pd} \sim E_l^2 L_l$, where E_x is the longitudinal electrostatic field in the snowplow regime and L_l is the length of laser pulse. Assuming that the group velocity of the laser pulse is close to c, we have

$$L_{pd} \sim L_l c / v_{\rm etch}, \tag{3}$$

where $v_{\rm etch}$ is the etching velocity of the laser pulse and it reads $v_{\rm etch} \simeq E_B^2 c/E_l^2 = 9\pi^2 n_e c/a_0 n_c$. Clearly it is proportional to n_e/a_0. Since the electron layer is formed by laser compression, its speed is from the beginning the same as that of the front of the laser pulse, namely $v_f = v_g - v_{\rm etch}$, where v_g is the laser group velocity.

The decoupling length of the PL is approximately $L_{dp} \sim l_{\rm inj}\ v_p/(v_p - v_f) = l_{\rm inj}\ v_p/[v_p - (v_g - v_{\rm etch})]$, where v_p is the proton velocity and $l_{\rm inj} \sim l_s/4$ from the simulation result. When v_g and v_p are both near the light speed, so that $v_{\rm etch} \gg v_p - v_g$, the proton decoupling length becomes

$$L_{dp} \sim \frac{1}{36\pi^2}\left(\frac{a_0 n_c}{n_e}\right)^{3/2} \lambda_l. \tag{4}$$

For $L_l > (a_0 n_c/n_e)^{1/2} \lambda_l/4$, the laser pump depletion length is larger than the decoupling length. So that the maximum energy the PL can achieve is

$$W_{\max} \sim (n_c/6n_e) m_e c^2 a_0^2, \tag{5}$$

which shows that the present two-phase scheme offers efficient proton acceleration. The acceleration length in the snowplow regime is proportional to $a_0^{3/2}$ and maximum proton energy scales with a_0^2. This scaling law indicates that a circularly polarized (CP) 10^{23} W/cm^2 116 fs laser pulse can generate a quasi-monoenergetic proton bunch with sub-TeV energy. In the following, we shall confirm this conclusion using PIC simulations, which amounts to following the trajectories of charged particles in self-consistent electromagnetic fields computed on a fixed mesh.

III. SIMULATION RESULTS AND DISCUSSION

For the simulation, we shall use the fully relativistic PIC code KLAP.[13,30] The $\lambda_l = 1\ \mu$m CP laser pulse has an axially trapezoidal profile with a $15\lambda_l$ flat top and a $20\lambda_l$ ramp on the rising side. The ramp is given by $a = a_0 sin^2(\pi t/40T_l)$ with $a_0 = 250$, where T_l is the laser period. At $t = 0$ it is normally incident from the left on a uniform, fully ionized hydrogen foil of thickness $d = 0.5\ \mu$m and normalized density $N = n_0/n_c = 20$. Behind the hydrogen foil is a tenuous heavy-ion-plasma of density $n_e = 0.01 n_c$ and charge-to-mass ratio 1/3. Here the simulation box is length 12 000λ_l in the x direction. The spatial resolution is 40 cells per λ_l. Each cell is filled with 20 000 macroparticles for the foil plasma and 10 macroparticles for the tenuous backside plasma. The initial electron and ion temperatures are both 1 keV.

We first consider the 1D problem, as in the analysis in the Sec. II. As shown in Fig. 2(a), at $t = 5.8T_l$, or the end of the laser-foil interaction when space-charge accelerated and compressed PL enter the backside plasma, the electron layer ahead of the laser pulse runs faster than the PL at near the light speed. No quasisteady double-layer can form because the RP force exerted on the electrons is larger than the space-charge force. Fig. 2(b) shows that in this stage the maximum energy of the PL is about 20 MeV and the speed is about 0.2c. The distance between the electron and proton layers continues to increase, as shown in the inset of Fig. 2(a). The laser pulse, now propagating in the backside plasma, snowplows the electrons in front of it and excites an intense electrostatic plasma wave in its wake, thus starting the second stage of proton acceleration in the present scheme.

Snapshots of the electron density, proton density, and electrostatic field are given in Figs. 2(c) and 2(d). We can see that the PL is continuously accelerated until the laser pulse is almost completely depleted at $t = 12\ 000T_l$. At that time, the

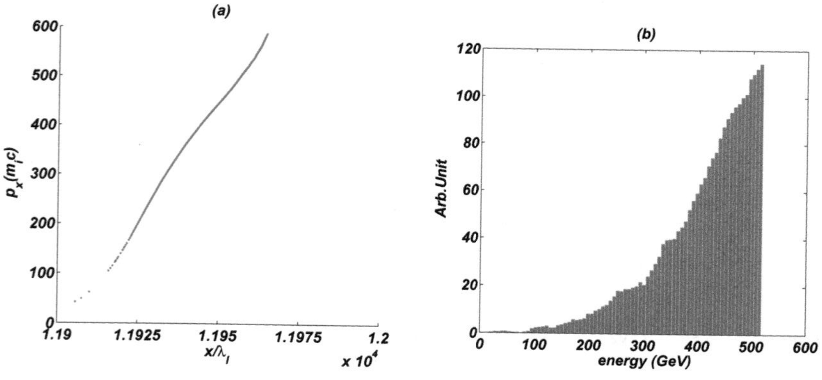

FIG. 3. (Color online) Simulation results at $t = 12\ 000T_l$. (a) Proton phase space p_x/m_pc *versus* x, and (b) energy spectrum of the trapped protons.

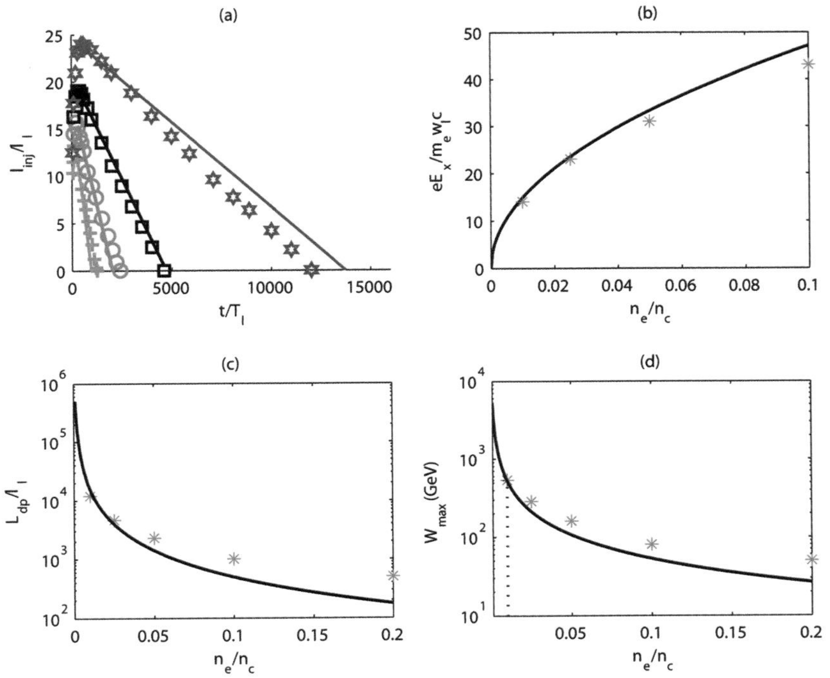

FIG. 4. (Color online) (a) Distance between proton beam and the laser pulse front *versus* time for different gas densities, where stars, diamonds, crosses, and squares correspond to the gas densities $0.01n_c$, $0.025n_c$, $0.05n_c$, and $0.1n_c$, respectively, (b) longitudinal electrostatic field $eE_x/m_e\omega_l c$, (c) decoupling length, (d) maximum proton energy. Here the theoretical curves are given as solid lines and the simulation results are shown as stars, diamonds, crosses, and squares.

PL approaches the laser front, leading to substantial reduction of the electrostatic field, as shown in Fig. 2(d). According to our analytical model, we have $E_x = 14.9E_0$, $l_{pd} = 11125\lambda_l$, $W_{max} = 532\,\text{GeV}$, $v_{\text{etch}} = 0.0036\,c$, and $v_f = 0.9964\,c$. These results agree well with that from the simulation. Fig. 3 shows that the energy spread (FWHM) of the beam is less than 20%. The maximum proton energy is 540 GeV and the averaged energy of the PL is 437 GeV. These are about 8 times that of the corresponding double-layer accelerating scheme.[26] About 1.72% of the protons are located in this energy window. The PL is compressed in the phase space and the length of the final quasi-monoenergetic layer is shorter than 100 μm. Such a layer can be used to excite a plasma wakefield that can in turn accelerate the plasma electrons to hundreds of GeV.[31]

As can be seen at the left edge of Fig. 4(a), the distance between the proton beam and the laser front increases up to

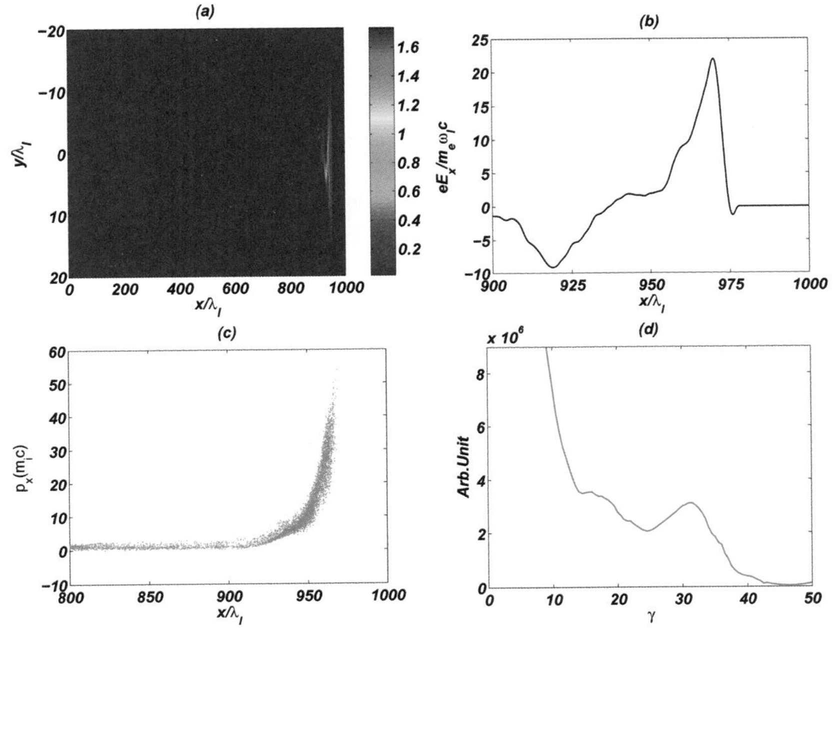

FIG. 5. (Color online) 2D simulation results at $t = 990T_l$. (a) Electron density profile, (b) electrostatic field on the axis, (c) proton phase space, and (d) proton spectrum.

a maximum (corresponding to l_{inj}). This distance then decreases due to the erosion of the laser pulse. The slopes of the curves correspond to the etching velocity of the laser pulse for the different backside-plasma densities. The maximum longitudinal-electrostatic field, decoupling length, and maximum proton energy *versus* the backside-plasma density are shown in Figs. 4(b)–4(d). We see that the maximum energy of the proton beam decreases with the plasma density. To the left of the dotted line (at $n_e < 0.01n_c$), the pulse depletion length is shorter than the decoupling length. Accordingly, to increase the proton energy, one can increase the laser intensity and reduce the backside-plasma density, so that both the decoupling length and pump depletion length are increased.

It is difficult to perform full scale 2D simulations (for obtaining sub-TeV protons) with our computing resources. We therefore consider a narrow simulation box $40 \times 1000\lambda_l^2$ with resolution of 5 cells/λ_l in the y plane and 20 cells/λ_l in the x plane. Each cell contains 4 particles for the backside plasma and 400 particles for the foil plasma. The foil parameters are the same as that in the 1D simulations, but the backside-plasma density is $0.2n_c$. The super-Gaussian laser pulse is given by $I = I_0 exp\{-(r/r_0)^4 - [(t-t_0)/\tau]^8\}$, where $r_0 = 10\,\mu$m, $t_0 = 55$ fs, and $\tau = 55$ fs. Fig. 5 shows that the snowplowed electron layer, the electrostatic field, and the phase space are similar to the 1D results. The maximum proton energy of about 50 GeV also agrees with the 1D simulation. Since the protons gain energy mainly from the electrostatic wake field,[26] the return current and magnetic field that appear in the 2D simulation do not affect proton acceleration in the snowplow regime. We also note that depreciation of the electron layer is compensated by the laser-snowplowed electrons from the backside plasma, so that the electron layer still exists at the end of our simulations. To suppress transverse instabilities such as Rayleigh-Taylor instability associated with laser-foil interactions,[13–20] an ultra-intense laser pulse and an ultra-thin foil are needed (just as these in the present acceleration scheme), since the instability growth becomes much slower due to relativistic effect in a moving frame. In fact, in the 2D case studied here, there is no evidence of these instabilities. Furthermore, as the pre-accelerated protons are detached from the initial foil, this transverse instability does not appear, since the main acceleration occurs in the stable LWFA stage, where the laser pulse interacts with a very low density underdense plasma.

IV. SUMMARY

In summary, a two-phase acceleration scheme using a foil and gas target driven by the ultra-intense laser pulse is presented. A 1D model that predicts the proton decoupling length, pump depletion length, and the maximum proton energy is given. The results are verified by 1D as well as 2D PIC simulations. It is shown that sub-TeV quasi-monoenergetic proton bunches can be generated by a centimeter-scale accelerator with a laser pulse of intensity 10^{23} W/cm^2 and duration 116 fs. The resulting proton beam length is shorter than 100 μm.

ACKNOWLEDGMENTS

We thank G. Mourou, C. Y. Zheng, H. Zhang, B. Liu, and H. C. Wu for useful discussions and help. This work was supported by the National Nature Science Foundation of China (Grants 10935002, 10974022, 11025523, 11175026, 11175029, 11105009) and the National Basic Research Program of China (Grant 2012CB80111). XQY would like to thank the Alexander von Humboldt Foundation for support.

[1] M. Borghesi, A. Schiavi, D. H. Campbell, M. G. Haines, O. Willi, A. J. Mackinnon, P. Patel, M. Galimberti, and L. A. Gizzi, Rev. Sci. Instrum. **74**, 1688 (2003).
[2] S. V. Bulanov, T. Zh. Esirkepovb, V. S. Khoroshkovc, A. V. Kuznetsovb, and F. Pegoraro, Phys. Lett. A **299**, 240 (2002); T. Esirkepov, S. V. Bulanov, K. Nishihara, T. Tajima, F. Pegoraro, V. S. Khoroshkov, K. Mima, H. Daido, Y. Kato, Y. Kitagawa, K. Nagai, and S. Sakabe, Phys. Rev. Lett. **89**, 175003 (2002).
[3] M. Roth, T. E. Cowan, M. H. Key, S. P. Hatchett, C. Brown, W. Fountain, J. Johnson, D. M. Pennington, R. A. Snavely, S. C. Wilks, K. Yasuike, H. Ruhl, F. Pegoraro, S. V. Bulanov, E. M. Campbell, M. D. Perry, and H. Powell, Phys. Rev. Lett. **86**, 436 (2001).
[4] K. W. D. Ledingham, P. McKenna, T. McCanny, S. Shimizu, J. M. Yang, L. Robson, J. Zweit, J. M. Gillies, J. Bailey, G. N. Chimon, R. J. Clarke, D. Neely, P. A. Norreys, J. L. Collier, R. P. Singhal, M. S. Wei, S. P. D. Mangles, P. Nilson, K. Krushelnick, and M. Zepf, J. Phys. D **36**, L79 (2003).
[5] J. W. Shearer, J. Garrison, J. Wong, and J. E. Swain, Phys. Rev. A **8**, 1582 (1973).
[6] S.C. Wilks, A. B. Langdon, T. E. Cowan, M. Roth, M. Singh, S. Hatchett, M. H. Key, D. Pennington, A. MacKinnon, and R. A. Snavely, Phys. Plasma **8**, 542 (2001).
[7] A. Macchi, F. Cattani, T. V. Liseykina, and F. Cornolti, Phys. Rev. Lett. **94**, 165003 (2005).
[8] X. Zhang, B. F. Shen, X. M. Li, Z. Y. Jin, F. C. Wang, and M. Wen, Phys. Plasmas **14**, 123108 (2007).
[9] X. Q. Yan, C. Lin, Z. M. Sheng, Z. Y. Guo, B. C. Liu, Y. R. Lu, J. X. Fang, and J. E. Chen, Phys. Rev. Lett. **100**, 135003 (2008).
[10] A. P. L. Robinson, M. Zepf, S. Kar, R. G. Evans, and C. Bellei, New J. Phys. **10**, 013021 (2008).
[11] C. S. Liu, V. K. Tripathi, and X. Shao, AIP Conf. Proc. **1061**, 246 (2008).
[12] A. Henig, S. Steinke, M. Schnüer, T. Sokollik, R. Hörlein, D. Kiefer, D. Jung, J. Schreiber, B. M. Hegelich, X. Q. Yan, J. Meyer-ter-Vehn, T. Tajima, P.V. Nickles, W. Sandner, and D. Habs, Phys. Rev. Lett. **103**, 245003 (2009).
[13] X. Q. Yan, H. C. Wu, Z. M. Sheng, J. E. Chen, and J. Meyer-ter-Vehn, Phys. Rev. Lett. **103**, 135001 (2009).
[14] T. Esirkepov, M. Borghesi, S. V. Bulanov, G. Mourou, and T. Tajima, Phys. Rev. Lett. **92**, 175003 (2004).
[15] B. Qiao, M. Zepf, M. Borghesi, and M. Geissler, Phys. Rev. Lett. **102**, 145002 (2009).
[16] M. Chen, A. Pukhov, and T. P. Yu, Phys. Rev. Lett. **103**, 024801 (2009).
[17] F. Pegoraro and S. V. Bulanov, Phys. Rev. Lett. **99**, 065002 (2007).
[18] L. Yin, B. J. Albright, B. M. Hegelich, K. J. Browers, K. A. Flippo, T. J. T. Kwan, and J. C. Fernandez, Phys. Plasmas **14**, 056706 (2007).
[19] Z. M. Zhang, X. T. He, Z. M. Sheng, and M. Y. Yu, Phys. Plasmas **17**, 043110 (2010).
[20] V. K. Tripathi, C. S. Liu, X. Shao, B. Eliasson, and R.Z. Sagdeev, Plasma Phys. Controlled Fusion **51**, 024014 (2009).
[21] T. Tajima and J. M. Dawson, Phys. Rev. Lett. **43**, 267 (1979).
[22] A. Pukhov and J. Meyer-ter-Vehn, Appl. Phys. B **74**, 355 (2002).
[23] W. P. Leemans, B. Nagler, A. J. Gonsalves, C. Toth, K. Nakamura, C. G. R. Geddes, E. Esarey, C. Schroeder, and S. M. Hooker, Nat. Phys. **2**, 696 (2006).
[24] W. Lu, C. Huang, M. Zhou and M. Tzoufras, Phys. Plasmas **13**, 056709 (2006).
[25] B. F. Shen, X. Zhang, Z. M. Sheng, M. Y. Yu, and J. Cary, Phys. Rev. ST Accel. Beams **12**, 121301 (2009).
[26] L. L. Yu, H. Xu, W. M. Wang, Z. M. Sheng, B. F. Shen, W. Yu, and J. Zhang, New J. Phys. 12, 045021 (2010).
[27] M. Ashour-Abdalla, J. N. Leboeuf, T. Tajima, J. M. Dawson, and C. F. Kennel, Phys. Rev. A **23**, 1906 (1981).
[28] X. Zhang, B. F. Shen, L. L. Ji, F. C. Wang, M. Wen, W. P. Wang, J. C Xu, and Y. H. Yu, Phys. Plasmas **17**, 123102 (2010).
[29] E. Esarey and M. Pilloff, Phys. Plasmas **2**, 1432 (1995).
[30] Z. M. Sheng, K. Mima, Y. Sentoku, K. Nishihara, and J. Zhang, Phys. Plasmas **9**, 3147 (2002).
[31] A. Caldwell, K. Lotov, A. Pukhov, and F. Simon, Nat. Phys. **5**, 363 (2009).

PHYSICS OF PLASMAS **20**, 013107 (2013)

Laser-driven collimated tens-GeV monoenergetic protons from mass-limited target plus preformed channel

F. L. Zheng,[1] S. Z. Wu,[1,2] H. C. Wu,[1] C. T. Zhou,[1,2] H. B. Cai,[1,2] M. Y. Yu,[3,4] T. Tajima,[5] X. Q. Yan,[1,6,a)] and X. T. He[1,2,b)]
[1]*Key Laboratory of HEDP of the Ministry of Education, CAPT, Peking University, Beijing 100871, China*
[2]*Institute of Applied Physics and Computational Mathematics, Beijing 100088, China*
[3]*Institute of Fusion Theory and Simulation, Zhejiang University, Hangzhou 310027, China*
[4]*Institut für Theoretische Physik I, Ruhr-Universität Bochum, D-44780 Bochum, Germany*
[5]*Fakultät f. Physik, LMU München, Garching D-85748, Germany,*
[6]*State Key Laboratory of Nuclear Physics and Technology, Peking University, Beijing 100871, China*

(Received 10 September 2012; accepted 27 December 2012; published online 11 January 2013)

Proton acceleration by ultra-intense laser pulse irradiating a target with cross-section smaller than the laser spot size and connected to a parabolic density channel is investigated. The target splits the laser into two parallel propagating parts, which snowplow the back-side plasma electrons along their paths, creating two adjacent parallel wakes and an intense return current in the gap between them. The radiation-pressure pre-accelerated target protons trapped in the wake fields now undergo acceleration as well as collimation by the quasistatic wake electrostatic and magnetic fields. Particle-in-cell simulations show that stable long-distance acceleration can be realized, and a 30 fs monoenergetic ion beam of >10 GeV peak energy and <2° divergence can be produced by a circularly polarized laser pulse at an intensity of about 10^{22} W/cm^2. © *2013 American Institute of Physics*. [http://dx.doi.org/10.1063/1.4775728]

I. INTRODUCTION

Laser-driven ion acceleration has been of much recent interest because of its broad scientific and technical applications. Protons with tens to hundreds MeV energies are useful for cancer therapy,[1] fast ignition in inertial confinement fusion,[2] high resolution radiography,[3–8] laser nuclear physics,[9] etc. Ions with still higher energies are relevant to high energy physics research.[10]

Radiation pressure acceleration (RPA) is promising for obtaining high-quality ion beams efficiently.[11–13] Existing studies have shown that GeV proton beams can be obtained by RPA using circularly polarized (CP) lasers with intensity above 10^{22} W/cm^2.[14–16] However, because of transverse instabilities and hole-boring by the laser pulse,[14–17] the acceleration length is rather limited and it is difficult to enhance proton energy without still higher laser intensity. Recently, it has been shown that tens GeV proton beams[18,19] can be obtained by sequential radiation pressure and bubble acceleration or a moving double-layer, or in a two-phase acceleration regime[20,21] by ultra-relativistic lasers. However, in classical RPA and laser wakefield acceleration (LWFA),[22] the accelerated ion bunch tends to diverge because of the ubiquitous intense space-charge field.[23–25] The resulting defocusing leads to rather low ion flux. Moreover, the effective laser-plasma interaction distance is also limited by laser diffraction. As a result, ions cannot remain collimated and be accelerated stably over a long distance.

In this paper, we propose a stable proton acceleration scheme using an ultra-intense CP laser pulse and a mass-limited target (MLT)[26] connected to an underdense parabolic plasma channel. The target splits the laser pulse into two separate but adjacent parts, which are then refractively guided by the preformed underdense plasma channel. The quasistatic longitudinal accelerating and transverse collimating fields associated with the electron return current at the center of the channel can then stably trap, collimate, and further accelerate the RPA target protons by LWFA over a long distance. In contrast to the existing acceleration schemes, both proton defocusing and laser diffraction do not occur here. Particle-in-cell (PIC) simulations show that ultrashort (30 fs) monoenergetic proton beams with peak energy >10 GeV and divergence angle less than 2° can be obtained with a $\sim 10^{22}$ W/cm^2 CP laser pulse.

II. PRODUCING HIGHLY COLLIMATED TENS-GeV MONOENERGETIC PROTONS IN THE TWIN-WAKES REGIME

A. PIC simulation modeling

To demonstrate the proposed scheme, we use the two-dimensional (2D) PIC code KLAP.[27] The CP laser pulse is of wavelength $\lambda = 1\,\mu$m and peak intensity $I_0 = 9.8 \times 10^{21}$ W/cm^2 (or the normalized laser parameter $a_0 = 60$). It has a super-Gaussian radial profile with spot size $2\sigma = 40\lambda$ and a $18T$ flattop envelope, where T is the laser period. The fully ionized uniform hydrogen MLT of thickness $D = 0.8\lambda$ and density of $N = 20n_c$, where n_c is the critical density, is initially located in $50\lambda \le x \le 50.8\lambda$, $-10\lambda \le y \le 10\lambda$. Behind the hydrogen MLT is an underdense tritium plasma channel with a parabolic density profile $n = n_1 + \Delta n(y^2/y_0^2)$, where $n_1 = 0.15n_c$, $\Delta n = 2n_c, y_0 = 50\lambda$, and it is located in $50.8\lambda < x \le 600\lambda, -40\lambda \le y \le 40\lambda$. The simulation box is

a)x.yan@pku.edu.cn.
b)xthe@iapcm.ac.cn.

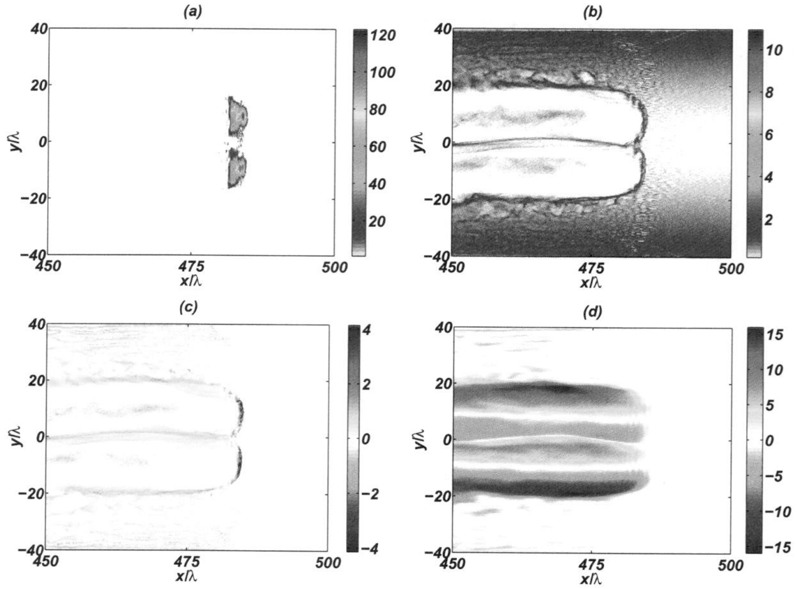

FIG. 1. (a) Square root of the laser intensity $\sqrt{E_y^2+E_z^2}$, normalized by $\mathcal{E}_0$. (b) Electron density normalized by n_c. (c) Current density j_x normalized by en_cc. (d) Spatial distribution of transverse quasistatic field $E_{ys}-B_{zs}$ (in the $\pm y$ directions), normalized by $\mathcal{E}_0$, at $t\sim 500T$.

of size $100\lambda\times 100\lambda$, corresponding to a moving window consisting of 3200×1000 cells, with each cell containing 800 macroparticles for the target and 6 for the underdense channel.

Figs. 1(a) and 1(b) show the square root of the laser intensity and electron density, respectively, at $t\sim 500T$. The transverse size $h=20\,\mu$m of the MLT is smaller than the laser spot size. The incoming laser pulse is split into two by the target. After passing the latter they continue to propagate forward in the preformed backside underdense plasma channel as adjacent twin pulses. The ponderomotive force of the latter expels the local plasma electrons transversely outward, creating two adjacent wakes, or twin wakes, which trap the RPA target protons in them. Fig. 1(c) shows the longitudinal current density j_x at $t\sim 500T$. One can see that there is an intense electron return-current sheet between the two wakes. Since the longitudinal velocity v_x of the trapped protons is near the light speed c, the transverse field experienced by the energetic protons can be written as $\mathcal{E}\hat{e}_y=E_{ys}\hat{e}_y+c^{-1}v_x\hat{e}_xc\times B_{zs}\hat{e}_z\sim(E_{ys}-B_{zs})\hat{e}_y$, where E_{ys} and B_{zs} are the self-generated electrostatic and magnetic fields, respectively. Fig. 1(d) shows the total transverse quasistatic field $\mathcal{E}=E_{ys}-B_{zs}$ normalized by $\mathcal{E}_0=m_e\omega c/e$, where m_e and $-e$ are the electron mass and charge, and ω and c are the laser frequency and light speed, at $t\sim 500T$. One can see that the twin wakes act as a moving potential well for confining, stabilizing, and further accelerating (now the second phase by LWFA) the RPA-protons for a long distance in the target-back-side underdense plasma. In addition, since the MLT electrons also move with the twin wakes, the expelled plasma electrons are prevented from rapidly refilling the back side of the channel, so that the latter can remain open for a long distance. The present acceleration mechanism therefore differs significantly from that of the classical bubble/blowout acceleration schemes.[22,28–30]

B. Guiding the twin wakes in the back-side underdense plasma channel

When the initial laser pulse is split by the MLT, the effective spot size σ_1 of each of the resulting twin light pulses is less than that of the original pulse. Similarly, the divergence angle $\theta_d=\sigma_1/Z_R$ is larger and the Rayleigh length $Z_R=\pi\sigma_1^2/\lambda$ smaller. Thus, without the underdense plasma channel behind the target the laser-plasma interaction distance would be rather limited. However, this undesirable scenario is avoided by the presence of a preformed parabolic density channel. To illustrate this effect, a plasma with the density profile $n=[n_1+\Delta n(y^2/y_0^2)]$ is placed at the back of the target. The dispersion relation for relativistic light waves is $\omega^2=c^2k^2+\omega_p^2/\gamma$, where k and $\omega_p/\gamma^{0.5}$ are the wave number and effective background plasma frequency, respectively, and $\gamma\sim\sqrt{1+a^2}$ is the relativistic factor associated with the electron quiver motion. For an ultra-intense CP laser pulse ($a\gg 1$), the index of refraction is $\eta(y)\sim\left[1-\frac{1}{a(y)n_c}\left(n_1+\Delta n\left(\frac{y^2}{y_0^2}\right)\right)\right]^{1/2}$, so that refractive light guiding (RLG) occurs if $d_y\eta<0$. Light diffraction can be cancelled by RLG if the maximum beam focusing angle is larger than the divergence angle θ_d, or

$$\frac{n_1}{2n_c}\left[\frac{1}{a(\sigma)}-\frac{1}{a(h/2)}\right]+\frac{\Delta n}{n_c}\left[\frac{\sigma^2}{a(\sigma)y_0^2}-\frac{(h/2)^2}{a(h/2)y_0^2}\right]\geq\left(\frac{\lambda}{\pi\sigma_1}\right)^2, \tag{1}$$

where the three terms in the inequality represent relativistic self-focusing guiding, preformed density channel RLG, and light diffraction, respectively. We note that this relation differs considerably from that of the $a\ll 1$ case. For the parameters under consideration, the self-focusing and RLG terms have the values $\sim$0.0021 and 0.013, respectively. The latter alone is already much larger than the diffraction term, which has the value 0.0041. One can also see that without RLG the proton bunch would diverge. Fig. 1(a) shows the profile of the light field at $t \sim 500T$. We see that RLG by the preformed parabolic density channel indeed results in long distance propagation of the twin light pulses and the corresponding twin wakes. In fact, the laser-plasma interaction distance is effectively increased to $\sim 7Z_R$. Moreover, the focusing effect of RLG also doubles the intensity of the twin pulses to $a_1 \sim 2a_0 = 120$, as shown in Fig. 1(a).

C. Scaling law for the highly collimated tens-GeV monoenergetic protons

We now consider the evolution of the pre-accelerated protons in more detail. From Figs. 2(a) and 2(c), one sees that the relativistic protons are confined along the original laser axis and they can propagate a long distance without divergence. It can be seen from Figs. 3(a) and 3(b) that these protons undergo further electrostatic acceleration in the twin-wakes along the original laser axis.

In the present twin wakes regime, the profile can be estimated by assuming that the transverse ponderomotive force of the laser $k_p \nabla a_1^2/\gamma \sim a_1/k_p R$, where R and k_p are the transverse radius and plasma wave number, is roughly balanced by the force of the ion channel $E_r \sim k_p R/2$.[30,31] Equating these two expressions yields $k_p R \sim \sqrt{2a_1}$. Moreover, the twin-wakes radius scales with the twin-pulses laser spot, $k_p R \sim k_p \sigma_1$. Through PIC simulation we have found that a more refined condition is $k_p R \sim k_p \sigma_1 \sim \pi\sqrt{a_1}$. Combining these equations, the maximum quasistatic field is thus given, $\mathcal{E}_{\max}(y)/\mathcal{E}_0 \sim (\pi/2)(a_1 n(y)/n_c)^{1/2}$. As a result, the maximum quasistatic transverse field at the center of the channel reads

$$\mathcal{E}_{\max} \sim (\pi/2)(a_1 n_1/n_c)^{1/2}\mathcal{E}_0. \qquad (2)$$

Fig. 1(d) shows that the maximum quasistatic transverse (or focusing) field $\mathcal{E}_{\max}$ induced by the charge separation, and the return current at the center of the channel is about $6\mathcal{E}_0$, agreeing well with the above theoretical scaling, where $a_1 \sim 2a_0 = 120$ is the laser front intensity due to the RLG effect. The angular distribution of the accelerated protons at $t \sim 600T$ is given in Fig. 2(b), which shows that the average emission angle for protons with energy greater than 8 GeV is less than 2°. That is, the output proton beam is well collimated by the transverse quasistatic field.

When the twin-pulses propagate in the underdense plasma, they etch back with $v_{\rm etch}$ due to local pump depletion, since the twin-pulses strongly interact with the front compressed electron layers because their rear parts are in a region of electron void. The laser will be depleted after a distance (pump depletion length), $L_{pd} \sim (c/v_{\rm etch})c\tau_{\rm FWHM}$, where $c\tau_{\rm FWHM}$ is laser pulse duration. Furthermore, the laser pump depletion length can be estimated by equating the laser pulse energy to the energy left behind in the wakefield, $\mathcal{E}^2_{y=\pm h/2} L_{pd} \sim \mathcal{E}_l^2 c\tau_{\rm FWHM}$, where $\mathcal{E}^2_{y=\pm h/2} \sim \pi(a_1 n_1/n_c)^{1/2}\mathcal{E}_0$ and $\mathcal{E}_l$ are the maximum quasistatic field and laser field of twin-pulses front, respectively. The laser etching velocity is thus $v_{\rm etch} \sim \pi^2 n_1 c/a_1 n_c$.

The distance that the trapped protons travel until they outrun the twin-wakes (dephasing length) is approximately $L_{dp} \sim l_{\rm inj} v_p/(v_p - v_f) = l_{\rm inj} v_p/[v_p - (v_g - v_{\rm etch})]$, where v_p and $v_f = v_g - v_{\rm etch}$ are the proton velocity and laser front velocity, v_g and $v_{\rm etch}$ are group velocity and etching velocity of laser pulse, $l_{\rm inj} \sim \frac{R}{2}$ from the simulation. When v_p and v_g are both near the light speed c, so that $v_{\rm etch} \gg v_p - v_g$, the proton dephasing length becomes

$$L_{dp} \sim \frac{R}{2}\frac{c}{v_{\rm etch}} \sim \frac{1}{4\pi^2}\left(\frac{a_1 n_c}{n_1}\right)^{3/2}\lambda. \qquad (3)$$

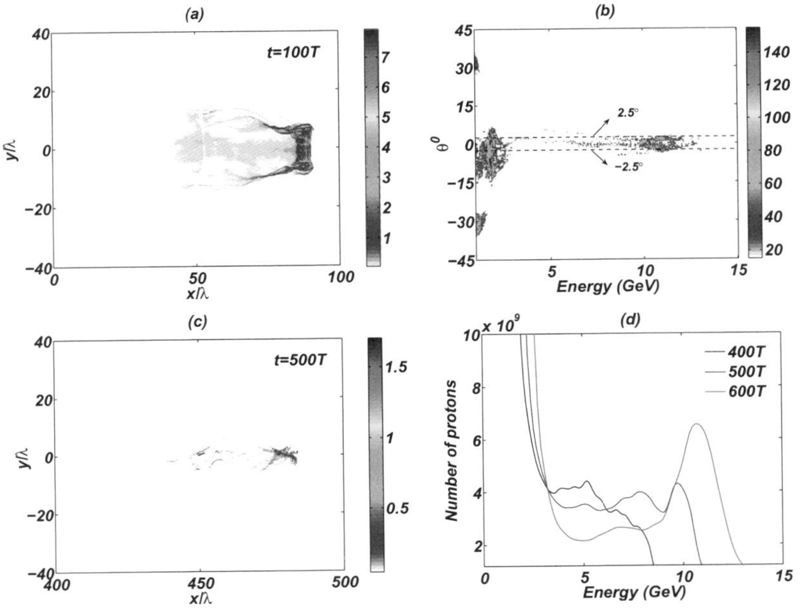

FIG. 2. Density of the accelerated protons in units of n_c at (a) $t \sim 100T$ and (c) $t \sim 500T$. (b) Angular distribution of the protons at $t \sim 600T$. The dashed lines correspond to the divergence angle ±2.5°. (d) Evolution of the proton energy spectrum.

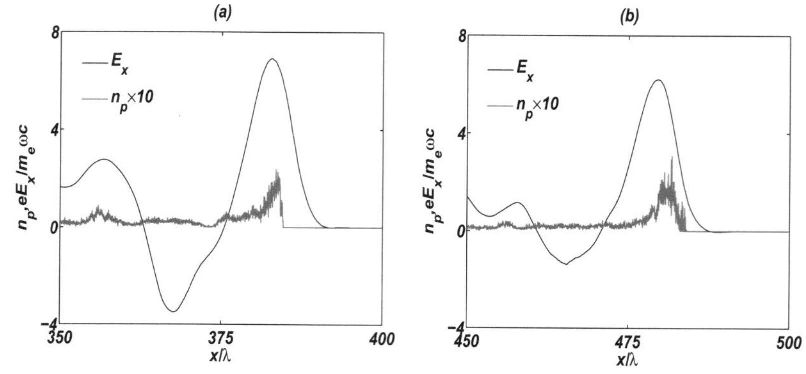

FIG. 3. The proton density n_p and longitudinal electrostatic field $eE_x/m_e\omega c$ along the original laser axis in the twin-wakes regime at (a) $t \sim 400T$ and (b) $t \sim 500T$, respectively.

Figure 4(a) for the temporal evolution of the maximum proton energy shows clearly the two stages, namely, RPA and LWFA, of the acceleration process. In the initial RPA regime, the maximum proton velocity reaches $0.945c$, or 1.93 GeV, at $t \sim 100T$. The pre-accelerated protons then undergo LWFA in the twin wakes until the light pulses are almost depleted at $t \sim 600T$, implying that pump depletion length is about $L_{pd} \simeq 600\,\mu$m. At that time, the trapped proton bunch approaches the front of the twin wakes. That is, the dephasing length L_{dp} is roughly the same as L_{pd}. The latter is also consistent with the theoretical value, as calculated by Eq. (3).

We also see that the maximum proton energy increases rapidly with time until $t \sim 600T$, when saturation occurs. Using the conditions for the stable twin wakes excitation, we can estimate the energy gain of the trapped proton beam by using the above equations,

$$\Delta E = eE_{\rm LW}L_{dp} \sim (a_1^2 n_c/n_1)m_e c^2/4, \tag{4}$$

where $E_{\rm LW} \sim (\pi/2)(a_1 n_1/n_c)^{1/2}\mathcal{E}_0$ is the average accelerating field due to the space-charge additive effect at the center of the twin wakes. Fig. 2(d) shows the evolution of energy spectrum. One can see that the latter improves with time and eventually achieves a sharp peak at about 12 GeV, which is highly consistent with the analytical model. Fig. 4(b) shows that the number of protons in the region $|y| \leq 3\lambda$ first decreases and then becomes constant at 1.5×10^{11}, indicating that the transverse spreading effect for the energetic protons is well under control in the back-side channel. Their total energy is above 500 J and the laser conversion efficiency is greater than 10%. The accelerated protons are well compressed in the phase space and the quasi-monoenergetic pulse has a duration of about 30 fs. The proton energy is plotted versus the foil thickness in Fig. 4(c). If the foil

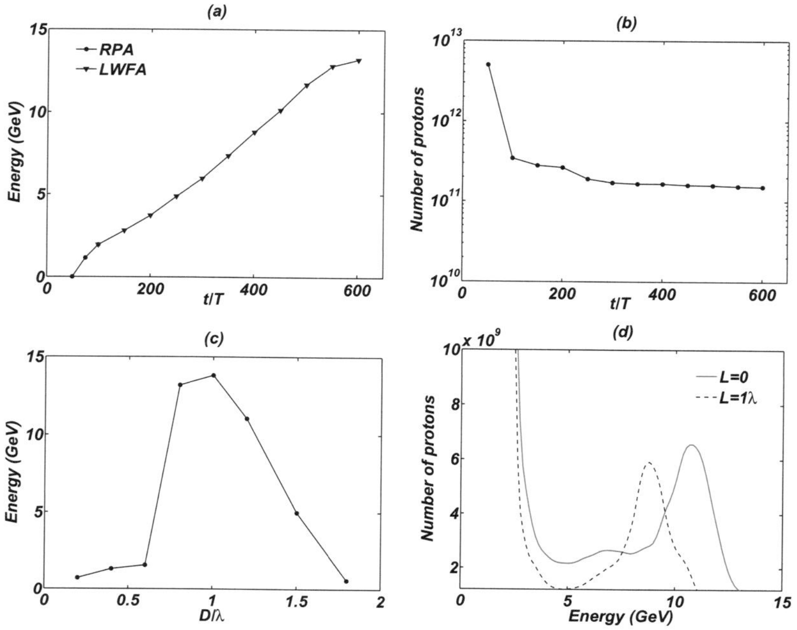

FIG. 4. Evolution of the (a) maximum proton energy, (b) number of protons in the axis region ($|y| \leq 3\lambda$), (c) proton energy versus the foil thickness D, (d) proton energy spectrum at $t = 600T$ with different preplasma density scale length from PIC simulations.

thickness were less than 0.8 μm, the foil protons would not be sufficiently pre-accelerated in the RPA regime and they would not be trapped by the twin wakes. On the other hand, if the foil thickness were greater than 1.2 μm, the dephasing length would be much lowered in the LWFA regime, resulting in sharp decrease of the final proton energies. The optimum proton spectra can be obtained with foil thickness in the range $1.0-1.2\,\mu$m.

D. Preplasma effect

At present, contrasts at about 10^{12} can be achieved experimentally[32] for amplified spontaneous emission (ASE) pulses at $I = 10^{10}\,\mathrm{W/cm^2}$. In order to model ASE and prepulse effects, we use a preplasma density profile obtained from hydrodynamic simulations.[33] Typically, the preplasma in the moderately overdense region has a density scale length $L = 1\lambda$. Compared to the $L = 0$ case, more laser energy is depleted in laser-foil interaction stage, so that the pump depletion length is reduced. However, the RPA pre-accelerated protons can be still trapped and collimated by the twin wakes. Fig. 4(d) shows that, because of the reduced pump depletion length, the peak proton energy is about 9 GeV, or smaller than that in the case of $L = 0$.

E. Single-wake case

A key issue in the proposed scheme is the formation of a stable proton focusing and collimating twin-wakes structure that moves at a relativistic speed with the target-modulated laser light. When the transverse size of target is too large comparing to the laser spot size, the initial laser pulse can fully interact with the target. The laser is not split into two parts and no twin wakes are formed. Without the transverse confinement in the underdense region, the RPA protons would diverge. This can be seen in the simulation by setting $h = 2\sigma = 40\,\mu$m, with the other laser and plasma parameters the same as in the preceding simulations. Instead of the stable twin wakes (as in Fig. 1(b)) a single wake is now generated by the laser pulse. Due to the defocusing effect of the space-charge field, the RPA protons quickly diverge in the wake, as can be seen in Fig. 5(b) for the emission angle (almost 20°) at $t \sim 600T$. Protons that deviate from the axis region do not experience LWFA. The maximum energy of the accelerated protons is only about 5 GeV, which is considerably lower than that in the stable twin-wakes case (Fig. 2(d)). The quality of the final proton bunch is thus relatively poor in terms of energy gain as well as emission angle.

F. Uniform back-side underdense plasma

The condition (1) for guiding the laser light in the underdense plasma is also important. If the maximum focusing angle is less than the beam divergence angle, RLG of the twin pulses in the density channel cannot occur. To see this, we set $\Delta n = 0$ (uniform plasma behind the target), with the other parameters unchanged. That is, the density-channel guiding effect is removed. For this case in consideration, the relativistic self-focusing term has the value ~ 0.0021, which is less than the beam diffraction term. It can not prevent the pulses from diffraction and the resulting acceleration process becomes unstable. Fig. 5(c) shows the density distribution of electrons in this case. We see that although a twin-wakes structure is still formed, but it does not survive for a long time. Only a fraction of the RPA protons is further accelerated in the much reduced distance than that in the preformed channel. In addition, the average divergence angle is considerably larger, namely 14°, as shown in Fig. 5(d).

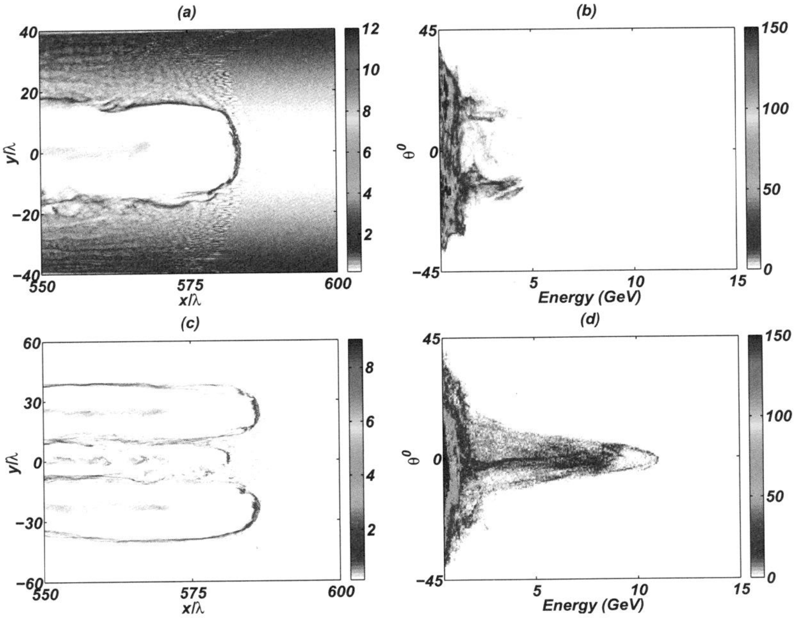

FIG. 5. Electron density at $t \sim 600T$ for the (a) single-wake (when the condition $h \leq \sigma$ is not satisfied) and (c) without guiding channel. The corresponding angular distributions of the protons are shown in (b) and (d), respectively.

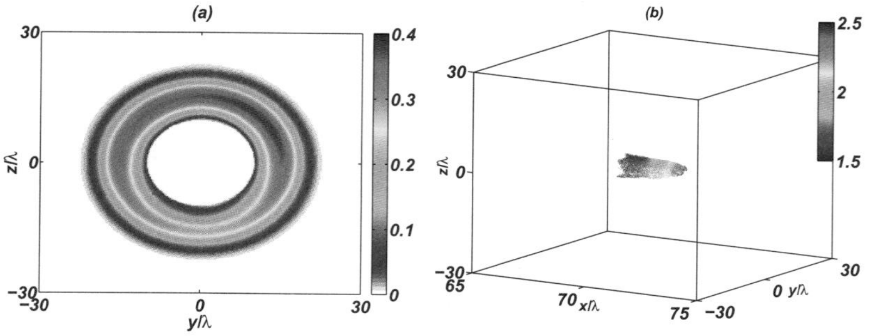

FIG. 6. (a) (z, y) section of the square root of the laser front intensity $\sqrt{E_y^2 + E_z^2}$, normalized by $\mathcal{E}_0$; (b) energy density of the accelerated protons (in units of GeV), at $t = 80T$.

III. 3D SIMULATION

To ensure that the obtained results are not 2D artifacts, we have also carried out 3D simulations. Because of limitation in the computing resources, we consider a narrow simulation box, namely, $60 \times 60 \times 100\lambda^3$. A circular target with density of $25n_c$ and radius $r = 3\lambda$ is used, while its initial position is moved to $x = 10\lambda$. The underdense plasma channel at the backside of the target, with density profile $n = 0.25n_c + 1.25n_c(r^2/r_0^2)$ is located in $11\lambda \le x \le 100\lambda$, $-10\lambda \le y \le 10\lambda$, $-10\lambda \le z \le 10\lambda$, where $r_0 = 20\lambda$. The laser spot size and duration are also reduced to be 8λ and $12T$, respectively. Fig. 6(a) displays the transverse intensity profile of laser front at $t = 80T$, where the center part of laser pulse is depleted. It evolves into a torus in 3D case instead of twin wakes. The trapped protons then undergo further acceleration as well as collimation in the torus, indicating the dynamical process of the accelerated protons is very similar to the 2D case. Fig. 6(b) shows that the accelerated protons are tightly compressed to a spotsize of less than 2.5 μm, and the maximum energy of protons reaches about 2.5 GeV at $t = 80\,T$.

IV. CONCLUSION

In conclusion, an efficient laser-driven proton acceleration scheme using a MLT with a preformed parabolic underdense channel is proposed. The laser pulse drives the target protons by RPA and is split by the MLT into two parts, generating behind them a twin-wakes structure in the channel. The pre-accelerated RPA target protons are efficiently collimated and further accelerated by LWFA in the twin wakes. With this scheme, both light diffraction and proton defocusing are avoided, making long-distance acceleration possible. The conditions and characteristics of proton acceleration from the PIC simulations agree well with that from the analytical estimates. It is shown that ultrashort (30 fs) highly collimated monoenergetic proton bunches with a peak energy of >10 GeV can be produced with a 60 fs CP laser pulse of intensity about 10^{22} W/cm^2. Such a short and collimated proton pulse can be used to excite a plasma wakefield that can in turn accelerate the plasma electrons.[34]

ACKNOWLEDGMENTS

This work is supported by the National Natural Science Foundation of China (Grant Nos. 10835003, 10935003, 10974022, 11175026, 11175029, 11272026), and the National Basic Research Program of China (Grant Nos. 2008CB717806, 2012CB801111).

[1]S. V. Bulanov, T. Zh. Esirkepov, V. S. Khoroshkov, A. V. Kuznetsov, and F. Pegoraro, Phys. Lett. A **299**, 240 (2002).
[2]M. Roth, T. E. Cowan, M. H. Key, S. P. Hatchett, C. Brown, W. Fountain, J. Johnson, D. M. Pennington, R. A. Snavely, S. C. Wilks, K. Yasuike, H. Ruhl, F. Pegoraro, S. V. Bulanov, E. M. Campbell, M. D. Perry, and H. Powell, Phys. Rev. Lett. **86**, 436 (2001).
[3]M. Borghesi, A. Schiavi, D. H. Campbell, M. G. Haines, O. Willi, A. J. Mackinnon, P. Patel, M. Galimberti, and L. A. Gizzi, Rev. Sci. Instrum. **74**, 1688 (2003).
[4]C. K. Li, F. H. Séguin, J. A. Frenje, M. Rosenberg, R. D. Petrasso, P. A. Amendt, J. A. Koch, O. L. Landen, H. S. Park, H. F. Robey, R. P. J. Town, A. Casner, F. Philippe, R. Betti, J. P. Knauer, D. D. Meyerhofer, C. A. Back, J. D. Kilkenny, and A. Nikroo, Science **327**, 1231 (2010).
[5]D. H. H. Hoffmann, A. Blazevic, P. Ni, O. Rosmej, M. Roth, N. A. Tahir, A. Tauschwitz, S. Udrea, D. Varentsov, K. Weyrich, and Y. Maron, Laser Part. Beams **23**, 47–53 (2005), available at http://journals.cambridge.org/action/displayAbstract?fromPage=online&aid=310085.
[6]P. M. Nilson, L. Willingale, M. C. Kaluza, C. Kamperidis, S. Minardi, M. S. Wei, P. Fernandes, M. Notley, S. Bandyopadhyay, M. Sherlock, R. J. Kingham, M. Tatarakis, Z. Najmudin, W. Rozmus, R. G. Evans, M. G. Haines, A. E. Dangor, and K. Krushelnick, Phys. Rev. Lett. **97**, 255001 (2006).
[7]C. K. Li, F. H. Séguin, J. A. Frenje, J. R. Rygg, R. D. Petrasso, R. P. J. Town, O. L. Landen, J. P. Knauer, and V. A. Smalyuk, Phys. Rev. Lett. **99**, 055001 (2007).
[8]Q. L. Dong, S. J. Wang, Q. M. Lu, C. Huang, D. W. Yuan, X. Liu, X. X. Lin, Y. T. Li, H. G. Wei, J. Y. Zhong, J. R. Shi, S. E. Jiang, Y. K. Ding, B. B. Jiang, K. Du, X. T. He, M. Y. Yu, C. S. Liu, S. Wang, Y. J. Tang, J. Q. Zhu, G. Zhao, Z. M. Sheng, and J. Zhang, Phys. Rev. Lett. **108**, 215001 (2012).
[9]M. Borghesi, J. Fuchs, S. V. Bulanov, A. J. Mackinnon, P. Patel, and M. Roth, Fusion Sci. Technol. **49**, 412 (2006), available at http://www.new.ans.org/pubs/journals/fst/a_1159.
[10]G. A. Mourou, T. Tajima, and S. V. Bulanov, Rev. Mod. Phys. **78**, 309–371 (2006).
[11]X. Q. Yan, C. Lin, Z. M. Sheng, Z. Y. Guo, B. C. Liu, Y. R. Lu, J. X. Fang, and J. E. Chen, Phys. Rev. Lett. **100**, 135003 (2008).
[12]O. Klimo, J. Psikal, J. Limpouch, and V. T. Tikhonchuk, Phys. Rev. ST Accel. Beams **11**, 031301 (2008).
[13]A. P. L. Robinson, M. Zepf, S. Kar, R. G. Evans, and C. Bellei, New J. Phys. **10**, 013021 (2008).
[14]X. Q. Yan, H. C. Wu, Z. M. Sheng, J. E. Chen, and J. Meyer-ter-Vehn, Phys. Rev. Lett. **103**, 135001 (2009).
[15]B. Qiao, M. Zepf, M. Borghesi, and M. Geissler, Phys. Rev. Lett. **102**, 145002 (2009).
[16]T. Esirkepov, M. Borghesi, S. V. Bulanov, G. Mourou, and T. Tajima, Phys. Rev. Lett. **92**, 175003 (2004).
[17]Z. M. Zhang, X. T. He, Z. M. Sheng, and M. Y. Yu, Phys. Plasmas **17**, 043110 (2010).
[18]B. F. Shen, X. Zhang, Z. M. Sheng, M. Y. Yu, and J. Cary, Phys. Rev. ST Accel. Beams **12**, 121301 (2009).
[19]L. L. Yu, H. Xu, W. M. Wang, Z. M. Sheng, B. F. Shen, W. Yu, and J. Zhang, New J. Phys. **12**, 045021 (2010).

[20]F. L. Zheng, H. Y. Wang, X. Q. Yan, T. Tajima, M. Y. Yu, and X. T. He, Phys. Plasmas **19**, 023111 (2012).
[21]F. L. Zheng, S. Z. Wu, C. T. Zhou, H. Y. Wang, X. Q. Yan and X. T. He, Europhys. Lett. **95**, 55005 (2011).
[22]T. Tajima and J. M. Dawson, Phys. Rev. Lett. **43**, 267 (1979).
[23]B. Qiao, M. Zepf, M. Borghesi, B. Dromey, M. Geissler, A. Karmakar, and P. Gibbon, Phys. Rev. Lett. **105**, 155002 (2010).
[24]S. V. Bulanov, E. Yu. Echkina, T. Zh. Esirkepov, I. N. Inovenkov, M. Kando, F. Pegoraro, and G. Korn, Phys. Rev. Lett. **104**, 135003 (2010).
[25]X. M. Zhang, B. F. Shen, L. L. Ji, F. C. Wang, M. Wen, W. P. Wang, J. C. Xu, and Y. H. Yu, Phys. Plasmas **17**, 123102 (2010).
[26]T. Sokollik, T. Paasch-Colberg, K. Gorling, U. Eichmann, M. Schnürer, S. Steinke, P. V. Nickles, A. Andreev, and W. Sandner, New J. Phys **12**, 11301 (2010), available at http://iopscience.iop.org/1367-2630/12/11/113013/fulltext/.
[27]Z. M. Sheng, K. Mima, Y. Sentoku, K. Nishihara, and J. Zhang, Phys. Plasmas **9**, 3147 (2002).
[28]A. Pukhov and J. Meyer-ter-Vehn, Appl. Phys. B **74**, 355 (2002).
[29]W. P. Leemans, B. Nagler, A. J. Gonsalves, C. Toth, K. Nakamura, C. G. R. Geddes, E. Esarey, C. Schroeder, and S. M. Hooker, Nature Phys. **2**, 696 (2006).
[30]W. Lu, C. Huang, M. Zhou, and M. Tzoufras, Phys. Plasmas **13**, 056709 (2006).
[31]W. Lu, M. Tzoufras, C. Joshi, F. S. Tsung, W. B. Mori, J. Vieira, R. A. Fonseca, and L. O. Silva, Phys. Rev. ST Accel. Beams **10**, 061301 (2007).
[32]A. Henig, D. Kiefer, K. Markey, D. C. Gautier, K. A. Flippo, S. Letzring, R. P. Johnson, T. Shimada, L. Yin, B. J. Albright, K. J. Bowers, J. C. Fernández, S. G. Rykovanov, H.-C. Wu, M. Zepf, D. Jung, V. Kh. Liechtenstein, J. Schreiber, D. Habs, and B. M. Hegelich, Phys. Rev. Lett. **103**, 045002 (2009).
[33]H. B. Cai, K. Mima, A. Sunahara, T. Johzaki, H. Nagatomo, S. P. Zhu, and X. T. He, Phys. Plasmas **17**, 023106 (2010).
[34]A. Caldwell, K. Lotov, A. Pukhov, and F. Simon, Nature Phys. **5**, 363 (2009).

PHYSICS OF PLASMAS 21, 123118 (2014)

Generation of high-energy mono-energetic heavy ion beams by radiation pressure acceleration of ultra-intense laser pulses

D. Wu,[1,2] B. Qiao,[1,2,a)] C. McGuffey,[1] X. T. He,[2] and F. N. Beg[1]
[1]*Center for Energy Research, University of California, San Diego, California 92093, USA*
[2]*School of Physics, Peking University, Beijing 100871, China*

(Received 9 September 2014; accepted 3 December 2014; published online 24 December 2014)

Generation of high-energy mono-energetic heavy ion beams by radiation pressure acceleration (RPA) of intense laser pulses is investigated. Different from previously studied RPA of protons or light ions, the dynamic ionization of high-Z atoms can stabilize the heavy ion acceleration. A self-organized, stable RPA scheme specifically for heavy ion beams is proposed, where the laser peak intensity is required to match with the large ionization energy gap when the successive ionization state passes the noble gas configurations [such as removing an electron from the helium-like charge state $(Z-2)^+$ to $(Z-1)^+$]. Two-dimensional particle-in-cell simulations show that a mono-energetic Al^{13+} beam with peak energy 1.0 GeV and energy spread of only 5% can be obtained at intensity of $7 \times 10^{20}\,\mathrm{W/cm^2}$ through the proposed scheme. A heavier, mono-energetic, ion beam (Fe^{26+}) can attain a peak energy of 17 GeV by increasing the intensity to $10^{22}\,\mathrm{W/cm^2}$. © *2014 AIP Publishing LLC*. [http://dx.doi.org/10.1063/1.4904402]

I. INTRODUCTION

Laser driven ion acceleration has become a highly active field of research over the past decade.[1,2] The applications[3] include tumor therapy, ultrafast radiography, isotope production, and fast ignition laser fusion. Most applications require a high-energy ion beam with large particle number, mono-energetic spectrum, and small divergence. Among a number of acceleration schemes identified to date, including target normal sheath acceleration (TNSA),[4] break-out afterburner acceleration,[5] shock acceleration,[6] and others, the newly emerged radiation pressure acceleration (RPA)[7–11] using circularly polarized (CP) lasers has been regarded as one of the most promising routes to obtaining such high-quality ion beams. However, all current studies are focused on protons and light ions. How to use RPA to produce high-energy mono-energetic heavy ion beams is still unknown. Meanwhile, heavy ion beams have a much broader range of applications related to medicine, materials, nuclear reactions,[12,13] quantum electrodynamics,[14] and high energy density plasmas.

In the idealistic RPA scheme, all ion species are co-accelerated and the heavy ion acceleration is not impeded by the light ion contaminants (hydrogen and carbon)[8,10,15,16] whereas this is a concern in the TNSA mechanism. Therefore, in principle, heavy ions can be efficiently accelerated to much higher energy due to their large mass. However, both simulations and experiments[17,18] show that in a realistic multidimensional case, the heavy ion species in RPA undergo rapid Coulomb explosion due to the deficiency of co-moving electrons and the transverse instabilities, which lead to very broad beam energy spectrum. Recently, two-dimensional (2D) particle-in-cell (PIC) simulation[19] indicated that a quasi-mono-energetic Fe^{24+} beam can be generated by using compound targets mixed with a heavier Au substrate. However, it relies on complicated target fabrication and the laser-ion conversion efficiency is extremely low.

In this paper, we investigate the RPA of heavy ion beams by self-consistently including the ionization dynamics. This investigation is different from previously studied RPA of protons or light ions, as it has been found that the dynamic ionization of high-Z atoms can self-organize and stabilize the acceleration of heavy ions. The central part of a Gaussian laser ionizes heavy ions to high charge state, while on the wings of the laser, the ionization charge states are low. Thus, the highly charged heavy ions only concentrate in the central part of the laser, self-organizing into a high density, high charge clump structure therein. The self-organized structure is due to its ionization property, which does not depend on the complicated target design or other external control. The stabilization is related to a large supplement of electrons produced from ionization, which keeps the accelerating plasma opaque to laser and helps neutralize the ion beam, suppressing transverse instabilities.

In this scheme, an appropriate matching condition between laser intensity, foil thickness, and the unique large gap feature in the ionization energies when the successive ionization of high-Z atoms reaches the noble gas configuration, such as removing an electron from the helium-like $(Z-2)^+$ to one-bound-electron charge states $(Z-1)^+$, needs to be satisfied. The condition of the regime has been analytically given and verified by 2D PIC simulations. Results show that a highly charged mono-energetic Al^{13+} beam with a total charge >2 nC, peak energy of 1 GeV, and energy spread of 5% is produced by CP lasers at intensity $7 \times 10^{20}\,\mathrm{W/cm^2}$. Similarly, a mono-energetic Fe^{26+} beam

a)Electronic mail: bqiao@ucsd.edu

with peak energy 17 GeV can be obtained if the laser intensity is increased to $4 \times 10^{22}\ \mathrm{W/cm^2}$.

II. THEORETICAL MODEL

In the RPA scheme, the laser ponderomotive force pushes the electrons forward, leaving the ions behind until the space charge electric field balances the ponderomotive force at a distance. These ions are therefore trapped by the combined forces of the electrostatic field of the electron sheath and the inertial force of the accelerating target. Together with the electron layer, they form a double layer and are collectively accelerated by the laser ponderomotive force longitudinally, leading to monoenergetic ion production. Although, the longitudinal velocity difference of ions and electrons is zero, the transverse velocity difference is very large. Electrons are rapidly oscillated by the strong transverse laser field with $p_e \sim a$. Heavy ions cannot respond to high frequent oscillations, remaining nearly fixed transversely. When an intense laser irradiates a solid high-Z target, a crucial, yet unexplored effect for heavy ion RPA is the importance of the successive ionization during acceleration, including both the field[20] and impact ionizations.[21] Considering the cross section of impact ionization,[22] which decreases rapidly with electron kinetic energy greater than 10 keV (states of Al), we can estimate that the impact ionization rate in our case is extremely low, as transverse kinetic energy of electrons can be as high as several MeV. For static high-Z atoms in an electric field, the ionization rate from charge states $(\zeta-1)^+$ to ζ^+ ($\zeta \le \mathrm{Z}$) can be calculated by the direct current Ammosov-Delone-Krainov (ADK) formula[23]

$$W_{l,m} = \omega_a C^2_{n^*l^*} \frac{(2l+1)(l+|m|)!}{2^{|m|}(|m|)!(l-|m|)!} \frac{U_\mathrm{i}}{2U_\mathrm{H}} \times \left[\frac{2E_\mathrm{H}}{E}\left(\frac{U_\mathrm{i}}{U_\mathrm{H}}\right)^{3/2}\right]^{2n^*-|m|-1} \exp\left[-\frac{2}{3}\frac{E_\mathrm{H}}{E}\left(\frac{U_\mathrm{i}}{U_\mathrm{H}}\right)^{3/2}\right], \tag{1}$$

where $\omega_a = \alpha^3 c/r_e = 4.13 \times 10^{-16}\mathrm{s}^{-1}$ is the atomic unit frequency, U_i is the ionization potential, and $U_\mathrm{H} = 1312\ \mathrm{kJ/mol}$ is the ionization potential of hydrogen (H) at fundamental state, $E_\mathrm{H} = m_e^2 e^5 h^{-4} = 5.14 \times 10^{11}\ \mathrm{V/m}$ is the corresponding electric field, m_e is electron mass, c is light speed, $\alpha = 1/137$ is the fine structure constant, $r_e = m_e/c^2$ is the classical electron radius, l, m are the electrons' orbital quantum number and its projection, respectively, $n^* = \zeta\sqrt{U_\mathrm{H}/U_\mathrm{i}}$ is the effective principal quantum number, $l^* = n_0^* - 1$ is the effective orbital number, and n_0^* is the effective principal quantum number of the ground state. The coefficients $C_{n^*l^*} = 2^{2n^*}/[n^*\Gamma(n^*+l^*+1)\Gamma(n^*-l^*)]$ in Eq. (1) can be calculated using semi-classical approximation as $C_{n^*l^*} \simeq 1/2\pi n^* \times [4e^2/(n^{*2}-l^{*2})]^{n^*} \times [(n^*-l^*)(n^*+l^*)]^{l^*+1/2}$. In the limit $l^* \ll n^*$, considering the fundamental state $l = m = 0$, we obtain the ionization rate as a function of the local electric field E for the charge state at ζ^+as,

$$W = \frac{\omega_a}{8\pi}\frac{E}{E_\mathrm{H}\zeta} \times \left(\frac{4\pi E_\mathrm{H}\zeta^3}{En^{*4}}\right)^{2n^*} \times \exp\left[-\frac{2}{3}\frac{E_\mathrm{H}}{E}\left(\frac{\zeta}{n^*}\right)^3\right]. \tag{2}$$

The ionization probability within a finite time δt is calculated as

$$I_p = 1 - \exp(-W\delta t). \tag{3}$$

From Eqs. (2) and (3), we see that the ionization dynamics heavily depend on the ionization potentials of different materials and charge states. Figures 1(a) and 1(b) plot the ionization energy potentials for different charge states of, respectively, Al and Fe based on the NIST database.[24] It can be clearly seen that a large gap in the ionization potentials exists when the successive ionization of high-Z atoms reaches the noble gas configuration, such as it passing the helium (He)-like configuration from Al^{11+} to Al^{12+} or from Fe^{24+} to Fe^{25+} [generally ζ^+ increases from $(Z-2)^+$ to $(Z-1)^+$]. In fact, such large ionization potential gap features are general for high-Z atoms because atoms/ions with noble-like configurations are always stable, which includes not only He- but also Neon (Ne)-like. The ionization potentials for He and Ne-like configurations of atoms/ions varying with the material atomic number Z can be also found from the NIST database. An empirical formula describing the dependence of the ionization potential of He-like ions on Z is given as[25]

$$U_\text{He-like} = \frac{65}{64}\left(Z^2 - \frac{4}{3}Z + \frac{4}{9}\right) \times U_\mathrm{H}. \tag{4}$$

Our stable heavy ion RPA regime is based on the above ionization gap feature of high-Z ultra thin foil targets during interactions with intense lasers. According to Eqs. (2)–(4),

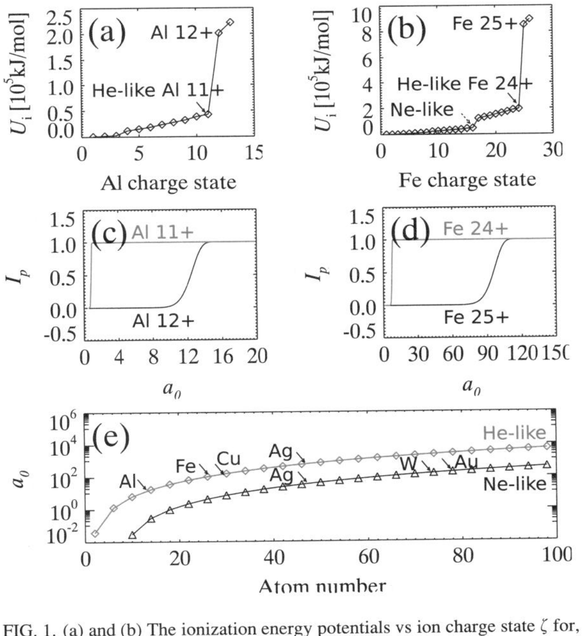

FIG. 1. (a) and (b) The ionization energy potentials vs ion charge state ζ for, respectively, Al and Fe. (c) and (d) The ionization probabilities of Al^{11+}, Al^{12+} and Fe^{24+}, Fe^{25+} vs the normalized laser amplitude a_0. (e) The required a_0 to achieve stable heavy ion RPA regime vs the atomic number Z of target materials, which is calculated from Eqs. (2)–(4) by matching the He-like ionization potential gaps (red). Those for matching the Ne-like gaps are also plotted (black).

the large gap in the ionization potentials indicates that a sharp increase of the electric field is required to achieve the highly charged state of heavy ions, such as $(Z-1)^+$ (or Z^+). Since laser electric field dominates during the ionization, Figures 1(c) and 1(d) plot the calculated ionization probability I_p of, respectively, Al^{11+}, Al^{12+} and Fe^{24+}, Fe^{25+} varying with the normalized laser electric field, i.e., amplitude $a_0 = eE/m_e c\omega_0$, where e is electron charge, ω_0 is laser frequency. δt is chosen to be the laser pulse duration τ and $\tau = 30T_0$ ($T_0 = 2\pi/\omega_0$) is assumed here. We see that [Fig. 1(c)] for a large range of $1 < a_0 < 12$, the dominant charge state of ions in the acceleration is Al^{11+}, while to achieve the highly charged state Al^{12+} (and/or Al^{13+}) dominating, a_0 has to be sharply increased from 12 to $a_0 > 16$. Similar feature can be seen in Fe in Fig. 1(d), where to achieve Fe^{25+}, $a_0 > 120$ is required.

To achieve stable RPA of heavy ions, one possible way is to choose the peak laser amplitude a_0 to be closely matching to that required to overcome the above ionization gap and reach the highly charged state $(Z-1)^+$ (or Z^+) of heavy ions. This will help to stabilize the acceleration in two ways. First, since the laser amplitude generally has transverse Gaussian distribution as $a = a_0 \times \exp(-r^2/r_0^2)$ (r_0 is the laser spot radius), only the ions near the laser axis can be ionized to extremely high charge state $(Z-1)^+$ (or Z^+), which undergo strong acceleration by the larger electric force due to their higher charge to mass ratio. The ions in the laser wing region are ionized to lower charge states due to dropping of laser field, undergoing weaker acceleration and rapid dispersion due to their lower charge to mass ratio and transverse instabilities. Eventually, a self-organized, transverse-size-limited ion beam near the laser axis is formed undergoing efficient acceleration, where the one-dimensional (1D) stable RPA dynamics are maintained. Second and most importantly, the high-order ionization process occurring near laser axis will produce a large number of supplementary electrons accompanying the heavy ion beam acceleration, which help to avoid Coulomb explosion and preserve stable RPA of the beam, as expected from existing results.[11,26] Therefore, a high-energy mono-energetic highly charged heavy ion beam can be obtained in the new regime. Note that both the self-organization and stabilization effects here relate to the dynamic ionization of heavy atoms, which cannot be achieved in RPA of protons or light ions. The required a_0 for this stable heavy ion RPA regime can be calculated from the coupled equations (2)–(4), which is shown in Fig. 1(e). The other condition is that the initial foil thickness l_0 should satisfy the required RPA condition,[10] $l_0/\lambda \simeq a_0 n_c/\pi Z_{eff} n$, that is, the radiation pressure balances the maximum charge separation field, where Z_{eff} is the effective charge state of targets during acceleration.

III. NUMERICAL RESULTS

In order to verify the new scheme, 2D PIC simulations are carried out, where the field ionization module is self-consistently integrated in the code. Multi-level ionization and energy conservation during the ionization process has also been taken into account.[27,28] An ultrathin Al foil with solid density $\rho_{Al} = 2.7\,g/cm^3$ ($6 \times 10^{22}\,cm^{-3}$) is chosen. According to the conditions discussed above [Fig. 1(e)], a CP laser pulse with $a_0 = 16$ (intensity $I_0 = 7 \times 10^{20}\,W/cm^2$) and wavelength $\lambda = 1\,\mu m$ is taken to be normally incident on the foil. The laser pulse has a transverse Gaussian profile with $r_0 = 3.0\,\mu m$ and a trapezoidal temporal profile of duration $\tau = 30T_0$, consisting of a plateau of $26T_0$ and rising and falling times of $2T_0$ each. The foil thickness is 20 nm satisfying the above optimal condition and the initial charge state is taken as Al^{3+} under temperature $T_e = T_{Al} = 10\,eV$. In the simulations, 8000 cells along x-axis and 1200 cells transversely along y-axis constitute a $8 \times 12\,\mu m$ simulation box. Each foil cell is filled with 900 electrons and 900 Al^{3+} ions. The foil is located at $x = 2.0\mu m$ and laser propagates from the left boundary $x = 0$. To more clearly identify the role of the ionization on the heavy ion RPA dynamics, simulations without the ionization effect

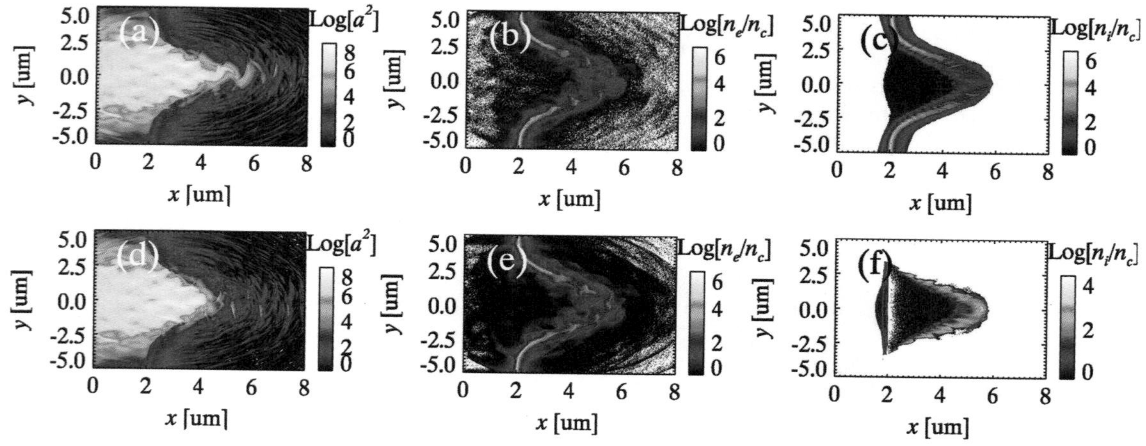

FIG. 2. 2D PIC simulation results: laser intensity [(a) and (d)] at $t = 25T_0$, electron [(b) and (e)] and ion [Al^{11+} in (c) and Al^{13+} in (f)] densities at $t = 30T_0$ for 20 nm Al foils irradiated by CP lasers at $7 \times 10^{20}\,W/cm^2$, where the ionization effect is not taken into account in (a)–(c) but self-consistently included in (d)–(f). Other parameters are shown in the text.

considered are also carried out for comparison, where the charge state is fixed to Al^{11+}.

Figure 2 shows the laser intensity distributions at time $t = 25T_0$ and the electron and Al ion density maps at $t = 30T_0$ for, respectively, without [(a)–(c)] and with [(d)–(f)] the ionization effect taken into account. We see that without ionization, the intense laser pulse easily penetrates through the accelerating plasma slab due to transverse instabilities [Fig. 2(a)], leading to rapid heating and dispersion of electrons in the latter [Fig. 2(b)], which eventually results in decompression and premature acceleration termination of the Al^{11+} beam. However, when the ionization is considered and the above conditions for our heavy ion RPA regime are satisfied, as expected, a self-organized highly mono-charged Al^{13+} beam near the laser axis is formed, which undergoes stable RPA of intense lasers [see Fig. 2(f)]. Furthermore, comparing Fig. 3(a) with Fig. 3(b), we clearly see that the ionization produces a large number of additional electrons as supplement to the accelerating slab, where the electron density [Fig. 3(b)] is much higher than the case without ionization [Fig. 3(a)]. This keeps the accelerating plasma slab opaque to intense laser and helps to stabilize the RPA of heavy Al^{13+} beam, as we mentioned above. The supplements of electrons (i.e., the two inner shell electrons) play an important role in stabilizing the RPA acceleration. Considering that the foil moves more and more rapidly, in its proper frame the incident wavelength increases, and correspondingly the critical density decreases. The two supplemental electrons from inner shell ionization provide a non-negligible increase in the free electron density allowing RPA to be sustained for a longer time duration.

The energy spectra of heavy Al ions for both cases at $t = 24$, 26, 28, and $30T_0$ are shown in Figs. 3(c) and 3(d), respectively. Clearly without the ionization, the Al^{11+} beam initially has a well-peaked energy spectrum at $t = 24T_0$ [the black line in Fig. 3(c)], but at later time it undergoes rapid instabilities and Coulomb explosion, leading to much broadened energy spectrum finally of the beam at $t = 30T_0$ [the blue line in Fig. 3(c)]. However, in our regime with the ionization included [Fig. 3(d)], the heavy Al ion beam experiences stable RPA, the pronounced peak in the energy spectrum is preserved till the laser is over at $t = 30T_0$. Because the ionization potential difference between Al^{12+} and Al^{13+} is very small, the ion beam at $t = 24$ and $26T_0$ [the black and red lines in Fig. 3(d)], having both Al^{12+} and Al^{13+} charge states, are further rapidly ionized to fully Al^{13+} later [the green and blue lines in Fig. 3(d)]. Eventually, when the laser pulse is over, a quasi-mono-energetic Al^{13+} beam with peak energy 40 MeV/u (1.08 GeV) and energy spread of only 5% is produced. The effective particle number in the beam is about 10^9, i.e., the total charge about 2 nC.

In order to identify the robustness of the regime, further simulations with Fe ultrathin foils are also carried out. To match the condition [Eqs. (2)–(4)], the incident CP laser is chosen to be $a_0 = 130$ ($I \simeq 4 \times 10^{22}\,\mathrm{W/cm^2}$) and $\tau = 14T_0$. The foil thickness at $7.9\,\mathrm{g/cm^3}$ is taken as 30 nm and the initial charge state is $\zeta_0 = 3$. All other parameters are the same as above. Figure 4 shows density maps and energy spectra of Fe^{26+} at $t = 12$ and $14T_0$. It can be clearly seen that a quasi-mono-energetic Fe^{26+} beam with peak energy 300 MeV/u (16.8 GeV) is also produced. For super heavy ions, such as W and Au, in order to obtain stable RPA of highly charged W^{73+} and Au^{78+} $[(Z-1)^+]$ beam, the corresponding laser intensities need to be $\sim 10^{25}\,\mathrm{W/cm^2}$ to match their He-like ionization potential gaps, which are unachievable so far. However, we can still carefully use their Ne-like ionization gaps to achieve stable RPA of lower-charge state such as W^{65+} and Au^{70+} $[(Z-9)^+]$, where the required intensity would drop to $\sim 10^{22}\,\mathrm{W/cm^2}$, reachable with current laser facilities.

Note that in the above simulations, the reason that the temporal profiles of laser pulses are chosen to be trapezoidal is only to confirm exactly the theoretical predictions from Eqs. (2)–(4). The regime can be applied for Gaussian or other temporal profiles as well by just more carefully

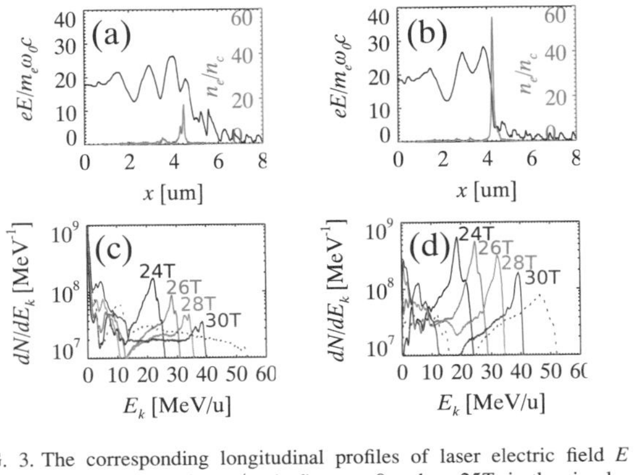

FIG. 3. The corresponding longitudinal profiles of laser electric field E (black) and electron density n_e/n_c (red) at $y = 0$ and $t = 25T_0$ in the simulation of Fig. 2 for the cases of, respectively, without (a) and with (b) the ionization effects. (c) and (d) The corresponding Al energy spectra at $t = 24$, 26, 28, and $30T_0$. Note that in (d) at $t = 24$ and $26T_0$, the spectra are calculated for a combination of both Al^{12+} and Al^{13+}, after which all ions are fully ionized to Al^{13+}. The dashed blue lines in (c) and (d) show the results when laser temporal profile changes to be $\sin^2$ function, as mentioned in the text.

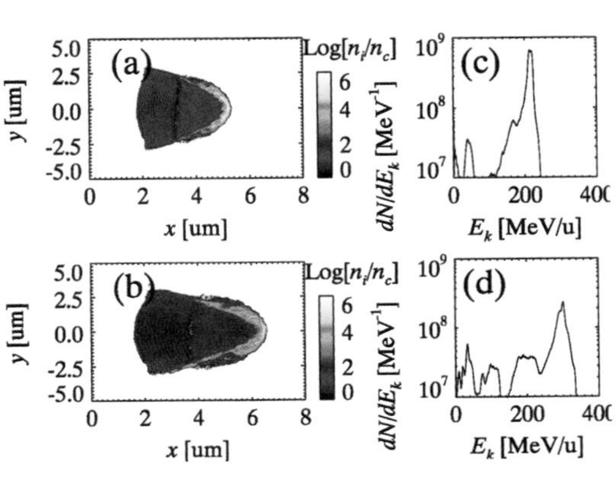

FIG. 4. 2D PIC simulation results for the case of Fe foils: ion density [(a) and (b)] and energy spectra [(c) and (d)] for Fe^{26+} at $t = 12$ and $14T_0$, where the foil thickness is 30 nm, $a_0 = 130$ ($4 \times 10^{22}\,\mathrm{W/cm^2}$), and $\tau = 14T_0$. Other parameters are the same as those in Fig. 2.

matching the condition. The simulations using an intense laser pulse with $\sin^2$ temporal profile as $\sin^2(\pi t/\tau)$ have also been run, where $a_0 = 24$ is taken and $\tau = 30T_0$ is chosen to keep the total laser energy the same as above. For this more realistic case, the final ion energy spectra are shown in Figs. 3(c) and 3(d) as the dashed lines, which show clearly in the regime, a high-energy Al^{13+} ion beam at peak energy 45 MeV/u is also obtained and the narrow-band energy spectral feature is preserved.

IV. DISCUSSIONS

There are additional practical considerations for this regime. Laser contrast in excess of 10^{10} is required, which can be achieved by using plasma mirrors. In a plasma, the free electrons constantly produce bremsstrahlung in collisions with the ions, especially heavy ions, due to their much higher coulomb force. To evaluate the role of this effect, we have considered a uniform plasma with thermal electrons distributed according to the Maxwell-Boltzmann distribution with the temperature Te (eV). The power density (power per volume) of the bremsstrahlung radiated is estimated to be[29] $P_{Br} = 1.69 \times 10^{-32} n_e T_e^{1/2} \sum[\zeta^2 n_\zeta]\ \mathrm{W/cm^3}$, where n_e ($\mathrm{cm^{-3}}$) is electron density number. However, for the RPA case, the thickness of the high-Z target is only 10s nm and the acceleration time is also only 10s laser periods, the total radiation losses are about $E_{Br} = P_{Br} \times r_0^2 \times l \times \tau \sim 3.0 \times 10^{-6}$ J, which is much smaller than the laser energy $E_{laser} = I \times r_0^2 \times \tau \sim 6.0$ J. Thus, the bremsstrahlung radiation loss can be neglected in our regime. An experiment was recently reported in which an intense positron beam was produced, due to the Bethe-Heitler (BH) process, out the back of Au targets when illuminated with ultra intense laser pulses.[30] However in their experiment, the target was of order mm thickness. Experiments also showed that the efficiency of positrons generated by the BH process heavily depends on the target thickness.[30] For our ultra-thin target with thickness of 10s nm, the energy losses due to pair-production can also be ignored.

However, there are also some concerns about the ADK ionization model used in this work, which does not take the Lorenz effect into consideration. For high speed atoms/ions with velocity close to light speed, the Lorenz ionization[31] needs to be taken into consideration. Our on-going and following work would focus on the more detailed atomic physics model and its influence on heavy ion acceleration.

V. CONCLUSIONS

In summary, we have proposed a novel RPA scheme specifically for generation of high-energy mono-energetic heavy ion beams from ultra thin foils irradiated by intense laser pulses. In this scheme, the dynamic ionization of high-Z atoms plays a key role in self-organization and stabilization of the heavy ion acceleration, different from previously studied RPA of protons or light ions. The ionization provides a large number of supplementary electrons that accompany and stabilize the RPA of heavy ion beams. By matching the laser intensity with the large ionization energy gap during the successive ionization when passing the noble gas configurations, a self-organized, stable RPA of highly charged heavy ion beam close to the laser axis is obtained. A clear physical picture of heavy ion RPA including their ionization dynamics has been given for the first time and verified numerically. Two-dimensional PIC simulations show a highly charged mono-energetic Al^{13+} beam with peak energy at 1 GeV and energy spread of only 5% is produced by ultra intense laser at intensity $7 \times 10^{20}\ \mathrm{W/cm^2}$. Similarly, a quasi-mono-energetic Fe^{26+} beam with peak energy 300 MeV/u (17 GeV) can be also obtained by increasing the intensity to $10^{22}\ \mathrm{W/cm^2}$. Both schemes could be tested experimentally with existing lasers.

ACKNOWLEDGMENTS

The calculations are performed on the extreme science and engineering discovery environment (XSEDE) supported by National Science Foundation. The work was partially supported by the Air Force Office of Scientific Research Grant (No. FA9550-14-1-0282).

[1]A. Macchi, M. Borghesi, and M. Passoni, Rev. Mod. Phys. **85**, 751 (2013).
[2]H. Daido, M. Nishiuchi, and A. S. Pirozhkov, Rep. Prog. Phys. **75**, 056401 (2012).
[3]M. Borghesi, J. Fuchs, S. V. Bulanov, A. J. MacKinnon, P. K. Patel, and M. Roth, Fusion Sci. Technol. **49**, 412 (2006).
[4]S. C. Wilks, A. B. Langdon, T. E. Cowan, M. Roth, M. Singh, S. Hatchett, M. H. Key, D. Pennington, A. MacKinnon, and R. A. Snavely, Phys. Plasmas **8**, 542 (2001).
[5]L. Yin, B. J. Albright, K. J. Bowers, D. Jung, J. C. Fernandez, and B. M. Hegelich, Phys. Rev. Lett. **107**, 045003 (2011).
[6]D. Haberberger, S. Tochitsky, F. Fiuza, C. Gong, R. A. Fonseca, L. O. Silva, W. B. Mori, and C. Joshi, Nat. Phys. **8**, 95 (2012).
[7]A. Macchi, F. Cattani, T. V. Liseykina, and F. Cornolti, Phys. Rev. Lett. **94**, 165003 (2005).
[8]A. P. L. Robinson, M. Zepf, S. Kar, R. G. Evans, and C. Bellei, New J. Phys. **10**, 013021 (2008).
[9]B. Qiao, M. Zepf, M. Borghesi, and M. Geissler, Phys. Rev. Lett. **102**, 145002 (2009).
[10]A. Macchi, S. Veghini, and F. Pegoraro, Phys. Rev. Lett. **103**, 085003 (2009).
[11]B. Qiao, M. Zepf, M. Borghesi, B. Dromey, M. Geissler, A. Karmakar, and P. Gibbon, Phys. Rev. Lett. **105**, 155002 (2010).
[12]F. N. Beg, K. Krushelnick, C. Gower, S. Torn, A. E. Dangor, A. Howard, T. Sumner, A. Bewick, V. Lebedenko, J. Dawson, D. Davidge, M. Joshi, and J. R. Gillespie, Appl. Phys. Lett. **80**, 3009 (2002).
[13]P. McKenna, K. W. D. Ledingham, J. M. Yang, L. Robson, T. McCanny, S. Shimizu, R. J. Clarke, D. Neely, K. Spohr, R. Chapman, R. P. Singhal, K. Krushelnick, M. S. Wei, and P. A. Norreys, Phys. Rev. E **70**, 036405 (2004).
[14]Th. Stohlker, H. Backe, H. F. Beyer, F. Bosch, A. Brauning-Demian, S. Hagmann, D. C. Ionescu, K. Jungmann, H.-J. Kluge, C. Kozhuharov, Th. Kuhl, D. Liesen, R. Mann, P. H. Mokler, and W. Quint, Nucl. Instrum. Methods Phys. Res., Sect. B **205**, 156 (2003).
[15]D. Wu, C. Y. Zheng, C. T. Zhou, X. Q. Yan, M. Y. Yu, and X. T. He, Phys. Plasmas **20**, 023102 (2013).
[16]D. Wu, C. Y. Zheng, B. Qiao, C. T. Zhou, X. Q. Yan, M. Y. Yu, and X. T. He, Phys. Rev. E **90**, 023101 (2014).
[17]B. Qiao, S. Kar, M. Geissler, P. Gibbon, M. Zepf, and M. Borghesi, Phys. Rev. Lett. **108**, 115002 (2012).
[18]S. Kar, K. F. Kakolee, B. Qiao, A. Macchi, M. Cerchez, D. Doria, M. Geissler, P. McKenna, D. Neely, J. Osterholz, R. Prasad, K. Quinn, B. Ramakrishna, G. Sarri, O. Willi, X. Y. Yuan, M. Zepf, and M. Borghesi, Phys. Rev. Lett. **109**, 185006 (2012).
[19]A. V. Korzhimanov, E. S. Efimenko, S. V. Golubev, and A. V. Kim, Phys. Rev. Lett. **109**, 245008 (2012).

[20]A. J. Kemp, R. E. W. Pfund, and J. Meyer-ter-Vehn, Phys. Plasmas **11**, 5648 (2004).
[21]Y. Sentoku and A. J. Kemp, J. Comput. Phys. **227**, 6846 (2008).
[22]G. M. Petrov, J. Davis, and Tz. Petrova, Plasma Phys. Controlled Fusion **51**, 095005 (2009).
[23]M. V. Ammosov, N. B. Delone, and V. P. Krainov, Sov. Phys. JETP **64**, 1191 (1986).
[24]See http://physics.nist.gov/PhysRefData/ASD/ionEnergy.html for evaluated data on ground states and ionization energies of atoms and atomic ions.
[25]H. Elo, Naturwissenschaften **94**, 779 (2007).
[26]B. Qiao, M. Zepf, P. Gibbon, M. Borghesi, B. Dromey, S. Kar, J. Schreiber, and M. Geissler, Phys. Plasmas **18**, 043102 (2011).
[27]M. Chen, E. Cormier-Michel, C. G. R. Geddes, D. L. Bruhwiler, L. L. Yu, E. Esarey, C. B. Schroeder, and W. P. Leemans, J. Comput. Phys. **236**, 220 (2013).
[28]R. Nuter, L. Gremillet, E. Lefebvre, A. Levy, T. Ceccotti, and P. Martin, Phys. Plasmas **18**, 033107 (2011).
[29]J. D. Huba, NRL Plasma Formulary, Naval Research Laboratory, Washington DC, 2002 revised, p. 57.
[30]H. Chen, S. C. Wilks, J. D. Bonlie, E. P. Liang, J. Myatt, D. F. Price, D. D. Meyerhofer, and P. Beiersdorfer, Phys. Rev. Lett. **102**, 105001 (2009).
[31]V. S. Popov, B. M. Karnakov, and V. D. Mur, J. Exp. Theor. Phys. **88**, 902 (1999).

IOP Publishing New J. Phys. 18 (2016) 093029 doi:10.1088/1367-2630/18/9/093029

New Journal of Physics
The open access journal at the forefront of physics

Deutsche Physikalische Gesellschaft DPG
IOP Institute of Physics

Published in partnership with: Deutsche Physikalische Gesellschaft and the Institute of Physics

CrossMark

OPEN ACCESS

RECEIVED
22 April 2016

REVISED
20 July 2016

ACCEPTED FOR PUBLICATION
1 August 2016

PUBLISHED
15 September 2016

PAPER

Generation of quasi-monoenergetic heavy ion beams via staged shock wave acceleration driven by intense laser pulses in near-critical plasmas

W L Zhang[1], B Qiao[1,2], X F Shen[1], W Y You[1], T W Huang[1], X Q Yan[1], S Z Wu[2,3], C T Zhou[1,3] and X T He[1,3]

[1] Center for Applied Physics and Technology, HEDPS, and State Key Laboratory of Nuclear Physics and Technology, School of Physics, Peking University, Beijing 100871, People's Republic of China
[2] Collaborative Innovation Center of Extreme Optics, Shanxi University, Taiyuan, Shanxi 030006, People's Republic of China
[3] Institute of Applied Physics and Computational Mathematics, Beijing 100094, People's Republic of China

E-mail: bqiao@pku.edu.cn

Keywords: staged shock waves, laser-driven heavy ion acceleration, adjustable energy spectrum, near-critical plasmas, shock transmission, shock wave acceleration

Abstract

Laser-driven ion acceleration potentially offers a compact, cost-effective alternative to conventional accelerators for scientific, technological, and health-care applications. A novel scheme for heavy ion acceleration in near-critical plasmas via staged shock waves driven by intense laser pulses is proposed, where, in front of the heavy ion target, a light ion layer is used for launching a high-speed electrostatic shock wave. This shock is enhanced at the interface before it is transmitted into the heavy ion plasmas. Monoenergetic heavy ion beam with much higher energy can be generated by the transmitted shock, comparing to the shock wave acceleration in pure heavy ion target. Two-dimensional particle-in-cell simulations show that quasi-monoenergetic C^{6+} ion beams with peak energy 168 MeV and considerable particle number 2.1×10^{11} are obtained by laser pulses at intensity of 1.66×10^{20} W cm^{-2} in such staged shock wave acceleration scheme. Similarly a high-quality Al^{10+} ion beam with a well-defined peak with energy 250 MeV and spread $\delta E/E_0 = 30\%$ can also be obtained in this scheme.

1. Introduction

High quality heavy ion beam generated from high-Z material can potentially be used for wide-ranging fields, including carbon ion-based fast ignition [1], injector for conventional heavy ion accelerators [2], nuclear reaction [3], study of exotic phenomenons in the interior of stars [4], and creation of quark-gluon plasma [5], etc. The advent of ultra-high intensity laser provides access to laboratory-sized compact ion accelerator via laser-plasma interactions. While the accelerations of protons and light ions have been extensively investigated, little has been reported on acceleration of heavy ions. In particular, the aforementioned applications [6] require the heavy ion beams have simultaneous high peak energy, high directionality, narrow energy spread, and high intensity flux, etc, which impose great challenges to the heavy ion acceleration.

Shock wave acceleration (SWA) [7–9] by ion reflection in near-critical plasmas is a promising candidate as a source of high quality ion beam. For the SWA in near-critical plasmas, the ions are continuously reflected to twice of the shock velocity during the shock wave stable propagation, resulting in production of intense quasi-monoenergetic ion beam with comparatively high particle number. These are in stark contrast to the target normal sheath acceleration [10–12], which typically produces an ion beam with exponential spectrum, large angular divergence, and low particle number. Furthermore, SWA can be achieved under a much relaxed experimental condition. This is also in contrast to the radiation pressure acceleration [13–18] which is exposed to various transverse instabilities and faces formidable experimental challenges. While for heavy ions with low Z/A, the shock velocity is slow due to the slow ion acoustic wave speed, i.e., $c_s = \sqrt{ZT_e/Am_p}$, where T_e is the electron temperature and m_p is the proton mass. In addition, the condition for ion reflection, i.e.,

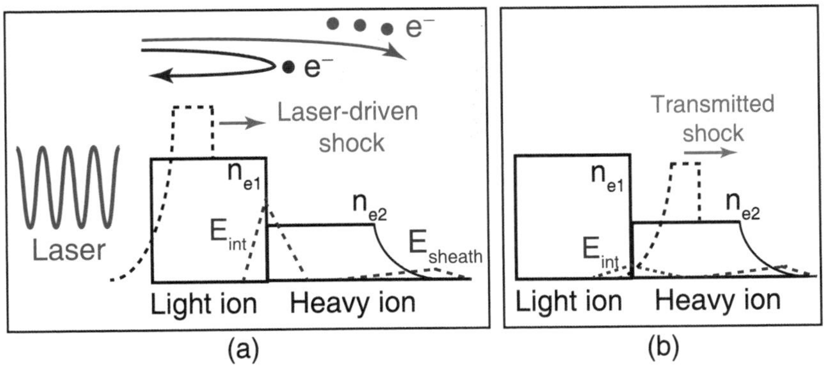

Figure 1. A schematic figure for the staged shock wave acceleration dynamics at moments (a) before the laser-driven shock arrives at the interface region and (b) after the transmitted shock is completely formed. The black solid lines represent the initial plasma density profile with electron density n_{e1} and n_{e2}, respectively. The blue dashed lines are the electrostatic fields induced at the interface region (E_{int}) and the rear side (E_{sheath}), respectively. The electric field associated with the shock front E_{sh} is not shown here. The black dashed line in (a) is the laser-compressed ion density profile acting as the downstream of the laser-driven shock. Similarly, the black dashed line in (b) represents the downstream of the transmitted shock. The hot electrons move toward the rear side with relativistic velocity and a portion of them will be turned back under the action of E_{int}.

$Ze\phi_s = 1/2Am_p v_s^2$, indicates the heavy ion reflection requires higher potential barrier given the same shock velocity, which limits the resultant beam density. Here e is the charge of electron, ϕ_s is the electrostatic potential of shock, and v_s denotes the shock velocity. The quick dissipation by the heavy ion reflection further drags the shock wave during its propagation, which will significantly broaden the resultant energy spread.

In this paper, a staged SWA scheme for production of high quality heavy ion beam is proposed, where, a light ion layer is used in front of the heavy ion target. When the light ion layer is irradiated by an intense laser, a high-speed collisionless electrostatic shock wave is firstly launched. Before this shock arrives at the interface between the light and heavy ion plasmas, a strong electrostatic field is induced at the interface, resulting in early recirculation of hot electrons in the light ion layer and pre-acceleration of the heavy ions. Both of these effects lead to a much enhanced transmitted shock wave in the heavy ion plasma. The transmitted shock can reflect heavy ions which gives rise to a monoenergetic ion beam. These features and the exciting enhancement on the beam quality compared with the typical SWA in pure heavy ion plasmas are confirmed by two-dimensional PIC simulations. The resultant energy spectrum is adjustable by varying the mass density jump between the two adjoining plasmas.

This paper is organized as follows. In section 2, the theoretical model for the staged SWA are depicted. In section 3, the 2D PIC simulation results are presented to confirm our theory. In section 4, further insights into the staged SWA dynamics are given and the parameter range for optimal acceleration is proposed. In section 5, discussion on the experimental feasibility of our proposed scheme is given. Section 6 is dedicated to the conclusion.

2. Theoretical model

A schematic figure for the staged SWA dynamics is shown in figure 1. The front light ion layer is irradiated by intense laser, which drives a high-speed shock wave under the conditions of appropriate density piling-up and electron heating [9]. This laser-driven shock (black dashed line in figure 1(a)) is hereinafter referred to as 'the first shock'. A shock has the nature of compression due to the acceleration of the post-shock medium. So when the first shock strikes the interface, the heavy ion plasma is compressed and piled up under the action of the shock field E_{sh}. A transmitted shock capable of reflecting heavy ions is thus formed (black dashed line in figure 1(b)). At the same time, either a shock wave or rarefaction wave (RW) is backscattered into the light ion layer. The character of the backscattered wave depends on the ion mass density jump between the two adjoining plasmas, i.e., $\rho_{i1}/\rho_{i2} = (A_1/Z_1)n_{e1}/(A_2/Z_2)n_{e2}$, and their respective thermal properties [19]. In addition, the velocity of the transmitted shock increases with ρ_{i1}/ρ_{i2}. Consequently, the resultant heavy ion spectrum in this staged SWA scheme can be adjusted by varying the density jump condition.

It is worth noting that there are some crucial merits in this staged SWA scheme. Before the first shock arrives at the interface where plasma density drops, the two plasmas expand into each other which induces an intense field E_{int}. On the other hand, hot electrons generated in the front laser-plasma interaction region move to the rear side and will enhance E_{int} by losing some kinetic energy when passing by the interface. A portion of the hot electrons even turn back under the condition $(\gamma_e - 1)m_e c^2 < e\phi_{int}$, where γ_e is the electron gamma factor, m_e is the mass of electron, c is the light speed in vacuum, and ϕ_{int} denotes the electrostatic potential across the interface. This early recirculation contributes to the uniform heating and accordingly a high temperature T_e in the light ion layer. The high temperature promises a low Mach number given the same radiation pressure drive. So the light ion reflection is minimized in the first shock, which greatly benefits our staged SWA scheme. On the contrary, if large number of light ions are reflected in the first shock, e.g. in the case using short pulse or thick light ion layer, the laser energy will be heavily depleted. This effect decreases the transmitted shock velocity and the final energy conversion efficiency from laser into the reflected heavy ions.

It is noted that the heavy ion reflection is assisted by the pre-acceleration of heavy ions by E_{int}, because the upstream ions will become less energetic in the shock-rest frame. When the mass of electron is neglected, E_{int} can be estimated as $E_{int} = -\nabla p_e/en_{e0}$, where the pressure gradient $\nabla p_e = T_e\nabla n_e = T_e(n_{e1} - n_{e2})/l_{int}$ and $n_{e0} \approx 1/2(n_{e1} + n_{e2})$. l_{int} is the scale thickness of the interface region which is assumed to have the same order of Debye length, i.e., $l_{int} = \delta\lambda_{De}$ with $\delta = O(1)$, $\lambda_{De} = \sqrt{\varepsilon_0 T_e/n_{e0}e^2}$, and ε_0 is permittivity constant of vacuum. After some algebra, E_{int} is given by

$$E_{int} = \sqrt{2n_{e1}T_e/\varepsilon_0}\,\frac{1 - n_{e2}/n_{e1}}{\delta\sqrt{1 + n_{e2}/n_{e1}}}. \tag{1}$$

This indicates that E_{int} increases with the front layer density n_{e1} and density jump n_{e1}/n_{e2} (or ρ_{i1}/ρ_{i2}). When n_{e1}/n_{e2} are sufficiently large, $E_{int} \approx \sqrt{2n_{e1}T_e/\varepsilon_0}/\delta$ which does not sensitively depend on ρ_{i1}/ρ_{i2}. So E_{int} has a weak scaling with ρ_{i1}/ρ_{i2} in the high ρ_{i1}/ρ_{i2} regime, which suggests that too large ρ_{i1}/ρ_{i2} will not help further boosting the first shock velocity via E_{int}.

3. Two-dimensional PIC simulation results

The proposed novel acceleration scheme is confirmed by 2D PIC EPOCH simulations [20]. In the simulations, the laser pulse with wavelength $\lambda = 1\ \mu$m is polarized along y direction and propagates along $+x$ direction. The transverse intensity distribution is a supergaussian ($I(r) \propto I_0\exp(-(r/r_0)^4)$) giving a full-width-at-half-maximum focal spot radius of 24 μm centered at $y = 30\ \mu$m. The peak intensity $I_0 = 1.66 \times 10^{20}$ W cm^{-2} corresponding to the normalized amplitude $a_0 = 11$. The laser has a flat-topped temporal profile (i.e., constant intensity) with duration $\tau = 267$ fs. The simulation box is $46 \times 60\ (\mu m)^2$ in the (x, y) plane. The light and heavy ions are chosen to be fully ionized proton (H^+) and carbon (C^{6+}) ions, respectively. In front of the heavy ion target is a 5 μm thick proton layer with electron density $n_{e1} = 10n_c$. Here n_c is the critical density $n_c = m_e\varepsilon_0\omega_0^2/e^2$, where ω_0 is the laser frequency. The carbon target consists of a 5 μm thick layer with density $n_{e2} = \frac{1}{4}n_{e1} = 2.5n_c$ followed by a rear plasma with an exponentially decaying profile. So the light and heavy ion plasmas have a ion mass density jump $\rho_{i1}/\rho_{i2} = 2/1$. The decaying C^{6+} plasma, with a scale length of $L_g = 8\ \mu$m, guarantees a suitably smooth density gradient which leads to a mitigated and constant sheath field at the rear side (i.e., E_{sheath} in figure 1). This effect benefits the generation of high quality ion beam in SWA [7, 9, 21]. The cell dimensions are resolved by the initial collisionless skin depth of light ion layer as $\Delta x = 0.24c/\omega_{pe}$ (12 nm) and $\Delta y = 0.4c/\omega_{pe}$ (20 nm), respectively, where $\omega_{pe} = \sqrt{n_{e1}e^2/m_e\varepsilon_0}$. There are 64 particles per cell per species. With higher number of particles, the simulation results are converged.

Figure 2 shows the evolutions of transverse electric field E_y, proton density $n(H^+)$, and carbon charge density $6n(C^{6+})$, respectively. In our simulation, when the laser interacts with the front hydrogen layer, the density piling-up due to the radiation pressure compression is obvious, see $n(H^+)$ at $t = 100$ fs (purple part in figure 2(a)). The piling-up ultimately evolves into a collisionless electrostatic shock wave whose front can be observed in figures 3(a) and (d). With the density jump, a strong electric field E_{int} (figure 3(d)) is induced at the interface as expected in theory. The simulation shows that $E_{int} \approx 1$ which is comparable to the first shock field $E_{sh} \approx 2$. On the other hand, the temperature is measured to be $T_e = 3.0$ MeV, so $\lambda_{De} = 0.15\ \mu$m. The scale length of the interface region is measured to be $l_{int} = 1.1\ \mu m \sim 7\lambda_{De}$. It suggests $E_{int} \approx 1.1$ from equation (1), which agrees well with the simulation result. E_{int} can accelerate both H^+ and C^{6+} before the arrival of the first shock. This pre-acceleration makes it more feasible for the C^{6+} ion reflection, because these C^{6+} ions will become less energetic in the shock-rest frame.

When the first shock is striking the interface, a superposition of E_{sh} and E_{int} is observed (figures 3(b) and (e)) which leads to a very strong field ($E \sim 3$). The C^{6+} layer is profoundly accelerated and compressed which shows the enhanced compression ability of the current shock. The large C^{6+} piling-up can be seen in figure 2(b). Consequently, the first shock is being transmitted into the C^{6+} plasma and the C^{6+} ion reflection starts (see its phase space in figures 3(e) and (f)). With the current simulation setup, a RW is backscattered into the proton layer resulting in a much rarefied hydrogen plasma (purple part in figure 2(b)).

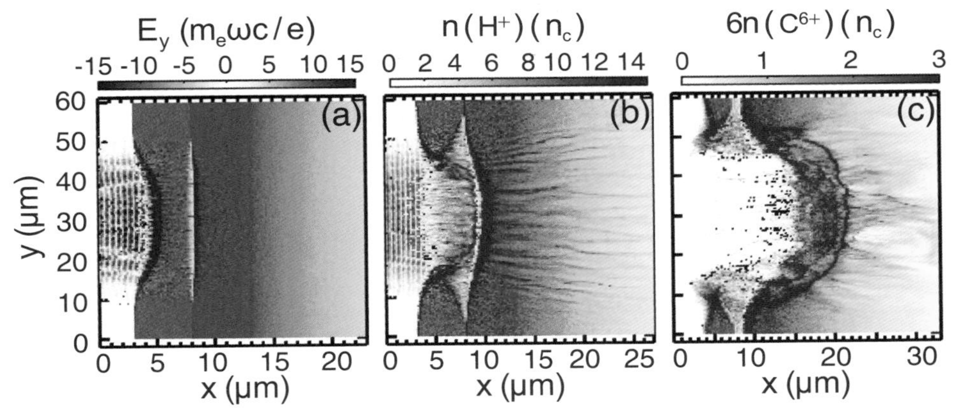

Figure 2. The spatial distributions of the transverse electric field E_y, proton density $n(\mathrm{H}^+)$, and carbon ion charge density $6n(\mathrm{C}^{6+})$ in the simulation in which staged shock waves are driven by intense laser in near-critical plasmas. The peak laser intensity is $I_0 = 1.66 \times 10^{20}$ W cm^{-2} and electron density in the front proton layer is $n_{e1} = 10n_c$. The electron density jump between the adjoining plasmas are $n_{e1}/n_{e2} = 4/1$ (i.e., $\rho_{i1}/\rho_{i2} = 2/1$). The three snapshots are taken at (a) $t = 100$ fs, (b) $t = 233$ fs, and (c) $t = 600$ fs, respectively. The electric field is normalized by $m_e\omega c/e$ and density by n_c.

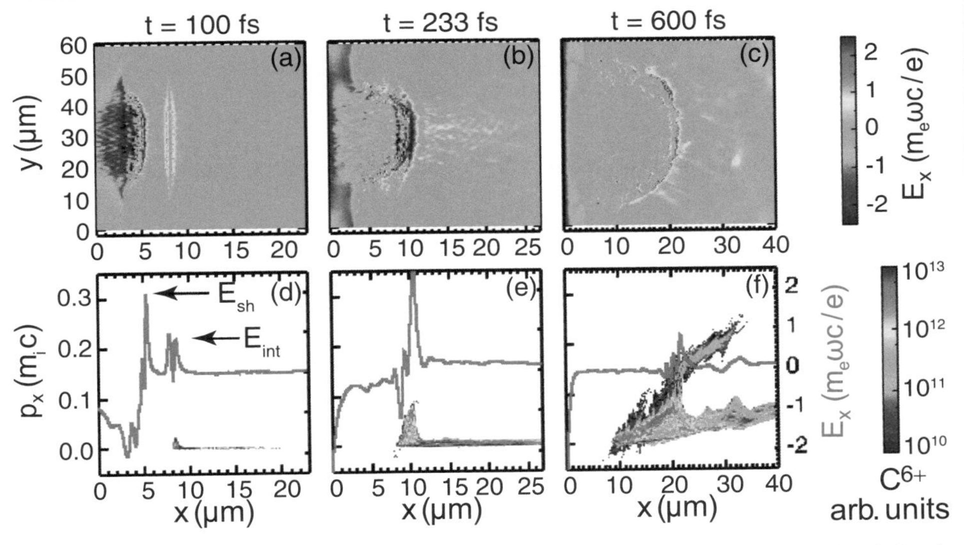

Figure 3. Results from the same simulation as in figure 2 at the same corresponding moments. (a)–(c) show the spatial distribution of longitudinal field E_x. (d)–(f) show the C^{6+} ion phase space ($p_x - x$). The red line represents the spatially averaged E_x (along a ± 10 μm average over y centered at $y = 30$ μm). The momentum is normalized by m_ic, where m_i is the mass of carbon ion.

After the first shock is completely transmitted into the C^{6+} plasma, the transmitted shock front is clearly shown in figures 2(c) and 3(c). The plateau in phase space (figure 3(f)) indicates a monoenergetic C^{6+} ion beam has been produced. For comparison with the staged SWA case, another PIC simulation, in which a target made up of pure C^{6+} plasmas with thickness 10 μm and electron density $10n_c$, are performed. Other parameters are kept unchanged. The energy spectra of both cases are compared in figure 4(a). In the staged SWA case (red line, $\rho_{i1}/\rho_{i2} = 2/1$), due to the lower density of the C^{6+} layer, the number of reflected C^{6+} ions (2.1×10^{11}) is about one half of that in pure carbon case (4.1×10^{11}, green line). The spectrum in pure carbon case barely shows a peak with energy $E_0 = 60$ MeV and spread up to be $\delta E/E_0 = 73\%$. In striking contrast, the spectrum in the staged SWA case shows a well-defined peak with $E_0 = 168$ MeV and $\delta E/E_0 = 31\%$. The strong adaptability of this staged SWA scheme is confirmed by replacing the carbon ions by heavier aluminum ($Z = +10, A = 27$, $\rho_{i1}/\rho_{i2} = 1.48/1$) and keeping other parameters unchanged. The results (figure 4(b)) also indicate a high quality

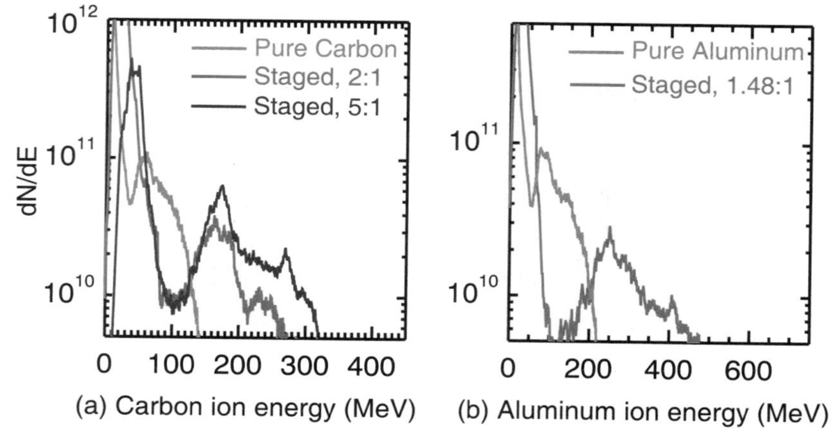

Figure 4. (a) Carbon ion spectra in the staged SWA cases with ion mass density jump of $\rho_{i1}/\rho_{i2} = 2/1$ (red line, corresponding to $n_{e1}/n_{e2} = 4/1$ as in figures 2 and 3) and 5/1 (blue line), respectively. The green line depicts the spectrum from the comparison case with pure carbon plasma. (b) Aluminum ion spectra in the staged SWA case with $\rho_{i1}/\rho_{i2} = 1.48/1$ (i.e., $n_{e1}/n_{e2} = 4/1$, red line) and in the comparison case with pure aluminum plasma (green line), respectively. The heavy ions for the above spectra are selected within a $\pm 10\ \mu$m spatial range over y direction centered at $y = 30\ \mu$m.

Al^{10+} ion beam with well-defined peak of $E_0 = 250$ MeV and $\delta E/E_0 = 30\%$ is generated in the staged SWA case. While in the comparison case with pure aluminum plasma, the spectrum shows a small peak with $E_0 = 90$ MeV and spread up to be $\delta E/E_0 = 71\%$. These simulations show that both the peak energy and spread, through our novel scheme, are greatly improved compared with the typical SWA in pure heavy ion plasmas.

4. Analysis of the acceleration dynamics and the parameters for optimal acceleration

The essential idea of the staged SWA scheme is that the formation of the shock wave in heavy ion plasmas is separated from the intense driving laser. Two merits based on this separation should be pointed out. First, we could apply very intense laser in the light ion layer to drive a high-speed shock. Having high charge-to-mass ratio Z/A, the light ions can respond to the electrostatic field quickly and accordingly have sufficient piling-up. This mechanism for remaining relativistically opaque by large piling-up and obtaining a high-speed shock wave is similar with the idea in reference [22]. Second, the heavy ion layer could have relatively low density which leads to a stable shock velocity with mild dissipation. While in the typical SWA scheme using pure heavy ion plasma with high density, the continuous reflection of heavy ions will impose a strong loading for the shock structure. This loading will drag the shock and diminish the corresponding potential ϕ_s dramatically. The declination in shock velocity (figure 5(b)) will result in a progressive broadening effects on the ion spectrum (green line in figure 4(a)).

The energy evolution in the whole simulation box offers further insights into the staged SWA mechanisms, see figure 5(a). The laser energy (blue line) is continuously injected into the system until it is switched off. The electron energy (black solid line) increases sharply with the effective heating by $\mathbf{J} \times \mathbf{B}$ effects, plasma waves excitation, etc. The H^+ population gains energy (green solid line) mainly through the laser-induced electrostatic field during the ponderomotive force compression stage and the first shock propagation. After $t \sim 200$ fs, the total energy of C^{6+} (red solid line) increases rapidly due to the acceleration of C^{6+} ions (figure 3(e)). The growth rate of the total C^{6+} energy in staged SWA case is higher than the pure carbon case (red dashed line), because the transmitted C^{6+} shock in staged SWA case is much faster. This is also consistent with their peak energy difference (figure 4(a)). The final value of the total C^{6+} energy is higher in staged SWA case which shows the enhancement of the energy conversion efficiency (blue line in figure 6(b)). After the laser is off (~267 fs), the total C^{6+} energy gain approaches a saturated level in the pure carbon case. This is because the shock will lose its momentum rapidly when it is reflecting large number of heavy C^{6+} ions. While in the staged SWA case, the transmitted C^{6+} shock has a relatively stable velocity with mild ion reflection when propagating in lower density plasma.

The property of the transmitted shock mainly depends on the ion mass density jump ρ_{i1}/ρ_{i2} [19], which lays the theoretical foundation for the adjustability of heavy ion spectrum by varying ρ_{i1}/ρ_{i2} in the staged SWA scheme. So a series of PIC simulations are performed to study the parameter range for optimal staged SWA.

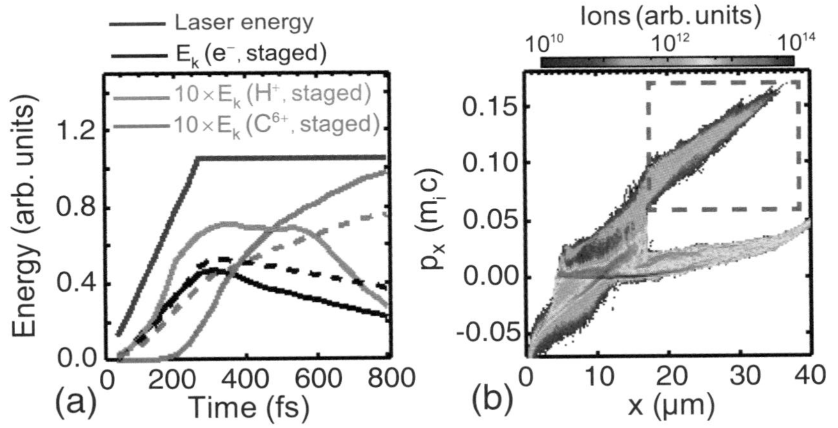

Figure 5. (a) The black, green, and red solid lines represent the total energy of all electrons, H^+ ions, and C^{6+} ions, respectively, in the staged SWA case (as in figures 2 and 3). The black and red dashed lines represent the total energy of all electrons and C^{6+} ions in the comparison case with pure carbon plasmas. Blue line depicts the laser energy for both cases. Ion energies are multiplied by a factor of 10. (b) The phase space of C^{6+} ions in the comparison case with pure carbon plasma at t = 800 fs. The dashed rectangle shows that the shock velocity is declining significantly with the dissipation by reflecting large number of C^{6+} ions.

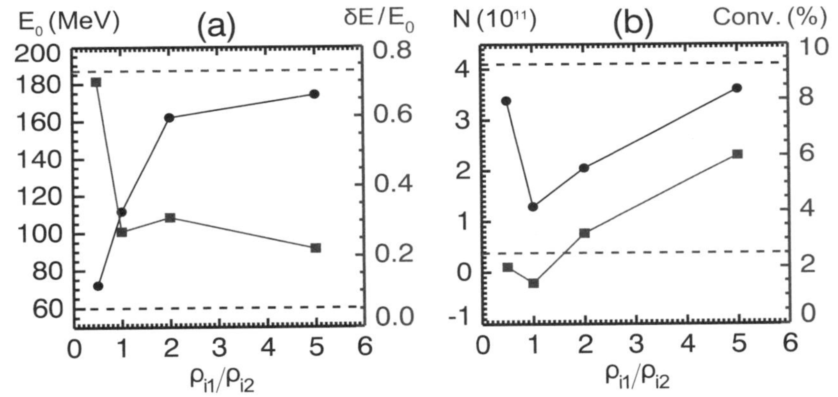

Figure 6. Results from a series of 2D PIC simulations for staged SWA scheme with the same laser intensity 1.66×10^{20} W cm^{-2} and front proton layer ($n_{e1} = 10n_c$). The heavy ion target is chosen to be C^{6+} plasmas. (a) The dependences of the peak energy E_0 (black solid circle) and the corresponding spread $\delta E/E_0$ (blue solid square) of the C^{6+} ion beams on the ion mass density jump ρ_{i1}/ρ_{i2}. The horizontal dashed lines are the results from comparison simulation with pure carbon plasma (black for E_0 and blue for $\delta E/E_0$). (b) Corresponding to (a), the dependences of the reflected C^{6+} number N (black solid circle) and the energy conversion efficiency (blue solid square) from laser to the reflected C^{6+} on ρ_{i1}/ρ_{i2}. The horizontal dashed lines are results from comparison simulation with pure carbon plasma (black for reflected C^{6+} number and blue for the energy conversion efficiency).

Figure 6(a) shows the dependences of peak energy E_0 (black solid circle) and spread $\delta E/E_0$ (blue solid square) of the C^{6+} ion beams on ρ_{i1}/ρ_{i2}. It indicates that E_0 in all of the staged SWA cases increase with ρ_{i1}/ρ_{i2} as expected, and is also superior to the comparison case (60 MeV, black dashed line). $\delta E/E_0$ shows the similar trend with ρ_{i1}/ρ_{i2} and advantage over the comparison case (73%, blue dashed line). It is noted that the improvements of E_0 and $\delta E/E_0$ still exist even without the help of ion mass density jump (see the cases with $\rho_{i1}/\rho_{i2} = 0.5$ and $\rho_{i1}/\rho_{i2} = 1$). In addition, in the case with $\rho_{i1}/\rho_{i2} = 0.5$ (i.e., $n_{e1}/n_{e2} = 1$), there is no interface-induced field E_{int} which could enhance the first shock velocity, but the peak energy (~72 MeV) is still enhanced. So the light ion layer used in staged SWA scheme does help in boosting the final heavy ion energy.

Figure 6 shows that the range of ρ_{i1}/ρ_{i2} for optimal staged SWA scheme lies in $1 \leqslant \rho_{i1}/\rho_{i2} \leqslant 5$. In this range, both E_0 and $\delta E/E_0$ are greatly improved compared with the SWA in pure carbon plasma. At the same time, the reflected C^{6+} ion number and the energy conversion efficiency increase with ρ_{i1}/ρ_{i2} dramatically (figure 6(b)).

For example, in the case with $\rho_{i1}/\rho_{i2} = 5$, E_0 is boosted to be 174 MeV and $\delta E/E_0$ drops to be only 22%. In addition, the ion beam has a considerable particle number of 3.6×10^{11} and high conversion efficiency of ~6%. It is noted that very intense $E_{\rm int}$ is established in this case due to large density jump (equation (1)). More hot electrons will be turned back by $E_{\rm int}$ which leads to high T_e in the front layer. Such high temperature and intense $E_{\rm int}$ give rise to fast expansion of the proton layer, so the laser will penetrate the target when the first shock is striking the interface. Large number of hot electrons will be generated and expand into vacuum. This leads to a strong sheath field which accelerates the upstream C^{6+} plasma. Such acceleration effects will help more upstream ions to be reflected by the transmitted shock, because the accelerated ions will become less energetic in the shock-rest frame. This explains the high particle number and high conversion efficiency in this case. But another two problems will arise for the high density jump regime ($\rho_{i1}/\rho_{i2} \geqslant 5$). First, the velocity of the reflected C^{6+} ions is given by $v_i = 2v_s - v_0$ in the non-relativistic limit, where v_0 is the expanding velocity of the upstream plasma. Consequently, the peak energy can no longer be boosted to a much higher value with larger ρ_{i1}/ρ_{i2}, as the expanding velocity v_0 cancels out part of the velocity gain by shock reflection. This explains the limited gain of the peak energy from the case with $\rho_{i1}/\rho_{i2} = 2$ to the case with $\rho_{i1}/\rho_{i2} = 5$ (figure 6(a)). Second, some reflected particles will obtain more energy under the action of the intense sheath field, which leads to the high energy tail in the spectrum (blue line in figure 4(a)). So, too large ρ_{i1}/ρ_{i2} is not suitable for the high quality ion beam generation.

5. Discussion on experimental feasibility

With the rapid progress of target fabrication technology, targets of relativistically near-critical density ($n_c \sim 10^{21}$ cm^{-3}, corresponding to laser with short wavelength) are not only makable but also controllable through various methods in experiments to date. Hydrogen target at density of $\sim 40n_c$, which has been produced either by extrusion from a liquid-helium-cooled cryostat [23] or by a cooling finger from a cryogenic system [24] in laser-plasma experiments, can be used as the low-Z layer in the staged SWA scheme. Foams with high hydrogen concentration can also provide well-characterized uniform density and temperature near-critical plasmas [25, 26]. As for the carbon target, carbon nanotube foams [27] can provide spatially well-defined carbon plasmas with electron densities at $\sim(3.4 \pm 1.7) \times 10^{21}$ cm^{-3}, which are very close to those used in the simulations of this paper. The exponentially decaying profile [8, 9, 28] at rear side of the target can be produced through the thermal expansion of the carbon target via either a low-power pre-heater pulse or a separate low-intensity ultraviolet (UV) laser [29], where the density scale length can be controlled via adjusting the expansion time of the carbon target, i.e., the delay time between the main pulse and the pre-heater pulse or the UV beam.

The proposed staged SWA scheme in this paper is rather robust. The basic physics of this scheme will not change even if the preplasma effect is take into account. The latter can influence the maximum piling-up of plasma density and the electron heating during the shock formation process [9]. The precise dependence of the resultant beam quality on the preplasma needs further systematic investigation. Another consideration is for the presence of non-planar interface between light and heavy ion plasmas. The non-planar interface may lead to spatially non-uniformity and modulate the transmission of the first shock into the heavy ion target, increasing the angular divergence of the shock-reflected heavy ions. In practical experiments, using a comparatively large spot size of the laser pulse, including the main pulse and the separate UV beam (for creating required plasma profile as stated above), can suppress this potential detrimental effect.

6. Conclusion

In this paper, a novel heavy ion acceleration scheme via staged shock waves is proposed in which a light ion layer is deployed in front of the heavy ion target. When the laser-driven shock in the light ion layer is transmitted into the heavy ion target, a quasi-monoenergetic heavy ion beam is generated. The resultant ion spectrum can be adjusted by varying the ion mass density jump between the adjoining plasmas. The interface-induced field $E_{\rm int}$ can enhance the first shock velocity and amplitude, which greatly benefits the ensuing heavy ion acceleration. 2D PIC simulations have confirmed our theory and show that a C^{6+} ion beam with peak energy of 168 MeV and considerable particle number of 2.1×10^{11} can be obtained in the staged SWA scheme with $\rho_{i1}/\rho_{i2} = 2/1$ at the laser intensity of 1.66×10^{20} W cm^{-2}. This acceleration scheme also applies to heavier ion species, such as Al^{10+} ion beam with well-defined peak of $E_0 = 250$ MeV and $\delta E/E_0 = 30\%$ can be generated in the staged SWA scheme with the same laser.

Acknowledgments

This work is supported by the National Natural Science Foundation of China, Nos. 11575298, 91230205, 11575031 and 11175026, the National Basic Research 973 Project, Nos. 2013CBA01500 and 2013CB834100, the National High-Tech 863 Project and the National Science Challenging Program. B Q acknowledges the support from Thousand Young Talents Program of China. The comments of the anonymous referees are appreciated. The computational resources are supported by the Special Program for Applied Research on Super Computation of the NSFC-Guangdong Joint Fund (the second phase).

References

[1] Honrubia J J, Fernandez J C, Temporal M, Hegelich B M and Meyer-ter-Vehn J 2009 *Phys. Plasmas* **16** 102701
[2] Krushelnick K *et al* 2000 *IEEE Trans. Plasma Sci.* **28** 1184
[3] McKenna P *et al* 2004 *Phys. Rev.* E **70** 036405
[4] Hoffmann D, Fortov V, Kuster M, Mintsev V, Sharkov B, Tahir N, Udrea S, Varentsov D and Weyrich K 2009 *Astrophys. Space Sci.* **322** 167
[5] Ludlam T and McLerran L 2003 *Phys. Today* **56** 48
[6] Macchi A, Borghesi M and Passoni M 2013 *Rev. Mod. Phys.* **85** 751
[7] Haberberger D, Tochitsky S, Fiuza F, Gong C, Fonseca R A, Silva L O, Mori W B and Joshi C 2012 *Nature Phys.* **8** 95
[8] Fiuza F, Stockem A, Boella E, Fonseca R A, Silva L O, Haberberger D, Tochitsky S, Gong C, Mori W B and Joshi C 2012 *Phys. Rev. Lett.* **109** 215001
[9] Zhang W L, Qiao B, Huang T W, Shen X F, You W Y, Yan X Q, Wu S Z, Zhou C T and He X T 2016 *Phys. Plasmas* **23** 073118
[10] Snavely R A *et al* 2000 *Phys. Rev. Lett.* **85** 2495
[11] Wilks S C, Langdon A B, Cowan T E, Roth M, Singh M, Hatchett S, Key M H, Pennington D, MacKinnon A and Snavely R A 2001 *Phys. Plasmas* **8** 542
[12] Xiao K D, Zhou C T, Qiao B and He X T 2015 *Phys. Plasmas* **22** 093112
[13] Esirkepov T, Borghesi M, Bulanov S V, Mourou G and Tajima T 2004 *Phys. Rev. Lett.* **92** 175003
[14] Robinson A P L, Zepf M, Kar S, Evans R G and Bellei C 2008 *New J. Phys.* **10** 013021
[15] Macchi A, Cattani F, Liseykina T V and Cornolti F 2005 *Phys. Rev. Lett.* **94** 165003
[16] Qiao B, Zepf M, Borghesi M and Geissler M 2009 *Phys. Rev. Lett.* **102** 145002
[17] Qiao B, Zepf M, Borghesi M, Dromey B, Geissler M, Karmakar A and Gibbon P 2010 *Phys. Rev. Lett.* **105** 155002
[18] Wu D, Qiao B, McGuffey C, He X T and Beg F N 2014 *Phys. Plasmas* **21** 123118
[19] Drake R P 2006 *High-Energy-Density Physics: Fundamentals, Inertial Fusion, and Experimental Astrophysics* (New York: Springer) ch 4
[20] Arber T D *et al* 2015 *Plasma Phys. Control. Fusion* **57** 113001
[21] Fiuza F, Stockem A, Boella E, Fonseca R A, Silva L O, Haberberger D, Tochitsky S, Gong C, Mori W B and Joshi C 2013 *Phys. Plasmas* **20** 056304
[22] Robinson A P L, Trines R M G M, Dover N P and Najmudin Z 2012 *Plasma Phys. Control. Fusion* **54** 115001
[23] van Kessel C G M and Sigel R 1974 *Phys. Rev. Lett.* **33** 1020
[24] Sigel R 1970 *Z. Nat.forsch.* A **25** 488
[25] Willingale L, Nilson P M, Thomas A G R, Bulanov S S, Maksimchuk A, Nazarov W, Sangster T C, Stoeckl C and Krushelnick K 2011 *Phys. Plasmas* **18** 056706
[26] Chen S N *et al* 2016 *Sci. Rep.* **6** 21495
[27] Bin J H *et al* 2015 *Phys. Rev. Lett.* **115** 064801
[28] Lecz Z and Andreev A 2015 *Phys. Plasmas* **22** 043103
[29] Haberberger D, Ivancic S, Hu S X, Boni R, Barczys M, Craxton R S and Froula D H 2014 *Phys. Plasmas* **21** 056304

PHYSICS OF PLASMAS **16**, 112703 (2009)

Enhancement of backward Raman scattering by electron-ion collisions

Z. J. Liu,[1,2] Shao-ping Zhu,[2] L. H. Cao,[2] C. Y. Zheng,[1,2] X. T. He,[1,2] and Yugang Wang[1]
[1]*State Key Laboratory of Nuclear Physics and Technology, Center for Applied Physics and Technology, Peking University, Beijing 100871, China*
[2]*Institute of Applied Physics and Computational Mathematics, Beijing 100094, China*

(Received 8 April 2009; accepted 15 October 2009; published online 4 November 2009)

The propagation of light waves in an underdense plasma is studied using one-dimensional Vlasov–Maxwell numerical simulation. It is found that the backward stimulated Raman scattering will be enhanced by electron-ion collisions. With appropriate electron-ion collision rate the Langmuir waves, driven via SRS, can be made to propagate for a long distance. © *2009 American Institute of Physics*. [doi:10.1063/1.3258839]

I. INTRODUCTION

Stimulated Raman scattering (SRS),[1,2] the resonant decay of an incident light wave into a scattered light wave and an electron plasma wave, has been widely studied for at least four decades.[3–13] This scattering process plays an important role in laser propagation in underdense plasmas. In laser driven inertial confinement fusion (ICF),[14] backward SRS leads to a net energy loss of the incident laser beam and to the generation of fast electrons that preheat the fusion fuel in the implosion compression, so that the backward SRS can be detrimental to the fusion. In the Raman laser amplification scheme,[15–18] SRS in plasma of a long pumping laser pulse is used to amplify a short, frequency downshifted seed pulse, in this instance SRS is beneficial. Many fundamental studies have been carried out experimentally,[5,7,8] with the aim of checking theoretical predictions[3,4] on the evolution of this backward SRS. Reference 7 reported that electron-ion collisions can inhibit SRS. However the collision rate depends on the electron temperature and density, as well as ion charge.[19] In high temperature plasmas the collision rate is usually less than the SRS growth rate and therefore SRS can develop. In collisionless plasmas, the burst behavior of SRS appears to be a common feature of nonlinear kinetic simulations.[11] It can be attributed to the increase in nonlinear Landau damping due to hot electron generation through the trapping, significant nonlinear frequency shift due to particle trapping in the electron plasma wave, and trapped-particle modulational instability that leads to phase-space vortices with long lifetimes. The bursts can be weakened by collisions, but the long-lived vortices can enhance the reflectivity of the SRS. Here we perform kinetic simulations with and without electron-ion collisions and show that collisional damping can enhance the Raman reflectivity under certain conditions. In Sec. II we introduce the physical model and simulation parameters. Moreover, in Sec. III we present and analysis the numerical results. Finally, conclusions are summarized in Sec. IV.

II. PHYSICAL MODEL

We shall use the one-dimensional (1D) Vlasov–Maxwell model.[9–13,20–23] The governing equations for the model are the Krook and Maxwell equations

$$\partial_t f_s + v_{xs}\partial_x f_s + \frac{q_s}{m_s}(E_x + v_{ys}B_z)\partial_{v_x} f_s = -\nu_{si}(f_s - f_{s0}), \quad (1)$$

$$(\partial_t \pm \partial_x)E^{\pm} = -J_y, \quad (2)$$

$$E^{\pm} \equiv E_y \pm B_z, \quad (3)$$

$$m_s\partial_t v_{ys} = q_s E_y, \quad (4)$$

$$J_y \equiv \sum_s q_s n_s v_{ys}, \quad (5)$$

$$\partial_t E_x = -J_x, \quad (6)$$

$$J_x \equiv \sum_s q_s n_s v_{xs}, \quad (7)$$

where f_s, q_s, m_s, and ν_{si} are the velocity distribution function, the charge, the mass, and the collision rate for species s, where $s=i,e$ for ions and electrons, f_{s0} is the initial equilibrium distribution function, E_x is the electrostatic field, E_y and B_z are the transverse electric and magnetic fields, respectively, v_{ys} is the quiver velocity in laser field, and n_s is the particle density. All the spatial variations and wave vectors only depend on x. The light wave is linearly polarized in the y direction. Equation (1) is solved by the splitting method.[24] The solver is split into separate spatial and velocity space updates. These updates are 1D and the constant velocity advections are carried out using the third order Van Leer method.[13] We shall use the incident light wavelength $\lambda_L = 0.526$ μm and the laser intensity 1.6×10^{16} W/cm^2, corresponding to the normalized amplitude $eE/m_e\omega_L c = 0.057$, where E is the incident light field, ω_L is the frequency, and c is the light speed in vacuum. The total simulation box length is 167 μm, and the plasma length is 100 μm, which located in the center with two sharp boundaries. The electron density is $0.03n_c$, where n_c is the critical density, and the electron temperature is 350 eV.[5] Under these conditions, the wave number of the Langmuir wave for backward SRS is $k_p\lambda_{De} = 0.27$, where k_p is the wave number of the Langmuir wave and λ_{De} is the Debye length.

It is well known that the Landau damping rate depends on the wave vector.[25] To include the electron ion collision

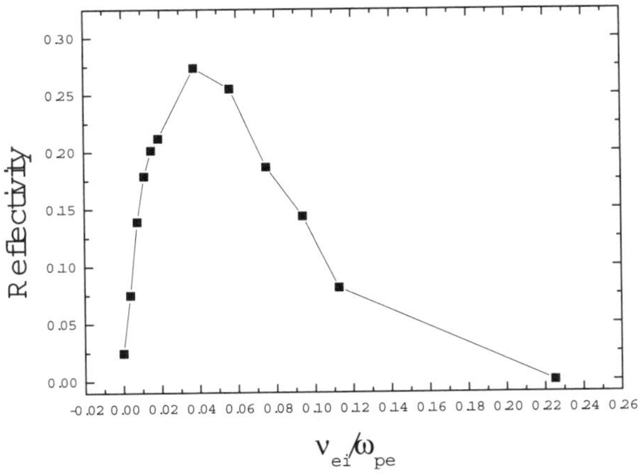

FIG. 1. The backward SRS fractions vs the collision rates. ν_{ei} is the electron-ion collision rate and ω_{pe} is the Langmuir frequency.

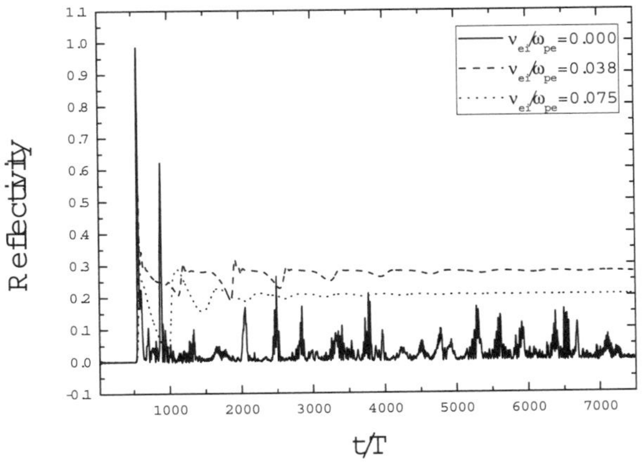

FIG. 2. The instantaneous reflectivity of SRS for three different cases. The time is normalized by the laser cycle T.

effects, we vary the collision rate[19] $\nu_{ei} \simeq 3 \times 10^{-6} \ln \Lambda n_e Z / T_e^{3/2}$ between 0 and $0.225\omega_{pe}$, where ω_{pe} is the plasma frequency, ln Λ is the Coulomb logarithm, and T_e is the electron temperature in eV, by varying the ion charge Z. The undamped growth rate[19] for the backward SRS is $k_p v_{os} \omega_{pe} / 4\sqrt{\omega_s \omega_{ek}} = 0.065\omega_{pe}$, where ω_s is the frequency of the scattering light and $\omega_{ek}^2 = \omega_{pe}^2 + 3k_p^2 v_{te}^2$ is the Langmuir wave frequency, and v_{os} is the electron quiver velocity in laser fields.

III. NUMERICAL RESULTS AND ANALYSIS

The average reflectivity for different collision rates is shown in Fig. 1. We can see that there is an optimal collision rate for enhancing the backward SRS. Higher or lower collision rates can both decrease the reflectivity. The average reflectivity of backward SRS with appropriate collision rate is more than 11 times that without collisions. Backward SRS may be detrimental or beneficial, depending on the application. For laser fusion, it should be avoided. However, for Raman amplification, the enhanced reflectivity may be desirable. Figure 1 also shows that SRS can be inhibited if the electron-ion collision rate is high.

In Fig. 1, we vary the collision rates ν_{ei}/ω_{pe} from 0 to 0.225. For our parameters, some collision rates, for example, $\nu_{ei}/\omega_{pe} = 0.094, 0.113, 0.225$ would require an ion charge in the range from 100 to 240. For low temperature plasmas or high density plasmas, these collision rates can be achieved. We include these results in order to demonstrate that stronger collision rates can indeed suppress SRS. For Raman amplification, the plasma temperature is relatively low and only several tens of eV, the collision rates can be achieved by much lower ionization charge. Reference 26 also used high collision rate to study SRS in the presence of trapped particle instability. In the simulations, we use laser intensity as 1.6×10^{16} W/cm^2. It is higher than the laser intensity that usually used in the ICF experiments. We also run simulations with laser intensity 4×10^{15} W/cm^2. The results are similar, but the collision rate can suppress SRS with ion charge less than 30.

Most kinetic simulations show that SRS exhibits bursting behavior, but with higher collision rate, it can be stabilized. In Fig. 2, we show the evolution of the backward SRS for three different collision rates. In all these cases, the SRS rises rapidly to a very high level at the beginning, and then it oscillates, followed by a few bursts. For $\nu_{ei}/\omega_{pe} = 0.038$ and 0.075, it reaches a stable state at $t = 3000T$, where T is the laser cycle. For $\nu_{ei}/\omega_{pe} = 0$, there is no stable state and the bursting behavior continues in time. The averaged reflectivity for $\nu_{ei}/\omega_{pe} = 0.038$ is much higher than that of the other two cases. For $\nu_{ei}/\omega_{pe} = 0$, the Langmuir wave may grow to very high amplitude and will eventually break. A large number of hot electrons are generated. The hot electrons can enhance the nonlinear Landau damping and quench the backward SRS.

The bursting behavior can be due to several effects, such as increase in nonlinear Landau damping by hot electrons generation, nonlinear frequency shift[27] by particle trapping in the Langmuir wave, and trapped-particle instabilities.[11] The reflectivity is normally small. However, for $\nu_{ei}/\omega_{pe} = 0.038$ it is 11 times that of $\nu_{ei}/\omega_{pe} = 0$ and the SRS reaches a steady state. The reason for this high level of reflectivity and appearance of a steady SRS state is that collisions can weaken the rapid growth of the Langmuir wave, preventing it from breaking up. As a result, the backward SRS reaches a steady state, where the damping and excitation of Langmuir waves reach a balance. For the same reason, at $\nu_{ei}/\omega_{pe} = 0.075$, the SRS also reaches a steady state, and the reflectivity is between the other two cases. Here the reflectivity is lower than that for $\nu_{ei}/\omega_{pe} = 0.038$ since the laser-Langmuir wave interaction region is shorter when the Langmuir wave is strongly damped. If the collision rate is sufficiently large, Langmuir waves cannot propagate in the plasma, and SRS does not occur, as shown in Fig. 1.

In Fig. 3, we show the electrostatic field distribution along the x direction at $t = 6700T$ and the corresponding frequency spectrum. The top panel in Fig. 3(a) is the electrostatic field for $\nu_{ei}/\omega_{pe} = 0$, and the middle and bottom panels for $\nu_{ei}/\omega_{pe} = 0.038$ and $\nu_{ei}/\omega_{pe} = 0.075$, respectively. Figure

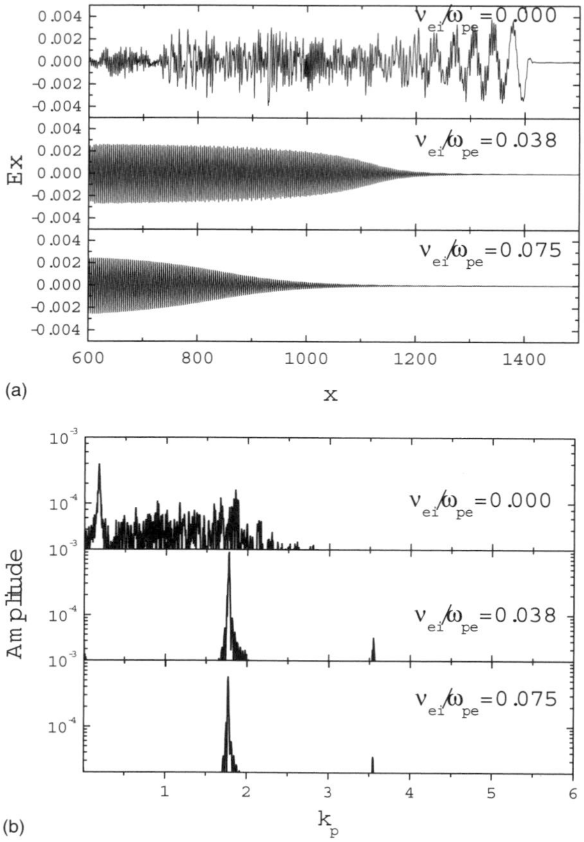

FIG. 3. Profile of the electrostatic field (a) and its spectra (b) for three different collision rates. The x in (a) is normalized by k_L^{-1}, where k_L is the incident light wave number in vacuum. The Langmuir wave number k_p is normalized by k_L and the amplitude is in relative units in (b).

3(b) shows the corresponding frequency spectrum. It is clearly seen that the electrostatic field is irregular for $\nu_{ei}/\omega_{pe}=0$, but it is very regular for other collision cases. For backward SRS to develop, the three-wave matching conditions must be satisfied and the interaction distance should be sufficiently long. We see that in the collisionless case (top panel) the spectrum between $k_p=0$ and 3 is a very broad but the amplitude relatively small. Although Langmuir waves exist, the backward SRS is weak because the amplitude of $k_p=1.775$ for backward SRS is small, as shown in Fig. 1. The amplitude at $k_p\sim 0.18$ is somewhat stronger. It corresponds to forward SRS, which begins at about $t=4000T$, but the scattering level (the ratio of forward SRS energy to the incident laser energy) is however very low. For $\nu_{ei}/\omega_{pe}=0.038$ (middle panel), the energy is concentrated in the backscattered Langmuir waves with $k_p=1.775$ (satisfying the three-wave-matching condition) and it is the strongest among the three cases. The SRS reflectivity is thus much higher than two other cases. For $\nu_{ei}/\omega_{pe}=0.075$, the three-wave matching conditions are also satisfied. However, because of collisional damping the amplitude is lower than that for $\nu_{ei}/\omega_{pe}=0.038$ and the propagating distance also shorter. The corresponding reflectivity is thus low.

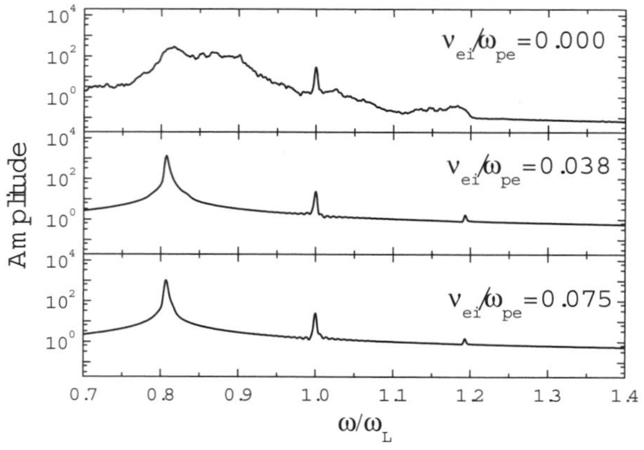

FIG. 4. The time integrated reflected light spectra for the entire run time. The field amplitude is in arbitrary units and the frequency ω is normalized by the incident frequency ω_L.

In Fig. 4 we show the reflected light spectrum. We can see that in the collisional cases the main peaks of the spectrum are in the form of intense spikes and their frequencies corresponding to the scattered-light frequency in the SRS. For the collisionless case, the scattered light has a broad spectrum, consistent with the corresponding broad electrostatic wave spectrum in Fig. 3. We have also run simulations with other plasma parameters and the results are similar.

IV. CONCLUSION

To summarize, backward SRS is studied using 1D Vlasov–Maxwell numerical simulation. Our results show that backward SRS can be enhanced when the collision rate is such that electron-ion collisions balance the excitation and damping of the excited Langmuir waves, so that the latter can propagate for a long distance. The enhanced SRS may be detrimental or beneficial, depending on the application. Furthermore, if sufficiently high collision rates can be achieved, these results suggest that collisional effects would act to suppress SRS.

ACKNOWLEDGMENTS

This work was supported by the National High-Tech ICF Committee of China, National Basic Research Program of China (973 Program Nos. 2007CB815101, 2007CB814802, and 2010CB832904) and the National Natural Science Foundation of China (Grant Nos. 10975023, 10835003, 10935003, and 10675024).

[1]M. V. Goldman and D. F. Dubois, Phys. Fluids **8**, 1404 (1965).
[2]G. Baym and R. W. Hellwarth, IEEE J. Quantum Electron. **1**, 309 (1965).
[3]C. S. Liu, M. N. Rosenbluth, and R. B. White, Phys. Fluids **17**, 1211 (1974).
[4]D. W. Forslund, J. M. Kindle, and E. L. Lindman, Phys. Fluids **18**, 1002 (1975).
[5]D. S. Montgomery, R. J. Focia, H. A. Rose, D. A. Russell, J. A. Cobble, J. C. Fernández, and R. P. Johnson, Phys. Rev. Lett. **87**, 155001 (2001); D. S. Montgomery, J. A. Cobble, J. C. Fernández, R. J. Focia, R. P.

Johnson, N. Renard-LeGalloudec, H. A. Rose, and D. A. Russell, Phys. Plasmas **9**, 2311 (2002).
[6]L. Yin, W. Daughton, B. J. Albright, B. Bezzerides, D. F. DuBois, J. M. Kindel, and H. X. Vu, Phys. Rev. E **73**, 025401(R) (2006).
[7]R. E. Turner, K. Estabrook, R. L. Kauffman, D. R. Bach, R. P. Drake, D. W. Phillion, B. F. Lasinski, E. M. Campbell, W. L. Kruer, and E. A. Williams, Phys. Rev. Lett. **54**, 189 (1985); R. P. Drake, E. A. Williams, P. E. Young, K. Estabrook, W. L. Kruer, H. A. Baldis, and T. W. Johnston, Phys. Rev. A **39**, 3536 (1989).
[8]H. A. Rose, D. F. DuBois, and B. Bezzerides, Phys. Rev. Lett. **58**, 2547 (1987).
[9]D. J. Strozzi, E. A. Williams, A. B. Langdon, and A. Bers, Phys. Plasmas **14**, 013104 (2007).
[10]M. Albrecht-Marc, A. Ghizzo, T. M. Johnston, T. Réveillé, D. Del Sarto, and P. Bertrand, Phys. Plasmas **14**, 072704 (2007).
[11]S. Brunner and E. J. Valeo, Phys. Rev. Lett. **93**, 145003 (2004).
[12]A. Ghizzo, P. Bertrand, M. Shoucri, T. W. Johnston, F. Fijalkow, and M. R. Feix, J. Comput. Phys. **90**, 431 (1990).
[13]A. Mangeney, F. Califano, C. Cavazzoni, and P. Travnicek, J. Comput. Phys. **179**, 495 (2002).
[14]J. D. Lindl, P. Amendt, R. L. Berger, S. G. Glendinning, S. H. Glenzer, S. W. Haan, R. L. Kauffman, O. L. Landen, and L. J. Suter, Phys. Plasmas **11**, 339 (2004); J. D. Lindl, *ibid.* **2**, 3933 (1995).
[15]N. A. Yampolsky and N. J. Fisch, Phys. Plasmas **16**, 072105 (2009).
[16]N. A. Yampolsky, N. J. Fisch, V. M. Malkin, E. J. Valeo, R. Lindberg, J. Wurtele, J. Ren, S. Li, A. Morozov, and S. Suckewer, Phys. Plasmas **15**, 113104 (2008).
[17]D. S. Clark and N. J. Fisch, Phys. Plasmas **10**, 4837 (2003).
[18]D. S. Clark and N. J. Fisch, Phys. Plasmas **10**, 3363 (2003).
[19]W. L. Kruer, *The Physics of Laser Plasma Interactions* (Addison-Wesley, Reading, MA, 1988).
[20]F. Califano, T. Cecchi, and C. Chiuderi, Phys. Plasmas **9**, 451 (2002).
[21]F. Califano and L. Galeotti, Phys. Plasmas **13**, 082102 (2006).
[22]N. J. Sircombe, T. D. Arber, and R. O. Dendy, Plasma Phys. Controlled Fusion **48**, 1141 (2006).
[23]M. Shoucri, Commun. Nonlinear Sci. Numer. Simul. **13**, 174 (2008).
[24]C. Z. Cheng and G. Knorr, J. Comput. Phys. **22**, 330 (1976).
[25]Z. J. Liu, J. Zheng, and C. X. Yu, Phys. Plasmas **9**, 1073 (2002).
[26]M. Mašek and K. Rohlena, Commun. Nonlinear Sci. Numer. Simul. **13**, 125 (2008).
[27]G. J. Morales and T. M. O'Neil, Phys. Rev. Lett. **28**, 417 (1972).

PHYSICS OF PLASMAS **16**, 093108 (2009)

The transition from plasma gratings to cavitons in laser-plasma interactions

Z. J. Liu,[1,2] X. T. He,[1,2] C. Y. Zheng,[1,2] and Y. G. Wang[1]
[1]*Center for Applied Physics and Technology, State Key Laboratory of Nuclear Physics and Technology, Peking University, Beijing 100871, China*
[2]*Institute of Applied Physics and Computational Mathematics, Beijing 100094, China*

(Received 8 June 2009; accepted 24 August 2009; published online 11 September 2009)

One-dimensional Vlasov–Maxwell simulations of laser-plasma interactions are presented. It is shown that plasma gratings and density cavitons are formed sequentially. There are strong electromagnetic fields in the cavitons and the electromagnetic structures are nearly standing and long-lived. The formation of gratings and cavitons can be explained by a nonlinear second-order differential equation. The electromagnetic fields trapped in cavitons have both subcycle and cycle structures. Plasma whose density is higher than the critical density can be formed around the cavitons. Gratings and high density plasmas can reflect light in a very high level. This may be detrimental to the inertial confinement fusion. © *2009 American Institute of Physics.* [doi:10.1063/1.3227647]

I. INTRODUCTION

Laser-plasma interaction is a very important issue of inertial confinement fusion (ICF). Parametric instabilities, such as stimulated Raman scattering[1] and stimulated Brillouin scattering (SBS),[2] may occur during the laser-plasma interactions. When a laser light incidents on plasma, whose density less than critical density $n_c = m_e\omega_L^2/4\pi e^2$, SBS can be developed. Where ω_L is the laser light angular frequency, m_e and $-e$ are the electron mass and charge. Many mechanisms for the saturation of SBS have been put forward. References 3–6 explain the low-level saturation of SBS is due to the creation of cavitons in plasmas. Cavitons can be formed in many cases and in many fields.[6–8]

Many literatures research the SBS using particle-in-cell methods,[3–5] but that method has some disadvantages, such as numerical noise. To better understand the laser-plasma interactions, here we developed a one-dimensional Vlasov–Maxwell code to simulate the laser-plasma interactions. Gratings, solitary waves, and cavitons can be developed in some cases during laser-plasma interactions. In the grating case, the incident light can be reflected in a high level. In the caviton case, the electromagnetic fields below the plasma frequency can be excited and related to transient solitonlike structures. The caviton is responsible for a strong local reduction in the reflectivity and goes along with an efficient but transient heating of the electrons and ions. For lower density plasma, once heating ceases, transmission starts to increase; local as well as global average reflectivities attain a very low value due to strong plasma density variations brought about by the caviton process.[3–5] However, for higher density plasma, the reflectivity may be enhanced for plasma density at the caviton edges higher than the critical density. We studied the gratings and cavitons using one-dimensional Vlasov–Maxwell simulations, the results show that the gratings and cavitons are formed sequentially and the reflectivity is very high for gratings and high density plasma formed.

II. PHYSICAL MODEL AND NUMERICAL SCHEME

The governing equations for the model are the Krook equation[9] and Maxwell equations

$$\partial_t f_s + v_{xs}\partial_x f_s + \frac{q_s}{m_s}(E_x + v_{ys}B_z)\partial_{v_x} f_s = -\nu_s(f_s - f_{s0}), \quad (1)$$

$$(\partial_t \pm \partial_x)E^{\pm} = -J_y, \quad (2)$$

$$E^{\pm} \equiv E_y \pm B_z, \quad (3)$$

$$m_s\partial_t v_{ys} = q_s E_y, \quad (4)$$

$$J_y \equiv \sum_s q_s n_s v_{ys}, \quad (5)$$

$$\partial_t E_x = -J_x, \quad (6)$$

$$J_x \equiv \sum_s q_s n_s v_{xs}, \quad (7)$$

where f_s, q_s, m_s, and ν_s are the velocity distribution function, the charge, the mass, and the collision rate for species s, where $s=i,e$ for ions and electrons, f_{s0} is the initial equilibrium velocity distribution function, E_x is the electrostatic field, and E_y and B_z are the light electric and magnetic fields, respectively. v_{ys} is the species quiver velocity in laser field and n_s is the number density. The light waves are linearly polarized in the y direction. Equation (1) is solved by the splitting method.[10] The solver is split into separate spatial, velocity space and collision updates. The spatial and velocity space updates are carried out using the third order Van Leer method[11] and collision update has exact solutions.

We inject the laser lights on both sides of boundaries, and both of the incident lights have wavelength 1 μm and intensity $I = 3.4\times 10^{15}$ W/cm^2, i.e., $eE_y/m_e\omega_L c = 0.05$, where E_y is the electric field of the incident light and c is the light speed in vacuum. The total simulation length is $1000k_L^{-1}$,

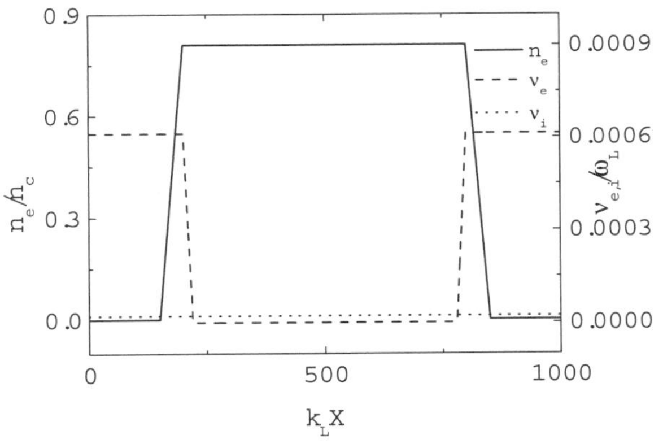

FIG. 1. The profiles of plasma density n_e and collision rates ν_e, ν_i for electrons and ions.

where $k_L=\omega_L/c$ is the light wave number in vacuum. The initial plasma is located in the center of the simulation box and the plasma length is $600k_L^{-1}$. The plasma density n_e is $0.81n_c$. The initial electron/ion velocity distribution functions are Maxwellian with temperature 1 keV/0.5 keV. To excluding the sheath effects on the plasma vacuum interfaces, we also set large ν_e on the boundaries. Figure 1 shows the profiles of initial plasma density and $\nu_{e,i}$. We take the mass ratio of ions to electrons as 100.

III. SIMULATION RESULTS

The evolutions of electron and ion density are shown in Fig. 2. The x-axis is normalized by k_L^{-1} and the y-axis is normalized by ω_L^{-1}, the color bar shows the density values, which is normalized by the critical density n_c. We can see from Fig. 2 that the electron and ion density profiles are very similar. Plasma gratings are formed very rapidly when the lasers incident on the plasma and cavitons are formed sequentially. The wave number of gratings is twice of that of incident light in plasmas. The formation of gratings can be easily understood. The two laser lights can form standing wave in plasmas, there are very strong ponderomotive forces between the wave node and antinode. The ponderomotive forces push the electrons to wave nodes and the light can be trapped between the neighboring wave nodes. To keep the charge neutral, ions will also move to the wave nodes, and then the gratings are formed. Plasma gratings can also modulate the light, the light with a certain frequency band can be stopped. There is a band gap that light can be completely reflected. The incident light is in the band gap,[12–14] and it will be reflected completely. The plasma gratings lifetime can be several tens of laser cycles. If the plasma has very low temperature and the mass ratio of ions to electrons is large enough, the gratings lifetime can be several hundreds of laser cycles.

Gratings are not the most stable state, they will develop into cavitons in some cases. If the electromagnetic field is strong enough, the ponderomotive force and the thermal pressure cannot get balance, the gratings will continued to develop till most electrons in the standing wave antinodes

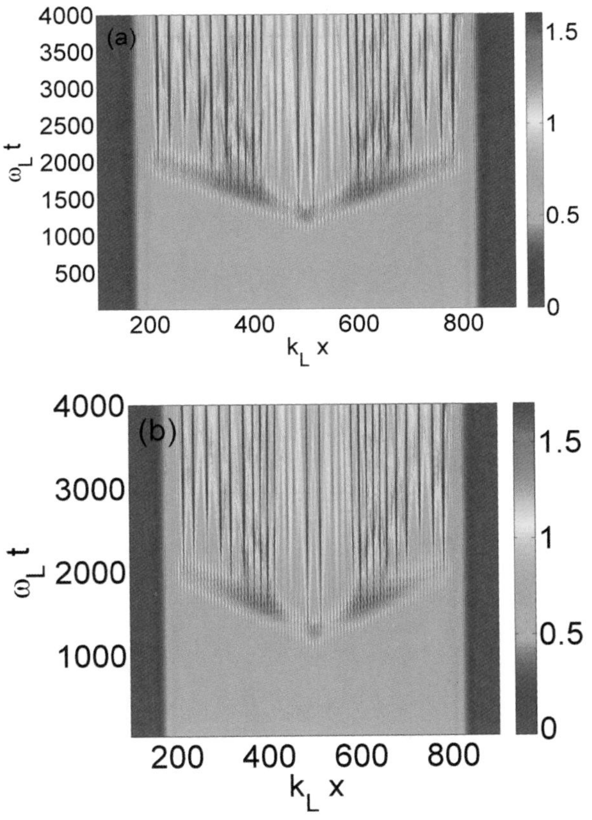

FIG. 2. (Color online) The evolution of electron (a) and ion density (b). The x-axis is normalized by incident light wave number k_L^{-1} and y-axis is normalized by light frequency ω_L^{-1}.

moved to the nodes. If the neighboring gratings have different plasma density, temperature or different electromagnetic field, the gratings may be destroyed and cavitons are formed. The cavitons are very easy to be formed, there are 28 cavitons for $\omega_L t=3000$ in the simulations box, the average plasma length between two cavitons is less than $600k_L^{-1}/28 \approx 21k_L^{-1}$, the length is too short for ion acoustic wave to develop. If the amplitude of incident light is small, the gratings can be formed and disappeared again and again, no caviton will be formed. For example, we did not observe the formation of caviton for $eE_y/m_e\omega_L c=0.01$. In this case, the ponderomotive force is too small and the density perturbation is only several percent. The laser intensity is very important during the transition from gratings to cavitons.

During the formation of cavitons, electrons and ions were heated to very high temperature, as shown in Fig. 3. The electrons and ions in the cavitons can be accelerated to tens of keV. For $\omega_L t=2000$, the electrons and ions begin to be pushed outside the cavitons. Moreover, for $\omega_L t=10\,000$, both low energy electrons and ions are trapped between the neighbored cavitons, while some high energy particles can pass through the cavitons.

The transition from gratings to cavitons can also be understood in physics. As we know, if the plasmas, which be-

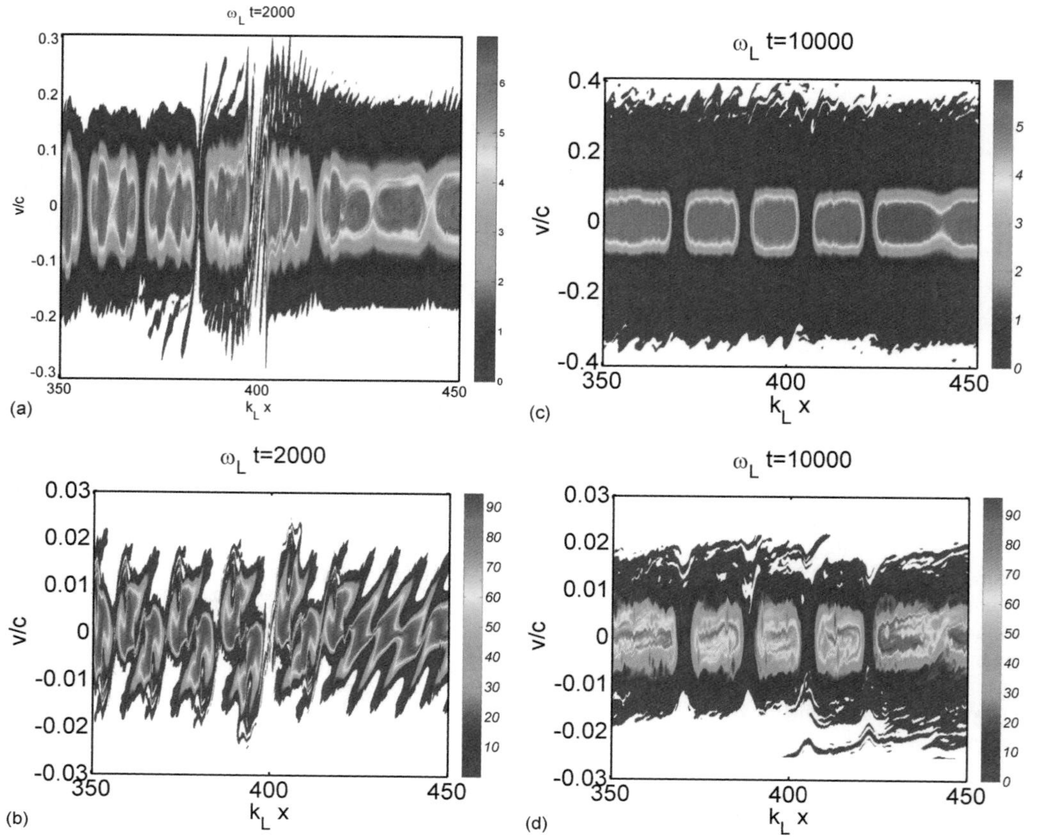

FIG. 3. (Color online) The electron and ion distribution functions at $\omega_L t=2000$ and $\omega_L t=10\,000$, (a) and (c) are electron distribution functions, and (b) and (d) are ion distribution functions. Velocity v is normalized by light speed c in vacuum.

tween two neighbored light wave packets, become warm and the two wave packets have different energy, the plasma will move to one side, and cavitons are formed finally. This process is a positive feedback process and the caviton width is larger than the grating width.

The electromagnetic fields that trapped in the cavitons have different frequencies from the incident light. The critical density for the trapped electromagnetic field must be less than the density of the surrounding plasmas. The spectral analysis of electric and magnetic fields in the caviton shows that there are some components which frequency is nearly half of the incident frequency, as shown in Fig. 4. The light frequency in the cavitons can be estimated by electromagnetic wave dispersion relation[15]

$$\omega^2 = \omega_{pe}^2 + k^2c^2, \tag{8}$$

where $\omega_{pe}^2 = 4\pi n_e e^2/m_e$ is the plasma frequency and ω and k are the electromagnetic wave frequency and wave number. For light trapped in the caviton, the plasma density can be taken as zero for electrons completed pushed to the edges and the wave number can be determined by the caviton width. We take the light wavelength twice of the caviton width. Under these assumptions, the light frequency is about $0.43\omega_L$, while the spectral analysis is showed that the frequency is between $0.48\omega_L$ and $0.62\omega_L$, $0.82\omega_L$ and $1.05\omega_L$ with broad range. The first peak in Fig. 4 can be explained by the caviton width slowly changing and the electron density profile in the caviton. The electron density is not strictly zero in the caviton, so the light spectra have a broad range and the

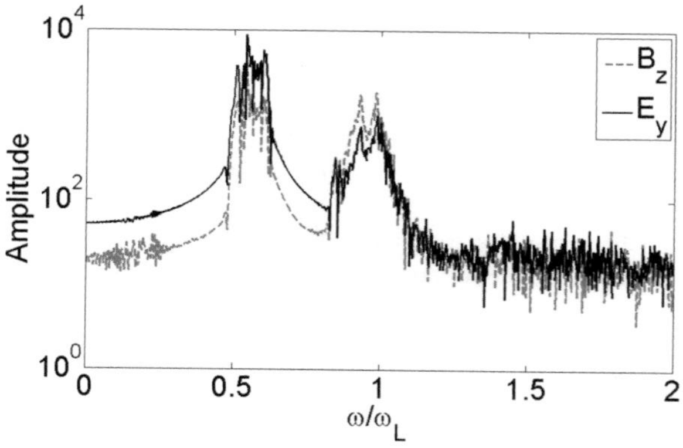

FIG. 4. (Color online) The frequency spectra of E_y and B_z at $x=243$ and $\omega_L t \in [2500, 5000]$, the black solid line is for E_y and the red dashed line is for B_z.

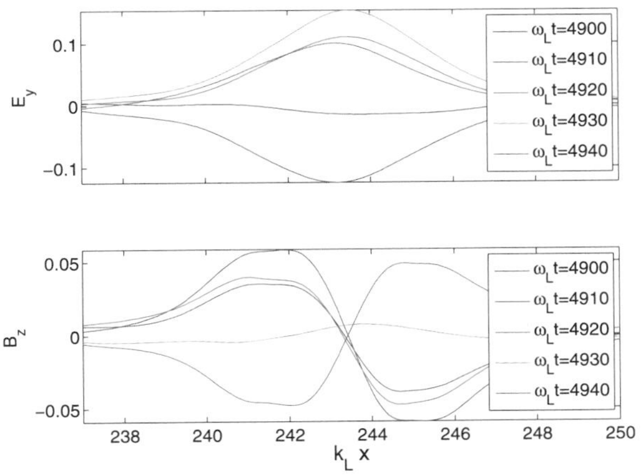

FIG. 5. (Color online) The profiles of E_y and B_z in cavitons. There are both cycle and half-cycle structures for E_y and B_z. For E_y, $\omega_L t=4910$ is a cycle structure and others are half-cycle structures. For B_z, $\omega_L t=4930$ is a half-cycle structure and others are cycle structures.

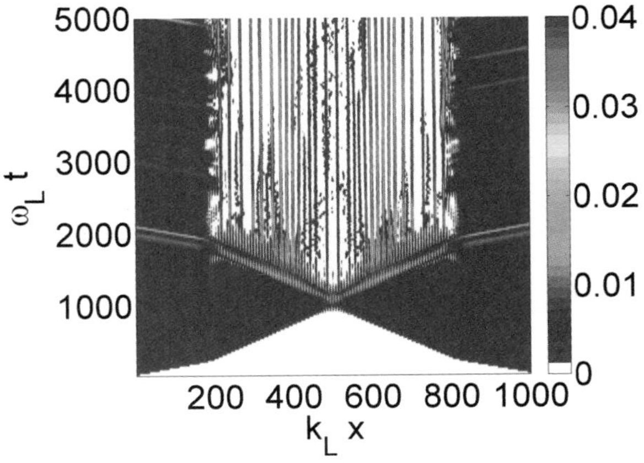

FIG. 6. (Color online) The electromagnetic field energy density for laser incident on both sides. The t is normalized by ω_L^{-1} and the x is normalized by k_L^{-1}.

analytical results have a slight different from the numerical results. For laser propagating in plasmas, the amplitude of magnetic field is less than that of electric field, just as shown in the first peak. However, for the second peak, the magnetic field amplitude is higher than electric field amplitude. In Fig. 5, we plot the profiles of E_y and B_z in a caviton. The electric and magnetic fields have both half-cycle and cycle structures. For the electric field, the half-cycle structure is dominated, and for the magnetic field, the cycle structure is dominated. The frequency analysis in Fig. 4 also shows the electric field is dominated in the half-cycle range and the magnetic field is dominated in the cycle range. We can call the half-cycle electromagnetic field as the ground state and the cycle electromagnetic field as the first excited state. It is very like an electron in a potential well in quantum mechanics. The frequency of the first excited state is not just twice of the ground state. For other higher excited states, the plasma density around the caviton is less than their critical densities, electromagnetic fields cannot be trapped in cavitons.

To show the light departing from plasma is reflected light, the field energy density $E_y^2+B_z^2$ is shown in Fig. 6. It is clearly show that there is no electromagnetic energy between the cavitons, the light departing from plasma is reflected light. The reflected light also shows bursting behaviors. The plasmas between the vacuum and caviton interact with the incident light and the reflected light from the overdense plasmas. These plasmas make the reflected light modulated in intensity. Moreover, the burst behavior may be resulted from the modified plasmas. It is detrimental to ICF. In the indirect-drive ICF, we wish the lights deliver energy to the *Hohlraum* wall, but the gratings and cavitons make the light reflected far from *Hohlraum* wall.

The characteristics of the cavitons are: (a) there are both subcycle and cycle electromagnetic entities trapped; (b) a very effective plasma depletion takes place in the regions where the electromagnetic energy is localized; (c) the radiation trapped inside the density cavitons has both lower and higher frequency components than the unperturbed plasma frequency, and some components are even higher than the incident light frequency; and (d) electron and ion densities remain quite close to each other so that the plasma can be considered quasineutral during its response over long times to the radiation. We also run for the mass ratio of ions to electrons is 1836, the results are similar, but the caviton formation time is later and gratings are formed and disappeared more than once. The gratings and cavitons change the plasma density profile and make the reflected light at high level. It should be avoided in ICF. The formation of gratings and cavitons needs the ponderomotive force to push the electron. Cavitons are easily formed with high intensity laser and high plasma density.

IV. THE ANALYTICAL MODEL

In Ref. 8, a one-dimensional model for the interaction of electromagnetic waves of relativistic amplitude was developed. The model holds for circularly polarized light and results in two coupled nonlinear second-order differential equations in the spatial coordinate x for the electrostatic potential $\phi(x)$ and the vector potential $a(x)$. With the quasineutral approximation, the relevant system reduces to one nonlinear second-order differential equation in $a(x)$, which can be easily solved numerically. The equation reads

$$a_{xx}+\omega^2 a(x)=G(a), \tag{9}$$

where

$$G(a)=a\left\{\frac{K_0(\gamma_{\perp e}/\lambda_e)}{K_1(1/\lambda_e)}\left[\frac{\gamma_{\perp i}K_1(1/\lambda_e)K_1(\gamma_{\perp i}/\rho\lambda_i)}{\gamma_{\perp e}K_1(1/\rho\lambda_i)K_1(\gamma_{\perp e}/\lambda_e)}\right]^{\alpha}+\rho Z\frac{K_0(\gamma_{\perp i}/\rho\lambda_i)}{K_1(1/\rho\lambda_i)}\left[\frac{\gamma_{\perp i}K_1(1/\lambda_e)K_1(\gamma_{\perp i}/\rho\lambda_i)}{\gamma_{\perp e}K_1(1/\rho\lambda_i)K_1(\gamma_{\perp e}/\lambda_e)}\right]^{\alpha-1}\right\},$$

and $\lambda_e=T_e/m_e$, $\lambda_i=T_i/m_e$, $\rho=m_e/m_i$, $\gamma_{\perp e}=\sqrt{1+a^2}$, $\gamma_{\perp i}=\sqrt{1+\rho^2Z^2a^2}$, Z is the ion charge, T_e and T_i are the temperatures of electrons and ions, and K_0 and K_1 are the zeroth

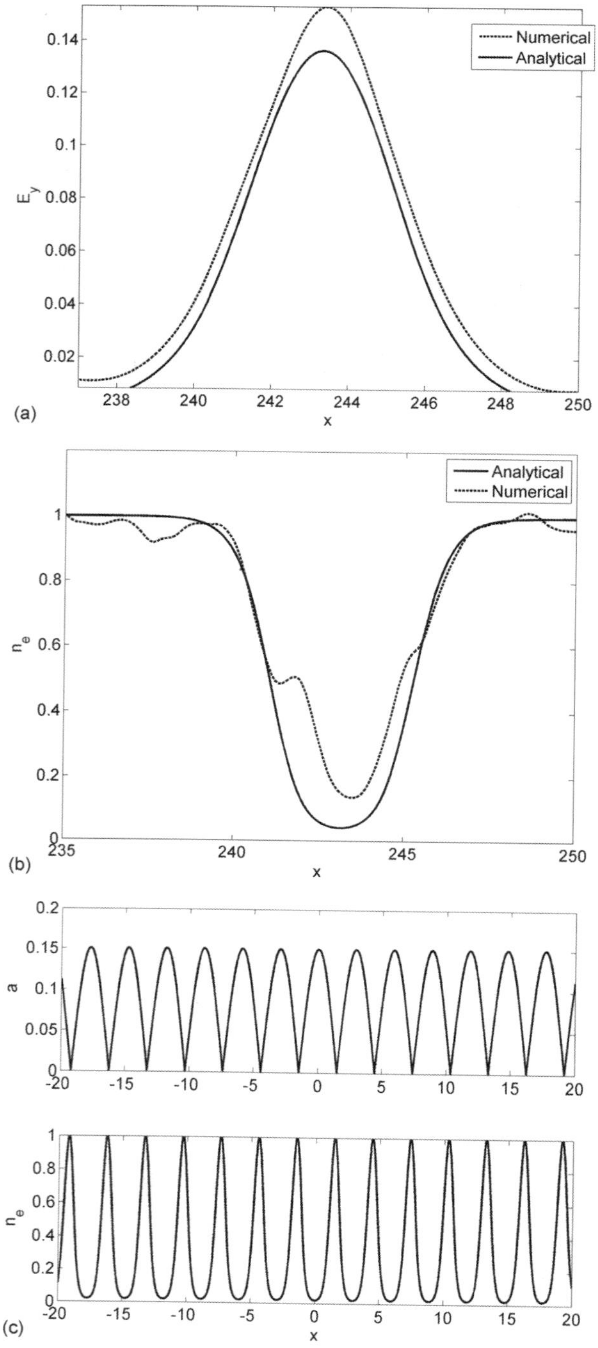

FIG. 7. The electric field (a) and electron density (b) in the caviton. Dashed lines are the simulation results and the solid lines are the analytical results with light frequency $0.6\omega_{pe}$. The vector potential and plasma density profiles (c) from the analytical results with light frequency $1.1\omega_{pe}$.

and first order of modified Bessel functions. For localized solutions, the electron and ion temperatures are taken as 1 and 0.5 keV, although they have high temperatures after the cavitons formed, and the light frequency is taken as $0.6\omega_{pe}$. The vector potential and the electron density profiles under these parameters are shown with solid lines in Fig. 7. For comparing, we also draw the results by the Vlasov–Maxwell simulation with dashed lines. We can see that the simulation results and the analytical results are very similar. The difference of vector potential in Fig. 7(a) maybe is that the analytical results hold only for circularly polarized light and our numerical results correspond to linearly polarized light. Equation (9) has both localized and oscillated solutions. We show an oscillated solution in Fig. 7(c) with the light frequency $1.1\omega_{pe}$. The profile is similar to the gratings. For unperturbed plasma, the incident light frequency is greater than the plasma frequency, but it may be less than the plasma frequency at some places for perturbed plasma. Spatially localized solutions of Eq. (9), which decrease exponentially for $x \to \pm\infty$, exist for[8]

$$\frac{K_0(1/\lambda_e)}{K_1(1/\lambda_e)} + \rho Z \frac{K_0(1/\rho\lambda_i)}{K_1(1/\rho\lambda_i)} - \frac{\omega^2}{\omega_{pe}^2} > 0. \tag{10}$$

For

$$\frac{K_0(1/\lambda_e)}{K_1(1/\lambda_e)} + \rho Z \frac{K_0(1/\rho\lambda_i)}{K_1(1/\rho\lambda_i)} - \frac{\omega^2}{\omega_{pe}^2} < 0, \tag{11}$$

an oscillated solution may exist. The density of part plasma can be twice of the initial density. For unperturbed plasma, Eq. (11) holds, as the gratings formed, Eq. (11) does not hold anymore and Eq. (10) becomes true. The transition from gratings to cavitons can be explained by Eqs. (9)–(11).

V. CONCLUSION

The gratings and cavitons are studied using one-dimensional Vlasov–Maxwell code. The results show that plasma gratings can evolve to cavitons. The lights in gratings push the plasma from high intensity region to low intensity region. If the plasma frequency at the low intensity region is greater than light frequency, the light will be trapped. The trapped lights have both cycle and subcycle structures. Cavitons can make plasma density higher than the critical density on somewhere. Moreover, the high density plasmas reflect the incident light in a very high level. In the indirect-drive ICF, lasers should deliver energy to the *Hohlraum* wall, but the gratings and cavitons can reflect light in a very high level in some cases. This may be detrimental to ICF.

ACKNOWLEDGMENTS

This work was supported by the National High-Tech ICF Committee of China, National Basic Research Program of China (973 Program Nos. 2007CB815101 and 2007CB814802) and the National Natural Science Foundation of China (Grant Nos. 10675024 and 10835003).

[1] M. V. Goldman and D. F. Dubois, Phys. Fluids **8**, 1404 (1965).
[2] R. W. Short and E. M. Epperlein, Phys. Rev. Lett. **68**, 3307 (1992).
[3] S. Weber, C. Riconda, and V. Tikhonchuk, Phys. Plasmas **12**, 043101 (2005).
[4] S. Weber, C. Riconda, and V. Tikhonchuk, Phys. Rev. Lett. **94**, 055005 (2005).

[5]S. Weber, M. Lontano, M. Passoni, C. Riconda, and V. Tikhonchuk, Phys. Plasmas **12**, 112107 (2005).
[6]M. Lontano, M. Passoni, C. Riconda, V. T. Tikhonchuk, and S. Weber, Laser Part. Beams **24**, 125 (2006).
[7]M. Lontano, S. V. Bulanov, J. Koga, M. Passoni, and T. Tajima, Phys. Plasmas **9**, 2562 (2002).
[8]M. Lontano, M. Passoni, and S. V. Bulanov, Phys. Plasmas **10**, 639 (2003).
[9]P. L. Bhatnagar, E. P. Gross, and M. Krook, Phys. Rev. **94**, 511 (1954).
[10]C. Z. Cheng and G. Knorr, J. Comput. Phys. **22**, 330 (1976).
[11]A. Mangeney, F. Califano, C. Cavazzoni, and P. Travnicek, J. Comput. Phys. **179**, 495 (2002).
[12]Z. M. Sheng, J. Zhang, and D. Umstadter, Appl. Phys. B: Lasers Opt. **77**, 673 (2003).
[13]H. C. Wu, Z. M. Sheng, Q. J. Zhang, Y. Cang, and J. Zhang, Phys. Plasmas **12**, 113103 (2005).
[14]Y. S. Kivshar and G. P. Agrawal, *Optical Solitons: From Fibers to Photonic Crystals* (Academic, New York, 2003).
[15]W. L. Kruer, *The Physics of Laser Plasma Interactions* (Addison-Wesley, Reading, MA, 1988).

PHYSICS OF PLASMAS **17**, 024502 (2010)

Excitation of coherent terahertz radiation by stimulated Raman scatterings

Z. J. Liu,[1,2] X. T. He,[1,2] C. Y. Zheng,[1,2] Shao-ping Zhu,[2] L. H. Cao,[2] and Yugang Wang[1]
[1]*Center for Applied Physics and Technology, State Key Laboratory of Nuclear Physics and Technology, Peking University, Beijing 100871, China*
[2]*Institute of Applied Physics and Computational Mathematics, Beijing 100094, China*

(Received 8 January 2010; accepted 5 February 2010; published online 26 February 2010)

The excitation of terahertz radiation by stimulated Raman scattering is considered. Vlasov–Maxwell numerical simulations show that powerful terahertz radiation can be produced by cascade Raman scatterings. © *2010 American Institute of Physics*. [doi:10.1063/1.3334999]

Terahertz spectroscopy is rapidly emerging as a new domain with rich implications in both fundamental and applied sciences. As a nonionizing electromagnetic radiation with submillimeter wavelength, it has promising applications in biological imaging, surface chemistry, and high-field condensed matter studies.[1] Unlike the microwave and optical frequency range, there are still no promising candidates for high efficiency terahertz generators.[2] Various methods, based on the principles of wake field excitation,[3,4] synchrotron radiation, transition radiation,[5] ionization-induced excitation,[6] and frequency up conversion of electromagnetic radiation[7] were proposed to generate terahertz radiation. Conventional sources of terahertz radiation using short pulse lasers rely on pulse generation in a solid and are generally limited to μJ/pulse.[8] Higher energies per pulse can be generated at accelerator facilities with intense bunched electron beams via synchrotron or transition radiation.[9] Recently, intense terahertz pulses with energies in the range 10–100 μJ/pulse have been generated as transition radiation by a laser generated and accelerated electron beam passing from plasma to vacuum.[8,10] Here we present a scheme for terahertz radiation generation that can provide tens of Joules energy.

The carbon dioxide (CO_2) gas laser which emits laser light with wavelength 10.6 μm can provide kilojoules energy. The frequency of the laser light, which is about 28.3 THz, is close to the terahertz range. Stimulated Raman scattering (SRS), a resonant decay of an incident light wave into a scattered light wave and an electron plasma wave, can occur in the CO_2-laser plasma interactions.[11,12] The frequency of the scattered light wave $\omega_s=\omega_L-\omega_{pe}$ is less than that of incident light wave, where ω_L is the frequency of the incident light and ω_{pe} is the frequency of electron plasma wave. If SRS occurs near quarter critical density, the frequency of the scattered light wave is $\omega_s \cong \omega_L/2$. The scattering process can occur again if the scattered light interacts with the plasma under proper conditions; the frequency of the second scattered light wave is $\omega_L/4$. An electromagnetic wave whose frequency falls in the terahertz range can thus be obtained. In this paper, a one dimensional Vlasov–Maxwell code[13] is used to study this process and the results show that terahertz radiation can indeed be produced by SRS.

The simulation parameters are as follows. The simulation box is one dimensional and all spatial variation is in the x direction. The CO_2 laser light propagates along the $-x$ direction, i.e., light propagates from right to left and is linearly polarized in y direction. The intensity of the incident CO_2-laser light is 1.1×10^{13} W/cm^2, corresponding to a normalized amplitude $eE/m_e\omega_L c=0.03$, where E is the incident light amplitude, m_e is the electron mass and c is the light speed in vacuum. The electron thermal velocity is $v_{te}=0.05c$, where $v_{te}=\sqrt{T_e/m_e}$ and T_e is the electron temperature. The initial electron density distribution is showed in Fig. 1. In Fig. 1, the ordinate is in unit of critical density n_c and the abscissa is in units of $1/k_L$, where k_L is the wave number of CO_2 laser light in vacuum. The three main plasma segment densities are 1/64, 1/16, and $1/4n_c$ with slight density gradients. Experimentally, the plasma layers can be produced by focusing a laser pulse onto gas jets that stream from nozzles in the y direction. The density gradients are not necessary. The Landau damping of plasma wave is relatively small and electron-ion collision damping is added such as Ref. 13. To exclude the stimulated Brillouin scattering (SBS), the effects of ions are not considered.

The incident light propagates from right to left in Fig. 1, and backward SRS occurs near the $1/4n_c$, the back scattered SRS light is reflected by the high density plasma at the right side. The incident and scattered lights propagation directions are showed with arrows in Fig. 1. If all the incident light is converted, according to the relations $\omega_s=\omega_L-\omega_{pe}$ and $\omega_s=\omega_{pe}$, the conversion efficiency to scattered light is about 0.50. In the simulations, the energy conversion efficiency is about 0.45. If the energy conversion efficiency is greater than 0.25, the normalized amplitude of scattered light is greater than that of incident light. The scattered light can also excite SRS near the density of $1/16n_c$, the second scattered light can also excite SRS near the density of $1/64n_c$. The second scattered light frequency is $\omega_L/4=7.1$ THz and the third is $\omega_L/8=3.5$ THz. The second and third scattered light propagation directions are also showed in Fig. 1 with arrows. The plasma density profiles can be modified to get scattered light with other frequencies.

Figure 2 is the light spectra at left boundary, the amplitude of the Fourier transform for E_y-B_z is in arbitrary units, and the frequency is normalized to the incident laser frequency, where E_y and B_z are the transverse electric and magnetic fields at left boundary. The main frequency of light in the left boundary is $1/8, 1/4, 1/2, 1\omega_L$. Part of ω_L light is the incident light passing through the plasma with very weak interactions during the SRS growth time. The amplitudes of the beat frequencies, such as $3/8, 3/4, 3/2\omega_L$, are small. Fig-

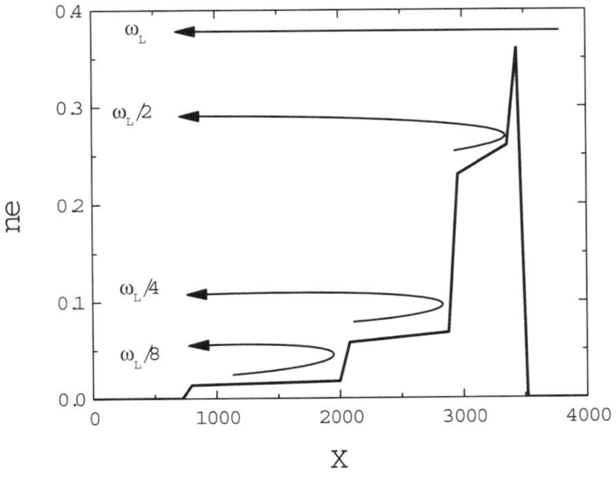

FIG. 1. Schematic of electron density distributions and lights propagation directions. The x-axis is in units of $1/k_L$, where k_L is the wave number of CO_2 laser light in vacuum and density in units of critical density. The arrows which marked with frequency show the propagation direction of corresponding wave.

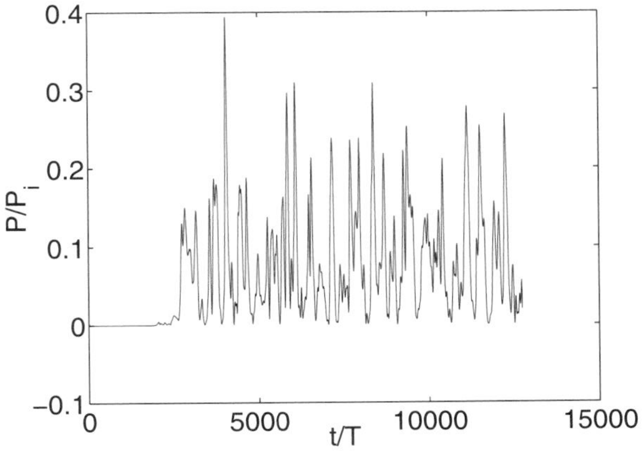

FIG. 3. Power of 3.5 THz radiation vs time, the time is normalized to the CO_2 laser cycle and the power is normalized to the incident laser light power. The average conversion efficiency is about 0.067 for the whole run and 0.085 for time between 2700 and 12 700.

ure 2 also shows that the scattered light has a very narrow spectral width. The spectral width of scattered light is related to the plasma and laser parameters. The nonlinear frequency shift of the plasma wave[14] and electron beam acoustic mode scattering[15] will broaden the spectrum of scattering light. Figure 3 is the power of 3.5 THz lights vary with time. The power of terahertz radiation has bursting behavior reminiscent of SRS in other kinetic simulations[13,16] and is related to the nonlinear saturation of the instability. The average 3.5 THz power in the whole run is 0.067 of the incident power.

The important point is the conversion efficiency. There are several methods for enhancing the SRS conversion efficiency, such as increasing the laser intensity, with proper electron-ion collision rates.[13] In the above simulations, the ions are fixed. When taking ion motion into account, the

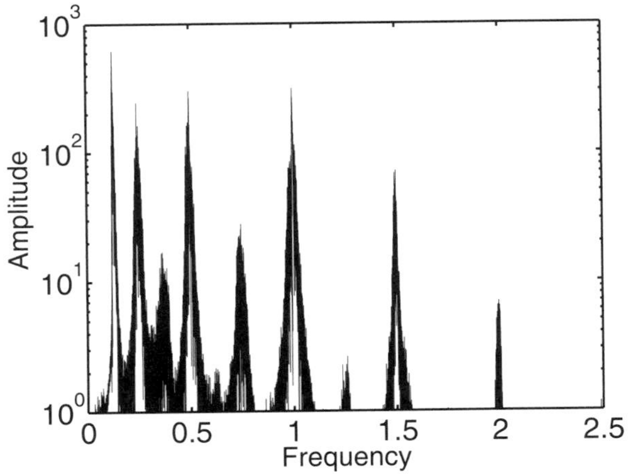

FIG. 2. Time integrated spectra of light in the left boundary. The amplitudes in arbitrary units and the frequency is normalized to the CO_2 laser light frequency.

maximum power of 3.5 THz radiation is about 0.0002 times the CO_2 laser power, while the maximum power of 7.1 THz radiation is 0.06 times the laser power. These radiation processes can last several hundreds of incident laser cycles. In laser plasma interactions, there are other instabilities that compete with SRS, such as SBS, the two plasmon decay instability, gratings, and cavitons formation.[17] These phenomena reduce the SRS conversion efficiency since they deplete the incident light or modify the plasma density. These processes should be avoided. To get a maximum conversion efficiency, the plasma and laser parameters can be adjusted, such as with two-ion species plasmas to increase the ion wave Landau damping or with plasmas doped high-Z elements to increase the collision damping. If scattering light near 10 THz is desired, the processes are relatively simple and the conversion efficiency is higher. For example, with plasma densities 0.12 and 0.09 n_c and an electron temperature 1.5 keV, the maximum conversion efficiency of 8.8 THz radiation can reach 0.05. If a scattered light frequency of 1 THz is desired, the processes are more complex and the conversion efficiency is lower.

In conclusion, we propose a new mechanism to generate powerful, coherent terahertz radiation. This mechanism can produce more energy and a longer pulse duration of terahertz radiation than that generated by femtosecond pulses, but relies on more precise control of plasma parameters.

This work was supported by the National High-Tech ICF Committee of China, National Basic Research Program of China (973 Program Nos. 2007CB815101, 2007CB814802, and 2010CB832904), and the National Natural Science Foundation of China (Grants Nos. 10975023, 10835003, 10935003, and 10974022).

[1] J. Orenstein and A. J. Millis, Science **288**, 468 (2000).
[2] J. H. Booske, Phys. Plasmas **15**, 055502 (2008).
[3] Z.-M. Sheng, H.-C. Wu, K. Li, and J. Zhang, Phys. Rev. E **69**, 025401(R) (2004).

[4]L.-H. Cao, W. Yu, H. Xu, C.-Y. Zheng, Z.-J. Liu, B. Li, and A. Bogaerts, Phys. Rev. E **70**, 046408 (2004).
[5]Y. Liu, A. Houard, B. Prade, S. Akturk, and A. Mysyrowicz, Phys. Rev. Lett. **99**, 135002 (2007).
[6]V. B. Gildenburg and N. V. Vvedenskii, Phys. Rev. Lett. **98**, 245002 (2007).
[7]R. L. Savage, C. Joshi, and W. B. Mori, Phys. Rev. Lett. **68**, 946 (1992).
[8]T. M. Antonsen, Jr. and J. Palastro, Phys. Plasmas **14**, 033107 (2007).
[9]B. E. Carlsten, K. A. Bishofberger, and R. J. Faehl, Phys. Plasmas **15**, 073101 (2008).
[10]W. P. Leemans, C. G. R. Geddes, J. Faure, Cs. Toth, J. van Tilborg, C. B. Schroeder, E. Esarey, G. Fubiani, D. Auerbach, B. Marcelis, M. A. Carnahan, R. A. Kaindl, J. Byrd, and M. C. Martin, Phys. Rev. Lett. **91**, 074802 (2003).
[11]H. A. Rose, D. F. DuBois, and B. Bezzerides, Phys. Rev. Lett. **58**, 2547 (1987).
[12]C. J. Walsh, D. M. Villeneuve, and H. A. Baldis, Phys. Rev. Lett. **53**, 1445 (1984).
[13]Z. J. Liu, S. P. Zhu, L. H. Cao, C. Y. Zheng, X. T. He, and Y. G. Wang, Phys. Plasmas **16**, 112703 (2009).
[14]G. J. Morales and T. M. O'Neil, Phys. Rev. Lett. **28**, 417 (1972).
[15]L. Yin, W. Daughton, B. J. Albright, B. Bezzerides, D. F. DuBois, J. M. Kindel, and H. X. Vu, Phys. Rev. E **73**, 025401(R) (2006).
[16]M. Albrecht-Marc, A. Ghizzo, T. W. Johnston, T. Réveillé, D. Del Sarto, and P. Bertrand, Phys. Plasmas **14**, 072704 (2007).
[17]Z. J. Liu, X. T. He, C. Y. Zheng, and Y. G. Wang, Phys. Plasmas **16**, 093108 (2009).

PHYSICS OF PLASMAS **18**, 032705 (2011)

Stimulated backward Brillouin scattering in two ion-species plasmas

Z. J. Liu,[1,2] C. Y. Zheng,[1,2] X. T. He,[1,2] and Yugang Wang[1]
[1]*State Key Laboratory of Nuclear Physics and Technology, Center for Applied Physics and Technology, Peking University, Beijing 100871, China*
[2]*Institute of Applied Physics and Computational Mathematics, Beijing 100094, China*

(Received 3 November 2010; accepted 7 March 2011; published online 24 March 2011)

Stimulated Brillouin back-scattering in mixed carbon and hydrogen plasmas is studied using one-dimensional Vlasov–Maxwell simulation. It is found that both the fast and slow ion acoustic waves can scatter the incident light. Carbon ions can be trapped in the slow ion acoustic wave, and the hydrogen ions can be trapped in both the fast and slow waves. The trapped ions tend to reduce the Landau damping of the ion acoustic waves, and both the fast and slow ion acoustic waves can be excited. From the time-integrated scattering spectra, the scattering peaks of the fast and slow ion acoustic waves can be clearly distinguished. © *2011 American Institute of Physics.*
[doi:10.1063/1.3570638]

I. INTRODUCTION

The stimulated Brillouin scattering (SBS), or resonant decay of a light wave into a scattered light wave and an ion acoustic wave, has been widely studied.[1] This scattering process plays an important role in laser propagation in underdense plasmas. Backward SBS leads to great energy loss by the incident laser and is detrimental to inertial confinement fusion (ICF).[2] Many methods have been proposed for reducing the SBS scattering level (the ratio of the scattered energy to the incident energy). These include laser smoothing[3–5] and increasing the ion acoustic damping. Theoretical results show that doping low-Z material with mid-Z or high-Z material can increase the Landau damping of the ion acoustic waves.[6,7] In two ion-species plasmas, there are two kinds of weakly damping acoustic waves, usually called fast and slow waves.[8–11] The general kinetic theory of ion acoustic waves in multispecies plasmas was developed by Fried, White, and Samec,[12] and the conditions under which one may or may not observe the two acoustic modes are given. The fast- and slow-wave dispersion relations are sensitive to the ion temperatures, a property that can be used to diagnose the plasma parameters.[9,13,14] Hydrocarbon (CH) plasmas have been widely studied experimentally.[15,16] But numerical simulations of SBS in CH plasmas are lacking.[17] Here we study SBS in CH plasmas using a one-dimensional Vlasov–Maxwell model.[18–20]

II. PHYSICAL MODEL

We assume that all spatial variations are in the x direction. The light wave is linearly polarized in the y direction, and the particles are collisionless. The incident light has a wavelength $\lambda_L=0.351$ μm and intensity 1.0×10^{16} W/cm^2, corresponding to the normalized amplitude $eE/m_e\omega_L c=0.03$, where E is the electric field and ω_L is the frequency of the light wave, c is the light speed in vacuum, $-e$ and m_e are the electron charge and mass, respectively. The simulation box length is 335 μm, and the plasma length is 200 μm, which is located at the center, and it has two sharp boundaries. To exclude stimulated Raman scattering, the electron density is chosen to be $0.3n_c$, where n_c is the critical density. The ions are composed of carbon and hydrogen ions with number density ratio 1:6. The temperatures T_e and T_i of the electrons and ions are both 5 keV. The space and time simulation steps are $0.2k_L^{-1}$ and $0.2\omega_L^{-1}$, where $k_L=2\pi/\lambda_L$ is the wavenumber in vacuum. The maximum velocities of the electrons, hydrogen ions, and carbon ions are $0.8c$, $0.02c$, and $0.005c$, respectively, at 513 intervals. Relativistic effects are not included. We seed a light field with intensity 1.0×10^{10} W/cm^2 and frequency $0.994\omega_L$. The incident laser propagates along the x axis with outgoing boundary conditions.

III. SIMULATION RESULTS

Figure 1 shows the instantaneous reflectivity. In the simulations, SBS does not reach a steady state and exhibits bursting behavior. The maximum transient reflectivity can be over 70% and the averaged reflectivity is as high as 18%. A small peak at $t/T_L=2000$ is observed. SBS bursting behavior has also been found in Ref. 21 by hybrid simulations.

The reflected energy spectra is shown in Fig. 2. It can be seen that there are two peaks located at $\sim 0.9940\omega_L$ and $\sim 0.9977\omega_L$, corresponding to the two branches of ion acoustic waves in the mixed plasma. Both ion acoustic waves can scatter the incident light. From the kinetic theory, we found that the frequencies of the fast and slow acoustic waves are $5.95\times 10^{-3}\omega_L$ and $2.78\times 10^{-3}\omega_L$, and the damping rates are $4.94\times 10^{-3}\omega_L$ and $0.85\times 10^{-3}\omega_L$, respectively. The damping rates are quite high. The resonantly excited ion waves can undergo wave steeping, harmonic generation, ion trapping, modulational and filamentation, as well as parametric scattering and decay.[22] The frequency discrepancy between the kinetic theory and the simulation results are mainly caused by the ion trapping, which can modify the ion wave frequencies. The nonlinear frequency shift is about +0.8% for the fast wave and is about -18% for the slow wave. In low-ZT_e/T_i plasmas, kinetic effects can be important. Particle trapping modifies the dispersion relation for the acoustic waves, and leads to a nonlinear frequency shift $\delta\Omega$ (Refs. 23–25)

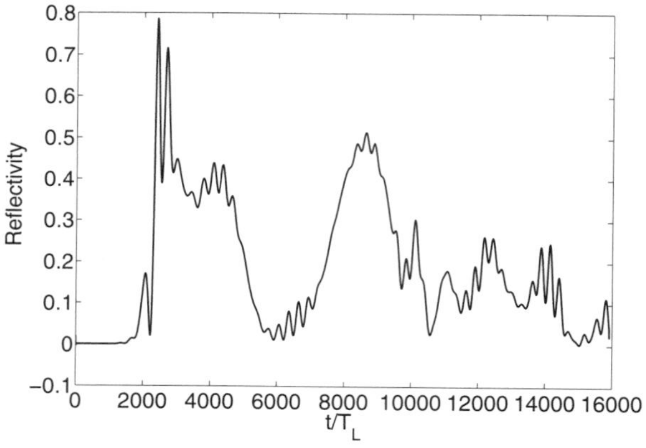

FIG. 1. The SBS reflected energy varies with time, where T_L is the laser period.

$$\frac{\delta\Omega}{\omega_a}=-\frac{1}{\sqrt{2\pi}}(v^4-v^2)e^{-v^2/2}\sqrt{\frac{ZT_e}{T_i}}\left(\frac{|\delta n|}{n_e}\right)^{1/2}, \tag{1}$$

where ω_a is the SBS acoustic wave frequency, $v=c_s/v_{\text{th}}$, $v_{\text{th}}=\sqrt{T_i/m_i}$ is the ions thermal velocity, and c_s is the phase velocity of the acoustic wave. The density perturbation for the slow wave is 10% if we use Zn_i instead n_e in Eq. (1). The simulation results show that the hydrogen-ion density perturbation is 2% and the carbon ion density is 10%. Equation (1) is in good agreement with that from the simulations for the slow wave, but it does not agree for the fast wave. This may be caused by hydrogen-ion trapping in the slow wave. The trapped hydrogen ions in the slow wave cause the distribution function to depart from Maxwellian and the hydrogen-ion temperature to increase. The resulting frequency shift is

$$\frac{\delta\Omega}{\omega_a}=\frac{3\,\delta T_i}{2(ZT_e+3T_i)}, \tag{2}$$

where δT_i is the change of the ion temperature and it is related to the amplitude of the slow wave. The trapped hydrogen ions in the slow wave lead to a positive nonlinear frequency shift, but the trapped ions in the fast wave lead to a negative frequency shift. The hydrogen density perturbation contains both the fast- and slow-wave perturbations, so that the value given by Eq. (1) is not accurate. Using Eq. (1) and summing over both the hydrogen and carbon ions, we obtain a 0.4% hydrogen density perturbation. The nonlinear frequency shift detunes the resonant coupling between the incident light and the driven ion acoustic wave, therefore saturating the SBS instability. The time-integrated scattering spectra are usually broadened for the heavily damped and nonlinearly frequency-shifted ion wave.

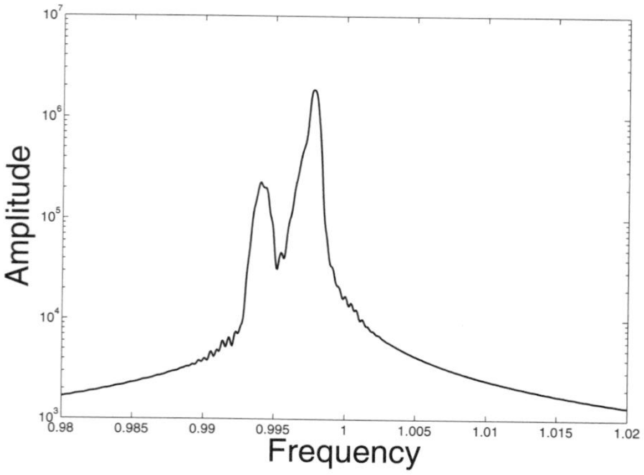

FIG. 2. The time-integrated scattering spectra. The frequency is normalized by the laser frequency, and the amplitude is in relative unit.

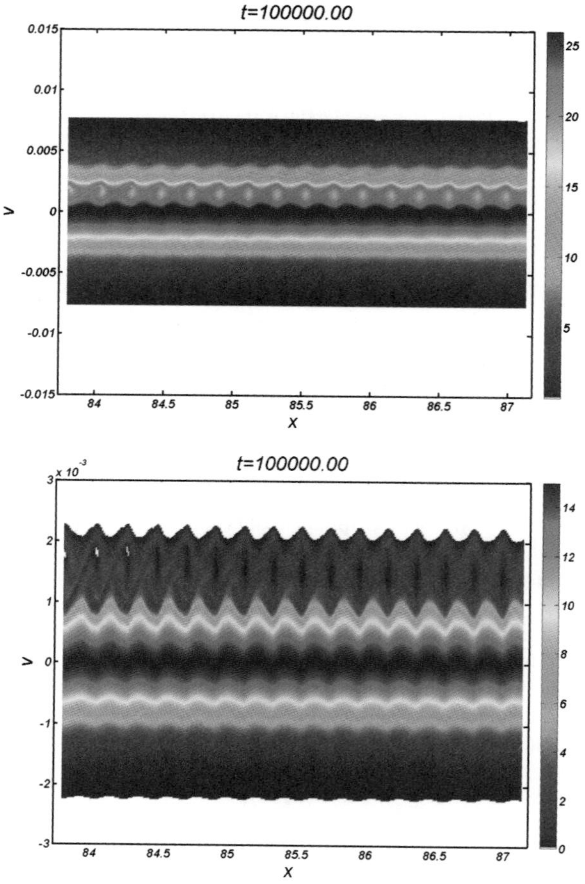

FIG. 3. (Color online) The velocity distribution functions for hydrogen (top) and carbon (bottom) ions at $\omega_L t=10^5$. The unit of x is μm and v is normalized by light speed in vacuum.

In larger-amplitude ion acoustic waves, more ions can be trapped. Figure 3 clearly shows the trapped ions. For the slow ion acoustic wave, electrons and (carbon and hydrogen) ions can be trapped. For the fast wave, electrons and hydrogen ions can be trapped, but carbon ions cannot be trapped. In the wave frame, the particles with energy $1/2m_s(v-v_p)^2 < |q_s\phi|$ can be trapped, where q_s and m_s are the charge and mass of species s. For the fast wave, the carbon ion is too heavy and cannot be trapped. The trapping width in phase

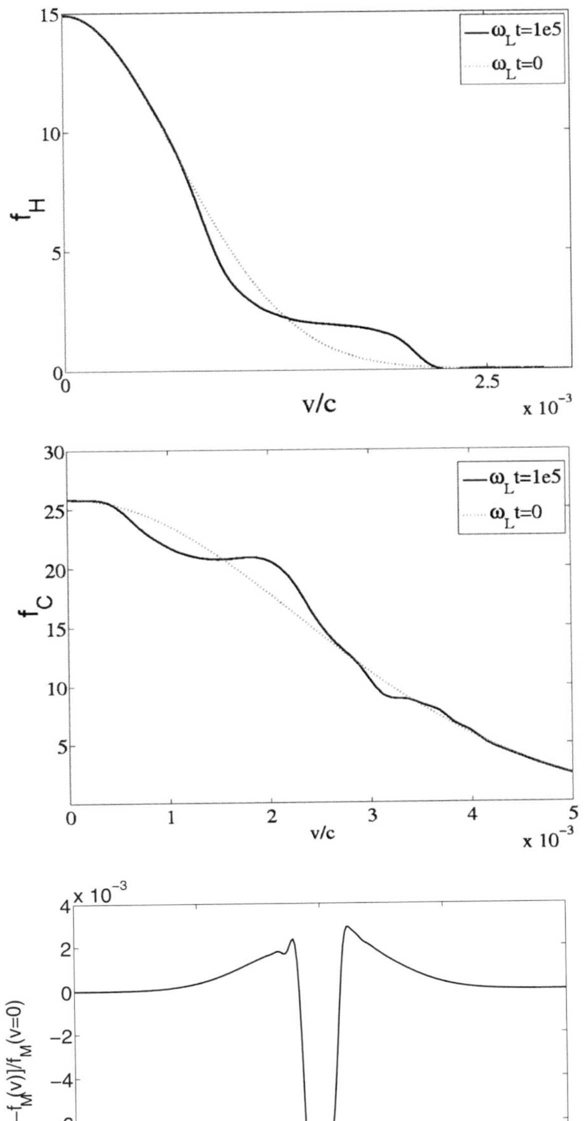

FIG. 4. (Color online) The spatial averaged distribution functions for hydrogen (top), and carbon (middle) ions, between 84 and 87 μm at $\omega_L t=10^5$ and $\omega_L t=0$, the difference between the electron distribution function and Maxwellian function (bottom), where f_s for $s=H,C,e$ are distribution functions for hydrogen, carbon and electron, $f_M=n_e/\sqrt{2\pi T_e/m_e}\times\exp(-m_e v^2/2T_e)$ is Maxwellian distribution function.

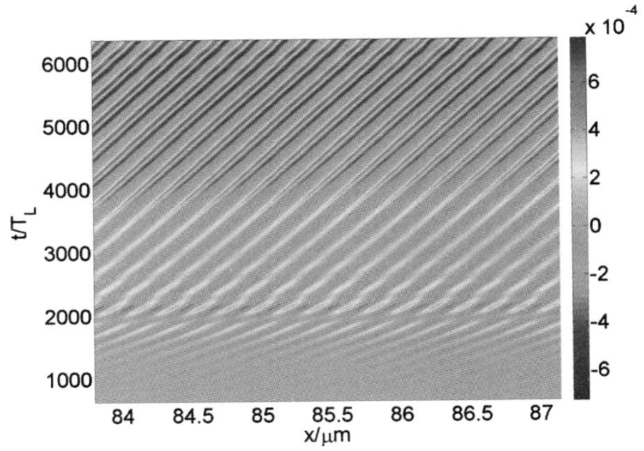

FIG. 5. (Color online) The electrostatic field.

space is related with the ion wave amplitude. The width of trapped hydrogen ions in fast wave is less than that in the slow wave, so the amplitude of fast wave is also less than that of slow wave. The SBS scattering level of the fast wave is less than that of the slow wave, just as shown in Fig. 2. The carbon ion trapping structures in the left part of Fig. 3 (bottom) and are different from that of the right part. The electrostatic field decreases and the kinematic energy of the carbon ions increases during the trapping structures like the left part of Fig. 3 (bottom) formed. The carbon ion density perturbation propagation velocity is also increased. We can consider the ion acoustic speed as the ion density perturbation velocity. The acoustic speed jump can reduce the spatial growth rate.[26]

Figure 4 shows the spatially averaged distribution functions at $\omega_L t=10^5$. We can clearly seen that hydrogen ion distribution function is distorted around the fast and slow wave phase velocities and the carbon ion distribution function is only distorted around the slow wave phase velocity. This is similar to the distortion of the electron distribution function in stimulated Raman scattering. The flattened distribution functions are due to particle trapping. The latter can significantly reduce the Landau damping after only a few bounce periods. Although electrons can be trapped by both fast and slow waves, the electron distribution function is not much distorted for $m_e \ll m_i$, where m_i is the ion mass. The electron bounce frequency $\omega_b=(ek_aE_x/m_e)^{1/2}$, where k_a and E_x are the wave number and electric field amplitude of ion acoustic wave, is too high comparing with the acoustic wave frequencies and there is no resonance between the electrons and the acoustic waves. Electrons perform only bounce motion, and little energy is exchanged between the electrons and waves. The spatially averaged electron distribution function is also shown in Fig. 4. One can see that it is close to Maxwellian.

Figure 5 shows the electrostatic field. For $t/T_L<2000$, the group velocity is about $3.2\times10^{-3}c$ and the strength of electrostatic field is relatively small. The group velocity is consistent with the fast wave in Fig. 2. Comparing Fig. 1 and Fig. 5, we can conclude that the first small peak in Fig. 1 is mainly caused by the strongly damped fast wave. For $t/T_L>2000$, the electrostatic field group velocity is about $1.5\times10^{-3}c$. It corresponds to the slow wave in Fig. 2. Its electrostatic field is stronger than that of the fast wave. The scattering level in Fig. 1 is dominated by the slow wave.

IV. CONCLUSION

In conclusion, we have studied the SBS process in a mixed carbon and hydrogen plasma at high ions temperature. We found that both fast and slow acoustic waves with heavy Landau damping can be excited by an intense laser. Both the fast and slow waves can scatter the incident light. In the SBS spectra, the fast- and slow-wave scattering peaks can be clearly distinguished. The trapping of hydrogen and carbon ions is also observed.

ACKNOWLEDGMENTS

This work was supported by Sci. & Tech. Funds of CAEP (Grant Nos. 2010B0102018 and 2010A0102004), National Basic Research Program of China (973 Program Nos. 2007CB815101, 2010CB832904, and 2007CB814802), and the National Natural Science Foundation of China (Grant Nos. 10975023, 10835003, and 10935003).

[1] W. L. Kruer, *The Physics of Laser Plasma Interactions* (Addison-Wesley, Reading, MA, 1988).
[2] J. D. Lindl, P. Amendt, R. L. Berger, S. G. Glendining, S. H. Glenzer, S. W. Hann, R. L. Kauffman, O. L. Landen, and L. Suter, Phys. Plasmas **11**, 339 (2004).
[3] Y. Kato, K. Mima, N. Miyanaga, S. Arinaga, Y. Kitagawa, M. Nakatsuka, and C. Yamanaka, Phys. Rev. Lett. **53**, 1057 (1984).
[4] S. Skupsky, R. W. Short, T. Kessler, R. S. Craxton, S. Letzring, and J. M. Soures, J. Appl. Phys. **66**, 3456 (1989).
[5] J. E. Rothenberg, J. Appl. Phys. **87**, 3654 (2000).
[6] V. Yu. Bychenkov, W. Rozmus, and V. T. Tikhonchuk, Phys. Rev. E **51**, 1400 (1995).
[7] H. X. Vu, J. M. Wallace, and B. Bezzerides, Phys. Plasmas **2**, 1682 (1995).
[8] E. A. Williams, R. L. Berger, R. P. Drake, A. M. Rubenchik, B. S. Bauer, D. D. Meyerhofer, A. C. Gaeris, and T. W. Johnston, Phys. Plasmas **2**, 129 (1995).
[9] Z. J. Liu, J. Zheng, and C. X. Yu, Phys. Plasmas **9**, 1073 (2002).
[10] R. L. Berger and E. J. Valeo, Phys. Plasmas **12**, 032104 (2005).
[11] H. X. Vu, J. M. Wallace, and B. Bezzerides, Phys. Plasmas **1**, 3542 (1994).
[12] B. D. Fried, R. B. White, and W. D. Samec, Phys. Fluids **14**, 2388 (1971).
[13] D. H. Froula, L. Divol, and S. H. Glenzer, Phys. Rev. Lett. **88**, 105003 (2002).
[14] S. H. Glenzer, C. A. Back, K. G. Estabrook, R. Wallace, K. Baker, B. J. MacGowan, B. A. Hammel, R. E. Cid, and J. S. De Groot, Phys. Rev. Lett. **77**, 1496 (1996).
[15] H. C. Bandulet, C. Labaune, K. Lewis, and S. Depierreux, Phys. Rev. Lett. **93**, 035002 (2004).
[16] P. Neumayer, R. L. Berger, L. Divol, D. H. Froula, R. A. London, B. J. MacGowan, N. B. Meezan, J. S. Ross, C. Sorce, L. J. Suter, and S. H. Glenzer, Phys. Rev. Lett. **100**, 105001 (2008).
[17] R. L. Berger, C. H. Still, E. A. Williams, and A. B. Langdon, Phys. Plasmas **5**, 4337 (1998).
[18] Z. J. Liu, X. T. He, C. Y. Zheng, and Y. G. Wang, Phys. Plasmas **16**, 093108 (2009).
[19] Z. J. Liu, Shao-ping Zhu, L. H. Cao, C. Y. Zheng, X. T. He, and Yugang Wang, Phys. Plasmas **16**, 112703 (2009).
[20] Z. J. Liu, X. T. He, C. Y. Zheng, Shao-ping Zhu, L. H. Cao, and Yugang Wang, Phys. Plasmas **17**, 024502 (2010).
[21] B. I. Cohen, L. Divol, A. B. Langdon, and E. A. Williams, Phys. Plasmas **12**, 052703 (2005).
[22] B. I. Cohen, B. F. Lasinski, A. B. Langdon, and E. A. Williams, Phys. Plasmas **4**, 956 (1997).
[23] L. Divol, R. L. Berger, B. I. Cohen, E. A. Williams, A. B. Langdon, B. F. Lasinski, D. F. Froula, and S. H. Glenzer, Phys. Plasmas **10**, 1822 (2003).
[24] G. J. Morales and T. M. O'Neil, Phys. Rev. Lett. **28**, 417 (1972).
[25] R. L. Dewar, Phys. Fluids **15**, 712 (1972).
[26] C. E. Clayton, C. Hoshi, and F. F. Chen, Phys. Rev. Lett. **51**, 1656 (1983).

PHYSICS OF PLASMAS 21, 032107 (2014)

Research on ponderomotive driven Vlasov–Poisson system in electron acoustic wave parametric region

C. Z. Xiao,[1] Z. J. Liu,[1,2] T. W. Huang,[1] C. Y. Zheng,[1,2] B. Qiao,[1,3] and X. T. He[1,2]
[1]*HEDPS, Center for Applied Physics and Technology, Peking University, Beijing 100871, China*
[2]*Institute of Applied Physics and Computational Mathematics, Beijing 100094, China*
[3]*School of Physics, Peking University, Beijing 100871, China*

(Received 26 October 2013; accepted 24 February 2014; published online 6 March 2014)

Theoretical analysis and corresponding 1D Particle-in-Cell (PIC) simulations of ponderomotive driven Vlasov–Poisson system in electron acoustic wave (EAW) parametric region are demonstrated. Theoretical analysis identifies that under the resonant condition, a monochromatic EAW can be excited when the wave number of the drive ponderomotive force satisfies $0.26 \lesssim k_d\lambda_D \lesssim 0.53$. If $k_d\lambda_D \lesssim 0.26$, nonlinear superposition of harmonic waves can be resonantly excited, called kinetic electrostatic electron nonlinear waves. Numerical simulations have demonstrated these wave excitation and evolution dynamics, in consistence with the theoretical predictions. The physical nature of these two waves is supposed to be interaction of harmonic waves, and their similar phase space properties are also discussed. © *2014 AIP Publishing LLC*. [http://dx.doi.org/10.1063/1.4867664]

I. INTRODUCTION

Excitation of electrostatic plasma waves is an essential feature of plasma physics that has aroused great interest for many years. The Langmuir and ion acoustic waves are the basic linear electrostatic waves. Nonlinear electrostatic waves have been found from the stationary solutions of Vlasov–Poisson equations, based on the Bernstein–Greene–Kruskal (BGK) technique,[1] such as solitary wave and shock wave. In 1991, Holloway and Dorning[2] noted that certain nonlinear wave structures can exist in plasmas, even at low amplitude, from a periodic BGK solution with infinitesimal potential. Since the dispersion relation of these waves is of the acoustic form, i.e., $\omega = 1.3kv_{the}$ (v_{the} is electron thermal velocity), they are called as electron acoustic waves (EAWs).

EAWs have been widely studied theoretically[3–6] and numerically. Unlike the ion acoustic waves, which are linear waves that including both electrons and ions, the EAWs are higher-frequency (between the ion acoustic wave frequency and the Langmuir wave frequency) nonlinear waves that involve only the electrons (the ions simply provide a uniform neutralized background). Within the linear theory, an EAW would be heavily Landau damped since its phase velocity is comparable to the electron thermal velocity. Therefore, particle trapping[7] plays a crucial role in the formation of nonlinear EAW. Due to the electron trapping, the distribution of electron velocities is effectively flat at the wave phase velocity, which turns off the Landau damping. The asymptotic state can be regarded as a nonlinear BGK mode. EAWs have also been observed in experiments with laser-produced plasmas,[8,9] pure electron plasmas,[10] pure ion plasmas,[11,12] and in related numerical simulations.[13–17]

To excite the electrostatic plasma waves, one possible way is to apply an external ponderomotive drive at a particular frequency and wave number into the collisionless plasmas. This can be achieved by the beating of two-counter-propagating laser electromagnetic waves. The evolution of nonlinear Langmuir waves excited by the stimulated Raman scattering in a ponderomotive driven plasma has been theoretically studied by Bénisti[18] and Lindberg.[19] However, how the ponderomotive driven plasma waves evolve in EAW parametric regions is still unclear.

In this paper, we systematically study the excitation and evolution dynamics of waves in a ponderomotive driven plasma in EAW parametric region from the Vlasov–Poisson equations by introducing an external electrostatic field in the form of the ponderomotive force. The so called ponderomotive driven Vlasov–Poisson equations are analyzed by means of Fourier–Laplace transform, and a series of one-dimensional (1D) particle-in-cell (PIC) simulations are carried out by varying the driven frequency and wave number. It shows that in the linear stage, the envelope of the electrostatic plasma wave in both the fundamental and second harmonic mode has a constant component plus a Langmuir term. When the ponderomotive drive wave number $k_d\lambda_D$ is large, the Langmuir term can be ignored, the constant wave amplitudes dominate. The dependence of electrostatic wave excitation on the driven wave number and frequency are investigated. It is found that monochromatic stable EAWs can be excited by nonlinear wave-particle interactions when the wave number of the drive ponderomotive force satisfies $0.26 \lesssim k_d\lambda_D \lesssim 0.53$. Nonlinear superposition of harmonic waves, so called kinetic electrostatic electron nonlinear (KEEN) waves, can be resonantly excited and undamped, when $k_d\lambda_D$ is less than 0.26. Furthermore, both EAWs and KEEN waves show the similiar phase space structure, so that they can be classified into a special set of BGK modes.

The paper is organized as follows: In Sec. II, analytical analysis of the ponderomotive driven Vlasov–Poisson system has been carried out, where the solutions of the fundamental and second harmonic modes are given. The condition for excitation of EAWs has also been identified. In Sec. III, 1D PIC simulation results show the evolution dynamics of such ponderomotive driven Vlasov–Poisson system in EAW parametric region and the excitations of EAWs and further nonlinear superposition of harmonic waves, i.e., KEEN waves. Summary and discussions are given in Sec. IV.

II. THEORETICAL ANALYSIS OF THE PONDEROMOTIVE DRIVEN VLASOV–POISSON SYSTEM

Laser-produced plasmas can be approximately described by the 1D electrostatic Vlasov–Poisson equations with an external electrostatic field, which represents the laser-driven ponderomotive force for a given time period. Considering ions as a uniform neutralized background, the electron distribution function obeys

$$\frac{\partial f}{\partial t}+v\frac{\partial f}{\partial x}-\frac{e}{m}(E+E_{ext})\frac{\partial f}{\partial v}=0, \tag{1}$$

$$\frac{\partial E}{\partial x}=4\pi n_0 e\left[1-\int_{-\infty}^{\infty} f dv\right], \tag{2}$$

where f is the normalized electron distribution function satisfying $\int_{-\infty}^{\infty} f dv=1$, E is the self-consistent electrostatic field in plasmas, and n_0, e, m refer to the background equilibrium density, electron charge and mass. E_{ext} is the external electrostatic field representing the ponderomotive drive. For simplicity, we assume E_{ext} to be in the form of a traveling wave with frequency ω_d, wave number k_d, and amplitude E_0 as

$$E_{ext}(x,t)=E_0\exp[i(k_d x-\omega_d t)], \tag{3}$$

where ω_d, k_d, and E_0 are related to the specific laser parameters and can be adjusted by the beating of two-counter-propagating laser waves in different frequencies and wave numbers. Note that here we use a sudden drive[21] to excite plasma waves instead of the adiabatic drives,[18,19] so it is expected that the plasma initially in equilibrium quickly turns to a forced oscillation state.

To seek the harmonic superposition solutions of Eqs. (1)–(3), we transform f and E in a Fourier series related to the external ponderomotive drive and assume the amplitude is only dependent on t

$$f(x,v,t)=f_0(v)+\sum_{n=1}^{\infty} f_n(v,t)\exp[in(k_d x-\omega_d t)], \tag{4}$$

$$E(x,t)=\sum_{n=1}^{\infty} E_n(t)\exp[in(k_d x-\omega_d t)], \tag{5}$$

where $f_n(v,t)$ and $E_n(t)$ are the envelopes of the excited electrostatic waves in each modes and f_0 is the Maxwellian distribution function of equilibrium plasmas. Substituting Eqs. (4) and (5) into Eqs. (1)–(3), we obtain that the electron distribution functions in various modes n evolve as

$$\begin{aligned}&\left\{\frac{\partial f_1}{\partial t}-i(\omega_d-k_d v)f_1-\frac{e}{m}[E_1+E_0]\frac{\partial f_0}{\partial v}\right\}e^{i(k_d x-\omega_d t)}\\&=-\sum_{n=2}^{\infty}\left[\frac{\partial f_n}{\partial t}-in(\omega_d-k_d v)f_n\right]e^{in(k_d x-\omega_d t)}\\&+\frac{e}{m}\sum_{n'=2}^{\infty}E_{n'}e^{in'(k_d x-\omega_d t)}\cdot\left\{\frac{\partial f_0}{\partial v}+\sum_{n=1}^{\infty}\frac{\partial f_n}{\partial v}e^{in(k_d x-\omega_d t)}\right\}\\&+\frac{e}{m}[E_1+E_0]\cdot\sum_{n=1}^{\infty}\frac{\partial f_n}{\partial v}e^{i(n+1)(k_d x-\omega_d t)},\end{aligned} \tag{6}$$

$$1+\frac{4\pi n_0 e}{ink_d}\int_{-\infty}^{\infty} dv\frac{f_n}{E_n}=0. \tag{7}$$

In Eqs. (6) and (7), the coefficients of a specific mode n should be identical on both sides, so that we can obtain the envelope evolution equations of each mode.

A. Solution of the fundamental mode

From Eqs. (6) and (7), we can easily get the envelope evolution equation of the fundamental mode as

$$\frac{\partial f_1}{\partial t}-i(\omega_d-k_d v)f_1=\frac{e}{m}(E_1+E_0)\frac{\partial f_0}{\partial v}, \tag{8}$$

and the corresponding Poisson equation as

$$1+\frac{4\pi n_0 e}{ik_d}\int_{-\infty}^{\infty} dv\frac{f_1}{E_1}=0. \tag{9}$$

Using the Laplace transform[22] and assuming initially $f_1(v,t=0)=E_1(t=0)=0$, we get that

$$E_1(p)=-\frac{{}^1\chi_e(p)}{1+{}^1\chi_e(p)}\cdot\frac{E_0}{2\pi ip}, \tag{10}$$

$$f_1(v,p)=\frac{eE_0}{m}\frac{\partial f_0}{\partial v}\cdot\frac{1}{2\pi ip[p-i(\omega_d-k_d v)][1+{}^1\chi_e(p)]}, \tag{11}$$

where p in normal mode is $-i\omega_e$ and ω_e is envelop frequency. ${}^1\chi_e(p)$ is the fundamental mode envelope susceptibility, which is defined as

$$\begin{aligned}{}^1\chi_e(p)={}^1\chi_e(\omega_e)&=\chi_e(\omega_d,k_d,\omega_e)\\&=-\frac{\omega_{pe}^2}{k_d^2}\int_{-\infty}^{\infty} dv\frac{\partial f_0/\partial v}{v-(\omega_d+\omega_e)/k_d},\end{aligned} \tag{12}$$

where $\omega_{pe}=\sqrt{4\pi n_0 e^2/m}$ is the electron plasma frequency. It is found that the fundamental mode envelope susceptibility is a shifted linear plasma susceptibility, i.e., $\chi_e(\omega_d,k_d,\omega_e)=\chi_l(\omega_e+\omega_d,k_d)$.

Further making the inverse-Laplace transform of Eqs. (10) and (11) through analytical continuation and contour integral, we obtain the envelop evolution equation of the fundamental mode as

$$E_1(t)=-\frac{{}^1\chi_l}{1+{}^1\chi_l}E_0+\sum_j\frac{1}{\omega_{ej}\left.\dfrac{\partial^1\chi_e(\omega_e)}{\partial\omega_e}\right|_{\omega_e=\omega_{ej}}}E_0 e^{-i\omega_{ej}t}, \tag{13}$$

$$\begin{aligned}f_1(v,t)=&\frac{ieE_0}{m(\omega_d-k_d v)}\frac{\partial f_0/\partial v}{1+{}^1\chi_l}+\frac{ieE_0}{m}\frac{\partial f_0}{\partial v}\\&\times\frac{1}{(k_d v-\omega_d)[1+{}^1\chi_e(\omega_e=k_d v-\omega_d)]}e^{-i(k_d v-\omega_d)t}\\&+\frac{ieE_0}{m}\frac{\partial f_0}{\partial v}\\&\times\sum_j\frac{1}{\omega_{ej}(\omega_{ej}+\omega_d-k_d v)\left.\dfrac{\partial^1\chi_e(\omega_e)}{\partial\omega_e}\right|_{\omega_e=\omega_{ej}}}e^{-i\omega_{ej}t},\end{aligned} \tag{14}$$

where ω_{ej} are the solutions of $1+{}^1\chi_e(\omega_e)=0$. These solutions include the shifted Landau normal mode corresponding to the smallest damping rate (Langmuir wave) and the higher-order modes with the larger damping rates. Among them, the higher-order modes vanish rapidly with the Landau normal mode left behind. Furthermore, the second term of Eq. (14) is the phase mixing term, which also tends to zero rapidly. Therefore, in the case of strong Landau damping (large $k_d\lambda_D$), only the first terms of Eqs. (13) and (14) remain, while for $k_d\lambda_D \ll 1$, the undamped Langmuir wave component needs to be taken into account. When both the higher-order and the phase mixing terms vanish, the envelope evolution equations of the fundamental mode for $k_d\lambda_D \ll 1$ can be simplified as

$$E_1(t) = -\frac{{}^1\chi_l}{1+{}^1\chi_l}E_0 + \left[\frac{1}{2}+\frac{\omega_d}{2(\omega_{ek}-\omega_d)}\right]E_0 e^{-i(\omega_{ek}-\omega_d)t}, \tag{15}$$

$$f_1(t) = \frac{ieE_0}{m(\omega_d - k_d v)}\frac{\partial f_0/\partial v}{1+{}^1\chi_l} + \frac{ieE_0}{m(\omega_{ek}-k_d v)}\frac{\partial f_0}{\partial v} \times \left[\frac{1}{2}+\frac{\omega_d}{2(\omega_{ek}-\omega_d)}\right]e^{-i(\omega_{ek}-\omega_d)t}, \tag{16}$$

where $\omega_{ek}=\sqrt{\omega_{pe}^2+3k_d^2v_{the}^2}$ is the Langmuir wave frequency.

B. Solution of the second harmonic wave

Second and high-order harmonic waves are excited through the mode coupling processes. In the linear stage, the envelope evolution equations of these harmonic waves can be obtained through Eqs. (6) and (7). The second harmonic wave envelope obeys

$$\frac{\partial f_2}{\partial t} - 2i(\omega_d - k_d v)f_2 = \frac{e}{m}E_2\frac{\partial f_0}{\partial v} + S_2, \tag{17}$$

where the term $S_2 = e(E_1+E_0)/m \times (\partial f_1/\partial v)$ is the source of the second harmonic wave. The other terms behave as a linear response. Using the Laplace transform and assuming the initial condition as $f_2(v,t=0)=E_2(t=0)=0$, we have

$$E_2(p) = -\frac{{}^2\chi_{nl}^{(2)}(p)}{1+{}^2\chi_e(p)}\cdot\frac{E_0}{2\pi i p}, \tag{18}$$

where ${}^2\chi_e(p) = {}^2\chi_e(\omega_e) = \chi_l(\omega_e+2\omega_d, 2k_d)$ is the envelope susceptibility of the second harmonic wave. The 2nd-order nonlinear susceptibility ${}^2\chi_{nl}^{(2)}$ associated with S_2 represents the nonlinear wave-wave interactions. For the EAW parametric region we studied, the Langmuir wave component in Eqs. (15) and (16) is negligible, so ${}^2\chi_{nl}^{(2)}$ can be expressed as

$${}^2\chi_{nl}^{(2)}(\omega_e) = -\frac{ieE_0}{m(1+{}^1\chi_l)^2}\frac{\omega_{pe}^2}{(2k_d)^2}\int_{-\infty}^{\infty}dv\frac{\frac{\partial}{\partial v}\left(\frac{\partial f_0/\partial v}{\omega_d-k_d v}\right)}{v-(\omega_e+2\omega_d)/2k_d}. \tag{19}$$

Making the inverse-Laplace transform of Eq. (18), we obtain the envelope evolution equation of the second harmonic wave as

$$E_2(t) = -\frac{{}^2\chi_{nl}^{(2)}(\omega_e=0)}{1+{}^2\chi_l}E_0 - \sum_{j'}\frac{{}^2\chi_{nl}^{(2)}(\omega_{ej'})}{\omega_{ej'}\left.\frac{\partial^2\chi_e(\omega_e)}{\partial\omega_e}\right|_{\omega_e=\omega_{ej'}}}E_0 e^{-i\omega_{ej'}t}, \tag{20}$$

where $\omega_{ej'}$ are the solutions of $1+{}^2\chi_e(\omega_e)=0$. As discussed above, when $2k_d\lambda_D$ is large, the second term of Eq. (20) vanishes due to strong damping. While for $2k_d\lambda_D \ll 1$, the Langmuir wave component is excited and can be written as

$$E_2(t) = -\frac{{}^2\chi_{nl}^{(2)}(\omega_e=0)}{1+{}^2\chi_l}E_0 - \left[\frac{1}{2}+\frac{\omega_d}{\omega_{ek2}-2\omega_d}\right]\cdot{}^2\chi_{nl}^{(2)}(\omega_{ek2} - 2\omega_d)E_0 e^{-i(\omega_{ek2}-2\omega_d)t}, \tag{21}$$

where $\omega_{ek2}=\sqrt{\omega_{pe}^2+3(2k_d)^2v_{the}^2}$ is the Langmuir wave frequency of the second harmonic wave. From Eq. (20), one can see that the amplitude of the second harmonic wave is about $E_2 \sim [E_0/(1+{}^1\chi_l)^2]E_1$. Similarly, the envelope evolutions of high-order harmonics can also be derived from Eqs. (6) and (7), but with lower amplitudes.

C. Waves excited in electron acoustic wave region and their nonlinear evolution

We analyze Eqs. (15) and (21) in electron acoustic wave parametric region by numerically calculating the linear plasma dielectric function, $\varepsilon_l = 1+\chi_l$, with an initial Maxwellian distribution. Figures 1(a) and 1(b) plot, respectively, the real part of ε_l and imaginary part of ε_l. In Fig. 1(a), the curve Re(ε_l) = 0 shows the well-known EAW dispersion relation with infinitesimal wave potential, where the upper and lower branches are, respectively, for the Langmuir wave and the EAW separated by the turning point $(k_{tp}\lambda_D, \omega_{tp}/\omega_{pe}) = (0.53, 1.14)$. The value of Re($\varepsilon_l$) is below zero inside the curve and above zero outside the curve, while Im(ε_l) is always positive for the whole parametric region. Moreover, both their absolute values increase when $k\lambda_D$ and ω/ω_{pe} decrease.

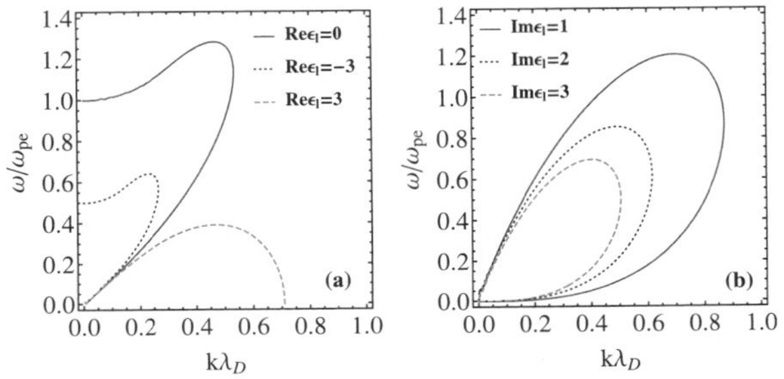

FIG. 1. The solutions of the real (a) and imaginary (b) parts of the linear plasma dielectric function ε_l. The initial distribution function f_0 is assumed to be Maxwellian.

If the combination of the wave number and frequency of the ponderomotive drive (k_d, ω_d) lies in the vicinity of the Langmuir wave dispersion relation $1 + \chi_l \approx 0$, both two terms in either Eqs. (15) or (21) are resonant, and the Langmuir wave is excited. However, in the parametric region of EAWs, since the imaginary part of $1 + \chi_l$ is large, the amplitude of fundamental mode is about E_0, and the amplitude of second harmonic wave is much smaller due to the factor $E_0/(1+{}^1\chi_l)^2$. Therefore, in the linear stage, no resonant waves are excited and the plasma experiences just a forced oscillation driven by the ponderomotive force.

When the driven plasma keeps evolving with time exceeding the characteristic bounce time $\tau_b = 2\pi/\omega_b$ $(\omega_b = \sqrt{ek_dE_0/|(1+{}^1\chi_l)|m})$ of this system, nonlinear wave-particle interactions become dominant, leading to excitation of fundamental and harmonic waves. The imaginary part of the wave envelope susceptibility decreases quickly due to the nonlinear Landau damping. And its real part is also slightly modified by the nonlinear frequency shift.[6,18,19,21,23–25] As a result, the absolute value of the dielectric function decreases, leading to rapid fundamental and harmonic mode growing, as shown in Eqs. (15) and (21). These wave modes become saturate when the imaginary part of the dielectric function turns to zero while the saturation level is determined by its real part. The nonlinear electrostatic wave modes can be resonantly excited when (k_d, ω_d) lies on the curve $\mathrm{Re}[1 + \chi_{nl}] = 0$, where χ_{nl} is the nonlinear envelope susceptibility due to nonlinear wave-particle interactions.

If assuming the nonlinear susceptibilities χ_{nl} Refs. 6, 18, 19, and 26 for fundamental mode and its harmonics are almost the same, and note that the nonlinear susceptibility is a modified linear EAW dispersion relation, then we plot the curve of $\mathrm{Re}[\varepsilon_{nl}] = 0$ in Figure 2. We see that when (k_d, ω_d) lies near the curve, the fundamental mode would be resonantly excited, and so do the second, third, and high-order harmonics. Since the curve of $\mathrm{Re}[\varepsilon_{nl}] = 0$ is bounded, we can estimate the number of excited harmonics using the linear curve's turning point $(k_{tp}\lambda_D, \omega_{tp}/\omega_{pe}) = (0.53, 1.14)$, which separates the Langmuir and EAW branches. The number of harmonics is finally estimated as

$$n \lesssim \frac{0.53}{k_d\lambda_D}. \tag{22}$$

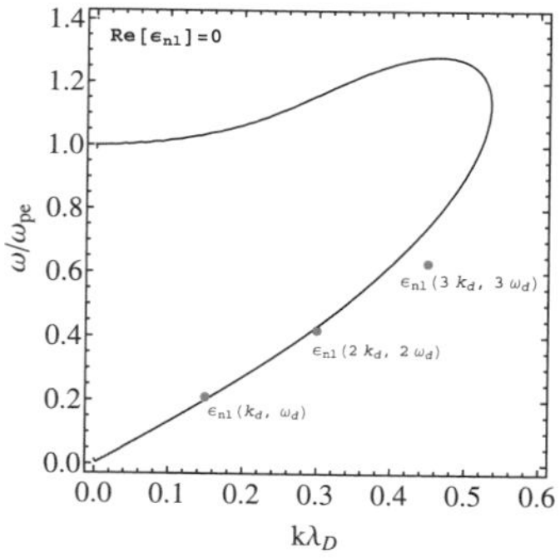

FIG. 2. Resonant excitation.

To excite a monochromatic EAW wave, the optimal condition of the ponderomotive drive should satisfy $0.26 \lesssim k_d\lambda_D \lesssim 0.53$.

III. 1D PIC SIMULATIONS CONFIRMING THE THEORY

To verify the above theoretical analysis, we carry out 1D PIC simulations to demonstrate the excitation and evolution dynamics of plasma waves in a ponderomotive driven plasma in EAW region. 5×10^6 total electrons are used for the simulation, where ions are simulated as a uniform neutralized background. The simulation box consisting of 256 grids with grid size $\Delta v = 0.0391 v_{the}$ covers the velocity space of $[-5v_{the}, 5v_{the}]$. To avoid nonlinear harmonic conversions, the spatial length of the simulation box is chosen to be exactly one ponderomotive drive wavelength, varying from $4\pi\lambda_D$ to $80\lambda_D$ for different driven wave numbers. Correspondingly, the spatial grid size Δx also changes, and its average value is about $0.078\lambda_D$. Time step of $0.1\omega_{pe}^{-1}$ is taken and periodic boundary condition is used. The external ponderomotive drive has the form

$$E_{ext} = E_0 P(t)\sin(k_d x - \omega_d t), \tag{23}$$

where $P(t)$ is the temporal envelope of the drive. Here we choose the same form as Ref. 27

$$P(t) = \begin{cases} \frac{1}{2}\left[1 + \tanh\left(t_p\left(\frac{2t}{t_r} - 1\right)\right)\right], & t < t_r \\ \frac{1}{2}\left[1 - \tanh\left(t_p\left(\frac{2(t-\tau)}{t_r} - 1\right)\right)\right], & t_r \le t < t_r + \tau \\ 0, & otherwise \end{cases}, \tag{24}$$

where τ is the duration. $t_r/2(\omega_{pe}^{-1})$ is the turn-on time of the ponderomotive drive, and the rising and down time is about $200/t_p(\omega_{pe}^{-1})$. $t_r = 100$ and $t_p = 20$ are taken in the simulations.

A. Overall picture of excited electrostatic waves

A series of simulations with the ponderomotive drive of different wave number and frequency (k_d, ω_d) are carried out.

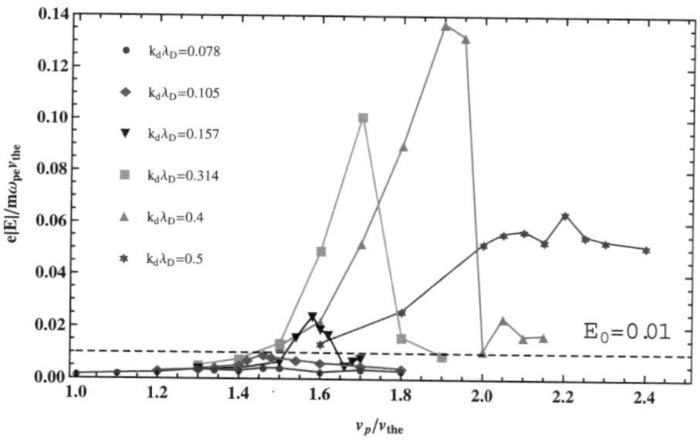

FIG. 3. Resonant curves for different driven wave numbers. The dashed line is the amplitude of the ponderomotive drive. v_p in the x-axis is the phase velocity of the ponderomotive drive as $v_p = \omega_d/k_d$.

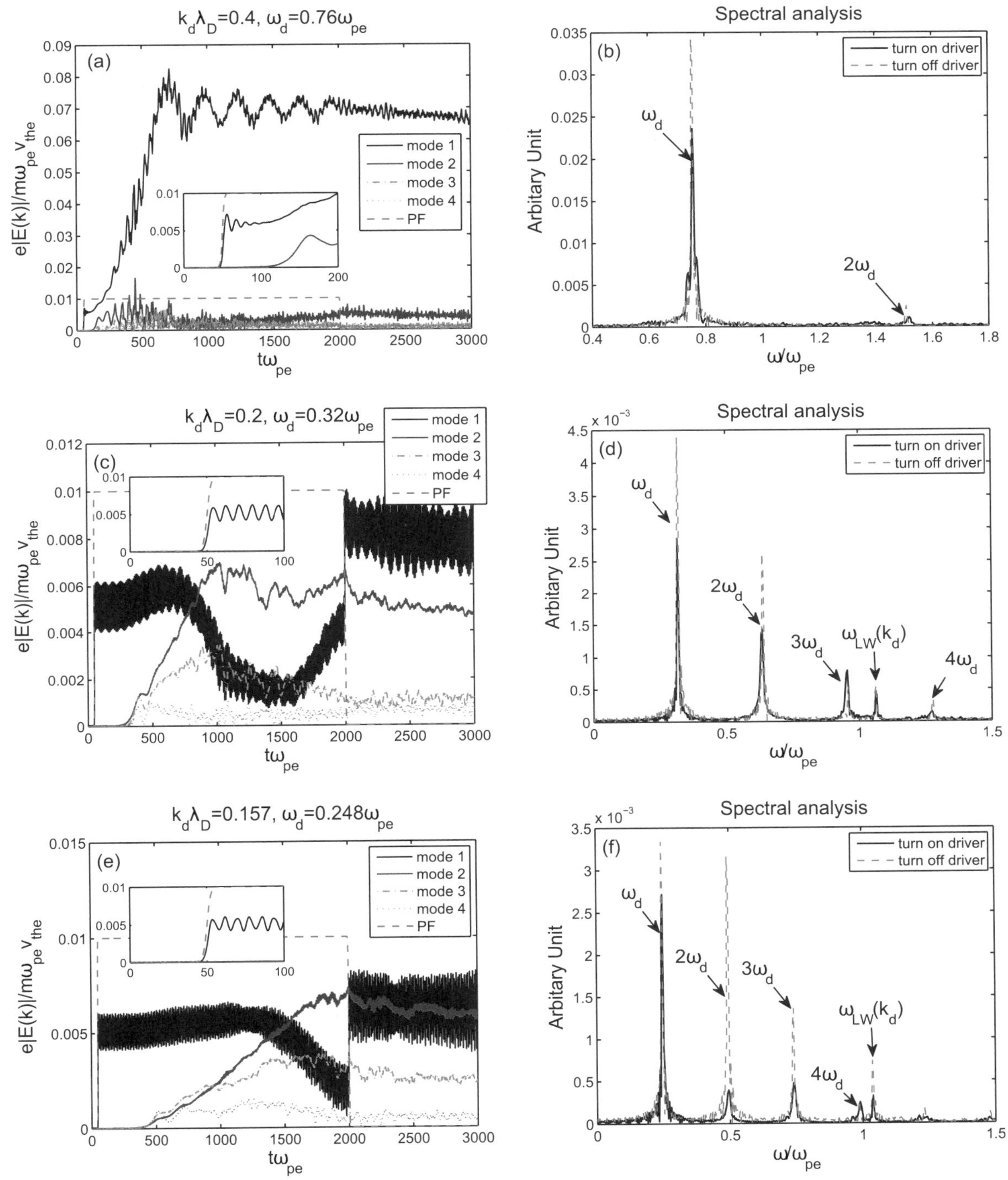

FIG. 4. Evolutions of the electric field amplitudes for plasma waves of the first 4 modes in a ponderomotive driven plasma. Three typical cases of $k_d\lambda_D = 0.4, 0.2$, and 0.157 are presented from top to bottom. (a), (c), and (e) show the amplitude evolution of spatial Fourier modes, where the modes 1–4 correspond to $k = k_d, 2k_d, 3k_d, 4k_d$, respectively, and the red dashed line is the amplitude of the ponderomotive force. The insets are the zoomed initial modes evolution. (b), (d), and (f) are the corresponding temporal spectral analysis, where the data of the turn-on drive (black solid line) are sampled from $t = 100$ to $1000\omega_{pe}^{-1}$, and the turn-off drive data (red dashed line) are sampled from $t = 2000$ to $t = 2900\omega_{pe}^{-1}$.

The amplitude of the drive E_0 is chosen as $0.01m\omega_{pe}v_{the}/e$. The duration is taken to be larger than the characteristic bounce time of the fundamental mode $\tau_b = 2\pi/\omega_b = 2\pi\sqrt{m|\varepsilon_l(k_d,\omega_d)|/ek_dE_0}$, so that nonlinear evolution of ponderomotive driven Vlasov–Poisson system can be observed. The drive is turned off after an appropriate duration

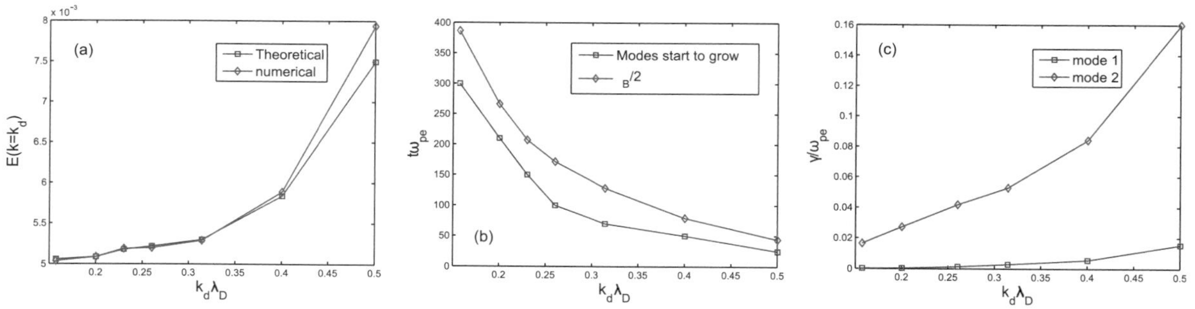

FIG. 5. (a) Comparison of theoretical and simulation results for the initial fundamental mode amplitude varying with the driver wave number $k_d\lambda_D$. (b) Relevance between modes initial growth time and τ_B. (c) Initial growth rates of the fundamental mode and second harmonic wave.

time, $\tau = 1950\omega_{pe}^{-1}$, and then the plasma evolves in a free Vlasov–Poisson system. The amplitude of the electric field $|E|$ becomes stable representing the resonant characters, which is shown in Fig. 3 for various (k_d, ω_d) cases.

Figure 3 clearly shows that the resonance cannot occur for all $k_d\lambda_D$ parametric region. For moderate $k_d\lambda_D$, especially for $k_d\lambda_D = 0.314$ and 0.4, a remarkable resonant peak is observed, while for small $k_d\lambda_D$ ($k_d\lambda_D \leq 0.157$), the amplitude increases slowly or even falls into a quite small value (lower than the drive amplitude) as the drive is off. In the near loss of resonance[6] region for $k_d\lambda_D = 0.5$, a broad plane of resonant region is observed, since the nonlinear dielectric function has nearly the same value. Through careful analysis, it is also found that the resonant point (k_d, ω_d) satisfies the nonlinear dispersion relation of EAWs, i.e., $\mathrm{Re}^1\varepsilon_{nl}(E) = 0$, especially in moderate $k_d\lambda_D$ region. Furthermore, the saturation level of these waves increases with $k_d\lambda_D$ and have a maximal value at about ten times of initial drive amplitude E_0 at $k_d\lambda_D = 0.4$.

B. Linear stage, growing and saturation processes of wave evolution

To clearly show the excited plasma wave evolution dynamics, the electric field amplitudes varying with time for the first 4 modes are plotted in Fig. 4 for three typical resonant cases as $(k_d\lambda_D, \omega_d/\omega_{pe}) = (0.4, 0.76), (0.2, 0.32)$, and (0.157, 0.248), respectively. Among them, Figs. 4(a), 4(c), and 4(e) show the amplitude evolution in spatial Fourier modes, where modes 1, 2, 3, and 4 represent fundamental mode, second harmonic wave, third harmonic wave, and fourth harmonic wave corresponding to $k = k_d, 2k_d, 3k_d, 4k_d$, and Figs. 4(b), 4(d), and 4(f) are the corresponding temporal Fourier analysis. It is clearly shown from Fig. 4 that the wave evolution behaviors in ponderomotive driven Vlasov–Poisson system in electron acoustic wave region include excitation of monochromatic wave and harmonic superposition due to various modes interactions.

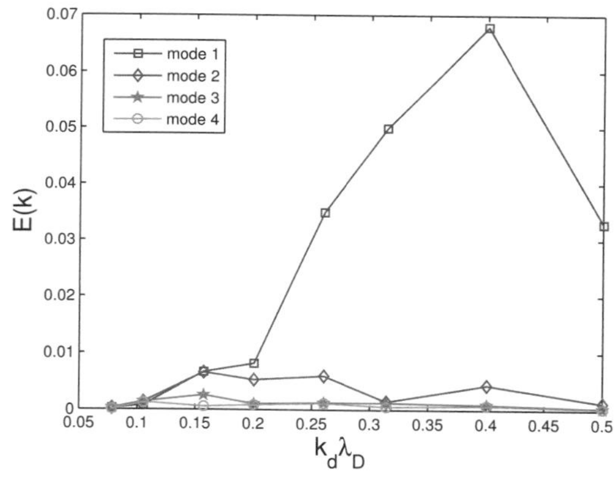

FIG. 6. The amplitude of mode 1–4 varying with $k_d\lambda_D$ when these modes are resonant excited at $t = 2100\omega_{pe}^{-1}$.

First, we look at the linear stage, growing and saturation processes of the wave evolutions. At $t = 50\omega_{pe}^{-1}$, the drive is turned on and its amplitude immediately increases to E_0. The insets of Figs. 4(a), 4(c), and 4(e) show clearly the initial evolution of modes 1 and 2. We can see that the amplitude of the fundamental mode quickly grows to the level of $E_0/2$ and then oscillates due to the plasma eigenmode (the Langmuir wave). This can be interpreted easily from our theoretical formula Eq. (13). The Langmuir component is also observed in the temporal spectral analysis. The simulation results of the fundamental mode amplitude evolution agree well with our theoretical analysis, as shown in Fig. 5(a).

When time scale reaches the characteristic bounce time τ_B, the initial forced oscillation changes, and the nonlinear modes start to grow. Comparison of the initial growth time of the nonlinear wave modes and half of the bounce time $\tau_B/2$ varying with the driven wave numbers $k_d\lambda_D$ is given in Fig. 5(b). It shows a positive correlation between two quantities, indicating that the modes do grow from their initial amplitudes due to nonlinear wave-particle interactions. The fundamental mode and its high-order harmonics start to grow at the same time but with different growth rates, which scales as $\left[\partial \mathrm{Im}(^n\chi_l)/\partial\omega\right]^{-1}$ [see Fig. 1(b)]. The initial growth rate of the fundamental mode (mode 1) and the second harmonics (mode 2) varying with $k_d\lambda_D$ is plotted in Fig. 5(c). We can see that the growth rate decreases when $k_d\lambda_D$ is reduced, and the growth rate of the fundamental model is smaller than that of the second harmonics, leading to different mode evolution behaviors. Specifically, at large $k_d\lambda_D$, the fundamental mode grows rapidly to a considerable level [see Fig. 4(a)], which suppresses the growth of the high-order harmonics though the latter has a large growth rate. Eventually, the high-order harmonics saturate at low levels. However, at small $k_d\lambda_D$, the fundamental mode has much smaller growth rate [see Figs. 4(c) and 4(e)] and spends much longer time to grow, during which the high-order harmonics, in particular, the second harmonics, grow almost freely to the ponderomotive force level.

After the initial growing stage, nonlinear wave-wave interaction becomes dominant, where the fundamental mode suppresses the growth of high-order harmonics at large $k_d\lambda_D$ region, and, in contrary, the high-order harmonics prevent the fundamental mode from growing up at small $k_d\lambda_D$ region. Then all modes start to saturate. Further, from the temporal spectral analysis of Figs. 4(b), 4(d), and 4(f), we see that in the ponderomotive driven plasma, the frequencies of harmonics are always an integer multiple of the drive frequency, which means that the harmonics propagate with the same velocity $v_p = \omega_d/k_d$.

C. EAWs and KEEN waves

The ponderomotive drive is turned off at $t = 2000\omega_{pe}^{-1}$ when the plasma wave modes saturate, then all waves propagate stably in the free Vlasov–Poisson system. These stable waves include the monochromatic waves [Fig. 4(a)] and the nonlinear waves of harmonic superposition [Figs. 4(c) and 4(e)], which are consistent with our theoretical analysis above. For $k_d\lambda_D = 0.4$, a monochromatic EAW wave is excited; for $k_d\lambda_D = 0.2$, the nonlinear wave structure of a combination of the fundamental mode plus the second harmonic wave is observed; and for $k_d\lambda_D = 0.157$, the nonlinear structure is a superposition of first three harmonics. Figure 6 plots the amplitude of mode 1 to 4 for various $k_d\lambda_D$ at time $t = 2100\omega_{pe}^{-1}$ when all modes are resonantly excited. It shows that the fundamental mode amplitude has a peak around $k_d\lambda_D = 0.4$ and quickly drops when $k_d\lambda_D$ decreases. Therefore, we can conclude that the monochromatic EAWs only exist in relatively large $k_d\lambda_D$ region, while nonlinear harmonic superposition structures dominate in small $k_d\lambda_D$ region.

These ponderomotive driven monochromatic waves first discovered by Valentini[16] in the same system and satisfying nonlinear EAW dispersion relation are what we called electron acoustic wave. EAWs excited in this system are nearly undamped, and they are proved to be BGK mode by many authors. Their phase space structures and the corresponding potential and electric field are plotted in Fig. 7 (for $k_d\lambda_D = 0.4$), which shows typical phase topology of the electron distribution in the sinusoidal wave potential and electric

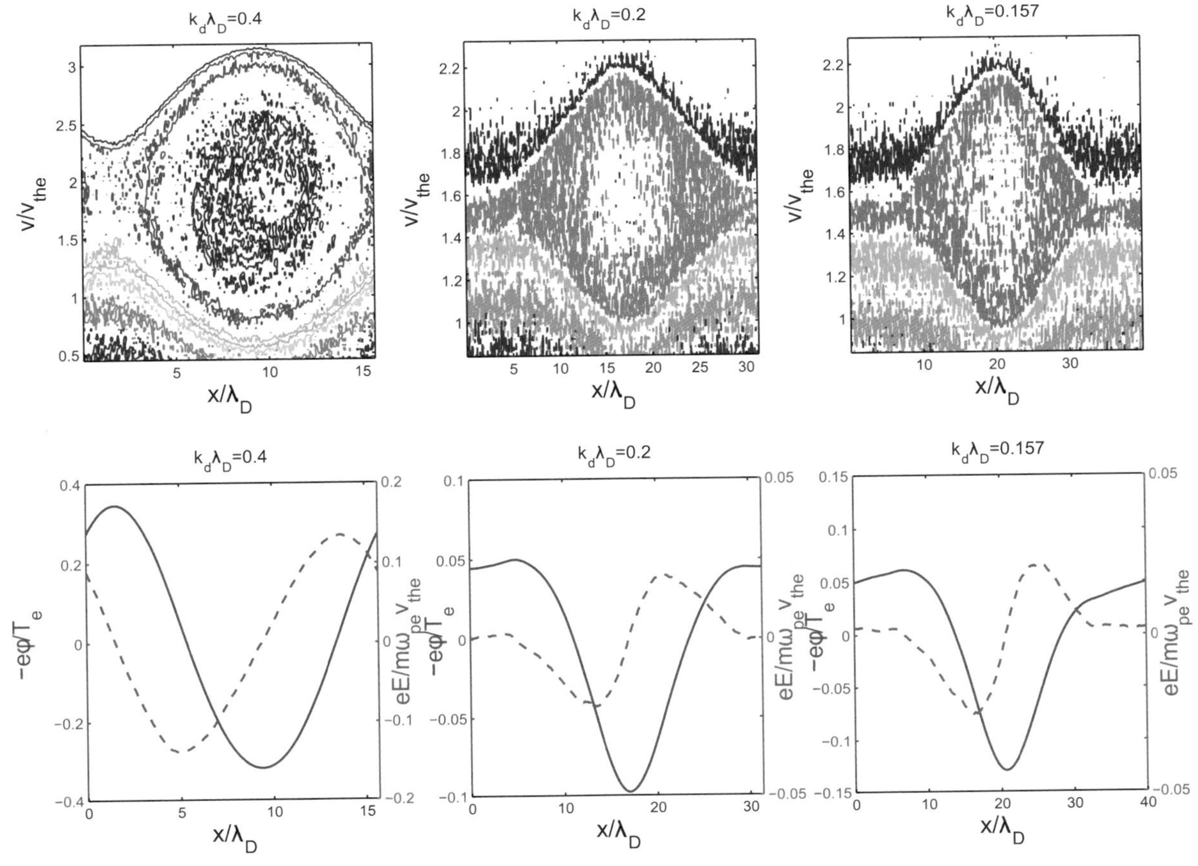

FIG. 7. Phase space images of electron distribution function and its corresponding potential and electric field distribution for $k_d\lambda_D = 0.4, 0.2, 0.157$, at $t = 2940, 2820, 2880\omega_{pe}^{-1}$, respectively, corresponding to the status in Fig. 4.

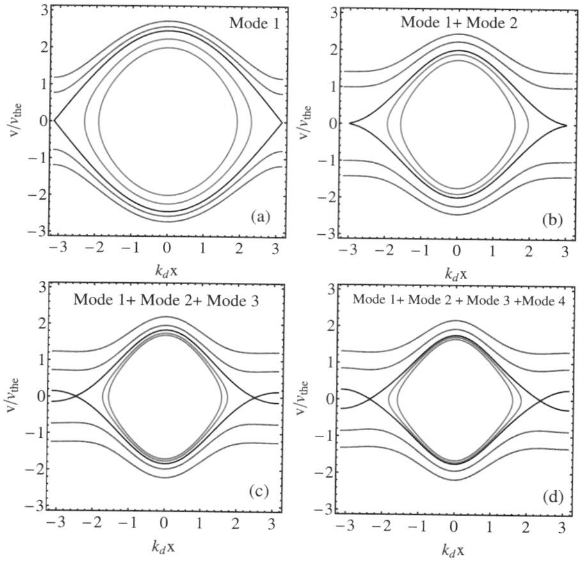

FIG. 8. Phase space structure of harmonic superposition in the wave frame. The separatrix (black line) delimits the trapping particles trajectories (red lines) from the passing particle trajectories (brown lines).

field. For the cases of $k_d\lambda_D = 0.2, 0.157$ in Fig. 7, the nonlinear waves of multiple harmonics superposition are a kind of resonantly excited KEEN waves,[17,20] whose phase space shows a superposition of harmonics with the same phase velocity and a well-organized mode amplitude.

To clarify this more clearly, we consider the electron motion under a simple electric field in the sinusoidal form, which consists of the first four harmonics as

$$E(x) = a\sin x + b\sin 2x + c\sin 3x + d\sin 4x, \qquad (25)$$

where a, b, c, and d are the amplitudes of different modes. The corresponding potential can be obtained from $E = -\nabla\phi$. Four cases of the harmonic waves with the same total electric field amplitude are taken as: (I) Mode 1: $a = 1.5, b = c = d = 0$; (II) mode 1+mode 2: $a = 1, b = 0.5, c = d = 0$; (III) mode 1+ mode 2+ mode 3: $a = 0.8, b = 0.6, c = 0.1, d = 0$; and (IV) mode 1+ mode 2+ mode 3+ mode 4: $a = 0.7, b = 0.6, c = 0.15, d = 0.05$. Electron trajectories under these potential cases are plotted in Fig. 8, showing similar structures as those in Fig. 7. This concludes that EAWs and KEEN waves are naturally of the same physics of harmonic superposition. Both of them are nearly undamped in Vlasov–Poisson system, indicating that harmonic superposition may be a set of solutions of BGK theory.

IV. CONCLUSIONS

We have presented theoretical and numerical study of ponderomotive driven Vlasov–Poisson system in electron acoustic wave parametric region. The plasma initially undergoes forced oscillation, including both the fundamental mode and high-order harmonic waves. However, at later time when the system evolves into the nonlinear wave-particle interaction stage, both the fundamental mode and high-order harmonics significantly increase and saturate. Analytical analysis and 1D PIC simulations identify that the excited nonlinear waves include monochromatic EAWs, which can be excited when the wave number of the drive ponderomotive force satisfies $0.26 \lesssim k_d\lambda_D \lesssim 0.53$, and kinetic electrostatic electron nonlinear (KEEN) waves, which is a nonlinear superposition of harmonic waves and can be resonantly excited when $k_d\lambda_D \lesssim 0.26$. These waves, which are supposed to be a specific set of BGK modes, have the same physical nature. Note that further strong nonlinear wave-wave interactions may lead to incoherent phase space structures, which will be studied in future.

ACKNOWLEDGMENTS

We thank Q. Jia, D. Wu, W. W. Wang, and S. L. Yang for useful discussions. This work was supported by the National Natural Science Foundation of China (Grant Nos. 11075025 and 11175029) and National Basic Research Program of China (Grant No. 2013CB834101).

[1] I. B. Bernstein, J. M. Greene, and M. D. Kruskal, Phys. Rev. **108**, 546 (1957).
[2] J. P. Holloway and J. J. Dorning, Phys. Rev. A **44**, 3856 (1991).
[3] V. B. Krapchev and A. K. Ram, Phys. Rev. A **22**, 1229 (1980).
[4] H. Schamel, Phys. Scr. **20**, 306 (1979).
[5] H. Schamel, Phys. Plasmas **7**, 4831 (2000).
[6] H. A. Rose and D. A. Russell, Phys. Plasmas **8**, 4784 (2001).
[7] T. M. O'Neil, Phys. Fluids **8**, 2255 (1965).
[8] D. S. Montgomery, R. J. Focia, H. A. Rose, D. A. Russell, J. A. Cobble, J. C. Fernandez, and R. P. Johnson, Phys. Rev. Lett. **87**, 155001 (2001).
[9] D. Turnbull, S. Li, A. Morozov, and S. Suckewer, Phys. Plasmas **19**, 083109 (2012).
[10] A. A. Kabantsev, F. Valentini, and C. F. Driscoll, AIP Conf. Proc. **862**, 13 (2006).
[11] F. Anderegg, C. F. Driscoll, D. H. Dubin, and T. M. O'Neil, Phys. Rev. Lett. **102**, 095001 (2009).
[12] F. Anderegg, C. F. Driscoll, D. H. Dubin, T. M. O'Neil, and F. Valentini, Phys. Plasmas **16**, 055705 (2009).
[13] Lj. Nikolić, M. M. Škorić, S. Ishiguro, and T. Sato, Phys. Rev. E **66**, 036404 (2002).
[14] N. J. Sircombe, T. D. Arber, and R. O. Dendy, Plasma Phys. Controlled Fusion **48**, 1141 (2006).
[15] L. Zhan-Jun, X. Jiang, Z. Chun-Yang, Z. Shao-Ping, C. Li-Hua, H. Xian-Tu, and W. Yu-Gang, Chin. Phys. B **19**, 075201 (2010).
[16] F. Valentini, T. M. O'Neil, and D. H. E. Dubin, Phys. Plasmas **13**, 052303 (2006).
[17] T. W. Johnston, Y. Tyshetskiy, A. Ghizzo, and P. Bertrand, Phys. Plasmas **16**, 042105 (2009).
[18] D. Bénisti and L. Gremillet, Phys. Plasmas **14**, 042304 (2007).
[19] R. R. Lindberg, A. E. Charman, and J. S. Wurtele, Phys. Plasmas **14**, 122103 (2007).
[20] B. Afeyan, K. Won, V. Savchenko, T. W. Johnston, A. Ghizzo, and P. Bertrand, "Kinetic electrostatic electron nonlinear (KEEN) waves and their interactions driven by the PF of crossing laser beams," in *Third International Conference on Inertial Fusion Sciences and Applications*, Monterey, CA, 7-12 September 2003, edited by B. Hammel, D. Meyerhofer, J. Meyer-ter-Vehn, and H. Azechi (American Nuclear Society, LaGrange Park, IL, 2004), Paper No. M034, p. 213.
[21] R. L. Berger, S. Brunner, T. Chapman, L. Divol, C. H. Still, and E. J. Valeo, Phys. Plasmas **20**, 032107 (2013).
[22] Laplace transform here, $g(p) = \frac{1}{2\pi i}\int_0^\infty e^{-pt}g(t)dt$, and inverse-Laplace transform $g(t) = \int_{\sigma-i\infty}^{\sigma+i\infty} e^{pt}g(p)dp$, where $\mathrm{Re}(p) = \sigma > \alpha$, α is growth index of $g(t)$.
[23] W. M. Manheimer and R. W. Flynn, Phys. Fluids **14**, 2393 (1971).
[24] G. J. Morales and T. M. O'Neil, Phys. Rev. Lett. **28**, 417 (1972).
[25] R. L. Dewar, Phys. Fluids **15**, 712 (1972).
[26] B. I. Cohen and A. N. Kaufman, Phys. Fluids **20**, 1113 (1977).
[27] J. W. Banks, R. L. Berger, S. Brunner, B. I. Cohen, and J. A. F. Hittinger, Phys. Plasmas **18**, 052102 (2011).

PHYSICS OF PLASMAS **22**, 052121 (2015)

Nonlinear evolution of stimulated Raman scattering near the quarter-critical density

C. Z. Xiao,[1] Z. J. Liu,[1,2] D. Wu,[1] C. Y. Zheng,[1,2,a)] and X. T. He[1,2]
[1]*HEDPS, Center for Applied Physics and Technology, Peking University, Beijing 100871, China*
[2]*Institute of Applied Physics and Computational Mathematics, Beijing 100084, China*

(Received 24 March 2015; accepted 8 May 2015; published online 26 May 2015)

Nonlinear evolution of stimulated Raman scattering (SRS) near the quarter-critical density is studied using one-dimensional (1D) and two-dimensional (2D) particle-in-cell simulations in homogeneous plasmas. In 1D configuration, with two-plasmon decay (TPD) instability excluded, the system evolves into two regimes distinguished by whether density cavities have been formed or not. At low temperatures, a cavity regime characterised by high absorption and low reflection, and at high temperatures nonlinear frequency shift regime due to particle trapping, are observed. Furthermore, a competition between SRS and TPD in 2D simulations reveals that the nonlinear SRS does play a significant role near the quarter-critical density, whose influences were mostly neglected before. © *2015 AIP Publishing LLC*. [http://dx.doi.org/10.1063/1.4921644]

I. INTRODUCTION

Laser plasma interactions (LPIs) are key issues in achieving the ignition goal in inertial confinement fusion (ICF). The interaction of nonrelativistic laser with the quarter-critical density plasma is an inevitable process in direct-drive ICF,[1] shock ignition,[2] and hybrid-drive ICF where indirect-drive and direct-drive processes are combined.[3,4] The physical complexity near the quarter-critical density is obvious since many instabilities can occur. In the past, the instability of most concern was the two-plasmon decay (TPD) instability, where an electromagnetic (EM) wave incident onto a quarter-critical density plasma decays into two Langmuir waves (LWs) (plasmons).[5] The probable energetic electrons produced from this instability can preheat the capsule which is hazardous to the ignition. Many theoretical[6] and experimental[7] works have been done in this field, recently.

However, stimulated Raman scattering (SRS) [the parametric decay of a laser into a scattered light wave and a Langmuir wave][8] is also a fairly significant instability in this region whose existence and effects are often underestimated. The reasons are as follows: First, SRS becomes an absolute instability near the quarter-critical density region. Second, the linear growth rates of SRS and TPD are comparable, although their inhomogeneous thresholds differ a lot, which makes TPD seemingly dominant in large scale length plasmas. Third, the SRS instability is more universal, while the TPD instability only happens in multi-dimensional configuration in a narrow density region near the quarter-critical. Moreover, the TPD instability occurs in the fixed plane consisting of the laser propagation direction and the laser polarization direction. Last, but the most important reason is the fact of the coexistence of SRS and TPD in many situations.[9–12] Above all, in this paper, we engage in revealing the significance of SRS near the quarter-critical density region.

Nonlinear saturation mechanisms of SRS, can include pump depletion, harmonic generation, particle trapping, Langmuir decay instability (LDI), cavity formation,[13,14] etc., we are most concerned with. In our circumstances, the cavity formation and the nonlinear frequency shift due to particle trapping are significant. The generation of density cavities with the corresponding self-trapped light was found in LPI experimentally[15,16] and numerically. These density cavities have been observed in various scenarios, including in the stimulated Brillouin scattering (SBS) instability at a density above the quarter-critical,[17–19] and in the oscillating two-stream instability (OTSI) at a density near the critical,[20] also, in ultra-short ultra-intense relativistic laser interaction with underdense plasmas.[21–23] The formation of the cavity was also reported near the quarter-critical density in inhomogeneous plasmas by using particle-in-cell (PIC) simulation briefly.[20,24] The density cavity prevents the SRS instability from developing, once cavities have been formed and it is an efficient energy absorption mechanism, as well. On the other hand, the kinetic effects in SRS, especially, the nonlinear frequency shift due to particle trapping,[25–28] are common for indirect-drive parametric regimes, however, they appear significant in our situation, as well.

Firstly, we restrict our consideration to an one-dimension, so that the effects of the SRS instability are studied solely. An 1D PIC code is used for analyzing the problem. In our paper, we analyze the cavitation process induced by SRS in detail near the quarter-critical density. We find that the cavity regime is robust in a large parameter region associated with the plasma density, laser intensity, and the electron and ion temperatures. If the laser intensity is high enough and/or temperatures are low enough, the system turns into the cavity regime in which the reflectivity is at low levels and absorptivity is pretty high. While, if the laser intensity is weaker and/or the temperatures are higher, there comes another regime. This regime, first found by Vu and Yin[26–28] in a lower density scenario, is dominated by the nonlinear frequency shift due to particle trapping, accompanied by the generation of the beam acoustic mode (BAM), which is similar to the electron acoustic wave[29] (EAW). The

a)zheng_chunyang@iapcm.ac.cn

reflectivity is enhanced in this regime. At the end of the paper, we conduct a realistic and simple study by two-dimensional (2D) PIC simulation in a homogeneous plasma near the quarter-critical density in order to verify the significance of SRS in the existence of the TPD instability.

The paper is organized as follows: In Sec. II, the physical model and simulation setup are discussed. Two different regimes distinguished by whether cavities are formed, are discussed in Secs. III and IV, in detail. In Sec. III, we present the regime of high laser intensity and/or low electron and ion temperatures when the density cavity can be generated by a strong SRS instability. The nonlinear evolution of SRS in that regime evolves to cavities. On the other hand, when laser intensity is low and/or temperatures are high, we find the nonlinear frequency shift of SRS which is discussed in Sec. IV. Some parametric analyses for these two regimes are presented in Sec. V. At last, a brief discussion about the importance of SRS in the two-dimensional configuration near the quarter-critical density is also given in Sec. VI. Section VII summarizes our main results.

II. PHYSICAL MODEL

Fully nonlinear relativistic PIC codes in one-dimension and two-dimensions are used for studying the problem. At first, in studies of the nonlinear evolution of SRS near the quarter-critical density, we utilize an 1D PIC code,[30] in order to eliminate the influence of the TPD instability which becomes important near the quarter-critical density. The importance of 2D effects in considering both SRS and TPD will be discussed briefly later, by using the 2D PIC code.[31] Throughout the paper, only homogeneous plasma cases are considered.

The simulation setup for 1D configuration is as follows. The initial homogeneous plasma density is $0.23n_c$, where n_c is the critical density for the incident laser light. The linearly polarized laser intensity in our simulations is $I = 2.4 \times 10^{15}\mathrm{W/cm^2}$ and the wavelength is $\lambda_0 = 1\,\mu\mathrm{m}$, the corresponding normalized vector potential is $a = 0.042$. This choice of laser intensity associates with the consideration that the higher laser intensity is used in direct-drive ICF or shock ignition than that used in indirect-drive. The quantity $I\lambda_0^2$ in our cases is $2.4 \times 10^{15}\mathrm{W}\mu\mathrm{m}^2/\mathrm{cm}^2$. The envelope of the laser is a flat-top, with an amplitude reaching a plateau in $5T_0$, where T_0 is the laser period, equals to 3.3 fs, and we use an infinite long pulse in our simulations. The simulation box in x-direction has a length of $120\,\mu\mathrm{m}$, and a plasma is uniformly distributed between $24\,\mu\mathrm{m}$ and $96\,\mu\mathrm{m}$. The plasma is isolated from both boundaries by a vacuum in order to study nonlinear evolution of plasmas without a boundary influence. The number of grids is 12 000 with a grid step $\Delta x = 0.01\lambda_0$, and the number of particles in each grid is 200. The simulation time is $10\,000T_0$, which corresponds to 33 ps with the time step, $\Delta t = 0.01T_0$. We use mobile ions with a realistic ions to electrons mass ratio, $m_i/m_e = 1836$. In the simulations, we use open boundary conditions for fields and absorbing boundary conditions for particles. As is well known that the condition of equilibrium between the plasma thermal pressure and the light pressure determines the formation of cavities, we choose two sets of temperature parameters representing two different regimes of nonlinear SRS evolution near the quarter-critical density. The low temperature regime of $T_e = 1\,\mathrm{keV}$ and $T_i = 0.2\,\mathrm{keV}$, where T_e and T_i are initial electron and ion temperatures, and the high temperature regime of $T_e = 3.5\,\mathrm{keV}$ and $T_i = 3.5\,\mathrm{keV}$.

For further discussion, in Sec. VI, we evaluate the SRS effects by using a 2D configuration. The parameters are similar to those of 1D simulations. The simulation boxes in longitudinal direction and transverse direction are $60\,\mu\mathrm{m}$ and $30\,\mu\mathrm{m}$, respectively. The temperatures of electrons and ions are $T_e = 1\,\mathrm{keV}$ and $T_i = 0.2\,\mathrm{keV}$ and we use a p-polarized plane electromagnetic wave with other parameters kept the same as in 1D case.

III. THE LOW TEMPERATURE REGIME: FORMATION OF CAVITIES

In this section, we present the case with $T_e = 1\,\mathrm{keV}$ and $T_i = 0.2\,\mathrm{keV}$. The boundary condition for particles in this process is chosen to be absorbing boundary condition, since we want to calculate the particle energy gain in this regime. However, it is a bit artificial since hot electrons would bounce back in the system due to the sheath potential at plasma-vacuum boundaries. In spite of this fact, the physical nature of this process remains.

A. Mechanism of cavity formation induced by SRS

The characteristics of SRS near the quarter-critical density are such that the wavenumber of the excited Langmuir wave is close to that of the original pump wave, while the wavenumber of the scattered wave is close to zero. In our case, the wavenumber of the pump in a homogeneous plasma ($n_0 = 0.23n_c$) is $k_0\lambda_D = 0.081$, where λ_D is Debye length. By applying the linear three-wave matching condition,[32] it is readily obtained that for stimulated Raman backscattering (SRBS): $k_{bLW}\lambda_D = 0.098$, $k_{bs}\lambda_D = -0.017$, and for stimulated Raman forward scattering (SRFS): $k_{fLW}\lambda_D = 0.063$, $k_{fs}\lambda_D = 0.018$ (the subscript *bs* and *fs* represent backward scatter and forward scatter, respectively; *bLW* and *fLW* represent the corresponding LWs, and the sign "−" denotes the direction opposite to the pump wave). Since the growth rates of these two instabilities are close[8] near the quarter-critical density, both scattered waves intensities are observed, as shown in Fig. 1. The SRBS occurs at $t = 200T_0$ then SRFS emerges sequentially. After continuous bursts about $700T_0$, the scatterings are suppressed due to the formation of cavities.

The mechanism of the formation of cavities near the quarter-critical density regime is simple. As Langdon[20] first found this phenomenon in 2D PIC simulation, we interpret the formation of cavity as the equilibrium between the plasma thermal pressure and the ponderomotive pressure of the scattered light. Since the density is near the quarter-critical, the laser frequency is almost twice the initial plasma frequency ω_{pe}, so that the scattered wave frequency is just a little bit above ω_{pe}. The scattered light can be readily trapped by the local density fluctuation. If the intensity of scattered wave is large enough, its ponderomotive force pushes the

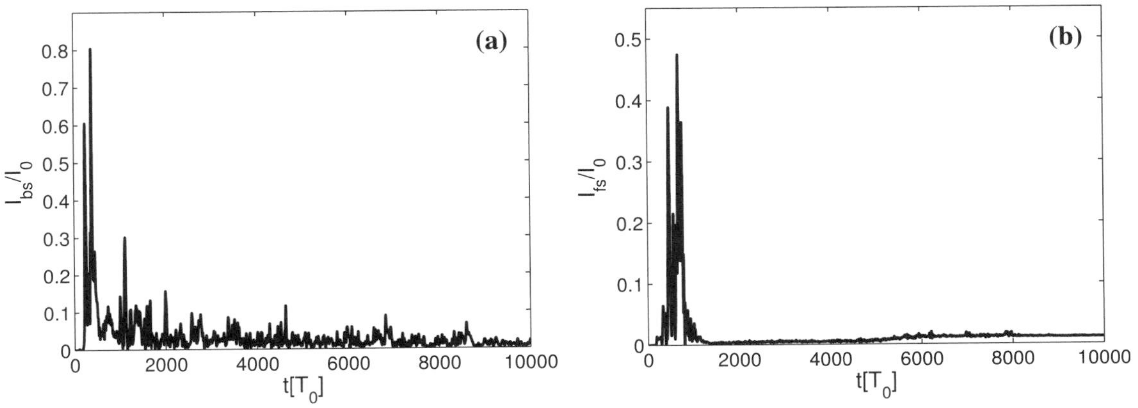

FIG. 1. The intensity variation of the backscattered (a) and forward scattered (b) light.

local density away and pile up around it. Then density cavity and the corresponding electromagnetic soliton form. This process plays a saturation and stagnation mechanism for SRS as well as SBS.[18] We thus summarise the process and write the cavity formation condition as

$$\frac{e^2E_s^2}{4m_e\omega_s^2} > \frac{|\delta n|}{n_0}(T_e + \gamma T_i), \tag{1}$$

where e, E_s, and ω_s are electron charge, scattered wave amplitude, and frequency, respectively; n_0 and δn are initial density and density fluctuation, respectively; and γ is the adiabatic index for ions. From Eq. (1), we know that the regime of the system is related to the laser intensity, plasma density, and temperatures. Therefore, when the laser intensity is high or temperatures are low, there appears the cavity regime, otherwise no cavity forms and the system will evolve into another regime, which is discussed in Sec. IV.

B. Laser-caviton interaction

Once the cavities are formed, nonlinear SRS evolution in this regime stops. The system from a parametric decay process turns into laser-caviton interaction in which the laser transfers its energy to particles, as shown in Fig. 2. Fig. 2(a) shows the spatio-temporal evolution of the ion density. The figure for electron density variation, which is not shown, is similar to the ion's, which means that while cavity forms, the quasi-neutrality condition is satisfied. Fig. 2(b) shows the Poynting flux in x-direction, the initial maximum Poynting flux of a laser pump is $1.8 \times 10^{-3} m_e^2\omega_0^2c^3/e^2$. Thus, we find that when the cavity has been formed, laser energy gets absorbed by particles which makes that the majority of the laser pulse cannot penetrate the plasma until the laser-caviton interaction is over. The cavitation process is similar to that induced by SBS.[18] We summarise this process in three aspects.

1. Cavity formation and initial acceleration of electrons and ions

We have picked up a representative cavitation process in Fig. 2, and its electron, ion density, and longitudinal field variation are shown in Fig. 3. When the cavity starts to form, the ponderomotive force of the scattered light first pushes away the electrons and then the charge separation forms a dipole electric field as we see in Fig. 3(c). This dipole field sustains for several hundreds of laser periods until the

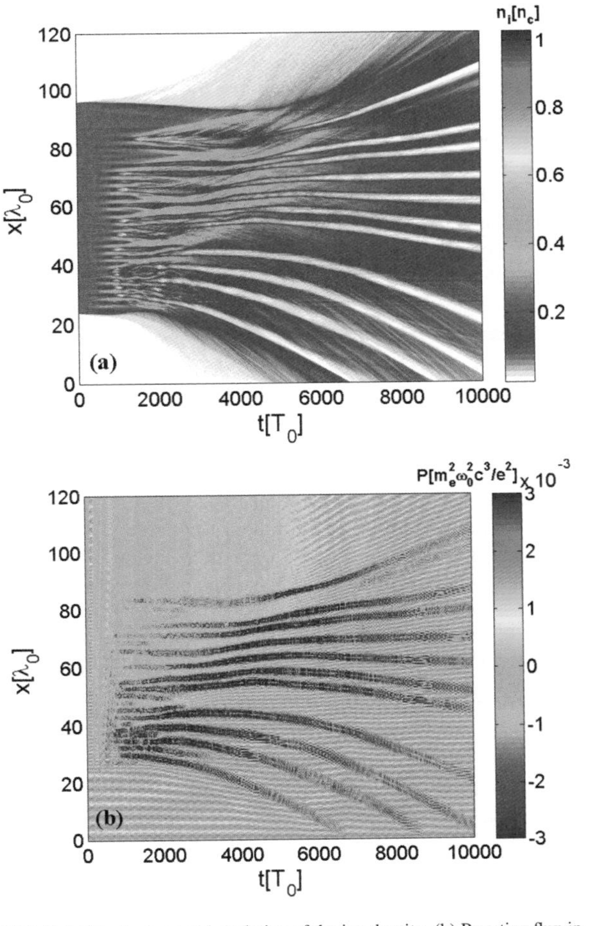

FIG. 2. (a) Spatio-temporal evolution of the ion density. (b) Poynting flux in x-direction.

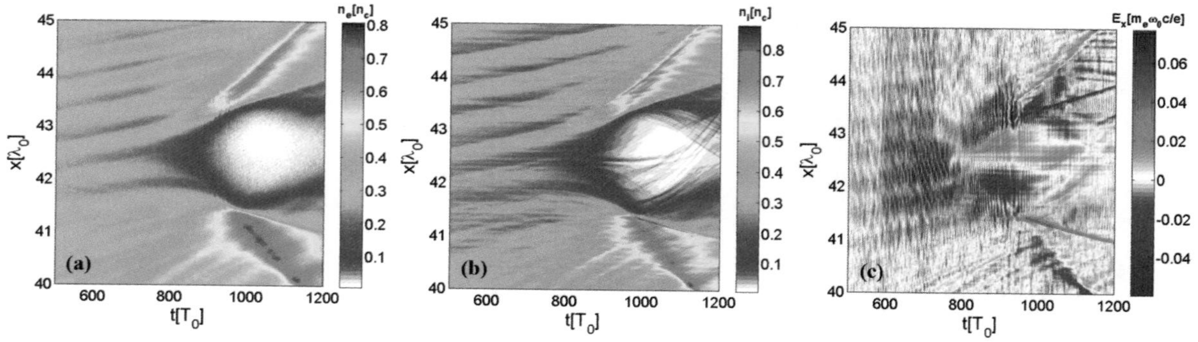

FIG. 3. Figures at the moment of caviton formation. (a) Electron density, (b) ion density, and (c) longitudinal field E_x.

cavities have been completely formed. The cavity dynamics is universal and some authors have predicted the distribution analytically.[33,34] Particles in the cavity gain energy from the ponderomotive force and the longitudinal field. It is not shown here that ions in the cavity are accelerated through collisionless shock wave in forward and backward directions, as discussed by Weber *et al.* in their paper.[18] It is also observed that four ion beams are accelerated in a cavity to the velocity of tens of the initial sound velocity. However, acceleration of electrons is not so obvious, since electrons have been already accelerated in the early SRS stage.

The size of the cavities is almost the same and they are little changing in our simulations. It means that the density cavities are rather stable in one-dimensional configuration which is different from the wave collapse model derived from Zakharov equation,[35] the reason is the dissipation by wave-particle interaction.[36] We observe that the scattered light trapped in the cavity is a half-cycle EM soliton with the group velocity close to zero. The structure of these EM solitons is reported both in nonrelativistic laser[37] and relativisitic laser cases.[38–40] The maximum value of the electric field of the EM soliton reaches about $0.4m_e\omega_0c/e$, so that it is relativistic EM soliton and the amplitude is very large as compared to the laser intensity. The size of EM solitons is about $2\pi c/\omega_{pe}$ so in our quarter-critical case, $\omega_{pe} \approx \omega_0/2$, the EM solitons, as well as the cavities, have the size of $2\lambda_0$. It is also found that the corresponding trapped frequency of the EM soliton downshifts from initial $\omega_0/2$ to around $\omega_0/4$ and the product of cavity size and trapped frequency is constant in the evolution, typical for the nonlinear Schrödinger type of EM solitons.[41]

2. Energy absorption due to laser-caviton interaction

In order to make clear where the laser energy goes, we analyze each energy component in the simulation. It is obvious that the system is energy conserved with the absorbing boundary condition. We show temporal evolutions of each energy component in Fig. 4. The total laser energy is divided into three components: backscattered light which escaped from the left boundary, transmitted through the right boundary and energy deposited in the simulation box, as shown in Fig. 4(a). The average reflectivity over $10\,000T_0$ is only 3.7%, while the average absorptivity reaches, as high as, 37.5%. Moreover, we note that after about $6000T_0$, the deposited energy does not grow anymore which means that the laser-caviton interaction is over. The high level of the laser energy deposited in the simulation box transforms into particle energy gain and field energy gain. Fig. 4(b) shows the average particle energy and Fig. 4(c) denotes the corresponding average field energy density. The initial mean

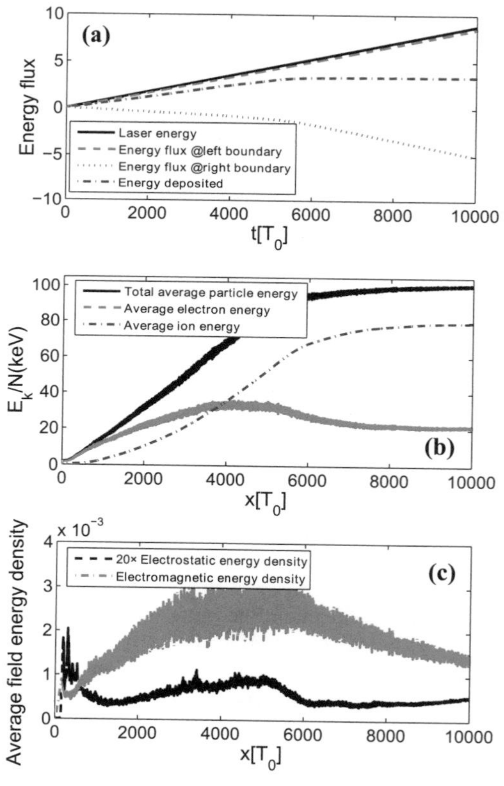

FIG. 4. (a) Laser energy into the simulation box: total energy (black solid), net energy into the left boundary (red dashed), net energy out of right boundary (blue dotted), and energy deposited (brown dashed-dotted). (b) Average particle energy of all particles (black solid), electrons (red dashed), and ions (blue dashed-dotted). (c) Average field energy density: electrostatic energy density (black dashed) which is amplified 20 times and electromagnetic energy density (red dashed dotted). Note that energy is conserved all the time.

electron and ion energy are 1.5 keV and 0.3 keV, respectively. During the SRS process ($100T_0$ to $900T_0$), the electrons gain energy from wave-particle interaction. Later, when cavities have been formed, both electrons and ions gain much energy from the laser, and the total particle energy augment increases linearly. In Fig. 4(c), we find that the average electromagnetic field energy density increases during the laser-caviton interaction due to increasing localized energy of self-trapped scattered light and so does the electrostatic field energy density.

It is worth noticing that more than 90% of the deposited energy converts into particles so that the formation of cavities is an efficient absorbing mechanism. Fig. 5(a) shows the electron energy spectra at three different times. The initial Raman heated electron spectrum (the black line at $t = 1000T_0$) has a classical bi-Maxwellian distribution, from which we found a hot electron tail with the effective temperature $T_h \approx 89.8$ keV. This temperature is a result of interaction between SRFS, which generates hot electrons with energy $m_e v_{fLW}^2/2 \approx 126$ keV, and SRBS, which generates hot electrons with energy $m_e v_{bLW}^2/2 \approx 52$ keV, where v_{fLW} and v_{bLW} are phase velocity of Langmuir wave of forward scattering and backward scattering, respectively. When the cavities have been formed in the system, laser energy is largely absorbed by particles. In the moment of $t = 5000T_0$, near the end of laser-caviton interaction, we find that a large number of electrons is accelerated by this mechanism. It is easy to obtain that the hot electron temperature at this moment is about 73.8 keV. Finally, at $t = 8000T_0$, the hot component is cooled down to a effective temperature of 47 keV. After that, the electron spectrum is unchanged. Meanwhile, it is found that the main bulk of the electron and ion velocity distribution functions is broadened during the laser-caviton interaction, as we see in the temporal evolution of electron and ion kinetic energy averaged over the low energy component (Figs. 5(b) and 5(c)), and then stop increasing after that. Unlike the resonant wave-particle interaction, the whole plasma is heated by this process. The physical nature of the heating mechanism is not discussed in this paper and will be reported elsewhere. According to above statistics, we can conclude that the laser-caviton interaction not only is an efficient heating mechanism but also generates super hot electrons.

3. Scale effects

In this subsection, we have shortened the simulation box length from 120 μm to 32.4 μm in order to study the scale effects in the cavitation regime. The ion density variation and the Poynting flux in x-direction are shown in Fig. 6. Compared with Fig. 2, we find that the cavitation processes with different scale lengths result in the same cavity size, $2\lambda_0$. Thus, with an infinite laser energy available to supply the trapped light, the plasma length basically determines a maximum number of cavities generated in a system. In the large scale case, our system generates nearly 20 cavities at the beginning and some cavities merge to form a larger and more stable cavity, so that the cavity number is 11, finally. While for a small scale case, only three cavities are generated which leads to different absorption efficiency. It is seen that the SRS instability is suppressed at $t = 1000T_0$ which makes that reflectivities at both scales are of the same order (6.7% for the large scale and 6.3% for the small scale, averaged over $6000T_0$). As seen in Fig. 6(b), it takes less time for a laser to totally penetrate through the plasma, so that the absorptivities in two cases differ a lot. The averaged absorptivity is 59.3% for the large scale and 8.5% for the small scale, however, the averaged absorptivities per cavity are of the same order: 5.4% for the large scale and 2.8% for the small scale. The reason of the factor of two may be as the cavity number in the large scale at the early stage is twice of that in the final stage.

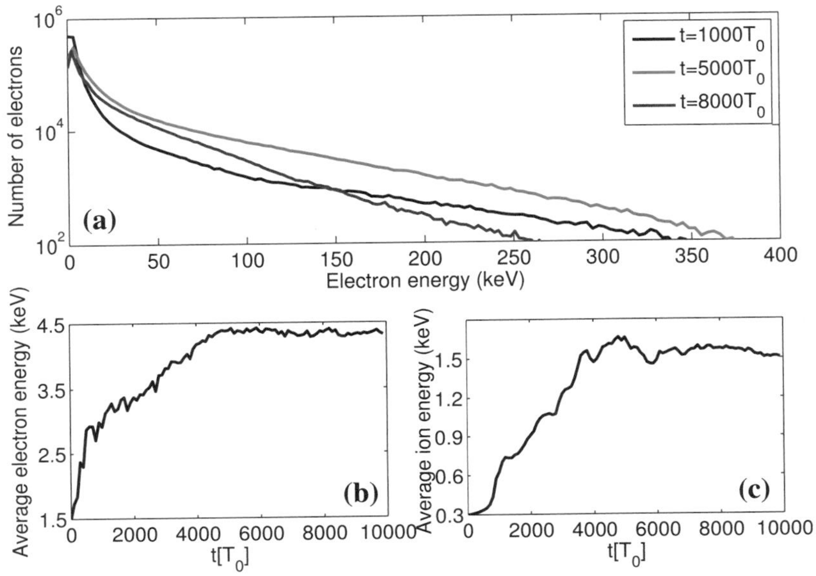

FIG. 5. Electron energy spectra at three different times (a): $t = 1000T_0$ (black line), $t = 5000T_0$ (red line), $t = 8000T_0$ (blue line). (b) Temporal evolution of electron energy averaged over electrons with the energy less than 12 keV. (c) Temporal evolution of ion energy averaged over ions with energy less than 5 keV.

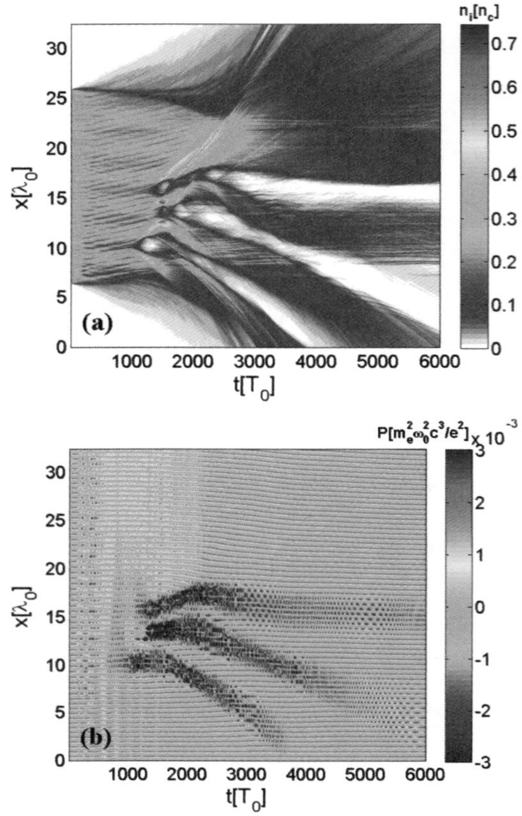

FIG. 6. Ion density variation (a) and Poynting flux in x-direction (b) in small scale simulation ($L_x = 32.4\,\mu$m) with other parameters kept the same as for the large scale.

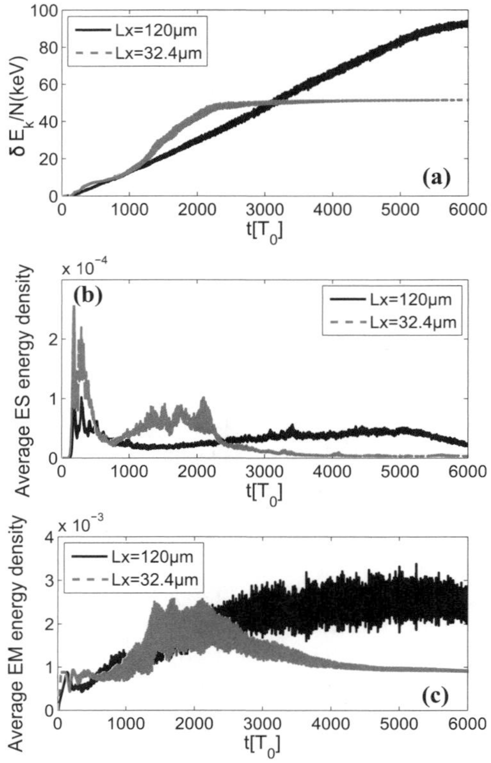

FIG. 7. The comparison of large scale ($L_x = 120\,\mu$m, black solid line) and small scale plasma ($L_x = 32.4\,\mu$m, red dashed line). (a) Total average particle energy gain. (b) Average electrostatic field energy density. (c) Average electromagnetic field energy density.

Furthermore, we present comparisons of the absorbed energy for particles, electrostatic energy density, and electromagnetic energy density for two different scales in Fig. 7. It is found that the fields with different scales are of the same order which makes that the particle gain the same order of the average energy. Thus, we conclude that under the same condition of infinite laser energy supply system, the plasma length basically determines the cavity number so that it will affect the energy absorption, eventually.

IV. THE HIGH TEMPERATURE REGIME: NONLINEAR FREQUENCY SHIFT DUE TO PARTICLE TRAPPING

In this section, we consider the case with $T_e = 3.5\,$keV and $T_i = 3.5\,$keV. As known from Eq. (1), the plasma thermal pressure can balance the ponderomotive force of the scattered wave so if the temperatures are high enough, no cavity forms during the initial SRS process. Therefore, we observe the nonlinear SRS evolution near the quarter-critical density at high temperature regime. Owing to the high ratio of the electron temperature to the ion temperature ($T_e/T_i = 1$), the ion acoustic wave is highly Landau damped so that SBS and the LDI are of less significance in this regime.

In an one-dimensional configuration, only forward scatter and backward scatter of SRS are allowed, thus we calculate the linear matching condition for the initial homogeneous density, $n_0 = 0.23n_c$. For the pump wave, the wavenumbers in plasma and the frequency are $k_0\lambda_D = 0.15$ and $\omega_0/\omega_{pe} = 2.085$; for SRBS, the wavenumbers of Langmuir wave and backscatter are $k_{bLW}\lambda_D = 0.175$ and $k_{bs}\lambda_D = -0.024$; for SRFS, the wave numbers of Langmuir wave and forward scatter are $k_{fLW}\lambda_D = 0.12$ and $k_{fs}\lambda_D = 0.03$. The fact that near the quarter-critical density the wavenumber of the scatter is much less than that of the pump wave makes that $k_{LW}\lambda_D$ solely depends on the electron temperature. Thus, it is estimated that the system turns into a kinetic regime ($k_{LW}\lambda_D \gtrsim 0.3$)[42] only when $T_e \gtrsim 10\,$keV.

The boundary conditions for particles in this regime are still chosen to be absorbing ones with vacuum at both boundaries. If we choose the thermal boundary condition, it will accumulate particles near each boundary, especially at the laser entrance boundary, which may break the balance condition and form cavities artificially. Other parameters are the same as in the low temperature case. The nonlinear evolution of SRS lasts for a long time, as seen in the temporal evolution of reflectivity in Fig. 8. The reflectivity of SRS shows continuous bursts from $t=0$ to $5000T_0$ then it stagnates for several hundreds of laser periods, followed by another continuous bursts period until it quenches at the moment $t = 9000T_0$. Two continuous bursts indicate that the nonlinear SRBS dominates in the whole simulation. The time averaged reflectivity in this regime reaches the value of 19.8%,

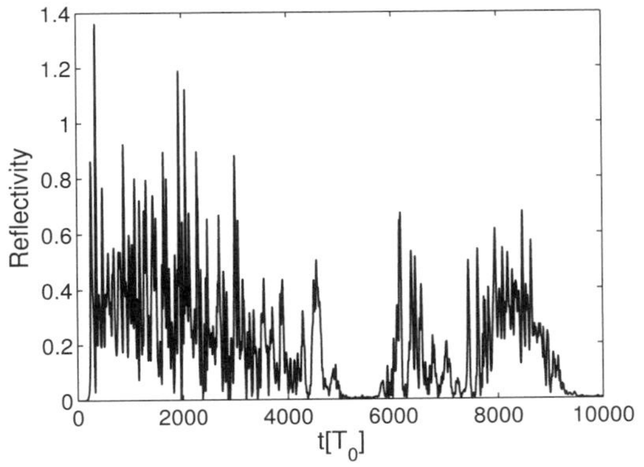

FIG. 8. Temporal evolution of the reflectivity.

as compared with that of cavity regime, which is just as low as 3.7%. It implies that strong intermittent nonlinear SRS occurs in this regime and then enhances the reflectivity, drastically. Meanwhile, 21.5% of the energy is absorbed by particles and fields, while the percentage is 37.5% in the cavity regime. The hot electron temperature heated by the nonlinear SRS in this regime is as high as 47 keV, eventually.

It is important to understand the nonlinear saturation mechanism of SRS in this regime. We realize that though $k_{LW}\lambda_D = 0.175$, LDI plays a negligible role in saturation due to strong Landau damping of the ion acoustic wave. Through the spectrum analysis, we interpret the saturation mechanism as the nonlinear frequency shift due to particle trapping which is shown in Fig. 9. We separate the process into two stages, the first bursts from $t = 0$ to $5000T_0$ and the second bursts from $t = 6000T_0$ to $9000T_0$. Figs. 9(a) and 9(b) depict the spectra of the longitudinal field E_x and the transverse field E_z, respectively. In Fig. 9(a), in addition to the LW dispersion relation, the linear LW excited by backward scattering ($k_{LW}\lambda_D = 0.175, \omega_{LW}/\omega_{pe} = 1.045$) suffers a frequency downshift sequentially, denoted by a "streak" along with acoustic like modes denoted by a "BAM." This phenomenon was first found in experiments,[43] Vu and Yin interpreted theoretically as a nonlinear frequency shift due to particle trapping and generation of BAMs with a flattening of the velocity distribution function in the context of lower density, lower temperatures, and lower laser intensity.[26–28] The difference between our and Vu's case is the kinetic quantity, $k_{LW}\lambda_D$, because of higher intensity, the required $k_{LW}\lambda_D$ for kinetic effect to be important is smaller than that required in the indirect-driven scenario. The corresponding scattered wave spectrum is shown in Fig. 9(b), where we find that the backscatter suffers a frequency upshift, while the forward scatter is of less significance in this regime.

As the LW frequency downshifts and the backscatter frequency upshifts, the reflectivity decreases gradually. After a short time break among the two bursts, the distribution function recovers to the Maxwellian and then another SRS bursts come about. Figs. 9(c) and 9(d) show the spectra of these bursts. Since the plasma temperatures are increased in the first stage, the initial reexcited LW has the wavenumber of $k'_{LW}\lambda_D = 0.15$ and the wave frequency of $\omega'_{LW}/\omega_{pe} = 1.14$ (red circle in Fig. 9(c)). Similarly, this new excited mode downshifts in frequency domain along with BAMs, however, the duration of the bursts becomes shorter and intensity becomes weaker. At the end of the bursts ($t = 9000T_0$), we again observe that a cavity has been formed in the density space, this phenomenon is also observed in the E_z spectrum denoted by "trapped light" which is again downshifted to the quarter of the laser frequency. The formation of the cavity stops the nonlinear

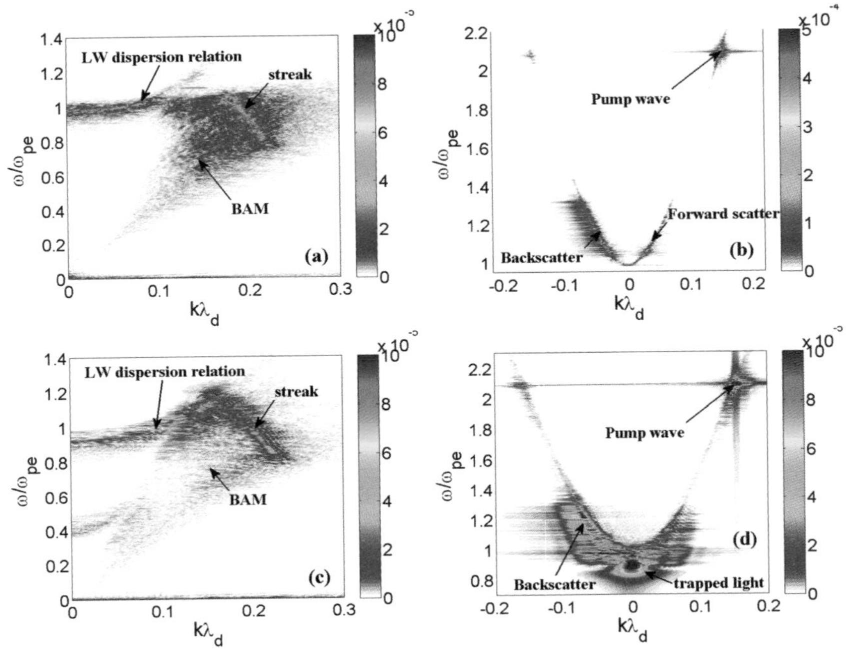

FIG. 9. Spectra of longitudinal field E_x ((a) and (c)) and transverse field E_z ((b) and (d)) average over $0 - 5000T_0$ ((a) and (b)) and $5000T_0 - 9000T_0$ ((c) and (d)).

SRS evolution then the system evolves along the way that we discussed in Sec. III.

V. PARAMETRIC STUDY

The condition of cavity formation, Eq. (1), determines the regime of the system. However, it is difficult to obtain an exact quantitative description for determining the cavity formation due to the complex evolution of E_s. Generally speaking, the condition of forming a cavity near the quarter-critical density should involve four aspects: (i) The initial density should not exceed quarter-critical density, since there is no SRS existing above that. (ii) The closer to the quarter-critical density the easier it is to form cavities. (iii) The electron and ion temperatures which relate to the plasma thermal pressure should be low. (iv) Since the amplitude of scattered wave has a positive correlation with the laser intensity, it is predictable that high intensity is beneficial to form cavities.

In this section, we present studies of two parameters: the temperatures and the laser intensity. The reflectivities and corresponding absorptivities vs temperatures are shown in Fig. 10 for $a = 0.042$ and $n_0 = 0.23n_c$ averaged over $6000T_0$. The cavity regime is characterised by the low reflectivity and high absorptivity. As seen in Figs. 10(a) and 10(b), the reflectivities at the level under 10% and the absorptivities above 50% belong to the cavity regime. In this regime, the reflectivity decreases as the temperatures increase since the Landau damping of the Langmuir wave and the ion acoustic wave is getting stronger, while the absorptivity changes a little due to a relative stable number of cavities. As the temperatures continue to increase, the plasma thermal pressure becomes larger than the ponderomotive force, so that we find, as the data in brown and black circle in Fig. 10 shows, that the reflectivity increases to about 20% and the absorptivity drops to 20% dramatically. The nonlinear frequency shift in SRS dominates in these high temperature parametric regions. It is also found that in the high temperature regime, the effects of ion acoustic wave are of less significance even when $T_e/T_i = 5$. The parameter determining the regimes only relates to the total thermal pressure of electrons and ions, i.e., $T_e + \gamma T_i$.

The laser intensity dependence of reflectivities and corresponding absorptivities is shown in Fig. 11 for $n_0 = 0.23n_c$. In Fig. 11(a), we find that all the data for $T_e = 1\,\text{keV}, T_i = 0.2\,\text{keV}$ have generated density cavities, and in which, as the laser intensity increases, the reflectivity reduces since it takes less time to form a cavity at higher

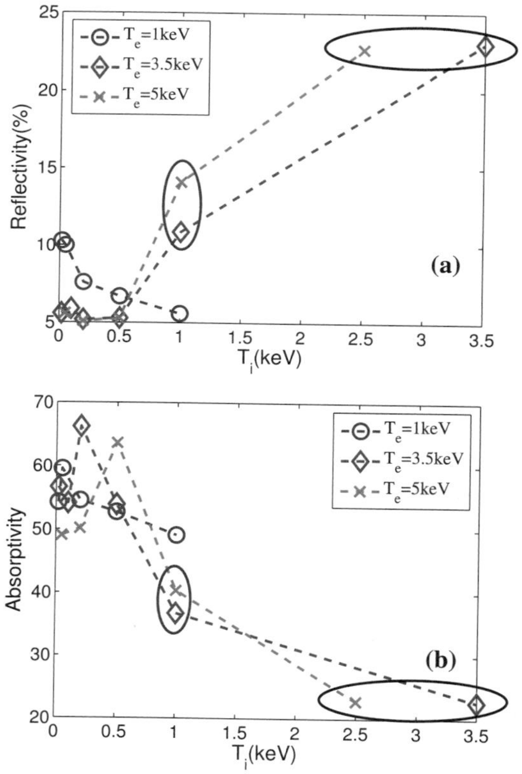

FIG. 10. Reflectivities and corresponding absorptivities vs temperatures at $a = 0.042$ and $n_0 = 0.23n_c$. The reflectivities and absorptivities are averaged over $6000T_0$. The data without circle represent the system forming cavities, while the brown circles represent that cavities are formed later than the former ones and the black circles represent that the system has evolving with no cavity.

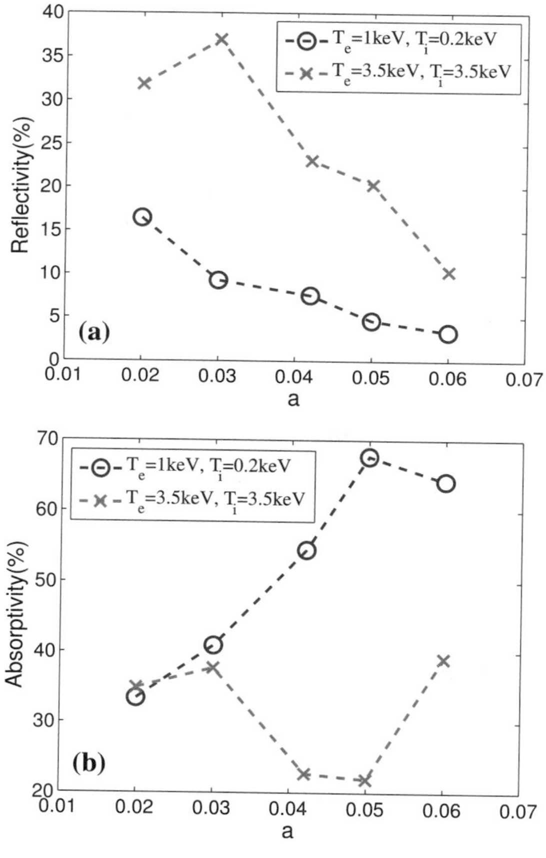

FIG. 11. Laser intensity dependence of reflectivities and corresponding absorptivities at $n_0 = 0.23n_c$ with two sets of temperatures: $T_e = 1\,\text{keV}$, $T_i = 0.2\,\text{keV}$ and $T_e = 3.5\,\text{keV}, T_i = 3.5\,\text{keV}$. The reflectivities and absorptivities are averaged over $6000T_0$.

intensities. However, in the set of $T_e = 3.5\,\mathrm{keV}, T_i = 3.5$ keV, only when $a = 0.06$, the cavity forms, the reflectivity also decreases with laser intensity due to the enhancement of the trapping effects. On the other hand, as we see in Fig. 11(b), the absorptivity increases with the laser amplitude when the system is in the cavity regime and saturates at the level of 60% due to the limitation of the size of the cavity and the system.

VI. 2D SIMULATION OF SRS AND TPD IN A HOMOGENEOUS PLASMA

We have studied above the nonlinear SRS evolution near the quarter-critical density in one-dimension, however, it would ignore the multidimensional effects which can be significant due to multi-instabilities, such as TPD and filamentation. In this section, we simulated a simple case in two-dimensions in order to understand the behavior of SRS and TPD simultaneously. When the laser is s-polarized, it is easy to observe a nonlinear SRS behavior of forming cavities as we discussed in Sec. III, since the TPD instability is excluded.[20] To include the TPD instability in our simulation, p-polarized laser (in z-y plane, where z is longitudinal direction) is chosen. The simulation box has the domain of $L_z = 60\,\mu\mathrm{m}$ and $L_y = 30\,\mu\mathrm{m}$ with the grid number $N_z = 2000$ and $N_y = 1000$. The plasma is uniformly distributed in $[5\,\mu\mathrm{m}, 55\,\mu\mathrm{m}] \times [-15\,\mu\mathrm{m}, 15\,\mu\mathrm{m}]$ with vacuum remaining on both sides. The plasma density is $0.23n_c$ and the temperatures are $T_e = 1\,\mathrm{keV}$ and $T_i = 0.2\,\mathrm{keV}$, with 80 particles per cell. The boundary condition is periodic in transverse and absorbing in longitudinal direction. We use a plane wave with a normalized amplitude $a = 0.042$ with the wavelength $\lambda_0 = 1\,\mu\mathrm{m}$. Other parameters are the same as for 1D simulation.

The longitudinal field energy variation is presented in Fig. 12. We find that in a homogeneous plasma, the TPD instability is first excited at $t = 100T_0$, the field energy increases from zero to the peak "a," where the TPD instability

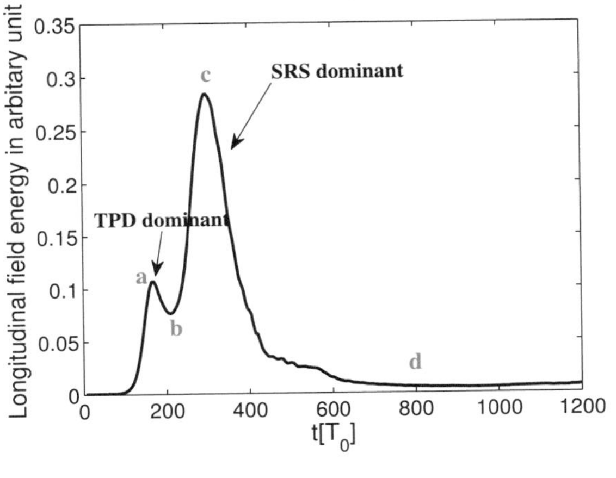

FIG. 12. The longitudinal field (E_z) energy variation in two-dimension homogeneous plasma simulation considering the effects of both SRS and TPD. The labels "a," "b," "c," and "d" represent the stage of TPD dominant, coexistence of TPD and SRS, SRS dominant and disappearance of both instabilities, respectively. No cavities are formed at the end of simulation.

gets saturated. The saturation mechanism is suppression by SRS. The corresponding longitudinal field spectrum at that moment is shown in Fig. 13(a), in which we find the signals of four excited LWs located along the maximum growth rate curves of the TPD, the black dashed hyperbola. So we can infer these modes to be Langmuir waves excited by TPD. In addition, in the center of the figure, there are weak signals of SRBS and SRFS with the wavenumber ($k_{bz}\lambda_D = 0.098, k_{by}\lambda_D = 0$) and ($k_{fz}\lambda_D = 0.063, k_{fy}\lambda_D = 0$). From Fig. 13(a), we realize that the scale of oblique density fluctuation (λ_{TPD}) is smaller than that of normal density fluctuation (λ_{SRS}) so that the SRS is easy to take control over the TPD. Fig. 13(b) indicates the coexistence of SRS and TPD corresponding to the moment labeled "b" in Fig. 12. The SRS signal is intensified while that of TPD weakens. Another peak in Fig. 12, point "c," represents that the system is dominated by SRS. In the corresponding spectrum, Fig. 13(c), we find both strong signals of SRBS and SRFS instead of TPD. The TPD oblique modes fade away due to the modification of density by the fluctuations in SRS. This behavior is natural for a homogeneous and the quarter-critical density region, as seen in other similar simulations, so that the effects of SRS in a homogeneous plasma are more significant than in large scale length system, as studied by many authors related to indirect-drive or shock ignition scenario.[6] While the time evolution continues, the SRS fades away as seen in Fig. 13(d) at $t = 800T_0$. This is perhaps due to the strong inhomogeneity of density fluctuations caused by the SRS and TPD together which then prevent the SRS instability from growing.

At the end of the simulation, we find no cavities formed which may be probably due to modification of transverse density fluctuation that no scattered light can be trapped in the local 2D density distribution. In this 2D simulation, although simple, it still reveals that both SRS and TPD instabilities occur in a homogeneous, near the quarter-critical density region, moreover, the TPD instability gets easily suppressed by the SRS instability, since the scale of SRS is large. This result reminds us to reconsider the effects of the SRS and its competition with the TPD instability.

VII. DISCUSSION AND CONCLUSIONS

The purpose of this paper was to draw the attention to reevaluate the significance of SRS near the quarter-critical density region. The influence of the TPD instability in generation of hot electrons in large scale length plasmas, which is related to the direct-drive ICF and the shock ignition is important; however, other instabilities, such as SRS, SBS, filamentation, are competing with TPD at the same time. Very recently, the multi-instability effects, including SRS, TPD, and SBS, were studied in the context of the shock ignition.[11] In other situations, such as homogeneous plasma or inhomogeneous plasma where the threshold of SRS is comparable with that of TPD, the SRS instability is probably more significant, as we see in our 2D homogeneous case.

To summarize our paper, we started from the 1D configuration. Although multi-dimension effects are excluded, the physical nature of the nonlinear SRS evolution has been mostly preserved. The regime of the system is distinguished

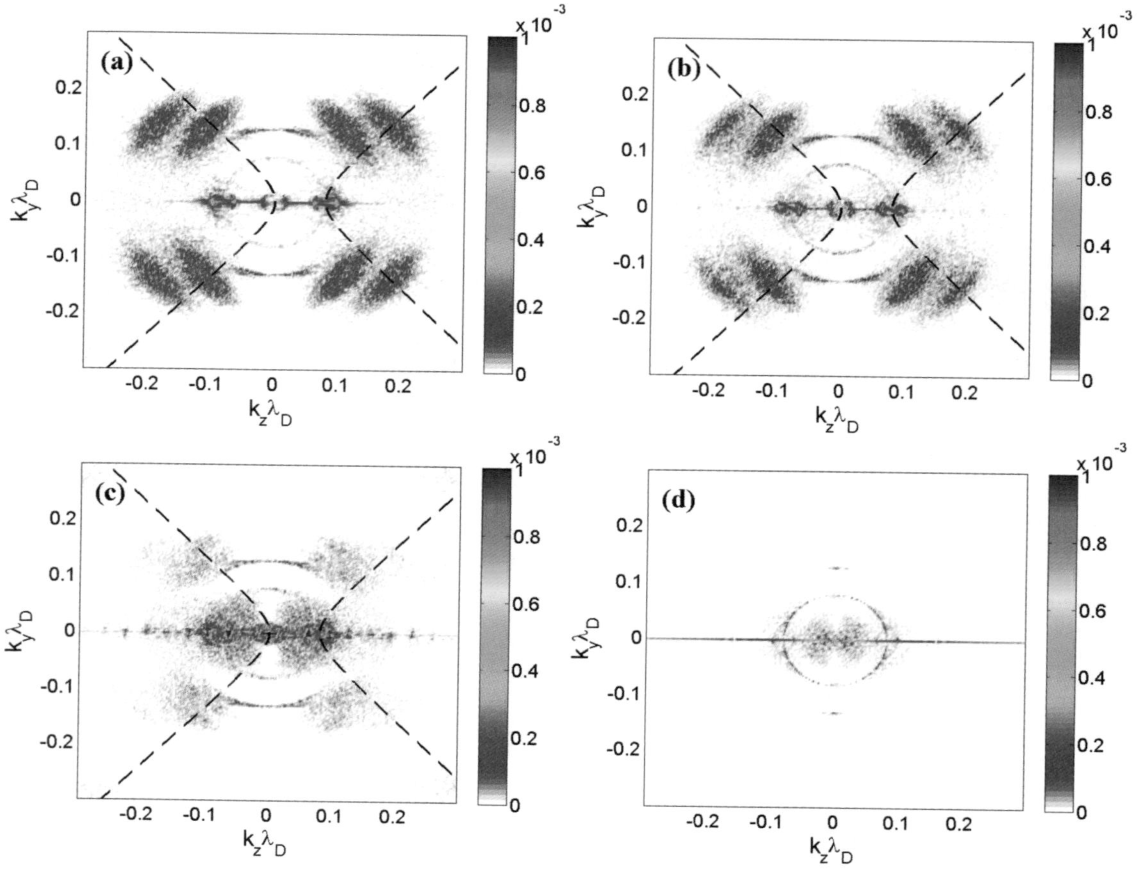

FIG. 13. The spectra of longitudinal electric field, $E_z(k_z, k_y)$, in k-space. (a)–(d) are snapshot at $t = 170T_0, 210T_0, 300T_0$, and $800T_0$ corresponding to the moment labeled "a, b, c, d" in Fig. 12, respectively. The black dashed hyperbola represents the location of maximum TPD growth rate.

by whether the cavity has been formed. One readily gets the general result from Eq. (1) that when the laser intensity is high and/or the temperatures are low, the system turns into the cavity regime, otherwise the nonlinear frequency shift due to particle trapping dominates. We have done 1D PIC simulations to verify this conjecture. In Sec. III, the low temperature regime of $T_e = 1\,\text{keV}, T_i = 0.2\,\text{keV}$ was discussed. We find that once cavities have been formed, the SRS instability stagnates with the reflectivity being at a low level. The laser-caviton interaction process transfers the laser energy to particle energy of both species. The rather large absorptivity implies that generated cavities do play a significant anomalous heating mechanism near the quarter-critical density, just like that one, near the critical surface. The absorbed energy is determined by a number of generated cavities which is related to the plasma length. In Sec. IV, the nonlinear frequency shift case is presented for $T_e = 3.5\,\text{keV}, T_i = 3.5$ keV. The initial SRBS mode suffers a frequency downshift due to the particle trapping with accompanying BAM modes due to flattening of the velocity distribution function. The reflectivity increases, drastically. Having experienced two long-duration bursts periods, the system generates a cavity, finally. In the studied parameter space, the saturation mechanisms of SRS should be the cavitation and the nonlinear frequency shift. Some parametric studies associated with the temperatures and laser intensity effects are accomplished in Sec. V.

We closed our paper with a simple homogeneous case that considers both SRS and TPD simultaneously by using 2D PIC simulation. The result is that TPD occurs first and soon SRS develops, suppressing the TPD instability. Finally, both instabilities disappear due to density modifications. This simple case reminds us of the significance of SRS existence in 2D configuration near the quarter-critical density. More competitions between SRS and TPD, and other instabilities, will be studied in multi-dimensions in the future.

ACKNOWLEDGMENTS

We are pleased to acknowledge useful discussions with Miloš M. Škorić. This research was supported by the National Natural Science Foundation of China (Grant Nos. 11175029, 11475030, and 11175030) and National Basic Research Program of China (Grant No. 2013CB834101).

[1]S. E. Bodner, D. G. Colombant, J. H. Gardner, R. H. Lehmberg, S. P. Obenschain, L. Phillips, A. J. Schmitt, J. D. Sethian, R. L. McCrory, W. Seka, C. P. Verdon, J. P. Knauer, B. B. Afeyan, and H. T. Powell, Phys. Plasmas **5**, 1901 (1998).
[2]R. Betti, C. D. Zhou, K. S. Anderson, L. J. Perkins, W. Theobald, and A. A. Solodov, Phys. Rev. Lett. **98**, 155001 (2007).
[3]Z. F. Fan, M. Chen, Z. S. Dai, H. B. Cai, S. P. Zhu, W. Y. Zhang, and X. T. He, e-print arXiv:1303.1252 [physics.plasm-ph].
[4]X. T. He, Z. F. Fan, J. W. Li, J. Liu, K. Lan, J. F. Wu, L. F. Wang, and W. H. Ye, e-print arXiv:1503.08979 [physics.plasm-ph].
[5]M. V. Goldman, Ann. Phys. N.Y. **38**, 117 (1966); C. S. Liu and M. N. Rosenbluth, Phys. Fluids **19**, 967 (1976); A. Simon, R. W. Short, E. A. Williams, and T. Dewandre, Phys. Fluids **26**, 3107 (1983); A. B. Langdon, B. F. Lasinski, and W. L. Kruer, Phys. Rev. Lett. **43**, 133 (1979); D. F. DuBois, D. A. Russell, and H. A. Rose, Phys. Rev. Lett. **74**, 3983 (1995); B. B. Afeyan and E. A. Williams, Phys. Plasmas **4**, 3827 (1997); D. A. Russell and D. F. DuBois, Phys. Rev. Lett. **86**, 428 (2001).
[6]R. Yan, A. V. Maximov, C. Ren, and F. S. Tsung, Phys. Rev. Lett. **103**, 175002 (2009); H. X. Vu, D. F. DuBois, D. A. Russell, and J. F. Myatt, Phys. Plasmas **17**, 072701 (2010); R. Yan, C. Ren, J. Li, A. V. Maximov, W. B. Mori, Z.-M. Sheng, and F. S. Tsung, Phys. Rev. Lett. **108**, 175002 (2012); H. X. Vu, D. F. DuBois, D. A. Russell, J. F. Myatt, and J. Zhang, Phys. Plasmas **21**, 042705 (2014); J. Zhang, J. F. Myatt, R. W. Short, A. V. Maximov, H. X. Vu, D. F. DuBois, and D. A. Russell, Phys. Rev. Lett. **113**, 105001 (2014).
[7]D. T. Michel, A. V. Maximov, R. W. Short, S. X. Hu, J. F. Myatt, W. Seka, A. A. Solodov, B. Yaakobi, and D. H. Froula, Phys. Rev. Lett. **109**, 155007 (2012); W. Seka, J. F. Myatt, R. W. Short, D. H. Froula, J. Katz, V. N. Goncharov, and I. V. Igumenshchev, Phys. Rev. Lett. **112**, 145001 (2014).
[8]W. L. Kruer, *The Physics of Laser-Plasma Interaction* (Addison-Wesley, New York, 1988).
[9]C. Riconda, S. Weber, V. Tikhonchuk, and A. Heron, Phys. Plasmas **18**, 092701 (2011).
[10]S. Weber, C. Riconda, O. Klimo, A. Heron, and V. Tikhonchuk, Phys. Rev. E **85**, 016403 (2012).
[11]S. Weber and C. Riconda, High Power Laser Sci. Eng. **3**, e6 (2015).
[12]H. Wen, A. V. Maximov, R. Yan, J. Li, C. Ren, and J. F. Myatt, in 56th Annual Meeting of the APS Division of Plasma Physics, New Orleans, LA, 2014, Abstract ID: BAPS.2014.DPP.PO4.5.
[13]D. A. Russell, D. F. Dubois, and H. A. Rose, Phys. Plasmas **6**, 1294 (1999).
[14]S. Brunner and E. J. Valeo, Phys. Rev. Lett. **93**, 145003 (2004).
[15]P. Y. Cheung, A. Y. Wong, C. B. Darrow, and S. Z. Qian, Phys. Rev. Lett. **48**, 1348 (1982).
[16]M. Borghesi, S. Bulanov, D. H. Campbell, R. J. Clarke, T. Z. Esirkepov, M. Galimberti, L. A. Gizzi, A. J. MacKinnon, N. M. Naumova, F. Pegoraro, H. Ruhl, A. Schiavi, and O. Willi, Phys. Rev. Lett. **88**, 135002 (2002).
[17]S. Weber, C. Riconda, and V. Tikhonchuk, Phys. Rev. Lett. **94**, 055005 (2005).
[18]S. Weber, M. Lontano, M. Passoni, C. Riconda, and V. Tikhonchuk, Phys. Plasmas **12**, 043101 (2005).
[19]Z. J. Liu, X. T. He, C. Y. Zheng, and Y. G. Wang, Phys. Plasmas **16**, 093108 (2009).
[20]A. Langdon and B. Lasinski, Phys. Fluids **26**, 582 (1983).
[21]S. Bulanov, T. Esirkepov, N. Naumova, F. Pegoraro, and V. Vshivkov, Phys. Rev. Lett. **82**, 3440 (1999).
[22]S. Bulanov, I. Inovenkov, V. Kirsanov, N. Naumova, and A. Sakharov, Phys. Fluids B **4**, 1935 (1992).
[23]D. Wu, C. Y. Zheng, X. Q. Yan, M. Y. Yu, and X. T. He, Phys. Plasmas **20**, 033101 (2013).
[24]O. Klimo, S. Weber, V. Tikhonchuk, and J. Limpouch, Plasma Phys. Controlled Fusion **52**, 055013 (2010).
[25]H. X. Vu, D. F. DuBois, and B. Bezzerides, Phys. Rev. Lett. **86**, 4306 (2001).
[26]H. X. Vu, L. Yin, D. F. DuBois, B. Bezzerides, and E. S. Dodd, Phys. Rev. Lett. **95**, 245003 (2005).
[27]L. Yin, W. Daughton, B. J. Albright, B. Bezzerides, D. F. DuBois, J. M. Kindel, and H. X. Vu, Phys. Rev. E **73**, 025401(R) (2006).
[28]L. Yin, W. Daughton, B. J. Albright, K. J. Bowers, D. S. Montgomery, J. L. Kline, J. C. Fernandez, and Q. Roper, Phys. Plasmas **13**, 072701 (2006).
[29]C. Z. Xiao, Z. J. Liu, T. W. Huang, C. Y. Zheng, B. Qiao, and X. T. He, Phys. Plasmas **21**, 032107 (2014).
[30]W. P. Yao, B. W. Li, L. H. Cao, F. L. Zheng, T. W. Huang, C. Z. Xiao, and M. M. Skoric, Laser Part. Beams **32**, 583 (2014).
[31]B. Liu, H. Y. Wang, J. Liu, L. B. Fu, Y. J. Xu, X. Q. Yan, and X. T. He, Phys. Rev. Lett. **110**, 045002 (2013).
[32]D. W. Forslund, J. M. Kindel, and E. L. Lindman, Phys. Fluids **18**, 1002 (1975); D. W. Forslund, J. M. Kindel, and E. L. Lindman, Phys. Fluids **18**, 1017 (1975).
[33]M. Lontano, S. V. Bulanov, J. Koga, M. Passoni, and T. Tajima, Phys. Plasmas **9**, 2562 (2002).
[34]M. Lontano, M. Passoni, C. Riconda, V. T. Tikhonchuk, and S. Weber, Laser Part. Beams **24**, 125 (2006).
[35]V. E. Zakharov, Zh. Eksp. Teor. Fiz. **62**, 1745 (1972) , [Sov. Phys. JETP **35**, 908 (1972)].
[36]T. M. O'Neil, Phys. Fluids **8**, 2255 (1965).
[37]S. Weber, C. Riconda, and V. Tikhonchuk, Phys. Plasmas **12**, 112107 (2005).
[38]P. K. Kaw, A. Sen, and T. Katsouleas, Phys. Rev. Lett. **68**, 3172 (1992).
[39]N. M. Naumova, S. V. Bulanov, T. Zh. Esirkepov, D. Farina, K. Nishihara, F. Pegoraro, H. Ruhl, and A. S. Sakharov, Phys. Rev. Lett. **87**, 185004 (2001).
[40]K. Mima, M. S. Jovanović, Y. Sentoku, Z. M. Sheng, M. M. Škorić, and T. Sato, Phys. Plasmas **8**, 2349 (2001).
[41]A. Mančić, Lj. Hadžievski, and M. M. Škorić, Phys. Plasmas **13**, 052309 (2006).
[42]L. Yin, B. J. Albright, H. A. Rose, D. S. Montgomery, J. L. Kline, R. K. Kirkwood, J. Milovich, S. M. Finnegan, B. Bergen, and K. J. Bowers, Phys. Plasmas **21**, 092707 (2014).
[43]J. L. Kline *et al.*, Phys. Rev. Lett. **94**, 175003 (2005).

PHYSICS OF PLASMAS **23**, 022704 (2016)

Competition between stimulated Raman scattering and two-plasmon decay in inhomogeneous plasma

C. Z. Xiao,[1] Z. J. Liu,[1,2] C. Y. Zheng,[1,2,3,a)] and X. T. He[1,2,3]
[1]*HEDPS, Center for Applied Physics and Technology, Peking University, Beijing 100871, China*
[2]*Institute of Applied Physics and Computational Mathematics, Beijing 100084, China*
[3]*Collaborative Innovation Center of IFSA (CICIFSA), Shanghai Jiao Tong University, Shanghai 200240, China*

(Received 20 October 2015; accepted 29 January 2016; published online 12 February 2016)

We demonstrate competitions between stimulated Raman scattering (SRS) and two-plasmon decay (TPD) in the laser polarization plane in inhomogeneous near quarter-critical density plasma by using linear convective gain analysis and two-dimensional (2D) particle-in-cell (PIC) simulations. Linear theoretical analysis implies that convective SRS occurs in a wider and lower density region than absolute SRS and has a shared occurrence region with convective TPD. This convective SRS prefers a parameter space with the laser intensity larger than the order of $10^{15}\,\mathrm{W/cm^2}$ and the density scale length about several hundreds microns, which may be common in large scale direct-drive scheme, shock ignition scheme, and hybrid-drive scheme. A convective nature and saturation mechanism under these parameter regions are identified to be Langmuir decay instability and strong pump depletion. The significance of this convective SRS is shown in our 2D PIC simulations that hot electrons are reduced through suppressing the electron staged acceleration by TPD in the lower density region due to its high phase velocity. Temperature induced competitions are also studied using a relativistic modification to the Langmuir wave dispersion relation when $T_e > 5\,\mathrm{keV}$. Both absolute and convective SRS are observed to be dominant in the simulations when the temperature is as high as 10 keV. *© 2016 AIP Publishing LLC.*
[http://dx.doi.org/10.1063/1.4941969]

I. INTRODUCTION

Inertial confinement fusion (ICF) is a promising approach to get ignition although we have met some difficulties. Several main approaches to ICF, including the well-known indirect-drive approach,[1] direct-drive approach,[2] and shock ignition,[3] are on their way. New ignition method has been proposed such as hybrid-drive approach[4,5] where indirect-drive and direct-drive processes are combined in order for better ignition symmetry and higher energy gain. The density scale length in hybrid-drive regime may be larger than the conventional approaches (about several hundreds microns) and the laser intensity is as high as $10^{15}\,\mathrm{W/cm^2}$. Laser plasma interactions (LPIs) under this scheme are what we are interested in.

In recent years, the two-plasmon decay (TPD) instability, where an incident laser decays into a pair of Langmuir waves, has been reconsidered due to its potential for generating energetic electrons near the quarter-critical density. Meanwhile, the counterpart one, stimulated Raman scattering (SRS) which decays into a scattering wave and a corresponding Langmuir wave, has a shared density region with TPD. It is interesting and expectable that the competition between SRS and TPD will occur in this near quarter-critical density regions, which is seldom studied by the previous researchers. For the experiments at OMEGA laser facility, the basic parameters for direct-drive ICF are that the laser intensity is of the order $10^{14}\,\mathrm{W/cm^2}$ and the density scale length is about a hundred micron. Under this scheme, the TPD threshold parameter $\eta \propto IL_n/T_e$ is just above unity, where I is laser intensity, L_n is density scale length, and T_e is electron temperature. In both the direct-drive experiments[6–10] and particle-in-cell (PIC) simulations,[11–13] we can hardly find the existence of SRS, except for some evidences of absolute SRS close to the quarter-critical density in PIC simulations. Wen[14] has studied the competition between this absolute SRS and absolute TPD based on the absolute SRS theory[15] by using a fluid code. Unfortunately, this narrow located absolute SRS is hardly observed in the present experiments, and it may thought to be unimportant to hot-electron generation. Other TPD studies, either using a Zakharov code[16,17] or a reduced particle code[18–20] with SRS effects excluding, concentrated themselves on the nonlinear evolution of TPD with multibeam effects because of the above reasons.

On the other hand, some authors have studied these problems under the scheme of shock ignition, in which the laser intensity can be as high as $10^{16}\,\mathrm{W/cm^2}$ in order to launch an intense shock. They observed that there are strong interactions between SRS and TPD when the intensity is so high or the electron temperature is high enough[21–23] through PIC simulations. Under this high intensity condition, actually, not only absolute SRS but also convective SRS can be observed in the simulations, which are ignored by the authors. Also in homogeneous near quarter-critical density plasma, the nonlinear evolution of SRS has been studied alone by virtue of an one-dimensional model.[24] Two-dimensional (2D) PIC simulations in homogenous plasma

[a)]zheng_chunyang@iapcm.ac.cn

also show a coexistence of absolute SRS and absolute TPD due to its infinite density scale length.

As for the instability natures of the three-wave parametric instabilities in inhomogeneous plasma, Rosenbluth's pioneering work[25] defined a gain factor, the so-called Rosenbluth gain factor, by which we can obtain the instability threshold and distinguish the convective and absolute instability by using general WKB methods. The physics near the quarter-critical density is complex just because it is the turning point of SRS and TPD where the WKB method is invalid. Absolute instabilities emerge here as theoretical works predicted.[15,26–28] Generally speaking, the threshold of absolute instability is much lower than that of convective instability. However, convective instability occurs in a much wider region and has the potential to trigger other effects.[29,30] For convective SRS, White[31] proved that the transition from Rosenbluth's theory to Drake and Lee's theory by using the WKB invalid condition is reasonable. For convective TPD, Yan[32] derived formulas for convective TPD by using the Rosenbluth's theory, but it cannot transit to the absolute TPD theory yet. These convective theories extent the instabilities occurrence region to lower density, which means that a wider interaction region between convective SRS and convective TPD will be found than that of absolute SRS and absolute TPD (shown in Fig. 1). Thus, in this paper, the near quarter-critical density region refers to the interaction region of both convective and absolute instability, and usually it refers from $0.2n_c$ to $0.25n_c$, where n_c is the critical density of the incident laser.

In this paper, we systematically studied the competition between SRS and TPD in inhomogeneous plasma near the quarter-critical density especially for the hybrid-drive ICF parameters. A generalized three-dimensional (3D) instability geometry is used for deriving the Rosenbluth convective gain for SRS and TPD near the quarter-critical density. The SRS side-scattering effects are also included. From this convective gain, we can easily observe an absolute side-scattering instability where the scattering light is perpendicular to the laser polarization plane.[33] While in the plane of laser polarization, competition between convective SRS and convective TPD occurs. We have compared the convective gains in both wavenumber space and real density space. In order for competitions at high temperatures ($T_e > 5\,\text{keV}$), a relativistic modified Langmuir wave dispersion relation is used. Thresholds for the instabilities near the quarter-critical density are also derived. In order for verifying the linear analysis, 2D PIC simulations are used to study the competitions in the linear and nonlinear stage. A convective nature under the hybrid-drive scheme and saturation mechanism are identified due to Langmuir decay instability (LDI) and strong pump depletion. Three cases are compared under the hybrid-drive parameters space. It is inspiring to find that the excitation of convective SRS will reduce the hot-electron generation through suppressing the staged acceleration by TPD in the lower density region due to its high phase velocity. At high temperature, SRS is the dominant instability, while TPD is suppressed by Landau damping.

The paper is organized as follow: In Sec. II, the linear theoretical analysis of the generalized convective gains is presented. Then in Sec. III, we performed 2D PIC simulations to show the linear verification and nonlinear evolution of the competition. Summary and conclusions are presented in Sec. IV.

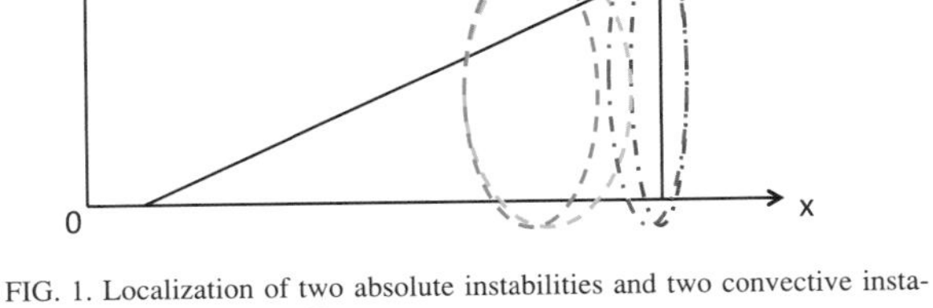

FIG. 1. Localization of two absolute instabilities and two convective instabilities in the density profile.

II. LINEAR THEORETICAL ANALYSIS OF SRS AND TPD CONVECTIVE GAIN

A. Generalized instability geometry for SRS

For a single laser beam normally incident into a near quarter-critical density plasma, several instabilities will occur, including SRS, stimulated Brillouin scattering (SBS), and TPD. In this theoretical section, we treat ion as fixed background so that no ion instabilities are discussed here. In order for a generalized scattering geometry including both backward scattering, forward scattering, and side-scattering, we draw the geometry of SRS and TPD in the wavenumber space, as shown in Fig. 2. The laser, which denoted by k_0, propagates along x axis and polarizes in y axis. TPD is mostly unstable in the x–y plane, and we denote two decay waves by k_1 and k_2, respectively. The geometry of SRS denotes by k_p and k_s for the Langmuir wave vector and the scattering wave vector, respectively, orientated by the angle θ (the angle between k_p and x axis) and ξ (the angle between k_p and $-z$ axis). Another auxiliary angle η, which is between k_s and E_0, is dependent on θ and ξ: $\cos^2\eta = k_p^2(\sin^2\theta - \cos^2\xi)/k_s^2$. The plasma density gradient is along x and has a local scale length $L_n = n_0(x)/[dn_0(x)/dx]$.

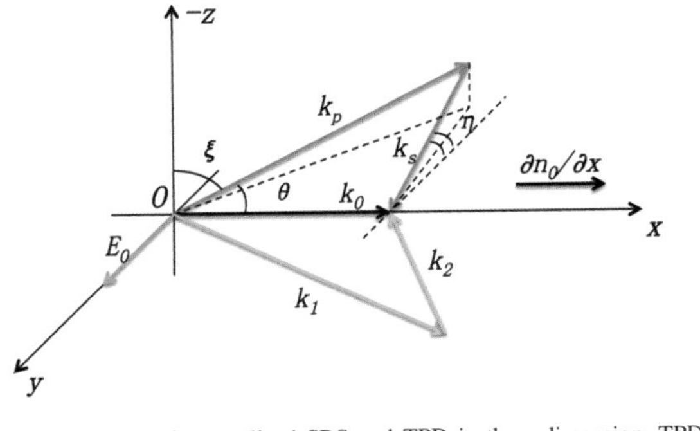

FIG. 2. Geometry of generalized SRS and TPD in three-dimension. TPD, denoted by k_1 and k_2, lies in the x–y plane. Generalized SRS, denoted by k_p and k_s, includes backward, forward, and side-scattering geometry.

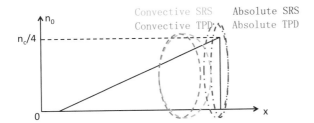

Next, we derive the Rosenbluth convective gain for this generalized SRS case near the quarter-critical density. The formula of the Rosenbluth convective gain is $G = 2\pi\gamma_0^2/\kappa'|V_{px}V_{sx}|$.[34] A linear density profile is assumed, $\omega_{pe}^2(x) = \omega_{pe0}^2(1 + x/L_n)$, where $\omega_{pe}^2(x) = 4\pi n_0(x)e^2/m_e$ is the local electron plasma frequency, e and m_e are electron charge and mass, respectively. The x component of the group velocities of decay waves are $V_{px} = 3k_{px}v_e^2/\omega_{LW}$ and $V_{sx} = k_{sx}c^2/\omega_s$, where $v_e = \sqrt{T_e/m_e}$ is the electron thermal velocity, c is the light speed, and ω_{LW} and ω_s are the frequencies of SRS Langmuir wave and scattering wave, respectively. κ is the wave number mismatch from the perfect resonant point, $\kappa(x) = k_0(x) - k_{px}(x) - k_{sx}(x)$. We can calculate κ' from the linear wave dispersion relations[35]

$$\kappa' = d\kappa/dx = \frac{\omega_{pe0}^2}{2L_n}\left(-\frac{1}{k_0c^2} + \frac{1}{k_{sx}c^2} + \frac{1}{3k_{px}v_e^2}\right) \approx \frac{\omega_{pe0}^2}{6k_{px}Lv_e^2}. \tag{1}$$

The homogeneous growth rate γ_0 is readily obtained from the coupling three-wave equations, and assume that the polarization direction of the scattering wave gives the maximum growth rate, then

$$\gamma_0 = \frac{k_pv_0}{4}\frac{\omega_{pe}}{\sqrt{\omega_{LW}\omega_s}}\sin\eta, \tag{2}$$

where v_0 is the electron quiver velocity, $v_0 = eE_0/m_e\omega_0c$, E_0 is the laser amplitude, and ω_0 is the laser frequency. Thus, we obtain the convective gain for SRS from the above formulas

$$G_{SRS} = \frac{2\pi}{8}\left(\frac{v_0}{c}\right)^2 L_n\frac{k_p^2}{|k_p\cos\theta - k_0|}\frac{k_s^2 - k_p^2(\sin^2\theta - \cos^2\xi)}{k_s^2}. \tag{3}$$

With the relation $k_s = \sqrt{k_p^2 - 2k_pk_0\cos\theta + k_0^2}$, the convective gain of SRS in wavenumber space is calculable for any scattering geometry.

We analysis Eq. (3) for some special cases. (i) Backward scattering (forward scattering) case, i.e., $\xi = \pi/2, \theta = 0$, the convective gain writes

$$G_{SRS} = \frac{2\pi}{8}\left(\frac{v_0}{c}\right)^2\frac{k_p^2}{|k_p - k_0|}L_n, \tag{4}$$

which returns to the common convective SRS threshold $(v_0/c)^2k_0L_n > 2$,[28] when $n_0 \ll n_c$. (ii) 90° side-scattering case, i.e., $k_{sx} = 0$. In this case, G_{SRS} is divergent, except for $\xi = \pi/2$, where it decreases to zero contrarily because of zero ponderomotive force. This divergence corresponds to an absolute instability.[33] (iii) When the density is very close to the quarter-critical density, the scattering wave vector k_s becomes very small. There is another absolute instability (the so-called absolute SRS) occurs, which can also be regarded as a limitation of 90° side-scattering.[28]

To obtain the density dependence of the convective gain, we need to solve the linear three-wave frequency matching condition, $\omega_0 = \omega_{LW} + \omega_s$. For the competition between SRS and TPD at high temperatures ($T_e > 5$ keV), we introduce a relativistic modification of Langmuir wave dispersion relation based on an initial relativistic Maxwellian, i.e., the Jüttner-Synge distribution function in the weak relativistic limit[36]

$$\omega_{LW}^2 \approx \omega_{pe}^2\left[1 - \frac{2}{5}\left(\frac{v_e}{c}\right)^2\right] + 3k_p^2v_e^2. \tag{5}$$

Further approximation is that $\omega_{LW} \approx \omega_{pe} + 3k_p^2v_e^2/2\omega_{pe} + \delta\omega$ when $k_p\lambda_D$ is not too large, where $\delta\omega = -\frac{5}{4}(v_e/c)^2\omega_{pe}$ is the linear frequency shift due to relativistic effects (λ_D is the Debye length, $\lambda_D = v_e/\omega_{pe}$). This temperature dependent linear frequency downshift is important for the cutoff of the instability at high temperature. Substitute Eq. (5) into the linear three-wave matching condition, consider only the backward/forward scattering case, and use the near the quarter-critical density condition, $k_s \ll k_0$, we obtain the Langmuir wave vector for SRS

$$k_p = \frac{1}{1 + 3(v_e/c)^2}\left\{k_0 \pm \sqrt{k_0^2 - [1 + 3(v_e/c)^2]\left[k_0^2 - \frac{(\omega_0 - 2\omega_{pe} - \delta\omega)2\omega_{pe}}{c^2}\right]}\right\}. \tag{6}$$

Combine Eqs. (4) and (6) we can get the SRS convective gain near the quarter-critical density, which will compare with the TPD convective gain in Subsection II C.

B. TPD convective gain and Landau damping effects

For the TPD convective gain, we still utilize the Rosenbluth's formula. It is readily to obtain the group velocities of TPD decay waves, $V_{1x} = 3k_{1x}v_e^2/\omega_1$ and $V_{2x} = 3k_{2x}v_e^2/\omega_2$, where ω_1 and ω_2 are the frequencies of two decay Langmuir waves. The wave number mismatch for TPD is then $\kappa' \approx k_0\omega_{pe0}^2/6v_e^2L_nk_{1x}k_{2x}$. The homogeneous growth rate for TPD is obtained from Yan's derivation, $\gamma_0^2 = -\frac{1}{16}k_y^2v_0^2\left(\frac{\omega_2}{\omega_1} - \frac{k_2^2}{k_1^2}\right)\left(\frac{\omega_1}{\omega_2} - \frac{k_1^2}{k_2^2}\right)$.[32] Thus, the TPD convective gain in two-dimension geometry is

$$G_{TPD} = \frac{2\pi}{24}\left(\frac{v_0}{v_e}\right)^2\frac{k_y^2L_n}{k_0}\left(\frac{\omega_2}{\omega_1} - \frac{k_2^2}{k_1^2}\right)\left(\frac{\omega_1}{\omega_2} - \frac{k_1^2}{k_2^2}\right), \tag{7}$$

where $k_y = |k_{1y}| = |k_{2y}|$.

The convective gain for TPD is much larger than that for SRS due to the small quantity v_e/c. However, since the wave vectors of the two decay wave increase quickly with decreasing plasma density, the Landau damping effects become an important cutoff regime for TPD. On the other hand, for SRS, only until the temperature is very high, Landau damping effects can greatly influence the instability, so that we ignore the damping effect in SRS case. To calculate the damping effects, we introduce an effective damping coefficient β, derived from the three-wave coupling theory. The damping coefficient is defined as, $\beta = L_0/L_a$,[37] where $L_0 = \sqrt{|V_{1x}V_{2x}|}/\gamma_0$ is the basic gain length, and $L_a = (\nu_1/V_{1x} + \nu_2/V_{2x})^{-1}$ is the absorption length. ν_1 and ν_2 are the usual Landau damping rate for Langmuir wave

$$\nu_i = \sqrt{\frac{\pi}{8}}\frac{\omega_{pe}}{(k_i\lambda_D)^3}\exp\left[-\frac{1}{2(k_i\lambda_D)^2} - \frac{3}{2}\right], \tag{8}$$

where $i = 1$, 2. The damping threshold for TPD is then $\beta < 2$. Substitute V_{ix} and γ_0 into the threshold condition, we obtain

$$\frac{1}{2}k_0 v_0 > \left(\frac{\nu_1}{k_{1x}} + \frac{\nu_2}{k_{2x}}\right)\frac{k_1 k_2\sqrt{k_{1x}k_{2x}}}{k_y(k_{1x}+k_{2x})}, \tag{9}$$

where we simplify γ_0 with $\omega_1 \approx \omega_2$.

We proceed with the same procedure as in Subsection II A to get the relativistic modified density dependence on the TPD convective gain. Solving the linear frequency matching condition, $\omega_0 = \omega_1 + \omega_2$, with the help of Eq. (5), it is readily to obtain the density dependent equation for TPD

$$\left(k_x - \frac{k_0}{2}\right)^2 + k_y^2 = \frac{1}{3}\frac{\omega_0 - 2\omega_{pe} - 2\delta\omega}{\lambda_D^2\omega_{pe}} - \left(\frac{k_0}{2}\right)^2, \tag{10}$$

where k_x is the x component of one of the TPD decay wave. This curve corresponds to a circle, valid only when $k_i\lambda_D$ is not too large. Further, we consider the TPD modes with the maximum growth rates (the well-known hyperbola curve)

$$\left(k_x - \frac{k_0}{2}\right)^2 - k_y^2 = \frac{k_0^2}{4}. \tag{11}$$

The convective gain for TPD, Eq. (7), is then simplified with the Taylor expansion

$$G_{TPD} = \frac{2\pi}{24}\left(\frac{v_0}{v_e}\right)^2 k_0 L_n\left[1 - \frac{9}{2}\left(\frac{v_e}{c}\right)^2 - 24\left(\frac{v_e}{c}\right)^2\left(\frac{k_y c}{\omega_0}\right)^2\right]. \tag{12}$$

Combine Eqs. (10) and (11), we obtain the relativistic modified linear TPD dispersion relation

$$k_x = \frac{k_0}{2} \pm \sqrt{\frac{(\omega_0 - 2\omega_{pe} - 2\delta\omega)\omega_{pe}}{6v_e^2}}, \tag{13}$$

$$k_y = \pm\sqrt{\frac{(\omega_0 - 2\omega_{pe} - 2\delta\omega)\omega_{pe}}{6v_e^2} - \left(\frac{k_0}{2}\right)^2}. \tag{14}$$

With Eqs. (12) and (14), we can get the density dependence of TPD convective gain.

C. Competition between SRS and TPD in the linear stage

The Rosenbluth convective gain provides the spatial amplification of initial perturbation as $I = I_0 e^G$, where I_0 is the nondriven, thermal, source intensity. Effective growth happens when the convective gain is larger than 2π. For instabilities with a shared occurrence region, their convective gain may determine the possibility of each instability. In order to compare the two electron instabilities near the quarter-critical density, we further transform Eq. (3) to wavenumber coordinate with the help of $k_x = k_p\cos\theta$, $k_y = k_p\cos\xi$. Eq. (3) then writes

$$G_{SRS}(k_x,k_y,k_z) = \frac{2\pi}{8}\left(\frac{v_0}{c}\right)^2 L_n\frac{\left(k_x^2+k_y^2+k_z^2\right)\left(k_0^2 - 2k_xk_0 + k_x^2 + k_z^2\right)}{|k_x - k_0|\left(k_x^2+k_y^2+k_z^2 - 2k_xk_0 + k_0^2\right)}. \tag{15}$$

The same procedure does for the TPD convective gain with the approximation that $\omega_{pe} \approx \omega_0/2$ in Eq. (7). Competitions between SRS and TPD in wavenumber space are shown in Fig. 3 with the laser intensity $I = 2.7\times10^{15}\mathrm{W/cm^2}$, the laser wavelength $\lambda_0 = 1/3\,\mu$m, and the density scale length $L_n = 245\,\mu$m. Figs. 3(a) and 3(b) show the low temperature case, $T_e = 2.5$ keV. The convective gain for SRS in the plane of polarization ($k_x - k_y$ plane, Fig. 3(a)) indicates that the most-likely modes lie close to k_0 ($k_0 \approx \sqrt{3}\omega_0/2c$ near the quarter-critical density). However, the value of its gain is much smaller than that of TPD mode in the plane and SRS side-scattering out of the plane ($k_x - k_z$ plane), as shown in Fig. 3(b). We plot the summation of G_{SRS} and G_{TPD} in Fig. 3(b) and find that the SRS convective gain is shallowed in the TPD convective gain in the plane, while the gain of 90° side-scattering out of the plane is comparable with the TPD convective gain, which means that strong interactions between SRS side-scattering and TPD will be observed. But in our paper, we mainly discuss the competition in the plane. As well-known that temperature will greatly suppress the growth of TPD, we plot the TPD convective gain at high temperature, $T_e = 10$ keV, shown in Fig. 3(c). The gain of SRS is almost independence of temperature, thus we find that at high temperature interactions between SRS and TPD in the plane will become severe.

In Fig. 4, the convective gains verses plasma density and temperature in the laser polarization plane are shown for the same parameters as in Fig. 3. The regions I and II represent spatial convective amplification region and absolute instability region, respectively. The narrow located absolute instabilities are very close to the upper cutoff curves,[38] so that in the lower and wider density region the competition between convective SRS and convective TPD is dominant.

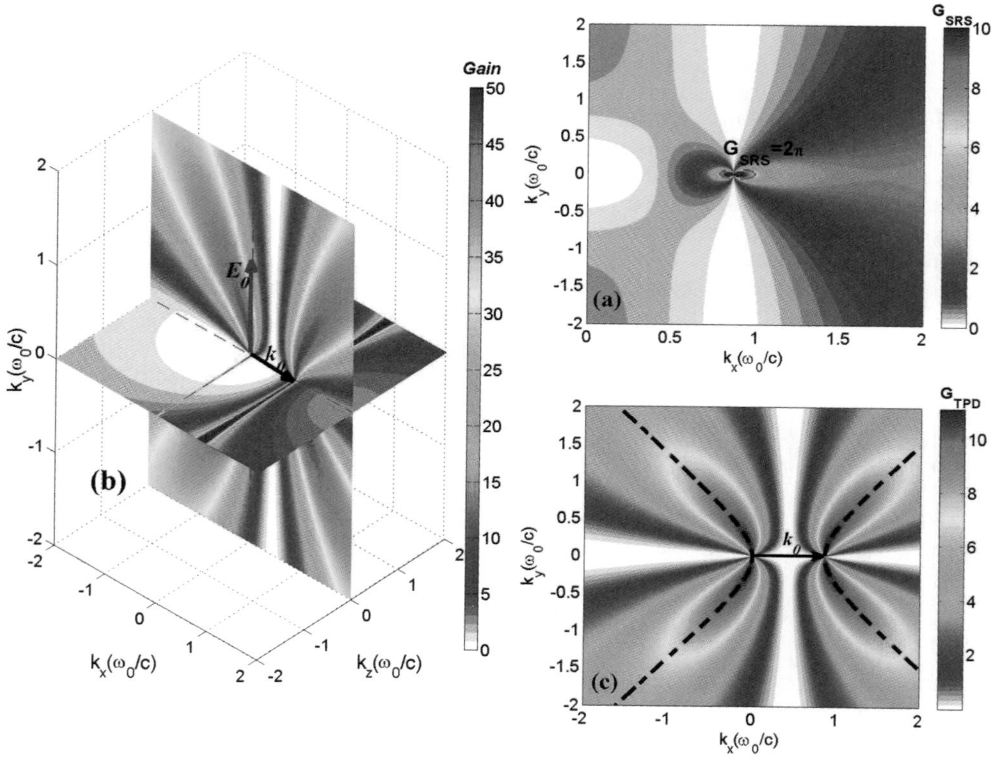

FIG. 3. Competition between SRS and TPD in the wavenumber space. Figs. 3(a) and 3(b) are the low temperature case, $T_e = 2.5\,\text{keV}$, while Fig. 3(c) is the high temperature case, $T_e = 10\,\text{keV}$. Figs. 3(a) and 3(c) are the convective gains for SRS and TPD, respectively. Fig. 3(b) presents a total effects of both SRS and TPD in 3D geometry. Other parameters, for hybrid-drive scheme, are the same: $I = 2.7 \times 10^{15}\,\text{W/cm}^2$, $\lambda_0 = 1/3\,\mu\text{m}$, and $L_n = 245\,\mu\text{m}$.

The weak SRS growth may compete with the strong TPD only when $G_{SRS} > 2\pi$ (the region above the curve $G_{SRS} = 2\pi$). This threshold curve is dropped when strengthening the laser intensity or enlarging the density scale length. In Fig. 4(b), Landau damping threshold is included, labeled "IV," which show us an increasing of the lower cutoff density with temperature increasing. So that it is naturally to expect that the SRS will dominate when temperature is high. Competition between convective SRS and convective TPD in the linear stage will affect the latter nonlinear evolution, which is presented in our PIC simulations in Section III.

We present the thresholds for the instabilities near the quarter-critical density in Fig. 5. Five basic instability natures are involved: (i) absolute SRS close to the cutoff density with the threshold, $k_0L_n(v_0/c)^{3/2} > 3^{-1/4}$;[28] (ii) absolute TPD close to the cutoff density with the threshold, $(v_0/v_e)^2 k_0 L_n > 16.54$;[27] (iii) absolute 90° side-scattering with the threshold, $(v_0/c)^2(\omega_0 L_n/c)^{4/3} > (\omega_{pe}^2/\omega_0^2)^{1/3}/(2 - 2\omega_{pe}/\omega_0 - \omega_{pe}^2/\omega_0^2)$;[33] (iv) convective SRS with the threshold, $G_{SRS} > 2\pi$; and (v) convective TPD with the threshold, $G_{TPD} > 2\pi$. In Fig. 5, thresholds for two different temperatures with a fixed plasma density $n_0 = 0.23n_c$ and laser wavelength $\lambda_0 = 1/3\,\mu\text{m}$ are presented. First, we come to the usual ignition temperature, $T_e = 2.5\,\text{keV}$ (Fig. 5(a)). For the direct-drive scheme at OMEGA laser facility, competition between absolute SRS, including side-scattering, and TPD (both convective and absolute) is likely to observed. When the ignition scale is enlarged (such as NIF target), or considering the multibeam effects, it is likely to observe the effects of convective SRS. For shock ignition, due to large laser intensity, all these instabilities are readily to occur. However, the physics will change when the temperature is higher, $T_e = 10\,\text{keV}$. The threshold of convective SRS is dropped while that of absolute and convective TPD is lifted, which makes it easier to trigger these instabilities as shown in Fig. 5(b). Three different cases related to the hybrid-drive scheme are shown in Fig. 5 and the corresponding linear and nonlinear evolutions are discussed in Section III by using PIC simulation method.

III. PARTICLE SIMULATIONS FOR THE HYBRID-DRIVE SCHEME

In this section, we present PIC simulations for the parameters under the hybrid-drive scheme. Three cases with different density scale lengths and electron temperatures are given. The different parameters are $L_n = 245\,\mu\text{m}, T_e = 2.5\,\text{keV}$ for case A; $L_n = 490\,\mu\text{m}, T_e = 2.5\,\text{keV}$ for case B; and $L_n = 245\mu\text{m}, T_e = 10\,\text{keV}$ for case C. Other parameters are the same for each case. The laser intensity $I = 2.7 \times 10^{15}\,\text{W/cm}^2$, the wavelength $\lambda_0 = 1/3\,\mu\text{m}$, and the ion ($H^+$) temperature $T_i = 0.5\,\text{keV}$. These simulation parameters are presented in Table I. From the threshold diagram, Fig. 5,

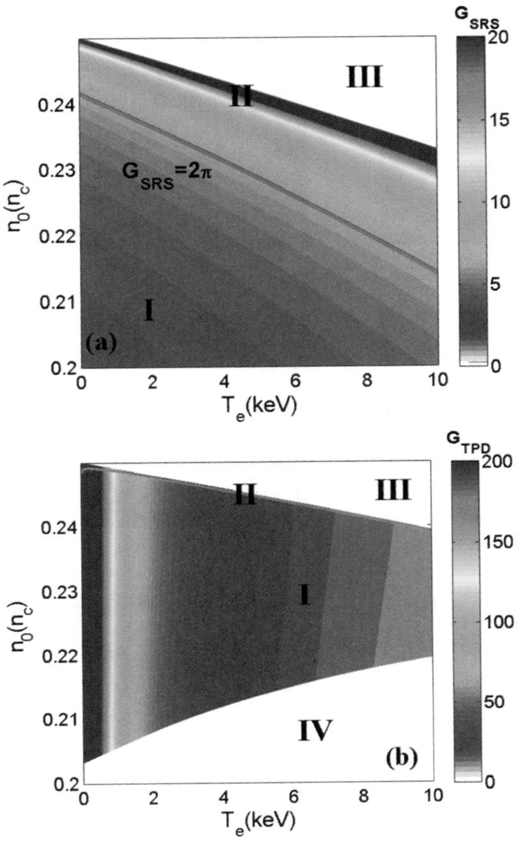

FIG. 4. Convective gains of SRS (a) and TPD (b) vs n_0 and T_e in the laser polarization plane. Other parameters, for hybrid-drive scheme, are the same as in Fig. 3: $I = 2.7 \times 10^{15}$ W/cm^2, $\lambda_0 = 1/3\,\mu$m, and $L_n = 245\,\mu$m.

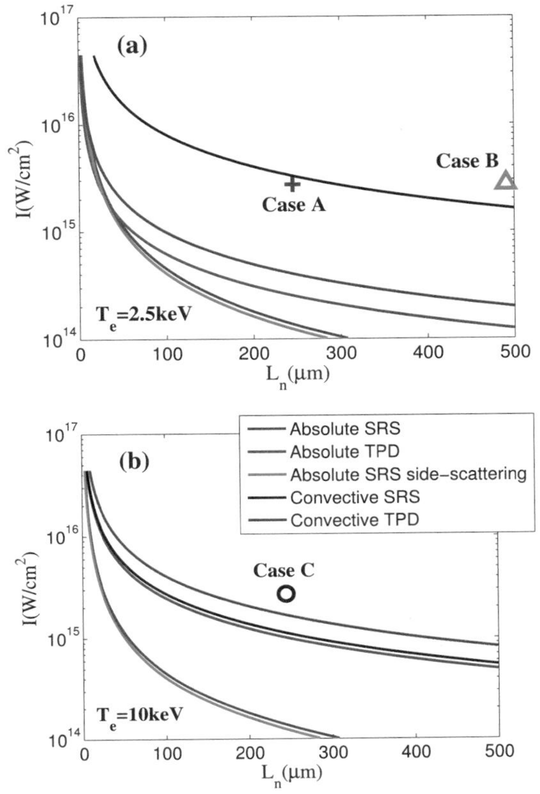

FIG. 5. Thresholds of the instabilities near the quarter-critical density for two different temperatures: (a) $T_e = 2.5$ keV and (b) $T_e = 10$ keV. Other parameters are $n_0 = 0.23n_c$ and $\lambda_0 = 1/3\,\mu$m.

we expect few convective SRS in case A, weak convective SRS in case B, and strong SRS in case C.

Our simulation setup is as follow. A 2D PIC code KLAP[39] is used to simulate the competition between convective SRS and convective TPD in the plane of polarization. Thus, a p-polarized plane electromagnetic wave propagates along x axis and oscillates in y axis. A linear density profile goes from $0.205n_c$ to $0.255n_c$. The simulation box is 60 μm in longitudinal and 10 μm in transverse for cases A and C, and 120 μm long for case B. The plasma longitudinally distributes in the center of the box with a 5 μm vacuum region in both sides for cases A and C, and 10 μm for case B. Periodic boundary condition is used in transverse direction, and absorbing boundary condition is used in longitudinal direction. The total simulation time is 4 ps, during which the physics reaches a steady state. Linear and nonlinear effects are mostly preserved in our simulations.

First, we compare our PIC simulations with linear analysis in the first picosecond. Fig. 6 shows the spectra of longitudinal field E_x for three different cases. In the first three figures, the time averaged spectra over 1 ps in the wavenumber space imply that convective SRS becomes stronger from case A to case C. Concretely, in Fig. 6(a), a common TPD spectrum is observed, with the maximum TPD growth rate (white dashed curve) concisely across the hyperbolic pattern. When the density scale length is elongated to 490 μm (case B, Fig. 6(b)), the TPD spectrum becomes weak, in addition, at $k_x \approx k_0, k_y = 0$, a signature of backward SRS is observed. Further, when the electron temperature increases to 10 keV (case C, Fig. 6(c)), strong SRS is dominant in the simulation accompany by the suppressed TPD in a narrower wavevector region. In addition to the signatures of SRS and TPD,

TABLE I. Simulation parameters, convective gains, and growth rates from theoretical analysis and the absolute instability growth rates from Drake's[15] and Simon's[27] theory.

	I (W/cm^2)	L_n (μm)	T_e (keV)	G_{TPD}	G_{SRS}	γ_{cTPD}/ω_0	γ_{cSRS}/ω_0	γ_{aTPD}/ω_0	γ_{aSRS}/ω_0
Case A	2.7×10^{15}	245	2.5	44.4	5.5	3.1×10^{-3}	3.7×10^{-3}	2.7×10^{-3}	2.9×10^{-3}
Case B	2.7×10^{15}	490	2.5	88.9	10.9	3.1×10^{-3}	3.7×10^{-3}	2.9×10^{-3}	3.1×10^{-3}
Case C	2.7×10^{15}	245	10	10.8	18.6	3.1×10^{-3}	3.3×10^{-3}	1.9×10^{-3}	2.9×10^{-3}

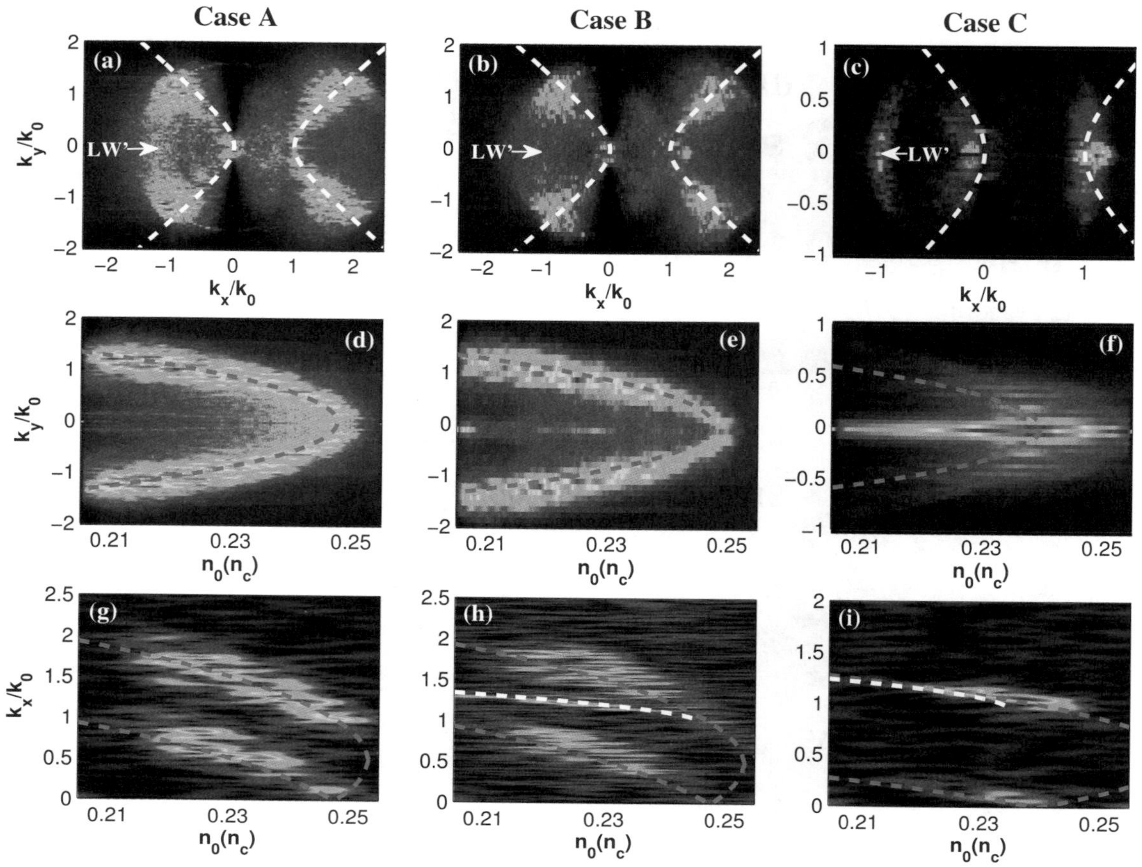

FIG. 6. Spectra of longitudinal field E_x for three different cases. (a)–(f) represent $E_x(k_x, k_y)$ and $E_x(n_0, k_y)$ averaged over 1 ps, respectively. (g), (h), and (i) are $E_x(n_0, k_x)$ plotted from the short-time Fourier transform (STFT) at a fixed time $t = 0.5$ ps and a fixed line in the center of the simulation box. The overlapped curves are plotted from our linear theoretical analysis, which show a good consistence with the simulation results.

secondary decay waves from initial LPI are also seen in these simulations, which are called LDI,[40] with corresponding ion acoustic wave signatures in ion density fluctuation (not shown here). Therefore, LDI may be a possible saturation mechanism under the hybrid-drive scheme.

The density dependences of the spectra are also studied. The second row represents the time averaged spectra vs n_0 and k_y. The convective SRS lies in $k_y = 0$, which is clearly seen in case B and case C. In case A, obscure image near the $n_c/4$ indicates the competition between absolute TPD and absolute SRS. The instability occurrence regions for all cases are concisely consistent with the linear analysis as shown in Fig. 4. The relativistic modified linear TPD dispersion relation curve (red dashed line) also shows a good agreement with the simulation results in case C. To diagnose the linear SRS dispersion relation, we present the spectra vs n_0 and longitudinal wave vector k_x in the third row by using the short-time Fourier transform (STFT) at a fixed time $t = 0.5$ ps and a fixed line in the center of the simulation box. The red dashed lines plotted from Eq. (13) are the TPD curves, and the white dashed lines plotted from Eq. (6) show the dispersion relation of backward SRS, which verifies the existence of convective SRS as well.

Further, we analysis the longitudinal field evolution. Fig. 7 shows the evolution of transversely averaged longitudinal field energy density $\langle E_x^2 \rangle$ and longitudinal field $\langle E_x \rangle$ for three different cases. The averaging process is defined as $\langle E_x^2 \rangle = \int E_x^2 dy / \int dy$, $\langle E_x \rangle = \int E_x dy / \int dy$. In case A, the averaged longitudinal field, Fig. 7(d), shows that absolute SRS close to the quarter-critical density is observed. This narrow located SRS growth starts from 0.3 ps, then saturates at about 0.6 ps. The linear growth rate of SRS and TPD (Table I) near the quarter-critical density is almost the same, but absolute SRS growth rate is a little bit larger. These absolute instabilities quickly saturate at maximum level of $E_{xmax}(absolute) \approx 0.034 m\omega_0 c/e$, causing strong pump depletion as shown in Fig. 8. The same way occurs for convective TPD. With a quite high convective gain, $G_{cTPD} = 44.4$, the convective TPD increases almost at the same time with the same growth rate with absolute instabilities, then saturates at a higher level of $E_{xmax}(cTPD) \approx 0.045 m\omega_0 c/e$. The noise level is about 10^{-4}, thus convective TPD saturates before it reaches the convective gain due pump depletion and LDI which is mentioned above. Convective SRS is not obvious in case A, since it is under the convective instability threshold. Thus, in the evolution of

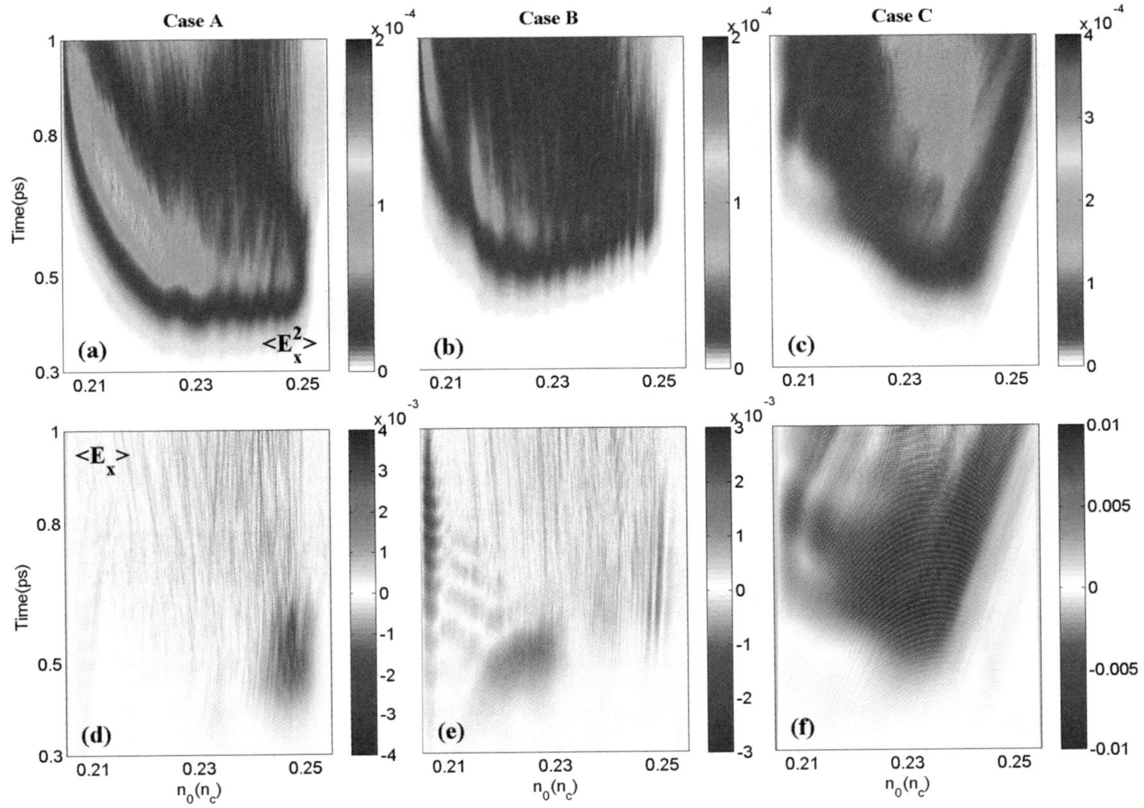

FIG. 7. Evolution of longitudinal field for three different cases. Transversely averaged field energy density, $\langle E_x^2 \rangle = \int E_x^2 dy / \int dy$, shows a total evolution of SRS and TPD. Transversely averaged field, $\langle E_x \rangle = \int E_x dy / \int dy$, eliminates the effects of TPD mode leaving the evolution of SRS alone.

total field energy density, Fig. 7(a), the TPD evolution indicates a typical convective nature and is dominant in this case. Low threshold absolute instabilities near the quarter-critical density are suppressed by strong pump depletion.

Take case B for comparison, Fig. 7(e) shows that a convective SRS growth starts from 0.3 ps at lower density region first. Its linear growth rate is larger than other instabilities presented in Table I. An exponential spatial growth of SRS toward the convective gain, $G_{cSRS} = 10.9$, saturates at $t = 0.5$ ps with $E_{xmax}(cSRS) \approx 0.026 m\omega_0 c/e$. Meanwhile,

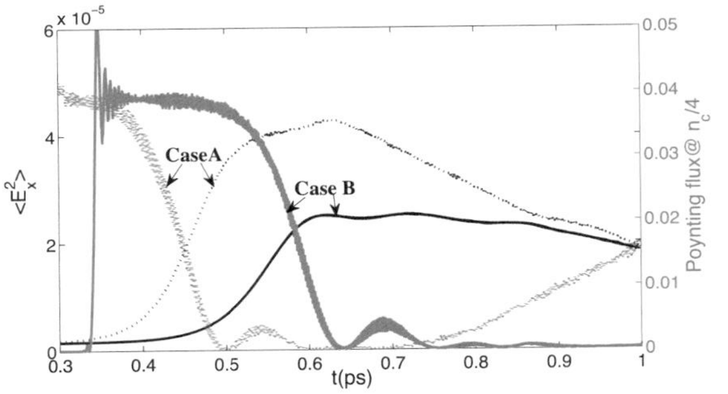

FIG. 8. Pump depletions in case A and case B in the first 1 ps. The black curves are the averaged longitudinal field energy $\langle E_x^2 \rangle$, and the red curves are the corresponding Poynting vector at quarter-critical density.

convective TPD grows along with convective SRS at lower density region but with smaller growth rate. The expected linear convective gain is the twice of case A, $G_{cTPD} = 88.9$, so at late time convective TPD is still the dominant instability over convective SRS. The saturation level of convective TPD is about $E_{xmax}(cTPD) \approx 0.036 m\omega_0 c/e$, which is smaller than that in case A, perhaps due to the existence of convective SRS and scale length effects.[41] Strong pump depletion is also a nonlinear saturation mechanism in case B due to large gains. As shown in Fig. 8, when the longitudinal field saturates, the Poynting flux at the quarter-critical density decreases to a small level, which indicates that laser is largely absorbed there, then it stagnates the absolute instabilities. The corresponding averaged longitudinal energy density (Fig. 7(b)) also shows convective natures of SRS and TPD.

As for the case C, the physics changes thoroughly. The strong SRS (Fig. 7(f)) occurs at 0.4 ps near $0.23 n_c$, then grows violently. In the linear theoretical analysis, both absolute and convective growth rate of SRS are larger than the TPD growth rate and they have larger convective gains. Thus, it seems that both convective SRS and absolute SRS exist in the simulation at the same time. In Fig. 7(c), we cannot distinguish the evolution of TPD from the total field energy density, and the total maximum longitudinal field is about $E_{xmax} \approx 0.065 m\omega_0 c/e$. Thus, in this high temperature

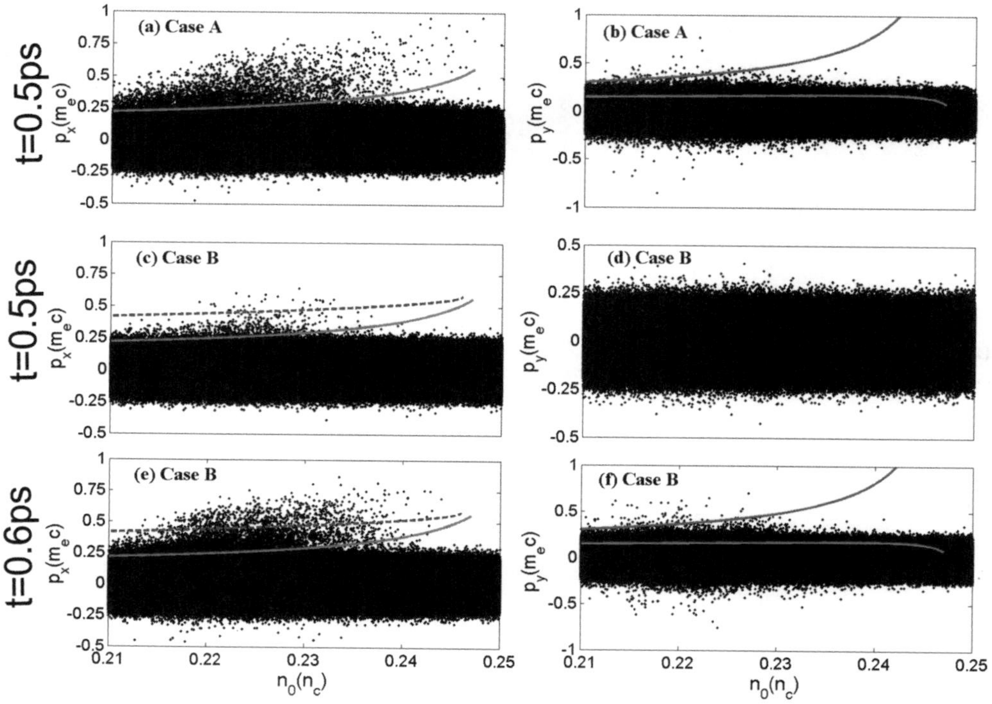

FIG. 9. Electron phase space of parallel $n_0 - p_x$ and perpendicular $n_0 - p_y$. (a) and (b) are from case A at $t = 0.5$ ps; (c) and (d) are from case B at $t = 0.5$ ps; and (e) and (f) are from case B at $t = 0.6$ ps. The red/green curve and blue dashed curve are the phase velocity of forward/backward propagating Langmuir wave generated by TPD and forward propagating Langmuir wave generated by backward SRS, respectively.

case, SRS is the dominant instability saturated by LDI, while TPD is highly suppressed by Landau damping effects.

The hot-electron generation near the quarter-critical density is one of the most important problems in the ignition issue. We compare the hot-electron generation in three different cases and find that hot electrons in case C are simply generated by the longitudinally propagating Langmuir wave generated from the backward SRS. The hot-electron temperature can be explained by the theory of SRS alone,[42,43] depending on the phase velocity of Raman-generated Langmuir wave. The significance of convective SRS on hot-electron generation competing with TPD is an unsolved and interesting problem, thus we present detailed phase space diagrams of case A and case B in Fig. 9. It is observed that through suppressing the staged acceleration by convective TPD in the lower density region, convective SRS can reduce hot electrons near the quarter-critical density, because of its high phase velocity.

Figs. 9(a) and 9(b) indicate a typical electron acceleration phase space through TPD staged acceleration mechanism[12] at $t = 0.5$ ps in the saturation phase of TPD in case A. The red curve is the phase velocity of forward propagating TPD Langmuir wave, while the phase velocity of backward propagating Langmuir wave is plotted in green. The corresponding wave potential depth in the momentum space, given by $\Delta p_x \approx \sqrt{2em_eE_{xmax}/k_0}$, is about $0.32m_ec$, so that a lot of electrons can be trapped by the forward TPD Langmuir wave in the low density region and staged accelerated to high phase velocity near the quarter-critical density as shown in Fig. 9(a). However, when convective SRS is dominant in the lower density region, for example, $t = 0.5$ ps in case B, its high phase velocity (the blue dashed line in Fig. 9(c)) can only trap few electrons with a narrow wave potential depth $\Delta p_x \approx 0.24m_ec$. The transverse phase space at that time, Fig. 9(d), implies that electrons have not been transversely accelerated by TPD yet. Late at $t = 0.6$ ps in case B, when convective TPD is dominant again, hot electrons trapped by TPD Langmuir wave with a width of $0.29m_ec$ result in a similar phase space as in case A.

This competition mechanism on hot-electron generation is efficient when convective SRS is strong compared with convective TPD, and the longer it maintains the more hot electrons can be reduced. However, in the first picosecond of case B, convective SRS is relative weak compared with TPD. Even we can hardly detect the scattering wave signature at the left boundary, it is perhaps since that near its critical density, scattering wave is almost reabsorbed by plasmas. From the hot-electron fraction evolution in the first picosecond (seen Fig. 11), the reduction of hot electrons by convective SRS is compensated by late convective TPD effects. But we can still expect that strong reduction of hot electrons can be observed, as long as the convective gain of SRS is greater.

To explore the nonlinear evolution of the system and further verify the significance of convective SRS, long time simulations to 4 ps are presented for case A and case B. The physics in the first picosecond determines the later nonlinear evolution, which is shown in Fig. 10. From the transversely

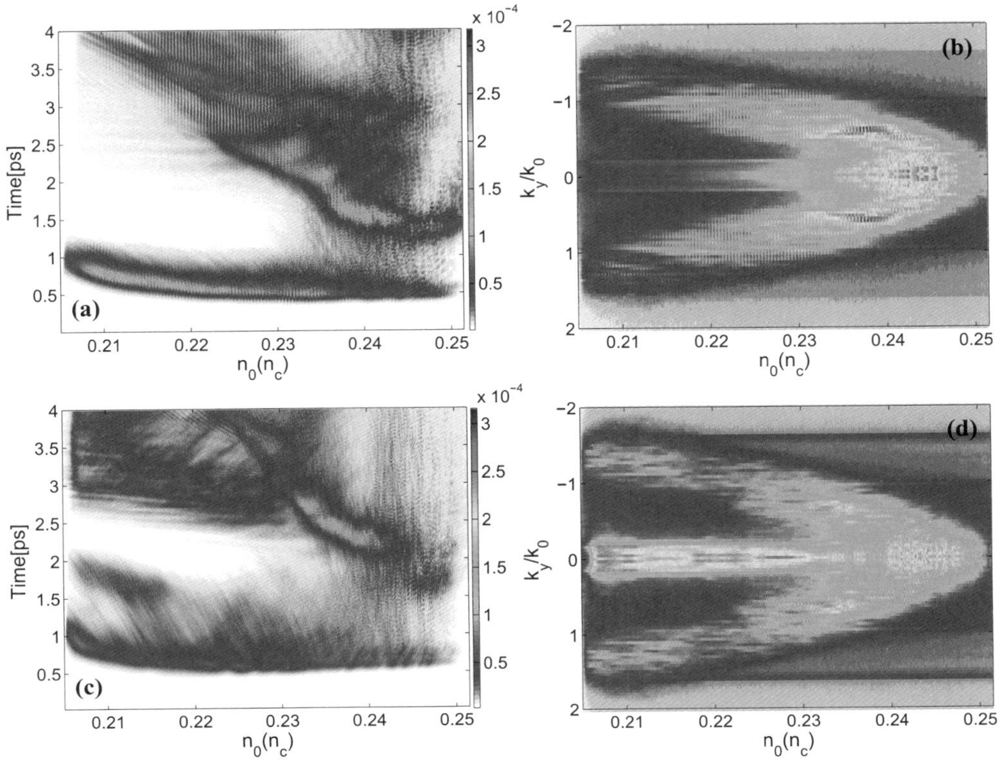

FIG. 10. Evolutions of transversely averaged longitudinal field energy density $\langle E_x^2 \rangle$ in 4 ps and corresponding time averaged spectra $E_x(n_0, k_y)$. (a) and (b) are from case A, and (c) and (d) are from case B.

averaged longitudinal field energy density evolutions, both case A (Fig. 10(a)) and case B (Fig. 10(c)) demonstrate an intermittent behaviors of LPI after the convective nature in the first picosecond. This phenomenon is reported by Yan,[13] but we record a quicker stagnation of LPI due to stronger convective gains resulting in stronger pump depletion. The convective nature is ubiquitous under the parameters far exceeding the thresholds of absolute instabilities, which is observed in the second burst of LPI as well. The recurrence of convective SRS occurs at $t \approx 2.5$ ps in case B with a stronger signature and dominates in the lower density region after that, since the increased electron temperature and generation of hot electrons from laser heating suppress the TPD instability. The competition is obvious in the time averaged spectra over 4 ps, shown in Figs. 10(b) and 10(d) depicting a convective TPD dominant regime and a convective SRS dominant regime, respectively, which is in good coincidence with the linear analysis.

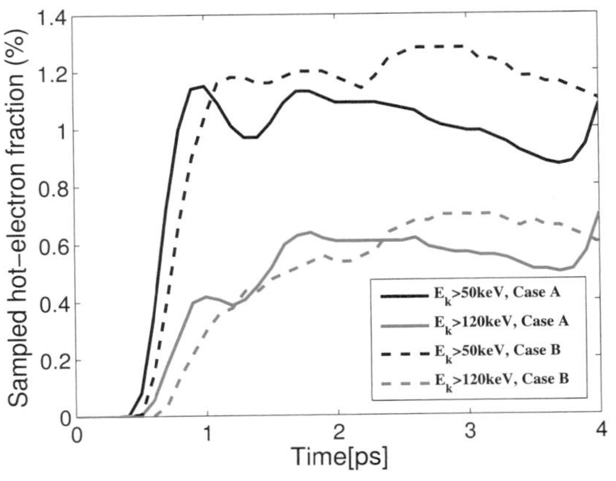

FIG. 11. Hot electron fraction from the sampled electrons. Solid: case A; dashed: case B; black: above 50 keV; red: above 120 keV.

Hot-electron production efficiencies in the total 4 ps for cases A and B are presented in Fig. 11. It is found that when convective SRS is excited for the first time, the hot-electron fraction in case B is lower than that in case A. However, later dominant TPD instability produces hot electrons to an equal level of case A. The recurrence of TPD makes a little increasement, then almost saturates in hot-electron fraction. As the TPD instability convectives to the lower density region, another sharp increasing at the end of simulation in case A (solid curves) raises the hot-electron fraction again. On the contrary, a descent in hot-electron fraction resulting from the recurrence of convective SRS dominated in the lower density after 3 ps is observed in case B. Though with a large density scale length, thus large linear gains, the hot-electron fraction is close to the small density scale length case, which means that a saturation in hot-electron fraction reaches under the hybrid-drive scheme. Even, if the convective SRS is strong compared with TPD, we would expect a decrease in hot electrons.

IV. CONCLUSIONS

In this paper, we present linear theoretical analysis and 2D PIC simulations of the competition between SRS and TPD, especially convective SRS and convective TPD in the plane of laser polarization. The linear theoretical analysis shows that the convective SRS prefers occurring in high laser intensity, large density scale length, and high temperatures situations. We can distinguish the occurrence region from absolute instabilities and convective instabilities and determine the threshold of convective SRS and TPD from the theory. In order to get a correct dependence of the density at high temperatures, the relativistic modified dispersion relation is used. The wider and lower density region of convective SRS and convective TPD makes it relevant to the hot electron generation. In our PIC simulations, three cases under the hybrid-drive scheme are presented. Great consistencies with our linear theoretical analysis are shown in our simulation results. A convective nature under the hybrid-drive scheme and saturation mechanism are identified due to LDI and strong pump depletion. Nonlinear evolution of the competition between SRS and TPD shows us the importance of the convective SRS that it can reduce the hot-electron generation through suppressing the staged acceleration by TPD in the lower density region due to its high phase velocity. Therefore, we conclude that the competition between convective SRS and convective TPD is important, and it may be observed in the future experiments under the hybrid-drive scheme, shock ignition scheme, and large scale direct-drive scheme as well.

ACKNOWLEDGMENTS

We thank L. H. Cao for useful discussion, and the anonymous referee for his suggestions that have improved this paper. This research was supported by the National Natural Science Foundation of China (Grant Nos. 11575035, 11175029, 11475030, and 11175030) and National Basic Research Program of China (Grant No. 2013CB834101).

[1]J. Lindl, Phys. Plasmas **2**, 3933 (1995).
[2]S. E. Bodner, D. G. Colombant, J. H. Gardner, R. H. Lehmberg, S. P. Obenschain, L. Phillips, A. J. Schmitt, J. D. Sethian, R. L. McCrory, W. Seka, C. P. Verdon, J. P. Knauer, B. B. Afeyan, and H. T. Powell, Phys. Plasmas **5**, 1901 (1998).
[3]R. Betti, C. D. Zhou, K. S. Anderson, L. J. Perkins, W. Theobald, and A. A. Solodov, Phys. Rev. Lett. **98**, 155001 (2007).
[4]Z. F. Fan, M. Chen, Z. S. Dai, H. B. Cai, S. P. Zhu, W. Y. Zhang, and X. T. He, e-print arXiv:1303.1252 [physics.plasm-ph].
[5]X. T. He, Z. F. Fan, J. W. Li, J. Liu, K. Lan, J. F. Wu, L. F. Wang, and W. H. Ye, e-print arXiv:1503.08979 [physics.plasm-ph].
[6]R. K. Follett, D. H. Edgell, R. J. Henchen, S. X. Hu, J. Katz, D. T. Michel, J. F. Myatt, J. Shaw, and D. H. Froula, Phys. Rev. E **91**, 031104 (2015).
[7]W. Seka, J. F. Myatt, R. W. Short, D. H. Froula, J. Katz, V. N. Goncharov, and I. V. Igumenshchev, Phys. Rev. Lett. **112**, 145001 (2014).
[8]D. T. Michel, A. V. Maximov, R. W. Short, J. A. Delettrez, D. Edgell, S. X. Hu, I. V. Igumenshchev, J. F. Myatt, A. A. Solodov, C. Stoeckl *et al.*, Phys. Plasmas **20**, 055703 (2013).
[9]S. X. Hu, D. T. Michel, D. H. Edgell, D. H. Froula, R. Follet, V. N. Goncharov, J. F. Myatt, S. Skupsky, and B. Yaakobi, Phys. Plasmas **20**, 032704 (2013).
[10]D. H. Froula *et al.*, Phys. Rev. Lett. **108**, 165003 (2012).
[11]R. Yan, A. V. Maximov, C. Ren, and F. S. Tsung, Phys. Rev. Lett. **103**, 175002 (2009).
[12]R. Yan, C. Ren, J. Li, A. V. Maximov, W. B. Mori, Z.-M. Sheng, and F. S. Tsung, Phys. Rev. Lett. **108**, 175002 (2012).
[13]R. Yan, J. Li, and C. Ren, Phys. Plasmas **21**, 062705 (2014).
[14]H. Wen, R. Yan, A. V. Maximov, and C. Ren, Phys. Plasmas **22**, 052704 (2015).
[15]J. F. Drake and Y. C. Lee, Phys. Rev. Lett. **31**, 1197 (1973).
[16]J. Zhang, J. F. Myatt, R. W. Short, A. V. Maximov, H. X. Vu, D. F. DuBois, and D. A. Russell, Phys. Rev. Lett. **113**, 105001 (2014).
[17]H. X. Vu, D. F. DuBois, D. A. Russell, J. F. Myatt, and J. Zhang, Phys. Plasmas **21**, 042705 (2014).
[18]H. X. Vu, D. F. DuBois, D. A. Russell, and J. F. Myatt, Phys. Plasmas **17**, 072701 (2010).
[19]H. X. Vu, D. F. DuBois, J. F. Myatt, and D. A. Russell, Phys. Plasmas **19**, 102703 (2012).
[20]H. X. Vu, D. F. DuBois, D. A. Russell, and J. F. Myatt, Phys. Plasmas **19**, 102708 (2012).
[21]S. Weber and C. Riconda, High Power Laser Sci. Eng. **3**, e6 (2015).
[22]S. Weber, C. Riconda, O. Klimo, A. Heron, and V. Tikhonchuk, Phys. Rev. E **85**, 016403 (2012).
[23]C. Riconda, S. Weber, V. Tikhonchuk, and A. Heron, Phys. Plasmas **18**, 092701 (2011).
[24]C. Z. Xiao, Z. J. Liu, D. Wu, C. Y. Zheng, and X. T. He, Phys. Plasmas **22**, 052121 (2015).
[25]M. N. Rosenbluth, Phys. Rev. Lett. **29**, 565 (1972).
[26]C. S. Liu and M. N. Rosenbluth, Phys. Fluids **19**, 967 (1976).
[27]A. Simon, R. W. Short, E. A. Williams, and T. Dewandre, Phys. Fluids **26**, 3107 (1983).
[28]C. S. Liu, M. N. Rosenbluth, and R. B. White, Phys. Fluids **17**, 1211 (1974).
[29]D. R. Nicholson and A. N. Kaufman, Phys. Rev. Lett. **33**, 1207 (1974).
[30]G. Laval, R. Pellat, and D. Pesme, Phys. Rev. Lett. **36**, 192 (1976).
[31]R. B. White, P. K. Kaw, D. Pesme, M. N. Rosenbluth, G. Laval, H. Huff, and R. Varma, Nucl. Fusion **14**, 45 (1974).
[32]R. Yan, A. V. Maximov, and C. Ren, Phys. Plasmas **17**, 052701 (2010).
[33]B. B. Afeyan and E. A. Williams, Phys. Fluids **28**, 3397 (1985).
[34]In the Rosenbluth's theory, if $V_{px}V_{sx} > 0$ (forward scattering), there is no unstable solution. However, in our analysis, this case is still considered here in order for the completeness of the theory.
[35]When $n_0 \rightarrow n_c/4$, $k_s \rightarrow 0$, so that the second term in Eq. (1) becomes non-negligible. However, this situation is valid only when n_0 is very close $n_c/4$. In the case considering convective SRS, we can ignore the first two terms.
[36]J. Bergman and B. Eliasson, Phys. Plasmas **8**, 1482 (2001).
[37]D. W. Forslund, J. M. Kindel, and E. L. Lindman, Phys. Fluids **18**, 1002 (1975).
[38]Actually, the TPD convective gain could not predict the location of absolute TPD, thus we plot the location of maximum absolute TPD growth rate with red curve in Fig. 4(b) from the Simon's work.
[39]B. Liu, H. Y. Wang, J. Liu, L. B. Fu, Y. J. Xu, X. Q. Yan, and X. T. He, Phys. Rev. Lett. **110**, 045002 (2013).
[40]C. Labaune, H. A. Baldis, B. S. Bauer, V. T. Tikhonchuk, and G. Laval, Phys. Plasmas **5**, 234 (1998).
[41]C. E. Max, E. M. Campbell, W. C. Mead, W. L. Kruer, D. W. Phillion, R. E. Turner, B. F. Lasinski, and K. G. Estabrook, "Scaling of laser-plasma interactions with laser wavelength and plasma size," in *Laser Interaction and Related Plasma Phenomena*, edited by H. Hora and G. H. Miley (Plenum Press, New York, 1984), Vol. 6, pp. 507–526.
[42]W. L. Kruer, K. Estabrook, B. F. Lasinski, and A. B. Langdon, Phys. Fluids **23**, 1326 (1980).
[43]K. Estabrook, W. L. Kruer, and B. F. Lasinski, Phys. Rev. Lett. **45**, 1399 (1980).

PHYSICAL REVIEW E **94**, 023205 (2016)

Fluid nonlinear frequency shift of nonlinear ion acoustic waves in multi-ion species plasmas in the small wave number region

Q. S. Feng,[1] C. Z. Xiao,[1,2] Q. Wang,[1] C. Y. Zheng,[1,3,4,*] Z. J. Liu,[1,3] L. H. Cao,[1,3,4] and X. T. He[1,3,4]
[1]*HEDPS, Center for Applied Physics and Technology, Peking University, Beijing 100871, China*
[2]*School of Physics and Electronics, Hunan University, Changsha 410082, People's Republic of China*
[3]*Institute of Applied Physics and Computational Mathematics, Beijing, 100094, China*
[4]*Collaborative Innovation Center of IFSA (CICIFSA), Shanghai Jiao Tong University, Shanghai, 200240, China*
(Received 2 June 2016; revised manuscript received 30 July 2016; published 24 August 2016)

The properties of the nonlinear frequency shift (NFS), especially the fluid NFS from the harmonic generation of the ion-acoustic wave (IAW) in multi-ion species plasmas, have been researched by Vlasov simulation. Pictures of the nonlinear frequency shift from harmonic generation and particle trapping are shown to explain the mechanism of NFS qualitatively. The theoretical model of the fluid NFS from harmonic generation in multi-ion species plasmas is given, and the results of Vlasov simulation are consistent with the theoretical result of multi-ion species plasmas. When the wave number $k\lambda_{De}$ is small, such as $k\lambda_{De} = 0.1$, the fluid NFS dominates in the total NFS and will reach as large as nearly 15% when the wave amplitude $|e\phi/T_e| \sim 0.1$, which indicates that in the condition of small $k\lambda_{De}$, the fluid NFS dominates in the saturation of stimulated Brillouin scattering, especially when the nonlinear IAW amplitude is large.

DOI: 10.1103/PhysRevE.94.023205

I. INTRODUCTION

Stimulated Brillouin scattering (SBS), or resonant decay of a light wave into a scattered light wave and an ion acoustic wave (IAW) [1], plays an important role in the successful ignition goal of inertial confinement fusion (ICF) [2,3]. Many methods have been used to reduce the SBS scattering level [4–7], including frequency detuning due to particle trapping [8–13], increasing linear Landau damping induced by kinetic ion heating [14,15], nonlinear damping due to wave breaking and trapping [16–19], or coupling with higher harmonics [20,21].

The nonlinear effect on the frequency and Landau damping rate of nonlinear ion acoustic waves have received renewed interest for their potential role in determining the saturation in stimulated Brillouin scattering of laser drivers in ICF [22–26]. Recently Albright *et al.* researched ion-trapping-induced and electron-trapping-induced IAW bowing and breakup; they found that SBS saturation was dominated by IAW bowing and breakup due to nonlinear frequency shift from particle trapping [27]. However, this paper will show if the wave number $k\lambda_{De}$ is small, the fluid nonlinear frequency shift from harmonic generation will dominate and should be considered especially when the wave amplitude is large.

The nonlinear behavior of an ion acoustic wave is a subject of fundamental interest to plasma physics. The total nonlinear frequency shift (NFS) comes from the harmonic generation and particle trapping. The harmonic nonlinearities in IAW were researched by Cohen *et al.* through the single specie cold ion fluid equations [22]. The fluid NFS due to harmonic nonlinearities was reinvestigated by Winjum *et al.*, although the research was about the electron plasma wave (EPW) [28]. The fluid frequency shift is positive and proportional to the square of the wave amplitude. Berger *et al.* derived the fluid NFS of IAW in single-ion specie plasmas from the isothermal cold ion fluid equations [29]. This paper will show the fluid NFS of IAW in multi-ion species plasmas.

The kinetic nonlinear frequency shift (KNFS) due to particle trapping was calculated four decades ago by Dewar [30] and Morales and O'Neil [31]. In their work, they assumed that the wave did not contain harmonics, and these derivations were for the case of an EPW and were applied to IAW in Ref. [29]. The KNFS is proportional to the square root of the wave amplitude. Recently Chapman *et al.* researched the kinetic nonlinear frequency shift in multi-ion species plasmas [32]. In their research, in the condition of the wave number $k\lambda_{De} = 1/3$, the KNFS dominated in the scope of wave amplitude researched. However, the fluid nonlinear frequency shift from harmonic generation made use of the result of single-ion specie plasmas that was derived in Ref. [29]. Therefore, it was clear that calculations of the KNFS matched Vlasov results well for low wave amplitude, but underestimated the total NFS at higher amplitudes where harmonic generation was expected to contribute a further positive frequency shift. For the system in Chapman's research, which is a multi-ion species system, the fluid NFS model should make use of the multi-ion species fluid NFS model. This paper will give a multi-ion species fluid NFS model to improve the single-ion specie fluid NFS model. And the multi-ion species fluid NFS model should have a further positive frequency shift than the single-ion specie fluid NFS model and will fit the Vlasov results better.

In this paper, when the wave number is small such as $k\lambda_{De} = 0.1$, the fluid NFS from harmonic generation is important, especially when the wave amplitude is large. And the multi-ion species fluid NFS model is given to calculate the fluid NFS. The Vlasov simulation results are verified to be consistent with the theoretical model of the total NFS including fluid NFS and kinetic NFS in multi-ion species plasmas. When $k\lambda_{De} = 0.1$, the total NFS can reach a value of nearly 15% when the nonlinear IAW amplitude $|e\phi/T_e|$ reaches about 0.1, which is much larger than the total NFS in the condition of

*zheng_chunyang@iapcm.ac.cn

$k\lambda_{De} = 1/3$ [32]. This indicates that when $k\lambda_{De}$ is small, the fluid NFS will reach a large level, especially when the wave amplitude is large, thus dominating in the total NFS so that the fluid NFS dominates in the saturation of SBS by IAW bowing and breakup [27].

II. THEORETICAL MODEL

A. Fluid theory: Nonlinear frequency shift from harmonic generation in multi-ion species plasmas

Taking the charge and the number fraction of the ion specie α to be Z_α and f_α, the electron density, n_e, and the total ion density, n_i, are related through the average charge, $\bar{Z}$, i.e., $n_e = n_i \bar{Z}$. The ion number densities are given by

$$n_\alpha = f_\alpha n_i = f_\alpha n_e/\bar{Z}, \quad \bar{Z} = \sum_\alpha f_\alpha Z_\alpha. \tag{1}$$

Here multi-ion species and isothermal Boltzmann electrons are considered in this cold ion fluid model, and the nonlinear frequency shift from harmonic generation is derived from the multispecies isothermal cold ions fluid equations:

$$\frac{\partial n_\alpha}{\partial t} + \frac{\partial n_\alpha v_\alpha}{\partial x} = 0,$$
$$\frac{\partial v_\alpha}{\partial t} + v_\alpha \frac{\partial v_\alpha}{\partial x} = -C_\alpha^2 \frac{\partial \tilde{\phi}}{\partial x}, \quad -\frac{T_e}{e}\frac{\partial^2 \tilde{\phi}}{\partial x^2} + 4\pi e n_{e0} \exp(\tilde{\phi}) = 4\pi \sum_\alpha q_\alpha n_\alpha, \tag{2}$$

where $C_\alpha = \sqrt{Z_\alpha T_e/m_\alpha}$ is the sound velocity of ion α, and $\tilde{\phi} = e\phi/T_e$ is the normalized electrostatic potential. The initial charge satisfies charge conservation, i.e., $\sum_\alpha q_\alpha n_{\alpha 0} = e n_{e0}$. Following the work in Ref. [29], one expands the variables in a Fourier series:

$$(\tilde{\phi}, n_\alpha, v_\alpha) = (\tilde{\phi}_0, n_{\alpha 0}, v_{\alpha 0}) + \frac{1}{2}\sum_{l \neq 0} (\tilde{\phi}_l, n_{\alpha l}, v_{\alpha l}) \exp[il(kx - \omega t)], \tag{3}$$

with the condition $(\tilde{\phi}_{-l}, n_{\alpha,-l}, v_{\alpha,-l}) = (\tilde{\phi}_l, n_{\alpha l}, v_{\alpha l})^*$ and keeping terms for $l = 0, \pm 1, \pm 2$ up to second order with $\exp\tilde{\phi} \simeq 1 + \tilde{\phi} + \frac{1}{2}\tilde{\phi}^2$. Retaining the same exponents in the Fourier series, for $v_0 = 0, n_{\alpha 0} = f_\alpha n_{i0} = f_\alpha n_{e0}/\bar{Z} = \text{const}$, one finds from Eq. (2) for $l = 0$

$$e n_{e0}\left(1 + \tilde{\phi}_0 + \frac{1}{2}\tilde{\phi}_0^2 + \frac{1}{4}|\tilde{\phi}_1|^2 + \frac{1}{4}|\tilde{\phi}_2|^2\right) = \sum_\alpha q_\alpha n_{\alpha 0}, \tag{4}$$

and by conservation of charge, $e n_{e0} = \sum_\alpha q_\alpha n_{\alpha 0}$, it thus obtains that

$$\tilde{\phi}_0 + \frac{1}{2}\tilde{\phi}_0^2 = -\frac{1}{4}|\tilde{\phi}_1|^2 - \frac{1}{4}|\tilde{\phi}_2|^2. \tag{5}$$

For $l = 1$, the equations are

$$-i\omega n_{\alpha 1} + ikn_{\alpha 0} v_{\alpha 1} = -i\frac{1}{2}kn_{\alpha,-1}v_{\alpha 2} - i\frac{1}{2}kn_{\alpha 2}v_{\alpha,-1},$$
$$-i\omega v_{\alpha 1} + ikC_\alpha^2 \tilde{\phi}_1 = -i\frac{1}{2}kv_{\alpha,-1}v_{\alpha 2},$$
$$\left[\frac{T_e}{e}k^2 + 4\pi e n_{e0}(1 + \tilde{\phi}_0)\right]\tilde{\phi}_1 - 4\pi \sum_\alpha q_\alpha n_{\alpha 1} = -\frac{1}{2}4\pi e n_{e0}\tilde{\phi}_2\tilde{\phi}_{-1}. \tag{6}$$

For $l = 2$, the corresponding equations are

$$-2i\omega n_{\alpha 2} + i2kn_{\alpha 0}v_{\alpha 2} + ikn_{\alpha 1}v_{\alpha 1} = 0,$$
$$-2i\omega v_{\alpha 2} = -i2k\left[C_\alpha^2\tilde{\phi}_2 + \frac{1}{4}v_{\alpha 1}^2\right],$$
$$\left[4\frac{T_e}{e}k^2 + 4\pi e n_{e0}(1 + \tilde{\phi}_0)\right]\tilde{\phi}_2 + \frac{1}{4}4\pi e n_{e0}\tilde{\phi}_1^2 = 4\pi \sum_\alpha q_\alpha n_{\alpha 2}. \tag{7}$$

Defining $C_{2s}^2 = \sum_\alpha \frac{Z_\alpha n_{\alpha 0}}{n_{e0}} C_\alpha^2$, $C_{4s}^4 = \sum_\alpha \frac{Z_\alpha n_{\alpha 0}}{n_{e0}} C_\alpha^4$, and $C_{sl}^2 = l^2k^2C_{2s}^2/(1 + l^2k^2\lambda_{De}^2)$, by keeping terms only to second order in $\tilde{\phi}_1$, then one obtains the relation between $\tilde{\phi}_2$ and $\tilde{\phi}_1$:

$$\tilde{\phi}_2 = A_{2\phi}\tilde{\phi}_1^2, \tag{8}$$

where

$$A_{2\phi} = \frac{1}{4}\frac{C_{s2}^2}{4\omega^2 - C_{s2}^2}\left[3\frac{k^2C_{4s}^4}{\omega^2 C_{2s}^2} - \frac{\omega^2}{k^2C_{2s}^2}\right]. \tag{9}$$

Using $\tilde{\phi}_0 \simeq -|\tilde{\phi}_1|^2/4$ and defining $C_{6s}^6 = \sum_\alpha Z_\alpha n_{\alpha 0} C_\alpha^6/n_{e0}$ correspondingly, retaining harmonic corrections to the linear dispersion relation again to the second order in $\tilde{\phi}_1$ and keeping the lowest nonlinear term, the nonlinear equation for the amplitude of $\tilde{\phi}_1$ is obtained as follows:

$$(\omega^2 - C_{s1}^2)\tilde{\phi}_1 = (A_{2\phi}C_{A_{2\phi}} + C_2)|\tilde{\phi}_1|^2\tilde{\phi}_1, \tag{10}$$

where

$$C_{A_{2\phi}} = \frac{C_{s1}^2}{2}\left[-\frac{\omega^2}{k^2C_{2s}^2} + 3\frac{k^2C_{4s}^4}{\omega^2C_{2s}^2}\right], \quad C_2 = \frac{C_{s1}^2}{8}\left[2\frac{\omega^2}{k^2C_{2s}^2} + 5\frac{k^4C_{6s}^6}{\omega^4C_{2s}^2}\right]. \tag{11}$$

Taking the effective fundamental IAW frequency after accounting for harmonic effects $\omega_{\text{harm}} = \omega_0 + \delta\omega_{\text{harm}}$ and the linear fundamental frequency of IAW $\omega_0 \simeq C_{s1} = kC_{2s}/\sqrt{1 + k^2\lambda_{De}^2} = kC_\alpha\sqrt{\sum_\alpha \frac{Z_\alpha n_{\alpha 0}}{n_{e0}}}/\sqrt{1 + k^2\lambda_{De}^2}$, then the frequency shift of the first harmonic due to the inclusion of the second harmonic terms in multispecies ion plasmas is given by

$$\frac{\delta\omega_{\text{harm}}}{\omega_0} = \frac{1}{2}\frac{\Delta}{\omega^2}|\tilde{\phi}_1|^2 = L|\tilde{\phi}_1|^2, \tag{12}$$

where $L = \frac{1}{2}\frac{\Delta}{\omega^2}$ is the fluid NFS coefficient and Δ is defined by

$$\Delta \equiv A_{2\phi}C_{A_{2\phi}} + C_2. \tag{13}$$

By combining Eqs. (11)–(13), then the final result of $\delta\omega_{\text{harm}}/\omega$ is related to the wave number $k\lambda_{De}$ and the second order in $\tilde{\phi}_1$:

$$\frac{\delta\omega_{\text{harm}}}{\omega_0} = L|\tilde{\phi}_1|^2 = \left[\frac{-2A_{2\phi}+1}{8(1+k^2\lambda_{De}^2)} + \frac{(1+k^2\lambda_{De}^2)^2}{(C_{2s}^2)^2} \times\left(\frac{6A_{2\phi}C_{4s}^4}{1+k^2\lambda_{De}^2} + 5\frac{C_{6s}^6}{C_{2s}^2}\right)\right]|\tilde{\phi_1}|^2, \quad (14)$$

where the second harmonic coefficient $A_{2\phi}$ is defined in Eq. (9).

As shown in Fig. 1(a), when $k\lambda_{De}$ is small, the second harmonic coefficient $A_{2\phi}$ is large and the contribution of $\tilde{\phi}_2$ is strong for the reason that the second harmonic becomes more resonant as $4\omega^2 - C_{s2}^2 = 4C_{s1}^2 - C_{s2}^2$ decreases strongly with $k\lambda_{De}$ decreasing. Figure 1(c) shows that with $k\lambda_{De}$ increasing, the fluid NFS coefficient first decreases quickly and then slowly increases, the turning point is $k\lambda_{De} \simeq 0.5$, and the SBS relevant wave number is usually $k\lambda_{De} < 0.5$. To research the fluid NFS in multi-ion species plasmas, CH (1:1) plasmas are taken as an typical multi-ion species plasmas. We can see that the second harmonic coefficient $A_{2\phi}$ of CH plasmas is slightly larger than that of the single-ion specie plasmas, and as a result, the fluid NFS coefficient L from the harmonic generation in CH plasmas is slightly higher than that of the single-ion specie plasmas as shown in Figs. 1(a) and 1(c). In particular, when $k\lambda_{De}$ is small, such as $k\lambda_{De} = 0.1$, the harmonic effect is obvious, and the deviation of the harmonic effect between the single-ion specie plasmas and the multi-ion species plasmas is also obvious, as shown in Figs. 1(b) and 1(d). When the number fraction $f_H = 0$ or $f_C = 0$, the results from the multi-ion species plasmas including $A_{2\phi}$ and L are the same with that of the single-ion specie plasmas. When the number fraction varies, the results from the multi-ion species plasmas including $A_{2\phi}$ and L varies correspondingly.

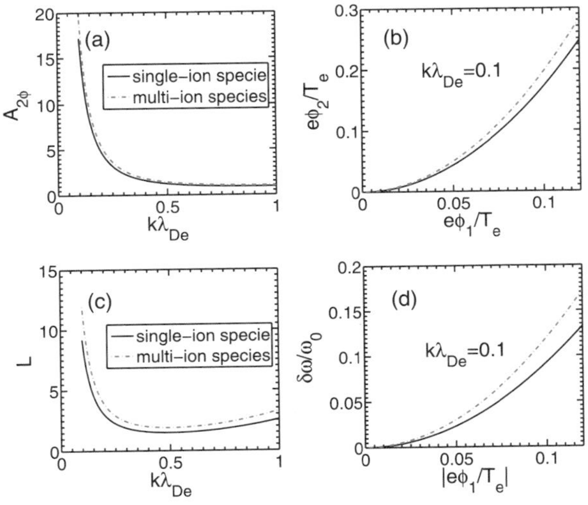

FIG. 1. (a) The coefficient of the second harmonic $A_{2\phi}$ and (c) the fluid NFS coefficient L as a function of the wave number $k\lambda_{De}$. In the condition of $k\lambda_{De} = 0.1$, (b) the second harmonic amplitude $e\phi_2/T_e$ and (d) the fluid NFS $\delta\omega_{\text{harm}}/\omega_0$ as a function of $|e\phi_1/T_e|$. The black solid line presents for single-ion specie plasmas and the red dot-dash line for CH (1:1) plasmas.

B. Kinetic theory: Nonlinear frequency shift from particles trapping in multi-ion species plasmas

The linear kinetic theory of ion-acoustic waves (IAWs) in homogeneous, nonmagnetized plasmas consisting of multi-species ions is considered. Assume that the same temperature of all ions equals T_i and the electron temperature equals T_e. The linear dispersion relation of the IAW in multi-ion species plasmas is given by

$$\epsilon_L(\omega,k) = 1 + \sum_j \frac{1}{(k\lambda_{Dj})^2}[1+\xi_j Z(\xi_j)] = 0, \quad (15)$$

where $\xi_j = \omega/(\sqrt{2}kv_{tj})$, and $Z(\xi_j)$ is the dispersion function, and j represents for electron, H ion, or C ion; $v_{tj} = \sqrt{T_j/m_j}$ is the thermal velocity of particle j; $\lambda_{Dj} = \sqrt{T_j/4\pi n_j Z_j^2 e^2}$ is the Debye length; and T_j, m_j, n_j, Z_j are the temperature, mass, density, and charge number of specie j, respectively.

The derivation of the kinetic nonlinear frequency shift (KNFS) of electrostatic waves resulting from particles trapping was originally presented in Refs. [30] and [31] and is applied to the case of IAWs in single ion specie plasmas (such as H or He ion) in Ref. [29]. The following KNFS in multi-ion species plasmas is given by Berger:

$$\delta\omega_{\text{NL}}^{\text{kin}} \simeq -\left[\frac{\partial\epsilon_L(\omega_L)}{\partial\omega}\right]^{-1} \sum_j \alpha_j \frac{\omega_{pj}^2}{k^2}\Delta v_{tr,j}\frac{d^2(f_0/N)}{dv^2}|_{v_\phi}, \quad (16)$$

where the initial distribution f_0 is assumed to be Maxwellian distribution, then

$$\frac{d^2(f_0/N)}{dv^2} = \frac{1}{\sqrt{2\pi}}\frac{1}{v_{th}^3}[(v/v_{th})^2 - 1]e^{-\frac{1}{2}(\frac{v}{v_{th}})^2},$$

and $\Delta v_{tr,j} = 2\sqrt{\frac{|q_j|\phi_0}{m_j}}$, where α is a constant which is dependent on how the wave was excited, $\alpha_j = \alpha_{\text{sud}} = 0.823$ for sudden excitation, $\alpha_j = \alpha_{ad} = 0.544$ for adiabatic excitation, and j represents an H ion or C ion; however, the excitation condition of electrons is adiabatic, i.e., $\alpha_e = \alpha_{ad} = 0.544$ [29]. In this paper, adiabatic excitation is used for it is more relevant to the conditions here and stimulated Brillouin scattering processes.

III. VLASOV SIMULATION

One dimension in space and velocity (1D1V) Vlasov-Poisson code [33,34] is taken to excite the nonlinear IAW and research the properties of the nonlinear frequency shift of the nonlinear IAW in CH plasmas. The split method [35,36] is taken to solve Vlasov equation: we split the time-stepping operator into free streaming in x and motion in v_x, and then we can get the advection equations. A third order Van Leer scheme (VL3) [37,38] is taken to solve the advection equations. A neutral, fully ionized, nonmagnetized CH plasmas (1:1 mixed) with the same temperature of all ion species ($T_H = T_C = T_i$) is considered. To solve the particles' behavior, the phase space domain is $[0,L_x] \times [-v_{\max},v_{\max}]$, where $L_x = 2\pi/k$ is the spatial scale discretized with $N_x = 128$ grid points in the spatial domain and $v_{\max} = 8v_{tj}$ (j represents electrons, H ions, or C ions) is the velocity scale with $N_v = 256$ grid points in the velocity domain. The periodic boundary condition is taken in the spatial domain, and the time step is $dt = 0.1\omega_{pe}^{-1}$. To generate a driven IAW, one considers the form of the external driving electric field (driver) as

$$\tilde{E}_d(x,t) = \tilde{E}_d(t)\sin(kx - \omega_d t), \tag{17}$$

where tilde presents for normalization $\tilde{E}_d = eE_d\lambda_{De}/T_e$. Here ω_d and k are the frequency and the wave number of the driver, respectively. $\tilde{E}_d(t)$ is the envelope of the driver only related with time

$$\tilde{E}_d(t) = \frac{\tilde{E}_d^{\max}}{1 + \left(\frac{t-t_0}{\frac{1}{2}t_0}\right)^{10}}, \tag{18}$$

where $\tilde{E}_d^{\max} = eE_d^{\max}\lambda_{De}/T_e$ is the maximum amplitude of the driver, and t_0 is the duration time of the peak driving electric filed. The form of the envelope of the driver can ensure that the waves are driven slowly by the driver up to a finite amplitude; thus a so-called "adiabatic" distribution of resonance particles is formed [29,30]. Otherwise, if the wave is initialized suddenly with a large amplitude, a so-called "sudden" distribution of resonance particles will be formed [29–31].

In this paper, the fundamental frequency of the IAW calculated by the linear dispersion relation is chosen as the driver frequency, i.e., $\omega_d = \omega_L$, which is close to the resonant frequency of the small-amplitude nonlinear IAW when $k\lambda_{De}$ is small. The amplitude of the driver $\tilde{E}_d^{\max}$ varies to excite different amplitude of nonlinear IAW. As shown in Fig. 2, the examples of the driver amplitude $\tilde{E}_d^{\max} = 0.03$ for the fast mode and $\tilde{E}_d^{\max} = 0.0533$ for the slow mode are taken to excite the nonlinear IAW with the amplitude $\tilde{E} \approx 0.01$. The external driving electric field is on from 0 to $2 \times 10^5\omega_{pe}^{-1}$ with duration time $t_0 = 1 \times 10^5\omega_{pe}^{-1}$, which is much longer than the bounce time of the particles $\tau_{bj} = 2\pi/\sqrt{q_jkE/m_j}$ so as to ensure the excitation of the nonlinear IAW. After the driver is off, the electric field oscillates at almost constant amplitude, with a value that depends on the the driver amplitude $\tilde{E}_d^{\max}$. We notice that impulsive spikes appear in the electric field and the electric signal appears to be composed of many frequencies. Fourier analysis (shown in Fig. 3) will reveal the development of the harmonics when the driver is on and off, and also the frequency shift due to the harmonic generation.

In Fig. 3 the frequency spectra analyses of the time period when the driver is on, i.e., $\omega_{pe}t \in [0,2 \times 10^5]$, are shown as the black dashed lines, which are labeled as ω_0. When the driver is off and the electric field of IAW reaches constant amplitude, i.e., $\omega_{pe}t \in [3 \times 10^5, 4.9 \times 10^5]$, the frequency spectra are shown as the red solid lines, which are labeled as $\omega_0 + \delta\omega$. When the driver is on, the amplitude of the fundamental frequency ω_0 is much larger than that of the harmonics. In this period the harmonic is weak and the frequency ω_0 is near the driver frequency ω_d. However, when the driver is off and the electric field of the nonlinear IAW reaches constant finite amplitude, the harmonics, especially the second harmonics, grow and reach an amplitude level comparable to that of the fundamental mode driven by the driver. In this period, the frequency of the fundamental modes will shift with a quantity of $\delta\omega$ relative to ω_0. The same process appears for the fast mode and the slow mode. These cases are taken to qualitatively show

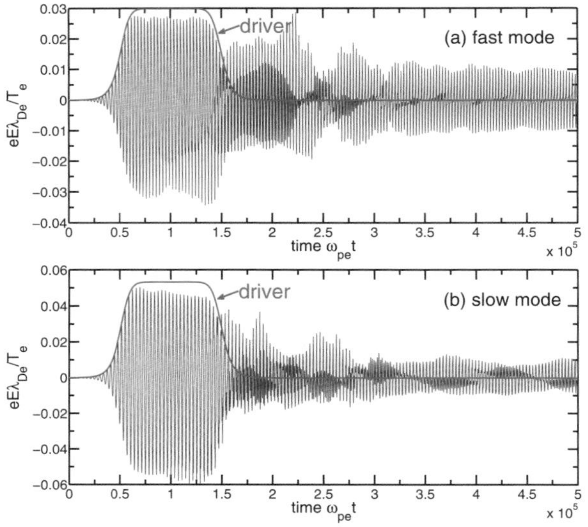

FIG. 2. Time evolution of the electric field, calculated at a fixed point $x_0 = 5\lambda_{De}$, for (a) $T_i/T_e = 0.1$, the fast IAW mode (the frequency of the driver $\omega_d = 1.965 \times 10^{-3}\omega_{pe}$, the maximum amplitude of the driver $eE_d^{\max}\lambda_{De}/T_e = 0.03$), and (b) $T_i/T_e = 0.5$, the slow IAW mode ($\omega_d = 1.75 \times 10^{-3}\omega_{pe}, eE_d^{\max}\lambda_{De}/T_e = 0.0533$) in the condition of $k\lambda_{De} = 0.1$. The red line is the envelope of the driver.

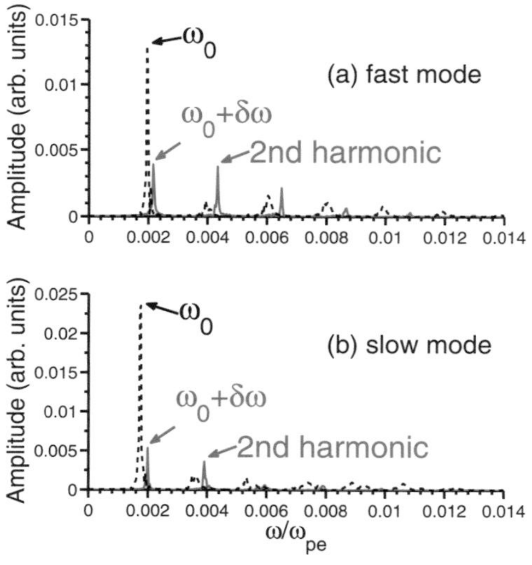

FIG. 3. The frequency spectra of the electric field at $x_0 = 5\lambda_{De}$ and $\omega_{pe}t \in [0,2 \times 10^5]$ (black dashed line), $\omega_{pe}t \in [3 \times 10^5, 4.9 \times 10^5]$ (red solid line) for (a) the fast mode and (b) the slow mode. This corresponds to Fig. 2.

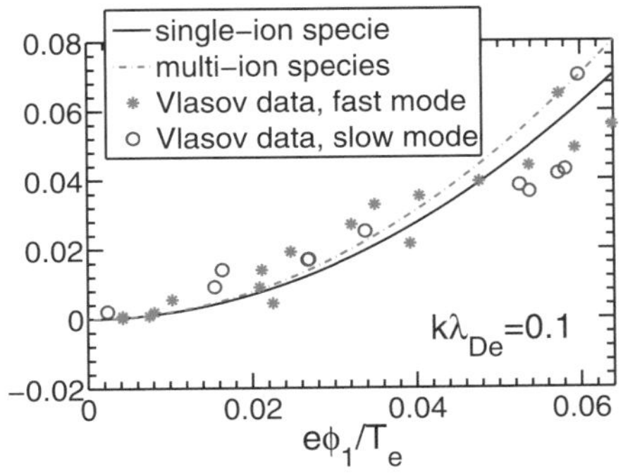

FIG. 4. The relation of the second harmonic amplitude and the first harmonic amplitude for the fast mode and the slow mode. The simulation results are compared to the fluid theory including the cold ion fluid model of single-ion specie and multi-ion species.

the nonlinear frequency shift of the IAW at the IAW electric field amplitude $\tilde{E} \approx 0.01$ corresponding to the condition of Fig. 2. The quantitatively calculations will be shown in Fig. 6. The quantity of the frequency shift $\delta\omega$ is related to not only the amplitude of the nonlinear IAW but also the wave number $k\lambda_{De}$, which will be discussed in Sec. IV.

Figure 4 shows that the second harmonic amplitude $e\phi_2/T_e$ varies with the fundamental mode amplitude $|\tilde{\phi}_1|$ (or the first harmonic amplitude, $\tilde{\phi}_1 = e\phi_1/T_e$, corresponding to the mode labeled $\omega_0 + \delta\omega$ in Fig. 3). With the fundamental mode amplitude $|\tilde{\phi}_1|$ increasing, the harmonics, especially the second harmonic, will increase correspondingly, which is consistent with the theoretical result of Fig. 1(b) from Eq. (8) in the appropriate range of $\tilde{\phi}_1$. However, if the fundamental mode amplitude $\tilde{\phi}_1$ is too large ($\tilde{\phi}_1 \gtrsim 0.06$), the theoretical curve from Eq. (8) will fail to fit the Vlasov simulation data well due to the large-amplitude higher-order harmonics such as the third even the fourth harmonics generation. Thus the multi-ion species fluid NFS model should be applied in the appropriate range of IAW amplitude. In the condition of the small wave number, such as $k\lambda_{De} = 0.1$, the harmonic effect is obvious, especially when the nonlinear IAW amplitude is large. As a result, the fluid NFS from harmonic generation will dominate in the total NFS. In our paper and also in Ref. [32], the multi-ion species plasmas are considered. Therefore, the multi-ion species fluid NFS should be taken rather than the single-ion specie fluid NFS that was taken in Chapman's research. Detailed discussion is given as shown in Sec. IV.

The total nonlinear frequency shift comes from the harmonic generation (fluid) and the particle trapping (kinetic). As shown in Fig. 5, the distributions and the phase pictures of particles for the fast mode and the slow mode are given to clarify the wave and particle interaction, and thus the nonlinear frequency shift from particles trapping. After a certain number of bounce times, a quasisteady, BGK-like state [39] is thus established. The phase velocities of both the fast mode and the slow mode are lower than the thermal velocity of electrons, thus the wave will gain energy from the electrons, and as a result, electrons will provide a positive frequency shift of the nonlinear IAW. The contribution of H ions and C ions to the NFS of the fast mode and the slow mode is different for the different phase velocity of the modes compared to the ions' thermal velocity. For the fast mode, the phase velocity is two to three times of the thermal velocity of H ions and much larger than that of C ions, thus the wave will give energy to H ions by trapping and nearly cannot trap C ions; as a result, H ions will provide a negative NFS, while C ions make no contribution to the NFS of the fast mode. However, for the slow mode, the phase velocity is close to the thermal velocity of H ions

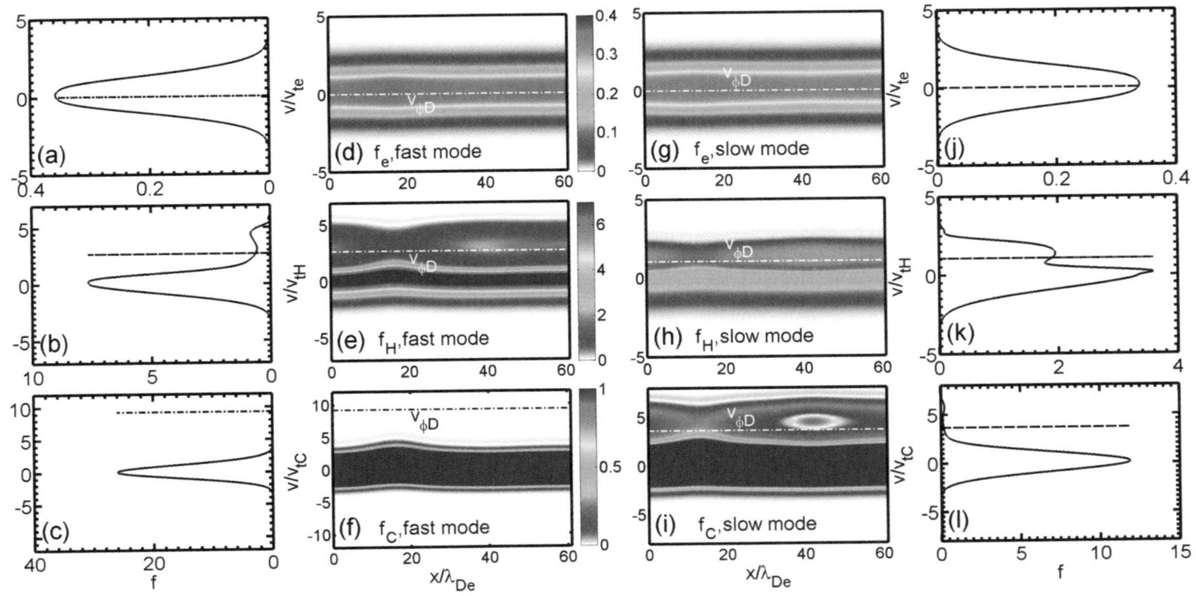

FIG. 5. The phase pictures and the corresponding distributions of electrons, H ions and C ions for the fast mode and the slow mode at the specific time $\omega_{pe}t_0 = 4 \times 10^5$, $\tilde{E} \approx 0.01$. $v_{\phi D}$ is the phase velocity of the driver.

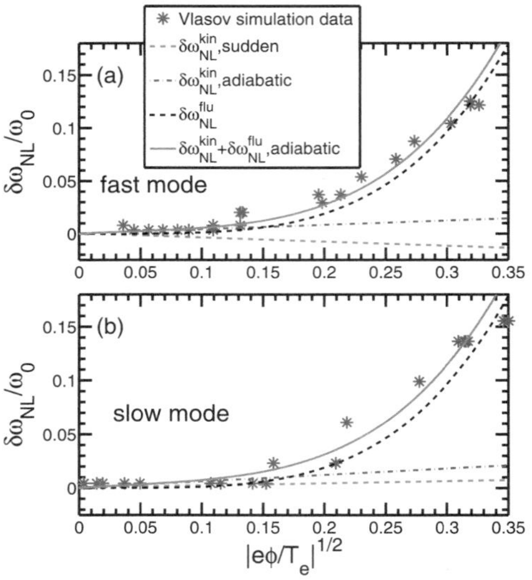

FIG. 6. The nonlinear frequency shift (NFS) of (a) the fast mode, $T_i/T_e = 0.1$ and (b) the slow mode, $T_i/T_e = 0.5$ compared to fluid theory in the multi-ion species plasmas (Sec. II A) and the kinetic theory (Sec. II B) in the condition of $k\lambda_{De} = 0.1$, where "sudden" and "adiabatic" represent for the excitation condition of the ions (C or H ions), and the excitation condition of the electrons is adiabatic, i.e., $\alpha_e = \alpha_{ad}$.

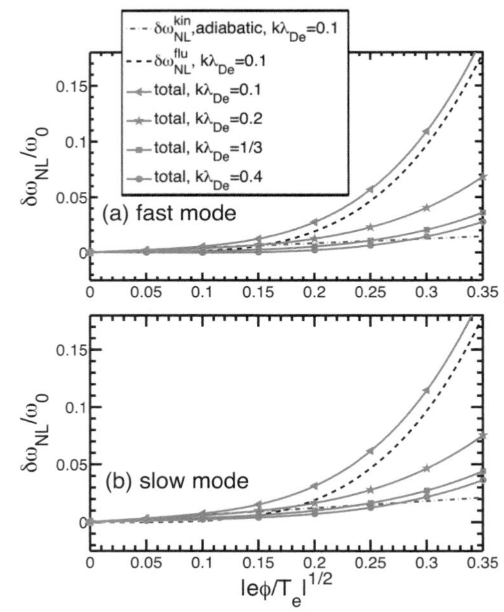

FIG. 7. The analytical results of the total nonlinear frequency shift of (a) the fast mode, $T_i/T_e = 0.1$ and (b) the slow mode, $T_i/T_e = 0.5$ for various values of $k\lambda_{De}$, where "adiabatic" is considered as the excitation condition of the ions (C or H ions). Here "total" represents the total NFS, i.e., the kinetic NFS plus the multi-ion species fluid NFS $\delta\omega_{\rm NL}^{\rm kin} + \delta\omega_{\rm NL}^{\rm flu}$, multi-ion species.

and three to four times of that of C ions, and the wave and H ions will nearly not exchange energy while the wave will give energy to C ions by trapping; thus H ions nearly make no contribution to the NFS while C ions provide a negative NFS of the slow mode. This is the explanation of KNFS from each species for both the fast mode and the slow mode in the view of energy exchange of the wave and particles, which is consistent with the explanation of KNFS in Ref. [32] through theoretical analysis.

In Fig. 6 the Vlasov simulation data of the NFS of nonlinear IAWs in the condition of $k\lambda_{De} = 0.1$ is given when the electric field amplitude of nonlinear IAW varies. The fundamental frequency of the nonlinear IAW ω_0 in the Vlasov data is chosen as the frequency calculated by the dispersion relation with no damping $\mathrm{Re}[\epsilon_L(\mathrm{Re}(\omega),k)] = 0$ [40]. In the condition of $k\lambda_{De} = 0.1$, the fluid NFS from harmonic generation is larger than the kinetic NFS from particle trapping when $|e\phi/T_e|^{1/2} \gtrsim 0.15$. And the total NFS, $\delta\omega_{\rm NL}/\omega_0$, can reach as large as nearly 15% when the nonlinear IAW amplitude $|e\phi/T_e|$ reaches 0.1, in which the Vlasov simulation data match the theory analytical curve ($\delta\omega_{\rm NL}^{\rm kin} + \delta\omega_{\rm NL}^{\rm flu}$) well. These results indicate the total NFS can reach a relative large value when the wave number $k\lambda_{De}$ is small, for the reason that the NFS from harmonic generation dominates and increases quickly with the electric field amplitude, which is much larger than that researched [32] in the condition of $k\lambda_{De} = 1/3$. A comparison of the total NFS of the modes in the condition of $k\lambda_{De} = 0.1$ and $k\lambda_{De} = 1/3$ are shown in Fig. 7. The Vlasov simulation data are consistent with the curve of the total NFS including fluid NFS and kinetic NFS, $\delta\omega_{\rm NL}^{\rm flu} + \delta\omega_{\rm NL}^{\rm kin}$, where adiabatic is taken as the excitation condition of the ions and the fluid NFS analysis $\delta\omega_{\rm NL}^{\rm flu}$ is given by the multispecies cold ions fluid model. This verifies that the theoretical model of the fluid NFS in multi-ion species plasmas is valid in appropriate scale of $\tilde{\phi}$ (not too large). In the research of Chapman, the theory analysis of NFS from harmonic generation used the single-specie cold ion fluid model, which was derived in Ref. [29]. It was clear that calculations of $\delta\omega_{\rm NL}^{\rm kin}$ from the kinetic NFS theory matched Vlasov data well for low $\tilde{\phi}$, in which the kinetic NFS dominated, but, underestimated the total NFS $\delta\omega_{\rm NL}$ at higher amplitudes of $\tilde{\phi}$ because harmonic generation in multi-ion species plasmas will contribute a further positive frequency shift than single-ion specie plasmas, as shown in Fig. 8. The comparison of the fluid NFS (or the total NFS) in the multi-ion species plasmas and the single-ion specie plasmas in the condition of Ref. [32] are shown in Fig. 8.

However, when the nonlinear IAW amplitude is too large, the higher-order harmonics such as the third and even the fourth harmonic will be excited resonantly to a large amplitude, thus the theoretical model of the fluid NFS from harmonic generation in multi-ion species plasmas should retain the higher order of the variables in a Fourier series.

IV. DISCUSSION

Figure 7 shows the total NFS (fluid plus kinetic) of CH plasmas (C:H=1:1) as a function of wave amplitude for various values of the wave number $k\lambda_{De}$. We can see that the total NFS in the condition of $k\lambda_{De} = 0.1$ is much larger than that in the

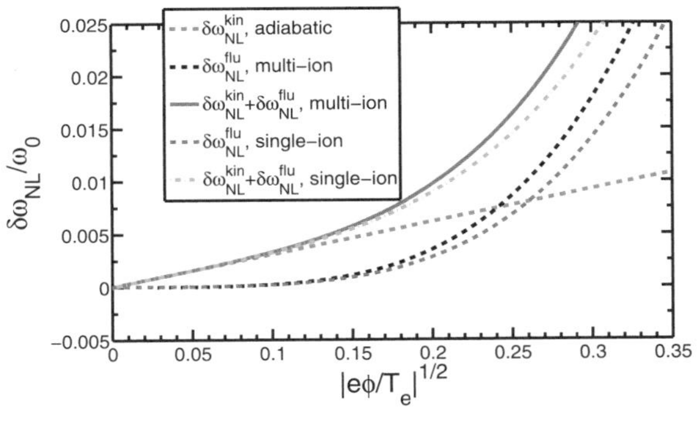

FIG. 8. The comparison of the multi-ion species fluid NFS (the total NFS: $\delta\omega_{\mathrm{NL}}^{\mathrm{kin}} + \delta\omega_{\mathrm{NL}}^{\mathrm{flu}}$, multi-ion) and the single-ion specie fluid NFS (the total NFS: $\delta\omega_{\mathrm{NL}}^{\mathrm{kin}} + \delta\omega_{\mathrm{NL}}^{\mathrm{flu}}$, single-ion) for the slow mode in the condition of Ref. [32], $T_i/T_e = 0.5, k\lambda_{De} = 1/3$, corresponding to Fig. 3(d) in Ref. [32].

condition of $k\lambda_{De} = 1/3$, which is the condition in Ref. [32]. The reason is that for small $k\lambda_{De}$, harmonic generation plays a prominent role in NFS. With $k\lambda_{De}$ increasing, the fluid NFS decreases obviously as shown in Fig. 1(c), and thus the total NFS decreases correspondingly.

To compare the multi-ion species fluid NFS and Ref. [32], the condition of the slow mode $T_i/T_e = 0.5, k\lambda_{De} = 1/3$, which is the same as Ref. [32], is taken as an example. The kinetic theory is the same as the research there (as shown in Sec. II B), and the excitation condition of the ions is "adiabatic". As shown in Fig. 8, especially when the IAW amplitude is large, the multi-ion species fluid NFS (Sec. II A) is larger than the single-ion specie fluid NFS, which was derived in Ref. [29] and made use of in Ref. [32]. Thus, the total NFS (kinetic NFS plus multi-ion species fluid NFS) is larger than the total NFS in Ref. [32] (kinetic NFS plus single-ion specie fluid NFS), especially when the IAW amplitude is large. This result of multi-ion species fluid NFS will make the total NFS fit the Vlasov data in Fig. 3(d) of Ref. [32] better, especially at higher IAW amplitude. This fluid NFS model of multi-ion species plasmas is a correction to Ref. [32], which just considered the single-ion specie fluid NFS model that was derived in Ref. [29].

When $f_C = 0, f_H = 1$, i.e., in the H plasmas, the fluid NFS from the multispecies cold ions fluid model is consistent with that from the single specie cold ion fluid model [29]. This indicates that the multi-ion species NFS model is valid and can also cover the single-ion specie NFS model if we take $f_C = 0$ or $f_H = 0$. We can find the single-ion specie fluid NFS model is independent of the ion specie. However, the multi-ion species fluid NFS model is related to the ion specie and the number fraction of the ion f_i. This paper takes $f_C = f_H = 0.5$ as an example; the fluid NFS of multi-ion species plasmas is larger than that of single-ion specie plasmas for the effect of mass and charge in multi-ion species plasmas, and thus harmonic generation in multi-ion species plasmas contributes a further positive frequency shift. This multi-ion species fluid NFS model gives the explanation for why the calculations of total NFS in Fig. 3(d) of Ref. [32] underestimate the real NFS $\delta\omega_{\mathrm{NL}}$ at higher IAW amplitudes.

V. CONCLUSIONS

In summary, the nonlinear frequency shift from harmonic generation derived from the multi-species cold ions fluid model has been given to calculate the fluid NFS and verified to be consistent with the Vlasov results for both the fast mode and the slow mode. And this multi-ion species fluid NFS model can be applied to many multi-ion species plasmas. Pictures of the NFS of nonlinear IAW from the harmonic generation and the particle trapping are shown to explain the NFS process. The fluid NFS from harmonic generation is related to not only the nonlinear IAW amplitude but also the wave number $k\lambda_{De}$. When the wave number $k\lambda_{De} = 0.1$, the fluid NFS from harmonic generation dominates in the scale of $|e\phi/T_e|^{1/2} \gtrsim 0.15$ and will reach as large as $\sim 15\%$ when $|e\phi/T_e|^{1/2} \sim 0.35$ in which the theory analytical calculations are consistent to the Vlasov simulation results. This indicates that the fluid NFS dominates in the saturation of SBS in the condition of small $k\lambda_{De}$, especially when the wave amplitude is large.

ACKNOWLEDGMENTS

We are pleased to acknowledge useful discussions with K. Q. Pan. This research was supported by the National Natural Science Foundation of China (Grant Nos. 11575035, 11475030, and 11435011) and National Basic Research Program of China (Grant No. 2013CB834101).

[1] W. L. Kruer, *The Physics of Laser Plasma Interactions* (Addison-Wesley, New York, 1998), pp. 87–95.

[2] J. D. Lindl, P. Amendt, R. L. Berger, S. G. Glendinning, S. H. Glenzer, S. W. Haan, R. L. Kauffman, O. L. Landen, and L. J. Suter, Phys. Plasmas **11**, 339 (2004).

[3] S. H. Glenzer, B. J. MacGowan, P. Michel, N. B. Meezan, L. J. Suter, S. N. Dixit, J. L. Kline, G. A. Kyrala, D. K. Bradley, D. A. Callahan, E. L. Dewald, L. Divol, E. Dzenitis, M. J. Edwards, A. V. Hamza, C. A. Haynam, D. E. Hinkel, D. H. Kalantar, J. D. Kilkenny, O. L. Landen, J. D. Lindl, S. LePape, J. D. Moody, A. Nikroo, T. Parham, M. B. Schneider, R. P. J. Town, P. Wegner, K. Widmann, P. Whitman, B. K. F. Young, B. Van Wonterghem, L. J. Atherton, and E. I. Moses, Science **327**, 1228 (2010).

[4] P. Neumayer, R. L. Berger, L. Divol, D. H. Froula, R. A. London, B. J. MacGowan, N. B. Meezan, J. S. Ross, C. Sorce, L. J. Suter, and S. H. Glenzer, Phys. Rev. Lett. **100**, 105001 (2008).

[5] L. Divol, B. I. Cohen, E. A. Williams, A. B. Langdon, and B. F. Lasinski, Phys. Plasmas **10**, 3728 (2003).

[6] R. L. Berger, C. H. Still, E. A. Williams, and A. B. Langdon, Phys. Plasmas **5**, 4337 (1998).

[7] K. Estabrook, W. L. Kruer, and M. G. Haines, Phys. Fluids B **1**, 1282 (1989).

[8] C. E. Clayton, C. Joshi, and F. F. Chen, Phys. Rev. Lett. **51**, 1656 (1983).

[9] H. Ikezi, K. Schwarzenegger, and A. L. Simons, Phys. Fluids **21**, 239 (1978).

[10] D. H. Froula, L. Divol, and S. H. Glenzer, Phys. Rev. Lett. **88**, 105003 (2002).
[11] A. A. Andreev and V. T. Tikhonchuk, Sov. Phys. JETP **68**, 1135 (1989).
[12] R. E. Giacone and H. X. Vu, Phys. Plasmas **5**, 1455 (1998).
[13] H. X. Vu, D. F. DuBois, and B. Bezzerides, Phys. Rev. Lett. **86**, 4306 (2001).
[14] C. J. Pawley, H. E. Huey, and N. C. Luhmann, Jr., Phys. Rev. Lett. **49**, 877 (1982).
[15] P. W. Rambo, S. C. Wilks, and W. L. Kruer, Phys. Rev. Lett. **79**, 83 (1997).
[16] W. L. Kruer, Phys. Fluids **23**, 1273 (1980).
[17] M. J. Herbst, C. E. Clayton, and F. F. Chen, Phys. Rev. Lett. **43**, 1591 (1979).
[18] J. Handke, S. A. H. Rizvi, and B. Kronast, Phys. Rev. Lett. **51**, 1660 (1983).
[19] J. E. Bernard and J. Meyer, Phys. Rev. Lett. **55**, 79 (1985).
[20] J. A. Heikkinen, S. J. Karttunen, and R. R. E. Salomaa, Phys. Fluids **27**, 707 (1984).
[21] W. Rozmus, M. Casanova, D. Pesme, A. Heron, and J.-C. Adam, Phys. Fluids B **4**, 576 (1992).
[22] B. I. Cohen, B. F. Lasinski, A. B. Langdon, and E. A. Williams, Phys. Plasmas **4**, 956 (1997).
[23] B. I. Cohen, B. F. Lasinski, A. B. Langdon, E. A. Williams, K. B. Wharton *et al.*, Phys. Plasmas **5**, 3408 (1998).
[24] D. H. Froula, L. Divol, D. G. Braun, B. I. Cohen, G. Gregori, A. Mackinnon, E. A. Williams, S. H. Glenzer, H. A. Baldis, D. S. Montgomery, and R. P. Johnson, Phys. Plasmas **10**, 1846 (2003).
[25] L. Divol, R. L. Berger, B. I. Cohen, E. A. Williams, A. B. Langdon, B. F. Lasinski, D. H. Froula, and S. H. Glenzer, Phys. Plasmas **10**, 1822 (2003).
[26] B. I. Cohen, L. Divol, A. B. Langdon, and E. A. Williams, Phys. Plasmas **12**, 052703 (2005).
[27] B. J. Albright, L. Yin, K. J. Bowers, and B. Bergen, Phys. Plasmas **23**, 032703 (2016).
[28] B. Winjum, J. Fahlen, and W. Mori, Phys. Plasmas **14**, 102104 (2007).
[29] R. L. Berger, S. Brunner, T. Chapman, L. Divol, C. H. Still, and E. J. Valeo, Phys. Plasmas **20**, 032107 (2013).
[30] R. L. Dewar, Phys. Fluids **15**, 712 (1972).
[31] G. J. Morales and T. O'Neil, Phys. Rev. Lett. **28**, 417 (1972).
[32] T. Chapman, R. L. Berger, S. Brunner, and E. A. Williams, Phys. Rev. Lett. **110**, 195004 (2013).
[33] Z. J. Liu, S. P. Zhu, L. H. Cao, C. Y. Zheng, X. T. He, and Y. Wang, Phys. Plasmas **16**, 112703 (2009).
[34] Z. J. Liu, X. T. He, C. Y. Zheng, and Y. G. Wang, Phys. Plasmas **16**, 093108 (2009).
[35] C. Z. Cheng and G. Knorr, J. Comp. Phys. **22**, 330 (1976).
[36] D. J. Strozzi, M. M. Shoucri, and A. Bers, Comput. Phys. Commun. **164**, 156 (2004).
[37] A. Mangeney, F. Califano, C. Cavazzoni, and P. Travnicek, J. Comput. Phys. **179**, 495 (2002).
[38] F. Califano and L. Galeotti, Phys. Plasmas **13**, 082102 (2006).
[39] I. B. Bernstein, J. M. Greene, and M. D. Kruskal, Phys. Rev. **108**, 546 (1957).
[40] Q. S. Feng *et al.*, Phys. Plasmas **23**, 082106 (2016).

APPLIED PHYSICS LETTERS **92**, 071502 (2008)

Intense-laser generated relativistic electron transport in coaxial two-layer targets

C. T. Zhou,[1,2,3,a)] X. T. He,[1,2,3] and M. Y. Yu[3]
[1]*Institute of Applied Physics and Computational Mathematics, P.O. Box 8009, Beijing 100088, People's Republic of China*
[2]*Center for Applied Physics and Technology, Peking University, Beijing 100871, People's Republic of China*
[3]*Institute for Fusion Theory and Simulation, Zhejiang University, Hangzhou 310027, People's Republic of China*

(Received 9 January 2008; accepted 30 January 2008; published online 21 February 2008)

The transport and heating of laser produced relativistic electron beam propagating through coaxial two-layer solid-density plasma are studied using two-dimensional hybrid fluid-particle-in-cell simulation. When the energetic electrons enter the plastic and copper target, stronger heating is observed in the plastic plasma. The heating of the plasma electrons and the formation of tens of million Gauss interface magnetic fields results from a large difference of the velocities of the returning electrons at the two sides of the interface. Moreover, the electron beam can have a completely different emittance behavior when the target consists of different-material layers. © *2008 American Institute of Physics*. [DOI: 10.1063/1.2884334]

The study of energetic electron transport through matter is important for fast ignition in inertial confinement fusion.[1,2] Ultrabright, multi-MeV, low-emittance electron beams also have applications in astrophysical plasmas, beam-plasma interactions, electron acceleration, etc.[3] Recently, advances in laser technology have pushed the maximum laser intensity to about 10^{22} W/cm^2,[3] thereby stimulating considerable interest in the interaction of intense laser with matter. By depositing its light energy at the critical density, a short-pulse high-intensity laser can generate relativistic electrons, which can heat the target material. There have been many experiments investigating the heating and transport of laser-produced relativistic electrons in metals and plastics, glass slabs, and dense plasmas. The transport of the energetic electrons is highly nonlinear and no analytical theory exists and some phenomena, such as beam heating and spreading, are still unclear. In this letter, we consider relativistic electron transport and heating in coaxial two-layer solid targets.

Although particle-in-cell (PIC) simulation[4] has been the main tool for studying laser-plasma coupling and transport of electrons in hot, mildly dense plasma near the relativistic critical density, fully explicit three-dimensional PIC simulation of beams interacting with overdense plasmas is expensive. The high target density severely constraints the explicit simulation time and space steps due to the limitations on the time and space scales that must be resolved with respect to the electron density. Thus, accurate representation of overdense plasmas requires special simulation techniques. One method for simulating overdense plasmas is to use a hybrid technique, which include the inertialess electron,[5] the implicit fluid moment,[6] and the direct-implicit fully kinetic PIC (Ref. 7) models and their variations.[8–10] The present work uses the fully electromagnetic implicit HFPIC code,[11,12] in particular to the third approach there. A direct implicit particle push and an implicit electromagnetic solver are employed. The Spitzer resistivity for the target electrons is also assumed.

We consider a two-layer envelope target consisting of two coaxial layers, labeled A and B in Fig. 1. The solid-density plasmas A and B are Cu^{+3} and $C^{+2}H^{+}$, respectively. The corresponding electron number densities are $n_{e,A}=2.55\times10^{23}$ and $n_{e,B}=2.883\times10^{22}$ cm^{-3}, respectively. The initial temperature of the plasma electrons and ions is 5 eV and the cylindrical envelops are each homogeneous in density. The simulation box (z,r) is 50×50 μm^2, the mesh resolution is 500×500 uniform cells, with two injected beam electrons and up to twelve plasma particles in each cell. The temporal resolution is $\delta t=0.02$ fs. The vacuum regions are 5 μm both in front of and behind the foil and 30 μm in the transverse directions. Moreover, outgoing-particle boundary conditions are used. The propagating electromagnetic fields are periodic in the transverse direction.

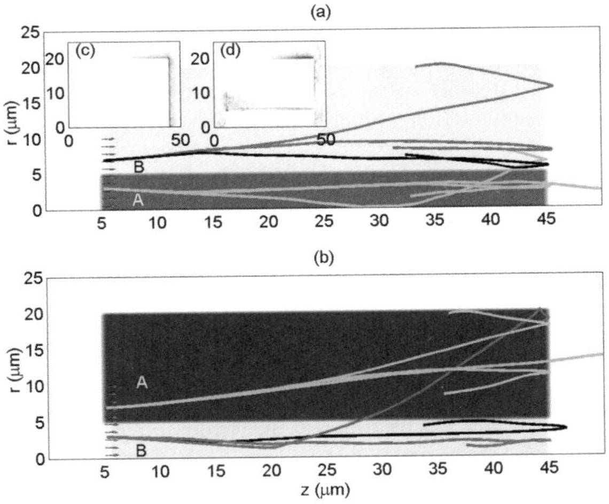

FIG. 1. (Color online) The coaxial two-layer target. A and B correspond to Cu^{+3} and $C^{+2}H^{+}$, respectively. [(a) and (b)] Trajectories of typical electrons (for $t=0-200$ fs) initially at $(z,r)=(5.2,3)$ and $(5.2,7)$ μm show the effects of electron acceleration, scatter, and reflection. (c) The electric field $(E_z^2+E_r^2+E_\theta^2)^{1/2}$. (d) The magnetic field $(B_z^2+B_r^2+B_\theta^2)^{1/2}$.

[a)]Electronic mail: zcangtao@iapcm.ac.cn.

We assume a parallel electron beam temperature 1.1 MeV and a transverse temperature 300 keV. The center of the injected hot electron beam is at $(z,r)=(5.2,0\ \mu\text{m})$. The injection current density is uniform, with radius $r_b = 10\ \mu\text{m}$ and duration 30 fs, and the total current is $I \approx 25$ MA, which is far beyond the critical Alfvén current limit of $I_A \approx 17.1\beta\gamma$ KA, where $\gamma=(1-\beta^2)^{-1/2}$ with $\beta=v/c$, where v is the quiver velocity of the plasma electrons, and c is the speed of light in vacuum. The injection electron density is $n_b \approx 1.7\times 10^{21}\ \text{cm}^{-3}$. Thus the hot electron temperature T_{hot} is[3] $T_{\text{hot}}\ (\text{MeV}) = 0.1[I_0\lambda_0^2/10^{17}\ (\text{W/cm}^2\ \mu\text{m}^2)]^{1/3}$, and the conversion efficiency is $\eta=0.000\,175\ [I_0\ (\text{W/cm}^2)]^{0.2661}$, our injected current density matches with the laser intensity $I_0 \approx 1.3 \times 10^{20}\ \text{W cm}^{-2}$ (with $\lambda_0 = 1\ \mu\text{m}$) and the laser power of $P\approx 500$ TW.

In order to compare the transport characteristics of energetic electrons in different target materials, we have considered two cases, shown in Figs. 1(a) and 1(b). In case I, a 10 μm diameter Cu^{+3} plasma is covered with a 15 μm $C^{+2}H^{+}$ foil and, in case II, the plasmas are interchanged. For existing experiments using aluminum, copper, or CH, direct collisional heating is not important and Joule heating, produced by the return current, dominates the heating of the target plasmas. When the hot electrons enter the plasmas, the resulting space-charge electric field allows the hot electrons to propagate by lowering the resistivity. The electron beam then propagates to the rear of the target or even escapes at an angle from the axial direction, as shown in Figs. 1(a) and 1(b). These electrons set up huge ($E_z \sim 10^{10}$ V/cm) space-charge electrostatic and magnetic fields at the surfaces of the slab, as shown in Figs. 1(c) and 1(d), so that the beam electrons experience multiple reflections, as seen in Figs. 1(a) and 1(b).

By comparing Figs. 1(c) and 1(d), it can be noted that strong magnetic fields ~30 MG, as also shown in Figs. 3(g) and 3(h), appear at the interface of Cu^{+3} and $C^{+2}H^{+}$ plasmas although the corresponding interface electric fields are small. Thus, some beam electrons [indicated by the black trajectories in Figs. 1(a) and 1(b)] can also be reflected by the intense interface magnetic fields. It is therefore of interest to see the difference in the beam electron motion for the two target configurations. Figures 2(a) and 2(b) show the phase space of the beam electrons at $t=140$ fs. In general, the beam electrons propagate to the right but some of them are angularly scattered out to the vacuum region in the radial direction. These escaping energetic electrons set up a large space charge field, which further bounds the low energy electrons in the region $r \leqslant 20\ \mu$m. Comparing Figs. 2(a) and 2(b), we note that in case I more beam electrons are concentrated in the region $r \leqslant 5\ \mu$m. But in Case II, most of the beam electrons are scattered to the region $10 \leqslant r \leqslant 20\ \mu$m. This behavior can also be observed in Fig. 4(b).

Beam electron propagation in overdense plasmas is accompanied by a return current. For MeV energy and MA electron currents, the beam will propagate with net currents at or above the Alfvén limit I_A. Assuming current neutralization, the fluid electron velocity can be estimated as $v_e \approx n_b v_b / n_e$, where the plasma electron density $n_e = n_A$ or n_B. For case I (II), because n_e in the region of $r \leqslant 5\ \mu$m is higher (lower) than in $5 \leqslant r \leqslant 20\ \mu$m, v_e in the former region should be slower (faster) than that in the latter region, as can be observed in Figs. 2(c) and 2(d).

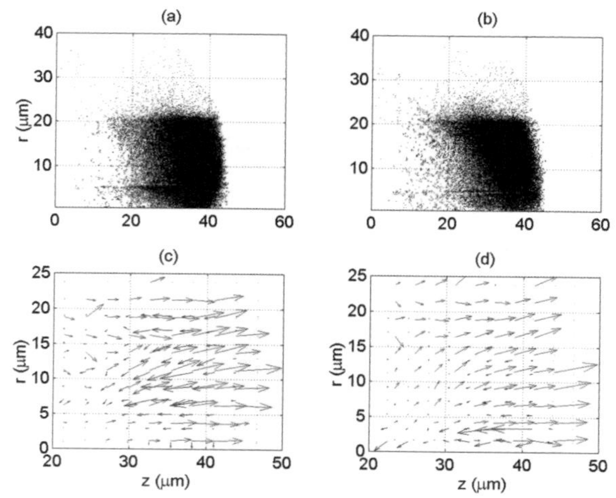

FIG. 2. (Color online) Behavior of the beam and target electrons at $t=140$ fs. [(a) and (b)] The phase-space trajectories of the beam electron. [(c) and (d)] The velocities of beam (red) and fluid plasma (blue) electrons.

We now analyze the heating process. The fluid electron temperature is given by[7] $dT_e/dt \sim T_e \nabla\cdot\mathbf{v}_e + Q_e/n_e + \cdots$, where the right-hand-side terms are the velocity change, ohmic heating rate, etc. For homogeneous plasmas, the averaged effect of the term $\nabla\cdot\mathbf{v}_e$ may be neglected. However, for both cases, Figs. 2(c) and 2(d) show that there is a velocity difference at the interface of the copper and plastic plasmas. Thus, the heating at the interface, as shown in Figs. 3(a) and 3(b), is associated with the velocity change process. Comparing Figs. 3(a)–3(d), we see that obvious heating by the return current, producing a resistive $\mathbf{E}=\mathbf{j}/\sigma \sim 10^8$ V/cm, where $\mathbf{j}$ is the current density and σ is the conductivity, for CH target. Furthermore, due to the filamentation instability and return current, local heating is increased. The phenomenon of hot electrons spreading out in a fanlike expansion can also be observed in the current density, shown in Figs. 3(e) and 3(f). The filaments are associated with the local magnetic field, as

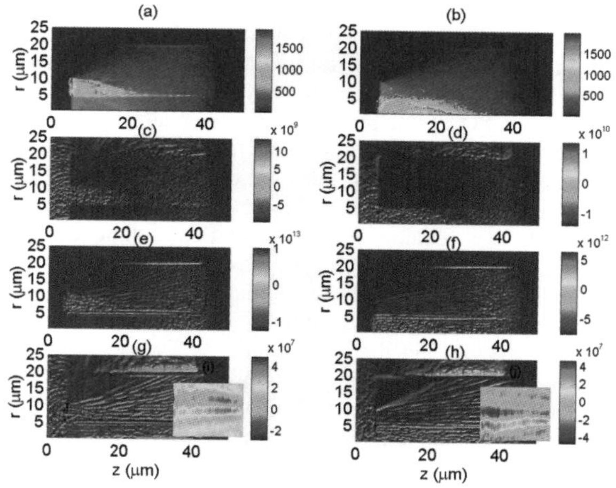

FIG. 3. (Color online) Snapshots of (a) and (b) the temperature T_e (eV) of the target electrons at $t=140$ fs. [(c) and (d)] The longitudinal electric field $E_z(z,r)$ (V/cm). [(e) and (f)] Current density $J_z(z,r)$ (A/cm^2). [(g) and (h)] The azimuthal magnetic field B_θ (Gauss). [(i) and (j)] An enlargement of (g) and (h) for $z\in[15,45]$ and $r\in[4,6]$ μm.

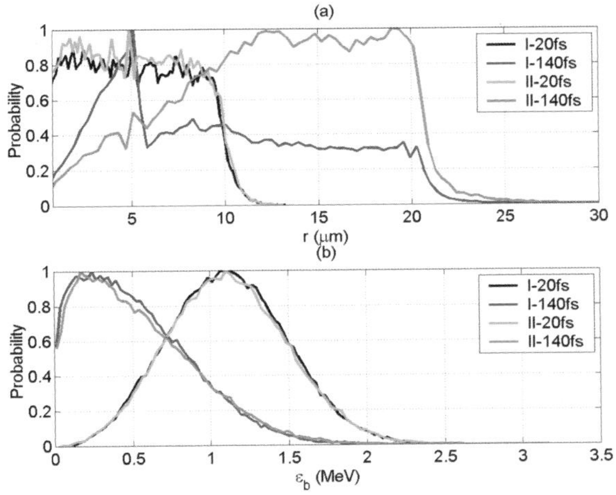

FIG. 4. (Color online) Comparison of the distributation of the number of beam electrons and their energy for cases I and II. (a) In the radial direction. (b) Energy spectra of the beam electrons.

shown in Figs. 3(g) and 3(h). For our two-coaxial-layer target, an important result is the strong magnetic fields appearing at the CH-copper interface, as shown in Figs. 3(g)–3(j). Comparing Figs. 3(i) and 3(j) with 2(c) and 2(d), we note that these interface magnetic fields originate from the difference in the speeds of the return electrons at the two sides of the interface. These strong magnetic fields then reflect some of the beam electrons by $\mathbf{v}\times\mathbf{B}$ drift, as shown in Figs. 1(a) and 1(b). Moreover, we observe in Figs. 2(a) and 2(b) that the interface magnetic field can also concentrate some beam electrons along the interface towards the right-hand side of the target.

In order to see the relative efficiency of the two cases for beam concentration, we count the number of the injected electrons in the radial direction at two different times. Figure 4(a) shows that at $t=20$ fs, the beam electrons are almost uniform with a radius of 10 μm for both cases. However, at $t=140$ fs, in case I a peak appears at $r=5$ μm, and the electrons in the region of $5<r<20$ μm have a flat profile. However in case II, more electrons have moved to the region $r>10$ μm. In other words, the injected relativistic electrons can diverge faster in case II than in case I. However, the beam energy loss and the energy distribution, given in Fig. 4(b), are essentially the same for both cases.

In conclusion, we have used a two-dimensional hybrid fluid-PIC code to investigate the transport and heating of targets by relativistic electron beams driven by intense laser pulses. Our two-layer target structure with two different target materials allows us to study the Joule heating by the return current. Our results illustrate for the first time that plasma electron heating at the interface of two different plasmas is due to the difference in velocities of the return electrons at the two sides of the interface. Furthermore, we also found that the strong interface magnetic field can influence the motion of the energetic electrons. In particular, the injected relativistic electrons can diverge faster if the inner plasma is CH and the outer is copper. The results here are useful to the realization of many applications[13,14] of energetic electrons, for example, in fast ignition, high-density materials, diagnostic physics, and other high energy density physics.

This work is supported by the the National Natural Science Foundation of China Grant Nos. 10575013, 10335020, and 10576007 and the National High-Tech ICF Committee and partially by the National Basic Research Program of China (973) (2007CB815101).

[1] S. Atzeni and J. Meyer-tar-Vehn, *Inertial Fusion—Beam Plasma Interaction, Hydrodynamic, Dense Plasma Physics* (Clarendon, Oxford, 2003).
[2] M. Tabak, J. Hammer, M. E. Glinsky, W. L. Kruer, S. C. Wilks, J. Woodworth, E. M. Campbell, M. D. Perry, and R. J. Mason, Phys. Plasmas **1**, 1626 (1994).
[3] P. Gibbon, *Short Pulse Laser Interactions with Matter: An Introduction* (Imperial College Press, London, 2005).
[4] O. Klimo and J. Limpouch, Laser Part. Beams **24**, 107 (2006).
[5] L. Gremillet, G. Bonnaud, and F. Amiranoff, Phys. Plasmas **9**, 941 (2002).
[6] J. R. Davies, Phys. Rev. E **68**, 056404 (2003).
[7] D. R. Welch, D. V. Rose, B. V. Oliver, and R. E. Clark, Nucl. Instrum. Methods Phys. Res. A **464**, 134 (2001).
[8] R. J. Mason, E. S. Doaa, and B. J. Albright, Phys. Rev. E **72**, 015401 (2005).
[9] J. J. Honrubia, M. Kaluza, J. Schreiber, G. D. Tsakiris, and J. Meyer-ter-Vehn, Phys. Plasmas **12**, 052708 (2005).
[10] R. B. Campbell, R. Kodama, K. A. Tanaka, and D. R. Welch, Phys. Rev. Lett. **94**, 055001 (2005).
[11] C. T. Zhou and X. T. He, Appl. Phys. Lett. **90**, 031503 (2007).
[12] C. T. Zhou, M. Y. Yu, and X. T. He, Phys. Plasmas **13**, 092109 (2006); J. Appl. Phys. **101**, 113302 (2007).
[13] H. Hora, Laser Part. Beams **23**, 441 (2005).
[14] H. Hora, Laser Part. Beams **25**, 37 (2007).

APPLIED PHYSICS LETTERS **92**, 151502 (2008)

Laser-produced energetic electron transport in overdense plasmas by wire guiding

C. T. Zhou,[1,2,a)] X. T. He,[1,2] and M. Y. Yu[2]
[1]*Institute of Applied Physics and Computational Mathematics, P.O. Box 8009, Beijing 100088, People's Republic of China and Center for Applied Physics and Technology, Peking University, Beijing 100871, People's Republic of China*
[2]*Institute for Fusion Theory and Simulation, Zhejiang University, Hangzhou 310027, People's Republic of China*

(Received 27 February 2008; accepted 24 March 2008; published online 16 April 2008)

Laser-driven energetic electron transport in a two-layered (Au and DT) ultrahigh density plasma is investigated. It is shown that the jump in the resistivity at the interface of the two plasmas plays an important role in the slowing down of the energetic beam electrons and heating of the plasmas. Furthermore, a thin gold wire in the DT plasma can further slow down the beam electrons and absorb a part of the beam energy. © *2008 American Institute of Physics*. [DOI: 10.1063/1.2908923]

As in conventional inertial confinement fusion, with the fast ignition scheme, the fuel is first compressed to ultrahigh density. The compressed fuel pellet is then illuminated by a petawatt short laser pulse. The energetic (MeV) electrons produced from the laser-plasma interaction are expected to ignite a small region inside the compressed fuel. A challenge[1,2] to the fast ignition is that the energetic electrons have to be transported from the region of relativistic critical density to a suitable ignition point inside the ultrahigh density plasma over a distance of hundreds of micrometers. Two approaches have been proposed for solving the problem of electron transport in ultrahigh density plasmas, namely, that of hole boring and cone guiding. In the latter, one can attach a hollow gold cone to the fuel shell. The igniter laser is sent through the cone whose tip is deep inside the latter. Some investigations showed that energetic electron flux can be two to three times higher than that without the cone. However, divergence of the electron beam is still large.

To narrow the beam divergence, a target design consisting of a hollow gold cone with a thin copper wire at its tip has been proposed.[3,4] Results show that the energetic electrons are guided by large electromagnetic fields around the wire and can propagate for a distance of several millimeters, together with more Ohmic heating. Most existing simulations of Ohmic heating of the wire plasma and enhancement of the return current have considered a plasma-vacuum interface. In practice, the guide cone and the wire are directly attached to the fuel so that the relevant interface is not plasma vacuum. Thus, it is quite important to study the beam electron transport in the high-density compressed fuel by using a thin gold wire in the plasma. In this letter, we present the first hybrid fluid-particle-in-cell (HFPIC) simulation of the propagation of high-energy electrons in the interface region of ultrahigh density Au-DT plasmas.

For the present problem traditional PIC techniques[5] may not be appropriate since the electron density n_e of a compressed target is much higher than critical. Accordingly, we employ the fully electromagnetic implicit HFPIC code,[6] which includes both PIC kinetic and fluid electrons to study in detail energetic-particle transport in high-density plasmas. Moreover, our code also involves particle-particle collisions that are described by the Spitzer collision frequency. Our two-ion-species target is shown in Fig. 1. The plasmas on the left- and right-hand sides are Au^{+20} and DT (with $\bar{Z}\sim 1$ and $A=2.5$), respectively. The same electron number density $n_e=1.2\times 10^{24}$ cm^{-3} is used for the two plasmas. The initial temperatures of the plasma electrons and ions are 100 eV. The DT plasma is initially homogeneous, but the gold plasma has a 5 μm long exponential rise on the front side in the beam propagation direction to the 3 μm long uniform density region, with the initial lowest density at $z=5$ μm being 8×10^{22} cm^{-3}. Two cases are considered. Case I [Fig. 1(a)] is without the gold wire. In case II [Fig. 1(b)], a gold-wire plasma with a radius of 2 μm is attached on the axis to the backside of the gold plasma. It extends to $z=50$ μm and is radially bounded by the DT plasma.

The simulation box (z,r) is 100×50 μm^2, the mesh resolution is 1000×1000 uniform cells, with two injected beam

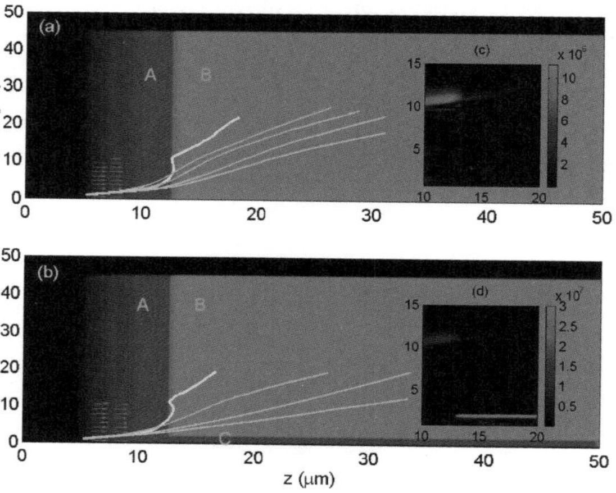

FIG. 1. (Color online) The two-layer target under consideration. (a) Without the gold wire. (b) With the gold wire in the DT plasma. A (also C, denoting the thin gold wire) and B represent the Au^{+20} and DT plasmas, respectively. The electron beam shortly after injection is schematically shown by the red arrows in the Au layer. The blue and yellow lines represent trajectories (from $t=0$ to 120 fs) of electrons initially at $(z,r)=(5.2,1)$ μm, showing electron acceleration, scattering, and pinching by the $\mathbf{J}\times\mathbf{B}$ force. The insets (c) and (d) show the magnetic field amplitude at $t=120$ fs. For clarity, the maximum amplitude ($\sim$80 MG) shown in (d) is limited to 30 MG.

[a)]Electronic mail: zcangtao@iapcm.ac.cn.

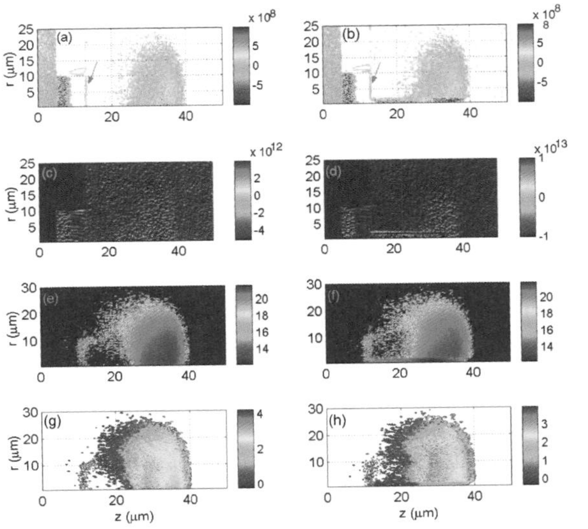

FIG. 2. (Color online) The field structures and behavior of the beam electrons for cases I (left column) and II (right column) at $t=120$ fs. (a) and (b) The longitudinal electric field $E_z(z,r)$ (V/cm). [(c) and (d)] The current density $J_z(z,r)$ (A/cm^2). [(e) and (f)] The beam density $\log_{10}[n_b[\mathrm{cm}^{-3}]]$. [(g) and (h)] The intensity distribution of the longitudinal velocity v_z of the beam electrons.

electrons and up to 16 plasma particles in each cell. The time step is 0.02 fs. The vacuum regions are 5 μm on each side. The electromagnetic fields are periodic in the transverse direction and propagating in the longitudinal direction. Outgoing-particle boundary conditions are used. To investigate the transport of multi-MeV electrons, we assume that the laser produced energetic electron beam is Gaussian, with parallel and transverse temperatures of 1.6 MeV and 240 keV, respectively. The beam radius is $r_b=10$ μm and its duration is 42 fs. The center of the beam is initially at (5.2,0 μm). The beam electron density is $n_b \sim 2.9 \times 10^{21}$ cm^{-3}. Its total current with the current density 1.4×10^{13} A cm^{-2} is far beyond the critical Alfvén limit $I_A \sim 17.1\beta\gamma$ K A, where $\gamma=(1-\beta^2)^{-1/2}$, $\beta=v/c$, v is the quiver velocity of the plasma electrons, and c is the speed of light in vacuum.

When the energetic electrons enter the relatively low-density gold plasma, the resulting electrostatic space-charge field initially prevents their propagation. The field energy is converted into heat energy of the plasmas. Thus, the resistivity as well as the space-charge field are reduced, allowing the energetic electrons to propagate further into the plasmas. For $n_e < 10^{23}$ cm^{-3}, collisional heating is unimportant since $\nu_{i,e}/\omega_{pe} < 0.025$,[7] where $\nu_{e,i}=4\times 10^{14} Z^2 (T_e/100\ \mathrm{eV})^{3/2}$ s^{-1} and ω_{pe} are the electron-ion collision and plasma oscillation frequencies, respectively. Instead, Joule heating by the return current dominates the heating of the low-density gold plasmas. The corresponding electric field is larger than 5×10^8 V/cm [see Figs. 2(a) and 2(b)]. Figures 2(c) and 2(d) show that filamentation instabilities associated with the self-magnetic fields and/or return current can further enhance the local plasma heating. It should be pointed out that in our simulation we have not resolved the local plasma Debye length and skin depth so that the observed large magnetic fields cannot be associated with the fine-scale filamentary structures from the Weibel instabilities. The trajectories of beam electrons shown in Figs. 1(a) and 1(b) reveal typical scattering behavior as they enter the (3 μm) high-density gold slab through the 5 μm slope. Figures 1(c) and 1(d) show the magnetic field at $t=120$ fs after the start of the interaction. There are two regions with intense magnetic fields. The one ($\sim$10 MG) at the edge of the electron beam inside the target can significantly change the direction of the beam because of the $\mathbf{j}\times\mathbf{B}$ force, as indicated by the yellow trajectories in Figs. 1(a) and 1(b).

At the interface, the discontinuity in the density and resistivity due to the Z^2 dependence in $\nu_{e,i}$ can give rise to discontinuities in the plasma and field quantities.[8,9] For the former case, Hora[8] showed that the longitudinal electric field, reaching value beyond 10^8 V/cm due to density gradient, can be observed by using a Gennine two fluid model and electric double layer. In the present problem, the resistivity jump between the Au and DT layers results in the appearance of large electric and magnetic fields at the interface A-B, as shown in Figs. 1(c), 1(d), 2(a), and 2(b). The enhanced magnetic fields can also bend the beam electrons toward the interface, as shown in Figs. 1(a) and 1(b) by the yellow trajectories. More importantly, the strong surface electric field [indicated in Figs. 2(a) and 2(b) by the red arrows] can stop the low-energy electrons [indicated in Figs. 2(e) and 2(f) by the red arrows]. Thus our simulation shows that the interface fields due to the jump in the resistivity and/or density plays an important role in the stopping and heating of the electrons near the interface, which is consistent with the theoretical predication.[10] Taking into account the higher (lower) resistivity of the gold (DT) plasma than that of the deuterated plastic (CD) plasma, it should be possible to lower the interface fields arising from the jump in the resistivity by inserting a CD layer between the Au and DT.

We now consider how the gold wire affects the beam electron transport in the DT plasma. Figures 2(e) and 2(f) show the beam density at $t=120$ fs for cases I and II, respectively. We note that the beam electron population in the region $0<r<2$ μm is very different for the two cases. In case I, most of the injected energetic electrons have entered the uniform DT plasma, with some lower-energy electrons propagating from the injection region as a diffuse beam with a spread angle of roughly 27°, which is consistent with experimental and simulation estimates. The electric field in the DT plasma is nonuniform. The beam electrons deposit their energy through the return current and thus heat DT plasma. In case II, the energy deposition processes is essentially the same as in case I. However, Figs. 2(b) and 1(d) show that the electric and magnetic fields in the Au wire is stronger than that in the DT plasma. The strongly fluctuating electromagnetic fields in and near the wire can cause the background electrons to execute complex motion that macroscopically enhances the plasma resistivity[11] and further slows down the energetic electrons. Comparing Figs. 2(g) and 2(h) for the beam electron velocity distribution, we can see that there is a clear slowdown of the beam electrons in the gold wire. This anomalous stopping results in a much higher electron population in the gold wire, as can be seen in Fig. 2(f).

In order to see the effect of the gold wire on beam concentration and energy loss, we count the number of the injected electrons in the radial direction at two different times and compare their energy conversion. Figure 3(a) shows that at $t=10$ fs, the beam electrons are almost uniform with a radius of 10 μm in both cases I and II. However, at

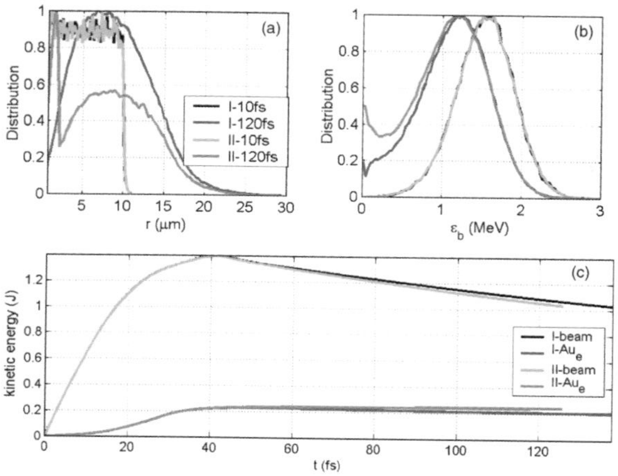

FIG. 3. (Color online) (a) The distribution of the number of beam electrons in the radial direction at 10 and 120 fs. (b) The energy spectra of the beam electrons at 10 and 120 fs. The legend in (a) also applies to (b). (c) The evolution of the kinetic energy of the beam and gold-plasma electrons.

$t=120$ fs, in case I more hot electrons have moved to the region $r>2$ μm. In case II, a peak appears at $r=2$ μm, and the electrons in the region $2<r<20$ μm are of the similar profile as in case I. That is, some injected relativistic electrons are guided by the wire. Figure 3(b) shows that the wire does not change the distribution of the high-energy (>0.7 MeV) electrons. However, the beam energy loss and the energy distribution of low-energy electrons in the two cases are clearly different. In particular, in case II, injected electrons with energy smaller than 0.3 MeV can be seen. By comparing Figs. 3(a) and 3(b) to Fig. 2(f), we see that the low-energy peak can be attributed to these electrons that are anomalously stopped by the microscopic turbulent fields in the wire. Figure 3(c) shows the energy conversion between the beam and background plasma electrons. For $t<42$ fs, the total kinetic energy of the injected electrons increases and is continuously being converted by the return currents into the energy of the background plasma. When $t>80$ fs, a clear difference in both the beam energy loss and the plasma electron energy gain can be observed in cases I and II. In Case II, the beam electrons can also deposit energy to the Au wire electrons because of Ohmic heating by the return current.

In conclusion, we have investigated the energetic electron propagation in a Au-DT plasma by using a HFPIC code. For a two-layer target, our simulation shows that the interface electric and magnetic fields arising from the jump in the resistivity of the two plasmas play an important role in the stopping of the energetic electrons and heating of the interface region by surface currents. We also show for the first time that a gold wire in DT plasmas not only enhances the magnetic field in a thin layer at the wire-plasma interface but also changes the characteristics of beam electron transport and energy deposition in the Au and DT plasmas. The slowdown and stopping of the beam electrons in the gold wire can be attributed to the higher resistivity of the Au plasma (compared to that of the DT plasma) is consistent with the results of most experiments and theories. Besides the fast ignition, wire guided electron beam transport at ultrahigh plasma and energy densities can have many applications[5,12,13] in laboratory astrophysics, medicine, material science, heavy-ion fusion, and other high-energy density physics.

This work is supported by the National Natural Science Foundation of China, Grant Nos. 10575013 and 10576007, and the National High-Tech ICF Committee and partially by the National Basic Research Program of China (973) (2007CB815101).

[1]H. Hora, J. Badziak, and M. N. Read, Phys. Plasmas **14**, 072701 (2007).
[2]M. H. Key, Phys. Plasmas **14**, 055502 (2007).
[3]R. Kodama, Y. Sentoku, Z. L. Chen, G. R. Kumar, S. P. Hatchett, Y. Toyama, T. E. Cowan, R. R. Freeman, J. Fuchs, Y. Izawa, M. H. Key, Y. Kitagawa, K. Kondo, T. Matsuoka, H. Nakamura, M. Nakatsutsumi, P. A. Norreys, T. Norimatsu, R. A. Snavely, R. B. Stephens, M. Tampo, K. A. Tanaka, and T. Yabuuchi, Nature (London) **432**, 1005 (2004).
[4]J. S. Green, K. L. Lancaster, K. U. Akli, C. D. Gregory, F. N. Beg, S. N. Chen, D. Clark, R. R. Freeman, S. Hawkes, C. Hernandez-Gomez, H. Habara, R. Heathcote, D. S. Hey, K. Highbarger, M. H. Key, R. Kodama, K. Krushelnick, I. Musgrave, H. Nakamura, M. Nakatsutsumi, N. Patel, R. Stephens, M. Storm, M. Tampo, W. Theobald, L. Van Woerkom, R. L. Weber, M. S. Wei, N. C. Woolsey, and P. A. Norreys, Nat. Phys. **3**, 853 (2007).
[5]A. F. Lifshitz, J. Faure, Y. Glinec, V. Malka, and P. Mora, Laser Part. Beams **24**, 255 (2006); O. Klimo and J. Limpouch, *ibid.* **24**, 107 (2006); A. Karmakar and A. Pukhov, *ibid.* **25**, 371 (2007).
[6]C. T. Zhou and X. T. He, Opt. Lett. **32**, 2444 (2007); Appl. Phys. Lett. **90**, 031503 (2007); **92**, 071502 (2008).
[7]C. Ran, M. Tzoufras, F. S. Tsung, W. B. Mori, S. Amorini, R. A. Fonseca, L. O. Silva, J. C. Adam, and A. Heron, Phys. Rev. Lett. **93**, 185004 (2004).
[8]H. Hora, *Plasmas at High Temperature and Density* (Springer, Heidelberg, 1991).
[9]R. G. Evans, High energy density physics **2**, 35 (2006).
[10]L. Cicchitelli, J. S. Elijah, S. Eliezer, A. K. Ghatak, M. P. Goldsworthy, H. Hora, and P. Lalousis, Laser Part. Beams **2**, 467 (1984).
[11]E. M. Campbell, R. R. Freeman, and K. A. Tanaka, Fusion Sci. Technol. **49**, 249 (2006).
[12]H. Hora, Laser Part. Beams **25**, 37 (2007).
[13]D. H. H. Hoffmann, K. Weyrich, H. Wahl, D. Gardés, R. Bimbot, and C. Fleurier, Phys. Rev. A **42**, 2313 (1990); S. Neef, R. Knobloch, D. H. H. Hoffmann, A. Tauschwitz, and S. S. Yu, Laser Part. Beams **24**, 71 (2006).

APPLIED PHYSICS LETTERS **97**, 201502 (2010)

Density effect on relativistic electron beams in a plasma fiber

C. T. Zhou,[1,2,3,a)] X. G. Wang,[3] S. Z. Wu,[1] H. B. Cai,[1,2] F. Wang,[2] and X. T. He[1,2,3]
[1]*Institute of Applied Physics and Computational Mathematics, Beijing 100094, People's Republic of China*
[2]*Center for Applied Physics and Technology, Peking University, Beijing 100871, People's Republic of China*
[3]*Institute for Fusion Theory and Simulation, Zhejiang University, Hangzhou 310027, People's Republic of China*

(Received 23 September 2010; accepted 2 November 2010; published online 18 November 2010)

Intense short-petawatt-laser driven relativistic electron beams in a hollow high-Z plasma fiber embedded in low-Z plasmas of different densities are studied. When the plasma is of lower density than the hollow fiber, resistive filamentation of the electron beam is observed. It is found that the electron motion and the magnetic field are highly correlated with tens of terahertz oscillation frequency. Depending on the material property around the hollow fiber and the plasma density, the beam electrons can be focused or defocused as it propagates in the plasma. Relativistic electron transport and target heating are also investigated. © *2010 American Institute of Physics*. [doi:10.1063/1.3518720]

Recently, lasers with intensities of two orders of magnitude beyond today's state of the art have been planned in the foreseeable future.[1,2] Such lasers allow important research and application of relativistic plasma physics issues,[3–12] ranging from compact particle and light sources, fast ignition in inertial confinement fusion, high energy density physics, etc. When a laser pulse interacts with a foil target, electrons produced at the relativistic critical density can be accelerated to relativistic energy by the ponderomotive force.[13] Nonlinear interaction between these beam electrons and the overdense plasma can lead to changes in the angular characteristics of the beam and nonuniform heating of the cold background plasma. Experiments and simulations[7,14] show that beam divergence can become large when laser intensity increases. To control the transverse divergence of the beam propagating in the target, several schemes for improving the beam collimation have been proposed recently. These include the use of special target configurations[6,12] such as the cone-wire, the cone-funnel, the cone-hollow, and the vacuum gap targets. In the plasma-vacuum interface based approaches,[6,7,15] the guiding of electrons by large surface electric fields generated around the target surface enables the electrons to propagate a long distance at a diameter similar to that of the plasma target. However, when the target is attached to a plasma, the interface is not plasma-vacuum, and recent studies[16,17] have shown that robust tens of megagauss interface magnetic fields are generated because of the resistivity and density gradients at the material interface. Such magnetic fields can considerably reduce the divergence of the electron beam. Although magnetic-field collimation has been intensively investigated,[6,7,16,17] it is not clear if these results remain valid for the hollow-guided scheme. In this Letter, we investigate the electron-beam transport and plasma heating when laser-driven relativistic electrons propagate in a hollow fiber immersed in a plasma. In particular, we are interested in the microturbulence behavior of the relativistic electrons.

As shown in Fig. 1(a), we consider that a hollow high-Z gold plasma fiber is attached to a thin disk (slab). The fiber is embedded in the low-Z plasma (here the deuterium-tritium plasma with $Z=1$ is considered). The two-dimensional model of the target structure is shown in Fig. 1. Both the slab and the hollow consist of gold plasma with a constant degree of ionization $Z=20$. To see how the hollow fiber controls and guides the relativistic electrons, we present the propagation dynamics of the injected electron beam for three cases. In case I, only the slab-hollow target is present. In cases II and III, the hollow is completely embedded in the uniform low-Z plasma. The densities of the low-Z plasma are $n_{\text{low-Z}}=n_{\text{Au}}/10$ and $10n_{\text{Au}}$, respectively, for the cases II and III, as shown in Figs. 3(d) and 3(g) by the thick green solid lines. We simulate the subsequent beam-plasma interaction by directly injecting a high-current electron beam from the left and follow its propagation and evolution. The beam is assumed to have a relativistic-Maxwellian distribution with an average temperature of 1.5 MeV along the x direction. Its width at $1/e^2$ maximum is 20 μm and its angular spread is 30° at $(z,x)=(20,0)$ [μm]. The beam rises to its maximum value in 50 fs and remains constant at 10^{13} A/cm^2. The simulation box (z,x) is 200 μm$\times$100 μm. In the present hybrid simulation,[16] the background plasma electrons and ions are treated as fluids and the injected beam electrons are treated by particle-in-cell simulation.

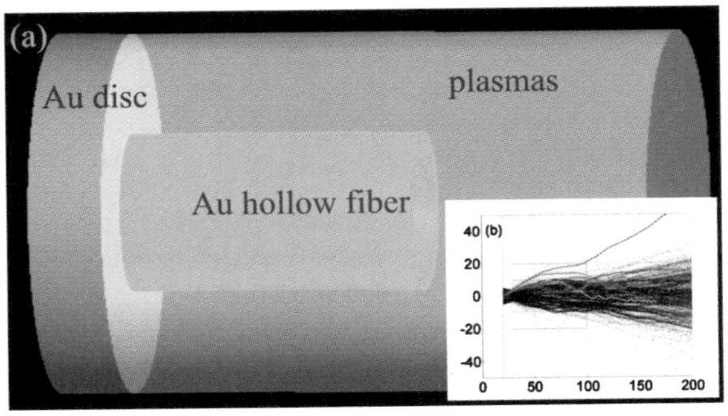

FIG. 1. (Color online) (a) Schematic of the gold-hollow plasma fiber around the low-density deuterium-tritium plasma. (b) The behavior of the beam electrons at $t=1$ ps. The thin dashed lines in (b) show the edges of the gold plasma. The trajectories of several typical electrons initially at $(x,z)=(0,20)$ show the propagation of the beam electrons in the plasma.

[a)]Electronic mail: zcangtao@iapcm.ac.cn.

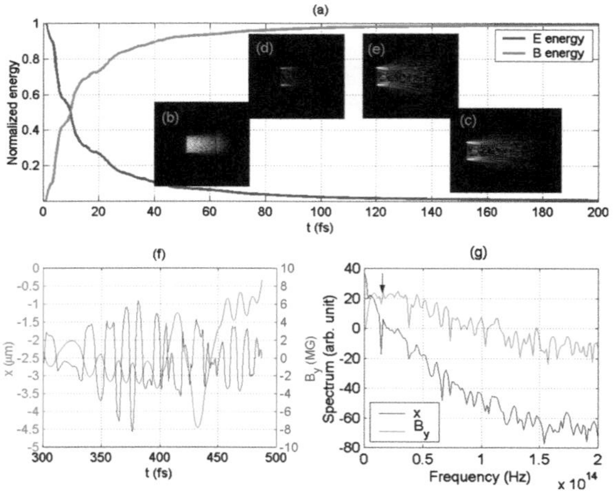

FIG. 2. (Color online) (a) Normalized energy of the electric and magnetic fields. The insets (b) and (c) show the electric field at $t=50$ and 200 fs, respectively. The insets (d) and (e) show the corresponding magnetic fields. (f) Evolution of the trajectory $x(t)$ of a typical beam electron and the magnetic field $B_y(t)$ around it and (g) the corresponding power spectra.

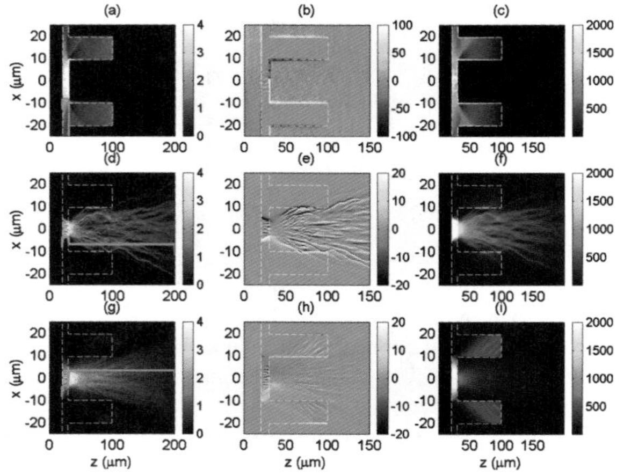

FIG. 3. (Color online) The beam-electron density (n_b/n_{cr}), the magnetic field component B_y (in megagauss), and the plasma electron temperature (in eV) at $t=1$ ps. The thin dashed lines in the subplots show the gold plasma and the thick solid lines in (a), (d), and (g) mark the electron density profile of the gold slab and the deuterium-tritium plasma in the beam propagation direction. The first, second, and third rows correspond to cases I, II, and III, respectively.

We first consider the relativistic electron transport for case II, as shown in Fig. 1. Initially, the relativistic electron propagation in the gold plasmas will set up a charge separation field, whose energy is then converted into the thermal energy of the plasma by Joule heating. The resistivity as well as the space-charge field is thus reduced, allowing the energetic electrons to propagate further into the plasmas. Since the beam current far exceeds the Alfvén current, the beam transport is possible only if sufficient current neutralization is effected by a return current of background plasma electrons. In the initial stage, the electron currents are in near equilibrium with each other and no resistive magnetic field is generated. As the electron beams propagate further into the target, nonuniform electric fields [see Figs. 2(b) and 2(c)] are induced. The latter can give rise to a magnetic field B_y [see Figs. 2(d) and 2(e)]. This stage leads to an exponential reduction of the electric field and growth of the magnetic field, as shown in Fig. 2(a). When the magnetic energy further increases, the B_y component of the magnetic field can reach several tens of megagauss and is strong enough to pinch the beam electrons. It finally saturates due to magnetic trapping of the beam electrons. The physical processes of exponential reduction of the electric field and growth of the magnetic field can be clearly observed in Figs. 2(a)–2(e).

When the electron beam further enters the gold slab, a strong edge magnetic field, as shown in Figs. 2(d) and 2(e), is generated by $\eta\nabla\times\mathbf{j}_f$, where $\mathbf{j}_f$ and η are the fast electron current and the Spitzer resistivity, respectively. This pushes the beam electrons toward the regions of higher current density and focuses the beam. The beam electrons deposit their energy to the background plasma through the return current. Joule heating of the cold background plasma in the return current [Fig. 3(f)] can be described by the cold-electron energy equation[6] $\partial T_e/\partial t\approx 2J_e^2/3\sigma n_e$, where σ and J_e are the electrical conductivity and the cold-electron current density, respectively. Since the resistivity decreases as the plasma is heated by the hot electrons, the resistivity gradient can appear at the beam-plasma edge. The source term $\nabla\eta\times\mathbf{j}_f$ therefore generates a magnetic field when there is a beam current perpendicular to a resistivity gradient. For a uniform plasma, this edge magnetic field hollows rather than focuses the beam. The unbalance between the focusing and the defocusing results in complex dynamics of the beam propagation. Figures 1(b) and 3(d) clearly indicate hosing-like instabilities of the beam in the overdense plasma. On the other hand, when the beam electrons propagate in the relatively low-density low-Z plasma, the high-current beam is also subject to a wide range of instabilities, such as the two-stream, Weibel, and filamentation instabilities. The two-stream instability results in the generation of Langmuir turbulence. The Weibel instability leads to the break up of the fast electron beam into small beamlets with a typical transverse dimension of the order of the collisionless skin depth or the Bennett radius[18] $\sqrt{8}(c/v)\lambda_b$, where v is the propagation velocity and λ_b is the Debye length of the electron beam obtained from the transverse temperature. Figure 1(b) exhibits a high degree of filamentation. The beamlets can also coalesce on a longer time scale. These collective effects can result in changes in the angular characteristics of the filamentation beam.

We now investigate how the microturbulent self-generated magnetic field affects the beam electron transport dynamics in the low-Z plasma. The resistive filamentation instability associated with the magnetic field is clearly observed in Figs. 3(d) and 3(e). In order to analyze the microturbulent dynamics of the beam electrons and the magnetic fields, we trace the trajectory $x(t)$ of one of the beam electrons, as shown in Fig. 1(b), and compare it with the magnetic field $B_y(t)$ around it, as shown in Fig. 2(f). We see that the magnetic field can reach several megagauss, which is strong enough to change the propagation direction of the beam electron. The electron motion and the magnetic field are highly correlated. The power spectra of $x(t)$ and $B_y(t)$ are given in Fig. 2(g), which shows a typical chaotic behavior. As indicated by the arrow in Fig. 2(g), $x(t)$ and $B_y(t)$ have the same fundamental frequency, $f_0\approx 16$ THz. This can correspond to very short period (therefore very short wavelength or spatial scale) plasma wave oscillations.

In order to have a better understanding of the beam electrons propagating in different density plasmas, we now compare their transport dynamics and heating for the three cases. Figure 3 summarizes the simulation results. For case I, the beam electrons first traverse the plasma slab and exit into the rear vacuum region, and generate large space-charge electrostatic fields at the front and back sides of the slab. These fields create across the plasma slab an electrostatic potential wall and result in multiple reflections of some beam electrons in the slab. By following the electron trajectories, we find that most beam electrons can propagate through the slab to the hollow fiber. A strong azimuthal surface magnetic field ($B_y \sim 100$ MG) [see Fig. 3(b)] and a strong transverse electric field ($E_x \sim 10^{12}$ V/m) surrounding the fiber plasma are created. The surface electric and magnetic fields in the inner and outer surfaces of the fiber confine and guide the hot electrons. The appearance of the thin enhanced return-current layers in both the inner and outer surfaces of the fiber plasma results in enhanced surface heating. The temperature in both surfaces can double that of the bulk of the fiber plasma, as shown in Fig. 3(c).

For cases II and III, the hollow gold plasma fiber is completely embedded in the deuterium-tritium plasma. The interface electric field (several times 10^9 V/m) of the gold slab and the plasma is not strong enough to stop the high-energy electrons. These beam electrons can propagate into the plasma, but their dynamic behaviors are quite different in cases II and III. The interface magnetic field between the gold fiber and the low-Z plasma is mainly from the density and resistivity gradients, etc. The steepest gradients of both at the interface discontinuity (between the different layers) lead to the largest interface magnetic fields. If we assume the Spitzer resistivity $\eta = 10^{-4} Z \ln \Lambda T^{-3/2}$, we can describe the interface magnetic field growth by $\partial \mathbf{B}/\partial t \sim \nabla \eta \times \mathbf{j}_f \sim T_{\text{low-}Z}^{3/2} - T_{\text{high-}Z}^{3/2}$. For case II, it can be seen from Fig. 3(f) that $T_{\text{low-}Z} > T_{\text{high-}Z}$ on both sides of the fiber's inner surface. Thus, on the inner surface, we have $B_y|_{x=10} < 0$ because the sign of $\mathbf{j}_f$ is negative. This interface magnetic field [>20 MG, shown in Fig. 3(e)] can significantly change the direction of the beam because of the $\mathbf{j} \times \mathbf{B}$ force. Figures 1(b) and 3(d) show that the collimating magnetic field focuses the beam electrons propagation into the deuterium-tritium plasmas inside the fiber. For case III, however, the inner surface magnetic field becomes $B_y|_{x=10} > 0$ because $T_{\text{low-}Z} < T_{\text{high-}Z}$. It is seen from Figs. 3(g) and 3(h) that the beam electrons can be attracted into the gold fiber. When the beam electrons further propagate into the high-density deuterium-tritium plasma, they can also be scattered into the gold fiber through collisions with the background electrons and ions. The enhanced magnetic potential around the fiber then bends the beam electrons such that they propagate along the gold fiber. Higher heating in the gold fiber than in the deuterium-tritium plasma is clearly observed in Fig. 3(i).

In summary, using two-dimensional hybrid simulation, we have investigated the microturbulent dynamics of relativistic electrons driven by petawatt-picosecond pulse lasers in a hollow-guided high-Z gold plasma fiber. We compared the transport and heating of the relativistic electrons propagating in a hollow fiber filled with low-Z deuterium-tritium plasmas of different densities. It is shown that with a hollow-vacuum target, the surface electric and magnetic fields in the inner and outer surfaces of the fiber confine and guide the hot electrons. The appearance of thin enhanced return-current layers in both the inner and outer surfaces of the fiber plasma results in enhanced surface heating. When the hollow is filled with different density plasmas, a different nonlinear behavior of the electron beam is observed. When the low-Z deuterium-tritium plasma has a lower density than the gold hollow fiber, the inner surface magnetic field due to the steep gradients of the resistivity and plasma density at the interface discontinuity behaves like a plasma mirror, and the resulting reflections of the electron trajectories control the beam electrons in the low-Z plasma. Our simulation results also indicate that the beam electrons and the self-generated microturbulent magnetic field are strongly correlated. However, when the low-Z deuterium-tritium plasma has higher density than the gold hollow fiber, the interface magnetic field in the inner surface of the hollow can change sign so that the relativistic electrons are attracted into the hollow fiber. The findings here may be useful in the design of fast-ignition targets and more efficient point light sources, as well as in other applications of intense laser-driven relativistic electrons.

This work is supported by the National Natural Science Foundation of China (Grant Nos. 10974022 and 10835003) and the National Basic Research Program of China (Grant No. 2007CB815101).

[1] V. Yanovsky, V. Chvykov, G. Kalinchenko, P. Rousseau, T. Planchon, T. Matsuoka, A. Maksimchuk, J. Nees, G. Cheriaux, G. Mourou, and K. Krushelnick, Opt. Express **16**, 2109 (2008).
[2] T. Feder, Phys. Today **63**(6), 20 (2010).
[3] J. Zhang, E. E. Fill, Y. Li, D. Schlögl, J. Steingruber, M. Holden, G. J. Tallents, A. Demir, P. Zeitoun, C. Danson, P. A. Norreys, F. Walsh, M. H. Key, C. L. Lewis, and A. G. McPhee, Opt. Lett. **21**, 1035 (1996).
[4] J. Meyer-ter-Vehn, Plasma Phys. Controlled Fusion **51**, 124001 (2009).
[5] H. Hora, B. Malekynia, M. Ghoranneviss, G. H. Miley, and X. He, Appl. Phys. Lett. **93**, 011101 (2008).
[6] C. T. Zhou, L. Y. Chew, and X. T. He, Appl. Phys. Lett. **97**, 051502 (2010), and references therein.
[7] J. S. Green, K. L. Lancaster, K. U. Akli, C. D. Gregory, F. N. Beg, S. N. Chen, D. Clark, R. R. Freeman, S. Hawkes, C. Hernandez-Gomez, H. Habara, R. Heathcote, D. S. Hey, K. Highbarger, M. H. Key, R. Kodama, K. Krushelnick, I. Musgrave, H. Nakamura, M. Nakatsutsumi, N. Patel, R. Stephens, M. Storm, M. Tampo, W. Theobald, L. Van Woerkom, R. L. Weber, M. S. Wei, N. C. Woolsey, and P. A. Norreys, Nat. Phys. **3**, 853 (2007).
[8] H. Hora, Laser Part. Beams **27**, 207 (2009).
[9] H. Hora, G. H. Miley, M. Ghoranneviss, B. Malekynia, N. Azizi, and X. He, Energy Environ. Sci. **3**, 479 (2010).
[10] A. A. Solodov, K. S. Anderson, R. Betti, V. Gotcheva, J. Myatt, J. A. Delettrez, S. Skupsky, W. Theobald, and C. Stoeckl, Phys. Plasmas **16**, 056309 (2009).
[11] V. Malka, J. Faure, Y. A. Gaduel, E. Lefebvre, A. Rousse, and K. T. Phuoc, Nat. Phys. **4**, 447 (2008).
[12] C. T. Zhou, X. G. Wang, S. C. Ruan, S. Z. Wu, L. Y. Chew, M. Y. Yu, and X. T. He, Phys. Plasmas **17**, 083103 (2010).
[13] S. C. Wilks and W. L. Kruer, IEEE J. Quantum Electron. **33**, 1954 (1997).
[14] O. Polomarov, I. Kaganovich, and G. Shvets, Phys. Rev. Lett. **101**, 175001 (2008).
[15] H. B. Cai, K. Mima, W. M. Zhou, T. Jozaki, H. Nagatomo, A. Sunahara, and R. J. Mason, Phys. Rev. Lett. **102**, 245001 (2009).
[16] C. T. Zhou, X. T. He, and M. Y. Yu, Appl. Phys. Lett. **92**, 151502 (2008); Appl. Phys. Lett. **92**, 071502 (2008).
[17] S. Z. Wu, C. T. Zhou, and S. P. Zhu, Phys. Plasmas **17**, 063103 (2010).
[18] W. H. Bennett, Phys. Rev. **98**, 1584 (1955).

PHYSICS OF PLASMAS 17, 083103 (2010)

Dynamics of relativistic electrons propagating in a funnel-guided target

C. T. Zhou,[1,2,3,a)] X. G. Wang,[3] S. C. Ruan,[4] S. Z. Wu,[1] L. Y. Chew,[5] M. Y. Yu,[3,6] and X. T. He[1,2,3]
[1]Institute of Applied Physics and Computational Mathematics, Beijing 100094, People's Republic of China
[2]Center for Applied Physics and Technology, Peking University, Beijing 100871, People's Republic of China
[3]Institute for Fusion Theory and Simulation, Zhejiang University, Hangzhou 310027, People's Republic of China
[4]Shenzhen University, Shenzhen 518060, People's Republic of China
[5]Division of Physics and Applied Physics, School of Physical and Mathematical Sciences, Nanyang Technological University, Nanyang Avenue, Singapore 639798
[6]Institut für Theoretische Physik I, Ruhr-Universität Bochum, Bochum D-44780, Germany

(Received 10 May 2010; accepted 29 June 2010; published online 12 August 2010)

Hot electron transport in a funnel-guided target is investigated using two-dimensional hybrid simulations. Relativistic electrons generated by a petawatt picosecond short-pulse laser interacting with a gold slab are guided by a funnel to the compressed core region of the fuel. It is shown that as the energetic electrons propagate into the compressed core, the interface magnetic fields in the inner and outer surfaces of the funnel can change sign so that the fast electrons are efficiently guided in the gold funnel to reach the dense core. It is also shown that with funnel guiding, fuel heating is enhanced compared to that without the funnel. The findings here may be useful in the design of more efficient fast-ignition schemes, as well as in other applications involving transport and heating of energetic electrons in targets. © 2010 American Institute of Physics. [doi:10.1063/1.3465905]

I. INTRODUCTION

Rapid development in laser technology has stimulated considerable interest in relativistic plasma physics,[1–10] such as laboratory astrophysics, compact particle accelerators, and fast ignition (FI) in inertial fusion. In the FI scheme,[3–5] a deuterium-tritium (DT) fuel capsule is first compressed to high density ($\sim 300\ \mathrm{g/cm^3}$) by long-pulse lasers. A petawatt picosecond (ps) short-pulse laser then generates an enormous current [$>10^{13}\ \mathrm{A/cm^2}$ of energetic (1–2 MeV) electrons with a beam diameter of $\sim 40\ \mu\mathrm{m}$], for igniting the fuel. To efficiently deposit the igniter energy in the compressed pellet plasma, two main schemes,[5] namely, "hole boring" and "cone-guiding," have been proposed. In the latter, a petawatt short-pulse laser of intensity of $10^{19}-10^{20}\ \mathrm{W/cm^2}$ interacts with both the wall and the tip of a gold cone to generate the required energetic electrons.

Cone-guided FI reduces both the laser energy and precision requirements. However, the collimation of the energetic electrons inside the target still remains a problem. The laser generated energetic electrons have an intrinsic angular spread or beam divergence ($>30°$) when the laser intensity is larger than $10^{18}\ \mathrm{W/cm^2}$.[11] Moreover, these energetic electrons have to propagate about a hundred or more microns from the tip of the cone to the dense core. Because the beam current is about 1 GA, which exceeds the Alfvén current more than 1000 times, the counterpropagating high-current electron streams are subject to two-stream, Weibel, filamentation, etc., instabilities.[8–12] Strong nonlinear interaction between the energetic electrons and the overdense plasma is also accompanied by propagation loss and deflection. Therefore, it is necessary to collimate or properly focus the energetic beam electrons.

Several schemes for improving the collimation of the electron beams have been proposed. These include the use of special target configurations such as the cone-wire,[13–16] the conical,[17] the two-layer or coaxial,[18,19] and the vacuum gap[7,9] targets. In the cone-wire scheme, a wire is attached to the tip of the gold cone. Experiments and simulations[14,15,20] showed that the large self-generated magnetic (SGM) fields around the wire can guide and collimate the energetic electrons. In the conical structure scheme, a small channel is attached to the outside tip of a structured cone target. It was found[17] that the electron beam formed at the inside tip of the cone can travel into the channel inside the tip for a long distance. In the two-layer structured target,[18,19] because of the resistivity and density gradients at the interface, the interface magnetic fields can considerably reduce the divergence of the electron beam. These schemes make use of the tens megagauss resistive SGM fields to control of the divergence of the beam. Thus Ohmic heating of the wire or channel plasma and enhancement of the return currents at plasma-vacuum interface have to be considered. In practice, the guide wire or channel should be directly attached to the fuel so that the relevant interfaces are not plasma vacuum. It is therefore important to investigate in more detail beam electron transport in a high-density thin wire, channel, or similar plasma. In this work, to illustrate that the energetic electrons generated by a short-pulse laser interacting with a traditional cone can be further guided by a funnel, we consider here the cone-target structure shown in Fig. 1(a). A tiny hollow cone, or funnel, is attached to a thin slab, which models the tip of the traditional cone.[4] The DT fuel of the target also fills the funnel so that

a)Electronic mail: zcangtao@iapcm.ac.cn.

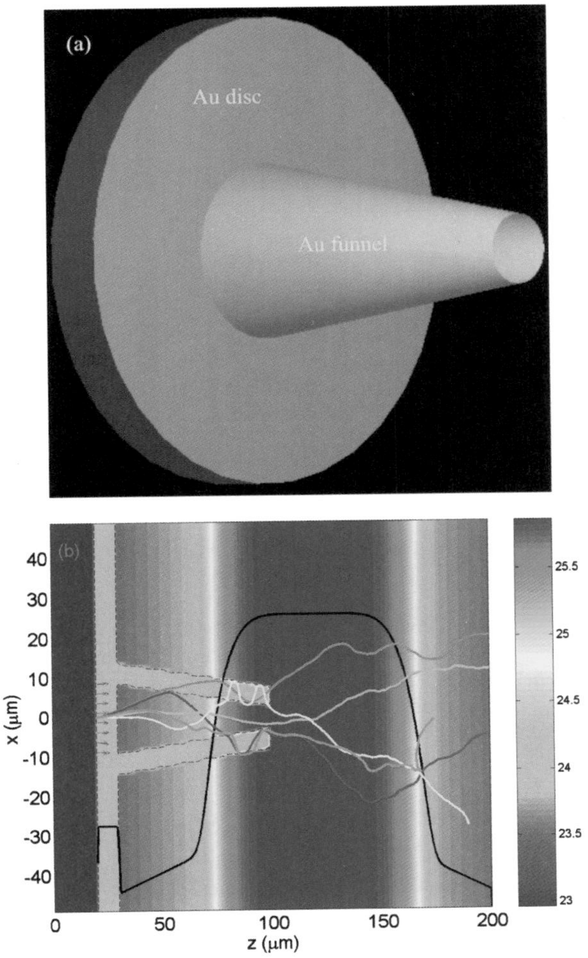

FIG. 1. (Color online) (a) Target configuration. Here the disk represents the tip of the traditional cone. (b) Simulation model. The cone tip or the disk in (a) is modeled by the slab on the left. Note that the front part of the attached funnel is immersed in the DT plasma. The thick solid line shows the electron density profile of both the gold slab and the DT fuel plasmas in the beam propagation axis. The electron beam shortly after injection is schematically shown by the short arrows in the slab in the Au slab. The trajectories of several typical electrons initially at $(x,z)=(0,20)$ by the $\mathbf{j}\times\mathbf{B}$ force, the surface magnetic fields around the funnel walls, as well as scattering between the hot electrons and the cold high-density plasma particles.

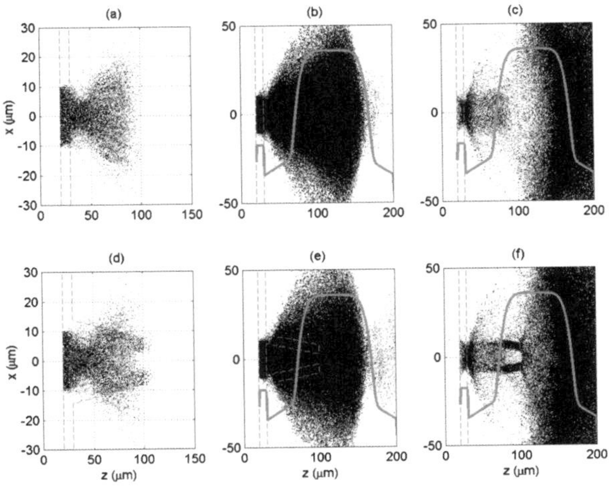

FIG. 2. (Color online) The behavior of the beam electrons at $t=800$ fs for cases I [(a)–(c)] and II [(d)–(f)]. (a) and (d) correspond to the beam electron energy $\epsilon_b>2$ MeV, (b) and (e) for $0.6<\epsilon_b<2$ MeV, and (d) and (f) for $\epsilon_b<0.6$ MeV. The thin dashed lines show the gold plasma, and the thick solid lines mark the electron density profile of the gold slab and DT plasmas in the beam propagation direction.

energy loss in heating the thin funnel walls can be reduced and most of the igniter energy can be deposited in a small region inside the compressed fuel.

II. TARGET CONFIGURATION AND ELECTRON DYNAMICS

The two-dimensional model of the target structure used in our simulation is shown in Fig. 1(b). The funnel is attached to the disk (modeled by the slab on the left). Both the slab and the funnel are made of gold with a constant degree of ionization $Z=20$. The funnel is completely embedded in the DT plasma, which is uniform in the transverse direction but has an exponential density profile rising from 1 to 300 $\mathrm{g/cm^3}$ in the beam propagation direction, as indicated in Fig. 1(b) by the black solid line.

Direct modeling of the laser-plasma interaction involving highly compressed targets by explicit particle-in-cell (PIC) simulations[21] is known to be difficult. For this reason, hybrid simulations[8–12] have often been used. Here we shall also use a hybrid simulation scheme, where the background plasma electrons and ions are treated as fluids and the injected beam electrons are treated as PIC particles. The simulation box (z,x) is $200\times 100\ \mu\mathrm{m}^2$. We shall not consider the initial laser-target interaction, which has been investigated by many authors.[5] Instead, we simulate the beam channel interaction by directly injecting a high-current electron beam from the left and follow its propagation and evolution. The beam is assumed to be relativistic-Maxwellian distributed with a temperature of 1.5 MeV along the z direction. Its width at $1/e^2$ maximum is 20 μm and its angular spread is 30° at $(z,x)=(20,0)$ (μm). The beam rises to the maximum value in 50 fs and then remains constant at $10^{13}\ \mathrm{A/cm^2}$.

To see how the funnel structure controls and guides the fast beam electrons, we shall compare the transport dynamics of the injected electron beam for two cases. In case I, only the traditional slab is present. In case II, a funnel is attached to the latter, as shown in Fig. 1(b). Figures 2–6 summarize the results of these two cases. Figures 2(a)–2(f) show the beam electron dynamics in cases I and II, respectively. Figure 2 shows the phase spaces of three different energy levels. For high (or low) energy beam electrons with $\epsilon_b>2$ MeV (or $\epsilon_b<0.6$ MeV), Figs. 2(a) and 2(d), or Figs. 2(c) and 2(f), clearly indicate that the beam electrons are guided by the funnel. Comparing Figs. 2(c) and 2(f) for the beam electron distributions, we can see clearly that there is stopping and scattering of the low-energy electrons at $z=30$ μm. This

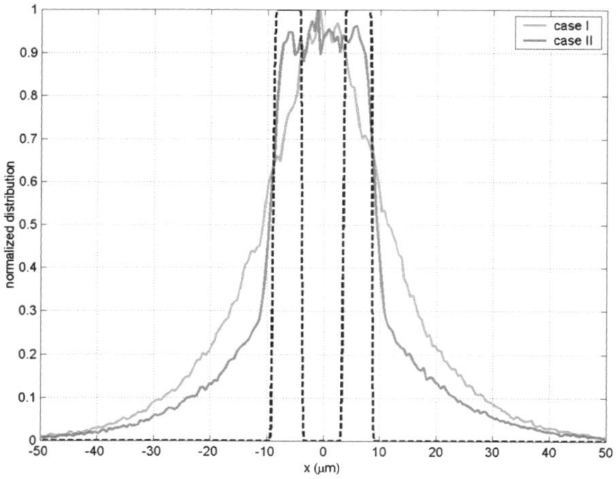

FIG. 3. (Color online) The distribution of the number of the beam electrons in the transverse direction at $t=800$ fs for the two cases. The thick dashed lines show the initial gold-funnel plasma.

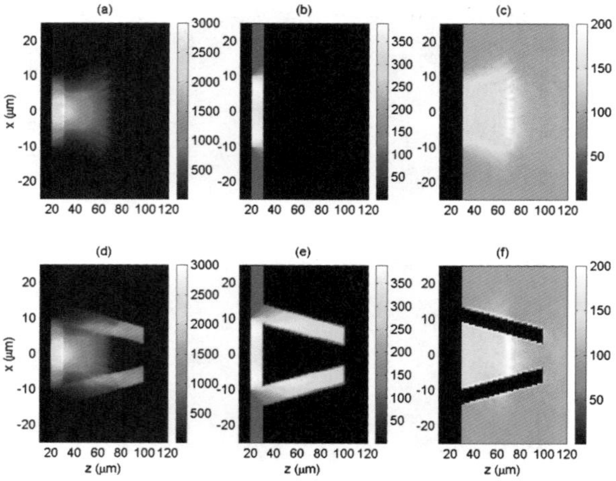

FIG. 5. (Color online) Snapshots of the plasma temperature (in eV) at $t=800$ fs. [(a) and (d)] Fluid electron temperature. [(b) and (e)] Au ion temperature. [(c) and (f)] DT ion temperature. The first and second rows correspond to cases I and II, respectively.

anomalous stopping because of the strong interface electric field, as can be seen in Figs. 4(c) and 4(f), results in a much higher electron population in the DT plasma close to the interface between the Au disk and the DT fuel, as shown in Figs. 4(a) and 4(d). For $0.6<\epsilon_b<2$ MeV, however, the control cannot be clearly observed from phase-space trajectories in Figs. 2(b) and 2(e). In order to see the relative efficiency of the two cases in terms of beam concentration, we count the number of the injected electrons in the transverse direction. The green and red lines in Fig. 3 are for cases I and II, respectively, at $t=0.8$ ps. In case I, the beam electrons are almost Gaussian, and 58% of the injected electrons is in $|x|<10$ μm. However in case II the beam electron distribution is of a flat-centered super-Gaussian profile, and when a funnel is used, 71% of the beam electrons is confined in $|x|<10$ μm. Thus, by attaching a funnel to the cone tip, one can considerably improved the collimation of the fast electrons.

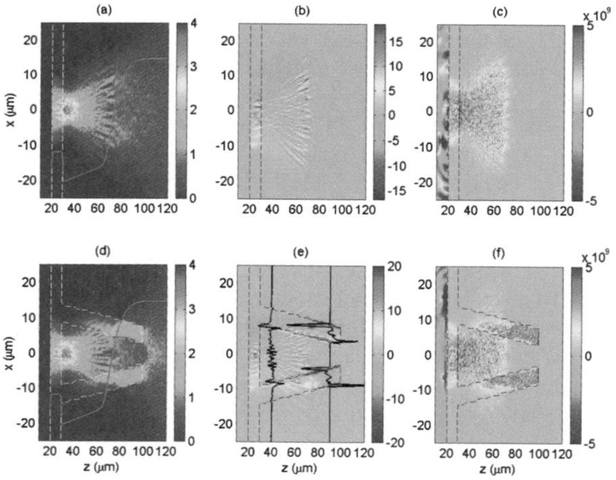

FIG. 4. (Color online) Snapshots of the hot-electron density and the magnetic and electric fields at $t=800$ fs. The thin dashed lines in all subplots show the gold plasma. [(a) and (d)] Hot-electron density (n_b/n_{cr}). The red line also marks the electron density profile in the beam propagation direction. [(b) and (e)] B_y component of magnetic field (in MG). Two thick solid lines gives B_y-profiles located at $z=40$ and 90 μm, respectively. [(c) and (f)] E_z component of electric field (in V/m). The first and second rows correspond to cases I and II, respectively.

III. TRANSPORT AND HEATING OF A FUNNEL-GUIDED TARGET

We now analyze the transport and heating of the beam electrons in the funnel. As the fast electrons enter the plasma, they create a charge separation field. In the initial stage, both the beam and the return currents are nearly in equilibrium with each other and no resistive magnetic field is generated. When the beam electrons propagate further into the plasma, the resulting space-charge field enhances their propagation since as the plasma is heated by the hot electrons, the plasma resistivity is lowered. We see in Figs. 4(c) and 4(f) that the electric field $\boldsymbol{E}$ can reach 10^9 V/m. The latter draws a return current of the background cold plasma electrons with current density larger than 10^{12} A/cm^2. Furthermore, the SGM fields grow in time according to Faraday's law $\partial\mathbf{B}/\partial t=-\nabla\times\mathbf{E}\approx\eta\nabla\times\mathbf{j}_f+\nabla\eta\times\mathbf{j}_f$, neglecting the displacement current. The SGM fields at the edge of the electron beam grow faster than that inside the beam because of the formation of a larger surface current at the beam-plasma interface. This edge SGM field (~10 MG) inside the target can significantly change the direction of the beam because of the $\mathbf{j}\times\mathbf{B}$ force.[11] As can be seen in Figs. 4(b) and 4(e), this process also occurs near the local peaks of the current density, leading to resistive filamentation.

Figures 4(a) and 4(d) show that the collimating SGM field focuses the beam in the slab representing the tip of the gold cone. Near the gold-DT interface discontinuity at $z=30$ μm, large interface electric and magnetic fields are generated because of the steep gradients in both the resistiv-

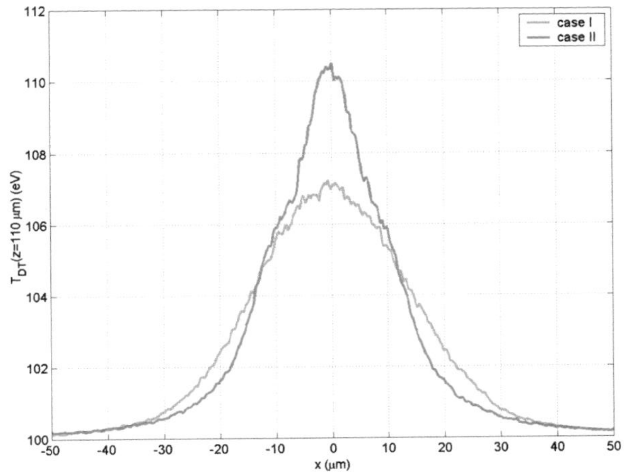

FIG. 6. (Color online) The DT ion temperature $T_{\mathrm{DT}}(z=110\ \mu\mathrm{m})$ (eV) for cases I and II.

ity and the plasma density. The strong interface electric field [$\sim 10^{10}$ V/cm, as seen in Figs. 4(c) and 4(f)] can prevent the beam electrons from propagating forward to the DT plasma. The slowdown and compression of the beam electrons close to the interface ($z=30\ \mu$m) can be clearly seen in Figs. 4(a) and 4(d), also shown in Figs. 2(c) and 2(f).

We now investigate how the gold funnel affects beam electron transport in the DT plasma. Figures 4(a) and 4(d) show the beam density at $t=0.8$ ps for cases I and II, respectively. We can see that the beam electron population in the region $30<z<100\ \mu$m for the two cases is very different. For case I, resistive filamentation of the beam electrons with large spread angle can be observed in Fig. 4(a). For case II, one finds in Fig. 4(e) large magnetic fields around the funnel wall. Since the source term $\nabla\eta\times\mathbf{j}_f$ can generate a magnetic field when there is a beam current perpendicular to a resistivity gradient, the discontinuities in the density and resistivity between the gold funnel and the DT plasma therefore lead to generation of large surface SGM fields on the interface. The energetic electrons in the region $30<z<40\ \mu$m are collimated by this resistive interface SGM field. However, one can see from Figs. 4(d) and 4(e) that the high-energy electrons can also reach the surface of the funnel. Because the gold plasma has a higher resistivity than the DT plasma at high temperature, it is seen from Fig. 4(d) that the energetic electrons can be attracted into the gold funnel.

In particular, near the dense region, the beam electrons can be scattered into the gold funnel through collisions with the DT electrons and ions, as shown in Fig. 1(b). As the beam further propagates into the funnel plasma, the beam electrons deposit their energy in the background plasma through the return current and result in heating of the plasma in the funnel, as shown in Fig. 5(d). Since the Spitzer resistivity $\eta=10^{-4}Z\ln\Lambda T^{-3/2}$ is used, the resistivity decreases as the plasma is heated by hot electrons. Thus the source term $\nabla\eta\times\mathbf{j}_f$ can change sign, as indicated by the solid black lines in Fig. 4(e), due to change of the net current at the interface as the DT plasma density increases in the propagation direction. This reversal of the interface magnetic fields appeared on both the inner and outer surfaces of the funnel, so the fields behave like a wall, which bends the beam electrons in the funnel into the axial direction. This bending process of the fast electrons by the magnetic potential around the funnel is clearly shown in Fig. 1(b).

The coupling of the beam electron energy to the DT fuel for the two cases is compared in Fig. 5 for the temperature. Figures 5(d) and 5(e) show that, because of Joule heating, the plasma temperature of the gold funnel is higher in the region $z>40\ \mu$m. In the region ($z<70\ \mu$m) of relatively low-density DT plasmas, the electric field in Figs. 4(c) and 4(f) is nonuniform. The beam electrons deposit their energy in the DT plasma through the return current and this results in relatively nonuniform heating. In case II, the energy deposition process is significantly different from that in case I. Since the electric and magnetic fields in and near the Au funnel are stronger than that in the DT plasma, they can cause the background electrons to execute complex motion that enhances the heating of the Au plasma [see Figs. 5(d) and 5(e)]. For relatively low-density DT plasmas, Figs. 5(a) and 5(d) show that the background (fluid) electron temperature can grow (due to Joule heating) with a faster rate than the thermal collisions. However, for higher DT density, the electric field is reduced and collisions between the hot electrons and the cold plasma particles become the dominant heating mechanism. This can be clearly observed in Figs. 5(c) and 5(f) for both cases. As the beam enters the dense region ($z=70\ \mu$m), it deposits its energy through scattering of pairs of relativistic electrons off the background plasmas. The DT plasma temperature in case II can grow faster than in case I. Furthermore, when the beam electrons reach the outer surface of the funnel, heating of the funnel plasma near the funnel surface is enhanced. Figure 6 shows that at $t=0.8$ ps the DT plasma ions near the head of the funnel ($z=110\ \mu$m) have been heated to a somewhat higher temperature in case II than in case I. That is, the Au funnel can indeed guide and collimate the hot-electron beam propagation in dense DT plasmas and enhance the heating of the DT fuel.

IV. CONCLUSION

In conclusion, the dynamics of energetic electrons driven by petawatt-picosecond pulse lasers in a new cone-funnel target configuration has been investigated using two-dimensional hybrid simulations. The hot-electron propagation through both the traditional cone-guided and cone-funnel-guided targets are compared. Our results show that hot electrons are first collimated between the inner walls of the funnel by the large interface magnetic fields appearing on the inner surface. However, since the interface magnetic fields in the inner and outer surfaces of the funnel eventually change sign due to fast electrons entering the gold funnel, it results in a change of the return currents near the interface between the funnel and the DT plasma. The resulting magnetic field around the funnel bends the electron beam along the gold funnel to reach the fuel core. Enhanced heating of the dense DT core for the funnel-guided target is also verified. Besides fast ignition, funnel-guided electron beam transport at ultrahigh plasma and energy densities can have applications in laboratory astrophysics, medicine, materials science, heavy-ion fusion, etc. We note that the reversal of the magnetic fields observed here may also be relevant to the study of anomalous magnetic fields in high energy density physics.[22–24]

ACKNOWLEDGMENTS

This work was supported by the National Natural Science Foundation of China (Grant Nos. 10974022 and 10835003), the National Basic Research Program of China (973) (Grant No. 2007CB815101), and partially by the National High-Tech ICF Committee.

[1]J. Meyer-ter-Vehn, Plasma Phys. Controlled Fusion **51**, 124001 (2009).
[2]W. Yu, V. Bychenkov, Y. Sentoku, M. Y. Yu, Z. M. Sheng, and K. Mima, Phys. Rev. Lett. **85**, 570 (2000).
[3]M. Tabak, J. Hammer, M. E. Glinsky, W. L. Kruer, S. C. Wilks, J. Woodworth, E. M. Campbell, M. D. Perry, and R. J. Mason, Phys. Plasmas **1**, 1626 (1994).
[4]R. Kodama, P. A. Norreys, K. Mima, A. E. Dangor, R. G. Evans, H. Fujita, Y. Kitagawa, K. Krushelnick, T. Miyakoshi, N. Miyanaga, T. Norimatsu, S. J. Rose, T. Shozaki, K. Shigemori, A. Sunahara, M. Tampo, K. A. Tanaka, Y. Toyama, T. Yamanaka, and M. Zepf, Nature (London) **412**, 798 (2001).
[5]S. Atzeni and J. Meyer-ter-Vehn, *Inertial Fusion—Beam Plasma Interaction, Hydrodynamic, Dense Plasma Physics* (Clarendon, Oxford, 2003).
[6]V. Malka, J. Faure, Y. A. Gauduel, E. Lefebvre, A. Rousse, and K. T. Phuoc, Nat. Phys. **4**, 447 (2008).
[7]H. B. Cai, K. Mima, W. M. Zhou, T. Jozaki, H. Nagatomo, A. Sunahara, and R. J. Mason, Phys. Rev. Lett. **102**, 245001 (2009).
[8]C. T. Zhou and X. T. He, Opt. Lett. **32**, 2444 (2007); Phys. Plasmas **15**, 123105 (2008).
[9]R. B. Campbell, R. Kodama, T. A. Mehlhorn, K. A. Tanaka, and D. R. Welch, Phys. Rev. Lett. **94**, 055001 (2005).
[10]D. R. Welch, D. V. Rose, B. V. Oliver, and R. E. Clark, Nucl. Instrum. Methods Phys. Res. A **464**, 134 (2001).
[11]A. A. Solodov, K. S. Anderson, R. Betti, V. Gotcheva, J. Myatt, J. A. Delettrez, S. Skupsky, W. Theobald, and C. Stoeckl, Phys. Plasmas **16**, 056309 (2009).
[12]J. J. Honrubia and J. Meyer-ter-Vehn, Nucl. Fusion **46**, L25 (2006).
[13]R. Kodama, Y. Sentoku, Z. L. Chen, G. R. Kumar, S. P. Hatchett, Y. Toyama, T. E. Cowan, R. R. Freeman, J. Fuchs, Y. Izawa, M. H. Key, Y. Kitagawa, K. Kondo, T. Matsuoka, H. Nakamura, M. Nakatsutsumi, P. A. Norreys, T. Norimatsu, R. A. Snavely, R. B. Stephens, M. Tampo, K. A. Tanaka, and T. Yabuuchi, Nature (London) **432**, 1005 (2004).
[14]J. S. Green, K. L. Lancaster, K. U. Akli, C. D. Gregory, F. N. Beg, S. N. Chen, D. Clark, R. R. Freeman, S. Hawkes, C. Hernandez-Gomez, H. Habara, R. Heathcote, D. S. Hey, K. Highbarger, M. H. Key, R. Kodama, K. Krushelnick, I. Musgrave, H. Nakamura, M. Nakatsutsumi, N. Patel, R. Stephens, M. Storm, M. Tampo, W. Theobald, L. Van Woerkom, R. L. Weber, M. S. Wei, N. C. Woolsey, and P. A. Norreys, Nat. Phys. **3**, 853 (2007).
[15]J. King, K. Akli, R. R. Freeman, J. Green, S. P. Hatchett, D. Hey, P. Jamangi, M. H. Norreys, P. K. Patel, T. Phillips, R. B. Stephen, R. B. Stephens, W. Theobald, R. J. Town, L. Van Woerkom, B. Zhang, and F. N. Beg, Phys. Plasmas **16**, 020701 (2009).
[16]A. G. MacPhee, L. Divol, A. J. Kemp, K. U. Akli, F. N. Beg, C. D. Chen, H. Chen, D. S. Hey, R. J. Fedosejevs, R. R. Freeman, M. Henesian, M. H. Key, S. Le Pape, A. Link, T. Ma, A. J. Mackinnon, V. M. Ovchinnikov, P. K. Patel, T. W. Phillips, R. B. Stephens, M. Tabak, R. Town, Y. Y. Tsui, L. D. Van Woerkom, M. S. Wei, and S. C. Wilks, Phys. Rev. Lett. **104**, 055002 (2010).
[17]N. Renard-Le Galloudec, E. d'Humiéres, B. I. Cho, J. Osterholz, Y. Sentoku, and T. Ditmire, Phys. Rev. Lett. **102**, 205003 (2009).
[18]S. Kar, A. P. L. Robinson, D. C. Carroll, O. Lundh, K. Markey, P. McKenna, P. Norreys, and M. Zepf, Phys. Rev. Lett. **102**, 055001 (2009).
[19]C. T. Zhou, X. T. He, and M. Y. Yu, Appl. Phys. Lett. **92**, 151502 (2008); **92**, 071502 (2008).
[20]A. P. L. Robinson and M. Sherlock, Phys. Plasmas **14**, 083105 (2007).
[21]C. K. Birdsall and A. B. Langdon, *Plasma Physics via Computer Simulation* (McGraw-Hill, New York, 1985).
[22]M. Honda, J. Meyer-ter-Vehn, and A. Pukhov, Phys. Plasmas **7**, 1302 (2000).
[23]R. P. Drake, *High-Energy-Density Physics* (Springer, Berlin, 2006).
[24]H. Hora, *Plasmas at High Temperature and Density* (Springer, Heidelberg, 1991).

APPLIED PHYSICS LETTERS **97**, 051502 (2010)

Propagation of energetic electrons in a hollow plasma fiber

C. T. Zhou,[1,2,a)] L. Y. Chew,[3] and X. T. He[1,2]
[1]*Institute of Applied Physics and Computational Mathematics, Beijing 100094, People's Republic of China*
[2]*Center for Applied Physics and Technology, Peking University, Beijing 100871, People's Republic of China*
[3]*Division of Physics and Applied Physics, School of Physical and Mathematical Sciences, Nanyang Technological University, Nanyang Avenue, Singapore 639798*

(Received 26 June 2010; accepted 14 July 2010; published online 4 August 2010)

Transport of energetic electrons in a hollow plasma fiber is investigated. The high-current electron beam induces in the fiber strong radial electric fields and azimuthal magnetic fields on the inner and outer surfaces of the hollow fiber. The hot electrons are pushed out by the surface magnetic field and returned into the fiber by the sheath electric field. Imbalance of the latter fields can drive chaotic oscillations of electrons around the fiber wall. Intense thin return-current layers inside both the inner and outer wall surfaces are observed. This enhances local joule heating around both surfaces by the return current. © *2010 American Institute of Physics*. [doi:10.1063/1.3475414]

The plasma fiber as a photonic device can have many applications[1–6] in high energy density physics, as well as for generating highly localized charged particles for medical treatment, isochoric heating for precise estimation of the equation of state, x-ray imaging, fast ignition in inertial fusion, etc.[3,7–10] Recently, a wire-guided plasma fiber was used to control the divergence of laser-driven megaelectronvolt electrons.[4] Simulations[5] and experiments[11] demonstrated that fast electrons generated by intense lasers can be guided by the large fields around a wire and propagate several millimeters along the wire. This effectively enhances the energy coupling between the laser and the wire. More recently, it has been shown that laser-driven hot electrons can be guided into a beam and transported over several tens of microns in a small channel.[12] The physical mechanisms to form a hollow plasma is obviously different from depletions of plasma from the axis of a laser beam due to ponderomotive self-focusing.[7,13,14]

In this paper we consider a hollow plasma fiber attached to the back of a target. The two-dimensional model of the target structure used in our simulation is shown in Fig. 1. The fiber is attached to the slab. Both the slab and the fiber consist of gold plasma with a constant degree of ionization $Z=20$. In our hybrid particle-in-cell (PIC) simulation scheme, the background plasma electrons and ions are treated as fluids and the injected beam electrons are treated as PIC particles.[6] The simulation box (z,x) is 200×100 μm^2. We shall not consider the initial laser-target interaction, which has been investigated by many authors. Instead, we simulate the follow up beam-plasma interaction by directly injecting a high-current electron beam from the left and follow its propagation and evolution. The beam is assumed to be relativistic-Maxwellian distributed with an average temperature of 1.5 MeV along the x direction. Its width at $1/e^2$ maximum is 20 μm and its angular spread is 30° at $(z,x)=(20,0)[\mu\text{m}]$. The beam rises to its maximum value in 50 fs, and remains constant at 10^{13} A/cm^2.

Figures 1–4 summarize our simulation results. At the very beginning of the beam-plasma interaction, the beam electrons traverse the plasma slab and exit into the rear vacuum region. Huge (several times 10^{12} V/m) space-charge electrostatic fields at the front and backsides of the slab are generated. These fields create across the plasma slab an electrostatic potential wall, as shown in Fig. 1 with the thick green solid line. The black and red electron trajectories in the slab of Fig. 1 indicate that some beam electrons can experience multiple reflections in the slab, as the mean free path of megaelectronvolt electrons is much larger than the slab thickness (10 μm). Furthermore, the blue arrows in Fig. 1 illustrate that most beam electrons propagate along the fiber interface, inducing a strong azimuthal surface magnetic field (B_y) surrounding the fiber plasma, as shown in Fig. 1 by the thick white solid line. A strong transverse electric field (E_x, in teravolt per meter) is also created [see Fig. 2(c)]. The fast electrons, which are pushed out by the magnetic field B_y,

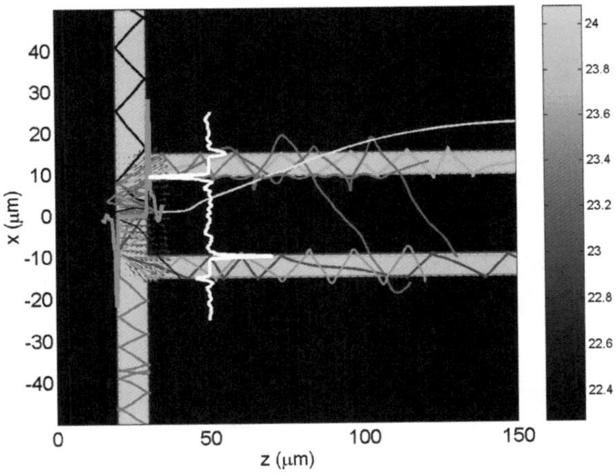

FIG. 1. (Color online) Two-dimensional model of a slab-and-hollow fiber plasma for the hybrid simulation. The electron beam shortly after injection is schematically represented by the blue arrows. The thick solid line around the surface of the slab gives E_z-profile at $x=0$ μm, and the thick white line gives the B_y-profile at $z=50$ μm. The trajectories of several typical electrons initially at $z=20$ μm show the effects of electron reflection in $20<z<30$ μm due to the longitudinal electric field (E_z), the guiding, pinching, and chaotic motion of the electrons due to both the azimuthal magnetic field (B_y) and the transverse electric field (E_x) around the plasma fiber.

a)Electronic mail: zcangtao@iapcm.ac.cn.

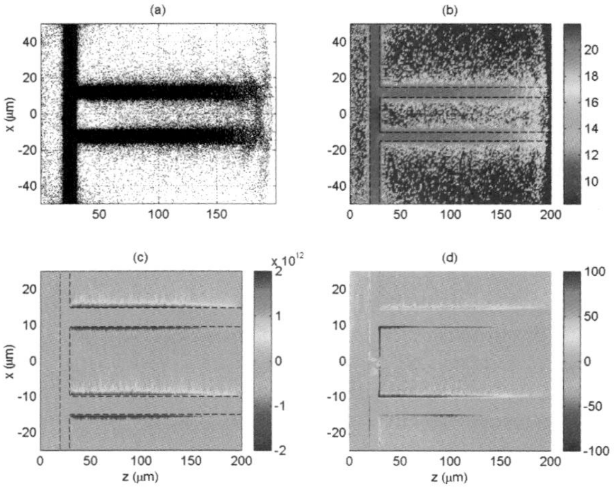

FIG. 2. (Color online) Snapshots of the beam-electron trajectories, beam-electron density, and electric fields at $t=0.7$ ps. (a) The phase-space behavior of the beam electrons. (b) Beam-electron density ($\log_{10} n_b$). (c) The transverse electric field component E_x (in volt per meter). (d) B_y component of the magnetic field (in megagauss).

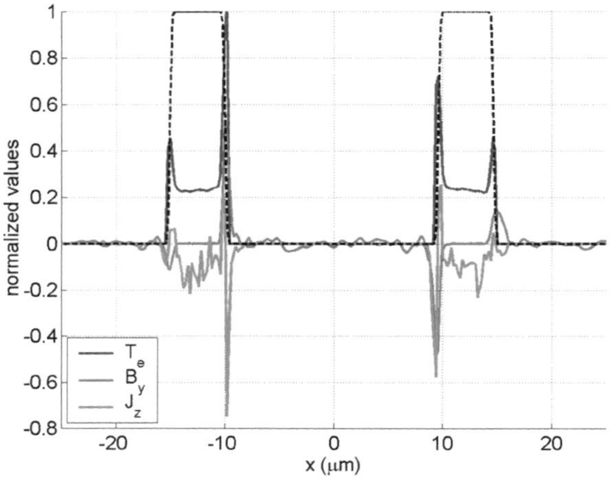

FIG. 4. (Color online) The normalized cold plasma-electron temperature T_e, the azimuthal magnetic field B_y, and the net current density J_z at $z=50$ μm and $t=0.7$ ps. The thick dashed lines show the initial gold-hollow plasma.

and are reflected into the fiber plasma by the transverse sheath field.[4] An imbalance between the azimuthal magnetic and the transverse electric field around the fiber plasma can drive chaotic oscillations of the electrons in the fiber wall, as shown in Fig. 1.

We now examine hot-electron transport in the slab-hollow fiber target. Hot-electron propagation in solid plasmas will set up a charge separation field, whose energy is then converted into the thermal energy of the plasma. Thus, the resistivity as well as the space-charge field is reduced, allowing the energetic electrons to propagate further into the plasmas. In the initial stage, these currents are nearly in equilibrium with each other and no resistive magnetic field is generated. The exponential reduction in the electric field and growth of the magnetic field can be clearly observed at the very beginning of the beam-plasma interaction. When the high-energy beam electrons exit into the back vacuum region, the strong sheath field formed at the backside of the slab reflects the beam electrons back into the front surface. Furthermore, the beam electrons exiting from the front surface form strong front-surface electric fields. Figure 1 clearly shows the strong sheath fields around the surface of the slab. This field bends the beam electrons in the slab layer, as seen in Figs. 2(a) and 2(b). The physical mechanism of bending of the hot elections in the slab plasma is well understood in connection to the study of the laser-driven electron and proton acceleration.[15]

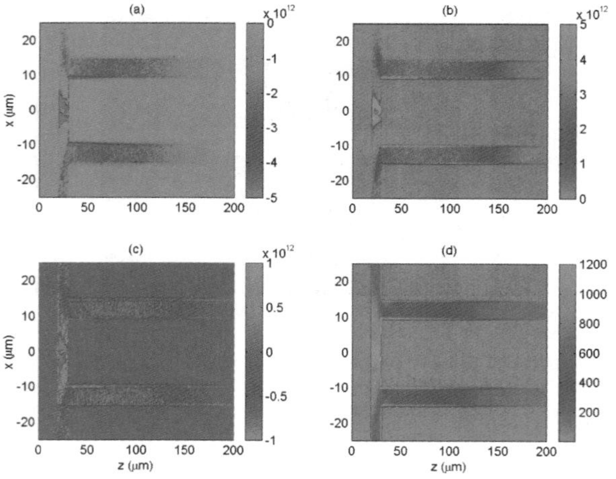

FIG. 3. (Color online) Snapshots of the current density J_z (in ampere per centimeter) and the azimuthal magnetic field at $t=0.7$ ps. [(a)–(c)] the forward hot electron, returning cold electron, and net current densities, respectively. (d) The background-plasma (fluid) electron temperature (in electronvolt).

Guiding and collimation of the energetic electrons along the plasma fiber can be deduced from Figs. 2(a) and 2(b). Tracing the electron trajectories shown in Fig. 1, we find that the beam electrons can also propagate through the slab to the hollow fiber. However the corresponding beam-electron trajectories are more complex than that for the simple slab target. On one hand, the beam electrons can exit both the inner and outer walls of the fiber plasma. In Fig. 2(c) huge radial space-charge electrostatic fields (reaching 2×10^{12} V/m) at the inner and outer surfaces of the fiber can be clearly observed. These fields create across the fiber plasma a transverse electrostatic potential wall. The transverse surface electric fields result in the beam electrons experiencing multiple reflections in the fiber wall. On the other hand, when the beam electrons propagate further into the fiber layer, the current of fast electrons exceeds the Alfvén current and beam transport is possible only if sufficient current neutralization is achieved by the return current of background plasma electrons. Figures 3(a) and 3(b) show that both the fast and return electron current density in the fiber wall can be larger than 5×10^{12} A/cm, and the corresponding net current density [Fig. 3(c)] exceeds 10^{12} A/cm. Figure 3(c) shows that a net current can exist inside the fiber bulk. In particular, thin return current layers inside the fiber surfaces can clearly be seen. Such a thin-layer return current was first reported in Ref. 5 for a wire target. In the present case, two thin layers can appear at the inner and outer surfaces. The azimuthal magnetic fields appearing in the inner and outer surfaces, as shown in Fig. 2(d), can be attributed to the net current along the fiber.

To see the energy coupling of the beam electrons to the fiber plasma, we now consider the fluid electron temperature T_e. Figure 3(d) shows nonuniform heating of the fiber plasma. In the bulk of the fiber, the average temperature is about 500 eV. The surface temperature in the outer surface layer of the fiber is about 1000 eV, i.e., double that of the bulk. Figure 4 shows the azimuthal magnetic field, the background plasma temperature, and the net current at $z=50\ \mu m$ and $t=700$ fs. Obviously, the fast-electron current is larger than the cold-electron return current in the inner surface. The direction of the azimuthal magnetic field on the inner surface is then determined by the net current. The magnetic field on the outside of the fiber is due to the return current at the fiber-vacuum interface. As can be seen in Figs. 3 and 4, the fast- and return-electron currents along the inner surface of the fiber are higher, thus the thin inner surface layer is heated more strongly. The corresponding temperature can be higher than 1500 eV. Heating for the cold background plasma in the fiber can be described by the cold-electron energy equation[16] $(3/2)n_e(\partial T_e/\partial t)=\nabla\cdot(\kappa\nabla T_e)+J_e^2/\sigma+Q_h$, where κ, σ, and J_e are the thermal conductivity, electrical conductivity, and the cold-electron current density, respectively. The drag of the hot electrons on the cold electrons is given by $Q_h=(3/2)T_h n_h/\tau_{ch}$, where $\tau_{ch}\equiv(3\sqrt{3}/8\pi)\times(m_e^{1/2}T_h^{3/2}/n_c e^4 \ln\Lambda)$, $\ln\Lambda$ is the usual Coulomb logarithm, and T_h and n_h are the hot electron temperature and density, respectively. Energy loss due to inelastic collisions and material effects is neglected. For relativistic beam electrons, the term ∇T_e can be neglected since it is smaller than the other terms by a factor T_e/T_h. One can also neglect the hot-electron drag since it is smaller than the joule heating term by a factor $(T_e/T_h)^{3/2}$. Thus joule heating J_e^2/σ dominates. Due to the enhanced return currents in the thin layers of both the inner and outer surfaces, the temperature in the surfaces of the fiber plasma is therefore higher, as can be seen in Figs. 3(d) and 4.

In conclusion, the propagation of hot electrons generated by a petawatt picosecond laser interacting with a gold slab-hollow fiber target is simulated using two-dimensional hybrid simulations. The hot electrons are guided by a hollow-fiber plasma behind the slab. It is shown that as the hot electrons propagate in the fiber plasma, the surface electric and magnetic fields in the inner and outer surfaces of the fiber confine and guide the hot electrons. The appearance of the thin enhanced return-current layers in both the inner and outer surfaces of the fiber plasma results in enhanced surface heating. The temperature in both surfaces is double that of the bulk of the fiber plasma. The results here should be useful in the design of more efficient point light sources, as well as in other applications of laser produced energetic electrons.

This work is supported by the National Natural Science Foundation of China, Grant Nos. 10974022 and 10835003, the National Basic Research Program of China (Grant No. 2007CB815101), and the National High-Tech ICF Committee.

[1]W. Yu, V. Bychenkov, Y. Sentoku, M. Y. Yu, Z. M. Sheng, and K. Mima, Phys. Rev. Lett. **85**, 570 (2000).
[2]H. Hora, B. Malekynia, M. Ghoranneviss, and G. H. Miley, Appl. Phys. Lett. **93**, 011101 (2008).
[3]V. Malka, J. Faure, Y. A. Gaduel, E. Lefebvre, A. Rousse, and K. T. Phuoc, Nat. Phys. **4**, 447 (2008).
[4]R. Kodama, Y. Sentoku, Z. L. Chen, G. R. Kumar, S. P. Hatchett, Y. Toyama, T. E. Cowan, R. R. Freeman, J. Fuchs, Y. Izawa, M. H. Key, Y. Kitagawa, K. Kondo, T. Matsuoka, H. Nakamura, M. Nakatsutsumi, P. A. Norreys, T. Norimatsu, R. A. Snavely, R. B. Stephens, M. Tampo, K. A. Tanaka, and T. Yabuuchi, Nature (London) **432**, 1005 (2004).
[5]J. S. Green, K. L. Lancaster, K. U. Akli, C. D. Gregory, F. N. Beg, S. N. Chen, D. Clark, R. R. Freeman, S. Hawkes, C. Hernandez-Gomez, H. Habara, R. Heathcote, D. S. Hey, K. Highbarger, M. H. Key, R. Kodama, K. Krushelnick, I. Musgrave, H. Nakamura, M. Nakatsutsumi, N. Patel, R. Stephens, M. Storm, M. Tampo, and W. Theobald, Nat. Phys. **3**, 853 (2007).
[6]C. T. Zhou, X. T. He, and M. Y. Yu, Appl. Phys. Lett. **92**, 151502 (2008); **92**, 071502 (2008).
[7]S. Atzeni and J. Meyer-tar-Vehn, *Inertial Fusion: Beam Plasma Interaction, Hydrodynamic, Dense Plasma Physics* (Clarendon, Oxford, 2003).
[8]W. M. Wang, Z. M. Sheng, and J. Zhang, Laser Part. Beams **27**, 3 (2009).
[9]L. Bakshi, S. Eliezer, Z. Henis, N. Nissim, L. Perelmutter, D. Moreno, M. Sudai, and M. Mond, Laser Part. Beams **27**, 79 (2009).
[10]S. Lourenco, N. Kowarsch, W. Scheid, and P. X. Wang, Laser Part. Beams **28**, 195 (2010).
[11]J. King, K. Akli, R. R. Freeman, J. Green, S. P. Hatchett, D. Hey, P. Jamangi, M. H. Norreys, P. K. Patel, T. Phillips, R. B. Stephen, R. B. Stephens, W. Theobald, R. J. Town, L. Van Woerkom, B. Zhang, and F. N. Beg, Phys. Plasmas **16**, 020701 (2009).
[12]N. Renard-Le Galloudec, E. d'Humiéres, B. I. Cho, J. Osterholz, Y. Sentoku, and T. Ditmire, Phys. Rev. Lett. **102**, 205003 (2009).
[13]H. Hora, Zeitschr. Physik **226**, 156 (1969).
[14]M. C. Richardson and A. J. Alcock, Appl. Phys. Lett. **18**, 357 (1971).
[15]C. T. Zhou and X. T. He, Opt. Lett. **32**, 2444 (2007); Appl. Phys. Lett. **90**, 031503 (2007).
[16]M. Glinsky, Phys. Plasmas **2**, 2796 (1995).

PHYSICS OF PLASMAS **18**, 022703 (2011)

Relativistic kinetic model for energy deposition of intense laser-driven electrons in fast ignition scenario

Sizhong Wu,[1,a)] Cangtao Zhou,[1,2] Shaoping Zhu,[1] Hua Zhang,[2,b)] and Xiantu He[1,2]
[1]*Institute of Applied Physics and Computational Mathematics, P.O. Box 8009, Beijing 100088, People's Republic of China*
[2]*Center for Applied Physics and Technology, Peking University, Beijing 100871, People's Republic of China*

(Received 23 November 2010; accepted 17 January 2011; published online 16 February 2011)

One of the most crucial steps for a fast ignition scenario is the energy deposition into the highly compressed deuterium-tritium core plasmas via intense laser-produced relativistic electrons. Based on fundamental principles, a kinetic model is developed by considering both binary collisions and the contribution due to collective process. The collision operator is exactly simplified by taking into account relativistic effects within the context of fast ignition. It is expressed in a differential form with the help of two analogous Rosenbluth potentials. The explicit formulation of a relativistic kinetic equation in three-dimensional momentum space is obtained by expanding the potential functions in terms of spherical harmonics, in which only simple differentiations and integrations are involved. Fast electron number is well conserved in this model. The range and penetration depth are also discussed. © *2011 American Institute of Physics*. [doi:10.1063/1.3553452]

I. INTRODUCTION

The fast ignition (FI) scenario[1] provides an alternative approach to realizing inertial confinement fusion (ICF),[2] in which the hot spot formation is separated from the capsule compression. This novel and time segmented scheme has drawn significant attention worldwide[3–7] ever since its inception due to its potential advantages over conventional ICF, notably in easing requirements of the implosion symmetry and the hydrodynamic instabilities, achieving higher gains, and reducing the driver energy.[3] The ignition of a compressed deuterium-tritium (DT) pellet generally falls into three steps. First, relativistic electrons (carrying a total current of ~GA with 1–3 MeV mean energy) are generated by an ultraintense ($\sim 10^{20}$ W cm^{-2}) short-pulse (~20 ps) laser interacting with solid targets.[8,9] Then, the electron beam will leave the interaction area and transport into the dense core. It will experience an extremely steep density gradient over a distance of about 100 μm, typically rising from the initial critical density ($\sim 10^{21}$ cm^{-3}) to the highly compressed core density ($\sim 10^{26}$ cm^{-3}).[10–12] Once reaching the edge of the core region, the beam will deposit its energy locally through collisions with background electrons and ions and heat the fuel to the ignition temperature, forming a hot spot and triggering a thermonuclear burn.[2]

The coupling efficiency of the petawatt laser energy to the hot spot is of primary importance in assessing the viability of a FI scenario. It requires a good understanding about the energy deposition of fast electrons in FI targets, which will ultimately determine the size and temperature of the hot spot and hence the thermonuclear burn dynamics. There have been many works contributed to this subject during the past two decades. In the original work introducing the general concepts of FI,[1] the range for ~MeV electrons in core plasmas was estimated based on the stopping power model. Later on, the stopping of fast electrons as well as their multiple scattering off target ions was properly treated.[13] This was further improved and extended by adding the equally significant multiple scattering contribution of target electrons[14,15] with the help of the multiple scattering theory.[16] Furthermore, a few numerical mistakes in the previous works were corrected.[17] The beam range, straggling, and blooming can be evaluated in a direct way. Most recently, various simulation techniques[18–22] have been employed to explore the essential characteristics of energy deposition.

The works above have helped to gain knowledge about energy deposition, such as the scattering effect, the pinching effect of the resistive magnetic field, etc. However, in order to make a comprehensive analysis about the energy efficiency in FI and propose integrated point design experiments, the dynamics of deposition must be investigated in more detail, especially its dependence on time, space, and momentum, which is mainly not included in the methods mentioned above. Therefore, the kinetic theory will be a complementary tool for this study since it describes the evolution of the microcosmic distribution function in phase space and can accurately reveal the dynamic behavior of fast electrons in dense DT core plasmas. In this paper, a relativistic kinetic model for fast electrons was established based on fundamental principles in the context of energy deposition in FI. The contributions from binary collisions with impact parameters less than Debye length as well as the contribution related to core plasmas' collective process are well included in our model. It is noticed that the first attempt to apply such kind of kinetic model in FI was made in Ref. 23, where an approximation was adopted in order to get a simple and linearized collision term. In our work, the relevant collision operator was introduced and simplified in an exact way. It was converted from its initial complicated integrodifferential

[a)]Electronic mail: wu_sizhong@iapcm.ac.cn.
[b)]Electronic mail: zhanghua01@gmail.com.

form to a differential expression with analogous Rosenbluth potential functions. The explicit expansion of potential functions in terms of spherical harmonics allowed us to rewrite the collision operator in a compact form where only simple integrations and differentiations are involved, leading to convenience for rapid numerical evaluation as well as analytic work.

The paper is organized as follows: Sec. II describes the kinetic model for energy deposition and the relevant collision operator is introduced and simplified. In Sec. III, the explicit form for the kinetic equation in three-dimensional momentum space is calculated with the help of spherical harmonics expansion. The fast electron number conservation is discussed analytically. Preliminary numerical results based on the derived relativistic kinetic model are presented in Sec. IV, as well as the discussions. The main conclusions are summarized in Sec. V. In the Appendices, we list the general expressions of the coefficients for the expansion of the collision operator and the derivatives needed for calculating these coefficients.

II. COLLISION OPERATOR

For a relativistic electron beam incident into a highly compressed plasma, its energy loss and deposition to the background particles are mainly due to collisions[15] and radiative processes[24] such as bremsstrahlung, pair creation, etc. For the fast electron energy range considered in FI, which is just within a few million electron volts, the contribution from radiative loss is negligible.[24] The collisions between fast electrons and background particles generally fall into two types, namely, the short-range and the long-range collisions, depending on the impact parameter l. The former refers to the binary Coulomb collisions with impact parameter less than Debye length ($l<\lambda_D$), which is well described by the Fokker–Planck (FP) collision operator. The latter is associated with the collective response of target plasmas[25] to the fast electrons ($l\geq\lambda_D$), which is often treated with the Balescu–Lenard (BL) collision term. The collisions between the fast electrons will not be considered in this paper since the beam density is far less then the target density in a FI scheme.

The relativistic kinetic equation for fast electrons can be expressed as[26,27]

$$\frac{\partial f_e}{\partial t}+\frac{\mathbf{p}}{\gamma m}\cdot\nabla f_e-e\left(\mathbf{E}+\frac{\mathbf{p}\times\mathbf{B}}{\gamma m}\right)\cdot\nabla_{\mathbf{p}}f_e=\sum_{b=be,bi}C(f_e,f_b), \tag{1}$$

where subscripts e and b refer to fast electrons and background plasma species (be for background electron and bi for background ion). f_e (or f_b) denotes the phase space distribution function for e (or species b) and $C(f_e,f_b)$ is the collision operator. $\mathbf{E}$ and $\mathbf{B}$ are, respectively, the electric and magnetic field. $\mathbf{p}$, γ, and m represent the momentum, the relativistic Lorentz factor, and the rest mass of fast electrons. The physical quantities for background plasma species b are all labeled with superscript $'$ throughout this paper. In order to deal with the relativistic effect involved in FI, the following formulae will be frequently recalled.

$$\mathbf{p}=\gamma m\mathbf{v},\quad \gamma=1/\sqrt{1-v^2/c^2}=\sqrt{1+p^2/m^2c^2}; \tag{2a}$$

$$\mathbf{p}'=\gamma' m'\mathbf{v}',\quad \gamma'=1/\sqrt{1-v'^2/c^2}=\sqrt{1+p'^2/m'^2c^2}, \tag{2b}$$

where v and v' are the velocities for the fast electron and background species b.

Generally, the collision operator can be written in a integrodifferential form,[27]

$$C(f_e,f_b)=\nabla_{\mathbf{p}}\cdot\int Q\cdot[f_b\nabla_{\mathbf{p}}f_e-f_e\nabla_{\mathbf{p}'}f_b]d^3\mathbf{p}', \tag{3}$$

where Q is the collision kernel. Furthermore, it can be cast into two parts, as mentioned, $Q=Q_{\mathrm{FP}}+Q_{\mathrm{BL}}$. Here, Q_{FP} and Q_{BL} represent the FP and BL kernels, respectively.

For binary collisions between fast electrons and background plasmas in FI, the fast electrons (with mean energy of a few million electron volts) are relativistic, while the background particles (about ~100 eV to ~keV) remain still nonrelativistic. Under this circumstance, the Landau kernel[28] will be appropriate for characterizing binary collisions,[29,30]

$$Q_{\mathrm{FP}}=2\pi e^2q'^2\ln\Lambda_{\mathrm{FP}}\frac{u^2\mathbf{I}-\mathbf{uu}}{u^3}, \tag{4}$$

where $\mathbf{u}=\mathbf{v}-\mathbf{v}'$ is the relative velocity of the two colliding particles, q' represents the charge for species b, and $\ln\Lambda_{\mathrm{FP}}$ is the corresponding Coulomb logarithm.[17] As for the BL kernel, it can be expressed as[26]

$$Q_{\mathrm{BL}}=2e^2q'^2\int_{k\leq\lambda_D^{-1}}\frac{\mathbf{kk}}{k^4}\frac{\delta(\mathbf{k}\cdot\mathbf{v}-\mathbf{k}\cdot\mathbf{v}')}{|\epsilon(\omega,\mathbf{k})|^2_{\omega=\mathbf{k}\cdot\mathbf{v}}}d^3\mathbf{k}, \tag{5}$$

where $\epsilon(\omega,\mathbf{k})$ is the dielectric function representing the effect of the collective oscillation of the background plasmas. ω and k denote the angular frequency and wave number of the oscillation, respectively. Equation (5) can be further reduced by employing $\epsilon(\omega,\mathbf{k})=1-\omega_p^2/\omega^2$ (ω_p is the background plasma frequency)

$$Q_{\mathrm{BL}}=2\pi e^2q'^2\ln\Lambda_{\mathrm{BL}}\frac{u^2\mathbf{I}-\mathbf{uu}}{u^3}, \tag{6}$$

where $\ln\Lambda_{\mathrm{BL}}$ is the related Coulomb logarithm with the expression

$$\ln\Lambda_{\mathrm{BL}}=\frac{1}{2}\ln|A-1|-\frac{1}{2}\frac{A}{A-1},\quad A=\frac{2}{3}\frac{p^2}{\gamma^2\lambda_D^2\omega_p^2}. \tag{7}$$

By substituting Eqs. (4) and (6) into Eq. (3) and after some algebra, we can re-express the collision operator as

$$C(f_e,f_b)=\Gamma\nabla_{\mathbf{p}}\cdot\int(\nabla_{\mathbf{v}}\nabla_{\mathbf{v}}u)\cdot[f_b\nabla_{\mathbf{p}}f_e-f_e\nabla_{\mathbf{p}'}f_b]d^3\mathbf{p}', \tag{8}$$

where $\Gamma=2\pi e^2q'^2(\ln\Lambda_{\mathrm{FP}}+\ln\Lambda_{\mathrm{BL}})$.

The first term on the right-hand side (RHS) of Eq. (8) (labeled as *RHS*.1) can be simplified as

$$RHS.1 = \Gamma \nabla_{\mathbf{p}} \cdot \left[\nabla_{\mathbf{v}} \nabla_{\mathbf{v}} \int u f_b d^3 \mathbf{p}' \right] \cdot \nabla_{\mathbf{p}} f_e. \tag{9}$$

The second term on the RHS (labeled as *RHS*.2) can be reduced through partial integration steps,

$$RHS.2 = \Gamma \nabla_{\mathbf{p}} \cdot \left[f_e \nabla_{\mathbf{v}} \int (\nabla_{\mathbf{p}'} \cdot \nabla_{\mathbf{v}} u) f_b d^3 \mathbf{p}' \right]. \tag{10}$$

By employing $\gamma' m' \nabla_{\mathbf{p}'} \cdot \nabla_{\mathbf{v}} u = -\gamma m \nabla_{\mathbf{p}} \cdot \nabla_{\mathbf{v}} u$, Eq. (10) is further expressed as

$$RHS.2 = -\Gamma \frac{m}{m'} \nabla_{\mathbf{p}} \cdot \left[f_e \nabla_{\mathbf{v}} \left(\gamma \nabla_{\mathbf{p}} \cdot \nabla_{\mathbf{v}} \int \frac{u f_b}{\gamma'} d^3 \mathbf{p}' \right) \right]. \tag{11}$$

Combining Eqs. (9) and (11), the collision operator in Eq. (3) is ready to be reformulated in a concise way,

$$C(f_e, f_b) = \Gamma \nabla_{\mathbf{p}} \cdot \left[(\nabla_{\mathbf{v}} \nabla_{\mathbf{v}} g) \cdot \nabla_{\mathbf{p}} f_e - \frac{m}{m'} f_e \nabla_{\mathbf{v}} (\gamma \nabla_{\mathbf{p}} \cdot \nabla_{\mathbf{v}} h) \right], \tag{12}$$

where we introduced

$$g = \int u f_b d^3 \mathbf{p}' \tag{13}$$

and

$$h = \int \frac{u}{\gamma'} f_b d^3 \mathbf{p}'. \tag{14}$$

It is noted that g and h contain the integration of background particle distribution function in momentum space and are somewhat analogous to the classical Rosenbluth potentials as defined in Refs. 31 and 32. Therefore, they will be referred to as potential functions hereafter. In such a form, the collision operator is now converted to a differential operator provided that the potential functions are explicitly given.

III. EXPLICIT FORM OF RELATIVISTIC KINETIC EQUATION

For the purpose of numerical calculation and analytical work, it is essential to express the kinetic equation explicitly.

A. Collision operator in three-dimensional momentum space

We are now ready to calculate the collision operator Eq. (12) in three-dimensional (3D) momentum space under the spherical coordinates (p, θ, ϕ). The first term on the RHS can be expanded using the algorithm for differential operator,

$$\begin{aligned}
\nabla_{\mathbf{p}} \cdot (\nabla_{\mathbf{v}} \nabla_{\mathbf{v}} g \cdot \nabla_{\mathbf{p}} f_e) = & \frac{1}{p^2} \frac{\partial}{\partial p} \left[p^2 \frac{\partial^2 g}{\partial v^2} \frac{\partial f_e}{\partial p} + p \frac{\partial}{\partial v} \left(\frac{1}{v} \frac{\partial g}{\partial \theta} \right) \frac{\partial f_e}{\partial \theta} + \frac{p}{\sin^2 \theta} \frac{\partial}{\partial v} \left(\frac{1}{v} \frac{\partial g}{\partial \phi} \right) \frac{\partial f_e}{\partial \phi} \right] + \frac{1}{p^2 \sin \theta} \frac{\partial}{\partial \theta} \\
& \times \left[p \sin \theta \frac{\partial}{\partial v} \left(\frac{1}{v} \frac{\partial g}{\partial \theta} \right) \frac{\partial f_e}{\partial p} + \sin \theta \left(\frac{1}{v} \frac{\partial g}{\partial v} + \frac{1}{v^2} \frac{\partial^2 g}{\partial \theta^2} \right) \frac{\partial f_e}{\partial \theta} + \frac{1}{v^2} \frac{\partial}{\partial \theta} \left(\frac{1}{\sin \theta} \frac{\partial g}{\partial \phi} \right) \frac{\partial f_e}{\partial \phi} \right] \\
& + \frac{1}{p^2 \sin \theta} \frac{\partial}{\partial \phi} \left[p \frac{\partial}{\partial v} \left(\frac{1}{v \sin \theta} \frac{\partial g}{\partial \phi} \right) \frac{\partial f_e}{\partial p} + \frac{1}{v^2} \frac{\partial}{\partial \theta} \left(\frac{1}{\sin \theta} \frac{\partial g}{\partial \phi} \right) \frac{\partial f_e}{\partial \theta} + \frac{1}{\sin \theta} \right. \\
& \left. \times \left(\frac{1}{v} \frac{\partial g}{\partial v} + \frac{\cot \theta}{v^2} \frac{\partial g}{\partial \theta} + \frac{1}{v^2 \sin^2 \theta} \frac{\partial^2 g}{\partial \phi^2} \right) \frac{\partial f_e}{\partial \phi} \right].
\end{aligned} \tag{15}$$

Letting $\lambda = \gamma \nabla_{\mathbf{p}} \cdot \nabla_{\mathbf{v}} h$, the second term is written as

$$\begin{aligned}
& -\frac{m}{m'} \nabla_{\mathbf{p}} \cdot [f_e \nabla_{\mathbf{v}} (\gamma \nabla_{\mathbf{p}} \cdot \nabla_{\mathbf{v}} h)] \\
& = -\frac{1}{p^2} \frac{\partial}{\partial p} \left(p^2 \frac{m}{m'} \frac{\partial \lambda}{\partial v} f_e \right) \\
& \quad - \frac{1}{p^2 \sin \theta} \frac{\partial}{\partial \theta} \left(\frac{m^2}{m'} \gamma \sin \theta \frac{\partial \lambda}{\partial \theta} f_e \right) \\
& \quad - \frac{1}{p^2 \sin \theta} \frac{\partial}{\partial \phi} \left(\frac{\gamma}{\sin \theta} \frac{m^2}{m'} \frac{\partial \lambda}{\partial \phi} f_e \right).
\end{aligned} \tag{16}$$

The relationship in Eq. (2) was used in obtaining the above equation.

Eventually, the collision operator in 3D momentum space can be formulated in a compact form by rearranging the terms in Eqs. (15) and (16)

$$\begin{aligned}
C(f_e, f_b) = & \frac{\Gamma}{p^2} \frac{\partial}{\partial p} \left(A_1 f_e + A_2 \frac{\partial f_e}{\partial p} + A_3 \frac{\partial f_e}{\partial \theta} + A_4 \frac{\partial f_e}{\partial \phi} \right) \\
& + \frac{\Gamma}{p^2 \sin \theta} \frac{\partial}{\partial \theta} \left(B_1 f_e + B_2 \frac{\partial f_e}{\partial p} + B_3 \frac{\partial f_e}{\partial \theta} + B_4 \frac{\partial f_e}{\partial \phi} \right) \\
& + \frac{\Gamma}{p^2 \sin \theta} \frac{\partial}{\partial \phi} \left(C_1 f_e + C_2 \frac{\partial f_e}{\partial p} + C_3 \frac{\partial f_e}{\partial \theta} \right. \\
& \left. + C_4 \frac{\partial f_e}{\partial \phi} \right).
\end{aligned} \tag{17}$$

where the corresponding coefficients $\{A_i, B_i, C_i\}$ $(i = 1, \ldots, 4)$

can be directly calculated from Eqs. (15) and (16). Their general forms are listed in Appendix A.

B. Potential functions

It is necessary to express the potential functions explicitly in order to get the differential form of the collision operator as mentioned in Sec. II. This can also be seen from Eq. (17), in which the coefficients are all related to the potential functions and their derivatives.

As a sophisticated method for calculating potential functions,[31,33,34] spherical harmonics are usually employed to expand the background distribution function. The basis function for expansion is chosen to be

$$Y_{lk}(\theta',\phi') = \sqrt{\frac{(l-k)!}{(l+k)!}}P_l^k(\cos\theta')e^{ik\phi'}, \tag{18}$$

where $P_l^k(\cos\theta')$ is the associated Legendre polynomial. The orthogonality relationship for the basis function is

$$\int_{\theta',\phi'} Y_{lk}^*(\theta',\phi')Y_{l'k'}(\theta',\phi')\sin\theta' d\theta' d\phi' = \frac{4\pi}{2l+1}\delta_{ll'}\delta_{kk'}, \tag{19}$$

where $\delta_{ll'}, \delta_{kk'}$ are the Kronecker delta.

Therefore, the background distribution function f_b can be expanded as

$$f_b = \sum_{l=0}^{\infty}\sum_{k=-l}^{l} F_{lk}Y_{lk}(\theta',\phi'), \tag{20}$$

where

$$F_{lk} = \frac{2l+1}{4\pi}\int_{\phi'=0}^{2\pi}\int_{\theta'=0}^{\pi} f_b Y_{lk}^*(\theta',\phi')\sin\theta' d\theta' d\phi'. \tag{21}$$

The relative velocity can be also expanded by making use of the expansion property of Legendre polynomials,

$$u = |\mathbf{v}-\mathbf{v}'| = \sum_{n=0}^{\infty}\left(\frac{(v_</v_>)^2}{2n+3} - \frac{1}{2n-1}\right)\frac{v_<^n}{v_>^{n-1}}P_n(\cos\alpha), \tag{22}$$

where $v_<$ and $v_>$ are, respectively, the smaller and the larger values of v and v', and α is the angle between $\mathbf{v}$ and $\mathbf{v}'$.

Having introduced the spherical harmonics expansion, we can now calculate the collision integration in potential function Eq. (13),

$$g = \sum_{l=0}^{\infty}\sum_{k=-l}^{l}\left\{\sum_{n=0}^{\infty}\int_{\theta',\phi'} Y_{lk}(\theta',\phi')P_n(\cos\alpha)\sin\theta' d\theta' d\phi' \times\left[\int_{v'<v} F_{lk}p'^2\left(\frac{1}{2n+3}\frac{v'^{n+2}}{v^{n+1}} - \frac{1}{2n-1}\frac{v'^n}{v^{n-1}}\right)dp' + \int_{v'>v} F_{lk}p'^2\left(\frac{1}{2n+3}\frac{v^{n+2}}{v'^{n+1}} - \frac{1}{2n-1}\frac{v^n}{v'^{n-1}}\right)dp'\right]\right\}. \tag{23}$$

The integration with respect to the angles θ' and ϕ' in Eq. (23) can be evaluated independently,

$$\int_{\phi'=0}^{2\pi}\int_{\theta'=0}^{\pi} Y_{lk}(\theta',\phi')P_n(\cos\alpha)\sin\theta' d\theta' d\phi' = \frac{4\pi}{2n+1}\delta_{nl}\sum_{k'=-n}^{n}[Y_{nk'}(\theta,\phi)\delta_{kk'}]. \tag{24}$$

Here, we have used the orthogonality relationship Eq. (19) and the addition theorem of Legendre polynomials

$$P_n(\cos\alpha) = \sum_{k'=-n}^{n} Y_{nk'}(\theta,\phi)Y_{nk'}^*(\theta',\phi'). \tag{25}$$

After some algebra, Eq. (23) can be further expressed as

$$g = \sum_{l=0}^{\infty}\sum_{k=-l}^{l}\frac{4\pi}{2l+1}Y_{lk}\left[\frac{1}{2l+3}\left(\int_0^{pm'/m}\frac{v'^{l+2}p'^2F_{lk}}{v^{l+1}}dp' + \int_{pm'/m}^{\infty}\frac{v^{l+2}p'^2F_{lk}}{v'^{l+1}}dp'\right) - \frac{1}{2l-1}\left(\int_0^{pm'/m}\frac{v'^l p'^2F_{lk}}{v^{l-1}}dp' + \int_{pm'/m}^{\infty}\frac{v^l p'^2F_{lk}}{v'^{l-1}}dp'\right)\right]. \tag{26}$$

Following the similar procedures, the potential function Eq. (14) can also be calculated,

$$h = \sum_{l=0}^{\infty}\sum_{k=-l}^{l}\frac{4\pi}{2l+1}Y_{lk}\left[\frac{1}{2l+3}\left(\int_0^{pm'/m}\frac{v'^{l+2}p'^2F_{lk}}{\gamma' v^{l+1}}dp' + \int_{pm'/m}^{\infty}\frac{v^{l+2}p'^2F_{lk}}{\gamma' v'^{l+1}}dp'\right) - \frac{1}{2l-1}\left(\int_0^{pm'/m}\frac{v'^l p'^2F_{lk}}{\gamma' v^{l-1}}dp' + \int_{pm'/m}^{\infty}\frac{v^l p'^2F_{lk}}{\gamma' v'^{l-1}}dp'\right)\right]. \tag{27}$$

The above expressions for potential functions can be further reduced by reorganizing the terms,

$$g = \sum_{l=0}^{\infty}\sum_{k=-l}^{l} G_{lk}Y_{lk}, \tag{28}$$

$$h = \sum_{l=0}^{\infty}\sum_{k=-l}^{l} H_{lk}Y_{lk}, \tag{29}$$

where

$$G_{lk}=\frac{4\pi}{2l+1}\left\{\frac{1}{2l+3}[v^{-(l+1)}M_{glk}+v^{l+2}N_{glk}]-\frac{1}{2l-1}[v^{-(l-1)}S_{glk}+v^{l}W_{glk}]\right\} \tag{30}$$

with

$$M_{glk}=\int_0^{pm'/m}v'^{l+2}p'^2F_{lk}dp', \tag{31a}$$

$$N_{glk}=\int_{pm'/m}^{\infty}\frac{p'^2}{v'^{l+1}}F_{lk}dp', \tag{31b}$$

$$S_{glk}=\int_0^{pm'/m}v'^{l}p'^2F_{lk}dp', \tag{31c}$$

$$W_{glk}=\int_{pm'/m}^{\infty}\frac{p'^2}{v'^{l-1}}F_{lk}dp' \tag{31d}$$

and

$$H_{lk}=\frac{4\pi}{2l+1}\left\{\frac{1}{2l+3}[v^{-(l+1)}M_{hlk}+v^{l+2}N_{hlk}]-\frac{1}{2l-1}[v^{-(l-1)}S_{hlk}+v^{l}W_{hlk}]\right\} \tag{32}$$

with

$$M_{hlk}=\int_0^{pm'/m}\frac{v'^{l+2}p'^2}{\gamma'}F_{lk}dp', \tag{33a}$$

$$N_{hlk}=\int_{pm'/m}^{\infty}\frac{p'^2}{\gamma'v'^{l+1}}F_{lk}dp', \tag{33b}$$

$$S_{hlk}=\int_0^{pm'/m}\frac{v'^{l}p'^2}{\gamma'}F_{lk}dp', \tag{33c}$$

$$W_{hlk}=\int_{pm'/m}^{\infty}\frac{p'^2}{\gamma'v'^{l-1}}F_{lk}dp'. \tag{33d}$$

It should be noticed that G_{lk} and H_{lk} are only related to the particle momentum or velocity while Y_{lk} is solely dependent on the angle variables. Therefore, the derivatives of potential functions with respect to velocity can be separated from those with respect to the angles,

$$\partial_v^s g=\sum_{l=0}^{\infty}\sum_{k=-l}^{l}(\partial_v^s G_{lk})Y_{lk}, \tag{34a}$$

$$\partial_\theta^s g=\sum_{l=0}^{\infty}\sum_{k=-l}^{l}G_{lk}(\partial_\theta^s Y_{lk}), \tag{34b}$$

$$\partial_\phi^s g=\sum_{l=0}^{\infty}\sum_{k=-l}^{l}G_{lk}(\partial_\phi^s Y_{lk}), \tag{34c}$$

$$\partial_v^s h=\sum_{l=0}^{\infty}\sum_{k=-l}^{l}(\partial_v^s H_{lk})Y_{lk}, \tag{34d}$$

$$\partial_\theta^s h=\sum_{l=0}^{\infty}\sum_{k=-l}^{l}H_{lk}(\partial_\theta^s Y_{lk}), \tag{34e}$$

$$\partial_\phi^s h=\sum_{l=0}^{\infty}\sum_{k=-l}^{l}H_{lk}(\partial_\phi^s Y_{lk}), \tag{34f}$$

where $\partial_x=\partial/\partial x(x=v,\theta,\phi)$ and the superscript s donates the order of derivatives.

C. Explicit formulation of relativistic kinetic equation

Having expanded the potential functions in 3D momentum space, we are now coming to get the relativistic kinetic equation,

$$\lambda=\sum_{l=0}^{\infty}\sum_{k=-l}^{l}\left[\frac{1}{\gamma^2 m}\partial_v^2H_{lk}+\frac{2}{mv}\partial_vH_{lk}-\frac{n(n+1)}{mv^2}H_{lk}\right]Y_{lk}. \tag{35}$$

The equation that the spherical harmonics satisfied is used for the above calculation

$$\frac{1}{\sin\theta}\frac{\partial}{\partial\theta}\left(\sin\theta\frac{\partial Y_{lk}}{\partial\theta}\right)+\frac{1}{\sin^2\theta}\frac{\partial^2 Y_{lk}}{\partial\phi^2}+l(l+1)Y_{lk}=0. \tag{36}$$

By substituting Eqs. (28) and (29) with Eqs. (34) and (35) into Eqs. (A1), the coefficients in the collision operator Eq. (17) can be well calculated,

$$A_1=-\frac{m^2}{m'}\sum_{l=0}^{\infty}\sum_{k=-l}^{l}\left\{v^2\partial_v^3H_{lk}+2v\partial_v^2H_{lk}-\gamma^2\times[2+l(l+1)]\partial_vH_{lk}+\frac{2\gamma^2}{v}l(l+1)H_{lk}\right\}Y_{lk}, \tag{37a}$$

$$A_2=\sum_{l=0}^{\infty}\sum_{k=-l}^{l}p^2(\partial_v^2G_{lk})Y_{lk}, \tag{37b}$$

$$A_3=\sum_{l=0}^{\infty}\sum_{k=-l}^{l}\left[\gamma m(\partial_vG_{lk})-\frac{\gamma^2m^2G_{lk}}{p}\right](\partial_\theta Y_{lk}), \tag{37c}$$

$$A_4=\sum_{l=0}^{\infty}\sum_{k=-l}^{l}\frac{1}{\sin^2\theta}\left[\gamma m(\partial_vG_{lk})-\frac{\gamma^2m^2G_{lk}}{p^2}\right](\partial_\theta Y_{lk}), \tag{37d}$$

$$B_1=-\frac{m}{m'}\gamma\sin\theta\sum_{l=0}^{\infty}\sum_{k=-l}^{l}\left[\frac{1}{\gamma^2}\partial_v^2H_{lk}+\frac{2}{v}\partial_vH_{lk}-\frac{l(l+1)}{v^2}H_{lk}\right](\partial_\theta Y_{lk}), \tag{37e}$$

$$B_2=\sum_{l=0}^{\infty}\sum_{k=-l}^{l}\sin\theta\left[\gamma m(\partial_vG_{lk})-\gamma^2m^2\frac{G_{lk}}{p}\right](\partial_\theta Y_{lk}), \tag{37f}$$

$$B_3=\sum_{l=0}^{\infty}\sum_{k=-l}^{l}\sin\theta\left[\frac{\gamma m(\partial_v G_{lk})}{p}Y_{lk}+\frac{\gamma^2 m^2 G_{lk}}{p^2}(\partial_\theta^2 Y_{lk})\right], \tag{37g}$$

$$B_4=\sum_{l=0}^{\infty}\sum_{k=-l}^{l}\frac{\gamma^2 m^2 G_{lk}}{p^2}\left[\frac{1}{\sin\theta}(\partial_{\theta\phi}^2 Y_{lk})-\frac{\cos\theta}{\sin^2\theta}(\partial_\phi Y_{lk})\right], \tag{37h}$$

$$C_1=-\frac{m}{m'}\frac{\gamma}{\sin\theta}\sum_{l=0}^{\infty}\sum_{k=-l}^{l}\left[\frac{1}{\gamma^2}\partial_v^2 H_{lk}+\frac{2}{v}\partial_v H_{lk}-\frac{l(l+1)}{v^2}H_{lk}\right]\times(\partial_\phi Y_{lk}), \tag{37i}$$

$$C_2=\sum_{l=0}^{\infty}\sum_{k=-l}^{l}\frac{1}{\sin\theta}\left[\gamma m(\partial_v G_{lk})-\gamma^2 m^2\frac{G_{lk}}{p}\right](\partial_\phi Y_{lk}), \tag{37j}$$

$$C_3=\sum_{l=0}^{\infty}\sum_{k=-l}^{l}\frac{\gamma^2 m^2 G_{lk}}{p^2}\left[\frac{1}{\sin\theta}(\partial_{\theta\phi}^2 Y_{lk})-\frac{\cos\theta}{\sin^2\theta}(\partial_\phi Y_{lk})\right], \tag{37k}$$

$$C_4=\sum_{l=0}^{\infty}\sum_{k=-l}^{l}\frac{1}{\sin\theta}\left[\frac{\gamma m(\partial_v G_{lk})}{p}Y_{lk}+\frac{\gamma^2 m^2 G_{lk}}{p^2}\cot\theta(\partial_\theta Y_{lk})+\frac{\gamma^2 m^2(\partial_\phi^2 Y_{lk})}{p^2\sin^2\theta}\right]. \tag{37l}$$

The derivatives appearing in Eqs. (37) can be found in Appendix B. Once the expansion for the background distribution function is completed, the full set of coefficients will be available directly.

Therefore, the relativistic kinetic equation for energy deposition of fast electrons can be explicitly expressed as

$$\left[\frac{\partial}{\partial t}+\frac{\mathbf{p}}{\gamma m}\cdot\nabla-e\left(\mathbf{E}+\frac{\mathbf{p}\times\mathbf{B}}{\gamma m}\right)\cdot\nabla_{\mathbf{p}}\right]f_e$$
$$=\sum_{b=be,bi}\left[\frac{\Gamma}{p^2}\frac{\partial}{\partial p}\left(A_1+A_2\frac{\partial}{\partial p}+A_3\frac{\partial}{\partial\theta}+A_4\frac{\partial}{\partial\phi}\right)f_e+\frac{\Gamma}{p^2\sin\theta}\frac{\partial}{\partial\theta}\left(B_1+B_2\frac{\partial}{\partial p}+B_3\frac{\partial}{\partial\theta}+B_4\frac{\partial}{\partial\phi}\right)f_e+\frac{\Gamma}{p^2\sin\theta}\frac{\partial}{\partial\phi}\left(C_1+C_2\frac{\partial}{\partial p}+C_3\frac{\partial}{\partial\theta}+C_4\frac{\partial}{\partial\phi}\right)f_e\right], \tag{38}$$

with the coefficients in Eq. (37). The kinetic equation in this form is convenient for rapid numerical calculation and analytical work since it involves only simple differentiations and integrations.

IV. PRELIMINARY NUMERICAL RESULTS AND ANALYSIS

Numerical investigations based on the derived kinetic model have been made in order to gain more insight into the relevant FI physics as well as the model in itself. The preliminary results and analysis are presented below. The particle number conservation, which is of vital importance in kinetic theory, is examined both analytically and numerically. The continuous range and penetration depth for fast electrons, which depict the profile of energy deposition region, are also calculated and compared with published works.

A. Particle number conservation

As for the fast electron beam incident into the dense DT core plasmas in FI, it will gradually lose the total moment and kinetic energy due to the dissipative collisions with background electrons and ions. However, its total particle number in real space will not decay with time since there are no particle annihilation or creation processes. The total electron number can be calculated from its distribution function,

$$N_e=\int_V\int_{p,\theta,\phi}f_e p^2 dp\sin\theta d\theta d\phi dV, \tag{39}$$

where V represents the real space. Its variation can be calculated by making use of the governing equation for f_e in Eq. (38),

$$\begin{aligned}\frac{dN_e}{dt}&=\int_V\int_{p,\theta,\phi}\frac{\partial f_e}{\partial t}p^2dp\sin\theta d\theta d\phi dV\\&=\int_V\int_{\theta,\phi}\sum_b\Gamma\left[\left(A_1f_e+A_2\frac{\partial f_e}{\partial p}+A_3\frac{\partial f_e}{\partial\theta}+A_4\frac{\partial f_e}{\partial\phi}\right)\Bigg|_{p=0}^{\infty}\right]\sin\theta d\theta d\phi dV\\&+\int_V\int_{p,\phi}\sum_b\Gamma\left[\left(B_1f_e+B_2\frac{\partial f_e}{\partial p}+B_3\frac{\partial f_e}{\partial\theta}+B_4\frac{\partial f_e}{\partial\phi}\right)\Bigg|_{\theta=0}^{\pi}\right]dpd\phi dV\\&+\int_V\int_{p,\theta}\sum_b\Gamma\left[\left(C_1f_e+C_2\frac{\partial f_e}{\partial p}+C_3\frac{\partial f_e}{\partial\theta}+C_4\frac{\partial f_e}{\partial\phi}\right)\Bigg|_{\phi=0}^{2\pi}\right]dpd\theta dV=0.\end{aligned} \tag{40}$$

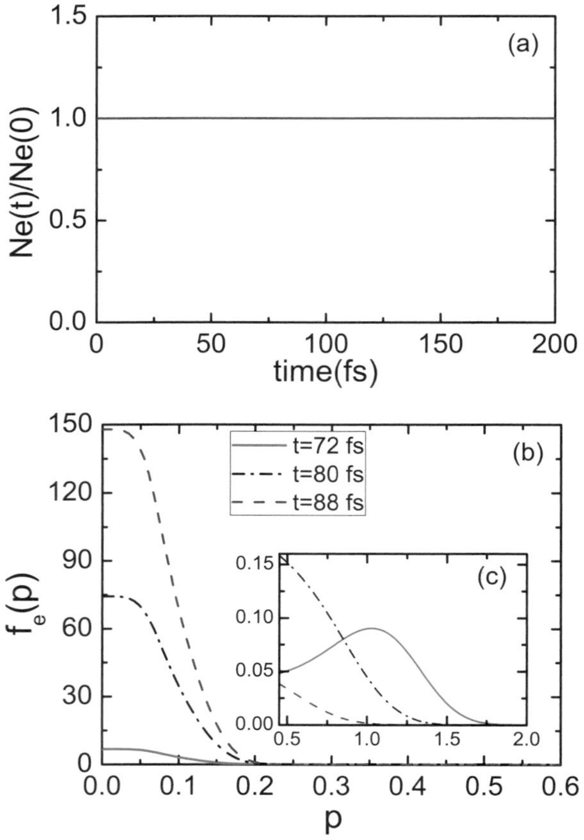

FIG. 1. (Color online) (a) Computed electron number (normalized to the initial injected number). [(b) and (c)] The normalized electron distribution function in momentum space at various times: 72 fs (red solid), 80 fs (black dash-dot), and 88 fs (blue dashed). The momentum is normalized to $m_e c$.

The physical boundaries where the distribution function and its derivatives become zero are used for the calculations above. Some coefficients vanish at the boundaries according to Eqs. (37).

In order to corroborate the number conservation in a more intuitional way, a typical case in FI is considered numerically.[17,20] A beam of 1.0 MeV monoenergetic electrons is incident into an initially homogeneous DT plasma with density $\rho=300\ \text{g/cm}^3$ and temperature $T=5$ keV. As shown in Fig. 1(a), the electron number does not vary with the beam-plasma interaction time, clearly indicating that the fast electron number is perfectly conserved in our kinetic model. This can be further explored by examining the electron distribution function (EDF) in momentum space at three different times, which is presented in Figs. 1(b) and 1(c). The EDF changes rapidly as time increases due to the collision operator. It is gradually flattened to zero in the high energy tail [indicated by subplot Fig. 2(c)] due to the advective term in model equation, but it will be raised up in the lower energy region [shown in Fig. 2(b)] as a result of the diffusion term. In this way, the particle number conservation is guaranteed, just as the classical case.[31] This is also consistent with the physical property of kinetic equation, as discussed in Refs. 26–28 and 31.

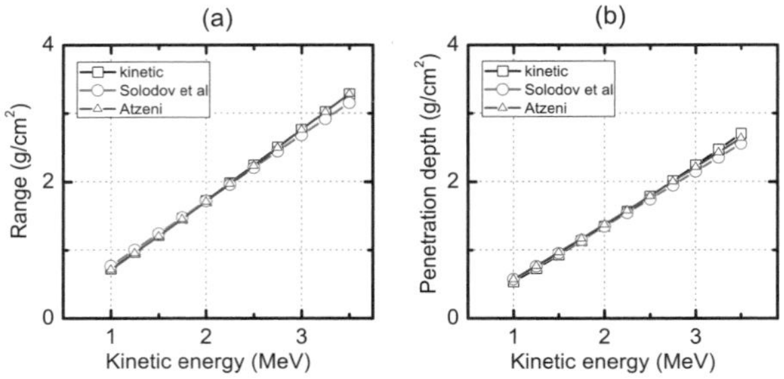

FIG. 2. (Color online) (a) The continuous range and (b) the penetration depth vs incident electron kinetic energy. The mass density and temperature for DT plasmas are $300\ \text{g}\cdot\text{cm}^{-3}$ and 5 keV, respectively.

B. Electron range and penetration depth

Generally, collisions impose two effects on the incident fast electrons in the context of FI. On one hand, it will cause the slowing down of electrons, leading to energy loss. On the other hand, it will scatter electrons to deviate their initial incident direction, resulting in beam blooming. Therefore, the continuous path length of electrons before they are stopped (also called continuous range) and the depth it can reach along the initial direction (called penetration depth) are both very important in evaluating the energy deposition profile and have been extensively studied by many authors.[13,14,17,20] Based on the derived kinetic model, it is straightforward to obtain these two quantities for specified parameters of fast electrons and core plasmas. In order to make a comparison with previous works, a monoenergetic beam is considered and the parameters for DT plasma is the same as the case considered in Sec. IV A. The initial kinetic energy of electrons are varied to see how the range and the penetration depth change. Particularly, we choose the energy range between 1.0 MeV and 3.5 MeV, which is very closely related to realistic FI ignition beam requirements.[3,7] Figures 2(a) and 2(b) plot the continuous range and the penetration depth as a function of initial energy. The lines with squares represent the results based on the kinetic model. The most recent published results by Solodov *et al.*[17] (line with circles) and Atzeni[20] (line with triangles) are also displayed in the same figures. It is clearly seen that the three models agree very well with each other in both the range and the penetration depth.

V. SUMMARY

To summarize, based on fundamental principles, we have established a relativistic kinetic model for energy deposition of fast electrons in highly dense core plasmas in a FI scenario. Both binary collisions and collective response are well included. The kinetic equation is explicitly formulated in 3D dimensional momentum space with the help of harmonic expansion technique. It is analytically proven that the derived relativistic kinetic equation conserves the particle number of fast electrons analytically and this is further verified by numerical results. Preliminary calculations on the range and the penetration depth for electrons with FI-relevant energies are presented and they agree very well with previously published works.

ACKNOWLEDGMENTS

This work is supported by the National Natural Science Foundation of China (Grant Nos. 10835003, 10935003, 10974022, and 10975023), and the National High-Tech ICF Committee and partially by the National Basic Research Program of China (973) (Program No. 2007CB815101).

APPENDIX A: COEFFICIENTS OF COLLISION OPERATOR

The general forms for the coefficients in collision operator Eq. (17) are listed below.

$$A_1 = -\frac{m}{m'} p^2 \frac{\partial \lambda}{\partial v}, \tag{A1a}$$

$$A_2 = p^2 \frac{\partial^2 g}{\partial v^2}, \tag{A1b}$$

$$A_3 = p\frac{\partial}{\partial v}\left(\frac{1}{v}\frac{\partial g}{\partial \theta}\right), \tag{A1c}$$

$$A_4 = \frac{p}{\sin^2\theta}\frac{\partial}{\partial v}\left(\frac{1}{v}\frac{\partial g}{\partial \phi}\right), \tag{A1d}$$

$$B_1 = -\frac{m^2}{m'}\gamma \sin\theta \frac{\partial \lambda}{\partial \theta}, \tag{A1e}$$

$$B_2 = p\sin\theta\frac{\partial}{\partial v}\left(\frac{1}{v}\frac{\partial g}{\partial \theta}\right), \tag{A1f}$$

$$B_3 = \sin\theta\left(\frac{1}{v}\frac{\partial g}{\partial v} + \frac{1}{v^2}\frac{\partial^2 g}{\partial \theta^2}\right), \tag{A1g}$$

$$B_4 = \frac{1}{v^2}\frac{\partial}{\partial \theta}\left(\frac{1}{\sin\theta}\frac{\partial g}{\partial \phi}\right), \tag{A1h}$$

$$C_1 = -\frac{m^2}{m'}\frac{\gamma}{\sin\theta}\frac{\partial \lambda}{\partial \phi}, \tag{A1i}$$

$$C_2 = p\frac{\partial}{\partial v}\left(\frac{1}{v\sin\theta}\frac{\partial g}{\partial \phi}\right), \tag{A1j}$$

$$C_3 = \frac{1}{v^2}\frac{\partial}{\partial \theta}\left(\frac{1}{\sin\theta}\frac{\partial g}{\partial \phi}\right), \tag{A1k}$$

$$C_4 = \frac{1}{\sin\theta}\left(\frac{1}{v}\frac{\partial g}{\partial v} + \frac{\cot\theta}{v^2}\frac{\partial g}{\partial \theta} + \frac{1}{v^2\sin^2\theta}\frac{\partial^2 g}{\partial \phi^2}\right). \tag{A1l}$$

APPENDIX B: THE DERIVATIVES OF POTENTIAL FUNCTIONS

The derivatives encountered in calculating the coefficients in Eq. (34) are provided. The derivatives of spherical harmonics with respect to angles θ, ϕ are not explicitly listed here, which can be easily calculated with the help of a textbook or mathematical handbook, such as in Ref. 35.

$$\partial_v G_{lk} = \sum_{l=0}^{\infty}\sum_{k=-l}^{l}\frac{4\pi}{2l+1}\left\{\frac{1}{2l+3}[-(l+1)v^{-(l+2)}M_{glk} + (l+2)v^{l+1}N_{glk}] - \frac{1}{2l-1}[-(l-1)v^{-l}S_{glk} + lv^{l-1}W_{glk}]\right\}, \tag{B1a}$$

$$\partial_v^2 G_{lk} = \sum_{l=0}^{\infty}\sum_{k=-l}^{l}\frac{4\pi}{2l+1}\left\{\frac{(l+2)(l+1)}{2l+3}[v^{-(l+3)}M_{glk} + v^{l}N_{glk}] - \frac{l(l-1)}{2l-1}[v^{-(l+1)}S_{glk} + v^{l-2}W_{glk}]\right\}, \tag{B1b}$$

$$\partial_v^3 G_{lk} = \sum_{l=0}^{\infty}\sum_{k=l}^{l}\frac{4\pi}{2l+1}\left\{\frac{(l+2)(l+1)}{2l+3}[-(l+3)v^{-(l+4)}M_{glk} + lv^{l-1}N_{glk}] - \frac{l(l-1)}{2l-1}\times[-(l+1)v^{-(l+2)}S_{glk} + (l-2)v^{l-3}W_{glk}]\right\}, \tag{B1c}$$

$$\partial_v H_{lk} = \sum_{l=0}^{\infty}\sum_{k=-l}^{l}\frac{4\pi}{2l+1}\left\{\frac{1}{2l+3}[-(l+1)v^{-(l+2)}M_{hlk} + (l+2)v^{l+1}N_{hlk}] - \frac{1}{2l-1}[-(l-1)v^{-l}S_{hlk} + lv^{l-1}W_{hlk}]\right\}, \tag{B1d}$$

$$\partial_v^2 H_{lk} = \sum_{l=0}^{\infty}\sum_{k=-l}^{l}\frac{4\pi}{2l+1}\left\{\frac{(l+2)(l+1)}{2l+3}[v^{-(l+3)}M_{hlk} + v^{l}N_{hlk}] - \frac{l(l-1)}{2l-1}[v^{-(l+1)}S_{hlk} + v^{l-2}W_{hlk}]\right\}, \tag{B1e}$$

$$\partial_v^3 H_{lk} = \sum_{l=0}^{\infty}\sum_{k=l}^{l}\frac{4\pi}{2l+1}\left\{\frac{(l+2)(l+1)}{2l+3}[-(l+3)v^{-(l+4)}M_{hlk} + lv^{l-1}N_{hlk}] - \frac{l(l-1)}{2l-1}\times[-(l+1)v^{-(l+2)}S_{hlk} + (l-2)v^{l-3}W_{hlk}]\right\}. \tag{B1f}$$

[1] M. Tabak, J. Hammer, M. E. Glinsky, W. L. Kruer, S. C. Wilks, J. Woodworth, E. M. Campbell, M. D. Perry, and R. J. Mason, Phys. Plasmas **1**, 1626 (1994).
[2] S. Atzeni and J. Meyer-ter-Vehn, *The Physics of Inertial Fusion* (Oxford University Press, London, 2004).
[3] S. Atzeni, Phys. Plasmas **6**, 3316 (1999).
[4] R. Kodama, P. A. Norreys, K. Mima, A. E. Dangor, R. G. Evans, H. Fujita, Y. Kitagawa, K. Krushelnick, T. Miyakoshi, N. Miyanaga, T. Norimatsu, S. J. Rose, T. Shozaki, K. Shigemori, A. Sunahara, M. Tampo, K. A. Tanaka, Y. Toyama, T. Yamanaka, and M. Zepf, Nature (London) **412**, 798 (2001); **418**, 933 (2002) and references therein.
[5] S. Atzeni and M. Tabak, Plasma Phys. Controlled Fusion **47**, B769 (2005).
[6] E. M. Campbell, R. R. Freeman, and K. A. Tanaka, Fusion Sci. Technol. **49**, 249 (2006) and references therein.
[7] M. H. Key, Phys. Plasmas **14**, 055502 (2007).
[8] S. C. Wilks, W. L. Kruer, M. Tabak, and A. B. Langdon, Phys. Rev. Lett. **69**, 1383 (1992).
[9] F. N. Beg, A. R. Bell, A. E. Dangor, C. N. Danson, A. P. Fews, M. E. Glinsky, B. A. Hammel, P. Lee, P. A. Norreys, and M. Tatarakis, Phys. Plasmas **4**, 447 (1997).
[10] R. R. Freeman, D. Batani, S. Baton, M. Key, and R. Stephens, Fusion Sci. Technol. **49**, 297 (2006).
[11] C. T. Zhou, L. Y. Chew, and X. T. He, Appl. Phys. Lett. **97**, 051502 (2010).
[12] S. Z. Wu, C. T. Zhou, and S. P. Zhu, Phys. Plasmas **17**, 063103 (2010).
[13] C. Deutsch, H. Furukawa, K. Mima, M. Murakami, and K. Nishihara, Phys. Rev. Lett. **77**, 2483 (1996); Phys. Rev. Lett. **85**, 1140(E) (2000).
[14] C. K. Li and R. D. Petrasso, Phys. Rev. E **70**, 067401 (2004).
[15] C. K. Li and R. D. Petrasso, Phys. Rev. E **73**, 016402 (2006); Phys. Plasmas **13**, 056314 (2006).
[16] H. W. Lewis, Phys. Rev. **78**, 526 (1950).
[17] A. A. Solodov and R. Betti, Phys. Plasmas **15**, 042707 (2008).
[18] B. Chrisman, Y. Sentoku, and A. J. Kemp, Phys. Plasmas **15**, 056309 (2008).
[19] J. J. Honrubia and J. Meyer-ter-Vehn, Plasma Phys. Controlled Fusion **51**, 014008 (2009).
[20] S. Atzeni, A. Schiavi, and J. R. Davies, Plasma Phys. Controlled Fusion **51**, 015016 (2009).
[21] A. A. Solodov, K. S. Anderson, R. Betti, V. Gotcheva, J. Myatt, J. A. Delettrez, S. Skupsky, W. Theobald, and C. Stoeckl, Phys. Plasmas **16**, 056309 (2009).
[22] T. Johzaki, Y. Nakao, and K. Mima, Phys. Plasmas **16**, 062706 (2009).
[23] T. Yokota, Y. Nakao, T. Johzaki, and K. Mima, Phys. Plasmas **13**, 022702 (2006).
[24] H. W. Koch and J. W. Motz, Rev. Mod. Phys. **31**, 920 (1959).
[25] D. Pines and D. Bohm, Phys. Rev. **85**, 338 (1952).
[26] N. A. Krall and A. W. Trivelpiece, *Principles of Plasma Physics* (McGraw-Hill, New York, 1973).
[27] Y. L. Klimontovich, *Kinetic Theory of Nonideal Gases and Nonideal Plasmas* (Pergamon, Oxford, 1982).
[28] L. Landau and E. Lifshitz, *Physical Kinetics* (Pergamon, New York, 1981).
[29] B. J. Braams and C. F. F. Karney, Phys. Rev. Lett. **59**, 1817 (1987).
[30] C. C. F. Karney, Comput. Phys. Rep. **4**, 183 (1986).
[31] M. N. Rosenbluth, W. M. MacDonald, and D. L. Judd, Phys. Rev. **107**, 1 (1957).
[32] B. A. Trubnikov, Sov. Phys. JETP **7**, 926 (1958).
[33] M. G. McCoy, A. A. Mirin, and J. Killeen, Comput. Phys. Commun. **24**, 37 (1981).
[34] M. R. Franz, University of California Report No. UCRL-96510, 1987.
[35] G. Arfken, *Mathematical Methods for Physicists*, 3rd ed. (Academic, Orlando, 1985).

PHYSICS OF PLASMAS 20, 072701 (2013)

Effects of the imposed magnetic field on the production and transport of relativistic electron beams

Hong-bo Cai,[1,2] Shao-ping Zhu,[1,a)] and X. T. He[1,2,b)]
[1]Institute of Applied Physics and Computational Mathematics, Beijing 100094, China
[2]Center for Applied Physics and Technology, Peking University, Beijing 100871, People's Republic of China

(Received 14 February 2013; accepted 30 May 2013; published online 1 July 2013)

The effects of the imposed uniform magnetic field, ranging from 1 MG up to 50 MG, on the production and transport of relativistic electron beams (REBs) in overdense plasmas irradiated by ultraintense laser pulse are investigated with two-dimensional particle-in-cell numerical simulations. This study gives clear evidence that the imposed magnetic field is capable of effectively confining the relativistic electrons in space even when the source is highly divergent since it forces the electrons moving helically. In comparison, the spontaneous magnetic fields, generated by the helically moving electrons interplaying with the current filamentation instability, are dominant in scattering the relativistic electrons. As the imposed magnetic field was increased from 1 MG to 50 MG, overall coupling from laser to the relativistic electrons which have the potential to heat the compressed core in fast ignition was found to increase from 6.9% to 21.3% while the divergence of the REB increases significantly from 64° to 90°. The simulations show that imposed magnetic field of the value of 3–30 MG could be more suitable to fast-ignition inertial fusion. © *2013 AIP Publishing LLC*. [http://dx.doi.org/10.1063/1.4812631]

I. INTRODUCTION

In the concept of fast ignition (FI),[1,2] a highly compressed deuterium-tritium pellet (10–20 μm at its core) is ignited by a ~10 ps, 10 kJ intense flux of MeV electrons (or ions). These high energy particles are generated by the absorption of an intense petawatt laser, at the edge of the pellet, which is usually ~50 μm away from the dense core. Many experiments have been carried out to study the feasibility of fast ignition with high energy particles.[3–6] One task is how to produce and transport enormous numbers of high-energy and high-quality charged particles. However, recent studies revealed that the relativistic electron beam (REB) produced from the interaction of an intense laser pulse with a solid-density target may be strongly divergent.[7,8] If the beam divergence is large, then the coupling efficiency could not be large and the required ignitor pulse energy can rapidly become unfeasibly large. State-of-the-art integrated simulations and experiments indicate that the coupling efficiency of short-pulse energy into the compressed deuterated-plastic (CD) core is only 3.5±1.0%.[4] Therefore, of particular importance is the possibility of collimating the REBs with the size of the compressed core.

To date, several mitigation works have been done regarding the control of the divergence of REBs, including self-generated magnetic field at the resistivity boundary in sandwich target,[9–11] high-resistivity-core-low-cladding targets,[12] and switchyard target,[13] magnetically guiding in cylindrically compressed matter,[14] two laser pulses,[15] ionization-driven resistive magnetic fields,[16] double cone target,[17] and so on. These works clearly showed the possibility for collimating fast electrons, but most of resistive collimation schemes also brought some complexity for the target design. Recent work by Strozzi *et al.* showed that imposed uniform axial magnetic fields recovered the 132 kJ ignition energy of artificially collimated source with hybrid PIC code Zuma.[18] Recently, Chang *et al.*[19] have produced 15 MG magnetic field by B-flux compression scheme with laser-driven implosion. Such strong imposed magnetic field has also been demonstrated with a capacitor-coil target and a ns-kJ laser without compression.[20–22] It seems that the imposed magnetic field raises great hopes by controlling the electron divergence and lowering the laser energy required to obtain the fuel burning for FI. However, there are still several physics issues have to be examined before a realistic assessment to FI can be made. First, how is the production and quality of the REB in the interaction region affected by the imposed magnetic field. Second, what is the role of spontaneous magnetic fields in collimating or scattering fast electrons.

In this paper, the interaction of an ultraintense laser pulse with overdense plasma with different imposed magnetic fields is investigated by 2D3V particle-in-cell (PIC) simulation code. The beneficial and detrimental effect of the imposed magnetic field on production and transport of REBs in overdense plasma are studied in details. Besides the imposed magnetic field, the spontaneous magnetic fields, generated by the interplaying of the helically electron currents and the current filamentation instability, play important role in the transport of the REBs. It is found that the imposed magnetic field is helpful for the confining of relativistic electrons while the spontaneous magnetic fields increase the divergence angle of REBs.

II. PIC SIMULATIONS ON THE GENERATION AND COLLIMATION OF REBs WITH IMPOSED MAGNETIC FIELDS

Here, we present some results from 2.5D simulations using the particle-in-cell (PIC) framework ASCENT.[17]

a)Electronic address: zhu_shaoping@iapcm.ac.cn
b)Electronic address: xthe@iapcm.ac.cn

Figure 1 is a sketch of the geometry of the simulations. The plasma consists of two species: electrons and carbon ions with an assumed charge state $Z_i = 6$ and $m_i/m_e = 22032$. The plasma density is $60n_c$. The p-polarized laser at $\lambda_0 = 1.06\ \mu$m wavelength and $3.45 \times 10^{19}\ \mathrm{W/cm^2}$ intensity irradiates the target from the left boundary. The intensity profile is Gaussian in the y direction with a spot size of $6.7\ \mu$m(FWHM). The laser rises in $10T_0$, where T_0 is the laser period, after which the laser amplitude is kept constant. A uniform axial magnetic field is imposed along the laser propagating direction. A typical simulation duration is $300T_0$, which corresponds to about 1 ps for $\lambda_0 = 1.06\ \mu$m.

Here, we use a grid size of $\Delta x = \Delta y = \lambda_0/64$ with 3840×3200 grid cells. The time step used is $0.01T_0$. 128 particles are used in one mesh, and the total number of particles is about 1.3×10^9. Both the field and particle boundary conditions are absorbing boundary conditions, either in the x or the y direction. Over the current simulation time scale (picoseconds), a large number of high-energy electrons escaping to the boundaries are reflected back by a large unphysical sheath field generated there. In order to eliminate this sheath field, we set cooling buffers at the boundaries in our simulation. Furthermore, in order to reduce the restrictions on the grid size compared with the Debye length, we used a fourth-order interpolation scheme to evaluate fields and currents.

Crucial issues for the fast ignition scheme are the collimation and transport of enormous numbers of high-energy electrons to the compressed core. The mechanism here proposed for transverse confinement of the high-energy electrons relies on the fact that the electrons move helically around the axis magnetic field. Strozzi *et al.*[18] studied the collimation effects of imposed magnetic field with coupled code of the hybrid-PIC code Zuma and radiation-hydrodynamics code Hydro. It was found that uniform axial magnetic fields ~50 MG are sufficient to mitigate the beam divergence and reduce the ignition beam energy. However, in their simulations, the short-pulse laser plasma interaction is not modeled directly and the artificial fast electron source

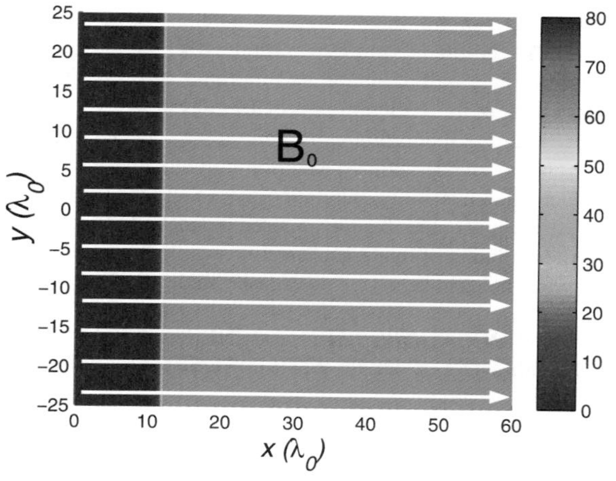

FIG. 1. Schematic of initial plasma density and imposed axial magnetic field.

is used regardless of the feedback of the transport on the laser plasma interaction and the influence of the imposed magnetic field.[18] Furthermore, the radial beam deviation, due to the laser ponderomotive force and propagation effects in preplasma, has been neglected in the fast electron transport calculations, overestimating the beam collimation by resistive magnetic fields.[23] Here, the laser plasma interaction process is included self-consistently in our simulations.

To understand the absorption mechanism under the imposed and spontaneous magnetic fields, we examine the equations of motion in prescribed fields as done by May.[25] We consider plane waves propagating along $\mathbf{e}_x$ with the $\mathbf{E}$ field polarized along $\mathbf{e}_y$, the imposed magnetic field $\mathbf{B}_0$ is along $\mathbf{e}_x$ and spontaneous magnetic fields $\mathbf{B}_s$ are along $\mathbf{e}_x$ and $\mathbf{e}_z$, for which the equations of motion are

$$\frac{dp_x}{dt} = -ev_y(B_z + B_{zs}), \tag{1}$$

$$\frac{dp_y}{dt} = -e\left[E_y + \frac{v_x}{c}(B_z + B_{zs})\right], \tag{2}$$

$$\frac{dp_z}{dt} = ev_y(B_0 + B_{xs}). \tag{3}$$

For simplicity, we assume 100% reflection and that the plasma acts like a perfect conductor, for which

$$\begin{aligned}\frac{e\mathbf{A}}{m_ec^2} &= [a_0\sin(kx-\omega t) + a_0\sin(kx+\omega t)]\mathbf{e}_y \\ &= 2a_0\sin(kx)\cos(\omega t)\mathbf{e}_y.\end{aligned} \tag{4}$$

We note that in the vacuum region ($x < 0$), the electric and magnetic fields are given by $\mathbf{E} = -\partial_t\mathbf{A}/c$ and $\mathbf{B} = \nabla \times \mathbf{A}$. Therefore, we have

$$\frac{eE_y}{m_e\omega c} = 2a_0\sin(kx)\sin(\omega t), \tag{5}$$

$$\frac{eB_z}{m_e\omega c} = 2a_0\cos(kx)\cos(\omega t). \tag{6}$$

Since the plasma is highly steepened by the laser pressure, a large fraction of the fast electrons are accelerated by this process:[25] First, the electrons moving with positive p_{y0} inside the plasma get deflected by the magnetic fields. Second, they leave normal to the surface and then the electric field E_y rapidly accelerates them in $\mathbf{e}_y$ direction. Finally, they are rotated forward into the target by the magnetic fields. Since the field is null inside, the kinetic energy given to the electrons that reenter the plasma can be considered as heating. This is a standing-wave pattern acceleration of the laser electromagnetic fields in front of the surface. The electric field E_y has a node at $z=0$, while B_x has an antinode at $z=0$. In the vacuum region, the energy gain can only come from E_y, so the electron must penetrate nearly $\frac{\lambda}{4}$ into the vacuum in order to gain significantly energy. However, if the electron leaves when $|\mathbf{B}| = |\mathbf{B}_z + \mathbf{B}_s|$ is a maximum ($eB_{z,\max}/m_ec\omega = 2a_0$, and B_s), then the Larmor radius will be small compared to $\frac{\lambda}{4}$ for high intensity lasers ($a_0 \gg 1$). Thus, the surface magnetic fields prevent significant heating

TABLE I. Fraction of the energy flux of the diagnosed fast electrons at different positions with respect to the input laser energy for different imposed magnetic fields. Here, only the electrons in the region $y=(-8\lambda_0, 8\lambda_0)$ and $t=(667, 1000)$ fs are diagnosed.

B_0 (MG)	Conversion efficiency (%)	$x=15\lambda_0$ (%)	$x=25\lambda_0$ (%)	$x=55\lambda_0$ (%)
0	39.8	37.6	23.9	7.1
1	40.3	36.2	23.3	6.9
3	38.5	43.7	22.8	20.1
5	38.3	43.7	33.0	20.1
10	38.0	45.7	36.4	23.4
20	38.1	47.9	36.9	24.2
30	37.9	51.2	34.6	22.2
40	38.0	47.4	34.4	22.1
50	37.7	57.6	31.1	21.3
50 (collisional)	38.5	54.7	35.4	17.1

of the electrons. From the Eqs. (1)–(3), one can see that the imposed axial magnetic field B_0 does not directly affect the forward and backward movement of the electrons, the surface magnetic field are laser field B_z and spontaneous magnetic field B_{zs}. As shown in Fig. 5, the magnitude and spatial distribution of the spontaneous magnetic fields are related to the imposed magnetic field B_0. Fortunately, the magnitude of B_{zs} is only about 20% of B_z. Therefore, the effect of B_{zs} to the laser conversion efficiency is limited. This has been confirmed in the second column of Table I, which shows the conversion efficiency is only slightly decreased as the increasing of the imposed magnetic fields.

In Fig. 2, the energy density distributions of electrons with energy between $0.5 \le E[\text{MeV}] \le 5.0$ are plotted. It is clearly seen that the produced fast electrons are very divergent for the case without imposed magnetic field. Alternatively, in the cases with imposed magnetic field, the electrons are collimated in space. The spot of the electron beam becomes smaller in the higher imposed magnetic field case. Especially, in the case $B_0 = 50$ MG [Fig. 2(d)], not only the electrons are confined very well but also much fewer electrons are scattered to two sides at the interaction region. It seems that the high energy electrons are confined much better in the higher imposed magnetic field case.

We next examine how the imposed magnetic field affects the forward fast electron flux. In Fig. 3, we plot the time-integrated electron energy observed at (a) $x=15\lambda_0$ and (b) $x=55\lambda_0$ as a function of y coordinate. One can see that the maximum value of the diagnosed electron energy increases with increasing magnetic field at $x=55\lambda_0$. Physically, the reason for this is the smaller gyroradius of the fast electrons can be, the more electron energy can be confined in the center. In comparison, at position $x=15\lambda_0$, the electron energy does not show much difference between the cases with $B_0 = 0, 5, 10$ MG, but there is a rapid increase for the case $B_0 = 50$ MG. Since $x=15\lambda_0$ is very near the interaction surface, the electron energy diagnosed at this position is linked to the absorption of the laser light. The absorption efficiency in Table I shows it only slightly influenced by the imposed magnetic field, leading to the similar electron energy flux at $x=15\lambda_0$ for $B_0 = 0, 5, 10$ MG. Here, we attribute the rapid increase for the case $B_0 = 50$ MG to the collective effects of imposed and spontaneous magnetic fields on the absorption and plasma dynamics at the surface. The guiding and collimation of MeV electrons under the imposed magnetic field are clearly seen in the electron trace images shown in Fig. 4. Here, the tracked MeV electrons are sampled randomly at the beginning near the laser plasma interface. One can see that the electrons propagate helically around the axial imposed magnetic field, by which the fast electrons are collimated in space. By comparing with the trajectories in Figs. 4(a), 4(b) and 4(c), 4(d), one can see that

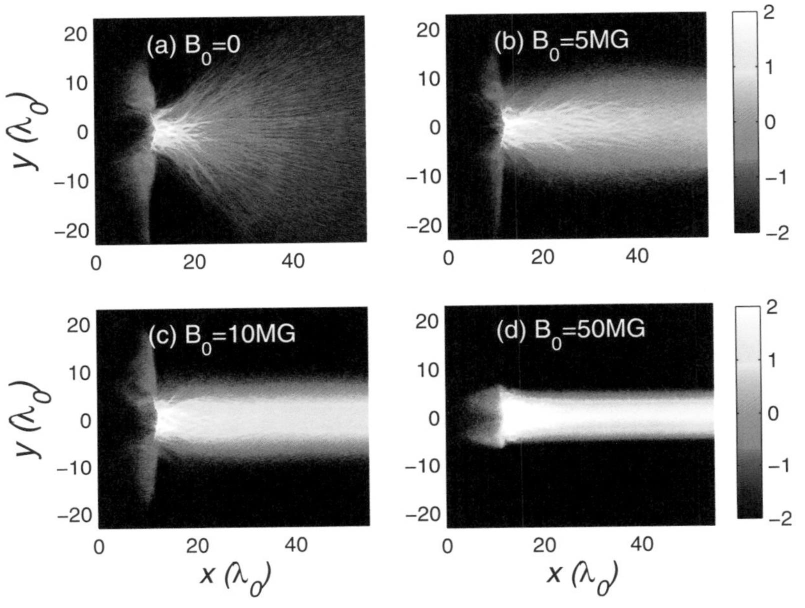

FIG. 2. The natural logarithm of the electron energy density for different imposed magnetic fields (a) $B_0 = 0$, (b) $B_0 = 5$ MG, (c) $B_0 = 10$ MG, and (d) $B_0 = 50$ MG at t = 1 ps. Here, the electron energy density is normalized by $m_e c^2 n_c$.

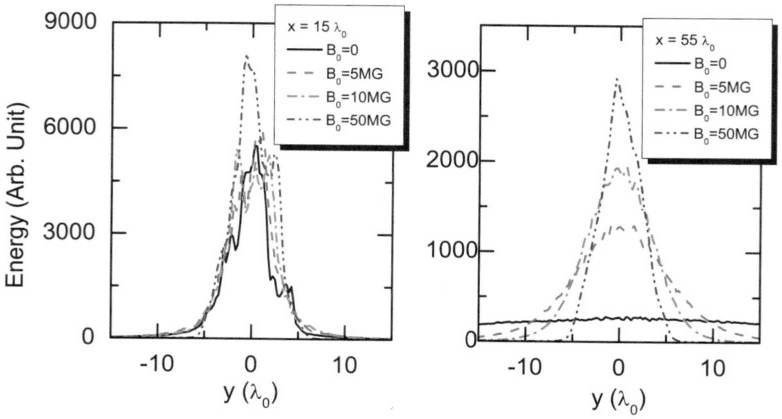

FIG. 3. The transverse distribution of the time-integrated energy of the fast electrons diagnosed at two different positions: (a) $x = 15\lambda_0$, and (b) $x = 55\lambda_0$. The electrons are diagnosed at time interval t = (667 fs, 1000 fs).

the gyroradius becomes much smaller for the case $B_0 = 50\,\text{MG}$. This will lead to the formation of a strong radial current during the transport of REB. Therefore, one can expect the generation of axial and azimuthal spontaneous magnetic fields inside the overdense plasma. Since the spontaneous magnetic fields can be as high as the imposed magnetic field, their effects on the transport of REBs should be taken into account. We will discuss them in detail in Sec. IV.

The fractions of energy flux of the diagnosed fast electrons f_e, defined as the ratio of the diagnosed electron kinetic energy from each position to the total input laser energy, are listed in Table I. Here, only the electrons ($0.5 \le E[\text{MeV}] \le 5.0$) in the region $y = (-8\lambda_0, 8\lambda_0)$ and $t = (667,\ 1000)$ fs are diagnosed. Note that the electrons diagnosed at $x = 15\lambda_0$ denote the fast electrons produced at the laser plasma interaction region, while the electrons diagnosed at $x = 55\lambda_0$ denote the electrons whose energy may contribute to heat the compress core in fast ignition scheme. Comparing the electrons diagnosed at $x = 55\lambda_0$, it is found that in the case without imposed magnetic field only 7.1% of the total absorbed laser energy is contained in the forward energy flux by fast electrons after they transport several tens of microns. When using imposed magnetic field, this fraction can be increase to about 20%. Noticed that there is a rapid increase of the diagnosed energy flux between the cases with imposed magnetic field 1 MG and 3 MG, while the energy flux does not increase significantly with the imposed magnetic field with $B_0 > 3\,\text{MG}$. The explanation for the rapid increase of the energy flux is quite simple. We can estimate the electron gyroradius as[24]

$$r_e = \frac{mc^2}{eB}(\gamma_0^2 - 1)^{1/2}(cgs) = 1.7 \times 10^3(\gamma_0^2 - 1)^{1/2}/B\,\text{cm}, \quad (7)$$

where γ_0 is the relativistic factor and B is the magnetic field in gauss (G). In the present simulations, the ultra-intense laser

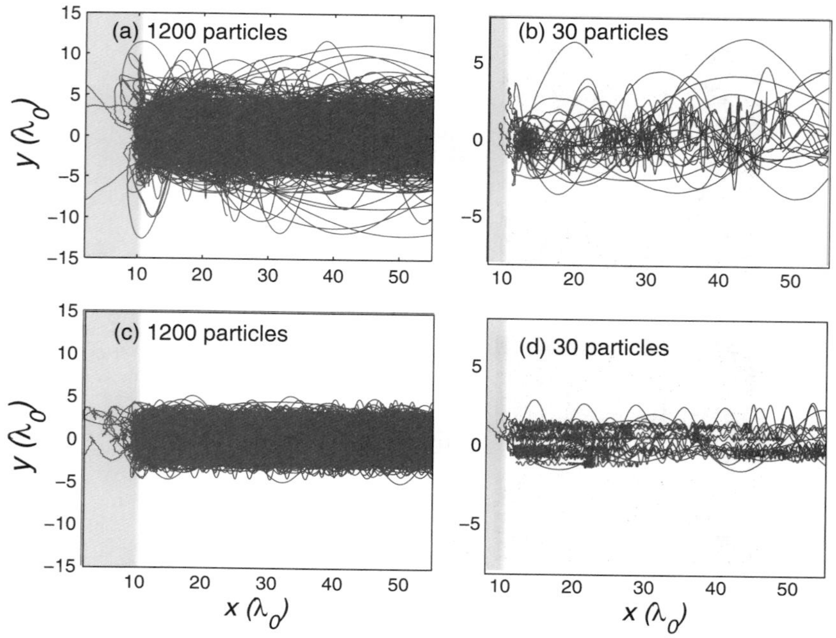

FIG. 4. Typical trajectories of the MeV electrons, showing the guiding and collimation of the electrons with the imposed magnetic fields.

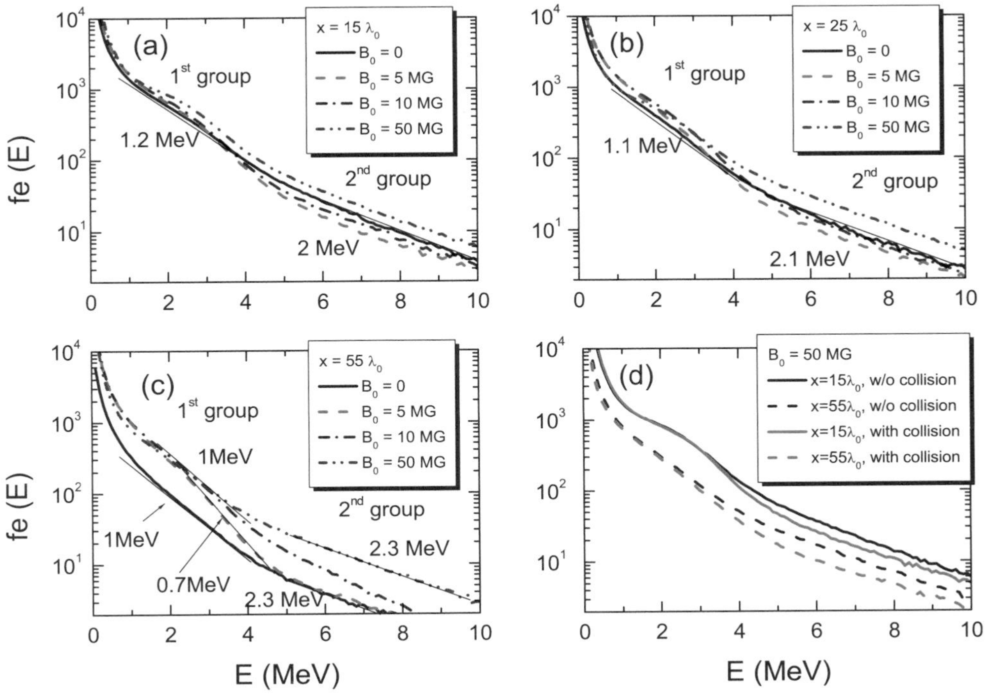

FIG. 5. The spectrum of the fast electrons diagnosed under different imposed magnetic fields at three positions: (a) $x = 15\lambda_0$, $x = 25\lambda_0$, and (c) $x = 55\lambda_0$. (d) The spectrum of the diagnosed fast electrons for the collision case. The electrons are diagnosed at time interval t = (667 fs, 1000 fs).

irradiates the overdense plasma surface normally, the relativistic electrons are mainly accelerated via two different mechanisms: (1) standing-wave pattern acceleration of the laser electromagnetic fields in front of the surface, and (2) $J \times B$ heating. Therefore, the temperature of the fast electrons can be estimated by: (1) the standing wave acceleration scaling law for the low-energy fast electrons,[25] which is typically a fraction of $\alpha \sim \frac{1}{3} - \frac{1}{2}$ of the ponderomotive temperature, $T_M = \alpha(\gamma_0 - 1) \times 0.511\,\text{MeV}$; (2) the ponderomotive scaling law by Wilks and Kruer[26] $T_W = (\gamma_0 - 1) \times 0.511\,\text{MeV}$ for high-energy fast electrons, where $\gamma_0 = (1 + a_0^2/2)^{1/2}$ and a_0 is the laser amplitude. Here, the temperatures of these two-group fast electrons are evaluated as $T_M \simeq 1\,\text{MeV}$ and $T_W \simeq 2\,\text{MeV}$, respectively. The maximum electron gyroradius can be evaluated as 47.3 μm, 15.8 μm, 9.4 μm, 4.7 μm, and 0.9 μm for the cases with imposed magnetic fields 1 MG, 3 MG, 5 MG, 10 MG, and 50MG, respectively. When the electron gyroradius is larger than the maximal electron transport distance d_0, it is not easy to confine it. Therefore, the rapid increase of diagnosed electron flux for the case with $B_0 \geq 3\,\text{MG}$ (electron gyroradius $r_e \simeq 15.8\,\mu\text{m} < d_0$) is understandable. From Eq. (7), we can also derive the minimal magnetic field for collimating fast electrons

$$B_{0,min} = \frac{m_e c^2}{e d_0}(\gamma_0^2 - 1)^{1/2}(cgs) = 1.7 \times 10^7 (\gamma_0^2 - 1)^{1/2}/d_0\,\text{G}, \tag{8}$$

where d_0 is the maximal transport distance for the fast electrons in unit of μm. On the other hand, when the electron gyroradius r_e is smaller than the laser wavelength, the induced spontaneous magnetic fields will result in large divergence angle of the fast electron beam, which is detrimental to the electron transport. Therefore, we can derive the maximal magnetic field for collimating fast electrons

$$B_{0,max} = \frac{m_e c^2}{e\lambda_0}(\gamma_0^2 - 1)^{1/2}(cgs) = 1.7 \times 10^7 (\gamma_0^2 - 1)^{1/2}/\lambda_0\,\text{G}. \tag{9}$$

Therefore, in order to collimate fast electrons, the suitable imposed magnetic field should be satisfied with

$$B_0 \in 1.7 \times 10^7 (\gamma_0^2 - 1)^{1/2}\left(\frac{1}{d_0}, \frac{1}{\lambda_0}\right)\,\text{G}. \tag{10}$$

As a numerical example, we evaluate Eq. (10) for the present simulation case. The temperature of the fast electrons is estimated by $T_M \simeq 1\,\text{MeV}$, which gives $\gamma_0 \simeq 3$. And the maximal transport distance of the fast electrons is about 40 μm, then the minimal and maximal magnetic fields are evaluated as $B_{0,min} = 1.2\,\text{MG}$, and $B_{0,max} = 47\,\text{MG}$.

We also checked the absorption fractions, which are calculated by subtracting the unabsorbed laser energy (laser energy escaped from the boundary and remained in the simulation system) from the total input energy in the simulations. At 1 ps, in the cases without imposed magnetic field and $B_0 = 1\,\text{MG}$, the absorption coefficients (second column in Table I) are about 39.8% and 40.3%, respectively. They are consistent with the fraction of energy flux f_e diagnosed at position $x = 15\lambda_0$. However, this absorption shows an unexpected slight decrease as the increasing of the imposed magnetic field. The results are not significantly modified if the ions are fixed. It means that the electron conversion rates will not increase with the imposed magnetic fields. In contrast, the fraction of energy flux f_e diagnosed at $x = 15\lambda_0$ and $x = 25\lambda_0$ shows a significant increase for larger imposed magnetic field for the cases with $B_0 \geq 3$ MeV. Notice that in these cases the forward energy flux diagnosed at $x = 15\lambda_0$ is much larger than the laser deposition. From a physical point of view, the explanation of the unusual energy flux is quite simple: the electrons are doing circular movement inside the magnetic field, that is to say, the effect of magnetic field tends to confine fast electrons to the target surface region. However, these piled electrons have no contribution to the heating of the core of the fast ignition pellet.

We next studied how the imposed magnetic fields affect the forward fast electron spectra. In Figs. 5(a)–5(c), we plot a series of spectra diagnosed at three positions under different imposed magnetic fields. Each spectrum is time-integrated from 667 fs to 1 ps. For reference, the case without imposed magnetic field is also shown, showing that the generation of two groups of fast electrons via distinct absorption mechanisms in our simulation cases. The first group of fast electrons has a temperature of 1.2 MeV in the interaction region ($x_0 = 15\lambda_0$), which is consistent with the scaling law of May:[25] $T_M = \alpha(\gamma_0 - 1)m_e c^2$. The second group of fast electrons has a temperature of 2 MeV, which is consistent with the ponderomotive scaling law of Wilks and Kruer. It also can be seen in Figs. 5(a)–5(c) the imposed magnetic fields (5 MG, 10 MG, 50 MG) do not significantly affect the spectra of the fast electrons, especially for these diagnosed at the interaction region [Fig. 5(a)]. In fact, the imposed axial magnetic field does not directly affect the standing-wave pattern acceleration of the laser electromagnetic fields and $\mathbf{J} \times \mathbf{B}$ heating, as discussed above. However, the imposed magnetic fields can affect the electron spectra during the transport, as shown in Fig. 5(c).

III. GENERATION OF THE SPONTANEOUS MAGNETIC FIELDS

Up to now, we have discussed the collimation effects of the imposed magnetic fields. However, when an ultraintense laser pulse with a finite spot size normally irradiating on an overdense plasma with 1 μm preplasma, REBs will be driven by the $\mathbf{J} \times \mathbf{B}$ heating at the laser plasma interface. The REBs interact with the dense plasma and generate huge magnetic fields oscillating in the transverse direction with very short wavelengths. The magnitude of these spontaneous generated magnetic fields can become hundreds of MG at high laser intensities. Various nonlinear mechanisms have been invoked to explain the generation of these magnetic fields,[28–32] including nonparallel temperature and density gradients in the ablated plasma,[28] the spatial gradients and temporal variations of the ponderomotive force exerted by the laser on the plasma electrons,[29] inverse Faraday effect,[30] the low-frequency drag current caused by the formation of the cavitation,[31] and current filamentation instability in the laser generated electron beams,[32] and so on. Here, at the plasma surface, the magnetic fields are produced via the spatial gradients and temporal variations of the ponderomotive force, while inside the overdense plasma, the magnetic fields are mainly due to the current filamentation instability. Note that the electron movement and the fast electron current are influenced by the imposed magnetic field; in practice, the imposed magnetic field plays an very important role in the nonlinear evolution of the spontaneous magnetic fields. The helically moving electrons interplay with the usual current filamentation instability to give rise to the spontaneous magnetic fields with very complex spatial structure. As pointed out in Refs. 23 and 27, the excitation of these spontaneous magnetic fields is considered as an important reason for the large divergence of REBs.[27] In this section, we will discuss the characteristics of the spontaneous magnetic fields and their effects on the transport of the REBs.

The importance of spontaneous generated magnetic fields $\langle B_z\rangle$ and $\langle B_x\rangle$ is clear in Figs. 6 and 7, where the spatial distributions of $\langle B_z\rangle$ and $\langle B_x\rangle$ have been plotted at time (a) (c) t = 500 fs and (b) (d) t = 1000 fs. Clearly, at t = 500 fs, both in the cases with $B_0 = 10$ MG and $B_0 = 50$ MG, short transverse filaments have been formed mainly due to the current filamentation instability in the overdense plasma region. The peak value of the magnetic filaments is larger than 100 MG, the distance between two filaments is about $\sim 0.5\,\lambda_0$ at the interface and $1\lambda_0$ just behind the interaction region. These spontaneous magnetic fields lead to the circular movements of the low energy fast electrons mentioned in the last section. In the case with $B_0 = 10$ MG, inside the overdense plasma the peak value of $\langle B_z\rangle$ drops to 10 MG and their size of the filaments is about $0.3\,\lambda_0$. Since two filament beams with currents in the same direction attract each other and tend to coalesce into a larger beam, the beams grow with time until saturation.[33] At time t = 1 ps, we can see that the size of the filaments becomes $1\lambda_0$. Fortunately, in this case, the filaments still peak along the beam propagation direction, which are helpful for the confinement of fast electrons in space.

In contrast, in the case with $B_0 = 50$ MG, the spontaneous fields $\langle B_z\rangle$ at the laser plasma interface are several times larger than those in the previous case. Notice that these magnetic fields induced by the current filamentation instability at the interface are considered as an important reason for the electron divergence angle,[23] their increase will certainly induce larger electron divergence angle. Furthermore, the helically moving fast electrons interplaying with the current filamentation instability result in the formation of fish-scale-shaped magnetic fields (both $\langle B_z\rangle$ and $\langle B_x\rangle$) inside the overdense plasma region. At t = 500 fs, the size of the fish-scale-shaped magnetic fields $\langle B_z\rangle$ is about $0.7\lambda_0$, which is about the size of gyroradius for electrons with energy $\sim$2 MeV. At time t = 1 ps, we also can see that the size of the magnetic structure becomes larger due to the coalescence of filaments. This fish-scale-shaped magnetic fields can also as high as several tens of MG. We may, therefore, assert that these fish-scale-shaped magnetic fields $\langle B_z\rangle$ play a role in scattering fast electrons.

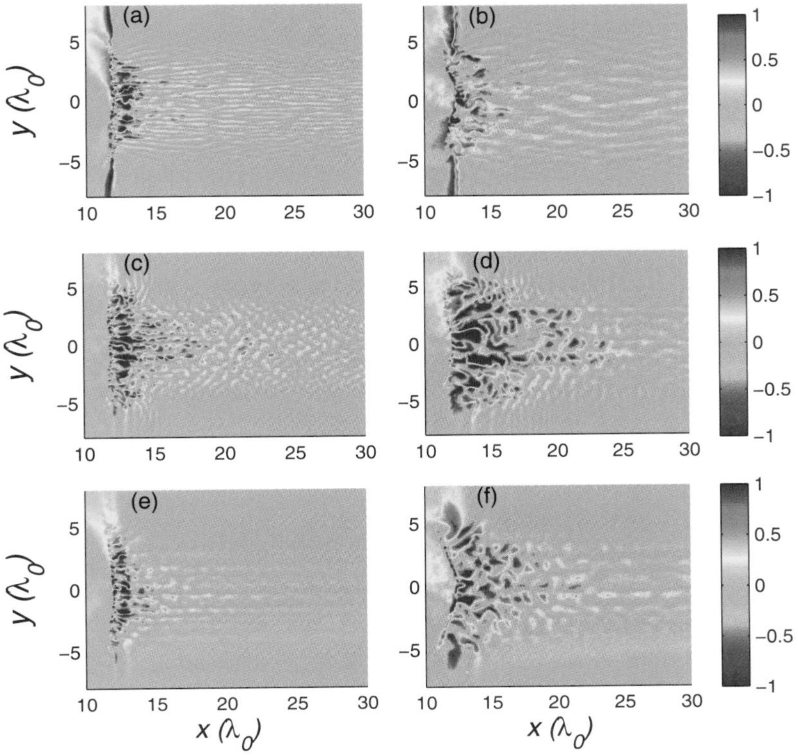

FIG. 6. The quasistatic magnetic fields $\langle B_z \rangle$ under imposed magnetic field (a)(b) $B_0 = 10\,\mathrm{MG}$ and (c)(d) $B_0 = 50\,\mathrm{MG}$ at different times: (a), (c) $t = 500$ fs and (b), (d) $t = 1000$ fs. (e), (f) The quasistatic magnetic fields $\langle B_z \rangle$ in the collisional case, other parameters are same as those in (c), (d). The unit of the electromagnetic fields is $m_e \omega_0 c/e$ (1 unit = 100 MG for magnetic field).

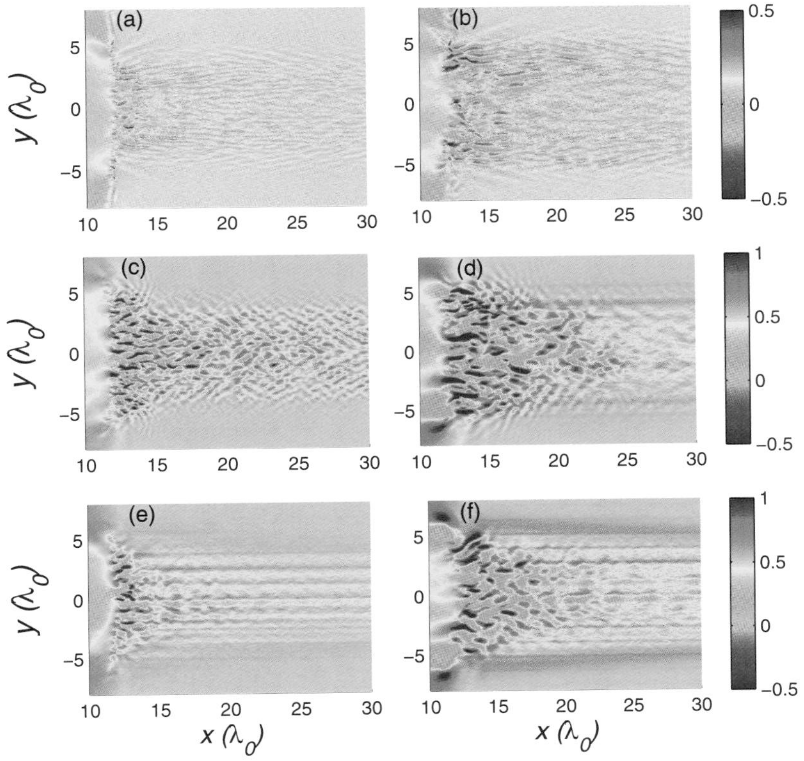

FIG. 7. The quasistatic magnetic fields $\langle B_x \rangle$ under imposed magnetic field (a), (b) $B_0 = 10\,\mathrm{MG}$ and (c), (d) $B_0 = 50\,\mathrm{MG}$ at different times: (a), (c) $t = 500$ fs and (b)(d) $t = 1000$ fs. (e), (f) The quasistatic magnetic fields $\langle B_x \rangle$ in the collisional case, other parameters are same as those in (c), (d). The unit of the electromagnetic fields is $m_e \omega_0 c/e$ (1 unit = 100 MG for magnetic field).

IV. EFFECTS OF THE IMPOSED AND SPONTANEOUS MAGNETIC FIELDS ON REB DIVERGENCE

We next check the effects of imposed and spontaneous magnetic fields on the divergence angle of the electrons. In our simulations, the laser plasma interaction region locates at $x = 11\text{–}13\lambda_0$, while the local angular distribution function of the fast electrons was extracted at the positions $15\lambda_0$, $25\lambda_0$, and $55\lambda_0$. The first diagnostic position is just after the zone where the current filamentation instability is excited. The second and third diagnostic positions locate $12\lambda_0$ and $42\lambda_0$ away from the interaction region. In order to distinguish the divergence angle of the fast electrons from that of the background electrons, only the electrons with energy higher than 200 keV and propagating through the diagnostic positions at time interval t = (667, 1000) fs were considered. The local dispersion has been characterized by a Gaussian function

$$f(\theta) = A\exp\left[-\frac{(\theta-\theta_r)^2}{\Delta\theta_0^2}\right], \tag{11}$$

where $\theta = \tan^{-1}[p_y/p_x]$, θ_r is the local electron mean propagation angle, and $\Delta\theta$ is the local divergence angle. As pointed out by Debayle *et al.*,[23] the local electron propagation angle is related to the beam transverse velocity, while the local divergence angle can be related to the transverse electron temperature. Here, we focus on the divergence angle.

Table II presents the divergence angle of fast electrons diagnosed at different positions for different imposed magnetic fields. We note the position $x = 15\lambda_0$ very close to the laser-plasma interaction region, allowing us to consider the divergence angle of the fast electrons diagnosed at this position as the divergence angle in the laser plasma interaction region. It is found for the case without imposed magnetic field, the divergence angle is about 63°, which is consistent with the simulation results in Ref. 23. In fact, when the imposed magnetic field is smaller than 30 MG, the divergence angle is measured as being $66(\pm 6)^\circ$. It means that for the cases with $B_0 \leq 30$ MG, the electron divergence angle is not significantly affected by the imposed magnetic field. However, when the imposed magnetic field B_x reaches as high as 50 MG, which is the same order of the spontaneous magnetic field due to the current filamentation instability in the interaction region, the divergence angle rapidly increases to 90°. The second diagnostic position locates at $x = 25\lambda_0$. Since the electrons with large divergence angle are already escaped from the central region, the divergence angle of the fast electrons is still slightly influenced by the imposed/spontaneous magnetic field at this distance during the transport. The third diagnostic position locates at $x = 55\lambda_0$. Simulation results show that the divergence angle increases with the imposed magnetic field. In the case $B_0 = 50$ MG, the divergence angle is about 60°, which is almost two times higher than the case without imposed magnetic field. That is to say, the imposed magnetic field can result in larger divergence angle after the electrons transport several tens of microns in the overdense plasma. Altogether, taking into consideration of the conversion efficiency (shown in Table I) and the electron divergence angle (shown in Table II), we may conclude that $B_0 = 3 \sim 30$ MG is more suitable to collimate the fast electrons.

TABLE II. Divergence angle of fast electrons diagnosed at different positions for different imposed magnetic fields. Here, only the electrons in the region $y = (-8\lambda_0, 8\lambda_0)$ and $t = (667, 1000)$ fs are diagnosed.

B_0 (MG)	$x = 15\lambda_0$ (deg)	$x = 25\lambda_0$ (deg)	$x = 55\lambda_0$ (deg)
0	63	59	33
1	64	61	33
3	72	66	36
5	64	53	46
10	67	50	51
20	60	56	56
30	71	60	59
40	76	61	61
50	90	61	60
50 (collisional)	61	64	57

V. COLLISIONAL EFFECT AND LASER INTENSITY

In overdense plasma, the spontaneous magnetic fields are mainly due to the development of the current-filamentation instability (CFI, sometimes also called as Weibel instability). The collisions cause detuning between the particle density perturbations and the reactive fields, which attenuate the CFI.[34] That is to say, in the present collisionless simulations, the CFI induced magnetic fields may be overestimated. In order to check the effect of collisions, we ran simulation of laser impinging on an overdense plasma with an electron density of $60n_c$, while the collision rate was calculated assuming solid carbon ($\sim$2.267 g/cm^{-3}).[23] This is clearly illustrated in Figs. 6 and 7. Figs. 6(e), 6(f), 7(e), and 7(f) show the quasistatic magnetic fields $\langle B_z \rangle$ and $\langle B_x \rangle$ under imposed magnetic field $B_0 = 50$ MG at different times. In comparison with the collisionless cases Figs. 6(c), 6(d) and Figs. 7(c), 7(d), one can see that both the spontaneous magnetic fields $\langle B_z \rangle$ and $\langle B_x \rangle$ are attenuated, especially for the region several microns away from the interaction region inside the overdense plasma.

The divergence angle is mainly due to the electron scattering in the small scale magnetic fields that appear in the laser-plasma interaction region due to the CFI.[23] From this point of view, the divergence angle will decrease for the collisional cases since the CFI is attenuated. Our simulation results show that the divergence angle at the interaction region ($x_0 = 15\lambda_0$) decreases from 90° in collisionless case to 61° in the collisional case ($a_0 = 5$ and $B_0 = 50$ MG). On the other hand, in solid density plasma, the resistive heating followed by diffusion and direct collisions between fast electrons and the background play an important roles for energy transport and transfer.[35] While collision is beneficial control of the CFI and the divergence angle in the interaction region, it decreases the quality of fast electrons by scattering and reducing the energy of electrons during the electron transport. Our simulations indicate that after the transport through several tens of microns of overdense plasma, the divergence angle of fast electron beam is still 57° at $x_0 = 55\lambda_0$ (shown in Table II), which is similar to the collisionless case. Table I also shows that the diagnosed energy flux decreases from 21.3% in the collisionless case to 17.1% in the collisional case. The collisional effect on the fast electron spectra is plotted in Fig. 5(d). It can be seen that in the interaction region ($x_0 = 15\lambda_0$), the number of the low-energy component of fast electrons is increased, while the high-energy component of fast electrons is decreased for the collisional case. Since the fast electrons lose energy during the transport because of collision, the numbers of both low-energy and high-energy component of the fast electrons diagnosed at $x = 55\lambda_0$ are decreased for the collisional case.

We have also studied how the current conclusions change for lower and higher laser intensities. The laser intensity varies from $5.5 \times 10^{18}\,\mathrm{W/cm^2}$ ($a = 2$) to $1.37 \times 10^{20}\,\mathrm{W/cm^2}$ ($a = 10$). For all laser intensities, we studied two different imposed magnetic fields: 10 MG and 50 MG. It is found that the absorption rate is increased significantly as the laser intensity increases from $5.5 \times 10^{18}\,\mathrm{W/cm^2}$ to $1.37 \times 10^{20}\,\mathrm{W/cm^2}$, which is consistent with the observation by Ping *et al.*[36] It is worth stressing that no significant difference of the two imposed magnetic fields (10 MG and 50 MG) for fixed laser intensity to the laser absorption and the fast electron flux diagnosed at $x_0 = 55\lambda_0$ is observed.

We also checked how the divergence angle changes for laser intensities. For the case with laser intensity $I_0 = 5.5 \times 10^{18}\,\mathrm{W/cm^2}$, the divergence angles of fast electrons diagnosed at the interaction region ($x_0 = 15\lambda_0$) are $\theta_{10} = 44°$ and $\theta_{50} = 60°$. Here, θ_{10} and θ_{50} denote the divergence angles for $B_0 = 10\,\mathrm{MG}$ and 50 MG, respectively. As the laser intensity becomes larger, both θ_{10} and θ_{50} increases significantly while the difference between θ_{10} and θ_{50} becomes smaller. In the case with laser intensity $1.37 \times 10^{20}\,\mathrm{W/cm^2}$, the divergence angles of fast electrons diagnosed at the interaction region are $\theta_{10} = 118°$ and $\theta_{50} = 120°$. We also checked the divergence angle of the forward fast electrons diagnosed at $x_0 = 55\lambda_0$. In the case $I_0 = 5.5 \times 10^{18}\,\mathrm{W/cm^2}$, the divergence angle is $\theta_{10} = 46°$ and $\theta_{50} = 48°$; in the case $I_0 = 8.8 \times 10^{19}\,\mathrm{W/cm^2}$ ($a = 8$), the divergence angle is $\theta_{10} = 73°$ and $\theta_{50} = 74°$; in the case $1.37 \times 10^{20}\,\mathrm{W/cm^2}$, the divergence angle is $\theta_{10} = 85°$ and $\theta_{50} = 85°$. The reason for the decrease of the divergence angles comparing with those measured at $x_0 = 15\lambda_0$ is quite simple: those electrons with large divergence angle have less possibility of transport several tens of microns.

VI. SUMMARY AND DISCUSSION

In summary, we performed 2.5D PIC simulations to study the collimation of fast electrons with the help of the imposed magnetic field in fast ignition scheme. It has been shown that in the case without imposed magnetic field, the fast electrons, only carried about 7% of the total input laser energy, can transport 42 μm from the interaction region to the boundary. In comparison, in the case with $B_0 > 3\,\mathrm{MG}$, the fast electrons, carried about 20% of the total input laser energy, can transport to the boundary. It is found that 3 MG imposed magnetic field is enough to collimate fast electrons and enhance the fast electron flux to the compressed core in FI scheme. However, the divergence angle of the fast electrons increases with the imposed magnetic field, in the case with $B_0 = 50\,\mathrm{MG}$ and $a = 5$, the divergence angle can be as large as 90° in the interaction region. It also found the divergence angle increases with the increasing of the laser intensity. Therefore, The imposed magnetic field with $B_0 = 3 \sim 30\,\mathrm{MG}$ may be an optimum value for collimating the fast electrons. This is consistent with our theoretical estimation, which indicates that when the magnitude of the magnetic field is satisfied with $B_0 \in 1.7 \times 10^7 (\gamma_0^2 - 1)^{1/2} (\frac{1}{d_0}, \frac{1}{\lambda_0})$ G, the fast electrons can be collimated in space. The results presented in the paper provide important information needed to design an optimized fast ignition target.

Here, the collimation effect of the uniformed imposed magnetic fields has been studied in detail. However, the structure of the magnetic field could be changed in the region close to the compressed core in FI. When there is an increase in the axial direction of the magnetic field, magnetic mirroring effect will reflect the electrons. In this case, the capability of magnetic confinement of the fast electrons will be reduced. Therefore, in FI experiments, we should pay more attention to the structure of these MG magnetic fields.

ACKNOWLEDGMENTS

We gratefully acknowledge C. T. Zhou, C. Y. Zheng, L. H. Cao, Z. J. Liu, T. Johzaki, and K. Mima for fruitful discussions. This work was supported by the National Natural Science Foundation of China (Grant Nos. 11275028, 11005011, 10935003, and 91230205).

[1] M. Tabak *et al.*, Phys. Plasmas **1**, 1626 (1994).

[2] S. Atzeni, A. Schiavi, J. J. Honrubia, X. Ribeyre, G. Schurtz, Ph. Nicolai, M. Olazabal-Loume, C. Bellei, R. G. Evans, and J. R. Davies, Phys. Plasmas **15**, 056311 (2008).

[3] R. Kodama, H. Shiraga, K. Shigemori, Y. Toyama, S. Fujioka, H. Azechi, H. Fujita, H. Habara, T. Hall, Y. Izawa, T. Jitsuno, Y. Kitagawa, K. M. Krushelnick, K. L. Lancaster, K. Mima, K. Nagai, M. Nakai, H. Nishimura, T. Norimatsu, P. A. Norreys, S. Sakabe, K. A. Tanaka, A. Youssef, M. Zepf, and T. Yamanaka, Nature **418**, 933 (2002); R. Kodama, P. A. Norreys, K. Mima, A. E. Dangor, R. G. Evans, H. Fujita, Y. Kitagawa, K. Krushelnick, T. Miyakoshi, N. Miyanaga, T. Norimatsu, S. J. Rose, T. Shozaki, K. Shigemori, A. Sunahara, M. Tampo, K. A. Tanakaka, Y. Toyama, T. Yamanaka, and M. Zepf, *ibid.* **412**, 798 (2001).

[4] W. Theobald, A. A. Solodov, C. Stoeckl, K. S. Anderson, R. Betti, T. R. Boehly, R. S. Craxton, J. A. Delettrez, C. Dorrer, J. A. Frenje, V. Yu. Glebov, H. Habara, K. A. Tanaka, J. P. Knauer, R. Lauck, F. J. Marshall, K. L. Marshall, D. D. Meyerhofer, P. M. Nilson, P. K. Patel, H. Chen, T. C. Sangster, W. Seka, N. Sinenian, T. Ma, F. N. Beg, E. Giraldez, and R. B. Stephens, Phys. Plasmas **18**, 056305 (2011).

[5] T. Ma, H. Sawada, P. K. Patel, C. D. Chen, L. Divol, D. P. Higginson, A. J. Kemp, M. H. Key, D. J. Larson, S. Le Pape, A. Link, A. G. MacPhee, H. S. McLean, Y. Ping, R. B. Stephens, S. C. Wilks, and F. N. Beg, Phys. Rev. Lett. **108**, 115004 (2012).

[6] J. J. Honrubia and J. Meyer-ter-Vehn, Plasma Phys. Controlled Fusion **51**, 014008 (2009).

[7] K. L. Lancaster, J. S. Green, D. S. Hey, K. U. Akli, J. R. Davies, R. J. Clarke, R. R. Freeman, H. Habara, M. H. Key, R. Kodama, K. Krushelnick, C. D. Murphy, M. Nakatsutsumi, P. Simpson, R. Stephens, C. Stoeckl, T. Yabuuchi, M. Zepf, and P. A. Norreys, Phys. Rev. Lett. **98**, 125002 (2007).

[8] J. S. Green, V. M. Ovchinnikov, R. G. Evans, K. U. Akli, H. Azechi, F. N. Beg, C. Bellei, R. R. Freeman, H. Habara, R. Heathcote, M. H. Key, J. A. King, K. L. Lancaster, N. C. Lopes, T. Ma, A. J. Mackinnon, K. Markey,

A. McPhee, Z. Najmudin, P. Nilson, R. Onofrei, R. Stephens, K. Takeda, K. A. Tanaka, W. Theobald, T. Tanimoto, J. Waugh, L. Van Woerkom, N. C. Woolsey, M. Zepf, J. R. Davies, and P. A. Norreys, Phys. Rev. Lett. **100**, 015003 (2008).
[9]S. Kar, A. P. L. Robinson, D. C. Carroll, O. Lundh, K. Markey, P. McKenna, P. Norreys, and M. Zepf, Phys. Rev. Lett. **102**, 055001 (2009).
[10]A. P. L. Robinson, M. Sherlock, and P. A. Norreys, Phys. Rev. Lett. **100**, 025002 (2008); A. P. L. Robinson and M. Sherlock, Phys. Plasmas **14**, 083105 (2007).
[11]H. B. Cai, S. P. Zhu, M. Chen, S. Z. Wu, X. T. He, and K. Mima, Phys. Rev. E **83**, 036408 (2011).
[12]B. Ramakrishna, S. Kar, A. P. L. Robinson, D. J. Adams, K. Markey, M. N. Quinn, X. H. Yuan, P. McKenna, K. L. Lancaster, J. S. Green, R. H. H. Scott, P. A. Norreys, J. Schreiber, and M. Zepf, Phys. Rev. Lett. **105**, 135001 (2010).
[13]A. P. L. Robinson, M. H. Key, and M. Tabak, Phys. Rev. Lett. **108**, 125004 (2012).
[14]F. Peärez, A. Debayle, J. Honrubia, M. Koenig, D. Batani, S. D. Baton, F. N. Beg, C. Benedetti, E. Brambrink, S. Chawla, F. Dorchies, C. Fourment, M. Galimberti, L. A. Gizzi, L. Gremillet, R. Heathcote, D. P. Higginson, S. Hulin, R. Jafer, P. Koester, L. Labate, K. L. Lancaster, A. J. MacKinnon, A. G. MacPhee, W. Nazarov, P. Nicolai, J. Pasley, R. Ramis, M. Richetta, J. J. Santos, A. Sgattoni, C. Spindloe, B. Vauzour, T. Vinci, and L. Volpe, Phys. Rev. Lett. **107**, 065004 (2011).
[15]R. H. H. Scott, C. Beaucourt, H.-P. Schlenvoigt, K. Markey, K. L. Lancaster, C. P. Ridgers, C. M. Brenner, J. Pasley, R. J. Gray, I. O. Musgrave, A. P. L. Robinson, K. Li, M. M. Notley, J. R. Davies, S. D. Baton, J. J. Santos, J.-L. Feugeas, Ph. Nicolai, G. Malka, V. T. Tikhonchuk, P. McKenna, D. Neely, S. J. Rose, and P. A. Norreys, Phys. Rev. Lett. **109**, 015001 (2012).
[16]Y. Sentoku, E. D'Humierres, L. Romagnani, P. Audebert, and J. Fuchs, Phys. Rev. Lett. **107**, 135005 (2011).
[17]H. B. Cai, K. Mima, W. M. Zhou *et al.*, Phys. Rev. Lett. **102**, 245001 (2009).
[18]D. J. Strozzi, M. Tabak, D. J. Larson, L. Divol, A. J. Kemp, C. Bellei, M. M. Marinak, and M. H. Key, Phys. Plasmas **19**, 072711 (2012).
[19]P. Y. Chang, G. Fiksel, M. Hohenberger, J. P. Knauer, R. Betti, F. J. Marshall, D. D. Meyerhofer, F. H. Seguin, and R. D. Petrasso, Phys. Rev. Lett. **107**, 035006 (2011).
[20]H. Daido, Y. Kato, K. Murai, S. Ninomiya, R. Kodama, G. Yuan, Y. Oshikane, M. Takagi, and H. Takabe, Phys. Rev. Lett. **75**, 1074 (1995).
[21]C. Courtois, A. D. Ash, D. M. Chambers, R. A. Grundy, and N. C. Woolsey, J. Appl. Phys. **98**, 054913 (2005).
[22]S. Fujioka, Z. Zhang, K. Ishihara, K. Shigemori, Y. Hironaka, T. Johzaki, A. Sunahare, N. Yamamoto, H. Nakashima, T. Watanabe, H. Shiraga, H. Nishimura, and H. Azechi, Sci. Rep. **3**, 1170 (2013); K. Mima, in Reports in 12th international workshop on fast ignition of fusion target, Napa Valley, CA, USA, 6–8 Nov. 2012.
[23]A. Debayle, J. J. Honrubia, E. d'Humieres, and V. T. Tikhonchuk, Phys. Rev. E **82**, 036405 (2010).
[24]J. D. Huba, *NRL Plasma Formulary* (Naval Research Laboratory, 2007).
[25]J. May, J. Tonge, F. Fiuza, R. A. Fonseca, L. O. Silva, C. Ren, and W. B. Mori, Phys. Rev. E **84**, 025401 (2011).
[26]S. C. Wilks, W. L. Kruer, M. Tabak, and A. B. Langdon, Phys. Rev. Lett. **69**, 1383 (1992).
[27]J. C. Adam, A. Heron, and G. Leval, Phys. Rev. Lett. **97**, 205006 (2006).
[28]J. A. Stamper, K. Papadopoulos, R. N. Sudan, S. O. Dean, E. A. McLean, and J. M. Dawson, Phys. Rev. Lett. **26**, 1012 (1971).
[29]R. Sudan, Phys. Rev. Lett. **70**, 3075 (1993).
[30]Z. M. Sheng and J. Meyer-ter-Vehn, Phys. Rev. E **54**, 1833 (1996).
[31]H. B. Cai, W. Yu, S. P. Zhu, and C. T. Zhou, Phys. Rev. E **76**, 036403 (2007).
[32]U. Wagner, M. Tatarakis, A. Gopal, F. N. Beg, E. L. Clark, A. E. Dangor, R. G. Evans, M. G. Haines, S. P. D. Mangles, P. A. Norreys, M.-S. Wei, M. Zepf, and K. Krushelnick, Phys. Rev. E **70**, 026401 (2004).
[33]T. N. Kato, Phys. Plasmas **12**, 080705 (2005).
[34]B. Hao, Z. M. Sheng, C. Ren, and J. Zhang, Phys. Rev. E **79**, 046409 (2009).
[35]A. Kemp, Y. Sentoku, V. Sotnikov, and S. C. Wilks, Phys. Rev. Lett. **97**, 235001 (2006).
[36]Y. Ping, R. Shepherd, B. F. Lasinski, M. Tabak, H. Chen, H. K. Chung, K. B. Fournier, S. B. hansen, A. Kemp, D. A. Liedahl, K. Widmann, S. C. Wilks, W. Rozmus, and M. Sherlock, Phys. Rev. Lett. **100**, 085004 (2008).

PHYSICAL REVIEW E **92**, 053106 (2015)

Mitigating the relativistic laser beam filamentation via an elliptical beam profile

T. W. Huang,[1,2] C. T. Zhou,[1,3,4,*] A. P. L. Robinson,[2,†] B. Qiao,[1,‡] H. Zhang,[3] S. Z. Wu,[3] H. B. Zhuo,[4] P. A. Norreys,[2,5] and X. T. He[1,3]

[1]*HEDPS, Center for Applied Physics and Technology and School of Physics, Peking University, Beijing 100871, People's Republic of China*
[2]*Central Laser Facility, STFC Rutherford-Appleton Laboratory, Didcot, OX11 0QX, United Kingdom*
[3]*Institute of Applied Physics and Computational Mathematics, Beijing 100094, People's Republic of China*
[4]*Science College, National University of Defense Technology, Changsha 410073, People's Republic of China*
[5]*Clarendon Laboratory, Department of Physics, University of Oxford, Parks Road, Oxford, OX1 3PU, United Kingdom*

(Received 15 May 2015; revised manuscript received 9 October 2015; published 16 November 2015)

It is shown that the filamentation instability of relativistically intense laser pulses in plasmas can be mitigated in the case where the laser beam has an elliptically distributed beam profile. A high-power elliptical Gaussian laser beam would break up into a regular filamentation pattern—in contrast to the randomly distributed filaments of a circularly distributed laser beam—and much more laser power would be concentrated in the central region. A highly elliptically distributed laser beam experiences anisotropic self-focusing and diffraction processes in the plasma channel ensuring that the unstable diffractive rings of the circular case cannot be produced. The azimuthal modulational instability is thereby suppressed. These findings are verified by three-dimensional particle-in-cell simulations.

DOI: 10.1103/PhysRevE.92.053106 PACS number(s): 52.38.Hb, 42.65.Sf, 52.35.Mw

I. INTRODUCTION

Nonlinear propagation of relativistically intense laser pulses through plasmas is of significant interest in laser plasma interactions (LPI) due to its relation to many potential applications [1–18], ranging from the acceleration of charged particles and efficient Raman amplification, to fast ignition in inertial confinement fusion. Particularly, for the laser channeling fast ignition, a stable plasma channel is required for a second laser pulse to stably propagate into the over-dense Deuterium-tritium (DT) plasma region. To ensure these applications to be feasible, the laser pulse interacting with the underdense plasmas must evolve in a robust way. However, high-intensity laser pulses suffer from many instabilities in underdense plasmas when the laser power exceeds the critical power $P_{\rm cr} \sim 17n_c/n_e$ GW (n_e is the plasma density and n_c is the critical plasma density). The laser energy inside the plasmas would be redistributed and the overall efficiency of the laser energy transport over long distances in plasmas would be strongly restricted because of the filamentation instability. This means that understanding how the filamentation instability can be suppressed is one of laser-plasma physics' highest priorities.

Self-focusing and filamentation instability of relativistic laser pulse in plasmas have been extensively investigated both experimentally and numerically [19–30]. The filamentation is essentially initiated by the beam noise and originates from the transverse modulational instability triggered by the nonlinear response of plasmas. Modulational instability can break up the high power laser beams into many small-scale filaments that each conveys a power of several times of $P_{\rm cr}$ and has a size on the order of the plasma skin depth [19–23]. If the laser power is relatively low, only one filament is created. For a wide laser beam conveying much higher powers, multiple filaments are produced. On the other hand, since the filamentation is a noise-induced process, the number and location of the filaments are unpredictable from shot to shot. The beam quality of the laser pulse would be severely degraded by the random filamentation process. For the nonrelativistic laser pulses propagating in a Kerr medium, several methods have been proposed to organize the filamentation process [31–33]. In particular, it has been shown experimentally that an elliptically distributed laser beam can dominate the noise effect and thus produce deterministic filamentation patterns [33]. However, for relativistically intense laser pulses propagating in plasmas, while the filamentation instability is always a crucial problem, it is not yet clear how one can reduce the filamentation instability.

In this work, we report a simple and efficient method for mitigating the filamentation instability of relativistically intense laser pulses in plasmas by imposing an appropriate ellipticity on the beam profile. It is shown that an initially circular (or slightly elliptical) laser beam produces a filamentation pattern that is the result of the azimuthal modulational instability developing on some diffractive rings. This leads to the laser energy being divided among several randomly distributed filaments. However, when a highly elliptical laser beam propagates in an underdense plasma, it breaks up into a regular filamentation pattern with most of the laser energy being concentrated in a single central filament.

II. THEORETICAL CONJECTURE

In order to investigate the filamentation instability of intense laser beams in plasmas, a circularly polarized laser pulse propagating in a homogeneous plasma was considered for simplicity. The duration of the laser pulse (τ) is assumed to satisfy $\tau_i \gg \tau \gg \tau_e$, where τ_i and τ_e are, respectively, the characteristic time of the ion and electron motion. In this case the ion motion can be neglected. Then under the paraxial approximation, one can obtain the relativistic nonlinear Schrödinger equation (RNSE) of the electric field

*zcangtao@iapcm.ac.cn
†alex.robinson@stfc.ac.uk
‡bqiao@pku.edu.cn

(E) of the laser pulse as [19–23]

$$2ik_L\frac{\partial}{\partial x}E+\nabla_\perp^2E+k_p^2\left(1-\frac{n}{\gamma}\right)E=0, \tag{1}$$

where x is the laser propagation direction and $\perp$ denotes the transverse direction. Here $\nabla_\perp^2=\partial_y^2+\partial_z^2$ is the Laplace operator, k_L is the laser wave vector, $k_p=\omega_p/c$, c is the vacuum light speed, ω_p is the relativistically corrected plasma frequency, $n=n_e/n_0$ is the normalized electron density, n_0 is the background plasma density, $\gamma=\sqrt{1+|a|^2}$ is the Lorentz factor of the electron, $a=eE/m_e\omega c$ is the normalized electric field, and m_e is the electron rest mass. To ensure the charge conservation condition, a finite electron temperature T_e is usually taken into account and then the electron density is given by $n=1+k_p^{-2}\nabla_\perp^2(\gamma+\alpha\ln n)$ [20], where $\alpha=T_e/m_ec^2$ is the normalized electron temperature. It is noted that both the relativistic effect and the ponderomotive expulsion of the electrons are considered in this physical model. In this case, the beam self-focusing and filamentation are induced by the combination of the relativistic and ponderomotive self-focusing effects. Using this one can develop two arguments for why ellipticity should reduce the filamentation.

In nature, filamentation instability is the nonlinear development of modulational instability. Particularly, for a circularly distributed Gaussian laser beam in plasmas, it has been observed that its filamentation pattern corresponds to some diffractive rings surrounding a central filament [23,24]. From Eq. (1), one can elucidate some important features of the filamentation instability. One can perform a perturbation analysis on a ring-shaped beam and consider the unstable modes emerging from this ring with a mean radius of $\bar{r}\equiv\sqrt{\int r^2|a|^2d\vec{r}/\int|a|^2d\vec{r}}$. Assuming a constant amplitude of a_0 for this ring beam, the diffraction term in Eq. (1) can be rewritten as $\nabla_\perp^2=\partial_s^2$, where $s=\bar{r}\phi$ and ϕ is the azimuthal angle. Let $\delta a(x,r,\phi)$ be a small perturbation on the ring and the corresponding plasma density perturbation is set as $\delta n(x,r,\phi)$. Considering the perturbations in the form of $e^{i\lambda x}\cos(M\phi)$, one can obtain $\lambda=\frac{k_p^2}{2k_L}F(a_0^2)\pm i\frac{M}{2k_L\bar{r}}\sqrt{2k_p^2(1+\gamma_0\mathrm{Re})F'(a_0^2)a_0^2-(\frac{M}{\bar{r}})^2}$, where $F(a_0^2)=1-1/\gamma_0$, $\gamma_0=\sqrt{1+a_0^2}$, $\mathrm{Re}=\frac{(\frac{M}{\bar{r}})^2}{k_p^2+\alpha(\frac{M}{\bar{r}})^2}$, and $F'(u)=\partial_uF$. It is clearly shown that the ring beams are unstable to the azimuthal perturbations when propagating in plasmas. This kind of instability is called as azimuthal modulational instability (AMI) in the context of nonlinear optics [34]. The growth rate of AMI in our case can be easily expressed as

$$\Gamma=\frac{1}{2k_L}\frac{M}{\bar{r}}\sqrt{k_p^2(1+\gamma_0\mathrm{Re})\frac{a_0^2}{\gamma_0^3}-\left(\frac{M}{\bar{r}}\right)^2}. \tag{2}$$

Here M refers to the integral number of modulations on the ring.

Figure 1(a) shows the growth rate of the AMI as a function of the normalized wave number $m=\frac{M}{k_L\bar{r}}$ at different electron temperatures. It is shown that with the increase of the electron temperature, the maximum growth rate of AMI is decreased and the fastest-growing modes are shifted to longer wavelengths [35]. In spite of this, it is noted that for these temperatures with $\alpha\leqslant0.1$ the maximum growth rate of AMI does not change significantly. The ratio of the thermal pressure (P_{th}) over the laser radiation pressure (P_{rad}) can be written as $P_{\mathrm{th}}/P_{\mathrm{rad}}=n_eT_e/a_0^2n_cm_ec^2$. For the parameters with $n_e\sim0.1n_c$ and $a_0\sim1$, the thermal pressure is only about several percents of the laser radiation pressure even for a hot plasma with $T_e=100$ keV. Thus, the thermal effect in our case does not affect the AMI significantly and one can consider the cold plasma for simplicity. For a cold plasma with $\alpha=0$, the maximum growth rate of AMI can be written as $\Gamma_{\max}\approx\frac{k_p^2}{4k_L}\frac{a_0^2}{\gamma_0^2}=k_L(\frac{n_e}{n_c})\frac{a_0^2}{4\gamma_0^3}$, which is similar to the growth rate of relativistic modulational instability obtained in Ref. [36]. The corresponding number of modulations can be estimated as $M_{\max}\approx\bar{r}k_pa_0/\sqrt{\gamma_0}=\bar{r}k_L\sqrt{\frac{n_e}{n_c}\frac{a_0}{\gamma_0}}$. Figure 1(b) shows the maximum growth rate of AMI as a function of the laser amplitudes at different plasma densities. This indicates that the filamentation instability mainly occurs for laser pulses with a moderate laser intensity propagating into the plasma with a relatively high density. In addition, it has been demonstrated that the laser beam would suffer from filamentation instability and break up into several filaments when both the normalized radius ($r_0\omega_p/c$) and normalized input power ($P_{\mathrm{in}}/P_{\mathrm{cr}}$) of the laser beam are large enough [23], where r_0 is the radius of the laser beam. It can also be seen from Fig. 1(c) that the maximum growth rate of AMI increases with these two normalized parameters. That is, the filamentation instability will be much easier to grow up when the two normalized parameters are large enough.

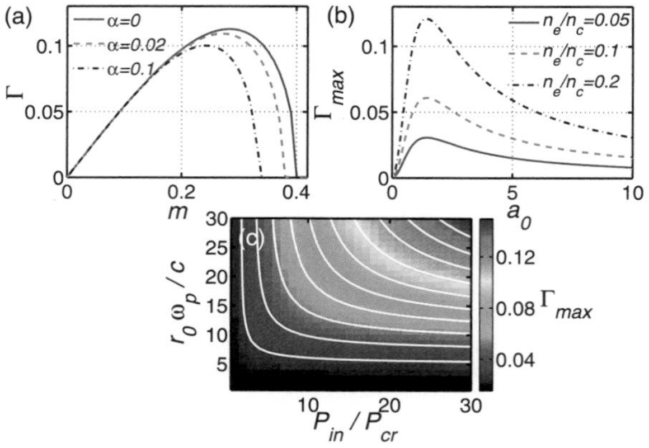

FIG. 1. (Color online) (a) The growth rate of the AMI as a function of the wave number at different initial temperatures, where $m=\frac{M}{k_L\bar{r}}$, $a_0=2$, and $n_e=0.2n_c$. (b) The maximum growth rate of AMI for different laser amplitudes and plasma densities. (c) The phase space of the maximum growth rate of AMI. Here, $P_{\mathrm{in}}/P_{\mathrm{cr}}=0.84\pi a_0^2\frac{r_0^2n_e}{\lambda_L^2n_c}$ and $r_0\omega_p/c=2\pi\frac{r_0}{\lambda_L}\sqrt{\frac{n_e}{n_c\gamma_0}}$, where λ_L is the laser wavelength and $\gamma_0=\sqrt{1+a_0^2}$ is the Lorentz factor. The beam radius r_0 is set as $8\lambda_L$ and the laser wavelength is 1 μm.

It is noted that the unstable rings essentially result from the cylindrical symmetry of the laser beam, which leads to an isotropic diffraction process. The characteristic length of the production of the unstable rings (x_r) is on the order of the

Rayleigh length of the self-focused laser beam that is inversely proportional to the plasma density. This means that the unstable rings are much easier to be generated when the plasma density is relatively high. However, if one introduces an anisotropy on the initial laser beam profile or degrade the symmetry of the laser beam, for example, using an elliptically distributed laser beam, its self-focusing and subsequent diffraction process would become anisotropic along the major and minor axes. It can be inferred that when the initial ellipticity of the laser beam is large enough, the difference value of the self-focusing distance (Δx_f) along the two transverse axes would exceed the value of x_r, i.e., $\Delta x_f > x_r$. In this case, the unstable ring modes cannot be produced due to the highly anisotropic diffraction process. Thereby the filamentation instability can be reduced and even suppressed. Here Δx_f is dependent on the plasma density and the laser intensity, as well as the initial beam ellipticity. However, it is impossible to get an analytical expression of Δx_f due to the nonlinearity of Eq. (1).

From another point of view, the critical power for self-focusing of an elliptically distributed laser beam can be considered, which also determines the threshold power for filamentation instability. To obtain this, Eq. (1) should be reduced into the nonrelativistic case, then the evolution of the laser spot $r^2 \equiv \int (y^2+z^2)|a|^2 dydz/P$ can be written as $\frac{d^2r^2}{dx^2} = \frac{2H}{k_L^2 P}$ with the laser power $P = \int |a|^2 dydz$ and the Hamiltonian $H = \int [|\nabla_\perp a|^2 - \frac{k_p^2}{4}|a|^4] dydz$ [37]. For an initial elliptical Gaussian beam with $a(0,y,z) = a_0 \exp[-y^2/r_1^2 - z^2/r_2^2]$, one obtains the mean-square radius of the laser beam as $r^2 = \frac{r_1 r_2}{2}[1 + (1 - \frac{P_{\rm in}}{P_{{\rm cr},h}})\frac{x^2}{x_R^2}]$, where $x_R = k_L r_1 r_2/2$ is the Rayleigh length, $P_{\rm in} = \pi r_1 r_2 a_0^2/2$ is the input laser power, $P_{{\rm cr},h} = P_{\rm cr} \cdot \frac{h^2+1}{2h}$ is the critical power, and $h = r_2/r_1 (\geqslant 1)$ is the ellipticity of the beam profile. It is noted that in this case self-focusing only occurs when $P_{\rm in} > P_{{\rm cr},h}$. This indicates that an initial beam ellipticity can increase the threshold power for self-focusing as well as the occurrence of filamentation instability. In addition, the self-focusing distance ($x_f \approx x_R/\sqrt{\frac{P_{\rm in}}{P_{{\rm cr},h}} - 1}$) of the whole laser beam is also enhanced with the increase of initial beam ellipticity. This also suggests that the filamentation process can be delayed for an elliptical laser beam. Thus, one concludes that it should be possible to manipulate the filamentation process of relativistic laser pulses in plasmas by invoking a proper beam ellipticity.

III. THREE-DIMENSIONAL PARTICLE-IN-CELL SIMULATIONS

In order to test this hypothesis, three-dimensional (3D) particle-in-cell (PIC) simulations were carried out using the code EPOCH [38]. It is emphasized that the self-focusing and filamentation process in 3D geometry are quite different from that in 2D geometry both quantitatively and qualitatively [27]. In this case, the role of 3D simulations cannot be underestimated. In the simulations, the filamentation of a circularly polarized laser pulse with different radial profiles in an uniform underdense plasma was considered. The simulation box is featured as $100 \times 30 \times 30\,\mu$m with a grid of $1000 \times 300 \times 300$ cells and eight particles of each species (electron and ion) per cell. The central laser wavelength is $\lambda = 1\,\mu$m, the field period is $T_L = \lambda/c$, the duration of the laser pulse is set as $160 T_L$ with a rising time of $5T_L$, and the plasma density is $n_e = 0.2 n_c$. The plasma is located in the region of $2\,\mu{\rm m} < x < 100\,\mu$m and the plasma density rises linearly from 2 to 6 μm then keeps as constant. The initial electron temperature is assumed to be 10 keV. Similar temperatures are expected in the coronal plasma of laser irradiated targets at $I\lambda_L^2 \sim 10^{14}\,{\rm W/cm^2}\mu{\rm m}^2$ [39]. The initial laser beam has an elliptical Gaussian profile $a(0,y,z) = a_0 \exp[-\frac{y^2}{r_0^2 * h} - \frac{z^2}{r_0^2/h}]$ with $a_0 = 0.8$, $r_0 = 8\,\mu$m, and $h = 1.0$, 1.2, and 1.5. This corresponds to a normalized input power of the laser beam as $P_{\rm in}/P_{\rm cr} \approx 22$ and a normalized radius as $r_0\omega_p/c \approx 20$. In this case, the filamentation instability would be much easier to grow up, as indicated in Fig. 1(c).

The simulation result is displayed in Fig. 2. It is shown from Fig. 2(a) that for a circular laser beam with $h = 1$ it experiences an initial self-focusing stage and then breaks up into four filaments. The filamentation process is clearly demonstrated by its corresponding transverse intensity patterns shown in Figs. 2(d), 2(g), 2(j), and 2(m). The laser beam first collapses into a focal spot with a size of about $2c/\omega_p$, which is in good agreement with the radius of the eigen mode in Eq. (1) [22]. During the initial self-focusing stage, electron cavitation gradually takes place and the nonlinear plasma focusing effect [the third term in Eq. (1)] is then reduced. On the contrary, the diffraction term ($\nabla_\perp^2 a$) gradually plays a dominant role and diffractive rings are generated due to the isotropic diffraction, as shown in Fig. 2(g). As illustrated above, the rings are unstable modes and they are sensitive to the azimuthal perturbations. The AMI can easily lead to disintegration of the rings into several filaments, as shown in Fig. 2(j). Finally, the laser power is redistributed into three filaments nucleated on an angular ring and a central filament, as shown in Fig. 2(m). It is interesting to note that the formation and break up of the ring structures seem to be a particular but universal route for the filamentation process of intense Gaussian laser beams propagating in nonlinear media [23,24,31–33,40].

When an initial ellipticity is imposed on the input laser beam profile, the filamentation process would be different from the circular case, as shown in Figs. 2(b) and 2(c). For a slightly elliptical beam with $h = 1.2$, the slight anisotropy, however, cannot suppress the generation of diffractive rings. The AMI is inevitable and the laser beam is broken up into several filaments, as shown in Figs. 2(k) and 2(n). However, when the beam ellipticity is large enough, the introduced anisotropy would take effect and a quite different filamentation process is occurred, as shown in Fig. 2(c). For a highly elliptical laser beam with $h = 1.5$, due to the anisotropic self-focusing process along its major and minor axes of the ellipse, its diffraction process in the plasma channel becomes asynchronous, as shown in Figs. 2(i), 2(l), and 2(o). The diffraction first occurs along its minor axis and then along its major axis. In this case the unstable ring structures cannot be produced and the AMI is suppressed. It can be seen from Fig. 2(o) that the final filamentation pattern consists of a strong central filament and some low intensity filaments along the major and minor axes. It is shown from Figs. 2(m) and 2(o) that the peak intensity of the central filament is about three

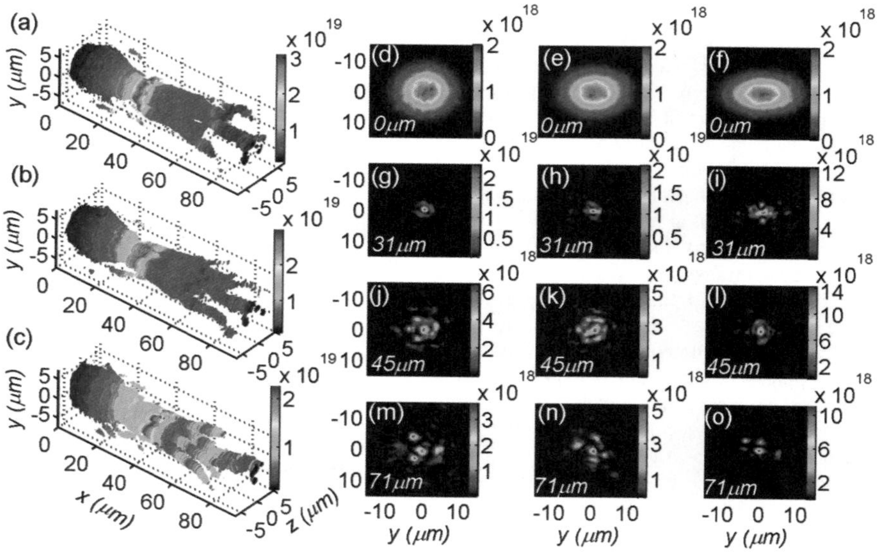

FIG. 2. (Color online) The 3D particle-in-cell simulation results. (a–c) The isosurface of laser intensity at $T = 400$ fs for different elliptical beams, where the isovalue is $1.2 \times 10^{18}\ \mathrm{W/cm^2}$ and the color depicts the maximum intensity. (d–o) The corresponding cross-sections of the laser intensity at different propagation distances. (a), (d), (g), (j), and (m) correspond to $h = 1.0$; (b), (e), (h), (k), and (n) correspond to $h = 1.2$; and (c), (f), (i), (l), and (o) correspond to $h = 1.5$. The simulation parameters correspond to the point C in Fig. 4(d).

times higher than that in the circular case. That is, in the highly elliptical case most of the laser power would be concentrated in a single central filament, while in the circular case the laser power is divided into several filaments.

To clearly demonstrate the filamentation stability in a mensurable way, one can define a normalized parameter as $\eta(x) = P_{\mathrm{cen}}(x)/P_{\mathrm{cen}}(0)$, where P_{cen} refers to the laser power concentrated in the central region with a radius about three times of the radius of the self-focused laser beam $(2c/\omega_p)$. Such a central region should cover most of the laser power of self-focused laser beam. When the filamentation instability occurs, the value of η will fall as the laser power is divided into multiple surrounding filaments. One can employ the averaged value of η to characterize the filamentation instability. To ascertain the stability boundary of filamentation instability, numerical simulations based on Eq. (1) [22] were also performed. In the simulations, the input laser beam is also modeled by an elliptical Gaussian profile. The normalized electron temperature is set as $\alpha = 0.02$. The central laser wavelength is assumed to be 1 μm and the initial beam waist is 8 μm, which is the same as that in Fig. 2. Figure 3 shows one numerical simulation result with the same parameters in Fig. 2(a) based on Eq. (1). It is clearly shown that the filamentation process for a circularly distributed laser beam in plasmas consists of the formation and break-up of a ring-shaped beam. The laser beam first experiences an initial self-focusing process and then breaks up into four filaments, as shown in Fig. 3(b). This is in good agreement with the 3D particle-in-cell simulation results shown in Figs. 2(d), 2(g), 2(j), and 2(m). It is noted that apart from some quantitative difference, the qualitative features about the filamentation process of relativistic laser beams in underdense plasmas can be well captured by the physical model.

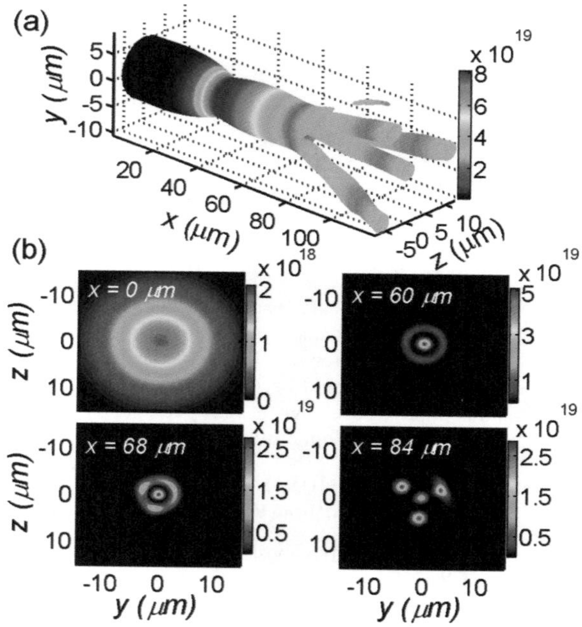

FIG. 3. (Color online) The numerical simulation results based on the physical model Eq. (1). (a) The isosurface of laser intensity for the initial conditions with $a_0 = 0.8$, $n_e = 0.2n_c$, $r_0 = 8\ \mu$m, and $h = 1$, where the isovalue is $1.2 \times 10^{18}\ \mathrm{W/cm^2}$, and the color depicts the maximum intensity. These parameters correspond to the point C in Fig. 4(d). (b) The corresponding cross-sections of the laser intensity at different propagation distances. The unit of laser intensity is $\mathrm{W/cm^2}$. The displayed space scale is $30 \times 30\ \mu$m, and the simulation box is $40 \times 40\ \mu$m.

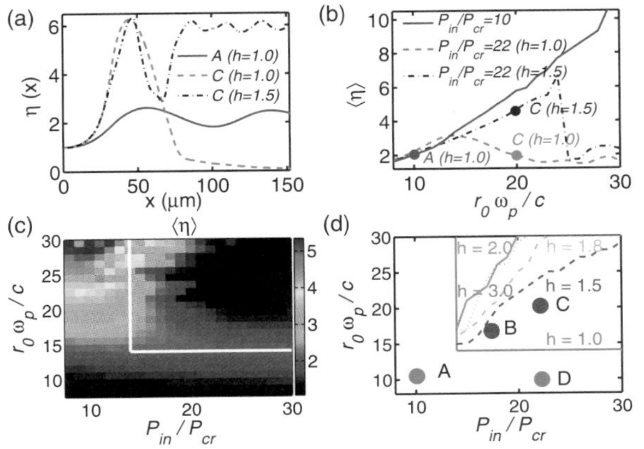

FIG. 4. (Color online) The numerical simulation results based on the physical model Eq. (1). (a) The evolutions of the normalized laser power concentrated in the central region for different initial parameters, where the solid blue line corresponds to the point A in (d), the red dashed line corresponds to the point C in (d) and also corresponds to the simulation result in Fig. 3, and the black dash-dotted line corresponds to the point C in (d) but with an initial ellipticity of $h = 1.5$. (b) The averaged value of η as a function of the normalized radius of the laser beam for different initial input powers and beam ellipticity. (c) The phase space of the averaged value of η, where the white line approximately depicts the stability boundary of filamentation instability and the initial beam ellipticity is $h = 1.0$. (d) Stability map of different elliptical laser beams propagating in underdense plasmas.

Figure 4(a) shows the evolutions of the normalized laser power concentrated in the central region $\eta(x)$ for different initial parameters. When the two normalized parameters of $r_0\omega_p/c$ and $P_{\rm in}/P_{\rm cr}$ are relatively small (A point), the laser beam would experience periodic self-focusing process and the value of η also displays periodic oscillation, as indicated by the blue solid line in Fig. 4(a). The red dashed line shown in Fig. 4(a) corresponds to the simulation result shown in Fig. 3. This shows that the value of η falls below its initial value when the filamentation instability occurs at about $x = 80\,\mu$m. In this case, the periodic oscillation does not occur. The laser power is divided into several filaments and the averaged value of η is only about 1.6. However, when an proper ellipticity is imposed on the initial beam profile, most of the laser power is concentrated in the central region and the averaged value of η is increased to be about 4.7, as shown in Figs. 4(a) and 4(b). Figure 4(b) shows the averaged value of η as a function of the normalized laser beam radius for different input laser powers. As the laser beam will self-focus into a small region with a radius about $2c/\omega_p$, the amplification factor of the peak laser intensity, which can be expressed as $I_f/I_0 = r_0^2/r_f^2 \approx (r_0\omega_p/2c)^2$, would increase with the value of $r_0\omega_p/c$. Here I_f refers to the peak laser intensity at the first self-focusing point and it is noted that the ratio of I_f/I_0 also determines the maximum value of η. For a fixed low-input power of $P_{\rm in} = 10P_{\rm cr}$, the filamentation instability cannot occur. In this case, the averaged value of η will also increase with the value of $r_0\omega_p/c$, as shown in Fig. 4(b). However, for a relatively high-input power of $P_{\rm in} = 22P_{\rm cr}$, when the initial value of $r_0\omega_p/c$ is large enough, the averaged value of η is much decreased. This indicates the occurrence of filamentation instability and the corresponding boundary value of $r_0\omega_p/c$ is about 14. Below this value it is noted that when the value of $r_0\omega_p/c$ is fixed, the averaged value of η is almost the same regardless of the input powers as it is only dependent on the value of $r_0\omega_p/c$ in this stable region. When the laser beam has an elliptically distributed profile, the boundary value of $r_0\omega_p/c$ is increased and the stable region is expanded, as shown in Fig. 4(b).

By employing the averaged value of η on the phase map of $P_{\rm in}/P_{\rm cr}$ and $r_0\omega_p/c$, one can ascertain the stability boundary of the occurrence of filamentation instability, as shown in Fig. 4(c). This indicates that the relativistic beam filamentation in plasmas only depends on the two normalized parameters of $r_0\omega_p/c$ and $P_{\rm in}/P_{\rm cr}$. When these two parameters are large enough, the beam filamentation would gradually take place and the averaged value of η would be much decreased in this unstable region. That is, the filamentation instability only occurs when the laser power is sufficiently large and the self-focusing process is sufficiently tight. In this case, the diffraction process becomes comparatively severe and the unstable rings are much easier to be generated. The stability boundaries for different elliptical laser beams are approximately drawn in Fig. 4(d). This indicates that with the increase of the initial beam ellipticity, the filamentation region in the circular case would be much compressed. These stability boundaries are also verified by conducting 3D particle-in-cell simulations, as shown in Fig. 5. It is shown from Fig. 5(a) that when the two parameters of $r_0\omega_p/c$ and $P_{\rm in}/P_{\rm cr}$ are located in the stable region, the value of η displays periodic oscillation. This suggests that the laser beam experiences periodic self-focusing process and the unstable rings are not

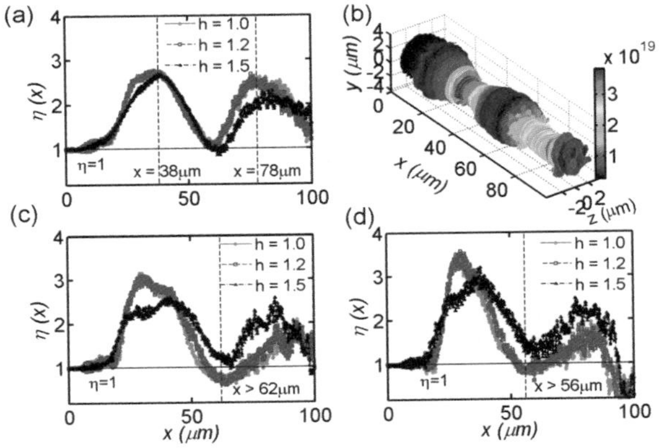

FIG. 5. (Color online) The 3D particle-in-cell simulation results at $T = 400\,fs$. (a), (c), and (d) The evolutions of the normalized laser power concentrated in the central region $[\eta(x)]$ for different elliptical cases and the simulation parameters are, respectively, corresponding to the A, B, and C points in Fig. 4(d). Panel (d) also corresponds to the result in Fig. 2. (b) Isosurface of the laser intensity corresponding to (a), where the isovalue is 2×10^{18} W/cm^2 and the initial beam ellipticity is $h = 1.0$. The dashed lines in (a) indicate the positions of the first and second self-focusing points. The dashed lines in (c) and (d) show the region of the occurrence of filamentation instability.

produced, as shown in Fig. 5(b). In this stable region, the beam profile would oscillate around the soliton-like state and the self-similar self-shaping process that is regardless of the initial beam profiles can occur [22]. Figures 5(c) and 5(d) show that when the two parameters of $r_0\omega_p/c$ and $P_{\rm in}/P_{\rm cr}$ are located in the unstable region, the periodic oscillation vanishes and the value of η also falls after the first self-focusing process. The value of η even falls below the unity and cannot return to its maximum value. This indicates the complete break-up of the whole laser beam, which can be clearly seen from Fig. 2. The subsequent increase of the value of η in the filamentation region, as shown in Figs. 5(c) and 5(d), is attributed to the interaction of different filaments [24]. This process is different from the stable case and clearly indicates the occurrence of the filamentation instability. For the laser beam with an appropriate ellipticity, it is shown that during the filamentation process the laser power concentrated in the central region is about two times larger than that in the circular and slightly elliptical cases, as shown in Figs. 5(c) and 5(d). For the parameters in Fig. 5(d), the averaged value of η can be increased from 1.20 to 1.75 when the circular beam profile is replaced by an elliptical beam profile with $h = 1.5$. This result is in good agreement with the stability boundary shown in Fig. 4(d). In addition, it can be seen from Figs. 5(c) and 5(d) that the self-focusing length is also increased by the initial beam ellipticity, which agrees well with our theoretical predictions.

IV. DISCUSSIONS

For simplicity we have only investigated the filamentation process of circularly polarized laser pulses in underdense plasmas. It is found from the 3D particle-in-cell simulations that the beam ellipticity can also reduce the filamentation instability of linearly polarized laser pulses in underdense plasmas, as shown in Fig. 6(a). During the filamentation process, in the elliptical case much more laser power can be concentrated in the central region compared with that in the circular case. Correspondingly, the averaged value of η in the elliptical case is also larger than that in the circular case, i.e., 2.32 versus 1.46. The production of the unstable rings essentially is a kind of geometry effect. An artificially introduced anisotropy can efficiently suppress the production of the unstable rings and thus reduce the filamentation instability. This is regardless of the polarizations of the laser pulse. Imposing an appropriate ellipticity on the beam profile should be an efficient way to reduce the filamentation instability. The beam ellipticity can be easily adjusted by using two cylindrical lens in the experiments.

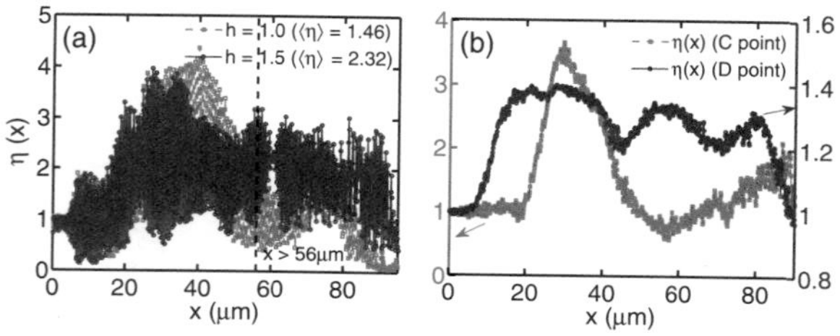

FIG. 6. (Color online) The 3D particle-in-cell simulation results at $T = 400\,\mathrm{fs}$. (a) and (b) The evolutions of the normalized laser power concentrated in the central region $[\eta(x)]$ for different initial parameters, where (a) corresponds to the C point in Fig. 4(d) but for a linearly polarized laser beam and (b) corresponds to the C and D points in Fig. 4(d) for a circularly polarized laser beam. The D point corresponds to the initial conditions with $a_0 = 1.6$, $r_0 = 4\,\mu\mathrm{m}$, and $n_e = 0.2n_c$. This gives a normalized radius as $r_0\omega_p/c \approx 8.2$ and a normalized input power of $P_{\rm in}/P_{\rm cr} = 22$, which keeps the same value of $P_{\rm in}/P_{\rm cr}$ as the C point. The dashed line in (a) shows the region of the occurrence of filamentation instability.

In addition, we suggest that experiments that employ the laser beam with higher intensity, or a smaller radius, will exhibit more stable propagation in underdense plasmas. This can be seen from Fig. 6(b). For the laser beam with a fixed input power, by increasing the laser intensity and decreasing the beam radius, the normalized parameter $r_0\omega_p/c = 2\pi\frac{r_0}{\lambda_L}\sqrt{\frac{n_e}{n_c\gamma_0}}$ can be much decreased. When the value of $r_0\omega_p/c$ is relatively small, the beam self-focusing process becomes smooth and the diffractive ring beams cannot be generated. In this case, the laser beam experiences periodic self-focusing process and the filamentation process does not occur, as shown in Fig. 6(b). On the other hand, the filamentation process should play an important role in electron acceleration, especially for the direct laser acceleration, a stable plasma channel ensures the resonance condition being well kept [6]. Once the filamentation process is manipulated, the corresponding electron dynamics would also be controlled. It is observed that the maximum electron energy can be increased and the divergence angle of the fast electrons can also be reduced when an appropriate ellipticity is imposed on the beam profile. In addition, for the laser beam with a fixed input power, the conversion efficiency of the laser energy into the energy of the fast electrons can be much increased by decreasing the value of $r_0\omega_p/c$. This shall be discussed in detail in a further work.

V. SUMMARY

In summary, we have proposed a simple and efficient way to mitigate the filamentation instability of relativistically intense laser beams in plasmas. It is found that the filamentation process can be organized and a high-quality laser beam can be obtained by imposing an appropriate ellipticity on the beam profile. Initially, this was a theoretical conjecture based on the properties of the filamentation instability; however, our 3D PIC simulations have clearly verified this idea. The stability map for the filamentation process of relativistically intense laser beams in plasmas is also given based on the numerical simulations. By imposing an appropriate ellipticity on the beam profile, the unstable region can be much compressed. Such a well-organized laser beam can be of many potential applications in relativistic high-energy density plasmas.

ACKNOWLEDGMENTS

This work is supported by the National Natural Science Foundation of China (Grants No. 91230205, No. 11575031, No. 11575298, and No. 11175026), the National Basic Research Program of China (973 Program: Grants No. 2013CBA01500 and No. 2013CB834100), and the National High-Tech 863 Project. B.Q. acknowledges the support from

Thousand Young Talents Program of China. T.W.H. acknowledges the SCARF computing clusters at Rutherford-Appleton Laboratory and the support from China Scholarship Council. We thank Max Tabak, Robbie Scott, Holger Schmitz, and Raoul Trines for their useful discussions and help. T.W.H. also thanks Dong Wu, Chengzhuo Xiao, Fanglan Zheng, Xueqing Yan at Peking University and Naren Ratan, Luke Ceurvorst, Muhammad Kasim, James Sadler, Jimmy Holloway, and Matthew Levy at Oxford University for their help and guidance in this work.

[1] M. Tabak, J. Hammer, M. E. Glinsky, W. L. Kruer, S. C. Wilks, J. Woodworth, E. M. Campbell, M. D. Perry, and R. J. Mason, Phys. Plasmas **1**, 1626 (1994).
[2] A. J. Kemp, F. Fiuza, A. Debayle, T. Johzaki, W. B. Mori, P. K. Patel, Y. Sentoku, and L. O. Silva, Nucl. Fusion **54**, 054002 (2014).
[3] P. A. Norreys, D. Batani, S. Baton, F. N. Beg, R. Kodama, P. M. Nilson, P. Patel, F. Pérez, J. J. Santos, R. H. H. Scott, V. T. Tikhonchuk, M. Wei, and J. Zhang, Nucl. Fusion **54**, 054004 (2014).
[4] P. A. Norreys, Nat. Photonics **3**, 423 (2009).
[5] E. Esarey, C. B. Schroeder, and W. P. Leemans, Rev. Mod. Phys. **81**, 1229 (2009).
[6] B. Liu, H. Y. Wang, J. Liu, L. B. Fu, Y. J. Xu, X. Q. Yan, and X. T. He, Phys. Rev. Lett. **110**, 045002 (2013).
[7] A. P. L. Robinson, A. V. Arefiev, and D. Neely, Phys. Rev. Lett. **111**, 065002 (2013).
[8] A. V. Arefiev, B. N. Breizman, M. Schollmeier, and V. N. Khudik, Phys. Rev. Lett. **108**, 145004 (2012).
[9] W. Yu, V. Bychenkov, Y. Sentoku, M. Y. Yu, Z. M. Sheng, and K. Mima, Phys. Rev. Lett. **85**, 570 (2000).
[10] A. P. L. Robinson, R. M. G. M. Trines, J. Polz, and M. Kaluza, Plasma Phys. Controlled Fusion **53**, 065019 (2011).
[11] J. Fuchs *et al.*, Phys. Rev. Lett. **80**, 1658 (1998).
[12] S. Y. Chen, G. S. Sarkisov, A. Maksimchuk, R. Wagner, and D. Umstadter, Phys. Rev. Lett. **80**, 2610 (1998).
[13] C. T. Zhou and X. T. He, Appl. Phys. Lett. **90**, 031503 (2007).
[14] F. L. Zheng, S. Z. Wu, H. Zhang, T. W. Huang, M. Y. Yu, C. T. Zhou, and X. T. He, Phys. Plasmas **20**, 123105 (2013).
[15] G. A. Mourou, T. Tajima, and S. V. Bulanov, Rev. Mod. Phys. **78**, 309 (2006).
[16] J. Fuchs *et al.*, Phys. Rev. Lett. **94**, 045004 (2005).
[17] R. M. G. M. Trines, F. Fiuza, R. Bingham, R. A. Fonseca, L. O. Silva, R. A. Cairns, and P. A. Norreys, Nat. Physics **7**, 87 (2011).
[18] A. G. MacPhee *et al.*, Phys. Rev. Lett. **104**, 055002 (2010).
[19] G. Z. Sun, E. Ott, Y. C. Lee, and P. Guzdar, Phys. Fluids **30**, 526 (1987); X. L. Chen and R. N. Sudan, Phys. Rev. Lett. **70**, 2082 (1993).
[20] M. D. Feit, A. M. Komashko, S. L. Musher, A. M. Rubenchik, and S. K. Turitsyn, Phys. Rev. E **57**, 7122 (1998); A. Kim, M. Tushentsov, F. Cattani, D. Anderson, and M. Lisak, *ibid.* **65**, 036416 (2002).
[21] B. Qiao, C. H. Lai, and C. T. Zhou, Appl. Phys. Lett. **91**, 221114 (2007); Phys. Plasmas **14**, 112301 (2007).
[22] T. W. Huang, C. T. Zhou, and X. T. He, Laser Part. Beams **33**, 347 (2015).
[23] A. B. Borisov, A. V. Borovskiy, O. B. Shiryaev, V. V. Korobkin, A. M. Prokhorov, J. C. Solem, T. S. Luk, K. Boyer, and C. K. Rhodes, Phys. Rev. A **45**, 5830 (1992); A. B. Borisov, A. V. Borovskiy, A. Mcpherson, K. Boyer, and C. K. Rhodes, Plasma Phys. Controlled Fusion **37**, 569 (1995).
[24] A. Pukhov and J. Meyer-ter-Vehn, Phys. Rev. Lett. **76**, 3975 (1996).
[25] Z. M. Sheng, K. Nishihara, T. Honda, Y. Sentoku, K. Mima, and S. V. Bulanov, Phys. Rev. E **64**, 066409 (2001).
[26] H. Y. Wang, C. Lin, Z. M. Sheng, B. Liu, S. Zhao, Z. Y. Guo, Y. R. Lu, X. T. He, J. E. Chen, and X. Q. Yan, Phys. Rev. Lett. **107**, 265002 (2011).
[27] K. A. Tanaka *et al.*, Phys. Rev. E **62**, 2672 (2000); A. Pukhov, Rep. Prog. Phys. **66**, 47 (2003).
[28] N. E. Andreev, L. M. Gorbunov, P. Mora, and R. R. Ramazashvili, Phys. Plasmas **14**, 083104 (2007).
[29] M. Borghesi, A. J. MacKinnon, L. Barringer, R. Gaillard, L. A. Gizzi, C. Meyer, O. Willi, A. Pukhov, and J. Meyer-ter-Vehn, Phys. Rev. Lett. **78**, 879 (1997).
[30] F. Sylla, A. Flacco, S. Kahaly, M. Veltcheva, A. Lifschitz, V. Malka, E. d'Humieres, I. Andriyash, and V. Tikhonchuk, Phys. Rev. Lett. **110**, 085001 (2013).
[31] G. Mechain, A. Couairon, M. Franco, B. Prade, and A. Mysyrowicz, Phys. Rev. Lett. **93**, 035003 (2004).
[32] A. Vincotte and L. Berge, Phys. Rev. Lett. **95**, 193901 (2005).
[33] A. Dubietis, G. Tamosauskas, G. Fibich, and B. Ilan, Opt. Lett. **29**, 1126 (2004); G. Fibich, S. Eisenmann, B. Ilan, and A. Zigler, *ibid.* **29**, 1772 (2004).
[34] D. V. Petrov and L. Torner, Opt. Quantum Electron. **29**, 1037 (1997); D. V. Petrov, L. Torner, J. Martorell, R. Vilaseca, J. P. Torres, and C. Cojocaru, Opt. Lett. **23**, 1444 (1998).
[35] K.-C. Tzeng and W. B. Mori, Phys. Rev. Lett. **81**, 104 (1998).
[36] C. D. Decker, W. B. Mori, K.-C. Tzeng, and T. Katsouleas, Phys. Plasmas **3**, 2047 (1996).
[37] P. Gibbon, *Short Pulse Laser Interactions with Matter: An Introduction* (Imperial College Press, London, 2005), pp. 97–106.
[38] C. S. Brady and T. D. Arber, Plasma Phys. Controlled Fusion **53**, 015001 (2011).
[39] S. Atzeni and J. Meyer-ter-Vehn, *The Physics of Inertial Fusion: Beam Plasma Interaction, Hydrodynamics, Hot Dense Matter* (Clarendon Press, Oxford, 2004).
[40] T. W. Huang, C. T. Zhou, and X. T. He, Phys. Rev. E **87**, 053103 (2013); T. W. Huang, C. T. Zhou, H. Zhang, and X. T. He, Phys. Plasmas **20**, 072111 (2013).

IOP Publishing
Plasma Physics and Controlled Fusion
Plasma Phys. Control. Fusion **57** (2015) 064003 (11pp)
doi:10.1088/0741-3335/57/6/064003

Physical studies of fast ignition in China

X T He[1,2,3], Hong-bo Cai[1,3], Si-zhong Wu[1], Li-hua Cao[1,3], Hua Zhang[1], Ming-qing He[1], Mo Chen[1], Jun-feng Wu[1], Cang-tao Zhou[1,3], Wei-Min Zhou[4], Lian-qiang Shan[4], Wei-wu Wang[4], Feng Zhang[4], Bi Bi[4], Zong-qing Zhao[4], Yu-qiu Gu[4], Bao-han Zhang[4], Wei Wang[5], Zhi-heng Fang[5], An-le Lei[5], Chen Wang[5], Wen-bing Pei[5] and Si-zu Fu[5]

[1] Institute of Applied Physics and Computational Mathematics, Beijing 100088, People's Republic of China
[2] National Hi-Tech Inertial Confinement Fusion Committee, Beijing 100088, People's Republic of China
[3] Center for Applied Physics and Technology, Peking University, Beijing 100871, People's Republic of China
[4] Laser Fusion Research Center, China Academy of Engineering Physics, Mianyang 621900, People's Republic of China
[5] Shanghai Institute of Laser Plasma, Shanghai 201800, People's Republic of China

E-mail: xthe@iapcm.ac.cn and cai_hongbo@iapcm.ac.cn

Received 17 October 2014, revised 15 January 2015
Accepted for publication 28 January 2015
Published 29 April 2015

Abstract

Fast ignition approach to inertial confinement fusion is one of the important goals today, in addition to central hot spot ignition in China. The SG-IIU and PW laser facilities are coupled to investigate the hot spot formation for fast ignition. The SG-III laser facility is almost completed and will be coupled with tens kJ PW lasers for the demonstration of fast ignition. In recent years, for physical studies of fast ignition, we have been focusing on the experimental study of implosion symmetry, M-band radiation preheating and mixing, advanced fast ignition target design, and so on. In addition, the modeling capabilities and code developments enhanced our ability to perform the hydro-simulation of the compression implosion, and the particle-in-cell (PIC) and hybrid-PIC simulation of the generation, transport and deposition of relativistic electron beams. Considerable progress has been achieved in understanding the critical issues of fast ignition.

Keywords: fast ignition, relativistic electron beam, advanced fast ignition target design

(Some figures may appear in colour only in the online journal)

1. Introduction

Fast ignition (FI) [1–4], which is an approach to inertial confinement fusion (ICF), is proposed to lower driver energy, relax symmetry requirements and fulfill promising high gains in inertial fusion energy power plants. In FI, a two-step ICF scheme, the fuel is pre-compressed to a high density >300 g cm^{-3} by a long-pulse (ns) driver (laser beams or x-rays), and then this highly compressed deuterium-tritium pellet ($10 \sim 20$ μm at its core) is ignited by a $\sim$ 10 ps, 10 kJ intense flux of particle beam. These high energy particles can be produced either by conventional accelerator or by lasers. But now the intensity of heavy ion beam produced by the conventional accelerator does not satisfy the ignitor requirement. The laser-driven ion beams have an excellent beam quality [5–7] and a characteristic maximum of energy deposition at the end of their range, which allows efficient heating of a localized volume in an ion fast ignition scheme. However, the conversion efficiency from laser energy into ion (proton) beam energy is still an important issue [8]. Nowdays, most of FI researches are focused on the generation and transport of high energy particles (especially high energy electrons) by the absorption of ultra-intense petawatt (PW) laser pulses at the edge of the compressed fuel [9–11]. Studies [12] showed that the electrons with energy of 1–3 MeV are optimal for FI. In order to accelerate electrons to this energy range, the ultra-intense PW laser needs an intensity of more than 10^{19} W cm^{-2}, a pulse duration of

(a) (b)

Figure 1. (*a*) Heating laser and (*b*) target chamber for the SG-IIU laser facility.

less than 20 ps, a spot size of less than 20 μm, and a laser intensity contrast of more than 10^8.

In order to study the FI related physics and to demonstrate the feasibility of the FI scheme, large scale ns lasers and PW lasers have been operated and/or are under construction in many countries, for example, an ultra-intense laser with an intensity larger than 10^{19} W cm^{-2}, pressures in excess of 1 Gbar and magnetic fields in excess of 100 MG. Recently, experiments at high power, high energy laser facilities and at large scale massively parallel numerical simulations have become available [13]. The current generation of laser facilities with kilojoule-class ultraintense short pulse laser beams are [14–17], the OMEGA facilities at the Laboratory for Laser Energetic (LLE) at Rochester University in USA, the FIREX-I facility at the Institute of Laser Engineering (ILE) at Osaka University in Japan, the Vulcan facilities at the Rutherford Appleton Laboratory in UK, and the Shenguang-II new (SG-IIU) laser facility at the Shanghai Institute of Optics and Fine Mechanics (SIOM) in China. Besides these, there are several other PW laser facilities available for research of the generation and transportation of high energy particles [18–22, 24], where ultra-intense laser pulses are used to test the feasibility of FI with these high energy particles [14, 17, 23, 24]. A couple of new facilities in construction or at design stage are the NIF-ARC at the Lawrence Livermore National Laboratory (LLNL) in US and the Fast Ignition Realization Experiment (FIREX) in Japan. The construction of these laser facilities provides an opportunity for the integrated FI experiments to test fast electron heating of a high-gain compressed fuel assembly in indirect-drive or polar direct-drive geometry.

In this paper, progress in the experimental research related to FI and the development of the integrated code toward the demonstration of ignition in China have been presented. We have been studying the FI scheme of laser fusion using the Shengguang (SG) laser facilities (SG-II, SG-IIU, SG-III prototype and SG-III laser facilities) and the Xingguang-III laser facility. We demonstrated the advanced target designs for FI and checked the survival of the cone-wire structure at the maximal compression time. The first phase of the FI program in China has been started based on a good symmetry of indirect-drive of cone-in-shell target implosion compression carried out on the SG-II laser facility. With the completion of the PW laser beam on the SG-IIU laser facility by the end of 2014, we will do proof-of-principle experiments at the sub-ignition scale to answer the crucial questions related to the feasibility of FI concept.

2. Up-to-date laser facilities for fast ignition

The FI approach to inertial confinement fusion is one of the important goals in addition to central hot spot ignition in China. Usually, the integrated FI experiments require both the large scale multi-beam long-pulse laser beams and the PW-ps laser beams. The large scale laser beams are well developed due to conventional ICF research. At present, the main challenge is the construction of PW lasers with sufficiently high energy. The SG-II laser facility has been completed since 2000, which has eight beams achieving a total energy of 3 kJ of 351 nm (3 ω) drive pulse, and a ninth beam of 1.2 kJ (3 ω, 2 ns) is used to provide an x-ray backlighter for the radiographic studies. Since then, we have carried out over 4000 shots related to the ICF experiments. We have also been studying the FI schemes using the SG-II laser facilities to validate modeling, exploring the trade-offs available and optimizing compression of the fuel, and designing advanced FI targets.

Now the SG-II laser facility has been upgraded to output larger laser energy and aims to carrying out integrated FI experiments. It has eight beams achieving a total energy of 24 kJ of 351 nm (3 ω) drive pulse (or 40 kJ of 526 nm (2 ω)), which are used to implode the deuterated-polystyrene (CD) shells. Besides, a PW heating laser beam has been constructed with the capablility of delivering an energy of 1 kJ energy in 1–10 ps, as shown in figure 1(*a*). This ultraviolet light enters a target chamber with a diameter of 2.6 m, where the beams are focused onto the inside of a hohlraum located at the center of the target chamber, as shown in figure 1(*b*). The new SG-IIU laser facility is designed to do proof-of-principle experiments for FI. The integrated FI experiments will be carried out in 2015. Moreover, the SG-III laser facility has almost been completed. Finally 48 beams will provide an output energy of 180 kJ (3 ω) in 2015. Four PW laser beams are scheduled at the SG-III laser facility, which can be used to make high-energy radiographs

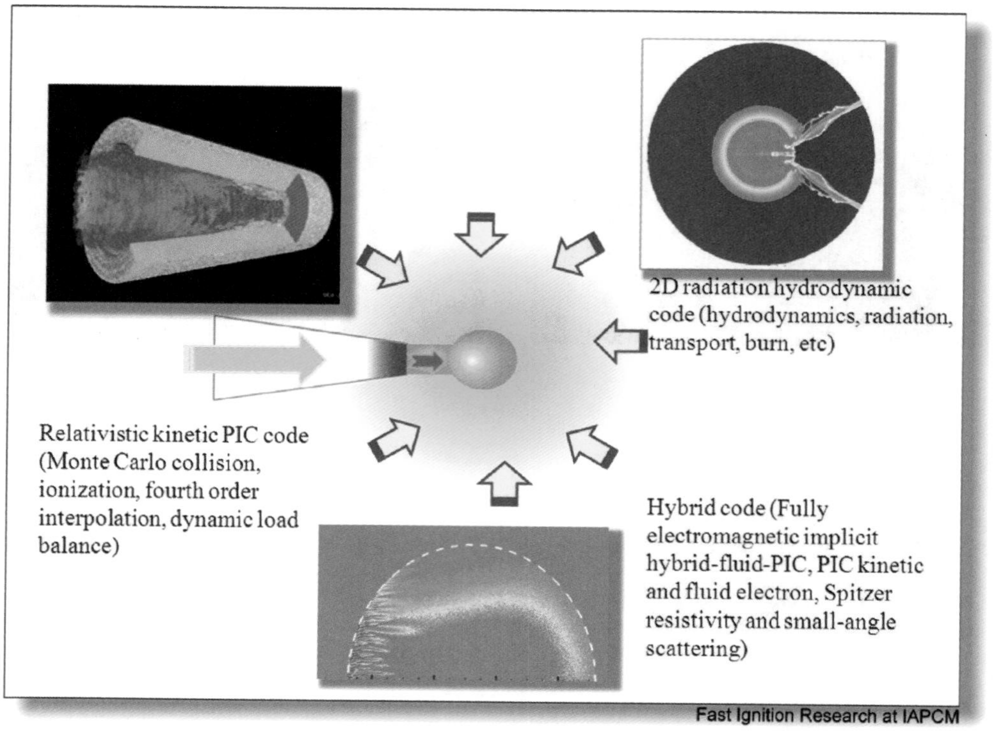

Figure 2. A schematic diagram of the ICFI at the IAPCM.

of the compressed core and produce high energy electrons or ions. These high energy particles can be used to probe material directly, or to produce efficiently high energy x-ray sources. Besides, the SG-III driver will be coupled with those PW lasers to make proof-of-principle experiments at the sub-ignition scale to answer some crucial questions related to the feasibility of FI.

3. Progress in numerical codes

In order to deal adequately with the complexities of FI physics, numerical simulations are of great importance in understanding the physics and exciting new ideas for FI. However, the physical processes in FI are very complex, particularly for the irradiation of the PW ultra-intense laser beam, such as nonlinear laser-plasma interaction, relativistic electron beam (REB) generation and transport [25–27], beam instabilities [28], and so on. Since FI physics have different time scales, large spatial ranges and multi-physics, it is impossible to simulate all the FI processes with one code. The detailed analysis of FI processes can only be carried out by integrated simulation, usually combining a radiation hydrodynamic code, a PIC code, a hybrid fluid-PIC code, or a Fokker–Planck code.

Nowadays, many universities and institutes have developed their integrated FI simulation codes to check the laser coupling efficiency and the feasibility of FI scheme. In ILE, the radiation hydrodynamic code PINOCO, the PIC code FISCOF, and the relativistic Fokker–Planck code FIBMET are integrated to investigate the core heating property of the FIREX-I experiments [15]. In LLNL, the radiation hydrodynamic code HYDRA, the PIC code PSC, and the hybrid code Zuma or LSP are coupled together to model FI processes with initial laser and target parameters [29]. In LLE, the radiation hydrodynamic code DRACO and the hybrid code LSP are integrated to estimate the coupling efficiency of fast electrons to the compressed CD target [14]. In Europe, the radiation hydrodynamic code, such as SARA and DEUD, the hybrid code PETRA and ZEPHYROS, the PIC code OSIRIS and the hybrid Vlasov–Fokker–Planck code VFP have been developed to study the physical issues of FI [3].

Recently, we have developed an 'Integrated Code for Fast Ignition' (ICFI) (shown in figure 2) at the IAPCM, which consists of 2D3V/3D3V (two/three dimensions in space and three dimensions in velocity) relativistic collisional PIC codes ASCENT [30] and LARED-P [31], 2D radiation hydrodynamic codes LARED-S [32, 33] and XRL2D [34], hybrid fluid-PIC codes EBT2D and 3D [35–37] and HFPIC [38–41].

The implosion dynamics of a cone-in-shell FI target is simulated by the LARED-S code, which is a 2D radiation hydrodynamics code including laser ray tracing, multi-group radiation diffusion and transport, electron and ion thermal conduction, atomic physics, nuclear reaction, alpha particle transport, and the quotidian equation of state. The laser plasma interaction (LPI) simulations are performed with the relativistic PIC codes ASCENT and LARED-P, which have been constructed on an infrastructure named JASMIN. JASMIN is aimed at structured or adaptive block-structured meshes for numerical simulations of complex systems on parallel computers [42]. With the support of JASMIN, our code

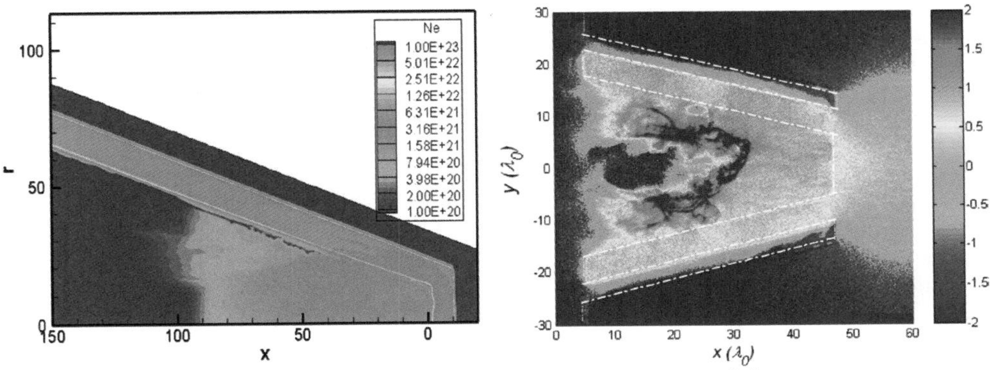

(a) (b)

Figure 3. (*a*) The radiation hydrodynamic simulation of the produced preplasma inside a cone target in irradiation of a 10^{12} W cm^{-2} ASE prepulse at time t = 2.4 ns, (*b*) the PIC simulation results of the electron energy density in irradiation of an ultraintense laser at time t = 1 ps. The initial double cone edge is drawn with the white dash–dotted lines.

can efficiently use thousands and even tens of thousands of CPU processors.

The transport and deposition of the high energy particles are studied by the hybrid code. The numerical algorithms of the hybrid code HFPIC are similar to those developed by Welch *et al* [43], in which both PIC kinetic and fluid electrons are included. The scheme uses a direct implicit particle push and an implicit electromagnetic solver, which relaxes the usual PIC restrictions on the temporal and spatial resolutions and still maintains a good energy conservation. In contrast, the numerical algorithms of the hybrid code EBT2D and 3D are different. The background target is treated as a cold and stationary fluid, while the fast electrons are described by the Fokker–Planck equation and solved by the PIC method [44]. Collisions between fast electrons are ignored and the background particles give only drag and random angular scattering terms. The Brown, Preston and Singleton (BPS) model [45] for the Coulomb logarithm including quantum and coupling effects for a wide range of plasma conditions has been supplemented in the code [46]. These algorithms are much simpler and enable us to do larger-scale simulations. Besides the ICFI codes, we developed a relativistic Fokker–Planck code to model the collisions between fast electrons and core plasma particles, details can be found in [47, 48].

4. Target design and physics researches

We carried out simulations and experiments to verify the key issues and increase the coupling efficiency for the electron-driven cone-in-shell FI scheme. There are several key issues that need to be verified in FI: M-band radiation preheating and mixing in indirect drive, reduction of the electron beam divergence during electron generation and transport, advanced FI target design, prepulse effects on electron energy spectrum and beam divergence, and so on. Recent results indicate that the coupling efficiency can be increased by two methods: (1) collimating REBs in specially engineered targets [22, 49, 50] with the spontaneously generated magnetic fields during its transport; (2) collimating fast electrons with an imposed magnetic field [51, 52].

4.1. Prepulse effects inside the cone target

An ultraintense PW laser pulse is usually accompanied by a long prepulse of amplified spontaneous emission (ASE) pedestal. The ASE prepulses typically have a pulse duration of 0.5–8 ns and an intensity contrast ratio of 10^{-10} to 10^{-6}. The prepulse usually creates a plasma corona with a size of hundreds of microns before the arrival of the main pulse. The size of preplasma is usually positively related to the prepulse energy on target. The main pulse of the ultraintense laser firstly interacts with this preplasma. Before the main pulse reaches the relativistic critical density, many complex physical processes, such as self-focusing, filamentation, beam instabilities, happen and affect the production of fast electrons. The spectrum and beam divergence of the produced high energy electrons in the interaction of ultraintense picoseconds laser pulses with high-Z solid targets are shown to be sensitive to the preplasma created by the prepulse and the ASE pedestal [53].

The radiation hydrodynamic code XRL2D was used to study the formation of the preplasma created by laser ablation from the inside wall of the cone due to the prepulse. Figure 3(*a*) shows the preplasma distribution inside a gold cone target at time t = 2.4 ns. Here, the gold cone has an inner full opening angle of 30°, an inner cone tip width of 20 μm and a side-wall thickness of 10 μm. It is irradiated by an ASE prepulse with a intensity of 10^{12} W cm^{-2} and a total energy of 10 mJ. The gold cone surface is ionized quickly and expands to form a preplasma with a scale of hundreds of microns (one-tenth of the critical density) inside the inner cone [54, 55]. In order to study the effects of the preplasma on the transport of the laser pulse and the production of the REBs, we carried out massive PIC simulations. Here, the initial conditions for the PIC simulations are imported from the hydrodynamic simulation to model the actual target geometry. Here, the cone target used in the PIC simulation is a double cone target, which has the merits of avoiding the side-escaping of fast electrons and improving the coupling efficiency [56, 57]. The p-polarized laser pulse with an intensity of 3×10^{19} W cm^{-2} irradiates the target from the left boundary. Figure 3(*b*) shows the natural logarithm of electron energy density at time $t = 1$ ps. One can see that the laser channels into the overdense plasma and the channeling deviates from the laser axis. The deviation of this laser channeling leads to a very different picture depending on the preplasma distribution and scale length. Detailed discussion of this preplasma effect will be presented in [58].

4.2. Effects of cone-funnel target on the collimation of fast electrons

Enhancing the transport efficiency of the REBs will increase the heating of the core plasma. The transport efficiency of the REBs in the overdense plasmas may be enhanced by the azimuthal magnetic field generated by the specially engineered targets. There are two kinds of targets that can produce magnetic fields to collimate the REBs with different generation mechanisms. One is the large resistivity gradient across the interfaces between different materials [25, 50, 59], the other is the nonparallel density gradient and fast electron current at the interfaces [30, 60, 62]. The former is studied with the hybrid-Vlasov–Fokker–Planck's simulation results [25, 50] and experimental results [22, 49], and the generation of magnetic fields is given by

$$\frac{\partial \mathbf{B}}{\partial t} = \eta \nabla \times \mathbf{j}_f + \nabla \eta \times \mathbf{j}_f, \qquad (1)$$

Where η is the resistivity, and $\mathbf{j}_f$ is the fast electron current density.

The second generation mechanism of magnetic field can be described by a simple analytical model. From the electron magneto-hydro-dynamic (EMHD) equations, the equation for the self-generated magnetic field during the propagation of REB through an overdense plasma is obtained

$$\frac{m_e c^2}{4\pi e^2} \nabla \times \left(\frac{\gamma_e \nabla \times \mathbf{B}}{n_{e0}}\right) + \mathbf{B} = -\frac{m_e c}{e} \nabla \times \left(\frac{\gamma_e n_h \mathbf{v}_h}{n_{e0}}\right), \qquad (2)$$

The term in the RHS of equation (2) is the source term of the magnetic field $\nabla \times \left(\frac{\gamma_e n_h \mathbf{v}_h}{n_{e0}}\right) \cong -\frac{\gamma_e}{n_{e0}^2} \nabla n_{e0} \times \mathbf{j}_h$. It means that the generation of a self-generated magnetic field is due to the nonparallel density gradient and fast electron current. We can solve this equation and obtain the spatial distribution of the quasi-static magnetic field. One can see that magnetic field tends to drive fast electrons into the lower density regions. At sufficiently strong density gradient, a huge magnetic field will be produced. Therefore, if one structures a target using two materials with different densities, in principle the lower density material should confine and guide the fast electrons as magnetic fields are produced at the interface between the two materials where the $\frac{\gamma_e}{n_{e0}^2} \nabla n_{e0} \times \mathbf{j}_h$ term will be large. We can solve equation (2) and obtain the relationship between the peak magnetic field as a function of the laser intensity and plasma density [30]:

$$B_{\max} \simeq 200 \frac{\eta}{f} \frac{I_{18}^{1/2} \lambda_{\mu \mathrm{m}}}{[(Z_2 n_{i2}) / n_c]^{1/2}} \frac{v_{h2}}{c} \mathrm{MG}. \qquad (3)$$

Here, $Z_2 n_{i2}$ denotes the plasma density of the lower density region and v_{h2} denotes the velocity of the fast electrons. Note that the peak magnetic field is proportional to the square root of the input laser intensity and inversely proportional to the square root of the plasma density in the lower density region. The PIC simulations proved the generation of the interface magnetic field by the nonparallel density gradients and fast-electron current near the interface. The generated quasistatic magnetic fields are as high as several tens of and even hundreds of MG [30]. The gyroradius of the relativistic electrons inside the magnetic field is given by

$$r_e = \frac{mc^2}{eB} (\gamma_0^2 - 1)^{1/2} \ (cgs) = 1.7 \times 10^3 (\gamma_0^2 - 1)^{1/2} / B \ \mathrm{cm}, \qquad (4)$$

Here, γ_0 is the relativistic factor of the electron and B is magnetic field in Gauss (G). Therefore, we can estimate that the gyroradius of 1MeV electron is 1.3 μm under for the case with a 100 MG magnetic field. Therefore, the generated interface magnetic field is strong enough to collimate the fast electrons.

We can apply this lower central density target to the advanced FI target design. In order to increase the fast electron transport efficiency, a variant target design containing a reentrant cone coupled with a wire (copper or gold wire) is introduced to guide the fast electrons [63, 39]. The wire attached to the tip of the gold cone is surrounded by low-Z materials, providing a high-resistivity-core-low-cladding target structure. Megagauss magnetic fields can be produced around the wire, which can collimate fast electrons and enhance the transport efficiency. As shown in figure 4, a funnel structure is attached on the tip of the cone target. The collimating effect of the fast electrons by this cone-funnel target is studied by the PIC code ASCENT. The cone and the funnel wall consist of the electrons and gold ions with an assumed charge state $Z_i = 40$ and $m_i/m_p = 195.4$, where $Z_i = 40$ corresponds to the average ionization state of the gold at 1 keV temperature. Both inside the funnel and outside of the cone target are fully ionized hydrogen plasma with density of $4n_c$. The p-polarized laser pulse at $\lambda_0 = 1.06$ μm wavelength and 1.2×10^{19} W cm^{-2} intensity irradiates the target from the left boundary. The intensity profile is Gaussian in the y direction with a spot size of 5 μm (FWHM). The laser rises within 20 laser periods, after which the laser amplitude is kept constant.

Figure 4(*b*) shows the spontaneous quasistatic magnetic field at time $t = 1$ ps. Clearly, a very large magnetic field has been generated at the interfaces of the funnel structure, which can play a role in collimating fast electrons. Figure 4(*c*) shows the energy density distributions of electrons with energies of $0.5 \leqslant E_e \leqslant 5$ MeV. It is clearly seen that the fast electrons are

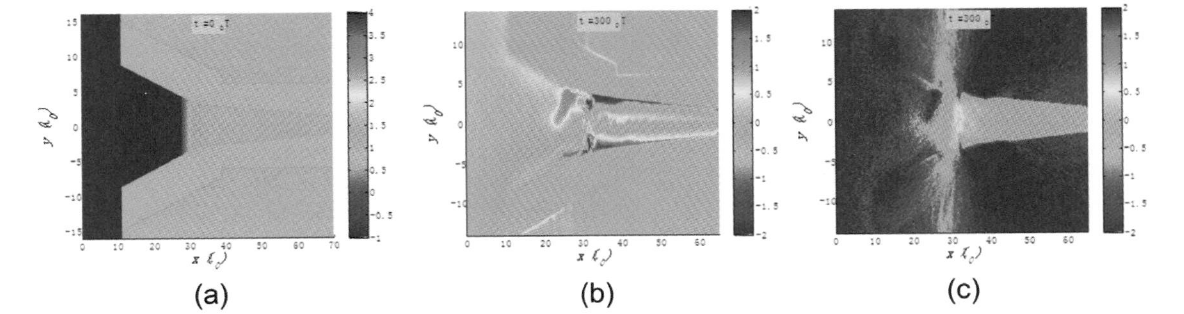

Figure 4. (*a*) The initial density profile of the cone-funnel structure target, (*b*) the spontaneous quasistatic magnetic field and (*c*) the electron energy density at time $t = 1$ ps. Here, (*a*) and (*c*) are on a logarithmic scale, the unit of the magnetic field is $m_e\omega_0 c/e$ (1unit = 100 MG), and the electron energy density is normalized by $m_e c^2 n_c$.

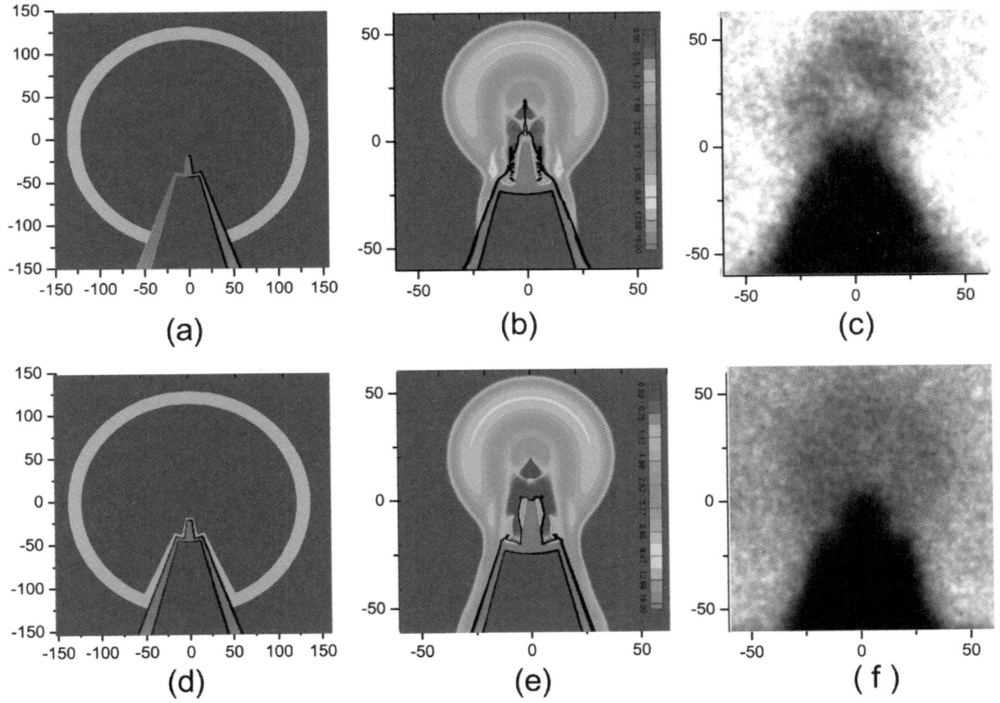

Figure 5. Hydro-simulation of the cone-wire target for the cases (*a*), (*b*) without a CH coating and (*d*), (*e*) with a CH coating at time $t = 1$ ns for the SG-II laser parameters. The experimental results for the cases (*c*) without a CH coating and (*f*) with a CH coating are presented (from Cai *et al* 2014 [68] and Zhou *et al* 2013 [69]. Reproduced with permission from Chinese Laser Press.).

well collimated inside the funnel structure due to the interface magnetic field.

4.3. Plastic (CH) coating on the outer surface of the cone-wire-in-target

Several key issues in cone-wire-in-shell FI should be confirmed before a relativistic assessment to FI can be made. First, can the cone-in-shell target survive at the maximal compression time? Second, how can we avoid the preheating of the gold cone and wire (or funnel structure) by the M-band radiation?

Firstly, we checked the implosion compression of the cone-wire-in-shell target on the SG-II laser facility. A gold wire is attached on the tip of the gold cone target. The wire diameter is 20 μm, wire length is 30–50 μm, and the distance from the wire tip to the pellet center is 20 μm. The CH coating on the outer surface of the cone and wire is about 3 μm. Hydrodynamic simulations [15] have shown that CH coating with a thickness of 0.5 μm on the cone tip reduced the mixing and increased the compressed areal density. Here, our experiments are carried out on the SG-II laser facility. Eight beams (260 J/beam, 3ω, 1 ns) enter into an 800 μm × 1350 μm hohlraum and results in a peak radiation temperature of 180 eV [69]. The implosion processes are diagnosed by a MCP channeling x-ray framing camera (XFC), XRD and a pinhole camera. As shown in figure 5(*f*) the XFC images clearly show that the cone-wire structure can survive at the maximal compression time for the case with the CH coating on the outer surface of the cone and wire structure.

Another important issue is the preheating of the cone-wire structure, which results in the expansion and mixing of the high-Z material. The mixing of ablated high-Z material into the central core is very detrimental to full scale FI. Even if small quantities of high-Z ions were mixed into the high density part of the assembled fuel, it would significantly stifle both ignition and widespread thermonuclear burning [65, 66]. Recent measurements of the M-band (2–5 keV) fraction in hohlraums driven similarly are in the 6–9% range at the peak [64]. This M-band radiation can penetrate the shell and ablates high-Z material from the gold cone and wire. This M-band is the main reason for the preheating of the gold cone and wire. The amount of high-Z ablated plasma is very sensitive to the amount of penetrating M-band radiation, but it is difficult to eliminate or shield this hard x-ray source.

Here, we studied the tamping effect of CH coating on preheating and expansion of the outer surface of the cone-wire target. The implosions of the cone-wire-in-shell target with and without the CH coating are also investigated by the LARED-S code. As shown in figure 5, the cone-wire structure survives at the maximal compression time. The black lines in (*b*) and (*e*) are the interface of the high-Z plasma. Figures 5(*c*) and (*f*) show the experimental x-ray radiographs at the maximal compression time. One can see that the high-Z gold is mixed and diffused into the core of the compressed fuel in the case without the CH coating. In contrast, figures 5(*d*)–(*f*) show the case with a several microns of CH tamper. In this case, the expansion and mixing of the high-Z plasma is effectively controlled by the CH tamper. However, the CH can expand vigorously under the influence of the softer drives, and then the carbon ion mixing is liable to be an issue [64, 65]. The compression implosion of cone-funnel-in-shell target are also carried out on the SG-II laser facility, and similar results are also found and discussed in another paper [67]. In the near future, we will study other tamping materials and heat the compressed cone-wire-in-shell target with a PW laser beam.

4.4. Effects of the imposed magnetic field on the production and transport of relativistic electron beams

The external magnetic field has been recognized as an important method to guide the fast electrons. Strozzi *et al* [51] shows that imposed uniform axial magnetic fields recovered the 132 kJ ignition energy of artificially collimated source with the hybrid PIC code Zuma. Fujioka *et al* [52] has demonstrated the generation of huge magnetic fields of the order of 10 MG with a capacitor-coil target irradiated by ns-kJ laser beams. Recently, we have also measured the spatial structure and temporal evolution of huge magnetic fields with the time-gated proton radiography method [70]. It seems that the imposed magnetic field has the advantages of controlling the electron divergence and lowering the laser energy required to obtain the fuel burning for FI. However, several questions still need to be confirmed: how large is the magnetic field needed? What is the effect of the magnetic field on absorption mechanism and the fast electron spectrum? In hybrid PIC simulations, the generation of fast electrons by the interaction of ultraintense laser pulse and plasmas is not modeled due to the large computational scale, and the artificial fast electron source is used regardless of the feedback of the transport on the laser plasma interaction and the influence of the imposed magnetic field. In our research [61], the interaction of an ultraintense laser pulse under different imposed magnetic fields is investigated by the collisional PIC code ASCENT. The beneficial and detrimental effects of the imposed magnetic fields on the production and transport of REBs are included self-consistently.

Here, we considered a p-polarized laser at $\lambda_0 = 1.06\ \mu$m wavelength and 3.45×10^{19} W cm^{-2} intensity which irradiates the carbon target whose plasma density is $60n_c$ from the left boundary. The intensity profile is Gaussian in the y direction with a spot size of 6.7 μm(FWHM). The laser rises in 10 T_0, where T_0 is the laser period, after which the laser amplitude is kept constant. A uniform axial magnetic field is imposed along the laser propagating direction. Simulations showed that the fast electrons are mainly accelerated by the 'not-so-resonant' $J \times B$ heating [72] or so-called $2a_0$ acceleration mechanism [73]. Fortunately, it is found that the external magnetic field does not directly affect the electron acceleration mechanism and the resulting electron energy spectrum [61]. Figure 6 shows the energy distribution of electrons with energies between $0.5 \leqslant E_e \leqslant 5$ MeV. It is clearly seen that the produced fast electrons are very divergent for the case without the imposed magnetic field. Alternatively, in the cases with the imposed magnetic field, the electrons are collimated in space. It seems that the fast electrons are confined much better in the higher imposed magnetic field case. When the imposed magnetic field is greater than 3 MG, the fast electrons are guided well by the imposed magnetic field, and the coupling efficiency from laser to forward fast electrons increases from 7% (without the imposed magnetic field) to 20% . Here, the fast electrons are diagnosed at position $x = 55\ \mu$m, which is about 40 μm away from the laser plasma interaction region. The physical explanation for the minimal magnetic field 3 MG is straightforward: the electrons move helically around the magnetic field lines, whose gyroradius are estimated by $r_e = \frac{mc^2}{eB}(\gamma_0^2 - 1)^{1/2}$ (cgs), where γ_0 is the relativistic factor and B is the magnetic field. If the electron gyroradius is smaller than the transport distance d_0 (it is usually 30 ~ 100 μm for fast ignition scheme), the electrons can be collimated by the magnetic field. Therefore, we can derive the minimal magnetic field for collimating fast electrons,

$$B_{0,\min} = \frac{m_e c^2}{e d_0}(\gamma_0^2 - 1)^{1/2}\ (cgs) = 1.7 \times 10^7(\gamma_0^2 - 1)^{1/2} / d_0 \quad \text{G}.$$

In the present case, the ultra-intense laser irradiates the overdense plasma surface normally, the relativistic electrons can be accelerated to a temperature of 1 ~ 3 MeV. Then we can estimate that the minimal magnetic field for collimating the fast electrons is about ~3 MG.

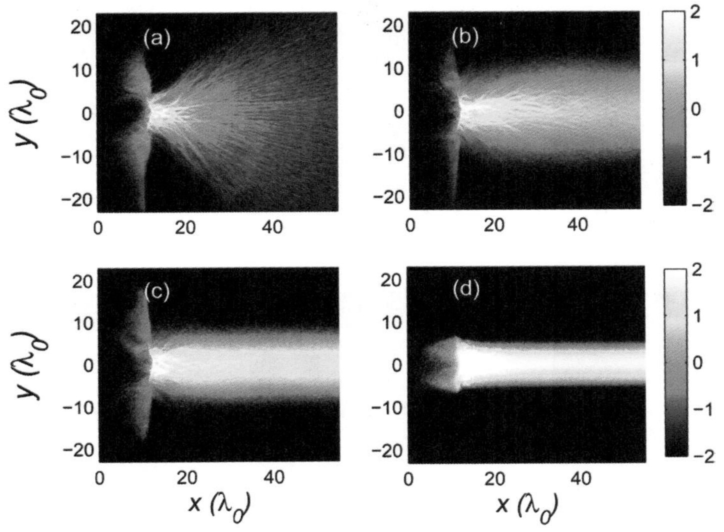

Figure 6. The natural logarithm of the electron energy density for different imposed magnetic fields (*a*) $B_0 = 0$, (*b*) $B_0 = 5$ MG, (*c*) $B_0 = 10$ MG and (*d*) $B_0 = 50$ MG at $t = 1$ ps. Here, the electron energy density is normalized by $m_ec^2n_c$ (Reproduced with permission from [71]. Copyright 2013, AIP Publishing LLC.).

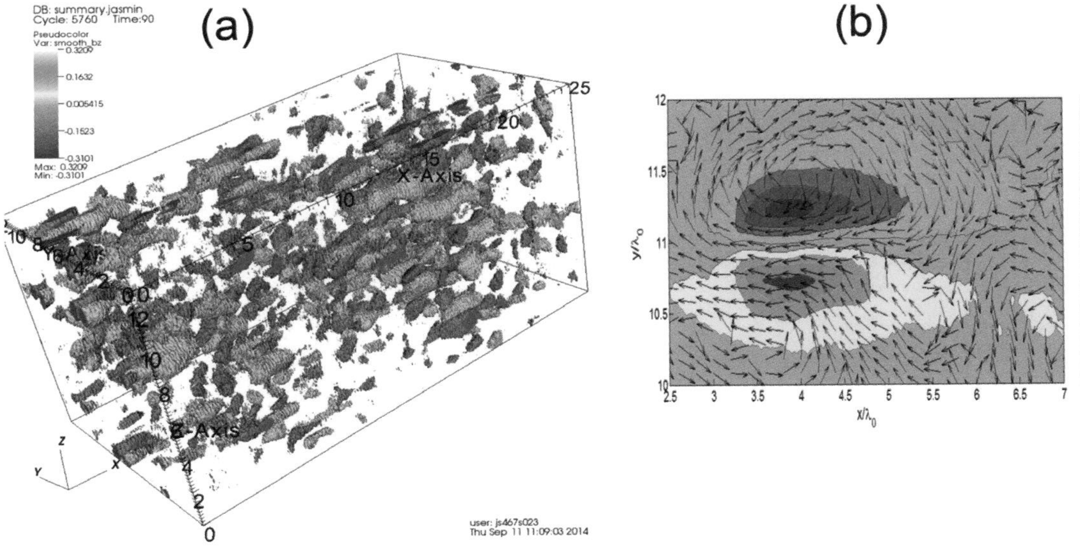

Figure 7. (*a*) Snapshots of the magnetic field Bz in beam-plasma system with an imposed magnetic field $B_x = 15$ MG at $t = 90T_0(T_0 = 2\pi c/\omega_p)$, (*b*) the detailed structure of the contours of a single magnetic dipole, and the arrows represent the magnetic field lines in the simulation plane.

However, the helically moving electrons under the magnetic field in the overdense plasma can result in a higher level of current filamentation instability and even magnetic dipoles near the interaction region [74]. Simulation results show that the beam divergence angle increases with the imposed magnetic field. In the case with $B_0 = 50$ MG and $a = 5$, the divergence angle can be as large as 90° near the interaction region. Taking into account both the conversion efficiency and the electron divergence angle, we may conclude that $B_0 = 3 \sim 30$ MG is more suitable to collimate the fast electrons [61, 76].

In order to clarify the formation mechanism of the magnetic dipoles and its effects on the REB divergence, we carried out a massive PIC simulation to study the transport of the REBs in the magnetized dense plasma and the kinetic processes of the beam instabilities and magnetic dipoles, see figure 7. Here, we assume a very long electron beam propagating in the $\boldsymbol{e}_x$ direction and there is no dependence on the coordinate x. The beam and plasma densities are n_b and n_p, respectively. The beam and background plasma electrons have velocities $v_{bx} = 0.979c$ (corresponding to 2 MeV fast electrons) and

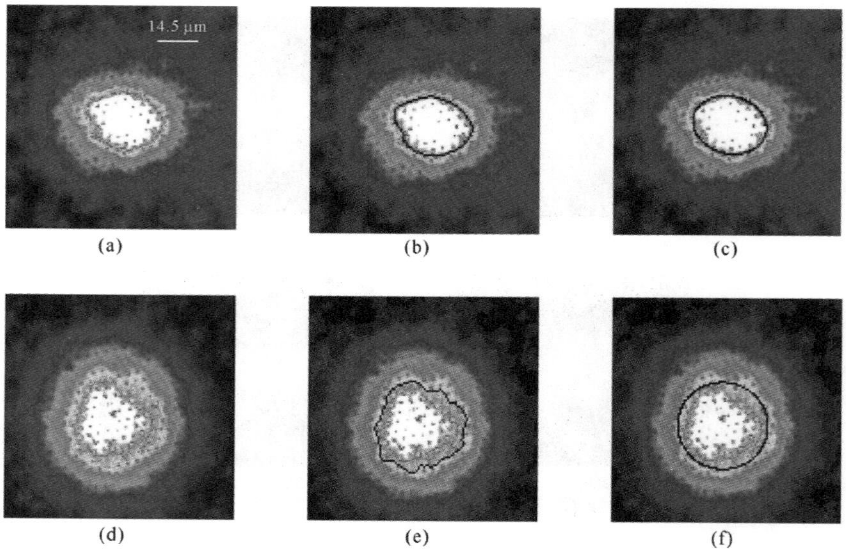

Figure 8. Curve fitting results for 50% peak x-ray emission intensity contour of CD shell targets for the cases (*a*)–(*c*) with a cone and (*d*)–(*f*) without a cone in hohlraums for a fixed length $L = 1700\ \mu$m near peak compression (from Bi *et al*, 2014 [77]). (*a*) Emission of cone target. (*b*) 50% peak intensity contour of cone target. (*c*) Ellipse fitting, $a/b = 0.768$. (*d*) Emission of target without cone. (*e*) 50% peak intensity contour of target without cone. (*f*) Ellipse fitting, $a/b = 0.967$.

$v_{px} = -\ 0.163c$. The current is neutralized: $n_b v_{bx} + n_p v_{px} = 0$. A uniform axial magnetic field is imposed along the $\boldsymbol{e}_x$ direction. It is found that the magnetic dipoles can cause the anomalous energy dissipation of the REB [74]. The random magnetic field induced by current filamentation instability (CFI) and magnetic dipoles scatters the fast electrons and results in a bigger divergence angle [75]. A parametric study of the beam instabilities is presented in the context of FI, and we found the a new way to suppress the CFI and its detrimental effects on the fast electron beam divergence [75].

4.5. Radiation symmetry of indirectly driven cone-in-shell targets

In China, the SG series laser facilities are configured to achieve laser based ICF using the indirect drive approach to ignition. Indirect drive ICF uses a small high-Z hohlraums to convert the incident laser energy into thermal x-rays. Here we have examined the implosion symmetry of an indirect driven cone-in-shell target. In order to study the influence of the guiding cone and hohlraum length on radiation symmetry of the compressed core, x-ray emission measurements in FI pre-compression were carried out on the SG-III prototype laser facility [77]. Here, the hohlraums have a dimensions of 1000 μm in diameter with laser entrance holes of 650 μm in diameter. The hohlraums have three kinds of lengths: 1500 μm, 1600 μm, and 1700 μm. The shell is 330 μm in diameter with a 15 μm thick CD wall. Firstly, we compared the symmetry for the cases with and without a gold cone for a fixed hohlraum length to check the effect of the gold cone on the implosion symmetry. Here, the gold cone is perpendicular to the hohlraum axis.

Figure 8 shows the snapshots of x-ray emission from the imploding capsule near peak compression. Here, the normal line of the image plane is perpendicular to both the hohlraum axis and the cone axis. The least-square ellipse fitting is applied to 50% peak intensity contour, thus the ellipse characters could be used to numerically analyze the core compression symmetry at the peak compression time. The x-ray emission shows that the compression symmetry is affected by the gold cone. Actually, this asymmetry is usually the P4 asymmetry, which is sensitive to the position of the spots relative to the capsule and can be tuned by changing the hohlraum length. Our experiments showed that shape deformation of the cone-in-shell target in hohlraum with a 1600 μm length is much better than that in the hohlraums with 1500 μm and 1700 μm lengths [77].

4.6. Integrated simulation of cone-in-shell FI

The FI scheme relies on the localized deposition of the particle energy in the core. Design studies have been performed for electrons with integrated simulations coupling the hydrodynamic code LARED-S with the hybrid PIC code EBT2D. The LARED-S code simulates the fuel assembly in a cylindrical geometry. The hybrid PIC code EBT2D simulates the propagation and energy deposition of the REBs in the pre-compressed cone-in-shell target. The contours of mass density (g cm^{-3}) and the temperature (MK) of imploded pre-compressed plasma are shown in figures 9(*a*) and (*b*), where the rectangle in white dashed lines indicates the calculating zone for the hybrid PIC code. Here, the compressed cone-shell target is irradiated by an ultraintense laser pulse with the intensity of $I = I_p \mathrm{e}^{-r^2/R_L{}^2} \mathrm{e}^{-4(t-t_L)^2/t_L{}^2}$ and the peak laser intensity of $I_p = 5 \times 10^{19}$ W cm^{-2}. Here, the spot size is $R_L = 5\ \mu$m, and the pulse duration is 1 ps with $t_L = 0.5$ ps. However, the laser-plasma interaction is not modeled actually. We assume

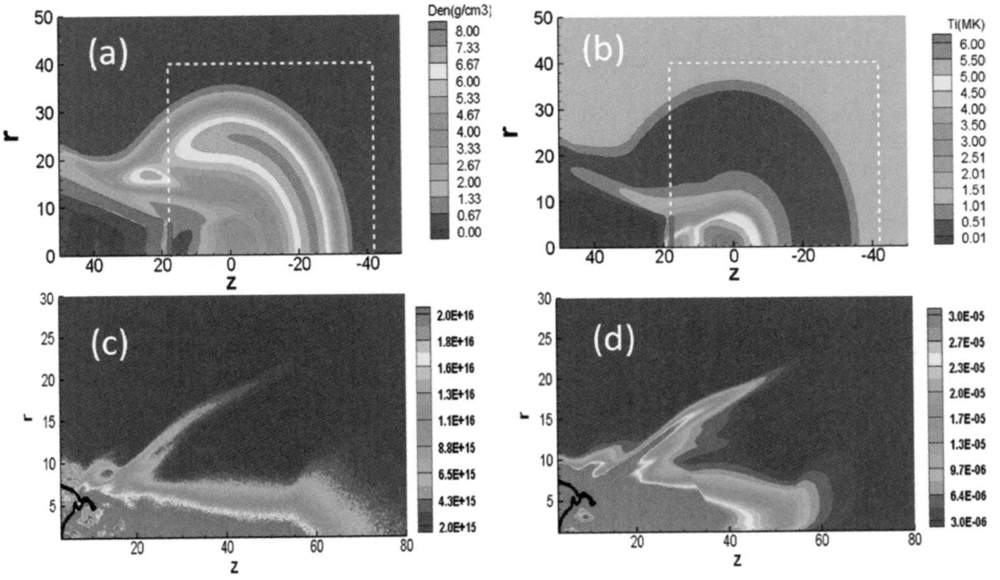

Figure 9. (*a*) The mass density (g cm^{-3}) and (*b*) temperature (MK) of the imploded pre-composed target obtained from hydro-simulations; (*c*) the current density (A m^{-2}) and (*d*) the energy deposition (*J*) of REB injected from the left side of the cone tip.

that the generated REBs have an exponential energy distribution with the average kinetic energy from the scaling law [44], $kT_f = 0.308[I_p(\mathrm{W/cm^2})\lambda(\mu\mathrm{m})^2]^{1/3}\mathrm{eV} \sim 1\,.14$ MeV, where $\lambda = 1$ μm is the laser wave length. The REB transport and deposition are calculated by the hybrid code. Figures 9(*c*) and (*d*) shows the distribution of current density and the energy deposition of the REB inside the compressed core. It is found that the REB can be scattered by the imploded plasma because of its high density. In the hybrid simulation, the energy deposition of the REBs is due to both the Ohmic heating and collisions. Further investigation of the cone-in-shell FI target will be carried out by the integrated simulation combing PIC, 2D radiation hydrodynamic code and hybrid code.

5. Summary

FI, which is considered to provide an alternative way to achieve ICF ignition with a considerably smaller laser energy, is being studied by many groups worldwide using ultra-intense short-pulse lasers. In the last ten years, it has opened new and very promising perspectives for the relativistic LPI, and has made significant progresses in understanding and improving ignition physics. However, many issues associated with FI still need to be verified, such as inefficient laser-target energy coupling. Recent research has shown that the REBs usually have a large divergence angle, therefore, the main challenge is how to collimate fast electrons during the transport progress so that relatively modest ignition can be achieved. Another problem is how to extend the present scaling laws to the ignition scale laser facilities. As we know, the present scaling laws and laser coupling efficiency are obtained from the experiments on the PW laser facilities with the output energy smaller than 1 kJ and a pulse duration smaller than 1 ps while in the ignition scale the laser beams have to deliver more than 200 kJ in 20 ps. Since the laser energy and pulse duration play an important role in the generation of fast electrons, it is still a challenge to extend the physics from these hundreds-of-Joule laser facilities to the thousands of times of larger one (200 kJ). Further efforts associated with this issue will be carried out in the future. Shock ignition [78, 79] provides attractive alternatives that may sidestep the present difficulties with laser-driven electron-beam fast ignition. However, the laser-plasma interaction, energy transport and target hydrodynamics still need further investigations [80].

In IAPCM, significant progresses in laser drives construction, code development and FI physics understanding have been achieved since the FI project was formally started in 2009. The SG series laser facilities together with the PW heating laser, have been built and upgraded to satisfy the requirement for the demonstration of the feasibility of the FI scheme. A lot of experiments have been carried out to study the compression symmetry, M-band radiating preheating and mixing, fast electron generation and transport and advanced FI target design. The reliable PIC code, hybrid-PIC code, and radiation hydrodynamic code have been developed and checked. The physics of pellet compression, and the production, transport, and deposition of fast electrons in overdense plasmas have been studied with those codes. Especially, a number of ideas to control the REB divergence have been described by our PIC code and hybrid-PIC codes. Furthermore, the FI target design has been carried out with our hydrodynamic code and demonstrated in the experiments on the SG series laser facilities. As the next step, the coupling efficiency of FI will be investigated with our integrated simulation codes and the SG-IIU laser facility.

Acknowledgments

Due to limit pages of this paper, we only introduced the research works of fast ignition at the Institute of Applied Physics and Computational Mathematics(IAPCM), Laser Fusion Research Center(LFRC), and Shanghai Institute of Laser Plasma(SLIP). In China, there are several other groups in universities and institutes also working on FI and contributing a lot to FI community in China. We would like to thank them very much for their contribution.

We gratefully acknowledge the target fabrication, laser operation groups at LFRC and Shanghai Institute of Optics and Fine Mechanics (SIOM), and J Miao from SIOM for providing the photos of the SG-IIU laser facility. This work was performed under the auspices of the National Hi-Tech Inertial Confinement Fusion Committee of China. HBC is supported by the National Natural Science Foundation of China (Grant No. 11275028), SZW is supported by the National Natural Science Foundation of China (Grant No. 11175026), and CTZ is supported by the National Natural Science Foundation of China (Grant No. 91230205).

References

[1] Tabak M *et al* 1994 *Phys. Plasmas* **1** 1626
[2] Atzeni S *et al* 2008 *Phys. Plasmas* **15** 056311
[3] Honrubia J J and Meyer-ter-Vehn J 2009 *Plasma Phys. Control. Fusion* **51** 014008
[4] Basov N G, Gus'kov S Yu and Feoktistov L P 1992 *J. Sov. Laser Res.* **13** 396
[5] Naumova N *et al* 2009 *Phys. Rev. Lett.* **102** 025002
[6] Wu D *et al* 2013 *Phys. Plasmas* **20** 023102
[7] Qiao B *et al* 2013 *Phys. Rev.* E **87** 013108
[8] Roth M 2009 *Plasma Phys. Control. Fusion* **51** 014004
[9] Kemp A J *et al* 2014 *Nucl. Fusion* **54** 054002
[10] Robinson A P L *et al* 2014 *Nucl. Fusion* **54** 054003
[11] Norreys P *et al* 2014 *Nucl. Fusion* **54** 054004
[12] Key M H *et al* 2008 *Phys. Plasmas* **15** 022701
[13] National Research Council 2013 *An Assessment of the Prospects for Inertial Fusion Energy* (Washington, DC: National Academies Press)
[14] Theobald W *et al* 2011 *Phys. Plasmas* **18** 056305
[15] Azechi H *et al* 2009 *Nucl. Fusion* **49** 104024
[16] Shiraga H *et al* 2014 *Nucl. Fusion* **54** 054005
[17] Norreys P A *et al* 2000 *Phys. Plasmas* **7** 3721
[18] Beg F N *et al* 1997 *Phys. Plasmas* **2** 447
[19] Chen C D *et al* 2009 *Phys. Plasmas* **16** 082705
[20] Li Y T *et al* 2006 *Phys. Rev. Lett.* **96** 165003
[21] Zhuo H B *et al* 2014 *Phys. Rev. Lett.* **112** 225001
[22] Ramakrishna B *et al* 2010 *Phys. Rev. Lett.* **105** 135001
[23] Kodama R *et al* 2002 *Nature* **418** 933
Kodama R *et al* 2001 *Nature* **412** 798
[24] Ma T *et al* 2012 *Phys. Rev. Lett.* **108** 115004
[25] Robinson A P L *et al* 2008 *Phys. Rev. Lett.* **100** 025002
[26] Debayle A, Honrubia J J, D'Humieres E and Tikhonchuk V T 2010 *Phys. Rev.* E **82** 036405
[27] Van Woerkom L *et al* 2008 *Phys. Plasmas* **15** 056304
[28] Bret A, Gremillet L, Benisti D and Lefebvre E 2008 *Phys. Rev. Lett.* **100** 205008
[29] Patel P 2010 *10th Workshop on Fast Ignition and Fusion Target* (*Shanghai, China, 17–21 October 2010*)
[30] Cai H B, Zhu S P, Chen M, Wu S Z, He X T and Mima K 2011 *Phys. Rev.* E **83** 036408
[31] Zheng C Y, Zhu S P and He X T 2005 *Phys. Plasmas* **12** 044505
[32] Pei W B *et al* 2007 *Commun. Comput. Phys.* **2** 255
[33] Wu J F *et al* 2014 *Phys. Plasmas* **21** 042707
[34] Zheng W D and Zhang G P 2007 *Chin. Phys.* **16** 2439
[35] Cao L H, Chang T Q, Pei W B, Liu Z J, Li M and Zheng C Y 2008 *Plasma Sci. Technol.* **10** 18
[36] Cao L H, Pei W B, Liu Z J, Chang T Q, Li B and Zheng C Y 2006 *Plasma Sci. Technol.* **8** 269
[37] Cao L H *et al* 2014 *Plasma Sci. Technol.* **16** 1007
[38] Zhou C T, He X T and Yu M Y 2008 *Appl. Phys. Lett.* **92** 071502
[39] Zhou C T and He X T 2008 *Phys. Plasmas* **15** 123105
[40] Zhou C T *et al* 2010 *Phys. Plasmas* **17** 083103
[41] Zhou C T *et al* 2012 *Laser Part. Beams* **30** 111
[42] Mo Z Y *et al* 2010 *Front. Comput. Sci. China* **4** 480
[43] Welch D R, Rose D V, Clark R E, Genoni T C and Hughes T P 2004 *Comput. Phys. Commun.* **164** 183
[44] Davies J R 2002 *Phys. Rev.* E **65** 026407
[45] Brown L S, Preston D L and Singleton R L Jr 2005 *Phys. Rep.* **410** 237
[46] Cao L H, Chen M, He X T, Yu W and Yu M Y 2012 *Phys. Plasmas* **19** 044503
[47] Wu S Z *et al* 2011 *Phys. Plasmas* **18** 022703
[48] Wu S Z, Zhang H, Zhou C T, Zhu S P and He X T 2013 *EPJ Web Conf.* **59** 05021
[49] Kar S *et al* 2009 *Phys. Rev. Lett.* **102** 055001
[50] Robinson A P L and Sherlock M 2007 *Phys. Plasmas* **14** 083105
[51] Strozzi D J, Tabak M, Larson D J, Divol L, Kemp A J, Bellei C, Marinak M M and Key M H 2012 *Phys. Plasmas* **19** 072711
[52] Fujioka S *et al* 2013 *Sci. Rep.* **3** 1170
Mima K 2012 *Reports in 12th Int. Workshop on Fast Ignition of Fusion Target* (*Napa Valley, CA, 4–8 November 2012*)
[53] Cai H B *et al* 2010 *Phys. Plasmas* **17** 023106
[54] Baton S D *et al* 2008 *Phys. Plasmas* **15** 042706
[55] MacPhee A G *et al* 2010 *Phys. Rev. Lett.* **104** 055002
[56] Cai H B *et al* 2009 *Phys. Rev. Lett.* **102** 245001
[57] Nakamura T *et al* 2007 *Phys. Plasmas* **14** 103105
[58] Wu S Z *et al* 2014 in preparation
[59] Robinson A P L, Key M H and Tabak M 2012 *Phys. Rev. Lett.* **108** 125004
[60] Cai H B *et al* 2011 *Phys. Plasmas* **18** 023106
[61] Cai H B, Zhu S P, He X T and Mima K 2013 *EPJ Web Conf.* **59** 17017
[62] Wu S Z, Zhou C T and Zhu S P 2010 *Phys. Plasmas* **17** 063103
[63] Akli K U *et al* 2012 *Phys. Rev.* E **86** 065402
[64] Stephens R B *et al* 2005 *Phys. Plasmas* **12** 056312
[65] Pasley J and Stephens R 2007 *Phys. Plasmas* **14** 054501
[66] Caruso A and Strangio C 2003 *JETP* **97** 948
[67] Zhou W M *et al* 2014 submitted
[68] Cai H B *et al* 2014 *High Power Laser Sci. Eng.* **2** 1
[69] Zhou W M *et al* 2013 *High Power Laser Part. Beams* **25** 3135
[70] Wang W W *et al* 2015 submitted
[71] Cai H B, Zhu S P and He X T 2013 *Phys. Plasmas* **20** 072701
[72] Cai H B, Yu W, Zhu S P and Zheng C Y 2006 *Phys. Plasmas* **13** 113105
[73] May J *et al* 2011 *Phys. Rev.* E **84** 025401
[74] Jia Q *et al* 2015 *Phys. Rev.* E at press
[75] Jia Q, Cai H B, Wang W W, Zhu S P, Sheng Z M and He X T 2013 *Phys. Plasmas* **20** 032113
[76] Wang W M *et al* 2015 *Phys. Rev. Lett.* **114** 015001
[77] Bi B *et al* 2014 *High Power Laser Part. Beams* **26** 092002
[78] Shcherbakov V A 1983 *Sov. J. Plasma Phys.* **9** 240
[79] Betti R *et al* 2007 *Phys. Rev. Lett.* **98** 155001
[80] Batani D *et al* 2014 *Nucl. Fusion* **54** 054009

流体力学不稳定性和可压缩流体湍流

流体力学不稳定性引起的界面物质混合在激光聚变和核武器内爆动力学以及天体超新星爆炸等问题中是十分重要的现象，发展后期会演化成可压缩流体湍流，通常还伴随着冲击波的作用。

贺贤土院士提出流体力学不稳定性与可压缩流体湍流耦合是影响界面多相流物质混合和能量传输的关键因素。他与合作者及学生在理论上解析证明了一个界面扰动的 Rayleigh-Taylor 不稳定性（Rayleigh-Taylor instability，RTI）增长能够反馈传播和诱导另一界面 RTI，即使后者不存在任何初始扰动；他们还发现了汇聚内爆界面 RTI 过程发生尖峰（spike）减速和气泡加速现象等。这可能用于解释为什么流体力学不稳定性过程会引发层破裂现象等等。

同时，贺贤土院士与合作者进行了深入的可压缩流体湍流及其与流体力学不稳定性耦合关系的研究，发现 RTI 引起的相干涡旋结构进入冲击波阵面后被破碎为大量无规运动小斑图，这被认为是可压缩流体湍流基元。此时，压缩产生的胀量与涡矢量耦合支配了可压缩湍流的演化。他们所做的大量三维模拟揭露了胀量和涡量等值面的复杂空间分布，并获得了高 Mach 数下的概率密度函数特性，为理解界面多相流特性和界面不同物质混合效应提供了重要信息。这些工作引起了国际同行的高度关注和评价。

第11卷 第5期
1999年10月
强 激 光 与 粒 子 束
HIGH POWER LASER AND PARTICLE BEAMS
Vol. 11, No. 5
Oct., 1999

文章编号:1001-4322(1999)05-0613-06

激光烧蚀瑞利-泰勒不稳定性数值研究*

叶文华,张维岩,陈光南,贺贤土
(北京应用物理与计算数学研究所,计算物理实验室,北京8009信箱,100088)

摘 要: 介绍了激光烧蚀流体不稳定性计算程序EUL3D,其计算结果与Takabe公式、FAST2D程序和LASNEX程序的结果以及日本大阪大学激光烧蚀瑞利—泰勒(RT)不稳定性实验,都较好符合,发现了横向电子热传导烧蚀在长波长扰动的非线性瑞利—泰勒不稳定性演变中起重要作用。在合理近似下,得到了烧蚀RT不稳定性线性增长率的预热致稳公式。此公式除包含了烧蚀对流致稳和密度梯度致稳因素外,还包含了Atwood数变小致稳因素,因此与各种情况的二维计算值都很好符合。

关键词: 瑞利-泰勒不稳定性;不稳定性致稳;高精度格式;活动网格
中图分类号:TN241　　**文献标识码:**A

为成功实现ICF的点火,通常要求烧蚀瑞利-泰勒不稳定性RTI有足够的致稳,即线性增长率降至经典值的一半左右[1]。目前美国国家点火装置(NIF)上的直接驱动点火靶设计已进入最后攻坚阶段,但烧蚀RT不稳定性的增长仍是点火靶设计中最不确定的因素。近年日本大阪大学和美国利弗莫尔实验室(LLNL)相继进行了激光烧蚀RT不稳定性线性增长率的实验测量[3,4],测量值与Takabe公式预言的数值明显不符,小于美国LASNEX程序Spitzer-Harm(SH)电子热传导[6]计算值的18%,这意味着直接驱动点火所要求的驱动激光器能量可大为减少。快电子的预热能较好解释致稳的原因。高能X光的预热对X光烧蚀RT不稳定性也同样产生致稳作用[7]。

我们采用高精度FCT(Flux Corrected-Transport)流体算法,解决了电子热传导和高精度FCT的耦合计算问题,研制了激光烧蚀RT不稳定性程序EUL3D[8,9]。其计算结果与国外的实验和计算结果都较好符合。通过深入分析预热致稳的计算结果,发现了Atwood变小的致稳作用,并得到了新的RTI线性增长率公式。

1 物理方程和计算方法

数值模拟使用非均匀和活动网格的激光烧蚀RTI程序EUL3D,它是以前EUL2D程序的三维版本。该程序采用分裂格式算法分开计算流体和热传导。流体计算采用六阶相位误差的FCT算法。多维热传导方程计算采用局部一维追赶的分裂格式算法。EUL3D程序考虑激光的逆轫致吸收和临界面附近的共振吸收。采用局域的SH电子热传导公式进行计算时,对电子热传导流进行限流。多数计算采用理想气体状态方程,全电离和单温近似。

流体不稳定性非线性发展产生大的剪切流,拉氏计算方法会遇到网格相交的严重困难。因此我们采用活动网格欧拉计算方法。不稳定性计算要求较高的网格分辨率,必须采用高精度欧拉算法。网格划分的不同对不稳定性计算结果有严重影响。我们用密度峰值处速度或靶心速

* 国家自然科学基金及国家863惯性约束聚变领域资助课题。
1999年6月15日收到原稿,1999年9月10日收到修改稿。第五届全国激光科学技术青年学术交流会优秀论文
叶文华,男,1959年6月出生,硕士,副研究员

度追踪烧蚀面，平移网格板块。烧蚀面附近均匀密分网格，以减少对流计算误差，两边网格逐渐放大，但放大因子不超过1.1。纵向计算边界尽量放得远些。激光吸收区和电子热传导区要保持合适的网格宽度。

2 与RTI线性增长率公式和国外数值结果的比较

与Takabe公式[5]、FAST2D程序[10]和LASNEX程序[11]计算结果比较。激光烧蚀RTI线性增长率的Takabe公式为$\gamma=0.9\sqrt{kg}-3kV_a$。式中$g$为加速度，$k$为波矢大小，$V_a$为烧蚀速度。图1(a)和图(b)给出了不同扰动波长的RTI线性增长率计算值与经典值($\gamma=\sqrt{kg}$)的比值。图1(a)为100μm CH靶，与文献[10]FAST2D程序计算结果比较；图1(b)为20μm CH靶，与文献[11]LASNEX程序计算结果比较。EUL2D程序计算的RTI线性增长率与Takabe公式、FAST2D程序和LASNEX程序的计算值较好符合。

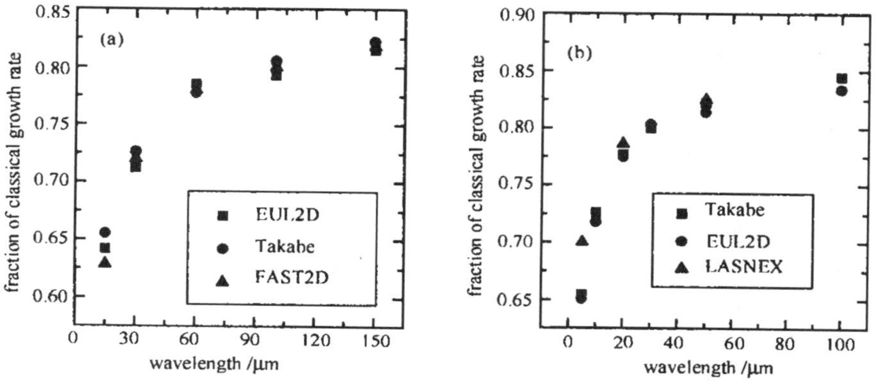

Fig. 1 Comparison of RTI growth rates. a. EUL2D、FAST2D and the Takabe formula. b. EUL2D. LASNEX and the Takabe formula

图1 不同扰动波长RTI线性增长率的计算值与经典值的比较

3 数值模拟大阪大学激光烧蚀RTI实验

日本大阪大学激光烧蚀RTI实验[12]条件：25μm CH靶，初始扰动振幅3μm，扰动波长100μm，入射激光波长0.53μm，峰值激光功率2×10^{14}W/cm^2。图2表明EUL2D程序的计算结果再现了大阪大学激光烧蚀RTI的实验结果。不仅在时间尺度上，而且在尖顶和气泡的空间尺度上，两者都较好符合。特别是当横向分点加密时，EUL2D程序计算得到了由于横向电子热传导烧蚀产生的尖刺图象，这在公开发表的文献中未见有过报道。这一结果表明横向电子热传导烧蚀在长波长扰动的不稳定性非线性演变中起着重要作用。

4 激光烧蚀RTI预热致稳研究

美国利弗莫尔实验室近来进行了激光烧蚀RT不稳定性实验，但线性增长率的LASNEX程序计算值平均大于实验值18%[4]。我们采用电子热传导公式$K_e=K_{nloc}f(T)$，数值模拟利弗莫尔的实验，式中$K_{nloc}=K_{SH}/[1+A\ln(T_{max}/T_0)]$[13]是近似非局域电子预热的热传导系数，$K_{SF}$是SH电子热传导系数，$A$取为1，$T_{max}$、$T_0$分别为靶最高温度和靶前温度。计算中$T_0$取为2.5eV。直接的电子FP方程的计算结果表明快电子能预热烧蚀面附近的区域，引入温度的分段阶梯函数$f(T)$，以增大低温电子热传导系数，从而数值模拟烧蚀面附近的RT不稳定性预热致稳效应。当计算中取 $f(T)=1$ 时，预热作用不大，线性增长率的二维计算值明显地高于实验值；当在低温区增大 $f(T)$ 时，靶密度分布与利弗莫尔实验室的 BGK 输运计算的结果接近，线性增长率的二维计算值与实验值较好符合。

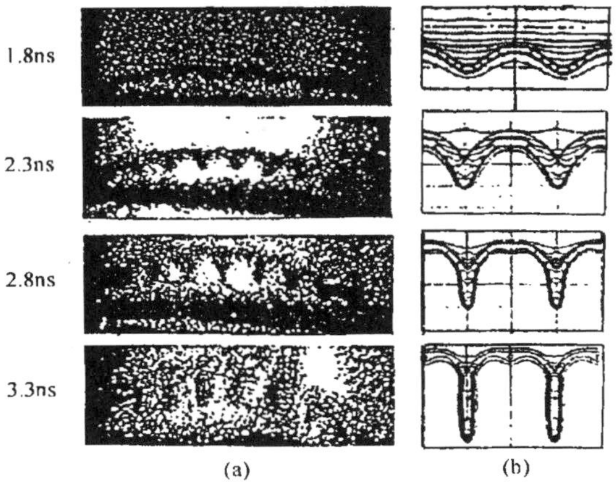

Fig. 2　Evolution of nonlinear RTI. (a) experiment of Osaka University, (b) simulation by the EUL2D code

图 2　激光烧蚀非线性 RTI 的演变

(a) 大阪大学实验室，(b) EUL2D 计算

在不可压缩、无粘和无热传导流体近似下，线性化流体方程寻求 $e^{ik_x x+ik_y y+\gamma t}$ 形式的解，可得到关于 RT 不稳定性线性增长率的二阶本征常微分方程。假定烧蚀面呈指数密度分布，密度梯度定标长度为 L。使用经典 RT 不稳定性的本征函数 $\exp(|k|x)$，可从二阶本征常微分方程直接得到线性增长率的解析表达式。考虑烧蚀对流致稳因素后，我们得到了预热情况的烧蚀 RT 不稳定性线性增长率公式 $\gamma=\sqrt{Akg/(1+AkL)}-2kV_a$。此公式包含了密度梯度致稳、Atwood 数变小致稳和烧蚀对流致稳，在 Atwood 数等于 1 时退化为 Lindl 公式，在陡烧蚀面情况与 Takabe 公式接近。Lindl 公式没有考虑 Atwood 数变小的致稳作用，所以只适用于 Atwood 数接近 1 的情况。

图 3 给出了增大低温电子热传导系数的二维计算结果(NLC 曲线)与实验测量值及不同公式拟合值的比较。Takabe 曲线和 Lindl 曲线是根据一维计算结果分别进行 Takabe 公式和 Lindl 公式拟合的结果，Measured 曲线是利弗莫尔实验室的实验测量结果，New form 曲线是我们新公式的拟合结果。NLC 曲线与利弗莫尔实验室的测量值较好符合，明显与 Takabe 曲线不符，与 Lindl 公式也有较大偏差。New form 曲线与 NLC 曲线相当好地符合。

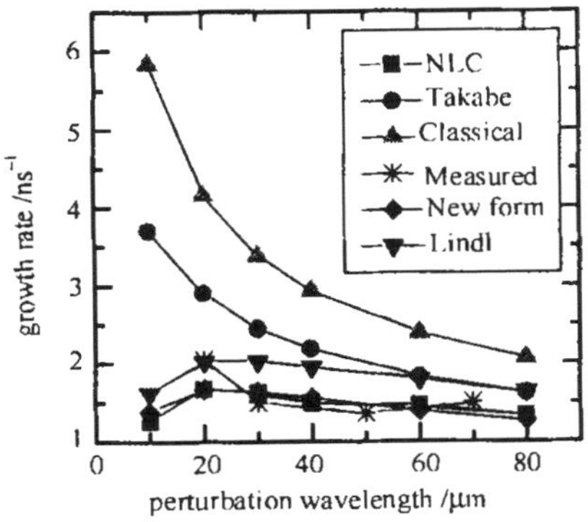

Fig. 3　Comparison of RT growth rates for 20μm DH foil

图 3　20μm CH 靶薄膜激光烧蚀 RT 不稳定性线性增长率比较

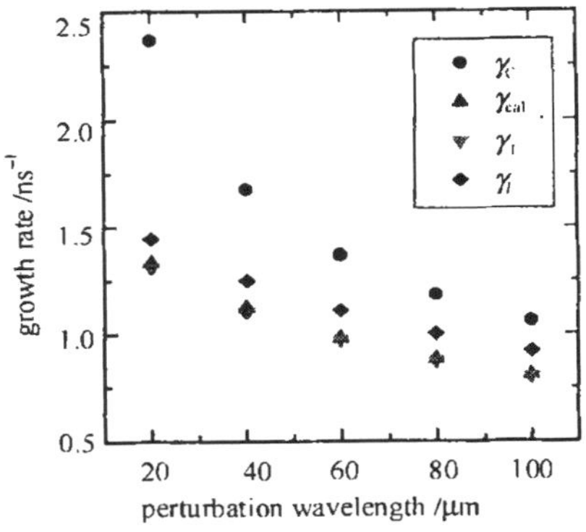

Fig. 4　Comparison of RT growth rates by 2D simulation and different formula

图 4　不同扰动波长的线性增长率二维计算值与不同公式拟合值的比较，$L=3.23\mu m$，$A=0.727$

我们数值研究不同预热情况的烧蚀 RT 不稳定性增长。计算条件为：100μm CH 靶，初始

密度 1.0g/cm^3，激光波长 0.53μm，激光功率线性上升 3ns 到峰值 3×10^{14}W/cm^2，然后维持峰值功率不变，扰动波长 40μm。先进行一维计算至 5ns 时刻，此时流体基本达稳态，然后记下一维计算结果，在烧蚀面附近加入密度扰动，进行二维烧蚀 RT 不稳定性的数值计算。

JJ1：

$$f(T)=1;$$

JJ2：

$$f(T)=\begin{cases}1.0, & T<43\text{eV}\\ 2.5, & 43\text{eV}<T<86\text{eV}\\ 1.5 & 86\text{eV}<T<172\text{eV}\\ 1.0, & T>172\text{eV}\end{cases}$$

JJ3 ：

$$f(T)=\begin{cases}1.0, & T<8.6\text{eV}\\ 1500, & 8.6\text{eV}<T<17.2\text{eV}\\ 300, & 17.2\text{eV}<T<43\text{eV}\\ 2.5, & 43\text{eV}<T<86\text{eV}\\ 1.5, & 86\text{eV}<T<172\text{eV}\\ 1.0, & T>172\text{eV}\end{cases}$$

JJ4：

$$f(T)=\begin{cases}1.0, & T<0.86\text{eV}\\ 10000, & 0.86\text{eV}<T<4.3\text{eV}\\ 4000, & 4.3\text{eV}<T<8.6\text{eV}\\ 1500, & 8.6\text{eV}<T<17.2\text{eV}\\ 300, & 17.2\text{eV}<T<43\text{eV}\\ 2.5, & 43\text{eV}<T<86\text{eV}\\ 1.5, & 86\text{eV}<T<172\text{eV}\\ 1.0, & T>172\text{eV}\end{cases}$$

JJ5：限流 SH 电子热传导，限流因子为 0.05；JJA：$f(T)=200/T^{3/2}\rho$，1eV$<T<$43eV 和 $r>r_p-12\mu$m，r_p 为密度峰值位置；JJB：$f(T)=3000/\rho$，1eV$<T<$43eV，$r>r_p-12\mu$m。

表 1 中 γ_C、γ_{cal}、γ_A 和 γ_L 分别为 RT 不稳定性线性增长率的经典值、二维计算值、新的线性增长率公式拟合值和 Lindl 公式拟合值。表 1 数据表明：JJ1、JJ2 和 JJ5 这三个模型的线性增长率都在经典值的 80%以上，只有较小的不稳定性致稳。这三个模型的密度峰值 ρ_2 都较高（大于 7g/cm^3），烧蚀面台阶下面的密度 ρ_1 很小，接近零，所以 Atwood 数 $A\approx1$。这三个模型的烧蚀面密度梯度定标长度 L 较小，因此密度梯度致稳作用较弱，由新公式或 Lindl 公式拟合的增长率在数值上接近，也与二维计算值符合。当逐渐增大低温电子热传导系数时，如从 JJ2 到 JJ3 和 JJ4，烧蚀 RT 不稳定性线性增长率二维计算值逐渐降低，由 Lindl 公式拟合的增长率与二维计算值逐渐发生偏离，但由新公式拟合的增长率仍与二维计算值很好符合。对于 JJ3、JJ4、JJA 和 JJB 这四个模型，线性增长率的 Lindl 公式拟合值与二维计算值出现较大偏差。出现偏差的原因不难理解：一方面，预热增大了靶的烧蚀，烧蚀面台阶底部的ρ_1值从不增大低温电子

表 1 不同预热情况的烧蚀 RT 不稳定性线性增长率比较

Table.1 Comparison of ablative RT growth rates at different preheat case

	L/μm	A	γ_C/ns^{-1}	γ_{cal}/ns^{-1}	γ_A/ns^{-1}	γ_L/ns^{-1}	γ_{cal}/γ_C	$\left\|\frac{\gamma_{cal}-\gamma_A}{\gamma_{cal}}\right\|$	$\left\|\frac{\gamma_{cal}-\gamma_L}{\gamma_{cal}}\right\|$
JJ1	1.32	0.974	1.626	1.415	1.381	1.397	0.870	0.024	0.013
JJ2	1.53	0.974	1.639	1.365	1.371	1.387	0.833	0.004	0.016
JJ3	2.78	0.757	1.735	1.228	1.207	1.346	0.708	0.017	0.096
JJ4	4.04	0.706	1.676	1.082	1.059	1.200	0.646	0.021	0.109
JJ5	0.96	0.974	1.827	1.656	1.585	1.604	0.906	0.043	0.031
JJA	3.23	0.727	1.672	1.125	1.107	1.249	0.673	0.016	0.110
JJB	2.27	0.625	1.662	1.032	1.059	1.298	0.621	0.026	0.258

热传导系数的接近零上升到 1g/cm^3 左右，另一方面，预热使烧蚀面内声速增大，靶的稀疏加快，靶峰值密度降低，所以烧蚀面 Atwood 数 A 明显变小，这四个模型的烧蚀面 Atwood 数都降到了 0.76 以下，出现了由界面 Atwood 数变小而产生的不稳定性致稳现象。因此线性增长率公式中必须考虑这种 Atwood 数变小的致稳效应。特别是 JJB 模型，虽然 L 仅为 2.27μm，密度梯度致稳不算大，但 A 降到了 0.625，结果线性增长率的二维计算值最小，Lindl 公式拟合值与二维计算值产生 25%的偏差。可见 Lindl 公式仅适用于 A 接近 1 的情况。

图 4 给出了不同扰动波长的线性增长率的二维计算值与不同拟合公式的比较。计算条件与 JJA 模型的相同。图中清楚表明各种波长扰动的线性增长率的 Lindl 公式拟合值与二维计算值有明显偏差，大于 10%，而新公式的拟合值与二维计算值相当好地符合。因此在 A 明显小于 1 的情况，除了要考虑密度梯度致稳作用外，还要考虑 Atwood 数变小的致稳作用。

5　结论

我们解决了活动欧拉网格下电子热传导和高精度 FCT 流体的耦合计算问题，研制了激光烧蚀流体不稳定性计算程序 EUL3D，其计算结果与 Takabe 公式、FAST2D 程序和 LASNEX 程序的结果较好符合。EUL3D 程序的计算再现了日本大阪大学激光烧蚀 RTI 的实验结果。我们发现横向电子热传导烧蚀在长波长扰动的不稳定性非线性演变中起着重要作用。当增大低温电子热传导系数时，烧蚀 RT 不稳定性线性增长率的二维计算值与利弗莫尔实验室的实验值较好符合。数值研究预热致稳作用，发现 Lindl 公式与二维计算值有明显偏差，Atwood 数变小产生的致稳作用是造成偏差的主要原因。在合理近似下，得到了烧蚀 RT 不稳定性线性增长率的预热致稳公式 $\gamma=\sqrt{Akg/(1+AkL)}-2kV$。此公式与各种情况的二维计算值都很好符合，在 Atwood 数等于 1 时退化为 Lindl 公式，在陡烧蚀面情况与 Takabe 公式接近。

参考文献

1 Kilkenny J, et al. *Phys Plasmas*, 1994, 1(5): 1379

2 Lindl J. *Phys Plasmas*, 1995, **2**(11): 3933

3 Shigemori K, et al. *Phys Rev Lett*, 1997, **78**: 250

4 Glendinning S, et al. *Phys Rev Lett*, 1997, **78**: 3318

5 Takabe H, et al. *Phys. Fluids*, 1983, **26**: 2299; Takabe H, et al. *ibid*, 1985, **28**: 3676

6 Spitzer L and Harm R. *Phys Rev*, 1953, **89**: 977

7 Weber S V, et al. *Phys Plasmas*, 1994, 1(11): 3652

8 叶文华等. 计算物理, 1998, **15**(3): 277

9 叶文华等. 强激光与粒子束, 1998, **10**(3): 403

10 Gardner J H. et al. *Phys Fluids*, 1991, **B3**: 1070

11 Tabak M. *Phys Fluids*, 1990, **B2**(5): 1070

12 Takabe H J. *Plasma Fusion Res*, 1997, **73**: 147～161. 313～329, 395～410

13 Xu Y and He X T. *Phys Rev*, 1994, **E50**: 443

NUMERICAL STUDY OF LASER ABLATIVE RAYLEIGH-TAYLOR INSTABILITY

YE Wen-hua, ZHANG Wei-yan, CHEN Guang-nan, and HE Xian-tu
Institute of Applied Physics and Computational Mathematics,
Laboratory of Computational Physics, P. O. Box 8009, *Beijing* 100088

ABSTRACT: Physical equations, numerical methods and slide grids of laser ablative hydrodynamic instability in the code EUL3D are introduced. Simulation results of the EUL3D code are good agreements with those of the Takabe formula, the FAST2D code and the LASNEX code, and the experimental result of laser ablative Rayleigh-Taylor instability in Osaka University. The lower Atwood stabilization due to the preheat is observed in simulations and a new growth rate formula is obtained. It includes the density gradient, the lower Atwood number and the ablative convection stabilizations, so it agrees well with all of 2D simulations.

KEY WORDS: Rayleigh-Taylor instability; instability stabilization; high-resolution scheme; slide grids

NOTES

Instability of relativistic electron beam with strong magnetic field

LIU Zhijing[1], H.L. Berk[2] & HE Xiantu[3]

1. Department of Astronomy and Applied Physics, University of Science and Technology of China, Hefei 230026, China;
2. Institute for Fusion Studies, University of Texas at Austin, Austin, TX 78712-1060, USA;
3. Institute for Applied Physics and Computational Mathematics, Beijing 100088, China

Abstract Stability criteria for a weak relativistic beam-plasma interaction in a strong magnetic field are found. Two beam modes occur, $\omega \approx k_z c$ and $\omega \approx k_z c - \omega_c$. The dispersion equation of electrostatic two-stream is exactly solved with analytical method.

Keywords: instability, relativistic electron beam, two-stream, strong magnetic field, dispersion relation.

Two-stream instability appears in magnetic confinement devices and laser fusion as well as relativistic electron beam fusion. It plays an important role in ion acceleration and ion compression as well as the ignition of target. In order to make full use of the instability and further control it, it is necessary to understand clearly the relationship between the instability and various parameters and the existent region of the instability.

The relativistic two-stream instability has been investigated theoretically in the limit of zero magnetic field[1], but there does not appear to be an analysis in the literature for this mode in a strong magnetic field. In this work we analyze the fluid equations for the beam and background electrons and delineate the stability parameters and growth rates. We find that electromagnetic effects are important over a wide range of parameters. We also discuss electrostatic two-stream instability excited by relativistic cold electron beam. Although the dispersion relation was discussed by many a researcher in the past, only a few of its approximative solutions were obtained[2—6]. In this note we also solve exactly the dispersion relation of electrostatic two-stream by using analytical method and calculate numerically the growth rate.

1 The dispersion relation in strong magnetic field

Both the beam and background plasma are described in the pressureless fluid limits where the beam speed is taken as v_b along the z directed magnetic field B_z while the background electron speed is zero. The self field of the beam is assumed to be much less than B_z and is neglected in this analysis. Similarly, toroidal effects are neglected. We assume ω, $\omega_{pe} \ll \gamma\omega_c$, where $\omega_{pe} \equiv (4\pi n_e e^2/m)^{1/2}$, $\gamma\omega_c \equiv eB_z/mc$, $\gamma = (1 - v_b^2/c^2)^{-1/2}$. n_e is the cold plasma density, m the electronic rest mass, $\gamma\omega_c$ the nonrelativistic cyclotron frequency, v_b the beam velocity. In this limit we can neglect the transverse dynamics of the cold plasma. However, we will include the approximation transverse dynamics of the beam, which is a consistent approximation if $\gamma \gg 1$. The perturbed equations are the form1).

1) Berk, H. L., Relativistic beam plasma instability in strong magnetic field, Report in Lawrence Livermor Lab., 1972, preprint 1—24.

$$\begin{cases}\left(\dfrac{\partial}{\partial t}+v_b\dfrac{\partial}{\partial z}\right)\delta n_b+\nabla\cdot(n_b\delta \boldsymbol{v}_b)=0,\\ \left(\dfrac{\partial}{\partial t}+v_b\dfrac{\partial}{\partial z}\right)(\gamma\delta \boldsymbol{v}_b)=\dfrac{e}{m}\left(\boldsymbol{E}+\dfrac{v_b}{c}\hat{z}\times\delta\boldsymbol{B}+\dfrac{\delta \boldsymbol{v}_b}{c}\times\boldsymbol{B}_z\right),\\ \dfrac{\partial}{\partial t}\delta n_e+\dfrac{\partial}{\partial z}(n_e\delta v_{ez})=0,\\ \dfrac{\partial}{\partial t}\delta v_{ez}=\dfrac{e}{m}E_z,\\ -\dfrac{1}{c}\dfrac{\partial}{\partial t}\delta\boldsymbol{B}=\nabla\times\boldsymbol{E},\\ \nabla\times\delta\boldsymbol{B}=\dfrac{1}{c}\dfrac{\partial}{\partial t}\boldsymbol{E}+4\pi e[(n_e\delta v_{ez}+\delta n_b v_b)\hat{z}+\delta n_b\delta \boldsymbol{v}_b].\end{cases} \tag{1}$$

We consider perturbations to be of the form $\hat{f}\exp[-i(\omega t-k_z z-k_x x)]$. From the perturbed equations of mass and motion and electromagnetic field we obtain the following dispersion relation:

$$\left(1-\frac{\omega_{pe}^2}{\omega^2}-\frac{\omega_b^2}{\gamma^2\tilde{\omega}^2}\right)(\omega^2-k_z^2c^2-\tilde{\omega}^2\Lambda)-k_x^2c^2(1-\Lambda/\gamma^2)=0, \tag{2}$$

where

$$\Lambda=\frac{\omega_B^2}{\tilde{\omega}^2-\omega_c^2}[(k^2c^2-\omega^2+\omega_b^2)/(k^2c^2-\omega^2+\omega_b^2\tilde{\omega}^2/(\tilde{\omega}^2-\tilde{\omega}_c^2))],$$

$$\tilde{\omega}=\omega-k_z v_b,\ \omega_b=(4\pi n_b e^2/m\gamma)^{1/2}.$$

2 Analysis of instability

To analyze the stability of eq. (2), we assume $\omega_b \ll \omega_{pe}$, ω_c. Therefore in eq. (2) we need only keep terms linear in ω_b^2. Hence we take $\Lambda=\omega_b^2/(\tilde{\omega}^2-\omega_c^2)$. The interesting regimes for stability then occur near the resonances where (A) $\tilde{\omega}=0$ and (B) $\tilde{\omega}^2=\omega_c^2$. For case (A) we can set $\Lambda=0$ and for case (B) we can set $\omega_b^2/\gamma^2\tilde{\omega}^2=0$.

(i) For case (A) $\tilde{\omega}\approx 0$. (2) can be approximated as

$$k_z^2=\omega_{pe}^2/\xi^2+\omega_b^2/[\gamma^2(\xi-v_b)^2]+k_x^2c^2/(\xi^2-c^2), \tag{3}$$

where $\xi=\omega/k_z$. If $k_x^2=0$, eq. (3) is unstable. Thus we have

$$1-\omega_{pe}^2/\omega^2-\omega_b^2/[\gamma^2(\omega-k_2 v_b)^2]=0. \tag{4}$$

This is a purely electrostatic case. It will be discussed in sec. 3 for cold plasma beam.

If we choose $\omega=\omega_{pe}+\delta\omega$ and $\omega_{pe}/k_z=v_b$, we then find from (2), that the unstable root is given by

$$\delta\omega=2^{-4/3}\frac{\omega_{pe}}{\gamma}\left(\frac{n_b}{n_e}\right)^{1/3}(-1+i\sqrt{3}). \tag{5}$$

If k_x^2 is sufficiently large, eq. (3) is stable (see fig. 1). In fig. 1 typical curves are shown. For $0<\xi<v_b$, two possibilities are evident, either the minimum occurs for $k_z^2>0$ (solid curve) or for $k_z^2<0$ (dotted curve).

Marginal stability corresponds to the minimum lying on the ξ axis. To find the stability let us assume $\omega_b \ll \omega_{pe}$, and $\gamma \gg 1$. Therefore $\xi_{min}\approx v_b\approx c$. We can write eq. (3) approximately as

$$k_z^2 = \omega_{pe}^2 / v_b^2 + \omega_b^2 / [\gamma^2 (\xi - v_b)^2] + k_x^2 c / [2(\xi - c)]. \quad (6)$$

Stability occurs for (a) $k_x > \frac{\omega_{pe}}{\gamma c}$, if $\gamma^2 \omega_b^2 < \omega_{pe}^2 / 8$, (b) $k_x > 2 \frac{\omega_{pe}}{\gamma^{3/4} c} \left(\frac{n_b}{n_e} \right)^{1/4}$, if $\gamma^2 \omega_b^2 >> \omega_{pe}^2 / 4$.

(ii) For case (B) $\tilde{\omega} \approx \omega_c, \omega_b << \gamma\omega_c$, eq. (2) can be approximated as

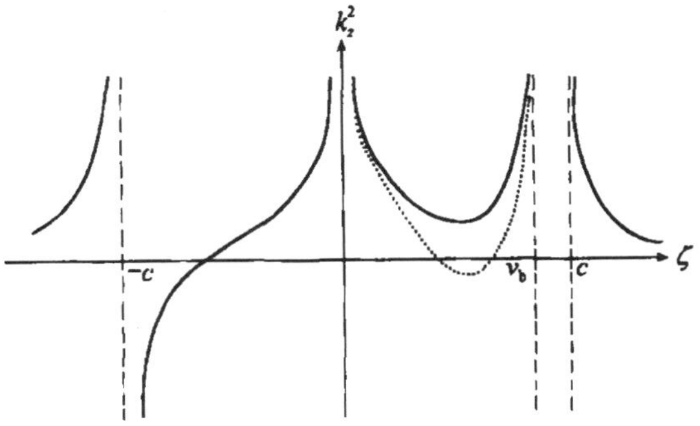

Fig. 1. The curves for k_z^2 vs. ξ.

$$\omega^2 - (\omega_{pe}^2 + k^2 c^2) + k_z^2 c^2 \omega_{pe}^2 / \omega^2 - \frac{\omega_b^2}{\tilde{\omega}^2 - \omega_c^2} \left[\tilde{\omega}^2 \left(1 - \frac{\omega_{pe}^2}{\omega^2} \right) - \frac{k_x^2 c^2}{\gamma^2} \right] = 0. \quad (7)$$

Let us choose ω_0 as the solution of eq. (7) when ω_b^2 is completely neglected. We then obtain

$$\omega_0^2 = \frac{1}{2}(k^2 c^2 + \omega_{pe}^2) \pm \frac{1}{2}[(k^2 c^2 + \omega_{pe}^2)^2 - 4\omega_{pe}^2 k_z^2 c^2]^{1/2}. \quad (8)$$

To simplify the results we use the following approximate relations for ω_0

$$\omega_0^2 = k_z^2 c^2 \omega_{pe}^2 / (\omega_{pe}^2 + k^2 c^2) \text{ and } \omega_0^2 = \omega_{pe}^2 + k^2 c^2. \quad (9)$$

They are pure oscillatory solutions. In the limit $\omega_{pe} << \omega_c$, it is clear that $\omega_0 << \omega_c$, $k_z v_b \approx \omega_c$. Since $kc > k_z v_b \approx \omega_c >> \omega_{pe}$, we have $\omega_0 = k_z \omega_{pe} / k$ which is an electrostatic oscillation.

Instability can be found for $k^2 > \omega_c^2 / v_b^2$. When $k v_b / \omega_c = \sqrt{3}$, the maximum growth rate ω_I is

$$\omega_I = 3^{-1/4} \left(\frac{1}{6} \right)^{1/2} \omega_b \left(\frac{\omega_{pe}}{\omega_c} \right)^{1/2}. \quad (10)$$

Now let us investigate the other limit $\omega_c < \omega_{pe} < \gamma\omega_c$. When $[1+(k^2c^2/\omega_{pe})]=4$, the maximum growth rate ω_I is then

$$\omega_I = \frac{3^{1/2}}{4} \omega_b. \quad (11)$$

If we combine eq. (10) and eq. (11), we find

$$\frac{\omega_I}{\omega_b} = \min \left[3^{-1/4} 6^{-1/2} \left(\frac{\omega_{pe}}{\omega_c} \right)^{1/2}, \frac{3^{1/2}}{4} \right]. \quad (12)$$

3 Electrostatic two-stream instability on cold relativistic electron beam

Let us set $u = v_b, \omega_{Be}^2 = \gamma\omega_b^2 = \frac{4\pi n_b e^2}{m}, \gamma_1 = \frac{1}{\gamma} = (1 - u^2/c^2)^{1/2}, k_z = k$. Thus, from eq. (4) we have

$$1 - \frac{\omega_{pe}^2}{\omega^2} - \frac{\gamma_1^3 \omega_{Be}^2}{(\omega - ku)^2} = 0. \quad (13)$$

Eq. (13) can be written as an algebraic equation of fourth order for ω

$$\omega^4 + a_1 \omega^3 + a_2 \omega^2 + a_3 \omega + a_4 = 0, \quad (14)$$

in which

$$a_1 = -2ku, a_2 = k^2u^2 - \omega_{pe}^2 - \gamma_1^3\omega_{Be}, a_3 = 2ku\omega_{pe}^2, a_4 = -k^2u^2\omega_{pe}^2. \tag{15}$$

According to the formulae of solving fourth order equation[7], we can obtain

$$\omega_j = x_j + \frac{1}{2}ku \qquad (j = 1, 2, 3, 4), \tag{16}$$

$$\begin{cases} x_1 = \frac{1}{2}\left\{-\left[(A+B) - \frac{2}{3}p\right]^{\frac{1}{2}} + 2\rho\cos\frac{\varphi}{2}\right\}, \\ x_2 = \frac{1}{2}\left\{\left[(A+B) - \frac{2}{3}p\right]^{\frac{1}{2}} - 2\rho\cos\frac{\varphi}{2}\right\}, \\ x_3 = \frac{1}{2}\left\{\left[(A+B) - \frac{2}{3}p\right]^{\frac{1}{2}} + \mathrm{i}2\rho\sin\frac{\varphi}{2}\right\}, \\ x_4 = \frac{1}{2}\left\{\left[(A+B) - \frac{2}{3}p\right]^{\frac{1}{2}} - \mathrm{i}2\rho\sin\frac{\varphi}{2}\right\}, \end{cases} \tag{17}$$

where

$$\begin{cases} A = -\frac{1}{3}\{[(\omega_{pe}^2 + \gamma_1^3\omega_{Be}^2 - k^2u^2)^3 + 54k^2u^2\omega_{pe}^2\gamma_1^3\omega_{Be}^2] \\ \qquad - 6\sqrt{3}ku\omega_{pe}\gamma_1^{3/2}\omega_{Be}[(\omega_{pe}^2 + \gamma_1^3\omega_{Be}^2 - k^2u^2)^3 + 27k^2u^2\omega_{pe}^2\gamma_1^3\omega_{Be}^2]^{1/2}\}^{1/3}, \\ B = -\frac{1}{3}\{[(\omega_{pe}^2 + \gamma_1^3\omega_{Be}^2 - k^2u^2)^3 + 54k^2u^2\omega_{pe}^2\gamma_1^3\omega_{Be}^2] \\ \qquad + 6\sqrt{3}ku\omega_{pe}\gamma_1^{3/2}\omega_{Be}[(\omega_{pe}^2 + \gamma_1^3\omega_{Be}^2 - k^2u^2)^3 + 27k^2u^2\omega_{pe}^2\gamma_1^3\omega_{Be}^2]^{1/2}\}^{1/3}, \\ \rho = \left[\left(\frac{1}{2}A + \frac{1}{2}B + \frac{2}{3}p\right)^2 + \frac{3}{4}(A-B)^2\right]^{1/4}, \quad p = -\frac{1}{2}k^2u^2 - \omega_{pe}^2 - \gamma_1^3\omega_{Be}^2, \\ \cos\varphi = -\left[\frac{1}{2}(A+B) + \frac{2}{3}p\right] \Big/ \rho^2, \quad \sin\varphi = \sqrt{\frac{3}{2}}(A-B) \Big/ \rho^2. \end{cases} \tag{18}$$

From the condition (15) satisfied by cubic equation with complex conjugate roots, it can be found that boundary of instability and damping is valid for

$$(\omega_{pe}^2 + \gamma_1^3\omega_{Be}^2 - k^2u^2)^3 + 27k^2u^2\omega_{pe}^2\gamma_1^3\omega_{Be}^2 = 0. \tag{19}$$

This is a cubic equation for k^2u^2. From Cadar formula we have found

$$\left(\frac{k^2u^2}{\omega_{pe}^2}\right)_c = 1 + \frac{\gamma_1^3\omega_{Be}^2}{\omega_{pe}^2} + 3\left[\frac{1}{2}\left(1 + \frac{\gamma^3\omega_{Be}^2}{\omega_{pe}^2}\right)\frac{\gamma^3\omega_{Be}^2}{\omega_{pe}^2}\right]^{1/3}\left\{\left[1 + \left(1 - \frac{4\gamma_1^3\omega_{Be}^2/\omega_{pe}^2}{(1+\gamma_1^3\omega_{Be}^2/\omega_{pe}^2)^2}\right)^{1/2}\right]^{1/3} + \left[1 - \left(1 - \frac{4\gamma_1^2\omega_{Be}^2/\omega_{pe}^2}{(1+\gamma_1^3\omega_{Be}^2/\omega_{pe}^2)^2}\right)^{1/2}\right]^{1/3}\right\}. \tag{20}$$

When $\gamma_1^3\omega_{Be}^2/\omega_{pe}^2 \ll 1$, we have

$$k_c \simeq \frac{\omega_{pe}}{u}\left[1+\frac{3}{2}\gamma_1\left(\frac{\omega_{Be}}{\omega_{pe}}\right)^{2/3}\right]. \tag{21}$$

Obviously, $\dfrac{ku}{\omega_{pe}} \le \left(\dfrac{ku}{\omega_{pe}}\right)_c$.

In the case of collision against each other for $n_{Be}= n_e$, we have

$$\left(\frac{ku}{\omega_{pe}}\right)_c \simeq (1+\gamma_1)^3.$$

For convenience's sake, write

$$u_1 = \frac{ku}{\omega_{pe}}, \quad u_2 = \frac{\omega_{Be}^2}{\omega_{pe}^2} = \frac{n_b}{n_e}. \tag{22}$$

Thus

$$\begin{cases} a = (1/\gamma_1^3 u_2 - u_1^2)^3 + 54u_1^2\gamma_1^3 u_2, \\ b = 6\sqrt{3}u_1\gamma_1^{3/2}u_2^{1/2}[(1+\gamma_1^3 u_2 - u_1^2)^3 + 27u_1^2\gamma_1^3 u_2]^{1/2}, \\ c = 2\left(1+\frac{1}{2}u_1^2+\gamma_1^3 u_2\right), \ \rho = 6^{-1/2}\omega_{pe}\rho_0, \\ \rho_0 = \{[(a+b)^{1/3}+(a-b)^{1/3}+2c]^2 + 3[(a+b)^{1/3}-(a-b)^{1/3}]^2\}^{1/4}, \\ \varphi = \mathrm{tg}^{-1}\dfrac{\sqrt{3}[(a+b)^{1/3}-(a-b)^{1/3}]}{(a+b)^{1/3}+(a-b)^{1/3}+2c}. \end{cases} \tag{23}$$

Hence we have found from (16)—(18), (22), (23) that

$$\omega_I/\omega_{pe} = I_m\omega/\omega_{pe} = \rho\sin\frac{\varphi}{2}/\omega_{pe}$$

$$= \left\{\rho_0^2 - [(a+b)^{\frac{1}{3}} + (a-b)^{\frac{1}{3}} + 2c]\right\}^{\frac{1}{3}} / 2\sqrt{3}, \tag{24}$$

$$\omega_\gamma/\omega_{pe} = R_e\omega/\omega_{pe}$$

$$= \left\{u_1 + 3^{-\frac{1}{2}}[c-((a+b)^{\frac{1}{3}}+(a-b)^{\frac{1}{3}})]^{\frac{1}{2}}\right\}/2. \tag{25}$$

Since electron energy $E=mc^2/\gamma_1$, $\gamma_1=mc^2/E=0.511/E$ [MeV].

The values of ω_I/ω_{pe} and $\omega_\gamma/\omega_{pe}$ can then be calculated from (16), (17), (23)—(25) as E, u_2 are given (see figs. (2—4)).

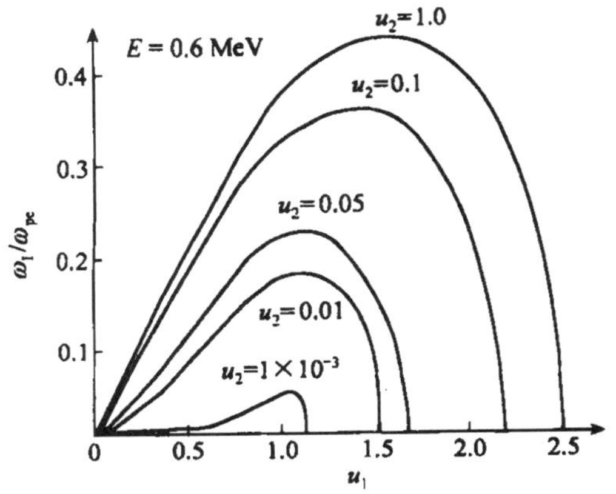

Fig. 2. The curves for ω_I/ω_{pe} vs. u_1 (as E=0.6 MeV, u_2 varies).

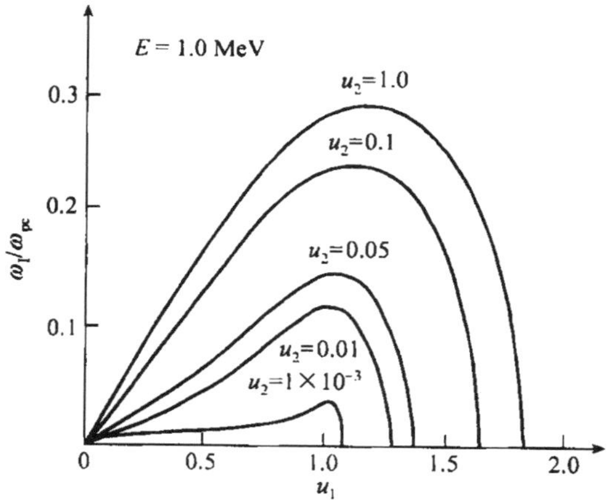

Fig. 3. The curves for ω_I/ω_{pe} vs. u_1 (as E=1.0 MeV, u_2 varies).

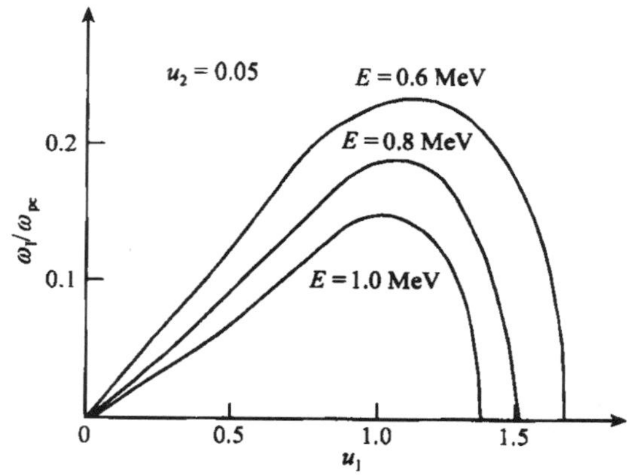

Fig. 4. The curves for ω_I/ω_{pe} vs. u_1 (as u_2=0.05, E varies).

4 Conclusion and discussion

Stability criteria for a weak relativistic beam-plasma interaction in a strong magnetic field are found. Two beam modes occur, $\omega \approx k_z c$ and $\omega \approx k_z c - \omega_c$. The former mode is stabilized if the k_x is sufficiently large. For electrostatic two-beam mode we have obtained analytic expressive forms (24) and (25). These solutions are general. The result of numerical calculation is expressed as figs. 2—4. The maximum of growth rate ω_I shifts to the region of $ku > \omega_{pe}$, the growth rate and the region of instability became larger with increasing density ratio n_{Be}/n_e, also became less with increasing energy, and effect of relativity can suppress electrostatic two beam instability, too. It can be seen that the higher the energy of relativistic electron beam, the larger the application range of approximation of weak stream.

References

1. Bludman, S. A., Watson, K. M., Rosenbluth, M.N., Statistical mechanics of relativistic streams Ⅰ, Ⅱ, Phys. Fluids, 1960, 3(5): 741.
2. Akhiezer, A. I., Akhiezer, I. A., Polovin, R. V. et al., Plasma Electrodynamics (Ⅰ), New York: Pergamon Press Ltd, 1975, Chapter 6, 339.
3. Breizman, B. N., Collective interaction of relativistic electron beams with plasmas, in Reviews of Plasma Phys. (ed. Kadomtsev, B. B.), New York: Consultants Bureau, 1990, 15: 61.
4. Buti, B., Relativistic effects on plasma oscillations and two-stream instability Ⅰ, Ⅱ, Phys. Fluids, 1963, 6(1): 89.
5. Thode, L. E., Sudan, R.N., Two-stream instability heating of plasma by relativistic electron beams, Phys. Rev. Lett., 1973, 30(16): 732.
6. Ferch, R.L., Sudan, R.N., Linear two-stream instability of warm relativistic electron beams, Plasma Phys., 1975, 17: 905.
7. Hua Luogeng, Introduction for Advanced Mathematics (in Chinese), Beijing: Science Press, 1974, 1(1): 34.

(Received October 8, 1998)

PHYSICAL REVIEW E, VOLUME 65, 057401

Stabilization of ablative Rayleigh-Taylor instability due to change of the Atwood number

Wenhua Ye,[1,2,*] Weiyan Zhang,[2] and X. T. He[1,2]

[1]*Department of Physics, Zhejiang University, Hangzhou 310028, People's Republic of China*

[2]*Institute of Applied Physics and Computational Mathematics, P.O. Box 8009, Beijing 100088, People's Republic of China*

(Received 29 March 2001; revised manuscript received 1 November 2001; published 21 May 2002)

Recent experiment [S.G. Glendinning *et al.*, Phys. Rev. Lett. **78**, 3318 (1997)] showed that the measured growth rate of laser ablative Rayleigh-Taylor (RT) instability with preheating is about 50% of the classic value and is reduced by about 18% compared with the simulated value obtained with the computer code LASNEX. By changing the temperature variation of the electron thermal conductivity at low temperatures, the density profile from the Bhatnagar-Gross-Krook approximation is recovered in the simulation, and the simulated RT growth rate is in good agreement with the experimental value from Glendinning *et al.* The preheated density profile on ablative RT stablization is studied numerically. A change of the Atwood number in the preheating case also leads to RT stabilization. The RT growth formula $\gamma=\sqrt{Akg/(1+AkL)}-2kV_a$ agrees well with experiment and simulation, and is appropriate for the preheating case.

DOI: 10.1103/PhysRevE.65.057401 PACS number(s): 52.35.Py, 81.15.Fg

Hydrodynamic instability places a fundamental limit on design parameters required for a capsule ignition of the inertial confinement fusion (ICF) [1]. In an ICF capsule implosion, the ablation front is Rayleigh-Taylor (RT) unstable. An accurate prediction of the RT growth is essential to the design of the ignition capsule, particularly for directly driven ICF [2]. The Takabe formula $\gamma_T=0.9\sqrt{kg}-\beta kV_a$ [3], which includes the ablation advection stabilization, is frequently used to discuss the linear growth rate of laser ablative RT instability, where k is the perturbed wave number, g is the acceleration, V_a is the ablation velocity, and $\beta\simeq3\sim4$ is expected in the direct-drive ablation. A modified version of the Takabe formula, which includes stabilization of the ablation advection and the density gradient scale length, was given by Lindl as $\gamma_{mT}=\sqrt{kg/(1+kL)}-\beta kV_a$ [1], where $L=[\rho/(d\rho/dz)]_{min}$ is the density gradient scale length and $\beta=2$ was given, for example, by Sanz [4]. In this paper we refer to this formula as the modified-by-Lindl Takabe formula or the Lindl formula. However, the reduction in growth of laser ablative RT instability in recent experiments [5–7], which deviates the modified-by-Lindl Takabe formula, is observed and has not fully been explained.

Recent experiments [5,6] have indicated that the growth rate of the areal density perturbation is significantly reduced as compared with the calculated results of the Spitzer-Harm (SH) electron thermal conductivity [8]. Glendinning *et al.* (GL) recognized that the difference between the experiment and the LASNEX simulation is due to a change in the longitudinal density profile resulting from preheating by energetic electrons. Honda *et al.* (HM) [9] calculated the lower RT growth rate using the two-dimensional (2D) full Fokker-Planck (FP) equation to compare with the SH electron transport coefficient, and showed that the density of the ablation front is changed from 2.0–2.4 g/cm^3 for SH and to 1.5–1.7 g/cm^3 for FP. GL simulated the foil acceleration in 1D geometry using the Bhatnagar-Gross-Krook (BGK) approximation [10] to the FP equation, and obtained a density distribution similar to the FP results given by HM. We do 2D simulations to further discuss the experiment result shown by GL. A similar change in the density profile is also seen by enhancing the electron thermal conductivity at low temperatures, and the RT growth rate in our simulation is in agreement with the GL's experimental value. It is found that there is a greater than 10% discrepancy in RT growth rate between our simulation and the modified-by-Lindl Takabe formula. We notice that this also occurs in Figs. 5 and 6 of Ref. [5]. Similarly, the FP result from HM was somewhat lower than the prediction of the modified-by-Lindl Takabe formula, using the BGK parameters. There may exist some physical reason that is not widely known. We see from the simulated density profile that the peak density of the ablation front drops considerably, so the Atwood number of the ablation front clearly decreases. Preheating not only enhances the stabilizations by the ablative advection and the finite density gradient scale length, but also leads to stabilization by the decreased Atwood number. The latter is another possible source for the reduction of ablative RT growth rates in the preheating case. Most of the heuristic formulas neglect the Atwood number because it takes a value of order unity in the less preheating case. Betti *et al.* gave a heuristic form $\gamma_{mL}=\sqrt{Akg/(1+AkL)}-\beta kV_a$ [11], here referred to as the modified Lindl formula. We find that the prediction of this formula is in closer agreement with GL's experimental value [5] and all our simulations. Preheating the ablation front is a useful way to stabilize the RT instability. Through effective preheating to the ablation front by x ray, the ablative RT growth and the laser imprint are suppressed in recent experiments [12,13]. In some designs of the National Ignition Facility (NIF) capsule, the RT stabilization by preheating of the shock, x ray etc., is important [14], so stabilization by decreased Atwood number can be a primary stabilization source.

The finite difference radiation hydrodynamic code LARED-S [15] is used in numerical simulations. LARED-S deals with compressible hydrodynamics, nonlinear electron conduction, multigroup radiation diffusion, ray tracing of laser absorbed by the inverse bremsstrahlung. The hydrodynamic equations

*Email address: Ye_wenhua@mail.iapcm.ac.cn

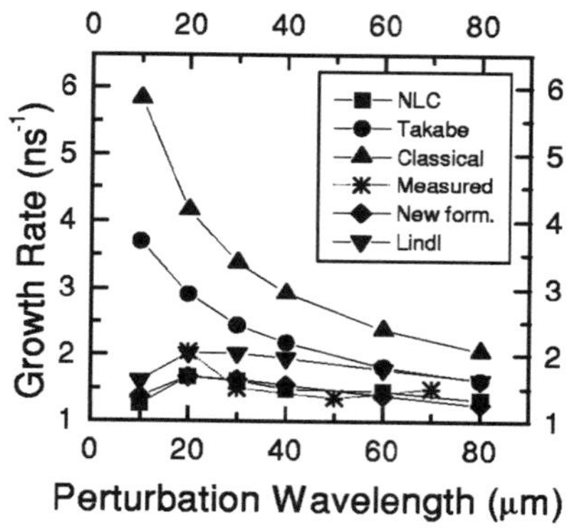

FIG. 1. Comparison of ablative growth rates for different formulas, the experiment and simulations for the 20-μm CH foil.

are integrated with the flux-corrected transport (FCT) algorithm [16] in space and the second-order Runge-Kutta method in time. The FCT with sixth-order accurate phase error is used and has second-order accuracy on the uniform part of the grid. The LARED-S code uses the slide zone version of FCT to trace the ablation surface and the fine uniform grids cluster near the steep ablation front. Simulations of the ablative RT instability in this paper are in two-dimensional Cartesian geometry.

The linear RT growth rate of the 2D simulation comes from the exponential fit of the Fourier amplitude of the areal density perturbation in the linear stage. The foil acceleration, the ablative velocity, the Atwood number, and the density gradient scale length are obtained by the parabolic fit to the one-dimensional simulation results during the same period. The Atwood number A is $(\rho_2-\rho_1)/(\rho_2+\rho_1)$, where ρ_2 is chosen as ρ_a, the peak density at the ablation front, and as in Ref. [17], ρ_1 is the blow-off density, located at $(2k)^{-1}$ from the position of peak density. The ablative RT growth experiment performed by GL under conditions of a 20-μm-thin CH foil, a drive laser pulse with a linear 1-ns ramp to a peak intensity 7×10^{13} W/cm^2 at the wavelength 0.53 μm, followed by a 2-ns flat section.

In our simulation, the electron thermal conductivity is considered in the form $\kappa=\kappa_{SH}f(T)h$, where $\kappa_{SH}\propto T^{5/2}$ is the SH formula and T is the electron temperature. h has a simple form of $1/[1+B\ln(T_{max}/T_0)]$ and is due to the fact that there exists a large temperature gradient near the critical surface, and the static potential effect has to be considered in electron thermal conduction [18], where T_{max} is the maximum electron temperature in the corona region, T_0 is the electron temperature at the critical surface, and B is an adjustable parameter, usually $B=1$. It was found that the electron thermal conductivity scales as 1/k at large k when there exists a nonnegligible electrostatic field, where $k=\ln(T_{max}/T_0)$. A function $f(T)$ is due to the preheating effect of energetic electrons, which is considered in a form of $f(T)=a$ when $T>43$ eV and $f(T)=b/T+c/T^{3/2}$ when 1 eV$<T\leqslant43$ eV to meet the experiment result of GL, where a, b, c are adjustable parameters. $f(T)$ is adjusted so that the density profile around the ablation front is close to the numerical solution for the 1D BGK [5] or for the FP equation of electron thermal transport [9,19], and it could be explained that the preheated ablation produces a lower temperature tongue at the head of the sharp front that is due to the SH electron thermal conduction. The tongue seems to be from an energetic electron contribution because energetic electrons have a longer mean free path.

The SH formula is rigorously valid only for fully ionized nondegenerate plasmas, i.e, for lower density and higher temperature. The application of the SH formula to dense plasmas can give erroneous results. The electron thermal conductivity calculated from the SH formula at a solid density and low temperature $T<10$ eV is incorrect by a large factor, which was shown by Tabak *et al.* [20]. The work of Lee and More gave some support to the arbitary change of the temperature variation of the electron thermal conductivity at lower temperatures [21]. The FP simulation showed that preheating by the electrons in the tail of the Maxwellian distribution is important for the density profile, and so more energy is transferred to the ablation front for FP than for SH. Our simulation shows that the density profile around the ablation front is very sensitive to the electron thermal conductivity at the head of the ablation front. In order to obtain the ablative RT growth rate in agreement with the experiment, it is possible to find a function $f(T)$ to fit the FP or the BGK density profile in the hydrodynamic simulation.

For $f(T)=1$, the ablation front is sharp due to less preheating, thus the linear RT growth rate from simulation agrees with the value of the Takabe formula. $f(T)$ is adjusted near the ablation front as follows:

$$f(T)=\begin{cases}1.0, & T<0.86\ \text{eV}\\ 10000, & 0.86\ \text{eV}\leqslant T<4.3\ \text{eV}\\ 4000, & 4.3\ \text{eV}\leqslant T<8.6\ \text{eV}\\ 1500, & 8.6\ \text{eV}\leqslant T<17.2\ \text{eV}\\ 300, & 17.2\ \text{eV}\leqslant T<43\ \text{eV}\\ 2.5, & 43\ \text{eV}\leqslant T<86\ \text{eV}\\ 1.5, & 86\ \text{eV}\leqslant T<172\ \text{eV}\\ 1.0, & T\geqslant172\ \text{eV},\end{cases}$$

which can be fit to $f(T)=186/T+11.5/T^{3/2}$ when 0.86 eV $<T<43$ eV, where T is in units of 100 eV as used in the simulation code, so the electron thermal conductivity versus temperature is $T^{5/2}f(T)=11.5T+186T^{3/2}$, which produces a slowly varying temperature profile as in the BGK or in the FP simulation. The density profile is similar to the BGK result [5], and L is increased from 1 μm at $f(T)=1$ to about 4 μm at the above $f(T)$. As shown in Fig. 1, the linear growth rate of RT instability from our simulation, as seen in the curve "NLC," agrees well with the measured value by GL, as shown by the curve "measured," but is lower than the prediction from the modified Takabe formula $\gamma_{mT}=\sqrt{kg/(1+kL)}-2kV_a$ (the curve "Lindl"). The Takabe formula $\gamma_T=0.9\sqrt{kg}-3kV_a$ shows greater value than the measured value.

There exists a discrepancy of over 10% between the Lindl formula and GL's experiment or our simulation. In order to

TABLE I. Comparison of ablative RT growth rates in different preheating cases for the 40-μm perturbation wavelength.

Model	L (μm)	A	g (μm/ns^2)	V_a (μm/ns)	γ_c (ns^{-1})	γ_{cal} (ns^{-1})	γ_{mL} (ns^{-1})	γ_{mT} (ns^{-1})	$\frac{\gamma_{cal}}{\gamma_c}$	$\left\|\frac{\gamma_{cal}-\gamma_{mL}}{\gamma_{cal}}\right\|$	$\left\|\frac{\gamma_{cal}-\gamma_{mT}}{\gamma_{cal}}\right\|$
Case 1	4.04	0.706	17.88	0.3540	1.676	1.082	1.059	1.200	0.646	0.021	0.109
Case 2	0.96	0.974	21.25	0.3144	1.827	1.656	1.585	1.604	0.906	0.043	0.031
Case 3	1.53	0.974	17.10	0.2703	1.639	1.365	1.371	1.387	0.833	0.004	0.016
Case 4	2.78	0.757	19.16	0.3246	1.735	1.228	1.207	1.346	0.708	0.017	0.096
Case 5	3.23	0.727	17.80	0.3540	1.672	1.125	1.107	1.249	0.673	0.016	0.110
Case 6	2.27	0.625	17.58	0.4099	1.662	1.032	1.059	1.298	0.621	0.026	0.258

understand the difference, an exponential density profile around the ablation front is assumed, and the RT growth rate $\gamma=\sqrt{Akg/(1+AkL)}$ is obtained at the approximation of linearity, incompressibility, no viscosity, and no heat transfer. For consideration of ablative advection stabilization, the ablative RT growth rate should be in the form $\gamma_{mL}=\sqrt{Akg/(1+AkL)}-2kV_a$ with the variable Atwood number that depends on the preheated density profile. This modified Lindl formula, i.e., the curve "new form," agrees very well with GL's experimental curve "measured" and our simulation with preheating. In the preheating case, the electron thermal flux at the head of the ablation front increases considerably, thus the temperature profile is not so steep that the power index of the electron thermal conductivity is much less than 5/2 in the SH model. This leads to the reduction of the peak density at the ablation front, so the density profile becomes smooth to compare with the SH model. As a result, we find that the modified Lindl formula with variable Atwood number may predict the preheating case. Rigorous simulation for preheating by energetic electrons or x rays is complicated, so $f(T)$ is changed to study the preheating effect on ablative RT growth, especially the Atwood number effect.

Since we see from the above discussion that a thin CH foil with 20 μm thickness is used in GL's experiment, the thin target easily expands so that preheating leads to the front density decreasing rapidly, and consequently the Atwood number is reduced. Therefore, one needs to further discuss whether or not the modified Lindl formula can better predict the thick target behavior when there exists preheating. We postulate a thicker CH foil, with a thickness of 100 μm and a density of 1.0 g/cm^3. The pulse duration of a 0.53 μm laser is shaped in the following manner: a linear 3 ns ramp to an intensity of 3×10^{14} W/cm^2, followed by an intensity that remains unchanged. At 5 ns, density perturbation with the wavelength 40 μm is superimposed on the one-dimensional density profile near the ablation front. The 2D simulation is performed in different adjusted $f(T)$ functions. Case 1: $f_1(T)$ is the same as $f(T)$. Case 2: $f_2(T)=1$, the Spitzer-Harm model with the limiter 0.05. Case 3: $f_3(T)=1$ is the same as the SH model when $T<43$ eV and when $T\geqslant172$ eV, but has a slight change when 43 eV$\leqslant T<172$ eV so as to show the SH effect on the RT growth rate. Case 4:

$$f_4(T)=\begin{cases}1.0, & T<8.6\ \text{eV}\\ 1500, & 8.6\ \text{eV}\leqslant T<17.2\ \text{eV}\\ 300, & 17.2\ \text{eV}\leqslant T<43\ \text{eV}\\ 2.5, & 43\ \text{eV}\leqslant T<86\ \text{eV}\\ 1.5, & 86\ \text{eV}\leqslant T<172\ \text{eV}\\ 1.0, & T\geqslant172\ \text{eV},\end{cases}$$

is chosen between both $f_3(T)$ and $f(T)$ for increasing preheating around the ablation front. Case 5: $f_5(T)=10/T^{3/2}$ (100 eV) when 1 eV$<T<$43 eV and $r>r_p-12$ μm so that the electron thermal conductivity versus T generates a less sharp ablation front for considering the thermal transport of energetic electrons, where r_p is the location of the peak density and the laser is incident on location r_∞. Preheating at the head of the ablation front is less than 12 μm because the length of the preheated tongue is usually shorter than 12 μm. Case 6: $f_6(T)=3000/\rho(\text{g/cm}^3)$ when 1 eV$<T$ <43 eV and $r>r_p-12$ μm. This form $f(T)$ considers the fact that the mean free path of energetic electrons is inversely proportional to the mass density.

In Table I, γ_c, γ_{cal}, and γ_{mT} are the RT growth rates for classic, two-dimensional simulation, and the modified Takabe formula, respectively. From Table I, we can see that the RT growth rates for cases 2 and 3 are greater than 80% of the classical value because of less preheating. However, as preheating increases from case 3 to case 4 and to case 1, the RT growth rate of the two-dimensional simulation decreases and the discrepancy increases between the value of the modified

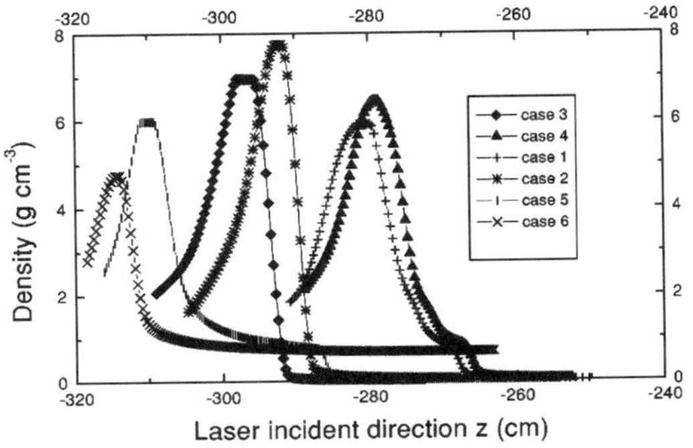

FIG. 2. The density profiles near the ablation front in different preheating cases.

Takabe formula and simulations. In the cases with more preheating, i.e., cases 5 and 6, similar results are shown. Particularly for case 6, the relative error of γ_{cal} and γ_{mT} is 26% though L is smaller—only 2.27 μm.

Figure 2 shows the density profile of different preheating cases. For cases 2 and 3, ρ_a is higher—over 7 g/cm^3—and ρ_1 is very low—about 0.2 g/cm^3, or below—so Atwood number A is close to 1 in these two less preheating cases. However, for cases 1, 5, and 6, ρ_a is lower, and ρ_1 increases to nearly 1 g/cm^3, so the Atwood number decreases to below 0.75. The density profiles of case 1 and case 5 are close; so are the Atwood numbers and the RT growth rates.

We see from the above results that even if a thick target is used in the preheating case, the modified Lindl formula (see Table I) given by γ_{mL} still agrees very well with the results of our simulation (see γ_{cal}). The density profile near the ablation front is also distinctly changed and the Atwood number of the ablation front is clearly less than 1, so the Atwood number stabilization is important, This formula is reduced to the modified Takabe formula at $A=1$, and is closed to the Takabe formula at the steep ablation front.

In summary, the linear RT growth rate in our simulation agrees well with GL's experiment by enhancing the electron thermal conductivity at a lower temperature and higher density. In the ablation region, preheating not only reduces the peak density, but also raises the density at the foot of the ablation front, thus the Atwood number of the ablation front decreases and contributes to the reduction of the ablation RT growth rate. The modified Lindl formula $\gamma_{mL} = \sqrt{Akg/(1+AkL)}-2kV_a$ is better suited for the preheating case. It agrees well with the experiment value given by GL and our simulation. We believe that this stabilization effect is also important in the x-ray preheating case for the x-ray ablative RT instability.

We acknowledge helpful discussions with Dr. Shaoping Zhu. This work was supported by the National High-Tech ICF Committee, partly supported by the National Natural Science Foundation of China under Grant Nos. 19932010 and 10135010, and the National Basic Research Project "Nonlinear Science" in China.

[1] J. Lindl, Phys. Plasmas **2**, 3933 (1995).
[2] C.P. Verdon, Bull. Am. Phys. Soc. **38**, 2010 (1993).
[3] H. Takabe *et al.*, Phys. Fluids **28**, 3676 (1985).
[4] J. Sanz, Phys. Rev. Lett. **73**, 2700 (1994).
[5] S.G. Glendinning *et al.*, Phys. Rev. Lett. **78**, 3318 (1997).
[6] K. Shigemori *et al.*, Phys. Rev. Lett. **78**, 250 (1997).
[7] D.L. Tubbs *et al.*, Phys. Plasmas **6**, 2095 (1999).
[8] L. Spitzer and R. Harm, Phys. Rev. **89**, 977 (1953).
[9] M. Honda, Ph.D. thesis, Osaka University, 1998.
[10] P.L. Bhatnagar *et al.*, Phys. Rev. **94**, 511 (1954).
[11] R. Betti *et al.*, Phys. Plasmas **5**, 1446 (1998).
[12] H. Azechi *et al.*, Phys. Plasmas **4**, 4079 (1997).
[13] C.J. Pawley *et al.*, Phys. Plasmas **6**, 565 (1999).
[14] S.E. Bodner *et al.*, Phys. Plasmas **5**, 1901 (1998).
[15] Wenhua Ye *et al.*, High power Laser Parti Beams **10**, 403 (1998), in Chinese.
[16] J.P. Boris and D.L. Book, J. Comput. Phys. **11**, 38 (1973).
[17] A.R. Piriz *et al.*, Phys. Plasmas **1**, 1117 (1997).
[18] Yan Xu and X.T. He, Phys. Rev. E **50**, 443 (1994).
[19] E.P. Epperlein, Laser Part. Beams **12**, 257 (1994).
[20] M. Tabak *et al.*, Phys. Fluids B **2**, 1007 (1990).
[21] Y.T. Lee and R.M. More, Phys. Fluids **27**, 1273 (1984).

第 52 卷 第 7 期 2003 年 7 月
1000-3290/2003/52(07)/1688-06
物 理 学 报
ACTA PHYSICA SINICA
Vol.52, No.7, July, 2003

二维不可压流体瑞利-泰勒不稳定性的非线性阈值公式*

吴俊峰[1)] 叶文华[2)3)] 张维岩[3)] 贺贤土[2)3)]

1)(中国工程物理研究院研究生部,北京 100088)

2)(浙江大学物理系,杭州 310028)

3)(北京应用物理与计算数学研究所,北京 100088)

(2002 年 6 月 21 日收到;2002 年 10 月 28 日收到修改稿)

分析了二维情况下平面几何、柱几何和球几何中瑞利-泰勒不稳定性发生非线性偏离的阈值问题,给出了三种几何中密度扰动振幅的定义,以及与界面扰动振幅的关系.由此得到了三种几何中密度扰动的非线性阈值公式,用高精度流体程序对三种几何中的不稳定性进行了数值模拟,验证了得到的非线性阈值公式.

关键词:瑞利-泰勒不稳定性,非线性阈值,密度振幅

PACC: 5253,5250J,4720

1. 引 言

当轻流体加速或减速重流体,或者在重力场中轻流体支撑重流体时,流体界面是不稳定的,产生瑞利-泰勒(RT)不稳定性[1]. RT 不稳定性在惯性约束聚变(ICF)、超新星爆炸、Z 箍缩(Z-pinch)等离子体、海洋混合层、晶体生长等方面具有重要作用. 为提高内爆效率,减少驱动器能量成本,ICF 中靶丸通常采用薄壳设计,内爆在加速和减速过程中 RT 不稳定,加速阶段的不稳定性严重影响靶丸的内爆压缩,导致烧蚀壳的破裂,而减速阶段的不稳定性导致高温中心热斑与周围冷物质的混合,大大降低中心区的温度,破坏 DT 的聚变点火[2]. 文献[3,4]总结和讨论了这方面很多工作.

众所周知,对于单模扰动,在线性阶段扰动振幅以指数形式增长,扰动界面是正弦或余弦形状,随着非线性的出现,模耦合作用增强,大量高次谐波被激发,界面逐渐变为不对称的气泡和尖钉形状,扰动振幅以常速度增长[5]. 在线性和非线性阶段,扰动以不同方式增长,因此确定其"衔接点",也就是非线性阈值,对于正确估计扰动增长有实际意义. 在线性和非线性阶段之间存在一个短暂的过渡期,此时模耦合产生的二次谐波与基模相比不可忽略,基模开始明显地偏离线性增长公式,可以用弱非线性理论描述这个过渡期. Catton 等人在文献[6]中用弱非线性理论对平面几何中 RT 不稳定性进行了分析,得到了基模、二阶模和三阶模的增长公式,如果取扰动增长偏离线性增长约 10% 作为非线性过程的开始,此时对应的基模扰动振幅约为 0.1λ. 实际应用中很多问题具有柱几何或球几何结构,例如 Z-pinch 内爆为柱几何结构,ICF 中内爆为球几何结构等等,因此讨论柱几何或球几何中的非线性阈值更有意义. 本文利用类比平面几何的方法,通过不稳定性在平面几何与柱几何或球几何之间的联系,得到了柱几何和球几何中单模扰动增长的非线性阈值公式,然后用数值模拟结果进行检验,同时比较平面几何、柱几何和球几何中不稳定性非线性阈值之间的差别.

2. RT 不稳定性的线性增长

在平面几何中,物理模型如图 1 所示,考虑两层理想不可压流体,其密度分别为 ρ_1 和 ρ_2,各占据 $y>0$ 和 $y<0$ 的区域,并且 $\rho_1>\rho_2$,重力加速度 $\boldsymbol{g}$ 平行于 y 轴,指向其负方向,为简单起见,这里只考虑

* 国家自然科学基金(批准号:19932010 和 10135010)和国家基础研究"非线性科学基金"资助的课题.

二维情况 x-y. 假设初始扰动为位移扰动 $y = A_0\cos kx$，其中 k 为扰动的波数，在线性近似下，可得方程[1]

$$\ddot{A}(t) - \alpha kgA(t) = 0, \qquad (1)$$

其中 $A(t)$为扰动振幅，$\alpha = \frac{\rho_1 - \rho_2}{\rho_1 + \rho_2}$为 Atwood 数.

在柱几何中，物理模型如图 2 所示，考虑两层理想不可压流体，外层流体的密度为 ρ_1，内层流体的密度为 ρ_2，并且 $\rho_1 > \rho_2$，界面到轴心的距离为 R，沿径向 r 存在向内的重力加速度 $\boldsymbol{g}$，这里仍然只考虑二维情况 r-φ. 假设初始扰动为位移扰动 $\Delta r = A_0\cos m\varphi$，其中 m 为扰动的模数，在线性近似下，可得方程[7]

$$\ddot{A}(t) - \alpha \frac{m}{R}gA(t) = 0. \qquad (2)$$

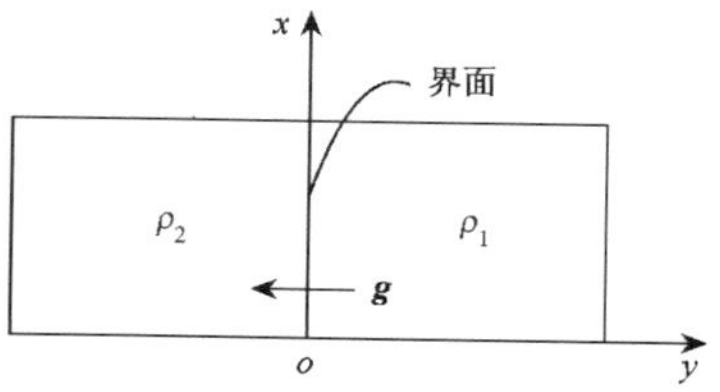

图 1　平面几何模型

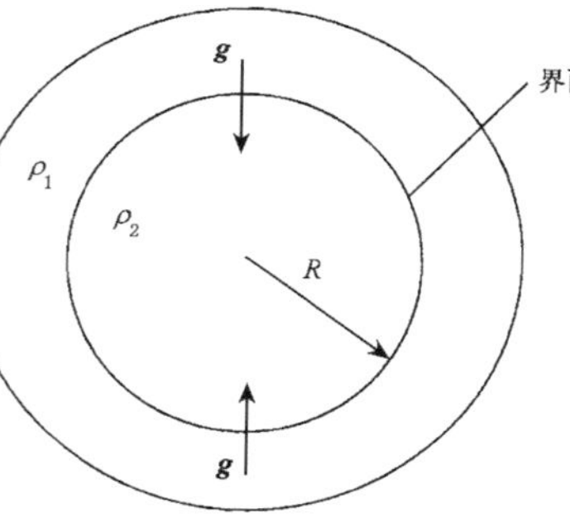

图 2　柱几何或球几何模型

在球几何中，物理模型也可以用图 2 表示，考虑两层理想不可压流体，外层流体的密度为 ρ_1，内层流体的密度为 ρ_2，并且 $\rho_1 > \rho_2$，而界面到球心的距离为 R，沿径向 r 存在向内的重力加速度 $\boldsymbol{g}$. 与前面情况类似，假设初始扰动为位移扰动 $\Delta r = A_0 Y_{lm}(\theta, \varphi)$，其中 $Y_{lm}(\theta, \varphi)$为归一化球谐函数，在线性近似下，可得方程

$$\ddot{A} - \frac{l(l+1)(\rho_1 - \rho_2)}{l\rho_1 + (l+1)\rho_2}\frac{g}{R}A = 0. \qquad (3)$$

3. 密度扰动幅度与非线性阈值

本文考虑单模的 RT 不稳定性增长. Catton 等人用展开至三阶的弱非线性理论对平面几何中单模扰动的 RT 不稳定性进行分析，随着基模扰动幅度的增长，各阶非线性项产生各次谐波，二阶非线性项产生二次谐波，三阶非线性项产生三次谐波，等等. Catton 三阶非线性理论表明：基模扰动进入非线性区，首先产生二次谐波，二次谐波的相位与基模相反，对基模增长起负反馈作用，二次谐波的增长导致基模增长变慢. 三次谐波的相位与基模相同，对基模增长起正反馈作用. 当基模的振幅达到 0.1λ 附近时，基模增长出现显著的非线性偏离，这时基模扰动偏离线性增长约 10%，二次谐波的振幅为基模的 30%，三次谐波的振幅为基模的 15%[8]. 而在柱几何或球几何中，特别是二维情况，各个模的不稳定性增长都有一个特征量，即为波长. 在界面到轴心或球心的平均距离趋于无限大极限时，柱几何或球几何中的扰动能够保持波长不变而退化到平面几何中. 几何效应对不稳定性增长的影响与界面到轴心或球心的平均距离有关. 在弱非线性阶段，对于界面到轴心或球心的平均距离和 Atwood 数不是非常小的情况，相对于不稳定性的非线性效应，几何效应的影响不是很大，因此柱几何或球几何的非线性阈值相对于平面几何即使可能有变化，也不会有较大改变. 我们在整个计算区域沿加速度方向进行密度积分，然后对获得的面密度分布进行傅里叶分析后得到密度扰动振幅. 这样做有两个原因：首先，ICF 激光器上背景光正面照相直接测量的是密度振幅[9]，为与实验结果比较，数值模拟也必须给出扰动的密度振幅；其次，在线性阶段和弱非线性阶段，不稳定性的界面扰动振幅较小，有时难以直接观测，而密度扰动振幅仍能给出定量的结果，因此研究密度扰动振幅与界面扰动振幅的关系，进而给出密度扰动振幅的非线性阈值很必要. 在这里界面扰动振幅的量纲为长度，密度扰动振幅的量纲与几何有关，在后面的讨论中可以看出，在不同的几何中，这两种振幅之间的关系不相同，我们将利用第二节给出的物理模型对不同的几何分别进行讨论.

平面几何情况如图 1 所示，在整个计算区域内对 t 时刻的密度分布 $\rho(x, y, t)$沿 y 轴方向进行积

分,得到只与坐标 x 有关的密度分布函数 $\tilde{\rho}(x,t)$,即

$$\tilde{\rho}(x,t)=\int_{-D/2}^{D/2}\rho(x,y,t)\mathrm{d}y, \quad (4)$$

其中 D 为计算区域在 y 轴方向的长度,然后对分布函数 $\tilde{\rho}(x,t)$进行傅里叶变换,得到基模的密度扰动振幅,即

$$\tilde{\tilde{\rho}}(t)=\frac{2}{L}\int_{-L/2}^{L/2}\tilde{\rho}(x,t)\cos kx\mathrm{d}x, \quad (5)$$

其中 $k=\frac{2\pi}{L}$为基模的波数,L 为计算区域在 x 轴方向的长度,也是基模的波长. 经过计算,得到界面扰动振幅 $A(t)$与密度扰动振幅 $\tilde{\tilde{\rho}}(t)$之间的关系为

$$\tilde{\tilde{\rho}}(t)=A(t)(\rho_1-\rho_2). \quad (6)$$

在平面几何中,密度扰动振幅的量纲为[质量/面积],与界面两侧流体的密度差成正比,因此不稳定性增长发生非线性偏离的阈值为

$$\tilde{\tilde{\rho}}_{\mathrm{th}}=0.1\lambda(\rho_1-\rho_2). \quad (7)$$

柱几何情况如图2所示,与平面几何情况类似,我们在整个计算区域内对密度分布 $\rho(r,\varphi,t)$沿 r 轴方向进行积分,得到只与坐标 φ 有关的密度分布函数 $\tilde{\rho}(\varphi,t)$,即

$$\tilde{\rho}(\varphi,t)=\int_{R-D/2}^{R+D/2}\rho(r,\varphi,t)r\mathrm{d}r, \quad (8)$$

其中 D 为计算区域在 r 轴方向的长度,然后对分布函数 $\tilde{\rho}(\varphi,t)$进行傅里叶变换,得到基模的密度扰动振幅,即

$$\tilde{\tilde{\rho}}(t)=\frac{1}{\pi}\int_{0}^{2\pi}\tilde{\rho}(\varphi,t)\cos m\varphi\mathrm{d}\varphi, \quad (9)$$

其中 m 为基模的模数. 经过计算,界面扰动振幅 $A(t)$与密度扰动振幅 $\tilde{\tilde{\rho}}(t)$之间的关系为

$$\tilde{\tilde{\rho}}(t)=A(t)R(\rho_1-\rho_2). \quad (10)$$

在柱几何中,密度扰动振幅的量纲为[质量/长度/弧度],与界面到对称轴的距离成正比,与界面两侧流体的密度差成正比,因此不稳定性发生非线性偏离的阈值为

$$\tilde{\tilde{\rho}}_{\mathrm{th}}=0.1\lambda R(\rho_1-\rho_2)=0.2\frac{2\pi}{m}R^2(\rho_1-\rho_2), \quad (11)$$

其中柱几何中单模扰动的波长为 $\lambda=\frac{2\pi R}{m}$.

球几何情况与柱几何情况类似,我们在整个计算区域内对密度分布 $\rho(r,\theta,\varphi,t)$沿 r 轴方向进行积分,得到只与坐标 θ 和 φ 有关的密度分布函数 $\tilde{\rho}(\theta,\varphi,t)$,即

$$\tilde{\rho}(\theta,\varphi,t)=\int_{R-D/2}^{R+D/2}\rho(r,\theta,\varphi,t)r^2\mathrm{d}r, \quad (12)$$

然后对分布函数 $\tilde{\rho}(\theta,\varphi,t)$进行球谐函数变换,得到基模的密度扰动振幅,即

$$\tilde{\tilde{\rho}}(t)=\int_{0}^{\pi}\int_{0}^{2\pi}\rho(\theta,\varphi,t)\mathrm{Y}_{lm}^{*}(\theta,\varphi)\sin\theta\mathrm{d}\theta\mathrm{d}\varphi, \quad (13)$$

其中 l 和 m 为基模的模数. 经过计算,界面扰动振幅 $A(t)$与密度扰动振幅 $\tilde{\tilde{\rho}}(t)$之间的关系为

$$\tilde{\tilde{\rho}}(t)=A(t)R^2(\rho_1-\rho_2). \quad (14)$$

在球几何中,密度扰动振幅的量纲为[质量/立体角],与界面到球心的距离平方成正比,与界面两侧流体的密度差成正比. 我们知道,球谐波不是"简单"的波,不可能对它们定义一个波长,但是利用拉普拉斯算子的本征值之间的对应关系,仿照平面情况,可以定义球谐波的波长[10]为

$$k^2=\left(\frac{2\pi}{\lambda}\right)^2=\frac{l(l+1)}{R^2}. \quad (15)$$

尤其当 $l\gg1$ 时,可以认为 l/R 为球谐波的波数. 另一方面,从第二节的讨论中可以看出,球谐波的线性增长率只与 l 有关,但是球谐波发生非线性偏离的阈值可能与 l 和 m 都有关系,不能简单地认为球谐波发生非线性偏离的阈值为 0.1λ,但在数值模拟中,我们使用的是二维流体力学程序,不稳定性的发展是方位角对称的,或者扰动的模数 $m=0$,可以近似地认为不稳定性发生非线性偏离的阈值为 0.1λ,因此对于二维球几何情况,不稳定性发生非线性偏离的阈值为

$$\begin{aligned}\tilde{\tilde{\rho}}_{\mathrm{th}}&=0.1\lambda R^2(\rho_1-\rho_2)\\&=0.1\frac{2\pi}{\sqrt{l(l+1)}}R^3(\rho_1-\rho_2).\end{aligned} \quad (16)$$

4. 数值模拟的结果

数值模拟中流体力学方程组为

$$\frac{\partial\rho}{\partial t}=-\nabla\cdot(\rho\boldsymbol{u}), \quad (17)$$

$$\frac{\partial(\rho\boldsymbol{u})}{\partial t}=-\nabla\cdot(\rho\boldsymbol{u}\boldsymbol{u})-\nabla p+\rho\boldsymbol{g}, \quad (18)$$

$$\frac{\partial\varepsilon}{\partial t}=-\nabla\cdot(\varepsilon\boldsymbol{u})-p\nabla\cdot\boldsymbol{u}, \quad (19)$$

分别为质量方程、动量方程和内能方程，其中 ρ 为质量密度，$\boldsymbol{u}$ 为流体速度，p 为压强，$\boldsymbol{g}$ 为重力场加速度，$\varepsilon = p/(\gamma - 1)$ 为单位体积内能，$\gamma = 10.0$ 为绝热指数．流体的可压缩性通过无量纲参数 $M^2 = \frac{g\lambda}{c_s^2}$ 确定[3]，其中 λ 为界面扰动的波长，c_s 为未加扰动时重流体中的声速，M 为重力波相速度与声速的比值，M 越小，流体越接近不可压缩．在下面的数值模拟中，M 的取值不超过 0.05，流体近似不可压缩．我们用六阶相位误差 FCT[11] 算法编制了二维平面几何、柱几何和球几何流体力学程序，对平面几何 x-y、柱几何 r-φ 和球几何 r-θ 中 RT 不稳定性进行了数值模拟．模拟结果与不稳定性在线性阶段的解析理论符合很好，同时利用前面讨论的结果，分析比较了数值模拟中各种几何下不稳定性增长的非线性阈值．

4.1. 平面几何中 RT 不稳定性

根据前面的讨论，平面几何中 RT 不稳定性在线性阶段具有指数增长形式，增长率为 $\gamma = \sqrt{\alpha k g}$．这里对两个例子进行了分析，数值模拟结果如图 3 所示．对于例 1，取 $\rho_1 = 40\mathrm{g\cdot cm^{-3}}$，$\rho_2 = 20\mathrm{g\cdot cm^{-3}}$，$\lambda = 174.5\mu\mathrm{m}$，$g = 2\times10^{14}\mathrm{cm\cdot s^{-2}}$；对于例 2，取 $\rho_1 = 35\mathrm{g\cdot cm^{-3}}$，$\rho_2 = 25\mathrm{g\cdot cm^{-3}}$，$\lambda = 174.5\mu\mathrm{m}$，$g = 2\times10^{14}\mathrm{cm\cdot s^{-2}}$．在这两个例子中，扰动有相同的波长，而界面两侧流体密度差不同．利用(7)式，可以得到两个例子发生非线性偏离的阈值分别为 0.03491 和 $0.01745\mathrm{g\cdot cm^{-2}}$，图 3 中用小圆圈标出，这时数值模拟结果约为线性理论结果的 90.7%．

4.2 柱几何中 RT 不稳定性

柱几何中 RT 不稳定性在线性阶段具有指数增长形式，增长率为 $\gamma = \sqrt{\alpha\frac{m}{R}g}$．这里对三个例子进行了分析，数值模拟结果如图 4 和图 5 所示．对于例 1，取 $\rho_1 = 20\mathrm{g\cdot cm^{-3}}$，$\rho_2 = 10\mathrm{g\cdot cm^{-3}}$，$R = 0.1\mathrm{cm}$，$m = 36$，$g = 10^{14}\mathrm{cm\cdot s^{-2}}$；对于例 2，取 $\rho_1 = 20\mathrm{g\cdot cm^{-3}}$，$\rho_2 = 10\mathrm{g\cdot cm^{-3}}$，$R = 0.1\mathrm{cm}$，$m = 24$，$g = 10^{14}\mathrm{cm\cdot s^{-2}}$；对于例 3，取 $\rho_1 = 20\mathrm{g\cdot cm^{-3}}$，$\rho_2 = 10\mathrm{g\cdot cm^{-3}}$，$R = 0.2\mathrm{cm}$，$m = 48$，$g = 10^{14}\mathrm{cm\cdot s^{-2}}$．图 4 中比较了例 1 与例 2，这时扰动只是模数不同．图 5 中比较了例 2 与例 3，这时扰动具有相同的波长，而界面到轴心的距离相差一倍．利用(11)式，可以得到上面三个例子发生非线性偏离的阈值分别为 0.001745，0.002618 和 $0.005236\mathrm{g\cdot cm^{-1}\cdot rad^{-1}}$，图 4 和图 5 中用小圆圈标出，这时数值模拟结果约为线性理论结果的 90.2%．

4.3. 球几何中 RT 不稳定性

球几何中 RT 不稳定性在线性阶段也具有指数增长形式，增长率为．$\gamma = \sqrt{\frac{l(l+1)(\rho_1 - \rho_2)}{l\rho_1 + (l+1)\rho_2}\frac{g}{R}}$．这里对两个例子进行了分析，数值模拟结果如图 6 所示．对于例 1，取 $\rho_1 = 20\mathrm{g\cdot cm^{-3}}$，$\rho_2 = 10\mathrm{g\cdot cm^{-3}}$，$R = 0.1\mathrm{cm}$，$l = 8$，$g = 10^{14}\mathrm{cm\cdot s^{-2}}$；对于例 2，取 $\rho_1 = 20\mathrm{g\cdot cm^{-3}}$，$\rho_2 = 10\mathrm{g\cdot cm^{-3}}$，$R = 0.1\mathrm{cm}$，$l = 5$，$g = 10^{14}\mathrm{cm\cdot s^{-2}}$．利用(16)式，可以得到上面两个例子发生非线性偏离的阈值分别为 0.0007405 和 $0.001147\mathrm{g\cdot sr^{-1}}$，图 6 中用小圆圈标出，这时数值模拟结果约为线性理论结果的 89.5%．

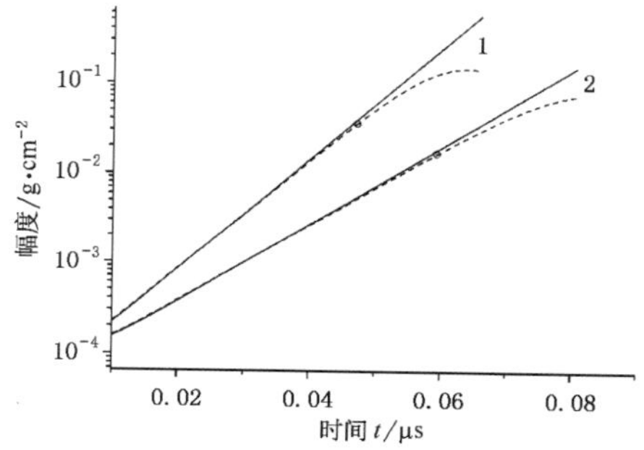

图 3　平面几何中密度扰动幅度随时间的变化　曲线 1 为例 1，曲线 2 为例 2．——为理论结果，……为数值模拟结果

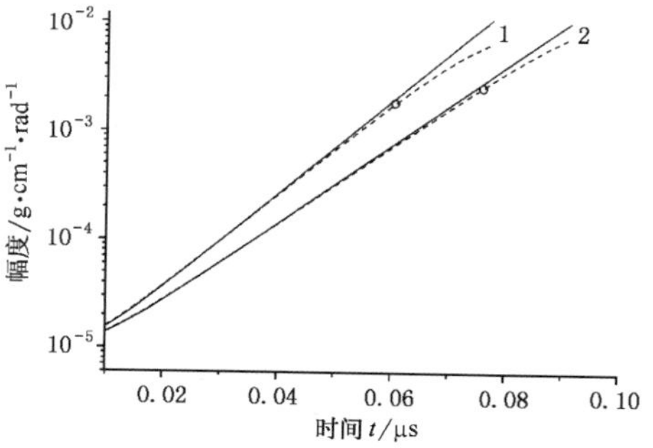

图 4　柱几何中密度扰动幅度随时间的变化(这时扰动只是模数不同)　曲线说明同图 3

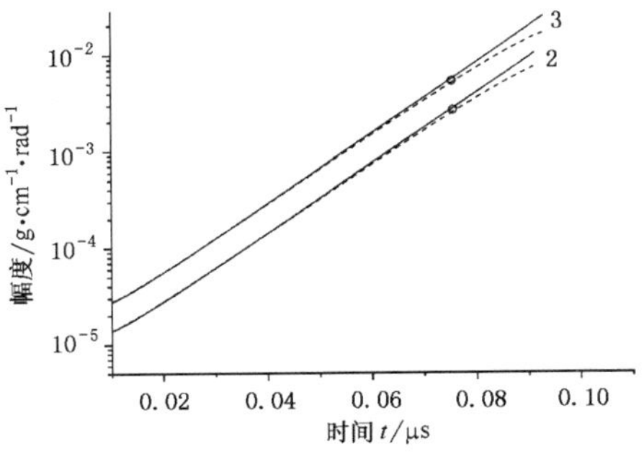

图5 柱几何中密度扰动幅度随时间的变化(这时扰动具有相同的波长,而界面到轴心的距离相差一倍) 曲线2为例2,曲线3为例3.——和……说明同图3

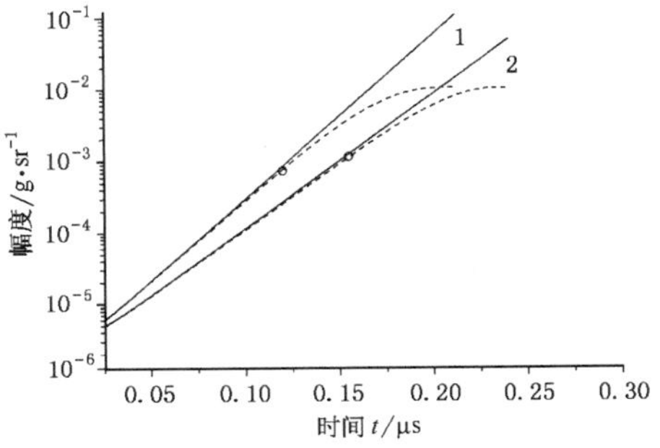

图6 球几何中密度扰动幅度随时间的变化 曲线1为例1,曲线2为例2.——和……说明同图3

5. 结 论

本文讨论了二维平面几何、柱几何和球几何中RT不稳定性发生非线性偏离的阈值问题,给出了三种几何中密度扰动振幅的非线性阈值公式,并且用数值模拟进行了检验. 这里取0.1λ为不稳定性界面扰动振幅的非线性阈值,数值模拟中基模增长偏离线性理论约为10%,但是,相对于平面几何情况,柱几何和球几何中的偏离稍微大一些,这表明柱几何和球几何中的模耦合机理可能与平面几何有些不同,存在几何效应的影响,这有待于进一步深入研究.

[1] Taylor G I 1950 *Proc. Roy. Soc. London* A **201** 192
[2] Ye W H, Zhang W Y and He X T 2000 *Acta Phys. Sin.* **49** 762(in Chinese)[叶文华、张维岩、贺贤土 2000 物理学报 **49** 762]
[3] Sharp D H 1984 *Physica* **12**D 3
[4] Lindl J 1995 *Phys. Plasmas* **2** 3933
[5] Dimonte G 2000 *Phys. Plasmas* **7** 2255
[6] Jacobs J W and Catton I 1988 *J. Fluid. Mech.* **187** 329
[7] Beck J B 1996 *Doctoral Dissertation* Purdue University
[8] Ofer D 1996 *Phys. Plasmas* **3** 3073
[9] Knauer J P *et al* 2000 *Phys. Plasmas* **7** 338
[10] Haan S W 1989 *Phys. Rev.* A **39** 5812
[11] Boris J P and Book D L 1973 *J. Comput. Phys.* **11** 38

Nonlinear threshold of two-dimensional Rayleigh-Taylor growth for incompressible liquid*

Wu Jun-Feng[1)] Ye Wen-Hua[2)3)] Zhang Wei-Yan[3)] He Xian-Tu[2)3)]
[1)] (*Graduate School, China Academy of Engineering Physics, Beijing* 100088, *China*)
[2)] (*Department of Physics, Zhejiang University, Hangzhou* 310028, *China*)
[3)] (*Institute of Applied Physics and Computational Mathematics, Beijing* 100088, *China*)
(Received 21 June 2002; revised manuscript received 28 October 2002)

Abstract

Nonlinear threshold of two-dimensional Rayleigh-Taylor(RT) instability is analyzed in planar and cylindrical and spherical geometries. Density amplitude is defined relating to instable interface and formulas of nonlinear threshold values for RT instability in three kinds of geometries are given, then high-order algorithm is used to simulate two-dimensional RT instability in these geometries, and the simulation results agree well with the formulas.

Keywords: Rayleigh-Taylor instability, nonlinear threshold, density amplitude
PACC: 5253, 5250J, 4720

* Project supported by the National Natural Science Foundation of China(Grant Nos. 19932010 and 10135010), and the National Basic Research Foundation for "Nonlinear Science" of China.

A LETTERS JOURNAL EXPLORING THE FRONTIERS OF PHYSICS
April 2009
www.epljournal.org
EPL, **86** (2009) 15002
doi: 10.1209/0295-5075/86/15002

Weakly nonlinear analysis on the Kelvin-Helmholtz instability

L. F. WANG[1,2], W. H. YE[3,4,1(a)], Z. F. FAN[1], Y. J. LI[2], X. T. HE[1,3,4] and M. Y. YU[5]

[1] *Laboratory of Computational Physics, Institute of Applied Physics and Computational Mathematics Beijing 100088, China*
[2] *School of Mechanics and Civil Engineering, China University of Mining and Technology Beijing 100083, China*
[3] *Department of Physics, Zhejiang University - Hangzhou 310027, China*
[4] *Center for Applied Physics and Technology, Peking University - Beijing 100871, China*
[5] *Institute for Fusion Theory and Simulation, Zhejiang University - Hangzhou 310027, China*

received 7 February 2009; accepted in final form 16 March 2009
published online 15 April 2009

PACS 52.35.Py – Macroinstabilities (hydromagnetic, *e.g.*, kink, fire-hose, mirror, balloning, tearing, trapped-particle, flute, Rayleigh-Taylor, etc.)
PACS 52.57.Fg – Implosion symmetry and hydrodynamic instability (Rayleigh-Taylor, Richtmyer-Meshkov, imprint, etc.)
PACS 47.20.Ft – Instability of shear flows (*e.g.*, Kelvin-Helmholtz)

Abstract – A weakly nonlinear model is proposed for the Kelvin-Helmholtz instability in two-dimensional incompressible fluids. The second- and third-harmonic generation effects of single-mode perturbation, as well as the nonlinear correction to the exponential growth of the fundamental modulation are analyzed. An important resonance in the mode-coupling process is found. The nonlinear saturation time depends on the initial perturbation amplitude and the density ratio of the two fluids, but the nonlinear saturation amplitude depends only on the initial perturbation amplitude. The weakly nonlinear result is supported by numerical simulation. The practical system of boundary layer containing thermal conductivity is analyzed. Their nonlinear saturation amplitude can be predicted by our weakly nonlinear model.

Introduction. – The Kelvin-Helmholtz (KH) instability occurs when two fluids flow with proximity to each other exhibiting a tangential velocity difference [1]. In this instability, small perturbations along the interface between two fluids undergo linear and nonlinear growth and may evolve into turbulent mixing through nonlinear interactions.

In an inviscid fluid, ignoring thermal conductivity, the unstable wave will grow linearly (the classical case) [2–4] until reaching large enough amplitudes when higher-harmonic generation sets on and the perturbation enters the nonlinear regime. Despite the progresses in research of the nonlinear evolution of the KH instability [5–8], understanding of the weakly nonlinear (WNL) regime of the KH instability is still unsatisfactory.

In this letter, a WNL model of the KH instability in a two-dimensional incompressible fluid is developed, adopting the standard velocity potential method. Analytical exhibiting of arbitrary Atwood numbers is derived for the generation of the second and third harmonics, and the nonlinear correction to the exponential growth of the fundamental modulation. We then include the thermal conductivity ablation by numerical simulation.

WNL model. – We introduce a Cartesian coordinate system where the x and y axes are along and normal to the undisturbed fluid interface. Consider two fluid layers, each with a finite velocity along the x-direction. An irrotational perturbation is assumed on each side of the interface between the fluids [9], so that we can write the velocity as $u_i = u_i^{(0)} + \nabla\phi_i$, where ϕ_i is the velocity potential of the perturbation. The physical quantities in the upper fluid are denoted by the subscript "A", in the lower fluid by the subscript "B". We analyze the growth of a small-amplitude perturbation $y = \eta(x,t)$ of the interface into the direction of the normal to the interface. The boundary condition that the pressure and normal velocity are both continuous at the interface can be written as

$$\begin{aligned} \eta_t + \eta_x\phi_{Ax} - \phi_{Ay}|_\eta &= 0, \\ \eta_t + \eta_x\phi_{Bx} - \phi_{By}|_\eta &= 0, \\ \left[\phi_{At} - \frac{1}{2}u_A^{(0)2} + \frac{1}{2}(\nabla\phi_A)^2\right]\bigg|_\eta &= \frac{\rho_B}{\rho_A}\left[\phi_{At} - \frac{1}{2}u_A^{(0)2} + \frac{1}{2}(\nabla\phi_A)^2\right]\bigg|_\eta, \end{aligned} \quad (1)$$

(a) E-mail: ye_wenhua@iapcm.ac.cn

where ρ is the density. The typical amplitude ε of the initial perturbation is assumed to be small (*i.e.*, smaller than its typical wavelength). The interface displacement η and perturbation velocity potential ϕ_i can be expanded into a power series in ε as

$$\eta(x,t)=\varepsilon\eta^{(1)}(x,t)+\varepsilon^2\eta^{(2)}(x,t)+\varepsilon^3\eta^{(3)}(x,t)+O(\varepsilon^4), \tag{2}$$

$$\phi_i(x,y,t)=\varepsilon\phi_i^{(1)}(x,y,t)+\varepsilon^2\phi_i^{(2)}(x,y,t) +\varepsilon^3\phi_i^{(3)}(x,y,t)+O(\varepsilon^4). \tag{3}$$

Substituting eqs. (2) and (3) into the boundary conditions (1) and collecting terms of the same power in ε, we will consider terms up to third order.

We consider a single-mode initial perturbation in the form $\eta^{(1)}(x,t=0)=\varepsilon\exp(ikx)+\text{c.c.}$, with c.c. being its complex conjugate. The first order gives the linear growth rate and frequency, respectively,

$$\gamma=k\frac{\sqrt{\left(u_A^{(0)}-u_B^{(0)}\right)r}}{1+r}, \tag{4}$$

$$\omega=k\frac{u_A^{(0)}+ru_B^{(0)}}{1+r}, \tag{5}$$

where $r=\rho_B/\rho_A$ is the density ratio.

In the WNL regime, mode coupling will soon generate higher harmonics. After a short linear growth time, the terms including $\exp(-\gamma t)$ can be ignored. In the second-order expansion, we write

$$\hat{\eta}^{(2)}(x,t)=\varepsilon^2(\hat{\eta}_{22}+t\hat{\eta}_{21})\exp(2nt+i2kx)+\text{c.c.}, \tag{6}$$

where $n=\gamma-i\omega$. The growth rate γ and frequency ω follow from linear stability analysis. The terms $\hat{\eta}_{22}(k)$ and $\hat{\eta}_{21}(k)$ originate from the second-harmonic generation.

The partial differential equation for the second harmonic is

$$\frac{\mathrm{d}^2\eta^{(2)}}{\mathrm{d}t^2}+\frac{4ik\left(u_A^{(0)}\rho_A+u_B^{(0)}\rho_B\right)}{\rho_A+\rho_B}\frac{\mathrm{d}\eta^{(2)}}{\mathrm{d}t} -\frac{4k^2\left(u_A^{(0)2}\rho_A+u_B^{(0)2}\rho_B\right)}{\rho_A+\rho_B}\eta^{(2)}= \varepsilon^2e^{2nt}\left[\frac{2k^3\left(u_A^{(0)2}\rho_A-u_B^{(0)2}\rho_B\right)}{\rho_A+\rho_B} -\frac{4ink^2\left(u_A^{(0)}\rho_A-u_B^{(0)}\rho_B\right)}{\rho_A+\rho_B}-2Akn^2\right], \tag{7}$$

where $A=(\rho_A-\rho_B)/(\rho_A+\rho_B)$. The left-hand side of eq. (7) are the second-harmonic generation terms. The right-hand side contains the coupling terms of two first-order quantities. A double root appears in eq. (7), because the growth rate of mode coupling is $2n$, which is equal to the growth rate of the second harmonic, *i.e.*, the process of second-harmonic generation is a resonant process. Therefore, we write the amplitude coefficient of the second-harmonic perturbation in the form $(\hat{\eta}_{22}+t\hat{\eta}_{21})$. Second-order analysis yields for the second-harmonic coefficients

$$\hat{\eta}_{22}=\frac{Ak}{4}+i\frac{\gamma}{2\left(u_A^{(0)}-u_B^{(0)}\right)}, \tag{8}$$

$$\hat{\eta}_{21}=-Ak\gamma-i\frac{2\gamma^2}{u_A^{(0)}-u_B^{(0)}}. \tag{9}$$

The partial differential equation for the third harmonic is relatively complex but similar to the second harmonic (7). For brevity, we will not write it here. We find that the third harmonic has also resonance in the mode-coupling process, which is written as $(\hat{\eta}_{33}+t\hat{\eta}_{32}+t^2\hat{\eta}_{31})$. For the same reason, the nonlinear correction coefficients are written as $(\hat{\eta}_{42}+t\hat{\eta}_{41})$. Thus, to the third-order expansion one has

$$\hat{\eta}^{(3)}(x,t) = \varepsilon^3(\hat{\eta}_{33}+t\hat{\eta}_{32}+t^2\hat{\eta}_{31})\exp(3nt+i3kx) -\varepsilon^3(\hat{\eta}_{42}+t\hat{\eta}_{41})\exp(3nt+ikx)+\text{c.c.}, \tag{10}$$

with third-harmonic coefficients,

$$\hat{\eta}_{33}=\frac{k^2(1-6r+r^2)}{9(1+r)^2}+i\frac{4Ak\gamma}{9\left(u_A^{(0)}-u_B^{(0)}\right)}, \tag{11}$$

$$\hat{\eta}_{32}=-\frac{2k^2\gamma(1-6r+r^2)}{3(1+r)^2}-i\frac{8Ak\gamma^2}{3\left(u_A^{(0)}-u_B^{(0)}\right)}, \tag{12}$$

$$\hat{\eta}_{31}=\frac{2k^4r\left(u_A^{(0)}-u_B^{(0)}\right)^2(1-6r+r^2)}{(1+r)^4} +i\frac{8Ark^3\gamma\left(u_A^{(0)}-u_B^{(0)}\right)}{(1+r)^2}, \tag{13}$$

and the nonlinear correction coefficients to the fundamental modulation,

$$\hat{\eta}_{42}=\frac{1}{4}k^2, \tag{14}$$

$$\hat{\eta}_{41}=\frac{k^2\gamma}{2}. \tag{15}$$

We introduce the following dimensionless parameters: $\rho_0=\rho_A$, $u_0=(u_A^{(0)}-u_B^{(0)})/2$, $l_0=\lambda$, $t_0=2\lambda/(u_A^{(0)}-u_B^{(0)})$, $P_0=1/4\rho_A(u_A^{(0)}-u_B^{(0)})^2$, where $\rho_A\geqslant\rho_B$, $u_A^{(0)}\geqslant u_B^{(0)}$, and we choose $(u_A^{(0)}+u_B^{(0)})/2$ as the reference velocity for normalization. Then the velocity, density and perturbation wave number can be written as $u_A^{(0)}=1$, $u_B^{(0)}=-1$, $\rho_A=1$,

$\rho_B = r$, and $k = 2\pi$. The WNL results can be rewritten as

$$\gamma = \frac{4\pi\sqrt{r}}{1+r},$$

$$\omega = \frac{2\pi(1-r)}{1+r},$$

$$\hat{\eta}_{22} = -\frac{\pi(\sqrt{r}-i)}{2(\sqrt{r}+i)},$$

$$\hat{\eta}_{21} = \frac{8\pi^2\sqrt{r}}{(\sqrt{r}+i)^2},$$

$$\hat{\eta}_{33} = \frac{4\pi^2(\sqrt{r}-i)^2}{9(\sqrt{r}+i)^2},$$

$$\hat{\eta}_{32} = -\frac{32\pi^3\sqrt{r}(\sqrt{r}-i)}{3(\sqrt{r}+i)^3},$$

$$\hat{\eta}_{31} = \frac{128\pi^4 r}{(\sqrt{r}+i)^4},$$

$$\hat{\eta}_{42} = \pi^2,$$

$$\hat{\eta}_{41} = \frac{8\pi^3\sqrt{r}}{1+r}.$$

The instability interface becomes

$$\begin{aligned}\eta(x,t) &= \varepsilon e^{\gamma t}\cos(kx-\omega t)\\ &\quad - \varepsilon^3\eta_4 e^{3\gamma t}\cos(kx-3\omega t)\\ &\quad + \varepsilon^2\eta_2 e^{2\gamma t}\cos[2(kx-\omega t)-\varphi_2]\\ &\quad + \varepsilon^3\eta_3 e^{3\gamma t}\cos[3(kx-\omega t)-\varphi_3], \qquad (16)\end{aligned}$$

where $\eta_4 = \eta_{4a} + \eta_{4b}t = 2\pi^2 + 16\pi^3 t\sqrt{r}/(1+r)$, $\eta_2 = \eta_{2a} + \eta_{2b}t = -\pi + 16\pi^2 t\sqrt{r}/(1+r)$, $\varphi_2 = \pi + \arctan[2\sqrt{r}/(1+r)]$, $\eta_3 = \eta_{3a} + \eta_{3b}t + \eta_{3c}t^2 = 8\pi^2/9 - 64\pi^3 t\sqrt{r}/[3(1+r)] + 256\pi^4 t^2 r/(1+r)^2$, and $\varphi_3 = \arctan[4(1-r)\sqrt{r}/(r^2-6r+1)]$ for $0 < r < 3-2\sqrt{2}$ and $r=1$, else $\varphi_3 = \pi + \arctan[4(1-r)\sqrt{r}/(r^2-6r+1)]$.

Instability interface. The instability interface from WNL theory at various times $t = 0$, 0.1, 0.2, 0.3, 0.4, 0.5, 0.6, 0.7, 0.8 are shown in fig. 1. The initial cosine perturbation grows exponentially and the phase changes like ωt at early stage (before 0.4). After that, the second and third harmonics cannot be ignored anymore. The interface in the two fluids becomes asymmetric. The "spike" in the heavy fluid becomes thin and the "bubble" in the light fluid becomes fat. The "windward side" of the interface broadens and the "leeward side" steepens. The amplitude in the heavy and light fluids become different in time. The amplitude in the heavy fluid will be larger than the light, which is apparent after $t = 0.6$. The calculations show that at early times the contribution of the resonant terms (*i.e.*, $t\eta_{2b}$, $t^2\eta_{3c}$, $t\eta_{4b}$) are very small, and the second and third harmonics grow exponentially. After a short linear period, the contribution of the resonance term increases significantly. The growth rates of the second and third harmonics become greater than 2γ and 3γ, respectively. From these considerations, we may conclude that it is the resonance that supports the evolution of the instability in the WNL regime of KH instability.

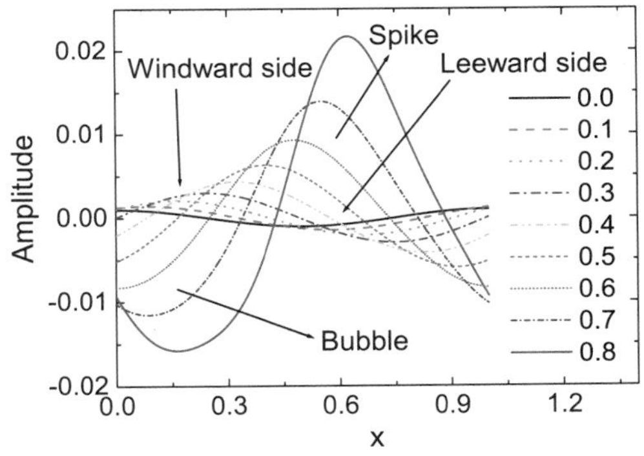

Fig. 1: (Color online) The perturbation interface at different times with initial perturbation $\varepsilon = 0.001$ and density ratio $r = 0.1$.

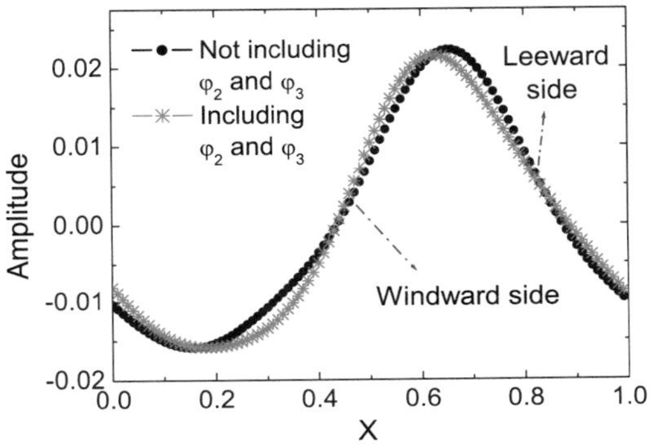

Fig. 2: (Color online) The interfaces with and without the second-order and third-order mode-coupling phases, φ_2 and φ_3. The density ratio is $r = 0.1$, the initial perturbation is $\varepsilon = 0.001$, and the time is $t = 0.8$.

Phase effect. The fundamental mode was assumed to be of zero phase. Performing the calculation one observes that the second and third harmonics develop second-order and third-order mode-coupling phases, φ_2 and φ_3, respectively, while the nonlinear correction to the fundamental mode lacks any mode-coupling phase. Figure 2 shows the interfaces with and without the second- order and third-order mode-coupling phases, φ_2 and φ_3. It is seen in fig. 2 that the mode-coupling phases φ_2 and φ_3 cause the "windward side" of the interface to steepen and smooth the "leeward side", which means that the mode-coupling phases φ_2 and φ_3 retard the deformation of the instability interface.

Nonlinear threshold. A relation between the nonlinear saturation time t_s, wave number k of the perturbation,

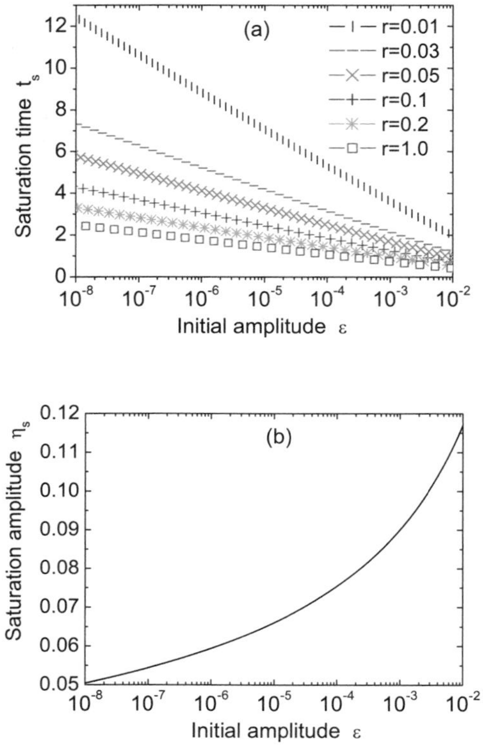

Fig. 3: The nonlinear threshold. (a) Nonlinear saturation time *vs.* initial perturbation amplitude at different density ratios; (b) Nonlinear saturation amplitude *vs.* initial perturbation amplitude.

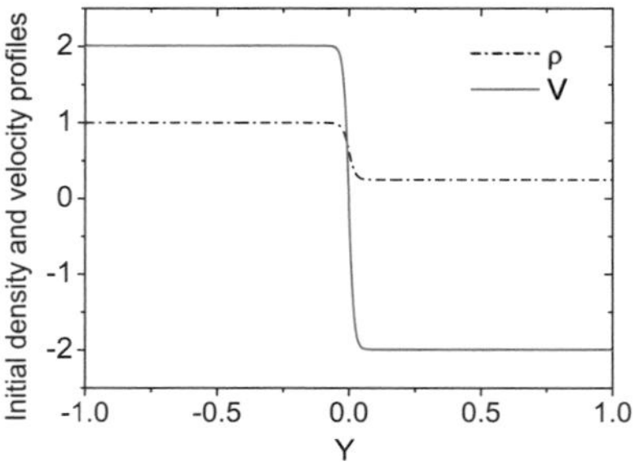

Fig. 4: (Color online) The initial density and velocity profiles used in the numerical simulations.

and the initial perturbation amplitude ε is obtained if we assume that the nonlinear process starts when the growth of fundamental mode deviates by 10% from linear growth:

$$\frac{1}{32}k^2+\frac{1}{16}\gamma k^2 t_\mathrm{s}=\frac{1}{10\eta_\mathrm{s}^2}, \tag{17}$$

where $\eta_\mathrm{s}=\varepsilon e^{\gamma t_\mathrm{s}}$ is the linear growth saturation amplitude of the fundamental mode. In normalized units this becomes

$$\frac{\pi^2}{8}+\frac{\pi^3\sqrt{r}}{1+r}t_\mathrm{s}=\frac{1}{10\varepsilon^2}\exp\left(-\frac{8\pi\sqrt{r}}{1+r}t_\mathrm{s}\right). \tag{18}$$

The relations between the initial perturbation amplitude and the nonlinear saturation time at the density ratios $r=0.01$, 0.03, 0.05, 0.1, 0.2, 1.0 are illustrated in fig. 3(a). The relation between the initial perturbation amplitude and the nonlinear saturation amplitude is plotted in fig. 3(b). For fixed density ratio, the nonlinear saturation time increases with initial perturbation amplitude. For fixed initial amplitude, the saturation time increases with decreasing density ratio. Equation (17) indicates that the value of γt_s depends only on the amplitude of the initial perturbation ε. That is to say, the nonlinear saturation amplitude depends only on the initial perturbation amplitude as shown in fig. 3(b).

Numerical simulation. – In this study, we assume that the fluid is inviscid. In plasma physics, the viscosity is so small that its contribution can be ignored. The basic equations are

$$\begin{aligned}\frac{\partial\rho}{\partial t}+\nabla\cdot(\rho\mathbf{V})&=0,\\ \frac{\partial(\rho\mathbf{V})}{\partial t}+p\nabla\cdot(\rho\mathbf{V}\mathbf{V}+p\mathbf{I})&=0,\\ \frac{\partial E}{\partial t}+\nabla\cdot(E+p)\mathbf{V}&=\nabla\cdot(\kappa\nabla T),\end{aligned} \tag{19}$$

where ρ, $\mathbf{V}$, E, p, and T represent the density, velocity, total energy, pressure, and temperature, respectively. For the thermal conductivity, the model of Spitzer-Harm (SH) is adopted [10], which is of the form $\kappa=\kappa_\mathrm{SH}\propto T^{5/2}$.

Figure 4 shows the boundary layer model used in this study. The initial density and velocity profiles are

$$u^{(0)}(y)=\frac{u_A^{(0)}+u_B^{(0)}}{2}-\frac{u_A^{(0)}-u_B^{(0)}}{2}\tanh\left(\frac{y}{L_u}\right), \tag{20}$$

$$\rho^{(0)}(y)=\frac{\rho_A^{(0)}+\rho_B^{(0)}}{2}-\frac{\rho_A^{(0)}-\rho_B^{(0)}}{2}\tanh\left(\frac{y}{L_\rho}\right), \tag{21}$$

where L_u and $L_\rho$are the respective half-thicknesses of the shear layer.

In the numerical simulations, we use the finite difference hydrodynamic code LARED-S [11]. We start with an interface perturbation $\eta(x)=\varepsilon\cos kx$. A velocity potential perturbation is also introduced at the beginning.

Classical case. L_u and L_ρ are assumed to be small ($\ll\lambda$). The LARED-S code was run in the nearly incompressible limit. For the second harmonic, it is difficult to distinguish the growth induced by the mode coupling and its proper growth, since it possesses a resonance. However, in the simulation the coefficients of t in the second-harmonic amplitude η_2 can be obtained. Because of the same reason, for the third harmonic and the nonlinear correction to the exponential growth of the fundamental

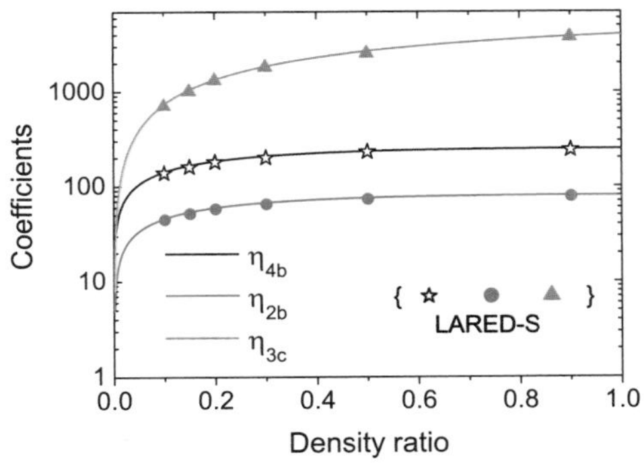

Fig. 5: (Color online) Results from the WNL theory and simulations.

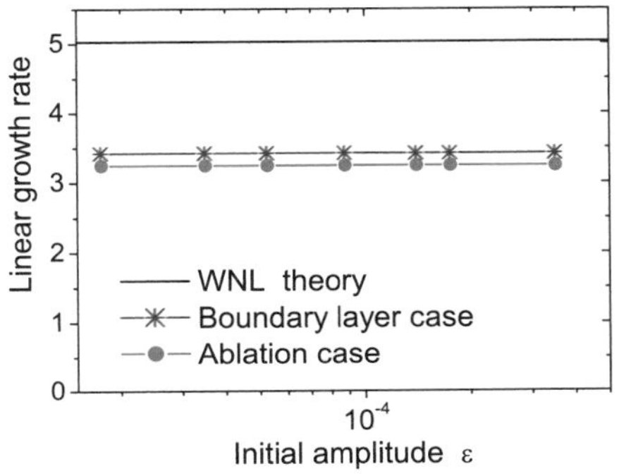

Fig. 6: (Color online) The linear growth rate *vs.* initial perturbation amplitude with density ratio $r = 0.25$.

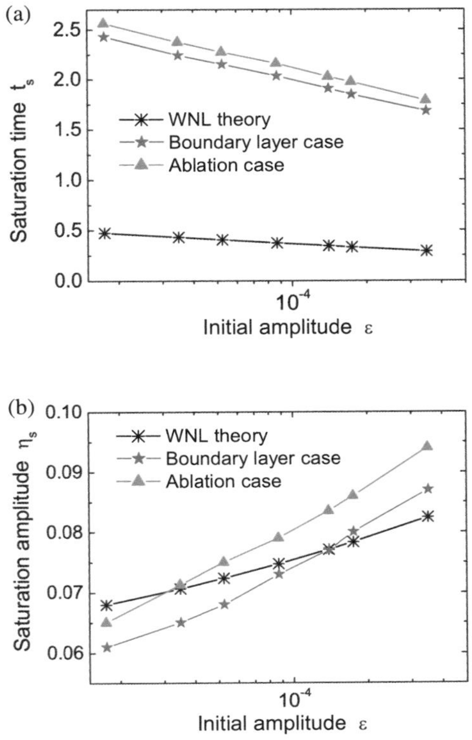

Fig. 7: (Color online) The nonlinear threshold. (a) Nonlinear saturation time *vs.* initial perturbation amplitude; (b) Nonlinear saturation amplitude *vs.* initial perturbation amplitude, given by the WNL model and simulations.

modulation, the coefficients of t^2 and t in η_3 and η_4 can be determined, respectively.

In fig. 5, we compare the results from our WNL model and the simulation. The vertical coordinate are the coefficients of time t, t^2, and t in η_2, η_3, and η_4, respectively. The simulation results are obtained from fitting the amplitudes of the second and third harmonics, and the nonlinear correction to the fundamental mode at different times. For the linear growth rate, the simulation gives almost the exact results. For η_{2b} and η_{3c}, our model agrees quite well with the simulation, and the deviation is within 2%. For η_{4b}, our model also agrees with the simulation, and the deviation is about 5%.

Similar results are obtained for other velocities, and densities of the flows or perturbation wavelengths. As can be seen from fig. 5, the values of η_{2b}, η_{3c}, and η_{4b} increase significantly with increasing density ratio. When the density ratio is < 0.5, the η_{2b}, η_{3c}, and η_{4b} are sensitive to the density ratio. When the density is > 0.5, the η_{2b}, η_{3c}, and η_{4b} are less sensitive to the density ratio and approach 77, 3700, and 240, respectively.

Boundary layer and ablation case. In practice, the interface is not exactly discontinuous and in some cases the thermal conductivity cannot be ignored. In this section, we study the linear growth rate and nonlinear saturation threshold at different perturbations. In the boundary layer case, L_u and L_ρ are equal to $1/3$ and the thermal conductivity term is ignored. In the ablation case, L_u and L_ρ are also equal to $1/3$ and the thermal conductivity was calculated. The initial static pressure is 2.5 in the above two cases.

Figure 6 shows the linear growth rate at different perturbations with density ratio of 0.25. The linear growth rate is 5.02653 given by the linear theory [1]. In the boundary and ablation cases, the linear growth rates are about 3.412 and 3.238, respectively, which is smaller than the classical case. The linear growth rate does not depend on the initial perturbations in the boundary or the ablation cases.

In fig. 7, we present the nonlinear saturation time and the nonlinear saturation amplitude at different initial perturbations with density ratio of 0.25. In the classical case, the nonlinear saturation time decreases with the increase of the initial perturbation, which also applies to the boundary and ablation cases.

As shown in fig. 7(b), the three cases have similar nonlinear saturation amplitudes at different initial perturbation. Our WNL model could predict the nonlinear saturation amplitude in the boundary layer and ablation cases. The nonlinear saturation amplitude in the ablation case is larger than that in the boundary case. When the initial perturbation is about $3.5 \times 10^{-5}(1.4 \times 10^{-4})$, the nonlinear saturation amplitude in the ablation (the boundary layer) case exceeds the classical case, respectively.

Conclusion. – In conclusion, a third-order WNL theory for the incompressible KH instability is developed. It agrees well with the simulation results and sheds new light on the nonlinear KH instability in practical system. In the WNL regime of KH instability, it is the resonance that enhances the nonlinear interaction. The nonlinear saturation time depends on the density ratio of the two fluids and the initial perturbation amplitude. The nonlinear saturation amplitude depends only on initial perturbation amplitude.

The ablative KH instability is analyzed by numerical simulation. Some nonlinear behavior in ablative KH instability can be explained by our WNL model. The trend of the nonlinear saturation time in practical system is similar to that in our WNL model. The nonlinear saturation amplitude can be previewed by our WNL model.

* * *

We wish to thank Guangcai Zhang and Aiguo Xu for helpful discussions. This work was supported by the National Basic Research Program of China (Grant No. 2007CB815100), the National Natural Science Foundation of China (Grant Nos. 10775020 and 10874242), and the Research Fund for the Doctoral Program of Higher Education of China (Grant No. 20070290008).

REFERENCES

[1] Chandrasekhar S., *Hydrodynamic and Hydromagnetic Stability* (Oxford University, London) 1961.
[2] Walker J. S., Talmage G., Brown S. H. and Sondergaard N. A., *Phys. Fluids A*, **5** (1993) 1466.
[3] Bau H. H., *Phys. Fluids*, **25** (1982) 1719.
[4] Miles J. W., *Phys. Fluids*, **23** (1980) 1719.
[5] Bodo G., Mignone A. and Rosner R., *Phys. Rev. E*, **70** (2004) 36304.
[6] Miura A., *Phys. Rev. Lett.*, **83** (1999) 1586.
[7] Horton W., Perez J. C., Carter T. and Bengtson R., *Phys. Plasmas*, **12** (2005) 22303.
[8] Blaauwgeers R., Eltsov V. B., Eska G., Finne A. P., Haley R. P., Krusius M., Ruohio J. J., Skrbek L. and Volovik G. E., *Phys. Rev. Lett.*, **89** (2002) 155301.
[9] Landau L. D. and Lifshitz E. M., *Fluid Mechanics* (Pergamon Press, Oxford) 1959.
[10] Spitzer L. and Harm R., *Phys. Rev.*, **89** (1953) 977.
[11] Ye W. H., Zhang W. Y. and He X. T., *Phys. Rev. E*, **65** (2002) 57401.

PHYSICS OF PLASMAS **17**, 122308 (2010)

Formation of large-scale structures in ablative Kelvin–Helmholtz instability

L. F. Wang,[1,2,3] W. H. Ye,[2,4,a)] Wai-Sun Don,[3] Z. M. Sheng,[5] Y. J. Li,[1] and X. T. He[2,4]
[1]*SMCE, China University of Mining and Technology, Beijing 100083, People's Republic of China*
[2]*CAPT, Peking University, Beijing 100871, People's Republic of China and LCP, Institute of Applied Physics and Computational Mathematics, Beijing 100088, People's Republic of China*
[3]*Department of Mathematics, Hong Kong Baptist University, Kowloon Tong, Hong Kong*
[4]*Department of Physics, Zhejiang University, Hangzhou 310027, People's Republic of China*
[5]*Beijing National Laboratory for Condensed Matter Physics, Institute of Physics, CAS, Beijing 100190, People's Republic of China and Department of Physics, Shanghai Jiao Tong University, Shanghai 200240, People's Republic of China*

(Received 22 June 2010; accepted 16 November 2010; published online 15 December 2010)

In this research, we studied numerically nonlinear evolutions of the Kelvin–Helmholtz instability (KHI) with and without thermal conduction, aka, the ablative KHI (AKHI) and the classical KHI (CKHI). The second order thermal conduction term with a variable thermal conductivity coefficient is added to the energy equation in the Euler equations in the AKHI to investigate the effect of thermal conduction on the evolution of large and small scale structures within the shear layer which separate the fluids with different velocities. The inviscid hyperbolic flux of Euler equation is computed via the classical fifth order weighted essentially nonoscillatory finite difference scheme and the temperature is solved by an implicit fourth order finite difference scheme with variable coefficients in the second order parabolic term to avoid severe time step restriction imposed by the stability of the numerical scheme. As opposed to the CKHI, fine scale structures such as the vortical structures are suppressed from forming in the AKHI due to the dissipative nature of the second order thermal conduction term. With a single-mode sinusoidal interface perturbation, the results of simulations show that the growth of higher harmonics is effectively suppressed and the flow is stabilized by the thermal conduction. With a two-mode sinusoidal interface perturbation, the vortex pairing is strengthened by the thermal conduction which would allow the formation of large-scale structures and enhance the mixing of materials. In summary, our numerical studies show that thermal conduction can have strong influence on the nonlinear evolutions of the KHI. Thus, it should be included in applications where thermal conduction plays an important role, such as the formation of large-scale structures in the high energy density physics and astrophysics.
 [doi:10.1063/1.3524550]

I. INTRODUCTION

The Kelvin–Helmholtz instability (KHI) occurs when there is a velocity shear layer between two fluids separated by a perturbed interface. It is considered as one of the important processes [other processes included but not limited to the well-known Rayleigh–Taylor instability (RTI) and Richtmyer–Meshkov instability] in the study of space plasma. Some of the well-known examples of the KHI in space plasma are the mass transfer in a planet's magnetopause[1–4] and the interaction between the solar wind and Earth's magnetosphere,[5] where large-scale structures of vortex formed by the KHI lead to large-scale plasma mixing across a shear layer. In the high energy density physics, examples of appearance of the KHI include high-Mach-number shocks[6–9] and jets, radioactive blast waves, and radioactively driven molecular clouds, gamma-ray bursts, and accretion black holes.[10–12] It has been observed that well collimated jets are formed in astrophysical objects[13–16] and jetlike long spikes are observed in direct-driven RTI experiments in inertial confinement fusion (ICF).[17–19]

In the classical KHI (CKHI) where thermal conduction is absent from the consideration, with a single-mode sinusoidal interface perturbation, the interface undergoes a linear and a weakly nonlinear growth[20] before a strongly nonlinear growth that leads to turbulent mixing. Recently, the effects of density gradient and velocity gradient across the perturbed interface on the linear growth of the KHI are also analyzed.[21,22] It is found that the increasing density gradient alone tends to enhance the KHI while the increasing velocity gradient alone tends to reduce it. Their combined effects tend to stabilize the flow.

In the direct-driven RTI experiments in ICF research, jetlike long spikes are formed while general characteristic vortical roll-up structures in the tips of the spikes are not observed.[17–19,23] In astrophysics, the observed astrophysical jets are long but stable.[6,13–16] It raises an interesting question about the absence of small-scale vortical roll-up structures in high energy density physics experiments and in astronomical observations mentioned above.[6,13–19,23] Additional physical mechanisms might have played a crucial role in suppressing the formation of the small-scale structures which is characteristic of the KHI. It has been suggested that the magnetic field, radiative cooling, and thermal conduction are three of probable physical mechanisms involved.[6,10,24–26] The effects of magnetic field[27–31] and radiative cooling[32–34] on the linear

[a)]Author to whom correspondence should be addressed. Electronic mail: ye_wenhua@iapcm.ac.cn.

and nonlinear behaviors of the flow have been investigated by several authors. However, the effect of thermal conduction on the KHI, or known as the ablative KHI (AKHI) in literature, is not well understood. Some recent numerical simulations of the AKHI (Ref. 35) indicated that, in a single-mode sinusoidal perturbed interface with a given wavelength λ and amplitude η_0, the growth rate of the amplitude of the fundamental (master) mode is linear, and in the mean time the amplitude of higher harmonics are reduced. Furthermore, a cutoff wavelength λ_c was observed where no higher harmonics exists when $\lambda<\lambda_c$. In a two-mode sinusoidal perturbed interface, vortex pairing or coalescence of vortices, i.e., inverse cascade, occurs. It is a well-known classical hydrodynamics phenomenon where the vortices coalesce into a larger vortex[36,37] which would enhance a large-scale mixing of materials.

In this study, we will focus on the numerical simulation of the AKHI in the nonlinear regime of the flow and its results will be compared with the results obtained from the simulations of the CKHI. The paper is organized as the follows. In Sec. II, the mathematical framework and numerical setup are presented. The governing Euler equations with an addition of the second order thermal conduction term to the energy equation, along with an auxiliary equation for the thermal conductivity, are presented. The numerical methods, namely, the classical fifth order weighted essentially nonoscillatory (WENO) finite difference scheme for the inviscid fluxes and fourth order implicit finite difference scheme for the temperature equation in the case of the AKHI, employed and initial setups of the problem in this study are briefly described. In Sec. III, numerical results from the simulation of a single-mode and two-mode sinusoidal perturbed interfaces with the CKHI and the AKHI are presented and discussed. The thermal conduction effects on the KHI are studied by comparing the nonlinear evolutions of the CKHI and the AKHI. It is found that the small-scale structures are suppressed while the vortex paring process is improved in the AKHI. The most obvious difference between the CKHI and the AKHI is that the more large-scale structures are observed in the AHKI in comparison with the CKHI. Concluding remarks are given in Sec. IV.

II. MATHEMATICAL FRAMEWORK AND NUMERICAL SETUP

A. Governing equations

In this study, we will model the physical system via the two dimensional inviscid compressible fluid[38,39] with thermal conduction. The governing equations for the single-fluid and one-temperature in two dimensional compact conservative form is

$$\mathbf{Q}_t + \nabla \cdot (\mathbf{F},\mathbf{G}) = \nabla \cdot (\mathbf{F}_v,\mathbf{G}_v), \tag{1}$$

where the conservative variable $\mathbf{Q}$, the inviscid fluxes vector $(\mathbf{F},\mathbf{G})$, and the thermal conduction fluxes vector $(\mathbf{F}_v,\mathbf{G}_v)$, in Cartesian coordinates. More explicitly, they are

$$\mathbf{Q} = (\rho,\rho u,\rho v,E)^T,$$

$$\mathbf{F} = [\rho u,\rho u^2+P,\rho uv,(E+P)u]^T, \quad \mathbf{F}_v = \left(0,0,0,\kappa\frac{\partial T}{\partial x}\right)^T, \tag{2}$$

$$\mathbf{G} = [\rho v,\rho uv,\rho v^2+P,(E+P)v]^T, \quad \mathbf{G}_v = \left(0,0,0,\kappa\frac{\partial T}{\partial y}\right)^T,$$

where ρ,u,v,E are the density, velocity in x, velocity in y, and total energy, respectively, and the pressure

$$P = (\gamma-1)\left[E - \frac{1}{2}\rho(u^2+v^2)\right], \tag{3}$$

$\gamma=\frac{5}{3}$ is the specific heat ratio. The equation of state closes the system of equations,

$$P = c_v(\gamma-1)\rho T, \tag{4}$$

where T is the temperature in megakelvins (MK), c_v is the specific heat at constant volume, and $c_v=86.2713$ $(\text{cm}/\mu\text{s})^2$ MK^{-1} for the CH material. For the thermal conductivity κ, we adopt the Spitzer–Härm electron thermal conductivity[40] $\kappa=\kappa_s T^n$ with $k_s=0.01$ and $n=\frac{5}{2}$ in the AKHI simulations and $\kappa=0$ in the CKHI simulations.

B. Numerical methods

To simplify the notation for discussion below, the system of partial differential equations can be expressed in the form of

$$\mathbf{Q}_t + L\mathbf{Q} = L_v\mathbf{Q}, \tag{5}$$

where L and L_v are the spatial differential operators for the first (inviscid, nonlinear, hyperbolic) and second order (linear, parabolic) terms, respectively.

The computational domain is rectangular in shape. The domain in x direction is $[-200\ \mu\text{m},\ 200\ \mu\text{m}]$ with length $L_x=400\ \mu\text{m}$. The domain in y direction is assumed to be periodical with length $L_y=100\ \mu\text{m}$. The free-stream boundary conditions are imposed at both ends of the domain in x since the flow is essentially unchanged far away from the center of the computational domain for a sufficiently large domain. Since the domain in y is periodical, periodical boundary condition is imposed on the upper and lower boundaries.

We approximate the inviscid flux $L\mathbf{Q}$ in Eq. (5), which is the Euler equations, via the well-known classical fifth order weighted essentially nonoscillatory conservative characteristicwise finite difference scheme (WENO) explicitly. Following Refs. 41–43 the hyperbolicity of the Euler equation (1) admits a complete set of right and left eigenvectors for the Jacobian of the system. The eigenvalues and eigenvectors are obtained via the linearized Riemann solver of Roe.[44] The first order Lax–Friedrichs flux is used as the low order building block for the high order reconstruction step of the WENO schemes. After projecting the positive and negative fluxes on the characteristic fields via the left eigenvectors, the high order WENO reconstruction step is applied to obtain the high order approximation at the cell boundaries using the surrounding cell-centered values, which are projected back

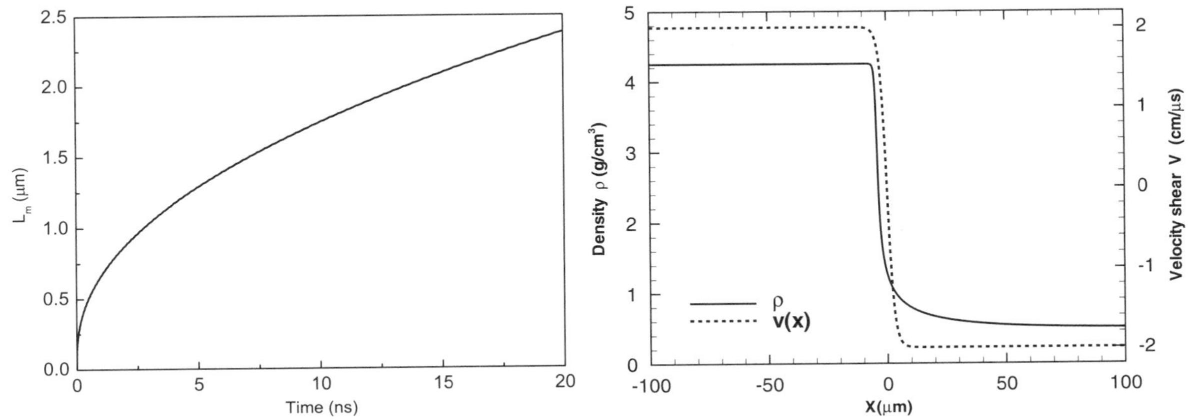

FIG. 1. (Left) The temporal evolution of the minimum density gradient length scale L_m. (Right) The profiles of the initial density $\rho(x,y,t=0)$ (solid line) and initial v-velocity $v(x,y,t=0)$ (dashed line) without interface perturbation at any given fixed location in y.

into the physical space via the right eigenvectors and added together to form a high order numerical flux at the cell-interfaces. The conservative difference of the reconstructed high order fluxes can then be computed for inviscid first order term $L\mathbf{Q}$. We refer to Ref. 43 for further details on the WENO algorithm for solving the hyperbolic conservation laws.

In the case that the thermal conductivity parameter $\kappa \neq 0$ as in the AKHI, the second order parabolic term involving the thermal conduction is solved via fourth order central implicit finite difference scheme to avoid the small time step restriction due to the severe stability condition imposed if solved by an explicit scheme.

In the simulations presented in Sec. III, the numbers of uniformly spaced grid points used in the x and y directions are $N_x=4000$ and $N_y=512$, respectively.

C. Initial condition setup

In the following discussion, we shall denote the properties of the fluid to the left of the interface by subscript l and to the right of the interface by subscript r unless stated otherwise (see Fig. 1 at time $t=0$).

One of the main result in our previous study in an ablative mixing between two fluids,[35] it is found that there is no absolute equilibrium (steady) flow when two fluids ablate with each other in the case of identical pressure but different densities. However, in the same study, it has been shown that a quasisteady density profile can be found by one dimensional calculation. In the case that the initial pressures of the two fluids are identical $P_l=P_r=40$ Mbar and the initial densities $\rho_l=4.25\ \mathrm{g/cm^3}$ and $\rho_r=0.5\ \mathrm{g/cm^3}$, the temporal evolution of the minimum density gradient scale length (typical thickness of the density transition layer) at the ablation surface, defined as $L_m=\min[|\rho/(\partial\rho/\partial x)|]$, is computed and shown in the left figure in Fig. 1. It can be observed that the rate of increase of L_m is decreasing, and after $t=5$ ns, the variation of L_m becomes small. Accordingly, we choose the ablative density profile at $t=14$ ns as an initial density profile and a shear vertical velocity in the form of

$$v(x)=\frac{1}{2}(v_l+v_r)-\frac{1}{2}(v_l-v_r)\tanh\left(\frac{y}{\delta}\right), \tag{6}$$

where v_l and v_r are the steady state fluid velocities in the left and right sides of the interface. The interface has a smooth hyperbolic tangent profile with an typical thickness $\delta=3\ \mu\mathrm{m}$. The velocity shear difference is $|v_l-v_r|=4\ \mathrm{cm}/\mu\mathrm{s}$ (see the right figure in Fig. 1).

The primitive variables (ρ,u,v,T) of the initial steady flow field are perturbed with a multimode sinusoidal interface perturbation at $x=x_0$ with wavelengths λ_i and amplitudes η_i in the y direction, namely,

$$S(x,y)=1+\sum_{i=1}^{m}\eta_i e^{-k_i|x-x_0|}\sin(k_i y), \tag{7}$$

where m is the number of perturbed modes, $x_0=0$ is the location of the interface, and $k_i=2\pi/\lambda_i$ are the wave numbers. The amplitudes are typically set as a fraction of wavelength $\eta_i=\alpha_i\lambda$ with $\frac{1}{10000}\le\alpha_i\le\frac{1}{1000}$.

III. NUMERICAL RESULTS AND ANALYSIS

In this work, we present the following two case studies.

(1) The first case is the comparison of the nonablative and the ablative KHI with a single-mode sinusoidal interface perturbation with wavelength $\lambda_1=50\ \mu\mathrm{m}$ and amplitude $\eta_1=0.0056\ \mu\mathrm{m}$.
(2) The second case is the comparison of the nonablative and the ablative KHI with a two-mode sinusoidal interface perturbation with wavelengths $\lambda_1=100\ \mu\mathrm{m}$ and $\lambda_2=50\ \mu\mathrm{m}$ and amplitudes $\eta_1=\eta_2=0.031999\ \mu\mathrm{m}$.

We note that the stability analysis introduced in Ref. 35 shows that the linear growth rate curve has a peak value which corresponds with the perturbation wavelength $\lambda=50\ \mu\mathrm{m}$ under the same physical conditions here.

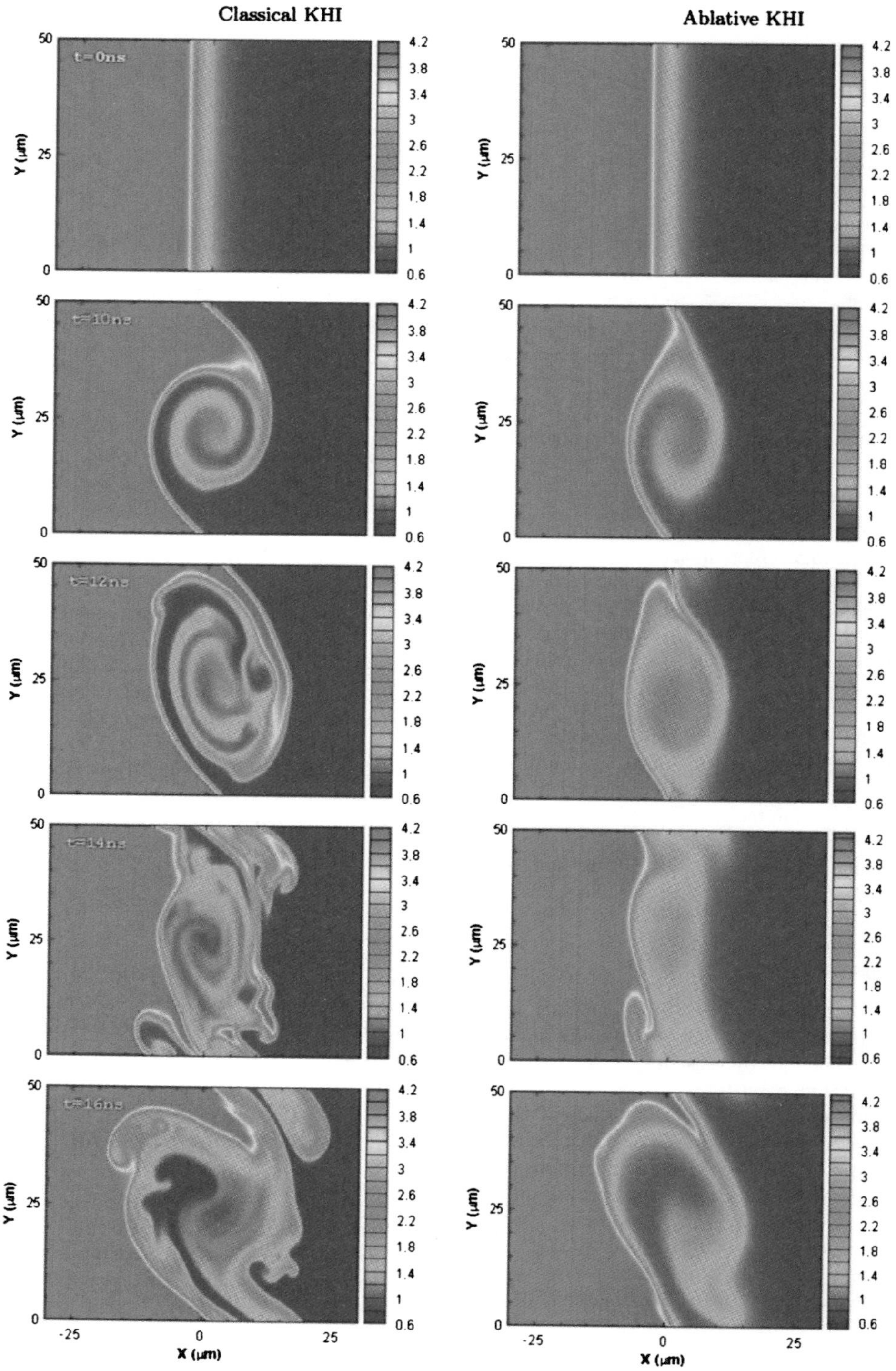

FIG. 2. (Color online) Density fields of the CKHI and the AKHI with a single-mode sinusoidal interface perturbation at times of $t=0,10,12,14,16$ ns. The perturbation wavelength is $\lambda_1=50$ μm and amplitude is $\eta_1=0.0056$ μm.

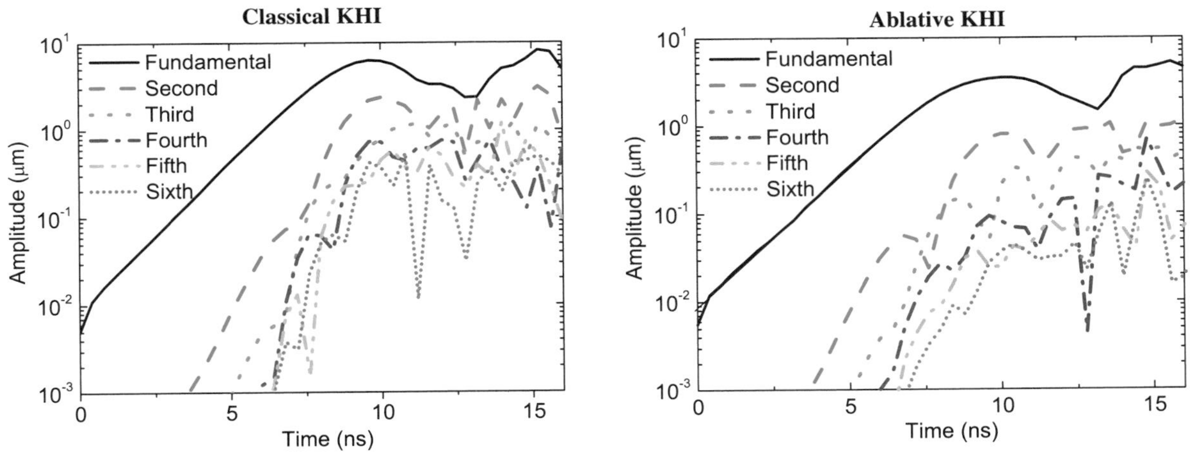

FIG. 3. (Color online) The temporal evolution of the amplitudes of the first six harmonics in the CKHI and the AKHI with a single-mode sinusoidal interface perturbation.

A. Single-mode sinusoidal perturbed interface

In Fig. 2, the density field with a single-mode sinusoidal interface perturbation at time $t=0,10,12,14,16$ ns without thermal conduction (CKHI) and with thermal conduction (AKHI) are shown. In this case, the most obvious difference between the CKHI and the AKHI is the rupture of the vortex arms and the lack of small-scale structures in the latter case. The ablative vortex structures are also narrower in comparison with the CKHI, which is prone to rupture and coalesce.

During the simulations, the x-averaged density field,[45]

$$\hat{\rho}(y,t) = \frac{1}{\rho_{\max}} \int \rho(x,y,t)dx, \quad (8)$$

where $\rho_{\max}=4.25$ g/cm^3 is the peak density, is computed. The Fourier analysis of the resulted x-averaged density field $\hat{\rho}(y,t)$, which is assumed to be periodical in y, is applied to obtain the time dependent amplitude of the k-harmonics of the field $|a_k(t)|, k=0,\ldots,N_y/2$, where $a_k(t)$ is the Fourier coefficients of the field $\hat{\rho}(y,t)$.

In Fig. 3, we show the temporal evolution of the amplitudes of the first six harmonics $|a_k(t)|$, $k=1,2,3,4,5,6$, for the CKHI and the AKHI with a single-mode interface perturbation with amplitude $\eta_1=0.0056$ μm. In the AKHI, the amplitudes of higher harmonics are strongly suppressed while the fundamental (master, $k=1$) mode has a strong growth. This is the main source of energy in the formation of large-scale structures such as the vortex roll-ups in the KHI. In astrophysics, many of the large-scale structures are observed.[46,47] Similar to the studies that showed a significant impact of radiative cooling on the collimation of the jets,[32–34] this study shows that the thermal conduction can significantly reduce the growth of the AKHI. In the nonlinear evolution of the ablative RTI, the strong shear flows are generated on both sides of the spike, especially at the tip of the

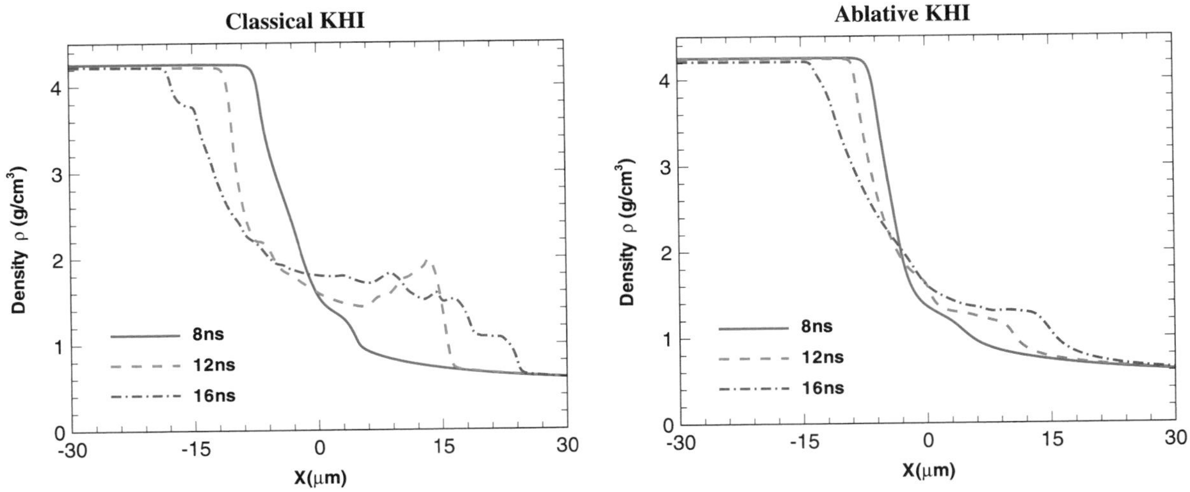

FIG. 4. (Color online) y-averaged density profile at times $t=8,12,16$ ns in the CKHI and the AKHI with a single-mode sinusoidal interface perturbation.

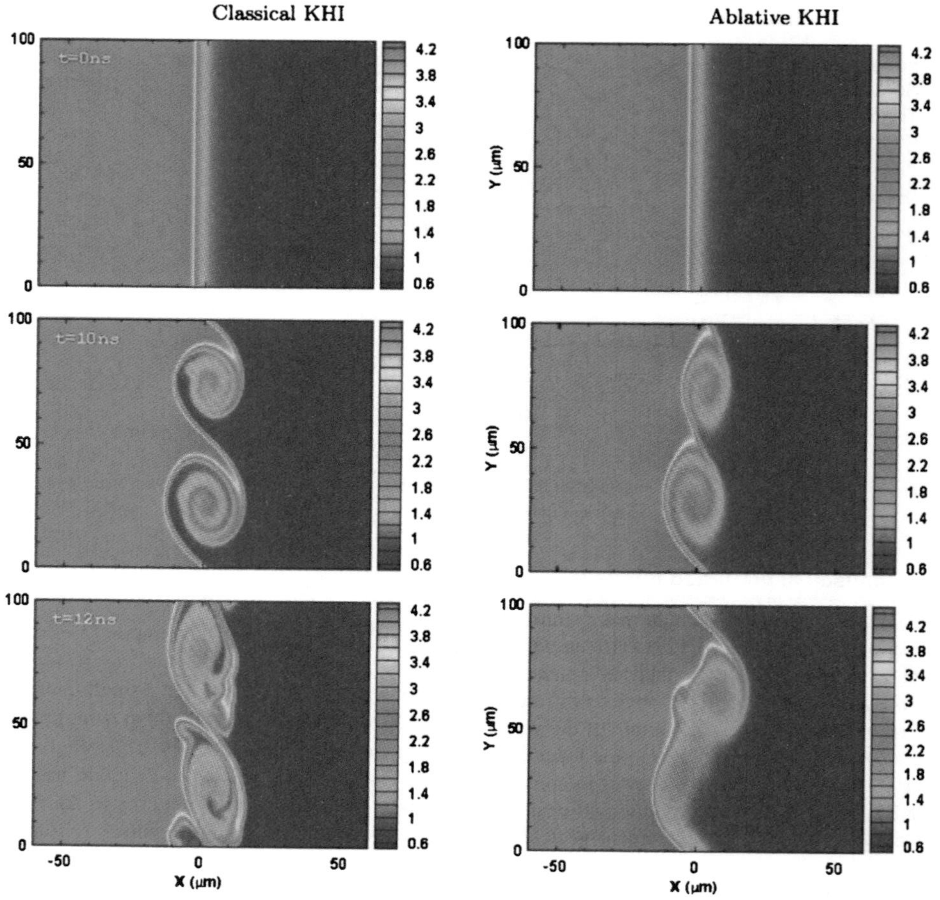

FIG. 5. (Color online) Density fields of the CKHI and the AKHI with a two-mode sinusoidal interface perturbation at times of $t=0,10,12$ ns. The perturbation wavelengths are $\lambda_1=100$ μm, $\lambda_2=50$ μm and amplitudes are $\eta_1=\eta_2=0.03199$ μm.

spike. When the preheating is considered,[24] no characteristic roll-ups (mushroom) structures are found at the spike head. That is, the AKHI is effectively suppressed. The reduction of growth of the amplitudes of the higher harmonics in the AKHI due to thermal conduction prevents the formation of the small-scale structures even without preheating. One can expect that the combined effect of the thermal conduction and preheating can be an effective mean to suppress the formation of small-scale structures and will be investigated in future study.

The y-averaged density profile of the CKHI and the AKHI with a single-mode sinusoidal interface perturbation at time $t=8,12,16$ ns are shown in Fig. 4. At time $t=16$ ns, we observe that, in the case of CKHI, the density profile has a thickness directly related to the size of the vortex, but in the case of AKHI, the initial profile is relatively well preserved.

B. Two-mode sinusoidal perturbed interface

In a two-mode sinusoidal interface perturbation, the velocity shear interface is perturbed with two sinusoidal modes with wavelength $\lambda_1=100$ μm and $\lambda_2=50$ μm and perturbation amplitudes $\eta_1=\eta_2=0.03199$ μm for the both modes. In Figs. 5 and 6, the density fields of the CKHI and the AKHI with a two-mode sinusoidal interface perturbation at time $t=0,10,12,14,16,18$ ns are shown. The similar vortex pairing occurs in both the CKHI and the AKHI. At the earlier time, before the development of the vortex pairing, the small-scale structures of the roll-up vortex are better developed in the CKHI than the AKHI. Comparing the two cases, the vortex pairing is strengthened and appeared earlier in the AKHI than the CKHI. Moreover, at the later time, after the development of the vortex pairing, the paired vortex arms are nearly destroyed in the CKHI. On the contrary, in the AKHI, at the earlier time ($t<12$ ns), before the development of the vortex pairing, the small-scale structures of the roll-up vortex are effectively suppressed. Moreover, after the development of the vortex pairing ($t>14$ ns), the vortex arms are still relatively well kept.

Similar to the single-mode case, the time dependent x-averaged density field is computed and analyzed by the Fourier analysis in y to obtain the amplitude of the harmon-

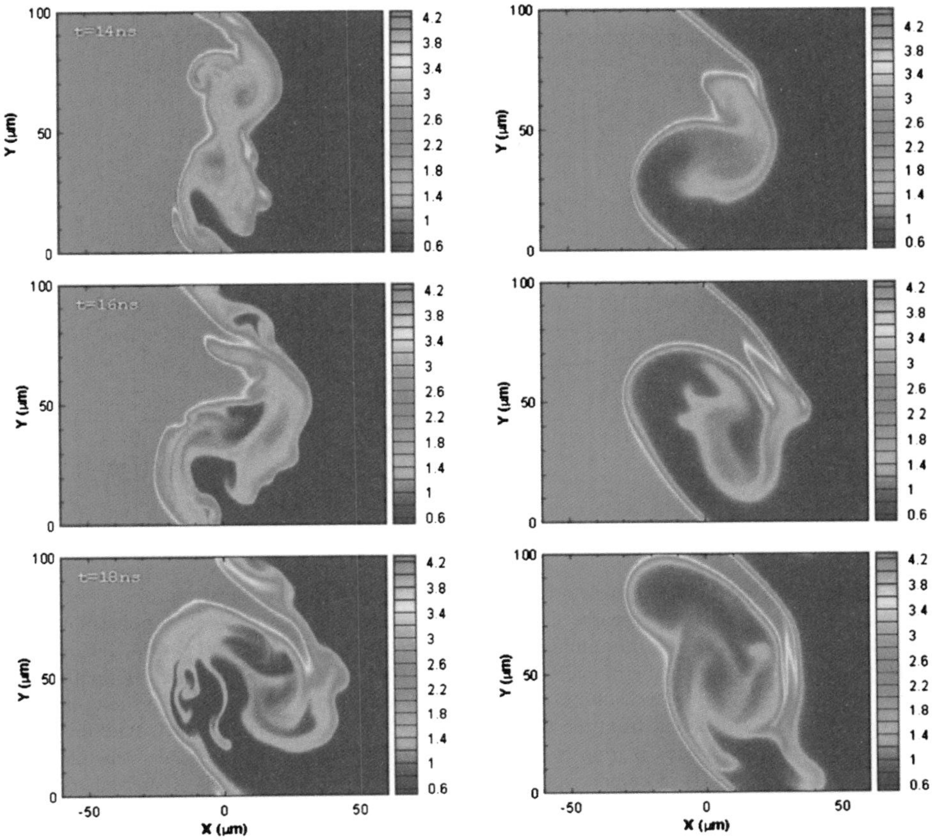

FIG. 6. (Color online) (Continuous from Fig. 5) Density fields of the CKHI and the AKHI with a two-mode sinusoidal interface perturbation at times of $t=14,16,18$ ns. The perturbation wavelengths are $\lambda_1=100$ μm, $\lambda_2=50$ μm and amplitudes are $\eta_1=\eta_2=0.03199$ μm.

ics of the field $|a_k(t)|, k=0,\ldots,N_y/2$. The temporal evolution of the amplitudes of the first four harmonics $|a_k(t)|$, $k=1,2,3,4$, is shown in Fig. 7 (the fundamental mode and second harmonic are drawn with thicker linewidth to emphasize the vortex pairing process). In the CKHI, before the fundamental mode's nonlinear growth exceeds the second harmonic's ($t<10$ ns), the second harmonic has a relatively large growth with maximum amplitude of about 6 μm. The fundamental mode begins to dominate over the other harmonics after $t>13$ ns. In the AKHI, on the other hand, the

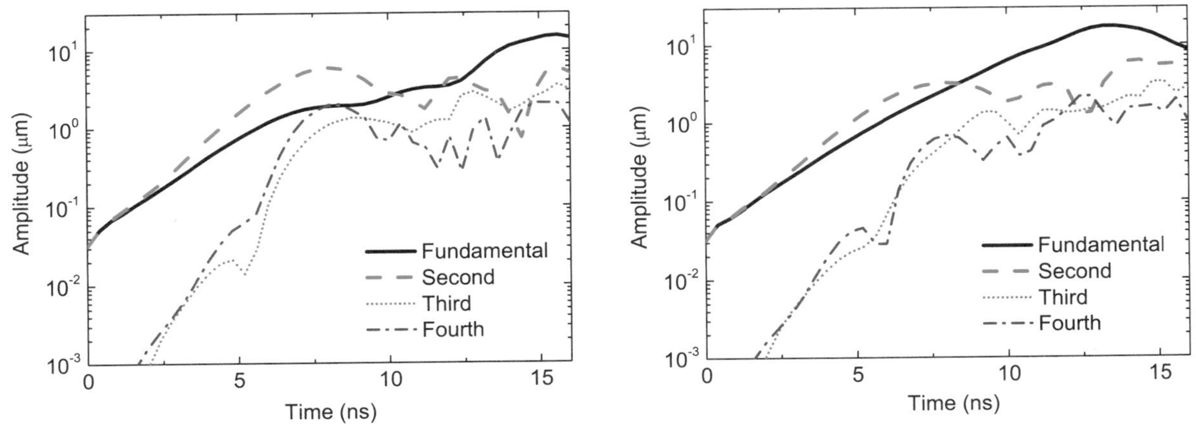

FIG. 7. (Color online) The temporal evolution of the amplitudes of the first four harmonics in the CKHI and the AKHI with a two-mode sinusoidal interface perturbation.

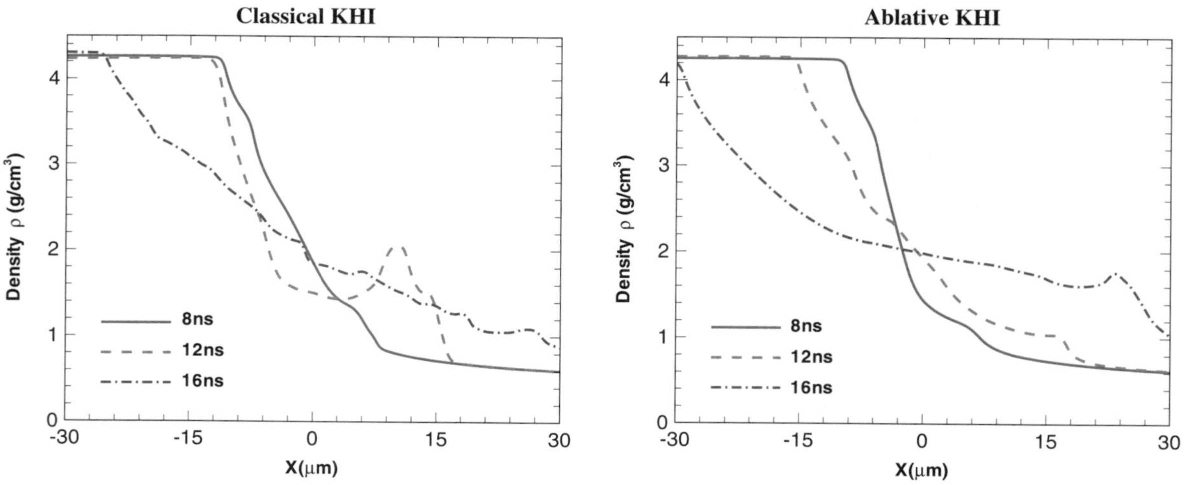

FIG. 8. (Color online) y-averaged density profile at time $t=8,12,16$ ns in the CKHI and the AKHI with a two-mode sinusoidal interface perturbation.

fundamental mode has an extended nonlinear growth regime which surpassing the second harmonic at the early stage of the flow development. As shown in Fig. 7, the fundamental mode's amplitude exceeds the second harmonic's after $t>8.5$ ns and becomes the dominant force in the flow development of the AKHI, which is different from the CKHI. Therefore, we can conclude that the vortex pairing is developed earlier and improved in the AKHI in comparison with the CKHI. Vortex pairing is believed to be a major process in the KHI which plays a crucial role in the interaction between the solar wind and Earth's magnetosphere and provides a mechanism by which the solar wind penetrates the Earth magnetosphere.[5] This study shows that, when the thermal conduction is included in the KHI process, the vortex pairing process is strengthened which can lead to large-scale material mixing.

The y-averaged density profiles of the CKHI and the AKHI with a two-mode sinusoidal interface perturbation at time $t=8,12,16$ ns are shown in Fig. 8. Here, one can see that the material mixing is increased by the vortex pairing in the AKHI. Furthermore, in the CKHI, a plateau appears near the central region of the initial shear layer $x=10$ at $t=12$ ns which is in general agreement with the results in Ref. 30 where the magnetic field effect is considered but its influence is weak in their studies since the magnetic field is perpendicular to the shear flows.

IV. SUMMARY AND CONCLUSION

In this research, we have conducted some preliminary numerical experiments in the simulation of the CKHI and the AKHI, which includes the effect of the thermal conduction in the process. The CKHI and the AKHI with interface being perturbed by a single-mode and a two-mode sinusoidal interface perturbation are simulated using the fifth order WENO finite difference scheme. We examined the temporal evolution of the density fields, the temporal evolution of the amplitude of the first few harmonics of x-averaged density field, and the y-averaged density profile. The thermal conduction effects on the nonlinear evolutions of the KHI are investigated by the comparison of the CKHI and the AKHI.

This study shows that the thermal conduction plays an important role in the spatial and temporal evolution of perturbed interface under the KHI. With the thermal conduction, the KHI growths of small-scale structures are reduced while the vortex pairing process is strengthened, which can result in the large-scale plasma mixing in systems relevant to the high energy density physics and the astrophysics. This effect might be partially responsible for the phenomenon in the stability of the astrophysical jets over long distance exceeding 10 jet diameters or more[13–16,46,47] and generic flows with jetlike long spikes in high energy density physics experiments.[17–19] Detailed research in these areas will be pursued and reported in our future work.

ACKNOWLEDGMENTS

The author (L.F.W.) would like to thank Professor S. Atzeni (Università di Roma "La Sapienza" and CNISM) for his helpful discussions, and the anonymous referee for suggestions that have improved the paper. He also would like to acknowledge the hospitality of the Department of Mathematics during his stay at Hong Kong Baptist University and supported by the RGC grant from Hong Kong Research Grants Council.

This research was supported by the National Basic Research Program of China (Grant No. 2007CB815100) and the National Natural Science Foundation of China (Grant Nos. 11075024, 10935003, 10905006, and 11074300). The author (W.S.D.) would like to thank the support provided by the RGC grant from Hong Kong Research Grants Council.

[1] A. Otto and D. H. Fairfield, J. Geophys. Res. **105**, 21175 (2000).
[2] D. H. Fairfield, A. Otto, T. Mukai, S. Kokubun, R. P. Lepping, J. T. Steinberg, A. J. Lazarus, and T. Yamamoto, J. Geophys. Res. **105**, 21159 (2000).

[3]S. Parhi, S. T. Suess, and M. Sulkanen, J. Geophys. Res. **104**, 14781 (1999).
[4]K. Nykyri and A. Otto, Geophys. Res. Lett. **28**, 3565 (2001).
[5]H. Hasegawa, M. Fujimoto, T.-D. Phan, H. Rème, A. Balogh, M. W. Duntop, C. Hashimoto, and R. TanDokoro, Nature (London) **430**, 755 (2004).
[6]B. A. Remington, R. P. Drake, and D. D. Ryutov, Rev. Mod. Phys. **78**, 755 (2006).
[7]E. C. Harding, J. F. Hansen, O. A. Hurricane, R. P. Drake, H. F. Robey, C. C. Kuranz, B. A. Remington, M. J. Bono, M. J. Grosskopf, and R. S. Gillespie, Phys. Rev. Lett. **103**, 045005 (2009).
[8]O. A. Hurricane, J. F. Hansen, H. F. Robey, B. A. Remington, M. J. Bono, E. C. Harding, R. P. Drake, and C. C. Kuranz, Phys. Plasmas **16**, 056305 (2009).
[9]E. C. Harding, R. P. Drake, Y. Aglitskiy, T. Plewa, A. L. Velikovich, R. S. Gillespie, J. L. Weaver, A. Visco, M. J. Grosskopf, and J. R. Ditmar, Phys. Plasmas **17**, 056310 (2010).
[10]R. P. Drake, *High-Energy-Density Physics: Fundamentals, Inertial Fusion and Experimental Astrophysics* (Springer, New York, 2006).
[11]S. Atzeni and J. Meyer-ter-Vehn, *The Physics of Inertial Fusion: Beam Plasma Interaction Hydrodynamics, Hot Dense Matter* (Oxford University, Oxford, 2004).
[12]Committee on High Density Plasma Physics Plasma Science and Committee Board on Physics and Astronomoy Division on Engineering and Physical Science, *Frontiers in High Energy Density Physics* (The National Academies, Washington, D.C., 2001).
[13]J. M. Foster, B. H. Wilde, P. A. Rosen, R. J. R. Williams, B. E. Blue, R. F. Coker, R. P. Drake, A. Frank, P. A. Keiter, A. M. Khokhlov, J. P. Knauer, and T. S. Perry, Astrophys. J. **634**, L77 (2005).
[14]S. V. Lebedev, J. P. Chittenden, F. N. Beg, S. N. Bland, A. Ciardi, D. Ampleford, S. Hughes, M. G. Haines, A. Frank, E. G. Blackman, and T. Gardiner, Astrophys. J. **564**, 113 (2002).
[15]P. E. Hardee, Astrophys. Space Sci. **293**, 117 (2004).
[16]M. Livio, Phys. Rep. **311**, 225 (1999).
[17]D. K. Bradley, D. G. Braun, S. G. Glendinning, M. J. Edwards, J. L. Milovich, C. M. Sorce, G. W. Collins, S. W. Haan, R. H. Page, and R. J. Wallace, Phys. Plasmas **14**, 056313 (2007).
[18]B. A. Remington, R. P. Drake, H. Takabe, and D. Arnett, Phys. Plasmas **7**, 1641 (2000).
[19]J. O. Kane, H. F. Robey, B. A. Remington, R. P. Drake, J. Knauer, D. D. Ryutov, H. Louis, R. Teyssier, O. Hurricane, D. Arnett, R. Rosner, and A. Calder, Phys. Rev. E **63**, 055401 (2001).
[20]L. F. Wang, W. H. Ye, Z. F. Fan, Y. J. Li, X. T. He, and M. Y. Yu, Europhys. Lett. **86**, 15002 (2009).
[21]L. F. Wang, C. Xue, W. H. Ye, and Y. J. Li, Phys. Plasmas **16**, 112104 (2009).
[22]L. F. Wang, W. H. Ye, and Y. J. Li, Phys. Plasmas **17**, 042103 (2010).
[23]S. Nakai, T. Yamanaka, Y. Izawa, Y. Kato, K. Nishihara, M. Nakatsuka, T. Sasaki, and H. Takabe, *et al.*, Plasma Physics and Controlled Fusion Research 1994, *15th International Conference Proceedings*, Seville, Spain, 28 September–1 October 1994 (IAEA, Vienna, 1995), Vol. 3, pp. 3–12.
[24]W. Ye and X. T. He, Proceeding of the Tenth International Workshop on the Physics of Compressible Turbulent Mixing, edited by M. Legrand and M. Vandenboomgaerde, Paris, France, 2006.
[25]C. D. Gregory, J. Howe, B. Loupias, S. Myers, M. M. Notley, Y. Sakawa, A. Oya, R. Kodama, M. Koenig, and N. C. Woolsey, Astrophys. J. **676**, 420 (2008).
[26]K. Yirak, A. Frank, and A. Cunningham, Astrophys. J. **672**, 996 (2008).
[27]A. Miura, Phys. Rev. Lett. **83**, 1586 (1999).
[28]A. Miura and P. L. Pritchett, J. Geophys. Res. **87**, 7431 (1982).
[29]T. K. M. Nakamura and M. Fujimoto, Phys. Rev. Lett. **101**, 165002 (2008).
[30]M. Faganello, F. Califano, and F. Pegoraro, Phys. Rev. Lett. **101**, 105001 (2008).
[31]D. A. Knoll and L. Chacón, Phys. Rev. Lett. **88**, 215003 (2002).
[32]J. M. Stone, N. Turner, K. Estabrook, R. Remington, D. Farley, S. G. Glendinning, and S. Glenzer, Astrophys. J., Suppl. Ser. **127**, 497 (2000).
[33]K. Shigemori, R. Kodama, D. R. Farley, T. Koase, K. G. Estabrook, B. A. Remington, D. D. Ryutov, Y. Ochi, H. Azechi, J. Stone, and N. Turner, Phys. Rev. E **62**, 8838 (2000).
[34]D. R. Farley, K. G. Estabrook, S. G. Glendinning, S. H. Glenzer, B. A. Remington, K. Shigemori, J. M. Stone, R. J. Wallace, G. B. Zimmerman, and J. A. Harte, Phys. Rev. Lett. **83**, 1982 (1999).
[35]L. F. Wang, W. H. Ye, and Y. J. Li, Europhys. Lett. **87**, 54005 (2009).
[36]C. D. Winant and F. K. Browand, J. Fluid Mech. **63**, 237 (1974).
[37]F. K. Browand and P. D. Weidman, J. Fluid Mech. **76**, 127 (1976).
[38]W. Ye, W. Zhang, and X. T. He, Phys. Rev. E **65**, 057401 (2002).
[39]L. F. Wang, W. H. Ye, and Y. J. Li, Phys. Plasmas **17**, 052305 (2010).
[40]L. Spitzer and R. Härm, Phys. Rev. **89**, 977 (1953).
[41]D. Balsara and C. W. Shu, J. Comput. Phys. **160**, 405 (2000).
[42]G. S. Jiang and C. W. Shu, J. Comput. Phys. **126**, 202 (1996).
[43]R. Borges, M. Carmona, B. Costa, and W. S. Don, J. Comput. Phys. **227**, 3191 (2008).
[44]P. L. Roe, J. Comput. Phys. **43**, 357 (1981).
[45]B. A. Remington, S. V. Weber, M. M. Marinak, S. W. Haan, J. D. Kilkenny, R. Wallace, and G. Dimonte, Phys. Rev. Lett. **73**, 545 (1994).
[46]B. Reipurth and J. Bally, Annu. Rev. Astron. Astrophys. **39**, 403 (2001).
[47]J. J. Hester, P. A. Scowen, R. Sankrit, T. R. Lauer, E. A. Ajhar, W. A. Baum, A. Code, D. G. Currie, G. E. Danielson, S. P. Ewald, S. M. Faber, C. J. Grillmair, E. J. Groth, J. A. Holtzman, D. A. Hunter, J. Kristian, R. M. Light, C. R. Lynds, D. G. Monet, E. J. O'Neil, Jr., E. J. Shaya, K. P. Seidelmann, and J. A. Westphal, Astron. J. **111**, 2349 (1996).

CHIN. PHYS. LETT. Vol. 27, No. 12 (2010) 125203

Jet-Like Long Spike in Nonlinear Evolution of Ablative Rayleigh–Taylor Instability *

YE Wen-Hua(叶文华)[1,2,3**], WANG Li-Feng(王立锋)[2,3,4], HE Xian-Tu(贺贤土)[1,2,3]

[1] *Department of Physics, Zhejiang University, Hangzhou 310027*

[2] *CAPT, Peking University, Beijing 100871*

[3] *LCP, Institute of Applied Physics and Computational Mathematics, Beijing 100088*

[4] *SMCE, China University of Mining and Technology, Beijing 100083*

(Received 5 July 2010)

We report the formation of jet-like long spike in the nonlinear evolution of the ablative Rayleigh–Taylor instability (ARTI) experiments by numerical simulations. A preheating model $\kappa(T) = \kappa_{SH}[1 + f(T)]$, where κ_{SH} is the Spitzer–Härm (SH) electron conductivity and $f(T)$ interprets the preheating tongue effect in the cold plasma ahead of the ablative front [Phys. Rev. E 65 (2002) 57401], is introduced in simulations. The simulation results of the nonlinear evolution of the ARTI are in general agreement with the experiment results. It is found that two factors, i.e., the suppressing of ablative Kelvin–Helmholtz instability (AKHI) and the heat flow cone in the spike tips, contribute to the formation of jet-like long spike in the nonlinear evolution of the ARTI.

PACS: 52. 57. Fg, 47. 20. Ma, 52. 35. Py DOI: 10.1088/0256-307X/27/12/125203

The ablative Rayleigh–Taylor instability (ARTI) is a critical issue for inertial confinement fusion (ICF).[1,2] The ablative front between the shell and the blow-off plasma in the acceleration phase and the inner interface between the shell and the hot spot in the deceleration stage are both prone to the ARTI. It can break up the implosion shell during the acceleration stage and destroy the central ignition hot-spot during the deceleration stage in ICF. Therefore, ICF targets must be designed to keep the ARTI growth at an acceptable level.[1,2]

It is well known that the mass ablation effects reduce the linear growth rate of the ARTI.[3−9] It is found the stabilizing mechanisms (i.e., the density gradient effect and ablation effect, etc.)[10−14] have double effects on the evolution of the ARTI. On the one hand, it definitely reduces its linear growth rate,[3−9] while, on the other hand, it can lead to the formation of a large amplitude of spike (i.e., the jet-like long spike) in the nonlinear evolution of the Rayleigh-Taylor instability (RTI).[15−19] The formation of jet-like long spike in the ARTI is dangerous for the ignition of ICF. Because the jet-like long spike would hinder the formation of the hot spot resulting in auto-ignition failure or destruction of the ignition hot spot during the deceleration stage in ICF. The preheating by the nonlocal electron transport, pre-pulse, shock, hot electron and hard x-rays, etc, can significantly reduce the growth of the ARTI, which is vital for ICF ignition since these preheating effects can dramatically enhance the ablation stabilization and the density gradient stabilization for the ARTI. However, the preheating effects mentioned above can also lead to the formation of jet-like long spike in the nonlinear evolution of the ARTI. Thus, understanding the formation of jet-like long spike is significant for the success of ICF.

In this Letter, we consider the nonlinear evolution of the ARTI including the preheating effect. The two-dimensional (2D) compressible fluid equations taking into account the laser absorption and the electron conduction are solved numerically using the LARED-S code,[20−26] which is widely used in studies on the ICF ignition target and hydrodynamic instabilities in the high energy density physics field. For simplification in simulation, the ideal equation of state (EOS) is used and the radiation effect is ignored. The governing equations are

$$\frac{\partial \rho}{\partial t} + \nabla \cdot (\rho \boldsymbol{u}) = 0, \tag{1a}$$

$$\frac{\partial (\rho \boldsymbol{u})}{\partial t} + \nabla \cdot (\rho \boldsymbol{u}\boldsymbol{u}) + \nabla p = 0, \tag{1b}$$

$$\frac{\partial E}{\partial t} + \nabla \cdot [(E + P)\boldsymbol{u}] = -\nabla \cdot \boldsymbol{q}(T) + S_L, \tag{1c}$$

where ρ, $\boldsymbol{u}$, T, p and E are the fluid density, velocity, temperature, pressure, and the total energy $E = \rho(\varepsilon + \boldsymbol{u}\cdot\boldsymbol{u}/2)$, respectively, ε is the specific internal energy, S_L is the energy source from laser absorption by inverse bremsstrahlung according to laser ray-tracing, $\boldsymbol{q}(T) = -\kappa \nabla T$ is the heat flow, $\kappa = h\kappa_{SH} f(T)$ is the electron heat conductivity, κ_{SH} is the Spitzer–Härm (SH) electron conductivity. The heat flux limitation is governed by the coefficient $h = (1 + |\boldsymbol{F}_D/\beta_e \boldsymbol{F}_L|)$,

*Supported by the National Basic Research Program of China under Grant No 2007CB815103, and the National Natural Science Foundation of China under Grant Nos 10935003, 10775020 and 11075024.

**To whom correspondence should be addressed. Email: ye_wenhua@iapcm.ac.cn where $\boldsymbol{F}_D = \kappa_{\rm SH} f(T)\nabla T$ is the diffusion electron flux, $\boldsymbol{F}_L$ is the free electron flux, and $\beta_e = 0.1$ is a weighting factor. Since the diffusion flux is much less than the free electron flux at the ablation front, we have $h \approx 1$. Here $f(T)$ is the preheating function.[20]

In the ARTI experiments of NOVA and the GEKKO XII, the linear ARTI growth rate is found to be significantly less than the predicted one. Fokker–Planck (FP) simulation shows that this reduction can be attributed to the preheating of the fluid by fast electrons.[3,4] In the nonlinear ARTI experiments of NOVA and the GEKKO XII, the observed formation of long spikes has been attributed to suppression of the harmonic growth by preheating from hard x-rays and fast electrons. Simulations show that the spike ruptures in the weak preheating case, such as the SH electron conduction, but we see jet-like long spikes in our recent experiments in Shengguang-II, which can also be seen in many other side-on experiments on the ICF laser facility.[15−19] Direct-drive experiments on the GEKKO XII[3] and the NOVA[4] have shown the large reduction of linear growth rate of the ARTI. The preheating calculated by the FP simulation of nonlocal electron heat transport gives good explanation for the reduced linear growth rate. To obtain accurate results of preheating (nonlocal energy transport) for the ARTI in ICF, one should solve the FP equation of electron and the transport equation of radiation. However, it costs very high to do this.

One knows that the growth of the ARTI is mainly dependent on the density and temperature profiles. For the steady ablative flow, the isobaric model is a good approximation, i.e., $p = \Gamma\rho T = $ const, where p is the pressure, ρ is the density and T is the temperature. For the ideal EOS, the thermal conductivity $\kappa(\rho, T) \approx \kappa(T)$, because ρ is not independent of T. From the simulation results of LARED-S code,[20] it is found that the profiles of density and temperature obtained from FP simulations can be reproduced by the preheating model $\kappa(T) = \kappa_{\rm SH}[1 + f(T)]$, where $\kappa_{\rm SH}$ is the SH electron conductivity and $f(T)$ interprets the preheating tongue effect in the cold plasma ahead of the ablative front.[20] In the blow-off region, the property of high temperature corresponding to $f(T) \ll 1$ leads to the conductivity $\kappa(T) \to \kappa_{\rm SH}$. Inside the ablation front, due to low temperature, we have $f(T) \gg 1$. Accordingly, ahead of the alation front, the conductivity $\kappa(T) \approx \kappa_{\rm SH} f(T) \gg \kappa_{\rm SH}$, which means the preheating effect is dominated in this zone. For temperatures greater than 0.10 MK, it has been shown that the choice $f(T) = \alpha T^{-1} + \beta T^{-3/2}$, where T is in units of MK, α and β are adjustable parameters, leads to ARTI growth rates as well as the fluid density and temperature profiles at the ablation front, in agreement well with the experiments.[20]

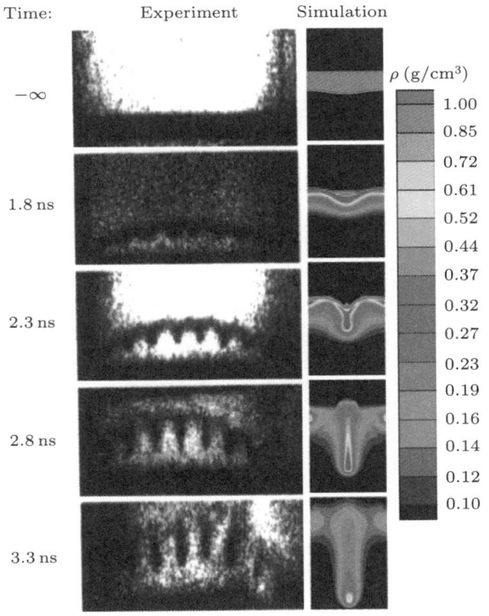

Fig. 1. (Color online) Comparison of nonlinear evolution of ATRI between the GEKKO XII experiment (from Ref. [18]) and the LARED-S simulation.

A beautiful example of the jet-like long spike in the ARTI experiment is from the laser facility of GEKKO XII.[18] It refers to a laser-accelerated 25-μm-thick plastic (CH) foil, with an imposed initial sinusoidal corrugation of amplitude of 1.5 μm and wavelength of 100 μm. The foil is irradiated by a 0.53 μm laser beam with a laser pulse of intensity $I = 2 \times 10^{14}\,{\rm W/cm^2}$.

In Fig. 1, we compare the results from LARED-S code using our preheating model with $\alpha = 6$, $\beta = 5.5$ and the GEKKO XII experiment.[18] General agreements between the experiment and the simulation are obtained, as shown in Fig. 1. We analyze the simulation result of the nonlinear evolution of the jet-like long spike in the GEKKO XII experiment. The density profiles and velocity fields are shown in Fig. 2.

As shown in Fig. 2(a), the spike grows by receiving feed from the surrounding fluid (the foil at the bottom of spike) before 2.5 ns. It can be seen that at 2.8 ns the fluid in the bottom of the spike is exhausted and the foil has been ruptured, which is more obvious in Figs. 2(c) and 2(d). When the foil at bottom of spike cannot feeds the growth of the spike, the foil begins to break up. After that, the spike grows longer and longer. The length of spike is about 89, 108 and 122 μm at time of 2.8, 3.0 and 3.3 ns, respectively.

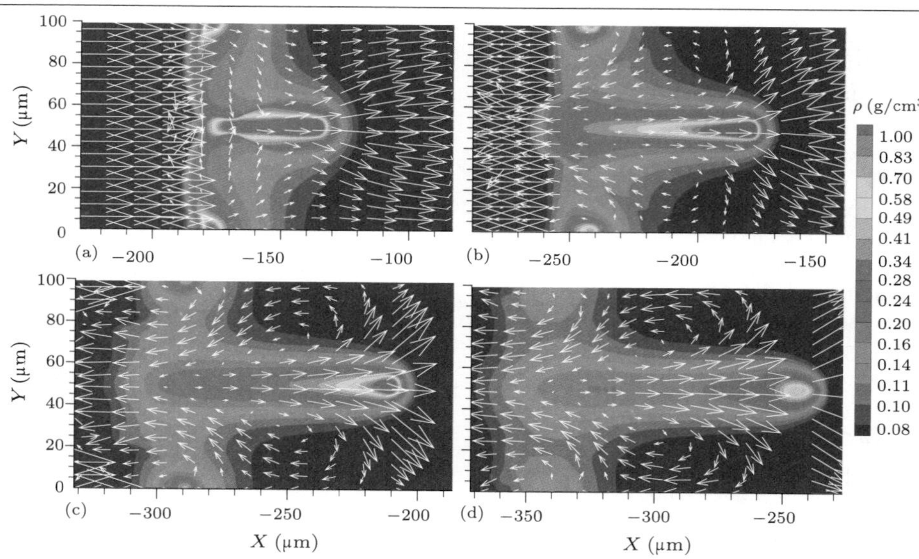

Fig. 2. (Color online) Density color map and velocity field of the spike at times $t = 2.5$ ns (a), 2.8 ns (b), 3.0 ns (c), and 3.3 ns (d).

The explicit shear flows are formed at the waist of the spike, as shown in Figs. 2(c) and 2(d). The direction of mass flow in the spike is from its bottom to head (the left to the right in Fig. 2) while in the surroundings it is from the right to the left. In order to further analyze the ablative Kelvin–Helmholtz instability (AKHI)[22] induced by the shear flows at both sides of the spike, we employ the profiles of pressure, density, temperature and velocity in the x direction, along the transverse y direction at different heights of the spike as the one-dimensional (1D) basic flow and introduced a 50-μm perturbation wavelength to calculate the linear growth rate of the AKHI. It is found that the growth of the AKHI is completely suppressed in our simulation.

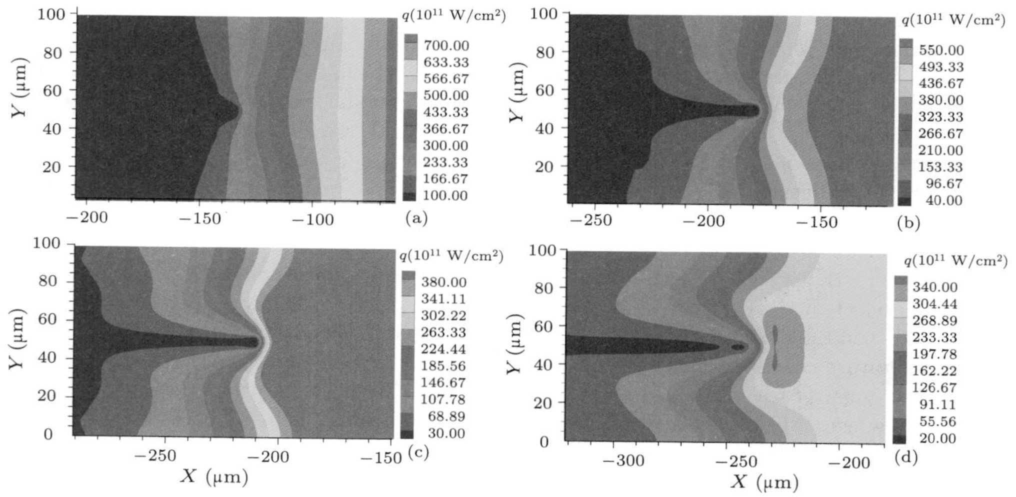

Fig. 3. (Color online) Heat flow color map of the spike at times $t = 2.5$ ns (a), 2.8 ns (b), 3.0 ns (c), and 3.3 ns (d).

We show the heat flows around the head of spike in Fig. 3. Here the heat flow is in the form $|\boldsymbol{q}(T)| = |-\kappa(T)\nabla T|$. The heat flow is modified by the spike, as shown in Fig. 3. At the spike tip, the heat flows form a cone, and the spike tip is surrounded by the cone. In surroundings of the high temperature heat flow cone, it is hard for the spike to become decentralized to form the classical Kelvin–Helmholtz (KH) mushroom structures. Detailed research in these areas will be pursued and reported in our future work.

In conclusion, we have presented the formation of jet-like long spike in the ablative Rayleigh–Taylor instability (ARTI) experiments by numerical simulations. The density profile, velocity field and heat flow at the spike tip are obtained. In our simulation of the ablative Kelvin–Helmholtz instability (AKHI), which introduces the transversal profiles at the different heights of the spike as the 1D basic flows, the perturbation with 50 μm wavelength does not grow. It is found that the heat flow cone at the spike tip, suppresses the growth of AKHI to form the classical Kelvin–Helmholtz (KH) mushroom structures.

References

[1] Lindl J D, Amendt P and Berger R L, Glendinning S G, Glenzer S H, Haan S W, Kauffan R L, Landen O L and Suter L J 2004 *Phys. Plasmas* **11** 339

[2] Atzeni S and Meyer-ter-Vehn J 2004 *The Physics of Inerial Fusion: Beam Plasma Interaction, Hydrodynamics, Hot Dense Mater* (OxfordOxford: University Press)

[3] Shigemori K, Azechi H, Nakai M, Honda M, Meguro K, Miyanaga N, Takabe H and Mima K 1997 *Phys. Rev. Lett.* **78** 250

[4] Glendinning S G, Dixit S N, Hammel B A, Kalantar D H, Key M H, Kilkenny J D, Knauer J P, Pennington D M, Remington B A, Wallace R J, and Weber S V 1997 *Phys. Rev. Lett.* **78** 3318

[5] Sakaiya T, Azechi H, Matsuoka M, Izumi N, Nakai M, Shigemori K, Shiraga H, SunaharaA, Takabe H and Yamanaka T 2002 *Phys. Rev. Lett.* **88** 145003

[6] Azechi H, Sakaiya T, Fujioka S, Tamari Y, Otani K, Shigemori K, Nakai M, Shiaga H, Miyanaga N and Mima K 2007 *Phys. Rev. Lett.* **98** 045002

[7] Sanz J, Ramírez J, Ramis R, Betti R and Town R P J 2002 *Phys. Rev. Lett.* **89** 195002

[8] Garnier J, Raviart P -A, Cherfils-Clèrouink C and Masse L 2003 *Phys. Rev. Lett.* **90** 185003

[9] Ikegawa T, Nishihara K 2002 *Phys. Rev. Lett.* **89** 115001

[10] Sanz J 2002 *Phys. Rev. Lett.* **73**(20) 2700

[11] Pirize A P, Sanz J and Ibañez L F 1997 *Phys. Plasmas* **4** 1117

[12] Goncharov V N, Betti R, McCrory R L, Sorotokin P and Verdon C P 1996 *Phys. Plasmas* **3** 1402

[13] Betti R, Goncharov V N, McCrory R L and Verdon C P 1998 *Phys. Plasmas* **5** 1446

[14] Masse L 2007 *Phys. Rev. Lett.* **98** 245001

[15] Remington B A, Haan S W, Glendinning S G, Kilkenny J D, Munro D H and Wallace R J 1992 *Phys. Fluids* B **4** 967

[16] Remington B A, Drake R P, Takabe H and Arnett D 2000 *Phys. Plasmas* **7** 1641

[17] Mason R J, Hollowell D E, Schappert G T and Batha S H 2001 *Phys. Plasmas* **8** 2338

[18] Nakai S, Yamanaka T, Izawa Y, Kato Y, Nishihara K, Nakatsuka M, Sasaki T, Takabe H, et al 1995 *Plasma Physics and Controlled Fusion Research 1994, Fifteen International Conference Proceedings* (Seville Spain 28 September–1 Octoober 1994) (IAEA: Vienna) vol 3 p 3

[19] Kane J O, Robey H F, Remington B A, Drake R P, Knauer J, Ryutov D D, Louis H, Teyssier R, Hurricane O, Arnett D, Rosner R and Calder A 2001 *Phys. Rev.* E **63** 055401

[20] Ye Wenhua, Zhang Weiyan and He X T 2002 *Phys. Rev.* E **65** 57401

[21] Wang L F, Ye W H, Fan Z F, Li Y J, He X T and Yu M Y 2009 *Europhys. Lett.* **86** 15002

[22] Wang L F, Ye W H and Li Y J 2009 *Europhys. Lett.* **87** 54005

[23] Wang L F, Ye W H and Li Y J 2010 *Phys. Plasmas* **17** 052305

[24] Wang L F, Xue C, Ye W H and Li Y J 2009 *Phys. Plasmas* **16** 112104

[25] Wang L F, Ye W H, Fan Z F and Li Y J 2010 *Europhys. Lett.* **89** 15001

[26] Wang L F, Ye W H and Li Y J 2010 *Chin. Phys. Lett.* **27** 025202

PHYSICS OF PLASMAS **17**, 122706 (2010)

Preheating ablation effects on the Rayleigh–Taylor instability in the weakly nonlinear regime

L. F. Wang,[1,2,3] W. H. Ye,[2,4,a)] Z. M. Sheng,[5] Wai-Sun Don,[3] Y. J. Li,[1] and X. T. He[2,4]
[1]*SMCE, China University of Mining and Technology, Beijing 100083, People's Republic of China*
[2]*CAPT, Peking University, Beijing 100871, People's Republic of China and LCP, Institute of Applied Physics and Computational Mathematics, Beijing 100088, People's Republic of China*
[3]*Department of Mathematics, Hong Kong Baptist University, Kowloon Tong, Hong Kong*
[4]*Department of Physics, Zhejiang University, Hangzhou 310027, People's Republic of China*
[5]*Beijing National Laboratory for Condensed Matter Physics, Institute of Physics, CAS, Beijing 100190, People's Republic of China and Department of Physics, Shanghai Jiao Tong University, Shanghai 200240, People's Republic of China*

(Received 22 June 2010; accepted 28 October 2010; published online 10 December 2010)

The two-dimensional Rayleigh–Taylor instability (RTI) with and without thermal conduction is investigated by numerical simulation in the weakly nonlinear regime. A preheat model $\kappa(T) = \kappa_{SH}[1+f(T)]$ is introduced for the thermal conduction [W. H. Ye, W. Y. Zhang, and X. T. He, Phys. Rev. E **65**, 057401 (2002)], where κ_{SH} is the Spitzer–Härm electron thermal conductivity coefficient and $f(T)$ models the preheating tongue effect in the cold plasma ahead of the ablation front. The preheating ablation effects on the RTI are studied by comparing the RTI with and without thermal conduction with identical density profile relevant to inertial confinement fusion experiments. It is found that the ablation effects strongly influence the mode coupling process, especially with short perturbation wavelength. Overall, the ablation effects stabilize the RTI. First, the linear growth rate is reduced, especially for short perturbation wavelengths and a cutoff wavelength is observed in simulations. Second, the second harmonic generation is reduced for short perturbation wavelengths. Third, the third-order negative feedback to the fundamental mode is strengthened, which plays a stabilization role. Finally, on the contrary, the ablation effects increase the generation of the third harmonic when the perturbation wavelengths are long. Our simulation results indicate that, in the weakly nonlinear regime, the ablation effects are weakened as the perturbation wavelength is increased. Numerical results obtained are in general agreement with the recent weakly nonlinear theories as proposed in [J. Sanz, J. Ramírez, R. Ramis *et al.*, Phys. Rev. Lett. **89**, 195002 (2002); J. Garnier, P.-A. Raviart, C. Cherfils-Clérouin *et al.*, Phys. Rev. Lett. **90**, 185003 (2003)]. © *2010 American Institute of Physics*. [doi:10.1063/1.3517606]

I. INTRODUCTION

A major goal of inertial confinement fusion (ICF) program is to initiate self-sustained fusion reaction at the core of a deuterium-tritium filled spherical capsule. It has long been recognized that the Rayleigh–Taylor instability (RTI), where an interface separating a heavier density material above a lighter density material under an acceleration (for example, gravity), poses severe challenges in achieving ICF autoignition and eventual self-sustained fusion reaction.[1,2] At the initial acceleration stage, when the imploding shell is accelerated inward by the low-density blow-off plasma, the outer shell surface becomes unstable to the RTI. Moreover, at the later stage, when the compressed fuel begins to decelerate the imploding pusher, the inner shell surface is prone to the instability.[1,2] This inherent physical instability can limit the implosion velocity and the drive energy and, in some cases, even break up the implosion shell during the acceleration stage. Furthermore, the instability would hinder the formation of hot spot resulting in the autoignition failure or destruction of ignition hot spot during the deceleration stage. Thus, the ICF target capsule should be designed in such a way that would minimize the growth of the instability at an acceptable level. In this report, the term instability refers to the RTI unless stated otherwise.

It is well-known in the literature that the density gradient effect reduces the RTI growth rate, which is approximated as $\gamma_L = \sqrt{Akg/(1+AkL_m)}$, where $A=(\rho_1-\rho_2)/(\rho_1+\rho_2)$ is the Atwood number, $\rho_1 > \rho_2$ are the densities of the fluids, $L_m = \min[|\rho/(\partial\rho/\partial x)|]$ is the minimum density gradient scale length, g is the acceleration, and $k=2\pi/\lambda$ is the perturbation wave number for a given perturbation wavelength λ at the interface of two fluids.[1,3–6]

When the thermal conduction is included in the modeling of the physical system, the instability is referred to as the ablative Rayleigh–Taylor instability (ARTI) as opposed to the classical Rayleigh–Taylor instability (CRTI) in the absence of the thermal conduction. Several analysis in the past studied the instability of the ablation front.[7–17] It has been shown that the ablation effect tends to stabilize the ARTI and a cutoff wavelength λ_c appears when the perturbation wavelength λ is sufficiently short. The linear growth rate derived by Bodner and Takabe,[10,11] $\gamma_T = \alpha\sqrt{kg/(1+kL_m)} - \beta k v_a$, generally agrees with the two-dimensional numerical

a)Author to whom correspondence should be addressed. Electronic mail: ye_wenhua@iapcm.ac.cn.

simulations[12] and experiments,[13] where L_m is the minimum density gradient scale length at the ablation surface, and v_a is the ablation velocity. The two parameters α and β depend on the flow parameters with $\alpha=0.9-0.95$ and $\beta=1-3$ typically.[13–15]

In direct-driven ICF experiments,[14,15] it has been shown that the growth rate of the areal density perturbation is significantly reduced when compared with the LASNEX numerical simulation results if the thermal conduction term using the Spitzer–Härm (SH) electron thermal conductivity[18] is included. Glendinning *et al.*[14] recognized that the difference between the experiment and the LASNEX simulation is due to a change in the longitudinal density profile resulting from preheat by energetic electrons originating in the plasma corona which penetrates beyond the ablation front. In addition, the experiments of Shigemori[15] and Sakaiya *et al.*[16] suggested that the nonlocal electron heat transport plays an important role in the suppression of ARTI at the ablation front. Subsequently, Ye *et al.*[17] found that the ARTI growth rate, $\gamma_{mL}=\sqrt{Akg/(1+AkL_m)}-2kv_a$, agrees well with the experiments and simulations and is a suitable model for the preheating case in direct-drive ICF experiments. The resulting density profile in direct-drive ICF experiments[19] and simulations[14,16] indicated that the estimation of the ablative density gradient length scale L_m is markedly improved when the preheating ablation effect is also taken into account. It should be noted that the first term in γ_T and γ_{mL} can be understood as the density gradient stabilization from the ablation effect, which plays an important role in one of the physical reasons for the stabilization mechanisms of the ICF system.

The physical origin of the dominant stabilizing effect of RTI has been attributed to the so-called ablation stabilization (the removal of modulated material by ablation),[10] overpressure stabilization,[20–23] anisotropic thermal diffusion stabilization,[24] etc. In previous studies where all physics were included have made it difficult to identify and isolate the relevant and dominant physical mechanism responsible for the stabilization effect. In this study, the influence of the ablation in the stabilization of the ARTI is isolated. For comparison, the results will be compared with those obtained from the simulations of the CRTI using the same density profile in the weakly nonlinear regime when the density maintains as a single value function along the direction of the acceleration.

The paper is organized as the following. In Sec. II, the mathematical framework and numerical setup are presented. The governing equations with an addition of the second order thermal conduction term to the energy equation, along with an auxiliary equation for the thermal conductivity are presented. The numerical methods, namely, the classical fifth order weighted essentially nonoscillatory (WENO) finite difference scheme for the inviscid fluxes and the fourth order implicit central finite difference scheme for the temperature equation in the case of ARTI, employed and initial setup of the problem in this study are briefly described. Section III presents some recent analytical study of evolution of the growth of the amplitude of the harmonics of the interface with increasing initial perturbation wavelengths in the weakly nonlinear regime. The harmonic generation coefficients of the first three harmonics are defined and their roles on the weakly nonlinear behaviors of the RTI will be examined in the next section. In Sec. IV, numerical results from simulations of a single-mode sinusoidal perturbed interface with different perturbation wavelengths are presented and discussed for both the ARTI and the CRTI in the weakly nonlinear regime. The thermal conduction effects on the RTI in the weakly nonlinear regime are presented in term of the evolutions of amplitude of the harmonics of the interface in the ARTI and the CRTI simulations. Concluding remarks are given in Sec. V.

II. NUMERICAL METHODS AND INITIAL CONDITION SETUP

A. Governing equations

In this study, the physical system is modeled via the two-dimensional inviscid one-temperature single-fluid with a bulk acceleration force in a constantly accelerating frame of reference.[25]

$$
\begin{aligned}
&\frac{\partial \rho}{\partial t} + \nabla \cdot (\rho \boldsymbol{u}) = 0, \\
&\frac{\partial (\rho \boldsymbol{u})}{\partial t} + \nabla \cdot (\rho \boldsymbol{u}\boldsymbol{u}) + \nabla P = \rho \boldsymbol{g}, \qquad (1)\\
&\frac{\partial E}{\partial t} + \nabla \cdot [(E+P)\boldsymbol{u}] = \nabla \cdot (\kappa \nabla T) + \rho \boldsymbol{u} \cdot \boldsymbol{g},
\end{aligned}
$$

where $\rho(\mathrm{g/cm^3})$, $\boldsymbol{u}(\mathrm{cm}/\mu\mathrm{s})$, $T(\mathrm{MK})$, and $\boldsymbol{g}(\mathrm{cm}/\mu\mathrm{s}^2)$ are, respectively, the density, velocity, temperature, and acceleration. $E=c_v\rho T+(1/2)\rho \boldsymbol{u}\cdot\boldsymbol{u}$ is the total energy, $P=\Gamma\rho T(\mathrm{Mbar})$ is the pressure, $c_v=\Gamma/(\bar{\gamma}-1)$ is the specific heat at constant volume. For the CH material used in this study, $\bar{\gamma}=5/3$ and $c_v=86.2713(\mathrm{cm}/\mu\mathrm{s})^2\mathrm{MK}^{-1}$ are used.

The term of $\nabla\cdot(\kappa\nabla T)$ is used to model the electron thermal conduction. The thermal conductivity coefficient $\kappa(T)=\kappa_{\mathrm{SH}}[1+f(T)]$ is introduced,[17] where κ_{SH} is the Spitzer–Härm electron thermal conductivity[18] and $f(T)$ models the preheating tongue effect in the cold plasma ahead of the ablation front. For temperatures $T>0.1$ MK, it has been shown that the ARTI growth rate, the density, and temperature profiles at the ablation front obtained from the $f(T)=\alpha T^{-1}+\beta T^{-3/2}$, agree well with the numerical simulations and the experiments.[17] In this paper, we choose a medium preheat case with parameters $\alpha=8.6$ and $\beta=1.6$. In simulating the CRTI, we set $\kappa_{\mathrm{SH}}=0$.

B. Numerical methods

The physical domain is rectangular. The domain in the x direction is $[x_l,x_r]=[-46.3\ \mu\mathrm{m},121.7\ \mu\mathrm{m}]$ with length $L_x=168\ \mu\mathrm{m}$. The domain in the y direction is assumed to be periodical with length $L_y=\lambda$, where λ is the perturbation wavelength of the single-mode sinusoidal interface perturbation. The free-stream boundary conditions are imposed at both ends of the domain in x since the flow is essentially unchanged near the boundaries of the computational domain

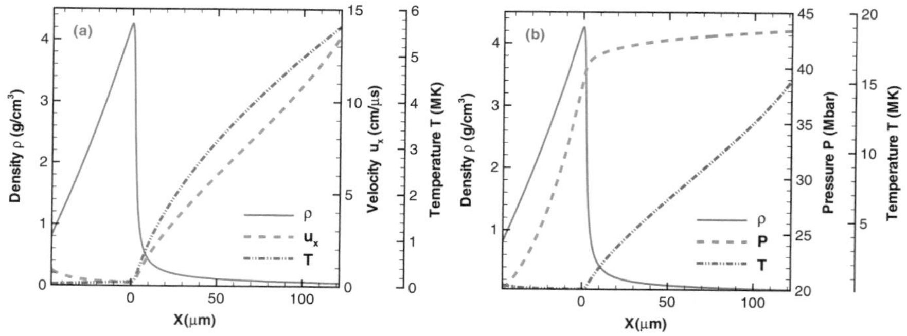

FIG. 1. (Color online) Distributions of the one-dimensional basic flows for the ARTI (a) and the CRTI (b). The two cases have the identical density profiles. The minimum scale length of density gradient at the ablation surface is $L_m = 1.83$ μm.

for a sufficiently large domain. Since the domain in y is periodical, periodical boundary conditions are imposed on the upper and lower boundaries.

To simplify the notation for discussion below, the system of partial differential equations can be expressed in the form of

$$\mathbf{Q}_t + L\mathbf{Q} = L_v\mathbf{Q} + \mathbf{S}, \tag{2}$$

where L and L_v are the spatial differential operators for the first (inviscid, nonlinear, hyperbolic) and second order (linear, parabolic) terms, respectively, and $\mathbf{S}$ is the bulk force source term.

We approximate the inviscid flux $L\mathbf{Q}$ in Eq. (2), which is the classical Euler equations, via the well-known classical fifth order characteristic-based weighted essentially nonoscillatory conservative finite difference scheme (WENO) explicitly. Following Refs. 26–28, the hyperbolicity of the Euler Eqs. (2) admits a complete set of right and left eigenvectors for the Jacobian of the system. The eigenvalues and eigenvectors are obtained via the Roe linearized Riemann solver.[29] The first order Lax–Friedrichs flux is used as the low order building block for the high order reconstruction step of the WENO scheme. After projecting the positive and negative fluxes on the characteristic fields via the left eigenvectors, the high order WENO reconstruction step is applied to obtain the high order approximation at the cell boundaries using the surrounding cell-centered values, which are then projected back into the physical space via the right eigenvectors and added together to form a high order numerical flux at the cell-interfaces. The conservative difference of the reconstructed high order fluxes can then be computed for inviscid first order term $L\mathbf{Q}$. We refer to Ref. 28 for further details on the WENO algorithm for solving the hyperbolic conservation laws.

In the case that the thermal conductivity coefficient $\kappa(T) \neq 0$ as in the ARTI, the second order parabolic term involving the thermal conduction is solved via a fourth order implicit central finite difference scheme to avoid the small time step restriction due to the severe stability condition imposed if solved by an explicit scheme. In the simulations presented in Sec. IV, the number of uniformly spaced grid points used in the x and y directions are $N_x = 1680$ and $N_y = 512$, respectively.

C. Initial condition setup

With the ARTI, the one-dimensional steady state flow field is obtained by integrating Eq. (1) in the accelerating reference frame of the ablation front from the peak density to the isothermal sonic points on both sides of the ablation front[30] (see the left figure of Fig. 1) using the averaged ablation parameters, the peak density $\rho_a = 4.3$ g/cm^3, the ablation velocity $v_a = 1.6$ μm/ns, the temperature at the peak density $T_a = 8.4 \times 10^{-2}$ MK, and the acceleration $g = 17$ μm/ns^2. With the CRTI, the density profile (see the right figure of Fig. 1) is identical to the one for the ARTI but the pressure profile is computed via the hydrostatic equation $P = P_0 + \int_{x_l}^{x_r} \rho g dx$ with hydrostatic pressure $P_0 = 20$ Mbar.

The primitive variables (ρ, u, v, T) of the initial steady flow field is perturbed with a single-mode sinusoidal interface perturbation at $x = x_0$ with wavelength λ and amplitudes η_0 in the y direction, namely,

$$S(x,y) = 1 + \eta_0 e^{-k_0|x-x_0|} \sin(k_0 y), \tag{3}$$

where $x_0 = 0$ is the location of the interface and $k_0 = 2\pi/\lambda$ is the wave number. The amplitude of the perturbation η_0 is typically set as a fraction of wavelength $\eta_0 = \alpha_0 \lambda$ with $1/10000 \le \alpha_0 \le 1/1000$. For example, the initial perturbed density field $\rho(x,y,t=0)$ becomes $\rho(x,y,0) = \rho_0(x,y) S(x,y)$, where $\rho_0(x,y)$ is the unperturbed initial steady density field.

III. ANALYTICAL HARMONIC ANALYSIS

The *linear* growth regime is defined by $k\eta(t)$ being small (typically $k\eta(t) < 0.628$), where $\eta(t)$ represents the typical spatial amplitude of the perturbed interface at time t. (The explicit dependent of η in time t will be dropped for simplicity sake.) After sufficient growth of the perturbation due to the instability, $k\eta$ is no longer small (typically $k\eta \sim 1$) and the flow enters the *nonlinear* regime. Before the strong nonlinear growth regimes, one has the *weakly nonlinear* growth regime.

In the weakly nonlinear growth regime, for a single-mode sinusoidal interface perturbation, within the framework of third-order perturbation theory,[1,25,31–34] the perturbation amplitudes of fundamental mode (first harmonic) η_1, second harmonic η_2, and third harmonic η_3 can be expressed as

$$\eta_1 = \eta_L - \tfrac{1}{16}(3A^2+1)k^2\eta_L^3,$$

$$\eta_2 = \tfrac{1}{2}Ak\eta_L^2, \tag{4}$$

$$\eta_3 = \tfrac{1}{2}\left(A^2 - \tfrac{1}{4}\right)k^2\eta_L^3,$$

where $\eta_L = \eta_0 e^{\gamma t}$ and γ are the spatial amplitude and growth rate in the linear regime, respectively. For problems with large Atwood number $A \to 1$, Eq. (4) reduce to[1,25,31–34]

$$\eta_1 = \eta_L - \tfrac{1}{4}k^2\eta_L^3,$$

$$\eta_2 = \tfrac{1}{2}k\eta_L^2, \tag{5}$$

$$\eta_3 = \tfrac{3}{8}k^2\eta_L^3.$$

It should be noted that the above potential expansion theories[31,32] are based on the assumptions that the interface between the two fluids is discontinuous and the thermal conduction is ignored. In ICF experiments, the thermal conduction is inevitable and a smooth mixing layer[19] is formed by the nonlocal electron heat transport or preheat effect of the hot electron, etc. It is, therefore, natural and important to investigate the growth of RTI with a smooth density profile. To investigate the effects of the smooth density transition layer and the ablation on the mode coupling process in the RTI, we generalize Eq. (5) by defining the second harmonic generation coefficient $C(2k)$, third harmonic generation coefficient $D(3k)$, and third-order feedback to the fundamental mode coefficient $E(k)$ and rewrite Eq. (4) as

$$\eta_1 = \eta_L - \tfrac{1}{4}E(k)k^2\eta_L^3,$$

$$\eta_2 = \tfrac{1}{2}C(2k)k\eta_L^2, \tag{6}$$

$$\eta_3 = \tfrac{3}{8}D(3k)k^2\eta_L^3.$$

For long perturbation wavelength ($\lambda \to \infty$), it is expected that Eq. (6) reduce to Eq. (5), or equivalently, one has $C(2k) \to A$, $D(3k) \to (4/3)[A^2-(1/4)]$, and $E(k) \to (1/4)(3A^2+1)$.

Another weakly nonlinear theory for the RTI up to the third-order was also developed by Berning *et al.*[35] which a slightly different form of the growths of the amplitudes of the harmonics was shown. The main reason for the difference between them is that the contribution from linear growth of the second harmonic and the third harmonic is also considered in Ref. 35. According to Eqs. (20) and (40) in Ref. 35, the growths of the second ($n=2$) and third harmonic ($n=3$) described above only consider the growth contribution from mode coupling,[1,25,31–34] i.e., $C_{nn}e^{n\gamma t}$, but drop their corresponding linear growths, i.e., $C_n e^{\sqrt{n}\gamma t}$, where C_{nn} and C_n are typical coefficients in the perturbation amplitudes formulas and $C_{nn} \sim C_n$.[35] In fact, for the growths of the second harmonic and the third harmonic, in a system with a single-mode sinusoidal interface perturbation, one can capture the main growth mechanism by considering the mode coupling growths and ignoring the contributions from their corresponding linear growth since there is $C_{nn}e^{n\gamma t} \gg C_n e^{\sqrt{n}\gamma t}$ within our focus of the weakly nonlinear regime which is typically in the range $0.005\lambda < \eta_3 < \eta_2 < \eta_1 < 0.5\lambda$, where λ is the perturbation wavelength. That is to say, in the present focused weakly nonlinear regime, the growth contribution of the amplitude from the linear growth of the second and third harmonics can be ignored in comparison with the mode coupling growth. Thus it is appropriate to use Eq. (6) in the investigation of RTI in the weakly nonlinear regime in the present research.

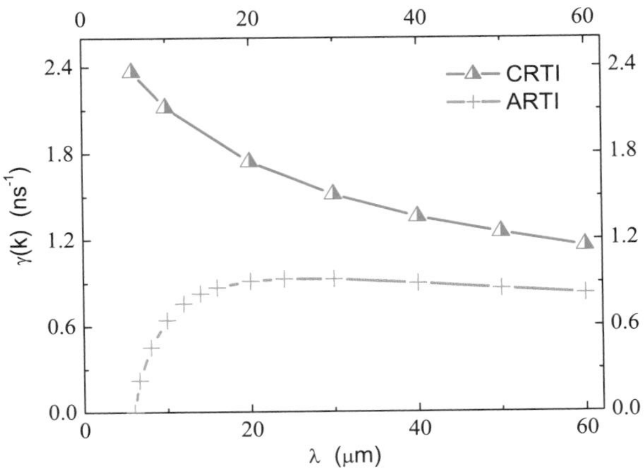

FIG. 2. (Color online) Linear growth rate curves $\gamma(k)$ for the ARTI and the CRTI.

Since the transition between the fluids is no longer sharp, we shall generate a reasonably defined interface in the fluid so that the growth rate of the amplitude of the perturbed interface can be tracked and analyzed by the Fourier analysis in the y direction. For this purpose, we shall define the interface as the location where the density gradient is the largest. In the weakly nonlinear regime where the density $\rho(x,y,t)$ remains as a single value function along the x direction, a function $f(y,t)$ where the density $\rho=2.4\ \mathrm{g/cm^3}$ can be uniquely defined and its gradient is the largest. Applying the Fourier analysis on the function $f(y,t)$, the absolute values of the Fourier coefficients $\{|a_n(t)|, n=1,\ldots,N_y/2\}$ yield the amplitudes of the nth harmonic of the interface in time t. The linear growth rate γ can then be obtained by fitting the amplitude of the fundamental mode within the linear growth regime. Similarly, the coefficients $\{E(k), C(2k), D(3k)\}$ can be obtained by fitting η_1, η_2, η_3, respectively, within the weakly nonlinear growth regime.

IV. NUMERICAL RESULTS AND DISCUSSIONS

The linear grow rate curves $\gamma(k)$ for the ARTI and the CRTI are plotted in Fig. 2. The result shows that the linear growth rate increases with the decreasing perturbation wavelength λ in the CRTI. However, in the ARTI, the linear growth rate is greatly reduced, especially for the short perturbation wavelength λ. The data also indicates that the cutoff wavelength is $\lambda_c \approx 6\ \mu$m in the ARTI simulation using our medium preheat model.

流体力学不稳定性和可压缩流体湍流 831

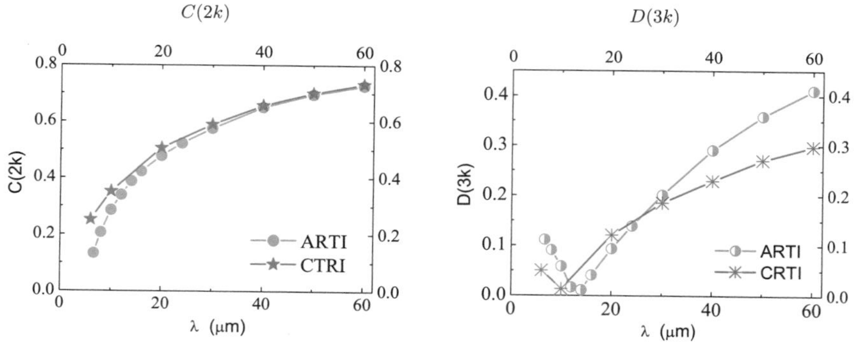

FIG. 3. (Color online) The second $C(2k)$ and third $D(3k)$ harmonic generation coefficients for the ARTI and the CRTI.

A. Ablation effects

To study the weakly nonlinear behaviors of the RTI, we examine the behavior of the second and third harmonic generation coefficients $C(2k)$ and $D(3k)$, respectively, as a function of wave numbers k for both the ARTI and the CRTI in Fig. 3. In the left figure, one can see that the ablation effect has an apparent influence on the coefficient $C(2k)$ of the second harmonic when $\lambda<20$ μm. The ablation effect is gradually weakened with an increasing perturbation wavelength λ. However, when $\lambda>40$ μm, the coefficient $C(2k)$ approaches almost the same value for both the ARTI and the CRTI. In the other words, the ablation essentially has no effect on $C(2k)$ for large λ. Moreover, as λ increases, $C(2k)\to A$, which is the classical predicted value for both the ARTI and the CRTI. In the right figure, one can see that the coefficient $D(3k)$ of the third harmonic first decreases and then increases with λ for both the ARTI and the CRTI.

This particular behavior of the coefficient $D(3k)$ has not been predicted by the classical third-order estimation.[31,32,35] To explain this behavior, we analyze the temporal evolution of the amplitudes of the first six harmonics at different perturbation wavelengths λ for both the ARTI and the CRTI. In the ARTI, similar to the CRTI,[25] it appears that:

- With wavelength $\lambda=10$ μm, the fourth harmonics ($n=4$) is generated earlier than the third harmonic ($n=3$);
- With wavelength $\lambda=14$ μm, the third and fourth harmonics are generated nearly at the same time, as illustrated in the left figure of Fig. 4; and
- With an increasing λ, the third harmonic in turn is generated earlier than the other higher harmonics ($n\geq 4$).

It is well known that the early generation of the higher harmonics ($n\geq 4$), especially the fourth harmonic, can greatly suppress the generation and the growth of the third harmonic. As noted above, for both the ARTI and the CRTI, the suppression of the third harmonic by the higher harmonics first decreases with wavelength and then increases with it. Accordingly, the $D(3k)$ curve first decreases with perturbation wavelength and then increases with it, as shown in the right figure of Fig. 3. Comparing the ARTI and the CRTI, for the large perturbation wavelength (30 μm$<\lambda<$60 μm) and small perturbation wavelengths ($\lambda<10$ μm), we also find that the suppression of the third harmonic by the higher harmonics is stronger in the CRTI than that in the ARTI. Therefore, $D(3k)$ in the ARTI is greater than that in the CRTI when $\lambda<10$ μm and 30 μm$<\lambda<$60 μm.

Figure 5 shows the coefficient $E(k)$ of the third-order feedback to the fundamental mode for the ARTI and the CRTI. For a short perturbation wavelength λ, $E(k)$ has very different behaviors in the ARTI and the CRTI. In the ARTI, $E(k)$ first decreases and then increases with increasing wave-

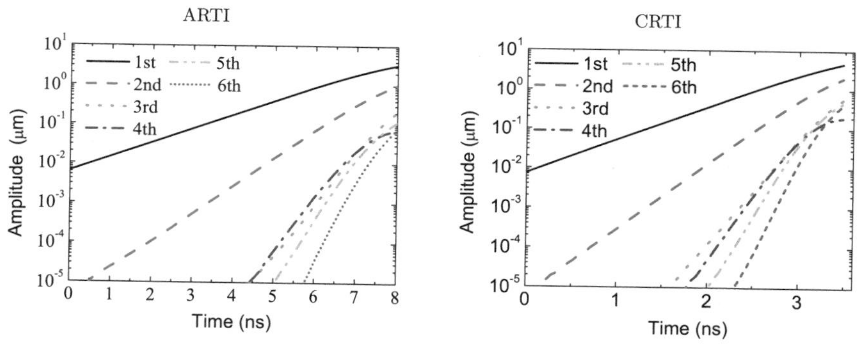

FIG. 4. (Color online) Temporal evolution of the amplitudes of the first six harmonics with a perturbation wavelength $\lambda=14$ μm for the ARTI and the CRTI using their respective initial conditions, as shown in Fig. 1.

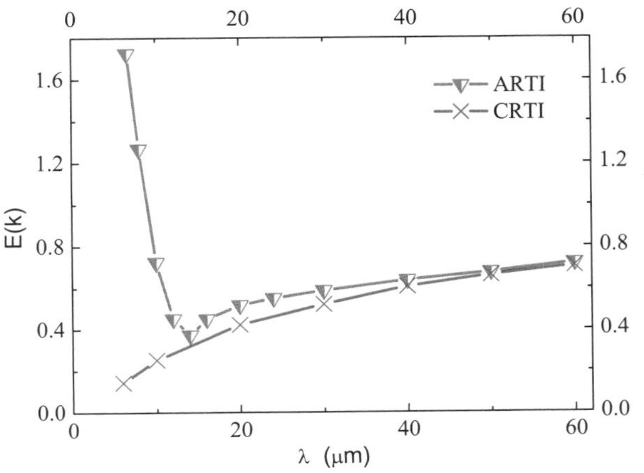

FIG. 5. (Color online) Coefficients of third-order feedback to the fundamental mode $E(k)$ for the ARTI and the CRTI.

length λ and a corresponding minimum wavelength $\lambda_{min} = 14$ μm. In the CRTI, however, $E(k)$ increases continuously with decreasing wavelength λ. Comparing the ARTI and the CRTI, we can conclude that the ablation plays a stabilization role since $E(k)$ in the ARTI is greater than that in the CRTI which resulted in the reduction of the amplitude of the fundamental harmonic η_1. In addition, it is noteworthy that for both the ARTI and the CRTI there is $E(k) < 1.0$, except for the ARTI when $\lambda \to \lambda_c$. Consequently, the nonlinear saturation amplitude[25] in both the ARTI and the CRTI should be increased and therefore exceeds the classical prediction 0.1λ,[31,32,35] which is harmful for the ICF ignition.

B. Transition layer thickness effects

In order to further identify which effect or effects, such as the nonlinearity, the transition layer thickness or the thermal conduction determines the counter-intuitive behavior, namely, the fourth harmonic growing faster than the third harmonic mentioned above, we present simulations of the CRTI with interface perturbation wavelength $\lambda = 14$ μm. Here, the one-dimensional steady flow of the density profile is in the form

$$\rho(x) = \frac{1}{2}(\rho_l + \rho_r) - \left[\frac{1}{2}(\rho_l - \rho_r)\right]\tanh\left(\frac{x}{L_D}\right),$$

where L_D is the typical half-thickness of the density transition layer, where $\rho_l = 4.25$ g/cm^3, $\rho_r = 0.5$ g/cm^3, with Atwood number $A = 0.7895$.[25] The CRTI simulations here are performed with four transition layer half-thicknesses $L_D = 0.4$, 0.6, 0.8, and 1.0 μm with the corresponding minimum density gradient scale lengths $L_m = 0.31$, 0.47, 0.63, and 0.79 μm. Please note that here the one-dimensional pressure profile is also computed via hydrostatic equation with again $P_0 = 20$ Mar and $g = 17$ μm/ns^2. The temporal evolutions of the amplitudes of the first six harmonics of the simulations are shown in Fig. 6.

For the $\lambda = 14$ μm perturbation wavelength, the counter-intuitive behavior of the fourth harmonic growing faster than the third harmonic mentioned above is observed in the ARTI

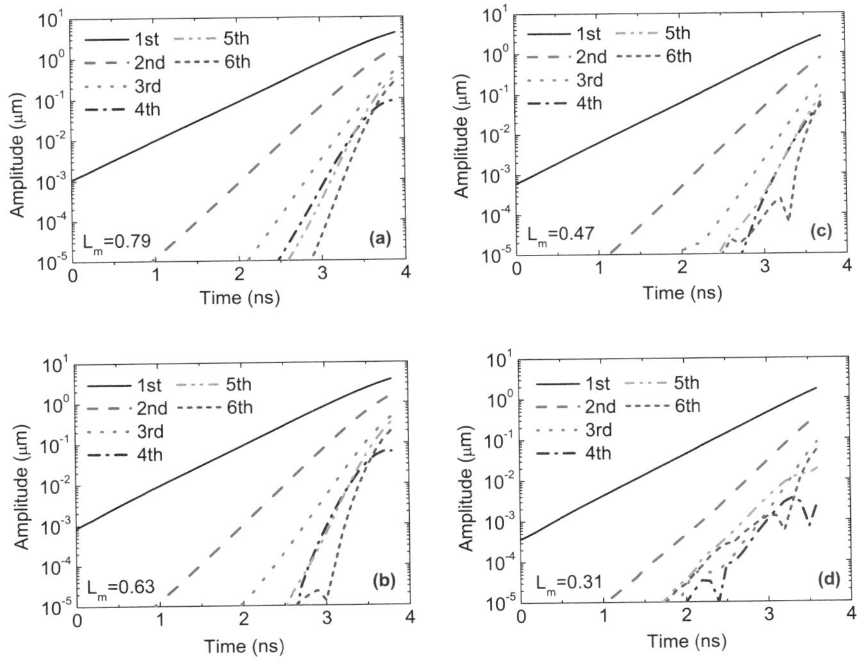

FIG. 6. (Color online) The temporal evolution of the amplitudes of the first six harmonics in the CRTI with a perturbation wavelength $\lambda = 14$ μm and the minimum density gradient scale lengths (a) $L_m = 0.79$ μm, (b) $L_m = 0.63$ μm, (c) $L_m = 0.47$ μm, and (d) $L_m = 0.31$ μm.

but not in the CRTI where the higher harmonics are generated one by one (see the left and right figure of Fig. 4, respectively). Therefore, one may conclude that the ablation effect strongly influences the mode coupling process and counterintuitive behaviors in the mode coupling process can occur when the effect of the thermal conduction is included. We would also like to point out that, as seen in Fig. 6, counterintuitive behavior with the lower harmonic growing faster than the higher one can also be observed when one decreases the transition layer thickness, L_D. Therefore, the transition layer thickness effect also strongly influences the mode coupling process in the absence of thermal conduction.[25] These results obtained above indicate that both the transition layer thickness L_D and the thermal conduction $\kappa(T)$ play central roles in the mode coupling process in the RTI. Detailed research in these areas will be pursued and reported in our future work.

C. Summary

In summary, for long perturbation wavelength, results obtained in this study are in general agreement with the recent weakly nonlinear theories of ARTI by Sanz and Ganier.[7,8] However, large differences appear as perturbation wavelength λ approaches the cutoff wavelength λ_c. The mode coupling coefficients given by Refs. 7 and 8 can be negative, especially for $E(k)$. This means that the third-order feedback strengthens the growth of the fundamental mode. That is, the fundamental mode cannot be saturated if one only considers the mode coupling process up to the third-order. Obviously, it is nonphysical. In our present work, contrarily, $E(k)$ is always positive. Moreover, the transition layer thickness also plays a key stabilization role in the ARTI growth, especially for the short perturbation wavelength λ in the weakly nonlinear regime. Note that the transition layer thickness effects were not sufficiently addressed in the work by Sanz or Garnier, but are studied in some detail here.

V. CONCLUSION

In this research, we have numerically investigated the two-dimensional RTI with and without thermal conduction (ARTI and CRTI, respectively) in the weakly nonlinear regime. A preheat model for the thermal conductivity is introduced to take into account of the preheating ablation effect of thermal conduction on the ARTI. The system of partial differential equations is solved using the high order weighted essentially non-oscillatory scheme and implicit central finite difference scheme. The preheating ablation effects and the ablation surface thickness effects on the RTI are studied by comparing the ARTI and the CRTI with an identical density profile relevant to inertial confinement fusion experiments.

It is found that the ablation effects dramatically influence the mode coupling process, especially for the short perturbation wavelengths. On the whole, the ablation effects stabilize the ARTI in the weakly nonlinear regime. First, the linear growth rate is reduced, especially for short perturbation wavelengths and a cutoff wavelength is observed in the simulations of the ARTI. Second, the second harmonic generation is reduced for the short perturbation wavelengths and the third-order negative feedback to the fundamental mode is strengthened. All of these factors noted above play a stabilization role on the ARTI in the weakly nonlinear regime. Finally, on the contrary, the ablation effects increase the generation of the third harmonic when the perturbation wavelengths are long. The counterintuitive behaviors with the lower harmonics growing faster than the higher ones are observed in both the ARTI and the CRTI and have been explained by analyzing the temporal evolution of the first six harmonics. Numerical results obtained in the present works are in general agreement with the recent weakly nonlinear theories.[7,8]

In direct-driven experiments in inertial confinement fusion, the jetlike long spikes are formed.[36–39] In the astrophysics, the astrophysical jets can remain well collimated over propagation distances exceeding ten jet diameters or more.[40] Our works show that when the thermal conduction with preheat is considered, the growth of the ARTI is stabilized in the weak nonlinear regime while the nonlinear saturation amplitude is improved, which is expected to improve the understanding of the stabilization factors for the astrophysical jet and of the formation mechanisms for the jetlike long spikes.

ACKNOWLEDGMENTS

The author (L.F.W.) would like to thank Professor S. Atzeni (Università di Roma "La Sapienza" and CNISM) and Professor K. Nishihara (Osaka University) for their helpful discussions and the anonymous referee for suggestions that have improved the paper. This research was supported by the National Basic Research Program of China (Grant No. 2007CB815100) and the National Natural Science Foundation of China (Grant Nos. 10935003, 11075024, 10905006, and 11074300). Both authors (W.S.D. and L.F.W.) would like to thank the support provided by the RGC grant from Hong Kong Research Grants Council. The author (L.F.W.) also would like to acknowledge the hospitality of the Department of Mathematics at Hong Kong Baptist University during his visit while conducting the research reported here.

[1] J. D. Lindl, P. Amendt, R. L. Berger, S. G. Glendinning, S. H. Glenzer, S. W. Haan, R. L. Kauffman, O. L. Landen, and L. J. Suter, Phys. Plasmas **11**, 339 (2004).
[2] S. Atzeni and J. Meyer-ter-Vehn, *The Physics of Inertial Fusion: Beam Plasma Interaction, Hydrodynamics, Hot Dense Mater* (Oxford Univeristy Press, Oxford, 2004).
[3] K. O. Mikaelian, Phys. Rev. Lett. **48**, 1365 (1982).
[4] K. O. Mikaelian, Phys. Rev. A **40**, 4801 (1989).
[5] A. B. Bud'ko and M. A. Liberman, Phys. Fluids B **4**, 3499 (1992).
[6] L. F. Wang, W. H. Ye, and Y. J. Li, Phys. Plasmas **17**, 042103 (2010).
[7] J. Sanz, J. Ramírez, R. Ramis, R. Betti, and R. P. J. Town, Phys. Rev. Lett. **89**, 195002 (2002).
[8] J. Garnier, P.-A. Raviart, C. Cherfils-Clérouin, and L. Masse, Phys. Rev. Lett. **90**, 185003 (2003).
[9] T. Ikegawa and K. Nishihara, Phys. Rev. Lett. **89**, 115001 (2002).
[10] S. Bodner, Phys. Rev. Lett. **33**, 761 (1974).
[11] H. Takabe, K. Mima, L. Montierth, and R. L. Morse, Phys. Fluids **28**, 3676 (1985).
[12] M. Tabak, D. H. Munro, and J. D. Lindl, Phys. Fluids B **2**, 1007 (1990).
[13] H. Azechi, T. Sakaiya, S. Fujioka, Y. Tamari, K. Otani, K. Shigemori, M. Nakai, H. Shiaga, N. Miyanaga, and K. Mima, Phys. Rev. Lett. **98**, 045002 (2007).

[14]S. G. Glendinning, S. N. Dixit, B. A. Hammer, D. H. Kalantat, M. H. Key, J. D. Kilkenny, J. P. Knauer, D. M. Pennington, B. A. Remington, R. J. Wallace, and S. V. Weber, Phys. Rev. Lett. **78**, 3318 (1997).
[15]K. Shigemori, H. Azechi, M. Nakai, M. Honda, K. Meguro, N. Miyanaga, H. Takabe, and K. Mima, Phys. Rev. Lett. **78**, 250 (1997).
[16]T. Sakaiya, H. Azechi, M. Matsuoka, N. Izumi, M. Nakai, K. Shigemori, H. Shiraga, A. Sunahara, H. Takabe, and T. Yamanaka, Phys. Rev. Lett. **88**, 145003 (2002).
[17]W. H. Ye, W. Y. Zhang, and X. T. He, Phys. Rev. E **65**, 057401 (2002).
[18]L. Spitzer and R. Härm, Phys. Rev. **89**, 977 (1953).
[19]S. Fujioka, H. Shiraga, M. Nishikino, K. Shigemori, A. Sunahara, M. Nakai, H. Azechi, K. Nishihara, and T. Yamanaka, Phys. Plasmas **10**, 4784 (2003).
[20]J. Sanz, Phys. Rev. Lett. **73**, 2700 (1994).
[21]A. P. Piriz, J. Sanz, and L. F. Ibañez, Phys. Plasmas **4**, 1117 (1997).
[22]V. N. Goncharov, R. Betti, R. L. McCrory, P. Sorotokin, and C. P. Verdon, Phys. Plasmas **3**, 1402 (1996).
[23]R. Betti, V. N. Goncharov, R. L. McCrory, and C. P. Verdon, Phys. Plasmas **5**, 1446 (1998).
[24]L. Masse, Phys. Rev. Lett. **98**, 245001 (2007).
[25]L. F. Wang, W. H. Ye, and Y. J. Li, Phys. Plasmas **17**, 052305 (2010).
[26]D. Balsara and C. W. Shu, J. Comput. Phys. **160**, 405 (2000).
[27]G. S. Jiang and C. W. Shu, J. Comput. Phys. **126**, 202 (1996).
[28]R. Borges, M. Carmona, B. Costa, and W. S. Don, J. Comput. Phys. **227**, 3191 (2008).
[29]P. L. Roe, J. Comput. Phys. **43**, 357 (1981).
[30]Z. F. Fan, J. S. Luo, and W. H. Ye, Phys. Plasmas **16**, 102104 (2009).
[31]J. W. Jacobs and I. Catton, J. Fluid Mech. **187**, 329 (1988).
[32]S. W. Haan, Phys. Fluids B **3**, 2349 (1991).
[33]B. A. Remington, S. V. Weber, M. M. Marinak, S. W. Haan, J. D. Kilkenny, R. J. Wallace, and G. Dimonte, Phys. Plasmas **2**, 241 (1995).
[34]V. A. Smalyuk, V. N. Goncharov, T. R. Boehly, J. P. Knauer, D. D. Meyerhofer, and T. C. Sangster, Phys. Plasmas **11**, 5038 (2004).
[35]M. Berning and A. M. Rubenchik, Phys. Fluids **10**, 1564 (1998).
[36]D. K. Bradley, D. G. Braun, S. G. Glendinning, M. J. Edwards, J. L. Milovich, C. M. Sorce, G. W. Collins, S. W. Haan, R. H. Page, and R. J. Wallace, Phys. Plasmas **14**, 056313 (2007).
[37]B. A. Remington, S. W. Haan, S. G. Glendinning, J. D. Kilkenny, D. H. Munro, and R. J. Wallace, Phys. Fluids B **4**, 967 (1992).
[38]B. A. Remington, R. P. Drake, H. Takabe, and D. Arnett, Phys. Plasmas **7**, 1641 (2000).
[39]J. O. Kane, H. F. Robey, B. A. Remington, R. P. Drake, J. Knauer, D. D. Ryutov, H. Louis, R. Teyssier, O. Hurricane, D. Arnett, R. Rosner, and A. Calder, Phys. Rev. E **63**, 055401 (2001).
[40]B. Reipurth and J. Bally, Annu. Rev. Astron. Astrophys. **39**, 403 (2001).

PHYSICS OF PLASMAS **17**, 122704 (2010)

Spike deceleration and bubble acceleration in the ablative Rayleigh–Taylor instability

W. H. Ye (叶文华),[1,2,a)] L. F. Wang (王立锋),[2,3,b)] and X. T. He (贺贤土)[1,2,c)]
[1]Department of Physics, Zhejiang University, Hangzhou 310027, People's Republic of China
[2]CAPT, Peking University, Beijing 100871, People's Republic of China and LCP, Institute of Applied Physics Computational Mathematics, Beijing 100088, People's Republic of China
[3]SMCE, China University of Mining and Technology, Beijing 100083, People's Republic of China

(Received 6 July 2010; accepted 14 September 2010; published online 7 December 2010)

The nonlinear evolutions of the Rayleigh–Taylor instability (RTI) with preheat is investigated by numerical simulation (NS). A new phase of the spike deceleration evolution in the nonlinear ablative RTI (ARTI) is discovered. It is found that nonlinear evolution of the RTI can be divided into the weakly nonlinear regime (WNR) and the highly nonlinear regime (HNR) according to the difference of acceleration velocities for the spike and the bubble. With respect to the classical RTI (i.e., without heat conduction), the bubble first accelerates in the WNR and then decelerates in the HNR while the spike holds acceleration in the whole nonlinear regime (NR). With regard to the ARTI, on the contrary, the spike first accelerates in the WNR and then decelerates in the HNR while the bubble keeps acceleration in the whole NR. The NS results indicate that it is the nonlinear overpressure effect at the spike tip and the vorticity accumulation inside the bubble that lead to, respectively, the spike deceleration and bubble acceleration, in the nonlinear ARTI. In addition, it is found that in the ARTI the spike saturation velocity increases with the perturbation wavelength.
 [doi:10.1063/1.3497006]

I. INTRODUCTION

In the inertial confinement fusion (ICF), the energy absorbed by the target material drives an ablation that leads to the mass off the shell outer surface and generates a low-density blow-off plasma, resulting in the expected shell acceleration.[1] Unfortunately the Rayleigh–Taylor instability (RTI) occurs during this regime when a less dense fluid (the ablation plasma) accelerates another fluid of higher density (the unabated material). On account of this instability, any perturbation either from the shell surfaces or from the laser beams would grow exponentially in time during the linear regime. After a short linear growth, the perturbation enters the nonlinear growth regime, where the "bubbles" of the lighter fluid rise through the denser fluid, and the "spikes" of the heavier fluid penetrates down through the lighter fluid. On the one hand, in the acceleration stage, the integrity of the dense shell is severely compromised by the RTI bubbles approaching the inner shell surface. On the other hand, in the deceleration stage, the growth of the hot spot is dramatically dependent on the cold spikes of the shell at the hot spot boundary. Therefore, the ICF targets must be designed to keep the ablative RTI (ARTI) growth at an acceptable level since it can break up the implosion shell during the acceleration stage and destroy the central ignition hot spot during the deceleration stage. The estimate for when these will happen naturally depends on the bubble (spike) velocity through the dense (light) fluid.

In the frame of the classical theory[1] [i.e., without heat conduction (HC)], when the heavier fluid of density ρ_h is superposed over the lighter fluid of density ρ_l in a gravitational field, $-g\hat{y}$, where g is the acceleration, a single-mode perturbation of wavelength λ with the perturbation amplitude η_0 [i.e., the perturbation is in the form $\eta(x)=\eta_0\cos(kx)$] on the interface between two fluids yields its linear growth rate[1] $\gamma_{cla}=\sqrt{A_T kg}$, where $k=2\pi/\lambda$ is the perturbation wave number, and $A_T=(\rho_h-\rho_l)/(\rho_h+\rho_l)$ is the Atwood number. It is well known that the density gradient effects (i.e., the interface width effects) reduce the RTI growth rate,[2] which can be approximated by the formula,[3–7] $\gamma_{cL}=\sqrt{A_T kg/(1+A_T kL_m)}$, where $L_m=\min[|\rho/(\partial\rho/\partial x)|]$ is the minimum density gradient scale length. When the ablation is included, the linear growth rate of the ARTI, $\gamma_{\rm abl}$, is remarkably reduced, especially for the short perturbation wavelength and a cut-off wavelength appears.[1,8–18] Due to the removal of modulated material by ablation, the reduction of $\gamma_{\rm abl}$ in comparison with γ_{cla}, is simply expressed by the Takabe–Bodner formula,[8,9] $\gamma_T=\alpha\sqrt{kg}-\beta kv_a$, where v_a is the ablation velocity (i.e., the rate at which the material is removed), and α and β are the numerical adjustable coefficients depending on the flow parameters. A modified version of this given by Lindl[10] including another stabilization factor, the width effect of the interface region, is in the form $\gamma_{mL}=\alpha\sqrt{kg/(1+kL_m)}-\beta kv_a$, where L_m is the minimum scale length of density gradient at the ablation surface. The direct-driven (DD) experiments[11,12] have shown that the growth rate of the areal density perturbation is significantly reduced as compared with the calculated results of the Spitzer–Härm (SH) electron thermal conductivity. Glendinning *et al.*[12] recognized that the difference between the experiment and the LASNEX simulation is due to a change in the longitudinal density profile resulting from preheat by energetic electrons originating in the plasma corona that penetrate beyond the ablation front. Furthermore,

[a)]Electronic mail: ye_wenhua@iapcm.ac.cn.
[b)]Electronic mail: wanglifengqq@126.com.
[c)]Electronic mail: xthe@iapcm.ac.cn.

the experiments of Shigemori and Sakaiya *et al.*[12,13] suggests that the nonlocal electron heat transport plays an important role in the suppression of RTI at the ablation front. Subsequently, Ye *et al.*[14] found that the ARTI growth rate formula, $\gamma_{mY}=\sqrt{Akg/(1+AkL_m)}-2kv_a$, agrees well with the experiments and simulations, and is appropriate for the preheating case in DD experiments in ICF. The observation of density profile in DD experiment[19] and simulation[12,13] indicates that L_m is markedly improved by the preheat ablation effect. In addition, the recent self-consistent linear theory[15–18] identified another physical mechanism of stabilization, which is expressed as $\gamma_{SC}=\sqrt{A_Tkg-A_T^2k^2v_a^2/r_D}-(1+A_T)kv_a$ for the case of large Froude number. Here, r_D is the effective density ratio of the blow-off plasma to the ablation front. The main difference between γ_{SC} and all the versions of the Takabe–Bodner formula (e.g., γ_T, γ_{mL}, and γ_{mY}) is the negative term under the square root.

Recently, a new phase of the nonlinear bubble evolution is discovered by Betti *et al.*,[20] which is attributed to the vorticity accumulation inside the bubble resulting from the mass ablation and vorticity convection off the ablation front. The focus of the present work investigates the single-mode velocities for the bubble and the spike including the interface thickness effects and the preheat ablation effects, which are both relevant to the typical ICF experiments by numerical simulation (NS). In order to separate the influence of the interface width and the ablation, we, therefore, consider situations with and without HC.

II. NUMERICAL METHODS AND BASIC FLOWS

Without considering the radiation, an ideal gas equation of state is used in our NS. The governing equations in a constantly accelerating reference frame for single-fluid and one-temperature are[21–24]

$$\frac{\partial \rho}{\partial t}+\nabla\cdot(\rho \boldsymbol{V})=0, \tag{1a}$$

$$\frac{\partial(\rho \boldsymbol{V})}{\partial t}+\nabla\cdot(\rho \boldsymbol{V}\boldsymbol{V})+\nabla p=\rho \boldsymbol{g}, \tag{1b}$$

$$\frac{\partial E}{\partial t}+\nabla\cdot[(E+P)\boldsymbol{V}]=\nabla\cdot(\kappa\nabla T)+\rho\boldsymbol{V}\cdot\boldsymbol{g}, \tag{1c}$$

where ρ, $\boldsymbol{V}$, T, and $\boldsymbol{g}$ are, respectively, the density, velocity, temperature, and acceleration. $E=C_V\rho T+\rho\boldsymbol{V}\cdot\boldsymbol{V}/2$ is the total energy, $p=\Gamma\rho T$ is the pressure, $C_V=\Gamma/(\gamma_h-1)$ is the specific heat at constant volume, and γ_h equals to 5/3. The term of $\nabla\cdot(\kappa\nabla T)$ represents the electron heat conduction, which is turned off when we consider the classical RTI (CRTI) (i.e., without HC). The preheat model $\kappa(T)=\kappa_{SH}[1+f(T)]$ is introduced in our NS,[14] where κ_{SH} is the SH electron heat conduction and $f(T)$ interprets the preheating tongue effect in the cold plasma ahead of the ablation front. For temperatures greater than 0.10 MK, it has been shown that the ARTI growth rates, the fluid density, and temperature profiles at the ablation front obtained from the choice $f(T)=\alpha T^{-1}+\beta T^{-3/2}$, where T is in the unit of megakelvin, α and β are adjustable parameters, agree well with the experiments.[14] In this paper, we choose two preheat cases, i.e., the weak preheat (WP) case with $\alpha=0.86$ and $\beta=0.24$, and the strong preheat (SP) case with $\alpha=8.6$ and $\beta=1.6$.

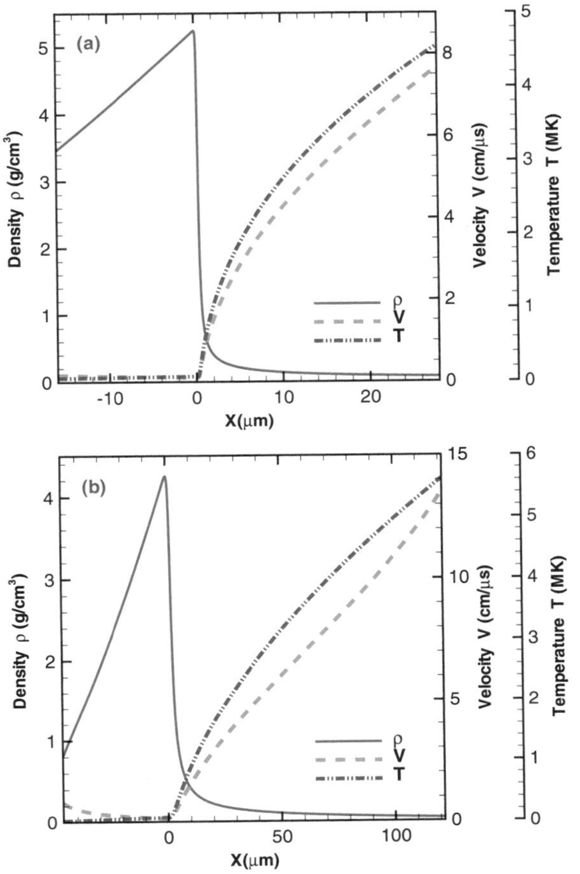

FIG. 1. (Color online) Basic flows for the WP (a) and SP (b) cases. The minimum scale length of density gradient at the ablation surface for the WP and SP cases are 0.23 and 1.83 μm, respectively.

With the ARTI, we obtain the steady state flow fields, as shown in Figs. 1(a) and 1(b) by integrating Eq. (1) in the accelerating reference frame of the ablation front from the peak density to the isothermal sonic points on both sides of the ablation front.[25] The averaged ablation parameters for the WP (SP) case are as follows: the peak density $\rho_a=5.25\ \mathrm{g/cm^3}$ ($4.25\ \mathrm{g/cm^3}$), the ablation velocity $v_a=1.19\ \mu\mathrm{m/ns}$ ($1.58\ \mu\mathrm{m/ns}$), the temperature at the peak density $T_a=8.23\times10^{-2}$ MK (8.37×10^{-2} MK), and the acceleration $g=17.85\ \mu\mathrm{m/ns^2}$ ($16.57\ \mu\mathrm{m/ns^2}$), which are obtained from the one-dimensional calculation,[25] which are adopted for the above integration. With the CRTI, the density profile is from the SP case, while the pressure is computed via the hydrostatic equation in the form $p=p_0+\int_l^r\rho g dx$, where $p_0=20$ Mbar is the static pressure, $g=16.57\ \mu\mathrm{m/ns^2}$ and the integral domain is from the left boundary to the right boundary within the calculation zone.

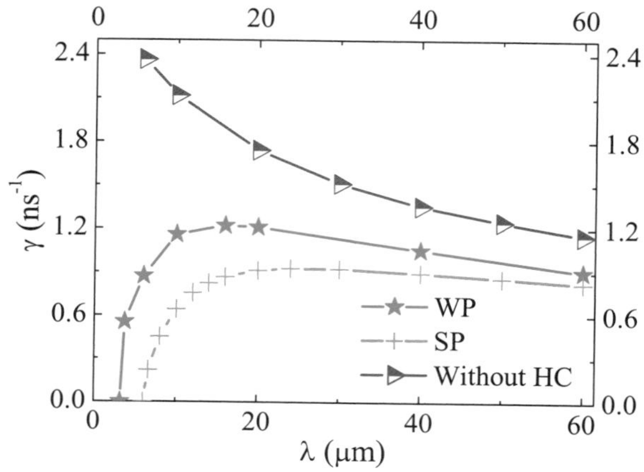

FIG. 2. (Color online) Linear growth rate curves $\gamma(k)$ for WP, SP, and classical cases. The cut-off wavelengths are 3.2 and 6.0 μm for the WP and SP cases, respectively.

III. SPIKE DECELERATION AND BUBBLE ACCELERATION

Since the transition between the fluids is no longer sharp, we shall generate a reasonably defined interface in the fluid so that the spatial evolution of the perturbed interface can be tracked and analyzed. For this purpose, we shall define the interface as the location where the density gradient is the largest in the one-dimensional steady density field. The linear growth rate can be obtained by fitting the amplitude of the fundamental mode within the linear growth regime. The linear grow rate curves $\gamma(k)$ are shown in Fig. 2. Considering the CRTI, the linear growth rate decreases with the perturbation wavelength λ. Regarding the ARTI, the linear growth rate is remarkably reduced in comparison with the CRTI, especially for the short perturbation wavelength and there is a cut-off perturbation wavelength that is about 3.2 μm (6.0 μm) in our WP (SP) preheat model.

The temporal evolutions of the normalized amplitudes η/λ, normalized velocities $v/\sqrt{g\lambda}$, and normalized accelerations a/g for the spike and the bubble with $\lambda=20$ μm and $\eta_0=0.0069$ μm are plotted in Fig. 3. In the CRTI, the spike length is always greater than the bubble's while the bubble length can surpass the spike's in the ARTI, as shown in Fig. 3(a). It is seen in Fig. 3(b) that in the nonlinear regime (NR) the bubble velocity in the ARTI can exceed the spike's, while this does not happen in the CRTI. As shown in Fig. 3(c), we can see that in the NR there is a spike (bubble) acceleration (deceleration) stage for the CRTI, while conversely there is a bubble (spike) acceleration (deceleration) phase in the ARTI. Because of the opposite behaviors of the acceleration velocities for bubble and spike in the ARTI and the CRTI, it is suggested that the NR of the RTI can be divided into the weakly nonlinear regime (WNR) and the highly nonlinear regime (HNR) according to the difference of acceleration velocities for the spike and the bubble. For the CRTI, the bubble first accelerates in the WNR and then decelerates in the HNR resulting in a bubble saturation velocity,[26,27] while

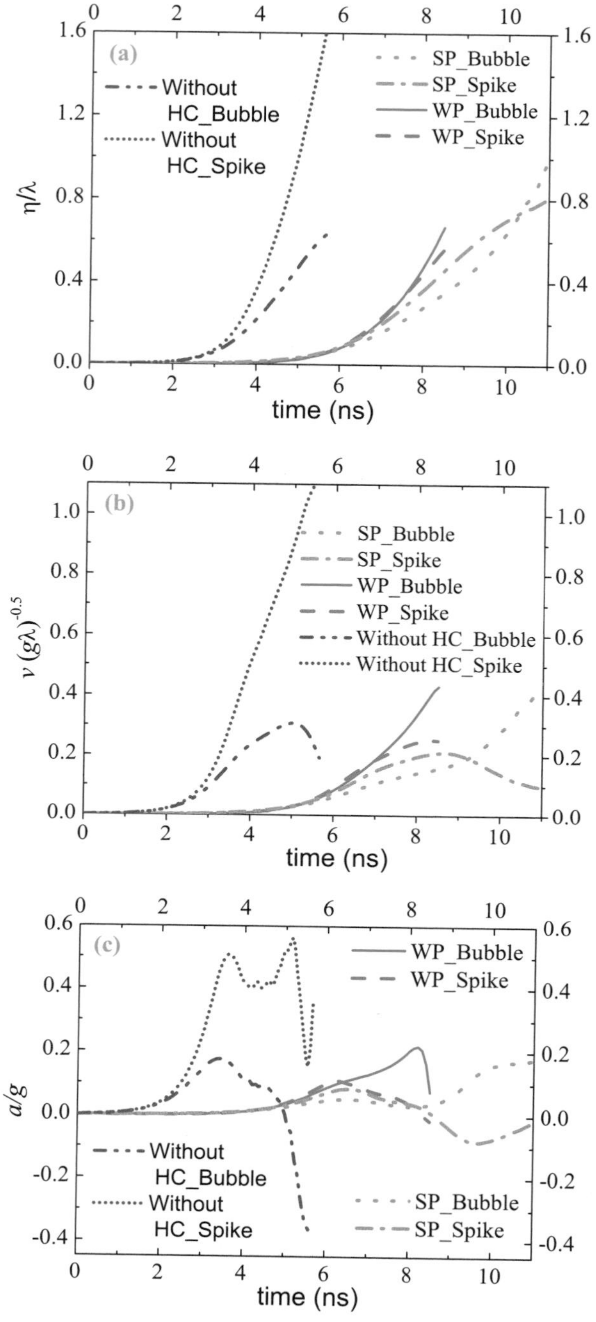

FIG. 3. (Color online) Normalized amplitudes (a), normalized velocities (b), and normalized accelerations (c) for the spike and the bubble vs time for a 20 μm perturbation wavelength with an initial amplitude 0.0069 μm. Solid, dotted, and dash dot dotted lines represent the bubbles for WP, SP, and classical cases, respectively. Dashed, dash dotted, and short dotted lines correspond to the spikes for WP, SP, and classical cases, respectively.

the spike keeps acceleration in the whole NR. Instead, for the ARTI, the spike first accelerates in the WNR and then decelerates in the HNR leading to a spike saturation velocity, while the bubble holds acceleration in the whole NR. The

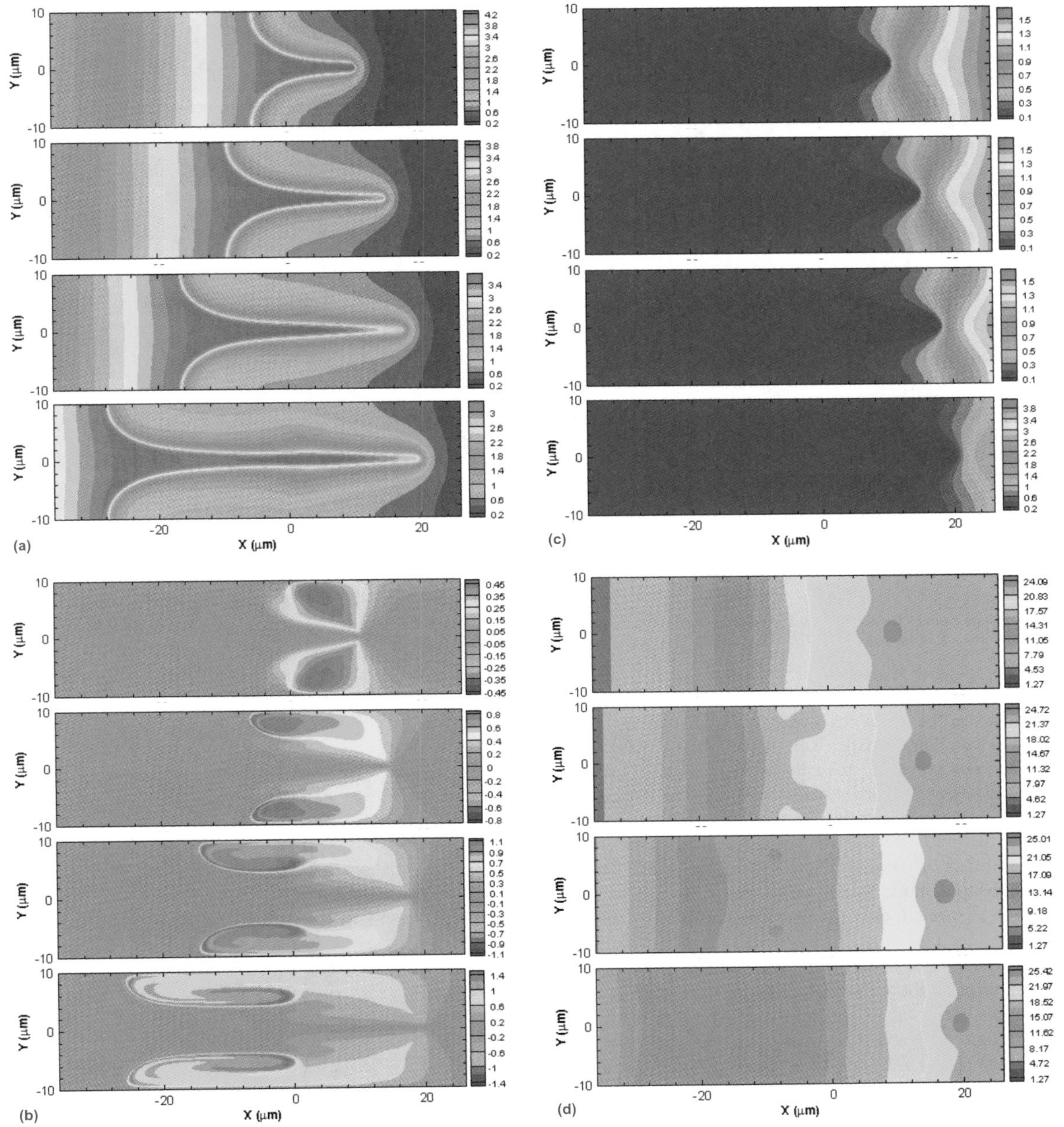

FIG. 4. (Color online) Density contours (a) and, respectively, the corresponding vorticity (b), absolute value of heat flux (c), and pressure (d) contours for the SP case of a 20 μm perturbation wavelength with an initial amplitude 0.0069 μm at $\gamma t=7.7$, 8.8, 9.8, and 11.0. The units of density, vorticity, heat flux, and pressure are g/cm^3, 10 ns^{-1}, 10^{13} W/cm^2, and megabar, respectively.

nonlinear behavior in the CRTI has been well understood[26,27] while the new highly nonlinear behaviors in the ARTI has not been fully understood. Here, we try to understand it by analyzing the nonlinear evolutions of the density, vorticity, absolute value of heat flux, and pressure.

Figure 4 shows the density contours and the corresponding contours of vorticity, absolute value of heat flux, and pressure for the SP case with $\lambda=20$ μm and $\eta_0=0.0069$ μm at $\gamma t=7.7$, 8.8, 9.8, and 11.0, where the vorticity is defined as $(\partial V_y/\partial x-\partial V_x/\partial y)$ and the absolute value of heat flux is defined as $|q|=\kappa\sqrt{(\partial T/\partial x)^2+(\partial T/\partial y)^2}$. The density contours and the corresponding vorticity contours in the CTRI for a 20 μm perturbation wavelength with $\eta_0=0.0069$ μm at $\gamma t=7.7$, 8.4, 9.0, 9.7, 10.4, and 11.1, are illustrated in Fig. 5.

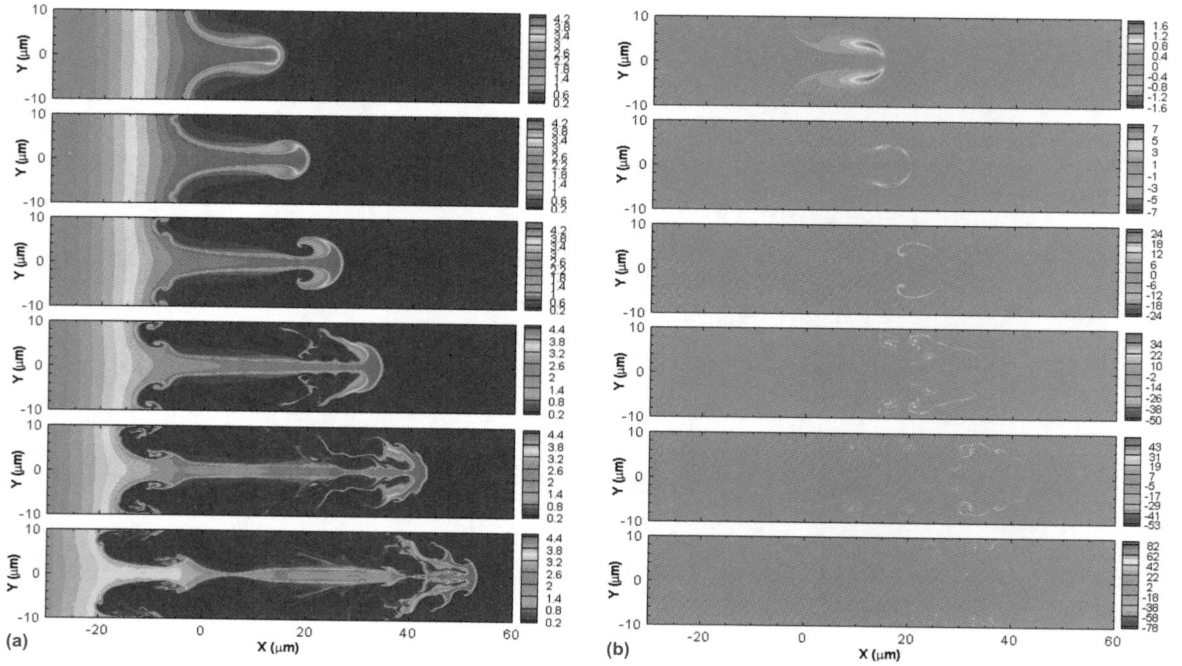

FIG. 5. (Color online) Density contours (a) and the corresponding vorticity contours (b) for the classical case of a 20 μm perturbation wavelength with an initial amplitude 0.0069 μm at $\gamma t=7.7$, 8.4, 9.0, 9.7, 10.4, and 11.1. The units of density and vorticity are g/cm^3 and 10 ns^{-1}, respectively.

Note that the density contours in the SP case are smooth showing no characteristic Kelvin–Helmholtz (KH) mushroom structures at the spike heads, as shown in Fig. 4(a), while the KH mushroom structures are fully developed in the classical case, as shown in Fig. 5(a). Accordingly, with and without the noticeable KH mushroom structures are the significant difference between the ARTI and the CRTI. The vortex climbs toward the bubble vertex in the SP case, as shown in Fig. 4(b), while it does not occur in the CRTI, as shown in Fig. 5(b). The phenomenon of the vorticity accumulation inside the bubble is first reported by Betti *et al.*,[20] which is used to explain the bubble acceleration in the HNR of the ARTI. Our results is in general agreement with their. Comparing Figs. 4(b) and 5(b), it is seen that the vorticity is strongly suppressed by the HC. The distribution of the heat flux round the spike tip forms a cone that encloses the spike tip, as shown in Fig. 4(c). One can expect that, in the surroundings of the high temperature heat flow cone, it is hard for the spike tip to become decentralized to form the classical KH mushroom structures. Detailed research in these areas will be pursued and reported in our future work.

We shows the density contours and the corresponding vorticity contours for the WP case with $\lambda=20$ μm and $\eta_0=0.0069$ μm at $\gamma t=8.4$, 9.0, 9.7, and 10.0, in Fig. 6. For the WP case, the characteristic KH mushroom structure appears in the NR regime, although these KH mushroom structures are not very typical, as shown in Fig. 6. Furthermore, it is found that the vorticity accumulation inside the bubble is reduced in comparison with the SP case, as shown in Fig. 4(b), but increased in comparison with the classical case, as

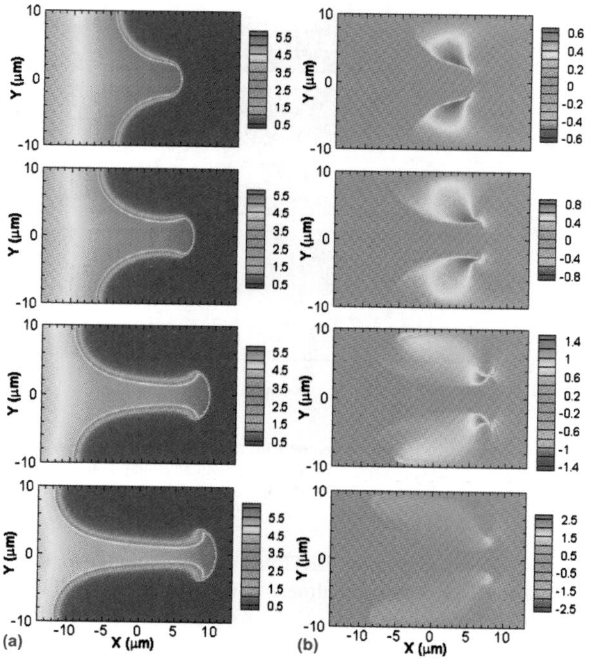

FIG. 6. (Color online) Density contours (a) and the corresponding vorticity contours (b) for the WP case of a 20 μm perturbation wavelength with an initial amplitude 0.0069 μm at $\gamma t=8.4$, 9.0, 9.7, and 10.0. The units of density and vorticity are g/cm^3 and 10 ns^{-1}, respectively.

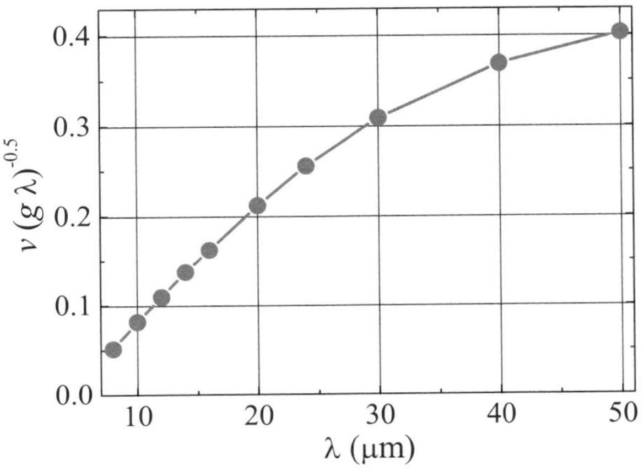

FIG. 7. (Color online) Normalized spike saturation velocities vs perturbation wavelengths for the SW case.

shown in Fig. 5(b). It is therefore concluded that the vorticity convection in the WP case is less pronounced in comparison with the SP case but more pronounced in comparison with the classical case.

As noted above, in the linear growth regime, the overpressure term (i.e., $A_T^2 k^2 v_a^2 / r_D$) can increase the ablative pressure at the spike tip providing an "restoring force." One can expect this effect can be strengthened in the NR. Because in the NR, the spike tip becomes more close to the hot laser absorption zone, where not only ∇T but also the heat conduction coefficient $\kappa(T)$ are improved. As can be seen in Fig. 4(d), the maximum pressure around the spike tip is about 24.09, 24.72, 25.01, and 25.42 Mbar for γt=7.7, 8.8, 9.8, and 11.0, respectively, while the initial pressure at the interface is only 22.05 Mbar, as shown in Fig. 1(b). It is seen that the restoring force $\propto k^2$ as noted above. Therefore, one can expect that the spike saturation velocity $\propto \lambda^2$. We show the normalized spike saturation velocities in Fig. 7. The saturation velocity increases with the perturbation wavelength, as shown in Fig. 7, which can be preliminarily understood as mentioned above. Detailed research in these areas will be pursued and reported in our future work.

IV. CONCLUSION

The nonlinear evolutions of the ARTI with preheat has been studied by NS in the present work. In order to better understand the ablation effect and ablation surface thickness effect, we also investigate the CRTI for comparison. We find a new regime of spike deceleration in the nonlinear evolution of the ARTI. Subsequently, it is therefore suggested that the nonlinear evolution of the RTI can be divided into the WNR and the HNR according to the difference of acceleration velocities for the spike and the bubble. On the one hand, for the CRTI, the bubble first accelerates in the WNR and then decelerates in the HNR while the spike holds acceleration in the whole NR. On the other hand, for the ARTI, instead, the spike first accelerates in the WNR and then decelerates in the HNR while the bubble keeps acceleration in the whole NR. Our NS results indicate that it is the nonlinear overpressure effect[15–18] at the spike tip and the vorticity accumulation inside the bubble[20] that lead to, respectively, the spike deceleration and bubble acceleration in the nonlinear ARTI. It is noteworthy that the effects of the interface width and the preheating ablation are not sufficiently included in the previous works, especially for the analytical investigations[15–18,20,26,27], which are fully considered in our numerical simulations presented here.

ACKNOWLEDGMENTS

This work was supported by the National Basic Research Program of China (Grant No. 2007CB815103) and the National Natural Science Foundation of China (Grant Nos. 10935003, 10775020, and 11075024).

[1] S. Atzeni and J. Meyer-ter-Vehn, *The Physics of Inerial Fusion: Beam Plasma Interaction, Hydrodynamics, Hot Dense Mater* (Oxford University Press, Oxford, 2004).
[2] K. O. Mikaelian, Phys. Rev. Lett. **48**, 1365 (1982).
[3] K. O. Mikaelian, Phys. Rev. A **33**, 1216 (1986).
[4] D. H. Munro, Phys. Rev. A **38**, 1433 (1988).
[5] K. O. Mikaelian, Phys. Rev. A **40**, 4801 (1989).
[6] A. B. Bud'ko and M. A. Liberman, Phys. Fluids B **4**, 3499 (1992).
[7] L. F. Wang, W. H. Ye, and Y. J. Li, Phys. Plasmas **17**, 042103 (2010).
[8] S. Bodner, Phys. Rev. Lett. **33**, 761 (1974).
[9] H. Takabe, K. Mima, L. Montierth, and R. L. Morse, Phys. Fluids **28**, 3676 (1985).
[10] J. D. Lindl, P. Amendt, R. L. Berger, S. G. Glendinning, S. H. Glenzer, S. W. Haan, R. L. Kauffman O. L. Landen, and L. J. Suter, Phys. Plasmas **11**, 339 (2004).
[11] K. Shigemori, H. Azechi, M. Nakai, M. Honda, K. Meguro, N. Miyanaga, H. Takabe, and K. Mima, Phys. Rev. Lett. **78**, 250 (1997).
[12] S. G. Glendinning, S. N. Dixit, B. A. Hammer, D. H. Kalantat, M. H. Key, J. D. Kilkenny, J. P. Knauer, D. M. Pennington, B. A. Remington, R. J. Wallace, and S. V. Weber, Phys. Rev. Lett. **78**, 3318 (1997).
[13] T. Sakaiya, H. Azechi, M. Matsuoka, N. Izumi, M. Nakai, K. Shigemori, H. Shiraga, A. Suna-hara, H. Takabe, and T. Yamanaka, Phys. Rev. Lett. **88**, 145003 (2002).
[14] W. H. Ye, W. Y. Zhang, and X. T. He, Phys. Rev. E **65**, 057401 (2002).
[15] J. Sanz, Phys. Rev. Lett. **73**, 2700 (1994).
[16] A. P. Piriz, J. Sanz, and L. F. Ibañez, Phys. Plasmas **4**, 1117 (1997).
[17] V. N. Goncharov, R. Betti, R. L. McCrory, P. Sorotokin, and C. P. Verdon, Phys. Plasmas **3**, 1402 (1996).
[18] R. Betti, V. N. Goncharov, R. L. McCrory, and C. P. Verdon, Phys. Plasmas **5**, 1446 (1998).
[19] S. Fujioka, H. Shiraga, M. Nishikino, K. Shigemori, A. Sunahara, M. Nakai, H. Azechi, K. Nishihara, and T. Yamanaka, Phys. Plasmas **10**, 4784 (2003).
[20] R. Betti and J. Sanz, Phys. Rev. Lett. **97**, 205002 (2006).
[21] L. F. Wang, X. Chuang, W. H. Ye, and Y. J. Li, Phys. Plasmas **16**, 112104 (2009).
[22] L. F. Wang, W. H. Ye, and Y. J. Li, Phys. Plasmas **17**, 052305 (2010).
[23] L. F. Wang, W. H. Ye, and Y. J. Li, Europhys. Lett. **87**, 54005 (2009).
[24] L. F. Wang, W. H. Ye, Z. F. Fan, and Y. J. Li, Europhys. Lett. **90**, 15001 (2010).
[25] Z. Fan, J. Luo, and W. Ye, Phys. Plasmas **16**, 102104 (2009).
[26] V. N. Goncharov, Phys. Rev. Lett. **88**, 134502 (2002).
[27] K. O. Mikaelian, Phys. Rev. E **67**, 026319 (2003).

PHYSICS OF PLASMAS **18**, 022704 (2011)

Competitions between Rayleigh–Taylor instability and Kelvin–Helmholtz instability with continuous density and velocity profiles

W. H. Ye (叶文华),[1,2,3,a)] L. F. Wang (王立锋),[2,3,4,b)] C. Xue (薛创),[3] Z. F. Fan (范征锋),[3] and X. T. He (贺贤土)[1,2,3]
[1]*Department of Physics, Zhejiang University, Hangzhou 310027, People's Republic of China*
[2]*CAPT, Peking University, Beijing 100871, People's Republic of China*
[3]*LCP, Institute of Applied Physics and Computational Mathematics, Beijing 100088, People's Republic of China*
[4]*SMCE, China University of Mining and Technology, Beijing 100083, People's Republic of China*

(Received 30 August 2010; accepted 9 January 2011; published online 17 February 2011)

In this research, competitions between Rayleigh–Taylor instability (RTI) and Kelvin–Helmholtz instability (KHI) in two-dimensional incompressible fluids within a linear growth regime are investigated analytically. Normalized linear growth rate formulas for both the RTI, suitable for arbitrary density ratio with continuous density profile, and the KHI, suitable for arbitrary density ratio with continuous density and velocity profiles, are obtained. The linear growth rates of pure RTI (γ_{RT}), pure KHI (γ_{KH}), and combined RTI and KHI (γ_{total}) are investigated, respectively. In the pure RTI, it is found that the effect of the finite thickness of the density transition layer (L_ρ) reduces the linear growth of the RTI (stabilizes the RTI). In the pure KHI, it is found that conversely, the effect of the finite thickness of the density transition layer increases the linear growth of the KHI (destabilizes the KHI). It is found that the effect of the finite thickness of the density transition layer decreases the "effective" or "local" Atwood number (A) for both the RTI and the KHI. However, based on the properties of $\gamma_{RT} \propto \sqrt{A}$ and $\gamma_{KH} \propto \sqrt{1-A^2}$, the effect of the finite thickness of the density transition layer therefore has a completely opposite role on the RTI and the KHI noted above. In addition, it is found that the effect of the finite thickness of the velocity shear layer (L_u) stabilizes the KHI, and for the most cases, the combined effects of the finite thickness of the density transition layer and the velocity shear layer ($L_\rho = L_u$) also stabilize the KHI. Regarding the combined RTI and KHI, it is found that there is a competition between the RTI and the KHI because of the completely opposite effect of the finite thickness of the density transition layer on these two kinds of instability. It is found that the competitions between the RTI and the KHI depend, respectively, on the Froude number, the density ratio of the light fluid to the heavy one, and the finite thicknesses of the density transition layer and the velocity shear layer. Furthermore, for the fixed Froude number, the linear growth rate ratio of the RTI to the KHI decreases with both the density ratio and the finite thickness of the density transition layer, but increases with the finite thickness of the velocity shear layer and the combined finite thicknesses of the density transition layer and the velocity shear layer ($L_\rho = L_u$). In summary, our analytical results show that the effect of the finite thickness of the density transition layer stabilizes the RTI and the overall combined effects of the finite thickness of the density transition layer and the velocity shear layer ($L_\rho = L_u$) also stabilize the KHI. Thus, it should be included in applications where the transition layer effect plays an important role, such as the formation of large-scale structures (jets) in high energy density physics and astrophysics and turbulent mixing. © *2011 American Institute of Physics*. [doi:10.1063/1.3552106]

I. INTRODUCTION

Hydrodynamic instabilities play key roles in astrophysics and inertial confinement fusion (ICF).[1–3] The Rayleigh–Taylor instability (RTI) occurs whenever fluid regions that vary in density experience a pressure gradient that opposes the density gradient. The Kelvin–Helmholtz instability (KHI) occurs when there is a velocity shear layer between two fluids separated by a perturbed interface.[4]

In ICF, the capsule shell undergoes the RTI both in the acceleration and the deceleration phases. It has long been recognized that the RTI poses severe challenges in achieving ICF autoignition and eventual self-sustained fusion reaction.[2,3] At the initial acceleration stage, when the imploding shell is accelerated inward by the low-density blow-off plasma, the "outer surface" between the outer shell and the ablated plasma becomes unstable to the RTI. Moreover, at the later stage when the compressed fuel begins to decelerate the imploding pusher, the "inner surface" between the hot spot and the inner shell is prone to the RTI. This inherent physical instability can limit the implosion velocity and the drive energy and, in some cases, even break up the implosion shell by the RTI bubble of the abated plasma during the

a)Electronic mail: ye_wenhua@iapcm.ac.cn.
b)Electronic mail: wanglifengqq@126.com and lif_wang@yahoo.com.

acceleration stage. Furthermore, the RTI would retard or hinder the formation of the hot spot by the cold RTI spike of the capsule shell, resulting in the destruction of the ignition hot spot or autoignition failure during the deceleration stage. Thus, the ICF target capsule should be designed in such a way that would minimize the growth of the instability at an acceptable level. It is considered that the KHI is one of the important processes [other processes included but not limited to the well known RTI and Richtmyer–Meshkov instability (RMI)] in the study of space plasma. Some of the well known examples of KHI in space plasma are the mass transfer in a planet's magnetopause[5–8] and the interaction between the solar wind and the Earth's magnetosphere[9] where large-scale structures of vortex formed by the KHI lead to large-scale plasma mixing across a shear layer. The RTI and the KHI also play central roles on the nonlinear evolutions of many other astrophysical phenomena, such as high-Mach-number shocks[10–13] and jets, radioactive blast waves and radioactively driven molecular clouds, gamma-ray bursts, and accretion black holes.[3,14,15] It has been observed that well collimated jets are formed in astrophysical objects[16–19] and jetlike long spikes are observed in direct-driven RTI experiments in ICF.[20–23]

In the nonlinear evolution of RTI, the KHI can be ignited due to the velocity shear at the spike tips, which is known as the secondary KHI in literature and forms the resulted classical Kelvin–Helmholtz (KH) mushroom structures.[24–32] Similarly, in the nonlinear evolution of KHI, the RTI occurs at this phase since the KH vortex rolls up, which generates a centrifugal force resulting in the secondary RTI.[9,33–35] In the nonlinear regime of RTI (KHI), the primary RTI (KHI) drives the whole dynamics and creates the conditions for the development of secondary KHI (RTI). Therefore, there is a kind of competition between a global nonlinear primary dynamics and a local linear (and nonlinear) secondary instability. In addition, for a more general case, there are both the RTI and the KHI initially in a real system, and, hence, there is a competition between the RTI and the KHI in the whole dynamics.

In our previous work, it is found that the density gradient effect (the effect of the finite thickness of the density transition layer) stabilizes the RTI but destabilizes the KHI while the velocity gradient effect (the effect of the finite thickness of the velocity shear layer) tends to stabilize the KHI.[36,37] Because of the definitely opposite influence of the density gradient effect on the RTI and on the KHI, one can expect that there is a competition between the RTI and the KHI when one changes the density profile or the velocity profile. Several studies have focused their attention on the pure RTI[38–41] or the pure KHI[33,42–46] but few on the combined RTI and KHI. A complete understanding of the relation between the RTI and the KHI is important in understanding the nonlinear evolutions of the RTI and the KHI and the final turbulent mixing. Thus, in the present work, we focus preliminarily on the initial competitions between the RTI and the KHI but not on a fully developed nonlinear primary instability and a local secondary instability, which will be pursued and reported in our future work.

The paper is organized as the following. In Sec. II, the density and velocity profiles used in this research, the eigenvalue equation governing the linear growth rates of the RTI and the KHI, and the resulted dispersion relation formula employing a suitable approximate eigenfunction considering both the RTI and the KHI in incompressible fluids are presented. The analytical expressions of the linear growth rate of the RTI, the KHI, and the total of RTI and KHI, are also shown in Sec. II. In Sec. III, the pure RTI is analyzed. It is found that the density gradient effect reduces the linear growth rate of the RTI. In Sec. IV, the pure KHI is investigated. It is found that the velocity gradient effect reduces the linear growth rate of the KHI but the density gradient effect increases it, while, in most cases, the overall combined effects of density and velocity gradients tend to stabilize the KHI. Sec. V presents the competitions between the RTI and the KHI. It is found that the competitions between the RTI and the KHI depend, respectively, on the Froude number, the density ratio, and the finite thicknesses of the density transition layer and the velocity shear layer. Concluding remarks and discussions are given in Sec. VI.

II. LINEAR DISPERSION

In the following discussion, we shall denote the properties of the fluid above the interface by the subscript h (the heavy fluid) and below the interface by the subscript l (the light fluid), unless stated otherwise. We consider here two fluids separated by an interface located at $y=0$ in an acceleration field, $-g\hat{y}$, where y is the height and g is the acceleration, respectively, referring to the following density and velocity profiles:[40]

$$\rho_0=\begin{cases}\rho_h-\dfrac{\Delta\rho}{2}\exp^{-y/L_\rho}, & y\geq 0\\[2ex] \rho_l+\dfrac{\Delta\rho}{2}\exp^{y/L_\rho}, & y<0\end{cases}, \tag{1}$$

$$u_0=\begin{cases}U_h-\dfrac{\Delta U}{2}\exp^{-y/L_u}, & y\geq 0\\[2ex] U_l+\dfrac{\Delta U}{2}\exp^{y/L_u}, & y<0\end{cases}, \tag{2}$$

where $\rho_h(\rho_l)$ and $U_h(U_l)$ are the density and the velocity in x direction away from the interface of the upper (lower) fluid, $\Delta\rho=\rho_h-\rho_l(\rho_h>\rho_l)$ is the density difference between the two fluids, $\Delta U=U_h-U_l$ is the velocity shear difference between the two fluids, and $L_\rho(L_u)$ is the typical half-thickness of the density transition layer (velocity shear layer). In this report, the term "thickness" refers to "typical half-thickness" and we suppose $\Delta U>0$ for simplicity's sake unless stated otherwise. The general eigenvalue equation for the combined RTI and KHI is the following:[36,37]

$$(n+iku_0)^2\left(\rho_0\frac{d^2\tilde{v}}{dy^2}+\frac{d\rho_0}{dy}\frac{d\tilde{v}}{dy}-k^2\rho_0\tilde{v}\right)-ik(n+iku_0)\times\left(\frac{d\rho_0}{dy}\frac{du_0}{dy}+\rho_0\frac{d^2u_0}{dy^2}\right)\tilde{v}+k^2g\frac{d\rho_0}{dy}\tilde{v}=0, \tag{3}$$

where $\tilde{v}(y)$ is the perturbed velocity eigenfunction, $i=\sqrt{-1}$ is the imaginary unit, k is the perturbation wave number for a given perturbation wavelength λ around the interface of two fluids, and n is the eigenvalue.

Substituting the initial density and velocity profiles [Eqs. (1) and (2)] into the eigenvalue equation [Eq. (3)] and adopting an approximate eigenfunction in the form[4,36,37]

$$\tilde{v}(y) = [n + iku_0(y)]e^{-k|y|}, \tag{4}$$

we can obtain the expected dispersion relation about the n,

$$\begin{aligned} 0 = {} & n^2 - k^2\frac{U_h^2 + rU_l^2}{1+r} + i2nk\frac{U_h + rU_l}{1+r} - ik^2 A\Delta U \\ & \times\left(n + ik\frac{U_h + U_l}{2}\right)\frac{1}{k + 1/L_\rho} - ik^2\Delta U \\ & \times\left(nA + ik\frac{U_h - rU_l}{1+r}\right)\frac{1}{k + 1/L_u} - \frac{1}{4}k^3\Delta U^2\frac{1}{k + 2/L_u} \\ & + ik^2 A\Delta U\left(n + ik\frac{U_h + U_l}{2}\right)\frac{1}{k + 1/L_\rho + 1/L_u} - \frac{Akg}{1 + kL_\rho}, \end{aligned} \tag{5}$$

where $r=\rho_l/\rho_h$ $(0\leq r\leq 1)$ is the density ratio of the light fluid to the heavy one and $A=(\rho_h-\rho_l)/(\rho_h+\rho_l)$ is the Atwood number of the instability system.

One can obtain the linear growth rate, γ, and the frequency, ω, by solving the above dispersion relation [Eq. (5)] about the eigenvalue $n=\gamma-i\omega$. The present work pays attention to the spatial amplitude competitions between the RTI and the KHI, and, therefore, we here only consider the linear growth rates of the RTI and the KHI but ignore the frequency of the KHI. We obtain the linear growth rate formulas for the RTI (γ_{RT}), the KHI (γ_{KH}), and the total of RTI and KHI (γ_{total}) by solving the dispersion relation [Eq. (5)],

$$\frac{\gamma_{RT}^2}{gk} = \frac{A}{1 + kL_\rho}, \tag{6}$$

$$\begin{aligned} \frac{\gamma_{KH}^2}{k^2\Delta U^2} = {} & \frac{r}{(1+r)^2} + \frac{1}{2}A^2\frac{1}{1 + 1/kL_\rho} - \frac{r}{(1+r)^2}\frac{2}{1 + 1/kL_u} + \frac{1}{4}\frac{1}{1 + 2/kL_u} - \frac{1}{2}A^2\frac{1}{1 + 1/kL_\rho + 1/kL_u} \\ & - \frac{1}{2}A^2\frac{1}{(1 + 1/kL_\rho)(1 + 1/kL_u)} + \frac{1}{2}A^2\frac{1}{(1 + 1/kL_\rho)(1 + 1/kL_\rho + 1/kL_u)} + \frac{1}{2}A^2\frac{1}{(1 + 1/kL_u)(1 + 1/kL_\rho + 1/kL_u)} \\ & - \frac{1}{4}A^2\frac{1}{(1 + 1/kL_\rho)^2} - \frac{1}{4}A^2\frac{1}{(1 + 1/kL_u)^2} - \frac{1}{4}A^2\frac{1}{(1 + 1/kL_\rho + 1/kL_u)^2}, \end{aligned} \tag{7}$$

$$\gamma_{total}^2 = \gamma_{RT}^2 + \gamma_{KH}^2, \tag{8}$$

where γ_{RT}^2/gk is the normalized square of linear growth rate of RTI and $\gamma_{KH}^2/k^2\Delta U^2$ is the normalized square of linear growth rate of KHI.

III. PURE RTI

In this section, we present the effect of the finite thickness of the density transition layer on the linear growth rate of pure RTI. It is well known that the "classical" linear growth rate of RTI is in the form[4]

$$\frac{\gamma_{RT_classical}^2}{kg} = A, \tag{9}$$

where the finite thickness of the density transition layer is zero. In the pure classical RTI, the Atwood number effect increases the linear growth rate of RTI because of the property $\gamma_{RT_classical} \propto A$. Let us now investigate the linear growth rate of pure RTI [Eq. (6)] including the finite thickness of density transition layer L_ρ. On the one hand, for long-wavelength modes ($kL_\rho \ll 1$) that are not affected by the finite thickness of the density transition layer and "see" or "feel" a sharp steplike density configuration, their linear growth rates become

$$\gamma_{RT}^2 \rightarrow \gamma_{RT_classical}, \quad kL_\rho \ll 1. \tag{10}$$

On the other hand, for short-wavelength modes ($kL_\rho \gg 1$) that are localized inside the finite thickness of the density transition layer and "feel" a quite homogeneous density configuration, the linear growth rates of RTI become[47]

$$\gamma_{RT}^2 \rightarrow \frac{Ag}{L_\rho}, \quad kL_\rho \gg 1, \tag{11}$$

which is an asymptotic value and not dependent on the perturbation wave number k any more.

It is noteworthy that if the density profile is smoothly varying between the two fluids and the minimum density gradient scale length in the form $L_{\rho_m}=\min[\rho/(d\rho/dy)]$ is finite [for example, the density profile as shown in Eq. (1)], then, the linear growth rate of RTI becomes

$$\frac{\gamma_{RT}^2}{kg} = \frac{A}{1 + AkL_{\rho_m}}, \tag{12}$$

which is another useful form widely used in computing the contribution from the finite density gradient to the linear growth of RTI around the ablation front when one considers the ablation effect under the context of ICF.[2,47,48]

We show the normalized linear growth rate of the RTI $\gamma_{RT}/\sqrt{gk}$ at different density ratios r and normalized thick-

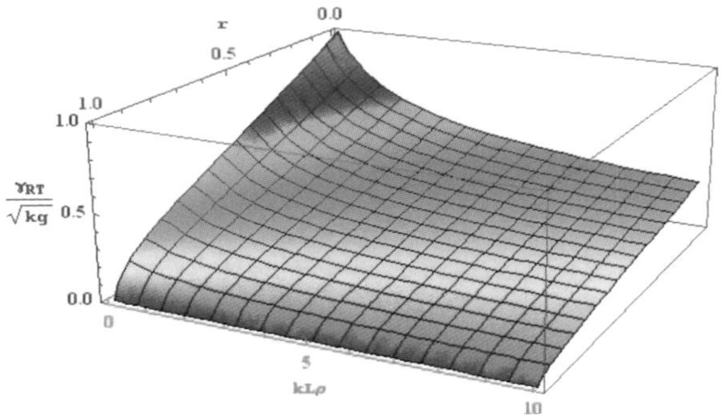

FIG. 1. (Color online) Normalized linear growth rates of RTI $\gamma_{RT}/\sqrt{gk}$ at different density ratios r and normalized thicknesses of density transition layer kL_ρ.

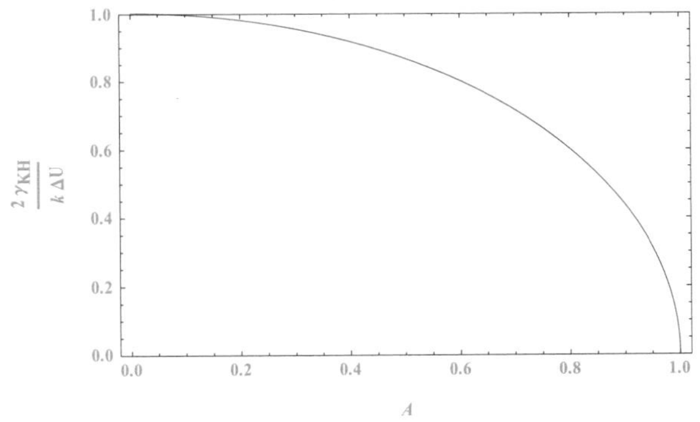

FIG. 2. (Color online) Normalized classical linear growth rates of KHI $2\gamma_{KH_classical}/(k\Delta U)$ at different Atwood numbers A.

nesses of density transition layer kL_ρ in Fig. 1. As can be seen in Fig. 1, the $\gamma_{RT}/\sqrt{gk}$ decreases with both the r and the kL_ρ, which has been well understood in the past 30 years.[2,36–41] Here, we try to understand the effect of the finite thickness of the density transition layer on the linear growth rate of RTI in the following way. The effect of the finite thickness of the density transition layer reduces the "local" or "effective" Atwood number of the instability system [i.e., from Atwood number=A in the classical RTI to the effective Atwood number=$A/(1+kL_\rho)$ in the case including the finite thickness of the density transition layer L_ρ], and hence, it reduces the linear growth rate of RTI since $\gamma_{RT}\propto$ Atwood number. In other words, for a fixed perturbation wavelength λ, when $kL_\rho \ll 1$, the perturbation velocity eigenfunction $\tilde{v}(y)$ "sees" or "feels" a steplike and quite inhomogeneous density configuration, while when $kL_\rho \gg 1$, the perturbation velocity eigenfunction $\tilde{v}(y)$ feels a nearly homogeneous one. Obviously, a homogeneous density configuration is more stable than an inhomogeneous one in the condition of RTI. That is to say, the buffer zone (the density transition layer) slows down the RTI, which has been widely employed in the ICF target design. However, we will see that this "situation" is completely opposite in the pure KHI in the next section.

IV. PURE KHI

In this section, we present the effects of the finite thickness of the density transition layer and the velocity shear layer on the linear growth rate of pure KHI.

It is well known that the "classical" linear growth rate of KHI is in the form[4]

$$\frac{4\gamma^2_{KH_classical}}{k^2\Delta U^2} = 1 - A^2, \tag{13}$$

where the finite thicknesses of the density transition layer and the velocity shear layer are both zero. In order to clearly understand the Atwood number effect on the linear growth of KHI, we show the normalized classical linear growth rate of KHI $2\gamma_{KH_classical}/(k\Delta U)$ at different Atwood numbers A in Fig. 2. It is easy to obtain that the Atwood number effect stabilizes the KHI because of $\gamma_{KH_classical}\propto\sqrt{1-A^2}$, as shown in Fig. 2, which is completely opposite to the "situation" in the classical RTI [Eq. (9)].

When the perturbation wavelength λ and the velocity shear difference ΔU between the two fluids are fixed, the maximum linear growth rate of KHI is in the form $\gamma_{KH_\max}=(1/2)k\Delta U$ with the thickness of velocity shear layer $L_u=0$ and the density ratio $r=1$ (the Atwood number $A=0$), which can be used to normalize the linear growth rate of KHI in the condition of continuous density and velocity profiles.

A. $L_\rho \neq 0$ and $L_u=0$

Let the thickness of the density transition layer $L_\rho \neq 0$ and the thickness of the velocity shear layer $L_u=0$, the linear growth rate of KHI [Eq. (7)] reduces to

$$\frac{4\gamma^2_{KH_den}}{k^2\Delta U^2} = 1 - \left(\frac{A}{1+kL_\rho}\right)^2, \tag{14}$$

which is dependent on the Atwood number A and the normalized thickness of density transition layer kL_ρ. The normalized linear growth rate of KHI $2\gamma_{KH_den}/(k\Delta U)$ in the condition of the thickness of velocity shear layer $L_u=0$ is illustrated in Fig. 3. As can be seen in Fig. 3, the $2\gamma_{KH_den}/(k\Delta U)$ decreases with the A but increases with the kL_ρ.

Similarly, like the effect of the finite thickness of the density transition layer in the RTI, we can conclude that in the KHI, the effect of the finite thickness of the density transition layer also reduces the "effective" or "local" Atwood number of the instability system by comparing Eqs. (13) and (14), because the Atwood number=A in Eq. (13) becomes the effective Atwood number=$A/(1+kL_\rho)$ in Eq. (14). However, the Atwood number effect reduces the KHI (i.e., $\gamma^2_{KH_classical}\propto 1-A^2$), which is completely opposite to that in

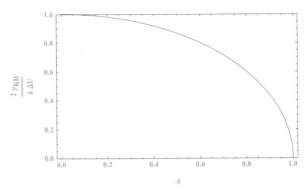

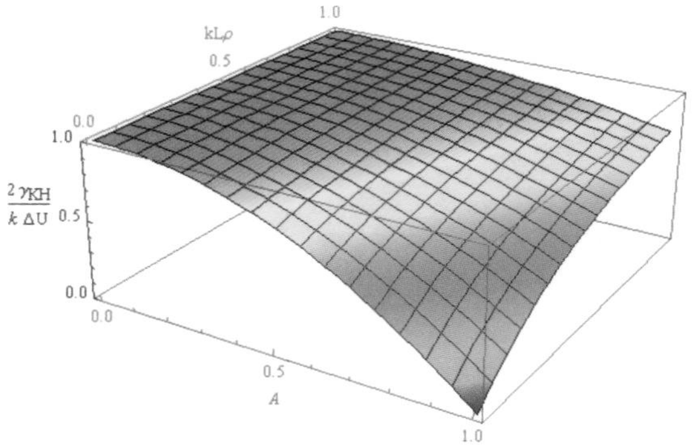

FIG. 3. (Color online) Normalized linear growth rates of KHI $2\gamma_{KH_den}/(k\Delta U)$ with the thickness of velocity shear layer $L_u=0$ at different Atwood numbers A and normalized thicknesses of density transition layer kL_ρ.

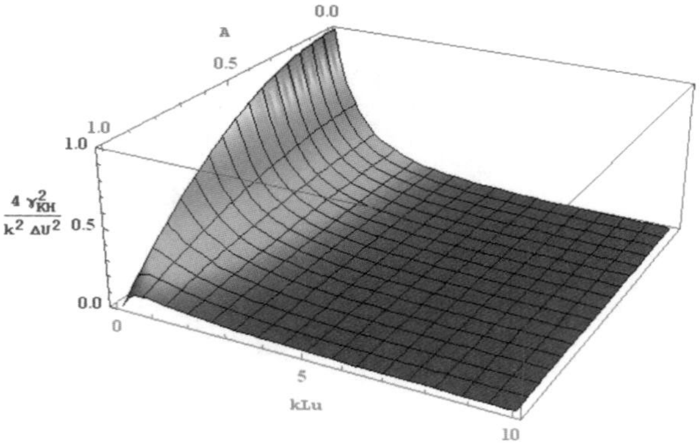

FIG. 4. (Color online) Normalized linear growth rates of KHI $4\gamma_{KH_vel}^2/(k\Delta U)^2$ with the thickness of density transition layer $L_\rho=0$ at different Atwood numbers A and normalized thicknesses of velocity shear layer kL_u.

the RTI (i.e., $\gamma_{RT_classical}^2 \propto A$). It is well known that the effect of the finite thickness of the density transition stabilizes the RTI. Therefore, on the contrary, the effect of the finite thickness of the density transition layer destabilizes the KHI.

On the one hand, for long-wavelength modes, the classical prediction,

$$\gamma_{KH_den}^2 \to \gamma_{KH_classical}^2, \quad kL_\rho \ll 1, \tag{15}$$

is recovered. On the other hand, for short-wavelength modes, we obtain

$$\gamma_{KH_den}^2 \to \frac{k^2\Delta U^2}{4}, \quad kL_\rho \gg 1, \tag{16}$$

which is the maximum linear growth rate of KHI.

B. $L_\rho=0$ and $L_u\neq 0$

Let the thickness of density transition layer $L_\rho=0$ and the thickness of velocity shear layer $L_u\neq 0$, the linear growth rate of KHI [Eq. (7)] reduces to

$$\frac{4\gamma_{KH_vel}^2}{k^2\Delta U^2} = \frac{2+2kL_u - A^2(2+kL_u)}{(1+kL_u)^2(2+kL_u)}, \tag{17}$$

which is dependent on the Atwood number A and the normalized finite thickness of velocity shear layer kL_u. We show the normalized linear growth rate of KHI $4\gamma_{KH_vel}^2/(k\Delta U)^2$, with the finite thickness of density transition layer $L_\rho=0$ at different Atwood numbers A and normalized thicknesses of velocity shear layer kL_u in Fig. 4. Overall, the $4\gamma_{KH_vel}^2/(k\Delta U)^2$ decreases with both the A and the kL_u, as shown in Fig. 4.

As can be seen in Eq. (17), on the one hand, for long-wavelength modes, the classical prediction,

$$\gamma_{KH_vel}^2 \to \gamma_{KH_classical}^2, \quad kL_u \ll 1, \tag{18}$$

is recovered. On the other hand, for short-wavelength modes, we obtain

$$\gamma_{KH_vel}^2 \to \frac{\Delta U^2}{4}\frac{2-A^2}{L_u^2}, \quad kL_u \gg 1, \tag{19}$$

which is an asymptotic value and not dependent on the perturbation wave number k any more.

C. $L_\rho=DL_U$

Let the thickness of density transition layer L_ρ be a fraction of the thickness of velocity shear layer L_u, namely, $L_\rho=DL_u=DL$ (In this subsection, there is always $L_u=L$ unless stated otherwise), the linear growth rate of KHI [Eq. (7)] reduces to,

$$\frac{4\gamma_{KH_LD}^2}{k^2\Delta U^2} = \frac{2(1+kL)[1+D+2DkL+kLD^2(1+kL)]^2 - A^2(2+kL)(1+D+2DkL)^2}{(1+kL)^2(2+kL)[1+D+2DkL+kLD^2(1+kL)]^2}, \tag{20}$$

which is dependent on the Atwood number A, the normalized finite thickness of velocity shear layer kL, and the coefficient D. Figure 5 refers to the normalized linear growth rates of KHI $4\gamma_{KH_LD}^2/(k\Delta U)^2$ [Eq. (20)] at different D and kL for the fixed Atwood numbers of $A=0.1$, 0.5, and 1.0, respectively. As shown in Fig. 5, the kL and the A influence the $4\gamma_{KH_LD}^2/(k\Delta U)^2$ strikingly and the coefficient D pronounces a significant sound only at the large A ($A=1.0$) and the small kL (about $kL<1.0$).

In most cases, the finite thickness of density transition

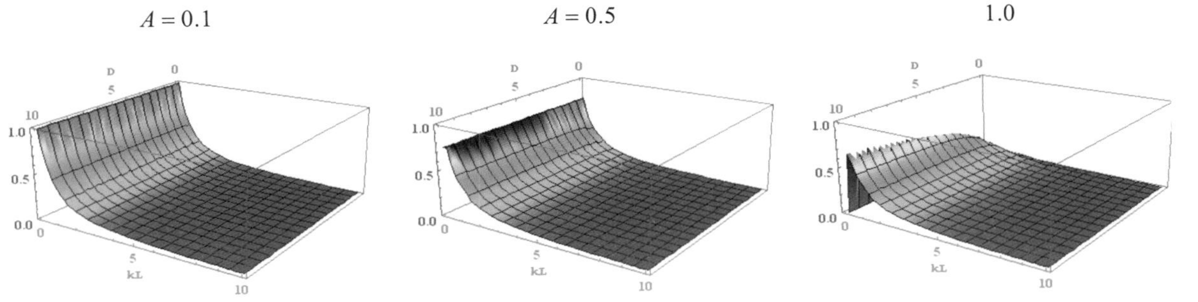

FIG. 5. (Color online) Normalized linear growth rates of KHI $4\gamma_{KH_LD}^2/(k\Delta U)^2$ at different coefficients D and normalized thicknesses of velocity shear layer kL for the fixed Atwood numbers of $A=0.1$, 0.5, and 1.0, respectively.

layer L_ρ and the finite thickness of velocity shear layer L_u have the same typical (dimensionless) value $L_\rho \approx L_u$.[33,49] Therefore, we here especially study the case of $L_\rho = L$. For the case $D=1$, Eq. (20) reduces to

$$\frac{4\gamma_{KH_L}^2}{k^2\Delta U^2} = \frac{2(2-2A^2+3kL+k^2L^2)}{(2+3kL+k^2L^2)^2}, \tag{21}$$

which is now only dependent on the Atwood number A and the thickness of density transition (velocity shear) layer kL. We illustrate the normalized linear growth rate of KHI $4\gamma_{KH_L}^2/(k\Delta U)^2$ [Eq. (21)] at different A and kL in Fig. 6. For the fixed A, when the A is small (about $A<0.80$), the $4\gamma_{KH_L}^2/(k\Delta U)^2$ decreases with the kL all the time but when the A is large (about $A>0.80$), the $4\gamma_{KH_L}^2/(k\Delta U)^2$ first increases with the kL and then decreases with the kL. For the fixed kL, when the kL is small (about $kL<0.50$), the $4\gamma_{KH_L}^2/(k\Delta U)^2$ increases with the A but when the kL is large (about $kL>0.50$), the $4\gamma_{KH_L}^2/(k\Delta U)^2$ is nearly not affected by the A. It is noteworthy that the kL plays a central role in the linear growth rate of KHI and the overall combined effects of the finite thickness of the density transition layer and the velocity shear layer stabilize the KHI, as shown in Fig. 6. It is well known that $\gamma_{KH_classical} \propto \sqrt{1-A^2}$ and the density gradient effect reduces the "effective" or "local" Atwood number like the density gradient effect in the RTI as noted above. One also knows that this "reduction" of the local A is distinct (or effective) in the condition of large A. Therefore, when the A is large enough, the destabilizing effect of L_ρ can play a dominant role, resulting in the γ_{KH_L} increasing with the kL while when the A is small, the stabilizing effect of L_u plays the dominant role, leading to the γ_{KH_L} decreasing with the kL. Overall, the effect of buffer zone (thickness) of velocity shear layer dominate over the effect of the buffer zone of the density transition layer, and hence, the combined effects of L_ρ and L_u, stabilize the KHI in most cases.

Now let us consider the asymptotic behavior of Eq. (21). On the one hand, for long-wavelength modes, the classical prediction

$$\gamma_{KH_L}^2 \to \gamma_{KH_classical}^2, \quad kL \ll 1, \tag{22}$$

is recovered. On the other hand, for short-wavelength modes, we obtain

$$\gamma_{KH_L}^2 \to \frac{\Delta U^2}{2L^2}, \quad kL \gg 1, \tag{23}$$

which is an asymptotic value and not dependent on the perturbation wave number k any more. Please note that the results obtained here are in contrast with what was found by Miura,[44] where a tanh profile considering the magnetic field effect is totally stable in the condition of KHI for $kL>1$.

D. Discussions

The normalized linear growth rate of KHI, $2\gamma_{KH}/(k\Delta U)$, is dependent on the normalized thicknesses of the density transition layer and the velocity shear layer, kL_ρ and kL_u, and the density ratio, r, as shown in Eq. (7). In this section, we discuss the linear growth rate of KHI at arbitrary density ratio with continuous density and velocity profiles in some more general cases.

Figure 7 shows the relations among the normalized linear growth rate of KHI $2\gamma_{KH}/(k\Delta U)$, the normalized thickness of density transition layer kL_ρ, and the normalized thickness of velocity shear layer kL_u with the density ratios of $r=0.01$, 0.1, and 1.0, respectively. At the fixed r and kL_ρ, the $2\gamma_{KH}/(k\Delta U)$ decreases with the kL_u while at the fixed r

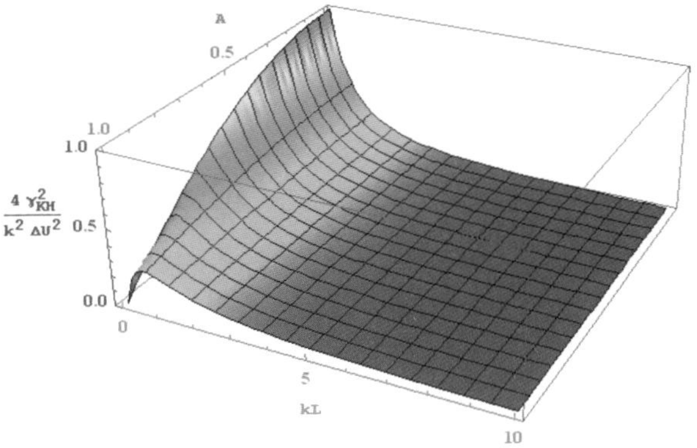

FIG. 6. (Color online) Normalized linear growth rates of KHI $4\gamma_{KH_L}^2/(k\Delta U)^2$ at different Atwood numbers A and normalized thicknesses of velocity shear (density transition) layer kL.

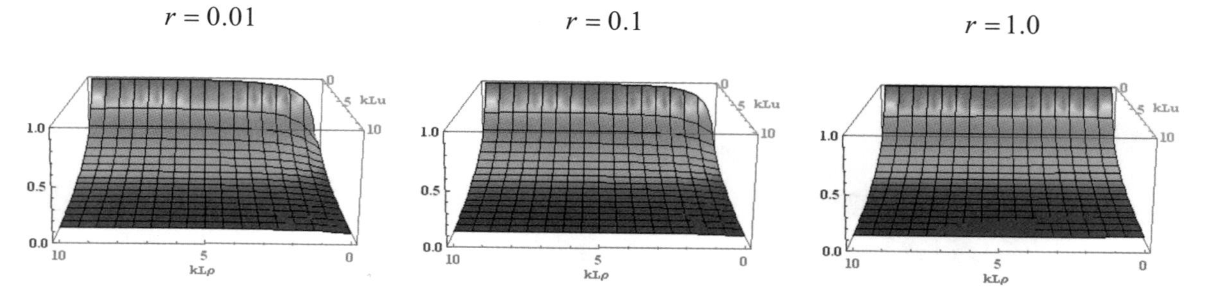

FIG. 7. (Color online) Normalized linear growth rates of KHI $2\gamma_{KH}/(k\Delta U)$, at different normalized thicknesses of density transition layer and velocity shear layer kL_ρ and kL_u for the fixed density ratios of r=0.01, 0.1, and 1.0, respectively.

and kL_u, the $2\gamma_{KH}/(k\Delta U)$ increases with the kL_ρ, especially when the kL_u is small, as shown in Fig. 7. At the small r (r=0.01), the $2\gamma_{KH}/(k\Delta U)$ is dependent strongly on the kL_ρ, as shown in the left figure of Fig. 7. One can expect that this dependence can be weakened with the increase of r, as shown in center figure of Fig. 7. Furthermore, when r=1, the kL_ρ therefore does not influence the $2\gamma_{KH}/(k\Delta U)$ at all, as shown in the right figure of Fig. 7

Figure 8 refers to the relations among the normalized linear growth rate of the KHI $2\gamma_{KH}/(k\Delta U)$, the normalized thickness of velocity shear layer kL_u, and the density ratio r for the fixed normalized thicknesses of the density transition layer of kL_ρ=0.0628, 0.628, and 6.28, respectively. As shown in Fig. 8, for the fixed kL_ρ, the $2\gamma_{KH}/(k\Delta U)$ decreases with the kL_u but increases with the r. At the small kL_ρ (kL_ρ=0.0628), the $2\gamma_{KH}/(k\Delta U)$ is dependent remarkably on the r, as shown in the left figure of Fig. 8. One can expect that this dependence can be slowed down with the increase of kL_ρ, as shown in the center figure of Fig. 8. When kL_ρ=6.28, the r nearly does not influence the $2\gamma_{KH}/(k\Delta U)$, as shown in the right figure of Fig. 8.

Figure 9 illustrates the relations among the normalized linear growth rate of KHI $2\gamma_{KH}/(k\Delta U)$, the normalized thickness of density transition layer kL_ρ, and the density ratio r for the fixed normalized thicknesses of the velocity shear layer of kL_u=0.0628, 0.628, and 6.28, respectively. As shown in Fig. 9, the $2\gamma_{KH}/(k\Delta U)$ increases with both the r and the kL_u. However, the striking character of Fig. 9 is the dramatic influence of kL_u on the KHI, namely, the kL_u strongly decreases the linear growth rate of KHI.

These phenomena can be understood as the following. The properties of $kL_\rho \gg 1$ and $kL_u \ll 1$ lead to the perturbation profile "seeing" or "feeling" a steplike configuration for the velocity but a quite homogeneous configuration for the density, which is obviously more unstable than an inhomogeneous configuration for the density ($kL_\rho \ll 1$). In other words, the effect of density transition layer (kL_ρ) destabilizes the KHI, as noted above. When the U_h and the U_l are fixed, the effect of the finite thickness of the velocity transition layer reduces the "effective" or the "local" velocity shear difference ΔU which is the "force" of the KHI. Thus, the effect of the finite thickness of the velocity transition layer (kL_u) stabilizes the KHI.

In order to further understand the relation between the linear growth rate of KHI γ_{KH} and the perturbation wave number k, we show the γ_{KH} at different k in Fig. 10. First, Fig. 10(a) shows the relation between γ_{KH} and k at different density ratios r with the thickness of density transition layer L_ρ=0.05, the thickness of velocity shear L_u=1.0, and the velocity shear difference ΔU=1.0. Second, Fig. 10(b) shows the relation between γ_{KH} and k at different L_ρ with r=0.1, L_u=1.0, and ΔU=1.0. Finally, Fig. 10(c) shows the relation between γ_{KH} and k at different L_u with r=1, L_ρ=1.0, and ΔU=2.0. As can be seen in Fig. 10, the γ_{KH} increases with the k. Furthermore, it is found that the r and the L_ρ play similar roles on the linear growth rate of KHI by comparing

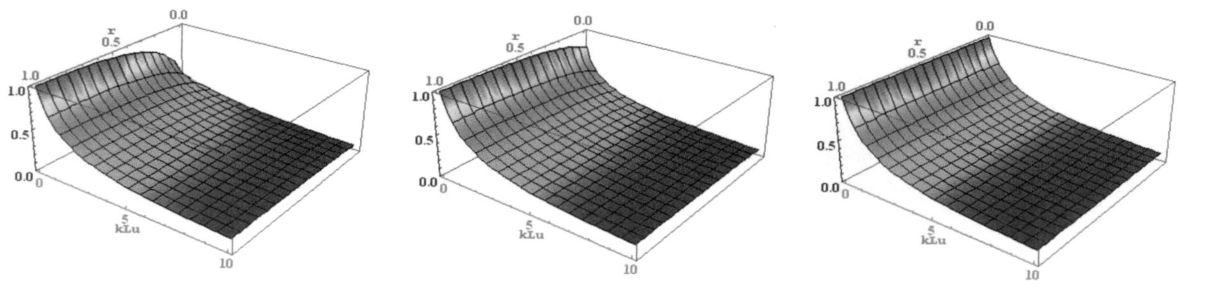

FIG. 8. (Color online) Normalized linear growth rates of KHI $2\gamma_{KH}/(k\Delta U)$, at different normalized thicknesses of velocity shear layer kL_u and density ratios r for the fixed thicknesses of density transition layer of kL_ρ=0.0628, 0.628, and 6.28, respectively.

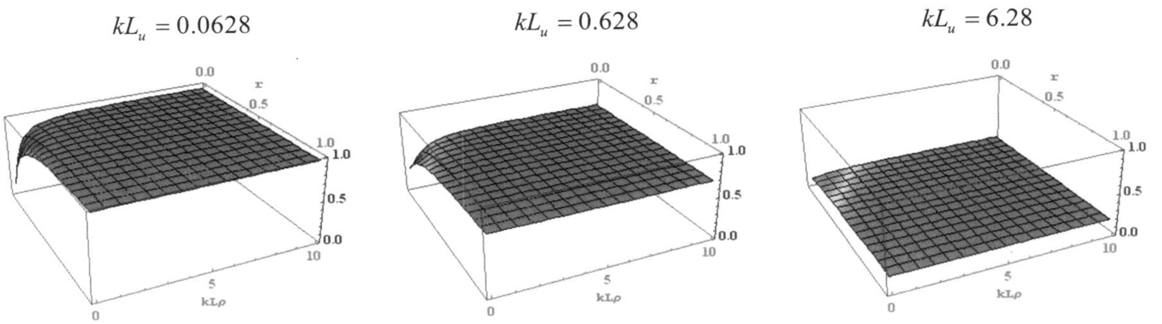

FIG. 9. (Color online) Normalized linear growth rates of KHI $2\gamma_{KH}/(k\Delta U)$ at different normalized thicknesses of density transition layer kL_ρ and density ratios r for the fixed normalized thicknesses of velocity shear layer of kL_u=0.0628, 0.628, and 6.28, respectively.

Figs. 10(a) and 10(b). In addition, it is found that the L_u plays a vital role on the linear growth rate of KHI, as shown in Fig. 10(c)

V. COMPETITIONS BETWEEN THE RTI AND THE KHI

In this section, we investigate the competition between the RTI and the KHI. We here define the Froude number as $Fr^2=\Delta U^2/(g\lambda)$, for the sake of normalization of the linear growth rate ratio of the RTI to the KHI.

A. $L_\rho \neq 0$ and $L_u=0$

In the condition of the thickness of density transition layer $L_\rho \neq 0$ and the thickness of velocity shear layer $L_u=0$, the square of linear growth rate ratio of the RTI to the KHI can be normalized as

$$Fr^2 \frac{\pi}{2} \frac{\gamma_{RT}^2}{\gamma_{KH}^2} = \frac{A}{1+kL_\rho} / \left[1 - \frac{A^2}{(1+kL_\rho)^2}\right], \tag{24}$$

which is dependent on the Atwood number A and the normalized thicknesses of density transition layer kL_ρ. The relations among the normalized linear growth rate ratio of the RTI to the KHI $Fr\sqrt{\pi}\gamma_{RT}/(\sqrt{2}\gamma_{KH})$, the Atwood number A, and the normalized thickness of density transition layer kL_ρ with the thickness of velocity shear layer $L_u=0$ are shown in Fig. 11. The $Fr\sqrt{\pi}\gamma_{RT}/(\sqrt{2}\gamma_{KH})$ decreases with the kL_ρ but increases with the A, as shown in Fig. 11.

B. $L_\rho=0$ and $L_u \neq 0$

In the condition of the thickness of density transition layer $L_\rho=0$ and the thickness of velocity shear layer $L_u \neq 0$, the normalized square of linear growth rate ratio of the RTI to the KHI can be written as

$$Fr^2 \frac{\pi}{2} \frac{\gamma_{RT}^2}{\gamma_{KH}^2} = A / \left[\frac{2(1+kL_u) - A^2(2+kL_u)}{(1+kL_u)^2(2+kL_u)}\right], \tag{25}$$

which is dependent on the Atwood number A and the normalized thickness of velocity shear layer kL_u. We show the normalized linear growth rate ratio of the RTI to the KHI $Fr\sqrt{\pi}\gamma_{RT}/(\sqrt{2}\gamma_{KH})$ at different Atwood numbers A and normalized thicknesses of velocity shear layer kL_u with the thickness of density transition layer $L_\rho=0$ in Fig. 12. For the fixed kL_u, the $Fr\sqrt{\pi}\gamma_{RT}/(\sqrt{2}\gamma_{KH})$ decreases with the A while for the fixed A, the $Fr\sqrt{\pi}\gamma_{RT}/(\sqrt{2}\gamma_{KH})$ increases with the kL_u except for the large A (about $A>0.95$) and small kL_u (about $kL_u<0.5$), as shown in Fig. 12.

C. $L_\rho=DL_u=DL$

Let the thickness of density transition layer L_ρ be a fraction of the thickness of velocity shear layer L_u, namely, $L_\rho=DL_u=DL$ (In this subsection, there is always $L_u=L$ unless stated otherwise), the square of linear growth rate ratio of the RTI to the KHI can be normalized as

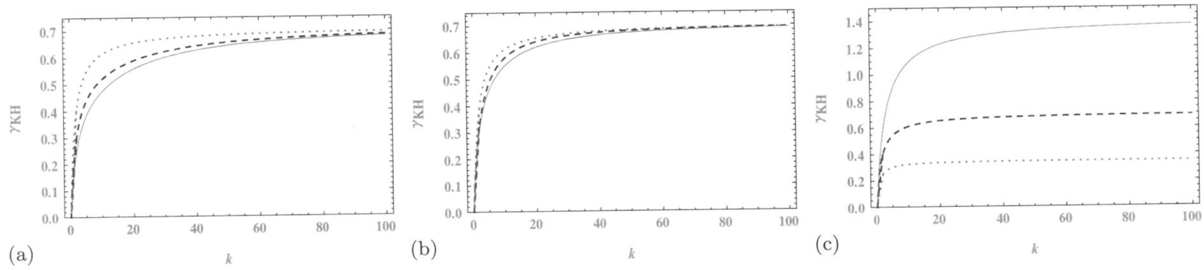

FIG. 10. (Color online) Linear growth rates of KHI γ_{KH} (a) with the thickness of density transition layer L_ρ=0.05, the thickness of velocity shear layer L_u=1.0, and the velocity shear difference ΔU=1.0 for the different density ratios of r=0.01 (solid), 0.1 (dashed), and 1.0 (dotted), respectively; (b) with r=0.1, L_u=1.0, and ΔU=1.0 for L_ρ=0.1 (solid), 0.2 (dashed), and 1.0 (dotted), respectively, and (c) with r=1, L_ρ=1.0, and ΔU=2.0 for L_u=0.5 (solid), 1.0 (dashed), and 2.0 (dotted), respectively.

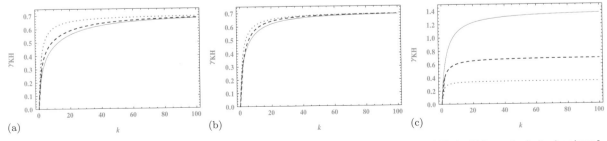

$$Fr^2\frac{\pi}{2}\frac{\gamma_{RT}^2}{\gamma_{KH}^2}=\left(\frac{A}{1+DkL}\right)\Bigg/\left[\frac{2(1+kL)(1+DkL)^2(1+D+DkL)^2-A^2(2+kL)(1+D+2DkL)^2}{(1+kL)^2(2+kL)(1+DkL)^2(1+D+DkL)^2}\right], \qquad (26)$$

which is dependent on the parameters of the Atwood number A, the normalized thickness of velocity shear layer kL, and the coefficient D.

Normalized linear growth rate ratios of the RTI to the KHI $Fr\sqrt{\pi}\gamma_{RT}/(\sqrt{2}\gamma_{KH})$ at different coefficients D and normalized thicknesses of velocity shear layer kL for the fixed Atwood numbers of $A=0.05$, 0.5, and 0.99, respectively, are illustrated in Fig. 13. In most cases, the D reduces the $Fr\sqrt{\pi}\gamma_{RT}/(\sqrt{2}\gamma_{KH})$ while the kL increases the $Fr\sqrt{\pi}\gamma_{RT}/(\sqrt{2}\gamma_{KH})$ except for the very small kL (about $kL<0.5$) in the condition of a large A ($A=0.99$), as shown in Fig. 13.

As noted above, in most cases, the thickness of density transition layer L_ρ and the thickness of velocity shear layer L_u have the same typical (dimensionless) value $L_\rho \approx L_u$.[33,49] Consequently, here, we also study the linear growth rate ratio of the RTI to the KHI in the condition of $L_\rho=L$.

For $D=1$, Eq. (26) reduces to

$$Fr^2\frac{\pi}{2}\frac{\gamma_{RT}^2}{\gamma_{KH}^2}=\left(\frac{A}{2+2kL}\right)\Bigg/\left[\frac{2-2A^2+3kL+kL^2}{(2+3kL+kL^2)^2}\right], \qquad (27)$$

which is only dependent on the Atwood number A and the normalized thickness of velocity shear layer (density transition layer) kL. Normalized linear growth rate ratios of the RTI to the KHI $Fr\sqrt{\pi}\gamma_{RT}/(\sqrt{2}\gamma_{KH})$ at different Atwood numbers A and normalized thicknesses of velocity shear (density transition) layer kL are shown in Fig. 14. The $Fr\sqrt{\pi}\gamma_{RT}/(\sqrt{2}\gamma_{KH})$ decreases with the kL but increases with the A, especially for the small kL ($kL<0.50$) and large A ($A>0.50$), as shown in Fig. 14. The property that the Atwood number effect increases the linear growth rate of RTI but decreases the linear growth rate of KHI leads to the $Fr\sqrt{\pi}\gamma_{RT}/(\sqrt{2}\gamma_{KH})$ increasing with the A. The fact that the $Fr\sqrt{\pi}\gamma_{RT}/(\sqrt{2}\gamma_{KH})$ decreases with the increase of kL, which is not very pronounced, means that the stabilizing effect of kL is more pronounced in the KHI than that in the RTI.

D. Discussions

As noted above, the effect of the buffer zone of the density transition layer stabilizes the RTI but destabilizes the KHI. Accordingly, there should be a competition between the RTI and the KHI when one changes the thickness of density transition layer L_ρ, the thickness of velocity shear layer L_u, or the density ratio of the two fluids r. In order to understand the competitions between the RTI and the KHI in some detail, we here again define the Froude number as $Fr=\Delta U/\sqrt{g\lambda}$. Then, the ratio of the normalized linear growth rate of the RTI, $\gamma_{RT}/\sqrt{gk}$, to the normalized linear growth rate of the KHI, $\gamma_{KH}/(k\Delta U)$, can be written as $\sqrt{2\pi}Fr\gamma_{RT}/\gamma_{KH}$.

Let us now consider the relations among the normalized linear growth rate ratio of the RTI to the KHI $\sqrt{2\pi}Fr\gamma_{RT}/\gamma_{KH}$, the normalized thickness of density transition layer kL_ρ, and the normalized thickness of velocity shear layer kL_u at the fixed density ratio r. Figure 15 shows the $\sqrt{2\pi}Fr\gamma_{RT}/\gamma_{KH}$ at the different kL_ρ and kL_u with the $r=0.1$, 0.5, and 0.99, respectively. As can be seen in Fig. 15, for the fixed kL_ρ, the $\sqrt{2\pi}Fr\gamma_{RT}/\gamma_{KH}$ increases with the kL_u while for the fixed kL_u, the $\sqrt{2\pi}Fr\gamma_{RT}/\gamma_{KH}$ decreases with the kL_ρ. It is noteworthy that the kL_u does not influence the RTI but stabilizes the KHI and the kL_ρ stabilizes the RTI but destabilizes the KHI. Therefore, the $\sqrt{2\pi}Fr\gamma_{RT}/\gamma_{KH}$ is reduced by

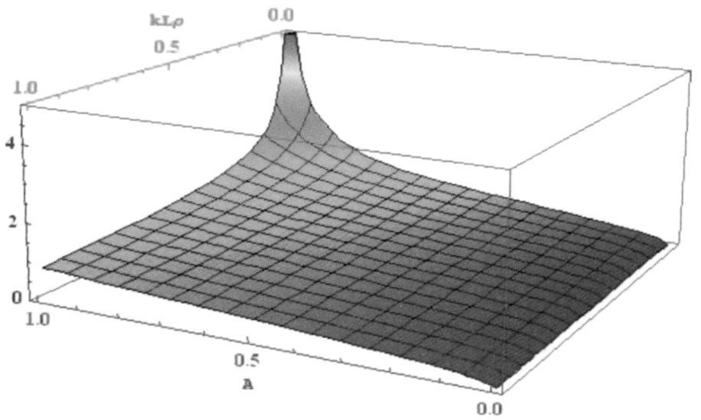

FIG. 11. (Color online) Normalized linear growth rate ratio of the RTI to the KHI $Fr\sqrt{\pi}\gamma_{RT}/(\sqrt{2}\gamma_{KH})$ at different Atwood numbers A and normalized thicknesses of density transition layer kL_ρ with the thickness of velocity shear layer $L_u=0$.

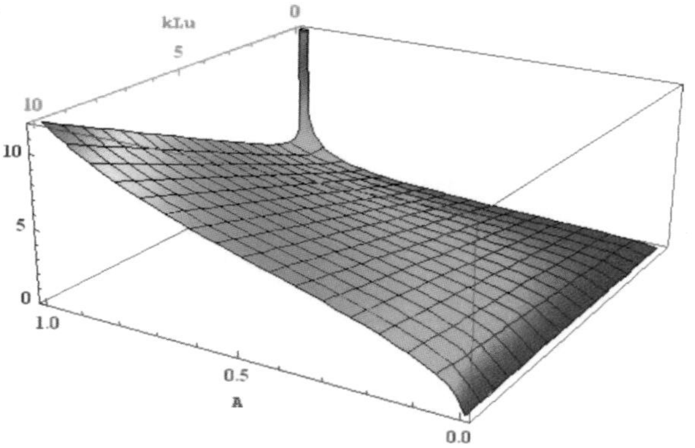

FIG. 12. (Color online) Normalized linear growth rate ratios of the RTI to the KHI $Fr\sqrt{\pi}\gamma_{RT}/(\sqrt{2}\gamma_{KH})$ at different Atwood numbers A and normalized thicknesses of velocity shear layer kL_u with the thickness of density transition layer $L_\rho=0$.

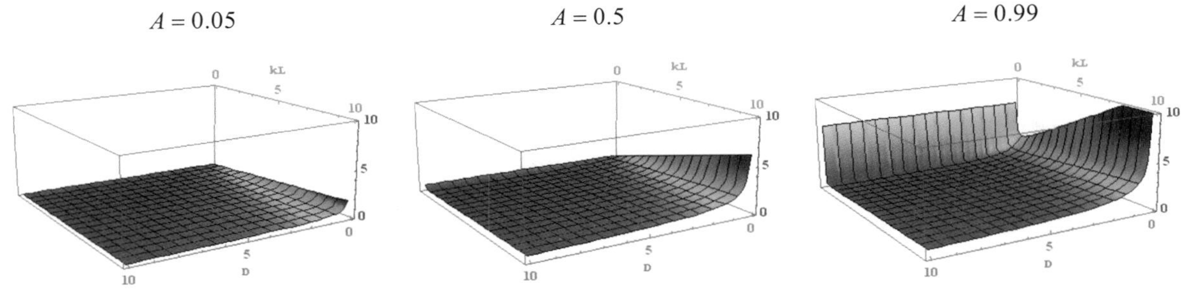

FIG. 13. (Color online) Normalized linear growth rate ratios of the RTI to the KHI $Fr\sqrt{\pi}\gamma_{RT}/(\sqrt{2}\gamma_{KH})$ at different coefficients D and normalized thicknesses of velocity shear layer kL for the fixed Atwood numbers $A=0.05$, 0.5, and 0.99, respectively.

the kL_ρ but increased by the kL_u. Furthermore, it is found that $\sqrt{2\pi}Fr\gamma_{RT}/\gamma_{KH}$ is dependent dramatically on the r, as shown in Fig. 15.

Figure 16 illustrates the normalized linear growth rate ratios of the RTI to the KHI, $Fr\sqrt{2\pi}\gamma_{RT}/\gamma_{KH}$, at different density ratios r and normalized thicknesses of velocity shear layer kL_u for the fixed normalized thicknesses of density transition layer of $kL_\rho=0.0628$, 0.628, and 6.28, respectively. The $Fr\sqrt{2\pi}\gamma_{RT}/\gamma_{KH}$ increases with the kL_u but decreases with the r, especially when the kL_ρ is small ($kL_\rho=0.0628$), as seen in Fig. 16. Comparing Fig. 16 at the different kL_ρ, one may conclude that the thickness of the density transition layer strongly influences the competitions between the RTI and the KHI. At the small kL_ρ, the RTI is increased while the KHI is effectively suppressed in comparison with the large kL_ρ, and, hence, the RTI dominates over the KHI, as shown in the left figure of Fig. 16. At the large kL_ρ, on the contrary, the RTI is effectively suppressed while the KHI is increased in comparison with the small kL_ρ, and, hence, the KHI dominates over the RTI, as shown in the right figure of Fig. 16.

Figure 17 refers to the normalized linear growth rate ratios of the RTI to the KHI, $Fr\sqrt{2\pi}\gamma_{RT}/\gamma_{KH}$, at different density ratios r and normalized thicknesses of density transition layer kL_ρ for the fixed normalized thicknesses of velocity shear layer of $kL_u=0.0628$, 0.628, and 6.28, respectively. The $Fr\sqrt{2\pi}\gamma_{RT}/\gamma_{KH}$ decreases with both the kL_ρ and

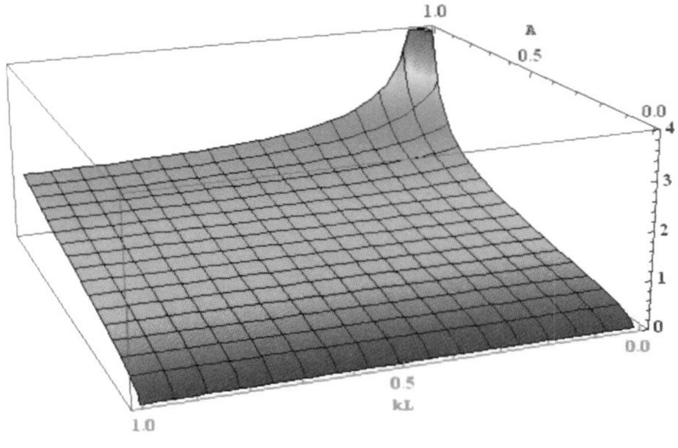

FIG. 14. (Color online) Normalized linear growth rate ratios of the RTI to the KHI $Fr\sqrt{\pi}\gamma_{RT}/(\sqrt{2}\gamma_{KH})$ at different Atwood numbers A and normalized thicknesses of velocity shear (density transition) layer kL.

the r, especially at the large kL_u ($kL_u=6.28$), as shown in the right figure of Fig. 17. It is noteworthy that the kL_u does not influence the linear growth rate of RTI though it plays a central role in the linear growth rate of KHI. One can expect that the larger kL_u corresponds to the larger $Fr\sqrt{2\pi}\gamma_{RT}/\gamma_{KH}$, as shown in Fig. 17.

The square of linear growth rates of the RTI (γ_{RT}^2), the KHI (γ_{KH}^2), and the total of RTI and KHI (γ_{total}^2) with the acceleration $g=1.0$, the thickness of density transition layer $L_\rho=1.0$, the thickness of velocity transition layer $L_u=1.0$, and the shear velocity difference $\Delta U=1.0$ for the fixed density ratios of $r=0.1$, 0.5, and 0.9, respectively, are plotted in Fig. 18. At the range of $0\leq k\leq 10$, when the r is small ($r=0.1$), the linear growth rate of KHI is always greater than the linear growth rate of RTI, while when the r is large ($r=0.9$), the linear growth rate of KHI is always smaller than the linear growth rate of RTI, as shown, respectively, in the left and the right figures of Fig. 18. Furthermore, at the medium r ($r=0.5$), there is a significant competition between the RTI and KHI. When $k<4$, there is $\gamma_{RT}>\gamma_{KH}$, while when $k>4$, there is $\gamma_{RT}<\gamma_{KH}$, as shown in the center figure of Fig. 18. In addition, with the increase of k, the KHI tends to dominate over the RTI. This can be well understood since there is $\gamma_{RT}\propto\sqrt{k}$ but $\gamma_{KH}\propto k$.

The square of linear growth rates of the RTI (γ_{RT}^2), the KHI (γ_{KH}^2), and the total of RTI and KHI (γ_{total}^2) with the acceleration $g=1.0$, the density ratio $r=0.5$, the thickness of velocity transition layer $L_u=1.0$, and the shear velocity difference $\Delta U=1.0$ for the fixed thickness of density transition layer of $L_\rho=0.1$, 1.0, and 10.0, respectively, are plotted in Fig. 19. At the range of $0\leq k\leq 10$, when the L_ρ is small ($L_\rho=0.1$), the linear growth rate of KHI is always greater than the linear growth rate of RTI, while when the L_ρ is large ($L_\rho=10.0$) the linear growth rate of KHI is always smaller than the linear growth rate of RTI, as shown, respectively, in the left and the right figures of Fig. 19. Furthermore, at the medium L_ρ ($L_\rho=1.0$), there is a significant competition between the RTI and KHI. When $k<4$, there is $\gamma_{RT}>\gamma_{KH}$, while when $k>4$, there is $\gamma_{RT}<\gamma_{KH}$, as shown in the center figure of Fig. 19. In addition, with the increase of k, the KHI tends to dominate over the RTI. This can be well understood since there is $\gamma_{RT}\propto\sqrt{k}$ but $\gamma_{KH}\propto k$ as mentioned above.

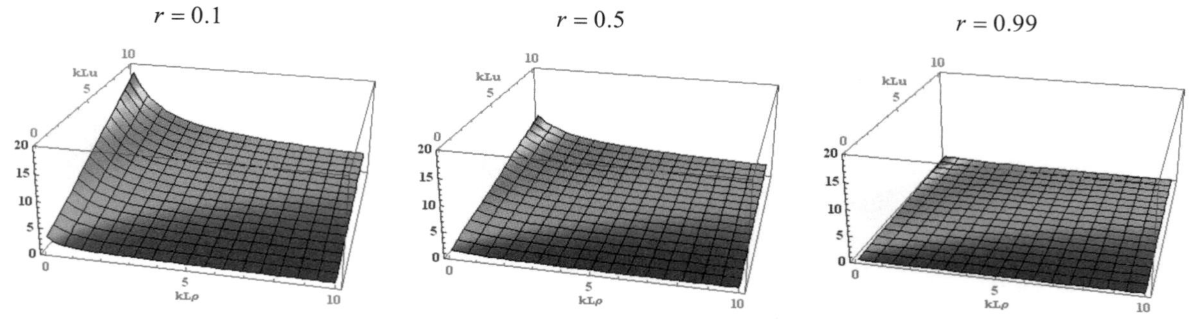

FIG. 15. (Color online) Normalized linear growth rate ratios of the RTI to the KHI $Fr\sqrt{2\pi}\gamma_{RT}/\gamma_{KH}$ at different normalized thicknesses of density transition layer and velocity shear layer kL_ρ and kL_u for the fixed density ratios of r=0.1, 0.5, and 0.99, respectively.

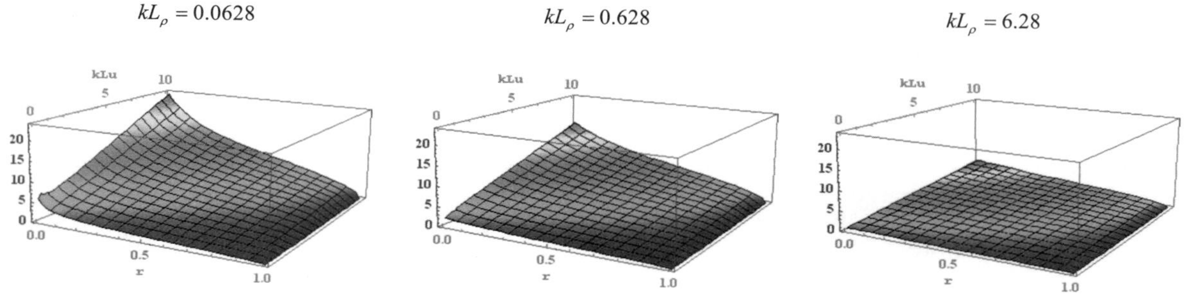

FIG. 16. (Color online) Normalized linear growth rate ratios of the RTI to the KHI $Fr\sqrt{2\pi}\gamma_{RT}/\gamma_{KH}$ at different density ratios r and normalized thicknesses of velocity shear layer kL_u for the fixed normalized thicknesses of density transition layer of kL_ρ=0.0628, 0.628, and 6.28, respectively.

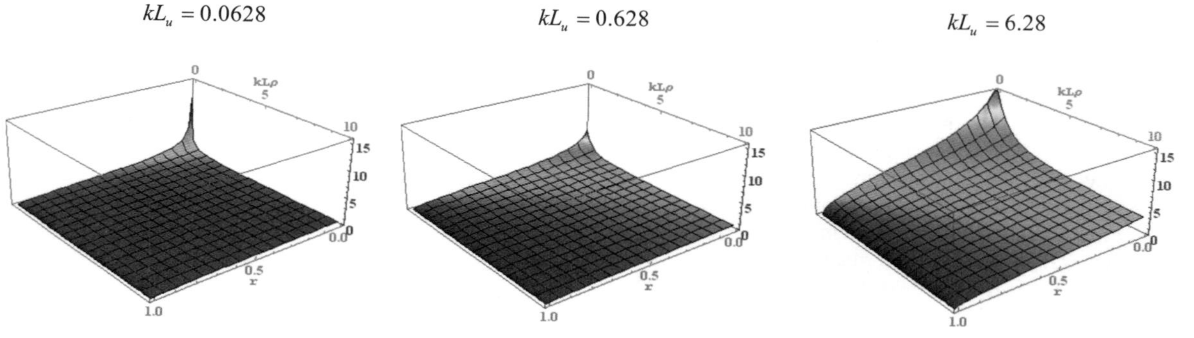

FIG. 17. (Color online) Normalized linear growth rate ratios of the RTI to the KHI $Fr\sqrt{2\pi}\gamma_{RT}/\gamma_{KH}$ at different density ratios r and normalized thicknesses of density transition layer kL_ρ for the fixed normalized thicknesses of velocity shear layer of kL_u=0.0628, 0.628, and 6.28, respectively.

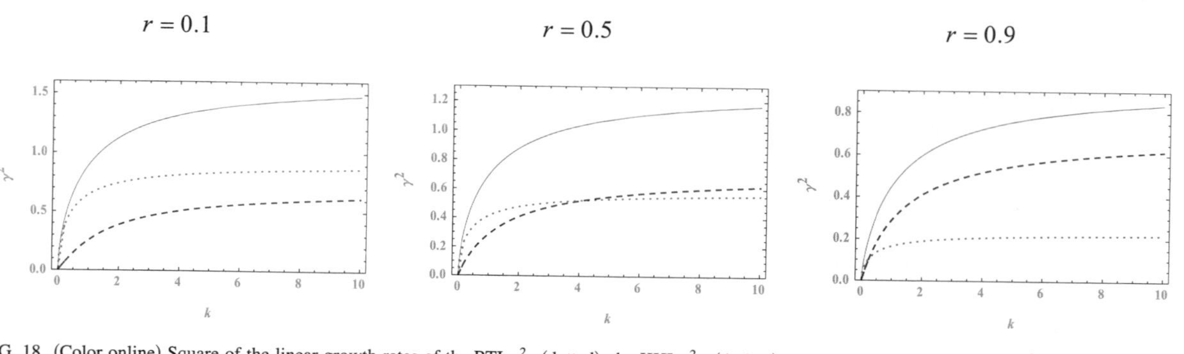

FIG. 18. (Color online) Square of the linear growth rates of the RTI γ_{RT}^2 (dotted), the KHI γ_{KH}^2 (dashed), and the total of RTI and KHI γ_{total}^2 (solid) with the acceleration g=1.0, thickness of density transition layer L_ρ=1.0, thickness of velocity transition layer L_u=1.0, and shear velocity difference ΔU=1.0 for the fixed density ratios of r=0.1, 0.5, and 0.9, respectively.

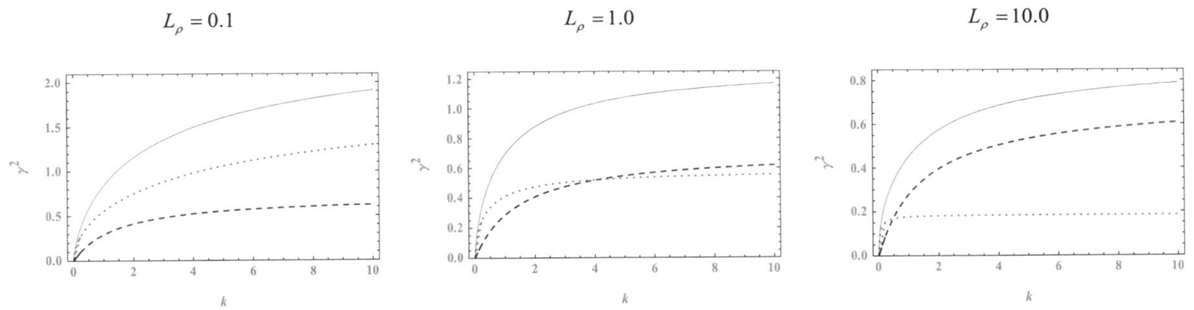

FIG. 19. (Color online) Square of the linear growth rates of the RTI γ_{RT}^2 (dotted), the KHI γ_{KH}^2 (dashed), and the total of RTI and KHI γ_{total}^2 (solid) with the acceleration $g=1.0$, density ratio $r=0.5$, thickness of velocity transition layer $L_u=1.0$, and shear velocity difference $\Delta U=1.0$ for the fixed thickness of density transition layer of $L_\rho=0.1$, 1.0, and 10.0, respectively.

VI. DISCUSSION AND CONCLUSION

In summary, in this research, we have studied analytically the competitions between the Rayleigh–Taylor instability (RTI) and the Kelvin–Helmholtz instability (KHI) in the condition of arbitrary density ratio and continuous density and velocity profiles. It is found that the competition between the RTI and the KHI is dependent on the Froude number, the density ratio of the two fluids, and the thicknesses of the density transition layer and the velocity shear layer. The analytical results are expected to improve the understanding of nonlinear evolutions of the RTI and the KHI and the final turbulent mixing.

In numerical simulations or experiments on the turbulent mixing driven by the RTI, it is found that the penetration of the two fluids grows at a similar rate with the density ratio of the two fluids.[24–28] Note that the density ratio here is defined as the light fluid to the heavy one. It is found that at a large density ratio, the interface rolls up early, i.e., the larger the density ratio is, the earlier the KHI ignition was and correspondingly the earlier Kelvin–Helmholtz (KH) mushroom structures appear at the tips of the spikes. As noted above, the density ratio effect stabilizes the RTI but destabilizes the KHI. Thus, for the same perturbation wavelength, the interface between the two fluids rolls up earlier at the fluids with larger density ratio in the simulations and the experiments.

It is worthy to note that there are continuous density and velocity profiles on both sides of the jet. In direct-driven or indirect-driven experiments in inertial confinement fusion (ICF), the jetlike long spikes are formed.[20–23] In astrophysics, the astrophysical jets can remain well collimated over propagation distances exceeding 10 jet diameters or more.[50] Our present works indicate that the density gradient effect stabilizes the RTI and, in most cases, the overall combined effects of density and velocity gradients also tend to stabilize the KHI, which can help us to improve the understanding of the stability for the astrophysical jets in astrophysics and of the formation mechanism for the jetlike long spikes in ICF.

ACKNOWLEDGMENTS

The author (L.F.W) wishes to thank the referee for his/her illuminating comments, which have helped improve the manuscript greatly. This work was supported by the National Basic Research Program of China (Grant No. 2007CB815100), the National Natural Science Foundation of China (Grant Nos. 10935003, 10905006, 11075024, 10974022 and 10835003), and the National High-Tech ICF Committee.

[1] B. A. Remington, D. Arnett, R. P. Drake, and H. Takabe, Science **284**, 1488 (1999).
[2] J. D. Lindl, P. Amendt, R. L. Berger, S. G. Glendinning, S. H. Glenzer, S. W. Haan, R. L. Kauffman, O. L. Landen, and L. J. Suter, Phys. Plasmas **11**, 339 (2004).
[3] S. Atzeni and J. Meyer-ter-Vehn, *The Physics of Inertial Fusion: Beam Plasma Interaction, Hydrodynamics, Hot Dense Mater* (Oxford University, London, 2004).
[4] S. Chandrasekhar, *Hydrodynamic and Hydromagnetic Stability* (Oxford University, London, 1968).
[5] A. Otto and D. H. Fairfield, J. Geophys. Res. **105**, 21175, doi:10.1029/1999JA000312 (2000).
[6] D. H. Fairfield, A. Otto, T. Mukai, S. Kokubun, R. P. Lepping, J. T. Steinberg, A. J. Lazarus, and T. Yamamoto, J. Geophys. Res. **105**, 21159, doi:10.1029/1999JA000316 (2000).
[7] S. Parhi, S. T. Suess, and M. Sulkanen, J. Geophys. Res. **104**, 14781, doi:10.1029/1999JA900041 (1999).
[8] K. Nykyri and A. Otto, Geophys. Res. Lett. **28**, 3565, doi:10.1029/2001GL013239 (2001).
[9] H. Hasegawa, M. Fujimoto, T.-D. Phan, H. Rème, A. Balogh, M. W. Dunlop, C. Hashimoto, and R. TanDokoro, Nature (London) **430**, 755 (2004).
[10] B. A. Remington, R. P. Drake, and D. D. Ryutov, Rev. Mod. Phys. **78**, 755 (2006).
[11] E. C. Harding, J. F. Hansen, O. A. Hurricane, R. P. Drake, H. F. Robey, C. C. Kuranz, B. A. Remington, M. J. Bono, M. J. Grosskopf, and R. S. Gillespie, Phys. Rev. Lett. **103**, 045005 (2009).
[12] O. A. Hurricane, J. F. Hansen, H. F. Robey, B. A. Remington, M. J. Bono, E. C. Harding, R. P. Drake, and C. C. Kuranz, Phys. Plasmas **16**, 056305 (2009).
[13] E. C. Harding, R. P. Drake, Y. Aglitskiy, T. Plewa, A. L. Velikovich, R. S. Gillespie, J. L. Weaver, A. Visco, M. J. Grosskopf, and J. R. Ditmar, Phys. Plasmas **17**, 056310 (2010).
[14] R. P. Drake, *High-Energy-Density Physics: Fundamentals, Inertial Fusion and Experimental Astrophysics* (Springer, New York, 2006).
[15] Committee on High Density Plasma Physics Plasma Science and Committee Board on Physics and Astronomoy Division on Engineering and Physical Science, *Frontiers in High Energy Density Physics* (The National

Academies, Washington, D.C., 2001).

[16]J. M. Foster, B. H. Wilde, P. A. Rosen, R. J. R. Williams, B. E. Blue, R. F. Coker, R. P. Drake, A. Frank, P. A. Keiter, A. M. Khokhlov, J. P. Knauer, and T. S. Perry, Astrophys. J. **634**, L77 (2005).

[17]S. V. Lebedev, J. P. Chittenden, F. N. Beg, S. N. Bland, A. Ciardi, D. Ampleford, S. Hughes, M. G. Haines, A. Frank, E. G. Blackman, and T. Gardiner, Astrophys. J. **564**, 113 (2002).

[18]P. E. Hardee, Astrophys. Space Sci. **293**, 117 (2004).

[19]M. Livio, Phys. Rep. **311**, 225 (1999).

[20]D. K. Bradley, D. G. Braun, S. G. Glendinning, M. J. Edwards, J. L. Milovich, C. M. Sorce, G. W. Collins, S. W. Haan, R. H. Page, and R. J. Wallace, Phys. Plasmas **14**, 056313 (2007).

[21]B. A. Remington, R. P. Drake, H. Takabe, and D. Arnett, Phys. Plasmas **7**, 1641 (2000).

[22]B. A. Remington, S. W. Haan, S. G. Glendinning, J. D. Kilkenny, D. H. Munro, and R. J. Wallace, Phys. Fluids B **4**, 967 (1992).

[23]J. O. Kane, H. F. Robey, B. A. Remington, R. P. Drake, J. Knauer, D. D. Ryutov, H. Louis, R. Teyssier, O. Hurricane, D. Arnett, R. Rosner, and A. Calder, Phys. Rev. E **63**, 055401 (2001).

[24]D. L. Youngs, Physica D **12**, 32 (1984).

[25]D. L. Youngs, Physica D **37**, 270 (1989).

[26]K. I. Read, Physica D **12**, 45 (1984).

[27]G. Dimonte and M. Schneider, Phys. Fluids **12**, 304 (2000).

[28]P. Ramaprabhu and G. Dimonte, Phys. Rev. E **71**, 036314 (2005).

[29]J. P. Wilkinson and J. W. Jacobs, Phys. Fluids **19**, 124102 (2007).

[30]A. R. Miles, D. G. Braun, M. J. Edwards, H. F. Robey, R. P. Drake, and D. R. Leibrandt, Phys. Plasmas **11**, 3631 (2004).

[31]A. R. Miles, M. J. Edwards, B. Blue, J. F. Hansen, H. F. Robey, R. P. Drake, C. Kuranz, and D. R. Leibrandt, Phys. Plasmas **11**, 5507 (2004).

[32]A. R. Miles, B. Blue, M. J. Edwards, J. A. Greenough, J. F. Hansen, H. F. Robey, R. P. Drake, C. Kuranz, and D. R. Leibrandt, Phys. Plasmas **12**, 056317 (2005).

[33]M. Faganello, F. Califano, and F. Pegoraro, Phys. Rev. Lett. **100**, 015001 (2008).

[34]Y. Matsumoto and M. Hoshino, Geophys. Res. Lett. **31**, L02807, doi:10.1029/2003GL018195 (2004).

[35]C. D. Winant and F. K. Browand, J. Fluid Mech. **63**, 237 (1974).

[36]L. F. Wang, C. Xue, W. H. Ye, and Y. J. Li, Phys. Plasmas **16**, 112104 (2009).

[37]L. F. Wang, W. H. Ye, and Y. J. Li, Phys. Plasmas **17**, 042103 (2010).

[38]K. O. Mikaelian, Phys. Rev. Lett. **48**, 1365 (1982).

[39]K. O. Mikaelian, Phys. Rev. A **40**, 4801 (1989).

[40]K. O. Mikaelian, Phys. Rev. A **33**, 1216 (1986).

[41]A. B. Bud'ko and M. A. Liberman, Phys. Fluids B **4**, 3499 (1992).

[42]A. Miura, Phys. Rev. Lett. **83**, 1586 (1999).

[43]A. Miura and P. L. Pritchett, J. Geophys. Res. **87**, 7431, doi:10.1029/JA087iA09p07431 (1982).

[44]A. Miura, Phys. Plasmas **4**, 2871 (1997).

[45]A. Miura, Phys. Plasmas **16**, 102107 (2009).

[46]T. K. M. Nakamura and M. Fujimoto, Phys. Rev. Lett. **101**, 165002 (2008).

[47]R. Betti, V. N. Goncharov, R. L. McCrory, and C. P. Verdon, Phys. Plasmas **5**, 1446 (1998).

[48]W. Ye, W. Zhang, and X. T. He, Phys. Rev. E **65**, 057401 (2002).

[49]H. F. Robey, Phys. Plasmas **11**, 4123 (2004).

[50]B. Reipurth and J. Bally, Annu. Rev. Astron. Astrophys. **39**, 403 (2001).

A LETTERS JOURNAL EXPLORING THE FRONTIERS OF PHYSICS

November 2011

EPL, **96** (2011) 35002
doi: 10.1209/0295-5075/96/35002

www.epljournal.org

Effect of preheating on the nonlinear evolution of the ablative Rayleigh-Taylor instability

W. H. YE[1,2], X. T. HE[1,2,3], W. Y. ZHANG[2] and M. Y. YU[1,4]

[1] *Institute for Fusion Theory and Simulation, Department of Physics, Zhejiang University Hangzhou 310027, PRC*
[2] *Institute of Applied Physics and Computational Mathematics - Beijing 100088, PRC*
[3] *Center for Applied Physics and Technology and Key Laboratory of High Energy Density Physics Simulation of the Ministry of Education, Peking University - Beijing 100871, PRC*
[4] *Theoretical Physics I, Ruhr University - D-44789 Bochum, Germany, EU*

received 29 April 2011; accepted in final form 13 September 2011
published online 18 October 2011

PACS 52.65.Kj – Magnetohydrodynamic and fluid equation
PACS 52.38.Mf – Laser ablation
PACS 52.57.Fg – Implosion symmetry and hydrodynamic instability (Rayleigh-Taylor, Richtmyer-Meshkov, imprint, etc.)

Abstract – The evolution of the ablative Rayleigh-Taylor instability in the presence of preheating of the cold pristine plasma ahead of the ablative front is investigated by two-dimensional numerical simulation. When the dependence of the linear ARTI growth rate on the excitation wavelength has a sharp maximum, a secondary spike-bubble structure at a harmonic of the excitation wavelength will be generated, leading to rupture of the original mode. On the other hand, for stronger preheating the linear growth rate has no sharp maximum and in the nonlinear evolution no competing secondary spike-bubble structure is generated. Instead, the original excitation evolves into an elongated jet. These results are in agreement with that from an analytical single-mode theory.

Introduction. – In intense-laser interaction with a target in inertial confinement fusion, a subsonic heat-conduction wave is excited at the laser-absorption spot on the target surface. The intense heat wave propagates into the cold dense plasma and generates a shock wave. As the latter propagates, it compresses, heats, and accelerates the local plasma. A rarefaction wave that decelerates the accelerating fluid is formed ahead of the heat front, where the Rayleigh-Taylor instability (RTI) can take place [1–3] The RTI is also common in gas combustion, the implosion phase of laser-target interaction, as well as in astrophysical gas clouds [1–6].Linear theory shows that ablation caused by the heat front can stabilize the RTI as well as the subsequent Kelvin-Helmholtz instability [7–13].

Most earlier studies of the ablative RTI (ARTI) used the fluid approach and invoked the Spitzer-Harm (SH) conductivity for the thermal transport. On the other hand, results from Fokker-Planck (FP) simulations showed [14,15] that under ICF conditions the SH theory is inadequate because of intense plasma heating by the laser-generated hot electrons, heat and shock waves, etc. [2,3], and the preheated region in front of the temperature gradient of the heat-wave front is often quite extended spatially. The heat transport process involved in the region is rather complex and several models have been introduced [2,3,15,16]. Simulations and analytical modeling have shown that the preheating can significantly affect the linear growth rate and cutoff wavelength of the ARTI, as well as its nonlinear development [14,17,18]. However, although the onset and initial evolution of ARTI have been widely investigated and well understood, a more detailed study of the later stages of the development is still lacking [3]. Here we investigate by two-dimensional numerical simulation the nonlinear evolution of the ARTI accounting for the preheating effect by numerical simulation and analytical modeling. It is found that if the dependence of the linear ARTI growth rate on the wavelength has a sharp maximum, then the excited mode will eventually rupture. Otherwise it will evolve into an elongated jet. The results agree with that from an analytical model invoking the single-mode theory.

Model for preheating. – The fluid conservation equations including laser absorption and electron conduction effects are solved numerically using the LARED-S code [19,20]. The energy equation is given by [7–9,19,20]

$$\partial_t E + \nabla \cdot [(E+P)\boldsymbol{u}] = -\nabla \cdot \boldsymbol{q}(T) + S_L, \tag{1}$$

where ρ, $\boldsymbol{u}$, T, P and E are the fluid density, velocity, temperature, pressure, and total energy $E=\rho(\varepsilon+u^2/2)$, respectively, ε is the specific internal energy, S_L is the energy source due to absorption of laser energy by inverse bremsstrahlung, and $\boldsymbol{q}=-\kappa\nabla T$ is the heat flow. To include preheating effects, we invoke a modified SH electron heat conductivity [13,19,21] $\kappa=\kappa_{\rm SH}(T)[1+f(T[\rm MK])]$, where $\kappa_{\rm SH}(T)$ is the classical SH conductivity, and $f(T)=\alpha T^{-1}+\beta T^{-3/2}$ takes into account preheating of the cold plasma ahead of the ablation front. The form (not unique) of the preheating function $f(T)$ is consistent with that of the steady-state Fokker-Planck theory for the given density and temperature profiles, and the free parameters α and β are to be determined empirically by comparing the simulated results of the problem of interest with that from the corresponding experiment, theory, and/or simulation [13]. One can see that in the high-temperature low-density blow-off region the SH conductivity $\kappa_{\rm SH}(T)$ is recovered since $f(T)\ll 1$, and in the low-temperature ablation front, where preheating is significant, we have $\kappa(T)\sim\kappa_{\rm SH}f(T)\gg\kappa_{\rm SH}$. It is convenient to define a characteristic preheating length $L=T(\partial_x T)^{-1}\sim\rho(\partial_x\rho)^{-1}=L_{\rm SH}[1+f(T)]$, where $L_{\rm SH}=\kappa_{\rm SH}/\rho_a v_a c_p$ is the classical SH length and ρ_a, v_a, c_p are the density, speed, and specific heat, respectively, of the plasma in the ablation front region. Other heat conduction models [15,22] can also be used, provided that preheating effects are included.

Simulation procedure and results. – For simulating the evolution of ARTI, we first obtain a quasistationary state by irradiating a $200\,\mu$m thick CH foil target with a $0.35\,\mu$m laser beam of intensity $I=I_0\sin^4(t/5\,\rm ns)/\sin^4(1.0)$ for $t<5\,$ns and $I=I_0=10^{15}\,\rm W/cm^2$ for $t\geqslant 5\,$ns. Such a self-consistent one-dimensional quasistationary state appears at $t\geqslant 7\,$ns, as shown in the left panel of fig. 1 for the density, temperature, and pressure near the ablation front. To initiate the ARTI on the ablation-front interface, we apply at this instant a small fluid displacement $x=\xi(y,0)=\delta\xi_0\cos(ky)$, where $\delta\xi_0$ is the amplitude and $\lambda=2\pi/k$ is the wavelength. In both the dense pristine plasma and the ablation regions, 1200 fine meshes of width 0.06–0.07 μm are used in the x-directions. The mesh width is gradually increased on both sides of the region. Up to 1400×200 meshes are used in the simulation.

The left panel of fig. 1 shows that in the ablation front the temperature and density are $\sim 0.2\,$MK and 5.6–6.5 $\rm g/cm^3$, respectively, and in the ablation layer they are $\sim 2\,$MK and $\sim 0.8\,\rm g/cm^3$, respectively. The right panel of fig. 1 shows that the linear ARTI growth rate depends significantly on L. It is thus convenient to define weak

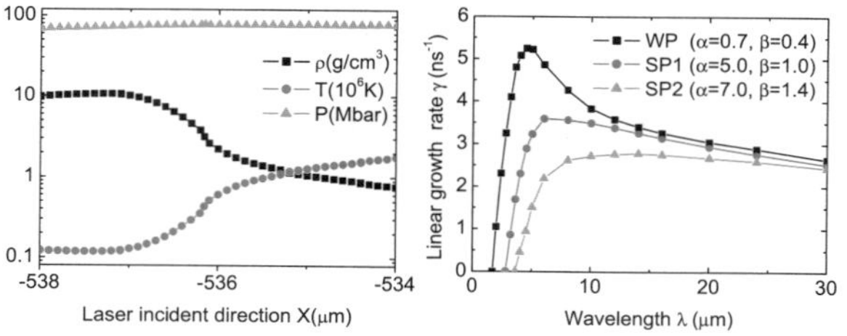

Fig. 1: (Color online) Left panel: profiles of the density (0.8–10.5 $\rm g/cm^3$), temperature (0.1–2 MK), and pressure ($\sim 70\,$Mb) for the WP regime near the ablation front at $t=9.09\,$ns and $x\sim -536.5\,\mu$m. The laser is incident from the right. The ARTI interface is near the density peak. Right panel: linear ARTI growth rate *vs.* the wavelength, from numerical simulations. The black squared curve is for $\alpha=0.7$ and $\beta=0.4$ (WP regime). The red circled curve is for $\alpha=5.0$ and $\beta=1.0$ (SP regime), where $\lambda_c\sim 3\,\mu$m, $L\sim 1.35\,\mu$m, $v_a\sim 0.80\,\mu$m/ns. The green triangled curve is for $\alpha=7.0$ and $\beta=1.4$ (also SP regime).

preheating (WP) by the condition $L/L_{\rm SH}=f(T)\ll f_0\sim 20$, which is the case indicated by the black squared curve (for $f(T)\sim 8$ at $T=0.2\,$MK). Here the linear growth rate has a sharp maximum at $\lambda\sim 5\,\mu$m, showing that a perturbation of the latter scale length is preferred. We also define strong preheating (SP) by the condition $L/L_{\rm SH}>f_0\sim 20$. It corresponds to the regime where the linear growth rate does not have a sharp maximum, as indicated in fig. 1 by the red circled and green triangled curves.

For $\alpha=0.7$ and $\beta=0.4$, L is about $0.58\,\mu$m, with $f_0\approx 8$. The initial perturbation is given by $\delta\xi_0=0.4\,\mu$m and $\lambda_1=12\,\mu$m. Figure 1 shows that the linear growth rate has a sharp maximum at $\lambda_m=0.375\lambda_1$ and a cutoff wavelength at $\lambda_c=0.14\lambda_1$. As expected, the linear growth rate agrees well with that from the modified Lindl-Takabe formula [11,19] $\gamma=[Akg/(1+\alpha_0 kL)]^{1/2}-\beta_0 v_a k$, where A is the Atwood number, g is the fluid acceleration, and the free parameters α_0 and β_0 are in the range 2–3. That is, in the short-wavelength regime the effect of ablation is significant and the linear growth rate decreases as L increases.

Our simulations show that besides the fundamental mode ($\lambda_1=12\,\mu$m), at least six harmonics having wavelengths larger than $\lambda_c=0.14\lambda_1$ appear. The evolution of the fundamental mode $\xi(y-\lambda_1/2,t)$ is shown in fig. 2 and that of the harmonic modes $\xi_n(t)$ at $y=\lambda_1/2$ for $n=1$–4 are shown in fig. 3. At $t=7.78\,$ns, the fundamental mode has evolved into a wide spatial structure including a bubble of size $\sim 0.24\lambda_1$. In the region preheated by the heat conduction wave, it grows at $>10^6\,$cm/s. Here the pressure remains high ($>70\,$Mb), the temperature increases sharply to $>1\,$MK, and the density decreases sharply to $<2\,\rm gm/cm^3$, as can be seen in the left panel of fig. 1. The front of the fundamental-mode spike becomes

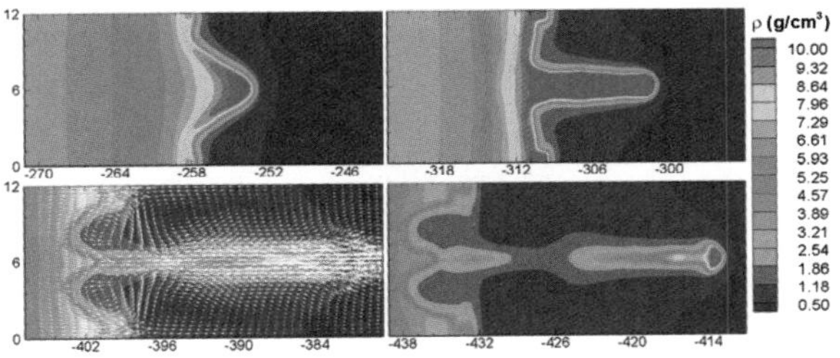

Fig. 2: (Color online) Amplitudes (of the ablation layer displacement) of the density spikes and bubbles for the WP regime ($\alpha = 0.7$ and $\beta = 0.4$). For clarity the center of the density pattern has been moved to $y = \lambda_1/2$. Upper left: the fundamental spike and bubble pattern at $t = 7.78$ ns. Upper right: the secondary density spikes (around $y = 0, 12\,\mu$m) at $t = 8.10$ ns. Lower left: the mass flux $\rho \boldsymbol{u}$ at $t = 8.56$ ns. Lower right: the fundamental spike ruptures at $t = 8.73$ ns.

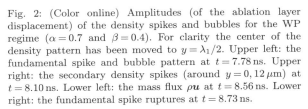

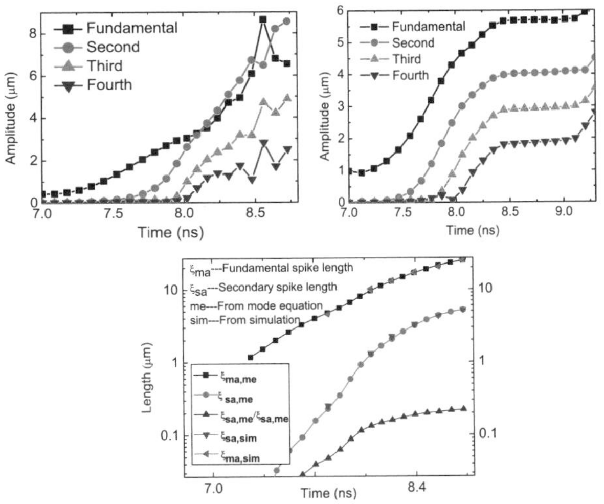

Fig. 3: (Color online) Amplitudes of the density spikes *vs.* time. Upper left: the fundamental mode and the second, third, and forth harmonics for the WP case. Upper right: the same for the SP1 case. Lower panel: lengths (displacements) of the fundamental and secondary spikes, and their ratio for the WP case.

flattened, but no mushroom structure appears. This can be attributed to the fact that there is enhanced energy loss due to increased electron heat conduction, so that velocity shear and fluid rotation needed for mushroom formation is suppressed [13].

One can see that at $t = 8.1$ ns the fundamental spike is still extending forward, although at a slower rate $\sim 2.5 \times 10^6$ cm/s. The displacement ξ of the layer is close to the modulation wavelength λ_1. That is, higher-harmonics and mode coupling effects become important, and the ARTI can be considered to have entered a new nonlinear stage. Here the boundary part of the high-density region is gradually ablated because of the high temperature there. In fact, the upper right panel of fig. 2 shows that the high-pressure ablation layer surrounding the spike is being compressed to a width of $\sim 2\,\mu$m and density of $\sim 10\,\mathrm{g/cm^3}$. The ablation also leads to increase of the bubble pressure due to fluid expansion around the boundaries of the region, as can be seen near the spike bottom, where the expansion (and ablation) speed is $\sim$2–5×10^5 cm/s. Consequently, the bubbles on both sides of the spike are extended and deformed, and the fundamental spike becomes even narrower. The bubbles are now nearly stationary.

The upper right panel of fig. 2 shows that secondary perturbations, or spikes, are excited in the center region of the fundamental bubble. The secondary spikes correspond to the second harmonic $\xi_2(t)$ with wavelength $\lambda_1/2$. Its rate of fluid displacement can be larger than that of the fundamental spike and thus compete with it for mass and energy feed. In fact, at $t > 8.1$ ns one finds in the left panel of fig. 3 that $\xi_2(t) \geqslant \xi_1(t)$, *i.e.*, the secondary spikes are overtaking the fundamental spike.

Single-mode model for ARTI evolution. – As already mentioned in the discussion of the simulation results, the evolution of the ARTI can be understood in terms of the nonlinear interaction among the fundamental and the secondary spikes associated with the harmonics of the fundamental excitation. For simplicity, we shall adopt the single-mode approach for considering the nonlinear RTI [2,3,18,23–29]. Accordingly, the excitation of higher harmonics and their interactions in the region of interest shall be investigated.

An initial finite-amplitude perturbation (the fundamental mode) of wave number $k_1 = 2\pi/\lambda_1$ can excite harmonic modes with wave numbers $k_n = nk_1$, where $n = 2, 3, \ldots$ [2,3,16]. The total surface displacement of the interface can be expressed as $\xi(y,t) = \sum_n \xi_n(y,t)\cos(k_n y)$. In the weakly nonlinear stage, $\xi(y,t)$ can be given by the first three harmonic modes [23]. As the instability continues, more modes must be included. Thus, the structure of the ARTI fluid displacement can be represented by $\xi(y - \lambda_1/2, t) = \sum_n \xi_n(y - \lambda_1/2, t)\cos[k_n(y - \lambda_1/2)]$, where $n = 1, 2, 3, \ldots$ with $n = 1$ being the fundamental mode. To maintain symmetry in the analysis, in this representation the y-coordinate has been shifted by $\lambda_1/2$. Accordingly, one obtains the displacement of the secondary spike $\xi_{sa} = \xi(\pm\lambda_1/2, t) - \xi_b = -\xi_1(t) + \xi_2(t) - \xi_3(t) + \xi_4 - \xi_b$, where $\xi_b = \xi(\pm\lambda_1/6, t) = [\xi_1(t) - \xi_2(t) - \xi_4(t)]/2 - \xi_3(t)$ is the leftmost of the bubble region, or $\xi_{sa} = -3[\xi_1(t) - \xi_2(t) - \xi_4(t)]/2$. Since $\xi_1 \sim \xi_2$, we have $\xi_{sa} \sim 3\xi_4(t)/2 > 0$. The displacement (distance from the tip of the fundamental spike to the edge of the bubble) of the fundamental spike is $\xi_{ma} = \xi(0,t) - \xi_b = [\xi_1(t) + 3\xi_2(t) + 3\xi_4(t)]/2$.

In fig. 3, the profiles of $\xi_n(t)$, $n = 1$–4, obtained by numerically solving the corresponding mode equations are shown. Using the value of ξ_4 at $t = 8.1\,\text{ns}$ from the upper left panel, we find $\xi_{sa} = 1.2\,\mu\text{m}$ (at the same instant) from the lower panel. This result agrees well with the corresponding value $1.3\,\mu\text{m}$ (for the secondary spikes in fig. 2, upper right panel) from the simulations. That is, by including four harmonics in the theory the obtained maximum displacement ξ_{sa} of the secondary spikes agrees well with that from the simulations. The lower panel of fig. 3 shows that the fundamental- and secondary-spike displacements $\xi_{ma,me}$ and $\xi_{sa,me}$, respectively, obtained from mode equations are in good agreement with those, $\xi_{ma,sim}$ and $\xi_{sa,sim}$, respectively, from the simulation.

For $\xi_{sa} < 0$ and $\xi_1 > \xi_2 > \ldots$, the distinct maximum in the linear ARTI growth rate becomes smoothed out, so is the secondary-spike excitation. In fact, the upper right panel of fig. 2 shows that at $t = 8.1\,\text{ns}$ a secondary spike-bubble structure forms in the center region of the fundamental bubble. The secondary bubble and the fundamental bubble coalesce into a wide and irregular bubble between the fundamental and the secondary spikes. The fundamental spike is further compressed and becomes still narrower. At $t = 8.56\,\text{ns}$, new eddies of the mass flux are generated between the fundamental and secondary spikes, as indicated by the arrows in the lower left panel of fig. 2. The mass flux pushes against the waist of the fundamental spike, such that the density of upper region of the fundamental spike increases to $\rho > 7.5$–$10\,\text{g/cm}^3$. The upper region is accelerated upward and becomes a jet. The secondary spike grows faster than the fundamental one. The ratio $R = \xi_{sa}/\xi_{ma} = \xi(\pm\lambda_1/6, t)/\xi(0, t)$, where $\xi(0, t) = \sum_n \xi_n(t)$ (n =1–4), is a measure of the displacement of the fundamental spike-bubble pattern at the center. It increases gradually, as shown by the red and green curves in the lower panel of fig. 3. That is, the secondary spikes acquire more fluid mass than the fundamental spike. The plasma at the foot of the fundamental spike are also fed into the secondary spikes, making them grow faster, as shown in the lower left panel of fig. 2, where the arrows indicate the direction and magnitude of the mass flux. Eventually, at $t = 8.73\,\text{ns}$ the fundamental spike ruptures into two parts. This scenario in agreement with that of the simulation (fig. 2, lower right panel). The latter also indicate that for the parameters considered the threshold above which the fundamental spike can ruptures is $R \sim 0.20$.

If the initial modulation wavelength λ_1 is larger than $12\,\mu\text{m}$, our analysis shows that harmonics higher than the fourth are excited. For $\lambda_1 = 24\,\mu\text{m}$, at least 12 higher harmonics are excited, and the first 10 modes have initial growth rates larger than that of the fundamental mode. That is, many high-harmonic modes can be excited simultaneously. The harmonics having linear growth rates near the sharp maximum grow rapidly, resulting in generation of harmonic spikes. From the simulations, we found that for $\lambda_1 > 12\,\mu\text{m}$ more short-wavelength (*i.e.*, high-harmonic) spikes appear in the fundamental bubble region. The fundamental spike eventually ruptures due to

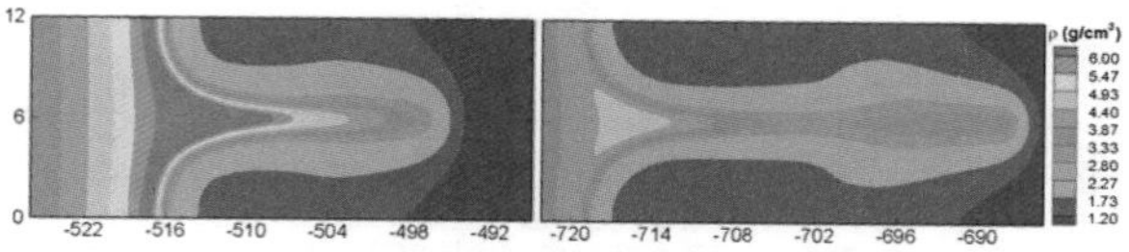

Fig. 4: (Color online) Evolution of the fundamental spike and bubble for the SP regime ($\alpha = 5.0$, $\beta = 1.0$). The initial modulation wavelength is $12\,\mu\text{m}$. Left: $t = 8.93\,\text{ns}$, right: $t = 9.67\,\text{ns}$.

lack of mass feed resulting from competition for the latter among the spikes.

As can also be seen in the right panel of fig. 1, the linear growth rate in the SP regime is smaller than that of the WP regime, and it has no sharp maximum. This is because the SP regime corresponds to larger L and lower Atwood number. The cutoff wavelength is about $2.8\,\mu\text{m}$ and the linear growth rate has no sharp maximum. For $\alpha = 7.0$ and $\beta = 1.4$ (strong preheating), the cutoff wavelength is about $3.6\,\mu\text{m}$. The corresponding growth rate *vs.* wavelength curve is relatively smooth and the maximum is not sharp. On the other hand, the right panel of fig. 3 shows that for $\alpha = 5.0$ and $\beta = 1.0$ (also strong preheating), except in the early linear growth phase, no harmonic mode grows faster than the fundamental mode. This can be attributed to the fact that energy and mass feed is spread over several modes, instead of only in the fastest growing mode, as in the WP case. In fact, in the SP regime the growth rates of the fundamental and secondary harmonics in the nonlinear stage is similar. As a result, the fundamental spike does not rupture. Instead, it eventually evolves into a long jet that enters into the higher-temperature underdense plasma at an average speed of $\sim 1.8 \times 10^6\,\text{cm/s}$, as shown in fig. 4. Such long jets have been frequently observed in experiments [3,29,30], as well as astrophysical clouds [4,29]. However, it should be cautioned that when several harmonics are excited, the single-mode and 2D assumptions can become invalid [3].

Summary. – In this letter, we have shown that the nonlinear evolution of ARTI in the presence of preheating can be interpreted in terms of the generation and interaction of harmonics of the fundamental perturbation. In agreement with that from existing numerical simulations [17,18,26], it is found that the initial growth of the fundamental spike-bubble system depends on the ablation length. However, the evolution also depends significantly on the preheating parameters. In fact, there can be two regimes of ARTI evolution. The WP regime is characterized by a distinct peak in the wavelength dependence of the linear growth rate and thus rapid excitation of the corresponding harmonic. In the nonlinear stage, the fundamental and the second harmonic compete with each other. Eventually the second harmonic dominates and the fundamental spike ruptures. In the SP regime, the dependence of the linear growth on the wavelength is not sharply peaked, so that many harmonics are excited. In the evolution the fundamental spike remains dominant and evolves into a elongated jet that spurts towards the high-temperature low-density blow-off region. Both scenarios of the ARTI evolution have often been observed in the laser-plasma compression experiments as well as in astrophysical plasmas [1,3,4,29,30].

* * *

We thank Gaimin Lu and L. F. Wang for useful discussions. This work was supported by the National High-Tech ICF Committee, the National Natural Science Foundation of China (11075024, 10835003, 10935003 and 10905006), and the National Basic Research Program of China (2007CB815100, 2007CB814802, 2008CB717806), and the ITER-CN Program (2009GB105005).

REFERENCES

[1] Kilkenny J. D. *et al.*, *Phys. Plasmas*, **1** (1994) 1379.
[2] Lindl J. D., *Inertial Confinement Fusion* (Springer, New York) 1998.
[3] Atzeni S. and Meyer-ter-Vehn J., *The Physics of Inertial Fusion* (Oxford University Press, New York) 2004.
[4] Zinnecker H. *et al.*, *Nature*, **394** (1998) 862.
[5] Cheng C. Z. and Johnson J. R., *J. Geophys. Res.*, **104** (1999) 41.
[6] Kanuer J. P. *et al.*, *Phys. Plasmas*, **7** (2000) 338.
[7] Bodner S. E., *Phys. Rev. Lett.*, **33** (1974) 761.
[8] Takabe H. *et al.*, *Phys. Fluids*, **26** (1983) 2299.
[9] Takabe H. *et al.*, *Phys. Fluids*, **28** (1985) 3676.
[10] Sanz J., *Phys. Rev. Lett.*, **73** (1994) 2700.
[11] Betti R. *et al.*, *Phys. Plasmas*, **5** (1998) 1446.
[12] Atzeni S. and Temporal M., *Phys. Rev. E*, **67** (2003) 057401.
[13] Wang L. F., Ye W. H. and Li Y. J., *EPL*, **87** (2009) 54005.
[14] Epperlein E. M. and Short R. W., *Phys. Fluids B*, **3** (1991) 3092.
[15] Weng S. M. *et al.*, *Phys. Rev. Lett.*, **100** (2008) 185001.
[16] Otani K. *et al.*, *Phys. Plasmas*, **17** (2010) 032702.
[17] Wang L. F. *et al.*, *EPL*, **90** (2010) 15001.
[18] Wang L. F. *et al.*, *Phys. Plasmas*, **17** (2010) 122706.
[19] Ye W. H., Zhang W. Y. and He X. T., *Phys. Rev. E*, **65** (2002) 057401.
[20] Pei W., *Commun. Comput. Phys.*, **2** (2007) 255.
[21] Ye W. H., Wang L. F. and He X. T., *Phys. Plasmas*, **17** (2010) 122704.
[22] Yu M. Y. *et al.*, *Phys. Fluids*, **23** (1980) 226.
[23] Sanz J., *Phys. Rev. Lett.*, **89** (2002) 195002.
[24] Garnier J. *et al.*, *Phys. Rev. Lett.*, **90** (2003) 185003.
[25] Ikegawa T. and Nishihara K., *Phys. Rev. Lett.*, **89** (2002) 115001.
[26] Wang F., *Phys. Plasmas*, **17** (2010) 052305.
[27] Robey H. F. *et al.*, *Phys. Plasmas*, **8** (2001) 2446.
[28] Kane J. O. *et al.*, *Phys. Rev. E*, **63** (2001) 055401R.
[29] Remington B. A. *et al.*, *Rev. Mod. Phys.*, **78** (2006) 755.
[30] Lei A. L. *et al.*, *Phys. Plasmas*, **16** (2009) 056307.

PHYSICS OF PLASMAS **19**, 012706 (2012)

Density gradient effects in weakly nonlinear ablative Rayleigh-Taylor instability

L. F. Wang (王立锋),[1,2,a)] W. H. Ye (叶文华),[1,2,b)] and X. T. He (贺贤土)[1,2,c)]
[1]*HEDPS and CAPT, Peking University, Beijing 100871, China*
[2]*LCP, Institute of Applied Physics and Computational Mathematics, Beijing 100088, China*

(Received 8 November 2011; accepted 15 December 2011; published online 26 January 2012)

In this research, density gradient effects (i.e., finite thickness of ablation front effects) in ablative Rayleigh-Taylor instability (ARTI), in the presence of preheating within the weakly nonlinear regime, are investigated numerically. We analyze the weak, medium, and strong ablation surfaces which have different isodensity contours, respectively, to study the influences of finite thickness of ablation front on the weakly nonlinear behaviors of ARTI. Linear growth rates, generation coefficients of the second and the third harmonics, and coefficients of the third-order feedback to the fundamental mode are obtained. It is found that the linear growth rate which has a remarkable maximum, is reduced, especially when the perturbation wavelength λ is short and a cut-off perturbation wavelength λ_c appears when the perturbation wavelength λ is sufficiently short, where no higher harmonics exists when $\lambda < \lambda_c$. The phenomenon of third-order positive feedback to the fundamental mode near the λ_c [J. Sanz *et al.*, Phys. Rev. Lett. **89**, 195002 (2002); J. Garnier *et al.*, Phys. Rev. Lett. **90**, 185003 (2003); J. Garnier and L. Masse, Phys. Plasmas **12**, 062707 (2005)] is confirmed in numerical simulations, and the physical mechanism of the third-order positive feedback is qualitatively discussed. Moreover, it is found that generations and growths of the second and the third harmonics are stabilized (suppressed and reduced) by the ablation effect. Meanwhile, the third-order negative feedback to the fundamental mode is also reduced by the ablation effect, and hence, the linear saturation amplitude (typically $\sim 0.2\lambda$ in our simulations) is increased significantly and therefore exceeds the classical prediction 0.1λ, especially for the strong ablation surface with a small perturbation wavelength. Overall, the ablation effect stabilizes the ARTI in the weakly nonlinear regime. Numerical results obtained are in general agreement with the recent weakly nonlinear theories and simulations as proposed [J. Sanz *et al.*, Phys. Rev. Lett. **89**, 195002 (2002); J. Garnier *et al.*, Phys. Rev. Lett. **90**, 185003 (2003); J. Garnier and L. Masse, Phys. Plasmas **12**, 062707 (2005)]. *© 2012 American Institute of Physics.* [doi:10.1063/1.3677821]

I. INTRODUCTION

The Rayleigh-Taylor instability (RTI) takes place wherever the perturbed fluid regions that differ in density experience a pressure gradient that opposes the density gradient.[1,2] It is of significant importance in inertial confinement fusion (ICF) (Refs. 3–6) and in astrophysics.[7–10] In ICF, the shell undergoes the RTI both in the acceleration and the deceleration phases. At the initial acceleration stage, when the imploding shell is accelerated inward by the low density blow-off plasma, the outer shell ablation front between the shell and the ablated plasma is unstable to the RTI, while at the later stage, when the compressed fuel begins to decelerate the imploding pusher, the inner shell surface between the hot spot and the shell is also prone to the RTI. This inherent physical instability can limit the implosion velocity and, in some cases, even break up the implosion shell. Furthermore, the instability would hinder the formation of hot spot resulting in the auto-ignition failure. (In this report, the term instability refers to the RTI unless stated otherwise.) Therefore, the ICF targets must be designed to keep the RTI growth at an acceptable level. In astrophysics, the RTI plays a central role in the evolutions of many astrophysical phenomena, such as supernova explosion,[11,12] high Mach number jets,[7,8] etc. Therefore, it is essential to understand and estimate the evolution of the RTI because of its significance in both fundamental research and engineering applications.

When the heavier fluid is superposed over the lighter fluid in a gravitational field, $-g\mathbf{e}_y$, where g is the acceleration, an initial single-mode perturbation of wavelength λ on an interface between two fluids of densities ρ_h and ρ_l ($\rho_h > \rho_l$) is the simplest case. Its initial growth is exponential with the linear growth rate $\gamma = \sqrt{A_T k g}$, where $k = 2\pi/\lambda$ is the perturbation wave number, and $A_T = (\rho_h - \rho_l)/(\rho_h + \rho_l)$ is the Atwood number. Fortunately, the density gradient effect (i.e., the effect of finite thickness of the perturbed interface) reduces the linear growth rate of the RTI, which can be approximated by the formula[4,13–16] $\gamma_L = \sqrt{A_T k g/(1 + A_T k L_m)}$, where $L_m = \min[|\rho/(\partial\rho/\partial x)|]$ is the minimum density gradient scale length. It has long been known that the mass ablation effects (i.e., the physical removal of modulated material through ablation and the ablation-driven convection of the vorticity off the ablation front) reduces the linear growth rate of the RTI.[17–28] When the ablation effect is considered, the linear growth rate of the ablative RTI (ARTI) is

a)Electronic mail: lif_wang@pku.edu.cn.
b)Electronic mail: ye_wenhua@iapcm.ac.cn.
c)Author to whom correspondence should be addressed. Electronic mail: xthe@iapcm.ac.cn.

remarkably reduced, especially for the short perturbation wavelength and a cut-off wavelength appears when the perturbation wavelength λ is sufficiently short, which can be simply expressed by the Takabe-Bodner formula[17–19] $\gamma_T = \alpha\sqrt{kg} - \beta k v_a$, where v_a is the ablation velocity. The two parameters α and β depend fundamentally on the flow parameters with $\alpha = 0.9 - 0.95$ and $\beta = 1 - 3$ typically.[20–22] A modified version of this including the effect of the finite thickness of the ablation front given by Lindl is in the form $\gamma_{mT} = \alpha\sqrt{kg/(1+kL_m)} - \beta k v_a$, where L_m is the minimum scale length of density gradient at the ablation surface.[4] Subsequently, Ye *et al.* found that the ARTI growth rate $\gamma_{mL} = \sqrt{A_T kg/(1+A_T kL_m)} - 2kv_a$, agrees well with the experiments and simulations, and is a suitable model for the preheating case in direct-drive ICF experiments.[24] In addition, the recent self-consistent linear theory[25–28] identified another physical mechanism of the stabilization (i.e., perturbed dynamic pressure), which can be expressed as $\gamma_{sc} = \sqrt{A_T kg - A_T^2 k^2 v_a^2/r_D} - (1+A_T)kv_a$ for the case of large Froude number, where r_D is the effective density ratio of the blow-off plasma to the ablation front. The main difference between γ_{sc} and all versions of the Takabe-Bodner formula is the negative term under the square root which is from the dynamic pressure (a restoring force).

In the past 40 years, most attentions were focused on the linear growth rate of the ARTI. However, only recently, the weakly nonlinear behaviors of the ARTI are studied by Sanz *et al.*[29] and Garnier *et al.*,[30,31] where the nonlinear harmonic generations up to the third-order is analyzed and the corresponding harmonic generation coefficients are obtained. However, the effect of the finite thickness (the density gradient effect) of the ablation front has not been well considered in both of their models. In ICF, gentle density profiles at the ablation front are generated by the preheating ablation effect (i.e., the nonlocal energy transport) from the corona of the heated plasma (for example, hot electrons, high-energy photons, and high-energy ions), the products of fusion with high-energy (for example, alpha particles and neutrons), and other particles with a long free path.[32–35] In fact, the preheating ablation effect plays a key role not only in the linear growth regime but also in the weakly nonlinear regime in the ARTI. The density gradient effects of the ablation surface in the ARTI, in the presence of preheating, have not been well understood. Thus, in the present work, we investigate the structure of the ablation surface in ARTI with preheating in the weakly nonlinear regime by numerical simulations. Our numerical results have been compared with the analytical theories by Sanz and Garnier, whose prediction of third-order positive feedback to the fundamental mode is reappeared in numerical simulations with a short perturbation wavelength around the cut-off wavelength.

The paper is organized as the following. In Sec. II, the mathematical framework and numerical setup are presented. Section III is devoted to some recent analytical study of evolution of the growth of the amplitude of the harmonics of the interface, with increasing initial perturbation wavelengths in the weakly nonlinear regime. The harmonic generation coefficients of the first three harmonics are defined and their roles on the weakly nonlinear behaviors of the RTI will be examined in Sec. IV. In Sec. IV, numerical results from simulations of a single-mode sinusoidal perturbed interface with different perturbation wavelengths are presented and discussed. Concluding remarks are given in Sec. V.

II. SIMULATION CODE AND BASIC FLOWS

A. Governing equations

In this study, the physical system is modeled via the two-dimensional inviscid one-temperature single-fluid with a bulk acceleration force in a constantly accelerating frame of Refs. 36 and 37. Without considering the radiation, an ideal gas equation of state is used. The governing equations are

$$\begin{aligned} &\frac{\partial \rho}{\partial t} + \nabla\cdot(\rho \boldsymbol{u}) = 0,\\ &\frac{\partial(\rho \boldsymbol{u})}{\partial t} + \nabla\cdot(\rho \boldsymbol{u}\boldsymbol{u}) + \nabla P = \rho \boldsymbol{g},\\ &\frac{\partial E}{\partial t} + \nabla\cdot[(E+P)\boldsymbol{u}] = \nabla\cdot(\kappa\nabla T) + \rho \boldsymbol{u}\cdot \boldsymbol{g}, \end{aligned} \tag{1}$$

where ρ, $\boldsymbol{u}$, T, and $\boldsymbol{g}$ are, respectively, the density, velocity, temperature, and acceleration. $E = c_v \rho T + (1/2)\rho \boldsymbol{u}\cdot\boldsymbol{u}$ is the total energy, $P = \Gamma \rho T$ is the pressure, and $c_v = \Gamma/(\bar{\gamma}-1)$ is the specific heat at constant volume. For the CH material used in this study, $\bar{\gamma} = 5/3$ and $c_v = 86.2713\ (\text{cm}/\mu\text{s})^2\ \text{MK}^{-1}$ are used.

The preheating function $\kappa(T) = \kappa_{SH}[1 + f(T)]$ is introduced in our numerical simulation, where κ_{SH} is the Spitzer-Härm (SH) electron heat conduction and $f(T)$ interprets the preheating tongue effect in the cold plasma ahead of the ablation front.[24] In the blow-off region, the property of high temperature corresponding to $f(T) \ll 1$, leads to the conductivity $\kappa(T) \to \kappa_{SH}$. Inner the ablation front, due to low temperature, there is $f(T) \gg 1$. Accordingly, ahead of the ablation front, the conductivity $\kappa(T) \approx \kappa_{SH} f(t) \gg \kappa_{SH}$, which means the preheating effect is dominated in this zone. For temperatures greater than 0.10 MK, it has been shown that the ARTI growth rates, the fluid density, and temperature profiles at the ablation front obtained from the choice $f(T) = aT^{-1} + bT^{3/2}$, where T is in the unit of MK, a and b are adjustable parameters, agree well with the experiments.[24] In the present research, we choose a weak preheating case with $a = 0.86$ and $b = 0.24$. It is noteworthy that the form (not unique) of the preheating function $f(T)$ is consistent with that of the steady-state Fokker-Planck theory[38] for the given density and temperature profiles, and the free parameters a and b are to be determined empirically by comparing the simulated results of the problem of interest with that from the corresponding experiment, theory, and/or simulation. In fact, other heat conduction models[39,40] can also be used, provided that preheating effects are included. More details about the preheating model used in our simulation can be found in Refs. 24 and 41.

B. Numerical methods

The physical domain is rectangular. The domain in x direction is $[-20\ \mu\text{m}, 30\ \mu\text{m}]$ with length $L_x = 50\ \mu\text{m}$. The domain in y direction is assumed to be periodical with length $L_y = \lambda$, where λ is the perturbation wavelength of the single-mode sinusoidal interface perturbation. The free-stream boundary conditions are imposed at both ends of the domain in x, since the flow is essentially unchanged near the boundaries of the computational domain for a sufficiently large domain, especially for the sharp interface case studied here. Since the domain in y is periodical, periodical boundary conditions are imposed on the upper and lower boundaries.

To simplify the notation for discussion below, the system of partial differential equations can be expressed in the form of

$$\mathbf{Q}_t + L\mathbf{Q} = L_v\mathbf{Q} + \mathbf{S}, \tag{2}$$

where L and L_v are the spatial differential operators for the first (inviscid, nonlinear, hyperbolic) and second order (linear, parabolic) terms, respectively, and $\mathbf{S}$ is the bulk force source term.

We approximate the inviscid flux $L\mathbf{Q}$ in Eq. (2), which is the classical Euler equations, via the well-known classical fifth order characteristic-based weighted essentially non-oscillatory conservative finite difference scheme (WENO) (Refs. 42, 43, and 44) explicitly. The second order parabolic term involving the thermal conduction is solved via a fourth order implicit central finite difference scheme, to avoid the small time step restriction due to the severe stability condition imposed if solved by an explicit scheme. More details about the numerical scheme used in our simulation can be referred in Ref. 36.

In the simulations presented in Sec. IV, the number of uniformly spaced grid points used in the x and y directions are $N_x = 2200$ and $N_y = 256$, respectively.

C. Initial condition setup

The one-dimensional steady state flow field is obtained by integrating Eq. (1) in the accelerating reference frame of the ablation front from the peak density to the isothermal sonic points on both sides of the ablation front (see Figure 1) using the averaged ablation parameters, the peak density $\rho_a = 5.25$ g/cm^3, the ablation velocity $v_a = 1.19$ μm/ns, the

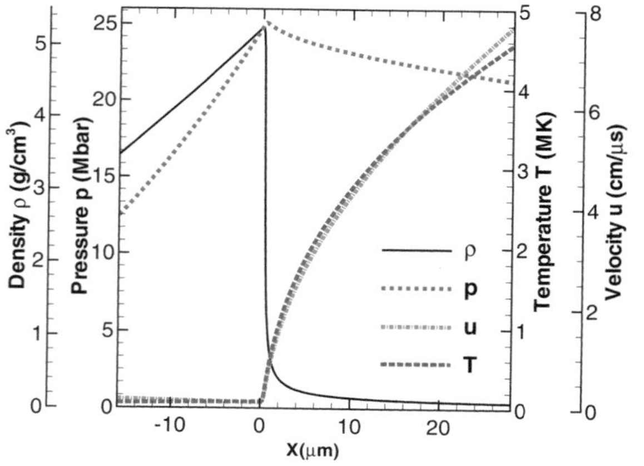

FIG. 1. (Color online) Distributions of the unperturbed one-dimensional basic flows. The minimum density gradient scale length of the ablation front is $L_m = 0.23$ μm in our present weak preheating model. In our numerical simulations, we employ very fine grid whose width is $\Delta x = 0.02$ μm to capture the sharp interface.

temperature at the peak density $T_a = 8.23 \times 10^{-2}$ MK, and the acceleration $g = 17.85$ μm/ns^2, which are obtained form one-dimensional laser-target interaction calculation.

The primitive variables (ρ, u, v, T) of the initial steady flow field is perturbed by a single-mode sinusoidal interface perturbation at $x = x_0$ with wavelength λ and amplitudes η_0 in the y direction, namely,

$$S(x,y) = 1 + \eta_0 e^{-k_0|x-x0|}\sin(k_0 y) \tag{3}$$

where $x_0 = 0$ is the location of the interface and $k_0 = 2\pi/\lambda$ is the wave number. The amplitude of the perturbation η_0 is typically set as a fraction of wavelength $\eta_0 = \alpha_0\lambda$ with $1/10\,000 \le \alpha_0 \le 5/10\,000$. For example, the initial perturbed density field $\rho(x,y,t=0)$ becomes $\rho(x,y,0) = \rho_0(x)S(x,y)$, where $\rho_0(x)$ is the unperturbed one-dimensional steady density field as shown in Figure 1.

III. ANALYTICAL METHOD AND TECHNOLOGY

This section is devoted to the detailed description of the analytical method and technology of the present paper. The *linear* growth regime is defined by $k\eta(t)$ being small (typically $k\eta(t) < 0.628$), where $\eta(t)$ represents the typical spatial amplitude of the perturbed interface at time t. (The explicit dependent of η in time t will be drop for simplicity sake.) After sufficient growth of the perturbation due to the instability, $k\eta$ is no longer small and the flow enters the *nonlinear* regime (typically $k\eta \sim 1$). Before the strong nonlinear growth regimes, one has the *weakly nonlinear* growth regime.

In the weakly nonlinear growth regime, for a single-mode sinusoidal interface perturbation, within the framework of third-order perturbation theory,[4,45–47] the perturbation amplitudes of fundamental mode (first harmonic) η_1, second harmonic η_2, and third harmonic η_3 can be expressed as

$$\begin{aligned}\eta_1 &= \eta_L - \frac{1}{16}(3A_T^2+1)k^2\eta_L^3,\\ \eta_2 &= \frac{1}{2}A_T k\eta_L^2,\\ \eta_3 &= \frac{1}{2}\left(A_T^2-\frac{1}{4}\right)k^2\eta_L^3,\end{aligned} \tag{4}$$

where $\eta_L = \eta_0 e^{\gamma t}$ and γ are the spatial amplitude and growth rate in the linear growth regime, respectively. For problems with large Atwood number $A_T \to 1$, Eq. (4) reduce to[4,45–49]

$$\begin{aligned}\eta_1 &= \eta_L - \frac{1}{4}k^2\eta_L^3,\\ \eta_2 &= \frac{1}{2}k\eta_L^2,\\ \eta_3 &= \frac{3}{8}k^2\eta_L^3\end{aligned} \tag{5}$$

As noted above, at the third-order, the growth of the fundamental mode is reduced by the nonlinear effect, i.e., third-order negative feedback to the fundamental mode. If we define the transition into the nonlinear regime to occur when the growth of the fundamental mode is reduced by 10% in comparison with its corresponding linear growth, then we have $\eta_s/\lambda = \sqrt{2}/\sqrt{5\pi^2(3A_T^2+1)}$, where η_s is the linear saturation amplitude.[47] Let $A_T = 1$, then $\eta_s \approx 0.1\lambda$ is recovered, which is the widely used linear saturation amplitude given by Layzer.[4,50]

It should be noted that the above potential expansion theories[45,46] are based on the assumptions that the interface between the two fluids is discontinuous and the thermal conduction is ignored. Furthermore, the direct-driven experiments[21,22] have shown that the growth rate of the areal density perturbation is significantly reduced as compared with the calculated results of the SH electron thermal conductivity. Glendinning *et al.*[21] recognized that the difference between the experiment and the LASNEX simulation is due to a change in the longitudinal density profile resulting from preheating by energetic electrons originating in the plasma corona which penetrate beyond the ablation front. Besides, the experiments of Shigemori[22] and Sakaiya *et al.*[23] suggest that the nonlocal electron heat transport plays an important role in the suppression of ARTI at the ablation front. The observation of density profile in direct-driven experiment[32] and simulation[21,22] indicates that L_m is markedly improved by the preheating ablation effect. It is, therefore, natural and important to investigate the growth of ARTI with a smooth density profile. To investigate the effect of the finite thickness of ablation front (the density gradient effects of the ablation surface) generated by the preheating on the mode coupling process in the ARTI, we generalize Eq. (5) by defining the second harmonic generation coefficient $C(2k)$, third harmonic generation coefficient $D(3k)$, and third-order feedback to the fundamental mode coefficient $E(k)$, and rewrite Eq. (5) as

$$\begin{aligned} \eta_1 &= \eta_L - \frac{1}{4}E(k)k^2\eta_L^3, \\ \eta_2 &= \frac{1}{2}C(2k)k\eta_L^2, \\ \eta_3 &= \frac{3}{8}D(3k)k^2\eta_L^3 \end{aligned} \tag{6}$$

For long perturbation wavelength ($\lambda \to \infty$), it is expected that Eq. (6) reduce to Eq. (5), or equivalently, one has $C(2k) \to A_T$, $D(3k) \to (4/3)(A_T^2 - 1/4)$, and $E(k) \to (1/4)(3A_T^2+1)$.

Since the transition between the fluid is no longer sharp (see Figure 2), we shall generate a reasonably defined interface in the fluid so that the growth of the amplitude of the perturbed interface can be tracked and analyzed by the Fourier analysis in the y direction. In the weakly nonlinear regime where the density $\rho(x,y,t)$ remains as a single value function along the x direction, a function $f(y,t)$ where the isodensity contour ρ_I can be uniquely defined. In the following, we will track the strong ($\rho_I = 1.2$ g/cm^3), medium ($\rho_I = 2.4$ g/cm^3), and weak ($\rho_I = 4.2$ g/cm^3) ablation surfaces, respectively, to investigate the preheating ablation effect on the growth of instability surfaces. Applying the Fourier analysis on the function $f(y,t)$, the absolute values of the Fourier coefficients $\{|a_n(t)|, n = 1, \cdots, N_y/2\}$ yield the amplitudes of the nth harmonic of the interface in time t. The linear growth rate γ can then be obtained by fitting the amplitude of the fundamental mode within the linear growth regime. Similarly, the coefficients $\{E(k), C(2k), D(3k)\}$ can be obtained by fitting η_1, η_2, η_3, respectively, within the weakly nonlinear growth regime. Section IV is devoted to the study of the temporal evolutions of the amplitudes of the first three harmonics.

IV. RESULTS AND DISCUSSIONS

The linear growth rate curves adopting our weak preheating model obtained from the strong ($\rho_I = 1.2$ g/cm^3), medium ($\rho_I = 2.4$ g/cm^3), and weak ($\rho_I = 4.2$ g/cm^3) ablation surfaces, respectively, are plotted in Figure 3. The results show that the linear growth rate does not increase with the decreasing perturbation wavelength all the time which is the case in the classical RTI.[51] The linear growth rate is strikingly reduced when the perturbation wavelength λ is less than the maximal growth wavelength $\lambda_{max} = 16.0$ μm, as shown in Figure 3. The data also indicates that the cutoff wavelength is $\lambda_c = 3.2$ μm using our weak preheating model. It should be pointed out that the three different isodensity ablation surfaces give the same linear growth rate

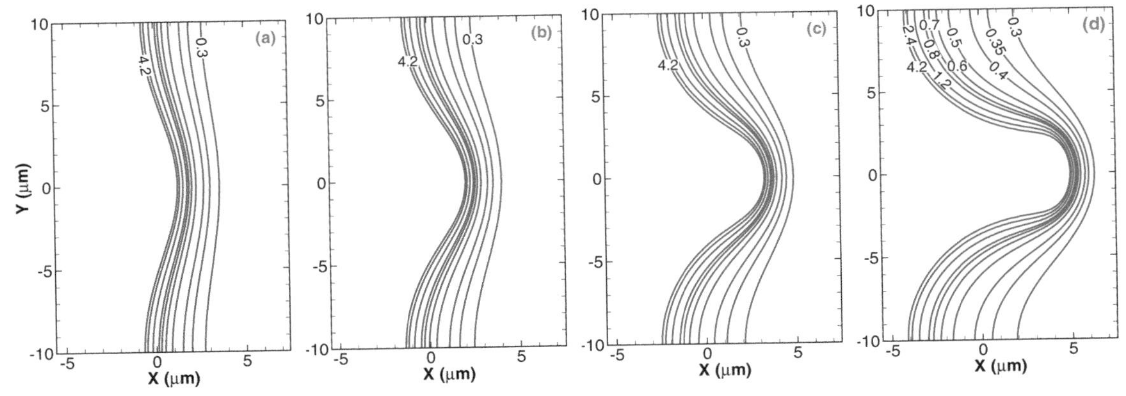

FIG. 2. (Color online) Isodensity contours for a 20 μm perturbation wavelength with an initial perturbation amplitude $\eta_0 = 0.00021$ μm at the time of (a) $t = 5.5$ ns, (b) $t = 6.0$ ns, (c) $t = 6.5$ ns, and (d) $t = 7.0$ ns. The isodensity contours of every figure from the right to the left are, respectively, $\rho_I = 0.3$, 0.35, 0.4, 0.5, 0.6, 0.7, 0.8, 1.2, 2.4, and 4.2 g/cm^3. It is found that the spike has a sharper gradient but the bubble has a shallower gradient.

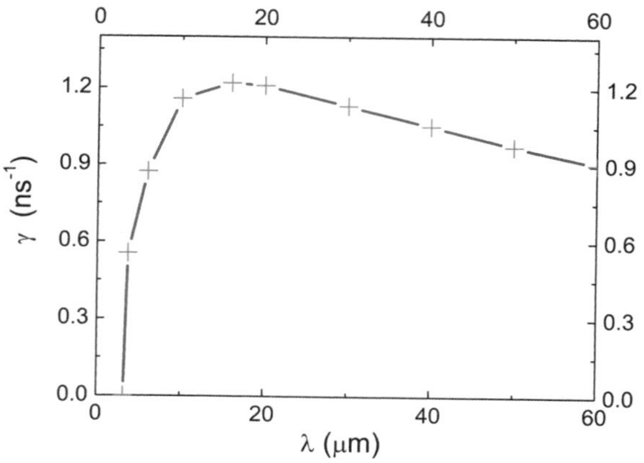

FIG. 3. (Color online) Linear growth rate curve obtained from the strong ($\rho_l = 1.2$ g/cm^3), medium ($\rho_l = 2.4$ g/cm^3), and weak ($\rho_l = 4.2$ g/cm^3) ablation surfaces, respectively. The cut-off wavelength is 3.2 μm and the maximal growth rate wavelength is 16.0 μm. The three different isodensity ablation surfaces show the same linear growth rate curve.

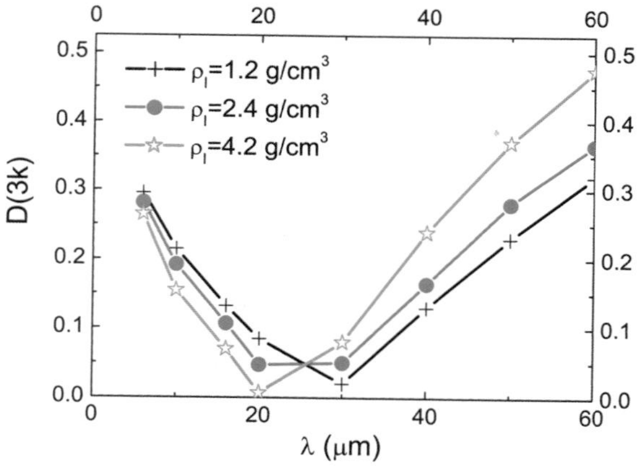

FIG. 5. (Color online) Third harmonic generation coefficients D(3k) obtained from the strong ($\rho_l = 1.2$ g/cm^3), medium ($\rho_l = 2.4$ g/cm^3), and weak ($\rho_l = 4.2$ g/cm^3) ablation surfaces, respectively.

curve. In other words, in the linear growth regime, the influence of preheating ablation on the growth of the instability surface is not pronounced. However, the preheating effect becomes more pronounced in the following weakly nonlinear regime as shown later.

We show the curves of second harmonic generation coefficient $C(2k)$ obtained from the strong ($\rho_l = 1.2$ g/cm^3), medium ($\rho_l = 2.4$ g/cm^3), and weak ($\rho_l = 4.2$ g/cm^3) ablation surfaces, respectively, in Figure 4. With the increasing perturbation wavelength λ, the $C(2k)$ tends to recover the classical value A_T. The ablation effect has a pronounced influence on the $C(2k)$ when $\lambda < 20$ μm, which is gradually weakened with the increasing λ. The $C(2k)$ is dependent significantly on the preheating ablation intensity, as shown in Figure 4, where the $C(2k)$ obtained from the weak ablation surface has the largest value for different perturbation wavelengths. That is to say, the ablation effect suppresses the second harmonic generation (stabilizes the second harmonic).

The curve of third harmonic generation coefficient $D(3k)$ obtained from the strong ($\rho_l = 1.2$ g/cm^3), medium ($\rho_l = 2.4$ g/cm^3), and weak ($\rho_l = 4.2$ g/cm^3) ablation surfaces, respectively, are illustrated in Figure 5. As seen in Figure 5, $D(3k)$ curves obtained from the three surface first decreases with the increasing perturbation wavelength and then increases with the increasing perturbation wavelength. This counterintuitive behavior cannot be predicted by the classical third-order estimation.[45,46] We had tried to understand this counterintuitive behavior in Refs. 36 and 51, which will not be further studied here. However, with the increasing perturbation wavelength λ, the $D(3k)$ has the tendency to recover its corresponding classical value $(4/3)(A_T^2 - 1/4)$. The inflection point wavelength [i.e., the corresponding perturbation wavelength with the smallest $D(3k)$], increases with the increasing preheating ablation intensity. For long perturbation wavelengths ($\lambda > 30$ μm), the strong ablation surface shows the smallest $D(3k)$, as can be seen in Figure 5. In other words, the ablation effect stabilizes the third harmonic's growth when the perturbation wavelengths are long.

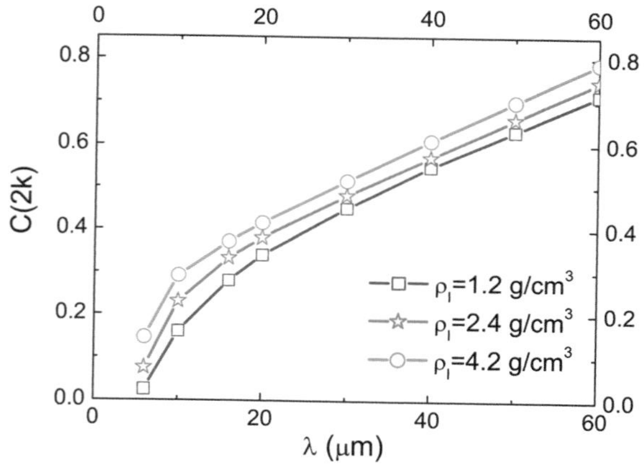

FIG. 4. (Color online) Second harmonic generation coefficients C(2k) obtained from the strong ($\rho_l = 1.2$ g/cm^3), medium ($\rho_l = 2.4$ g/cm^3), and weak ($\rho_l = 4.2$ g/cm^3) ablation surfaces, respectively.

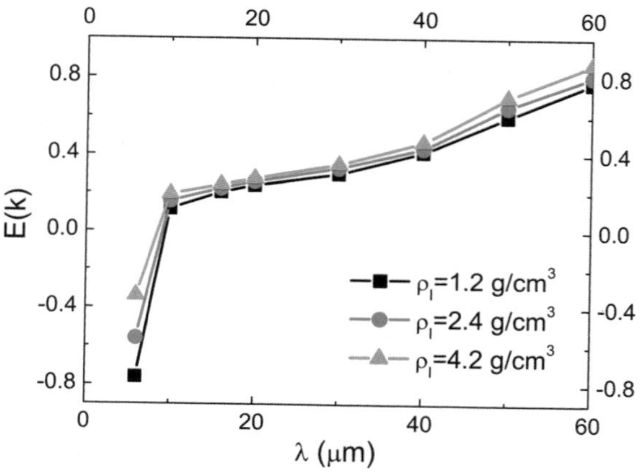

FIG. 6. (Color online) Coefficients of the third-order feedback to the fundamental mode E(k) obtained from the strong ($\rho_l = 1.2$ g/cm^3), medium ($\rho_l = 2.4$ g/cm^3), and weak ($\rho_l = 4.2$ g/cm^3) ablation surfaces, respectively.

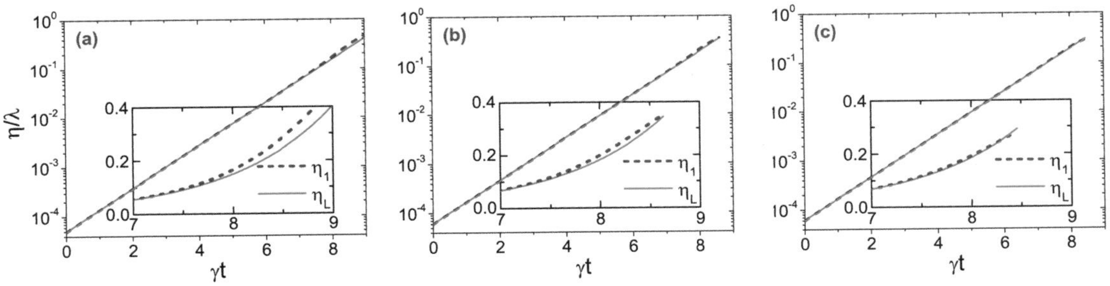

FIG. 7. (Color online) Temporal evolutions of the amplitudes of the fundamental mode of a 6.0 μm perturbation wavelength with an initial perturbation amplitude $\eta_0 = 0.00021$ μm obtained from the numerical simulation (η_1) and the linear growth theory (η_L) for (a) strong ($\rho_I = 1.2$ g/cm^3), (b) medium ($\rho_I = 2.4$ g/cm^3), and (c) weak ($\rho_I = 4.2$ g/cm^3) ablation surfaces, respectively. The third-order positive feedback to the fundamental mode becomes pronounced after $\gamma t > 7.0$.

Figure 6 shows the third-order feedback coefficient to the fundamental mode $E(k)$ curves obtained from the strong ($\rho_I = 1.2$ g/cm^3), medium ($\rho_I = 2.4$ g/cm^3), and weak ($\rho_I = 4.2$ g/cm^3) ablation surfaces, respectively. The most counterintuitive behavior is the third-order positive feedback to the fundamental mode (i.e., $E(k) < 0$) for the short perturbation wavelength around the cut-off wavelength λ_c, as shown in Figure 6. The phenomenon of third-order positive feedback to the fundamental is hard to understand since the fundamental mode cannot be saturated if one considers the mode coupling process within the third-order. Anyway, it is worth emphasizing that when the λ is long enough, the $E(k)$ also has the tendency to recover its corresponding classical prediction $(1/4)(3A_T^2 + 1)$. In order to further understand this interesting phenomenon, we will analyze the temporal evolutions of the amplitudes of the fundament mode and the corresponding isodensity contours at the duration of third-order positive feedback to the fundamental mode.

Figure 7 shows the temporal evolutions of the amplitudes of the fundamental mode obtained from the numerical simulation, and the linear growth theory for the strong ($\rho_I = 1.2$ g/cm^3), medium ($\rho_I = 2.4$ g/cm^3), and weak ($\rho_I = 4.2$ g/cm^3) ablation surfaces, respectively. The three different isodensity contours, at the period of the third-order positive feedback to the fundamental mode, are shown in Figure 8. The third-order positive feedback to the fundamental mode becomes pronounced after $\gamma t > 7.0$, as shown in Figure 7. The third-order positive feedback to the fundamental mode is more significant in the temporal evolutions of the amplitudes of the fundamental mode obtained from the strong ablation surface, which is in consistent with Figure 6 where the coefficients of the third-order feedback to the fundamental mode E(k) obtained from the strong ablation surface have the largest negative value. Consequently, we can conclude that it is the effect of the ablation that leads to this counterintuitive phenomenon of the third-order positive feedback to the fundamental mode. However, the growth of the fundamental mode can surely be saturated after these counterintuitive durations in our simulations. Thus, it is the nonlinearity higher than the third-order that eventually saturate the linear growth of the fundamental mode. Comparing the three different ablation surfaces in Figure 8, it is found that the strong ablation surface has a more blunt spike, which corresponds to the strong positive third-order feedback to the fundamental mode. Regarding the strong ablation surface, the generation and the growth of higher harmonics ($n \geq 2$) are effectively suppressed, and hence, the fundamental mode has an elongated linear growth phase. In other words, this shows that the preheating ablation, on the one hand, reduces the linear growth rate, but on the other hand, it also increases the duration of the linear growth stage.

Here, it is worthwhile to point out that there are always $0 < E(3k) < 1$ for $\lambda > 10$ μm, as shown in Figure 6. Consequently, the linear saturation amplitude mentioned above should be increased and therefore exceeds the classical prediction 0.1λ,[4,45,47,50] which is harmful for the ICF ignition since the large linear saturation amplitude can lead to the formation of jet-like spikes resulting in hot spot formation hard or even impossible. We plot the normalized linear saturation amplitude η_s/λ at different perturbation wavelengths obtained from the strong ($\rho_I = 1.2$ g/cm^3), medium ($\rho_I = 2.4$ g/cm^3), and weak ($\rho_I = 4.2$ g/cm^3) ablation surfaces, respectively, in Figure 9. As expected, the strong ablation surface shows the largest linear saturation amplitude, and the linear saturation amplitude decreases with the

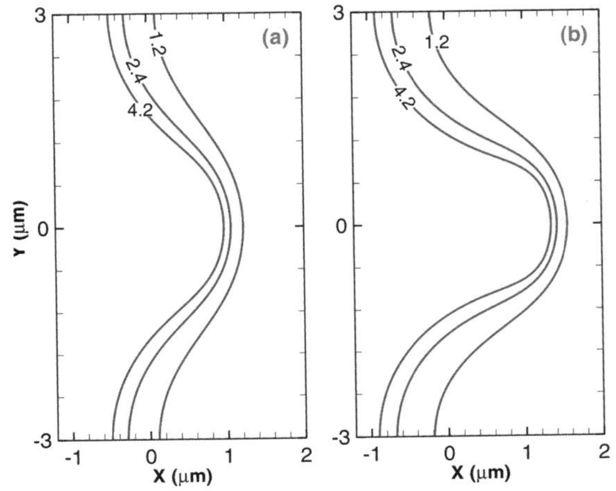

FIG. 8. (Color online) Isodensity contours for a 6.0 μm perturbation wavelength with an initial perturbation amplitude $\eta_0 = 0.00021$ μm at (a) $\gamma t = 7.8$ and (b) $\gamma t = 8.3$ within the period of third-order positive feedback to the fundamental mode. The isodensity contours from the right to the left are, respectively, $\rho_I = 1.2$, 2.4, and 4.2 g/cm^3.

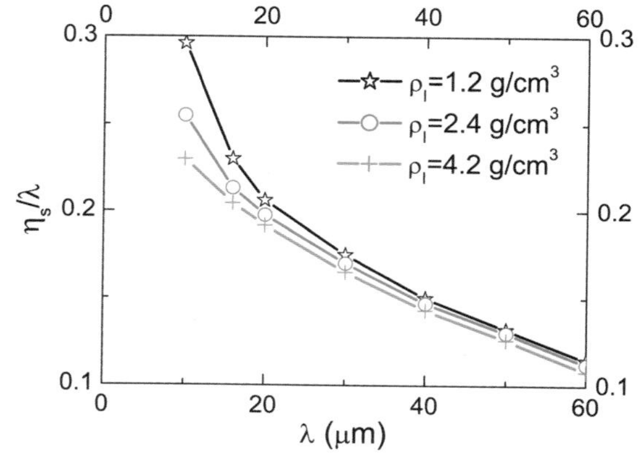

FIG. 9. (Color online) Normalized linear saturation amplitude η_s/λ at different perturbation wavelengths obtained from the strong ($\rho_l = 1.2$ g/cm^3), medium ($\rho_l = 2.4$ g/cm^3), and weak ($\rho_l = 4.2$ g/cm^3) ablation surfaces, respectively.

increasing perturbation wavelengths. One can understand this in the following way. The preheating ablation effects remarkably influence the short perturbation wavelength and the strong ablation surface.

In order to further understand the effects of the preheating ablation and the finite thickness of the ablation front, we show the mode coupling coefficients given by Sanz *et al.*[29] in Figure 10, where the index number of thermal conduction coefficient is $n = 2.5$ and the cut-off wavelength is $\lambda_c = 3.2$ μm. Note that Ganier[30] shows a similar mode coupling coefficients just like Figure 10. As can be seen in Figure 10, both the second harmonic generation coefficient and the coefficient of third-order feedback to the fundamental mode increase with the increasing perturbation wavelength, while the third harmonic generation coefficient first decreases with the increasing perturbation wavelength and then increases with it. Our simulation results are in generally consistent with these trends, but C(2k) and D(3k) are always positive in our simulations where the effect of the finite thickness of ablation front has been well considered.

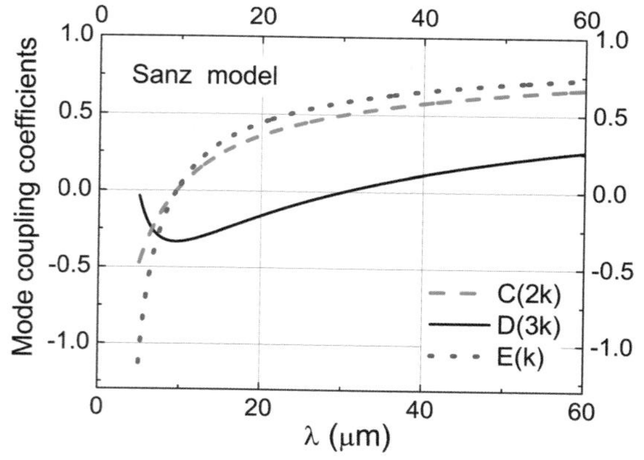

FIG. 10. (Color online) Mode coupling coefficients C(2k), D(3k), and E(k) given by Sanz *et al.*[29] with the index number of the coefficient of thermal conduction $n = 2.5$ and the cut-off wavelength $\lambda_c = 3.2$ μm.

In summary, for long perturbation wavelengths, results obtained in this study are in general agreement with the recent weakly nonlinear theories and simulations of the ARTI by Sanz and Ganier.[29,30] However, large differences appear as perturbation wavelength λ approach the cut-off wavelength λ_c The mode coupling coefficients, $\{E(k), C(2k), D(3k)\}$, given by Refs. 29 and 30 can be negative. Obviously, it is nonphysical in practical system where the effects of finite thickness of ablation front should be fully considered. In the present work, contrarily, the $C(2k)$ and the $D(3k)$ are always positive, except for the $E(k)$ near the cut-off wavelength where there is a third-order positive feedback to the fundamental mode. Moreover, it is found that the effects of finite thickness of ablation front play a key role of stabilization in the ARTI growth not only in the linear regime but also in the weakly nonlinear regime, especially for the short perturbation wavelength. Note that the effects of finite thickness of ablation front were not sufficiently addressed in the work by Sanz or Garnier, but are studied in some detail here.

V. CONCLUSION

In this article, we have investigated the structure of the ablation surface in the ablative Rayleigh-Taylor instability in the presence of preheating within the weakly nonlinear regime. It is found that the spikes have sharper gradients, but the bubbles have shallower gradients. The density gradient effect (the ablation front width effect) on ARTI is obtained by analyzing the three different isodensity contours of the ablation surfaces.

The simulation results show that the preheating ablation effect crucially influences the mode coupling process in the ARTI. First, the linear growth rate is reduced, especially when the perturbation wavelength λ is short, and a cut-off perturbation wavelength λ_c appears when the perturbation wavelength λ is sufficiently short where no higher harmonics exists when $\lambda < \lambda_c$. Second, the generations and growths of the second and third harmonics are suppressed and reduced. Third, the ablation effect suppresses the third-order negative feedback to the fundamental mode, except for the short perturbation wavelengths near the λ_c where the feedback can be positive, and hence, leads to a larger linear saturation amplitude which can obviously exceed its corresponding classical prediction 0.1λ. Moreover, the simulation results indicate that there is a positive third-order feedback to the fundamental mode for perturbations near the cut-off wavelength. Furthermore, our numerical results obtained are in general agreement with the recent weakly nonlinear theories and simulations.[29–31]

In direct-driven experiments and simulations in inertial confinement fusion, the jet-like spikes are formed.[52–59] In astrophysics, the astrophysical jets can remain well collimated over propagation distances exceeding 10 jet diameters or more.[60] Our numerical results show that when the preheating ablation effect is considered, the growths of the second and the third harmonics of ARTI are suppressed, while the linear saturation amplitude of the fundamental mode is increased, which is expected to improve the understanding of the stabilization factors for the astrophysical jet and of the formation mechanisms for the jet-like spikes in ICF.

ACKNOWLEDGMENTS

The author (L.F.W.) would like to thank the anonymous referee for suggestions that have improved the paper. This work was supported by the National Basic Research Program of China (Grant Nos. 2007CB815101 and 2007CB815105) and the National Natural Science Foundation of China (Grant Nos. 10935003, 10905006, 11075024, 10974022, and 10835003), the National High-Tech ICF Committee. L.F.W. would like to acknowledge the support from China Postdoctoral Science Foundation.

[1]L. Rayleigh, *Scientific Papers II* (Cambridge University Press, Cambridge, England, 1900).
[2]G. I. Taylor, Proc. R. Soc. London, Ser. A **201**, 192 (1950).
[3]J. H. Nuckolls, J. Wood, A Thiessen, and G. Zimmerman, Nature **239**, 139 (1972).
[4]J. D. Lindl, P. Amendt, R. L. Berger, S. G. Glendinning, S. H. Glenzer, S. W. Haan, R. L. Kauffman, O. L. Landen, and L. J. Suter, Phys. Plasmas **11**, 339 (2004).
[5]S. Atzeni and J. Meyer-ter-Vehn, *The Physics of Inertial Fusion: Beam Plasma Interaction Hydrodynamics, Hot Dense Mater* (Oxford University, Oxford, 2004).
[6]X. T. He and W. Y. Zhang, Eur. Phys. J. D **44**, 227 (2007).
[7]B. A. Remington, D. Arnett, R. P. Drake, and H. Takabe, Science **284**, 1488 (1999).
[8]B. A. Remington, R. P. Drake, and D. D. Ryutov, Rev. Mod. Phys. **78**, 755 (2006).
[9]K. Nomoto, K. Iwanoto, and N. Kishimoto, Science **276**, 1378 (1997).
[10]V. N. Gamezo, A. M. Khokhlov, E. S. Oran, A. Y. Chtchelkanova, and R. O. Roseenberg, Science **299**, 3 (2003).
[11]T. Ebisuzaki, T. Shigeyama, and K. Nomoto, Astrophys. J. **344**, L65 (1989).
[12]I. Hachisu, T. Matsuda, K. Nomoto, and T. Shigeyama, Astrophys. J. **358**, L57 (1990).
[13]K. O. Mikaelian, Phys. Rev. Lett. **48**(19), 1365 (1982).
[14]K. O. Mikaelian, Phys. Rev. E **54**(4), 3676 (1996).
[15]A. B. Bud'ko, and M. A. Liberman, Phys. Fluids B **4**, 3499 (1992).
[16]L. F. Wang, W. H. Ye, and Y. J. Li, Phys. Plasmas **17**, 042103 (2010).
[17]S. Bodner, Phys. Rev. Lett. **33**, 761 (1974).
[18]H. Takabe, K. Mima, L. Montierth, and R. L. Morse, Phys. Fluids **28**, 3676 (1985).
[19]M. Tabak, D. H. Munro, and J. D. Lindl, Phys. Fluids B **2**, 1007 (1990).
[20]H. Azechi, T. Sakaiya, S. Fujioka, Y. Tamari, K. Otani, K. Shigemori, M. Nakai, H. Shiraga, N. Miyanaga, and K. Mima, Phys. Rev. Lett. **98**, 045002 (2007).
[21]S. G. Glendinning, S. N. Dixit, B. A. Hammel, D. H. Kalantar, M. H. Key, J. D. Kilkenny, J. P. Knauer, D. M. Pennington, B. A. Remington, R. J. Wallace, and S. V. Weber, Phys. Rev. Lett. **78**, 3318 (1997).
[22]K. Shigemori, H. Azechi, M. Nakai, M. Honda, K. Meguro, N. Miyanaga, H. Takabe, and K. Mima, Phys. Rev. Lett. **78**, 250 (1997).
[23]T. Sakaiya, H. Azechi, M. Matsuoka, N. Izumi, M. Nakai, K. Shigemori, H. Shiraga, A. Sunahara, H. Takabe, and T. Yamanaka, Phys. Rev. Lett. **88**, 145003 (2002).
[24]W. H. Ye, W. Y. Zhang, and X. T. He, Phys. Rev. E **65**, 57401 (2002).
[25]J. Sanz, Phys. Rev. Lett. **73**(20), 2700 (1994).
[26]A. R. Piriz, J. Sanz, and L. F. Ibañez, Phys. Plasmas **4**(4), 1117 (1997).
[27]V. N. Goncharov, R. Betti, R. L. McCrory, P. Sorotokin, and C. P. Verdon, Phys. Plasmas **3**(4), 1402 (1996).
[28]R. Betti, V. N. Goncharov, R. L. McCrory, and C. P. Verdon, Phys. Plasmas **5**(5), 1446 (1998).
[29]J. Sanz, J. Ramírez, R. Ramis, R. Betti, and R. P. J. Town, Phys. Rev. Lett. **89**, 195002 (2002).
[30]J. Garnier, P.-A. Raviart, C. Cherfils-Clérouin, and L. Masse, Phys. Rev. Lett. **90**, 185003 (2003).
[31]J. Garnier and L. Masse, Phys. Plasmas **12**, 062707 (2005).
[32]S. Fujioka, H. Shiraga, M. Nishikino, K. Shigemori, A. Sunahara, M. Nakai, H. Azechi, K. Nishihara, and T. Yamanaka, Phys. Plasmas **10**, 4784 (2003).
[33]K. Otani, K. Shigemori, T. Kadono, Y. Hironaka, M. Nakai, H. Shiraga, H. Azechi, K. Mima, N. Ozaki, T. Kimura, K. Miyanishi, R. Kodama, T. Sakaiya, and A. Sunahara, Phys. Plasmas **17**, 032702 (2010).
[34]R. E. Olson, R. J. Leeper, A. Nobile, and J. A. Oertel, Phys. Rev. Lett. **91**(23), 235002 (2003).
[35]V. A. Smalyuk, D. Shvarts, R. Betti, J. A. Delettrez, D. H. Edgell, V. Yu. Glebov, V. N. Goncharov, R. L. McCrory, D. D. Meyerhofer, P. B. Radha, S. P. Regan, T. C. Sangster, W. Seka, S. Skupsky, C. Stoeckl, and B. Yaakobi, Phys. Rev. Lett. **100**, 185005 (2008).
[36]L. F. Wang, W. H. Ye, Z. M. Sheng, W. S. Don, Y. J. Li, and X. T. He, Phys. Plasmas **17**, 122706 (2010).
[37]W. H. Ye, L. F. Wang, and X. T. He, Phys. Plasmas **17**, 122704 (2010).
[38]J. Li, B. Zhao, H. Li, and J. Zheng, Plasma Phys. Controlled Fusion **52**, 085008 (2010).
[39]D. G. Colombant, W. M. Manheimer, and M. Busquet, Phys. Plasmas **12**, 072702 (2005).
[40]W. Manheimer, D. Colombant, and V. Goncharov, Phys. Plasma **15**, 083103 (2008).
[41]W. H. Ye, L. F. Wang, and X. T. He, Chin. Phys. Lett. **27**, 125203 (2010).
[42]D. Balsara and C. W. Shu, J. Comput. Phys. **160**, 405 (2000).
[43]G. S. Jiang and C. W. Shu, J. Comput. Phys. **126**, 202 (1996).
[44]R. Borges, M. Carmona, B. Costa, and W. S. Don, J. Comput. Phys. **227**, 3191 (2008).
[45]J. W. Jacobs and I. Catton, J. Fluid Mech. **187**, 329 (1988).
[46]S. W. Haan, Phys. Fluids B **3**, 2349 (1991).
[47]L. F. Wang, W. H. Ye, and Y. J. Li, Chin. Phys. Lett. **27**, 025203 (2010).
[48]B. A. Remington, S. V. Weber, M. M. Marinak, S. W. Haan, J. D. Kilkenny, R. J. Wallace, and G. Dimonte, Phys. Plasmas **2**(1), 241 (1995).
[49]V. A. Smalyuk, V. N. Goncharov, T. R. Boehly, J. P. Knauer, D. D. Meyerhofer, and T. C. Sangster, Phys. Plasmas **11**(11), 5083 (2004).
[50]D. Layzer, Astrophys. J. **122**, 1 (1955).
[51]L. F. Wang, W. H. Ye, and Y. J. Li, Phys. Plasmas **17**, 052305 (2010).
[52]D. K. Bradley, D. G. Braun, S. G. Glendinning, M. J. Edwards, J. L. Milovich, C. M. Sorce, G. W. Collins, S. W. Haan, R. H. Page, and R. J. Wallace, Phys. Plasmas **14**, 056313 (2007).
[53]B. A. Remington, S. W. Haan, S. G. Glendinning, J. D. Kilkenny, D. H. Munro, and R. J. Wallace, Phys. Fluids B **4**, 967 (1992).
[54]B. A. Remington, R. P. Drake, H. Takabe, and D. Arnett, Phys. Plasmas **7**, 1641 (2000).
[55]J. O. Kane, H. F. Robey, B. A. Remington, R. P. Drake, J. Knauer, D. D. Ryutov, H. Louis, R. Teyssier, O. Hurricane, D. Arnett, R. Rosner, and A. Calder, Phys. Rev. E **63**, 055401 (2001).
[56]R. J. Mason, D. E. Hollowell, G. T. Schappert, and S. H. Batha, Phys. Plasmas **8**, 2338 (2001).
[57]S. Nakai, T. Yamanaka, Y. Izawa, Y. Kato, K. Nishihara, M. Nakatsuka, T. Sasaki, H. Takabe *et al.*, in *Proceedings of the 15th International Conference on Plasma Physics and Controlled Fusion Research 1994, Seville, Spain, 28 September-1 October 1994* (IAEA, Vienna, 1995), Vol. 3, pp. 3–12.
[58]W. H. Ye and X. T. He, in *Proceeding of the Tenth International Workshop on the Physics of Compressible Turbulent Mixing*, edited by M. Legrand and M. Vandenboomgaerde (Paris, France, 2006).
[59]W. H. Ye, X. T. He, W. Y. Zhang, and M. Y. Yu, EPL **96**, 35002 (2011).
[60]B. Reipurth and J. Bally, Annu. Rev. Astron. Astrophys. **39**, 403 (2001).

PHYSICS OF PLASMAS **19**, 100701 (2012)

Formation of jet-like spikes from the ablative Rayleigh-Taylor instability

L. F. Wang (王立锋),[1,2,a)] W. H. Ye (叶文华),[1,2,b)] X. T. He (贺贤土),[1,2,c)] W. Y. Zhang (张维岩),[2] Z. M. Sheng (盛政明),[3,4] and M. Y. Yu (郁明阳)[5,6]
[1]*HEDPS and CAPT, Peking University, Beijing 100871, China*
[2]*Institute of Applied Physics and Computational Mathematics, Beijing 100088, China*
[3]*Key Laboratory for Laser Plasmas (MoE) and Department of Physics, Shanghai Jiaotong University, Shanghai 200240, China*
[4]*Beijing National Laboratory of Condensed Matter Physics, Institute of Physics, CAS, Beijing 100190, China*
[5]*Institute for Fusion Theory and Simulation, Zhejiang University, Hangzhou 310027, China*
[6]*Institute for Theoretical Physics I, Ruhr University, Bochum D-44780, Germany*

(Received 27 May 2012; accepted 2 October 2012; published online 11 October 2012)

The mechanism of jet-like spike formation from the ablative Rayleigh-Taylor instability (ARTI) in the presence of preheating is reported. It is found that the preheating plays an essential role in the formation of the jet-like spikes. In the early stage, the preheating significantly increases the plasma density gradient, which can reduce the linear growth of ARTI and suppress its harmonics. In the middle stage, the preheating can markedly increase the vorticity convection and effectively reduce the vorticity intensity resulting in a broadened velocity shear layer near the spikes. Then the growth of ablative Kelvin-Helmholtz instability is dramatically suppressed and the ARTI remains dominant. In the late stage, nonlinear bubble acceleration further elongates the bubble-spike amplitude and eventually leads to the formation of jet-like spikes. © *2012 American Institute of Physics*. [http://dx.doi.org/10.1063/1.4759161]

Control of the ablative Rayleigh-Taylor instability (ARTI) is crucially important for the success of inertial confinement fusion (ICF).[1] This inherent physical instability can limit the implosion velocity and the size of the hot spot, and in some cases even break up the implosion shell or make hot spot formation impossible, resulting in the auto-ignition failure. So far, nonlinear evolution of ARTI is still not very clear, especially the formation of jet-like structures, which are widely observed in high energy density hydrodynamic instability experiments and simulations in ICF research.[2–4]

It is well documented that in the linear growth regime the ablation stabilizes the ARTI and a cutoff wavelength appears when the perturbation wavelength is sufficiently short.[5–10] The linear growth rate of Bodner-Takabe-Lindl $\gamma_{BTL} = \alpha\sqrt{kg/(1+kL_m)} - \beta k v_a$ approximately fits the two-dimensional numerical simulations and experiments, where k is the wave number, g is the acceleration, L_m is the minimum scale length of density gradient at the ablation surface, and v_a is the ablation velocity.[6] The parameters α and β depends basically on the flow parameters, and usually $\alpha = 0.9 - 0.95$ and $\beta = 1 - 3$. A self-consistent theory[7] shows that the linear growth rate of the ARTI takes the form

$$\gamma = \sqrt{\hat{A}_T g k + \delta^2 k^4 L_0^2 v_a^2 + \left(\omega^2 - \frac{1}{\xi_l}\right) k^2 v_a^2} - \delta k^2 L_0 v_a - \hat{\beta} k v_a,$$

where the parameters are given by Eq. (8) in Ref. 7, which is an asymptotic formula valid for arbitrary Froude numbers. The self-consistent theory contains the effects of the dynamic overpressure due to the thermal conduction, of the damping due to the fire polishing and the vorticity convection, and of the density gradients of the ablation surface. The physical explanation of the stabilization of the ARTI has been described clearly by Piriz *et al.*,[8] and references therein. The recent experiments[9,10] suggest that the preheating by energetic electrons and the nonlocal electron heat transport plays an important role in the suppression of ARTI at the ablation front. The observed density profiles in the direct-drive experiments[11] and simulations[9,10] indicate that the L_m is significantly improved by the preheating.

Without considering radiation, an ideal gas equation of state is used in our numerical simulation.[12] For the heat flow, a preheating mode[5] $\kappa(T) = \kappa_{SH}[1 + f(T)]$ is introduced in our numerical simulation to include the preheating effects, where κ_{SH} is the Spitzer-Härm electron heat conduction and $f(T)$ interprets the preheating tongue effect in the cold plasma ahead of the ablation front. For temperatures greater than 0.10 MK, it has been shown that the ARTI growth rates, the fluid density profiles, and temperature profiles at the ablation front obtained from the choice $f(T) = aT^{-1} + bT^{-3/2}$, where T is in the unit of MK, a and b are adjustable parameters, agree well with the experiments.[5] In this paper, we choose a preheating profile with $a = 8.6$ and $b = 1.6$. It should be mentioned here that the form of the preheating function $f(T)$ is consistent with that of Fokker-Planck calculations,[13] and the free parameters a and b are determined empirically by comparing the simulation results with that from the corresponding experiment, theory, and/or simulation. Other heat-conduction models with preheating effects can also be used.

We obtain the steady state flow fields, as shown in the left figure of Fig. 1, by integrating the governing equations from

[a)]Electronic mail: lif_wang@pku.edu.cn.
[b)]Electronic mail: ye_wenhua@iapcm.ac.cn.
[c)]Electronic mail: xthe@iapcm.ac.cn.

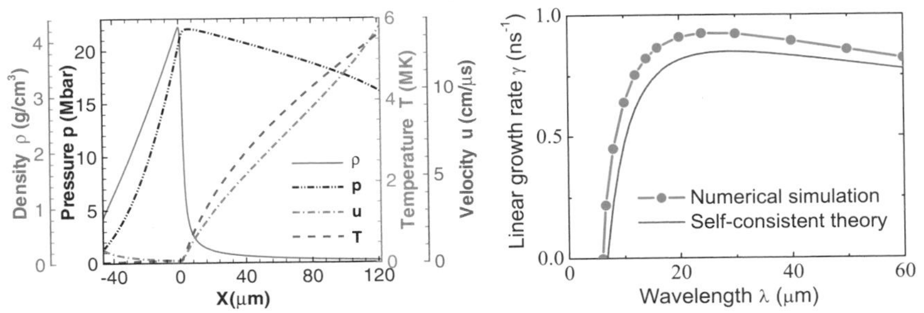

FIG. 1. Left panel: distributions of the unperturbed one-dimensional basic flows with $L_m = 1.83\,\mu$m. Right panel: linear growth rate curves from the numerical simulation and the self-consistent theory.

the peak density to the isothermal sonic points on both sides of the ablation front. The averaged ablation parameters, i.e. the peak density $\rho_a = 4.25\,\mathrm{g/cm^3}$, the ablation velocity $v_a = 1.58\,\mu\mathrm{m/ns}$, the temperature at the peak density $T_a = 8.37 \times 10^{-2}\,\mathrm{MK}$, and the acceleration $g_a = 16.57\,\mu\mathrm{m/ns^2}$, are adopted for the integration. The four parameters are obtained from a one-dimensional (1D) calculation. In 1D simulation, a 200 μm thick CH foil target with a density of $1.0\,\mathrm{g/cm^3}$ is irradiated by a 0.35 μm laser beam of intensity $I = I_0(t/4\,\mathrm{ns})$ for $t < 4\,\mathrm{ns}$ and $I = I_0 = 10^{14}\,\mathrm{W/cm^2}$ for $t \geq 4\,\mathrm{ns}$. The 1D quasistationary state appears at $t \geq 7\,\mathrm{ns}$.

Since the transition between the fluid is no longer sharp, we shall generate a reasonably defined interface in the fluid so that the perturbed interface can be tracked and analyzed by the Fourier analysis in the y direction. For this purpose, we shall define the interface as the location where the density gradient is the largest. The right figure of Fig. 1 displays that the linear growth rate γ has a maximum at $\lambda_m \simeq 24\,\mu$m and a cutoff wavelength at $\lambda_c \simeq 6\,\mu$m. It can be seen that there is agreement between the numerical simulation and the self-consistent theory [Eq. (8) in Ref. 7].

Two-dimensional simulation results indicated that the nonlinear evolution of the ARTI is substantially dependent on the preheating intensity and the perturbation wavelength. The preheating intensity directly influences the profile of the linear growth rate curve of the ARTI. On the one hand, for the weak preheating case which has a sharp peak in its linear growth rate curve, the nonlinear evolution of the single-mode ARTI exhibits three regions with different characteristics related to the number of excited Fourier modes. Regarding region I (i.e. short-wavelength perturbations near the cutoff), the growth of a single-mode perturbation has a periodic behavior (i.e. growth-evaporation-growth). Medium-wavelength perturbations near the peak (i.e. region II) evolve into jet-like structure. In region III, the spike of a long-wavelength perturbation (typically $\lambda \sim 2\lambda_m$) can be ruptured due to the excessive growth of its harmonics around the bubble vertex. On the other hand, there is not a sharp maximum in the linear growth rate curve of the strong preheating case. Instead, the long-wavelength perturbations grow into elongated spike and would not be ruptured. The wavelength dependent nonlinear evolution of ARTI including the preheating intensity effect will be reported elsewhere.

This research focus on the formation of jet-like spikes with the present preheating profile. Typically, we study the evolution of a 24 μm-wavelength perturbation in the following. Fig. 2 illustrates the temporal evolution of density, vorticity, absolute value of heat flux, and pressure for a 24 μm-wavelength perturbation at $\gamma t = 6.67$, 8.34, and 10.0, where the vorticity is defined as $\omega = (\partial u_y/\partial x - \partial u_x/\partial y)$ and the absolute value of heat flux is defined as $|Q| = \kappa\sqrt{(\partial T/\partial x)^2 + (\partial T/\partial y)^2}$.

Note that the density contours are smooth, with no characteristic Kelvin-Helmholtz instability (KHI) mushroom structures at the spike heads, as shown in Fig. 2(a), while the KHI mushroom structures are fully developed in the classical RTI (CRTI).[14] Accordingly, with and without any noticeable

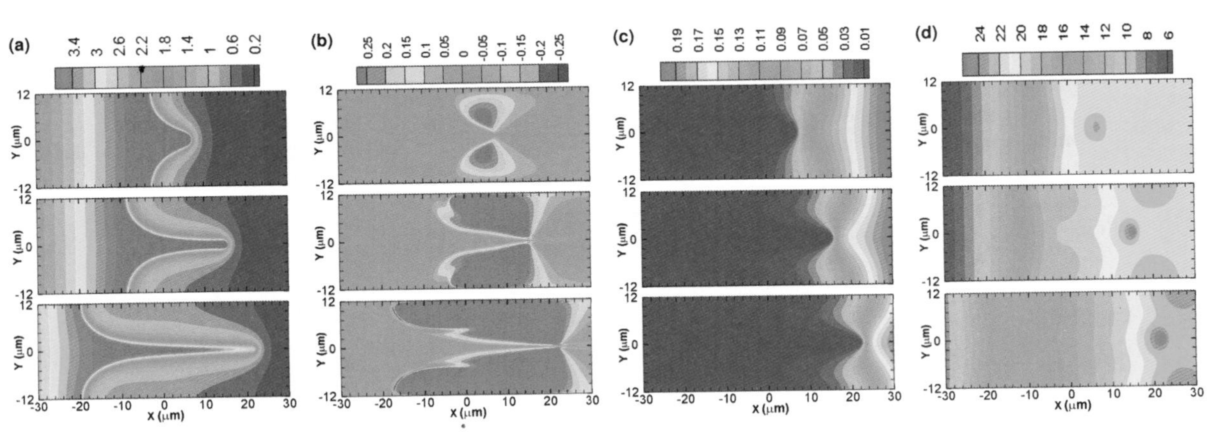

FIG. 2. Density (a), vorticity (b), absolute value of heat flux (c), and pressure (d) contours for a 24 μm-wavelength perturbation with initial 0.007 μm-amplitude at $\gamma t = 6.67$, 8.34, and 10.0, respectively. The units of density, vorticity, heat flux, and pressure are $\mathrm{g/cm^3}$, $10\,\mathrm{ns^{-1}}$, $10^{14}\,\mathrm{W/cm^2}$, and Mbar, respectively.

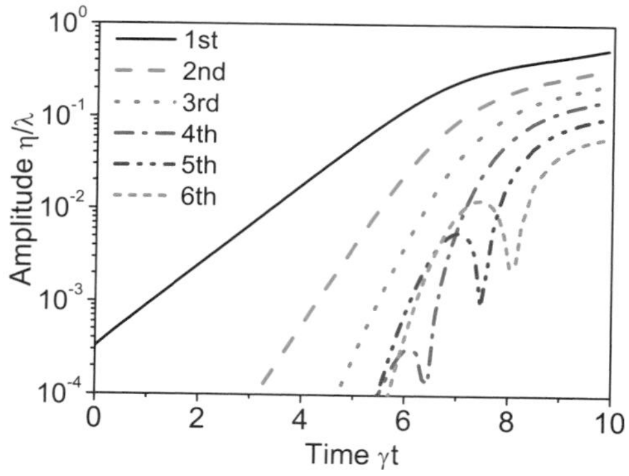

FIG. 3. Temporal evolution of the normalized amplitudes of the first six harmonics for a 24 μm-wavelength perturbation with an initial 0.007 μm-amplitude.

KHI mushroom structures is the most prominent difference between the nonlinear evolution of the ARTI and the CRTI. Furthermore, Fig. 2(c) shows that the distribution of the heat flux around the spike tip features a cone which encloses the spike tip. In addition, we note that the evolution of jet-like spike in ARTI in the presence of preheating consists of three stages, as shall be discussed below.

The temporal evolution of the amplitudes of the first six harmonics for the $\lambda_m = 24\,\mu$m-wavelength perturbation is plotted in Fig. 3. As can be seen in Fig. 3 at $\gamma t = 6$, the amplitudes of the fundamental mode, the second harmonic, and the third harmonic are $\eta_1 = 0.124\lambda_m$, $\eta_2 = 0.025\lambda_m$, and $\eta_3 = 0.004\lambda_m$, respectively. However, within the framework of third-order expansion theory,[12] when the amplitude of the fundamental mode is $\eta_1 \simeq 0.1\lambda$ which is its saturation amplitude, there are $\eta_2 \simeq 0.032\lambda$ and $\eta_3 \simeq 0.014\lambda$, respectively. Comparison with the classical theory, the growth of the higher harmonics is strongly suppressed but the fundamental mode has an extended linear (nonlinear) growth regime, as shown obviously in Fig. 3. Therefore, it can be concluded that in the early stages the fundamental mode dominates over the higher harmonics. In fact, we have studied the interface width[15] and the thermal conduction[12] effects on the weakly nonlinear RTI growth. It was found that the effect of the thickness of the ablation front from the preheating can effectively suppress the generation of the higher harmonics, and hence increase the saturation amplitude of the fundamental mode. It has indicated that the preheating would lead to a reduction of the effective Atwood number and the density ahead the ablation front with the consequence of an increasing of the ablation velocity, resulting in the decrease of the linear growth rate of the ARTI.[5] With regard to the formation of the jet-like spikes, the effect of preheating in the early stage is primarily to increase the ablation surface width, and hence, damp the growth of the high harmonics.

With the growth of the instability, shear flows are formed around the spike tip. However, in the presence of preheating the ablative KHI (AKHI)[16] is fully suppressed, with no mushroom structure, as shown in Fig. 2(a). It is found that the vorticity intensity near the spike interface is significantly reduced but the vorticity convection inside the bubble is strengthened in the ARTI, as compared with the CRTI. Now let us consider the growth of AKHI. Fig. 4 shows the y direction profiles of the density and the shear velocity (velocity in the x direction) around the spike tips at $\gamma t = 6.67$ and 8.5. As can be seen evidently in Fig. 4, the density has a sharp profile while the shear velocity has a gentle profile. The sharp density profile is good for the RTI growth but bad for the KHI growth.[17] On the other hand, the gentle shear velocity profile is bad for the KHI growth. Consequently, with the present profiles of density and shear velocity, it is difficult for the spike tip to generate the KHI and form mushroom structures. In addition, it is found that the mass ablation can further suppress the KHI growth.[16] Comparison with the CRTI and the ARTI, and with the ARTI in the presence of different preheating,[14] it is found that preheating can effectively suppress the KHI growth and mushroom formation at the spike tips. That is, in the middle stage, the AKHI is dramatically suppressed by the ablation in the present of preheating.

When the bubble-spike amplitude is comparable to its perturbation wavelength, the instability enters the highly nonlinear regime. Fig. 5 depicts the temporal evolution of the normalized amplitude η/λ_m, normalized acceleration a/g_a, and normalized pressure gradient $(\Delta p/\Delta x)/(\Delta p_0/\Delta x_0)$ of the spike and the bubble versus the time for 24 μm-wavelength perturbation. Here, $x_{ref} = -30\,\mu$m is the reference point where the pressure is $p_{ref} = 5.23$ Mbar at $t = 0$ for calculating the pressure gradient, and $\Delta p_0/\Delta x_0 = 0.51$ Mbar/μm is the pressure gradient of the unperturbed interface.

In the CRTI, the spike amplitude is always greater than the bubble's while the bubble amplitude can surpass the spike's in the ARTI. It is well known that there is a spike (bubble) acceleration (deceleration) stage in the CRTI, while conversely there is a bubble (spike) acceleration (deceleration) phase in the ARTI as shown in Fig. 5. It is found that

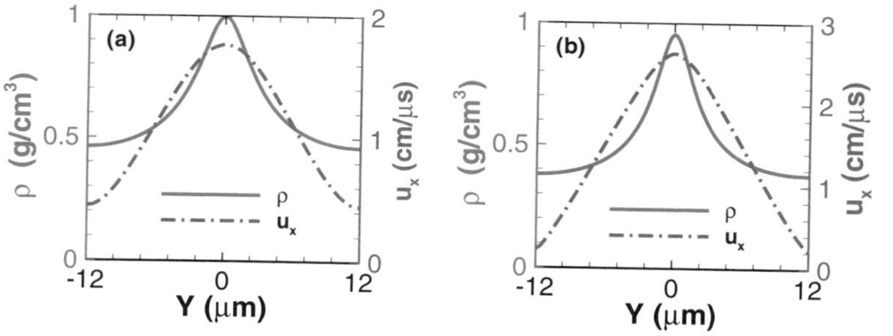

FIG. 4. y-direction profiles of the density and the shear velocity (velocity in the x-direction) at spike tips for a 24 μm-wavelength perturbation with an initial 0.007 μm-amplitude. (a) $\gamma t = 6.67$, $x = 8.5\,\mu$m and (b) $\gamma t = 8.34$, $x = 16.4\,\mu$m.

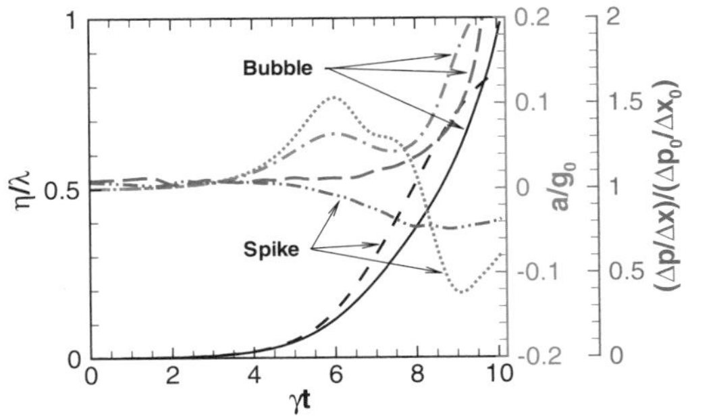

FIG. 5. Normalized amplitudes η/λ, normalized accelerations a/g_a, and normalized pressure gradients $(\Delta p/\Delta x)/(\Delta p_0/\Delta x_0)$ for the spike and the bubble versus the time for a 24 μm-wavelength perturbation with an initial 0.007 μm-amplitude. Solid, dashed dotted, and long dashed lines represent the amplitude, acceleration, and pressure gradient, respectively, for bubbles. Dashed, dotted, and dashed dot dotted lines correspond the amplitude, acceleration, and pressure gradient, respectively, for spike.

the overpressure effect at the spike tip and the vorticity accumulation inside the bubble that can lead to, respectively, the spike deceleration and bubble acceleration. As mentioned above, the ARTI is stabilized by the dynamic overpressure[7,8] which leads to a restoring force (linear with the perturbation amplitude), generated by steepening (flattening) of the temperature gradient at the perturbation peaks (valleys) of the ablation front. One can expect this effect can be strengthened in the nonlinear regime. Since in the nonlinear regime the spike tip becomes closer to the hot laser light absorption zone, there not only ∇T but also the heat conduction coefficient $\kappa(T)$ are increased. As can be seen in Fig. 2(d), the maximum pressure around the spike tip is about 22.45, 24.28, and 25.11 Mbar for $\gamma t = 6.67$, 8.34, and 10.0, respectively, while the pressure at the unperturbed interface is only 22.05 Mbar, as shown in Fig. 1(b). It can be seen from Fig. 2(b) that the vortex climbs towards the bubble vertex, while it does not occur in the CRTI. The phenomenon of the vorticity accumulation inside the bubble has been reported by Betti and Sanz[18] to explain the bubble acceleration in the highly nonlinear ARTI. Here, we consider the equivalent force (pressure gradients) for the bubble and the spike. The normalized pressure gradients $(\Delta p/\Delta x)/(\Delta p_0/\Delta x_0)$ for the bubble increases with the time while that for the spike decreases with the time, as shown in Fig. 5. In the late stage, the spike velocity is decreased but the bubble acceleration can cover this spike deceleration. Moreover, the combined effect of spike deceleration and bubble acceleration is increasing the bubble-spike amplitude as shown in Fig. 5, and finally results in the formation of jet-like spikes.

In conclusion, we have presented a mechanism of jet-like spike formation in the ARTI via numerical simulation. It is demonstrated that preheating plays an essential role in the formation of jet-like spikes. In the early stage, the preheating contributes to increase the density gradient stabilization of RTI. Thus, the fundamental mode dominates over other high harmonics. In the middle stage, due to the strengthened vorticity convection and the reduced intensity of vorticity, the velocity shear layer around the spike is broadened resulting in the stabilization of the KHI, and the RTI consequently dominates over the KHI. Accordingly, there are no clear mushroom structures at the spike tips. In other words, the suppression of the KHI in the middle stage is a necessary condition for the formation of the jet-like spikes. In the late stage, the nonlinear bubble acceleration owing to the vorticity accumulation towards the bubble vertex further elongates the bubble-spike amplitude and finally leads to formation of the jet-like spikes in the ARTI with preheating.

The author (L.F.W) would like to thank the three anonymous referees for suggestions that have greatly improved the paper. This work was supported by the National Natural Science Foundation of China (Grant Nos. 10935003, 11275031, 10835003, and 11274026). L.F.W would also like to acknowledge the support from China Postdoctoral Science Foundation (No. 2012T50018).

[1]S. Atzeni and J. Meyer-ter-Vehn, *The Physics of Inertial Fusion: Beam Plasma Interaction Hydrodynamics, Hot Dense Matter* (Oxford University Press, Oxford, 2004).
[2]J. O. Kane, H. F. Robey, B. A. Remington, R. P. Drake, J. Knauer, D. D. Ryutov, H. Louis, R. Teyssier, O. Hurricane, D. Arnett, R. Rosner, and A. Calder, Phys. Rev. E **63**, 055401 (2001).
[3]R. J. Mason, D. E. Hollowell, G. T. Schappert, and S. H. Batha, Phys. Plasmas **8**(5), 2338 (2001).
[4]A. Casner, V. A. Smalyuk, L. Masse, I. Igumenshchev, S. Liberatore, L. Jacquet, C. Chicanne, P. Loiseau, O. Poujade, D. K. Bradley, H. S. Park, and B. A. Remington, Phys. Plasmas **19**, 082708 (2012).
[5]Wenhua Ye, Weiyan Zhang, and X. T. He, Phys. Rev. E **65**, 57401 (2002).
[6]M. Tabak, D. H. Munro, and J. D. Lindl, Phys. Fluids B **2**, 1007 (1990).
[7]R. Betti, V. N. Goncharov, R. L. McCrory, and C. P. Verdon, Phys. Plasmas **5**(5), 1446 (1998), and references therein.
[8]S. A. Piriz, A. R. Piriz, and N. A. Tahir, Phys. Plasmas **16**, 082706 (2009).
[9]S. G. Glendinning, S. N. Dixit, B. A. Hammer, D. H. Kalantat, M. H. Key, J. D. Kilkenny, J. P. Knauer, D. M. Pennington, B. A. Remington, R. J. Wallace, and S. V. Weber, Phys. Rev. Lett. **78**, 3318 (1997).
[10]T. Sakaiya, H. Azechi, M. Matsuoka, N. Izumi, M. Nakai, K. Shigemori, H. Shiraga, A. Sunahara, H. Takabe, and T. Yamanaka, Phys. Rev. Lett. **88**, 145003 (2002).
[11]S. Fujioka, H. Shiraga, M. Nishikino, K. Shigemori, A. Sunahara, M. Nakai, H. Azechi, K. Nishihara, and T. Yamanaka, Phys. Plasmas **10**, 4784 (2003).
[12]L. F. Wang W. H. Ye, Z. M. Sheng, W.-S. Don, Y. J. Li, and X. T. He, Phys. Plasma **17**, 122706 (2010).
[13]J. Li, B. Zhao, H. Li, and J. Zheng, Plasma Phys. Controlled Fusion **52**, 085008 (2010).
[14]W. H. Ye, L. F. Wang, and X. T. He, Phys. Plasma **17**, 122704 (2010).
[15]L. F. Wang, W. H. Ye, and Y. J. Li, Phys. Plasma **17**, 052305 (2010).
[16]L. F. Wang, W. H. Ye, W.-S. Don, Z. M. Sheng, Y. J. Li, and X. T. He, Phys. Plasma **17**, 122308 (2010).
[17]L. F. Wang, C. Xue, W. H. Ye, and Y. J. Li, Phys Plasmas **16**, 112104 (2009).
[18]R. Betti and J. Sanz, Phys. Rev. Lett. **97**, 205002 (2006).

Scaling and Statistics in Three-Dimensional Compressible Turbulence

Jianchun Wang,[1] Yipeng Shi,[1] Lian-Ping Wang,[2] Zuoli Xiao,[1] X. T. He,[1] and Shiyi Chen[1,*]
[1]*SKLTCS and CAPT, College of Engineering, Peking University, Beijing 100871, China*
[2]*Department of Mechanical Engineering, University of Delaware, Newark, Delaware 19716, USA*
(Received 9 November 2011; published 25 May 2012)

The scaling and statistical properties of three-dimensional compressible turbulence are studied using high-resolution numerical simulations and a heuristic model. The two-point statistics of the solenoidal component of the velocity field are found to be not significantly different from those of incompressible turbulence, while the scaling exponents of the velocity structure function for the compressive component become saturated at high orders. Both the simulated flow and the heuristic model reveal the presence of a power-law tail in the probability density function of negative velocity divergence (high compression regime). The power-law exponent is different from that in Burgers turbulence, and this difference is shown to have a major contribution from the pressure effect, which is absent in the Burgers turbulence.

DOI: 10.1103/PhysRevLett.108.214505 PACS numbers: 47.27.−i, 47.40.−x, 47.53.+n

Compressible three-dimensional (3D) fluid turbulence is of great importance to a large number of industrial applications and natural phenomena, including high-temperature reactive flows, transonic and hypersonic aircrafts, interplanet space exploration, and star-forming clouds in galaxies [1]. An accurate description (e.g., subgrid-scale stress modeling [2]) of small-scale compressible turbulence is desired when modeling complex compressible turbulence. In this Letter, we study the effects of compressibility and shock discontinuities on the statistics of fluid turbulence, with specific attention to the similarity and difference between 3D compressible turbulence and one-dimensional Burgers turbulence.

Since it is difficult to analyze 3D compressible turbulence, Burgers [3] first systematically studied a nonlinear model of fluid turbulence; i.e., the one-dimensional Burgers equation. Since then, the one-dimensional Burgers turbulence has frequently been investigated theoretically and numerically [4–13]. According to multifractal theory [4], isolated shocks connected by smooth ramps lead to bifractal scaling exponents of the velocity structure function in the Burgers turbulence. Mitra *et al.* [5] performed simulations of one-dimensional Burgers turbulence with up to 2^{20} mesh points to study the multiscaling of velocity structure functions. They found that scaling exponents asymptotically saturate to one with increasing orders. Much effort has also been made to exploit the asymptotic behavior at the tail of the probability density function (PDF) of the negative velocity derivative in Burgers turbulence [7–13]. E *et al.* [10] predicted that the large negative velocity gradients stem mainly from preshocks, leading to the $-7/2$ power-law tail in the PDF of negative velocity gradients (provided that preshocks do not cluster). Bec [11] verified this result using a novel particle and shock tracking numerical method. Boldyrev, Linde, and Polyakov [12] carried out simulations of random forced Burgers turbulence using a standard shock capturing scheme. They obtained a power-law exponent of about -3.4, very close to the theoretical value of -3.5.

Compared to incompressible turbulence, 3D compressible turbulence is more complex due to nonlinear interactions between solenoidal and compressive modes of velocity fluctuations and couplings between the velocity field and pressure field [14]. Besides the vortex-filament induced intermittency observed in the incompressible turbulence, shock waves in the compressible turbulence add an intermittent dissipation mechanism of different topological structures [15]. Schmidt, Federrath, and Klessen [16] studied the two-point velocity statistics of compressible turbulence at a root-mean-square (rms) Mach number of 5.5 by numerical simulations of Euler equations. A universal scaling has been recovered by reformulation of the refined similarity hypothesis in terms of the mass-weighted velocity $\rho^{1/3}\boldsymbol{u}$. They also reported that the most intermittent dissipative structures were shocks, due to extreme compressibility of the flow field. Galtier and Banerjee [17] derived an exact relation for correlation functions in compressible isothermal turbulence that mimics the Kolmogorov 4/5 law in the incompressible isotropic turbulence. Consequently, they theoretically revealed the effect of dilation and compression on the local energy transfer. By dimensional arguments, they further obtained a $k^{-5/3}$ spectrum of the density-weighted velocity $\rho^{1/3}\boldsymbol{u}$.

Here, a forced compressible turbulence is simulated in a cubic box with periodic conditions at 1024^3 grid resolution, using a novel hybrid approach described in [18] [also see Supplemental Material [19]]. A total of 20 flow fields at the statistical stationary stage spanning $2.68 \le t/T_e \le 4.63$ are extracted to analyze the flow statistics, where t is time and T_e is the large eddy turnover time defined by $T_e = \sqrt{3}L/u'$, where L is the integral length scale and u' is the rms velocity magnitude. The turbulent Mach number is $M_t = u'/\langle c\rangle = 1.03$, where $\langle c\rangle$ is the average sound speed, and the Taylor microscale Reynolds number is $R_\lambda = 254$. The rms velocity divergence is found to be $\theta' = 0.35\omega'$, where ω' is the rms vorticity magnitude. It is also found (not shown in the Letter) that power spectra for velocity $\mathbf{u}$, and density-weighted velocities $\rho^{1/3}\mathbf{u}$ and $\rho^{1/2}\mathbf{u}$ almost overlap, indicating a minor effect of density fluctuations on the velocity spectrum.

In order to reveal the underlying physics in the compressible turbulence, we employ the Helmholtz decomposition [14]; namely, the fluid velocity $\mathbf{u}$ is decomposed into a solenoidal component $\mathbf{u}^s$ and a compressive component $\mathbf{u}^c$: $\mathbf{u} = \mathbf{u}^s + \mathbf{u}^c$, where $\nabla \cdot \mathbf{u}^s = 0$ and $\nabla \times \mathbf{u}^c = 0$. In our simulated flow, the ratio of rms fluctuations, $u^{c\prime}/u^{s\prime}$, is found to be 0.22, comparable to 0.18 reported in Porter, Pouquet, and Woodward [20] at a similar turbulent Mach number.

Figure 1 shows the normalized PDFs of the longitudinal increments, $\Delta u^s(r) \equiv \Delta \mathbf{u}^s(\mathbf{r}) \cdot \mathbf{r}/r$, of the solenoidal velocity component at different separations, where $\mathbf{r}$ is the separation vector and $r = |\mathbf{r}|$. The PDFs exhibit stretched exponential tails at small spatial separations and approach Gaussian as the separation increases. These trends are very similar to those found in incompressible turbulence [21]. Figure 2 shows the normalized PDFs of the longitudinal increments, $\Delta u^c(r) \equiv \Delta \mathbf{u}^c(\mathbf{r}) \cdot \mathbf{r}/r$, of the compressive velocity component. The shapes of the PDFs are highly skewed at small separations and always have a longer tail than Gaussians for all separations. In addition, at the one-grid-length separation the PDF has a power-law tail with an exponent of -2.5 on the left side. We note that, due to the power-law behavior of the PDF tail, when the viscosity asymptotically approaches zero, special care is required since the variance of the velocity increment may become unbounded. Similar results have been reported for Burgers turbulence. In the random-force driven Burgers turbulence, the PDFs of velocity increments have an algebraic tail on the left side, leading to strong intermittency and bifractality of the scaling exponents [8].

We plot the normalized PDFs of the pressure increments Δp in Fig. 3. The shapes of these PDFs are nearly symmetric and have longer tails at small spatial separations than those in the incompressible turbulence at the same Taylor Reynolds number (also see [22]). The pressure changes drastically in the high compression regime, leading to intensive pressure increments at small separations. Power-law tails are also found in the pressure-increment PDF at the one-grid-length separation, with a power-law exponent of -3.

The scaling exponents for the longitudinal structure functions in the inertial subrange are defined as

$$S^L_{s,n}(r) \equiv \langle |\Delta u^s(r)|^n \rangle \sim r^{\zeta^L_{s,n}}, \tag{1}$$

$$S^L_{c,n}(r) \equiv \langle |\Delta u^c(r)|^n \rangle \sim r^{\zeta^L_{c,n}}, \tag{2}$$

where $\zeta^L_{s,n}$ and $\zeta^L_{c,n}$ are the scaling exponents for the two velocity components, respectively. Our results show that $\zeta^L_{s,n}$ agrees well with those from the incompressible turbulence [23,24]. This implies that at $M_t \sim 1$, the two-point statistics of the solenoidal velocity component are insensitive to the presence of shocks. In addition, similar to results in [20,25], we find that the scaling exponents of the full velocity $\mathbf{u}$ are also the same as those in the incompressible turbulence. However, the scaling exponents of the compressive velocity component, shown in the inset of Fig. 4, is drastically different. A saturation of $\zeta^L_{c,n}$ is observed for $n \geq 5$, and the saturated value is estimated to be $\zeta^L_{c,\infty} \approx 0.7$. The compensated longitudinal structure functions at orders 5 and 6 are shown in Fig. 4, where the separation is normalized by the Kolmogorov length η. According to the multifractal theory, the saturation of exponents is caused by the domination of frontlike structures [25]. Benzi *et al.* [25] observed that the density field displays frontlike structures, leading to saturation of the scaling exponents for density structure functions in a weakly compressible turbulence at rms Mach number 0.3. However, for the velocity field, they did not find any difference between weakly compressible turbulence and incompressible

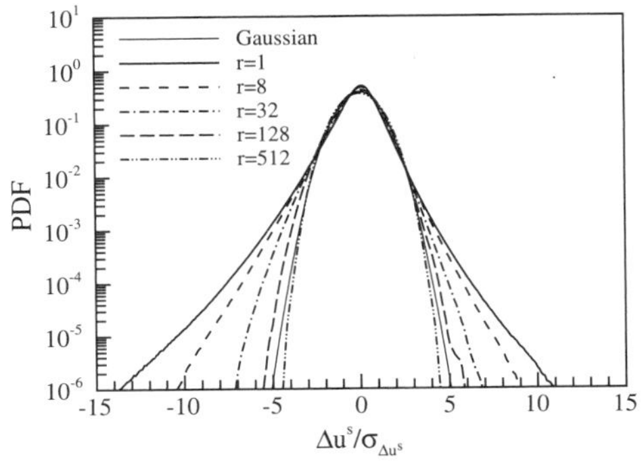

FIG. 1. Normalized probability density functions of Δu^s.

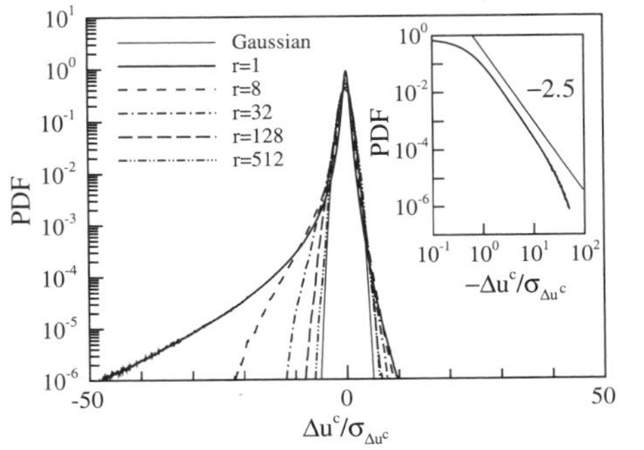

FIG. 2. Normalized probability density functions of Δu^c. Inset: log-log plot of the same PDF for the negative Δu^c at the one-grid-length separation.

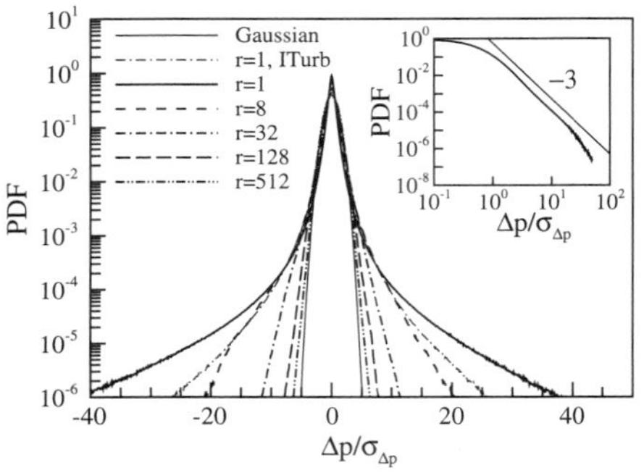

FIG. 3. Normalized probability density functions of Δp. Thin dash-dotted line is the corresponding PDF at the one-grid-length separation for incompressible turbulence (ITurb) at the same Taylor Reynolds number. Inset: log-log plot of the PDF for positive Δp in compressible turbulence at the one-grid-length separation.

turbulence. On the other hand, due to a relatively higher turbulent Mach number in our simulated flow, shocks produce a significant number of frontlike structures in the compressive velocity field, causing saturation of scaling exponents $\zeta_{c,n}^L$. Furthermore, substantial uncertainties in the exponents at high orders are associated with large temporal fluctuations, which ascribe to instability of intermittency structures [26].

The PDF of velocity divergence (Fig. 5) exhibits a power-law tail for large negative divergences and a superexponential tail for large positive divergences, both qualitatively similar to the PDF of the velocity derivative in Burgers turbulence [13]. In Burgers turbulence with infinitesimal viscosity, the large negative velocity gradients stem mainly from preshocks, leading to the power-law tail in the PDF of negative velocity gradients [10]. In the

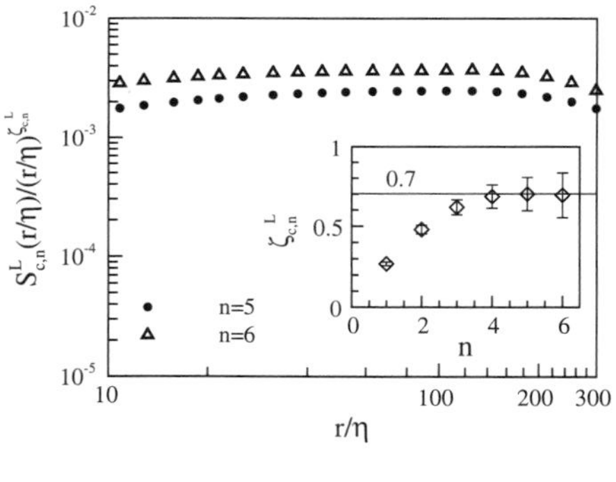

FIG. 4. Compensated structure functions for $n = 5$ and $n = 6$ as a function of normalized separation. Inset: scaling exponents $\zeta_{c,n}^L$ as a function of n. The separation r is normalized by the Kolmogorov length scale $\eta = [\langle \nu \rangle^3/\epsilon]^{1/4}$, where ν is the kinematic viscosity and ϵ is the average kinetic energy dissipation rate [14].

compressible turbulence, through studying the contours of velocity divergence (not shown here), we have identified that the power-law regime of velocity divergence has a major contribution from preshocks and weak shocklets rather than strong shock waves. Otherwise, the power-law exponent for the PDF of velocity divergence is -2.5, the same as that for the PDF of the longitudinal increments of the compressive velocity component at the one-grid-length separation but qualitatively different from the power-law exponent (-3.5) for the PDF of the velocity gradient in one-dimensional Burgers turbulence. To understand this difference, we write down the equation of the velocity derivative in Burgers turbulence [9]:

$$\frac{\partial \xi}{\partial t} + u\frac{\partial \xi}{\partial x} = -\xi^2 + \nu\frac{\partial^2 \xi}{\partial x^2} + \frac{\partial f}{\partial x}, \tag{3}$$

where $u(x, t)$ is the velocity and $\xi(x, t)$ is the velocity gradient. The forcing $f(x, t)$ is used to maintain the Burgers turbulence to be statistically stationary. In contrast, the governing equation for velocity divergence in 3D Navier-Stokes flow is

$$\frac{\partial \theta}{\partial t} + u_j\frac{\partial \theta}{\partial x_j} = -\frac{\partial u_j}{\partial x_i}\frac{\partial u_i}{\partial x_j} - \frac{1}{\gamma M^2}\frac{\partial}{\partial x_i}\left(\frac{1}{\rho}\frac{\partial p}{\partial x_i}\right) + \frac{4\nu_0}{3Re}\frac{\partial^2 \theta}{\partial x_i^2}. \tag{4}$$

To simplify the discussions, we have neglected the effect of density fluctuations on the viscous term. It is seen that there are some similarities between the term $\frac{\partial u_j}{\partial x_i}\frac{\partial u_i}{\partial x_j}$ in Eq. (4) and ξ^2 in Eq. (3), the former causing strong skewness of the PDF of velocity divergence provided that the turbulent Mach number is larger than 0.3 [27].

Below, we demonstrate how the pressure term alters the power-law exponent of the PDF of velocity divergence in the high compression regime. Following the similar procedure provided by Gotoh and Kraichnan [6], we can derive the Liouville equation for the PDF of velocity divergence $P(\theta)$ as follows (see Supplemental Material [19]):

$$\frac{\partial P}{\partial t} - \frac{\partial(\theta^2 P)}{\partial \theta} - \theta P + D = F, \tag{5}$$

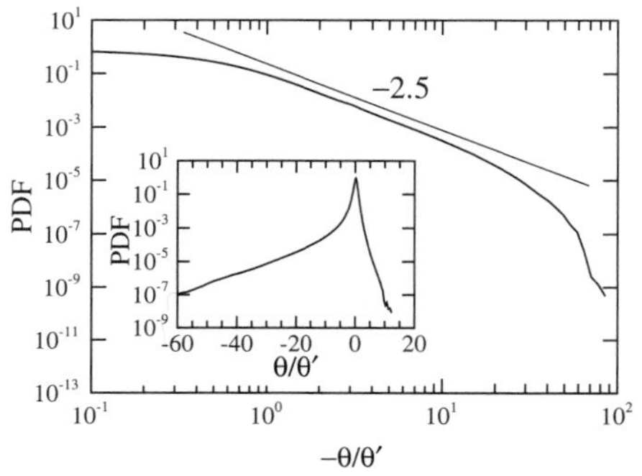

FIG. 5. Probability density function of velocity divergence.

where the dissipation term is

$$D(\theta) = \frac{4\nu_0}{3Re}\frac{\partial}{\partial\theta}\left[\left\langle\frac{\partial^2\theta}{\partial x_i^2}\,\middle|\,\theta\right\rangle P\right] \qquad (6)$$

and the *forcing* term, including effects of pressure and anisotropic straining, is

$$F = \frac{\partial}{\partial\theta}\left[\left\langle\frac{1}{\gamma M^2}\frac{\partial}{\partial x_i}\left(\frac{1}{\rho}\frac{\partial p}{\partial x_i}\right) + \left(\frac{\partial u_j}{\partial x_i}\frac{\partial u_i}{\partial x_j} - \theta^2\right)\,\middle|\,\theta\right\rangle P\right]. \qquad (7)$$

This equation is written formally the same as the PDF equation of velocity derivative in Burgers turbulence but with a different dissipation term and forcing term [13]. It has been argued previously [7–13] that Burgers turbulence has a power-law tail in the PDF of ξ for the high compression regime. If Burgers turbulence is driven by large scale white-in-time random forcing, this tail is believed to be universal with an exponent -3.5 in the limit of vanishing viscosity. Other exponents are possible if Burgers turbulence is driven by forcing with prescribed power-law spectra [8].

In Fig. 6, we plot the average normalized straining $\frac{1}{\theta'^2} \times \frac{\partial u_j}{\partial x_i}\frac{\partial u_i}{\partial x_j}$ conditioned on the velocity divergence as a function of velocity divergence. For the compression regime (i.e., $\theta < 0$), we find that the straining term can be well approximated by $\frac{\theta^2}{\theta'^2}$, implying that the effect of stretching-tilting dynamics on the PDF of velocity divergence is small. In addition, its compressive component $\frac{1}{\theta'^2}\frac{\partial u_j^c}{\partial x_i}\frac{\partial u_i^c}{\partial x_j}$ dominates the overall straining term. These interesting approximations are consistent with the numerical simulations that show that intensive velocity gradients are dominated by variations in one spatial direction during compression in

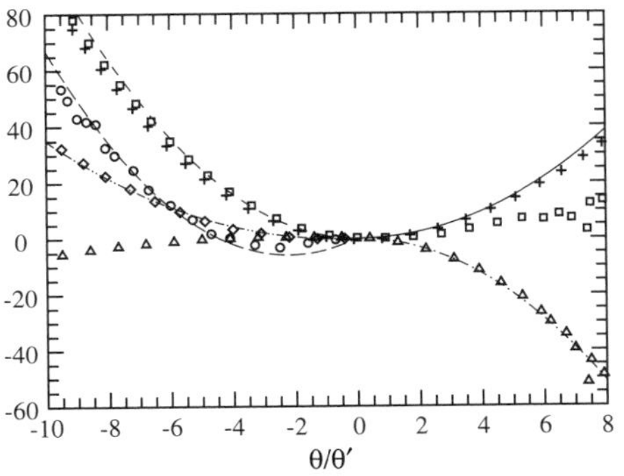

FIG. 6. Conditional average of $\langle\frac{1}{\theta'^2}\frac{\partial u_j}{\partial x_i}\frac{\partial u_i}{\partial x_j}|\frac{\theta}{\theta'}\rangle$ (squares), $\langle\frac{1}{\theta'^2} \times \frac{\partial u_j^c}{\partial x_i}\frac{\partial u_i^c}{\partial x_j}|\frac{\theta}{\theta'}\rangle$ (plusses), $\langle\frac{1}{\theta'^2}\frac{\partial u_j}{\partial x_i}\frac{\partial u_i}{\partial x_j} - (\frac{\theta}{\theta'})^2|\frac{\theta}{\theta'}\rangle$ (triangles), $\langle\frac{1}{\theta'^2}\frac{4\nu_0}{3Re} \times \frac{\partial^2\theta}{\partial x_i^2}|\frac{\theta}{\theta'}\rangle$ (diamonds) and $\langle\frac{1}{\theta'^2}\frac{1}{\gamma M^2}\frac{\partial}{\partial x_i}(\frac{1}{\rho}\frac{\partial p}{\partial x_i})|\frac{\theta}{\theta'}\rangle$ (circles). The lines represent $\alpha(\frac{\theta}{\theta'})^2 + \beta\frac{\theta}{\theta'}$ with $(\alpha, \beta) = (1.0, 0.0)$ (dashed line), $(\alpha, \beta) = (0.6, 0.0)$ (solid line), $(\alpha, \beta) = (-0.8, 0.0)$ (dash-dotted line), $(\alpha, \beta) = (0.4, 0.5)$ (dash-dot-dotted line) and $(\alpha, \beta) = (1.2, 5.3)$ (long dashed line).

the shock waves [14,28]. Therefore, the solenoidal velocity component has very limited contribution to the PDF equation of velocity divergence in the compression regime. In contrast, in the expansion regime (i.e., $\theta > 0$), the straining term is no longer close to the square of the divergence. Based on dimensional analysis, the anisotropic straining term $[\frac{1}{\theta'^2}\frac{\partial u_j}{\partial x_i}\frac{\partial u_i}{\partial x_j} - (\frac{\theta}{\theta'})^2]$ can be represented by the square of the divergence multiplied by a nondimensional constant. Figure 6 shows that this nondimensional constant is around -0.8 in the expansion regime instead of almost zero in the compression regime. The conditional straining is always positive, implying that the magnitude of the velocity divergence is increased by the straining when $\theta < 0$, but it is decreased by the straining when $\theta > 0$. However, the magnitude of the conditional straining for $\theta > 0$ is only about $1/5$ of the values for $\theta < 0$, indicating that the effect of straining on the velocity divergence is substantially weaker in the expansion regime. Therefore, the role of the straining term in Eq. (4) is similar to that of the ξ^2 term in Eq. (3) in the compression regime due to shock structures, but in the expansion regime it is weakened by relaxation of multidirectional advection and by the solenoidal velocity dynamics, where the flow is dominated by vortex structures.

We now focus our discussion on the PDF equation in the strong compression regime where the power-law tail appears. In Fig. 6, we show average values of the viscous term and pressure term conditioned on θ in the PDF equation. As also shown in the figure, the anisotropic staining effect is small compared with the other two terms and may be neglected. The viscous term is well fitted by a parabola $\alpha_\nu(\frac{\theta}{\theta'})^2 + \beta_\nu\frac{\theta}{\theta'}$ with $\alpha_\nu = 0.4$ and $\beta_\nu = 0.5$. A similar procedure was suggested by Gotoh and Kraichnan [9] based on an analysis of the viscous term inside an equilibrium single shock in Burgers turbulence. The pressure term can also be approximated by a parabola with coefficients $\alpha_p = 1.2$ and $\beta_p = 5.3$. With these approximations, we obtain the following solution at the stationary stage:

$$P(\theta) = C_0\theta^{-1}\left[\theta + \frac{(\beta_\nu - \beta_p)}{1 + \alpha_p - \alpha_\nu}\right]^{-1-[1/(1+\alpha_p-\alpha_\nu)]}. \qquad (8)$$

It follows that, for large negative θ, $P(\theta) \propto \theta^{-q}$, where $q = 2 + \frac{1}{1+\alpha_p-\alpha_\nu}$. Using the fitting values for α_p and α_ν, we then obtain $q = 2.56$. This is very close to the value of 2.5 shown in Fig. 5. An exponent of -3 is obtained without considering the pressure term and viscous term, which is consistent with the case of Burgers turbulence [13]. The viscous term enlarges the exponent as in Burgers turbulence [13]. The key difference here is the modification of the power-law exponent from the pressure term. We note that the role of the pressure is opposite to that of the viscosity in determining this exponent, and the effect of pressure predominates.

Finally, we emphasize that the viscosity is small, but finite, in our discussion on the PDF of velocity divergence.

Several issues require further investigations, including the relative contributions of preshocks and weak shocklets to the power-law tail of the PDF at higher Reynolds numbers, the asymptotic behavior of the power-law exponent in the limit of vanishing viscosity, and the effect of Mach number on the power-law behavior.

We thank Qionglin Ni, Zhenhua Xia, and Zhou Jiang for many useful discussions. This work was supported by the National Natural Science Foundation of China (Grants No. 10921202 and No. 91130001) and the National Science and Technology Ministry under a subproject of the 973 program (Grant No. 2009CB724101). Simulations were done on a cluster computer at the Center for Computational Science and Engineering at Peking University, China and on Bluefire at NCAR, USA through CISL-35751014 and CISL-35751015. L.-P. W. acknowledges support from the National Science Foundation under Grants No. ATM-0730766 and OCI-0904534.

---

*syc@pku.edu.cn

[1] P. Padoan and A. Norlund, Astrophys. J. **576**, 870 (2002).
[2] B. Kosovic, D. I. Pullin, and R. Samtaney, Phys. Fluids **14**, 1511 (2002).
[3] J.M. Burgers, in *Advances in Applied Mechanics* (Academic, New York, 1948), Vol. 1, p. 171.
[4] E. Aurell, U. Frisch, J. Lutsko, and M. Vergassola, J. Fluid Mech. **238**, 467 (1992).
[5] D. Mitra, J. Bec, R. Pandit, and U. Frisch, Phys. Rev. Lett. **94**, 194501 (2005).
[6] T. Gotoh and R. H. Kraichnan, Phys. Fluids A **5**, 445 (1993).
[7] A. M. Polyakov, Phys. Rev. E **52**, 6183 (1995).
[8] A. Chekhlov and V. Yakhot, Phys. Rev. E **52**, 5681 (1995); V. Yakhot and A. Chekhlov, Phys. Rev. Lett. **77**, 3118 (1996).
[9] T. Gotoh and R.H. Kraichnan, Phys. Fluids **10**, 2859 (1998).
[10] W. E, K. Khanin, A. Mazel, and Y. Sinai, Phys. Rev. Lett. **78**, 1904 (1997).
[11] J. Bec, Phys. Rev. Lett. **87**, 104501 (2001).
[12] S. Boldyrev, T. Linde, and A. Polyakov, Phys. Rev. Lett. **93**, 184503 (2004).
[13] J. Bec and K. Khanin, Phys. Rep. **447**, 1 (2007).
[14] R. Samtaney, D. I. Pullin, and B. Kosovic, Phys. Fluids **13**, 1415 (2001).
[15] L. Pan, P. Padoan, and A. G. Kritsuk, Phys. Rev. Lett. **102**, 034501 (2009).
[16] W. Schmidt, C. Federrath, and R. Klessen, Phys. Rev. Lett. **101**, 194505 (2008).
[17] S. Galtier and S. Banerjee, Phys. Rev. Lett. **107**, 134501 (2011).
[18] J. Wang, L.-P. Wang, Z. Xiao, Y. Shi, and S. Chen, J. Comput. Phys. **229**, 5257 (2010).
[19] See Supplemental Material at http://link.aps.org/supplemental/10.1103/PhysRevLett.108.214505 for a brief description of the hybrid simulation approach used in the Letter and a derivation of the PDF equation governing velocity divergence.
[20] D. Porter, A. Pouquet, and P. Woodward, Phys. Rev. E **66**, 026301 (2002).
[21] S. Chen, G. D. Doolen, R. H. Kraichnan, and Z. She, Phys. Fluids A **5**, 458 (1993).
[22] N. Cao, S. Chen, and G. D. Doolen, Phys. Fluids **11**, 2235 (1999).
[23] Z. S. She and E. Lévêque, Phys. Rev. Lett. **72**, 336 (1994).
[24] N. Cao, S. Chen, and Z. S. She, Phys. Rev. Lett. **76**, 3711 (1996).
[25] R. Benzi, L. Biferale, R. T. Fisher, L. P. Kadanoff, D. Q. Lamb, and F. Toschi, Phys. Rev. Lett. **100**, 234503 (2008).
[26] S. Chen and N. Cao, Phys. Rev. Lett. **78**, 3459 (1997).
[27] S. Pirozzoli and F. Grasso, Phys. Fluids **16**, 4386 (2004).
[28] J. Wang, Y. Shi, L.-P. Wang, Z. Xiao, X. He, and S. Chen, Phys. Fluids **23**, 125103 (2011).

PHYSICS OF PLASMAS 19, 072704 (2012)

Stabilization of the Rayleigh-Taylor instability in quantum magnetized plasmas

L. F. Wang (王立锋),[1,2,a)] B. L. Yang (羊本林),[2,3] W. H. Ye (叶文华),[1,2,b)] and X. T. He (贺贤土)[1,2,c)]
[1]HEDPS and CAPT, Peking University, Beijing 100871, China
[2]Institute of Applied Physics and Computational Mathematics, Beijing 100088, China
[3]Graduate School, China Academy of Engineering Physics, Beijing 100088, China

(Received 10 May 2012; accepted 20 June 2012; published online 18 July 2012)

In this research, stabilization of the Rayleigh-Taylor instability (RTI) due to density gradients, magnetic fields, and quantum effects, in an ideal incompressible plasma, is studied analytically and numerically. A second-order ordinary differential equation (ODE) for the RTI including quantum corrections, with a continuous density profile, in a uniform external magnetic field, is obtained. Analytic expressions of the linear growth rate of the RTI, considering modifications of density gradients, magnetic fields, and quantum effects, are presented. Numerical approaches are performed to solve the second-order ODE. The analytical model proposed here agrees with the numerical calculation. It is found that the density gradients, the magnetic fields, and the quantum effects, respectively, have a stabilizing effect on the RTI (reduce the linear growth of the RTI). The RTI can be completely quenched by the magnetic field stabilization and/or the quantum effect stabilization in proper circumstances leading to a cutoff wavelength. The quantum effect stabilization plays a central role in systems with large Atwood number and small normalized density gradient scale length. The presence of external transverse magnetic fields beside the quantum effects will bring about more stability on the RTI. The stabilization of the linear growth of the RTI, for parameters closely related to inertial confinement fusion and white dwarfs, is discussed. Results could potentially be valuable for the RTI treatment to analyze the mixing in supernovas and other RTI-driven objects. © *2012 American Institute of Physics*. [http://dx.doi.org/10.1063/1.4737162]

I. INTRODUCTION

Rayleigh-Taylor[1–3] instability (RTI) is an inherent physical problem in inertial confinement fusion (ICF), occurring when a lighter fluid is accelerated against a heavier one. Control of the growth of the RTI is extremely essential for the success of standard central ignition ICF.[4–7] This instability can limit the implosion velocity and the hot spot size, hinder (delay) the hot spot formation, reduce the fusion gain, and in some cases even break up the implosion shell or make the hot spot formation impossible resulting in the auto-ignition failure. Thus, the RTI must be controlled at an accepted level in ICF. On the other hand, there are a number of astrophysical systems, in which the RTI is expected to be important,[8–13] such as the large-scale mixing that takes place in the explosion of a supergiant star.[14,15] In a core collapse supernova, the sudden release of huge amounts of energy at the star's core drives a blast wave propagating out through layers of less dense material. As the interfaces between these layers is decelerated subsequently, the density gradients in an opposing direction to the pressure gradients around the interface are established, and then the RTI occurs.[16–18] It is widely accepted that the most probable mechanism to mix the supernova (SN) ejecta, for example SN 1987A in the Large Magellanic Cloud, is the RTI. The present research focuses on the RTI and in this report term instability refers to the RTI unless stated otherwise.

On the one hand, the performance of an ICF target is limited by the growth of the RTI. On the other hand, this instability is closely related to stellar structure and evolution. Hence, it is important to understand the physical mechanisms that can affect such instability, especially its stabilization as well as the details of this instability, because of its significance in both fundamental researches and engineering applications. Such knowledge will aid our understanding of the origin of star formation, such as white dwarfs and neutron stars.

When a heavier fluid is superposed over a lighter one, in a gravitational field $-ge_y$, where g is acceleration, a single-mode perturbation, with wave number $k = 2\pi/\lambda$, where λ is the perturbation wavelength, on the interface between two fluids of densities ρ^h and ρ^l ($\rho^h > \rho^l$), is a simplest case. According to the classical linear theory,[1–3] a cosinusoidal initial interface perturbation with small amplitude η_0 ($\eta_0 \ll \lambda$) grows exponentially in time t, $\eta_L = \varepsilon e^{\gamma_c t}$, where $\gamma_c = \sqrt{Akg}$ is the linear growth rate with $A = (\rho^h - \rho^l)/\ (\rho^h + \rho^l)$ being the Atwood number. When the typical perturbation amplitude can compared to its wavelength, higher harmonics are excited, and then the RTI enters the nonlinear regime.[19–24]

There are several physical mechanisms influencing the RTI, such as the density gradients[25–29] (i.e., the finite thickness of the perturbed interface), and the mass ablation[30–40] (i.e., the restoring force due to the thermal conduction, the physical removal of modulated material through ablation and the ablation-driven convection of the vorticity off the ablation

[a)]Electronic mail: lif_wang@pku.edu.cn.
[b)]Electronic mail: ye_wenhua@iapcm.ac.cn.
[c)]Electronic mail: xthe@iapcm.ac.cn.

front). In addition, it is found that the magnetic field plays a vital role in the RTI which can reduce its linear growth rate.[3] In ICF, the temperature gradients and the density gradients of laser induced plasmas can generate large-scale circulating magnetic fields.[41] Furthermore, the Weibel instability, the resonant absorption, and the motion of super-thermal electrons can also generate such magnetic fields.[42] The magnetic RTI is expected to be of enormous importance for many astrophysical systems, such as buoyant bubbles generated by radio jets in clusters of galaxies.[43] Moreover, magnetic fields observed in some astrophysical jets[44] and numerical modelings[45] strongly suggest that magnetohydrodynamics (MHD) plays a crucial role in governing the jet formation and collimation.[46] Several studies have focused their attentions on the effects of magnetic field on the RTI with a uniform magnetic field or a sharp magnetic field interface, which indicate that the magnetic field plays a prominent role in damping the linear growth of the RTI, and if the strength of magnetic field is strong enough, the RTI can even be quenched.[3] Besides, the magnetic field gradients on the RTI have been investigated in our previous work.[47] It is shown that the linear growth rate of the RTI increases with increasing thickness of the magnetic field transition layer, especially for the case of small thickness of the magnetic field transition layer. When the magnetic field transition layer width is large enough, the linear growth rate of the RTI can be saturated. Thus when one increases the thickness of the magnetic field transition layer, the linear growth rate of the RTI increases only in a certain range, which depends on the strength of magnetic field.

Quantum plasma is a new emerging and a rapidly growing field of plasma whose constitutes can be electrons, positrons, and ions. It is well-known that a quantum plasma has the properties of high-particle number densities and low-temperatures, in comparison to the classical plasma. It plays a significant role in dense astrophysical environments, particularly in the interior of Jupiter, white dwarfs, and neutron stars. For the behavior of the charged plasma particles, quantum effects become pronounced when the de Broglie wavelength of the charge carriers (e.g., electrons and positrons) is comparable to the dimension of the plasma system. Recently, plasmas with quantum modifications have been extensively studied.[48–56] There are two well-known mathematical approaches used for the quantum plasmas. The first one is the Wigner model which describes the statistical behavior of plasmas based on the Wigner-Poisson system. The other one is the Hartree model which describes the hydrodynamic behavior of plasmas based on the Schrödinger-Poisson system. The quantum hydrodynamic (QHD) model consists of a set of equations describing the transport of charge density, momentum, and energy in plasmas. It worth noting that the QHD model has been introduced to astrophysics.[57,58] Moreover, the quantum magnetohydrodynamic (QMHD) model has been obtained by Haas[53] with the help of QHD. In the present work, we will study the stabilization of the linear growth of the RTI due to the density gradients, the magnetic fields, and the quantum effects, in ideal incompressible plasmas by virtue of the QMHD model.

The rest of the paper is organized as follows. Section II is devoted to the detailed description of the theoretical framework of the present paper. A second-order ordinary differential equation (ODE) for the RTI, including the quantum effects, with a continuous density profile, in a uniform external transverse magnetic field, is obtained in Sec. II. Numerical methods solving the resulting ODE are described in Sec. III. The analytic results of the linear growth rate of the RTI, with the corrections of the density gradients, the magnetic fields, and the quantum effects, are presented in Sec. IV. We investigate the stabilization of the RTI from the density gradients, the magnetic fields, and the quantum effects, as well as the combined contributions and the competitions among them, in Sec. V. The stabilization of the linear growth of the RTI due to magnetic fields and quantum effects, with parameters closely related to ICF and white dwarfs, is discussed in some details in Sec. VI. Finally, some concluding remarks are given in Sec. VII.

II. THEORETICAL FRAMEWORK

In this section, an ideal incompressible quantum magnetized plasma whose constituent is a fluid of electrons and singly charged ions is considered. For mathematical simplicity, the dissipation effects arising from resistance and viscosity are ignored in our study. In a Cartesian coordinate, the governing equations (i.e., QMHD) for this system[53] are

$$\frac{d\rho}{dt} + \rho(\nabla \cdot \boldsymbol{u}) = 0, \tag{1}$$

$$\rho\frac{d\boldsymbol{u}}{dt} = -\nabla p + \rho\boldsymbol{g} + \frac{1}{\mu_0}(\nabla \times \boldsymbol{B}) \times \boldsymbol{B} + \frac{\hbar^2}{2m_e m_i}\rho\nabla\left(\frac{\nabla^2\sqrt{\rho}}{\sqrt{\rho}}\right), \tag{2}$$

$$\nabla \times \boldsymbol{E} = -\frac{\partial \boldsymbol{B}}{\partial t}, \tag{3}$$

$$\boldsymbol{E} = -\boldsymbol{u} \times \boldsymbol{B} + \frac{\hbar^2}{2em_e}\nabla\left(\frac{\nabla^2\sqrt{\rho}}{\sqrt{\rho}}\right), \tag{4}$$

here ρ, $\boldsymbol{u}$, p, g, $\boldsymbol{E}$, and $\boldsymbol{B}$ are, respectively, density, velocity, thermal pressure, acceleration, electronic field, and magnetic field. Here, μ_0 is vacuum permeability, $\hbar$ is the Planck's constant divided by 2π, m_e is the mass of electrons, m_i is the mass of ions, and e is the magnitude of the electron charge. The last $\hbar$-dependent term in the right-hand side of Eq. (2), which is the so-called Bohm potential, is due to quantum modifications in the density fluctuation. Quantum corrections are clearly contained within the Bohm potential, which is the only quantum contribution in comparison with standard fluid models for charged particle systems. In dense astrophysical objects, such as white dwarfs and neutron stars, quantum corrections to magnetohydrodynamics can play a pivotal role. However, the quantum effects can be negligible if the density profile of the plasma system is slowly varying in comparison with its typical length scale, due to the presence of a third-order derivative of density in the Bohm potential. Note that the classical magnetohydrodynamic equations can be recovered in the limit of $\hbar = 0$. Since the fluid is assumed to be incompressible, we have

$$\nabla \cdot \boldsymbol{u} = 0. \tag{5}$$

We consider small perturbations of the equilibrium state such that every fluid variable φ (representing $\boldsymbol{u}$, $\boldsymbol{B}$, ρ, and p) has the form

$$\varphi = \varphi^{(0)} + \varphi^{(1)}, \tag{6}$$

where $\varphi^{(0)}$ is the zero-order/unperturbed quantity and $\varphi^{(1)} \ll \varphi^{(0)}$ is the first-order/perturbed value. In the following discussion, we shall denote the properties of unperturbed quantities by the superscript (0) and the properties of perturbed quantities by the superscript (1) unless otherwise stated. The first-order linearized system of equations derived from Eqs. (1)–(5) becomes

$$0 = \frac{\partial \rho^{(1)}}{\partial t} + \boldsymbol{u}^{(0)} \cdot \nabla \rho^{(1)} + \boldsymbol{u}^{(1)} \cdot \nabla \rho^{(0)}, \tag{7}$$

$$\begin{aligned} 0 = {} & \rho^{(0)}\left(\frac{\partial \boldsymbol{u}^{(1)}}{\partial t} + \boldsymbol{u}^{(0)} \cdot \nabla \boldsymbol{u}^{(1)} + \boldsymbol{u}^{(1)} \cdot \nabla \boldsymbol{u}^{(0)}\right) \\ & + \rho^{(1)}\left(\frac{\partial \boldsymbol{u}^{(0)}}{\partial t} + \boldsymbol{u}^{(0)} \cdot \nabla \boldsymbol{u}^{(0)}\right) \\ & - \rho^{(1)}\boldsymbol{g} + \nabla\left(p^{(1)} + \frac{\boldsymbol{B}^{(0)} \cdot \boldsymbol{B}^{(1)}}{\mu_0}\right) \\ & - \frac{1}{\mu_0}\left(\boldsymbol{B}^{(0)} \cdot \nabla \boldsymbol{B}^{(1)} + \boldsymbol{B}^{(1)} \cdot \nabla \boldsymbol{B}^{(0)}\right) - \boldsymbol{Q}^{(1)}, \end{aligned} \tag{8}$$

$$\begin{aligned} 0 = {} & -\frac{\partial \boldsymbol{B}^{(1)}}{\partial t} + \boldsymbol{B}^{(0)} \cdot \nabla \boldsymbol{u}^{(1)} + \boldsymbol{B}^{(1)} \cdot \nabla \boldsymbol{u}^{(0)} \\ & - \boldsymbol{u}^{(0)} \cdot \nabla \boldsymbol{B}^{(1)} - \boldsymbol{u}^{(1)} \cdot \nabla \boldsymbol{B}^{(0)}, \end{aligned} \tag{9}$$

$$0 = \nabla \cdot \boldsymbol{u}^{(1)}, \tag{10}$$

where the last term in the right-hand side of Eq. (8) $\boldsymbol{Q}^{(1)}$ is the first-order perturbation of Bohm potential term, with the expression being

$$\boldsymbol{Q}^{(1)} = \frac{\hbar^2}{2m_e m_i}\boldsymbol{T}^{(1)},$$

where

$$\begin{aligned} \boldsymbol{T}^{(1)} = {} & \frac{\rho^{(1)}}{2\left(\rho^{(0)}\right)^2}\nabla^2\rho^{(0)}\nabla\rho^{(0)} - \frac{\rho^{(1)}}{(\rho^{(0)})^3}\left(\nabla\rho^{(0)}\right)^2\nabla\rho^{(0)} \\ & + \frac{\rho^{(1)}}{4\left(\rho^{(0)}\right)^2}\nabla\left(\nabla\rho^{(0)}\right)^2 + \frac{(\nabla\rho^{(0)})^2}{2\left(\rho^{(0)}\right)^2}\nabla\rho^{(1)} \\ & - \frac{\nabla^2\rho^{(0)}}{2\rho^{(0)}}\nabla\rho^{(1)} + \frac{\nabla\rho^{(0)} \cdot \nabla\rho^{(1)}}{\left(\rho^{(0)}\right)^2}\nabla\rho^{(0)} + \frac{1}{2}\nabla\left(\nabla^2\rho^{(1)}\right) \\ & - \frac{\nabla^2\rho^{(1)}}{2\rho^{(0)}}\nabla\rho^{(0)} - \frac{1}{2\rho^{(0)}}\nabla\left(\nabla\rho^{(0)} \cdot \nabla\rho^{(1)}\right). \end{aligned}$$

For further analysis, the first-order vector equations (7)–(10) are converted to scalar equations. In the present work, the distributions of the initial/unperturbed flows are as follows: $\boldsymbol{u}^{(0)} = u^{(0)}(y)e_x$, $\boldsymbol{B}^{(0)} = B^{(0)}(y)e_x$, and $\boldsymbol{g} = -ge_y$. The perturbed quantities of $\boldsymbol{B}^{(1)}$ and $\boldsymbol{u}^{(1)}$ are assumed in the forms $\boldsymbol{u}^{(1)} = u^{(1)}e_x + v^{(1)}e_y$ and $\boldsymbol{B}^{(1)} = B^{(1)}e_x + C^{(1)}e_y$. Substituting the unperturbed and perturbed quantities into Eqs. (7)–(10), six scalar equations are obtained

$$0 = \frac{\partial \rho^{(1)}}{\partial t} + v^{(1)}\frac{d\rho^{(0)}}{dy} + u^{(0)}\frac{\partial \rho^{(1)}}{\partial x}, \tag{11}$$

$$\begin{aligned} & \rho^{(0)}\left(\frac{\partial u^{(1)}}{\partial t} + u^{(0)}\frac{\partial u^{(1)}}{\partial x} + v^{(1)}\frac{du^{(0)}}{dy}\right) \\ & = -\frac{\partial p^{(1)}}{\partial t} + \frac{C^{(1)}}{\mu_0}\frac{dB^{(0)}}{dy} + Q_x^{(1)}, \end{aligned} \tag{12}$$

$$\begin{aligned} & \rho^{(0)}\left(\frac{\partial v^{(1)}}{\partial t} + u^{(0)}\frac{\partial v^{(1)}}{\partial x}\right) \\ & = \frac{1}{\mu_0}\left(B^{(0)}\frac{\partial C^{(1)}}{\partial x} - B^{(0)}\frac{\partial B^{(1)}}{\partial y} - B^{(1)}\frac{dB^{(0)}}{dy}\right) \\ & - \frac{\partial p^{(1)}}{\partial y} - \rho^{(1)}g + Q_y^{(1)}, \end{aligned} \tag{13}$$

$$\frac{\partial B^{(1)}}{\partial t} = B^{(0)}\frac{\partial u^{(1)}}{\partial x} + C^{(1)}\frac{du^{(0)}}{dy} - u^{(0)}\frac{\partial B^{(1)}}{\partial x} - v^{(1)}\frac{dB^{(0)}}{dy}, \tag{14}$$

$$\frac{\partial C^{(1)}}{\partial t} = B^{(0)}\frac{\partial v^{(1)}}{\partial x} - u^{(0)}\frac{\partial C^{(1)}}{\partial x}, \tag{15}$$

$$0 = \frac{\partial u^{(1)}}{\partial x} + \frac{\partial v^{(1)}}{\partial y}, \tag{16}$$

where the linearized part of quantum modification in x direction [i.e., the last term in Eq. (12)] is

$$Q_x^{(1)} = \frac{\hbar^2}{4m_e m_i}T_x^{(1)},$$

$$\begin{aligned} T_x^{(1)} = {} & \frac{1}{(\rho^{(0)})^2}\left(\frac{d\rho^{(0)}}{dy}\right)^2\frac{\partial \rho^{(1)}}{\partial x} - \frac{1}{\rho^{(0)}}\frac{d^2\rho^{(0)}}{dy^2}\frac{\partial \rho^{(1)}}{\partial x} \\ & - \frac{1}{\rho^{(0)}}\frac{d\rho^{(0)}}{dy}\frac{\partial^2\rho^{(1)}}{\partial x\partial y} + \frac{\partial^3\rho^{(1)}}{\partial x^3} + \frac{\partial^3\rho^{(1)}}{\partial x\partial y^2}, \end{aligned}$$

and that in y direction [i.e., the last term in Eq. (13)] is

$$Q_y^{(1)} = \frac{\hbar^2}{2m_e m_i}T_y^{(1)},$$

$$\begin{aligned} T_y^{(1)} = {} & \frac{\rho^{(1)}}{(\rho^{(0)})^2}\frac{d^2\rho^{(0)}}{dy^2}\frac{d\rho^{(0)}}{dy} - \frac{\rho^{(1)}}{(\rho^{(0)})^3}\left(\frac{d\rho^{(0)}}{dy}\right)^2\frac{d\rho^{(0)}}{dy} \\ & - \frac{1}{\rho^{(0)}}\frac{d^2\rho^{(0)}}{dy^2}\frac{\partial \rho^{(1)}}{\partial y} + \frac{3}{2(\rho^{(0)})^2}\left(\frac{d\rho^{(0)}}{dy}\right)^2\frac{\partial \rho^{(1)}}{\partial y} \\ & - \frac{1}{2\rho^{(0)}}\frac{d\rho^{(0)}}{dy}\frac{\partial^2\rho^{(1)}}{\partial x^2} - \frac{1}{\rho^{(0)}}\frac{d\rho^{(0)}}{dy}\frac{\partial^2\rho^{(1)}}{\partial y^2} \\ & + \frac{1}{2}\frac{\partial^3\rho^{(1)}}{\partial x^2\partial y} + \frac{1}{2}\frac{\partial^3\rho^{(1)}}{\partial y^3}. \end{aligned}$$

The Fourier analysis (normal mode expansion) is performed on the perturbed quantities of $\rho^{(1)}$, $u^{(1)}$, $v^{(1)}$, $p^{(1)}$, $B^{(1)}$, and $C^{(1)}$ by letting

$$\{\rho^{(1)}, u^{(1)}, v^{(1)}, p^{(1)}, B^{(1)}, C^{(1)}\} = \{\tilde{\rho}(y), \tilde{u}(y), \tilde{v}(y), \tilde{p}(y), \tilde{B}(y), \tilde{C}(y)\}e^{ikx+nt}, \quad (17)$$

where $n = \gamma - i\omega$ is the eigenvalue with γ and ω being the linear growth rate and frequency, respectively. The tilde quantities of $\tilde{\rho}$, $\tilde{u}$, $\tilde{v}$, $\tilde{p}$, $\tilde{B}$, and $\tilde{C}$ are, respectively, the y-dependent part of the perturbed fluid density, velocities in x and y directions, pressure, and magnetic fields in x and y directions, representing the perturbation profiles.

Substituting Eq. (17) into the linearized equations (11)–(16), a second-order ODE about the y-dependent part of the perturbed fluid velocity $\tilde{v}(y)$ is yielded as the following

$$\left(n^2\rho^{(0)} + k^2\frac{B_0^2}{\mu_0} + k^2A^*\right)\frac{d^2\tilde{v}}{dy^2} + \left(n^2\frac{d\rho^{(0)}}{dy} - k^2B^*\right)\frac{d\tilde{v}}{dy} - k^2\left[n^2\rho^{(0)} + k^2\frac{B_0^2}{\mu_0} - (g - k^2C^*)\frac{d\rho^{(0)}}{dy}\right]\tilde{v} = 0, \quad (18)$$

where

$$A^* = \frac{\hbar^2}{4m_em_i}\frac{1}{\rho^{(0)}}\left(\frac{d\rho^{(0)}}{dy}\right)^2,$$

$$B^* = \frac{\hbar^2}{4m_em_i}\left[\frac{1}{(\rho^{(0)})^2}\left(\frac{d\rho^{(0)}}{dy}\right)^2 - \frac{2}{\rho^{(0)}}\frac{d^2\rho^{(0)}}{dy^2}\right]\frac{d\rho^{(0)}}{dy},$$

$$C^* = \frac{\hbar^2}{4m_em_i}\frac{1}{\rho^{(0)}}\frac{d\rho^{(0)}}{dy}.$$

It is worth noting that we have assumed that the initial magnetic field $B^{(0)} = B_0$ is a constant and the initial velocity $u^{(0)} = 0$, for mathematical simplicity. Equation (18) combined with appropriate boundary conditions comprises an eigenvalue problem, where $\tilde{v}(y)$ is the eigenfunction associated with the eigenvalue n. It is the standard form of the RTI, with continuous density profiles, in a uniform external magnetic field, including the quantum corrections.

Ignoring the quantum effects, Eq. (18) is reduced to

$$\left(n^2\rho^{(0)} + k^2\frac{B_0^2}{\mu_0}\right)\frac{d^2\tilde{v}}{dy^2} + n^2\frac{d\rho^{(0)}}{dy}\frac{d\tilde{v}}{dy} - k^2\left(n^2\rho^{(0)} + k^2\frac{B_0^2}{\mu_0}\right)\tilde{v} = 0, \quad (19)$$

which is the characteristic equation used to analyze the magnetic field gradient effects on the linear growth of the RTI.[47] If we let $B_0 = 0$ and ignore the quantum effects, then Eq. (18) can be simplified to read

$$n^2\rho^{(0)}\frac{d^2\tilde{v}}{dy^2} + n^2\frac{d\rho^{(0)}}{dy}\frac{d\tilde{v}}{dy} - k^2n^2\rho^{(0)}\tilde{v} = 0, \quad (20)$$

which has been analyzed by Mikaelian[25] and Munro[27] to research the linear growth of the RTI with continuous density profiles. Section III is devoted to describe the numerical methods solving Eq. (18).

III. NUMERICAL METHODS

The second-order ODE (18) about the y-dependent part of the perturbed fluid velocity $\tilde{v}(y)$ can be rewritten as

$$A_0(n,k)\frac{d^2\tilde{v}}{dy^2} + B_0(n,k)\frac{d\tilde{v}}{dy} + C_0(n,k)\tilde{v} = 0, \quad (21)$$

where

$$A_0(n,k) = n^2\rho^{(0)} + k^2\frac{B_0^2}{\mu_0} + k^2A^*,$$

$$B_0(n,k) = n^2\frac{d\rho^{(0)}}{dy} - k^2B^*,$$

$$C_0(n,k) = -k^2\left[n^2\rho^{(0)} + k^2\frac{B_0^2}{\mu_0} - (g - k^2C^*)\frac{d\rho^{(0)}}{dy}\right].$$

The first-order and the second-order derivatives can be solved by the central difference scheme with a second-order precision. The calculation zone ranges from 0 to $N+1$, with step length h. The boundary condition is $\tilde{v}_0 = \tilde{v}_{N+1} = 0$. For numerical simplicity, we set $A = 1$, $B = B_0/A_0$, and $C = C_0/A_0$, and then the differential equation for node j can be expressed as

$$A_j\frac{\tilde{v}_{j+1} - 2\tilde{v}_j + \tilde{v}_{j-1}}{h^2} + B_j\frac{\tilde{v}_{j+1} - \tilde{v}_{j-1}}{2h} + C_j\tilde{v}_j = 0, \quad j = 1, 2, 3, ..., N. \quad (22)$$

In matrix form, Eq. (22) reads $M\tilde{V} = 0$, where $\tilde{V} = (\tilde{v}_1, \tilde{v}_2, ..., \tilde{v}_N)^T$ is an eigencolumn with N components, and

$$M(k,n) = \begin{pmatrix} h^2C_1 - 2A_1 & A_1 + 0.5hB_1 & 0 & \cdots & 0 \\ A_2 - 0.5hB_2 & h^2C_2 - 2A_2 & A_2 + 0.5hB_2 & \cdots & 0 \\ \vdots & \vdots & \vdots & \vdots & \vdots \\ 0 & \cdots & 0 & A_N - 0.5hB_N & h^2C_N - 2A_N \end{pmatrix}$$

is an $N \times N$ tridiagonal matrix. It is known that only if the determinant $\|M(k,n)\|_{N\times N} = 0$, can the homogeneous linear equations (22) have nonzero roots.

In our numerical calculation, we use the following recursion formula to calculate the determinant $\|M\|_{N\times N} = T_N$

$$\begin{aligned} T_0 &= 1, \\ T_1 &= h^2 C_1 - 2A_1, \\ T_j &= (h^2 C_j - 2A_j)T_{j-1} \\ &\quad - (A_{j-1} + 0.5hB_{j-1})(A_j - 0.5hB_j)T_{j-2}, \quad j = 2,3,\ldots,N. \end{aligned} \tag{23}$$

For a fixed wave number k, the eigenvalue n^*, with which the determinant $\|M(k,n^*)\|_{N\times N}$ approximately equals to zero, can be obtained, by the iterative method. The main idea of this method is described as the following. The relationship between the determinant $\|M(k,n)\|$ and the eigenvalue n is expressed by a quadratic polynomial

$$\|M(k,n)\| = d\cdot n^2 + e\cdot n + f, \tag{24}$$

where d, e, and f are coefficients for the quadratic polynomial. In order to start the iteration, three approximate values are estimated, for example n_1, $n_2 = 1.1n_1$, and $n_3 = 0.9n_1$, where n_1 can be estimated within a suitable range, which do not make any differences to the calculation results. It is worth emphasizing that referring to the analytical formulas of the linear growth rate of the RTI is a good method to estimate n_1. Here, the initial n_1 is set to be $n_1^2 = Akg/(1+kL_\rho) - 2B_0^2k^2/[\mu_0(\rho_h+\rho_l)]$. The value of the determinants $m_1 = \|M(k,n_1)\|$, $m_2 = \|M(k,n_2)\|$, and $m_3 = \|M(k,n_3)\|$ can be calculated by the recursion formula equation (23). Substituting n_i, m_i, $(i = 1, 2, 3)$ into Eq. (24), three linear equations about d, e, f are obtained

$$\begin{pmatrix} n_1^2 & n_1 & 1 \\ n_2^2 & n_2 & 1 \\ n_3^2 & n_3 & 1 \end{pmatrix}\begin{pmatrix} d \\ e \\ f \end{pmatrix} = \begin{pmatrix} m_1 \\ m_2 \\ m_3 \end{pmatrix}. \tag{25}$$

Solving Eq. (25), coefficients for the quadratic polynomial are obtained as $d = \Delta_1/\Delta$, $e = \Delta_2/\Delta$, $f = \Delta_3/\Delta$, where

$$\begin{aligned} \Delta &= (n_1 - n_2)\cdot(n_1 - n_3)\cdot(n_2 - n_3), \\ \Delta_1 &= (n_2 - n_3)m_1 + (n_3 - n_1)m_2 + (n_1 - n_2)m_3, \\ \Delta_2 &= (n_3^2 - n_2^2)m_1 + (n_1^2 - n_3^2)m_2 + (n_2^2 - n_1^2)m_3, \\ \Delta_3 &= n_2n_3(n_2 - n_3)m_1 + n_1n_3(n_3 - n_1)m_2 \\ &\quad + n_1n_2(n_1 - n_2)m_3, \end{aligned}$$

are Vandermonde determinants. Now we can re-substitute the coefficients d, e, f into Eq. (24) and let

$$d\cdot n_4^2 + e\cdot n_4 + f = 0. \tag{26}$$

Then, a new eigenvalue $n_4 = -e/2d + \sqrt{(e/2d)^2 - f/d}$ is obtained. To update the older eigenvalue n_1, n_2, and n_3, we adopt the following strategy, $n_3 = n_2$, $n_2 = n_1$, and $n_1 = n_4$. Repeat the iteration proceeding until the precision is acceptable, such as $\|M(k,n_4)\| < 10^{-6}$ or $|n_4 - n_1| < 10^{-6}$, and then let $n^* = n_4$. Here, we obtain the eigenvalue n^* for Eq. (21) within an acceptable precision.

Once the eigenvalue n^* is obtained, the coefficients of Eq. (22) are completely fixed. The y-dependent part of the perturbed fluid velocity $\tilde{v}_j(y)$ for node j $(j = 2,3,4,\ldots,N)$ can therefore be calculated, by setting an appropriate value to $\tilde{v}_1$, for example $\tilde{v}_1 = 0.001$. Normalizing the resulting $\tilde{v}_j(y)$ $(j = 2,3,4,\ldots,N)$, the spatial distribution of the y-dependent part of the perturbed fluid velocity $\tilde{v}(y)$ is then obtained.

IV. ANALYTICAL DEVELOPMENT AND DISCUSSIONS

It is well-known that the characteristic equation (20), with a continuous density profile, has been analyzed by Mikaelian[25,26] to research the effects of finite thickness of density transition layer on the linear growth of the RTI in the past decades. In this section, we will follow his analytical method to solve Eq. (18) including the quantum corrections, with a continuous density profile, in a uniform external transverse magnetic field.

A. Analytic expressions of the linear growth rate

Since the limitation of the y-dependent part of the perturbed fluid velocity $\tilde{v}(y)$ is zero when y tends to be infinity [i.e., $\lim_{y\to\pm\infty}\tilde{v}(y) = 0$], Eq. (18) can be integrated over $-\infty < y < +\infty$ by parts, resulting in a simple form

$$\begin{aligned} &n^2\int_{-\infty}^{\infty}\rho^{(0)}\tilde{v}dy + k^2\frac{B_0^2}{\mu_0}\int_{-\infty}^{\infty}\tilde{v}dy - g\int_{-\infty}^{\infty}\frac{d\rho^{(0)}}{dy}\tilde{v}dy \\ &+ k^2\int_{-\infty}^{\infty}C^*\frac{d\rho^{(0)}}{dy}\tilde{v}dy = 0. \end{aligned} \tag{27}$$

Note that given a density profile $\rho^{(0)}(y)$ and an appropriate eigenfunction $\tilde{v}(y)$, the eigenvalue n can then be obtained by integrating equation (27). Fortunately, it is a good approximation to adopt the eigenfunction[25] as follows:

$$\tilde{v}(y) = \tilde{v}_{max}e^{-k|y|} = \begin{cases} \tilde{v}_{max}e^{-ky}, & y > 0 \\ \tilde{v}_{max}e^{ky}, & y < 0 \end{cases}, \tag{28}$$

where $\tilde{v}_{max}$ is the maximum of the y-dependent part of the perturbed fluid velocity $\tilde{v}(y)$. Here, we chose the constant plus exponential density profile suggested in Refs. 26, 59–61

$$\rho^{(0)} = \begin{cases} \rho_h - \dfrac{\rho_h - \rho_l}{2}e^{-y/L_\rho}, & y > 0 \\ \rho_l + \dfrac{\rho_h - \rho_l}{2}e^{y/L_\rho}, & y < 0 \end{cases}, \tag{29}$$

where L_ρ is called the density gradient scale length, and ρ_h (ρ_l) is the density away from the interface of the upper (lower) fluid. The density profile is discontinuous at $y = 0$ when $L_\rho = 0$, while it is homogeneous when $L_\rho = \infty$.

Note that the eigenvalue n in Eq. (27) is a real number. Therefore, in the following discussion, we shall use the linear growth rate γ to replace the eigenvalue n unless stated otherwise. Substituting the density profile (29) together with the approximate eigenfunction (28) into the integral equation (27), the following dispersion relation about the linear growth rate is obtained

$$\gamma^2 = \gamma_d^2 - \Gamma_m^2 - \Gamma_q^2, \tag{30}$$

where $\gamma_d^2 = Akg/[(kL_\rho + 1)]$, $\Gamma_m^2 = 2k^2B_0^2/[\mu_0(\rho_h + \rho_l)]$, and $\Gamma_q^2 = k^3Q(k)/(\rho_h + \rho_l)$. Here

$$Q(k) = \frac{\hbar^2}{4m_em_i}\left(\frac{\rho_h-\rho_l}{2L_\rho}\right)^2 \int_{-\infty}^{0} \frac{1}{\rho_l + \dfrac{\rho_h-\rho_l}{2}e^{y/L_\rho}} e^{(k+2/L_\rho)y}dy + \frac{\hbar^2}{4m_em_i}\left(\frac{\rho_h-\rho_l}{2L_\rho}\right)^2 \times \int_{0}^{\infty} \frac{1}{\rho_h - \dfrac{\rho_h-\rho_l}{2}e^{-y/L_\rho}} e^{-(k+2/L_\rho)y}dy. \tag{31}$$

Owing to $Q(k) > 0$, it is concluded that the quantum effects tend to stabilize the RTI (reduce the linear growth rate of the RTI). It is known that the quantum effects are introduced by the mass density fluctuation and represented by the Bohm potential term in Eq. (2). As can be seen, the Bohm potential term provides a force in the same direction of the density gradient, which is in the direction against the gravity. Clearly, the quantum effects play a stabilizing role in the linear growth of the RTI similar to the magnetic fields in some way.

As can be seen in Eq. (30), the square of the linear growth rate of the RTI, γ^2, consists of three terms. The first term in the right-hand side of Eq. (30), γ_d^2, is the square of the linear growth rate of the RTI only considering the density gradient stabilization. The second term in the right-hand side of Eq. (30), Γ_m^2, represents the uniform external transverse magnetic field stabilization. The last term in the right-hand side of Eq. (30), Γ_q^2, represents the quantum effect stabilization.

It is seen obviously that the integration of $Q(k)$ is hard to be calculated analytically. Fortunately, when $\rho_l < \rho_h < 3\rho_l$ $(0 < A < 0.5)$, the integration equation (31) can be expressed as a summation of three infinite series. The result takes the form

$$Q(k) = \left[\frac{1}{\rho_l}\sum_{p=0}^{\infty}\frac{\left(\dfrac{\rho_h/\rho_l-1}{2}\right)^{2p}}{k+(2p+2)/L_\rho} - \frac{1}{\rho_l}\sum_{p=0}^{\infty}\frac{\left(\dfrac{\rho_h/\rho_l-1}{2}\right)^{2p+1}}{k+(2p+3)/L_\rho} + \frac{1}{\rho_h}\sum_{p=0}^{\infty}\frac{\left(\dfrac{1-\rho_l/\rho_h}{2}\right)^{p}}{k+(p+2)/L_\rho}\right] \times \frac{\hbar^2}{4m_em_i}\left(\frac{\rho_h-\rho_l}{2L_\rho}\right)^2. \tag{32}$$

For $\rho_h \geq 3\rho_l$ $(0.5 \leq A < 1.0)$, we shall calculate the integration of $Q(k)$ numerically in the present work. It is worth pointing out that the $Q(k)$ in Eq. (32) is composed of three sum terms. Now let us analyze the convergence of the summation of three infinite series. It is found that keeping the first finite several terms in the summation can result in the solution of $Q(k)$ with very high accuracy. The square of the linear growth rate of the RTI can be expressed approximately as γ_p^2 by setting $p = 0, 1, 2, \ldots, q$. The $Q(k)$ can also be expressed approximately as $Q_p(k)$ by setting $p = 0, 1, 2, \ldots, q$. It is found that when the number (q) of the first several terms in the summation of the three infinite series is large enough, the discrepancy between γ_q^2 and γ_∞^2 [$Q_q(k)$ and $Q_\infty(k)$] is acceptable. Fig. 1 shows the γ^2 and the $Q(k)$ as a function of the k with $q = 1$, 2, 4, 8, and ∞, respectively. It clearly demonstrates the trend of convergence of γ^2 and $Q(k)$ with the increasing q. Keeping $q = 8$ for the summation in Eq. (32), γ_8^2 [$Q_8(k)$] gives the solution within 99.5% accuracy in comparison to γ_∞^2 [$Q_\infty(k)$], as shown in Fig. 1. Therefore, in the following discussion, for $0 < A < 0.5$, we just use γ_8^2 [$Q_8(k)$] to represent γ_∞^2 [$Q_\infty(k)$], for mathematical simplicity, unless stated otherwise.

In fact, we can use the square of the linear growth rate of the RTI only considering the density gradient stabilization, $\gamma_d^2 = A_Tkg/(kL_\rho + 1)$, to normalize Eq. (30), and then the normalized one becomes

$$\frac{\gamma^2}{\gamma_d^2} = 1 - \Gamma_M - \Gamma_Q, \tag{33}$$

where the normalized Γ_M (Γ_Q) represents the magnetic field (quantum effect) stabilization. Here, the expression of Γ_M is

$$\Gamma_M = kL_M(kL_\rho + 1), \tag{34}$$

where $L_M = B_0^2/[(\rho_h - \rho_l)g\mu_0/2]$, and the expression of Γ_Q is

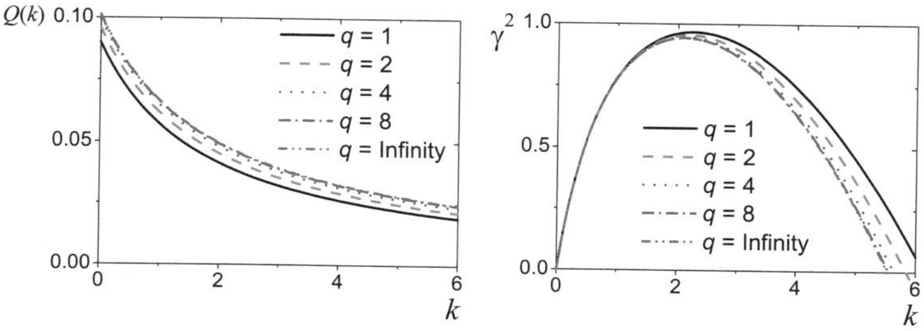

FIG. 1. The integration of $Q(k)$ and the square of the linear growth rate of the RTI γ^2 as a function of wave number k for different number (q) of the first several terms in the summation [Eq. (32)], $q = 1$ (solid line), 2 (dashed line), 4 (dotted line), 8 (dashed-dotted line), and infinity (dashed-double-dotted line), respectively, with $\mu_0 = 1.0$, $\hbar = 1.0$, $m_em_i = 1.0$, $g = 3.5$, $B_0 = 0.1$, $\rho_l = 1.0$, $\rho_h = 2.7$, and $L_\rho = 1.0$.

$$\Gamma_Q = k^3 L_Q^3 \frac{kL_\rho + 1}{kL_\rho}\left(\int_{-\infty}^{0} \frac{e^{(kL_\rho+2)\zeta}}{(1-A)/A + e^{\zeta}} d\zeta + \int_{0}^{\infty} \frac{e^{-(kL_\rho+2)\zeta}}{(1+A)/A - e^{-\zeta}} d\zeta\right), \quad (35)$$

where $L_Q^3 = \hbar^2/(8gm_e m_i)$. Regarding $0 < A < 0.5$, Eq. (35) can be written as

$$\Gamma_Q \simeq k^3 L_Q^3 Q_8(A) \frac{kL_\rho + 1}{kL_\rho}, \quad (36)$$

where

$$Q_8(A) = \sum_{p=0}^{8} \frac{[A/(1-A)]^{2p+1}}{kL_\rho + 2p + 2} - \sum_{p=0}^{8} \frac{[A/(1-A)]^{2p+2}}{kL_\rho + 2p + 3} + \sum_{p=0}^{8} \frac{[A/(1+A)]^{p+1}}{kL_\rho + p + 2}.$$

B. Comparisons with numerical solutions

Using the numerical method introduced in Sec III, the second-order ODE (18) can also be solved numerically. To compare the analytical results with the numerical solutions, we have depicted the square of the linear growth rate of the RTI γ^2 as a function of wave number k, from the analytical model and the numerical calculation, in Fig. 2. It can be seen that there is qualitative agreement between the analytical results and the numerical solutions. The same trend of the linear growth rate of the RTI with the wave number from the analytical results and the numerical solutions can be noticed in Fig. 2. The γ^2 first increases quickly with the k, reaching a maximum and then drops slowly to zero. Nevertheless, the numerical solutions and the analytical results do not exactly coincide with each other. When the wave number $k = 2.1$, the linear growth rate of the RTI from both the numerical solutions and the analytical results reach their maximums with $\gamma_{analytical} = 0.51$ and $\gamma_{numerical} = 0.46$, respectively. The numerical solution is 9.8% less than the analytical result at their maximums. Which mechanism induces the discrepancy between the numerical solutions and the analytical results? The underlying reason is that the assumption of the eigenfunction equation (28) is an approximate estimate. This can be seen as follows.

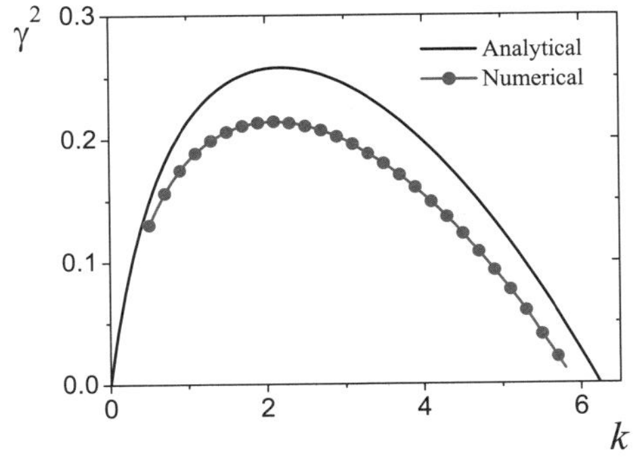

FIG. 2. The square of the linear growth rate of the RTI γ^2 as a function of wave number k from the analytical model and the numerical solutions with $\mu_0 = 1.0$, $\hbar = 1.0$, $m_e m_i = 1.0$, $g = 3.5$, $B_0 = 0.1$, $\rho_l = 1.3$, $\rho_h = 1.7$, and $L_\rho = 1.1$. The analytical results are from Eq. (30). The spatial distributions of the normalized y-dependent part of the perturbed fluid velocity of the wave number $k = 2$ and 4 are plotted in Figure 3.

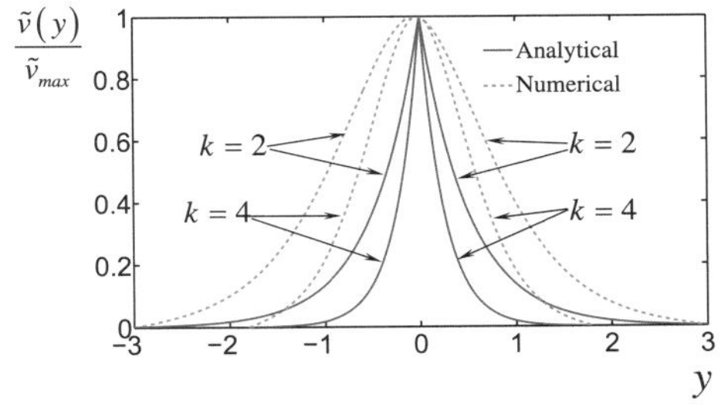

FIG. 3. Spatial distribution of the normalized y-dependent part of the perturbed fluid velocity $\tilde{v}(y)/\tilde{v}_{max}$ for two wave number $k = 2$ and $k = 4$ with parameters same as Figure 2. The analytical results are from Eq. (28).

Before integrating the eigenvalue equation (27), we have to assume an analytic eigenfunction $\tilde{v}(y)$, for example Eq. (28), and substitute the constructed eigenfunction into the integral equation as mentioned above. Since the assumption of the analytic eigenfunction is just an simple approximation which in fact is the "exact" eigenfunction for the case with a sharp density discontinuity,[3] it will certainly bring some errors to describe the spatial distribution of the y-dependent part of the perturbed fluid velocity for the case with a continuous density profile, and hence, affect the linear growth rate of the RTI. To see this more clearly, we have studied the spatial distribution of the normalized y-dependent part of the perturbed fluid velocity $\tilde{v}(y)/\tilde{v}_{max}$ from the analytical assumptions and the numerical solutions, respectively. It is found that the spatial distribution of $\tilde{v}(y)/\tilde{v}_{max}$ is strongly dependent on k. Typically, in Fig. 3, we plot the $\tilde{v}(y)/\tilde{v}_{max}$ with $k = 2.0$ and $k = 4.0$. It can be seen that for a fixed k, the analytic assumptions cannot preciously estimate the $\tilde{v}(y)/\tilde{v}_{max}$. For the analytic assumptions, they have a very sharp profile and peak in the $y = 0$. On the other hand, the numerical solutions are much more smoother at their maximum and peak in the $y < 0$ region. We see that the eigenfunctions from both the numerical solutions and the analytic assumptions become more localized with increasing k. And for a fixed y, the value of the $\tilde{v}(y)/\tilde{v}_{max}$ from numerical solutions are always larger than that from the analytic assumptions. This explains why the linear growth rates of the RTI from the analytical results are always larger than that from the numerical solutions, as shown in Fig. 2. Notwithstanding the $\tilde{v}(y)/\tilde{v}_{max}$ from the adopted assumptions does not exactly coincide with that from numerical solutions, they do exhibit the main characteristic in common. For example, the profiles of spatial distribution of $\tilde{v}(y)/\tilde{v}_{max}$ from the analytic assumption and the numerical solutions have the same behavior (i. e., their trends with k are identical). Therefore, the approximation of the eigenfunction (28) can physically represent the spatial distribution of the y-dependent part of the perturbed fluid velocity. Generally speaking, the numerical solutions agree well with the analytical results. In short, the assumption of Eq. (28) is not an exact eigenfunction but at least an acceptable one.

C. Cutoff and maximum linear growth rate wave numbers

Now let us consider the dependence of the linear growth rate on the perturbation wave number. If the last two terms on the right-hand side of Eq. (30) are large enough, the instability would be totally restrained. In other words, when the perturbation wave number become large enough, the magnetic field stabilization and/or the quantum effect stabilization become/becomes large enough, resulting in the quenching of the instability. On the other hand, it is well-known that the linear growth rate of the RTI increases monotonically with the wave number under the classical condition, for example only considering the density gradient stabilization.

As mentioned above, we can therefore define two useful wave numbers, k_c and k_{max}. Here, k_c is the cutoff wave number where the linear growth rate of the RTI becomes zero and k_{max} is called the maximum linear growth rate wave number leading to the square of linear growth rate of the RTI reaching the maximum. The k_c satisfies the following

$$\frac{Ak_c g}{k_c L_\rho + 1} - \frac{2k_c^2 B_0^2}{(\rho_h + \rho_l)\mu_0} - \frac{k_c^3 Q(k_c)}{\rho_h + \rho_l} = 0. \tag{37}$$

In the case of $k > k_c$, the instability is fully quenched and becomes a steady wave. The RTI would occur only for the long-wavelength perturbation when the magnetic field stabilization and/or the quantum effect stabilization are/is considered. The RTI appearing in the short-wavelength perturbation region can be restrained by the magnetic field stabilization and/or the quantum effect stabilization. When $k < k_{max}$, the γ^2 increase monotonically with increasing k while when $k > k_{max}$, the γ^2 starts to decrease with increasing k monotonically. To obtain the k_{max}, we have to solve the equation as follows:

$$\frac{d\gamma^2}{dk}\Big|_{k=k_{max}} = 0. \tag{38}$$

Note that the equations that k_c and k_{max} satisfied are cubic equations. Generally speaking, these equations cannot be solved analytically. Here, the k_c and the k_{max} are obtained numerically by the method of bisection.

Fig. 4 exhibits the maximum linear growth rate wave number k_{max} and the cut off wave number k_c as a function of magnetic field B_0. It is seen evidently that when the strength of magnetic field B_0 is small, k_{max} and k_c decrease with increasing B_0 quickly, and then they decrease slowly for a stronger magnetic field. This implies that the magnetic fields decrease the maximum linear growth rate wave number and the cut off wave number.

Figs. 5 and 6 show the cut off wave number k_c and the maximum linear growth rate wave number k_{max} at different density gradient scale lengths L_ρ and densities of the heavy fluid ρ_h, respectively. We find that in most cases for a fixed L_ρ, both k_{max} and k_c increase with increasing ρ_h monotonously. However, the density gradient scale length plays a more complicated role in the k_{max} and the k_c. For a large ρ_h, typically $\rho_h > 3.0$ in Figs. 5 and 6, the k_{max} and the k_c first

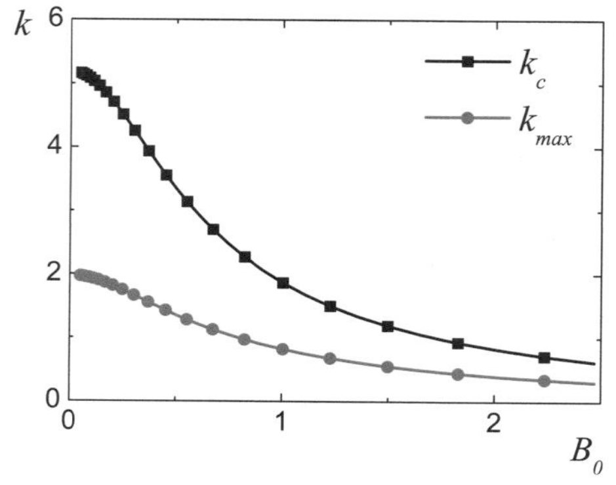

FIG. 4. Cutoff and maximum linear growth rate wave numbers k_c and k_{max} against the strength of magnetic field B_0 with $\mu_0 = 1.0$, $\hbar = 1.0$, $m_e m_i = 1.0$, $g = 3.5$, $\rho_l = 1.0$, $\rho_h = 4.7$, and $L_\rho = 1.1$.

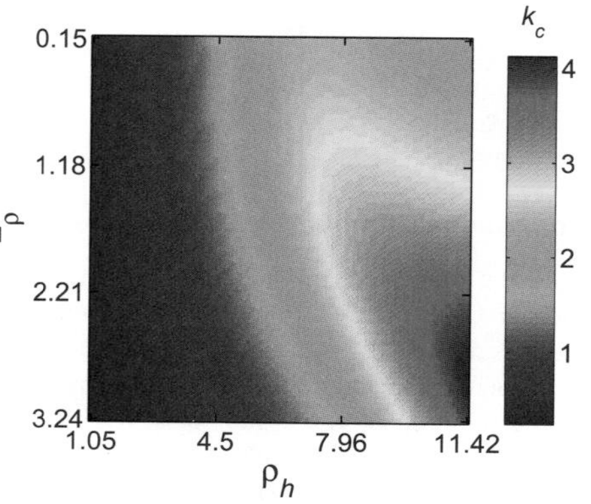

FIG. 5. Cut off wave numbers k_c at different density gradient scale lengths L_ρ and densities of the heavy fluid ρ_h with $\mu_0 = 1.0$, $\hbar = 1.0$, $m_e m_i = 1.0$, $g = 3.5$, $B_0 = 0.5$, and $\rho_l = 1.0$.

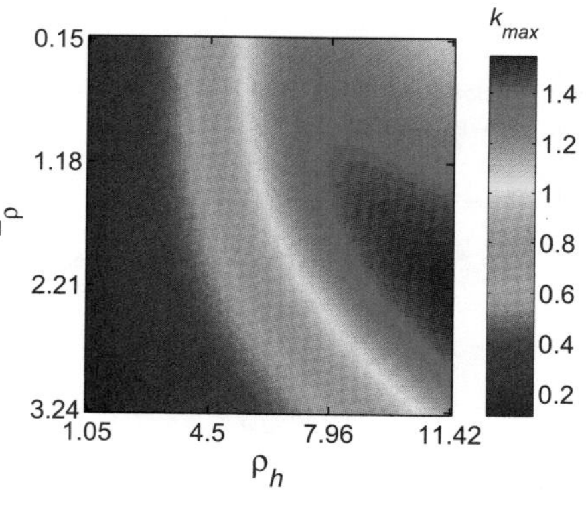

FIG. 6. Maximum linear growth rate wave numbers k_{max} at different density gradient scale lengths L_ρ and densities of the heavy fluid ρ_h with parameters same as Fig. 5.

increase with increasing L_ρ, reaching their maximum and then decrease gradually.

V. EFFECTS OF DENSITY GRADIENT, QUANTUM MECHANISM, AND MAGNETIC FIELD

In this section, we will perform a theoretical analysis of the stabilization of the instability due to the density gradients, the magnetic fields, and the quantum effects, as well as the combined contributions and the competitions among them. As mentioned in Sec. IV, the result of $Q(k)$ can be expressed as the series form for $0 < A < 0.5$, whereas to the best of our knowledge, only numerical solution of $Q(k)$ can be obtained for $0.5 \le A < 1$. In order to obtain a clear physical understanding, our discussions here about the quantum effects are within $0 < A < 0.5$ and adopt the series solution of $Q(k)$ [Eq. (32)].

A. Density gradient stabilization

The linear growth rate of the RTI only considering the density gradient stabilization can be expressed as

$$\frac{\gamma_d^2}{kg} = \frac{A}{1 + kL_\rho}. \tag{39}$$

On the one hand, the long-wavelength perturbations ($kL_\rho \ll 1$) are not influenced by the finite thickness of density transition layer and "feel" a sharp step-like density configuration. Their linear growth rates reduce to

$$\gamma_d^2 \rightarrow \gamma_c^2, \quad kL_\rho \ll 1. \tag{40}$$

On the other hand, the short-wavelength perturbations ($kL_\rho \gg 1$) are localized inside the density transition layer and "feel" a quite homogeneous density configuration. Their linear growth rates become

$$\gamma_d^2 \rightarrow \frac{Ag}{L_\rho}, \quad kL_\rho \gg 1. \tag{41}$$

It is worth noting that, if the density is smoothly varying between the two fluids and the minimum density gradient scale length in the form $L_{\rho(min)} = \min[\rho/(d\rho/dy)]$ is finite, e.g., the density profile as shown in Eq. (29), then the linear growth rate of RTI can be written as

$$\frac{\gamma_d^2}{kg} = \frac{A}{1 + AkL_{\rho(min)}}, \tag{42}$$

which is a useful estimation widely used in computing the contribution of the density gradient stabilization from the ablation in the context of ICF.

B. Magnetic field stabilization

The linear growth rate of the RTI only considering the magnetic field stabilization reads as

$$\frac{\gamma_m^2}{kg} = A(1 - kL_B). \tag{43}$$

Note that if the strength of magnetic field is strong enough (i.e., $kL_B \ge 1$), the RTI can be completely suppressed. The main influence of the magnetic field concerns the reduction in the linear growth rate of the RTI compared to the classical case γ_c and the stabilization of the fluid interface with a larger wave number than the cut-off one.

C. Quantum effect stabilization

The linear growth rate of the RTI only considering the quantum effect stabilization takes the form

$$\frac{\gamma_q^2}{kg} \simeq A\left[1 - \frac{k^3 L_B^3 Q_8(A)}{kL_\rho}\right]. \tag{44}$$

As can be seen from Eq. (44) that similar to the magnetic fields, the primary result of the quantum corrections also concerns the reduction in the linear growth rate of the RTI compared to the classical case γ_c and the stabilization of the fluid interface with a wave number larger than the cut-off one.

From Eq. (36), it can be seen that the quantum effect stabilization term Γ_Q is the function of A, kL_ρ, and kL_Q. To see how the A, kL_ρ, and kL_Q affect the Γ_Q clearly, we plot the Γ_Q as a function of A and kL_ρ at the fixed $kL_Q = 1.0$ in Fig. 7. The Γ_Q increases with increasing A, which means that quantum effects play a pronounced role when the A is large. This is because of the Bohm potential term in Eq. (2) which is due to the quantum modification in the mass density fluctuation. It is well-known that the density fluctuation is remarkable when the A is large. As can be seen in Fig. 7, the Γ_Q is remarkable only for the small normalized density gradient scale length. That is to say, the quantum effect stabilization of the RTI is need to be well considered only if the kL_ρ is small enough.

D. Combined contributions and stabilization competitions

As can be seen from Eq. (30), both the Γ_q and the Γ_m are positive, which means that the Γ_q and the Γ_m play a role in decreasing the linear growth of the RTI. To see this

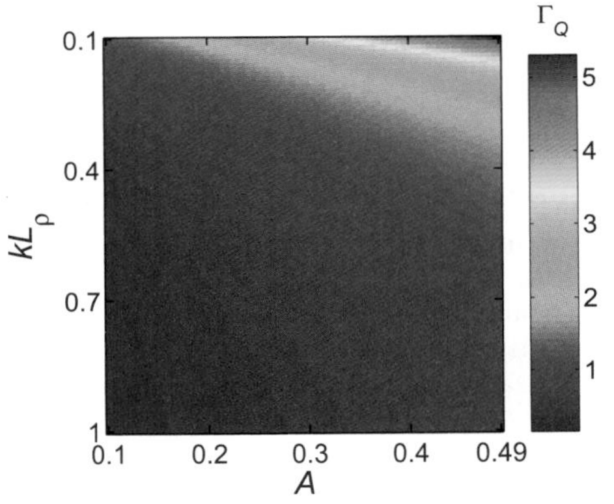

FIG. 7. Quantum effects Γ_Q as a function of A and kL_ρ with $kL_Q = 1.0$.

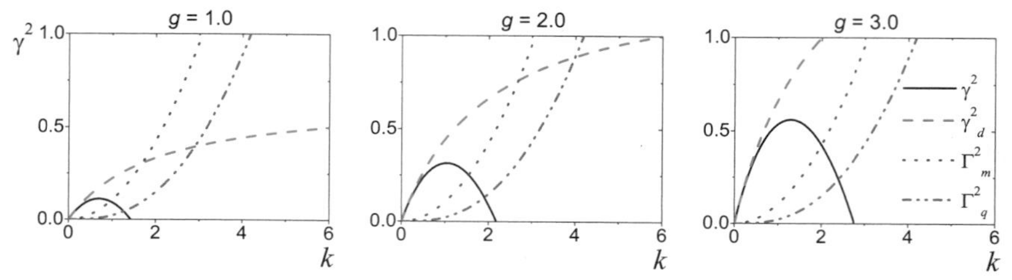

FIG. 8. The γ_d^2 (dashed line), Γ_m^2 (dotted line), Γ_q^2 (dashed-double-dotted line), and γ^2 (solid line) against the k with $\mu_0 = 1.0$, $\hbar = 1.0$, $m_e m_i = 1.0$, $B_0 = 0.4$, $\rho_l = 1.0$, $\rho_h = 2.0$, and $L_\rho = 0.5$, for the fixed $g = 1.0$, 2.0, and 3.0, respectively.

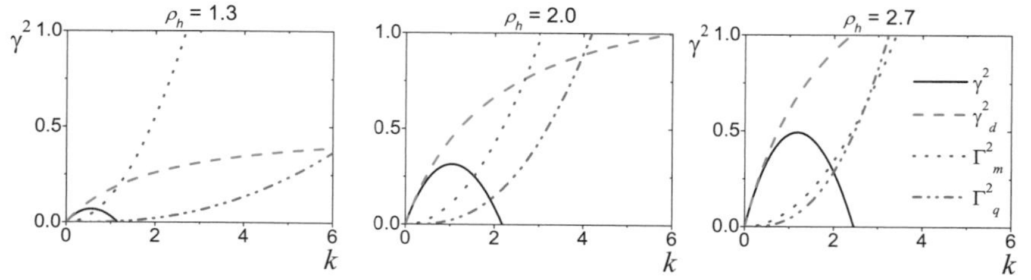

FIG. 9. The γ_d^2 (dashed line), Γ_m^2 (dotted line), Γ_q^2 (dashed-double-dotted line), and γ^2 (solid line) against the k with $\mu_0 = 1.0$, $\hbar = 1.0$, $m_e m_i = 1.0$, $g = 2.0$, $B_0 = 0.4$, $\rho_l = 1.0$, and $L_\rho = 0.5$, for the fixed $\rho_h = 1.3$, 2.0, and 2.7, respectively.

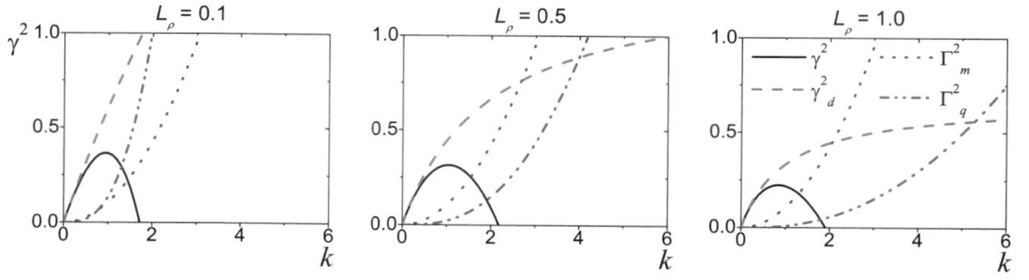

FIG. 10. The γ_d^2 (dashed line), Γ_m^2 (dotted line), Γ_q^2 (dashed-double-dotted line), and γ^2 (solid line) against the k with $\mu_0 = 1.0$, $\hbar = 1.0$, $m_e m_i = 1.0$, $g = 2.0$, $B_0 = 0.4$, $\rho_l = 1.0$, and $\rho_h = 2.0$, for the fixed $L_\rho = 0.1$, 0.5, and 1.0, respectively.

clearly, we analyze the relationship between the γ^2 (γ_d, Γ_m, and Γ_q) and the k.

Fig. 8 displays the square of the linear growth rate of the RTI only considering the density gradient stabilization (γ_d^2), the magnetic field stabilization term (Γ_m^2), the quantum effect stabilization term (Γ_q^2), and the square of the linear growth rate of the RTI considering the combined stabilization (γ^2), as a function of wave number (k) with the fixed acceleration $g = 1.0$, 2.0, and 3.0, respectively. As can be seen in Fig. 8, the curves of γ_d^2, Γ_m^2, Γ_q^2 increases with increasing k, while the curve of γ^2 first increases with k researching a maximum k_{max} and then decay to zero at the cutoff wave number k_c. We can see that for the fixed k, the larger the g is, the larger the γ_d^2, γ^2, k_{max} and k_c are.

Fig. 9 presents the γ_d^2, Γ_m^2, Γ_q^2, and γ^2 as a function of k with the fixed $\rho_h = 1.3$, 2.0, and 2.7, respectively. As can be seen in Fig. 9, the curves of γ_d^2, Γ_m^2, Γ_q^2 increases with increasing k, while the curve of γ^2 first increases with k researching a maximum k_{max} and then decay to zero at the cutoff wave number k_c. For the fixed k, the larger the ρ_h is, the larger the γ_d^2, Γ_q^2, γ^2, k_{max}, and k_c are, but the smaller the Γ_m^2 is. This implies that the Atwood number effect tends to strengthen the density gradient stabilization and the quantum effect stabilization of the RII, but the sum of the densities of two fluids weakens the magnetic field stabilization of the instability.

Fig. 10 exhibits the γ_d^2, Γ_m^2, Γ_q^2, and γ^2 as a function of k with the fixed $L_\rho = 0.1$, 0.5, and 1.0, respectively. As can be seen in Fig. 10, the curves of γ_d^2, Γ_m^2, Γ_q^2 increases with increasing k, while the curve of γ^2 first increases with k researching a maximum k_{max} and then decay to zero at the cutoff wave number k_c. For the fixed k, the larger the L_ρ is, the smaller the γ_d^2, Γ_q^2, γ^2, and k_{max} are, as shown in Fig. 10. This means that the density gradient scale length effect plays a role in strengthening the density gradient stabilization and the quantum effect stabilization and hence damps the RTI growth.

Fig. 11 depicts the γ_d^2, Γ_m^2, Γ_q^2, and γ^2 as a function of k with the fixed $B_0 = 0.1$, 0.4, and 1.0, respectively. As can be seen in Fig. 11, the curves of γ_d^2, Γ_m^2, Γ_q^2 increases with increasing k, while the curve of γ^2 first increases with k researching a maximum k_{max} and then decay to zero at the cutoff wave number k_c. For the fixed k, the larger the B_0 is, the larger the γ_m^2 is, but the smaller the γ^2, k_{max} and k_c are, as shown in Fig. 11. This shows the magnetic field stabilization of the RTI.

Now let us discuss the relation between the square of the normalized linear growth rate of the RTI γ^2/γ_c^2 and the normalized density gradient scale length kL_ρ in Eq. (33). It is known that the magnetic field stabilization of the instability is remarkable for large kL_ρ, while the quantum effect stabilization plays a pronounced role at the small kL_ρ. Thus, there is a competition between them when one changes the kL_ρ. When the kL_ρ is small, the Γ_Q plays a more substantial role than the Γ_M. As mentioned above, the increasing kL_ρ would lead the Γ_Q to decrease, and hence, the γ^2/γ_c^2 increases with increasing

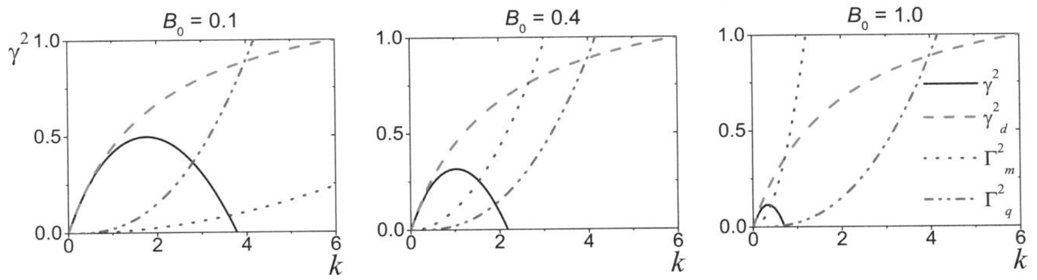

FIG. 11. The γ_d^2 (dashed line), Γ_m^2 (dotted line), Γ_q^2 (dashed-double-dotted line), and γ^2 (solid line) against the k with $\mu_0 = 1.0$, $\hbar = 1.0$, $m_e m_i = 1.0$, $g = 2.0$, $\rho_l = 1.0$, $\rho_h = 2.0$, and $L_\rho = 0.5$, for the fixed $B_0 = 0.1$, 0.4, and 1.0, respectively.

kL_ρ at first when the kL_ρ is small. With increasing kL_ρ, the Γ_Q becomes smaller and smaller (i.e., the quantum effect stabilization does not play a dominant role). Meanwhile, the Γ_M [which is proportional to $kL_\rho + 1$ as shown in Eq. (34)] begins to play a major role at the large kL_ρ. Therefore, the γ^2/γ_c^2 is reduced, for the further increasing kL_ρ. In a word, for the small kL_ρ the RTI is suppressed by the quantum effect stabilization, while for the large kL_ρ the RTI is damped by the magnetic field stabilization, and hence, for the medium kL_ρ the RTI would reach its maximum linear growth rate.

VI. APPLICATIONS IN ICF AND ASTROPHYSICS

This section is devoted to the detailed analysis of the stabilization of the linear growth of the RTI due to the magnetic fields and the quantum effects, for parameters closely related to ICF (Refs. 62 and 63) and white dwarfs,[64,65] respectively. In order to discuss physical problems more explicitly, we study the iso-contour of $\Gamma_Q = 0.1$ and 1.0, and $\Gamma_M = 0.1$ and 1.0, respectively, in the $k - L_\rho$ planes. The quantum effect (magnetic field) stabilization can be neglected if $\Gamma_Q < 0.1$ ($\Gamma_M < 0.1$) and the instability is fully quenched when $\Gamma_Q \geq 1$ ($\Gamma_M \geq 1$). We should carefully consider the quantum effect (magnetic field) stabilization when $0.1 < \Gamma_Q < 1.0$ ($0.1 < \Gamma_M < 1.0$).

Fig. 12 illustrates the stabilization of the linear growth of the RTI due to the magnetic fields and the quantum effects, for parameters closely related to ICF, in the presence of magnetic field with strength $B_0 = 1.0$, 10^2, and 10^4 T, respectively. As can be seen, in the first figure of Fig. 12 for $B_0 = 1.0$ T, there are five regimes with difference physical mechanisms. In regime (a), there are $\Gamma_Q \geq 1.0$ or $\Gamma_M \geq 1.0$, and hence, the instability is completely damped by the magnetic field stabilization and/or the quantum effect stabilization. In regime (b), there are $0.1 < \Gamma_Q < 1.0$ and $0.1 < \Gamma_M < 1.0$. Therefore, in regime (b), both the quantum effect stabilization and the magnetic field stabilization play a central role in the linear growth of the RTI. In regime (c), there are $0.1 < \Gamma_M < 1.0$ and $\Gamma_Q < 0.1$. Accordingly, in regime (c), one can only consider the magnetic field stabilization but ignore the quantum effect stabilization. In regime (d), there are $0.1 < \Gamma_Q < 1.0$ and $\Gamma_M < 0.1$. In regime (d), instead, one should carefully consider the quantum effect stabilization but ignore the magnetic field stabilization. In regime (e), there are $\Gamma_Q \leq 0.1$ or $\Gamma_M \leq 0.1$. This implies in regime (e) that both the quantum effect stabilization and the magnetic field stabilization can be ignored and the density gradient stabilization dominates the linear growth of the RTI.

Regarding the quantum effect stabilization within the context of ICF, under considerations $L_\rho = 10^{-9} - 10^{-3}$ m, one consider it only when $k \gtrsim 2 \times 10^8$ m^{-1}, as shown in Fig. 12. In other words, in ICF, the quantum effect stabilization plays a pronounced role only when the wave length of a perturbation can comparable to the diameter of an atom. Considering the parameters directly related to the present ICF experiments and designs with $k \sim 10^4 - 10^6$ m^{-1} and $L_\rho \sim 10^{-7} - 10^{-5}$ m, it is found that when $B_0 < 10^2$ T, both the magnetic field stabilization and the quantum effect stabilization, can be ignored, as shown in the first two figures of Fig. 12; when $10^2 < B_0 < 10^4$ T, the magnetic field stabilization plays a significant role in the RTI growth and should be included carefully; when $B_0 > 10^4$ T, the instability is completely suppressed by the magnetic field stabilization, as shown in the third figure of Fig. 12.

Fig. 13 displays the stabilization of the linear growth of the RTI due to the magnetic fields and the quantum effects, for parameters closely related to white dwarfs, in the presence of magnetic field with strength $B_0 = 10$, 10^3, and 10^5

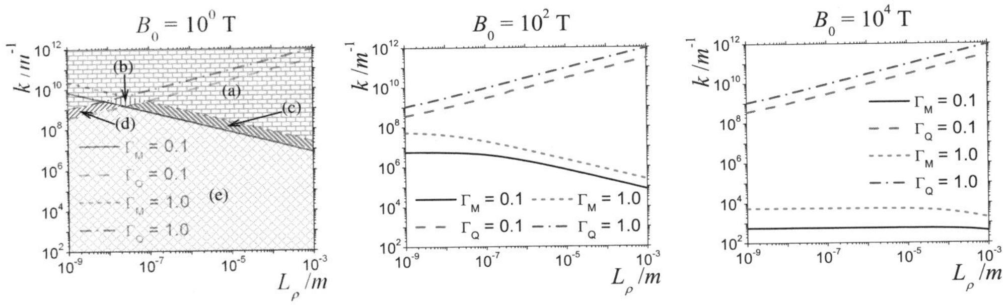

FIG. 12. Stabilization of the linear growth of the RTI due to the magnetic fields and the quantum effects, for parameters closely related to ICF, in the presence of the magnetic field with strength $B_0 = 1.0$, 10^2, and 10^4 T, respectively. The parameters are $\mu_0 \simeq 1.26 \times 10^{-6}$ H/m, $\hbar \simeq 1.05 \times 10^{-34}$ J s, $m_e \simeq 9.11 \times 10^{-31}$kg, $m_i = 13m_p \simeq 2.17 \times 10^{-26}$ kg, $g = 1.0 \times 10^{14}$ m/s^2, $\rho_l = 1.0 \times 10^3$ kg/m^3, and $\rho_h = 1.0 \times 10^4$ kg/m^3.

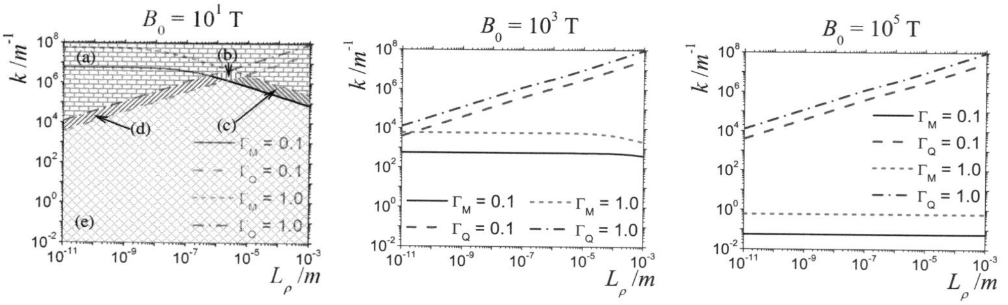

FIG. 13. Stabilization of the linear growth of the RTI due to the magnetic fields and the quantum effects, for parameters closely related to white dwarfs, in the presence of the magnetic field with strength $B_0 = 10$, 10^3, and 10^5 T, respectively. The parameters are $\mu_0 \simeq 1.26 \times 10^{-6}$ H/m, $\hbar \simeq 1.05 \times 10^{-34}$ J s, $m_e \simeq 9.11 \times 10^{-31}$ kg, $m_i = 12m_p \simeq 2.01 \times 10^{-26}$ kg, $g = 1.0 \times 10^6$ m/s^2, $\rho_l = 1.0 \times 10^9$ kg/m^3, and $\rho_h = 1.0 \times 10^{10}$ kg/m^3.

T, respectively. As can be seen in the first figure of Fig. 13 for $B_0 = 10$ T, there are five regimes with difference physical mechanisms, similar to the first figure of Fig. 12.

As can be seen in Fig. 13, under considerations $L_\rho = 10^{-11} - 10^{-3}$ m, the quantum effect stabilization begins to play a pronounced role only when the $k \gtrsim 2 \times 10^4$ m^{-1}. However, when the magnetic field with strength achieves $\sim 10^3$ T, the quantum effect stabilization can be ignored, as shown in the second and the third figures of Fig. 13. In the presence of magnetic field with strength up to $\sim 10^3$ T, the instability with $k \gtrsim 5 \times 10^3$ m^{-1} is entirely quenched; the instability with $k \sim 5 \times 10^2 - 5 \times 10^3$ m^{-1} should carefully consider the magnetic field stabilization; the instability with $k \lesssim 5 \times 10^2$ m^{-1} could ignore the magnetic field stabilization but only consider the density gradient stabilization. In the presence of magnetic field with strength up to $\sim 10^5$, the instability with $k \gtrsim 5 \times 10^{-1}$ m^{-1} is completely suppressed; the instability with $k \sim 5 \times 10^{-2} - 5 \times 10^{-1}$ m^{-1} should include the magnetic field stabilization; the instability with $k \lesssim 5 \times 10^{-2}$ m^{-1} could ignore the magnetic field stabilization but only consider the density gradient stabilization.

VII. SUMMARY AND CONCLUSION

In this study, based on the QMHD model, we have analytically as well as numerically investigated the stabilization of the linear growth of the RTI from density gradients, magnetic fields, and quantum effects, in an ideal incompressible magnetized plasma. A second-order ordinary different equation (eigenvalue equation) for the RTI considering corrections of density gradients, magnetic fields, and quantum effects are obtained. Using a constant plus exponential density distribution, analytic expressions of the linear growth rate of the RTI, with modifications of density gradients, magnetic fields, and quantum effects, are presented. Numerical approaches are performed to solve the second-order eigenvalue equation resulting in the linear growth rate of the RTI. The analytical model proposed here agrees with the numerical calculation.

Our results demonstrate that the density gradient (magnetic field) has a stabilizing effect on the RTI (reduce the linear growth of the RTI) and the RTI is affected significantly by the quantum effects. The quantum effect stabilization is remarkable when the density gradient scale length is sufficiently short. Moreover, the Atwood number effect plays a role in strengthening the quantum effect stabilization of the instability. The presence of the external transverse magnetic fields beside the quantum effects will bring about more stability on the RTI. The RTI can be completely quenched by the magnetic field stabilization and/or the quantum effect stabilization in proper circumstances resulting in a cutoff wavelength.

The stabilization of the linear growth of the RTI for parameters closely related to ICF and white dwarfs is discussed in some details. It is found that the magnetic field stabilization always plays a more pronounced role in the linear growth of the RTI, in comparison to the quantum effect stabilization. The magnetic field stabilization should be well considered when it reaches as high as 10^2 T and the RTI disappears for a magnetic field with strength up to 10^4 T, in the context of the present ICF experiments and designs. Regarding perturbations with wave number greater than 5×10^{-1} m^{-1}, the RTI is completely quenched in a white dwarf when its magnetic field strength is up to 10^5 T. Our results are expected to cause further understanding of mixing in supernovas and other RTI-driven objects such as ICF ignition targets.

We would like to point out that our regular theory as shown above is based on the hypothesis of ideal incompressible plasmas and the Bohm potential is the only quantum correction. In fact, in real systems, there are still other factors contributing to quantum effects, e.g., the equation of state (EoS) of a non-ideal gas and the compressibility, which will be pursued in our future research.

ACKNOWLEDGMENTS

This work was supported by the National Natural Science Foundation of China (Grant Nos. 10935003, 10905006, 11075024, 11074300, and 10835003) and the National High-Tech ICF Committee.

[1]L. Rayleigh, Proc. London Math. Soc. **14**, 170 (1883).
[2]G. Taylor, Proc. R. Soc. London, Ser. A **201**, 192 (1950).
[3]S. Chandrasekhar, *Hydrodynamic and Hydromagnetic Stability* (Oxford University, London, 1961).
[4]J. H. Nuckolls, J. Wood, A. Thiessen, and G. Zimmerman, Nature **239**, 139 (1972).

[5]J. D. Lindl, P. Amendt, R. L. Berger, S. G. Glendinning, S. H. Glenzer, S. W. Haan, R. L. Kauffman, O. L. Landen, and L. J. Suter, Phys. Plasmas **11**, 339 (2004).
[6]S. Atzeni and J. Meyer-ter-Vehn, *The Physics of Inertial Fusion: Beam Plasma Interaction Hydrodynamics, Hot Dense Mater* (Oxford University, Oxford, 2004).
[7]X. T. He and W. Y. Zhang, Eur. Phys. J. D **44**, 227 (2007).
[8]J. O. Kane, H. F. Robey, B. A. Remington, R. P. Drake, J. Knauer, D. D. Ryutov, H. Louis, R. Teyssier, O. Hurricane, D. Arnett, R. Rosner, and A. Calder, Phys. Rev. E **63**, 055401 (2001).
[9]D. Arnett, Astrophys. J., Suppl. Ser. **127**, 213 (2000).
[10]D. H. Kalantar, B. A. Remington, E. A. Chandler, J. D. Colvin, D. M. Gold, K. O. Mikaelian, S. V. Weber, L. G. Wiley, J. S. Wark, A. Loveridge, A. Hauer, B. H. Failor, M. A. Meyers, and G. Ravichandran, Astrophys. J. Suppl. Ser. **127**, 357 (2000).
[11]J. Kane, D. Arnett, B. A. Remington, S. G. Glendinning, G. Bazan, R. P. Drake, and B. A. Fryxell, Astrophys. J. Suppl. Ser. **127**, 365 (2000).
[12]F. Nakamura, C. F. Mckee, R. I. Klein, and R. T. Fisher, Astrophys. J. Suppl. Ser. **164**, 477 (2006).
[13]H. Azechi, K. Shigemori, M. Nakai, N. Miyanaga, and H. Takabe, Astrophys. J. Suppl. Ser. **127**, 219 (2000).
[14]V. N. Gamezo, A. M. Khokhlov, E. S. Oran, A. Y. Chtchelkanova, and R. O. Rosenberg, Science **299**, 77 (2003).
[15]A. Burrows, Nature **403**, 727 (2000).
[16]I. Hachisu, T. Matsuda, K. Nomoto, and T. Shigeyama, Astrophys. J **358**, L57 (1990).
[17]E. Müller, B. Fryxell, and D. Arnett, Astron. Astrophys. **251**, 505 (1991).
[18]R. P. Drake, *High-Energy-Density Physics: Fundamentals, Inertial Fusion and Experimental Astrophysics* (Springer, New York, 2006).
[19]Q. Zhang, Phys. Rev. Lett. **81**, 3391 (1998).
[20]V. N. Goncharov, Phys. Rev. Lett. **88**, 134502 (2002).
[21]K. O. Mikaelian, Phys. Rev. E **67**, 026319 (2003).
[22]S.-I. Sohn, Phys. Rev. E **67**, 026301 (2003).
[23]P. Ramaprabhu and G. Dimonte, Phys. Rev. E **71**, 036314 (2005).
[24]P. Ramaprabhu, G. Dimonte, Y. N. Young, A. C. Calder, and B. Fryxell, Phys. Rev. E **74**, 066308 (2006).
[25]K. O. Mikaelian, Phys. Rev. Lett. **48**, 1365 (1982).
[26]K. O. Mikaelian, Phys. Rev. A **33**, 1216 (1986).
[27]D. H. Munro, Phys. Rev. A **38**, 1433 (1988).
[28]D. Ucer and V. V. Mirnov, Astrophys. J. Suppl. Ser. **127**, 509 (2000).
[29]L. F. Wang, W. H. Ye, and Y. J. Li, Phys. Plasmas **17**, 052305 (2010).
[30]S. Bodner, Phys. Rev. Lett. **33**, 761 (1974).
[31]M. Tabak, D. H. Munro, and J. D. Lindl, Phys. Fluids B **2**, 1007 (1990).
[32]H. Takabe, K. Mima, L. Montierth, and R. L. Morse, Phys. Fluids **28**, 3676 (1985).
[33]J. Sanz, Phys. Rev. E **53**, 4026 (1996).
[34]A. R. Piriz, J. Sanz, and L. F. Ibañez, Phys. Plasmas **4**(4), 1117 (1997).
[35]V. N. Goncharov, R. Betti, R. L. McCrory, P. Sorotokin, and C. P. Verdon, Phys. Plasmas **3**(4), 1402 (1996).
[36]R. Betti, V. N. Goncharov, R. L. McCrory, and C. P. Verdon, Phys. Plasmas **5**(5), 1446 (1998).
[37]W. Ye, W. Zhang, and X. T. He, Phys. Rev. E **65**, 057401 (2002).
[38]L. F. Wang, W. H. Ye, Z. M. Sheng, W.-S. Don, Y. J. Li, and X. T. He, Phys. Plasmas **17**, 122706 (2010).
[39]L. F. Wang, W. H. Ye, and X. T. He, Phys. Plasmas **19**, 012706 (2012).
[40]W. H. Ye, L. F. Wang, and X. T. He, Phys. Plasmas **17**, 122704 (2010).
[41]R. J. Mason and M. Tabak, Phys. Rev. Lett. **80**, 524 (1998).
[42]M. K. Srivastava, S. V. Lawande, M. Khan, D. Chandra, and B. Chakrborty, Phys. Fluids B **4**, 4086 (1992).
[43]K. Robinson, L. J. Dursi, P. M. Ricker, R. Rosner, A. C. Calder, M. Zingale, J. W. Truran, T. Linde, A. Caceres, B. Fryxell, K. Olson, K. Riley, A. Siegel, and N. Vladimirova, Astrophys. J. **601**, 621 (2004).
[44]W. H. T. Vlemmings, P. J. Diamond, and H. Imai, Nature **400**, 58 (2006).
[45]Y. Kato, Astrophys. Space Sci. **307**, 11 (2007).
[46]P. M. Bellan, Phys. Plasmas **12**, 058301 (2005).
[47]B. L. Yang, L. F. Wang, W. H. Ye, and C. Xue, Phys. Plasmas **18**, 072111 (2011).
[48]S. Ali, W. M. Moslem, P. K. Shukla, and R. Schlickeiserd, Phys. Plasmas **14**, 082307 (2007).
[49]D. Anderson, B. Hall, M. Lisak, and M. Marklund, Phys. Rev. E **65**, 046417 (2002).
[50]A. Bret, Phys. Plasmas **14**, 084503 (2007).
[51]L. G. Garcia, F. Haas, L. P. L. de Oliveira, and J. Goedert, Phys. Plasmas **12**, 012302 (2005).
[52]F. Haas, G. Manfredi, and M. Feix, Phys. Rev. E **62**, 2763 (2000).
[53]F. Haas, Phys. Plasmas **12**, 062117 (2005).
[54]A. Luque, H. Schamel, and R. Fedele, Phys. Lett. A **324**, 185 (2004).
[55]B. Shokri and A. A. Rukhadze, Phys. Plasmas **6**, 4467 (1999).
[56]P. K. Shukla and S. Ali, Phys. Plasmas **13**, 082101 (2006).
[57]M. Marklund, G. Brodin, and L. Stenflo, Phys. Rev. Lett. **91**, 163601 (2003).
[58]M. Marklund and P. K. Shukla, Rev. Mod. Phys. **78**, 455 (2006).
[59]L. F. Wang, C. Xue, W. H. Ye, and Y. J. Li, Phys. Plasmas **16**, 112104 (2009).
[60]L. F. Wang, W. H. Ye, and Y. J. Li, Phys. Plasmas **17**, 042103 (2010).
[61]W. H. Ye, L. F. Wang, C. Xue, Z. F. Fan, and X. T. He, Phys. Plasmas **18**, 022704 (2011).
[62]H. B. Cai, S. P. Zhu, M. Chen, S. Z. Wu, X. T. He, and K. Mima, Phys. Rev. E **83**, 036408 (2011).
[63]R. J. Mason, E. S. Dodd, and B. J. Albright, Phys. Rev. E **72**, 015401 (2005).
[64]M. Zingale, S. E. Woosley, J. B. Bell, M. S. Day, and C. A. Rendleman, J. Phys: Conf. S. **16**, 405 (2005).
[65]V. V. Bychkov and M. A. Liberman, Astrophys. Space Sci. **233**, 287 (1995).

Acceleration of Passive Tracers in Compressible Turbulent Flow

Yantao Yang, Jianchun Wang, Yipeng Shi,* Zuoli Xiao, X. T. He, and Shiyi Chen†
State Key Laboratory for Turbulence and Complex Systems, Center for Applied Physics and Technology, and Key Laboratory of High Energy Density Physics Simulation (Ministry of Education), College of Engineering, Peking University, Beijing, 100871, China
(Received 6 September 2012; published 7 February 2013)

In compressible turbulence at high Reynolds and Mach numbers, shocklets emerge as a new type of flow structure in addition to intense vortices as in incompressible turbulence. Using numerical simulation of compressible homogeneous isotropic turbulence, we conduct a Lagrangian study to explore the effects of shocklets on the dynamics of passive tracers. We show that shocklets cause very strong intermittency and short correlation time of tracer acceleration. The probability density function of acceleration magnitude exhibits a -2.5 power-law scaling in the high compression region. Through a heuristic model, we demonstrate that this scaling is directly related to the statistical behavior of strong negative velocity divergence, i.e., the local compression. Tracers experience intense acceleration near shocklets, and most of them are decelerated, usually with large curvatures in their trajectories.

DOI: 10.1103/PhysRevLett.110.064503 PACS numbers: 47.27.E−, 47.40.−x

Compressible fluid turbulence has proven to be a key phenomenon in many natural environments and engineering flows, including star formation from interstellar gas [1], inertial confinement fusion [2], and hypersonic aircraft. One interesting problem is the motion of particles in those flows, which is crucial to understanding transport and mixing in turbulence [3–5]. The Lagrangian investigation has attracted increasing interest, and it has significantly advanced our knowledge of turbulence. For incompressible flow, it is now possible to track large numbers of particles simultaneously; see Ref. [6] and references therein. Experimental and numerical results have revealed that particle acceleration exhibits strong intermittency that is primarily caused by intense vortex filaments [7–9], and up to 40 times the root mean square (rms) value has been observed [10]. The probability density function (PDF) of acceleration magnitude obeys a log-normal distribution [11].

However, it is very difficult to track particles in experiments with compressible turbulence. Numerical simulation serves as the major tool for Lagrangian study. Some efforts have been made to develop the Lagrangian numerical simulation [12–14] and to explore the mass distribution in isothermal flow [15,16] and the alignment dynamics for isentropic compressible magnetohydrodynamics [17]. Recently, the passive tracer has been used in a compressible turbulence simulation to investigate pair dispersion and mixing in the cosmic intracluster medium [18]. The Lagrangian investigation of passive tracers has also been used to explore the difference between solenoidal and compressive forcing for an isothermal ideal gas flow [19]. In this Letter, we focus on the effects of shocklets on the dynamics of passive tracers, particularly their acceleration, using three-dimensional compressible viscous turbulent flow.

We simulate three-dimensional compressible turbulence in a periodic box of width 2π. A novel hybrid numerical method with a resolution of 512^3 is used [20]. After the flow reaches a statistically steady state, a million passive tracers are seeded uniformly into the flow domain. The dynamic equations for a passive tracer in dimensionless form are given by

$$\dot{\mathbf{x}}(t) = \mathbf{u}(\mathbf{x}, t), \tag{1}$$

$$\dot{\mathbf{u}} = \mathbf{a} = -\frac{1}{\rho\gamma M^2}\nabla p + \frac{1}{\rho \mathrm{Re}}\nabla\cdot\sigma + \mathcal{F}, \tag{2}$$

where $\mathbf{x}$, $\mathbf{u}$, and $\mathbf{a}$ denote the location, velocity, and acceleration of the tracer, respectively; ρ is local density; p is pressure; σ is the viscous stress tensor; and γ is the ratio of specific heat at constant pressure to that at constant volume. A dot stands for the Lagrangian or material derivative. The Mach number $M = U/c$ is defined as the ratio of velocity to the speed of sound. The Reynolds number Re is equal to $\rho UL/\mu$ with L being the length and μ the dynamic viscosity. The three terms on the right-hand side of (2) represent the accelerations due to the pressure gradient, viscous stress, and external driving force, respectively.

In order to track the tracer, Eq. (1) is solved by using the second-order Runge-Kutta method. The physical quantities of the tracer are calculated by a trilinear interpolation to avoid unphysical oscillations near shocklets. Recent studies suggest that the Kolmogorov phenomenology can be extended to compressible turbulence [21–23]. Thus we may define the viscous scale $\eta = (\nu^3/\epsilon)^{1/4}$ and the Kolmogorov time $\tau_\eta = (\nu/\epsilon)^{1/2}$, where ν is the kinematic viscosity and ϵ is the energy-dissipation rate per unit mass. To accurately track tracers, we set the grid size to be $dx \approx 0.96\eta$ and update locations every $\tau_\eta/100$. The total integration time is about $10T^E$, where T^E is the large eddy turnover time. We compute the Lagrangian statistics during the last $8T^E$ with a sampling interval of $\tau_\eta/10$. The resulting turbulent Mach number $M_T = u_{\rm rms}/\bar{c} \approx 1.03$. Hereafter, the subscript "rms" denotes the root mean square value, and the overline represents the average over all particles and time. The Reynolds number based on the Taylor microscale Re_λ is about 153. The Lagrangian integral time scale T^L, which is defined by the integral of the velocity autocorrelation, is about 1.07. For comparison, T^E is about 1.1 and τ_η about 0.063.

We use the dilatation $\theta = \nabla \cdot \mathbf{u}$ to identify the flow region with different local compression. A large negative θ indicates strong compression. In order to isolate the compression effect, we decompose the entire flow domain into three regions: (a) the compression region with $\theta < -\theta^n_{\rm rms}$, (b) the smooth region with $-\theta^n_{\rm rms} < \theta < \theta^p_{\rm rms}$, and (c) the expansion region with $\theta > \theta^p_{\rm rms}$. Here $\theta^n_{\rm rms}$ and $\theta^p_{\rm rms}$ are the rms values computed from negative and positive θ of the Eulerian field, respectively.

A major difference between compressible and incompressible flow is the appearance of shocklets [20,24]. In Fig. 1(a), we show instantaneous volume renderings of two different types of flow structures, i.e., sheetlike shocklets and tubelike vortices. In the same figure, we also place several typical particle trajectories near those flow structures. An immediate observation is that tracers change their directions abruptly when they cross the shocklets, as shown in Fig. 1(b). Away from the shocklets, tracers may rotate around the intense vortices, as shown in Fig. 1(c). Thus, for compressible turbulence the shocklet is a new source of extreme acceleration other than intense vortices.

The two different types of tracer motion caused by shocklets and vortices together determine the time dynamics of acceleration. Figure 2 shows the Lagrangian autocorrelations of acceleration magnitude a and three Cartesian components a_i with $i = 1, 2, 3$, which are defined as $R = \overline{a'(t)a'(t+\tau)}/a^2_{\rm rms}$. Here, $a'(t) = a(t) - \bar{a}$. The curves of the incompressible flow from Ref. [11] are also shown for comparison. Autocorrelations in the compressible case decay much faster than those in the incompressible case, which may be attributed to the shocklets.

Furthermore, we have calculated the autocorrelation coefficients near different flow structures. For shocklets, $R^s = \langle a'(0)a'(\tau)\rangle/(\sqrt{\langle a'(0)^2\rangle}\sqrt{\langle a'(\tau)^2\rangle})$. Here, $\langle\cdot\rangle$ represents the ensemble average over all possible time intervals $0 < \tau < \tau_\eta$, during which θ has a local minimum that is smaller than $-\theta^n_{\rm rms}$. R^v is calculated with the same formula for all time intervals $0 < \tau < 5\tau_\eta$, during which tracers stay in the smooth vortical region with $-\theta^n_{\rm rms} < \theta < \theta^p_{\rm rms}$. Acceleration decorrelates much faster near shocklets than near vortices. When tracers move through shocklets, they experience extremely high acceleration during a short period, as shown in inset (a) in Fig. 3. The peak value is much bigger than the average over the entire domain. Naturally, the autocorrelation coefficients will quickly decay when τ is larger than the width of the acceleration burst. Figure 3 indicates that R^s of magnitude drops to around 0.5 at $\tau = 0.1\tau_\eta$. R^s of magnitude decays slightly faster than those of the components, since shocklets induce only large acceleration normal to the shock surface and have little effect on the tangential component. The curves in the vortical region are very similar to those of incompressible flow reported in Ref. [11]. R^v of components

FIG. 1 (color online). (a) Instantaneous rendering of shocklets (dark brown sheetlike surface) and vortices (light gray tubelike surface), with typical tracer trajectories. (b) Tracers changing direction near a shocklet. (c) Tracers rotating around a vortex. Trajectories start from the near (blue) ends and last for a time period of $10\tau_\eta$.

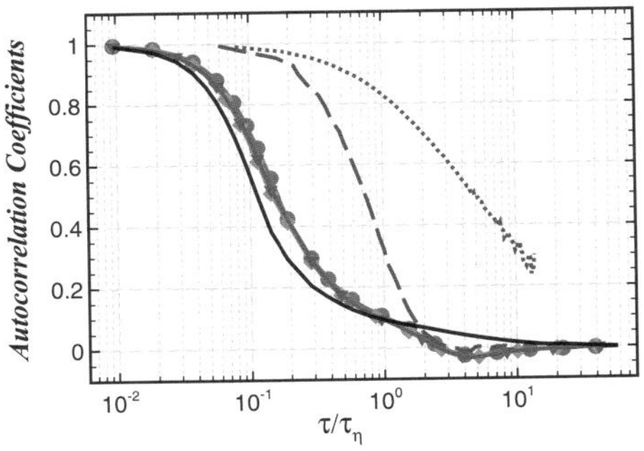

FIG. 2 (color online). The autocorrelation coefficients of acceleration magnitude a (black line) and components a_i (lines with symbols), compared to the autocorrelations of magnitude (dotted line) and one component (dashed line) of an incompressible flow (reproduced from Ref. [11]).

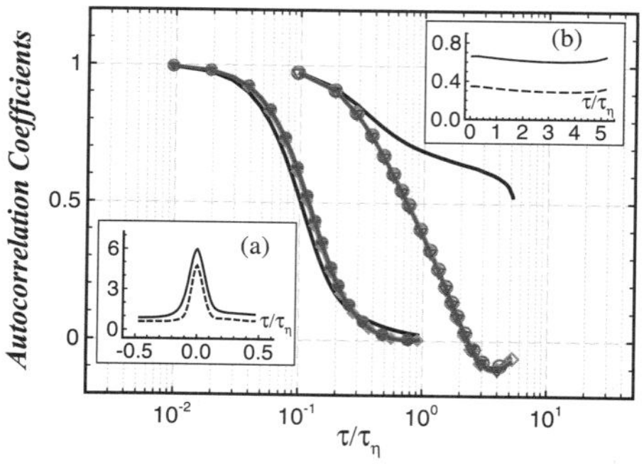

FIG. 3 (color online). Autocorrelation coefficients R^s (left bunch) and R^v (right bunch), with the same line style as Fig. 2. The insets show $\langle a(\tau)\rangle/\bar{a}$ and $\sqrt{\langle a'(\tau)^2\rangle}/a_{\rm rms}$ with solid and dashed lines, respectively. (a) Near shocklets and (b) trapped by vortices.

cross zero at around $2.3\tau_\eta$. The ensemble average and rms of a remain nearly constant when tracers are trapped by vortices, as shown in inset (b) in Fig. 3.

In Fig. 4, we show the total PDF of acceleration magnitude and the PDFs in the three different regions. For large a, the total PDF curve overlaps with that of the compression region, which implies that the extreme acceleration events are all detected near the shocklets. For small a, the smooth and expansion regions contribute to the major portion of the total PDF. The expansion region, which is often located closely downstream from the shocklets, does not induce very large acceleration. As shown in inset (a) in Fig. 4, the PDF curve of the smooth region matches the log-normal distribution with a variance of 1, as proposed for incompressible turbulence by Refs. [11,25]. This implies that, in the low-dilatation region of compressible turbulence, the tracer acceleration behaves similarly to that in incompressible turbulence.

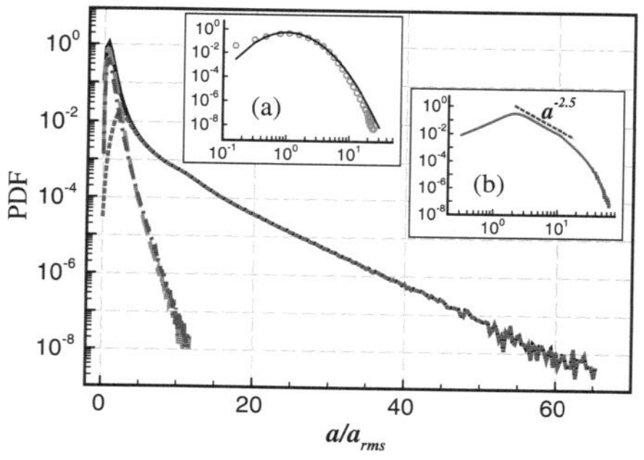

FIG. 4 (color online). Log-linear plots of the PDFs of acceleration magnitude a in three flow regions, normalized by the total number of data points. Black solid line: all tracers; red dotted line: the compression region; green dashed line: the smooth region; and blue dash-dotted line: the expansion region. Inset (a), Log-normal distribution in the smooth region; and inset (b), -2.5 power-law distribution in the compression region.

The curve of the compression region develops a $a^{-2.5}$ power law in the PDF tail; see inset (b) in Fig. 4. It has been found that in the shocklet region, the θ PDF also follows the -2.5 power law [26]. Figure 5 shows the joint PDF of a and θ, $P(a, \theta)$. A distinct ridge exists in the negative θ region. We also plotted the PDF quotient $Q = \frac{P(a,\theta)}{P(a)P(\theta)}$ [27], which measures the correlation between a and θ. Q equals unity if a and θ are independent. $Q > 1$ ($Q < 1$) represents positive (negative) correlation. The contours immediately suggest that a large acceleration is very likely accompanied by a large magnitude of θ, which usually happens near shocklets.

The strong correlation between the large acceleration and strong compression may be explained as follows. Let us denote the three principle axes of the strain-rate tensor by $\mathbf{e}_i$ with $i = 1, 2, 3$, which are associated with the eigenvalues sorted in descending order. Near strong shocklets, the compression occurs mainly in the $\mathbf{e}_3$ direction. Without losing generality, we set $\mathbf{e}_3$ to point from upstream of the shocklet to downstream. Then, the velocity divergence can be approximated by $\theta \approx (w_2 - w_1)/dl$, where w_1 and w_2 are the velocity components in the $\mathbf{e}_3$ direction upstream and downstream of shocklet, respectively, and dl is the shock thickness. It is reasonable to assume that the time scale of shocklet evolution is much longer than the time it takes a tracer to cross the shock region. Therefore, tracer acceleration can be calculated approximately as $a \approx |w_2 - w_1|/dt$, where dt is the time for a tracer to travel through a shocklet. Immediately, one has $a \approx |\theta| dl/dt = |\theta| u_n$, where u_n can be treated as the normal velocity with which the tracer moves through a shocklet.

Using the above model, we can compute the PDF of a from that of θ. In shocklets region the θ PDF has power-law scaling at $\theta < 0$, i.e., $f_{|\theta|} \sim |\theta|^\zeta$ with $\zeta \approx -2.5$ [26]. Our numerical results have indicated that u_n is statistically independent of θ and has a quasi-Gaussian distribution, which we denote as f_{u_n}. Then, by the algebra of random variables, we have

$$f_a(a) \sim \int f_{u_n}(u_n)(a/u_n)^\zeta u_n^{-1} du_n = Ca^\zeta. \tag{3}$$

Thus, the a PDF must have the same power-law scaling as the θ PDF, which is exactly what we found in our simulation.

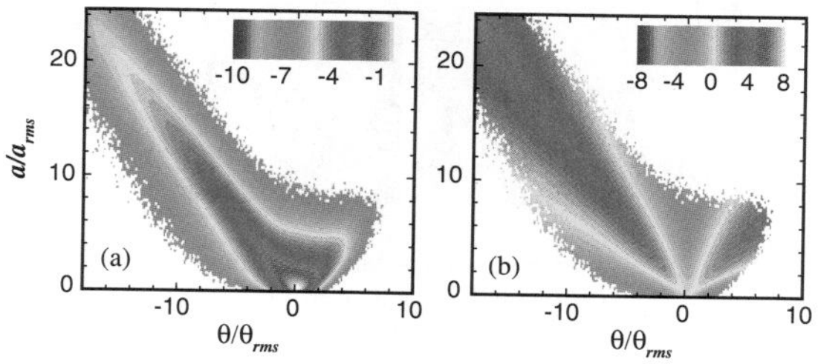

FIG. 5 (color online). (a) Logarithm of the joint PDF $P(a, \theta)$. (b) Logarithm of the PDF quotient $Q(a, \theta)$.

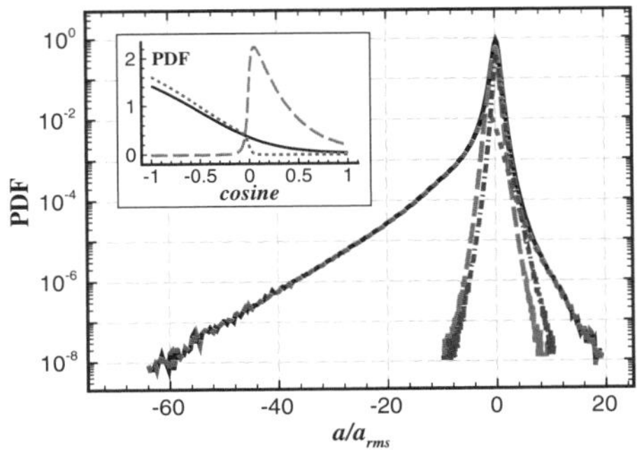

FIG. 6 (color online). PDFs of a_L in different regions with the same line styles as in Fig. 4, normalized by the total number of data points. Inset: PDFs of $\cos\alpha$ in the compression region, normalized separately. Black solid line: sampled over all tracers; red dotted line: decelerated tracers with $a_L < 0$; and green dashed line: accelerated tracers with $a_L > 0$.

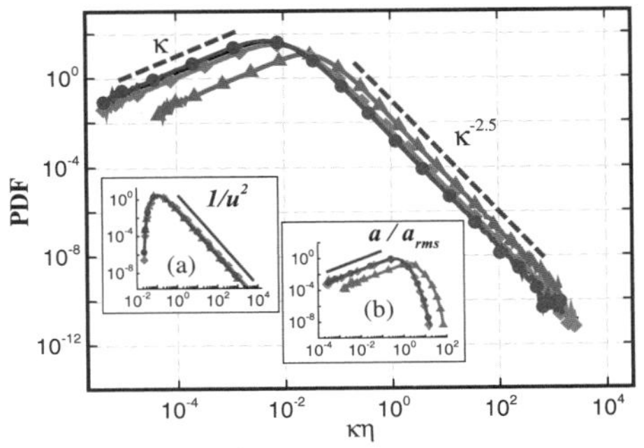

FIG. 7 (color online). Log-log plot of PDF distribution of normalized curvature $\kappa\eta$ of the whole field (black solid line), the compression region (red triangle), the smooth region (greed diamond), and the expansion region (blue circle). Inset (a): the $-5/2$ scaling of $1/u^2$ as $u^2 \to 0$. Inset (b): the linear scaling of a_T as $a_T \to 0$.

To further understand the tracer motion near shocklets, we decompose the acceleration into a longitudinal component $a_L = \mathbf{a} \cdot \mathbf{u}/|\mathbf{u}|$ and a transverse component $\mathbf{a}_T = \mathbf{a} - a_L\mathbf{u}/|\mathbf{u}|$. Evidently, a_L determines the time-changing rate of the velocity magnitude. $\mathbf{a}_T$ is directly linked to the curvature of tracer trajectory, which will be discussed later. The PDFs of a_L are plotted in Fig. 6. The curve of the smooth region is symmetric about $a_L = 0$, which is the same as in our incompressible results (not shown here). The PDF of the expansion region shows slight asymmetry, and the peak is located at a very small positive value. The PDF of the compression region has very strong asymmetry, and the left tail is much longer than the right one. In the compression region, the negative pressure gradient $-\nabla p$ is basically perpendicular to the shocklet surface pointing upstream and in the direction of acceleration according to (2). We investigated the cosine of the angle α between the velocity $\mathbf{u}$ and negative pressure gradient $-\nabla p$. The PDF of $\cos\alpha$ in the compression region is plotted in the inset in Fig. 6. It has a peak at $\cos\alpha = -1$ (solid line), which means that tracers are most likely to hit shocklets vertically from upstream. For most decelerated tracers, $\cos\alpha < 0$ (dotted line), and they move against the shocklets. Some tracers are accelerated near shocklets with $\cos\alpha > 0$ (dashed line). Basically, these tracers are overtaken by shocklets. Numerical results show that the ratio of the numbers of decelerated tracers to accelerated ones is 7:1 for the present case.

We now focus on the curvature of tracer trajectory, which is directly related to the transverse acceleration $\mathbf{a}_T$ by $\kappa \equiv |\mathbf{a}_T|/u^2$ [27]. Near shocklets, the total acceleration is very likely perpendicular to the shock surface. When a tracer moves obliquely through such structures, it undergoes strong transverse acceleration, and its trajectory may have a very large curvature in these regions. The PDFs of κ in different regions are plotted in Fig. 7. At the present Reynolds and Mach numbers, the peak of total PDF is located at $\kappa\eta \approx 5.8 \times 10^{-3}$, which is very close to that of incompressible flow [27]. The two tails of all PDF curves exhibit the same power laws as the incompressible case [27]. This is to be expected, because the extremely small and large curvatures are associated with the statistical behavior of a_T as $a_T \to 0$ and $1/u^2$ as $u \to 0$. The insets show that the $1/u^2$ PDF is almost the same for the three different regions, and the compressibility only shifts the PDF of a_T toward a larger value. Consequently, the κ PDF of the compression region also shifts toward a larger curvature value. The peak of PDF in the compression region is located at $\kappa\eta \approx 2.4 \times 10^{-2}$. We believe that, for a higher Mach number, this PDF curve would move even farther into the larger curvature region.

In conclusion, we present several significant effects of shocklets on the dynamics of passive tracers in compressible turbulence. Shocklets induce very large acceleration in a very short time period and therefore quickly decorrelate the acceleration magnitude and components. The PDF of acceleration magnitude in the compression region obeys a -2.5 power-law scaling, which is directly related to the PDF of negative θ. Most tracers move through shocklets from upstream, during which they experience obvious deceleration and strong lateral acceleration due to the large adverse pressure gradient. Meanwhile, large curvatures appear in their trajectories. It is also found that a small portion of tracers may be accelerated as they are overtaken by shocklets. Even though the present study is based on one single set of parameters, we infer that the main conclusions are still qualitatively applicable for higher Mach and Reynolds numbers.

The authors appreciate the valuable comments of the anonymous referees. This work was supported partially by the National Natural Science Foundation of China (NSFC Grants No. 10902002, No. 10921202, and No. 91130001) and the National Science and Technology Ministry under a subproject of the 973 program (Grant No. 2009CB724101). Y. Y. acknowledges partial support by NSFC Grant No. 11274026 and China Postdoctoral Foundation Grant No. 2011M500194. The simulations were done on the TH-1A supercomputer at Tianjin, China.

*ypshi@coe.pku.edu.cn
†syc@pku.edu.cn

[1] M.-M. Mac Low and R. S. Klessen, Rev. Mod. Phys. **76**, 125 (2004).
[2] V. A. Thomas and R. J. Kares, Phys. Rev. Lett. **109**, 075004 (2012).
[3] S. Pope, Annu. Rev. Fluid Mech. **26**, 23 (1994).
[4] B. I. Shraiman and E. D. Siggia, Nature (London) **405**, 639 (2000).
[5] P. Yeung, Annu. Rev. Fluid Mech. **34**, 115 (2002).
[6] F. Toschi and E. Bodenschatz, Annu. Rev. Fluid Mech. **41**, 375 (2009).
[7] L. Biferale, G. Boffetta, A. Celani, A. Lanotte, and F. Toschi, Phys. Fluids **17**, 021701 (2005).
[8] F. Toschi, L. Biferale, G. Boffetta, A. Celani, B. Devenish, and A. Lanotte, J. Turbul. **6**, N15 (2005).
[9] A. Reynolds, N. Mordant, A. Crawford, and E. Bodenschatz, New J. Phys. **7**, 58 (2005).
[10] A. La Porta, G. A. Voth, A. M. Crawford, J. Alexander, and E. Bodenschatz, Nature (London) **409**, 1017 (2001).
[11] N. Mordant, A. M. Crawford, and E. Bodenschatz, Phys. Rev. Lett. **93**, 214501 (2004).
[12] H. Schamel, Phys. Rep. **392**, 279 (2004).
[13] S. Kitsionas, C. Federrath, R. Klessen, W. Schmidt, D. Price, L. Dursi, M. Gritschneder, S. Walch, R. Piontek, J. Kim, A. Jappsen, P. Ciecielag, and M. Mac Low, Astron. Astrophys. **508**, 541 (2009).
[14] V. Springel, Annu. Rev. Astron. Astrophys. **48**, 391 (2010).
[15] C. Beetz, C. Schwarz, J. Dreher, and R. Grauer, Phys. Lett. A **372**, 3037 (2008).
[16] C. Schwarz, C. Beetz, J. Dreher, and R. Grauer, Phys. Lett. A **374**, 1039 (2010).
[17] J. Gibbon and D. Holm, New J. Phys. **9**, 292 (2007).
[18] F. Vazza, C. Gheller, and G. Brunetti, Astron. Astrophys. **513**, A32 (2010).
[19] L. Konstandin, C. Federrath, R. Klessen, and W. Schmidt, J. Fluid Mech. **692**, 183 (2011).
[20] J. Wang, L. P. Wang, Z. Xiao, Y. Shi, and S. Chen, J. Comput. Phys. **229**, 5257 (2010).
[21] G. Falkovich, I. Fouxon, and Y. Oz, J. Fluid Mech. **644**, 465 (2010).
[22] H. Aluie, Phys. Rev. Lett. **106**, 174502 (2011).
[23] H. Aluie, S. Li, and H. Li, Astrophys. J. Lett. **751**, L29 (2012).
[24] R. Samtaney, D. Pullin, and B. Kosović, Phys. Fluids **13**, 1415 (2001).
[25] C. Beck, Phys. Rev. Lett. **98**, 064502 (2007).
[26] J. Wang, Y. Shi, L. P. Wang, Z. Xiao, X. T. He, and S. Chen, Phys. Rev. Lett. **108**, 214505 (2012).
[27] H. Xu, N. T. Ouellette, and E. Bodenschatz, Phys. Rev. Lett. **98**, 050201 (2007).

Cascade of Kinetic Energy in Three-Dimensional Compressible Turbulence

Jianchun Wang, Yantao Yang, Yipeng Shi,* Zuoli Xiao, X. T. He, and Shiyi Chen†

State Key Laboratory for Turbulence and Complex Systems, Center for Applied Physics and Technology, Key Laboratory of High Energy Density Physics Simulation, College of Engineering, Peking University, Beijing 100871, China

(Received 26 January 2013; published 21 May 2013)

The conservative cascade of kinetic energy is established using both Fourier analysis and a new exact physical-space flux relation in a simulated compressible turbulence. The subgrid scale (SGS) kinetic energy flux of the compressive mode is found to be significantly larger than that of the solenoidal mode in the inertial range, which is the main physical origin for the occurrence of Kolmogorov's $-5/3$ scaling of the energy spectrum in compressible turbulence. The perfect antiparallel alignment between the large-scale strain and the SGS stress leads to highly efficient kinetic energy transfer in shock regions, which is a distinctive feature of shock structures in comparison with vortex structures. The rescaled probability distribution functions of SGS kinetic energy flux collapse in the inertial range, indicating a statistical self-similarity of kinetic energy cascades.

DOI: 10.1103/PhysRevLett.110.214505 PACS numbers: 47.27.E−, 47.40.−x, 47.53.+n

Compressible turbulence is of fundamental importance in a wide range of engineering flows and natural phenomena including high-temperature reactive flows, hypersonic aircrafts, and star formation in galaxies [1,2]. However, the basic processes occurring in compressible turbulence are less understood as compared to those occurring in incompressible turbulence. It is crucial to clarify the underlying physics of the interscale transfer of kinetic energy in compressible turbulence. In this Letter, we describe an investigation of Kolmogorov's $-5/3$ scaling of the kinetic energy spectrum and the conservative cascade of kinetic energy in the presence of large-scale shock waves in compressible turbulence, with a specific focus on the drastic enhancement in kinetic energy flux through shock structures.

Fully developed three-dimensional (3D) incompressible turbulence exhibits Kolmogorov's $-5/3$ energy spectrum in the inertial range that is generated and maintained by a nonequilibrium process, viz., the conservative cascade of kinetic energy from large scales to small scales [3,4]. In compressible turbulence, there are nonlinear interactions between solenoidal and compressive modes of velocity fluctuations. Moreover, quasi-2D shock waves add a new type of flow structures in addition to quasi-1D intense vortices as in incompressible turbulence, which further complicates the kinetic energy transfer process in compressible turbulence. Interestingly, the Kolmogorov's $-5/3$ energy spectrum of velocity has been reported in supersonic motions of interstellar media [5–7]. High resolution numerical simulations for supersonic isothermal turbulence showed the $-5/3$ spectrum of the density-weighted velocity $\mathbf{v} = \rho^{1/3}\mathbf{u}$, where ρ is the density and $\mathbf{u}$ is the velocity [8,9]. Aluie [10,11] proved that the kinetic energy cascades conservatively in compressible turbulence, provided that the pressure-dilatation cospectrum decays at a sufficiently rapid rate. This analysis was further supported by numerical simulations of both forced and decaying compressible turbulence [12].

Here, we utilize a novel hybrid numerical method [13] to simulate a compressible turbulence in a cubic box at a 1024^3 grid resolution. We apply large-scale force to both solenoidal and compressive components of the velocity field [14–16]. The flow is evolved to about $6T_e$, where the large-eddy turn over time $T_e = \sqrt{3}L/u^{\text{rms}}$, where L is the integral length scale and u^{rms} is the rms velocity magnitude. Various statistics of the simulated flow are averaged over the time interval $3.6 < t/T_e < 6.0$. The turbulent Mach number $M_t = u^{\text{rms}}/\langle c\rangle = 0.62$, where $\langle c\rangle$ is the average sound speed, and the Taylor microscale Reynolds number $R_\lambda = 160$. The highest local Mach number exceeds 2.5, and sheetlike shock waves of large-scale size are generated in the simulated flow.

Helmholtz decomposition of the density weighted velocity $\mathbf{w} = \sqrt{\rho}\mathbf{u}$ yields a solenoidal component $\mathbf{w}^s$ and a compressive component $\mathbf{w}^c$: $\mathbf{w} = \mathbf{w}^s + \mathbf{w}^c$, where $\nabla \cdot \mathbf{w}^s = 0$ and $\nabla \times \mathbf{w}^c = 0$. The average kinetic energy $E_k = \langle w^2/2\rangle$ can be decomposed by $E_k = E_k^s + E_k^c$, where $E_k^s = \langle (w^s)^2/2\rangle$ and $E_k^c = \langle (w^c)^2/2\rangle$. In our simulation, $E_k^c/E_k^s = 2.7$; i.e., the compressive kinetic energy is significantly larger than its solenoidal counterpart. Figure 1(a) shows that the kinetic energy spectrum exhibits a $-5/3$ scaling and its compressive component displays a -2 scaling in the inertial range of $4 \le k \le 20$ (see Supplemental Material [17]). To clarify the k^{-2} compressive spectrum, we consider the equation of the velocity potential Ψ defined by $\mathbf{u}^c = -\nabla\Psi$ (where $\mathbf{u}^c$ is the compressive velocity component, see Supplemental Material [17]):

$$\partial_t\Psi - \frac{1}{2}\nabla\Psi \cdot \nabla\Psi = \frac{4\nu}{3}\nabla^2\Psi + N_p + N_{\text{sol}} + F, \quad (1)$$

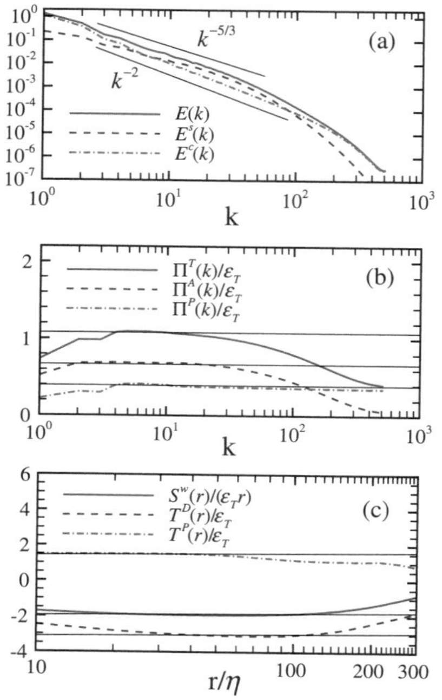

FIG. 1 (color online). (a) Spectra for kinetic energy and its two components; (b) Fourier-space kinetic energy flux; (c) structure functions in the physical-space kinetic energy flux.

where, $N_p = \nabla^{-2}[\nabla \cdot (\nabla p/\rho)]$ and p is the pressure. $N_{\rm sol}$ represents the impact of the solenoidal velocity component. ν is the coefficient of viscosity, and F is the potential of external force. According to the theory of 3D Burgers turbulence [18], large-scale shock waves can be generated through the effect of potential gradient square $\nabla\Psi \cdot \nabla\Psi$, leading to a k^{-2} spectrum of $\mathbf{u}^c$ at moderate and high Mach numbers. Moreover, we find that the two spectra for $\mathbf{u}^c$ and $\mathbf{w}^c$ are almost identical, indicating the negligible effect of density in our simulated flow.

In Fig. 1(b), we present the kinetic energy flux due to the advection and the pressure, namely,

$$\Pi^T(k) = \Pi^A(k) + \Pi^P(k), \tag{2}$$

where $\Pi^A(k) = \sum_{|\mathbf{k}|<k}\langle{\rm Re}[\hat{\mathbf{w}}^*(\mathbf{k}) \cdot (\widehat{\mathbf{u} \cdot \nabla\mathbf{w} + \mathbf{w}\theta/2})(\mathbf{k})]\rangle$ and $\Pi^P(k) = \sum_{|\mathbf{k}|<k}\langle{\rm Re}[\hat{\mathbf{w}}^*(\mathbf{k}) \cdot (\widehat{\nabla p/\sqrt{\rho}})(\mathbf{k})]\rangle$ [19,20]. The energy flux has been normalized by the total dissipation $\epsilon_T = -\langle p\theta\rangle + \epsilon_0$ [21], i.e., the total conversion rate of kinetic energy into internal energy by the pressure-dilatation work $-\langle p\theta\rangle$ and the viscous dissipation ϵ_0. The kinetic energy flux is approximately constant in the inertial range of $4 \le k \le 20$. Particularly, the ratio of $\Pi^A(k)$ to $\Pi^P(k)$ is about 0.65:0.35 in the inertial range, indicating the remarkable effect of pressure on the kinetic energy transfer.

We now introduce the physical-space kinetic energy flux defined by [3,22]

$$\epsilon(\mathbf{r}) \equiv -\partial_t\langle \mathbf{w}(\mathbf{x})\mathbf{w}(\mathbf{x}+\mathbf{r})/2\rangle|_{\rm NL}, \tag{3}$$

where, $\partial_t()|_{\rm NL}$ denotes the time-rate-of-change by the nonlinear terms including advection and pressure. Specifically, the flux for homogeneous isotropic turbulence (see Supplemental Material [17]) is $\epsilon(\mathbf{r}) = [-\nabla_{\mathbf{r}} \cdot (S^w(r)\mathbf{r}/r) + T^D(r) + T^P(r)]/4$, where $S^w(r) = \langle(\delta\mathbf{w})^2\delta\mathbf{u}\cdot\mathbf{r}/r\rangle$ is a third-order structure function of velocity. $T^D(\mathbf{r}) = \langle\delta\mathbf{w}\cdot(\theta\mathbf{w}' - \theta'\mathbf{w})\rangle$ and $T^P(\mathbf{r}) = 2\langle\mathbf{w}'\cdot(\nabla p/\sqrt{\rho}) + \mathbf{w}\cdot(\nabla' p'/\sqrt{\rho'})\rangle$, respectively, represent the impact of dilatational motion and pressure. Here, increments of $\mathbf{w}$ and $\mathbf{u}$ are defined by $\delta\mathbf{w} = \mathbf{w}' - \mathbf{w}$ and $\delta\mathbf{u} = \mathbf{u}' - \mathbf{u}$, respectively, and the prime is used to denote variables at the point $\mathbf{x}' = \mathbf{x} + \mathbf{r}$. We plot the rescaled third-order structure function $S^w(r)/r$ as a function of the separation r in Fig. 1(c). The separation r has been normalized by the Kolmogorov length scale η. The function $S^w(r)/r$ is nearly constant in an inertial range, such that $S^w \approx -C_0^w\epsilon_T r$, similar to the case of incompressible turbulence, but the coefficient $C_0^w = 1.95$ is substantially larger than $C_0 = 4/3$ found in incompressible turbulence [23]. Moreover, Fig. 1(c) shows that $T^D(r)/\epsilon_T \approx -C_0^D$ and $T^P(r)/\epsilon_T \approx C_0^P$ in the inertial range, where $C_0^D = 3.15$ and $C_0^P = 1.45$. Consequently, the physical-space kinetic energy flux via advection [i.e., the first and second terms in $\epsilon(\mathbf{r})$] is about $0.68\epsilon_T$, and that via pressure is about $0.36\epsilon_T$.

We suppose that the dissipation acts mainly at the smallest scales of turbulence [10,11], leading to $\epsilon(\mathbf{r}) = \epsilon_T$ for the separation $\mathbf{r}$ in the inertial range in compressible turbulence driven by a proper large-scale force. This yields a new flux relation in compressible isotropic turbulence [22–24], which can be expressed by

$$\nabla_{\mathbf{r}} \cdot \left(S^w(r)\frac{\mathbf{r}}{r}\right) = -4\epsilon_T + T^D(r) + T^P(r). \tag{4}$$

Particularly, $C_0^w = (4 + C_0^D - C_0^P)/3$ in the inertial range, in good agreement with the numerical results. The effect of dilatation is to enlarge the coefficient C_0^w, while the effect of pressure is to decrease C_0^w. We note that the impact of external forcing is small, considering that the variation in the density field is not very strong in our simulated flow compared to that in the supersonic isothermal turbulence at the turbulent root-mean-square Mach number 6 [24]. In the limit of incompressible turbulence, this relation reduces to

$$\nabla_{\mathbf{r}} \cdot \langle(\delta\mathbf{u})^2\delta\mathbf{u}\rangle = -4\epsilon_0, \tag{5}$$

which is equivalent to Kolmogorov's 4/5 law in isotropic incompressible turbulence [23].

We now employ a "coarse-graining" approach [10] to study the cascade of kinetic energy. We begin with the definition of a classically filtered field $\bar{\mathbf{a}}(\mathbf{x}) \equiv \int d^3\mathbf{r}G_l(\mathbf{r})\mathbf{a}(\mathbf{x}+\mathbf{r})$, where $G_l(\mathbf{r}) \equiv l^{-3}G(\mathbf{r}/l)$ is the kernel and $G(\mathbf{r})$ is a normalized window function. The Favre filtered field is defined as $\tilde{\mathbf{a}} \equiv \overline{\rho\mathbf{a}}/\bar{\rho}$. The filtered equation for the kinetic energy reads

$$\partial_t \frac{\bar{\rho}|\tilde{\mathbf{u}}|^2}{2} + \nabla \cdot \mathbf{J}_l = -\Pi_l - \Lambda_l - \Phi_l - D_l + \varepsilon^{\mathrm{inj}}, \quad (6)$$

where, the kinetic energy flux due to the subgrid scale (SGS) stress is $\Pi_l = -\nabla\tilde{\mathbf{u}}:[\bar{\rho}(\widetilde{\mathbf{u}\mathbf{u}} - \tilde{\mathbf{u}}\tilde{\mathbf{u}})]$, the energy flux due to the turbulent mass flux $\Lambda_l = \nabla\bar{p} \cdot (\overline{\rho\mathbf{u}} - \bar{\rho}\,\tilde{\mathbf{u}})/\bar{\rho}$, and the large scale pressure-dilatation $\Phi_l = -\bar{p}\theta_l$ [10]. $\mathbf{J}_l$ is the spatial transport flux of large-scale kinetic energy, D_l the viscous dissipation term, and $\varepsilon^{\mathrm{inj}}$ the energy injected by forcing. We present the spatial average of the SGS energy flux and the pressure-dilatation in Fig. 2(a). The filter width l has been normalized by the Kolmogorov length scale η. The pressure-dilatation Φ_l is nearly constant at scales $l/\eta < 100$, qualitatively consistent with the result reported by Aluie *et al.* [12]. The average of both Π_l and Λ_l are nearly constant over an inertial range $20 < l/\eta < 100$, indicating the existence of a conservative cascade of kinetic energy through the inertial range [10–12].

Helmholtz decomposition on the SGS energy flux yields $\Pi_l = \Pi_l^s + \Pi_l^c$, where $\Pi_l^s = -\nabla(\tilde{\mathbf{w}}^s/\sqrt{\bar{\rho}}):[\bar{\rho}(\widetilde{\mathbf{u}\mathbf{u}} - \tilde{\mathbf{u}}\tilde{\mathbf{u}})]$ and $\Pi_l^c = -\nabla(\tilde{\mathbf{w}}^c/\sqrt{\bar{\rho}}):[\bar{\rho}(\widetilde{\mathbf{u}\mathbf{u}} - \tilde{\mathbf{u}}\,\tilde{\mathbf{u}})]$, that is, the turbulent stress acting against the large-scale strain which is due to the solenoidal or compressive component of the density-weighted velocity, where $\tilde{\mathbf{w}}^s$ and $\tilde{\mathbf{w}}^c$ are, respectively, the solenoidal and compressive components of $\tilde{\mathbf{w}} = \sqrt{\bar{\rho}}\tilde{\mathbf{u}}$. We note that the average of Π_l^s (Π_l^c) is exactly equal to the average flux of $\tilde{E}^s$ (or $\tilde{E}^c$) from large scales to subgrid scales, owing to SGS stress, where $\tilde{E}^s = \langle(\tilde{w}^s)^2/2\rangle$ and $\tilde{E}^c = \langle(\tilde{w}^c)^2/2\rangle$. Figure 2(b) shows that the average SGS energy flux of both solenoidal and compressive modes is nearly constant in the inertial interval, namely, $\langle\Pi_l^s\rangle \approx 0.075\epsilon_T$ and $\langle\Pi_l^c\rangle \approx 0.62\epsilon_T$. Thus, the compressive component of the kinetic energy cascade is significantly faster than its solenoidal counterpart, leading to the dominance of the solenoidal kinetic energy spectrum at high wave numbers in the inertial range. Therefore, both the overall kinetic energy spectrum and its solenoidal component exhibit Kolmogorov's $-5/3$ scaling at high wave numbers in the inertial range. This is further confirmed by numerical simulation of Euler equations with an eighth-order hyperviscosity [13], at the same turbulent Mach number $M_t = 0.62$ and 512^3 grid resolution [see Fig. 2(c)]. The power-law exponents are found to be -1.66 and -2.02, respectively for the solenoidal and compressive kinetic energy spectra in the range of $4 \leq k \leq 30$ by least square estimation (see Supplemental Material [17]). Since the inertial-range statistics are independent of the dissipation mechanism [25,26], both the overall kinetic energy spectrum and its solenoidal component should have a clear $-5/3$ scaling at higher Reynolds numbers in compressible turbulence with normal viscosity.

Figure 3 shows a 3D visualization for the isosurfaces of the solenoidal component Π_l^s and the compressive component Π_l^c of kinetic energy flux. Only the 512^3 mesh points are shown. $(\Pi_l^s)^{\mathrm{rms}}$ and $(\Pi_l^c)^{\mathrm{rms}}$ are the rms values of Π_l^s and Π_l^c, respectively. The figure reveals a dramatic difference between the spatial structures of Π_l^s and Π_l^c. The isosurfaces of Π_l^c have large-scale sheetlike structures, whereas the isosurfaces of Π_l^s tend to exhibit coarse-grained blocklike structures that are similar to those in incompressible turbulence [27]. This observation indicates a significantly greater intermittency of the compressive

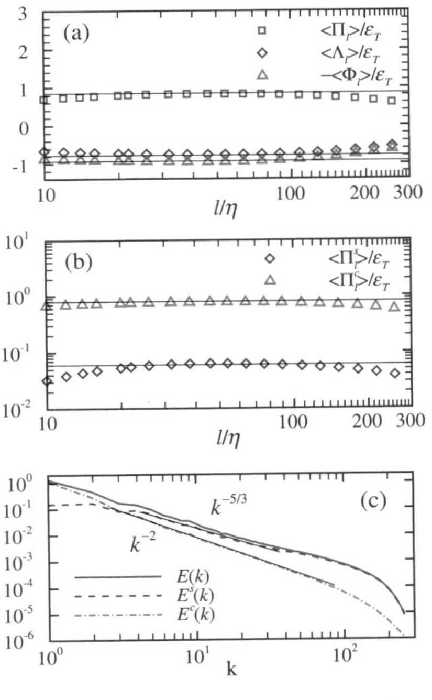

FIG. 2 (color online). (a) SGS kinetic energy flux and large scale pressure-dilatation work; (b) solenoidal and compressive components of SGS kinetic energy flux Π_l; (c) spectra for kinetic energy and its two components in the hyperviscosity simulation.

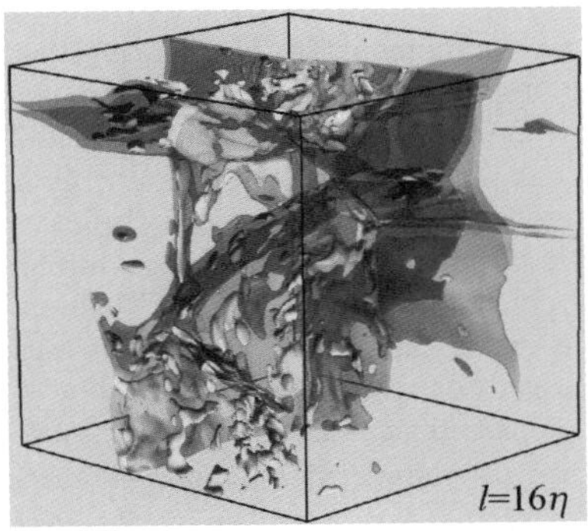

FIG. 3 (color). Isosurfaces for the solenoidal and compressive components of SGS kinetic energy flux Π_l in a 512^3 subdomain (the filter width $l = 16\eta$). (1) Yellow surfaces for $\Pi_l^s = 12.0(\Pi_l^s)^{\mathrm{rms}}$, and (2) red surfaces for $\Pi_l^c = 12.0(\Pi_l^c)^{\mathrm{rms}}$.

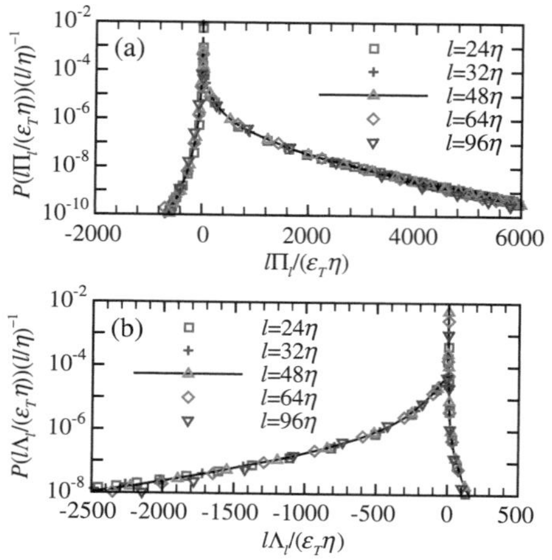

FIG. 4 (color online). Rescaled probability distribution functions (a) $P[l\Pi_l/(\epsilon_T\eta)](l/\eta)^{-1}$ and (b) $P[l\Lambda_l/(\epsilon_T\eta)](l/\eta)^{-1}$.

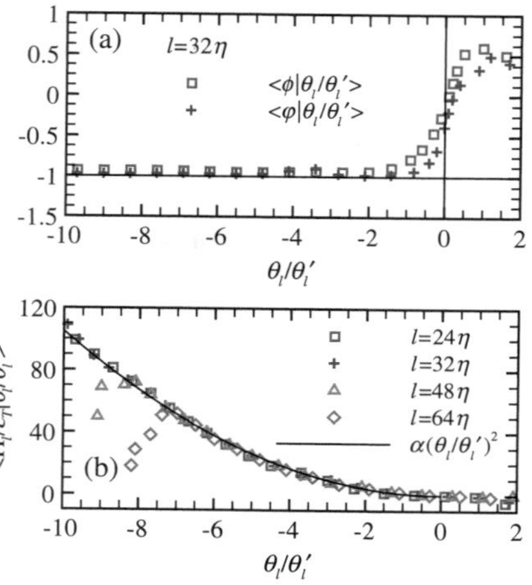

FIG. 5 (color online). (a) Conditional average of ψ and ϕ; (b) conditional average of Π_l.

energy flux component compared to its solenoidal counterpart.

The rescaled probability distribution functions (PDFs) of the SGS kinetic energy flux are depicted in Fig. 4. The rescaled PDFs of Π_l exhibit a strong skewness toward the positive side, showing a substantial difference from the case of incompressible turbulence [27], wherein the PDFs of the SGS energy flux have a moderate skewness. Therefore, most of the kinetic energy transfer due to the SGS stress is from large scales to small scales. Meanwhile, the rescaled PDFs of Λ_l display a strong skewness toward the negative side, indicating that most of the kinetic energy transfer due to the SGS mass flux is from small scales to large scales. We note that the extended and long PDF tails of the SGS kinetic energy flux have a major contribution from shock regions. Moreover, the tails of the rescaled PDFs of the SGS kinetic energy flux collapse to the same distribution for all l in the inertial range. According to the multifractal theory [26], the scaling exponents for the moments of $l\Pi_l$ and $l\Lambda_l$ should saturate at the value of $z_\infty = 1$ for high orders, namely, $\langle |l\Pi_l|^n\rangle \sim l^{z_\infty}$ and $\langle |l\Lambda_l|^n\rangle \sim l^{z_\infty}$ for large n. This prediction reveals a statistical scale-invariant property of energy flux in the inertial range and demonstrates the similarity between compressible turbulence and 1D Burgers turbulence [28].

To further clarify the impact of shock waves on kinetic energy transfer, we apply the average of various quantities conditioned on the filtered dilatation $\theta_l = \nabla\cdot\tilde{\mathbf{u}}$. Figure 5(a) depicts the conditional average of ψ and ϕ for $l=32\eta$, where $\psi=\tilde{\mathbf{S}}:\tilde{\tau}/(|\tilde{\mathbf{S}}||\tilde{\tau}|)$ and $\phi=\nabla\bar{p}\cdot(\overline{\rho\mathbf{u}}-\bar{\rho}\tilde{\mathbf{u}})/(|\nabla\bar{p}||(\overline{\rho\mathbf{u}}-\bar{\rho}\tilde{\mathbf{u}})|)$. Here, the large-scale strain is $\tilde{\mathbf{S}} = (\nabla\tilde{\mathbf{u}} + \nabla\tilde{\mathbf{u}}^T)/2$ and the SGS stress is $\tilde{\tau} = [\bar{\rho}(\widetilde{\mathbf{u}\mathbf{u}} - \tilde{\mathbf{u}}\,\tilde{\mathbf{u}})]$. Both the averages ψ and ϕ are close to -1 for $\theta_l/\theta_l' < -2$, indicating antiparallel alignments between the large-scale strain and the SGS stress, and between the pressure gradient and the turbulent mass flux vector in shock regions. The perfect alignments maximize the SGS kinetic energy flux, which is a distinctive feature of shock structures in comparison to vortex structures. In 3D incompressible turbulence, vortex dynamics induces a considerably weaker alignment between the large-scale strain and the SGS stress. For strong positive filtered dilatation, there is a tendency of parallel alignment between the large-scale strain and the SGS stress, indicating that an inverse cascade of kinetic energy due to SGS stress occurs locally in strong expansion regions. We plot the average of the SGS kinetic energy flux conditioned on the normalized filtered dilatation in Fig. 5. The conditional average of normalized Π_l is well approximated by $\alpha(\theta_l/\theta_l')^2$ in compression regions, where $\alpha = 1.05$ for filter widths l in the inertial range. This observation is consistent with the kinetic energy cascade scenario [10–12], as well as the fact that the local viscous dissipation of kinetic energy increases linearly with the square of dilatation in shock regions [21].

Finally, we emphasize that the $k^{-5/3}$ spectrum of the overall kinetic energy and the k^{-2} spectrum of the compressive kinetic energy are revealed at a moderate turbulent Mach number. As Mach number increases, the interactions between the solenoidal and compressive modes will become stronger, probably leading to a tendency of energy equipartition between the two modes [29]. Previous studies on numerical simulations of Euler equations showed that the spectrum exponents for both the solenoidal and compressive velocity are very close to -2 at high Mach numbers [8,16]. However, the simulations might be substantially contaminated by the excessive numerical dissipation of low-order schemes in these works. Therefore, several issues require further investigations through high-order numerical simulations, including the spectrum of solenoidal velocity and the effect of density variations on the kinetic energy spectrum at high turbulent Mach numbers.

The authors appreciate the valuable discussions with professor L.-P. Wang. This work was supported by the National Natural Science Foundation of China (Grants No. 11221061 and No. 91130001) and the National Science and Technology Ministry under a subproject of the 973 Program (Grant No. 2009CB724101). The simulations were done on the TH-1A super computer at Tianjin, China. Y. Y. is also supported partially by NSFC Grant No. 11274026 and China Postdoctoral Foundation Grant No. 2011M500194.

*ypshi@coe.pku.edu.cn
†syc@pku.edu.cn

[1] P. Sagaut and C. Cambon, *Homogeneous Turbulence Dynamics* (Cambridge University Press, Cambridge, England, 2008).
[2] P. Padoan and A. Norlund, Astrophys. J. **576**, 870 (2002).
[3] U. Frisch, *Turbulence: The Legacy of A. N. Kolmogorov* (Cambridge University Press, Cambridge, England, 1995).
[4] J. Cardy, G. Falkovich, and K. Gawedzki, *Non-equilibrium Statistical Mechanics and Turbulence* (Cambridge University Press, Cambridge, England, 2008).
[5] J.W. Armstrong, B.J. Rickett, and S.R. Spangler, Astrophys. J. **443**, 209 (1995).
[6] N. Kevlahan and R. Pudritz, Astrophys. J. **702**, 39 (2009).
[7] H. Xu, H. Li, D.C. Collins, S. Li, and M.L. Norman, Astrophys. J. **725**, 2152 (2010).
[8] A. G. Kritsuk, M. L. Norman, P. Padoan, and R. Wagner, Astrophys. J. **665**, 416 (2007).
[9] W. Schmidt, C. Federrath, and R. Klessen, Phys. Rev. Lett. **101**, 194505 (2008).
[10] H. Aluie, Phys. Rev. Lett. **106**, 174502 (2011).
[11] H. Aluie, Physica (Amsterdam) **247**, 54 (2013).
[12] H. Aluie, S. Li, and H. Li, Astrophys. J. Lett. **751**, L29 (2012).
[13] J. Wang, L.-P. Wang, Z. Xiao, Y. Shi, and S. Chen, J. Comput. Phys. **229**, 5257 (2010).
[14] S. Kida and S. A. Orszag, J. Sci. Comput. **5**, 85 (1990).
[15] M. R. Petersen and D. Livescu, Phys. Fluids **22**, 116101 (2010).
[16] C. Federrath, J. Roman-Duval, R. S. Klessen, W. Schmidt, and M.-M. Mac Low, Astron. Astrophys. **512**, A81 (2010).
[17] See Supplemental Material at http://link.aps.org/supplemental/10.1103/PhysRevLett.110.214505 for a brief description of large versions of Figs. 1(a) and 2(c) and also contains derivations of the governing equations for the velocity potential and physical-space flux relation.
[18] J. Bec and K. Khanin, Phys. Rep. **447**, 1 (2007).
[19] H. Miura and S. Kida, Phys. Fluids **7**, 1732 (1995).
[20] B. Thornber and Y. Zhou, Phys. Rev. E **86**, 056302 (2012).
[21] R. Samtaney, D. I. Pullin, and B. Kosovic, Phys. Fluids **13**, 1415 (2001).
[22] G. Falkovich, I. Fouxon, and Y. Oz, J. Fluid Mech. **644**, 465 (2010).
[23] S. Galtier and S. Banerjee, Phys. Rev. Lett. **107**, 134501 (2011).
[24] R. Wagner, G. Falkovich, A. G. Kritsuk, and M. L. Norman, J. Fluid Mech. **713**, 482 (2012).
[25] N. Cao, S. Chen, and Z. S. She, Phys. Rev. Lett. **76**, 3711 (1996).
[26] R. Benzi, L. Biferale, R. T. Fisher, L. P. Kadanoff, D. Q. Lamb, and F. Toschi, Phys. Rev. Lett. **100**, 234503 (2008).
[27] Q. Chen, S. Chen, G. L. Eyink, and D. D. Holm, Phys. Rev. Lett. **90**, 214503 (2003).
[28] J. Wang, Y. Shi, L.-P. Wang, Z. Xiao, X. T. He, and S. Chen, Phys. Rev. Lett. **108**, 214505 (2012).
[29] R. H. Kraichnan, J. Acoust. Soc. Am., **27**, 438 (1955).

PHYSICAL REVIEW E **87**, 013001 (2013)

Nonlinear Rayleigh-Taylor instability of rotating inviscid fluids

J. J. Tao,* X. T. He, and W. H. Ye
SKLTCS and CAPT, Department of Mechanics and Aerospace Engineering, College of Engineering, Peking University, Beijing 100871, China

F. H. Busse
Physikalisches Institut der Universität Bayreuth, D-95440 Bayreuth, Germany
(Received 23 October 2012; published 3 January 2013)

It is demonstrated theoretically that the nonlinear stage of the Rayleigh-Taylor instability can be retarded at arbitrary Atwood numbers in a rotating system with the axis of rotation normal to the acceleration of the interface between two uniform inviscid fluids. The Coriolis force provides an effective restoring force on the perturbed interface, and the uniform rotation will always decrease the nonlinear saturation amplitude of the interface at any disturbance wavelength.

DOI: 10.1103/PhysRevE.87.013001 PACS number(s): 47.20.Cq, 47.20.Ma, 47.20.Ky

I. INTRODUCTION

The instability of an interface, which separates two fluids and is accelerated into the heavier fluid, is known as the Rayleigh-Taylor instability (RTI) in honor of Rayleigh, who studied the unstable mode in a gravitational field [1], and Taylor, who considered the linear stability of unstable surface waves under acceleration [2]. RTI is of interest in a wide variety of fields, such as deep convection in oceans, inertial-confinement fusion (ICF), supernovae explosions, plasma physics, and vortex control. There are three stages in the RTI development. Interface disturbances in the linear stage of the classical RTI grow exponentially with a growth rate:

$$\gamma = \sqrt{gkA}, \quad A = \frac{\rho_1 - \rho_2}{\rho_1 + \rho_2},$$

where g is the interface acceleration, k is the wave number in the plane normal to g, and A is the Atwood number. ρ_1 and ρ_2 are the densities of the denser and lighter fluids, respectively. In the second stage, the interface is twisted into bubbles of lighter fluid and spikes of the heavier one penetrating into the lighter fluid along with the growth of perturbations. The bubble velocity saturates later on and the growth of the amplitude turns from exponential to linear in time. Such a transition is commonly referred to as a nonlinear saturation. At the last stage, the strong shear near the interface may cause vortices due to Kelvin-Helmholtz instability and then extensive interfacial mixing ensues. In order to study the nonlinear behavior of bubbles and spikes, weakly nonlinear theory has been developed [3–5] for irrotational flows.

Since RTI may dramatically reduce the performance of inertial confinement fusion (ICF) at both the initial implosion acceleration stage and the later deceleration stage by RT mixing [6], RTI should be suppressed as completely as possible [7–9]. For inviscid and irrotational uniform fluids, it has been shown that a uniform rotation with its axis parallel to the interface acceleration decreases the growth rate of normal mode disturbances [10]. Following this pioneering work a series of linear RTI analyses have been carried out to study the effect of viscosity [11], general rotation [12], and surface tension [13]. For exponentially stratified incompressible and viscous fluids, it is found that rotation tends to inhibit RTI through viscous effects [14]. Recent numerical simulations demonstrated that rotation with viscosity and diffusion, at least in the Boussinesq approximation or for low Atwood number flow, could retard RTI [15]. In the above study the rotation axis is parallel to the gravity, and the RTI suppression is explained by analyzing the vorticity field in the mixing zone. In other words, some studies of rotating RTI mainly focused on the case where the rotating vector is parallel to the interface acceleration, and hence the nonlinear influence of rotation about an axis normal to the direction of the acceleration on the bubble-spike formation was not studied theoretically. This is the motivation of the present paper.

II. PHYSICAL MODEL AND NONLINEAR THEORY

We consider an inner cylindrical fluid domain surrounded by an outer fluid. The inviscid fluids are immiscible and rotate around the z axis with the constant angular velocity Ω. y points in the radial direction and x is in the azimuthal direction. The acceleration g is in the same direction as y, and the densities of inner and outer fluids near the interface are assumed to be constant values ρ_1 and ρ_2, respectively. The interface at $r = R$ is accelerated to the inner fluid due to a disturbance as shown in Fig. 1(a). The disturbed flow near the interface can be described in the rectangular coordinate system fixed at the undisturbed interface as shown in Fig. 1(b) when the interface displacement η and the azimuthal wavelength are much smaller than R.

The two-dimensional potential flow of fluids is described by the hydrodynamical potentials ϕ_j and stream functions ψ_j; hence the disturbing velocities are

$$u_j = \frac{\partial \phi_j}{\partial x} = \frac{\partial \psi_j}{\partial y}, \quad v_j = \frac{\partial \phi_j}{\partial y} = -\frac{\partial \psi_j}{\partial x}, \tag{1}$$

where the subscript $j = 1, 2$ indicate inner and outer fluids, respectively. We assume that the instability is restricted to the neighborhood of the interface and hence apply the same boundary conditions as in the case of the traditional RTI. The

*jjtao@pku.edu.cn

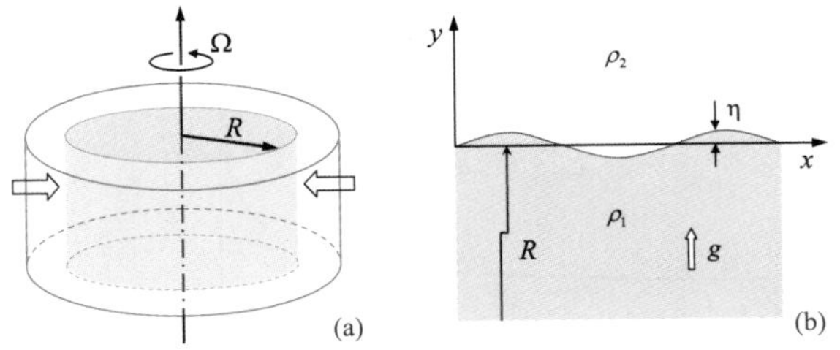

FIG. 1. (Color) (a) Schematic of the rotating system and (b) the coordinates used to study the interface instability. It is assumed that the interface displacement $\eta \ll R$, the radius of the interface.

governing equations and boundary conditions are

$$\nabla^2\psi_j = \nabla^2\phi_j = 0, \quad \begin{matrix} \nabla\psi_1 = \nabla\phi_1 = 0, & y \to -\infty, \\ \nabla\psi_2 = \nabla\phi_2 = 0, & y \to \infty, \end{matrix} \tag{2}$$

$$\partial_t \eta + \nabla\phi_j \cdot \nabla\eta = \phi_{j,y}, \quad y = \eta, \tag{3}$$

$$\begin{aligned} &\rho_1 \left\{\phi_{1,t} + \tfrac{1}{2}(\nabla\phi_1)^2 - gy + 2\Omega\psi_1\right\} \\ &= \rho_2 \left\{\phi_{2,t} + \tfrac{1}{2}(\nabla\phi_2)^2 - gy + 2\Omega\psi_2\right\}, \quad y = \eta. \end{aligned} \tag{4}$$

The above governing equations are the same as those of the classical RTI except for the Coriolis force terms. It is noted that the centrifugal effect brought by rotation is included in the acceleration g and will be discussed at the end of this paper. The interface displacement η and the disturbance variables ψ_j and ϕ_j can be expanded in powers of ϵ,

$$\begin{aligned} \eta(x,t) &= \epsilon\eta^1(x,t) + \epsilon^2\eta^2(x,t) + \epsilon^3\eta^3(x,t) + O(\epsilon^4), \\ \phi_j(x,y,t) &= \epsilon\phi_j^1(x,y,t) + \epsilon^2\phi_j^2(x,y,t) + \epsilon^3\phi_j^3(x,y,t) \\ &\quad + O(\epsilon^4), \quad j = 1,2, \\ \psi_j(x,y,t) &= \epsilon\psi_j^1(x,y,t) + \epsilon^2\psi_j^2(x,y,t) + \epsilon^3\psi_j^3(x,y,t) \\ &\quad + O(\epsilon^4), \quad j = 1,2. \end{aligned}$$

We insert this ansatz into Eqs. (2) to (4), which are expanded in a Taylor series about the undisturbed interfacial position ($\eta = 0$). By collecting the terms with the same powers in ϵ, we obtain the systems of order 1, 2, and 3, governing the linear instability and the second- and third-order corrections. In the first order of ϵ we obtain the equations

$$\nabla^2\psi_j^1 = \nabla^2\phi_j^1 = 0, \quad \begin{matrix} \nabla\psi_1^1 = \nabla\phi_1^1 = 0, & y \to -\infty, \\ \nabla\psi_2^1 = \nabla\phi_2^1 = 0, & y \to \infty, \end{matrix} \tag{5}$$

$$\partial_t \eta^1 = \phi_{j,y}^1, \quad y = 0, \tag{6}$$

$$\rho_1\left(\phi_{1,t}^1 - g\eta^1 + 2\Omega\psi_1^1\right) = \rho_2\left(\phi_{2,t}^1 - g\eta^1 + 2\Omega\psi_2^1\right), \quad y = 0. \tag{7}$$

Introducing normal mode of η^1, ψ_j^1, and ϕ_j^1 in the form of $\sim e^{i(kx+\omega t)}$, where k is the wave number in the x direction, we solve (5) to (7) and obtain the frequency for unstable mode $\omega = -i\gamma - \Omega A$, where $\gamma = \sqrt{gkA - A^2\Omega^2}$ is the linear growth rate and the Atwood number is $A = (\rho_1 - \rho_2)/(\rho_1 + \rho_2)$. For the case of a basic-mode initial disturbance of the form $\epsilon\cos(kx)$, the corresponding growing solution η is obtained up to the third order as follows:

$$\begin{aligned} \eta = &\ \epsilon\cos(kx - A\Omega t)e^{\gamma t} + \epsilon^2 C_2\cos(2kx - 2A\Omega t)e^{2\gamma t} \\ &+ \epsilon^3\left[C_{31}^c\cos(kx - A\Omega t) + C_{31}^s\sin(kx - A\Omega t)\right. \\ &\left.+ C_3\cos(3kx - 3A\Omega t)\right]e^{3\gamma t}, \end{aligned} \tag{8}$$

where

$$\begin{aligned} C_2 &= \frac{kA}{2}, \\ C_{31}^c &= -\frac{k^2}{16\gamma^2}(A^4\Omega^2 + A^2\Omega^2 + 3A^2\gamma^2 + \gamma^2), \\ C_{31}^s &= \frac{k^2 A\Omega}{8\gamma}(A^2 + 1), \\ C_3 &= \frac{k^2}{8}(4A^2 - 1). \end{aligned} \tag{9}$$

Different from the traditional RTI, the unstable modes are not stationary but traveling waves, though the phase velocity is much smaller than the rotation velocity ΩR. When the angular velocity Ω decreases to zero, the linear growth rate γ approaches the traditional value $\gamma_0 = \sqrt{gkA}$, and with $A = 1$ and $\Omega = 0$ the previous nonlinear solution of the interface position η is recovered [4].

The uniform rotation brings two additional forces, the Coriolis force and the centrifugal force. In order to analyze the Coriolis effect on the RTI, we first assume that the total acceleration of the interface g is a constant. According to the expression of the linear growth rate $\gamma = \sqrt{gkA - A^2\Omega^2}$, the Coriolis force always diminishes the growth of the RTI. It is shown in Fig. 2(a) that this retardation effect becomes stronger when the Atwood number A is increased. The reciprocal of the dimensionless angular velocity $\Omega^* = \Omega/\sqrt{gk}$ is the Rossby number. Secondly, since only the product gA appears in γ, the Coriolis suppression effect on the linear RTI works for both an acceleration stage ($g > 0$ and $A > 0$) and a deceleration stage ($g < 0$ and $A < 0$).

According to Eq. (8), the nonlinear evolution of the fundamental mode up to the third order can be expressed as $\eta_N\cos(kx - A\Omega t)$, where the nonlinear amplitude $\eta_N = (\eta_L + \eta_L^3 C_{31}^c)$ and the linear amplitude $\eta_L = \epsilon e^{\gamma t}$. Since $C_{31}^c < 0$ [see Eq. (9)], the third-order modulation always diminishes the growth of the fundamental mode. Such a nonlinear effect is described by the nonlinear suppression factor S:

$$S = \frac{\eta_N - \eta_L}{\eta_L^3 k^2} = -\frac{1}{16\gamma^2}(3A^2\gamma^2 + \gamma^2 + A^4\Omega^2 + A^2\Omega^2). \tag{10}$$

Apparently, S is a negative value; hence the larger $-S$, the stronger the nonlinear suppression exerted on the basic mode. It is shown in Fig. 2(b) that larger rotation velocity and Atwood number will cause stronger suppression of the nonlinear evolution of the RTI. The shape variation of spikes and bubbles including higher-order harmonics along with the uniform rotation are shown in Figs. 2(c) and 2(d).

The Coriolis suppression mechanism is straightforward and may be explained as follows. In first-order approximation, the disturbance velocity in the x (azimuthal) direction at

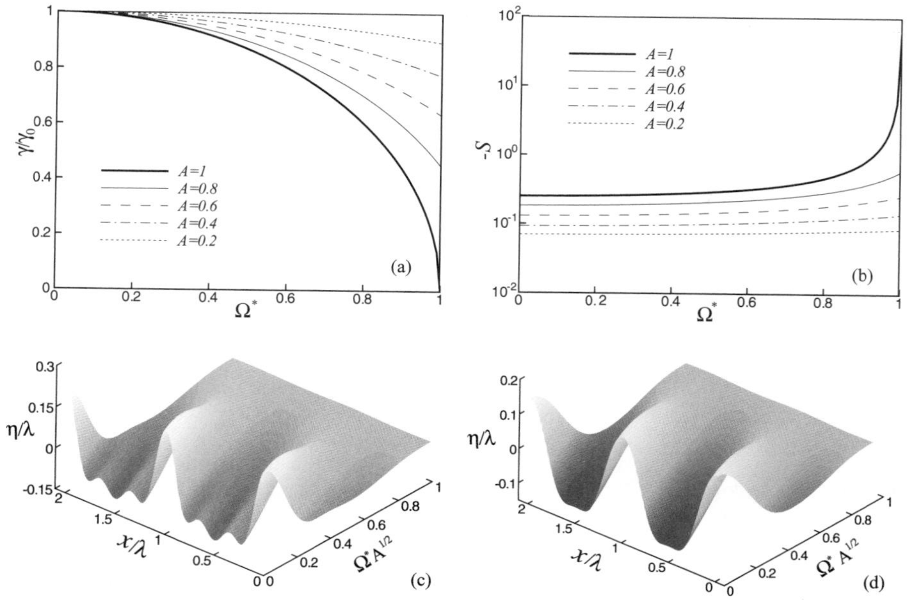

FIG. 2. (Color) (a) Linear growth rate γ and (b) the nonlinear suppression factor S of the fundamental mode as functions of the dimensionless rotation frequency. (c), (d) Interface displacements η at $\gamma_0 t = 5$ for $\epsilon k = 0.001$ in the cases $A = 1$ and $A = 0.4$, respectively. The definitions $\gamma_0 = \sqrt{gkA}$, $\Omega^* = \Omega/\sqrt{gk}$, and the wavelength $\lambda = 2\pi/k$ have been used.

the interface can be expressed as $u_1 \simeq -u_2 \simeq \epsilon e^{\gamma t}\cos(kx - A\Omega t + \theta)$, where θ is the rotation-induced phase difference between u and the interface perturbation η. $\cos\theta = A\Omega/\gamma_0 = \sqrt{A}\Omega^*$. It is shown in Fig. 3(a) that $\theta = \pi/2$ for the classical RTI case without rotation. However, a uniform rotation normal to the acceleration of the interface will change θ and cause a normal Coriolis force $-2\Omega u$. Since u_1 and u_2 share the same absolute value but with opposite directions at the interface, the direction of the effective normal Coriolis force F_n is the same as that of the heavier side. Consequently, the integrated F_n over every half wavelength behaves as a restoring force exerted on the distorted interface. When Ω^* increases to $1/\sqrt{A}$ or $\theta = 0$, u_1 is in the same phase as the interface displacement, and the effective normal Coriolis force, shown as arrows in Fig. 3(b), becomes a restoring force everywhere on the interface. It is easy to check that when $\Omega < 0$ the suppression mechanism of the Coriolis force works as well.

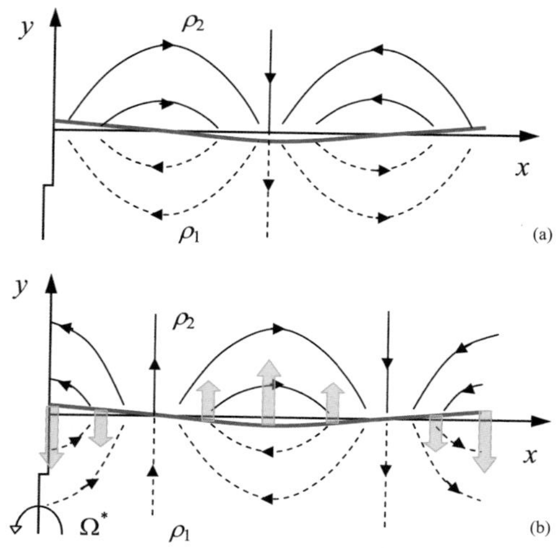

FIG. 3. (Color) Schematics illustrating the streamlines of inner (dashed lines) and outer (solid lines) fluids near the interface at (a) $\Omega^* = 0$ and (b) $\Omega^* A = 1$. The thick blue lines indicates the interface, and the arrows in (b) represent the effective Coriolis force.

In reality, the rotation induced centrifugal force changes the acceleration of the interface. In this paper we assume that the amplitude η and wavelength λ of the interface displacement are much smaller than R, and it is easy to find that Eq. (8) remains the same for cases including the centrifugal effect except that the interface acceleration $g = g_0 + R\Omega^2$ is used, where g_0 is the value without centrifugal effect. Consequently, the linear growth rate, including both the Coriolis effect and the centrifugal effect, is

$$\gamma = \sqrt{g_0 kA + R\Omega^2 kA - A^2\Omega^2}. \tag{11}$$

Since $Rk = 2\pi R/\lambda > 2\pi > A$, the effect of uniform rotation will enhance the linear RTI by increasing γ in the acceleration stage ($A > 0, g_0 > 0$) and retard the RTI by decreasing γ in the deceleration stage ($A < 0, g_0 < 0$). The perturbation enters the nonlinear regime when the amplitude is larger than a certain value. The shape of the interface changes from the sinusoidal form to broad thin bubbles and narrow thick spikes. It is convenient to define the transition into the nonlinear regime as occurring when the nonlinear saturation amplitude η_s is reached, where the amplitude of the fundamental mode is reduced by 10% due to nonlinear effects, i.e., $\eta_s = \eta_N = 0.9\eta_L$. From Eq. (8) we get

$$\frac{\eta_s}{\lambda} = \frac{1}{\pi}\sqrt{\frac{2}{5(3A^2+1)+5(A^4\Omega^2+A^2\Omega^2)/\gamma^2}}. \tag{12}$$

The appearance of an additional term $5(A^4\Omega^2 + A^2\Omega^2)/\gamma^2$ in the above equation is caused by the Coriolis force, and the centrifugal effect only changes the linear growth rate γ. Because this additional term is always positive, the surprising conclusion can be drawn that the uniform rotation will always decrease the nonlinear saturation amplitude at an arbitrary Atwood number and wavelength, during the acceleration stage as well as the deceleration stage. Bubbles and spikes will grow to a larger length scale before entering the nonlinear regime if the Atwood number and the rotation angular velocity are small. When $A = 1$ and $\Omega = 0$ the ratio $\eta_s/\lambda \simeq 0.1$, which is the widely used threshold for nonlinearity [16].

III. DISCUSSION

When the uniform rotation is significantly strong, i.e., $R\Omega^2 = |g_0|$, the RTI will be completely suppressed by rotation at the deceleration stage ($A < 0$ and $g_0 < 0$). For the acceleration stage ($A > 0$ and $g_0 > 0$), the additional term in Eq. (12) may be simplified as

$$\begin{aligned} 5(A^4\Omega^2 + A^2\Omega^2)/\gamma^2 &\simeq 5(A^4\Omega^2 + A^2\Omega^2)/(2g_0kA) \\ &= 5(A^3 + A)/(2Rk) \\ &\ll 5(A^3 + A)/4\pi < 1; \end{aligned}$$

hence the nonlinear saturation amplitude η_s is only slightly retarded by rotation. It is noted that the interface of the present model is a circular cylinder; hence the current stabilization mechanism is applicable to the equatorial region but not the polar regimes of a compressible rotating sphere. However, numerical simulations [15] have shown that when the gravity is parallel to the rotation, just as what happens at the polar regimes, the mixing zone generated by the RTI can be retarded by the effect of the Coriolis force on the vorticity field. Though these two mechanisms are different, the one proposed in this paper is applicable for potential flows and the other one is found in rotational flows; their cooperation makes it possible to suppress the Rayleigh-Taylor instability near a spherical interface by using rotation, and may have potential applications in the field of inertial confinement fusion, which includes both the acceleration and the deceleration stages.

ACKNOWLEDGMENTS

This work has been supported by the NSFC Grants (No. 10972007, No. 10921202, and No. 11225209).

[1] L. Rayleigh, Proc. London Math. Soc. **14**, 170 (1883).
[2] G. I. Taylor, Proc. R. Soc. London A **201**, 192 (1950).
[3] A. H. Nayfeh, J. Fluid Mech. **38**, 619 (1969).
[4] J. W. Jacobs and I. Catton, J. Fluid Mech. **187**, 329 (1988).
[5] V. N. Goncharov, Phys. Rev. Lett. **88**, 134502 (2002).
[6] J. D. Lindl, *Inertial Confinement Fusion: The Quest for Ignition and Energy Gain Using Indirect Drive* (Springer-Verlag, New York, 1998).
[7] J. Garnier, P. A. Raviart, C. Cherfils-Clérouin, and L. Masse, Phys. Rev. Lett. **90**, 185003 (2003).
[8] J. Sanz, R. Betti, R. Ramis, and J. Ramírez, Plasma Phys. Control. Fusion **46**, B367 (2004).
[9] W. H. Ye, L. F. Wang, and X. T. He, Phys. Plasmas **17**, 122704 (2010).
[10] S. Chandrasekhar, *Hydrodynamic and Hydromagnetic Stability* (Oxford University Press, London, 1961).
[11] P. K. Bhatia and R. P. S. Chhonkar, Astrophys. Space Sci. **114**, 271 (1985).
[12] L. A. Dávalos-Orozco, Dynam. Atmos. Oceans **23**, 247 (1996).
[13] N. F. El-Ansary, G. A. Hoshoudy, A. S. Abd-Elrady, and A. H. Ayyad, Phys. Chem. Chem. Phys. **4**, 1464 (2002).
[14] R. Hide, Quart. J. Mech. Applied Math. **9**, 22 (1956).
[15] G. F. Carnevale, P. Orlandi, Y. Zhou, and R. C. Kloosterziel, J. Fluid Mech. **457**, 181 (2002).
[16] D. Layzer, Astrophys. J. **122**, 1 (1955).

PHYSICS OF PLASMAS 20, 042708 (2013)

Weakly nonlinear incompressible Rayleigh-Taylor instability growth at cylindrically convergent interfaces

L. F. Wang (王立锋),[1,2,a)] J. F. Wu (吴俊峰),[1] W. H. Ye (叶文华),[1,2,3,b)] W. Y. Zhang (张维岩),[1] and X. T. He (贺贤土)[1,2,b)]
[1]Institute of Applied Physics and Computational Mathematics, Beijing 100094, China
[2]HEDPS, Center for Applied Physics and Technology, Peking University, Beijing 100871, China
[3]Department of Physic, Zhejiang University, Hangzhou 310027, China

(Received 1 March 2013; accepted 15 April 2013; published online 29 April 2013)

A weakly nonlinear (WN) model has been developed for the incompressible Rayleigh-Taylor instability (RTI) in cylindrical geometry. The transition from linear to nonlinear growth is analytically investigated via a third-order solutions for the cylindrical RTI initiated by a single-mode velocity perturbation. The third-order solutions can depict the early stage of the interface asymmetry due to the bubble-spike formation, as well as the saturation of the linear (exponential) growth of the fundamental mode. The WN results in planar RTI [Wang *et al.*, Phys. Plasmas **19**, 112706 (2012)] are recovered in the limit of high-mode number perturbations. The difference between the WN growth of the RTI in cylindrical geometry and in planar geometry is discussed. It is found that the interface of the inward (outward) development spike/bubble is extruded (stretched) by the additional inertial force in cylindrical geometry compared with that in planar geometry. For interfaces with small density ratios, the inward growth bubble can grow fast than the outward growth spike in cylindrical RTI. Moreover, a reduced formula is proposed to describe the WN growth of the RTI in cylindrical geometry with an acceptable precision, especially for small-amplitude perturbations. Using the reduced formula, the nonlinear saturation amplitude of the fundamental mode and the phases of the Fourier harmonics are studied. Thus, it should be included in applications where converging geometry effects play an important role, such as the supernova explosions and inertial confinement fusion implosions.
[http://dx.doi.org/10.1063/1.4803067]

I. INTRODUCTION

The Rayleigh-Taylor instability (RTI) occurs at a perturbed interface between two fluids where a less dense fluid accelerates a more dense fluid.[1–3] The mechanism of RTI for an incompressible fluid in a gravitational field was originally explored by Rayleigh[1] in the 1880s and later generalized to all accelerated fluids by Taylor.[2] Understanding the material mixing caused by the RTI is important to a wide variety of applications including, but not limited to, supernova explosions[4–6] and inertial confinement fusion (ICF) implosions.[7–9] In the standard central hot-spot ignition ICF scheme, a spherical capsule consisting of a ablator-fuel shell filled with a deuterium-tritium gas is imploded by directly irradiating lasers or indirectly irradiating x-rays. It is essential to maintain the integrity of the compressed shell during the entire implosion. However, any imperfections in the target manufacture or non-uniform driver sources can grow rapidly during the implosion due to the inherent RTI. It can mix not only the fuel materials with the ablator in the acceleration phase but also the hot core with the cold shell in the deceleration phase, destroying fuel compression symmetry, or even rupturing the shell (if the shell is too thin) and quenching the hot-spot (if the low-temperature spike of the shell around the hot-spot boundary has an excessive growth). In short, the RTI plays a critical role in determining the success and performance of ICF. On the other hand, the RTI is believed to occur in core-collapse supernova explosions where a blast wave moving outward crosses a density drop at a perturbed interface creating the situation (i.e., a density gradient in an opposing direction to the pressure gradient) for the RTI. Furthermore, it is expected that the RTI plays a main role in the occurrence of material mixing and the evolution of supernova explosions.

The initial growth of a single-mode cosinusoidal interface perturbation with wavelength λ on a surface between two incompressible, inviscid, irrotational, and immiscible fluids with densities $\rho^{\rm H}$ and $\rho^{\rm L}$ ($\rho^{\rm H} > \rho^{\rm L}$ and the heavy fluid is superposed over the light one) in an acceleration field $\mathbf{g} = -g\hat{\mathbf{e}}_z$, is exponential, $\eta(t) = \eta_0 e^{\gamma_{cl} t}$ with a *linear* growth rate $\gamma_{cl} = \sqrt{A_T g k}$, where η_0 is the initial amplitude of the perturbation ($\eta_0 \ll \lambda$), t is the time, $A_T = (\rho^{\rm H} - \rho^{\rm L})/(\rho^{\rm H} + \rho^{\rm L})$ is the Atwood number, and $k = 2\pi/\lambda$ is perturbation wave number.[1–3] As the amplitude of the perturbation grows, the nonlinear effects from mode-coupling become pronounced, generating higher harmonics and causing the saturation of the exponential growth of the Fourier mode's amplitude, and hence the original symmetry shape of the cosinusoidal interface begins to lose. This *weakly nonlinear* (WN) growth regime has been extensively studied by the method of successive approximations.[10–17] Hydrodynamical equations of the

a)Electronic addresses: wang_lifeng@iapcm.ac.cn and lif_wang@pku.edu.cn.
b)Authors to whom correspondence should be addressed. Electronic addresses: ye_wenhua@iapcm.ac.cn and xthe@iapcm.ac.cn

RTI for two fluids of constant density up to arbitrary orders are derived and is actually sought up to the second-order by Ingraham in the 1950s.[10] Subsequently, considering the fluid-vacuum interface ($A_T = 1$) and the surface tension effects, the WN growth of the RTI up to the third-order is analyzed.[11,12] It is worth mentioning that up to the third-order correction, a feedback from mode-coupling to the fundamental mode and secular terms can be noted.[11,12] Moreover, an extended three-dimensional version for various initial perturbations has been given by Jacobes and Catton in 1988.[13] For the multi-mode perturbations, a second-order solution is obtained by Haan[14] and the third-order nonlinearity in the presence of surface tension has been studied by Garnier using statistical analysis.[15] In addition, the RTI up to the third-order corrections for arbitrary Atwood numbers ignoring the surface tension effects has also been explored.[16,17] It is worth noting that the series solution from the method of successive approximations tends to diverge for any positive time no matter how small the initial amplitude parameter is.[10] However, the solutions up to the first few corrections beyond the linear approximation can provide a qualitative understanding of the growth in the WN regime.[18–21] The *nonlinear* growth regime of the RTI is typically described by the classic model from Layzer where bubbles of light fluid rise into the heavy fluid at the linear-in-time growth determined by the bubble radius and the acceleration, $\eta(t) \sim U_b t$, where $U_b = \sqrt{g/(3k)}$ is the bubble velocity.[22] The Layzer's theory has been extended to spikes by Zhang[23] and for arbitrary Atwood numbers by Goncharov.[24] However, these models are strictly valid only for planar RTI and hence ignore the effects introduced by converging geometries which is of great interest in ICF and astrophysics. The work of Bell[25] and Plesset[26] introduced the effects of spherical/cylindrical convergence on RTI growth but only for the linear growth regime.

As mentioned earlier, incompressibility has been commonly assumed in the theoretical study of the classical RTI (CRTI). But for the ablative RTI (ARTI) where the thermal conduction effects are considered, the assumption of the constant fluid density is therefore not applicable. Nevertheless, it is well known that incompressibility can be satisfied in a generalized way with less restrictions which only assumes that the pressure variation is negligibly small compared with the density and temperature variations (i.e., isobaric approximation).[27] In fact, the incompressible model has been widely used in the analytical treatment of the ARTI at the ablation front over the past decades.[27–35] Additionally, it has been reported by several authors that the linear growth of the ARTI is obviously suppressed comparison to that of the CRTI, especially for short-wavelength perturbations. The dominant stabilizing factors have been attributed to the damping due to the fire-polishing and the vorticity convection,[28,36] the dynamic overpressure due to the thermal conduction,[37] the anisotropic thermal diffusion,[38,39] and the finite-thickness of the ablation surface.[31,40] Recently, WN models for the ARTI have been developed.[41–44] The main result is that the WN growth of the ARTI is mitigated by the thermal conduction effects.[19–21,45–47] However, those studies are performed primarily in planar geometry. While the RTI has been studied experimentally and theoretically at spherical interfaces,[48–55] very little research has been performed at cylindrical interfaces.[56,57] The RTI growth at cylindrical interfaces can not only display the convergence effects but also give a direct line-of-sight diagnostic access (i.e., the amount of perturbation growth of the RTI in cylindrical geometry can be measured visibly by looking down the z-axis). In the present work, a WN model is proposed for the CRTI in a cylindrically convergent geometry, based on the potential flow description for incompressible fluids.

The remainder of this paper is organized as follows. Section II contains the theoretical formulation of the potential model, the initial conditions, and the resulting WN perturbation solutions up the third-order accuracy. In Sec. III, we analyze the temporal evolution of the interface position in a cylindrically convergent geometry, compare the WN growth of the RTI in cylindrical geometry with its plane counterpart, and propose a reduced formula. Using the reduced formula, the nonlinear saturation amplitude (NSA) of the fundamental mode and the phases of the second and the third harmonics are investigated. Moreover, the precision of the reduced formula is also studied in Sec. III. Section IV summarizes the major results of the present study and includes some concluding remarks.

II. THEORETICAL MODEL AND MATHEMATICAL FORMULATION

This section is devoted to a brief description of the theoretical framework of the present work. In Sec. II A, we will demonstrate the governing equation of the potential model and the solution procedure of successive approximations. Section II B focuses on the initial conditions for cylindrical RTI. Analytic expressions for the temporal evolution of the amplitude of the first several harmonics are given in Sec. II C.

A. Basic equations

A cylindrical coordinate system is established, where the θ and the r are, respectively, along and normal to the undisturbed interface between the two fluids. The perturbed interface between the two fluids is parameterized as $r(\theta,t) = R_0 + \eta(\theta,t)$, where R_0 is the radius of the unperturbed interface and $\eta(\theta,t)$ is the perturbation displacement. In the following discussion, we shall denote the properties of the exterior fluid ($R_0 < r < \infty$) by the superscript "ex" and that of the interior fluid ($0 < r < R_0$) by the superscript "in". Considering two layer fluids with $\rho^{ex} > \rho^{in}$ (ρ is the density) in a gravitational field $\mathbf{g} = -g\hat{\mathbf{e}}_r$ (e.g., the hot-spot/fuel interface during the deceleration phase of the ICF implosion) to be, incompressible, inviscid, irrotational, and immiscible, the governing equations for this system are

$$\nabla^2 \phi^{ex} = \nabla^2 \phi^{in} = 0, \tag{1}$$

$$\frac{\partial \eta}{\partial t} + \frac{1}{r^2}\frac{\partial \eta}{\partial \theta}\frac{\partial \phi^{ex}}{\partial \theta} - \frac{\partial \phi^{ex}}{\partial r} = 0 \quad \text{at} \quad r = R_0 + \eta, \tag{2}$$

$$\frac{\partial \eta}{\partial t} + \frac{1}{r^2}\frac{\partial \eta}{\partial \theta}\frac{\partial \phi^{in}}{\partial \theta} - \frac{\partial \phi^{in}}{\partial r} = 0 \quad \text{at} \quad r = R_0 + \eta, \tag{3}$$

$$\rho^{\text{ex}}\left\{\frac{\partial\phi^{\text{ex}}}{\partial t}+\frac{1}{2}\left[\left(\frac{\partial\phi^{\text{ex}}}{\partial r}\right)^2+\frac{1}{r^2}\left(\frac{\partial\phi^{\text{ex}}}{\partial\theta}\right)^2\right]+gr\right\}$$
$$-\rho^{\text{in}}\left\{\frac{\partial\phi^{\text{in}}}{\partial t}+\frac{1}{2}\left[\left(\frac{\partial\phi^{\text{in}}}{\partial r}\right)^2+\frac{1}{r^2}\left(\frac{\partial\phi^{\text{in}}}{\partial\theta}\right)^2\right]+gr\right\}$$
$$=f(t)\quad\text{at}\quad r=R_0+\eta,\tag{4}$$

where ϕ^{ex} and ϕ^{in} are the velocity potentials which are related to the fluid velocities and $f(t)$ is an arbitrary function of time. The Laplace equation [Eq. (1)] describes the potential flow of the two fluids. Equations (2) and (3) represent the continuity condition of the normal component of the velocity for each fluid across the interface, where $\partial\eta/\partial t$ is the normal velocity of the interface between two fluids. Equation (4) means the continuity condition of the pressure across the interface (i.e., Bernoulli equation).

The perturbation displacement $\eta(\theta,t)$ and the velocity potentials $\phi^{\text{ex}}(r,\theta,t)$ and $\phi^{\text{in}}(r,\theta,t)$ can be expanded into a power series in a formal parameter ε

$$\eta(\theta,t)=\sum_{n=1}^{\infty}\varepsilon^n\eta^n(\theta,t)$$
$$=\sum_{n=1}^{\infty}\varepsilon^n\sum_{l=0}^{\lfloor\frac{n}{2}\rfloor}\eta_{n,n-2l}(t)\cos[(n-2l)m\theta],\tag{5}$$

$$\phi^{\text{ex}}(r,\theta,t)=\sum_{n=1}^{\infty}\varepsilon^n(\phi^{\text{ex}})^n(r,\theta,t)$$
$$=\sum_{n=1}^{\infty}\varepsilon^n\sum_{l=0}^{\lfloor\frac{n}{2}\rfloor}\phi^{\text{ex}}_{n,n-2l}(t)r^{-(n-2l)m}\cos[(n-2l)m\theta],\tag{6}$$

$$\phi^{\text{in}}(r,\theta,t)=\sum_{n=1}^{\infty}\varepsilon^n(\phi^{\text{in}})^n(r,\theta,t)$$
$$=\sum_{n=1}^{\infty}\varepsilon^n\sum_{l=0}^{\lfloor\frac{n}{2}\rfloor}\phi^{\text{in}}_{n,n-2l}(t)r^{(n-2l)m}\cos[(n-2l)m\theta],\tag{7}$$

where $\eta^n(\theta,t)$ is the nth-order component of the perturbation displacement (n is a natural number), and $(\phi^{\text{ex}})^n(r,\theta,t)$ and $(\phi^{\text{ex}})^n(r,\theta,t)$ are the nth-order component of the velocity potential. Here, m is the mode number of the instability (the number of wavelengths of the perturbation in one circumference), and the Gauss's symbol $\lfloor n/2\rfloor$ denotes the maximum integer that is less than or equal to $n/2$. The $\eta_{n,n-2l}(t)$ $[\phi^{\text{ex}}_{n,n-2l}(t),\ \phi^{\text{in}}_{n,n-2l}(t)]$ is the amplitude of the perturbation displacement (of the velocity potential) of the $(n-2l)$th Fourier mode in the nth-order of ε ($n=1,\ 2,\ \cdots;$ $l=0,\ 1,\ \cdots,\ \lfloor n/2\rfloor$). For $l=0$, quantities represent the nth harmonic generation; for $l\geq 1$, they represent the nth feedback to the $(n-2l)$th harmonic. Note that velocity potentials $\phi^{\text{ex}}(r,\theta,t)$ and $\phi^{\text{in}}(r,\theta,t)$ have satisfied the Laplace equation (1) and the conditions $\nabla\phi^{\text{ex}}(r,\theta,t)|_{r\to\infty}=0$ and $\nabla\phi^{\text{in}}(r,\theta,t)|_{r\to 0}=0\ (m\geq 2)$. In fact, the amplitudes of the $(n-2l)$th Fourier mode in the nth-order of ε [i.e., $\eta_{n,n-2l}(t)$, $\phi^{\text{ex}}_{n,n-2l}(t)$ and $\phi^{\text{in}}_{n,n-2l}(t)$, $n=1,\ 2,\ \cdots;$ $l=0,\ 1,\ \cdots,\ \lfloor n/2\rfloor$], are the unknowns that we always want to obtain.

As mentioned above, the boundary conditions are based on the interface. First, we substitute Eqs. (5)–(7) to Eqs. (2)–(4) replacing r with $R_0+\eta(\theta,t)$. Second, we collect terms of the same power in ε to construct the successive equations. Based on the Taylor series theory, the resulting boundary condition equations that hold at the interface directly can be expressed as power series in the formal expansion parameter ε. That is to say, the nth-order governing equations containing ε^n are obtained. Finally, the amplitudes of nth Fourier harmonics can be obtained by successively solving the resulting nth-order equations. In fact, the nth-order equations are the second-order ordinary differential equation (ODE). Setting $\varepsilon=1$, up to the pth-order approximation, the amplitude of the qth harmonic and the perturbation displacement can be expressed as $\eta^p_q(t)=\sum_{l=0}^{\lfloor\frac{p-q}{2}\rfloor}\eta_{q+2l,q}(t)$ and $\eta(\theta,t)=\sum_{q=1}^{p}\eta^p_q(t)\cos(qm\theta)$, respectively, where $\lfloor\frac{p-q}{2}\rfloor$ is the Gauss' symbol, and $l=0,\ 1,\ \cdots,\ \lfloor\frac{p-q}{2}\rfloor;\ q=1,\ 2,\ \cdots,\ p$.

B. Initial conditions

To solve the resultant second-order ODE at each order, one need the initial conditions of the perturbed interface. Since the fluids are assumed to be incompressible, the area surrounded by the interface between the two fluids should be preserved. This means that the area surrounded by the perturbed interface equals to the initial area, namely,

$$\frac{1}{2}\int_0^{2\pi}R_0^2d\theta=\frac{1}{2}\int_0^{2\pi}[R_0+\eta(\theta,t)]^2d\theta.\tag{8}$$

The nonzero parts of Eq. (8) are from the terms with even order of ε. We obtain at the second-order $O(\varepsilon^2)$

$$\eta_{2,0}(t)=-\frac{1}{4R_0}\eta^2_{1,1}(t),\tag{9}$$

where $\eta_{2,0}(t)$ is the second-order feedback to the zeroth harmonic and $\eta_{1,1}(t)$ is the amplitude of the linear growth of the fundamental mode. Since $\eta_{2,0}(t=0)=0$, we get $\eta_{1,1}(t=0)=0$. This indicates that the initial amplitude of the interface perturbation is zero [i.e., $\eta(\theta,t=0)=0$]. In fact, if one imposes a cosinusoidal interface perturbation in the form $r(\theta,t)=R_0+\eta_0\cos(m\theta)$, the area surrounded by the perturbed interface becomes

$$\pi\tilde{R_0}^2=\frac{1}{2}\int_0^{2\pi}[R_0+\eta_0\cos(m\theta)]^2d\theta=\pi\left(R_0^2+\frac{\eta_0^2}{2}\right),\tag{10}$$

where $\tilde{R_0}=\sqrt{R_0^2+\eta_0^2/2}$ is the "equivalent radius" of the perturbed interface. In other words, after imposing a cosinusoidal interface perturbation for an unperturbed interface with a radius of R_0 in incompressible fluids, the equivalent radius of the unperturbed interface becomes $\tilde{R_0}=\tilde{R_0}(R_0,\eta_0)$.

It is well-known that in most real systems (e.g., ICF implosions or supernova explosions), the interface will have both velocity and interface perturbations when the RTI starts, since the surface imperfections or other nonuniformities have undergone the Richtmyer-Meshkov instability and/or Landau-Darrieus instability before the steady acceleration (the RTI growth) stage.[58,59] However, as mentioned in our previous work[60] when ignoring the terms decaying exponentially in time and keep only the dominant terms, an initial interface perturbation in the form $\eta(x,t=0)=\eta_0\cos(kx)$ can be regarded as an equivalent initial velocity perturbation in the form $v(x,t=0)=\eta_0\gamma_{cl}\cos(kx)$. Namely, for the seeding of the RTI, the interface (velocity) perturbations are almost a duplication of the velocity (interface) perturbations. Thus, in order to keep the zero-order quantity (equilibrium state), here we impose only a single-mode velocity perturbation in the following form:

$$\eta_{n,n-2l}|_{t=0}=0, \tag{11}$$

$$\left.\frac{\partial\eta_{n,n-2l}}{\partial t}\right|_{t=0}=a_0\gamma\delta_{1n}\delta_{1(n-2l)}, \tag{12}$$

where a_0 is the equivalent interface perturbation amplitude for a velocity perturbation amplitude $v_0=a_0\gamma$, $\gamma=\sqrt{Amg/R_0}$ is the linear growth rate of the RTI in cylindrical geometry, δ_{1n} and $\delta_{1(n-2l)}$ are Kronecker delta function, $a_0/(2\pi R_0/m)\ll 1$, $n=1,\ 2,\ \cdots$, and $l=0,\ 1,\ \cdots,\ \lfloor n/2\rfloor$. Here, $A=(\rho^{ex}-\rho^{in})/(\rho^{ex}+\rho^{in})$ is the Atwood number in cylindrical geometry.

C. WN solutions

At the first-order $O(\varepsilon)$, the governing equation for the fundamental mode generation reduces to the well-known equation

$$\frac{d^2\eta_{1,1}}{dt^2}-\gamma^2\eta_{1,1}=0, \tag{13}$$

which is a second-order homogeneous ODE with constant coefficients. With the initial conditions $\eta_{1,1}(t=0)=0$ and $\partial\eta_{1,1}(t=0)/\partial t=a_0\gamma$ as mentioned in Sec. II B, we obtain in the linear approximation

$$\eta_{1,1}=\overbrace{a_0\sinh(\gamma t)}^{(\mathbf{k})}, \tag{14}$$

which depicts the temporal evolution of the amplitude of the linear growth of the fundamental mode in cylindrical RTI initiated by a single-mode cosinusoidal velocity perturbation.

At the second-order $O(\varepsilon^2)$, the second harmonic generation equation takes the form:

$$\frac{d^2\eta_{2,2}}{dt^2}-2\gamma^2\eta_{2,2}=-\frac{a_0^2\gamma^2}{2R_0}[(Am+1)\cosh(2\gamma t)+Am], \tag{15}$$

which is a second-order nonhomogeneous ODE with constant coefficients, and the equation of the second-order feedback to the zero harmonic reads

$$\frac{d^2\eta_{2,0}}{dt^2}=-\frac{a_0^2\gamma}{4R_0}\sinh(2\gamma t). \tag{16}$$

Note that the $\eta_{2,0}$ vanishes in planar RTI. The existence of the feedback to the zeroth harmonic is important in cylindrical geometry since it guarantees the conservation of the area (mass) surrounded by the interface between the two fluids. By imposing the initial conditions $\eta_{2,2}(t=0)=0$ and $d\eta_{2,2}(t=0)/dt=0$; $\eta_{2,0}(t=0)=0$ and $d\eta_{2,0}(t=0)/dt=0$, the second-order solutions can be expressed as

$$\eta_{2,2}=\overbrace{-\frac{a_0^2}{2}\left(\frac{m}{R_0}\right)A\sinh^2(\gamma t)}^{(\mathbf{k},\mathbf{k})}-\frac{a_0^2}{4}\left(\frac{1}{R_0}\right)\left[\overbrace{\sinh^2(\gamma t)+\cosh^2(\gamma t)}^{(\mathbf{k},\mathbf{k})}-\overbrace{\cosh(\sqrt{2}\gamma t)}^{(2\mathbf{k})}\right], \tag{17}$$

$$\eta_{2,0}=\overbrace{-\frac{a_0^2}{4}\left(\frac{1}{R_0}\right)\sinh^2(\gamma t)}^{(\mathbf{k},-\mathbf{k})}, \tag{18}$$

where the explicit m/R_0 and $1/R_0$ dependencies of $\eta_{2,2}$ and explicit $1/R_0$ dependence of $\eta_{2,0}$ are displayed clearly. Note that the second term in the right-hand side of Eq. (17) and the $\eta_{2,0}$ given by Eq. (18) are implicitly dependent on A, m, and g. On the one hand, the second harmonic generation is from both the mode-coupling of the wave vectors (**k**, **k**) and the linear growth of wave vector (2**k**). On the other hand, the second-order feedback to the zero harmonic is only from the mode-coupling of the wave vectors (**k**, –**k**).

At the third-order $O(\varepsilon^3)$, the equation for the third harmonic generation can be written as

$$\begin{aligned}\frac{d^2\eta_{3,3}}{dt^2}-3\gamma^2\eta_{3,3}&=\frac{a_0^3\gamma^2}{8R_0^2}[6(4A^2m^2-m^2+7Am+3)\cosh^2(\gamma t)\sinh(\gamma t)\\&-3\cosh(\sqrt{2}\gamma t)\sinh(\gamma t)-2\sqrt{2}(3Am+1)\\&\times\sinh(\sqrt{2}\gamma t)\cosh(\gamma t)-(6Am+5)\sinh(\gamma t)],\end{aligned} \tag{19}$$

and the equation of the third-order feedback to the fundamental mode takes the form:

$$\begin{aligned}\frac{d^2\eta_{3,1}}{dt^2}-\gamma^2\eta_{3,1}&=-\frac{a_0^3\gamma^2}{8R_0^2}[4(3A^2m^2+m^2-Am-9)\cosh^2(\gamma t)\sinh(\gamma t)\\&-(2Am-3)\cosh(\sqrt{2}\gamma t)\sinh(\gamma t)\\&-2\sqrt{2}(Am-1)\sinh(\sqrt{2}\gamma t)\cosh(\gamma t)\\&-(4A^2m^2+2m^2-2Am-11)\sinh(\gamma t)].\end{aligned} \tag{20}$$

Using the initial conditions $\eta_{3,3}(t=0)=0$ and $d\eta_{3,3}(t=0)/dt=0$; $\eta_{3,1}(t=0)=0$ and $d\eta_{3,1}(t=0)/dt=0$, the third-order solutions reads

$$\eta_{3,3} = \overbrace{\frac{a_0^3}{8}\left(\frac{m}{R_0}\right)^2 (4A^2-1)\sinh^3(\gamma t)}^{(\mathbf{k},\mathbf{k},\mathbf{k})} + \frac{a_0^3}{8}\left(\frac{m}{R_0^2}\right)A\left[\overbrace{4\sinh^3(\gamma t)+3\cosh^2(\gamma t)\sinh(\gamma t)}^{(\mathbf{k},\mathbf{k},\mathbf{k})} - \overbrace{3\cosh(\sqrt{2}\gamma t)\sinh(\gamma t)}^{(2\mathbf{k},\mathbf{k})}\right]$$
$$+\frac{a_0^3}{32}\left(\frac{1}{R_0^2}\right)\left[\overbrace{2\sinh^3(\gamma t)+10\cosh^2(\gamma t)\sinh(\gamma t)}^{(\mathbf{k},\mathbf{k},\mathbf{k})} - \overbrace{3\sqrt{2}\sinh(\sqrt{2}\gamma t)\cosh(\gamma t) - 4\cosh(\sqrt{2}\gamma t)\sinh(\gamma t)}^{(2\mathbf{k},\mathbf{k})}\right], \quad (21)$$

$$\eta_{3,1} = -\frac{a_0^3}{16}\left(\frac{m}{R_0}\right)^2\left[\overbrace{(3A^2+1)\sinh^3(\gamma t)}^{(\mathbf{k},\mathbf{k},-\mathbf{k})} + \overbrace{(A^2+1)\sinh(\gamma t) - \underbrace{\gamma t(A^2+1)\cosh(\gamma t)}_{\text{secular term}}}^{(\mathbf{k},-\mathbf{k},\mathbf{k})}\right]$$
$$+\frac{a_0^3}{16}\left(\frac{m}{R_0^2}\right)A\left[\overbrace{\sinh^3(\gamma t)}^{(\mathbf{k},\mathbf{k},-\mathbf{k})} + \overbrace{2\cosh(\sqrt{2}\gamma t)\sinh(\gamma t)}^{(2\mathbf{k},-\mathbf{k})} - \overbrace{\sinh(\gamma t) - \underbrace{\gamma t\cosh(\gamma t)}_{\text{secular term}}}^{(\mathbf{k},-\mathbf{k},\mathbf{k})}\right]$$
$$+\frac{a_0^3}{16}\left(\frac{1}{R_0^2}\right)\left[\overbrace{4\sinh^3(\gamma t)+5\cosh^2(\gamma t)\sinh(\gamma t)}^{(\mathbf{k},\mathbf{k},-\mathbf{k})} - \overbrace{\cosh(\sqrt{2}\gamma t)\sinh(\gamma t) - \sqrt{2}\sinh(\sqrt{2}\gamma t)\cosh(\gamma t)}^{(2\mathbf{k},-\mathbf{k})} - \overbrace{\underbrace{2\gamma t\cosh(\gamma t)}_{\text{secular term}}}^{(\mathbf{k},-\mathbf{k},\mathbf{k})}\right], \quad (22)$$

where the explicit m^2/R_0^2, m/R_0^2, and $1/R_0$ dependencies of the $\eta_{3,3}$ ($\eta_{3,1}$) are displayed clearly. The third harmonic generation is caused by the mode-coupling of the wave vectors (**k**, **k**, **k**) and (2**k**, **k**). The third-order feedback to the fundamental mode is caused by mode-coupling of the wave vectors (**k**, **k**, –**k**), (**k**, –**k**, **k**), and (2**k**, –**k**). In addition, the secular term appears in the mode-coupling of the wave vectors (**k**, – **k**, **k**) in the expression of $\eta_{3,1}$.

It is well-known that the amplitudes of the Fourier harmonics should converge to that in planar geometry for a quite short-wavelength perturbation compared to the initial radius R_0, namely, for the equivalent perturbation wavelength $\tilde{\lambda}_0 = 2\pi R_0/m \ll R_0$. Note that the term explicitly proportional to m/R_0 in the second-order amplitude given by Eq. (17) and terms explicitly proportional to m^2/R_0^2 in the third-order amplitude given by Eqs. (21) and (22), play the dominant role in each order for high-mode number perturbations. If we define the equivalent wave number of the perturbation in cylindrical geometry as $\tilde{k}_0 = m/R_0$, the WN results in planar RTI[60] can be recovered by keeping only the dominant terms (i.e., term explicitly including m/R_0 in $\eta_{2,2}$ and terms explicitly including m^2/R_0^2 in $\eta_{3,3}$ and $\eta_{3,1}$). In other words, the WN behavior of the RTI in the cylindrical case is similar to that in the planar case for high-mode number perturbations. Thus, the convergence effects play a pronounced role for low-mode number perturbations as shall be discussed below.

It is worth pointing out that the appearance of the explicit Atwood number independence terms in the second harmonic generation amplitude [i.e., the second term in the right-hand side of Eq. (17)], in the third harmonic generation [i.e., the third term in the right-hand side of Eq. (21)], and in the third-order feedback to the fundamental mode [i.e., the third term in the right-hand side of Eq. (22)], plays an important role in the WN growth of the RTI in a cylindrically converging geometry, which can further affect the development of the RTI with decreasing R_0. This is a main difference between the RTI in cylindrical geometry and in planar geometry. For example, assuming a limit case where $\rho^{ex} \simeq \rho^{in}$ and keeping Ag fixed, the amplitude of the second harmonic in the planar case tends to be zero while the explicit Atwood number independence terms in the cylindrical case is kept. Namely, the effects of cylindrical geometry on the RTI will play a more pronounced role with the decreasing A.

It is interesting to note that the above WN results [i.e., Eqs. (14), (17), (18), (21), and (22)] can be directly applied to the other density configuration with $\rho^{ex} < \rho^{in}$ in the presence of $\mathbf{g} = g\hat{\mathbf{e}}_r$ (e.g., the ablation surface during the acceleration phase of the ICF implosion) by substituting $g \to -g$ and $A \to -A$, since the combination Ag appears everywhere. Analysis of such system, which is of great interest in the supernova explosions and ICF implosions, will also be considered in detail here.

III. WEAKLY NONLINEAR ANALYSIS

In Subsections III A–III C, we focus on the WN growth of the cylindrical RTI and compare the difference between the WN growth of the RTI in cylindrical geometry and in its plane counterpart. Furthermore, a reduced formula is proposed by keeping only the dominant terms in the WN results. Applying the reduced formula, the nonlinear saturation amplitude of the fundamental mode and the phases of the Fourier harmonics are studied. Besides, the precision of the reduced formula will also be discussed.

A. WN growth

We address in this section the temporal evolution of the interface position of the cylindrical RTI in the WN regime.

In Fig. 1 (Fig. 2), we illustrate the WN development of the cylindrical RTI growth for Atwood numbers $A > 0\,(A < 0)$, using the perturbation solutions up to the third-order approximation. As the typical amplitude of the perturbation becomes similar in size to the equivalent perturbation wavelength $\tilde{\lambda}_0$, the shape of the interface begins to develop considerable outside-to-inside asymmetry, exhibiting the bubble-spike features. On the one hand, for $A > 0$, the wave crests broaden into bubbles and the wave troughs thin into spikes as shown in Fig. 1. On the other hand, for $A < 0$, instead, the wave crests thin into spikes and the wave troughs broaden into bubbles as shown in Fig. 2. The bubble-spike asymmetry can be seen markedly, especially for the case with a large Atwood number. With the time proceeding, the bubble tips are broadened but the spike heads become narrow, leading to the growing asymmetry of the interface. As can also be seen in Fig. 1 (Fig. 2), when the Atwood number approaches to $+1$ (-1), the spike becomes more narrow and grows faster while the bubble becomes more round and grows slower. It means that the bubble-spike asymmetry in both shape and penetration is negligible when $A \to 0$, and becomes pronounced as $A \to \pm 1$.

In Fig. 3, we plot the temporal evolution of the amplitudes of bubble (η_b) and spike (η_s) for Atwood numbers $A > 0$ and $A < 0$ using the data of Figs. 1 and 2. As can be seen, with the time proceeding, the amplitudes of bubble and spike increase rapidly. For a fixed γt, the spike amplitude increases with increasing absolute value of the Atwood number but the bubble amplitude decreases with increasing absolute value of the Atwood number. For the most A, at a fixed γt, the spike amplitude is greater than the bubble's, except for $A = -0.1$, as indicated in Fig. 3. Generally speaking, in cylindrical RTI for small density variations with $A < 0$ the bubble can grow fast than the spike, which is not observed in planar RTI.

Due to the excitation of the zeroth harmonic, the averaged radius of the perturbed interface $\bar{R}$ will be corrected by the nonlinear mode-coupling, namely, $\bar{R}(t) = R_0 + \eta_{2,0}(t)$. Figure 4 exhibits the $\bar{R}(t)$ decreases with increasing γt. This also means that the effective wavelength of the perturbation tends to become short but the linear growth rate tends to become large since $\tilde{\lambda}(t) = 2\pi\bar{R}(t)/m$ and $\gamma(t) = \sqrt{Amg/\bar{R}(t)}$. This property vanishes in planar RTI where the λ and the γ are fixed.

B. Comparison with planar geometry

Now let us consider the difference between the WN growth of the RTI in cylindrical geometry and in planar geometry. Figure 5 displays the temporal evolution of the amplitude of the fundamental mode ($\eta_1 = \eta_{1,1} + \eta_{3,1}$) for Atwood numbers $A > 0$ and $A < 0$ in the cylindrical case and in the planar case, respectively. The Atwood number dependence of the temporal evolution of amplitude of the

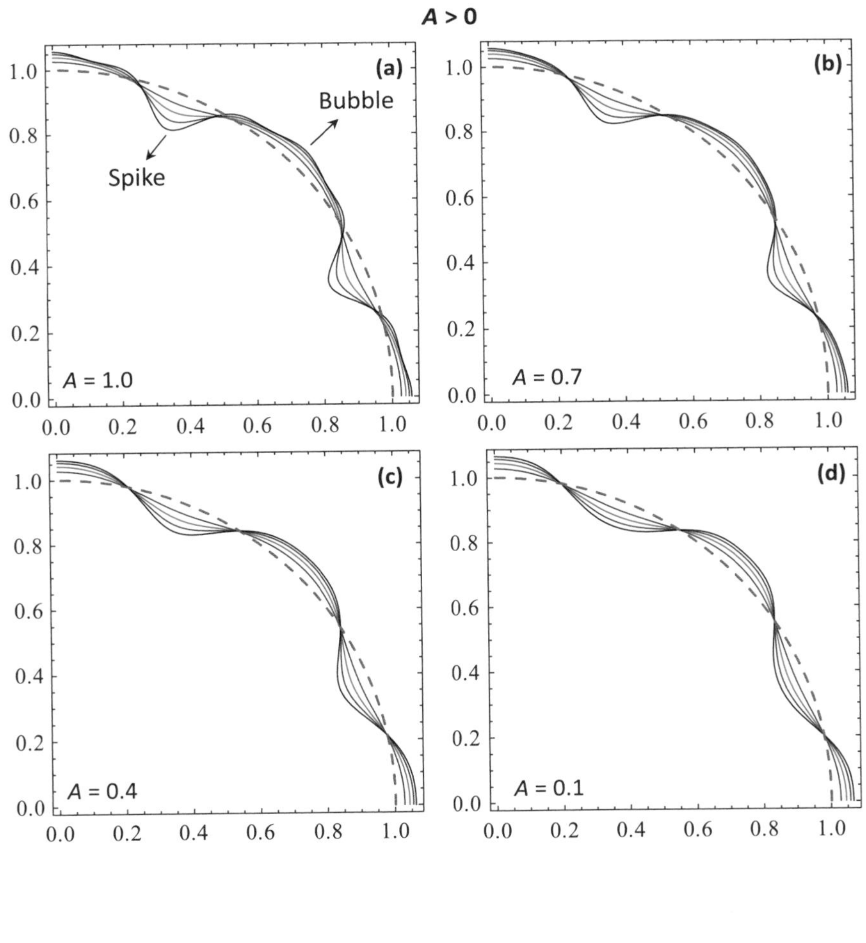

FIG. 1. Temporal evolution of the interface position $r(\theta, t)$ for Atwood numbers $A = 1.0$ (a), 0.7 (b), 0.4 (c), and 0.1 (d) at normalized time $\gamma t = 4.3$, 4.8, 5.1, and 5.3, respectively. The parameters are $a_0 = 0.001\tilde{\lambda}_0$, $R_0 = 1.0$, $m = 8$. The interface develops outside-to-inside asymmetry with the wave crests broadening into bubbles and the troughs thinning into spikes (e.g., the hot-spot/fuel interface during the deceleration phase of the ICF implosion).

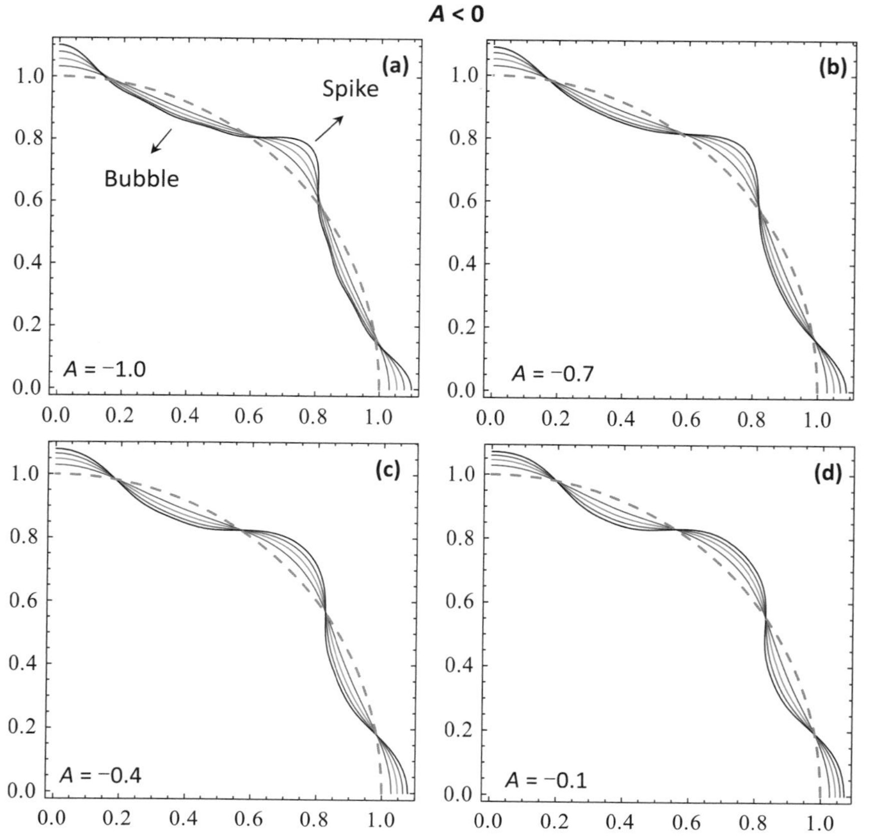

FIG. 2. Temporal evolution of the interface position $r(\theta, t)$ for Atwood numbers $A = -1.0$ (a), -0.7 (b), -0.4 (c), and -0.1 (d) at normalized time $\gamma t = 4.3$, 4.8, 5.1, and 5.3, respectively. The parameters are the same as the data of Fig. 1. The interface develops outside-to-inside asymmetry with the wave crests thinning into spikes and the troughs broadening into bubbles (e.g., the ablation surface during the acceleration phase of the ICF implosion).

fundamental mode can be seen noticeably in Fig. 5. In the initial linear growth regime, the amplitude of the fundamental mode increases exponentially with increasing γt, until its growth is alleviated and saturated by the third-order negative feedback. In other words, an inflexion (maximum) can be found in the temporal evolution of the amplitude of the fundamental mode. Since the decreasing amplitude in the theoretical model is invalid, the scope of application of the expansion theory is before the fundamental mode exceeding the maximum amplitude. As illustrated in Fig. 5, the maximum amplitude of the fundamental mode is inversely proportional to the absolute value of the Atwood number. The variation of the Atwood number extremely influences the amplitude of the fundamental mode. At a fixed γt, the smaller the Atwood number is, the larger the amplitude of the fundamental mode is. Note that the for a fixed A and γt, the amplitude of the fundamental mode in cylindrical RTI is larger than that in planar RTI.

Based on Fourier analysis, when the second harmonic is considered, the bubble-spike interface can be formed. Moreover, the phase and the amplitude of the second harmonic play different roles, which can be considered separately. On the one hand, the distribution of the bubble-spike asymmetry is determined by the phase shift between the second harmonic and the fundamental mode. On the other hand, the specific shape and penetration of bubble-spike asymmetry are affected by the amplitude of the second harmonic, namely, the larger the amplitude of the second harmonic is, the more rounded (shortened) the bubble tip (amplitude) becomes, but more narrowed (elongated) the spike tip (amplitude) becomes. As can be seen in Eqs. (17) and (21), the amplitudes of the second and the third harmonics can be negative or positive which is determined by the Atwood numbers. The negative amplitude means the phase of the Fourier harmonic is shifted by π against the fundamental mode and the positive amplitude means that the phase of the Fourier harmonic is the same as the fundamental mode. In the following discussions, for the sake of simplicity, the sign of the amplitude of the Fourier harmonic will be ignored when we study the amplitude unless otherwise stated.

Temporal evolution of the amplitudes of the second harmonic ($\eta_2 = \eta_{2,2}$) for Atwood numbers $A > 0$ and $A < 0$, are illustrated in Fig. 6. It is well-known that in planar RTI the phase of the second harmonic is shifted by π against the fundamental mode for $A > 0$ and the phase of the second harmonic is the same as the fundamental mode for $A < 0$.[10,13,14,16,17,60] As can be seen from Eq. (17) in cylindrical RTI, for $A > 0$, the phase of the second harmonic is shifted by π against the fundamental mode while for $A < 0$ with small density ratios, the phase of the second harmonic can still be shifted by π against the fundamental mode, as shown in Fig. 6 for $A = -0.1$. Note that for a fixed A and γt, the amplitude of the second harmonic in cylindrical RTI is larger than that in planar RTI when $A > 0$, as illustrated in

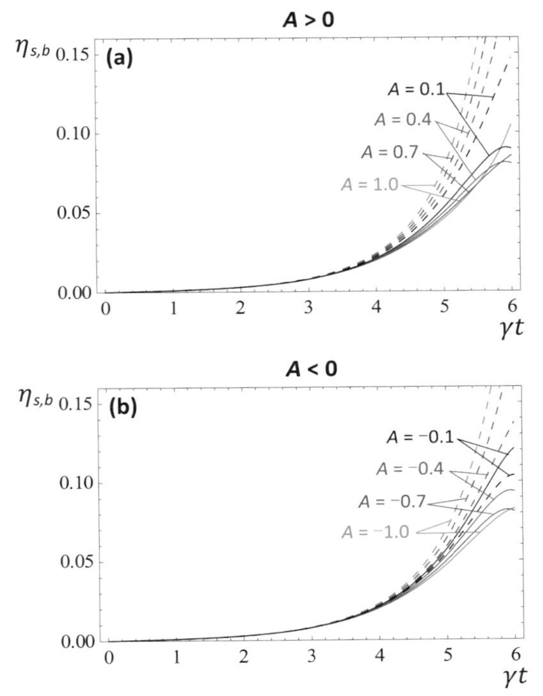

FIG. 3. Temporal evolution of the amplitudes of bubble (solid line) and spike (dashed line) for Atwood number $A = 1.0$, 0.7, 0.4, and 0.1 (a), and $A = -1.0$, -0.7, -0.4, and -0.1 (b), respectively. The parameters are the same as the data of Fig. 1. The bubble amplitude is larger than the spike's for $A = -0.1$.

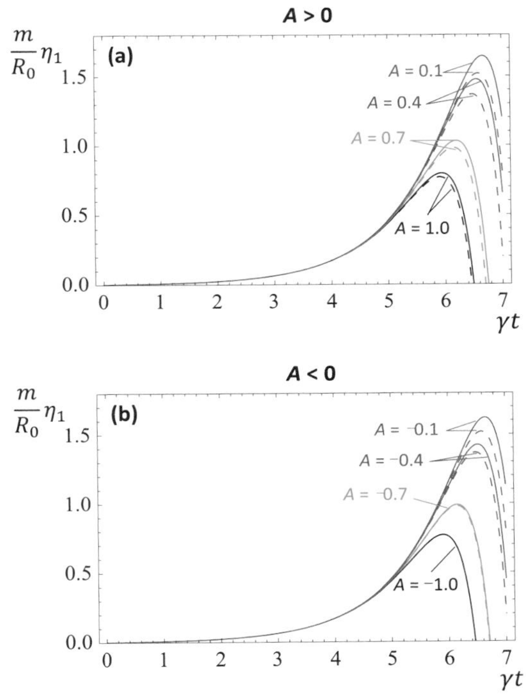

FIG. 5. Temporal evolution of the normalized amplitudes of the fundamental mode $m\eta_1/R_0$ for Atwood numbers $A = 1.0$, 0.7, 0.4, and 0.1 (a), and $A = -1.0$, -0.7, -0.4, and -0.1 (b), respectively, in cylindrical geometry (solid line) and in planar geometry (dashed line). The parameters are the same as the data of Fig. 1.

Fig. 6(a). Instead, for a fixed A and γt, the amplitude of the second harmonic in cylindrical RTI is less than that in planar RTI when $A < 0$, as illustrated in Fig. 6(b).

We present the temporal evolution of the amplitudes of the third harmonic ($\eta_3 = \eta_{3,3}$) for Atwood numbers $A > 0$ and $A < 0$ in Fig. 7. It can be seen that the phase and the amplitude of the third harmonic is essentially dependent on the Atwood number in both cylindrical RTI and planar RTI. As can be seen, when the phase of the third harmonic is the same as the fundamental mode, the amplitude of the third harmonic in cylindrical RTI is larger than in planar RTI for $A > 0$ but the amplitude of the third harmonic in cylindrical RTI is less than in planar RTI for $A < 0$. On the other hand, when the phase of the third harmonic is shifted by π against the fundamental mode, the amplitude of the third harmonic in cylindrical RTI is less than in planar RTI for $A > 0$ but the amplitude of the third harmonic in cylindrical RTI is larger than in planar RTI for $A < 0$.

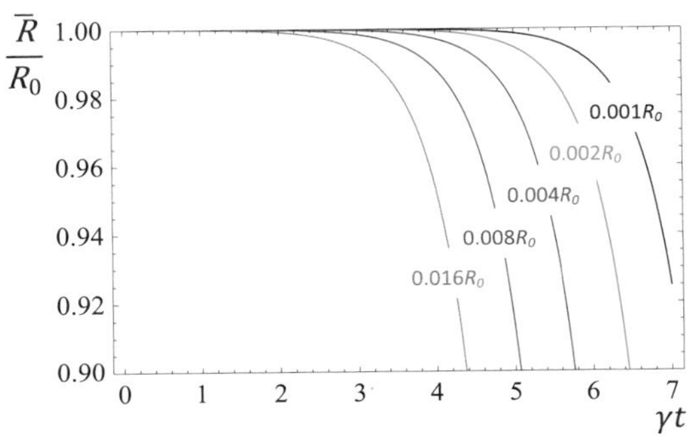

FIG. 4. Temporal evolution of the normalized averaged radius of the perturbed interface $\bar{R}(t)/R_0$ at different perturbation amplitudes of $a_0 = 0.001R_0$, $0.002R_0$, $0.004R_0$, $0.008R_0$, and $0.016R_0$. The parameters are $R_0 = 1.0$.

A simple physical interpretation of the above phenomena of the difference between the WN growth in cylindrical RTI and in planar RTI can be given, from the perspective of momentum equation. The two-dimensional momentum equations in cylindrical geometry are

$$\frac{\partial(\rho v_r)}{\partial t} = -\frac{1}{r}\frac{\partial}{\partial r}(r\rho v_r^2) - \frac{1}{r}\frac{\partial}{\partial \theta}(\rho v_r v_\theta) - \frac{\partial p}{\partial r} + \frac{\rho v_\theta^2}{r} - \rho g, \quad (23)$$

$$\frac{\partial(\rho v_\theta)}{\partial t} = -\frac{1}{r}\frac{\partial}{\partial r}(r\rho v_r v_\theta) - \frac{1}{r}\frac{\partial}{\partial \theta}(\rho v_\theta^2) - \frac{1}{r}\frac{\partial p}{\partial r} - \frac{\rho v_r v_\theta}{r}, \quad (24)$$

where additional terms of inertial forces $\rho v_\theta^2/r$ and $\rho v_r v_\theta/r$ compared to the planar case are included. Here, v_r is the velocity in r direction and v_θ in θ direction, and p is the pressure. In the WN growth regime, these two terms are the second-order components and can be ignored when $R_0 \to +\infty$. Considering the motion of fluids on both sides of the interface, after a simple force analysis, it is found that the resultant force of $\rho v_\theta^2/r$ and $\rho v_r v_\theta/r$ for the case of $A > 0$

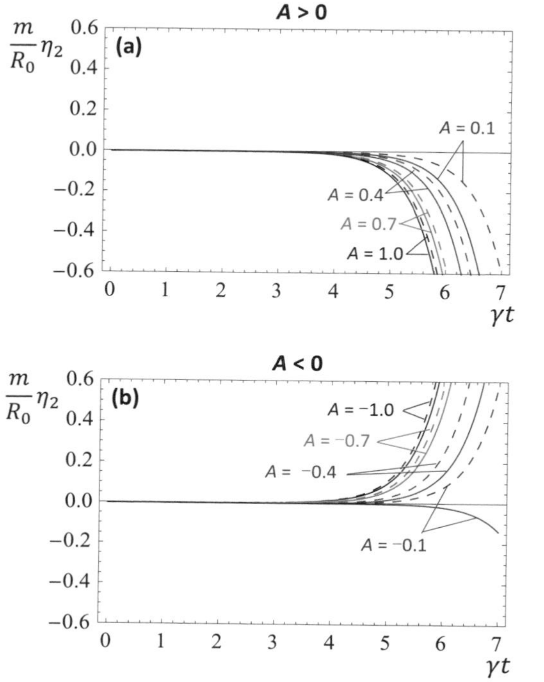

FIG. 6. Temporal evolution of the normalized amplitudes of the second harmonic $m\eta_2/R_0$ for Atwood numbers $A=1.0$, 0.7, 0.4, and 0.1 (a), and $A=-1.0$, -0.7, -0.4, and -0.1 (b), respectively, in cylindrical geometry (solid line) and in planar geometry (dashed line). The parameters are the same as the data of Fig. 1.

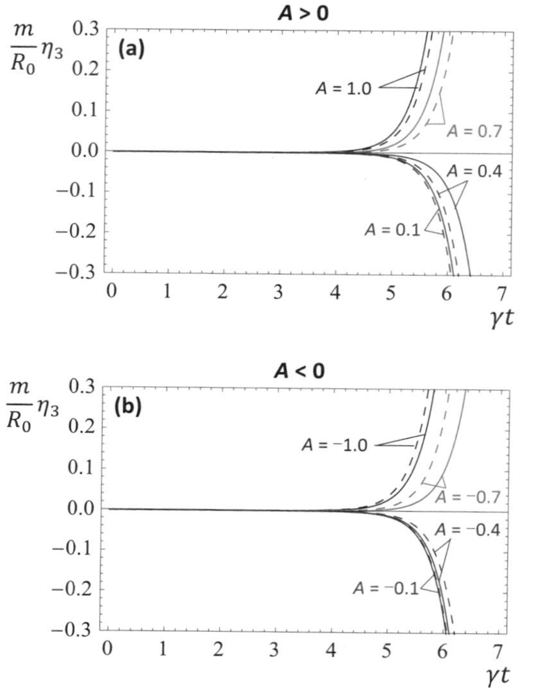

FIG. 7. Temporal evolution of the normalized amplitudes of the third harmonic $m\eta_3/R_0$ for Atwood numbers $A=1.0$, 0.7, 0.4, and 0.1 (a), and $A=-1.0$, -0.7, -0.4, and -0.1 (b), respectively, in cylindrical geometry (solid line) and in planar geometry (dashed line). The parameters are the same as the data of Fig. 1.

(as shown in Fig. 1) tends to extrude the spike head but stretch the bubble tip, while the resultant force of $\rho v_\theta^2/r$ and $\rho v_r v_\theta/r$ for the case of $A<0$ (as shown in Fig. 2) stretches the spike head but extrudes the bubble tip, which coincides quite well with Figs. 5–7. In other words, the inward development interface in cylindrical RTI is extruded while the outward development interface is stretched by the additional inertial forces compared with that in planar RTI. These effects are from the effects of cylindrically converging geometry.

C. Reduced formulas

If we drop the terms decaying exponentially in time and keep only the dominant terms at each order (i.e., terms including $e^{\gamma t}$, $e^{2\gamma t}$, or $e^{3\gamma t}$), then the amplitudes of the fundament mode $\eta_1(t)$, the second harmonic $\eta_2(t)$, and the third harmonic $\eta_3(t)$, up to the third-order correction for $m\geq 2$, reduce to

$$\eta_1=\left\{1-\frac{1}{16R_0^2}[m^2(3A^2+1)-Am-9]\eta_L^2\right\}\eta_L, \tag{25}$$

$$\eta_2=-\frac{1}{2R_0}(Am+1)\eta_L^2, \tag{26}$$

$$\eta_3=\frac{1}{8R_0^2}[m^2(4A^2-1)+7Am+3]\eta_L^3, \tag{27}$$

and the averaged radius of the perturbed interface $\bar{R}(t)$ becomes

$$\bar{R}=R_0-\frac{1}{4R_0}\eta_L^2, \tag{28}$$

where $\eta_L=\frac{1}{2}(v_0/\gamma)e^{\gamma t}$ is the linear regime spatial amplitude with v_0 being the amplitude of a cosinusoidal single-mode initial velocity perturbation in the form $v(x,t=0)=v_0\cos(m\theta)$.

As can be seen in Eq. (25) up to the third-order correction, the exponential growth of the fundamental mode is alleviate by the nonlinear mode-coupling effects, namely, the third-order negative feedback to the fundamental mode. If we define the transition from the linear to the nonlinear growth regime to occur when the growth of the fundamental mode (η_1) is reduced by 10% in comparison to its corresponding linear (exponential) growth (η_L),[61] namely, $(\eta_L-\eta_1)/\eta_L=0.1$, then we obtain

$$\frac{m}{R_0}\eta_{1s}^{(3)}=2\sqrt{\frac{2}{5}}\frac{1}{\sqrt{(3A^2+1)-A/m-9/m^2}}, \tag{29}$$

for $m\geq 2$ [$A>(1+\sqrt{61})/12$ if $m=2$; $A>1/9$ if $m=3$], where $\eta_{1s}^{(3)}$ is the NSA of the fundamental mode with corrections up to the third-order accuracy. Let $A=1$, then $\tilde{k}_0\eta_{1s}^{(3)}\simeq 0.628\tilde{\lambda}_0$ is recovered when $m\to\infty$, which is a typical threshold for nonlinearity widely used in planar RTI.[9]

The dependence of the normalized NSA (NNSA) of the fundamental mode $m\eta_{1s}^{(3)}/R_0$ on the Atwood number A and the mode number m is distinctly described by Eq. (29). In Fig. 8, we exhibit the NNSA of the fundamental mode at different mode numbers for several Atwood numbers. On the one hand, the NNSA of the fundamental mode decreases with increasing m for a fixed A. On the other hand, the NNSA of the fundamental mode decreases with increasing absolute value of A for a fixed m, as depicted in Fig. 8, which is similar to that in planar case.[61]

Equation (26) indicates that the phase of the second harmonic is shifted by π against the fundament mode for $A > 0$ or $-1/m < A < 0$ and the phase of the second harmonic is the same as the fundamental mode for $A < -1/m$. Additionally, for $A = -1/m$, the second harmonic disappears. The dependence of phase on the Atwood number and the mode number becomes more complicated for the third harmonic. There is a critical Atwood number A^c for the phase of third harmonic. For the case of $A > 0$, when $A > A^c$, the third harmonic has the same phase as the fundamental mode; when $A < A^c$, its phase is shift by π against the fundament mode; and for $A = A^c$, the third harmonic disappears. For the case of $A < 0$, when $A < A^c$, the third harmonic has the same phase as the fundamental mode; when $A > A^c$, its phase is shift by π against the fundament mode; and for $A = A^c$, the third harmonic disappears. Here, the critical Atwood number for the phase of the third harmonic can be expressed as

$$A^c = \begin{cases} \dfrac{-7m + m\sqrt{1+16m^2}}{8m^2} & A > 0, \\ \dfrac{-7m - m\sqrt{1+16m^2}}{8m^2} & A < 0. \end{cases} \qquad (30)$$

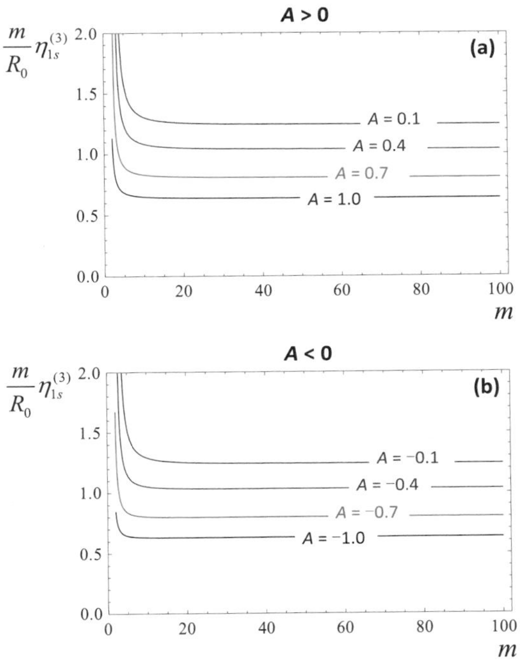

FIG. 8. The normalized nonlinear saturation amplitude of the fundamental mode $m\eta_{1s}^{(3)}/R_0$ versus the mode number m for Atwood numbers $A = 1.0$, 0.7, 0.4, and 0.1 (a), and $A = -1.0$, -0.7, -0.4, and -0.1 (b), respectively.

Figure 9 is a plot of the critical Atwood number of the phase of the third harmonic at different mode numbers. As can be seen obviously, A^c decreases with increasing m and tends to $A^c = \pm\frac{1}{2}$ as $m \to \infty$.

We now analyze the precision of the reduced formula [Eqs. (25)–(28)]. The temporal evolution of the normalized amplitudes of the fundamental mode, initiated with different amplitudes of the velocity perturbation for several mode numbers, is plotted in Fig. 10 for $A = 1$ and Fig. 11 for $A = -1$, respectively.

As can be seen, the amplitude of the fundamental first raises gradually, reaching a maximum value and then drops rapidly to zero. Namely, one can recognize a point of inflexion in the curve. In the early stage, the fundamental mode grows exponentially. With increasing amplitude, the growth of fundamental mode begins to fall below the linear growth due to the nonlinear mode-coupling effects. This continues until the maximum amplitude is reached. Exceeding the maximum amplitude, the amplitude decreases sharply, which in fact is always invalid. The maximum amplitude predicted by the "exact formula" [Eqs. (14) and (22)] slightly varies with the amplitude of the initial velocity perturbation. However, the maximum amplitude given by the "reduced formula" [Eq. (25)] shows the same value for different amplitudes of the initial perturbation. Note that the maximum amplitude here is larger than its NSA counterpart mentioned above.

For a fixed γt, the amplitude of the fundamental mode from the exact formula first less than and then larger than that from the reduced formula for $A = 1$, as can be seen from comparison of the corresponding cases with the same mode numbers and initial velocity perturbation amplitudes in Fig. 10. However, for $A = -1$ at a fixed γt, the amplitude of the fundamental mode from the exact formula is less than that from the reduced formula, as can be seen in Fig. 11. Nevertheless, the reduced formula Eq. (25) can better describe the temporal evolution of the amplitude of the fundamental mode initiated by a high-mode, small-amplitude initial velocity perturbations. For $m > 8$ and $a_0 < 0.01\tilde{\lambda}_0$, the reduced formula Eq. (25) can nearly depict accurately the temporal evolution of the amplitudes of the fundamental mode in cylindrical RTI, as shown in Figs. 10(c) and 10(d) and Figs. 11(c) and 11(d), respectively.

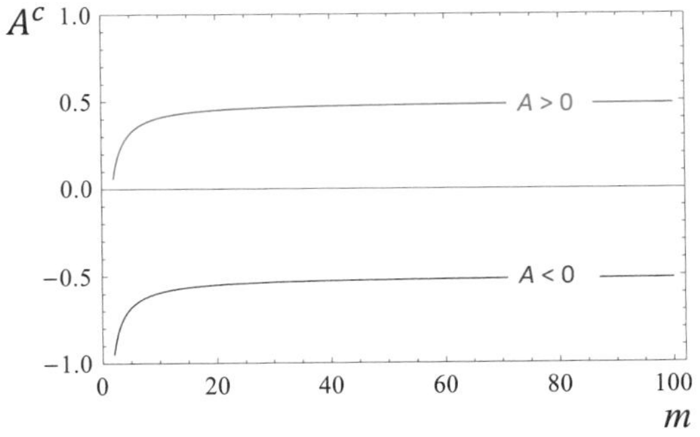

FIG. 9. The critical Atwood number of the phase of the third harmonic A^c versus the mode number m.

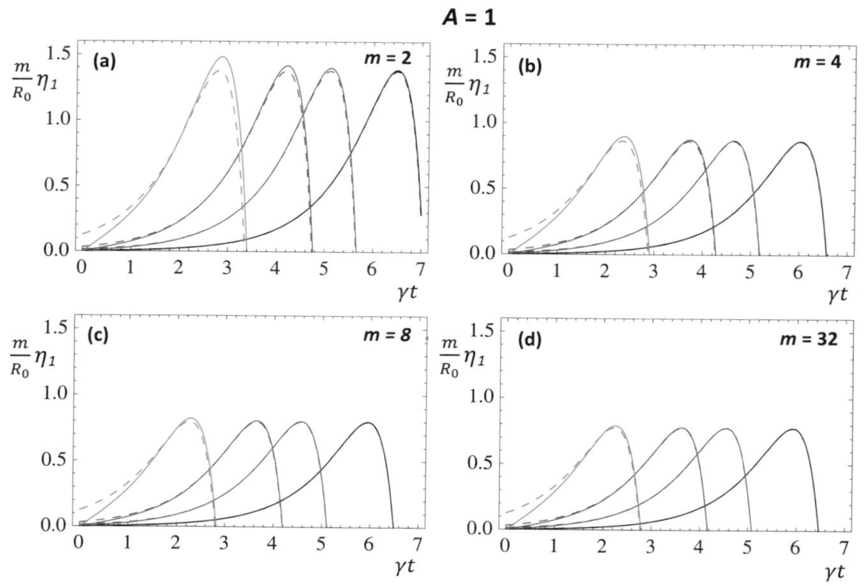

FIG. 10. Temporal evolution of the normalized amplitude of the fundamental mode $m\eta_1/R_0$ initiated with different velocity perturbation amplitudes of $a_0 = 0.001\tilde{\lambda}_0$, $0.004\tilde{\lambda}_0$, $0.01\tilde{\lambda}_0$, and $0.04\tilde{\lambda}_0$ for mode numbers $m = 2$ (a), 4 (b), 8 (c), and 32 (d), respectively. The parameters are $A = 1.0$ and $R_0 = 1.0$. The solid curves give the results of Eqs. (14) and (22), whereas the dashed curves are the results of Eq. (25).

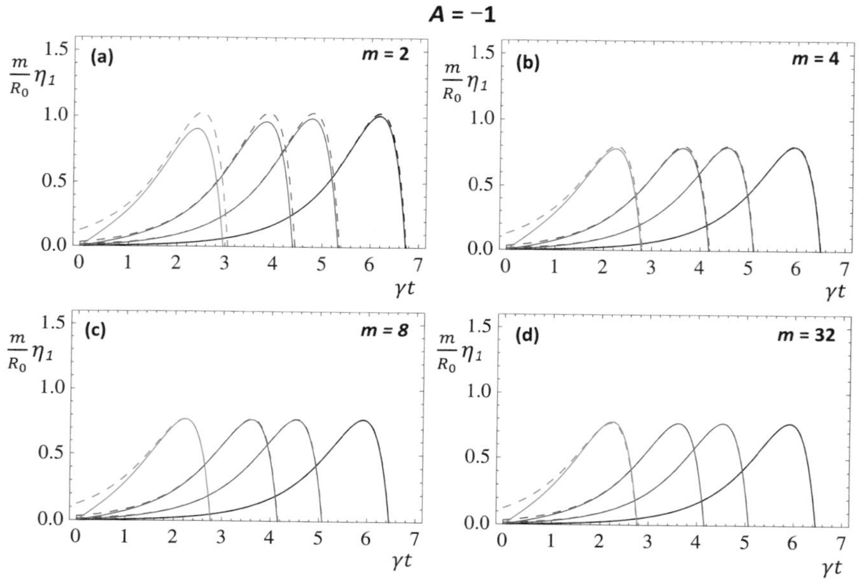

FIG. 11. Temporal evolution of the normalized amplitude of the fundamental mode $m\eta_1/R_0$ initiated with different velocity perturbation amplitudes of $a_0 = 0.001\tilde{\lambda}_0$, $0.004\tilde{\lambda}_0$, $0.01\tilde{\lambda}_0$, and $0.04\tilde{\lambda}_0$ for mode numbers $m = 2$ (a), 4 (b), 8 (c), and 32 (d), respectively. The parameters are $A = -1.0$ and $R_0 = 1.0$. The solid curves give the results of Eqs. (14) and (22), whereas the dashed curves are the results of Eq. (25).

IV. CONCLUSIONS

In this article, based on the potential flow theory for ideal incompressible fluids, a WN model is porposed for the RTI in cylindrical geometry. Due to the area preservative property in incompressible fluids, the RTI initiated by a single-mode velocity perturbation is studied analytically. The nonlinear solutions up to the third-order approximation are obtained. It is shown that the second-order feedback to the zeroth harmonic is appeared in cylindrical RTI. The existence of the second-order feedback to the zeroth harmonic is essential for the incompressible fluids in the cylindrically convergent geometry, which can preserve the area (mass) surrounded by the perturbed interface.

Our third-order nonlinear model can describe the early stage of the formation of bubble-spike structure and the saturation of the linear growth of the fundament mode, for arbitrary Atwood numbers ($A > 0$ or $A < 0$). The bubble (spike) becomes more round (narrow) and grows slower (faster) when the Atwood number approaches ± 1. The effective wavelength of the perturbation tends to become short in cylindrical geometry due to the decrease of the averaged radius of the perturbed interface. The effects of convergence on the RTI play a significant role in the case with low-mode number perturbations. We have compared the WN growth of the RTI in cylindrical geometry with that in planar geometry, and provide a physical explanation for the interface of "extruded inward growth" and "stretched outward growth" in cylindrical RTI compared to its planar counterpart. It is interesting to point out that the inward development bubble can grows fast than the outward development spike in cylindrical RTI with a small density variation. Dropping the terms decaying exponentially in time and keeping only the dominant terms, a reduce formula for the WN growth of the RTI is proposed which is very suitable for small-amplitude initial velocity perturbations. Applying the reduced formula, the NSA of the fundamental mode and the phases of the second and the third harmonics are discussed. We find that the NNSA decreases with increasing mode numbers. The phases of the second and the third harmonics are strongly dependent on Atwood numbers and mode numbers.

It is interesting to point out that the cylindrical RTI initiated by a single-mode interface perturbation or by a single-mode both interface and velocity perturbations can also be studied by the present WN solutions, since an initial interface perturbation can be treated as an equivalent velocity perturbation within an acceptable approximation.[60] Moreover, it is worthwhile to stress that the WN analysis of our model can be performed up to arbitrary orders without any more assumptions with the help of a computerized algebraical calculation (e.g., MATHEMATICA or MAPLE software). The present model is expected to cause further understanding of the convergence effects on the WM growth of the RTI in cylindrical geometry and to be probably important for the assessment of the RTI in supernova explosions and ICF implosions.

ACKNOWLEDGMENTS

The author (L.F.W.) acknowledges valuable discussions with Dr. Jianfa Gu, Dr. Zhengfeng Fan, Dr. Zhensheng Dai, Mr. Chuang Xue, and Mr. Wanhai Liu. The authors are also indebted to the referee for his/her useful comments and suggestions. This work was supported by the National Natural Science Foundation of China (Grant Nos. 10935003, 11275031, 11075024, and 11274026), and the National Basic Research Program of China (Grant No. 2013CB834100).

[1]L. Rayleigh, Proc. London Math. Soc. **14**, 170 (1883).
[2]G. Taylor, Proc. R. Soc. London, Ser. A **201**, 192 (1950).
[3]S. Chandrasekhar, *Hydrodynamic and Hydromagnetic Stability* (Oxford University, London, 1961).
[4]V. N. Gamezo, A. M. Khokhlov, E. S. Oran, A. Y. Chtchelkanova, and R. O. Rosenberg, Science **299**, 77 (2003).
[5]T. Ebisuzaki, T. Shigeyama, and K. Nomoto, Astrophys. J. **344**, L65 (1989).
[6]I. Hachisu, T. Matsuda, K. Nomoto, and T. Shigeyama, Astrophys. J. **358**, L57 (1990).
[7]J. Nuckolls, L. Wood, A. Thiessen, and G. Zimmerman. Nature **239**, 139 (1972).
[8]S. Atzeni and J. Meyer-ter-Vehn, *The Physics of Inertial Fusion: Beam Plasma Interaction Hydrodynamics, Hot Dense Mater* (Oxford University, Oxford, 2004).
[9]J. D. Lindl, P. Amendt, R. L. Berger, S. G. Glendinning, S. H. Glenzer, S. W. Haan, R. L. Kauffman, O. L. Landen, and L. J. Suter, Phys. Plasmas **11**(2), 339 (2004).
[10]R. L. Ingraham, Proc. Phys. Soc. London, Ser. B **67**, 748 (1954).
[11]C. T. Chang, Phys. Fluids **2**, 656 (1959).
[12]H. W. Emmons, C. T. Chang, and B. C. Watson, J. Fluid Mech. **7**, 177 (1960).
[13]J. W. Jacobs and I. Catton, J. Fluid Mech. **187**, 329 (1988).
[14]S. W. Haan, Phys. Fluids B **3**, 2349 (1991).
[15]J. Garnier, C. Cherfils-Clérouin, and P.-A. Holstein, Phys. Rev. E **68**, 036401 (2003).
[16]M. Berning and A. M. Rubenchik, Phys. Fluids **10**, 1564 (1998).
[17]L.-F. Wang, W.-H. Ye, and Y.-J. Li, Chin. Phys. Lett. **27**, 025203 (2010).
[18]J. W. Jacobs and I. Catton, J. Fluid Mech. **187**, 353 (1988).
[19]B. A. Remington, S. W. Haan, S. G. Glendinning, J. D. Kilkenny, D. H. Munro, and R. J. Wallace, Phys. Fluids B **4**(4), 967 (1992).
[20]B. A. Remington, S. V. Weber, M. M. Marinak, S. W. Haan, J. D. Kilkenny, R. J. Wallace, and G. Dimonte, Phys. Plasmas **2**(1), 241 (1995).
[21]V. A. Smalyuk, V. N. Goncharov, T. R. Boehly, J. P. Knauer, D. D. Meyerhofer, and T. C. Sangster, Phys. Plasmas **11**(11), 5038 (2004).
[22]D. Layzer, Astrophys. J **122**, 1 (1955).
[23]Q. Zhang, Phys. Rev. Lett. **81**(16), 3391 (1998).
[24]V. N. Goncharov, Phys. Rev. Lett. **88**(13), 134502 (2002).
[25]G. I. Bell, Los Alamos National Laboratory Report No. LA-1321, 1951.
[26]M. S. Plesset, J. Appl. Phys. **25**, 96 (1954).
[27]H. J. Kull, Phys. Rep. **206**(5), 197 (1991).
[28]S. Bodner, Phys. Rev. Lett. **33**, 761 (1974).
[29]D. L. Book, Plasma Phys. Controlled Fusion **34**, 737 (1992).
[30]K. Mikaelian, Phys. Rev. A **46**, 6621 (1992).
[31]M. Tabak, D. H. Munro, and J. D. Lindl, Phys. Fluids B **2**(5), 1007 (1990).
[32]R. Betti, R. L. McCrory, and C. P. Verdon, Phys. Rev. Lett. **71**, 3131 (1993).
[33]R. Betti, V. Goncharov, R. L. McCrory, E. Tnrano, and C. P. Verdon, Phys. Rev. E **50**, 3968 (1994).
[34]J. G. Wouchuk and A. R. Piriz, Phys. Plasmas **2**, 493 (1995).
[35]V. N. Goncharov, R. Betti, R. L. McCrory, P. Sorotokin, and C. P. Verdon, Phys. Plasmas **3**(4), 1402 (1996).
[36]Z. Fan, S. Zhu, W. Pei, W. Hua, M. Li, X. Xu, J. Wu, Z. Dai, and L. Wang, EPL **99**, 65003 (2012).
[37]A. R. Piriz, J. Sanz, and L. F. Ibañez, Phys. Plasmas **4**(4), 1117 (1997).
[38]L. Masse, Phys. Rev. Lett. **98**, 245001 (2007).
[39]G. Huser, A. Casner, L. Masse, S. Liberatore, D. Galmiche, L. Jacquet, and M. Theobald, Phys. Plasmas **18**, 012706 (2011).
[40]L. F. Wang, W. H. Ye, and Y. J. Li, Phys. Plasmas **17**, 042103 (2010).
[41]J. Sanz, J. Ramìrez, R. Ramis, R. Betti, and R. P. J. Town, Phys. Rev. Lett. **89**, 195002 (2002).
[42]T. Ikegawa and K. Nishihara, Phys. Rev. Lett. **89**, 115001 (2002).
[43]J. Garnier, P.-A. Raviart, C. Cherfils-Clérouin, and L. Masse, Phys. Rev. Lett. **90**, 185003 (2003).
[44]J. Garnier and L. Masse, Phys. Plasmas **12**, 062707 (2005).
[45]Z. Fan, J. Luo, and W. Ye, Phys. Plasmas **16**, 102104 (2009).
[46]L. F. Wang, W. H. Ye, Z. M. Sheng, W.-S. Don, Y. J. Li, and X. T. He, Phys. Plasmas **17**, 122706 (2010).
[47]L. F. Wang, W. H. Ye, and X. T. He, Phys. Plasmas **19**, 012706 (2012).
[48]C. Cherfils, S. G. Glendinning, D. Galmiche, B. A. Remington, A. L. Richard, S. Haan, R. Wallace, N. Dague, and D. H. Kalantar, Phys. Rev. Lett. **83**(26), 5507 (1999).
[49]S. G. Glendinning, J. Colvin, S. Haan, D. H. Kalantar, O. L. Landen, M. M. Marinak, B. A. Remington, R. Wallace, C. Cherfils, N. Dague, L. Divol, D. Galmiche, and A. L. Richard, Phys. Plasmas **7**, 2033 (2000).
[50]K. O. Mikaelian, Phys. Rev. A **42**(6), 3400 (1990).
[51]D. S. Clark and M. Tabak, Phys. Rev. E **71**, 055302(R) (2005).
[52]D. S. Clark and M. Tabak, Phys. Rev. E **72**, 056308 (2005).
[53]K. O. Mikaelian, Phys. Fluids **17**, 094105 (2005).
[54]P. Amendt, J. D. Colvin, J. D. Ramshaw, H. F. Robey, and O. L. Landen, Phys. Plasmas **10**(3), 820 (2003).
[55]P. Amendt, Phys. Plasmas **13**, 042702 (2006).
[56]W. W. Hsing, C. W. Barnes, J. B. Beck, N. M. Hoffman, D. Galmiche, A. Richard, J. Edwards, P. Graham, S. Rothman, and B. Thomas, Phys. Plasmas **4**(5), 1832 (1997).
[57]S. T. Weir, E. A. Chandler, and B. T. Goodwin, Phys. Rev. Lett. **80**, 3763 (1998).
[58]A. R. Piriz and R. F. Portugues, Phys. Plasmas **10**(6), 2449 (2003).
[59]V. N. Goncharov, Phys. Rev. Lett. **82**(10), 2091 (1999).
[60]L. F. Wang, J. F. Wu, Z. F. Fan, W. H. Ye, X. T. He, W. Y. Zhang, J. F. Gu, Z. S. Dai, and C. Xue, Phys. Plasmas **19**, 112706 (2012).
[61]W. H. Liu, L. F. Wang, W. H. Ye, and X. T. He, Phys. Plasmas **19**, 042705 (2012).

PHYSICS OF PLASMAS 21, 042707 (2014)

Indirect-drive ablative Rayleigh-Taylor growth experiments on the Shenguang-II laser facility

J. F. Wu (吴俊峰),[1] W. Y. Miao (缪文勇),[2] L. F. Wang (王立锋),[1,3] Y. T. Yuan (袁永腾),[2] Z. R. Cao (曹柱荣),[2] W. H. Ye (叶文华),[1,3,a)] Z. F. Fan (范征锋),[1] B. Deng (邓博),[2] W. D. Zheng (郑无敌),[1] M. Wang (王敏),[1] W. B. Pei (裴文兵),[1] S. P. Zhu (朱少平),[1] S. E. Jiang (江少恩),[2] S. Y. Liu (刘慎业),[2] Y. K. Ding (丁永坤),[2] W. Y. Zhang (张维岩),[1] and X. T. He (贺贤土)[1,3]
[1]*Institute of Applied Physics and Computational Mathematics, Beijing 100094, China*
[2]*Research Center of Laser Fusion, China Academy of Engineering Physics, Mianyang 621900, China*
[3]*HEDPS, Center for Applied Physics and Technology, Peking University, Beijing 100871, China*

(Received 13 March 2014; accepted 6 April 2014; published online 22 April 2014)

In this research, a series of single-mode, indirect-drive, ablative Rayleigh-Taylor (RT) instability experiments performed on the Shenguang-II laser facility [X. T. He and W. Y. Zhang, Eur. Phys. J. D **44**, 227 (2007)] using planar target is reported. The simulation results from the one-dimensional hydrocode for the planar foil trajectory experiment indicate that the energy flux at the hohlraum wall is obviously less than that at the laser entrance hole. Furthermore, the non-Planckian spectra of x-ray source can strikingly affect the dynamics of the foil flight and the perturbation growth. Clear images recorded by an x-ray framing camera for the RT growth initiated by small- and large-amplitude perturbations are obtained. The observed onset of harmonic generation and transition from linear to nonlinear growth regime is well predicted by two-dimensional hydrocode simulations. © *2014 AIP Publishing LLC*. [http://dx.doi.org/10.1063/1.4871721]

I. INTRODUCTION

In laser-driven inertial confinement fusion (ICF), to create the fusion hotspot, the shell must maintain its integrity throughout the entire implosion to prevent excessive mixing between the cold shell and the hot core, which would lead to hotspot quenching prior to achieving fusion relevant temperatures (pressures).[1,2] In ICF implosions, the Richtmyer-Meshkov (RM) instability[3,4] is initiated early during the shock transit time prior to the target steady acceleration.[5] In fact, the RM instability (RMI) acts as a seed for the faster growing Rayleigh-Taylor (RT) instability[6,7] that occurs after the target begins to be accelerated. During the acceleration phase of the ICF implosion, small perturbations in the shell's areal density (ρR) seeded by manufacturing defects and/or laser non-uniformities which have undergone the RM growth grow at the ablation-front and ablator-fuel interface due to the RT instability (RTI) and Bell-Plesset (BP) Effects.[8–10] As the target shell converts its kinetic energy into hotspot internal energy (i.e., the deceleration phase), the ρR perturbations at the shell's inner surface grow due to RTI, RMI, and BP effects, forming the jet-like cold spikes,[11,12] which would hinder the hotspot formation and shrink the size of the hotspot. Additionally, the material mixing from the ablator to the hotspot due to ablation-front instability has been identical as the main factor leading to the performance degrading of the implosion.[13,14]

Because of its fundamental importance to the success of ICF, a considerable amount of theoretical research has been devoted to study of the linear and nonlinear growths of the ablative RTI (ARTI) and its application to implosions.[15–27] Various formulas have been developed for the linear growth rate of the ARTI.[15–22] Typically, a widely used formula considering the stabilization of density-gradient of the ablation-front[2] is given by

$$\gamma = \sqrt{\frac{kg}{1+kL_m}} - \beta k v_a, \tag{1}$$

where $k = 2\pi/\lambda$ is the perturbation wave number with λ being perturbation wavelength, g is the acceleration, L_m is the density-gradient scale length of the ablation-front, v_a is the ablation velocity, β depends fundamentally on the flow parameters and varies in numerical simulations, usually it is 3–4. The linear stability theory for the ARTI valid for arbitrary Froude numbers[19] has been developed by Goncharov *et al.* This self-consistent linear growth formula contains the effects of density gradient, restoring force, and damping.[18] More physical explanation of the stabilization of the ARTI can be found in Ref. 21. The linear growth rate of ARTI in the presence of preheating has been investigated by Ye *et al.* considering a change of the Atwood number of the ablation-front.[22] After sufficient linear growth, the typical amplitude of the perturbation becomes no longer small in size to the perturbation wavelength; the shape of the interface begins to develop considerable top-to-bottom asymmetry with the wave crests thinning into spikes and the troughs broadening into bubbles. The weakly nonlinear (WN) models for the ARTI have been developed.[23–25] It is found that the ablation effects strongly suppress the mode-coupling, which is in general agreement with our previous simulation results.[26,27]

A number of ARTI experiments have been conducted, especially for planar targets.[28–37] In experiments on Nova,[28] the linear growth rate of the ARTI for a 50 μm-wavelength

[a)]Electronic mail: ye_wenhua@iapcm.ac.cn

perturbation in indirect-drive is about 60% of its classical counterpart. The Nova RT experiments have been well modeled by Weber *et al.* considering the effects of opacity, equation of state, and radiation drive spectrum.[29] Multimode (up to eight modes) ARTI experiments clearly show the transition from the linear growth to the nonlinear growth regime in the form of the appearance of "beat modes" and the formation of "bubble-spike" pattern.[30] Three-dimensional (3-D) ARTI has been measured by Marinak *et al.* in order to study the effect of 3-D perturbation shape on the RT growth and saturation.[31] The experimental comparison of the ARTI and the classical RTI has remarkably exhibited the ablative stabilization effects on the short-wavelengths perturbation.[32] The dispersion curve of the indirect-drive ARTI is constructed by minimizing the influence of the shock using numerical experiments.[33] In direct-drive ARTI, both the experiment of Shigemori *et al.*[34] and that of Glendinning *et al.*[35] suggest that the discrepancy between the measured growth rates and the hydrocode simulations using Spitzer-Härm electron thermal conductivity can most likely be explained by the nonlocal electron heat transport, especially for the short-wavelength perturbations.[36] The ARTI using laser-intensity as high as $\sim 1.5 \times 10^{15}\ \mathrm{W/cm^2}$ has been studied by Smalyuk *et al.* showing strong preheating effects from both the nonlocal electron transport and hot electrons.[37] Recently, the ARTI experiment with rugby hohlraums on OMEGA laser facility is reported[38] and the experiment design aiming at the transition from WN growth to highly nonlinear regimes at the National Ignition Facility has been proposed[39] by Casner *et al.*

This study is devoted to planar indirect-drive ARTI experiments for single-mode perturbations with initial small and large amplitudes using brominated polystyrene targets. The paper is organized as follows. In Sec. II, we address the experimental setup. Section III contains the drive characterization. We present the results of the acceleration of the foil in Sec. IV. The discussion of single-mode ablative RT growth results with initial small and large amplitudes is described in Sec. V. Section VI summarizes the present investigation and includes some concluding remarks.

II. EXPERIMENTAL CONFIGURATION

The experiments are performed on the Shenguang-II laser facility.[40] Single-mode perturbation with a 50 μm-wavelength is imposed on one side of CH(Br) foil ($C_{50}H_{47}Br_3$). The foil is mounted across a diagnostic hole with a 500 μm diameter on the wall of a cylindrical Au hohlraum with 1700 μm long and 1000 μm diameter. The foil is ablated and accelerated by x-rays as schematically illustrated in Fig. 1 for face-on experiments. The setup for side-on experiments is similar, but the hohlraum is rotated about its axis of symmetry by 90°. In face-on experiments, the growth of areal-mass perturbations in the foil which is diagnosed by the transmitted backlighter x-rays is recorded as a function of time with an x-ray framing camera (XFC) (see Fig. 8 or Fig. 10). In the side-on experiments, the trajectory of the foil is obtained by imaging the transmitted x-rays through the foil on an x-ray streak camera (XSC) in the form of an edge-on shadow (see Fig. 6).

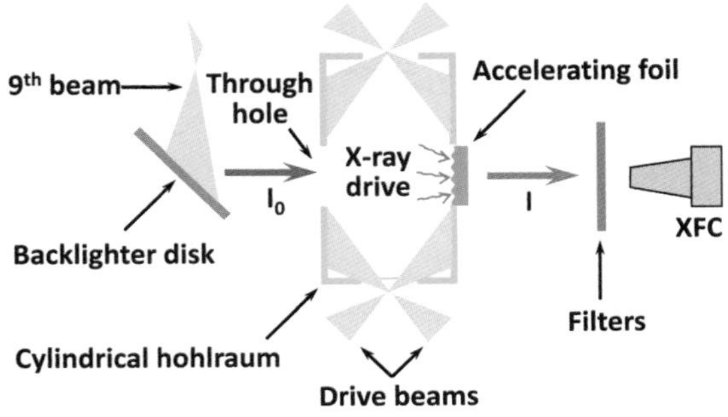

FIG. 1. Schematic of the experimental setup (not to scale) for the indirect-drive RT experiments. The foil is mounted on the right wall of a cylindrical gold hohlraum with an inward surface perturbation. Eight drive beams enter the ends of the hohlraum resulting in a thermal x-ray drive. When the foil is accelerated by the x-ray ablation towards the XFC, the ninth beam striking a Mo disk generates a back x-ray illumination, which translates through the perturbed foil leading to the exposure at the XFC.

The spatial resolution function $R(x-x_0)$ is measured by streaking a one-dimensional image of a backlit grid. We approximate $R(x-x_0)$ as a weighted sum of two Gaussians

$$R(x-x_0) = \frac{1}{1+\alpha}\left[e^{-\frac{(x-x_0)^2}{2\sigma_1^2}} + \alpha e^{-\frac{(x-x_0)^2}{2\sigma_2^2}}\right], \qquad (2)$$

where $\alpha = 0.05$, $\sigma_1 = 3\ \mu$m, and $\sigma_2 = 30\ \mu$m. In practice, the spatial resolution of the imaging system can be conveniently expressed by the modulation transfer function (MTF), which is defined as the ratio of observed contrast to the actual one and is calculated from the Fourier transform of $R(x-x_0)$, as can be seen in Figure 2. In our experiments, the MTF at the first three harmonics for the 50 μm-wavelength perturbation is about 0.81, 0.50, and 0.23, respectively.

III. DRIVE CHARACTERIZATION

Eight Shenguang-II laser beams are focused into the hohlraum through the laser entrance hole (LEH), which then can be converted to the expected thermal x-rays as shown in

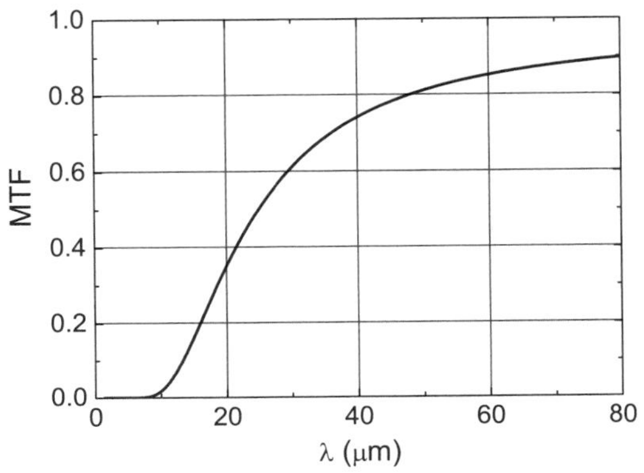

FIG. 2. The instrument MTF of the XFC versus the perturbation wavelength.

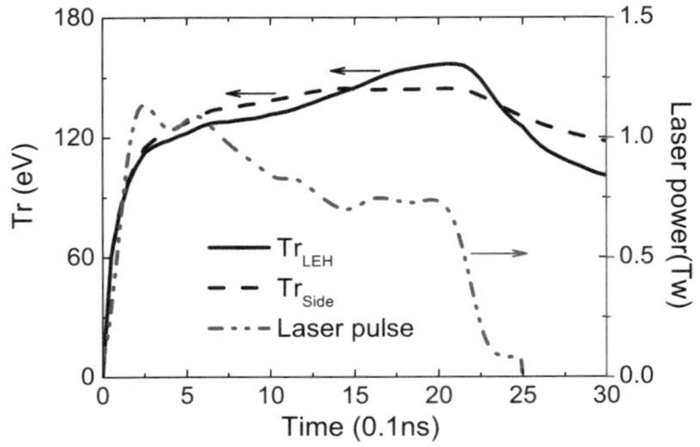

FIG. 3. Laser power, computed radiation flux temperature $T_{r,side}$ at the hohlraum wall, and radiation temperature inferred from the Dante measurement $T_{r,LEH}$ at the LEH versus time.

Fig. 1, and generate an ~140 eV radiation drive (see Fig. 3). Each of the 0.35 μm-wavelength beam is 2 ns in duration with an energy output of ~0.3 kJ.

Figure 3 shows a measured laser power profile and the x-ray flux temporal history, which is characterized by the radiation temperature, T_r. Here, the x-ray flux can be expressed as $F_x = \sigma T_r^4$ with σ being the Stefan-Boltzmann constant. A Dante measures the x-ray spectrum emitted by a section of hohlraum around the LEH. The radiation temperature nearby the LEH, $T_{r,LEH}$, is inferred from the Dante measurement. The radiation temperature at the hohlraum wall (i.e., the temperature at the location of the foil mounted), $T_{r,side}$, results from LARED-H[41] which is a two-dimensional (2-D) hohlraum simulation code. As can be seen in Fig. 3, there is evident difference between the curve of $T_{r,LEH}$ and that of $T_{r,side}$. The peak of the curve of $T_{r,LEH}$ is 158 eV but that of $T_{r,side}$ is 145 eV. The difference between them is about 9% corresponding to x-ray flux about 40%.

In Fig. 4, we illustrate a typical drive x-ray spectrum near the peak laser power. In fact, the drive spectra vary with the time throughout the drive pulse. As can be seen in Fig. 4, the drive spectrum is "harder" than a Planckian spectrum, in part because of the Au M-band emission from the laser spots. Though we use the Br-doped CH foil, this x-ray preheating effect can still dramatically influence the foil acceleration and the perturbation growth, as shall be discussed below.

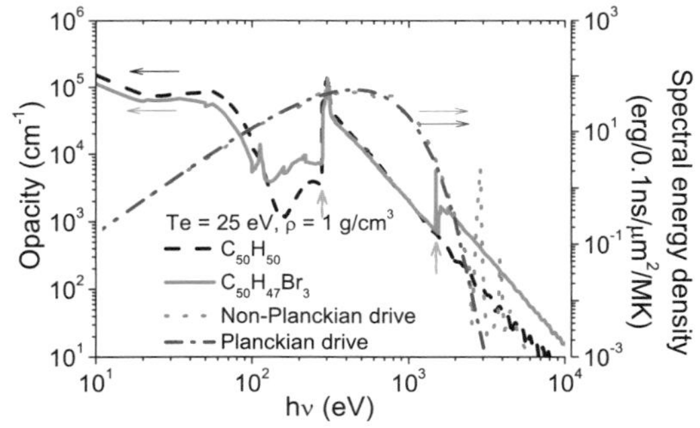

FIG. 4. Representative drive spectra during the peak power and opacity for CH and CH(Br) versus photon energy. The opacity is calculated for a representative ablation-front condition (temperature 25 eV, density 1 g/cm^3).

The opacity from the relativistic Hartree-Fock-Slater self-consistent average atom model OPINCH[42] for representative ablation-front temperature and density (25 eV, 1 g/cm^3) is depicted in Fig. 4. The CH and CH(Br) have similar opacity in the 0.27–1.5 keV band. However, obviously higher opacity of CH(Br) in the 1.5–10 keV band can better shield the preheating from Au M-band than the pure CH foil.

The backlighter material Mo (Z = 42) is used in the experiments according to the total optical depth of the CH (Br) foils. The calculation result of x-ray spectrum of a Mo disk radiated by a 1.2 kJ, 2 ns-square, 0.35 μm-wavelength backlighter beam is shown in Fig. 5. The Mo x-ray spectrum is dominated by the L-band emission at about 2.4–2.8 keV, which can transmit through the thick foils in the face-on experiments in order to observe an extended acceleration stage before the foil burnthrough.

IV. ACCELERATION MEASUREMENTS

The acceleration of the foil can be measured and recorded by the XSC in the side-on experiments. Figure 6 displays the measured position of a 25 μm thick CH(Br) foil. A 50 μm diameter tungsten filament is used to scale the flight distance of the foil, as shown in Fig. 6.

Figure 7 compares the experimental foil trajectory from two shots (2012-139 and 2012-141) with the results from foil-only radiation hydrodynamic code LARED-S[22,43,44] using $T_{r,LEH}$ with Planckian spectrum, $T_{r,side}$ with Planckian spectrum, and $T_{r,side}$ with non-Planckian spectrum, respectively. The result of shot 2012-139 can be repeated by that of shot 2012-141 showing excellent repeatability. The back edge of the foil does not begin to move until the shock breakout at the rear (~1.2 ns). The foil then accelerates during the interval ~1.2–2.3 ns. After the drive is turned off (~2.3 ns), the foil begins to coast. The three curves obtained from simulations give visibly different foil position. The flight distance using $T_{r,LEH}$ with Planckian spectrum is much larger than the experiment results, since the radiation temperature measured at the LEH had larger peak value than that at the wall (see Fig. 3), which directly

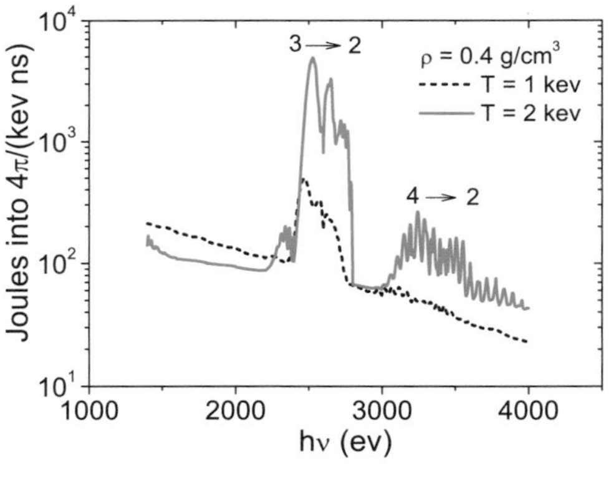

FIG. 5. The x-ray emission spectra resulted from a 2 ns square, 1.2 kJ, 0.35 μm-wavelength Shenguang-II 9th beam irradiating a backlighter disk of Mo at temperature of 1 keV and 2 keV, respectively. The $n = 3 \rightarrow 2$ L-band transitions are marked.

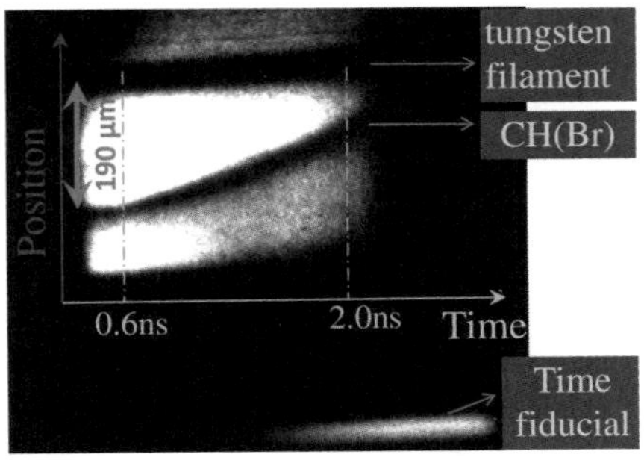

FIG. 6. Trajectory of the CH(Br) foil observed in the side-on backlighting streaked image of the CH(Br) plane foil showing the target acceleration properties.

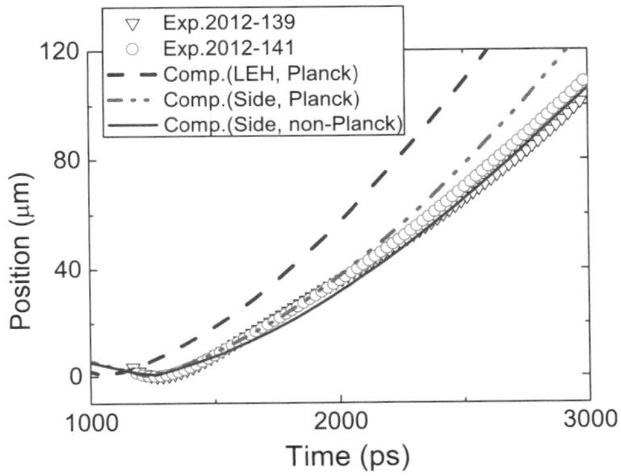

FIG. 7. Comparison of the foil position from the side-on measured radiography and the LARED-S 1-D hydrodynamic simulation.

drives the foils mounted on the side of the hohlraum. As can be seen, visible difference exists between the flight distances using $T_{r,side}$ with non-Planckian and that with Planckian spectra. Detailed simulation results indicate that the absorbed energy of the CH(Br) foil adopting the non-Planckian spectrum is less than that using Planckian spectrum, because of a larger ratio of high-energy photos in a tail of the non-Planckian spectrum (see Fig. 4). The simulation foil trajectory using $T_{r,side}$ with non-Planckian spectrum is quantitatively consistent with the experimental position.

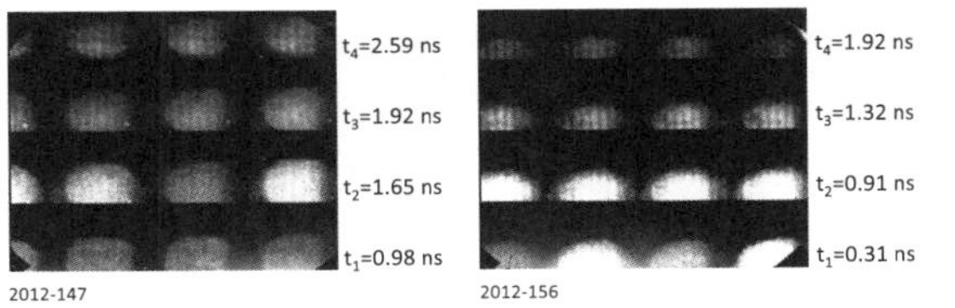

FIG. 8. Face-on backlighting image from framing camera in the small-amplitude perturbation experiments for a 50 μm-wavelength perturbation with a 0.5 μm initial amplitude. The four strips (strips 1 to 4) are timed from bottom to top. Every strip consists of 4 images (images 1 to 4), which are timed from right to left. The value of t_n reflects the trigger time of the strip relative to the start of the backlighting pulse. Each image in the strip is separated by 56 ps.

V. RT GROWTH MEASUREMENTS

The RT growth at the ablation-front is recorded by the XFC in the face-on experiments. The 24 μm-thick CH(Br) foil imposed a 50 μm-wavelength cosinusoidal perturbation is used in these experiments. The principal data from these experiments are the radiographic images obtained at selected times by the XFC, which can produce up to 16 images separated in time by a total of $\sim$1.8 ns. The temporal resolution of framing camera is 50 ps.

A. Small-amplitude perturbations

In order to obtain the linear-growth-factor (e-folding) of the ablative RT growth, the ratio of the initial amplitude of the perturbation to its wavelength, η_0/λ, should be enough small (at least $\eta_0/\lambda < 0.1$). In the series of small-amplitude perturbation experiments, we set $\eta_0/\lambda = 0.01$.

Figure 8 presents the data from the XFC for two separate shots (2012-147 and 2012-156). The early-time amplitude of the foil is small, and hence the "contrast," namely, the optical-depth modulation, ΔOD, is small. As the perturbation grows, the contrast becomes large. After sufficient linear growth, the perturbation enters the nonlinear growth regime, exhibiting the characteristic "bubble-spike" pattern. Later in time, the contrast decreases rapidly after the bubbles break out at the foil rear.

Since the perturbation growth in contrast (ΔOD) has been measured as a function of time (see Fig. 8), the Fourier transform coefficients of each harmonic can then be obtained as a function of time. Figure 9(a) compares the experimentally

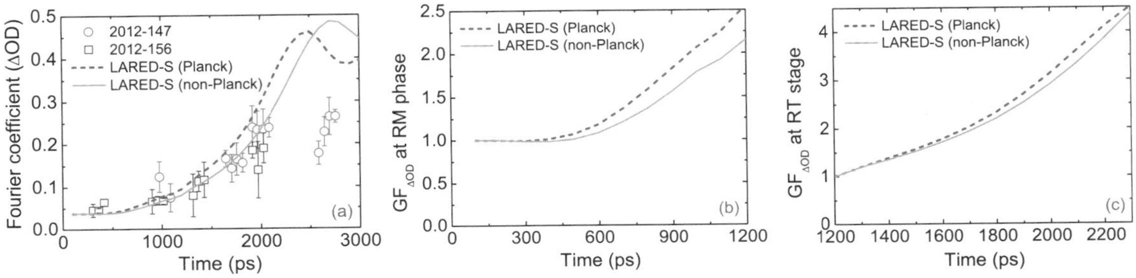

FIG. 9. (a) Comparison of the full time dependence of the Fourier coefficient of ΔOD from the experimental measurement as shown in Figure 8 and that from the predictions of the LARED-S 2-D hydrodynamic simulations. Temporal evolution of the growth factor of the contrast, $GF_{\Delta OD}$, within the early RM phase (b) and later RT growth stage (c), resulted from numerical simulations adopting the Planckian and non-Planckian spectra, respectively.

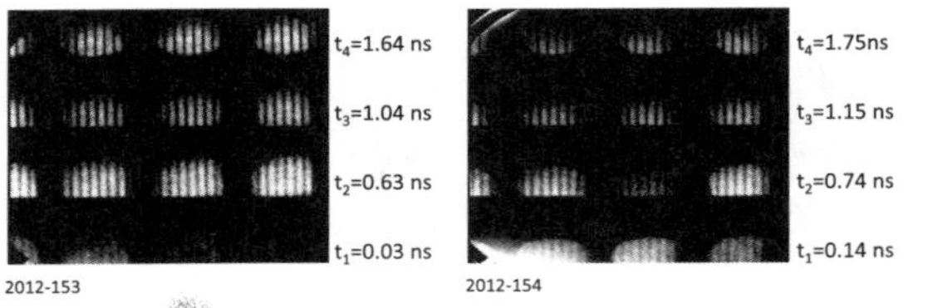

FIG. 10. Face-on backlighting image from framing camera in the large-amplitude perturbation experiments for a 50 μm-wavelength perturbation with a 3.0 μm initial amplitude. The four strips (strips 1 to 4) are timed from bottom to top. Every strip consists of 4 images (images 1 to 4), which are timed from right to left. The value of t_n reflects the trigger time of the strip relative to the start of the backlighting pulse. Each image in the strip is separated by 56 ps.

measured Fourier coefficient and the predictions of the LARED-S. It can be seen that the agreement between experiments and simulations is considerably favorable before the drive is closed. The simulations adopting the non-Planckian spectrum lead to better agreement with experiments, as shown in Fig. 9(a). That is to say, the preheating from high-energy photos can still affect the perturbation growth, though we have applied the Br-doped CH ($C_{50}H_{47}Br_3$) shielding. After the x-ray driver is closed, the foil begins to decompress and the contrast becomes obscure, and hence the simulation results gradually deviate the experimental observations.

Figure 9(b) shows the temporal evolution of the RM growth during the initial shock traversal phase ($\lesssim$1.2 ns) from LARED-S simulations using Planckian and non-Planckian spectra, respectively. Here, the optical-depth growth factor, $GF_{\Delta OD}$, is defined as the ratio of final to initial contrast. Figure 9(c) shows the temporal evolution of the RT growth during the steady acceleration stage (~1.2–2.3 ns) from LARED-S simulations using Planckian and non-Planckian spectra, respectively. As can be seen in Figure 9(b), preheating from the tail of the non-Planckian spectrum strikingly suppresses the RM growth. The seeding of the ARTI by RM growth in simulations using Planckian and non-Planckian spectra is $GF_{\Delta OD} \sim 2.5$ and 2.1, respectively. The $GF_{\Delta OD}$ of RT growth at the end of steady acceleration stage (~2.3 ns) in simulations using Planckian and non-Planckian spectra is ~4.5 and ~4.4, respectively. By comparing the results plotted in Figure 9(b) with those in Figure 9(c), we find that the preheating effects play a more pronounced role in the RM growth than that in RT growth in our experiments. This kind of preheating effect on the growth of ablative RM and RT instabilities, especially on the WN growth of ARTI, will be pursued experimentally in the future.

B. Large-amplitude perturbations

A series of large-amplitude perturbations with $\eta_0/\lambda = 3/50$ have been performed to observe onset of harmonic generation. The "raw" XFC images for large-amplitude face-on experiments from two separate shots (2012-153 and 2012-154) are exhibited in Fig. 10. As can be seen, at the early shock transit stage, the contrast (ΔOD) was small and the light and dark streaks have nearly the same widths. After the foil is accelerated, the contrast increases gradually. The dark streak (which corresponds to the spike) narrows, and the light one (which corresponds to the bubble) broadens, showing characteristic WN interface asymmetry pattern.

Figure 11 compares the first three Fourier coefficients of ΔOD from the experiments with that from LARED-S numerical simulations. The experimental measurement from the two shots gives similar results, namely, the repeatability of the data is acceptable. As can be seen, the simulation results are shown in reasonable agreement with experiments. The onset of harmonic generation and the transition from linear into nonlinear growth regime can be markedly observed in Fig. 11. The Δ OD of fundamental mode reaches its peak at ~1.7 ns and then enters the nonlinear growth regime. The ratio of maximum to initial contrast in the fundamental mode is $GF_{max,\Delta OD} \sim 3$. When the growth of ΔOD of the fundamental mode has arrived at its peak, that of the second and the third harmonic still have a quick growth, as shown in Fig. 11. The $GF_{max,\Delta OD}$ in the second (third) harmonic is about 7 (5). As can be seen in Figure 11, the simulated Fourier coefficients with the non-Planckian drive are a little smaller than that with Planckian drive, since the extra preheating from the M-band energy of non-Planckian drive has reduced the RM growth before the shock breakout (steady acceleration), as discussed previously. Moreover, at the early stage, there is a small discrepancy between the simulated and experimental Fourier coefficients, since the amplitude of the harmonics is too small compared with the noise level of the measured system. The contrast of noise is $<\sim 0.03$ in our experiments.

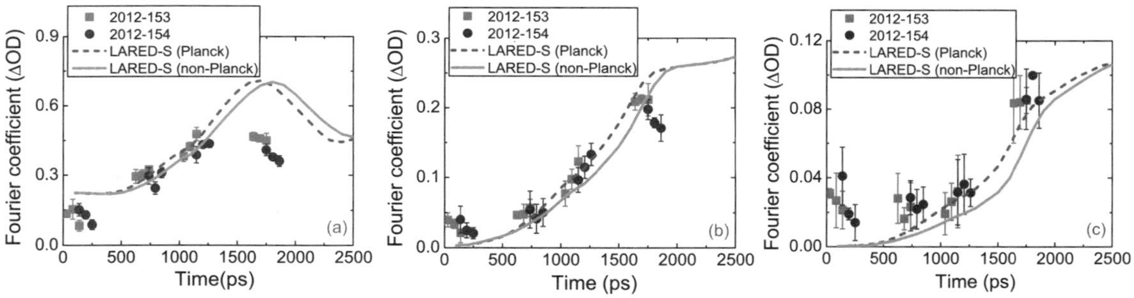

FIG. 11. Comparison of full time dependence of the first three Fourier coefficients of ΔOD from experimental data in Figure 10 and that from the predictions of the LARED-S 2-D hydrodynamic simulations. (a) Fundamental mode, (b) the second harmonic, and (c) the third harmonic.

VI. SUMMARY

The Shenguang-II laser facility has been used to investigate the ablation-front RT growth. We have conducted a series of experiments and simulations to study the growth of single-mode perturbation with initial small and large amplitudes. The comparison of foil acceleration with 1-D hydrodynamic simulation shows well modeled radiation drive. In the face-on experiments, the linear growth and transition into the nonlinear regime are recorded by XFC and well predicted by 2-D hydrodynamic simulations. The onset of harmonic generation and the growth of harmonics have been well captured in the experiments and predicted by simulations.

In the flight acceleration experiments, we find that the simulation foil trajectory using non-Planckian spectrum is more consistent with the experimental position. Simulation results indicate that since a larger ratio of high-energy photos in a tail of the non-Planckian spectrum, the absorbed energy of the CH (Br) foil adopting the Planckian spectrum is more than that using the non-Planckian spectrum, and hence leads to a larger flight distance. In the small-amplitude experiments, it is found that the preheating from high-energy photos reduces both the seeding of the ARTI by RM growth during the initial shock traversal phase and the linear growth of the ARTI. In the large-amplitude experiments, it is seen that the preheating from high-energy photos delays the generation and growth of the second and third harmonics. In a word, our experiment results suggest that high-energy photos in the tail of the non-Planckian spectrum (preheatings) play a pronounced role in the foil flight and perturbation growth. These ablative RT growth measurements provide a useful benchmark for our LARED-S hydrocode and give us confidence in our predictive capability as we design ARTI experiments on the Shenguang-III prototype and Shenguang-III laser facilities with more drive energy (the Shenguang-III has a laser energy output of ~150 kJ).

ACKNOWLEDGMENTS

We gratefully acknowledge technical supports from the Shenguang-II operation group for assistance and patience during these experiments, and the supports of the technical staff at Research Center of Laser Fusion CAEP for target fabrication and plasma diagnostics. This work was supported by the National Natural Science Foundation of China under Grant Nos. 11275031, 11105013, 11205017, 11274026, and the National Basic Research Program of China under Grant No. 2013CB834100.

[1]S. Atzeni and J. Meyer-ter-Vehn, *The Physics of Inertial Fusion: Beam Plasma Interaction Hydrodynamics, Hot Dense Matter* (Oxford University, Oxford, 2004).
[2]J. D. Lindl, P. Amendt, R. L. Berger, S. G. Glendinning, S. H. Glenzer, S. W. Haan, R. L. Kauffman, O. L. Landen, and L. J. Suter, Phys. Plasmas **11**(2), 339 (2004).
[3]R. D. Richtmyer, Commun. Pure. Appl. Math. **13**, 297 (1960).
[4]E. E. Meshkov, Fluid Dyn. **4**, 101 (1969).
[5]V. N. Goncharov, Phys. Rev. Lett. **82**(10), 2091 (1999).
[6]L. Rayleigh, Proc. London Math. Soc. **14**, 170 (1883).
[7]G. Taylor, Proc. R. Soc. London, Ser. A **201**, 192 (1950).
[8]G. I. Bell, LANL Report No. LA-1321, 1951.
[9]M. S. Plesset, J. Appl. Phys. **25**, 96 (1954).
[10]H.-Y. Guo, X.-J. Yu, L.-F. Wang, W.-H. Ye, J.-F. Wu, and Y.-J. Li, Chin. Phys. Lett. **31**(4), 044702 (2014).
[11]W. H. Ye, X. T. He, W. Y. Zhang, and M. Y. Yu, EPL **96**, 35002 (2011).
[12]L. F. Wang, W. H. Ye, X. T. He, W. Y. Zhang, Z. M. Sheng, and M. Y. Yu, Phys. Plasmas **19**, 100701 (2012).
[13]V. N. Goncharov and O. A. Hurricane, LLNL Report No. LLNL-TR-562014, 2012.
[14]S. P. Regan, R. Epstein, B. A. Hammel, L. J. Suter, H. A. Scott, M. A. Barrios, D. K. Bradley, D. A. Callahan, C. Cerjan, G. W. Collins, S. N. Dixit, T. Döppner, M. J. Edwards, D. R. Farley, K. B. Fournier, S. Glenn, S. H. Glenzer, I. E. Golovkin, S. W. Haan, A. Hamza, D. G. Hicks, N. Izumi, O. S. Jones, J. D. Kilkenny, J. L. Kline, G. A. Kyrala, O. L. Landen, T. Ma, J. J. MacFarlane, A. J. MacKinnon, R. C. Mancini, R. L. McCrory, N. B. Meezan, D. D. Meyerhofer, A. Nikroo, H.-S. Park, J. Ralph, B. A. Remington, T. C. Sangster, V. A. Smalyuk, P. T. Springer, and R. P. J. Town, Phys. Rev. Lett. **111**, 045001 (2013).
[15]J. D. Lindl and W. C. Mead, Phys. Rev. Lett. **34**(20), 1273 (1975).
[16]S. Bodner, Phys. Rev. Lett. **33**, 761 (1974).
[17]H. Takabe, K. Mima, L. Montierth, and R. L. Morse, Phys. Fluids **28**, 3676 (1985).
[18]A. R. Piriz, J. Sanz, and L. F. Ibañez, Phys. Plasmas **4**(4), 1117 (1997).
[19]V. N. Goncharov, R. Betti, R. L. McCrory, and C. P. Verdon, Phys. Plasmas **3**(12), 4665 (1996).
[20]R. Betti, V. N. Goncharov, R. L. McCrory, and C. P. Verdon, Phys. Plasmas **5**(5), 1446 (1998).
[21]S. A. Piriz, A. R. Piriz, and N. A. Tahir, Phys. Plasmas **16**, 082706 (2009).
[22]W. Ye, W. Zhang, and X. T. He, Phys. Rev. E **65**, 57401 (2002).
[23]J. Sanz, J. Ramìrez, R. Ramis, R. Betti, and R. P. J. Town, Phys. Rev. Lett. **89**, 195002 (2002).
[24]J. Garnier and L. Masse, Phys. Plasmas **12**, 062707 (2005).
[25]T. Ikegawa and K. Nishihara, Phys. Rev. Lett. **89**, 115001 (2002).
[26]L. F. Wang, W. H. Ye, Z. M. Sheng, W.-S. Don, Y. J. Li, and X. T. He, Phys. Plasmas **17**, 122706 (2010).
[27]L. F. Wang, W. H. Ye, and X. T. He, Phys. Plasmas **19**, 012706 (2012).
[28]B. A. Remington, S. W. Haan, S. G. Glendinning, J. D. Kilkenny, D. H. Munro, and R. J. Wallace, Phys. Fluids B **4**(4), 967 (1992).
[29]S. V. Weber, B. A. Remington, S. W. Haan, B. G. Wilson, and J. K. Nash, Phys. Plasmas **1**(11), 3652 (1994).
[30]B. A. Remington, S. V. Weber, M. M. Marinak, S. W. Haan, J. D. Kilkenny, R. J. Wallace, and G. Dimonte, Phys. Plasmas **2**(1), 241 (1995).
[31]M. M. Marinak, S. G. Glendinning, R. J. Wallace, B. A. Remington, K. S. Budil, and S. W. Haan, Phys. Rev. Lett. **80**(20), 4426 (1998).
[32]K. S. Budil, B. A. Remington, T. A. Peyser, K. O. Mikaelian, P. L. Miller, N. C. Woolsey, W. M. Wood-Vasey, and A. M. Rubenchik, Phys. Rev. Lett. **76**(24), 4536 (1996).
[33]K. S. Budil, B. Lasinski, M. J. Edwards, A. S. Wan, B. A. Remington, S. V. Weber, S. G. Glendinning, L. Suter, and P. E. Stry, Phys. Plasmas **8**(5), 2344 (2001).
[34]K. Shigemori, H. Azechi, M. Nakai, M. Honda, K. Meguro, N. Miyanaga, H. Takabe, and K. Mima, Phys. Rev. Lett. **78**, 250 (1997).
[35]S. G. Glendinning, S. N. Dixit, B. A. Hammer, D. H. Kalantat, M. H. Key, J. D. Kilkenny, J. P. Knauer, D. M. Pennington, B. A. Remington, R. J. Wallace, and S. V. Weber, Phys. Rev. Lett. **78**, 3318 (1997).
[36]T. Sakaiya, H. Azechi, M. Matsuoka, N. Izumi, M. Nakai, K. Shigemori, H. Shiraga, A. Sunahara, H. Takabe, and T. Yamanaka, Phys. Rev. Lett. **88**, 145003 (2002).
[37]V. A. Smalyuk, S. X. Hu, V. N. Goncharov, D. D. Meyerhofer, T. C. Sangster, C. Stoeckl, and B. Yaakobi, Phys. Plasmas **15**, 082703 (2008).
[38]A. Casner, D. Galmiche, G. Huser, and J.-P. Jadaud, Phys. Plasmas **16**, 092701 (2009).
[39]A. Casner, V. A. Smalyuk, L. Masse, I. Igumenshchev, S. Liberatore, L. Jacquet, C. Chicanne, P. Loiseau, O. Poujade, D. K. Bradley, H. S. Park, and B. A. Remington, Phys. Plasmas **19**, 082708 (2012).
[40]X. T. He and W. Y. Zhang, Eur. Phys. J. D **44**, 227 (2007).
[41]Q. Duan, T. Chang, W. Zhang, G. Wang, and C. Wang, Chin. J. Comput. Phys. **16**(6), 610 (1999) (in Chinese).
[42]F. J. D. Serduke, E. Minguez, S. J. Davidson, and C. A. Iglesias, J. Quant. Spectrosc. Radiat. Transfer **65**, 527 (2000).
[43]W. Ye, W. Zhang, and G. Cheng, High Power Laser Part. Beams **10**(3), 403 (1998) (in Chinese).
[44]Z. Fan, S. Zhu, W. Pei, W. Ye, M. Li, X. Xu, J. Wu, Z. Dai, and L. Wang, EPL **99**, 65003 (2012).

PHYSICS OF PLASMAS 21, 122710 (2014)

Weakly nonlinear Rayleigh-Taylor instability of a finite-thickness fluid layer

L. F. Wang (王立锋),[1,2,a)] H. Y. Guo (郭宏宇),[3] J. F. Wu (吴俊峰),[1,b)] W. H. Ye (叶文华),[1,2,c)] Jie Liu (刘杰),[1,2] W. Y. Zhang (张维岩),[1] and X. T. He (贺贤土)[1,2]
[1]Institute of Applied Physics and Computational Mathematics, Beijing 100094, China
[2]HEDPS, Center for Applied Physics and Technology, Peking University, Beijing 100871, China
[3]Graduate School, China Academy of Engineering Physics, Beijing 100088, China

(Received 12 September 2014; accepted 26 November 2014; published online 22 December 2014)

A weakly nonlinear (WN) model has been developed for the Rayleigh-Taylor instability of a finite-thickness incompressible fluid layer (slab). We derive the coupling evolution equations for perturbations on the (upper) "linearly stable" and (lower) "linearly unstable" interfaces of the slab. Expressions of temporal evolutions of the amplitudes of the perturbation first three harmonics on the upper and lower interfaces are obtained. The classical feedthrough (interface coupling) solution obtained by Taylor [Proc. R. Soc. London A **201**, 192 (1950)] is readily recovered by the first-order results. Our third-order model can depict the WN perturbation growth and the saturation of linear (exponential) growth of the perturbation fundamental mode on both interfaces. The dependence of the WN perturbation growth and the slab distortion on the normalized layer thickness (kd) is analytically investigated via the third-order solutions. Comparison is made with Jacobs-Catton's formula [J. W. Jacobs and I. Catton, J. Fluid Mech. **187**, 329 (1988)] of the position of the "linearly unstable" interface. Using a reduced formula, the saturation amplitude of linear growth of the perturbation fundamental mode is studied. It is found that the finite-thickness effects play a dominant role in the WN evolution of the slab, especially when $kd < 1$. Thus, it should be included in applications where the interface coupling effects are important, such as inertial confinement fusion implosions and supernova explosions.
[http://dx.doi.org/10.1063/1.4904363]

I. INTRODUCTION

Of great importance to a successful inertial confinement fusion (ICF)[1] is the control of hydrodynamic instabilities,[2,3] such as Rayleigh-Taylor (RT),[4,5] Richtmyer-Meshkov (RM),[6,7] or Kelvin-Helmholtz (KH) instabilities,[8] which can cause any imperfections on the capsule surfaces or material volume defects in the ablator and/or ice to grow, destroying the fuel compression symmetry, injecting ablator dopants into the hot-spot, mixing the cold fuel with the hotspot, or even completely rupturing the capsule in extreme cases. In the smoothly accelerated ICF capsules, the RT instability (RTI) is believed to be the dominant mechanism for the total perturbation growth[3] and is the focus of the present work. Moreover, it is expected that the RTI plays a key role in the material mixing and nonlinear evolution of supernova explosions.[9–11]

The typical ICF capsule consists of two layers, namely, an ablator layer on the outside and a fuel layer on the inside. Therefore, there are at least three surfaces, namely, the ablation front, the fuel-ablator interface, and the hot-spot boundary in ICF implosions. After the shell is accelerated, the RTI caused by a low-density fluid pushing on a high-density fluid results in perturbation growth on the ablation front. Also, preheating due to hard x-rays and hot electrons can lead to the ablator density that is less than the fuel density during the shell acceleration. There is then a fuel-ablator interface, which is prone to the RTI. A low-density, high-temperature hot-spot is produced while the shock arrives at the capsule center. This light hot fuel begins to push outward on the converging heavy cold one, eventually stagnating the compression. The RTI now grows during this deceleration regime on the hot-spot boundary. Additionally, in the capsule with a graded-doped ablator, any of the internal interfaces between doped regions could also cause the RT growth.

The classical RTI refers to a semi-infinite fluid of density ρ_L accelerating the other semi-infinite fluid of density ρ_H ($\rho_H > \rho_L$).[4,5] Small amplitude perturbations on the interface of wave number $k = 2\pi/\lambda$ grow exponentially in time, $\eta(t) \sim \eta_0 e^{\gamma t}$, with a linear growth rate $\gamma = \sqrt{A_T kg}$, where λ is the perturbation wavelength, η_0 ($\eta_0 \ll \lambda$) is the initial perturbation amplitude, t is the time, $A_T = (\rho_H - \rho_L)/(\rho_H + \rho_L)$ is the Atwood number, and g is the acceleration. Most studies deal with this classical configuration of semi-infinite fluids.[12–23] However, in fact, we want to accelerate a relatively thin shell in ICF implosions, as mentioned above. When a fluid layer of a finite-thickness is accelerated in vacuum (Taylor's case[5]), the lower interface is subjected to the RTI while the upper interface is intuitively "linearly stable." How perturbations on one interface feed through to another interface are crucial for ICF implosions.[2,3] Most earlier work[5,24–26] on finite-thickness layers has focused primarily on the linear growth regime. The stability of a fluid slab accelerated by uniform pressure is reported in Taylor's original work.[5] The linear growth of RTI at the interface of any number of stratified fluids is discussed by Mikaelian.[24] The physics of "interface coupling" (or "feedthrough") which

a)Electronic mail: wang_lifeng@iapcm.ac.cn
a)Electronic mail: wu_junfeng@iapcm.ac.cn
c)Electronic mail: ye_wenhua@iapcm.ac.cn

means the evolution of the perturbation at one interface depends not only on its initial perturbation but also on the perturbation at the adjacent interfaces, can be clearly seen in Mikaelian's linear model for finite-thickness fluid layers.[24–26] Furthermore, the author has extended his linear model in plane geometry to that in spherical[27] and cylindrical[28] geometries. Beyond a certain amplitude, when $k\eta$ becomes no longer small, the perturbations enter the nonlinear growth regime. In fact, in the low-adiabat ICF implosions[29–32] and even in the recent high-foot high-adiabat ICF implosions,[33] the perturbation amplitudes will reach levels where the linear growth theory is unlikely to be valid. Though the RT growth will slow with increasing amplitude ($k\eta \sim 1$), it is critical that the behavior be better understood, at least in the weakly nonlinear (WN) regime, so perturbation amplitudes and the shell distortion can be predicted. While the linear RT growth of small perturbations on the interface of finite-thickness layers (shells) has been relatively well understood, the nonlinear stage[34] is still not well understood. In this work, we treat the RTI of a finite-thickness fluid layer in the WN regime up to the third-order.

The remainder of this article is organized as follows. Section II contains the theoretical formulation of the potential model, the WN solutions up the third-order correction, and a reduced formula. In Sec. III, we discuss the normalized layer thickness dependence of the WN perturbation growth on the "linearly stable" and "linearly unstable" interfaces, and the saturation amplitude of the linear (exponential) growth of the fundamental mode. Finally, the concluding remarks and some applications are given in Sec. IV.

II. THEORETICAL MODEL AND MATHEMATICAL FORMULATION

In this section, we will demonstrate a brief description of the theoretical framework of the present work. Section II A is devoted to the governing equation of the potential model. Section II B gives the coupling evolution equations and the analytic expressions for the temporal evolution of the amplitudes of the perturbation first three harmonics on the upper and lower interfaces of the layer. Finally, a reduced formula is developed by keeping only the dominant terms in Sec. II C.

A. Derivation of general equations

This study treats a finite-thickness, incompressible, inviscid, irrotational, and immiscible fluid slab, of density ρ and thickness d, with semi-infinite vacuum on each side, immersing in a gravitational field $\mathbf{g} = -g\mathbf{e}_y$, as shown in Figure 1. A rectangular coordinate system is established. The x-direction is along the upper undisturbed interface and the y-direction is normal to the undisturbed interfaces. In the following discussion, physical quantities of the upper interface shall be denoted by the superscript "u" and that of the lower interface by the superscript "l," unless otherwise stated.

The motion equations of the upper and lower interfaces are parameterized, respectively, as

$$F^u(x,y,t) = y - \eta^u(x,t), \tag{1}$$

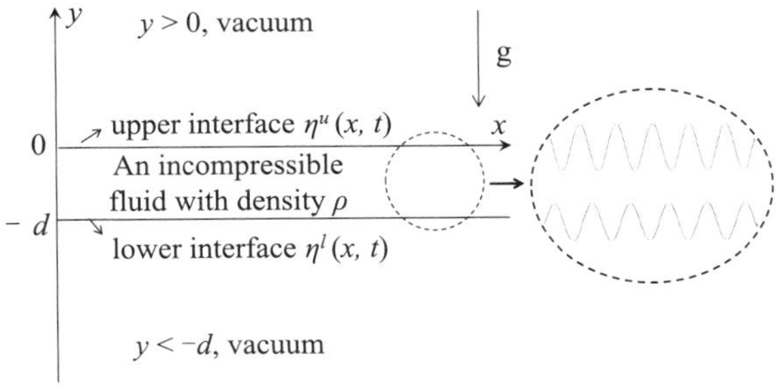

FIG. 1. Schematic drawing of the finite-thickness fluid slab considered in the present work.

$$F^l(x,y,t) = y + d - \eta^l(x,t), \tag{2}$$

where $\eta^u(x,t)$ and $\eta^l(x,t)$ denote the perturbation displacements of the upper and lower interfaces, respectively. Let $\Phi^{(0)}(t)$ be the velocity potential of the initial unperturbed fluid, the velocity potential of the perturbed fluid can then be expressed as

$$\Phi(x,y,t) = \Phi^{(0)}(t) + \varphi(x,y,t), \tag{3}$$

where φ is the additional velocity potential due to the perturbed interface and $\nabla^2\varphi = 0$. According to the Bernoulli equation, the pressures at the upper and lower interfaces can be written, respectively, as

$$p^u = -\rho\left[\frac{\partial\Phi}{\partial t} + \frac{1}{2}(\nabla\Phi)^2 + g\eta^u + f(t)\right], \tag{4}$$

$$p^l = -\rho\left[\frac{\partial\Phi}{\partial t} + \frac{1}{2}(\nabla\Phi)^2 - gd + g\eta^l + f(t)\right], \tag{5}$$

where $f(t)$ is an arbitrary function of time. For the unperturbed case, they reduce to

$$p^{u(0)} = -\rho\left[\frac{\partial\Phi^{(0)}}{\partial t} + f(t)\right], \tag{6}$$

$$p^{l(0)} = -\rho\left[\frac{\partial\Phi^{(0)}}{\partial t} - gd + f(t)\right], \tag{7}$$

where the pressure equilibrium condition $p^{l(0)} - p^{u(0)} = \rho g d$ has been satisfied. Therefore, the perturbed pressure at the two interfaces should be zero, i.e.,

$$\rho\left[\frac{\partial\varphi}{\partial t} + \frac{1}{2}(\nabla\varphi)^2 + g\eta^u\right] = 0 \text{ at } y = \eta^u, \tag{8}$$

$$\rho\left[\frac{\partial\varphi}{\partial t} + \frac{1}{2}(\nabla\varphi)^2 + g\eta^l\right] = 0 \text{ at } y = \eta^l. \tag{9}$$

According to kinematic boundary conditions, the normal components of the interface velocities $[-(\partial F/\partial t)/|\nabla F|]$ should equal to the normal components of the fluid velocities ($\nabla\Phi \cdot \mathbf{n}$, where $\mathbf{n} = \nabla F/|\nabla F|$ is the unit normal vector to the interface), namely,

$$-\frac{\dfrac{\partial F^u}{\partial t}}{|\nabla F^u|} = \nabla\Phi \cdot \frac{\nabla F^u}{|\nabla F^u|} \text{ at } y = \eta^u, \tag{10}$$

$$-\frac{\frac{\partial F^l}{\partial t}}{|\nabla F^l|} = \nabla\Phi \cdot \frac{\nabla F^l}{|\nabla F^l|} \text{ at } y = \eta^l. \tag{11}$$

Substituting Eqs. (1) and (2), respectively, into Eqs. (10) and (11), we obtain

$$\frac{\partial \eta^u}{\partial t} = -\frac{\partial \varphi}{\partial x}\frac{\partial \eta^u}{\partial x} + \frac{\partial \varphi}{\partial y} \text{ at } y = \eta^u, \tag{12}$$

$$\frac{\partial \eta^l}{\partial t} = -\frac{\partial \varphi}{\partial x}\frac{\partial \eta^l}{\partial x} + \frac{\partial \varphi}{\partial y} \text{ at } y = \eta^l. \tag{13}$$

The governing equations of the system are Eqs. (8), (9), (12), and (13).

We expand η^u, η^l, and φ into a power series in a formal expansion parameter ϵ

$$\begin{aligned}\eta^u(x,t) = {} & \epsilon\eta^u_{1,1}(t)\cos(kx) + \epsilon^2[\eta^u_{2,0}(t) + \eta^u_{2,2}(t)\cos(2kx)] \\ & + \epsilon^3[\eta^u_{3,1}(t)\cos(kx) + \eta^u_{3,3}(t)\cos(3kx)] + \ldots,\end{aligned} \tag{14}$$

$$\begin{aligned}\eta^l(x,t) = {} & \epsilon\eta^l_{1,1}(t)\cos(kx) + \epsilon^2[\eta^l_{2,0}(t) + \eta^l_{2,2}(t)\cos(2kx)] \\ & + \epsilon^3[\eta^l_{3,1}(t)\cos(kx) + \eta^l_{3,3}(t)\cos(3kx)] + \ldots,\end{aligned} \tag{15}$$

$$\begin{aligned}\varphi(x,y,t) = {} & \epsilon[e^{ky}A_{1,1}(t) + e^{-ky}B_{1,1}(t)]\cos(kx) \\ & + \epsilon^2\{A_{2,0}(t) + [e^{2ky}A_{2,2}(t) \\ & + e^{-2ky}B_{2,2}(t)]\cos(2kx)\} \\ & + \epsilon^3\{[e^{ky}A_{3,1}(t) + e^{-ky}B_{3,1}(t)]\cos(kx) \\ & + [e^{3ky}A_{3,3}(t) + e^{-3ky}B_{3,3}(t)]\cos(3kx)\} + \ldots,\end{aligned} \tag{16}$$

where $\eta^i_{n,n-2m}$, $A_{n,n-2m}$, and $B_{n,n-2m}$ are the amplitudes of the $(n-2m)$ th Fourier mode in the nth-order of ϵ ($i = u, l$; $n = 1, 2, 3, \ldots$; $m = 0, \ldots \lfloor n/2 \rfloor$). Here, the Gauss's symbol $\lfloor n/2 \rfloor$ represents the maximum integer less than or equal to $n/2$.

We first substitute Eqs. (14)–(16) into the governing equations. Then, we generate a power series expansion for the governing equations about the point $y = 0$ or $y = -d$. After that, the nth-order governing equations containing ϵ^n are obtained. Finally, the amplitudes of the $(n-2m)$th Fourier mode in the nth-order of $\epsilon(n = 1, 2, 3, \ldots; m = 0, \ldots \lfloor n/2 \rfloor)$ can be obtained by successively solving the resulted nth-order equations, as shall be discussed below. More details of the employed recursive procedure for obtaining the nonlinear solutions can be found in our previous publications.[17] The solutions of the interface position can then be obtained by setting $\epsilon = 1$.

It is worthwhile to point out that we truncate the formal power series in Eqs. (14)–(16), keeping only the first three orders (third-order WN models).[12–17] That is to say, the results here are now valid up to third-order accuracy in the perturbation amplitude. The generation of second and third harmonics, as well as the fundamental mode feedback can be investigated with the present third-order results. Furthermore, the developed third-order WN model can be extended to higher orders, if needed. However, as the amplitude increase, the amplitude of the fundamental mode obtained from the WN solution first begins to fall below the linear prediction, reaching a maximum amplitude, and then decreases rapidly (see Figure 2). Since the amplitude decrease is clearly unrealistic, the "inflexion" point can be viewed as the validity limit of the WN solution. Since the asymptotic series quickly break down when $k\eta(t) \sim 1$, it is a reasonable approximation to neglect the fourth- or higher-order effects.

B. Weakly nonlinear solutions

At the first-order $O(\epsilon)$, the coupled second-order ordinary differential equations (ODEs) for $\eta^u_{1,1}$ and $\eta^l_{1,1}$ are

$$\frac{d^2\eta^u_{1,1}}{dt^2} - \frac{2e^{-\xi}}{1+e^{-2\xi}}\frac{d^2\eta^l_{1,1}}{dt^2} + \gamma^2\frac{1-e^{-2\xi}}{1+e^{-2\xi}}\eta^u_{1,1} = 0, \tag{17}$$

$$\frac{d^2\eta^l_{1,1}}{dt^2} - \frac{2e^{-\xi}}{1+e^{-2\xi}}\frac{d^2\eta^u_{1,1}}{dt^2} - \gamma^2\frac{1-e^{-2\xi}}{1+e^{-2\xi}}\eta^l_{1,1} = 0, \tag{18}$$

where $\gamma^2 = gk$ is the square of linear growth rate of the classical RTI ($A_T = 1$) and $\xi = kd$ is the normalized slab

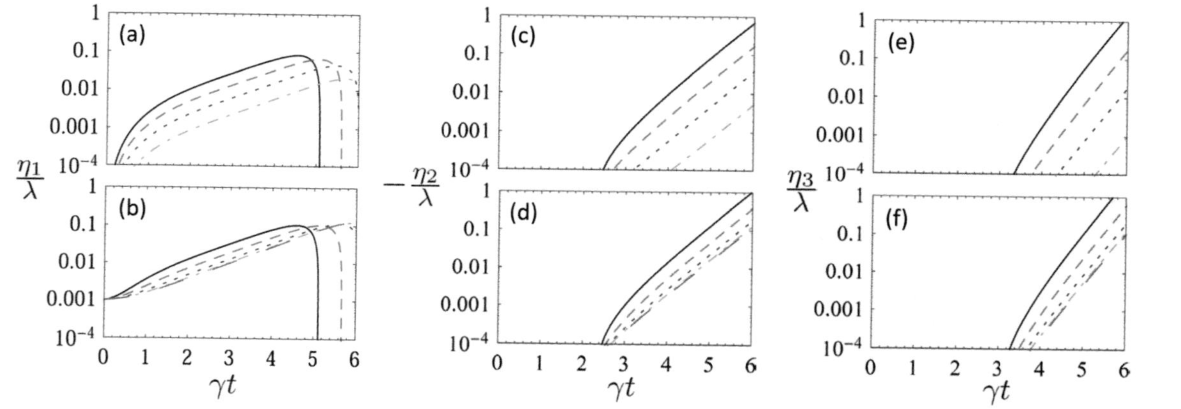

FIG. 2. Temporal evolution of normalized amplitudes of the perturbation fundamental mode [(a) and (b)], second [(c) and (d)] and third [(e) and (f)] harmonics on the upper [(a), (c), (e)] and lower [(b), (d), (f)] interfaces initiated by only the lower interface perturbation with $kd = 0.2$ (solid lines), 0.4 (dashed lines), 0.8 (dotted lines), and 1.6 (dashed-dotted lines), respectively. The long-dashed lines represent the corresponding amplitudes calculated by Jacobs-Catton's formula. The initial conditions are $k = 1$, $a = 0$ and $b = 0.001\lambda$.

thickness. For the most common initial conditions (the layer initially at rest with the same wavelength perturbations on the upper and lower interfaces), namely, $\eta^u_{1,1}(t=0)=a$, $d\eta^u_{1,1}(t=0)/dt=0$, $\eta^l_{1,1}(t=0)=b$, and $d\eta^l_{1,1}(t=0)/dt=0$, we obtain

$$\eta^u_{1,1}(t)=\Xi[\tilde{a}\cos(\Gamma)+e^{-\xi}\tilde{b}\cosh(\Gamma)], \tag{19}$$

$$\eta^l_{1,1}(t)=\Xi[e^{-\xi}\tilde{a}\cos(\Gamma)+\tilde{b}\cosh(\Gamma)], \tag{20}$$

where $\Xi=1/(1-e^{-2\xi})$, $\Gamma=\gamma t$, $\tilde{a}=a-be^{-\xi}$, and $\tilde{b}=b-ae^{-\xi}$. This linear results are identical to that of Ref. 5.

From the linear solutions [Eqs. (19) and (20)], it is seen that the two interfaces of a finite-thickness fluid layer are coupled, especially for $\xi<1$, and the physics of interface coupling (feed-through) is represented by the factors of Ξ and $e^{-\xi}$. Note that the Ξ decreases with increasing ξ with an lower limit $\Xi\to 1$ when $\xi\to\infty$. That is to say, the linear RT growth of a "thin" shell is more severe than a "thick" shell, while the other conditions are kept constant. Thus, to reduce this effect, the ICF capsule needs a thicker ablator and/or thicker fuel.[35,36] Moreover, the interface coupling is quite clear in the linear solutions where the $\eta^u_{1,1}$ and the $\eta^l_{1,1}$ depend on both a and b. Even when $a=0$ $(b=0)$, $\eta^u_{1,1}(t)\neq 0$ $(\eta^l_{1,1}(t)\neq 0)$ since perturbations feed through from the lower (upper) interface to the upper (lower) interface.

Note that in Eqs. (19) and (20) the linear solutions contain a growing $\cosh(\Gamma)$ mode and an oscillating $\cos(\Gamma)$ mode. If one sets $b=ae^{-\xi}$, the growing modes on both interfaces are completely suppressed and interfaces become stable, except for harmless oscillations. Likewise, if $a=be^{-\xi}$, the oscillatory modes cancel. Moreover, it is shown that cancelling any one of the modes at one interface will automatically cancel that same mode at the other interface.

The effect of one interface to the other interface is decayed by a factor of $e^{-\xi}$. For $\xi\gg 1$ (i.e., the case of interface decouple), we have

$$\eta^u_{1,1}(t)=a\cos(\Gamma), \tag{21}$$

$$\eta^l_{1,1}(t)=b\cosh(\Gamma), \tag{22}$$

which are, respectively, the linear solution of the upper and lower interfaces for the case of infinite-thickness fluid (i.e., the classical RTI results[4]). Regarding $a=0$ and $b\neq 0$, the linear solutions reduce to

$$\eta^u_{1,1}(t)=\Xi[-e^{-\xi}b\cos(\Gamma)+e^{-\xi}b\cosh(\Gamma)], \tag{23}$$

$$\eta^l_{1,1}(t)=\Xi[-e^{-2\xi}b\cos(\Gamma)+b\cosh(\Gamma)], \tag{24}$$

where the effect of the lower interface on the upper is denoted by the terms in Eq. (23) containing $e^{-\xi}b$ as a factor, and the effect of upper interface on the low is represented by the terms in Eq. (24) containing $e^{-2\xi}b$ as a factor. Regarding $a\neq 0$ and $b=0$, the linear solutions become

$$\eta^u_{1,1}(t)=\Xi[a\cos(\Gamma)-e^{-2\xi}a\cosh(\Gamma)], \tag{25}$$

$$\eta^l_{1,1}(t)=\Xi[e^{-\xi}a\cos(\Gamma)-e^{-\xi}a\cosh(\Gamma)], \tag{26}$$

where the effect of the upper interface on the low is denoted by the terms in Eq. (26) containing $e^{-\xi}a$ as a factor, and the effect of low interface on the upper is represented by the terms in Eq. (25) containing $e^{-2\xi}a$ as a factor.

At the second-order $O(\epsilon^2)$, the coupled second-order ODEs for $\eta^u_{2,2}$ and $\eta^l_{2,2}$ are

$$\begin{aligned}&\frac{d^2\eta^u_{2,2}}{dt^2}-\frac{2e^{-2\xi}}{1+e^{-4\xi}}\frac{d^2\eta^l_{2,2}}{dt^2}+2\gamma^2\frac{1-e^{-4\xi}}{1+e^{-4\xi}}\eta^u_{2,2}\\&+\gamma^2k\left[\frac{1}{2}\tilde{a}^2C_1\cos(2\Gamma)-\frac{1}{2}e^{-2\xi}C_1\tilde{b}^2\cosh(2\Gamma)\right.\\&-\frac{2}{e^{2\xi}+e^{-2\xi}}C_{ab}\cos(\Gamma)\cosh(\Gamma)\\&\left.+2e^{-2\xi}C_1C_{ab}\sin(\Gamma)\sinh(\Gamma)+\frac{e^{\xi}+e^{3\xi}}{1+e^{4\xi}}a\hat{b}\right]=0,\end{aligned} \tag{27}$$

and

$$\begin{aligned}&\frac{d^2\eta^l_{2,2}}{dt^2}-\frac{2e^{-2\xi}}{1+e^{-4\xi}}\frac{d^2\eta^u_{2,2}}{dt^2}-2\gamma^2\frac{1-e^{-4\xi}}{1+e^{-4\xi}}\eta^l_{2,2}\\&+\gamma^2k\left[\frac{1}{2}\tilde{b}^2C_1\cosh(2\Gamma)-\frac{1}{2}e^{-2\xi}C_1\tilde{a}^2\cos(2\Gamma)\right.\\&-\frac{2}{e^{2\xi}+e^{-2\xi}}C_{ab}\cos(\Gamma)\cosh(\Gamma)\\&\left.-2e^{-2\xi}C_1C_{ab}\sin(\Gamma)\sinh(\Gamma)+\frac{e^{\xi}+e^{3\xi}}{1+e^{4\xi}}b\hat{a}\right]=0,\end{aligned} \tag{28}$$

where $\hat{a}=a-b(e^{-\xi}+e^{\xi})/2$, $\hat{b}=b-a(e^{-\xi}+e^{\xi})/2$, $C_1=(1+e^{-2\xi})/(1-e^{-2\xi}+e^{-4\xi}-e^{-6\xi})$, and $C_{ab}=a^2+b^2-ab(e^{-\xi}+e^{\xi})$. For initial conditions $\eta^u_{2,2}(t=0)=0$, $d\eta^u_{2,2}(t=0)/dt=0$, $\eta^l_{2,2}(t=0)=0$, and $d\eta^l_{2,2}(t=0)/dt=0$, we obtain

$$\begin{aligned}\eta^u_{2,2}(t)=&\frac{k}{2}\Xi^2\left[-e^{-2\xi}\tilde{b}^2\cosh^2(\Gamma)+\tilde{a}^2\cos^2(\Gamma)\right.\\&-\frac{e^{-2\xi}}{1+e^{-2\xi}}D_1\tilde{b}\cosh\left(\sqrt{2}\Gamma\right)-\frac{1}{1+e^{-2\xi}}D_2\tilde{a}\cos\left(\sqrt{2}\Gamma\right)\\&\left.-2e^{-\xi}\frac{1-e^{-2\xi}}{1+e^{-2\xi}}\tilde{a}\tilde{b}\cos(\Gamma)\cosh(\Gamma)\right],\end{aligned} \tag{29}$$

and

$$\begin{aligned}\eta^l_{2,2}(t)=&\frac{k}{2}\Xi^2\left[-\tilde{b}^2\cosh^2(\Gamma)+e^{-2\xi}\tilde{a}^2\cos^2(\Gamma)-\frac{1}{1+e^{-2\xi}}D_1\tilde{b}\cosh(\sqrt{2}\Gamma)-\frac{e^{-2\xi}}{1+e^{-2\xi}}D_2\tilde{a}\cos(\sqrt{2}\Gamma)\right.\\&\left.+2e^{-\xi}\frac{1-e^{-2\xi}}{1+e^{-2\xi}}\tilde{a}\tilde{b}\cos(\Gamma)\cosh(\Gamma)\right],\end{aligned} \tag{30}$$

贺贤土论文选集

926

where $D_1 = ae^{-\xi}(3+e^{-2\xi}) - b(1+3e^{-2\xi})$ and $D_2 = a(1+3e^{-2\xi}) - be^{-\xi}(3+e^{-2\xi})$. Additionally, due to the principles of conservation of mass, there is no excitation of a zeroth harmonic, namely, $\eta^u_{2,0}(t) = 0$ and $\eta^l_{2,0}(t) = 0$.

At the third-order $O(\epsilon^3)$, the coupled second-order ODEs for $\eta^u_{3,1}$ and $\eta^l_{3,1}$ are

$$\frac{d^2\eta^u_{3,1}}{dt^2} - \frac{2e^{-\xi}}{1+e^{-2\xi}}\frac{d^2\eta^l_{3,1}}{dt^2} + \gamma^2\frac{1-e^{-2\xi}}{1+e^{-2\xi}}\eta^u_{3,1} + \gamma^2k^2\Big[-\tilde{a}E_1E_{ab1}\cos(\Gamma) - e^{-\xi}\tilde{b}E_1E_{ab2}\cosh(\Gamma) - \tilde{a}^3E_1(8+3e^{-2\xi})\cos^3(\Gamma)$$
$$- e^{-\xi}\tilde{b}^3E_1(8e^{-2\xi}+3)\cosh^3(\Gamma) - E_2\tilde{a}^2D_2\sin(\Gamma)\sin(\sqrt{2}\Gamma) - e^{-\xi}\tilde{a}\tilde{b}D_2E_2\sin(\sqrt{2}\Gamma)\sinh(\Gamma)$$
$$- e^{-2\xi}\tilde{a}\tilde{b}D_1E_2\sin(\Gamma)\sinh(\sqrt{2}\Gamma) - e^{-3\xi}\tilde{b}^2D_1E_2\sinh(\Gamma)\sinh(\sqrt{2}\Gamma) + \tilde{a}^2D_2E_3\cos(\Gamma)\cos(\sqrt{2}\Gamma)$$
$$+ e^{-\xi}\tilde{b}^2D_1E_3\cosh(\Gamma)\cosh(\sqrt{2}\Gamma) + e^{-\xi}\tilde{a}^2\tilde{b}E_4\cos(2\Gamma)\cosh(\Gamma) - e^{-2\xi}\tilde{a}\tilde{b}^2E_5\cos(\Gamma)\cosh(2\Gamma)$$
$$+ \tilde{a}^3E_6\cos(\Gamma)\sin(\Gamma)^2 - \tilde{b}^3E_7\cosh(\Gamma)\sinh^2(\Gamma) + \frac{3}{4}\sqrt{2}\tilde{a}^3E_2\sin(\Gamma)\sin(2\Gamma) - \frac{3}{4}\sqrt{2}e^{-5\xi}\tilde{b}^3E_2\sinh(\Gamma)\sinh(2\Gamma)$$
$$+ \tilde{a}^2\tilde{b}E_8\sin(2\Gamma)\sinh(\Gamma) - e^{-\xi}\tilde{a}\tilde{b}^2E_8\sin(\Gamma)\sinh(2\Gamma)\Big] = 0, \tag{31}$$

and

$$\frac{d^2\eta^l_{3,1}}{dt^2} - \frac{2e^{-\xi}}{1+e^{-2\xi}}\frac{d^2\eta^u_{3,1}}{dt^2} - \gamma^2\frac{1-e^{-2\xi}}{1+e^{-2\xi}}\eta^l_{3,1} - \gamma^2k^2\Big[-\tilde{a}e^{-\xi}E_1E_{ab3}\cos(\Gamma) - \tilde{b}E_1E_{ab4}\cosh(\Gamma)$$
$$- \tilde{a}^3e^{-\xi}(3+8e^{-2\xi})E_1\cos^3(\Gamma) - \tilde{b}^3(3e^{-2\xi}+8)E_1\cosh^3(\Gamma) - E_2\tilde{b}^2D_1\sinh(\Gamma)\sinh(\sqrt{2}\Gamma) - e^{-\xi}E_2\tilde{a}\tilde{b}D_1\sin(\Gamma)\sinh(\sqrt{2}\Gamma)$$
$$- e^{-2\xi}E_2\tilde{a}\tilde{b}D_2\sin(\sqrt{2}\Gamma)\sinh(\Gamma) - e^{-3\xi}E_2\tilde{a}^2D_2\sin(\Gamma)\sin(\sqrt{2}\Gamma) - e^{-\xi}\tilde{a}^2D_2E_3\cos(\Gamma)\cos(\sqrt{2}\Gamma)$$
$$- \tilde{b}^2D_1E_3\cosh(\Gamma)\cosh(\sqrt{2}\Gamma) + e^{-\xi}\tilde{a}\tilde{b}^2E_4\cos(\Gamma)\cosh(2\Gamma) - e^{-2\xi}\tilde{a}^2\tilde{b}E_5\cos(2\Gamma)\cosh(\Gamma)$$
$$+ \tilde{a}^3E_7\cos(\Gamma)\sin^2(\Gamma) + e^{-\xi}\tilde{a}^2\tilde{b}E_8\sin(2\Gamma)\sinh(\Gamma) - \tilde{b}^3E_6\cosh(\Gamma)\sinh^2(\Gamma)$$
$$- \tilde{a}\tilde{b}^2E_8\sin(\Gamma)\sinh(2\Gamma) - \frac{3\sqrt{2}}{4}\tilde{b}^3E_2\sinh(\Gamma)\sinh(2\Gamma) + \frac{3\sqrt{2}}{4}e^{-5\xi}\tilde{a}^3E_2\sin(\Gamma)\sin(2\Gamma)\Big] = 0, \tag{32}$$

where

$$E_1 = \frac{(1+e^{-2\xi})}{16(1-e^{-4\xi})^2},$$
$$E_2 = \frac{1}{\sqrt{2}(1-e^{-4\xi})^2},$$
$$E_3 = \frac{(1-e^{-2\xi})}{2(1-e^{-4\xi})^2},$$
$$E_4 = \frac{(3-13e^{-2\xi})}{8(1-e^{-4\xi})^2},$$
$$E_5 = \frac{(13-3e^{-2\xi})}{8(1-e^{-4\xi})^2},$$
$$E_6 = \frac{3e^{-2\xi}(3e^{-2\xi}+11)}{16(1-e^{-4\xi})^2},$$
$$E_7 = \frac{3e^{-\xi}(11e^{-2\xi}+3)}{16(1-e^{-4\xi})^2},$$
$$E_8 = \frac{3e^{-\xi}(1+e^{-2\xi})}{4(1-e^{-4\xi})^2},$$
$$E_{ab1} = -2abe^{-\xi}(7e^{-2\xi}+4) + b^2e^{-2\xi}(e^{-2\xi}+10) + a^2(6e^{-4\xi}+e^{-2\xi}+4),$$
$$E_{ab2} = b^2(10e^{-2\xi}+1) - 2abe^{-\xi}(4e^{-2\xi}+7) + a^2(4e^{-4\xi}+e^{-2\xi}+6),$$
$$E_{ab3} = a^2(10e^{-2\xi}+1) - 2abe^{-\xi}(4e^{-2\xi}+7) + b^2(4e^{-4\xi}+e^{-2\xi}+6),$$
$$E_{ab4} = -2abe^{-\xi}(7e^{-2\xi}+4) + a^2e^{-2\xi}(e^{-2\xi}+10) + b^2(6e^{-4\xi}+e^{-2\xi}+4).$$

For initial conditions $\eta^u_{3,1}(t=0)=0$, $d\eta^u_{3,1}(t=0)/dt=0$, $\eta^l_{3,1}(t=0)=0$, and $d\eta^l_{3,1}(t=0)/dt=0$, we obtain

$$\begin{aligned}\eta^u_{3,1}=&-\frac{1}{8}k^2\Xi^3\Big[F_1\tilde{a}^3\cos^3(\Gamma)+F_2\tilde{b}^3\cosh^3(\Gamma)-F_3\tilde{a}^2\tilde{b}\cos^2(\Gamma)\cosh(\Gamma)-F_4\tilde{a}\tilde{b}^2\cos(\Gamma)\cosh^2(\Gamma)+2e^{-2\xi}F_5\tilde{a}^2\tilde{b}\sin(2\Gamma)\sinh(\Gamma)\\&-2e^{-\xi}F_5\tilde{a}\tilde{b}^2\sin(\Gamma)\sinh(2\Gamma)+F_6D_1\tilde{b}^2\cosh(\Gamma)\cosh\big(\sqrt{2}\Gamma\big)-F_7D_2\tilde{a}^2\cos(\Gamma)\cos\big(\sqrt{2}\Gamma\big)-2e^{-\xi}D_1F_5\tilde{a}\tilde{b}\cos(\Gamma)\cosh\big(\sqrt{2}\Gamma\big)\\&+2D_2F_5\tilde{a}\tilde{b}\cos\big(\sqrt{2}\Gamma\big)\cosh(\Gamma)-F_5F_{ab1}\tilde{b}\cos(\Gamma)+e^{-\xi}F_5F_{ab2}\tilde{a}\cosh(\Gamma)-e^{-\xi}F_5^{-1}\tilde{a}^3\Gamma\sin(\Gamma)+e^{-2\xi}F_5^{-1}\tilde{b}^3\Gamma\sinh(\Gamma)\Big],\end{aligned}\tag{33}$$

and

$$\begin{aligned}\eta^l_{3,1}=&-\frac{1}{8}k^2\Xi^3\Big[F_2\tilde{a}^3\cos^3(\Gamma)+F_1\tilde{b}^3\cosh^3(\Gamma)-F_4\tilde{a}^2\tilde{b}\cos^2(\Gamma)\cosh(\Gamma)-F_3\tilde{a}\tilde{b}^2\cos(\Gamma)\cosh^2(\Gamma)-2F_5D_1\tilde{a}\tilde{b}\cos(\Gamma)\cosh\big(\sqrt{2}\Gamma\big)\\&+2e^{-\xi}F_5\tilde{a}^2\tilde{b}\sin(2\Gamma)\sinh(\Gamma)-2e^{-2\xi}F_5\tilde{a}\tilde{b}^2\sin(\Gamma)\sinh(2\Gamma)+D_1F_7\tilde{b}^2\cosh(\Gamma)\cosh\big(\sqrt{2}\Gamma\big)-D_2F_6\tilde{a}^2\cos(\Gamma)\cos\big(\sqrt{2}\Gamma\big)\\&+2e^{-\xi}\tilde{a}\tilde{b}D_2F_5\cos\big(\sqrt{2}\Gamma\big)\cosh(\Gamma)-e^{-\xi}F_5F_{ab1}\tilde{b}\cos(\Gamma)+F_5F_{ab2}\tilde{a}\cosh(\Gamma)-e^{-2\xi}F_5^{-1}\tilde{a}^3\Gamma\sin(\Gamma)+e^{-\xi}F_5^{-1}\tilde{b}^3\Gamma\sinh(\Gamma)\Big],\end{aligned}\tag{34}$$

where

$$\begin{aligned}F_1&=2+3e^{-2\xi},\\F_2&=e^{-\xi}(3+2e^{-2\xi}),\\F_3&=\frac{e^{-\xi}(3+7e^{-2\xi})}{1+e^{-2\xi}},\\F_4&=\frac{e^{-2\xi}(7+3e^{-2\xi})}{1+e^{-2\xi}},\\F_5&=\frac{e^{-\xi}}{1+e^{-2\xi}},\\F_6&=\frac{2e^{-\xi}(2+e^{-2\xi})}{1+e^{-2\xi}},\\F_7&=\frac{2(1+2e^{-2\xi})}{1+e^{-2\xi}},\\F_{ab1}&=-b^2(1-3e^{-4\xi}-6e^{-2\xi})-8abe^{-\xi}(1+e^{-2\xi})+a^2(3-e^{-4\xi}+6e^{-2\xi}),\\F_{ab2}&=a^2(1-3e^{-4\xi}-6e^{-2\xi})+8abe^{-\xi}(1+e^{-2\xi})-b^2(3-e^{-4\xi}+6e^{-2\xi}).\end{aligned}$$

At the third-order $O(\epsilon^3)$, the second-order ODEs for $\eta^u_{3,3}$ and $\eta^l_{3,3}$ are

$$\begin{aligned}&\frac{d^2\eta^u_{3,3}}{dt^2}-\frac{2e^{-3\xi}}{1+e^{-6\xi}}\frac{d^2\eta^l_{3,3}}{dt^2}+3\gamma^2\frac{1-e^{-6\xi}}{1+e^{-6\xi}}\eta^u_{3,3}+\gamma^2k^2[\tilde{a}^2D_2G_1\sin(\Gamma)\sin\big(\sqrt{2}\Gamma\big)+e^{-\xi}\tilde{a}\tilde{b}D_2G_1\sin\big(\sqrt{2}\Gamma\big)\sinh(\Gamma)\\&+e^{-2\xi}\tilde{a}\tilde{b}D_1G_1\sin(\Gamma)\sinh\big(\sqrt{2}\Gamma\big)+e^{-3\xi}\tilde{b}^2D_1G_1\sinh(\Gamma)\sinh\big(\sqrt{2}\Gamma\big)-\tilde{a}\tilde{b}D_1G_2\cos(\Gamma)\cosh\big(\sqrt{2}\Gamma\big)\\&-\tilde{a}\tilde{b}D_2G_3\cos\big(\sqrt{2}\Gamma\big)\cosh(\Gamma)-\tilde{a}^3G_4\cos(\Gamma)\sin^2(\Gamma)+e^{-3\xi}\tilde{b}^3G_4\cosh(\Gamma)\sinh^2(\Gamma)+G_5\tilde{a}^2\tilde{b}\cos^2(\Gamma)\cosh(\Gamma)\\&-G_6\tilde{a}\tilde{b}^2\cos(\Gamma)\cosh^2(\Gamma)-G_7\tilde{a}\tilde{b}^2\cos(\Gamma)\sinh^2(\Gamma)+G_8\tilde{a}^2\tilde{b}\cosh(\Gamma)\sin^2(\Gamma)-G_9\tilde{a}^2\tilde{b}\sin(2\Gamma)\sinh(\Gamma)\\&+G_{10}\tilde{a}\tilde{b}^2\sin(\Gamma)\sinh(2\Gamma)-G_{11}\tilde{a}^3\sin(\Gamma)\sin(2\Gamma)+e^{-3\xi}\tilde{b}^3G_{11}\sinh(\Gamma)\sinh(2\Gamma)]=0,\end{aligned}\tag{35}$$

and

$$\begin{aligned}&\frac{d^2\eta^l_{3,3}}{dt^2}-\frac{2e^{-3\xi}}{1+e^{-6\xi}}\frac{d^2\eta^u_{3,3}}{dt^2}-3\gamma^2\frac{1-e^{-6\xi}}{1+e^{-6\xi}}\eta^l_{3,3}-\gamma^2k^2[\tilde{b}^2D_1G_1\sinh(\Gamma)\sinh\big(\sqrt{2}\Gamma\big)+e^{-\xi}\tilde{a}\tilde{b}D_1G_1\sin(\Gamma)\sinh\big(\sqrt{2}\Gamma\big)\\&+e^{-2\xi}\tilde{a}\tilde{b}D_2G_1\sin\big(\sqrt{2}\Gamma\big)\sinh(\Gamma)+e^{-3dk}\tilde{a}^2D_2G_1\sin(\Gamma)\sin\big(\sqrt{2}\Gamma\big)+\tilde{a}\tilde{b}D_2G_2\cos\big(\sqrt{2}\Gamma\big)\cosh(\Gamma)\\&+\tilde{a}\tilde{b}D_1G_3\cos(\Gamma)\cosh\big(\sqrt{2}\Gamma\big)+\tilde{b}^3G_4\cosh(\Gamma)\sinh^2(\Gamma)-e^{-3\xi}\tilde{a}^3G_4\cos(\Gamma)\sin^2(\Gamma)+\tilde{a}\tilde{b}^2G_5\cos(\Gamma)\cosh^2(\Gamma)\\&-\tilde{a}^2\tilde{b}G_6\cos^2(\Gamma)\cosh(\Gamma)+\tilde{a}^2\tilde{b}G_7\cosh(\Gamma)\sin^2(\Gamma)-\tilde{a}\tilde{b}^2G_8\cos(\Gamma)\sinh^2(\Gamma)+\tilde{a}\tilde{b}^2G_9\sin(\Gamma)\sinh(2\Gamma)\\&-\tilde{a}^2\tilde{b}G_{10}\sin(2\Gamma)\sinh(\Gamma)+\tilde{b}^3G_{11}\sinh(\Gamma)\sinh(2\Gamma)-e^{-3\xi}\tilde{a}^3G_{11}\sin(\Gamma)\sin(2\Gamma)]=0,\end{aligned}\tag{36}$$

where

$$G_1 = \frac{3(1+e^{-2\xi}+e^{-4\xi})}{\sqrt{2}(1-e^{-4\xi})^2(1-e^{-2\xi}+e^{-4\xi})},$$

$$G_2 = \frac{3e^{-2\xi}(e^{-2\xi}+2)}{2(-e^{-10\xi}+e^{-6\xi}-e^{-4\xi}+1)},$$

$$G_3 = \frac{3e^{-\xi}(2e^{-2\xi}+1)}{2(-e^{-10\xi}+e^{-6\xi}-e^{-4\xi}+1)},$$

$$G_4 = \frac{9}{4(1-e^{-2\xi})^2(1+e^{-2\xi})},$$

$$G_5 = \frac{3e^{-\xi}(6e^{-4\xi}+11e^{-2\xi}+1)}{4(-e^{-10\xi}+e^{-6\xi}-e^{-4\xi}+1)},$$

$$G_6 = \frac{3e^{-2\xi}(e^{-4\xi}+11e^{-2\xi}+6)}{4(-e^{-10\xi}+e^{-6\xi}-e^{-4\xi}+1)},$$

$$G_7 = \frac{3e^{-2\xi}(5e^{-6\xi}+2e^{-4\xi}+8e^{-2\xi}+3)}{4(1-e^{-4\xi})^2(1-e^{-2\xi}+e^{-4\xi})},$$

$$G_8 = \frac{3e^{-\xi}(3e^{-6\xi}+8e^{-4\xi}+2e^{-2\xi}+5)}{4(1-e^{-4\xi})^2(1-e^{-2\xi}+e^{-4\xi})},$$

$$G_9 = \frac{3e^{-\xi}(3e^{-6\xi}+4e^{-4\xi}+10e^{-2\xi}+1)}{4(1-e^{-4\xi})^2(1-e^{-2\xi}+e^{-4\xi})},$$

$$G_{10} = \frac{3e^{-2\xi}(e^{-6\xi}+10e^{-4\xi}+4e^{-2\xi}+3)}{4(1-e^{-4\xi})^2(1-e^{-2\xi}+e^{-4\xi})},$$

$$G_{11} = \frac{9e^{-2\xi}}{4(1-e^{-2\xi})^2(1+e^{-6\xi})}.$$

For initial conditions $\eta^u_{3,3}(t=0)=0$, $d\eta^u_{3,3}(t=0)/dt=0$, $\eta^l_{3,3}(t=0)=0$, and $d\eta^l_{3,3}(t=0)/dt=0$, we obtain

$$\begin{aligned}\eta^u_{3,3} = \frac{3}{8}k^2\Xi^3[&\tilde{a}^3\cos^3(\Gamma) + e^{-3\xi}\tilde{b}^3\cosh^3(\Gamma) + 2e^{-2\xi}\tilde{b}^2 D_1F_5\cosh(\Gamma)\cosh(\sqrt{2}\Gamma) - 2e^{\xi}\tilde{a}^2D_2F_5\cos(\Gamma)\cos(\sqrt{2}\Gamma)\\ &-\tilde{a}^2\tilde{b}H_1\cos^2(\Gamma)\cosh(\Gamma) + \tilde{a}\tilde{b}^2H_2\cos(\Gamma)\cosh^2(\Gamma) - \tilde{a}^2\tilde{b}H_3\cosh(\Gamma)\sin^2(\Gamma) + \tilde{a}^2\tilde{b}H_3\sin(2\Gamma)\sinh(\Gamma)\\ &+e^{-3\xi}\tilde{a}\tilde{b}^2H_3\cos(\Gamma)\sinh^2(\Gamma) - e^{-3\xi}\tilde{a}\tilde{b}^2H_3\sin(\Gamma)\sinh(2\Gamma) + \tilde{a}\tilde{b}D_1H_4\cos(\Gamma)\cosh(\sqrt{2}\Gamma) + \tilde{a}\tilde{b}D_2H_5\cos(\sqrt{2}\Gamma)\cosh(\Gamma)\\ &+\tilde{a}H_{ab1}H_6\cos(\sqrt{3}\Gamma) + e^{-3\xi}\tilde{b}H_{ab2}H_6\cosh(\sqrt{3}\Gamma)],\end{aligned} \tag{37}$$

and

$$\begin{aligned}\eta^l_{3,3} = \frac{3}{8}k^2\Xi^3\Big[&e^{-3\xi}\tilde{a}^3\cos^3(\Gamma) + \tilde{b}^3\cosh^3(\Gamma) + 2e^{\xi}\tilde{b}^2D_1F_5\cosh(\Gamma)\cosh(\sqrt{2}\Gamma) - 2e^{-2\xi}\tilde{a}^2D_2F_5\cos(\Gamma)\cos(\sqrt{2}\Gamma)\\ &+\tilde{a}^2\tilde{b}H_2\cos^2(\Gamma)\cosh(\Gamma) - \tilde{a}\tilde{b}^2H_1\cos(\Gamma)\cosh^2(\Gamma) + \tilde{a}\tilde{b}^2H_3\cos(\Gamma)\sinh^2(\Gamma) - \tilde{a}\tilde{b}^2H_3\sin(\Gamma)\sinh(2\Gamma)\\ &-e^{-3\xi}\tilde{a}^2\tilde{b}H_3\cosh(\Gamma)\sin^2(\Gamma) + e^{-3\xi}\tilde{a}^2\tilde{b}H_3\sin(2\Gamma)\sinh(\Gamma) - \tilde{a}\tilde{b}D_1H_5\cos(\Gamma)\cosh(\sqrt{2}\Gamma) - \tilde{a}\tilde{b}D_2H_4\cos(\sqrt{2}\Gamma)\cosh(\Gamma)\\ &+e^{-3\xi}\tilde{a}H_{ab1}H_6\cos(\sqrt{3}\Gamma) + \tilde{b}H_{ab2}H_6\cosh(\sqrt{3}\Gamma)],\end{aligned} \tag{38}$$

where

$$H_1 = \frac{e^{-\xi}(2-3e^{-6\xi}-e^{-4\xi}+8e^{-2\xi})}{(1+e^{-4\xi}+e^{-2\xi})(1+e^{-2\xi})},$$

$$H_2 = \frac{e^{-2\xi}(3-2e^{-6\xi}-8e^{-4\xi}+e^{-2\xi})}{(1+e^{-4\xi}+e^{-2\xi})(1+e^{-2\xi})},$$

$$H_3 = \frac{e^{-\xi}(1-e^{-2\xi})^2}{(1+e^{-4\xi}+e^{-2\xi})(1+e^{-2\xi})},$$

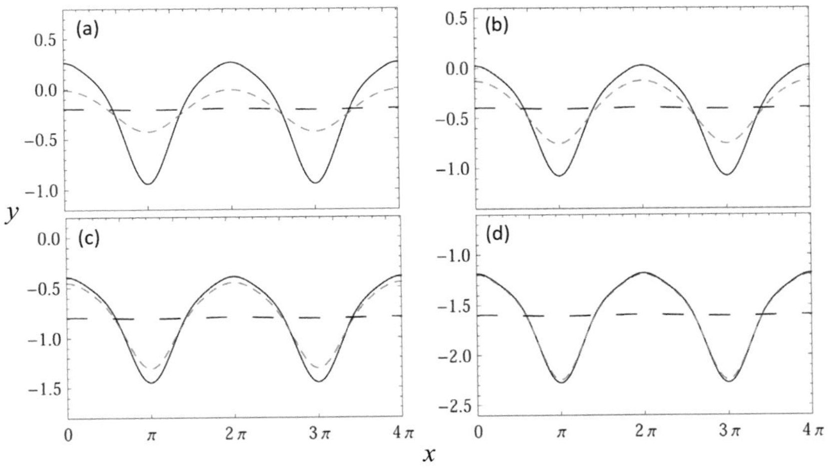

FIG. 3. Comparison of the position of the lower (linearly unstable) interface obtained from the present model (solid lines) with that calculated by the Jacobs-Catton's formula (dashed lines) initiated only lower interface perturbation with $kd=0.2$ at $\gamma t=4.2$ (a), $kd=0.4$ at $\gamma t=4.6$ (b), $kd=0.8$ at $\gamma t=4.9$ (c), and $kd=1.6$ at $\gamma t=5.1$ (d), respectively, with initial conditions being the same as in Figure 2.

$$H_4=\frac{2e^{-2\xi}(1-e^{-4\xi}-e^{-2\xi})}{(1+e^{-4\xi}+e^{-2\xi})(1+e^{-2\xi})},$$

$$H_5=\frac{2e^{-\xi}(1-e^{-4\xi}+e^{-2\xi})}{(1+e^{-4\xi}+e^{-2\xi})(1+e^{-2\xi})},$$

$$H_6=\frac{1}{(1+e^{-4\xi}+e^{-2\xi})(1+e^{-2\xi})},$$

$$\begin{aligned}H_{ab1}&=-6abe^{-\xi}(1+e^{-2\xi})^3+a^2(1+8e^{-6\xi}+9e^{-4\xi}\\&\quad+6e^{-2\xi})+b^2e^{-2\xi}(8+e^{-6\xi}+6e^{-4\xi}+9e^{-2\xi}),\end{aligned}$$

$$\begin{aligned}H_{ab2}&=-6abe^{-\xi}(1+e^{-2\xi})^3+b^2(1+8e^{-6\xi}+9e^{-4\xi}\\&\quad+6e^{-2\xi})+a^2e^{-2\xi}(8+e^{-6\xi}+6e^{-4\xi}+9e^{-2\xi}).\end{aligned}$$

C. Reduced formulas

If we drop the terms decaying exponentially in time (i.e., terms including $e^{-\Gamma}$, $e^{-2\Gamma}$, $e^{-3\Gamma}$) and the oscillation terms (i.e., terms including $\cos\Gamma$, $\cos^2\Gamma$, $\cos^3\Gamma$), keeping only the dominant terms (i.e., terms including e^{Γ}, $e^{2\Gamma}$, $e^{3\Gamma}$) at each order, then the amplitudes of the fundamental mode η_1, second η_2, and third η_3 harmonics up to the third-order for the upper and low interfaces, respectively, reduce to

$$\begin{cases}\eta_1^u=\left[1-\dfrac{1}{4}k^2\left(\eta_L^u e^{\xi}\right)^2\left(\dfrac{3}{2}+e^{-2\xi}\right)\right]\eta_L^u,\\ \eta_2^u=-\dfrac{1}{2}k(\eta_L^u)^2,\\ \eta_3^u=\dfrac{3}{8}k^2(\eta_L^u)^3,\end{cases}\tag{39}$$

and

$$\begin{cases}\eta_1^l=\left[1-\dfrac{1}{4}k^2\left(\eta_L^l\right)^2\left(1+\dfrac{3}{2}e^{-2\xi}\right)\right]\eta_L^l,\\ \eta_2^l=-\dfrac{1}{2}k\left(\eta_L^l\right)^2,\\ \eta_3^l=\dfrac{3}{8}k^2\left(\eta_L^l\right)^3,\end{cases}\tag{40}$$

where $\eta_L^u=\Xi\tilde{b}e^{-\xi}(e^{\Gamma}/2)$ and $\eta_L^l=\Xi\tilde{b}(e^{\Gamma}/2)$.

Note that when $\xi\to\infty$ Eq. (40) can be reduced to the Jacobs-Catton's formula[12]

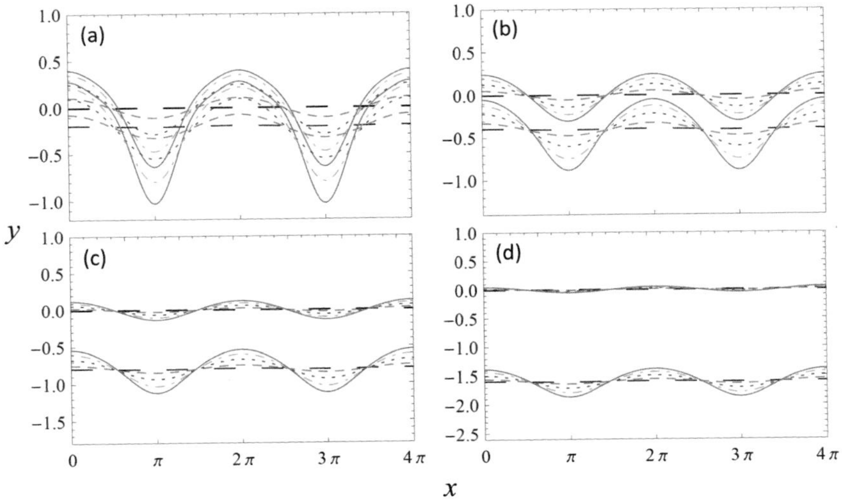

FIG. 4. Temporal evolution of positions of the upper and lower interfaces initiated by only the lower interface perturbation at $\gamma t=0$ (long-dashed lines), 2.5 (dashed lines), 3.5 (dotted lines), 4.0 (dashed-dotted lines), and 4.3 (solid lines) for $kd=0.2$ (a), 0.4 (b), 0.8 (c), and 1.6 (d), respectively, with initial conditions being the same as in Figure 2.

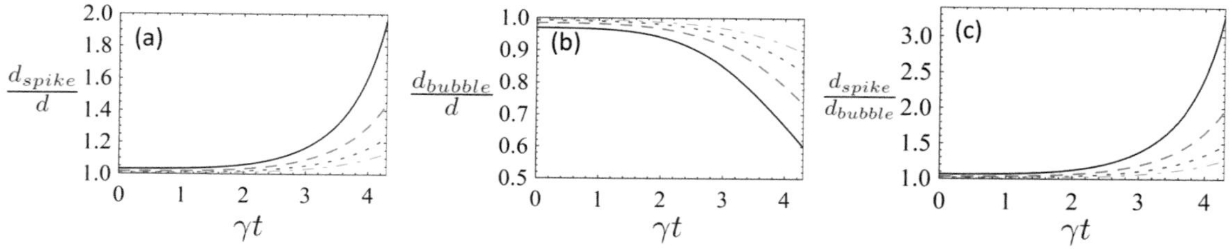

FIG. 5. Temporal evolution of the normalized layer thicknesses at the spike tip (a) and the bubble vertex (b), and the ratios of layer thicknesses at the spike tip and that at the bubble vertex (c) initiated by only the lower interface perturbation for $kd = 0.2$ (solid lines), 0.4 (dashed lines), 0.8 (dotted lines), and 1.6 (dashed-dotted lines), respectively, with initial conditions being the same as in Figure 2.

$$\begin{cases} \eta_1^{JC} = \left[1 - \frac{1}{4}k^2\left(\eta_L^{JC}\right)^2\right]\eta_L^{JC}, \\ \eta_2^{JC} = -\frac{1}{2}k\left(\eta_L^{JC}\right)^2, \\ \eta_3^{JC} = \frac{3}{8}k^2\left(\eta_L^{JC}\right)^3, \end{cases} \tag{41}$$

where $\eta_L^{JC} = \eta_0(e^{\Gamma}/2)$.

III. WEAKLY NONLINEAR ANALYSIS

In this section, we will study the WN perturbation growth on the upper and low interfaces. The WN growth of the RTI initiated with the lower interface perturbation alone, the upper interface perturbation alone, and both the lower and upper interface perturbations, are analyzed in Secs. III A, III B, and III C, respectively. We address the saturation of linear growth of the fundamental mode in Sec. III D.

A. Lower interface perturbations

First, let us consider the WN growth of the RTI initiated by only the lower interface perturbation. In Figure 2, we illustrate the temporal evolution of normalized amplitudes of the perturbation first three harmonics ($\eta_1/\lambda, -\eta_2/\lambda$, and η_3/λ) on the upper and lower interfaces initiated by only the lower interface perturbation with different normalized slab thicknesses. Also plotted in Figures 2(b), 2(d), and 2(f) are the corresponding amplitudes calculated by the Jacobs-Catton's formula [Eq. (41)].

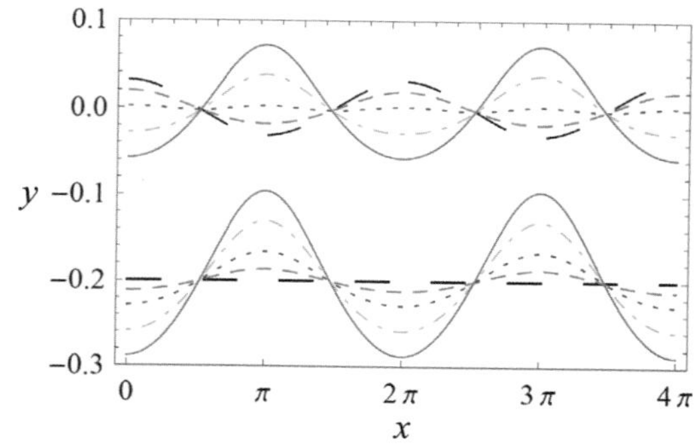

FIG. 6. Temporal evolution of positions of the upper and lower interfaces initiated by only the upper interface perturbation for $kd = 0.2$ at $\gamma t = 0$ (long-dashed lines), 0.4 (dashed lines), 0.63 (dotted lines), 0.9 (dashed-dotted lines), and 1.1 (solid lines), respectively. The initial conditions are $k = 1$, $a = 0.005\lambda$, and $b = 0$.

In the early stage ($1 < \gamma t < 4$), the perturbation fundamental modes on both interfaces grow exponentially. With increasing amplitude, the amplitude growth of fundamental mode begins to fall below the linear (exponential) growth due to the third-order negative feedback (nonlinear mode-coupling effects). This continues until a "maximum amplitude" is reached. Exceeding that maximum, the amplitude decreases sharply, which in fact is invalid in the theoretical model. This means the range of validity of the present model is within the maximum of the fundamental mode. Furthermore, the saturation of linear growth of the fundamental mode will be discussed in detail in Sec. III D. It can be seen from Figure 2 that at a fixed γt the normalized amplitudes of the perturbation first three harmonics increase with decreasing kd. The Jacobs-Catton's formula can only relatively well describe the temporal evolution of the amplitudes of the perturbation first three harmonics for a large kd case.

Figure 3 shows a comparison of the position of the lower interface obtained from the present model with that calculated by the Jacobs-Catton's formula initiated by only lower interface perturbation with different kd. Significant differences in interface positions between the present model and the Jacobs-Catton's formula can be observed in Figure 3, especially when kd is small. Jacobs-Catton's formula underestimates the perturbation growth of the lower interface. For $kd = 0.2$, the bubble-spike amplitude calculated by the Jacobs-Catton's formula is only $\sim$1/3 of that obtained from our model. The finite-thickness effects become important with decreasing kd. Consequently, when kd is small, to predict the WN perturbation growth, one should carefully consider the finite-thickness effects.

In Figure 4, the kd dependence of the temporal evolution of positions of the upper and lower interfaces can be seen noticeably. For a small, the upper and lower interfaces are strongly coupled (see Figure 4(a)). The perturbation amplitude of the upper (linearly stable) interface is approximately the same as that of the lower (linearly unstable) interface. With increasing kd, the finite-thickness effects become weak. For a large kd, as can be seen in Figure 4(d), the perturbation growth of the upper (linearly stable) interface are negligibly small in comparison to that of the lower (linearly unstable) interface. On the other hand, for $kd = 0.2$ at $\gamma t = 4.3$, the bubble-spike asymmetry can be seen markedly, as shown in Figure 4(a). However, for $kd = 1.6$ at the same

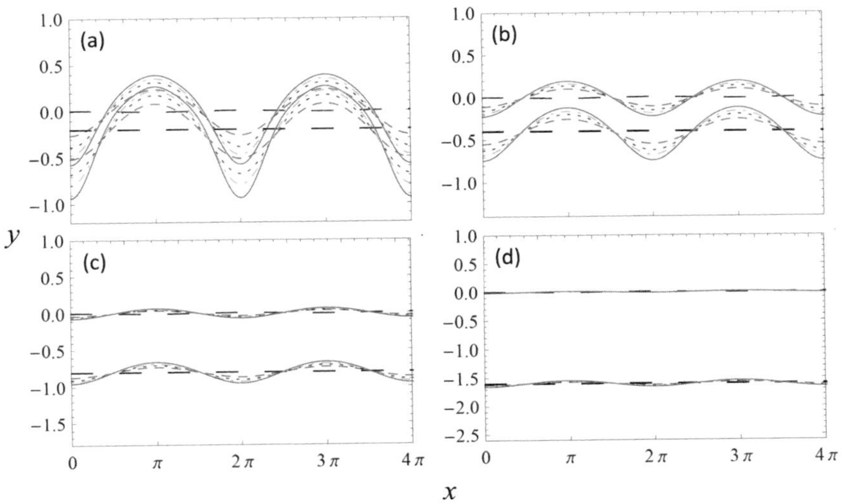

FIG. 7. Temporal evolution of positions of the upper and lower interfaces initiated by only the upper interface perturbation at $\gamma t = 0$ (long-dashed lines), 3.6 (dashed lines), 4.0 (dotted lines), 4.2 (dashed-dotted lines), and 4.4 (solid lines) for $kd = 0.2$ (a), 0.4 (b), 0.8 (c), and 1.6 (d), respectively. The initial conditions are $k = 1$, $a = 0.001\lambda$, and $b = 0$.

time, the symmetry interface structures still hold, as shown in Figure 4(d).

In order to study the slab distortion, we analyze the layer thickness at the spike tip (d_{spike}) and that at the bubble vertex (d_{bubble}). Figure 5 shows the temporal evolution of the normalized layer thicknesses at the spike tip and the bubble vertex initiated by only the lower interface perturbation for different kd. The d_{spike} increases with increasing γt but the d_{bubble} decreases with increasing γt. That is to say, the slab thickness variations are amplified by the RTI of a finite-thickness fluid layer. At a fixed γt, the d_{spike} decreases with increasing kd but the d_{bubble} increases with increasing kd. This means the interface coupling effects increase the slab thickness variations. The slab thickness variations (i.e., the slab distortion) can be well represented by the normalized parameter d_{spike}/d_{bubble}. The kd dependence of the slab distortion can be obviously seen in Fig. 5(c). At a fixed γt, the d_{spike}/d_{bubble} increases with decreasing kd, namely, the "thin slab" results in a more severe distortion. Thus, the "thin shell" effects should be included in predictions of shell distortion in ICF implosions.

B. Upper interface perturbations

Second, we impose only the upper interface perturbation. As can be seen from Eq. (26), the phase of the lower interface perturbation is shifted by π in comparison to that of the upper initial interface perturbation. This can be clearly seen in Figure 6 that depicts the positions of the upper and lower interfaces initiated by only the upper interface perturbation for $kd = 0.2$ at different normalized times.

Initially, a perturbation is imposed on the upper (linearly stable) interface and the lower linearly unstable interface is kept unperturbed. The perturbation amplitude of the upper interface first decreases with increasing γt until to zero at $\gamma t \sim 0.6$, and then increases with increasing γt but change its phase. On the other hand, the perturbation amplitude of the lower interface increases with increasing γt all the time but

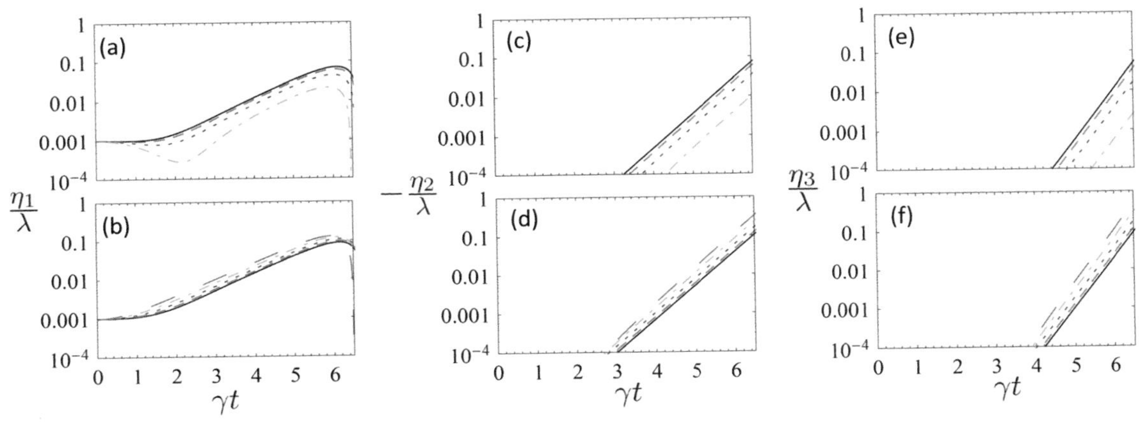

FIG. 8. Temporal evolution of normalized amplitudes of the perturbation fundamental mode [(a) and (b)], second [(c) and (d)] and third [(e) and (f)] harmonics on the upper [(a), (c), (e)] and lower [(b), (d), (f)] interfaces initiated by both the upper and lower interface perturbations with $kd = 0.2$ (solid lines), 0.4 (dashed lines), 0.8 (dotted lines), and 1.6 (dashed-dotted lines), respectively. The long-dashed lines represent the corresponding amplitudes calculated by Jacobs-Catton's formula. The initial conditions are $k = 1$ and $a = b = 0.001\lambda$ (in-phase case).

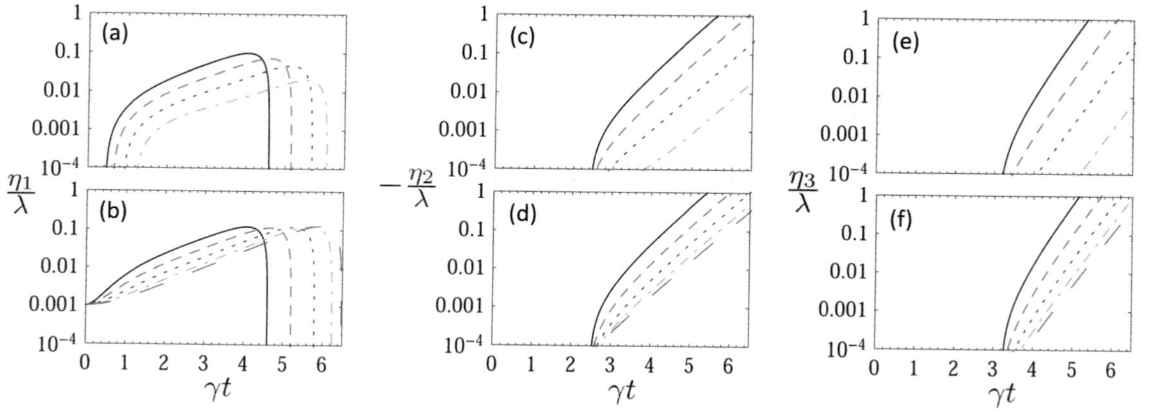

FIG. 9. Temporal evolution of normalized amplitudes of the perturbation fundamental mode [(a) and (b)], second [(c) and (d)] and third [(e) and (f)] harmonics on the upper [(a), (c), (e)] and lower [(b), (d), (f)] interfaces initiated by both the upper and lower interface perturbations with $kd = 0.2$ (solid lines), 0.4 (dashed lines), 0.8 (dotted lines), and 1.6 (dashed-dotted lines), respectively. The long-dashed lines represent the corresponding amplitudes calculated by Jacobs-Catton's formula. The initial conditions are $k = 1$ and $a = -b = -0.001\ \lambda$ (out-phase case).

with a negative phase. This is because the perturbation phase of lower (linearly unstable) interface felt or seen will be changed when the upper (linearly stable) interface perturbation feeds through to the lower (linearly unstable) interface. After the RTI on the lower interface has been initiated, the perturbation will be feeds through to the upper interface. Finally, the perturbation phase of the upper interface will be the same as that of the lower interface, as can be seen in Figure 6.

It is notable that the potential influences of compressibility are not included throughout this paper. When the compressibility effects[37] are considered, the extra degrees of freedom from sound waves and compressibility would likely become important. Considerations of compressibility effects are beyond the scope of this work.

Figure 7 is the plot of positions of the upper and lower interfaces initiated by only the upper interface perturbation at different normalized times with different normalized layer thicknesses. The dependence of WN evolution of the RTI at the two interfaces on the normalized layer thickness has been quite clearly depicted in Figure 7. For a small kd, the initial perturbation of the upper (linearly stable) interface well feeds through to the lower (linearly unstable) interface, resulting in a quick RT growth at the lower interface. Then, the RT growth at the lower interface rapidly feeds through to the upper interface, leading to a great distortion of the fluid layer, as shown in Figure 7(a). On the other hand, the feedthrough process is mitigated with increasing kd, as can be seen from comparison of perturbation amplitude at the same γt with different kd in Figure 7.

C. Both lower and upper interface perturbations

More generally, we address the RTI initiated by both the upper and lower interface perturbations. The initial perturbation phase of the upper and lower interfaces can be different in our model. The in-phase and out-phase cases are studied here.

Figure 8 presents the temporal evolution of normalized amplitudes of the perturbation first three harmonics on the

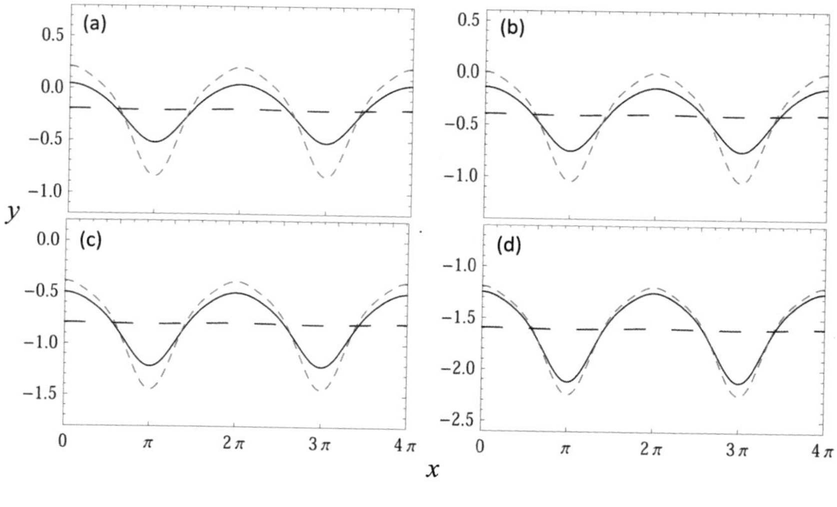

FIG. 10. Comparison of the position of the lower (linearly unstable) interface obtained from the present model (solid lines) with that calculated by the Jacobs-Catton's formula (dashed lines) initiated by both the upper and lower interface perturbations at $\gamma t = 5.1$ with $kd = 0.2$ (a), 0.4 (b), 0.8 (c), and 1.6 (d), respectively, with initial conditions being the same as in Figure 8 (in-phase case).

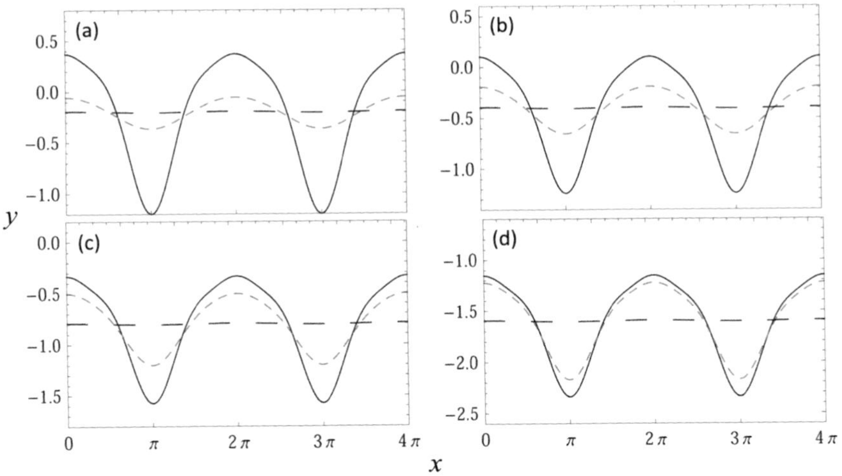

FIG. 11. Comparison of the position of the lower (linearly unstable) interface obtained from the present model (solid lines) with that calculated by the Jacobs-Catton's formula (dashed lines) initiated by both the upper and lower interface perturbations with $kd = 0.2$ at $\gamma t = 3.9$ (a), $kd = 0.4$ at $\gamma t = 4.3$ (b), $kd = 0.8$ at $\gamma t = 4.7$ (c), and $kd = 1.6$ at $\gamma t = 5.0$ (d), respectively, with initial conditions being the same as in Figure 9 (out-phase case).

upper and lower interfaces initiated by both the upper and lower interface perturbations with initial amplitudes $a = b = 0.001\lambda$ (in-phase case). The amplitudes of the perturbation first three harmonics on the lower interface calculated by the Jacobs-Catton's formula are also plotted for comparison purpose. Figure 9 shows the corresponding out-phase case.

It is found that the amplitude growth of the perturbation first three harmonics in the in-phase case (Figure 8) is slowed down in comparison to that in the out-phase case (Figure 9). On the other hand, the Jacobs-Catton's formula overestimates the amplitudes of the perturbation first three harmonics on the lower interface in the in-phase case but underestimates that in the out-phase case. This is because the phase difference of the upper interface perturbation in two cases. When one considers the interface coupling effects, the "effective" initial perturbation amplitude driving the RT growth of the lower interface becomes $\eta_0^{RT} = b - ae^{-kd}$. For in-phase case ($a = b$), the η_0^{RT} is reduced to $\sim 0.18b$, $0.33b$, $0.55b$, and $0.80b$, respectively, for $kd = 0.2$, 0.4, 0.8, and 1.6. However, for out-phase case ($a = -b$), the η_0^{RT} is increased to $\sim 1.82b$, $1.67b$, $1.45b$, and $1.20b$, respectively, for $kd = 0.2$, 0.4, 0.8, and 1.6.

The difference of positions of the lower interface obtained from the present model and the Jacobs-Catton's formula can be visibly seen in Figure 10 (for in-phase case) and Figure 11 (for out-phase case). Again, the Jacobs-Catton's formula overestimates the perturbation growth for the in-phase case (see Figure 10) but severely underestimates that for the out-phase case (see Figure 11). In fact, both cases can occur in ICF implosions. Note that the perturbation growth of the out-phase case in a finite-thickness slab is more severe than prediction by semi-infinite fluid models, e.g., Jacobs-Catton's formula[12] or Haan model.[38] Therefore, the "thin shell" effects should be well considered in ICF implosions when predicting the perturbation growth and shell distortion.

D. Saturation amplitudes

Now let us analyze the saturation amplitude (SA) of linear growth of the perturbation fundamental mode[3,16,18,38] with corrections up to the third-order. The SA of the fundamental mode, η_{1s}^i, can be defined as the linear growth amplitude of the fundamental mode at the saturation time when the growth of the fundamental mode (η_1^i) is reduced by 10% in comparison to its corresponding linear growth η_L^i, namely, $(\eta_L^i - \eta_1^i)/\eta_L^i = 1/10$, where $i = u$, l. Then, the saturation amplitudes (SAs) of the fundamental mode for the upper and low interfaces (i.e., η_{1s}^u and η_{1s}^l) with corrections up to the third-order are obtained. The η_{1s}^u and η_{1s}^l normalized by its perturbation wavelength λ can be expressed as

$$\begin{cases} \dfrac{\eta_{1s}^u}{\lambda} = \dfrac{e^{-\xi}}{\pi\sqrt{5}\sqrt{3 + 2e^{-2\xi}}}, \\ \dfrac{\eta_{1s}^l}{\lambda} = \dfrac{1}{\pi\sqrt{5}\sqrt{2 + 3e^{-2\xi}}}. \end{cases} \tag{42}$$

For problems with a large normalized layer thickness ($\xi \to \infty$), Eq. (42) reduces to

$$\begin{cases} \dfrac{\eta_{1s}^u}{\lambda} = 0, \\ \dfrac{\eta_{1s}^l}{\lambda} = \dfrac{1}{\pi\sqrt{10}}, \end{cases} \tag{43}$$

which is the classical SA of 0.1λ.[3,15,16,18]

Figure 12 displays the normalized layer thickness dependence of the normalized SAs of the perturbation fundamental mode on the upper and lower interfaces. As can be seen in Figure 12, the finite-thickness effects (as $kd \to 0$) are to decrease the SA of the lower (linearly unstable) interface (η_{1s}^l) from $\sim 0.1\lambda$ to $\sim 0.064\lambda$, but to increase the SA of the upper (linearly stable) interface (η_{1s}^u) from 0 to $\sim 0.064\lambda$. That is to say, in the real ICF implosion, the SA of the fundamental mode on the unstable interface (e.g., the ablation front and the ablator-fuel interface in the acceleration regime, or the hot-spot boundary in the deceleration regime) can be much less than the classical prediction 0.1λ. It should be noted that the smaller SA means the earlier nonlinear growth. We would like to point out that the finite-thickness effects are not included in the Youngs' model[39,40] or the Haan saturation model[38] that has been used to predict mix

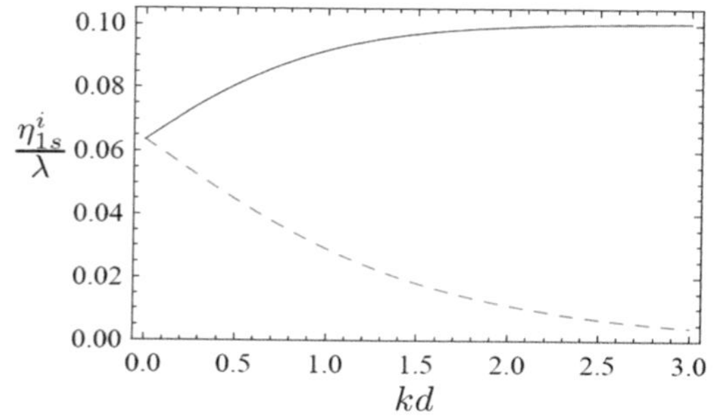

FIG. 12. Normalized SAs (η_{1s}^i/λ) of the perturbation fundamental mode on the upper "linearly stable" (dashed lines) and lower "linearly unstable" (solid lines) interfaces versus the normalized layer thicknesses (kd).

depth by post-processing one-dimensional calculations from ICF codes.[41] On the other hand, it is reported that when interface coupling effects are ignored, ablation effects[42–46] and/or interface with effects[47,48] tend to increase the SA, especially for short-wavelength perturbations. The SA of linear growth of the perturbation fundamental mode of the RTI includes the interface coupling, ablation and interface width effects will be described in the future.

IV. SUMMARY AND CONCLUSIONS

In this article, based on potential flow description for ideal incompressible fluids, a WN model up to the third order has been proposed for the single-mode RTI of a finite-thickness fluid layer (slab). Intuitively, the lower interface of the slab is prone to the RTI while the upper interface of that slab is "linearly stable." The WN growth of the RTI initiated by only the linearly unstable interface perturbation, only the linearly stable interface perturbation, and both linearly unstable and linearly stable interface perturbations with different phases, are studied analytically. How the perturbation on the linearly unstable interface feeds through to the linearly stable interface and vice versa (i.e., interface coupling effects) have been discussed in detail in this work. It is found that the interface coupling effects become important when the fluid is of a thickness on the order of a perturbation wavelength. It is notable that the theory based on the semi-infinite fluid assumption, e.g., Jacobs-Catton' model[12] and Haan's saturation model,[38] always give invalid predictions, especially when the interface coupling (feedthrough) is important. Even if the perturbation amplitude of the linearly unstable interface is initially zero (i.e., only the linearly stable interface perturbation), the RTI would still grow due to the feedthrough, also leading to severe slab distortion.

Our third-order nonlinear model can describe the kd dependence of the WN perturbation growth on the two interfaces (i.e., the slab distortion) and the saturation of the exponential growth of the perturbation fundamental mode. When kd is small, the interface coupling effects play an important role in the WN perturbation growth on both interfaces and hence the distortion of the slab. On the other hand, the WN perturbation growth of the lower (linearly unstable) interface becomes less dependent on the perturbations of the upper (linearly stable) interface with increasing kd, and vice versa. It is shown that the felt or seen perturbation phase of the linearly unstable interface is shifted by π in comparison to the initial perturbation phase of the linearly stable interface when the linearly stable interface perturbation feeds through to the linearly unstable interface. Thus, when the perturbations on the linearly stable and linearly unstable interfaces with different phases, one should carefully consider the finite-thickness effects. Moreover, the finite-thickness effects (as $kd \rightarrow 0$) are to decrease the SA of the linearly unstable interface perturbation, but to increase that of the linearly stable interface perturbation.

In ICF implosions, the source of perturbation at the hot-spot boundary can be either initial perturbations on shell's inner surface or perturbations that feedthrough from the outer capsule surface (e.g., the ablator front or the fuel-ablator interface), isolated defects in the ablator and/or ice, and the fill-tube and tent perturbations, etc. Feedthrough can be a severe problem in ICF implosions, which contributes pronounced perturbation growth to the inner surface.[2,3,35,36,49–52] The simulation results discussed in Ref. 53 indicate that feedthrough tends to be the most prominent candidate for the performance degradation in double-shell ignition targets. Furthermore, in beryllium target designs for the National Ignition Facility, simulation results have obviously shown that the capsule yield robustness depends significantly on the shell thickness since a thicker shell leads to an enhanced feedthrough damping.[35,36] Thus, the fuel and ablator thicknesses should be well optimized in ICF target design. Our model can be used to predict how much of the outside perturbation feeds through to hot-spot interface and the most dangerous perturbation wavelengths. Meanwhile, it is hoped that this model can shed some light on ways to know which interface should be paid special attention and how to fabricate the target when imperfections on a particular surface cannot to fabricate to the required standard.

It should be pointed out that our WN theory has neglected compressibility and converging geometry effects.[54,55] However, they are quite important and should be well considered in ICF implosions. Research in these areas will be pursued in the future.

ACKNOWLEDGMENTS

We would like to thank the referee for useful comments and suggestions that improved this paper. This work was supported by the Foundation of President of Chinese Academy of Engineering Physics (Grant No. 2014-1-040), the foundation of CAEP (Grant No. 2014A0202010), the National Natural Science Foundation of China (Grant Nos. 11275031, 11475034, 11475032, and 11274026), and National Basic Research Program of China (Grant No. 2013CB834100).

[1]J. Nuckolls, L. Wood, A. Thiessen, and G. Zimmerman, Nature **239**, 139 (1972).
[2]S. Atzeni and J. Meyer-ter-Vehn, *The Physics of Inertial Fusion: Beam Plasma Interaction Hydrodynamics, Hot Dense Mater* (Oxford University, Oxford, 2004).
[3]J. D. Lindl, P. Amendt, R. L. Berger, S. G. Glendinning, S. H. Glenzer, S. W. Haan, R. L. Kauffman, O. L. Landen, and L. J. Suter, Phys. Plasmas **11** (2), 339 (2004).

[4]L. Rayleigh, Proc. London Math. Soc. **14**, 170 (1883).
[5]G. Taylor, Proc. R. Soc. London A **201**, 192 (1950).
[6]R. D. Richtmyer, Commun. Pure Appl. Math. **13**, 297 (1960).
[7]E. E. Meshkov, Fluid Dyn. **4**, 101 (1969).
[8]S. Chandrasekhar, *Hydrodynamic and Hydromagnetic Stability* (Oxford University, London, 1961).
[9]V. N. Gamezo, A. M. Khokhlov, E. S. Oran, A. Y. Chtchelkanova, and R. O. Rosenberg, Science **299**, 77 (2003).
[10]E. Müller, B. Fryxell, and D. Arnett, Astron. Astrophys. **251**, 505 (1991); available at http://adsabs.harvard.edu/abs/1991A%26A...251..505M.
[11]I. Hachisu, T. Matsuda, K. Nomoto, and T. Shigeyama, Astrophys. J. **358**, L57 (1990) and references therein.
[12]J. W. Jacobs and I. Catton, J. Fluid Mech. **187**, 329 (1988).
[13]S. W. Haan, Phys. Fluids B **3**, 2349 (1991).
[14]M. Berning and A. M. Rubenchik, Phys. Fluids **10**, 1564 (1998).
[15]L.-F. Wang, W.-H. Ye, and Y.-J. Li, Chin. Phys. Lett. **27**, 025203 (2010).
[16]W. H. Liu, L. F. Wang, W. H. Ye, and X. T. He, Phys. Plasmas **19**, 042705 (2012).
[17]L. F. Wang, J. F. Wu, Z. F. Fan, W. H. Ye, X. T. He, W. Y. Zhang, Z. S. Dai, J. F. Gu, and C. Xue, Phys. Plasmas **19**, 112706 (2012).
[18]D. Layzer, Astrophys. J. **122**, 1 (1955).
[19]Q. Zhang, Phys. Rev. Lett. **81**, 3391 (1998).
[20]V. N. Goncharov, Phys. Rev. Lett. **88**, 134502 (2002).
[21]K. O. Mikaelian, Phys. Rev. E **67**, 026319 (2003).
[22]S. I. Sohn, Phys. Rev. E **67**, 026301 (2003).
[23]W. H. Liu, L. F. Wang, W. H. Ye, and X. T. He, Phys. Plasmas **20**, 062101 (2013).
[24]K. O. Mikaelian, Phys. Rev. A **26**, 2140 (1982).
[25]K. O. Mikaelian, Phys. Rev. A **28**, 1637 (1983).
[26]K. O. Mikaelian, Phys. Fluids **7**(4), 888 (1995).
[27]K. O. Mikaelian, Phys. Rev. A **42**(6), 3400 (1990).
[28]K. O. Mikaelian, Phys. Fluids **17**, 094105 (2005).
[29]B. A. Hammel, S. W. Hann, D. S. Clark, M. J. Edwards, S. H. Langer, M. M. Marinak, M. V. Patel, J. D. Salmonson, and H. A. Scott, High Energy Density Phys. **6**, 171 (2010).
[30]D. S. Clark, D. E. Hinkel, D. C. Eder, O. S. Jones, S. W. Haan, B. A. Hammel, M. M. Marinak, J. L. Milovich, H. F. Robey, L. J. Suter, and R. P. J. Town, Phys. Plasmas **20**, 056318 (2013).
[31]J. Lindl, O. Landen, J. Edwards, and NIC Team, Phys. Plasmas **21**, 020501 (2014).
[32]K. S. Raman, V. A. Smalyuk, D. T. Casey, S. W. Haan, D. E. Hoover, O. A. Hurricane, J. J. Kroll, A. Nikroo, J. L. Peterson, B. A. Remington, H. F. Robey, D. S. Clark, B. A. Hammel, O. L. Landen, M. M. Marinak, D. H. Munro, K. J. Peterson, and J. Salmonson, Phys. Plasmas **21**, 072710 (2014).
[33]O. A. Hurricane, D. A. Callahan, D. T. Casey, E. L. Dewald, T. R. Dittrich, T. Doppner, M. A. Barrios Garcia, D. E. Hinkel, L. F. Berzak Hopkins, P. Kervin, J. L. Kline, S. Le Pape, T. Ma, A. G. MacPhee, J. L. Milovich, J. Moody, A. E. Pak, P. K. Patel, H.-S. Park, B. A. Remington, H. F. Robey, J. D. Salmonson, P. T. Springer, R. Tommasini, L. R. Benedetti, J. A. Caggiano, P. Celliers, C. Cerjan, R. Dylla-Spears, D. Edgell, M. J. Edwards, D. Fittinghoff, G. P. Grim, N. Guler, N. Izumi, J. A. Frenje, M. Gatu Johnson, S. Haan, R. Hatarik, H. Herrmann, S. Khan, J. Knauer, B. J. Kozioziemski, A. L. Kritcher, G. Kyrala, S. A. Maclaren, F. E. Merrill, P. Michel, J. Ralph, J. S. Ross, J. R. Rygg, M. B. Schneider, B. K. Spears, K. Widmann, and C. B. Yeamans, Phys. Plasmas **21**, 056314 (2014).
[34]E. Ott, Phys. Rev. Lett. **29**, 1429 (1972).
[35]A. N. Simakov, D. C. Wilson, S. A. Yi, J. L. Kline, D. S. Clark, J. L. Milovich, J. D. Salmonson, and S. H. Batha, Phys. Plasmas **21**, 022701 (2014).
[36]S. A. Yi, A. N. Simakov, D. C. Wilson, R. E. Olson, J. L. Kline, D. S. Clark, B. A. Hammel, J. L. Milovich, J. D. Salmonson, B. J. Kozioziemski, and S. H. Batha, Phys. Plasmas **21**, 092701 (2014).
[37]A. R. Piriz, Phys. Plasmas **8**(12), 5268 (2001).
[38]S. W. Haan, Phys. Rev. A **39**, 5812 (1989).
[39]D. Youngs, Physica D **12**, 32 (1984).
[40]D. Youngs, Physica D **37**, 270 (1989).
[41]L. Welser-Sherrill, R. C. Mancini, D. A. Haynes, S. W. Haan, I. E. Golovkin, J. J. MacFarlane, P. B. Radha, J. A. Delettrez, S. P. Regan, J. A. Koch, N. Izumi, R. Tommasini, and V. A. Smalyuk, Phys. Plasmas **14**, 072705 (2007).
[42]J. Garnier and L. Masse, Phys. Plasmas **12**, 062707 (2005).
[43]J. Sanz, J. Ramírez, R. Ramis, R. Betti, and R. P. J. Town, Phys. Rev. Lett. **89**(19), 195002 (2002).
[44]L. F. Wang, W. H. Ye, Z. M. Sheng, W.-S. Don, Y. J. Li, and X. T. He, Phys. Plasmas **17**, 122706 (2010).
[45]L. F. Wang, W. H. Ye, and X. T. He, Phys. Plasmas **19**, 012706 (2012).
[46]L. F. Wang, W. H. Ye, X. T. He, W. Y. Zhang, Z. M. Sheng, and M. Y. Yu, Phys. Plasmas **19**, 100701 (2012).
[47]L. F. Wang, W. H. Ye, and Y. J. Li, Phys. Plasmas **17**, 052305 (2010).
[48]L. F. Wang, W. H. Ye, Z. F. Fan, and Y. J. Li, EPL **90**, 15001 (2010).
[49]W. W. Hsing and N. M. Hoffman, Phys. Rev. Lett. **78**(20), 3876 (1997).
[50]S. T. Weir, E. A. Chandler, and B. T. Goodwin, Phys. Rev. Lett. **80**(17), 3763 (1998).
[51]K. Shigemori, H. Azechi, M. Nakai, T. Endo, T. Nagaya, and T. Yamanaka, Phys. Rev. E **65**, 045401(R) (2002).
[52]S. P. Regan, J. A. Delettrez, V. N. Goncharov, F. J. Marshall, J. M. Soures, V. A. Smalyuk, P. B. Radha, B. Yaakobi, R. Epstein, V.Yu. Glebov, P. A. Jaanimagi, D. D. Meyerhofer, T. C. Sangster, W. Seka, S. Skupsky, and C. Stoeckl, Phys. Rev. Lett. **92**(18), 185002 (2004).
[53]J. L. Milovich, P. Amendt, M. Marinak, and H. Robey, Phys. Plasmas **11** (4), 1552 (2004).
[54]R. Epstein, Phys. Plasmas **11**(11), 5114 (2004).
[55]P. Amendt, J. D. Colvin, J. D. Ramshaw, H. F. Robey, and O. L. Landen, Phys. Plasmas **10**(3), 820 (2003).

THE ASTROPHYSICAL JOURNAL, 815:46 (6pp), 2015 December 10 doi:10.1088/0004-637X/815/1/46

MODELING SUPERSONIC-JET DEFLECTION IN THE HERBIG–HARO 110-270 SYSTEM WITH HIGH-POWER LASERS

DAWEI YUAN[1], JUNFENG WU[2], YUTONG LI[1,3], XIN LU[1], JIAYONG ZHONG[4,5], CHUANLEI YIN[1], LUNING SU[1], GUOQIAN LIAO[1], HUIGANG WEI[4], KAI ZHANG[4], BO HAN[4], LIFENG WANG[2], SHAOEN JIANG[6], KAI DU[6], YONGKUN DING[6], JIANQIANG ZHU[7], XIANTU HE[2], GANG ZHAO[4], AND JIE ZHANG[1,3,8]

[1] National Laboratory for Condensed Matter Physics, Institute of Physics, Chinese Academy of Sciences, Beijing 100080, China; ytli@iphy.ac.cn, jzhang@sjtu.edu.cn
[2] Institute of Applied Physics and Computational Mathematics, Beijing 100088, China
[3] IFSA Collaborative Innovation Center, Shanghai Jiaotong University, Shanghai 200240, China
[4] Key Laboratory of Optical Astronomy, National Astronomical Observatories, Chinese Academy of Sciences, Beijing 100012, China
[5] Department of Astronomy, Beijing Normal University, Beijing 100875, China
[6] Research Center for Laser Fusion, China Academy of Engineering Physics, Mianyang 621900, China
[7] National Laboratory on High Power Lasers and Physics, Shanghai, 201800, China
[8] Key Laboratory for Laser Plasmas (MoE) and Department of Physics, Shanghai Jiao Tong University, Shanghai 200240, China
Received 2015 March 29; accepted 2015 November 1; published 2015 December 8

ABSTRACT

Herbig–Haro (HH) objects associated with newly born stars are typically characterized by two high Mach number jets ejected in opposite directions. However, HH 110 appears to only have a single jet instead of two. Recently, Kajdi et al. measured the proper motions of knots in the whole system and noted that HH 110 is a continuation of the nearby HH 270. It has been proved that the HH 270 collides with the surrounding mediums and is deflected by 58°, reshaping itself as HH 110. Although the scales of the astrophysical objects are very different from the plasmas created in the laboratory, similarity criteria of physical processes allow us to simulate the jet deflection in the HH 110/270 system in the laboratory with high power lasers. A controllable and repeatable laboratory experiment could give us insight into the deflection behavior. Here we show a well downscaled experiment in which a laser-produced supersonic-jet is deflected by 55° when colliding with a nearby orthogonal side-flow. We also present a two-dimensional hydrodynamic simulation with the Euler program, LARED-S, to reproduce the deflection. Both are in good agreement. Our results show that the large deflection angle formed in the HH 110/270 system is probably due to the ram pressure from a flow–flow collision model.

Key words: Herbig–Haro objects – ISM: jets and outflows – plasmas – shock waves

1. INTRODUCTION

Jet deflection (De Young 1991; Reipurth et al. 1996; Mendoza & Longair 2001) is mysterious in astronomical phenomena. In young stellar objects (YSOs; Bally et al. 1996; Mendoza & Longair 2001) and active galactic nuclei (AGNs; O'Dea & Owen 1986; Balsara & Norman 1992), collimated jets are found to propagate away from their initial trajectories. The HH 110/270 system is one of the most enigmatic cases among them. Observations of the proper motions of knots in the whole system support that HH 110 is a continuation of the nearby HH 270. HH 270 might interact with the surrounding mediums and was deflected by 58°, reshaping itself as HH 110. However, what causes the deflection remains unclear. Two mechanisms are proposed to explain the deflection phenomena. One is collisionless, caused by ambient magnetic fields (Hurka et al. 1999), and the other one is collisional, caused by ram pressure (Fiedler & Henriksen 1984; Mendoza & Longair 2001; Kajdic et al. 2012). Simulation results (Hurka et al. 1999) show that the collisionless mechanism is not suitable for the HH 110/270 system. Additionally, the enhancement of special line emission and a cloud observed nearby the HH 110/270 system (Choi 2001; Kajdic et al. 2012) also support the collisional explanation.

With the development of high-power pulse technology, laboratory study becomes a new complementary way to investigate astrophysical phenomena and verify the physical model (Remington et al. 1999, 2006; Bellan et al. 2009; Morita et al. 2010; Zhong et al. 2010; Liu et al. 2011; Dong et al. 2012; Kugland et al. 2012; Yuan et al. 2013; Albertazzi et al. 2014; Huntington et al. 2015). Recently, jet deflection has been investigated in laboratories by applying the scaling law of hydrodynamics (Ryutov et al. 1999). A flow-cloud model, in which HH 270 collides obliquely with a dense cloud and gets defected, was first proposed (De Young 1991). Lebedev et al. (2004) studied the jet deflection behavior occurring in YSOs and AGNs using the Z-pinch facility. A jet with a deflection angle of 21° was observed in their experiment. Lebedev et al. (2004) also predicted that the deflection angle would be larger if the ram pressure was enhanced ($p = \rho v^2$). Hartigan et al. (2009) carried out another deflection experiment at the Omega laser facility and conducted the comparison of dimensionless parameters of HH 110 with experimental. Both experiments used the flow-cloud model, where the cloud was produced by extreme-UV and soft X-radiation from the flow.

New observational evidence shows that HH 270 may collide with another flow instead of a dense cloud. A recently identified flow, named as HH 451, is presented near the dense cloud (Kajdic et al. 2012), whose knot motions are following HH 110. This indicates that the flow–flow collision may be an alternative interpretation to the deflection. Here we present our results of modeling the jet deflection with a flow–flow collision model, in which two orthogonal flows are generated by individual laser beams respectively. In our experiment, the supersonic-jet is deflected by 55° when it collides with the orthogonal side-flow. The down-scaled experiment results and the scaling relation show that the deflection behavior of the HH 110/270 system can be reproduced using the flow–flow model in the laboratory.

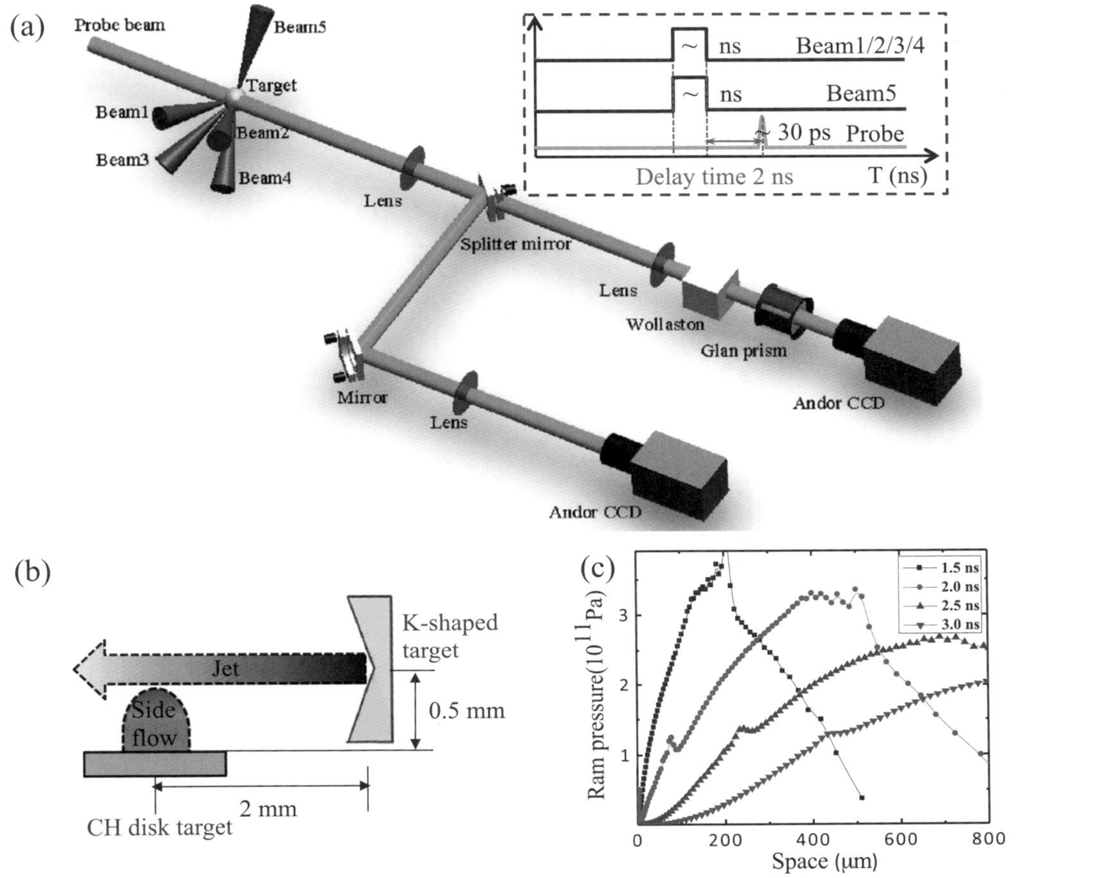

Figure 1. (a) Schematic view of the experimental setup. (b) The configuration of targets. (c) The evolution of the ram pressure dependent on spatial-temporal scale.

2. EXPERIMENT SETUP

The experiment was performed with the SG-II laser facility, which can output eight main laser beams (Beam 1–8) and deliver a total energy of 2.0 kJ in 1 ns at a wavelength of 351 nm. The schematic view of the experimental setup is shown in Figure 1(a). Four laser beams (Beam1–4) were divided into two bunches (Bunch 1 and Bunch 2), each of which consisted of two laser beams. The interaction of the jet with the side-flow was measured by a Nomarski interferometer with a magnification factor of three. Another laser beam with a wavelength of 527 nm and duration of 30 ps was used as a probe. The timing of the laser beams was shown in the inset at the right top of Figure 1(a). Beam1–4 for the jet generation were synchronized. The delay time of the probe with respect to the falling edge of Beam1 was kept as 2 ns. The timing of the side-flow generation was determined by the time separation between Beam1 and Beam5.

Figure 1(b) shows the schematic diagram of the generation and interaction of two orthogonal flows. Note that it is possible that the astronomical collision is a three-dimensional interaction with a significant velocity out of the plane of the HH 110/270 system. To catch the main physical picture, we only investigate the interaction with a two-dimensional orthogonal geometry in the laboratory. The whole target assembly consisted of a K-shaped Cu target with an open angle of 120° and a plane CH disk target. The two bunches were symmetrically focused on the interior surfaces of the K-shaped Cu target with a separation of 400 μm. The collision of the two plasma bubbles produced from the interior surfaces led to a collimated plasma jet. The diameter of each focal spot was 150 μm full width at half maximum (FWHM), giving an incident laser intensity of $\sim 2.5 \times 10^{15}$ W cm^{-2}. The side-flow was synchronously generated by using an individual Beam5 to irradiate the CH disk target. To obtain a low-density and quasi-one-dimensional expanding flow, the diameter of the focal spot of Beam5 was set to be as large as $\sim$600 μm. The laser energy was 50 J, giving an intensity of $\sim 1.8 \times 10^{13}$ W cm^{-2}. For such a large spot, the lateral expansion along the target surface was small because of the weak temperature gradient and density gradient. Since the collimated jet and the side-flow were generated separately, it was easy to control their characteristics by changing the parameters of the driven laser pulses.

Two issues of targets should be considered. The first is the horizontal distance between the K-shaped target and the plane CH disk target. We set the horizontal distance to be 2 mm,

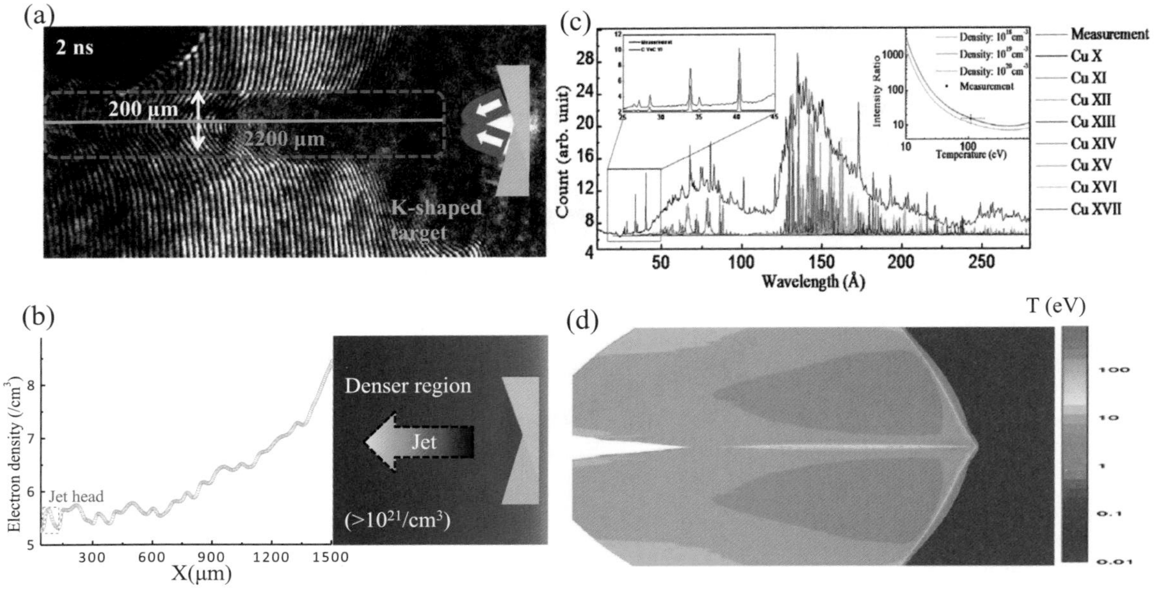

Figure 2. (a) The interferogram of a jet freely propagating into vacuum, taken at a delay time of 2 ns. (b) The electron density profile along the propagating direction obtained by Abel inversion. (c) The characteristic spectral lines obtained by the X-ray flat-field spectrometer. (d) The plasma temperature distribution obtained by the MULTI code.

considering the similarity criteria limitation and the spatial location of the laser beams (this will be discussed below). The second issue is the vertical distance, which decides the strength of the ram pressure, i.e., the deflection angle. According to Lebedev et al.'s (2004) prediction, we should enhance the ram pressure of the side-flow to obtain a significant deflection angle observed in the HH 110/270 system. Figure 1(c) shows the spatial-temporal evolution of the side-flow ram pressure, which is calculated by the one-dimensional Medusa code (Yan et al. 2006). The peak of ram pressure drops slowly depending on time and changes sharply depending on space. We choose the vertical distance to be 500 μm, where the ram pressure from the side-flow has the maximum value (~3.5E11 Pa) at 2 ns. It also ensures that the collimated jet from the K-shaped target can reach the side-flow and has enough time to interact with it.

3. EXPERIMENTAL RESULTS AND DISCUSSIONS

Figure 2(a) shows the interferogram of a jet freely propagating into a vacuum, taken at a delay time of 2 ns. Figure 2(b) is the density distribution of the jet by Abel inversing the interferogram. The jet generated by the K-shaped target remains well collimated. The length of the jet is about 2.2 mm and the width is about 0.2 mm, respectively. This indicates that the longitudinal propagation velocity ($v^{\rm jet}$) is about 1100 ± 100 km s^{-1} and the lateral expansion velocity is about 100 km s^{-1}. The average electron density ($n_e^{\rm jet}$) at the jet head is about 5.5×10^{19} cm^{-3}. Figure 2(c) shows the characteristic spectral lines measured by an X-ray flat-field spectrometer. Because of the thin layer of CH contaminants on the Cu target surface, the emission lines include both Cu and C ions. The average temperature of the jet is determined by the line intensity ratio of C v and C vi, as shown in the inset at the left top of Figure 2(c). Two lines originate from the transitions $1s^{2\,1}S_0$–$1s\,2p\,^3P_2$ and $1s^{2\,1}S_0$–$1s\,3d\,^1P_1$, respectively. The line blending from other transitions is also considered in our fitting. We find that the line ratio is not sensitive to the electron density below 10^{19} cm^{-3}, as shown in the inset at the right top of Figure 2(c). The average temperature is estimated to be about 110 eV with the ratio. We also use the temperature to fitting the Cu emission lines and find that most lines are from the higher ionization Cu x–xvii. Considering that the temperature ~110 eV obtained by the spectrometer is temporally integrated, we also use the two-dimensional MULTI code to calculate the plasma expansion. Figure 2(d) shows the calculated plasma temperature distribution at 2 ns. The temperature is in the range of 150–200 eV. Therefore, we take the average temperature of the jet to be ~170 eV.

To model the jet deflection relevant to the HH 110/270 system in the laboratory, we should confirm that both jets share similar dimensionless parameters according to Ryutov's theory. The Reynolds number (Re) and the Peclet number (Pe) are required to be much larger than unity, i.e., ${\rm Re} = R^{\rm jet} v^{\rm jet}/\nu \gg 1$ and ${\rm Pe} = R^{\rm jet} v^{\rm jet}/\sigma \gg 1$, where ν and σ are the viscosity and the thermal diffusivity, respectively. These values can ensure that the viscosity and heat conduction in the laboratory and in the astrophysical system are unimportant. In order to satisfy the behavior as a fluid, particles in the jet should be localized, that is to say, the ratio of the ion–ion mean free path (MFP), $\lambda_{ii}^{\rm jet\text{-}jet}$, to the jet radius, $R^{\rm jet}$, should be much less than unity, i.e., $\zeta = \lambda_{ii}^{\rm jet\text{-}jet}/R^{\rm jet} \ll 1$. Then, the similarity between the laboratory and astrophysical jet is determined by the other three scaling parameters, the Euler number (Courtois et al. 2004), ${\rm Eu} = v(\rho/p)^{1/2}$, which ensures that the Euler equations are invariant under Euler transformation and where p, and ρ, are the jet pressure and the mass density, the Mach number, $M = v^{\rm jet}/c_s = v^{\rm jet}/\left(\left(\bar{Z}^{\rm jet} T_e^{\rm jet} + T_i^{\rm jet}\right)/m_i\right)^{1/2}$, which ensures

Table 1
Comparison of the Dimensionless Parameters of the HH Objects with the Experimental Jet at 2 ns

Parameters	HH Objects[a]	Experiment
Collisionality, ζ	3E-7	~4E-4
Mach number, M	>5	~10–15
Peclet number, Pe	1E7	~3E2
Reynolds number, Re	7E8	~5E6
Euler number, Eu	22	~9
Radiation cooling parameter,	0.1–10	0.2–0.3

Reference. (a) Hartigan et al. (2009).

that the jet is supersonic, and the cooling parameter (Loupias et al. 2007), $\chi = \tau_{\rm rad}/\tau_{\rm hydro}$, which determines whether the radiation cooling effect plays an important role in the jet evolution and where $\tau_{\rm rad}$, and $\tau_{\rm hydro}$, are the cooling time and the hydrodynamic one ($\tau_{\rm hydro} = R^{\rm jet}/c_s$). In astrophysical HH objects, the radiation cooling effect is strong and the value of the cooling parameter is less than unity. This means the jet is "radiative" when the radiation cooling time is much smaller than the hydrodynamic time.

The radiation cooling parameter is estimated using a steady-state coronal equilibrium model (Post et al. 1977). The cooling time can be expressed as (Shigemori et al. 2000), $\tau_{\rm rad} = \left[(3/2)(\bar{Z}^{\rm jet} + 1)n_i^{\rm jet}T^{\rm jet}\right]/\left(\Gamma \bar{Z}^{\rm jet} n_i^{\rm jet2}\right)$, where $\bar{Z}^{\rm jet}$ is the average ionization of the jet, $T^{\rm jet}$ is the temperature of the jet, $n_i^{\rm jet}$ is the ionic density, and Γ is the cooling function. Assuming quasi neutrality is satisfied in our experiment, that is, $\bar{Z}^{\rm jet}n_i^{\rm jet} = n_e^{\rm jet} = 5.5 \times 10^{19}\,{\rm cm}^{-3}$, and a single temperature for the ions and electrons, $T_e^{\rm jet} = T_i^{\rm jet} = T^{\rm jet} = 170\,{\rm eV}$. Since $\bar{Z}^{\rm jet} \approx 10$ is relatively large, obtained by the 1D MEDUSA simulation and estimated by Post's model, the radiative cooling function is given as $\Gamma \sim \frac{2 \times 10^{-18}\,{\rm erg\,cm^3}}{\rm s} = 1.2 \times 10^{-6}\,{\rm eV\,cm^{-3}\,s^{-1}}$. With those parameters, χ is estimated to be 0.2–0.3. This value indicates that the jet in our experiment is "radiative."

Table 1 shows the comparison of the dimensionless parameters of the HH objects with the laboratory jet at 2 ns. One can see that they share hydrodynamic similarities with similar dimensionless parameters. This indicates that it is reasonable to use the laboratory jet to model the astrophysical jet in the HH 110/270 system.

Figure 3(a) illustrates a whole morphology of the HH 110/270 system by combining the images of Hα, [S II] and H_2 emission lines, recently observed with large view field ground-based and high resolution space-based telescopes (Kajdic et al. 2012). The deflection angle between HH 110 and HH 270 is about 58°. The newly identified HH 451 is located at the deflection point of the HH 110/270 system. Figure 3(b) shows the interferogram taken at 2 ns in our experiment. The free propagating jet to the left is deflected by the side-flow and reforms itself as a new well-collimated jet. The deflected part becomes narrower and denser than the free-propagating part, due to the transverse confining pressure from the side-flow (Lebedev et al. 2004). The right edge of the deflected jet is sharp without perturbation, which is similar to the astro-observation (Kajdic et al. 2012). The deflection angle, θ, is as large as about 55°. The whole morphology of the experiment is comparable to the structure of the HH 110/270 system and HH 451 shown in Figure 3(a).

To understand the experiment, we have used a hydrodynamic code, LARED-S (Ye et al. 2002, 2010), to reproduce the interaction of the supersonic-jet with the side-flow. The LARED-S is a two Eulerian code, based on an orthogonal structured mesh. A numerical algorithm with a conservative second-order Godunov method in Eulerian form is used to compute gas dynamics. Electron and ion conduction are treated using an operator splitting procedure. Conductivities and electron–ion coupling coefficients are computed with Lee and Moore's model. The energy deposition through ray tracing and inverse bremsstrahlung is also included in the code. In the calculations, the Quotidian Equation of State (QEOS) is used for the equations of state of all the materials. The QEOS model uses a combined EOS with a modified Cowan model for the ion EOS and with a scaled Thomas–Fermi table for the electron EOS. To do this, the supersonic-jet itself, the side-flow itself, and the supersonic-jet interacting with the side-flow should meet the localization conditions (Ryutov et al. 1999).

For the supersonic-jet, the plasma temperature ($T^{\rm jet}$) is about 170 eV, the ionization state ($\bar{Z}^{\rm jet}$) is 10, and the electron density ($n_e^{\rm jet}$) is $5.5 \times 10^{19}\,{\rm cm}^{-3}$. The ion–ion MFP of the jet is expressed as (Zheng 2009) $\lambda_{ii}^{\rm jet-jet} = 3\sqrt{\pi}/2(k_B T^{\rm jet})^2/(e^4(\bar{Z}^{\rm jet})^4 n_i^{\rm jet}\,{\rm Ln}\Lambda^{\rm jet})$, where ${\rm Ln}\Lambda^{\rm jet}$ is the Coulomb logarithm. The ion–ion MFP is estimated to be 4×10^{-6} cm. It is significantly smaller than the jet radius ($R^{\rm jet} \sim 10^{-2}$ cm). The electron density and electron temperature distributions of the CH side-flow are simulated by the 1D-Medusa code (Yan et al. 2006). Figure 4(a) shows the simulated distributions in the target normal direction at 2 ns. The plasma temperature ($T^{\rm flow}$) is about 470 eV and the electron density ($n_e^{\rm flow}$) is about $7 \times 10^{20}\,{\rm cm}^{-3}$ at 400 μm from the CH target. Assuming the maximum ionization of CH plasma ($\bar{Z}^{\rm flow}$) is six. From Figure 3(b), the average speed of the sideways flow ($v^{\rm flow}$) is estimated to be larger than $2.5 \times 10^7\,{\rm cm\,s}^{-1}$. With those parameters, the ion–ion MFP of the sideways flow can be calculated as $\lambda_{ii}^{\rm flow-flow} = 3\sqrt{\pi}/2(k_B T^{\rm flow})^2/(e^4(\bar{Z}^{\rm flow})^4 n_i^{\rm flow}\,{\rm Ln}\Lambda^{\rm flow}) = 1.5 \times 10^{-4}$ cm, which is much less than the diameter of the disk ($D^{\rm flow} = 0.1$ cm). For the impact of the side-flow on the supersonic-jet, the ion–ion MFP is expressed as (Chenais-Popovics et al. 1997) $\lambda_{ii}^{\rm jet-flow} = \bar{m}_i^2 (v^{\rm jet-flow})^4/(4\pi e^4 \bar{Z}^4 \bar{n}_i^{\rm jet-flow}\,{\rm Ln}\Lambda^{\rm jet-flow})$, where $\bar{m}_i$ is the average ion mass, $\bar{Z}$ is the average ionization state, $\bar{n}_i^{\rm jet-wind}$ is the average ion density, and ${\rm Ln}\Lambda^{\rm jet-flow}$ is the Coulomb logarithm. With the above parameters, $\lambda_{ii}^{\rm jet-flow}$ is calculated as $\lambda_{ii}^{\rm jet-flow} = 3.7 \times 10^{-3}$ cm, $\lambda_{ii}^{\rm jet-flow}/R^{\rm jet} = 10^{-2}$–$10^{-1}$. From the above discussions, we can see that all of the localization conditions are satisfied.

Figure 3(c) shows the simulated 2D electron density contour of the jet–flow interaction. The parameters are initialized with the properties measured experimentally. The electron density is in units of 1E24 cm^{-3}. The simulated pattern reproduces the experimental and the astronomical observation. The deflection angle is as large as 55°, which agrees well with the experimental. The sharp density jump at the interaction interface in Figures 3(b) and (c) is a collisional shock (Hartigan et al. 2009), which is the boundary separating the supersonic-jet from the flow. Figure 4(b) shows the electron density distribution crossing the collisional shock in Figures 3(b) and (c).

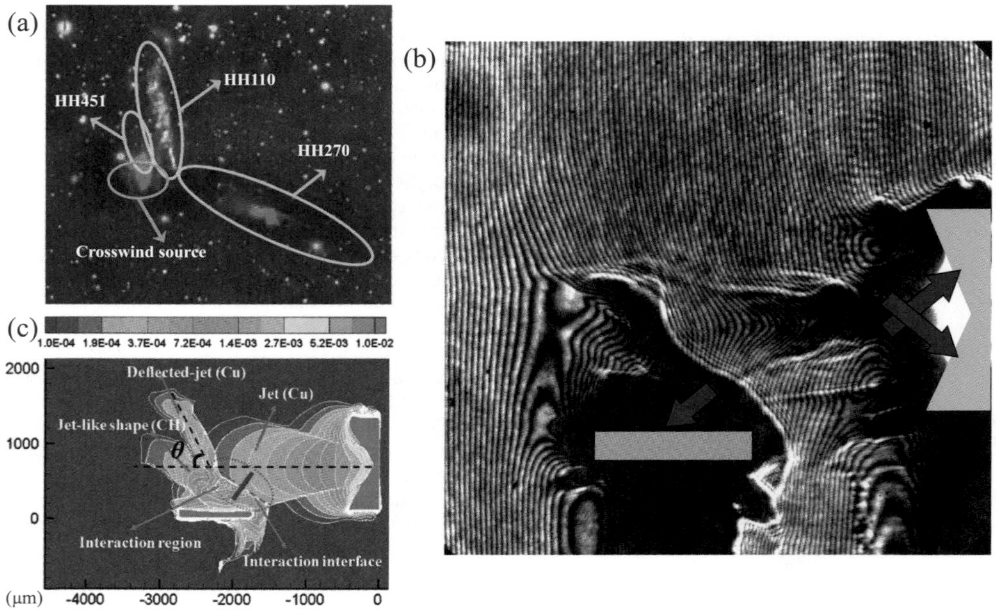

Figure 3. Images illustrating the jet deflection: (a) the combination image obtained by ground-based observations and the space-based telescopes[5]; (b) the experimental interferogram taken at a delay time of 2 ns; (c) the two-dimensional density map of the interaction simulated with the LARED-S code. The color bar stands for the electron density in units of $1E24/cm^3$.

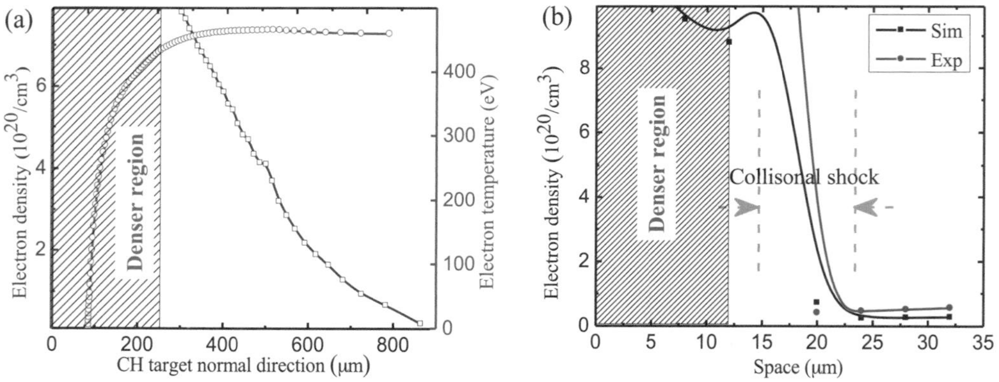

Figure 4. (a) The initial electron density and temperature distributions of the side-flow in the CH target normal direction at 2 ns, obtained with the Medusa code[27]. (b) Comparison of the electron density profiles along the cyan lines, as shown in Figures 3(b) and (c). The sharp density gradient is known as a collisional shock, which separates the supersonic-jet from the side-flow.

The shock plays an important role during the interaction. It survives on the collisions and can be regarded as the momentum exchange surface (Hartigan et al. 2009). The ram pressure from the side-flow deflects the incoming supersonic-jet with an angle of 55° and compresses the deflected jet, forming a sharp rim on one side while no perturbation on the other side. The ram pressure from the supersonic-jet also compresses the side-flow and forces it to follow the motion of the deflected jet. This may explain the observed new features that HH 110 flow has a sharp edge and some knots of HH 110 lies along the well-defined axis of the HH 451 flow (Kajdic et al. 2012).

In addition, the deflection angle θ can be expressed as (De Young 1991), $\theta = \tan^{-1}\left(\sqrt{\frac{R_p}{R_{bend}}}\right)$, where R_p is the scale over which the ram pressure acts on the supersonic-jet deflection, and R_{bend} is the radius of curvature of the deflection of the jet. Using the bending equation (De Young 1991), $\frac{\rho^{jet}(v^{jet})^2}{R_{bend}} = \frac{\rho^{flow}(v^{flow})^2}{R_p}$, θ can be expressed as $\theta = \tan^{-1}\left(\sqrt{[(n_e^{flow}/\bar{Z}^{flow})m_i^{flow}(v^{flow})^2]\Big/\Big[(n_e^{jet}/\bar{Z}^{jet})m_i^{jet}(v^{jet})^2\Big]}\right)$, where $\rho^{flow}(v^{flow})^2 = (n_e^{flow}/\bar{Z}^{flow})m_i^{flow}(v^{flow})^2$ is the ram pressure from the side-flow and $\rho^{jet}(v^{jet})^2 =$

Table 2
Similarity of the HH 110/270 System to the Laser-produced Plasmas, with $a = 5 \times 10^{15}$, $b = 4 \times 10^{-17}$, $c = 1.6 \times 10^{-19}$

Parameters	HH 110/270[a]	Laser-produced Plasmas	HH 110/270 (Scaled)
Length (cm)	1e15	2e-1	~2e-1
Mass density (g cm^{-3})	2e-20	5e-4	~5e-4
Pressure (Pa)	1e-9	6e9	~6e9
Time (s)	2e8	2e-9	~2.5e-9

Reference. (a) Hartigan et al. (2009).

$(n_e^{\rm jet}/\bar{Z}^{\rm jet})m_i^{\rm jet}(v^{\rm jet})^2$ is the ram pressure from the supersonic-jet. Obviously, the deflection angle is dependent on the ram pressure of the side-flow, whose characteristic parameters are shown in Figure 4(a). Taking the above experimental and the simulation parameters into the above equation, we estimate the deflection angle to be $52\overset{\circ}{.}50 \pm 4°$, which is also consistent with the experimental and simulated.

We do not take the magnetic field effect on the deflection into account. The possible mechanism for magnetic field generation in our experiment is the Biermann battery effect. For the side-flow from the disk target, the focal spot of the driven laser is as large as about 600 μm. Such a large spot will greatly suppress the self-generated magnetic field due to low density gradient and temperature gradient (Fox et al. 2013). In addition, the lifetime of the magnetic field under similar experimental conditions is found to be shorter than 2 ns (Li et al. 2007; Gao et al. 2015). In our experiment the timing zero is defined with respect to the falling edge of the laser pulses. Our delay of 2 ns is actually equal to their 3 ns if using the same timing definition as theirs. Therefore, the B field will be very weak during the flow–flow collision. For the supersonic-jet, we use two laser bunches focused on the interior surfaces of the K-shaped target, and then two plasma bubbles were generated. The collision between the two bubbles would pinch out a collimated jet. Recent experimental results with similar conditions (Li et al. 2013) show that both bubbles would carry spontaneous magnetic fields from the laser focal spots with them, due to the Biermann battery effect. Magnetic reconnection (MR) would occur when two plasma bubbles collided at the center. After MR, the magnetic fields would have azimuthal components around the jet axis. The field strength was estimated as ~30 T by proto radiography. The azimuthal magnetic field would collimate the jet by the $J \times B$ force, rather than deflect it. Therefore, the self-generation B field would not be important for the jet deflection in our experiment.

Although the space and timescales of laboratory plasmas are different from the astrophysical, similarity criteria allows a direct scaling between the HH 110/270 system and the laboratory plasmas, in which the variables of the systems remain invariant under the following transformation, $r = ar_1$, $\rho = b\rho_1$, $p = cp_1$, $t = a\sqrt{b/c}\,t_1$, $v = \sqrt{c/b}\,v_1$, where a, b, c are transformation coefficients, r is the characteristic length, ρ is the mass density, p is the pressure, t is the characteristic time, and v is the velocity. The parameters with subscript 1 represent ones in laboratory. Table 2 shows the similarity of the HH 110/270 system and laser-produced plasmas, with the transformation coefficients $a = 5 \times 10^{15}$, $b = 4 \times 10^{-17}$, and $c = 4 \times 10^{-19}$. The scaled parameters of the HH 110/270 system in the third column are very similar to those of the laser-produced plasmas in the second column. Our experimental results and the simulation results show that the deflection in the HH 110/270 system can be well-modeled in the laboratory.

4. CONCLUSION

The similarity criteria allow us to investigate supersonic-jet deflection in the HH 110/270 system in the laboratory. A large deflection angle $\theta \sim 55°$ is demonstrated. The 2D LARED-S code is used to simulate the experimental results. The numerical simulations and theoretical analysis show that the deflection jet is induced by the ram pressure from the side-flow. A collisional shock provides the momentum exchange and plays an important role in the jet deflection. The well-modeling experiments support that the HH 110/270 system is shaped by the ram pressure from a flow–flow collision model. Our experiments show that laser-produced jets could provide a test-bed to study astrophysical phenomena, such as the jet propagation and deflection of HH objects.

The authors thank the staff of the SG-II laser facility for operating the laser and target area. This work is supported by the National Basic Research Program of China (grant No. 2013CBA01501) and the National Natural Science Foundation of China (grant Nos. 11135012 and 11220101002).

REFERENCES

Albertazzi, B., Ciardi, A., Nakatsutsumi, M., et al. 2014, Sci, 346, 325
Bally, J., Devine, D., & Alten, V. 1996, ApJ, 473, 921
Balsara, D., & Norman, M. 1992, ApJ, 393, 631
Bellan, P. M., Livio, M., Kato, Y., et al. 2009, PhPl, 16, 041005
Chenais-Popovics, C., Renaudin, P., Rancu, O., et al. 1997, PhPl, 4, 190
Choi, M. 2001, ApJ, 550, 817
Courtois, C., Grundy, R. A. D., Ash, A. D., et al. 2004, PhPl, 11, 3386
De Young, D. S. 1991, ApJ, 371, 69
Dong, Q. L., Wang, S. J., Lu, Q. M., et al. 2012, PhRvL, 108, 215001
Fiedler, R., & Henriksen, R. N. 1984, ApJ, 281, 554
Fox, W., Fiksel, G., Bhattacharjee, A., et al. 2013, PhRvL, 111, 225002
Gao, L., Nilson, P. M., Igumenshchev, I. V., et al. 2015, PhRvL, 114, 215003
Hartigan, P., Foster, J. M., Wilde, B. H., et al. 2009, ApJ, 705, 1073
Huntington, C. M., Fiuza, F., Ross, J. S., et al. 2015, NatPh, 11, 173
Hurka, J. D., Schmid-Burgk, J., & Hardee, P. E. 1999, A&A, 343, 558
Kajdic, P., Reipurth, B., Raga, A. C., et al. 2012, AJ, 143, 106
Kugland, N. L., Ryutov, D. D., Chang, P.-Y., et al. 2012, NatPh, 8, 809
Lebedev, S. V., Ampleford, D., Ciardi, A., et al. 2004, ApJ, 616, 988
Li, C. K., Ryutov, D. D., Hu, S. X., et al. 2013, PhRvL, 111, 235003
Li, C. K., Séguin, F. H., & Frenje, J. A. 2007, PhRvL, 99, 055001
Liu, X., Zhang, Y., Zhong, J. Y., et al. 2011, NJPh, 13, 093001
Loupias, B., Koenig, M., Falize, E., et al. 2007, PhRvL, 99, 265001
Mendoza, S., & Longair, M. S. 2001, MNRAS, 324, 149
Morita, T., Sakawa, Y., Kuramitsu, Y., et al. 2010, PhPl, 17, 122702
O'Dea, C. P., & Owen, F. N. 1986, ApJ, 301, 841
Post, D. E., Jensen, R. V., Tarter, C. B., et al. 1977, ADNDT, 20, 397
Reipurth, B., Raga, A. C., & Heathcote, S. 1996, A&A, 311, 989
Remington, B. A., Arnett, D., Drake, R. P., et al. 1999, Sci, 284, 1488
Remington, B. A., Drake, R. P., & Ryutov, D. D. 2006, RvMP, 78, 755
Ryutov, D., Drake, R. P., Kane, J., et al. 1999, ApJ, 518, 821
Shigemori, K., Kodama, R., Farley, D. R., et al. 2000, PhRvE, 62, 8838
Yan, F., Zhang, J., Dong, Q. L., et al. 2006, PhRvA, 73, 023812
Ye, W. H., Wang, L. F., & He, X. T. 2010, ChPhL, 27, 125203
Ye, W. H., Zhang, W. Y., & He, X. T. 2002, PhRvE, 65, 057401
Yuan, D. W., Li, Y. T., Liu, X., et al. 2013, HEDP, 9, 239
Zheng, C. K. 2009, Plasma Physics (1st ed.; Beijing: Peking Univ. Press)
Zhong, J. Y., Li, Y. T., Wang, X. G., et al. 2010, NatPh, 6, 984

PHYSICS OF PLASMAS 22, 082702 (2015)

Weakly nonlinear Bell-Plesset effects for a uniformly converging cylinder

L. F. Wang (王立锋),[1,2,a)] J. F. Wu (吴俊峰),[1] H. Y. Guo (郭宏宇),[3] W. H. Ye (叶文华),[1,2] Jie Liu (刘杰),[1,2] W. Y. Zhang (张维岩),[1] and X. T. He (贺贤土)[1,2]
[1]Institute of Applied Physics and Computational Mathematics, Beijing 100094, China
[2]HEDPS, Center for Applied Physics and Technology, Peking University, Beijing 100871, China
[3]Graduate School, China Academy of Engineering Physics, Beijing 100088, China

(Received 30 March 2015; accepted 10 July 2015; published online 5 August 2015)

In this research, a weakly nonlinear (WN) model has been developed considering the growth of a small perturbation on a cylindrical interface between two incompressible fluids which is subject to arbitrary radial motion. We derive evolution equations for the perturbation amplitude up to third order, which can depict the linear growth of the fundamental mode, the generation of the second and third harmonics, and the third-order (second-order) feedback to the fundamental mode (zero-order). WN solutions are obtained for a special uniformly convergent case. WN analyses are performed to address the dependence of interface profiles, amplitudes of inward-going and outward-going parts, and saturation amplitudes of linear growth of the fundamental mode on the Atwood number, the mode number (m), and the initial perturbation. The difference of WN evolution in cylindrical geometry from that in planar geometry is discussed in some detail. It is shown that interface profiles are determined mainly by the inward and outward motions rather than bubbles and spikes. The amplitudes of inward-going and outward-going parts are strongly dependent on the Atwood number and the initial perturbation. For low-mode perturbations, the linear growth of fundamental mode cannot be saturated by the third-order feedback. For fixed Atwood numbers and initial perturbations, the linear growth of fundamental mode can be saturated with increasing m. The saturation amplitude of linear growth of the fundamental mode is typically 0.2λ–0.6λ for $m < 100$, with λ being the perturbation wavelength. Thus, it should be included in applications where Bell-Plesset [G. I. Bell, Los Alamos Scientific Laboratory Report No. LA-1321, 1951; M. S. Plesset, J. Appl. Phys. **25**, 96 (1954)] converging geometry effects play a pivotal role, such as inertial confinement fusion implosions. © *2015 AIP Publishing LLC*. [http://dx.doi.org/10.1063/1.4928088]

I. INTRODUCTION

Converging geometry hydrodynamic instabilities are of critical importance in fields of inertial confinement fusion (ICF)[1,2] and astrophysics.[3] It was realized that the nonlinear growth of hydrodynamic instabilities plays a crucial role in the performance degradation of central ignition of ICF implosions.[4–12] In ICF ignition or high-gain target design,[1,2] substantial compression of the fuel is required, and a spherical convergence-ratio C_r (ratio of initial outer capsule radius to the final compressed hot fuel radius) of $\sim$30–40 is necessary to achieve the required conditions. Generally, the dynamic of the shells in ICF implosions consists of four stages:[13,14] shock transit, acceleration, coasting, and deceleration. In the earliest period of time, the initial outer-surface perturbations are susceptible to the ablative Richtmyer-Meshkov (RM) instability which does not significantly amplify the perturbations. For a typical ICF implosion, the shell is accelerated up to $C_r \sim 2$–3. Then, the final compression is achieved during the coasting and deceleration phase. Significant perturbation growth of the interface occurs throughout the acceleration phase due to Rayleigh-Taylor (RT) instability and during the coasting phase due to Bell-Plesset (BP)[15,16] related convergence effects. After the convergent shock reflects from center and returns to the shell, several (typically twice) impulsive deceleration (RM growth) is followed before a continuous deceleration (RT growth).[17] BP convergent effects are magnified during the deceleration phase because of the large C_r.

The effect of convergence on the fluid stability was first introduced by Bell[15] in 1951. Bell considers the growth of small perturbation on a cylindrical or spherical interface in arbitrary radial motion for incompressible and compressible fluids. It is found that when considering perturbed surface in curved geometry, additional effects appear even in the absence of acceleration, which have no equivalent on a plane fluid interface. Later, Plesset[16] investigates the stability conditions for a spherical interface between two incompressible fluids in radial motion and obtains analogous results. This effect is now referred to collectively as the BP effects. Recently, the BP-type treatment was used to assess the instability growth with different flow arrangements in both cylindrical and spherical geometries.[18–20] Mikaelian studies the instability growth of an arbitrary number of cylindrical[18] or spherical[19] concentric shells. Amendt[20] assesses the effects of contiguous density gradients at an accelerating spherical surface. A more general extension of Bell's[15] work on the effect of fluid compressibility for free surface was developed by Amendt[21] which studies time-dependent uniform densities on both sides of a surface in spherical geometry. Epstein[22] investigated the BP effects in planar, cylindrical, and spherical geometries and clarified a clear distinction

a)Electronic mail: wang_lifeng@iapcm.ac.cn

between the BP effect and RT instability. These analyses considered small perturbation in both cylindrical and spherical geometries for compressible and incompressible fluids but addressed only the linear growth regime of BP effects. However, to the best of our knowledge, no study of the contribution to the BP growth due to weakly nonlinear (WN) effects has been performed. In fact, the hydrodynamic instability in the present ICF implosions has entered the highly nonlinear growth regime.[9–11] Thus, it is practically important to construct even a WN model, which can help us at least to better understand the physics of onset of nonlinearity (i.e., mode-coupling process), the results from experiments, the reliability of numerical simulations, and the applicability of phenomenological models for the highly nonlinear regime.

This paper studies the WN effects on BP perturbation growth for semi-infinite incompressible fluids in cylindrical geometry up to third order in amplitude. Focuses are on the WN growth of spatial perturbations on a surface of uniformity compressing cylinder. The remainder of this work is organized as follows. Section II introduces the first-, second-, and third-order governing equations for the single-mode perturbation growth. Section III provides the analytical WN solutions. Section IV discusses the temporal evolution of the perturbed interface, the differences between the amplitude growth of inward-going and outward-moving parts, and the saturation amplitude (SA) of linear growth of the fundamental mode. Finally, conclusions and remarks are given in Section V.

II. THEORETICAL MODEL AND MATHEMATICAL FORMULATION

A cylindrical coordinate (r, θ) is introduced in this work. We consider the problem of the stability of two incompressible, inviscid, irrotational, and immiscible fluids in arbitrary radial motion. It can be stated as follows. A fluid of density ρ^i is contained within a cylinder of radius $R(t)$. The other fluid of density ρ^e occupies the region exterior to this cylinder. The perturbed interface between the two fluids is parameterized as $r(\theta, t) = R(t) + \eta(\theta, t)$, where $\eta(\theta, t)$ is the perturbation displacement. The initial and boundary conditions have cylindrical symmetry, and we want to obtain the equation of motion for the perturbation displacement. We introduce a velocity potential ϕ so that the local fluid velocity generally follows the form $\mathbf{v} = \nabla\phi$. A source or sink of mass at the origin is introduced, since the divergent behavior of ϕ at $r = 0$ [see Equation (6)]. In the following discussion, we shall denote the properties of the exterior fluid ($R < r < \infty$) by the superscript "e" and that of the interior fluid ($0 < r < R$) by the superscript "i." The governing equations for this system are

$$\nabla^2\phi^e = \nabla^2\phi^i = 0, \tag{1}$$

$$\frac{\partial\eta}{\partial t} + \frac{1}{r^2}\frac{\partial\eta}{\partial\theta}\frac{\partial\phi^e}{\partial\theta} - \frac{\partial\phi^e}{\partial r} = 0 \quad \text{at} \quad r = R + \eta, \tag{2}$$

$$\frac{\partial\eta}{\partial t} + \frac{1}{r^2}\frac{\partial\eta}{\partial\theta}\frac{\partial\phi^i}{\partial\theta} - \frac{\partial\phi^i}{\partial r} = 0 \quad \text{at} \quad r = R + \eta, \tag{3}$$

$$\rho^e\left\{\frac{\partial\phi^e}{\partial t} + \frac{1}{2}\left[\left(\frac{\partial\phi^e}{\partial r}\right)^2 + \frac{1}{r^2}\left(\frac{\partial\phi^e}{\partial\theta}\right)^2\right]\right\} - \rho^i\left\{\frac{\partial\phi^i}{\partial t} + \frac{1}{2}\left[\left(\frac{\partial\phi^i}{\partial r}\right)^2 + \frac{1}{r^2}\left(\frac{\partial\phi^i}{\partial\theta}\right)^2\right]\right\} = f(t) \quad \text{at} \quad r = R + \eta, \tag{4}$$

where ϕ^e and ϕ^i are the velocity potentials relating to the fluid velocities, and $f(t)$ is an arbitrary function of time. The potential flow of the two fluids is described by the Laplace equation [Eq. (1)]. Equations (2) and (3) represent the kinematic boundary conditions, namely, the normal component of the velocity for each fluid across the interface should equal to the normal velocity of the interface between two fluids. Equation (4) corresponds to the continuity condition of the pressure across the interface (i.e., Bernoulli equation).

The perturbation displacement and the velocity potentials can be expanded into a power series in a formal parameter ε

$$\eta(\theta, t) = \sum_{n=1}^{\infty}\varepsilon^n\eta^{(n)}(\theta, t) = \sum_{n=1}^{\infty}\varepsilon^n\sum_{l=0}^{\lfloor\frac{n}{2}\rfloor}\eta_{n,n-2l}(t)\cos[(n-2l)m\theta], \tag{5}$$

$$\phi^e(r, \theta, t) = \sum_{n=1}^{\infty}\varepsilon^n(\phi^e)^{(n)}(r, \theta, t) = \sum_{n=1}^{\infty}\varepsilon^n\sum_{l=0}^{\lfloor\frac{n}{2}\rfloor}\phi^e_{n,n-2l}(t)r^{-(n-2l)m}\cos[(n-2l)m\theta], \tag{6}$$

$$\phi^i(r, \theta, t) = \sum_{n=1}^{\infty}\varepsilon^n\left(\phi^i\right)^{(n)}(r, \theta, t) = \sum_{n=1}^{\infty}\varepsilon^n\sum_{l=0}^{\lfloor\frac{n}{2}\rfloor}\phi^i_{n,n-2l}(t)r^{(n-2l)m}\cos[(n-2l)m\theta], \tag{7}$$

where $\eta^{(n)}$ is the nth-order component of the perturbation displacement, $(\phi^e)^{(n)}$ is the nth-order component of the velocity potential of the exterior fluid, and $(\phi^i)^{(n)}$ is the nth-order component of the velocity potential of the interior fluid (n is a natural number). The $\eta_{n,n-2l}$, $\phi^e_{n,n-2l}$, and $\phi^i_{n,n-2l}$ are amplitudes of the $(n-2l)$th Fourier mode in the nth-order of ε ($n = 1, 2, \cdots; l = 0, 1, \cdots, \lfloor n/2\rfloor$). Here, m is the mode number of the perturbation, and the Gauss's symbol $\lfloor n/2\rfloor$ denotes the maximum integer that is less than or equal to $n/2$. For $l = 0$, quantities represent the nth harmonic generation; for $l \geq 1$, they represent the nth-order feedback to the $(n-2l)$th-order. Amplitudes of the $(n-2l)$th Fourier mode in the nth-order of ε are in fact the unknowns that we want to obtain.

As mentioned previously, the boundary conditions [Eqs. (2)–(4)] are based on the perturbed interface. Thus, we first substitute Eqs. (5)–(7) to Eqs. (2)–(4). Second, we generate a power series expansion for resulted governing equations about the point $r = R(t)$ and replace r with $R(t) + \eta(\theta, t)$. Then, we collect terms of the same power in ε to construct the nth-order governing equations. Finally, nth-order governing equations are expressed as a second-order ordinary differential equation (ODE) for the $\eta_{n,n-2l}(t)$ ($n = 1, 2, \cdots$; $l = 0, 1, \cdots, \lfloor n/2 \rfloor$).

At the first-order $O(\varepsilon)$, the second-order ODE for $\eta_{1,1}$ is

$$\ddot{\eta}_{1,1} + 2\frac{\dot{R}}{R}\dot{\eta}_{1,1} + (1 - Am)\frac{\ddot{R}}{R}\eta_{1,1} = 0, \tag{8}$$

where $A = (\rho^e - \rho^i)/(\rho^e + \rho^i)$ is the Atwood number. The first-order equation (8), as far as we know, was first derived by Bell[15] considering RT growth in cylindrical geometry for two specific cases $A = 1$ and -1. In addition, governing equation (8) for linear growth BP effects has been used by Mikaelian considering RT and RM instability in stratified cylindrical shells.[19] It is convenient to make the following substitution:

$$\eta_{1,1}(t) = C_r a(t), \tag{9}$$

where $C_r = R_0/R(t)$ is the convergent ratio with R_0 being the value of $R(t)$ at some fixed time t_0. Equation (8) then reduces to

$$\ddot{a} - \gamma^2 a = 0, \tag{10}$$

where $\gamma^2 = A\frac{m}{R}\ddot{R}$. This equation is easily solvable, which in fact is analogous to the resulting equation in planar geometry.[23]

At the second-order $O(\varepsilon^2)$, the second-order ODE for $\eta_{2,2}$ can be written as

$$\begin{aligned}\ddot{\eta}_{2,2} &+ 2\frac{\dot{R}}{R}\dot{\eta}_{2,2} + (1 - 2Am)\frac{\ddot{R}}{R}\eta_{2,2} \\ &+ Am\eta_{1,1}\frac{1}{2R^2}\left[4\dot{R}\dot{\eta}_{1,1} + \ddot{R}\eta_{1,1}\right] \\ &+ \frac{1}{2R}\left[\ddot{\eta}_{1,1}\eta_{1,1} + (1 + 2Am)\dot{\eta}_{1,1}^2\right] = 0,\end{aligned} \tag{11}$$

which is the second harmonic generation equation. On the other hand, the governing equation for the second-order feedback to the zero harmonic takes the form

$$\dot{\eta}_{2,0} + \frac{\dot{R}}{R}\eta_{2,0} + \frac{1}{2R}\eta_{1,1}\dot{\eta}_{1,1} = 0. \tag{12}$$

Note that the $\eta_{2,0}$ does not appear in the planar geometry.[23]

At the third-order $O(\varepsilon^3)$, the second-order ODE for $\eta_{3,3}$ can be expressed as

$$\begin{aligned}\ddot{\eta}_{3,3} &+ 2\frac{\dot{R}}{R}\dot{\eta}_{3,3} + (1 - 3Am)\frac{\ddot{R}}{R}\eta_{3,3} - \left[(Am - 3m^2)\frac{\dot{R}^2}{4R^4} + Am\frac{\ddot{R}}{4R^3}\right]\eta_{1,1}^3 \\ &- (Am - m^2)\frac{3\dot{R}}{2R^3}\eta_{1,1}^2\dot{\eta}_{1,1} + \frac{3m^2}{4R^2}\eta_{1,1}\dot{\eta}_{1,1}^2 + Am\frac{3\ddot{R}}{2R^2}\eta_{1,1}\eta_{2,2} + 3Am\frac{\dot{R}}{R^2}\eta_{1,1}\dot{\eta}_{2,2} \\ &+ \frac{1}{2R}\eta_{1,1}\ddot{\eta}_{2,2} + 3Am\frac{\dot{R}}{R^2}\dot{\eta}_{1,1}\eta_{2,2} + \frac{1}{2R}\ddot{\eta}_{1,1}\eta_{2,2} + (1 + 3Am)\frac{1}{R}\dot{\eta}_{1,1}\dot{\eta}_{2,2} = 0,\end{aligned} \tag{13}$$

which governs the third harmonic generation. On the other hand, the second-order ODE for $\eta_{3,1}$ reads

$$\begin{aligned}\ddot{\eta}_{3,1} &+ 2\frac{\dot{R}}{R}\dot{\eta}_{3,1} + (1 - Am)\frac{\ddot{R}}{R}\eta_{3,1} + \left((Am - 3m^2)\frac{\dot{R}^2}{4R^4} + m^2\frac{\ddot{R}}{4R^3}\right)\eta_{1,1}^3 \\ &+ \frac{\dot{R}}{2R^3}\eta_{1,1}^2\dot{\eta}_{1,1} + (m^2 + Am - 2)\frac{1}{4R^2}\eta_{1,1}^2\ddot{\eta}_{1,1} + (m^2 + 2Am - 4)\frac{1}{4R^2}\eta_{1,1}\dot{\eta}_{1,1}^2 \\ &+ (Am - 1)\left(\frac{\ddot{R}}{R^2} - \frac{\dot{R}^2}{R^3}\right)\eta_{1,1}\eta_{2,0} + (Am - 1)\frac{\dot{R}}{R^2}(\eta_{1,1}\dot{\eta}_{2,0} + \dot{\eta}_{1,1}\eta_{2,0}) \\ &+ (1 + Am)\frac{1}{R}\dot{\eta}_{1,1}\dot{\eta}_{2,0} + \frac{1}{R}\ddot{\eta}_{1,1}\eta_{2,0} - Am\frac{\dot{R}}{R^2}(\eta_{1,1}\dot{\eta}_{2,2} + \dot{\eta}_{1,1}\eta_{2,2}) - Am\frac{\ddot{R}}{2R^2}\eta_{1,1}\eta_{2,2} \\ &+ \frac{1}{2R}\eta_{1,1}\ddot{\eta}_{2,2} + (1 - 2Am)\frac{1}{2R}\ddot{\eta}_{1,1}\eta_{2,2} + (1 - Am)\frac{1}{R}\dot{\eta}_{1,1}\dot{\eta}_{2,2} = 0,\end{aligned} \tag{14}$$

which describes the third-order feedback to the fundamental mode.

In the present work, we truncate the formal power series in Equations (5)–(7) at the third-order. Nevertheless, the linear growth of the fundamental mode, the generation of second harmonic, the second-order feedback to the zeroth-order (i.e., equilibrium radius R), the generation of the third harmonics, and the third-order feedback to the fundamental mode can be analyzed, respectively, by governing equations (8), (11), (12), (13), and (14). Moreover, the third-order WN model can be extended to higher orders, if needed. However, since the asymptotic series will break down when the interface distortions are significant, it is a reasonable approximation keeping only the first three orders.

III. WN SOLUTIONS

Evolution equations for the growth of a small perturbation on a cylindrical fluid surface which is subject to arbitrary radial motion up to third order are obtained in Section II. We now consider the analytic solutions of the above governing equations. As far as we know, there is no general solution for an arbitrary implosion (explosion) history $R(t)$. Here, we therefore consider a special case $\ddot{R}=0$ and $\dot{R}=V_0=\text{const}$. In other words, we study the case of "undriven growth" in the absence of force that drives the RT instability or accelerationless BP growth. Accordingly, the history $R(t)$ can be expressed as $R(t)=R_0+V_0t$.

Generally, we impose both interface and velocity perturbations in the following form:

$$\eta_{n,n-2l}|_{t=0}=\eta_0\delta_{1n}\delta_{1(n-2l)}, \tag{15}$$

$$\dot{\eta}_{n,n-2l}|_{t=0}=\dot{\eta}_0\delta_{1n}\delta_{1(n-2l)}, \tag{16}$$

where δ_{1n} and $\delta_{1(n-2)}$ are the Kronecker delta function, $n=1, 2, 3$ and $l=0, \ldots, \lfloor n/2 \rfloor$. In fact, we initiate the fundamental mode perturbation alone.

Substituting the expression of $R(t)$ and using the initial conditions, we obtain the linear solution at the first order $O(\varepsilon)$

$$\eta_{1,1}=\eta_0+\dot{\eta}_0tC_r, \tag{17}$$

which is identical to that of Ref. 19 [Equation (A12)]. Note that the first-order solution is independent of the Atwood number. The second-order $O(\varepsilon^2)$ solutions are

$$\eta_{2,2}=\eta_0\dot{\eta}_0t\left(A\frac{m}{R}-\frac{1}{2}\frac{1}{R}\right)(C_r-1)+\dot{\eta}_0^2t^2\left[\left(\frac{1}{6}A\frac{m}{R}-\frac{1}{4}\frac{1}{R}\right)C_r^2-\frac{2}{3}A\frac{m}{R}C_r\right], \tag{18}$$

$$\eta_{2,0}=-\left(\eta_0\dot{\eta}_0t\frac{1}{2}\frac{1}{R}C_r+\dot{\eta}_0^2t^2\frac{1}{4}\frac{1}{R}C_r^2\right), \tag{19}$$

where the explicit $1/R$ and C_r dependencies of $\eta_{2,0}$ and explicit A, m/R, $1/R$, and C_r dependencies of $\eta_{2,2}$ are displayed clearly. At the third-order $O(\varepsilon^3)$, we have

$$\begin{aligned}\eta_{3,3}=&-\eta_0^3\frac{1}{8}\left(3\frac{m^2}{R^2}-A\frac{1}{R}\frac{m}{R}\right)\left(1-2\frac{1}{C_r}+\frac{1}{C_r^2}\right)\\&+\eta_0^2\dot{\eta}_0\frac{1}{8}t\left\{\left[(8A^2-1)\frac{m^2}{R^2}-9A\frac{1}{R}\frac{m}{R}+\frac{2}{R^2}\right]C_r-\left[(16A^2-8)\frac{m^2}{R^2}-8A\frac{1}{R}\frac{m}{R}+\frac{2}{R^2}\right]+\left[(8A^2-7)\frac{m^2}{R^2}+A\frac{1}{R}\frac{m}{R}\right]\frac{1}{C_r}\right\}\\&+\eta_0\dot{\eta}_0^2\frac{1}{24}t^2\left\{\left(9A^2\frac{m^2}{R^2}-26A\frac{1}{R}\frac{m}{R}+\frac{9}{R^2}\right)C_r^2-2\left[(27A^2-3)\frac{m^2}{R^2}-19A\frac{1}{R}\frac{m}{R}+\frac{3}{R^2}\right]C_r\left[5(3A^2-1)\frac{m^2}{R^2}-2A\frac{1}{R}\frac{m}{R}\right]\right\}\\&+\dot{\eta}_0^3t^3\left\{\left(\frac{1}{40}A^2\frac{m^2}{R^2}-\frac{11}{60}A\frac{1}{R}\frac{m}{R}+\frac{1}{8}\frac{1}{R^2}\right)C_r^3-\left(\frac{3}{10}A^2\frac{m^2}{R^2}-\frac{8}{15}A\frac{1}{R}\frac{m}{R}\right)C_r^2+\left[\left(\frac{31}{40}A^2-\frac{1}{8}\right)\frac{m^2}{R^2}+\frac{3}{20}A\frac{1}{R}\frac{m}{R}\right]C_r\right\},\end{aligned} \tag{20}$$

$$\begin{aligned}\eta_{3,1}=&\eta_0^3\frac{1}{8}\left(3\frac{m^2}{R^2}-A\frac{1}{R}\frac{m}{R}\right)\left(1-2\frac{1}{C_r}+\frac{1}{C_r^2}\right)\\&+\eta_0^2\dot{\eta}_0\frac{1}{24}t\left\{-\left[(8A^2-7)\frac{m^2}{R^2}+19A\frac{1}{R}\frac{m}{R}-18\frac{1}{R^2}\right]C_r+2\left[(8A^2-10)\frac{m^2}{R^2}+7A\frac{1}{R}\frac{m}{R}-\frac{3}{R^2}\right]-\left[(8A^2-13)\frac{m^2}{R^2}-5A\frac{1}{R}\frac{m}{R}+\frac{12}{R^2}\right]\frac{1}{C_r}\right\}\\&+\eta_0\dot{\eta}_0^2\frac{1}{24}t^2\left\{\left[(A^2+2)\frac{m^2}{R^2}-22A\frac{1}{R}\frac{m}{R}+27\frac{1}{R^2}\right]C_r^2+2\left[(5A^2-5)\frac{m^2}{R^2}+13A\frac{1}{R}\frac{m}{R}-3\frac{1}{R^2}\right]C_r-\left[(11A^2-5)\frac{m^2}{R^2}-2A\frac{1}{R}\frac{m}{R}\right]\right\}\\&+\dot{\eta}_0^3t^3\left\{\left(\frac{1}{120}A^2\frac{m^2}{R^2}-\frac{11}{60}A\frac{1}{R}\frac{m}{R}+\frac{3}{8}\frac{1}{R^2}\right)C_r^3-\left[\left(\frac{1}{60}A^2+\frac{1}{12}\right)\frac{m^2}{R^2}-\frac{11}{30}A\frac{1}{R}\frac{m}{R}\right]C_r^2-\left[\left(\frac{19}{120}A^2-\frac{1}{24}\right)\frac{m^2}{R^2}+\frac{1}{60}A\frac{1}{R}\frac{m}{R}\right]C_r\right\},\end{aligned} \tag{21}$$

where the explicit A, A^2, m/R, $m^2/R^2\,1/R$, $1/R^2$, $1/C_r$, $1/C_r^2$, C_r, C_r^2, and C_r^3 dependencies of the $\eta_{3,3}$ ($\eta_{3,1}$) are displayed clearly. Note that the third-order feedback to the fundamental mode can be positive, especially for small mode numbers. Now, up to the third-order correction, the interfacial profile can be expressed as

$$r(\theta,t)=\bar{R}(t)+\bar{\eta}(\theta,t), \tag{22}$$

where $\bar{R}(t)=R(t)+\eta_{2,0}$ is the averaged radius of the interface, and $\bar{\eta}(\theta,t)=\eta_1(t)\cos(m\theta)+\eta_{2,2}(t)\cos(2m\theta)+\eta_{3,3}(t)\cos(3m\theta)$ with $\eta_1(t)=\eta_{1,1}(t)+\eta_{3,1}(t)$ being the amplitude of the fundamental mode including the third-order correction. In the following discussion, the perturbation wavelength is defined as $\lambda = 2\pi\bar{R}/m$.

We make a specific study for the case when $R(t) = R_0$ (i.e., ignoring convergence effects). Let $C_r = 1$ (i.e., $V_0 = 0$), the third-order WN solutions reduce to

$$\eta_{1,1} = \eta_0 + \dot{\eta}_0 t, \tag{23}$$

$$\eta_{2,2} = -\frac{1}{4R_0}\dot{\eta}_0^2 t^2 (2Am + 1), \tag{24}$$

$$\eta_{2,0} = -\frac{1}{4R_0}\dot{\eta}_0 t(\dot{\eta}_0 t + 2\eta_0), \tag{25}$$

$$\eta_{3,3} = \frac{1}{8R_0^2}\dot{\eta}_0^2 t^2 \{\dot{\eta}_0 t\left[(4A^2 - 1)m^2 + 4Am + 1\right] - \eta_0(3m^2 - 2Am - 1)\}, \tag{26}$$

$$\eta_{3,1} = -\frac{1}{24R_0^2}\dot{\eta}_0^2 t^2 \{\dot{\eta}_0 t[(4A^2 + 1)m^2 - 4Am - 9] + 3\eta_0(m^2 - 2Am - 7)\}, \tag{27}$$

which are identical to the corresponding ones from Ref. 27 for cylindrical geometry effects on RM instability. Moreover, let $\dot{\eta}_0 = 1$, the main results in Ref. 28 are recovered, which study on a vortex sheet due to incompressible RM instability in cylindrical geometry.

IV. WEAKLY NONLINEAR ANALYSIS

In Section III, we have obtained the WN solutions for the growth of a small perturbation on a cylindrical fluid surface which is subject to a uniformly radial motion (collapsing or expanding). Our interest is primarily in ICF implosions. Here, we focus on the uniformly cylindrical convergent case, and hence the value of V_0 is set to be negative in the following discussions. That is to say, we study the growth of perturbation caused purely by converging geometry effects in this section.

For simplicity, we will consider the WN perturbation growth due to pure BP effects initiated by only velocity perturbation. Temporal evolution of position of the perturbed interfaces is given in Section IV A. Section IV B provides the temporal evolution of the amplitudes of the inward-going and outward-going parts. Section IV C presents the saturation amplitude of linear growth of the perturbation fundamental mode.

A. Interfacial profiles

Now we analyse the temporal evolution of the interface position of the BP perturbation growth in the WN regime. In Figures 1 and 2, we illustrate the WN development of the BP growth with Atwood numbers $A = \pm 1$ for various modes $m = 1, 2, 3, 4, 5, 6, 8$, and 16. The average radius of the interface $\bar{R}(t)$ is also plotted for comparison purpose. For $A = 1$ ($A = -1$), the inward-going and outward-moving parts correspond to the spike (bubble) and bubble (spike), respectively. In planar geometry, the WN growth of bubbles and spikes is quite different from each other, where the bubble tip is broadened but the spike head becomes narrow.[23] However, as shown in Figures 1 and 2, the profile of the ingoing spike (outgoing bubble) for $A = 1$ is not so different from those of the ingoing bubble (outgoing spike) for $A = -1$. In cylindrical geometry, the inward development part is extruded becoming narrow head, while the outward development part is stretched resulting broadened tip. Thus, interface profile is determined mainly by the inward or outward going rather than by bubbles or spikes, especially for low-mode perturbations. This is similar to our previous investigation considering the RT growth on a fixed cylindrical interface.[24]

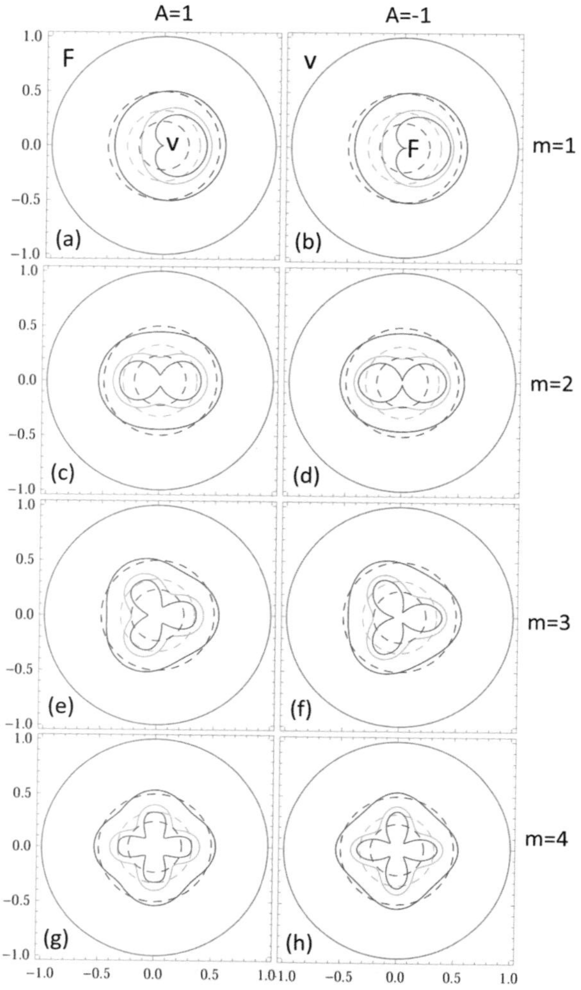

FIG. 1. Interfacial profiles for modes $m = 1$ [(a), (b)], $m = 2$ [(c), (d)], $m = 3$ [(e), (f)], and $m = 4$ [(g), (h)] with $A = 1$ [(a), (c), (e), (g)] and $A = -1$ [(b), (d), (f), (h)], initiated by only velocity perturbation, at $C_r = 1$, 2, 3, and 4, respectively. The initial velocity perturbation amplitudes for $m = 1$, 2, 3, and 4 are $\dot{\eta}_0 = -0.55V_0$, $-0.50V_0$, $-0.45V_0$, and $-0.40V_0$, respectively. Note that the colors (red, blue, cyan, and purple) correspond to the values of convergence ($C_r = 1$, 2, 3, and 4). The letters "F" and "V" represent fluid and vacuum, respectively. The average radius of the interface $\bar{R}(t)$ is also plotted by dashed lines.

B. Amplitudes of inward-going and outward-moving parts

This section is devoted to the temporal evolution of amplitudes of inward-going and outward-going parts. The

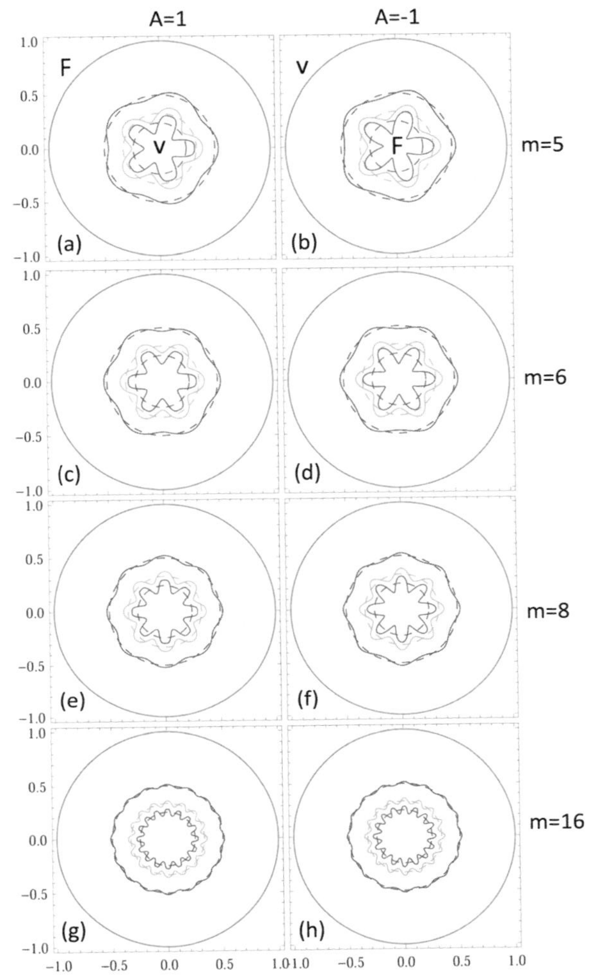

FIG. 2. (Continuation from Figure 1) Interfacial profiles for modes $m = 5$ [(a), (b)], $m = 6$ [(c), (d)], $m = 8$ [(e), (f)], and $m = 16$ [(g), (h)] with $A = 1$ [(a), (c), (e), (g)] and $A = -1$ [(b), (d), (f), (h)], initiated by only velocity perturbation, at $C_r = 1$, 2, 3, and 4, respectively. The initial velocity perturbation amplitudes for $m = 5$, 6, 8, and 16 are $\dot{\eta}_0 = -0.35V_0$, $-0.30V_0$, $-0.25V_0$, and $-0.15V_0$, respectively. Note that the colors (red, blue, cyan, and purple) correspond to the values of convergence ($C_r = 1$, 2, 3, and 4). The letters "F" and "V" represent fluid and vacuum, respectively. The average radius of the interface $\bar{R}(t)$ is also plotted by dashed lines.

amplitudes of outward-going and inward-going parts are $\bar{\eta}(\theta, t)|_{\theta=2l\pi/m}$ and $\bar{\eta}(\theta, t)|_{\theta=(2l+1)\pi/m}$ $(l = 0, 1, \cdots, m-1)$, respectively.

Figure 3 displays the dependence of normalized amplitudes of inward-going parts (η_{in}/λ) and outward-going parts (η_{out}/λ) on the Atwood number (A) and the initial velocity perturbation amplitude ($\dot{\eta}_0$). As can be seen in Figure 3(a), for $m = 1$, the ingoing (spike) amplitude is larger than the outgoing (bubble) amplitude; for $m = 2$–64, the ingoing (spike) amplitude becomes smaller than the outgoing (bubble) amplitude; for $m = 200$, the ingoing (spike) amplitude is larger than the outgoing (bubble) amplitude. The amplitude growth of outgoing and ingoing parts is determined by the growth of the second and third harmonics. The positive amplitude of the second (third) harmonic increases the amplitude of outward-going parts but decreases that of inward-going parts. For $m = 1$ and $m = 200$, the amplitudes of the second and third harmonics are negative, as shown in Figure 4(a), and hence the amplitude of inward-going part is larger than that of outward-going part. Moreover, as shown in Figure 4(a), for $m = 2$–64, the absolute value of the amplitude of the second harmonic (positive) is larger than that of the second harmonic (negative) at fixed C_r, and hence the amplitude of outward-going part grows quicker than that of inward-going part. As can be seen in Figure 3(b), the amplitudes of the inward-going part are larger than that of the outward-going part, which can be explained by the amplitudes of the second and third harmonics as shown in Figure 4(b).

As can be seen in Figure 3(c), the amplitudes of inward-going parts are larger than that of outward-going parts for $m = 1$–64. With increasing $\dot{\eta}_0$, only for $m = 1$–4 the amplitudes of inward-going parts are larger than that of outward-going parts, as shown in Figure 3(d).

As we know from previous work,[23] the amplitude of spike grows quicker than that of bubble in planar geometry. As shown in Figures 3(a) and 3(b), the amplitude of ingoing spike is larger than that of outgoing bubble for $m = 200$. On the other hand, as shown in Figures 3(c) and 3(d), the amplitude of outgoing spike is larger than that of ingoing bubble for $m = 200$. That is to say, the effects of BP growth on the temporal evolution of amplitudes of bubbles and spikes become weakened with increasing mode number, and the planar approximation may be suitable for the analysis of the physics when $m \to \infty$.

In a word, as for low-mode spatial perturbations of a uniformity compressing cylindrical shell, at the outer surface the ingoing bubbles grow faster than the outgoing spikes, but at the inner surface the ingoing spikes grow faster than the outgoing bubbles. Since the excessive growth of bubbles at the outer surface of the shell can rupture the shell, and the severe growth of spikes at the inner surface of the shell can delay the hot-spot formation or even quench the hot-spot, resulting in ignition failure. Therefore, the growth of low-mode perturbations should be strictly controlled during ICF implosions.

C. Saturation amplitude

In this section, we study the SA[1,2,23,24] of linear growth of the perturbation fundamental mode considering the third-order correction. The SA of the fundamental mode, η_s, can be defined as the linear growth amplitude of the fundamental mode at the saturation time, when the growth of the fundamental mode η_1 is reduced by 10% in comparison to its corresponding linear growth η_L, where $\eta_L = \eta_{1,1}$. The saturation time t_s can be obtained by solving $(\eta_L - \eta_1)/\eta_L = 1/10$. Then, substituting t_s into $\eta_L(t)$, the η_s is obtained as $\eta_L(t_s)$. The analytic solution of t_s is always hard to obtain. In fact, we obtain η_s numerically.

Figure 5 shows the initial velocity perturbation amplitude ($\dot{\eta}_0$) dependence of the normalized SA (η_s/λ) at fixed Atwood numbers $A = \pm 1$. As mentioned above, the third-order feedback to the fundamental mode can be positive for

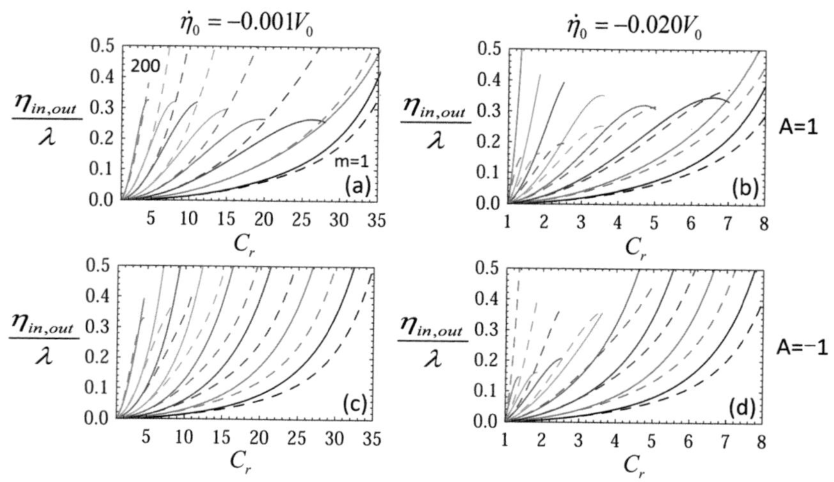

FIG. 3. Normalized amplitudes of inward-going parts η_{in}/λ (solid lines) and outward-going parts η_{out}/λ (dashed lines) for Atwood numbers $A = 1$ (a, b) and -1 (c, d) with initial velocity perturbation amplitudes $\dot{\eta}_0 = -0.001V_0$ (a, c) and $-0.020V_0$ (b, d) at mode numbers $m = 1$ (black), 2 (red), 4 (blue), 8 (magenta), 16 (cyan), 32 (brown), 64 (pink), and 200 (green), respectively.

small mode numbers [see Equation (21)]. As can be seen from Figure 5(a) for $A = 1$, the linear growth of the fundamental mode of perturbation mode number $m = 1$ and $m = 2$ cannot be saturated by the third-order feedback which in fact is positive at different initial velocity perturbation amplitudes ($\dot{\eta}_0$) [see Figures 6(a) and 6(b)]. For small $\dot{\eta}_0$ (e.g., $\dot{\eta}_0 = -0.001V_0$), the η_s/λ first decreases with increasing m (arriving at a minimum), then increases with increasing m (arriving at a maximum), and finally decreases with increasing m. With increasing $\dot{\eta}_0$ (e.g., $\dot{\eta}_0 = -0.02V_0$), the m dependence of η_s/λ becomes simple, namely, the η_s/λ decreases with increasing m. For a fixed m, the η_s/λ decreases with increasing $\dot{\eta}_0$. In the limit $m \to \infty$, there are $\eta_s/\lambda \to 0.11$.

For $A = -1$, the η_s/λ decreases with increasing m at different $\dot{\eta}_0$, as shown in Figure 5(b). However, the mode number of perturbations with positive third-order feedback is dependent on the $\dot{\eta}_0$ [see Figures 6(c) and 6(d)]. The smallest m can be saturated by the third-order feedback is $m = 86$, 50, 27, 18, and 12 for $\eta_0 = -0.001V_0$, $-0.002V_0$, $-0.005V_0$, $-0.01V_0$, and $-0.02V_0$, respectively. There are also $\eta_s/\lambda \to 0.11$ when $m \to \infty$.

In short, the linear growth of fundamental mode for low-mode perturbations of the surface in a uniformity compressing cylindrical shell cannot be saturated by the third-order feedback. It is generally recognized that the linear (exponential) growth of single-mode RT perturbation begins to saturate when its amplitude is $\sim 0.1\lambda$.[23–25] The mixing models (e.g., Youngs' model[29,30] or Haan saturation model[31]) are usually based on the notion that the after short linear (exponential) growth of imposed perturbations, the nonlinear saturation occurs entering a constant terminal velocity phase.[32,33] However, the growth due to BP effects in converging geometry is in a different way. We can say, in some sense, that the growth of low-mode perturbations in ICF implosions (especially for high convergence-ratio target) is severe and hard to control, since the growth of low-mode perturbations due to BP effects cannot be saturated. This may be consistent with the simulation results of the recent high-foot implosions on the National Ignition Facility (NIF).[4] In high-foot implosions on NIF, the ablation-front instability has been controlled effectively but the low-mode perturbations at the inner surface still has a large growth

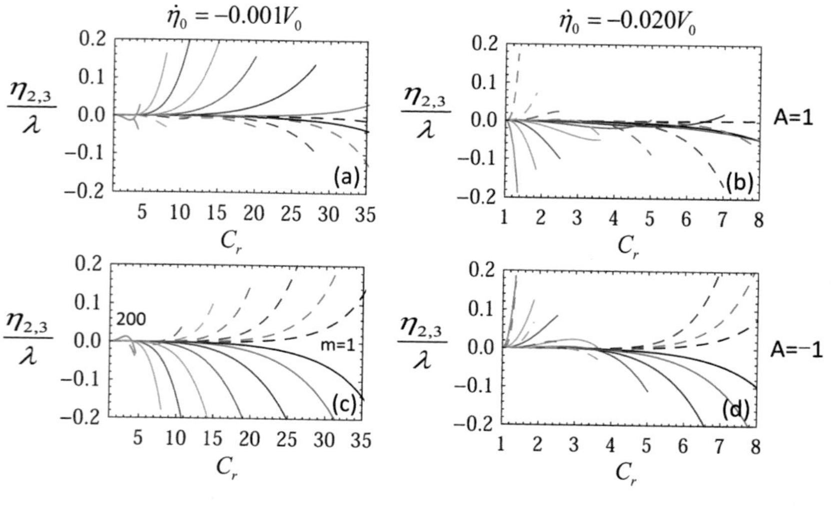

FIG. 4. Normalized amplitudes of the second harmonic η_2/λ (solid lines) and the third harmonic η_3/λ (dashed lines) for Atwood numbers $A = 1$ (a, b) and -1 (c, d) with initial velocity perturbation amplitudes $\dot{\eta}_0 = -0.001V_0$ (a, c) and $-0.020V_0$ (b, d) at mode numbers $m = 1$ (black), 2 (red), 4 (blue), 8 (magenta), 16 (cyan), 32 (brown), 64 (pink), and 200 (green), respectively.

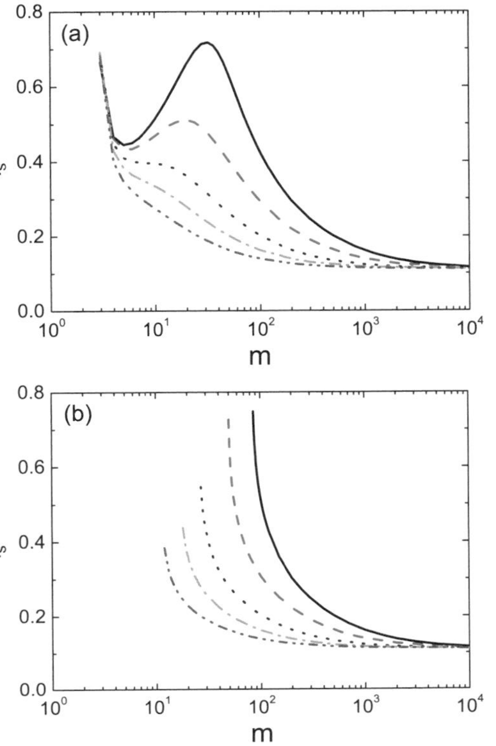

FIG. 5. The normalized saturation amplitude (η_s/λ) of linear growth of the perturbation fundamental mode for Atwood number $A = 1$ (a) and -1 (b) with initial velocity perturbation amplitudes $\dot{\eta}_0 = -0.001V_0$ (solid lines), $-0.002V_0$ (dashed lines), $-0.005V_0$ (dotted lines), $-0.01V_0$ (dashed-dotted lines), and $-0.02V_0$ (dashed-dotted-dotted lines).

leading to severe hot-spot interface distortion.[34] As for the ICF ignition target design, it is worthwhile to point out that a particular attention must be given to the control of the growth of low-mode perturbations, even for the design of high-adiabat implosions.

V. SUMMARY AND CONCLUSIONS

In this article, a WN model for BP perturbation growth has been developed in cylindrical geometry under assumptions that the motion is irrotational and incompressible. This model includes nonlinearity up to the third order in amplitude, which can be used to describe the saturation of linear growth of the fundamental mode and to determine the WN growth of the perturbation. WN solutions for a special uniformly convergent case (i.e., purely converging geometry effects) are obtained. The WN growth of the interfacial profile due to BP effects is determined mainly by the inward or outward going rather than bubbles or spikes. The amplitudes of inward-going and outward-going parts are strongly dependent on the initial perturbations, the mode number, and the Atwood number. Moreover, the linear growth of fundamental mode for low-mode perturbations cannot be saturated by the third-order corrections.

Our interest is primarily in converging geometry hydrodynamic instabilities during ICF implosion. Although ICF capsules are spherical shell, the main convergence effects are captured in cylindrical geometry at reduced levels.[19] However, the exterior fluid in our model has been assumed as semi-infinite, and the velocity potential implies a source or sink at the origin. Critical points about the direct applications of the present analysis to hydrodynamic instabilities in ICF implosions are assumptions that the perturbation wavelength is small compared to the thickness of the shell and that the densities of fluids bordering the shell are negligible small. In fact, when the perturbation wavelength is of the same magnitude as the shell thickness, we expect some modifications in amplitude growth. In our previous work, this problem for a plane slab has been analyzed in detail.[25] Additionally, compressibility effects are ignored in the present model, which plays an important role in converging geometry hydrodynamic instabilities during ICF implosions.[26] In fact, the effect of compressibility becomes much more pronounced in the deceleration stage ($C_r > 3$) of ICF implosions. The effect of compressibility is predicted to dramatically mitigate the linear

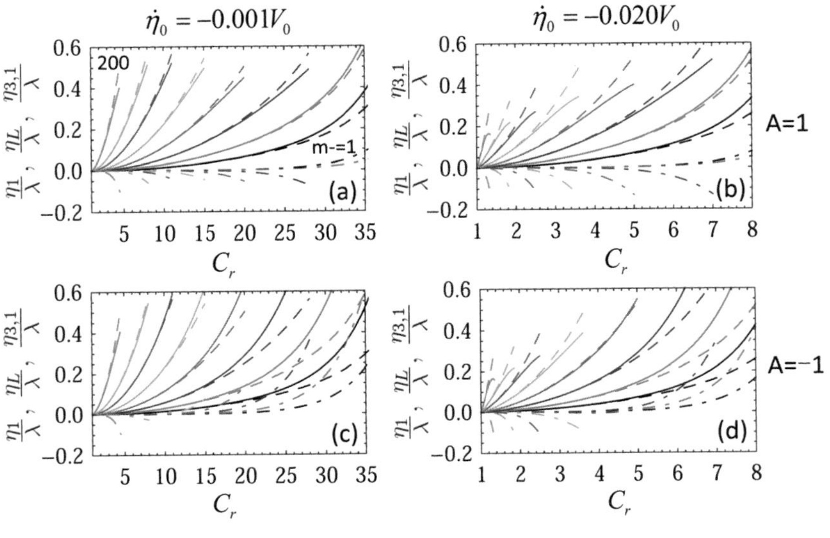

FIG. 6. Normalized amplitudes of the fundamental mode including the third-order correction η_1/λ (solid lines), the linear growth of the fundamental mode η_L/λ (dashed lines), and the third-order feedback to the fundamental mode $\eta_{3,1}/\lambda$ (dashed-dotted lines) for Atwood numbers $A = 1$ (a, b) and -1 (c, d) with initial velocity perturbation amplitudes $\dot{\eta}_0 = -0.001V_0$ (a, c) and $-0.020V_0$ (b, d) at mode numbers $m = 1$ (black), 2 (red), 4 (blue), 8 (magenta), 16 (cyan), 32 (brown), 64 (pink), and 200 (green), respectively.

growth of BP growth.[21] Moreover, the influence of compressibility on the nonlinear mode coupling due to BP growth will be pursued in our future research.

ACKNOWLEDGMENTS

We are grateful to an anonymous referee for useful comments and suggestions that improved this paper. This work was supported by the foundation of CAEP (No. 2014A0202010), the Foundation of President of Chinese Academy of Engineering Physics (Grant No. 2014-1-040), the National Natural Science Foundation of China (Grant Nos. 11275031, 11475034, 11475032, and 11274026), and National Basic Research Program of China (Grant No. 2013CB834100).

[1] J. D. Lindl, P. Amendt, R. L. Berger, S. G. Glendinning, S. H. Glenzer, S. W. Haan, R. L. Kauffman, O. L. Landen, and L. J. Suter, Phys. Plasmas **11**(2), 339 (2004).
[2] S. Atzeni and J. Meyer-ter-Vehn, *The Physics of Inertial Fusion: Beam Plasma Interaction Hydrodynamics, Hot Dense Mater* (Oxford University, Oxford, 2004).
[3] B. A. Remington, R. P. Drake, and D. D. Ryutov, Rev. Mod. Phys. **78**, 755 (2006).
[4] O. A. Hurricane, D. A. Callahan, D. T. Casey, E. L. Dewald, T. R. Dittrich, T. Döppner, M. A. Barrios Garcia, D. E. Hinkel, L. F. Berzak Hopkins, P. Kervin, J. L. Kline, S. L. Pape, T. Ma, A. G. MacPhee, J. L. Milovich, J. Moody, A. E. Pak, P. K. Patel, H.-S. Park, B. A. Remington, H. F. Robey, J. D. Salmonson, P. T. Springer, R. Tommasini, L. R. Benedetti, J. A. Caggiano, P. Celliers, C. Cerjan, R. Dylla-Spears, D. Edgell, M. J. Edwards, D. Fittinghoff, G. P. Grim, N. Guler, N. Izumi, J. A. Frenje, M. Gatu Johnson, S. Haan, R. Hatarik, H. Herrmann, S. Khan, J. Knauer, B. J. Kozioziemski, A. L. Kritcher, G. Kyrala, S. A. Maclaren, F. E. Merrill, P. Michel, J. Ralph, J. S. Ross, J. R. Rygg, M. B. Schneider, B. K. Spears, K. Widmann, and C. B. Yeamans, Phys. Plasmas **21**, 056314 (2014).
[5] V. N. Goncharov, T. C. Sangster, R. Betti, T. R. Boehly, M. J. Bonino, T. J. B. Collins, R. S. Craxton, J. A. Delettrez, D. H. Edgell, R. Epstein, R. K. Follett, C. J. Forrest, D. H. Froula, V. Yu. Glebov, D. R. Harding, R. J. Henchen, S. X. Hu, I. V. Igumenshchev, R. Janezic, J. H. Kelly, T. J. Kessler, T. Z. Kosc, S. J. Loucks, J. A. Marozas, F. J. Marshall, A. V. Maximov, R. L. McCrory, P. W. McKenty, D. D. Meyerhofer, D. T. Michel, J. F. Myatt, R. Nora, P. B. Radha, S. P. Regan, W. Seka, W. T. Shmayda, R. W. Short, A. Shvydky, S. Skupsky, C. Stoeckl, B. Yaakobi, J. A. Frenje, M. Gatu-Johnson, R. D. Petrasso, and D. T. Casey, Phys. Plasmas **21**, 056315 (2014).
[6] D. T. Casey, V. A. Smalyuk, K. S. Raman, J. L. Peterson, L. Berzak Hopkins, D. A. Callahan, D. S. Clark, E. L. Dewald, T. R. Dittrich, S. W. Haan, D. E. Hinkel, D. Hoover, O. A. Hurricane, J. J. Kroll, O. L. Landen, A. S. Moore, A. Nikroo, H.-S. Park, B. A. Remington, H. F. Robey, J. R. Rygg, J. D. Salmonson, R. Tommasini, and K. Widmann, Phys. Rev. E **90**, 011102 (2014).
[7] V. A. Smalyuk, D. T. Casey, D. S. Clark, M. J. Edwards, S. W. Haan, A. Hamza, D. E. Hoover, W. W. Hsing, O. Hurricane, J. D. Kilkenny, J. Kroll, O. L. Landen, A. Moore, A. Nikroo, L. Peterson, K. Raman, B. A. Remington, H. F. Robey, S. V. Weber, and K. Widmann, Phys. Rev. Lett. **112**, 185003 (2014).
[8] K. S. Raman, V. A. Smalyuk, D. T. Casey, S. W. Haan, D. E. Hoover, O. A. Hurricane, J. J. Kroll, A. Nikroo, J. L. Peterson, B. A. Remington, H. F. Robey, D. S. Clark, B. A. Hammel, O. L. Landen, M. M. Marinak, D. H. Munro, K. J. Peterson, and J. Salmonson, Phys. Plasmas **21**, 072710 (2014).
[9] C. R. Weber, D. S. Clark, A. W. Cook, D. C. Eder, S. W. Haan, B. A. Hammel, D. H. Hinkel, O. S. Jones, M. M. Marinak, J. L. Milovich, P. K. Patel, H. F. Robey, J. D. Salmonson, S. M. Sepke, and C. A. Thomas, Phys. Plasmas **22**, 032702 (2015).
[10] D. S. Clark, M. M. Marinak, C. R. Weber, D. C. Eder, S. W. Haan, B. A. Hammel, D. E. Hinkel, O. S. Jones, J. L. Milovich, P. K. Patel, H. F. Robey, J. D. Salmonson, S. M. Sepke, and C. A. Thomas, Phys. Plasmas **22**, 022703 (2015).
[11] D. S. Clark, D. E. Hinkel, D. C. Eder, O. S. Jones, S. W. Haan, B. A. Hammel, M. M. Marinak, J. L. Milovich, H. F. Robey, L. J. Suter, and R. P. J. Town, Phys. Plasmas **20**, 056318 (2013).
[12] V. A. Smalyuk, S. V. Weber, D. T. Casey, D. S. Clark, J. E. Field, S. W. Haan, B. A. Hammel, A. V. Hamza, D. E. Hoover, O. L. Landen, A. Nikroo, H. F. Robey, and C. R. Weber, Phys. Plasmas **22**, 072704 (2015).
[13] P. B. Radha, V. N. Goncharov, T. J. B. Collins, J. A. Delettrez, Y. Elbaz, V. Yu. Glebov, R. L. Keck, D. E. Keller, J. P. Knauer, J. A. Marozas, F. J. Marshall, P. W. McKenty, D. D. Meyerhofer, S. P. Regan, T. C. Sangster, D. Shvarts, S. Skupsky, Y. Srebro, R. P. J. Town, and C. Stoeckl, Phys. Plasmas **12**, 032702 (2005).
[14] L. Welser-Sherrill, R. C. Mancini, D. A. Haynes, S. W. Haan, I. E. Golovkin, J. J. MacFarlane, P. B. Radha, J. A. Delettrez, S. P. Regan, J. A. Koch, N. Izumi, R. Tommasini, and V. A. Smalyuk, Phys. Plasmas **14**, 072705 (2007).
[15] G. I. Bell, Los Alamos Scientific Laboratory Report No. LA-1321, 1951.
[16] M. S. Plesset, J. Appl. Phys. **25**, 96 (1954).
[17] M. Temporal, S. Jaouen, L. Masse, and B. Canaud, Phys. Plasmas **13**, 122701 (2006).
[18] K. O. Mikaelian, Phys. Rev. A **42**, 3400 (1990).
[19] K. O. Mikaelian, Phys. Fluids **17**, 094105 (2005).
[20] P. Amendt, Phys. Plasmas **13**, 042702 (2006).
[21] P. Amendt, J. D. Colvin, J. D. Ramshaw, H. F. Robey, and O. L. Landen, Phys. Plasmas **10**(3), 820 (2003).
[22] R. Epstein, Phys. Plasmas **11**(11), 5114 (2004).
[23] L. F. Wang, J. F. Wu, Z. F. Fan, W. H. Ye, X. T. He, W. Y. Zhang, Z. S. Dai, J. F. Gu, and C. Xue, Phys. Plasmas **19**, 112706 (2012).
[24] L. F. Wang, J. F. Wu, W. H. Ye, W. Y. Zhang, and X. T. He, Phys. Plasmas **20**, 042708 (2013).
[25] L. F. Wang, H. Y. Guo, J. F. Wu, W. H. Ye, J. Liu, W. Y. Zhang, and X. T. He, Phys. Plasmas **21**, 122710 (2014).
[26] V. N. Goncharov, P. McKenty, S. Skupsky, R. Betti, R. L. McCrory, and C. Cherfile-Clérouin, Phys. Plasmas **7**(12), 5118 (2000).
[27] W. H. Liu, X. T. He, and C. P. Yu, Phys. Plasmas **19**, 072108 (2012).
[28] C. Matsuoka and K. Nishihara, Phys. Rev. E **74**, 066303 (2006).
[29] D. Youngs, Physica D **12**, 32 (1984).
[30] D. Youngs, Physica D **37**, 270 (1989).
[31] S. W. Haan, Phys. Rev. A **39**, 5812 (1989).
[32] Q. Zhang, Phys. Rev. Lett. **81**(16), 3391 (1998).
[33] V. N. Goncharov, Phys. Rev. Lett. **88**(13), 134502 (2002).
[34] T. R. Dittrich, O. A. Hurricane, D. A. Callahan, E. L. Dewald, T. Döppner, D. E. Hinkel, L. F. Berzak Hopkins, S. LePape, T. Ma, J. Milovich, J. C. Moreno, P. K. Patel, H.-S. Park, B. A. Remington, and J. Salmonson, Phys. Rev. Lett. **112**, 055002 (2014).

J. Fluid Mech. (2016), *vol.* 786, R6, doi:10.1017/jfm.2015.681

JFM RAPIDS
journals.cambridge.org/rapids

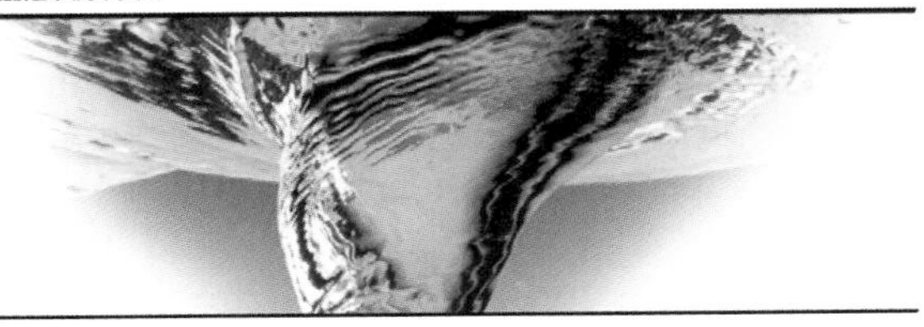

Intermittency caused by compressibility: a Lagrangian study

Yantao Yang[1,2], **Jianchun Wang**[1], **Yipeng Shi**[1,†], **Zuoli Xiao**[1], **X. T. He**[1] **and Shiyi Chen**[1,3,4]

[1]State Key Laboratory for Turbulence and Complex System, HEDPS and Center for Applied Physics and Technology, College of Engineering, Peking University, Beijing 100871, PR China

[2]Physics of Fluids Group, Faculty of Science and Technology, MESA+ Research Institute, and J. M. Burgers Centre for Fluid Dynamics, University of Twente, PO Box 217, 7500 AE Enschede, The Netherlands

[3]South University of Science and Technology of China, Shenzhen 518055, PR China

[4]Collaborative Innovation Center of Advanced Aero-Engine, Beijing 100191, PR China

(Received 10 September 2015; revised 16 October 2015; accepted 12 November 2015; first published online 7 December 2015)

We investigate how compressibility affects the turbulent statistics from a Lagrangian point of view, particularly in the parameter range where the flow transits from the incompressible type to a state dominated by shocklets. A series of three-dimensional simulations were conducted for different types of driving and several Mach numbers. For purely solenoidal driving, as the Mach number increases a new self-similar region first emerges in the Lagrangian structure functions at sub-Kolmogorov time scale and gradually extends to larger time scale. In this region the relative scaling exponent saturates and the saturated value decreases as the compressibility becomes stronger, which can be attributed to the shocklets. The scaling exponent for the inertial range is still very close to that of incompressible turbulence for small Mach number, and discrepancy becomes visible when the Mach number is high enough. When the driving force is dominated by the compressive component the shocklet-induced self-similar region occupies a much wider range of time scales than that in the purely solenoidal driving case. Regardless of the type of driving force, the probability density functions of the velocity increment collapse onto one another for the time scales in the new self-similar region after proper normalization.

Key words: compressible turbulence, homogeneous turbulence, intermittency

† Email address for correspondence: ypshi@coe.pku.edu.cn

1. Introduction

Compressible turbulence is of great relevance in many natural phenomena and engineering applications, such as in supersonic interstellar turbulence (Padoan & Nordlund 2002; Scalo & Elmegreen 2004; McKee & Ostriker 2007; Federrath *et al.* 2010; Hennebelle & Falgarone 2012) and in the flow around hypersonic aircraft. While incompressible turbulent flows are characterized by the nonlinear interactions among vortices within a wide range of scales (Frisch 1995), compressible turbulence is even more complex due to the appearance of the shocklet structures (Samtaney, Pullin & Kosović 2001; Pan, Padoan & Kritsuk 2009; Wang *et al.* 2011). Substantial progress has been made recently in better understanding compressible turbulence, such as the scaling and statistics (Kritsuk *et al.* 2007; Schmidt, Federrath & Klessen 2008; Wang *et al.* 2012), and two-point correlations (Galtier & Banerjee 2011; Wagner *et al.* 2012; Banerjee & Galtier 2013). It is now believed that an inertial energy cascade also exists in compressible turbulence (Aluie 2011; Aluie, Li & Li 2012; Kritsuk, Wagner & Norman 2013; Wang *et al.* 2013*a*).

Turbulent statistics are known to be intermittent, and the structure functions exhibit anomalous scaling (Frisch 1995), which has also been observed in compressible turbulence (Kritsuk *et al.* 2007; Schmidt *et al.* 2008; Galtier & Banerjee 2011; Wagner *et al.* 2012). For incompressible turbulence, Lagrangian study, i.e. investigation of statistical properties by following particles advected by turbulence, has provided many interesting and valuable results in the past few years (Toschi & Bodenschatz 2009). For very weakly compressible turbulence with Mach number of order 0.3, both the Eulerian and the Lagrangian statistics are very close to those of the incompressible flow (Benzi *et al.* 2008, 2010). Recent research at much higher Mach number, however, has revealed that the compressibility has a pronounced influence on the statistics (Konstandin *et al.* 2012; Yang *et al.* 2013).

The statistics of compressible turbulence depends not only on the flow parameters such as the Mach number and the Reynolds number, but also on the nature of the external driving force. Federrath (2013) conducted numerical simulations at very high Mach number of order 17 for both solenoidal (divergence-free) driving and compressive (curl-free) driving. Compressive driving produces a much more intermittent density distribution than solenoidal driving. Similar behaviour has been reported at lower Mach number of order 5 in Konstandin *et al.* (2012), in which the authors discovered that the structure functions have more intermittent scalings for compressive driving than solenoidal driving. More interestingly, the results revealed that the Lagrangian statistics show stronger intermittency than the Eulerian ones but have weaker dependence on different types of forcing. While for ideal gas turbulence purely driven by a solenoidal force at a Mach number of approximately 1 the structure functions of the whole velocity are very close to those of the incompressible case, the structure functions of the compressive velocity component exhibit much stronger intermittent scalings (Wang *et al.* 2012).

All of the above results provide useful knowledge about how the compressibility affects the intermittency in turbulent flow. However, it is still not clear how the turbulent statistics respond as the Mach number approaches unity from a smaller value. Since this regime corresponds to the transition from an incompressible flow to a flow state where the compressibility starts to affect the flow properties, investigation in this regime will provide valuable insights into the effects of compressibility on turbulent statistics. In the present study we conduct a series of numerical simulations in the aforementioned regime with different driving forces. We focus on the Lagrangian

Case	f^s:f^c	M_t	R_λ	τ_η	T^E
S045	1:0	0.45	159.7	0.060	1.087
S075	1:0	0.75	159.3	0.062	1.093
S104	1:0	1.04	152.9	0.063	1.107
C069	1:4	0.69	130.4	0.058	1.218

TABLE 1. Parameters of the numerical simulations. Columns from left to right: case name, the ratio of the solenoidal force to the compressible force, the turbulent Mach number, the Taylor microscale Reynolds number, the viscous time scale and the integral time scale. In the case names the leading letter indicates purely solenoidal driving (S) or the driving force dominated by the compressive component (C), and the following three digits indicate the Mach number.

structure functions (LSFs) by tracking passive tracers advected by turbulence. The pth order LSF is defined as $S_p(\tau) = \langle [\delta u(\tau)]^p \rangle$, where $\delta u(\tau) = u(t+\tau) - u(t)$ is the increment of a Cartesian component of tracer velocity over a time lag τ.

Before we proceed to the results, we would like to point out that recently there were debates on the physical meaning of single-particle LSFs and the existence of inertial scalings in those LSFs (Falkovich *et al.* 2012; Lanotte *et al.* 2013). Some alternatives to traditional LSFs have been proposed in the past few years (e.g. Huang *et al.* 2013; Lévêque & Naso 2014). Nevertheless, LSFs have been extensively used in the Lagrangian study of turbulence and have provided useful information about the intermittent statistics of turbulence, e.g. see the review of Toschi & Bodenschatz (2009) and numerous references therein. Furthermore, in the current study LSFs will serve as a tool to highlight the difference between compressible and incompressible flows, and the effects of the strength of the compressibility.

This paper is organized as follows. In § 2 we will briefly describe the numerical simulations and flow fields. In § 3 the intermittency will be discussed by using Lagrangian statistics. Finally, conclusions will be given in § 4.

2. Numerical simulations

Three-dimensional compressible turbulence for an ideal gas is simulated in a periodic box of $(2\pi)^3$ by using a hybrid scheme (Wang *et al.* 2010). Namely, for the convective term an eighth-order compact scheme is used in the smooth region away from the shocklets and a seventh-order weighted essentially non-oscillatory scheme is used near the shocklets. For time advancing we employ an explicit low-storage second-order Runge–Kutta method. The flow is driven by a large-scale force which can be applied to both solenoidal and compressive components of the velocity field. More details about the numerical method are reported in Wang *et al.* (2013*a*). Four cases are simulated with different driving forces and turbulent Mach numbers $M_t = U_{rms}/c$, where U_{rms} is the root-mean-square (r.m.s.) value of the velocity magnitude and c is the mean speed of sound. The resolution is fixed at 512^3. The parameters for the simulations are listed in table 1. Cases S045, S075 and S104 are driven by purely solenoidal forces. They have different values of M_t and similar Taylor microscale Reynolds numbers R_λ. Case C069 is driven by a combination of compressive and solenoidal forces with a ratio of 4:1.

A Lagrangian module has been developed for particle tracking (Yang *et al.* 2013). For each case, after the flow reaches the statistically steady state, a million

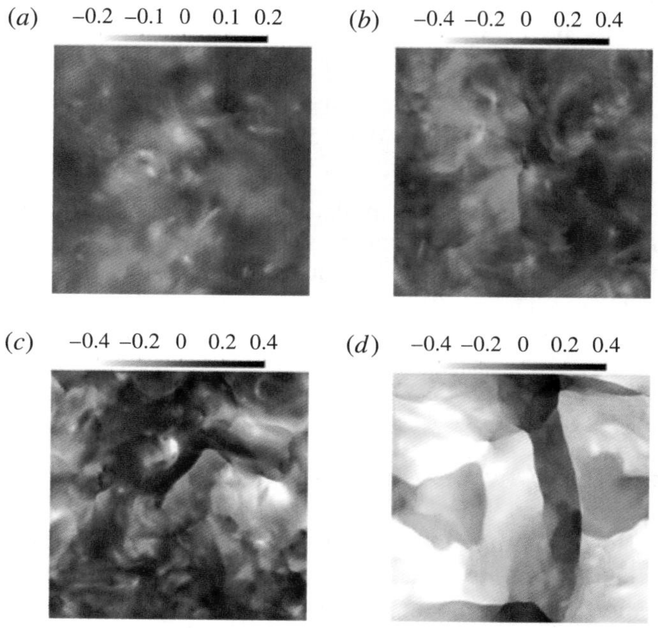

FIGURE 1. Contours of $\log(\rho/\rho_0)$ on the midplane of the domain for (*a*) case S045, (*b*) case S075, (*c*) case S104 and (*d*) case C069. Due to the small density variation in case S045, the colour map in (*a*) has half of the value range compared with the others.

passive tracers are seeded uniformly and advected by the background flow. The flow information at the exact locations of particles is obtained by a trilinear interpolation of the values at surrounding grid points. The particle locations are then updated by using the same time-marching scheme as the Eulerian solver. The trilinear interpolation is adequate in our simulations because (1) our mesh size is slightly smaller than the Kolmogorov scale and (2) linear interpolation may avoid any unphysical oscillation near shocklets where the flow quantities change abruptly. The flow then evolves for a time period of approximately $11T^E$. Here, $T^E = \sqrt{3}L/u_{rms}$ is the large-eddy turnover time, where L is the integral length scale. During the last $8T^E$, two sets of tracer data are stored. One set of data covers the whole $8T^E$ period with a sampling rate of approximately $0.1\tau_\eta$, where $\tau_\eta = (\nu/\epsilon)^{1/2}$ denotes the Kolmogorov time scale, with ν being the mean kinematic viscosity and ϵ the energy-dissipation rate per unit mass. Another set of data consists of several segments, all of which last a time period of $0.4\tau_\eta$ and are separated by $T^E/3$ from one another. Within each segment the sampling rate is approximately $0.01\tau_\eta$. From the former dataset we calculate the Lagrangian statistics for large time scales, which are usually averaged over a time period of approximately $5T^E$ and include in total 4×10^7 data points. The statistics of small time scales are calculated from the latter dataset, which covers a time period of approximately $3T^E$ and contains 10^7 data points.

In figure 1 we show the contours of the logarithm of density $\log(\rho/\rho_0)$ on one slice, where ρ_0 is the mean density. For case S045 with solenoidal driving at the smallest Mach number, the density is non-uniform but no distinct shocklet is observed. As the Mach number increases, the inhomogeneity of the density is enhanced and the shocklets become stronger. These shocklets are still localized in space, i.e. the size of the shocklets is much smaller than the domain size. For case C069 with hybrid driving, the morphology of the shock structures is totally different from those with solenoidal force. Large-scale shocks develop and extend through the entire domain. The flow field is divided into blocks by these large shock structures, and within each block the variation of density is relatively weak.

3. Lagrangian statistics

Different force schemes have profound influences on the behaviour of the LSFs, as illustrated in figure 2, where we plot the LSFs up to the sixth order. For case S045, the LSFs are quite similar to those of the incompressible turbulence, since at such a low Mach number the compressibility effect is negligible, which has also been confirmed by simulations with different numerical schemes (Benzi *et al.* 2008). As M_t increases, the LSFs at $\tau > 10\tau_\eta$ show no visible changes. However, a new self-similar region emerges below the Kolmogorov time scale and expands as M_t increases. For case C069, the LSFs exhibit a self-similar region in the range $0.2\tau_\eta < \tau < 8\tau_\eta$ and saturate for larger τ. The shape is very similar to those at higher M_t (Konstandin *et al.* 2012). All of these results indicate that for the Mach number considered here a purely solenoidal driving only has notable effects on the LSFs at short time scale, while a compressive driving can change the behaviour of the LSFs from the sub-Kolmogorov time scale to much larger scales.

We performed an extended self-similarity (ESS) analysis (Benzi *et al.* 1993) and calculated the local slope $\chi_p = \mathrm{d}\log(S_p(\tau))/\mathrm{d}\log(S_2(\tau))$. In figure 3 we plot the curves of χ_p for $p = 4$ and 6. All of the curves approach the dimensional non-intermittent value $p/2$ as τ decreases to zero. This confirms that our Lagrangian simulation has sufficient time resolution. For case S045, the curve shares similar characteristics to those for incompressible turbulence (Arnèodo *et al.* 2008; Biferale *et al.* 2008; Toschi & Bodenschatz 2009). A dip appears at approximately $\tau \approx 3\tau_\eta$, which is caused by small-scale vortex filaments (Biferale *et al.* 2005; Bec *et al.* 2006). We refer to this as the vortex dip. Interestingly, a new dip emerges at sub-Kolmogorov time scale and becomes deeper when M_t increases, as shown by the curves of cases S075 and S104. Meanwhile the vortex dip becomes shallower and eventually invisible in the curve of case S104. Thus, a new type of intermittency develops at small time scale when the compressibility is strong enough. For case C069, the situation is totally different. As τ increases from small time scale, χ_p quickly drops to near unity and stays at this low value over a relatively wide range of time scales, and then slowly increases before the LSFs saturate. This new dip, starting from sub-Kolmogorov time scale, will be referred to as the shocklet dip. We will show below that it can be related to shocklets.

Our results indicate that two different types of self-similarities may exist at different time scales. Thus, two relative scaling exponents can be extracted for both self-similar regions. For case S045, only one relative scaling exponent ζ_p is calculated by a linear fitting of $(\log(S_2(\tau))\ \log(S_p(\tau)))$ at $8 < \tau/\tau_\eta < T_{max} = 20$ with T_{max} comparable to T^E. This region corresponds to the plateau at the scales above the vortex dip in the χ_p curves, as shown in figure 3. Another relative scaling exponent ξ_p is calculated by a linear fitting at $0.2 < \tau/\tau_\eta < 0.7$ for case S075 and $0.2 < \tau/\tau_\eta < 1.0$ for case S104. For case C069, we only calculate ξ_p over the region $0.2 < \tau/\tau_\eta < 8.0$. These regions are marked by grey shade in figure 2. The exponents are shown in figure 4. One may expect that ζ_p will recover the value of the incompressible flow as M_t decreases to zero. This is exactly the case in our results, as shown in figure 4, where ζ_p from our simulations is compared with the values reported in Benzi *et al.* (2010). For cases S045 and S075 with lower M_t, ζ_p takes a value very close to that of the

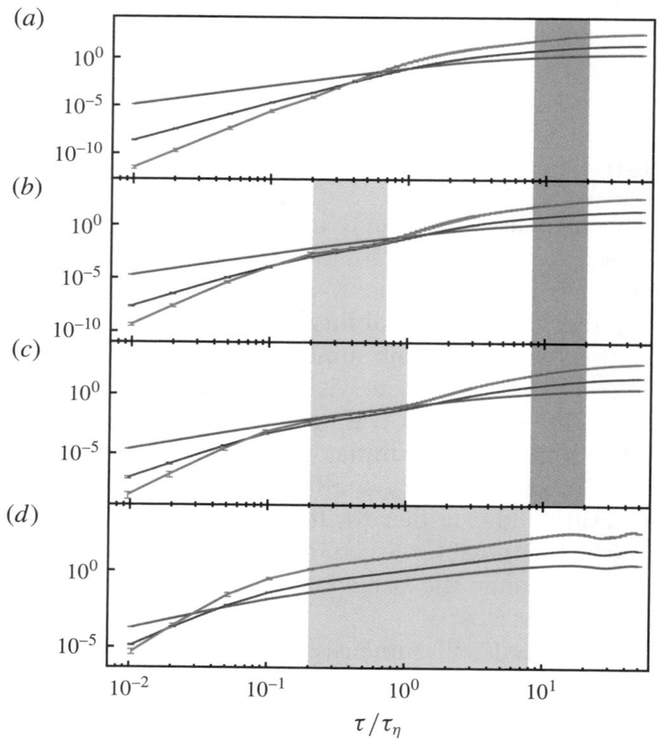

FIGURE 2. Even-order LSFs for (*a*) case S045, (*b*) case S075, (*c*) case S104 and (*d*) case C069: green, $p=2$; blue, $p=4$; red, $p=6$. The error bar is estimated by the standard deviation of the temporary fluctuation. The dark grey shade marks the range where the relative scaling exponent ζ_p is calculated and the light grey shade marks the range where ξ_p is calculated.

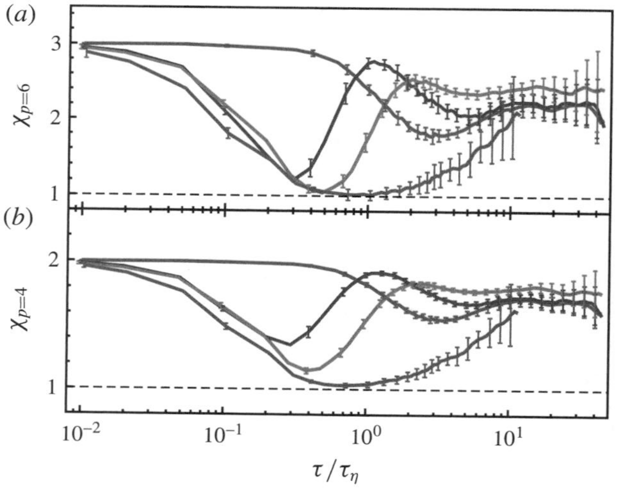

FIGURE 3. Extended self-similarity local slope χ_p for $p=4$ and 6: green, case S045; blue, case S075; red, case S104; grey, case C069. The error bar is estimated by the standard deviation of the temporary fluctuation and is only shown at every other data point for clarity. For case C069, χ_p is computed in the range of $\tau < 12\tau_\eta$ before the LSFs saturate.

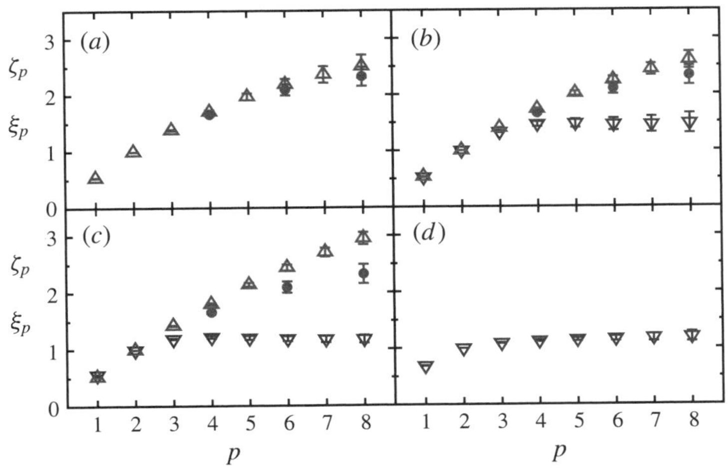

FIGURE 4. Extended self-similarity relative scaling exponents up to $p = 8$ for ζ_p (red upward triangles) and ξ_p (blue downward triangles) for (*a*) case S045, (*b*) case S075, (*c*) case S104 and (*d*) case C069. The grey circles in the plots for cases S045, S075 and S104 mark the values for the incompressible case with $Re_\lambda \sim 600$ (Benzi *et al.* 2010). For cases S045 and C069, only the applicable exponents are shown.

incompressible case, although our simulations have smaller Reynolds numbers. For higher M_t such as case S104, ζ_p is considerably higher than the incompressible value at high order. This indicates that compressibility starts to affect the overall turbulent statistics over a wide range of time scales.

The relative exponent ξ_p calculated in the new self-similar region shows totally different behaviour compared with ζ_p. It is slightly larger than ζ_p for $p = 1$ and smaller than ζ_p for $p > 2$, indicating a much stronger intermittency. It saturates when $p > 4$ for cases S075 and S104. The saturated exponent ξ_∞, which is calculated by averaging ξ_p over $p = 5$–8, is approximately 1.47 for case S075 and 1.19 for case S104. For case C069, ξ_p does not saturate completely but increases very slowly when $p > 4$. We define $\xi_\infty = \xi_8 \approx 1.19$ for case C069. Based on multi-fractal theory, the saturation of the exponent suggests that the tails of the probability density function (PDF) $P(\delta u, \tau)$ should collapse for different values of τ in the self-similar region when normalized by τ^{ξ_∞} (Benzi *et al.* 2008; Wang *et al.* 2013*b*). We plot in figure 5 the normalized PDF $\tau^{-\xi_\infty} P$ for case S104 within $\tau/\tau_\eta \in (0.2, 1.0)$ and for case C069 within $\tau/\tau_\eta \in (0.2, 8.0)$, i.e. the region where ξ_p is evaluated. Indeed, the tails of the normalized PDFs collapse. It is remarkable that for case C069 the peak of the normalized PDFs at $\delta u = 0$ decreases for several orders as τ/τ_η increases, while the tails are still very close to one another for the entire range of $\tau/\tau_\eta \in (0.2, 8.0)$.

It has been revealed that the exponents of the Eulerian structure functions saturate due to the appearance of the front-like or shocklet structures (Benzi *et al.* 2008; Wang *et al.* 2013*b*). From the Lagrangian point of view, when a passive tracer moves through a shocklet it gains a velocity increment which is determined by the strength of the shocklet. In a homogeneous flow, a reasonable assumption is that the probability of a tracer encountering a shocklet is proportional to the time lag τ, which gives $\delta u \sim \text{constant} \cdot \tau$. Therefore, the saturated value of the exponent should be very close to 1 if the shocklets in the flow have similar strength and tracers only penetrate shocklets once in the time interval considered. This argument is consistent

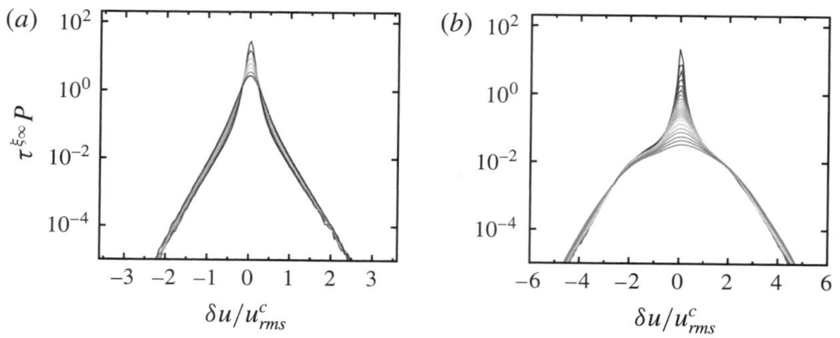

FIGURE 5. Probability density functions of $\delta u/u^c_{rms}$ normalized by τ^{ξ_∞}, where u^c_{rms} is the r.m.s. value of the velocity component: (*a*) $\tau/\tau_\eta \in (0.2, 1.0)$ for case S104 and (*b*) $\tau/\tau_\eta \in (0.2, 8)$ for case C069. The colour changes from blue to red as τ increases.

with our numerical results. For a short time interval, it is likely that tracers encounter shocklets at most once, and thus the local relative exponent should be close to 1. Indeed, in figure 3 we observe that with solenoidal driving force the local minimum of χ_p within the shocklet dip approaches 1 as M_t increases, while the χ_p of case C069 equals 1 at approximately $\tau = \tau_\eta$. For the three cases with solenoidal driving force, the major portion of the flow field is dominated by vortical structures, and the LSFs are similar to those of the incompressible cases at large time scales. However, the effects of compressibility do emerge at higher M_t. With hybrid driving, however, shocks are more pronounced and dramatically change the LSFs even at large time lags.

Now it is clear that the shocklet dip in the local slope curves is related to the shocklet structures in flow. It is worth pointing out that the vortex dip and the shocklet dip have different shapes and respond differently to their relevant control parameters. For incompressible flow, the location and shape of the vortex dip are universal and have minor dependences on the Reynolds number (Arnèodo *et al.* 2008). The shocklet dip reported here, on the other hand, expands over time scales as M_t increases or shocklets become dominant. This is caused by the different natures of the extreme velocity differences induced by vortex filaments and shocklets. For a tracer trapped by a vortex filament, the largest velocity difference occurs at the time lag during which the tracer travels half of the circle around the vortex filament, regardless of the strength of the vortex filament. However, when a tracer penetrates a shocklet, the velocity differences over a time lag are predominantly determined by the strength of the shocklet, provided that the tracer is located on different sides of the shocklet at the beginning and end of that time lag. As the shocklet becomes stronger, its influence lasts for a longer time period.

4. Conclusions

In this work we reported several interesting results about the intermittent statistics and the anomalous scaling of LSFs for compressible homogeneous turbulence, which provide new insights into the compressibility origin of turbulent intermittency. A series of 3D numerical simulations were conducted for different turbulent Mach numbers M_t and different types of driving force. When the flow is purely driven by a solenoidal force, the LSFs are very close to those of incompressible turbulence at small M_t. For large M_t, the LSFs exhibit a new self-similar region at the sub-Kolmogorov time scale, which is characterized by a saturated relative scaling exponent and is attributed to shocklet structures. This region expands to larger time scales and the saturated exponent decreases towards 1 as M_t increases. The behaviour of the LSFs is still very close to that for incompressible flow at larger time scales. When M_t is high enough, say approximately 1, compressibility starts to affect the statistics at both small and large time scales.

The nature of the driving force also has profound influences on the LSFs. When a driving force dominated by a compressive component is applied, very strong shock structures emerge at relatively smaller M_t compared with the flow solely driven by a solenoidal force, and the LSFs only have a single self-similar region ranging from the sub-Kolmogorov time scale to much larger time scales before the LSFs saturate. This self-similar region is caused by strong shocks, and the relative exponent saturates for high-order LSFs. We also demonstrated that the saturation of the relative exponent in the compressibility-related self-similar region can be explained by a simple tracer–shocklet interaction model. Our results indicated that for time lags in this region the PDFs of the velocity increments collapse after proper normalization, which is consistent with multi-fractal theory; a similar phenomenon has been reported for Eulerian statistics.

Another issue that is not discussed in the present study is the effect of the Reynolds number, since all of our simulations have similar Re_λ. Some conjectures may be made based on the current results and their differences compared with incompressible turbulence. For solenoidal driving with high enough Mach number, the new self-similar region at small time scales is dominated by compressible effects, since for incompressible turbulence one would expect no self-similar behaviour at such small time scales. Thus, the Reynolds number should have a minor influence on the existence of this region. For fixed Re, as the Mach number increases the new self-similar region starts to affect the statistics at large time scales. While higher Re indicates wider scale separation in the flow, the change of the LSFs at large time scales may only appear at higher Mach number as Re becomes larger. When the driving force is dominated by the compressive component, the morphology of the shocklet is totally different from the solenoidal driving case, and compressibility changes the LSFs from sub-Kolmogorov scales to much larger scales. The effect of Re in such flow is not clear at this stage. More simulations with different values of Re are needed to fully understand the effects of the Reynolds number in the future.

Acknowledgements

The authors appreciate the useful comments from the anonymous referees. This work was supported partially by the National Natural Science Foundation of China (NSFC grants 11221061 and 91130001) and the National Basic Research Program of China (grant 2013CB834100). Y.Y. appreciates the partial support by NSFC grant 11274026. Y.S. also acknowledges the support by NSFC grant U1330107. The simulations were carried out at the Tianhe-1A supercomputing facility at the National Supercomputer Center in Tianjin, China.

References

ALUIE, H. 2011 Compressible turbulence: the cascade and its locality. *Phys. Rev. Lett.* **106**, 174502.

ALUIE, H., LI, S. & LI, H. 2012 Conservative cascade of kinetic energy in compressible turbulence. *Astrophys. J. Lett.* **751** (2), L29.

ARNÈODO, A., BENZI, R., BERG, J., BIFERALE, L., BODENSCHATZ, E., BUSSE, A., CALZAVARINI, E., CASTAING, B., CENCINI, M., CHEVILLARD, L., FISHER, R. T., GRAUER, R., HOMANN, H., LAMB, D., LANOTTE, A. S., LÉVÈQUE, E., LÜTHI, B., MANN, J., MORDANT, N., MÜLLER, W.-C., OTT, S., OUELLETTE, N. T., PINTON, J.-F., POPE, S. B., ROUX, S. G., TOSCHI, F., XU, H. & YEUNG, P. K. 2008 Universal intermittent properties of particle trajectories in highly turbulent flows. *Phys. Rev. Lett.* **100**, 254504.

BANERJEE, S. & GALTIER, S. 2013 Exact relation with two-point correlation functions and phenomenological approach for compressible magnetohydrodynamic turbulence. *Phys. Rev.* E **87**, 013019.

BEC, J., BIFERALE, L., CENCINI, M., LANOTTE, A. S. & TOSCHI, F. 2006 Effects of vortex filaments on the velocity of tracers and heavy particles in turbulence. *Phys. Fluids* **18** (8), 081702.

BENZI, R., BIFERALE, L., FISHER, R., LAMB, D. Q. & TOSCHI, F. 2010 Inertial range Eulerian and Lagrangian statistics from numerical simulations of isotropic turbulence. *J. Fluid Mech.* **653**, 221–244.

BENZI, R., BIFERALE, L., FISHER, R. T., KADANOFF, L. P., LAMB, D. Q. & TOSCHI, F. 2008 Intermittency and universality in fully developed inviscid and weakly compressible turbulent flows. *Phys. Rev. Lett.* **100**, 234503.

BENZI, R., CILIBERTO, S., TRIPICCIONE, R., BAUDET, C., MASSAIOLI, F. & SUCCI, S. 1993 Extended self-similarity in turbulent flows. *Phys. Rev.* E **48**, R29–R32.

BIFERALE, L., BODENSCHATZ, E., CENCINI, M., LANOTTE, A. S., OUELLETTE, N. T., TOSCHI, F. & XU, H. 2008 Lagrangian structure functions in turbulence: a quantitative comparison between experiment and direct numerical simulation. *Phys. Fluids* **20** (6), 065103.

BIFERALE, L., BOFFETTA, G., CELANI, A., LANOTTE, A. & TOSCHI, F. 2005 Particle trapping in three-dimensional fully developed turbulence. *Phys. Fluids* **17** (2), 021701.

FALKOVICH, G., XU, H., PUMIR, A., BODENSCHATZ, E., BIFERALE, L., BOFFETTA, G., LANOTTE, A. S. & TOSCHI, F. 2012 On Lagrangian single-particle statistics. *Phys. Fluids* **24** (5), 055102.

FEDERRATH, C. 2013 On the universality of supersonic turbulence. *Mon. Not. R. Astron. Soc.* **436** (2), 1245–1257.

FEDERRATH, C., ROMAN-DUVAL, J., KLESSEN, R. S., SCHMIDT, W. & MAC LOW, M. M. 2010 Comparing the statistics of interstellar turbulence in simulations and observations. *Astron. Astrophys.* **512**, A81.

FRISCH, U. 1995 *Turbulence: The Legacy of A. N. Kolmogorov*. Cambridge University Press.

GALTIER, S. & BANERJEE, S. 2011 Exact relation for correlation functions in compressible isothermal turbulence. *Phys. Rev. Lett.* **107**, 134501.

HENNEBELLE, P. & FALGARONE, E. 2012 Turbulent molecular clouds. *Astron. Astrophys. Rev.* **20** (1), 55.

HUANG, Y., BIFERALE, L., CALZAVARINI, E., SUN, C. & TOSCHI, F. 2013 Lagrangian single-particle turbulent statistics through the Hilbert–Huang transform. *Phys. Rev.* E **87**, 041003.

KONSTANDIN, L., FEDERRATH, C., KLESSEN, R. S. & SCHMIDT, W. 2012 Statistical properties of supersonic turbulence in the Lagrangian and Eulerian frameworks. *J. Fluid Mech.* **692**, 183–206.

KRITSUK, A. G., NORMAN, M. L., PADOAN, P. & WAGNER, R. 2007 The statistics of supersonic isothermal turbulence. *Astrophys. J.* **665** (1), 416–431.

KRITSUK, A. G., WAGNER, R. & NORMAN, M. L. 2013 Energy cascade and scaling in supersonic isothermal turbulence. *J. Fluid Mech.* **729**, R1.

LANOTTE, A. S., BIFERALE, L., BOFFETTA, G. & TOSCHI, F. 2013 A new assessment of the second-order moment of Lagrangian velocity increments in turbulence. *J. Turbul.* **14** (7), 34–48.

LÉVÊQUE, E. & NASO, A. 2014 Introduction of longitudinal and transverse Lagrangian velocity increments in homogeneous and isotropic turbulence. *Europhys. Lett.* **108** (5), 54004.

MCKEE, C. F. & OSTRIKER, E. C. 2007 Theory of star formation. *Annu. Rev. Astron. Astrophys.* **45** (1), 565–687.

PADOAN, P. & NORDLUND, Å. 2002 The stellar initial mass function from turbulent fragmentation. *Astrophys. J.* **576** (2), 870.

PAN, L., PADOAN, P. & KRITSUK, A. G. 2009 Dissipative structures in supersonic turbulence. *Phys. Rev. Lett.* **102**, 034501.

SAMTANEY, R., PULLIN, D. I. & KOSOVIĆ, B. 2001 Direct numerical simulation of decaying compressible turbulence and shocklet statistics. *Phys. Fluids* **13** (5), 1415–1430.

SCALO, J. & ELMEGREEN, B. G. 2004 Interstellar turbulence. Part II: implications and effects. *Annu. Rev. Astron. Astrophys.* **42**, 275–316.

SCHMIDT, W., FEDERRATH, C. & KLESSEN, R. 2008 Is the scaling of supersonic turbulence universal? *Phys. Rev. Lett.* **101**, 194505.

TOSCHI, F. & BODENSCHATZ, E. 2009 Lagrangian properties of particles in turbulence. *Annu. Rev. Fluid Mech.* **41** (1), 375–404.

WAGNER, R., FALKOVICH, G., KRITSUK, A. G. & NORMAN, M. L. 2012 Flux correlations in supersonic isothermal turbulence. *J. Fluid Mech.* **713**, 482–490.

WANG, J., SHI, Y., WANG, L., XIAO, Z., HE, X. T. & CHEN, S. 2011 Effect of shocklets on the velocity gradients in highly compressible isotropic turbulence. *Phys. Fluids* **23** (12), 125103.

WANG, J., SHI, Y., WANG, L., XIAO, Z., HE, X. T. & CHEN, S. 2012 Scaling and statistics in three-dimensional compressible turbulence. *Phys. Rev. Lett.* **108**, 214505.

WANG, J., WANG, L., XIAO, Z., SHI, Y. & CHEN, S. 2010 A hybrid numerical simulation of isotropic compressible turbulence. *J. Comput. Phys.* **229**, 5257–5279.

WANG, J., YANG, Y., SHI, Y., XIAO, Z., HE, X. T. & CHEN, S. 2013*a* Cascade of kinetic energy in three-dimensional compressible turbulence. *Phys. Rev. Lett.* **110**, 214505.

WANG, J., YANG, Y., SHI, Y., XIAO, Z., HE, X. T. & CHEN, S. 2013*b* Statistics and structures of pressure and density in compressible isotropic turbulence. *J. Turbul.* **14** (6), 21–37.

YANG, Y., WANG, J., SHI, Y., XIAO, Z., HE, X. T. & CHEN, S. 2013 Acceleration of passive tracers in compressible turbulent flow. *Phys. Rev. Lett.* **110**, 064503.

PHYSICS OF PLASMAS **23**, 122702 (2016)

Main drive optimization of a high-foot pulse shape in inertial confinement fusion implosions

L. F. Wang (王立锋),[1,2,3,a)] W. H. Ye (叶文华),[1,2,3] J. F. Wu (吴俊峰),[1] Jie Liu (刘杰),[1,2,3] W. Y. Zhang (张维岩),[1] and X. T. He (贺贤土)[1,2,3,4]
[1]Institute of Applied Physics and Computational Mathematics, Beijing 100094, China
[2]HEDPS, Center for Applied Physics and Technology, Peking University, Beijing 100871, China
[3]IFSA Collaborative Innovation Center of MoE, Shanghai Jiao-Tong University, Shanghai 200240, China
[4]Institute of Fusion Theory and Simulation, Zhejiang University, Hangzhou 310027, China

(Received 8 July 2016; accepted 3 November 2016; published online 6 December 2016)

While progress towards hot-spot ignition has been made achieving an alpha-heating dominated state in high-foot implosion experiments [Hurricane *et al.*, Nat. Phys. **12**, 800 (2016)] on the National Ignition Facility, improvements are needed to increase the fuel compression for the enhancement of the neutron yield. A strategy is proposed to improve the fuel compression through the recompression of a shock/compression wave generated by the end of the main drive portion of a high-foot pulse shape. Two methods for the peak pulse recompression, namely, the decompression-and-recompression (DR) and simple recompression schemes, are investigated and compared. Radiation hydrodynamic simulations confirm that the peak pulse recompression can clearly improve fuel compression without significantly compromising the implosion stability. In particular, when the convergent DR shock is tuned to encounter the divergent shock from the capsule center at a suitable position, not only the neutron yield but also the stability of stagnating hot-spot can be noticeably improved, compared to the conventional high-foot implosions [Hurricane *et al.*, Phys. Plasmas **21**, 056314 (2014)]. *Published by AIP Publishing.* [http://dx.doi.org/10.1063/1.4971237]

I. INTRODUCTION

The high-foot implosion experiments on the National Ignition Facility (NIF) have achieved an encouraging progress towards indirectly driven Inertial Confinement Fusion (ICF) ignition[1,2] with a yield about 26 kJ and stagnation pressure about 220 Gbar[3–6] although they are still short of the pressure required for the hot-spot ignition conditions ($\sim$350 Gbar). Compared to the previous low-foot implosions during the National Ignition Campaign (NIC)[7] on the NIF, the high-foot implosions[8–11] employ a higher radiation temperature during the first stage of the ignition pulse shape to drive a stronger first-shock into the target. The high-foot designs have been successful in controlling implosion stability at the ablation front, and hence have experimentally led to an increase in neutron yield by an order of magnitude achieving an alpha-heating dominated state,[3] but at the cost of reduced fuel compressibility that of course is needed to achieve the ignition conditions. Additionally, the improved performance of high-foot implosions, albeit in the presence of severe low-mode hot-spot asymmetries, strongly indicates that controlling instability growth, as the high-foot implosion does by increasing the shell adiabat, is critical to achieving ICF ignition.[12]

Note that the improved neutron yield and stability property of the high-foot implosions come at the expense of a reduced fuel areal density (compression). It is well known that high fuel compression is a crucial requirement for ICF ignition and burn-up, which can increase the rate of fusion reactions. Therefore, in order to further increase the neutron yield, the high-foot design has to improve the final fuel compression, especially at the energy scale available on the NIF. An important question then is whether it is possible to develop an implosion to further improve the fuel compression of the conventional high-foot implosion but without sacrificing the implosion hydrodynamic stability. This paper attempts to tackle this problem. Specifically, this paper describes a general design study aimed at surveying the peak pulse recompression effects on implosion performance.

Many detailed design studies of the foot of the ignition pulse shape have been performed previously, especially on the "picket" and "trough" of the early time portion of the laser pulse,[13–21] namely, "adiabat-shaping" technology. Specifically, a design space survey of the foot (the picket and trough) of the laser pulse is performed in reference.14 In fact, the concept of adiabat-shaping technology was proposed by Gardner *et al.*[22] and Bodner *et al.*[23] and subsequently explored in direct-drive implosions.[24–27] The present work focuses on the main drive portion of the ignition pulse shape.

Any promising avenues for improving the fuel compression of conventional high-foot implosions deserve serious investigation. This paper, following the conception of controlling implosion stability of conventional high-foot design, presents a scheme modifying the end of the main drive portion of the high-foot pulse shape to generate a shock/compression wave to compress the fuel again. Two schemes creating the shock/compression wave are studied and compared. In the simple recompression (SR) scheme, the peak temperature of the main drive is directly improved to a higher one when the shell is accelerated to a certain velocity (typically 260–320 km/s). On the other hand, the peak temperature of the main drive is first reduced to a lower one

a)Electronic mail: wang_lifeng@iapcm.ac.cn

when the shell is accelerated to a certain velocity (typically 150–240 km/s), and after a short-distance flight (typically ~ 100-μm) it is improved to a higher one in the decompression-and-recompression (DR) scheme.[28] The implosion performance and implosion stability of the SR and DR schemes are compared with the conventional high-foot design.

The rest of this paper is organized as follows. Section II briefly describes the simulation model used in the present work. Section III presents a comparison of the recompression effects of peak pulse on the one-dimensional (1-D) implosion performance. Section IV outlines the two-dimensional (2-D) stability analysis on each modified peak pulses, and the paper concludes with Section V.

II. MODELING AND SIMULATION

This study uses the plastic (CH) ablator point design target.[29] The outer radius of the CH-ablator is 1140-μm. The 190-μm-thick CH ablator has a density of 1.0 g/cm^3. The layer of deuterium-tritium (DT) ice has an initial thickness of 80 μm with a density of 0.25 g/cm^3. The interior of the DT-ice layer is filled with DT-gas with a density of 3×10^{-4} g/cm^3. We perform highly resolved 1-D and 2-D simulations using the capsule-only detailed radiation-hydrodynamic code LARED-S,[30] which has been widely used in the field of ICF hydrodynamics instabilities.[31–38] The present calculations were carried out using the version of LARED-S that is parallelized in the scheme of message-passing interface (MPI). It is a multi-dimensional Eulerian mesh based code with multi-group diffusion of radiation transport, flux-limited Spitzer-Harm̈ thermal conduction for electrons and ions, and multi-group diffusion of charged particles for alpha-particles. The opacity used in the calculation of radiation transport is from tables generated using the modified relativistic Hartree-Fock-Slater self-consistent average atom model.[39] A modified tabular quotidian equation of state (QEOS)[40] is used to model the CH ablator and DT fuel.

The grids in the radial direction can move and be nonuniformly spaced. The spacing and motion of the radial grids are chosen to provide the maximum resolution for the compressed shell throughout the time of implosion. The radial meshes near the critical regions of the target, for example, the ablation front, the ablator-fuel interface, and the inner shell surface, are locally refined. In the simulation of this work, the maximum spatial resolution is 0.05-μm in the radial direction. The 2-D simulations use a $(r-\theta)$ geometry mesh. The number of grid points used in the radial direction is $N_r = 2000$ and that in the poloidal direction is $N_\theta = 40$.

To study the recompression effects, we consider two ways of achieving main drive recompression, namely, the decompression-and-recompression (DR) high-foot pulse shape and the simple recompression (SR) high-foot pulse shape. The main difference between the DR and SR high-foot pulse shapes is the main drive portion. Figure 1 depicts the radiation temperature versus time ($T_r(t)$) for the high-foot capsule described above with different peak pulses under consideration. Two variants of DR and SR high-foot pulse shapes are investigated separately. The recompression is earlier in DR (SR) high-foot A than that in DR (SR) high-foot B, as shown in Figure 1. The peak driving temperature of conventional high-foot pulse shape is fixed at 300-eV. In DR high-foot implosions, the driving temperature in the *decompression phase* is decreased from 300-eV to 270-eV and that in the *recompression phase* is increased to 320-eV. In SR high-foot implosions, the driving temperature is directly increased from 300-eV to 320-eV in the *recompression phase*. In our calculations, the time integral of $T_r^4(t)$ is held fixed for all of these pulse shapes. It is interesting to note that the capsule absorbed energy is not constrained in the five models. However, it does not vary quite much for the same target and integral of $T_r^4(t)$, in the presence of a similar main drive temperature.[41]

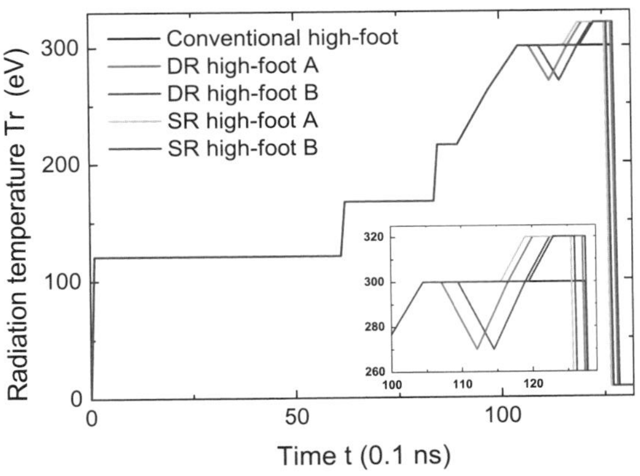

FIG. 1. The radiation temperature versus time $T_r(t)$ for the conventional high-foot implosion, DR high-foot implosions, and SR high-foot implosions. The inset is a magnified plot of the main drive portion of the pulse shape. The integral of $T_r^4(t)$ is held fixed.

The 2-D cylindrical hohlraum simulation results from the LARED-JC code[30] indicate that the SR and DR high-foot pulse shapes can be generated by a laser facility with energy $E_{laser} \sim 1.8$-MJ (including 15% energy loss of the incident energy) and maximum power $P_{laser} \sim 520$-TW. In the simulations, the depleted uranium (DU) cylindrical hohlraum is 9.40-mm long and 5.75-mm in diameter, and the diameter of the laser entrance hole (LEH) is 3.10-mm. Additionally, LARED-JC simulation results indicate that a spherical hohlraum with 6-LEH[42] using similar requirements for energy and power can also generate the SR and DR high-foot pulse shapes.

In the following discussions, we will compare the implosion performance driven by these different peak pulses with the conventional three-shock high-foot implosions.[8] Subsequently, recompression effects of the peak pulse of the drive on implosion performance are then obtained.

III. 1-D IMPLOSION PERFORMANCE

In this section, a series of 1-D calculations are performed to assess the 1-D implosion performance of the peak pulse recompression. The effects of DR and SR high-foot on the 1-D implosion performance are compared with that from the conventional high-foot implosion.

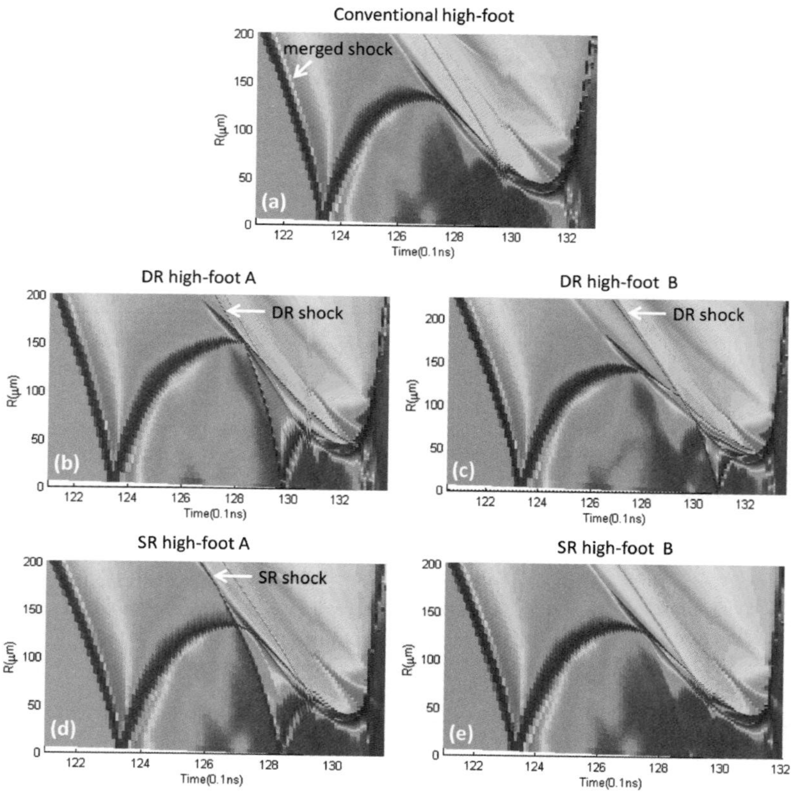

FIG. 2. Comparison of shock trajectories close to the stagnation time in the time-radius plane for the conventional high-foot implosion (a), DR high-foot implosions ((b) and (c)), and SR high-foot implosions ((d) and (e)).

Figure 2 shows the plot of the radial velocity gradient versus time and target radius for conventional high-foot implosions, DR high-foot implosions, and SR high-foot implosions, which distinctly visualizes the shock trajectory. In conventional high-foot implosions, three-shock merged into a single one before reaching the capsule center, as shown in Figure 2(a). The merged shock is reflected off the center of the target (return shock) and subsequently interacts with the incoming inner shell surface. The shell is slowed down in an impulsive manner when interacting with the return shock and generates a new convergent shock. A series of shocks are then reflected off the center and the inner shell surface increasing the pressure of the low-density DT. Eventually, the hot-spot is formed and the DT inside the hot-spot reaches a pressure large enough to continuously slow down the shell up to the stagnation point. In the implosion of DR high-foot A, a DR shock is generated by the end of the main drive and collides with the return shock around the foot of the inner shell surface, as shown in Figure 2(b). In contrast, the DR shock encounters the return shock near the waist of the inner shell surface in the DR high-foot B implosion,[28] as can be seen in Figure 2(c). In the SR high-foot A implosion, a SR shock is generated by the end of the main drive, which collides with the return shock near the foot of the inner shell surface, as illustrated in Figure 2(d). Note that the end of the main drive in the implosion of SR high-foot B generates a compression wave, which cannot be clearly seen in Figure 2(e). However, recompression effects of the compression wave in SR high-foot B implosions can be easily seen in Figure 3.

The consequence of peak pulse recompression can be immediately seen in Figure 3, which plots the in-flight dense shell thickness ΔR during the acceleration regime for the conventional high-foot implosion, DR high-foot implosions, and SR high-foot implosions. Here, the ΔR is defined as the distance between the inner shell surface and the radiation ablation front. As can be seen clearly in Figure 3, the DR

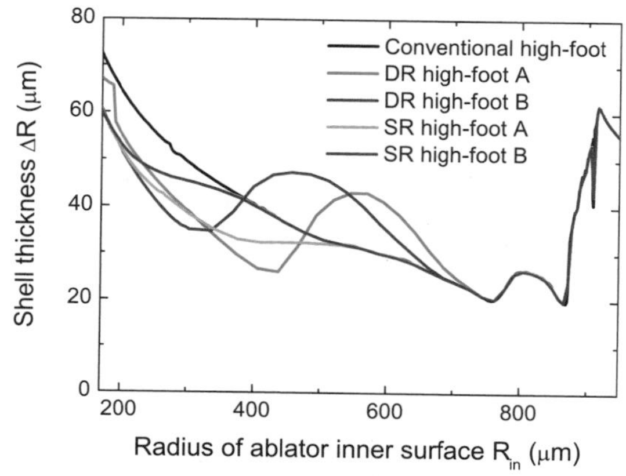

FIG. 3. The dense shell thickness ΔR versus the ablator inner radius R_{in} for the conventional high-foot implosion, DR high-foot implosions, and SR high-foot implosions.

TABLE I. Comparison of simulated ideal 1-D performance metrics with different peak pulses.

	Conventional high-foot	DR high-foot A	DR high-foot B	SR high-foot A	SR high-foot B
Yield (MJ)	6.7	8.9	14.1	13.1	14.8
Fuel V_{imp} (km/s)	385	385	370	415	394
Fuel adiabat	2.6	2.4	2.4	2.5	2.5
Fuel ρR (g/cm^2)	1.17	1.13	1.20	1.20	1.25
P_{stag} (Gbar)[a]	390	395	415	495	480
CR[b]	37	35	36	35	36

[a]Alpha particle deposition off.
[b]R_{abl} (at initial time)/$R_{hotspot}$(at stagnation time).

high-foot implosion increases the ΔR by the decompression pulse in the middle of the acceleration but decreases it by the recompression pulse in the late of the acceleration. By contrast, the SR high-foot implosion just directly decreases the ΔR by the recompression pulse in the late of the acceleration. It turns out that the ΔR is a critical parameter in the design of ICF capsule since the Rayleigh-Taylor growth and the feed-through are directly related to the ratio $R_{in}/\Delta R$,[29] where R_{in} is the ablator inner radius. Furthermore, the influence of thickness on the Rayleigh-Taylor instability of an incompressible fluid layer has been discussed in Reference 43.

As shown in Figure 2, the implosion dynamic of these five pulse shapes is noticeably different. The end of the main drive portion of the pulse shape can evidently affect the in-flight thickness of the shell (see Figure 3) and the dynamics of the deceleration phase. Moreover, the main drive recompression effects on the 1-D implosion performance metrics and the implosion stability (see Section IV) will be addressed in the following.

Table I summarizes the comparison of various ideal 1-D performance metrics for the conventional high-foot implosion, DR high-foot implosions, and the SR high-foot implosions. Here, the fuel adiabat is defined as $\alpha = P/P_{cold}$, where P_{cold} is the minimum pressure as $1000\,\mathrm{g/cm^3}$ from the DT equation of state.[44] The yield of conventional high-foot implosion is 6.7-MJ. Compared to the conventional high-foot implosion, the DR and SR high-foot implosions can significantly improve the yield. The SR high-foot implosions can slightly increase the fuel peak implosion velocity (V_{imp}), and the DR high-foot implosions can potentially influence the peak implosion velocity. Note that the implosion of DR high foot B has the smallest peak implosion velocity that is 370 km/s. The fuel adiabat of conventional high-foot implosion is 2.6, which is larger than the fuel adiabat of 80-eV foot design but less than that of 100-eV foot design as described in Ref. 9. The DR and SR high-foot implosions can slightly reduce the fuel adiabat. The SR high-foot implosions can improve somewhat the fuel areal density (ρR) but the effects of DR high-foot implosions on the fuel ρR are dependent on the time of DR. Compared to the conventional high-foot implosion, the DR and SR high-foot implosions can effectively improve the stagnation pressure (P_{stag}), especially for the SR high-foot implosions. The DR and SR high-foot implosions cannot appreciably change the convergence ratio (CR), compared to the conventional high-foot implosion.

Figure 4 shows the density and pressure profiles from 1-D simulation of the five cases: conventional high-foot, DR high-foot A, DR high-foot B, SR high-foot A, and SR high-foot B. All are shown at a radius corresponding to about 1/3 of the initial radius. The properties of the ablation front for the five cases are similar, including the density-gradient scale-length. Note that the stabilization effect from the density-gradient scale-length plays a crucial role in reducing the ablation-front Rayleigh-Taylor growth, especially for the long-wavelength (typically $\ell < 10$, where ℓ is the mode number) nonuniformities that lead to an overall shell deformation and intermediate-wavelength (typically $10 < \ell < 50$) perturbations that would lead to shell breakup. Therefore, all these designs would have similar acceleration-phase Rayleigh-Taylor growth (see Section IV). The implosion of SR high-foot A is clearly unstable at the interface between the CH ablator and the DT fuel. This implosion would be severely susceptible to intermediate- and short-wavelength (typically $\ell > 50$) Rayleigh-Taylor growth. By contrast, the implosion of DR high-foot B provides a better match of ablator and fuel densities, which would reduce the intermediate- and short-wavelength Rayleigh-Taylor growth at ablator-fuel interface during the acceleration regime (see Section IV).

Figure 5 presents the radial mass density profiles at peak implosion velocity for the conventional high-foot implosion, DR high-foot implosions, and SR high-foot implosions. As

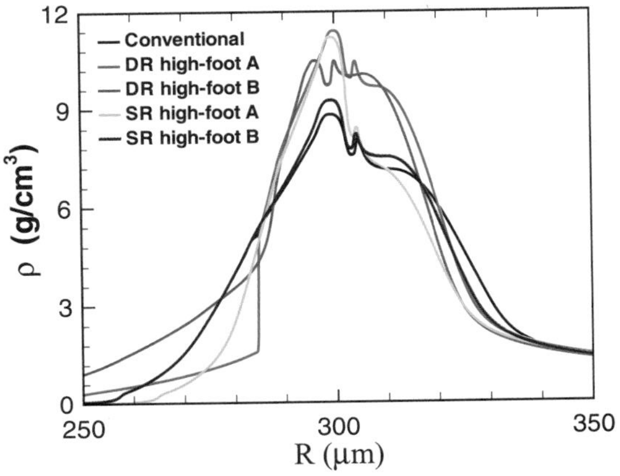

FIG. 4. Radial profiles of density, in-flight at a radius below half initial ($R_{in} \sim 300\,\mu$m) for the conventional high-foot implosion, DR high-foot implosions, and SR high-foot implosions.

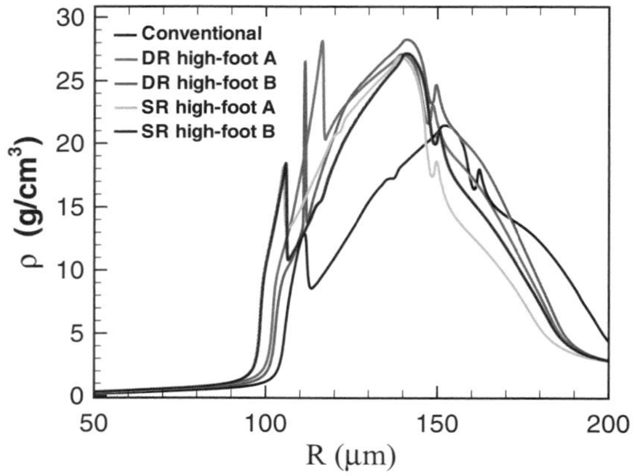

FIG. 5. Comparison of shell density profiles at peak implosion velocity resulting from different peak pulses.

can be seen in Figure 5, the recompression of the main drive portion of the pulse (i.e., peak pulse) can markedly improve the peak density of the shell at peak implosion velocity. Note that increasing the shell density is an effective way to improve the shell dynamic pressure, which is approximately equal to the final hot-spot pressure,[45] especially for the case of limited peak implosion velocity.

IV. 2-D STABILITY ANALYSIS

In this section, we present stability analysis results from a series of 2-D simulations. The computational domain in the poloidal direction is $[\pi/2 - \pi/\ell, \pi/2 + \pi/\ell]$, where ℓ is the mode number. We initiate single-mode surface perturbation in the form $\delta R = A_0 \cos(\ell\theta)$ in calculations, where A_0 is the initial perturbation amplitude.

Linear stability analysis of the implosions driven by the five different pulse shapes has been performed and the linear growth factor (GF) results are obtained. In fact, the A_0 in our calculation has been well adjusted to keep the evolution of perturbations within the linear growth regime. Here, the GF is measured by comparing the mode interface amplitude η at a chosen time to that at the other chosen time. The GF during the acceleration stage is defined as $\eta_{peak\,vel.}/\eta_{acc.\,beg.}$ and the GF at peak implosion velocity is $\eta_{peak\,vel.}/\eta_{init.}$, where $\eta_{peak\,vel.}$ is the perturbation amplitude at peak implosion velocity, $\eta_{acc.\,beg.}$ is the perturbation amplitude at the beginning of shell acceleration, and $\eta_{init.}$ is the initial perturbation amplitude. Moreover, the hot-spot surface GF at stagnation is measured by comparing the interface amplitude at the stagnation time to its initial value.

Figures 6(a) and 6(b) show the single-mode growth factors measured at the ablation front during the acceleration stage and at peak implosion velocity, respectively, for the five different main drive pulse shapes. The difference of ablation front GF for the various main drive pulse shapes is not very dramatic. This is because the density-gradient scale-length of the ablation front is very similar (see Figure 4). However, the difference of the main drive temperature and the peak implosion velocity can still alter ablation-front GF to some extent. The highest peak implosion velocity of SR high-foot A corresponds to the largest ablation-front GF. The ablation front GF during the acceleration stage is from the acceleration Rayleigh-Taylor growth, which can be expressed as $\exp\left(\int \gamma(t)dt\right)$. Here, the $N_e = \int \gamma(t)dt$ is the so-called number of e-folding. The N_e is proportional to the peak implosion velocity.[46] The high implosion velocity results in large GF. The conventional high-foot implosion and DR high-foot B implosion have a low ablation front GF. The lower peak implosion velocity but higher compression makes the ablation front GF from the implosion of DR high-foot B nearly the same as that from the conventional high-foot implosion.

Figures 6(c) and 6(d) plot the GF measured at the fuel-ablator interface during the acceleration stage and at peak implosion velocity for the five cases. The implosion of SR

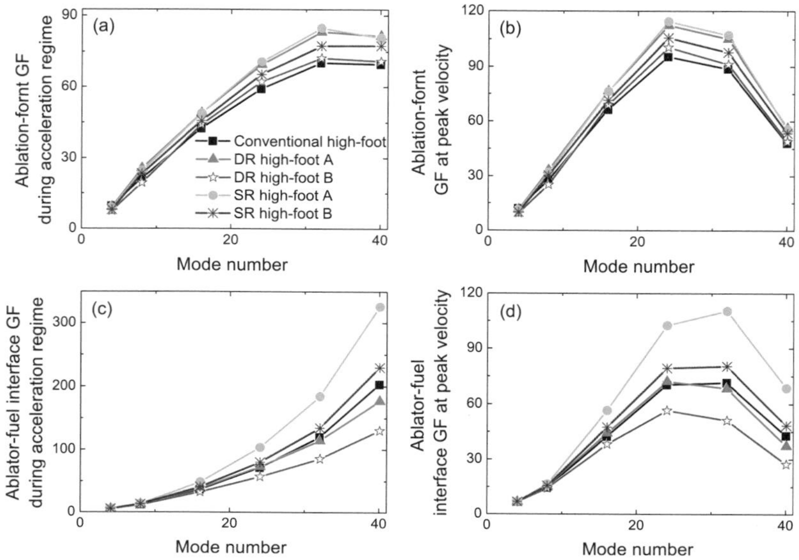

FIG. 6. Comparison of the ablation front ((a) and (b)) and ablator-fuel interface ((c) and (d)) growth factor spectrum versus mode number with perturbations initially seeded on the ablator outer surface for implosions of conventional high-foot (black), DR high-foot A (red), DR high-foot B (blue), SR high-foot A (cyan), and SR high-foot B (wine).

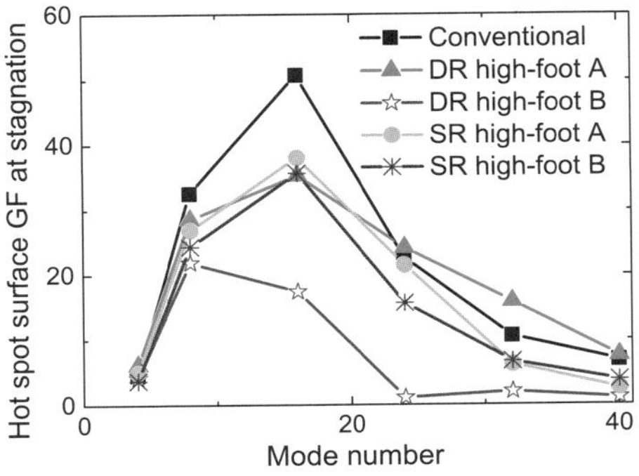

FIG. 7. Comparison of the hot-spot surface growth factor spectrum versus mode number at stagnation time with perturbations initially seeded on the ablator outer surface for the implosion with different peak pulses.

(DR) high-foot A (B) has the highest (lowest) GF at the fuel-ablator interface, which is consistent with the density profiles as shown in Figure 4. The DR pulse can effectively alleviate the fuel-ablator interface Rayleigh-Taylor growth. But the SR pulse aggravates the Rayleigh-Taylor growth at the fuel-ablator interface. As seen in Figures 6(c) and 6(d), the GF of

TABLE II. Comparison of bubble-and-spike amplitudes at stagnation time with different peak pulses for perturbation $\ell = 16$ initially seeded on the inner DT-ice surface.

	Conventional high-foot	DR high-foot A	DR high-foot B	SR high-foot A	SR high-foot B
$A_0 = 1\ \mu m$	5.1	4.2	1.9	3.3	3.6
$A_0 = 2\ \mu m$	9.5	8.5	4.2	7.0	7.2

the moderate-mode can be reduced by approximately a factor of two in DR high-foot implosions in comparison with that in SR high-foot implosions. Thus, a DR pulse would indeed help to suppress the fuel-ablator interface instability growth.

It is important to point out that the x-ray spectrum of radiation temperature $T_r(t)$ is approximated by a Planckian distribution and the ablator is not doped in the present research. Indeed, it is well known that the preheating of hard x-ray (mainly contributed by the hohlraum wall Au M-band radiation) cannot be ignored and hence the ablator should be doped to adjust the preheating intensity balancing the instability growth at the ablation front and that at the ablator-fuel interface. The non-Planckian spectrum of radiation temperature

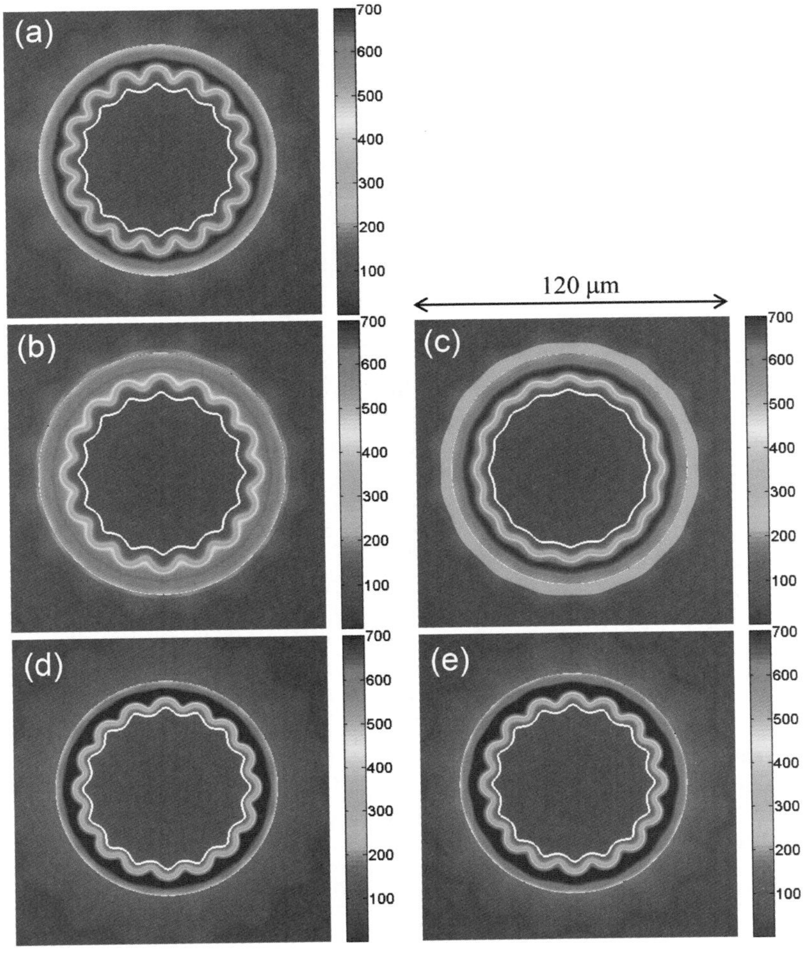

FIG. 8. Comparison of density contours at stagnation time for the conventional high-foot implosion (a), the DR high-foot A implosion (b), the DR high-foot B implosion (c), the SR high-foot A implosion (d), and the SR high-foot B implosion (e) with the perturbation $\ell = 16$ and $A_0 = 1.0$-μm initially seeded on the inner DT-ice surface. Scales are the same for all plots. The white line marks the ion temperature contour of 4.3-keV.

and silicon (Si)-doped ablator will be included in the future work.

The hot-spot surface GF is an important metric for measuring the robustness of capsules to hydrodynamic instabilities. Figure 7 compares the hot-spot GF at stagnation time for the high-foot implosions with different peak pulses. As seen in Figure 7, the maximum value of the hot-spot GF is approximately 22 for the DR high-foot B implosion, 50 for the conventional high-foot implosion, and 36 for other three implosions. Therefore, the recompression of the main drive can indeed help to suppress the hot-spot instability growth, especially the DR high-foot B implosion. It is found that the deceleration-phase Rayleigh-Taylor growth is retarded by shock collision near the waist of the inner shell surface in the implosion of DR high-foot B.[28]

When the amplitude of the perturbation becomes comparable to its wavelength, it enters the nonlinear growth regime. Then, the characteristic bubble-and-spike structure can be immediately seen. Table II summarizes a comparison of bubble-and-spike amplitudes at stagnation time among implosions with different peak pulses for the initial inner DT-ice surface perturbation $\ell = 16$ with $A_0 = 1.0$-μm and 2.0-μm, respectively. As expected, the peak pulse recompression can effectively improve the hot-spot stability, which is consistent with the GF results in Figure 7. Again, the bubble-and-spike amplitude from the DR high-foot B implosion is smallest for a fixed initial perturbation amplitude.

A more visual comparison of the inner hot-spot surface stability for implosions with different peak pulses is illustrated in Figure 8. This figure compares the stagnation time density contours among the conventional high-foot implosions, the DR high-foot implosions, and the SR high-foot implosions, for the perturbation $\ell = 16$ and $A_0 = 1.0$-μm initially seeded on the inner DT-ice surface. As can be seen obviously in Figure 8, the shell integrity and the hot-spot distortion are noticeably improved in the recompression implosions compared to the conventional high-foot implosions, especially the DR high-foot B implosion. By the way, note that the improved stagnation-stage stability was also found in shock-ignition simulations.[47–49]

V. CONCLUSIONS

In this article, we have studied the main drive optimization of a high-foot pulse shape in indirect-drive ICF implosions. The goal is to identify a peak pulse shape improving the final fuel compression without serious degradation of the implosion stability. In the pulse shape design, we follow the concept of high-foot implosions taking advantage of its stability properties but further improve the fuel compression by modifying the end of the main drive portion to generate a shock/compression wave. There are two ways of achieving the peak pulse recompression, namely, the DR[28] and SR schemes. The SR and DR high-foot implosions are compared to the conventional high-foot implosion[8] for the 1-D implosion performance and 2-D implosion stability. It is found that the peak pulse recompression (SR and DR high-foot implosions) can considerably improve the 1-D implosion performance including the fuel areal density and neutron yield. More importantly, the implosions of peak pulse recompression can substantially improve the stability of the stagnating hot-spot surface, especially the DR high-foot B implosion.

It is well known that the implosion hydrodynamic stability is a key issue for achieving ICF hot-spot ignition, especially the stability of stagnating hot-spot.[50–54] The 2-D stability analysis indicates that implosions of peak pulse recompression can improve the stability of the stagnating hot-spot surface. However, the slightly improved stagnating hot-spot surface stability of SR high-foot implosions is at the expense of reduced the shell stability during the acceleration stage, in particular the stability of the fuel-ablator interface. Radiation hydrodynamic simulations indicate that the DR high-foot B implosion can remarkably improve the 1-D performance without a loss of implosion stability, compared to the conventional high-foot implosion. In fact, it can particularly improve the stability of the hot-spot surface in the deceleration phase. Thus, the implosion of DR high-foot B deserves to be considered by the experimentalist and it can be of help in the path to ignition in ICF. Further robustness studies of DR high-foot B implosion remain a subject for the future work.

Finally, it is worth mentioning that the conclusions obtained from these simulations are of course dependent on the physics model, the parameters of equation of state and opacity, and even the numerical method used in simulations. Predicting the implosion performance via numerical simulations requires accurate modeling of many physical process involved, which is extremely challenging, especially the stagnation phase of the implosion. Additionally, the current simulation approaches does not capture all the implosion physics of indirect-drive ICF. Thus, the recompression effect of the peak pulse in a high-foot implosion is not without uncertainties. These uncertainties needed to be directly verified by the future implosion experiments.

ACKNOWLEDGMENTS

The authors would like to thank Guoli Ren, Wenyi Huo, and Ke Lan at IAPCM for their hohlraum simulations. We also thank the anonymous referee for insightful comments and suggestions that led to notable improvement of this paper. This work was supported by the Foundation of President of Chinese Academy of Engineering Physics (Grant No. 2014-1-040), the National Natural Science Foundation of China (Grant Nos. 11275031, 11675026, 11475034, 11575033, and 11274026), and National Basic Research Program of China (Grant No. 2013CB834100).

[1] J. D. Lindl, P. Amendt, R. L. Berger, S. G. Glendinning, S. H. Glenzer, S. W. Haan, R. L. Kauffman, O. L. Landen, and L. J. Suter, Phys. Plasmas **11**(2), 339 (2004).
[2] S. Atzeni and J. Meyer-ter-Vehn, *The Physics of Inertial Fusion: Beam Plasma Interaction Hydrodynamics, Hot Dense Mater* (Oxford University, Oxford, 2004).
[3] O. A. Hurricane, D. A. Callahan, D. T. Casey, E. L. Dewald, T. R. Dittrich, T. Döppner, S. Haan, D. E. Hinkel, L. F. Berzak Hopkins, O. Jones, A. L. Kritcher, S. L. Pape, T. Ma, A. G. MacPhee, J. L. Milovich, J. Moody, A. Pak, H.-S. Park, P. K. Patel, J. E. Ralph, H. F. Robey, J. S. Ross, J. D. Salmonson, B. K. Spears, P. T. Springer, R. Tommasini, F. Albert, L. R. Benedetti, R. Bionta, E. Bond, D. K. Bradley, J. Caggiano, P.

M. Celliers, C. Cerjan, J. A. Church, R. Dylla-Spears, D. Edgell, M. J. Edwards, D. fittinghoff, M. A. Barrios Garcia, A. Hamza, R. Hatarik, H. Herrmann, M. Hohenberger, D. Hoover, J. L. Kline, G. Kyrala, B. Kozioziemski, G. Grim, J. E. Field, J. Frenje, N. Izumi, M. Gatu Johnson, S. F. Khan, J. Knauer, T. Kohut, O. Landen, F. Merrill, P. Michel, A. Moore, S. R. Nagel, A. Nikroo, T. Parhan, R. R. Rygg, D. Sayre, M. Schneider, D. Shaughnessy, D. Strozzi, R. P. J. Town, D. Turnbull, P. Volegov, A. Wan, K. Widmann, C. Wilde, and C. Yeamans, "Inertially confined fusion plasmas dominated by alpha-particle self-heating," Nat. Phys. **12**, 800 (2016).

[4]T. Ma, O. A. Hurricane, D. A. Callahan, M. A. Barrios, D. T. Casey, E. L. Dewald, T. R. Dittrich, T. Döppner, S. W. Haan, D. E. Hinkel, L. F. Berzak Hopkins, S. Le Pape, A. G. MacPhee, A. Pak, H.-S. Park, P. K. Patel, B. A. Remington, H. F. Robey, J. D. Salmonson, P. T. Springer, R. Tommasini, L. R. Benedetti, R. Bionta, E. Bond, D. K. Bradley, J. Caggiano, P. Celliers, C. J. Cerjan, J. A. Church, S. Dixit, R. Dylla-Spears, D. Edgell, M. J. Edwards, J. Field, D. N. Fittinghoff, J. A. Frenje, M. Gatu Johnson, G. Grim, N. Guler, R. Hatarik, H. W. Herrmann, W. W. Hsing, N. Izumi, O. S. Jones, S. F. Khan, J. D. Kilkenny, J. Knauer, T. Kohut, B. Kozioziemski, A. Kritcher, G. Kyrala, O. L. Landen, B. J. MacGowan, A. J. Mackinnon, N. B. Meezan, F. E. Merrill, J. D. Moody, S. R. Nagel, A. Nikroo, T. Parham, J. E. Ralph, M. D. Rosen, J. R. Rygg, J. Sater, D. Sayre, M. B. Schneider, D. Shaughnessy, B. K. Spears, R. P. J. Town, P. L. Volegov, A. Wan, K. Widmann, C. H. Wilde, and C. Yeamans, Phys. Rev. Lett. **114**, 145004 (2015).

[5]T. Döeppner, D. A. Callahan, O. A. Hurricane, D. E. Hinkel, T. Ma, H.-S. Park, L. F. Berzak Hopkins, D. T. Casey, P. Celliers, E. L. Dewald, T. R. Dittrich, S. W. Haan, A. L. Kritcher, A. MacPhee, S. Le Pape, A. Pak, P. K. Patel, P. T. Springer, J. D. Salmonson, R. Tommasini, L. R. Benedetti, E. Bond, D. K. Bradley, J. Caggiano, J. Church, S. Dixit, D. Edgell, M. J. Edwards, D. N. Fittinghoff, J. Frenje, M. Gatu Johnson, G. Grim, R. Hatarik, M. Havre, H. Hermann, N. Izumi, S. F. Khan, J. L. Kline, J. Knauer, G. A. Kyrala, O. L. Landen, F. E. Merrill, J. Moody, A. S. Moore, A. Nikroo, J. E. Ralph, B. A. Remington, H. F. Robey, D. Sayre, M. Schneider, H. Streckert, R. Town, D. Turnbull, P. L. Volegov, A. Wan, K. Widmann, C. H. Wilde, and C. Yeamans, Phys. Rev. Lett. **115**, 055001 (2015).

[6]D. A. Callahan, O. A. Hurricane, D. E. Hinkel, T. Döeppnerppner, T. Ma, H.-S. Park, M. A. Barrios Garcia, L. F. Berzak Hopkins, D. T. Casey, C. J. Cerjan, E. L. Dewald, T. R. Dittrich, M. J. Edwards, S. W. Haan, A. V. Hamza, J. L. Kline, J. P. Knauer, A. L. Kritcher, O. L. Landen, S. LePape, A. G. MacPhee, J. L. Milovich, A. Nikroo, A. E. Pak, P. K. Patel, J. R. Rygg, J. E. Ralph, J. D. Salmonson, B. K. Spears, P. T. Springer, R. Tommasini, L. R. Benedetti, R. M. Bionta, E. J. Bond, D. K. Bradley, J. A. Caggiano, J. E. Field, D. N. Fittinghoff, J. Frenje, M. Gatu Johnson, G. P. Grim, R. Hatarik, F. E. Merrill, S. R. Nagel, N. Izumi, S. F. Khan, R. P. J. Town, D. B. Sayre, P. Volegov, and C. H. Wilde, Phys. Plasmas **22**, 056314 (2015).

[7]M. J. Edwards, P. K. Patel, J. D. Lindl, L. J. Atherton, S. H. Glenzer, S. W. Haan, J. D. Kilkenny, O. L. Landen, E. I. Moses, A. Nikroo, R. Petrasso, T. C. Sangster, P. T. Springer, S. Batha, R. Benedetti, L. Bernstein, R. Betti, D. L. Bleuel, T. R. Boehly, D. K. Bradley, J. A. Caggiano, D. A. Callahan, P. M. Celliers, C. J. Cerjan, K. C. Chen, D. S. Clark, G. W. Collins, E. L. Dewald, L. Divol, S. Dixit, T. Deoppner, D. H. Edgell, J. E. Rai, M. Farrell, R. J. Fortner, J. Frenje, M. B. Gatu Johnson, E. Giraldez, V. Yu. Glebov, G. Grim, B. A. Hammel, A. V. Hamza, D. R. Harding, S. P. Hatchett, N. Hein, H. W. Herrmann, D. Hicks, D. E. Hinkel, M. Hoppe, W. W. Hsing, N. Izumi, B. Jacoby, O. S. Jones, D. Kalantar, R. Kauffman, J. L. Kline, J. P. Knauer, J. A. Koch, B. J. Kozioziemski, G. Kyrala, K. N. LaFortune, S. Le Pape, R. J. Leeper, R. Lerche, T. Ma, B. J. MacGowan, A. J. MacKinnon, A. Macphee, E. R. Mapoles, M. M. Marinak, M. Mauldin, P. W. McKenty, M. Meezan, P. A. Michel, J. Milovich, J. D. Moody, M. Moran, D. H. Munro, C. L. Olson, K. Opachich, A. E. Pak, T. Parham, H.-S. Park, J. E. Ralph, S. P. Regan, B. Remington, H. Rinderknecht, H. F. Robey, M. Rosen, S. Ross, J. D. Salmonson, J. Sater, D. H. Schneider, F. H. Séguin, S. M. Sepke, D. A. Shaughnessy, V. A. Smalyuk, B. K. Spears, C. Stoeckl, W. Stoeffl, L. Suter, C. A. Thomas, R. Tommasini, R. P. Town, S. V. Weber, P. J. Webner, K. Widman, M. Wilke, D. C. Wilson, C. B. Yeamans, and A. Zylstra, Phys. Plasmas **20**, 070501 (2013).

[8]O. A. Hurricane, D. A. Callahan, D. T. Casey, P. M. Celliers, C. Cerjan, E. L. Dewald, T. R. Dittrich, T. Doeppner, D. E. Hinkel, L. F. Berzak Hopkins, J. L. Kline, S. LePape, T. Ma, A. G. MacPhee, J. L. Milovich, A. Pak, H.-S. Park, P. K. Patel, B. A. Remington, J. D. Salmonson, P. T. Springer, and R. Tommasini, Nature **506**, 343 (2014).

[9]O. A. Hurricane, D. A. Callahan, D. T. Casey, E. L. Dewald, T. R. Dittrich, T. Döppner, M. A. Barrios Garcia, D. E. Hinkel, L. F. Berzak Hopkins, P. Kervin, J. L. Kline, S. L. Pape, T. Ma, A. G. MacPhee, J. L. Milovich, J. Moody, A. E. Pak, P. K. Patel, H.-S. Park, B. A. Remington, H. F. Robey, J. D. Salmonson, P. T. Springer, R. Tommasini, L. R. Benedetti, J. A. Caggiano, P. Celliers, C. Cerjan, R. Dylla-Spears, D. Edgell, M. J. Edwards, D. Fittinghoff, G. P. Grim, N. Guler, N. Izumi, J. A. Frenje, M. Gatu Johnson, S. Haan, R. Hatarik, H. Herrmann, S. Khan, J. Knauer, B. J. Kozioziemski, A. L. Kritcher, G. Kyrala, S. A. Maclaren, F. E. Merrill, P. Michel, J. Ralph, J. S. Ross, J. R. Rygg, M. B. Schneider, B. K. Spears, K. Widmann, and C. B. Yeamans, Phys. Plasmas **21**, 056314 (2014).

[10]T. R. Dittrich, O. A. Hurricane, D. A. Callahan, E. L. Dewald, T. Döppner, D. E. Hinkel, L. F. Berzak Hopkins, S. LePape, T. Ma, J. Milovich, J. C. Moreno, P. K. Patel, H.-S. Park, B. A. Remington, and J. Salmonson, Phys. Rev. Lett. **112**, 055002 (2014).

[11]H.-S. Park, O. A. Hurricane, D. A. Callahan, D. T. Casey, E. L. Dewald, T. R. Dittrich, T. Döppner, D. E. Hinkel, L. F. Berzak Hopkins, S. Le Pape, T. Ma, P. K. Patel, B. A. Remington, H. F. Robey, and J. D. Salmonson, Phys. Rev. Lett. **112**, 055001 (2014).

[12]D. S. Clark, C. R. Weber, J. L. Milovich, J. D. Salmonson, A. L. Kritcher, S. W. Haan, B. A. Hammel, D. E. Hinkel, O. A. Hurricane, O. S. Jones, M. M. Marinak, P. K. Patel, H. F. Robey, S. M. Sepke, and M. J. Edwards, Phys. Plasmas **23**, 056302 (2016).

[13]J. L. Peterson, D. S. Clark, L. P. Masse, and L. J. Suter, Phys. Plasmas **21**, 092710 (2014).

[14]D. S. Clark, J. L. Milovich, D. E. Hinkel, J. D. Salmonson, J. L. Peterson, L. F. Berzak Hopkins, D. C. Eder, S. W. Haan, O. S. Jones, M. M. Marinak, H. F. Robey, V. A. Smalyuk, and C. R. Weber, Phys. Plasmas **21**, 112705 (2014).

[15]A. G. MacPhee, J. L. Peterson, D. T. Casey, D. S. Clark, S. W. Haan, O. S. Jones, O. L. Landen, J. L. Milovich, H. F. Robey, and V. A. Smalyuk, Phys. Plasmas **22**, 080702 (2015).

[16]D. T. Casey, V. A. Smalyuk, K. S. Raman, J. L. Peterson, L. Berzak Hopkins, D. A. Callahan, D. S. Clark, E. L. Dewald, T. R. Dittrich, S. W. Haan, D. E. Hinkel, D. Hoover, O. A. Hurricane, J. J. Kroll, O. L. Landen, A. S. Moore, A. Nikroo, H.-S. Park, B. A. Remington, H. F. Robey, J. R. Rygg, J. D. Salmonson, R. Tommasini, and K. Widmann, Phys. Rev. E **90**, 011102 (2014).

[17]D. T. Casey, J. L. Milovich, V. A. Smalyuk, D. S. Clark, H. F. Robey, A. Pak, A. G. MacPhee, K. L. Baker, C. R. Weber, T. Ma, H.-S. Park, T. Döppner, D. A. Callahan, S. W. Haan, P. K. Patel, J. L. Peterson, D. Hoover, A. Nikroo, C. B. Yeamans, F. E. Merrill, P. L. Volegov, D. N. Fittinghoff, G. P. Grim, M. J. Edwards, O. L. Landen, K. N. Lafortune, B. J. MacGowan, C. C. Widmayer, D. B. Sayre, R. Hatarik, E. J. Bond, S. R. Nagel, L. R. Benedetti, N. Izumi, S. Khan, B. Bachmann, B. K. Spears, C. J. Cerjan, M. Gatu Johnson, and J. A. Frenje, Phys. Rev. Lett. **115**, 105001 (2015).

[18]K. L. Baker, H. F. Robey, J. L. Milovich, O. S. Jones, V. A. Smalyuk, D. T. Casey, A. G. MacPhee, A. Pak, P. M. Celliers, D. S. Clark, O. L. Landen, J. L. Peterson, L. F. Berzak-Hopkins, C. R. Weber, S. W. Haan, T. D. Döppner, S. Dixit, E. Giraldez, A. V. Hamza, K. S. Jancaitis, J. J. Kroll, K. N. Lafortune, B. J. MacGowan, J. D. Moody, A. Nikroo, C. C. Widmayer *et al.*, Phys. Plasmas **22**, 052702 (2015).

[19]J. L. Milovich, H. F. Robey, D. S. Clark, K. L. Baker, D. T. Casey, C. Cerjan, J. Field, A. G. MacPhee, A. Pak, P. K. Patel, J. L. Peterson, V. A. Smalyuk, and C. R. Weber, Phys. Plasmas **22**, 122702 (2015).

[20]V. A. Smalyuk, H. F. Robey, T. Döppner, O. S. Jones, J. L. Milovich, B. Bachmann, K. L. Baker, L. F. Berzak Hopkins, E. Bond, D. A. Callahan, D. T. Casey, P. M. Celliers, C. Cerjan, D. S. Clark, S. N. Dixit, M. J. Edwards, E. Giraldez, S. W. Haan, A. V. Hamza, M. Hohenberger, D. Hoover, O. A. Hurricane, K. S. Jancaitis, J. J. Kroll, K. N. Lafortune, O. L. Landen, B. J. MacGowan, A. G. MacPhee, A. Nikroo, A. Pak, P. K. Patel, J. L. Peterson, C. R. Weber, C. C. Widmayer, and C. Yeamans, Phys. Plasmas **22**, 080703 (2015).

[21]H. F. Robey, V. A. Smalyuk, J. L. Milovich, T. Döppner, D. T. Casey, K. L. Baker, J. L. Peterson, B. Bachmann, L. F. Berzak Hopkins, E. Bond, J. A. Caggiano, D. A. Callahan, P. M. Celliers, C. Cerjan, D. S. Clark, S. N. Dixit, M. J. Edwards, N. Gharibyan, S. W. Haan, B. A. Hammel, A. V. Hamza, R. Hatarik, O. A. Hurricane, K. S. Jancaitis, O. S. Jones, G. D. Kerbel, J. J. Kroll, K. N. Lafortune, O. L. Landen, T. Ma, M. M. Marinak, B. J. MacGowan, A. G. MacPhee, A. Pak, M. Patel, P. K. Patel, L. J.

Perkins, D. B. Sayre, S. M. Sepke, B. K. Spears, R. Tommasini, C. R. Weber, C. C. Widmayer, C. Yeamans, E. Giraldez, D. Hoover, A. Nikroo, M. Hohenberger, and M. Gatu Johnson, Phys. Plasmas **23**, 056303 (2016).
[22]J. H. Gardner, S. E. Bodner, and J. P. Dahlburg, Phys. Fluids B **3**, 1070 (1991).
[23]S. E. Bodner, D. G. Colombant, J. H. Gardner, R. H. Lehmberg, S. P. Obenschain, L. Phillips, A. J. Schmitt, J. D. Sethian, R. L. McCrory, W. Seka, C. P. Verdon, J. P. Knauer, B. B. Afeyan, and H. T. Powell, Phys. Plasmas **5**, 1901 (1998).
[24]V. N. Goncharov, J. P. Knauer, P. W. McKenty, P. B. Radha, T. C. Sangster, S. Skupsky, R. Betti, R. L. McCrory, and D. D. Meyerhofer, Phys. Plasmas **10**, 1906 (2003).
[25]K. Anderson and R. Betti, Phys. Plasmas **10**, 4448 (2003).
[26]K. Anderson and R. Betti, Phys. Plasmas **11**, 5 (2004).
[27]R. Betti, K. Anderson, J. Knauer, T. J. B. Collins, R. L. McCrory, P. W. McKenty, and S. Skupsky, Phys. Plasmas **12**, 042703 (2005).
[28]L. F. Wang, W. H. Ye, J. F. Wu, J. Liu, W. Y. Zhang, and X. T. He, Phys. Plasmas **23**, 052713 (2016).
[29]J. D. Lindl, Phys. Plasmas **2**, 3933 (1995).
[30]X. T. He and W. Y. Zhang, Eur. Phys. J. D **44**, 227 (2007).
[31]W.-H. Ye, W.-H. Zhang, and G.-H. Chen, High Power Laser Part. Beams **10**, 403 (1998) (in Chinese).
[32]W. Ye, W. Zhang, and X. T. He, Phys. Rev. E **65**, 57401 (2002).
[33]L. F. Wang, C. Xue, W. H. Ye, and Y. J. Li, Phys. Plasmas **16**, 112104 (2009).
[34]L. F. Wang, W. H. Ye, Z. M. Sheng, W.-S. Don, Y. J. Li, and X. T. He, Phys. Plasmas **17**, 122706 (2010).
[35]L. F. Wang, W. H. Ye, W.-S. Don, Z. M. Sheng, Y. J. Li, and X. T. He, Phys. Plasmas **17**, 122308 (2010).
[36]L. F. Wang, W. H. Ye, X. T. He, W. Y. Zhang, Z. M. Sheng, and M. Y. Yu, Phys. Plasmas **19**, 100701 (2012).
[37]X. T. He, J. W. Li, Z. F. Fan, L. F. Wang, J. Liu, K. Lan, J. F. Wu, and W. H. Ye, Phys. Plasmas **23**, 082706 (2016).
[38]L. F. Wang, W. H. Ye, W. Y. Zhang, and X. T. He, Phys. Scr. **T155**, 014018 (2013).
[39]F. J. D. Serduke, E. Minguez, S. J. Davidson, and C. A. Iglesias, J. Quantum Spectrosc. Radiat. Transfer **65**, 527 (2000).
[40]R. M. More, K. H. Warren, D. A. Young, and G. B. Zimmerman, Phys. Fluids **31**(10), 3059 (1988).
[41]S. W. Haan, M. C. Herrmann, T. R. Dittrich, A. J. Fetterman, M. M. Marinak, D. H. Munro, S. M. Pollaine, J. D. Salmonson, G. L. Strobel, and L. J. Suter, Phys. Plasmas **12**, 056316 (2005).
[42]K. Lan, J. Liu, Z. Li, X. Xie, W. Huo, Y. Chen, G. Ren, C. Zheng, D. Yang, S. Li, Z. Yang, L. Guo, S. Li, M. Zhang, X. Han, C. Zhai, L. Hou, Y. Li, K. Deng, Z. Yuan, X. Zhang, F. Wang, G. Yuan, H. Zhang, B. Jiang, L. Huang, W. Zhang, K. Du, K. Zhao, P. Li, W. Wang, J. Su, X. Deng, D. Hu, W. Zhou, H. Jia, Y. Ding, W. Zhang, and X. He, Matter Radiat. Extremes **1**, 8 (2016).
[43]L. F. Wang, H. Y. Guo, J. F. Wu, W. H. Ye, J. Liu, W. Y. Zhang, and X. T. He, Phys. Plasmas **21**, 122710 (2014).
[44]S. W. Haan, J. D. Lindl, D. A. Callahan, D. S. Clark, J. D. Salmonson, B. A. Hammel, L. J. Atherton, R. C. Cook, M. J. Edwards, S. Glenzer, A. V. Hamza, S. P. Hatchett, D. E. Hinkel, D. D. Ho, O. S. Jones, O. L. Landen, B. J. MacGowan, M. M. Marinak, J. L. Milovich, E. I. Moses, D. H. Munro, S. M. Pollaine, J. E. Ralph, H. F. Robey, B. K. Spears, P. T. Springer, L. J. Suter, C. A. Thomas, R. P. Town, S. V. Weber, D. C. Wilson, G. Kyrala, M. C. Herrmann, R. E. Olson, K. Peterson, R. Vesey, D. D. Meyerhofer, A. Nikroo, H. L. Wilkens, H. Huang, and K. A. Moreno, Phys. Plasmas **18**, 051001 (2011).
[45]K. Nishihara, M. Murakami, H. Azechi, T. Jitsuno, T. Kanabe, M. Katayama, N. Miyanaga, M. Nakai, M. Nakatsuka, K. Tsubakimoto, and S. Nakai, Phys. Plasmas **1**, 1653 (1994).
[46]C. D. Zhou and R. Betti, Phys. Plasmas **14**, 072703 (2007).
[47]X. Ribeyre, G. Schurtz, M. Lafon, S. Galera, and S. Weber, Plasma Phys. Controlled Fusion **51**, 015013 (2009).
[48]S. Atzeni, X. Ribeyre, G. Schurtz, A. J. Schmitt, B. Canaud, R. Betti, and L. J. Perkins, Nucl. Fusion **54**, 054008 (2014).
[49]S. Atzeni, A. Marocchino, and A. Schiavi, Plasma Phys. Controlled Fusion **57**, 014022 (2015).
[50]S. W. Haan, S. M. Pollaine, J. D. Lindl, L. J. Suter, R. L. Berger, L. V. Powers, W. E. Alley, P. A. Amendt, J. A. Futterman, W. Kirk Levedahl, M. D. Rosen, D. P. Rowley, R. A. Sacks, A. I. Shestakov, G. L. Strobe, M. Tabak, S. V. Weber, and G. B. Zimmerman, Phys. Plasmas **2**, 2480 (1995).
[51]M. C. Herrmann, M. Tabak, and J. D. Lindl, Phys. Plasmas **8**, 2296 (2001).
[52]R. Betti, K. Anderson, V. N. Goncharov, R. L. Mccrory, D. D. Meyerhofer, S. Skupsky, and R. P. J. Town, Phys. Plasmas **9**, 2277 (2002).
[53]S. Atzeni, A. Schiavi, and M. Temporal, Plasma Phys. Controlled Fusion **46**, B111 (2004).
[54]B. Srinivasan and X.-Z. Tang, Phys. Plasmas **21**, 102704 (2014).

温稠密物质

温稠密物质（warm dense matter，WDM）是高能量密度状态下的一种十分重要的物质，是介于通常凝聚态物质和经典等离子体之间的一种物质。它是高压和温度 T 小于费米温度 T_F 状态下电子、离子（部分电离）、原子、分子和它们的聚束共存并相互作用的多成分集合体，极化（电荷屏蔽）和自能移动效应起着十分重要作用。

WDM 是 ICF 和武器内爆过程中以及很多天体行星中存在的高密度物质，是具有部分电离和量子效应的重要高压物质。长期以来，没有一种理论能有效地预言 WDM 的性质。其困难是：随着温度升高，原子（离子）在电离边界附近产生了近无限多激发能级。很长时间内，国际上没有一种准确方法能在有限时间内大量仔细计算这些激发能级，给出诸如 WDM 状态方程和输运系数等信息。

贺贤土院士在 WDM 特性研究方面取得了重要进展。他给出了 WDM 部分简并状态的近似边界，揭示了电荷屏蔽产生的相互作用能对束缚能级合并、加宽和向电离边界移动的显著影响。同时他领导的WDM研究小组发现，在电离边界附近，绝大部分电离电子动能可以很好地满足 Born 近似条件，因此密度泛函理论中薛定锷方程解就可用平面波近似解。这一关键发现使得长期的难题基本得到了解决。基于这一发现，贺贤土院士领导的 WDM 研究小组改造了传统的量子分子动力学（quantum molecular dynamics，QMD）程序，新的“Extended QMD”程序比传统的节省了几千倍的计算时间，使得 ICF 和核武器内爆过程物质以及很多星体内部物质的状态方程和输运系数的数据库精确计算成为可能，引起了国际同行的关注。

Classical Aspects in Above-Threshold Ionization with a Midinfrared Strong Laser Field

W. Quan,[1] Z. Lin,[1] M. Wu,[1] H. Kang,[1] H. Liu,[1] X. Liu,[1,*] J. Chen,[2,3,†] J. Liu,[2,3] X. T. He,[2,3] S. G. Chen,[3] H. Xiong,[4] L. Guo,[4] H. Xu,[4] Y. Fu,[4] Y. Cheng,[4,‡] and Z. Z. Xu[4,§]

[1]*State Key Laboratory of Magnetic Resonance and Atomic and Molecular Physics, Wuhan Institute of Physics and Mathematics, Chinese Academy of Sciences, Wuhan 430071, China*

[2]*Center for Applied Physics and Technology, Peking University, Beijing 100084, China*

[3]*Institute of Applied Physics and Computational Mathematics, P.O. Box 8009, Beijing 100088, China*

[4]*State Key Laboratory of High Field Laser Physics, Shanghai Institute of Optics and Fine Mechanics, Chinese Academy of Sciences, P.O. Box 800-211, Shanghai 201800, China*

(Received 5 January 2009; revised manuscript received 26 March 2009; published 24 August 2009; publisher error corrected 31 August 2009)

We present high resolution photoelectron energy spectra of noble gas atoms from high intensity above-threshold ionization (ATI) at midinfrared wavelengths. An unexpected structure at the very low-energy portion of the spectra, in striking contrast to the prediction of the simple-man theory, has been revealed. A semiclassical model calculation is able to reproduce the experimental feature and suggests the prominent role of the Coulomb interaction of the outgoing electron with the parent ion in producing the peculiar structure in long wavelength ATI spectra.

DOI: 10.1103/PhysRevLett.103.093001 PACS numbers: 32.80.Rm, 32.80.Wr, 42.50.Hz

When exposed to a strong laser field, atoms may absorb more photons than necessary for ionization, leading to a series of peaks separated by one photon energy in the photoelectron spectrum. This process, coined as above-threshold ionization (ATI), was discovered almost 30 years ago [1]. Since then, ATI has been continuously advancing our understanding of strong-field physics [2].

One of the most prominent features in ATI, in the long wavelength, high intensity limit, is the rapid decay of the ionization rate with the electron energy in the low-energy part of the photoelectron spectrum, followed by a high-energy plateau [3] extending up to $10U_p$ ($U_p = I/4\omega^2$, denotes the ponderomotive energy, where I is the laser intensity and ω the frequency). The physical insight therein was found to be rooted in a rather simple semiclassical simple-man description of the ionization process [4], based on the pioneering work by Keldysh [5]. Here, a pivotal role is given to the Keldysh parameter $\gamma = \sqrt{I_p/2U_p}$, where I_p is the ionization potential. Under the condition of $\gamma \ll 1$, the simple-man picture of the ionization dynamics may be verified. Therein, the motion of the active electron, initially tunnel ionized, is entirely determined by its interaction with the laser electric field, while the effect of the binding potential is neglected.

Despite the great success and pivotal role of this intuitive semiclassical picture in understanding ATI processes of atoms in the strong field, many puzzles still exist. For example, the photoelectron spectra of argon show a resonancelike structure at intermediate electron energy [6], which is beyond the scope of the semiclassical picture and still under debate [7]. Furthermore, recent experiments using the sophisticated cold target recoil-ion momentum spectroscopy (COLTRIMS) show a clear minimum at zero in the electron momentum distribution along the laser polarization direction [8], obviously deviating from the prediction of the simple-man theory [4]. Theoretical studies with the semiclassical approach [9,10] have been focused on this peculiar low electron momentum structure, suggesting a prominent role of the Coulomb potential. However, further COLTRIMS experiments [11,12] reveal many resonancelike peaks in the electron momentum distribution, a characteristic of the multiphoton ionization (MPI). As these experiments were mostly performed at 800 nm wavelength, one difficulty one may encounter is to reliably stay in the regime in which the semiclassical picture is valid (i.e., $\gamma \ll 1$).

In principle, the semiclassical regime could be easily reached by increasing the laser wavelength, compared to the intensity, considering the fact that $U_p \propto I\lambda^2$. Thanks to the advances in ultrafast laser technology, tunable long wavelength lasers with focusable peak intensities around 10^{14} W/cm^2 have become available recently, providing a potential for going deeply into the tunnel regime without saturated ionization of target atoms or molecules. Indeed, a pioneering experiment with midinfrared wavelengths has been performed very recently and a λ^2 scaling of the classical behavior both in ATI and closely related high harmonic generation was observed [13], enabling the classical limit to be approached with unprecedented power.

In this Letter, we present the ATI spectra of noble gas atoms in ultrashort intense laser pulses at midinfrared wavelengths. In particular, the application of long wavelength provides the opportunity to go deep in the semiclassical regime, exploring the classical aspects in strong-field physics more closely. Indeed, our data at long wavelengths show a pronounced structure in the very low-energy part of photoelectron spectra. While this finding is in striking contrast to the prediction of the simple-man theory, a simulation based on a semiclassical model is able to reproduce this feature qualitatively and suggests it as a classical aspect in ATI at long wavelengths.

In our experiments, wavelength-tunable midinfrared femtosecond laser pulses are generated by an optical parametric amplifier (OPA, TOPAS-C, Light Conversion, Inc.) pumped by a commercial Ti:sapphire laser system (Legend, Coherent, Inc.) [14]. The pulse energy from OPA is variable, before focused into the vacuum chamber, by means of an achromatic half wave plate followed by a polarizer. The photoelectron energy spectra are measured with a field-free time-of-flight spectrometer. By means of a cryopump, the base pressure of the spectrometer is maintained below 10^{-8} mbar. Photoionized electrons are detected by microchannel plates that subtend a 5° angle at the light-atom interaction point, and are recorded by a multihit time-to-digital converter with an ultimate resolution of 25 ps (TDC8HP, Roentdek Handels GmbH). An energy resolution of about 20 meV for 1 eV electrons has been measured for this spectrometer.

Figures 1(a), 1(c), and 1(e) show the low-energy part of the measured photoelectron spectra of xenon at different laser wavelengths and intensities. The data at longer wavelengths, e.g., 1500 and 2000 nm, reveal a pronounced structure at very low electron energy. This structure seemingly consists of two humps at higher laser intensities [see Figs. 1(c) and 1(e)]. The first hump lies at electron energy below 1 eV, while the second broader one at slightly higher electron energy exhibits a strong dependence on both the laser wavelength and intensity. At the shortest wavelength of 800 nm, the latter is hardly seen, while distinct resonant structures within each ATI order, an indication that the MPI process is dominant, become prominent. We note that this feature is also present for the other noble gas species we have studied, e.g., krypton and argon, and thus seems to be a universal phenomenon for high intensity ATI spectra at long wavelengths.

This peculiar low-energy structure, which becomes more significant at 2000 nm, is in striking contrast to the prediction of the simple-man theory, while, with which, the overall evolution of the ATI spectra, e.g., the appearance of the well-known plateau [3] and an approximately quadratically increase of the 10 U_p cutoff with the laser wavelength λ, is well predicted [see the inset in Fig. 1(a)]. In the remaining part of this Letter, we will focus on this feature at long wavelengths, with emphasis on the second hump.

One major question concerns whether the second hump that well develops at long wavelengths, e.g., 2000 nm, is a classical aspect in the ATI spectra, as for the well-established high-energy plateau. We address this question by comparing the experimental data with the results of the theoretical simulations based on a semiclassical description of the ionization process.

Details of the semiclassical approach have been described elsewhere [15]. In brief, we consider a xenon

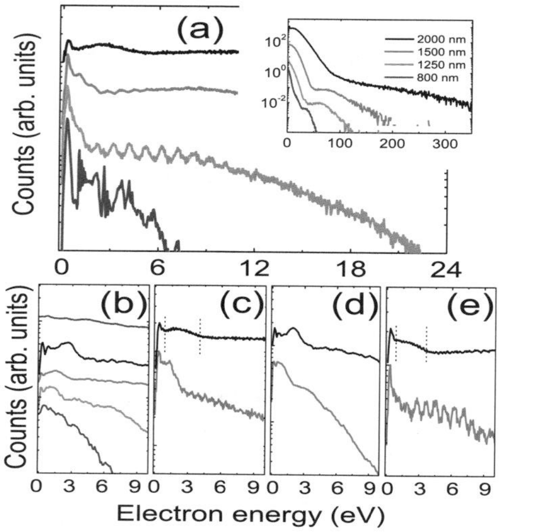

FIG. 1 (color online). Experimental [(a), (c), and (e)] and calculated [(b),(d)] photoelectron spectra of xenon. (a) $I = 8.0 \times 10^{13}$ W/cm^2, $\lambda = 800$, 1250, 1500, and 2000 nm from bottom to top, respectively. The complete spectra are shown in the inset. The laser pulse durations are ~40 fs at 800 nm, ~30 fs at 1250 and 1500 nm, while ~90 fs at 2000 nm. (b) $I = 8.0 \times 10^{13}$ W/cm^2 and $\lambda = 800$, 1250, 1500, and 2000 nm, with Coulomb potential for the curves from bottom to top, respectively. While the uppermost curve is for $I = 8.0 \times 10^{13}$ W/cm^2 and $\lambda = 2000$ nm without Coulomb potential. (c),(d) $\lambda = 2000$ nm, $I = 3.2$, 6.4×10^{13} W/cm^2 for the lower and upper curves, respectively. (e) $\lambda = 1500$ nm, $I = 4.0$, 10.0×10^{13} W/cm^2 for the lower and upper curves, respectively. In (c) and (e), the boundaries of the second hump are indicated by the dashed lines for higher intensities. The curves are shifted in the vertical direction for visual convenience.

atom (with hydrogenlike potential of $V = -Z_{\text{eff}}/r$ where Z_{eff} is the effective nuclear charge) interacting with an external laser field. The laser field $\epsilon(t) = f(t)E_0 \cos\omega t$ has a constant amplitude for the first ten cycles and is ramped off within 3-cycles. For simplification, the electron is released within the time interval $-\pi/2 \le \omega t \le \pi/2$ through tunneling with a rate given by the ADK formula [16]. The subsequent evolution of the electron is determined by Newton's equation of motion. This model has achieved great success in explaining various strong-field ionization phenomena [15] under the circumstances that the semiclassical picture can be verified, which should be the case for our long wavelength data (e.g., $\gamma = 0.45$ at 2000 nm and 8×10^{13} W/cm^2).

In Figs. 1(b) and 1(d), the calculated photoelectron spectra of xenon at various wavelengths and intensities are shown. To compare directly with the data, in the calculation, we select out only those electrons emitted along the field direction. In general, the calculation reproduces qualitatively the overall features in the data at long wavelengths, i.e., a pronounced low-energy structure, the separation into two humps, and the significance of the second hump with the increase of the intensity and wavelength. For shorter wavelengths, the discrepancy between the data and simulation becomes non-negligible. This may not be a surprise, considering the fact that the MPI mechanism, which is beyond the scope of the semiclassical treatment, becomes prominent. Note that the first hump can also be reproduced to some extent at long wavelengths in the simulation. However, the persistence of this hump structure in the data even at, e.g., 800 nm, indicates that other mechanisms may play a role at shorter wavelengths, which is still under investigation. Upon the reasonable agreement at long wavelengths, one may conclude that the peculiar low-energy structure is a pure classical aspect in strong-field single ionization. Moreover, the calculated result without including the Coulomb potential, as shown in Fig. 1(b), shows that the structure completely disappears. In the following, we will further demonstrate that the structure is a consequence of the Coulomb interaction between the electron and the parent ion [17]. Note that the influence of Coulomb potential on single ionization of atoms has been studied at laser wavelength of 800 nm recently [9,10,18].

In order to facilitate the analysis, we present the calculated density plot of the two-dimensional electron momentum distributions, and classify the electron trajectories that contribute to the distributions into two categories, i.e., direct electrons and the indirect ones, by their initial tunneling ionization times. In the context of simple-man theory, the direct electron is tunnel ionized in the rising edge, i.e., $\omega t \in [-\pi/2, 0]$, of the oscillating laser electric field, and leaves directly without further interaction with the ion core. In contrast, the indirect one may return and collide with the core after initially ionized in the falling edge ($\omega t \in [0, \pi/2]$). It is well known that the latter one, if backscattered upon the ion core, produces the high-energy plateau in ATI spectra [3]. Moreover, to reveal the distinct role of the Coulomb potential for direct and indirect electrons, respectively, we compare the results with and without including the Coulomb potential.

Figures 2(a) and 2(c) depict the two-dimensional momentum distributions for electrons initially tunnel ionized in the rising and falling edges of the oscillating laser electric field, respectively, without including the Coulomb potential. Not surprisingly, the results show that most electrons are distributed around momentum of zero, in both parallel and perpendicular directions with respect to the laser polarization, in accordance with the simple-man model.

In Figs. 2(b) and 2(d), the calculations are made with the Coulomb interaction included. It is obvious that the overall momentum distribution in the direction perpendicular to the laser electric field is much narrower than that without considering the Coulomb potential and, in particular, the electrons even concentrate in the field direction in the low momentum region. For both the rising and falling edges, one sees that the momentum distributions clearly shift in the laser polarization direction. This shift can be understood as the overall influence of the Coulomb potential on the wave packet's longitudinal momentum of the electron as it leaves the saddle point after tunneling. Moreover, for the rising edge, the ionized electrons are accumulated in a small region which is close to, but deviates from, zero momentum. The energy position of this highly distributed region is in accordance with the first hump in the photoelectron spectra. For the falling edge, the electrons are accumulated in two small regions. One is close to zero while its center position is in between the two main humps in the calculated photoelectron spectra, which is, however, not distinguishable in the data. The other one is wider in the direction of the field and centers around 0.42 a.u., which corresponds to the kinetic energy of 2.4 eV, approximately the center position of the observed second hump.

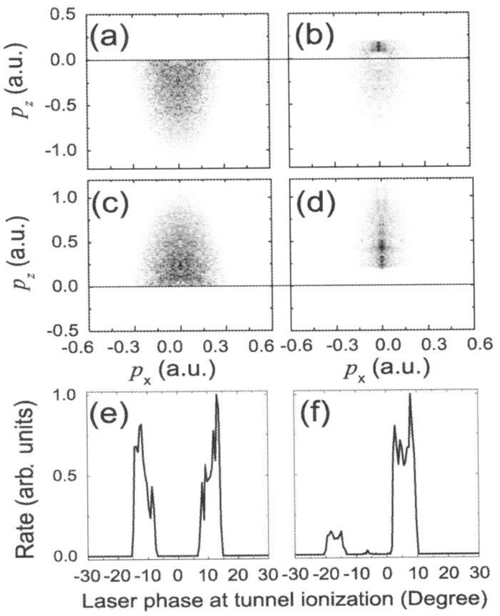

FIG. 2. (a)–(d) Calculated electron momentum distributions (z is the laser polarization direction). $I = 8 \times 10^{13}$ W/cm^2 and $\lambda = 2000$ nm. See the text for the details. (e),(f) Electron yields as a function of the laser phase at the moment of tunnel ionization, without and with the Coulomb potential considered, respectively. Here the statistics are only collected for the electron trajectories that contribute to the energy range of the second hump structure (i.e., $1.0 < E_{kin} < 3.7$ eV, where E_{kin} is the electron energy) in Fig. 1(b) at $I = 8 \times 10^{13}$ W/cm^2.

From Fig. 2(b), it indicates that, in our semiclassical treatment, the first hump comes from the electrons ionized very close to the crest of the field in the rising edge. The trajectories of these electrons, which are expected to leave the ion directly according to the simple-man picture, are significantly changed by the ion potential and hence the electrons may return to the core and interact with it strongly, leading to the intense concentration in the momentum and energy spectra. For the falling edge shown in Fig. 2(d), the electrons returning to the ion with small kinetic energies will be influenced by the attractive ion potential strongly. Therefore, not only the electrons' transverse momenta but also the final energies of part of these electrons will be changed, resulting in the first concentration region in Fig. 2(d). When the phase of the ionization time increases in the falling edge, the electron returns to the ion with larger kinetic energy and therefore the core potential can no longer change its final momentum in the field direction significantly. However, the same as in the first concentration region, the momenta of these electrons transverse to the field direction are also significantly changed by the ion potential, resulting in electron's accumulation in the field direction. This phenomenon related to the indirect electron can be referred to as the Coulomb focusing effect, which has been studied in the single ionization process [19] and manifests its important role in the double ionization process [20].

To further understand how the Coulomb potential affects the indirect electron dynamics and produces the second hump in the electron spectra, we collect all trajectories that constitute this hump structure and trace back their tunneling ionization times. Without considering the Coulomb potential, the tunneling times are confined in two regions, one in the rising edge and the other in the falling edge, which are symmetric with respect to the origin [Fig. 2(e)]. This is consistent with the simple-man prediction, in which the final electron energy is solely determined by the laser phase when the electron tunnels out. In contrast to this, the distribution of the tunneling time with Coulomb potential included [Fig. 2(f)] shows that most electrons are emitted in the falling edge, i.e., the contributions are mainly from the indirect electron. Moreover, both phase distributions in Fig. 2(f) shift to the left compared to that in Fig. 2(e).

One more intriguing point is that the right boundary of the right phase distribution in Fig. 2(f) is very close to the phase of 12°, the maximum phase at which the tunneled electron may experience multiple returns, in the context of simple-man theory. This agrees with the above analysis that the second hump is due to the Coulomb focusing effect. More details will be given elsewhere.

In conclusion, we have performed an experimental study on ATI of noble gas atoms at midinfrared wavelengths. An unexpected structure has been revealed in the very low-energy part of the measured photoelectron spectra. Resorting to a semiclassical model, which should be more reliable at long wavelengths, we are able to reproduce this peculiar structure and suggest the origin of this structure as a classical Coulomb effect in long wavelength strong-field physics.

We are grateful to Professor M. S. Zhan for many useful discussions. This work is supported by NNSF of China (No. 10674153 and No. 10574019), the National Basic Research Program of China (No. 2006CB806000), and the CAEP Foundation (No. 2006z0202 and No. 2008B0102007).

Note added.—Very recently, a similar experimental observation has been presented by Blaga *et al.* [21].

*xjliu@wipm.ac.cn
†chen_jing@iapcm.ac.cn
‡ycheng-45277@hotmail.com
§zzxu@mail.shcnc.ac.cn

[1] P. Agostini *et al.*, Phys. Rev. Lett. **42**, 1127 (1979).
[2] For a review, see, e.g., W. Becker *et al.*, Adv. At. Mol. Opt. Phys. **48**, 35 (2002).
[3] G. G. Paulus *et al.*, Phys. Rev. Lett. **72**, 2851 (1994).
[4] P. B. Corkum, N. H. Burnett, and F. Brunel, Phys. Rev. Lett. **62**, 1259 (1989); K. J. Schafer *et al.*, *ibid.* **70**, 1599 (1993); P. B. Corkum, *ibid.* **71**, 1994 (1993); G. G. Paulus *et al.*, J. Phys. B **27**, L703 (1994).
[5] L. V. Keldysh, Sov. Phys. JETP **20**, 1307 (1965).
[6] P. Hansch, M. A. Walker, and L. D. Van Woerkom, Phys. Rev. A **55**, R2535 (1997); M. P. Hertlein, P. H. Bucksbaum, and H. G. Muller, J. Phys. B **30**, L197 (1997); G. G. Paulus *et al.*, Phys. Rev. A **64**, 021401 (2001).
[7] See, for example, H. G. Muller and F. C. Kooiman, Phys. Rev. Lett. **81**, 1207 (1998); R. Kopold, W. Becker, M. Kleber, and G. G. Paulus, J. Phys. B **35**, 217 (2002).
[8] R. Moshammer *et al.*, Phys. Rev. Lett. **91**, 113002 (2003).
[9] J. Chen and C. H. Nam, Phys. Rev. A **66**, 053415 (2002).
[10] K. Dimitriou *et al.*, Phys. Rev. A **70**, 061401(R) (2004).
[11] A. Rudenko *et al.*, J. Phys. B **37**, L407 (2004).
[12] A. S. Alnaser *et al.*, J. Phys. B **39**, L323 (2006).
[13] P. Colosimo *et al.*, Nature Phys. **4**, 386 (2008).
[14] Y. Fu *et al.*, Phys. Rev. A **79**, 013802 (2009).
[15] B. Hu, J. Liu, and S. G. Chen, Phys. Lett. A **236**, 533 (1997); J. Chen, J. Liu, and W. M. Zheng, Phys. Rev. A **66**, 043410 (2002); D. F. Ye, X. Liu, and J. Liu, Phys. Rev. Lett. **101**, 233003 (2008).
[16] M. V. Ammosov, N. B. Delone, and V. P. Krainov, Sov. Phys. JETP **64**, 1191 (1986).
[17] Very recently, an *S*-matrix theory with strong-field approximation [H. R. Reiss, Phys. Rev. Lett. **102**, 143003 (2009)] has been employed to account for experimental data at an even longer wavelength of 10.6 μm [W. Xiong *et al.*, J. Phys. B **21**, L159 (1988)], in which a low-energy peak and another broad one at rather high energy were observed. However, the dependence of the two peaks on the specific atom species therein indicates a different mechanism and is not relevant to our case.
[18] D. G. Arbó *et al.*, Phys. Rev. Lett. **96**, 143003 (2006); Z. Chen *et al.*, Phys. Rev. A **74**, 053405 (2006); M. Abu-samha *et al.*, J. Phys. B **41**, 245601 (2008).
[19] A. Rudenko *et al.*, J. Phys. B **38**, L191 (2005).
[20] T. Brabec, M. Y. Ivanov, and P. B. Corkum, Phys. Rev. A **54**, R2551 (1996); V. R. Bhardwaj *et al.*, Phys. Rev. Lett. **86**, 3522 (2001).
[21] C. I. Blaga *et al.*, Nature Phys. **5**, 335 (2009).

JOURNAL OF APPLIED PHYSICS **108**, 044909 (2010)

Hugoniot of shocked liquid deuterium up to 300 GPa: Quantum molecular dynamic simulations

Cong Wang,[1] Xian-Tu He,[1,2] and Ping Zhang[1,2,a)]
[1]*LCP, Institute of Applied Physics and Computational Mathematics, P.O. Box 8009, Beijing 100088, People's Republic of China*
[2]*Center for Applied Physics and Technology, Peking University, Beijing 100871, People's Republic of China*

(Received 27 April 2010; accepted 29 June 2010; published online 23 August 2010)

Quantum molecular dynamic (QMD) simulations are introduced to study the thermophysical properties of liquid deuterium under shock compression. The principal Hugoniot is determined from the equation of states, where contributions from molecular dissociation and atomic ionization are also added onto the QMD data. At pressures below 100 GPa, our results show that the local maximum compression ratio of 4.5 can be achieved at 40 GPa, which is in good agreement with magnetically driven flyer and convergent-explosive experiments; At the pressure between 100 and 300 GPa, the compression ratio reaches a maximum of 4.95, which agrees well with recent high power laser-driven experiments. In addition, the nonmetal-metal transition and optical properties are also discussed. © *2010 American Institute of Physics*. [doi:10.1063/1.3467969]

I. INTRODUCTION

Recent studies of materials under extreme conditions, which require improved understandings of the thermophysical properties in the new and complex regions, have gained much scientific interest.[1] The combination of high temperature and high density defines "warm dense matter" (WDM)-a strongly correlated state, which is characterized by partially dissociated, ionized and degenerated states, and the modeling of the dynamical, electronic, and optical properties for such system is rather challenging. Due to their simplicity, hydrogen and its isotopes (deuterium and tritium) have been studied intensively,[2] where the relative pressure and temperature have reached megabar range and several electron-volt. Specially, as one of the target materials in the inertial confinement fusion experiments,[3] deuterium has been extensively investigated through experimental measurements and theoretical models. Gas gun,[4] converging explosive,[5] magnetically driven flyer,[6] and high power laser-driven[7–9] experiments have been applied to probe the physical properties of deuterium during single or multiple dynamic compression. Theoretically, approximations have been introduced to simulate warm dense deuterium (hydrogen), such as, interatomic potential method,[10] linear mixing model,[11] chemical model fluid variational theory (FVT),[12] path integral Monte Carlo (PIMC),[13–15] and quantum molecular dynamics (QMDs).[16,17]

Till date, although a number of explanatory and predictive results in some cases have already been provided by experimental and theoretical studies, however, many fundamental questions of deuterium under extreme conditions are still yet to be clarified. The equation of states (EOS), especially the Hugoniot curve, are essential in this context. Since five to sixfold the initial densities have been detected by laser-driven experiments at megabar pressure regime[7–9] and supported by PIMC simulations,[13,15] considerable controversies in the deuterium EOS have been raised. Meanwhile, converging explosives[5] and magnetically driven flyer[6] experiments indicate that the compression ratio (η) shows a maximum close to 4.3, which is in good agreement with QMD results.[16,17] Furthermore, in adiabatic and isentropic compressions, the nonmetal-metal transition of deuterium (hydrogen), which is accompanied by the change of optical spectroscopies,[18] has been a major issue recently. The links between nonmetal-metal transition and dissociation (ionization) under dynamic compression are of particular significance.[2]

The chemical pictures of deuterium under extreme conditions could be briefly described as two processes: (i) partial dissociation of molecules, $D_2 \rightleftarrows 2D$, and (ii) a subsequent ionization of atoms, $D \rightleftarrows e + D^+$. QMD simulations, where electrons are modeled by quantum theory, are convinced to be a powerful tool to describe the chemical reactions, such as dissociation and recombination of molecules. Meanwhile, the dynamical, electrical, and optical properties of WDM have already been proved to be successfully investigated by QMD simulations.[19,20] However, the ionization of atoms is not well defined in the framework of density functional theory (DFT). Considering these facts, thus, in this paper we applied the corrected QMD simulations to shock compressed deuterium, and the calculated compression ratio is substantially increased according to the ionization of atoms in the warm dense fluid.

II. COMPUTATIONAL METHOD

We have performed simulations for deuterium by employing the Vienna Ab-initio Simulation Package (VASP).[21,22] A fixed volume supercell of N atoms, which is repeated periodically throughout the space, forms the elements of the calculation. By involving Born–Oppenheimer approximation, electrons are fully quantum mechanically treated through plane-wave, finite-temperature DFT,[16] and the electronic states are populated according to the Fermi–Dirac distribution at temperature T_e. The exchange correlation func-

[a)]Electronic mail: zhang_ping@iapcm.ac.cn.

tional is determined by generalized gradient approximation with the parametrization of Perdew–Wang 91.[23] The ion-electron interactions are represented by a projector augmented wave pseudopotential.[24] The system is calculated with the isokinetic ensemble (NVT), where the ionic temperature T_i is kept constant every time step by velocity scaling, and the system is kept in local thermodynamical equilibrium by setting the electron (T_e) and ion (T_i) temperatures to be equal.

The plane-wave cutoff energy is selected to be 600.0 eV, so that the pressure is converged within 5% accuracy. Γ point is employed to sample the Brillouin zone in molecular dynamic simulations, because EOS (conductivity) can only be modified within 5% (15%) for the selection of higher number of **k** points. A total number of 128 atoms (64 deuterium molecules) is included in a cubic cell, and over 300 (densities and temperatures) points are calculated. The densities adopted in our simulations range from 0.167 to 0.9 g/cm^3 and temperatures between 20 and 50 000 K, which highlight the regime of principal Hugoniot. All the dynamic simulations are lasted for 4–6 ps, and the time steps for the integrations of atomic motion are 0.5–2 fs according to different densities (temperatures).[25] Then, the subsequent 1 ps simulations are used to calculate EOS as running averages.

III. THE EOS

In QMD simulations, zero point vibration energy ($1/2h\nu_{\mathrm{vib}}$) and ionization energy (13.6 eV/atom) are excluded, thus, the internal energy and pressure should be corrected as follows:

$$E = E_{\mathrm{QMD}} + \frac{1}{2}N(1-\alpha)E_{\mathrm{vib}} + N\beta E_{\mathrm{ion}}, \tag{1}$$

$$P = P_{\mathrm{QMD}} + (1+\beta)\frac{\rho k_B T}{m_D}, \tag{2}$$

where N is the total number of atoms for the present system, and m_D presents the mass of deuterium atom. The density and temperature are denoted by ρ and T, respectively, and k_B stands for Boltzmann constant. α and β are the dissociation degree and ionization degree, while E_{vib} and E_{ion} correspond to the zero point vibration energy and ionization energy, respectively. E_{QMD} and P_{QMD} are calculated from VASP. Various corrections to QMD simulations have already been applied to model WDM,[16,17] but contributions from atomic ionization, which are particularly important at high pressure, are still in absence. Ionization degree of aluminum under extreme conditions has been successfully quantified through Drude model,[26] however, the simple metallic model is not suitable for the present system. Here, a new and effective method in accounting for contributions to the EOS from molecular dissociation and atomic ionization has been demonstrated.

The dissociation degree is important in determining the internal energy, from which EOS can be derived, especially at low temperatures and the initial state on the Hugoniot curve. The dissociation degree could be evaluated through the coordination number

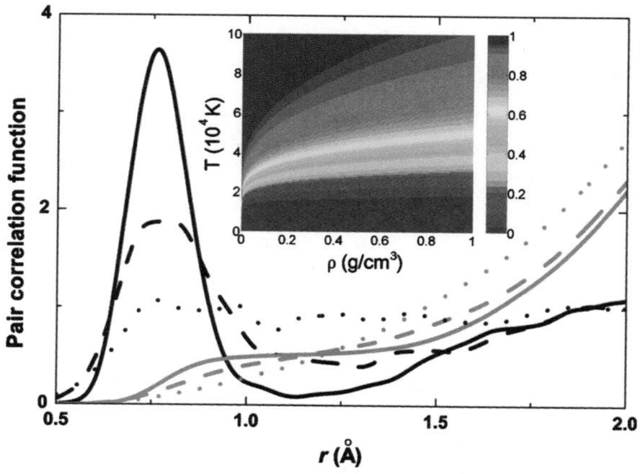

FIG. 1. (Color online) Calculated PCF (black line) and coordination number (red line) at temperatures of 3000 K (solid line), 5000 K (dashed line), and 10 000 K (dotted line) for $\rho=0.6$ g/cm^3. Inset is the contour plot of the ionization degree as a function of density and temperature.

$$K(r) = \frac{N-1}{\Omega}\int_0^r 4\pi r'^2 g(r')dr', \tag{3}$$

where Ω is the volume of the supercell. The coordination number is a weighted integral over the pair correlation function (PCF) $g(r)$ of the ions. The doubled value of K at the maximum of $g(r)$ ($r=0.75$ Å), is equal to the fraction of atoms forming molecules in the supercell. The sampled PCF and $K(r)$ are labeled in Fig. 1. Fast dissociation of molecules emerges at the temperature between 4000 and 7000 K, and a region featured with $(\partial P/\partial T)_V<0$, which is not presented here, is observed. Our results show that molecular deuterium can be neglected above 15 000 K due to thermal dissociation.

The ionization degree of the system can be evaluated through Saha equation

$$\frac{\beta^2}{1-\beta} = \frac{2\Omega}{\lambda^3}\exp\left(-\frac{E_{\mathrm{ion}}}{k_B T}\right), \tag{4}$$

$$\lambda = \sqrt{\frac{h^2}{2\pi m_e k_B T}}, \tag{5}$$

where only one level of ionization process is considered. In the present formula, m_e stands for the electron mass. As shown in the inset in Fig. 1, the ionization of deuterium could be neglected below 10 000 K. At the temperature between 10 000 and 50 000 K, where the modeling of the principal Hugoniot of deuterium is rather difficult, partially ionized warm dense fluid is formed, and the ionization of atoms is of predominance in determining the EOS.

Following Lenosky *et al.*,[27] Beule *et al.*,[28] and Holst *et al.*,[17] we fit the internal energy and pressure by expansions in terms of density (gram per cubic centimeter) and temperature (10^3 K). The corrected QMD data for internal energy (electron-volt per atom) can be expanded as follows:

$$E = \sum_{i=0}^{4} A_i(T)\rho^i, \tag{6}$$

TABLE I. Coefficients a_{ik} in expansion for the internal energy E.

i	a_{i0}	a_{i1}	a_{i2}	a_{i3}	a_{i4}
0	4.1065	−0.1111	13.4393	−6.7345	0.2532
1	2.7497	−0.5432	2.6361	−9.6771	1.5115
2	12.8700	1.6686	15.9184	8.8453	−3.3917
3	−37.4966	−18.1714	30.6954	2.8396	3.1274
4	2.4465	3.1352	3.6421	5.3229	−1.2843

$$A_i(T) = a_{i0} \exp\left[-\left(\frac{T-a_{i1}}{a_{i2}}\right)^2\right] + a_{i3} + a_{i4}T. \quad (7)$$

The total pressure given in gigapascals can be similarly expanded as E with the expansion coefficients b_{jk}. The expansion coefficients a_{ik} and b_{jk} for E and P (accuracy better than 5%) are summarized in Tables I and II, respectively.

Based on the EOS, the principal Hugoniot curve can be derived from the following equation:

$$(E_0 - E_1) = \frac{1}{2}\left(\frac{1}{\rho_0} - \frac{1}{\rho_1}\right)(P_0 + P_1), \quad (8)$$

where the subscripts 0 and 1 refer to the initial and shocked states. In our present simulations, the initial density ρ_0 is $0.167\ \mathrm{g/cm^3}$ with the respective internal energy $E_0 = -3.28\ \mathrm{eV/atom}$ at $T_0 = 20\ \mathrm{K}$. The pressure P_0 of the starting point on the Hugoniot can be neglected compared to high pressures of shocked states.

The principal Hugoniot is shown in Fig. 2, where previous theoretical and experimental results are also provided for comparison. At pressures below 100 GPa, our results indicate that the principal Hugoniot experiences a local maximum compression ratio of 4.5 around 40 GPa, which can be attributed to the dissociation of molecules. The present Hugoniot agrees well with previous experiments, such as gas gun,[4] magnetically launched flyer plates,[6] and converging explosives.[5] At the pressure between 40 and 100 GPa, the Hugoniot curve shows a stiff behavior, and the compression ratio lies between 4.25 and 4.5. Meanwhile, the ionization of atoms increases remarkably at $P > 50$ GPa (see the inset in Fig. 3) and consequently intenerates the fluid. Thus, the combined effect of the molecular dissociation and atomic ionization results in a local minimum of η (4.25) along the Hugoniot.

Recent high power laser-driven experiments[8,9] suggest that deuterium is stiff ($\eta_{max} \approx 4.2$) below 100 GPa and become softer ($\eta \approx 4.5-5.5$) above 110 GPa, which can be described by the present simulations. Our results indicate that molecular deuterium can be neglected at this stage, and the atomic ionization dominates the characteristic of the Hugoniot with η lies between 4.5 and 4.95 (maximum is reached at 200 GPa), which is in accord with recent experiments.[29] The wide-range behavior of the Hugoniot is characterized by two stage transitions—dissociation under low pressure and ionization at higher pressure, and the present results show excellent agreement with experimental ones. The Hugoniots from mere QMD simulations hardly exceed 100 GPa (Ref. 16) except for that of Holst *et al.*,[17] but η does not exceed 4.5 (at $P > 100$ GPa). Although some PIMC simulations[13,15] show five to sixfold compressions, the simulated data are not yet comparable with experiments. Due to the intrinsic approximations, no consistency has been detected between our results and those of linear mixing model[11] and chemical model FVT.[12]

IV. DYNAMIC CONDUCTIVITY AND OPTICAL PROPERTIES

Let us turn now to see the nonmetal-metal transition by studying the optical and conductive behaviors of the warm dense deuterium. The real part of dynamic conductivity $\sigma_1(\omega)$ can be evaluated through the following Kubo–Greenwood formula

$$\sigma_1(\omega) = \frac{2\pi}{3\omega\Omega}\sum_{\mathbf{k}} w(\mathbf{k})\sum_{j=1}^{N}\sum_{i=1}^{N}\sum_{\alpha=1}^{3}[f(\epsilon_i,\mathbf{k}) - f(\epsilon_j,\mathbf{k})] \times |\langle\Psi_{j,\mathbf{k}}|\nabla_\alpha|\Psi_{i,\mathbf{k}}\rangle|^2\delta(\epsilon_{j,\mathbf{k}} - \epsilon_{i,\mathbf{k}} - \hbar\omega), \quad (9)$$

where i and j summations range over N discrete bands included in the calculation. The α sum is over the three spatial directions. $f(\epsilon_i,\mathbf{k})$ describes the occupation of the ith band, with the corresponding energy $\epsilon_{i,\mathbf{k}}$ and the wave function $\Psi_{i,\mathbf{k}}$ at $\mathbf{k}$. $w(\mathbf{k})$ is the $\mathbf{k}$-point weighting factor.

From the calculated dynamic conductivity along the Hugoniot curve, the dc conductivity σ_{dc}, which follows the static limit $\omega \to 0$ of $\sigma_1(\omega)$, is then extracted and plotted in Fig. 3 as a function of the Hugoniot pressure. As shown in Fig. 3, σ_{dc} increases rapidly with pressure up to 40 GPa toward the formation of metallic state of deuterium, which is in agreement with the experimental measurements.[4] Similar

TABLE II. Coefficients b_{ik} in expansion for the total pressure P.

j	b_{j0}	b_{j1}	b_{j2}	b_{j3}	b_{j4}
0	41.9168	−1.2676	−8.5830	−55.5348	5.4594
1	101.2582	6.0739	15.4631	−29.1378	−8.2198
2	60.8838	−0.6898	4.6233	−124.8318	24.2147
3	277.4649	−10.0840	46.9646	−233.0212	−5.4740
4	8.0324	−3.8221	−1.0263	30.4031	−0.3011

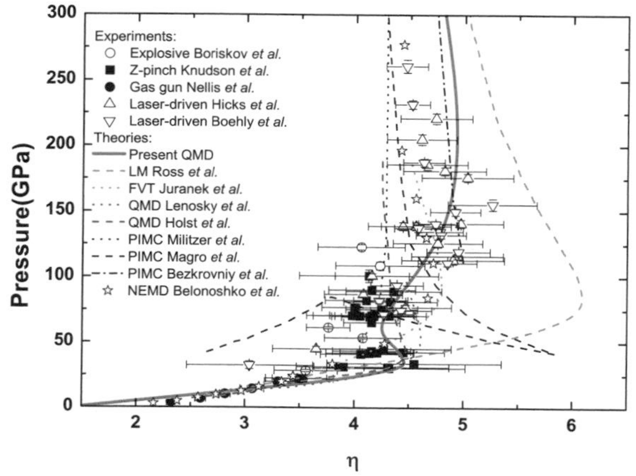

FIG. 2. (Color online) Simulated principal Hugoniot curve (solid red curve). Previous data are also shown for comparison. Theories: QMD results by Lenosky *et al.* (Ref. 16) (dotted blue curve) and Holst *et al.* (Ref. 17) (dashed blue curve); restricted PIMC simulations by Magro *et al.* (Ref. 13) (dashed black curve), Militzer *et al.* (Ref. 14) (dotted black curve), and direct PIMC by Bezkrovniy *et al.* (Ref. 15) (dashed dot black curve); modeling nonequilibrium MD (NEMD) process by interatomic potential (Ref. 10) (open stars), the linear mixing model of Ross (Ref. 11) (dashed green curve), and the chemical model FVT (Ref. 12) (dotted green curve). Experiments: gas gun by Nellis *et al.* (Ref. 4) (solid circle), Z-pinch by Knudson *et al.* (Ref. 6) (solid square), explosives of Boriskov *et al.* (Ref. 5) (open circle), laser-driven by Hicks *et al.* (Ref. 9) (up open triangle), and Boehly *et al.* (Ref. 8) (down open triangle).

tendency has also been found in the QMD simulations of warm dense hydrogen.[17] When further increasing the Hugoniot pressure, one finds from Fig. 3 that σ_{dc} keeps almost invariant and the warm dense deuterium maintains its metallic behavior. Here we address that the nonmetal-metal transition is induced by gradual dissociation of molecules and thermal activation of electronic states, instead of atomic ionization, which is not observed until 50 GPa according to the charge density distribution in the QMD simulations. Quantitative analysis can be clarified through plotting the ionization along the Hugoniot as shown in the inset in Fig. 3. Meanwhile, optical reflectivity, with the respective wavelength of 808 nm, is shown along the principal Hugoniot in Fig. 4, where good agreement has been achieved between our present work and previous experiments.[18] The increase in reflectance (from 0.05 to 0.6) is observed, and this can be interpreted as a gradual transition from a molecular insulating fluid to a partially dissociated and metallic fluid at above 40 GPa.

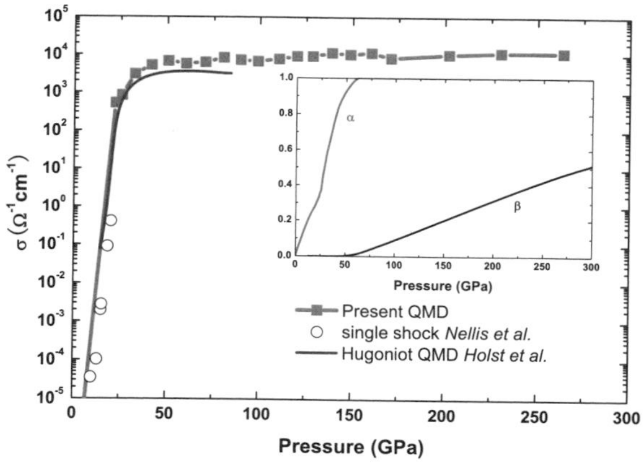

FIG. 3. (Color online) Calculated dc conductivity along the Hugoniot curve (solid squares). Previous data (Refs. 4 and 17) are also shown for comparison. Inset is the degree of dissociation α (red line) and ionization β (black line) along the Hugoniot.

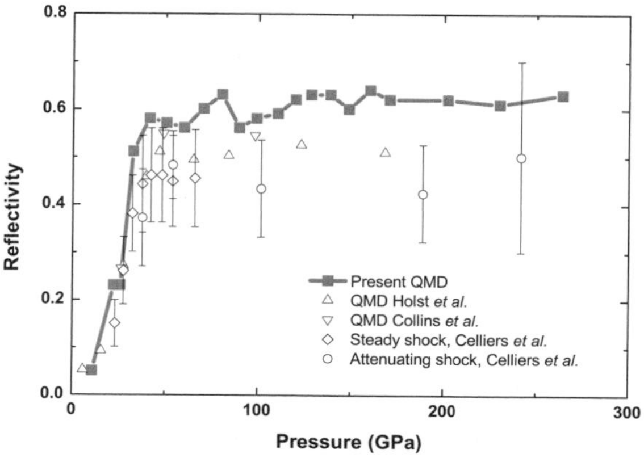

FIG. 4. (Color online) Calculated optical reflectivity of wavelength 808 nm along the Hugoniot. Previous data (Refs. 17, 18, and 30) are also plotted for comparison.

V. CONCLUSION

In summary, we have performed QMD simulations to study the thermophysical properties of deuterium under extreme conditions. The Hugoniot EOS has been evaluated through QMD calculations and corrected by taking into account the molecular dissociation described by the coordination number $K(r)$ and the atomic ionization described by Saha equation. The corrected Hugoniot has shown good agreement with the experimental data in a wide range of shock conditions, which thus indicates the importance of physical picture of a two-stage transition, i.e., dissociation and ionization. The principal Hugoniot reveals a local maximum compression ratio of 4.5 at 40 GPa. With the increase in pressure, η reaches a maximum of 4.95 at about 200 GPa, and the contribution from atomic ionization demonstrates softened character of the Hugoniot. Smooth transition from a molecular insulating fluid to an partially dissociated and metallic fluid are observed at 40 GPa. Our calculated optical constants along the Hugoniot have shown excellent agreement with experiments. In addition, smooth fit functions constructed in the present paper for the internal energy and total pressure are expected to be useful for the future studies of warm dense deuterium.

ACKNOWLEDGMENTS

This work was supported by NSFC.

[1]R. Ernstorfer, M. Harb, C. T. Hebeisen, G. Sciaini, T. Dartigalongue, and R. J. D. Miller, Science **323**, 1033 (2009).
[2]W. J. Nellis, Rep. Prog. Phys. **69**, 1479 (2006).
[3]F. Philippe, A. Casner, T. Caillaud, O. Landoas, M. C. Monteil, S. Libera-

tore, H. S. Park, P. Amendt, H. Robey, C. Sorce, C. K. Li, F. Seguin, M. Rosenberg, and R. Petrasso, Phys. Rev. Lett. **104**, 035004 (2010).
[4]W. J. Nellis, A. C. Mitchell, M. van Thiel, G. J. Devine, and R. J. Trainor, J. Chem. Phys. **79**, 1480 (1983).
[5]G. V. Boriskov, A. I. Bykov, R. Ilkaev, V. D. Selemir, G. V. Simakov, R. F. Trunin, V. D. Urlin, A. N. Shuikin, and W. J. Nellis, Phys. Rev. B **71**, 092104 (2005).
[6]M. D. Knudson, D. L. Hanson, J. E. Bailey, C. A. Hall, J. R. Asay, and C. Deeney, Phys. Rev. B **69**, 144209 (2004).
[7]G. W. Collins, L. B. Da Silva, P. Celliers, D. M. Gold, M. E. Foord, R. J. Wallace, A. Ng, S. V. Weber, K. S. Budil, and R. Cauble, Science **281**, 1178 (1998).
[8]T. R. Boehly, D. G. Hicks, P. M. Celliers, T. J. B. Collins, R. Earley, J. H. Eggert, D. Jacobs-Perkins, S. J. Moon, E. Vianello, D. D. Meyerhofer, and G. W. Collins, Phys. Plasmas **11**, L49 (2004).
[9]D. G. Hicks, T. R. Boehly, P. M. Celliers, J. H. Eggert, S. J. Moon, D. D. Meyerhofer, and G. W. Collins, Phys. Rev. B **79**, 014112 (2009).
[10]A. B. Belonoshko, A. Rosengren, N. V. Skorodumova, S. Bastea, and B. Johansson, J. Chem. Phys. **122**, 124503 (2005).
[11]M. Ross, Phys. Rev. B **58**, 669 (1998).
[12]H. Juranek and R. Redmer, J. Chem. Phys. **112**, 3780 (2000).
[13]W. R. Magro, D. M. Ceperley, C. Pierleoni, and B. Bernu, Phys. Rev. Lett. **76**, 1240 (1996).
[14]B. Militzer and D. M. Ceperley, Phys. Rev. Lett. **85**, 1890 (2000).
[15]V. Bezkrovniy, V. S. Filinov, D. Kremp, M. Bonitz, M. Schlanges, W. D. Kraeft, P. R. Levashov, and V. E. Fortov, Phys. Rev. E **70**, 057401 (2004).
[16]T. Lenosky, S. Bickham, J. Kress, and L. Collins, Phys. Rev. B **61**, 0163 (2000).
[17]B. Holst, R. Redmer, and M. P. Desjarlais, Phys. Rev. B **77**, 184201 (2008).
[18]P. M. Celliers, G. W. Collins, L. B. DaSilva, D. M. Gold, R. Cauble, R. J. Wallace, M. E. Foord, and B. A. Hammel, Phys. Rev. Lett. **84**, 5564 (2000).
[19]A. Kietzmann, R. Redmer, M. P. Desjarlais, and T. R. Mattsson, Phys. Rev. Lett. **101**, 070401 (2008).
[20]W. Lorenzen, B. Holst, and R. Redmer, Phys. Rev. Lett. **102**, 115701 (2009).
[21]G. Kresse and J. Hafner, Phys. Rev. B **47**, R558 (1993).
[22]G. Kresse and J. Furthmüller, Phys. Rev. B **54**, 11169 (1996).
[23]J. P. Perdew, *Electronic Structure of Solids* (Akademie Verlag, Berlin, 1991), p. 11.
[24]P. E. Blöchl, Phys. Rev. B **50**, 17953 (1994).
[25]The time steps have been taken as $\Delta t = a/20\sqrt{k_B T/m_D}$, where $a = (3/4\pi n_i)^{1/3}$ is the ionic sphere radius (n_i is the ionic number density), $k_B T$ presents the kinetic energy, and m_D is the ionic mass. For lower densities and temperatures, where $\Delta t > 2$ fs according to the equation, the time steps have been rounded to 2 fs, and the results have been tested.
[26]S. Mazevet, M. P. Desjarlais, L. A. Collins, J. D. Kress, and N. H. Magee, Phys. Rev. E **71**, 016409 (2005).
[27]T. J. Lenosky, J. D. Kress, and L. A. Collins, Phys. Rev. B **56**, 5164 (1997).
[28]D. Beule, W. Ebeling, A. Föster, H. Juranek, S. Nagel, R. Redmer, and G. Röpke, Phys. Rev. B **59**, 14177 (1999).
[29]M. D. Knudson and M. P. Desjarlais, Phys. Rev. Lett. **103**, 225501 (2009).
[30]L. A. Collins, S. R. Bickham, J. D. Kress, S. Mazevet, T. J. Lenosky, N. J. Troullier, and W. Windl, Phys. Rev. B **63**, 184110 (2001).

贺贤土论文选集

982

IOP Publishing | Journal of Physics B: Atomic, Molecular and Optical Physics

J. Phys. B: At. Mol. Opt. Phys. **43** (2010) 235601 (3pp) doi:10.1088/0953-4075/43/23/235601

Intensity determination of superintense laser pulses via ionization fraction in the relativistic tunnelling regime

D F Ye[1,2], G G Xin[2,3], J Liu[1,2,3] and X T He[1,2]

[1] Center for Applied Physics and Technology, Peking University, Beijing 100084, People's Republic of China

[2] Institute of Applied Physics and Computational Mathematics, PO Box 8009 (28), Beijing 100088, People's Republic of China

[3] College of Physics and Information, Beijing Institute of Technology, Beijing 100081, People's Republic of China

E-mail: liu-jie@iapcm.ac.cn

Received 22 April 2010
Published 19 November 2010
Online at stacks.iop.org/JPhysB/43/235601

Abstract

We address the quantum tunnelling (QT) effect in the scheme that retrieves the laser intensity via the ionization fraction of multiply charged ions proposed recently. Our discussions are based on both the analytical tunnelling rate formula and numerical calculations using the classical trajectory Monte Carlo + tunnelling (CTMC-T) technique. Both approaches show that the correction from the QT effect is up to one order of magnitude.

(Some figures in this article are in colour only in the electronic version)

Recent advances in optical technology have made it possible to obtain laser pulses with peak intensity up to 10^{22} W cm^{-2} [1]. The next generation of laser systems is envisaged to reach even higher intensities in the near future. The availability of superintense laser pulses opens a window to the physical phenomena occurring in the relativistic and ultra-relativistic domain (see [2] and [3] for two recent reviews). In order to take full advantage of such pulses, it is crucial to stabilize and measure the peak laser intensity.

The traditional measurement of less powerful lasers simply involves recording the output pulse's energy, the pulse duration and the beam spot size [4]. For ultrahigh laser intensities, however, this technique becomes less viable because the mirrors involved in the measurement process may not sustain such intensities anymore.

Instead of direct measurements, recently, some theoretical works have shed light on this long-standing unsettled issue by proposing to retrieve the laser intensity via the ionization fraction of multiply charged ions [5]. The theoretical benchmark data in the procedure were obtained by the classical trajectory Monte Carlo (CTMC) simulation [6, 7] where the quantum tunnelling (QT) effect is ignored. As an important complement to the above method, we address the QT effect in this brief report. Our discussions are based on both the analytic tunnelling rate obtained in [8] and numerical calculations using the classical trajectory Monte Carlo + tunnelling (CTMC-T) technique [9] that has been extended to the relativistic regime by us. Both approaches have demonstrated the correction from the QT effect is up to one order of magnitude.

We start our detailed discussions with a brief review of the method presented in [5]. It applies relativistic classical trajectory simulations to calculate the ionization fraction of various high-Z hydrogen-like ions under a wide range of laser intensities. For a fixed Z, there is a most sensitive (MS) point on the calculated ionization curve where the ionization fraction sharply rises for a slight increase of the laser intensity. These MS points are pivotal for the adequate determination of laser intensity. One reads off all the MS points from the ionization curves and draws a diagram that contains two reference curves showing the variation of ionization fraction and laser intensity corresponding to the MS points versus the ionic charge Z. Then, for a given laser beam, one must start from $Z_{\min}$ where the ionization fraction is near one, and increase the ionic charge until $Z_{\max}$ with an ionization fraction of nearly zero is measured. The crossing of the measured curve and the reference curve finally determines the applied laser intensity.

In the above procedure, the two reference curves are obtained based on the CTMC simulation, for which only over-the-barrier ionization is considered, and the tunneling ionization is totally neglected. However, for a practical laser pulse, the electric field needs to increase gradually from zero to the maximum, and the ions may be completely depleted at the ascending edge of the laser pulse before it enters the over-the-barrier regime. In this case, the ionization dynamics, which is dominated by tunneling ionization, cannot be described by the CTMC simulation. On the other hand, a complete quantum-mechanical solution of the full-dimensional Dirac equation for an electron in both a Coulomb and a laser field is still a demanding computational problem. To address the QT effect, we resort to two different semiclassical approaches: the analytic tunneling rate obtained from a semiclassical solution of the Dirac equation [8] and numerical calculations using the CTMC-T technique [9].

Quantum tunneling calculation. We restrict our following discussions to the quasistatic tunneling regime where the relativistic generalized Keldysh parameter [10] $\gamma = (\omega c/|\mathbf{E}_0|)\sqrt{1-[(c^2-Z^2/2)/c^2]^2}$ is far below one (atomic units ($\hbar = m_e = |e| = 1$) are used throughout this paper). Here, $|\mathbf{E}_0|$ is the electric field strength, ω is the angular frequency, and $c = 137$ is the speed of light. The tunneling rate for the linearly polarized laser field in this regime has been obtained by Milosevic, Krainov and Brabec from a semiclassical solution of the Dirac equation [8]:

$$w_0(t_0) = \frac{c^2[|\mathbf{E}(\mathbf{r}_0,t_0)|/c^3]^{1-2\varepsilon}}{2\sqrt{3}\xi\Gamma(2\varepsilon+1)}\sqrt{\frac{3-\xi^2}{3+\xi^2}}\left[\frac{4\xi^3(3-\xi^2)^2}{\sqrt{3}(1+\xi^2)}\right]^{2\varepsilon} \times \exp\left[6\mu\arcsin\frac{\xi}{\sqrt{3}} - \frac{2\sqrt{3}c^3\xi^3}{|\mathbf{E}(\mathbf{r}_0,t_0)|(1+\xi^2)}\right]. \quad (1)$$

Here, $\mu = Z/c$, $\varepsilon = \sqrt{1-\mu^2}$, $\xi = \sqrt{1-\varepsilon(\sqrt{\varepsilon^2+8}-\varepsilon)/2}$ and $\Gamma()$ is the standard Gamma function. $\mathbf{E}(\mathbf{r}_0, t_0)$ denotes the instantaneous laser electric field when the electron escapes the Coulomb barrier. In our following calculations, we adopt the rectangular Cartesian coordinate system and define axes Ox, Oy and Oz as the magnetic, propagation and polarization directions, respectively. The ion ensemble is placed at the origin so that $\mathbf{r}_0 \approx 0$. The laser field is a plane electromagnetic wave represented by the vector potential:

$$\mathbf{A}(\mathbf{r},t) = \frac{|\mathbf{E}_0|}{\omega} f(\eta)\sin(\omega\eta+\phi_0)\,\mathbf{e}_z, \quad (2)$$

with $f(\eta)$ the envelope function, $\eta = t - y/c$ the proper time, ϕ_0 the carrier envelope phase and $\mathbf{e}_z$ the unit vector in the z direction (the same for $\mathbf{e}_x$ and $\mathbf{e}_y$). The electromagnetic field can be obtained by means of

$$\mathbf{E} = -\frac{\partial\mathbf{A}}{\partial t} = -\frac{\partial A}{\partial t}\mathbf{e}_z, \quad (3)$$

$$\mathbf{B} = \nabla\times\mathbf{A} = \frac{\partial A}{\partial y}\mathbf{e}_x. \quad (4)$$

The ionization fraction is determined by integrating $w_0(t_0)$ over the time duration $[t_{\text{initial}}, t_{\text{final}}]$:

$$\chi = 1 - \exp\left[-\int_{t_{\text{initial}}}^{t_{\text{final}}} w_0(t_0)\,dt_0\right]. \quad (5)$$

CTMC-T calculation. The calculation is in the same spirit as that in [5]. What is new is that tunneling is now included via the Wentzel–Kramers–Brillouin (WKB) approximation. In the calculation, the motion of electrons is governed by the following relativistic Newton–Lorentz equations:

$$\frac{d\mathbf{r}}{dt} = \frac{1}{\gamma}\mathbf{p}, \qquad \frac{d\mathbf{p}}{dt} = -\left[\mathbf{E}(\mathbf{r},t) + \frac{Z\mathbf{r}}{r^3} + \frac{1}{\gamma}\mathbf{p}\times\mathbf{B}(\mathbf{r},t)\right], \quad (6)$$

with a relativistic factor $\gamma = \sqrt{1+|\mathbf{p}|^2/c^2}$. The integration is performed separately over an ensemble of at least 1000 particles with energy equal to the ground state of high-Z ions $E_g = c^2\sqrt{1-(Z/c)^2}$. The ensemble is prepared with the procedure proposed in [11], which has been generalized to the high-Z ions by us as follows.

We first generate a planar ensemble in the x–y plane, then choose arbitrary sets of Euler angles to rotate the planar ensemble into three dimensions. It has been proved that all possible initial states can be reached by choosing a parameter t_r randomly from the interval $[0, \Delta t = 2\pi/c^3W^{3/2}]$ [11]. The parameter t_r fixes the eccentric angle u in the Kepler orbit through the expression $u = c^3W^{3/2}t_r + W^2c^4\bar{R}\Delta R\sin u$, which can be immediately obtained with iterative numerical technique. The eccentric angle determines the radius r and polar angle φ by the expressions $r = \bar{R} - \Delta R\cos u$ and $\varphi = 2\sqrt{\frac{s}{s-1}}\arctan\left[\sqrt{\frac{W}{s-1}}c^2(\bar{R}+\Delta R)\tan\frac{u}{2}\right]$. Transforming back to the Cartesian coordinate, one obtains the initial positions $(x_0, y_0) = (r\cos u, r\sin u)$. The corresponding momenta (p_{x_0}, p_{y_0}) can be obtained by solving the equations for the ground state energy E_g and angular momentum L:[4] $E_g = \sqrt{c^4+(p_{x_0}^2+p_{y_0}^2)c^2} - Z/\sqrt{x_0^2+y_0^2}$, $L = x_0p_{y_0} - y_0p_{x_0}$. In the above expressions, we have used the abbreviations $W = 1 - E_g^2/c^4$, $s = c^2L^2/Z^2$, $\bar{R} = E_g/Wc^4$ and $\Delta R = \sqrt{1-sW}/c^2W$. The rotation of the above planar ensemble into three-dimensional space has been discussed in [12]. Once the ensemble has been prepared, each ion is traced by numerically integrating equation (6) with the standard fourth to fifth Runge–Kutta algorithm. The QT effect is included semiclassically by allowing the bound electron to tunnel through the potential barrier whenever it reaches the outer turning point, where $p_{i,z} = 0$ and $z_i\epsilon(t) < 0$, with a tunneling probability P_i^{tul} given by the WKB approximation:

$$P_i^{\text{tul}} = \exp\left[-2\sqrt{2}\int_{z_i^{\text{in}}}^{z_i^{\text{out}}}\sqrt{V(z_i)-V(z_i^{\text{in}})}\,dz_i\right]. \quad (7)$$

Here, z_i^{in} and z_i^{out} are the two roots ($|z_i^{\text{out}}| > |z_i^{\text{in}}|$) of the equation for z_i, $V(z_i) = -Z/r_i + z_i\epsilon(t) = -Z/r_i^{\text{in}} + z_i^{\text{in}}\epsilon(t)$ [9].

[4] The possible value of $|L|$ is between the interval $[Z/c, 1]$. However, for the hydrogen atom ($Z = 1$), it has been shown that the CTMC-T results are very close to the accurate quantum-mechanical results if L is restricted to a small value very close to zero (exactly, $L = 0.01$). The reason for doing so is that zero is the quantum-mechanical angular momentum number of the s state [9]. For high-Z ions, however, $|L| \approx 0$ is classically forbidden, we thus choose $|L|$ to be the smallest allowed value (Z/c) instead.

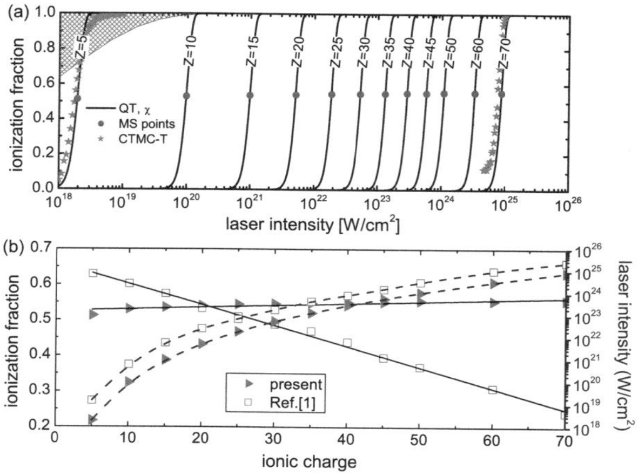

Figure 1. (a) The ionization fraction for several different hydrogen-like ions with nuclear charge *Z* as a function of the peak laser intensity. The ionization fraction is calculated at the end of a single-cycle planar pulse ($f(\eta) = 1$ if $0 < \eta < 2\pi/\omega$, otherwise $f(\eta) = 0$) of wavelength $\lambda = 1054$ nm and $\phi_0 = 0$. The hatched area denotes the over-the-barrier ionization regime, the other part is the tunnelling regime. For a given ionic charge *Z*, the intensity threshold for over-the-barrier ionization can be calculated from equation (8). The connection of these critical points for different *Z* determines the boundary of the hatched area. The full circles are the most sensitive points. Evidently, they all fall into the tunnelling regime. (b) The solid line with full triangles defines the most sensitively measured ionization fraction (left axis), whereas the dashed line with full triangles shows the corresponding laser intensity (right axis) as a function of the respective optimal *Z*. The lines are fits and the triangles indicate the deduced points from (a). The two lines with open squares are the corresponding results taken from [5] for comparison.

Results and discussions. Figure 1(a) shows the ionization fraction for several different hydrogen-like ions with nuclear charge *Z* as a function of the peak laser intensity. It clearly shows that the CTMC-T calculations agree well with that from the tunnelling formula of [8]. For both calculations, each ionization curve exhibits a common feature: a flat profile near zero followed by a rather steep rise ending up with a plateau of complete ionization, the same as that demonstrated in [5]. However, all our curves shift towards the low intensity regime by about one order of magnitude. This result indicates that a laser field far from entering the over-the-barrier regime is indeed strong enough to significantly ionize or even completely deplete the high-*Z* ions through QT. As we have expected, the MS points, which are the benchmark data for determining the peak laser intensity, all fall into the tunnelling regime. This is clearly illustrated by the full circles in figure 1(a). Here, the hatched area denotes the over-the-barrier ionization regime, the other part being the tunnelling regime. The boundary between these two regimes is estimated by

$$|\mathbf{E}_{\text{bs}}| = \frac{c^4}{4Z}(1 - \sqrt{1 - Z^2/c^2})^2, \tag{8}$$

that is the relativistic generalization of the barrier suppression field strength at which the maximum of the effective Coulomb barrier $-2\sqrt{Z|\mathbf{E}_{\text{bs}}|}$ (i.e. the local maximum of $f(x) = -Z/x - |\mathbf{E}_{\text{bs}}|x$) equals the binding energy $c^2(\sqrt{1 - Z^2/c^2} - 1)$.

The ionization fraction and laser intensity corresponding to the MS point for each curve in figure 1(a) are read off. The results are shown in figure 1(b) along with the data taken from [5] for comparison. The curves of intensity versus ionic charge exhibit an analogous tendency, however, the present results are up to one eighth lower than the pure classical calculations of [5]. On the other hand, the two curves for ionization fraction versus ionic charge manifest a complete difference: the curve from [5] exhibits a rapid decrease as the ionic charge increases, whereas the present calculations predict almost a straight line with slight ascent. The differences between the reference curves will unambiguously lead to divergent gauge of the laser intensity. For instance, following the procedure of [5], suppose Ca^{19+} ($Z = 20$) is selected as the optimal ion for determining the laser intensity; then the benchmark data from [5] predict a laser intensity of 2.5×10^{22} W cm^{-2} while the benchmark data that include QT effect predict 5.1×10^{21} W cm^{-2}. The former overestimates the laser intensity by five times.

In conclusion, measuring the intensity of ultra-strong laser pulses is an experimentally challenging task. The work of [5] opened a promising route to this long-standing unsettled issue. However, in the classical CTMC treatment, the QT effect is totally ignored despite the fact that the parameters are deep in the tunnelling regime. In this work, it is shown that the results based on the Milosevic formula and a newly developed CTMC-T method agree and both demonstrate the significant role played by QT. Our result should be an important complement to [5], and of importance for possible implementation in practical experiments.

Acknowledgments

This work is supported by the National Fundamental Research Programme of China (Contact no 2007CB815103) and the National Natural Science Foundation of China (Contact no 10725521).

References

[1] Yanovsky V *et al* 2008 *Opt. Express* **16** 2109
[2] Salamina Y I *et al* 2006 *Phys. Rep.* **427** 41
[3] Mourou G A, Tajima T and Bulanov S V 2006 *Rev. Mod. Phys.* **78** 309
[4] See, for example, Albert O *et al* 2000 *Opt. Lett.* **25** 1125
[5] Hetzheim H G and Keitel C H 2009 *Phys. Rev. Lett.* **102** 083003
[6] Abrines R and Percival I C 1966 *Proc. Phys. Soc. Lond.* **88** 861
[7] Leopold J G and Percival I C 1979 *J. Phys. B: At. Mol. Opt. Phys.* **12** 709
[8] Milosevic N, Krainov V P and Brabec T 2002 *Phys. Rev. Lett.* **89** 193001
[9] Cohen J S 2001 *Phys. Rev.* A **64** 043412
[10] Keldysh V 1965 *Sov. Phys. JETP* **20** 1307
[11] Schmitz H, Boucke K and Kull H J 1998 *Phys. Rev.* A **57** 46
[12] Cohen J S 1982 *Phys. Rev.* A **26** 3008

Ab Initio Simulations of Dense Helium Plasmas

Cong Wang,[1] Xian-Tu He,[1,2] and Ping Zhang[1,2,*]
[1]LCP, Institute of Applied Physics and Computational Mathematics, P.O. Box 8009, Beijing 100088, People's Republic of China
[2]Center for Applied Physics and Technology, Peking University, Beijing 100871, People's Republic of China
(Received 1 December 2010; published 8 April 2011)

We study the thermophysical properties of dense helium plasmas by using quantum molecular dynamics and orbital-free molecular dynamics simulations, where densities are considered from 400 to 800 g/cm^3 and temperatures up to 800 eV. Results are presented for the equation of state. From the Kubo-Greenwood formula, we derive the electrical conductivity and electronic thermal conductivity. In particular, with the increase in temperature, we discuss the change in the Lorenz number, which indicates a transition from strong coupling and degenerate state to moderate coupling and partial degeneracy regime for dense helium.

DOI: 10.1103/PhysRevLett.106.145002 PACS numbers: 52.27.Gr, 52.25.Fi, 52.65.Yy

Pressure induced physical properties of hot dense helium plasmas are of crucial interest for inertial confinement fusion (ICF) and astrophysics [1–3]. Complete burning of a deuterium-tritium (DT) capsule follows the fusion reaction $D + T \rightarrow n + \text{He} + 17.6$ MeV; thus, accurate knowledge of the equation of states (EOS) and the relative transport coefficients for helium are essential in typical ICF designs. In the direct-drive scheme, thermal transports in ICF plasmas play a central role in predicting laser absorption [4], shock timing [5], and Rayleigh-Taylor instabilities grown at the fuel-ablator interface or at the hot spot–fuel interface [6–8], while in the indirect-drive ICF, the efficiency of the x-ray conversion is also determined by thermal conduction [9]. Accurate modeling of electrical conductivity is important for precisely determining interactions between electrons and plasmas, because high-energy electron beams have been considered to be the most suitable source for igniting the hot spot much smaller than the dense DT core in the fast ignitor [10]. Because of these important items mentioned above, the EOS and electronic transport properties for hot dense helium are highly recommended to be presented and understood. From the theoretical point of view, various assumptions have been used to predict the electrical conductivity and electronic thermal conductivity for weakly coupled and strongly degenerated plasmas [11–14]. These classic methods widely disagree at high density and are incorrect to model nonlinear screening caused by electronic polarization. Quantum molecular dynamic (QMD) simulation [15,16], which is free of adjustable parameters or empirical interionic potentials, has been proven to be ideally suited for studying warm dense matter. However, large numbers of occupied electronic states and overlap between pseudocores limit the use of this method at high temperatures and high densities.

In the present work, direct QMD simulations based on plane-wave density functional theory (DFT) have been adopted to study helium at both high temperatures and high densities, along the 400 to 800 g/cm^3 isochore and temperatures up to 800 eV. The EOS data have been determined for a wide range of densities and temperatures. We apply Kubo-Greenwood formula as a starting point for the evaluation of the dynamic conductivity $\sigma(\omega)$ from which the dc conductivity (σ_{dc}) and electronic thermal conductivity (K) can be extracted.

We introduce the *ab initio* plane-wave code ABINIT [17–19] to perform QMD simulations. A series of volume-fixed supercells including N atoms, which are repeated periodically throughout the space, form the elements of our calculations. After Born-Oppenheimer approximation, electrons are quantum mechanically treated through plane-wave, finite-temperature DFT, where the electronic states are populated according to Fermi-Dirac distributions. The exchange-correlation functional is determined by local density approximation (LDA) with Teter-Pade parametrization [20], and the temperature dependence of exchange-correlation functional, which is convinced to be as small as negligible, is not taken into account. The selection of pseudopotential approximations, which separate core electrons from valence electrons, prevents general QMD simulations from high-density region with the convenience of saving computational cost. Because of the density-induced overlap of the pseudopotential cutoff radius and delocalization of core electrons, this frozen-core approximation only works at moderate densities. As a consequence, a Coulombic pseudopotential with a cutoff radius $r_s = 0.001$ a.u., which is used to model dense helium up to 800 g/cm^3, has been built to overcome the limitations. The plane-wave cutoff energy is set to 200.0 a.u., because large basis set is necessary in modelling wave functions near the core. Sufficient occupational band numbers are included in the overall calculations (the occupation down to 10^{-6} for electronic states are considered). Γ point and $3 \times 3 \times 3$ Monkhorst-Pack scheme **k** points are used to sample the Brillouin zone in molecular dynamics simulations and electronic structure calculations, respectively, because EOS (transport coefficients) can only be modified within 5% (15%) for the selection of higher number of **k** points. A total number of 64 helium atoms are used in the cubic box. Isokinetic ensemble is adopted in present simulations, and local equilibrium is kept through setting the electronic (T_e) and ionic (T_i) temperatures to be equal. Each dynamics simulation is lasted for 6000 steps, and the time steps for the integrations of atomic motion are selected according to different densities (temperatures) [21]. Then, EOS is averaged over the subsequent 1000 step simulations.

QMD simulations have only been run for temperatures up to 300 eV, for higher temperatures, the thermal excited electronic states increase dramatically, and are currently numerically intractable. For temperatures between 10 and 1000 eV, the EOS of dense helium are also obtained from orbital-free molecular dynamic (OFMD) simulations for comparison with the QMD results. In this scheme, orbital-free functional is derived from the semiclassical development of the Mermin functional [22], which leads to the finite-temperature Thomas-Fermi expression for the kinetic part. OFMD simulations are numerically available for temperatures up to 1000 eV, where QMD simulations became prohibitive. The electronic density, which is the only variable in OFMD simulations, is Fourier transformed during the calculation, and numerical convergence in terms of the mesh size has been insured for all the simulations.

Two nondimensional parameters are used to characterize the state of plasmas, namely, the ion-ion coupling and the electron degeneracy parameters. The former one is commonly defined as $\Gamma_{\rm ii} = Z^{*2}/(k_B T a)$, which describes the ratio of the mean electrostatic potential energy and the mean kinetic energy of the ions. Here, Z^* is the average ionization degree, which is equal to 2 in the present system and a is the ionic sphere radius. The degeneracy parameter $\theta = T/T_F$ is the ratio of the temperature to the Fermi temperature $T_F = (3/\pi^2 n_e)^{2/3}/3$. Both of the parameters are summarized in Table I for the studied densities and temperatures.

The accurate determination of the transport coefficients depends on a precise description of the EOS. The wide-range EOS are obtained from both quantum and semiclassical molecular dynamic simulations. As have been shown in Table I, our simulations start from strongly coupled and highly degenerate states, then reach moderate coupling and partial degeneracy states. To examine our results, we also calculate the EOS of hydrogen along 80 g/cm^3 isochore up to 1000 eV, and the data are in accordance with previous theoretical predictions [23,24], as shown in Fig. 1. Good agreements are also shown between our QMD and OFMD results, due to the ideal metallization of helium in the hot dense regime. The simulated EOS along temperature for dense helium shows systematic behavior, and smooth functions ($P = \sum A_{ij}\rho^i T^j$), which can be simply used in hydrodynamic simulations for hot dense helium and astrophysical applications, have been constructed (coefficients A_{ij} have been listed in Table II).

TABLE I. Ion-ion coupling parameter ($\Gamma_{\rm ii}$) and electron degeneracy parameter θ.

ρ (g/cm^3)	400		480		600		800	
T (eV)	$\Gamma_{\rm ii}$	θ	$\Gamma_{\rm ii}$	θ	$\Gamma_{\rm ii}$	θ	$\Gamma_{\rm ii}$	θ
10	28.85	0.02	30.66	0.02	33.03	0.01	36.35	0.01
20	14.43	0.04	15.33	0.03	16.51	0.03	18.18	0.02
50	5.77	0.09	6.13	0.08	6.61	0.07	7.27	0.06
100	2.89	0.18	3.07	0.16	3.30	0.14	3.64	0.11
200	1.44	0.36	1.53	0.32	1.65	0.27	1.82	0.22
300	0.96	0.54	1.02	0.47	1.10	0.41	1.21	0.34
400	0.72	0.71	0.77	0.63	0.83	0.54	0.91	0.45
500	0.58	0.89	0.61	0.79	0.66	0.68	0.73	0.56
800	0.36	1.43	0.38	1.26	0.41	1.09	0.45	0.90

The linear response of hot dense helium to external electrical field and temperature gradient can be characterized by the electrical and heat current densities. The key to evaluate these linear-response transport properties is the kinetic coefficients based on the Kubo-Greenwood formula

$$\hat{\sigma}(\epsilon) = \frac{1}{\Omega}\sum_{k,k'}|\langle\psi_k|\hat{v}|\psi_{k'}\rangle|^2\delta(\epsilon_k - \epsilon_{k'} - \epsilon), \quad (1)$$

where $\langle\psi_k|\hat{v}|\psi_{k'}\rangle$ are the velocity matrix elements, Ω is the volume of the supercell, and ϵ_k are the electronic eigenvalues. The kinetic coefficients $\mathcal{L}_{ij}$ in the Chester-Thellung version [25] are given by

$$\mathcal{L}_{ij} = (-1)^{i+j}\int d\epsilon\hat{\sigma}(\epsilon)(\epsilon - \mu)^{(i+j-2)}\left(-\frac{\partial f(\epsilon)}{\partial\epsilon}\right), \quad (2)$$

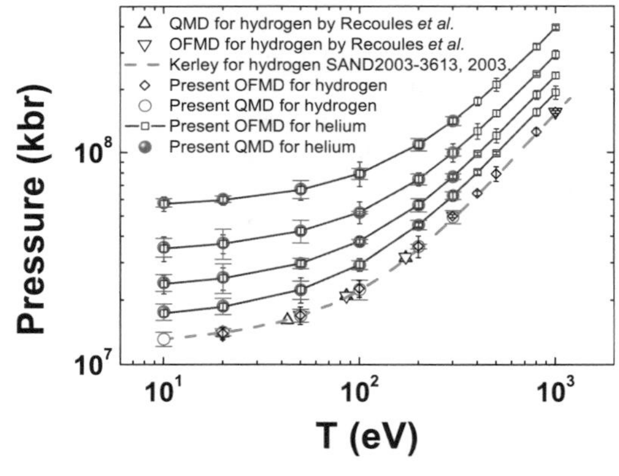

FIG. 1 (color online). Calculated EOS (pressure versus temperature) for dense helium. The QMD and OFMD results, where the error bars are shown by using wide and narrow caps, are labeled as red solid circles and blue open squares, respectively. Previous theoretical predictions obtained by Recoules *et al.* (up and down triangles) [23] and Kerley (green dashed line) [24], where the EOS of dense hydrogen along the 80 g/cm^3 isochore, are also shown for comparison.

TABLE II. Pressure (kbar) expansion coefficients A_{ij} in terms of density (g/cm^3) and temperature (eV).

i	A_{i0}	A_{i1}	A_{i2}
0	26.77	2710.10	−7.02
1	10 819.04	398.98	0.08
2	69.92	−0.07	0.00

with μ being the chemical potential and $f(\epsilon)$ the Fermi-Dirac distribution function. We obtain the electrical conductivity σ

$$\sigma = \mathcal{L}_{11}, \tag{3}$$

and electronic thermal conductivity K is

$$K = \frac{1}{T}\left(\mathcal{L}_{22} - \frac{\mathcal{L}_{12}^2}{\mathcal{L}_{11}}\right), \tag{4}$$

where T is the temperature. Eqs. (3) and (4) are energy-dependent, then the electrical conductivity and electronic thermal conductivity are obtained through extrapolating to zero energy. Those formulation are implemented in the ABINIT code, and have lead to good results for liquid aluminum [26] and hot dense hydrogen [23], where Troullier-Martins potential and Coulombic potential were used, respectively. Within the framework of finite-temperature DFT, chemical potential is evaluated through fitting the set of occupation numbers corresponding to the set of eigenvalues with the usual functional form for the Fermi-Dirac distribution. The δ function in Eq. (1) is broaden in the calculations by a Gaussian function, which has been tested to obtain smooth curves.

The Lorentz number L is defined as

$$L = \frac{K}{\sigma T} = \gamma \frac{e^2}{k_B^2}. \tag{5}$$

The rule of force responsible for the electronic scattering characterizes γ. From the Wiedemann-Franz law, γ is $\pi^2/3$ and L is 2.44×10^{-8} in the degenerate regime (low temperature). For the nondegenerate case (high temperature), γ is 1.5966 and L is 1.18×10^{-8}. No classical assumptions are available for γ value in the intermediate region. As a consequence, the electronic thermal conductivity can not be deduced from the electrical conductivity by using Wiedemann-Franz law. Direct predictions of electrical and thermal conductivity from QMD simulations are then rather interesting.

In order to get converged transport coefficients, ten independent snapshots, which are selected during one molecular dynamics simulation at given conditions, are picked up to calculate electrical conductivity and electronic thermal conductivity as running averages. For temperatures below 300 eV, atomic configurations are directly extracted from QMD simulations, while they are taken from OFMD simulations for higher temperatures. Let us stress here that, one-body electronic states are populated according to a Fermi-Dirac distribution in QMD simulations, and a large number of occupied orbitals are introduced at high temperatures. Thus this method is rather time consuming.

In Fig. 2 we plot the electrical conductivity and electronic thermal conductivity as functions of temperature. While at relatively low temperatures up to 100 eV the electrical conductivity changes slowly, it turns to smoothly increase at higher temperatures. The electronic thermal conductivity shows a sustaining increase with temperature, as in Fig. 2 (lower panel). As temperatures arise from 10 to 800 eV, electronic thermal conductivity is of great importance in ICF applications, where these thermodynamic conditions are encountered in the pellet target fusion. In general ICF designs, several models are used to evaluate thermal conductivity, such as Hubbard and Lee-More models [12,14]. Previous theoretical predictions for dense hydrogen have demonstrated that the electronic thermal conductivity obtained by QMD simulations are in fairly good agreement with the Hubbard model. For very high temperatures, the results for Lee-More model merge into the Spitzer thermal conductivity [11], which are accordant with QMD simulations [23]. As a comparison, we also plot the thermal conductivity for dense hydrogen in Fig. 2. It is indicated that at temperatures below 200 eV, electronic thermal conductivity for hydrogen plasma is larger than that of helium. However, they tend to merge with each other for higher temperatures.

Using the calculated electrical and thermal conductivity, the Lorentz ratio is extracted and shown in Fig. 3. In the strong coupling and degenerate region, the computed Lorentz number vibrates around ideal value of 2.44×10^{-8}, which is predicted by the nearly free electron model. This model is valid in the case where Born approximation is applicable because scattering of the electrons by the ions is sufficiently weak. With the increase of

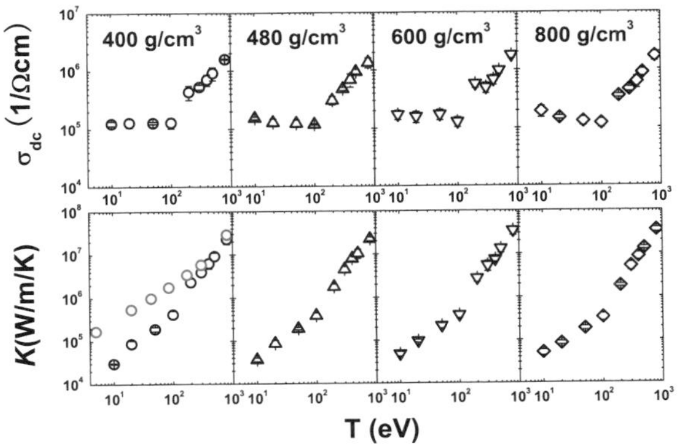

FIG. 2 (color online). Calculated electrical conductivity (upper panels) and electronic thermal conductivity (lower panels) as functions of temperature at four densities of 400, 480, 600, and 800 g/cm^3. For comparison, electronic thermal conductivity for dense hydrogen is also shown as red open circles.

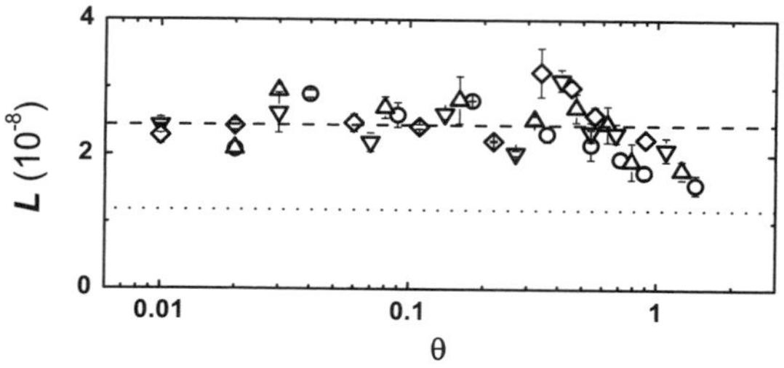

FIG. 3. Lorenz number as a function of electron degeneracy parameter. To distinguish different densities, open circle, up triangle, down triangle, and open diamond are used to denote densities of 400, 480, 600, 800 g/cm^3, respectively. The relative degeneracy parameter can be read from Table I. The dashed line is the value of the degenerate limit, and the dotted line is the value of nondegenerate case.

temperature, where dense helium enters moderate coupling and a partial degeneracy regime, Lorenz number tends to decrease and approach to the nondegenerate value of 1.18×10^{-8}.

In summary, we have performed *ab initio* QMD simulations to study the thermophysical properties of dense helium plasmas under extreme conditions as reached in ICF experiments. As a result, highly converged EOS data (pressures up to 10^8 kbar), for which Coulombic potential with a very small cutoff radius is adopted to model ultradense helium, have been obtained. We have constructed smooth functions to fit the QMD data for the pressure, which is applicable for astrophysics and ICF designs. Using Kubo-Greenwood formula, the electrical conductivity and electronic thermal conductivity have been systematically determined and carefully discussed. The examination of Lorenz number, as well as that of the calculated plasmas parameters (Table I), has indicated a gradual transition from a strong coupling and degenerate state to a moderate coupling and partial degeneracy regime for dense helium. We expect the present simulated results provide a guiding line in the practical ICF hydrodynamical simulations with the helium plasma participated.

This work was supported by NSFC under Grants No. 11005012 and No. 51071032, by the National Basic Security Research Program of China, and by the National High-Tech ICF Committee of China.

*Corresponding author.
zhang_ping@iapcm.ac.cn

[1] S. Atzeni and J. Meyer-ter-Vehn, *The Physics of Inertial Fusion: Beam Plasma Interaction, Hydrodynamics, Hot Dense Matter*, International Series of Monographs on Physics (Clarendon Press, Oxford, 2004).
[2] J. D. Lindl, *Inertial Confinement Fusion: The Quest for Ignition and Energy Gain Using Indirect Drive* (Springer-Verlag, New York, 1998).
[3] J. Nuckolls *et al.*, Nature (London) **239**, 139 (1972).
[4] W. Seka *et al.*, Phys. Plasmas **15**, 056312 (2008).
[5] T. R. Boehly *et al.*, Phys. Plasmas **13**, 056303 (2006).
[6] H. Azechi *et al.*, Phys. Plasmas **4**, 4079 (1997).
[7] S. G. Glendinning *et al.*, Phys. Rev. Lett. **78**, 3318 (1997).
[8] V. Lobatchev and R. Betti, Phys. Rev. Lett. **85**, 4522 (2000).
[9] J. D. Lindl *et al.*, Phys. Plasmas **11**, 339 (2004).
[10] H. Cai *et al.*, Phys. Rev. Lett. **102**, 245001 (2009).
[11] L. Spitzer *et al.*, Phys. Rev. **89**, 977 (1953).
[12] W. B. Hubbard *et al.*, Astrophys. J. **146**, 858 (1966).
[13] H. Brysk *et al.*, Plasma Phys. **17**, 473 (1975).
[14] Y. T. Lee *et al.*, Phys. Fluids **27**, 1273 (1984).
[15] A. Kietzmann *et al.*, Phys. Rev. Lett. **101**, 070401 (2008).
[16] W. Lorenzen, B. Holst, and R. Redmer, Phys. Rev. Lett. **102**, 115701 (2009).
[17] The ABINIT code, which is a common project of the Université Catholique de Louvain, Corning Incorporated, Commissariat à l'Energie Atomique, Université de Liège, Mitsubishi Chemical Corp. and other contributions, is available at http://www.abinit.org.
[18] X. Gonze *et al.*, Comput. Mater. Sci. **25**, 478 (2002).
[19] F. Bottin *et al.*, Comput. Mater. Sci. **42**, 329 (2008).
[20] S. Goedecker, M. Teter, and J. Hutter, Phys. Rev. B **54**, 1703 (1996).
[21] The time steps have been taken as $\Delta t = a/20\sqrt{k_B T/m_{\mathrm{He}}}$, where $a = (3/4\pi n_i)^{1/3}$ is the ionic sphere radius (n_i is the ionic number density), $k_B T$ presents the kinetic energy, and m_{He} is the ionic mass. A few hundred of plasma periods are included in this choice of simulations.
[22] M. Brack and R. K. Bhaduri, *Semiclassical Physics* (Westview Press, Boulder, CO, 2003), ISBN 0813340845.
[23] V. Recoules *et al.*, Phys. Rev. Lett. **102**, 075002 (2009).
[24] G. I. Kerley, Sandia National Laboratory Tech. Rep. Report No. SAND2003-3613, 2003.
[25] G. V. Chester and A. Thellung, Proc. Phys. Soc. London **77**, 1005 (1961).
[26] V. Recoules and J. P. Crocombette, Phys. Rev. B **72**, 104202 (2005).

PHYSICS OF PLASMAS **18**, 082707 (2011)

Thermophysical properties for shock compressed polystyrene

Cong Wang,[1] Xian-Tu He,[1,2] and Ping Zhang[1,2,3,a)]
[1]*LCP, Institute of Applied Physics and Computational Mathematics, P.O. Box 8009, Beijing 100088, People's Republic of China*
[2]*Center for Applied Physics and Technology, Peking University, Beijing 100871, People's Republic of China*
[3]*National Institute for Fusion Science, Toki-shi 509-5292, Japan*

(Received 29 April 2011; accepted 21 July 2011; published online 23 August 2011)

We have performed quantum molecular dynamic simulations for warm dense polystyrene at high pressures. The principal Hugoniot up to 770 GPa is derived from wide range equation of states. The optical conductivity is calculated via the Kubo-Greenwood formula, from which the dc electrical conductivity and optical reflectivity are determined. The nonmetal-to-metal transition is identified by gradual decomposition of the polymer. Our results show good agreement with recent high precision laser-driven experiments. © *2011 American Institute of Physics*. [doi:10.1063/1.3625273]

I. INTRODUCTION

In high energy density ($E/V \geq 10^{11}\,\mathrm{J/m^3}$) physics (HEDP), pressure-induced response of materials, which can be probed through shock wave experiments,[1–4] is of crucial interest for inertial confinement fusion (ICF), chemical (or explosive) process. Fundamental experiments in typical ICF designs, such as the investigation of hydrodynamic instabilities, have introduced plastic compounds as basic candidates to achieve high gain.[5] Furthermore, in recent fast ignitor experiments, the electrical properties of polymers have been often implicated for the transport of relativistic electron beams in solid targets as compared to conducting materials.[6] Understanding the behavior of polymers at several megabar regime brings better insight into physical processes in ICF.

Some of the ablators in the indirect-drive mode of national ignition facility (NIF) will be made of glow-discharge polymer (GDP) with various levels of germanium doping (Ge-GDP).[7] Due to the fact that no high pressure data exist for Ge-GDP, polystyrene (CH), which is closest in structure, is considered as a coarse indicator for shock timing simulations of NIF targets involving such ablators.[8] The absolute equation of states (EOS) of polystyrene along the principal Hugoniot curve has been studied by using gas gun up to 50 GPa (Refs. 9–12) and high energy laser-driven shock waves up to 4000 GPa (Refs. 13–15). High energy dynamic compressions produce the so-called warm dense matter (WDM), where simultaneous dissociations, ionizations, and degenerations make the transition between condensed matter physics to plasma physics. The EOS and the relative properties, such as the Hugoniot curve, are important features in this context. Moreover, electrical properties such as the dielectric function are closely related to the dynamic conductivity, which determines the dc electrical conductivity in the static limit. Then optical reflectance can also be extracted. All these quantities are indispensable in characterizing the unique behavior of shock compressed polystyrene, especially for high pressures at, or exceeding, 1 Mbar. Recently, the development and application of quantum molecular dynamic (QMD) simulation[16] techniques for WDM are available due to the enormous progress in computer capacity. QMD simulations, where quantum effects and correlations are systematically treated, have conducted highly predictive results to describe WDM, as demonstrated by recent simulations[17] on two CH2 polymers, polyethylene, and poly 4-methyl-1-pentene (PMP).

In the present work, QMD simulations are applied to calculate a broad spectrum of thermophysical properties for shock compressed polystyrene. The self-consistent electronic structure calculation within density functional theory (DFT) yields the charge density distribution in the simulation supercell at every time step. The ion-ion pair correlation function (PCF), which is important for characterizing and identifying phase transitions, can be given by molecular dynamics run. The Hugoniot curve, which is derived from EOS data for a wide region of densities and temperatures, is determined and compared with available experimental and theoretical results. As a starting point, we use the Kubo-Greenwood formula to evaluate the dynamic conductivity $\sigma(\omega)$, from which the dc conductivity, the dielectric function $\varepsilon(\omega)$, and the optical reflectivity can be settled.

II. COMPUTATIONAL METHOD

We introduce *ab initio* plane wave code VASP (Refs. 18 and 19) to perform QMD simulations. The elements of our calculations consist of a series of volume-fixed supercells including N atoms, which are repeated periodically throughout the space. By involving Born-Oppenheimer approximation, electrons are quantum mechanically treated through plane-wave, finite-temperature DFT (Ref. 20), where the electronic states are populated according to Fermi-Dirac distributions. Sufficient occupational bands are included in the overall calculations (the occupational number down to 10^{-6} for electronic states are considered). The exchange-correlation functional is determined by generalized gradient approximation (GGA) with the parametrization of Perdew-Wang 91 (Ref. 21). The ion-electron interactions are represented by a projector augmented wave (PAW) pseudopotential.[22] Isokinetic ensemble (NVT) is adopted in the present simulations,

[a)]Author to whom correspondence should be addressed. Electronic mail: zhang_ping@iapcm.ac.cn.

where the ionic temperature T_i is controlled by Nośe thermostat,[23] and the system is kept in local equilibrium by setting the electron (T_e) and ion (T_i) temperatures to be equal.

The simulations have been done for 128 atoms (namely, eight C_8H_8 units) in a supercell. A plane-wave cutoff energy of 600.0 eV is selected, so that a convergence of better than 3% is secured. The Brillouin zone is sampled by Γ point and $3 \times 3 \times 8$ Monkhorst-Pack scheme[24] **k** points in molecular dynamic simulations and electronic structure calculations, respectively, because EOS (conductivity) can only be modified within 5% (15%) for the selection of higher number of **k** points. The densities adopted in our simulations range from 1.05 g/cm^3 to 3.50 g/cm^3 and temperatures between 300 and 80 000 K, which highlight the regime of principal Hugoniot. All the dynamic simulations are lasted 6000 steps, and the time steps for the integrations of atomic motion are selected according to different densities (temperatures).[25] Then, the subsequent 1000 step simulations are used to calculate EOS as running averages.

III. RESULTS AND DISCUSSION

High precision EOS data are essential for understanding the electrical and optical properties of polystyrene under extreme conditions. The present EOS have been examined theoretically through the Rankine-Hugoniot (RH) equations, which follow from conservation of mass, momentum, and energy across the front of shock waves. The locus of points in (*E, P,* and *V*)-space described by RH equations satisfy the following relations:

$$E_1 - E_0 = \frac{1}{2}(P_1 + P_0)(V - V_0), \tag{1}$$

$$P_1 - P_0 = \rho_0 u_s u_p, \tag{2}$$

$$V_1 = V_0(1 - u_p/u_s), \tag{3}$$

where subscripts 0 and 1 present the initial and shocked state, and *E*, *P*, and *V* denote internal energy, pressure, and volume, respectively. The u_s and u_p correspond to the shock and mass velocities of the material behind the shock front. As the starting point along the Hugoniot, the initial density is $\rho_0 = 1.05$ g/cm^3 and the internal energy is $E_0 = -87.44$ kJ/g at a temperature of 300 K. Compared to the high pressure of shocked states along the Hugoniot, the initial pressure P_0 can be treated approximately as zero. We use smooth functions to fit the internal energy and pressure in terms of temperature at sampled density and derive Hugoniot point from Eq. (1). The calculated Hugoniot data are listed in Table I.

TABLE I. Polystyrene Hugoniot results from QMD simulations.

ρ (g/cm^3)	P (GPa)	T (eV)	u_s (km/s)	u_p (km/s)
2.04	22.07	0.19	6.58	3.19
2.10	24.64	0.20	6.85	3.43
2.43	53.36	0.31	9.46	5.37
2.59	73.22	0.37	10.83	6.44
2.79	126.51	0.73	13.90	8.67
3.01	208.43	1.24	17.46	11.37
3.08	277.85	1.82	20.04	13.21
3.18	361.07	2.15	22.66	15.18
3.29	592.61	4.25	28.79	19.60
3.39	766.46	5.43	32.52	22.45

The principal Hugoniot curve is plotted in Fig. 1, where previous theoretical and experimental results are also shown for comparison. The EOS of polystyrene has been previously probed by gas gun experiments, which have the advantage of high precision, but the pressure hardly exceeds 50 GPa. The present QMD results for a CH polymer and recent QMD results[17] for CH2 polymers all agree well with the gas gun data. Recently, high energy laser-driven experiments have detected pressures up to 4000 GPa. Meanwhile, however, large error bars were introduced at above 100 GPa (Refs. 13, 14), and thus the use of low-precision EOS of the polymer to predict the behavior of NIF Ge-GDP ablator materials provides an unacceptable uncertainty. Then, high precision EOS (up to 1000 GPa) for polystyrene was obtained by using α-quartz as an impedance-matching (IM) standard.[8] As shown in Fig. 1, previous data by Ozaki *et al.*[14] (IM with an

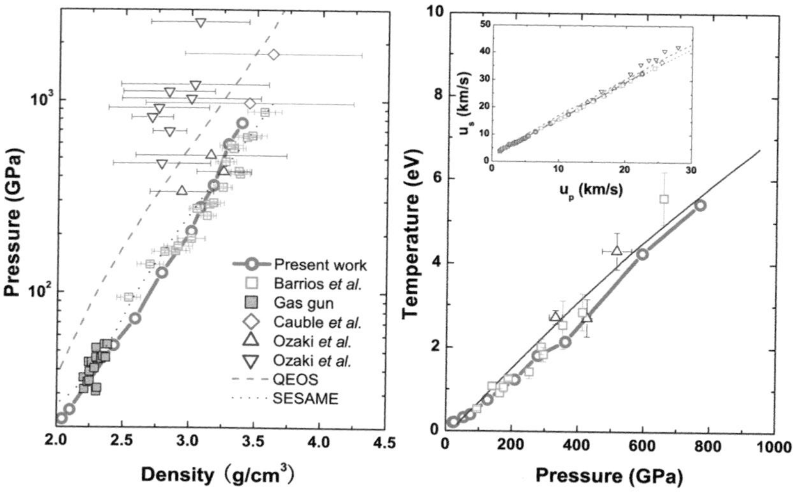

FIG. 1. (Color online) The *P-V* (left panel) and *T-P* (right panel) Hugoniot curves of shocked polystyrene. Inset is the (u_s, u_p) diagram. Previous results are also shown for comparison. Experiments: gas-gun results (the gray filled squares);[9,12] absolute measurements on NOVA (magenta diamonds);[13] and high energy laser-driven measurements by Ozaki *et al.*[14,15] (blue upward and downward triangles) and Barrios *et al.* (orange squares).[8] Theories: QEOS and SESAME (Ref. 26) results are denoted by green dashed line and purple dotted line, respectively.

aluminum standard) and Cauble *et al.*[13] (absolute data) show clearly stiffer behavior compared to those results with quartz standard[8,15] and SESAME model.[26] The possible reason for this stiff behavior is likely due to x-ray preheating of the samples, as has been stated by those authors. Then, thicker pushers and low-Z ablators were used in the newly measured data from Ozaki *et al.*[15] to reduce pre-heating of the samples. The experiments, where IM with a quartz standard were also used, show results that are closer to those by Barrios *et al.* Our simulated results show good agreements with those experiments where quartz standard was used.

Temperature, as has been focused as one of the most important parameters in experiments, is difficult to be measured because of the uncertainty in determining the optical-intensity loss for ultraviolet part of the spectrum in adiabatic or isentropic shock compressions, especially for the temperature exceeding several eV.[27] QMD simulations can provide efficient predictions for shock temperatures. As has been shown in Fig. 1 (right panel), our calculated Hugoniot temperatures are accordant with experimental results up to 2 eV and discrepancy emerges at higher temperatures. Beyond 2 eV, the predicted temperatures from both experiments and SESAME model at a given pressure are slightly higher compared to our calculations.

We also examine the structural transition of polystyrene under shock compressions by using PCF, which is evaluated at equilibrium during the molecular dynamics simulations. Along the Hugoniot, four points of PCF are shown in Fig. 2. At the initial state [see Fig. 2(a)], the ideal condensed polymer phase is indicated by one peak for C-H bond, meanwhile, two peaks (π-type and sp^3-type) for C-C bond. With the increase of pressure ($\sim$20 GPa), phenyl decomposes, which is indicated by that the two C-C peaks merge together (a transverse from π bond to sp^2, sp^3 like bond). At this stage, diamondlike carbon nanoparticles (with defects) are formed. The present phenomenon agrees well with the behavior of shock compressed benzene.[28,29] Further, increase of pressure produces molecular and atomic hydrogen, which is indicated by the peak in H-H PCF around 0.75 Å [Fig. 2(b)]. Decomposition of hydrocarbons continues until 126 GPa, where PCF shows no obvious evidence for the existence of C-H bond. Whereas, for higher pressures, the PCFs do not show visible difference and are thus not presented here. SESAME model suggested that the softening behavior of Hugoniot is caused by the chemical decomposition of the polymer at 200$\sim$400 GPa. Whereas, our calculations show that the decomposition pressure lies between 20 and 126 GPa, which is much lower than that of SESAME.

Having clarified the EOS, let us turn now to study the electrical and optical properties for shocked polystyrene. Based on Kubo-Greenwood formula, the electrical and optical properties can be obtained as described in Ref. 28. In order to get converged results, 30 independent snapshots, which are selected during one molecular dynamics simulation at given conditions, are picked up to calculate dynamic conductivity as running averages. The dc conductivity σ_{dc} which follows from the static limit $\omega \rightarrow 0$ of $\sigma_1(\omega)$ is then evaluated and plotted in Fig. 3 (left panel) as a function of shock velocity. Initially, σ_{dc} increases rapidly with shock velocity up to 14 km/s (126 GPa) towards the formation of metallic state of polystyrene. Then, one can find from Fig. 3 that σ_{dc} keeps almost invariant and the shock compressed polystyrene maintains its metallic behavior. The onset of metallization ($\sigma_{dc} > 10^3 \Omega^{-1}$ cm^{-1}) is observed at around 10 km/s ($\sim$50 GPa), where dissociation of the polymer and formations of hydrogen (molecular and atomic species) govern the characteristics of warm dense polystyrene. We stress here that the nonmetal to metal transition is induced by gradual chemical decompositions and thermal activations of the electronic state. In the same pressure range (20$\sim$200 GPa),

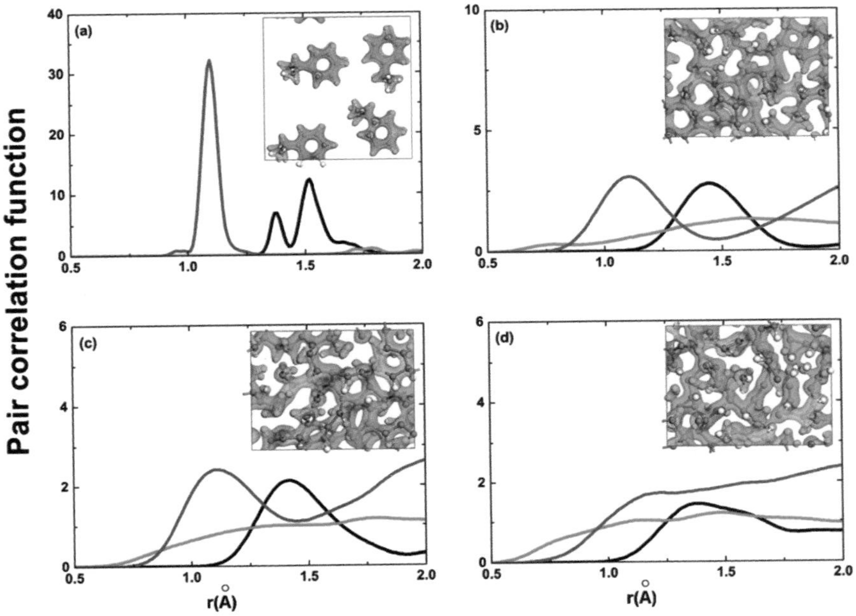

FIG. 2. (Color online) Pair correlation functions for four different points along the principal Hugoniot curve. (a)$\sim$(d) correspond to the starting point, 53.36 GPa, 73.22 GPa, and 126.51 GPa, respectively. C-H, C-C, and H-H bonds are denoted by blue, black, and red lines, respectively. Insets are the sampled atomic structure and charge density distribution at each state (gray balls for carbon and white balls for hydrogen).

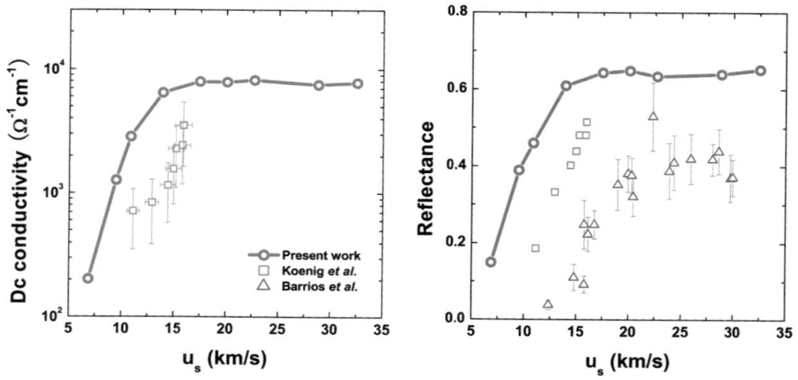

FIG. 3. (Color online) Dc conductivity (left panel) and optical reflectivity at the wavelength of 532 nm (right panel). Previous measurements by Koenig *et al.*[30] (green squares) and Barrios *et al.*[8] (upward triangles) are also plotted.

we observe larger σ_{dc} compared to experimental measurement.[30] This overestimation of σ_{dc} values could be attributed to the use of DFT-based molecular dynamics, which is known to underestimate band gaps in many systems.

Optical reflectance, from which emissivity can be derived, is of great interest in experiment for determining shock temperature. Along the Hugoniot, we show in Fig. 3 (right panel) the variation of optical reflectance at 532 nm as a function of shock velocity. Optical properties of polystyrene have been experimentally studied by Koenig *et al.*[30] with u_s of 11$\sim$16 km/s (80$\sim$170 GPa), and steadily increasing reflectivities reaching values up to $\sim$50% were observed. The results by Ozaki *et al.* indicated reflectivity from 16% to 42% in u_s range of 22$\sim$27 km/s (300$\sim$500 GPa).[15] Recent experiments by Barrios *et al.*[8] suggested a drastic increase in reflectivity around the Hugoniot pressure of 100 GPa, followed by saturated value of 40% around 250$\sim$300 GPa. Discrepancies in reflectivities among experiments have been claimed to be arisen from probe-beam stability or from differences in diagnostic configurations. Along the Hugoniot, our QMD calculations provide the general feature for the optical reflectance—steep increase (from 15% to 60%) followed by saturation (at $\sim$126 GPa). The change in optical reflectivity with pressure can be interpreted as the gradual insulator-conductor transition at above 20 GPa, where the atoms strongly fluctuate with neighbors to dense, partially ionized plasma at high pressures above 126 GPa.

IV. CONCLUSION

In summary, we have carried out QMD simulations to study the thermophysical properties for warm dense polystyrene. The Hugoniot EOS data of polystyrene up to $\sim$770 GPa, which is in agreement with dynamic experiments in a wide range of shock conditions, have been evaluated through QMD calculations. Contribution from chemical decomposition demonstrates the steep increase of σ_{dc} and optical reflectance observed at 20$\sim$126 GPa.

ACKNOWLEDGMENTS

This work was supported by NSFC under Grant Nos. 11005012 and 51071032, by the National Basic Security Research Program of China, by the National High-Tech ICF Committee of China, and by the Core University Program.

[1]M. D. Knudson and M. P. Desjarlais, Phys. Rev. Lett. **103**, 225501 (2009).
[2]W. J. Nellis, Rep. Prog. Phys. **69**, 1479 (2006).
[3]D. G. Hicks, T. R. Boehly, P. M. Celliers, D. K. Bradley, J. H. Eggert, R. S. McWilliams, R. Jeanloz, and G. W. Collins, Phys. Rev. B **78**, 174102 (2008).
[4]D. G. Hicks, T. R. Boehly, P. M. Celliers, J. H. Eggert, S. J. Moon, D. D. Meyerhofer, and G. W. Collins, Phys. Rev. B **79**, 014112 (2009).
[5]J. D. Lindl, *Inertial Confinement Fusion: The Quest for Ignition and Energy Gain Using Indirect Drive* Springer-Verlag, New York, 1998.
[6]H. Cai, K. Mima, W. Zhou, T. Jozaki, H. Nagatomo, A. Sunahara, and R. J. Mason, Phys. Rev. Lett. **102**, 245001 (2009).
[7]S. W. Haan, M. C. Herrmann, J. D. Salmonson, P. A. Amendt, D. A. Callahan, T. R. Dittrich, M. J. Edwards, O. S. Jones, M. M. Marinak, D. H. Munro, S. M. Pollaine, B. K. Spears, and L. J. Suter, Eur. Phys. J. D **44**, 249 (2007).
[8]M. A. Barrios, D. G. Hicks, T. R. Boehly, D. E. Fratanduono, J. H. Eggert, P. M. Celliers, G. W. Collins, and D. D. Meyerhofer, Phys. Plasmas **17**, 056307 (2010).
[9]*LASLShock Hugoniot Data*, Los Alamos Series on Dynamic Material Properties, edited by S. P. Marsh University of California Press, Berkeley, 1980.
[10]R. G. McQueen, S. P. Marsh, J. W. Taylor, J. N. Fritz, and J. W. Carter, in *High-Velocity Impact Phenomena*, edited by R. Kinslow Academic, New York, 1970 Chap. VII, Sec. II, p. 293.
[11]M. V. Thiel, J. Shaner, and E. Salinas, Lawrence Livermore National Laboratory Report No. UCRL-50108, 1977.
[12]A. V. Bushman, I. V. Lomonosov, V. E. Fortov, K. V. Khishchenko, M. V. Zhernokletov, and Yu. N. Sutulov, JETP **82**, 895 (1996). Available at http://www.jetp.ac.ru/cgi-bin/index/e/82/5/p895?a=list.
[13]R. Cauble, L. B. Da Silva, T. S. Perry, D. R. Bach, K. S. Budil, P. Celliers, G. W. Collins, A. Ng, T. W. Barbee, Jr., B. A. Hammel, N. C. Holmes, J. D. Kilkenny, R. J. Wallace, G. Chiu, and N. C. Woolsey, Phys. Plasmas **4**, 1857 (1997).
[14]N. Ozaki, T. Ono, K. Takamatsu, K. A. Tanaka, M. Nakano, T. Kataoka, M. Yoshida, K. Wakabayashi, M. Nakai, K. Nagai, K. Shigemori, T. Yamanaka, and K. Kondo, Phys. Plasmas **12**, 124503 (2005).
[15]N. Ozaki, T. Sano, M. Ikoma, K. Shigemori, T. Kimura, K. Miyanishi, T. Vinci, F. H. Ree, H. Azechi, T. Endo, Y. Hironak, Y. Hori, A. Iwamoto, T. Kadono, H. Nagatomo, M. Nakai, T. Norimatsu, T. Okuchi, K. Otani, T. Sakaiya, K. Shimizu, A. Shiroshita, A. Sunahara, H. Takahashi, and R. Kodama, Phys. Plasmas **16**, 062702 (2009).
[16]A. Kietzmann, R. Redmer, M. P. Desjarlais, and T. R. Mattsson, Phys. Rev. Lett. **101**, 070401 (2008).
[17]T. R. Mattsson, J. Matthew, D. Lane, K. R. Cochrane, M. P. Desjarlais, A. P. Thompson, F. Pierce, and G. S. Grest, Phys. Rev. B **81**, 054103 (2010).
[18]G. Kresse and J. Hafner, Phys. Rev. B **47**, R558 (1993).
[19]G. Kresse and J. Furthmüller, Phys. Rev. B **54**, 11169 (1996).
[20]V. Recoules, F. Lambert, A. Decoster, B. Canaud, and J. Clérouin, Phys. Rev. Lett. **102**, 075002 (2009).
[21]J. P. Perdew, *Electronic Structure of Solids* (Akademie Verlag, Berlin, 1991).
[22]P. E. Blöchl, Phys. Rev. B **50**, 17953 (1994).
[23]S. Nosé, J. Chem. Phys. **81**, 511 (1984).
[24]H. J. Monkhorst and J. D. Pack, Phys. Rev. B **13**, 5188 (1976).
[25]The time steps have been taken as $\Delta t = a/20\sqrt{k_BT/m}$, where $a = (3/4\pi n_i)^{1/3}$ is the ionic sphere radius (n_i is the ionic number density), k_BT presents the kinetic energy, and m is the ionic mass.

[26]S. P. Lyon and J. D. Johnson, Los Alamos National Laboratory Report No. LA-CP-98-100, (1998).
[27]Q. F. Chen, private communication (2010).
[28]C. Wang and P. Zhang, J. Appl. Phys. **107**, 083502 (2010).
[29]W. J. Nellis, D. C. Hamilton, and A. C. Mitchell, J. Chem. Phys. **115**, 1015 (2001).
[30]M. Koenig, F. Philippe, A. Benuzzi-Mounaix, D. Batani, M. Tomasini, E. Henry, and T. Hall, Phys. Plasmas **10**, 3026 (2003).

Characteristic Spectrum of Very Low-Energy Photoelectron from Above-Threshold Ionization in the Tunneling Regime

C. Y. Wu,* Y. D. Yang, Y. Q. Liu, and Q. H. Gong†
State Key Laboratory for Mesoscopic Physics, Department of Physics, Peking University, Beijing 100871, China

M. Wu and X. Liu
State Key Laboratory of Magnetic Resonance and Atomic and Molecular Physics, Wuhan Institute of Physics and Mathematics, Chinese Academy of Sciences, Wuhan 430071, China

X. L. Hao and W. D. Li
Institute of Theoretical Physics and Department of Physics, Shanxi University, 030006 Taiyuan, China

X. T. He and J. Chen‡
HEDPS, Center for Applied Physics and Technology, Peking University, Beijing 100084, China
Institute of Applied Physics and computational Mathematics, P.O. Box 8009, Beijing 100088, China
(Received 18 August 2011; revised manuscript received 13 January 2012; published 23 July 2012)

We report an experimental and theoretical study of very low-energy photoelectrons in tunneling ionization process from noble gas atoms interacting with ultrashort intense infrared laser pulses. A universal peak structure with electron energy well below 1 eV in the photoelectron spectrum, corresponding to the double-hump structure in the longitudinal momentum distribution, is identified experimentally for all atomic species. Our quantum and semiclassical analysis reveal the role of long-range Coulomb potential in the production of this very low-energy peak structure.

DOI: 10.1103/PhysRevLett.109.043001 PACS numbers: 33.80.Rv, 33.80.Wz, 42.50.Hz

Since its first observation in 1979 [1], above-threshold ionization (ATI) has been a fundamental process that plays an essential role in understanding the intense laser-matter interaction [2]. In the tunneling regime where the Keldysh parameter $\gamma = \sqrt{I_p/2U_p} \ll 1$ (where I_p is the ionization potential and $U_p = F_0^2/(2\omega)^2$ the ponderomotive energy, ω the laser angular frequency and F_0 the peak amplitude of the laser field) [3], the ATI spectrum was expected to be consistent with the Simpleman's picture originating from the semiclassical viewpoint of the atom-laser interaction [4]. The overall spectrum drops fast from the origin until $2U_p$ where a flat plateau emerges and extends to a cutoff at $10U_p$, which can be well explained by a rescattering process [5,6].

In contrast to the prediction of the standard tunneling picture, it was discovered that low-energy photoelectrons in the ATI process possess nontrivial behavior. For example, the low-energy momentum distribution of photoelectrons shows some intriguing structures deviating obviously from the Simpleman's perspective [7,8], which has been intensively investigated recently and its mechanism is still in debate [9–14]. More recently, a peculiar low-energy structure (LES) has been found in ATI spectra in the tunneling regime [15,16], which is also in striking contrast to the prediction of the Simpleman's model. Semiclassical model [16–18] has been applied to explore this low-energy structure and revealed the essential role of the long range Coulomb potential in its production. Moreover, significant effect of the Coulomb potential on the low-energy photoelectron dynamics has also been explored in near-threshold high harmonic generation [19].

At present, there are still several puzzles in the low-energy ATI spectrum. One is that although the two experimental observations agree on the LES with electron energies above 1 eV [15,16], Ref. [16] reports a sharp peak structure, or very-low-energy structure (VLES), which is well below 1 eV and depends hardly on the laser wavelength and intensity. In contrast, this VLES was not found in Ref. [15]. The other one concerns the relationship between the LES and the particular structure in the momentum distribution [8,12,13], as both of them are related to the dynamics of low-energy photoelectron. Therefore, the dynamics of photoelectron in ATI process, especially the low-energy electron, is still far from being fully understood, even though ATI process has been studied for more than thirty years since its first observation [1]. In this Letter, we aim to resolve these puzzles and perform a comprehensive analysis on the dynamics of photoelectron with very low energy (< 1 eV) in the tunneling regime.

The experiments were performed using a newly built cold target recoil-ion momentum spectroscopy (COLTRIMS) [20]. A commercial multipass femtosecond amplifier (Femtolasers GmbH) delivers laser pulses with a central wavelength of ~800 nm and pulse duration of 24 fs at a repetition rate of 3 kHz. An optical parametric amplifier (OPA, TOPAS-C, Light Conversion, Inc.) is used to generate wavelength-tunable midinfrared femtosecond laser pulses. The pulse durations are ~30 fs at 1320 nm and ~60 fs at 1800 nm. The laser beam was then tightly focused on the collimated supersonic gas jet in an ultrahigh vacuum chamber (~ 5×10^{-11} mbar). Electrons and ions thus generated were accelerated by applying weak homogenous electric (3 V/cm) and magnetic (~ 5 G) fields and recorded by two opposite position-sensitive delay-line detectors (Roentdek Handels GmbH). From the time of flight and position on the detectors the momentum vectors of electrons and ions were calculated. The momentum resolution is ~0.02 a.u. along the laser polarization direction. In particular, we employ the electron-ion momentum conservation in the data analysis to ensure that the electrons are generated from the single ionization of the gas targets.

Figs. 1(a)–1(c) show the low-energy part of the measured photoelectron spectra (collected in a solid angle of 5–8 degrees in the field direction) of noble gas atoms at different laser wavelengths and intensities. The parameters chosen here guarantee that the process is in the tunneling regime ($\gamma < 1$) and $I_p/\omega \gg 1$ to suppress the resonance channel in the ionization process. Several ATI peaks can be clearly seen in the spectrum for wavelength of 800 nm. With increasing wavelength, the ATI peaks become less pronounced and can be hardly distinguished in the spectrum for 1800 nm. On the other hand, the LES [15,16] becomes noticeable in the long-wavelength spectra. We will denote this as the high-energy low-energy structure (HLES) throughout this Letter, in order to distinguish it from the very-low-energy structure. (For visual convenience, the HLES and the VLES are marked by dashed and solid arrows, respectively, in all figures of energy spectra in this Letter). It is intriguing that all spectra show a VLES below 1 eV, which is located at about 0.5 eV for 800 nm and 0.1–0.2 eV for 1320 nm and 1800 nm, respectively.

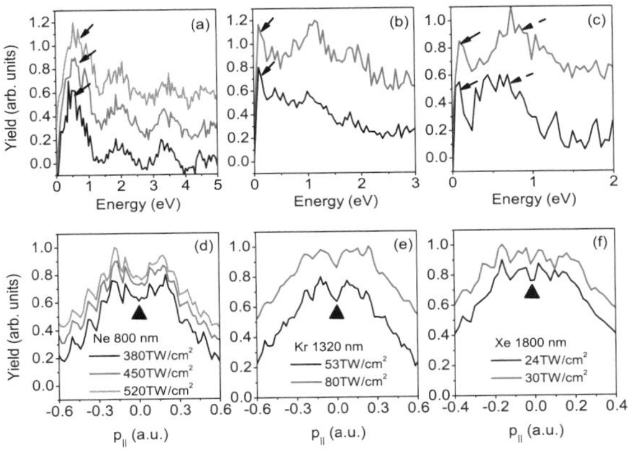

FIG. 1 (color online). Low-energy photoelectron energy spectra and longitudinal momentum distributions of Ne [(a),(d)], Kr [(b),(e)], and Xe [(c),(f)] in infrared laser fields. For visual convenience, the VLES and HLES are marked by solid and dashed arrows in all figures of energy spectrum, respectively (see text).

For comparison, longitudinal momentum distributions (LMD) corresponding to the ATI spectra, which are obtained by integrating over all transverse momenta of the photoelectrons, are depicted in Figs. 1(d)–1(f). It is noteworthy that all distributions show the profile of a double-hump structure (DHS): a central minimum with a pronounced hump on each side that consists of a series of peaks. For 800 nm, this structure reproduces the result presented by Rudenko *et al.* [12]. In addition, it can be found that the humps on either side of the central minimum correspond to the VLES in the energy spectrum. For 800 nm, for example, the hump in the momentum distribution locates at about 0.2 a.u., corresponding to the peak at about 0.5 eV in the energy spectrum. This DHS in the LMD can be qualitatively reproduced in the semiclassical simulation [7,9], *S*-matrix calculation [11,14], and numerical solution of three-dimensional time-dependent Schrödinger equation (TDSE) [10,21]. However, its underlying mechanism has not been fully explored.

To reveal the underlying mechanism of these structures, we calculate the spectrum of Xe atom using a three-dimensional TDSE code [22] (In our calculation, the laser parameters are the same as those in the experiment and a Gaussian envelop is assumed for the pulse shape). As shown in Figs. 2(a) and 2(b), TDSE calculation with focal average qualitatively reproduces the experimental observation: a VLES well below 1 eV in the ATI spectrum and a DHS in the LMD. For comparison, calculations with a short-range ion potential and a hydrogen-like potential are also presented in Fig. 2. It is noteworthy that, for short-range potential, the structures in the ATI spectrum, including both the HLES and the VLES, disappear and accordingly the LMD shows a central peak distribution [see Fig. 2(b)], indicating that the long-range Coulomb potential plays an essential role in the dynamics of low-energy photoelectron in ATI [10,21]. On the other

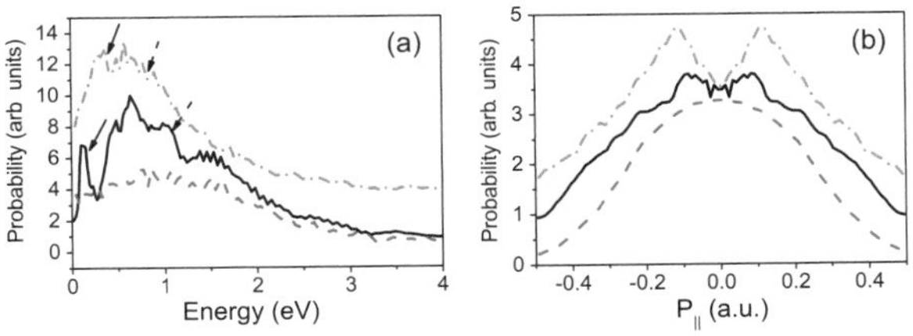

FIG. 2 (color online). Low-energy photoelectron energy spectrum (a) and longitudinal momentum distribution (b) calculated using numerical solution of TDSE for Xe (black solid line) and short-range potential (red dashed line). Calculated result using a hydrogen-like potential with the same ground state energy of Xe is also shown (green dash-dotted line) (see text). $I = 3 \times 10^{13}$ W/cm^2 and $\lambda = 1800$ nm.

side, the results with a hydrogen-like potential are qualitatively consistent with the experimental observation. However, a more pronounced central minimum in the momentum distribution [Fig. 2(b)] has been found in the calculations than that in the experimental data, which will be further discussed later.

To gain further insight into these peculiar structures, we compare the experimental data with the result of the theoretical simulation based on a semiclassical description of the ionization process. Details of the semiclassical approach have been described elsewhere [23]. In brief, we consider model atoms with hydrogen-like potentials of $V = -Z_{\rm eff}/r$ (where $Z_{\rm eff}$ is the effective nuclear charge to give ground state energy equal to that of corresponding atoms) interacting with an external laser field. The laser field $\epsilon(t) = f(t)E_0 \cos\omega t$ has a constant amplitude for the first ten cycles and is ramped off within 3 cycles. For simplification, the electron is released within the time interval $-\pi/2 \leq \omega t \leq \pi/2$ through tunneling with a rate given by the Ammosov-Delone-Krainov formula [24]. The subsequent evolution of the electron is determined by the Newton's equation of motion.

The simulation results are shown in Fig. 3 and exhibit the following main features: all energy spectra show a VLES below 1 eV and a HLES, which develops and shifts with intensity; the position of the VLES locates at about 0.5 eV and shifts slightly with intensity for 800 nm, while for 1320 and 1800 nm, it locates at about 0.1 eV independently of intensity; all LMDs show a DHS and, on each side, the hump consists of one pronounced peak at low intensity or two or more peaks at high intensity; the peaks beside the minimum in the LMD correspond to the VLES in the ATI spectrum. Clearly, the main features shown in the semiclassical simulation are qualitatively consistent with the experimental results.

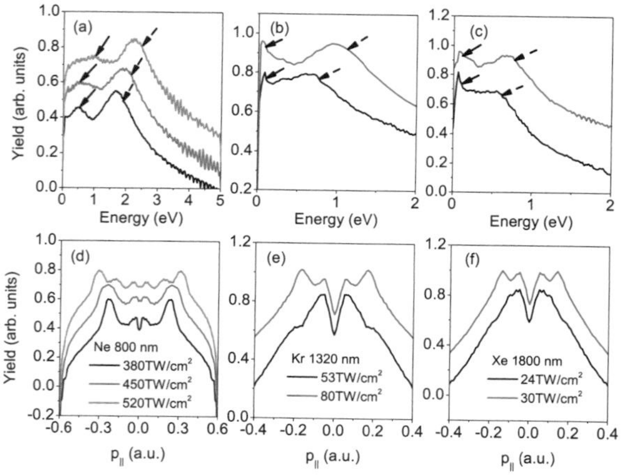

FIG. 3 (color online). Low-energy photoelectron energy spectra and longitudinal momentum distributions of atoms calculated using semiclassical model. (a),(d) Ne. $\lambda = 800$ nm; (b),(e) Kr. $\lambda = 1320$ nm; (c),(f) Xe. $\lambda = 1800$ nm.

In order to shed more light on the physical picture behind, we further calculate the two-dimensional electron momentum distributions (2dMD). As shown in Figs. 4(a) and 4(b), besides a shift of both the direct electron and indirect electron [25] to the opposite direction of the tunneling exit of the electron due to the Coulomb potential [16], the total 2dMD [Fig. 4(c)] shows some more particular structures. A dense lump structure (DLS) locating below $p_{\parallel} = 0.2$ a.u. [region inside a black solid pane in Fig. 4(c)] followed by a hatlike structure [region inside a blue dotted pane in Fig. 4(c)] appears near the z axis (field direction). In addition, two strips, with electron momenta $p_{\parallel}$ of roughly 0.07 and 0.17 a.u., extend out in the direction almost perpendicular to the field direction. The DLS and strip structures in 2dMD [Fig. 4(c)] give rise to DHS in LMD [Fig. 3(e)]. Moreover, the VLES and HLES in Fig. 3(b) correspond to the DLS and hat-like structure in Fig. 4(c), respectively.

Comparing Fig. 4(c) with Figs. 4(a) and 4(b), one can see that the position of the DLS corresponds to the positions of the lower boundary of the indirect electrons and the upper boundary of the direct electrons in the

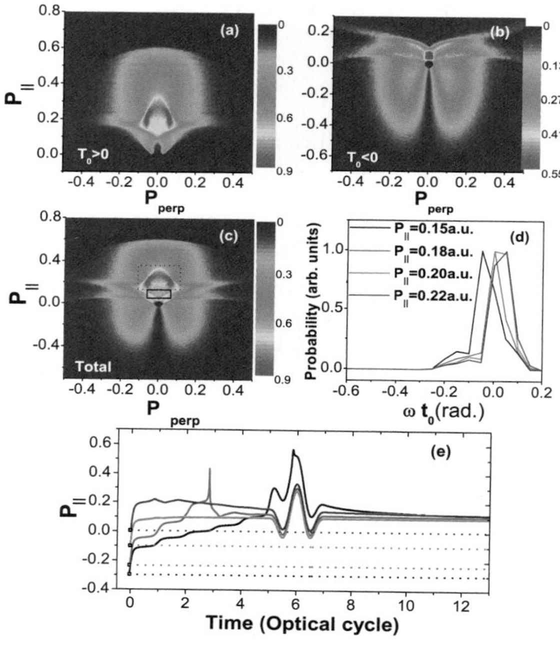

FIG. 4 (color online). (a)–(c) Calculated two-dimensional electron momentum distributions for Kr (z is the laser polarization direction). $I = 8 \times 10^{13}$ W/cm^2 and $\lambda = 1320$ nm. See the text for the details. (d) Distributions of the initial tunnel-ionization phase (ωt_0) of the electron with the corresponding final drift momentum in the field direction. (e) Temporal evolution of drift momenta of four trajectories with different initial conditions with (solid line) and without (dotted line) Coulomb potential.

2dMD, indicating that the DHS is closely related to the shift due to the Coulomb potential. This is partly consistent with the analysis of Dimitriou *et al.* [9]. Moreover, as shown in Figs. 1 and 3, the prominent HLES in the energy spectrum is hardly distinguished while the VLES still manifests in the LMD. This is because the HLES is mainly due to the Coulomb focusing effect [16–18]; integration over the transverse momentum mostly smears out this effect. In contrast, the DHS will not be spoiled by the integration since it partly comes from the shift due to the Coulomb potential.

Moreover, it deserves noting that the DLS and other structures, e.g., the strips shown in Fig. 4(c) cannot be only attributed to Coulomb distortion since distortion can only give rise to a smooth distribution with a shifted maximum in the LMD. This point can be clearly seen in Fig. 4(d) that shows the distributions of the initial tunnel-ionization phase (ωt_0) of the electron with the corresponding final drift momentum in the field direction. In general, the phase increases with momentum and the averaged phase of the electron with $P_{\parallel}$ in the range from 0.15 to 0.22 a.u. is at about $\omega t_0 = 0$, which is in agreement with the distortion picture [9]. However, the phase distributions exhibit a distinct overlap for different longitudinal momenta, indicating that the interaction between the electron and the ion core significantly redistributes the electron momentum. This can be further demonstrated by Fig. 4(e), which shows the evolution of the longitudinal momenta of electron trajectories with different initial conditions (i.e., with different tunnel ionization time and transverse momenta). Without the Coulomb potential, the electrons' final longitudinal momenta are solely determined by the initial tunnel ionization time t_0 [$p = -A(t_0)$ where $A(t)$ is the vector potential of the laser field] and are quite different. In contrast, all of them develop an identical momentum and contribute to the DLS at about 0.1–0.2 a.u. in the 2dMD [Fig. 4(c)] with Coulomb potential considered. Moreover, this picture also applies to the strip structures in Fig. 4(c). In Fig. 4(e), abrupt changes of the drift momenta indicate strong interaction of the electrons with the Coulomb potential when they return to the core. Note that this strong interaction does not require a very close approaching of the electron to the ion [26] since the electron with initial phase $\omega t \sim 0$ considered here returns to the core with very low kinetic energy. It is worth mentioning that the mechanism revealed here for the VLES is similar to an electron energy bunching process discussed in Ref. [27].

It is noteworthy that, by comparing Fig. 3 with Fig. 4, the VLES in the energy spectrum for long wavelengths (1320 nm and 1800 nm) more likely corresponds to the first hump beside the minimum but not the second hump in the LMD since the energy spectrum is obtained by integrating over a small solid angle (5–8 degree) around the field direction and the LMD is obtained by integrating the total transverse momentum. However, for the 800 nm case with much higher intensities, the most apparent humps locate at larger momenta [about ± 0.2 a.u., see Fig. 3(d)] in the momentum distribution. As a consequence, the VLES shifts to a higher energy position. All these features are consistent with the experimental observations.

A closer inspection, however, reveals some discrepancies between the semiclassical and experimental spectra. This can be attributed to the quantum nature of the process, which is obviously neglected in the semiclassical model. One is that the ATI structure shown in Figs. 1(a) and 1(b), corresponding to absorption of integral number of photons, is absent in the semiclassical spectrum. In fact, these ATI peaks tend to smear out the classical effects shown in Fig. 3. The HLES in the energy spectrum [Fig. 3(a)] is hardly distinguishable in Fig. 1(a) [28]. In addition, the fine structure (multiple humps developed at high intensity) in the LMD is also hardly distinguishable in the experimental result. However, the DHS in Fig. 3(d) is qualitatively consistent with the experiment in spite of disturbance from the ATI peaks visible in Fig. 1(d). Furthermore, the energy peak at about 0.5 eV, which appears as an ATI peak in Fig. 1(a) and corresponds to the DHS in the LMD, can be considered to correspond to the VLES shown in Fig. 3(a) since these structures are absent in both TDSE and semiclassical simulations for short-range potential (not shown here). Moreover, this quantum nature of the process, which tends to spoil the visibility of the classical effect, diminishes with increasing wavelength, which is clearly shown by the fact that the experimental result is more consistent with the semiclassical simulation at longer wavelengths.

The other discrepancy is that the minimum in the semiclassical LMD is more pronounced than the experimental data and TDSE simulation for Kr and Xe, which can be attributed to two effects. One is that the classical treatment overestimates the probability of electrons with very low return energy being trapped by the Coulomb potential, resulting in the suppression of the electron spectrum near the origin [7]. Another one is that Freeman-resonance effect, which is absent in the semiclassical simulation, still play a noticeable role here. It has been demonstrated that the resonance process will largely affect the ATI spectrum, especially the longitudinal momentum distribution, even in the tunneling regime [13,21]. Under the laser intensities achieved in our experiment at longer wavelengths, the resonance channel can still have an important contribution due to abundant excited states in Kr and Xe atoms [29], resulting in a less prominent DHS shown in Figs. 1(e) and 1(f). This point can be seen by the fact that the spectra, especially the momentum distribution, are much more close to the experimental result when a model potential, which faithfully reproduces the energy-level structure of Xe, is used in the TDSE calculations (see Fig. 2), comparing with that of a hydrogen-like potential, which shows a deep central minimum in the momentum distribution.

In summary, we report experimental and theoretical studies on very low-energy ATI photoelectron in the tunneling ionization process. The COLTRIMS measurement reveals a universal peak structure below 1 eV in the energy spectrum, which corresponds to a double-hump structure in the longitudinal momentum distribution for all noble gas atomic species. These structures can be qualitatively reproduced in quantum and semiclassical simulations. Our analysis reveals the role of the long-range Coulomb potential in the origin of these particular structures.

This work was partially supported by NNSFC (Nos. 11074026, 11134001, 11121091, 10934004, and 11074155) and the National Basic Research Program of China (Nos. 2011CB8081002 and 2012CB921603).

*cywu@pku.edu.cn
†qhgong@pku.edu.cn
‡chen_jing@iapcm.ac.cn

[1] P. Agostini, F. Fabre, G. Mainfray, G. Petite, and N. K. Rahman, Phys. Rev. Lett. **42**, 1127 (1979).
[2] For a review, see, e.g., W. Becker, F. Grasbon, R. Kopold, D. B. Miloševi, G. G. Paulus, and H. Walther, Adv. At. Mol. Opt. Phys. **48**, 35 (2002).
[3] L. V. Keldysh, Zh. Eksp. Teor. Fiz. **47**, 1945 (1964); [Sov. Phys. JETP **20**, 1307 (1965)].
[4] H. B. van Linden van den Heuvell and H. G. Muller, *Multiphoton Processes*, edited by S. J. Smith and P. L. Knight (Cambridge University Press, Cambridge, England, 1988), p. 25.
[5] P. B. Corkum, N. H. Burnett, and F. Brunel, Phys. Rev. Lett. **62**, 1259 (1989); P. B. Corkum, Phys. Rev. Lett. **71**, 1994 (1993).
[6] G. G. Paulus, W. Nicklich, H. Xu, P. Lambropoulos, and H. Walther, Phys. Rev. Lett. **72**, 2851 (1994).
[7] J. Chen and C. H. Nam, Phys. Rev. A **66**, 053415 (2002).
[8] R. Moshammer *et al.*, Phys. Rev. Lett. **91**, 113002 (2003).
[9] K. I. Dimitriou, D. G. Arbo, S. Yoshida, E. Persson, and J. Burgdorfer, Phys. Rev. A **70**, 061401(R) (2004).
[10] Z. Chen, T. Morishita, A.-T. Le, M. Wickenhauser, X.-M. Tong, and C. D. Lin, Phys. Rev. A **74**, 053405 (2006).
[11] L. Guo, J. Chen, J. Liu, and Y. Q. Gu, Phys. Rev. A **77**, 033413 (2008).
[12] A. Rudenko, K. Zrost, C. D. Schröter, V. L. B. de Jesus, B. Feuerstein, R. Moshammer, and J. Ullrich, J. Phys. B **37**, L407 (2004).
[13] A. S. Alnaser, C. M. Maharjan, P. Wang, and I. V. Litvinyuk, J. Phys. B **39**, L323 (2006).
[14] F. H. M. Faisal and G. Schlegel, J. Phys. B **38**, L223 (2005); J. Mod. Opt. **53**, 207 (2006).
[15] C. I. Blaga, F. Catoire, P. Colosimo, G. G. Paulus, H. G. Muller, P. Agostini, and L. F. DiMauro, Nature Phys. **5**, 335 (2009).
[16] W. Quan *et al.*, Phys. Rev. Lett. **103**, 093001 (2009).
[17] C. P. Liu and K. Z. Hatsagortsyan, Phys. Rev. Lett. **105**, 113003 (2010).
[18] T. Yan, S. V. Popruzhenko, M. J. J. Vrakking, and D. Bauer, Phys. Rev. Lett. **105**, 253002 (2010).
[19] D. C. Yost, T. R. Schibli, J. Ye, J. L. Tate, J. Hostetter, M. B. Gaarde, and K. J. Schafer, Nature Phys. **5**, 815 (2009); E. P. Power, A. M. March, F. Catoire, E. Sistrunk, K. Krushelnick, P. Agostini, and L. F. DiMauro, Nature Photon. **4**, 352 (2010).
[20] Y. Liu, X. Liu, Y. Deng, C. Wu, H. Jiang, and Q. Gong, Phys. Rev. Lett. **106**, 073004 (2011).
[21] M. Abu-samha and L. B. Madsen, J. Phys. B **41**, 151001 (2008); M. Abu-samha, D. Dimitrovski, and L. B. Madsen, J. Phys. B **41**, 245601 (2008).
[22] J. Chen, B. Zeng, X. Liu, Y. Cheng, and Z. Z. Xu, New J. Phys. **11**, 113021 (2009). In our calculation under single-active-electron approximation for Xe, a model potential that can faithfully reproduce the energy-level structure of Xe is used.
[23] B. Hu, J. Liu, and S. G. Chen, Phys. Lett. A **236**, 533 (1997); J. Chen, J. Liu, and S. G. Chen, Phys. Rev. A **61**, 033402 (2000); J. Chen, J. Liu, and W. M. Zheng, Phys. Rev. A **66**, 043410 (2002); D. F. Ye, X. Liu, and J. Liu, Phys. Rev. Lett. **101**, 233003 (2008).
[24] N. B. Delone and V. P. Krainov, J. Opt. Soc. Am. B **8**, 1207 (1991).
[25] Here direct and indirect electrons imply the electrons those are tunnel-ionized in the rising edge ($\omega t \in [-\pi/2, 0]$) and the falling edge ($\omega t \in [0, \pi/2]$), respectively. To reduce the calculations, we have restricted the initial phase intervals to the interval $\omega t \in [-\phi, 0]$ and $[0, \phi]$ where ϕ is given a value such that the final kinetic energy of the electron ionized in these initial phase intervals satisfies $E_k < 2$ eV when the Coulomb potential is absent.
[26] The distance between the electron and the ion core is usually $\gtrsim 10$ a.u. according to our calculation.
[27] A. Kästner, U. Saalmann, J. M. Rost, Phys. Rev. Lett. **108**, 033201 (2012).
[28] The HLES can be roughly distinguished in large scale energy spectrum of He for 800 nm at intensity of 4–6 $\times 10^{14}$ W/cm^2. P. Agostini (private communication).
[29] C. M. Maharjan, A. S Alnaser, I. Litvinyuk, P. Ranitovic, and C. L. Cocke, J. Phys. B **39**, 1955 (2006).

Equation of State for Shock Compressed Xenon in the Ionization Regime: ab Initio Study[*]

WANG Cong (王聪),[1] GU Yun-Jun (顾云军),[2] CHEN Qi-Feng (陈其峰),[2] HE Xian-Tu (贺贤土),[1,3] and ZHANG Ping (张平)[1,3,†]

[1]LCP, Institute of Applied Physics and Computational Mathematics, P.O. Box 8009, Beijing 100088, China
[2]Laboratory for Shock Wave and Detonation Physics Research, Institute of Fluid Physics, P.O. Box 919-102, Mianyang 621900, China
[3]Center for Applied Physics and Technology, Peking University, Beijing 100871, China

(Received December 14, 2011; revised manuscript received April 20, 2012)

Abstract *Quantum molecular dynamic (QMD) simulations have been applied to study the thermophysical properties of liquid xenon under dynamic compressions. The equation of state (EOS) obtained from QMD calculations are corrected according to Saha equation, and contributions from atomic ionization, which are of predominance in determining the EOS at high temperature and pressure, are considered. For the pressures below 160 GPa, the necessity in accounting for the atomic ionization has been demonstrated by the Hugoniot curve, which shows excellent agreement with previous experimental measurements, and three levels of ionization have been proved to be sufficient at this stage.*

PACS numbers: 65.20.De, 64.30.Jk, 31.15.xv
Key words: equation of states, quantum molecular dynamics, xenon

1 Introduction

Accurate understandings of the thermodynamic properties of materials under extreme conditions are of great scientific interest.[1] The relative high temperatures (several eV) and densities (several g/cm^3) produce the so-called "warm dense matter" (WDM), which provides an active research platform by combining the traditional plasma physics and condensed matter physics. WDM usually appears in shock or laser heated solids, inertial confinement fusion, and giant planetary interiors, where simultaneous dissociations, ionizations, and degenerations make modelling of the dynamical, electrical, and optical properties of WDM extremely challenging.

The study of xenon in the warm dense region has long been focused because of the potential applications in power engineering and the significance in astrophysics.[2] Various experimental measurements[3−9] and theoretical models[10−14] have been applied to probe the EOS of xenon. Since the ionization equilibrium are not interfered with the dissociation equilibrium between molecules and atoms, xenon, which is characterized by monoatomic molecule and close shell electronic structure, is particularly suitable for the investigation of the high pressure behavior under extreme conditions. Despite its simple atomic structure, modelling the EOS of xenon under dynamic compression, especially the Hugoniot curve, is an essential issue. For instance, 0-K isotherm, which is important in determining the physical properties of solid xenon under static pressure, has been modelled by augmented-plane-wave (APW) electron band theory method.[12] While, the EOS of xenon under dynamic compression has been calculated through the fluid perturbation theory (FPT), and compared with experimental data.[12] On the other hand, the chemical picture of fluid xenon can be simply described through the concept of plasma physics, deep insight into the shock compressed xenon has been extracted by the self-consistent fluid variational theory (SFVT).[14]

However, the electronic structure has been convinced to be the key role in determining the thermophysical properties of WMD.[15] Due to the intrinsic approximations of these classical methods, fully quantum mechanical description of xenon under shock compressions still remains to be presented and understood. Meanwhile, QMD simulations, where quantum effects are considered by the combinations of classical molecular dynamics for the ions and density functional theory (DFT) for electrons, have already been proved to be successful in describing WDM.[16−17]

In the present paper, QMD simulations have been used to calculate the thermophysical properties of warm dense xenon. The EOS are determined for a wide range of densities and temperatures, and compared with experimental measurements and different theoretical models. The Hugoniot curves are then derived from the EOS. The rest of the paper is organized as follows. The simulation details are briefly described in Sec. 2; Corrections to QMD data and the EOS are presented in Sec. 3. Finally, we close our paper with a summary of our main results.

---

[*]Supported by the Foundation for Development of Science and Technology of China Academy of Engineering Physics under Grant No. 2009B0301037
[†]Corresponding author, E-mail: zhang_ping@iapcm.ac.cn
© 2011 Chinese Physical Society and IOP Publishing Ltd

http://www.iop.org/EJ/journal/ctp   http://ctp.itp.ac.cn

2 Computational Method

In this work, we have performed simulations for shock compressed xenon by employing the Vienna Ab-initio Simulation Package (VASP), which was developed at the Technical University of Vienna.[18–19] N atoms in a fixed volume supercell, repeated throughout the space periodically, form the elements of simulations. After introducing Born–Oppenheimer approximation, the thermo-activated electrons are fully quantum mechanically treated through plane-wave, finite-temperature DFT.[20] The electronic states are populated according to the Fermi-Dirac distribution at temperature T_e. The exchange correlation functional is determined by generalized gradient approximation (GGA) with the parametrization of Perdew–Wang 91.[21] Projector augmented wave (PAW) pseudopotential[22] are used to present the ion-electron interactions. The electronic structure is obtained from a series of given ionic positions, which are subsequently varied according to the forces calculated within the framework of DFT via the Hellmann–Feynman theorem at each molecular dynamics step. The simulations are executed with the isokinetic ensemble (NVT), where the ionic temperature (T_i) is controlled by Nosé thermostat,[23] and the system is kept in local thermo-dynamical equilibrium by setting the electron temperature T_e equal to T_i.

The bulk of the QMD simulations are performed using Γ point to sample the Brillouin zone in molecular dynamic simulations with the plane-wave cutoff of 600.0 eV, because EOS (pressure and energy) can only be modified within 5% for the selection of higher number of $\boldsymbol{k}$ points and larger cutoff energy. A number of 64 atoms is included in the simulation box. To cover typical Hugoniot points with respect to the data obtained from experiments, the densities adopted in our simulations range from 2.965 to 10.0 g/cm^3 and temperatures between 165 and 50000 K. Each density and temperature point is typically simulated for 4 $\sim$ 6 ps, and the time step selected for the integrations of atomic motion lies between 0.5 and 2.0 fs with respect to different conditions. Then, the subsequent 1 ps at equilibrium are used to calculate EOS as running averages.

3 Results and Discussion

3.1 *Effects of Electrons in Warm Dense Xenon*

Internal energy for shock compressed materials can be expressed as:

$$E_{\text{tot}} = E_{\text{ion}} + T_{\text{ion}} + \int \mathrm{d}r V_{\text{ext}}(r) n(r) + \frac{1}{2}\int \mathrm{d}r V_H(r) n(r) + E_{\text{xc}} + T_{\text{electron}} + \triangle E_I\,, \quad (1)$$

where E_{ion} and T_{ion} correspond to the interaction energy and the kinetic energy of the bare nuclear. The third term comes from the fixed external potential $V_{\text{ext}}(r)$ (in most cases the potential is due to the classical nuclei), in which the electrons move. The fourth term is the classical electrostatic energy of the electronic density, which can be obtained from the Hartree potential. E_{xc} and T_{electron} denote exchange-correlation energy and the kinetic energy of electrons. Finally, the ionization energy is added in the last term, which stands for the energy to generate free electrons and should be seriously considered in the warm dense region. In QMD simulations, the last term is excluded, and the simulated results should be corrected. The total internal energy is then expressed as:

$$E_{\text{tot}} = E_{\text{QMD}} + N\sum_i \alpha_i E_i\,,$$

where N is the total number of atoms for the present system, and α_i is the i-th level ionization degree, E_i is the energy required to remove i electrons from a neutral atom, creating an i-level ion. Here, E_i equals to $\sum_i \triangle\epsilon_i$, and $\triangle\epsilon_i$ is the i-th ionization energy.

Free electrons can be introduced at high temperatures and pressures by subsequent ionization of atoms. The ionization precess can be revealed from the change of the charge density distribution with increasing density and temperature, as shown in Fig. 1. In lower density and temperature region, the localization of electrons represents the atomic fluid. While, metallic behaviors of fluid xenon are indicated by delocalization of electrons, which can be attributed to the thermo-activation of electronic states. Similar phenomena have already been found for another nobel gas helium[24–25] and the expanded fluid alkali metal Li.[16]

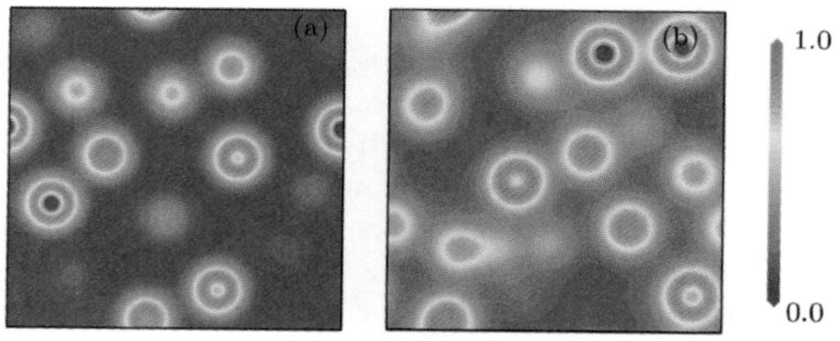

Fig. 1 (Color online) Charge density distribution for different conditions: $\rho = 5.0$ g/cm^3, $T = 1000$ K (a); and $\rho = 9.5$ g/cm^3, $T = 40000$ K (b).

Although ionizations of electrons can be qualitative described in the framework of finite-temperature DFT, contributions to pressure from noninteracting electrons are still lacking. Thus, the total pressure should be corrected as:

$$P_{\text{tot}} = P_{\text{QMD}} + \sum_i \mathrm{i}\alpha_i \frac{\rho k_B T}{m}\,, \quad (2)$$

where m presents the mass of xenon atom. The density and temperature are denoted by ρ and T respectively, and k_B stands for Boltzmann constant. E_{QMD} and P_{QMD} are obtained from VASP. Since, it is difficult to quantify the ionization degree for i-th level from direct QMD results, as an alternative selection, we use Saha equation to verify the degree of each ionization level:

$$\frac{n_{i+1}n_e}{n_i} = \frac{2\Omega}{\lambda^3}\frac{g_{i+1}}{g_i}\exp\Big(-\frac{\triangle\epsilon_i}{k_B T}\Big)\,, \quad (3)$$

$$\lambda = \sqrt{\frac{h^2}{2\pi m_e k_B T}}\,, \tag{4}$$

where n_i is the density of atoms in the i-th state of ionization, that is with i electrons removed, g_i is the degeneracy of states for the i-ions, n_e is the electron density, λ is the thermal de Broglie wavelength of an electron, m_e is the electron mass, and Ω is the volume of the supercell. Here, α_i equals to $n_i\Omega/N$. Similar approximations have already been used to study the EOS of liquid deuterium under shock compressions,[26] where soften behavior of the Hugoniot curve has been found at high pressure, and the simulated results are consist with available experiments.

In the present work, three levels of ionization have been considered (the relative ionization energies are 12.1, 21.2, and 32.1 eV respectively), and the relative electron number (equals to $\sum_i i\alpha_i$) as a function of density and temperature are labelled in Fig. 2. Xenon under extreme conditions is considered to be partially ionized plasma, where ionization equilibrium: $Xe^{(i-1)+} \leftrightarrows Xe^{i+} + e$, stands for the changes in the electronic structure under high temperatures and pressures. Three levels of atomic ionization have been taken into account in the present calculations. As an example, the composition for density of 6.0 g/cm^3 is plotted in Fig. 3. In the case of the 3$^{\rm rd}$ ionization, the fraction of charged ions Xe^+, and Xe^{2+} increase to maximum, then decrease, and the number of Xe^{3+} and electrons increase monotonously with temperature, while the number of neutral atom Xe decrease with the increase of temperature. The Hugoniot of shocked compressed liquid xenon can be effectively modified with the consideration of atomic ionizations, and the detailed discussion can be found in Subsec. 3.2.

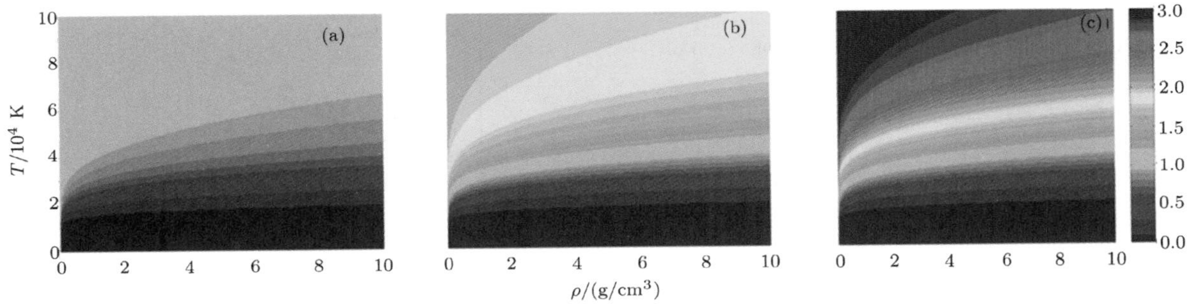

Fig. 2 (Color online) The ionized electron number considering the 1$^{\rm st}$ (a), 2$^{\rm nd}$ (b), and 3$^{\rm rd}$ (c) ionization as a function of the density and temperature.

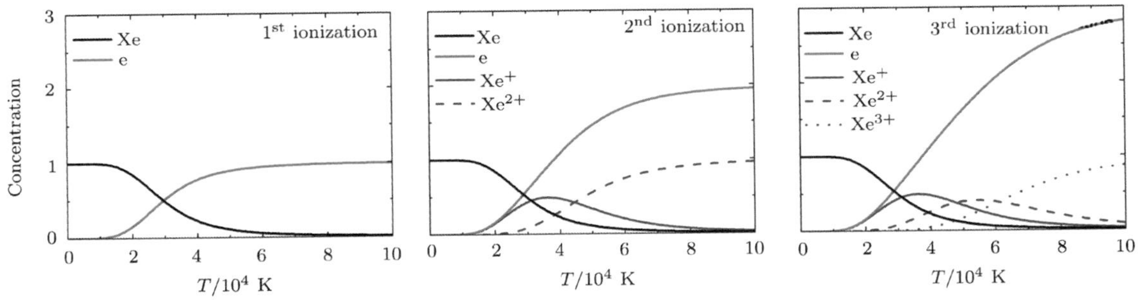

Fig. 3 (Color online) Component of plasma xenon (considering the 1$^{\rm st}$, 2$^{\rm nd}$, and 3$^{\rm rd}$ ionizations) as a function of temperature at $\rho = 6.0$ g/cm^3.

3.2 *The Equation of State*

Let us turn now to see the Hugoniot by investigating the corrected EOS of warm dense xenon. Following Lenosky *et al.*, Beule *et al.*, and Holst *et al.*,[27–29] smooth functions are used to fit the internal energies and pressures by expansions in terms of density (g/cm^3) and temperature (10^3 K). The corrected QMD data for internal energy (eV/atom) can be expanded as follows:

$$E = \sum_{i=0}^{2} A_i(T)\rho^i\,, \tag{5}$$

$$A_i(T) = a_{i0}\exp\Big[-\Big(\frac{T-a_{i1}}{a_{i2}}\Big)^2\Big] + a_{i3} + a_{i4}T\,. \tag{6}$$

The total pressure given in GPa can be similarly expanded

as:

$$P = \sum_{j=0}^{2} B_j(T)\rho^j\,, \tag{7}$$

$$B_j(T) = b_{j0}\exp\Big[-\Big(\frac{T-b_{j1}}{b_{j2}}\Big)^2\Big] + b_{j3} + b_{j4}T\,. \tag{8}$$

The expansion coefficients a_{ik} and b_{jk} for E and P (accuracy better than 5%) are summarized in Tables 1 and 2, respectively.

Table 1 Coefficients a_{ik} in expansion for the internal energy E.

i	a_{i0}	a_{i1}	a_{i2}	a_{i3}	a_{i4}
0	57.3976	−17.3142	28.2460	−38.0864	1.7336
1	11.2512	13.0002	−67.2259	−11.4532	0.0057
2	−0.7180	41.0407	73.9168	0.6057	0.0060

Table 2 Coefficients b_{ik} in expansion for the total pressure P.

j	b_{j0}	b_{j1}	b_{j2}	b_{j3}	b_{j4}
0	−38.9090	−15.8020	72.3514	63.7065	−0.0607
1	0.3299	26.0136	−5.7763	−12.5064	−0.0294
2	0.7302	1.0722	19.2058	0.9256	0.0441

A crucial measurement of the EOS data of xenon under shock conditions is the Hugoniot, which can be derived from the following equation:

$$(E_0 - E_1) + \frac{1}{2}(V_0 - V_1)(P_0 + P_1) = 0\,, \tag{9}$$

where E is the internal energy, P is the pressure, V is the volume, and the subscripts 0 and 1 denote the initial and shocked state, respectively. The Hugoniot relation, which describes the locus of points in (E, P, V)-space, follows from conservation of mass, momentum, and energy for an isolated system compressed by a pusher at a constant velocity. In our simulations, the initial density ρ_0 is 2.965 g/cm^3, the relative internal energy $E_0 = 0.01$ eV/atom at $T_0 = 165$ K. The initial pressure P_0 of the start point on the Hugoniot can be approximated to be zero compared to high pressure of shocked states.

The Hugoniot curve, where results from our simulations and previous predictions from both experiments and theories are labelled for comparison, is shown in Fig. 4. In order to clarify the influence of atomic ionization, direct QMD simulated EOS (EOS_D) and those corrected EOS (EOS_C) by Saha equation are also provided in the present work. Agreements can be found between experimental measurements and results from EOS_D only below 70 GPa. However, at higher pressures ($P > 70$ GPa), the Hugoniot curve from EOS_D shows an abrupt increase, which suggests a stiff behavior, and unphysical results are obtained here because of the ignorance of atomic ionization. Considerable softening of the Hugoniot can be achieved by the EOS_C, and reliable results are gradually revealed through accounting for the first (EOS_C1), second (EOS_C2), and third (EOS_C3) level of ionization. Here, the wide-range EOS for shock compressed liquid xenon can be well described by EOS_C3 below 160 GPa, where good agreements are detected between our simulation results and experiments.[3–4,7–8] Theoretically, no Hugoniot points by previous FPT method[12] are available to be used to compare with experiments at pressures higher than 60 GPa, and the Hugoniot curve from SFVT[14] seems to be inconsistent with the results from Nellis *et al.*[4]

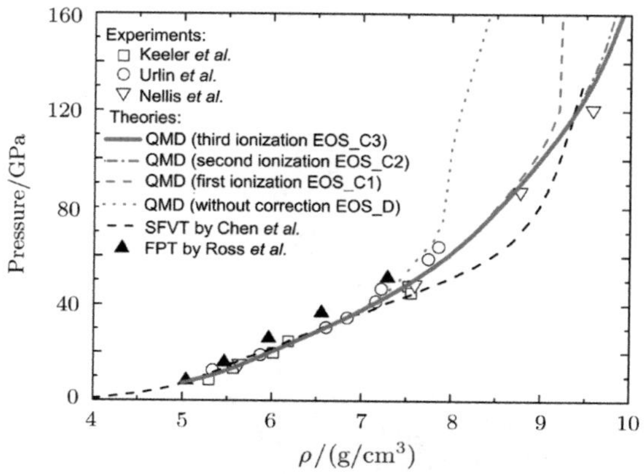

Fig. 4 (Color online) Hugoniot for shocked liquid xenon are compared with previous experimental[3–4,7–8] and theoretical results.[12,14] Specially, the results, which are corrected according to different ionization level, are also provided.

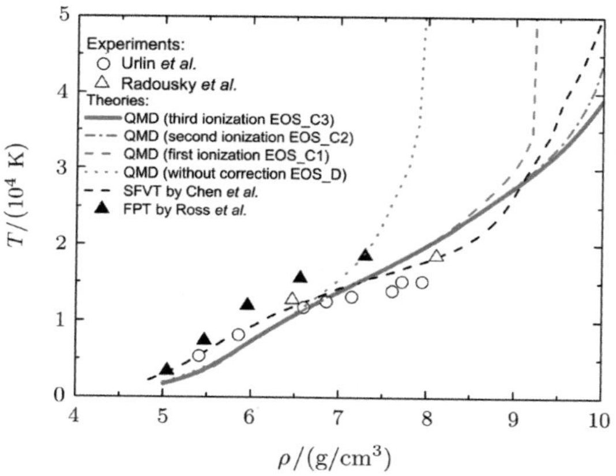

Fig. 5 (Color online) Temperature is labelled as a function of density along the Hugoniot. Previous experimental[8,10] and theoretical[12,14] results are also included.

Temperature, which is focused as one of the most important parameters in experiments, is difficult to be measured because of the uncertainty in determining the optical-intensity loss for ultraviolet part of the spectrum in adiabatic or isentropic shock compressions, especially for the temperature exceeding 20000 K. QMD simulations provide powerful tools to predict shock temperature. Our calculated Hugoniot temperature is shown in Fig. 5 as a function of the density. For comparison, the previous simulated (FPT[12] and SFVT[14]) and experimental (by Radousky *et al.*[10] and Urlin *et al.*[8]) results are also shown in Fig. 5. One can find that the temperatures from EOS_D and EOS_C3 begin to depart at about 10000 K (the respective density is 6.5 g/cm^3), which highlights the start point of ionization on the Hugoniot. Discontinuous change in temperature, which is similar to the behavior of Hugoniot, has been found in the EOS_D results, while, smooth variations of temperature along the Hugoniot are predicted by the EOS_C3 results. The shock temperatures from EOS_C3 are accordant with the available experiments.

4 Conclusion

In summary, QMD simulations, where contributions from ionized electrons are taken into account, are introduced to investigate the EOS of shock compressed liquid xenon, and ionization equilibrium: $\mathrm{Xe}^{(i-1)+} \leftrightarrows \mathrm{Xe}^{i+} + e$ is described by Saha equation. Here, three levels of ionization are considered, and it has been demonstrated that the accuracy of the calculated EOS can be effectively improved through the present approximations. The corrected EOS (EOS_C3) data are fitted by smooth functions, from which Hugoniot curve are then derived. Good agreements have been achieved between our calculated results and those from experimental measurements.

References

[1] R. Ernstorfer, M. Harb, C.T. Hebeisen, G. Sciaini, T. Dartigalongue, and R.J.D. Miller, Science **323** (2009) 1033.

[2] R. Paul Drake, *High Energy Density Physics: Fundamental Inertial Fusion, and Experimental Astrophysics*, Springer, Berlin (2006) p. 5.

[3] R.N. Keeler, M. van Thiel, and B.J. Alder, Physica (Utrecht) **31** (1965) 1437.

[4] W.J. Nellis, M. van Thiel, and A.C. Mitchell, Phys. Rev. Lett. **48** (1982) 816.

[5] V.E. Fortov, A.A. Leontev, A.N. Dremin, and V.K. Gryaznov, Zh. Eksp. Teor. Fiz. **71** (1976) 225 (in Russian).

[6] V.K. Gryaznov, M.V. Zhernokletov, V.N. Zubarev, I.L. Iosilevsky, and V.E. Fortov, Zh. Eksp. Teor. Fiz. **78** (1980) 573 (in Russian).

[7] V.D. Urlin, M.A. Mochalov, and O.L. Mikhailova, Mat. Model. **3** (1991) 42 (in Russian).

[8] V.D. Urlin, M.A. Mochalov, and O.L. Mikhailova, High Press. Res. **8** (1991) 595.

[9] M.I. Eremets, E.A. Gregoryanz, V.V. Struzhkin, H. Mao, R.J. Hemley, N. Mulders, and N.M. Zimmerman, Phys. Rev. Lett. **85** (2000) 2797.

[10] H.B. Radousky and M. Ross, Phys. Lett. A **129** (1988) 43.

[11] S. Kuhlbrodt, R. Redmer, H. Reinholz, G. Ropke, B. Holst, V.B. Mintsev, V.K. Grayaznov, N.S. Shilkin, and V.E. Fortov, Contrib. Plasma Phys. **45** (2005) 61.

[12] M. Ross and A.K. McMahan, Phys. Rev. B **21** (1980) 1658.

[13] V. Schwarz, H. Juranek, and R. Redmer, Phys. Chem. Chem. Phys. **7** (2005) 1990.

[14] Q.F. Chen, L.C. Cai, Y.J. Gu, and Y. Gu, Phys. Rev. E **79** (2009) 016409.

[15] C. Wang and P. Zhang, J. Chem. Phys. **132** (2010) 154307.

[16] A. Kietzmann, R. Redmer, M.P. Desjarlais, and T.R. Mattsson, Phys. Rev. Lett. **101** (2008) 070401.

[17] W. Lorenzen, B. Holst, and R. Redmer, Phys. Rev. Lett. **102** (2009) 115701.

[18] G. Kresse and J. Hafner, Phys. Rev. B **47** (1993) R558.

[19] G. Kresse and J. Furthmüller, Phys. Rev. B **54** (1996) 11169.

[20] T. Lenosky, S. Bickham, J. Kress, and L. Collins, Phys. Rev. B **61** (2000) 1.

[21] J.P. Perdew, *Electronic Structure of Solids*, Akademie Verlag, Berlin (1991).

[22] P.E. Blöchl, Phys. Rev. B **50** (1994) 17953.

[23] S. Nosé, J. Chem. Phys. **81** (1984) 511.

[24] A. Kietzmann, B. Holst, R. Redmer, M.P. Desjarlais, and T.R. Mattsson, Phys. Rev. Lett. **98** (2007) 190602.

[25] C. Wang, X.T. He, and P. Zhang, Phys. Rev. Lett. **106** (2011) 145002.

[26] C. Wang, X.T. He, and P. Zhang, J. App. Phys. **108** (2010) 044909.

[27] T.J. Lenosky, J.D. Kress, and L.A. Collins, Phys. Rev. B **56** (1997) 5164.

[28] D. Beule, W. Ebeling, A. Föster, H. Juranek, S. Nagel, R. Redmer, and G. Röpke, Phys. Rev. B **59** (1999) 14177.

[29] B. Holst, R. Redmer, and M.P. Desjarlais, Phys. Rev. B **77** (2008) 184201.

Commun. Comput. Phys.
doi: 10.4208/cicp.290411.121211a

Vol. **12**, No. 4, pp. 1121-1128
October 2012

First-Principles Calculations of Shocked Fluid Helium in Partially Ionized Region

Cong Wang[1], Xian-Tu He[1,2] and Ping Zhang[1,2,*]

[1] *LCP, Institute of Applied Physics and Computational Mathematics, P.O. Box 8009, Beijing 100088, China.*
[2] *Center for Applied Physics and Technology, Peking University, Beijing 100871, China.*

Received 29 April 2011; Accepted (in revised version) 12 December 2011

Communicated by Xingao Gong

Available online 17 April 2012

Abstract. Quantum molecular dynamic simulations have been employed to study the equation of state (EOS) of fluid helium under shock compressions. The principal Hugoniot is determined from EOS, where corrections from atomic ionization are added onto the calculated data. Our simulation results indicate that principal Hugoniot shows good agreement with gas gun and laser driven experiments, and maximum compression ratio of 5.16 is reached at 106 GPa.

AMS subject classifications: 81Q05, 82D10, 85A04

Key words: Equation of state, warm dense matter, ab initio simulations.

1 Introduction

High-pressure introduced response of materials, which requires accurate understandings of the thermophysical properties into new and complex region, has gained much scientific interest recently [1]. The relative high temperature and high density are usually referred as the so-called "warm dense matter" (WDM)—a strongly correlated state, where simultaneous dissociations, ionizations, and degenerations make modelling of the dynamical, electrical, and optical properties of WDM extremely challenging [2]. WMD, which provides an active research platform by combining the traditional plasma physics

*Corresponding author. *Email addresses:* wang_cong@iapcm.ac.cn (C. Wang), xthe@iapcm.ac.cn (X.-T. He), zhang_ping@iapcm.ac.cn (P. Zhang)

and condensed matter physics, usually appears in shock or laser heated targets [3], inertial confinement fusion [4], and giant planetary interiors [5].

Next to hydrogen, helium is the most abundant element in the universe, and physical properties of warm dense helium, especially the EOS, are critical for astrophysics [6,7]. For instance, the structure and evolution of stars, White Dwarfs, and Giant Planets [8–11], and therefore the understanding of their formation, depends sensitively on the EOS of hydrogen and helium at several megabar regime. For all planetary models, accurate EOS data are essential in solving the hydrostatic equation. As a consequence, a series of experimental measurements and theoretical approaches have been applied to investigate the EOS of helium. Liquid helium was firstly single shocked to 15.6 GPa using two stage light gas gun by Nellis *et al.*, then double shocked to 56 GPa, and the calculated temperature are 12000 and 22000 K respectively [12]. Maximum compression ratio ($\eta_{max}\approx 6$) was achieved by laser driven shock experiments with the crossover pressure around 100 GPa [13], However, soon after that, Knudson *et al.* have modified η_{max} to be 5.1 [14]. Theoretically, since the ionization equilibrium is not interfered with the dissociation equilibrium between molecules and atoms, helium, which is characterized by monoatomic molecule and close shell electronic structure, is particularly suitable for the investigation of the high pressure behavior under extreme conditions. Free energy based chemical models by Ross *et al.* [15], Chen *et al.* [16], and Kowalski *et al.* [17] have been used to investigate the principal Hugoniot of liquid helium, and the results are accordant with gas gun experiments. However, considerable controversies have been raised at megabar pressure regime, especially since the data was probed by laser shock wave experiments. Interatomic potential method predict η_{max}=4 for shock compressed helium [15], whereas, the EOS used by the astrophysical community from Saumon *et al.* (SCVH) [8], path-integral Monte Carlo (PIMC) [18], and activity expansion (ACTEX) [19] calculations provide an increase in compressibility at the beginning of ionization. For the initial density of ρ_0=0.1233 g/cm^3, SCVH and ACTEX simulation results indicate the maximum compression ratio lies around 6 at 300 GPa and 100 GPa respectively [8,19], while PIMC calculations suggest η_{max}=5.3 near 360 GPa [18].

On the other hand, quantum molecular dynamic (QMD) simulations, where quantum effects are considered by the combinations of classical molecular dynamics for the ions and density functional theory (DFT) for electrons, have already been proved to be successful in describing thermophysical properties of materials at complex conditions [20,21]. However, the DFT based molecular dynamic simulations (with or without accounting for excited electrons) do not provide reasonable results at the ionization region, mainly because the atomic ionization is not well defined in the framework of DFT. Considering these facts mentioned above, thus, in the present work, we applied the corrected QMD simulations to study shock compressed helium, and the EOS, which is compared with experimental measurements and different theoretical models, are determined for a wide range of densities and temperatures. The calculated compression ratio is substantially increased according to the ionization of atoms in the warm dense fluid.

2 Computational method

The Vienna Ab-initio Simulation Package (VASP) [22, 23], which was developed at the Technical University of Vienna, has been employed to perform simulations for helium. The elements of our calculations consist of a series of volume-fixed supercells including N atoms, which are repeated periodically throughout the space. By involving Born-Oppenheimer approximation, electrons are quantum mechanically treated through plane-wave, finite-temperature (FT) DFT [24], where the electronic states are populated according to Fermi-Dirac distributions at temperature T_e. The exchange-correlation functional is determined by generalized gradient approximation (GGA) with the parametrization of Perdew-Wang 91 [25]. The ion-electron interactions are represented by a projector augmented wave (PAW) pseudopotential [26]. Isokinetic ensemble (NVT) is adopted in present simulations, where the ionic temperature T_i is controlled by Nośe thermostat [27], and the system is kept in local equilibrium by setting the electron (T_e) and ion (T_i) temperatures to be equal.

The plane-wave cutoff energy is selected to be 700.0 eV so that the pressure is converged within 3% accuracy. Γ point is used to sample the Brillouin zone in molecular dynamics simulations, because EOS can only be modified within 5% for the selection of higher number of **k** points. 64 helium atoms are included in the cubic supercell. The densities selected in our simulations range from 0.1233 to 0.8 g/cm^3 and temperatures between 4 and 50000 K, which highlight the regime of the principal Hugoniot. All the dynamic simulations are lasted for 6000 steps, and the time steps for the integrations of atomic motion are selected according to different densities (temperatures) [28]. Then, the subsequent 1000 steps of simulation are used to calculate EOS as running averages.

3 Results and discussion

Different corrections to QMD simulations have already been used to model the thermophysical properties of WDM, such as the zero point vibrational energy modification for hydrogen and deuterium [29]. However, quantitative descriptions of the ionization of atoms in the frame of DFT is still lacking, except for the Drude model for aluminium introduced by Mazevet *et al.* [30], but the simple metallic model is not suitable for studying warm dense helium. Here, we adopt similar approximations as described in [31], where the effect of atomic ionization are considered, to study the EOS of fluid helium. In this work, the EOS are expressed as follows [31]:

$$E = E_{QMD} + N\beta E_{ion}, \tag{3.1}$$

$$P = P_{QMD} + (1+\beta)\frac{\rho k_B T}{m_{He}}, \tag{3.2}$$

where E_{QMD} and P_{QMD} are data obtained from regular QMD calculations, and N is the total number of atoms considered in supercells. m_{He} and k_B represent the mass of helium atom and Boltzmann constant. The density and temperature are denoted by ρ and T, respectively. β stands for the ionization degree, and E_{ion} is the ionization energy. Van der waals interactions are treated as small as negligible in the present work.

In the present work, only the first ionization level is considered, and the ionization degree of helium can be evaluated through Saha equation:

$$\frac{\beta^2}{1-\beta}=\frac{2\Omega}{\lambda^3}\exp(-\frac{E_{ion}}{k_BT}), \tag{3.3}$$

$$\lambda=\sqrt{\frac{h^2}{2\pi m_e k_B T}}, \tag{3.4}$$

where, m_e and Ω present the mass of electron and volume of the supercell. The ionization energy E_{ion} is determined by $E_{ion}=E^{He+}(\rho,T)-E^{He}(\rho,T)$, where $E^{He+}(\rho,T)$ and $E^{He}(\rho,T)$ are total energy of He^+ and He at the relative density and temperature, respectively. These energies are calculated from FT-DFT.

Based on the approximations mentioned above, the calculated EOS was examined theoretically along the principal Hugoniot, the locus of states that satisfy the Rankine-Hugoniot (RH) equations, which are derived from conservation of mass, momentum, and energy across the front of shock waves. The RH equations describe the locus of states in (E, P, V)-space satisfying the following relation:

$$E_1-E_0=\frac{1}{2}(P_1+P_0)(V-V_0), \tag{3.5}$$

$$P_1-P_0=\rho_0 u_s u_p, \tag{3.6}$$

$$V_1=V_0(1-u_p/u_s), \tag{3.7}$$

where E, P, V present internal energy, pressure, volume, and subscripts 0 and 1 present the initial and shocked state, respectively. In Eqs. (3.6) and (3.7), u_s is the velocity of the shock wave and u_p corresponds to the mass velocity of the material behind the shock front. In our present simulations, the initial density for helium is ρ_0=0.1233 g/cm^3 and the liquid specimen is controlled at a temperature of 4 K, where the internal energy is E_0=−0.02 eV/atom. The initial pressure P_0 can be treated approximately as zero compared to the high pressure of shocked states along the Hugoniot. The Hugoniot points are obtained as follows: (i) smooth functions are used to fit the internal energy and pressure in terms of temperature at sampled density; (ii) then, Hugoniot points are derived from Eq. (3.5).

The principal Hugoniot curve for helium is shown in Fig. 1, where previous experimental measurements and theoretical predictions are also plotted for comparison. For pressures below 30 GPa, our results show good agreements with the data detected by gas gun experiments [12]. At higher pressures, the principal Hugoniot curve shows a maximum compression ratio $\eta_{max}\approx 6$ with the crossover pressure around 100 GPa, as have been reported by laser driven experiments [13]. Soon after that, the use of quartz

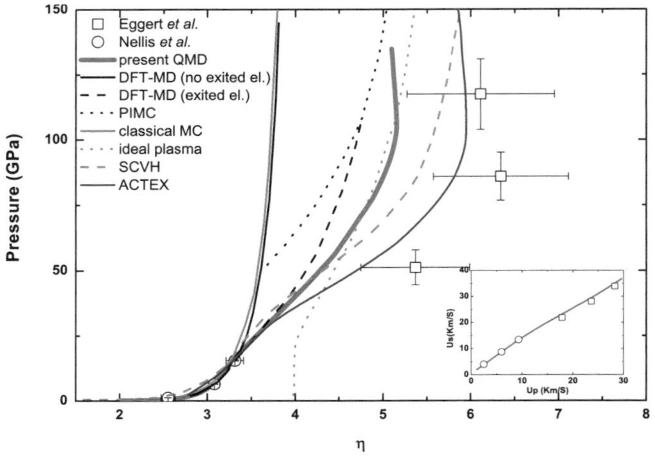

Figure 1: (Color online) Principal Hugoniot up to 130 GPa is shown for the present work (red line), where previous theoretical predictions and experimental measurements are also shown for comparison. Experiments: Two stage light gas gun data [12] is denoted by open circle; Laser driven experimental results [13] are labeled by open square. Theories: (i) Classical MC simulation results [32] are plotted as green solid line; (ii) ideal plasma's Hugoniot curve [18] is shown as green dotted line; (iii) SCVH method [8] is labeled as green dashed line; (iv) ACTEX results [19] are blue solid line; (v) Previous DFT simulation results [18] (with and without accounting for electron excitation) are shown as black dashed and solid line, respectively; (vi) PIMC results [18] are labeled as black dotted line. Inset is the plot of shock wave velocity verse mass velocity.

as a shock wave standard by Knudson *et al.* has improve the shock data, and η_{max} has been reduced to be 5.1 [14]. Our corrected QMD simulation results indicate that, the atomic ionization dominates the characteristic of the EOS at $P>30$, and the principal Hugoniot shows soften behavior with the increase of pressure, then reach its maximum (η_{max}=5.16) with the corresponding pressure of 106 GPa. Then, stiff behavior has been found at $P>110$ GPa along the Hugoniot. As has been shown in Fig. 2, with the increase

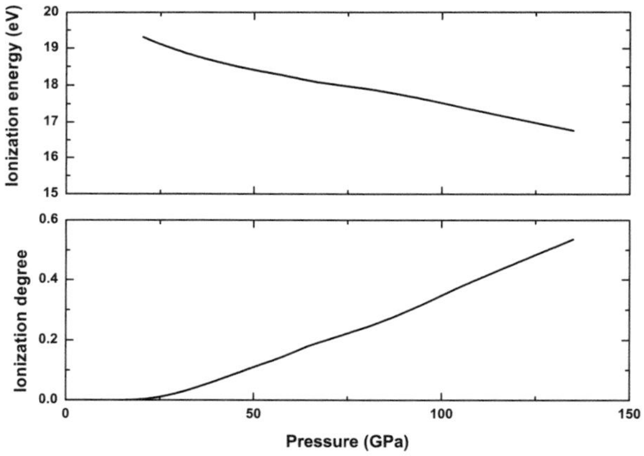

Figure 2: Ionization energy and ionization degree are plotted as functions of pressure along the principal Hugoniot in the upper and lower panels, respectively.

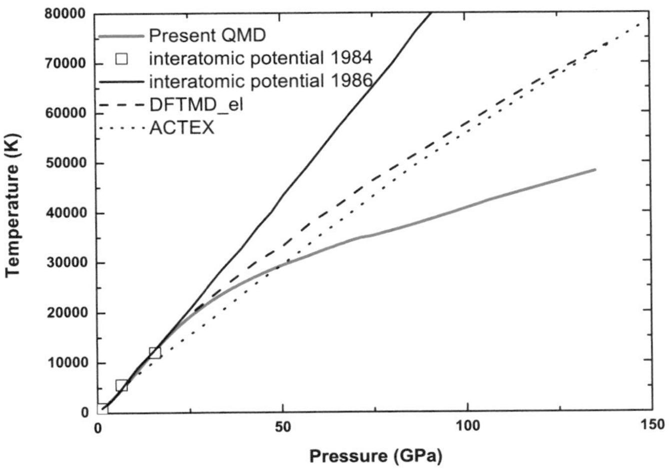

Figure 3: (Color online) Comparison between our calculated shock temperature and previous theoretical predictions. Interatomic potential results are shown as open square [12] and solid line [15]; DFT [18] and ACTEX [19] results are plotted as dashed and dotted line, respectively.

of pressure along the Hugoniot, ionization energy decreases, and continuous increase of atomic ionization has become considerable in determining the EOS of fluid helium. Maximum compression ratio is achieved with the ionization degree of 40%. The wide-range behavior of our simulated principal Hugoniot for helium shows excellent agreement with experimental ones.

Concerning other theoretical models, classical MC results show stiff behavior along the Hugoniot, and η_{max} lies around 4 at 1000 GPa [32]. Direct DFT simulation results show stiff behavior up to 110 GPa, while, by combining PIMC simulations, η_{max} was reported to be 5.3 at 360 GPa [18], but the predicted pressure is still too high compared with laser driven experiments. ACTEX [19] and SCVH [8] calculations predict η_{max} to be 6 near 100 GPa and 300 GPa respectively, but the results do not agree with those of Knudson *et al.* [14].

Temperature, which is focused as one of the most important parameters in experiments, is difficult to be measured because of the uncertainty in determining the optical-intensity loss for ultraviolet part of the spectrum in adiabatic or isentropic shock compressions, especially for the temperature exceeding several electron-volt [33]. QMD simulations provide powerful tools to predict shock temperature. The calculated Hugoniot temperature has been shown in Fig. 3 as a function of pressure along the Hugoniot. Previous theoretical predictions, such as interatomic potential [12,15], DFT-MD (excited electron) [18], and ACTEX [19] are also provided for comparison. Disagreements have been found to begin at around 30 GPa with the relative temperature of 20000 K, which highlights the starting point of ionization on the Hugoniot, and the predicted temperatures by those models are higher at a given pressure compared with our calculations.

4 Conclusion

In summary, We have performed DFT based QMD simulations to study the thermophysical properties of helium under extreme conditions. The Hugoniot EOS has been evaluated through QMD calculations and corrected by taking into account the atomic ionization described by Saha equation, where only first ionization is considered. The corrected Hugoniot has been proved accord well with the experimental data in a wide range of shock conditions, which thus indicates the importance of atomic ionization. Maximum compression ratio of 5.16 reveals at round 106 GPa along the principal Hugoniot, and the softened characteristic of the Hugoniot has been demonstrated by the contributions from

Acknowledgments

This work was supported by NSFC under Grant No. 11005012, by the National Basic Security Research Program of China, and by the National High-Tech ICF Committee of China.

References

[1] R. Ernstorfer, M. Harb, C. T. Hebeisen, G. Sciaini, T. Dartigalongue, and R. J. D. Miller, The formation of warm dense matter: Experimental evidence for electronic bond hardening in gold, Science, 323 (2009), 1033-1037.
[2] W. J. Nellis, Dynamic compression of materials: metallization of fluid hydrogen at high pressures, Rep. Prog. Phys., 69 (2006), 1479.
[3] D. G. Hicks, T. R. Boehly, P. M. Celliers, J. H. Eggert, S. J. Moon, D. D. Meyerhofer,and G. W. Collins, Laser-driven single shock compression of fluid deuterium from 45 to 220 GPa, Phys. Rev. B, 79 (2009), 014112.
[4] F. Philippe, A. Casner, T. Caillaud, O. Landoas, M. C. Monteil, S. Liberatore, H. S. Park, P. Amendt, H. Robey, C. Sorce, C. K. Li, F. Seguin, M. Rosenberg, R. Petrasso, V. Glebov, and C. Stoeckl, Experimental demonstration of X-ray drive enhancement with rugby-shaped hohlraums, Phys. Rev. Lett., 104 (2010), 035004.
[5] W. Lorenzen, B. Holst, and R. Redmer, Demixing of hydrogen and helium at megabar pressures, Phys. Rev. Lett., 102 (2009), 115701.
[6] D. J. Stevenson and E. E. Salpeter, The phase diagram and transport properties for hydrogen-helium fluid planets, Astrophys. J. Suppl., 35 (1977), 221-237.
[7] D. J. Stevenson and E. E. Salpeter, The dynamics and helium distribution in hydrogen-helium fluid planets, Astrophys. J. Suppl., 35 (1977), 239-261.
[8] D. Saumon, G. Chabrier, and H. M. Van Horn, An equation of state for low-mass stars and giant planets, Astrophys. J. Suppl. Ser., 99 (1995), 713-41.
[9] V.Ya. Ternovoi, S. V. Kvitov, A. A. Pyalling, A. S. Filimonov and V. E. Fortov, Experimental determination of the conditions for the transition of Jupiters atmosphere to the conducting state, JETP Lett., 79 (2004), 6-9.
[10] J. Vorberger, I. Tamblyn, B. Militzer, and S. A. Bonev, Hydrogen-helium mixtures in the interiors of giant planets, Phys. Rev. B., 75 (2007), 024206.

[11] M. A. C. Perryman, Extra-solar planets, Rep. Prog. Phys., 63 (2000), 1209-1272.
[12] W. J. Nellis, N. C. Holmes, A. C. Mitchell, R. J. Trainor, G. K. Governo, M. Ross, and D. A. Young, Shock Compression of liquid helium to 56 GPa (560 kbar), Phys. Rev. Lett., 53 (1984), 1248.
[13] J. Eggert, S. Brygoo, P. Loubeyre, R. S. McWilliams, P. M. Celliers, D. G. Hicks, T. R. Boehly, R. Jeanloz, and G.W. Collins, Hugoniot data for helium in the ionization regime, Phys. Rev. Lett., 100 (2008), 124503.
[14] M. D. Knudson and M. P. Desjarlais, Shock compression of quartz to 1.6 TPa: Redefining a pressure standard, Phys. Rev. Lett., 103 (2009), 225501.
[15] M. Ross and D. A. Young, Helium at high density, Phys. Lett. A, 118 (1986), 463-466.
[16] Q. F. Chen, Y. Zhang, L. C. Cai, Y. J. Gu, and F. Q. Jing, Self-consistent variational calculation of the dense fluid helium in the region of partial ionization, Phys. Plasma., 14 (2007), 012703.
[17] P. M. Kowalski, S. Mazevet, D. Saumon, and M. Challacombe, Equation of state and optical properties of warm dense helium, Phys. Rev. B, 76 (2007), 075112.
[18] B. Militzer, First principles calculations of shock compressed fluid helium, Phys. Rev. Lett., 97 (2006), 175501.
[19] M. Ross, F. Rogers, N. Winter, and G. Collins, Activity expansion calculation of shock-compressed helium: The liquid Hugoniot, Phys. Rev. B, 76 (2007), 020502(R).
[20] C. Wang and P. Zhang, Ab initio study of shock compressed oxygen, J. Chem. Phys., 132 (2010), 154307.
[21] C. Wang and P. Zhang, The equation of state and nonmetal-metal transition of benzene under shock compression, J. Appl. Phys., 107 (2010), 083502.
[22] G. Kresse and J. Hafner, Ab initio molecular dynamics for liquid metals, Phys. Rev. B, 47 (1993), R558.
[23] G. Kresse and J. Furthmüller, Efficient iterative schemes for ab initio total-energy calculations using a plane-wave basis set, Phys. Rev. B, 54 (1996), 11169.
[24] T. Lenosky, S. Bickham, J. Kress, and L. Collins, Density-functional calculation of the Hugoniot of shocked liquid deuterium, Phys. Rev. B, 61 (2000), 1.
[25] J. P. Perdew, Electronic Structure of Solids, Akademie Verlag, Berlin (1991).
[26] P. E. Blöchl, Projector augmented-wave method, Phys. Rev. B, 50 (1994), 17953.
[27] S. Nosé, A unified formulation of the constant temperature molecular dynamics methods, J. Chem. Phys., 81 (1984), 511.
[28] The time steps have been taken as $\triangle t = a/20\sqrt{k_B T/m_{He}}$, where $a = (3/4\pi n_i)^{1/3}$ is the ionic sphere radius (n_i is the ionic number density), $k_B T$ presents the kinetic energy, and m_{He} is the ionic mass.
[29] B. Holst, R. Redmer, and M. P. Desjarlais, Thermophysical properties of warm dense hydrogen using quantum molecular dynamics simulations, Phys. Rev. B, 77 (2008), 184201.
[30] S. Mazevet, M. P. Desjarlais, L. A. Collins, J. D. Kress, and N. H. Magee, Simulations of the optical properties of warm dense aluminum, Phys. Rev. E, 71 (2005), 016409.
[31] C. Wang, X. T. He, and P. Zhang, Hugoniot of shocked liquid deuterium up to 300 GPa: Quantum molecular dynamic simulations, J. Appl. Phys., 108 (2010), 044909.
[32] R. A. Aziz, A. R. Janzen, and M. R. Moldover, Ab initio calculations for helium: A standard for transport property measurements, Phys. Rev. Lett., 74 (1995), 1586.
[33] Q. F. Chen, private communication, (2010).

PRL 109, 093001 (2012) PHYSICAL REVIEW LETTERS week ending 31 AUGUST 2012

Low Yield of Near-Zero-Momentum Electrons and Partial Atomic Stabilization in Strong-Field Tunneling Ionization

Hong Liu,[1] Yunquan Liu,[1,*] Libin Fu,[2,3] Guoguo Xin,[4,5] Difa Ye,[2] Jie Liu,[2,3,4,†] X. T. He,[2,3] Yudong Yang,[1] Xianrong Liu,[1] Yongkai Deng,[1] Chengyin Wu,[1] and Qihuang Gong[1,‡]

[1]*Department of Physics and State Key Laboratory for Mesoscopic Physics, Peking University, Beijing 100871, China*
[2]*National Key Laboratory of Science and Technology on Computation Physics, Institute of Applied Physics and Computational Mathematics, 100088 Beijing, China*
[3]*Center for Applied Physics and Technology, Peking University, 100084 Beijing, China*
[4]*School of Science, Beijing Institute of Technology, Beijing 100081, China*
[5]*Department of Physics, Northwest University, Xian, Shaanxi 710069, China*

(Received 22 November 2011; published 27 August 2012)

We measure photoelectron angular distributions of single ionization of krypton and xenon atoms by laser pulses at 1320 nm, 0.2–1.0 $\times 10^{14}$ W/cm^2, and observe that the yield of near-zero-momentum electrons in the strong-field tunneling ionization regime is significantly suppressed. Semiclassical simulations indicate that this local ionization suppression effect can be attributed to a fraction of the tunneled electrons that are released in a certain window of the initial field phase and transverse velocity are ejected into Rydberg elliptical orbits with a frequency much smaller than that of the laser; i.e., the corresponding atoms are stabilized. These electrons with high-lying atomic orbits are thus prevented from ionization, resulting in the substantially reduced near-zero-momentum electron yield. The refined transition between the Rydberg states of the stabilized atoms has implication on the THz radiation from gas targets in strong laser fields.

DOI: 10.1103/PhysRevLett.109.093001 PACS numbers: 32.80.Rm, 32.80.Wr

Atomic stabilization in superintense high-frequency fields has been studied theoretically for decades (see, e.g., [1] for a review) and has led to a wealth of profound and intriguing topics in strong-field physics [2–4]. It is usually discussed in the Kramers-Henneberger (KH) frame; i.e., the moving coordinate frame of a free electron responding to the laser field. In the KH frame, the ground-state wave function of the atom splits into two nonoverlapping peaks and the atom becomes stabilized against ionization when the laser frequency is much higher than the bound state frequency of the atom [5]. Recent theoretical investigations further reveal that this concept is not exclusively associated with high frequencies, as widely assumed [6]. The stabilization can also occur when the field frequency is sufficiently large compared to some typical atomic excitation energy. Lacking experimental evidence, this theory is not yet solidified. On the other hand, in an intense low-frequency laser field (e.g., in the infrared regime) the tunneling limit of multiphoton ionization is more appropriately described by the Keldysh theory [7]. With the picture of the Keldysh theory, an effect of ionization suppression associated with tunneling was predicted at some specific field strengths [8]. The controversial issue in the deduction however has been argued recently [9], in which the dependence of the ionization rate on the laser field is found to be monotonic. In contrast to the high-frequency multiphoton ionization, the low-frequency atomic stabilization in the tunneling regime is a more subtle and unsettled question that needs further experimental and theoretical investigation.

In this Letter we present experimental observation that the yield of near-zero-momentum electrons from single ionization of xenon (Xe) and krypton (Kr) atoms produced by a linearly polarized infrared laser is much suppressed in the deep tunneling ionization regime, where the Keldysh parameter γ is much less than unity. Here $\gamma = \sqrt{I_p/2U_p}$, I_p is the ionization potential, U_p is the ponderomotive potential, $U_p = \varepsilon^2/4\omega^2$, ε is the field amplitude, ω is the field frequency, and atomic units (a.u.) are used unless otherwise specified. In the experiment, we did not observe the traditional atomic stabilization in the strict sense that the total ionization yield decreases or at least cease to increase as the laser intensity increases. Instead, we observed that, as the intensity of the field increases, the relative contribution of low-energy photoelectrons to the total ionization yield decreases. With a three-dimensional semiclassical electron ensemble model, we have reproduced this pronounced local ionization suppression effect and uncovered the underlying physics as partial atomic stabilization. We found that, when the electrons initially tunnel in a certain time window at the rising front of a laser cycle, a fraction of atoms can be finally excited into the Rydberg states with high-lying elliptical orbits. Because the frequency associated with the orbit is much smaller than that of the laser, the effect of the fast laser-driven electron oscillations is averaged out on the time scale of the orbital motion and the electrons can remain in the elliptical orbits after the laser pulse is switched off. Those stabilized atoms can make up the suppressed yield of the low-energy ionized electrons. Our analysis provides an intuitive picture for the low-frequency stabilization in strong-field tunneling ionization that largely extends the ADK (Ammosov-Delone-Krainov) theory [10] and enriches the rescattering scenario [11].

We used linearly polarized 1320 nm radiation generated by an optical parametric amplification system that was pumped by 25 fs, 795 nm pulses from a Ti:Sa laser system with 3 kHz repetition rate, amplified pulse energy up to 0.8 mJ. The estimated pulse duration at 1320 nm was 35–40 fs. The laser intensity was controlled with a pair of polarizers and was calibrated with the electron energy cutoff of 10 U_p.

We measured the photoelectron angular distributions (PADs) with a reaction microscope [12] (for the principle, see [13]). The spectrometer has the photoelectron momentum resolution ~0.02 a.u. along the time-of-flight direction and ~0.05 a.u. along the transverse direction. Ions and electrons were measured with two position-sensitive channel plate detectors, respectively. We applied a weak electric (~ 3 V/cm) and magnetic (~ 5 G) field applied along the time-of-flight axis. From the time-of-flight and position on the detectors, the full momentum vectors of particles were reconstructed. In the offline analysis, the photoelectrons were selected in coincidence with their singly charged parent atomic ions. We have carefully adjusted the magnetic field and electrical field along the spectrometer to avoid the poor resolution of the transverse momentum for the low-energy electrons. The laser polarization direction was along the time-of-flight axis.

Figure 1 shows two-dimensional PADs in momentum space (P_z, $P_\perp$) of Xe and Kr in the intensities of $0.2\text{–}1\times10^{14}\ \mathrm{W/cm^2}$ at 1320 nm. The P_z and $P_\perp$($P_\perp = \sqrt{P_x^2+P_y^2}$) represent the momentum parallel and perpendicular to the laser polarization axis, respectively. One may first observe the "fanout" structures on the PADs. Indeed, there is a large body of theoretical literature on the origin of resonant structures [14] in the multiphoton ionization regime with both the classical and quantum models [15,16], which is beyond the scope of this Letter. The striking finding in Fig. 1 is that the relative yield of near-threshold electrons is more suppressed when the laser intensity increases.

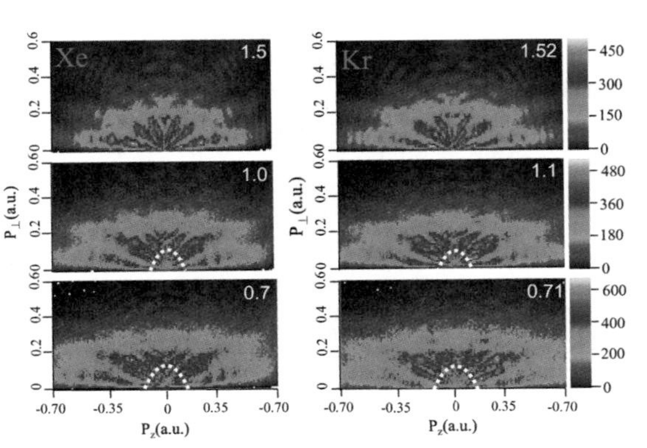

FIG. 1 (color online). Experimental two-dimensional PADs in momentum space (P_z, $P_\perp$) of Kr and Xe in the intensities of $0.2\text{–}1\times10^{14}\ \mathrm{W/cm^2}$ at 1320 nm. The Keldysh parameters are labeled at the top right corner.

In order to achieve deep insight into the ionization dynamics behind the striking PADs, we have performed three-dimensional semiclassical electron ensemble simulations including the tunneling effect (for the details see, e.g., [17,18]). Briefly, in the model the electron initial position along the laser polarization direction is derived from the Landau effective potential theory [19]. The tunneled electrons have initially zero longitudinal velocity and a Gaussian-like transverse velocity distribution. Each electron trajectory is weighed by the ADK ionization rate $\varpi(t_0, v_\perp{}^i) = \varpi_0(t_0)\varpi_1(v_\perp{}^i)$ [10]. $\varpi_1(v_\perp{}^i) \propto v_\perp{}^i \exp\times[\sqrt{2I_p}(v_\perp{}^i)^2/|\varepsilon(t_0)|]$ is the distribution of initial transverse velocity, and $\varpi_0(t_0) = |(2I_p)^2/\varepsilon(t_0)|^{2/\sqrt{2I_p}-1}\exp\times[-2(2I_p)^{3/2}/|3\varepsilon(t_0)|]$ depends on the field phase ωt_0 at the instant of ionization as well as the ionization potential I_p.

After tunneling, the electron evolution in the combined oscillating laser field and Coulomb field is solved via the Newtonian equation, $\ddot{\vec{r}} = -\vec{r}/r^3 - \varepsilon\cos\omega t\vec{e}_z$, where the polarization direction is along the z axis, r is the distance between the electron and nucleus, ε and ω are the amplitude and frequency of the laser field, respectively. The dressed energy of the electron in an oscillating laser field takes the form of $E=\frac{1}{2}(\dot{\vec{r}}+\frac{\varepsilon}{\omega}\sin\omega t\vec{e}_z)^2-\frac{1}{r}$. In the simulation, the laser field has a constant amplitude with ten cycles and is ramped off within three laser cycles.

The simulation results of two-dimensional momentum distributions of (P_z, $P_\perp$) for Xe atoms at the intensity of $6\times10^{13}\ \mathrm{W/cm^2}$ and the wavelength of 1300 nm are presented in Fig. 2(a). Generally, the simulation results agree with the experimental results qualitatively except for the interference patterns. In particular, it reproduces the phenomenon of local ionization suppression in PADs at the origin.

In order to trace the suppressed events, we illustrate the energy distribution of all electrons that tunnel from the first half of the laser cycle $\omega t\in(-\pi/2, \pi/2)$, regardless of whether the final energy is positive or negative, in Fig. 2(c). The figure exhibits that, after the laser pulse there is a large number of tunneled electrons with the final negative energies within (− 0.01, 0) a.u., which means that those electrons are finally bounded by the atomic potential. The binding energies of those Rydberg states ($E < 0.01$ a.u.) are much smaller than the photon energy of the low-frequency light (for 1320 nm, the photon energy is ~0.04 a.u.). It has been found experimentally that a large

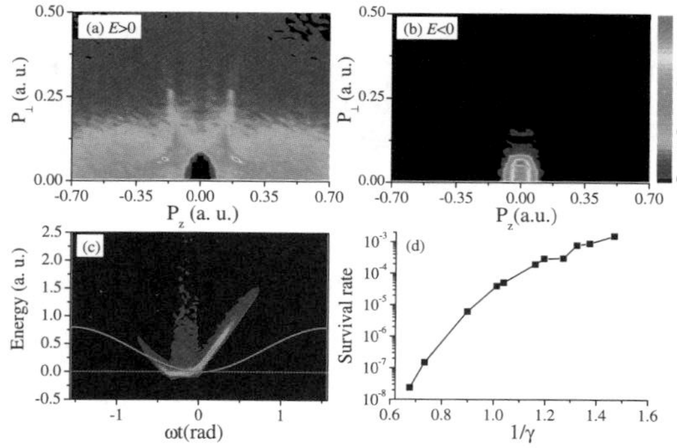

FIG. 2 (color online). Simulation results of two-dimensional momentum distribution (P_z, $P_\perp$) of the ionized electrons (a) and the un-ionized electrons (b) for Xe at the intensity of 6×10^{13} W/cm^2 at 1300 nm. (c) The energy distribution of all electrons vs the initial phase. The solid curve shows the prediction of the simple-man model. (d) The survival rate with respect to the inverse Keldysh parameter.

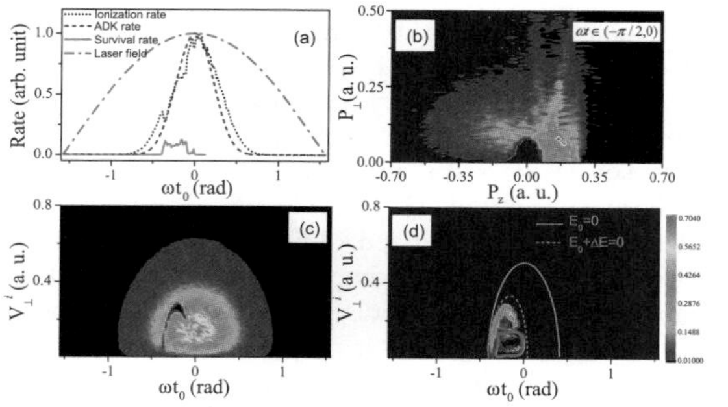

FIG. 3 (color online). (a) The ionization (or survival) rate vs the initial phase (see the text for details). (b) Two-dimensional momentum distribution (P_z, $P_\perp$) that is contributed from the rising front of the first half laser cycle. (c),(d) show the initial transverse velocities of tunneled electrons with respect to the laser phase that contribute to ionization and stabilization process, respectively. In (d), the solid and dashed curves indicate the theoretical predictions for the boundary for the un-ionized window (see text for details).

number of excited neutral atoms can survive in strong laser fields [20]. In Fig. 2(b), we show the two-dimensional momentum distribution of those electrons with the negative energy (the survival events), which can really make up the suppressed yield of those low-energy ionized electrons. The fact that a substantial part of the tunneled electrons end up in the bound states should affect the momentum spectrum of ionized electrons, particularly the low-energy part. Moreover, the survival rate increases with the inverse Keldysh parameter as presented in Fig. 2(d). This is consistent with the experimental observation that the relative yield of near-zero-momentum electrons is more suppressed in deep tunneling regime.

As shown in Fig. 2(c), in the presence of the Coulomb field, the energies of the electrons released at the rising front of the laser cycle are depressed and the energies of the electrons released at the descending front are enhanced, as compared with the prediction by the simple-man model [21] (the solid curve in Fig. 2(c)]. Some electrons tunneled from the phase region slightly before the field maximum can achieve much higher energy than 2 U_p because of the electron chaotic motions [17].

In order to shed more light on the local ionization suppression effect, we now consider the subcycle ionization dynamics. In the combined Coulomb and laser field, the ionization (survival) rate from the first half of the laser cycle is shown in Fig. 3(a). Without the Coulomb field, electrons released from the rising front of the laser cycle will drift away from their parent ions and never return to the vicinity of the nucleus. However, if the Coulomb potential is present, the final ionization rate deviates dramatically from the ADK rate [dashed curve in Fig. 3(a)]. The suppressed ionization results from several distinct phase regions at the rising front of a laser cycle, leading to the survival rate [solid curve in Fig. 3(a)]. On closer inspection, two-dimensional momentum distribution of (P_z, $P_\perp$) is shown when the electrons are launched within the rising front of the first half laser cycle $\omega t \in (-\pi/2, 0)$ in Fig. 3(b).

To see why those electrons released from those time windows remain un-ionized, we consider the role of the initial transverse velocity $v^i_\perp$ of the tunneled electrons. We show the distributions of the initial transverse velocities of the tunneled electrons that contribute to the ionization rate and survival rate with respect to the field phase in Figs. 3(c) and 3(d), respectively. The launch window of the laser phase and initial transverse velocities for those negative energy electrons consists of a regular hippocampuslike structure (the bright regime in Fig. 3(d)], and an irregular regime [encircled in Fig. 3(d)]. Those electrons ionized slightly before the field maximum with a small transverse velocity are strongly affected by the attraction of the Coulomb field, especially moving with an initially small drift velocity along the laser field.

The physics behind the un-ionized window can be understood by the following approximate theoretical analysis. Considering the Coulomb potential, the final energy of the tunneled electron at the birth time of t_0 can be given by $E_0 = \frac{1}{2}(v^i{}_\perp \hat{e}_\perp + \frac{\varepsilon}{\omega} \sin\omega t_0 \hat{e}_z)^2 - \frac{1}{|z_0|}$, where $z_0 = \left(I_p + \sqrt{I_p^2 - 4\varepsilon \cos\omega t_0}\right)/2\varepsilon \cos\omega t_0$ is the tunnel exit point. The energy value can be smaller than zero, indicating tunneling without ionization. In Fig. 3(d), the boundary expressed by $E_0 = 0$ in t_0 and $v^i{}_\perp$ plane is plotted with the solid curve, below which the initial energy of the tunneling electron is negative. Nevertheless, the tunneled electron will be accelerated in the consequent scattering mediated by the Coulomb and light fields. The energy gain can be expressed analytically by $\Delta E = -\int_{t_0}^{\infty} \frac{z\varepsilon}{\omega r^3} \sin\omega t dt$. Ignoring the Coulomb focusing effect, we approximate the electron orbit to be $z = \frac{\varepsilon \sin\omega t_0}{\omega}(t - t_0) + \frac{\varepsilon}{\omega^2}(\cos\omega t - \cos\omega t_0) - z_0$, $r = \sqrt{[v^i_\perp(t - t_0)]^2 + z^2}$, and then the energy change can be obtained by evaluating the above integral numerically. We plot the line of $E_0 + \Delta E = 0$ in Fig. 3(d) (labeled by the dashed line), which gives a good approximation for the boundary that confines the un-ionized window. However, since some electrons inside the *arc* region might experience multiple forward and backward scatterings, their orbits are essentially chaotic [17] and cannot be described by the approximate orbits. During the chaotic scattering, those electrons can acquire additional energy and are eventually ionized, thus making the boundary of the un-ionized window irregular.

Figures 4(a) and 4(b) show the typical trajectories associated with the regular hippocampuslike structure and the irregular structure, respectively, in the specific launching window. The electrons born with a certain field phase and transverse velocity at the rising front of a laser cycle can finally be launched into the elliptical orbits that have the negative energy. In the first optical cycle after tunneling, the electron obtains or releases energy depending on the instantaneous field phases, and then is pumped into the Rydberg elliptical orbits. The difference between those two typical trajectories is that the electron in Fig. 4(a) is ejected directly into an elliptical orbit without collision with the atomic ion. While the tunneled electron in Fig. 4(b) experiences the hard collision with the core during its launching process. Our statistics indicate that the first type orbits constitute ~90% and the second type orbits contribute the residual ~10% of the total un-ionized electrons, respectively.

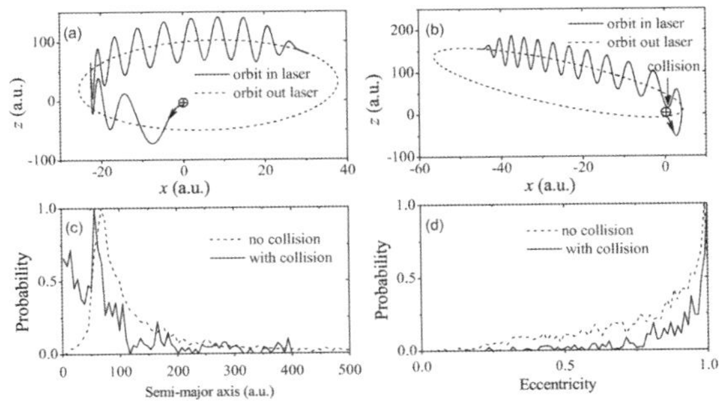

FIG. 4. (a),(b) indicate the typical trajectories in the hippocampuslike structure regime and irregular regime (encircled with solid curve) in Figs. 3(d), respectively. The first type trajectory represents the tunneled electron ejected into the elliptical orbits without collisions, and the second one experiences collisions with nucleus. (c),(d) are the statistical analysis on the semimajor axis and eccentricity distributions corresponding to those two typical trajectories, respectively.

We illustrate the statistic analysis on the semimajor axis and the eccentricity of the elliptic orbits in Figs. 4(c) and 4(d). Those electrons with collisions prefer to move in the elliptic orbits with the smaller semimajor axes and larger eccentricities. The energy of an electron in the elliptic obit is $E = -1/2a$, where a is the semimajor axis. One can find that the distribution of semimajor axis mostly falls into the regime $1/2a \ll \omega$; i.e., the binding energies of the Rybderg states are much smaller than the photon energy. Because the classical elliptic orbit frequencies ($\sim 1/a^{3/2}$) are much smaller than the laser frequency, the fast oscillating motion driven by laser field can be safely averaged out and the electrons will finally remain on the elliptical orbits [see Figs. 4(a) and 4(b)]. This is analogous to the stabilization condition for the Rydberg atoms in the low-frequency light field [5]. However, in our case, the electrons are released from the ground state through tunneling rather than prepared in the Rydberg states directly and thus the stringent atomic stabilization cannot be observed.

With the semiclassical model, we have identified the local ionization suppression as a signature of partial atomic stabilization in the strong-field tunneling ionization regime. The underlying dynamics is that the electrons are dynamically mediated by the Coulomb and light potential in multiple forward and backward scattering processes. Since only the tunneling effect is considered in our semiclassical model, the quantum transition between those Rydberg states is ignored. However, it will efficiently produce the radiation from the allowed transition in the atomic stabilization process. The transition frequency between the Rydberg states is well in THz regime, and thus strong-field atomic stabilization has evident implications on the THz radiation from gas targets in strong laser fields [22], e.g., femtosecond laser filamentation [23].

Note that the above finding differs from the low-energy hump structure of electron spectra observed in the laser polarization direction [24,25], which was attributed as electron multiple forward scattering due to overall influence of the Coulomb potential on the wave packet's longitudinal momentum of the electron [26,27]. This effect results in the enhancement of low-energy photoelectrons in the laser polarization plane. With fully-differential measurement on PADs, we observe that the yield of near-threshold electrons is suppressed in all solid angles.

In conclusion, we have investigated the partial atomic stabilization effect in strong-field tunneling ionization of atoms both experimentally and theoretically. We identify that a fraction of tunneled electrons released in a certain window of the field phase and transverse velocity can be successively launched into two types of elliptical orbits and is finally stabilized there. This leads to the suppressed yield of near-zero-momentum photoelectrons, i.e., a phenomenon that is universal and is very essential in strong-field tunneling ionization. The identification provides a deep intuition for further quantum control using an attosecond vacuum ultraviolet pulse to excite an atom at a proper time window in a strong few-cycle low-frequency light field.

The semiclassical simulation does not replicate all of the experimental details, e.g., the interference structures. For tunneling ionization at the long wavelength limit, the PADs of low-energy photoelectrons show more or less constant interference structure and we do not observe multiphoton channel closing or channel shift effect. On the other hand, in deep tunneling ionization ($\gamma < 1$) the interference structure in the PADs deviates from the radial distribution. This is very different from the fan structure in multiphoton regime [13,28] (see also Fig. 1 for the largest Keldysh parameter), and a dedicated quantum model is required. Similar to the ionization of the metastable xenon atoms in 7.5 μm far-infrared laser fields [29], these interference structures can be associated with photoelectron holography.

We would like to thank Dr. T. Morishita and Dr. Ming-Yang Yu for helpful discussions. This work is financially supported by the NCET in University from the Ministry of Education, the NSFC (11125416, 61078025, 11121091, 11134001, 91021021), and the 973 program (2011CB921503).

*yunquan.liu@pku.edu.cn
†liu_jie@iapcm.ac.cn
‡qhgong@pku.edu.cn

[1] M. Gavrila, J. Phys. B **35**, R147 (2002).
[2] Q. Su and J. H. Eberly, J. Opt. Soc. Am. B **7**, 564 (1990); J. H. Eberly and K. C. Kulander, Science **262**, 1229 (1993).
[3] M. Dörr, R. M. Potvliege, and R. Shakeshaft, Phys. Rev. Lett. **64**, 2003 (1990).
[4] K. C. Kulander, K. J. Schafer, and J. L. Krause, Phys. Rev. Lett. **66**, 2601 (1991); M. P. de Boer, J. H. Hoogenraad, R. B. Vrijen, L. D. Noordam, and H. G. Muller, Phys. Rev. Lett. **71**, 3263 (1993).
[5] The traditional atomic stabilization requires that the laser frequency is higher than or at least comparable with the bound energy. Recently, the stabilization issue is extended to the Rydberg atoms, e.g., M. V. Fedorov and A. M. Movsesian, J. Phys. B **21**, L155 (1988); B. Piraux and R. M. Potvliege, Phys. Rev. A **57**, 5009 (1998); M. Pont and R. Shakeshaft, Phys. Rev. A **44**, R4110 (1991); S. Askeland, S. A. Sørngård, I. Pilskog, R. Nepstad, and M. Førre, Phys. Rev. A **84**, 033423 (2011).
[6] M. Gavrila, I. Simbotin, and M. Stroe, Phys. Rev. A **78**, 033404 (2008); M. Stroe, I. Simbotin, and M. Gavrila, Phys. Rev. A **78**, 033405 (2008).
[7] L. V. Keldysh, Zh. Eksp. Teor. Fiz. **47**, 1945 (1964) [Sov. Phys. JETP **20**, 1307 (1965)].
[8] R. V. Kulyagin and V. D. Taranukhin, Laser Phys. **3**, 644 (1993).
[9] V. P. Gavrilenk and E. Oks, Can. J. Phys. **89**, 849 (2011).
[10] A. M. Perelomov, V. S. Popov, and V. M. Teren'ev, Zh. Eksp. Teor. Fiz. **52**, 514 (1967) [Sov. Phys. JETP **25**, 336 (1967)]; M. V. Ammosov, N. B. Delone, and V. P. Krainov, Zh. Eksp. Teor. Fiz. **91**, 2008 (1986) [Sov. Phys. JETP **64**, 1191 (1986)]; N. B. Delone and V. P. Krainov, J. Opt. Soc. Am. B **8**, 1207 (1991).
[11] K. J. Schafer, B. Yang, L. F. DiMauro, and K. C. Kulander, Phys. Rev. Lett. **70**, 1599 (1993); P. B. Corkum, Phys. Rev. Lett. **71**, 1994 (1993); P. B. Corkum, Phys. Today **64**, 36 (2011).
[12] Y. Liu, X. Liu, Y. Deng, C. Wu, H. Jiang, and Q. Gong, Phys. Rev. Lett. **106**, 073004 (2011); Y. Liu *et al.*Phys. Rev. Lett.**104**, 173002 (2010); Y. Deng, Y. Liu, X. Liu, H. Liu, Y. Yang, C. Wu, and Q. Gong, Phys. Rev. A **84**, 065405 (2011).
[13] J. Ullrich, R. Moshammer, A. Dorn, R. Dörner, L. Ph. H. Schmidt, and H. Schmidt-Böcking, Rep. Prog. Phys. **66**, 1463 (2003).
[14] A. Rudenko, K. Zrost, C. D. Schröter, V. L. B. de Jesus, B. Feuerstein, R. Moshammer, and J. Ullrich, J. Phys. B **37**, L407 (2004).
[15] D. G. Arbó, S. Yoshida, E. Persson, K. I. Dimitriou, and J. Burgdörfer, Phys. Rev. Lett. **96**, 143003 (2006).
[16] Z. Chen, T. Morishita, A.-T. Le, M. Wickenhauser, X. M. Tong, and C. D. Lin, Phys. Rev. A **74**, 053405 (2006).
[17] B. Hu, J. Liu, and S. G. Chen, Phys. Lett. A **236**, 533 (1997).
[18] D. Ye and J. Liu, Phys. Rev. A **81**, 043402 (2010).
[19] L. D. Landau and E. M. Lifshitz, *Quantum Mechanics* (Pergamon, New York, 1977).
[20] T. Nubbemeyer, K. Gorling, A. Saenz, U. Eichmann, and W. Sandner, Phys. Rev. Lett. **101**, 233001 (2008).
[21] In the simple-man model, the electron is supposed to release at origin with zero velocity and the subsequent dynamics is described classically without considering Coulomb potential. For details, refer to H. B. van Linden van den Heuvell and H. G. Muller, in *Multiphoton Processes*, edited by S. J. Smith and P. L. Knight (Cambridge University Press, Cambridge, England, 1988); T. F. Gallagher, Phys. Rev. Lett. **61**, 2304 (1988); P. B. Corkum, N. H. Burnett, and F. Brunel, Phys. Rev. Lett. **62**, 1259 (1989).
[22] C. D'Amico, A. Houard, M. Franco, B. Prade, A. Mysyrowicz, A. Couairon, and V. T. Tikhonchuk, Phys. Rev. Lett. **98**, 235002 (2007).
[23] P. Béjot, J. Kasparian, S. Henin, V. Loriot, T. Vieillard, E. Hertz, O. Faucher, B. Lavorel, and J.-P. Wolf, Phys. Rev. Lett. **104**, 103903 (2010).
[24] C. I. Blaga, F. Catoire, P. Colosimo, G. G. Paulus, H. G. Muller, P. Agostini, and L. F. DiMauro, Nature Phys. **5**, 335 (2009); F. Catoire, C. I. Blaga, E. Sistrunk, H. G. Muller, P. Agostini, and L. F. DiMauro, Laser Phys. **19**, 1574 (2009).
[25] W. Quan *et al.*, Phys. Rev. Lett. **103**, 093001 (2009); C. Wu, Y. Yang, Y. Liu, Q. Gong, M. Wu, X. Liu, X. Hao, W. Li, X. He, and J. Chen, Phys. Rev. Lett. **109**, 043001 (2012).
[26] C. Liu and K. Z. Hatsagortsyan, Phys. Rev. Lett. **105**, 113003 (2010).
[27] T. -M. Yan, S. V. Popruzhenko, M. J. J. Vrakking, and D. Bauer, Phys. Rev. Lett. **105**, 253002 (2010).
[28] M. Li, Y. Liu, H. Liu, Y. Yang, J. Yuan, X. Liu, Y. Deng, C. Wu, and Q. Gong, Phys. Rev. A **85**, 013414 (2012).
[29] Y. Huisman *et al.*, Science **331**, 61 (2010).

PHYSICS OF PLASMAS 19, 042702 (2012)

The equation of state, electronic thermal conductivity, and opacity of hot dense deuterium-helium plasmas

Cong Wang,[1] Xian-Tu He,[1,2] and Ping Zhang[1,2,a)]
[1]LCP, Institute of Applied Physics and Computational Mathematics, P.O. Box 8009, Beijing 100088, People's Republic of China
[2]Center for Applied Physics and Technology, Peking University, Beijing 100871, People's Republic of China

(Received 30 January 2012; accepted 16 March 2012; published online 4 April 2012)

Thermophysical properties of dense deuterium-helium plasmas along the 160 g/cm^3 isochore with temperatures up to 800 electron-volt are reported. From Kubo-Greenwood formula, the electronic thermal conductivity and Rosseland mean opacity are determined by means of quantum molecular dynamics (QMD) simulations. Equation of states is obtained by QMD and orbital free molecular dynamics. The electronic heat conductance is compared with several models currently used in inertial confinement fusion designs. Our results indicate that only in the weak coupling regime, the opacity is sensitive to the concentration of helium. © *2012 American Institute of Physics*. [http://dx.doi.org/10.1063/1.3699536]

I. INTRODUCTION

Inertial confinement fusion (ICF) has been pursued for decades due to the potential way to generate clean energy.[1–3] In typical ICF designs, a deuterium-tritium (DT) capsule is driven to implode either directly by high-power laser pulses or indirectly by x rays in a hohlraum with the productions of helium (He) and neutron (n), and the complete burning of the fuel follows the fusion reaction $D + T \rightarrow n + He + 17.6$ MeV. The unique behaviors of D-He mixtures in high energy density physics (HEDP), which is understood to be the thermodynamic regime for energy density exceeding 10^{11} J/m^3, are of crucial interest for understanding the fundamental physical process in ICF. During the imploding process, D-T fuels and the relative productions, which are dynamically compressed by shocks and/or adiabatic compression waves driven by laser ablation,[4] could reach the plasma conditions at densities up to $10^2 \sim 10^3$ g/cm^3 and at temperatures varying from a few to several hundreds of electron-volts (eV).[1] In this context, high precision wide range equation of states (EOS) and electronic heat conduction for dense D-He mixtures are essential. First, they are key ingredients to determine laser absorption,[5] shock timing,[6] and all kinds of hydrodynamical instabilities grown at the fuel-ablator[7,8] or hot spot-fuel interface.[9,10] Then, the accurate knowledge of thermophysical properties of dense D-He mixtures is important for the correct description of the propagation of the thermonuclear burn wave and the evolution of the central hot spot against the dense cold fuel.[11] Thus, the high pressure responses for D-He mixtures are highly recommended to be present and understood.

At the present stage, to create and properly characterize HEDP conditions are difficult in laboratory, and measurements are scarce. Theoretical approximations used to predict the thermophysical properties, such as the EOS and transport coefficients in a self-consistent way, open a challenging field of research. Classical methods for modelling weakly coupled and strongly degenerated plasmas[12–15] widely disagree at high density and are incorrect to determine nonlinear screening caused by electronic polarization. Quantum molecular dynamics (QMD) simulations,[16] where electrons are fully quantum mechanically treated according to finite-temperature density-functional theory (FT-DFT), provide a powerful method to compute the electronic structure and ionic configuration in local thermodynamic equilibrium. However, the application of this method is limited by large number of occupied electronic states and overlap between pseudocores in the hot dense region. Recently, the use of short cutoff radius Coulombic pseudopotential has overcome these shortcomings and affords efficient ways to study dense hydrogen[17] and helium.[18] Here, we present for the first time QMD simulations of a broad spectrum of thermophysical properties for D-He plasma mixtures in both high temperatures and high densities.

Our QMD simulations on D-He plasmas have been performed at several temperatures from $k_BT = 2$ eV to 800 eV along the 160 g/cm^3 isochore, which highlight the HEDP conditions realized in the imploding process. The self-consistent electronic structure calculation within FT-DFT yields the charge density distribution in the simulation cubic cell at every time step. Then, the EOS is determined at local thermodynamical equilibrium conditions. We use Kubo-Greenwood formula as a starting point for evaluation of the kinetic coefficients, from which the electronic thermal conductivity can be extracted, and then the Rosseland mean opacity (RMO) is obtained.

II. COMPUTATIONAL METHODS

The ionic trajectories and the EOS in thermodynamical equilibrium are generated through two methods. On one hand, *ab initio* plane wave code ABINIT (Ref. 19) has been applied to perform QMD simulations for temperatures up to 200 eV. With N atoms included, a series of volume-fixed cubic cells, which are repeated periodically throughout the space, form the elements of our calculations. We use the

[a)]Author to whom correspondence should be addressed. Electronic mail: zhang_ping@iapcm.ac.cn.

local density approximation (LDA) exchange-correlation functional with Teter-Pade parametrization.[20] The temperature dependence of exchange-correlation functional, which is considered to be as small as negligible, is not taken into account. With the convenience of saving computational cost, the application of pseudopotential approximations, which separate core electrons from valence electrons, prevents general QMD simulations from high-density region. This frozen-core approximation only works at moderate densities, due to the density-induced overlap of the pseudocores and delocalization of core electrons. In this paper, Coulombic pseudopotentials for both D and He with a cutoff radius $r_s = 0.001$ AU, which are used to model dense D, He, and their mixtures at 160 g/cm^3, have been built to overcome the limitations. We have chosen a plane-wave cutoff E_{cut} at 200 AU, because large basis set is necessary in modelling wavefunctions near the core and the pressure is converged within 2% accuracy. In the overall calculations, we have considered sufficient occupational band numbers (the occupation down to 10^{-6} for electronic states are ensured). Γ point is used to sample the Brillouin zone in molecular dynamics simulations, because EOS can only be modified within 2% for the selection of higher number of **k** points.

A total number of 128 electrons are enclosed in the simulation box, that is, 128 atoms for pure D, 64 atoms for pure He, respectively. As for the mixtures, the substitutional rate is defined as $\alpha = 2N_{He}/(N_D + 2N_{He})$, where N_D and N_{He} are the number of atoms for D and He, have been set to be 25%, 50%, and 75%, and the density of the supercell is kept invariant during all of the simulations. Isokinetic ensemble is adopted, and local equilibrium is secured through setting the electronic (T_e) and ionic (T_i) temperatures to be equal to the sampled temperatures (T). Each selected point has been calculated for 6000 steps, and the time steps for the integrations of atomic motion are determined according to different thermodynamic conditions.[21] Then, the subsequent 1000 molecular dynamics steps are used to calculate EOS as running averages.

We stress here that in general QMD simulations, the high temperatures commonly introduce large number of one-body electronic states, which are populated through the Fermi-Dirac distribution, and make the calculations numerically intractable. However, we consider the plasmas in a very dense region, where the ratio between temperature and Fermi energy stays less than the order of unity, even up to 200 eV, and the degeneracy effects effectively reduce the thermal excitations. The long period QMD simulations become prohibitive at temperatures exceeding 300 eV due to the dramatic increase of the thermally excited electronic states. As a consequence, we introduce orbital-free molecular dynamic (OFMD) (Refs. 22–24) to evaluate the EOS, which is numerically extremely efficient and allows calculations up to 800 eV. In this scheme, the semiclassical development of the Mermin functional,[25] which leads to the finite-temperature Thomas-Fermi expression for the kinetic part, determines the orbital-free functional. The only variable in OFMD simulations—the electronic density, is Fourier transformed during calculations, and numerical convergence in terms of the mesh size has been insured for all the simulations. As for the previous QMD simulations, computational parameters are not changed, and the same number of particles has been propagated in a cubic simulation cell with periodic boundary conditions.

III. RESULTS AND DISCUSSION

To characterize the high pressure behavior of the present D-He plasmas, we introduce two commonly used nondimensional parameters, that is, the ion-ion coupling parameter $\Gamma_{ii} = Z^{*2}/(k_B T a)$ and the electron degeneracy parameter $\theta = T/T_F$. Here, $Z^* = n_e/n_i$ represents the average ionization degree, with n_e and n_i denoting the electronic and ionic number densities respectively, $a = (3/4\pi n_i)^{1/3}$ is the ionic sphere radius, and $T_F = (3\pi^2 n_e)^{2/3}/2$ is the Fermi temperature. The ion-ion coupling and the electron degeneracy parameters for the studied dense plasmas are shown in Fig. 1. Γ_{ii} for the mixtures lie between the values of the pure D and He. Here, the electron density is kept constant for the calculated supercells, thus, the degeneracy parameter for all the systems are equal at a given temperature.

The simulated matter goes from strongly coupled ($\Gamma_{ii} \gg 1$) and highly degenerate states ($\theta \ll 1$), then reaches moderate coupling ($\Gamma_{ii} \sim 10^{-1}$) and partial degeneracy states ($\theta \sim 1$). The EOS for the pure and mixed systems, which are restricted here to the pressure–temperature relationship, are obtained from both QMD and OFMD simulations, as have been shown in Fig. 2. In order to distinguish the data from these methods, we have shown the EOS in three temperature regimes (see Fig. 2). It is indicated that the discrepancy between QMD and OFMD results lies within 2% for temperatures below 20 eV, whereas, beyond this value, the difference can be treated as small as negligible. The excellent accordance between the present QMD and OFMD (the reference at high temperatures) shows the capacity of quantum simulations to deal, not only with cold and dense system as in Ref. 22 but also with high temperatures in the extremely dense regime. By examining the EOS data, we have found that the pressure obtained from both QMD and OFMD follows a simple linear mixing rule under the present thermodynamic equilibrium conditions. At given temperatures, the pressure for dense D-He plasmas $P(\alpha, T)$ with mixing

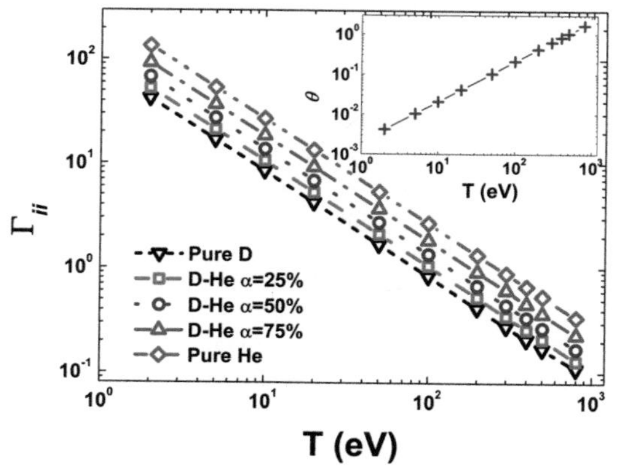

FIG. 1. The ion-ion coupling parameter (Γ_{ii}) for dense Deuterium, Helium, and their mixtures at 160 g/cm^3 are shown. Inset is the electron degeneracy parameter (θ).

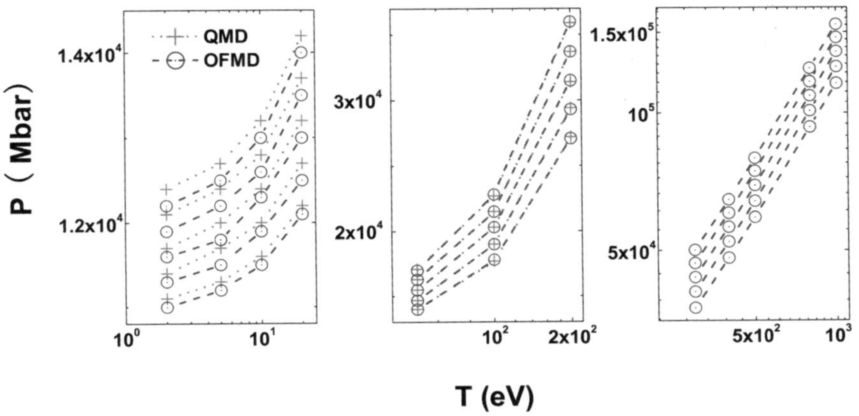

FIG. 2. Calculated EOS (pressure versus temperature) for dense deuterium, helium, and their mixtures at 160 g/cm^3. The QMD and OFMD results are labelled as red crosses and blue circles, respectively. These curves from lower to upper region shown in the figure correspond to pure helium, mixtures (α = 25%, 50%, and 75%), and pure deuterium. The error bars are included in the symbols.

parameter α can be expressed as $P(\alpha,T) = (1-\alpha)P(0,T) + \alpha P(1,T)$.

Having clarified the EOS for the pure and mixed deuterium-helium plasmas, let us turn now to examine the transport coefficients. The heat currents characterize the high-pressure linear response of hot dense plasmas to external temperature gradient. The key to evaluate electronic thermal conductivity is the kinetic coefficients based on the Kubo-Greenwood formula[26–28]

$$\hat{\sigma}(\epsilon) = \frac{1}{\Omega}\sum_{k,k'} |\langle\psi_k|\hat{v}|\psi_{k'}\rangle|^2 \delta(\epsilon_k - \epsilon_{k'} - \epsilon), \qquad (1)$$

where Ω is the volume of the supercell, $\langle\psi_k|\hat{v}|\psi_{k'}\rangle$ are the velocity matrix elements, and ϵ_k are the electronic eigenvalues. In the calculations, a Gaussian function, which has been tested to obtain smooth curves, has been used to broaden the δ function in Eq. (1). In the Chester-Thellung version,[29] the kinetic coefficients $\mathcal{L}_{ij}$ are described as

$$\mathcal{L}_{ij} = (-1)^{i+j}\int d\epsilon\hat{\sigma}(\epsilon)(\epsilon-\mu)^{(i+j-2)}\left(-\frac{\partial f(\epsilon)}{\partial \epsilon}\right), \qquad (2)$$

with $f(\epsilon)$ being the Fermi-Dirac distribution function and μ the chemical potential. Within the framework of FT-DFT, chemical potential is evaluated through fitting the set of occupation numbers corresponding to the set of eigenvalues with the usual functional form for the Fermi-Dirac distribution. Then, the electronic thermal conductivity K is given by

$$K = \frac{1}{T}\left(\mathcal{L}_{22} - \frac{\mathcal{L}_{12}^2}{\mathcal{L}_{11}}\right). \qquad (3)$$

The present formulation used to evaluate K has already been implemented in the ABINIT code and readily leads to good results for warm-dense aluminum.[30] Note that a $3\times3\times3$ Monkhorst-Pack ***k***-points mesh has been adopted in calculating the transport coefficients, since compared to the EOS, they are much more sensitive to the precise electronic quantum states. At given conditions, we have selected ten independent snapshots during the relative molecular dynamics simulation to calculate transport coefficients, so as to get statistically converged results. The ionic configurations for temperatures at and below 200 eV are directly extracted from QMD simulations, while the atomic positions are taken from OFMD simulations for higher temperatures.

In Fig. 3, the electronic thermal conductivity is shown as a function of temperature, where error bars are smaller than the size of the symbols. Semiclassical methods currently used in hydrodynamical simulations of ICF and QMD results for hydrogen along the isochore of 80 g/cm^3 up to 800 eV are also plotted as a comparison with the previous theoretical reports.[17,31] Hubbard model,[13] where independent plane waves are used to treat the electronic states, assumes the Born approximation and a complete Fermi degeneracy. Then, the atomic configurations are supposed to be Debye or one-component plasma like. Whereas the thermal conditions considered in our calculations range from temperatures well below T_F to partial degeneracy regime, which goes beyond the hypothesis of Hubbard model. The Lee-More model,[15] which is not applicable in the degenerate range, uses different formulas for the electron collision time in solid, liquid, and plasma. At high temperatures beyond T_F, QMD and

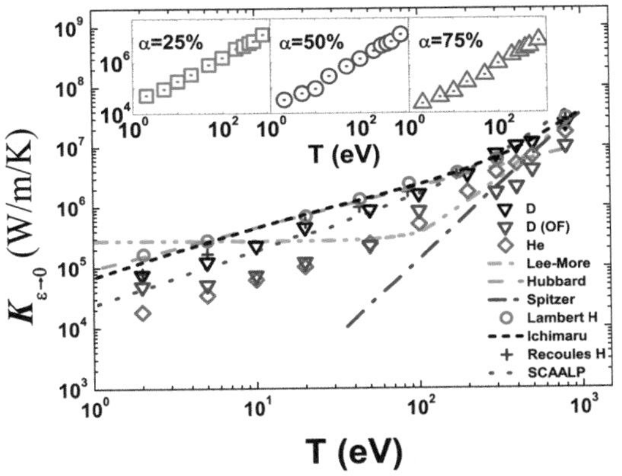

FIG. 3. Electronic heat conduction for deuterium and helium are shown as a function of temperature at 160 g/cm^3. Previous theoretical predictions for hydrogen along the 80 g/cm^3 (equivalent deuterium mass density of 160 g/cm^3) isochore, such as Hubbard,[13] Lee-More,[15] Spitzer,[12] QMD,[17,31] Ichimaru,[32,33] and SCAALP (Ref. 34) thermal conductivity models are also provided for comparison. Insets are results for D-He mixtures at 160 g/cm^3.

Lee-More model merge into the Spitzer thermal conductivity.[12] The Ichimaru model[32,33] is widely used in astrophysics, but it is restricted to moderate coupling region, namely, $\Gamma < 2$. Remarkably, our QMD data of the electronic heat conductance for pure deuterium show reasonable consistence with recent developed self-consistent average-atom model (SCAALP) (Ref. 34) for hydrogen in a wide range. This is another (besides the agreement with the OFMD result for the wide-range EOS data) good hint to suggest that the thermophysical properties of the D-He plasma in the extreme HEDP conditions can be effectively studied by using the QMD approach with appropriate pseudopotentials. Meanwhile, we have taken orbital free charge density as input to simulate the electronic thermal conductivity, which is labelled as D (OF) in Fig. 3, and the results are smaller compared with *ab initio* data at sampled temperatures.

Because of the difficulty in characterizing the electronic structure and pair distribution functions, semiclassical theoretical models in predicting the electronic heat conduction for D-He mixtures are rare, especially in the hot dense region. Due to their importance in ICF hydrodynamical simulations, QMD results are provided. The present data for mixtures are plotted in the insets of Fig. 3, so as to avoid overlaps with those of D and He. As temperatures rise from 10 to 800 eV, electronic heat conduction for mixture shows a sustaining increase with temperature. These values lie within the region expanded by the results from pure D and He. With the increase of He substitution, the electronic thermal conductivity of the mixtures decreases at a given temperature. This phenomenon can be understood as the reduction of electron activity deduced by continuously incremental binding from He nucleus.

In Fig. 4, we display the RMO (κ_R) as a function of Γ_{ii}. The RMO is given as[35]

$$\frac{1}{\kappa_R}=\frac{\int_0^\infty n^2(\omega)[\partial B(\omega,T)/\partial T]/\alpha(\rho,\omega,T)d\omega}{\int_0^\infty n^2(\omega)B(\omega,T)d\omega}, \quad (4)$$

where $\partial B(\omega,T)/\partial T$ is the derivative of the normalized Planck function with peaks around $4k_BT$. The evaluation of the absorption coefficient $\alpha(\rho,\omega,T)$ and index of refraction $n(\omega)$ can be found in Refs. 30, 35, and 36. The integral of Eq. (4), as well as the calculation of $\alpha(\omega)$ and $n(\omega)$, has been performed in the energy boundary of 0.06 to 6000 eV. As shown in Fig. 4, a clear tendency emerges between the strong and the weak coupling states. As for the former, where $\Gamma_{ii} > 10$ and $\theta \sim 10^{-2}$, the opacity is insensitive with the He concentration increases. On the other hand, for the latter, ion-ion couplings are effectively weaken due to the thermal activation, and the opacity drops six orders of magnitude as temperatures rise from 20 eV to 800 eV, where the slope of the opacity at high energy is 3.56. This can be attributed to the fact that the evaluation of RMO is sensitive to the data of the absorption coefficient and index of refraction at around $4k_BT$. For instance, we have selected two sampled temperature points to show the absorption coefficient for hot dense deuterium, where the result of 800 eV drops five orders of magnitude compared to the value of 20 eV, as is indicated in Fig. 5. However, the difference in index of refraction appears at low energy (Fig. 5 lower panel), thus does not affect RMO at high temperatures. In order to clarify the effect introduced by $n(\omega)$, we have shown the results for RMO with the selection of setting the index of refraction to be equal to one for deuterium (Fig. 4), and divergency only exists at temperatures below 20 eV. Let us stress here, the effect on RMO due to the participation of He only appears at high temperatures.

IV. SUMMARY

In conclusion, we have adopted quantum molecular dynamics calculations to study the EOS, electronic heat conduction, and RMO for dense D, He, and D-He mixtures under extreme conditions as reached in ICF designs. The simulated matter exhibits from a strong coupling and degenerate state to a moderate coupling and partial degeneracy regime. By introducing a very small cutoff radius Coulombic pseudopotential, high precision EOS (pressures up to 10^8 kbar) for the studied species have been obtained, and the EOS of mixtures have been demonstrated to follow a simple linear mixing

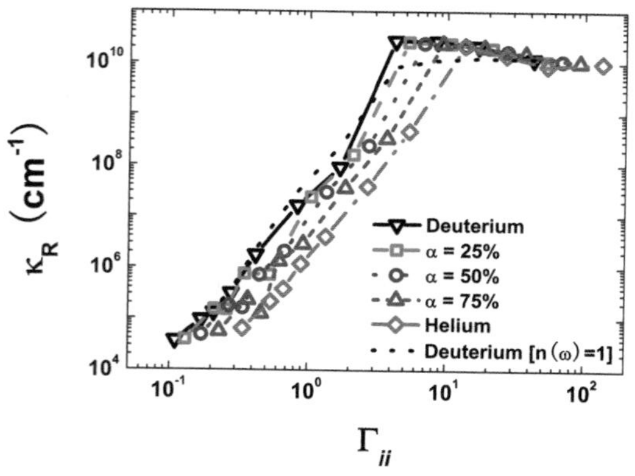

FIG. 4. Comparison of Rosseland mean opacity as a function of coupling parameter (Γ_{ii}) at 160 g/cm^3.

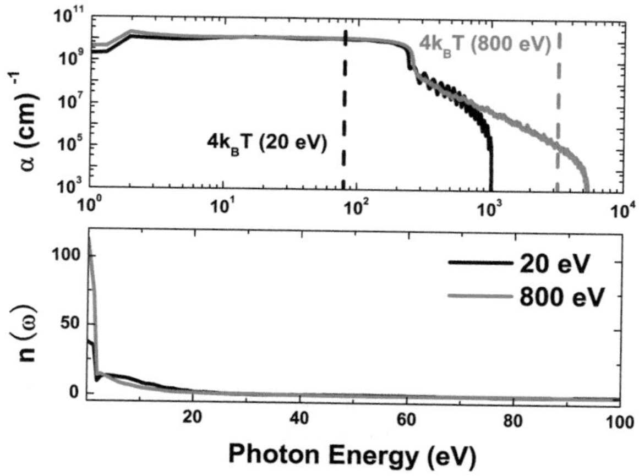

FIG. 5. The absorption coefficient and index of refraction are shown for hot dense deuterium (160 g/cm^3) at the upper and lower panel, respectively.

rule along the fusion reaction path. The electronic thermal conductivity, for which Kubo-Greenwood formula is adopted, has been systematically determined and carefully discussed. Meanwhile, the high pressure behavior of the opacity is closely related to the ion-ion coupling parameters. Our present results offer a guiding line towards the practical ICF hydrodynamical simulations with D-He mixture involved.

ACKNOWLEDGMENTS

This work was supported by NSFC under Grant Nos. 11005012 and 51071032 and by the National Basic Security Research Program of China.

[1]S. Atzeni and J. Meyer-ter-Vehn, *The Physics of Inertial Fusion: Beam Plasma Interaction, Hydrodynamics, Hot Dense Matter*, International Series of Monographs on Physics (Clarendon Press, Oxford, 2004).
[2]J. D. Lindl, *Inertial Confinement Fusion: The Quest for Ignition and Energy Gain Using Indirect Drive* (Springer-Verlag, New York, 1998).
[3]A. H. Boozer, Rev. Mod. Phys. **76**, 1071 (2004).
[4]S. X. Hu, B. Militzer, V. N. Goncharov, and S. Skupsky, Phys. Rev. Lett. **104**, 235003 (2010).
[5]W. Seka, D. H. Edgell, J. P. Knauer, J. F. Myatt *et al.*, Phys. Plasmas **15**, 056312 (2008).
[6]T. R. Boehly, E. Vianello, J. E. Miller, R. S. Craxton *et al.*, Phys. Plasmas **13**, 056303 (2006).
[7]V. A. Smalyuk, D. Shvarts, R. Betti, J. A. Delettrez *et al.*, Phys. Rev. Lett. **100**, 185005 (2008).
[8]B. A. Hammel, M. J. Edwards, S. W. Haan, M. M. Marinak *et al.*, J. Phys.: Conf. Ser. **112**, 022007 (2008).
[9]V. Lobatchev and R. Betti, Phys. Rev. Lett. **85**, 4522 (2000).
[10]M. Temporal, S. Jaouen, L. Masse, and B. Canaud, Phys. Plasmas **13**, 122701 (2006).
[11]P. Gauthier, F. Chaland, and L. Masse, Phys. Rev. E **70**, 055401(R), (2004).
[12]L. Spitzer and R. Härm, Phys. Rev. **89**, 977 (1953).
[13]W. B. Hubbard, Astrophys. J. **146**, 858 (1966).
[14]H. Brysk, P. M. Campbell, and P. Hammerling, Plasma Phys. **17**, 473 (1975).
[15]Y. T. Lee and R. M. More, Phys. Fluids **27**, 1273 (1984).
[16]W. Lorenzen, B. Holst, and R. Redmer, Phys. Rev. Lett. **102**, 115701 (2009).
[17]V. Recoules, F. Lambert, A. Decoster, B. Canaud, and J. Clérouin, Phys. Rev. Lett. **102**, 075002 (2009).
[18]C. Wang, X. T. He, and P. Zhang, Phys. Rev. Lett. **106**, 145002 (2011).
[19]See http://www.abinit.org for details of ABINIT code.
[20]S. Goedecker, M. Teter, and J. Hutter, Phys. Rev. B **54**, 1703 (1996).
[21]The time steps have been taken as $\Delta t = a/20\sqrt{k_B T/m_D}$, where m_D is the ionic mass of deuterium. A few hundred of plasma periods are included in this choice of simulations.
[22]S. Mazevet, F. Lambert, F. Bottin, G. Zérah, and J. Clérouin, Phys. Rev. E **75**, 056404 (2007).
[23]J. F. Danel, L. Kazandjian, and G. Zérah, Phys. Plasmas **13**, 092701 (2006).
[24]F. Lambert, J. Clerouin, and G. Zérah, Phys. Rev. E **73**, 016403 (2006).
[25]M. Brack and R. K. Bhaduri, *Semiclassical Physics* (Westview, Boulder, CO, 2003).
[26]V. Recoules and J. P. Crocombette, Phys. Rev. B **72**, 104202 (2005).
[27]S. Mazevet, M. Torrent, V. Recoules, and F. Jollet, High Energy Density Phys. **6**, 84 (2010).
[28]M. Pozzo, M. P. Desjarlais, and D. Alf, Phys. Rev. B **84**, 054203 (2011).
[29]G. V. Chester and A. Thellung, Proc. Phys. Soc. London **77**, 1005 (1961).
[30]S. Mazevet, M. P. Desjarlais, L. A. Collins *et al.*, Phys. Rev. E **71**, 016409 (2005).
[31]F. Lambert, V. Recoules, A. Decoster, J. Clérouin, and M. Desjarlais, Phys. Plasmas **18**, 056306 (2011).
[32]S. Ichimaru and S. Tanaka, Phys. Rev. A **32**, 1790 (1985).
[33]H. Kitamura and S. Ichimaru, Phys. Rev. E **51**, 6004 (1995).
[34]G. Faussurier, C. Blancard, P. Cossé, and P. Renaudin, Phys. Plasmas **17**, 052707 (2010).
[35]F. Perrot, Laser Part. Beams **14**, 731 (1996).
[36]S. Mazevet, L. A. Collins, N. H. Magee, J. D. Kress, and J. J. Keady, Astron. Astrophys. **405**, L5 (2003).

PHYSICAL REVIEW E **87**, 043105 (2013)

Equations of state and transport properties of warm dense beryllium: A quantum molecular dynamics study

Cong Wang,[1,2] Yao Long,[1] Ming-Feng Tian,[1] Xian-Tu He,[1,2] and Ping Zhang[1,2,*]
[1] *Institute of Applied Physics and Computational Mathematics, P.O. Box 8009, Beijing 100088, People's Republic of China*
[2] *Center for Applied Physics and Technology, Peking University, Beijing 100871, People's Republic of China*
(Received 14 February 2013; published 12 April 2013)

We have calculated the equations of state, the viscosity and self-diffusion coefficients, and electronic transport coefficients of beryllium in the warm dense regime for densities from 4.0 to 6.0 g/cm^3 and temperatures from 1.0 to 10.0 eV by using quantum molecular dynamics simulations. The principal Hugoniot curve is in agreement with underground nuclear explosive and high-power laser experimental results up to ~20 Mbar. The calculated viscosity and self-diffusion coefficients are compared with the one-component plasma model, using effective charges given by the average-atom model. The Stokes-Einstein relationship, which connects viscosity and self-diffusion coefficients, is found to hold fairly well in the strong coupling regime. The Lorenz number, which is the ratio between thermal and electrical conductivities, is computed via Kubo-Greenwood formula and compared to the well-known Wiedemann-Franz law in the warm dense region.

DOI: 10.1103/PhysRevE.87.043105 PACS number(s): 52.25.Fi, 52.65.Yy, 64.30.Jk, 72.15.Cz

I. INTRODUCTION

The nature of compressed matter is of considerable interest for many fields of modern physics, including astrophysics [1], inertial confinement fusion (ICF) [2,3], and other related fields [4]. Materials under a pressure greater than a few Mbar can be driven into a strongly coupled, partially ionized fluid state, which is defined as the so-called warm dense matter (WDM). Theoretical modeling and experimental detection of the high-pressure behavior of WDM are of great challenge and are being intensively investigated. Among various kinds of WDM, warm dense beryllium (Be) is of particular current interest. The equations of state (EOS) and transport properties of Be are very important in ICF, due to its appearance in the ablator of the deuterium-tritium (D-T) capsule. The compressibility of the capsule, laser absorption, and instability growth at the fuel-ablator interface sensitively depend on the thermophysical properties of Be [5].

The EOS of Be at shock Hugoniot up to ~18 Mbar have been accessed by strong shock waves generated by underground nuclear explosives [6,7]. Then, it is also possible to probe similar pressure range in the laboratory by high-intensity laser [8]. Despite the success of these techniques in detecting wide range EOS of typical matters, one should note that after several decades, only seven Hugoniot points are available for Be, and more experimental data are desired for building successful theoretical models, such as interatomic potentials or chemical models currently used in SESAME EOS [9]. Apart from the EOS, the atomic diffusion coefficients and fluid viscosity are key ingredients to control hydrodynamic instabilities near interfaces [10–12]. The electronic dynamic conductivity, from which dielectric function can be obtained, determines a series of interactions between laser and matters [13,14]. Due to these key issues to be addressed, therefore, the thermophysical properties of Be in the warm dense region are highly recommended to be understood in a systematic and self-consistent way.

In the present work, quantum molecular dynamics (QMD) simulations [15,16], where electrons are fully quantum mechanically treated by finite-temperature density functional theory (FT-DFT), have been introduced to study warm dense Be. The EOS have been extracted from a series of simulations, which were performed for a canonical ensemble (NVT) with temperature, volume of the simulation box, and particle number therein as given quantities. Then, the Hugoniot curve is calculated from the Rankine-Hugoniot relation. The self-diffusion coefficient and viscosity have been computed from the trajectory by the velocity and the stress tensor autocorrelation function. The dynamic conductivity, from which the DC conductivity and electronic thermal conductance are derived, has been obtained from Kubo-Greenwood formula. The rest of the paper is organized as follows. In Sec. II, we briefly describe the QMD simulations and computational method in determining the atomic transport properties and electronic dynamic conductivity. In Sec. III, discussions are presented for the EOS and transport properties. In Sec. IV, we present our conclusions.

II. COMPUTATIONAL METHOD

A. Quantum molecular dynamics

Our QMD simulations employed the Vienna *ab initio* Simulation Package (VASP) [17,18]. A series of volume fixed supercells including N atoms, which are repeated periodically throughout the space, form the elements of our calculations. Introducing Born-Oppenheimer approximation, electrons are quantum mechanically treated through plane-wave FT-DFT. The interaction between electron and ion is presented by a projector augmented wave (PAW) pseudopotential. The exchange-correlation functional is determined within Perdew-Burke-Ernzerhof generalized gradient approximation. The ions move classically according to the forces from the electron density and the ion-ion repulsion. The system is kept in local thermodynamic equilibrium with equal T_e and T_i. The electronic temperature is kept fixed through Fermi-Dirac

*Corresponding author: zhang_ping@iapcm.ac.cn

distribution of the electronic states, and the ion temperature is secured through Nosé-Hoover thermostat [19].

We have chosen 216 atoms in the unit cell with periodic boundary condition. A range of densities from $\rho = 1.84$ g/cm^3 to 6.0 g/cm^3 and temperatures from $T = 300$ K to $T = 120\,000$ K are selected to highlight main Hugoniot regions. The convergence of the thermodynamic quantities plays an important role in the accuracy of QMD simulations. In the present work, a plane-wave cutoff energy of 800 eV is employed in all simulations so that the pressure is converged within 2%. We have also checked out the convergence with respect to a systematic enlargement of the **k**-point set in the representation of the Brillouin zone. The correction of higher-order **k** points on the EOS data is slight and negligible. In the molecular dynamics simulations, only the Γ point of the Brillouin zone is included, while $4 \times 4 \times 4$ Monkhorst-Pack scheme grid points are used in the electronic structure calculations. The dynamic simulations have lasted $10 \sim 20$ ps with time steps of $0.5 \sim 1.0$ fs according to different conditions. For each pressure and temperature, the system is equilibrated within $1 \sim 2$ ps. The EOS data are obtained by averaging over the final 5 ps molecular dynamic simulations.

B. Transport properties

The self-diffusion coefficient D can either be calculated from the trajectory by the mean-square displacement

$$D = \frac{1}{6t}\langle |R_i(t) - R_i(0)|^2 \rangle, \quad (1)$$

or by the velocity autocorrelation function

$$D = \frac{1}{3}\int_0^\infty \langle V_i(t)\cdot V_i(0)\rangle dt, \quad (2)$$

where R_i is the position and V_i is the velocity of the ith nucleus. Only in the long-time limit, these two formulas of D are formally equivalent. Sufficient length of trajectories has been generated to secure a null contribution to the integral, and the mean-square displacement away from the origin consistently fits to a straight line. The diffusion coefficients obtained from these two approaches lie within 1% accuracy of each other, here, we report the results from velocity autocorrelation function.

The viscosity

$$\eta = \lim_{t\to\infty} \bar{\eta}(t), \quad (3)$$

has been computed from the autocorrelation function of the off-diagonal component of the stress tensor [20]

$$\bar{\eta}(t) = \frac{V}{k_B T}\int_0^t \langle P_{12}(0)P_{12}(t')\rangle dt'. \quad (4)$$

The results are averaged from the five independent off-diagonal components of the stress tensor P_{xy}, P_{yz}, P_{zx}, $(P_{xx} - P_{yy})/2$, and $(P_{yy} - P_{zz})/2$.

Different from the self-diffusion coefficient, which involves single-particle correlations and attains significant statistical improvement from averaging over the particles, the viscosity depends on the entire system and thus needs very long trajectories to reach statistical accuracy. To shorten the length of the trajectory, we use empirical fits [21] to the integrals of the autocorrelation functions. Thus, extrapolation of the fits to $t \to \infty$ can more effectively determine the basic dynamical properties. Both of the D and $\bar{\eta}$ have been fit to the functional in the form of $A[1 - \exp(-t/\tau)]$, where A and τ are free parameters. Reasonable approximation to the viscosity can be produced from the finite time fitting procedure, which also serves to damp the long-time fluctuations.

The fractional statistical error in calculating a correlation function C for molecular-dynamics trajectories [22] can be given by

$$\frac{\triangle C}{C} = \sqrt{\frac{2\tau}{T_{\rm traj}}}, \quad (5)$$

where τ is the correlation time of the function, and $T_{\rm traj}$ is the length of the trajectory. In the present work, we generally fitted over a time interval of $[0, 4\tau - 5\tau]$.

C. Dynamic conductivity

The key to evaluate the electrical transport properties is the kinetic coefficients. They are calculated using the following Kubo-Greenwood formulation:

$$\hat{\sigma}(\epsilon) = \frac{1}{\Omega}\sum_{k,k'} |\langle \psi_k|\hat{v}|\psi_{k'}\rangle|^2 \delta(\epsilon_k - \epsilon_{k'} - \epsilon), \quad (6)$$

where $\langle \psi_k|\hat{v}|\psi_{k'}\rangle$ are the velocity matrix elements, Ω is the volume of the supercell, and ϵ_k are the electronic eigenvalues. The kinetic coefficients $\mathcal{L}_{ij}$ in the Chester-Thellung version [23] are given by

$$\mathcal{L}_{ij} = (-1)^{i+j}\int d\epsilon\,\hat{\sigma}(\epsilon)(\epsilon - \mu)^{(i+j-2)}\left(-\frac{\partial f(\epsilon)}{\partial \epsilon}\right), \quad (7)$$

where μ is the chemical potential and $f(\epsilon)$ is the Fermi-Dirac distribution function. The electrical conductivity σ is obtained as

$$\sigma = \mathcal{L}_{11}, \quad (8)$$

and electronic thermal conductivity K is

$$K = \frac{1}{T}\left(\mathcal{L}_{22} - \frac{\mathcal{L}_{12}^2}{\mathcal{L}_{11}}\right), \quad (9)$$

where T is the temperature. Equations (8) and (9) are energy dependent, then the electrical conductivity and electronic thermal conductivity are obtained through extrapolating to zero energy. In order to get converged transport coeÀfficients, ten independent snapshots, which are selected during one molecular dynamic simulation at given conditions, are selected to calculate electrical conductivity and electronic thermal conductivity as running averages.

III. RESULTS AND DISCUSSION

A. The equations of state

Wide range EOS (ρ from 4.0 to 6.0 g/cm^3 and temperatures of $1 \sim 10$ eV) have been calculated according to QMD simulations, and the internal energy E (eV/atom) and pressure P (GPa) are fitted by expansions in terms of density (g/cm^3)

TABLE I. Expansion coefficients A_{ij} for the internal energy E (eV/atom).

$A_{i,j}$	$j=0$	$j=1$	$j=2$
$i=0$	0.4369	0.5386	0.3149
$i=1$	−0.7316	1.1936	−0.1149
$i=2$	0.3215	−0.1241	0.0117

and temperature (eV) as follows:

$$E = \sum A_{ij}\rho^i T^j, \tag{10}$$

$$P = \sum B_{ij}\rho^i T^j. \tag{11}$$

The fitted coefficient for $A_{i,j}$and $B_{i,j}$ are summarized in Tables I and II. Here, the internal energy E at 1.84 g/cm^3 and $T = 300$ K has been taken as zero.

Based on the fitted EOS, the principal Hugoniot curve can be derived from the Rankine-Hugoniot equation, which is the locus of points in (E,P,V) space satisfying the condition

$$(E_0 - E_1) + \tfrac{1}{2}(V_0 - V_1)(P_0 + P_1) = 0, \tag{12}$$

where the subscripts 0 and 1 denote the initial and shocked state, respectively. This relation follows from conservation of mass, momentum, and energy for an isolated system compressed by a pusher at a constant velocity. In the canonical (NVT) ensemble in which both E and P are temperature dependent, the locus of states which satisfies Eq. (12) is the so-called principal Hugoniot, which describes the shock adiabat between the initial and final states.

The Hugoniot curve is shown in Fig. 1, where previous theoretical and experimental results are also provided for comparison. In the warm dense region, the nature of the continuous transition from condensed matter to dense plasma remains an outstanding and interesting issue in high-pressure physics. One way to address this issue is to measure EOS in the range of 0.1 to 5 TPa. The high shock pressures required to span this range have been reached traditionally with strong shock waves generated by underground nuclear explosives [6,7], then accessed by high intensity lasers [8]. Good agreement is found from Fig. 1 between our QMD-determined EOS and those obtained experimentally. The present QMD EOS indicates a smooth transition from condensed matter to plasma at temperatures from 1.0 eV to 10.0 eV, where we do not find any signs that suggest a first-order plasma phase transition. Theoretically, SESAME EOS [9] also shows an overall accordance with our results. On the contrast, the Hugoniot curve by multiphase EOS is softer at $\rho \sim 6$ g/cm^3, which documents an inaccurate matching of mixed theoretical models [24].

TABLE II. Expansion coefficients B_{ij} for pressure P (GPa).

$B_{i,j}$	$j=0$	$j=1$	$j=2$
$i=0$	−17.9930	10.2237	−0.7659
$i=1$	−62.4972	24.4101	0.3885
$i=2$	38.7063	−0.2107	−0.0218

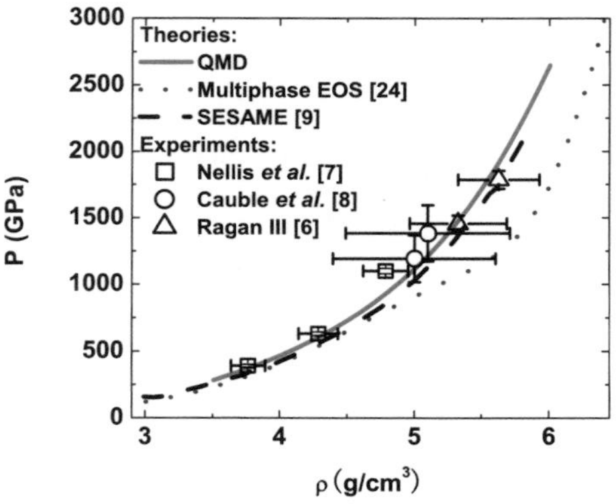

FIG. 1. (Color online) Hugoniot curve computed by QMD simulations (solid red line) are compared with previous results. Underground nuclear explosive experiments by Nellis *et al.* [7] and Ragan III [6] are labeled as open square and triangle. High power laser results by Cauble *et al.* [8] are shown as open circles. SESAME [9] and multiphase EOS [24] are shown as dashed black line and dotted blue lines, respectively.

B. Diffusion and viscosity

We have performed *ab initio* quantum-mechanical simulations with the FT-DFT method to benchmark the dynamic properties of Be in the WDM regime. An example of the QMD results for the self-diffusion coefficient and viscosity of 5.0 g/cm^3 at temperatures of 1.0, 5.0, and 10.0 eV are displayed with their fits in Fig. 2. The current simulations have the trajectory of 10 ∼ 20 ps and correlation times between 100 and 200 fs. The computed error lies within 10% for the viscosity. Due to the fitting procedure and extrapolation to infinite time a total uncertainty of ∼20% is estimated by experience. Since the particle average gives an additional $\frac{1}{\sqrt{N}}$ advantage, the error in the self-diffusion coefficient is less than 1%.

Idealized models, such as the one-component plasma (OCP) model, concern the interaction through the Coulomb potential within a neutralizing background of electrons. A large number of molecular dynamics and Monte Carlo simulations based on the OCP model [25–30] have demonstrated that physical properties like diffusion and viscosity can be represented in terms of coupling coefficient, which is defined by the ratio of the potential to kinetic energy

$$\Gamma = \frac{Z^2 e^2}{a k_B T}, \tag{13}$$

where Ze is the ion charge, and $a = (3/4\pi n_i)^{1/3}$ is the ion-sphere radius with $n_i = \rho/M$ the number density. A memory function has been used by Hansen *et al.* [30] to analyze the velocity autocorrelation function to obtain the diffusion coefficient for the classical OCP

$$\frac{D}{\omega_p a^2} = 2.95\Gamma^{-1.34}, \tag{14}$$

with $\omega_p = (4\pi n_i/M)^{1/2} Ze$ being the ion plasma frequency.

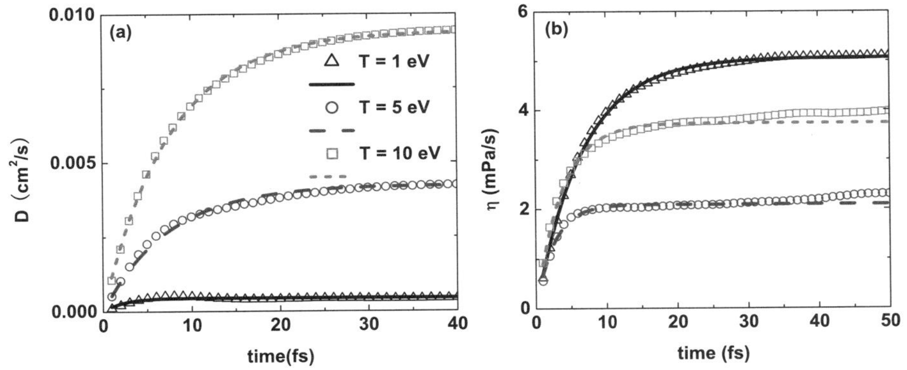

FIG. 2. (Color online) Self-diffusion coefficient (a) and viscosity (b) as a function of time at a density of 5.0 g/cm^3 and different temperatures of 1.0 (triangles), 5.0 (circles), and 10 eV (squares). The fits (lines) are performed for a sample window of [0,4τ-5τ].

Bastea [25] has performed classical molecular-dynamics simulations of the OCP and fits his results to the form

$$\frac{\eta}{n_i M \omega_p a^2} = A\Gamma^{-2} + B\Gamma^{-s} + C\Gamma, \tag{15}$$

with $s = 0.878$, $A = 0.482$, $B = 0.629$, and $C = 0.00188$. As the OCP model is restricted to a fully ionized plasma, determination of the ionization degree for WDM permits an extension of the OCP formulas to cooler systems. A reasonable choice is to replace Z in Eq. (13) with an effective charge $\bar{Z}$. As a consequence, we have introduced average-atom (AA) model [31], which solves the Hartree-Fock-Slater equation in a self-consistent field approximation assuming a finite temperature. In the densities and temperatures we explore, the effective charge $\bar{Z}$ is 2.0, which corresponds to the case of full ionization of the 2s electrons.

QMD and OCP results for the self-diffusion coefficient, at the density of 5.0 g/cm^3, are shown in the left panel in Fig. 3. The tendency for the self-diffusion coefficient with respect to temperature is similar for QMD simulations and OCP model. However, the OCP model predicts a larger value of the diffusion coefficient compared to QMD results ($\sim$1.5 times bigger). Results for the viscosity, which consist of contributions.from interatomic potential and kinetic motion of particles, are plotted in the right panel of Fig. 3. The minimum of viscosity along temperature can be attributed to a combined effect, that is, contribution from interatomic potential, which decreases with temperature, while contribution from kinetic motion increases with temperature. The OCP model indicates the local minimum around 2.0 eV, and QMD suggests the location around 3.0 eV.

The Stokes-Einstein relation gives a connection between the diffusion and shear viscosity

$$F_{\mathrm{SE}}[D,\eta] = \frac{D\eta}{k_B T n_i^{1/3}} = C_{\mathrm{SE}}, \tag{16}$$

where F_{SE} is a shorthand notation for the relationship between the transport coefficients and C_{SE} is a constant. Several prescriptions [32] for determining C_{SE} are available. Chisolm and Wallace [33] have provided an empirical value of 0.18 ± 0.02 from a theory of liquids near melting. On the other hand, C_{SE}, which has been derived based on the motion of a test particle through a solvent, was assumed to range from $1/6\pi$ [34] to $1/4\pi$ [35] depending on the limits of the slip coefficient from infinity (stick) to zero (slip). Here, we have examined the behavior of Be over the various regimes we explore. As

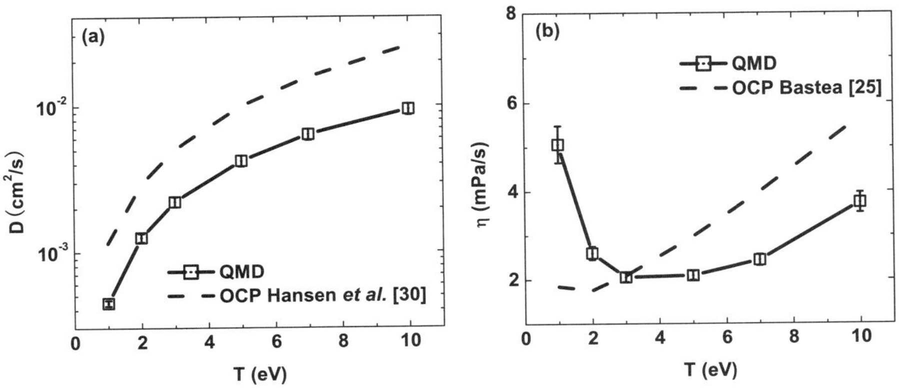

FIG. 3. Self-diffusion coefficient (a) and viscosity (b) as a function of temperature at a density of 5.0 g/cm^3. Only statistical error has been considered here.

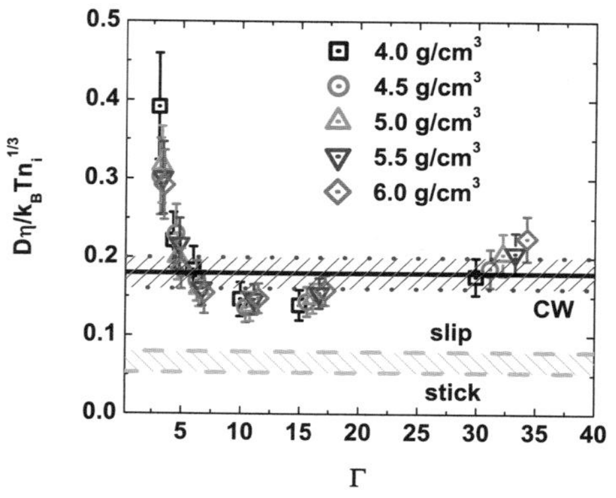

FIG. 4. (Color online) Examination of the Stokes-Einstein relation along the coupling parameter in the warm dense region. Predictions by Chisolm and Wallace [33] are shown as dotted blue region and denoted as CW in the figure. The flat dashed cyan lines show the constant values of C_{SE} for stick and slip boundary conditions [34,35].

shown in Fig. 4, we have plotted the Stokes-Einstein expression $F_{\mathrm{SE}}(D,\eta)$ as a function of the coupling parameter Γ using the diffusion coefficients and viscosities from QMD simulations at densities of $4 \sim 6$ g/cm^3 and temperatures from $1 \sim 10$ eV. As the expected fitting error of $\sim$20% for the viscosity, the QMD results are bounded by the classical values of C_{SE} from below (slip limit) and the Chisolm-Wallace liquid metal value from above at temperatures below 7.0 eV (corresponding to $\Gamma > 4.0$). The function $F_{\mathrm{SE}}[D,\eta]$ at the higher temperature evinces a sharp increase with temperature. The near-linear rise of the diffusion coefficient with temperature in the whole region basically cancels the temperature dependence of the denominator. As a consequence, the behavior of $F_{\mathrm{SE}}[D,\eta]$ is dominated by the viscosity, which rises sharply at high temperatures as shown in Fig. 3. As mentioned above, this abrupt bend in the viscosity with temperature reflects a change from potential- to kinetics-dominated regimes.

C. Lorenz number

We have also computed the Lorenz number defined as

$$L = \frac{K}{\sigma T} = \gamma \frac{e^2}{k_B^2}, \tag{17}$$

where K and σ are the thermal and electrical conductivities, respectively. γ depends on the screened potential and corresponds to the scattering of the electrons [23]. In a degenerate ($\theta \leqslant 1$) and coupled plasma ($\Gamma \geqslant 1$), L or γ is a constant ($\pi^2/3$), reaching the ideal Sommerfeld number, which is the value valid for metals, and the Wiedemann-Franz law is recovered in an elastically interacting electron system. As temperature is very high, the WDM enters into a nondegenerate case ($\theta \gg 1$ and $\Gamma \ll 1$), the Lorenz number reaches the value of kinetic matter (4 or 1.5966 depending on the e-e collisions). In the intermediate region, no assumptions can be found for predicting the Lorenz number and one cannot deduce the thermal conductivity from the electrical conductivity by using Wiedemann-Franz law.

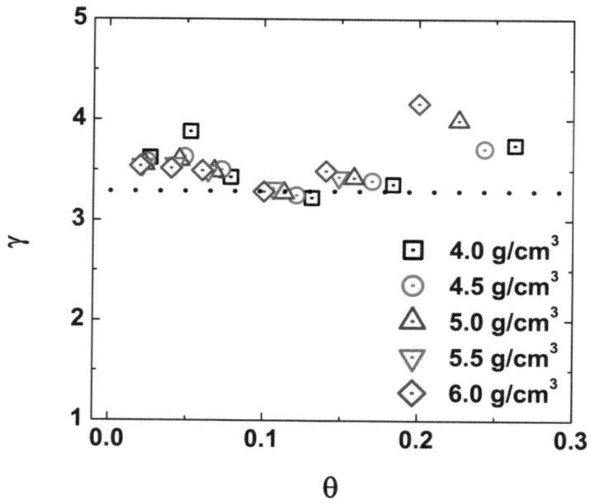

FIG. 5. (Color online) Calculated Lorenz number γ as a function of degenerate parameter θ. The ideal Sommerfeld number is plotted as dotted line in the figure.

In Fig. 5 we show the behavior of γ as a function of the degeneracy parameter θ ($\theta = k_B T/E_F$, with Fermi energy $E_F = \hbar^2 \frac{(3\pi^2 n_e)^{2/3}}{2m_e}$ and electron density n_e). Here, we should stress that in QMD simulations, the electrical conductivity σ and electronic thermal conductance K can be directly evaluated without using any assumption of Lorenz number, which is highly dependent on the two nondimensional parameters θ and Γ. In the present warm dense regime, the Lorenz number oscillates around the Sommerfeld limit at low temperatures, and as θ increases, a departure of the Lorenz number from the ideal value can be observed from Fig. 5.

IV. CONCLUSION

In the present work, clear chains have been demonstrated in investigating the thermophysical properties of warm dense Be. The EOS has been calculated through *ab initio* molecular dynamic simulations, and smooth functions have been constructed to fit the QMD wide range EOS data, which show good agreement with the underground nuclear explosive and high pulsed laser experimental results. Based on Green-Kubo relation, the self-diffusion coefficients and viscosity have then been determined, and as a reference, the OCP model with an effective charge determined from the average-atom model has also been employed. A Stokes-Einstein relation between the viscosity and diffusion coefficient holds the general feature of liquids as predicted by Chisolm and Wallace in the strong coupling region ($\Gamma > 4$) while a sharp increase has been observed as temperature rises. The Kubo-Greenwood formula provides an efficient way to study the electrical conductivity and electronic heat conductance in the warm dense regime. Through QMD simulations we have shown that the Wiedemann-Franz law is satisfied for the degenerate regime. Our present results are expected to shed light on the hydrodynamic modeling of target implosions in ICF design.

ACKNOWLEDGMENTS

This work was supported by the National Natural Science Foundation of China (NSFC) under Grants No.

11275032, No. 11005012, and No. 51071032, by the National Basic Security Research Program of China, and by the National High-Tech ICF Committee of China.

[1] D. Saumon, G. Chabrier, and H. M. Van Horn, Astrophys. J. Suppl. Ser. **99**, 713 (1995).

[2] S. Atzeni and J. Meyer-ter-Vehn, *The Physics of Inertial Fusion: Beam Plasma Interaction, Hydrodynamics, Hot Dense Matter*, International Series of Monographs on Physics (Clarendon Press, Oxford, 2004).

[3] J. D. Lindl, *Inertial Confinement Fusion: The Quest for Ignition and Energy Gain Using Indirect Drive* (Springer-Verlag, New York, 1998).

[4] V. E. Fortov *et al.*, Phys. Rev. Lett. **99**, 185001 (2007).

[5] J. D. Lindl, P. Amendt, R. L. Berger, S. G. Glendinning, S. H. Glenzer, S. W. Haan, R. L. Kauffman, O. L. Landen, and L. J. Suter, Phys. Plasmas **11**, 339 (2004).

[6] C. E. Ragan III, Phys. Rev. A **25**, 3360 (1982).

[7] W. J. Nellis, J. A. Moriarty, A. C. Mitchell, and N. C. Holmes, J. Appl. Phys. **82**, 2225 (1997).

[8] R. Cauble, T. S. Perry, D. R. Bach, K. S. Budil, B. A. Hammel, G. W. Collins, D. M. Gold, J. Dunn, P. Celliers, L. B. Da Silva, M. E. Foord, R. J. Wallace, R. E. Stewart, and N. C. Woolsey, Phys. Rev. Lett. **80**, 1248 (1998).

[9] S. P. Lyon and J. D. Johnson, Los Alamos National Laboratory Report No. LA-UR-92-3407, 1992 (unpublished).

[10] H. F. Robey, Y. Zhou, A. C. Buckingham, P. Keiter, B. A. Remington, and R. P. Drake, Phys. Plasmas **10**, 614 (2003).

[11] R. Betti, V. N. Goncharov, R. L. McCrory, and C. P. Verdon, Phys. Plasmas **5**, 1446 (1998).

[12] B. A. Hammel, M. J. Edwards, S. W. Haan, M. M. Marinak, M. Patel, H. Robey, and J. Salmonson, J. Phys.: Conf. Ser. **112**, 022007 (2008).

[13] W. Seka *et al.*, Phys. Plasmas **15**, 056312 (2008).

[14] T. R. Boehly *et al.*, Phys. Plasmas **13**, 056303 (2006).

[15] V. Bezkrovniy, V. S. Filinov, D. Kremp, M. Bonitz, M. Schlanges, W. D. Kraeft, P. R. Levashov, and V. E. Fortov, Phys. Rev. E **70**, 057401 (2004).

[16] T. Lenosky, S. Bickham, J. Kress, and L. Collins, Phys. Rev. B **61**, 1 (2000).

[17] G. Kresse and J. Hafner, Phys. Rev. B **47**, R558 (1993).

[18] G. Kresse and J. Furthmüller, Phys. Rev. B **54**, 11169 (1996).

[19] P. H. Hünenberger, Adv. Polym. Sci. **173**, 105 (2005); D. J. Evans and B. L. Holian, J. Chem. Phys. **83**, 4069 (1985).

[20] M. P. Allen and D. J. Tildesley, *Computer Simulation of Liquids* (Oxford University Press, New York, 1987).

[21] J. D. Kress, James S. Cohen, D. P. Kilcrease, D. A. Horner, and L. A. Collins, Phys. Rev. E **83**, 026404 (2011).

[22] R. Zwanzig and N. K. Ailawadi, Phys. Rev. **182**, 280 (1969).

[23] G. V. Chester and A. Thellung, Proc. Phys. Soc. (London) **77**, 1005 (1961).

[24] L. X. Benedict, T. Ogitsu, A. Trave, C. J. Wu, P. A. Sterne, and E. Schwegler, Phys. Rev. B **79**, 064106 (2009).

[25] S. Bastea, Phys. Rev. E **71**, 056405 (2005).

[26] F. Lambert, Ph.D. thesis, Université Paris XI. Commissariatà l' Énergie Atomique, 2007 (unpublished).

[27] B. Bernu and P. Vieillefosse, Phys. Rev. A **18**, 2345 (1978).

[28] Z. Donkó, B. Nyíri, L. Szalai, and S. Holló, Phys. Rev. Lett. **81**, 1622 (1998); Z. Donkó and B. Nyíri, Phys. Plasmas **7**, 45 (2000).

[29] J. Daligault, Phys. Rev. Lett. **96**, 065003 (2006); **103**, 029901(E) (2009).

[30] J. P. Hansen, I. R. McDonald, and E. L. Pollock, Phys. Rev. A **11**, 1025 (1975).

[31] B. F. Rozsnyai, Phys. Rev. A **5**, 1137 (1972).

[32] M. Cappelezzo, C. A. Capellari, S. H. Pezzin, and A. F. Coehlo, J. Chem. Phys. **126**, 224516 (2007).

[33] E. Chisolm and D. Wallace, in *Shock Compression of Condensed Matter* - 2005, AIP Conf. Proc. No. 845, edited by M. D. Furnish, M. Elert, T. P. Russell, and C. T. White (AIP, New York, 2006).

[34] A. Einstein, Z. Elektrochem. **14**, 235 (1908).

[35] W. Sutherland, Philos. Mag. **9**, 781 (1905).

PHYSICAL REVIEW E 88, 033106 (2013)

Thermophysical properties of hydrogen-helium mixtures: Re-examination of the mixing rules via quantum molecular dynamics simulations

Cong Wang,[1,2] Xian-Tu He,[1,2] and Ping Zhang[1,2,*]
[1]Institute of Applied Physics and Computational Mathematics, P.O. Box 8009, Beijing 100088, People's Republic of China
[2]Center for Applied Physics and Technology, Peking University, Beijing 100871, People's Republic of China
(Received 23 June 2013; published 23 September 2013)

Thermophysical properties of hydrogen, helium, and hydrogen-helium mixtures have been investigated in the warm dense matter regime at electron number densities ranging from $6.02 \times 10^{29} \sim 2.41 \times 10^{30}$ m^{-3} and temperatures from 4000 to 20 000 K via quantum molecular dynamics simulations. We focus on the dynamical properties such as the equation of states, diffusion coefficients, and viscosity. Mixing rules (density matching, pressure matching, and binary ionic mixing rules) have been validated by checking composite properties of pure species against that of the fully interacting mixture derived from quantum molecular dynamics simulations. These mixing rules reproduce pressures within 10% accuracy, while it is 75% and 50% for the diffusion and viscosity, respectively. The binary ionic mixing rule moves the results into better agreement. Predictions from one component plasma model are also provided and discussed.

DOI: 10.1103/PhysRevE.88.033106 PACS number(s): 52.27.Cm, 31.15.A−, 51.30.+i, 96.15.Nd

I. INTRODUCTION

The study of the equation of state (EOS), transport properties, and mixing rules of hydrogen (H) and helium (He) under extreme conditions of high pressure and temperature is not only of fundamental interest, but also of essential practical applications for astrophysics [1]. For instance, giant planets such as Jupiter and Saturn require accurate EOS as the basic input into the respective interior models in order to solve the hydrostatic equation and investigate the solubility of the rocky core [2,3]. On the other side, the evolution of stars and the design of the thermal protection system are assisted by high precision transport coefficients of H-He mixtures at high pressures [4]. In addition, the viscosity and mutual diffusion coefficients are also important input properties for hydrodynamic simulations in modeling the stability of the hot spot-fuel interfaces and the degree of fuel contamination in inertial confinement fusion (ICF) [5,6].

Since direct experimental access such as shock wave experiments is limited in the Mbar regime [7], the states deep in the interior of Jupiter ($\sim$ 45 Mbar) and Saturn ($\sim$ 10 Mbar) [8] can not be duplicated in the laboratory. As a consequence, theoretical modeling provides most of the insight into the internal structure of giant planets. The EOS of H-He mixtures has been treated by a linear mixing (LM) of the individual EOS via fluid perturbation theory [9] and Monte Carlo simulations [10]. Recently, several attempts have been made to calculate the EOS of H-He mixtures by means of quantum molecular dynamics (QMD) simulations. Klepeis *et al.* [11] applied the local density approximation of density functional theory (LDA-DFT) calculations for solid H-He mixtures, implying demixing for Jupiter and Saturn at 15 000 K for a He fraction of $x = N_{\rm He}/(N_{\rm He} + N_{\rm H}) = 0.07$. Vorberger *et al.* [12], Lorenzen *et al.* [8], and Militzer [13] performed QMD simulations by using generalized gradient approximation (GGA) instead of the LDA in order to evaluate the accuracy of the LM approximation and study the demixing of H-He at Mbar pressures. Pfaffenzeller *et al.* [14] introduced Car-Parrinello molecular dynamics (CPMD) simulations to calculate the excess Gibbs free energy of mixing at a lower temperature compared to that set in the work of Lorenzen *et al.* [8] and Militzer [13].

Considering the transport properties, QMD [15] and orbital-free molecular dynamics (OFMD) [16] simulations have been introduced to study hydrogen and its isotropic deuterium (D) and tritium (T). Self-diffusion coefficients in the pure H system and mutual diffusion for D-T mixtures were determined for temperatures T =1 to 10 eV and equivalent H mass densities 0.1 to 8.0 g/cm^3 [17–23]. QMD and OFMD simulations of self-diffusion, mutual diffusion, and viscosity have recently been performed on heavier elements (Fe, Au, Be) [24–26] and on mixtures of Li and H [27].

This work selects the H-He mixture as a representative system and examines some of the standard mixing rules with respect to the EOS and transport properties (viscosity, self-diffusion, and mutual diffusion coefficients) in the warm dense regime that covers standard extreme conditions as reached in the interiors of Jupiter and Saturn. The thermophysical properties of the full mixture and the individual species have been derived from QMD simulations, where the electrons are quantum mechanically treated through finite-temperature (FT) DFT and ions move classically. In the next section, we present the formalism for QMD and for determining the static and transport properties. Then, EOS, viscosity, and diffusion coefficients for H-He mixtures are presented, and the QMD results are compared with the results from reduced models and QMD-based linear mixing models. Finally, concluding remarks are given.

II. FORMALISM

In this section, a brief description of the fundamental formalism employed to investigate H-He mixtures is introduced. The basic quantum mechanical density functional theory forms the basis of our simulations. The implementation of schemes in determining diffusion and viscosity is discussed. Mixing

*Corresponding author: zhang_ping@iapcm.ac.cn

rules that combine pure species quantities to form composite properties are also presented.

A. Quantum molecular dynamics

QMD simulations have been performed for H-He mixtures by using the Vienna *ab initio* simulation package (VASP) [28,29]. In these simulations, the electrons are treated fully quantum mechanically by employing a plane-wave FT-DFT description, where the electronic states follow the Fermi-Dirac distribution. The ions move classically according to the forces from the electron density and the ion-ion repulsion. Simulations have been performed in the NVT (canonical) ensemble where the number of particles N and the volume are fixed. The system was assumed to be in local thermodynamic equilibrium with the electron and ion temperatures being equal ($T_e = T_i$). In these calculations, the electronic temperature has been kept constant according to Fermi-Dirac distribution, and ion temperature is controlled by the Nośe thermostat [30].

At each step during MD simulations, a set of electronic state functions $[\Psi_{i,k}(r,t)]$ for each **k**-point are determined within the Kohn-Sham construction by

$$H_{\rm KS}\Psi_{i,k}(r,t) = \epsilon_{i,k}\Psi_{i,k}(r,t), \quad (1)$$

with

$$H_{\rm KS} = -\frac{1}{2}\nabla^2 + V_{\rm ext} + \int \frac{n(r')}{|r-r'|}dr' + v_{xc}(r), \quad (2)$$

in which the four terms, respectively, represent the kinetic contribution, the electron-ion interaction, the Hartree contribution, and the exchange-correlation term. The electronic density is obtained by

$$n(r) = \sum_{i,k} f_{i,k}|\Psi_{i,k}(r,t)|^2. \quad (3)$$

Then, by applying the velocity Verlet algorithm, based on the force from interactions between ions and electrons, a new set of positions and velocities are obtained for ions.

All simulations are performed with 256 atoms and 128 atoms for pure species of H and He, and as for the case of the H-He mixture, a total number of 245 atoms (234 H atoms and 11 He atoms) for a mixing ratio $x = \frac{2N_{\rm He}}{N_{\rm H}+2N_{\rm He}} = 8.59\%$ (corresponding to the H-He immiscible region determined in the work of Morales *et al.* [31]) has been adopted, where a cubic cell of length L (volume $V = L^3$) is periodically repeated. The simulated densities range from 1.0 to 4.0 g/cm^3 for the pure H system. As for the pure He and H-He mixtures, the size of the supercell is chosen to be the same as that for pure H to secure a constant electron number density (in the range $6.02\times10^{29} \sim 2.41\times10^{30}$ m^{-3}). The temperature from 4000 to 20 000 K has been selected to highlight the conditions in the interiors of Jupiter and Saturn. The convergence of the thermodynamic quantities plays an important role in the accuracy of QMD simulations. In this work, a plane-wave cutoff energy of 1200 eV is employed in all simulations so that the pressure is converged within 2%. We have also checked out the convergence with respect to a systematic enlargement of the **k**-point set in the representation of the Brillouin zone. In the molecular dynamic simulations, only the Γ point of the Brillouin zone is included. The dynamic simulation is lasted 20 000 steps with time steps of $0.2 \sim 0.7$ fs according to different densities and temperatures. For each pressure and temperature, the system is equilibrated within $0.5 \sim 1$ ps. The EOS data are obtained by averaging over the final $1 \sim 3$ ps molecular dynamic simulations.

B. Transport properties

The self-diffusion coefficient D can be calculated either from the trajectory by the mean-square displacement

$$D = \frac{1}{6t}\langle|\boldsymbol{R}_i(t) - \boldsymbol{R}_i(0)|^2\rangle \quad (4)$$

or by the velocity autocorrelation function

$$D = \frac{1}{3}\int_0^\infty \langle \boldsymbol{V}_i(t)\cdot\boldsymbol{V}_i(0)\rangle dt, \quad (5)$$

where R_i is the position and V_i is the velocity of the ith nucleus. Only in the long-time limit are these two formulas of D formally equivalent. We have generated trajectories of sufficient length such that the velocity autocorrelation function becomes zero and contributes no further to the integral, and the mean mean-square displacement away from the origin consistently fits to a straight line. The diffusion coefficient obtained from these two approaches lies within 1% accuracy of each other. Here, we report the results from the velocity autocorrelation function.

We have also computed the mutual-diffusion coefficient

$$D_{\alpha\beta} = \lim_{t\to\infty}\overline{D_{\alpha\beta}}(t) \quad (6)$$

from the autocorrelation function

$$\overline{D_{\alpha\beta}}(t) = \frac{Q}{3Nx_\alpha x_\beta}\int_0^t \langle A(0)A(t')\rangle dt' \quad (7)$$

with

$$A(t) = x_\beta\sum_{i=1}^{N_\alpha}\boldsymbol{V}_i(t) - x_\alpha\sum_{j=1}^{N_\beta}V_j(t), \quad (8)$$

where the concentration and particle number of species α are denoted by x_α and N_α, respectively, and the total number of particles in the simulation box $N = \sum_\alpha N_\alpha$. The quantity Q is the thermodynamic factor related to the second derivation of the Gibbs free energy with respect to concentrations [32]. In the present simulations, the Q value has been adopted equal to unity since studies with Leonard-Jones and other model potentials have shown that for dissimilar constituents the Q factor departs from unity by about 10% [33].

The viscosity

$$\eta = \lim_{t\to\infty}\bar{\eta}(t) \quad (9)$$

has been computed from the autocorrelation function of the off-diagonal component of the stress tensor [34]

$$\bar{\eta}(t) = \frac{V}{k_BT}\int_0^t \langle P_{12}(0)P_{12}(t')\rangle dt'. \quad (10)$$

The results are averaged from the five independent off-diagonal components of the stress tensor P_{xy}, P_{yz}, P_{zx}, $(P_{xx}-P_{yy})/2$, and $(P_{yy}-P_{zz})/2$.

Different from the self-diffusion coefficient, which involves single-particle correlations and attains significant statistical improvement from averaging over the particles, the viscosity depends on the entire system and thus needs very long trajectories so as to gain statistical accuracy. To shorten the length of the trajectory, we use empirical fits [35] to the integrals of the autocorrelation functions. Thus, extrapolation of the fits to $t \to \infty$ can more effectively determine the basic dynamical properties. Both of the D and $\bar{\eta}$ have been fit to the functional in the form of $A[1-\exp(-t/\tau)]$, where A and τ are free parameters. Reasonable approximation to the viscosity can be produced from the finite time fitting procedure, which also serves to damp the long-time fluctuations.

The fractional statistical error in calculating a correlation function C for molecular dynamics trajectories [36] can be given by

$$\frac{\triangle C}{C} = \sqrt{\frac{2\tau}{T_{\mathrm{traj}}}}, \tag{11}$$

where τ is the correlation time of the function and T_{traj} is the length of the trajectory. In this work, we generally fitted over a time interval of $[0, 4\tau - 5\tau]$.

C. Mixing rules

Here, we examine two representative mixing rules, namely, the density-matching rule (MRd) with the inspiration of a two-species ideal gas, and the pressure-matching rule (MRp), which follows from two interacting immiscible fluids.

In the MRd, the volume of the individual species is set equal to that of the mixture ($V_{\mathrm{H}} = V_{\mathrm{He}} = V_{\mathrm{H\text{-}He}}$), and QMD simulations are performed for H at a density of $N_{\mathrm{H}}/V_{\mathrm{H\text{-}He}}$ and He at $N_{\mathrm{He}}/V_{\mathrm{H\text{-}He}}$ at a temperature T. Then, pressure predicted by MRd is determined by simply adding the individual pressures from the pure species H and He simulations. Other transport coefficients, such as mutual diffusion and viscosity, follow the same prescription and are summarized as

$$\begin{aligned} V_{\mathrm{H\text{-}He}} &= V_{\mathrm{H}} = V_{\mathrm{He}}, \\ P^{d}_{\mathrm{H\text{-}He}} &= P_{\mathrm{H}} + P_{\mathrm{He}}, \\ D^{d}_{\mathrm{H\text{-}He}} &= D_{\mathrm{H}} + D_{\mathrm{He}}, \\ \eta^{d}_{\mathrm{H\text{-}He}} &= \eta_{\mathrm{H}} + \eta_{\mathrm{He}}. \end{aligned} \tag{12}$$

The superscript is used to denote values predicted from MRd. The derived pressure based on the density mixing rule generally follows from the ideal noninteracting H and He gas in a volume $V_{\mathrm{H\text{-}He}}$.

The MRp has a more complicated construction compared to MRd. MRp can be characterized by the following prescription:

$$\begin{aligned} V_{\mathrm{H\text{-}He}} &= V_{\mathrm{H}} + V_{\mathrm{He}}, \\ P^{p}_{\mathrm{H\text{-}He}} &= P_{\mathrm{H}} = P_{\mathrm{He}}, \\ D^{p}_{\mathrm{H\text{-}He}} &= \nu_{\mathrm{H}} D_{\mathrm{H}} + \nu_{\mathrm{He}} D_{\mathrm{He}}, \\ \eta^{p}_{\mathrm{H\text{-}He}} &= \nu_{\mathrm{H}} \eta_{\mathrm{H}} + \nu_{\mathrm{He}} \eta_{\mathrm{He}}. \end{aligned} \tag{13}$$

In this case, we have performed a series of QMD simulations on the individual species H and He, where the volumes change under a constraint ($V_{\mathrm{H\text{-}He}} = V_{\mathrm{H}} + V_{\mathrm{He}}$) until the individual pressures equal to each other ($P_{\mathrm{H}} = P_{\mathrm{He}}$). The total pressure becomes the predicted value. Here, we use the excess or electronic pressure P_e to evaluate this MRp mixing rule. Composite properties such as mutual diffusion and viscosity are evaluated by combining the individual species results via volume fractions ($\nu_\alpha = V_\alpha / V_{\mathrm{H\text{-}He}}$).

Finally, we also derive properties of the mixture from a slightly more complex mixing rule [37], as the so-called binary ionic mixture (BIM):

$$\sum_i \nu_i \frac{\gamma_i - \gamma_m}{\gamma_i + \frac{3}{2}\gamma_m} = 0 \tag{14}$$

with γ the predicted mutual-diffusion coefficient or the viscosity. The subscript m denotes the mixture and i the pure species.

III. RESULTS AND DISCUSSION

In this section, the wealth of information derived from QMD calculations is mainly presented through figures, and the general trends of the EOS as well as transport coefficients are concentrated in the text. It is, therefore, interesting to explore not only to get insight into the interior physical properties of giant gas planets, but also to examine a series of mixing rules for hydrogen and helium. Additionally, one can consider the influence of helium on the EOS and transport coefficients of mixing.

A. Equation of state

High precision EOS data of hydrogen and helium are essential for understanding the evolution of Jupiter and target implosion in ICF. Experimentally, the EOS of hydrogen and helium in the fluid regime have been studied through gas gun [38], chemical explosive [39], magnetic driven plate flyer [40], and high power laser [41–43]. Since these experiments were limited by the conservation of mass, momentum, and energy, the explored density of warm dense matter (WDM) was limited within $5 \sim 6$ times of the initial density. Recently, a new technique combining a diamond anvil cell (DAC) and high intensity laser pulse has successfully been proved to provide visible ways to generate shock Huguniot data of hydrogen over a significantly broader density-temperature regime than previous experiments [44]. However, the density therein was still restricted within 1 g/cm^3.

In our simulations, wide range EOS for H, He, and H-He mixtures have been determined according to the QMD method. The EOS can be divided into two parts. That is, contributions from the noninteracting motion of ions (P_i) and the electronic term (P_{el}),

$$P_{\mathrm{total}} = P_i + P_{el}, \tag{15}$$

where P_{el} is calculated directly through DFT. In Fig. 1(a), we have compared our results of P_{el} with that of Holst *et al.* [45], where the electronic pressure is expressed as a smooth function in terms of density and temperature. The results agree with each other with a very slight difference (accuracy within 5%), which can be attributed to the fitting errors. In the simulated density and temperature regime, we do not find any signs for liquid-liquid phase transition ($\frac{\partial P}{\partial T}|_V < 0$) and plasma phase transition ($\frac{\partial P}{\partial V}|_T > 0$), which are characterized by molecular dissociation and electronic

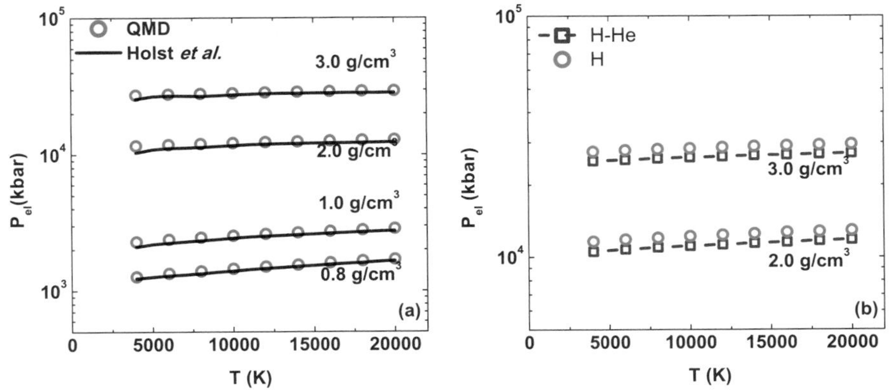

FIG. 1. (Color online) The present simulated QMD EOS have been shown as a function of temperature (with error bar smaller than the symbolic size). (a) Comparison with the results of Holst *et al.* [45] for hydrogen; (b) the variation in EOS after considering the mixing of helium.

ionization, respectively. With considering the mixing of He into H, the electronic pressure is effectively reduced, as has been shown in Fig. 1(b).

MRd accounts for contributions from noninteracting H and He subsystems in the volume of the mixtures. In MRd, the pressure contributed from noninteracting ions is the same as that of the mixture, but the electronic pressure is much lower due to the low electronic density in the pure species simulations. It is indicated that the electronic pressure is underestimated by the MRd model at about 8% $\sim$ 9% (see Fig. 2). For the MRp model, we have first performed a series of pure species simulations at a wider density (temperature) regime compared to H-He mixtures. Then, the simulated EOS data are fitted into smooth functions in terms of density and temperature. Under the constraint of $V_{\text{H-He}} = V_{\text{H}} + V_{\text{He}}$, we have predicted the electronic pressures $P_{\text{H-He}} = P_{\text{H}} = P_{\text{He}}$ according to the MRp model by solving the pure species EOS function at certain densities and temperatures, as shown in Fig. 2. It is indicated that the MRp model agrees better with direct QMD simulations (accuracy within 3%); the difference mainly comes from the ionic interactions between H and He species after mixing.

B. Diffusion and viscosity

QMD simulations have been performed within the framework of FT-DFT to benchmark the dynamic properties of H, He, and H-He mixtures in the WDM regime. Illustrations for the self-diffusion coefficients and viscosity (for H and He at densities of 2.0 and 8.0 g/cm^3, respectively) at a temperature 12 000 K as well as their fits are shown in Fig. 3. The trajectory of the present simulations lasts 4.0 $\sim$ 14.0 ps, and correlation times lie between 1.0 and 15.0 fs. As a consequence, the computational error for the viscosity lies within 10%. After accounting for the fitting error and extrapolation to infinite time, a total uncertainty of $\sim$20% can be estimated. The uncertainty in the self-diffusion coefficients is smaller than 1%, due to the additional $\frac{1}{\sqrt{N}}$ advantage given by particle average.

Dynamic properties of WDM are generally governed by two dimensionless quantities, namely, ionic coupling (Γ) and electronic degenerate (θ) parameters. The former is defined as the ratio of the potential to kinetic energy $\Gamma = \frac{(Ze)^2}{ak_BT}$ with ionic charge Z and ion-sphere radius $a = (3/4\pi n_i)^{1/3}$ (n_i is the number density), while the latter is defined as $\theta = T/T_F$, where T_F is the Fermi temperature. It has been reported that dynamic properties such as diffusion coefficients and viscosity can be represented purely in terms of ionic coupling parameter Γ according to molecular dynamics or Monte Carlo simulations based on the one-component plasma (OCP) model [37,46–50], where ions move classically in a neutralizing background of electrons. For instance, Hansen *et al.* [50] introduced a memory function to analyze the velocity autocorrelation function, and obtained the diffusion coefficient $D = 2.95\Gamma^{-1.34}\omega_p a^2$ with the plasma frequency

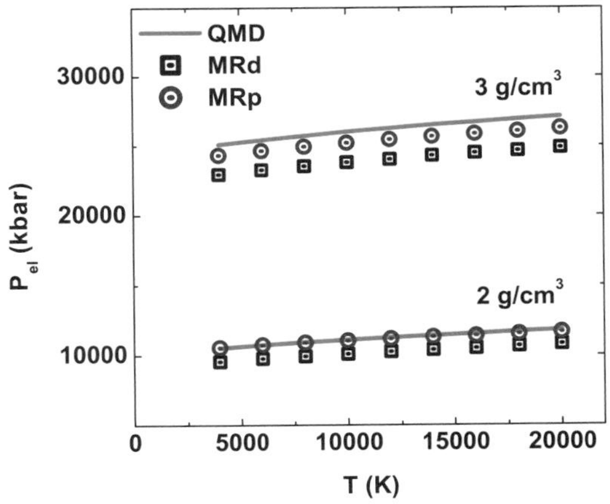

FIG. 2. (Color online) Comparison of electronic pressure between direct QMD simulation results and those from mixing rules (error bar being smaller than symbol size). QMD results are denoted by red solid lines, while the MRd and MRp data are presented by open squares and open circles, respectively.

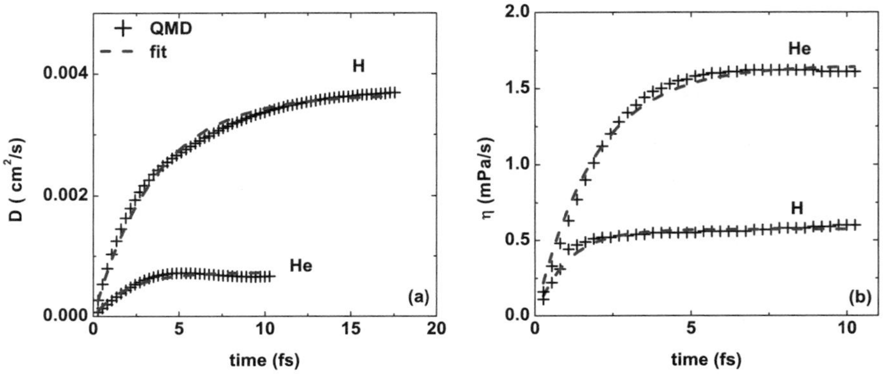

FIG. 3. (Color online) Transport properties are shown for H and He at 12 000 K with densities of 2 and 8 g/cm^3, respectively. (a) Self-diffusion coefficient; (b) viscosity. The direct QMD simulated results are presented by black crosses, while the fitted results are denoted by blue dashed lines.

$\omega_p = (4\pi n_i/M)^{1/2}Ze$. Based on classical molecular dynamics simulations, Bastea [37] has fitted the viscosity into the following form:

$$\eta = (A\Gamma^{-2} + B\Gamma^{-s} + C\Gamma)n_i M\omega_p a^2, \tag{16}$$

with $s = 0.878$, $A = 0.482$, $B = 0.629$, and $C = 0.00188$.

Since the OCP model is restricted to a fully ionized plasma, we use $Z = 1.0$ (or 2.0) for hydrogen (or helium) to compute the self-diffusion coefficient and viscosity. In Fig. 4(a), we show comparison between the QMD and OCP models [50] for hydrogen and helium at densities of 2 and 8 g/cm^3. The general tendency for the self-diffusion coefficient with respect to temperature is similar for the QMD and OCP models, however, the difference up to ~60% is observed between the two results. For the viscosity [Fig. 4(b)], the OCP model [37] predicts smaller (larger) values for hydrogen (helium) compared to QMD simulations. The viscosity is governed by interactions between particles and ionic motions; contribution from the former one decreases with the increase of temperature, while it increases for the latter one. As a consequence, the viscosity may have a local minimum along temperature. For hydrogen, the local minimum locates around 10 000 and 14 000 K indicated by the QMD and OCP models, respectively, while in the case of helium, we do not observe any signs of the local minimum in the simulated regime. The difference of the transport coefficients between QMD and OCP is large. Here, we would like to stress that OCP is a classical model in treating fully ionized plasmas, where ions move in a neutralized background of electrons. As a consequence, OCP is not suitable for describing WDM, where partial ionization and dissociation coexists.

In Fig. 5, we have shown the mutual diffusion coefficient $D_{\text{H-He}}$ and viscosity $\eta_{\text{H-He}}$ for the H-He mixture with an electron number density of 1.204×10^{30} m^{-3}, and results from mixing models are also provided. The transport coefficients predicted by MRd can be directly evaluated through Eq. (12). For the MRp model, we have first fitted the self-diffusion coefficients and viscosity in terms of density and temperature, after determining the volume for each species under the constraint of $P_{\text{H}} = P_{\text{He}}$, the transport coefficients are then obtained. In the BIM model, we have used $\nu_{\text{H}} = 91.41\%$ and $\nu_{\text{He}} = 8.59\%$, then, the transport coefficients are determined

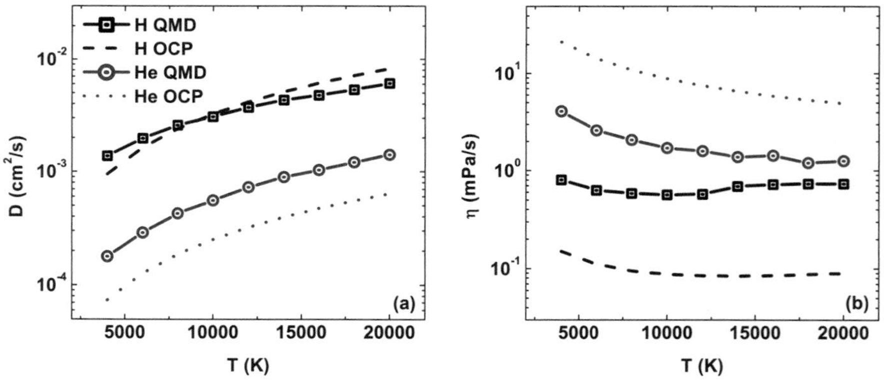

FIG. 4. (Color online) Comparison between QMD simulations and the OCP model (error bars are smaller than symbol size). (a) Self-diffusion coefficient; (b) viscosity. The sampled density for hydrogen is 2 g/cm^3. For helium, the density is 8 g/cm^3. The electron number density for pure species (H and He) is 1.204×10^{30} m^{-3}.

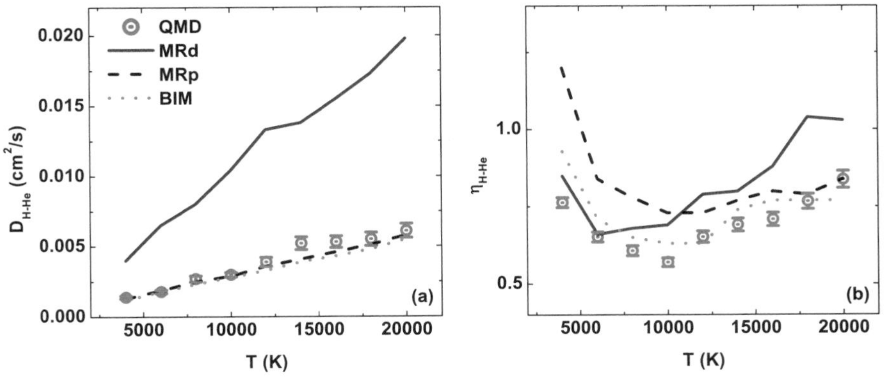

FIG. 5. (Color online) Comparison between QMD simulations and mixing rules. (a) Mutual-diffusion coefficient; (b) viscosity. The sampled mixture has a electron number density of 1.204×10^{30} m^{-3} and temperature from 4000 ~ 20 000 K.

by Eq. (14). Here, we would like to stress that in some mixture studies based on average atom models [51], the properties of pure species are derived from perturbed-atom models, where boundary conditions are introduced from the surrounding medium by treating a single atom within a cell. In this work, the dynamic properties of different mixing rules originate from QMD calculations of the individual species. Despite being divorced of the H-He interactions, the pure species calculations still contain complex intra-atomic interactions based on large samples of atoms.

The mutual diffusion coefficient of the H-He mixture shows a linear increase with temperature, as indicated in Fig. 5(a). The data from MRp and BIM models have a better agreement with QMD simulations compared to that of the MRd model, where ion densities are reduced and results in a larger diffusion coefficient. The viscosity of the H-He mixture has a more complex behavior than pure species under extreme condition. As shown in Fig. 5(b), the MRd is valid at low temperature, while the MRp works at higher temperature. The BIM rule moves the results into better agreement with the H-He mixture, leaving within 30% or better for the simulated conditions.

IV. CONCLUSIONS

In summary, we have performed systematic QMD simulations of H, He, and H-He mixtures in the warm dense regime. This study concentrated on thermophysical properties such as the EOS, diffusion coefficient, and viscosity, which are of crucial interest in astrophysics and ICF. Various mixing rules have been tested against full QMD simulations. We have found that the predicted electronic pressures are in good agreement with the direct QMD simulation results when using the MRp. For the mutual-diffusion coefficients and viscosity, whereas, we have found that compared to MRd and MRp, BIM leads to more improved results for the mixture.

ACKNOWLEDGMENT

This work was supported by NSFC under Grants No. 11275032, No. 11005012, and No. 51071032, by the foundation for Development of Science and Technology of China Academy of Engineering Physics under Grant No. 2013B0102019, by the National Basic Security Research Program of China, and by the National High-Tech ICF Committee of China.

[1] W. B. Hubbard, Science **214**, 145 (1981).
[2] H. F. Wilson and B. Militzer, Astrophys. J. **745**, 54 (2012).
[3] H. F. Wilson and B. Militzer, Phys. Rev. Lett. **108**, 111101 (2012).
[4] D. Bruno, C. Catalfamo, M. Capitelli, G. Colonna, O. De Pascale, P. Diomede, C. Gorse, A. Laricchiuta, S. Longo, D. Giordano, and F. Pirani, Phys. Plasmas **17**, 112315 (2010).
[5] S. Atzeni and J. Meyer-ter-Vehn, *The Physics of Inertial Fusion: Beam Plasma Interaction, Hydrodynamics, Hot Dense Matter*, International Series of Monographs on Physics (Clarendon, Oxford, 2004).
[6] J. D. Lindl, *Inertial Confinement Fusion: The Quest for Ignition and Energy Gain Using Indirect Drive* (Springer, New York, 1998).
[7] M. D. Knudson and M. P. Desjarlais, Phys. Rev. Lett. **103**, 225501 (2009).
[8] W. Lorenzen, B. Holst, and R. Redmer, Phys. Rev. Lett. **102**, 115701 (2009).
[9] D. J. Stevenson, Phys. Rev. B **12**, 3999 (1975).
[10] W. B. Hubbard and H. E. DeWitt, Astrophys. J. **290**, 388 (1985).
[11] J. E. Klepeis, K. J. Schafer, T. W. Barbee III, and M. Ross, Science **254**, 986 (1991).
[12] J. Vorberger, I. Tamblyn, B. Militzer, and S. A. Bonev, Phys. Rev. B **75**, 024206 (2007).
[13] B. Militzer, Phys. Rev. B **87**, 014202 (2013).
[14] O. Pfaffenzeller, D. Hohl, and P. Ballone, Phys. Rev. Lett. **74**, 2599 (1995).
[15] C. Wang, X.-T. He, and P. Zhang, Phys. Rev. Lett. **106**, 145002 (2011).
[16] V. Recoules, F. Lambert, A. Decoster, B. Canaud, and J. Clérouin, Phys. Rev. Lett. **102**, 075002 (2009).

[17] L. A. Collins, J. D. Kress, D. L. Lynch, and N. Troullier, J. Quant. Spectrosc. Radiat. Transfer **51**, 65 (1994).
[18] I. Kwon, J. D. Kress, and L. A. Collins, Phys. Rev. B **50**, 9118 (1994).
[19] L. Collins, I. Kwon, J. Kress, N. Troullier, and D. Lynch, Phys. Rev. E **52**, 6202 (1995).
[20] J. G. Clérouin and S. Bernard, Phys. Rev. E **56**, 3534 (1997).
[21] I. Kwon, L. Collins, J. Kress, and N. Troullier, Europhys. Lett. **29**, 537 (1995).
[22] J. Clérouin and J.-F. Dufrêche, Phys. Rev. E **64**, 066406 (2001).
[23] J. D. Kress, J. S. Cohen, D. A. Horner, F. Lambert, and L. A. Collins, Phys. Rev. E **82**, 036404 (2010).
[24] F. Lambert, J. Clérouin, and S. Mazevet, Europhys. Lett. **75**, 681 (2006).
[25] F. Lambert, J. Clérouin, and G. Zérah, Phys. Rev. E **73**, 016403 (2006).
[26] C. Wang, Y. Long, M. F. Tian, X.-T. He, and P. Zhang, Phys. Rev. E **87**, 043105 (2013).
[27] D. A. Horner, F. Lambert, J. D. Kress, and L. A. Collins, Phys. Rev. B **80**, 024305 (2009).
[28] G. Kresse and J. Hafner, Phys. Rev. B **47**, 558 (1993).
[29] G. Kresse and J. Furthmüller, Phys. Rev. B **54**, 11169 (1996).
[30] P. H. Hünenberger, Adv. Polym. Sci. **173**, 105 (2005); D. J. Evans and B. L. Holian, J. Chem. Phys. **83**, 4069 (1985).
[31] M. A. Morales, C. Pierleoni, E. Schwegler, and D. M. Ceperley, Proc. Natl. Acad. Sci. USA **106**, 1324 (2009).
[32] J. P. Hansen and I. R. McDonald, *The Theory of Simple Liquids*, 2nd ed. (Academic, New York, 1986), Chap. 8.
[33] M. Schoen and C. Hoheisel, Mol. Phys. **52**, 33 (1984); **52**, 1029 (1984).
[34] M. P. Allen and D. J. Tildesley, *Computer Simulation of Liquids* (Oxford University Press, New York, 1987).
[35] J. D. Kress, James S. Cohen, D. P. Kilcrease, D. A. Horner, and L. A. Collins, Phys. Rev. E **83**, 026404 (2011).
[36] R. Zwanzig and N. K. Ailawadi, Phys. Rev. **182**, 280 (1969).
[37] S. Bastea, Phys. Rev. E **71**, 056405 (2005).
[38] W. J. Nellis, Rep. Prog. Phys. **69**, 1479 (2006).
[39] G. V. Boriskov, A. I. Bykov, R. Ilkaev, V. D. Selemir, G. V. Simakov, R. F. Trunin, V. D. Urlin, A. N. Shuikin, and W. J. Nellis, Phys. Rev. B **71**, 092104 (2005).
[40] M. D. Knudson, D. L. Hanson, J. E. Bailey, C. A. Hall, J. R. Asay, and C. Deeney, Phys. Rev. B **69**, 144209 (2004).
[41] G. W. Collins, L. B. Da Silva, P. Celliers, D. M. Gold, M. E. Foord, R. J. Wallace, A. Ng, S. V. Weber, K. S. Budil, and R. Cauble, Science **281**, 1178 (1998).
[42] T. R. Boehly, D. G. Hicks, P. M. Celliers, T. J. B. Collins, R. Earley, J. H. Eggert, D. Jacobs-Perkins, S. J. Moon, E. Vianello, D. D. Meyerhofer, and G. W. Collins, Phys. Plasmas **11**, L49 (2004).
[43] D. G. Hicks, T. R. Boehly, P. M. Celliers, J. H. Eggert, S. J. Moon, D. D. Meyerhofer, and G. W. Collins, Phys. Rev. B **79**, 014112 (2009).
[44] P. Loubeyre, S. Brygoo, J. Eggert, P. M. Celliers, D. K. Spaulding, J. R. Rygg, T. R. Boehly, G. W. Collins, and R. Jeanloz, Phys. Rev. B **86**, 144115 (2012).
[45] B. Holst, R. Redmer, and M. P. Desjarlais, Phys. Rev. B **77**, 184201 (2008).
[46] F. Lambert, Ph.D. thesis, Université Paris XI Commissariatà l' Énergie Atomique, 2007.
[47] B. Bernu and P. Vieillefosse, Phys. Rev. A **18**, 2345 (1978).
[48] Z. Donkó, B. Nyíri, L. Szalai, and S. Holló, Phys. Rev. Lett. **81**, 1622 (1998); Z. Donkó and B. Nyíri, Phys. Plasmas **7**, 45 (2000).
[49] J. Daligault, Phys. Rev. Lett. **96**, 065003 (2006); **103**, 029901(E) (2009).
[50] J. P. Hansen, I. R. McDonald, and E. L. Pollock, Phys. Rev. A **11**, 1025 (1975).
[51] R. M. More, Adv. At. Mol. Phys. **21**, 305 (1985).

PHYSICAL REVIEW E **88**, 013106 (2013)

Transport properties of dense deuterium-tritium plasmas

Cong Wang,[1,2] Yao Long,[1] Xian-Tu He,[1,2] Jun-Feng Wu,[1] Wen-Hua Ye,[1,2] and Ping Zhang[1,2,*]
[1]*Institute of Applied Physics and Computational Mathematics, P. O. Box 8009, Beijing 100088, People's Republic of China*
[2]*Center for Applied Physics and Technology, Peking University, Beijing 100871, People's Republic of China*
(Received 9 February 2012; revised manuscript received 6 June 2013; published 29 July 2013)

Consistent descriptions of the equation of states and information about the transport coefficients of the deuterium-tritium mixture are demonstrated through quantum molecular dynamic (QMD) simulations (up to a density of 600 g/cm^3 and a temperature of 10^4 eV). Diffusion coefficients and viscosity are compared to the one-component plasma model in different regimes from the strong coupled to the kinetic one. Electronic and radiative transport coefficients, which are compared to models currently used in hydrodynamic simulations of inertial confinement fusion, are evaluated up to 800 eV. The Lorentz number is discussed from the highly degenerate to the intermediate region. One-dimensional hydrodynamic simulation results indicate that different temperature and density distributions are observed during the target implosion process by using the Spitzer model and *ab initio* transport coefficients.

DOI: 10.1103/PhysRevE.88.013106 PACS number(s): 52.25.Fi, 51.30.+i, 51.20.+d, 52.65.Yy

I. INTRODUCTION

Inertial confinement fusion (ICF) has been a long-desired goal since the implosion of a capsule of cryogenic deuterium-tritium (DT) covered with a plastic ablator is a visible way to generate a virtually unlimited source of energy [1,2]. In direct-drive ICF, nominally identical Megajoule-class laser beams illuminate the frozen capsule. To trigger the ignition and maximize the thermonuclear energy gain, high compression of the DT fuel and a high temperature of the hot spot should be achieved. In the process of compression, the DT shell accelerates and starts to converge since the laser beams ablate the capsule. The DT shell would experience a wide range of thermodynamical conditions at densities up to several hundreds of a gram per cubic centimeters (g/cm^3) and temperatures from a few to hundreds of electron volts (eV).

Demonstrations of the laser-driven ICF ignition and fusion energy gain in the laboratory are now prepared by the National Ignition Facility (NIF) [3,4]. Understanding and controlling the high-pressure behaviors of DT fuel are of crucial interest for the success of ignition experiments. The equation of states (EOS) of DT fuel are essential for ICF designs because the low adiabat compressibility of the material is dominated by the EOS [5]. During the imploding process, impurities from ablators or hohlraum are inadvertently mixed into the fuel, which strongly influence the burn efficiency and the instability growth near the interface. Thus, the viscosity and diffusion coefficients of the DT mixture are important parameters in hydrodynamic modeling treating interfaces instabilities [6], as well as in further investigating the multiphysical effects of high-Z dopants. Finally, accurate knowledge of the electronic and radiative transport coefficients in dense DT plasmas are also important for precisely predicting all kinds of hydrodynamical instabilities grown at the ablation front, at the fuel-ablator interface, or at the hot-spot–fuel interface, and play central roles in describing the evolution of the central hot spot against the cold fuel [7].

Despite the importance, the thermophysical properties are poorly known, particularly in the hot dense regime. QMD simulations have proven to be powerful tools to study warm dense matter, but are limited in terms of temperatures and densities. For instance, the EOS of hydrogen-helium mixtures were studied at densities between 0.19 and 0.66 g/cm^3 and temperatures from 500 to 8000 K by Vorberger *et al.* [8], then extended to a dense regime ($\rho \sim 6$ g/cm^3) [9]. Morales *et al.* [10] modified the Kerley 2003 EOS through QMD and coupled electron-ion Monte Carlo (CEIMC) simulations, however, the EOS at higher density and temperature still need to be addressed. Recently, the high-density EOS of hydrogen [11] was built by using a two-fluid model combining the quantum Green's function technique for the electrons and a classical description for the ions, and the results agree with QMD calculations in the high density limit. The electronic conductivity was determined for temperatures from 1000 to 50 000 K and equivalent H mass densities of $0.05 \sim 20.0$ g/cm^3 [12,13]. In this work, we focus on the thermophysical properties of the DT mixture at the equivalent H mass density from 80 to 240 g/cm^3 and temperatures up to 10^4 eV for the EOS (800 eV for the electronic transport coefficients).

II. COMPUTATIONAL METHOD

The thermophysical properties of the DT mixture are determined by combinations of first-principles molecular dynamics (FPMD) and orbital-free molecular dynamics (OFMD) [14] using the ABINIT package [15]. In FPMD, electrons are quantum mechanically treated through plane-wave finite temperature density functional theory (FT-DFT) after the introduction of the Born-Oppenheimer approximation. As for OFMD simulations, the orbital-free functional is derived from the semiclassical development of the Mermin functional [16], which leads to the finite-temperature Thomas-Fermi expression for the kinetic part.

Considering the DT mixture at high densities, the Coulombic pseudopotential with short cutoff radius ($r_{cut} = 0.001$ a.u.) has been built to avoid the overlap between pseudocores [17].

*Corresponding author: zhang_ping@iapcm.ac.cn

As a consequence, the large cutoff energy $E_{\mathrm{cut}} = 200$ Hartree are used. In this paper, we report QMD calculations for DT mixtures along the 200 to 600 g/cm^3 isochore with Γ point to represent the Brillouin zone and 128 particles in a cubic cell, where FPMD and OFMD are performed up to 100 and 10 000 eV, respectively. We have checked the simulations with respect to a systematic enlargement of the cutoff energy, number of particles, as well as the **k**-point set, and the convergence is better than 5%. Dynamic simulations are carried out from 8000 to 200 000 steps in the isokinetic ensemble [18]. Due to the computational limits of the finite temperature density-functional theory approach, the electronic structure is calculated up to 800 eV, and a $4 \times 4 \times 4$ Monkhorst-Pack k-point mesh is used. In characterizing the plasma states, the ionic coupling and degeneracy parameters have been used. The definition of the first one is $\Gamma_{ii} = Z^{*2}/(k_B T a)$ (the ratio between the electrostatic potential and the kinetic energy), and the later one is $\theta = T/T_F$ (the ratio of temperature to Fermi temperature). The presently studied plasmas goes from strongly coupled ($\Gamma_{ii} \sim 60$) and degenerated states ($\theta \sim 0.01$) to the kinetic states ($\Gamma_{ii} \sim 0.01$ and $\theta \sim 20$).

III. RESULTS AND DISCUSSION

A. Equation of states

Concerning the EOS, the difference between FPMD and OFMD lies within a 2% accuracy below 20 eV, however, as temperature increases, the difference becomes negligible. We have constructed the EOS as polynomial expansions (error within 2%) of the density (200–600 g/cm^3) and temperature (1–10^4 eV) as follows:

$$P = \sum A_{ij} \rho^i T^j, \tag{1}$$

which can be conveniently used in hydrodynamic simulations in ICF or astrophysics (Table I). The present fitted results are only valid at the density-temperature region mentioned above.

Figure 1 shows the calculated EOS as a function of temperature at sampled densities. QMD results are compared to both the ideal and Debye-Hückel models [19]. The pressure of the ideal model (P_{id}) can be viewed as the combination of noninteracting classical ions and fermionic electrons. The self-consistent solution of the Poisson equation for a screened charges plasmas leads to the well-known Debye-Hückel model, where pressure can be explicitly expressed as $P_D = P_{id} - \frac{k_B T}{24\pi\lambda_D^3}$. Here λ_D is the Debye length. At a density of 249.45 g/cm^3, the QMD pressures are 9–40% smaller than those given by the ideal model, as temperature lies below 100.0 eV. However, the Debye-Hückel model underestimates the QMD results by 60% at the lowest temperature (2.0 eV) considered. For very high temperatures, beyond T_F, QMD

TABLE I. Pressure (kbar) expansion coefficients A_{ij} for Eq. (1) in terms of density (g/cm^3) and temperature (eV).

i	A_{i0}	A_{i1}	A_{i2}
0	671.1939	816.8465	−0.1286
1	12645.2534	753.1419	0.0020
2	151.1943	−0.0401	0.0000

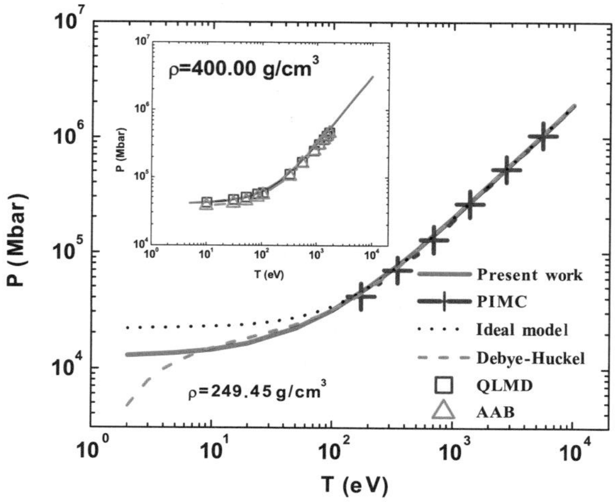

FIG. 1. (Color online) EOS as a function of temperature at 249.45 g/cm^3 (inset is the EOS for $\rho = 400$ g/cm^3). The present QMD results are plotted as red solid lines. Previous results, including the ideal model (black dotted line) and Debye-Hückel model (green dashed line), as well as PIMC (blue cross) [20], QLMD (blue open square) [21], and AAB (green triangle) [22] numerical simulations, are also shown for comparison.

simulation results and the two classical methods merge together into the noninteracting classical ions and electrons (with a difference smaller than 1%). It is clear that the oversimplified ideal or Debye-Hückel model fails to describe plasmas at high density and low temperature. Furthermore, good agreement in EOS is achieved between the present results and other simulation methods, such as the path integral Monte Carlo (PIMC), quantum Langevin molecular dynamics (QLMD), and average atom model with energy-level broadening (AAB) [20–22].

B. Diffusion coefficient and viscosity

Previous QMD simulations in determining the self-diffusion coefficients and viscosity were limited at low densities (equivalent hydrogen mass density up to 8 g/cm^3) and temperatures (10 eV) [23–26]. The present work substantially extends the study on the transport coefficients into a hot dense regime, and comprehensive comparisons with the one-component plasma (OCP) model are described. Here, the self-diffusion coefficient (D) is derived via particle velocity according to the Green-Kubo relation. The viscosity (η) is computed from the autocorrelation function of the five off-diagonal components of the stress tensor: P_{xy}, P_{yz}, P_{zx}, $(P_{xx} - P_{yy})/2$, and $(P_{yy} - P_{zz})/2$. To reduce the length of the trajectory, empirical functions $C[1 - \exp(-t/\tau)]$, where C and τ are free parameters, have been used to fit the integrals of the autocorrelation functions [26]. The fitting procedure is effective in damping the variation and produces reasonable approximations to η (the computed error lies within 10%). Due to the fitting procedure and extrapolation to infinite time, the total error of 20% has been estimated. The uncertainty in the self-diffusion coefficient, where an additional $1/\sqrt{N}$ advantage is secured by particle averages, lies within 2%.

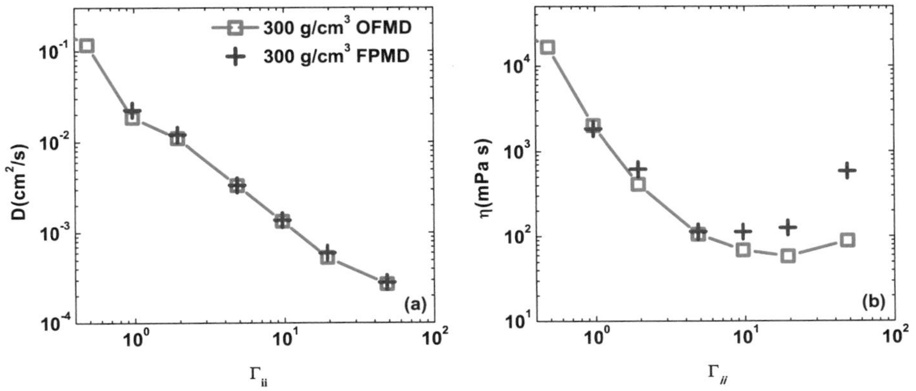

FIG. 2. (Color online) Transport coefficients at a density of 300 g/cm^3 are shown along with coupling parameter $\Gamma_{ii} = \frac{(Ze)^2}{ak_BT}$ (Ze is the ion charge and a the ion-sphere radius). (a) The diffusion coefficients. (b) The viscosity. Results from FPMD and OFMD are denoted by blue cross and red open square, respectively.

The self-diffusion coefficient and viscosity for the DT mixture calculated by FPMD and OFMD as a function of coupling parameter Γ_{ii} at a sampled density of 300 g/cm^3 are shown in Figs. 2(a) and 2(b), respectively. As indicated in the figures, FPMD and OFMD results for self-diffusion coefficients are in generally good agreement (error within 5% or better). As for the viscosity, the divergence between FPMD and OFMD could reach up to ~70% at low temperatures, however, as temperature increases, they tend to merge together. To compare these results with OCP simulations conveniently, the self-diffusion coefficient D and viscosity η are reduced to a dimensionless form: $D^* = D/\omega_p a^2$ and $\eta^* = \eta/n_i M\omega_p a^2$, where $\omega_p = (4\pi n_i/M)^{1/2}Ze$ is the plasma frequency for ions of mass ($M = 2.5$ amu.). A memory-function analysis of the velocity autocorrelation function has been used to obtain the diffusion coefficient of OCP ($D^* = 2.95\Gamma^{-1.34}$), however, this result is not accurate at $\Gamma \leqslant 4$ [27]. Recently, a more accurate fit (valid at $0.5 \leqslant \Gamma \leqslant 200$) to the OCP simulations has been provided by Daligault [28]. Yukawa models have also been applied to simulate the transport properties for particles moving through the screened Coulomb potential in the strongly coupled region ($\Gamma > 1$) [29,30]. In Fig. 3(a), we compare OFMD data with those OCP and Yukawa (with $\kappa = 0.1$) [31] simulations. The general feature of our simulation result agrees with that of Hansen *et al.* [27] and Ohta *et al.* [29], but visible divergence (3–50%) is still observed. The accordance of the Daligault's fit with the present results is better at low temperatures. With increasing temperature, the Daligault's OCP fitting overestimates the OFMD results by ~40% at $\Gamma_{ii} = 0.5$. The present results for the reduced viscosity are shown in Fig. 3(b), where the results from OCP and Yukawa simulations are also provided for comparison. By performing classical molecular dynamics of OCP, Bastea [32] fit η^* to the form $\eta^* = 0.482\Gamma_{ii}^{-2} + 0.629\Gamma_{ii}^{-0.878} + 0.00188\Gamma_{ii}$. The

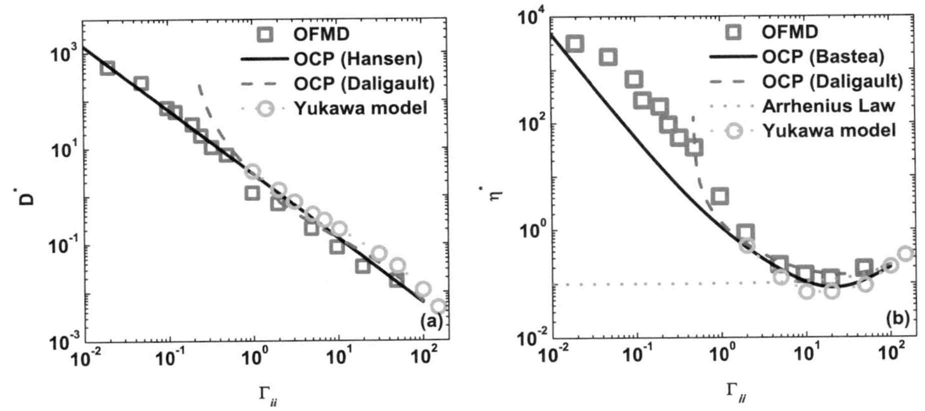

FIG. 3. (Color online) The reduced transport coefficients are shown along with coupling parameter Γ_{ii}. (a) The reduced diffusion coefficients: Present OFMD simulation results at a density of 300 g/cm^3 are plotted as red open squares. OCP results obtained by Hansen *et al.* [27] and Daligault [28] are denoted by black solid line and magenta dashed line, respectively. The Yukawa [31] diffusion coefficients are shown as cyan open circles. (b) The reduced viscosity, where results from OCP (Bastea [32]) and Arrhenius law are plotted as black solid line and orange dotted line, respectively. Other symbols in (b) have the same meaning as those in (a).

present simulation results for viscosity indicate a similar behavior between dense plasma and normal fluid, that is, both of the movement of particles and the action of interparticle forces contribute to the transport of momentum. At large coupling, the viscosity increases with decreasing temperature (or adding density), where an Arrhenius-type relation ($\eta^* = 0.1 \times e^{0.008\Gamma_{ii}}$) is observed. At the intermediate region, $10 \leqslant \Gamma_{ii} \leqslant 50$, contributions from the two mechanisms vary with similar magnitude, leading to a shallow minimum. OCP simulations indicate that the minimum lies at around $\Gamma_{ii} = 25$ [28], while our FPMD and OFMD results suggest the minimum near $\Gamma_{ii} = 11$ and $\Gamma_{ii} = 15$, which are accordant with Yukawa simulations [30]. When $\Gamma_{ii} < 10$, as in a gas, the movement of particles is predominant and the viscosity increases with temperature. Classical models such as OCP and Arrhenius law are not expected to work well for hot dense plasmas.

C. Electronic and radiative transport coefficients

Apart from the EOS and ionic transport coefficients, the electronic transport coefficients are derived using QMD in the following steps. Ten snapshots are directly taken from FPMD equilibrium trajectories up to 100 eV, however, beyond this value, the structures are extracted from OFMD simulations. Then the dynamical conductivity $\sigma(\omega) = \sigma_1(\omega) + i\sigma_2(\omega)$ is evaluated through the Kubo-Greenwood formula as averages of the selected configurations. The dc conductivity ($\sigma_{\rm dc}$) follows from the static limit $\omega \to 0$ of $\sigma_1(\omega)$. In the Chester-Thellung version [33], the kinetic coefficients $\mathcal{L}_{ij}$ are described as

$$\mathcal{L}_{ij} = (-1)^{i+j} \int d\epsilon \hat{\sigma}(\epsilon)(\epsilon - \mu)^{(i+j-2)} \left(-\frac{\partial f(\epsilon)}{\partial \epsilon} \right), \quad (2)$$

with $f(\epsilon)$ being the Fermi-Dirac distribution function and μ the chemical potential. The electronic thermal conductivity K_e is given by

$$K_e = \frac{1}{T}\left(\mathcal{L}_{22} - \frac{\mathcal{L}_{12}^2}{\mathcal{L}_{11}} \right). \quad (3)$$

The electronic heat conduction as a function of temperature along the isochore of 200 g/cm^3 up to 800 eV is shown in Fig. 4(a). The results from theoretical models, which are currently used in hydrodynamics simulations for ICF, as well as previous *ab initio* simulations, are provided for comparison. The Hubbard model [34] is valid for a highly degenerate electron system, where the electronic states are treated by using independent plane waves. The interactions between nuclei are not screened by electrons, and atomic configurations are assumed to be Debye- or OCP-like. Unfortunately, thermal conditions in ICF often enter the partially degenerate regime, which clearly goes beyond the hypothesis of the Hubbard model. The Lee-More model [35], which is not quite accurate in the degenerate state, uses different formulas for the electron collision time in a solid, liquid, and plasma. The Spitzer model [36], which exhibits a power law on the temperature ($T^{5/2}$), has been dedicated to kinetic plasmas, and is not suitable for the dense plasmas considered in this work. At high temperatures, our QMD results and Lee-More model merged together into the Spitzer conductivity. From the quantum Boltzmann equation coupled with Ziman theory, Ichimaru [37] derived hydrogen transport coefficients restricted to the moderate coupling $\Gamma < 2$. From Fig. 4, clearly, the Ichimaru model results overestimate the present data (up to ~30%) at the strong coupled regime, and then the difference reduces to 10%. Remarkably, our QMD data of the electronic thermal conductance for DT mixtures show reasonable consistency with the recently developed self-consistent average-atom model (SCAALP) [38] and previous QMD simulations for hydrogen [7,39] in a wide range.

The results for $\sigma_{\rm dc}$ are presented and compared to those obtained by classical models or *ab initio* simulations in Fig. 4(b). It has been demonstrated that the Hubbard model recovered the Wiedemann-Franz low in the degenerate limit, and thus the electrical conductance are computed from the thermal conductance. For the Spitzer model, the electrical conductivity is obtained by using the Lorenz number at the lower limit of kinetic matter including *e-e* collisions. Our QMD results agree well with the Ichimaru model [37] and data by Lambert *et al.* [14] in a wide region between the degenerate state and the kinetic state.

The thermal conductance and the dynamic conductivities are calculated so as to examine the Lorenz number defined as

$$L = \frac{K_e}{\sigma_{\rm dc} T} = \gamma \frac{k_B^2}{e^2}. \quad (4)$$

The nature of the screened potential, which is responsible for the scattering of the electrons, determines γ. L is constant in the degenerate and coupled state (γ is equal to $\pi^2/3$), reducing to the ideal Sommerfeld number. In the nondegenerate case, γ reaches 1.5966 considering *e-e* collisions. In the intermediate degenerate region, it is difficult to deduce the electronic thermal conductivity from electrical conductance by Wiedemann-Franz law due to the fact that there exists no assumption on the γ value. In Fig. 5(a) we show the γ value as a function of the degeneracy parameter θ for different densities. The simulation results indicate that the Lorenz number oscillates around the Sommerfeld limit in the degenerate region. As θ increases, the departure of the Lorenz ratio from the ideal value towards the lower limit is observed.

Another useful coefficient, which plays an important role in the energy transport formulation of ICF and astrophysics, is the radiative thermal conductivity (K_r). As usual, it is connected to the Rosseland mean path length ($l_R = \frac{1}{\rho\kappa_R}$ with κ_R the Rosseland mean opacity, RMO) by the following relation [1]:

$$K_r = \tfrac{16}{3}\sigma_B l_R(\rho,T)T^3, \quad (5)$$

where σ_B is the Stefan-Boltzmann constant. In particular, the RMO can be in the form

$$\frac{1}{\kappa_R} = \int_0^\infty \frac{B'(\omega)}{\alpha(\omega)} d\omega, \quad (6)$$

where $B'(\omega)$ is the derivative with respect to the temperature of the normalized Plank's function. $\alpha(\omega) = \frac{4\pi}{n(\omega)}\sigma_1(\omega)$ is the absorption coefficient with $n(\omega)$ the real part of the index of refraction. A comparison for K_r is made in Fig. 5(b) between our QMD simulations and those given by fully ionized plasma and the Dyson limit. Zeldovich and Raizer have proven that [40] the RMO of fully ionized plasma can be obtained as $\kappa_R^{\rm ideal} = 0.014(Z^3/A^2)\rho(k_B T)^{-7/2}$ cm^2/g. In this relation, Z is the atomic number, A is the atomic

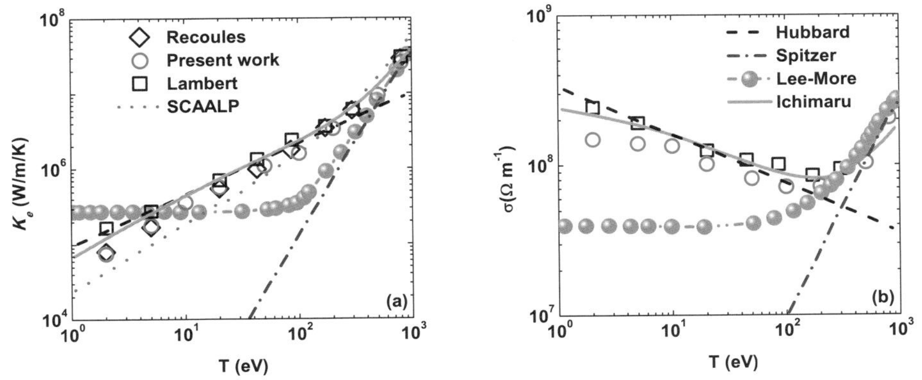

FIG. 4. (Color online) (a) Electronic thermal conductivity of the present work (red open circle) has been compared with previous QMD simulations from Recoules *et al.* (black open diamond) and Lambert *et al.* (black open square) [7,39], and the results from classical models are also shown, such as Hubbard (black dashed line) [34], Spitzer (blue dash dotted line) [36], Lee-More (green solid circle) [35], Ichimaru model (orange solid line) [37], and SCAALP (magenta dotted line) [38]. (b) Electrical conductivity, where the present results are shown with red open circle [other symbols and lines are the same with those in (a)].

weight, and k_BT is expressed in kilo-eV. The upper limit of RMO [$\kappa_R \leqslant 6.0 \times 10^3 Z/(Ak_BT)$ cm^2/g] has been reported by Bernstein and Dyson via applying Schwartz's inequality to a particular factoring of the integrand in the definition of RMO [41]. Our QMD simulation results indicate that K_e governs the energy transport process in the fully degenerate state. As the matter enters the moderate degenerate case, K_r rises about four orders of magnitude as the temperature increases from $\sim 10^2$ eV to $\sim 10^3$ eV, and becomes dominant when θ exceeds ~ 0.5.

D. Applications to ICF

Since the behavior of the electronic thermal conductivity from *ab initio* simulations is quite different from those models, such as the Hubbard model and Spitzer model, as a consequence, one fitted formula for K_e in terms of ρ (in g/cm^3) and T (eV) has been built and applied to investigating ICF target implosions:

$$K_e(\rho,T) = (P_1 + P_2\rho)T^{5/2} + \left(P_3 + P_4\Gamma + P_5 \ln(T)\frac{1}{1+\exp(\theta)}\right)\rho T, \quad (7)$$

with $P_1 = 0.5135$, $P_2 = 0.0006$, $P_3 = 522.4466$, $P_4 = 8.1557$, and $P_5 = -23.5396$. The current K_e model considers a Hubbard-like behavior at the electronic degenerate region and merges to the Spitzer model at an extremely high temperature. In the partially degenerate region, ionic coupling and electronic degenerate parameters are also introduced in the formula to present the complex behavior of the nonideal plasmas.

To study the influence introduced by K_e in ICF implosions simplified one-dimensional single temperature hydrodynamic simulations are performed, where the results are compared by using K_e from QMD simulations and the Spitzer model. Using radiation-hydrodynamics codes (one-dimensional

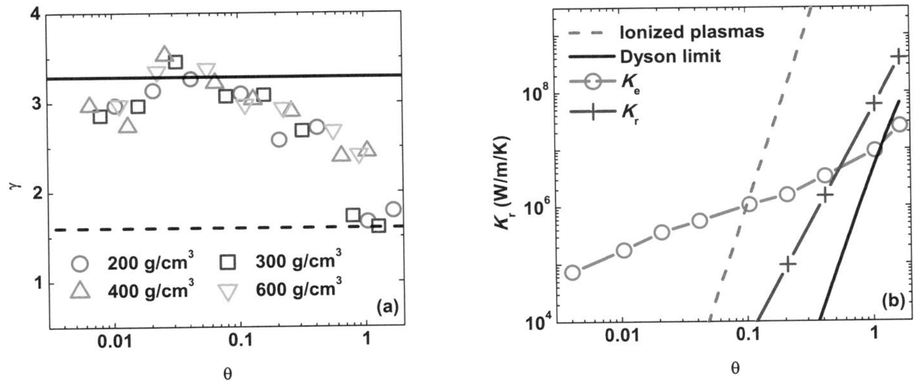

FIG. 5. (Color online) (a) At densities of 200 (red open circle), 300 (blue open square), 400 (green up triangle), and 600 (cyan down triangle) g/cm^3, Lorentz numbers (γ) have been shown as a function of the degenerate parameter θ. (b) Radiation thermal conductivity (K_r) along θ: QMD results are denoted by blue cross; results from Dyson limit and fully ionized plasmas are plotted as black solid line and magenta dashed line, respectively; The electronic thermal conductivity is presented with red open circle.

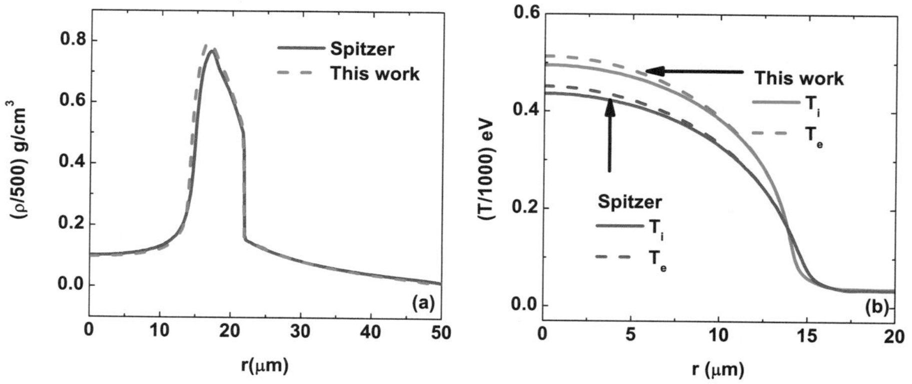

FIG. 6. (Color online) Results for the (a) density and (b) temperature distribution along the radius of the DT capsule [comparison between the Spitzer model and the present *ab initio* model Eq. (7)].

LARED-S [42]) for simulations of target implosions, we have explored the implications of the *ab initio* electronic transport coefficient for designing and understanding the ICF implosions. In the present hydrodynamic simulations, the standard flux-limited ($f = 0.06$) thermal transport model is used and a direct-drive implosion is considered, where the DT capsule is exposed to be illuminated through the laser pulse shapes consisting of a prepulse (with the power of 0.75 TW for 1.6 ns) and the main pulse (with the power of 33.0 TW for 0.75 ns). The cryogenic DT target (360-μm radius) has a 10-μm ablator (hydrocarbon) and 100-μm DT ice (with a DT gas of 0.3 mg/cm^3 inside). By applying the Spitzer model and present fitted model for K_e, the density (normalized by 500 g/cm^3) and temperature (normalized by 1000 eV) profiles at the stagnation stage (~31 ns after the main pulse) are shown in Fig. 6. Prominently, it is indicated that the density of the DT fuel is more compressed (by about 3%), and the temperature in the hot-spot region is even more enhanced (by about 13%) with our present model [Eq. (7)]. This is physically reasonable by the fact shown in Fig. 3(a) that the classical Spitzer model fundamentally underestimates the electronic thermal conductivity in the partially degenerate regime, which turns out to substantially influence different kinds of heat transport during the implosion, as shown in Fig. 6.

IV. CONCLUSION

In summary, we have determined the EOS and transport coefficients of DT plasmas within QMD simulations in the hot dense regime as reached in future ICF experiments. The wide range EOS has been built from the coupled to the kinetic regime to describe pressures up to 10^6 Mbar. A clear chain of simulations in computing the thermophysical properties of hot dense plasma has been demonstrated. The ionic diffusion coefficient and viscosity have been simulated and compared to various OCP model simulations. The electronic and radiative thermal transport coefficients have also been determined, clearly showing different weights when crossing different pressure-temperature regimes. The ability to simulate these parameters in a self-consistent way shown by our results opens aa alternative way to validate the classical theoretical models currently used in hydrodynamical simulations for ICF.

ACKNOWLEDGMENTS

This work was supported by the NSFC under Grants No. 11275032, No. 11005012, and No. 51071032; by the National Basic Security Research Program of China; and by the National High-Tech ICF Committee of China.

[1] S. Atzeni and J. Meyer-ter-Vehn, *The Physics of Inertial Fusion: Beam Plasma Interaction, Hydrodynamics, Hot Dense Matter*, International Series of Monographs on Physics (Clarendon, Oxford, 2004).
[2] J. D. Lindl, *Inertial Confinement Fusion: The Quest for Ignition and Energy Gain Using Indirect Drive* (Springer-Verlag, New York, 1998).
[3] S. H. Glenzer, B. J. MacGowan, P. Michel, N. B. Meezan *et al.*, Science **327**, 1228 (2010).
[4] S. H. Glenzer, B. J. MacGowan, N. B. Meezan, P. A. Adams *et al.*, Phys. Rev. Lett. **106**, 085004 (2011).
[5] S. X. Hu, B. Militzer, V. N. Goncharov, and S. Skupsky, Phys. Rev. Lett. **104**, 235003 (2010).
[6] J. L. Milovich, P. Amendt, M. Marinak, and H. Robey, Phys. Plasmas **11**, 1552 (2004).
[7] V. Recoules, F. Lambert, A. Decoster, B. Canaud, and J. Clérouin, Phys. Rev. Lett. **102**, 075002 (2009).
[8] J. Vorberger, I. Tamblyn, B. Militzer, and S. A. Bonev, Phys. Rev. B **75**, 024206 (2007).
[9] W. Lorenzen, B. Holst, and R. Redmer, Phys. Rev. Lett. **102**, 115701 (2009).
[10] M. A. Morales, L. X. Benedict, D. S. Clark, Eric Schwegler *et al.*, High Energy Density Phys. **8**, 5 (2012).
[11] J. Vorberger, D. O. Gericke, and W.-D. Kraeft, High Energy Density Phys. **9**, 448 (2013).

[12] B. Holst, R. Redmer, and M. P. Desjarlais, Phys. Rev. B **77**, 184201 (2008).
[13] B. Holst, M. French, and R. Redmer, Phys. Rev. B **83**, 235120 (2011).
[14] F. Lambert, J. Clerouin, and G. Zérah, Phys. Rev. E **73**, 016403 (2006).
[15] For details of ABINIT code, please refer to http://www.abinit.org.
[16] M. Brack and R. K. Bhaduri, *Semiclassical Physics* (Westview, Boulder, CO, 2003).
[17] C. Wang, X. T. He, and P. Zhang, Phys. Rev. Lett. **106**, 145002 (2011).
[18] The time steps have been taken as $\Delta t = a/20\sqrt{k_B T/m_i}$, where m_i is the ionic mass and a the mean ionic sphere radius written as $a = (3/4\pi n_i)^{1/3}$ with n_i the ionic number density.
[19] P. Debye and E. Hückel, Z. Phys. **24**, 185 (1923).
[20] S. X. Hu, B. Militzer, V. N. Goncharov, and S. Skupsky, Phys. Rev. B **84**, 224109 (2011).
[21] J.-Y. Dai, Y. Hou, and J.-M. Yuan, Astrophys. J. **721**, 1158 (2010).
[22] Y. Hou, F. Jin, and J.-M. Yuan, Phys. Plasmas **13**, 093301 (2006).
[23] I. Kwon, J. D. Kress, and L. A. Collins, Phys. Rev. B **50**, 9118 (1994).
[24] L. Collins, I. Kwon, J. Kress, N. Troullier, and D. Lynch, Phys. Rev. E **52**, 6202 (1995).
[25] J. G. Clérouin and S. Bernard, Phys. Rev. E **56**, 3534 (1997).
[26] J. D. Kress, J. S. Cohen, D. A. Horner, F. Lambert, and L. A. Collins, Phys. Rev. E **82**, 036404 (2010).
[27] J. P. Hansen, I. R. McDonald, and E. L. Pollock, Phys. Rev. A **11**, 1025 (1975).
[28] J. Daligault, Phys. Rev. Lett. **96**, 065003 (2006).
[29] H. Ohta and S. Hamaguchi, Phys. Plasmas **7**, 4506 (2000).
[30] T. Saigo and S. Hamaguchi, Phys. Plasmas **9**, 1210 (2002).
[31] $\kappa = k_D a$ is a dimensionless parameter with k_D^{-1} the screening length and a the interparticle distance.
[32] S. Bastea, Phys. Rev. E **71**, 056405 (2005).
[33] G. V. Chester and A. Thellung, Proc. Phys. Soc. London **77**, 1005 (1961).
[34] W. B. Hubbard, Astrophys. J. **146**, 858 (1966).
[35] Y. T. Lee and R. M. More, Phys. Fluids **27**, 1273 (1984).
[36] L. Spitzer and R. Härm, Phys. Rev. **89**, 977 (1953).
[37] S. Ichimaru and S. Tanaka, Phys. Rev. A **32**, 1790 (1985).
[38] G. Faussurier, C. Blancard, P. Cossé, and P. Renaudin, Phys. Plasmas **17**, 052707 (2010).
[39] F. Lambert, V. Recoules, A. Decoster, J. Clérouin, and M. Desjarlais, Phys. Plasmas **18**, 056306 (2011).
[40] Y. B. Zeldovich and Y. P. Raizer, *Physics of Shock Waves and High Temperature Hydrodynamic Phenomena* (Academic, New York, 1998).
[41] B. H. Armstrong, Astrophys. J. **136**, 309 (1962).
[42] W. Pei, Commun. Comput. Phys. **2**, 255 (2007).

PHYSICS OF PLASMAS 22, 062708 (2015)

Applications of deuterium-tritium equation of state based on density functional theory in inertial confinement fusion

Cong Wang,[1,2] Zheng-Feng Fan,[1] Xian-Tu He,[1,2] Wen-Hua Ye,[1,2] and Ping Zhang[1,2,a)]
[1]Institute of Applied Physics and Computational Mathematics, P.O. Box 8009, Beijing 100088, People's Republic of China
[2]Center for Applied Physics and Technology, Peking University, Beijing 100871, People's Republic of China

(Received 26 February 2015; accepted 4 June 2015; published online 24 June 2015)

An accurate equation of state for deuterium-tritium mixture is of crucial importance in inertial confinement fusion. The equation of state can determine the compressibility of the imploding target and the energy deposited into the fusion fuel. In the present work, a new deuterium-tritium equation of state, which is calculated according to quantum molecular dynamic and orbital free molecular dynamic simulations, has been used to study the target implosion hydrodynamics. The results indicate that the peak density predicted by the new equation of state is ~10% higher than the quotidian equation of state data. During the implosion, the areal density and neutron yield are also discussed. © *2015 AIP Publishing LLC*. [http://dx.doi.org/10.1063/1.4922900]

I. INTRODUCTION

Inertial confinement fusion (ICF), which is considered as one of the potential ways to generate ultimate clean energy, has been pursued over decades.[1–4] In ICF design, a cryogenic deuterium-tritium (DT) capsule can be driven to implosion either directly by the laser beams or indirectly by **X**-rays created in a hohlraum,[5–7] and the DT fuel should be compressed into high density and high temperature[8] to achieve ignition. In order to reduce the driving energy required for ignition, the capsule temperature should be kept as low as possible, and this can be attained by multiple shock waves or ramp compression waves created by specially shaped laser pulses.[9–11] During the implosion, two parameters, labeled as the first shock pressure and the peak implosion velocity, are important in characterizing the target performance. The former one (α), which is defined as the ratio between the shock induced-pressure and Fermi pressure $\alpha = P/P_F$, sets the adiabat parameter of the compression path in the low foot or the high foot ICF designs.[3,8,11,12] The latter one is proportional to the energy deposited into the DT fuel.

Accurate knowledge about the thermodynamic property of dense plasma is essential in ICF design.[13–16] Among the numerous parameters that condition the success of future ICF experiments, the equation of state (EOS) of the capsule material under high-energy-density regime plays a central role.[17,25] For instance, the EOS of the target material determines the compressibility of the capsule and the shock adiabat α, and is essential for designing ICF ignition target and predicting fusion energy production. Exactly because of this importance, the EOS of ICF-relevant material has been extensively investigated in recent years.[18–25] In order to perform ICF hydrodynamic simulations, the pressure and energy of target material in a wide density and temperature range should be provided in terms of analytical formula or table. Both of SESAME and quotidian EOS (QEOS) are widely used EOS tables[26,27] in recent ICF hydrodynamic simulations. The former one introduces a chemical model,[28–31] which describes physical property of hydrogen in terms of well-defined chemical species (molecules, atoms, electrons, and protons), and is expected to be valid in the weak coupling regime. In QEOS, the electronic contribution to EOS is given by Thomas-Fermi (TF) model, and Cowan model is used to determine the ionic contribution. It is clear that the TF theory in terms of the spherical-cell model for hot dense plasma used in QEOS ignores the bonding effect, the electronic orbit, and the exchange correlation effect. As a consequence, the EOS of ICF material in strong coupling and degenerate state, where many-body effect is remarkable, cannot be well described in terms of the spherical-cell model, and more accurate methods are highly recommended to be used. Recently, quantum molecular dynamics (QMD) method, where the electrons are fully quantum mechanically treated by finite-temperature density functional theory (FT-DFT) and ions are treated classically, has already been proved to be a powerful tool to study the thermophysical properties of matters under extreme condition.[15] Other theoretical methods such as path-integral Monte Carlo (PIMC) and coupled electron-ion Monte Carlo (CEIMC) have also been employed to study ICF materials.[32,33] Although the DT EOS has been pursued over decades, there are still some fundamental issues should be addressed. For instance, the first order liquid-liquid phase transition and plasma phase transition in hydrogen and its isotopes are still under intense discussions. As a result, a wide range DT EOS table is highly recommended to be constructed in a more accurate and self consistent manner.

In this work, a DT DFT-EOS table,[15] which is obtained from QMD and orbital-free molecular dynamics (OFMD)[34–37] simulations, has been used to study ICF target implosion. In QMD, the electrons are treated quantum mechanically through FT-DFT with the only approximation of the exchange-correlation functional. Due to a large number of partially occupied electronic states at extremely high

[a)]Author to whom correspondence should be addressed. Electronic mail: zhang_ping@iapcm.ac.cn

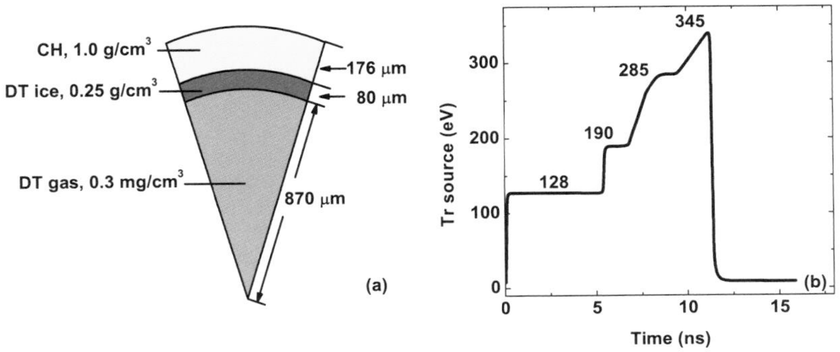

FIG. 1. The capsule model used in current hydrodynamic simulations is shown in the left panel (a). The radiation temperature (b) driving the capsule.

temperatures, QMD simulations will meet the computational bottleneck. To overcome this limitation, OFMD simulations, where the electronic kinetic energy is expressed as the functional of the electron density, have been used to calculate high temperature EOS instead of QMD. In the present work, indirect-drive ignition target design has been performed, and the results obtained from DFT-EOS and QEOS have been compared and discussed. It will be shown that the density distribution and the energy gain of the target sensitively depend on the EOS. The rest of this paper is organized as follows: A brief description of the target and simulation method is presented in Sec. II. In Sec. III, we analyze and discuss our main results. The paper is summarized in Sec. IV.

II. SIMULATION METHOD

This work gives a description of an indirect-drive implosion. Hydrodynamic simulations using the multidimensional radiation-hydrodynamics code LARED-S have been performed, and only its one-dimensional version is adopted. The standard flux-limited ($f = 0.06$) Spitzer thermal transport model, multigroup (20 groups) radiative transfer, and multigroup fusion-product transport are included in this code.

The geometry of the capsule is shown in Fig. 1(a), and the spherical pellet with a 1126 μm outer radius (including a 176 μm thick ablator and 80 μm DT ice). The ablator density is 1.0 g/cm^3, and the DT ice density is 0.25 g/cm^3 (with 0.3 mg/cm^3 DT gas). In our design, the shock adiabat is largely set by the first and second shock, and the third and fourth shock adds a relatively small amount of entropy to the target [the pulse shape is shown in Fig. 1(b)]. Two hydrodynamic simulations, where QEOS and DFT-EOS are implemented in LARED-S code, have been performed in the present work. First, the radiation temperature shape is optimized by hydrodynamic simulations using QEOS. Second, hydrodynamic simulations under the same driven mode are performed by using DFT-EOS.

III. RESULTS AND DISCUSSION

A DT DFT-EOS table (100 × 100 density-temperature points) has been built based on QMD and OFMD calculations in a wide density ($\rho_{DT} = 2.5 \times 10^{-3}$–$3.3 \times 10^3$ g/cm^3) and temperature range ($T = 0.086$–4300 eV).[15] At higher-temperatures ($T > 4300$ eV), an ideal gas model is used. The principal Hugoniot for D given by DFT-EOS table is shown in Fig. 2, and previous results as obtained from SESAME,[26] QEOS,[27] and PIMC simulations[17] are also shown for comparison. The present results indicate that the compression ratio (η) shows a maximum of 4.5 at about 40 GPa, where the fluid is governed by the gradual molecular dissociation. At higher pressure, η decreases and reaches a value of 4.23 at ~950 GPa, which indicating the formation of monoatomic fluid. In the fully ionized regime, η shrinks into 4.0 at a pressure exceeding 10^3 GPa. The QEOS results[27] shown in Fig. 2 indicate that the maximum compression ratio locus at $\eta = 4.9$, and the relative pressure is 250 GPa, which is not agreed with experiments (one can finds more experimental results in Ref. 15). The SESAME Hugoniot[26] gives a maximum η at around 4.5, but predicts a much higher pressure than DFT-EOS. Previous PIMC simulations results[17] are accordance with our DFT-EOS results at a pressure >50 GPa, however, PIMC results have failed to reproduce the experimental ones at pressure below 50 GPa.

Comparison between DFT-EOS and QEOS at high densities is shown in Fig. 3. The electronic pressures (P_e), given by QEOS, overestimate the DFT-EOS results up to ~20% at a density of 12.5 g/cm^3, because the TF model used in QEOS ignores the exchange-correlation effect. The pressure difference between QEOS and DFT-EOS is gradually reduced as the density increase, e.g., at 53.75 g/cm^3 in Fig. 3(b). It is

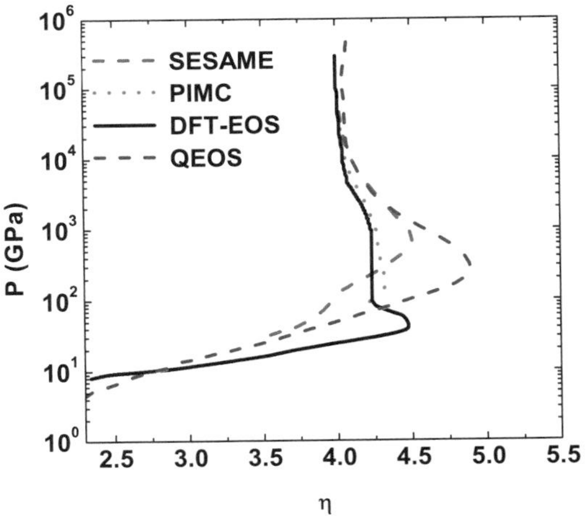

FIG. 2. The principal Hugoniot given by DFT-EOS is compared with PIMC,[17] QEOS,[27] and SESAME[26] data.

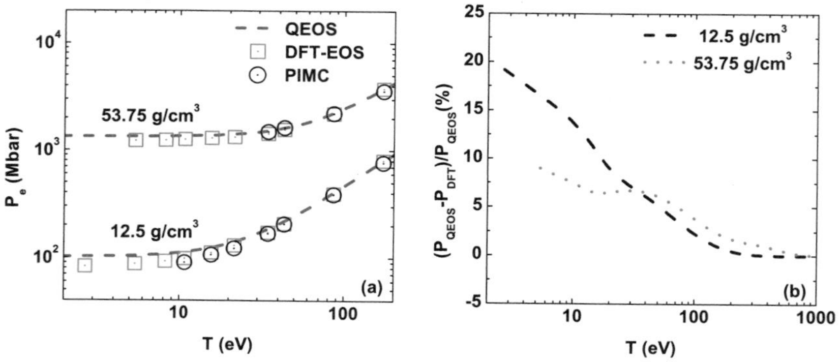

FIG. 3. The electronic pressures (P_e) of dense DT plasmas given by DFT-EOS, PIMC, and QEOS are shown in the left panel (a). The difference between DFT-EOS and QEOS has been shown in the right panel (b).

also indicated that the pressure given by DFT-EOS is in good agreement with PIMC calculations[17] at densities concerned in the present work.

After comparisons between DFT-EOS and QEOS, we now investigate the influence of different EOS on the implosion hydrodynamics. In this section, we can explore how the DFT-EOS may affect ICF implosions by comparing the results with QEOS. After the first and the second laser pulse illuminated plastic CH ablator, shock waves have been launched into the target, and the DT shell is compressed to the warm dense or hot dense plasma regime. Then, the capsule is driven to be compressed by the main pulse. In order to quantify how the different EOS table affects the imploding shell, the density and temperature profiles of the capsule are plotted in Fig. 4, where two different time snap shots at t = 9.8 and 11.3 ns during the implosion are selected. The blue dashed lines represent the density/temperature distribution of QEOS results, and the red solid lines denote DFT-EOS data. Figures 4(a) and 4(b) plot the in-flight plasma conditions at t = 9.8 ns, which is just ahead of the stagnation stage, and the mass density and electronic temperature are plotted as a function of the target radius. At this earlier time, the DT shell of the target has converged to a radius of ~240 μm from its original value of 950 μm, and the CH layer has not been fully ablated, which can be identified by the density peak in Fig. 4(a). The peak density and average temperature (for r lies between 200 and 240 μm) are $\rho \approx 14.6\,\mathrm{g/cm^3}$ and $T_e \approx 0.4$ MK by DFT-EOS simulation, and QEOS predicts $\rho \approx 12.5\,\mathrm{g/cm^3}$ and $T_e \approx 0.36$ MK. It is indicated that the DFT-EOS simulation predicted a 16.8% higher peak density and an 11.1% higher temperature compared with QEOS data. As shown in Fig. 3, the DFT-EOS shows a softer behavior than QEOS at a given density and temperature, and this can explain the higher peak density, as

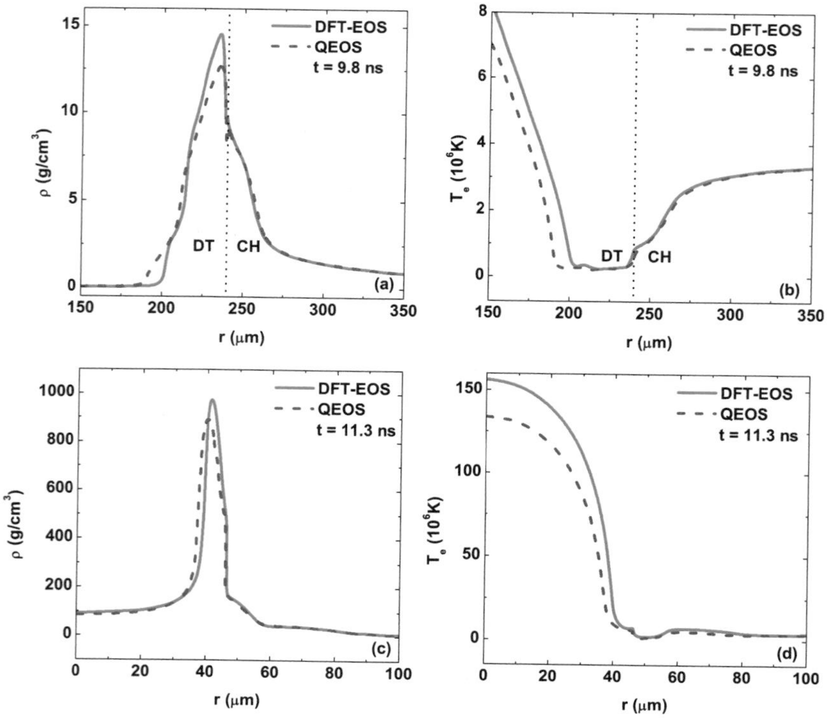

FIG. 4. The mass density and temperature profile as a function of the target radius at t = 9.8 ns are shown in (a) and (b), respectively, density and temperature distribution at 11.3 ns are shown in (c) and (d). The red solid and blue dashed lines represent the DFT-EOS and QEOS simulations, respectively.

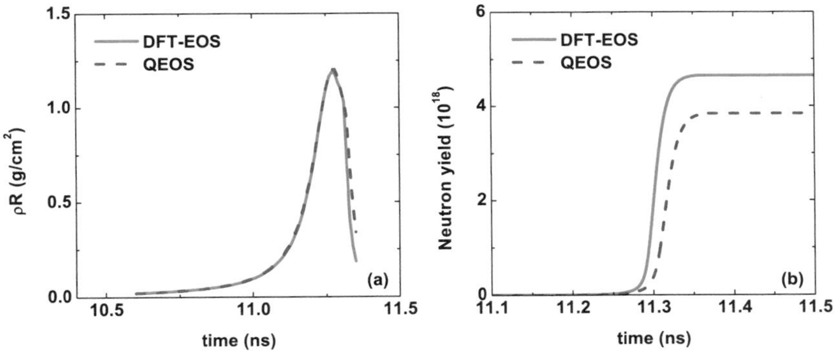

FIG. 5. The areal density (ρR) and neutron yield as a function of time. The red solid and blue dashed lines represent the DFT-EOS and QEOS simulations, respectively.

shown in Fig. 4(a). The 11.1% higher temperature obtained from DFT-EOS simulations can be attributed to the lower internal energy. In QEOS, TF theory in terms of the spherical-cell model is used, and the bonding effect, the shell effect, and the exchange correlation effect are neglected. Due to these limitations mentioned above, the internal energy given by QEOS is higher than that predicted by DFT-EOS at a given density. During the implosion, the laser illumination transfers a certain amount of energy to the target, which is heated and driven to implode. If the target is compressed to the same density, a lower increase in the internal energy means more driven energy is left and is consequently partitioned to heat the target, which results in a higher temperature. The change in the density and temperature will certainly influence the target performance. In Figs. 4(c) and 4(d), we show the density and temperature profile at 11.3 ns, which represents the peak compression. At the stagnation stage, the CH layer has been fully ablated, and the radius of the target has been compressed to $\sim 46\,\mu$m. It is shown in Fig. 3 that the peak density predicted by DFT-EOS ($\rho \approx 984\,$g/cm^3) is $\sim$10%, higher than that given by QEOS simulation ($\rho \approx 894\,$g/cm^3), and the DFT-EOS temperature ($T_e \sim 3.0$ MK) is $\sim$20%, higher than QEOS temperature ($T_e \sim 2.5$ MK).

The evolution of the areal density (ρR) and neutron yield as a function of time has been shown in Figs. 5(a) and 5(b), respectively. It is indicated that the areal density distribution does not sensitively depend on the two EOS tables used in the present indirect-driven scheme, and only a maximum difference of 2% is observed. One can find that the neutron yield is increased by about 21% when we use DFT-EOS instead of QEOS. The absolute neutron yield given by DFT-EOS is $\sim 4.65 \times 10^{18}$, and it is $\sim 3.84 \times 10^{18}$ for QEOS. As a consequence, the energy gain increases from 11.0 MJ (QEOS) to 13.1 MJ (DFT-EOS). The current results could be attributed to the different density-temperature trajectories directed by the two EOS tables used in the present work.

IV. SUMMARY

In conclusion, a DT DFT-EOS table, which is applied in hydrodynamic simulations, has been built based on QMD and OFMD simulations. This DFT-EOS table covers a wide density-temperature range, including the warm dense and hot dense regime. Comparison has been made between DFT-EOS and QEOS. For a given density and temperature, DFT-EOS predicts a lower electronic pressure than QEOS. Radiation-hydrodynamics simulations of implosions have been performed by using both of DFT-EOS and QEOS tables. Simulation results indicate that the density/temperature distribution of the target, and the neutron yield change dramatically according to different EOS. Compared with QEOS, DFT-EOS predicts a $\sim$10% higher peak density, a $\sim$20% larger average electronic temperature, and a $\sim$21% higher thermonuclear energy gain.

ACKNOWLEDGMENTS

This work was supported by the foundation of President of Chinese Academy of Engineering Physics (Grant No. 2014-1-036), by the National Natural Science Foundation (NSFC) of China under Grant Nos. 11475031, 11275032, 11005012, 11475032, and 91130002, by the foundation for Development of Science and Technology of China Academy of Engineering Physics under Grant No. 2013B0102019, by the National High Technology Research and Development Program of China (Grant No. 2012AA01A303), by the National Basic Security Research Program of China, and by the National High-Tech ICF Committee of China.

[1]S. Atzeni and J. Meyer-ter-Vehn, *The Physics of Inertial Fusion: Beam Plasma Interaction, Hydrodynamics, Hot Dense Matter* (Clarendon, Oxford, 2004).

[2]J. D. Lindl, *Inertial Confinement Fusion: The Quest for Ignition and Energy Gain Using Indirect Drive* (Springer-Verlag, New York, 1998).

[3]T. R. Dittrich, O. A. Hurricane, D. A. Callahan, E. L. Dewald, T. Döppner, D. E. Hinkel, L. F. Berzak Hopkins, S. Le Pape, T. Ma, J. L. Milovich, J. C. Moreno, P. K. Patel, H.-S. Park, B. A. Remington, and J. D. Salmonson, Phys. Rev. Lett. **112**, 055002 (2014).

[4]O. A. Hurricane, D. A. Callahan, D. T. Casey, P. M. Celliers, C. Cerjan, E. L. Dewald, T. R. Dittrich, T. Döppner, D. E. Hinkel, L. F. B. Hopkins, J. L. Kline, S. Le Pape, T. Ma, A. G. MacPhee, J. L. Milovich, A. Pak, H.-S. Park, P. K. Patel, B. A. Remington, J. D. Salmonson, P. T. Springer, and R. Tommasini, Nature **506**, 343–348 (2014).

[5]S. W. Haan, J. D. Lindl, D. A. Callahan, D. S. Clark, J. D. Salmonson, B. A. Hammel, L. J. Atherton, R. C. Cook, M. J. Edwards, S. Glenzer, A. V. Hamza, S. P. Hatchett, M. C. Herrmann, D. E. Hinkel, D. D. Ho, H. Huang, O. S. Jones, J. Kline, G. Kyrala, O. L. Landen, B. J. MacGowan, M. M. Marinak, D. D. Meyerhofer, J. L. Milovich, K. A. Moreno, E. I. Moses, D. H. Munro, A. Nikroo, R. E. Olson, K. Peterson, S. M. Pollaine, J. E. Ralph, H. F. Robey, B. K. Spears, P. T. Springer, L. J. Suter, C. A. Thomas, R. P. Town, R. Vesey, S. V. Weber, H. L. Wilkens, and D. C Wilson, Phys. Plasmas **18**, 051001 (2011).

[6]D. D. Meyerhofer, R. L. McCrory, R. Betti, T. R. Boehly, D. T. Casey, T. J. B. Collins, R. S. Craxton, J. A. Delettrez, D. H. Edgell, R. Epstein, K. A. Fletcher, J. A. Frenje, Y. Yu. Glebov, V. N. Goncharov, D. R. Harding, S. X. Hu, I. V. Igumenshchev, J. P. Knauer, C. K. Li, J. A. Marozas, F. J. Marshall, P. W. McKenty, P. M. Nilson, S. P. Padalino, R. D. Petrasso, P. B. Radha, S. P. Regan, T. C. Sangster, F. H. Séguin, W. Seka, R. W. Short, D. Shvarts, S. Skupsky, J. M. Soures, C. Stoeckl, W. Theobald, and B. Yaakobi, Nucl. Fusion **51**, 053010 (2011).

[7]R. L. McCrory, R. Betti, T. R. Boehly, D. T. Casey, T. J. B. Collins, R. S. Craxton, J. A. Delettrez, D. H. Edgell, R. Epstein, J. A. Frenje, D. H. Froula, M. Gatu-Johnson, V. Yu. Glebov, V. N. Goncharov, D. R. Harding, M. Hohenberger, S. X. Hu, I. V. Igumenshchev, T. J. Kessler, J. P. Knauer, C. K. Li, J. A. Marozas, F. J. Marshall, P. W. McKenty, D. D. Meyerhofer, D. T. Michel, J. F. Myatt, P. M. Nilson, S. J. Padalino, R. D. Petrasso, P. B. Radha, S. P. Regan, T. C. Sangster, F. H. Séguin, W. Seka, R. W. Short, A. Shvydky, S. Skupsky, J. M. Soures, C. Stoeckl, W. Theobald, B. Yaakobi, and J. D. Zuegel, Nucl. Fusion **53**, 113021 (2013).

[8]H.-S. Park, O. A. Hurricane, D. A. Callahan, D. T. Casey, E. L. Dewald, T. R. Dittrich, T. Döppner, D. E. Hinkel, L. F. Berzak Hopkins, S. Le Pape, T. Ma, P. K. Patel, B. A. Remington, H. F. Robey, J. D. Salmonson, and J. L. Kline, Phys. Rev. Lett. **112**, 055001 (2014).

[9]R. Betti and C. Zhou, Phys. Plasmas **12**, 110702 (2005).

[10]R. Betti, K. Anderson, T. R. Boehly, T. J. B. Collins, R. S. Craxton, J. A. Delettrez, D. H. Edgell, R. Epstein, V. Yu. Glebov, V. N. Goncharov, D. R. Harding, R. L. Keck, J. H. Kelly, J. P. Knauer, S. J. Loucks, J. A. Marozas, F. J. Marshall, A. V. Maximov, D. N. Maywar, R. L. McCrory, P. W. McKenty, D. D. Meyerhofer, J. Myatt, P. B. Radha, S. P. Regan, C. Ren, T. C. Sangster, W. Seka, S. Skupsky, A. A. Solodov, V. A. Smalyuk, J. M. Soures, C. Stoeck, W. Theobald, B. Yaakobi, C. Zhou, J. D. Zuegel, J. A. Frenje, C. K. Li, R. D. Petrasso, and F. H. Seguin, Plasma Phys. Controlled Fusion **48**, B153 (2006).

[11]V. N. Goncharov, T. C. Sangster, T. R. Boehly, S. X. Hu, I. V. Igumenshchev, F. J. Marshall, R. L. McCrory, D. D. Meyerhofer, P. B. Radha, W. Seka, S. Skupsky, C. Stoeckl, D. T Casey, J. A. Frenje, and R. D. Petrasso, Phys. Rev. Lett. **104**, 165001 (2010).

[12]M. J. Edwards, J. D. Lindl, B. K. Spears, S. V. Weber, L. J. Atherton, D. L. Bleuel, D. K. Bradley, D. A. Callahan, C. J. Cerjan, D Clark, G. W. Collins, J. E. Fair, R. J. Fortner, S. H. Glenzer, S. W. Haan, B. A. Hammel, A. V. Hamza, S. P. Hatchett, N. Izumi, B. Jacoby, O. S. Jones, J. A. Koch, B. J. Kozioziemski, O. L. Landen, R. Lerche, B. J. MacGowan, A. J. MacKinnon, E. R. Mapoles, M. M. Marinak, M. Moran, E. I. Moses, D. H. Munro, D. H. Schneider, S. M. Sepke, D. A. Shaughnessy, P. T. Springer, R. Tommasini, L. Bernstein, W. Stoeffl, R. Betti, T. R. Boehly, T. C. Sangster, V. Yu. Glebov, P. W. McKenty, S. P. Regan, D. H. Edgell, J. P. Knauer, C. Stoeckl, D. R. Harding, S. Batha, G. Grim, H. W. Herrmann, G. Kyrala, M. Wilke, D. C. Wilson, J. Frenje, R. Petrasso, K. Moreno, H. Huang, K. C. Chen, E. Giraldez, J. D. Kilkenny, M. Mauldin, N. Hein, M. Hoppe, A. Nikroo, and R. J. Leeper, Phys. Plasmas **18**, 051003 (2011).

[13]V. Recoules, F. Lambert, A. Decoster, B. Canaud, and J. Clérouin, Phys. Rev. Lett. **102**, 075002 (2009).

[14]C. Wang, X.-T. He, and P. Zhang, Phys. Rev. Lett. **106**, 145002 (2011).

[15]C. Wang and P. Zhang, Phys. Plasmas **20**, 092703 (2013).

[16]L. Caillabet, B. Canaud, G. Salin, S. Mazevet, and P. Loubeyre, Phys. Rev. Lett. **107**, 115004 (2011).

[17]S. X. Hu, B. Militzer, V. N. Goncharov, and S. Skupsky, Phys. Rev. B **84**, 224109 (2011).

[18]W. J. Nellis, Rep. Prog. Phys. **69**, 1479 (2006).

[19]G. V. Boriskov, A. I. Bykov, R. Ilkaev, V. D. Selemir, G. V. Simakov, R. F. Trunin, V. D. Urlin, A. N. Shuikin, and W. J. Nellis, Phys. Rev. B **71**, 092104 (2005).

[20]M. D. Knudson, D. L. Hanson, J. E. Bailey, C. A. Hall, J. R. Asay, and C. Deeney, Phys. Rev. B **69**, 144209 (2004).

[21]T. R. Boehly, D. G. Hicks, P. M. Celliers, T. J. B. Collins, R. Earley, J. H. Eggert, D. Jacobs-Perkins, S. J. Moon, E. Vianello, D. D. Meyerhofer, and G. W. Collins, Phys. Plasmas **11**, L49 (2004).

[22]D. G. Hicks, T. R. Boehly, P. M. Celliers, J. H. Eggert, S. J. Moon, D. D. Meyerhofer, and G. W. Collins, Phys. Rev. B **79**, 014112 (2009).

[23]B. Militzer and D. M. Ceperley, Phys. Rev. Lett. **85**, 1890 (2000).

[24]B. Holst, R. Redmer, and M. P. Desjarlais, Phys. Rev. B **77**, 184201 (2008).

[25]L. Caillabet, S. Mazevet, and P. Loubeyre, Phys. Rev. B **83**, 094101 (2011).

[26]G. I. Kerley, Phys. Earth Planet. Inter. **6**, 78 (1972).

[27]R. M. More, K. H. Warren, D. A. Young, and G. B. Zimmerman, Phys. Fluids **31**, 3059 (1988).

[28]D. Saumon and G. Chabrier, Phys. Rev. A **46**, 2084 (1992).

[29]M. Ross, Phys. Rev. B **58**, 669 (1998).

[30]F. J. Rogers, Contrib. Plasma Phys. **41**, 179 (2001).

[31]H. Juranek, R. Redmer, and Y. Rosenfeld, J. Chem. Phys. **117**, 1768 (2002).

[32]C. Pierleoni, D. M. Ceperley, and M. Holzmann, Phys. Rev. Lett. **93**, 146402 (2004).

[33]W. R. Magro, D. M. Ceperley, C. Pierleoni, and B. Bernu, Phys. Rev. Lett. **76**, 1240 (1996).

[34]F. Lambert, J. Clérouin, and S. Mazevet, Europhys. Lett. **75**, 681 (2006).

[35]S. Mazevet, F. Lambert, F. Bottin, G. Zérah, and J. Clérouin, Phys. Rev. E **75**, 056404 (2007).

[36]F. Lambert, J. Clérouin, J.-F. Danel, L. Kazandjian, and G. Zérah, Phys. Rev. E **77**, 026402 (2008).

[37]W. Pei, Commun. Comput. Phys. **2**, 255 (2007).

PHYSICS OF PLASMAS 22, 062707 (2015)

First-principles calculation of principal Hugoniot and K-shell X-ray absorption spectra for warm dense KCl

Shijun Zhao,[1,2] Shen Zhang,[1,2] Wei Kang,[1,2,a)] Zi Li,[3] Ping Zhang,[1,3] and Xian-Tu He[1,2,3,b)]
[1]*HEDPS, Center for Applied Physics and Technology, Peking University, Beijing 100871, China*
[2]*College of Engineering, Peking University, Beijing 100871, China*
[3]*Institute of Applied Physics and Computational Mathematics, Beijing 100088, China*

(Received 6 March 2015; accepted 28 May 2015; published online 17 June 2015)

Principal Hugoniot and K-shell X-ray absorption spectra of warm dense KCl are calculated using the first-principles molecular dynamics (FPMD) method. Evolution of electronic structures as well as the influence of the approximate description of ionization on pressure (caused by the underestimation of the energy gap between conduction bands and valence bands) in the first-principles method are illustrated by the calculation. It is shown that approximate description of ionization in FPMD has small influence on Hugoniot pressure due to mutual compensation of electronic kinetic pressure and virial pressure. The calculation of X-ray absorption spectra shows that the band gap of KCl persists after the pressure ionization of the $3p$ electrons of Cl and K taking place at lower energy, which provides a detailed understanding to the evolution of electronic structures of warm dense matter. © *2015 AIP Publishing LLC*. [http://dx.doi.org/10.1063/1.4922672]

I. INTRODUCTION

Warm dense matter (WDM) is of particular interest to astrophysics, geophysics, and inertial confinement fusion (ICF).[1–3] Recent developments of high-power laser facilities[4,5] and corresponding diagnosing techniques[6,7] greatly facilitate its investigation, making WDM accessible to systematical laboratory studies. Laser-driven shock compression is a typical way to generate warm dense states in large-scale laser facilities, such as Shen-Guang IIU.[8,9] Among materials used in shock compression experiments, KCl is a good choice for X-ray absorption measurements[8,10] considering its large intrinsic band gap of 8.69 eV,[11] which efficiently reduces the preheating of diagnosing X-ray in the early stages of the experiments.[8,10] As a typical ionic crystalline material (under ambient conditions) composed of heterogeneous species, KCl is also beneficial for illustrating the evolution of electronic structures (e.g., pressure ionization, energy level shifts, and metalization) of heterogeneous ionic materials under warm dense conditions. Those properties can now be readily detected by their X-ray absorption spectra (XAS). In particular, the variation of the intrinsic band gap under compression has a significant influence on the degree of ionization, which is a critical parameter in several commonly used models of equation of state (EOS), e.g., the quotidian equation of state (QEOS) model.[12]

From the methodological point of view, KCl is also an illustrating example to display the influence of the approximate description of ionization, caused by the underestimation of the energy gap between conduction bands and valence bands, in the first-principles molecular dynamics (FPMD) method, now an influential theoretical tool in the study of WDM.[13] The FPMD method, usually implemented with local density approximation (LDA) or the Perdew-Burke-Ernzerhof (PBE) version of generalized gradient approximation (GGA)[14] to the exchange-correlation interaction, has been known to underestimate the energy gap by 40%–60% for nonmetals,[15] which causes a serious overestimation in the ionization, and was thus postulated to bring about sizable deviations in the calculation of thermal properties of WDM. However, this conjecture is not well supported by recent studies. The deviation turns out to be small. FPMD calculations on a large variety of materials, including hydrogen,[16–18] helium,[19] aluminum,[20,21] and iron[22] display a good agreement with measured Hugoniot data. These calculations cover a large range of temperature and pressure, at which electrons in the outermost shell and inner shell can be ionized. How to understand this small effect remains an open problem. The answer to it, however, is important to the improvement of the FPMD method. The ionization of KCl takes place under a condition relatively easy to realize through current WDM experimental facilities, which makes it a favorable example to examine this effect.

Using the FPMD method, we investigate the variation of electronic structures and the influence of the approximate description of ionization to the calculation of pressure in warm dense KCl. The variation of electronic structures is illustrated through the calculation of XAS. Our results show that the energy gap, originally between the occupied valence band and unoccupied conduction band under ambient conditions, persists after the occurrence of pressure ionization of the $3p$ electrons of Cl and K taking place at a lower energy. Pressure ionization and thermal smearing are shown as the major factors to prevent the deviation of pressure caused by the approximate description of ionization from global accumulation along the Hugoniot. In addition, cancellation between electronic kinetic pressure and virial pressure further reduces the deviation, leading to a small influence of the approximate description of ionization in the FPMD method, in line with the trend illustrated in previous calculations.[16–22]

a)weikang@pku.edu.cn
b)xthe@iapcm.ac.cn

The rest of the article is organized as follows. In Sec. II, methods and numerical details used in the calculation of principal Hugoniots and K-shell XAS are summarized. In Sec. III, the influence of the approximate description of ionization on the calculation of pressure is discussed through the calculation of principle Hugoniot. In Sec. IV, the variation of electronic structures of KCl in warm dense states is displayed together with the calculation of XAS. We conclude the article with a short summary in Sec. V.

II. COMPUTATIONAL METHODS AND NUMERICAL DETAILS

A. Calculation of principal Hugoniots

Principal Hugoniot of shock-compressed KCl is calculated through FPMD simulations consisting of 54 atoms, i.e., 27 formula units of KCl. It should be noted that for the calculated Hugoniot results, the errors are on the order of 5% for the obtained temperature and pressure due to the statistical fluctuations originated from FPMD. Finite-temperature density functional theory (FTDFT)[23] is employed in the simulation to describe the statistics of electrons. The calculation is performed using the `Quantum-Espresso` package.[24] The PBE version of GGA[14] is used to account for the exchange-correlation interaction. Projector augmented wave (PAW) pseudopotentials[25] generated by the `ATOMPAW` program[26] are employed to model the electron-ion interactions. The plane-wave cutoff is set to be 50 Ry, which controls the deviation within 0.01 eV in the calculation of total energy. Only the Γ point is used to sample the Brillouin zone in the FPMD simulations.

For a given density, a series of simulations are carried out at different ionic temperatures, which are controlled by the Nośe-Hoover thermostat.[27] The electronic states are populated according to the Fermi-Dirac distribution and thermodynamical equilibrium is maintained by setting the electronic temperature (T_e) the same as the ionic temperature (T_i). In some cases, a T_e different from the value of T_i is used to illustrate the influence of the approximate description of ionization. Typical dynamic simulations last for 3–5 ps with an appropriate time step of 0.4–1.0 fs, varying with temperature and density. Data presented in the article are collected from the last 2000 steps.

The principal Hugoniot points are determined by solving the Hugoniot equation[28]

$$E_1 - E_0 = \frac{1}{2}(P_1 + P_0)(V_0 - V_1), \tag{1}$$

where E represents the average internal energy, V is the average volume, and P is the pressure. Subscripts 0 and 1 refer to initial and shocked states, respectively. Kinetic pressure of ions, equal to k_BT/V, is added manually in the calculation to accurately account for the total pressure. The reference state (E_0, V_0) is determined from a face-centered cubic (fcc) structure under ambient conditions. P_0 is neglected in the calculation as it is several orders smaller than P_1. An auxiliary cubic polynomial interpolating function is employed in solving Eq. (1) to reduce the numerical error less than 1%.

Note that KCl undergoes a polymorphic phase transition from a fcc lattice (B1, NaCl structure) to a body-centered cubic (bcc) lattice (B2, CsCl structure) during the shock compression.[29–32] The transition pressure varies from 1.97 to 2.5 GPa, depending on experimental conditions.[29–32] The calculation of Hugoniot starts from a B2 structure at the pressure around 3.5 GPa, and phase transition points along the Hugoniot are avoided.

B. Calculation of XAS

The XAS of warm dense KCl is calculated via first-principles methods based on FPMD, similar to those used in previous studies.[33,34] Theoretically, X-ray absorption is characterized by its cross-section $\sigma(\omega)$, which is calculated in a perturbative way as[35]

$$\frac{\sigma(\omega)}{4\pi^2\alpha\hbar\omega} = \sum_f |\langle f|\hat{\epsilon}\cdot \boldsymbol{r}|i\rangle|^2(1-\mathcal{F}(E_f))\delta(E_f - E_i - \hbar\omega), \tag{2}$$

where letters f and i represent final and initial states, respectively, $\hat{\epsilon}$ is a unit vector denoting the polarization of incident light, α is the fine structure constant, and $\mathcal{F}(E_f)$ is the Fermi-Dirac distribution of final states at finite temperature. Calculated using Eq. (2), K-shell XAS of shock-compressed Al have displayed a good agreement with experimental measurements.[34,36] Information on electronic structures and thermodynamical properties of WDM can be further derived through these calculations.[34,36]

The XAS calculation is performed using the `XSPECTRA` package[35,37] supplied with the `Quantum-Espresso` distribution.[24] A minor modification is made to account for the Fermi-Dirac distribution of electrons at finite temperature. Wave functions of the final states used in the XAS calculation are prepared by a density functional theory (DFT) calculation on the atomic configurations generated in the previous Hugoniot calculation. The plane-wave energy cutoff is 50 Ry, and the hole in the 1s state of the excited Cl atom is manually included in the core of the pseudopotential with the help of the GIPAW pseudopotential[38] to approximately account for electron-hole interactions. Other atoms in the system are presented using the Troullier-Martins pseudopotentials.[39] The exchange-correlation functional is kept the same as that used in the Hugoniot calculation. Here, we have used $4\times4\times4$ Monkhorst-Pack grids for the sampling of Brillouin zone. This parameter has been tested to ensure the convergence of the spectra calculation. With the wave functions prepared, $\sigma(\omega)$ is then calculated through Eq. (2). For each thermodynamical condition, a total of 8 XAS are calculated on selected snapshots along the trajectory of ions, separated by 200 time steps. The final results of XAS are averages of the spectra of each snapshot.

III. PRINCIPAL HUGONIOT AND THE INFLUENCE OF THE APPROXIMATE DESCRIPTION OF IONIZATION

Calculated principal Hugoniot of KCl is tabulated in Table I and also presented in Fig. 1(a) together with available experimental measurements.[30–32] Theoretical principal

TABLE I. Calculated Hugoniot data for KCl from QMD in this work.

Density (g/cm^3)	Temperature (K)	Energy (eV/formula)	Pressure (GPa)
2.5	242	0.16	3.9
3.2	1953	1.61	22.3
3.6	3304	3.90	45.7
3.9	7510	6.59	69.2
4.2	9791	9.07	89.5
4.6	14 778	14.57	132.4
4.9	19 632	19.39	167.8
5.5	29 408	30.13	244.3
5.8	38 348	40.63	316.8
6.3	50 104	56.12	422.2

Hugoniots derived from the QEOS model[12] (calculated using the `mpqeos` program included in the `MULTI` package[40]) as well as Altshuler's model[29] are also displayed for comparison. Corresponding temperature along the Hugoniot is displayed in Fig. 1(b). Fig. 1(a) shows that the FPMD Hugoniot generally agrees well with experimental data. A small overestimation about 25 GPa in pressure is observed at densities higher than 5 g/cm^3 in comparison with the measurements of Kormer *et al.*[30]

The origin of this deviation is not clear so far. On the theoretical side, it is possibly the reflection of the overestimation of ionization in the FPMD method. Calculation on LiF[41] displayed a similar trend compared with the measurements of Kormer *et al.*. On the other hand, uncertainties in experimental measurements could also be a cause of the deviation. Fig. 1(a) shows that the deviation starts from a sharp discontinuity taking place around 5 g/cm^3 in the measured Hugoniot. This discontinuity does not belong to any of the first-order transitions recognized so far, as have been summarized by Duvall and Graham in Ref. 42. The corresponding pressure (150 GPa) of the discontinuity is far beyond the pressure of the polymorphic B1–B2 phase transition (~2 GPa (Ref. 32)) and the pressure of the solid-liquid phase transition (33 GPa–48 GPa).[42]

The QEOS Hugoniot lies above the experimental measurements as well as the FPMD Hugoniot. The overestimation is derived from a much higher estimation to the temperature, as displayed in Fig. 1(b). It is quite unusual that the QEOS has a poor performance on a typical ionic crystalline material in the calculation of Hugoniot. It works reasonably well for other typical ionic crystalline materials like NaCl[12] and LiF.[41] The Altshuler's model has a similar performance as the QEOS model in the calculation of pressure. However, it gives an even higher estimation to the temperature and tends to overestimate the pressure on further compression. The large variation of temperatures displayed in Fig. 1(b) highlights the necessity to take temperature into account as a critical reference in building theoretical EOS models.

Calculated degree of ionization along the Hugoniot is displayed in Fig. 2(b). Using the FPMD method, the statistics of electron is described by Mermin's functional,[23] which assigns an average occupation number to each electronic state according to the Fermi-Dirac distribution. By this method, the ionization ratio $R(T)$ is estimated as[43]

$$R(T) = \frac{1}{N_i \bar{Z}} \int_{\mu}^{\infty} \frac{D(E)dE}{e^{(E-\mu)/k_B T} + 1}, \tag{3}$$

where N_i is the total number of ions, $\bar{Z}$ is the averaged charge number for each ion, $D(E)$ is the density of state, and μ is the chemical potential of the system.

As a result of the mass-action law,[44] the ionization ratio $R(T)$ is roughly proportional to $\exp[-(E_g/2k_B T)]$ at modest

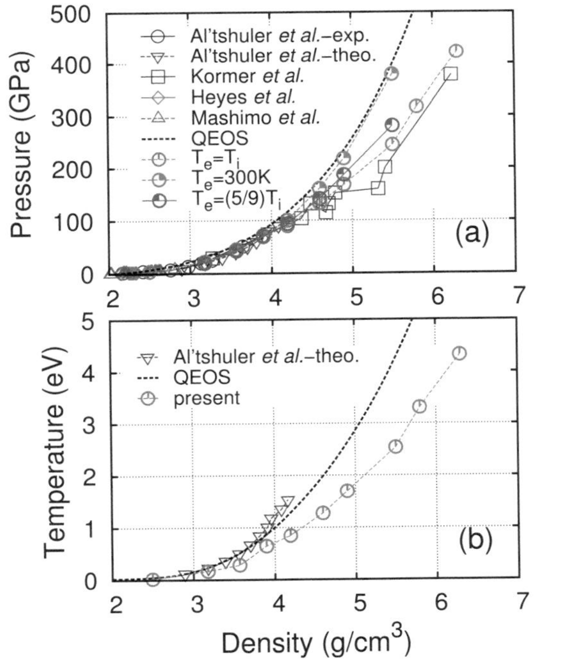

FIG. 1. Shocked principal Hugoniots of KCl in (a) P-ρ diagram and (b) T-ρ diagram. The experimental results from Altshuler *et al.*,[29] Kormer *et al.*,[30] Heyes *et al.*,[31] and Mashimo *et al.*[32] as well as the theoretical results from Altshuler *et al.*[29] and QEOS[12] are presented for comparison.

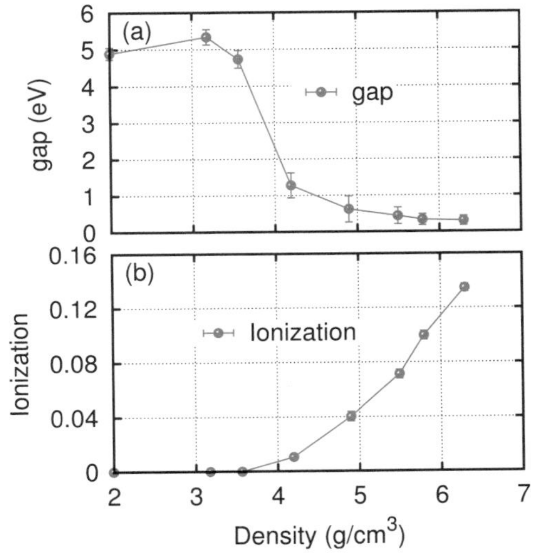

FIG. 2. (a) Energy gap E_g and (b) ionization ratio along the principal Hugoniot calculated using the FPMD simulations.

temperatures, where E_g is the energy of the gap. For temperature much higher than $E_g/2$, μ would be much lower than the energy of electrons at the bottom of the shell, as the consequence of finite number of electrons. Accordingly, $R(T)$, as estimated by Eq. (3), would approach to 100%. For commonly used LDA and GGA exchange-correlation functionals in the FPMD simulations, E_g is generally underestimated by 40%–60% for nonmetals,[15] which implies a sizable overestimation to the $R(T)$ of KCl.

The variation of E_g along the Hugoniot is also an important factor affecting the estimation of $R(T)$. When temperature and density keep increasing along the Hugoniot, E_g approaches to zero as the result of pressure ionization and thermal smearing. The vanishing of the band gap along the Hugoniot is broadly observed in materials of finite energy gap under ambient conditions, among which LiF[41] is a recent example under extensive studies. Fig. 2(a) displays the variation of band gap along the Hugoniot, determined through Eq. (3).[43] In practice, it is more convenient to calculate the band gap along FPMD trajectories. In the simulation, the Fermi-Dirac distribution of electrons was determined at each FPMD step. When the energy of electronic state is greater than the chemical potential, those states are taken as ionized states. The final results presented in Fig. 2 are averaged from the last 500 steps in the trajectory. It shows that the energy gap vanishes around 6.3 g/cm^3, where the $3p$ electrons of Cl are hybridized with the $3d$ electrons of K. This value can also be derived from the variation of density of states (DOS), as displayed in Fig. 3(d). The DOS is obtained from an additional calculation on the atomic configurations generated by the FPMD simulation. The electronic smearing parameters are the same. It provides a threshold indicating when the system becomes completely metallic and the inaccuracy caused by the gap becomes negligibly small. Beyond the threshold, the reliability of the FPMD method has been well demonstrated by the calculations on typical metallic materials, such as Al and Fe.[20–22,34,45] An accuracy of several percentages in both pressure and electronic structures can be achieved in the warm dense region (before the localized electrons in the lower shell are thermally excited). These results imply that the inaccuracy of ionization in the FPMD method is constrained by the pressure ionization and thermal smearing, and it will not accumulate globally, i.e., increase unboundedly, along the Hugoniot. This is, in indeed, what displays in Fig. 1(a) and in the previous calculations.[16–22] Extending this observation to a much higher temperature region (above 100 eV) strongly depends on the temperature effect of exchange-correlation functionals. Recent progress on this topic can be found elsewhere.[13,46]

To quantitatively estimate how an approximate description of ionization in FPMD affects the Hugoniot result, we manually adjust the temperature of electrons. Besides the Hugoniot calculated at $T_e = T_i$, Fig. 1(a) also displays the Hugoniots of $T_e = 300$ K and $T_e = (5/9)T_i$. The fraction of 5/9 gives an approximate correction to the overestimation of ionization caused by the underestimated band gap for the

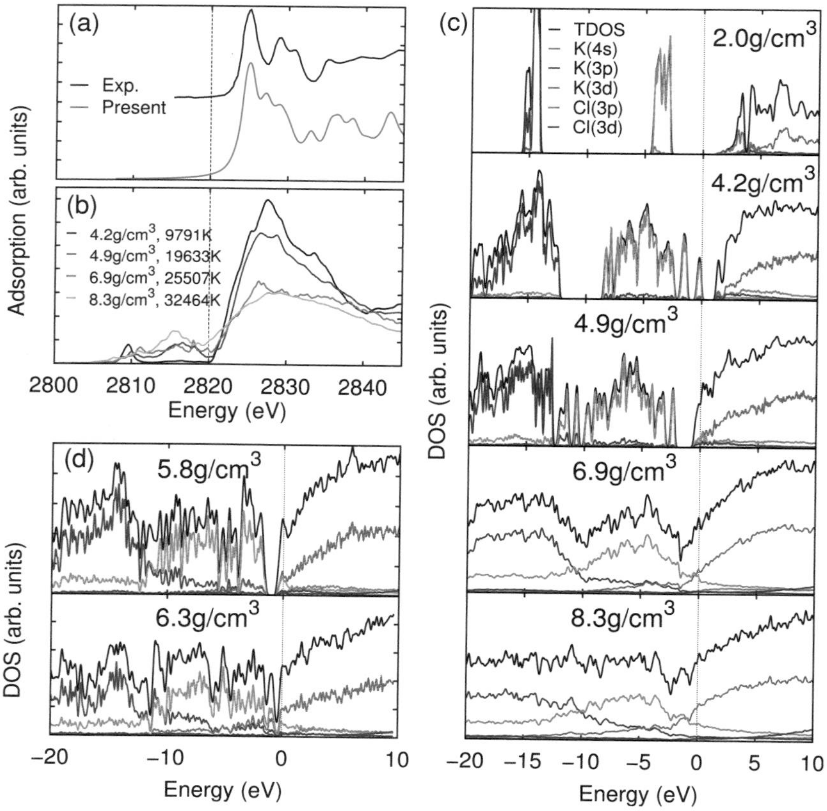

FIG. 3. K-XAS of Cl and corresponding PDOS for KCl. (a) XAS calculated under ambient conditions and corresponding experimental result;[47] and (b) XAS under typical warm dense conditions. (c) Corresponding PDOS for (a) and (b), where an energy shift of 2820 eV is applied to move the chemical potential to the vicinity of the origin point. (d) PDOS before and after the vanishing of energy gap.

TABLE II. Contributions of ionic kinetic pressure $P_{i,k}$, electronic kinetic pressure $P_{e,k}$, and virial pressure P_{pot} to the total pressure (in GPa) at selected electronic temperature and ionic temperature for thermal states $\rho = 3.9\,g/cm^3$ and $4.9\,g/cm^3$ corresponding to typical ionization states along the principal Hugoniot.

		3.9 g/cm^3				4.9 g/cm^3			
		$P_{i,k}$	$P_{e,k}$	P_{pot}	Total	$P_{i,k}$	$P_{e,k}$	P_{pot}	Total
$T_i = 0.5\,eV$	$T_e = T_i$	5.07	1262.82	−1204.83	63.06				
	$T_e = (5/9)T_i$	5.04	1268.42	−1205.98	67.47				
	$T_e = 300\,K$	5.06	1267.07	−1205.67	66.46				
$T_i = 1.0\,eV$	$T_e = T_i$	10.10	1281.90	−1210.36	81.64				
	$T_e = (5/9)T_i$	10.06	1279.66	−1209.00	80.72				
	$T_e = 300\,K$	10.10	1280.92	−1208.90	82.12				
$T_i = 2.0\,eV$	$T_e = T_i$	20.19	1305.98	−1218.50	107.67	25.37	1699.98	−1547.22	178.13
	$T_e = (5/9)T_i$	20.20	1299.31	−1214.45	105.06	25.51	1694.25	−1541.46	178.29
	$T_e = 300\,K$	20.22	1296.38	−1212.94	103.66	25.39	1691.45	−1538.74	178.10
$T_i = 3.0\,eV$	$T_e = T_i$					38.02	1735.11	−1563.49	209.63
	$T_e = (5/9)T_i$					38.09	1714.11	−1550.48	201.72
	$T_e = 300\,K$					37.98	1705.38	−1543.42	199.94
$T_i = 4.0\,eV$	$T_e = T_i$					50.81	1779.85	−1582.51	248.16
	$T_e = (5/9)T_i$					50.78	1735.46	−1558.39	227.86
	$T_e = 300\,K$					50.72	1722.14	−1547.44	225.42

$\rho < 4.0\,g/cm^3$ ($T_e < E_g/2$) part in the Hugoniot curve. The calculation at $T_e = T_i$ gives a better estimation to the pressure for the $\rho > 6.3\,g/cm^3$ part, when the energy gap vanishes and the material is completely metallic. Using these two calculations as references, the deviation of the calculation can be estimated as the smallest difference from the $T_e = T_i$ or $T_e = (5/9)T_i$ curves to the experimental measurements. As displayed in Fig. 1, the largest deviation in the calculation takes place in the region between $\rho = 4\,g/cm^3$ and $\rho = 6\,g/cm^3$. Below that, the $T_e = 5/9\ T_i$ curve only has a small correction of 1–2 GPa to the $T_e = T_i$ curve because of the small magnitude of *R*(*T*), as displayed in Fig. 2(b). The largest deviation is about 50 GPa (~25%) at $\rho \sim 5.3\,g/cm^3$, including experimental uncertainties. This gives an estimation to the worst scenario in the Hugoniot calculation.

To clarify the underlying mechanism of the small influence of the approximate ionization, a detailed decomposition of pressure is given in Table II for selected electronic and ionic temperatures, corresponding to Hugoniot points at $\rho = 3.9$ and $4.9\,g/cm^3$. As displayed in Fig. 2(a), they represent the starting and ending points of the sharp transition of the band gap along the Hugoniot. In Table II, the pressure is divided into three parts according to their origins as $P(T,\rho) = P_{e,k}(T,\rho) + P_{i,k}(T,\rho) + P_{pot}(T,\rho)$. Subscripts *e*, *k* and *i*, *k* represent kinetic pressures of electrons and ions, respectively, and subscript *pot* refers to virial pressures contributed by ion-ion, electron-electron, and ion-electron interactions. According to the virial theorem,[15] P_k is proportional to $2E_k$, and P_{pot} is approximately proportional to E_{pot}.

Table II displays the change in P_{pot} and $P_{e,k}$ brought about by different ionizations. Generally, $P_{e,k}$ grows with the increase of ionization, whereas P_{pot} goes to the opposite direction. The fluctuation at low T_e ($T_e < 0.6\,eV$ in the table) is caused by electronic excitations from localized electronic states to non-localized states near chemical potential. This is a characteristic of ionic materials. For KCl, the fluctuation is caused by the excitation of the localized 3*p* electrons of Cl to the non-localized 4*s* and 3*d* electrons of K. Most of the changes are brought by the variation of $P_{e,k}$. The direction of the change depends on the competition between the average kinetic energies of the localized and non-localized electrons.

The cancellation between the increase of $P_{e,k}$ and decrease of P_{pot} leads to a much smaller change in the total pressure. For the Hugoniot point at $\rho = 3.9\,g/cm^3$ ($T_i \sim 1.0\,eV$), the $T_e = (5/9)T_i$ calculation gives a better estimation to the ionization. Its correction to the total pressure is less than 1 GPa, which is about 1% of the FPMD result. For the Hugoniot point at $\rho = 4.9\,g/cm^3$ ($T_i \sim 2.0\,eV$), the correction calculated with $T_e = (5/9)T_i$ is less than 0.5 GPa. The real correction should be much smaller than that, because the energy gap is now close to zero, as displayed in Fig. 2(a).

IV. K-SHELL XAS OF Cl AND EVOLUTION OF ELECTRONIC STRUCTURES

XAS provide information on both ionization and electronic structures. Owing to the thermal excitation of electrons in WDM, electronic structures 10–20 eV below the chemical potential can also be detected by XAS. This is quite different from the low temperature cases, where only electronic structures above or close to the chemical potential are available for X-ray transitions. Accordingly, terminologies commonly used in XAS of condensed matter should be used cautiously to avoid discrepancies. Here, we do not distinguish subcategories of XAS and focus on absorption properties near the beginning of K-absorption spectra (usually called the edge region) for Cl ions in the shock-generated warm dense KCl. Note that the absorption edge in WDM could be far below (or above in some extreme cases) the chemical potential, as will be illustrated in the following part of this section.

The K-absorption spectra under ambient conditions are compared with experimental measurements[47] in Fig. 3(a). It serves as a reference illustrating how the calculated spectra can be interpreted. Fig. 3(a) shows that the calculation can reproduce the general feature of the spectra with small difference in details. The energy gap issue is paid special attention in the calculation. A correction of 3.7 eV obtained from a separate G_0W_0 calculation[48] is added to the FPMD energy gap ($\sim$5 eV), so that the calculated gap value agrees with the experimental measurement of 8.69 eV.[11] Hybrid functional HSE06[49] only slightly improves the gap to about 6 eV, and therefore is not helpful to the gap issue. Note that the G_0W_0 correction is not included in the XAS calculation of warm dense states, because the accuracy of the method relies on the description of ionization in the FPMD method, which is generally overestimated before the vanishing of the gap. Some kind of self-consistent GW approach[50] is necessary to get a theoretically consistent result.

Fig. 3(b) displays calculated K-absorption spectra of warm dense KCl. The prominent feature of the spectra is the persistence of the energy gap (around 2820 eV) under compression, disappearing at a high density of $\sim$6.3 g/cm^3. The figure also shows that the position of the chemical potential, around 2820 eV, is insensitive to the variation of density and temperature, which is similar to what was found in warm dense Al.[34]

Projected density of states (PDOS) in Figs. 3(c) and 3(d) show that the electronic states across the gap have different origins. The majority of the electronic states below the gap consist of the 3*p* states of Cl and K; whereas those above the gap mainly come from the 4*s* and 3*d* states of K. The figures also show a strong hybridization between the 3*p* orbitals of Cl and K under compression. This trend is further illustrated in Fig. 3(b) by the increase of band width in the sub-band below the gap (below 2820 eV). According to the theory of pressure ionization,[51,52] this hybridization of electronic states of different ions is an evidence of the pressure ionization, i.e., a transition from localized states to non-localized states. This transition is also confirmed by the variation of wave functions of the 3*p* orbitals at $E = 2816$ eV, as displayed in Fig. 4 for warm dense states $\rho = 2.0$ g/cm^3 ($T = 300$ K) and $\rho = 5.8$ g/cm^3 ($T = 3.3$ eV). Here, 2816 eV is the energy of the maximum PDOS for the 3*p* states of Cl under ambient conditions, and the thermal state of $\rho = 5.8$ g/cm^3 is the last state in the calculation just before the closing of the gap. The PDOS after the closing of gap is also displayed in Fig. 3(d). These results show that the 3*p* states of both Cl and K are part of the continuum (non-localized states), and the gap around $E = 2820$ eV is in the middle of unbounded states above the bottom of the continuum.

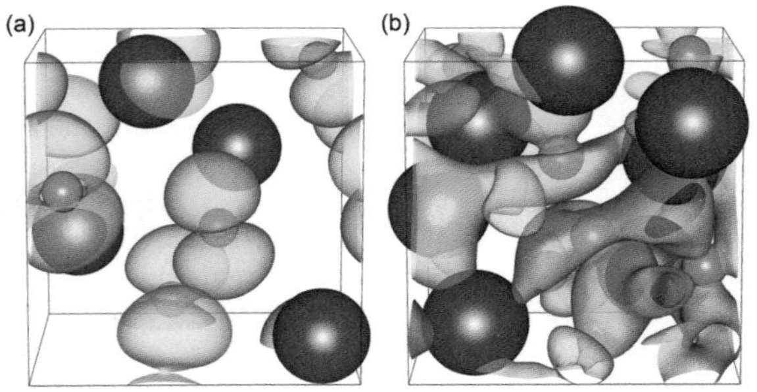

FIG. 4. Probability distribution of the 3*p* wave functions of KCl at $E = 2816$ eV. (a) The 3*p* wave functions of Cl under ambient conditions. (b) The hybrid 3*p* wave functions of Cl and K at 5.8 g/cm^3. The K and Cl ions are represented by purple and green balls, respectively. The isovalue is set to be 0.001.

Besides the absorption cross section σ, the transmission (or attenuation) of the material[8,10] was also measured in experiments. On many occasions, it was used to determine the absorption edge positions.[8,10] The transmission is proportional to $\exp(-\sigma\tilde{h})$, where $\tilde{h}$ is the areal number density. Fig. 5 displays the dependency of transmission spectra (corresponding to the K-absorption of Cl) on density and temperature. The $\tilde{h}$ here takes the value of 790 Å^{-2}, derived from recent experiment of Zhao *et al.*[8] Fig. 5(a) displays the variation of transmission as a function of temperature, at a fixed density of 5.5 g/cm^3. It shows that the absorption edge decreases with the growth of temperature, which is mainly caused by the thermal broadening of depopulated electronic states below the chemical potential. Fig. 5(b) presents the transmission spectra at fixed temperature of 3 eV, showing that the absorption edge is insensitive to the variation of density. Similar trend has also been found in warm dense Al,[34] indicating that the K-absorption edge is useful to diagnose the temperature of warm dense state for a variety of materials. The feature corresponding to the energy gap appears as a peak around 2820 eV in Fig. 5, which gives an explanation to the "bump" feature observed in the time-resolved transmission spectra measured by Bradley *et al.*[10]

To compare with the experiments of Zhao *et al.*,[8] in which the intensity of transmitted X-ray was measured, the intensity $I(E)$ is further derived from σ according to $I(E) = I_0(E)\exp[-\sigma(E)\tilde{h}]$, with $I_0(E)$ the intensity of backlighter. Fig. 6 supplies a comparison between calculated and measured $I(E)$. The background is deducted according to the left part of the measured spectra below the absorption edge. The intensity of backlighter[8] measured in a separate experiment is also included as references, and the energy scale is shifted

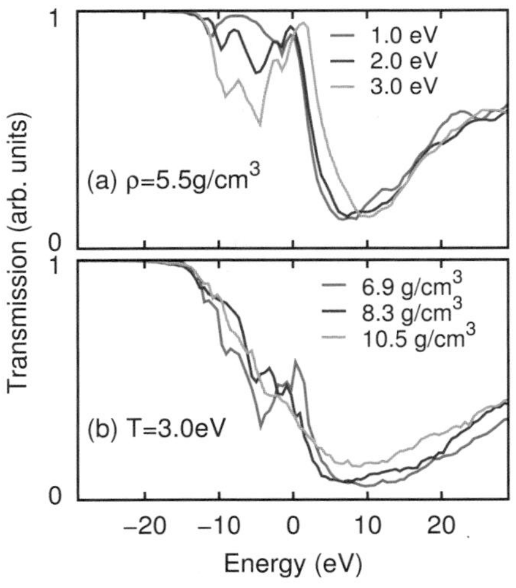

FIG. 5. Transmission spectra of warm dense KCl (a) at fixed density $\rho = 5.5$ g/cm^3; and (b) at fixed temperature $T = 3.0$ eV.

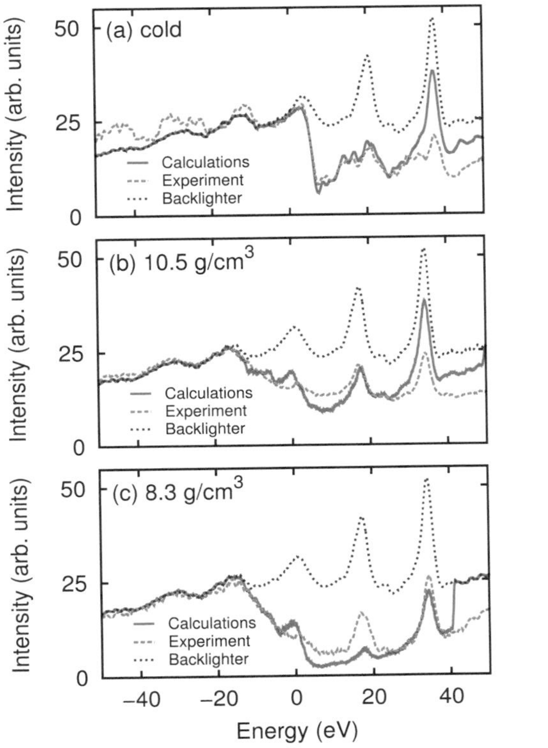

FIG. 6. Transmitted X-ray Intensity of KCl for (a) cold samples; (b) 10.5 g/cm^3; and (c) 8.3 g/cm^3. Experimental measurements and intensity of backlight are also presented[8] for comparison. The energy scale is shifted by 2820 eV to put the chemical potential at the vicinity of the origin point.

by 2820 eV so that the position of the chemical potential is located at the vicinity of the origin point. Fig. 6(a) displays the calculation of KCl under ambient conditions. It well reproduces the XAS up to ~30 eV above the absorption edge. Figs. 6(b) and 6(c) show that our calculation is in reasonable agreement with the experimental measurements up to the chemical potential. For the next 20 eV above that, the calculated intensity is lower than the experimental measurements by about 50%, which suggests that a refined theoretical treatment is necessary to deal with the spectra above the energy gap in warm dense materials.

V. SUMMARY

In summary, we investigate the variation of electronic structures and the influence of the approximate description of ionization to the calculation of pressure through first-principles calculation of principal Hugoniot and XAS for warm dense KCl. We show that pressure ionization and thermal smearing are the major factors to limit the deviation in the calculation of pressure. They prevent global accumulation of such deviations along the Hugoniot. In addition, the cancellation between kinetic pressure of electrons and virial pressure of interactions further reduces the error. The calculation of XAS shows that the band gap of KCl persists after the occurrence of the pressure ionization of the $3p$ electrons of Cl and K, which provide a detailed understanding to the evolution of electronic structures for WDM.

ACKNOWLEDGMENTS

We appreciate useful discussions with Jiamin Yang and Yang Zhao on the details of experiments. This work was financially supported by the NSFC (Grant No. 11274019).

[1]N. Nettelmann, A. Becker, B. Holst, and R. Redmer, Astrophys. J **750**, 52 (2012).

[2]J. D. Lindl, P. Amendt, R. L. Berger, S. G. Glendinning, S. H. Glenzer, S. W. Haan, R. L. Kauffman, O. L. Landen, and L. J. Suter, Phys. Plasmas **11**, 339 (2004).

[3]J. Lindl, Phys. Plasmas **2**, 3933 (1995).

[4]O. A. Hurricane, D. A. Callahan, D. T. Casey, P. M. Celliers, C. Cerjan, E. L. Dewald, T. R. Dittrich, T. Doppner, D. E. Hinkel, L. F. B. Hopkins, J. L. Kline, S. Le Pape, T. Ma, A. G. MacPhee, J. L. Milovich, A. Pak, H.-S. Park, P. K. Patel, B. A. Remington, J. D. Salmonson, P. T. Springer, and R. Tommasini, Nature **506**, 343 (2014).

[5]W. Y. Zhang and X. T. He, J. Phys.: Conf. Ser. **112**, 032001 (2008).

[6]L. M. Barker and R. E. Hollenbach, J. Appl. Phys. **43**, 4669 (1972).

[7]J. E. Miller, T. R. Boehly, A. Melchior, D. D. Meyerhofer, P. M. Celliers, J. H. Eggert, D. G. Hicks, C. M. Sorce, J. A. Oertel, and P. M. Emmel, Rev. Sci. Instrum. **78**, 034903 (2007).

[8]Y. Zhao, J. Yang, J. Zhang, G. Yang, M. Wei, G. Xiong, T. Song, Z. Zhang, L. Bao, B. Deng *et al.*, Phys. Rev. Lett. **111**, 155003 (2013).

[9]J. Zhang, H. Li, Y. Zhao, G. Xiong, Z. Yuan, H. Zhang, G. Yang, J. Yang, S. Liu, S. Jiang, Y. Ding, B. Zhang, Z. Zheng, Y. Xu, X. Meng, and J. Yan, Phys. Plasmas **19**, 113302 (2012).

[10]D. Bradley, J. Kilkenny, S. Rose, and J. Hares, Phys. Rev. Lett. **59**, 2995 (1987).

[11]D. Roessler and W. Walker, Phys. Rev. **166**, 599 (1968).

[12]R. M. More, K. H. Warren, D. A. Young, and G. B. Zimmerman, Phys. Fluids **31**, 3059 (1988).

[13]R. R. S. T. Frank Graziani, *Frontiers and Challenges in Warm Dense Matter*, Lecture Notes in Computational Science and Engineering Vol. 96, edited by M. Desjarlais (Springer, 2014).

[14]J. P. Perdew, K. Burke, and M. Ernzerhof, Phys. Rev. Lett. **77**, 3865 (1996).

[15]R. Martin, *Electronic Structure: Basic Theory and Practical Methods* (Cambridge University Press, 2008).

[16]D. Beule, W. Ebeling, A. Förster, H. Juranek, S. Nagel, R. Redmer, and G. Röpke, Phys. Rev. B **59**, 14177 (1999).

[17]B. Holst, R. Redmer, and M. P. Desjarlais, Phys. Rev. B **77**, 184201 (2008).

[18]T. J. Lenosky, S. R. Bickham, J. D. Kress, and L. A. Collins, Phys. Rev. B **61**, 1 (2000).

[19]A. Kietzmann, B. Holst, R. Redmer, M. P. Desjarlais, and T. R. Mattsson, Phys. Rev. Lett. **98**, 190602 (2007).

[20]S. Mazevet and G. Zérah, Phys. Rev. Lett. **101**, 155001 (2008).

[21]D. Minakov, P. Levashov, K. Khishchenko, and V. Fortov, J. Appl. Phys. **115**, 223512 (2014).

[22]D. Alfe, G. Price, and M. Gillan, Phys. Rev. B **65**, 165118 (2002).

[23]N. D. Mermin, Phys. Rev. **137**, A1441 (1965).

[24]P. Giannozzi, S. Baroni, N. Bonini, M. Calandra, R. Car, C. Cavazzoni, D. Ceresoli, G. L. Chiarotti, M. Cococcioni, I. Dabo *et al.*, J. Phys.: Condens. Matter **21**, 395502 (2009).

[25]P. E. Blöchl, Phys. Rev. B **50**, 17953 (1994).

[26]N. Holzwarth, A. Tackett, and G. Matthews, Comput. Phys. Commun. **135**, 329 (2001).

[27]S. Nosé, J. Chem. Phys. **81**, 511 (1984).

[28]I. Zel'dovich, Y. Zel'ovich, and Y. Raizer, *Physics of Shock Waves and High-Temperature Hydrodynamic Phenomena*, Dover Books on Physics (Dover Publications, 2002).

[29]V. Al'tshuler, M. Pavlovskii, L. Kuleshova, and G. Simakov, Sov. Phys.: Solid State **5**, 203 (1963).

[30]S. Kormer, M. Sinitsyn, A. Funtikov, V. Urlin, and A. Blinov, Sov. Phys. JETP **20**, 811 (1965) [J. Exptl. Theoret. Phys. (U.S.S.R.) **47**, 1202 (1964) (in Russian)].

[31]D. Hayes, J. Appl. Phys. **45**, 1208 (1974).

[32]T. Mashimo, K. Nakamura, K. Tsumoto, Y. Zhang, S. Ando, and H. Tonda, J. Phys.: Condens. Matter **14**, 10783 (2002).

[33]V. Recoules and S. Mazevet, Phys. Rev. B **80**, 064110 (2009).

[34]S. Zhang, S. Zhao, W. Kang, P. Zhang, and X.-T. He, e-print arXiv:1502.06082 [physics.plasm-ph].
[35]M. Taillefumier, D. Cabaret, A.-M. Flank, and F. Mauri, Phys. Rev. B **66**, 195107 (2002).
[36]A. Benuzzi-Mounaix, F. Dorchies, V. Recoules, F. Festa, O. Peyrusse, A. Levy, A. Ravasio, T. Hall, M. Koenig, N. Amadou *et al.*, Phys. Rev. Lett. **107**, 165006 (2011).
[37]C. Gougoussis, M. Calandra, A. P. Seitsonen, and F. Mauri, Phys. Rev. B **80**, 075102 (2009).
[38]C. J. Pickard and F. Mauri, Phys. Rev. B **63**, 245101 (2001).
[39]N. Troullier and J. L. Martins, Phys. Rev. B **43**, 1993 (1991).
[40]R. Ramis, R. Schmalz, and J. Meyer-Ter-Vehn, Comput. Phys. Commun. **49**, 475 (1988).
[41]J. Clérouin, Y. Laudernet, V. Recoules, and S. Mazevet, Phys. Rev. B **72**, 155122 (2005).
[42]G. E. Duvall and R. A. Graham, Rev. Mod. Phys. **49**, 523 (1977).
[43]P. Kowalski, S. Mazevet, D. Saumon, and M. Challacombe, Phys. Rev. B **76**, 075112 (2007).
[44]W. Ebeling, Physica **38**, 378 (1968).
[45]J. Dai, D. Kang, Z. Zhao, Y. Wu, and J. Yuan, Phys. Rev. Lett. **109**, 175701 (2012).
[46]T. Sjostrom and J. Daligault, Phys. Rev. B **90**, 155109 (2014).
[47]J. Trischka, Phys. Rev. **67**, 318 (1945).
[48]M. S. Hybertsen and S. G. Louie, Phys. Rev. B **34**, 5390 (1986).
[49]J. Heyd, G. E. Scuseria, and M. Ernzerhof, J. Chem. Phys. **118**, 8207 (2003).
[50]M. van Schilfgaarde, T. Kotani, and S. Faleev, Phys. Rev. Lett. **96**, 226402 (2006).
[51]D. Saltzman, *Atomic Physics in Hot Plasmas* (Oxford University Press, 1998).
[52]D. Kremp, M. Schlanges, and T. Bornath, *Quantum Statistics of Nonideal Plasmas, Atomic, Optical, and Plasma Physics* (Springer, 2005).

www.nature.com/scientificreports

SCIENTIFIC REPORTS

OPEN

Received: 08 October 2015
Accepted: 08 January 2016
Published: 08 February 2016

First-Principles Investigation to Ionization of Argon Under Conditions Close to Typical Sonoluminescence Experiments

Wei Kang[1,2], Shijun Zhao[1,2], Shen Zhang[1,2], Ping Zhang[1,3], Q. F. Chen[4] & Xian-Tu He[1,2,3]

Mott effect, featured by a sharp increase of ionization, is one of the unique properties of partially ionized plasmas, and thus of great interest to astrophysics and inertial confinement fusion. Recent experiments of single bubble sonoluminescence (SBSL) revealed that strong ionization took place at a density two orders lower than usual theoretical expectation. We show from the perspective of electronic structures that the strong ionization is unlikely the result of Mott effect in a pure argon plasma. Instead, first-principles calculations suggest that other ion species from aqueous environments can energetically fit in the gap between the continuum and the top of occupied states of argon, making the Mott effect possible. These results would help to clarify the relationship between SBSL and Mott effect, and further to gain an better understanding of partially ionized plasmas.

A sharp rising in ionization is a remarkable feature of Mott effect[1] in partially degenerate plasmas. The effect is also known as Mott transition[2], pressure ionization[3], or plasma phase transition (PPT)[4] on various occasions. It indicates the disintegration of bounded states and the formation of extended states with reduced ionization potential (IP). Usually, a companion structural transition is also expected together with the steep change of ionization[5]. It is arguably the most prominent nonideality character of partially ionized plasmas of strong coupling[2], a focus of current studies of warm dense matter (WDM). Mott effect is therefore of particular interest to astrophysics[6,7] as well as inertial confinement fusion[8,9].

The physical origin of Mott effect, however, has been under debate since it was first noticed in 1940's[10]. It is considered an result either derived from the competition between geometric confinement and the spatial extension of wave functions, i.e., the picture of pressure ionization[2,3], or from a balance between screened Coulomb interaction and quantum repulsion, i.e., the picture of PPT[4,11]. These two pictures are closely related but have subtle differences. Nevertheless, both theories suggest a transition taking place at such density that atomic distance is comparable with the diameter of the outer-most electronic orbital. This prediction was well supported by recent experiments and first-principles calculations on dense hydrogen systems[12–16], where a phase transition with abrupt increasing in ionization was observed around $\rho \sim 1.0\,\mathrm{g/cm^3}$, in line with the theoretical expectation.

However, in single bubble sonoluminescence (SBSL) experiments[17–19], where noble gas, e.g., argon (Ar), was compressed by the collapse of the cavity in aqueous solutions[20], a steep increase of ionization was observed at a number density $\rho_N \sim 10^{21}/\mathrm{cm^3}$, i.e., $\sim 0.07\,\mathrm{g/cm^3}$, and at a temperature $T \sim 1\,\mathrm{eV}$, i.e., 11000 K. The ρ_N corresponds to an average atomic distance of 10 Å, which is much larger than the length scale (~2 Å) of the outer-most $3p$ electronic orbital of Ar atoms. A theory developed recently by Kappus *et al.*[21] suggested that sonoluminescence was essentially the result of Mott effect in noble-gas-element plasmas. They proposed a classical chemical model with Debye-Hückle screening, which showed that the lowering of IP was greatly enhanced by the ionization process, making a sharp increase in ionization possible at a much lower density than that predicted by prior theories[22]. If so, this would lead to a significant revision to the physical picture of Mott effect, and consequently an updated understanding of the ionization mechanism of partially degenerate plasmas.

[1]HEDPS, Center for Applied Physics and Technology, Peking University, Beijing 100871, China. [2]College of Engineering, Peking University, Beijing 100871, China. [3]Institute of Applied Physics and Computational Mathematics, Beijing 100088, China. [4]National Key Laboratory of Shock Wave and Detonation Physics, Institute of Fluid Physics, P. O. Box 919-102, Mianyang, Sichuan, China. Correspondence and requests for materials should be addressed to W.K. (email: weikang@pku.edu.cn) or X.-T.H. (email: xthe@iapcm.ac.cn)

However, as was pointed out by Redmer and Holst[1], the appearance of Mott effect in a theory is sensitively relied on the details of the effective interaction at the distance comparable with the size of an atom. It would be desirable if a method can take as much as possible the details of the interaction into consideration. Following Knudson *et al.*[16], Morales *et al.*[12], Lorenzen *et al.*[23], and Scandolo[24], one can alternatively use first-principles method to detect Mott effect. Their method probes the change in pressure induced by the companion structural transition. The method works well as long as its resolution overcomes the finite-size effect[12,23,24], which, as they revealed, could be attained in a calculation composed of 400–500 atoms. However, it should be noted that the method itself is difficult to exclude the possibility of a transition beyond its resolution. Under this condition, information provided by other properties is then critical to pin down the problem. Electronic structures are a suitable candidate to serve this purpose. They are expected to go through a remarkable change in order to accommodate the sharp increase of ionization when Mott effect takes place. As a consequence, the effective IP will be greatly reduced, which can be seen through a simple estimation with Saha's equation[3]. More importantly, electronic structures are less constrained by the finite-size effect. They are locally determined near the chemical potential (or Fermi level at zero temperature) in an amorphous system because of the randomness of ionic positions. This localization was first illustrated by Anderson[25] and was observed in electronic structure calculations on a variety of WDM systems[26–28]. It greatly reduces the system size in a calculation from over 400 atoms to ~40 atoms without notable changes in electronic structures.

In this work, we concentrate on electronic structures of Ar at equilibrium, calculated using the first-principles molecular dynamics (FPMD) method. In the calculation, ions are described via classical Newtonian molecular dynamics (MD), whereas electrons are treated quantum-mechanically through density functional theory (DFT). Two levels of quantum descriptions are used in the calculation. Mean-field effects of excited electrons are accounted for through finite-temperature DFT (FTDFT) formula proposed by Mermin[29], which assigns the population of excited electrons according to the Fermi-Dirac distribution. Dynamical effects associated with electronic excitations[30], are taken into consideration through real-time time-dependent DFT (TDDFT) method[31], which explicitly describes the evolution of electronic wave functions with reasonable accuracy[32,33]. In both methods, local density approximation (LDA)[34] is used to describe the exchange-correlation interaction between electrons. These two FPMD methods are referred as FTDFT+MD and TDDFT+MD respectively hereinafter. A detailed account can be found in the methods section.

The inaccuracy brought about by the methodology is also considered. It has been well known that LDA underestimates the energy gap between occupied and unoccupied states at low temperature, usually 40–60% for semiconductors and insulators[27,35]. Electronic structures thus calculated favor the occurrence of Mott effect as a result of severe overestimation to the ionization. We show that even taking the overestimation into account, the sharp increase of ionization in SBSL experiments is unlikely the result of Mott effect in a pure Ar plasma system. Instead, ions and cations originated from dissociated water molecules are energetically located in the energy gap. They possibly provide the necessary energy levels to essentially reduce the effective IP, which can further trigger the Mott effect.

Results and Discussion

Thermodynamical Properties. In a typical SBSL experiment containing Ar, a sharp increase of ionization takes place with strong visible light emitted at a pressure over 4000 bar and at a temperature ~1.5 eV (17000 K)[20,36]. This temperature is far beyond the liquid-gas critical temperature ~150 K[37]. The number density of Ar ion n_i is approximately $4 \times 10^{21}/\mathrm{cm}^3$, i.e., 0.25 g/cm^3, about half of the number density at the liquid-gas critical point. However, surprisingly, the number density of electrons n_e is roughly the same as or even higher than n_i, making the degree of ionization $\alpha \geq 1$. The coupling constant[2] Γ is then greater than 2, which indicates the Ar plasma is strongly coupled.

Testing calculation on experimental conditions shows that electronic structures and IP are essentially the same as those of an isolated Ar atom. The interaction between Ar atoms is thus negligible according to the quantum perturbation theory. Compared with experimental value of 15.7 eV, the calculation (10.3 eV) underestimates the IP by 33%. α is therefore overestimated by about 5 times, as estimated using $\alpha \sim \exp(-I_0/2T)$[38] with I_0 the magnitude of IP. Even with this overestimation, the calculated α is less than 3%, which is about twice the value of 1.4 % given by the self-consistent fluid variational theory(SCFVT)[39] and essentially implies no Mott effect.

However, the absence of Mott effect is possibly caused by the drift of temperature and density predicted by the calculation, which is inevitable for the FTDFT+MD method due to various approximations involved in the calculation, as has been noticed in the calculation of hydrogen[12,16]. A detailed discussion on the influence of the approximations, including the semi-local approximation to the exchange-correlation functional and the neglecting of nuclear quantum effects, is carefully presented in the recent work of Knudson *et al.*[16]. To further exclude this possibility, a series of FTDFT+MD calculations are carried out surrounding the experimental density and temperature. Based on previous theoretical estimations[4,11,22], which predicted that Mott effect took place at a length scale comparable to the diameter of the outer-most electronic orbitals, we consider the density between 1 g/cm^3 and 8.34 g/cm^3. The upper limit of 8.34 g/cm^3 corresponds to an average atomic distance ~2 Å, slightly larger than the diameter $d = 1.94$ Å of the $3p$ orbital for an isolated Ar atom. The diameter of the $3p$ orbital is calculated as $2\sqrt{\langle 3p|\hat{r}^2|3p\rangle}$, where $\hat{r}$ is the radius operator, and the wave function $|3p\rangle$ is extracted from the all electron atomic DFT-LDA calculation of Ar atom. This definition follows Atzeni and Meyer-ter-Vehn in the discussion of the lowering of ionization potential[40]. The lower bound corresponds to the density at which interactions between atoms start to make observable changes in their electronic structures. The temperature concerned is between 10^3 K and 2×10^4 K. Higher temperature is not considered because both theories and experiments showed that Mott effect prefers lower temperature. It can only take place below some critical temperature[4,11–13,21,22].

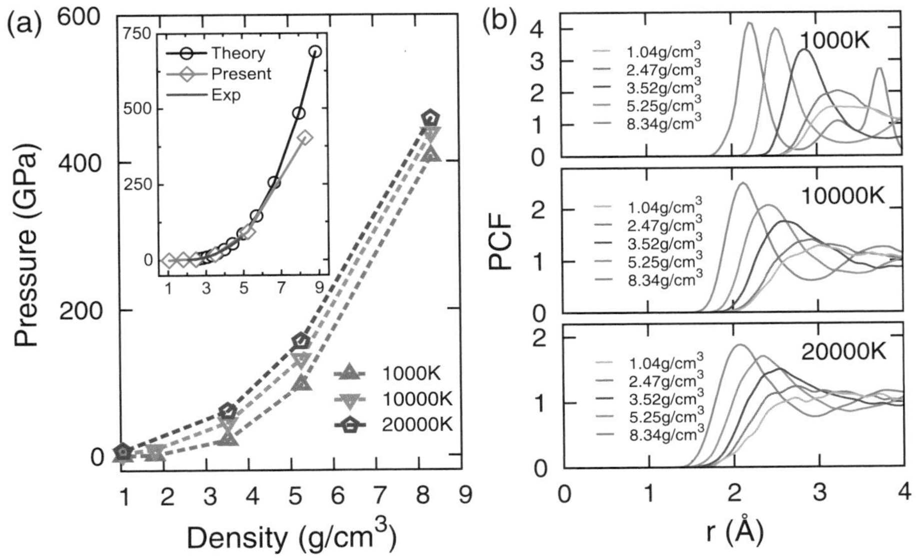

Figure 1. (**a**) Pressure isotherms of Ar calculated using the FTDFT+MD method at $T = 10^3$ K, 10^4 K, and 2×10^4 K. Displayed in the inset is the pressure isotherm at room temperature, compared with experimental measurements of Shimizu *et al.*[41], and Monte Carlo calculation of Ross *et al.*[42] with an empirical pair interaction. (**b**) Pair correlation function $g(r)$ along the isotherms displayed in (**a**), which indicates short-range structures of Ar plasma.

Figure 1(a) displays pressure isotherms of Ar calculated using the FTDFT+MD method at $T = 10^3$ K, 10^4 K, and 2×10^4 K, where isotherms of $T = 10^3$ K and $T = 2 \times 10^4$ K define the boundary of our calculations, and the isotherm of $T = 10^4$ K indicates the median of the temperature measured in SBSL experiments. To give an estimation to the reliability of the FTDFT+MD method, also displayed in the inset is the calculated isotherm at room temperature, compared with experimental results measured by Brillouin scattering[41] and theoretical results calculated using Monte Carlo method together with an empirical pair interaction[42]. The calculated pressure agrees well with the experimental result, but deviates from the Monte Carlo calculation at densities higher than 8 g/cm^3. This deviation is expected because the atomic distance at that density is only slightly larger than the diameter (1.94 Å) of the $3p$ orbital of Ar atom, where tunneling effect cannot be well accounted for by the empirical pair potential fitted from experimental data at low densities. In Fig. 1(b), pair correlation functions $g(r)$ along the isotherms are presented to reveal structural changes. When a first-order phase transition takes place, $g(r)$ usually undergoes a significant change. This, however, is only observed at $\rho = 8.34$ g/cm^3 and $T = 10^3$ K in Fig. 1(b), which is an extreme high density and low temperature condition compared with that in SBSL experiments[20,36]. Except that case, $g(r)$ and isotherms in Fig. 1 display a smooth change with respect to T and ρ. It shows that structural transition is not a prominent feature near the experimental conditions, which is consistent with the usual understanding of Mott effect[2,4,5], and is supported by FTDFT+MD calculations on warm dense helium[43].

Electronic Structures. In addition to thermodynamical properties, the electronic structure of Ar plasma is also calculated using the FTDFT+MD method. A strong reduction (over 75 %) in IP is expected according to the estimation of Kappus *et al.*[44] on experimental results of Xe. The IP, estimated as the energy gap across the chemical potential in the calculation, can then be used as an indication to the occurrence of Mott effect. In the FTDFT+MD method, the IP lowering is assessed with the help of Ewald summation technique[34], and the screened interaction between Ar ions is calculated using DFT through Hellmann-Feynman theorem[34] or Ehrenfest theorem[30]. These treatments are more sophisticated than the Debye-Hückle screening model and the IP lowering model (i.e., $\Delta(IP) = e^2/\sqrt{T/8\pi n_e e^2}$). In particular, it self-consistently includes the enhancement of ionization to the lowering of IP. Calculations on hydrogen systems[12,16,23,24] have shown that they are capable to predict the existence of Mott effect. However the calculated transition point sensitively depends on the details of the calculation, e.g., the choice of density functional and the inclusion of nuclear quantum effect, as has been carefully discussed by Knudson *et al.*[16].

Total density of electronic states (TDOS) of Ar plasma along the $T = 10^4$ K isotherm is presented in Fig. 2. It is a convenient way to display the electronic structure and the IP. Note that the position of chemical potential in the figure has been shifted to the origin point of the energy scale. In Fig. 2(a–c), the TDOS have three separate parts associated with different atomic orbitals: The band below -15 eV comes from the $3s$ states, the segment in the middle is from the $3p$ states, and the continuum above the chemical potential is formed by the bottom of conduction bands. The $3s$ and $3p$ states in the figure are broadened significantly due to thermal effects and

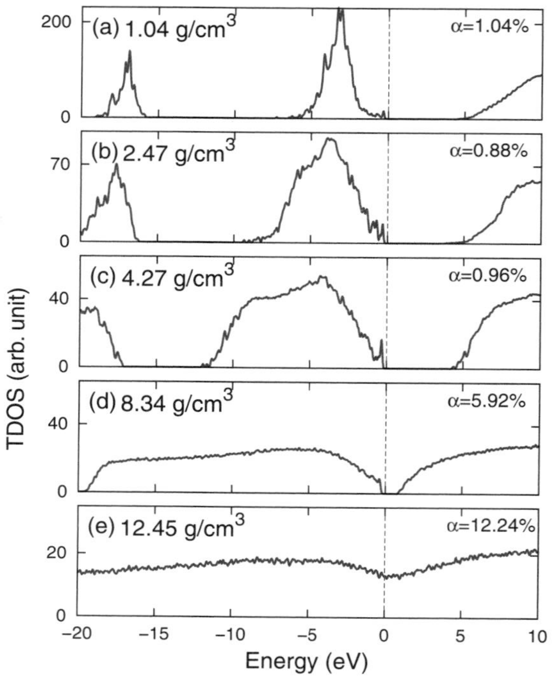

Figure 2. Total density of states(TDOS) of Ar plasma along the $T = 10^4$ K isotherm, calculated using the FTDFT+MD method. The degree of ionization α is also displayed together with the TDOS.

tunneling effects. The calculated IP is ~5 eV in a large range of density until $\rho \sim 4.3\,\text{g/cm}^3$, from which the IP starts to shrink to zero. This tendency is similar to that observed in KCl under shock compression[27]. Figure 2(d) displays the TDOS at $\rho = 8.34\,\text{g/cm}^3$, corresponding to an atomic distance comparable with the average diameter of the outer-most 3p orbital of Ar. The small IP (~0.8 eV) suggests that the Ar plasma is close to a nonmetal-metal transition point. The merging of the 3s and 3p states also implies a remarkable change in the electronic structure. Figure 2(e) shows the TDOS of a metallic state at $\rho = 12.45\,\text{g/cm}^3$. The disappearing of energy gap leads to an α over 12%, which is twice of that for $\rho = 8.34\,\text{g/cm}^3$ in Fig. 2(d). If one takes the nonmetal-metal transition as the indication of Mott effect, the variation of electronic structures displayed in Fig. 2 agrees with the usual understanding of Mott effect, i.e., the effect takes place where atomic distance is comparable with the diameter of the outer-most electronic orbital. No evidence is found to support Mott effect at lower density.

The influence of temperature is presented in Fig. 3, where TDOS of selected temperatures along the isochore of $\rho = 1\,\text{g/cm}^3$ are displayed. When temperature increases from $T = 10^3$ K to $T = 2 \times 10^4$ K, the 3s and 3p TDOS are broadened by about twice, and the IP decreases smoothly from ~7 eV to ~4 eV. No abrupt change of IP is observed in the temperature range. As has been pointed out by the PPT model[2] and the pressure ionization model[4], Mott effect takes place below a critical temperature, at which thermal energy (effectively repulsive) and screened Coulomb energy (effectively attractive) arrive at a balance. So, when there is no indication of Mott effect at low temperature, the effect is less likely to be found at higher temperature. Even lower temperature gives rise to a negligible α, which is unlikely to reach an $\alpha \sim 1$ as SBSL experiments suggested.

Influence of Dynamical Screening. To understand the role of dynamical screening effect in determining electronic structures, which is not sufficiently accounted for in the FTDFT+MD method, we further calculate the ionization ratio of Ar under typical conditions with the influence of dynamical screening included using the TDDFT method. It gives an explicit description to the real-time evolution of excited electrons by solving the time-dependent Kohn-Sham equation using adiabatic local density approximation (ALDA), and has been applied to a few occasions, e.g., atomic collisions[45], laser-solid interactions[33], and optical responses of solids[46], with reasonable results. Note that, because of the locality of ALDA[30], long range excitation process is not fully described in the TDDFT+MD method. To take this process into consideration, a long range exchange correlation functional is necessary. In the TDDFT+MD method, only the temperature of ions are controlled by a Nosé-Hoover heat reservoir[47], the electrons are thermalized through ion-electron interaction. It has been noticed that[48], without external time-dependent sources, such as photons, it is extremely inefficient for the TDDFT+MD method to thermalize electrons from their ground states to a desired temperature through the ion-electron interaction[48]. A more rigorous discussion on the thermalization is provided by Modine and Hatcher[48]. Alternatively, as adopted in the work for Ar with a large band gap, one can boost the thermalization process by increasing the ion temperature

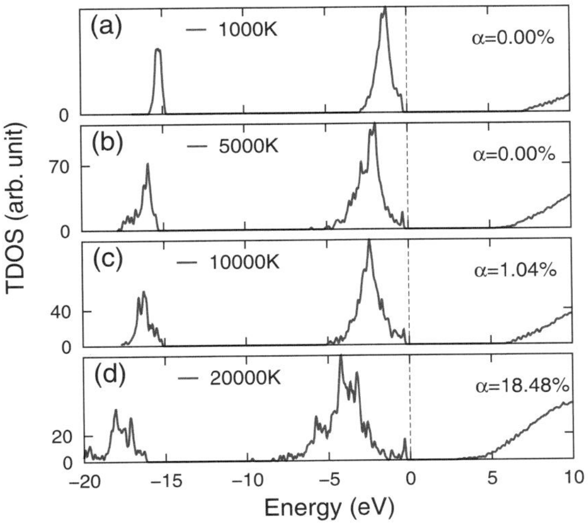

Figure 3. Total density of states (TDOS) of Ar plasma along the $\rho = 1\,g/cm^3$ isochore at selected temperatures, calculated using the FTDFT+MD method. α is also displayed to show the variation of ionization.

artificially (10–20 times of the desired temperature in our work) at the early stage of the simulation. When the electron system arrives at a temperature close to the desired one, the ion temperature is switched back to the desired temperature. The electrons can then reach the desired state much faster. In the actual calculation, the optimal timing to switch the ion temperature is determined by a few of trials case by case. Details of the method are referred to the methods section.

Figure 4 displays the evolution of α at two representative densities $\rho = 3.6\,g/cm^3$ and $\rho = 10.5\,g/cm^3$. Figure 4(a) is calculated with an artificial boost of ionization at the beginning to accelerate the calculation. It shows that the system has an α of 17.2% at $T = 2 \times 10^4\,K$ for $\rho = 3.6\,g/cm^3$, which is similar to the value calculated using the SCFVT method[39] but much lower than the value ($\alpha \geq 1$) observed in the experiments. An indication of typical Mott effect is illustrated in Fig. 4(b) at $\rho = 10.5\,g/cm^3$. When the energy gap closes and the system becomes metallic, stepwise multi-ionization processes are observed, and α reaches 12% at 0.48 ps (20000 a.u.) for $T = 5 \times 10^3\,K$ without artificial boost to the ionization. Further increase in α is observed when temperature rises, and the ionization process does not stop in our longest calculation at $T = 2 \times 10^4\,K$ after 0.72 ps (30000 a.u.) In general, Fig. 4 shows that dynamical screening included in the TDDFT+MD calculation does not induce a sharp transition at low density, and therefore, the physical picture presented by the FTDFT+MD calculation is qualitatively reliable.

Influences of Other Ion Species. From the perspective of electronic structures, we have shown that the strong ionization observed in the Ar SBSL experiments is quite unlikely the result of Mott effect of pure Ar plasma. However, it is still possible to obtain an effective IP small enough to meet the requirement of Mott effect when a mixture of plasmas composed of several ion species are considered. Energetically, this can be achieved by inserting energy levels of other ion species into the gap of Ar plasma. Indeed, as pointed out by chemical reaction models[49] and experiments[20], ionized water and other ion species from the solution can also exist inside the collapsing bubble in a SBSL experiment. Figure 5 shows the effect of water mixed with Ar plasma at $\rho = 1.7\,g/cm^3$, $T = 10^4\,K$. This state has a density close to the lowest density bound in our calculation at typical SBSL temperatures, and is expected to have the least possibility to present the Mott effect according to the trend illustrated in the FTDFT+MD and TDDFT+MD calculations. To distinguish the contribution of different ion species, the density of states is decomposed into projected density of states (PDOS) according to the composing elements (Ar, O, and H) and their electronic orbitals. For pure Ar plasma, as displayed in Fig. 5(a), a large IP of 5.3 eV is observed. When H_2O at the same number density ($N_{H_2O} : N_{Ar} = 9:9$) is introduced, as displayed in Fig. 5(b), the density of states near the chemical potential is then dominated by the p orbital of O, originated from OH^- radicals or H_2O molecules. At this temperature ($T = 10^4\,K$), OH^- does not have enough energy to decompose into O^{2-} and H^+. So, the contribution of O^{2-} is negligible. As a result, the energy gap in Ar plasma is filled by energy levels associated with ionized water, and the effective IP of the mixed Ar and H_2O plasma is substantially reduced, making the strong ionization possible.

In summary, we show from the perspective of electronic structures that the strong ionization observed in the SBSL experiments is unlikely the result of Mott effect in a pure Ar plasma. Instead, a mixture of plasmas composed of Ar and water energetically favors the occurrence of strong ionization around $T = 10^4\,K$. Our first-principles investigation supports the traditional understanding of Mott effect, i.e., the Mott effect happens at the density where atomic distance is comparable with the diameter of the outer-most electronic orbitals. Although the

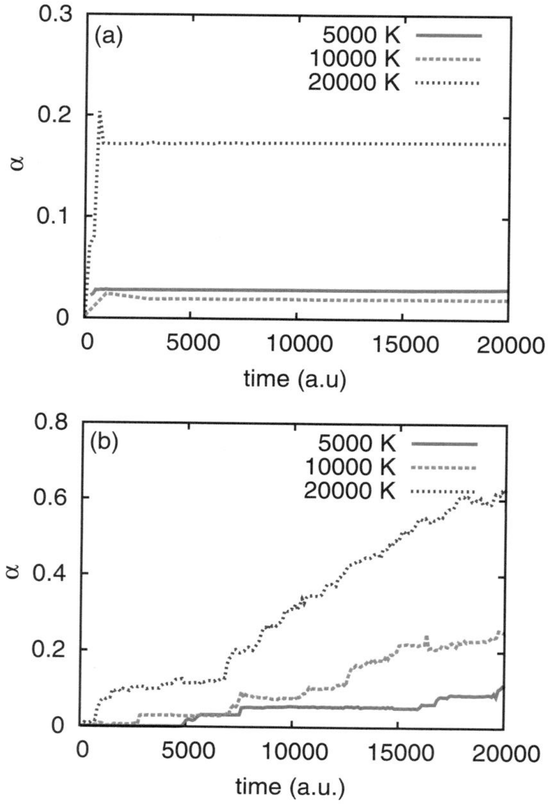

Figure 4. Degree of ionization α at $\rho = 3.6\,g/cm^3$ (**a**) and $\rho = 10.5\,g/cm^3$ (**b**), calculated using TDDFT+MD method. Degree of ionizations at various temperatures are distinguished by colors. They represent the evolution of ionization for typical non-metallic and metallic systems respectively. Note that different scales of degree of ionization are used in (**a**,**b**), and an artificial boost of ionization is employed at the early stage in the calculation of (**a**) to make the ionization reaches its equilibrium faster.

detailed mechanism of Mott effect is still an open question, this investigation would help to have a better understanding of Mott effect as well as its relationship to SBSL experiments.

Methods

FTDFT+MD simulations are carried out by the Vienna ab initio simulation package (VASP)[50], where Perdew-Zunger parametrization of local density approximation (LDA) is used for the exchange and correlation potential[34]. The 3s and 3p electrons are explicitly treated as valence in the calculation. The population of electronic states is assigned according to Fermi-Dirac distribution, and electronic temperature is set the same as that of ions to ensure local thermodynamical equilibrium[29]. Electron-ion interactions are modeled using the projector augmented wave (PAW) pseudo-potentials[51]. Wave functions are expanded up to a cutoff energy of 400 eV. Through the FTDFT+MD simulations, periodic boundary conditions are enforced and the Brillouin zone is sampled by the Γ point. Note that in the calculation of density of states, a $4 \times 4 \times 4$ Monkhorst-Pack k-point grid is used to ensure the accuracy. For pure Ar plasma, 64 atoms are included in the FTDFT+MD simulation, while for the mixture of H_2O and Ar, a total number of 36 atoms are used. The time step varies from 0.4–1 fs depending on the density and temperature. Each simulation lasts for 8000–20000 time steps, and the results presented are averaged over the last 2000 steps with a time interval of 200 steps. Degree of ionization is calculated following ref. 27, and the IP is calculated as the energy difference between the two adjacent Kohn-Sham orbitals across the chemical potential at finite temperature, or as the energy difference between the lowest unoccupied and the highest occupied Kohn-Sham orbital at $T = 0$. Note that the Kohn-Sham energy level is interpreted as an approximation to the quasiparticle energy following the convention of electronic structure calculation for extended systems[34], in order to keep in line with the many-body quantum-statistical theory of plasmas[2].

TDDFT+MD calculations are carried out using the OCTOPUS code[31]. ALDA functional is used to describe the exchange-correlation interaction. A total number of 8 Ar atoms are treated in the calculation through norm-conserving Troullier-Martins pseudopotential[52], with each having eight valence electrons. Periodic

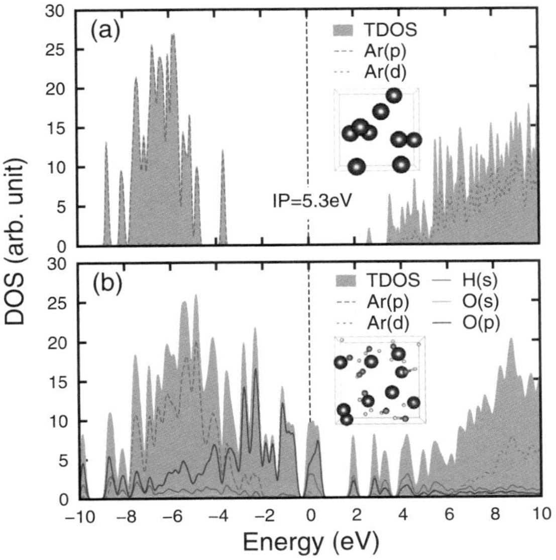

Figure 5. Total and projected density of states for pure Ar at $\rho = 1.7\,g/cm^3$ (**a**) and Ar mixed with H_2O of the same number density (**b**). Both are calculated at $T = 10^4\,K$ using the FTDFT+MD method. Ion configurations are also displayed in the insets, where golden spheres represent Ar, small white spheres are H, and red spheres are O.

boundary conditions are used throughout, and wave functions are represented using a three-dimensional spatial grid with 20 points on each direction (approximately equivalent to an energy cutoff of 50 Ry). The Brillouin zone is sampled by a $2 \times 2 \times 2$ Monkhorst-Pack shifted grid. In the calculation, ionic temperature is controlled through the Nosé-Hoover heat reservoir[47], while wave functions of electrons propagate according to the time-dependent Kohn-Sham equation with a time step about 0.1 a.u. (1 a.u. = 0.024 fs) up to 30000 a.u. To ensure the stability of this large time step, the propagator is calculated using the a Lanczos method[53] up to an order of 30, and the error of the propagator is kept less than 10^{-8} in each step.

In the framework of TDDFT, the degree of ionization $\alpha(t)$ is calculated as the charge excited from the ground state per atom. By projecting the time-dependent wave functions onto the ground-state Kohn-Sham wave functions at that instant, $\alpha(t)$ can be evaluated as

$$\alpha(t) = \frac{n_{ex}(t)}{N_a} = \frac{1}{N_a}\sum_{nn'k} w_k\left(\delta_{nn'} - |\langle\phi_{nk}^{gs}(t)|\varphi_{n'k}^{td}(t)\rangle|^2\right) \tag{1}$$

where N_a is the number of atoms, n_{ex} is the average number of excited electrons, w_k is the weight of each k-point in the summation of Brillouin zone, $\phi_{nk}^{gs}(t)$ is the ground state Kohn-Sham wave functions calculated using the ion configuration at instant t, and $\varphi_{n'k}^{td}(t)$ is the time-dependent electronic wave functions.

References

1. Redmer, R., Hensel, F. & Holst, B. *Metal-to-Nonmetal Transitions*. Springer Series in Materials Science (Springer Berlin Heidelberg, 2010).
2. Kremp, D., Schlanges, M. & Bornath, T. *Quantum Statistics of Nonideal Plasmas*. Atomic, Optical, and Plasma Physics (Springer, 2005).
3. Saltzman, D. *Atomic Physics in Hot Plasmas* (Oxford University Press, 1998).
4. Norman, G. E. Plasma phase transition. *Contrib. Plasma Phys.* **41,** 127 (2001).
5. Silvera, I. The insulator-metal transition in hydrogen. *Proc. Am. Thorac. Soc.* **107,** 12743 (2010).
6. Nettelmann, N., Becker, A., Holst, B. & Redmer, R. Jupiter models with improved ab initio hydrogen equation of state. *The Astrophysical Journal* **750,** 52 (2012).
7. Ternovoi, V., Kvitov, S., Pyalling, A., Filimonov, A. & Fortov, V. Experimental determination of the conditions for the transition of jupiter's atmosphere to the conducting state. *Journal of Experimental and Theoretical Physics Letters* **79,** 6–9 (2004).
8. Lindl, J. Development of the indirect drive approach to inertial confinement fusion and the target physics basis for ignition and gain. *Phys. Plasmas* **2,** 3933–4024 (1995).
9. Zhang, W. Y. & He, X. T. Recent progress and future prospects for ife in china. *Journal of Physics: Conference Series* **112,** 032001 (2008).
10. Zeldovich, Y. B. & Landau, L. D. On the relation between the liqquid and the gaseous states of metals. *Zh. Eksp. Teor. Fiz.* **18,** 194 (1943).
11. Ebeling, W. & Norman, G. Coulombic phase transitions in dense plasmas. *Journal of Statistical Physics* **110,** 861 (2003).
12. Morales, M. A., Pierleoni, C., Schwegler, E. & Ceperley, D. M. Evidence for a first-order liquid-liquid transition in high-pressure hydrogen from ab initio simulations. *Proc. Natl. Acad. Sci. USA* **107,** 12799–12803 (2010).

13. Dzyabura, V., Zaghoo, M. & Silvera, I. F. Evidence of a liquid–liquid phase transition in hot dense hydrogen. *Proc. Natl. Acad. Sci. USA* **110,** 8040–8044 (2013).
14. Weir, S., Mitchell, A. & Nellis, W. Metallization of fluid molecular hydrogen at 140 gpa (1.4 mbar). *Phys. Rev. Lett.* **76,** 1860 (1996).
15. Fortov, V. E. *et al.* Phase transition in a strongly nonideal deuterium plasma generated by quasi-isentropical compression at megabar pressures. *Phys. Rev. Lett.* **99,** 185001 (2007).
16. Knudson, M. D. *et al.* Direct observation of an abrupt insulator-to-metal transition in dense liquid deuterium. *Science* **348,** 1455–1460 (2015).
17. Brenner, M. P., Hilgenfeldt, S. & Lohse, D. Single-bubble sonoluminescence. *Rev. Mod. Phys.* **74,** 425–484 (2002).
18. Gaitan, D. F., Crum, L. A., Church, C. C. & Roy, R. A. Sonoluminescence and bubble dynamics for a single, stable, cavitation bubble. *J. Acoust. Soc. Am.* **91,** 3166 (1992).
19. Barber, B. P. & Putterman, S. J. Light scattering measurements of the repetitive supersonic implosion of a sonoluminescencing bubble. *Phys. Rev. Lett.* **69,** 3839 (1992).
20. Flannigan, D. J. & Suslick, K. S. Plasma formation and temperature measurement during single-bubble cavitation. *Nature* **434,** 52 (2005).
21. Kappus, B., Khalid, S., Chakravarty, A. & Putterman, S. Phase transition to an opaque plasma in a sonoluminescing bubble. *Phys. Rev. Lett.* **106,** 234302 (2011).
22. Quan, W. L. *et al.* Equations of state, transport properties, and compositions of argon plasma: Combination of self-consistent fluid variation theory and linear response theory. *Phys. Rev. E* **91,** 023106 (2015).
23. Lorenzen, W., Holst, B. & Redmer, R. First-order liquid-liquid phase transition in dense hydrogen. *Phys. Rev. B* **82,** 195107 (2010).
24. Scandolo, S. Liquid-liquid phase transition in compressed hydrogen from first-principles simulations. *Proc. Natl. Acad. Sci. USA* **100,** 3051–3053 (2003).
25. Anderson, P. W. Absence of diffusion in certain random lattices. *Phys. Rev.* **109,** 1492 (1958).
26. Zhang, S., Zhao, S., Kang, W., Zhang, P. & He, X.-T. Link between k-absorption edges and thermodynamic properties of warm-dense plasmas established by improved first-principles method. *arXiv preprint arXiv:1502.06082* (2015).
27. Zhao, S. *et al.* First-principles calculation of principal hugoniot and k-shell x-ray absorption spectra for warm dense kcl. *Phys. Plasmas* **22,** 062707 (2015).
28. Dai, J., Kang, D., Zhao, Z., Wu, Y. & Yuan, J. Dynamic ionic clusters with flowing electron bubbles from warm to hot dense iron along the hugoniot curve. *Phys. Rev. Lett.* **109,** 175701 (2012).
29. Mermin, N. D. Thermal properties of the inhomogeneous electron gas. *Phys. Rev.* **137,** A1441 (1965).
30. Graziani, F., Desjarlais, M. P., Redmer, R. & Trickey, S. B. (eds.) *Frontiers and Challenges in Warm Dense Matter* (Springer, 2014).
31. Marques, M. A., Castro, A., Bertsch, G. F. & Rubio, A. octopus: a first-principles tool for excited electron–ion dynamics. *Computer Physics Communications* **151,** 60–78 (2003).
32. Correa, A. A., Kohanoff, J., Artacho, E., Sánchez-Portal, D. & Caro, A. Nonadiabatic forces in ion-solid interactions: the initial stages of radiation damage. *Phys. Rev. Lett.* **108,** 213201 (2012).
33. Otobe, T. *et al.* First-principles electron dynamics simulation for optical breakdown of dielectrics under an intense laser field. *Phys. Rev. B* **77,** 165104 (2008).
34. Martin, R. *Electronic Structure: Basic Theory and Practical Methods* (Cambridge University Press, 2008).
35. Hybertsen, M. S. & Louie, S. G. Electron correlation in semiconductors and insulators: Band gaps and quasiparticle energies. *Phys. Rev. B* **34,** 5390–5413 (1986).
36. Suslick, K. S. & Flannigan, D. J. Inside a collapsing bubble: Sonoluminescence and the conditions during cavitation. *Ann. Rev. Phys. Chem.* **59,** 659 (2008).
37. Haynes, W. (ed.) *CRC handbook of Chemistry and Physics* (CRC Press, Boca Raton, FL, 2011), 92nd edn.
38. Ashcroft, N. W. & Mermin, N. D. *Solid State Physics* (Thomson Learning, 1976).
39. Chen, Q. F., Zheng, J., Gu, Y. J., Chen, Y. L. & Cai, L. C. Equation of state of partially ionized argon plasma. *Phys. Plasmas* **18,** 112704 (2011).
40. Atzeni, S. & Meyer-ter Vehn, J. *The Physics of Inertial Fusion: BeamPlasma Interaction, Hydrodynamics, Hot Dense Matter: BeamPlasma Interaction, Hydrodynamics, Hot Dense Matter*, vol. 125 (Oxford University Press, 2004).
41. Shimizu, H., Tashiro, H., Kume, T. & Sasaki, S. High-pressure elastic properties of solid argon to 70 gpa. *Phys. Rev. Lett.* **86,** 4568 (2001).
42. Ross, M., Mao, H., Bell, P. & Xu, J. The equation of state of dense argon: a comparison of shock and static studies. *The Journal of chemical physics* **85,** 1028–1033 (1986).
43. Kietzmann, A., Holst, B., Redmer, R., Desjarlais, M. P. & Mattsson, T. R. Quantum molecular dynamics simulations for the nonmetal-to-metal transition in fluid helium. *Phys. Rev. Lett.* **98,** 190602 (2007).
44. Kappus, B., Bataller, A. & Putterman, S. J. Energy balance for a sonoluminescence bubble yields a measure of ionization potential lowering. *Phys. Rev. Lett.* **111,** 234301 (2013).
45. Barragán, P., Errea, L. F., Méndez, L. & Rabadán, I. Isotope effect in ion-atom collisions. *Phys. Rev. A* **82,** 030701 (2010).
46. Onida, G., Reining, L. & Rubio, A. Electronic excitations: density-functional versus many-body green's-function approaches. *Rev. Mod. Phys.* **74,** 601–659 (2002).
47. Nosé, S. A unified formulation of the constant temperature molecular dynamics methods. *The Journal of Chemical Physics* **81,** 511–519 (1984).
48. Modine, N. A. & Hatcher, R. M. Representing the thermal state in time-dependent density functional theory. *J Chem. Phys.* **142,** 204111 (2015).
49. An, Y. Mechanism of single-bubble sonoluminescence. *Phys. Rev. E* **74,** 026304 (2006).
50. Kresse, G. & Hafner, J. Ab initio molecular dynamics for liquid metals. *Phys. Rev. B* **47,** 558–561 (1993).
51. Blöchl, P. E. Projector augmented-wave method. *Phys. Rev. B* **50,** 17953–17979 (1994).
52. Troullier, N. & Martins, J. L. Efficient pseudopotentials for plane-wave calculations. *Phys. Rev. B* **43,** 1993–2006 (1991).
53. M. Hochbruck, C. L. On krylov subspace approximations to the matrix exponential operator. *SIAM J. Numer. Anal.* 34, 1911 (1997).

Acknowledgements
This work is financially supported by the NSFC (Grant No. 11274019) and NSFC-NSAF (Grant No. U1530113).

Author Contributions
W.K. and X.-T.H. plan the project, W.K., S-J.Z. and S.Z. carry out the numerical calculation, W.K., P.Z., Q.F.C. and X.-T.H. theoretically analysis the results, all authors contributed to the writing of the manuscript.

Additional Information
Competing financial interests: The authors declare no competing financial interests.

How to cite this article: Kang, W. *et al.* First-Principles Investigation to Ionization of Argon Under Conditions Close to Typical Sonoluminescence Experiments. *Sci. Rep.* **6**, 20623; doi: 10.1038/srep20623 (2016).

贺贤土论文选集

PHYSICAL REVIEW B **93**, 115114 (2016)

Link between K absorption edges and thermodynamic properties of warm dense plasmas established by an improved first-principles method

Shen Zhang,[1,2] Shijun Zhao,[1,2] Wei Kang,[1,2,*] Ping Zhang,[1,3] and Xian-Tu He[1,4,†]
[1]*HEDPS, Center for Applied Physics and Technology, Peking University, Beijing 100871, China*
[2]*College of Engineering, Peking University, Beijing 100871, China*
[3]*LCP, Institute of Applied Physics and Computational Mathematics, Beijing 100088, China*
[4]*Institute of Applied Physics and Computational Mathematics, Beijing 100088, China*

(Received 31 October 2014; revised manuscript received 21 February 2016; published 7 March 2016)

A precise calculation that translates shifts of x-ray K absorption edges to variations of thermodynamic properties allows quantitative characterization of interior thermodynamic properties of warm dense plasmas by x-ray absorption techniques, which provides essential information for inertial confinement fusion and other astrophysical applications. We show that this interpretation can be achieved through an improved first-principles method. Our calculation shows that the shift of K edges exhibits selective sensitivity to thermal parameters and thus would be a suitable temperature index to warm dense plasmas. We also show with a simple model that the shift of K edges can be used to detect inhomogeneity inside warm dense plasmas when combined with other experimental tools.

DOI: 10.1103/PhysRevB.93.115114

I. INTRODUCTION

Warm dense matter (WDM) generally refers to a state of matter between solids and ideal plasmas. A typical WDM material usually has a density comparable to solids and a temperature from several eV to tens of eV [1], which are conditions fuel materials experience in the early stages of inertial confinement fusion (ICF) [2]. In addition, WDM broadly exists in various astronomical objects such as giant planets and brown dwarfs [3], as well as in the core part of the earth [4]. Understanding the property of WDM is thus of particular interest to the investigation of these systems.

With its large penetration depth and high resolution in time and space [5–7], x-ray absorption is ideal for diagnosing interior properties of WDM, where x-ray absorption measurements provide information on electronic structures. Variations in thermodynamic properties are obtained by detecting induced changes in electronic structures. The position of the K absorption edge (K edge) is defined by the transition between a K-shell electronic state and the lowest unoccupied electronic state. Since Bradley *et al.* [8], much effort has been spent trying to use the shift of the K edge to quantitatively characterize thermodynamic properties in a region well beneath the surface of WDM, which is of great interest but not yet well understood. The effectiveness of this approach depends not only on accurate measurement of the x-ray absorption spectra but also on the precision of calculations translating the shifts of K-edge energies into variations in thermal states.

Recent years have witnessed a substantial improvement in x-ray diagnostic techniques [5–7]. By x-ray absorption techniques, the K edge can now be determined with a temporal resolution less than 10 ps [6]. Theoretical methods based on first-principles molecular dynamics (FPMD), i.e., a combination of density functional theory (DFT) for electrons and classical molecular dynamics for ions, have been established to be effective in calculating thermodynamic properties for a variety of materials in their warm dense states [9–12]. Determining K edges of WDM using first-principles methods is more complicated. Unlike K-edge calculation for a crystalline structure [13], where only a limited number of ions in a primitive cell have to be considered due to translational symmetry, WDM K-edge calculation involves a large number of ions. Moreover, substantial influence of core electrons has to be taken into account properly. These two factors, if not well handled, could cause unpredictable computational costs. The challenge is to find an appropriate treatment of core electrons in a system of a large number of ions to keep computational costs within the limit of current computational resources while maintaining the theoretical accuracy required. The serial work of Mazevet and Zérah, Recoules and Mazevet, and Benuzzi-Mounaix *et al.* [14–16] on warm dense aluminum lays the foundation for accurate calculation of WDM K edges. By simplifying the treatment of core electrons using a pseudopotential method within the framework of FPMD, they were able to obtain K edges close to those measured below a density of 5 g/cm^3 along the principal Hugoniot of Al in shock experiments. However, when further compressed, their calculation [14–16] generally overestimates the magnitude of K-edge shifts by more than 30% (which is far beyond the error bars of experimental data), and the overestimation tends to increase with the Hugoniot compression. Essential improvements to the calculation have to be made for quantitative characterization of thermal states of WDM.

In this work, we provide an improved FPMD calculation of K edges for an extensively studied WDM material of shock-generated warm dense Al [14–22]. The calculated K edges display excellent agreement with recent experimental data, as long as both K- and L-shell core electrons of Al are properly described. This allows a reliable translation between shifts in K-edge energies and variations in thermodynamic properties inside WDM. Our results also reveal that the shift of the K edge is more sensitive to the change of temperature than to the change of density, which indicates that the K-edge

*weikang@pku.edu.cn
†xthe@pku.edu.cn

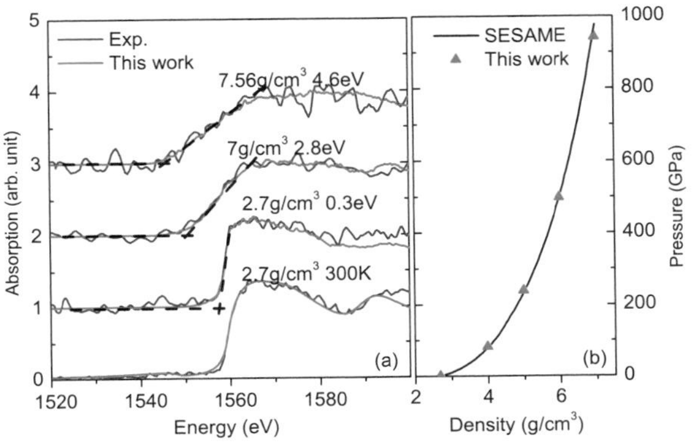

FIG. 1. (a) X-ray absorption structures (XAS) of K-shell electrons calculated (smooth curves) at selected thermal states of Al, compared with experimental measurements (undulating curves). The XAS are presented in an arbitrary unit and shifted vertically with equal space for different thermal parameters. The slope of K edges and corresponding abscissa are shown by dashed lines. (b) Pressures calculated (triangles) along the principal Hugoniot. The solid line is the principal Hugoniot derived from the SESAME 3700 equation-of-state table [24], which is an accurate representation to the experimental Hugoniot of Al.

shift is a good index for interior temperature of WDM. In addition, the calculation suggests that when combined with other temperature-measuring techniques, e.g., streaked optical pyrometers (SOP) [23], the shift of the K edge can be used to detect inhomogeneity inside WDM, which could provide further insights into the interior of WDM.

II. METHODOLOGY AND NUMERICAL DETAILS

Our calculation consists of three consecutive steps: (i) Atomic trajectories are generated at given thermal states of warm dense Al, using the FPMD method together with an appropriately designed pseudopotential including both M- and L-shell electrons, which precisely accounts for electronic structures and ion-ion interactions under high pressure but still at a reasonable computational cost. The inclusion of L-shell electrons is revealed to be one of the crucial factors for getting accurate K-edge energies. It contributes more than 2/3 of the improvement, especially at high temperature. (ii) Averaged x-ray absorption spectra (XAS) are calculated on atomic configurations uniformly sampled along the generated trajectory, and the K-edge position $E_{K,\mu}$ with respect to the chemical potential μ is then determined directly from the XAS as the intersection of its slope and the abscissa, as illustrated in Fig. 1(a). (iii) The energy of $1s$ states E_{1s} with respect to μ is determined by an all-electron DFT calculation on selected atomic configurations generated in the first step, which is the most challenging part of our calculations. In order to account for the temperature effect on $1s$ states with enough accuracy, more than 400 electrons have to be explicitly included in the calculation with a spatial resolution less than 0.2 bohr for wave functions. It should be noted that the temperature effect has a substantial contribution to E_{1s} but was not considered in the previous calculations [14–16]. Consequently, the K-edge energy E_K is determined as the difference between E_{1s} and $E_{k,\mu}$. Here, the chemical potential is a convenient choice of intermediate energy reference in the calculation. It does not appear in the final results of XAS but serves as a common energy origin to align the energies in the three calculation steps. The energy presented by the calculation is the transition energy from a $1s$ electron to an empty state, the availability of which is described by the Fermi-Dirac distribution. The edge E_K is then the lowest energy among those transitions. This is exactly what the experiments measured [14–16].

Our calculations are carried out using the QUANTUM ESPRESSO package [25]. The XAS are calculated using the XSPECTRA program [13] included in the package with a minor modification to describe the high-temperature electron distribution of WDM. Note that the original code was designed for zero-temperature calculations, where the electron distribution was simplified as a Heaviside function. It is modified as a Fermi-Dirac function to describe electrons at finite temperature in our calculation. A Perdew-Burke-Ernzerhof (PBE) type of exchange-correlation functional [26] is used throughout.

The first two steps of our calculation are similar to those employed in the previous studies [14–16] but with a home-made pseudopotential including both M-shell and L-shell electrons (i.e., $2s^2 2p^6 3s^2 3p^1$) as valence electrons, which essentially improves the accuracy of the electronic structure and ion-ion interaction. In the calculation, we adopt a plane-wave-type FPMD together with the Born-Oppenheimer approximation, as implemented in the QUANTUM ESPRESSO package. The pseudopotential takes the ultrasoft form [27] with a core cutoff radius of 1.4 bohrs, so that a plane-wave cutoff energy of 30 Ry and a shifted $2 \times 2 \times 2$ k-point mesh can be used to further reduce computational costs. The atomic trajectories are generated in a canonical system, i.e., a system of constant NVT, consisting of 32 Al atoms in a cubic box with periodic boundary conditions assumed. A time step of 1 fs is used, and atomic configurations in the last 1 ps are kept for the XAS calculation after the system evolves for more than 1 ps. The XAS is averaged on eight snapshots uniformly picked from the trajectory of the last 1 ps. For each snapshot, the XAS is calculated following the established method in Ref. [13], which approximately includes electron-hole interactions by putting a hole state in the K shell via a pseudopotential of the Gauge Including Projector Augmented Waves (GIPAW) format [28]. A shifted $4 \times 4 \times 4$ k-point mesh and 400 electronic states are used in the XAS calculation together with a plane-wave cutoff of 50 Ry to guarantee its accuracy up to 40 eV above K edges. As shown in Fig. 1(a), the position of K edge $E_{K,\mu}$ is measured directly from the XAS as the intersection of the K-edge slope and the abscissa.

In the third step, E_{1s} is determined by an all-electron plane-wave DFT calculation from the configurations used in the XAS calculation. The calculation is performed using a projector augmented-wave (PAW) pseudopotential [29] together with a plane-wave cutoff of 400 Ry and a core radius cutoff of 0.15 bohr. Benchmark calculations on atomic Al show that, using this pseudopotential, the energy of the $1s$ state can be converged within 0.5 eV ($<0.1\%$) to the reference result obtained by any atomic all-electron code used to generate pseudopotentials. To compare with experimental results directly, a constant energy shift of 63.0 eV is added to

K-edge energies in order to compensate the underestimation to E_{1s} caused by the DFT method itself. Because each atom in the snapshots has different environments, their E_{1s} values are slightly different from each other. We take the average of all atoms at selected snapshots as the calculated value of E_{1s} in this work. The uncertainty brought about by the averaging process varies from less than 0.5 eV at the uncompressed state to $\sim$1.5 eV at 8 g/cm^3, which is much smaller than the uncertainty of experimental measurements.

III. RESULTS

A. Calibration of the results

Figure 1 displays a comparison between our results and some available experimental measurements. It gives an estimation of the confidence of our method in reproducing thermodynamic properties and electronic structures of warm dense Al. Selected XAS for typical thermal conditions are displayed in Fig. 1(a) as smooth curves. As a comparison, experimental XAS (undulating curves) for the same thermal parameters are also displayed. A close match between these two XAS suggests that $E_{K,\mu}$ can be determined numerically within $\sim$1 eV of experimental values. The results are calculated with 32 atoms included in the calculation and using the home-made pseudopotential. The convergence is systematically examined with the number of atoms included, which shows that even with 8 atoms, the calculated x-ray absorption can be well compared with experimental results in Fig. 1. Increasing the number of atoms only makes the fluctuation smaller. For an additional check, we also reproduce the K-edge calculation in Refs. [14–16] with 32 atoms. A very small difference is revealed.

Figure 1(b) shows the calculated pressure of Al along the principal Hugoniot. Also displayed is the Hugoniot derived from the SESAME 3700 equation-of-state table [24], which gives an accurate account for the experimental measurements of shocked Al [30]. Our calculation agrees well with the experiments, with an overall deviation less than 2%. The convergence of pressure with an increasing number of atoms is displayed in Fig. 2 for typical densities along the principal Hugoniot. It shows that when the atom number increases from 8 to 32, the calculated error in pressure decreases from 6.5% to 1.9% compared with the SESAME [30] value.

B. K edges of warm dense Al

In Fig. 3, we present calculated K-edge shifts of Al referring to its uncompressed state $\rho = 2.7$ g/cm^3 and $T = 300$ K. Both calculated and experimentally measured K-edge shifts along the principal Hugoniot are displayed in Fig. 3(a). The calculated results, shown as solid curves with diamonds, reproduce well the experimental results of Hall *et al.*, displayed as solid dots in Fig. 3(a) [21]. Earlier experimental results of DaSilva *et al.* [22], however, exhibit an observable deviation $\sim$3 eV from our calculation and Hall *et al.*'s measurements at $\rho = 6.0$ g/cm^3 as the result of low temporal resolution and insufficient characterization to plasma states in the earlier experiments [16]. Figure 3(b) shows that K-edge shifts under reshocked and unloading conditions can also be well described. The majority of our results are well located inside

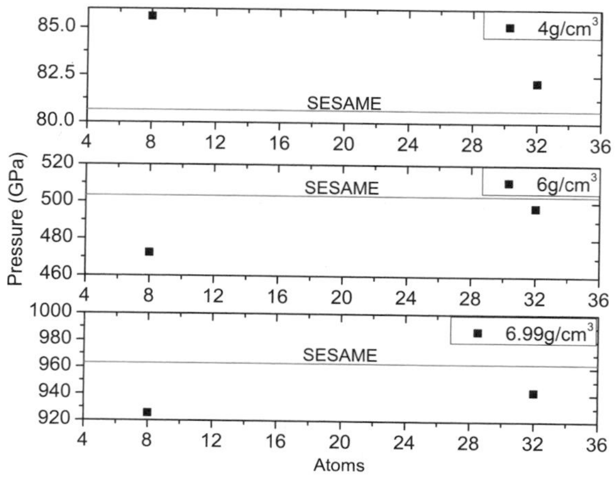

FIG. 2. Convergence of pressure along the principal Hugoniot for typical densities. The results are calculated using the pseudopotential generated for the calculation. The red line represents experimental measurements taken from the SESAME table [30].

the experimental error bars, except two of them (highlighted by dotted circles) that have slightly larger deviations. Since no systematic trends of these deviations are observed, they are probably caused by fluctuations in thermal parameters or by the inhomogeneity of plasma states. Figure 3(b) also shows that K-edge shifts sensitively depend on thermal parameters of plasma states. The K-edge shifts calculated with instant thermal parameters have distinguishable differences from those (displayed as solid curves with squares) calculated with subtly different thermal parameters derived from a hydrodynamic code [16]. The latter, representing K-edge shifts under ideal reshock conditions, approximately cross the center region of the experimental data.

C. Sources of improvement

The agreement between our calculation and experiments is attributed to a much improved estimation of E_K. There are two sources for the improvement. One is the temperature effect of E_{1s}, which was considered to be small and neglected in the previous calculations [14–16]. However, as we show in Fig. 4(a), this effect is substantial at a high compressing ratio when both ρ and T are high. The temperature effect goes into the correction indirectly. It first induces a spatial redistribution of ions and electrons at high temperature, which in turn causes a correction to the Coulomb potential energy part of E_{1s}. The other is the correction of L-shell electrons included in the FPMD calculation, which gives a better account of wave functions at high density and causes less blueshift in μ, as illustrated by Fig. 4(b). Figure 4(c) displays the net effect of these two contributions. It shows that the major contribution to the improvement comes from the correction to μ, at all compressing ratios. The correction of E_{1s} contributes less than 1/3 of the improvement but increases with further compression. Both corrections have significant contributions at a large compressing ratio.

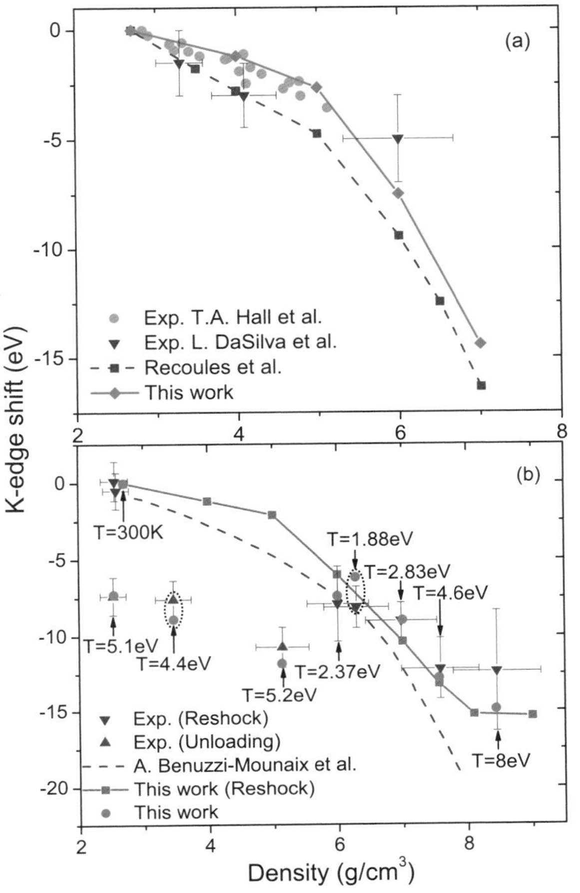

FIG. 3. Calculated K-edge shifts of Al referring to its uncompressed solid state, $\rho = 2.7$ g/cm^3 and $T = 300$ K, compared with experimental measurements and previous calculations of similar methods [14–16]. (a) K edges calculated along the principal Hugoniot. Experimental data are taken from Hall *et al.* [21] and DaSilva *et al.* [22]. The previous calculation is taken from Recoules and Mazevet [15]. (b) K edges under reshocked and unloading conditions, calculated with two subtly different sets of thermal parameters. Solid dots are calculated with thermal parameters measured by experiments, which are explicitly indicated in the figure [16]. Solid curve with squares is calculated with parameters derived from a hydrodynamic simulation for the reshock condition [16]. The two points slightly outside the experimental error bar are marked with dotted circles. Calculations from Benuzzi-Mounaix *et al.* [16] are also displayed for comparison.

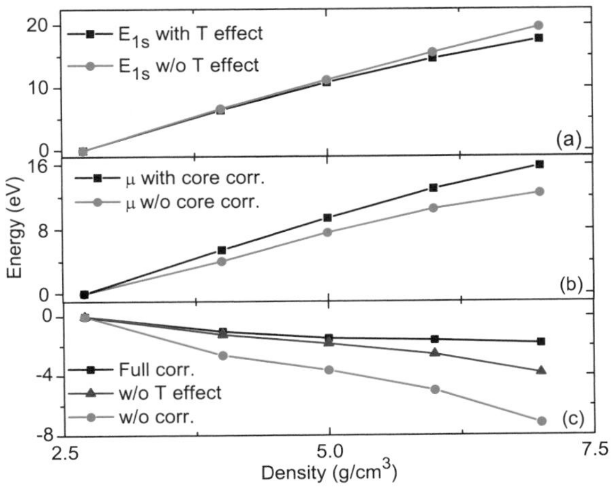

FIG. 4. Decomposition of contributions to the improvement of K-edge calculation along the principal Hugoniot of Al, referring to the standard solid state of Al. (a) Improvement to the calculation of E_{1s} caused by the temperature effect. (b) Improvement to the calculation of μ as a result of including L-shell electrons explicitly in generating atomic trajectories. (c) Calculated energy of $\mu - E_{1s}$ with both corrections in (a) and (b), compared with the result before improvements are taken into account. An intermediate result (displayed as a solid curve with triangles) including corrections to μ is displayed to visualize the fraction of contribution for each correction.

D. Selective sensitivity to thermal parameters

To quantitatively characterize the relation between plasma states of Al and K-edge shifts, a systematic examination is presented in Figs. 5(a) and 5(b), covering a variety states from solids to WDM. K-edge shifts at different densities are displayed in Fig. 5(a) as a function of temperature. At low temperature, the K edges decrease linearly at a slope of 2.9 ± 0.2, which is almost independent of the variation of density. When temperature further increases, a turning point occurs somewhere between $T = 2.5$ eV and $T = 5$ eV, depending on the density of WDM. Figure 5(b) displays K edges at different temperatures with respect to density. The flat shape of K edges at low temperatures, as illustrated by the $T = 0.5$ eV and $T = 2.5$ eV curves, confirms the insensitivity of K-edge shifts to the variation of density. These results suggest that the K edge has selective sensitivity to the variation of thermal parameters. In the parameter range investigated, the K edge is reasonably sensitive to the change of temperature and thus would be useful as a temperature index. Compared to the SOP technique [23], which detects temperatures on the surface, K-edge shifts reflect the temperature inside WDM, which is a feature attractive to the study of bulk WDM.

Alternatively, the temperature of bulk WDM can be probed by other x-ray diagnostic methods, for example, the imaging x-ray Thomson scattering (imaging-XRTS) spectroscopy [31,32], which takes advantage of the relation between the electronic temperature and the shape of the Thomson scattering spectrum profile. Since it is an essentially different temperature-measuring mechanism from that used in the x-ray absorption edge method, these two methods can be regarded as independent of each other. A combination of them can thus afford a stronger diagnostic of the bulk WDM.

There have also been attempts which propose to determine the temperature of WDM directly from the slope of the K edge [33]. Because this method strongly relies on the condition that the material is very close to a homogeneous electron gas, the range of its applicability is severely constrained by the underlying assumption compared to the imaging-XRTS

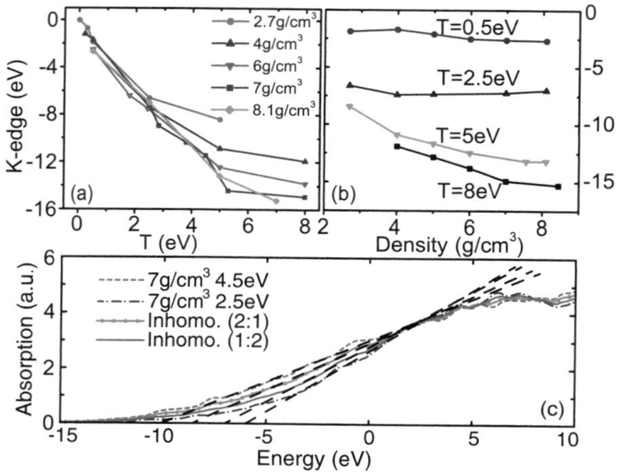

FIG. 5. (a) K-edge shift as a function of temperature at fixed densities. (b) K-edge shift with respect to density calculated at constant temperatures. (c) XAS for model inhomogeneous warm dense Al systems comprising two parts of the same density but having different temperatures and volume ratios. XAS for the two homogeneous parts are also displayed as dashed curves for reference. The K-edge position $E_{K,\mu}$ is measured as the intersection of the slopes and the abscissa. The slopes are displayed as dashed lines.

method and the shift of the K edge, which take all the properties of the real material into consideration.

E. Potential inhomogeneity detection

Additionally, we show with a simplified model that K-edge shifts, when combined with SOP, allow the probing of inhomogeneity in bulk WDM. XAS of model inhomogeneous warm dense Al systems are displayed in Fig. 5(c). The model system comprises two homogeneous parts of the same density $\rho = 7.0$ g/cm^3. The two homogeneous parts have different temperatures of $T = 4.5$ eV and $T = 2.5$ eV. They are put together with different mass ratios of 2:1 and 1:2, respectively. The effective spectra are obtained by the Beer-Lambert law to mix two homogeneous systems, which does not need a simulation of a heterogeneous system. The mass weight comes from the fact that the two components have different thicknesses and masses, although they have the same density. Dashed curves in Fig. 5(c) represent the XAS of the two homogeneous parts. Our calculation shows that the XAS of the inhomogeneous system is a mass-weighted average of these two homogeneous systems, and the K edge of the inhomogeneous system is different from those of the two homogeneous systems. Since the shift of the K edges is less sensitive to the change of density in the warm dense region, the density inhomogeneity is taken into account as the mass weight in the average. A real inhomogeneous WDM system is composed of a large number of such small homogeneous parts along the path of the x ray. According to our calculation, the measured K edge of a real inhomogeneous warm dense system is different from that calculated with the temperature measured by SOP at the surface. Except for extreme cases where only density inhomogeneity exists, a deviation between these two K edges thus indicates the appearance of inhomogeneity.

IV. SUMMARY

In summary, we show that an accurate estimation of the K-edge shift of warm dense Al can be achieved by an improved first-principles calculation when the effect of core electrons is carefully taken into account. A calculation of such accuracy would open a new possibility for an x-ray absorption technique to quantitatively characterize internal plasma states of WDM, which is of particular interest to a variety of fields, including ICF, astrophysics, and geophysics.

ACKNOWLEDGMENT

This work is financially supported by the NSFC (Grant No. 11274019) and NSAF (Grant No. U1530113).

[1] B. Barbrel, M. Koenig, A. Benuzzi-Mounaix, E. Brambrink, C. Brown, D. O. Gericke, B. Nagler, M. R. Le Gloahec, D. Riley, C. Spindloe *et al.*, Phys. Rev. Lett. **102**, 165004 (2009).

[2] S. Atzeni and J. Meyer-ter Vehn, *The Physics of Inertial Fusion: Beam Plasma Interaction, Hydrodynamics, Hot Dense Matter*, International Series of Monographs on Physics Vol. 125 (Oxford University Press, Oxford, 2004).

[3] T. Guillot, Science **286**, 72 (1999).

[4] G. Huser, M. Koenig, A. Benuzzi-Mounaix, E. Henry, T. Vinci, B. Faral, M. Tomasini, B. Telaro, and D. Batani, Phys. Plasmas **12**, 060701 (2005).

[5] B. Yaakobi, T. R. Boehly, T. C. Sangster, D. D. Meyerhofer, B. A. Remington, P. G. Allen, S. M. Pollaine, H. E. Lorenzana, K. T. Lorenz, and J. A. Hawreliak, Phys. Plasmas **15**, 062703 (2008).

[6] A. Levy, F. Dorchies, C. Fourment, M. Harmand, S. Hulin, J. J. Santos, D. Descamps, S. Petit, and R. Bouillaud, Rev. Sci. Instrum. **81**, 063107 (2010).

[7] Y. Zhao, J. Yang, J. Zhang, G. Yang, M. Wei, G. Xiong, T. Song, Z. Zhang, L. Bao, B. Deng, Y. Li, X. He, C. Li, Y. Mei, R. Yu, S. Jiang, S. Liu, Y. Ding, and B. Zhang, Phys. Rev. Lett. **111**, 155003 (2013).

[8] D. K. Bradley, J. Kilkenny, S. J. Rose, and J. D. Hares, Phys. Rev. Lett. **59**, 2995 (1987).

[9] M. P. Desjarlais, Phys. Rev. B **68**, 064204 (2003).

[10] A. Kietzmann, R. Redmer, M. P. Desjarlais, and T. R. Mattsson, Phys. Rev. Lett. **101**, 070401 (2008).

[11] S. A. Bonev, B. Militzer, and G. Galli, Phys. Rev. B **69**, 014101 (2004).

[12] C. Wang, X.-T. He, and P. Zhang, Phys. Rev. Lett. **106**, 145002 (2011).

[13] M. Taillefumier, D. Cabaret, A.-M. Flank, and F. Mauri, Phys. Rev. B **66**, 195107 (2002); C. Gougoussis, M. Calandra, A. P. Seitsonen, and F. Mauri, *ibid.* **80**, 075102 (2009).

[14] S. Mazevet and G. Zérah, Phys. Rev. Lett. **101**, 155001 (2008).

[15] V. Recoules and S. Mazevet, Phys. Rev. B **80**, 064110 (2009).

[16] A. Benuzzi-Mounaix, F. Dorchies, V. Recoules, F. Festa, O. Peyrusse, A. Levy, A. Ravasio, T. Hall, M. Koenig, N. Amadou, E. Brambrink, and S. Mazevet, Phys. Rev. Lett. **107**, 165006 (2011).
[17] M. P. Desjarlais, J. D. Kress, and L. A. Collins, Phys. Rev. E **66**, 025401 (2002).
[18] U. Zastrau *et al.*, Laser Part. Beams **30**, 45 (2012).
[19] S. M. Vinko *et al.*, Nature **482**, 59 (2012).
[20] S. Vinko, O. Ciricosta, and J. Wark, Nat. Commun. **5**, 3533 (2014).
[21] T. A. Hall, J. Al-Kuzee, A. Benuzzi, M. Koenig, J. Krishnan, N. Grandjouan, D. Batani, S. Bossi, and S. Nicolella, Europhys. Lett. **41**, 495 (1998).
[22] L. DaSilva, A. Ng, B. K. Godwal, G. Chiu, F. Cottet, M. C. Richardson, P. A. Jaanimagi, and Y. T. Lee, Phys. Rev. Lett. **62**, 1623 (1989).
[23] J. E. Miller, T. R. Boehly, A. Melchior, D. D. Meyerhofer, P. M. Celliers, J. H. Eggert, D. G. Hicks, C. M. Sorce, J. A. Oertel, and P. M. Emmel, Rev. Sci. Instrum. **78**, 034903 (2007); Z. Chen, L. Hao, W. Zhebin, J. Xiaohua, Z. Huige, L. Yonggang, L. Zhichao, L. Sanwei, Y. Dong, D. Yongkun, Z. Bin, H. Guangyue, and Z. Jian, Plasma Sci. Technol. **16**, 571 (2014).
[24] Los Alamos National Laboratory, SESAME, Technical Report No. LA-UR-92-3407, 1992.
[25] P. Giannozzi, S. Baroni, N. Bonini, M. Calandra, R. Car, C. Cavazzoni, D. Ceresoli, G. L. Chiarotti, M. Cococcioni, I. Dabo *et al.*, J. Phys.: Condens. Matter **21**, 395502 (2009).
[26] J. P. Perdew, K. Burke, and M. Ernzerhof, Phys. Rev. Lett. **77**, 3865 (1996).
[27] D. Vanderbilt, Phys. Rev. B **41**, 7892 (1990).
[28] C. J. Pickard and F. Mauri, Phys. Rev. B **63**, 245101 (2001).
[29] P. E. Blöchl, Phys. Rev. B **50**, 17953 (1994).
[30] P. Celliers, G. Collins, D. Hicks, and J. Eggert, J. Appl. Phys. **98**, 113529 (2005).
[31] E. J. Gamboa, C. M. Huntington, M. R. Trantham, P. A. Keiter, R. P. Drake, D. S. Montgomery, J. F. Benage, and S. A. Letzring, Rev. Sci. Instrum. **83**, 10E108 (2012).
[32] K. Falk, E. J. Gamboa, G. Kagan, D. S. Montgomery, B. Srinivasan, P. Tzeferacos, and J. F. Benage, Phys. Rev. Lett. **112**, 155003 (2014).
[33] F. Dorchies, F. Festa, V. Recoules, O. Peyrusse, A. Benuzzi-Mounaix, E. Brambrink, A. Levy, A. Ravasio, M. Koenig, T. Hall, and S. Mazevet, Phys. Rev. B **92**, 085117 (2015).

编辑说明

本书是贺贤土院士自 1982 年至 2017 年的科研论文的汇集，在较长的时间跨度上较为全面地反映了中国理论物理的发展历程，可作为相关科学领域重要的参考资料。本书收录的文章是贺贤土院士的部分代表性作品。因写作时间跨度长达三十余年，且文章在不同刊物上发表，因此文章的风格和格式有所不同。本着尊重原文的原则，我们只对文章做了影印出版，其中难免有瑕疵，恳请读者原谅并予以指正。

2017 年 5 月

图书在版编目（CIP）数据

贺贤土论文选集：英文 /《贺贤土论文选集》编委会编. — 杭州：浙江大学出版社，2017.5
ISBN 978-7-308-16958-5

Ⅰ. ①贺… Ⅱ. ①贺… Ⅲ. ①核物理学—文集—英文②理论物理学—文集—英文 Ⅳ. ① O57-53 ② O41-53

中国版本图书馆 CIP 数据核字 (2017) 第 095558 号

贺贤土论文选集

《贺贤土论文选集》编委会　编

策　　划　许佳颖
责任编辑　金佩雯
责任校对　许佳颖　侯鉴峰
封面设计　程　晨
出版发行　浙江大学出版社
（杭州市天目山路148号　邮政编码310007）
（网址：http://www.zjupress.com）
排　　版　杭州兴邦电子印务有限公司
印　　刷　浙江海虹彩色印务有限公司
开　　本　889mm × 1194mm　1/16
印　　张　67.75
字　　数　370 千
版 印 次　2017 年 5 月第 1 版　2017 年 5 月第 1 次印刷
书　　号　ISBN 978-7-308-16958-5
定　　价　398.00 元